Algebra and Trigonometry

Sixth Edition

▸ **Ron Larson**

▸ **Robert P. Hostetler**

The Pennsylvania State University
The Behrend College

▸ **With the assistance of David C. Falvo**

The Pennsylvania State University
The Behrend College

Houghton Mifflin Company Boston New York

Publisher : Jack Shira
Managing Editor: Cathy Cantin
Development Manager: Maureen Ross
Development Editor: Laura Wheel
Assistant Editor: Jennifer King
Assistant Editor: James Cohen
Supervising Editor: Karen Carter
Senior Project Editor: Patty Bergin
Production Technology Supervisor: Gary Crespo
Senior Marketing Manager: Danielle Potvin
Marketing Associate: Nicole Mollica
Senior Manufacturing Coordinator: Jane Spelman
Composition and Art: Meridian Creative Group
Cover Design Manager: Diana Coe

Cover Image: © Ralph Mercer

Printed in the U.S.A.

Library of Congress Catalog Card Number: 2002115582

ISBN: 0-618-31782-1

7 8 9–DOW–10 09 08 07 06

Arithmetic Operations:

$ab + ac = a(b + c)$ $\qquad \dfrac{a}{b} + \dfrac{c}{d} = \dfrac{ad + bc}{bd}$ $\qquad \dfrac{a + b}{c} = \dfrac{a}{c} + \dfrac{b}{c}$ $\qquad \dfrac{\left(\frac{a}{b}\right)}{\left(\frac{c}{d}\right)} = \dfrac{ad}{bc}$

$a\left(\dfrac{b}{c}\right) = \dfrac{ab}{c}$ $\qquad \dfrac{a - b}{c - d} = \dfrac{b - a}{d - c}$ $\qquad \dfrac{ab + ac}{a} = b + c, \; a \neq 0$

$\dfrac{\left(\frac{a}{b}\right)}{c} = \dfrac{a}{bc}$ $\qquad \dfrac{a}{\left(\frac{b}{c}\right)} = \dfrac{ac}{b}$

Exponents and Radicals:

$a^0 = 1, \; a \neq 0$ $\qquad \dfrac{a^x}{a^y} = a^{x-y}$ $\qquad \left(\dfrac{a}{b}\right)^x = \dfrac{a^x}{b^x}$ $\qquad \sqrt[n]{a^m} = a^{m/n} = \left(\sqrt[n]{a}\right)^m$

$a^{-x} = \dfrac{1}{a^x}$ $\qquad (a^x)^y = a^{xy}$ $\qquad \sqrt{a} = a^{1/2}$ $\qquad \sqrt[n]{ab} = \sqrt[n]{a}\sqrt[n]{b}$

$a^x a^y = a^{x+y}$ $\qquad (ab)^x = a^x b^x$ $\qquad \sqrt[n]{a} = a^{1/n}$ $\qquad \sqrt[n]{\left(\dfrac{a}{b}\right)} = \dfrac{\sqrt[n]{a}}{\sqrt[n]{b}}$

Algebraic Errors to Avoid:

$\dfrac{a}{x + b} \neq \dfrac{a}{x} + \dfrac{a}{b}$ (To see this error, let $a = b = x = 1$.)

$\sqrt{x^2 + a^2} \neq x + a$ (To see this error, let $x = 3$ and $a = 4$.)

$a - b(x - 1) \neq a - bx - b$ [Remember to distribute negative signs. The equation should be $a - b(x - 1) = a - bx + b$.]

$\dfrac{\left(\frac{x}{a}\right)}{b} \neq \dfrac{bx}{a}$ [To divide fractions, invert and multiply. The equation should be $\dfrac{\left(\frac{x}{a}\right)}{b} = \dfrac{\left(\frac{x}{a}\right)}{\left(\frac{b}{1}\right)} = \left(\dfrac{x}{a}\right)\left(\dfrac{1}{b}\right) = \dfrac{x}{ab}$.]

$\sqrt{-x^2 + a^2} \neq -\sqrt{x^2 - a^2}$ (The negative sign cannot be factored out of the square root.)

$\dfrac{\cancel{a} + bx}{\cancel{a}} \neq 1 + bx$ (This is one of many examples of incorrect cancellation. The equation should be $\dfrac{a + bx}{a} = \dfrac{a}{a} + \dfrac{bx}{a} = 1 + \dfrac{bx}{a}$.)

$\dfrac{1}{x^{1/2} - x^{1/3}} \neq x^{-1/2} - x^{-1/3}$ (This error is a more complex version of the first error.)

$(x^2)^3 \neq x^5$ [This equation should be $(x^2)^3 = x^2x^2x^2 = x^6$.]

Conversion Table:

1 centimeter ≈ 0.394 inch	1 joule ≈ 0.738 foot-pound	1 mile ≈ 1.609 kilometers
1 meter ≈ 39.370 inches	1 gram ≈ 0.035 ounce	1 gallon ≈ 3.785 liters
≈ 3.281 feet	1 kilogram ≈ 2.205 pounds	1 pound ≈ 4.448 newtons
1 kilometer ≈ 0.621 mile	1 inch ≈ 2.540 centimeters	1 foot-lb ≈ 1.356 joules
1 liter ≈ 0.264 gallon	1 foot ≈ 30.480 centimeters	1 ounce ≈ 28.350 grams
1 newton ≈ 0.225 pound	≈ 0.305 meter	1 pound ≈ 0.454 kilogram

Contents

Chapter 11 Sequences, Series, and Probability 800

A Word from the Authors

Welcome to *Algebra and Trigonometry*, Sixth Edition. In this revision we continue to focus on promoting student success, while providing an accessible text that offers flexible teaching and learning options.

In keeping with our philosophy that students learn best when they know what they are expected to learn, we have retained the thematic study thread from the Fifth Edition. We first introduce this study thread in the Chapter Opener. Each chapter begins with a study guide that contains a comprehensive overview of the chapter concepts (*What you should learn*), a list of *Important Vocabulary* integral to learning the chapter concepts, a list of additional chapter-specific *Study Tools*, and additional text-specific resources. The study guide allows students to get organized and prepare for the chapter. Then, each section opens with a a set of learning objectives outlining the concepts and skills students are expected to learn (*What you should learn*), followed by an interesting real-life application used to illustrate why it is important to learn the concepts in that section (*Why you should learn it*). *Study Tips* at point-of-use provide support as students read through the section. And finally, to provide study support and a comprehensive review of the chapter, each chapter concludes with a chapter summary (*What did you learn?*), which reinforces the section objectives, and chapter *Review Exercises*, which are correlated to the chapter summary.

In addition to providing in-text study support, we have taken care to write a text for the student. We paid careful attention to the presentation, using precise mathematical language and clear writing, to create an effective learning tool. We are committed to providing a text that makes the mathematics within it accessible to all students. In the Sixth Edition, we have revised and improved upon many text features designed for this purpose. The *Technology*, *Exploration* features have been expanded. *Chapter Tests*, which gave students an opportunity for self-assessment, are included in every chapter. We have retained the *Synthesis* exercises, which check students' conceptual understanding, and the *Review* exercises, which reinforce skills learned in previous sections within each section exercise set. Also, students have access to several media resources that offer additional text-specific resources to enhance the learning process.

From the time we first began writing in the early 1970s, we have always viewed part of our authoring role as that of providing instructors with flexible teaching programs. The optional features within the text allow instructors with different pedagogical approaches to design their course to meet both their instructional needs and the needs of their students. Instructors who stress applications and problem solving, or exploration and technology, and more traditional methods will be able to use this text successfully. We hope you enjoy the Sixth Edition.

Ron Larson

Robert P. Hostetler

Acknowledgments

We would like to thank the many people who helped us at various stages of this project. Their encouragement, criticisms, and suggestions have been invaluable to us.

Sixth Edition Reviewers

Ahmad Abusaid, Southern Polytechnic University; Catherine Banks, Texas Woman's College; Jared Burch, College of the Sequoias; Dr. Michelle R. DeDeo, University of North Florida; Brian Hickey, East Central College; Gangadhar R. Hiremath, Miles College; Erick Hofacker, University of Wisconsin-River Falls; Dr. Kevin W. Hopkins, Southwest Baptist University; Charles W. Johnson, South Georgia College; Gary S. Kersting, North Central Michigan College; Namyong Lee, Minnesota State University; Mary Leeseberg, Manatee Community College; Tristan Londré, Blue River Community College; Bruce N. Lundberg, University of Southern Colorado; Dr. Carl V. Lutzer, Rochester Institute of Technology; Rudy Maglio, Oakton Community College; James Miller, West Virginia University; Steve O'Donnell, Rogue Community College; Armando I. Perez, Laredo Community College; Rita Randolfi, Brevard Community College; David Ray, The University of Tennessee at Martin; Miguel San Miguel Gonzalez, Texas A&M International University; Scott Satake, North Idaho College; Jed Soifer, Atlantic Cape Community College; Dr. Roy N. Tucker, Palo Alto College and The University of Texas at San Antonio; Karen Villarreal, Xavier University of Louisiana; Carol Walker, Hinds Community College; J. Lewis Walston, Methodist College; Jun Wang, Alabama State University; Ibrahim Wazir, American International School; Robert Wylie, Carl Albert State College.

Previous Edition Reviewers

James Alsobrook, Southern Union State Community College; Sherry Biggers, Clemson University; Charles Biles, Humboldt State University; Randall Boan, Aims Community College; Jeremy Carr, Pensacola Junior College; D. J. Clark, Portland Community College; Donald Clayton, Madisonville Community College; Linda Crabtree, Metropolitan Community College; David DeLatte, University of North Texas; Gregory Dlabach, Northeastern Oklahoma A & M College; Joseph Lloyd Harris, Gulf Coast Community College; Jeff Heiking, St. Petersburg Junior College; Celeste Hernandez, Richland College; Heidi Howard, Florida Community College at Jacksonville; Wanda Long, St. Charles County Community College; Wayne F. Mackey, University of Arkansas; Rhonda MacLeod, Florida State University; M. Maheswaran, University of Wisconsin–Marathon County; Valerie Miller, Georgia State University; Katharine Muller, Cisco Junior College; Bonnie Oppenheimer, Mississippi University for Women; James Pohl, Florida Atlantic University; Hari Pulapaka, Valdosta State University; Michael Russo, Suffolk County Community College; Cynthia Floyd Sikes, Georgia Southern University; Susan Schindler, Baruch College–CUNY; Stanley Smith, Black Hills State University.

We would like to extend a special thanks to all of the instructors who took time to participate in our phone interviews.

We would like to thank the staff of Larson Texts, Inc. and the staff of Meridian Creative Group, who assisted in proofreading the manuscript, preparing and proofreading the art package, and typesetting the supplements.

We are grateful to our wives, Deanna Gilbert Larson and Eloise Hostetler, for their love, patience, and support. Also, a special thanks goes to R. Scott O'Neil.

If you have suggestions for improving this text, please feel free to write to us. Over the years we have received many useful comments from both instructors and students, and we value these comments very much.

Ron Larson
Robert P. Hostetler

How can this book help you

Support for Student Success

- Larson provides clear, easy-to-read examples that include all the steps needed to understand a new concept.
- Numerous examples are provided throughout the book that correspond to the exercise sets, giving students support with the key concepts in their homework assignments.
- Additional resources are also available, such as SMARTHINKING's live, one-on-one online tutoring service. This enables students to receive tutorial help from the comfort and privacy of their own home.
- Key course material is also presented on a DVD by a qualified instructor, making it easy to review content or material missed due to an absence.

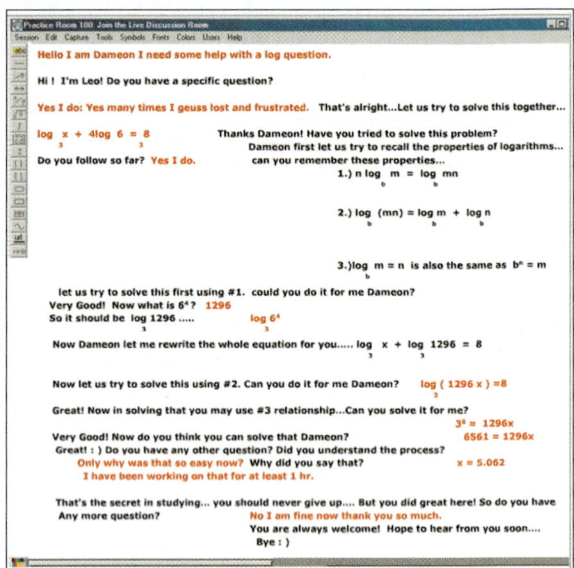

Options for Students and Instructors

- Concepts are presented through examples, applications, technology, or explorations to adapt the course to the curriculum needs or student learning styles.
- A variety of exercises that increase in difficulty allows professors the flexibility to assign homework to students with various learning styles. Exercise options include skills, technology, critical thinking, writing, applications, modeling data, true/false, proofs, and theoretical questions.
- The P.S. Problem Solving section at the end of every chapter offers more challenging exercises for advanced students.
- This text provides a solid mathematical foundation by foreshadowing concepts that will be used in future courses. Topics that will be especially helpful to students in Calculus are labeled with an "Algebra of Calculus" icon.

Exploration

Graph each of the functions with a graphing utility. Determine whether the function is *even*, *odd*, or *neither*.

Technology

You can use a graphing utility to determine the domain of a composition of functions. For the composition in Example 5, enter the function

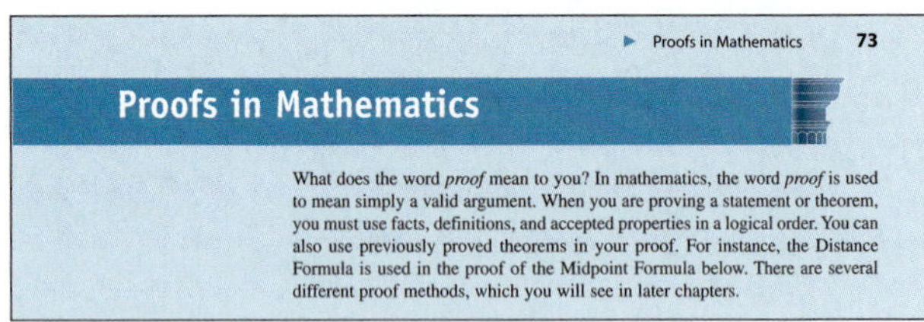
Proofs in Mathematics 73

Proofs in Mathematics

What does the word *proof* mean to you? In mathematics, the word *proof* is used to mean simply a valid argument. When you are proving a statement or theorem, you must use facts, definitions, and accepted properties in a logical order. You can also use previously proved theorems in your proof. For instance, the Distance Formula is used in the proof of the Midpoint Formula below. There are several different proof methods, which you will see in later chapters.

succeed in your math course?

For more information, see pages xii–xxv.

Applications That Motivate Students

- Applications in the exposition, examples, and exercises use real life data for students to see the relevance of what they are learning.
- Interesting topics are included throughout the book to help students see the practical, as well as theoretical, side of mathematics.
- Sourced data sets are included throughout the text, allowing students the opportunity to generate mathematical models that represent real data.

83. ***Height of a Ball*** The height y (in feet) of a ball thrown by a child is

$$y = -\frac{1}{12}x^2 + 2x + 4$$

where x is the horizontal distance (in feet) from the point at which the ball is thrown (see figure).

(a) How high is the ball when it leaves the child's hand? (*Hint:* Find y when $x = 0$.)

(b) What is the maximum height of the ball?

(c) How far from the child does the ball strike the ground?

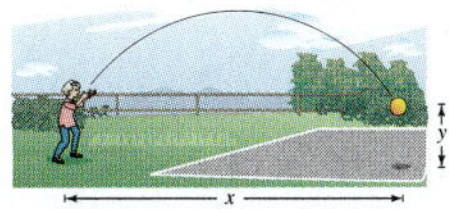

84. ***Path of a Diver*** The path of a diver is

$$y = -\frac{4}{9}x^2 + \frac{24}{9}x + 12$$

where y is the height (in feet) and x is the horizontal distance from the end of the diving board (in feet). What is the maximum height of the diver?

85. ***Graphical Analysis*** From 1960 to 2000, the per capita consumption C of cigarettes by Americans (age 18 and older) can be modeled by

$$C = 4258 + 6.5t - 1.62t^2, \quad 0 \le t \le 40$$

where t is the year, with $t = 0$ corresponding to

▶ **Model It**

86. ***Data Analysis*** The numbers y (in thousands) of hairdressers and cosmetologists in the United States for the years 1995 through 2000 are shown in the table. (Source: U.S. Bureau of Labor Statistics)

Year	Number of hairdressers and cosmetologists, y
1995	750
1996	737
1997	748
1998	763
1999	784
2000	820

(a) Use a graphing utility to create a scatter plot of the data. Let x represent the year, with $x = 5$ corresponding to 1995.

(b) Use the *regression* feature of a graphing utility to find a quadratic model for the data.

(c) Use a graphing utility to graph the model in the same viewing window as the scatter plot. How well does the model fit the data?

(d) Use the *trace* feature of the graphing utility to approximate the year in which the number of hairdressers and cosmetologists was the least.

(e) Use the model to predict the number of hairdressers and cosmetologists in 2005.

Readable and Understandable Text for Students

- Examples, explanations, and proofs begin and end on the same page to allow students to see concepts as a whole, without page- turning distractions. This unique design is one more example of the carefully developed texts created by the Larson Team.
- Examples include detailed solutions that show all steps to make it easy for students to understand the material being presented.
- Many examples include numerical, algebraic, and/or graphical presentations to provide students an opportunity to see the solution represented in a way that is most clear to them.

Technology

You can use a graphing utility to determine the domain of a composition of functions. For the composition in Example 5, enter the function composition as

$$y = \left(\sqrt{9 - x^2}\right)^2 - 9.$$

You should obtain the graph shown below. Use the *trace* feature to determine that the x-coordinates of points on the graph extend from -3 to 3. So, the domain of $(f \circ g)(x)$ is $-3 \le x \le 3$.

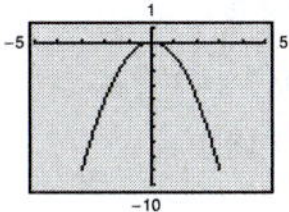

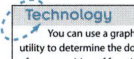

Example 5 Finding the Domain of a Composite Function

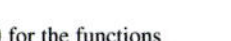

Find the composition $(f \circ g)(x)$ for the functions

$$f(x) = x^2 - 9 \quad \text{and} \quad g(x) = \sqrt{9 - x^2}.$$

Then find the domain of $(f \circ g)$.

Solution

$$\begin{aligned}(f \circ g)(x) &= f(g(x)) \\ &= f\left(\sqrt{9 - x^2}\right) \\ &= \left(\sqrt{9 - x^2}\right)^2 - 9 \\ &= 9 - x^2 - 9 \\ &= -x^2\end{aligned}$$

From this, it might appear that the domain of the composition is the set of all real numbers. Because the domain of f is the set of all real numbers and the domain of g is $-3 \le x \le 3$, the domain of $(f \circ g)$ is $-3 \le x \le 3$.

In Examples 4 and 5, you formed the composition of two given functions. In calculus, it is also important to be able to identify two functions that make up a given composite function. For instance, the function h given by

$$h(x) = (3x - 5)^3$$

is the composition of f with g, where $f(x) = x^3$ and $g(x) = 3x - 5$. That is,

$$h(x) = (3x - 5)^3 = [g(x)]^3 = f(g(x)).$$

Basically, to "decompose" a composite function, look for an "inner" function and an "outer" function. In the function h above, $g(x) = 3x - 5$ is the inner function and $f(x) = x^3$ is the outer function.

Example 6 Finding Components of Composite Functions

Express the function $h(x) = \dfrac{1}{(x-2)^2}$ as a composition of two functions.

Solution

One way to write h as a composition of two functions is to take the inner function to be $g(x) = x - 2$ and the outer function to be

$$f(x) = \frac{1}{x^2} = x^{-2}.$$

Then you can write

$$h(x) = \frac{1}{(x-2)^2} = (x-2)^{-2} = f(x-2) = f(g(x)).$$

Textbook Highlights

Student Success Tools

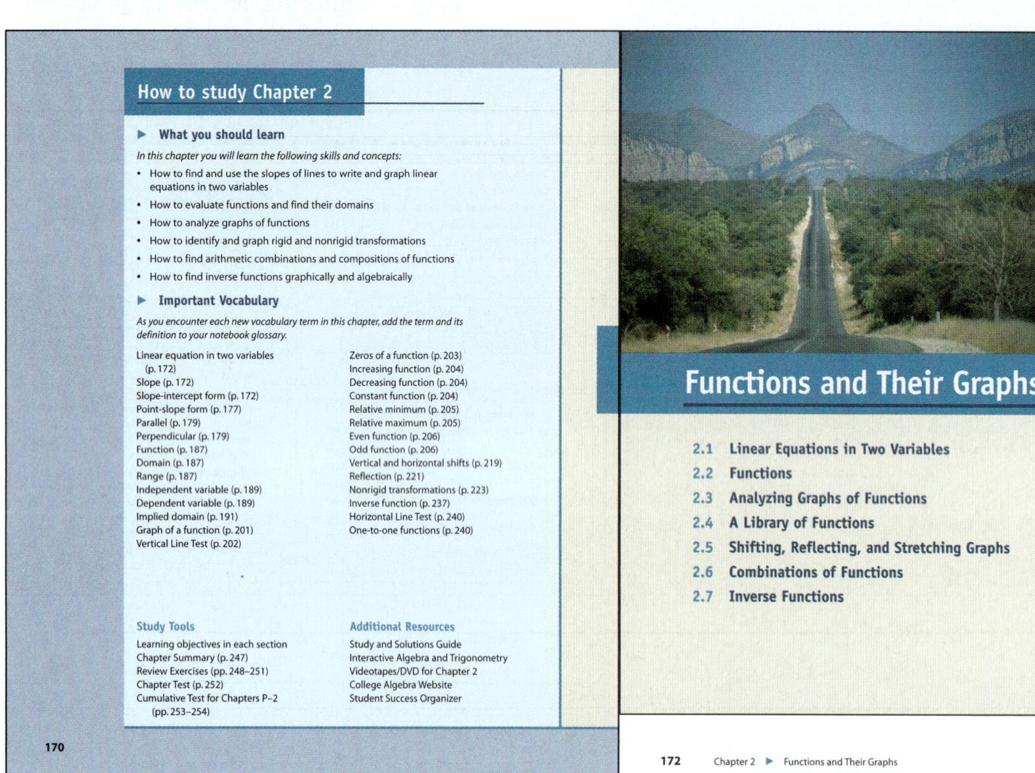

How to study Chapter 2

What you should learn

In this chapter you will learn the following skills and concepts:

- How to find and use the slopes of lines to write and graph linear equations in two variables
- How to evaluate functions and find their domains
- How to analyze graphs of functions
- How to identify and graph rigid and nonrigid transformations
- How to find arithmetic combinations and compositions of functions
- How to find inverse functions graphically and algebraically

Important Vocabulary

As you encounter each new vocabulary term in this chapter, add the term and its definition to your notebook glossary.

Linear equation in two variables (p. 172)
Slope (p. 172)
Slope-intercept form (p. 172)
Point-slope form (p. 177)
Parallel (p. 179)
Perpendicular (p. 179)
Function (p. 187)
Domain (p. 187)
Range (p. 187)
Independent variable (p. 189)
Dependent variable (p. 189)
Implied domain (p. 191)
Graph of a function (p. 201)
Vertical Line Test (p. 202)
Zeros of a function (p. 203)
Increasing function (p. 204)
Decreasing function (p. 204)
Constant function (p. 204)
Relative minimum (p. 205)
Relative maximum (p. 205)
Even function (p. 206)
Odd function (p. 206)
Vertical and horizontal shifts (p. 219)
Reflection (p. 221)
Nonrigid transformations (p. 223)
Inverse function (p. 237)
Horizontal Line Test (p. 240)
One-to-one functions (p. 240)

Study Tools

Learning objectives in each section
Chapter Summary (p. 247)
Review Exercises (pp. 248–251)
Chapter Test (p. 252)
Cumulative Test for Chapters P–2 (pp. 253–254)

Additional Resources

Study and Solutions Guide
Interactive Algebra and Trigonometry
Videotapes/DVD for Chapter 2
College Algebra Website
Student Success Organizer

170

Andreas Strenberg/Getty Images

2

Functions and Their Graphs

2.1 Linear Equations in Two Variables
2.2 Functions
2.3 Analyzing Graphs of Functions
2.4 A Library of Functions
2.5 Shifting, Reflecting, and Stretching Graphs
2.6 Combinations of Functions
2.7 Inverse Functions

"How to Study This Chapter"

The chapter-opening study guide includes: *What you should learn*, an objective-based overview of the main concepts of the chapter, *Important Vocabulary*, key mathematical terms integral to learning the concepts outlined in *What you should learn*, a list of *Study Tools*, additional study resources within the text chapter, and *Additional Resources*, text-specific supplemental resources available for each chapter.

Section Openers include: "What you should learn"

A list of section objectives outlining the main concepts to help students focus while reading through the section.

"Why you should learn it"

A real-life application or a reference to other branches of mathematics illustrates the relevance of the section's content. The real-life application is showcased in *Model It* found in the section exercise set.

172 Chapter 2 ▶ Functions and Their Graphs

2.1 Linear Equations in Two Variables

What you should learn

- How to use slope to graph linear equations in two variables
- How to find slopes of lines
- How to write linear equations in two variables
- How to use slope to identify parallel and perpendicular lines
- How to use linear equations in two variables to model and solve real-life problems

Why you should learn it

Linear equations in two variables can be used to model and solve real-life problems. For instance, in Exercise 119 on page 185, a linear equation is used to model the average monthly cellular phone bills for subscribers in the United States.

Dick Luria/Getty Images

Using Slope

The simplest mathematical model for relating two variables is the **linear equation in two variables** $y = mx + b$. The equation is called *linear* because its graph is a line. (In mathematics, the term *line* means *straight line.*) By letting $x = 0$, you can see that the line crosses the y-axis at $y = b$, as shown in Figure 2.1. In other words, the y-intercept is $(0, b)$. The steepness or slope of the line is m.

$$y = mx + b$$

Slope — y-Intercept

The **slope** of a nonvertical line is the number of units the line rises (or falls) vertically for each unit of horizontal change from left to right, as shown in Figure 2.1 and Figure 2.2.

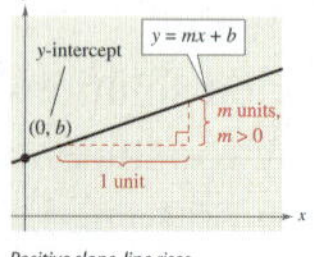

Positive slope, line rises.
FIGURE 2.1

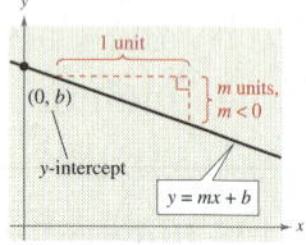

Negative slope, line falls.
FIGURE 2.2

A linear equation that is written in the form $y = mx + b$ is said to be written in **slope-intercept form.**

The Slope-Intercept Form of the Equation of a Line

The graph of the equation

$$y = mx + b$$

is a line whose slope is m and whose y-intercept is $(0, b)$.

Exploration

Use a graphing utility to compare the slopes of the lines $y = mx$ where $m = 0.5$, 1, 2, and 4. Which line rises most quickly? Now, let $m = -0.5$, -1, -2, and -4. Which line falls most quickly? Use a square setting to obtain a true geometric perspective. What can you conclude about the slope and the "rate" at which the line rises or falls?

The icon identifies examples and concepts related to features of the Learning Tools CD-ROM and the *Interactive* and *Internet* versions of this text. For more details see the chart on pages *xix–xxiii*.

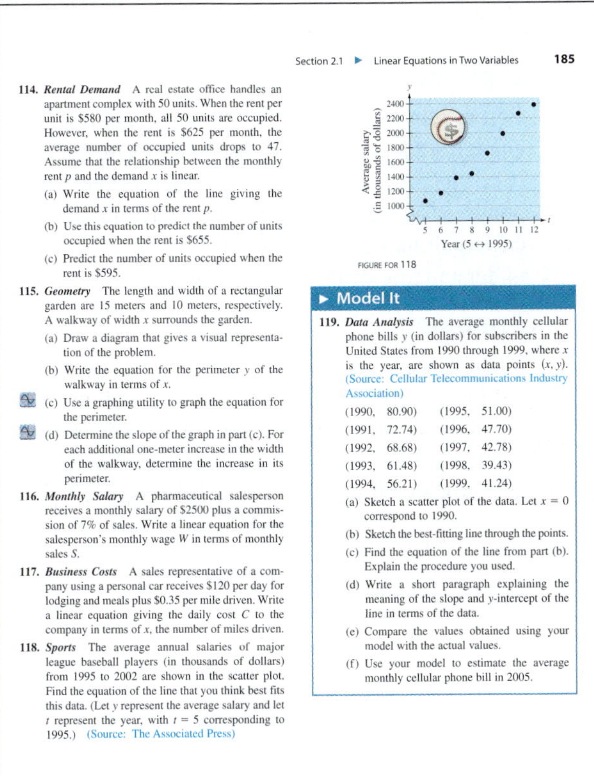

114. *Rental Demand* A real estate office handles an apartment complex with 50 units. When the rent per unit is \$580 per month, all 50 units are occupied. However, when the rent is \$625 per month, the average number of occupied units drops to 47. Assume that the relationship between the monthly rent p and the demand x is linear.

(a) Write the equation of the line giving the demand x in terms of the rent p.

(b) Use this equation to predict the number of units occupied when the rent is \$655.

(c) Predict the number of units occupied when the rent is \$595.

115. *Geometry* The length and width of a rectangular garden are 15 meters and 10 meters, respectively. A walkway of width x surrounds the garden.

(a) Draw a diagram that gives a visual representation of the problem.

(b) Write the equation for the perimeter y of the walkway in terms of x.

(c) Use a graphing utility to graph the equation for the perimeter.

(d) Determine the slope of the graph in part (c). For each additional one-meter increase in the width of the walkway, determine the increase in its perimeter.

116. *Monthly Salary* A pharmaceutical salesperson receives a monthly salary of \$2500 plus a commission of 7% of sales. Write a linear equation for the salesperson's monthly wage W in terms of monthly sales S.

117. *Business Costs* A sales representative of a company using a personal car receives \$120 per day for lodging and meals plus \$0.35 per mile driven. Write a linear equation giving the daily cost C to the company in terms of x, the number of miles driven.

118. *Sports* The average annual salaries of major league baseball players (in thousands of dollars) from 1995 to 2002 are shown in the scatter plot. Find the equation of the line that you think best fits this data. (Let y represent the average salary and let t represent the year, with $t = 5$ corresponding to 1995.) (Source: The Associated Press)

FIGURE FOR 118

▶ Model It

119. *Data Analysis* The average monthly cellular phone bills y (in dollars) for subscribers in the United States from 1990 through 1999, where x is the year, are shown as data points (x, y). (Source: Cellular Telecommunications Industry Association)

(1990, 80.90) (1995, 51.00)
(1991, 72.74) (1996, 47.70)
(1992, 68.68) (1997, 42.78)
(1993, 61.48) (1998, 39.43)
(1994, 56.21) (1999, 41.24)

(a) Sketch a scatter plot of the data. Let $x = 0$ correspond to 1990.

(b) Sketch the best-fitting line through the points.

(c) Find the equation of the line from part (b). Explain the procedure you used.

(d) Write a short paragraph explaining the meaning of the slope and y-intercept of the line in terms of the data.

(e) Compare the values obtained using your model with the actual values.

(f) Use your model to estimate the average monthly cellular phone bill in 2005.

Chapter Summary

▶ *What* did you learn?

	Review Exercises
Section 2.1	
☐ How to use slope to graph linear equations in two variables	1–14
☐ How to find slopes of lines	15–18
☐ How to write linear equations in two variables	19–26
☐ How to use slope to identify parallel and perpendicular lines	27, 28
☐ How to use linear equations in two variables to model and solve real-life problems	29–32
Section 2.2	
☐ How to determine whether relations between two variables are functions	33–38
☐ How to use function notation and evaluate functions	39–42
☐ How to find the domains of functions	43–48
☐ How to find the difference quotients	49, 50
☐ How to use functions to model and solve real-life problems	51–54
Section 2.3	
☐ How to use the Vertical Line Test for functions	55–58
☐ How to find the zeros of functions	59–62
☐ How to determine intervals on which functions are increasing or decreasing	63, 64
☐ How to identify even and odd functions	65–68
Section 2.4	
☐ How to identify and graph linear, squaring, cubic, square root, reciprocal, step, and other piecewise-defined functions	69–82
☐ How to recognize graphs of common functions	83, 84
Section 2.5	
☐ How to use vertical and horizontal shifts to sketch graphs of functions	85–88
☐ How to use reflections to sketch graphs of functions	89–94
☐ How to use nonrigid transformations to sketch graphs of functions	95–98
Section 2.6	
☐ How to add, subtract, multiply, and divide functions	99, 100
☐ How to find the composition of one function with another function	101–104
☐ How to use combinations of functions to model and solve real-life problems	105, 106
Section 2.7	
☐ How to find inverse functions informally and verify that two functions are inverse functions of each other	107, 108
☐ How to use graphs to determine whether functions have inverse functions	109, 110
☐ How to use the Horizontal Line Test to determine if functions are one-to-one	111–114
☐ How to find inverse functions algebraically	115–120

NEW! Model It

Often involving real-life data, these multi-part applications, referenced in *Why you should learn it*, offer students the opportunity to generate and analyze mathematical models.

"What did you learn?" Chapter Summary

The chapter summary provides a concise, section-by-section review of the section objectives. These objectives are correlated to the chapter Review Exercises allowing students to identify sections and concepts needing further review and study.

Review Exercises

Following the chapter summary, the Review Exercises provide additional practice and review of chapter concepts. The Review Exercises are organized by section and keyed directly to the section objectives listed in the chapter summary.

Additional Student Success Tools include point-of-use *Study Tips* and *Chapter* and *Cumulative Tests*.

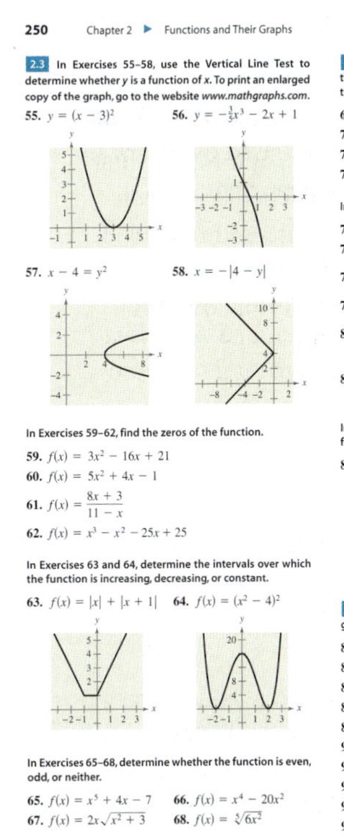

2.3 In Exercises 55–58, use the Vertical Line Test to determine whether y is a function of x. To print an enlarged copy of the graph, go to the website *www.mathgraphs.com*.

55. $y = (x - 3)^2$ **56.** $y = -\frac{3}{5}x^3 - 2x + 1$

57. $x - 4 = y^2$ **58.** $x = -|4 - y|$

In Exercises 59–62, find the zeros of the function.

59. $f(x) = 3x^2 - 16x + 21$

60. $f(x) = 5x^2 + 4x - 1$

61. $f(x) = \dfrac{8x + 3}{11 - x}$

62. $f(x) = x^3 - x^2 - 25x + 25$

In Exercises 63 and 64, determine the intervals over which the function is increasing, decreasing, or constant.

63. $f(x) = |x| + |x + 1|$ **64.** $f(x) = (x^2 - 4)^2$

In Exercises 65–68, determine whether the function is even, odd, or neither.

65. $f(x) = x^5 + 4x - 7$ **66.** $f(x) = x^4 - 20x^2$

67. $f(x) = 2x\sqrt{x^2 + 3}$ **68.** $f(x) = \sqrt[5]{6x^2}$

2.4 In Exercises 69–72, write the linear function f such that it has the indicated function values. Sketch a graph of the function.

69. $f(2) = -6,\ f(-1) = 3$

70. $f(0) = -5,\ f(4) = -8$

71. $f(-\frac{4}{5}) = 2,\ f(\frac{11}{5}) = 7$

72. $f(3.3) = 5.6,\ f(-4.7) = -1.4$

In Exercises 73–82, graph the function.

73. $f(x) = 3 - x^2$ **74.** $h(x) = x^3 - 2$

75. $f(x) = -\sqrt{x}$ **76.** $f(x) = \sqrt{x + 1}$

77. $g(x) = \dfrac{3}{x}$ **78.** $g(x) = \dfrac{1}{x + 5}$

79. $f(x) = [\![x]\!] - 2$ **80.** $g(x) = [\![x + 4]\!]$

81. $f(x) = \begin{cases} 5x - 3, & x \ge -1 \\ -4x + 5, & x < -1 \end{cases}$

82. $f(x) = \begin{cases} x^2 - 2, & x < -2 \\ 5, & -2 \le x \le 0 \\ 8x - 5, & x > 0 \end{cases}$

In Exercises 83 and 84, identify the transformed common function shown in the graph.

83. **84.**

2.5 In Exercises 85–98, identify the transformation of the graph of f and sketch the graph of h.

85. $f(x) = x^2,\quad h(x) = x^2 - 9$

86. $f(x) = x^3,\quad h(x) = (x - 2)^3 + 2$

87. $f(x) = \sqrt{x},\quad h(x) = \sqrt{x - 7}$

88. $f(x) = |x|,\quad h(x) = |x + 3| - 5$

89. $f(x) = x^2,\quad h(x) = -(x + 3)^2 + 1$

90. $f(x) = x^3,\quad h(x) = -(x - 5)^3 - 5$

91. $f(x) = [\![x]\!],\quad h(x) = -[\![x]\!] + 6$

92. $f(x) = \sqrt{x},\quad h(x) = -\sqrt{x + 1} + 9$

93. $f(x) = |x|,\quad h(x) = -|-x + 4| + 6$

94. $f(x) = x^2,\quad h(x) = -(x + 1)^2 - 3$

FEATURES

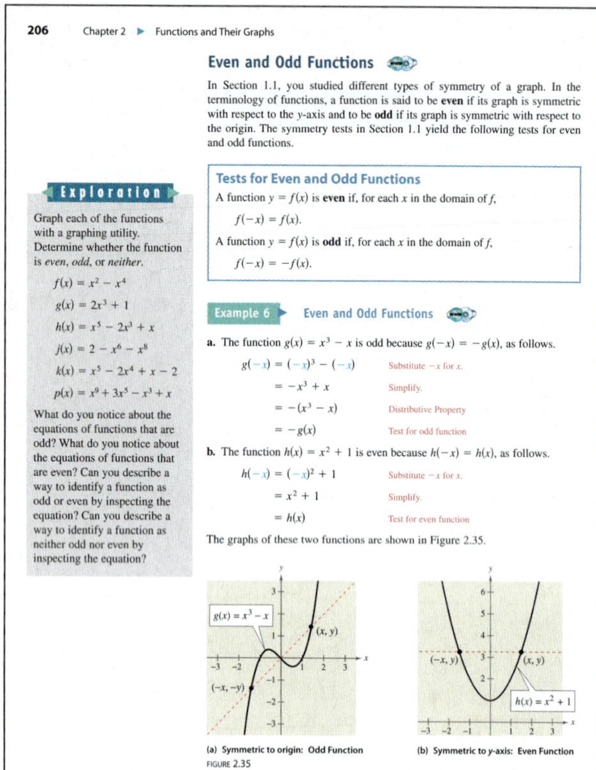
206 Chapter 2 ▶ Functions and Their Graphs

Even and Odd Functions

In Section 1.1, you studied different types of symmetry of a graph. In the terminology of functions, a function is said to be **even** if its graph is symmetric with respect to the y-axis and to be **odd** if its graph is symmetric with respect to the origin. The symmetry tests in Section 1.1 yield the following tests for even and odd functions.

Exploration

Graph each of the functions with a graphing utility. Determine whether the function is *even, odd,* or *neither.*

$f(x) = x^2 - x^4$

$g(x) = 2x^3 + 1$

$h(x) = x^5 - 2x^3 + x$

$j(x) = 2 - x^6 - x^8$

$k(x) = x^5 - 2x^4 + x - 2$

$p(x) = x^9 + 3x^5 - x^3 + x$

What do you notice about the equations of functions that are odd? What do you notice about the equations of functions that are even? Can you describe a way to identify a function as odd or even by inspecting the equation? Can you describe a way to identify a function as neither odd nor even by inspecting the equation?

Tests for Even and Odd Functions

A function $y = f(x)$ is **even** if, for each x in the domain of f,

$f(-x) = f(x)$.

A function $y = f(x)$ is **odd** if, for each x in the domain of f,

$f(-x) = -f(x)$.

Example 6 Even and Odd Functions

a. The function $g(x) = x^3 - x$ is odd because $g(-x) = -g(x)$, as follows.

$g(-x) = (-x)^3 - (-x)$ Substitute $-x$ for x.

$= -x^3 + x$ Simplify.

$= -(x^3 - x)$ Distributive Property

$= -g(x)$ Test for odd function

b. The function $h(x) = x^2 + 1$ is even because $h(-x) = h(x)$, as follows.

$h(-x) = (-x)^2 + 1$ Substitute $-x$ for x.

$= x^2 + 1$ Simplify.

$= h(x)$ Test for even function

The graphs of these two functions are shown in Figure 2.35.

$g(x) = x^3 - x$; (x, y); $(-x, -y)$

(a) Symmetric to origin: Odd Function

$h(x) = x^2 + 1$; $(-x, y)$; (x, y)

(b) Symmetric to y-axis: Even Function

FIGURE 2.35

Exploration

Before introducing selected topics, *Explorations* engage students in active discovery of mathematical concepts and relationships, often through the power of technology, while strengthening their critical thinking skills and developing an intuitive understanding of theoretical concepts.

Examples

Each example was carefully chosen to illustrate a particular mathematical concept or problem solving skill. Every example contains step-by-step solutions, most with side-by-side explanations that lead students through the solution process.

Technology

Point-of-use instructions for graphing utilities appear in the margin. Emphasis is placed on using technology as a tool for visualizing mathematical concepts, for verifying solutions, and for facilitating mathematical computation. The use of technology is optional and this feature and related exercises, identified by the icon , can be omitted without loss of continuity in coverage of topics.

Algebra of Calculus

Special emphasis is given to the algebraic techniques used in calculus. Algebra of Calculus examples and exercises are integrated throughout the text and are identified by the symbol ∫.

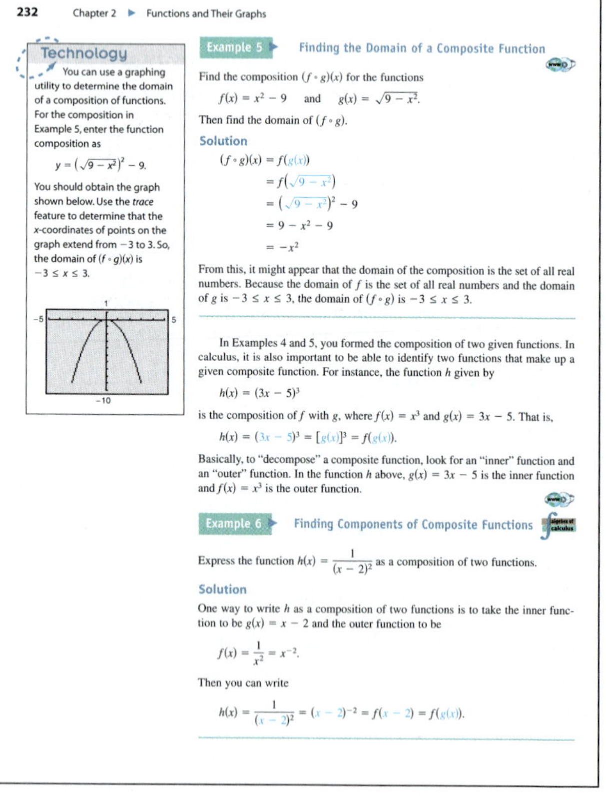
232 Chapter 2 ▶ Functions and Their Graphs

Technology

You can use a graphing utility to determine the domain of a composition of functions. For the composition in Example 5, enter the function composition as

$y = (\sqrt{9 - x^2})^2 - 9.$

You should obtain the graph shown below. Use the *trace* feature to determine that the x-coordinates of points on the graph extend from -3 to 3. So, the domain of $(f \circ g)(x)$ is $-3 \le x \le 3$.

Example 5 Finding the Domain of a Composite Function

Find the composition $(f \circ g)(x)$ for the functions

$f(x) = x^2 - 9$ and $g(x) = \sqrt{9 - x^2}$.

Then find the domain of $(f \circ g)$.

Solution

$(f \circ g)(x) = f(g(x))$

$= f(\sqrt{9 - x^2})$

$= (\sqrt{9 - x^2})^2 - 9$

$= 9 - x^2 - 9$

$= -x^2$

From this, it might appear that the domain of the composition is the set of all real numbers. Because the domain of f is the set of all real numbers and the domain of g is $-3 \le x \le 3$, the domain of $(f \circ g)$ is $-3 \le x \le 3$.

In Examples 4 and 5, you formed the composition of two given functions. In calculus, it is also important to be able to identify two functions that make up a given composite function. For instance, the function h given by

$h(x) = (3x - 5)^3$

is the composition of f with g, where $f(x) = x^3$ and $g(x) = 3x - 5$. That is,

$h(x) = (3x - 5)^3 = [g(x)]^3 = f(g(x))$.

Basically, to "decompose" a composite function, look for an "inner" function and an "outer" function. In the function h above, $g(x) = 3x - 5$ is the inner function and $f(x) = x^3$ is the outer function.

Example 6 Finding Components of Composite Functions

Express the function $h(x) = \dfrac{1}{(x - 2)^2}$ as a composition of two functions.

Solution

One way to write h as a composition of two functions is to take the inner function to be $g(x) = x - 2$ and the outer function to be

$f(x) = \dfrac{1}{x^2} = x^{-2}$.

Then you can write

$h(x) = \dfrac{1}{(x - 2)^2} = (x - 2)^{-2} = f(x - 2) = f(g(x))$.

P.S. Problem Solving

1. As a salesperson, you receive a monthly salary of \$2000, plus a commission of 7% of sales. You are offered a new job at \$2300 per month, plus a commission of 5% of sales.

(a) Write a linear equation for your current monthly wage W_1 in terms of your monthly sales S.

(b) Write a linear equation for the monthly wage W_2 of your new job offer in terms of the monthly sales S.

(c) Use a graphing utility to graph both equations in the same viewing window. Find the point of intersection. What does it signify?

(d) You think you can sell \$20,000 per month. Should you change jobs? Explain.

2. For the numbers 2 through 9 on a telephone keypad (see figure), create two relations: one mapping numbers onto letters, and the other mapping letters onto numbers. Are both relations functions? Explain.

3. What can be said about the sum and difference of each of the following?

(a) Two even functions

(b) Two odd functions

(c) An odd function and an even function

4. The two functions

$f(x) = x$ and $g(x) = -x$

are their own inverse functions. Graph each function and explain why this is true. Graph other linear functions that are their own inverse functions. Find a general formula for a family of linear functions that are their own inverse functions.

5. Prove that a function of the following form is even.

$y = a_{2n}x^{2n} + a_{2n-2}x^{2n-2} + \cdots + a_2x^2 + a_0$

6. A miniature golf professional is trying to make a hole-in-one on the miniature golf green shown. A coordinate plane is placed over the golf green. The golf ball is at the point (2.5, 2) and the hole is at the point (9.5, 2). The professional wants to bank the ball off the side wall of the green at the point (x, y). Find the coordinates of the point (x, y). Then write an equation for the path of the ball.

7. At 2:00 P.M. on April 11, 1912, the *Titanic* left Cobh, Ireland, on her voyage to New York City. At 11:40 P.M. on April 14, the *Titanic* struck an iceberg and sank, having covered only about 2100 miles of the approximately 3400-mile trip.

(a) What was the total length of the *Titanic's* voyage in hours?

(b) What was the *Titanic's* average speed in miles per hour?

(c) Write a function relating the *Titantic's* distance from New York City and the number of hours traveled. Find the domain and range of the function.

(d) Graph the function from part (c).

8. Consider the functions $f(x) = 4x$ and $g(x) = x + 6$.

(a) Find $(f \circ g)(x)$.

(b) Find $(f \circ g)^{-1}(x)$.

(c) Find $f^{-1}(x)$ and $g^{-1}(x)$.

(d) Find $(g^{-1} \circ f^{-1})(x)$ and compare the result with that of part (b).

(e) Repeat parts (a) through (d) for $f(x) = x^3 + 1$ and $g(x) = 2x$.

(f) Write two one-to-one functions f and g, and repeat parts (a) through (d) for these functions.

(g) Make a conjecture about $(f \circ g)^{-1}(x)$ and $(g^{-1} \circ f^{-1})(x)$.

256

9. You are in a boat 2 miles from the nearest point on the coast. You are to travel to a point Q, 3 miles down the coast and 1 mile inland (see figure). You can row at 2 miles per hour and walk at 4 miles per hour.

(a) Write the total time T of the trip as a function of x.

(b) Determine the domain of the function.

(c) Use a graphing utility to graph the function. Be sure to choose an appropriate viewing window.

(d) Use the *zoom* and *trace* features to find the value of x that minimizes T.

(e) Write a brief paragraph interpreting these values.

10. The Heaviside function $H(x)$ is widely used in engineering applications. (See figure.) To print an enlarged copy of the graph, go to the website *www.mathgraphs.com*.

$$H(x) = \begin{cases} 1, & x \ge 0 \\ 0, & x < 0 \end{cases}$$

Sketch the graph of each function by hand.

(a) $H(x) - 2$ (b) $H(x - 2)$ (c) $-H(x)$

(d) $H(-x)$ (e) $\frac{1}{2}H(x)$ (f) $-H(x - 2) + 2$

11. Let $f(x) = \dfrac{1}{1 - x}$.

(a) What are the domain and range of f?

(b) Find $f(f(x))$. What is the domain of this function?

(c) Find $f(f(f(x)))$. Is the graph a line? Why or why not?

12. Show that the Associative Property holds for compositions of functions—that is,

$(f \circ (g \circ h))(x) = ((f \circ g) \circ h)(x)$.

13. Consider the graph of the function f shown in the figure. Use this graph to sketch the graph of each function. To print an enlarged copy of the graph, go to the website *www.mathgraphs.com*.

(a) $f(x + 1)$ (b) $f(x) + 1$ (c) $2f(x)$ (d) $f(-x)$

(e) $-f(x)$ (f) $|f(x)|$ (g) $f(|x|)$

14. Use the graphs of f and f^{-1} to complete each table of function values.

(a)

x	$f(f^{-1}(x))$
-4	
-2	
0	
4	

(b)

x	$(f + f^{-1})(x)$
-3	
-2	
0	
1	

(c)

x	$(f \cdot f^{-1})(x)$
-3	
-2	
0	
1	

(d)

x	$\lvert f^{-1}(x) \rvert$
-4	
-3	
0	
4	

257

FEATURES

NEW! P.S. Problem Solving

Each chapter concludes with a collection of thought-provoking and challenging exercises that further explore and expand upon the chapter concepts. These exercises have unusual characteristics that set them apart from traditional text exercises.

NEW! Proofs in Mathematics

At the end of every chapter, Proofs in Mathematics emphasizes the importance of proofs in mathematics. Proofs of important mathematical properties and theorems are presented as well as discussions of various proof techniques.

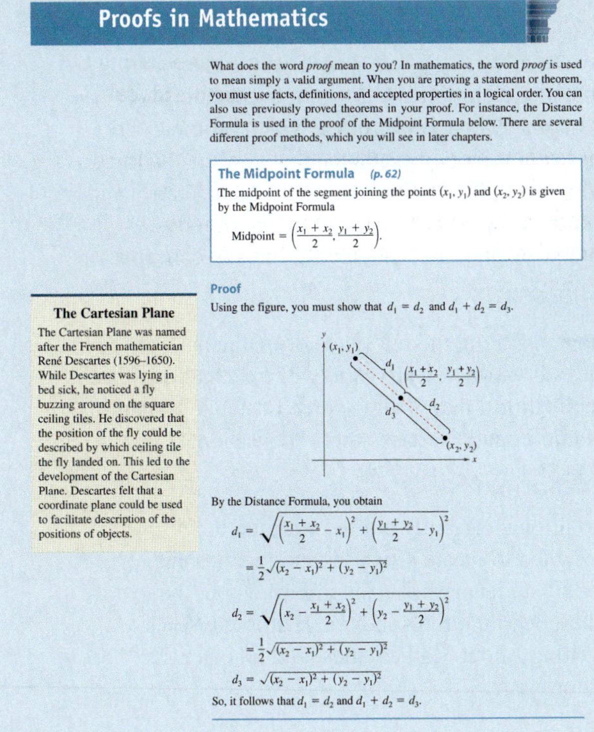

▶ Proofs in Mathematics 73

Proofs in Mathematics

What does the word *proof* mean to you? In mathematics, the word *proof* is used to mean simply a valid argument. When you are proving a statement or theorem, you must use facts, definitions, and accepted properties in a logical order. You can also use previously proved theorems in your proof. For instance, the Distance Formula is used in the proof of the Midpoint Formula below. There are several different proof methods, which you will see in later chapters.

The Midpoint Formula *(p. 62)*

The midpoint of the segment joining the points (x_1, y_1) and (x_2, y_2) is given by the Midpoint Formula

$$\text{Midpoint} = \left(\frac{x_1 + x_2}{2}, \frac{y_1 + y_2}{2}\right).$$

The Cartesian Plane

The Cartesian Plane was named after the French mathematician René Descartes (1596–1650). While Descartes was lying in bed sick, he noticed a fly buzzing around on the square ceiling tiles. He discovered that the position of the fly could be described by which ceiling tile the fly landed on. This led to the development of the Cartesian Plane. Descartes felt that a coordinate plane could be used to facilitate description of the positions of objects.

Proof

Using the figure, you must show that $d_1 = d_2$ and $d_1 + d_2 = d_3$.

By the Distance Formula, you obtain

$$d_1 = \sqrt{\left(\frac{x_1 + x_2}{2} - x_1\right)^2 + \left(\frac{y_1 + y_2}{2} - y_1\right)^2}$$

$$= \frac{1}{2}\sqrt{(x_2 - x_1)^2 + (y_2 - y_1)^2}$$

$$d_2 = \sqrt{\left(x_2 - \frac{x_1 + x_2}{2}\right)^2 + \left(y_2 - \frac{y_1 + y_2}{2}\right)^2}$$

$$= \frac{1}{2}\sqrt{(x_2 - x_1)^2 + (y_2 - y_1)^2}$$

$$d_3 = \sqrt{(x_2 - x_1)^2 + (y_2 - y_1)^2}$$

So, it follows that $d_1 = d_2$ and $d_1 + d_2 = d_3$.

3.2 Exercises

In Exercises 1–8, match the polynomial function with its graph. [The graphs are labeled (a), (b), (c), (d), (e), (f), (g), and (h).]

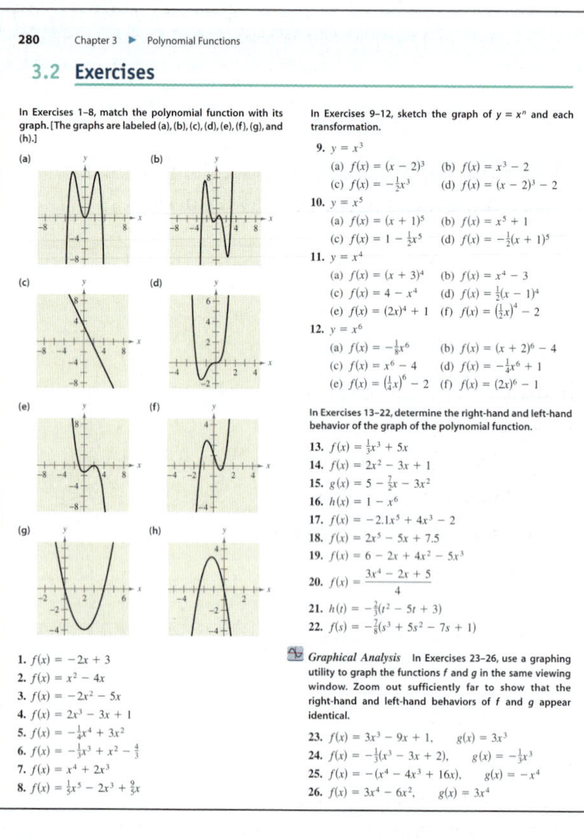

1. $f(x) = -2x + 3$
2. $f(x) = x^2 - 4x$
3. $f(x) = -2x^2 - 5x$
4. $f(x) = 2x^3 - 3x + 1$
5. $f(x) = -\frac{1}{4}x^4 + 3x^2$
6. $f(x) = -\frac{1}{3}x^3 + x^2 - \frac{4}{3}$
7. $f(x) = x^4 + 2x^3$
8. $f(x) = \frac{1}{5}x^5 - 2x^3 + \frac{9}{5}x$

In Exercises 9–12, sketch the graph of $y = x^n$ and each transformation.

9. $y = x^3$
   (a) $f(x) = (x - 2)^3$ (b) $f(x) = x^3 - 2$
   (c) $f(x) = -\frac{1}{2}x^3$ (d) $f(x) = (x - 2)^3 - 2$
10. $y = x^5$
   (a) $f(x) = (x + 1)^5$ (b) $f(x) = x^5 + 1$
   (c) $f(x) = 1 - \frac{1}{2}x^5$ (d) $f(x) = -\frac{1}{2}(x + 1)^5$
11. $y = x^4$
   (a) $f(x) = (x + 3)^4$ (b) $f(x) = x^4 - 3$
   (c) $f(x) = 4 - x^4$ (d) $f(x) = \frac{1}{2}(x - 1)^4$
   (e) $f(x) = (2x)^4 + 1$ (f) $f(x) = \left(\frac{1}{2}x\right)^4 - 2$
12. $y = x^6$
   (a) $f(x) = -\frac{1}{8}x^6$ (b) $f(x) = (x + 2)^6 - 4$
   (c) $f(x) = x^6 - 4$ (d) $f(x) = -\frac{1}{4}x^6 + 1$
   (e) $f(x) = \left(\frac{1}{4}x\right)^6 - 2$ (f) $f(x) = (2x)^6 - 1$

In Exercises 13–22, determine the right-hand and left-hand behavior of the graph of the polynomial function.

13. $f(x) = \frac{1}{3}x^3 + 5x$
14. $f(x) = 2x^2 - 3x + 1$
15. $g(x) = 5 - \frac{7}{2}x - 3x^2$
16. $h(x) = 1 - x^6$
17. $f(x) = -2.1x^5 + 4x^3 - 2$
18. $f(x) = 2x^5 - 5x + 7.5$
19. $f(x) = 6 - 2x + 4x^2 - 5x^3$
20. $f(x) = \dfrac{3x^4 - 2x + 5}{4}$
21. $h(t) = -\frac{2}{3}(t^2 - 5t + 3)$
22. $f(s) = -\frac{7}{8}(s^3 + 5s^2 - 7s + 1)$

Graphical Analysis In Exercises 23–26, use a graphing utility to graph the functions f and g in the same viewing window. Zoom out sufficiently far to show that the right-hand and left-hand behaviors of f and g appear identical.

23. $f(x) = 3x^3 - 9x + 1, \quad g(x) = 3x^3$
24. $f(x) = -\frac{1}{3}(x^3 - 3x + 2), \quad g(x) = -\frac{1}{3}x^3$
25. $f(x) = -(x^4 - 4x^3 + 16x), \quad g(x) = -x^4$
26. $f(x) = 3x^4 - 6x^2, \quad g(x) = 3x^4$

Synthesis

True or False? In Exercises 93–95, determine whether the statement is true or false. Justify your answer.

93. A fifth-degree polynomial can have five turning points in its graph.
94. It is possible for a sixth-degree polynomial to have only one solution.
95. The graph of the function

$$f(x) = 2 + x - x^2 + x^3 - x^4 + x^5 + x^6 - x^7$$

rises to the left and falls to the right.

96. ***Graphical Analysis*** Describe a polynomial function that could represent the graph. (Indicate the degree of the function and the sign of its leading coefficient.)

(a) (b)

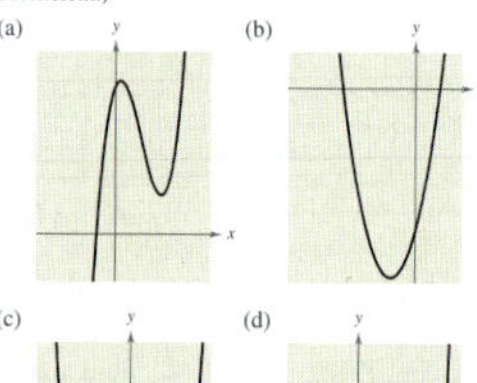

(c) (d)

97. ***Graphical Reasoning*** Sketch a graph of the function $f(x) = x^4$. Explain how the graph of g differs (if it does) from the graph of f. Determine whether g is odd, even, or neither.
   (a) $g(x) = f(x) + 2$ (b) $g(x) = f(x + 2)$
   (c) $g(x) = f(-x)$ (d) $g(x) = -f(x)$
   (e) $g(x) = f\left(\frac{1}{2}x\right)$ (f) $g(x) = \frac{1}{2}f(x)$
   (g) $g(x) = f\left(x^{3/4}\right)$ (h) $g(x) = (f \circ f)(x)$

98. ***Exploration*** Explore the transformations of the form $g(x) = a(x - h)^5 + k$.
   (a) Use a graphing utility to graph the functions

$$y_1 = -\frac{1}{3}(x - 2)^5 + 1$$

and

$$y_2 = \frac{3}{5}(x + 2)^5 - 3.$$

Determine whether the graphs are increasing or decreasing. Explain.

   (b) Will the graph of g always be increasing or decreasing? If so, is this behavior determined by a, h, or k? Explain.
   (c) Use a graphing utility to graph the function

$$H(x) = x^5 - 3x^3 + 2x + 1.$$

Use the graph and the result of part (b) to determine whether H can be written in the form $H(x) = a(x - h)^5 + k$. Explain.

Review

In Exercises 99–102, solve the equation by factoring.

99. $2x^2 - x - 28 = 0$
100. $3x^2 - 22x - 16 = 0$
101. $12x^2 + 11x - 5 = 0$
102. $x^2 + 24x + 144 = 0$

In Exercises 103–106, solve the equation by completing the square.

103. $x^2 - 2x - 21 = 0$
104. $x^2 - 8x + 2 = 0$
105. $2x^2 + 5x - 20 = 0$
106. $3x^2 + 4x - 9 = 0$

In Exercises 107–110, factor the expression completely.

107. $5x^2 + 7x - 24$
108. $6x^3 - 61x^2 + 10x$
109. $4x^4 - 7x^3 - 15x^2$
110. $y^3 + 216$

In Exercises 111–116, describe the transformation from a common function that occurs in the function. Then sketch its graph.

111. $f(x) = (x + 4)^2$
112. $f(x) = 3 - x^2$
113. $f(x) = \sqrt{x + 1} - 5$
114. $f(x) = 7 - \sqrt{x - 6}$
115. $f(x) = 2[\![x]\!] + 9$
116. $f(x) = 10 - \frac{1}{3}[\![x + 3]\!]$

Exercise

A hallmark feature of the text, the exercise sets contain a variety of computational, conceptual, and applied problems. Each section exercise set contains *Synthesis* exercises, which promote further exploration of mathematical concepts, critical thinking skills, and writing about mathematics and *Review* exercises, which provide continuous review of previously learned skills and concepts.

Applications

Demonstrating the relevance of mathematics to the real world, a wide variety of practical, real-life applications, many with sourced data, are found in examples and exercises throughout the text.

Additional Features

Additional carefully crafted learning tools designed to create a rich learning environment for all students can be found throughout the text. These learning tools include Historical Notes, Writing About Mathematics, and an extensive art program.

83. ***Height of a Ball*** The height y (in feet) of a ball thrown by a child is

$$y = -\frac{1}{12}x^2 + 2x + 4$$

where x is the horizontal distance (in feet) from the point at which the ball is thrown (see figure).
   (a) How high is the ball when it leaves the child's hand? (*Hint:* Find y when $x = 0$.)
   (b) What is the maximum height of the ball?
   (c) How far from the child does the ball strike the ground?

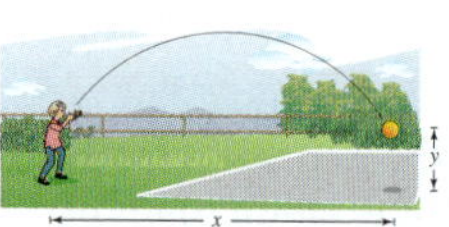

84. ***Path of a Diver*** The path of a diver is

$$y = -\frac{4}{9}x^2 + \frac{24}{9}x + 12$$

where y is the height (in feet) and x is the horizontal distance from the end of the diving board (in feet). What is the maximum height of the diver?

85. ***Graphical Analysis*** From 1960 to 2000, the per capita consumption C of cigarettes by Americans (age 18 and older) can be modeled by

$$C = 4258 + 6.5t - 1.62t^2, \quad 0 \le t \le 40$$

where t is the year, with $t = 0$ corresponding to 1960. (Source: *Tobacco Situation and Outlook Yearbook*)
   (a) Use a graphing utility to graph the model.
   (b) Use the graph of the model to approximate the maximum average annual consumption. Beginning in 1966, all cigarette packages were required by law to carry a health warning. Do you think the warning had any effect? Explain.
   (c) In 1990, the U.S. population (age 18 and over) was 185,105,441. Of those, about 63,423,167 were smokers. What was the average annual cigarette consumption *per smoker* in 1990? What was the average daily cigarette consumption *per smoker*?

▶ Model It

86. ***Data Analysis*** The numbers y (in thousands) of hairdressers and cosmetologists in the United States for the years 1995 through 2000 are shown in the table. (Source: U.S. Bureau of Labor Statistics)

Year	Number of hairdressers and cosmetologists, y
1995	750
1996	737
1997	748
1998	763
1999	784
2000	820

   (a) Use a graphing utility to create a scatter plot of the data. Let x represent the year, with $x = 5$ corresponding to 1995.
   (b) Use the *regression* feature of a graphing utility to find a quadratic model for the data.
   (c) Use a graphing utility to graph the model in the same viewing window as the scatter plot. How well does the model fit the data?
   (d) Use the *trace* feature of the graphing utility to approximate the year in which the number of hairdressers and cosmetologists was the least.
   (e) Use the model to predict the number of hairdressers and cosmetologists in 2005.

87. ***Wind Drag*** The number of horsepower y required to overcome wind drag on an automobile is approximated by

$$y = 0.002s^2 + 0.005s - 0.029, \quad 0 \le s \le 100$$

where s is the speed of the car (in miles per hour).
   (a) Use a graphing utility to graph the function.
   (b) Graphically estimate the maximum speed of the car if the power required to overcome wind drag is not to exceed 10 horsepower. Verify your estimate analytically.

Program Components

Algebra and Trigonometry, Student Edition
Algebra and Trigonometry, Instructor's Annotated Edition

Interactive Algebra and Trigonometry 3.0 CD-ROM (can be used alone or with the printed textbook)
Internet Algebra and Trigonometry 3.0 (can be used alone or with the printed textbook)

Additional Resources

Student Resources

Student Success Organizer

Study and Solutions Guide
by Dianna L. Zook, (Indiana University/Purdue University–Fort Wayne)

Student Technology Resources

Instructional Videotapes for Graphing Calculators
by Dana Mosely

Learning Tools Student CD-ROM

Smarthinking™.com live online tutoring

Instructional DVDs by Dana Mosely

Instructional Videotapes for Graphing Calculators
by Dana Mosely

Interactive Algebra and Trigonometry 3.0 CD-ROM

Internet Algebra and Trigonometry 3.0

HM eduSpace website

BlackBoard Course Cartridge
WebCT e-pack

Textbook website (math.college.hmco.com)

Instructor Resources

Instructor Success Organizer

Complete Solutions Guide
by Dianna L. Zook, (Indiana University/Purdue University–Fort Wayne)

Instructor's Annotated Edition

Test Item File

Instructor Technology Resources

HMClassPrep™ Instructor's CD-ROM

HM Testing 6.03

PowerPoint Presentations

Instructional Videotapes by Dana Mosely
(ideal for libraries and resource centers)

Interactive Algebra and Trigonometry 3.0 CD-ROM

Internet Algebra and Trigonometry 3.0

HM eduSpace website

BlackBoard Course Cartridge
WebCT e-pack

Textbook website (math.college.hmco.com)

For more information on these and other resources available, visit our website at **math.college.hmco.com.**

Algebra and Trigonometry, Sixth Edition
Learning Tools Student CD-ROM

The Learning Tools Student CD-ROM that accompanies the text provides students with an unprecedented quantity of support materials and resources that help bring mathematics to life with motion and sound. These electronic learning tools are separated into three components described below. The CD-ROM also provides access to MathGraphs, ACE Practice Tests, and SMARTHINKING, the online tutoring center.

Study the Lesson

The Glossary of Terms provides a comprehensive list of important mathematical terms for each chapter with a short definition of each term.

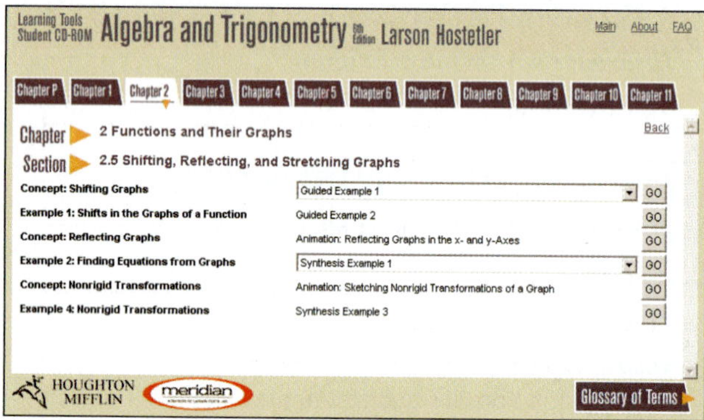

Review and Practice

- **Guided Examples** provide a full range of support by walking students step-by-step through problems that relate to a specific concept in the text.
- **Synthesis Examples** require the use of more than one concept from a section and encourage students to work through a solution of a problem one step at a time.

Visualize and Extend the Concepts

- **Animations** use motion and sound to explain concepts and can be played, paused, stopped, and replayed as many times as the student desires.
- **Simulations** encourage students to explore mathematical concepts experimentally.
- **Editable Graph Explorations** engage students in active discovery of mathematical concepts and relationships through the use of technology.

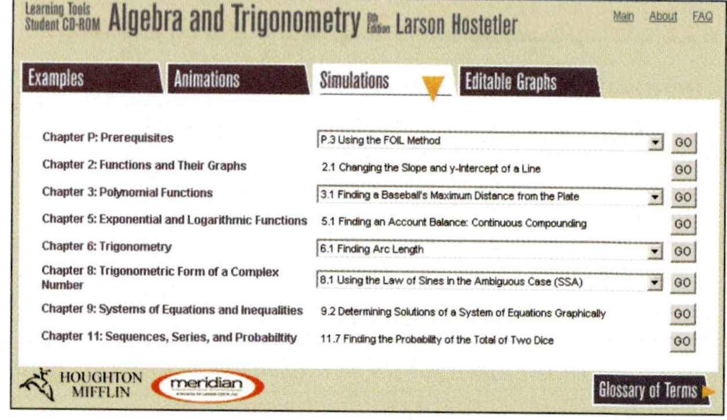

Algebra and Trigonometry, Sixth Edition
Learning Tools Student CD-ROM

Selected examples and concepts throughout the text are identified by the Learning Tools Student CD-ROM icon. The chart on this and the following pages indicates the feature(s) of the CD—Guided Example, Synthesis Example, Animation, Simulation, and Editable Graph Exploration—that corresponds to the example or concept.

Chapter	Section	Example/Concept	Guided Example	Synthesis Example	Animation	Simulation	Editable Graph
P	1	Real Numbers	✓				
P	1	Ordering Real Numbers	✓				
P	1	Example 1	✓		✓		
P	1	Example 2	✓				
P	1	Absolute Value and Distance		✓			
P	1	Examples 4, 5	✓	✓			
P	1	Algebraic Expressions	✓		✓		
P	1	Basic Rules of Algebra	✓				
P	2	Examples 1, 2	✓	✓			
P	2	Scientific Notation	✓				
P	2	Examples 3–6, 9–15	✓				
P	2	Example 8		✓			
P	3	Examples 1, 2, 4, 6–8	✓				
P	3	Operations with Polynomials		✓			
P	3	Example 3	✓			✓	
P	3	Example 5	✓	✓			
P	3	Example 9				✓	
P	4	Examples 2–4, 7, 9–10	✓				
P	4	Example 5		✓			
P	4	Example 8	✓	✓	✓		
P	5	Examples 1–6, 8	✓				
P	5	Example 7	✓	✓			
P	6	Algebraic Errors to Avoid	✓				
P	6	Examples 3–6	✓				
P	7	Examples 1, 3, 4, 7	✓				
P	7	Examples 2, 8			✓		
P	7	Example 6		✓			
1	1	The Graph of an Equation	✓				
1	1	Example 1					✓
1	1	Example 2			✓		
1	1	Example 3	✓				✓
1	1	Symmetry			✓		
1	1	Examples 4, 7	✓				
1	1	Example 5		✓	✓		
1	1	Example 6		✓			✓
1	2	Equations and Solutions of Equations	✓	✓			
1	2	Examples 2–4	✓				
1	2	Equations That Lead to Linear Equations		✓			
1	3	Introduction to Problem Solving	✓				
1	3	Examples 1, 2	✓	✓			
1	3	Examples 5, 7	✓				

RESOURCES

Chapter	Section	Example/Concept	Guided Example	Synthesis Example	Animation	Simulation	Editable Graph
1	3	Common Formulas		✓			
1	4	Factoring	✓				
1	4	Examples 1, 2	✓	✓			
1	4	Example 3	✓	✓	✓		
1	4	The Quadratic Formula	✓				
1	4	Example 4	✓		✓		
1	4	Examples 5–7	✓				
1	5	The Imaginary Unit *i*	✓				
1	5	Examples 1, 3, 4, 6	✓				
1	5	Examples 2, 5	✓	✓			
1	6	Examples 1–3, 5–7	✓				
1	6	Example 4	✓	✓			
1	6	Applications	✓				
1	7	Examples 2–4	✓	✓			
1	7	Example 5	✓				
1	8	Example 1	✓	✓			
1	8	Examples 2, 4–6	✓				
2	1	Using Slope	✓			✓	
2	1	Example 1		✓			
2	1	Examples 4, 5	✓	✓			
2	1	Examples 6, 7	✓				
2	1	Parallel and Perpendicular Lines	✓				
2	2	Examples 2, 6, 7, 9	✓				
2	2	Examples 3, 5	✓	✓			
2	3	Examples 1, 2	✓				
2	3	Examples 4, 6	✓	✓			
2	3	Even and Odd Functions				✓	
2	4	Example 2	✓				
2	4	Example 3		✓			
2	5	Shifting Graphs	✓		✓		
2	5	Example 1	✓				
2	5	Reflecting Graphs			✓		
2	5	Examples 2, 4		✓			
2	5	Nonrigid Transformations			✓		
2	6	Arithmetic Combinations of Functions	✓		✓		
2	6	Examples 2, 4, 5	✓				
2	6	Composition of Functions			✓		
2	6	Example 6		✓			
2	7	Examples 1, 2, 7	✓				
2	7	The Graph of an Inverse Function			✓		
2	7	Examples 5, 6	✓	✓			
2	7	Finding Inverse Functions Algebraically	✓				
3	1	Example 2					✓
3	1	Example 3		✓			✓
3	1	Example 4	✓				
3	1	Example 5	✓			✓	
3	2	Examples, 3, 4, 7	✓				
3	2	Example 5		✓			
3	2	Example 6	✓	✓			
3	3	Example 1	✓	✓			
3	3	Example 2		✓	✓		

Chapter	Section	Example/Concept	Guided Example	Synthesis Example	Animation	Simulation	Editable Graph
3	3	Example 4	✓		✓		
3	3	Example 6	✓				
3	4	Examples 1, 5, 6, 8, 9	✓				
3	4	Examples 3, 7		✓			
3	4	Other Tests for Zeros of Polynomials	✓				
3	4	Example 10	✓	✓	✓		
3	5	Examples 2, 4, 5	✓	✓			
3	5	Example 3	✓				
3	5	Example 6				✓	
4	1	Example 2	✓	✓			
4	2	Examples 1–4	✓				
4	2	Example 5	✓	✓			✓
4	2	Example 6		✓			
4	3	Example 1	✓	✓			
4	3	Examples 2, 3	✓				
4	3	Example 4		✓			
4	4	Introduction			✓		
4	4	Parabolas	✓				
4	4	Examples 2, 3, 5, 7	✓				
4	4	Ellipses			✓		✓
4	4	Example 4	✓	✓			
4	4	Example 6	✓	✓			
4	5	Examples 1, 5	✓				
4	5	Equations of Conics in Standard Form	✓				
4	5	Example 2	✓	✓			
4	5	Example 3	✓				✓
4	5	Example 4	✓		✓		✓
5	1	Example 2		✓			✓
5	1	Example 3	✓				✓
5	1	Example 4	✓	✓	✓		
5	1	Examples 5, 7, 8	✓				
5	1	Applications				✓	
5	2	Example 1	✓	✓			
5	2	Examples 2, 7, 8, 10	✓				
5	2	Example 4			✓		
5	2	Example 6	✓	✓	✓		
5	3	Examples 1, 2, 6	✓				
5	3	Example 4	✓	✓			
5	3	Example 5		✓			
5	4	Examples 1, 2, 4–8, 10	✓				
5	4	Example 3	✓	✓			
5	5	Example 1	✓	✓			
5	5	Examples 2, 3, 5, 6	✓				
6	1	Degree Measure	✓				
6	1	Examples 1, 2, 4, 5, 8	✓				
6	1	Radian Measure		✓			
6	1	Example 3			✓		
6	1	Example 6	✓			✓	
6	2	The Trigonometric Functions			✓		
6	2	Examples 1, 5, 7	✓	✓			
6	2	Example 3			✓		

Chapter	Section	Example/Concept	Guided Example	Synthesis Example	Animation	Simulation	Editable Graph
6	2	Trigonometric Identities	✓				
6	2	Examples 6, 8	✓				
6	3	Examples 1, 2, 4, 5, 8	✓				
6	3	Reference Angles				✓	
6	3	Example 6		✓			
6	4	Amplitude and Period	✓				
6	4	Example 2		✓			✓
6	4	Example 4					✓
6	4	Example 5	✓		✓		
6	4	Example 6	✓				✓
6	4	Mathematical Modeling	✓				
6	5	Examples 1, 3-6	✓				
6	5	Example 2			✓		
6	5	Graphs of Reciprocal Functions					✓
6	6	Example 2			✓		
6	6	Other Inverse Trigonometric Functions		✓	✓		
6	6	Examples 3, 5, 6	✓				
6	6	Examples 4, 7	✓	✓			
6	7	Examples 1, 3, 5, 6	✓				
6	7	Example 4	✓	✓			
7	1	Examples 1-3, 5, 8	✓				
7	1	Example 4	✓	✓			
7	1	Example 6		✓			
7	2	Introduction			✓		
7	2	Examples 1, 3, 5	✓				
7	3	Examples 3, 4, 6, 8	✓				
7	3	Equations of Quadratic Type	✓				
7	3	Example 5		✓			
7	3	Multiple Angles		✓			
7	4	Sum and Difference Formulas		✓			
7	4	Examples 1, 3, 4, 8	✓				
7	4	Example 2		✓			
7	5	Multiple-Angle Formulas			✓		
7	5	Examples 1, 9	✓				
7	5	Example 3		✓			
7	5	Power-Reducing Formulas			✓		
7	5	Example 5	✓	✓			
7	5	Half-Angle Formulas			✓		
7	5	Example 6	✓	✓			
7	5	Product-To-Sum Formulas	✓				
8	1	Examples 1, 3, 5–7	✓				
8	1	The Ambiguous Case (SSA)				✓	
8	2	Example 1	✓	✓			
8	2	Examples 2, 4, 5	✓				
8	2	Example 3		✓			
8	3	Examples 2, 9	✓				
8	3	Vector Operations			✓		
8	3	Example 3	✓	✓	✓		
8	3	Unit Vectors	✓				
8	3	Examples 6, 7		✓			
8	3	Direction Angles		✓			
8	4	Examples 1, 2, 6, 8	✓				

Chapter	Section	Example/Concept	Guided Example	Synthesis Example	Animation	Simulation	Editable Graph
8	4	Example 4	✓	✓			
8	4	Finding Vector Components				✓	
8	5	Examples 1-3, 6, 8	✓				
8	6	Examples 4, 5	✓	✓			
8	6	Roots of Complex Numbers		✓			
9	1	Examples 1, 5–7	✓				
9	1	Example 3	✓	✓	✓		
9	1	Example 4			✓		
9	2	Examples 1, 5, 6, 9	✓				
9	2	Example 3	✓	✓			
9	2	Example 4			✓		
9	2	Applications	✓				
9	3	Examples 3–5, 7–10	✓				
9	3	Nonsquare Systems	✓				
9	3	Applications	✓				
9	4	The Graph of an Inequality	✓				
9	4	Example 1					✓
9	4	Example 4	✓	✓	✓		
9	4	Examples 5, 9	✓				
9	5	Example 1			✓		
9	5	Examples 3–6	✓				
10	1	Examples 1, 2, 5, 7–9	✓				
10	1	Example 6	✓	✓	✓		
10	2	Examples 1, 8	✓				
10	2	Example 3	✓	✓			
10	2	Examples 7, 11	✓		✓		
10	3	Examples 1, 4, 5	✓				
10	3	Example 3	✓		✓		
10	4	The Determinant of a 2 x 2 Matrix		✓			
10	4	Example 1	✓				
10	4	Example 3	✓	✓			
10	4	Example 4		✓			
10	5	Examples 1, 3–5, 7, 8	✓				
10	5	Example 2	✓	✓	✓		
11	1	Example 3		✓			
11	1	Factorial Notation		✓			
11	1	Examples 5–7	✓				
11	1	Summation Notation		✓			
11	2	Examples 1, 6–8	✓				
11	2	Examples 2, 3		✓			
11	3	Examples 1–4, 6, 8	✓				
11	3	Geometric Series			✓		
11	3	Example 7		✓			
11	4	Introduction		✓			
11	4	Examples 1, 4	✓				
11	5	Examples 2, 4, 7	✓				
11	5	Example 5		✓	✓		
11	6	Examples 4–9	✓				
11	7	Examples 1, 2, 5, 8, 9, 11	✓				
11	7	Example 3	✓			✓	

Interactive Algebra and Trigonometry 3.0 CD-ROM and Internet Algebra and Trigonometry 3.0

To accommodate a wide variety of teaching and learning styles, *Algebra and Trigonometry* is also available as *Interactive Algebra and Trigonometry* 3.0 on an interactive CD-ROM and *Internet Algebra and Trigonometry* 3.0. Students using the interactive CD-ROM or those with internet access will benefit from a wide range of compelling, interactive pedagogy, plus solutions to all odd exercises in the text. For instructors who conduct part of their course online, the internet version is an ideal solution, offering the additional advantages of online interaction with instructors and course management tools.

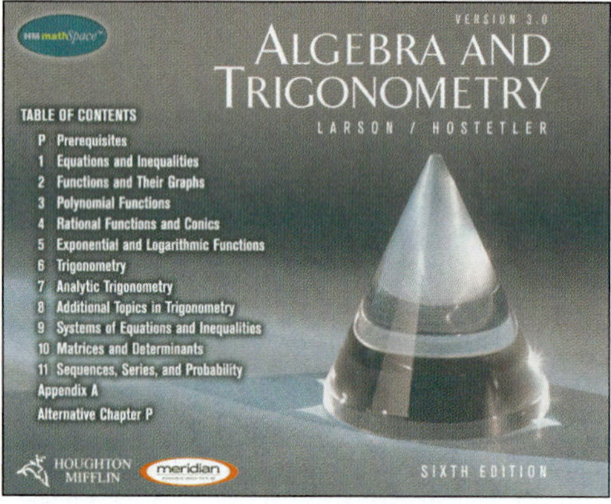

Hands-on Interaction

- The graphing calculator emulator provides students with an onscreen graphing utility that can be used for computation and exploration.
- The animations, simulations, and editable graph explorations make mathmatical concepts come alive.
- Guided and synthesis examples are designed to have students work through a solution one step at a time.
- Section quizzes require students to enter free-response answers, click and drag answers into place, or click on correct answers.

Classroom and Syllabus Management Systems

All of the content of the Sixth Edition—a wealth of applications, exercises, worked-out examples, and detailed explanations—is included in *Interactive Algebra and Trigonometry* 3.0 on CD-ROM and *Internet Algebra and Trigonometry* 3.0. Instructors have the flexibility of customizing content and interactive features for students as desired. Instructors may simply add dates to a default syllabus or may modify the order of topics. Either way, a customized syllabus is easy to distribute electronically and update instantly. This tool is particularly useful for managing distance learning courses.

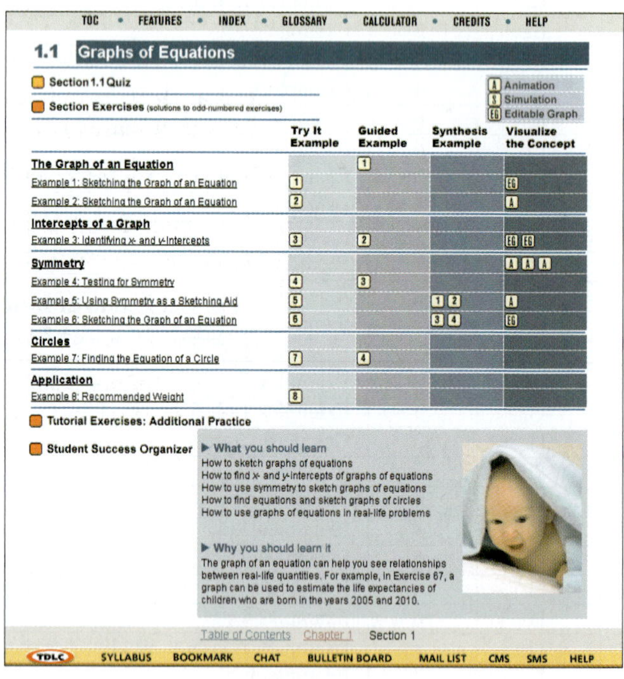

Features

Exercises with full worked-out solutions to all of the odd exercises in the text provide immediate feedback for students.

Try Its allow students to try problems similar to the examples and to check their work using the worked-out solutions provided.

Guided Examples provide a full range of support by walking students step-by-step through problems that relate to a specific concept in the text.

Synthesis Examples require the use of more than one concept from a section and encourage students to work through a solution of a problem one step at a time.

Animations, which use motion and sound to explain concepts, can be played, paused, stopped, and replayed as many times as the student desires.

Simulations are interactive activities that encourage exploration and hands-on use of mathematical concepts.

Editable Graphs encourage students to explore concepts by graphing "editable" graphs. Students can also change the viewing window and use *zoom* and *trace* features.

MathGraphs are enlarged, printable versions of graphs from exercises in the book in which students are asked to draw on the graphs.

Tutorial Exercises are *additional* exercises that furnish students with much needed guided practice and refer back to a **Guided Example** for help if necessary.

Graphing Calculator Emulator is a powerful tool built into the program for convenient computation and exploration, and is also useful for working exercises that require the use of a graphing calculator.

Section Quizzes with responses and **Chapter Tests** with answers help students assess their mastery of the material.

Chapter Pre-Tests and **Post-Tests** offer added practice and assessment opportunities.

Glossary of Terms provides a comprehensive list of important mathematical terms, which students can quickly and easily access at any time.

Index and **Features Index** facilitate cross-referencing by providing complete searchable text-specific content.

Syllabus Builder enables instructors to save administrative time and to convey important information online.

How to study Chapter P

▶ What you should learn

In this chapter you will learn the following skills and concepts:

- How to represent and order real numbers and use inequalities
- How to evaluate algebraic expressions using the basic rules of algebra
- How to use properties of exponents and radicals to simplify and evaluate expressions
- How to add, subtract, multiply, and factor polynomials
- How to determine the domains of algebraic expressions and simplify rational expressions
- How to avoid common algebraic errors and use algebraic techniques common in calculus
- How to plot points in the coordinate plane and use the Distance and Midpoint Formulas

▶ Important Vocabulary

As you encounter each new vocabulary term in this chapter, add the term and its definition to your notebook glossary.

Real numbers (p. 2)
Real number line (p. 2)
Inequality (p. 3)
Absolute value (p. 5)
Variables (p. 6)
Algebraic expressions (p. 6)
Coefficient (p. 6)
Evaluate (p. 6)
Additive inverse (p. 6)
Multiplicative inverse (p. 6)
Factors (p. 8)
Exponential form (p. 12)
Scientific notation (p. 14)
Principal *n*th root (p. 15)
Index (p. 15)
Radicand (p. 15)
Simplest form (p. 17)
Conjugate (p. 18)
Polynomial (p. 24)
Degree (of a polynomial) (p. 24)
Domain (p. 41)
Equivalent (expressions) (p. 41)
Rational expression (p. 41)
Complex fractions (p. 45)
Rectangular coordinate system (p. 58)
Ordered pair (p. 58)
Distance Formula (p. 60)
Midpoint Formula (p. 62)

Study Tools

Learning objectives in each section
Chapter Summary (p. 68)
Review Exercises (pp. 69–71)
Chapter Test (p. 72)

Additional Resources

Study and Solutions Guide
Interactive College Algebra
Videotapes/DVD for Chapter P
Algebra and Trigonometry Website
Student Success Organizer

Craig Tuttle/Corbis

P

Prerequisites

An expanded version of Sections P.1–P.4 is available on the text-specific website at *college.hmco.com* and on the *Interactive* version of this text. This expanded version contains the following sections.

Operations with Real Numbers; Properties of Real Numbers; Algebraic Expressions; Operations with Polynomials; Factoring Polynomials; Factoring Trinomials

P.1 Review of Real Numbers and Their Properties

▶ **What you should learn**

- How to represent and classify real numbers
- How to order real numbers and use inequalities
- How to find the absolute values of real numbers and find the distance between two real numbers
- How to evaluate algebraic expressions
- How to use the basic rules and properties of algebra

▶ **Why you should learn it**

Real numbers are used to represent many real-life quantities. For example, in Exercise 69 on page 10, you will use real numbers to represent the federal deficit.

Doug Armand/Getty Images

Real Numbers 

Real numbers are used in everyday life to describe quantities such as age, miles per gallon, container size, and population. Real numbers are represented by symbols such as

$$-5, 9, 0, \frac{4}{3}, 0.666 \ldots, 28.21, \sqrt{2}, \pi, \text{ and } \sqrt[3]{-32}.$$

Here are some important subsets of the real numbers.

$\{1, 2, 3, 4, \ldots\}$ Set of natural numbers

$\{0, 1, 2, 3, 4, \ldots\}$ Set of whole numbers

$\{\ldots, -3, -2, -1, 0, 1, 2, 3, \ldots\}$ Set of integers

A real number is **rational** if it can be written as the ratio p/q of two integers, where $q \neq 0$. For instance, the numbers

$$\frac{1}{3} = 0.3333 \ldots = 0.\overline{3}, \frac{1}{8} = 0.125, \text{ and } \frac{125}{111} = 1.126126 \ldots = 1.\overline{126}$$

are rational. The decimal representation of a rational number either repeats $\left(\text{as in } \frac{173}{55} = 3.1\overline{45}\right)$ or terminates $\left(\text{as in } \frac{1}{2} = 0.5\right)$. A real number that cannot be written as the ratio of two integers is called **irrational.** Irrational numbers have infinite nonrepeating decimal representations. For instance, the numbers

$$\sqrt{2} \approx 1.4142136 \quad \text{and} \quad \pi \approx 3.1415927$$

are irrational. (The symbol $\approx$ means "is approximately equal to.")

Real numbers are represented graphically by a **real number line.** The point 0 on the real number line is the **origin.** Numbers to the right of 0 are positive, and numbers to the left of 0 are negative, as shown in Figure P.1. The term **nonnegative** describes a number that is either positive or zero.

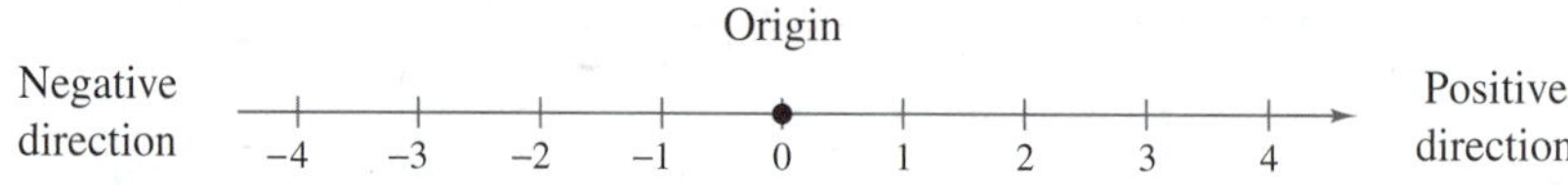

FIGURE P.1 *The Real Number Line*

As illustrated in Figure P.2, there is a *one-to-one correspondence* between real numbers and points on the real number line.

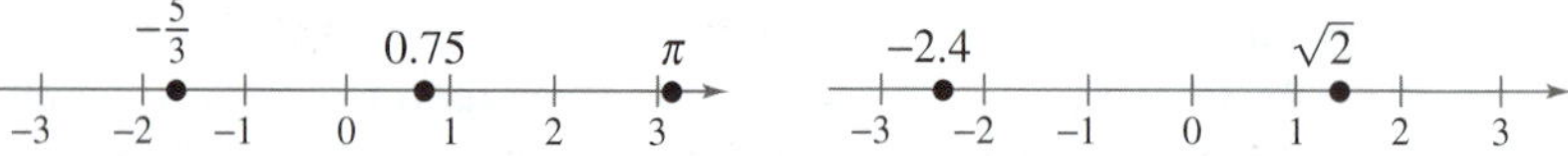

Every real number corresponds to exactly one point on the real number line.

Every point on the real number line corresponds to exactly one real number.

FIGURE P.2 *One-to-One Correspondence*

The icon identifies examples and concepts related to features of the Learning Tools CD-ROM and the *Interactive* and *Internet* versions of this text. For more details see the chart on pages *xix-xxiii*.

Ordering Real Numbers 

One important property of real numbers is that they are *ordered*.

> ### Definition of Order on the Real Number Line
>
> If a and b are real numbers, a is less than b if $b - a$ is positive. The **order** of a and b is denoted by the **inequality**
>
> $$a < b.$$
>
> This relationship can also be described by saying that b is *greater than* a and writing $b > a$. The inequality $a \le b$ means that a is *less than or equal to* b, and the inequality $b \ge a$ means that b is *greater than or equal to* a. The symbols $<$, $>$, $\le$, and $\ge$ are *inequality symbols*.

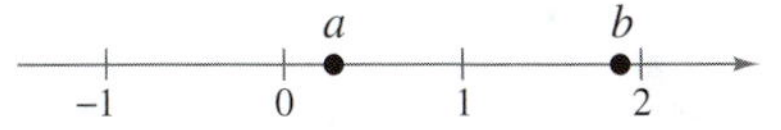

FIGURE P.3 *$a < b$ if and only if a lies to the left of b.*

Geometrically, this definition implies that $a < b$ if and only if a lies to the *left* of b on the real number line, as shown in Figure P.3.

Example 1 ▶ Interpreting Inequalities 

Describe the subset of real numbers represented by each inequality.

a. $x \le 2$ **b.** $-2 \le x < 3$

Solution

FIGURE P.4

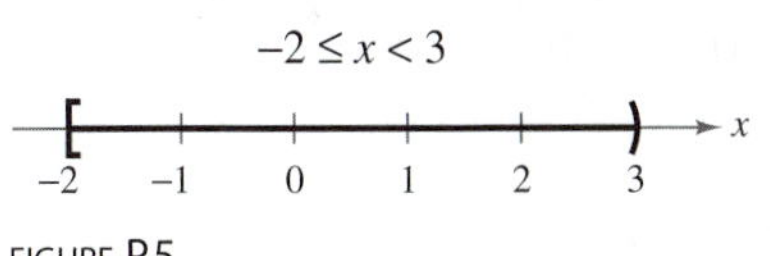

FIGURE P.5

a. The inequality $x \le 2$ denotes all real numbers less than or equal to 2, as shown in Figure P.4.

b. The inequality $-2 \le x < 3$ means that $x \ge -2$ and $x < 3$. This "double inequality" denotes all real numbers between -2 and 3, including -2 but not including 3, as shown in Figure P.5.

Inequalities can be used to describe subsets of real numbers called **intervals.** In the bounded intervals below, the real numbers a and b are the **endpoints** of each interval.

The *Interactive* CD-ROM and *Internet* versions of this text offer a Try It for each example in the text.

Bounded Intervals on the Real Number Line

Notation	*Interval Type*	*Inequality*	*Graph*
$[a, b]$	Closed	$a \le x \le b$	
(a, b)	Open	$a < x < b$	
$[a, b)$		$a \le x < b$	
$(a, b]$		$a < x \le b$	

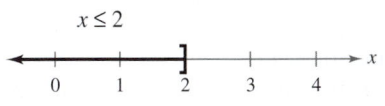

The symbols ∞, **positive infinity,** and $-\infty$, **negative infinity,** do not represent real numbers. They are simply convenient symbols used to describe the unboundedness of an interval such as $(1, \infty)$ or $(-\infty, 3]$.

Unbounded Intervals on the Real Number Line

Notation	*Interval Type*	*Inequality*	*Graph*
$[a, \infty)$		$x \geq a$	
(a, ∞)	Open	$x > a$	
$(-\infty, b]$		$x \leq b$	
$(-\infty, b)$	Open	$x < b$	
$(-\infty, \infty)$	Entire real line	$-\infty < x < \infty$	

Example 2 ▶ Using Inequalities to Represent Intervals

Use inequality notation to describe each of the following.

a. c is at most 2.

b. m is at least -3.

c. All x in the interval $(-3, 5]$

Solution

a. The statement "c is at most 2" can be represented by $c \leq 2$.

b. The statement "m is at least -3" can be represented by $m \geq -3$.

c. "All x in the interval $(-3, 5]$" can be represented by $-3 < x \leq 5$.

Example 3 ▶ Interpreting Intervals

Give a verbal description of each interval.

a. $(-1, 0)$ **b.** $[2, \infty)$ **c.** $(-\infty, 0)$

Solution

a. This interval consists of all real numbers that are greater than -1 and less than 0.

b. This interval consists of all real numbers that are greater than or equal to 2.

c. This interval consists of all real numbers that are less than zero (the negative real numbers).

The **Law of Trichotomy** states that for any two real numbers a and b, *precisely* one of three relationships is possible:

$a = b, \quad a < b, \quad \text{or} \quad a > b.$ Law of Trichotomy

Absolute Value and Distance 

The **absolute value** of a real number is its *magnitude*, or the distance between the origin and the point representing the real number on the real number line.

Definition of Absolute Value

If a is a real number, then the absolute value of a is

$$|a| = \begin{cases} a, & \text{if } a \geq 0 \\ -a, & \text{if } a < 0. \end{cases}$$

Notice in this definition that the absolute value of a real number is never negative. For instance, if $a = -5$, then $|-5| = -(-5) = 5$. The absolute value of a real number is either positive or zero. Moreover, 0 is the only real number whose absolute value is 0. So, $|0| = 0$.

Exploration

Absolute value expressions can be evaluated on a graphing utility. When an expression such as $|3 - 8|$ is evaluated, parentheses should surround the expression, as in abs(3 − 8). Evaluate each expression below. What can you conclude?

a. $|6|$ **b.** $|-1|$
c. $|5 - 2|$ **d.** $|2 - 5|$

Example 4 ▶ Evaluating the Absolute Value of a Number 

Evaluate $\dfrac{|x|}{x}$ for (a) $x > 0$ and (b) $x < 0$.

Solution

a. If $x > 0$, then $|x| = x$ and $\dfrac{|x|}{x} = \dfrac{x}{x} = 1$.

b. If $x < 0$, then $|x| = -x$ and $\dfrac{|x|}{x} = \dfrac{-x}{x} = -1$.

Properties of Absolute Values

1. $|a| \geq 0$ **2.** $|-a| = |a|$

3. $|ab| = |a||b|$ **4.** $\left|\dfrac{a}{b}\right| = \dfrac{|a|}{|b|}, \quad b \neq 0$

Absolute value can be used to define the distance between two points on the real number line. For instance, the distance between -3 and 4 is

$$|-3 - 4| = |-7|$$
$$= 7$$

as shown in Figure P.6.

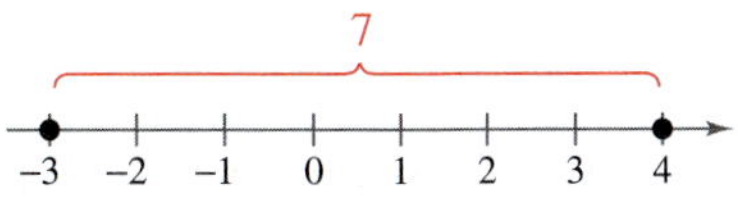

FIGURE P.6 *The distance between* -3 *and 4 is 7.*

Distance Between Two Points on the Real Line

Let a and b be real numbers. The **distance between a and b** is

$$d(a, b) = |b - a| = |a - b|.$$

Algebraic Expressions 

One characteristic of algebra is the use of letters to represent numbers. The letters are **variables,** and combinations of letters and numbers are **algebraic expressions.** Here are a few examples of algebraic expressions.

$$5x, \qquad 2x - 3, \qquad \frac{4}{x^2 + 2}, \qquad 7x + y$$

> **Definition of an Algebraic Expression**
>
> An **algebraic expression** is a collection of letters **(variables)** and real numbers **(constants)** combined using the operations of addition, subtraction, multiplication, division, and exponentiation.

The **terms** of an algebraic expression are those parts that are separated by *addition*. For example,

$$x^2 - 5x + 8 = x^2 + (-5x) + 8$$

has three terms: x^2 and $-5x$ are the **variable terms** and 8 is the **constant term.** The numerical factor of a variable term is the **coefficient** of the variable term. For instance, the coefficient of $-5x$ is -5, and the coefficient of x^2 is 1.

To **evaluate** an algebraic expression, substitute numerical values for each of the variables in the expression. Here are two examples.

The *Interactive* CD-ROM and *Internet* versions of this text offer a Quiz for every section of the text.

Expression	*Value of Variable*	*Substitute*	*Value of Expression*
$-3x + 5$	$x = 3$	$-3(3) + 5$	$-9 + 5 = -4$
$3x^2 + 2x - 1$	$x = -1$	$3(-1)^2 + 2(-1) - 1$	$3 - 2 - 1 = 0$

When an algebraic expression is evaluated, the **Substitution Principle** is used. It states that "If $a = b$, then a can be replaced by b in any expression involving a." In the first evaluation shown above, for instance, 3 is *substituted* for x in the expression $-3x + 5$.

Basic Rules of Algebra 

There are four arithmetic operations with real numbers: *addition*, *multiplication*, *subtraction*, and *division*, denoted by the symbols $+$, $\times$ or $\cdot$, $-$, and $\div$. Of these, addition and multiplication are the two primary operations. Subtraction and division are the inverse operations of addition and multiplication, respectively.

Subtraction: Add the opposite. *Division:* Multiply by the reciprocal.

$$a - b = a + (-b) \qquad \text{If } b \neq 0, \text{ then } a/b = a\left(\frac{1}{b}\right) = \frac{a}{b}.$$

In these definitions, $-b$ is the **additive inverse** (or opposite) of b, and $1/b$ is the **multiplicative inverse** (or reciprocal) of b. In the fractional form a/b, a is the **numerator** of the fraction and b is the **denominator.**

Because the properties of real numbers on page 7 are true for variables and algebraic expressions as well as for real numbers, they are often called the **Basic Rules of Algebra.**

> **STUDY TIP**
>
> Try to formulate a verbal description of each property. For instance, the first property states that *the order in which two real numbers are added does not affect their sum.*

Basic Rules of Algebra

Let a, b, and c be real numbers, variables, or algebraic expressions.

Property		*Example*
Commutative Property of Addition:	$a + b = b + a$	$4x + x^2 = x^2 + 4x$
Commutative Property of Multiplication:	$ab = ba$	$(4 - x)x^2 = x^2(4 - x)$
Associative Property of Addition:	$(a + b) + c = a + (b + c)$	$(x + 5) + x^2 = x + (5 + x^2)$
Associative Property of Multiplication:	$(ab)c = a(bc)$	$(2x \cdot 3y)(8) = (2x)(3y \cdot 8)$
Distributive Properties:	$a(b + c) = ab + ac$	$3x(5 + 2x) = 3x \cdot 5 + 3x \cdot 2x$
	$(a + b)c = ac + bc$	$(y + 8)y = y \cdot y + 8 \cdot y$
Additive Identity Property:	$a + 0 = a$	$5y^2 + 0 = 5y^2$
Multiplicative Identity Property:	$a \cdot 1 = a$	$(4x^2)(1) = 4x^2$
Additive Inverse Property:	$a + (-a) = 0$	$5x^3 + (-5x^3) = 0$
Multiplicative Inverse Property:	$a \cdot \dfrac{1}{a} = 1, \quad a \neq 0$	$(x^2 + 4)\left(\dfrac{1}{x^2 + 4}\right) = 1$

Because subtraction is defined as "adding the opposite," the Distributive Properties are also true for subtraction. So, the first Distributive Property can be applied to an expression of the form $a(b - c)$ as follows.

$$a(b - c) = ab - ac$$

Properties of Negation

Let a and b be real numbers, variables, or algebraic expressions.

Property	*Example*
1. $(-1)a = -a$	$(-1)7 = -7$
2. $-(-a) = a$	$-(-6) = 6$
3. $(-a)b = -(ab) = a(-b)$	$(-5)3 = -(5 \cdot 3) = 5(-3)$
4. $(-a)(-b) = ab$	$(-2)(-x) = 2x$
5. $-(a + b) = (-a) + (-b)$	$-(x + 8) = (-x) + (-8)$
	$= -x - 8$

STUDY TIP

Be sure you see the difference between the *opposite of a number* and a *negative number*. If a is already negative, then its opposite, $-a$, is positive. For instance, if $a = -5$, then

$$-a = -(-5) = 5.$$

Properties of Equality

Let a, b, and c be real numbers, variables, or algebraic expressions.

1. If $a = b$, then $a + c = b + c$. — Add c to each side.
2. If $a = b$, then $ac = bc$. — Multiply each side by c.
3. If $a + c = b + c$, then $a = b$. — Subtract c from each side.
4. If $ac = bc$ and $c \neq 0$, then $a = b$. — Divide each side by c.

STUDY TIP

The "or" in the Zero-Factor Property includes the possibility that either or both factors may be zero. This is an **inclusive or,** and it is the way the word "or" is generally used in mathematics.

Properties of Zero

Let a and b be real numbers, variables, or algebraic expressions.

1. $a + 0 = a$ and $a - 0 = a$ **2.** $a \cdot 0 = 0$

3. $\frac{0}{a} = 0, \quad a \neq 0$ **4.** $\frac{a}{0}$ is undefined.

5. Zero-Factor Property: If $ab = 0$, then $a = 0$ or $b = 0$.

Properties and Operations of Fractions

Let a, b, c, and d be real numbers, variables, or algebraic expressions such that $b \neq 0$ and $d \neq 0$.

1. Equivalent Fractions: $\frac{a}{b} = \frac{c}{d}$ if and only if $ad = bc$.

2. Rules of Signs: $-\frac{a}{b} = \frac{-a}{b} = \frac{a}{-b}$ and $\frac{-a}{-b} = \frac{a}{b}$

3. Generate Equivalent Fractions: $\frac{a}{b} = \frac{ac}{bc}, \quad c \neq 0$

4. Add or Subtract with Like Denominators: $\frac{a}{b} \pm \frac{c}{b} = \frac{a \pm c}{b}$

5. Add or Subtract with Unlike Denominators: $\frac{a}{b} \pm \frac{c}{d} = \frac{ad \pm bc}{bd}$

6. Multiply Fractions: $\frac{a}{b} \cdot \frac{c}{d} = \frac{ac}{bd}$

7. Divide Fractions: $\frac{a}{b} \div \frac{c}{d} = \frac{a}{b} \cdot \frac{d}{c} = \frac{ad}{bc}, \quad c \neq 0$

STUDY TIP

In Property 1 of fractions, the phrase "if and only if" implies two statements. One statement is: If $a/b = c/d$, then $ad = bc$. The other statement is: If $ad = bc$, where $b \neq 0$ and $d \neq 0$, then $a/b = c/d$.

Example 5 ▶ Properties and Operations of Fractions 

a. Equivalent fractions: $\frac{x}{5} = \frac{3 \cdot x}{3 \cdot 5} = \frac{3x}{15}$ **b.** Divide fractions: $\frac{7}{x} \div \frac{3}{2} = \frac{7}{x} \cdot \frac{2}{3} = \frac{14}{3x}$

c. Add fractions with unlike denominators: $\frac{x}{3} + \frac{2x}{5} = \frac{5 \cdot x + 3 \cdot 2x}{3 \cdot 5} = \frac{11x}{15}$

If a, b, and c are integers such that $ab = c$, then a and b are **factors** or **divisors** of c. A **prime number** is an integer that has exactly two positive factors:— itself and 1—such as 2, 3, 5, 7, and 11. The numbers 4, 6, 8, 9, and 10 are **composite** because they can be written as the product of two or more prime numbers. The number 1 is neither prime nor composite. The **Fundamental Theorem of Arithmetic** states that every positive integer greater than 1 can be written as the product of prime numbers in precisely one way (disregarding order). For instance, the *prime factorization* of 24 is $24 = 2 \cdot 2 \cdot 2 \cdot 3$.

P.1 Exercises

The *Interactive* CD-ROM and *Internet* versions of this text contain step-by-step solutions to all odd-numbered exercises. They also provide Tutorial Exercises for additional help.

In Exercises 1–6, determine which numbers are (a) natural numbers, (b) integers, (c) rational numbers, and (d) irrational numbers.

1. $-9, -\frac{7}{2}, 5, \frac{2}{3}, \sqrt{2}, 0, 1, -4, 2, -11$
2. $\sqrt{5}, -7, -\frac{7}{3}, 0, 3.12, \frac{5}{4}, -3, 12, 5$
3. $2.01, 0.666 \ldots, -13, 0.010110111 \ldots, 1, -6$
4. $2.3030030003 \ldots, 0.7575, -4.63, \sqrt{10}, -75, 4$
5. $-\pi, -\frac{1}{3}, \frac{6}{3}, \frac{1}{2}\sqrt{2}, -7.5, -1, 8, -22$
6. $25, -17, -\frac{12}{5}, \sqrt{9}, 3.12, \frac{1}{2}\pi, 7, -11.1, 13$

In Exercises 7–10, use a calculator to find the decimal form of the rational number. If it is a nonterminating decimal, write the repeating pattern.

7. $\frac{5}{8}$
8. $\frac{1}{3}$
9. $\frac{41}{333}$
10. $\frac{6}{11}$

In Exercises 11 and 12, approximate the numbers and place the correct symbol (< or >) between them.

11. 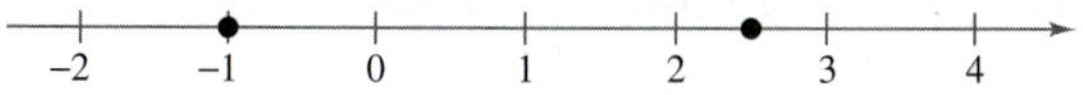

12. 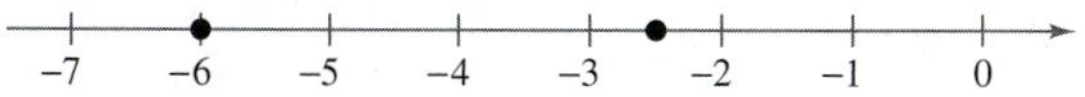

In Exercises 13–18, plot the two real numbers on the real number line. Then place the appropriate inequality symbol (< or >) between them.

13. $-4, -8$
14. $-3.5, 1$
15. $\frac{3}{2}, 7$
16. $1, \frac{16}{3}$
17. $\frac{5}{6}, \frac{2}{3}$
18. $-\frac{8}{7}, -\frac{3}{7}$

In Exercises 19–28, verbally describe the subset of real numbers represented by the inequality. Then sketch the subset on the real number line. State whether the interval is bounded or unbounded.

19. $x \le 5$
20. $x \ge -2$
21. $x < 0$
22. $x > 3$
23. $x \ge 4$
24. $x < 2$
25. $-2 < x < 2$
26. $0 \le x \le 5$
27. $-1 \le x < 0$
28. $0 < x \le 6$

In Exercises 29–36, use inequality notation to describe the set.

29. All x in the interval $(-2, 4]$
30. All y in the interval $[-6, 0)$
31. y is nonnegative.
32. y is no more than 25.
33. t is at least 10 and at most 22.
34. k is less than 5 but no less than -3.
35. The dog's weight W is more than 65 pounds.
36. The annual rate of inflation r is expected to be at least 2.5% but no more than 5%.

In Exercises 37–40, give a verbal description of the interval.

37. $[0, 8)$
38. $[-5, 7]$
39. $(-6, \infty)$
40. $(-\infty, 4]$

In Exercises 41–50, evaluate the expression.

41. $|-10|$
42. $|0|$
43. $|3 - 8|$
44. $|4 - 1|$
45. $|-1| - |-2|$
46. $-3 - |-3|$
47. $\dfrac{-5}{|-5|}$
48. $-3|-3|$
49. $\dfrac{|x + 2|}{x + 2}, \quad x < -2$
50. $\dfrac{|x - 1|}{x - 1}, \quad x > 1$

In Exercises 51–56, place the correct symbol (<, >, or =) between the pair of real numbers.

51. $|-3| \quad \square \quad -|-3|$
52. $|-4| \quad \square \quad |4|$
53. $-5 \quad \square \quad -|5|$
54. $-|-6| \quad \square \quad |-6|$
55. $-|-2| \quad \square \quad -|2|$
56. $-(-2) \quad \square \quad -2$

In Exercises 57–64, find the distance between *a* and *b*.

57. 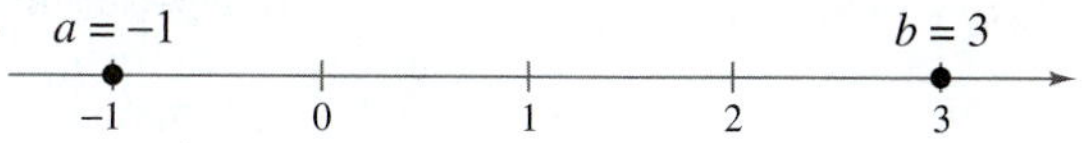

58. 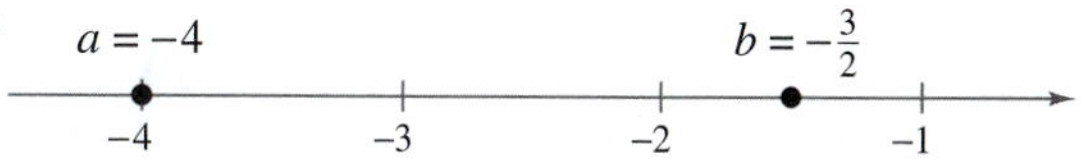

59. $a = 126, b = 75$
60. $a = -126, b = -75$
61. $a = -\frac{5}{2}, b = 0$
62. $a = \frac{1}{4}, b = \frac{11}{4}$
63. $a = \frac{16}{5}, b = \frac{112}{75}$
64. $a = 9.34, b = -5.65$

Budget Variance **In Exercises 65–68, the accounting department of a sports drink bottling company is checking to see whether the actual expenses of a department differ from the budgeted expenses by more than $500 or by more than 5%. Fill in the missing parts of the table, and determine whether each actual expense passes the "budget variance test."**

	Budgeted Expense, b	*Actual Expense, a*	$\|a - b\|$	$0.05b$
65. Wages	$112,700	$113,356		
66. Utilities	$9,400	$9,772		
67. Taxes	$37,640	$37,335		
68. Insurance	$2,575	$2,613		

▶ Model It

69. ***Federal Deficit*** The bar graph shows the federal government receipts (in billions of dollars) for selected years from 1960 through 2000. (Source: U.S. Office of Management and Budget)

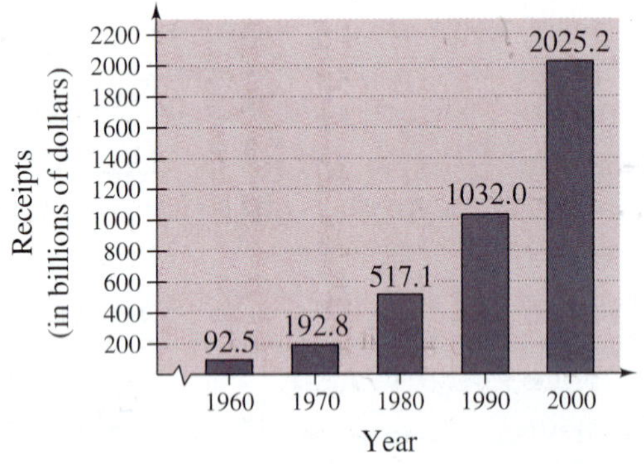

(a) Complete the table. (*Hint:* Find |Receipts – Expenditures|.)

Year	Expenditures (in billions)	Surplus or deficit (in billions)
1960	$92.2	
1970	$195.6	
1980	$590.9	
1990	$1253.2	
2000	$1788.8	

(b) Use the table in part (a) to construct a bar graph showing the magnitude of the surplus or deficit for each year.

70. ***Veterans*** The table shows the number of surviving spouses of deceased veterans of United States wars (as of May 2001). Construct a circle graph showing the percent of surviving spouses for each war as a fraction of the total number of surviving spouses of deceased war veterans. (Source: Department of Veteran Affairs)

War	Number of surviving spouses
Civil War	1
Indian Wars	0
Spanish-American War	386
Mexican Border War	181
World War I	25,573
World War II	272,793
Korean War	63,579
Vietnam War	114,514
Gulf War	6,261

In Exercises 71–78, use absolute value notation to describe the situation.

71. While traveling on the Pennsylvania Turnpike, you pass milepost 57 near Pittsburgh, then milepost 236 near Gettysburg. How far do you travel during that time period?

72. While traveling on the Pennsylvania Turnpike, you pass milepost 326 near Valley Forge, then milepost 351 near Philadelphia. How far do you travel during that time period?

73. The temperature in Bismarck, North Dakota, was 60° at noon, then 23° at midnight. What was the change in temperature over the 12-hour period?

74. The temperature in Chicago, Illinois was 48° last night at midnight, then 82° at noon today. What was the change in temperature over the 12-hour period?

75. The distance between x and 5 is no more than 3.

76. The distance between x and -10 is at least 6.

77. y is at least six units from 0.

78. y is at most two units from a.

In Exercises 79–84, identify the terms. Then identify the coefficients of the variable terms of the expression.

79. $7x + 4$

80. $6x^3 - 5x$

81. $\sqrt{3}x^2 - 8x - 11$

82. $3\sqrt{3}x^2 + 1$

83. $4x^3 + \frac{x}{2} - 5$ **84.** $3x^4 - \frac{x^2}{4}$

In Exercises 85–90, evaluate the expression for each value of x. (If not possible, state the reason.)

	Expression	*Values*	
85.	$4x - 6$	(a) $x = -1$	(b) $x = 0$
86.	$9 - 7x$	(a) $x = -3$	(b) $x = 3$
87.	$x^2 - 3x + 4$	(a) $x = -2$	(b) $x = 2$
88.	$-x^2 + 5x - 4$	(a) $x = -1$	(b) $x = 1$
89.	$\frac{x+1}{x-1}$	(a) $x = 1$	(b) $x = -1$
90.	$\frac{x}{x+2}$	(a) $x = 2$	(b) $x = -2$

In Exercises 91–100, identify the rule(s) of algebra illustrated by the statement.

91. $x + 9 = 9 + x$ **92.** $2\left(\frac{1}{2}\right) = 1$

93. $\frac{1}{h+6}(h + 6) = 1, \quad h \neq -6$

94. $(x + 3) - (x + 3) = 0$

95. $2(x + 3) = 2x + 6$ **96.** $(z - 2) + 0 = z - 2$

97. $1 \cdot (1 + x) = 1 + x$

98. $x + (y + 10) = (x + y) + 10$

99. $x(3y) = (x \cdot 3)y = (3x)y$

100. $\frac{1}{7}(7 \cdot 12) = \left(\frac{1}{7} \cdot 7\right)12 = 1 \cdot 12 = 12$

In Exercises 101–108, perform the operation(s). (Write fractional answers in simplest form.)

101. $\frac{3}{16} + \frac{5}{16}$ **102.** $\frac{6}{7} - \frac{4}{7}$

103. $\frac{5}{8} - \frac{5}{12} + \frac{1}{6}$ **104.** $\frac{10}{11} + \frac{6}{33} - \frac{13}{66}$

105. $12 \div \frac{1}{4}$ **106.** $-\left(6 \cdot \frac{4}{8}\right)$

107. $\frac{2x}{3} - \frac{x}{4}$ **108.** $\frac{5x}{6} \cdot \frac{2}{9}$

109. (a) Use a calculator to complete the table.

n	1	0.5	0.01	0.0001	0.000001
$5/n$					

(b) Use the result from part (a) to make a conjecture about the value of $5/n$ as n approaches 0.

110. (a) Use a calculator to complete the table.

n	1	10	100	10,000	100,000
$5/n$					

(b) Use the result from part (a) to make a conjecture about the value of $5/n$ as n increases without bound.

Synthesis

True or False? **In Exercises 111 and 112, determine whether the statement is true or false. Justify your answer.**

111. If $a < b$, then $\frac{1}{a} < \frac{1}{b}$, where $a \neq b \neq 0$.

112. Because $\frac{a+b}{c} = \frac{a}{c} + \frac{b}{c}$, then $\frac{c}{a+b} = \frac{c}{a} + \frac{c}{b}$.

113. ***Exploration*** Consider $|u + v|$ and $|u| + |v|$.

(a) Are the values of the expressions always equal? If not, under what conditions are they unequal?

(b) If the two expressions are not equal for certain values of u and v, is one of the expressions always greater than the other? Explain.

114. ***Think About It*** Is there a difference between saying that a real number is positive and saying that a real number is nonnegative? Explain.

115. ***Think About It*** Because every even number is divisible by 2, is it possible that there exist any even prime numbers? Explain.

116. ***Writing*** Describe the differences among the sets of natural numbers, integers, rational numbers, and irrational numbers.

In Exercises 117 and 118, use the real numbers A, B, and C shown on the number line. Determine the sign of each expression.

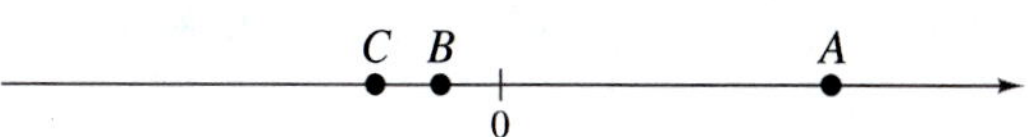

117. (a) $-A$ (b) $B - A$

118. (a) $-C$ (b) $A - C$

119. ***Writing*** You may hear it said that to take the absolute value of a real number you simply remove any negative sign and make the number positive. Can it ever be true that $|a| = -a$ for a real number a? Explain.

P.2 Exponents and Radicals

▶ **What you should learn**

- How to use properties of exponents
- How to use scientific notation to represent real numbers
- How to use properties of radicals
- How to simplify and combine radicals
- How to rationalize denominators and numerators
- How to use properties of rational exponents

▶ **Why you should learn it**

Real numbers and algebraic expressions are often written with exponents and radicals. For instance, in Exercise 113 on page 23, you will use an expression involving rational exponents to find the time required for a funnel to empty for different water heights.

Integer Exponents

Repeated *multiplication* can be written in **exponential form.**

Repeated Multiplication	*Exponential Form*
$a \cdot a \cdot a \cdot a \cdot a$	a^5
$(-4)(-4)(-4)$	$(-4)^3$
$(2x)(2x)(2x)(2x)$	$(2x)^4$

In general, if a is a real number and n is a positive integer, then

$$a^n = \underbrace{a \cdot a \cdot a \cdots a}_{n \text{ factors}}$$

where n is the **exponent** and a is the **base.** The expression a^n is read "a to the nth **power.**" In Property 3 below, be sure you see how to use a negative exponent.

Properties of Exponents

Let a and b be real numbers, variables, or algebraic expressions, and let m and n be integers. (All denominators and bases are nonzero.)

Property	*Example*
1. $a^m a^n = a^{m+n}$	$3^2 \cdot 3^4 = 3^{2+4} = 3^6 = 729$
2. $\dfrac{a^m}{a^n} = a^{m-n}$	$\dfrac{x^7}{x^4} = x^{7-4} = x^3$
3. $a^{-n} = \dfrac{1}{a^n} = \left(\dfrac{1}{a}\right)^n$	$y^{-4} = \dfrac{1}{y^4} = \left(\dfrac{1}{y}\right)^4$
4. $a^0 = 1, \quad a \neq 0$	$(x^2 + 1)^0 = 1$
5. $(ab)^m = a^m b^m$	$(5x)^3 = 5^3x^3 = 125x^3$
6. $(a^m)^n = a^{mn}$	$(y^3)^{-4} = y^{3(-4)} = y^{-12} = \dfrac{1}{y^{12}}$
7. $\left(\dfrac{a}{b}\right)^m = \dfrac{a^m}{b^m}$	$\left(\dfrac{2}{x}\right)^3 = \dfrac{2^3}{x^3} = \dfrac{8}{x^3}$
8. $\lvert a^2 \rvert = \lvert a \rvert^2 = a^2$	$\lvert(-2)^2\rvert = \lvert -2 \rvert^2 = (2)^2 = 4$

It is important to recognize the difference between expressions such as $(-2)^4$ and -2^4. In $(-2)^4$, the parentheses indicate that the exponent applies to the negative sign as well as to the 2, but in $-2^4 = -(2^4)$, the exponent applies only to the 2. So,

$$(-2)^4 = 16 \quad \text{and} \quad -2^4 = -16.$$

The properties of exponents listed on the preceding page apply to *all* integers m and n, not just to positive integers. For instance, by Property 2, you can write

$$\frac{3^4}{3^{-5}} = 3^{4-(-5)} = 3^{4+5} = 3^9.$$

Example 1 ▶ Using Properties of Exponents 

Use the properties of exponents to simplify each expression.

a. $(-3ab^4)(4ab^{-3})$ **b.** $(2xy^2)^3$ **c.** $3a(-4a^2)^0$ **d.** $\left(\frac{5x^3}{y}\right)^2$

Solution

a. $(-3ab^4)(4ab^{-3}) = (-3)(4)(a)(a)(b^4)(b^{-3}) = -12a^2b$

b. $(2xy^2)^3 = 2^3(x)^3(y^2)^3 = 8x^3y^6$

c. $3a(-4a^2)^0 = 3a(1) = 3a, \quad a \neq 0$

d. $\left(\frac{5x^3}{y}\right)^2 = \frac{5^2(x^3)^2}{y^2} = \frac{25x^6}{y^2}$

Example 2 ▶ Rewriting the Positive Exponents 

Rewrite each expression with positive exponents.

a. x^{-1} **b.** $\frac{1}{3x^{-2}}$ **c.** $\frac{12a^3b^{-4}}{4a^{-2}b}$ **d.** $\left(\frac{3x^2}{y}\right)^{-2}$

Solution

a. $x^{-1} = \frac{1}{x}$ Property 3

b. $\frac{1}{3x^{-2}} = \frac{1(x^2)}{3} = \frac{x^2}{3}$ The exponent -2 does not apply to 3.

c. $\frac{12a^3b^{-4}}{4a^{-2}b} = \frac{12a^3 \cdot a^2}{4b \cdot b^4}$ Property 3

$= \frac{3a^5}{b^5}$ Property 1

d. $\left(\frac{3x^2}{y}\right)^{-2} = \frac{3^{-2}(x^2)^{-2}}{y^{-2}}$ Properties 5 and 7

$= \frac{3^{-2}x^{-4}}{y^{-2}}$ Property 6

$= \frac{y^2}{3^2x^4}$ Property 3

$= \frac{y^2}{9x^4}$ Simplify.

STUDY TIP

Rarely in algebra is there only one way to solve a problem. Don't be concerned if the steps you use to solve a problem are not exactly the same as the steps presented in this text. The important thing is to use steps that you understand and, of course, steps that are justified by the rules of algebra. For instance, you might prefer the following steps for Example 2(d).

$$\left(\frac{3x^2}{y}\right)^{-2} = \left(\frac{y}{3x^2}\right)^2 = \frac{y^2}{9x^4}$$

Note how Property 3 is used in the first step of this solution. The fractional form of this property is

$$\left(\frac{a}{b}\right)^{-m} = \left(\frac{b}{a}\right)^m.$$

Technology

You can use a calculator to evaluate expressions with exponents. For instance, evaluate -3^{-2} as follows.

Scientific:

3 [+/−] [y^x] 2 [+/−] [=]

Graphing:

[(−)] 3 [^] [(−)] 2

The display will be as follows.

−.1111111111

Historical Note
The French mathematician Nicolas Chuquet (ca. 1500) wrote *Triparty en la science des nombres*, in which a form of exponent notation was used. Our expressions $6x^3$ and $10x^2$ were written as $.6.^3$ and $.10.^2$. Zero and negative exponents were also represented, so x^0 would be written as $.1.^0$ and $3x^{-2}$ as $.3.^{2m}$. Chuquet wrote that $.72.^1$ divided by $.8.^3$ is $.9.^{2m}$. That is, $72x \div 8x^3 = 9x^{-2}$.

Scientific Notation 

Exponents provide an efficient way of writing and computing with very large (or very small) numbers. For instance, there are about 359 billion billion gallons of water on Earth—that is, 359 followed by 18 zeros.

$$359{,}000{,}000{,}000{,}000{,}000{,}000$$

It is convenient to write such numbers in **scientific notation.** This notation has the form $c \times 10^n$, where $1 \le c < 10$ and n is an integer. So, the number of gallons of water on Earth can be written in scientific notation as

$$3.59 \times 100{,}000{,}000{,}000{,}000{,}000{,}000 = 3.59 \times 10^{20}.$$

The *positive* exponent 20 indicates that the number is *large* (10 or more) and that the decimal point has been moved 20 places. A *negative* exponent indicates that the number is *small* (less than 1). For instance, the mass (in grams) of one electron is approximately

$$9.0 \times 10^{-28} = 0.0000000000000000000000000009.$$

28 decimal places

Example 3 ▶ Scientific Notation 

Write the number in scientific notation.

a. 0.0000782 **b.** 836,100,000

Solution

a. $0.0000782 = 7.82 \times 10^{-5}$

b. $836{,}100{,}000 = 8.361 \times 10^8$

Example 4 ▶ Decimal Notation 

Write the number in decimal notation.

a. 9.36×10^{-6} **b.** 1.345×10^2

Solution

a. $9.36 \times 10^{-6} = 0.00000936$ **b.** $1.345 \times 10^2 = 134.5$

Technology

Most calculators automatically switch to scientific notation when they are showing large (or small) numbers that exceed the display range.

To *enter* numbers in scientific notation, your calculator should have an exponential entry key labeled

[EE] or [EXP].

Consult the user's guide for your calculator for instructions on keystrokes and how numbers in scientific notation are displayed.

Radicals and Their Properties

A **square root** of a number is one of its two equal factors. For example, 5 is a square root of 25 because 5 is one of the two equal factors of 25. In a similar way, a **cube root** of a number is one of its three equal factors, as in $125 = 5^3$.

Definition of *n*th Root of a Number

Let a and b be real numbers and let $n \geq 2$ be a positive integer. If

$$a = b^n$$

then b is an ***n*th root of *a*.** If $n = 2$, the root is a **square root.** If $n = 3$, the root is a **cube root.**

Some numbers have more than one nth root. For example, both 5 and -5 are square roots of 25. The *principal square root* of 25, written as $\sqrt{25}$, is the positive root, 5. The **principal *n*th root** of a number is defined as follows.

Principal *n*th Root of a Number

Let a be a real number that has at least one nth root. The **principal *n*th root of *a*** is the nth root that has the same sign as a. It is denoted by a **radical symbol**

$$\sqrt[n]{a}. \qquad \text{Principal } n\text{th root}$$

The positive integer n is the **index** of the radical, and the number a is the **radicand.** If $n = 2$, omit the index and write $\sqrt{a}$ rather than $\sqrt[2]{a}$. (The plural of index is *indices.*)

A common misunderstanding is that the square root sign implies both negative and positive roots. This is not correct. The square root sign implies only a positive root. When a negative root is needed, you must use the negative sign with the square root sign.

Incorrect: $\sqrt{4} = \pm 2$ (crossed out) *Correct:* $-\sqrt{4} = -2$ *and* $\sqrt{4} = 2$

Example 5 ▶ Evaluating Expressions Involving Radicals

a. $\sqrt{36} = 6$ because $6^2 = 36$.

b. $-\sqrt{36} = -6$ because $6^2 = 36$.

c. $\sqrt[3]{\dfrac{125}{64}} = \dfrac{5}{4}$ because $\left(\dfrac{5}{4}\right)^3 = \dfrac{5^3}{4^3} = \dfrac{125}{64}$.

d. $\sqrt[5]{-32} = -2$ because $(-2)^5 = -32$.

e. $\sqrt[4]{-81}$ is not a real number because there is no real number that can be raised to the fourth power to produce -81.

Here are some generalizations about the nth roots of real numbers.

Generalizations About nth Roots of Real Numbers

Real number a	Integer n	Root(s) of a	Example
$a > 0$	$n > 0$, is even.	$\sqrt[n]{a},\ -\sqrt[n]{a}$	$\sqrt[4]{81} = 3,\ -\sqrt[4]{81} = -3$
$a > 0$ or $a < 0$	n is odd.	$\sqrt[n]{a}$	$\sqrt[3]{-8} = -2$
$a < 0$	n is even.	No real roots	$\sqrt{-4}$ is not a real number.
$a = 0$	n is even or odd.	$\sqrt[n]{0} = 0$	$\sqrt[5]{0} = 0$

Integers such as 1, 4, 9, 16, 25, and 36 are called **perfect squares** because they have integer square roots. Similarly, integers such as 1, 8, 27, 64, and 125 are called **perfect cubes** because they have integer cube roots.

Properties of Radicals

Let a and b be real numbers, variables, or algebraic expressions such that the indicated roots are real numbers, and let m and n be positive integers.

Property	*Example*
1. $\sqrt[n]{a^m} = \left(\sqrt[n]{a}\right)^m$	$\sqrt[3]{8^2} = \left(\sqrt[3]{8}\right)^2 = (2)^2 = 4$
2. $\sqrt[n]{a} \cdot \sqrt[n]{b} = \sqrt[n]{ab}$	$\sqrt{5} \cdot \sqrt{7} = \sqrt{5 \cdot 7} = \sqrt{35}$
3. $\dfrac{\sqrt[n]{a}}{\sqrt[n]{b}} = \sqrt[n]{\dfrac{a}{b}}, \quad b \neq 0$	$\dfrac{\sqrt[4]{27}}{\sqrt[4]{9}} = \sqrt[4]{\dfrac{27}{9}} = \sqrt[4]{3}$
4. $\sqrt[m]{\sqrt[n]{a}} = \sqrt[mn]{a}$	$\sqrt[3]{\sqrt{10}} = \sqrt[6]{10}$
5. $\left(\sqrt[n]{a}\right)^n = a$	$\left(\sqrt{3}\right)^2 = 3$
6. For n even, $\sqrt[n]{a^n} = \lvert a \rvert$.	$\sqrt{(-12)^2} = \lvert -12 \rvert = 12$
For n odd, $\sqrt[n]{a^n} = a$.	$\sqrt[3]{(-12)^3} = -12$

A common special case of Property 6 is $\sqrt{a^2} = \lvert a \rvert$.

Example 6 ▶ Using Properties of Radicals 

Use the properties of radicals to simplify each expression.

a. $\sqrt{8} \cdot \sqrt{2}$ **b.** $\left(\sqrt[3]{5}\right)^3$ **c.** $\sqrt[3]{x^3}$ **d.** $\sqrt[6]{y^6}$

Solution

a. $\sqrt{8} \cdot \sqrt{2} = \sqrt{8 \cdot 2} = \sqrt{16} = 4$

b. $\left(\sqrt[3]{5}\right)^3 = 5$

c. $\sqrt[3]{x^3} = x$

d. $\sqrt[6]{y^6} = \lvert y \rvert$

Simplifying Radicals

An expression involving radicals is in **simplest form** when the following conditions are satisfied.

1. All possible factors have been removed from the radical.
2. All fractions have radical-free denominators (accomplished by a process called *rationalizing the denominator*).
3. The index of the radical is reduced.

To simplify a radical, factor the radicand into factors whose exponents are multiples of the index. The roots of these factors are written outside the radical, and the "leftover" factors make up the new radicand.

> **STUDY TIP**
>
> When you simplify a radical, it is important that both expressions are defined for the same values of the variable. For instance, in Example 7(b), $\sqrt{75x^3}$ and $5x\sqrt{3x}$ are both defined only for nonnegative values of x. Similarly, in Example 7(c), $\sqrt[4]{(5x)^4}$ and $5|x|$ are both defined for all real values of x.

Example 7 ▶ Simplifying Even Roots

a. $\sqrt[4]{48} = \sqrt[4]{16 \cdot 3} = \sqrt[4]{2^4 \cdot 3} = 2\sqrt[4]{3}$ (16: Perfect 4th power; 3: Leftover factor)

b. $\sqrt{75x^3} = \sqrt{25x^2 \cdot 3x}$ (25x²: Perfect square; 3x: Leftover factor)   Find largest square factor.

$= \sqrt{(5x)^2 \cdot 3x}$

$= 5x\sqrt{3x}$ Find root of perfect square.

c. $\sqrt[4]{(5x)^4} = |5x| = 5|x|$

In Example 7(b), the expression $\sqrt{75x^3}$ makes sense only for nonnegative values of x.

Example 8 ▶ Simplifying Odd Roots

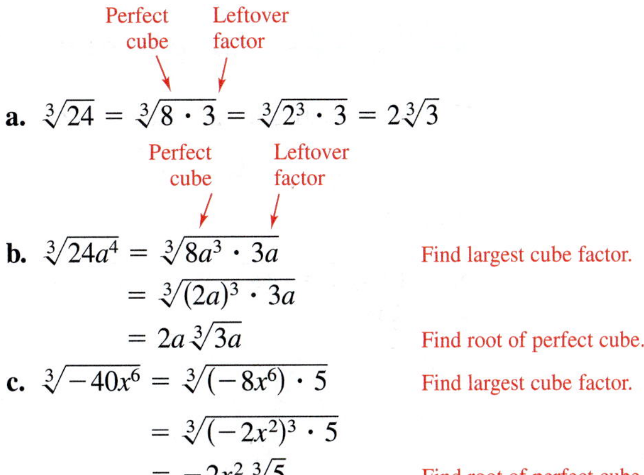

a. $\sqrt[3]{24} = \sqrt[3]{8 \cdot 3} = \sqrt[3]{2^3 \cdot 3} = 2\sqrt[3]{3}$ (8: Perfect cube; 3: Leftover factor)

b. $\sqrt[3]{24a^4} = \sqrt[3]{8a^3 \cdot 3a}$ (8a³: Perfect cube; 3a: Leftover factor)   Find largest cube factor.

$= \sqrt[3]{(2a)^3 \cdot 3a}$

$= 2a\sqrt[3]{3a}$ Find root of perfect cube.

c. $\sqrt[3]{-40x^6} = \sqrt[3]{(-8x^6) \cdot 5}$ Find largest cube factor.

$= \sqrt[3]{(-2x^2)^3 \cdot 5}$

$= -2x^2\sqrt[3]{5}$ Find root of perfect cube.

Radical expressions can be combined (added or subtracted) if they are **like radicals**—that is, if they have the same index and radicand. For instance, $\sqrt{2}$, $3\sqrt{2}$, and $\frac{1}{2}\sqrt{2}$ are like radicals, but $\sqrt{3}$ and $\sqrt{2}$ are unlike radicals. To determine whether two radicals can be combined, you should first simplify each radical.

Example 9 ▶ Combining Radicals 

a. $2\sqrt{48} - 3\sqrt{27} = 2\sqrt{16 \cdot 3} - 3\sqrt{9 \cdot 3}$ Find square factors.

$= 8\sqrt{3} - 9\sqrt{3}$ Find square roots.

$= (8 - 9)\sqrt{3}$ Combine like terms.

$= -\sqrt{3}$ Simplify.

b. $\sqrt[3]{16x} - \sqrt[3]{54x^4} = \sqrt[3]{8 \cdot 2x} - \sqrt[3]{27 \cdot x^3 \cdot 2x}$ Find cube factors.

$= 2\sqrt[3]{2x} - 3x\sqrt[3]{2x}$ Find cube roots.

$= (2 - 3x)\sqrt[3]{2x}$ Combine like terms.

Rationalizing Denominators and Numerators

To rationalize a denominator or numerator of the form $a - b\sqrt{m}$ or $a + b\sqrt{m}$, multiply both numerator and denominator by a **conjugate:** $a + b\sqrt{m}$ and $a - b\sqrt{m}$ are conjugates of each other. If $a = 0$, then the rationalizing factor for $\sqrt{m}$ is itself, $\sqrt{m}$. For cube roots, choose a rationalizing factor that generates a perfect cube.

Example 10 ▶ Rationalizing Single-Term Denominators 

Rationalize the denominator of each expression.

a. $\dfrac{5}{2\sqrt{3}}$ **b.** $\dfrac{2}{\sqrt[3]{5}}$

Solution

a. $\dfrac{5}{2\sqrt{3}} = \dfrac{5}{2\sqrt{3}} \cdot \dfrac{\sqrt{3}}{\sqrt{3}}$ $\sqrt{3}$ is rationalizing factor.

$= \dfrac{5\sqrt{3}}{2(3)}$

$= \dfrac{5\sqrt{3}}{6}$

b. $\dfrac{2}{\sqrt[3]{5}} = \dfrac{2}{\sqrt[3]{5}} \cdot \dfrac{\sqrt[3]{5^2}}{\sqrt[3]{5^2}}$ $\sqrt[3]{5^2}$ is rationalizing factor.

$= \dfrac{2\sqrt[3]{5^2}}{\sqrt[3]{5^3}}$

$= \dfrac{2\sqrt[3]{25}}{5}$

Example 11 ▶ Rationalizing a Denominator with Two Terms

$$\frac{2}{3+\sqrt{7}} = \frac{2}{3+\sqrt{7}} \cdot \frac{3-\sqrt{7}}{3-\sqrt{7}}$$ Multiply numerator and denominator by conjugate of denominator.

$$= \frac{2(3-\sqrt{7})}{3(3)+3(-\sqrt{7})+\sqrt{7}(3)-(\sqrt{7})(\sqrt{7})}$$ Use Distributive Property.

$$= \frac{2(3-\sqrt{7})}{(3)^2-(\sqrt{7})^2}$$ Simplify.

$$= \frac{2(3-\sqrt{7})}{9-7}$$ Square terms of denominator.

$$= \frac{2(3-\sqrt{7})}{2} = 3-\sqrt{7}$$ Simplify.

Sometimes it is necessary to rationalize the numerator of an expression. For instance, in Section P.5 you will use the technique shown in the next example to rationalize the numerator of an expression from calculus.

STUDY TIP

Do not confuse the expression $\sqrt{5}+\sqrt{7}$ with the expression $\sqrt{5+7}$. In general, $\sqrt{x+y}$ does not equal $\sqrt{x}+\sqrt{y}$. Similarly, $\sqrt{x^2+y^2}$ does not equal $x+y$.

Example 12 ▶ Rationalizing a Numerator

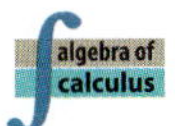

$$\frac{\sqrt{5}-\sqrt{7}}{2} = \frac{\sqrt{5}-\sqrt{7}}{2} \cdot \frac{\sqrt{5}+\sqrt{7}}{\sqrt{5}+\sqrt{7}}$$ Multiply numerator and denominator by conjugate of numerator.

$$= \frac{(\sqrt{5})^2-(\sqrt{7})^2}{2(\sqrt{5}+\sqrt{7})}$$ Simplify.

$$= \frac{5-7}{2(\sqrt{5}+\sqrt{7})}$$ Square terms of numerator.

$$= \frac{-2}{2(\sqrt{5}+\sqrt{7})} = \frac{-1}{\sqrt{5}+\sqrt{7}}$$ Simplify.

Rational Exponents

Definition of Rational Exponents

If a is a real number and n is a positive integer such that the principal nth root of a exists, then $a^{1/n}$ is defined as

$$a^{1/n} = \sqrt[n]{a}, \text{ where } 1/n \text{ is the } \textbf{rational exponent} \text{ of } a.$$

Moreover, if m is a positive integer that has no common factor with n, then

$$a^{m/n} = (a^{1/n})^m = (\sqrt[n]{a})^m \quad \text{and} \quad a^{m/n} = (a^m)^{1/n} = \sqrt[n]{a^m}.$$

The symbol $\int$ indicates an example or exercise that highlights algebraic techniques specifically used in calculus.

STUDY TIP

Rational exponents can be tricky, and you must remember that the expression $b^{m/n}$ is not defined unless $\sqrt[n]{b}$ is a real number. This restriction produces some unusual-looking results. For instance, the number $(-8)^{1/3}$ is defined because $\sqrt[3]{-8} = -2$, but the number $(-8)^{2/6}$ is undefined because $\sqrt[6]{-8}$ is not a real number.

The numerator of a rational exponent denotes the *power* to which the base is raised, and the denominator denotes the *index* or the *root* to be taken.

Power
Index

$$b^{m/n} = \left(\sqrt[n]{b}\right)^m = \sqrt[n]{b^m}$$

When you are working with rational exponents, the properties of integer exponents still apply. For instance,

$$2^{1/2}2^{1/3} = 2^{(1/2)+(1/3)} = 2^{5/6}.$$

Example 13 ▶ Changing from Radical to Exponential Form 

a. $\sqrt{3} = 3^{1/2}$

b. $\sqrt{(3xy)^5} = \sqrt[2]{(3xy)^5} = (3xy)^{(5/2)}$

c. $2x\sqrt[4]{x^3} = (2x)(x^{3/4}) = 2x^{1+(3/4)} = 2x^{7/4}$

Example 14 ▶ Changing from Exponential to Radical Form 

a. $(x^2 + y^2)^{3/2} = \left(\sqrt{x^2 + y^2}\right)^3 = \sqrt{(x^2 + y^2)^3}$

b. $2y^{3/4}z^{1/4} = 2(y^3z)^{1/4} = 2\sqrt[4]{y^3z}$

c. $a^{-3/2} = \dfrac{1}{a^{3/2}} = \dfrac{1}{\sqrt{a^3}}$

d. $x^{0.2} = x^{1/5} = \sqrt[5]{x}$

Technology

There are four methods of evaluating radicals on most graphing calculators. For square roots, you can use the square root key $\boxed{\sqrt{\ }}$. For cube roots, you can use the cube root key $\boxed{\sqrt[3]{\ }}$. For other roots, you can first convert the radical to exponential form and then use the exponential key $\boxed{\wedge}$, or you can use the *n*th root key $\boxed{\sqrt[x]{\ }}$.

Rational exponents are useful for evaluating roots of numbers on a calculator, for reducing the index of a radical, and for simplifying expressions in calculus.

Example 15 ▶ Simplifying with Rational Exponents 

a. $(-32)^{-4/5} = \left(\sqrt[5]{-32}\right)^{-4} = (-2)^{-4} = \dfrac{1}{(-2)^4} = \dfrac{1}{16}$

b. $(-5x^{5/3})(3x^{-3/4}) = -15x^{(5/3)-(3/4)} = -15x^{11/12}, \quad x \neq 0$

c. $\sqrt[9]{a^3} = a^{3/9} = a^{1/3} = \sqrt[3]{a}$ Reduce index.

d. $\sqrt[3]{\sqrt{125}} = \sqrt[6]{125} = \sqrt[6]{(5)^3} = 5^{3/6} = 5^{1/2} = \sqrt{5}$

e. $(2x - 1)^{4/3}(2x - 1)^{-1/3} = (2x - 1)^{(4/3)-(1/3)}$

$= 2x - 1, \quad x \neq \dfrac{1}{2}$

f. $\dfrac{x - 1}{(x - 1)^{-1/2}} = \dfrac{x - 1}{(x - 1)^{-1/2}} \cdot \dfrac{(x - 1)^{1/2}}{(x - 1)^{1/2}}$

$= \dfrac{(x - 1)^{3/2}}{(x - 1)^0}$

$= (x - 1)^{3/2}, \quad x \neq 1$

P.2 Exercises

In Exercises 1–4, write the expression as a repeated multiplication problem.

1. 8^5
2. $(-2)^7$
3. -0.4^6
4. 11.3^4

In Exercises 5–8, write the expression using exponential notation.

5. $(4.9)(4.9)(4.9)(4.9)(4.9)(4.9)$
6. $(2\sqrt{5})(2\sqrt{5})(2\sqrt{5})(2\sqrt{5})$
7. $(-10)(-10)(-10)(-10)(-10)$
8. $-(\frac{3}{2} \times \frac{3}{2} \times \frac{3}{2} \times \frac{3}{2})$

In Exercises 9–16, evaluate each expression.

9. (a) $3^2 \cdot 3$ (b) $3 \cdot 3^3$
10. (a) $\dfrac{5^5}{5^2}$ (b) $\dfrac{3^2}{3^4}$
11. (a) $(3^3)^2$ (b) -3^2
12. (a) $(2^3 \cdot 3^2)^2$ (b) $\left(-\frac{3}{5}\right)^3\left(\frac{5}{3}\right)^2$
13. (a) $\dfrac{3 \cdot 4^{-4}}{3^{-4} \cdot 4^{-1}}$ (b) $32(-2)^{-5}$
14. (a) $\dfrac{4 \cdot 3^{-2}}{2^{-2} \cdot 3^{-1}}$ (b) $(-2)^0$
15. (a) $2^{-1} + 3^{-1}$ (b) $(2^{-1})^{-2}$
16. (a) $3^{-1} + 2^{-2}$ (b) $(3^{-2})^2$

In Exercises 17–20, use a calculator to evaluate the expression. (If necessary, round your answer to three decimal places.)

17. $(-4)^3(5^2)$
18. $(8^{-4})(10^3)$
19. $\dfrac{3^6}{7^3}$
20. $\dfrac{4^3}{3^{-4}}$

In Exercises 21–28, evaluate the expression for the given value of x.

	Expression	*Value*
21.	$-3x^3$	2
22.	$7x^{-2}$	4
23.	$6x^0$	10
24.	$5(-x)^3$	3
25.	$2x^3$	-3
26.	$-3x^4$	-2
27.	$4x^2$	$-\frac{1}{2}$
28.	$5(-x)^3$	$\frac{1}{3}$

In Exercises 29–34, simplify each expression.

29. (a) $(-5z)^3$ (b) $5x^4(x^2)$
30. (a) $(3x)^2$ (b) $(4x^3)^2$
31. (a) $6y^2(2y^4)^2$ (b) $\dfrac{3x^5}{x^3}$
32. (a) $(-z)^3(3z^4)$ (b) $\dfrac{25y^8}{10y^4}$
33. (a) $\dfrac{7x^2}{x^3}$ (b) $\dfrac{12(x+y)^3}{9(x+y)}$
34. (a) $\dfrac{r^4}{r^6}$ (b) $\left(\dfrac{4}{y}\right)^3\left(\dfrac{3}{y}\right)^4$

In Exercises 35–40, rewrite each expression with positive exponents and simplify.

35. (a) $(x + 5)^0, \quad x \neq -5$ (b) $(2x^2)^{-2}$
36. (a) $(2x^5)^0, \quad x \neq 0$ (b) $(z + 2)^{-3}(z + 2)^{-1}$
37. (a) $(-2x^2)^3(4x^3)^{-1}$ (b) $\left(\dfrac{x}{10}\right)^{-1}$
38. (a) $(4y^{-2})(8y^4)$ (b) $\left(\dfrac{x^{-3}y^4}{5}\right)^{-3}$
39. (a) $3^n \cdot 3^{2n}$ (b) $\left(\dfrac{a^{-2}}{b^{-2}}\right)\left(\dfrac{b}{a}\right)^3$
40. (a) $\dfrac{x^2 \cdot x^n}{x^3 \cdot x^n}$ (b) $\left(\dfrac{a^{-3}}{b^{-3}}\right)\left(\dfrac{a}{b}\right)^3$

In Exercises 41–44, write the number in scientific notation.

41. Land area of Earth: 57,300,000 square miles
42. Light year: 9,460,000,000,000 kilometers
43. Relative density of hydrogen: 0.0000899 gram per cubic centimeter
44. One micron (millionth of a meter): 0.00003937 inch

In Exercises 45–48, write the number in decimal notation.

45. Worldwide daily consumption of Coca-Cola: 4.568×10^9 servings (Source: The Coca-Cola Company)

46. Interior temperature of the sun: 1.5×10^7 degrees Celsius

47. Charge of an electron: 1.602×10^{-19} coulomb

48. Width of a human hair: 9.0×10^{-5} meter

In Exercises 49 and 50, evaluate each expression without using a calculator.

49. (a) $\sqrt{25 \times 10^8}$ (b) $\sqrt[3]{8 \times 10^{15}}$

50. (a) $(1.2 \times 10^7)(5 \times 10^{-3})$ (b) $\dfrac{(6.0 \times 10^8)}{(3.0 \times 10^{-3})}$

In Exercises 51–54, use a calculator to evaluate each expression. (Round your answer to three decimal places.)

51. (a) $750\left(1 + \dfrac{0.11}{365}\right)^{800}$

(b) $\dfrac{67{,}000{,}000 + 93{,}000{,}000}{0.0052}$

52. (a) $(9.3 \times 10^6)^3(6.1 \times 10^{-4})$

(b) $\dfrac{(2.414 \times 10^4)^6}{(1.68 \times 10^5)^5}$

53. (a) $\sqrt{4.5 \times 10^9}$ (b) $\sqrt[3]{6.3 \times 10^4}$

54. (a) $(2.65 \times 10^{-4})^{1/3}$ (b) $\sqrt{9 \times 10^{-4}}$

In Exercises 55–66, fill in the missing form of the expression.

	Radical Form	*Rational Exponent Form*
55.	$\sqrt{9}$	
56.	$\sqrt[3]{64}$	
57.		$32^{1/5}$
58.		$-(144^{1/2})$
59.		$196^{1/2}$
60.	$\sqrt[3]{614.125}$	
61.	$\sqrt[3]{-216}$	
62.		$(-243)^{1/5}$
63.		$27^{2/3}$
64.	$(\sqrt[4]{81})^3$	
65.	$\sqrt[4]{81^3}$	
66.		$16^{5/4}$

In Exercises 67–74, evaluate each expression without using a calculator.

67. (a) $\sqrt{9}$ (b) $\sqrt[3]{8}$

68. (a) $\sqrt{49}$ (b) $\sqrt[3]{\frac{27}{8}}$

69. (a) $(\sqrt[3]{-125})^3$ (b) $27^{1/3}$

70. (a) $\sqrt[4]{562^4}$ (b) $36^{3/2}$

71. (a) $32^{-3/5}$ (b) $\left(\frac{16}{81}\right)^{-3/4}$

72. (a) $100^{-3/2}$ (b) $\left(\frac{9}{4}\right)^{-1/2}$

73. (a) $\left(-\dfrac{1}{64}\right)^{-1/3}$ (b) $\left(\dfrac{1}{\sqrt{32}}\right)^{-2/5}$

74. (a) $\left(-\dfrac{125}{27}\right)^{-1/3}$ (b) $-\left(\dfrac{1}{125}\right)^{-4/3}$

In Exercises 75–78, use a calculator to approximate the number. (Round your answer to three decimal places.)

75. (a) $\sqrt{57}$ (b) $\sqrt[5]{-27^3}$

76. (a) $\sqrt[3]{45^2}$ (b) $\sqrt[6]{125}$

77. (a) $(-12.4)^{-1.8}$ (b) $(5\sqrt{3})^{-2.5}$

78. (a) $\dfrac{7 - (4.1)^{-3.2}}{2}$ (b) $\left(\dfrac{13}{3}\right)^{-3/2} - \left(-\dfrac{3}{2}\right)^{13/3}$

In Exercises 79–84, simplify by removing all possible factors from each radical.

79. (a) $\sqrt{8}$ (b) $\sqrt[3]{24}$

80. (a) $\sqrt[3]{\frac{16}{27}}$ (b) $\sqrt{\frac{75}{4}}$

81. (a) $\sqrt{72x^3}$ (b) $\sqrt{\dfrac{18^2}{z^3}}$

82. (a) $\sqrt{54xy^4}$ (b) $\sqrt{\dfrac{32a^4}{b^2}}$

83. (a) $\sqrt[3]{16x^5}$ (b) $\sqrt{75x^2y^{-4}}$

84. (a) $\sqrt[4]{(3x^2)^4}$ (b) $\sqrt[5]{96x^5}$

In Exercises 85–88, perform the operations and simplify.

85. $\dfrac{(2x^2)^{3/2}}{2^{1/2}x^4}$

86. $\dfrac{x^{4/3}y^{2/3}}{(xy)^{1/3}}$

87. $\dfrac{x^{-3} \cdot x^{1/2}}{x^{3/2} \cdot x^{-1}}$

88. $\dfrac{5^{-1/2} \cdot 5x^{5/2}}{(5x)^{3/2}}$

In Exercises 89–92, rationalize the denominator of the expression. Then simplify your answer.

89. $\dfrac{1}{\sqrt{3}}$

90. $\dfrac{5}{\sqrt{10}}$

91. $\dfrac{2}{5 - \sqrt{3}}$

92. $\dfrac{3}{\sqrt{5} + \sqrt{6}}$

In Exercises 93–96, rationalize the numerator of the expression. Then simplify your answer.

93. $\dfrac{\sqrt{8}}{2}$ **94.** $\dfrac{\sqrt{2}}{3}$

95. $\dfrac{\sqrt{5}+\sqrt{3}}{3}$ **96.** $\dfrac{\sqrt{7}-3}{4}$

In Exercises 97 and 98, reduce the index of each radical.

97. (a) $\sqrt[4]{3^2}$ (b) $\sqrt[6]{(x+1)^4}$

98. (a) $\sqrt[6]{x^3}$ (b) $\sqrt[4]{(3x^2)^4}$

In Exercises 99 and 100, write each expression as a single radical. Then simplify your answer.

99. (a) $\sqrt{\sqrt{32}}$ (b) $\sqrt{\sqrt[4]{2x}}$

100. (a) $\sqrt{\sqrt{243(x+1)}}$ (b) $\sqrt{\sqrt[3]{10a^7b}}$

In Exercises 101–106, simplify each expression.

101. (a) $2\sqrt{50}+12\sqrt{8}$ (b) $10\sqrt{32}-6\sqrt{18}$

102. (a) $4\sqrt{27}-\sqrt{75}$ (b) $\sqrt[3]{16}+3\sqrt[3]{54}$

103. (a) $5\sqrt{x}-3\sqrt{x}$ (b) $-2\sqrt{9y}+10\sqrt{y}$

104. (a) $8\sqrt{49x}-14\sqrt{100x}$

(b) $-3\sqrt{48x^2}+7\sqrt{75x^2}$

105. (a) $3\sqrt{x+1}+10\sqrt{x+1}$

(b) $7\sqrt{80x}-2\sqrt{125x}$

106. (a) $-\sqrt{x^3-7}+5\sqrt{x^3-7}$

(b) $11\sqrt{245x^3}-9\sqrt{45x^3}$

In Exercises 107–110, complete the statement with <, =, or >.

107. $\sqrt{5}+\sqrt{3}$ ☐ $\sqrt{5+3}$

108. $\sqrt{\dfrac{3}{11}}$ ☐ $\dfrac{\sqrt{3}}{\sqrt{11}}$

109. 5 ☐ $\sqrt{3^2+2^2}$ **110.** 5 ☐ $\sqrt{3^2+4^2}$

111. ***Period of a Pendulum*** The period T (in seconds) of a pendulum is

$$T = 2\pi\sqrt{\frac{L}{32}}$$

where L is the length of the pendulum (in feet). Find the period of a pendulum whose length is 2 feet.

The symbol $\int$ indicates an example or exercise that highlights algebraic techniques specifically used in calculus.

The symbol [graphing utility icon] indicates an exercise or parts of an exercise in which you are instructed to use a graphing utility.

112. ***Erosion*** A stream of water moving at the rate of v feet per second can carry particles of size $0.03\sqrt{v}$ inches. Find the size of the largest particle that can be carried by a stream flowing at the rate of $\frac{3}{4}$ foot per second.

▶ Model It

113. ***Mathematical Modeling*** A funnel is filled with water to a height of h centimeters. The time t (in seconds) for the funnel to empty is

$$t = 0.03[12^{5/2} - (12-h)^{5/2}], \quad 0 \le h \le 12.$$

(a) Use the *table* feature of a graphing utility to find the times required for the funnel to empty for water heights of $h = 0$, $h = 1$, $h = 2, \ldots h = 12$ centimeters.

(b) Is there a limiting value of time required for the water to empty as the height of the water becomes closer to 12 centimeters? Explain.

114. ***Speed of Light*** The speed of light is approximately 11,180,000 miles per minute. The distance from the sun to Earth is approximately 93,000,000 miles. Find the time for light to travel from the sun to Earth.

Synthesis

True or False? **In Exercises 115 and 116, determine whether the statement is true or false. Justify your answer.**

115. $\dfrac{x^{k+1}}{x} = x^k$ **116.** $(a^n)^k = a^{n^k}$

117. Verify that $a^0 = 1$, $a \neq 0$. (*Hint:* Use the property of exponents $a^m/a^n = a^{m-n}$.)

118. Explain why each of the following pairs is not equal.

(a) $(3x)^{-1} \neq \dfrac{3}{x}$ (b) $y^3 \cdot y^2 \neq y^6$

(c) $(a^2b^3)^4 \neq a^6b^7$ (d) $(a+b)^2 \neq a^2+b^2$

(e) $\sqrt{4x^2} \neq 2x$ (f) $\sqrt{2}+\sqrt{3} \neq \sqrt{5}$

119. ***Exploration*** List all possible digits that occur in the units place of the square of a positive integer. Use that list to determine whether $\sqrt{5233}$ is an integer.

120. ***Think About It*** Square the real number $2/\sqrt{5}$ and note that the radical is eliminated from the denominator. Is this equivalent to rationalizing the denominator? Why or why not?

P.3 Polynomials and Special Products

▶ What you should learn

- How to write polynomials in standard form
- How to add, subtract, and multiply polynomials
- How to use special products to multiply polynomials
- How to use polynomials to solve real-life problems

▶ Why you should learn it

Polynomials can be used to model and solve real-life problems. For instance, in Exercise 104 on page 32, a polynomial is used to model the stopping distance of an automobile.

Nicholas DeVore/Tony Stone Images

Polynomials

The most common type of algebraic expression is the **polynomial.** Some examples are

$$2x + 5, \quad 3x^4 - 7x^2 + 2x + 4, \quad \text{and} \quad 5x^2y^2 - xy + 3.$$

The first two are *polynomials in x* and the third is a *polynomial in x and y*. The terms of a polynomial in x have the form ax^k, where a is the **coefficient** and k is the **degree** of the term. For instance, the polynomial

$$2x^3 - 5x^2 + 1 = 2x^3 + (-5)x^2 + (0)x + 1$$

has coefficients 2, -5, 0, and 1.

Definition of a Polynomial in x

Let $a_0, a_1, a_2, \ldots, a_n$ be real numbers and let n be a nonnegative integer. A polynomial in x is an expression of the form

$$a_nx^n + a_{n-1}x^{n-1} + \cdots + a_1x + a_0$$

where $a_n \neq 0$. The polynomial is of **degree** n, a_n is the **leading coefficient,** and a_0 is the **constant term.**

Polynomials with one, two, and three terms are called **monomials, binomials,** and **trinomials,** respectively. In **standard form,** a polynomial is written with descending powers of x.

Example 1 ▶ Writing Polynomials in Standard Form 

	Polynomial	*Standard Form*	*Degree*
a.	$4x^2 - 5x^7 - 2 + 3x$	$-5x^7 + 4x^2 + 3x - 2$	7
b.	$4 - 9x^2$	$-9x^2 + 4$	2
c.	8	$8 \quad (8 = 8x^0)$	0

A polynomial that has all zero coefficients is called the zero polynomial, denoted by 0. No degree is assigned to this particular polynomial. For polynomials in more than one variable, the degree of a *term* is the sum of the exponents of the variables in the term. The degree of the *polynomial* is the degree of the highest-degree term. The leading coefficient of the polynomial is the coefficient of the highest-degree term. Expressions such as the following are not polynomials.

$x^3 - \sqrt{3x} = x^3 - (3x)^{1/2}$ The exponent "1/2" is not an integer.

$x^2 + 5x^{-1}$ The exponent "-1" is not a nonnegative integer.

Operations with Polynomials

You can add and subtract polynomials in much the same way you add and subtract real numbers. Simply add or subtract the *like terms* (terms having the same variables to the same powers) by adding their coefficients. For instance, $-3xy^2$ and $5xy^2$ are like terms and their sum is

$$\begin{aligned}-3xy^2 + 5xy^2 &= (-3 + 5)xy^2\\ &= 2xy^2.\end{aligned}$$

> **STUDY TIP**
>
> A common mistake is to fail to change the sign of each term inside parentheses preceded by a negative sign. For instance, note that
>
> $$\begin{aligned}&-(x^2 - x + 3)\\ &\quad = -x^2 + x - 3\end{aligned}$$
>
> and
>
> $$\begin{aligned}&-(x^2 - x + 3)\\ &\quad \neq -x^2 - x + 3.\end{aligned}$$

Example 2 ▶ Sums and Differences of Polynomials 

a. $(5x^3 - 7x^2 - 3) + (x^3 + 2x^2 - x + 8)$

$= (5x^3 + x^3) + (2x^2 - 7x^2) - x + (8 - 3)$ — Group like terms.

$= 6x^3 - 5x^2 - x + 5$ — Combine like terms.

b. $(7x^4 - x^2 - 4x + 2) - (3x^4 - 4x^2 + 3x)$

$= 7x^4 - x^2 - 4x + 2 - 3x^4 + 4x^2 - 3x$ — Distributive Property

$= (7x^4 - 3x^4) + (4x^2 - x^2) + (-3x - 4x) + 2$ — Group like terms.

$= 4x^4 + 3x^2 - 7x + 2$ — Combine like terms.

To find the product of two polynomials, use the left and right Distributive Properties. For example, if you treat $5x + 7$ as a single quantity, you can multiply $3x - 2$ by $5x + 7$ as follows.

$$\begin{aligned}(3x - 2)(5x + 7) &= 3x(5x + 7) - 2(5x + 7)\\ &= (3x)(5x) + (3x)(7) - (2)(5x) - (2)(7)\\ &= 15x^2 + 21x - 10x - 14\end{aligned}$$

Product of First terms; Product of Outer terms; Product of Inner terms; Product of Last terms

$$= 15x^2 + 11x - 14$$

Note in this **FOIL Method** that for binomials the outer (O) and inner (I) terms are like terms and can be combined.

Example 3 ▶ Using the FOIL Method 

Use the FOIL Method to find the product of $2x - 4$ and $x + 5$.

Solution

F O I L

$$\begin{aligned}(2x - 4)(x + 5) &= 2x^2 + 10x - 4x - 20\\ &= 2x^2 + 6x - 20\end{aligned}$$

When multiplying two polynomials, be sure to multiply each term of one polynomial by *each* term of the other. A vertical arrangement is helpful.

Example 4 ▶ A Vertical Arrangement for Multiplication

Multiply $x^2 - 2x + 2$ by $x^2 + 2x + 2$ using a vertical arrangement.

Solution

$x^2 - 2x + 2$	Write in standard form.
$\times\ x^2 + 2x + 2$	Write in standard form.
$x^4 - 2x^3 + 2x^2$	$x^2(x^2 - 2x + 2)$
$2x^3 - 4x^2 + 4x$	$2x(x^2 - 2x + 2)$
$2x^2 - 4x + 4$	$2(x^2 - 2x + 2)$
$x^4 + 0x^3 + 0x^2 + 0x + 4 = x^4 + 4$	Combine like terms.

So,

$$(x^2 - 2x + 2)(x^2 + 2x + 2) = x^4 + 4.$$

Special Products

Some binomial products have special forms that occur frequently in algebra.

Special Products

Let u and v be real numbers, variables, or algebraic expressions.

Special Product	*Example*
Sum and Difference of Same Terms	
$(u + v)(u - v) = u^2 - v^2$	$(x + 4)(x - 4) = x^2 - 4^2$ $= x^2 - 16$
Square of a Binomial	
$(u + v)^2 = u^2 + 2uv + v^2$	$(x + 3)^2 = x^2 + 2(x)(3) + 3^2$ $= x^2 + 6x + 9$
$(u - v)^2 = u^2 - 2uv + v^2$	$(3x - 2)^2 = (3x)^2 - 2(3x)(2) + 2^2$ $= 9x^2 - 12x + 4$
Cube of a Binomial	
$(u + v)^3 = u^3 + 3u^2v + 3uv^2 + v^3$	$(x + 2)^3 = x^3 + 3x^2(2) + 3x(2^2) + 2^3$ $= x^3 + 6x^2 + 12x + 8$
$(u - v)^3 = u^3 - 3u^2v + 3uv^2 - v^3$	$(x - 1)^3 = x^3 - 3x^2(1) + 3x(1^2) - 1^3$ $= x^3 - 3x^2 + 3x - 1$

Example 5 ▶ Sum and Difference of Same Terms 

Find the product of $5x + 9$ and $5x - 9$.

Solution

The product of a sum and a difference of the *same* two terms has no middle term and takes the form $(u + v)(u - v) = u^2 - v^2$.

$$\begin{aligned}(5x + 9)(5x - 9) &= (5x)^2 - 9^2 \\ &= 25x^2 - 81\end{aligned}$$

STUDY TIP

When squaring a binomial, note that the resulting middle term is always *twice* the product of the two terms.

Example 6 ▶ Square of a Binomial 

Find $(6x - 5)^2$.

Solution

The square of a binomial has the form $(u - v)^2 = u^2 - 2uv + v^2$.

$$\begin{aligned}(6x - 5)^2 &= (6x)^2 - 2(6x)(5) + 5^2 \\ &= 36x^2 - 60x + 25\end{aligned}$$

Example 7 ▶ Cube of a Binomial 

Find $(3x + 2)^3$.

Solution

The cube of a binomial has the form

$$(u + v)^3 = u^3 + 3u^2v + 3uv^2 + v^3.$$

Note the *decrease* of powers of u and the *increase* of powers of v.

$$\begin{aligned}(3x + 2)^3 &= (3x)^3 + 3(3x)^2(2) + 3(3x)(2^2) + 2^3 \\ &= 27x^3 + 54x^2 + 36x + 8\end{aligned}$$

Example 8 ▶ The Product of Two Trinomials 

Find the product of $x + y - 2$ and $x + y + 2$.

Solution

By grouping $x + y$ in parentheses, you can write the product of the trinomials as a special product.

$$\begin{aligned}(x + y - 2)(x + y + 2) &= [(x + y) - 2][(x + y) + 2] \\ &= (x + y)^2 - 2^2 \\ &= x^2 + 2xy + y^2 - 4\end{aligned}$$

Application

Example 9 ▶ Volume of a Box

An open box is made by cutting squares from the corners of a piece of metal that is 16 inches by 20 inches, as shown in Figure P.7. The edge of each cut-out square is x inches. Find the volume of the box when $x = 1$, $x = 2$, and $x = 3$.

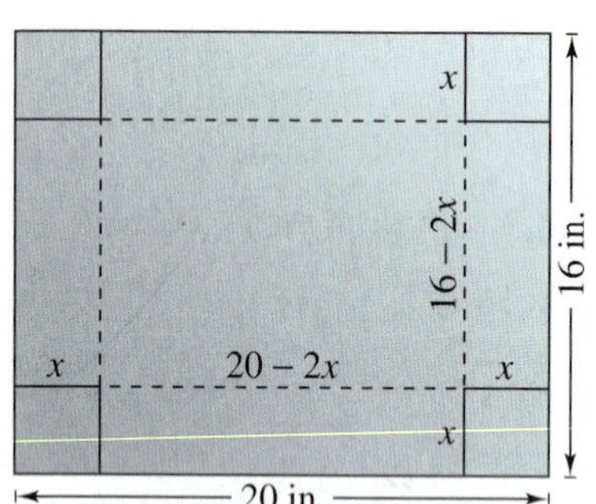

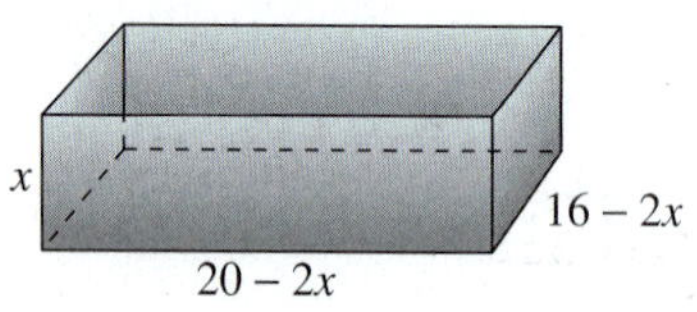

FIGURE P.7

Solution

The volume of a rectangular box is equal to the product of its length, width, and height. From the figure, the length is $20 - 2x$, the width is $16 - 2x$, and the height is x. So, the volume of the box is

$$\begin{aligned}\text{Volume} &= (20 - 2x)(16 - 2x)(x)\\ &= (320 - 72x + 4x^2)(x)\\ &= 320x - 72x^2 + 4x^3.\end{aligned}$$

When $x = 1$ inch, the volume of the box is

$$\begin{aligned}\text{Volume} &= 320(1) - 72(1)^2 + 4(1)^3\\ &= 252 \text{ cubic inches.}\end{aligned}$$

When $x = 2$ inches, the volume of the box is

$$\begin{aligned}\text{Volume} &= 320(2) - 72(2)^2 + 4(2)^3\\ &= 384 \text{ cubic inches.}\end{aligned}$$

When $x = 3$ inches, the volume of the box is

$$\begin{aligned}\text{Volume} &= 320(3) - 72(3)^2 + 4(3)^3\\ &= 420 \text{ cubic inches.}\end{aligned}$$

Writing ABOUT MATHEMATICS

Mathematical Experiment In Example 9, the volume of the open metal box is given by

$$\text{Volume} = 320x - 72x^2 + 4x^3.$$

You want to create a box that has as much volume as possible. From Example 9, you know that by cutting one-, two-, and three-inch squares from the corners, you can create boxes whose volumes are 252, 384, and 420 cubic inches, respectively. What are the possible values of x that make sense in the problem? Write your answer as an interval. Try several other values of x to find the size of the square that should be cut from the corners to produce a box that has maximum volume. Write a summary of your findings.

P.3 Exercises

In Exercises 1–6, match the polynomial with its description. [The polynomials are labeled (a), (b), (c), (d), (e), and (f).]

(a) $3x^2$ (b) $1 - 2x^3$

(c) $x^3 + 3x^2 + 3x + 1$ (d) 12

(e) $-3x^5 + 2x^3 + x$ (f) $\frac{2}{3}x^4 + x^2 + 10$

1. A polynomial of degree 0
2. A trinomial of degree 5
3. A binomial with leading coefficient -2
4. A monomial of positive degree
5. A trinomial with leading coefficient $\frac{2}{3}$
6. A third-degree polynomial with leading coefficient 1

In Exercises 7–10, write a polynomial that fits the description. (There are many correct answers.)

7. A third-degree polynomial with leading coefficient -2
8. A fifth-degree polynomial with leading coefficient 6
9. A fourth-degree binomial with a negative leading coefficient
10. A third-degree binomial with an even leading coefficient

In Exercises 11–18, find the degree and leading coefficient of the polynomial.

11. $2x^2 - x + 1$
12. $-3x^4 + 2x^2 - 5$
13. $x^5 - 1$
14. 3
15. $1 - x + 6x^4 - 4x^5$
16. $3 + 2x$
17. $4x^3y - 3xy^2 + x^2y^3$
18. $-x^5y + 2x^2y^2 + xy^4$

In Exercises 19–24, is the expression a polynomial? If so, write the polynomial in standard form.

19. $2x - 3x^3 + 8$
20. $2x^3 + x - 3x^{-1}$
21. $\dfrac{3x + 4}{x}$
22. $\dfrac{x^2 + 2x - 3}{2}$
23. $y^2 - y^4 + y^3$
24. $\sqrt{y^2 - y^4}$

In Exercises 25–42, perform the operation and write the result in standard form.

25. $(6x + 5) - (8x + 15)$
26. $(2x^2 + 1) - (x^2 - 2x + 1)$
27. $-(x^3 - 2) + (4x^3 - 2x)$
28. $-(5x^2 - 1) - (-3x^2 + 5)$
29. $(15x^2 - 6) - (-8.3x^3 - 14.7x^2 - 17)$
30. $(15.2x^4 - 18x - 19.1) - (13.9x^4 - 9.6x + 15)$
31. $5z - [3z - (10z + 8)]$
32. $(y^3 + 1) - [(y^2 + 1) + (3y - 7)]$
33. $3x(x^2 - 2x + 1)$
34. $y^2(4y^2 + 2y - 3)$
35. $-5z(3z - 1)$
36. $(-3x)(5x + 2)$
37. $(1 - x^3)(4x)$
38. $-4x(3 - x^3)$
39. $(2.5x^2 + 3)(3x)$
40. $(2 - 3.5y)(2y^3)$
41. $-4x\left(\frac{1}{8}x + 3\right)$
42. $2y\left(4 - \frac{7}{8}y\right)$

In Exercises 43–52, perform the operation.

43. Add $7x^3 - 2x^2 + 8$ and $-3x^3 - 4$.
44. Add $2x^5 - 3x^3 + 2x + 3$ and $4x^3 + x - 6$.
45. Subtract $x - 3$ from $5x^2 - 3x + 8$.
46. Subtract $-t^4 + 0.5t^2 - 5.6$ from $0.6t^4 - 2t^2$.
47. Multiply $-6x^2 + 15x - 4$ and $5x + 3$.
48. Multiply $4x^4 + x^3 - 6x^2 + 9$ and $x^2 + 2x + 3$.
49. $(x^2 + 9)(x^2 - x - 4)$
50. $(x - 2)(x^2 + 2x + 4)$
51. $(x^2 - x + 1)(x^2 + x + 1)$
52. $(x^2 + 3x - 2)(x^2 - 3x - 2)$

In Exercises 53–88, multiply or find the special product.

53. $(x + 3)(x + 4)$
54. $(x - 5)(x + 10)$
55. $(3x - 5)(2x + 1)$
56. $(7x - 2)(4x - 3)$
57. $(2x + 3)^2$
58. $(4x + 5)^2$
59. $(2x - 5y)^2$
60. $(5 - 8x)^2$
61. $(x + 10)(x - 10)$
62. $(2x + 3)(2x - 3)$
63. $(x + 2y)(x - 2y)$
64. $(2x + 3y)(2x - 3y)$
65. $[(m - 3) + n][(m - 3) - n]$
66. $[(x + y) + 1][(x + y) - 1]$
67. $[(x - 3) + y]^2$
68. $[(x + 1) - y]^2$
69. $(2r^2 - 5)(2r^2 + 5)$

70. $(3a^3 - 4b^2)(3a^3 + 4b^2)$

71. $(x + 1)^3$ **72.** $(x - 2)^3$

73. $(2x - y)^3$ **74.** $(3x + 2y)^3$

75. $(4x^3 - 3)^2$ **76.** $(8x + 3)^2$

77. $\left(\frac{1}{2}x - 3\right)^2$ **78.** $\left(\frac{2}{3}t + 5\right)^2$

79. $\left(\frac{1}{3}x - 2\right)\left(\frac{1}{3}x + 2\right)$ **80.** $\left(2x + \frac{1}{5}\right)\left(2x - \frac{1}{5}\right)$

81. $(1.2x + 3)^2$ **82.** $(1.5y - 3)^2$

83. $(1.5x - 4)(1.5x + 4)$

84. $(2.5y + 3)(2.5y - 3)$

85. $5x(x + 1) - 3x(x + 1)$

86. $(2x - 1)(x + 3) + 3(x + 3)$

87. $(u + 2)(u - 2)(u^2 + 4)$

88. $(x + y)(x - y)(x^2 + y^2)$

In Exercises 89–92, find the product. The expressions are not polynomials, but the formulas can still be used.

89. $(\sqrt{x} + \sqrt{y})(\sqrt{x} - \sqrt{y})$ **90.** $(5 + \sqrt{x})(5 - \sqrt{x})$

91. $(x - \sqrt{5})^2$ **92.** $(x + \sqrt{3})^2$

93. ***Cost, Revenue, and Profit*** An electronics manufacturer can produce and sell x radios per week. The total cost C (in dollars) for producing x radios is

$$C = 73x + 25{,}000$$

and the total revenue R (in dollars) is

$$R = 95x.$$

Find the profit P obtained by selling 5000 radios per week.

94. ***Cost, Revenue, and Profit*** An artist can produce and sell x craft items per month. The total cost C (in dollars) for producing x craft items is

$$C = 460 + 12x$$

and the total revenue R (in dollars) is

$$R = 36x.$$

Find the profit P obtained by selling 42 craft items per month.

95. ***Compound Interest*** After 2 years, an investment of \$500 compounded annually at an interest rate r will yield an amount of

$$500(1 + r)^2.$$

(a) Write this polynomial in standard form.

(b) Use a calculator to evaluate the polynomial for the values of r shown in the table.

r	$2\frac{1}{2}\%$	3%	4%	$4\frac{1}{2}\%$	5%
$500(1 + r)^2$					

(c) What conclusion can you make from the table?

96. ***Compound Interest*** After 3 years, an investment of \$1200 compounded annually at an interest rate r will yield an amount of $1200(1 + r)^3$.

(a) Write this polynomial in standard form.

(b) Use a calculator to evaluate the polynomial for the values of r shown in the table.

r	2%	3%	$3\frac{1}{2}\%$	4%	$4\frac{1}{2}\%$
$1200(1 + r)^3$					

(c) What conclusion can you make from the table?

97. ***Volume of a Box*** A take-out fast-food restaurant is constructing an open box by cutting squares from the corners of a piece of cardboard that is 18 centimeters by 26 centimeters (see figure). The edge of each cut-out square is x centimeters. Find the volume when $x = 1$, $x = 2$, and $x = 3$.

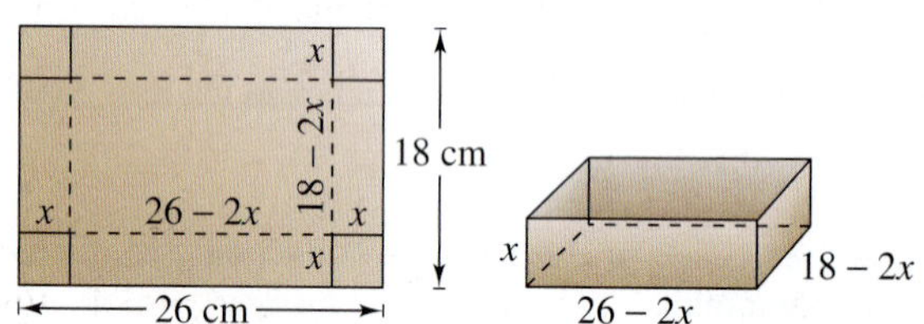

98. ***Volume of a Box*** An overnight shipping company is designing a closed box by cutting along the solid lines and folding along the broken lines on the rectangular piece of corrugated cardboard shown in the figure. The length and width of the rectangle are 45 centimeters and 15 centimeters, respectively. Find the volume of the shipping box in terms of x. Find the volume when $x = 3$, $x = 5$, and $x = 7$.

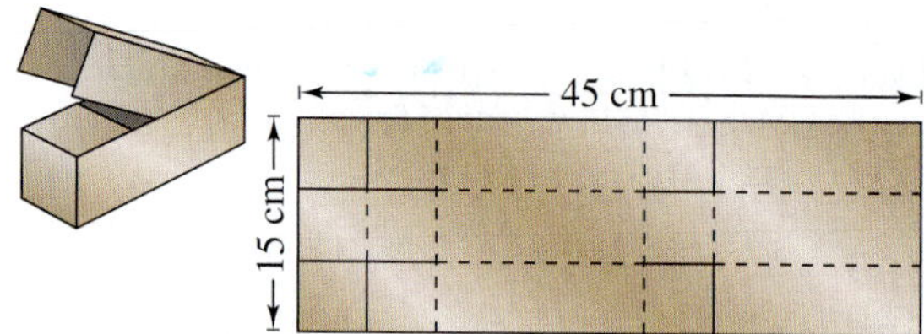

99. *Geometry* Find the area of the shaded region in each figure. Write your result as a polynomial in standard form.

(a)
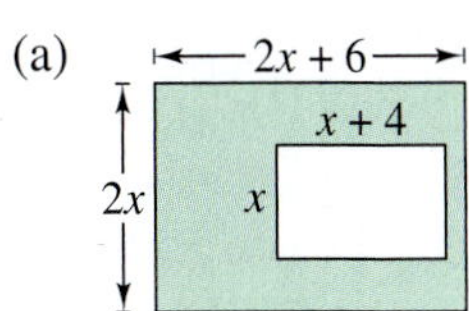

(b)
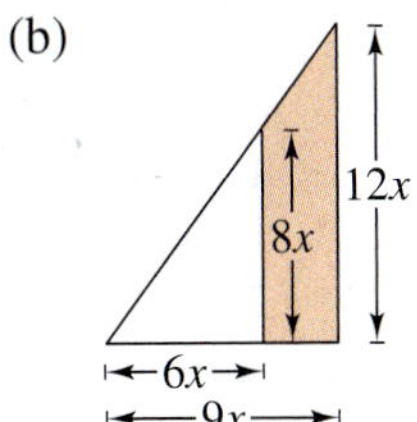

(c)
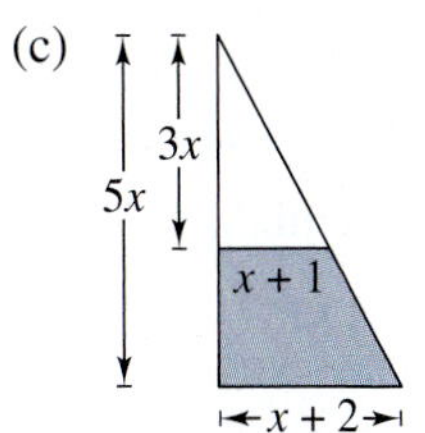

(d)
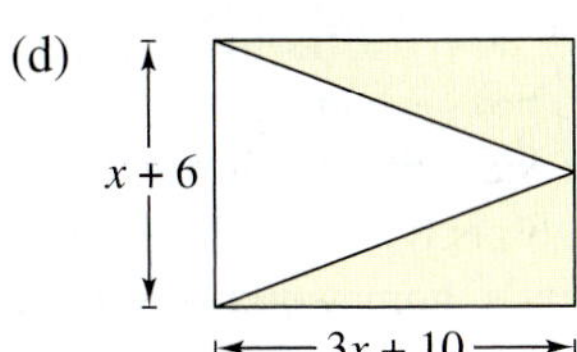

100. *Geometry* Find the area of the shaded region in each figure. Write your result as a polynomial in standard form.

(a) (b)
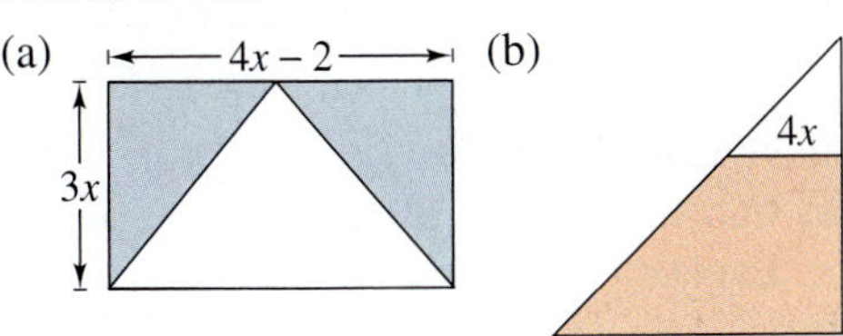

(c)
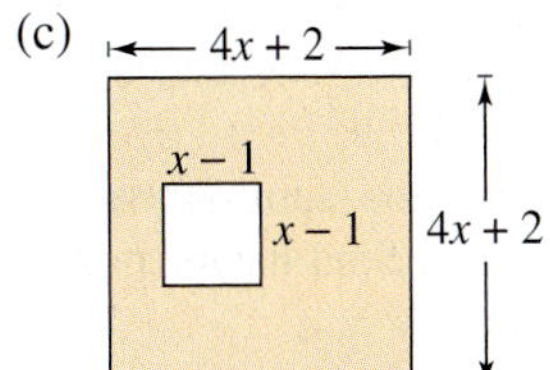

(d)
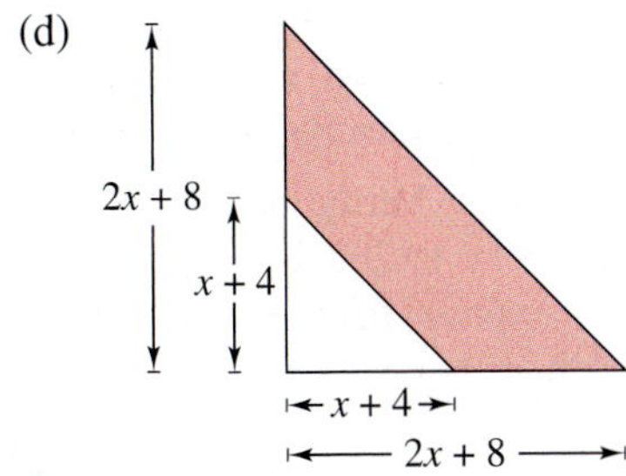

Geometry **In Exercises 101 and 102, find a polynomial that represents the total number of square feet for the floor plan shown in the figure.**

101.
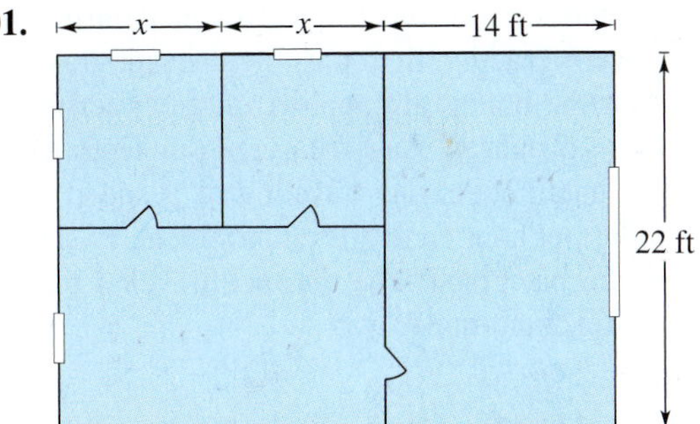

102.
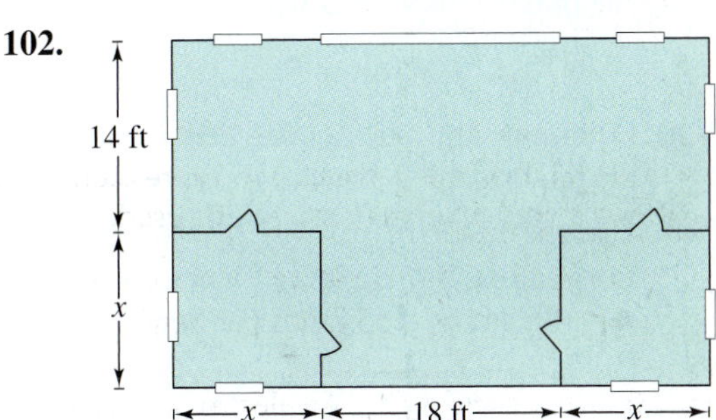

103. *Engineering* A uniformly distributed load is placed on a one-inch-wide steel beam. When the span of the beam is x feet and its depth is 6 inches, the safe load S is approximated by

$$S_6 = (0.06x^2 - 2.42x + 38.71)^2.$$

When the depth is 8 inches, the safe load is approximated by

$$S_8 = (0.08x^2 - 3.30x + 51.93)^2.$$

(a) Use the bar graph to estimate the difference in the safe loads for these two beams when the span is 12 feet.

(b) How does the difference in safe load change as the span increases?

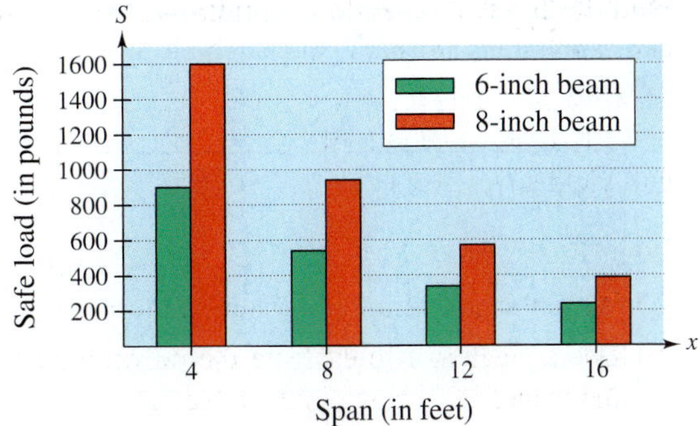

▶ Model It

104. ***Stopping Distance*** The stopping distance of an automobile is the distance traveled during the driver's reaction time plus the distance traveled after the brakes are applied. In an experiment, these distances were measured (in feet) when the automobile was traveling at a speed of x miles per hour on dry, level pavement, as shown in the bar graph. The distance traveled during the reaction time R was

$$R = 1.1x$$

and the braking distance B was

$$B = 0.0475x^2 - 0.001x + 0.23.$$

(a) Determine the polynomial that represents the total stopping distance T.

(b) Use the result of part (a) to estimate the total stopping distance when $x = 30$, $x = 40$, and $x = 55$ miles per hour.

(c) Use the bar graph to make a statement about the total stopping distance required for increasing speeds.

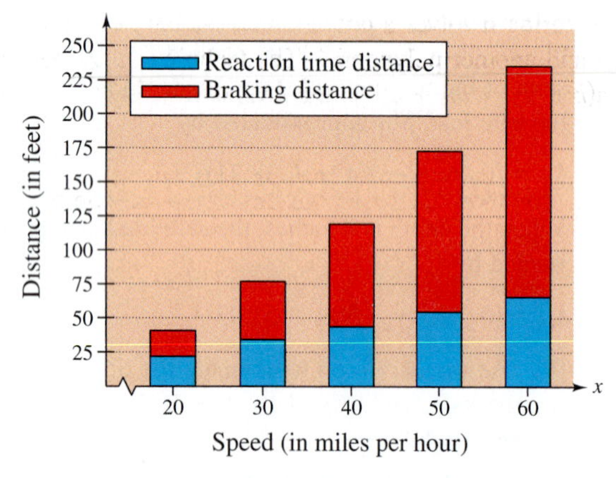

Geometry **In Exercises 105 and 106, use the area model to write two different expressions for the area. Then equate the two expressions and name the algebraic property that is illustrated.**

105.

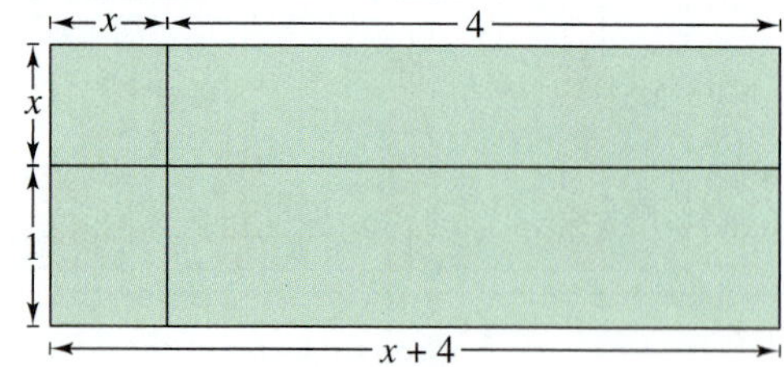

106.

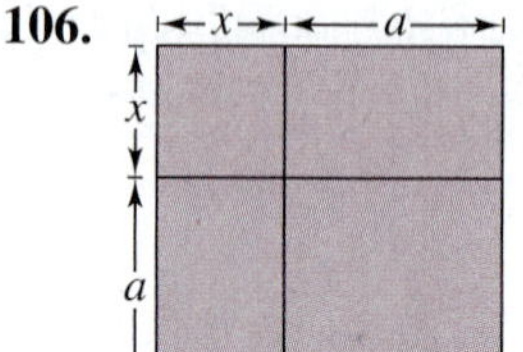

Synthesis

True or False? **In Exercises 107 and 108, determine whether the statement is true or false. Justify your answer.**

107. The product of two binomials is always a second-degree polynomial.

108. The sum of two binomials is always a binomial.

109. Find the degree of the product of two polynomials of degrees m and n.

110. Find the degree of the sum of two polynomials of degrees m and n if $m < n$.

111. ***Writing*** A student's homework paper included the following.

$$(x - 3)^2 = x^2 + 9$$

Write a paragraph fully explaining the error and give the correct method for squaring a binomial.

112. A third-degree polynomial and a fourth-degree polynomial are added.

(a) Can the sum be a fourth-degree polynomial? Explain or give an example.

(b) Can the sum be a second-degree polynomial? Explain or give an example.

(c) Can the sum be a seventh-degree polynomial? Explain or give an example.

113. ***Think About It*** Must the sum of two second-degree polynomials be a second-degree polynomial? If not, give an example.

114. ***Think About It*** When the polynomial $-x^3 + 3x^2 + 2x - 1$ is subtracted from an unknown polynomial, the difference is $5x^2 + 8$. If it is possible, find the unknown polynomial.

115. ***Logical Reasoning*** Verify that $(x + y)^2$ is not equal to $x^2 + y^2$ by letting $x = 3$ and $y = 4$ and evaluating both expressions. Are there any values of x or y for which $(x + y)^2 = x^2 + y^2$? Explain.

P.4 Factoring

▶ What you should learn

- How to remove common factors from polynomials
- How to factor special polynomial forms
- How to factor trinomials as the product of two binomials
- How to factor polynomials by grouping

▶ Why you should learn it

Polynomial factoring can be used to solve real-life problems. For instance, in Exercise 136 on page 40, factoring is used to develop an alternative formula for the volume of concrete used to make a cylindrical storage tank.

Arnulf Husmo/Tony Stone Images

Polynomials with Common Factors

The process of writing a polynomial as a product is called **factoring.** It is an important tool for solving equations and for simplifying rational expressions.

Unless noted otherwise, when you are asked to factor a polynomial, you can assume that you are looking for factors with integer coefficients. If a polynomial cannot be factored using integer coefficients, then it is **prime** or **irreducible over the integers.** For instance, the polynomial $x^2 - 3$ is irreducible over the integers. Over the *real numbers*, this polynomial can be factored as

$$x^2 - 3 = (x + \sqrt{3})(x - \sqrt{3}).$$

A polynomial is **completely factored** when each of its factors is prime. For instance

$$x^3 - x^2 + 4x - 4 = (x - 1)(x^2 + 4) \qquad \text{Completely factored}$$

is completely factored, but

$$x^3 - x^2 - 4x + 4 = (x - 1)(x^2 - 4) \qquad \text{Not completely factored}$$

is not completely factored. Its complete factorization is

$$x^3 - x^2 - 4x + 4 = (x - 1)(x + 2)(x - 2).$$

The simplest type of factoring involves a polynomial that can be written as the product of a monomial and another polynomial. The technique used here is the Distributive Property, $a(b + c) = ab + ac$, in the *reverse* direction.

$$ab + ac = a(b + c) \qquad a \text{ is a common factor.}$$

Removing (factoring out) a common factor is the first step in completely factoring a polynomial.

Example 1 ▶ Removing Common Factors

Factor each expression.

a. $6x^3 - 4x$

b. $-4x^2 + 12x - 16$

c. $(x - 2)(2x) + (x - 2)(3)$

Solution

a. $6x^3 - 4x = 2x(3x^2) - 2x(2)$ — $2x$ is a common factor.

$= 2x(3x^2 - 2)$

b. $-4x^2 + 12x - 16 = -4(x^2) + (-4)(-3x) + (-4)4$ — -4 is a common factor.

$= -4(x^2 - 3x + 4)$

c. $(x - 2)(2x) + (x - 2)(3) = (x - 2)(2x + 3)$ — $x - 2$ is a common factor.

Factoring Special Polynomial Forms

Some polynomials have special forms that arise from the special product forms on page 26. You should learn to recognize these forms so that you can factor such polynomials easily.

Factoring Special Polynomial Forms

Factored Form	*Example*
Difference of Two Squares	
$u^2 - v^2 = (u + v)(u - v)$	$9x^2 - 4 = (3x)^2 - 2^2 = (3x + 2)(3x - 2)$
Perfect Square Trinomial	
$u^2 + 2uv + v^2 = (u + v)^2$	$x^2 + 6x + 9 = x^2 + 2(x)(3) + 3^2 = (x + 3)^2$
$u^2 - 2uv + v^2 = (u - v)^2$	$x^2 - 6x + 9 = x^2 - 2(x)(3) + 3^2 = (x - 3)^2$
Sum or Difference of Two Cubes	
$u^3 + v^3 = (u + v)(u^2 - uv + v^2)$	$x^3 + 8 = x^3 + 2^3 = (x + 2)(x^2 - 2x + 4)$
$u^3 - v^3 = (u - v)(u^2 + uv + v^2)$	$27x^3 - 1 = (3x)^3 - 1^3 = (3x - 1)(9x^2 + 3x + 1)$

One of the easiest special polynomial forms to factor is the difference of two squares. Think of this form as follows.

$$u^2 - v^2 = (u + v)(u - v)$$

Difference   Opposite signs

To recognize perfect square terms, look for coefficients that are squares of integers and variables raised to *even powers*.

STUDY TIP

In Example 2, note that the first step in factoring a polynomial is to check for a common factor. Once the common factor is removed, it is often possible to recognize patterns that were not immediately obvious.

Example 2 ▶ Removing a Common Factor First

$$\begin{aligned} 3 - 12x^2 &= 3(1 - 4x^2) && \text{3 is a common factor.} \\ &= 3[1^2 - (2x)^2] \\ &= 3(1 + 2x)(1 - 2x) && \text{Difference of two squares} \end{aligned}$$

Example 3 ▶ Factoring the Difference of Two Squares

a. $(x + 2)^2 - y^2 = [(x + 2) + y][(x + 2) - y]$
$= (x + 2 + y)(x + 2 - y)$

b. $$\begin{aligned} 16x^4 - 81 &= (4x^2)^2 - 9^2 \\ &= (4x^2 + 9)(4x^2 - 9) && \text{Difference of two squares} \\ &= (4x^2 + 9)[(2x)^2 - 3^2] \\ &= (4x^2 + 9)(2x + 3)(2x - 3) && \text{Difference of two squares} \end{aligned}$$

A perfect square trinomial is the square of a binomial, and it has the following form.

$$u^2 + 2uv + v^2 = (u + v)^2 \quad \text{or} \quad u^2 - 2uv + v^2 = (u - v)^2$$

Like signs   Like signs

Note that the first and last terms are squares and the middle term is twice the product of u and v.

Example 4 ▶ Factoring Perfect Square Trinomials 

Factor each trinomial.

a. $x^2 - 10x + 25$

b. $16x^2 + 8x + 1$

Solution

a. $x^2 - 10x + 25 = x^2 - 2(x)(5) + 5^2 = (x - 5)^2$

b. $16x^2 + 8x + 1 = (4x)^2 + 2(4x)(1) + 1^2 = (4x + 1)^2$

Exploration

Rewrite $u^6 - v^6$ as the difference of two squares. Then find a formula for completely factoring $u^6 - v^6$. Use your formula to factor $x^6 - 1$ and $x^6 - 64$ completely.

The next two formulas show the sums and differences of cubes. Pay special attention to the signs of the terms.

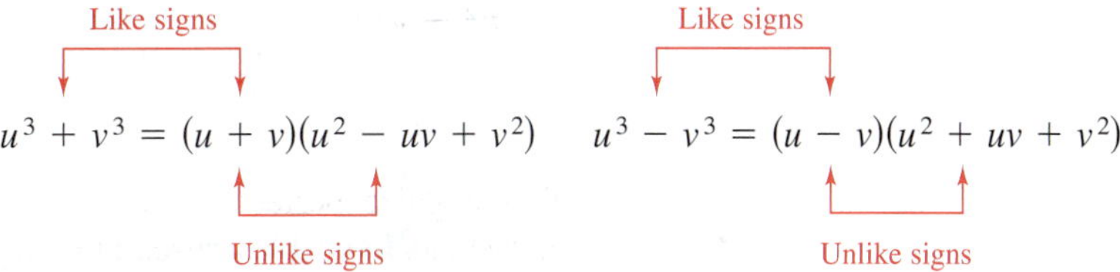

Example 5 ▶ Factoring the Difference of Cubes 

Factor $x^3 - 27$.

Solution

$x^3 - 27 = x^3 - 3^3$ Rewrite 27 as 3^3.

$= (x - 3)(x^2 + 3x + 9)$ Factor.

Example 6 ▶ Factoring the Sum of Cubes

a. $y^3 + 8 = y^3 + 2^3$ Rewrite 8 as 2^3.

$= (y + 2)(y^2 - 2y + 4)$ Factor.

b. $3(x^3 + 64) = 3(x^3 + 4^3)$ Rewrite 64 as 4^3.

$= 3(x + 4)(x^2 - 4x + 16)$ Factor.

Trinomials with Binomial Factors

To factor a trinomial of the form $ax^2 + bx + c$, use the following pattern.

Factors of a

$$ax^2 + bx + c = (\square x + \square)(\square x + \square)$$

Factors of c

The goal is to find a combination of factors of a and c such that the outer and inner products add up to the middle term bx. For instance, in the trinomial $6x^2 + 17x + 5$, you can write

$$\begin{aligned}(2x + 5)(3x + 1) &= \overset{F}{6x^2} + \overset{O}{2x} + \overset{I}{15x} + \overset{L}{5} \\ &= 6x^2 + \underset{O + I}{17x} + 5.\end{aligned}$$

Note that the outer (O) and inner (I) products add up to $17x$.

Example 7 ▶ Factoring a Trinomial: Leading Coefficient Is 1

Factor $x^2 - 7x + 12$.

Solution

The possible factorizations are

$$(x - 2)(x - 6), \quad (x - 1)(x - 12), \quad \text{and} \quad (x - 3)(x - 4).$$

Testing the middle term, you will find the correct factorization to be

$$x^2 - 7x + 12 = (x - 3)(x - 4).$$

Example 8 ▶ Factoring a Trinomial: Leading Coefficient Is Not 1

Factor $2x^2 + x - 15$.

Solution

The eight possible factorizations are as follows.

$$(2x - 1)(x + 15) \qquad (2x + 1)(x - 15)$$
$$(2x - 3)(x + 5) \qquad (2x + 3)(x - 5)$$
$$(2x - 5)(x + 3) \qquad (2x + 5)(x - 3)$$
$$(2x - 15)(x + 1) \qquad (2x + 15)(x - 1)$$

Testing the middle term, you will find the correct factorization to be

$$2x^2 + x - 15 = (2x - 5)(x + 3). \qquad \text{O + I} = 6x - 5x = x$$

STUDY TIP

If the original trinomial has no common monomial factor, its binomial factors cannot have common monomial factors. For instance, when factoring $4x^2 - 3x - 10$, you do not have to test factors, such as $(2x + 2)$, that have a common factor of 2.

Factoring by Grouping

Sometimes polynomials with more than three terms can be factored by a method called **factoring by grouping.** It is not always obvious which terms to group, and sometimes several different groupings will work.

Example 9 ▶ Factoring by Grouping 

Use factoring by grouping to factor $x^3 - 2x^2 - 3x + 6$.

Solution

$$\begin{aligned} x^3 - 2x^2 - 3x + 6 &= (x^3 - 2x^2) - (3x - 6) && \text{Group terms.} \\ &= x^2(x - 2) - 3(x - 2) && \text{Factor groups.} \\ &= (x - 2)(x^2 - 3) && \text{Distributive Property} \end{aligned}$$

STUDY TIP

Another way to factor the polynomial in Example 9 is to group the terms as follows.

$$\begin{aligned} &x^3 - 2x^2 - 3x + 6 \\ &= (x^3 - 3x) - (2x^2 - 6) \\ &= x(x^2 - 3) - 2(x^2 - 3) \\ &= (x^2 - 3)(x - 2) \end{aligned}$$

As you can see, you obtain the same result as in Example 9.

Factoring a trinomial can involve quite a bit of trial and error. Some of this trial and error can be lessened by using factoring by grouping. The key to this method of factoring is knowing how to rewrite the middle term. In general, to factor a trinomial $ax^2 + bx + c$ by grouping, choose factors of the product ac that add up to b and use these factors to rewrite the middle term. This technique is illustrated in Example 10.

Example 10 ▶ Factoring a Trinomial by Grouping 

Use factoring by grouping to factor $2x^2 + 5x - 3$.

Solution

In the trinomial $2x^2 + 5x - 3$, $a = 2$ and $c = -3$, which implies that the product ac is -6. Now, -6 factors as $(6)(-1)$ and $6 - 1 = 5 = b$. So, you can rewrite the middle term as $5x = 6x - x$. This produces the following.

$$\begin{aligned} 2x^2 + 5x - 3 &= 2x^2 + 6x - x - 3 && \text{Rewrite middle term.} \\ &= (2x^2 + 6x) - (x + 3) && \text{Group terms.} \\ &= 2x(x + 3) - (x + 3) && \text{Factor groups.} \\ &= (x + 3)(2x - 1) && \text{Distributive Property} \end{aligned}$$

So, the trinomial factors as $2x^2 + 5x - 3 = (x + 3)(2x - 1)$.

Guidelines for Factoring Polynomials

1. Factor out any common factors using the Distributive Property.
2. Factor according to one of the special polynomial forms.
3. Factor as $ax^2 + bx + c = (mx + r)(nx + s)$.
4. Factor by grouping.

P.4 Exercises

In Exercises 1–4, find the greatest common factor of the expressions.

1. 90, 300
2. 36, 84, 294
3. $12x^2y^3$, $18x^2y$, $24x^3y^2$
4. $15(x+2)^3$, $42x(x+2)^2$

In Exercises 5–12, factor out the common factor.

5. $3x + 6$
6. $5y - 30$
7. $2x^3 - 6x$
8. $4x^3 - 6x^2 + 12x$
9. $x(x-1) + 6(x-1)$
10. $3x(x+2) - 4(x+2)$
11. $(x+3)^2 - 4(x+3)$
12. $(3x-1)^2 + (3x-1)$

In Exercises 13–18, find the greatest common factor such that the remaining factors have only integer coefficients.

13. $\frac{1}{2}x + 4$
14. $\frac{1}{3}y + 5$
15. $\frac{1}{2}x^3 + 2x^2 - 5x$
16. $\frac{1}{3}y^4 - 5y^2 + 2y$
17. $\frac{2}{3}x(x-3) - 4(x-3)$
18. $\frac{4}{5}y(y+1) - 2(y+1)$

In Exercises 19–28, factor the difference of two squares.

19. $x^2 - 36$
20. $x^2 - 49$
21. $16y^2 - 9$
22. $49 - 9y^2$
23. $16x^2 - \frac{1}{9}$
24. $\frac{4}{25}y^2 - 64$
25. $(x-1)^2 - 4$
26. $25 - (z+5)^2$
27. $9u^2 - 4v^2$
28. $25x^2 - 16y^2$

In Exercises 29–38, factor the perfect square trinomial.

29. $x^2 - 4x + 4$
30. $x^2 + 10x + 25$
31. $4t^2 + 4t + 1$
32. $9x^2 - 12x + 4$
33. $25y^2 - 10y + 1$
34. $36y^2 - 108y + 81$
35. $9u^2 + 24uv + 16v^2$
36. $4x^2 - 4xy + y^2$
37. $x^2 - \frac{4}{3}x + \frac{4}{9}$
38. $z^2 + z + \frac{1}{4}$

In Exercises 39–46, factor the sum or difference of cubes.

39. $x^3 - 8$
40. $x^3 - 27$
41. $y^3 + 64$
42. $z^3 + 125$
43. $8t^3 - 1$
44. $27x^3 + 8$
45. $u^3 + 27v^3$
46. $64x^3 - y^3$

In Exercises 47–50, factor a negative real number from the polynomial and then write the polynomial factor in standard form.

47. $25 - 5x^2$
48. $5 + 3y^2 - y^3$
49. $-2t^3 + 4t + 6$
50. $-3x^5 - 3x^2 + 6x + 9$

In Exercises 51–64, factor the trinomial.

51. $x^2 + x - 2$
52. $x^2 + 5x + 6$
53. $s^2 - 5s + 6$
54. $t^2 - t - 6$
55. $20 - y - y^2$
56. $24 + 5z - z^2$
57. $x^2 - 30x + 200$
58. $x^2 - 13x + 42$
59. $3x^2 - 5x + 2$
60. $2x^2 - x - 1$
61. $5x^2 + 26x + 5$
62. $12x^2 + 7x + 1$
63. $-9z^2 + 3z + 2$
64. $-5u^2 - 13u + 6$

In Exercises 65–72, factor by grouping.

65. $x^3 - x^2 + 2x - 2$
66. $x^3 + 5x^2 - 5x - 25$
67. $2x^3 - x^2 - 6x + 3$
68. $5x^3 - 10x^2 + 3x - 6$
69. $6 + 2x - 3x^3 - x^4$
70. $x^5 + 2x^3 + x^2 + 2$
71. $6x^3 - 2x + 3x^2 - 1$
72. $8x^5 - 6x^2 + 12x^3 - 9$

In Exercises 73–78, factor the trinomial by grouping.

73. $3x^2 + 10x + 8$
74. $2x^2 + 9x + 9$
75. $6x^2 + x - 2$
76. $6x^2 - x - 15$
77. $15x^2 - 11x + 2$
78. $12x^2 - 13x + 1$

In Exercises 79–112, completely factor the expression.

79. $6x^2 - 54$
80. $12x^2 - 48$
81. $x^3 - 4x^2$
82. $x^3 - 9x$

83. $x^2 - 2x + 1$

84. $16 + 6x - x^2$

85. $1 - 4x + 4x^2$

86. $-9x^2 + 6x - 1$

87. $2x^2 + 4x - 2x^3$

88. $2y^3 - 7y^2 - 15y$

89. $9x^2 + 10x + 1$

90. $13x + 6 + 5x^2$

91. $\frac{1}{81}x^2 + \frac{2}{9}x - 8$

92. $\frac{1}{8}x^2 - \frac{1}{96}x - \frac{1}{16}$

93. $3x^3 + x^2 + 15x + 5$

94. $5 - x + 5x^2 - x^3$

95. $x^4 - 4x^3 + x^2 - 4x$

96. $3u - 2u^2 + 6 - u^3$

97. $\frac{1}{4}x^3 + 3x^2 + \frac{3}{4}x + 9$

98. $\frac{1}{5}x^3 + x^2 - x - 5$

99. $(t - 1)^2 - 49$

100. $(x^2 + 1)^2 - 4x^2$

101. $(x^2 + 8)^2 - 36x^2$

102. $2t^3 - 16$

103. $5x^3 + 40$

104. $4x(2x - 1) + (2x - 1)^2$

105. $5(3 - 4x)^2 - 8(3 - 4x)(5x - 1)$

106. $2(x + 1)(x - 3)^2 - 3(x + 1)^2(x - 3)$

107. $7(3x + 2)^2(1 - x)^2 + (3x + 2)(1 - x)^3$

108. $7x(2)(x^2 + 1)(2x) - (x^2 + 1)^2(7)$

109. $3(x - 2)^2(x + 1)^4 + (x - 2)^3(4)(x + 1)^3$

110. $2x(x - 5)^4 - x^2(4)(x - 5)^3$

111. $5(x^6 + 1)^4(6x^5)(3x + 2)^3 + 3(3x + 2)^2(3)(x^6 + 1)^5$

112. $\frac{x^2}{2}(x^2 + 1)^4 - (x^2 + 1)^5$

Geometric Modeling **In Exercises 113–116, match the factoring formula with the correct "geometric factoring model." [The models are labeled (a), (b), (c), and (d).] For instance, a factoring model for**

$$2x^2 + 3x + 1 = (2x + 1)(x + 1)$$

is shown in the figure.

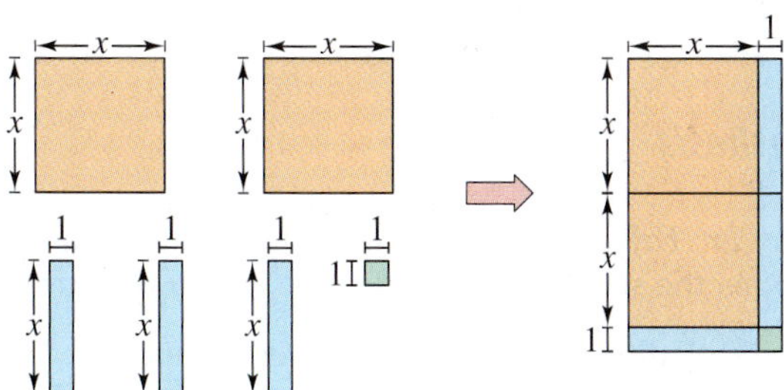

(a)

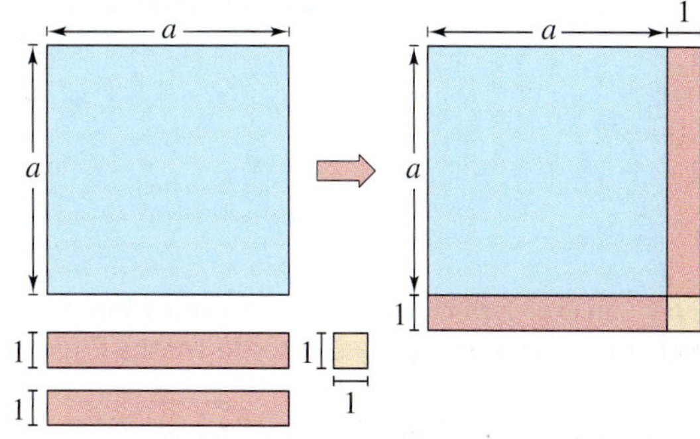

(b)

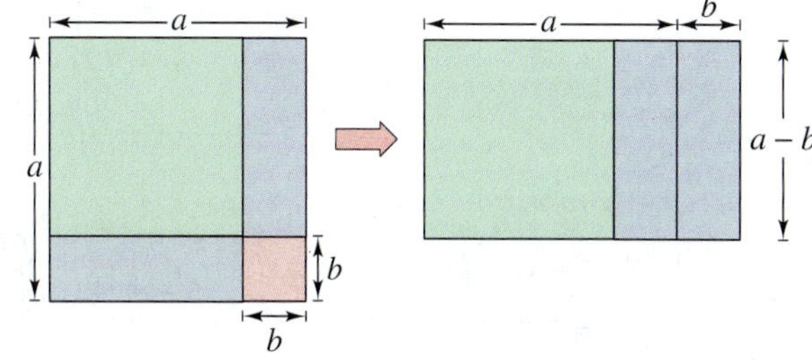

(c)

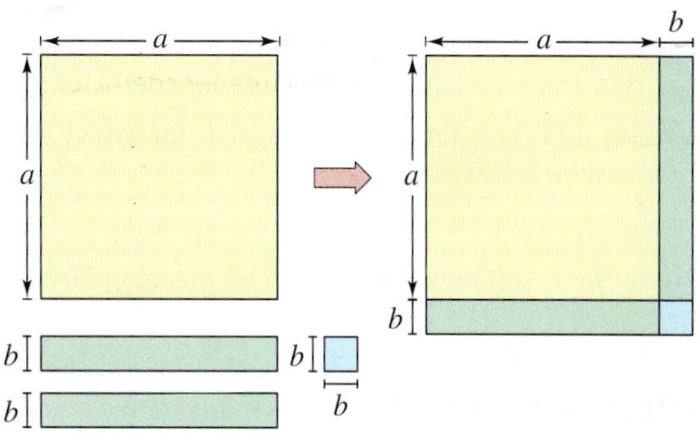

(d)

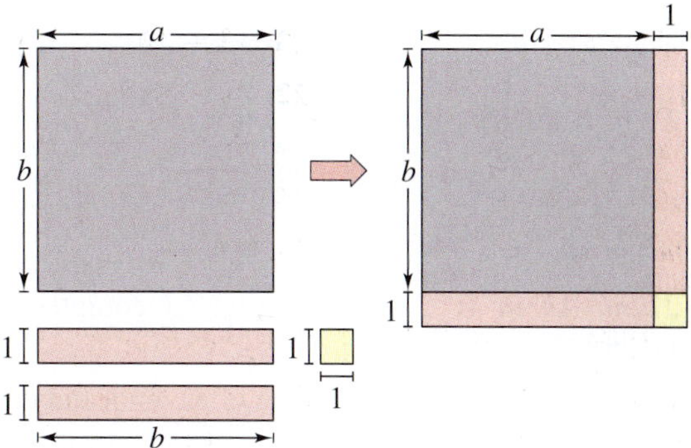

113. $a^2 - b^2 = (a + b)(a - b)$

114. $a^2 + 2ab + b^2 = (a + b)^2$

115. $a^2 + 2a + 1 = (a + 1)^2$

116. $ab + a + b + 1 = (a + 1)(b + 1)$.

Geometric Modeling **In Exercises 117–120, draw a "geometric factoring model" to represent the factorization.**

117. $3x^2 + 7x + 2 = (3x + 1)(x + 2)$

118. $x^2 + 4x + 3 = (x + 3)(x + 1)$

119. $2x^2 + 7x + 3 = (2x + 1)(x + 3)$

120. $x^2 + 3x + 2 = (x + 2)(x + 1)$

Geometry **In Exercises 121–124, write an expression in factored form for the area of the shaded portion of the figure.**

121.

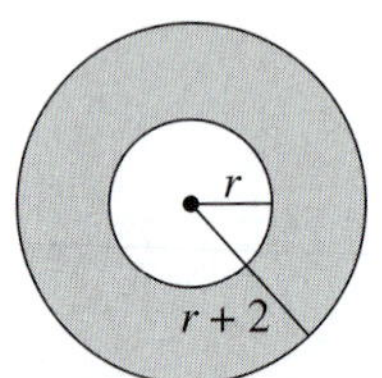

122.

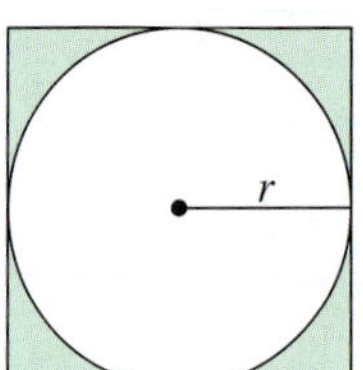

123.

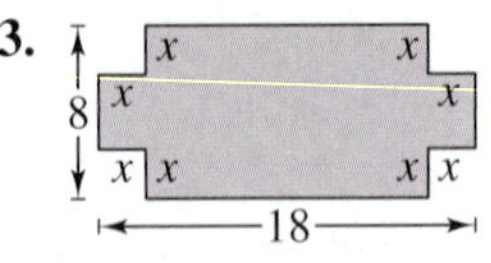

124.

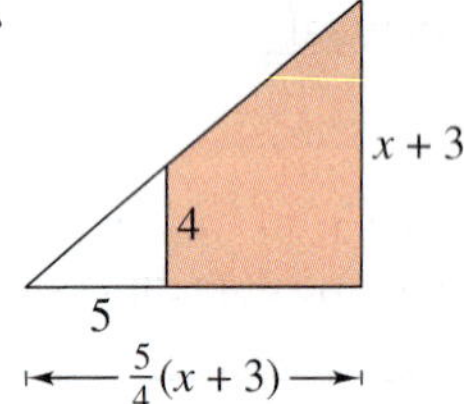

In Exercises 125–128, find all values of *b* for which the trinomial can be factored.

125. $x^2 + bx - 15$

126. $x^2 + bx + 50$

127. $x^2 + bx - 12$

128. $x^2 + bx + 24$

In Exercises 129–132, find two integer values of *c* such that the trinomial can be factored. (There are many correct answers.)

129. $2x^2 + 5x + c$

130. $3x^2 - 10x + c$

131. $3x^2 - x + c$

132. $2x^2 + 9x + c$

133. ***Error Analysis*** Describe the error.

$$9x^2 - 9x - 54 = (3x + 6)(3x - 9)$$
$$= 3(x + 2)(x - 3)$$

134. ***Think About It*** Is $(3x - 6)(x + 1)$ completely factored? Explain.

135. ***Chemistry*** The rate of change of an autocatalytic chemical reaction is $kQx - kx^2$, where Q is the amount of the original substance, x is the amount of substance formed, and k is a constant of proportionality. Factor the expression.

▶ Model It

136. ***Geometry*** The volume V of concrete used to make the cylindrical concrete storage tank shown in the figure is

$$V = \pi R^2 h - \pi r^2 h$$

where R is the outside radius, r is the inside radius, and h is the height of the storage tank.

(a) Factor the expression for the volume.

(b) From the result of part (a), show that the volume of concrete is

2π(average radius)(thickness of the tank)h.

(c) An 80-pound bag of concrete mix yields $\frac{3}{5}$ cubic foot of concrete. Find the number of bags required to construct a concrete storage tank having the following dimensions.

Outside radius, $R = 4$ feet

Inside radius, $r = 3\frac{2}{3}$ feet

Height, h feet

(d) Use the *table* feature of a graphing utility to create a table showing the number of bags of concrete required to construct the storage tank in part (c) with heights of $h = \frac{1}{2}, h = 1, h = \frac{3}{2}, h = 2, . . . , h = 6$ feet.

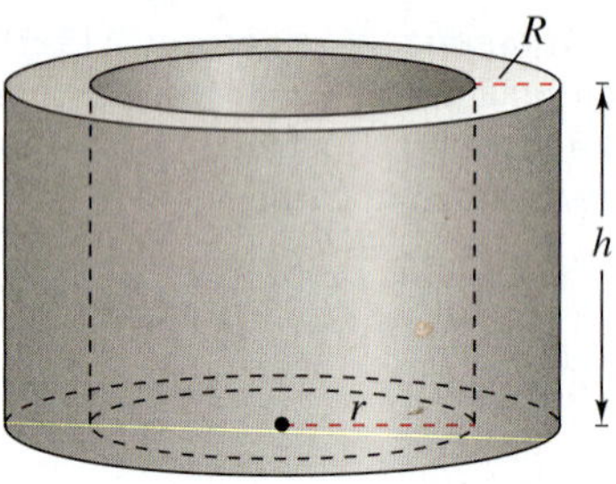

Synthesis

True or False? **In Exercises 137 and 138, determine whether the statement is true or false. Justify your answer.**

137. The difference of two perfect squares can be factored as the product of conjugate pairs.

138. The sum of two perfect squares can be factored as the binomial sum squared.

139. Factor $x^{2n} - y^{2n}$ completely.

140. Factor $x^{3n} + y^{3n}$ completely.

141. Factor $x^{3n} - y^{2n}$ completely.

P.5 Rational Expressions

▶ **What you should learn**

- How to find domains of algebraic expressions
- How to simplify rational expressions
- How to add, subtract, multiply, and divide rational expressions
- How to simplify complex fractions

▶ **Why you should learn it**

Rational expressions can be used to solve real-life problems. For instance, in Exercise 78 on page 50, a rational expression is used to model the cost per ounce of precious metals from 1994 through 1999.

Patrick Cocklin/Tony Stone Images

Domain of an Algebraic Expression

The set of real numbers for which an algebraic expression is defined is the **domain** of the expression. Two algebraic expressions are **equivalent** if they have the same domain and yield the same values for all numbers in their domain. For instance, $(x + 1) + (x + 2)$ and $2x + 3$ are equivalent because

$$\begin{aligned}(x + 1) + (x + 2) &= x + 1 + x + 2\\ &= x + x + 1 + 2\\ &= 2x + 3.\end{aligned}$$

Example 1 ▶ Finding the Domain of an Algebraic Expression

a. The domain of the polynomial

$$2x^3 + 3x + 4$$

is the set of all real numbers. In fact, the domain of any polynomial is the set of all real numbers, unless the domain is specifically restricted.

b. The domain of the radical expression

$$\sqrt{x - 2}$$

is the set of real numbers greater than or equal to 2, because the square root of a negative number is not a real number.

c. The domain of the expression

$$\frac{x + 2}{x - 3}$$

is the set of all real numbers except $x = 3$, which would produce an undefined division by zero.

The quotient of two algebraic expressions is a *fractional expression*. Moreover, the quotient of two *polynomials* such as

$$\frac{1}{x}, \quad \frac{2x - 1}{x + 1}, \quad \text{or} \quad \frac{x^2 - 1}{x^2 + 1}$$

is a **rational expression.** Recall that a fraction is in simplest form if its numerator and denominator have no factors in common aside from ± 1. To write a fraction in simplest form, divide out common factors.

$$\frac{a \cdot \not{c}}{b \cdot \not{c}} = \frac{a}{b}, \quad c \neq 0$$

The key to success in simplifying rational expressions lies in your ability to *factor* polynomials.

Simplifying Rational Expressions

When simplifying rational expressions, be sure to factor each polynomial completely before concluding that the numerator and denominator have no factors in common.

Example 2 ▶ **Simplifying a Rational Expression** 

Write $\dfrac{x^2 + 4x - 12}{3x - 6}$ in simplest form.

Solution

$$\frac{x^2 + 4x - 12}{3x - 6} = \frac{(x + 6)\cancel{(x - 2)}}{3\cancel{(x - 2)}} \qquad \text{Factor completely.}$$

$$= \frac{x + 6}{3}, \quad x \neq 2 \qquad \text{Divide out common factors.}$$

Note that the original expression is undefined when $x = 2$ (because division by zero is undefined). To make sure that the simplified expression is *equivalent* to the original expression, you must restrict the domain of the simplified expression by excluding the value $x = 2$.

STUDY TIP

In Example 2, do not make the mistake of trying to simplify further by dividing out terms.

$$\frac{x + 6}{3} = \frac{x + \cancel{6}}{\cancel{3}} = x + 2$$

Remember that to simplify fractions, divide out common *factors*, not terms.

Sometimes it may be necessary to change the sign of a factor to simplify a rational expression, as shown in Example 3(b).

Example 3 ▶ **Simplifying Rational Expressions** 

Write each expression in simplest form.

a. $\dfrac{x^3 - 4x}{x^2 + x - 2}$ **b.** $\dfrac{12 + x - x^2}{2x^2 - 9x + 4}$

Solution

a. $$\frac{x^3 - 4x}{x^2 + x - 2} = \frac{x(x^2 - 4)}{(x + 2)(x - 1)}$$

$$= \frac{x\cancel{(x + 2)}(x - 2)}{\cancel{(x + 2)}(x - 1)} \qquad \text{Factor completely.}$$

$$= \frac{x(x - 2)}{(x - 1)}, \quad x \neq -2 \qquad \text{Divide out common factors.}$$

b. $$\frac{12 + x - x^2}{2x^2 - 9x + 4} = \frac{(4 - x)(3 + x)}{(2x - 1)(x - 4)} \qquad \text{Factor completely.}$$

$$= \frac{-\cancel{(x - 4)}(3 + x)}{(2x - 1)\cancel{(x - 4)}} \qquad (4 - x) = -(x - 4)$$

$$= -\frac{3 + x}{2x - 1}, \quad x \neq 4 \qquad \text{Divide out common factors.}$$

Operations with Rational Expressions

To multiply or divide rational expressions, use the properties of fractions discussed in Section P.1. Recall that to divide fractions, you invert the divisor and multiply.

Example 4 ▶ Multiplying Rational Expressions 

$$\frac{2x^2 + x - 6}{x^2 + 4x - 5} \cdot \frac{x^3 - 3x^2 + 2x}{4x^2 - 6x} = \frac{(2x - 3)(x + 2)}{(x + 5)(x - 1)} \cdot \frac{x(x - 2)(x - 1)}{2x(2x - 3)}$$

$$= \frac{(x + 2)(x - 2)}{2(x + 5)}, \qquad x \neq 0, x \neq 1, x \neq \tfrac{3}{2}$$

In this text, when performing operations with rational expressions, the convention of listing *by the simplified expression* all values of x that must be specifically excluded from the domain in order to make the domains of the simplified and original expressions agree is followed. In Example 4, for instance, the restrictions $x \neq 0$, $x \neq 1$, and $x \neq \frac{3}{2}$ are listed with the simplified expression in order to make the two domains agree. Note that the value $x = -5$ is excluded from both domains, so it is not necessary to list this value.

Example 5 ▶ Dividing Rational Expressions 

$$\frac{x^3 - 8}{x^2 - 4} \div \frac{x^2 + 2x + 4}{x^3 + 8} = \frac{x^3 - 8}{x^2 - 4} \cdot \frac{x^3 + 8}{x^2 + 2x + 4} \qquad \text{Invert and multiply.}$$

$$= \frac{(x - 2)(x^2 + 2x + 4)}{(x + 2)(x - 2)} \cdot \frac{(x + 2)(x^2 - 2x + 4)}{x^2 + 2x + 4}$$

$$= x^2 - 2x + 4, \qquad x \neq \pm 2 \qquad \text{Divide out common factors.}$$

To add or subtract rational expressions, you can use the LCD (least common denominator) method or the basic definition

$$\frac{a}{b} \pm \frac{c}{d} = \frac{ad \pm bc}{bd}, \qquad b \neq 0, d \neq 0. \qquad \text{Basic definition}$$

This definition provides an efficient way of adding or subtracting *two* fractions that have no common factors in their denominators.

Example 6 ▶ Subtracting Rational Expressions 

$$\frac{x}{x - 3} - \frac{2}{3x + 4} = \frac{x(3x + 4) - 2(x - 3)}{(x - 3)(3x + 4)} \qquad \text{Basic definition}$$

$$= \frac{3x^2 + 4x - 2x + 6}{(x - 3)(3x + 4)} \qquad \text{Distributive Property}$$

$$= \frac{3x^2 + 2x + 6}{(x - 3)(3x + 4)} \qquad \text{Combine like terms.}$$

For three or more fractions, or for fractions with a repeated factor in the denominators, the LCD method works well. Recall that the least common denominator of several fractions consists of the product of all prime factors in the denominators, with each factor given the highest power of its occurrence in any denominator. Here is a numerical example.

$$\frac{1}{6}+\frac{3}{4}-\frac{2}{3}=\frac{1\cdot 2}{6\cdot 2}+\frac{3\cdot 3}{4\cdot 3}-\frac{2\cdot 4}{3\cdot 4}$$

The LCD is 12.

$$=\frac{2}{12}+\frac{9}{12}-\frac{8}{12}$$

$$=\frac{3}{12}$$

$$=\frac{1}{4}$$

Sometimes the numerator of the answer has a factor in common with the denominator. In such cases the answer should be simplified. For instance, in the example above, $\frac{3}{12}$ was simplified to $\frac{1}{4}$.

Example 7 ▶ Combining Rational Expressions: The LCD Method

Perform the operations and simplify.

$$\frac{3}{x-1}-\frac{2}{x}+\frac{x+3}{x^2-1}$$

Solution

Using the factored denominators $(x-1)$, x, and $(x+1)(x-1)$, you can see that the LCD is $x(x+1)(x-1)$.

$$\frac{3}{x-1}-\frac{2}{x}+\frac{x+3}{(x+1)(x-1)}$$

$$=\frac{3(x)(x+1)}{x(x+1)(x-1)}-\frac{2(x+1)(x-1)}{x(x+1)(x-1)}+\frac{(x+3)(x)}{x(x+1)(x-1)}$$

$$=\frac{3(x)(x+1)-2(x+1)(x-1)+(x+3)(x)}{x(x+1)(x-1)}$$

$$=\frac{3x^2+3x-2x^2+2+x^2+3x}{x(x+1)(x-1)}$$

Distributive Property

$$=\frac{3x^2-2x^2+x^2+3x+3x+2}{x(x+1)(x-1)}$$

Group like terms.

$$=\frac{2x^2+6x+2}{x(x+1)(x-1)}$$

Combine like terms.

$$=\frac{2(x^2+3x+1)}{x(x+1)(x-1)}$$

Factor.

Complex Fractions

Fractional expressions with separate fractions in the numerator, denominator, or both are called **complex fractions.** Here are two examples.

$$\frac{\left(\frac{1}{x}\right)}{x^2+1} \quad \text{and} \quad \frac{\left(\frac{1}{x}\right)}{\left(\frac{1}{x^2+1}\right)}$$

A complex fraction can be simplified by combining the fractions in its numerator into a single fraction and then combining the fractions in its denominator into a single fraction. Then invert the denominator and multiply.

Example 8 ▶ Simplifying a Complex Fraction

$$\frac{\left(\frac{2}{x}-3\right)}{\left(1-\frac{1}{x-1}\right)} = \frac{\left[\frac{2-3(x)}{x}\right]}{\left[\frac{1(x-1)-1}{x-1}\right]}$$ Combine fractions.

$$= \frac{\left(\frac{2-3x}{x}\right)}{\left(\frac{x-2}{x-1}\right)}$$ Simplify.

$$= \frac{2-3x}{x} \cdot \frac{x-1}{x-2}$$ Invert and multiply.

$$= \frac{(2-3x)(x-1)}{x(x-2)}, \quad x \neq 1$$

Another way to simplify a complex fraction is to multiply its numerator and denominator by the LCD of all fractions in its numerator and denominator. This method is applied to the fraction in Example 8 as follows.

$$\frac{\left(\frac{2}{x}-3\right)}{\left(1-\frac{1}{x-1}\right)} = \frac{\left(\frac{2}{x}-3\right)}{\left(1-\frac{1}{x-1}\right)} \cdot \frac{x(x-1)}{x(x-1)}$$ LCD is $x(x-1)$.

$$= \frac{\left(\frac{2-3x}{\cancel{x}}\right) \cdot \cancel{x}(x-1)}{\left(\frac{x-2}{\cancel{x-1}}\right) \cdot x(\cancel{x-1})}$$

$$= \frac{(2-3x)(x-1)}{x(x-2)}, \quad x \neq 1$$

The next three examples illustrate some methods for simplifying rational expressions involving negative exponents and radicals. These types of expressions occur frequently in calculus.

To simplify an expression with negative exponents, one method is to begin by factoring out the common factor with the smaller exponent. Remember that when factoring, you subtract exponents. For instance, in $3x^{-5/2} + 2x^{-3/2}$ the smaller exponent is $-\frac{5}{2}$ and the common factor is $x^{-5/2}$.

$$\begin{aligned} 3x^{-5/2} + 2x^{-3/2} &= x^{-5/2}[3(1) + 2x^{-3/2-(-5/2)}] \\ &= x^{-5/2}(3 + 2x^1) \\ &= \frac{3 + 2x}{x^{5/2}} \end{aligned}$$

Example 9 ▶ Simplifying an Expression

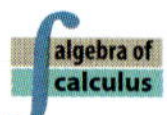

Simplify the following expression containing negative exponents.

$$x(1 - 2x)^{-3/2} + (1 - 2x)^{-1/2}$$

Solution

Begin by factoring out the common factor with the *smaller exponent*.

$$\begin{aligned} x(1 - 2x)^{-3/2} + (1 - 2x)^{-1/2} &= (1 - 2x)^{-3/2}[x + (1 - 2x)^{(-1/2)-(-3/2)}] \\ &= (1 - 2x)^{-3/2}[x + (1 - 2x)^1] \\ &= \frac{1 - x}{(1 - 2x)^{3/2}} \end{aligned}$$

A second method for simplifying an expression with negative exponents is shown in the next example.

Example 10 ▶ Simplifying a Complex Fraction

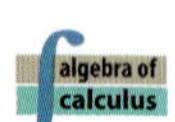

$$\begin{aligned} &\frac{(4 - x^2)^{1/2} + x^2(4 - x^2)^{-1/2}}{4 - x^2} \\ &\qquad = \frac{(4 - x^2)^{1/2} + x^2(4 - x^2)^{-1/2}}{4 - x^2} \cdot \frac{(4 - x^2)^{1/2}}{(4 - x^2)^{1/2}} \\ &\qquad = \frac{(4 - x^2)^1 + x^2(4 - x^2)^0}{(4 - x^2)^{3/2}} \\ &\qquad = \frac{4 - x^2 + x^2}{(4 - x^2)^{3/2}} \\ &\qquad = \frac{4}{(4 - x^2)^{3/2}} \end{aligned}$$

Example 11 ▶ Rewriting a Difference Quotient

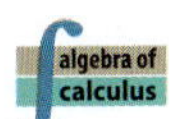

The expression from calculus

$$\frac{\sqrt{x+h}-\sqrt{x}}{h}$$

is an example of a *difference quotient*. Rewrite this expression by rationalizing its numerator.

Solution

$$\begin{aligned}\frac{\sqrt{x+h}-\sqrt{x}}{h} &= \frac{\sqrt{x+h}-\sqrt{x}}{h}\cdot\frac{\sqrt{x+h}+\sqrt{x}}{\sqrt{x+h}+\sqrt{x}}\\ &= \frac{\left(\sqrt{x+h}\right)^2-\left(\sqrt{x}\right)^2}{h\left(\sqrt{x+h}+\sqrt{x}\right)}\\ &= \frac{h}{h\left(\sqrt{x+h}+\sqrt{x}\right)}\\ &= \frac{1}{\sqrt{x+h}+\sqrt{x}}, \qquad h \neq 0\end{aligned}$$

Notice that the original expression is undefined when $h = 0$. So, you must exclude $h = 0$ from the domain of the simplified expression so that the expressions are equivalent.

Difference quotients, such as that in Example 11, occur frequently in calculus. Often, they need to be rewritten in an equivalent form that can be evaluated when $h = 0$. Note that the equivalent form is not simpler than the original form, but it has the advantage that it is defined when $h = 0$.

Writing ABOUT MATHEMATICS

Comparing Domains of Two Expressions Complete the table by evaluating the expressions

$$\frac{x^2-3x+2}{x-2} \quad \text{and} \quad x-1$$

for the given values of x. If you have a graphing utility with a *table* feature, use it to help create the table. Write a short paragraph describing the equivalence or nonequivalence of the two expressions.

x	-3	-2	-1	0	1	2	3
$\frac{x^2-3x+2}{x-2}$							
$x-1$							

P.5 Exercises

In Exercises 1–8, find the domain of the expression.

1. $3x^2 - 4x + 7$
2. $2x^2 + 5x - 2$
3. $4x^3 + 3, \quad x \ge 0$
4. $6x^2 - 9, \quad x > 0$
5. $\dfrac{1}{x - 2}$
6. $\dfrac{x + 1}{2x + 1}$
7. $\sqrt{x + 1}$
8. $\sqrt{6 - x}$

In Exercises 9 and 10, find the missing factor in the numerator such that the two fractions are equivalent.

9. $\dfrac{5}{2x} = \dfrac{5(\quad)}{6x^2}$
10. $\dfrac{3}{4} = \dfrac{3(\quad)}{4(x + 1)}$

In Exercises 11–28, write the rational expression in simplest form.

11. $\dfrac{15x^2}{10x}$
12. $\dfrac{18y^2}{60y^5}$
13. $\dfrac{3xy}{xy + x}$
14. $\dfrac{2x^2y}{xy - y}$
15. $\dfrac{4y - 8y^2}{10y - 5}$
16. $\dfrac{9x^2 + 9x}{2x + 2}$
17. $\dfrac{x - 5}{10 - 2x}$
18. $\dfrac{12 - 4x}{x - 3}$
19. $\dfrac{y^2 - 16}{y + 4}$
20. $\dfrac{x^2 - 25}{5 - x}$
21. $\dfrac{x^3 + 5x^2 + 6x}{x^2 - 4}$
22. $\dfrac{x^2 + 8x - 20}{x^2 + 11x + 10}$
23. $\dfrac{y^2 - 7y + 12}{y^2 + 3y - 18}$
24. $\dfrac{x^2 - 7x + 6}{x^2 + 11x + 10}$
25. $\dfrac{2 - x + 2x^2 - x^3}{x^2 - 4}$
26. $\dfrac{x^2 - 9}{x^3 + x^2 - 9x - 9}$
27. $\dfrac{z^3 - 8}{z^2 + 2z + 4}$
28. $\dfrac{y^3 - 2y^2 - 3y}{y^3 + 1}$

In Exercises 29 and 30, complete the table. What can you conclude?

29.

x	0	1	2	3	4	5	6
$\dfrac{x^2 - 2x - 3}{x - 3}$							
$x + 1$							

30.

x	0	1	2	3	4	5	6
$\dfrac{x - 3}{x^2 - x - 6}$							
$\dfrac{1}{x + 2}$							

31. ***Error Analysis*** Describe the error.

$$\frac{5x^3}{2x^3 + 4} = \frac{5x^3}{2x^3 + 4} = \frac{5}{2 + 4} = \frac{5}{6}$$

32. ***Error Analysis*** Describe the error.

$$\frac{x^3 + 25x}{x^2 - 2x - 15} = \frac{x(x^2 + 25)}{(x - 5)(x + 3)} = \frac{x(x - 5)(x + 5)}{(x - 5)(x + 3)} = \frac{x(x + 5)}{x + 3}$$

Geometry In Exercises 33 and 34, find the ratio of the area of the shaded portion of the figure to the total area of the figure.

33.

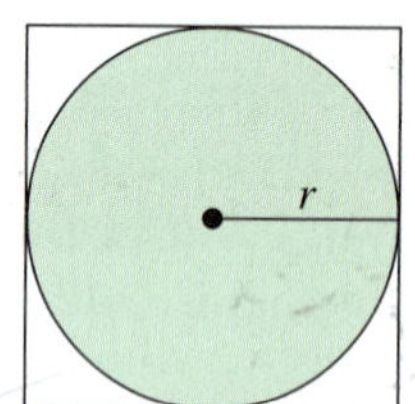

34.

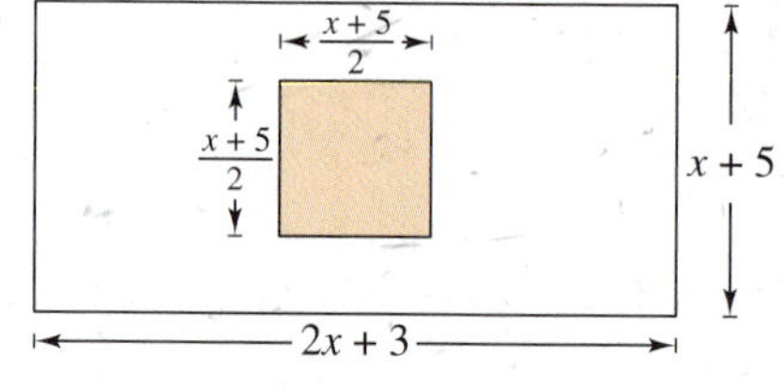

In Exercises 35–42, perform the multiplication or division and simplify.

35. $\dfrac{5}{x - 1} \cdot \dfrac{x - 1}{25(x - 2)}$
36. $\dfrac{x + 13}{x^3(3 - x)} \cdot \dfrac{x(x - 3)}{5}$
37. $\dfrac{r}{r - 1} \cdot \dfrac{r^2 - 1}{r^2}$
38. $\dfrac{4y - 16}{5y + 15} \cdot \dfrac{2y + 6}{4 - y}$
39. $\dfrac{t^2 - t - 6}{t^2 + 6t + 9} \cdot \dfrac{t + 3}{t^2 - 4}$

40. $\dfrac{x^2 + xy - 2y^2}{x^3 + x^2y} \cdot \dfrac{x}{x^2 + 3xy + 2y^2}$

41. $\dfrac{x^2 - 36}{x} \div \dfrac{x^3 - 6x^2}{x^2 + x}$

42. $\dfrac{x^2 - 14x + 49}{x^2 - 49} \div \dfrac{3x - 21}{x + 7}$

In Exercises 43–52, perform the addition or subtraction and simplify.

43. $\dfrac{5}{x - 1} + \dfrac{x}{x - 1}$

44. $\dfrac{2x - 1}{x + 3} + \dfrac{1 - x}{x + 3}$

45. $6 - \dfrac{5}{x + 3}$

46. $\dfrac{3}{x - 1} - 5$

47. $\dfrac{3}{x - 2} + \dfrac{5}{2 - x}$

48. $\dfrac{2x}{x - 5} - \dfrac{5}{5 - x}$

49. $\dfrac{1}{x^2 - x - 2} - \dfrac{x}{x^2 - 5x + 6}$

50. $\dfrac{2}{x^2 - x - 2} + \dfrac{10}{x^2 + 2x - 8}$

51. $-\dfrac{1}{x} + \dfrac{2}{x^2 + 1} + \dfrac{1}{x^3 + x}$

52. $\dfrac{2}{x + 1} + \dfrac{2}{x - 1} + \dfrac{1}{x^2 - 1}$

In Exercises 53–58, factor the expression by removing the common factor with the smaller exponent.

53. $x^5 - 2x^{-2}$

54. $x^5 - 5x^{-3}$

55. $x^2(x^2 + 1)^{-5} - (x^2 + 1)^{-4}$

56. $2x(x - 5)^{-3} - 4x^2(x - 5)^{-4}$

57. $2x^2(x - 1)^{1/2} - 5(x - 1)^{-1/2}$

58. $4x^3(2x - 1)^{3/2} - 2x(2x - 1)^{-1/2}$

Error Analysis **In Exercises 59 and 60, describe the error.**

59. $$\frac{x + 4}{x + 2} - \frac{3x - 8}{x + 2} = \frac{x + 4 - 3x - 8}{x + 2} = \frac{-2x - 4}{x + 2} = \frac{-2(x + 2)}{x + 2} = -2$$

60. $$\frac{6 - x}{x(x + 2)} + \frac{x + 2}{x^2} + \frac{8}{x^2(x + 2)} = \frac{x(6 - x) + (x + 2)^2 + 8}{x^2(x + 2)} = \frac{6x - x^2 + x^2 + 4 + 8}{x^2(x + 2)} = \frac{6(x + 2)}{x^2(x + 2)} = \frac{6}{x^2}$$

In Exercises 61–70, simplify the complex fraction.

61. $\dfrac{\left(\dfrac{x}{2} - 1\right)}{(x - 2)}$

62. $\dfrac{(x - 4)}{\left(\dfrac{x}{4} - \dfrac{4}{x}\right)}$

63. $\dfrac{\left[\dfrac{x^2}{(x + 1)^2}\right]}{\left[\dfrac{x}{(x + 1)^3}\right]}$

64. $\dfrac{\left(\dfrac{x^2 - 1}{x}\right)}{\left[\dfrac{(x - 1)^2}{x}\right]}$

65. $\dfrac{\left[\dfrac{1}{(x + h)^2} - \dfrac{1}{x^2}\right]}{h}$

66. $\dfrac{\left(\dfrac{x + h}{x + h + 1} - \dfrac{x}{x + 1}\right)}{h}$

67. $\dfrac{\left(\sqrt{x} - \dfrac{1}{2\sqrt{x}}\right)}{\sqrt{x}}$

68. $\dfrac{\left(\dfrac{t^2}{\sqrt{t^2 + 1}} - \sqrt{t^2 + 1}\right)}{t^2}$

69. $\dfrac{3x^{1/3} - x^{-2/3}}{3x^{-2/3}}$

70. $\dfrac{-x^3(1 - x^2)^{-1/2} - 2x(1 - x^2)^{1/2}}{x^4}$

In Exercises 71 and 72, rationalize the numerator of the expression.

71. $\dfrac{\sqrt{x + 2} - \sqrt{x}}{2}$

72. $\dfrac{\sqrt{z - 3} - \sqrt{z}}{3}$

Probability **In Exercises 73 and 74, consider an experiment in which a marble is tossed into a box whose base is shown in the figure. The probability that the marble will come to rest in the shaded portion of the box is equal to the ratio of the shaded area to the total area of the figure. Find the probability.**

73.

74.

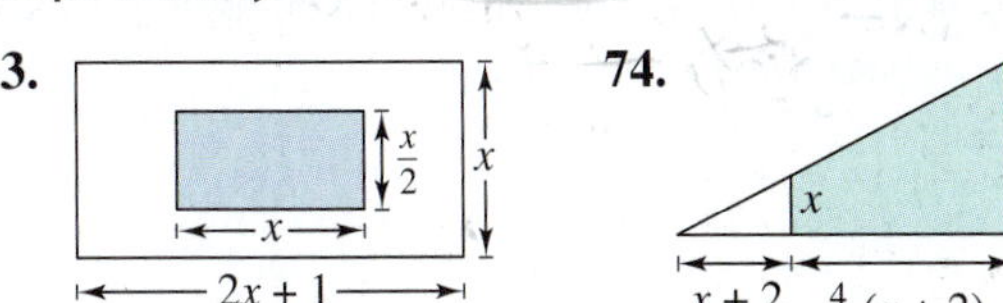

75. *Rate* A photocopier copies at a rate of 16 pages per minute.

(a) Find the time required to copy one page.

(b) Find the time required to copy x pages.

(c) Find the time required to copy 60 pages.

76. *Finance* The formula that approximates the annual interest rate r of a monthly installment loan is given by

$$r = \frac{\left[\frac{24(NM - P)}{N}\right]}{\left(P + \frac{NM}{12}\right)}$$

where N is the total number of payments, M is the monthly payment, and P is the amount financed.

(a) Approximate the annual interest rate for a four-year car loan of \$16,000 that has monthly payments of \$400.

(b) Simplify the expression for the annual interest rate r, and then rework part (a).

77. *Refrigeration* When food (at room temperature) is placed in a refrigerator, the time required for the food to cool depends on the amount of food, the air circulation in the refrigerator, the original temperature of the food, and the temperature of the refrigerator. The model that gives the temperature of food that has an original temperature of 75°F and is placed in a 40°F refrigerator is

$$T = 10\left(\frac{4t^2 + 16t + 75}{t^2 + 4t + 10}\right)$$

where T is the temperature (in degrees Fahrenheit) and t is the time (in hours).

(a) Complete the table.

t	0	2	4	6	8	10
T						

t	12	14	16	18	20	22
T						

(b) What value of T does the mathematical model appear to be approaching?

▶ Model It

78. *Precious Metals* The costs per fine ounce of gold and per troy ounce of platinum for the years 1994 through 1999 are shown in the table. (Source: U.S. Bureau of Mines, U.S. Geological Survey)

Year, t	Gold	Platinum
1994	\$385	\$411
1995	\$386	\$425
1996	\$389	\$398
1997	\$332	\$397
1998	\$295	\$373
1999	\$285	\$365

Mathematical models for this data are

$$\text{Cost of gold} = \frac{6.79t^2 - 95.6t + 356}{0.0205t^2 - 0.278t + 1}$$

and

$$\text{Cost of platinum} = \frac{-148.2t + 192}{-0.46t + 1}$$

where $t = 4$ corresponds to the year 1994.

(a) Create a table using the models to estimate the costs of the two metals for the given years.

(b) Compare the estimates given by the models with the actual costs.

(c) Determine a model for the ratio of the cost of gold to the cost of platinum.

(d) Use the model from part (c) to find the ratio over the given years. Over this period of time, did the cost of gold increase or decrease relative to the cost of platinum?

Synthesis

True or False? **In Exercises 79 and 80, determine whether the statement is true or false. Justify your answer.**

79. $\dfrac{x^{2n} - 1^{2n}}{x^n - 1^n} = x^n + 1^n$

80. $\dfrac{x^2 - 3x + 2}{x - 1} = x - 2$ for all values of x.

81. *Think About It* How do you determine whether a rational expression is in simplest form?

P.6 Errors and the Algebra of Calculus

▶ **What you should learn**

- How to avoid common algebraic errors
- How to recognize and use algebraic techniques that are common in calculus

▶ **Why you should learn it**

An efficient command of algebra is critical in the study of calculus.

Algebraic Errors to Avoid 

This section contains five lists of common algebraic errors: errors involving parentheses, errors involving fractions, errors involving exponents, errors involving radicals, and errors involving dividing out. Many of these errors are made because they seem to be the *easiest* things to do. For instance, the operations of subtraction and division are often believed to be commutative and associative. The following examples illustrate the fact that subtraction and division are neither commutative nor associative.

Not commutative	*Not associative*
$4 - 3 \neq 3 - 4$	$8 - (6 - 2) \neq (8 - 6) - 2$
$15 \div 5 \neq 5 \div 15$	$20 \div (4 \div 2) \neq (20 \div 4) \div 2$

Errors Involving Parentheses

Potential Error	*Correct Form*	*Comment*
~~$a - (x - b) = a - x - b$~~	$a - (x - b) = a - x + b$	Change all signs when distributing minus sign.
~~$(a + b)^2 = a^2 + b^2$~~	$(a + b)^2 = a^2 + 2ab + b^2$	Remember the middle term when squaring binomials.
~~$\left(\frac{1}{2}a\right)\left(\frac{1}{2}b\right) = \frac{1}{2}(ab)$~~	$\left(\frac{1}{2}a\right)\left(\frac{1}{2}b\right) = \frac{1}{4}(ab) = \frac{ab}{4}$	$\frac{1}{2}$ occurs twice as a factor.
~~$(3x + 6)^2 = 3(x + 2)^2$~~	$(3x + 6)^2 = [3(x + 2)]^2 = 3^2(x + 2)^2$	When factoring, apply exponents to all factors.

Errors Involving Fractions

Potential Error	*Correct Form*	*Comment*
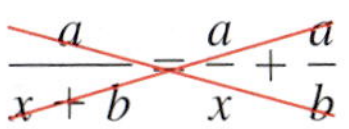	Leave as $\frac{a}{x + b}$.	Do not add denominators when adding fractions.
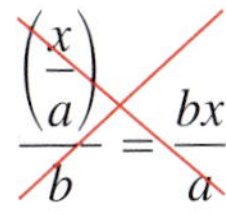	$\frac{\left(\frac{x}{a}\right)}{b} = \left(\frac{x}{a}\right)\left(\frac{1}{b}\right) = \frac{x}{ab}$	Multiply by the reciprocal when dividing fractions.
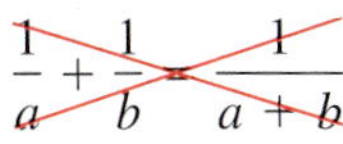	$\frac{1}{a} + \frac{1}{b} = \frac{b + a}{ab}$	Use the property for adding fractions.
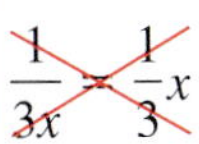	$\frac{1}{3x} = \frac{1}{3} \cdot \frac{1}{x}$	Use the property for multiplying fractions.
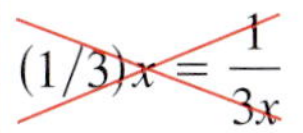	$(1/3)x = \frac{1}{3} \cdot x = \frac{x}{3}$	Be careful when using a slash to denote division.
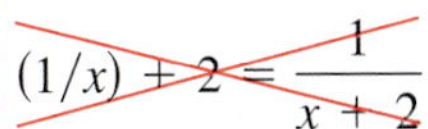	$(1/x) + 2 = \frac{1}{x} + 2 = \frac{1 + 2x}{x}$	Be careful when using a slash to denote division.

Errors Involving Exponents

Potential Error	*Correct Form*	*Comment*
$(x^2)^3 = x^5$	$(x^2)^3 = x^{2 \cdot 3} = x^6$	Multiply exponents when raising a power to a power.
$x^2 \cdot x^3 = x^6$	$x^2 \cdot x^3 = x^{2+3} = x^5$	Add exponents when multiplying powers with like bases.
$2x^3 = (2x)^3$	$2x^3 = 2(x^3)$	Exponents have priority over coefficients.
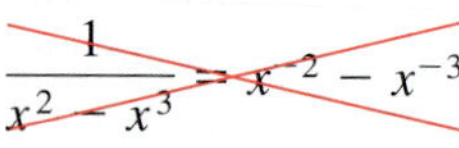 $\frac{1}{x^2 - x^3} = x^{-2} - x^{-3}$	Leave as $\dfrac{1}{x^2 - x^3}$.	Do not move term-by-term from denominator to numerator.

Errors Involving Radicals

Potential Error	*Correct Form*	*Comment*
$\sqrt{5x} = 5\sqrt{x}$	$\sqrt{5x} = \sqrt{5}\sqrt{x}$	Radicals apply to every factor inside the radical.
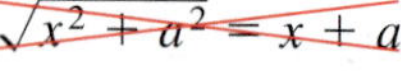 $\sqrt{x^2 + a^2} = x + a$	Leave as $\sqrt{x^2 + a^2}$.	Do not apply radicals term-by-term.
$\sqrt{-x + a} = -\sqrt{x - a}$	Leave as $\sqrt{-x + a}$.	Do not factor minus signs out of square roots.

Errors Involving Dividing Out

Potential Error	*Correct Form*	*Comment*
 $\frac{a + bx}{a} = 1 + bx$	$\dfrac{a + bx}{a} = \dfrac{a}{a} + \dfrac{bx}{a} = 1 + \dfrac{b}{a}x$	Divide out common factors, not common terms.
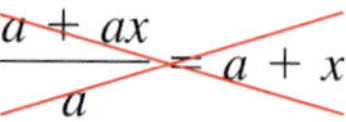 $\frac{a + ax}{a} = a + x$	$\dfrac{a + ax}{a} = \dfrac{a(1 + x)}{a} = 1 + x$	Factor before dividing out.
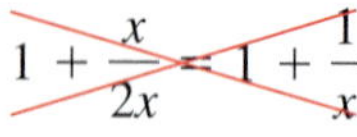 $1 + \frac{x}{2x} = 1 + \frac{1}{x}$	$1 + \dfrac{x}{2x} = 1 + \dfrac{1}{2} = \dfrac{3}{2}$	Divide out common factors.

A good way to avoid errors is to work *slowly*, *write neatly*, and *talk to yourself*. Each time you write a step, ask yourself why the step is algebraically legitimate. You can justify the step below because *dividing the numerator and denominator by the same nonzero number produces an equivalent fraction.*

$$\frac{2x}{6} = \frac{\cancel{2} \cdot x}{\cancel{2} \cdot 3} = \frac{x}{3}$$

Example 1 ▶ Using the Property for Adding Fractions

Describe and correct the error. $\frac{1}{2x} + \frac{1}{3x} = \frac{1}{5x}$

Solution

When adding fractions, use the property for adding fractions: $\dfrac{1}{a} + \dfrac{1}{b} = \dfrac{b + a}{ab}$.

$$\frac{1}{2x} + \frac{1}{3x} = \frac{3x + 2x}{6x^2} = \frac{5x}{6x^2} = \frac{5}{6x}$$

Some Algebra of Calculus

In calculus it is often necessary to take a simplified algebraic expression and "unsimplify" it. See the following lists, taken from a standard calculus text.

Unusual Factoring

Expression	*Useful Calculus Form*	*Comment*
$\frac{5x^4}{8}$	$\frac{5}{8}x^4$	Write with fractional coefficient.
$\frac{x^2 + 3x}{-6}$	$-\frac{1}{6}(x^2 + 3x)$	Write with fractional coefficient.
$2x^2 - x - 3$	$2\left(x^2 - \frac{x}{2} - \frac{3}{2}\right)$	Factor out the leading coefficient.
$\frac{x}{2}(x + 1)^{-1/2} + (x + 1)^{1/2}$	$\frac{(x + 1)^{-1/2}}{2}[x + 2(x + 1)]$	Factor out factor with lowest power.

Writing with Negative Exponents

Expression	*Useful Calculus Form*	*Comment*
$\frac{9}{5x^3}$	$\frac{9}{5}x^{-3}$	Move the factor to the numerator and change the sign of the exponent.
$\frac{7}{\sqrt{2x - 3}}$	$7(2x - 3)^{-1/2}$	Move the factor to the numerator and change the sign of the exponent.

Writing a Fraction as a Sum

Expression	*Useful Calculus Form*	*Comment*
$\frac{x + 2x^2 + 1}{\sqrt{x}}$	$x^{1/2} + 2x^{3/2} + x^{-1/2}$	Divide each term by $x^{1/2}$.
$\frac{1 + x}{x^2 + 1}$	$\frac{1}{x^2 + 1} + \frac{x}{x^2 + 1}$	Rewrite the fraction as the sum of fractions.
$\frac{2x}{x^2 + 2x + 1}$	$\frac{2x + 2 - 2}{x^2 + 2x + 1}$	Add and subtract the same term.
	$= \frac{2x + 2}{x^2 + 2x + 1} - \frac{2}{(x + 1)^2}$	Rewrite the fraction as the difference of fractions.
$\frac{x^2 - 2}{x + 1}$	$x - 1 - \frac{1}{x + 1}$	Use long division. (See Section 3.3.)
$\frac{x + 7}{x^2 - x - 6}$	$\frac{2}{x - 3} - \frac{1}{x + 2}$	Use the method of partial fractions. (See Section 4.3.)

Inserting Factors and Terms

Expression	*Useful Calculus Form*	*Comment*
$(2x - 1)^3$	$\frac{1}{2}(2x - 1)^3(2)$	Multiply and divide by 2.
$7x^2(4x^3 - 5)^{1/2}$	$\frac{7}{12}(4x^3 - 5)^{1/2}(12x^2)$	Multiply and divide by 12.
$\frac{4x^2}{9} - 4y^2 = 1$	$\frac{x^2}{9/4} - \frac{y^2}{1/4} = 1$	Write with fractional denominators.
$\frac{x}{x+1}$	$\frac{x+1-1}{x+1} = 1 - \frac{1}{x+1}$	Add and subtract the same term.

The next five examples demonstrate many of the steps in the preceding lists.

Example 2 ▶ Factors Involving Negative Exponents

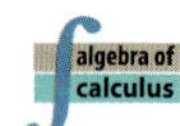

Factor $x(x + 1)^{-1/2} + (x + 1)^{1/2}$.

Solution

When multiplying factors with like bases, you add exponents. When factoring, you are undoing multiplication, and so you *subtract* exponents.

$$\begin{aligned} x(x + 1)^{-1/2} + (x + 1)^{1/2} &= (x + 1)^{-1/2}[x(x + 1)^0 + (x + 1)^1] \\ &= (x + 1)^{-1/2}[x + (x + 1)] \\ &= (x + 1)^{-1/2}(2x + 1) \end{aligned}$$

Here is another way to simplify the expression in Example 2.

$$\begin{aligned} x(x + 1)^{-1/2} + (x + 1)^{1/2} &= x(x + 1)^{-1/2} + (x + 1)^{1/2} \cdot \frac{(x + 1)^{1/2}}{(x + 1)^{1/2}} \\ &= \frac{x(x + 1)^0 + (x + 1)^1}{(x + 1)^{1/2}} = \frac{2x + 1}{\sqrt{x + 1}} \end{aligned}$$

Example 3 ▶ Inserting Factors in an Expression

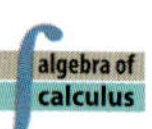

Insert the required factor: $\frac{x + 2}{(x^2 + 4x - 3)^2} = (\quad)\frac{1}{(x^2 + 4x - 3)^2}(2x + 4)$.

Solution

The expression on the right side of the equation is twice the expression on the left side. To make both sides equal, insert a factor of $\frac{1}{2}$.

$$\frac{x + 2}{(x^2 + 4x - 3)^2} = \left(\frac{1}{2}\right)\frac{1}{(x^2 + 4x - 3)^2}(2x + 4)$$

Right side is multiplied and divided by 2.

Example 4 ▶ Rewriting Fractions

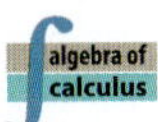

Explain the following.

$$\frac{4x^2}{9} - 4y^2 = \frac{x^2}{9/4} - \frac{y^2}{1/4}$$

Solution

To write the expression on the left side of the equation in the form given on the right side, multiply the numerators and denominators of both terms by $\frac{1}{4}$.

$$\begin{aligned}\frac{4x^2}{9} - 4y^2 &= \frac{4x^2}{9}\left(\frac{1/4}{1/4}\right) - 4y^2\left(\frac{1/4}{1/4}\right)\\ &= \frac{x^2}{9/4} - \frac{y^2}{1/4}\end{aligned}$$

Example 5 ▶ Rewriting with Negative Exponents

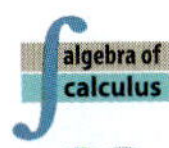

Rewrite each expression using negative exponents.

a. $\dfrac{-4x}{(1 - 2x^2)^2}$ **b.** $\dfrac{2}{5x^3} - \dfrac{1}{\sqrt{x}} + \dfrac{3}{5(4x)^2}$

Solution

a. $\dfrac{-4x}{(1 - 2x^2)^2} = -4x(1 - 2x^2)^{-2}$

b. Begin by writing the second term in exponential form.

$$\begin{aligned}\frac{2}{5x^3} - \frac{1}{\sqrt{x}} + \frac{3}{5(4x)^2} &= \frac{2}{5x^3} - \frac{1}{x^{1/2}} + \frac{3}{5(4x)^2}\\ &= \frac{2}{5}x^{-3} - x^{-1/2} + \frac{3}{5}(4x)^{-2}\end{aligned}$$

Example 6 ▶ Writing a Fraction as a Sum of Terms

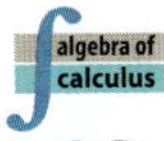

Rewrite each fraction as the sum of three terms.

a. $\dfrac{x^2 - 4x + 8}{2x}$ **b.** $\dfrac{x + 2x^2 + 1}{\sqrt{x}}$

Solution

a. $$\begin{aligned}\frac{x^2 - 4x + 8}{2x} &= \frac{x^2}{2x} - \frac{4x}{2x} + \frac{8}{2x}\\ &= \frac{x}{2} - 2 + \frac{4}{x}\end{aligned}$$

b. $$\begin{aligned}\frac{x + 2x^2 + 1}{\sqrt{x}} &= \frac{x}{x^{1/2}} + \frac{2x^2}{x^{1/2}} + \frac{1}{x^{1/2}}\\ &= x^{1/2} + 2x^{3/2} + x^{-1/2}\end{aligned}$$

P.6 Exercises

In Exercises 1–18, describe and correct the error.

1. $2x - (3y + 4) = 2x - 3y + 4$
2. $5z + 3(x - 2) = 5z + 3x - 2$
3. $\dfrac{4}{16x - (2x + 1)} = \dfrac{4}{14x + 1}$
4. $\dfrac{1 - x}{(5 - x)(-x)} = \dfrac{x - 1}{x(x - 5)}$
5. $(5z)(6z) = 30z$
6. $x(yz) = (xy)(xz)$
7. $a\left(\dfrac{x}{y}\right) = \dfrac{ax}{ay}$
8. $(4x)^2 = 4x^2$
9. $\sqrt{x + 9} = \sqrt{x} + 3$
10. $\sqrt{25 - x^2} = 5 - x$
11. $\dfrac{2x^2 + 1}{5x} = \dfrac{2x + 1}{5}$
12. $\dfrac{6x + y}{6x - y} = \dfrac{x + y}{x - y}$
13. $\dfrac{1}{a^{-1} + b^{-1}} = \left(\dfrac{1}{a + b}\right)^{-1}$
14. $\dfrac{1}{x + y^{-1}} = \dfrac{y}{x + 1}$
15. $(x^2 + 5x)^{1/2} = x(x + 5)^{1/2}$
16. $x(2x - 1)^2 = (2x^2 - x)^2$
17. $\dfrac{3}{x} + \dfrac{4}{y} = \dfrac{7}{x + y}$
18. $\dfrac{1}{2y} = (1/2)y$

In Exercises 19–38, insert the required factor in the parentheses.

19. $\dfrac{3x + 2}{5} = \dfrac{1}{5}(\square)$
20. $\dfrac{7x^2}{10} = \dfrac{7}{10}(\square)$
21. $\frac{2}{3}x^2 + \frac{1}{3}x + 5 = \frac{1}{3}(\square)$
22. $\frac{3}{4}x + \frac{1}{2} = \frac{1}{4}(\square)$
23. $x^2(x^3 - 1)^4 = (\square)(x^3 - 1)^4(3x^2)$
24. $x(1 - 2x^2)^3 = (\square)(1 - 2x^2)^3(-4x)$
25. $\dfrac{4x + 6}{(x^2 + 3x + 7)^3} = (\square)\dfrac{1}{(x^2 + 3x + 7)^3}(2x + 3)$
26. $\dfrac{x + 1}{(x^2 + 2x - 3)^2} = (\square)\dfrac{1}{(x^2 + 2x - 3)^2}(2x + 2)$
27. $\dfrac{3}{x} + \dfrac{5}{2x^2} - \dfrac{3}{2}x = (\square)(6x + 5 - 3x^3)$
28. $\dfrac{(x - 1)^2}{169} + (y + 5)^2 = \dfrac{(x - 1)^3}{169(\square)} + (y + 5)^2$
29. $\dfrac{9x^2}{25} + \dfrac{16y^2}{49} = \dfrac{x^2}{(\square)} + \dfrac{y^2}{(\square)}$
30. $\dfrac{3x^2}{4} - \dfrac{9y^2}{16} = \dfrac{x^2}{(\square)} - \dfrac{y^2}{(\square)}$
31. $\dfrac{x^2}{1/12} - \dfrac{y^2}{2/3} = \dfrac{12x^2}{(\square)} - \dfrac{3y^2}{(\square)}$
32. $\dfrac{x^2}{4/9} + \dfrac{y^2}{7/8} = \dfrac{9x^2}{(\square)} + \dfrac{8y^2}{(\square)}$
33. $x^{1/3} - 5x^{4/3} = x^{1/3}(\square)$
34. $3(2x + 1)x^{1/2} + 4x^{3/2} = x^{1/2}(\square)$
35. $(1 - 3x)^{4/3} - 4x(1 - 3x)^{1/3} = (1 - 3x)^{1/3}(\square)$
36. $\dfrac{1}{2\sqrt{x}} + 5x^{3/2} - 10x^{5/2} = \dfrac{1}{2\sqrt{x}}(\square)$
37. $\dfrac{1}{10}(2x + 1)^{5/2} - \dfrac{1}{6}(2x + 1)^{3/2} = \dfrac{(2x + 1)^{3/2}}{15}(\square)$
38. $\dfrac{3}{7}(t + 1)^{7/3} - \dfrac{3}{4}(t + 1)^{4/3} = \dfrac{3(t + 1)^{4/3}}{28}(\square)$

In Exercises 39–44, write the fraction as a sum of two or more terms.

39. $\dfrac{16 - 5x - x^2}{x}$
40. $\dfrac{x^3 - 5x^2 + 4}{x^2}$
41. $\dfrac{4x^3 - 7x^2 + 1}{x^{1/3}}$
42. $\dfrac{2x^5 - 3x^3 + 5x - 1}{x^{3/2}}$
43. $\dfrac{3 - 5x^2 - x^4}{\sqrt{x}}$
44. $\dfrac{x^3 - 5x^4}{3x^2}$

In Exercises 45–56, simplify the expression.

45. $\dfrac{-2(x^2 - 3)^{-3}(2x)(x + 1)^3 - 3(x + 1)^2(x^2 - 3)^{-2}}{[(x + 1)^3]^2}$
46. $\dfrac{x^5(-3)(x^2 + 1)^{-4}(2x) - (x^2 + 1)^{-3}(5)x^4}{(x^5)^2}$
47. $\dfrac{(6x + 1)^3(27x^2 + 2) - (9x^3 + 2x)(3)(6x + 1)^2(6)}{[(6x + 1)^3]^2}$
48. $\dfrac{(4x^2 + 9)^{1/2}(2) - (2x + 3)\left(\frac{1}{2}\right)(4x^2 + 9)^{-1/2}(8x)}{[(4x^2 + 9)^{1/2}]^2}$
49. $\dfrac{(x + 2)^{3/4}(x + 3)^{-2/3} - (x + 3)^{1/3}(x + 2)^{-1/4}}{[(x + 2)^{3/4}]^2}$
50. $(2x - 1)^{1/2} - (x + 2)(2x - 1)^{-1/2}$
51. $\dfrac{2(3x - 1)^{1/3} - (2x + 1)\left(\frac{1}{3}\right)(3x - 1)^{-2/3}(3)}{(3x - 1)^{2/3}}$
52. $\dfrac{(x + 1)\left(\frac{1}{2}\right)(2x - 3x^2)^{-1/2}(2 - 6x) - (2x - 3x^2)^{1/2}}{(x + 1)^2}$
53. $\dfrac{1}{(x^2 + 4)^{1/2}} \cdot \dfrac{1}{2}(x^2 + 4)^{-1/2}(2x)$

54. $\dfrac{1}{x^2 - 6}(2x) + \dfrac{1}{2x + 5}(2)$

55. $(x^2 + 5)^{1/2}(\frac{3}{2})(3x - 2)^{1/2}(3) +$
$(3x - 2)^{3/2}(\frac{1}{2})(x^2 + 5)^{-1/2}(2x)$

56. $(3x + 2)^{-1/2}(3)(x - 6)^{1/2}(1) +$
$(x - 6)^3(-\frac{1}{2})(3x + 2)^{-3/2}(3)$

▶ Model It

57. ***Athletics*** A triathlete has set up a course for training as part of her regimen in preparation for an upcoming triathlon. She is dropped off by a boat 2 miles from the nearest point on shore. The finish line is 4 miles down the coast and 2 miles inland (see figure). She can swim 2 miles per hour and run 6 miles per hour. The time t (in hours) required for her to reach the finish line can be approximated by the model

$$t = \frac{\sqrt{x^2 + 4}}{2} + \frac{\sqrt{(4 - x)^2 + 4}}{6}$$

where x is the distance down the coast (in miles) to which she swims and then leaves the water to start her run.

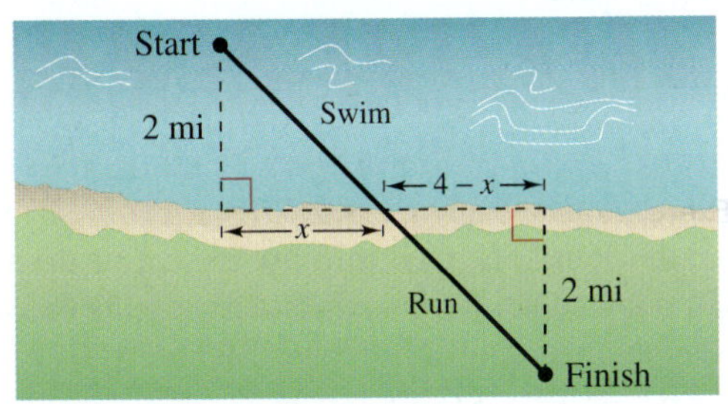

(a) Find the time required for the triathlete to finish when she swims to the points $x = 0.5$, $x = 1.0$, $x = 1.5, \ldots, x = 3.5$, and $x = 4.0$ miles down the coast.

(b) Use your results from part (a) to determine the distance down the coast that will yield the minimum amount of time required for the triathlete to reach the finish line.

(c) The expression below was obtained using calculus. It can be used to find the minimum amount of time required for the triathlete to reach the finish line. Simplify the expression.

$$\frac{1}{2}x(x^2 + 4)^{-1/2} + \frac{1}{6}(x - 4)(x^2 - 8x + 20)^{-1/2}$$

58. (a) Verify that $y_1 = y_2$ analytically.

$$y_1 = x^2\left(\frac{1}{3}\right)(x^2 + 1)^{-2/3}(2x) + (x^2 + 1)^{1/3}(2x)$$

$$y_2 = \frac{2x(4x^2 + 3)}{3(x^2 + 1)^{2/3}}$$

(b) Complete the table and demonstrate the equality in part (a) numerically.

x	-2	-1	$-\frac{1}{2}$	0	1	2	$\frac{5}{2}$
y_1							
y_2							

Synthesis

True or False? **In Exercises 59–62, determine whether the statement is true or false. Justify your answer.**

59. $x^{-1} + y^{-2} = \dfrac{y^2 + x}{xy^2}$

60. $\dfrac{1}{x^{-2} + y^{-1}} = x^2 + y$

61. $\dfrac{1}{\sqrt{x} + 4} = \dfrac{\sqrt{x} - 4}{x - 16}$

62. $\dfrac{x^2 - 9}{\sqrt{x} - 3} = \sqrt{x} + 3$

In Exercises 63–66, find and correct any errors.

63. $x^n \cdot x^{3n} = x^{3n^2}$

64. $(x^n)^{2n} + (x^{2n})^n = 2x^{2n^2}$

65. $x^{2n} + y^{2n} = (x^n + y^n)^2$

66. $\dfrac{x^{2n} \cdot x^{3n}}{x^{3n} + x^2} = \dfrac{x^{5n}}{x^{3n} + x^2}$

67. ***Think About It*** You are taking a course in calculus, and for one of the homework problems you obtain the following answer.

$$\frac{1}{10}(2x - 1)^{5/2} + \frac{1}{6}(2x - 1)^{3/2}$$

The answer in the back of the book is

$$\frac{1}{15}(2x - 1)^{3/2}(3x + 1).$$

Are these two answers equivalent? If so, show how the second answer can be obtained from the first.

P.7 Graphical Representation of Data

▶ What you should learn

- How to plot points in the Cartesian plane
- How to use the Distance Formula to find the distance between two points
- How to use the Midpoint Formula to find the midpoint of a line segment
- How to use a coordinate plane to model and solve real-life problems

▶ Why you should learn it

The Cartesian plane can be used to represent relationships between two variables. For instance, in Exercise 27 on page 65, a graph represents the minimum wage in the United States from 1950 to 2000.

David Young-Wolff/PhotoEdit

The Cartesian Plane

Just as you can represent real numbers by points on a real number line, you can represent ordered pairs of real numbers by points in a plane called the **rectangular coordinate system,** or the **Cartesian plane,** named after the French mathematician René Descartes (1596–1650).

The Cartesian plane is formed by using two real number lines intersecting at right angles, as shown in Figure P.8. The horizontal real number line is usually called the ***x*-axis,** and the vertical real number line is usually called the ***y*-axis.** The point of intersection of these two axes is the **origin,** and the two axes divide the plane into four parts called **quadrants.**

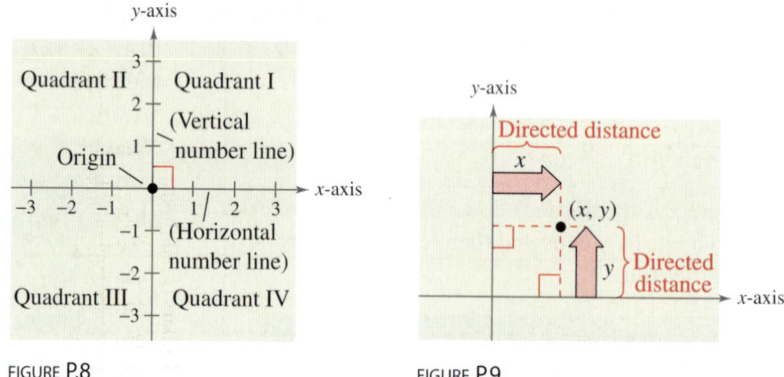

FIGURE P.8

FIGURE P.9

Each point in the plane corresponds to an **ordered pair** (x, y) of real numbers x and y, called **coordinates** of the point. The ***x*-coordinate** represents the directed distance from the y-axis to the point, and the ***y*-coordinate** represents the directed distance from the x-axis to the point, as shown in Figure P.9.

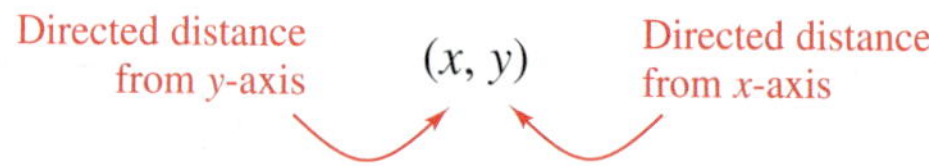

The notation (x, y) denotes both a point in the plane and an open interval on the real number line. The context will tell you which meaning is intended.

Example 1 ▶ Plotting Points in the Cartesian Plane 

Plot the points $(-1, 2)$, $(3, 4)$, $(0, 0)$, $(3, 0)$, and $(-2, -3)$.

Solution

To plot the point $(-1, 2)$, imagine a vertical line through -1 on the x-axis and a horizontal line through 2 on the y-axis. The intersection of these two lines is the point $(-1, 2)$. The other four points can be plotted in a similar way, as shown in Figure P.10.

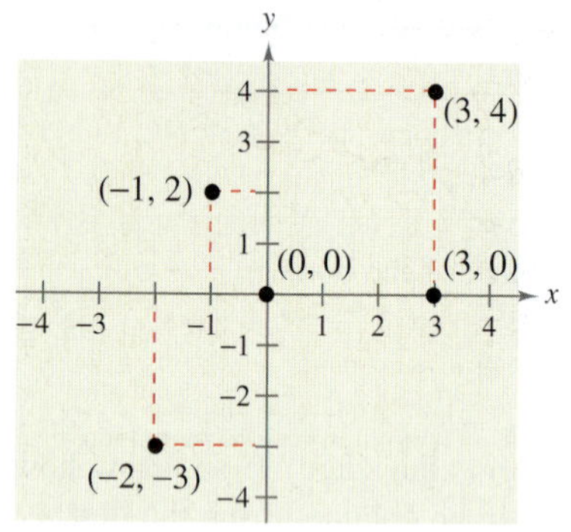

FIGURE P.10

The beauty of a rectangular coordinate system is that it allows you to *see* relationships between two variables. It would be difficult to overestimate the importance of Descartes's introduction of coordinates *in* the plane. Today, his ideas are in common use in virtually every scientific and business-related field.

Example 2 ▶ Sketching a Scatter Plot

Year, t	Amount, A
1990	475
1991	577
1992	521
1993	569
1994	609
1995	562
1996	707
1997	723
1998	718
1999	739

From 1990 through 1999, the amount A (in millions of dollars) spent on skiing equipment in the United States is shown in the table, where t represents the year. Sketch a scatter plot of the data. (Source: National Sporting Goods Association)

Solution

To sketch a *scatter plot* of the data shown in the table, you simply represent each pair of values by an ordered pair (t, A) and plot the resulting points, as shown in Figure P.11. For instance, the first pair of values is represented by the ordered pair (1990, 475). Note that the break in the t-axis indicates that the numbers between 0 and 1990 have been omitted.

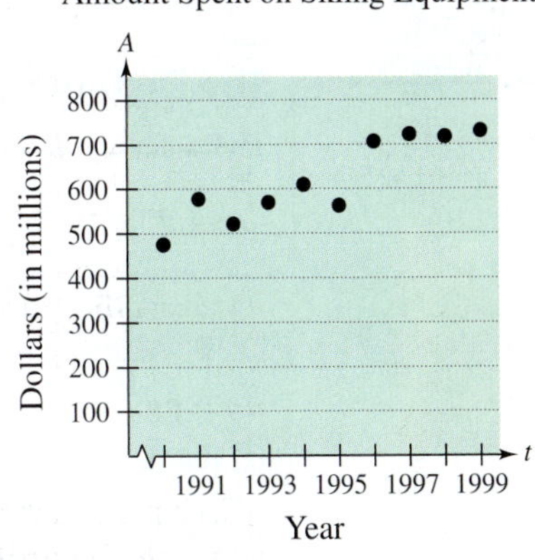

FIGURE P.11

STUDY TIP

In Example 2, you could have let $t = 1$ represent the year 1990. In that case, the horizontal axis would not have been broken, and the tick marks would have been labeled 1 through 10 (instead of 1990 through 1999).

Technology

The scatter plot in Example 2 is only one way to represent the data graphically. Two other techniques are shown at the right. The first is a bar graph and the second is a line graph. All three graphical representations were created with a computer. If you have access to a graphing utility, try using it to represent graphically the data given in Example 2.

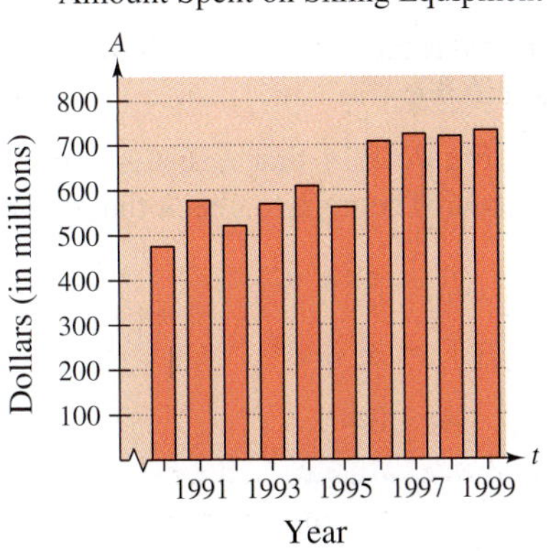

Bar Graph

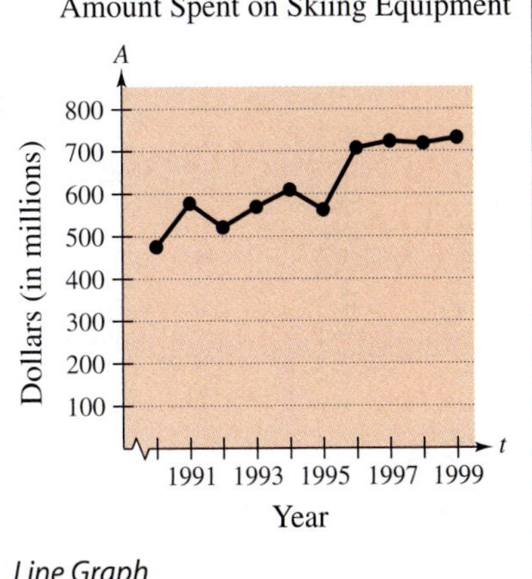

Line Graph

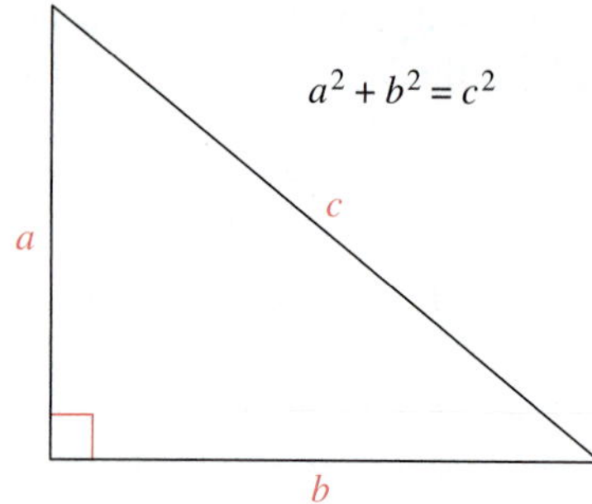

FIGURE P.12

The Distance Formula

Recall from the Pythagorean Theorem that, for a right triangle with hypotenuse of length c and sides of lengths a and b, you have

$$a^2 + b^2 = c^2 \qquad \text{Pythagorean Theorem}$$

as shown in Figure P.12. (The converse is also true. That is, if $a^2 + b^2 = c^2$, then the triangle is a right triangle.)

Suppose you want to determine the distance d between two points (x_1, y_1) and (x_2, y_2) in the plane. With these two points, a right triangle can be formed, as shown in Figure P.13. The length of the vertical side of the triangle is $|y_2 - y_1|$, and the length of the horizontal side is $|x_2 - x_1|$. By the Pythagorean Theorem, you can write

$$d^2 = |x_2 - x_1|^2 + |y_2 - y_1|^2$$
$$d = \sqrt{|x_2 - x_1|^2 + |y_2 - y_1|^2}$$
$$d = \sqrt{(x_2 - x_1)^2 + (y_2 - y_1)^2}.$$

This result is the **Distance Formula.**

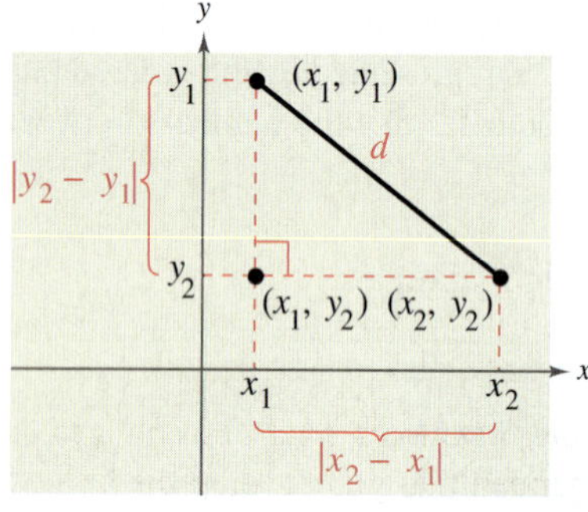

FIGURE P.13

The Distance Formula

The distance d between the points (x_1, y_1) and (x_2, y_2) in the plane is

$$d = \sqrt{(x_2 - x_1)^2 + (y_2 - y_1)^2}.$$

Example 3 ▶ Finding a Distance 

Find the distance between the points $(-2, 1)$ and $(3, 4)$.

Solution

Let $(x_1, y_1) = (-2, 1)$ and $(x_2, y_2) = (3, 4)$. Then apply the Distance Formula.

$$d = \sqrt{(x_2 - x_1)^2 + (y_2 - y_1)^2} \qquad \text{Distance Formula}$$
$$= \sqrt{[3 - (-2)]^2 + (4 - 1)^2} \qquad \text{Substitute for } x_1, y_1, x_2, \text{ and } y_2.$$
$$= \sqrt{(5)^2 + (3)^2} \qquad \text{Simplify.}$$
$$= \sqrt{34} \qquad \text{Simplify.}$$
$$\approx 5.83 \qquad \text{Use a calculator.}$$

Note in Figure P.14 that a distance of 5.83 looks about right. You can use the Pythagorean Theorem to check that the distance is correct.

$$d^2 \stackrel{?}{=} 3^2 + 5^2 \qquad \text{Pythagorean Theorem}$$
$$(\sqrt{34})^2 \stackrel{?}{=} 3^2 + 5^2 \qquad \text{Substitute for } d.$$
$$34 = 34 \qquad \text{Distance checks. ✓}$$

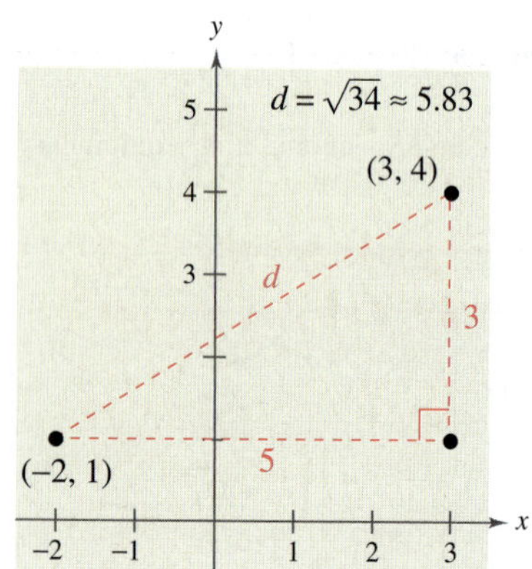

FIGURE P.14

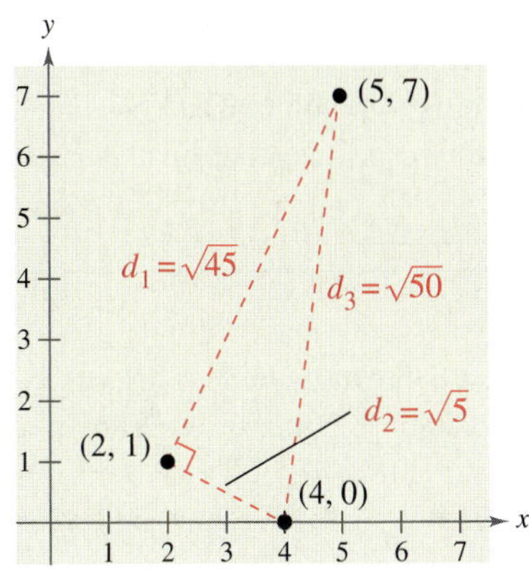

FIGURE P.15

Example 4 ▶ Verifying a Right Triangle 

Show that the points (2, 1), (4, 0), and (5, 7) are vertices of a right triangle.

Solution

The three points are plotted in Figure P.15. Using the Distance Formula, you can find the lengths of the three sides as follows.

$$d_1 = \sqrt{(5-2)^2 + (7-1)^2} = \sqrt{9+36} = \sqrt{45}$$
$$d_2 = \sqrt{(4-2)^2 + (0-1)^2} = \sqrt{4+1} = \sqrt{5}$$
$$d_3 = \sqrt{(5-4)^2 + (7-0)^2} = \sqrt{1+49} = \sqrt{50}$$

Because

$$(d_1)^2 + (d_2)^2 = 45 + 5$$
$$= 50$$
$$= (d_3)^2$$

you can conclude that the triangle must be a right triangle.

The figures provided with Examples 3 and 4 were not really essential to the solution. Nevertheless, it is strongly recommended that you develop the habit of including sketches with your solutions—even if they are not required.

Example 5 ▶ Finding the Length of a Pass 

During the first quarter of the 2002 Orange Bowl, Brock Berlin, the quarterback for the University of Florida, threw a pass from the 37-yard line, 40 yards from the sideline. The pass was caught by the wide receiver Taylor Jacobs on the 3-yard line, 20 yards from the same sideline, as shown in Figure P.16. How long was the pass?

Solution

You can find the length of the pass by finding the distance between the points (40, 37) and (20, 3).

$$d = \sqrt{(40-20)^2 + (37-3)^2}$$ Distance Formula

$$= \sqrt{400 + 1156}$$ Simplify.

$$= \sqrt{1556}$$ Simplify.

$$\approx 39$$ Use a calculator.

So, the pass was about 39 yards long.

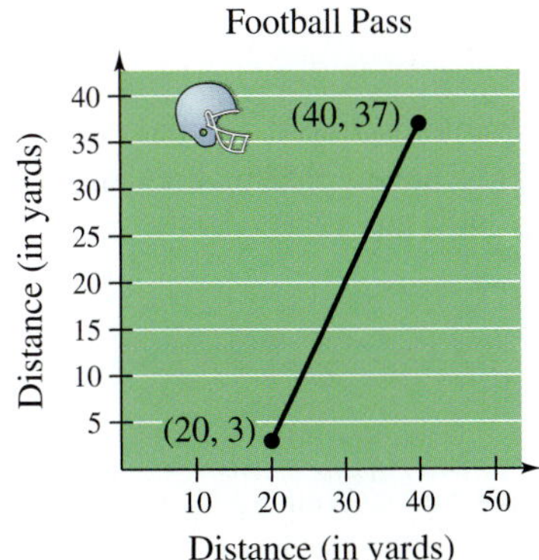

FIGURE P.16

In Example 5, the scale along the goal line does not normally appear on a football field. However, when you use coordinate geometry to solve real-life problems, you are free to place the coordinate system in any way that is convenient for the solution of the problem.

The Midpoint Formula

To find the **midpoint** of the line segment that joins two points in a coordinate plane, you can simply find the average values of the respective coordinates of the two endpoints using the **Midpoint Formula.**

> **The Midpoint Formula**
>
> The midpoint of the line segment joining the points (x_1, y_1) and (x_2, y_2) is given by the Midpoint Formula
>
> $$\text{Midpoint} = \left(\frac{x_1 + x_2}{2}, \frac{y_1 + y_2}{2}\right).$$

For a proof of the Midpoint Formula, see Proofs in Mathematics on page 73.

Example 6 ▶ Finding a Line Segment's Midpoint 

Find the midpoint of the line segment joining the points $(-5, -3)$ and $(9, 3)$, as shown in Figure P.17.

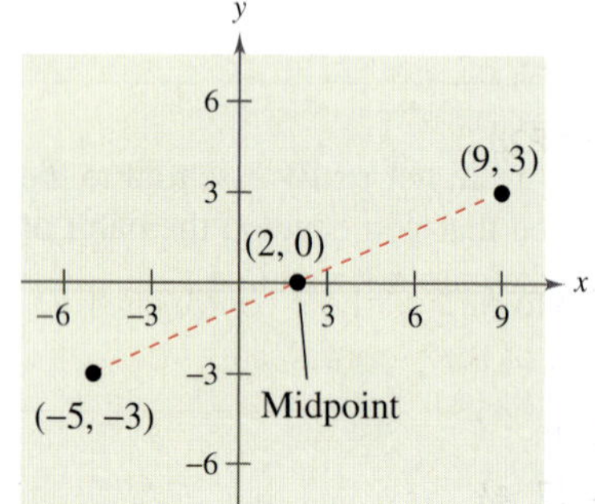

FIGURE P.17

Solution

Let $(x_1, y_1) = (-5, -3)$ and $(x_2, y_2) = (9, 3)$.

$$\text{Midpoint} = \left(\frac{x_1 + x_2}{2}, \frac{y_1 + y_2}{2}\right) \quad \text{Midpoint Formula}$$

$$= \left(\frac{-5 + 9}{2}, \frac{-3 + 3}{2}\right) \quad \text{Substitute for } x_1, y_1, x_2, \text{ and } y_2.$$

$$= (2, 0) \quad \text{Simplify.}$$

Example 7 ▶ Estimating Annual Revenue  

The United Parcel Service had annual revenues of \$24.8 billion in 1998 and \$29.8 billion in 2000. Without knowing any additional information, what would you estimate the 1999 revenue to have been? (Source: United Parcel Service of America Corp.)

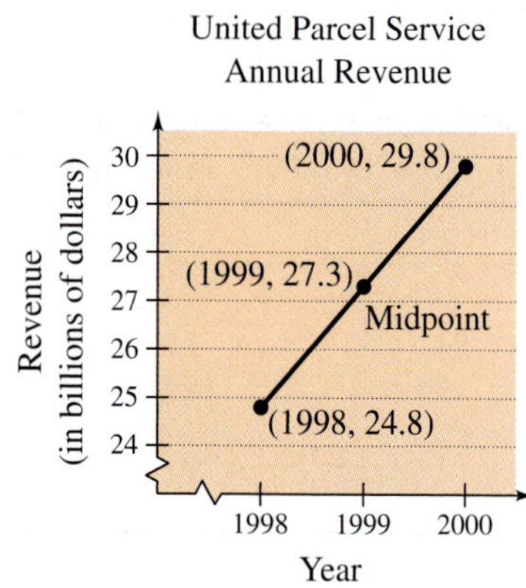

FIGURE P.18

Solution

One solution to the problem is to assume that revenue followed a linear pattern. With this assumption, you can estimate the 1999 revenue by finding the midpoint of the line segment connecting the points (1998, 24.8) and (2000, 29.8).

$$\text{Midpoint} = \left(\frac{1998 + 2000}{2}, \frac{24.8 + 29.8}{2}\right)$$

$$= (1999, 27.3)$$

So, you would estimate the 1999 revenue to have been about \$27.3 billion, as shown in Figure P.18. (The actual 1999 revenue was \$27.1 billion.)

Application

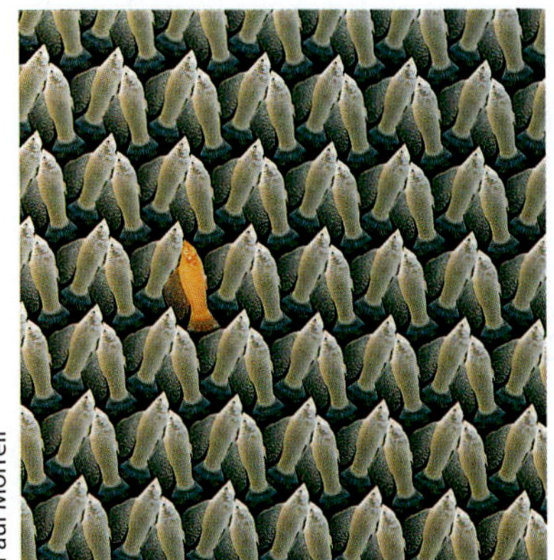

Paul Morrell

Much of computer graphics, including this computer-generated goldfish tessellation, consists of transformations of points in a coordinate plane. One type of transformation, a translation, is illustrated in Example 8. Other types include reflections, rotations, and stretches.

Example 8 ▶ Translating Points in the Plane

The triangle in Figure P.19 has vertices at the points $(-1, 2)$, $(1, -4)$, and $(2, 3)$. Shift the triangle three units to the right and two units upward and find the vertices of the shifted triangle, as shown in Figure P.20.

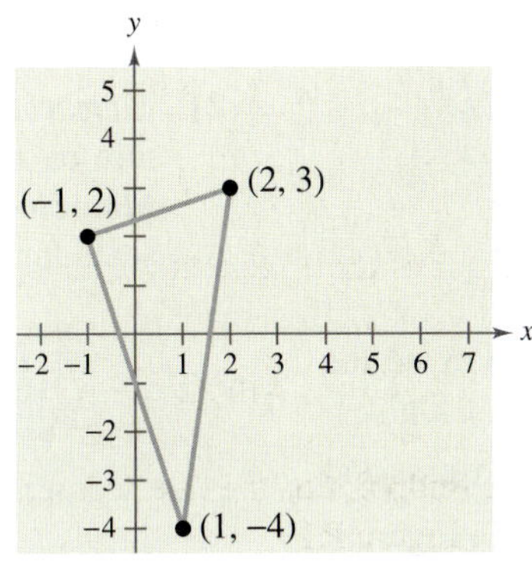

FIGURE P.19

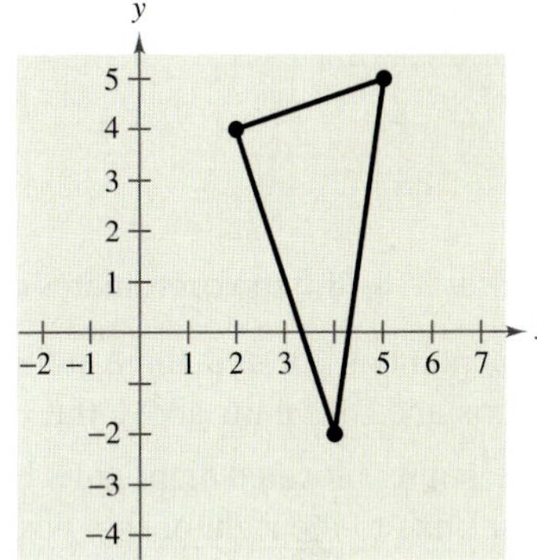

FIGURE P.20

Solution

To shift the vertices three units to the right, add 3 to each of the x-coordinates. To shift the vertices two units upward, add 2 to each of the y-coordinates.

Original Point	*Translated Point*
$(-1, 2)$	$(-1 + 3, 2 + 2) = (2, 4)$
$(1, -4)$	$(1 + 3, -4 + 2) = (4, -2)$
$(2, 3)$	$(2 + 3, 3 + 2) = (5, 5)$

Writing ABOUT MATHEMATICS

Extending the Example Example 8 shows how to translate points in a coordinate plane. Write a short paragraph describing how each of the following transformed points is related to the original point.

Original Point	*Transformed Point*
(x, y)	$(-x, y)$
(x, y)	$(x, -y)$
(x, y)	$(-x, -y)$

P.7 Exercises

In Exercises 1 and 2, approximate the coordinates of the points.

1.

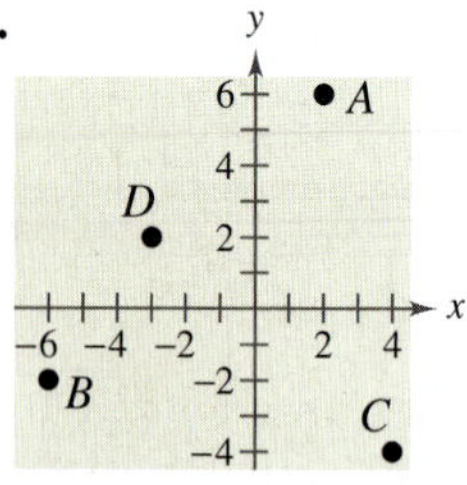

2.

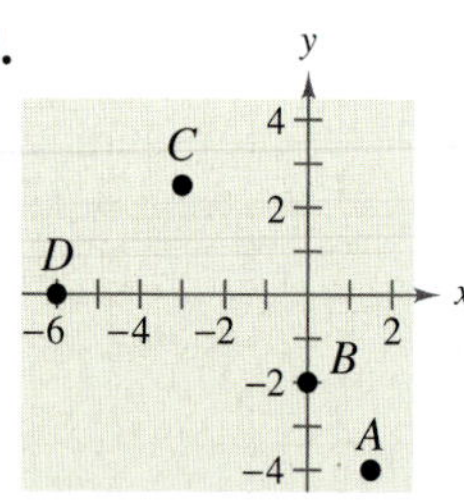

In Exercises 3–6, find the coordinates of the point.

3. The point is located three units to the left of the y-axis and four units above the x-axis.

4. The point is located eight units below the x-axis and four units to the right of the y-axis.

5. The point is located five units below the x-axis and the coordinates of the point are equal.

6. The point is on the x-axis and 12 units to the left of the y-axis.

In Exercises 7–16, determine the quadrant(s) in which (x, y) is located so that the condition(s) is (are) satisfied.

7. $x > 0$ and $y < 0$

8. $x < 0$ and $y < 0$

9. $x = -4$ and $y > 0$

10. $x > 2$ and $y = 3$

11. $y < -5$

12. $x > 4$

13. $(x, -y)$ is in the second quadrant.

14. $(-x, y)$ is in the fourth quadrant.

15. $xy > 0$

16. $xy < 0$

In Exercises 17–20, the polygon is shifted to a new position in the plane. Find the coordinates of the vertices of the polygon in its new position.

17.

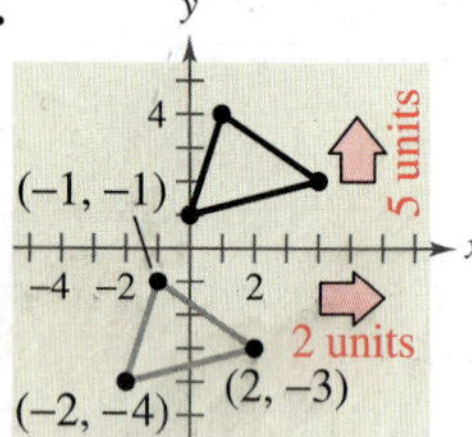

18.

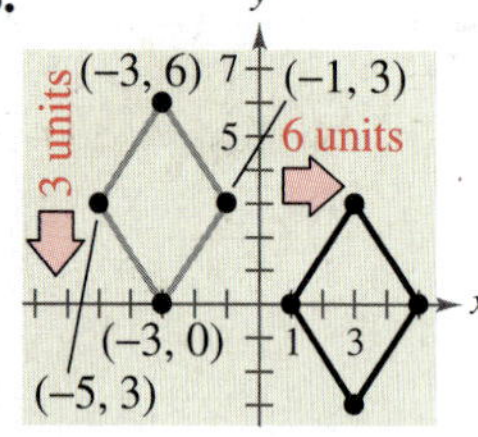

19. Original coordinates of vertices:

$(-7, -2)$, $(-2, 2)$, $(-2, -4)$, $(-7, -4)$

Shift: eight units upward, four units to the right

20. Original coordinates of vertices:

$(5, 8)$, $(3, 6)$, $(7, 6)$, $(5, 2)$

Shift: 6 units downward, 10 units to the left

In Exercises 21 and 22, sketch a scatter plot of the data given in the table.

21. ***Meteorology*** The table shows the lowest temperature on record y (in degrees Fahrenheit) in Duluth, Minnesota, for each month x, where $x = 1$ represents January. (Source: NOAA)

Month, x	Temperature, y
1	−39
2	−33
3	−29
4	−5
5	17
6	27
7	35
8	32
9	22
10	8
11	−23
12	−34

22. ***Number of Stores*** The table shows the number y of Wal-Mart stores for each year x from 1993 through 2000. (Source: Wal-Mart Stores, Inc.)

Year, x	Number of stores, y
1993	2440
1994	2759
1995	2943
1996	3054
1997	3406
1998	3599
1999	3985
2000	4190

Retail Price **In Exercises 23 and 24, use the graph below, which shows the average retail price of 1 pound of butter from 1993 to 1999.** (Source: U.S. Bureau of Labor Statistics)

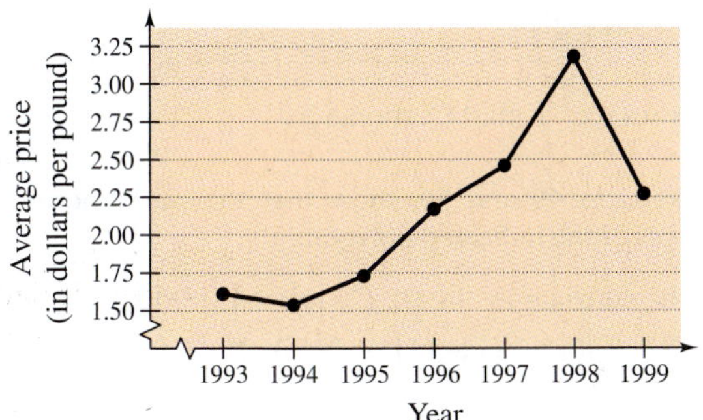

23. Approximate the highest price of a pound of butter shown in the graph. When did this occur?

24. Approximate the percent change in the price of butter from the price in 1994 to the highest price shown in the graph.

Advertising **In Exercises 25 and 26, use the graph below, which shows the cost of a 30-second television spot (in thousands of dollars) during the Super Bowl from 1989 to 2001.** (Source: USA Today Research)

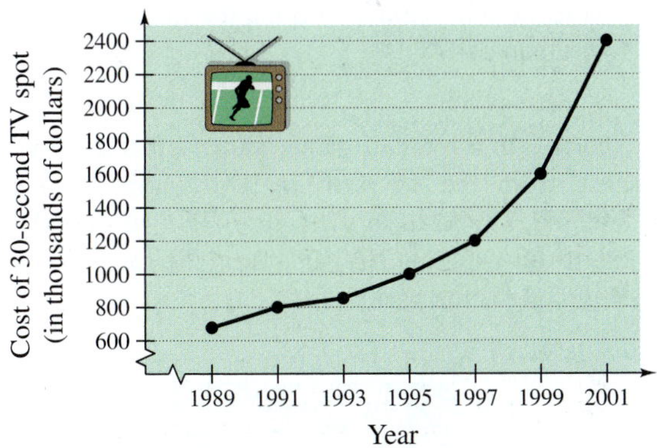

25. Approximate the percent increase in the cost of a 30-second spot from Super Bowl XXIII in 1989 to Super Bowl XXXV in 2001.

26. Estimate the percent increase in the cost of a 30-second spot (a) from Super Bowl XXIII in 1989 to Super Bowl XXVII in 1993 and (b) from Super Bowl XXVII in 1993 to Super Bowl XXXV in 2001.

▶ Model It

27. ***Labor Force*** Use the graph below, which shows the minimum wage in the United States (in dollars) from 1950 to 2000. (Source: U.S. Employment Standards Administration)

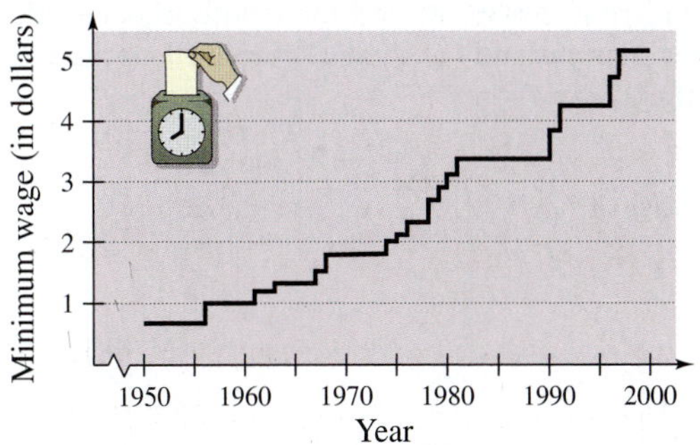

(a) Which decade shows the greatest increase in minimum wage?

(b) Approximate the percent increases in the minimum wage from 1990 to 1995 and from 1995 to 2000.

(c) Use the percent increase from 1995 to 2000 to predict the minimum wage in 2005.

(d) Do you believe that your prediction in part (c) is reasonable? Explain.

28. ***Data Analysis*** Use the table below, which shows the mathematics entrance test scores x and the final examination scores y in an algebra course for a sample of 10 students.

x	22	29	35	40	44
y	53	74	57	66	79

x	48	53	58	65	76
y	90	76	93	83	99

(a) Sketch a scatter plot of the data shown in the table.

(b) Find the entrance exam score of any student with a final exam score in the 80s.

(c) Does a higher entrance exam score imply a higher final exam score? Explain.

In Exercises 29–32, find the distance between the points. (*Note:* In each case, the two points lie on the same horizontal or vertical line.)

29. $(6, -3), (6, 5)$

30. $(1, 4), (8, 4)$

31. $(-3, -1), (2, -1)$

32. $(-3, -4), (-3, 6)$

In Exercises 33–36, (a) find the length of each side of the right triangle, and (b) show that these lengths satisfy the Pythagorean Theorem.

33. 

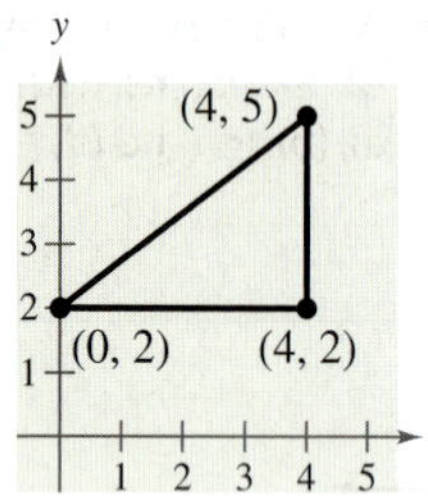

34. 

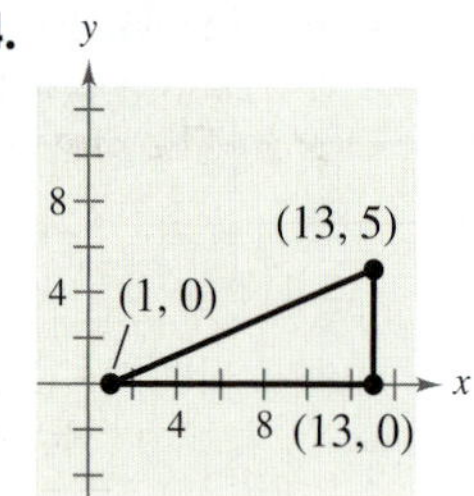

35.

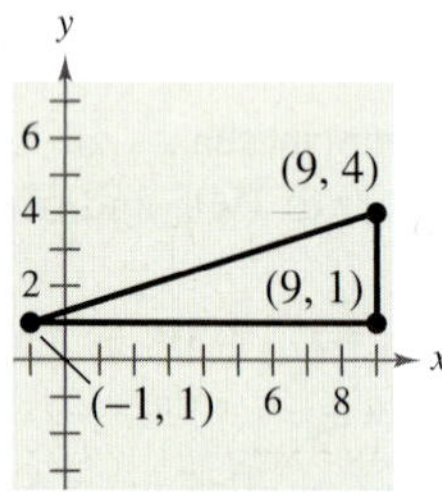

36.

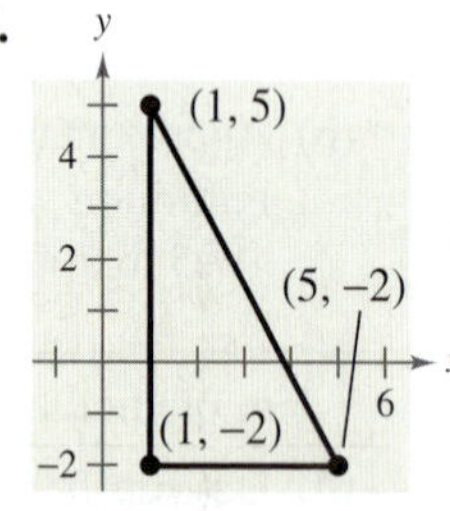

In Exercises 37–46, (a) plot the points, (b) find the distance between the points, and (c) find the midpoint of the line segment joining the points.

37. $(1, 1), (9, 7)$

38. $(1, 12), (6, 0)$

39. $(-4, 10), (4, -5)$

40. $(-7, -4), (2, 8)$

41. $(-1, 2), (5, 4)$

42. $(2, 10), (10, 2)$

43. $\left(\frac{1}{2}, 1\right), \left(-\frac{5}{2}, \frac{4}{3}\right)$

44. $\left(-\frac{1}{3}, -\frac{1}{3}\right), \left(-\frac{1}{6}, -\frac{1}{2}\right)$

45. $(6.2, 5.4), (-3.7, 1.8)$

46. $(-16.8, 12.3), (5.6, 4.9)$

Sales **In Exercises 47 and 48, use the Midpoint Formula to estimate the sales of Target Corporation and Kmart Corporation in 1998, given the sales in 1996 and 2000. Assume that the sales followed a linear pattern.**

47. *Target*

Year	Sales (in millions)
1996	\$25,371
2000	\$36,903

(Source: Target Corporation)

48. *Kmart*

Year	Sales (in millions)
1996	\$31,437
2000	\$37,028

(Source: Kmart Corporation)

In Exercises 49 and 50, show that the points form the vertices of the indicated polygon.

49. Right triangle: $(4, 0), (2, 1), (-1, -5)$

50. Isosceles triangle: $(1, -3), (3, 2), (-2, 4)$

51. A line segment has (x_1, y_1) as one endpoint and (x_m, y_m) as its midpoint. Find the other endpoint (x_2, y_2) of the line segment in terms of x_1, y_1, x_m, and y_m.

52. Use the result of Exercise 51 to find the coordinates of the endpoint of a line segment if the coordinates of the other endpoint and midpoint are, respectively,

(a) $(1, -2), (4, -1)$ and (b) $(-5, 11), (2, 4)$.

53. Use the Midpoint Formula three times to find the three points that divide the line segment joining (x_1, y_1) and (x_2, y_2) into four parts.

54. Use the result of Exercise 53 to find the points that divide the line segment joining the given points into four equal parts.

(a) $(1, -2), (4, -1)$ (b) $(-2, -3), (0, 0)$

55. ***Sports*** In a football game, a quarterback throws a pass from the 15-yard line, 10 yards from the sideline, as shown in the figure. The pass is caught on the 40-yard line, 45 yards from the same sideline. How long is the pass?

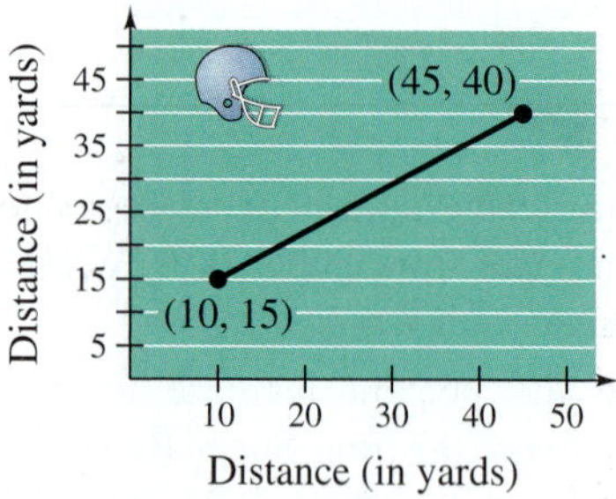

56. ***Flying Distance*** A jet plane flies from Naples, Italy in a straight line to Rome, Italy, which is 120 kilometers west and 150 kilometers north of Naples. How far does the plane fly?

57. ***Make a Conjecture*** Plot the points $(2, 1)$, $(-3, 5)$, and $(7, -3)$ on a rectangular coordinate system. Then change the sign of the x-coordinate of each point and plot the three new points on the same rectangular coordinate system. Make a conjecture about the location of a point when each of the following occurs.

(a) The sign of the x-coordinate is changed.

(b) The sign of the y-coordinate is changed.

(c) The signs of both the x- and y-coordinates are changed.

58. ***Music*** The graph shows the numbers of recording artists who were elected to the Rock and Roll Hall of Fame from 1986 to 2001.

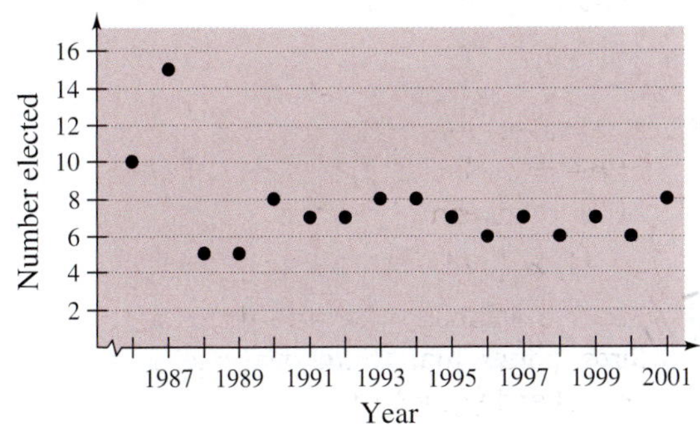

(a) Describe any trends in the data. From these trends, predict the number of artists elected in 2004.

(b) Why do you think the numbers elected in 1986 and 1987 were greater than in other years?

59. ***Revenue*** Polo Ralph Lauren Corp. had annual revenues of \$1713.1 million in 1998 and \$2225.8 million in 2000. Use the Midpoint Formula to estimate the revenue in 1999. (Source: Polo Ralph Lauren Corp.)

60. ***Revenue*** Zale Corp. had annual revenues of \$1428.9 million in 1999 and \$2068.2 million in 2001. Use the Midpoint Formula to estimate the revenue in 2000. (Source: Zale Corp.)

Synthesis

True or False? **In Exercises 61 and 62, determine whether the statement is true or false. Justify your answer.**

61. In order to divide a line segment into 16 equal parts, you would have to use the Midpoint Formula 16 times.

62. The points $(-8, 4)$, $(2, 11)$, and $(-5, 1)$ represent the vertices of an isosceles triangle.

63. ***Think About It*** What is the y-coordinate of any point on the x-axis? What is the x-coordinate of any point on the y-axis?

64. ***Think About It*** When plotting points on the rectangular coordinate system, is it true that the scales on the x- and y-axes must be the same? Explain.

In Exercises 65–68, use the plot of the point (x_0, y_0) in the figure. Match the transformation of the point with the correct plot. [The plots are labeled (a), (b), (c), and (d).]

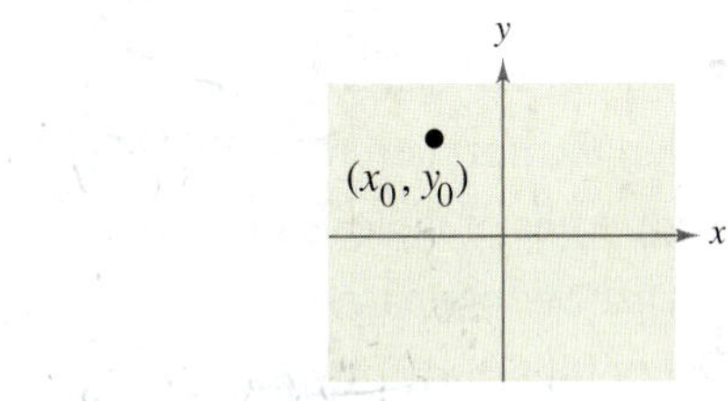

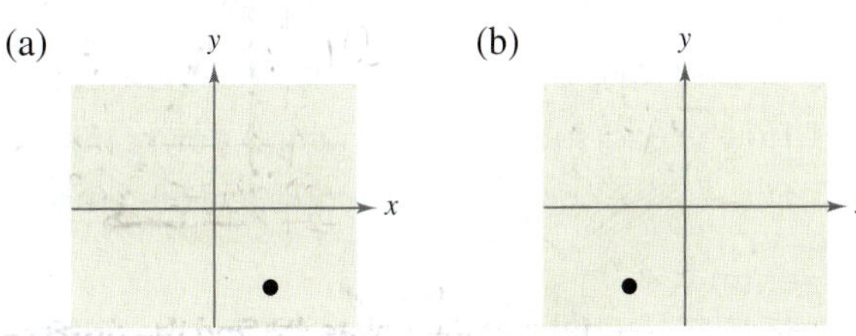

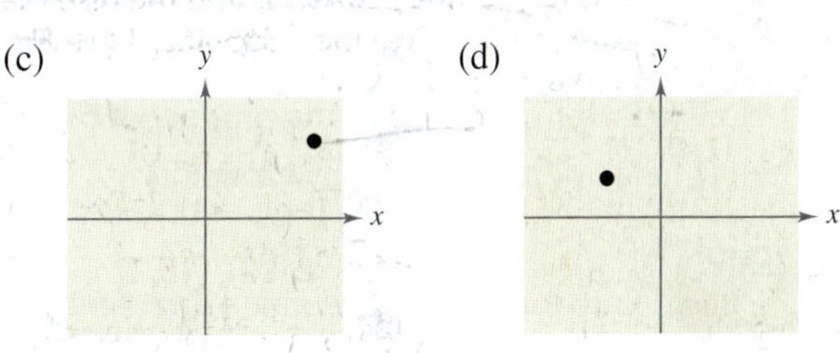

65. $(x_0, -y_0)$

66. $(-2x_0, y_0)$

67. $\left(x_0, \frac{1}{2}y_0\right)$

68. $(-x_0, -y_0)$

69. ***Proof*** Prove that the diagonals of the parallelogram in the figure intersect at their midpoints.

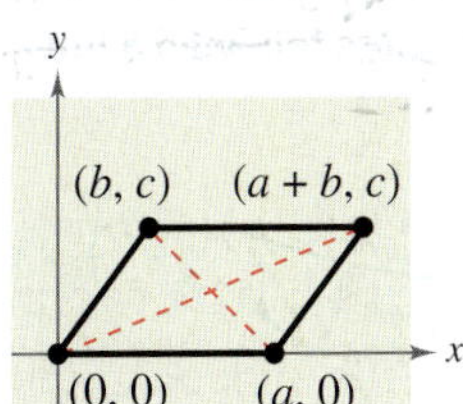

Chapter Summary

▶ *What* did you learn?

Section P.1	Review Exercises
☐ How to represent, classify, and order real numbers and use inequalities	1–6
☐ How to find the absolute values of real numbers and find the distance between two real numbers	7–12
☐ How to evaluate algebraic expressions	13–16
☐ How to use the basic rules and properties of algebra	17–26
Section P.2	
☐ How to use properties of exponents	27–30
☐ How to use scientific notation to represent real numbers	31–34
☐ How to use properties of radicals to simplify and combine radicals	35–42
☐ How to rationalize denominators and numerators	43–46
☐ How to use properties of rational exponents	47–50
Section P.3	
☐ How to write polynomials in standard form	51–54
☐ How to add, subtract, and multiply polynomials	55–64
☐ How to use polynomials to solve real-life problems	65, 66
Section P.4	
☐ How to remove common factors from polynomials	67, 68
☐ How to factor special polynomial forms	69–72
☐ How to factor trinomials as the product of two binomials	73, 74
☐ How to factor polynomials by grouping	75, 76
Section P.5	
☐ How to find domains of algebraic expressions	77, 78
☐ How to simplify rational expressions	79, 80
☐ How to add, subtract, multiply, and divide rational expressions	81–84
☐ How to simplify complex fractions	85, 86
Section P.6	
☐ How to avoid common algebraic errors	87–92
☐ How to recognize and use algebraic techniques that are common in calculus	93–100
Section P.7	
☐ How to plot points in the Cartesian plane	101–104
☐ How to use the Distance Formula to find the distance between two points	105, 106
☐ How to use the Midpoint Formula to find the midpoint of a line segment	107, 108
☐ How to use a coordinate plane to model and solve real-life problems	109, 110

Review Exercises

P.1 In Exercises 1 and 2, determine which numbers in the set are (a) natural numbers, (b) integers, (c) rational numbers, and (d) irrational numbers.

1. $\{11, -14, -\frac{8}{9}, \frac{5}{2}, \sqrt{6}, 0.4\}$
2. $\{\sqrt{15}, -22, -\frac{10}{3}, 0, 5.2, \frac{3}{7}\}$

In Exercises 3 and 4, use a calculator to find the decimal form of each rational number. If it is a nonterminating decimal, write the repeating pattern. Then plot the numbers on the real number line and place the appropriate inequality sign (< or >) between them.

3. (a) $\frac{5}{6}$ (b) $\frac{7}{8}$
4. (a) $\frac{9}{25}$ (b) $\frac{5}{7}$

In Exercises 5 and 6, give a verbal description of the subset of real numbers represented by the inequality, and sketch the subset on the real number line.

5. $x \leq 7$
6. $x > 1$

In Exercises 7 and 8, find the distance between *a* and *b*.

7. $a = -92, b = 63$
8. $a = -112, b = -6$

In Exercises 9–12, use absolute value notation to describe the expression.

9. The distance between x and 7 is at least 4.
10. The distance between x and 25 is no more than 10.
11. The distance between y and -30 is less than 5.
12. The distance between z and -16 is greater than 8.

In Exercises 13–16, evaluate the expression for each value of *x*. (If not possible, state the reason.)

	Expression	*Values*	
13.	$12x - 7$	(a) $x = 0$	(b) $x = -1$
14.	$x^2 - 6x + 5$	(a) $x = -2$	(b) $x = 2$
15.	$-x^2 + x - 1$	(a) $x = 1$	(b) $x = -1$
16.	$\frac{x}{x-3}$	(a) $x = -3$	(b) $x = 3$

In Exercises 17–20, identify the rule of algebra illustrated by the equation.

17. $2x + (3x - 10) = (2x + 3x) - 10$
18. $(t + 4)(2t) = (2t)(t + 4)$
19. $0 + (a - 5) = a - 5$
20. $\frac{2}{y+4} \cdot \frac{y+4}{2} = 1, \quad y \neq -4$

In Exercises 21–26, perform the operation without using a calculator.

21. $|-3| + 4(-2) - 6$
22. $\frac{|-10|}{-10}$
23. $\frac{5}{18} \div \frac{10}{3}$
24. $(16 - 8) \div 4$
25. $6[4 - 2(6 + 8)]$
26. $-4[16 - 3(7 - 10)]$

P.2 In Exercises 27 and 28, simplify each expression.

27. (a) $3x^2(4x^3)^3$ (b) $\frac{5y^6}{10y}$
28. (a) $(-2z)^2(8z^3)$ (b) $\frac{36x^5}{9x^{10}}$

In Exercises 29 and 30, rewrite each expression with positive exponents and simplify.

29. (a) $\frac{6^2u^3v^{-3}}{12u^{-2}v}$ (b) $\frac{3^{-4}m^{-1}n^{-3}}{9^{-2}mn^{-3}}$
30. (a) $(x + y^{-1})^{-1}$ (b) $\left(\frac{x^{-3}}{y}\right)\left(\frac{x}{y}\right)^{-1}$

In Exercises 31 and 32, write the number in scientific notation.

31. *Sales of Tommy Hilfiger Corporation in 2000:* \$1,880,900,000 (Source: Tommy Hilfiger Corporation)
32. *Number of meters in 1 foot:* 0.3048

In Exercises 33 and 34, write the number in decimal notation.

33. *Distance between the sun and Jupiter:* 4.836×10^8 miles
34. *Ratio of day to year:* 2.74×10^{-3}

In Exercises 35–38, simplify each expression.

35. (a) $\sqrt[3]{27^2}$ (b) $\sqrt{49^3}$
36. (a) $\sqrt[3]{\frac{64}{125}}$ (b) $\sqrt{\frac{81}{100}}$
37. (a) $\left(\sqrt[3]{216}\right)^3$ (b) $\sqrt[4]{32^4}$
38. (a) $\sqrt[3]{\frac{2x^3}{27}}$ (b) $\sqrt[5]{64x^6}$

In Exercises 39 and 40, simplify each expression.

39. (a) $\sqrt{50} - \sqrt{18}$ (b) $2\sqrt{32} + 3\sqrt{72}$

40. (a) $\sqrt{8x^3} + \sqrt{2x}$ (b) $\sqrt{18x^5} - \sqrt{8x^3}$

41. ***Writing*** Explain why $\sqrt{5u} + \sqrt{3u} \neq 2\sqrt{2u}$.

42. ***Engineering*** The rectangular cross section of a wooden beam cut from a log of diameter 24 inches (see figure) will have a maximum strength if its width w and height h are $w = 8\sqrt{3}$ and $h = \sqrt{24^2 - (8\sqrt{3})^2}$. Find the area of the rectangular cross section and express the answer in simplest form.

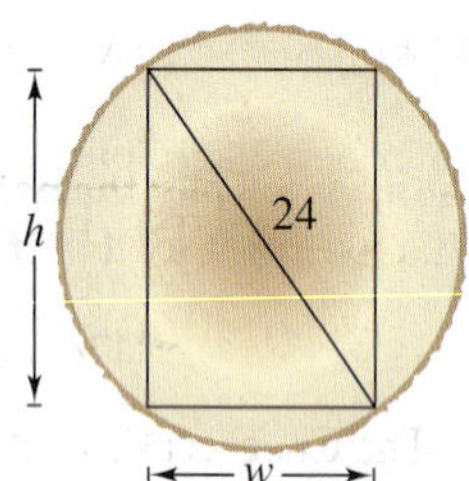

In Exercises 43 and 44, rewrite the expression by rationalizing the denominator. Simplify your answer.

43. $\dfrac{1}{2 - \sqrt{3}}$ **44.** $\dfrac{1}{\sqrt{5} - 1}$

In Exercises 45 and 46, rewrite the expression by rationalizing the numerator. Simplify your answer.

45. $\dfrac{\sqrt{7} - 1}{2}$ **46.** $\dfrac{\sqrt{2} - \sqrt{11}}{3}$

In Exercises 47–50, simplify the expression.

47. $(16)^{3/2}$ **48.** $(64)^{-2/3}$

49. $(3x^{2/5})(2x^{1/2})$ **50.** $(x - 1)^{1/3}(x - 1)^{-1/4}$

P.3 **In Exercises 51–54, write the polynomial in standard form. Identify the degree and leading coefficient.**

51. $3 - 11x^2$ **52.** $3x^3 - 5x^5 + x - 4$

53. $-4 - 12x^2$ **54.** $12x - 7x^2 + 6$

In Exercises 55–58, perform the operation and write the result in standard form.

55. $-(3x^2 + 2x) + (1 - 5x)$

56. $8y - [2y^2 - (3y - 8)]$

57. $2x(x^2 - 5x + 6)$

58. $(3x^3 - 1.5x^2 + 4)(-3x)$

In Exercises 59–64, find the product.

59. $(3x - 6)(5x + 1)$ **60.** $\left(x - \dfrac{1}{x}\right)(x + 2)$

61. $(2x - 3)^2$ **62.** $(6x + 5)(6x - 5)$

63. $(3\sqrt{5} + 2)(3\sqrt{5} - 2)$ **64.** $(x - 4)^3$

65. ***Surface Area*** The surface area S of a right circular cylinder is $S = 2\pi r^2 + 2\pi rh$.

(a) Draw a right circular cylinder of radius r and height h. Use the figure to explain how the surface area formula was obtained.

(b) Find the surface area when the radius is 6 inches and the height is 8 inches.

66. ***Geometry*** Find a polynomial that represents the total number of square feet for the floor plan shown in the figure.

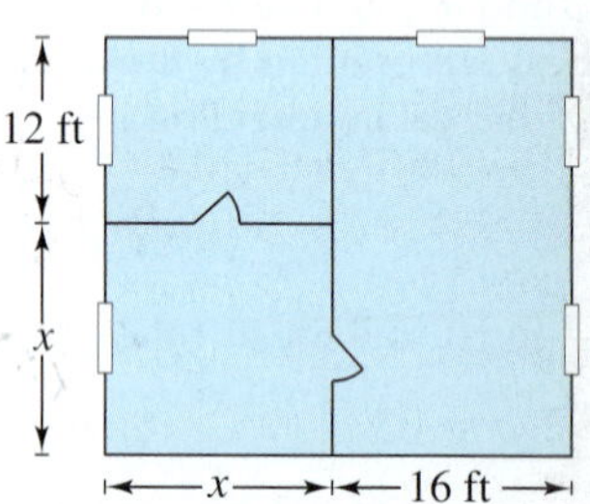

P.4 **In Exercises 67–76, factor completely.**

67. $x^3 - x$

68. $x(x - 3) + 4(x - 3)$

69. $25x^2 - 49$

70. $x^2 - 12x + 36$

71. $x^3 - 64$ **72.** $8x^3 + 27$

73. $2x^2 + 21x + 10$ **74.** $3x^2 + 14x + 8$

75. $x^3 - x^2 + 2x - 2$

76. $x^3 - 4x^2 + 2x - 8$

P.5 **In Exercises 77 and 78, find the domain of the expression.**

77. $\dfrac{1}{x + 6}$ **78.** $\sqrt{x + 4}$

In Exercises 79 and 80, write the rational expression in simplest form.

79. $\dfrac{x^2 - 64}{5(3x + 24)}$ **80.** $\dfrac{x^3 + 27}{x^2 + x - 6}$

In Exercises 81–84, perform the operation and simplify.

81. $\dfrac{x^2-4}{x^4-2x^2-8}\cdot\dfrac{x^2+2}{x^2}$

82. $\dfrac{4x-6}{(x-1)^2}\div\dfrac{2x^2-3x}{x^2+2x-3}$

83. $\dfrac{1}{x-1}+\dfrac{1-x}{x^2+x+1}$

84. $\dfrac{3x}{x+2}-\dfrac{4x^2-5}{2x^2+3x-2}$

In Exercises 85 and 86, simplify the complex fraction.

85. $\dfrac{\left[\dfrac{3a}{(a^2/x)-1}\right]}{\left(\dfrac{a}{x}-1\right)}$

86. $\dfrac{\left(\dfrac{1}{2x-3}-\dfrac{1}{2x+3}\right)}{\left(\dfrac{1}{2x}-\dfrac{1}{2x+3}\right)}$

P.6 In Exercises 87–92, describe and correct the error.

87. $10(4\cdot 7)=40\cdot 70$

88. $\left(\frac{1}{3}x\right)\left(\frac{1}{3}y\right)=\frac{1}{3}xy$

89. $(3^4)^4=3^8$

90. $\sqrt{3^2+4^2}=3+4$

91. $(5+8)^2=5^2+8^2$

92. $\dfrac{7+5(x+3)}{x+3}=12$

In Exercises 93–96, insert the missing factor.

93. $\frac{2}{3}x^4-\frac{3}{8}x^3+\frac{5}{6}x^2=\frac{1}{24}x^2(\quad)$

94. $\dfrac{t}{\sqrt{t+1}}-\sqrt{t+1}=\dfrac{1}{\sqrt{t+1}}(\quad)$

95. $2x(x^2-3)^{1/3}-5(x^2-3)^{4/3}=(x^2-3)^{1/3}(\quad)$

96. $y(y-1)^{5/4}-y^2(y-1)^{1/4}=y(y-1)^{1/4}(\quad)$

In Exercises 97 and 98, write the fraction as a sum of two or more terms.

97. $\dfrac{x^3+5x^2+7}{x}$

98. $\dfrac{2x^3-x^2+4}{x^{1/2}}$

In Exercises 99 and 100, simplify the expression.

99. $\dfrac{x(x+2)^{-1/2}+(x+2)^{1/2}}{(x+2)^{3/2}}$

100. $\dfrac{\frac{2}{3}x(4+x)^{-1/2}-\frac{2}{15}(4+x)^{1/2}}{(4+x)^{5/2}}$

P.7 In Exercises 101 and 102, determine the quadrant(s) in which (x, y) is located so that the condition(s) is (are) satisfied.

101. $x>0$ and $y=-2$

102. $y>0$

In Exercises 103 and 104, the polygon is shifted to a new position in the plane. Find the coordinates of the vertices of the polygon in its new position.

103. Original coordinates of vertices:
$(4, 8), (6, 8), (4, 3), (6, 3)$
Shift: three units downward, two units to the left

104. Original coordinates of vertices:
$(0, 1), (3, 3), (0, 5), (-3, 3)$
Shift: five units upward, four units to the left

In Exercises 105 and 106, (a) plot the points and (b) find the distance between the points.

105. $(-3, 8), (1, 5)$

106. $(5.6, 0), (0, 8.2)$

In Exercises 107 and 108, (a) plot the points and (b) find the midpoint of the line segment joining the points.

107. $(-2, 6), (4, -3)$

108. $(0, -1.2), (-3.6, 0)$

109. ***Revenue*** The Cheesecake Factory had annual revenues of \$347.5 million in 1999 and \$539.1 million in 2001. Use the Midpoint Formula to estimate the revenue in 2000. (Source: The Cheesecake Factory, Inc.)

110. ***Meteorology*** The apparent temperature is a measure of relative discomfort to a person from heat and high humidity. The table shows the actual temperatures x (in degrees Fahrenheit) versus the apparent temperatures y (in degrees Fahrenheit) for a relative humidity of 75%.

x	70	75	80	85	90	95	100
y	70	77	85	95	109	130	150

(a) Sketch a scatter plot of the data shown in the table.

(b) Find the change in the apparent temperature when the actual temperature changes from 70°F to 100°F.

Synthesis

True or False? In Exercises 111 and 112, determine whether the statement is true or false. Justify your answer.

111. A binomial sum squared is equal to the sum of the terms squared.

112. x^n-y^n factors as conjugates for all values of n.

Chapter Test

The *Interactive* CD-ROM and *Internet* versions of this text offer Chapter Pre-Tests and Chapter Post-Tests, both of which have randomly generated exercises with diagnostic capabilities.

Take this test as you would take a test in class. When you are finished, check your work against the answers given in the back of the book.

1. Place < or > between the real numbers $-\frac{10}{3}$ and $-|-4|$.
2. Find the distance between the real numbers -5.4 and $3\frac{3}{4}$.
3. Identify the rule of algebra illustrated by $(5 - x) + 0 = 5 - x$.

In Exercises 4 and 5, evaluate each expression without using a calculator.

4. (a) $27\left(-\frac{2}{3}\right)$ (b) $\frac{5}{18} \div \frac{15}{8}$ (c) $\left(-\frac{3}{5}\right)^3$ (d) $\left(\frac{3^2}{2}\right)^{-3}$
5. (a) $\sqrt{5} \cdot \sqrt{125}$ (b) $\frac{\sqrt{72}}{\sqrt{2}}$ (c) $\frac{5.4 \times 10^8}{3 \times 10^3}$ (d) $(3 \times 10^4)^3$

In Exercises 6 and 7, simplify each expression.

6. (a) $3z^2(2z^3)^2$ (b) $(u - 2)^{-4}(u - 2)^{-3}$ (c) $\left(\frac{x^{-2}y^2}{3}\right)^{-1}$
7. (a) $9z\sqrt{8z} - 3\sqrt{2z^3}$ (b) $(4x^{3/5})(x^{1/3})$ (c) $\sqrt[3]{\frac{16}{v^5}}$
8. Write the polynomial $x + 4x^4 - 5 - 3x^2$ in standard form. Identify the degree and leading coefficient.

In Exercises 9–12, perform the operation and simplify.

9. $(x^2 + 3) - [3x + (8 - x^2)]$
10. $(x + \sqrt{5})(x - \sqrt{5})$
11. $\frac{8x}{x - 3} + \frac{24}{3 - x}$
12. $\frac{\left(\frac{2}{x} - \frac{2}{x + 1}\right)}{\left(\frac{4}{x^2 - 1}\right)}$
13. Factor (a) $2x^4 - 3x^3 - 2x^2$ and (b) $x^3 + 2x^2 - 4x - 8$ completely.
14. Rationalize each denominator. (a) $\frac{16}{\sqrt[3]{16}}$ (b) $\frac{6}{1 - \sqrt{3}}$
15. Find the domain of $\frac{1 - x}{4 - x}$.
16. Multiply: $\frac{y^2 + 8y + 16}{2y - 4} \cdot \frac{8y - 16}{(y + 4)^3}$.
17. A T-shirt company can produce and sell x T-shirts per day. The total cost C (in dollars) for producing x T-shirts is $C = 1480 + 6x$, and the total revenue R (in dollars) is $R = 15x$. Find the profit obtained by selling 225 T-shirts per day.
18. Plot the points $(-2, 5)$ and $(6, 0)$. Find the coordinates of the midpoint of the line segment joining the points and the distance between the points.
19. Write an expression for the area of the shaded region in the figure at the left, and simplify the result.

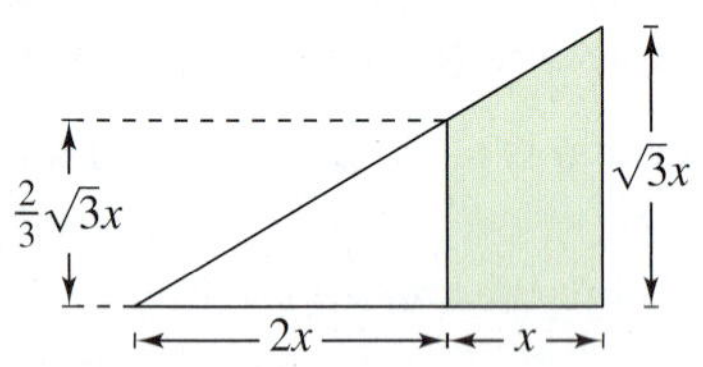

FIGURE FOR 19

Proofs in Mathematics

What does the word *proof* mean to you? In mathematics, the word *proof* is used to mean simply a valid argument. When you are proving a statement or theorem, you must use facts, definitions, and accepted properties in a logical order. You can also use previously proved theorems in your proof. For instance, the Distance Formula is used in the proof of the Midpoint Formula below. There are several different proof methods, which you will see in later chapters.

The Midpoint Formula *(p. 62)*

The midpoint of the segment joining the points (x_1, y_1) and (x_2, y_2) is given by the Midpoint Formula

$$\text{Midpoint} = \left(\frac{x_1 + x_2}{2}, \frac{y_1 + y_2}{2}\right).$$

The Cartesian Plane

The Cartesian Plane was named after the French mathematician René Descartes (1596–1650). While Descartes was lying in bed sick, he noticed a fly buzzing around on the square ceiling tiles. He discovered that the position of the fly could be described by which ceiling tile the fly landed on. This led to the development of the Cartesian Plane. Descartes felt that a coordinate plane could be used to facilitate description of the positions of objects.

Proof

Using the figure, you must show that $d_1 = d_2$ and $d_1 + d_2 = d_3$.

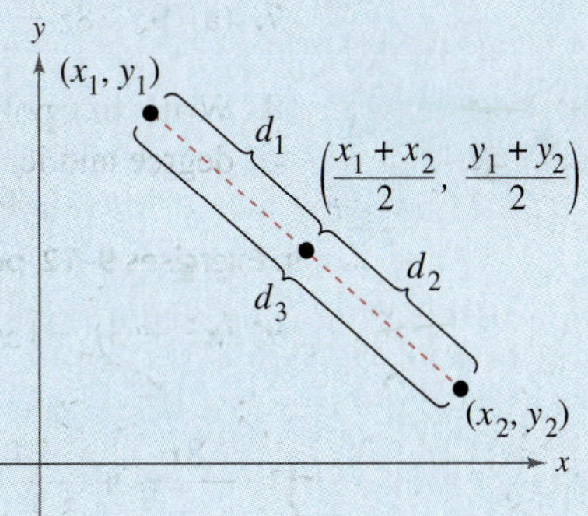

By the Distance Formula, you obtain

$$d_1 = \sqrt{\left(\frac{x_1 + x_2}{2} - x_1\right)^2 + \left(\frac{y_1 + y_2}{2} - y_1\right)^2}$$

$$= \frac{1}{2}\sqrt{(x_2 - x_1)^2 + (y_2 - y_1)^2}$$

$$d_2 = \sqrt{\left(x_2 - \frac{x_1 + x_2}{2}\right)^2 + \left(y_2 - \frac{y_1 + y_2}{2}\right)^2}$$

$$= \frac{1}{2}\sqrt{(x_2 - x_1)^2 + (y_2 - y_1)^2}$$

$$d_3 = \sqrt{(x_2 - x_1)^2 + (y_2 - y_1)^2}$$

So, it follows that $d_1 = d_2$ and $d_1 + d_2 = d_3$.

P.S. Problem Solving

1. The NCAA states that the men's and women's shot for track and field competition must comply with the following specifications. (Source: NCAA)

	Men's	Women's
Weight (minimum)	7.26 kg	4.0 kg
Diameter (minimum)	110 mm	95 mm
Diameter (maximum)	130 mm	110 mm

(a) Find the maximum and minimum volumes of both the men's and women's shots.

(b) The density of an object is an indication of how heavy the object is. To find the density of an object, divide its mass (weight) by its volume. Find the maximum and minimum densities of both the men's and women's shots.

(c) A shot is usually made out of iron. If a ball of cork has the same volume as an iron shot, do you think they would have the same density? Explain your reasoning.

2. Find an example for which $|a - b| > |a| - |b|$, and an example for which $|a - b| = |a| - |b|$. Then prove that $|a - b| \geq |a| - |b|$ for all a, b.

3. A major feature of Epcot Center in Disney World is called Spaceship Earth. The building is shaped as a sphere and weighs 1.6×10^7 pounds, which is equal in weight to 1.58×10^8 golf balls. Use these values to find the approximate weight (in pounds) of one golf ball. Then convert the weight to ounces. (Source: Disney.com)

4. The average life expectancy at birth in 1999 for men and women were 73.9 years and 79.4 years, respectively. Assuming an average healthy heart rate of 70 beats per minute, find the number of beats in a lifetime for a man and a woman. (Source: National Center for Health Statistics)

5. The accuracy of an approximation to a number is related to how many significant digits there are in the approximation. Write a definition of significant digits and illustrate the concept with examples.

6. The table shows the census population (in millions) of the United States from 1950 to 2000. (Source: U.S. Census Bureau)

Year	Population
1950	151.33
1960	179.32
1970	203.30
1980	226.54
1990	248.72
2000	281.42

(a) Find the increase in population from each census year to the next.

(b) Over which decade did the population increase the most? the least?

(c) Find the percent increase in population from each census year to the next.

(d) Over which decade was the percent increase the greatest? the least?

7. Find the annual depreciation rate r from the bar graph below. To find r by the declining balances method, use the formula

$$r = 1 - \left(\frac{S}{C}\right)^{1/n}$$

where n is the useful life of the item (in years), S is the salvage value (in dollars), and C is the original cost (in dollars).

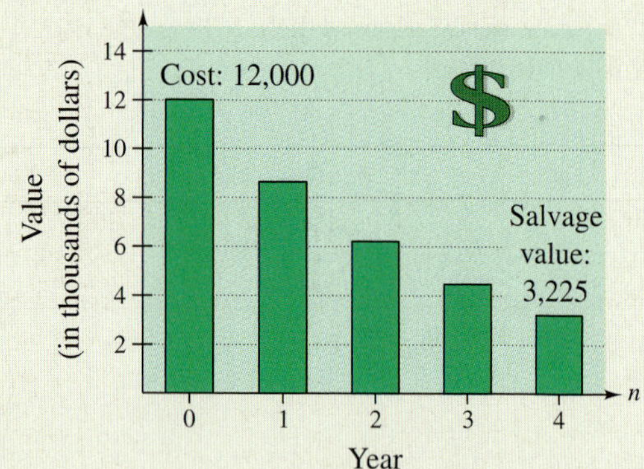

8. Johannes Kepler (1571–1630), a well-known German astronomer, discovered a relationship between the average distance of a planet from the sun and the time (or period) it takes the planet to orbit the sun. People then knew that planets that are closer to the sun take less time to complete an orbit than planets that are farther from the sun. Kepler discovered that the distance and period are related by an exact mathematical formula.

The table shows the average distance x (in astronomical units) and period y (in years) for the five planets that are closest to the sun. By completing the table, can you rediscover Kepler's relationship? Write a paragraph that summarizes your conclusions.

Planet	Mercury	Venus	Earth	Mars	Jupiter
x	0.387	0.723	1.000	1.524	5.203
$\sqrt{x}$					
y	0.241	0.615	1.000	1.881	11.862
$\sqrt[3]{y}$					

9. A stained glass window is designed in the shape of a rectangle with a semicircular arch (see figure). The width of the window is 2 feet and the perimeter is approximately 13.14 feet. Find the smallest amount of glass required to construct the window.

FIGURE FOR 9

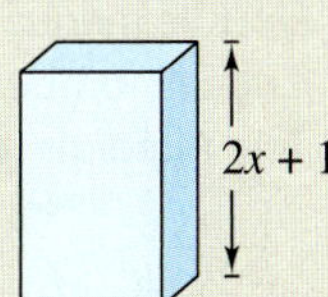

FIGURE FOR 10

10. The volume V of the box (in cubic inches) shown in the figure is modeled by

$$V = 2x^3 + x^2 - 8x - 4$$

where x is measured in inches. Find an expression for the surface area of the box. Then find the surface area when $x = 6$ inches.

11. Verify that $y_1 \neq y_2$ by letting $x = 0$ and evaluating y_1 and y_2.

$$y_1 = 2x\sqrt{1 - x^2} - \frac{x^3}{\sqrt{1 - x^2}}$$

$$y_2 = \frac{2 - 3x^2}{\sqrt{1 - x^2}}$$

Change y_2 so that $y_1 = y_2$.

12. Prove that

$$\left(\frac{2x_1 + x_2}{3}, \frac{2y_1 + y_2}{3}\right)$$

is one of the points of trisection of the line segment joining (x_1, y_1) and (x_2, y_2). Find the midpoint of the line segment joining

$$\left(\frac{2x_1 + x_2}{3}, \frac{2y_1 + y_2}{3}\right)$$

and (x_2, y_2) to find the second point of trisection.

13. Use the results of Exercise 12 to find the points of trisection of the line segment joining the following points.

(a) $(1, -2), (4, 1)$ (b) $(-2, -3), (0, 0)$

14. Although graphs can help visualize relationships between two variables, they can also be used to mislead people. The graphs shown below represent the same data points.

(a) Which of the two graphs is misleading, and why? Discuss other ways in which graphs can be misleading.

(b) Why would it be beneficial for someone to use a misleading graph?

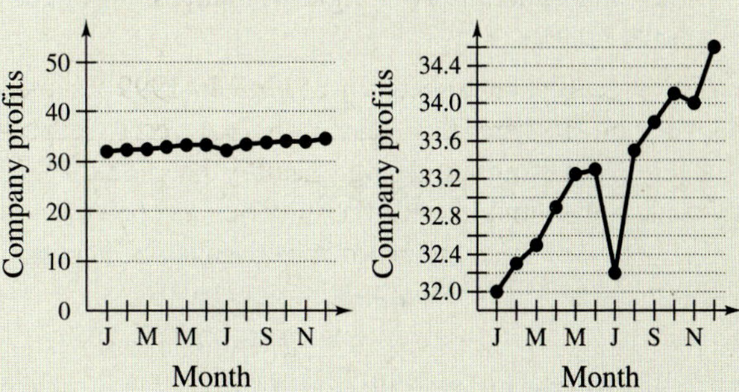

How to study Chapter 1

▶ What you should learn

In this chapter you will learn the following skills and concepts:

- How to sketch the graph of an equation
- How to solve linear equations, quadratic equations, polynomial equations, radical equations, and absolute value equations
- How to perform operations with complex numbers
- How to solve linear inequalities, polynomial inequalities, rational inequalities, and inequalities involving absolute value

▶ Important Vocabulary

As you encounter each new vocabulary term in this chapter, add the term and its definition to your notebook glossary.

Equation in two variables (p. 78)
Solution of equation in two variables (p. 78)
Graph of an equation (p. 78)
Intercepts (p. 80)
Symmetry (p. 80)
Circle (p. 83)
Equation in one variable (p. 88)
Solution of equation in one variable (p. 88)
Identity equation (p. 88)
Conditional equation (p. 88)
Linear equation in one variable (p. 88)
Equivalent equations (p. 89)
Extraneous solution (p. 91)
Quadratic equation (p. 109)
Quadratic Formula (p. 112)
Discriminant (p. 113)
Position equation (p. 115)
Complex number (p. 123)
Imaginary number (p. 123)
Pure imaginary number (p. 123)
Complex conjugates (p. 126)
Principal square root of a negative number (p. 127)
Polynomial equation (p. 130)
Solution of an inequality (p. 141)
Graph of an inequality (p. 141)
Linear inequality in one variable (p. 143)
Double inequality (p. 144)
Critical numbers (p. 151)
Test intervals (p. 151)

Study Tools

Learning Objectives in each section
Chapter Summary (p. 161)
Review Exercises (pp. 162–165)
Chapter Test (p. 166)

Additional Resources

Study and Solutions Guide
Interactive College Algebra
Videotapes/DVD for Chapter 1
College Algebra Website
Student Success Organizer

1

Equations and Inequalities

1.1 Graphs of Equations

▶ **What you should learn**

- How to sketch graphs of equations
- How to find x- and y-intercepts of graphs of equations
- How to use symmetry to sketch graphs of equations
- How to find equations and sketch graphs of circles
- How to use graphs of equations in solving real-life problems

▶ **Why you should learn it**

The graph of an equation can help you see relationships between real-life quantities. For example, in Exercise 67 on page 87, a graph can be used to estimate the life expectancies of children who are born in the years 2005 and 2010.

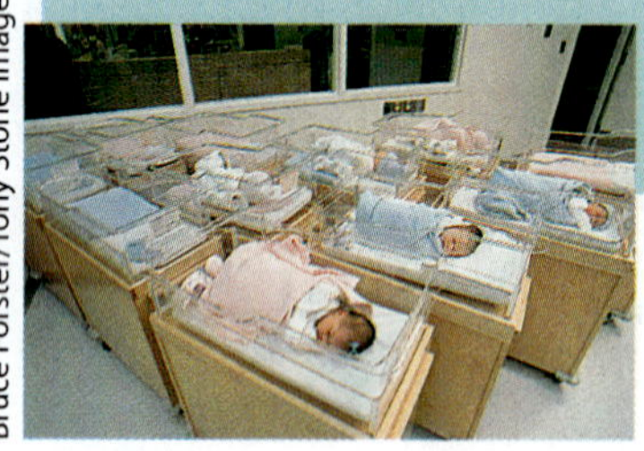

Bruce Forster/Tony Stone Images

The Graph of an Equation 

In Section P.7, you used a coordinate system to represent graphically the relationship between two quantities. There, the graphical picture consisted of a collection of points in a coordinate plane.

Frequently, a relationship between two quantities is expressed as an **equation in two variables.** For instance, $y = 7 - 3x$ is an equation in x and y. An ordered pair (a, b) is a **solution** or **solution point** of an equation in x and y if the equation is true when a is substituted for x and b is substituted for y. For instance, $(1, 4)$ is a solution of $y = 7 - 3x$ because $4 = 7 - 3(1)$ is a true statement.

In this section you will review some basic procedures for sketching the graph of an equation in two variables. The **graph of an equation** is the set of all points that are solutions of the equation.

Example 1 ▶ Sketching the Graph of an Equation 

Sketch the graph of $y = 7 - 3x$.

Solution

The simplest way to sketch the graph of an equation is the *point-plotting method.* With this method, you construct a table of values that consists of several solution points of the equation. For instance, when $x = 0$,

$$y = 7 - 3(0)$$
$$= 7$$

which implies that $(0, 7)$ is a solution point of the graph.

x	$y = 7 - 3x$	(x, y)
0	7	(0, 7)
1	4	(1, 4)
2	1	(2, 1)
3	−2	(3, −2)
4	−5	(4, −5)

From the table, it follows that

$$(0, 7), (1, 4), (2, 1), (3, -2), \text{ and } (4, -5)$$

are solution points of the equation. After plotting these points, you can see that they appear to lie on a line, as shown in Figure 1.1. The graph of the equation is the line that passes through the five plotted points.

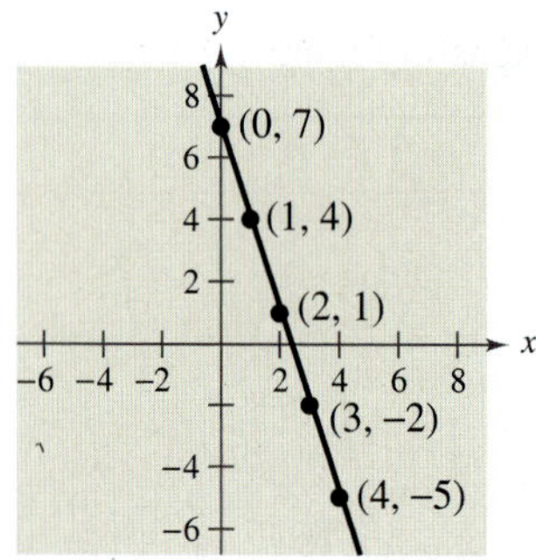

FIGURE 1.1

The icon [www] identifies examples and concepts related to features of the Learning Tools CD-ROM and the *Interactive* and *Internet* versions of this text. For more details see the chart on pages *xix-xxiii*.

The *Interactive* CD-ROM and *Internet* versions of this text offer a Try It for each example in the text.

STUDY TIP

One of your goals in this course is to learn to classify the basic shape of a graph from its equation. For instance, you will learn that the *linear equation* in Example 1 has the form

$$y = mx + b$$

and its graph is a line. Similarly, the *quadratic equation* in Example 2 has the form

$$y = ax^2 + bx + c$$

and its graph is a parabola.

Example 2 ▶ Sketching the Graph of an Equation 

Sketch the graph of

$$y = x^2 - 2.$$

Solution

Begin by constructing a table of values.

x	-2	-1	0	1	2	3
$y = x^2 - 2$	2	-1	-2	-1	2	7
(x, y)	$(-2, 2)$	$(-1, -1)$	$(0, -2)$	$(1, -1)$	$(2, 2)$	$(3, 7)$

Next, plot the points given in the table, as shown in Figure 1.2. Finally, connect the points with a smooth curve, as shown in Figure 1.3.

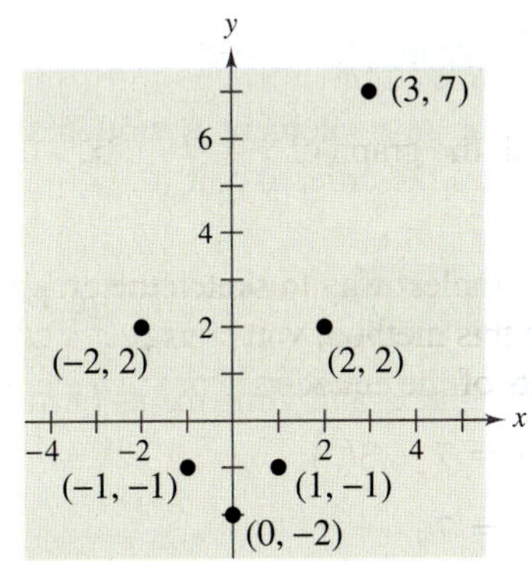

FIGURE 1.2

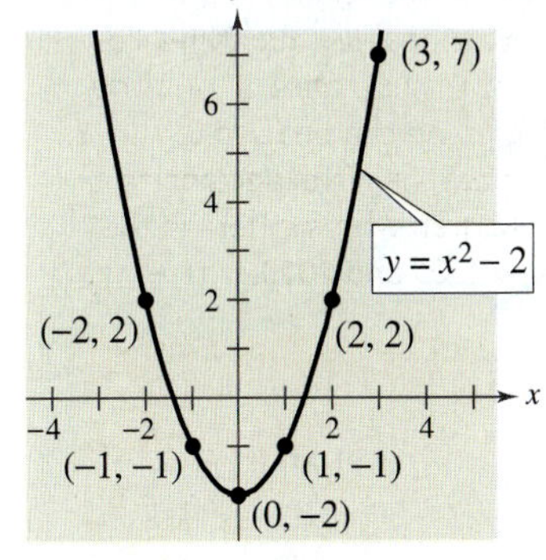

FIGURE 1.3

The point-plotting technique demonstrated in Examples 1 and 2 is easy to use, but it has some shortcomings. With too few solution points, you can misrepresent the graph of an equation. For instance, if only the four points

$$(-2, 2), (-1, -1), (1, -1), \text{ and } (2, 2)$$

in Figure 1.2 were plotted, any one of the three graphs in Figure 1.4 would be reasonable.

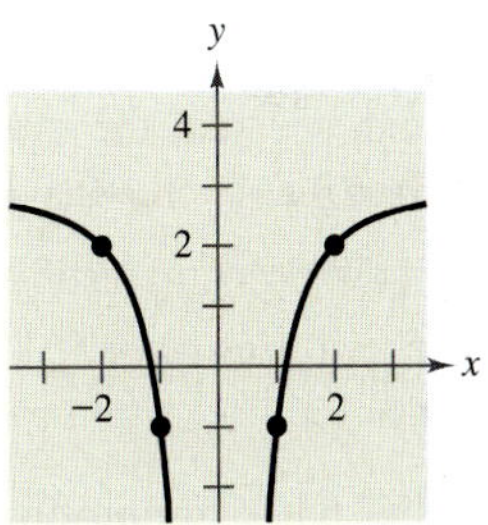

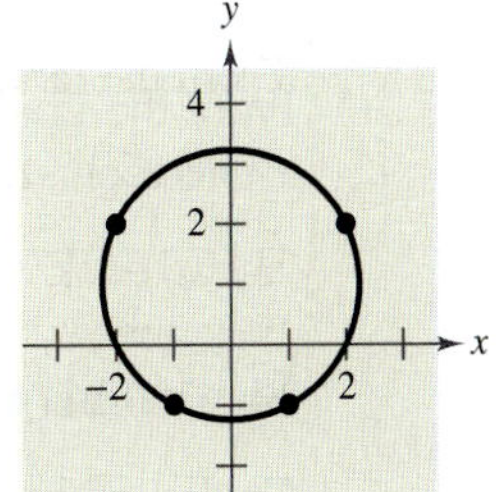

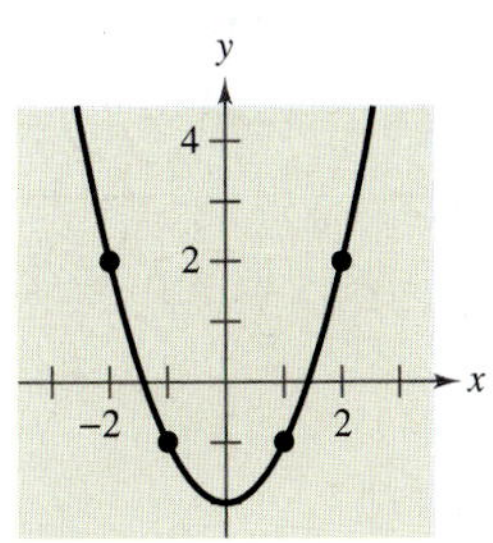

FIGURE 1.4

Intercepts of a Graph

It is often easy to determine the solution points that have zero as either the x-coordinate or the y-coordinate. These points are called **intercepts** because they are the points at which the graph intersects the x- or y-axis. It is possible for a graph to have no intercepts, one intercept, or several intercepts, as shown in Figure 1.5.

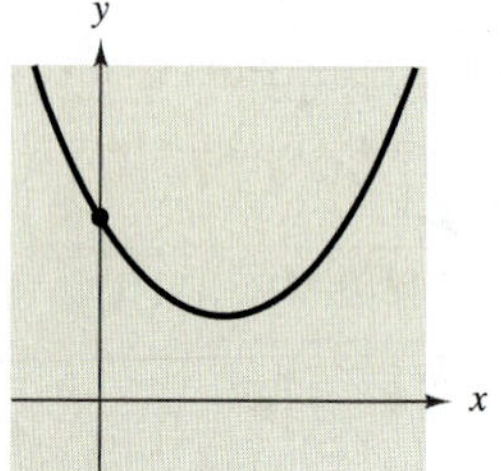

No x-intercept
One y-intercept

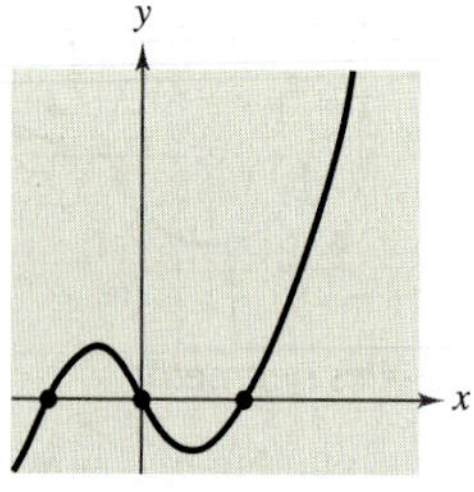

Three x-intercepts
One y-intercept

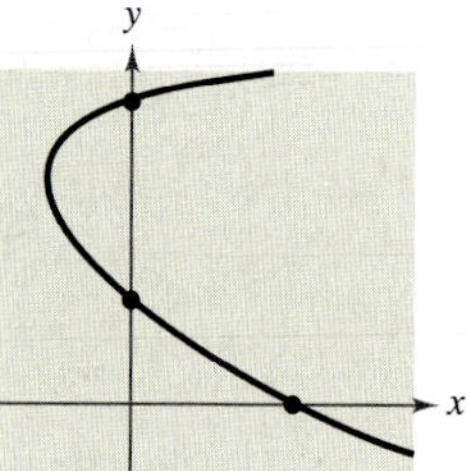

One x-intercept
Two y-intercepts

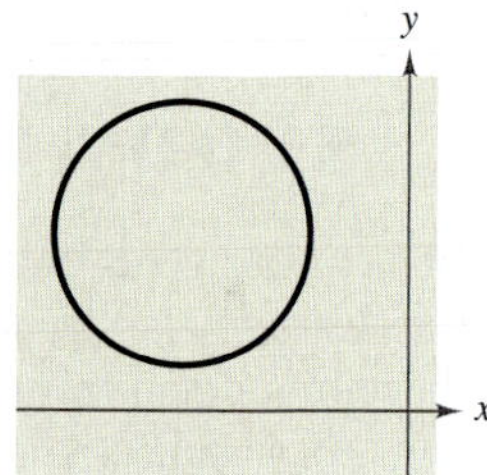

No intercepts

FIGURE 1.5

Note that an x-intercept is written as the ordered pair $(x, 0)$ and a y-intercept is written as the ordered pair $(0, y)$.

Example 3 ▶ Identifying x- and y-Intercepts 

Identify the x- and y-intercepts of the graph of $y = x^3 + 1$ shown in Figure 1.6.

The *Interactive* CD-ROM and *Internet* versions of this text offer a Quiz for every section of the text.

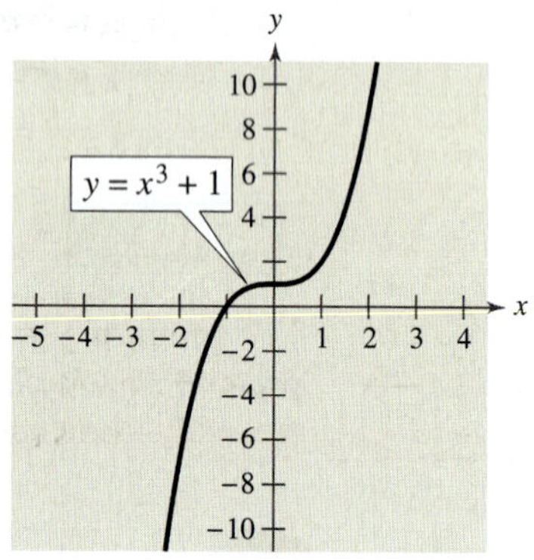

FIGURE 1.6

Solution

From the graph, you can see that the graph of the equation $y = x^3 + 1$ has an x-intercept (where y is zero) at $(-1, 0)$ and a y-intercept (where x is zero) at $(0, 1)$.

Symmetry

Graphs of equations can have **symmetry** with respect to one of the coordinate axes or with respect to the origin. Symmetry with respect to the x-axis means that if the Cartesian plane were folded along the x-axis, the portion of the graph above the x-axis would coincide with the portion below the x-axis. Symmetry with respect to the y-axis or the origin can be described in a similar manner, as shown in Figure 1.7.

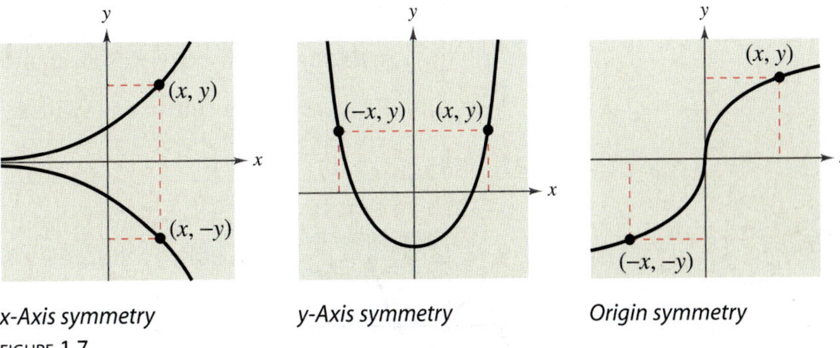

FIGURE 1.7

Knowing the symmetry of a graph *before* attempting to sketch it is helpful, because then you need only half as many solution points to sketch the graph. There are three basic types of symmetry, described as follows.

Graphical Tests for Symmetry

1. A graph is **symmetric with respect to the x-axis** if, whenever (x, y) is on the graph, $(x, -y)$ is also on the graph.
2. A graph is **symmetric with respect to the y-axis** if, whenever (x, y) is on the graph, $(-x, y)$ is also on the graph.
3. A graph is **symmetric with respect to the origin** if, whenever (x, y) is on the graph, $(-x, -y)$ is also on the graph.

Example 4 ▶ Testing for Symmetry 

The graph of $y = x^2 - 2$ is symmetric with respect to the y-axis because the point $(-x, y)$ is also on the graph of $y = x^2 - 2$. (See Figure 1.8.) The table at the left confirms that the graph is symmetric with respect to the y-axis.

x	y	(x, y)
-3	7	$(-3, 7)$
-2	2	$(-2, 2)$
-1	-1	$(-1, -1)$
1	-1	$(1, -1)$
2	2	$(2, 2)$
3	7	$(3, 7)$

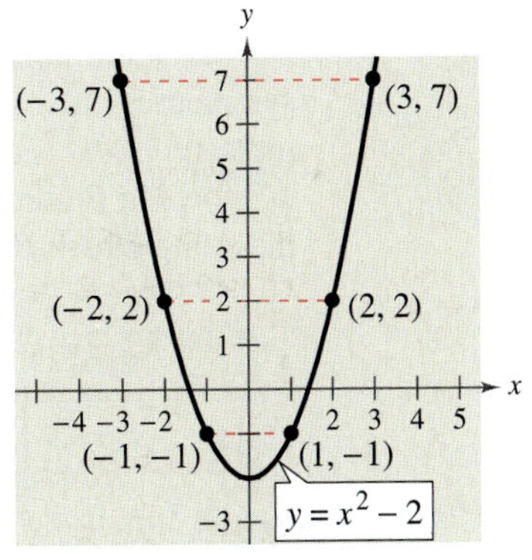

FIGURE 1.8 *y-Axis symmetry*

Algebraic Tests for Symmetry

1. The graph of an equation is symmetric with respect to the x-axis if replacing y with $-y$ yields an equivalent equation.
2. The graph of an equation is symmetric with respect to the y-axis if replacing x with $-x$ yields an equivalent equation.
3. The graph of an equation is symmetric with respect to the origin if replacing x with $-x$ and y with $-y$ yields an equivalent equation.

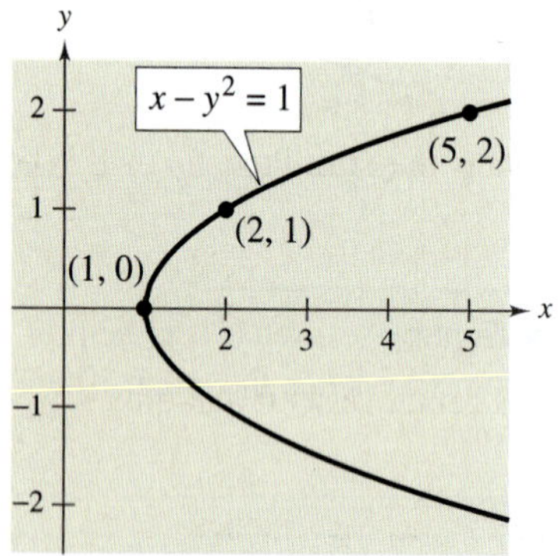

FIGURE 1.9

Example 5 ▶ Using Symmetry as a Sketching Aid 

Use symmetry to sketch the graph of

$$x - y^2 = 1.$$

Solution

Of the three tests for symmetry, the only one that is satisfied is the test for x-axis symmetry because $x - (-y)^2 = 1$ is equivalent to $x - y^2 = 1$. So, the graph is symmetric with respect to the x-axis. Using symmetry, you need only to find the solution points above the x-axis and then reflect them to obtain the graph, as shown in Figure 1.9.

y	$x = y^2 + 1$	(x, y)
0	1	(1, 0)
1	2	(2, 1)
2	5	(5, 2)

STUDY TIP

Notice that when creating the table in Example 5, it is easier to choose y-values and then find the corresponding x-values of the ordered pairs.

Example 6 ▶ Sketching the Graph of an Equation 

Sketch the graph of

$$y = |x - 1|.$$

Solution

This equation fails all three tests for symmetry and consequently its graph is not symmetric with respect to either axis or to the origin. The absolute value sign indicates that y is always nonnegative. Create a table of values and plot the points as shown in Figure 1.10. From the table, you can see that $x = 0$ when $y = 1$. So, the y-intercept is $(0, 1)$. Similarly, $y = 0$ when $x = 1$. So, the x-intercept is $(1, 0)$.

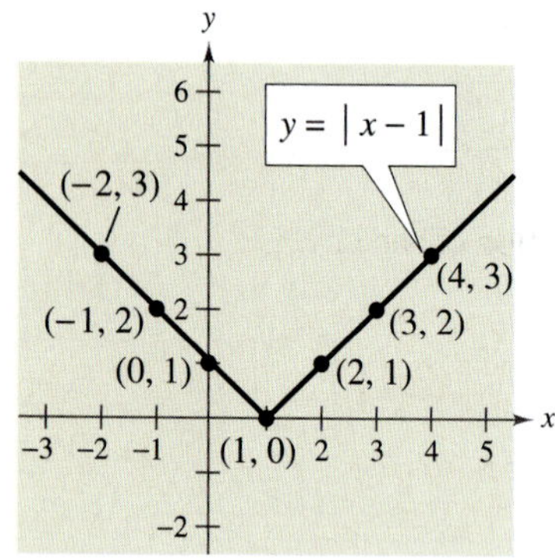

FIGURE 1.10

x	−2	−1	0	1	2	3	4
$y = \|x - 1\|$	3	2	1	0	1	2	3
(x, y)	(−2, 3)	(−1, 2)	(0, 1)	(1, 0)	(2, 1)	(3, 2)	(4, 3)

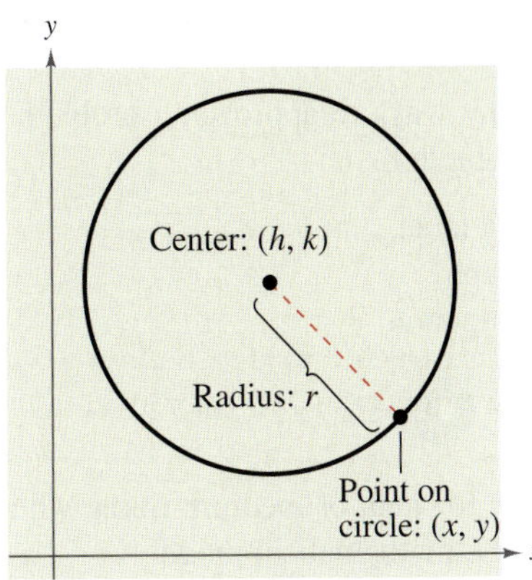

FIGURE 1.11

Throughout this course, you will learn to recognize several types of graphs from their equations. For instance, you will learn to recognize that the graph of a second-degree equation of the form

$$y = ax^2 + bx + c$$

is a parabola (see Example 2). Another easily recognized graph is that of a **circle.**

Circles

Consider the circle shown in Figure 1.11. A point (x, y) is on the circle if and only if its distance from the center (h, k) is r. By the Distance Formula,

$$\sqrt{(x - h)^2 + (y - k)^2} = r.$$

By squaring each side of this equation, you obtain the **standard form of the equation of a circle.**

Standard Form of the Equation of a Circle

The point (x, y) lies on the circle of radius r and center (h, k) if and only if

$$(x - h)^2 + (y - k)^2 = r^2.$$

STUDY TIP

To find the correct h and k, it may be helpful to rewrite the quantities $(x + 1)^2$ and $(y - 2)^2$, using subtraction.

$(x + 1)^2 = [x - (-1)]^2,$

$h = -1$

$(y - 2)^2 = [y - (2)]^2,$

$k = 2$

From this result, you can see that the standard form of the equation of a circle *with its center at the origin,* $(h, k) = (0, 0)$, is simply

$x^2 + y^2 = r^2.$ Circle with center at origin

Example 7 ▶ Finding the Equation of a Circle 

The point $(3, 4)$ lies on a circle whose center is at $(-1, 2)$, as shown in Figure 1.12. Write the standard form of the equation of this circle.

Solution

The radius of the circle is the distance between $(-1, 2)$ and $(3, 4)$.

$r = \sqrt{(x - h)^2 + (y - k)^2}$ Distance Formula

$r = \sqrt{[3 - (-1)]^2 + (4 - 2)^2}$ Substitute for x, y, h, and k.

$= \sqrt{4^2 + 2^2}$ Simplify.

$= \sqrt{16 + 4}$ Simplify.

$= \sqrt{20}$ Radius

Using $(h, k) = (-1, 2)$ and $r = \sqrt{20}$, the equation of the circle is

$(x - h)^2 + (y - k)^2 = r^2$ Equation of circle

$[x - (-1)]^2 + (y - 2)^2 = (\sqrt{20})^2$ Substitute for h, k, and r.

$(x + 1)^2 + (y - 2)^2 = 20.$ Standard form

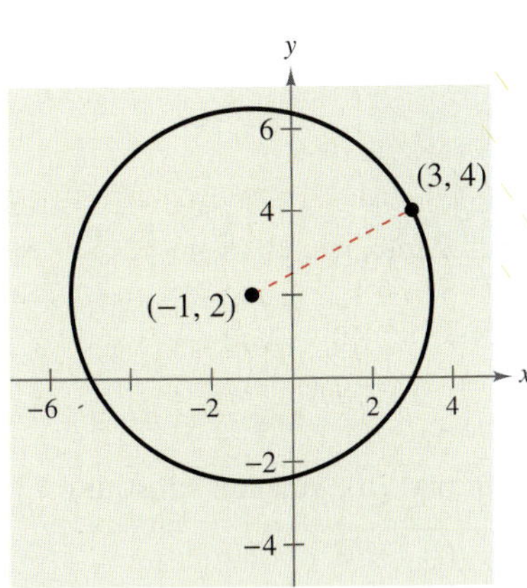

FIGURE 1.12

STUDY TIP

You should develop the habit of using at least two approaches to solve every problem. This helps build your intuition and helps you check that your answer is reasonable.

Application

In this course, you will learn that there are many ways to approach a problem. Three common approaches are illustrated in Example 8.

A *Numerical Approach:* Construct and use a table.
A *Graphical Approach:* Draw and use a graph.
An *Analytical Approach:* Use the rules of algebra.

Example 8 ▶ Recommended Weight 

The median recommended weight y (in pounds) for men of medium frame who are 25 to 59 years old can be approximated by the mathematical model

$$y = 0.073x^2 - 6.99x + 289.0, \quad 62 \le x \le 76$$

where x is the man's height in inches. (Source: Metropolitan Life Insurance Company)

a. Construct a table of values that shows the median recommended weights for men with heights of 62, 64, 66, 68, 70, 72, 74, and 76 inches.

b. Use the table of values to sketch a graph of the model. Then use the graph to estimate *graphically* the median recommended weight for a man whose height is 71 inches.

c. Use the model to confirm *analytically* the estimate you found in part (b).

Solution

Height, x	Weight, y
62	136.2
64	140.6
66	145.6
68	151.2
70	157.4
72	164.2
74	171.5
76	179.4

a. You can use a calculator to complete the table, as shown at the left.

b. The table of values can be used to sketch the graph of the function, as shown in Figure 1.13. From the graph, you can estimate that a height of 71 inches corresponds to a weight of about 161 pounds.

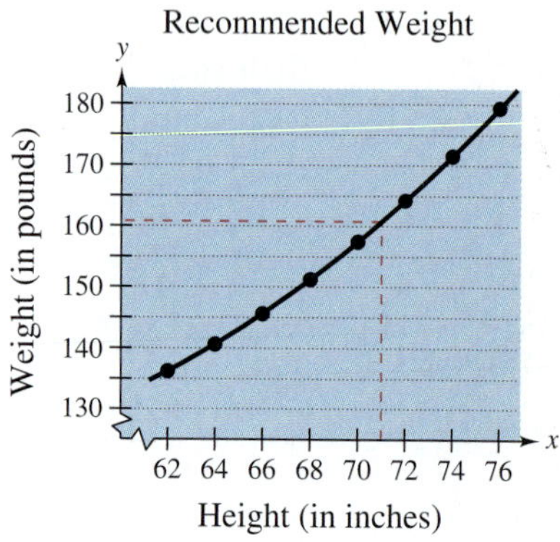

FIGURE 1.13

c. To confirm algebraically the estimate found in part (b), you can substitute 71 for x in the model.

$$y = 0.073(71)^2 - 6.99(71) + 289.0 \approx 160.70$$

So, the graphical estimate of 161 pounds is fairly good.

1.1 Exercises

The *Interactive* CD-ROM and *Internet* versions of this text contain step-by-step solutions to all odd-numbered exercises. They also provide Tutorial Exercises for additional help.

In Exercises 1–4, determine whether each point lies on the graph of the equation.

	Equation	*Points*	
1.	$y = \sqrt{x+4}$	(a) $(0, 2)$	(b) $(5, 3)$
2.	$y = x^2 - 3x + 2$	(a) $(2, 0)$	(b) $(-2, 8)$
3.	$y = 4 - \lvert x - 2 \rvert$	(a) $(1, 5)$	(b) $(6, 0)$
4.	$y = \frac{1}{3}x^3 - 2x^2$	(a) $\left(2, -\frac{16}{3}\right)$	(b) $(-3, 9)$

In Exercises 5–8, complete the table. Use the resulting solution points to sketch the graph of the equation.

5. $y = -2x + 5$

x	-1	0	1	2	$\frac{5}{2}$
y					
(x, y)					

6. $y = \frac{3}{4}x - 1$

x	-2	0	1	$\frac{4}{3}$	2
y					
(x, y)					

7. $y = x^2 - 3x$

x	-1	0	1	2	3
y					
(x, y)					

8. $y = 5 - x^2$

x	-2	-1	0	1	2
y					
(x, y)					

In Exercises 9–12, find the *x*- and *y*-intercepts of the graph of the equation.

9. $y = 16 - 4x^2$

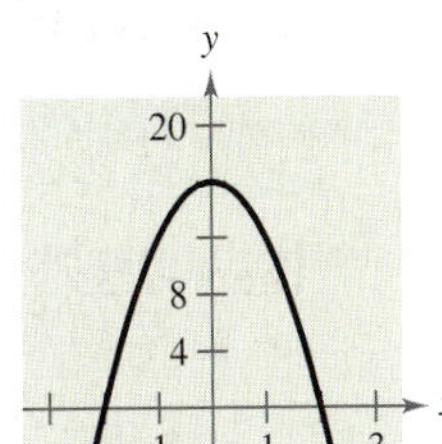

10. $y = (x + 3)^2$

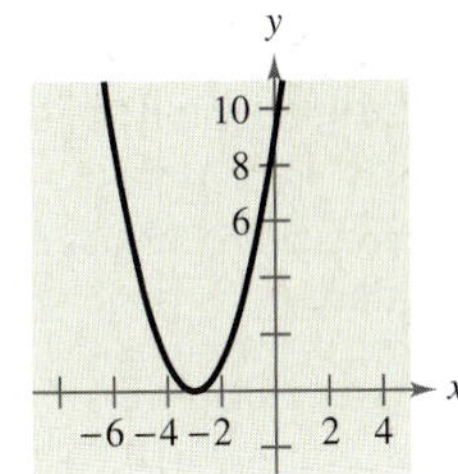

11. $y = 2x^3 - 4x^2$

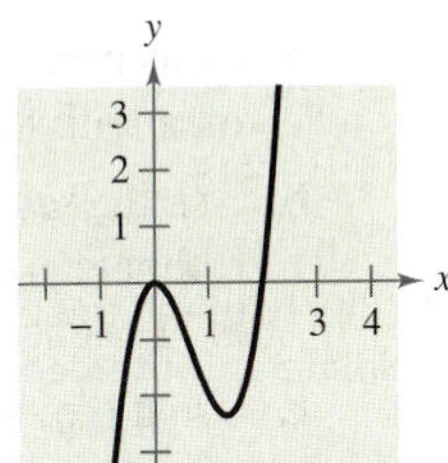

12. $y^2 = x + 1$

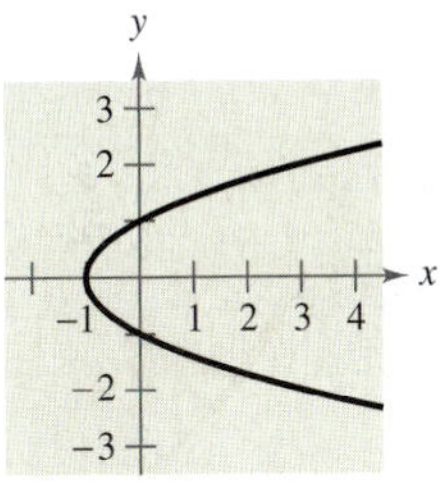

In Exercises 13–20, use the algebraic tests to check for symmetry with respect to both axes and the origin.

13. $x^2 - y = 0$

14. $x - y^2 = 0$

15. $y = x^3$

16. $y = x^4 - x^2 + 3$

17. $y = \dfrac{x}{x^2 + 1}$

18. $y = \dfrac{1}{x^2 + 1}$

19. $xy^2 + 10 = 0$

20. $xy = 4$

In Exercises 21–24, assume that the graph has the indicated type of symmetry. Sketch the complete graph of the equation. To print an enlarged copy of the graph, go to the website *www.mathgraphs.com*.

21.

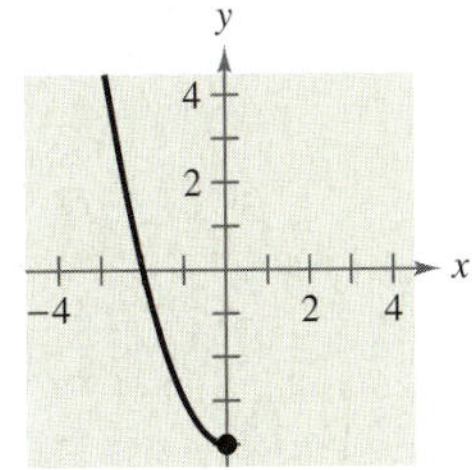

y-Axis symmetry

22.

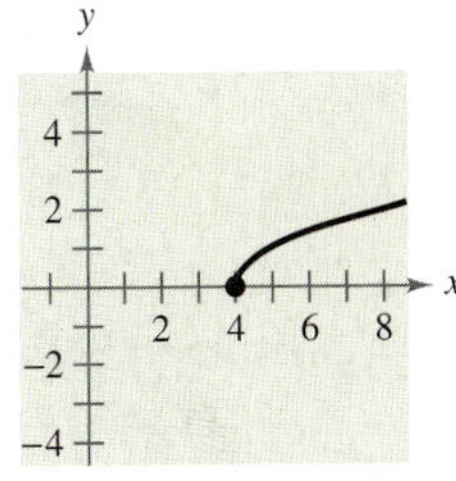

x-Axis symmetry

23. 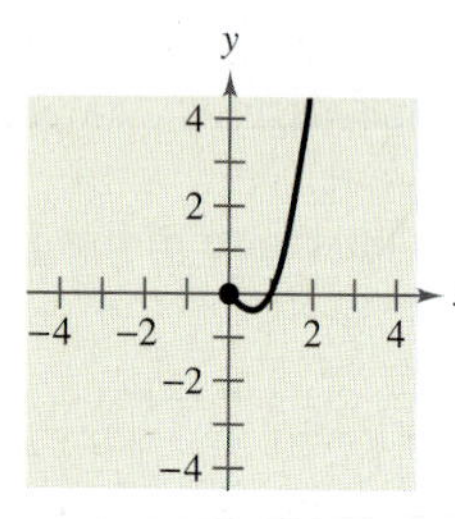

Origin symmetry

24. 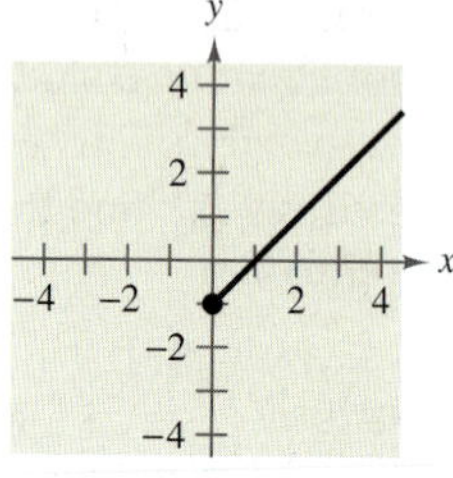

y-Axis symmetry

In Exercises 25–36, use symmetry to sketch the graph of the equation.

25. $y = -3x + 1$
26. $y = 2x - 3$
27. $y = x^2 - 2x$
28. $y = -x^2 - 2x$
29. $y = x^3 + 3$
30. $y = x^3 - 1$
31. $y = \sqrt{x - 3}$
32. $y = \sqrt{1 - x}$
33. $y = |x - 6|$
34. $y = 1 - |x|$
35. $x = y^2 - 1$
36. $x = y^2 - 5$

 **In Exercises 37–48, use a graphing utility to graph the equation. Use a standard setting. Approximate any intercepts.**

37. $y = 3 - \frac{1}{2}x$
38. $y = \frac{2}{3}x - 1$
39. $y = x^2 - 4x + 3$
40. $y = x^2 + x - 2$
41. $y = \dfrac{2x}{x - 1}$
42. $y = \dfrac{4}{x^2 + 1}$
43. $y = \sqrt[3]{x}$
44. $y = \sqrt[3]{x + 1}$
45. $y = x\sqrt{x + 6}$
46. $y = (6 - x)\sqrt{x}$
47. $y = |x + 3|$
48. $y = 2 - |x|$

In Exercises 49–56, write the standard form of the equation of the specified circle.

49. Center: $(0, 0)$; radius: 4
50. Center: $(0, 0)$; radius: 5
51. Center: $(2, -1)$; radius: 4
52. Center: $(-7, -4)$; radius: 7
53. Center: $(-1, 2)$; solution point: $(0, 0)$
54. Center: $(3, -2)$; solution point: $(-1, 1)$
55. Endpoints of a diameter: $(0, 0)$, $(6, 8)$
56. Endpoints of a diameter: $(-4, -1)$, $(4, 1)$

In Exercises 57–62, find the center and radius of the circle, and sketch its graph.

57. $x^2 + y^2 = 25$
58. $x^2 + y^2 = 16$
59. $(x - 1)^2 + (y + 3)^2 = 9$
60. $x^2 + (y - 1)^2 = 1$
61. $\left(x - \frac{1}{2}\right)^2 + \left(y - \frac{1}{2}\right)^2 = \frac{9}{4}$
62. $(x - 2)^2 + (y + 1)^2 = 3$

63. ***Depreciation*** A manufacturing plant purchases a new molding machine for \$225,000. The depreciated value y after t years is

$$y = 225{,}000 - 20{,}000t, \qquad 0 \le t \le 8.$$

Sketch the graph of the equation.

64. ***Consumerism*** You purchase a jet ski for \$8100. The depreciated value y after t years is

$$y = 8100 - 929t, \qquad 0 \le t \le 6.$$

Sketch the graph of the equation.

65. ***Geometry*** A rectangle of length x and width w has a perimeter of 12 meters.
   (a) Draw a rectangle that gives a visual representation of the problem. Use the specified variables to label the sides of the rectangle.
   (b) Show that the width of the rectangle is $w = 6 - x$ and its area is $A = x(6 - x)$.
    (c) Use a graphing utility to graph the area equation.
    (d) From the graph in part (c), estimate the dimensions of the rectangle that yield a maximum area.

66. ***Geometry*** A rectangle of length x and width w has a perimeter of 22 yards.
   (a) Draw a rectangle that gives a visual representation of the problem. Use the specified variables to label the sides of the rectangle.
   (b) Show that the width of the rectangle is $w = 11 - x$ and its area is $A = x(11 - x)$.
   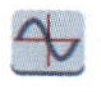 (c) Use a graphing utility to graph the area equation. Be sure to adjust your window settings.
   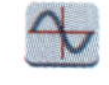 (d) From the graph in part (c), estimate the dimensions of the rectangle that yield a maximum area.

The symbol indicates an exercise or parts of an exercise in which you are instructed to use a graphing utility.

▶ Model It

67. ***Population Statistics*** The table shows the life expectancy of a child (at birth) in the United States for selected years from 1920 to 2000. (Source: U.S. National Center for Health Statistics, U.S. Census Bureau)

Year, t	Life expectancy, y
1920	54.1
1930	59.7
1940	62.9
1950	68.2
1960	69.7
1970	70.8
1980	73.7
1990	75.4
2000	77.1

A model for the life expectancy during this period is

$$y = -0.0025t^2 + 0.572t + 44.31$$

where y represents the life expectancy and t is the time in years, with $t = 20$ corresponding to 1920.

(a) Sketch a scatter plot of the data.

(b) Graph the model for the data and compare the scatter plot and the graph.

(c) Use the graph of the model to estimate the life expectancy of a child for the years 2005 and 2010.

(d) Do you think this model can be used to predict the life expectancy of a child 50 years from now? Explain.

68. ***Federal Debt*** The per capita federal debt y (in dollars) of the United States from 1950 through 2000 can be approximated by the model

$$y = 0.047t^3 + 9.23t^2 - 206.3t + 1984$$

where t is the time in years, with $t = 0$ corresponding to 1950. Use a graphing utility to graph the model and estimate the per capita federal debt for the year 2005. (Source: U. S. Census Bureau and U.S. Department of the Treasury)

69. ***Electronics*** The resistance y (in ohms) of 1000 feet of solid copper wire at 68 degrees Fahrenheit can be approximated by the model

$$y = \frac{10{,}770}{x^2} - 0.37, \qquad 5 \le x \le 100$$

where x is the diameter of the wire in mils (0.001 in.). Use a graphing utility to graph the model and estimate the resistance when $x = 50$. (Source: American Wire Gage)

Synthesis

True or False? **In Exercises 70 and 71, determine whether the statement is true or false. Justify your answer.**

70. In order to find the y-intercepts of the graph of an equation, let $y = 0$ and solve the equation for x.

71. The graph of a linear equation of the form $y = mx + b$ has one y-intercept.

72. ***Think About It*** Suppose you correctly enter an expression for the variable y on a graphing utility. However, no graph appears on the display when you graph the equation. Give a possible explanation and the steps you could take to remedy the problem. Illustrate your explanation with an example.

73. ***Think About It*** Find a and b if the graph of $y = ax^2 + bx^3$ is symmetric with respect to (a) the y-axis and (b) the origin. (There are many correct answers.)

74. In your own words, explain how the display of a graphing utility changes if the maximum setting for x is changed from 10 to 20.

Review

75. Identify the terms: $9x^5 + 4x^3 - 7$.

76. Rewrite the expression using exponential notation.

$$-(7 \times 7 \times 7 \times 7)$$

In Exercises 77–82, simplify the expression.

77. $\sqrt{18x} - \sqrt{2x}$

78. $\sqrt[4]{x^5}$

79. $\dfrac{70}{\sqrt{7x}}$

80. $\dfrac{55}{\sqrt{20} - 3}$

81. $\sqrt[6]{t^2}$

82. $\sqrt[3]{\sqrt{y}}$

1.2 Linear Equations in One Variable

▶ What you should learn

- How to identify different types of equations
- How to solve linear equations in one variable
- How to solve equations that lead to linear equations
- How to find x- and y-intercepts of graphs of equations algebraically.
- How to use linear equations to model and solve real-life problems

▶ Why you should learn it

Linear equations are used in many real-life applications. For example, in Exercise 95 on page 95, linear equations can be used to model the relationship between the length of a thigh bone and the height of a person, helping researchers learn about ancient cultures.

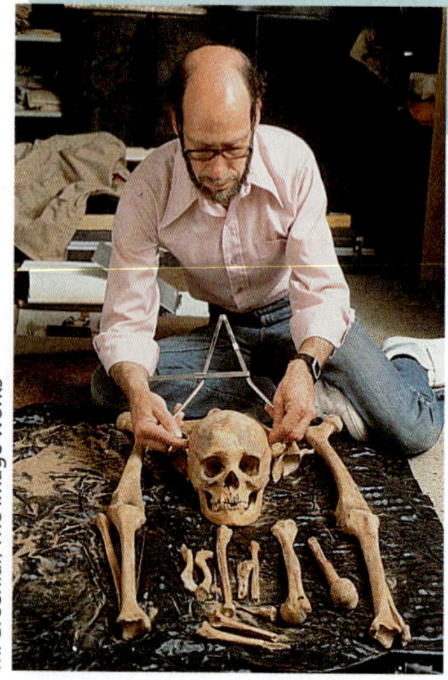
M. Greenlar/The Image Works

Equations and Solutions of Equations 

An **equation** in x is a statement that two algebraic expressions are equal. For example

$$3x - 5 = 7,\ x^2 - x - 6 = 0, \text{ and } \sqrt{2x} = 4$$

are equations. To **solve** an equation in x means to find all values of x for which the equation is true. Such values are **solutions.** For instance, $x = 4$ is a solution of the equation

$$3x - 5 = 7$$

because $3(4) - 5 = 7$ is a true statement.

The solutions of an equation depend on the kinds of numbers being considered. For instance, in the set of rational numbers, $x^2 = 10$ has no solution because there is no rational number whose square is 10. However, in the set of real numbers, the equation has the two solutions $\sqrt{10}$ and $-\sqrt{10}$.

An equation that is true for *every* real number in the domain of the variable is called an **identity.** For example

$$x^2 - 9 = (x + 3)(x - 3) \qquad \text{Identity}$$

is an identity because it is a true statement for any real value of x, and

$$\frac{x}{3x^2} = \frac{1}{3x} \qquad \text{Identity}$$

where $x \neq 0$, is an identity because it is true for any nonzero real value of x.

An equation that is true for just *some* (or even none) of the real numbers in the domain of the variable is called a **conditional equation.** For example, the equation

$$x^2 - 9 = 0 \qquad \text{Conditional equation}$$

is conditional because $x = 3$ and $x = -3$ are the only values in the domain that satisfy the equation. The equation $2x - 4 = 2x + 1$ is conditional because there are no real values of x for which the equation is true. Learning to solve conditional equations is the primary focus of this chapter.

Linear Equations in One Variable

Definition of a Linear Equation

A **linear equation in one variable** x is an equation that can be written in the standard form

$$ax + b = 0$$

where a and b are real numbers with $a \neq 0$.

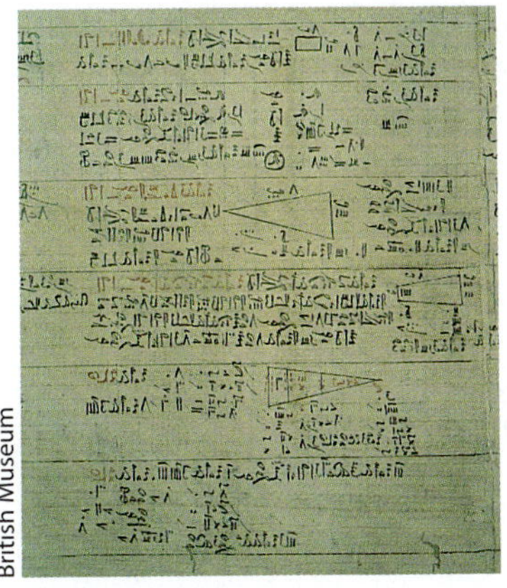

British Museum

Historical Note

This ancient Egyptian papyrus, discovered in 1858, contains one of the earliest examples of mathematical writing in existence. The papyrus itself dates back to around 1650 B.C., but it is actually a copy of writings from two centuries earlier. The algebraic equations on the papyrus were written in words. Diophantus, a Greek who lived around A.D. 250, is often called the Father of Algebra. He was the first to use abbreviated word forms in equations.

A linear equation has exactly one solution. To see this, consider the following steps. (Remember that $a \neq 0$.)

$ax + b = 0$	Write original equation.
$ax = -b$	Subtract b from each side.
$x = -\dfrac{b}{a}$	Divide each side by a.

To solve a conditional equation in x, isolate x on one side of the equation by a sequence of **equivalent** (and usually simpler) **equations,** each having the same solution(s) as the original equation. The operations that yield equivalent equations come from the Substitution Principle and the Properties of Equality studied in Chapter P.

Generating Equivalent Equations

An equation can be transformed into an *equivalent equation* by one or more of the following steps.

	Given Equation	*Equivalent Equation*
1. Remove symbols of grouping, combine like terms, or simplify fractions on one or both sides of the equation.	$2x - x = 4$	$x = 4$
2. Add (or subtract) the same quantity to (from) *each* side of the equation.	$x + 1 = 6$	$x = 5$
3. Multiply (or divide) *each* side of the equation by the same *nonzero* quantity.	$2x = 6$	$x = 3$
4. Interchange the two sides of the equation.	$2 = x$	$x = 2$

Exploration

Use a graphing utility to graph the equation $y = 3x - 6$. Use the result to estimate the x-intercept of the graph. Explain how the x-intercept is related to the solution of the equation $3x - 6 = 0$, as shown in Example 1(a).

Example 1 ▶ Solving a Linear Equation

a. $3x - 6 = 0$	Original equation
$3x = 6$	Add 6 to each side.
$x = 2$	Divide each side by 3.
b. $5x + 4 = 3x - 8$	Original equation
$2x + 4 = -8$	Subtract $3x$ from each side.
$2x = -12$	Subtract 4 from each side.
$x = -6$	Divide each side by 2.

After solving an equation, you should check each solution in the original equation. For instance, you can check the solution to Example 1(a) as follows.

$3x - 6 = 0$	Write original equation.
$3(2) - 6 \stackrel{?}{=} 0$	Substitute 2 for x.
$0 = 0$	Solution checks. ✓

Try checking the solution to Example 1(b).

Some linear equations have no solutions because all the x-terms sum to zero and a contradictory (false) statement such as $0 = 5$ or $12 = 7$ is obtained. For instance, the linear equation

$$x = x + 1$$

has no solution. Watch for this type of linear equation in the exercises.

Example 2 ▶ Solving a Linear Equation 

Solve

$$6(x - 1) + 4 = 3(7x + 1).$$

Solution

$6(x - 1) + 4 = 3(7x + 1)$	Write original equation.
$6x - 6 + 4 = 21x + 3$	Distributive Property
$6x - 2 = 21x + 3$	Simplify.
$6x = 21x + 5$	Add 2 to each side.
$-15x = 5$	Subtract $21x$ from each side.
$x = -\dfrac{1}{3}$	Divide each side by -15.

Check

Check this solution by substituting $-\frac{1}{3}$ for x in the original equation.

$6(x - 1) + 4 = 3(7x + 1)$	Write original equation.
$6\left(-\frac{1}{3} - 1\right) + 4 \stackrel{?}{=} 3\left[7\left(-\frac{1}{3}\right) + 1\right]$	Substitute $-\frac{1}{3}$ for x.
$6\left(-\frac{4}{3}\right) + 4 \stackrel{?}{=} 3\left[-\frac{7}{3} + 1\right]$	Simplify.
$6\left(-\frac{4}{3}\right) + 4 \stackrel{?}{=} 3\left(-\frac{4}{3}\right)$	Simplify.
$-\frac{24}{3} + 4 \stackrel{?}{=} -\frac{12}{3}$	Multiply.
$-8 + 4 \stackrel{?}{=} -4$	Simplify.
$-4 = -4$	Solution checks. ✓

So, the solution is $x = -\frac{1}{3}$.

Technology

You can use a graphing utility to check that a solution is reasonable. One way to do this is to graph the left side of the equation, then graph the right side of the equation, and determine the point of intersection. For instance, in Example 2, if you graph the equations

$y_1 = 6(x - 1) + 4$ The left side

$y_2 = 3(7x + 1)$ The right side

in the same viewing window, they should intersect when $x = -\frac{1}{3}$, as shown in the graph below.

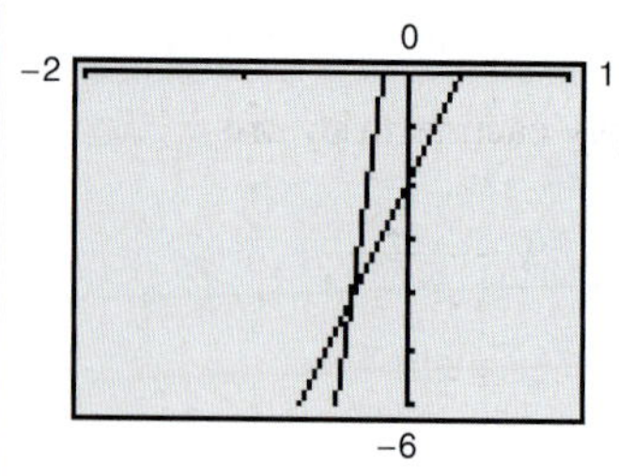

Equations That Lead to Linear Equations

To solve an equation involving fractional expressions, find the least common denominator (LCD) of all terms and multiply every term by the LCD.

Example 3 ▶ An Equation Involving Fractional Expressions

Solve $\frac{x}{3} + \frac{3x}{4} = 2$.

Solution

$$\frac{x}{3} + \frac{3x}{4} = 2 \qquad \text{Write original equation.}$$

$$(12)\frac{x}{3} + (12)\frac{3x}{4} = (12)2 \qquad \text{Multiply each term by the LCD of 12.}$$

$$4x + 9x = 24 \qquad \text{Divide out and multiply.}$$

$$13x = 24 \qquad \text{Combine like terms.}$$

$$x = \frac{24}{13} \qquad \text{Divide each side by 13.}$$

The solution is $x = \frac{24}{13}$. Check this in the original equation.

STUDY TIP

An equation with a *single fraction* on each side can be cleared of denominators by **cross multiplying,** which is equivalent to multiplying by the LCD and then dividing out.

$$\frac{a}{b} = \frac{c}{d} \qquad \text{LCD is } bd.$$

$$\frac{a}{b} \cdot bd = \frac{c}{d} \cdot bd \qquad \text{Multiply by LCD.}$$

$$ad = cb \qquad \text{Divide out common factors.}$$

By comparing the last equation with the original equation, you can see that the left numerator was multiplied by the right denominator and the right numerator was multiplied by the left denominator. Try cross multiplying the following equation to clear the equation of denominators.

$$\frac{2}{x-3} = \frac{3}{x+1}$$

When multiplying or dividing an equation by a *variable* quantity, it is possible to introduce an extraneous solution. An **extraneous solution** is one that does not satisfy the original equation.

Example 4 ▶ An Equation with an Extraneous Solution

Solve $\frac{1}{x-2} = \frac{3}{x+2} - \frac{6x}{x^2-4}$.

Solution

The LCD is $x^2 - 4$, or $(x + 2)(x - 2)$. Multiply each term by this LCD.

$$\frac{1}{x-2}(x+2)(x-2) = \frac{3}{x+2}(x+2)(x-2) - \frac{6x}{x^2-4}(x+2)(x-2)$$

$$x + 2 = 3(x - 2) - 6x, \qquad x \neq \pm 2$$

$$x + 2 = 3x - 6 - 6x$$

$$x + 2 = -3x - 6$$

$$4x = -8$$

$$x = -2 \qquad \text{Extraneous solution}$$

In the original equation, $x = -2$ yields a denominator of zero. So, $x = -2$ is an extraneous solution, and the original equation has *no solution*.

Finding Intercepts Algebraically

In Section 1.1 you learned to find x- and y-intercepts using a graphical approach. You can also use an algebraic approach to find x- and y-intercepts, as follows.

> **Finding Intercepts Algebraically**
>
> **1.** To find x-intercepts, set y equal to zero and solve the equation for x.
>
> **2.** To find y-intercepts, set x equal to zero and solve the equation for y.

Here is an example.

$$y = 4x + 1 \Rightarrow 0 = 4x + 1 \Rightarrow -1 = 4x \Rightarrow -\tfrac{1}{4} = x$$

$$y = 4x + 1 \Rightarrow y = 4(0) + 1 \Rightarrow y = 1$$

So, the x-intercept of $y = 4x + 1$ is $\left(-\frac{1}{4}, 0\right)$ and the y-intercept is $(0, 1)$.

Application

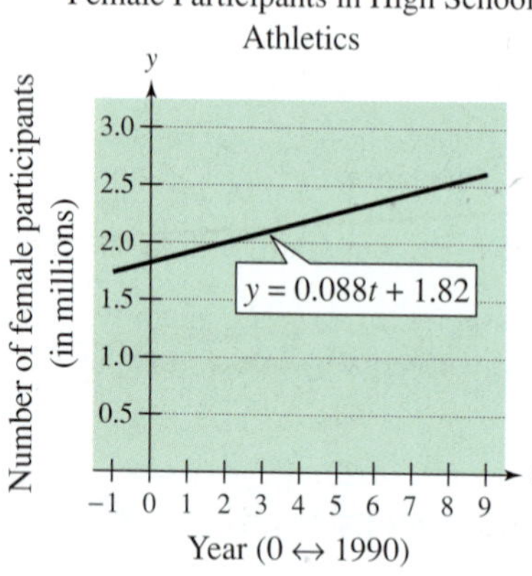

FIGURE 1.14

Example 5 ▶ Female Participants in Athletic Programs 

The number y of female participants in high school athletic programs (in millions) in the United States from 1989 to 1999 can be approximated by the linear model

$$y = 0.088t + 1.82, \qquad -1 \le t \le 9$$

where $t = 0$ represents 1990. (a) Find the y-intercept of the graph of the linear model shown in Figure 1.14 algebraically. (b) Assuming that this linear pattern continues, find the year in which there will be 3.14 million female participants. (Source: National Federation of State High School Associations)

Solution

a. To find the y-intercept, let $t = 0$ and solve for y as follows.

$$y = 0.088t + 1.82 \qquad \text{Write original equation.}$$

$$= 0.088(0) + 1.82 \qquad \text{Substitute 0 for } t.$$

$$= 1.82 \qquad \text{Simplify.}$$

So, the y-intercept is $(0, 1.82)$.

b. Let $y = 3.14$ and solve the equation $3.14 = 0.088t + 1.82$ for t.

$$3.14 = 0.088t + 1.82 \qquad \text{Write original equation.}$$

$$1.32 = 0.088t \qquad \text{Subtract 1.82 from each side.}$$

$$15 = t \qquad \text{Divide each side by 0.088.}$$

Because $t = 0$ represents 1990, $t = 15$ must represent 2005. So, from this model, there will be 3.14 million female participants in 2005.

1.2 Exercises

In Exercises 1–10, determine whether each value of x is a solution of the equation.

Equation	Values	
1. $5x - 3 = 3x + 5$	(a) $x = 0$	(b) $x = -5$
	(c) $x = 4$	(d) $x = 10$
2. $7 - 3x = 5x - 17$	(a) $x = -3$	(b) $x = 0$
	(c) $x = 8$	(d) $x = 3$
3. $3x^2 + 2x - 5$	(a) $x = -3$	(b) $x = 1$
$= 2x^2 - 2$	(c) $x = 4$	(d) $x = -5$
4. $5x^3 + 2x - 3$	(a) $x = 2$	(b) $x = -2$
$= 4x^3 + 2x - 11$	(c) $x = 0$	(d) $x = 10$
5. $\frac{5}{2x} - \frac{4}{x} = 3$	(a) $x = -\frac{1}{2}$	(b) $x = 4$
	(c) $x = 0$	(d) $x = \frac{1}{4}$
6. $3 + \frac{1}{x+2} = 4$	(a) $x = -1$	(b) $x = -2$
	(c) $x = 0$	(d) $x = 5$
7. $\sqrt{3x - 2} = 4$	(a) $x = 3$	(b) $x = 2$
	(c) $x = 9$	(d) $x = -6$
8. $\sqrt[3]{x - 8} = 3$	(a) $x = 2$	(b) $x = -5$
	(c) $x = 35$	(d) $x = 8$
9. $6x^2 - 11x - 35 = 0$	(a) $x = -\frac{5}{3}$	(b) $x = -\frac{2}{7}$
	(c) $x = \frac{7}{2}$	(d) $x = \frac{5}{3}$
10. $10x^2 + 21x - 10 = 0$	(a) $x = \frac{2}{5}$	(b) $x = -\frac{5}{2}$
	(c) $x = -\frac{1}{3}$	(d) $x = -2$

In Exercises 11–20, determine whether the equation is an identity or a conditional equation.

11. $2(x - 1) = 2x - 2$

12. $3(x + 2) = 5x + 4$

13. $-6(x - 3) + 5 = -2x + 10$

14. $3(x + 2) - 5 = 3x + 1$

15. $4(x + 1) - 2x = 2(x + 2)$

16. $-7(x - 3) + 4x = 3(7 - x)$

17. $x^2 - 8x + 5 = (x - 4)^2 - 11$

18. $x^2 + 2(3x - 2) = x^2 + 6x - 4$

19. $3 + \frac{1}{x+1} = \frac{4x}{x+1}$

20. $\frac{5}{x} + \frac{3}{x} = 24$

In Exercises 21 and 22, justify each step of the solution.

21.
$$\begin{aligned} 4x + 32 &= 83 \\ 4x + 32 - 32 &= 83 - 32 \\ 4x &= 51 \\ \frac{4x}{4} &= \frac{51}{4} \\ x &= \frac{51}{4} \end{aligned}$$

22.
$$\begin{aligned} 3(x - 4) + 10 &= 7 \\ 3x - 12 + 10 &= 7 \\ 3x - 2 &= 7 \\ 3x - 2 + 2 &= 7 + 2 \\ 3x &= 9 \\ \frac{3x}{3} &= \frac{9}{3} \\ x &= 3 \end{aligned}$$

In Exercises 23–38, solve the equation and check your solution.

23. $x + 11 = 15$

24. $7 - x = 19$

25. $7 - 2x = 25$

26. $7x + 2 = 23$

27. $8x - 5 = 3x + 20$

28. $7x + 3 = 3x - 17$

29. $2(x + 5) - 7 = 3(x - 2)$

30. $3(x + 3) = 5(1 - x) - 1$

31. $x - 3(2x + 3) = 8 - 5x$

32. $9x - 10 = 5x + 2(2x - 5)$

33. $\frac{5x}{4} + \frac{1}{2} = x - \frac{1}{2}$

34. $\frac{x}{5} - \frac{x}{2} = 3 + \frac{3x}{10}$

35. $\frac{3}{2}(z + 5) - \frac{1}{4}(z + 24) = 0$

36. $\frac{3x}{2} + \frac{1}{4}(x - 2) = 10$

37. $0.25x + 0.75(10 - x) = 3$

38. $0.60x + 0.40(100 - x) = 50$

In Exercises 39–42, solve the equation in two ways. Then explain which way is easier.

39. $3(x - 1) = 4$

40. $4(x + 3) = 15$

41. $\frac{1}{3}(x + 2) = 5$

42. $\frac{3}{4}(z - 4) = 6$

 ***Graphical Analysis*** **In Exercises 43–48, use a graphing utility to graph the equation and approximate any x-intercepts. Set $y = 0$ and solve the resulting equation. Compare the results with the graph's x-intercepts.**

43. $y = 2(x - 1) - 4$
44. $y = \frac{4}{3}x + 2$
45. $y = 20 - (3x - 10)$
46. $y = 10 + 2(x - 2)$
47. $y = -38 + 5(9 - x)$
48. $y = 6x - 6\left(\frac{16}{11} + x\right)$

In Exercises 49–58, find the x- and y-intercepts of the graph of the equation algebraically.

49. $y = 12 - 5x$
50. $y = 16 - 3x$
51. $y = -3(2x + 1)$
52. $y = 5 - (6 - x)$
53. $2x + 3y = 10$
54. $4x - 5y = 12$
55. $\frac{2x}{5} + 8 - 3y = 0$
56. $\frac{8x}{3} + 5 - 2y = 0$
57. $4y - 0.75x + 1.2 = 0$
58. $3y + 2.5x - 3.4 = 0$

In Exercises 59–80, solve the equation and check your solution. (If not possible, explain why.)

59. $x + 8 = 2(x - 2) - x$
60. $8(x + 2) - 3(2x + 1) = 2(x + 5)$
61. $\frac{100 - 4x}{3} = \frac{5x + 6}{4} + 6$
62. $\frac{17 + y}{y} + \frac{32 + y}{y} = 100$
63. $\frac{5x - 4}{5x + 4} = \frac{2}{3}$
64. $\frac{10x + 3}{5x + 6} = \frac{1}{2}$
65. $10 - \frac{13}{x} = 4 + \frac{5}{x}$
66. $\frac{15}{x} - 4 = \frac{6}{x} + 3$
67. $3 = 2 + \frac{2}{z + 2}$
68. $\frac{1}{x} + \frac{2}{x - 5} = 0$
69. $\frac{x}{x + 4} + \frac{4}{x + 4} + 2 = 0$
70. $\frac{7}{2x + 1} - \frac{8x}{2x - 1} = -4$
71. $\frac{2}{(x - 4)(x - 2)} = \frac{1}{x - 4} + \frac{2}{x - 2}$
72. $\frac{4}{x - 1} + \frac{6}{3x + 1} = \frac{15}{3x + 1}$
73. $\frac{1}{x - 3} + \frac{1}{x + 3} = \frac{10}{x^2 - 9}$
74. $\frac{1}{x - 2} + \frac{3}{x + 3} = \frac{4}{x^2 + x - 6}$
75. $\frac{3}{x^2 - 3x} + \frac{4}{x} = \frac{1}{x - 3}$
76. $\frac{6}{x} - \frac{2}{x + 3} = \frac{3(x + 5)}{x^2 + 3x}$
77. $(x + 2)^2 + 5 = (x + 3)^2$
78. $(x + 1)^2 + 2(x - 2) = (x + 1)(x - 2)$
79. $(x + 2)^2 - x^2 = 4(x + 1)$
80. $(2x + 1)^2 = 4(x^2 + x + 1)$

In Exercises 81–88, solve for x.

81. $4(x + 1) - ax = x + 5$
82. $4 - 2(x - 2b) = ax + 3$
83. $6x + ax = 2x + 5$
84. $5 + ax = 12 - bx$
85. $19x + \frac{1}{2}ax = x + 9$
86. $-5(3x - 6b) + 12 = 8 + 3ax$
87. $-2ax + 6(x + 3) = -4x + 1$
88. $\frac{4}{5}x - ax = 2\left(\frac{2}{5}x - 1\right) + 10$

In Exercises 89–92, solve the equation for x. (Round your solution to three decimal places.)

89. $0.275x + 0.725(500 - x) = 300$
90. $2.763 - 4.5(2.1x - 5.1432) = 6.32x + 5$
91. $\frac{2}{7.398} - \frac{4.405}{x} = \frac{1}{x}$
92. $\frac{3}{6.350} - \frac{6}{x} = 18$

93. ***Geometry*** The surface area S of the circular cylinder shown in the figure is

$$S = 2\pi(25) + 2\pi(5h)$$

Find the height h of the cylinder if the surface area is 471 square feet. Use 3.14 for π.

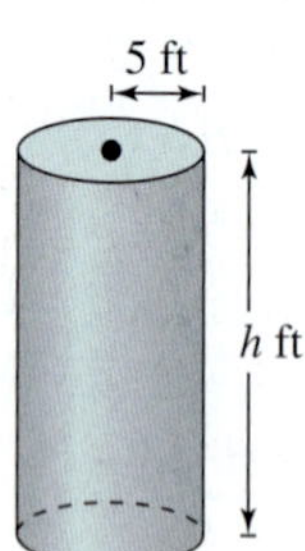

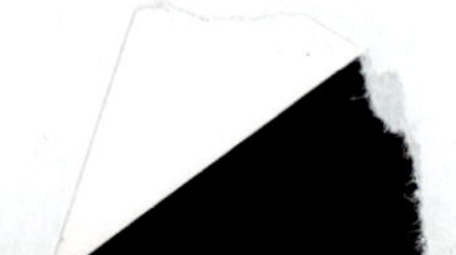

94. ***Geometry*** The surface area S of the rectangular solid in the figure is $S = 2(24) + 2(4x) + 2(6x)$. Find the length x of the box if the surface area is 248 square centimeters.

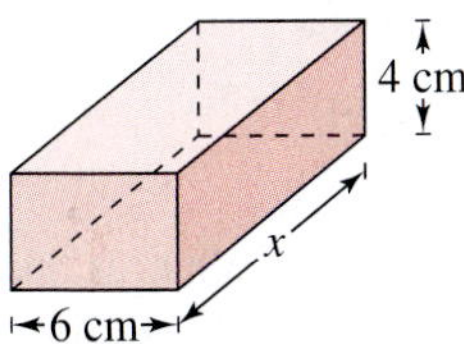

▶ Model It

95. ***Anthropology*** The relationship between the length of an adult's femur (thigh bone) and the height of the adult can be approximated by the linear equations

$y = 0.432x - 10.44$ Female

$y = 0.449x - 12.15$ Male

where y is the length of the femur in inches and x is the height of the adult in inches (see figure).

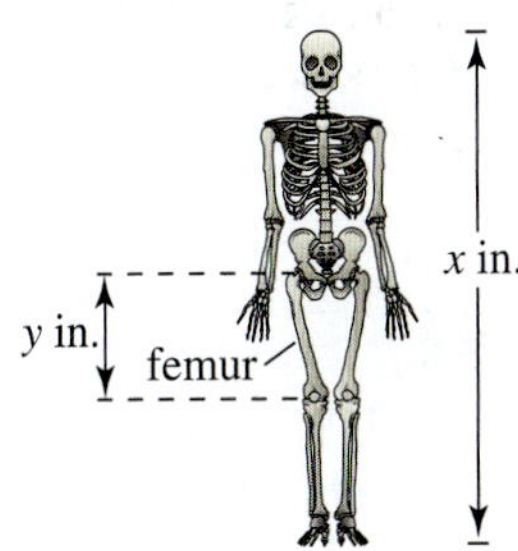

(a) An anthropologist discovers a femur belonging to an adult human female. The bone is 16 inches long. Estimate the height of the female.

(b) From the foot bones of an adult human male, an anthropologist estimates that the person's height was 69 inches. A few feet away from the site where the foot bones were discovered, the anthropologist discovers a male adult femur that is 19 inches long. Is it likely that both bones came from the same person?

(c) Complete the table to determine if there is a height of an adult for which an anthropologist would not be able to determine whether the femur belonged to a male or a female.

▶ Model It *(continued)*

Height, x	Female femur length, y	Male femur length, y
60		
70		
80		
90		
100		
110		

(d) Solve part (c) algebraically by setting the two equations equal to each other and solving for x. Compare your solutions. Do you believe an anthropologist would ever have the problem of not being able to determine whether a femur belonged to a male or female? Why or why not?

96. ***Tax Credits*** Use the following information about a possible tax credit for a family consisting of two adults and two children (see figure).

Earned income: E

Subsidy: $S = 10{,}000 - \frac{1}{2}E, \quad 0 \le E \le 20{,}000$

Total income: $T = E + S$

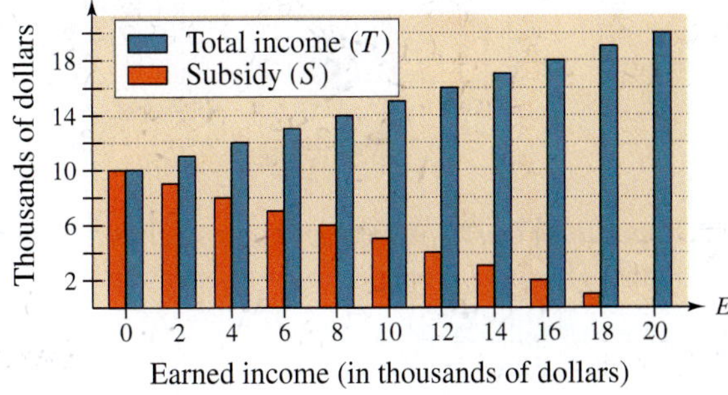

(a) Express the total income T in terms of E.

(b) Find the earned income E if the subsidy is \$6600.

(c) Find the earned income E if the total income is \$13,800.

(d) Find the subsidy S if the total income is \$12,500.

97. ***Consumerism*** The number of light trucks sold y (in millions) in the United States from 1992 to 1999 can be approximated by the model

$$y = 0.451t + 3.81$$

where $t = 2$ represents 1992. (Source: U.S. Bureau of Economic Analysis)

(a) Use the model to create a line graph of the number of light trucks sold from 1992 to 1999.

(b) Use the graph to determine the year during which the number of light trucks sold reached 6 million.

98. ***Labor Statistics*** The number of married women y (in millions) in the civilian work force in the United States from 1990 to 1999 (see figure) can be approximated by the model

$$y = 0.41t + 30.9$$

where $t = 0$ represents 1990. According to this model, during which year did the number reach 33 million? Explain how to answer this question graphically and algebraically. (Source: U.S. Bureau of Labor Statistics)

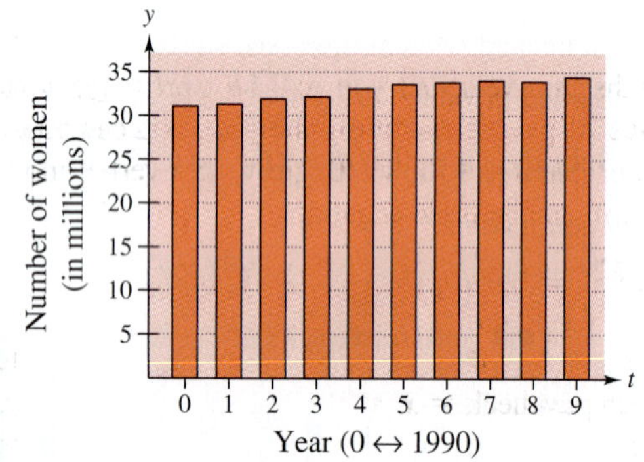

99. ***Operating Cost*** A delivery company has a fleet of vans. The annual operating cost C per van is

$$C = 0.32m + 2500$$

where m is the number of miles traveled by a van in a year. What number of miles will yield an annual operating cost of $10,000?

100. ***Flood Control*** A river has risen 8 feet above its flood stage. The water begins to recede at a rate of 3 inches per hour. Write a mathematical model that shows the number of feet above flood stage after t hours. If the water continually recedes at this rate, when will the river be 1 foot above its flood stage?

Synthesis

True or False? **In Exercises 101 and 102, determine whether the statement is true or false. Justify your answer.**

101. The equation $x(3 - x) = 10$ is a linear equation.

102. The equation $x^2 + 9x - 5 = 4 - x^3$ has no real solution.

103. ***Think About It*** What is meant by "equivalent equations"? Give an example of two equivalent equations.

104. ***Writing*** Describe the steps used to transform an equation into an equivalent equation.

105. ***Exploration***

(a) Complete the table.

x	-1	0	1	2	3	4
$3.2x - 5.8$						

(b) Use the table in part (a) to determine the interval in which the solution to the equation $3.2x - 5.8 = 0$ is located. Explain your reasoning.

(c) Complete the table.

x	1.5	1.6	1.7	1.8	1.9	2.0
$3.2x - 5.8$						

(d) Use the table in part (c) to determine the interval in which the solution to the equation $3.2x - 5.8 = 0$ is located. Explain how this process can be used to approximate the solution to any desired degree of accuracy.

106. ***Exploration*** Use the procedure in Exercise 105 to approximate the solution to the equation $0.3(x - 1.5) - 2 = 0$ accurate to two decimal places.

Review

In Exercises 107 and 108, simplify the expression.

107. $\dfrac{x^2 + 5x - 36}{2x^2 + 17x - 9}$ **108.** $\dfrac{x^2 - 49}{x^3 + x^2 + 3x - 21}$

In Exercises 109–112, sketch the graph of the equation.

109. $y = 3x - 5$ **110.** $y = -\frac{1}{2}x - \frac{9}{2}$

111. $y = -x^2 - 5x$ **112.** $y = \sqrt{5 - x}$

1.3 Modeling with Linear Equations

▶ What you should learn

- How to use a verbal model in a problem-solving plan
- How to write and use mathematical models to solve real-life problems
- How to solve mixture problems
- How to use common formulas to solve real-life problems

▶ Why you should learn it

You can use linear equations to determine the percents of incomes and of expenses of the federal government that come from various sources. See Exercise 44 on page 105.

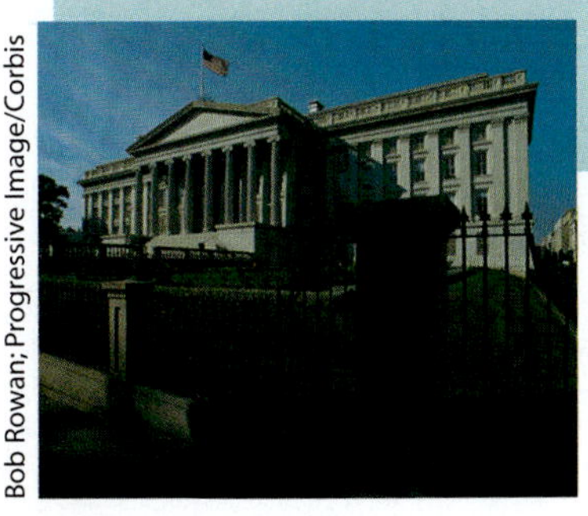

Bob Rowan; Progressive Image/Corbis

Introduction to Problem Solving 

In this section you will learn how algebra can be used to solve problems that occur in real-life situations. The process of translating phrases or sentences into algebraic expressions or equations is called **mathematical modeling.** A good approach to mathematical modeling is to use two stages. Begin by using the verbal description of the problem to form a *verbal model*. Then, after assigning labels to the quantities in the verbal model, form a *mathematical model* or *algebraic equation*.

Verbal Description  Verbal Model  Algebraic Equation

When you are trying to construct a verbal model, it is helpful to look for a *hidden equality*—a statement that two algebraic expressions are equal.

Example 1 ▶ Using a Verbal Model  

You have accepted a job for which your annual salary will be \$27,236. This salary includes a year-end bonus of \$500. You will be paid twice a month. What will your gross pay be for each paycheck?

Solution

Because there are 12 months in a year and you will be paid twice a month, it follows that you will receive 24 paychecks during the year. You can construct an algebraic equation for this problem as follows. Begin with a verbal model, then assign labels, and finally form an algebraic equation.

Verbal Model: Income for year $=$ 24 paychecks $+$ Bonus

Labels: Income for year $= 27{,}236$ (dollars)
Amount of each paycheck $= x$ (dollars)
Bonus $= 500$ (dollars)

Equation: $27{,}236 = 24x + 500$

The algebraic equation for this problem is a *linear equation* in the variable x, which you can solve as follows.

$$27{,}236 = 24x + 500 \qquad \text{Write original equation.}$$

$$27{,}236 - 500 = 24x + 500 - 500 \qquad \text{Subtract 500 from each side.}$$

$$26{,}736 = 24x \qquad \text{Simplify.}$$

$$\frac{26{,}736}{24} = \frac{24x}{24} \qquad \text{Divide each side by 24.}$$

$$1114 = x \qquad \text{Simplify.}$$

So, your gross pay for each paycheck will be \$1114.

A fundamental step in writing a mathematical model to represent a real-life problem is translating key words and phrases into algebraic expressions and equations. The following list gives several examples.

Translating Key Words and Phrases

Key Words and Phrases	*Verbal Description*	*Algebraic Expression or Equation*
Equality: Equals, equal to, is, are, was, will be, represents	• The sale price S is \$10 less than the list price L.	$S = L - 10$
Addition: Sum, plus, greater than, increased by, more than, exceeds, total of	• The sum of 5 and x • Seven more than y	$5 + x$ or $x + 5$ $7 + y$ or $y + 7$
Subtraction: Difference, minus, less than, decreased by, subtracted from, reduced by, the remainder	• The difference of 4 and b • Three less than z	$4 - b$ $z - 3$
Multiplication: Product, multiplied by, twice, times, percent of	• Two times x • Three percent of t	$2x$ $0.03t$
Division: Quotient, divided by, ratio, per	• The ratio of x to 8	$\frac{x}{8}$

Using Mathematical Models

Example 2 ▶ Finding the Percent of a Raise  

You have accepted a job that pays \$8 an hour. You are told that after a two-month probationary period, your hourly wage will be increased to \$9 an hour. What percent raise will you receive after the two-month period?

Solution

Verbal Model: Raise = Percent · Old wage

Labels:
Old wage = 8 (dollars per hour)
New wage = 9 (dollars per hour)
Raise = 9 − 8 = 1 (dollars per hour)
Percent = r (percent in decimal form)

Equation:

$$1 = r \cdot 8$$

$$\frac{1}{8} = r \quad \text{Divide each side by 8.}$$

$$0.125 = r \quad \text{Rewrite fraction as a decimal.}$$

You will receive a raise of 0.125 or 12.5%.

STUDY TIP

Writing the unit for each label in a real-life problem helps you determine the unit for the answer. This is called *unit analysis*. When the same unit of measure occurs in the numerator and denominator of an expression, you can divide out the unit. For instance, unit analysis verifies that the unit for time in the formula below is hours.

$$\text{Time} = \frac{\text{distance}}{\text{rate}} = \frac{\text{miles}}{\frac{\text{miles}}{\text{hour}}} = \cancel{\text{miles}} \cdot \frac{\text{hours}}{\cancel{\text{miles}}} = \text{hours}$$

Example 3 ▶ Finding the Percent of Monthly Expenses 

Your family has an annual income of \$57,000 and the following monthly expenses: mortgage (\$1100), car payment (\$375), food (\$295), utilities (\$240), and credit cards (\$220). The total value of the monthly expenses represents what percent of your family's annual income?

Solution

The total amount of your family's monthly expenses is \$2230. The total monthly expenses for 1 year are \$26,760.

Verbal Model: Monthly expenses = Percent · Income

Labels: Income = 57,000 (dollars)
Monthly expenses = 26,760 (dollars)
Percent = r (in decimal form)

Equation: $26{,}760 = r \cdot 57{,}000$

$\frac{26{,}760}{57{,}000} = r$ Divide each side by 57,000.

$0.469 \approx r$ Use a calculator.

Your family's monthly expenses are approximately 0.469 or 46.9% of your family's annual income.

Example 4 ▶ Finding the Dimensions of a Room 

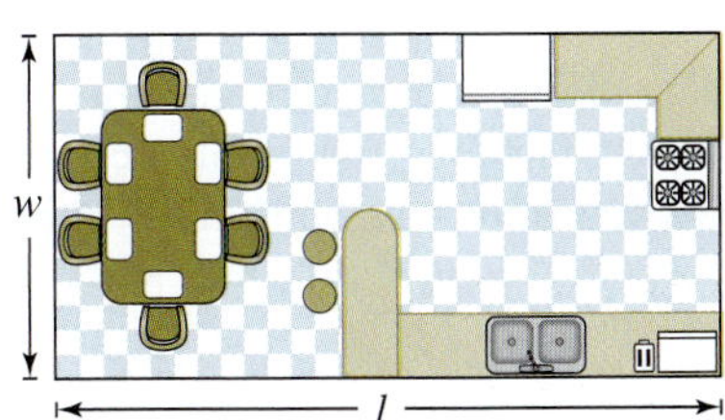

FIGURE 1.15

A rectangular kitchen is twice as long as it is wide, and its perimeter is 84 feet. Find the dimensions of the kitchen.

Solution

For this problem, it helps to sketch a diagram, as shown in Figure 1.15.

Verbal Model: 2 · Length + 2 · Width = Perimeter

Labels: Perimeter = 84 (feet)
Width = w (feet)
Length = $l = 2w$ (feet)

Equation: $2(2w) + 2w = 84$

$6w = 84$ Group like terms.

$w = 14$ Divide each side by 6.

Because the length is twice the width, you have

$l = 2w$ Length is twice width.

$= 2(14)$ Substitute.

$= 28.$ Simplify.

So, the dimensions of the room are 14 feet by 28 feet.

Example 5 ▶ A Distance Problem

A plane is flying nonstop from Atlanta to Portland, a distance of about 2700 miles, as shown in Figure 1.16. After 1.5 hours in the air, the plane flies over Kansas City (a distance of 820 miles from Atlanta). Estimate the time it will take the plane to fly from Atlanta to Portland.

Solution

Verbal Model: Distance = Rate · Time

Labels:
Distance = 2700 (miles)
Time = t (hours)
Rate = $\dfrac{\text{distance to Kansas City}}{\text{time to Kansas City}} = \dfrac{820}{1.5}$ (miles per hour)

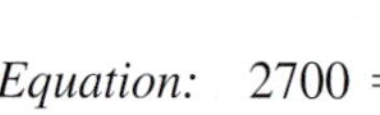

Equation:

$$2700 = \frac{820}{1.5}t$$

$$4050 = 820t$$

$$\frac{4050}{820} = t$$

$$4.94 \approx t$$

The trip will take about 4.94 hours, or about 4 hours and 57 minutes.

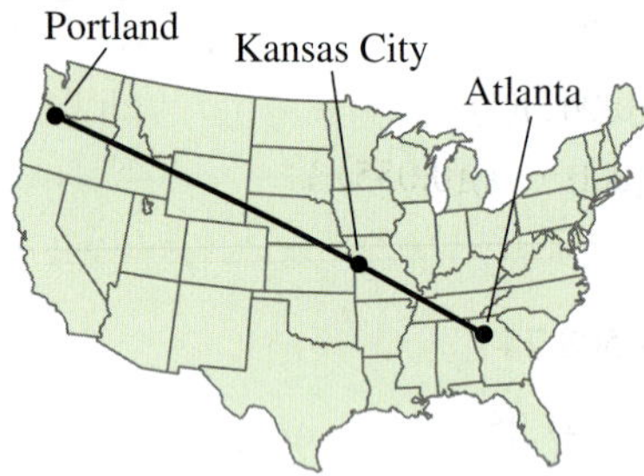

FIGURE 1.16

Example 6 ▶ An Application Involving Similar Triangles

To determine the height of the Aon Center Building (in Chicago), you measure the shadow cast by the building and find it to be 142 feet long, as shown in Figure 1.17. Then you measure the shadow cast by a four-foot post and find it to be 6 inches long. Estimate the building's height.

Solution

To solve this problem, you use a result from geometry that states that the ratios of corresponding sides of similar triangles are equal.

Verbal Model: $\dfrac{\text{Height of building}}{\text{Length of building's shadow}} = \dfrac{\text{Height of post}}{\text{Length of post's shadow}}$

Labels:
Height of building = x (feet)
Length of building's shadow = 142 (feet)
Height of post = 4 feet = 48 inches (inches)
Length of post's shadow = 6 (inches)

Equation:

$$\frac{x}{142} = \frac{48}{6}$$

$$x = 1136$$

So, the Aon Center Building is about 1136 feet high.

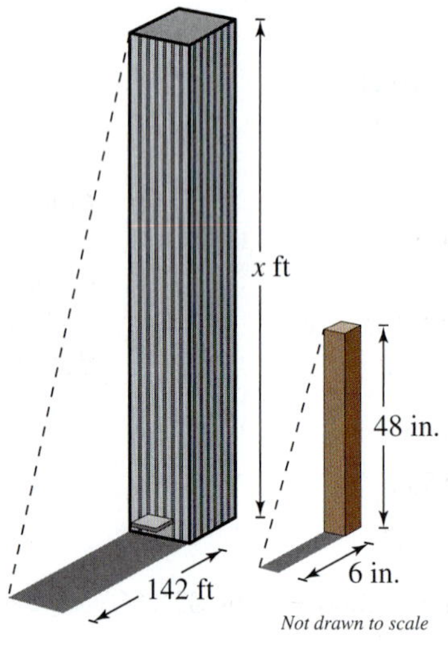

FIGURE 1.17

Mixture Problems

Problems that involve two or more rates are called mixture problems. They are not limited to mixtures of chemical solutions, as shown in Examples 7 and 8.

Example 7 ▶ A Simple Interest Problem  

STUDY TIP

Example 7 uses the simple interest formula $I = Prt$, where I is the interest, P is the principal, r is the annual interest rate (in decimal form), and t is the time in years.

You invested a total of \$10,000 at $4\frac{1}{2}\%$ and $5\frac{1}{2}\%$ simple interest. During 1 year, the two accounts earned \$508.75. How much did you invest in each?

Solution

Verbal Model: Interest from $4\frac{1}{2}\%$ + Interest from $5\frac{1}{2}\%$ = Total interest

Labels:
Amount invested at $4\frac{1}{2}\% = x$ (dollars)
Amount invested at $5\frac{1}{2}\% = 10{,}000 - x$ (dollars)
Interest from $4\frac{1}{2}\% = Prt = (x)(0.045)(1)$ (dollars)
Interest from $5\frac{1}{2}\% = Prt = (10{,}000 - x)(0.055)(1)$ (dollars)
Total interest $= 508.75$ (dollars)

Equation:

$$0.045x + 0.055(10{,}000 - x) = 508.75$$

$$-0.01x = -41.25$$

$$x = 4125$$

So, \$4125 was invested at $4\frac{1}{2}\%$ and $10{,}000 - x$ or \$5875 was invested at $5\frac{1}{2}\%$.

Example 8 ▶ An Inventory Problem 

A store has \$30,000 of inventory in 13-inch and 19-inch color televisions. The profit on a 13-inch set is 22% and the profit on a 19-inch set is 40%. The profit for the entire stock is 35%. How much was invested in each type of television?

Solution

Verbal Model: Profit from 13-inch sets + Profit from 19-inch sets = Total profit

Labels:
Inventory of 13-inch sets $= x$ (dollars)
Inventory of 19-inch sets $= 30{,}000 - x$ (dollars)
Profit from 13-inch sets $= 0.22x$ (dollars)
Profit from 19-inch sets $= 0.40(30{,}000 - x)$ (dollars)
Total profit $= 0.35(30{,}000) = 10{,}500$ (dollars)

Equation:

$$0.22x + 0.40(30{,}000 - x) = 10{,}500$$

$$-0.18x = -1500$$

$$x \approx 8333.33$$

So, \$8333.33 was invested in 13-inch sets and $30{,}000 - x$ or \$21,666.67 was invested in 19-inch sets.

Common Formulas 

Many common types of geometric, scientific, and investment problems use ready-made equations called **formulas.** Knowing these formulas will help you translate and solve a wide variety of real-life applications.

Common Formulas for Area *A*, Perimeter *P*, Circumference *C*, and Volume *V*

Square

$A = s^2$

$P = 4s$

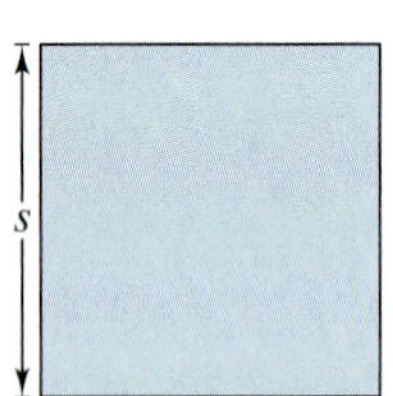

Rectangle

$A = lw$

$P = 2l + 2w$

Circle

$A = \pi r^2$

$C = 2\pi r$

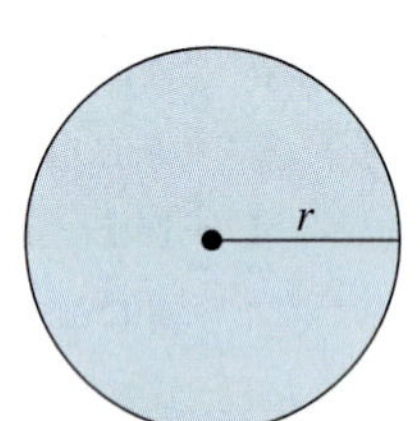

Triangle

$A = \frac{1}{2}bh$

$P = a + b + c$

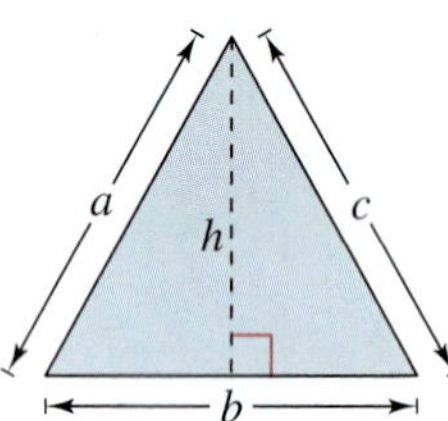

Cube

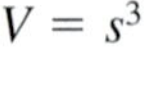

$V = s^3$

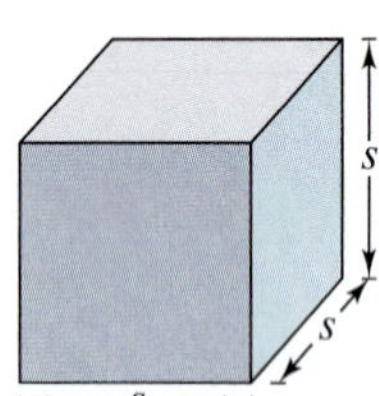

Rectangular Solid

$V = lwh$

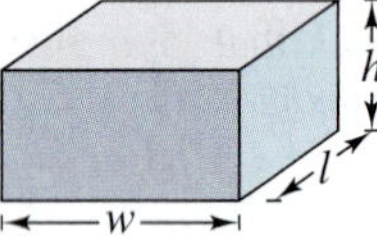

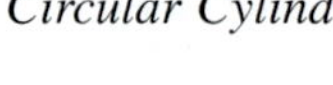

Circular Cylinder

$V = \pi r^2 h$

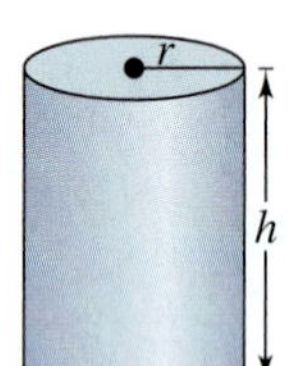

Sphere

$V = \frac{4}{3}\pi r^3$

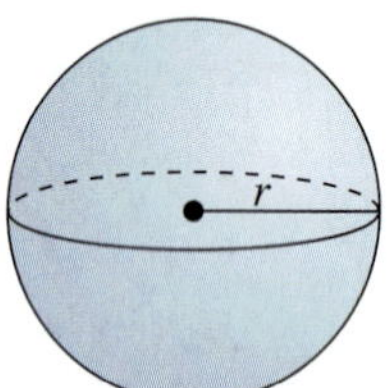

Miscellaneous Common Formulas

Temperature:

$F = \frac{9}{5}C + 32$ — F = degrees Fahrenheit, C = degrees Celsius

Simple Interest:

$I = Prt$ — I = interest, P = principal, r = annual interest rate, t = time in years

Compound Interest:

$A = P\left(1 + \frac{r}{n}\right)^{nt}$ — A = balance, P = principal, r = annual interest rate, n = compoundings per year, t = time in years

Distance:

$d = rt$ — d = distance traveled, r = rate, t = time

When working with applied problems, you often need to rewrite one of the common formulas. For instance, the formula for the perimeter of a rectangle, $P = 2l + 2w$, can be rewritten or solved for w as $w = \frac{1}{2}(P - 2l)$.

Example 9 ▶ Using a Formula 

A cylindrical can has a volume of 200 cubic centimeters (cm³) and a radius of 4 centimeters (cm), as shown in Figure 1.18. Find the height of the can.

FIGURE 1.18

Solution

The formula for the *volume of a cylinder* is $V = \pi r^2 h$. To find the height of the can, solve for h.

$$h = \frac{V}{\pi r^2}$$

Then, using $V = 200$ and $r = 4$, find the height.

$$h = \frac{200}{\pi(4)^2} \quad \text{Substitute 200 for } V \text{ and 4 for } r.$$

$$= \frac{200}{16\pi} \quad \text{Simplify denominator.}$$

$$\approx 3.98 \quad \text{Use a calculator.}$$

You can use unit analysis to check that your answer is reasonable.

$$\frac{200 \text{ cm}^3}{16\pi \text{ cm}^2} \approx 3.98 \text{ cm}$$

Writing ABOUT MATHEMATICS

Translating Algebraic Formulas Most people use algebraic formulas every day—sometimes without realizing it because they use a verbal form or think of an often-repeated calculation in steps. Translate each of the following verbal descriptions into an algebraic formula, and demonstrate the use of each formula.

a. ***Designing Billboards*** "The letters on a sign or billboard are designed to be readable at a certain distance. Take half the letter height in inches and multiply by 100 to find the readable distance in feet."—Thos. Hodgson, Hodgson Signs (Source: *Rules of Thumb* by Tom Parker)

b. ***Percent of Calories from Fat*** "To calculate percent of calories from fat, multiply grams of total fat per serving by 9, then divide by the number of calories per serving." (Source: *Good Housekeeping*)

c. ***Building Stairs*** "A set of steps will be comfortable to use if two times the height of one riser plus the width of one tread is equal to 26 inches."—Alice Lukens Bachelder, gardener (Source: *Rules of Thumb* by Tom Parker)

1.3 Exercises

In Exercises 1–10, write a verbal description of the algebraic expression without using the variable.

1. $x + 4$
2. $t - 10$
3. $\dfrac{u}{5}$
4. $\dfrac{2}{3}x$
5. $\dfrac{y - 4}{5}$
6. $\dfrac{z + 10}{7}$
7. $-3(b + 2)$
8. $\dfrac{-5(x - 1)}{8}$
9. $12x(x - 5)$
10. $\dfrac{(q + 4)(3 - q)}{2q}$

In Exercises 11–22, write an algebraic expression for the verbal description.

11. The sum of two consecutive natural numbers
12. The product of two consecutive natural numbers
13. The product of two consecutive odd integers, the first of which is $2n - 1$
14. The sum of the squares of two consecutive even integers, the first of which is $2n$
15. The distance traveled in t hours by a car traveling at 50 miles per hour
16. The travel time for a plane traveling at a rate of r kilometers per hour for 200 kilometers
17. The amount of acid in x liters of a 20% acid solution
18. The sale price of an item that is discounted 20% of its list price L
19. The perimeter of a rectangle with a width x and a length that is twice the width
20. The area of a triangle with base 20 inches and height h inches
21. The total cost of producing x units for which the fixed costs are \$1200 and the cost per unit is \$25
22. The total revenue obtained by selling x units at \$3.59 per unit

In Exercises 23–26, translate the statement into an algebraic expression or equation.

23. Thirty percent of the list price L
24. The amount of water in q quarts of a liquid that is 35% water
25. The percent of 500 that is represented by the number N
26. The percent change in sales from one month to the next if the monthly sales are S_1 and S_2, respectively

In Exercises 27 and 28, write an expression for the area of the region in the figure.

27. 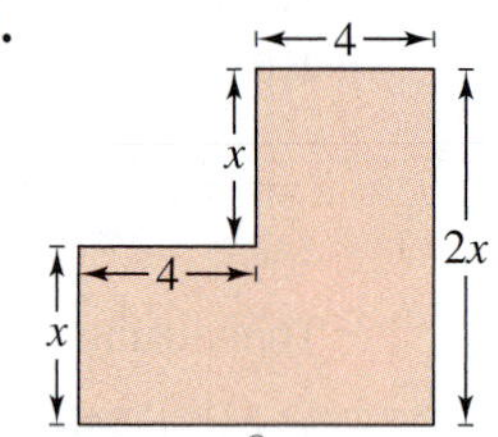

28. 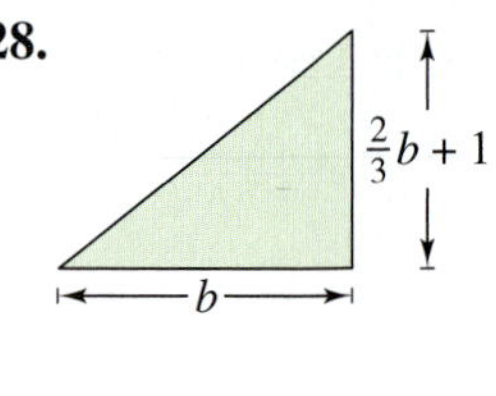

Number Problems **In Exercises 29–34, write a mathematical model for the number problem and solve.**

29. The sum of two consecutive natural numbers is 525. Find the numbers.
30. The sum of three consecutive natural numbers is 804. Find the numbers.
31. One positive number is 5 times another number. The difference between the two numbers is 148. Find the numbers.
32. One positive number is $\frac{1}{5}$ of another number. The difference between the two numbers is 76. Find the numbers.
33. Find two consecutive integers whose product is 5 less than the square of the smaller number.
34. Find two consecutive natural numbers such that the difference of their reciprocals is $\frac{1}{4}$ the reciprocal of the smaller number.

In Exercises 35–40, solve the percent equation.

35. What is 30% of 45?
36. What is 175% of 360?
37. 432 is what percent of 1600?
38. 459 is what percent of 340?
39. 12 is $\frac{1}{2}$% of what number?
40. 70 is 40% of what number?

41. ***Finance*** A family has annual loan payments equaling 58.6% of their annual income. During the year, their loan payments total \$13,077.75. What is their annual income?

42. ***Finance*** A salesperson's weekly paycheck is 15% less than her coworker's paycheck. The two paychecks total \$645. Find the amount of each paycheck.

43. ***Discount*** The price of a swimming pool has been discounted 16.5%. The sale price is \$1210.75. Find the original list price of the pool.

▶ Model It

44. ***Government*** The tables show the sources of income (in billions of dollars) and expenses (in billions of dollars) for the federal government in 1999. (Source: United States Office of Management and Budget)

Source of income	Income
Corporation taxes	184.68
Income tax	879.48
Social Security	611.83
Other	151.46

Source of expenses	Expense
Interest on debt	229.74
Health and human services	1058.89
Defense department	274.87
Other	139.54

(a) Find the percent of the total income for each category. Then use these percents to label the circle graph. To print an enlarged copy of the graph, refer to *www.mathgraphs.com*.

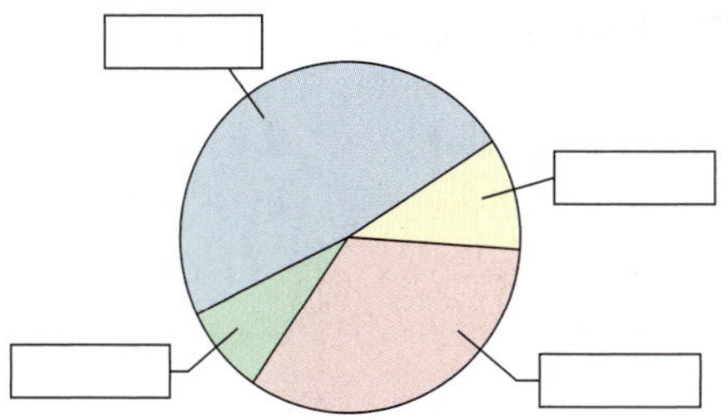

(b) Find the percent of the total expenses for each category. Then use these percents to label the circle graph. To print an enlarged copy of the graph, refer to *www.mathgraphs.com*.

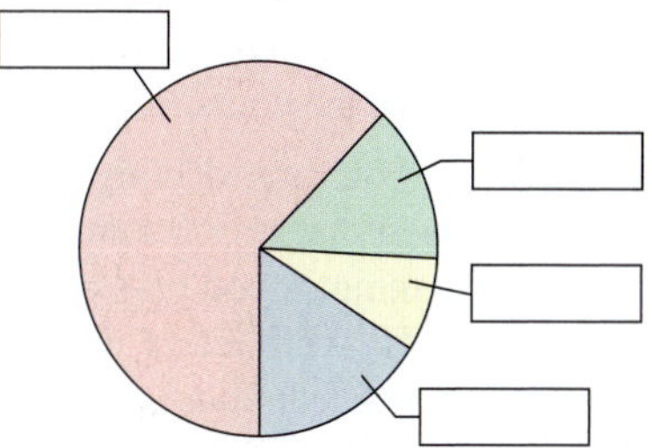

(c) Compare the total income and total expenses. How much of a surplus or deficit is there?

In Exercises 45–48, the values or prices of various items are given for 1980 and 2000. Find the percent change for each item. (Source: 2001 Statistical Abstract of the U.S.)

	Item	*1980*	*2000*
45.	Gallon of diesel fuel	\$0.82	\$0.94
46.	Cable TV monthly basic rate	\$7.69	\$30.08
47.	An ounce of gold	\$613.00	\$280.00
48.	New one-family home	\$64,600	\$169,000

49. ***Dimensions of a Room*** A room is 1.5 times as long as it is wide, and its perimeter is 25 meters.

(a) Draw a diagram that represents the problem. Identify the length as l and the width as w.

(b) Write l in terms of w and write an equation for the perimeter in terms of w.

(c) Find the dimensions of the room.

50. ***Dimensions of a Picture Frame*** A picture frame has a total perimeter of 2 meters. The height of the frame is 0.62 times its width.

(a) Draw a diagram that represents the problem. Identify the width as w and the height as h.

(b) Write h in terms of w and write an equation for the perimeter in terms of w.

(c) Find the dimensions of the picture frame.

51. ***Course Grade*** To get an A in a course, you must have an average of at least 90 on four tests of 100 points each. The scores on your first three tests were 87, 92, and 84. What must you score on the fourth test to get an A for the course?

52. ***Course Grade*** You are taking a course that has four tests. The first three tests are 100 points each and the fourth test is 200 points. To get an A in the course, you must have an average of at least 90% on the four tests. Your scores on the first three tests were 87, 92, and 84. What must you score on the fourth test to get an A for the course?

53. ***Travel Time*** You are driving on a Canadian freeway to a town that is 300 kilometers from your home. After 30 minutes you pass a freeway exit that you know is 50 kilometers from your home. Assuming that you continue at the same constant speed, how long will it take for the entire trip?

54. ***Travel Time*** Two cars start at an interstate interchange and travel in the same direction at average speeds of 40 miles per hour and 55 miles per hour. How much time must elapse before the two cars are 5 miles apart?

55. ***Travel Time*** On the first part of a 317-mile trip, a salesperson averaged 58 miles per hour. He averaged only 52 miles per hour on the last part of the trip because of an increased volume of traffic. The total time of the trip was 5 hours and 45 minutes. Find the amount of time at each of the two speeds.

56. ***Travel Time*** Students are traveling in two cars to a football game 135 miles away. The first car leaves on time and travels at an average speed of 45 miles per hour. The second car starts $\frac{1}{2}$ hour later and travels at an average speed of 55 miles per hour. How long will it take the second car to catch up to the first car? Will the second car catch up to the first car before the first car arrives at the game?

57. ***Travel Time*** Two families meet at a park for a picnic. At the end of the day one family travels east at an average speed of 42 miles per hour and the other travels west at an average speed of 50 miles per hour. Both families have approximately 160 miles to travel.

(a) Find the time it takes each family to get home.

(b) Find the time that will have elapsed when they are 100 miles apart.

(c) Find the distance the eastbound family has to travel after the westbound family has arrived home.

58. ***Average Speed*** A truck driver traveled at an average speed of 55 miles per hour on a 200-mile trip to pick up a load of freight. On the return trip (with the truck fully loaded), the average speed was 40 miles per hour. What was the average speed for the round trip?

59. ***Wind Speed*** An executive flew in the corporate jet to a meeting in a city 1500 kilometers away. After traveling the same amount of time on the return flight, the pilot mentioned that they still had 300 kilometers to go. The air speed of the plane was 600 kilometers per hour. How fast was the wind blowing? (Assume that the wind direction was parallel to the flight path and constant all day.)

60. ***Physics*** Light travels at the speed of 3.0×10^8 meters per second. Find the time in minutes required for light to travel from the sun to Earth (a distance of 1.5×10^{11} meters).

61. ***Radio Waves*** Radio waves travel at the same speed as light, 3.0×10^8 meters per second. Find the time required for a radio wave to travel from Mission Control in Houston to NASA astronauts on the surface of the moon 3.86×10^8 meters away.

62. ***Height of a Tree*** To obtain the height of a tree (see figure), you measure the tree's shadow and find that it is 8 meters long. You also measure the shadow of a two-meter lamppost and find that it is 75 centimeters long. How tall is the tree?

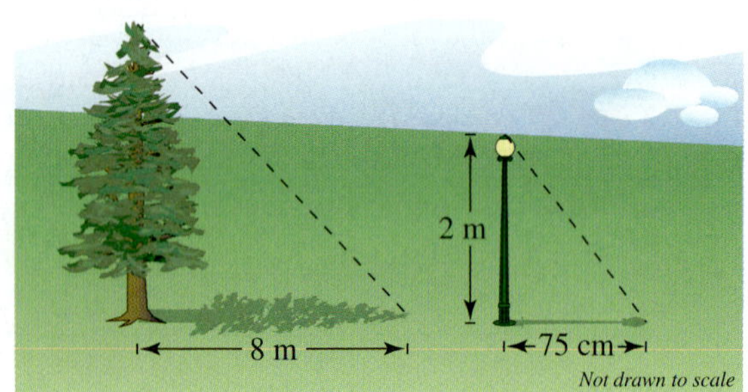

63. ***Height of a Building*** To obtain the height of the Chrysler Building in New York (see figure), you measure the building's shadow and find that it is 87 feet long. You also measure the shadow of a four-foot stake and find that it is 4 inches long. How tall is the Chrysler building?

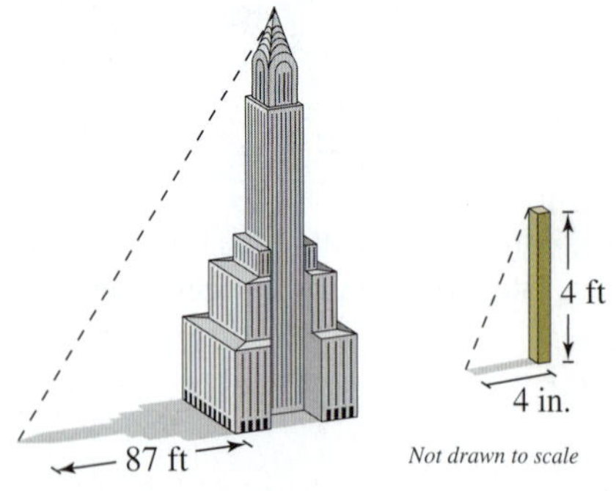

64. ***Flagpole Height*** A person who is 6 feet tall walks away from a flagpole toward the tip of the shadow of the flagpole. When the person is 30 feet from the flagpole, the tips of the person's shadow and the shadow cast by the flagpole coincide at a point 5 feet in front of the person. Find the height of the flagpole.

65. ***Shadow Length*** A person who is 6 feet tall walks away from a 50-foot silo toward the tip of the silo's shadow. At a distance of 32 feet from the silo, the person's shadow begins to emerge beyond the silo's shadow. How much farther must the person walk to be completely out of the silo's shadow?

66. ***Investment*** You plan to invest \$12,000 in two funds paying $4\frac{1}{2}\%$ and 5% simple interest. (There is more risk in the 5% fund.) Your goal is to obtain a total annual interest income of \$580 from the investments. What is the smallest amount you can invest in the 5% fund and still meet your objective?

67. ***Investment*** You plan to invest \$25,000 in two funds paying 3% and $4\frac{1}{2}\%$ simple interest. (There is more risk in the $4\frac{1}{2}\%$ fund.) Your goal is to obtain a total annual interest income of \$1000 from the investments. What is the smallest amount you can invest in the $4\frac{1}{2}\%$ fund and still to meet your objective?

68. ***Business*** A nursery has \$20,000 of inventory in dogwood trees and red maple trees. The profit on a dogwood tree is 25% and the profit on a red maple is 17%. The profit for the entire stock is 20%. How much was invested in each type of tree?

69. ***Business*** An automobile dealer has \$600,000 of inventory in compact cars and midsize cars. The profit on a compact car is 24% and the profit on a midsize car is 28%. The profit for the entire stock is 25%. How much was invested in each type of car?

70. ***Mixture Problem*** Using the values in the table, determine the amounts of solutions 1 and 2, respectively, needed to obtain the desired amount and concentration of the final mixture.

	Concentration			Amount of final solution
	Solution 1	Solution 2	Final solution	
(a)	10%	30%	25%	100 gal
(b)	25%	50%	30%	5 L
(c)	15%	45%	30%	10 qt
(d)	70%	90%	75%	25 gal

71. ***Mixture Problem*** A 100% concentrate is to be mixed with a mixture having a concentration of 40% to obtain 55 gallons of a mixture with a concentration of 75%. How much of the 100% concentrate will be needed?

72. ***Mixture Problem*** A forester mixes gasoline and oil to make 2 gallons of mixture for his two-cycle chain-saw engine. This mixture is 32 parts gasoline and 1 part two-cycle oil. How much gasoline must be added to bring the mixture to 40 parts gasoline and 1 part oil?

73. ***Mixture Problem*** A grocer mixes peanuts that cost \$2.49 per pound and walnuts that cost \$3.89 per pound to make 100 pounds of a mixture that costs \$3.19 per pound. How much of each kind of nut is put into the mixture?

74. ***Company Costs*** An outdoor furniture manufacturer has fixed costs of \$10,000 per month and average variable costs of \$8.50 per unit manufactured. The company has \$85,000 available to cover the monthly costs. How many units can the company manufacture? (*Fixed costs* are those that occur regardless of the level of production. *Variable costs* depend on the level of production.)

75. ***Company Costs*** A plumbing supply company has fixed costs of \$10,000 per month and average variable costs of \$9.30 per unit manufactured. The company has \$85,000 available to cover the monthly costs. How many units can the company manufacture? (*Fixed costs* are those that occur regardless of the level of production. *Variable costs* depend on the level of production.)

76. ***Water Depth*** A trough is 12 feet long, 3 feet deep, and 3 feet wide (see figure). Find the depth of the water when the trough contains 70 gallons (1 gallon $\approx$ 0.13368 cubic foot).

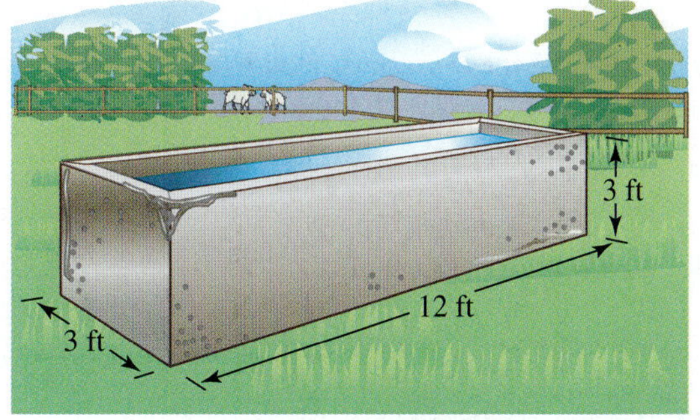

Physics **In Exercises 77 and 78, you have a uniform beam of length L with a fulcrum x feet from one end (see figure). Objects with weights W_1 and W_2 are placed at opposite ends of the beam. The beam will balance when $W_1x = W_2(L - x)$. Find x such that the beam will balance.**

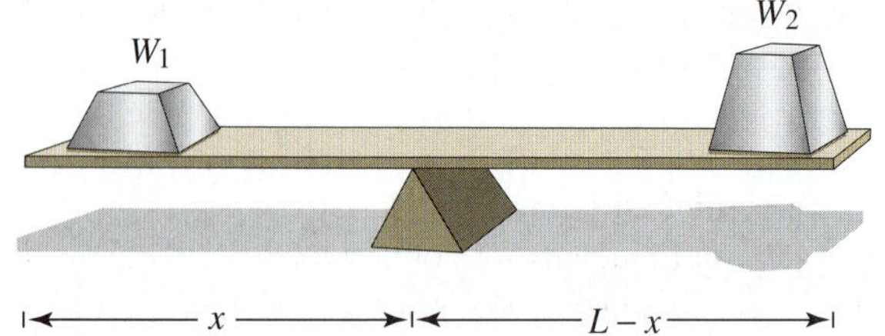

77. Two children weighing 50 pounds and 75 pounds are playing on a seesaw that is 10 feet long.

78. A person weighing 200 pounds is attempting to move a 550-pound rock with a bar that is 5 feet long.

In Exercises 79–90, solve for the indicated variable.

79. ***Area of a Triangle***

Solve for h: $A = \frac{1}{2}bh$

80. ***Volume of a Right Circular Cylinder***

Solve for h: $V = \pi r^2 h$

81. ***Markup*** Solve for C: $S = C + RC$

82. ***Investment at Simple Interest***

Solve for r: $A = P + Prt$

83. ***Volume of an Oblate Spheroid***

Solve for b: $V = \frac{4}{3}\pi a^2 b$

84. ***Volume of a Spherical Segment***

Solve for r: $V = \frac{1}{3}\pi h^2(3r - h)$

85. ***Free-Falling Body*** Solve for a: $h = v_0 t + \frac{1}{2}at^2$

86. ***Lensmaker's Equation***

Solve for R_1: $\frac{1}{f} = (n - 1)\left(\frac{1}{R_1} - \frac{1}{R_2}\right)$

87. ***Capacitance in Series Circuits***

Solve for C_1: $C = \dfrac{1}{\dfrac{1}{C_1} + \dfrac{1}{C_2}}$

88. ***Arithmetic Progression***

Solve for a: $S = \frac{n}{2}[2a + (n - 1)d]$

89. ***Arithmetic Progression***

Solve for n: $L = a + (n - 1)d$

90. ***Geometric Progression*** Solve for r: $S = \dfrac{rL - a}{r - 1}$

91. ***Volume of a Billiard Ball*** A billiard ball has a volume of 5.96 cubic inches. Find the radius of a billiard ball.

92. ***Length of a Tank*** The diameter of a cylindrical propane gas tank is 4 feet. The total volume of the tank is 603.2 cubic feet. Find the length of the tank.

Synthesis

True or False? **In Exercises 93 and 94, determine whether the statement is true or false. Justify your answer.**

93. "8 less than z cubed divided by the difference of z squared and 9" can be written as

$$\frac{z^3 - 8}{(z - 9)^2}.$$

94. The volume of a cube with a side of length 9.5 inches is greater than the volume of a sphere with a radius of 5.9 inches.

95. Consider the linear equation $ax + b = 0$.

(a) What is the sign of the solution if $ab > 0$?

(b) What is the sign of the solution if $ab < 0$?

In each case, explain your reasoning.

96. Write a linear equation that has the solution $x = -3$. (There are many correct answers.)

Review

In Exercises 97–100, simplify the expression.

97. $(5x^4)(25x^2)^{-1}$, $x \neq 0$

98. $\sqrt{150s^2t^3}$

99. $\dfrac{3}{x - 5} + \dfrac{2}{5 - x}$

100. $\dfrac{5}{x} + \dfrac{3x}{x^2 - 9} - \dfrac{10}{x + 3}$

In Exercises 101–104, rationalize the denominator.

101. $\dfrac{10}{7\sqrt{3}}$

102. $\dfrac{6}{\sqrt{10} - 2}$

103. $\dfrac{5}{\sqrt{6} + \sqrt{11}}$

104. $\dfrac{14}{3\sqrt{10} - 1}$

1.4 Quadratic Equations

▶ **What you should learn**

- How to solve quadratic equations by factoring
- How to solve quadratic equations by extracting square roots
- How to solve quadratic equations by completing the square
- How to use the Quadratic Formula to solve quadratic equations
- How to use quadratic equations to model and solve real-life problems

▶ **Why you should learn it**

Quadratic equations can be used to model and solve real-life problems. For instance, in Exercise 118 on page 120, you will use a quadratic equation to model the time it takes an object to fall from the top of the CN Tower.

Chris Thomaidis/Getty Images

Factoring 

A **quadratic equation** in x is an equation that can be written in the general form

$$ax^2 + bx + c = 0$$

where a, b, and c are real numbers with $a \neq 0$. A quadratic equation in x is also known as a **second-degree polynomial equation** in x.

In this section, you will study four methods for solving quadratic equations: *factoring*, *extracting square roots*, *completing the square*, and the *Quadratic Formula*. The first method is based on the Zero-Factor Property from Section P.1.

If $ab = 0$, then $a = 0$ or $b = 0$. Zero-Factor Property

To use this property, write the left side of the general form of a quadratic equation as the product of two linear factors. Then find the solutions of the quadratic equation by setting each linear factor equal to zero.

Example 1 ▶ Solving a Quadratic Equation by Factoring 

a. $2x^2 + 9x + 7 = 3$ Original equation

$2x^2 + 9x + 4 = 0$ Write in general form.

$(2x + 1)(x + 4) = 0$ Factor.

$2x + 1 = 0 \Rightarrow x = -\frac{1}{2}$ Set 1st factor equal to 0.

$x + 4 = 0 \Rightarrow x = -4$ Set 2nd factor equal to 0.

The solutions are $x = -\frac{1}{2}$ and $x = -4$. Check these in the original equation.

b. $6x^2 - 3x = 0$ Original equation

$3x(2x - 1) = 0$ Factor.

$3x = 0 \Rightarrow x = 0$ Set 1st factor equal to 0.

$2x - 1 = 0 \Rightarrow x = \frac{1}{2}$ Set 2nd factor equal to 0.

The solutions are $x = 0$ and $x = \frac{1}{2}$. Check these in the original equation.

Be sure you see that the Zero-Factor Property works *only* for equations written in general form (in which the right side of the equation is zero). So, all terms must be collected on one side *before* factoring. For instance, in the equation

$$(x - 5)(x + 2) = 8$$

it is *incorrect* to set each factor equal to 8. Try to solve this equation correctly.

Technology

You can use a graphing utility to check graphically the real solutions of a quadratic equation. Begin by writing the equation in general form. Then set y equal to the left side and graph the resulting equation. The x-intercepts of the equation represent the real solutions of the original equation. For example, to check the solutions of $6x^2 - 3x = 0$, graph $y = 6x^2 - 3x$, as shown below. Notice that the x-intercepts occur at $x = 0$ and $x = \frac{1}{2}$, as found in Example 1(b).

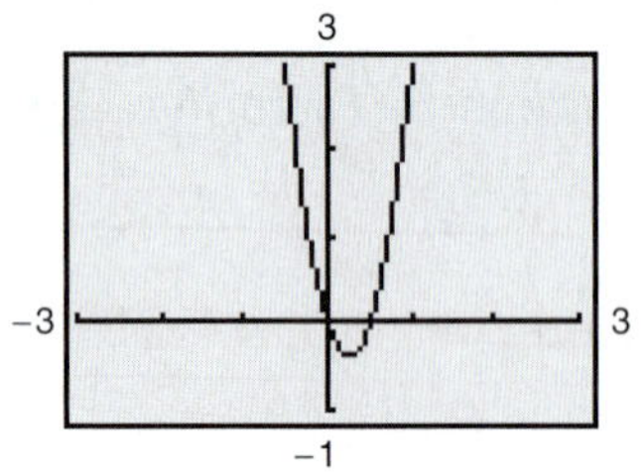

Extracting Square Roots

There is a nice shortcut for solving quadratic equations of the form $u^2 = d$, where $d > 0$ and u is an algebraic expression. By factoring, you can see that this equation has two solutions.

$$u^2 = d \qquad \text{Write original equation.}$$

$$u^2 - d = 0 \qquad \text{Write in general form.}$$

$$(u + \sqrt{d})(u - \sqrt{d}) = 0 \qquad \text{Factor.}$$

$$u + \sqrt{d} = 0 \quad \Rightarrow \quad u = -\sqrt{d} \qquad \text{Set 1st factor equal to 0.}$$

$$u - \sqrt{d} = 0 \quad \Rightarrow \quad u = \sqrt{d} \qquad \text{Set 2nd factor equal to 0.}$$

Because the two solutions differ only in sign, you can write the solutions together, using a "plus or minus sign," as

$$u = \pm\sqrt{d}.$$

This form of the solution is read as "u is equal to plus or minus the square root of d." Solving an equation of the form $u^2 = d$ without going through the steps of factoring is called **extracting square roots.**

Extracting Square Roots

The equation $u^2 = d$, where $d > 0$, has exactly two solutions:

$$u = \sqrt{d} \qquad \text{and} \qquad u = -\sqrt{d}.$$

These solutions can also be written as

$$u = \pm\sqrt{d}.$$

Example 2 ▶ Extracting Square Roots 

Solve each equation by extracting square roots.

a. $4x^2 = 12$ **b.** $(x - 3)^2 = 7$

Solution

a. $4x^2 = 12$ Write original equation.

$x^2 = 3$ Divide each side by 4.

$x = \pm\sqrt{3}$ Extract square roots.

The solutions are $x = \sqrt{3}$ and $x = -\sqrt{3}$. Check these in the original equation.

b. $(x - 3)^2 = 7$ Write original equation.

$x - 3 = \pm\sqrt{7}$ Extract square roots.

$x = 3 \pm \sqrt{7}$ Add 3 to each side.

The solutions are $x = 3 \pm \sqrt{7}$. Check these in the original equation.

Completing the Square

The equation in Example 2(b) was given in the form $(x - 3)^2 = 7$ so that you could find the solution by extracting square roots. Suppose, however, that the equation had been given in the general form $x^2 - 6x + 2 = 0$. Because this equation is equivalent to the original, it has the same two solutions, $x = 3 \pm \sqrt{7}$. However, the left side of the equation is not factorable, and you cannot find its solutions unless you rewrite the equation by **completing the square.**

Completing the Square

To **complete the square** for the expression $x^2 + bx$, add $(b/2)^2$, which is the square of half the coefficient of x. Consequently,

$$x^2 + bx + \left(\frac{b}{2}\right)^2 = \left(x + \frac{b}{2}\right)^2.$$

When solving quadratic equations by completing the square, you must add $(b/2)^2$ to *each side* in order to maintain equality. If the leading coefficient is *not* 1, you must divide each side of the equation by the leading coefficient *before* completing the square, as shown in Example 3.

Example 3 ▶ Completing the Square 

Solve $3x^2 - 4x - 5 = 0$ by completing the square.

Solution

$$3x^2 - 4x - 5 = 0 \qquad \text{Write original equation.}$$

$$3x^2 - 4x = 5 \qquad \text{Add 5 to each side.}$$

$$x^2 - \frac{4}{3}x = \frac{5}{3} \qquad \text{Divide each side by 3.}$$

$$x^2 - \frac{4}{3}x + \left(-\frac{2}{3}\right)^2 = \frac{5}{3} + \left(-\frac{2}{3}\right)^2 \qquad \text{Add } \left(-\tfrac{2}{3}\right)^2 \text{ to each side.}$$

$\left(\text{half of } -\frac{4}{3}\right)^2$

$$x^2 - \frac{4}{3}x + \frac{4}{9} = \frac{19}{9} \qquad \text{Simplify.}$$

$$\left(x - \frac{2}{3}\right)^2 = \frac{19}{9} \qquad \text{Perfect square trinomial.}$$

$$x - \frac{2}{3} = \pm\frac{\sqrt{19}}{3} \qquad \text{Extract square roots.}$$

$$x = \frac{2}{3} \pm \frac{\sqrt{19}}{3} \qquad \text{Add } \tfrac{2}{3} \text{ to each side.}$$

The solutions are $x = \dfrac{2 \pm \sqrt{19}}{3}$. Check these in the original equation.

The Quadratic Formula

Often in mathematics you are taught the long way of solving a problem first. Then, the longer method is used to develop shorter techniques. The long way stresses understanding and the short way stresses efficiency.

For instance, you can think of completing the square as a "long way" of solving a quadratic equation. When you use completing the square to solve a quadratic equation, you must complete the square for *each* equation separately. In the following derivation, you complete the square *once* in a general setting to obtain the **Quadratic Formula**—a shortcut for solving a quadratic equation.

$$ax^2 + bx + c = 0 \qquad \text{Write in general form, } a \neq 0.$$

$$ax^2 + bx = -c \qquad \text{Subtract } c \text{ from each side.}$$

$$x^2 + \frac{b}{a}x = -\frac{c}{a} \qquad \text{Divide each side by } a.$$

$$x^2 + \frac{b}{a}x + \left(\frac{b}{2a}\right)^2 = -\frac{c}{a} + \left(\frac{b}{2a}\right)^2 \qquad \text{Complete the square.}$$

$\left(\text{half of } \frac{b}{a}\right)^2$

$$\left(x + \frac{b}{2a}\right)^2 = \frac{b^2 - 4ac}{4a^2} \qquad \text{Simplify.}$$

$$x + \frac{b}{2a} = \pm\sqrt{\frac{b^2 - 4ac}{4a^2}} \qquad \text{Extract square roots.}$$

$$x = -\frac{b}{2a} \pm \frac{\sqrt{b^2 - 4ac}}{2|a|} \qquad \text{Solutions}$$

Note that because $\pm 2|a|$ represents the same numbers as $\pm 2a$, you can omit the absolute value sign. So, the formula simplifies to

$$x = \frac{-b \pm \sqrt{b^2 - 4ac}}{2a}.$$

STUDY TIP

You can solve every quadratic equation by completing the square or using the Quadratic Formula.

The Quadratic Formula

The solutions of a quadratic equation in the general form

$$ax^2 + bx + c = 0, \qquad a \neq 0$$

are given by the **Quadratic Formula**

$$x = \frac{-b \pm \sqrt{b^2 - 4ac}}{2a}.$$

The Quadratic Formula is one of the most important formulas in algebra. You should learn the verbal statement of the Quadratic Formula:

"Negative b, plus or minus the square root of b squared minus $4ac$, *all* divided by $2a$."

Exploration

From each graph, can you tell whether the discriminant is positive, zero, or negative? Explain your reasoning. Find each discriminant to verify your answers.

a. $x^2 - 2x = 0$

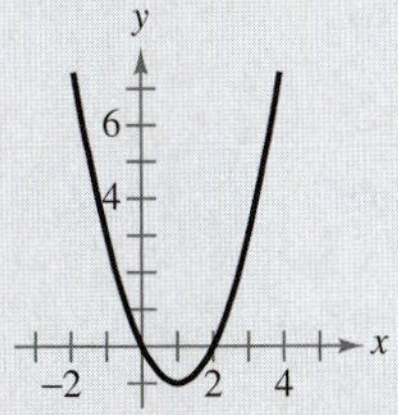

b. $x^2 - 2x + 1 = 0$

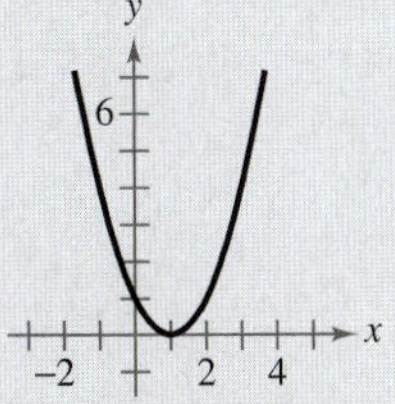

c. $x^2 - 2x + 2 = 0$

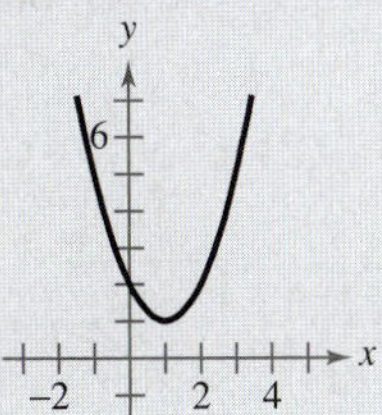

How many solutions would part (c) have if the linear term was $2x$? If the constant was -2?

In the Quadratic Formula, the quantity under the radical sign, $b^2 - 4ac$, is called the **discriminant** of the quadratic expression $ax^2 + bx + c$. It can be used to determine the nature of the solutions of a quadratic equation.

Solutions of a Quadratic Equation

The solutions of a quadratic equation $ax^2 + bx + c = 0$, $a \neq 0$, can be classified as follows. If the discriminant $b^2 - 4ac$ is

1. *positive*, then the quadratic equation has *two* distinct real solutions and its graph has *two* x-intercepts.
2. *zero*, then the quadratic equation has *one* repeated real solution and its graph has *one* x-intercept.
3. *negative*, then the quadratic equation has *no* real solutions and its graph has *no* x-intercepts.

If the discriminant of a quadratic equation is negative, as in case 3 above, then its square root is imaginary (not a real number) and the Quadratic Formula yields two complex solutions. You will study complex solutions in Section 1.5.

When using the Quadratic Formula, remember that *before* the formula can be applied, you must first write the quadratic equation in general form.

Example 4 ▶ The Quadratic Formula: Two Distinct Solutions

Use the Quadratic Formula to solve $x^2 + 3x = 9$.

Solution

The general form of the equation is $x^2 + 3x - 9 = 0$. The discriminant is $b^2 - 4ac = 9 + 36 = 45$, which is positive. So, the equation has two real solutions. You can solve the equation as follows.

$$x^2 + 3x - 9 = 0 \qquad \text{General form}$$

$$x = \frac{-b \pm \sqrt{b^2 - 4ac}}{2a} \qquad \text{Quadratic Formula}$$

$$x = \frac{-3 \pm \sqrt{(3)^2 - 4(1)(-9)}}{2(1)} \qquad \text{Substitute } a = 1, b = 3, \text{ and } c = -9.$$

$$x = \frac{-3 \pm \sqrt{45}}{2} \qquad \text{Simplify.}$$

$$x = \frac{-3 \pm 3\sqrt{5}}{2} \qquad \text{Simplify.}$$

The two solutions are:

$$x = \frac{-3 + 3\sqrt{5}}{2} \quad \text{and} \quad x = \frac{-3 - 3\sqrt{5}}{2}.$$

Check these in the original equation.

Applications

Quadratic equations often occur in problems dealing with area. Here is a simple example. "A square room has an area of 144 square feet. Find the dimensions of the room." To solve this problem, let x represent the length of each side of the room. Then, by solving the equation

$$x^2 = 144$$

you can conclude that each side of the room is 12 feet long. Note that although the equation $x^2 = 144$ has two solutions, $x = -12$ and $x = 12$, the negative solution does not make sense in the context of the problem, so you choose the positive solution.

Example 5 ▶ Finding the Dimensions of a Room  

A bedroom is 3 feet longer than it is wide (see Figure 1.19) and has an area of 154 square feet. Find the dimensions of the room.

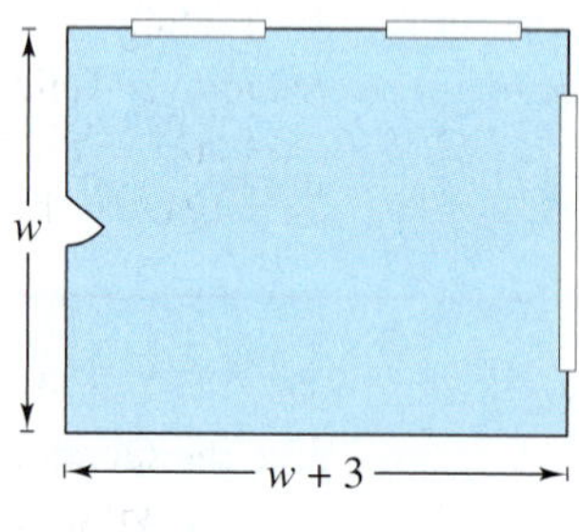

FIGURE 1.19

Solution

Verbal Model:	Width of room	·	Length of room	=	Area of room

Labels: Width of room $= w$ (feet)
Length of room $= w + 3$ (feet)
Area of room $= 154$ (square feet)

Equation:

$$w(w + 3) = 154$$

$$w^2 + 3w - 154 = 0$$

$$(w - 11)(w + 14) = 0$$

$$w - 11 = 0 \quad \Rightarrow \quad w = 11$$

$$w + 14 = 0 \quad \Rightarrow \quad w = -14$$

Choosing the positive value, you find that the width is 11 feet and the length is $w + 3$, or 14 feet. You can check this solution by observing that the length is 3 feet longer than the width *and* that the product of the length and width is 154 square feet.

Another common application of quadratic equations involves an object that is falling (or projected into the air). The general equation that gives the height of such an object is called a **position equation,** and on the *Earth's* surface it has the form

$$s = -16t^2 + v_0 t + s_0.$$

In this equation, s represents the height of the object (in feet), v_0 represents the initial velocity of the object (in feet per second), s_0 represents the initial height of the object (in feet), and t represents the time (in seconds).

Example 6 ▶ Falling Time  

A construction worker on the 24th floor of a building (see Figure 1.20) accidentally drops a wrench and yells "Look out below!" Could a person at ground level hear this warning in time to get out of the way?

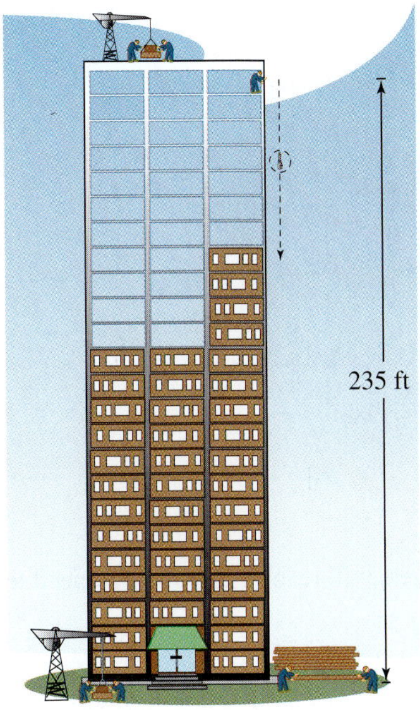

235 ft

FIGURE 1.20

Solution

Assume that each floor of the building is 10 feet high, so that the wrench is dropped from a height of 235 feet (the construction worker's hand is 5 feet below the ceiling of the 24th floor). Because sound travels at about 1100 feet per second, it follows that a person at ground level hears the warning within 1 second of the time the wrench is dropped. To set up a mathematical model for the height of the wrench, use the position equation

$$s = -16t^2 + v_0 t + s_0.$$

Because the object is dropped rather than thrown, the initial velocity is $v_0 = 0$ feet per second. Moreover, because the initial height is $s_0 = 235$ feet, you have the following model.

$$s = -16t^2 + (0)t + 235 = -16t^2 + 235$$

After falling for 1 second, the height of the wrench is $-16(1)^2 + 235 = 219$ feet. After falling for 2 seconds, the height of the wrench is $-16(2)^2 + 235 = 171$ feet. To find the number of seconds it takes the wrench to hit the ground, let the height s be zero and solve the equation for t.

$$s = -16t^2 + 235$$ Write position equation.

$$0 = -16t^2 + 235$$ Substitute 0 for height.

$$16t^2 = 235$$ Add $16t^2$ to each side.

$$t^2 = \frac{235}{16}$$ Divide each side by 16.

$$t = \frac{\sqrt{235}}{4}$$ Extract positive square root.

$$t \approx 3.83$$ Use a calculator.

The wrench will take about 3.83 seconds to hit the ground. If the person hears the warning 1 second after the wrench is dropped, the person still has almost 3 seconds to get out of the way.

STUDY TIP

The position equation used in Example 6 ignores air resistance. This implies that it is appropriate to use the position equation only to model falling objects that have little air resistance and that fall over short distances.

Example 7 ▶ Quadratic Modeling: Number of Lawyers  

From 1983 to 1999, the number of lawyers L (in thousands) in the United States can be modeled by the quadratic equation

$$L = 0.008t^2 + 19.59t + 552.8, \quad 3 \le t \le 19$$

where t is the time in years, with $t = 3$ corresponding to 1983. The number of lawyers is shown graphically in Figure 1.21. According to this model, in which year will the number of lawyers reach or surpass 1 million? (Source: U.S. Bureau of Labor Statistics)

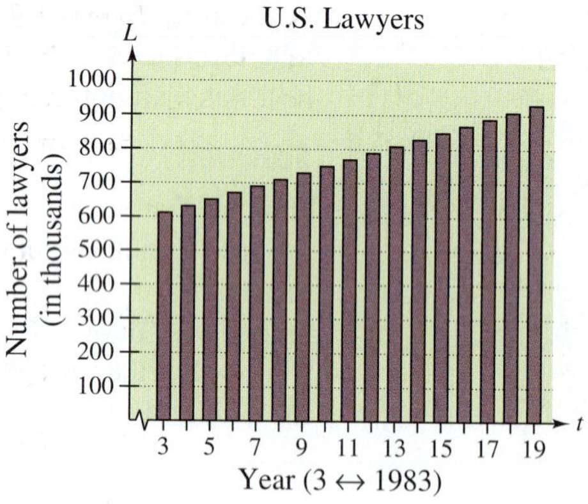

FIGURE 1.21

Solution

To find the year in which the number of lawyers will reach 1 million, you need to solve the equation

$$0.008t^2 + 19.59t + 552.8 = 1000.$$

To begin, write the equation in general form.

$$0.008t^2 + 19.59t - 447.2 = 0$$

Then apply the Quadratic Formula.

$$t = \frac{-19.59 \pm \sqrt{(19.59)^2 - 4(0.008)(-447.2)}}{2(0.008)}$$

Choosing the positive solution, you find that

$$t = \frac{-19.59 + \sqrt{(19.59)^2 - 4(0.008)(-447.2)}}{2(0.008)}$$

$$= \frac{-19.59 + \sqrt{398.08}}{0.016}$$

$$\approx 22.62.$$

Because $t = 3$ corresponds to 1983, it follows that $t \approx 22.62$ must correspond to 2002. So, the number of lawyers should have reached 1 million during the year 2002.

A fourth type of application that often involves a quadratic equation is one dealing with the hypotenuse of a right triangle. These types of applications often use the Pythagorean Theorem, which states that

$$a^2 + b^2 = c^2 \qquad \text{Pythagorean Theorem}$$

where a and b are the legs of a right triangle and c is the hypotenuse.

Example 8 ▶ An Application Involving the Pythagorean Theorem

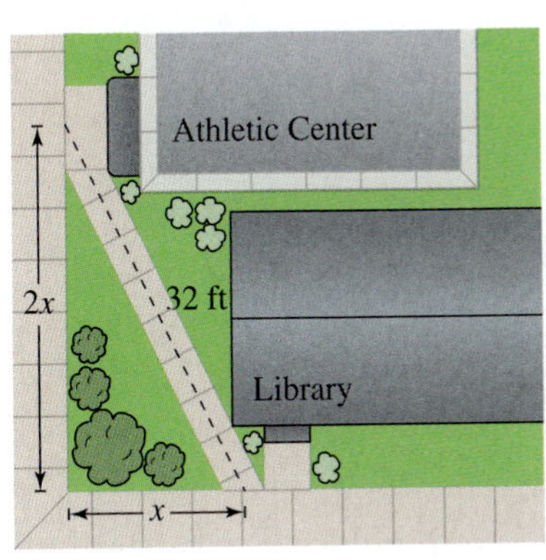

FIGURE 1.22

An L-shaped sidewalk from the athletic center to the library on a college campus is shown in Figure 1.22. The sidewalk was constructed so that the length of one sidewalk forming the L is twice as long as the other. The length of the diagonal sidewalk that cuts across the grounds between the two buildings is 32 feet. How many feet does a person save by walking on the diagonal sidewalk?

Solution

Using the Pythagorean Theorem, you have

$$x^2 + (2x)^2 = 32^2 \qquad \text{Pythagorean Theorem}$$

$$5x^2 = 1024 \qquad \text{Combine like terms.}$$

$$x^2 = 204.8 \qquad \text{Divide each side by 5.}$$

$$x = \sqrt{204.8} \qquad \text{Extract positive square root.}$$

The total distance covered by walking on the L-shaped sidewalk is

$$\begin{aligned} x + 2x &= 3x \\ &= 3\sqrt{204.8} \\ &\approx 42.9 \text{ feet.} \end{aligned}$$

Walking on the diagonal sidewalk saves a person about $42.9 - 32 = 10.9$ feet.

Writing ABOUT MATHEMATICS

Comparing Solution Methods In this section, you studied four algebraic methods for solving quadratic equations. Solve each of the quadratic equations below in several different ways. Write a short paragraph explaining which method(s) you prefer. Does your preferred method depend on the equation?

a. $x^2 - 4x - 5 = 0$

b. $x^2 - 4x = 0$

c. $x^2 - 4x - 3 = 0$

d. $x^2 - 4x - 6 = 0$

1.4 Exercises

In Exercises 1–6, write the quadratic equation in general form.

1. $2x^2 = 3 - 8x$
2. $x^2 = 16x$
3. $(x - 3)^2 = 3$
4. $13 - 3(x + 7)^2 = 0$
5. $\frac{1}{5}(3x^2 - 10) = 18x$
6. $x(x + 2) = 5x^2 + 1$

In Exercises 7–20, solve the quadratic equation by factoring.

7. $6x^2 + 3x = 0$
8. $9x^2 - 1 = 0$
9. $x^2 - 2x - 8 = 0$
10. $x^2 - 10x + 9 = 0$
11. $x^2 + 10x + 25 = 0$
12. $4x^2 + 12x + 9 = 0$
13. $3 + 5x - 2x^2 = 0$
14. $2x^2 = 19x + 33$
15. $x^2 + 4x = 12$
16. $-x^2 + 8x = 12$
17. $\frac{3}{4}x^2 + 8x + 20 = 0$
18. $\frac{1}{8}x^2 - x - 16 = 0$
19. $x^2 + 2ax + a^2 = 0$
20. $(x + a)^2 - b^2 = 0$

In Exercises 21–34, solve the equation by extracting square roots. List both the exact solution and the decimal solution rounded to two decimal places.

21. $x^2 = 49$
22. $x^2 = 169$
23. $x^2 = 11$
24. $x^2 = 32$
25. $3x^2 = 81$
26. $9x^2 = 36$
27. $(x - 12)^2 = 16$
28. $(x + 13)^2 = 25$
29. $(x + 2)^2 = 14$
30. $(x - 5)^2 = 30$
31. $(2x - 1)^2 = 18$
32. $(4x + 7)^2 = 44$
33. $(x - 7)^2 = (x + 3)^2$
34. $(x + 5)^2 = (x + 4)^2$

In Exercises 35–44, solve the quadratic equation by completing the square.

35. $x^2 - 2x = 0$
36. $x^2 + 4x = 0$
37. $x^2 + 4x - 32 = 0$
38. $x^2 - 2x - 3 = 0$
39. $x^2 + 6x + 2 = 0$
40. $x^2 + 8x + 14 = 0$
41. $9x^2 - 18x = -3$
42. $9x^2 - 12x = 14$
43. $8 + 4x - x^2 = 0$
44. $4x^2 - 4x - 99 = 0$

In Exercises 45–50, rewrite the quadratic portion of the algebraic expression as the sum or difference of two squares by completing the square.

45. $\dfrac{1}{x^2 + 2x + 5}$
46. $\dfrac{1}{x^2 - 12x + 19}$
47. $\dfrac{4}{x^2 + 4x - 3}$
48. $\dfrac{5}{x^2 + 25x + 11}$
49. $\dfrac{1}{\sqrt{6x - x^2}}$
50. $\dfrac{1}{\sqrt{16 - 6x - x^2}}$

Graphical Analysis In Exercises 51–58, use a graphing utility to graph the equation. Use the graph to approximate any x-intercepts of the graph. Set $y = 0$ and solve the resulting equation. Compare the result with the x-intercepts of the graph.

51. $y = (x + 3)^2 - 4$
52. $y = (x - 4)^2 - 1$
53. $y = 1 - (x - 2)^2$
54. $y = 9 - (x - 8)^2$
55. $y = -4x^2 + 4x + 3$
56. $y = 4x^2 - 1$
57. $y = x^2 + 3x - 4$
58. $y = x^2 - 5x - 24$

In Exercises 59–66, use the discriminant to determine the number of real solutions of the quadratic equation.

59. $2x^2 - 5x + 5 = 0$
60. $-5x^2 - 4x + 1 = 0$
61. $2x^2 - x - 1 = 0$
62. $x^2 - 4x + 4 = 0$
63. $\frac{1}{3}x^2 - 5x + 25 = 0$
64. $\frac{4}{7}x^2 - 8x + 28 = 0$
65. $0.2x^2 + 1.2x - 8 = 0$
66. $9 + 2.4x - 8.3x^2 = 0$

In Exercises 67–90, use the Quadratic Formula to solve the equation.

67. $2x^2 + x - 1 = 0$
68. $2x^2 - x - 1 = 0$
69. $16x^2 + 8x - 3 = 0$
70. $25x^2 - 20x + 3 = 0$
71. $2 + 2x - x^2 = 0$
72. $x^2 - 10x + 22 = 0$
73. $x^2 + 14x + 44 = 0$
74. $6x = 4 - x^2$
75. $x^2 + 8x - 4 = 0$
76. $4x^2 - 4x - 4 = 0$
77. $12x - 9x^2 = -3$
78. $16x^2 + 22 = 40x$
79. $9x^2 + 24x + 16 = 0$
80. $36x^2 + 24x - 7 = 0$
81. $4x^2 + 4x = 7$
82. $16x^2 - 40x + 5 = 0$
83. $28x - 49x^2 = 4$
84. $3x + x^2 - 1 = 0$
85. $8t = 5 + 2t^2$
86. $25h^2 + 80h + 61 = 0$

87. $(y - 5)^2 = 2y$

88. $(z + 6)^2 = -2z$

89. $\frac{1}{2}x^2 + \frac{3}{8}x = 2$

90. $\left(\frac{5}{7}x - 14\right)^2 = 8x$

In Exercises 91–98, use the Quadratic Formula to solve the equation. (Round your answer to three decimal places.)

91. $5.1x^2 - 1.7x - 3.2 = 0$

92. $2x^2 - 2.50x - 0.42 = 0$

93. $-0.067x^2 - 0.852x + 1.277 = 0$

94. $-0.005x^2 + 0.101x - 0.193 = 0$

95. $422x^2 - 506x - 347 = 0$

96. $1100x^2 + 326x - 715 = 0$

97. $12.67x^2 + 31.55x + 8.09 = 0$

98. $-3.22x^2 - 0.08x + 28.651 = 0$

In Exercises 99–108, solve the equation using any convenient method.

99. $x^2 - 2x - 1 = 0$

100. $11x^2 + 33x = 0$

101. $(x + 3)^2 = 81$

102. $x^2 - 14x + 49 = 0$

103. $x^2 - x - \frac{11}{4} = 0$

104. $x^2 + 3x - \frac{3}{4} = 0$

105. $(x + 1)^2 = x^2$

106. $a^2x^2 - b^2 = 0$

107. $3x + 4 = 2x^2 - 7$

108. $4x^2 + 2x + 4 = 2x + 8$

109. ***Floor Space*** The floor of a one-story building is 14 feet longer than it is wide. The building has 1632 square feet of floor space.

(a) Draw a diagram that gives a visual representation of the floor space. Represent the width as w and show the length in terms of w.

(b) Write a quadratic equation in terms of w.

(c) Find the length and width of the floor of the building.

110. ***Dimensions of a Corral*** A rancher has 100 meters of fencing to enclose two adjacent rectangular corrals (see figure). The rancher wants the enclosed area to be 350 square meters. What dimensions should the rancher use to obtain this area?

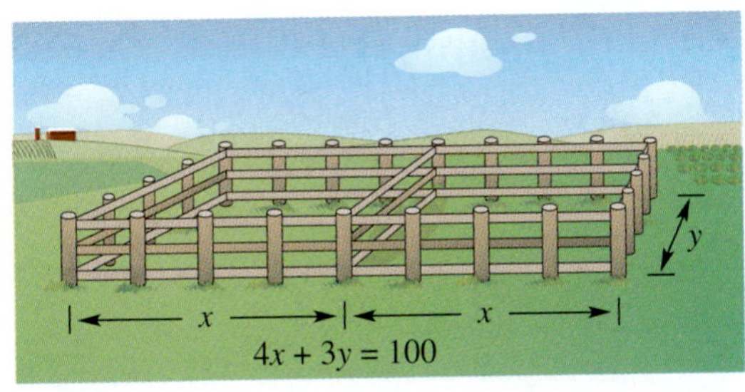

111. ***Packaging*** An open box with a square base (see figure) is to be constructed from 84 square inches of material. The height of the box is 2 inches. What are the dimensions of the box? (*Hint:* The surface area is $S = x^2 + 4xh$.)

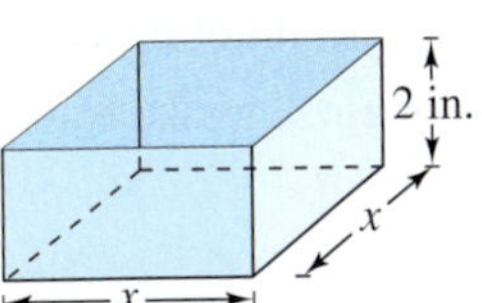

112. ***Packaging*** An open box is to be made from a square piece of material by cutting two-centimeter squares from the corners and turning up the sides (see figure). The volume of the finished box is to be 200 cubic centimeters. Find the size of the original piece of material.

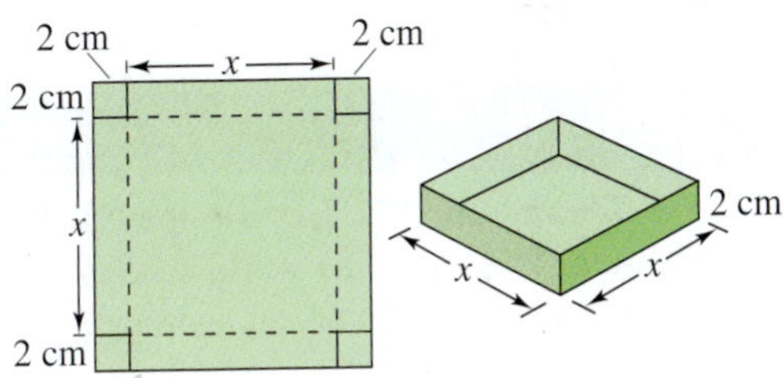

113. ***Mowing the Lawn*** Two landscapers must mow a rectangular lawn that measures 100 feet by 200 feet. Each wants to mow no more than half of the lawn. The first starts by mowing around the outside of the lawn. How wide a strip must the first landscaper mow on each of the four sides in order to mow no more than half of the lawn? The mower has a 24-inch cut. Approximate the required number of trips around the lawn.

114. ***Seating*** A rectangular classroom seats 72 students. If the seats were rearranged with three more seats in each row, the classroom would have two fewer rows. Find the original number of seats in each row.

In Exercises 115–118, use the position equation given in Example 6 as the model for the problem.

115. ***Military*** A B-52 stratofortress flying at 32,000 feet over level terrain drops a 500-pound bomb.

(a) How long will it take until the bomb strikes the ground?

(b) The plane is flying at 600 miles per hour. How far will the bomb travel horizontally during its descent?

116. ***Eiffel Tower*** You drop a coin from the top of the Eiffel Tower in Paris. The building has a height of 984 feet.

(a) Use the position equation to write a mathematical model for the height of the coin.

(b) Find the height of the coin after 4 seconds.

(c) How long will it take before the coin strikes the ground?

117. ***Sports*** You throw a baseball straight up into the air at a velocity of 45 feet per second. You release the baseball at a height of 5.5 feet and catch it when it falls back to a height of 6 feet.

(a) Use the position equation to write a mathematical model for the height of the baseball.

(b) What is the height of the baseball after 0.5 second?

(c) How many seconds is the baseball in the air?

▶ Model It

118. ***CN Tower*** At 1815 feet tall, the CN Tower in Toronto, Ontario is the world's tallest self-supporting structure. An object is dropped from the top of the tower.

(a) Use the position equation to write a model for the height of the object.

(b) Complete the table.

Time, t	0	2	4	6	8	10	12
Height, s							

(c) From the table in part (b), determine the time interval during which the object reaches the ground. Numerically approximate the time it takes the object to reach the ground.

(d) Find the time it takes the object to reach the ground algebraically. How close was your numerical approximation?

(e) Use a graphing utility with the appropriate viewing window to verify your answer to parts (c) and (d).

119. ***Geometry*** The hypotenuse of an isosceles right triangle is 5 centimeters long. How long are its sides?

120. ***Geometry*** An equilateral triangle has a height of 10 inches. How long is one of its sides? (*Hint:* Use the height of the triangle to partition the triangle into two congruent right triangles.)

121. ***Flying Speed*** Two planes leave simultaneously from Chicago's O'Hare Airport, one flying due north and the other due east (see figure). The northbound plane is flying 50 miles per hour faster than the eastbound plane. After 3 hours, the planes are 2440 miles apart. Find the speed of each plane.

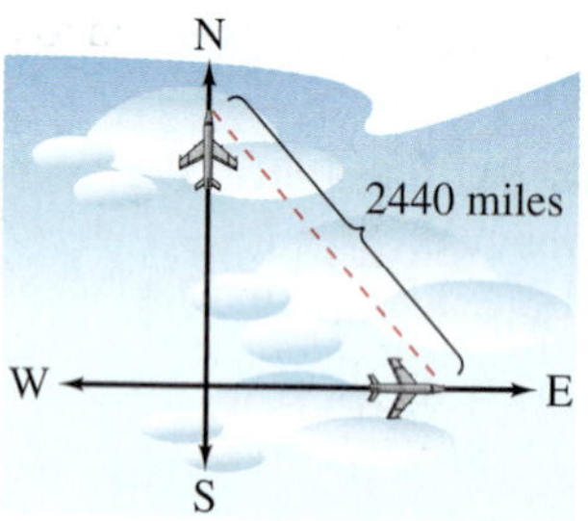

122. ***Boating*** A winch is used to tow a boat to a dock. The rope is attached to the boat at a point 15 feet below the level of the winch (see figure).

(a) Use the Pythagorean Theorem to write an equation giving the relationship between l and x.

(b) Find the distance from the boat to the dock when there is 75 feet of rope out.

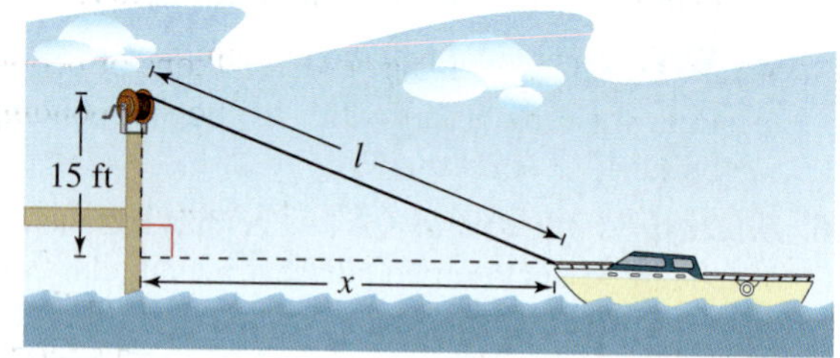

123. *Demand* The demand equation for a product is $p = 20 - 0.0002x$, where p is the price per unit and x is the number of units sold. The total revenue for selling x units is

$$\text{Revenue} = xp = x(20 - 0.0002x).$$

How many units must be sold to produce a revenue of $500,000?

124. *Demand* The demand equation for a product is $p = 60 - 0.0004x$, where p is the price per unit and x is the number of units sold. The total revenue for selling x units is

$$\text{Revenue} = xp = x(60 - 0.0004x).$$

How many units must be sold to produce a revenue of $220,000?

Cost **In Exercises 125–128, use the cost equation to find the number of units x that a manufacturer can produce for the given cost C. Round your answer to the nearest positive integer.**

125. $C = 0.125x^2 + 20x + 500$ $\quad C = \$14{,}000$

126. $C = 0.5x^2 + 15x + 5000$ $\quad C = \$11{,}500$

127. $C = 800 + 0.04x + 0.002x^2$ $\quad C = \$1680$

128. $C = 800 - 10x + \dfrac{x^2}{4}$ $\quad C = \$896$

129. *Population Statistics* The population of the United States from 1800 to 1890 can be approximated by the model

$$\text{Population} = 0.68522x^2 + 0.0871x + 6.047$$

where the population is given in millions of people and t represents time, with $t = 0$ corresponding to 1800, $t = 1$ corresponding to 1810, and so on. If this model had continued to be valid up through the present time, when would the population of the United States have reached 250 million? Judging from the graph, would you say this model was a good representation of the population through 1890? (Source: U.S. Bureau of the Census)

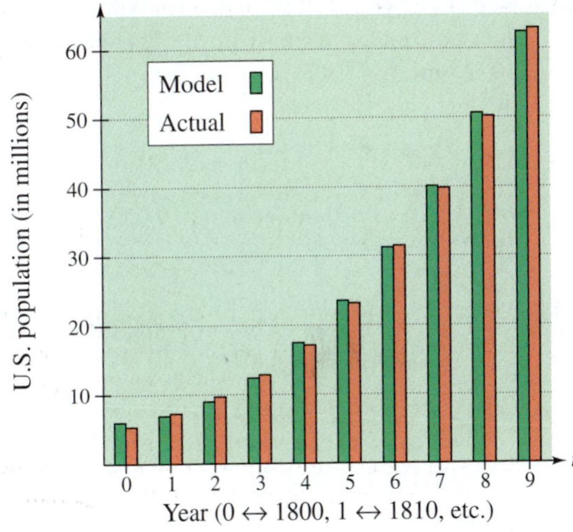

FIGURE FOR 129

130. *Biology* The metabolic rate of an ectothermic organism increases with increasing temperature within a certain range. The graph shows experimental data for the oxygen consumption C (in microliters per gram per hour) of a beetle at certain temperatures. This data can be approximated by the model

$$C = 0.45x^2 - 1.65x + 50.75, \qquad 10 \le x \le 25$$

where x is the air temperature in degrees Celsius.

(a) The oxygen consumption is 150 microliters per gram per hour. What is the air temperature?

(b) The temperature is increased from 10°C to 20°C. The oxygen consumption is increased by approximately what factor?

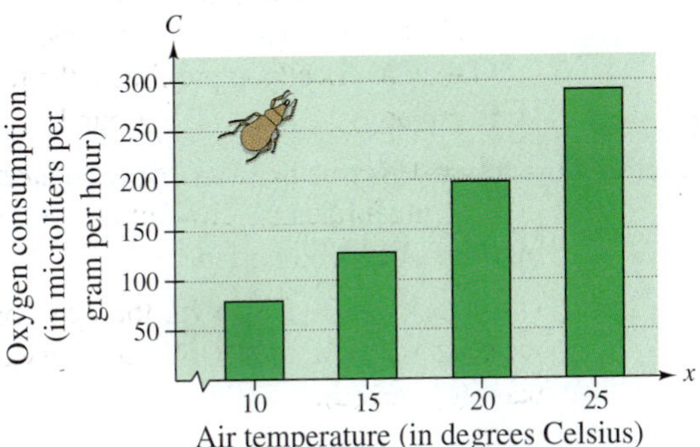

131. ***Boating*** The total amount S (in billions of dollars) spent on pleasure boats in the United States from 1992 through 1999 can be approximated by the model

$$S = -0.0959t^2 + 1.770t + 2.29$$

where t is the time, with $t = 2$ corresponding to 1992. (Source: National Sporting Goods Association)

(a) Use a graphing utility to graph the model over the interval $2 \le t \le 9$.

(b) If the model is used to forecast future sales, will sales ever exceed 12 billion dollars? If so, estimate the year.

132. ***Flying Distance*** A commercial jet flies to three cities whose locations form the vertices of a right triangle (see figure). The total flight distance (from Oklahoma City to Austin to New Orleans and back to Oklahoma City) is approximately 1348 miles. It is 560 miles between Oklahoma City and New Orleans. Approximate the other two distances.

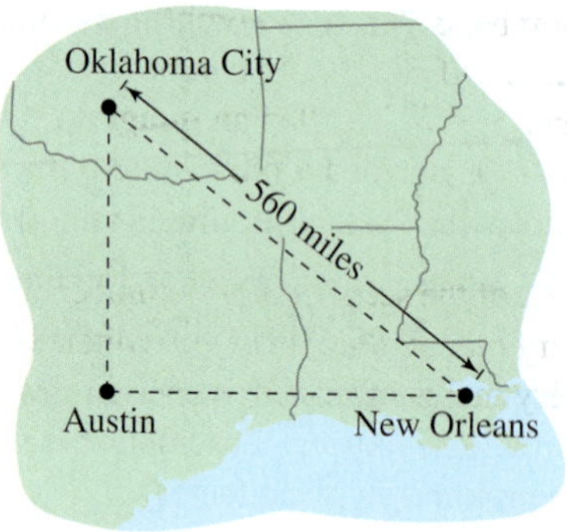

Synthesis

True or False? **In Exercises 133 and 134, determine whether the statement is true or false. Justify your answer.**

133. The quadratic equation $-3x^2 - x = 10$ has two real solutions.

134. If $(2x - 3)(x + 5) = 8$, then either $2x - 3 = 8$ or $x + 5 = 8$.

135. To solve the equation

$$2x^2 + 3x = 15x$$

a student divides each side by x and solves the equation $2x + 3 = 15$. The resulting solution $(x = 6)$ satisfies the original equation. Is there an error? Explain.

136. The graphs show the solutions of equations plotted on the real number line. In each case, determine whether the solution(s) is (are) for a linear equation, a quadratic equation, both, or neither. Explain.

(a) 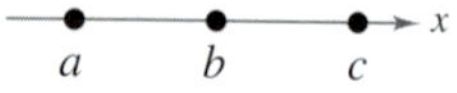

(b) 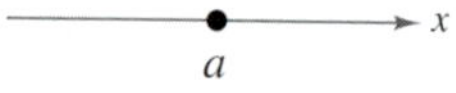

(c) 

(d) 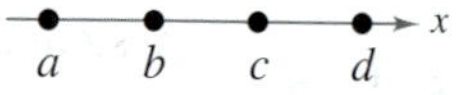

137. Solve $3(x + 4)^2 + (x + 4) - 2 = 0$ in two ways.

(a) Let $u = x + 4$, and solve the resulting equation for u. Then solve the u-solution for x.

(b) Expand and collect like terms in the equation, and solve the resulting equation for x.

(c) Which method is easier? Explain.

138. Solve the equations, given that a and b are not zero.

(a) $ax^2 + bx = 0$

(b) $ax^2 - ax = 0$

Think About It **In Exercises 139–142, write a quadratic equation that has the given solutions. (There are many correct answers.)**

139. -3 and 6

140. -4 and -11

141. 8 and 14

142. $\frac{1}{6}$ and $-\frac{2}{5}$

Review

In Exercises 143–146, identify the rule of algebra being demonstrated.

143. $(10x)y = 10(xy)$

144. $-4(x - 3) = -4x + 12$

145. $7x^4 + (-7x^4) = 0$

146. $(x + 4) + x^3 = x + (4 + x^3)$

In Exercises 147–150, find the product.

147. $(x + 3)(x - 6)$

148. $(x - 8)(x - 1)$

149. $(x + 4)(x^2 - x + 2)$

150. $(x + 9)(x^2 - 6x + 4)$

1.5 Complex Numbers

▶ What you should learn

- How to use the imaginary unit i to write complex numbers
- How to add, subtract, and multiply complex numbers
- How to use complex conjugates to write the quotient of two complex numbers in standard form
- How to find complex solutions of quadratic equations

▶ Why you should learn it

You can use complex numbers to model and solve real-life problems in electronics. For instance, in Exercise 84 on page 129, you will learn how to use complex numbers to find the impedance of an electrical circuit.

Chris Hamilton/Corbis

The Imaginary Unit *i* 

In Section 1.4, you learned that some quadratic equations have no real solutions. For instance, the quadratic equation

$$x^2 + 1 = 0 \qquad \text{Equation with no real solution}$$

has no real solution because there is no real number x that can be squared to produce -1. To overcome this deficiency, mathematicians created an expanded system of numbers using the **imaginary unit *i*,** defined as

$$i = \sqrt{-1} \qquad \text{Imaginary unit}$$

where $i^2 = -1$. By adding real numbers to real multiples of this imaginary unit, the set of **complex numbers** is obtained. Each complex number can be written in the **standard form $a + bi$.** The real number a is called the **real part** of the **complex number $a + bi$,** and the number bi (where b is a real number) is called the **imaginary part** of the complex number.

Definition of a Complex Number

If a and b are real numbers, the number $a + bi$ is a **complex number,** and it is said to be written in **standard form.** If $b = 0$, the number $a + bi = a$ is a real number. If $b \neq 0$, the number $a + bi$ is called an **imaginary number.** A number of the form bi, where $b \neq 0$, is called a **pure imaginary number.**

The set of real numbers is a subset of the set of complex numbers, as shown in Figure 1.23. This is true because every real number a can be written as a complex number using $b = 0$. That is, for every real number a, you can write $a = a + 0i$.

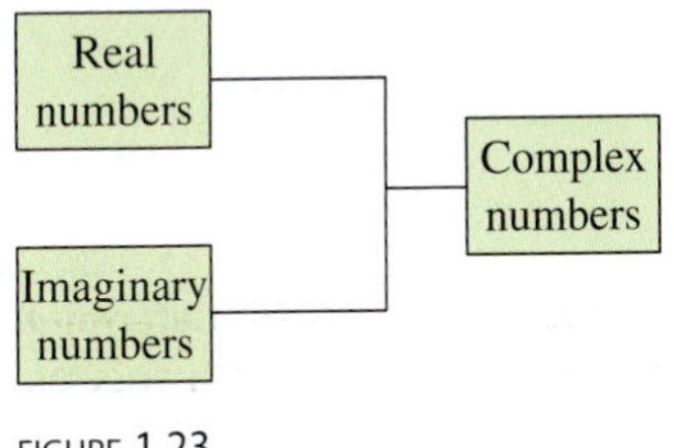

FIGURE 1.23

Equality of Complex Numbers

Two complex numbers $a + bi$ and $c + di$, written in standard form, are equal to each other

$$a + bi = c + di \qquad \text{Equality of two complex numbers}$$

if and only if $a = c$ and $b = d$.

Operations with Complex Numbers

To add (or subtract) two complex numbers, you add (or subtract) the real and imaginary parts of the numbers separately.

Addition and Subtraction of Complex Numbers

If $a + bi$ and $c + di$ are two complex numbers written in standard form, their sum and difference are defined as follows.

Sum: $(a + bi) + (c + di) = (a + c) + (b + d)i$

Difference: $(a + bi) - (c + di) = (a - c) + (b - d)i$

The **additive identity** in the complex number system is zero (the same as in the real number system). Furthermore, the **additive inverse** of the complex number $a + bi$ is

$-(a + bi) = -a - bi.$ Additive inverse

So, you have

$(a + bi) + (-a - bi) = 0 + 0i = 0.$

Example 1 ▶ **Adding and Subtracting Complex Numbers** 

a. $(4 + 7i) + (1 - 6i) = 4 + 7i + 1 - 6i$ Remove parentheses.

$= (4 + 1) + (7i - 6i)$ Group like terms.

$= 5 + i$ Write in standard form.

b. $(1 + 2i) - (4 + 2i) = 1 + 2i - 4 - 2i$ Remove parentheses.

$= (1 - 4) + (2i - 2i)$ Group like terms.

$= -3 + 0$ Simplify.

$= -3$ Write in standard form.

c. $3i - (-2 + 3i) - (2 + 5i) = 3i + 2 - 3i - 2 - 5i$

$= (2 - 2) + (3i - 3i - 5i)$

$= 0 - 5i$

$= -5i$

d. $(3 + 2i) + (4 - i) - (7 + i) = 3 + 2i + 4 - i - 7 - i$

$= (3 + 4 - 7) + (2i - i - i)$

$= 0 + 0i$

$= 0$

Note in Examples 1(b) and 1(d) that the sum of two complex numbers can be a real number.

Exploration

Complete the following.

$i^1 = i$	$i^7 =$ ▢
$i^2 = -1$	$i^8 =$ ▢
$i^3 = -i$	$i^9 =$ ▢
$i^4 = 1$	$i^{10} =$ ▢
$i^5 =$ ▢	$i^{11} =$ ▢
$i^6 =$ ▢	$i^{12} =$ ▢

What pattern do you see? Write a brief description of how you would find i raised to any positive integer power.

Many of the properties of real numbers are valid for complex numbers as well. Here are some examples.

Associative Properties of Addition and Multiplication

Commutative Properties of Addition and Multiplication

Distributive Property of Multiplication Over Addition

Notice below how these properties are used when two complex numbers are multiplied.

$$(a + bi)(c + di) = a(c + di) + bi(c + di) \quad \text{Distributive Property}$$
$$= ac + (ad)i + (bc)i + (bd)i^2 \quad \text{Distributive Property}$$
$$= ac + (ad)i + (bc)i + (bd)(-1) \quad i^2 = -1$$
$$= ac - bd + (ad)i + (bc)i \quad \text{Commutative Property}$$
$$= (ac - bd) + (ad + bc)i \quad \text{Associative Property}$$

Rather than trying to memorize this multiplication rule, you should simply remember how the Distributive Property is used to multiply two complex numbers.

Example 2 ▶ Multiplying Complex Numbers 

a. $4(-2 + 3i) = 4(-2) + 4(3i)$ Distributive Property
$= -8 + 12i$ Simplify.

b. $(2 - i)(4 + 3i) = 2(4 + 3i) - i(4 + 3i)$ Distributive Property
$= 8 + 6i - 4i - 3i^2$ Distributive Property
$= 8 + 6i - 4i - 3(-1)$ $i^2 = -1$
$= (8 + 3) + (6i - 4i)$ Group like terms.
$= 11 + 2i$ Write in standard form.

c. $(3 + 2i)(3 - 2i) = 3(3 - 2i) + 2i(3 - 2i)$ Distributive Property
$= 9 - 6i + 6i - 4i^2$ Distributive Property
$= 9 - 6i + 6i - 4(-1)$ $i^2 = -1$
$= 9 + 4$ Simplify.
$= 13$ Write in standard form.

d. $(3 + 2i)^2 = (3 + 2i)(3 + 2i)$ Square of a binomial
$= 3(3 + 2i) + 2i(3 + 2i)$ Distributive Property
$= 9 + 6i + 6i + 4i^2$ Distributive Property
$= 9 + 6i + 6i + 4(-1)$ $i^2 = -1$
$= 9 + 12i - 4$ Simplify.
$= 5 + 12i$ Write in standard form.

Complex Conjugates

Notice in Example 2(c) that the product of two complex numbers can be a real number. This occurs with pairs of complex numbers of the form $a + bi$ and $a - bi$, called **complex conjugates.**

$$\begin{aligned}(a + bi)(a - bi) &= a^2 - abi + abi - b^2i^2 \\ &= a^2 - b^2(-1) \\ &= a^2 + b^2\end{aligned}$$

Example 3 ▶ Multiplying Conjugates 

Multiply each complex number by its complex conjugate.

a. $1 + i$ **b.** $4 - 3i$

Solution

a. The complex conjugate of $1 + i$ is $1 - i$.

$$(1 + i)(1 - i) = 1^2 - i^2 = 1 - (-1) = 2$$

b. The complex conjugate of $4 - 3i$ is $4 + 3i$.

$$(4 - 3i)(4 + 3i) = 4^2 - (3i)^2 = 16 - 9i^2 = 16 - 9(-1) = 25$$

To write the quotient of $a + bi$ and $c + di$ in standard form, where c and d are not both zero, multiply the numerator and denominator by the complex conjugate of the *denominator* to obtain

$$\begin{aligned}\frac{a + bi}{c + di} &= \frac{a + bi}{c + di}\left(\frac{c - di}{c - di}\right) \\ &= \frac{(ac + bd) + (bc - ad)i}{c^2 + d^2}. \qquad \text{Standard form}\end{aligned}$$

Example 4 ▶ Writing a Quotient of Complex Numbers in Standard Form 

$$\begin{aligned}\frac{2 + 3i}{4 - 2i} &= \frac{2 + 3i}{4 - 2i}\left(\frac{4 + 2i}{4 + 2i}\right) && \text{Multiply numerator and denominator by complex conjugate of denominator.} \\ &= \frac{8 + 4i + 12i + 6i^2}{16 - 4i^2} && \text{Expand.} \\ &= \frac{8 - 6 + 16i}{16 + 4} && i^2 = -1 \\ &= \frac{2 + 16i}{20} && \text{Simplify.} \\ &= \frac{1}{10} + \frac{4}{5}i && \text{Write in standard form.}\end{aligned}$$

Complex Solutions of Quadratic Equations

When using the Quadratic Formula to solve a quadratic equation, you often obtain a result such as $\sqrt{-3}$, which you know is not a real number. By factoring out $i = \sqrt{-1}$, you can write this number in standard form.

$$\sqrt{-3} = \sqrt{3(-1)} = \sqrt{3}\sqrt{-1} = \sqrt{3}i$$

The number $\sqrt{3}i$ is called the *principal square root* of -3.

STUDY TIP

The definition of principal square root uses the rule

$$\sqrt{ab} = \sqrt{a}\sqrt{b}$$

for $a > 0$ and $b < 0$. This rule is not valid if *both* a and b are negative. For example,

$$\begin{aligned}\sqrt{-5}\sqrt{-5} &= \sqrt{5(-1)}\sqrt{5(-1)}\\ &= \sqrt{5}i\sqrt{5}i\\ &= \sqrt{25}i^2\\ &= 5i^2 = -5\end{aligned}$$

whereas

$$\sqrt{(-5)(-5)} = \sqrt{25} = 5.$$

To avoid problems with square roots of negative numbers, be sure to convert to standard form *before* multiplying.

Principal Square Root of a Negative Number

If a is a positive number, the **principal square root** of the negative number $-a$ is defined as

$$\sqrt{-a} = \sqrt{a}i.$$

Example 5 ▶ Writing Complex Numbers in Standard Form

a. $\sqrt{-3}\sqrt{-12} = \sqrt{3}i\sqrt{12}i = \sqrt{36}i^2 = 6(-1) = -6$

b. $\sqrt{-48} - \sqrt{-27} = \sqrt{48}i - \sqrt{27}i = 4\sqrt{3}i - 3\sqrt{3}i = \sqrt{3}i$

c. $$\begin{aligned}(-1 + \sqrt{-3})^2 &= (-1 + \sqrt{3}i)^2\\ &= (-1)^2 - 2\sqrt{3}i + (\sqrt{3})^2(i^2)\\ &= 1 - 2\sqrt{3}i + 3(-1)\\ &= -2 - 2\sqrt{3}i\end{aligned}$$

Example 6 ▶ Complex Solutions of a Quadratic Equation

Solve (a) $x^2 + 4 = 0$ and (b) $3x^2 - 2x + 5 = 0$.

Solution

a. $x^2 + 4 = 0$ — Write original equation.

$x^2 = -4$ — Subtract 4 from each side.

$x = \pm 2i$ — Extract square roots.

b. $3x^2 - 2x + 5 = 0$ — Write original equation.

$x = \dfrac{-(-2) \pm \sqrt{(-2)^2 - 4(3)(5)}}{2(3)}$ — Quadratic Formula

$= \dfrac{2 \pm \sqrt{-56}}{6}$ — Simplify.

$= \dfrac{2 \pm 2\sqrt{14}i}{6}$ — Write $\sqrt{-56}$ in standard form.

$= \dfrac{1}{3} \pm \dfrac{\sqrt{14}}{3}i$ — Write in standard form.

1.5 Exercises

In Exercises 1–4, find real numbers *a* and *b* such that the equation is true.

1. $a + bi = -10 + 6i$

2. $a + bi = 13 + 4i$

3. $(a - 1) + (b + 3)i = 5 + 8i$

4. $(a + 6) + 2bi = 6 - 5i$

In Exercises 5–16, write the complex number in standard form.

5. $4 + \sqrt{-9}$

6. $3 + \sqrt{-16}$

7. $2 - \sqrt{-27}$

8. $1 + \sqrt{-8}$

9. $\sqrt{-75}$

10. $\sqrt{-4}$

11. 8

12. 45

13. $-6i + i^2$

14. $-4i^2 + 2i$

15. $\sqrt{-0.09}$

16. $\sqrt{-0.0004}$

In Exercises 17–26, perform the addition or subtraction and write the result in standard form.

17. $(5 + i) + (6 - 2i)$

18. $(13 - 2i) + (-5 + 6i)$

19. $(8 - i) - (4 - i)$

20. $(3 + 2i) - (6 + 13i)$

21. $\left(-2 + \sqrt{-8}\right) + \left(5 - \sqrt{-50}\right)$

22. $\left(8 + \sqrt{-18}\right) - \left(4 + 3\sqrt{2}i\right)$

23. $13i - (14 - 7i)$

24. $22 + (-5 + 8i) + 10i$

25. $-\left(\frac{3}{2} + \frac{5}{2}i\right) + \left(\frac{5}{3} + \frac{11}{3}i\right)$

26. $(1.6 + 3.2i) + (-5.8 + 4.3i)$

In Exercises 27–40, perform the operation and write the result in standard form.

27. $\sqrt{-6} \cdot \sqrt{-2}$

28. $\sqrt{-5} \cdot \sqrt{-10}$

29. $\left(\sqrt{-10}\right)^2$

30. $\left(\sqrt{-75}\right)^2$

31. $(1 + i)(3 - 2i)$

32. $(6 - 2i)(2 - 3i)$

33. $6i(5 - 2i)$

34. $-8i(9 + 4i)$

35. $\left(\sqrt{14} + \sqrt{10}i\right)\left(\sqrt{14} - \sqrt{10}i\right)$

36. $\left(3 + \sqrt{-5}\right)\left(7 - \sqrt{-10}\right)$

37. $(4 + 5i)^2$

38. $(2 - 3i)^2$

39. $(2 + 3i)^2 + (2 - 3i)^2$

40. $(1 - 2i)^2 - (1 + 2i)^2$

In Exercises 41–48, write the complex conjugate of the complex number. Then multiply the number by its complex conjugate.

41. $6 + 3i$

42. $7 - 12i$

43. $-1 - \sqrt{5}i$

44. $-3 + \sqrt{2}i$

45. $\sqrt{-20}$

46. $\sqrt{-15}$

47. $\sqrt{8}$

48. $1 + \sqrt{8}$

In Exercises 49–58, write the quotient in standard form.

49. $\dfrac{5}{i}$

50. $-\dfrac{14}{2i}$

51. $\dfrac{2}{4 - 5i}$

52. $\dfrac{5}{1 - i}$

53. $\dfrac{3 + i}{3 - i}$

54. $\dfrac{6 - 7i}{1 - 2i}$

55. $\dfrac{6 - 5i}{i}$

56. $\dfrac{8 + 16i}{2i}$

57. $\dfrac{3i}{(4 - 5i)^2}$

58. $\dfrac{5i}{(2 + 3i)^2}$

In Exercises 59–62, perform the operation and write the result in standard form.

59. $\dfrac{2}{1 + i} - \dfrac{3}{1 - i}$

60. $\dfrac{2i}{2 + i} + \dfrac{5}{2 - i}$

61. $\dfrac{i}{3 - 2i} + \dfrac{2i}{3 + 8i}$

62. $\dfrac{1 + i}{i} - \dfrac{3}{4 - i}$

In Exercises 63–72, use the Quadratic Formula to solve the quadratic equation.

63. $x^2 - 2x + 2 = 0$

64. $x^2 + 6x + 10 = 0$

65. $4x^2 + 16x + 17 = 0$

66. $9x^2 - 6x + 37 = 0$

67. $4x^2 + 16x + 15 = 0$

68. $16t^2 - 4t + 3 = 0$

69. $\frac{3}{2}x^2 - 6x + 9 = 0$

70. $\frac{7}{8}x^2 - \frac{3}{4}x + \frac{5}{16} = 0$

71. $1.4x^2 - 2x - 10 = 0$

72. $4.5x^2 - 3x + 12 = 0$

In Exercises 73–80, simplify the complex number and write it in standard form.

73. $-6i^3 + i^2$

74. $4i^2 - 2i^3$

75. $-5i^5$

76. $(-i)^3$

77. $(\sqrt{-75})^3$

78. $(\sqrt{-2})^6$

79. $\dfrac{1}{i^3}$

80. $\dfrac{1}{(2i)^3}$

81. Cube each complex number.

(a) 2 (b) $-1 + \sqrt{3}i$ (c) $-1 - \sqrt{3}i$

82. Raise each complex number to the fourth power.

(a) 2 (b) -2 (c) $2i$ (d) $-2i$

83. Express each of the powers of i as i, $-i$, 1, or -1.

(a) i^{40} (b) i^{25} (c) i^{50} (d) i^{67}

▶ Model It

84. ***Impedance*** The opposition to current in an electrical circuit is called its impedance. The impedance z in a parallel circuit with two pathways satisfies the equation

$$\frac{1}{z} = \frac{1}{z_1} + \frac{1}{z_2}$$

where z_1 is the impedance (in ohms) of pathway 1 and z_2 is the impedance of pathway 2.

(a) The impedance of each pathway in a parallel circuit is found by adding the impedances of all components in the pathway. Use the table to find z_1 and z_2.

(b) Find the impedance z.

	Resistor	Inductor	Capacitor
Symbol	$a\Omega$	$b\Omega$	$c\Omega$
Impedance	a	bi	$-ci$

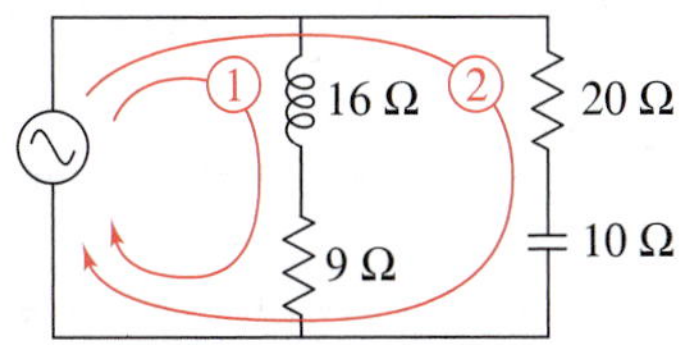

Synthesis

True or False? **In Exercises 85–87, determine whether the statement is true or false. Justify your answer.**

85. There is no complex number that is equal to its complex conjugate.

86. $-i\sqrt{6}$ is a solution of $x^4 - x^2 + 14 = 56$.

87. $i^{44} + i^{150} - i^{74} - i^{109} + i^{61} = -1$

88. ***Error Analysis*** Describe the error.

$$\sqrt{-6}\sqrt{-6} = \sqrt{(-6)(-6)} = \sqrt{36} = 6$$

89. ***Proof*** Prove that the complex conjugate of the product of two complex numbers $a_1 + b_1 i$ and $a_2 + b_2 i$ is the product of their complex conjugates.

90. ***Proof*** Prove that the complex conjugate of the sum of two complex numbers $a_1 + b_1 i$ and $a_2 + b_2 i$ is the sum of their complex conjugates.

Review

In Exercises 91–94, perform the operation and write the result in standard form.

91. $(4 + 3x) + (8 - 6x - x^2)$

92. $(x^3 - 3x^2) - (6 - 2x - 4x^2)$

93. $\left(3x - \frac{1}{2}\right)(x + 4)$

94. $(2x - 5)^2$

In Exercises 95–98, solve the equation and check your solution.

95. $-x - 12 = 19$

96. $8 - 3x = -34$

97. $4(5x - 6) - 3(6x + 1) = 0$

98. $5[x - (3x + 11)] = 20x - 15$

99. ***Volume of an Oblate Spheroid***

Solve for a: $V = \frac{4}{3}\pi a^2 b$

100. ***Newton's Law of Universal Gravitation***

Solve for r: $F = \alpha \dfrac{m_1 m_2}{r^2}$

101. ***Mixture Problem*** A five-liter container contains a mixture with a concentration of 50%. How much of this mixture must be withdrawn and replaced by 100% concentrate to bring the mixture up to 60% concentration?

1.6 Other Types of Equations

▶ **What you should learn**

- How to solve polynomial equations of degree three or greater
- How to solve equations involving radicals
- How to solve equations involving fractions or absolute values
- How to use polynomial equations and equations involving radicals to model and solve real-life problems

▶ **Why you should learn it**

Polynomial equations, radical equations, and absolute value equations can be used to model and solve real-life problems. For instance, in Exercise 102 on page 140, a radical equation can be used to model the total cost of a power line project.

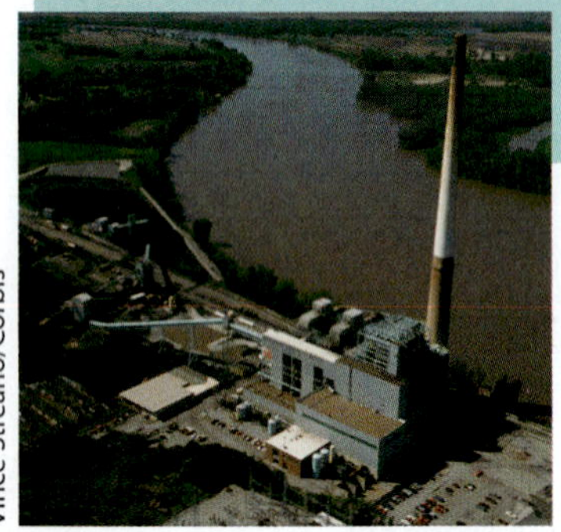

Vince Streano/Corbis

Polynomial Equations

In this section you will extend the techniques for solving equations to nonlinear and nonquadratic equations. At this point in the text, you have only four basic methods for solving nonlinear equations—*factoring*, *extracting square roots*, *completing the square*, and the *Quadratic Formula*. So the main goal of this section is to learn to *rewrite* nonlinear equations in a form to which you can apply one of these methods.

Example 1 shows how to use factoring to solve a **polynomial equation,** which is an equation that can be written in the general form

$$a_nx^n + a_{n-1}x^{n-1} + \cdot \cdot \cdot + a_2x^2 + a_1x + a_0 = 0.$$

Example 1 ▶ **Solving a Polynomial Equation by Factoring** 

Solve $3x^4 = 48x^2$.

Solution

First write the polynomial equation in general form with zero on one side, factor the other side, and then set each factor equal to zero and solve.

$3x^4 = 48x^2$		Write original equation.
$3x^4 - 48x^2 = 0$		Write in general form.
$3x^2(x^2 - 16) = 0$		Factor out common factor.
$3x^2(x + 4)(x - 4) = 0$		Write in factored form.
$3x^2 = 0$	⇨ $x = 0$	Set 1st factor equal to 0.
$x + 4 = 0$	⇨ $x = -4$	Set 2nd factor equal to 0.
$x - 4 = 0$	⇨ $x = 4$	Set 3rd factor equal to 0.

You can check these solutions by substituting in the original equation, as follows.

Check

$3(0)^4 = 48(0)^2$	0 checks. ✓
$3(-4)^4 = 48(-4)^2$	-4 checks. ✓
$3(4)^4 = 48(4)^2$	4 checks. ✓

So, you can conclude that the solutions are $x = 0$, $x = -4$, and $x = 4$.

A common mistake that is made in solving an equation such as that in Example 1 is to divide each side of the equation by the variable factor x^2. This loses the solution $x = 0$. When solving an equation, always write the equation in general form, then factor the equation and set each factor equal to zero. Do not divide each side of an equation by a variable factor in an attempt to simplify the equation.

Technology

You can use a graphing utility to check graphically the solutions of the equation in Example 2. To do this, graph the equation

$$y = x^3 - 3x^2 + 3x - 9.$$

As shown below, the x-intercept of the graph occurs at the *real* zero of the function, $x = 3$, confirming the result found in Example 2.

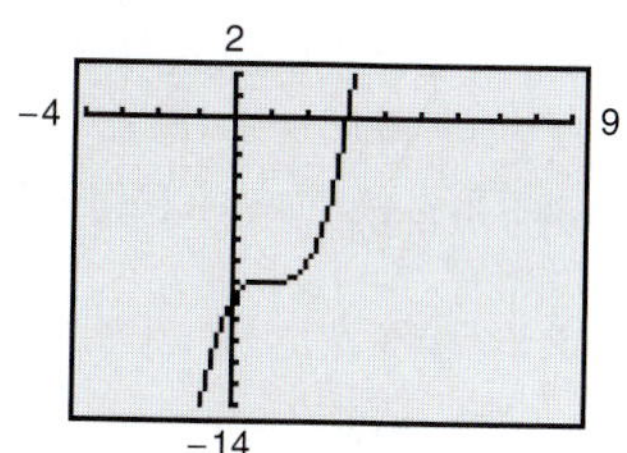

Try using a graphing utility to check the solutions found in Example 3.

For a review of factoring special polynomial forms, see Section P.4.

Example 2 ▶ Solving a Polynomial Equation by Factoring 

Solve $x^3 - 3x^2 + 3x - 9 = 0$.

Solution

$$x^3 - 3x^2 + 3x - 9 = 0 \quad \text{Write original equation.}$$

$$x^2(x - 3) + 3(x - 3) = 0 \quad \text{Factor by grouping.}$$

$$(x - 3)(x^2 + 3) = 0 \quad \text{Distributive Property}$$

$$x - 3 = 0 \Rightarrow x = 3 \quad \text{Set 1st factor equal to 0.}$$

$$x^2 + 3 = 0 \Rightarrow x = \pm\sqrt{3}i \quad \text{Set 2nd factor equal to 0.}$$

The solutions are $x = 3$, $x = \sqrt{3}i$, and $x = -\sqrt{3}i$.

Occasionally, mathematical models involve equations that are of quadratic type. In general, an equation is of quadratic type if it can be written in the form

$$au^2 + bu + c = 0$$

where $a \neq 0$ and u is an algebraic expression.

Example 3 ▶ Solving an Equation of Quadratic Type 

Solve $x^4 - 3x^2 + 2 = 0$.

Solution

This equation is of quadratic type with $u = x^2$.

$$(x^2)^2 - 3(x^2) + 2 = 0$$

To solve this equation, you can factor the left side of the equation as the product of two second-degree polynomials.

$$x^4 - 3x^2 + 2 = 0 \quad \text{Write original equation.}$$

$$\overbrace{(x^2)^2}^{u^2} - \overbrace{3(x^2)}^{3u} + 2 = 0 \quad \text{Quadratic form}$$

$$(x^2 - 1)(x^2 - 2) = 0 \quad \text{Partially factor.}$$

$$(x + 1)(x - 1)(x^2 - 2) = 0 \quad \text{Factor completely.}$$

$$x + 1 = 0 \Rightarrow x = -1 \quad \text{Set 1st factor equal to 0.}$$

$$x - 1 = 0 \Rightarrow x = 1 \quad \text{Set 2nd factor equal to 0.}$$

$$x^2 - 2 = 0 \Rightarrow x = \pm\sqrt{2} \quad \text{Set 3rd factor equal to 0.}$$

The solutions are $x = -1$, $x = 1$, $x = \sqrt{2}$, and $x = -\sqrt{2}$. Check these in the original equation.

Equations Involving Radicals

The steps involved in solving the remaining equations in this section will often introduce *extraneous solutions*, as discussed in Section 1.2. Operations such as squaring each side of an equation, raising each side of an equation to a rational power, and multiplying each side of an equation by a variable quantity all can introduce extraneous solutions. So, when you use any of these operations, checking is crucial.

Example 4 ▶ Solving Equations Involving Radicals 

a. $\sqrt{2x+7} - x = 2$ — Original equation

$\sqrt{2x+7} = x + 2$ — Isolate radical.

$2x + 7 = x^2 + 4x + 4$ — Square each side.

$0 = x^2 + 2x - 3$ — Write in general form.

$0 = (x+3)(x-1)$ — Factor.

$x + 3 = 0 \Rightarrow x = -3$ — Set 1st factor equal to 0.

$x - 1 = 0 \Rightarrow x = 1$ — Set 2nd factor equal to 0.

By checking these values, you can determine that the only solution is $x = 1$.

b. $\sqrt{2x-5} - \sqrt{x-3} = 1$ — Original equation

$\sqrt{2x-5} = \sqrt{x-3} + 1$ — Isolate $\sqrt{2x-5}$.

$2x - 5 = x - 3 + 2\sqrt{x-3} + 1$ — Square each side.

$2x - 5 = x - 2 + 2\sqrt{x-3}$ — Combine like terms.

$x - 3 = 2\sqrt{x-3}$ — Isolate $2\sqrt{x-3}$.

$x^2 - 6x + 9 = 4(x-3)$ — Square each side.

$x^2 - 10x + 21 = 0$ — Write in general form.

$(x-3)(x-7) = 0$ — Factor.

$x - 3 = 0 \Rightarrow x = 3$ — Set 1st factor equal to 0.

$x - 7 = 0 \Rightarrow x = 7$ — Set 2nd factor equal to 0.

The solutions are $x = 3$ and $x = 7$. Check these in the original equation.

> **STUDY TIP**
>
> The essential operations in Example 4 are isolating the square root and squaring each side. In Example 5, this is equivalent to isolating the factor with the rational exponent and raising each side to the *reciprocal power.*

Example 5 ▶ Solving an Equation Involving a Rational Exponent

$(x-4)^{2/3} = 25$ — Original equation

$x - 4 = 25^{3/2}$ — Raise each side to the $\frac{3}{2}$ power.

$x - 4 = 125$ — Simplify.

$x = 129$ — Add 4 to each side.

The solution is $x = 129$. Check this in the original equation.

Equations with Fractions or Absolute Values

To solve an equation involving fractions, multiply each side of the equation by the least common denominator (LCD) of all terms in the equation. This procedure will "clear the equation of fractions." For instance, in the equation

$$\frac{2}{x^2 + 1} + \frac{1}{x} = \frac{2}{x}$$

you can multiply each side of the equation by $x(x^2 + 1)$. Try doing this and solve the resulting equation. You should obtain one solution: $x = 1$.

Example 6 ▶ Solving an Equation Involving Fractions 

Solve $\frac{2}{x} = \frac{3}{x - 2} - 1$.

Solution

For this equation, the least common denominator of the three terms is $x(x - 2)$, so you begin by multiplying each term of the equation by this expression.

$$\frac{2}{x} = \frac{3}{x - 2} - 1 \qquad \text{Write original equation.}$$

$$x(x - 2)\frac{2}{x} = x(x - 2)\frac{3}{x - 2} - x(x - 2)(1) \qquad \text{Multiply each term by the LCD.}$$

$$2(x - 2) = 3x - x(x - 2) \qquad \text{Simplify.}$$

$$2x - 4 = -x^2 + 5x \qquad \text{Simplify.}$$

$$x^2 - 3x - 4 = 0 \qquad \text{Write in general form.}$$

$$(x - 4)(x + 1) = 0 \qquad \text{Factor.}$$

$$x - 4 = 0 \implies x = 4 \qquad \text{Set 1st factor equal to 0.}$$

$$x + 1 = 0 \implies x = -1 \qquad \text{Set 2nd factor equal to 0.}$$

Check $x = 4$

$$\frac{2}{x} = \frac{3}{x - 2} - 1$$

$$\frac{2}{4} \stackrel{?}{=} \frac{3}{4 - 2} - 1$$

$$\frac{1}{2} \stackrel{?}{=} \frac{3}{2} - 1$$

$$\frac{1}{2} = \frac{1}{2} \checkmark$$

Check $x = -1$

$$\frac{2}{x} = \frac{3}{x - 2} - 1$$

$$\frac{2}{-1} \stackrel{?}{=} \frac{3}{-1 - 2} - 1$$

$$-2 \stackrel{?}{=} -1 - 1$$

$$-2 = -2 \checkmark$$

So, the solutions are $x = 4$ and $x = -1$.

To solve an equation involving an absolute value, remember that the expression inside the absolute value signs can be positive or negative. This results in *two* separate equations, each of which must be solved. For instance, the equation

$$|x - 2| = 3$$

results in the two equations $x - 2 = 3$ and $-(x - 2) = 3$, which implies that the equation has two solutions: $x = 5$ and $x = -1$.

Example 7 ▶ Solving an Equation Involving Absolute Value

Solve $|x^2 - 3x| = -4x + 6$.

Solution

Because the variable expression inside the absolute value signs can be positive or negative, you must solve the following two equations.

First Equation

$$x^2 - 3x = -4x + 6 \qquad \text{Use positive expression.}$$

$$x^2 + x - 6 = 0 \qquad \text{Write in general form.}$$

$$(x + 3)(x - 2) = 0 \qquad \text{Factor.}$$

$$x + 3 = 0 \Rightarrow x = -3 \qquad \text{Set 1st factor equal to 0.}$$

$$x - 2 = 0 \Rightarrow x = 2 \qquad \text{Set 2nd factor equal to 0.}$$

Second Equation

$$-(x^2 - 3x) = -4x + 6 \qquad \text{Use negative expression.}$$

$$x^2 - 7x + 6 = 0 \qquad \text{Write in general form.}$$

$$(x - 1)(x - 6) = 0 \qquad \text{Factor.}$$

$$x - 1 = 0 \Rightarrow x = 1 \qquad \text{Set 1st factor equal to 0.}$$

$$x - 6 = 0 \Rightarrow x = 6 \qquad \text{Set 2nd factor equal to 0.}$$

Check

$$|(-3)^2 - 3(-3)| \stackrel{?}{=} -4(-3) + 6 \qquad \text{Substitute } -3 \text{ for } x.$$

$$18 = 18 \qquad -3 \text{ checks. } \checkmark$$

$$|(2)^2 - 3(2)| \stackrel{?}{=} -4(2) + 6 \qquad \text{Substitute 2 for } x.$$

$$2 \neq -2 \qquad \text{2 does not check.}$$

$$|(1)^2 - 3(1)| \stackrel{?}{=} -4(1) + 6 \qquad \text{Substitute 1 for } x.$$

$$2 = 2 \qquad \text{1 checks. } \checkmark$$

$$|(6)^2 - 3(6)| \stackrel{?}{=} -4(6) + 6 \qquad \text{Substitute 6 for } x.$$

$$18 \neq -18 \qquad \text{6 does not check.}$$

The solutions are $x = -3$ and $x = 1$.

Applications 

It would be impossible to categorize the many different types of applications that involve nonlinear and nonquadratic models. However, from the few examples and exercises that are given, you will gain some appreciation for the variety of applications that can occur.

Example 8 ▶ Reduced Rates

A ski club chartered a bus for a ski trip at a cost of \$480. In an attempt to lower the bus fare per skier, the club invited nonmembers to go along. After five nonmembers joined the trip, the fare per skier decreased by \$4.80. How many club members are going on the trip?

Solution

Begin the solution by creating a verbal model and assigning labels.

Verbal Model: Cost per skier · Number of skiers = Cost of trip

Labels:
Cost of trip = 480 (dollars)
Number of ski club members = x (people)
Number of skiers = $x + 5$ (people)
Original cost per member = $\dfrac{480}{x}$ (dollars per person)
Cost per skier = $\dfrac{480}{x} - 4.80$ (dollars per person)

Equation:

$$\left(\frac{480}{x} - 4.80\right)(x + 5) = 480$$

$$\left(\frac{480 - 4.8x}{x}\right)(x + 5) = 480 \qquad \text{Write } \left(\frac{480}{x} - 4.80\right) \text{ as a fraction.}$$

$$(480 - 4.8x)(x + 5) = 480x \qquad \text{Multiply each side by } x.$$

$$480x + 2400 - 4.8x^2 - 24x = 480x \qquad \text{Multiply.}$$

$$-4.8x^2 - 24x + 2400 = 0 \qquad \text{Subtract } 480x \text{ from each side.}$$

$$x^2 + 5x - 500 = 0 \qquad \text{Divide each side by } -4.8.$$

$$(x + 25)(x - 20) = 0 \qquad \text{Factor.}$$

$$x + 25 = 0 \quad \Rightarrow \quad x = -25$$

$$x - 20 = 0 \quad \Rightarrow \quad x = 20$$

Choosing the positive value of x, you can conclude that 20 ski club members are going on the trip. Check this in the original statement of the problem, as follows.

$$\left(\frac{480}{20} - 4.80\right)(20 + 5) \stackrel{?}{=} 480 \qquad \text{Substitute 20 for } x.$$

$$(24 - 4.80)25 \stackrel{?}{=} 480 \qquad \text{Simplify.}$$

$$480 = 480 \qquad \text{20 checks. ✓}$$

Interest in a savings account is calculated by one of three basic methods: simple interest, interest compounded n times per year, and interest compounded continuously. The next example uses the formula for interest that is compounded n times per year.

$$A = P\left(1 + \frac{r}{n}\right)^{nt}$$

In this formula, A is the balance in the account, P is the principal (or original deposit), r is the annual interest rate (in decimal form), n is the number of compoundings per year, and t is the time in years. In Chapter 5, you will study a derivation of the formula above for interest compounded continuously.

Example 9 ▶ Compound Interest 

When your cousin was born, your grandparents deposited \$5000 in a long-term investment in which the interest was compounded quarterly. Today, on your cousin's 25th birthday, the value of the investment is \$25,062.59. What is the annual interest rate for this investment?

Solution

Formula: $A = P\left(1 + \frac{r}{n}\right)^{nt}$

Labels:

Balance $= A = 25{,}062.59$	(dollars)
Principal $= P = 5000$	(dollars)
Time $= t = 25$	(years)
Compoundings per year $= n = 4$	(compoundings per year)
Annual interest rate $= r$	(percent in decimal form)

Equation: $25{,}062.59 = 5000\left(1 + \frac{r}{4}\right)^{4(25)}$

$\frac{25{,}062.59}{5000} = \left(1 + \frac{r}{4}\right)^{100}$ Divide each side by 5000.

$5.0125 \approx \left(1 + \frac{r}{4}\right)^{100}$ Use a calculator.

$(5.0125)^{1/100} = 1 + \frac{r}{4}$ Raise each side to reciprocal power.

$1.01625 \approx 1 + \frac{r}{4}$ Use a calculator.

$0.01625 = \frac{r}{4}$ Subtract 1 from each side.

$0.065 = r$ Multiply each side by 4.

The annual interest rate is about 0.065, or 6.5%. Check this in the original statement of the problem.

1.6 Exercises

In Exercises 1–24, find all solutions of the equation. Check your solutions in the original equation.

1. $4x^4 - 18x^2 = 0$

2. $20x^3 - 125x = 0$

3. $x^4 - 81 = 0$

4. $x^6 - 64 = 0$

5. $x^3 + 216 = 0$

6. $27x^3 - 512 = 0$

7. $5x^3 + 30x^2 + 45x = 0$

8. $9x^4 - 24x^3 + 16x^2 = 0$

9. $x^3 - 3x^2 - x + 3 = 0$

10. $x^3 + 2x^2 + 3x + 6 = 0$

11. $x^4 - x^3 + x - 1 = 0$

12. $x^4 + 2x^3 - 8x - 16 = 0$

13. $x^4 - 4x^2 + 3 = 0$

14. $x^4 + 5x^2 - 36 = 0$

15. $4x^4 - 65x^2 + 16 = 0$

16. $36t^4 + 29t^2 - 7 = 0$

17. $x^6 + 7x^3 - 8 = 0$

18. $x^6 + 3x^3 + 2 = 0$

19. $\dfrac{1}{x^2} + \dfrac{8}{x} + 15 = 0$

20. $6\left(\dfrac{x}{x+1}\right)^2 + 5\left(\dfrac{x}{x+1}\right) - 6 = 0$

21. $2x + 9\sqrt{x} = 5$

22. $6x - 7\sqrt{x} - 3 = 0$

23. $3x^{1/3} + 2x^{2/3} = 5$

24. $9t^{2/3} + 24t^{1/3} + 16 = 0$

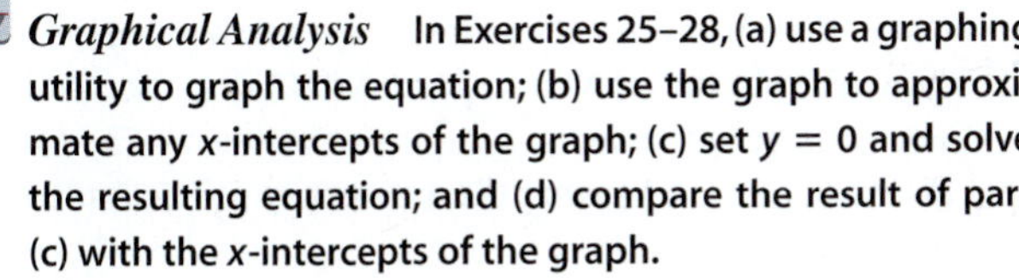

Graphical Analysis **In Exercises 25–28, (a) use a graphing utility to graph the equation; (b) use the graph to approximate any x-intercepts of the graph; (c) set $y = 0$ and solve the resulting equation; and (d) compare the result of part (c) with the x-intercepts of the graph.**

25. $y = x^3 - 2x^2 - 3x$

26. $y = 2x^4 - 15x^3 + 18x^2$

27. $y = x^4 - 10x^2 + 9$

28. $y = x^4 - 29x^2 + 100$

In Exercises 29–52, find all solutions of the equation. Check your solutions in the original equation.

29. $\sqrt{2x} - 10 = 0$

30. $4\sqrt{x} - 3 = 0$

31. $\sqrt{x - 10} - 4 = 0$

32. $\sqrt{5 - x} - 3 = 0$

33. $\sqrt[3]{2x + 5} + 3 = 0$

34. $\sqrt[3]{3x + 1} - 5 = 0$

35. $-\sqrt{26 - 11x} + 4 = x$

36. $x + \sqrt{31 - 9x} = 5$

37. $\sqrt{x + 1} = \sqrt{3x + 1}$

38. $\sqrt{x + 5} = \sqrt{x - 5}$

39. $\sqrt{x} - \sqrt{x - 5} = 1$

40. $\sqrt{x} + \sqrt{x - 20} = 10$

41. $\sqrt{x + 5} + \sqrt{x - 5} = 10$

42. $2\sqrt{x + 1} - \sqrt{2x + 3} = 1$

43. $\sqrt{x + 2} - \sqrt{2x - 3} = -1$

44. $4\sqrt{x - 3} - \sqrt{6x - 17} = 3$

45. $(x - 5)^{3/2} = 8$

46. $(x + 3)^{3/2} = 8$

47. $(x + 3)^{2/3} = 8$

48. $(x + 2)^{2/3} = 9$

49. $(x^2 - 5)^{3/2} = 27$

50. $(x^2 - x - 22)^{3/2} = 27$

51. $3x(x - 1)^{1/2} + 2(x - 1)^{3/2} = 0$

52. $4x^2(x - 1)^{1/3} + 6x(x - 1)^{4/3} = 0$

Graphical Analysis **In Exercises 53–56, (a) use a graphing utility to graph the equation; (b) use the graph to approximate any x-intercepts of the graph; (c) set $y = 0$ and solve the resulting equation; and (d) compare the result of part (c) with the x-intercepts of the graph.**

53. $y = \sqrt{11x - 30} - x$

54. $y = 2x - \sqrt{15 - 4x}$

55. $y = \sqrt{7x + 36} - \sqrt{5x + 16} - 2$

56. $y = 3\sqrt{x} - \dfrac{4}{\sqrt{x}} - 4$

In Exercises 57–70, find all solutions of the equation. Check your solutions in the original equation.

57. $x = \frac{3}{x} + \frac{1}{2}$

58. $\frac{4}{x} - \frac{5}{3} = \frac{x}{6}$

59. $\frac{1}{x} - \frac{1}{x+1} = 3$

60. $\frac{4}{x+1} - \frac{3}{x+2} = 1$

61. $\frac{20-x}{x} = x$

62. $4x + 1 = \frac{3}{x}$

63. $\frac{x}{x^2-4} + \frac{1}{x+2} = 3$

64. $\frac{x+1}{3} - \frac{x+1}{x+2} = 0$

65. $|2x - 1| = 5$

66. $|3x + 2| = 7$

67. $|x| = x^2 + x - 3$

68. $|x^2 + 6x| = 3x + 18$

69. $|x + 1| = x^2 - 5$

70. $|x - 10| = x^2 - 10x$

Graphical Analysis **In Exercises 71–74, (a) use a graphing utility to graph the equation; (b) use the graph to approximate any *x*-intercepts of the graph; (c) set $y = 0$ and solve the resulting equation; and (d) compare the result of part (c) with the *x*-intercepts of the graph.**

71. $y = \frac{1}{x} - \frac{4}{x-1} - 1$

72. $y = x + \frac{9}{x+1} - 5$

73. $y = |x + 1| - 2$

74. $y = |x - 2| - 3$

In Exercises 75–78, find the real solutions of the equation analytically. (Round your answers to three decimal places.)

75. $3.2x^4 - 1.5x^2 - 2.1 = 0$

76. $7.08x^6 + 4.15x^3 - 9.6 = 0$

77. $1.8x - 6\sqrt{x} - 5.6 = 0$

78. $4x^{2/3} + 8x^{1/3} + 3.6 = 0$

Think About It **In Exercises 79–86, find an equation that has the given solutions. (There are many correct answers.)**

79. $-2, 5$

80. $0, 3, 5$

81. $-\frac{7}{3}, \frac{6}{7}$

82. $-\frac{1}{8}, -\frac{4}{5}$

83. $\sqrt{3}, -\sqrt{3}, 4$

84. $2\sqrt{7}, -\sqrt{7}$

85. $-1, 1, i, -i$

86. $4i, -4i, 6, -6$

87. ***Chartering a Bus*** A college charters a bus for \$1700 to take a group to a museum. When six more students join the trip, the cost per student drops by \$7.50. How many students were in the original group?

88. ***Renting an Apartment*** Three students are planning to rent an apartment for a year and share equally in the cost. By adding a fourth person, each person could save \$75 a month. How much is the monthly rent?

89. ***Airspeed*** An airline runs a commuter flight between Portland, Oregon and Seattle, Washington, which are 145 miles apart. If the average speed of the plane could be increased by 40 miles per hour, the travel time would be decreased by 12 minutes. What airspeed is required to obtain this decrease in travel time?

90. ***Average Speed*** A family drove 1080 miles to their vacation lodge. Because of increased traffic density, their average speed on the return trip was decreased by 6 miles per hour and the trip took $2\frac{1}{2}$ hours longer. Determine their average speed on the way to the lodge.

91. ***Mutual Funds*** A deposit of \$2500 in a mutual fund reaches a balance of \$3052.49 after 5 years. What annual interest rate on a certificate of deposit compounded monthly would yield an equivalent return?

92. ***Mutual Funds*** A sales representative for a mutual funds company describes a "guaranteed investment fund" that the company is offering to new investors. You are told that if you deposit \$10,000 in the fund you will be guaranteed a return of at least \$25,000 after 20 years. (Assume the interest is compounded quarterly.)

(a) What is the annual interest rate if the investment only meets the minimum guaranteed amount?

(b) After 20 years, you receive \$32,000. What is the annual interest rate?

93. ***Saturated Steam*** The temperature T (in degrees Fahrenheit) of saturated steam increases as pressure increases. This relationship is approximated by the model

$$T = 75.82 - 2.11x + 43.51\sqrt{x}, \quad 5 \le x \le 40$$

where x is the absolute pressure (in pounds per square inch).

(a) Use the model to complete the table.

Absolute pressure, x	Temperature, T
5	
10	
15	
20	
25	
30	
35	
40	

(b) The temperature of steam at sea level is 212°F. Use the table in part (a) to approximate the absolute pressure at this temperature.

(c) Solve part (b) algebraically.

(d) Use a graphing utility to verify your solutions for part (b) and part (c).

94. ***Airline Passengers*** An airline offers daily flights between Chicago and Denver. The total monthly cost C (in millions of dollars) of these flights is $C = \sqrt{0.2x + 1}$, where x is the number of passengers (in thousands). The total cost of the flights for June is 2.5 million dollars. How many passengers flew in June?

95. ***Demand*** The demand equation for a video game is modeled by $p = 40 - \sqrt{0.01x + 1}$, where x is the number of units demanded per day and p is the price per unit. Approximate the demand when the price is \$37.55.

96. ***Demand*** The demand equation for a coffee maker is modeled by $p = 40 - \sqrt{0.0001x + 1}$, where x is the number of units demanded per day and p is the price per unit. Approximate the demand when the price is \$34.70.

97. ***Baseball*** A baseball diamond has the shape of a square in which the distance from home plate to second base is approximately $127\frac{1}{2}$ feet. Approximate the distance between the bases.

98. ***Meteorology*** A meteorologist is positioned 100 feet from the point where a weather balloon is launched. When the balloon is at height h, the distance d (in feet) between the meteorologist and the balloon is $d = \sqrt{100^2 + h^2}$.

(a) Use a graphing utility to graph the equation. Use the *trace* feature to approximate the value of h when $d = 200$.

(b) Complete the table. Use the table to approximate the value of h when $d = 200$.

h	160	165	170	175	180	185
d						

(c) Find h algebraically when $d = 200$.

(d) Compare the results of each method. In each case, what information did you gain that wasn't apparent in another solution method?

99. ***Geometry*** You construct a cone with a base radius of 8 inches. The surface area S of the cone can be represented by the equation

$$S = 8\pi\sqrt{64 + h^2}$$

where h is the height of the cone.

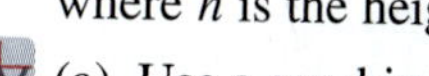

(a) Use a graphing utility to graph the equation. Use the *trace* feature to approximate the value of h when $S = 350$ square inches.

(b) Complete the table. Use the table to approximate the value of h when $S = 350$.

h	8	9	10	11	12	13
S						

(c) Find h algebraically when $S = 350$.

(d) Compare the results of each method. In each case, what information did you gain that wasn't apparent in another solution method?

100. ***Labor*** Working together, two people can complete a task in 8 hours. Working alone, one person takes 2 hours longer than the other to complete the task. How long would it take for each person to complete the task?

101. ***Labor*** Working together, two people can complete a task in 12 hours. Working alone, one person takes 3 hours longer than the other to complete the task. How long would it take for each person to complete the task?

▶ Model It

102. ***Power Line*** A power station is on one side of a river that is $\frac{3}{4}$ mile wide, and a factory is 8 miles downstream on the other side of the river, as shown in the figure. It costs $24 per foot to run power lines overland and $30 per foot to run them underwater.

(a) Write the total cost C of the project as a function of x (see figure).

(b) Find the total cost when $x = 3$.

(c) Find the length x when $C = \$1{,}098{,}662.40$.

(d) Use a graphing utility to graph the function from part (a).

(e) Use your graph from part (d) to find the value of x that minimizes the cost.

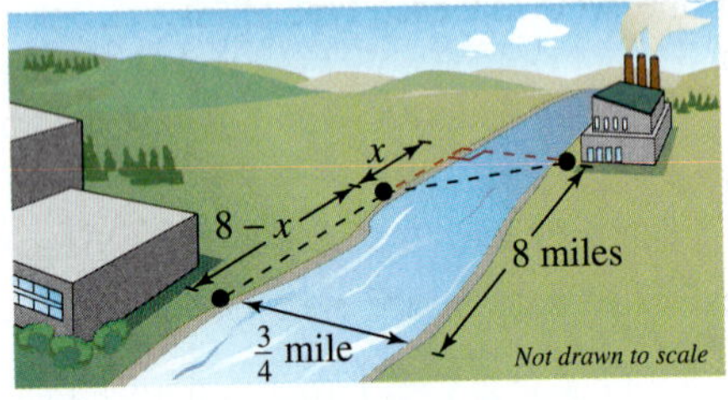

In Exercises 103 and 104, solve for the indicated variable.

103. ***A Person's Tangential Speed in a Rotor***

Solve for g: $v = \sqrt{\dfrac{gR}{\mu s}}$

104. ***Inductance***

Solve for Q: $i = \pm\sqrt{\dfrac{1}{LC}}\sqrt{Q^2 - q}$

Synthesis

True or False? In Exercises 105 and 106, determine whether the statement is true or false. Justify your answer.

105. An equation can never have more than one extraneous solution.

106. When solving an absolute value equation, you will always have to check more than one solution.

In Exercises 107 and 108, find x such that the distance between the given points is 13. Explain your results.

107. $(1, 2), (x, -10)$

108. $(-8, 0), (x, 5)$

In Exercises 109 and 110, find y such that the distance between the given points is 17. Explain your results.

109. $(0, 0), (8, y)$

110. $(-8, 4), (7, y)$

In Exercises 111 and 112, consider an equation of the form $x + |x - a| = b$, where a and b are constants.

111. Find a and b when the solution to the equation is $x = 9$. (There are many correct answers.)

112. ***Writing*** Write a short paragraph listing the steps required to solve this equation involving absolute values.

In Exercises 113 and 114, consider an equation of the form $x + \sqrt{x - a} = b$, where a and b are constants.

113. Find a and b when the solution to the equation is $x = 20$. (There are many correct answers.)

114. ***Writing*** Write a short paragraph listing the steps required to solve this equation involving radicals.

Review

In Exercises 115–118, perform the operation and simplify.

115. $\dfrac{8}{3x} + \dfrac{3}{2x}$

116. $\dfrac{2}{x^2 - 4} - \dfrac{1}{x^2 - 3x + 2}$

117. $\dfrac{2}{z + 2} - \left(3 - \dfrac{2}{z}\right)$

118. $25y^2 \div \dfrac{xy}{5}$

In Exercises 119 and 120, find all real solutions of the equation.

119. $x^2 - 22x + 121 = 0$

120. $x(x - 20) + 3(x - 20) = 0$

1.7 Linear Inequalities in One Variable

▶ **What you should learn**

- How to represent solutions of linear inequalities in one variable.
- How to solve linear inequalities in one variable.
- How to solve inequalities involving absolute values
- How to use inequalities to model and solve real-life problems

▶ **Why you should learn it**

Inequalities can be used to model and solve real-life problems. For instance, in Exercise 98 on page 149, you will use a linear inequality to analyze data about the maximum weight a weightlifter can bench press.

Brian Smith/Stock Boston

Introduction

Simple inequalities were reviewed in Section P.1. There, you used the inequality symbols $<$, $\leq$, $>$, and $\geq$ to compare two numbers and to denote subsets of real numbers. For instance, the simple inequality

$$x \geq 3$$

denotes all real numbers x that are greater than or equal to 3.

In this section you will expand your work with inequalities to include more involved statements such as

$$5x - 7 < 3x + 9$$

and

$$-3 \leq 6x - 1 < 3.$$

As with an equation, you **solve an inequality** in the variable x by finding all values of x for which the inequality is true. Such values are **solutions** and are said to **satisfy** the inequality. The set of all real numbers that are solutions of an inequality is the **solution set** of the inequality. For instance, the solution set of

$$x + 1 < 4$$

is all real numbers that are less than 3.

The set of all points on the real number line that represent the solution set is the **graph of the inequality.** Graphs of many types of inequalities consist of intervals on the real number line. See Section P.1 to review the nine basic types of intervals on the real number line. Note that each type of interval can be classified as *bounded or unbounded.*

Example 1 ▶ Intervals and Inequalities

Write an inequality to represent each interval, and state whether the interval is bounded or unbounded.

a. $(-3, 5]$
b. $(-3, \infty)$
c. $[0, 2]$
d. $(-\infty, \infty)$

Solution

a. $(-3, 5]$ corresponds to $-3 < x \leq 5$. Bounded
b. $(-3, \infty)$ corresponds to $-3 < x$. Unbounded
c. $[0, 2]$ corresponds to $0 \leq x \leq 2$. Bounded
d. $(-\infty, \infty)$ corresponds to $-\infty < x < \infty$. Unbounded

Properties of Inequalities

The procedures for solving linear inequalities in one variable are much like those for solving linear equations. To isolate the variable, you can make use of the **Properties of Inequalities.** These properties are similar to the properties of equality, but there are two important exceptions. When each side of an inequality is multiplied or divided by a negative number, the direction of the inequality symbol must be reversed. Here is an example.

$$-2 < 5 \quad \text{Original inequality}$$

$$(-3)(-2) > (-3)(5) \quad \text{Multiply each side by } -3 \text{ and reverse inequality.}$$

$$6 > -15 \quad \text{Simplify.}$$

Two inequalities that have the same solution set are **equivalent.** For instance, the inequalities

$$x + 2 < 5$$

and

$$x < 3$$

are equivalent. To obtain the second inequality from the first, you can subtract 2 from each side of the inequality. The following list describes the operations that can be used to create equivalent inequalities.

Properties of Inequalities

Let a, b, c, and d be real numbers.

1. Transitive Property

$a < b$ and $b < c$ ⇒ $a < c$

2. Addition of Inequalities

$a < b$ and $c < d$ ⇒ $a + c < b + d$

3. Addition of a Constant

$a < b$ ⇒ $a + c < b + c$

4. Multiplication by a Constant

For $c > 0$, $a < b$ ⇒ $ac < bc$

For $c < 0$, $a < b$ ⇒ $ac > bc$

Each of the properties above is true if the symbol $<$ is replaced by $\le$ and the symbol $>$ is replaced by $\ge$. For instance, another form of the multiplication property would be as follows.

For $c > 0$, $a \le b$ ⇒ $ac \le bc$

For $c < 0$, $a \le b$ ⇒ $ac \ge bc$

Solving a Linear Inequality in One Variable

The simplest type of inequality is a **linear inequality** in one variable. For instance, $2x + 3 > 4$ is a linear inequality in x.

In the following examples, pay special attention to the steps in which the inequality symbol is reversed. Remember that when you multiply or divide by a negative number, you must reverse the inequality symbol.

Example 2 ▶ Solving Linear Inequalities 

Solve each inequality.

a. $5x - 7 > 3x + 9$

b. $1 - \dfrac{3x}{2} \geq x - 4$

Solution

a. $5x - 7 > 3x + 9$ Write original inequality.

$2x - 7 > 9$ Subtract $3x$ from each side.

$2x > 16$ Add 7 to each side.

$x > 8$ Divide each side by 2.

The solution set is all real numbers that are greater than 8, which is denoted by $(8, \infty)$. The graph of this solution set is shown in Figure 1.24.

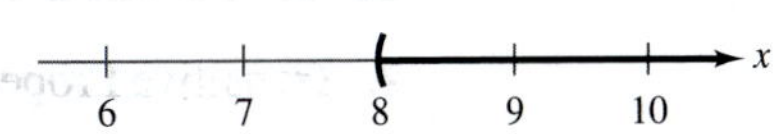

Solution interval: $(8, \infty)$

FIGURE 1.24

b. $1 - \dfrac{3x}{2} \geq x - 4$ Write original inequality.

$2 - 3x \geq 2x - 8$ Multiply each side by 2.

$2 - 5x \geq -8$ Subtract $2x$ from each side.

$-5x \geq -10$ Subtract 2 from each side.

$x \leq 2$ Divide each side by -5 and reverse the inequality.

The solution set is all real numbers that are less than or equal to 2, which is denoted by $(-\infty, 2]$. The graph of this solution set is shown in Figure 1.25.

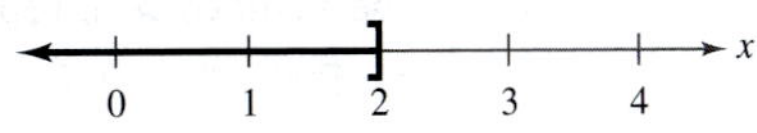

Solution interval: $(-\infty, 2]$

FIGURE 1.25

STUDY TIP

Checking the solution set of an inequality is not as simple as checking the solutions of an equation. You can, however, get an indication of the validity of a solution set by substituting a few convenient values of x.

Sometimes it is possible to write two inequalities as a **double inequality.** For instance, you can write the two inequalities $-4 \le 5x - 2$ and $5x - 2 < 7$ more simply as

$$-4 \le 5x - 2 < 7.$$

This form allows you to solve the two inequalities together, as demonstrated in Example 3.

Example 3 ▶ Solving a Double Inequality 

To solve a double inequality, you can isolate x as the middle term.

$$-3 \le 6x - 1 < 3 \qquad \text{Original inequality}$$

$$-3 + 1 \le 6x - 1 + 1 < 3 + 1 \qquad \text{Add 1 to each part.}$$

$$-2 \le 6x < 4 \qquad \text{Simplify.}$$

$$\frac{-2}{6} \le \frac{6x}{6} < \frac{4}{6} \qquad \text{Divide each part by 6.}$$

$$-\frac{1}{3} \le x < \frac{2}{3} \qquad \text{Simplify.}$$

The solution set is all real numbers that are greater than or equal to $-\frac{1}{3}$ and less than $\frac{2}{3}$, which is denoted by $\left[-\frac{1}{3}, \frac{2}{3}\right)$. The graph of this solution set is shown in Figure 1.26.

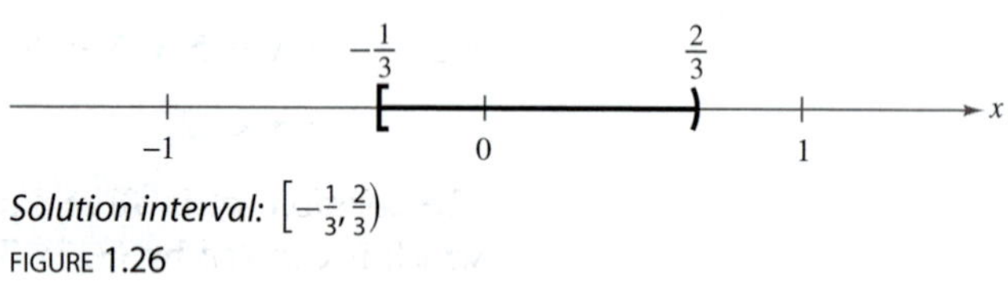

Solution interval: $\left[-\frac{1}{3}, \frac{2}{3}\right)$

FIGURE 1.26

The double inequality in Example 3 could have been solved in two parts as follows.

$$-3 \le 6x - 1 \quad \text{and} \quad 6x - 1 < 3$$

$$-2 \le 6x \qquad\qquad 6x < 4$$

$$-\frac{1}{3} \le x \qquad\qquad x < \frac{2}{3}$$

The solution set consists of all real numbers that satisfy *both* inequalities. In other words, the solution set is the set of all values of x for which

$$-\frac{1}{3} \le x < \frac{2}{3}.$$

When combining two inequalities to form a double inequality, be sure that the inequalities satisfy the Transitive Property. For instance, it is *incorrect* to combine the inequalities $3 < x$ and $x \le -1$ as $3 < x \le -1$. This "inequality" is wrong because 3 is not less than -1.

Technology

A graphing utility can be used to identify the solution set of the graph of an inequality. For instance, to find the solution set of $|x - 5| < 2$ (see Example 4), rewrite the inequality as $|x - 5| - 2 < 0$ and enter

Y1 = abs (X − 5) − 2

and press the graph key. The graph should look like the one shown below.

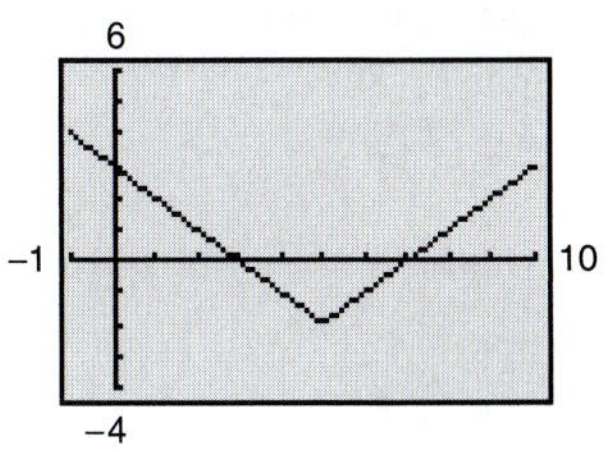

Notice that the graph is below the x-axis on the interval (3, 7).

Inequalities Involving Absolute Values

Solving an Absolute Value Inequality

Let x be a variable or an algebraic expression and let a be a real number such that $a \geq 0$.

1. The solutions of $|x| < a$ are all values of x that lie between $-a$ and a.

   $|x| < a$ if and only if $-a < x < a$.

2. The solutions of $|x| > a$ are all values of x that are less than $-a$ or greater than a.

   $|x| > a$ if and only if $x < -a$ or $x > a$.

These rules are also valid if < is replaced by ≤ and > is replaced by ≥.

Example 4 ▶ **Solving an Absolute Value Inequality** 

Solve each inequality.

a. $|x - 5| < 2$ **b.** $|x + 3| \geq 7$

Solution

a.

$\lvert x - 5\rvert < 2$	Write original inequality.
$-2 < x - 5 < 2$	Write equivalent inequalities.
$-2 + 5 < x - 5 + 5 < 2 + 5$	Add 5 to each part.
$3 < x < 7$	Simplify.

The solution set is all real numbers that are greater than 3 and less than 7, which is denoted by (3, 7). The graph of this solution set is shown in Figure 1.27.

b.

$\lvert x + 3\rvert \geq 7$			Write original inequality.
$x + 3 \leq -7$	or	$x + 3 \geq 7$	Write equivalent inequalities.
$x + 3 - 3 \leq -7 - 3$		$x + 3 - 3 \geq 7 - 3$	Subtract 3 from each side.
$x \leq -10$		$x \geq 4$	Simplify.

The solution set is all real numbers that are less than or equal to -10 *or* greater than or equal to 4. The interval notation for this solution set is $(-\infty, -10] \cup [4, \infty)$. The symbol $\cup$ is called a *union* symbol and is used to denote the combining of two sets. The graph of this solution set is shown in Figure 1.28.

STUDY TIP

Note that the graph of the inequality $|x - 5| < 2$ can be described as all real numbers *within* two units of 5, as shown in Figure 1.27.

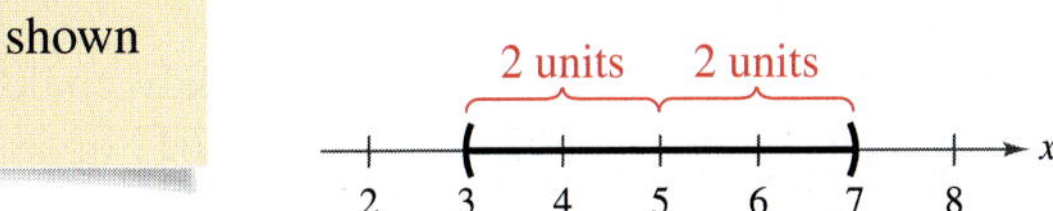

$|x - 5| < 2$: *Solutions lie inside* (3, 7)
FIGURE 1.27

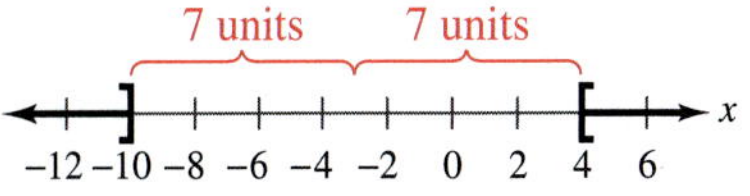

$|x + 3| \geq 7$: *Solutions lie outside* (−10, 4)
FIGURE 1.28

Applications

The problem-solving plan described in Section 1.3 can be used to model and solve real-life problems that involve inequalities, as illustrated in Example 5.

Example 5 ▶ Comparative Shopping  

A subcompact car can be rented from Company A for \$180 per week with no extra charge for mileage. A similar car can be rented from Company B for \$100 per week plus 20 cents for each mile driven. How many miles must you drive in a week in order for the rental fee for Company B to be more than that for Company A?

Solution

Verbal Model: Weekly cost for Company B > Weekly cost for Company A

Labels: Miles driven in one week $= m$ (miles)
Weekly cost for Company A $= 180$ (dollars)
Weekly cost for Company B $= 100 + 0.20m$ (dollars)

Inequality: $100 + 0.2m > 180$

$$0.2m > 80$$

$$m > 400 \text{ miles}$$

If you drive more than 400 miles in a week, Company B costs more.

Example 6 ▶ Accuracy of a Measurement 

You go to a candy store to buy chocolates that cost \$9.89 per pound. The scale that is used in the store has a state seal of approval that indicates the scale is accurate to within half an ounce. According to the scale, your purchase weighs one-half pound and costs \$4.95. How much might you have been undercharged or overcharged as a result of inaccuracy in the scale?

Solution

Let x represent the *true* weight of the candy. Because the scale is accurate to within half an ounce (or $\frac{1}{32}$ of a pound), the difference between the exact weight (x) and the scale weight $\left(\frac{1}{2}\right)$ is less than or equal to $\frac{1}{32}$ of a pound. That is, $\left|x - \frac{1}{2}\right| \le \frac{1}{32}$. You can solve this inequality as follows.

$$-\frac{1}{32} \le x - \frac{1}{2} \le \frac{1}{32}$$

$$\frac{15}{32} \le x \le \frac{17}{32}$$

$$0.46875 \le x \le 0.53125$$

In other words, your "one-half pound" of candy could have weighed as little as 0.46875 pound (which would have cost \$4.64) or as much as 0.53125 pound (which would have cost \$5.25). So, you could have been overcharged by as much as \$0.31 or undercharged by as much as \$0.30.

1.7 Exercises

In Exercises 1–6, write an inequality that represents the interval, and state whether the interval is bounded or unbounded.

1. $[-1, 5]$ **2.** $(2, 10]$
3. $(11, \infty)$ **4.** $[-5, \infty)$
5. $(-\infty, -2)$ **6.** $(-\infty, 7]$

In Exercises 7–12, match the inequality with its graph. [The graphs are labeled (a), (b), (c), (d), (e), and (f).]

(a) 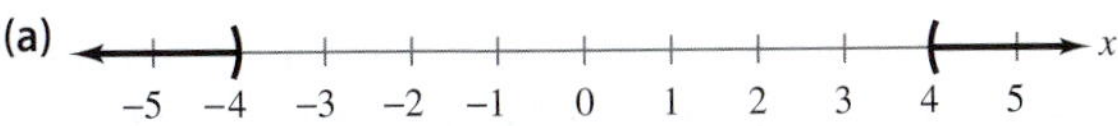

(b) 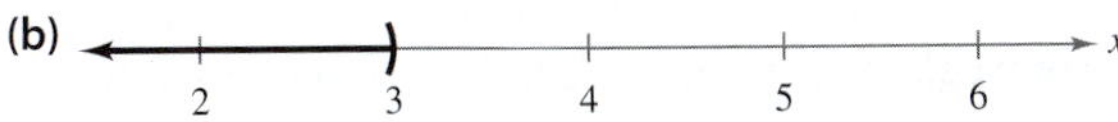

(c) 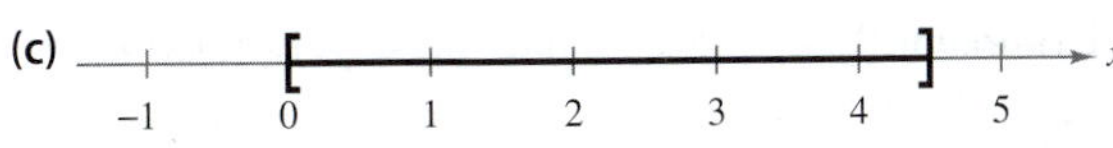

(d) 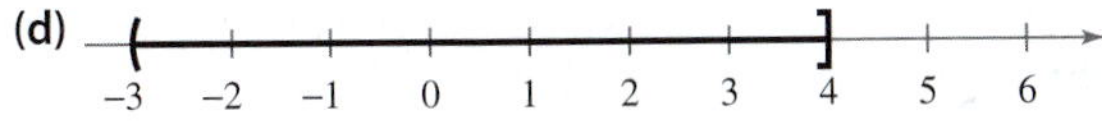

(e)

(f)

7. $x < 3$ **8.** $x \geq 5$
9. $-3 < x \leq 4$ **10.** $0 \leq x \leq \frac{9}{2}$
11. $|x| < 3$ **12.** $|x| > 4$

In Exercises 13–18, determine whether each value of x is a solution of the inequality.

	Inequality	*Values*	
13.	$5x - 12 > 0$	(a) $x = 3$	(b) $x = -3$
		(c) $x = \frac{5}{2}$	(d) $x = \frac{3}{2}$
14.	$2x + 1 < -3$	(a) $x = 0$	(b) $x = -\frac{1}{4}$
		(c) $x = -4$	(d) $x = -\frac{3}{2}$
15.	$0 < \dfrac{x-2}{4} < 2$	(a) $x = 4$	(b) $x = 10$
		(c) $x = 0$	(d) $x = \frac{7}{2}$
16.	$-1 < \dfrac{3-x}{2} \leq 1$	(a) $x = 0$	(b) $x = -5$
		(c) $x = 1$	(d) $x = 5$
17.	$\lvert x - 10\rvert \geq 3$	(a) $x = 13$	(b) $x = -1$
		(c) $x = 14$	(d) $x = 9$
18.	$\lvert 2x - 3\rvert < 15$	(a) $x = -6$	(b) $x = 0$
		(c) $x = 12$	(d) $x = 7$

In Exercises 19–44, solve the inequality and sketch the solution on the real number line. (Some equalities have no solutions.)

19. $4x < 12$ **20.** $10x < -40$
21. $-2x > -3$ **22.** $-6x > 15$
23. $x - 5 \geq 7$ **24.** $x + 7 \leq 12$
25. $2x + 7 < 3 + 4x$ **26.** $3x + 1 \geq 2 + x$
27. $2x - 1 \geq 1 - 5x$ **28.** $6x - 4 \leq 2 + 8x$
29. $4 - 2x < 3(3 - x)$ **30.** $4(x + 1) < 2x + 3$
31. $\frac{3}{4}x - 6 \leq x - 7$ **32.** $3 + \frac{2}{7}x > x - 2$
33. $\frac{1}{2}(8x + 1) \geq 3x + \frac{5}{2}$ **34.** $9x - 1 < \frac{3}{4}(16x - 2)$
35. $3.6x + 11 \geq -3.4$
36. $15.6 - 1.3x < -5.2$
37. $1 < 2x + 3 < 9$
38. $-8 \leq -(3x + 5) < 13$
39. $-4 < \dfrac{2x - 3}{3} < 4$ **40.** $0 \leq \dfrac{x + 3}{2} < 5$
41. $\frac{3}{4} > x + 1 > \frac{1}{4}$ **42.** $-1 < 2 - \dfrac{x}{3} < 1$
43. $3.2 \leq 0.4x - 1 \leq 4.4$
44. $4.5 > \dfrac{1.5x + 6}{2} > 10.5$

In Exercises 45–60, solve the inequality and sketch the solution on the real number line. (Some inequalities have no solution.)

45. $|x| < 6$ **46.** $|x| > 4$
47. $\left|\dfrac{x}{2}\right| > 1$ **48.** $\left|\dfrac{x}{5}\right| > 3$
49. $|x - 5| < -1$ **50.** $|x - 7| < -5$
51. $|x - 20| \leq 6$ **52.** $|x - 8| \geq 0$
53. $|3 - 4x| \geq 9$ **54.** $|1 - 2x| < 5$
55. $\left|\dfrac{x - 3}{2}\right| \geq 4$ **56.** $\left|1 - \dfrac{2x}{3}\right| < 1$
57. $|9 - 2x| - 2 < -1$ **58.** $|x + 14| + 3 > 17$
59. $2|x + 10| \geq 9$ **60.** $3|4 - 5x| \leq 9$

Graphical Analysis **In Exercises 61–68, use a graphing utility to graph the inequality and identify the solution set.**

61. $6x > 12$

62. $3x - 1 \leq 5$

63. $5 - 2x \geq 1$

64. $3(x + 1) < x + 7$

65. $|x - 8| \leq 14$

66. $|2x + 9| > 13$

67. $2|x + 7| \geq 13$

68. $\frac{1}{2}|x + 1| \leq 3$

Graphical Analysis **In Exercises 69–74, use a graphing utility to graph the equation. Use the graph to approximate the values of x that satisfy each inequality.**

	Equation	*Inequalities*	
69.	$y = 2x - 3$	(a) $y \geq 1$	(b) $y \leq 0$
70.	$y = \frac{2}{3}x + 1$	(a) $y \leq 5$	(b) $y \geq 0$
71.	$y = -\frac{1}{2}x + 2$	(a) $0 \leq y \leq 3$	(b) $y \geq 0$
72.	$y = -3x + 8$	(a) $-1 \leq y \leq 3$	(b) $y \leq 0$
73.	$y = \lvert x - 3 \rvert$	(a) $y \leq 2$	(b) $y \geq 4$
74.	$y = \left\lvert \frac{1}{2}x + 1 \right\rvert$	(a) $y \leq 4$	(b) $y \geq 1$

In Exercises 75–80, find the interval(s) on the real number line for which the radicand is nonnegative (greater than or equal to zero).

75. $\sqrt{x - 5}$

76. $\sqrt{x - 10}$

77. $\sqrt{x + 3}$

78. $\sqrt{3 - x}$

79. $\sqrt[4]{7 - 2x}$

80. $\sqrt[4]{6x + 15}$

81. ***Think About It*** The graph of $|x - 5| < 3$ can be described as all real numbers within 3 units of 5. Give a similar description of $|x - 10| < 8$.

82. ***Think About It*** The graph of $|x - 2| > 5$ can be described as all real numbers more than 5 units from 2. Give a similar description of $|x - 8| > 4$.

In Exercises 83–90, use absolute value notation to define the interval (or pair of intervals) on the real number line.

83.

84.

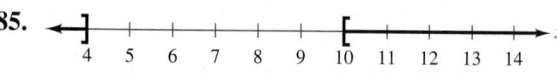

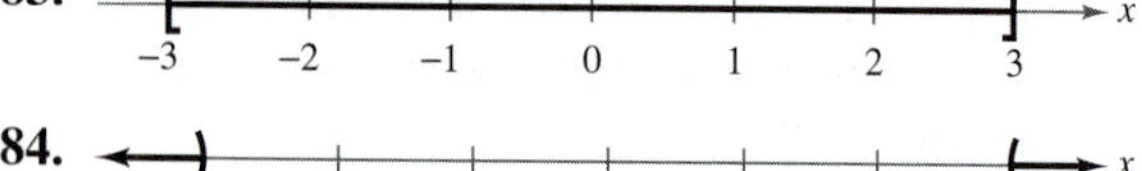

85.

86.

87. All real numbers within 10 units of 12

88. All real numbers at least 5 units from 8

89. All real numbers more than 5 units from -3

90. All real numbers no more than 7 units from -6

91. ***Car Rental*** You can rent a midsize car from Company A for \$250 per week with unlimited mileage. A similar car can be rented from Company B for \$150 per week plus 25 cents for each mile driven. How many miles must you drive in a week in order for the rental fee for Company B to be greater than that for Company A?

92. ***Copying Costs*** Your department sends its copying to the photocopy center of your company. The center bills your department \$0.10 per page. You have investigated the possibility of buying a departmental copier for \$3000. With your own copier, the cost per page would be \$0.03. The expected life of the copier is 4 years. How many copies must you make in the four-year period to justify buying the copier?

93. ***Investment*** In order for an investment of \$1000 to grow to more than \$1062.50 in 2 years, what must the annual interest rate be? $[A = P(1 + rt)]$

94. ***Investment*** In order for an investment of \$750 to grow to more than \$825 in 2 years, what must the annual interest rate be? $[A = P(1 + rt)]$

95. ***Cost, Revenue, and Profit*** The revenue for selling x units of a product is $R = 115.95x$. The cost of producing x units is

$$C = 95x + 750.$$

To obtain a profit, the revenue must be greater than the cost. For what values of x will this product return a profit?

96. ***Cost, Revenue, and Profit*** The revenue for selling x units of a product is $R = 24.55x$. The cost of producing x units is

$$C = 15.4x + 150{,}000.$$

To obtain a profit, the revenue must be greater than the cost. For what values of x will this product return a profit?

97. ***Data Analysis*** The admissions office of a college wants to determine whether there is a relationship between IQ scores x and grade-point averages y after the first year of school. An equation that models the data the admissions office obtained is

$$y = 0.067x - 5.638.$$

(a) Use a graphing utility to graph the model.

(b) Use the graph to estimate the values of x that predict a grade-point average of at least 3.0.

▶ Model It

98. ***Data Analysis*** You want to determine whether there is a relationship between an athlete's weight x (in pounds) and the athlete's maximum bench-press weight y (in pounds). The table shows a sample of data from 12 athletes.

Athlete's weight, x	Bench-press weight, y
165	170
184	185
150	200
210	255
196	205
240	295
202	190
170	175
185	195
190	185
230	250
160	155

(a) Use a graphing utility to plot the data.

(b) A model for this data is $y = 1.3x - 36$. Use a graphing utility to graph the model in the same viewing window used in part (a).

(c) Use the graph to estimate the values of x that predict a maximum bench-press weight of at least 200 pounds.

(d) Verify the estimate from part (c) algebraically.

▶ Model It *(continued)*

(e) Use the graph to write a statement about the accuracy of the model. If you think the graph indicates that an athlete's weight is not a particularly good indicator of the athlete's maximum bench-press weight, list other factors that might influence an individual's maximum bench-press weight.

99. ***Teachers' Salaries*** The average salary S (in thousands of dollars) for elementary and secondary teachers in the United States from 1980 to 2000 is approximated by the model

$$S = 1.33t + 16.8, \qquad 0 \le t \le 19$$

where $t = 0$ represents 1980. According to this model, when will the average teacher's salary exceed \$45,000? (Source: National Education Association)

100. ***Egg Production*** The number of eggs E (in billions) produced in the United States from 1990 to 1999 can be modeled by $E = 1.55t + 67.5$, where $t = 0$ represents 1990. According to the model, when will the number of eggs produced exceed 88 billion? (Source: U.S. Department of Agriculture)

101. ***Geometry*** The side of a square is measured as 10.4 inches with a possible error of $\frac{1}{16}$ inch. Using these measurements, determine the interval containing the possible areas of the square.

102. ***Geometry*** The side of a square is measured as 24.2 centimeters with a possible error of 0.25 centimeter. Using these measurements, determine the interval containing the possible areas of the square.

103. ***Accuracy of Measurement*** You buy a bag of oranges for \$0.95 per pound. The weight that is listed on the bag is 4.65 pounds. The scale that weighed the bag is accurate to within 1 ounce. How much might you have been undercharged or overcharged?

104. ***Accuracy of Measurement*** You buy six T-bone steaks that cost \$3.98 per pound. The weight that is listed on the package is 5.72 pounds. The scale that weighed the package is accurate to within $\frac{1}{2}$ ounce. How much might you have been undercharged or overcharged?

105. ***Time Study*** A time study was conducted to determine the length of time required to perform a particular task in a manufacturing process. The times required by approximately two-thirds of the workers in the study satisfied the inequality

$$\left|\frac{t - 15.6}{1.9}\right| < 1$$

where t is time in minutes. Determine the interval on the real number line in which these times lie.

106. ***Height*** The heights h of two-thirds of the members of a population satisfy the inequality

$$\left|\frac{h - 68.5}{2.7}\right| \le 1$$

where h is measured in inches. Determine the interval on the real number line in which these heights lie.

107. ***Meteorology*** An electronic device is to be operated in an environment with relative humidity h in the interval defined by $|h - 50| \le 30$. What are the minimum and maximum relative humidities for the operation of this device?

108. ***Music*** Michael Kasha of Florida State University used physics and mathematics to design a new classical guitar. He used the model for the frequency of the vibrations on a circular plate

$$v = \frac{2.6t}{d^2}\sqrt{\frac{E}{\rho}}$$

where v is the frequency (in vibrations per second), t is the plate thickness (in millimeters), d is the diameter of the plate, E is the elasticity of the plate material, and ρ is the density of the plate material. For fixed values of d, E, and ρ, the graph of the equation is a line (see figure).

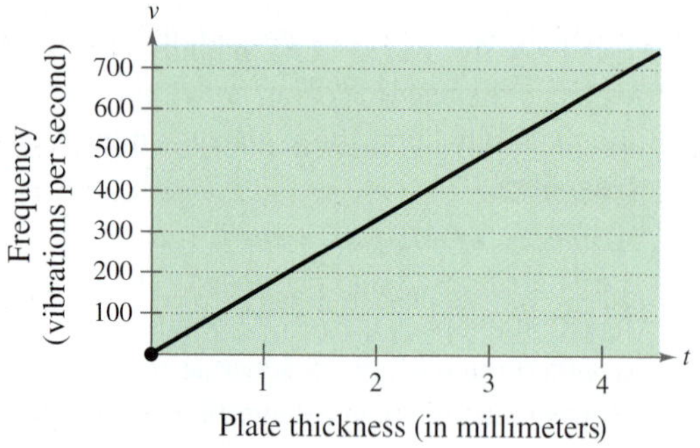

(a) Estimate the frequency when the plate thickness is 2 millimeters.

(b) Estimate the plate thickness when the frequency is 600 vibrations per second.

(c) Approximate the interval for the plate thickness when the frequency is between 200 and 400 vibrations per second.

(d) Approximate the interval for the frequency when the plate thickness is less than 3 millimeters.

Synthesis

True or False? **In Exercises 109 and 110, determine whether the statement is true or false. Justify your answer.**

109. If a, b, and c are real numbers, and $a \le b$, then $ac \le bc$.

110. If $-10 \le x \le 8$, then $-10 \ge -x$ and $-x \ge -8$.

111. Identify the graph of the inequality $|x - a| \ge 2$.

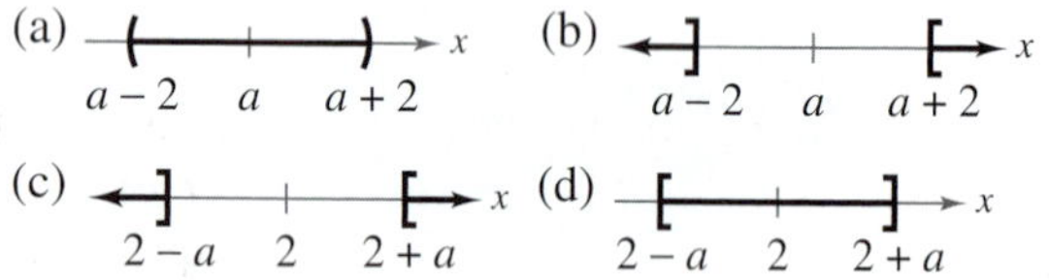

112. Find sets of values of a, b, and c such that $0 \le x \le 10$ is a solution of the inequality $|ax - b| \le c$.

Review

In Exercises 113–116, find the distance between each pair of points. Then find the midpoint of the line segment joining the points.

113. $(-4, 2), (1, 12)$

114. $(1, -2), (10, 3)$

115. $(3, 6), (-5, -8)$

116. $(0, -3), (-6, 9)$

In Exercises 117–124, solve the equation.

117. $3(x - 1) = 30$

118. $8x - 5(x + 4) = -19$

119. $-6(2 - x) - 12 = 36$

120. $4(x + 7) - 9 = -6(-x - 1)$

121. $2x^2 - 19x - 10 = 0$

122. $3x^2 - x - 10 = 0$

123. $14x^2 + 5x - 1 = 0$

124. $x^3 + 5x^2 - 4x - 20 = 0$

125. Find the coordinates of the point located 3 units to the left of the y-axis and 10 units above the x-axis.

126. Determine the quadrant(s) in which the point (x, y) could be located if $y > 0$.

1.8 Other Types of Inequalities

▶ **What you should learn**

- How to solve polynomial inequalities
- How to solve rational inequalities
- How to use inequalities to model and solve real-life problems

▶ **Why you should learn it**

Inequalities can be used to model and solve real-life problems. For instance, in Exercise 71 on page 159, a polynomial inequality is used to model the percent of households owning a television and having cable in the United States.

Stephen Ferry/Getty Images

Polynomial Inequalities

To solve a polynomial inequality such as $x^2 - 2x - 3 < 0$, you can use the fact that a polynomial can change signs only at its zeros (the x-values that make the polynomial equal to zero). Between two consecutive zeros, a polynomial must be entirely positive or entirely negative. This means that when the real zeros of a polynomial are put in order, they divide the real number line into intervals in which the polynomial has no sign changes. These zeros are the **critical numbers** of the inequality, and the resulting intervals are the **test intervals** for the inequality. For instance, the polynomial above factors as

$$x^2 - 2x - 3 = (x + 1)(x - 3)$$

and has two zeros, $x = -1$ and $x = 3$. These zeros divide the real number line into three test intervals:

$$(-\infty, -1), \quad (-1, 3), \quad \text{and} \quad (3, \infty). \qquad \text{(See Figure 1.29.)}$$

So, to solve the inequality $x^2 - 2x - 3 < 0$, you need only test one value from each of these test intervals to determine whether the value satisfies the original inequality. If so, you can conclude that the interval is a solution of the inequality.

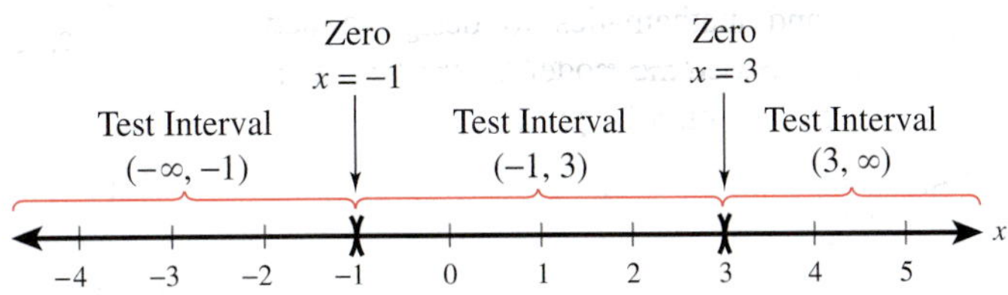

FIGURE 1.29 *Three test intervals for* $x^2 - 2x - 3$

You can use the same basic approach to determine the test intervals for any polynomial.

Finding Test Intervals for a Polynomial

To determine the intervals on which the values of a polynomial are entirely negative or entirely positive, use the following steps.

1. Find all real zeros of the polynomial, and arrange the zeros in increasing order (from smallest to largest). These zeros are the critical numbers of the polynomial.
2. Use the critical numbers of the polynomial to determine its test intervals.
3. Choose one representative x-value in each test interval and evaluate the polynomial at that value. If the value of the polynomial is negative, the polynomial will have negative values for every x-value in the interval. If the value of the polynomial is positive, the polynomial will have positive values for every x-value in the interval.

Example 1 ▶ Solving a Polynomial Inequality 

Solve

$$x^2 - x - 6 < 0.$$

Solution

By factoring the polynomial as

$$x^2 - x - 6 = (x + 2)(x - 3)$$

you can see that the critical numbers are $x = -2$ and $x = 3$. So, the polynomial's test intervals are

$$(-\infty, -2), \quad (-2, 3), \quad \text{and} \quad (3, \infty). \qquad \text{Test intervals}$$

In each test interval, choose a representative x-value and evaluate the polynomial.

Interval	*x-Value*	*Polynomial Value*	*Conclusion*
$(-\infty, -2)$	$x = -3$	$(-3)^2 - (-3) - 6 = 6$	Positive
$(-2, 3)$	$x = 0$	$(0)^2 - (0) - 6 = -6$	Negative
$(3, \infty)$	$x = 4$	$(4)^2 - (4) - 6 = 6$	Positive

From this you can conclude that the inequality is satisfied for all x-values in $(-2, 3)$. This implies that the solution of the inequality $x^2 - x - 6 < 0$ is the interval $(-2, 3)$, as shown in Figure 1.30.

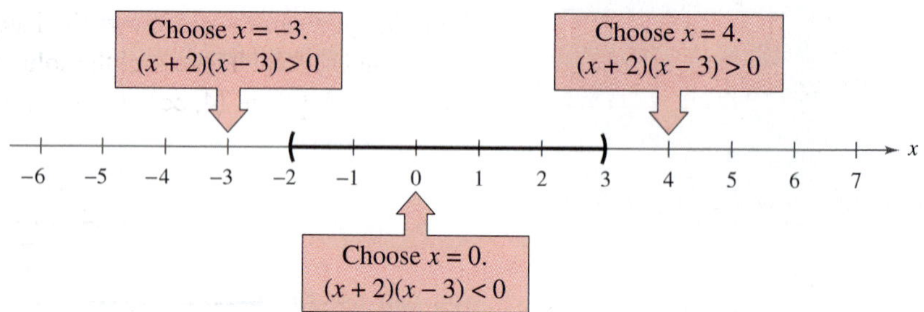

FIGURE 1.30

As with linear inequalities, you can check the reasonableness of a solution by substituting x-values into the original inequality. For instance, to check the solution found in Example 1, try substituting several x-values from the interval $(-2, 3)$ into the inequality

$$x^2 - x - 6 < 0.$$

Regardless of which x-values you choose, the inequality should be satisfied.

You can also use a graph to check the result of Example 1. Sketch the graph of $y = x^2 - x - 6$, as shown in Figure 1.31. Notice that the graph is below the x-axis on the interval $(-2, 3)$.

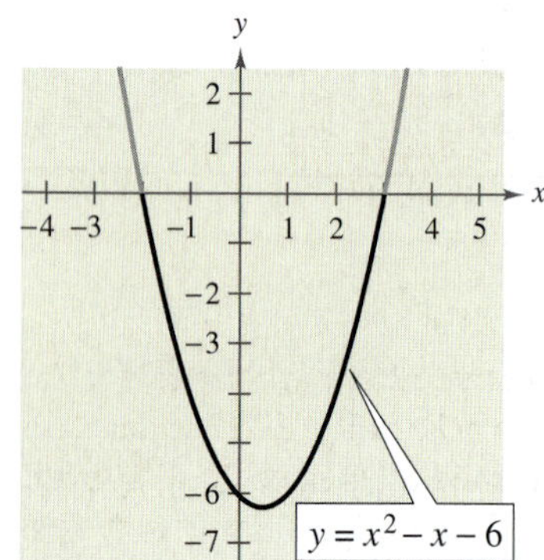

FIGURE 1.31

In Example 1, the polynomial inequality was given in general form (with the polynomial on one side and zero on the other). Whenever this is not the case, you should begin the solution process by writing the inequality in general form.

Example 2 ▶ Solving a Polynomial Inequality 

Solve $2x^3 - 3x^2 - 32x > -48$

Solution

Begin by writing the inequality in general form.

$$2x^3 - 3x^2 - 32x > -48 \qquad \text{Write original inequality.}$$

$$2x^3 - 3x^2 - 32x + 48 > 0 \qquad \text{Write in general form.}$$

$$(x - 4)(x + 4)(2x - 3) > 0 \qquad \text{Factor.}$$

The critical numbers are $x = -4$, $x = \frac{3}{2}$, and $x = 4$, and the test intervals are $(-\infty, -4)$, $\left(-4, \frac{3}{2}\right)$, $\left(\frac{3}{2}, 4\right)$, and $(4, \infty)$.

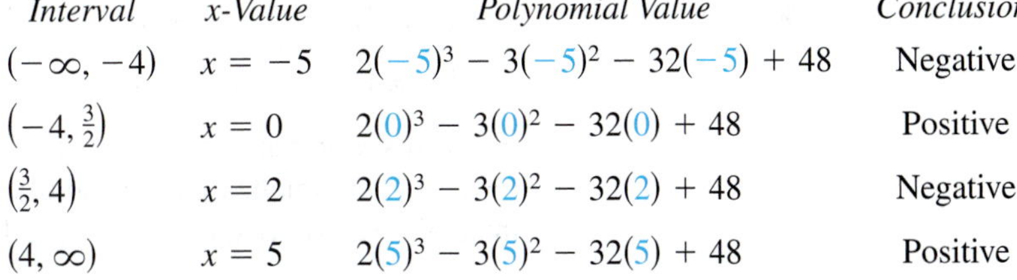

Interval	*x-Value*	*Polynomial Value*	*Conclusion*
$(-\infty, -4)$	$x = -5$	$2(-5)^3 - 3(-5)^2 - 32(-5) + 48$	Negative
$\left(-4, \frac{3}{2}\right)$	$x = 0$	$2(0)^3 - 3(0)^2 - 32(0) + 48$	Positive
$\left(\frac{3}{2}, 4\right)$	$x = 2$	$2(2)^3 - 3(2)^2 - 32(2) + 48$	Negative
$(4, \infty)$	$x = 5$	$2(5)^3 - 3(5)^2 - 32(5) + 48$	Positive

From this you can conclude that the inequality is satisfied on the open intervals $\left(-4, \frac{3}{2}\right)$ and $(4, \infty)$. Therefore, the solution set consists of all real numbers in the intervals $\left(-4, \frac{3}{2}\right)$ and $(4, \infty)$, as shown in Figure 1.32.

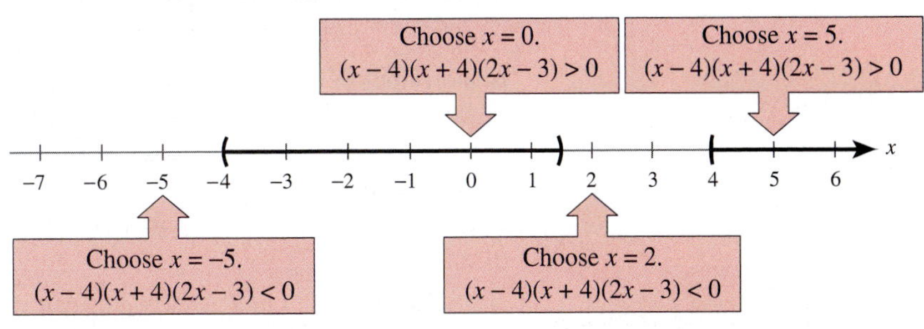

FIGURE 1.32

STUDY TIP

You may find it easier to determine the sign of a polynomial from its *factored* form. For instance, in Example 2, if the test value $x = 2$ is substituted into the factored form

$$(x - 4)(x + 4)(2x - 3)$$

you can see that the sign pattern of the factors is

$$(-)(+)(+)$$

which yields a negative result. Try using the factored form of the polynomial to determine the sign of the polynomial in the test intervals of the other examples in this section.

When solving a polynomial inequality, be sure you have accounted for the particular type of inequality symbol given in the inequality. For instance, in Example 2, note that the original inequality contained a "greater than" symbol and the solution consisted of two open intervals. If the original inequality had been

$$2x^3 - 3x^2 - 32x \geq 48$$

the solution would have consisted of the closed interval $\left[-4, \frac{3}{2}\right]$ and the interval $[4, \infty)$.

Each of the polynomial inequalities in Examples 1 and 2 has a solution set that consists of a single interval or the union of two intervals. When solving the exercises for this section, watch for unusual solution sets, as illustrated in Example 3.

Example 3 ▶ Unusual Solution Sets

a. The solution set of the following inequality consists of the entire set of real numbers, $(-\infty, \infty)$.

$$x^2 + 2x + 4 > 0$$

b. The solution set of the following inequality consists of the single real number $\{-1\}$.

$$x^2 + 2x + 1 \leq 0$$

c. The solution set of the following inequality is empty.

$$x^2 + 3x + 5 < 0$$

d. The solution set of the following inequality consists of all real numbers except $x = 2$.

$$x^2 - 4x + 4 > 0$$

Exploration

You can use a graphing utility to verify the results in Example 3. For instance, the graph of

$$y = x^2 + 2x + 4$$

is shown below. Notice that the y-values are greater than 0 for all values of x, as stated in Example 3(a). Use the graphing utility to graph the following:

$$y = x^2 + 2x + 1 \qquad y = x^2 + 3x + 5 \qquad y = x^2 - 4x + 4$$

Explain how you can use the graphs to verify the results of parts (b), (c), and (d) of Example 3.

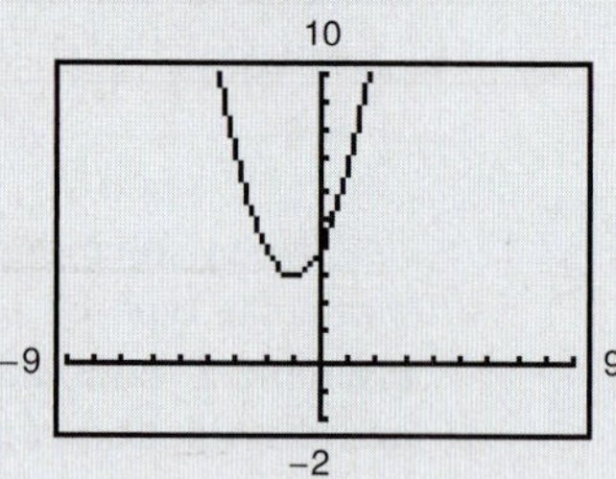

Rational Inequalities

The concepts of critical numbers and test intervals can be extended to rational inequalities. To do this, use the fact that the value of a rational expression can change sign only at its *zeros* (the x-values for which its numerator is zero) and its *undefined values* (the x-values for which its denominator is zero). These two types of numbers make up the *critical numbers* of a rational inequality.

Example 4 ▶ Solving a Rational Inequality 

Solve $\frac{2x-7}{x-5} \leq 3$.

Solution

Begin by writing the rational inequality in general form.

$$\frac{2x-7}{x-5} \leq 3 \qquad \text{Write original inequality.}$$

$$\frac{2x-7}{x-5} - 3 \leq 0 \qquad \text{Write in general form.}$$

$$\frac{2x-7-3x+15}{x-5} \leq 0 \qquad \text{Add fractions.}$$

$$\frac{-x+8}{x-5} \leq 0 \qquad \text{Simplify.}$$

Critical numbers: $x = 5, x = 8$ Zeros and undefined values of rational expression

Test intervals: $(-\infty, 5), (5, 8), (8, \infty)$

Test: Is $\frac{-x+8}{x-5} \leq 0$?

After testing these intervals, as shown in Figure 1.33, you can see that the inequality is satisfied on the open intervals $(-\infty, 5)$ and $(8, \infty)$. Moreover, because $(-x+8)/(x-5) = 0$ when $x = 8$, you can conclude that the solution set consists of all real numbers in the intervals $(-\infty, 5) \cup [8, \infty)$. (Be sure to use a closed interval to indicate that x can equal 8.)

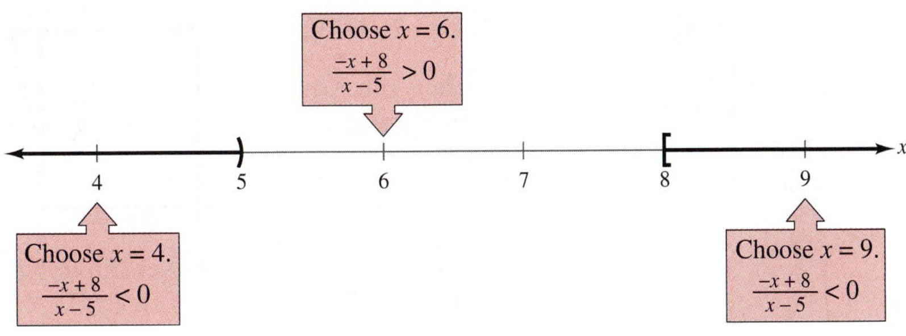

FIGURE 1.33

Applications

One common application of inequalities comes from business and involves profit, revenue, and cost. The formula that relates these three quantities is

$$\text{Profit} = \text{Revenue} - \text{Cost}$$

$$P = R - C.$$

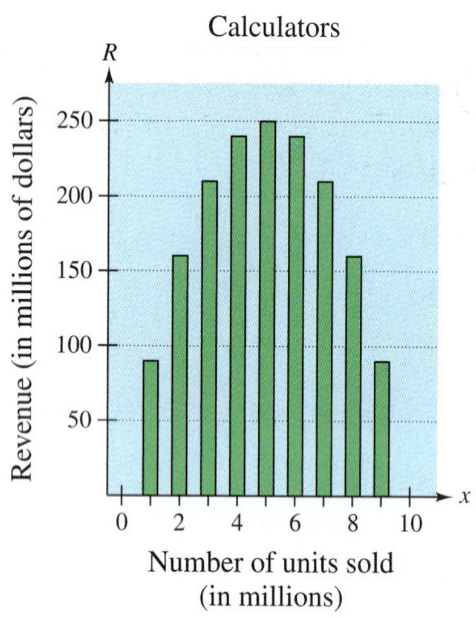

FIGURE 1.34

Example 5 ▶ Increasing the Profit for a Product  

The marketing department of a calculator manufacturer has determined that the demand for a new model of calculator is

$$p = 100 - 0.00001x, \qquad 0 \le x \le 10{,}000{,}000 \qquad \text{Demand equation}$$

where p is the price per calculator (in dollars) and x represents the number of calculators sold. (If this model is accurate, no one would be willing to pay \$100 for the calculator. At the other extreme, the company couldn't sell more than 10 million calculators.) The revenue for selling x calculators is

$$R = xp = x(100 - 0.00001x) \qquad \text{Revenue equation}$$

as shown in Figure 1.34. The total cost of producing x calculators is \$10 per calculator plus a development cost of \$2,500,000. So, the total cost is

$$C = 10x + 2{,}500{,}000. \qquad \text{Cost equation}$$

What price should the company charge per calculator to obtain a profit of at least \$190,000,000?

Solution

Verbal Model: $\text{Profit} = \text{Revenue} - \text{Cost}$

Equation: $P = R - C$

$$P = 100x - 0.00001x^2 - (10x + 2{,}500{,}000)$$

$$P = -0.00001x^2 + 90x - 2{,}500{,}000$$

To answer the question, solve the inequality

$$P \ge 190{,}000{,}000$$

$$-0.00001x^2 + 90x - 2{,}500{,}000 \ge 190{,}000{,}000.$$

When you write the inequality in general form, find the critical numbers and the test intervals, and then test a value in each test interval, you can find the solution to be

$$3{,}500{,}000 \le x \le 5{,}500{,}000$$

as shown in Figure 1.35. Substituting the x-values in the original price equation shows that prices of

$$\$45.00 \le p \le \$65.00$$

will yield a profit of at least \$190,000,000.

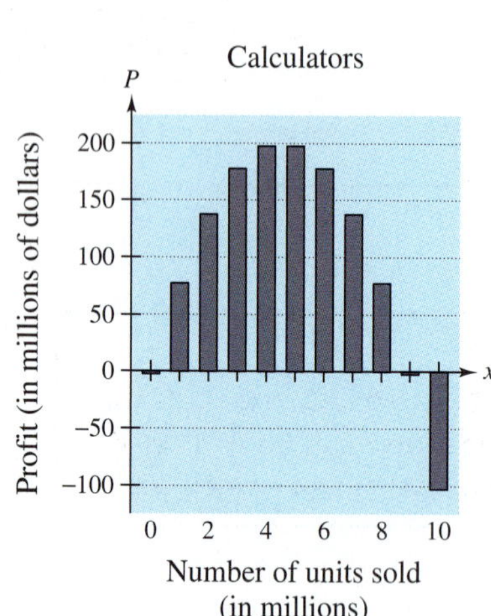

FIGURE 1.35

Another common application of inequalities is finding the domain of an expression that involves a square root, as shown in Example 6.

Example 6 ▶ Finding a Domain of an Expression 

Find the domain of $\sqrt{64 - 4x^2}$.

Solution

Remember that the domain of an expression is the set of all x-values for which the expression is defined. Because $\sqrt{64 - 4x^2}$ is defined (has real values) only if $64 - 4x^2$ is nonnegative, the domain is given by $64 - 4x^2 \geq 0$.

$$64 - 4x^2 \geq 0 \quad \text{Write in general form.}$$

$$16 - x^2 \geq 0 \quad \text{Divide each side by 4.}$$

$$(4 - x)(4 + x) \geq 0 \quad \text{Write in factored form.}$$

So, the inequality has two critical numbers: $x = -4$ and $x = 4$. You can use these two numbers to test the inequality as follows.

Critical numbers: $x = -4, x = 4$

Test intervals: $(-\infty, -4), (-4, 4), (4, \infty)$

Test: Is $(4 - x)(4 + x) \geq 0$?

A test shows that the inequality is satisfied in the *closed interval* $[-4, 4]$. So, the domain of the expression $\sqrt{64 - 4x^2}$ is the interval $[-4, 4]$, as shown in Figure 1.36.

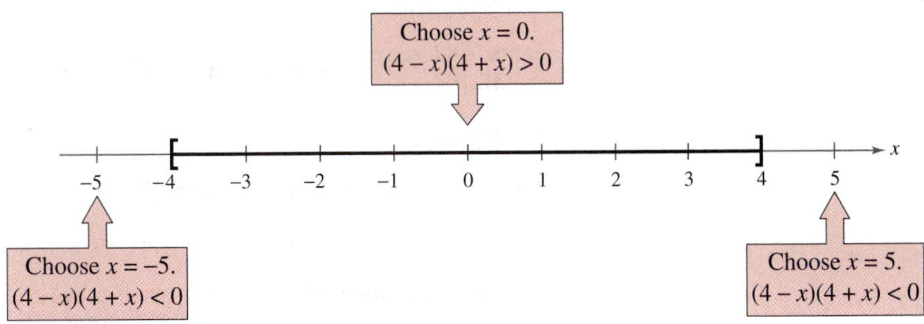

FIGURE 1.36

Writing ABOUT MATHEMATICS

Profit Analysis Consider the relationship $P = R - C$ described on page 156. Write a paragraph discussing why it might be beneficial to solve $P < 0$ if you owned a business. Use the situation described in Example 5 to illustrate your reasoning.

1.8 Exercises

In Exercises 1–4, determine whether each value of x is a solution of the inequality.

Inequality	Values	
1. $x^2 - 3 < 0$	(a) $x = 3$	(b) $x = 0$
	(c) $x = \frac{3}{2}$	(d) $x = -5$
2. $x^2 - x - 12 \geq 0$	(a) $x = 5$	(b) $x = 0$
	(c) $x = -4$	(d) $x = -3$
3. $\dfrac{x+2}{x-4} \geq 3$	(a) $x = 5$	(b) $x = 4$
	(c) $x = -\frac{9}{2}$	(d) $x = \frac{9}{2}$
4. $\dfrac{3x^2}{x^2+4} < 1$	(a) $x = -2$	(b) $x = -1$
	(c) $x = 0$	(d) $x = 3$

In Exercises 5–8, find the critical numbers.

5. $2x^2 - x - 6$

6. $9x^3 - 25x^2$

7. $2 + \dfrac{3}{x-5}$

8. $\dfrac{x}{x+2} - \dfrac{2}{x-1}$

In Exercises 9–24, solve the inequality and graph the solution on the real number line.

9. $x^2 \leq 9$

10. $x^2 < 36$

11. $(x+2)^2 < 25$

12. $(x-3)^2 \geq 1$

13. $x^2 + 4x + 4 \geq 9$

14. $x^2 - 6x + 9 < 16$

15. $x^2 + x < 6$

16. $x^2 + 2x > 3$

17. $x^2 + 2x - 3 < 0$

18. $x^2 - 4x - 1 > 0$

19. $x^2 + 8x - 5 \geq 0$

20. $-2x^2 + 6x + 15 \leq 0$

21. $x^3 - 3x^2 - x + 3 > 0$

22. $x^3 + 2x^2 - 4x - 8 \leq 0$

23. $x^3 - 2x^2 - 9x - 2 \geq -20$

24. $2x^3 + 13x^2 - 8x - 46 \geq 6$

In Exercises 25–30, solve the inequality and write the solution set in interval notation.

25. $4x^3 - 6x^2 < 0$

26. $4x^3 - 12x^2 > 0$

27. $x^3 - 4x \geq 0$

28. $2x^3 - x^4 \leq 0$

29. $(x-1)^2(x+2)^3 \geq 0$

30. $x^4(x-3) \leq 0$

Graphical Analysis In Exercises 31–34, use a graphing utility to graph the equation. Use the graph to approximate the values of x that satisfy each inequality.

Equation	Inequalities	
31. $y = -x^2 + 2x + 3$	(a) $y \leq 0$	(b) $y \geq 3$
32. $y = \frac{1}{2}x^2 - 2x + 1$	(a) $y \leq 0$	(b) $y \geq 7$
33. $y = \frac{1}{8}x^3 - \frac{1}{2}x$	(a) $y \geq 0$	(b) $y \leq 6$
34. $y = x^3 - x^2 - 16x + 16$	(a) $y \leq 0$	(b) $y \geq 36$

In Exercises 35–48, solve the inequality and graph the solution on the real number line.

35. $\dfrac{1}{x} - x > 0$

36. $\dfrac{1}{x} - 4 < 0$

37. $\dfrac{x+6}{x+1} - 2 < 0$

38. $\dfrac{x+12}{x+2} - 3 \geq 0$

39. $\dfrac{3x-5}{x-5} > 4$

40. $\dfrac{5+7x}{1+2x} < 4$

41. $\dfrac{4}{x+5} > \dfrac{1}{2x+3}$

42. $\dfrac{5}{x-6} > \dfrac{3}{x+2}$

43. $\dfrac{1}{x-3} \leq \dfrac{9}{4x+3}$

44. $\dfrac{1}{x} \geq \dfrac{1}{x+3}$

45. $\dfrac{x^2+2x}{x^2-9} \leq 0$

46. $\dfrac{x^2+x-6}{x} \geq 0$

47. $\dfrac{5}{x-1} - \dfrac{2x}{x+1} < 1$

48. $\dfrac{3x}{x-1} \leq \dfrac{x}{x+4} + 3$

Graphical Analysis In Exercises 49–52, use a graphing utility to graph the equation. Use the graph to approximate the values of x that satisfy each inequality.

Equation	Inequalities	
49. $y = \dfrac{3x}{x-2}$	(a) $y \leq 0$	(b) $y \geq 6$
50. $y = \dfrac{2(x-2)}{x+1}$	(a) $y \leq 0$	(b) $y \geq 8$
51. $y = \dfrac{2x^2}{x^2+4}$	(a) $y \geq 1$	(b) $y \leq 2$
52. $y = \dfrac{5x}{x^2+4}$	(a) $y \geq 1$	(b) $y \leq 0$

In Exercises 53–58, find the domain of x in the expression.

53. $\sqrt{4 - x^2}$ **54.** $\sqrt{x^2 - 4}$

55. $\sqrt{x^2 - 7x + 12}$ **56.** $\sqrt{144 - 9x^2}$

57. $\sqrt{\dfrac{x}{x^2 - 2x - 35}}$ **58.** $\sqrt{\dfrac{x}{x^2 - 9}}$

In Exercises 59–64, solve the inequality. (Round your answers to two decimal places.)

59. $0.4x^2 + 5.26 < 10.2$ **60.** $-1.3x^2 + 3.78 > 2.12$

61. $-0.5x^2 + 12.5x + 1.6 > 0$

62. $1.2x^2 + 4.8x + 3.1 < 5.3$

63. $\dfrac{1}{2.3x - 5.2} > 3.4$ **64.** $\dfrac{2}{3.1x - 3.7} > 5.8$

65. ***Physics*** A projectile is fired straight upward from ground level with an initial velocity of 160 feet per second.

(a) At what instant will it be back at ground level?

(b) When will the height exceed 384 feet?

66. ***Physics*** A projectile is fired straight upward from ground level with an initial velocity of 128 feet per second.

(a) At what instant will it be back at ground level?

(b) When will the height be less than 128 feet?

67. ***Geometry*** A rectangular playing field with a perimeter of 100 meters is to have an area of at least 500 square meters. Within what bounds must the length of the rectangle lie?

68. ***Geometry*** A rectangular parking lot with a perimeter of 440 feet is to have an area of at least 8000 square feet. Within what bounds must the length of the rectangle lie?

69. ***Investment*** P dollars, invested at interest rate r compounded annually, increases to an amount

$$A = P(1 + r)^2$$

in 2 years. An investment of \$1000 is to increase to an amount greater than \$1100 in 2 years. The interest rate must be greater than what percent?

70. ***Cost, Revenue, and Profit*** The revenue and cost equations for a product are

$$R = x(50 - 0.0002x) \quad \text{and} \quad C = 12x + 150{,}000$$

where R and C are measured in dollars and x represents the number of units sold. How many units must be sold to obtain a profit of at least \$1,650,000?

▶ Model It

71. ***Cable Television*** The percent C of households in the United States that owned a television and had cable from 1970 to 2000 can be modeled by

$$C = 0.0002t^4 - 0.018t^3 + 0.48t^2 - 1.5t + 8$$

where t is the year, with $t = 0$ corresponding to 1970. (Source: Nielsen Media Research)

 (a) Use a graphing utility to graph the equation.

(b) Complete the table to determine the year in which the percent of households that own a television and have cable will exceed 72%.

t	30	32	34	36	38	40
C						

 (c) Use the *trace* feature of a graphing utility to verify your answer to part (b).

(d) Complete the table to determine the years during which the percent of households that own a television and have cable will be between 72% and 100%.

t	40	41	42	43	44	45	46	47
C								

 (e) Use the *trace* feature of a graphing utility to verify your answer to part (d).

(f) Explain why the model may have values greater than 100% even though such values are not reasonable.

72. ***Safe Load*** The maximum safe load uniformly distributed over a one-foot section of a two-inch-wide wooden beam is approximated by the model

$$\text{Load} = 168.5d^2 - 472.1$$

where d is the depth of the beam.

(a) Evaluate the model for $d = 4$, $d = 6$, $d = 8$, $d = 10$, and $d = 12$. Use the results to create a bar graph.

(b) Determine the minimum depth of the beam that will safely support a load of 2000 pounds.

73. ***Resistors*** When two resistors of resistance R_1 and R_2 are connected in parallel (see figure), the total resistance R satisfies the equation

$$\frac{1}{R} = \frac{1}{R_1} + \frac{1}{R_2}.$$

Find R_1 for a parallel circuit in which $R_2 = 2$ ohms and R must be at least 1 ohm.

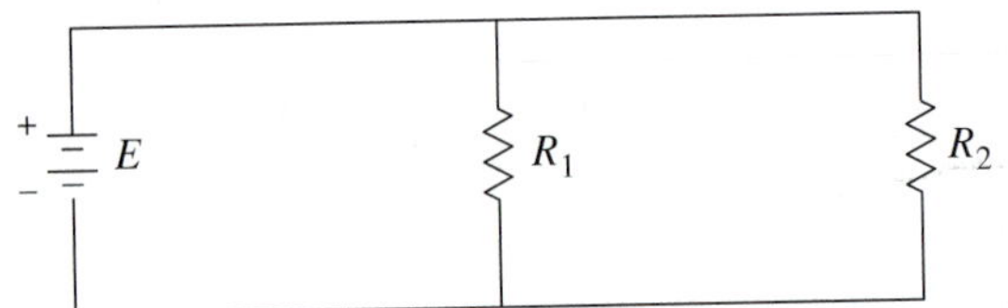

74. ***Education*** The percent P of the U.S. population that had completed 4 years of college or more from 1960 to 1999 is approximated by the model

$$P = 0.0006t^2 + 0.441t + 7.21$$

where t is the year, with $t = 0$ corresponding to 1960. According to this model, during what year will more than 27% of the population be college graduates? (Source: U.S. Census Bureau)

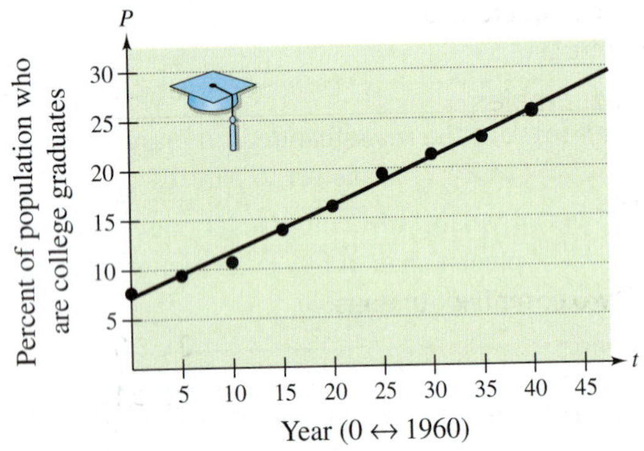

Synthesis

True or False? **In Exercises 75 and 76, determine whether the statement is true or false. Justify your answer.**

75. The zeros of the polynomial $x^3 - 2x^2 - 11x + 12 \geq 0$ divide the real number line into four test intervals.

76. The solution set of the inequality $\frac{3}{2}x^2 + 3x + 6 \geq 0$ is the entire set of real numbers.

Exploration **In Exercises 77–80, find the interval for b such that the equation has at least one real solution.**

77. $x^2 + bx + 4 = 0$

78. $x^2 + bx - 4 = 0$

79. $3x^2 + bx + 10 = 0$

80. $2x^2 + bx + 5 = 0$

81. (a) Write a conjecture about the interval for b in Exercises 77–80. Explain your reasoning.

(b) What is the center of the interval for b in Exercises 77–80?

82. Consider the polynomial $(x - a)(x - b)$ and the real number line shown below.

(a) Identify the points on the line at which the polynomial is zero.

(b) In each of the three subintervals of the line, write the sign of each factor and the sign of the product.

(c) For what x-values does the polynomial change signs?

Review

In Exercises 83–86, factor the expression completely.

83. $4x^2 + 20x + 25$

84. $(x + 3)^2 - 16$

85. $x^2(x + 3) - 4(x + 3)$

86. $2x^4 - 54x$

In Exercises 87 and 88, write an expression for the area of the region.

87. **88.**

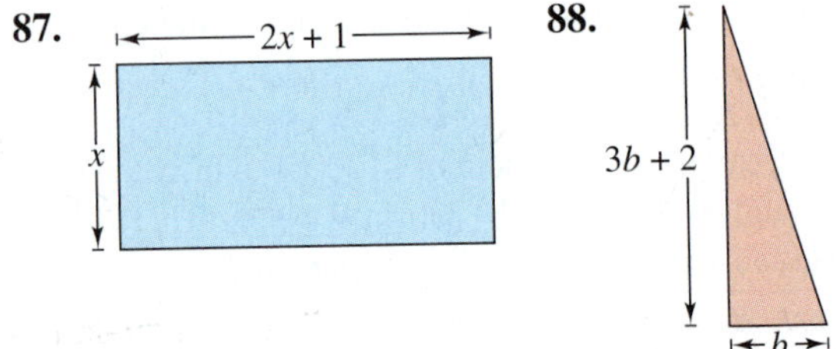

Chapter Summary

▶ *What* did you learn?

Section 1.1	Review Exercises
☐ How to sketch graphs of equations and find x- and y- intercepts of graphs of equations	1–12
☐ How to use symmetry to sketch graphs of equations	13–20
☐ How to find equations and sketch graphs of circles	21–28
☐ How to use graphs of equations in solving real-life problems	29, 30
Section 1.2	
☐ How to identify equations and solve linear equations in one variable	31–38
☐ How to solve equations that lead to linear equations	39–42
☐ How to find x- and y-intercepts of graphs of equations algebraically	43–50
☐ How to use linear equations to model and solve real-life problems	51, 52
Section 1.3	
☐ How to use a verbal model in a problem-solving plan	53, 54
☐ How to write and use mathematical models to solve real-life problems	55–58
☐ How to solve mixture problems and use common formulas to solve real-life problems	59–64
Section 1.4	
☐ How to solve quadratic equations by factoring, by extracting square roots, by completing the square, and by using the Quadratic Formula	65–74
☐ How to use quadratic equations to model and solve real-life problems	75, 76
Section 1.5	
☐ How to use the imaginary unit i to write complex numbers	77–80
☐ How to add, subtract, and multiply complex numbers	81–86
☐ How to use complex conjugates to write the quotient of two complex numbers in standard form	87–90
☐ How to find complex solutions of quadratic equations	91–94
Section 1.6	
☐ How to solve polynomial equations of degree three or greater and solve equations involving radicals, fractions, or absolute values	95–112
☐ How to use different types of equations to model and solve real-life problems	113, 114
Section 1.7	
☐ How to represent solutions of linear inequalities in one variable, solve linear inequalities in one variable, and solve inequalities involving absolute values	115–128
☐ How to use inequalities to model and solve real-life problems	129, 130
Section 1.8	
☐ How to solve polynomial and rational inequalities	131–138
☐ How to use inequalities to model and solve real-life problems	139, 140

Review Exercises

1.1 In Exercises 1–4, complete a table of values. Use the solution points to sketch the graph of the equation.

1. $y = 3x - 5$
2. $y = -\frac{1}{2}x + 2$
3. $y = x^2 - 3x$
4. $y = 2x^2 - x - 9$

In Exercises 5–10, sketch the graph *by hand*.

5. $y - 2x - 3 = 0$
6. $3x + 2y + 6 = 0$
7. $y = \sqrt{5 - x}$
8. $y = \sqrt{x + 2}$
9. $y + 2x^2 = 0$
10. $y = x^2 - 4x$

In Exercises 11 and 12, find the x- and y-intercepts of the graph of the equation.

11. $y = (x - 3)^2 - 4$
12. $y = |x + 1| - 3$

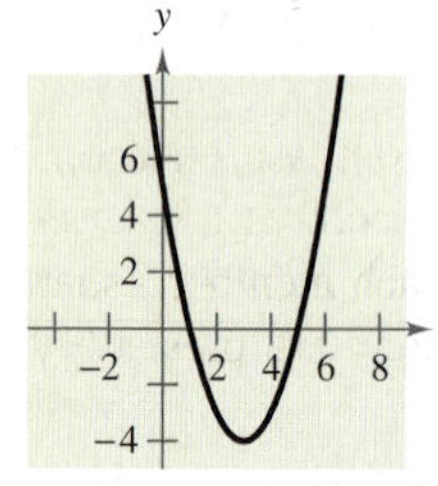

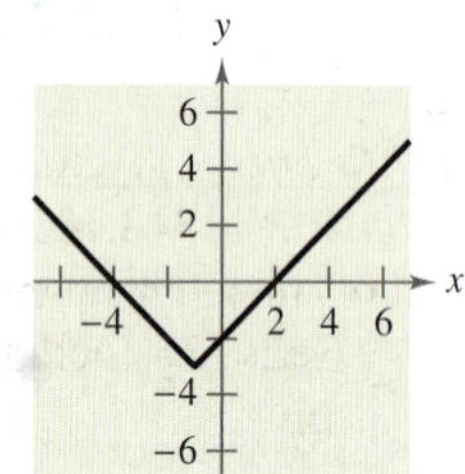

In Exercises 13–20, use the algebraic tests to check for symmetry with respect to both axes and the origin. Then sketch the graph.

13. $y = -4x + 1$
14. $y = 5x - 6$
15. $y = 5 - x^2$
16. $y = x^2 - 10$
17. $y = x^3 + 3$
18. $y = -6 - x^3$
19. $y = \sqrt{x + 5}$
20. $y = |x| + 9$

In Exercises 21–26, find the center and radius of the circle and sketch its graph.

21. $x^2 + y^2 = 9$
22. $x^2 + y^2 = 4$
23. $(x + 2)^2 + y^2 = 16$
24. $x^2 + (y - 8)^2 = 81$
25. $\left(x - \frac{1}{2}\right)^2 + (y + 1)^2 = 36$
26. $(x + 4)^2 + \left(y - \frac{3}{2}\right)^2 = 100$

27. Find the standard form of the equation of the circle for which the endpoints of a diameter are $(0, 0)$ and $(4, -6)$.

28. Find the standard form of the equation of the circle for which the endpoints of a diameter are $(-2, -3)$ and $(4, -10)$.

29. ***Physics*** The force F (in pounds) required to stretch a spring x inches from its natural length (see figure) is

$$F = \frac{5}{4}x, \quad 0 \le x \le 20.$$

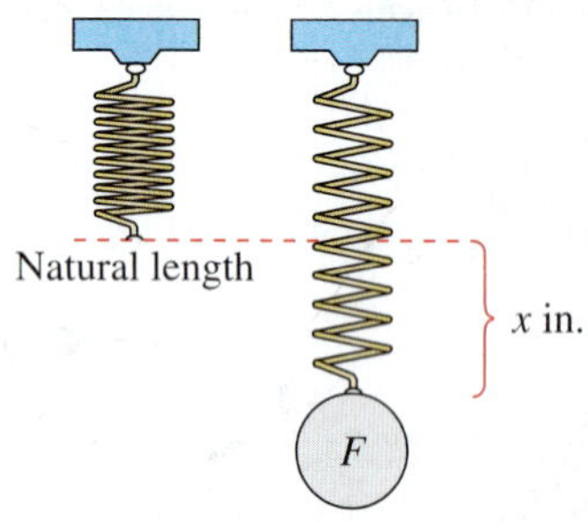

(a) Use the model to complete the table.

x	0	4	8	12	16	20
Force, F						

(b) Sketch a graph of the model.

(c) Use the graph to estimate the force necessary to stretch the spring 10 inches.

30. ***Number of Stores*** The number N of Home Depot stores from 1993 to 2000 can be approximated by the model

$$N = 9.53t^2 + 162$$

where t is the time (in years), with $t = 3$ corresponding to 1993. (Source: Home Depot, Inc.)

(a) Sketch a graph of the model.

(b) Use the graph to estimate the year in which the number of stores will be 2000.

1.2 In Exercises 31–34, determine whether the equation is an identity or a conditional equation.

31. $6 - (x - 2)^2 = 2 + 4x - x^2$
32. $3(x - 2) + 2x = 2(x + 3)$
33. $-x^3 + x(7 - x) + 3 = x(-x^2 - x) + 7(x + 1) - 4$
34. $3(x^2 - 4x + 8) = -10(x + 2) - 3x^2 + 6$

In Exercises 35–42, solve the equation (if possible) and check your solution.

35. $3x - 2(x + 5) = 10$ **36.** $4x + 2(7 - x) = 5$

37. $4(x + 3) - 3 = 2(4 - 3x) - 4$

38. $\frac{1}{2}(x - 3) - 2(x + 1) = 5$

39. $\frac{x}{5} - 3 = \frac{2x}{2} + 1$ **40.** $\frac{4x - 3}{6} + \frac{x}{4} = x - 2$

41. $\frac{18}{x} = \frac{10}{x - 4}$ **42.** $\frac{5}{x - 2} = \frac{13}{2x - 3}$

In Exercises 43–50, find the *x*- and *y*-intercepts of the graph of the equation algebraically.

43. $y = 3x - 1$ **44.** $y = -5x + 6$

45. $y = 2(x - 4)$ **46.** $y = 4(7x + 1)$

47. $y = -\frac{1}{2}x + \frac{2}{3}$ **48.** $y = \frac{3}{4}x - \frac{1}{4}$

49. $3.8y - 0.5x + 1 = 0$ **50.** $1.5y + 2x - 1.2 = 0$

51. ***Geometry*** The surface area of the cylinder shown in the figure is approximated by

$$S = 2(3.14)(3)^2 + 2(3.14)(3)h.$$

The surface area is 244.92 square inches. Find the height h of the cylinder.

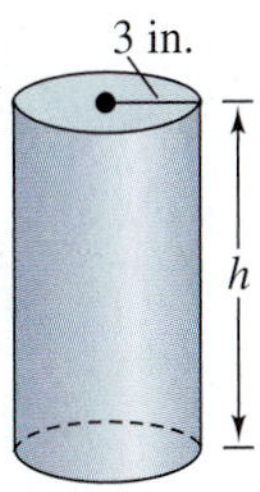

52. ***Temperature*** The Fahrenheit and Celsius temperature scales are related by the equation

$$C = \frac{5}{9}F - \frac{160}{9}.$$

Find the Fahrenheit temperature that corresponds to 100° Celsius.

1.3 **53.** ***Profit*** In October, a greeting card company's total profit was 12% more than it was in September. The total profit for the two months was \$689,000. Write a verbal model, assign labels, and write an algebraic equation to find the profit for each month.

54. ***Discount*** The price of a television set has been discounted \$85. The sale price is \$340. Write a verbal model, assign labels, and write an algebraic equation to find the percent discount.

55. ***Shadow Length*** A person who is 6 feet tall walks away from a streetlight toward the tip of the streetlight's shadow. When the person is 15 feet from the streetlight, the tip of the person's shadow and the shadow cast by the streetlight coincide at a point 5 feet in front of the person (see figure). How tall is the streetlight?

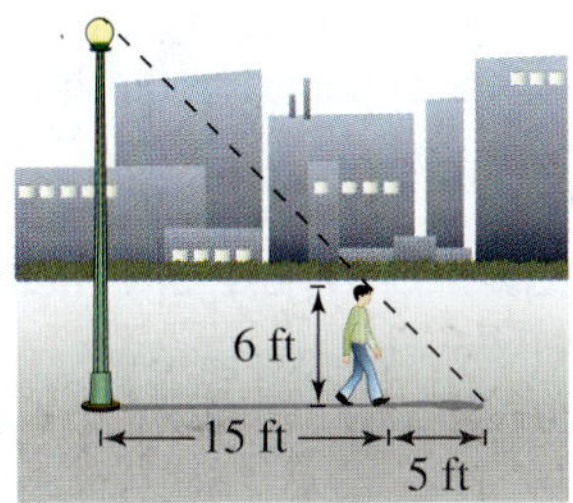

56. ***Finance*** A group agrees to share equally in the cost of a \$48,000 piece of machinery. If it can find two more group members, each member's share will decrease by \$4000. How many are presently in the group?

57. ***Business Venture*** You are planning to start a small business that will require an investment of \$90,000. You have found some people who are willing to share equally in the venture. If you can find three more people, each person's share will decrease by \$2500. How many people have you found so far?

58. ***Average Speed*** You commute 56 miles one way to work. The trip to work takes 10 minutes longer than the trip home. Your average speed on the trip home is 8 miles per hour faster. What is your average speed on the trip home?

59. ***Mixture Problem*** A car radiator contains 10 liters of a 30% antifreeze solution. How many liters will have to be replaced with pure antifreeze if the resulting solution is to be 50% antifreeze?

60. ***Investment*** You invested \$6000 at $4\frac{1}{2}\%$ and $5\frac{1}{2}\%$ simple interest. During the first year, the two accounts earned \$305. How much did you invest in each?

In Exercises 61 and 62, solve for the indicated variable.

61. ***Volume of a Cone***

Solve for h: $V = \frac{1}{3}\pi r^2 h$

62. ***Kinetic Energy***

Solve for m: $E = \frac{1}{2}mv^2$

63. ***Travel Time*** Two cars start at a given time and travel in the same direction at average speeds of 40 miles per hour and 55 miles per hour. How much time will elapse before the two cars are 10 miles apart?

64. ***Geometry*** The volume of a circular cylinder is 81π cubic feet. The cylinder's radius is 3 feet. What is the height of the cylinder?

1.4 In Exercises 65–74, use any method to solve the quadratic equation.

65. $15 + x - 2x^2 = 0$ **66.** $2x^2 - x - 28 = 0$

67. $6 = 3x^2$ **68.** $16x^2 = 25$

69. $(x + 4)^2 = 18$ **70.** $(x - 8)^2 = 15$

71. $x^2 - 12x + 30 = 0$ **72.** $x^2 + 6x - 3 = 0$

73. $-2x^2 - 5x + 27 = 0$

74. $-20 - 3x + 3x^2 = 0$

75. ***Simply Supported Beam*** A simply supported 20-foot beam supports a uniformly distributed load of 1000 pounds per foot. The bending moment M (in foot-pounds) x feet from one end of the beam is given by $M = 500x(20 - x)$.

(a) Where is the bending moment zero?

(b) Use a graphing utility to graph the equation.

(c) Use the graph to determine the point on the beam where the bending moment is the greatest.

76. ***Physics*** A ball is thrown upward with an initial velocity of 30 feet per second from a point that is 24 feet above the ground. The height h (in feet) of the ball at time t (in seconds) after it is thrown is

$$h = -16t^2 + 30t + 24.$$

Find the time when the ball hits the ground.

1.5 In Exercises 77–80, write the complex number in standard form.

77. $6 + \sqrt{-4}$ **78.** $3 - \sqrt{-25}$

79. $i^2 + 3i$ **80.** $-5i + i^2$

In Exercises 81–86, perform the operation and write the result in standard form.

81. $(7 + 5i) + (-4 + 2i)$

82. $\left(\frac{\sqrt{2}}{2} - \frac{\sqrt{2}}{2}i\right) - \left(\frac{\sqrt{2}}{2} + \frac{\sqrt{2}}{2}i\right)$

83. $5i(13 - 8i)$ **84.** $(1 + 6i)(5 - 2i)$

85. $(10 - 8i)(2 - 3i)$ **86.** $i(6 + i)(3 - 2i)$

In Exercises 87 and 88, write the quotient in standard form.

87. $\dfrac{6 + i}{4 - i}$ **88.** $\dfrac{3 + 2i}{5 + i}$

In Exercises 89 and 90, perform the operation and write the result in standard form.

89. $\dfrac{4}{2 - 3i} + \dfrac{2}{1 + i}$ **90.** $\dfrac{1}{2 + i} - \dfrac{5}{1 + 4i}$

In Exercises 91–94, find all solutions of the equation.

91. $3x^2 + 1 = 0$ **92.** $2 + 8x^2 = 0$

93. $x^2 - 2x + 10 = 0$ **94.** $6x^2 + 3x + 27 = 0$

1.6 In Exercises 95–112, find all solutions of the equation. Check your solutions in the original equation.

95. $5x^4 - 12x^3 = 0$ **96.** $4x^3 - 6x^2 = 0$

97. $x^4 - 5x^2 + 6 = 0$

98. $9x^4 + 27x^3 - 4x^2 - 12x = 0$

99. $\sqrt{x + 4} = 3$ **100.** $\sqrt{x - 2} - 8 = 0$

101. $\sqrt{2x + 3} + \sqrt{x - 2} = 2$

102. $5\sqrt{x} - \sqrt{x - 1} = 6$

103. $(x - 1)^{2/3} - 25 = 0$ **104.** $(x + 2)^{3/4} = 27$

105. $(x + 4)^{1/2} + 5x(x + 4)^{3/2} = 0$

106. $8x^2(x^2 - 4)^{1/3} + (x^2 - 4)^{4/3} = 0$

107. $\dfrac{5}{x} = 1 + \dfrac{3}{x + 2}$ **108.** $\dfrac{6}{x} + \dfrac{8}{x + 5} = 3$

109. $|x - 5| = 10$ **110.** $|2x + 3| = 7$

111. $|x^2 - 3| = 2x$ **112.** $|x^2 - 6| = x$

113. ***Demand*** The demand equation for a hair dryer is

$$p = 42 - \sqrt{0.001x + 2}$$

where x is the number of units demanded per day and p is the price per unit. Find the demand if the price is set at \$29.95.

114. ***Data Analysis*** The amount C of chlorofluorocarbon gases (CFCs) in thousands of metric tons emitted in the United States from 1993 to 1999 can be approximated by the model

$$C = 2.60t^2 - 48.7t + 269$$

where $t = 3$ represents 1993. The actual amounts emitted are shown in the table. (Source: U.S. Energy Information Administration)

Year, t	CFCs emitted, C
1993	148
1994	109
1995	102
1996	67
1997	51
1998	49
1999	41

(a) Use a graphing utility to compare the data with the model.

(b) Use the graph in part (a) to estimate the amount of CFCs emitted in 2005.

(c) Use the model to verify algebraically the estimate from part (b).

1.7 **In Exercises 115–118, write an inequality that represents the interval and state whether the interval is bounded or unbounded.**

115. $(-7, 2]$ **116.** $(4, \infty)$

117. $(-\infty, -10]$ **118.** $[-2, 2]$

In Exercises 119–128, solve the inequality.

119. $9x - 8 \le 7x + 16$ **120.** $\frac{15}{2}x + 4 > 3x - 5$

121. $4(5 - 2x) \le \frac{1}{2}(8 - x)$

122. $\frac{1}{2}(3 - x) > \frac{1}{3}(2 - 3x)$

123. $-19 < 3x - 17 \le 34$ **124.** $-3 \le \dfrac{2x - 5}{3} < 5$

125. $|x| \le 4$ **126.** $|x - 2| < 1$

127. $|x - 3| > 4$ **128.** $\left|x - \frac{3}{2}\right| \ge \frac{3}{2}$

129. ***Geometry*** The side of a square is measured as 19.3 centimeters with a possible error of 0.5 centimeter. Using these measurements, determine the interval containing the area of the square.

130. ***Cost, Revenue, and Profit*** The revenue for selling x units of a product is $R = 125.33x$. The cost of producing x units is $C = 92x + 1200$. To obtain a profit, the revenue must be greater than the cost. Determine the smallest value of x for which this product returns a profit.

1.8 **In Exercises 131–138, solve the inequality.**

131. $x^2 - 6x - 27 < 0$ **132.** $x^2 - 2x \ge 3$

133. $6x^2 + 5x < 4$ **134.** $2x^2 + x \ge 15$

135. $\dfrac{2}{x + 1} \le \dfrac{3}{x - 1}$ **136.** $\dfrac{x - 5}{3 - x} < 0$

137. $\dfrac{x^2 + 7x + 12}{x} \ge 0$ **138.** $\dfrac{1}{x - 2} > \dfrac{1}{x}$

139. ***Investment*** P dollars invested at interest rate r compounded annually increases to an amount $A = P(1 + r)^2$ in 2 years. An investment of \$5000 is to increase to an amount greater than \$5500 in 2 years. The interest rate must be greater than what percent?

140. ***Population of a Species*** A biologist introduces 200 ladybugs into a crop field. The population P of the ladybugs is approximated by the model

$$P = \frac{1000(1 + 3t)}{5 + t}$$

where t is the time in days. Find the time required for the population to increase to at least 2000 ladybugs.

Synthesis

True or False? **In Exercises 141 and 142, determine whether the statement is true or false. Justify your answer.**

141. $\sqrt{-18}\sqrt{-2} = \sqrt{(-18)(-2)}$

142. The equation $325x^2 - 717x + 398 = 0$ has no solution.

143. Explain why it is important to check your solutions to certain types of equations.

144. ***Error Analysis*** What is wrong with the following solution?

$$|11x + 4| \ge 26$$

$$11x + 4 \le 26 \quad \text{or} \quad 11x + 4 \ge 26$$

$$11x \le 22 \qquad\qquad 11x \ge 22$$

$$x \le 2 \qquad\qquad x \ge 2$$

Chapter Test

The *Interactive* CD-ROM and *Internet* versions of this text offer Chapter Pre-Tests and Chapter Post-Tests, both of which have randomly generated exercises with diagnostic capabilities.

Take this test as you would take a test in class. When you are finished, check your work against the answers given in the back of the book.

In Exercises 1–6, check for symmetry with respect to both axes and the origin. Then sketch the graph of the equation. Identify any *x*- and *y*-intercepts.

1. $y = 4 - \frac{3}{4}x$ **2.** $y = 4 - \frac{3}{4}|x|$ **3.** $y = 4 - (x - 2)^2$

4. $y = x - x^3$ **5.** $y = \sqrt{3 - x}$ **6.** $(x - 3)^2 + y^2 = 9$

In Exercises 7–12, solve the equation (if possible).

7. $\frac{2}{3}(x - 1) + \frac{1}{4}x = 10$ **8.** $(x - 3)(x + 2) = 14$

9. $\dfrac{x - 2}{x + 2} + \dfrac{4}{x + 2} + 4 = 0$ **10.** $x^4 + x^2 - 6 = 0$

11. $2\sqrt{x} - \sqrt{2x + 1} = 1$ **12.** $|3x - 1| = 7$

In Exercises 13–16, solve the inequality. Sketch the solution on the real number line.

13. $-3 \le 2(x + 4) < 14$ **14.** $\dfrac{2}{x} > \dfrac{5}{x + 6}$

15. $2x^2 + 5x > 12$ **16.** $|x - 15| \ge 5$

17. Perform each operation and write the result in standard form.

(a) $10i - \left(3 + \sqrt{-25}\right)$ (b) $\left(2 + \sqrt{3}i\right)\left(2 - \sqrt{3}i\right)$

18. Write the quotient in standard form: $\dfrac{5}{2 + i}$.

19. The sales y (in billions of dollars) for Gateway, Inc. from 1993 to 2000 can be approximated by the model

$$y = 1.16t - 1.9 \qquad 3 \le t \le 10$$

where t is the time (in years), with $t = 3$ corresponding to 1993. (Source: Gateway, Inc.)

(a) Sketch a graph of the model.

(b) Assuming that the pattern continues, use the graph in part (a) to estimate the sales in 2004.

(c) Use the model to verify algebraically the estimate from part (b).

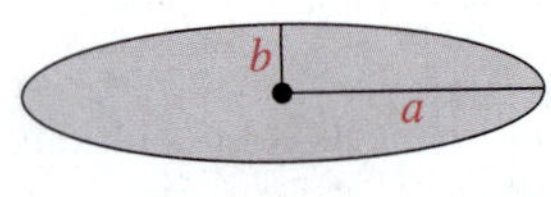

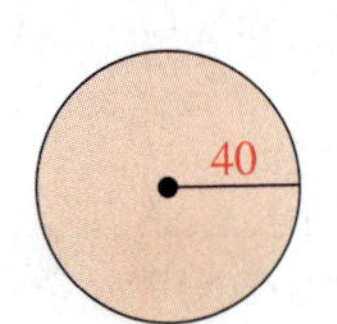

FIGURE FOR 21

20. On the first part of a 350-kilometer trip, a salesperson travels 2 hours and 15 minutes at an average speed of 100 kilometers per hour. The salesperson needs to arrive at the destination in another hour and 20 minutes. Find the average speed required for the remainder of the trip.

21. The area of the ellipse in the figure at the left is $A = \pi ab$. If a and b satisfy the constraint $a + b = 100$, find a and b such that the area of the ellipse equals the area of the circle.

Proofs in Mathematics

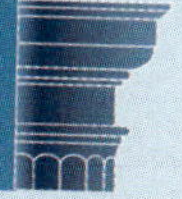

Conditional Statements

Many theorems are written in the **if-then form** "if p, then q," which is denoted by

$p \rightarrow q$ Conditional statement

where p is the **hypothesis** and q is the **conclusion.** Here are some other ways to express the conditional statement $p \rightarrow q$.

p implies q. p, only if q. p is sufficient for q.

Conditional statements can be either true or false. The conditional statement $p \rightarrow q$ is false only when p is true and q is false. To show that a conditional statement is true, you must prove that the conclusion follows for all cases that fulfill the hypothesis. To show that a conditional statement is false, you need only to describe a single **counterexample** that shows that the statement is not always true.

For instance, $x = -4$ is a counterexample that shows that the following statement is false.

If $x^2 = 16$, then $x = 4$.

The hypothesis "$x^2 = 16$" is true because $(-4)^2 = 16$. However, the conclusion "$x = 4$" is false. This implies that the given conditional statement is false.

For the conditional statement $p \rightarrow q$, there are three important associated conditional statements.

1. The **converse** of $p \rightarrow q$: $q \rightarrow p$
2. The **inverse** of $p \rightarrow q$: $\sim p \rightarrow \sim q$
3. The **contrapositive** of $p \rightarrow q$: $\sim q \rightarrow \sim p$

The symbol $\sim$ means the **negation** of a statement. For instance, the negation of "The engine is running" is "The engine is not running."

Example **Writing the Converse, Inverse, and Contrapositive**

Write the converse, inverse, and contrapositive of the conditional statement "If I get a B on my test, then I will pass the course."

Solution

a. *Converse:* If I pass the course, then I got a B on my test.

b. *Inverse:* If I do not get a B on my test, then I will not pass the course.

c. *Contrapositive:* If I do not pass the course, then I did not get a B on my test.

In the example above, notice that neither the converse nor the inverse is logically equivalent to the original conditional statement. On the other hand, the contrapositive *is* logically equivalent to the original conditional statement.

P.S. Problem Solving

1. Let x represent the time (in seconds) and let y represent the distance (in feet) between you and a tree. Sketch a possible graph that shows how x and y are related if you are walking toward the tree.

2. (a) Find the following sums

$1 + 2 + 3 + 4 + 5 =$ ▭

$1 + 2 + 3 + 4 + 5 + 6 + 7 + 8 =$ ▭

$1 + 2 + 3 + 4 + 5 + 6 + 7 + 8 + 9 + 10 =$ ▭

(b) Use the following formula for the sum of the first n natural numbers to verify your answers to part (a).

$$1 + 2 + 3 + \cdot \cdot \cdot + n = \frac{1}{2}n(n + 1)$$

(c) Use the formula in part (b) to find n if the sum of the first n natural numbers is 210.

3. The area of an ellipse is given by $A = \pi ab$ (see figure). For a certain ellipse, it is required that $a + b = 20$.

(a) Show that $A = \pi a(20 - a)$.

(b) Complete the table.

a	4	7	10	13	16
A					

(c) Find two values of a such that $A = 300$.

(d) Use a graphing utility to graph the area equation.

(e) Find the a-intercepts of the graph of the area equation. What do these values represent?

(f) What is the maximum area? What values of a and b yield the maximum area?

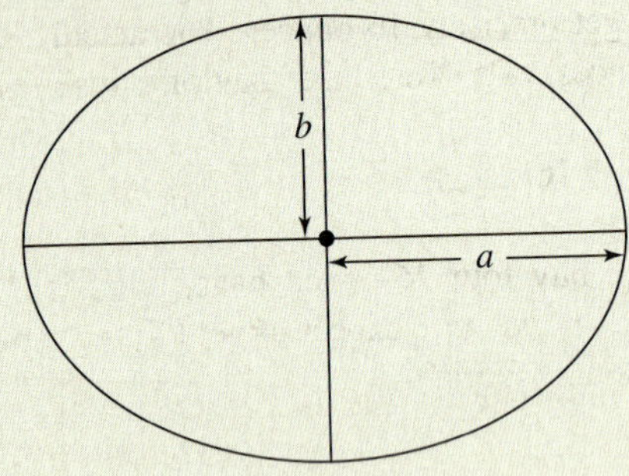

4. A building code requires that a building be able to withstand a certain amount of wind pressure. The pressure P (in pounds per square foot) from wind blowing at s miles per hour is given by

$$P = 0.00256s^2.$$

(a) A two-story library is designed. Buildings this tall are often required to withstand wind pressure of 20 pounds per square foot. Under this requirement, how fast can the wind be blowing before it produces excessive stress on the building?

(b) To be safe, the library is designed so that it can withstand wind pressure of 40 pounds per square foot. Does this mean that the library can survive wind blowing at twice the speed you found in part (a)? Justify your answer.

(c) Use the pressure formula to explain why even a relatively small increase in the wind speed could have potentially serious effects on a building.

5. For a bathtub with a rectangular base, Toricelli's Law implies that the height h of water in the tub t seconds after it begins draining is given by

$$h = \left(\sqrt{h_0} - \frac{2\pi d^2\sqrt{3}}{lw}t\right)^2$$

where l and w are the tub's length and width, d is the diameter of the drain, and h_0 is the water's initial height. (All measurements are in inches.) You completely fill a tub with water. The tub is 60 inches long by 30 inches wide by 25 inches high and has a drain with a two-inch diameter.

(a) Find the time it takes for the tub to go from being full to half-full.

(b) Find the time it takes for the tub to go from being half-full to empty.

(c) Based on your results in parts (a) and (b), what general statement can you make about the speed at which the water drains.

6. (a) Consider the sum of squares $x^2 + 9$. If the sum can be factored, then there are integers m and n such that $x^2 + 9 = (x + m)(x + n)$. Write two equations relating the sum and the product of m and n to the coefficients in $x^2 + 9$.

(b) Show that there are no integers m and n that satisfy both equations you wrote in part (a). What can you conclude?

7. A Pythagorean Triple is a group of three integers, such as 3, 4, and 5, that could be the lengths of the sides of a right triangle.
   (a) Find two other Pythagorean Triples.
   (b) Notice that $3 \cdot 4 \cdot 5 = 60$. Is the product of the three numbers in each Pythagorean Triple evenly divisible by 3? by 4? by 5?
   (c) Write a conjecture involving Pythagorean Triples and divisibility by 60.

8. Determine the solutions x_1 and x_2 of each quadratic equation. Use the values of x_1 and x_2 to fill in the boxes.

Equation	x_1, x_2	$x_1 + x_2$	$x_1 \cdot x_2$
(a) $x^2 - x - 6 = 0$			
(b) $2x^2 + 5x - 3 = 0$			
(c) $4x^2 - 9 = 0$			
(d) $x^2 - 10x + 34 = 0$			

9. Consider a general quadratic equation

   $ax^2 + bx + c = 0$

   whose solutions are x_1 and x_2. Use the results of Exercise 8 to determine a relationship among the coefficients a, b, and c and the sum $x_1 + x_2$ and the product $x_1 \cdot x_2$ of the solutions.

10. (a) The principal cube root of 125, $\sqrt[3]{125}$, is 5. Evaluate the expression x^3 for each value of x.

   (i) $x = \dfrac{-5 + 5\sqrt{3}i}{2}$

   (ii) $x = \dfrac{-5 - 5\sqrt{3}i}{2}$

   (b) The principal cube root of 27, $\sqrt[3]{27}$, is 3. Evaluate the expression x^3 for each value of x.

   (i) $x = \dfrac{-3 + 3\sqrt{3}i}{2}$

   (ii) $x = \dfrac{-3 - 3\sqrt{3}i}{2}$

   (c) Use the results of parts (a) and (b) to list possible cube roots of (i), 1, (ii) 8, and (iii) 64. Verify your results algebraically.

11. The multiplicative inverse of z is a complex number z_m such that $z \cdot z_m = 1$. Find the multiplicative inverse of each complex number.

   (a) $z = 1 + i$ (b) $z = 3 - i$ (c) $z = -2 + 8i$

12. Prove that the product of a complex number $a + bi$ and its complex conjugate is a real number.

13. A **fractal** is a geometric figure that consists of a pattern that is repeated infinitely on a smaller and smaller scale. The most famous fractal is called the **Mandelbrot Set**, named after the Polish-born mathematician Benoit Mandelbrot. To draw the Mandelbrot Set, consider the following sequence of numbers.

   $c, c^2 + c, (c^2 + c)^2 + c, [(c^2 + c)^2 + c]^2 + c, . . .$

   The behavior of this sequence depends on the value of the complex number c. If the sequence is bounded (the absolute value of each number in the sequence, $|a + bi| = \sqrt{a^2 + b^2}$, is less than some fixed number N), the complex number c is in the Mandelbrot Set, and if the sequence is unbounded (the absolute value of the terms of the sequence become infinitely large), the complex number c is not in the Mandelbrot Set. Determine whether the complex number c is in the Mandelbrot Set.

   (a) $c = i$ (b) $c = 1 + i$ (c) $c = -2$

14. Use the equation

   $4\sqrt{x} = 2x + k$

   to find three different values of k such that the equation has two solutions, one solution, and no solution. Describe the process you used to find the equations.

15. Use the graph of $y = x^4 - x^3 - 6x^2 + 4x + 8$ to solve the inequality $x^4 - x^3 - 6x^2 + 4x + 8 > 0$.

16. When you buy a 16-ounce bag of chips, you expect to get precisely 16 ounces. The actual weight w (in ounces) of a "16-ounce" bag of chips is given by

   $|w - 16| \le \frac{1}{2}.$

   You buy four 16-ounce bags, what is the greatest amount you can expect to get? What is the smallest amount? Explain.

How to study Chapter 2

▶ What you should learn

In this chapter you will learn the following skills and concepts:

- How to find and use the slopes of lines to write and graph linear equations in two variables
- How to evaluate functions and find their domains
- How to analyze graphs of functions
- How to identify and graph rigid and nonrigid transformations
- How to find arithmetic combinations and compositions of functions
- How to find inverse functions graphically and algebraically

▶ Important Vocabulary

As you encounter each new vocabulary term in this chapter, add the term and its definition to your notebook glossary.

Linear equation in two variables (p. 172)
Slope (p. 172)
Slope-intercept form (p. 172)
Point-slope form (p. 177)
Parallel (p. 179)
Perpendicular (p. 179)
Function (p. 187)
Domain (p. 187)
Range (p. 187)
Independent variable (p. 189)
Dependent variable (p. 189)
Implied domain (p. 191)
Graph of a function (p. 201)
Vertical Line Test (p. 202)
Zeros of a function (p. 203)
Increasing function (p. 204)
Decreasing function (p. 204)
Constant function (p. 204)
Relative minimum (p. 205)
Relative maximum (p. 205)
Even function (p. 206)
Odd function (p. 206)
Vertical and horizontal shifts (p. 219)
Reflection (p. 221)
Nonrigid transformations (p. 223)
Inverse function (p. 237)
Horizontal Line Test (p. 240)
One-to-one functions (p. 240)

Study Tools

Learning objectives in each section
Chapter Summary (p. 247)
Review Exercises (pp. 248–251)
Chapter Test (p. 252)
Cumulative Test for Chapters P–2 (pp. 253–254)

Additional Resources

Study and Solutions Guide
Interactive College Algebra
Videotapes/DVD for Chapter 2
College Algebra Website
Student Success Organizer

2

Functions and Their Graphs

2.1 Linear Equations in Two Variables

▶ What you should learn

- How to use slope to graph linear equations in two variables
- How to find slopes of lines
- How to write linear equations in two variables
- How to use slope to identify parallel and perpendicular lines
- How to use linear equations in two variables to model and solve real-life problems

▶ Why you should learn it

Linear equations in two variables can be used to model and solve real-life problems. For instance, in Exercise 119 on page 185, a linear equation is used to model the average monthly cellular phone bills for subscribers in the United States.

Dick Luria/Getty Images

Using Slope

The simplest mathematical model for relating two variables is the **linear equation in two variables** $y = mx + b$. The equation is called *linear* because its graph is a line. (In mathematics, the term *line* means *straight line*.) By letting $x = 0$, you can see that the line crosses the y-axis at $y = b$, as shown in Figure 2.1. In other words, the y-intercept is $(0,\ b)$. The steepness or slope of the line is m.

$$y = mx + b$$

Slope ↑ (m) ↑ (b) y-Intercept

The **slope** of a nonvertical line is the number of units the line rises (or falls) vertically for each unit of horizontal change from left to right, as shown in Figure 2.1 and Figure 2.2.

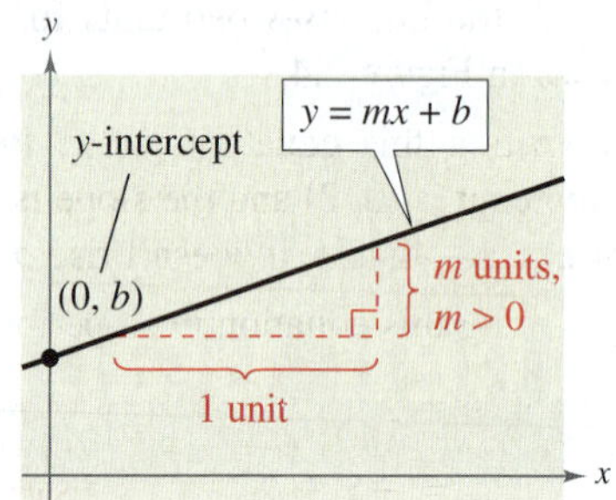

Positive slope, line rises.

FIGURE 2.1

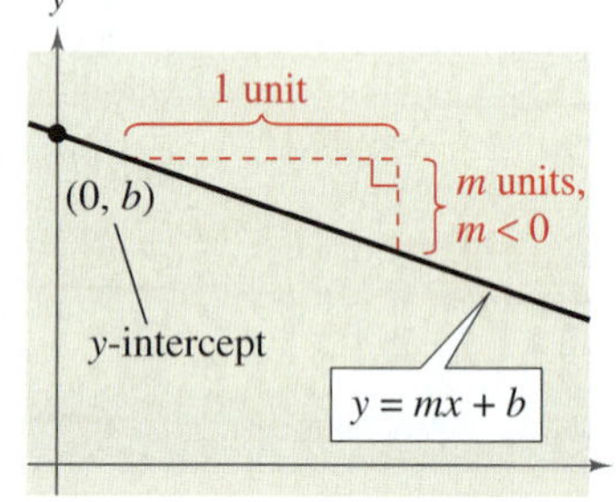

Negative slope, line falls.

FIGURE 2.2

A linear equation that is written in the form $y = mx + b$ is said to be written in **slope-intercept form.**

The Slope-Intercept Form of the Equation of a Line

The graph of the equation

$$y = mx + b$$

is a line whose slope is m and whose y-intercept is $(0, b)$.

Exploration

Use a graphing utility to compare the slopes of the lines $y = mx$ where $m = 0.5$, 1, 2, and 4. Which line rises most quickly? Now, let $m = -0.5$, -1, -2, and -4. Which line falls most quickly? Use a square setting to obtain a true geometric perspective. What can you conclude about the slope and the "rate" at which the line rises or falls?

The icon identifies examples and concepts related to features of the Learning Tools CD-ROM and the *Interactive* and *Internet* versions of this text. For more details see the chart on pages *xix-xxiii*.

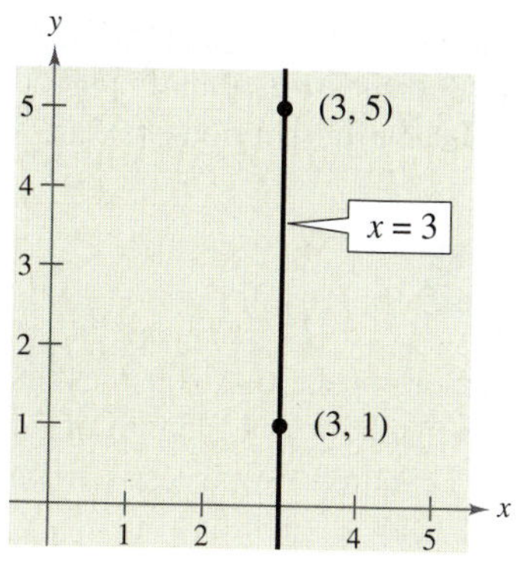

FIGURE 2.3 *Slope is undefined.*

Once you have determined the slope and the y-intercept of a line, it is a relatively simple matter to sketch its graph. In the next example, note that none of the lines is vertical. A vertical line has an equation of the form

$x = a.$ Vertical line

The equation of a vertical line cannot be written in the form $y = mx + b$ because the slope of a vertical line is undefined, as indicated in Figure 2.3.

Example 1 ▶ Graphing a Linear Equation

Sketch the graph of each linear equation.

a. $y = 2x + 1$

b. $y = 2$

c. $x + y = 2$

Solution

a. Because $b = 1$, the y-intercept is $(0, 1)$. Moreover, because the slope is $m = 2$, the line *rises* two units for each unit the line moves to the right, as shown in Figure 2.4.

b. By writing this equation in the form $y = (0)x + 2$, you can see that the y-intercept is $(0, 2)$ and the slope is zero. A zero slope implies that the line is horizontal—that is, it doesn't rise *or* fall, as shown in Figure 2.5.

c. By writing this equation in slope-intercept form

$x + y = 2$ Write original equation.

$y = -x + 2$ Subtract x from each side.

$y = (-1)x + 2$ Write in slope-intercept form.

you can see that the y-intercept is $(0, 2)$. Moreover, because the slope is $m = -1$, the line *falls* one unit for each unit the line moves to the right, as shown in Figure 2.6.

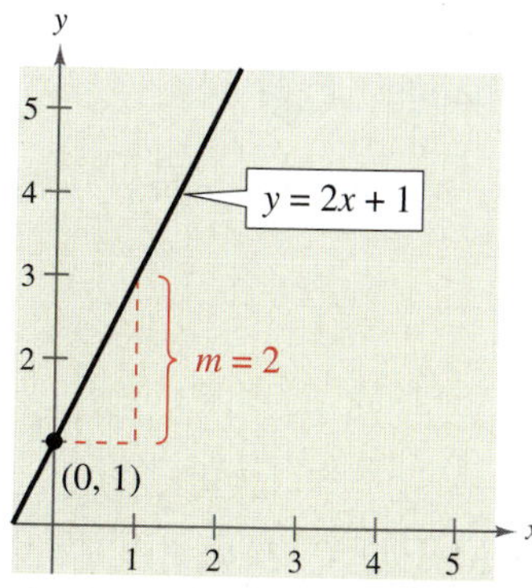

When m is positive, the line rises.
FIGURE 2.4

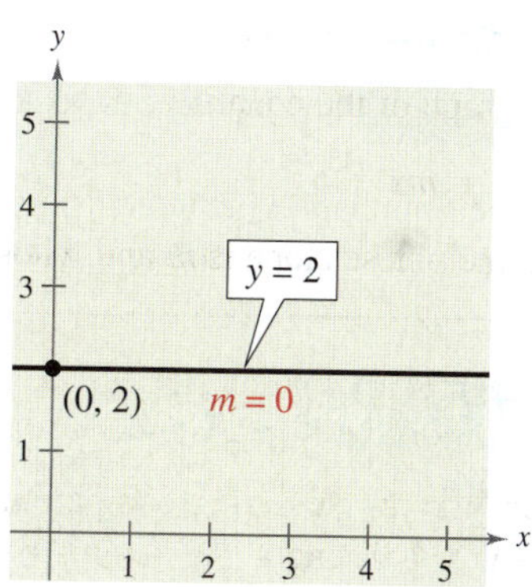

When m is 0, the line is horizontal.
FIGURE 2.5

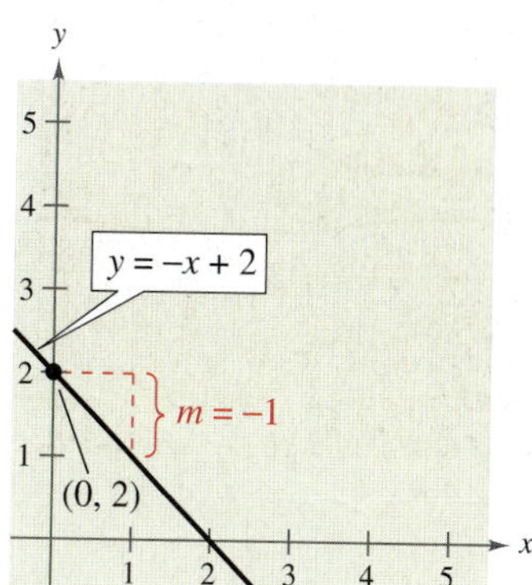

When m is negative, the line falls.
FIGURE 2.6

In real-life problems, the slope of a line can be interpreted as either a *ratio* or a *rate*. If the x-axis and y-axis have the same unit of measure, then the slope has no units and is a **ratio.** If the x-axis and y-axis have different units of measure, then the slope is a **rate** or **rate of change.**

Example 2 ▶ Using Slope as a Ratio 

The *Interactive* CD-ROM and *Internet* versions of this text offer a Try It for each example in the text.

The maximum recommended slope of a wheelchair ramp is $\frac{1}{12}$. A business is installing a wheelchair ramp that rises 22 inches over a horizontal length of 24 feet. Is the ramp steeper than recommended? (Source: Americans with Disabilities Act Handbook)

Solution

The horizontal length of the ramp is 24 feet or $12(24) = 288$ inches, as shown in Figure 2.7. So, the slope of the ramp is

$$\begin{aligned}\text{Slope} &= \frac{\text{vertical change}}{\text{horizontal change}}\\ &= \frac{22 \text{ in.}}{288 \text{ in.}}\\ &\approx 0.076.\end{aligned}$$

Because $\frac{1}{12} \approx 0.083$, the slope of the ramp is not steeper than recommended.

FIGURE 2.7

Example 3 ▶ Using Slope as a Rate of Change 

A kitchen appliance manufacturing company determines that the total cost in dollars of producing x units of a blender is

$$C = 25x + 3500. \quad \text{Cost equation}$$

Describe the practical significance of the y-intercept and slope of this line.

Solution

The y-intercept $(0, 3500)$ tells you that the cost of producing zero units is \$3500. This is the *fixed cost* of production—it includes costs that must be paid regardless of the number of units produced. The slope of $m = 25$ tells you that the cost of producing each unit is \$25, as shown in Figure 2.8. Economists call the cost per unit the *marginal cost*. If the production increases by one unit, then the "margin," or extra amount of cost, is \$25. So, the cost increases at a rate of \$25 per unit.

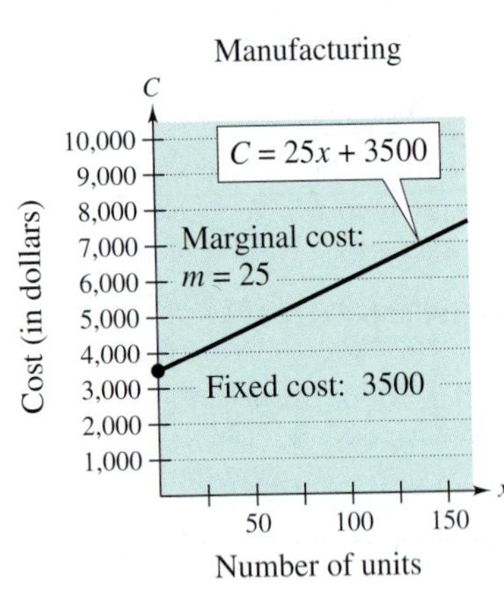

FIGURE 2.8 *Production cost*

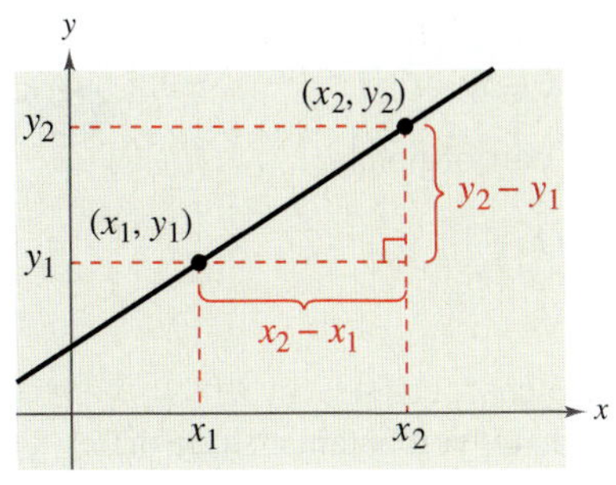

FIGURE 2.9

Finding the Slope of a Line

Given an equation of a line, you can find its slope by writing the equation in slope-intercept form. If you are not given an equation, you can still find the slope of a line. For instance, suppose you want to find the slope of the line passing through the points (x_1, y_1) and (x_2, y_2), as shown in Figure 2.9. As you move from left to right along this line, a change of $(y_2 - y_1)$ units in the vertical direction corresponds to a change of $(x_2 - x_1)$ units in the horizontal direction.

$$y_2 - y_1 = \text{the change in } y = \text{rise}$$

and

$$x_2 - x_1 = \text{the change in } x = \text{run}$$

The ratio of $(y_2 - y_1)$ to $(x_2 - x_1)$ represents the slope of the line that passes through the points (x_1, y_1) and (x_2, y_2).

$$\begin{aligned} \text{Slope} &= \frac{\text{change in } y}{\text{change in } x} \\ &= \frac{\text{rise}}{\text{run}} \\ &= \frac{y_2 - y_1}{x_2 - x_1} \end{aligned}$$

The Slope of a Line Passing Through Two Points

The **slope** m of the nonvertical line through (x_1, y_1) and (x_2, y_2) is

$$m = \frac{y_2 - y_1}{x_2 - x_1}$$

where $x_1 \neq x_2$.

When this formula is used for slope, the *order of subtraction* is important. Given two points on a line, you are free to label either one of them as (x_1, y_1) and the other as (x_2, y_2). However, once you have done this, you must form the numerator and denominator using the same order of subtraction.

$$m = \frac{y_2 - y_1}{x_2 - x_1} \qquad m = \frac{y_1 - y_2}{x_1 - x_2} \qquad m = \frac{y_2 - y_1}{x_1 - x_2}$$

Correct Correct Incorrect

For instance, the slope of the line passing through the points (3, 4) and (5, 7) can be calculated as

$$m = \frac{7 - 4}{5 - 3} = \frac{3}{2}$$

or, reversing the subtraction order in both the numerator and denominator, as

$$m = \frac{4 - 7}{3 - 5} = \frac{-3}{-2} = \frac{3}{2}.$$

The *Interactive* CD-ROM and *Internet* versions of this text offer a Quiz for every section of the text.

Example 4 ▶ Finding the Slope of a Line Through Two Points

Find the slope of the line passing through each pair of points.

a. $(-2,\ 0)$ and $(3, 1)$ **b.** $(-1,\ 2)$ and $(2, 2)$

c. $(0, 4)$ and $(1, -1)$ **d.** $(3, 4)$ and $(3, 1)$

Solution

a. Letting $(x_1,\ y_1) = (-2,\ 0)$ and $(x_2, y_2) = (3, 1)$, you obtain a slope of

$$m = \frac{y_2 - y_1}{x_2 - x_1} = \frac{1 - 0}{3 - (-2)} = \frac{1}{5}.$$

See Figure 2.10.

b. The slope of the line passing through $(-1, 2)$ and $(2, 2)$ is

$$m = \frac{2 - 2}{2 - (-1)} = \frac{0}{3} = 0.$$

See Figure 2.11.

c. The slope of the line passing through $(0, 4)$ and $(1, -1)$ is

$$m = \frac{-1 - 4}{1 - 0} = \frac{-5}{1} = -5.$$

See Figure 2.12.

d. The slope of the line passing through $(3, 4)$ and $(3, 1)$ is

$$m = \frac{1 - 4}{3 - 3} = \frac{-3}{0}.$$

See Figure 2.13.

Because division by 0 is undefined, the slope is undefined and the line is vertical.

STUDY TIP

In Figures 2.10 to 2.13, note the relationships between slope and the orientation of the line.

a. Positive slope; line rises from left to right

b. Zero slope; line is horizontal

c. Negative slope; line falls from left to right

d. Undefined slope; line is vertical

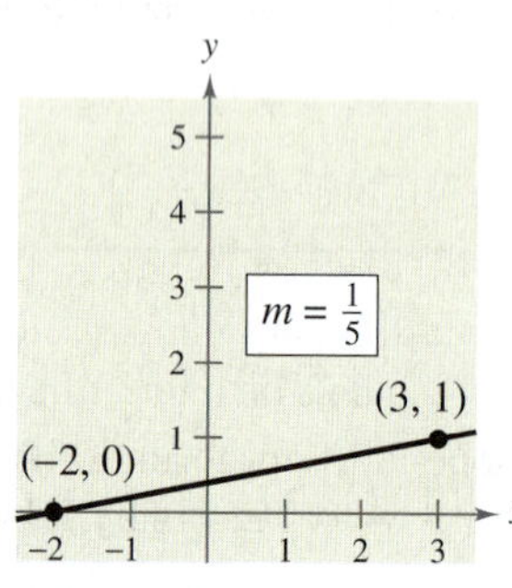

FIGURE 2.10

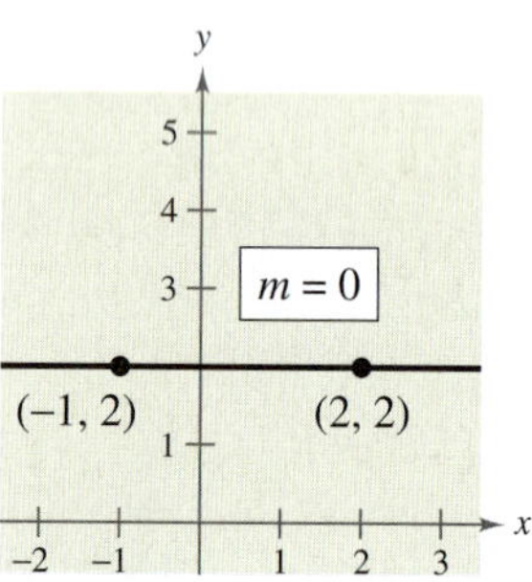

FIGURE 2.11

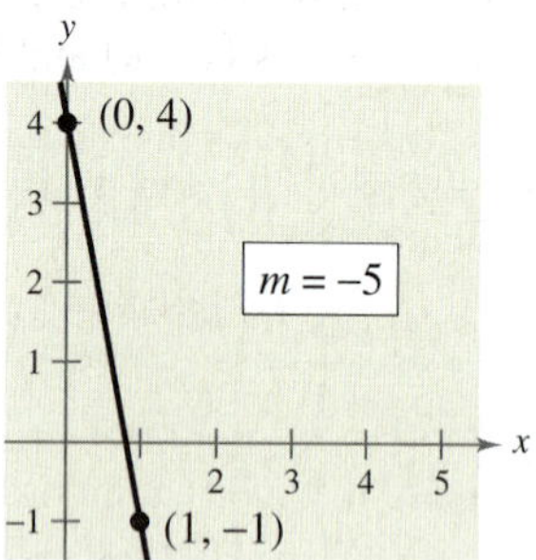

FIGURE 2.12

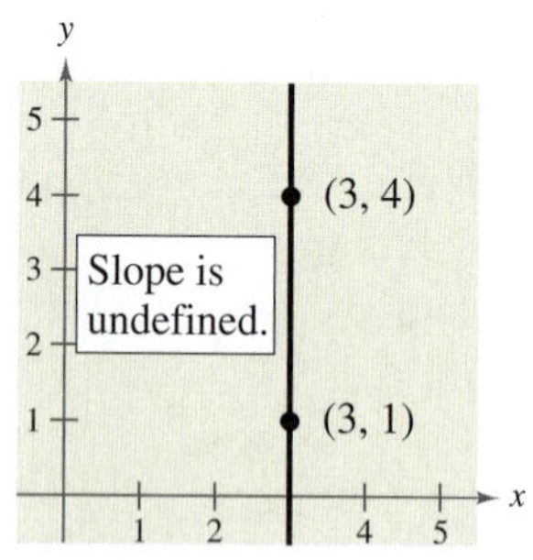

FIGURE 2.13

Writing Linear Equations in Two Variables

If (x_1, y_1) is a point on a line of slope m and (x, y) is *any other* point on the line, then

$$\frac{y - y_1}{x - x_1} = m.$$

This equation, involving the variables x and y, can be rewritten in the form

$$y - y_1 = m(x - x_1)$$

which is the **point-slope form** of the equation of a line.

Point-Slope Form of the Equation of a Line

The equation of the line with slope m passing through the point (x_1, y_1) is

$$y - y_1 = m(x - x_1).$$

The point-slope form is most useful for *finding* the equation of a line. You should remember this form.

Example 5 ▶ Using the Point-Slope Form 

Find the slope-intercept form of the equation of the line that has a slope of 3 and passes through the point $(1, -2)$.

Solution

Use the point-slope form with $m = 3$ and $(x_1, y_1) = (1, -2)$.

$$y - y_1 = m(x - x_1) \qquad \text{Point-slope form}$$

$$y - (-2) = 3(x - 1) \qquad \text{Substitute for } m, x_1, \text{ and } y_1.$$

$$y + 2 = 3x - 3 \qquad \text{Simplify.}$$

$$y = 3x - 5 \qquad \text{Write in slope-intercept form.}$$

The slope-intercept form of the equation of the line is $y = 3x - 5$. The graph of this line is shown in Figure 2.14.

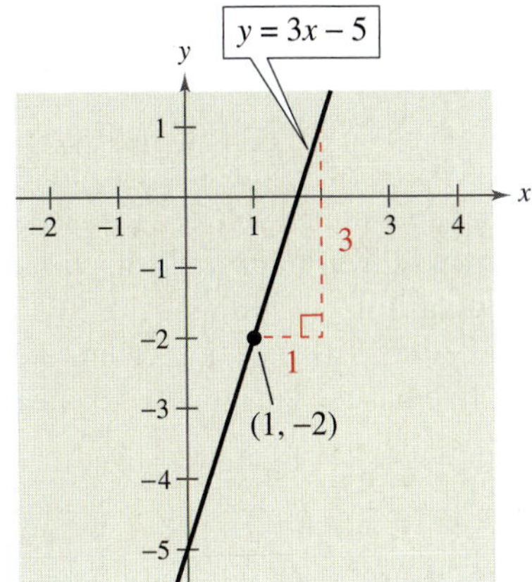

FIGURE 2.14

The point-slope form can be used to find an equation of the line passing through two points (x_1, y_1) and (x_2, y_2). To do this, first find the slope of the line

$$m = \frac{y_2 - y_1}{x_2 - x_1}, \qquad x_1 \neq x_2$$

and then use the point-slope form to obtain the equation

$$y - y_1 = \frac{y_2 - y_1}{x_2 - x_1}(x - x_1). \qquad \text{Two-point form}$$

This is sometimes called the **two-point form** of the equation of a line.

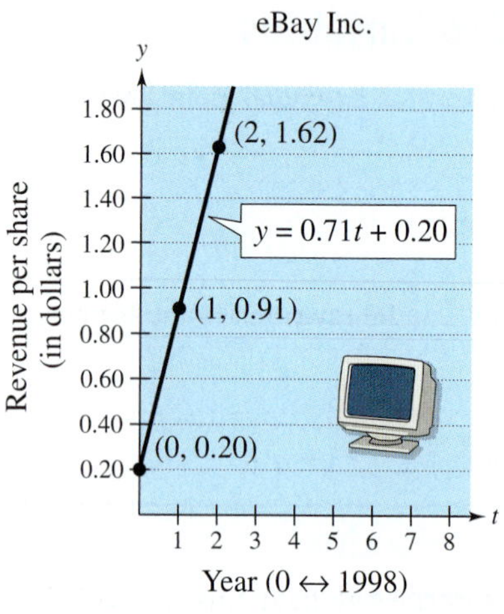

FIGURE 2.15

Example 6 ▶ Predicting Revenue per Share

The revenue per share for eBay Inc. was \$0.20 in 1998 and \$0.91 in 1999. Using only this information, write a linear equation that gives the revenue per share in terms of the year. Then predict the revenue per share for 2000. (Source: eBay Inc.)

Solution

Let $t = 0$ represent 1998. Then the two given values are represented by the data points $(0, 0.20)$ and $(1, 0.91)$. The slope of the line through these points is

$$m = \frac{0.91 - 0.20}{1 - 0}$$

$$= 0.71.$$

Using the point-slope form, you can find the equation that relates the revenue per share y and the year t to be

$$y - 0.20 = 0.71(t - 0) \qquad \text{Write in point-slope form.}$$

$$y = 0.71t + 0.20. \qquad \text{Write in slope-intercept form.}$$

According to this equation, the revenue per share in 2000 was \$1.62, as shown in Figure 2.15. (In this case, the prediction is quite good—the actual revenue per share in 2000 was \$1.60.)

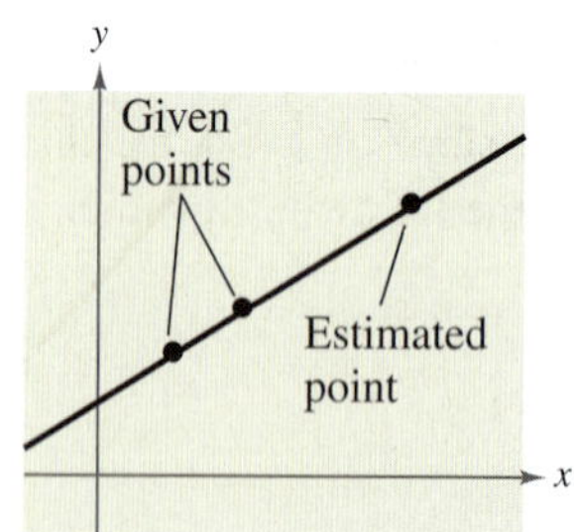

Linear extrapolation
FIGURE 2.16

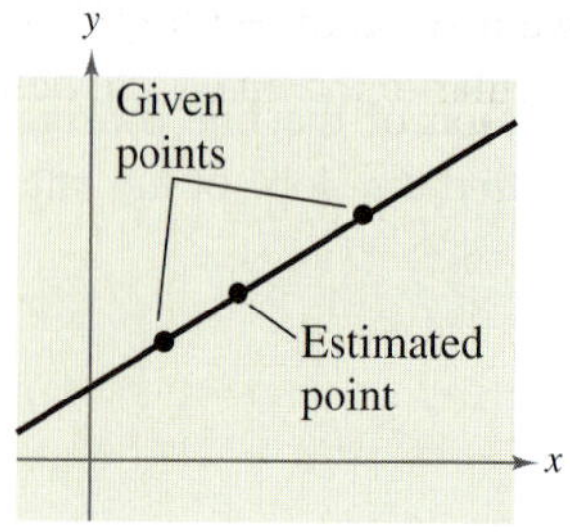

Linear interpolation
FIGURE 2.17

The prediction method illustrated in Example 6 is called **linear extrapolation.** Note in Figure 2.16 that an extrapolated point does not lie between the given points. When the estimated point lies between two given points, as shown in Figure 2.17, the procedure is called **linear interpolation.**

Because the slope of a vertical line is not defined, its equation cannot be written in slope-intercept form. However, every line has an equation that can be written in the **general form**

$$Ax + By + C = 0 \qquad \text{General form}$$

where A and B are not both zero. For instance, the vertical line given by $x = a$ can be represented by the general form $x - a = 0$.

Equations of Lines

1. General form: $Ax + By + C = 0$
2. Vertical line: $x = a$
3. Horizontal line: $y = b$
4. Slope-intercept form: $y = mx + b$
5. Point-slope form: $y - y_1 = m(x - x_1)$
6. Two-point form: $y - y_1 = \dfrac{y_2 - y_1}{x_2 - x_1}(x - x_1)$

Exploration

Find d_1 and d_2 in terms of m_1 and m_2, respectively (see figure). Then use the Pythagorean Theorem to find a relationship between m_1 and m_2.

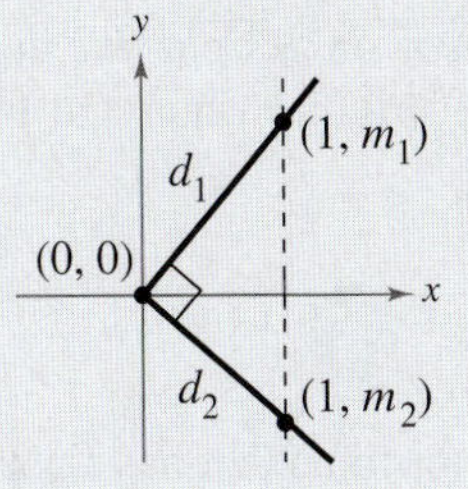

Parallel and Perpendicular Lines 

Slope can be used to decide whether two nonvertical lines in a plane are parallel, perpendicular, or neither.

Parallel and Perpendicular Lines

1. Two distinct nonvertical lines are **parallel** if and only if their slopes are equal. That is, $m_1 = m_2$.
2. Two nonvertical lines are **perpendicular** if and only if their slopes are negative reciprocals of each other. That is, $m_1 = -1/m_2$.

Example 7 ▶ Finding Parallel and Perpendicular Lines 

Find the slope-intercept forms of the equations of the lines that pass through the point $(2, -1)$ and are (a) parallel to and (b) perpendicular to the line $2x - 3y = 5$.

Solution

By writing the equation of the given line in slope-intercept form

$2x - 3y = 5$ Write original equation.

$-3y = -2x + 5$ Subtract $2x$ from each side.

$y = \frac{2}{3}x - \frac{5}{3}$ Write in slope-intercept form.

you can see that it has a slope of $m = \frac{2}{3}$, as shown in Figure 2.18.

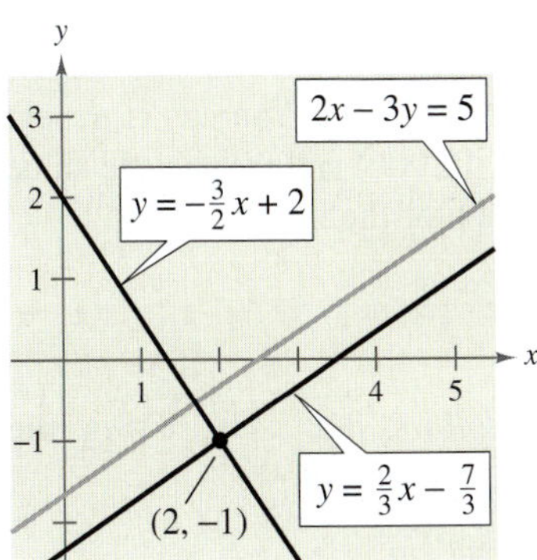

FIGURE 2.18

a. Any line parallel to the given line must also have a slope of $\frac{2}{3}$. So, the line through $(2, -1)$ that is parallel to the given line has the following equation.

$y - (-1) = \frac{2}{3}(x - 2)$ Write in point-slope form.

$3(y + 1) = 2(x - 2)$ Multiply each side by 3.

$3y + 3 = 2x - 4$ Distributive Property

$2x - 3y - 7 = 0$ Write in general form.

$y = \frac{2}{3}x - \frac{7}{3}$ Write in slope-intercept form.

b. Any line perpendicular to the given line must have a slope of $-1/(2/3)$ or $-3/2$. So, the line through $(2, -1)$ that is perpendicular to the given line has the following equation.

$y - (-1) = -\frac{3}{2}(x - 2)$ Write in point-slope form.

$2(y + 1) = -3(x - 2)$ Multiply each side by 2.

$2y + 2 = -3x + 6$ Distributive Property

$3x + 2y - 4 = 0$ Write in general form.

$y = -\frac{3}{2}x + 2$ Write in slope-intercept form.

Technology

On a graphing utility, lines will not appear to have the correct slope unless you use a viewing window that has a square setting. For instance, try graphing the lines in Example 7 using the standard setting $-10 \le x \le 10$ and $-10 \le y \le 10$. Then reset the viewing window with the square setting $-9 \le x \le 9$ and $-6 \le y \le 6$. On which setting do the lines $y = \frac{2}{3}x - \frac{5}{3}$ and $y = -\frac{3}{2}x + 2$ appear perpendicular?

Notice in Example 7 how the slope-intercept form is used to obtain information about the graph of a line, whereas the point-slope form is used to write the equation of a line.

Application

Most business expenses can be deducted in the same year they occur. One exception is the cost of property that has a useful life of more than 1 year. Such costs must be *depreciated* over the useful life of the property. If the *same amount* is depreciated each year, the procedure is called *linear* or *straight-line depreciation.* The *book value* is the difference between the original value and the total amount of depreciation accumulated to date.

Example 8 ▶ Straight-Line Depreciation 

STUDY TIP

In many real-life applications, the two data points that determine the line are often given in a disguised form. Note how the data points are described in Example 8.

Your publishing company has purchased a $12,000 machine that has a useful life of 8 years. The salvage value at the end of 8 years is $2000. Write a linear equation that describes the book value of the machine each year.

Solution

Let V represent the value of the machine at the end of year t. You can represent the initial value of the machine by the data point (0, 12,000) and the salvage value of the machine by the data point (8, 2000). The slope of the line is

$$m = \frac{2000 - 12{,}000}{8 - 0} = -\$1250$$

which represents the annual depreciation in *dollars per year*. Using the point-slope form, you can write the equation of the line as follows.

$$V - 12{,}000 = -1250(t - 0) \qquad \text{Write in point-slope form.}$$

$$V = -1250t + 12{,}000 \qquad \text{Write in slope-intercept form.}$$

The table shows the book value at the end of each year, and the graph of the equation is shown in Figure 2.19.

Useful Life of a Machine

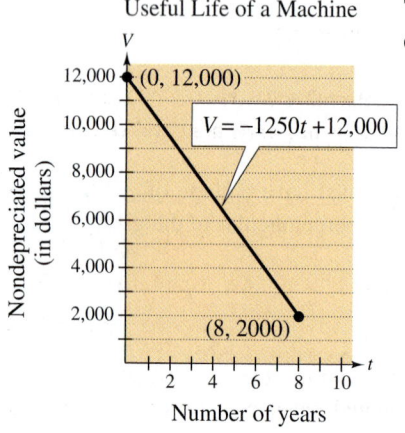

FIGURE 2.19 *Straight-line depreciation*

Year, t	Value, V
0	12,000
1	10,750
2	9,500
3	8,250
4	7,000
5	5,750
6	4,500
7	3,250
8	2,000

2.1 Exercises

The *Interactive* CD-ROM and *Internet* versions of this text contain step-by-step solutions to all odd-numbered exercises. They also provide Tutorial Exercises for additional help.

In Exercises 1 and 2, identify the line that has each slope.

1. (a) $m = \frac{2}{3}$
   (b) m is undefined.
   (c) $m = -2$

2. (a) $m = 0$
   (b) $m = -\frac{3}{4}$
   (c) $m = 1$

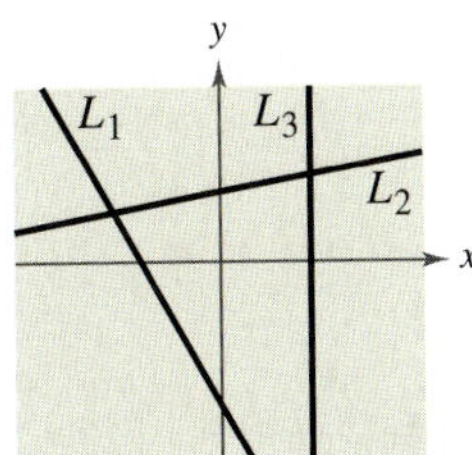

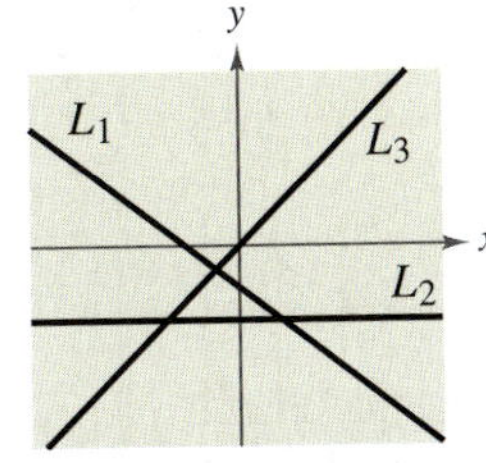

In Exercises 3 and 4, sketch the lines through the point with the indicated slopes on the same set of coordinate axes.

	Point	*Slopes*
3.	(2, 3)	(a) 0 (b) 1 (c) 2 (d) −3
4.	(−4, 1)	(a) 3 (b) −3 (c) $\frac{1}{2}$ (d) Undefined

In Exercises 5–8, estimate the slope of the line.

5.

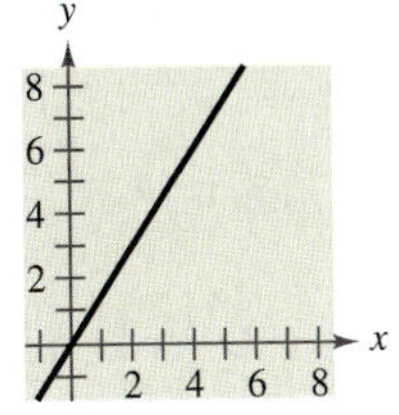

6.

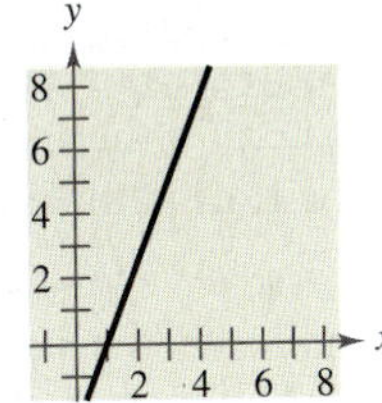

7.

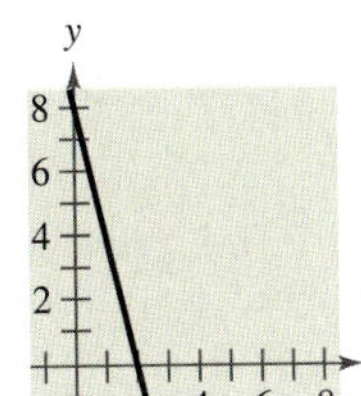

8.

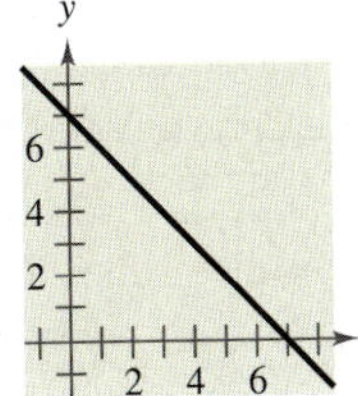

In Exercises 9–20, find the slope and *y*-intercept (if possible) of the equation of the line. Sketch the line.

9. $y = 5x + 3$
10. $y = x - 10$
11. $y = -\frac{1}{2}x + 4$
12. $y = -\frac{3}{2}x + 6$
13. $5x - 2 = 0$
14. $3y + 5 = 0$
15. $7x + 6y = 30$
16. $2x + 3y = 9$
17. $y - 3 = 0$
18. $y + 4 = 0$
19. $x + 5 = 0$
20. $x - 2 = 0$

In Exercises 21–28, plot the points and find the slope of the line passing through the pair of points.

21. $(-3, -2), (1, 6)$
22. $(2, 4), (4, -4)$
23. $(-6, -1), (-6, 4)$
24. $(0, -10), (-4, 0)$
25. $\left(\frac{11}{2}, -\frac{4}{3}\right), \left(-\frac{3}{2}, -\frac{1}{3}\right)$
26. $\left(\frac{7}{8}, \frac{3}{4}\right), \left(\frac{5}{4}, -\frac{1}{4}\right)$
27. $(4.8, 3.1), (-5.2, 1.6)$
28. $(-1.75, -8.3), (2.25, -2.6)$

In Exercises 29–38, use the point on the line and the slope of the line to find three additional points through which the line passes. (There are many correct answers.)

	Point	*Slope*
29.	(2, 1)	$m = 0$
30.	(−4, 1)	m is undefined.
31.	(5, −6)	$m = 1$
32.	(10, −6)	$m = -1$
33.	(−8, 1)	m is undefined.
34.	(−3, −1)	$m = 0$
35.	(−5, 4)	$m = 2$
36.	(0, −9)	$m = -2$
37.	(7, −2)	$m = \frac{1}{2}$
38.	(−1, −6)	$m = -\frac{1}{2}$

In Exercises 39–42, determine whether the lines L_1 and L_2 passing through the pairs of points are parallel, perpendicular, or neither.

39. L_1: $(0, -1), (5, 9)$
    L_2: $(0, 3), (4, 1)$
40. L_1: $(-2, -1), (1, 5)$
    L_2: $(1, 3), (5, -5)$
41. L_1: $(3, 6), (-6, 0)$
    L_2: $(0, -1), \left(5, \frac{7}{3}\right)$
42. L_1: $(4, 8), (-4, 2)$
    L_2: $(3, -5), \left(-1, \frac{1}{3}\right)$

43. ***Sales*** The following are the slopes of lines representing annual sales y in terms of time x in years. Use the slopes to interpret any change in annual sales for a one-year increase in time.
    (a) The line has a slope of $m = 135$.
    (b) The line has a slope of $m = 0$.
    (c) The line has a slope of $m = -40$.

44. ***Revenue*** The following are the slopes of lines representing daily revenues y in terms of time x in days. Use the slopes to interpret any change in daily revenues for a one-day increase in time.

(a) The line has a slope of $m = 400$.

(b) The line has a slope of $m = 100$.

(c) The line has a slope of $m = 0$.

45. ***Earnings per Share*** The graph shows the earnings per share of stock for Auto Zone, Inc. for the years 1991 through 2001. (Source: Auto Zone, Inc.)

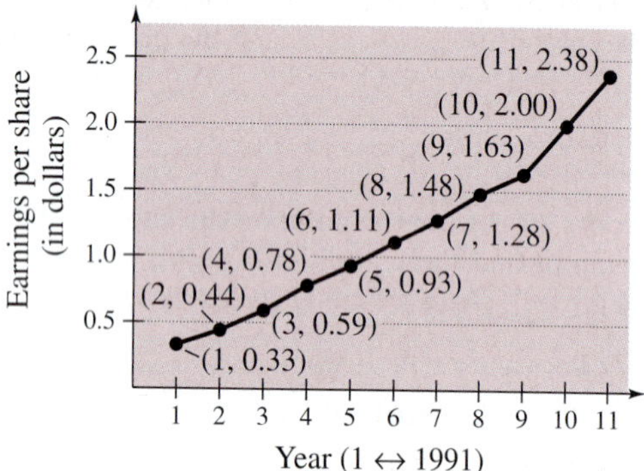

(a) Use the slopes to determine the years in which the earnings per share showed the greatest increase and the smallest increase.

(b) Find the slope of the line segment connecting the years 1991 and 2001.

(c) Interpret the meaning of the slope in part (b) in the context of the problem.

46. ***Net Profit*** The graph shows the net profit (in millions of dollars) for Outback Steakhouse for the years 1991 through 2001. (Source: Outback Steakhouse, Inc.)

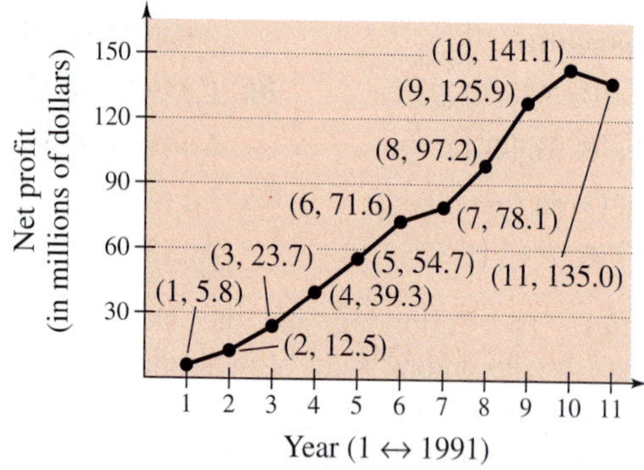

(a) Use the slopes to determine the years in which the net profit showed the greatest increase and the smallest increase.

(b) Find the slope of the line segment connecting the years 1991 and 2001.

(c) Interpret the meaning of the slope in part (b) in the context of the problem.

47. ***Road Grade*** From the top of a mountain road, a surveyor takes several horizontal measurements x and several vertical measurements y, as shown in the table (x and y are measured in feet).

x	y
300	-25
600	-50
900	-75
1200	-100
1500	-125
1800	-150
2100	-175

(a) Sketch a scatter plot of the data.

(b) Use a straightedge to sketch the best-fitting line through the points.

(c) Find an equation for the line you sketched in part (b).

(d) Interpret the meaning of the slope of the line in part (c) in the context of the problem.

(e) The surveyor needs to put up a road sign that indicates the steepness of the road. For instance, a surveyor would put up a sign that states "8% grade" on a road with a downhill grade that has a slope of $-\frac{8}{100}$. What should the sign state for the road in this problem?

48. ***Road Grade*** You are driving on a road that has a 6% uphill grade. This means that the slope of the road is $\frac{6}{100}$. Approximate the amount of vertical change in your position if you drive 200 feet.

Rate of Change **In Exercises 49 and 50, you are given the dollar value of a product in 2003 and the rate at which the value of the product is expected to change during the next 5 years. Use this information to write a linear equation that gives the dollar value V of the product in terms of the year t. (Let $t = 3$ represent 2003.)**

	2003 Value	*Rate*
49.	\$2540	\$125 increase per year
50.	\$156	\$4.50 increase per year

Graphical Interpretation **In Exercises 51–54, match the description of the situation with its graph. Also determine the slope of each graph and interpret the slope in the context of the situation. [The graphs are labeled (a), (b), (c), and (d).]**

(a)
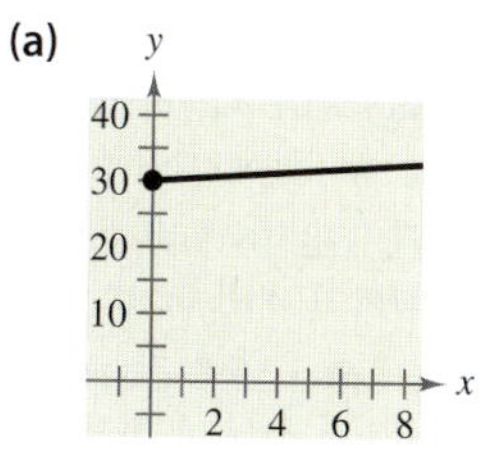

(b)
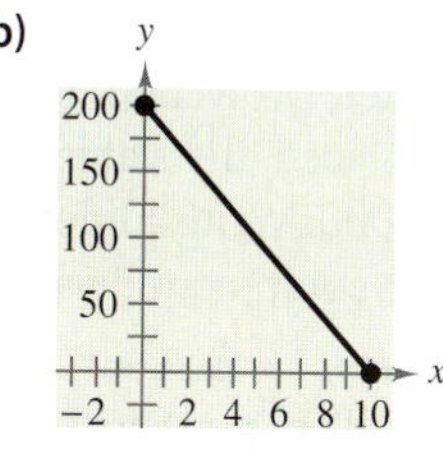

(c)
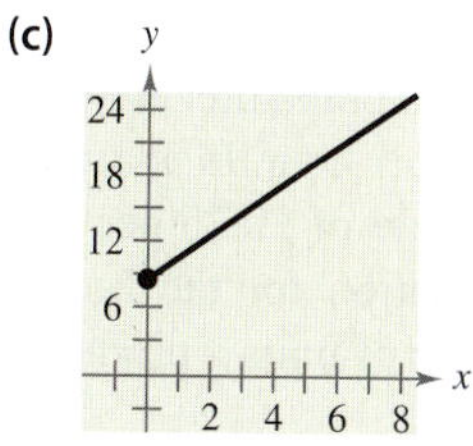

(d)
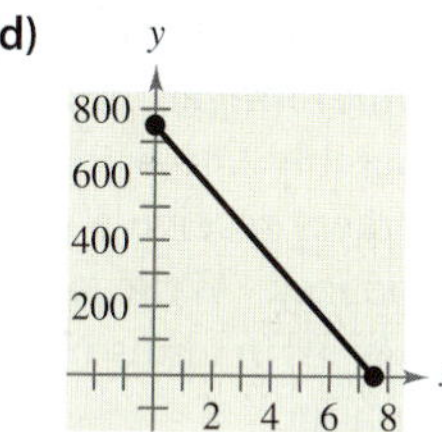

51. A person is paying \$20 per week to a friend to repay a \$200 loan.

52. An employee is paid \$8.50 per hour plus \$2 for each unit produced per hour.

53. A sales representative receives \$30 per day for food plus \$0.32 for each mile traveled.

54. A word processor that was purchased for \$750 depreciates \$100 per year.

In Exercises 55–66, find the slope-intercept form of the equation of the line that passes through the given point and has the indicated slope. Sketch the line.

	Point	*Slope*
55.	$(0, -2)$	$m = 3$
56.	$(0, 10)$	$m = -1$
57.	$(-3, 6)$	$m = -2$
58.	$(0, 0)$	$m = 4$
59.	$(4, 0)$	$m = -\frac{1}{3}$
60.	$(-2, -5)$	$m = \frac{3}{4}$
61.	$(6, -1)$	m is undefined.
62.	$(-10, 4)$	m is undefined.
63.	$\left(4, \frac{5}{2}\right)$	$m = 0$
64.	$\left(-\frac{1}{2}, \frac{3}{2}\right)$	$m = 0$
65.	$(-5.1, 1.8)$	$m = 5$
66.	$(2.3, -8.5)$	$m = -\frac{5}{2}$

In Exercises 67–80, find the slope-intercept form of the equation of the line passing through the points. Sketch the line.

67. $(5, -1), (-5, 5)$

68. $(4, 3), (-4, -4)$

69. $(-8, 1), (-8, 7)$

70. $(-1, 4), (6, 4)$

71. $\left(2, \frac{1}{2}\right), \left(\frac{1}{2}, \frac{5}{4}\right)$

72. $(1, 1), \left(6, -\frac{2}{3}\right)$

73. $\left(-\frac{1}{10}, -\frac{3}{5}\right), \left(\frac{9}{10}, -\frac{9}{5}\right)$

74. $\left(\frac{3}{4}, \frac{3}{2}\right), \left(-\frac{4}{3}, \frac{7}{4}\right)$

75. $(1, 0.6), (-2, -0.6)$

76. $(-8, 0.6), (2, -2.4)$

77. $(2, -1), \left(\frac{1}{3}, -1\right)$

78. $\left(\frac{1}{5}, -2\right), (-6, -2)$

79. $\left(\frac{7}{3}, -8\right), \left(\frac{7}{3}, 1\right)$

80. $(1.5, -2), (1.5, 0.2)$

In Exercises 81–86, use the *intercept form* to find the equation of the line with the given intercepts. The intercept form of the equation of a line with intercepts $(a, 0)$ and $(0, b)$ is

$$\frac{x}{a} + \frac{y}{b} = 1, \quad a \neq 0, \quad b \neq 0.$$

81. x-intercept: $(2, 0)$
y-intercept: $(0, 3)$

82. x-intercept: $(-3, 0)$
y-intercept: $(0, 4)$

83. x-intercept: $\left(-\frac{1}{6}, 0\right)$
y-intercept: $\left(0, -\frac{2}{3}\right)$

84. x-intercept: $\left(\frac{2}{3}, 0\right)$
y-intercept: $(0, -2)$

85. Point on line: $(1, 2)$
x-intercept: $(c, 0)$
y-intercept: $(0, c), \quad c \neq 0$

86. Point on line: $(-3, 4)$
x-intercept: $(d, 0)$
y-intercept: $(0, d), \quad d \neq 0$

In Exercises 87–96, write the slope-intercept forms of the equations of the lines through the given point (a) parallel to the given line and (b) perpendicular to the given line.

	Point	*Line*
87.	$(2, 1)$	$4x - 2y = 3$
88.	$(-3, 2)$	$x + y = 7$
89.	$\left(-\frac{2}{3}, \frac{7}{8}\right)$	$3x + 4y = 7$
90.	$\left(\frac{7}{8}, \frac{3}{4}\right)$	$5x + 3y = 0$
91.	$(-1, 0)$	$y = -3$
92.	$(4, -2)$	$y = 1$
93.	$(2, 5)$	$x = 4$
94.	$(-5, 1)$	$x = -2$
95.	$(2.5, 6.8)$	$x - y = 4$
96.	$(-3.9, -1.4)$	$6x + 2y = 9$

Graphical Interpretation **In Exercises 97–100, identify any relationships that exist among the lines, and then use a graphing utility to graph the three equations in the same viewing window. Adjust the viewing window so that the slope appears visually correct—that is, so that parallel lines appear parallel and perpendicular lines appear to intersect at a right angle.**

97. (a) $y = 2x$ (b) $y = -2x$ (c) $y = \frac{1}{2}x$

98. (a) $y = \frac{2}{3}x$ (b) $y = -\frac{3}{2}x$ (c) $y = \frac{2}{3}x + 2$

99. (a) $y = -\frac{1}{2}x$ (b) $y = -\frac{1}{2}x + 3$ (c) $y = 2x - 4$

100. (a) $y = x - 8$ (b) $y = x + 1$ (c) $y = -x + 3$

In Exercises 101–104, find a relationship between x and y such that (x, y) is equidistant from the two points.

101. $(4, -1), (-2, 3)$

102. $(6, 5), (1, -8)$

103. $\left(3, \frac{5}{2}\right), (-7, 1)$

104. $\left(-\frac{1}{2}, -4\right), \left(\frac{7}{2}, \frac{5}{4}\right)$

105. ***Cash Flow per Share*** The cash flow per share for Timberland Co. was \$0.18 in 1995 and \$3.65 in 2000. Write a linear equation that gives the cash flow per share in terms of the year. Let $t = 0$ represent 1995. Then predict the cash flows for the years 2005 and 2010. (Source: Timberland Co.)

106. ***Number of Stores*** In 1996 there were 3927 J.C. Penney stores and in 2000 there were 3800 stores. Write a linear equation that gives the number of stores in terms of the year. Let $t = 0$ represent 1996. Then predict the numbers of stores for the years 2005 and 2010. (Source: J.C. Penney Co.)

107. ***Annual Salary*** A jeweler's salary was \$28,500 in 1998 and \$32,900 in 2000. The jeweler's salary follows a linear growth pattern. What will the jeweler's salary be in 2005?

108. ***College Enrollment*** Ohio University had 27,913 students in 1999 and 28,197 students in 2001. The enrollment appears to follow a linear growth pattern. How many students will Ohio University have in 2005? (Source: Ohio University)

109. ***Depreciation*** A sub shop purchases a used pizza oven for \$875. After 5 years, the oven will have to be replaced. Write a linear equation giving the value V of the equipment during the 5 years it will be in use.

110. ***Depreciation*** A school district purchases a high-volume printer, copier, and scanner for \$25,000. After 10 years, the equipment will have to be replaced. Its value at that time is expected to be \$2000. Write a linear equation giving the value V of the equipment during the 10 years it will be in use.

111. ***Sales*** A discount outlet is offering a 15% discount on all items. Write a linear equation giving the sale price S for an item with a list price L.

112. ***Hourly Wage*** A microchip manufacturer pays its assembly line workers \$11.50 per hour. In addition, workers receive a piecework rate of \$0.75 per unit produced. Write a linear equation for the hourly wage W in terms of the number of units x produced per hour.

113. ***Cost, Revenue, and Profit*** A roofing contractor purchases a shingle delivery truck with a shingle elevator for \$36,500. The vehicle requires an average expenditure of \$5.25 per hour for fuel and maintenance, and the operator is paid \$11.50 per hour.

(a) Write a linear equation giving the total cost C of operating this equipment for t hours. (Include the purchase cost of the equipment.)

(b) Assuming that customers are charged \$27 per hour of machine use, write an equation for the revenue R derived from t hours of use.

(c) Use the formula for profit $(P = R - C)$ to write an equation for the profit derived from t hours of use.

(d) Use the result of part (c) to find the break-even point—that is, the number of hours this equipment must be used to yield a profit of 0 dollars.

114. *Rental Demand* A real estate office handles an apartment complex with 50 units. When the rent per unit is $580 per month, all 50 units are occupied. However, when the rent is $625 per month, the average number of occupied units drops to 47. Assume that the relationship between the monthly rent p and the demand x is linear.

(a) Write the equation of the line giving the demand x in terms of the rent p.

(b) Use this equation to predict the number of units occupied when the rent is $655.

(c) Predict the number of units occupied when the rent is $595.

115. *Geometry* The length and width of a rectangular garden are 15 meters and 10 meters, respectively. A walkway of width x surrounds the garden.

(a) Draw a diagram that gives a visual representation of the problem.

(b) Write the equation for the perimeter y of the walkway in terms of x.

(c) Use a graphing utility to graph the equation for the perimeter.

(d) Determine the slope of the graph in part (c). For each additional one-meter increase in the width of the walkway, determine the increase in its perimeter.

116. *Monthly Salary* A pharmaceutical salesperson receives a monthly salary of $2500 plus a commission of 7% of sales. Write a linear equation for the salesperson's monthly wage W in terms of monthly sales S.

117. *Business Costs* A sales representative of a company using a personal car receives $120 per day for lodging and meals plus $0.35 per mile driven. Write a linear equation giving the daily cost C to the company in terms of x, the number of miles driven.

118. *Sports* The average annual salaries of major league baseball players (in thousands of dollars) from 1995 to 2002 are shown in the scatter plot. Find the equation of the line that you think best fits this data. (Let y represent the average salary and let t represent the year, with $t = 5$ corresponding to 1995.) (Source: The Associated Press)

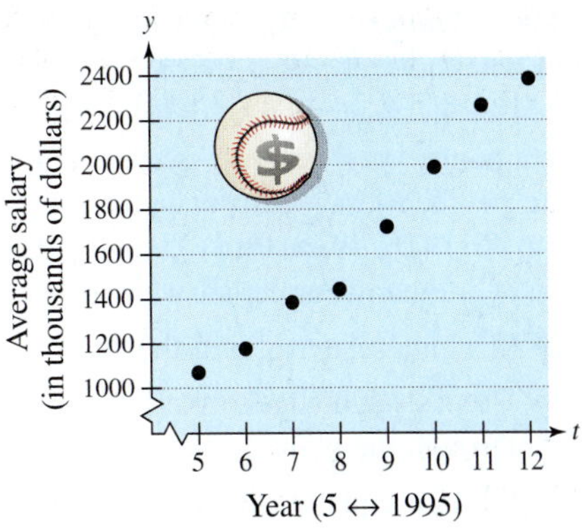

FIGURE FOR 118

▶ Model It

119. *Data Analysis* The average monthly cellular phone bills y (in dollars) for subscribers in the United States from 1990 through 1999, where x is the year, are shown as data points (x, y). (Source: Cellular Telecommunications Industry Association)

(1990, 80.90) (1995, 51.00)
(1991, 72.74) (1996, 47.70)
(1992, 68.68) (1997, 42.78)
(1993, 61.48) (1998, 39.43)
(1994, 56.21) (1999, 41.24)

(a) Sketch a scatter plot of the data. Let $x = 0$ correspond to 1990.

(b) Sketch the best-fitting line through the points.

(c) Find the equation of the line from part (b). Explain the procedure you used.

(d) Write a short paragraph explaining the meaning of the slope and y-intercept of the line in terms of the data.

(e) Compare the values obtained using your model with the actual values.

(f) Use your model to estimate the average monthly cellular phone bill in 2005.

120. ***Data Analysis*** An instructor gives regular 20-point quizzes and 100-point exams in an algebra course. Average scores for six students, given as data points (x, y) where x is the average quiz score and y is the average test score, are $(18, 87)$, $(10, 55)$, $(19, 96)$, $(16, 79)$, $(13, 76)$, and $(15, 82)$. [*Note:* There are many correct answers for parts (b)–(d).]

(a) Sketch a scatter plot of the data.

(b) Use a straightedge to sketch the best-fitting line through the points.

(c) Find an equation for the line sketched in part (b).

(d) Use the equation in part (c) to estimate the average test score for a person with an average quiz score of 17.

(e) The instructor adds 4 points to the average test score of everyone in the class. Describe the changes in the positions of the plotted points and the change in the equation of the line.

Synthesis

True or False? **In Exercises 121 and 122, determine whether the statement is true or false. Justify your answer.**

121. A line with a slope of $-\frac{5}{7}$ is steeper than a line with a slope of $-\frac{6}{7}$.

122. The line through $(-8, 2)$ and $(-1, 4)$ and the line through $(0, -4)$ and $(-7, 7)$ are parallel.

123. Explain how you could show that the points $A(2, 3)$, $B(2, 9)$, and $C(4, 3)$ are the vertices of a right triangle.

124. Explain why the slope of a vertical line is said to be undefined.

125. With the information given in the graphs, is it possible to determine the slope of each line? Is it possible that the lines could have the same slope? Explain.

(a)

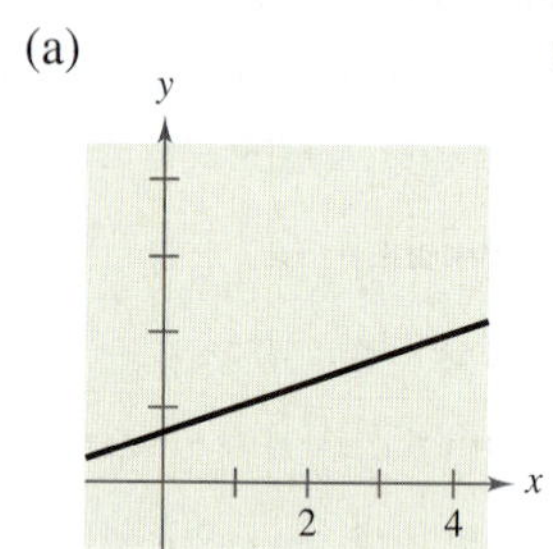

(b)

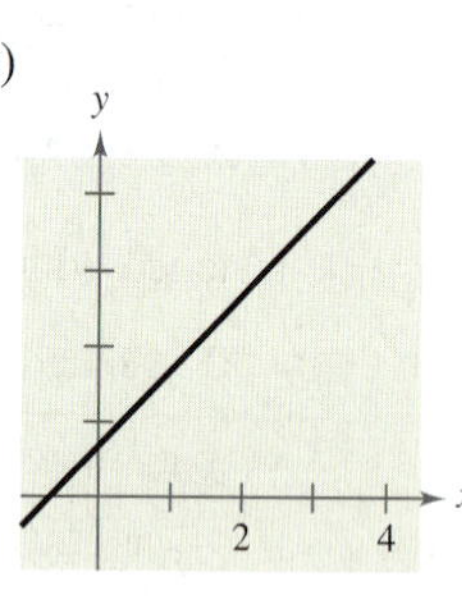

126. The slopes of two lines are -4 and $\frac{5}{2}$. Which is steeper? Explain.

127. The value V of a molding machine t years after it is purchased is

$$V = -4000t + 58{,}500, \quad 0 \le t \le 5.$$

Explain what the V-intercept and slope measure.

128. ***Think About It*** Is it possible for two lines with positive slopes to be perpendicular? Explain.

Review

In Exercises 129–132, match the equation with its graph. [The graphs are labeled (a), (b), (c), and (d).]

(a)

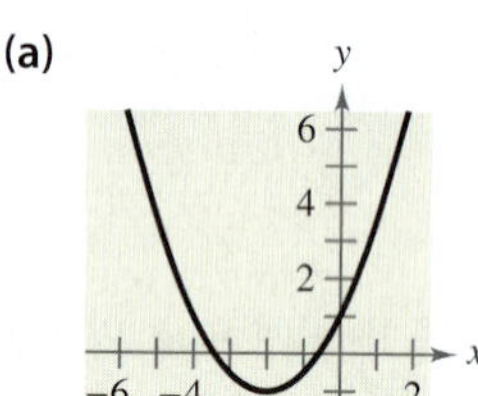

(b)

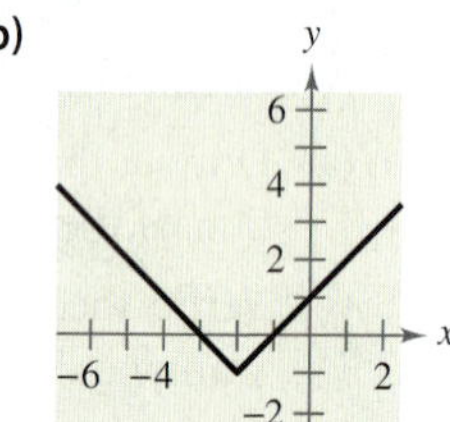

(c)

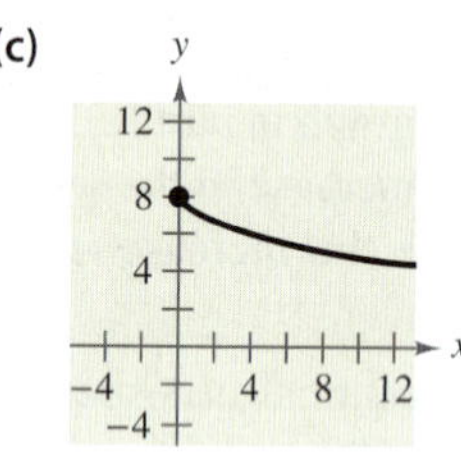

(d)

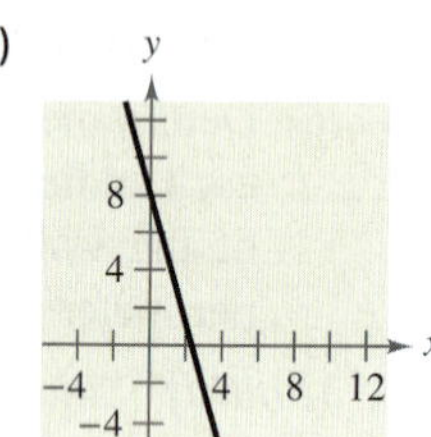

129. $y = 8 - 3x$

130. $y = 8 - \sqrt{x}$

131. $y = \frac{1}{2}x^2 + 2x + 1$

132. $y = |x + 2| - 1$

In Exercises 133–138, find all the solutions of the equation. Check your solution(s) in the original equation.

133. $-7(3 - x) = 14(x - 1)$

134. $\dfrac{8}{2x - 7} = \dfrac{4}{9 - 4x}$

135. $2x^2 - 21x + 49 = 0$

136. $x^2 - 8x + 3 = 0$

137. $\sqrt{x - 9} + 15 = 0$

138. $3x - 16\sqrt{x} + 5 = 0$

2.2 Functions

▶ What you should learn

- How to determine whether relations between two variables are functions
- How to use function notation and evaluate functions
- How to find the domains of functions
- How to use functions to model and solve real-life problems

▶ Why you should learn it

Functions can be used to model and solve real-life problems. For instance, in Exercise 99 on page 200, you will use a function to find the number of threatened and endangered fish in the world.

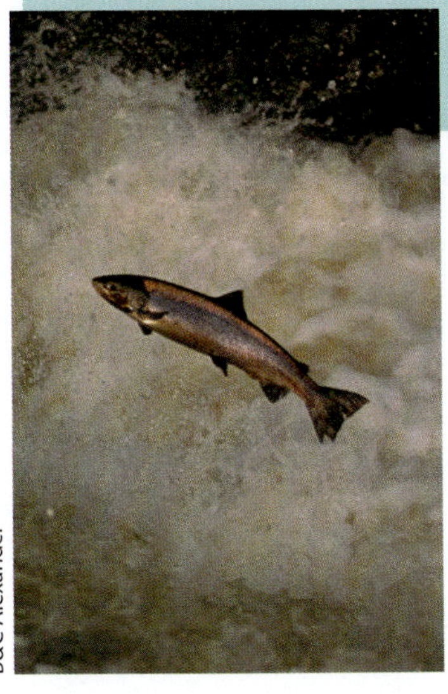

B&C Alexander

Introduction to Functions

Many everyday phenomena involve two quantities that are related to each other by some rule of correspondence. The mathematical term for such a rule of correspondence is a **relation.** In mathematics, relations are often represented by mathematical equations and formulas. For instance, the simple interest I earned on \$1000 for 1 year is related to the annual interest rate r by the formula $I = 1000r$.

The formula $I = 1000r$ represents a special kind of relation that matches each item from one set with exactly one item from a different set. Such a relation is called a **function.**

Definition of a Function

A **function** f from a set A to a set B is a relation that assigns to each element x in the set A exactly one element y in the set B. The set A is the **domain** (or set of inputs) of the function f, and the set B contains the **range** (or set of outputs).

To help understand this definition, look at the function that relates the time of day to the temperature in Figure 2.20.

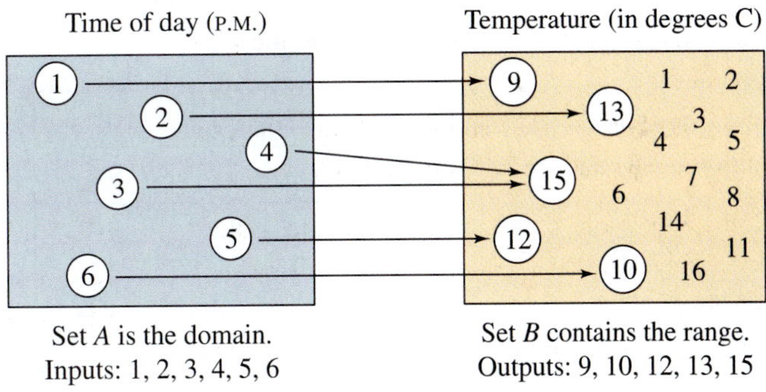

FIGURE 2.20

This function can be represented by the following ordered pairs, in which the first coordinate is the input and the second coordinate is the output.

$$\{(1, 9°), (2, 13°), (3, 15°), (4, 15°), (5, 12°), (6, 10°)\}$$

Characteristics of a Function from Set A to Set B

1. Each element in A must be matched with an element in B.
2. Some elements in B may not be matched with any element in A.
3. Two or more elements in A may be matched with the same element in B.
4. An element in A (the domain) cannot be matched with two different elements of B.

Functions are commonly represented in four ways.

> **Four Ways to Represent a Function**
>
> 1. *Verbally* by a sentence that describes how the input variable is related to the output variable
> 2. *Numerically* by a table or a list of ordered pairs that matches input values with output values
> 3. *Graphically* by points on a graph in a coordinate plane in which the input values are represented by the horizontal axis and the output values are represented by the vertical axis
> 4. *Algebraically* by an equation in two variables

To determine whether or not a relation is a function, you must decide whether each input value is matched with exactly one output value. If any input value is matched with two or more output values, the relation is not a function.

Example 1 ▶ Testing for Functions

Determine whether the relation represents y as a function of x.

a. The input value x is the number of representatives from a state, and the output value y is the number of senators.

b.

Input x	Output y
2	11
2	10
3	8
4	5
5	1

c.

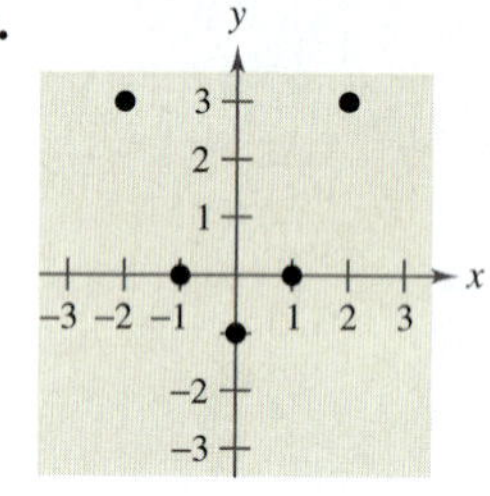

FIGURE 2.21

Solution

a. This verbal description *does* describe y as a function of x. Regardless of the value of x, the value of y is always 2. Such functions are called *constant functions*.

b. This table *does not* describe y as a function of x. The input value 2 is matched with two different y-values.

c. The graph in Figure 2.21 *does* describe y as a function of x. Each input value is matched with exactly one output value.

Representing functions by sets of ordered pairs is common in *discrete mathematics*. In algebra, however, it is more common to represent functions by equations or formulas involving two variables. For instance, the equation

$$y = x^2 \qquad y \text{ is a function of } x.$$

represents the variable y as a function of the variable x. In this equation, x is

The Granger Collection

Historical Note

Leonhard Euler (1707–1783), a Swiss mathematician, is considered to have been the most prolific and productive mathematician in history. One of his greatest influences on mathematics was his use of symbols, or notation. The function notation $y = f(x)$ was introduced by Euler.

the **independent variable** and y is the **dependent variable.** The domain of the function is the set of all values taken on by the independent variable x, and the range of the function is the set of all values taken on by the dependent variable y.

Example 2 ▶ Testing for Functions Represented Algebraically

Which of the equations represent(s) y as a function of x?

a. $x^2 + y = 1$ **b.** $-x + y^2 = 1$

Solution

To determine whether y is a function of x, try to solve for y in terms of x.

a. Solving for y yields

$x^2 + y = 1$ Write original equation.

$y = 1 - x^2.$ Solve for y.

To each value of x there corresponds exactly one value of y. So, y is a function of x.

b. Solving for y yields

$-x + y^2 = 1$ Write original equation.

$y^2 = 1 + x$ Add x to each side.

$y = \pm\sqrt{1 + x}.$ Solve for y.

The $\pm$ indicates that to a given value of x there correspond two values of y. So, y is not a function of x.

Function Notation

When an equation is used to represent a function, it is convenient to name the function so that it can be referenced easily. For example, you know that the equation $y = 1 - x^2$ describes y as a function of x. Suppose you give this function the name "f." Then you can use the following **function notation.**

Input	*Output*	*Equation*
x	$f(x)$	$f(x) = 1 - x^2$

The symbol $f(x)$ is read as *the value of f at x* or simply *f of x*. The symbol $f(x)$ corresponds to the y-value for a given x. So, you can write $y = f(x)$. Keep in mind that f is the *name* of the function, whereas $f(x)$ is the *value* of the function at x. For instance, the function

$$f(x) = 3 - 2x$$

has *function values* denoted by $f(-1)$, $f(0)$, $f(2)$, and so on. To find these values, substitute the specified input values into the given equation.

For $x = -1$, $f(-1) = 3 - 2(-1) = 3 + 2 = 5.$

For $x = 0$, $f(0) = 3 - 2(0) = 3 - 0 = 3.$

For $x = 2$, $f(2) = 3 - 2(2) = 3 - 4 = -1.$

STUDY TIP

In Example 3, note that $g(x + 2)$ is not equal to $g(x) + g(2)$. In general, $g(u + v) \neq g(u) + g(v)$.

Although f is often used as a convenient function name and x is often used as the independent variable, you can use other letters. For instance,

$$f(x) = x^2 - 4x + 7, \quad f(t) = t^2 - 4t + 7, \quad \text{and} \quad g(s) = s^2 - 4s + 7$$

all define the same function. In fact, the role of the independent variable is that of a "placeholder." Consequently, the function could be described by

$$f(\quad) = (\quad)^2 - 4(\quad) + 7.$$

Example 3 ▶ Evaluating a Function 

Let $g(x) = -x^2 + 4x + 1$. Find

a. $g(2)$ **b.** $g(t)$ **c.** $g(x + 2)$.

Solution

a. Replacing x with 2 in $g(x) = -x^2 + 4x + 1$ yields the following.

$$g(2) = -(2)^2 + 4(2) + 1 = -4 + 8 + 1 = 5$$

b. Replacing x with t yields the following.

$$g(t) = -(t)^2 + 4(t) + 1 = -t^2 + 4t + 1$$

c. Replacing x with $x + 2$ yields the following.

$$\begin{aligned} g(x + 2) &= -(x + 2)^2 + 4(x + 2) + 1 \\ &= -(x^2 + 4x + 4) + 4x + 8 + 1 \\ &= -x^2 - 4x - 4 + 4x + 8 + 1 \\ &= -x^2 + 5 \end{aligned}$$

A function defined by two or more equations over a specified domain is called a **piecewise-defined function.**

Example 4 ▶ A Piecewise-Defined Function

Evaluate the function when $x = -1$, 0, and 1.

$$f(x) = \begin{cases} x^2 + 1, & x < 0 \\ x - 1, & x \geq 0 \end{cases}$$

Solution

Because $x = -1$ is less than 0, use $f(x) = x^2 + 1$ to obtain

$$f(-1) = (-1)^2 + 1 = 2.$$

For $x = 0$, use $f(x) = x - 1$ to obtain

$$f(0) = (0) - 1 = -1.$$

For $x = 1$, use $f(x) = x - 1$ to obtain

$$f(1) = (1) - 1 = 0.$$

Technology

Use a graphing utility to graph $y = \sqrt{4 - x^2}$. What is the domain of this function? Then graph $y = \sqrt{x^2 - 4}$. What is the domain of this function? Do the domains of these two functions overlap? If so, for what values?

The Domain of a Function

The domain of a function can be described explicitly or it can be *implied* by the expression used to define the function. The **implied domain** is the set of all real numbers for which the expression is defined. For instance, the function

$$f(x) = \frac{1}{x^2 - 4}$$ Domain excludes x-values that result in division by zero.

has an implied domain that consists of all real x other than $x = \pm 2$. These two values are excluded from the domain because division by zero is undefined. Another common type of implied domain is that used to avoid even roots of negative numbers. For example, the function

$$f(x) = \sqrt{x}$$ Domain excludes x-values that result in even roots of negative numbers.

is defined only for $x \geq 0$. So, its implied domain is the interval $[0, \infty)$. In general, the domain of a function *excludes* values that would cause division by zero *or* that would result in the even root of a negative number.

Example 5 ▶ Finding the Domain of a Function 

Find the domain of each function.

a. f: $\{(-3, 0), (-1, 4), (0, 2), (2, 2), (4, -1)\}$ **b.** $g(x) = \dfrac{1}{x + 5}$

c. Volume of a sphere: $V = \frac{4}{3}\pi r^3$ **d.** $h(x) = \sqrt{4 - x^2}$

Solution

a. The domain of f consists of all first coordinates in the set of ordered pairs.

$$\text{Domain} = \{-3, -1, 0, 2, 4\}$$

b. Excluding x-values that yield zero in the denominator, the domain of g is the set of all real numbers $x \neq -5$.

c. Because this function represents the volume of a sphere, the values of the radius r must be positive. So, the domain is the set of all real numbers r such that $r > 0$.

d. This function is defined only for x-values for which

$$4 - x^2 \geq 0.$$

Using the methods described in Section 1.8, you can conclude that $-2 \leq x \leq 2$. So, the domain is the interval $[-2, 2]$.

In Example 5(c), note that the domain of a function may be implied by the physical context. For instance, from the equation

$$V = \tfrac{4}{3}\pi r^3$$

you would have no reason to restrict r to positive values, but the physical context implies that a sphere cannot have a negative radius.

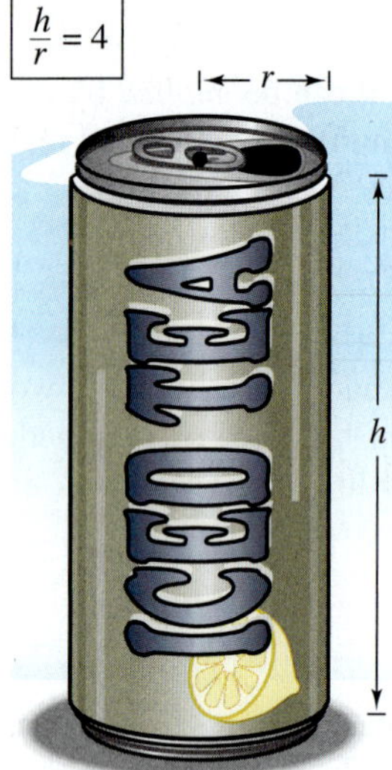

FIGURE 2.22

Applications

Example 6 ▶ The Dimensions of a Container

You work in the marketing department of a soft-drink company and are experimenting with a new can for iced tea that is slightly narrower and taller than a standard can. For your experimental can, the ratio of the height to the radius is 4, as shown in Figure 2.22.

a. Write the volume of the can as a function of the radius r.

b. Write the volume of the can as a function of the height h.

Solution

a. $V(r) = \pi r^2 h = \pi r^2(4r) = 4\pi r^3$ — Write V as a function of r.

b. $V(h) = \pi\left(\frac{h}{4}\right)^2 h = \frac{\pi h^3}{16}$ — Write V as a function of h.

Example 7 ▶ The Path of a Baseball

A baseball is hit at a point 3 feet above ground at a velocity of 100 feet per second and an angle of 45°. The path of the baseball is given by the function

$$f(x) = -0.0032x^2 + x + 3$$

where y and x are measured in feet, as shown in Figure 2.23. Will the baseball clear a 10-foot fence located 300 feet from home plate?

Solution

When $x = 300$, the height of the baseball is

$$f(300) = -0.0032(300)^2 + 300 + 3$$
$$= 15 \text{ feet.}$$

So, the ball will clear the fence.

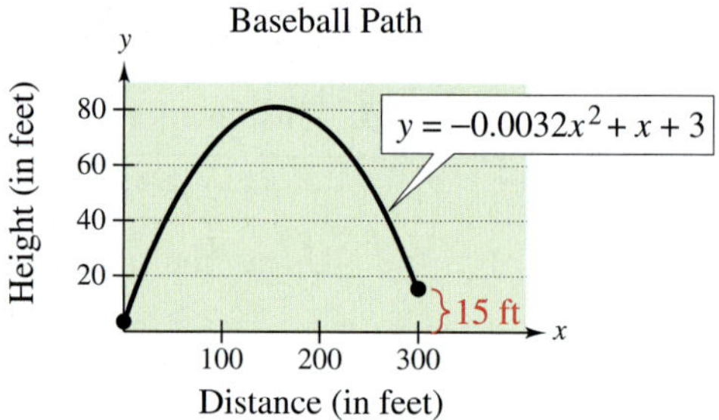

FIGURE 2.23

In the equation in Example 7, the height of the baseball is a function of the distance from home plate.

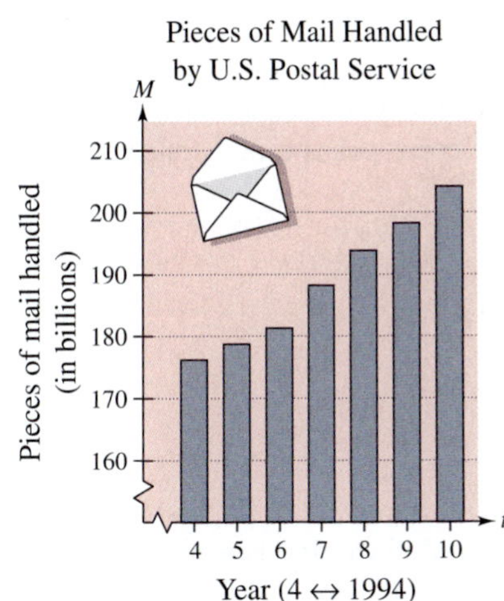

FIGURE 2.24

Example 8 ▶ Pieces of Mail Handled 

The number M (in billions) of pieces of mail handled by the U.S. Postal Service increased in a linear pattern from 1994 to 1996, as shown in Figure 2.24. Then, in 1997, the number handled took a jump and, until 2000, increased in a *different* linear pattern. These two patterns can be approximated by the function

$$M(t) = \begin{cases} 167.2 + 2.70t, & 4 \le t \le 6 \\ 152.0 + 5.57t, & 7 \le t \le 10 \end{cases}$$

where $t = 4$ represents 1994. Use this function to approximate the total number of pieces of mail handled from 1994 to 2000. (Source: U.S. Postal Service)

Solution

From 1994 to 1996, use $M(t) = 167.2 + 2.70t$.

$$\underbrace{178.0,}_{1994} \quad \underbrace{180.7,}_{1995} \quad \underbrace{183.4}_{1996}$$

From 1997 to 2000, use $M(t) = 152.0 + 5.57t$.

$$\underbrace{191.0,}_{1997} \quad \underbrace{196.6,}_{1998} \quad \underbrace{202.1,}_{1999} \quad \underbrace{207.7}_{2000}$$

The total of these seven amounts is 1339.5, which implies that the total number of pieces of mail handled was approximately 1.3 trillion.

One of the basic definitions in calculus employs the ratio

$$\frac{f(x+h) - f(x)}{h}, \qquad h \neq 0.$$

This ratio is called a **difference quotient,** as illustrated in Example 9.

Example 9 ▶ Evaluating a Difference Quotient 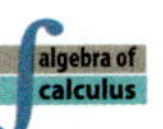 

For $f(x) = x^2 - 4x + 7$, find $\dfrac{f(x+h) - f(x)}{h}$.

Solution

$$\frac{f(x+h) - f(x)}{h} = \frac{[(x+h)^2 - 4(x+h) + 7] - (x^2 - 4x + 7)}{h}$$

$$= \frac{x^2 + 2xh + h^2 - 4x - 4h + 7 - x^2 + 4x - 7}{h}$$

$$= \frac{2xh + h^2 - 4h}{h} = \frac{h(2x + h - 4)}{h} = 2x + h - 4, \ h \neq 0$$

The symbol ∫ indicates an example or exercise that highlights algebraic techniques specifically used in calculus.

Summary of Function Terminology

Function: A **function** is a relationship between two variables such that to each value of the independent variable there corresponds exactly one value of the dependent variable.

Function Notation: $y = f(x)$

f is the *name* of the function.

y is the **dependent variable.**

x is the **independent variable.**

$f(x)$ is the *value of the function at x.*

Domain: The **domain** of a function is the set of all values (inputs) of the independent variable for which the function is defined. If x is in the domain of f, f is said to be *defined* at x. If x is not in the domain of f, f is said to be *undefined* at x.

Range: The range of a function is the set of all values (outputs) assumed by the dependent variable (that is, the set of all function values).

Implied Domain: If f is defined by an algebraic expression and the domain is not specified, the **implied domain** consists of all real numbers for which the expression is defined.

Writing ABOUT MATHEMATICS

Everyday Functions In groups of two or three, identify common real-life functions. Consider everyday activities, events, and expenses, such as long distance telephone calls and car insurance. Here are two examples.

a. The statement, "Your happiness is a function of the grade you receive in this course" *is not* a correct mathematical use of the word "function." The word "happiness" is ambiguous.

b. The statement, "Your federal income tax is a function of your adjusted gross income" *is* a correct mathematical use of the word "function." Once you have determined your adjusted gross income, your income tax can be determined.

Describe your functions in words. Avoid using ambiguous words. Can you find an example of a piecewise-defined function?

2.2 Exercises

In Exercises 1–4, is the relationship a function?

1. *Domain* *Range*

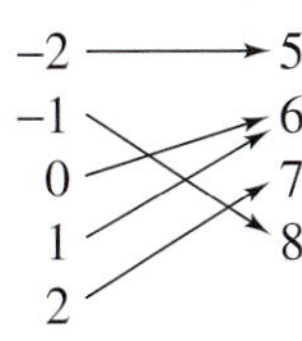

2. *Domain* *Range*

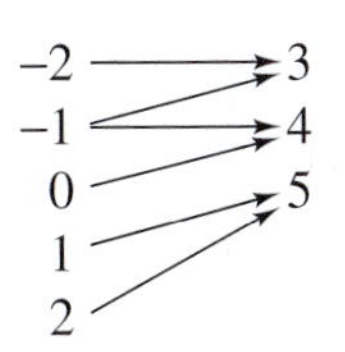

3. *Domain* *Range*

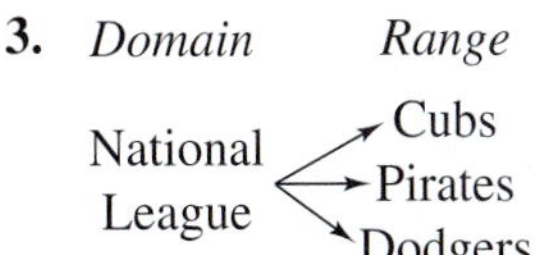

American League — Orioles, Yankees, Twins

4. *Domain* *Range*

(Year) (Number of North Atlantic tropical storms and hurricanes)

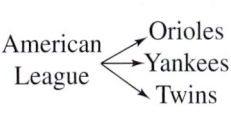

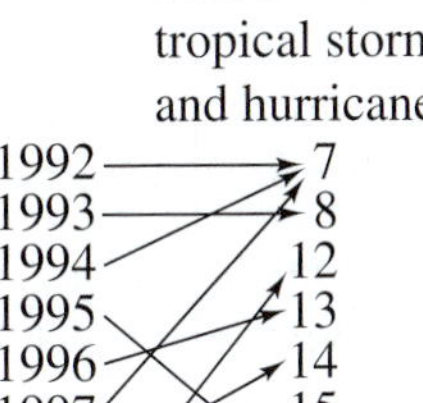

In Exercises 5–8, does the table describe a function? Explain your reasoning.

5.

Input value	-2	-1	0	1	2
Output value	-8	-1	0	1	8

6.

Input value	0	1	2	1	0
Output value	-4	-2	0	2	4

7.

Input value	10	7	4	7	10
Output value	3	6	9	12	15

8.

Input value	0	3	9	12	15
Output value	3	3	3	3	3

In Exercises 9 and 10, which sets of ordered pairs represent functions from *A* to *B*? Explain.

9. $A = \{0, 1, 2, 3\}$ and $B = \{-2, -1, 0, 1, 2\}$

(a) $\{(0, 1), (1, -2), (2, 0), (3, 2)\}$

(b) $\{(0, -1), (2, 2), (1, -2), (3, 0), (1, 1)\}$

(c) $\{(0, 0), (1, 0), (2, 0), (3, 0)\}$

(d) $\{(0, 2), (3, 0), (1, 1)\}$

10. $A = \{a, b, c\}$ and $B = \{0, 1, 2, 3\}$

(a) $\{(a, 1), (c, 2), (c, 3), (b, 3)\}$

(b) $\{(a, 1), (b, 2), (c, 3)\}$

(c) $\{(1, a), (0, a), (2, c), (3, b)\}$

(d) $\{(c, 0), (b, 0), (a, 3)\}$

Circulation of Newspapers **In Exercises 11 and 12, use the graph, which shows the circulation (in millions) of daily newspapers in the United States.** (Source: Editor & Publisher Company)

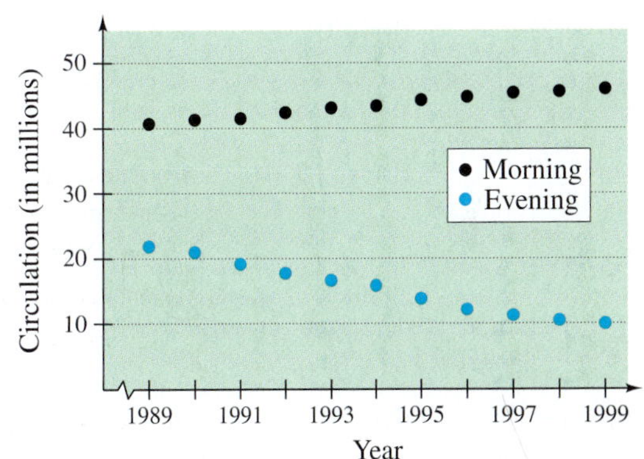

11. Is the circulation of morning newspapers a function of the year? Is the circulation of evening newspapers a function of the year? Explain.

12. Let $f(x)$ represent the circulation of evening newspapers in year x. Find $f(1998)$.

In Exercises 13–22, determine whether the equation represents *y* as a function of *x*.

13. $x^2 + y^2 = 4$

14. $x = y^2$

15. $x^2 + y = 4$

16. $x + y^2 = 4$

17. $2x + 3y = 4$

18. $(x - 2)^2 + y^2 = 4$

19. $y^2 = x^2 - 1$

20. $y = \sqrt{x + 5}$

21. $y = |4 - x|$

22. $|y| = 4 - x$

In Exercises 23–36, evaluate the function at each specified value of the independent variable and simplify.

23. $f(x) = 2x - 3$

(a) $f(1)$ (b) $f(-3)$ (c) $f(x - 1)$

24. $g(y) = 7 - 3y$

(a) $g(0)$ (b) $g\left(\frac{7}{3}\right)$ (c) $g(s + 2)$

25. $V(r) = \frac{4}{3}\pi r^3$

(a) $V(3)$ (b) $V\left(\frac{3}{2}\right)$ (c) $V(2r)$

26. $h(t) = t^2 - 2t$

(a) $h(2)$ (b) $h(1.5)$ (c) $h(x + 2)$

27. $f(y) = 3 - \sqrt{y}$

(a) $f(4)$ (b) $f(0.25)$ (c) $f(4x^2)$

28. $f(x) = \sqrt{x + 8} + 2$

(a) $f(-8)$ (b) $f(1)$ (c) $f(x - 8)$

29. $q(x) = \dfrac{1}{x^2 - 9}$

(a) $q(0)$ (b) $q(3)$ (c) $q(y + 3)$

30. $q(t) = \dfrac{2t^2 + 3}{t^2}$

(a) $q(2)$ (b) $q(0)$ (c) $q(-x)$

31. $f(x) = \dfrac{|x|}{x}$

(a) $f(2)$ (b) $f(-2)$ (c) $f(x - 1)$

32. $f(x) = |x| + 4$

(a) $f(2)$ (b) $f(-2)$ (c) $f(x^2)$

33. $f(x) = \begin{cases} 2x + 1, & x < 0 \\ 2x + 2, & x \geq 0 \end{cases}$

(a) $f(-1)$ (b) $f(0)$ (c) $f(2)$

34. $f(x) = \begin{cases} x^2 + 2, & x \leq 1 \\ 2x^2 + 2, & x > 1 \end{cases}$

(a) $f(-2)$ (b) $f(1)$ (c) $f(2)$

35. $f(x) = \begin{cases} 3x - 1, & x < -1 \\ 4, & -1 \leq x \leq 1 \\ x^2, & x > 1 \end{cases}$

(a) $f(-2)$ (b) $f\left(-\frac{1}{2}\right)$ (c) $f(3)$

36. $f(x) = \begin{cases} 4 - 5x, & x \leq -2 \\ 0, & -2 < x \leq 2 \\ x^2 + 1, & x > 2 \end{cases}$

(a) $f(-3)$ (b) $f(4)$ (c) $f(-1)$

In Exercises 37–42, complete the table.

37. $f(x) = x^2 - 3$

x	$f(x)$
-2	
-1	
0	
1	
2	

38. $g(x) = \sqrt{x - 3}$

x	$g(x)$
3	
4	
5	
6	
7	

39. $h(t) = \frac{1}{2}|t + 3|$

t	$h(t)$
-5	
-4	
-3	
-2	
-1	

40. $f(s) = \dfrac{|s - 2|}{s - 2}$

s	$f(s)$
0	
1	
$\frac{3}{2}$	
$\frac{5}{2}$	
4	

41. $f(x) = \begin{cases} -\frac{1}{2}x + 4, & x \leq 0 \\ (x - 2)^2, & x > 0 \end{cases}$

x	$f(x)$
-2	
-1	
0	
1	
2	

42. $h(x) = \begin{cases} 9 - x^2, & x < 3 \\ x - 3, & x \geq 3 \end{cases}$

x	$h(x)$
1	
2	
3	
4	
5	

In Exercises 43–50, find all real values of x such that $f(x) = 0$.

43. $f(x) = 15 - 3x$

44. $f(x) = 5x + 1$

45. $f(x) = \dfrac{3x - 4}{5}$

46. $f(x) = \dfrac{12 - x^2}{5}$

47. $f(x) = x^2 - 9$

48. $f(x) = x^2 - 8x + 15$

49. $f(x) = x^3 - x$

50. $f(x) = x^3 - x^2 - 4x + 4$

In Exercises 51–54, find the value(s) of x for which $f(x) = g(x)$.

51. $f(x) = x^2 + 2x + 1, \quad g(x) = 3x + 3$

52. $f(x) = x^4 - 2x^2, \quad g(x) = 2x^2$

53. $f(x) = \sqrt{3x} + 1, \quad g(x) = x + 1$

54. $f(x) = \sqrt{x} - 4, \quad g(x) = 2 - x$

In Exercises 55–68, find the domain of the function.

55. $f(x) = 5x^2 + 2x - 1$

56. $g(x) = 1 - 2x^2$

57. $h(t) = \dfrac{4}{t}$

58. $s(y) = \dfrac{3y}{y + 5}$

59. $g(y) = \sqrt{y - 10}$

60. $f(t) = \sqrt[3]{t + 4}$

61. $f(x) = \sqrt[4]{1 - x^2}$

62. $f(x) = \sqrt[4]{x^2 + 3x}$

63. $g(x) = \dfrac{1}{x} - \dfrac{3}{x + 2}$

64. $h(x) = \dfrac{10}{x^2 - 2x}$

65. $f(s) = \dfrac{\sqrt{s - 1}}{s - 4}$

66. $f(x) = \dfrac{\sqrt{x + 6}}{6 + x}$

67. $f(x) = \dfrac{x - 4}{\sqrt{x}}$

68. $f(x) = \dfrac{x - 5}{\sqrt{x^2 - 9}}$

In Exercises 69–72, assume that the domain of f is the set $A = \{-2, -1, 0, 1, 2\}$. Determine the set of ordered pairs that represents the function f.

69. $f(x) = x^2$

70. $f(x) = x^2 - 3$

71. $f(x) = |x| + 2$

72. $f(x) = |x + 1|$

Exploration **In Exercises 73–76, match the data with one of the following functions**

$f(x) = cx, \; g(x) = cx^2, \; h(x) = c\sqrt{|x|}, \text{ and } r(x) = \dfrac{c}{x}$

and determine the value of the constant c that will make the function fit the data in the table.

73.

x	y
-4	-32
-1	-2
0	0
1	-2
4	-32

74.

x	y
-4	-1
-1	$-\frac{1}{4}$
0	0
1	$\frac{1}{4}$
4	1

75.

x	y
-4	-8
-1	-32
0	Undef.
1	32
4	8

76.

x	y
-4	6
-1	3
0	0
1	3
4	6

In Exercises 77–84, find the difference quotient and simplify your answer.

77. $f(x) = x^2 - x + 1, \quad \dfrac{f(2 + h) - f(2)}{h}, h \neq 0$

78. $f(x) = 5x - x^2, \quad \dfrac{f(5 + h) - f(5)}{h}, h \neq 0$

79. $f(x) = x^3 + 3x, \quad \dfrac{f(x + h) - f(x)}{h}, h \neq 0$

80. $f(x) = 4x^2 - 2x, \quad \dfrac{f(x + h) - f(x)}{h}, h \neq 0$

81. $g(x) = \dfrac{1}{x^2}, \quad \dfrac{g(x) - g(3)}{x - 3}, x \neq 3$

82. $f(t) = \dfrac{1}{t - 2}, \quad \dfrac{f(t) - f(1)}{t - 1}, t \neq 1$

83. $f(x) = \sqrt{5x}, \quad \dfrac{f(x) - f(5)}{x - 5}, x \neq 5$

84. $f(x) = x^{2/3} + 1, \quad \dfrac{f(x) - f(8)}{x - 8}, x \neq 8$

The symbol ∫ indicates an example or exercise that highlights algebraic techniques specifically used in calculus.

85. ***Geometry*** Write the area A of a square as a function of its perimeter P.

86. ***Geometry*** Write the area A of a circle as a function of its circumference C.

87. ***Maximum Volume*** An open box of maximum volume is to be made from a square piece of material 24 centimeters on a side by cutting equal squares from the corners and turning up the sides (see figure).

(a) The table shows the volume V (in cubic centimeters) of the box for various heights x (in centimeters). Use the table to estimate the maximum volume.

Height, x	Volume, V
1	484
2	800
3	972
4	1024
5	980
6	864

(b) Plot the points (x, V). Does the relation defined by the ordered pairs represent V as a function of x?

(c) If V is a function of x, write the function and determine its domain.

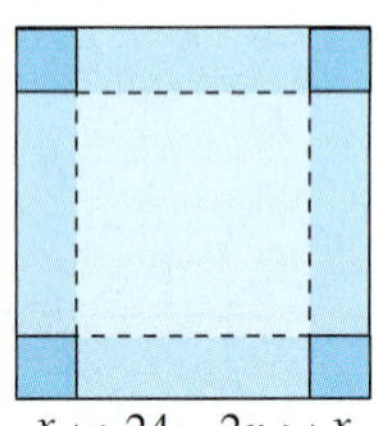

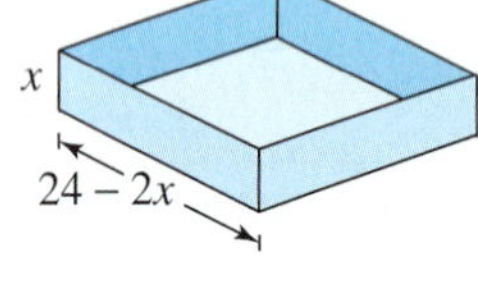

88. ***Maximum Profit*** The cost per unit in the production of a radio model is \$60. The manufacturer charges \$90 per unit for orders of 100 or less. To encourage large orders, the manufacturer reduces the charge by \$0.15 per radio for each unit ordered in excess of 100 (for example, there would be a charge of \$87 per radio for an order size of 120).

(a) The table shows the profit P (in dollars) for various numbers of units ordered, x. Use the table to estimate the maximum profit.

Units, x	Profit, P
110	3135
120	3240
130	3315
140	3360
150	3375
160	3360
170	3315

(b) Plot the points (x, P). Does the relation defined by the ordered pairs represent P as a function of x?

(c) If P is a function of x, write the function and determine its domain.

89. ***Geometry*** A right triangle is formed in the first quadrant by the x- and y-axes and a line through the point $(2, 1)$ (see figure). Write the area A of the triangle as a function of x, and determine the domain of the function.

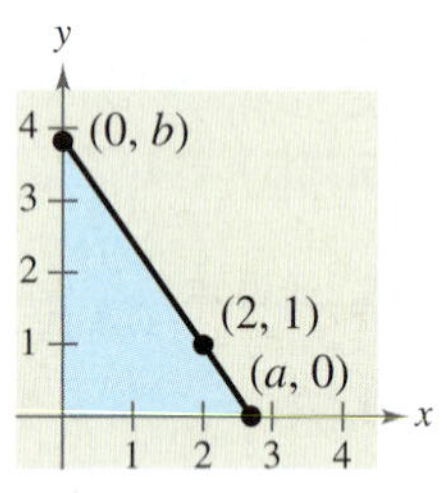

FIGURE FOR 89

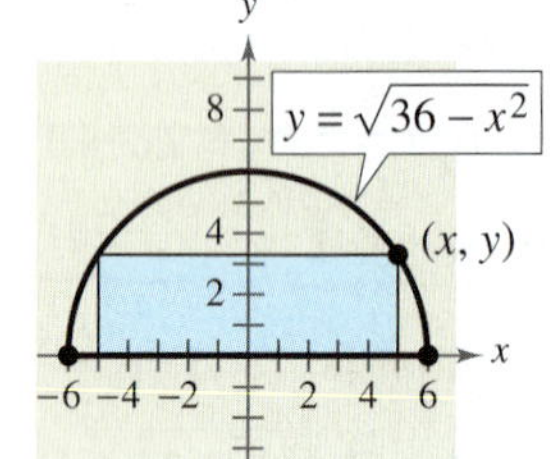

FIGURE FOR 90

90. ***Geometry*** A rectangle is bounded by the x-axis and the semicircle $y = \sqrt{36 - x^2}$ (see figure). Write the area A of the rectangle as a function of x, and determine the domain of the function.

91. ***Average Price*** The average price p (in thousands of dollars) of a new mobile home in the United States from 1990 to 1999 (see figure) can be approximated by the model

$$p(t) = \begin{cases} 0.543t^2 - 0.75t + 27.8, & 0 \le t \le 4 \\ 1.89t + 27.1, & 5 \le t \le 9 \end{cases}$$

where $t = 0$ represents 1990. Use this model to find the average prices of a mobile home in 1990, 1994, 1996, and 1999. (Source: U.S. Census Bureau)

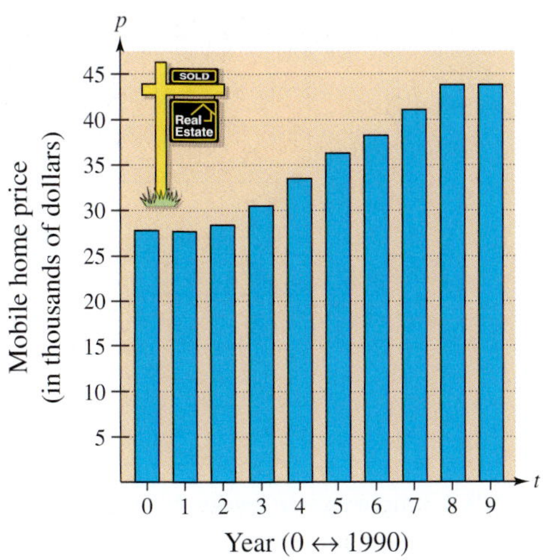

FIGURE FOR 91

92. ***Postal Regulations*** A rectangular package to be sent by the U.S. Postal Service can have a maximum combined length and girth (perimeter of a cross section) of 108 inches (see figure).

(a) Write the volume V of the package as a function of x. What is the domain of the function?

 (b) Use a graphing utility to graph your function. Be sure to use the appropriate window setting.

(c) What dimensions will maximize the volume of the package? Explain your answer.

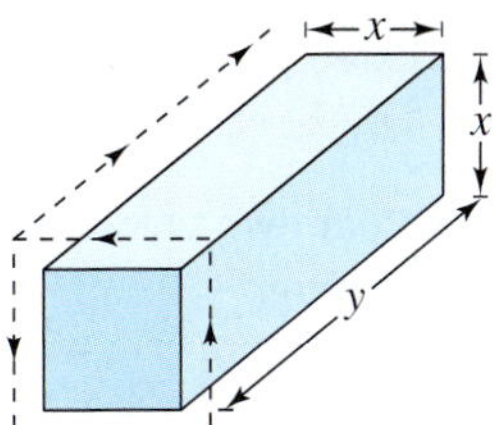

93. ***Cost, Revenue, and Profit*** A company produces a product for which the variable cost is \$12.30 per unit and the fixed costs are \$98,000. The product sells for \$17.98. Let x be the number of units produced and sold.

(a) The total cost for a business is the sum of the variable cost and the fixed costs. Write the total cost C as a function of the number of units produced.

(b) Write the revenue R as a function of the number of units sold.

(c) Write the profit P as a function of the number of units sold. (*Note:* $P = R - C$.)

94. ***Average Cost*** The inventor of a new game believes that the variable cost for producing the game is \$0.95 per unit and the fixed costs are \$6000. The inventor sells each game for \$1.69. Let x be the number of games sold.

(a) The total cost for a business is the sum of the variable cost and the fixed costs. Write the total cost C as a function of the number of games sold.

(b) Write the average cost per unit $\overline{C} = C/x$ as a function of x.

95. ***Transportation*** For groups of 80 or more people, a charter bus company determines the rate per person according to the formula

$$\text{Rate} = 8 - 0.05(n - 80), \qquad n \geq 80$$

where the rate is given in dollars and n is the number of people.

(a) Write the revenue R for the bus company as a function of n.

(b) Use the function in part (a) to complete the table. What can you conclude?

n	90	100	110	120	130	140	150
$R(n)$							

96. ***Physics*** The force F (in tons) of water against the face of a dam is estimated by the function $F(y) = 149.76\sqrt{10}y^{5/2}$, where y is the depth of the water in feet.

(a) Complete the table. What can you conclude from the table?

y	5	10	20	30	40
$F(y)$					

(b) Use the table to approximate the depth at which the force against the dam is 1,000,000 tons.

(c) Find the depth at which the force against the dam is 1,000,000 tons algebraically.

97. ***Height of a Balloon*** A balloon carrying a transmitter ascends vertically from a point 3000 feet from the receiving station.

(a) Draw a diagram that gives a visual representation of the problem. Let h represent the height of the balloon and let d represent the distance between the balloon and the receiving station.

(b) Write the height of the balloon as a function of d. What is the domain of the function?

98. *Path of a Ball* The height y (in feet) of a baseball thrown by a child is

$$y = -\frac{1}{10}x^2 + 3x + 6$$

where x is the horizontal distance (in feet) from where the ball was thrown. Will the ball fly over the head of another child 30 feet away trying to catch the ball? (Assume that the child who is trying to catch the ball holds a baseball glove at a height of 5 feet.)

▶ Model It

99. *Wildlife* The graph shows the number of threatened and endangered fish species in the world from 1996 through 2001. Let $f(t)$ represent the number of threatened and endangered fish species in the year t. (Source: U.S. Fish and Wildlife Service)

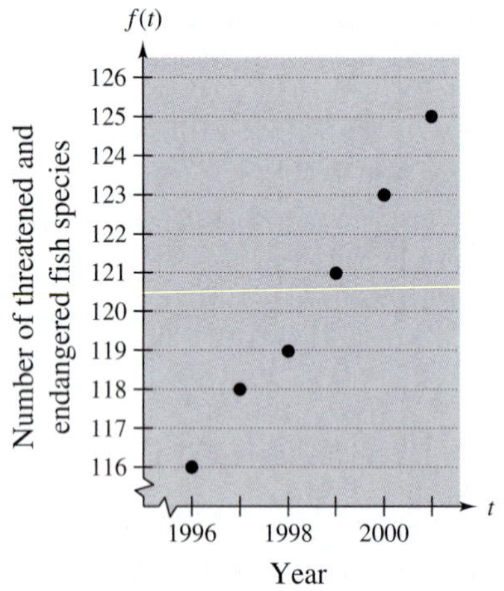

(a) Find

$$\frac{f(2001) - f(1996)}{2001 - 1996}$$

and interpret the result in the context of the problem.

(b) Find a linear model for the data algebraically. Let N represent the number of threatened and endangered fish species and let $x = 6$ correspond to 1996.

(c) Use the model found in part (b) to complete the table.

▶ Model It *(continued)*

x	N
6	
7	
8	
9	
10	
11	

(d) Compare your results from part (c) with the actual data.

(e) Use a graphing utility to find a linear model for the data. Let $x = 6$ correspond to 1996. How does the model you found in part (b) compare with the model given by the graphing utility?

Synthesis

True or False? **In Exercises 100 and 101, determine whether the statement is true or false. Justify your answer.**

100. The domain of the function $f(x) = x^4 - 1$ is $(-\infty, \infty)$, and the range of $f(x)$ is $(0, \infty)$.

101. The set of ordered pairs $\{(-8, -2), (-6, 0), (-4, 0), (-2, 2), (0, 4), (2, -2)\}$ represents a function.

102. *Writing* In your own words, explain the meanings of *domain* and *range*.

Review

In Exercises 103–106, solve the equation.

103. $\frac{t}{3} + \frac{t}{5} = 1$

104. $\frac{3}{t} + \frac{5}{t} = 1$

105. $\frac{3}{x(x + 1)} - \frac{4}{x} = \frac{1}{x + 1}$

106. $\frac{12}{x} - 3 = \frac{4}{x} + 9$

In Exercises 107–110, find the equation of the line passing through the pair of points.

107. $(-2, -5), (4, -1)$

108. $(10, 0), (1, 9)$

109. $(-6, 5), (3, -5)$

110. $\left(-\frac{1}{2}, 3\right), \left(\frac{11}{2}, -\frac{1}{3}\right)$

2.3 Analyzing Graphs of Functions

▶ **What you should learn**

- How to use the Vertical Line Test for functions
- How to find the zeros of functions
- How to determine intervals on which functions are increasing or decreasing
- How to identify even and odd functions

▶ **Why you should learn it**

Graphs of functions can help you visualize relationships between variables in real life. For instance, in Exercise 76 on page 209, you will use the graph of a function to represent visually the merchandise trade balance for the United States.

The Graph of a Function

In Section 2.2, you studied functions from an algebraic point of view. In this section, you will study functions from a graphical perspective.

The **graph of a function** f is the collection of ordered pairs $(x, f(x))$ such that x is in the domain of f. As you study this section, remember that

$$x = \text{the directed distance from the } y\text{-axis}$$

$$f(x) = \text{the directed distance from the } x\text{-axis}$$

as shown in Figure 2.25.

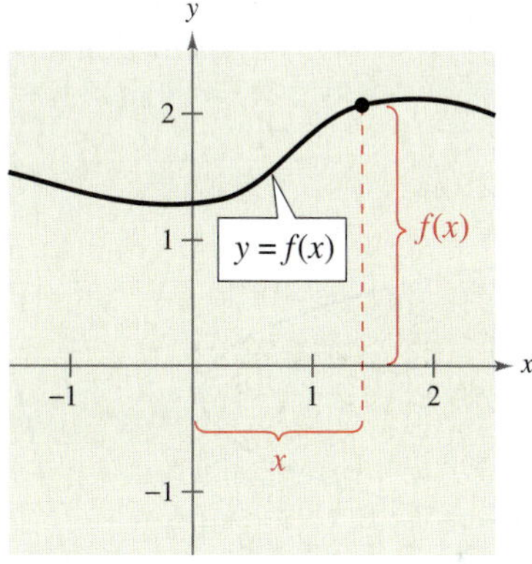

FIGURE 2.25

Example 1 ▶ **Finding the Domain and Range of a Function** 

Use the graph of the function f, shown in Figure 2.26, to find (a) the domain of f, (b) the function values $f(-1)$ and $f(2)$, and (c) the range of f.

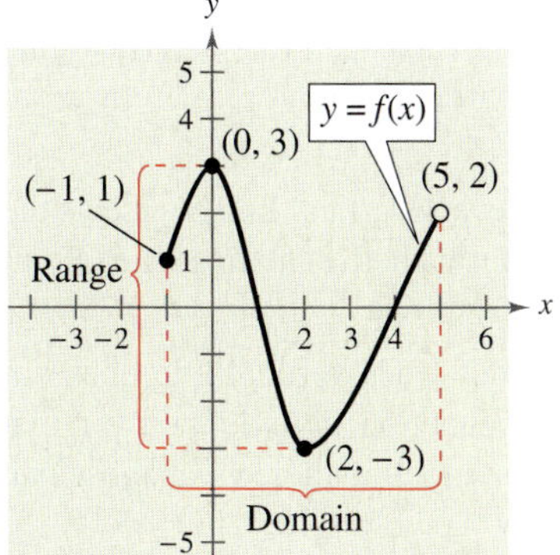

FIGURE 2.26

Solution

a. The closed dot at $(-1, 1)$ indicates that $x = -1$ is in the domain of f, whereas the open dot at $(5, 2)$ indicates $x = 5$ is not in the domain. So, the domain of f is all x in the interval $[-1, 5)$.

b. Because $(-1, 1)$ is a point on the graph of f, it follows that $f(-1) = 1$. Similarly, because $(2, -3)$ is a point on the graph of f, it follows that $f(2) = -3$.

c. Because the graph does not extend below $f(2) = -3$ or above $f(0) = 3$, the range of f is the interval $[-3, 3]$.

The use of dots (open or closed) at the extreme left and right points of a graph indicates that the graph does not extend beyond these points. If no such dots are shown, assume that the graph extends beyond these points.

By the definition of a function, at most one y-value corresponds to a given x-value. This means that the graph of a function cannot have two or more different points with the same x-coordinate, and no two points on the graph of a function can be vertically above and below each other. It follows, then, that a vertical line can intersect the graph of a function at most once. This observation provides a convenient visual test called the **Vertical Line Test** for functions.

> ### Vertical Line Test for Functions
> A set of points in a coordinate plane is the graph of y as a function of x if and only if no *vertical* line intersects the graph at more than one point.

Example 2 ▶ Vertical Line Test for Functions 

Use the Vertical Line Test to decide whether the graphs in Figure 2.27 represent y as a function of x.

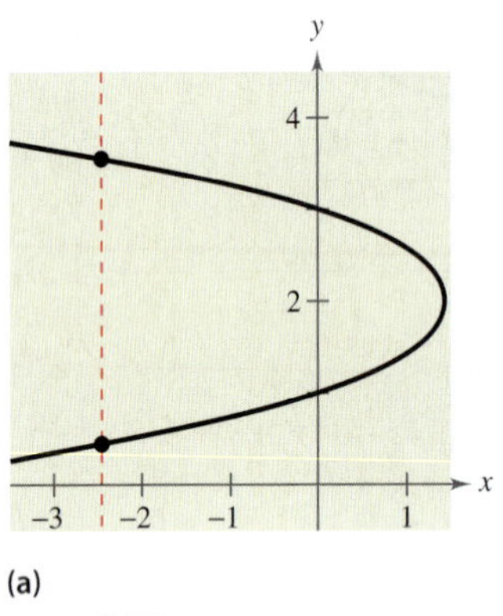

(a)

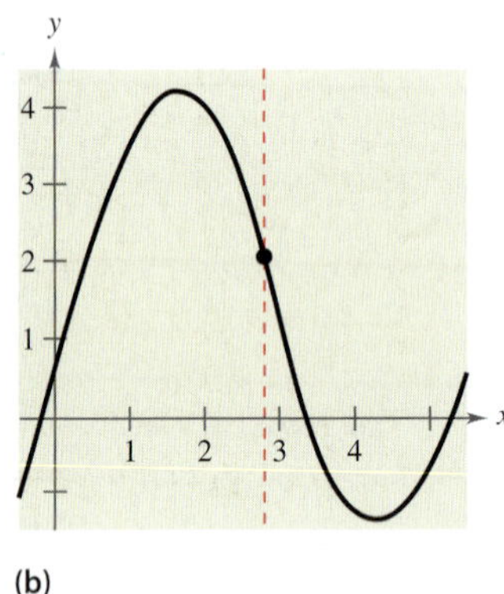

(b)

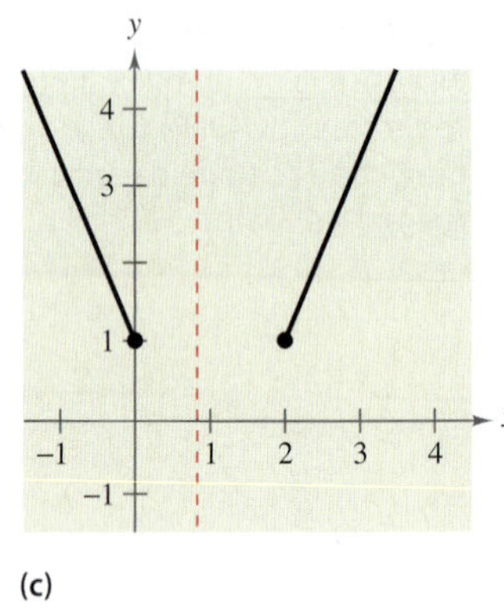

(c)

FIGURE 2.27

Solution

a. This *is not* a graph of y as a function of x, because you can find a vertical line that intersects the graph twice. That is, for a particular input x, there is more than one output y.

b. This *is* a graph of y as a function of x, because every vertical line intersects the graph at most once. That is, for a particular input x, there is at most one output y.

c. This *is* a graph of y as a function of x. (Note that if a vertical line does not intersect the graph, it simply means that the function is undefined for that particular value of x.) That is, for a particular input x, there is at most one output y.

Zeros of a Function

If the graph of a function of x has an x-intercept at $(a, 0)$, then a is a **zero** of the function.

> **Zeros of a Function**
>
> The **zeros of a function** f of x are the x-values for which $f(x) = 0$.

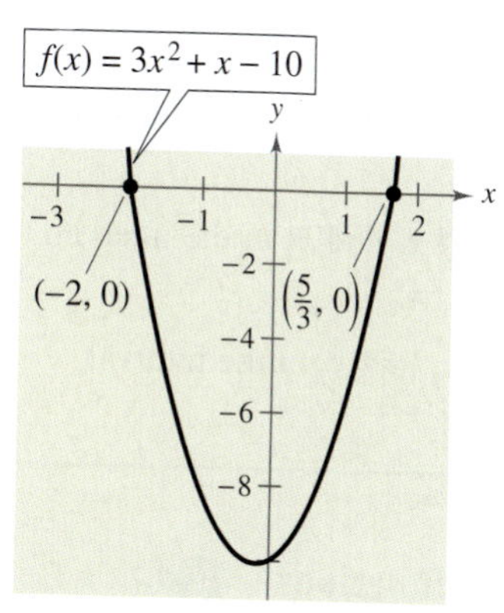

Zeros of f: $x = -2, x = \frac{5}{3}$

FIGURE 2.28

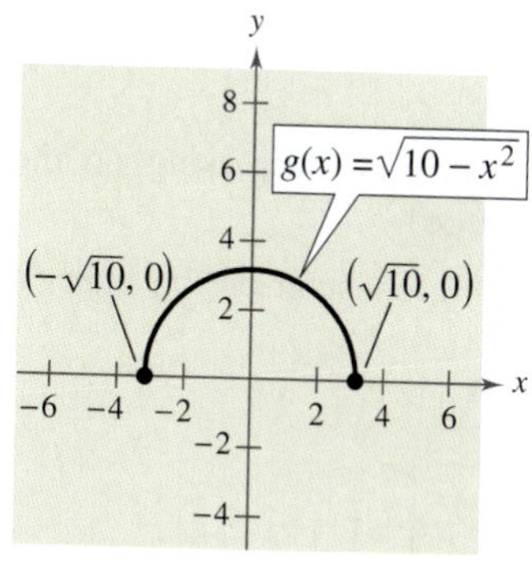

Zeros of g: $x = \pm\sqrt{10}$

FIGURE 2.29

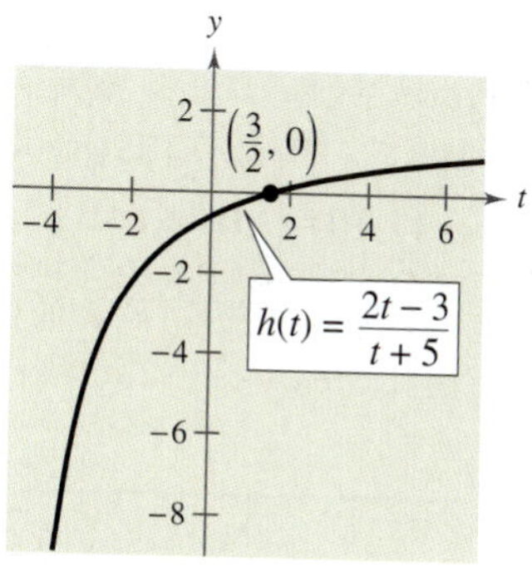

Zero of h: $t = \frac{3}{2}$

FIGURE 2.30

Example 3 ▶ Finding the Zeros of a Function

Find the zeros of each function.

a. $f(x) = 3x^2 + x - 10$ **b.** $g(x) = \sqrt{10 - x^2}$ **c.** $h(t) = \dfrac{2t - 3}{t + 5}$

Solution

To find the zeros of a function, set the function equal to zero and solve for the independent variable.

a.

$$3x^2 + x - 10 = 0 \quad \text{Set } f(x) \text{ equal to 0.}$$

$$(3x - 5)(x + 2) = 0 \quad \text{Factor.}$$

$$3x - 5 = 0 \Rightarrow x = \tfrac{5}{3} \quad \text{Set 1st factor equal to 0.}$$

$$x + 2 = 0 \Rightarrow x = -2 \quad \text{Set 2nd factor equal to 0.}$$

The zeros of f are $x = \frac{5}{3}$ and $x = -2$. In Figure 2.28, note that the graph of f has $\left(\frac{5}{3}, 0\right)$ and $(-2, 0)$ as its x-intercepts.

b.

$$\sqrt{10 - x^2} = 0 \quad \text{Set } g(x) \text{ equal to 0.}$$

$$10 - x^2 = 0 \quad \text{Square each side.}$$

$$10 = x^2 \quad \text{Add } x^2 \text{ to each side.}$$

$$\pm\sqrt{10} = x \quad \text{Extract square root.}$$

The zeros of g are $x = -\sqrt{10}$ and $x = \sqrt{10}$. In Figure 2.29, note that the graph of g has $\left(-\sqrt{10}, 0\right)$ and $\left(\sqrt{10}, 0\right)$ as its x-intercepts.

c.

$$\frac{2t - 3}{t + 5} = 0 \quad \text{Set } h(t) \text{ equal to 0.}$$

$$2t - 3 = 0 \quad \text{Multiply each side by } t + 5.$$

$$2t = 3 \quad \text{Add 3 to each side.}$$

$$t = \frac{3}{2} \quad \text{Divide each side by 2.}$$

The zero of h is $t = \frac{3}{2}$. In Figure 2.30, note that the graph of h has $\left(\frac{3}{2}, 0\right)$ as its t-intercept.

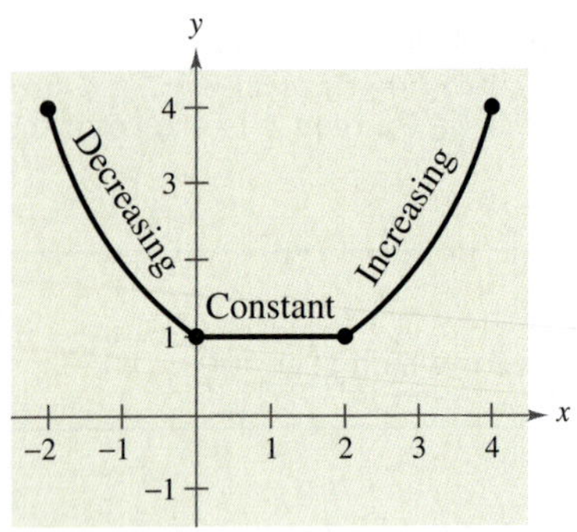

FIGURE 2.31

Increasing and Decreasing Functions

The more you know about the graph of a function, the more you know about the function itself. Consider the graph shown in Figure 2.31. As you move from *left to right*, this graph decreases, then is constant, and then increases.

Increasing, Decreasing, and Constant Functions

A function f is **increasing** on an interval if, for any x_1 and x_2 in the interval, $x_1 < x_2$ implies $f(x_1) < f(x_2)$.

A function f is **decreasing** on an interval if, for any x_1 and x_2 in the interval, $x_1 < x_2$ implies $f(x_1) > f(x_2)$.

A function f is **constant** on an interval if, for any x_1 and x_2 in the interval, $f(x_1) = f(x_2)$.

Example 4 ▶ Increasing and Decreasing Functions 

In Figure 2.32, use the graphs to describe the increasing or decreasing behavior of each function.

Solution

a. This function is increasing over the entire real line.

b. This function is increasing on the interval $(-\infty, -1)$, decreasing on the interval $(-1, 1)$, and increasing on the interval $(1, \infty)$.

c. This function is increasing on the interval $(-\infty, 0)$, constant on the interval $(0, 2)$, and decreasing on the interval $(2, \infty)$.

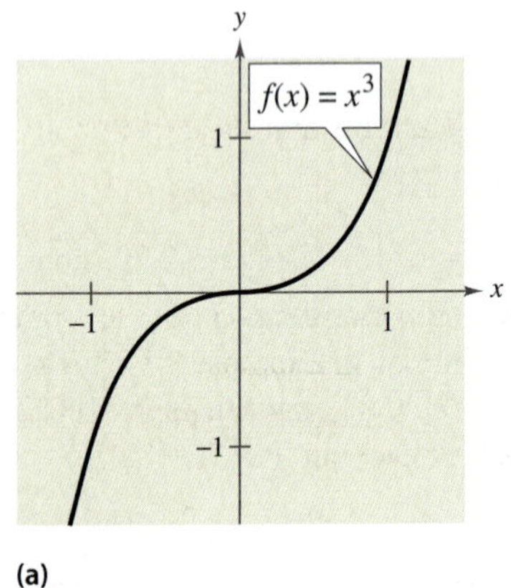

(a)

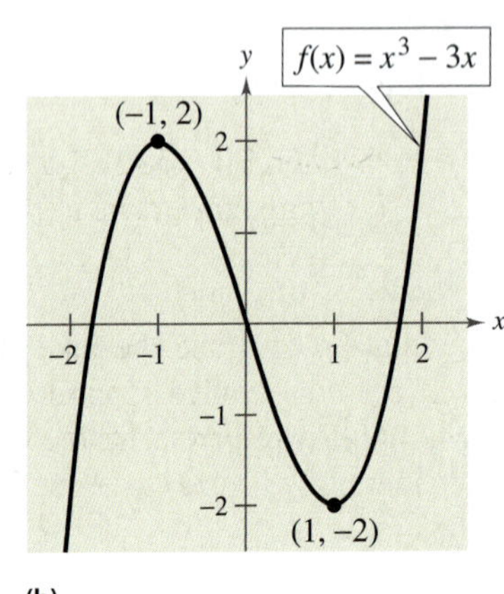

(b)

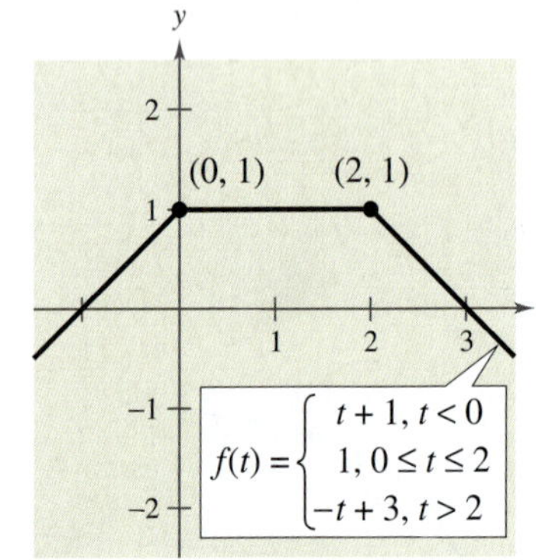

(c)

FIGURE 2.32

To help you decide whether a function is increasing, decreasing, or constant on an interval, you can evaluate the function for several values of x. However, calculus is needed to determine, for certain, all intervals on which a function is increasing, decreasing, or constant.

STUDY TIP

A relative minimum or relative maximum is also referred to as a *local* minimum or *local* maximum.

The points at which a function changes its increasing, decreasing, or constant behavior are helpful in determining the **relative minimum** or **relative maximum** values of the function.

Definition of Relative Minimum and Relative Maximum

A function value $f(a)$ is called a **relative minimum** of f if there exists an interval (x_1, x_2) that contains a such that

$$x_1 < x < x_2 \quad \text{implies} \quad f(a) \le f(x).$$

A function value $f(a)$ is called a **relative maximum** of f if there exists an interval (x_1, x_2) that contains a such that

$$x_1 < x < x_2 \quad \text{implies} \quad f(a) \ge f(x).$$

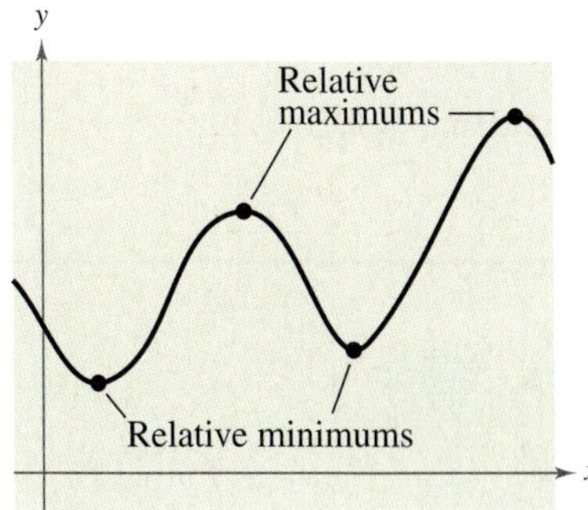

FIGURE 2.33

Figure 2.33 shows several different examples of relative minima and relative maxima. In Section 3.1, you will study a technique for finding the *exact point* at which a second-degree polynomial function has a relative minimum or relative maximum. For the time being, however, you can use a graphing utility to find reasonable approximations of these points.

Example 5 ▶ Approximating a Relative Minimum

Use a graphing utility to approximate the relative minimum of the function $f(x) = 3x^2 - 4x - 2$.

Solution

The graph of f is shown in Figure 2.34. By using the *zoom* and *trace* features of a graphing utility, you can estimate that the function has a relative minimum at the point

$(0.67, -3.33)$. Relative minimum

Later, in Section 3.1, you will be able to determine that the exact point at which the relative minimum occurs is $\left(\frac{2}{3}, -\frac{10}{3}\right)$.

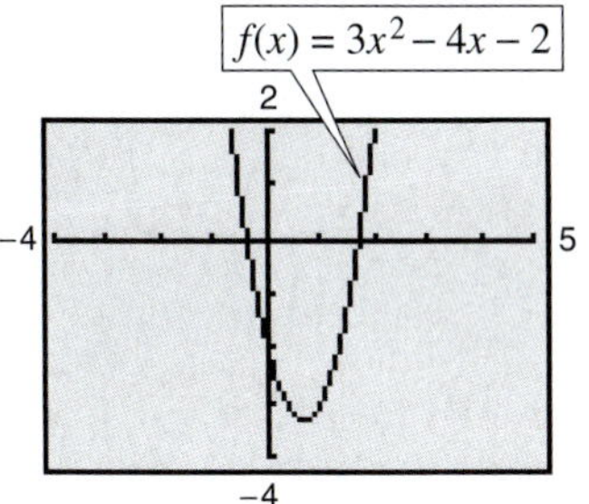

FIGURE 2.34

You can also use the *table* feature of a graphing utility to approximate numerically the relative minimum of the function in Example 5. Using a table that begins at 0.6 and increments the value of x by 0.01, you can approximate that the minimum of $f(x) = 3x^2 - 4x - 2$ occurs at the point $(0.67, -3.33)$.

Technology

If you use a graphing utility to estimate the x- and y-values of a relative minimum or relative maximum, the *automatic zoom* feature will often produce graphs that are nearly flat. To overcome this problem, you can manually change the vertical setting of the viewing window. The graph will stretch vertically if the values of Ymin and Ymax are closer together.

Even and Odd Functions

In Section 1.1, you studied different types of symmetry of a graph. In the terminology of functions, a function is said to be **even** if its graph is symmetric with respect to the y-axis and to be **odd** if its graph is symmetric with respect to the origin. The symmetry tests in Section 1.1 yield the following tests for even and odd functions.

Tests for Even and Odd Functions

A function $y = f(x)$ is **even** if, for each x in the domain of f,

$$f(-x) = f(x).$$

A function $y = f(x)$ is **odd** if, for each x in the domain of f,

$$f(-x) = -f(x).$$

Exploration

Graph each of the functions with a graphing utility. Determine whether the function is *even*, *odd*, or *neither*.

$f(x) = x^2 - x^4$

$g(x) = 2x^3 + 1$

$h(x) = x^5 - 2x^3 + x$

$j(x) = 2 - x^6 - x^8$

$k(x) = x^5 - 2x^4 + x - 2$

$p(x) = x^9 + 3x^5 - x^3 + x$

What do you notice about the equations of functions that are odd? What do you notice about the equations of functions that are even? Can you describe a way to identify a function as odd or even by inspecting the equation? Can you describe a way to identify a function as neither odd nor even by inspecting the equation?

Example 6 ▶ Even and Odd Functions

a. The function $g(x) = x^3 - x$ is odd because $g(-x) = -g(x)$, as follows.

$g(-x) = (-x)^3 - (-x)$ Substitute $-x$ for x.

$= -x^3 + x$ Simplify.

$= -(x^3 - x)$ Distributive Property

$= -g(x)$ Test for odd function

b. The function $h(x) = x^2 + 1$ is even because $h(-x) = h(x)$, as follows.

$h(-x) = (-x)^2 + 1$ Substitute $-x$ for x.

$= x^2 + 1$ Simplify.

$= h(x)$ Test for even function

The graphs of these two functions are shown in Figure 2.35.

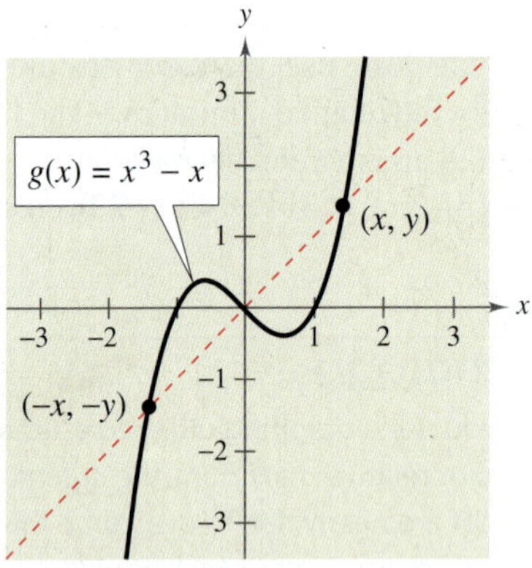

(a) Symmetric to origin: Odd Function

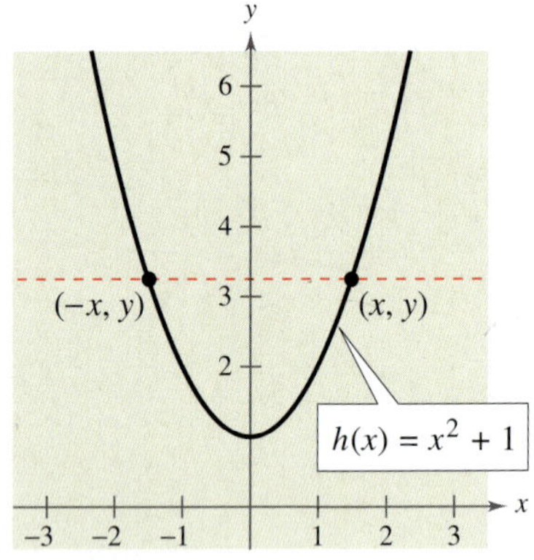

(b) Symmetric to y-axis: Even Function

FIGURE 2.35

2.3 Exercises

In Exercises 1–4, use the graph of the function to find the domain and range of *f*.

1.

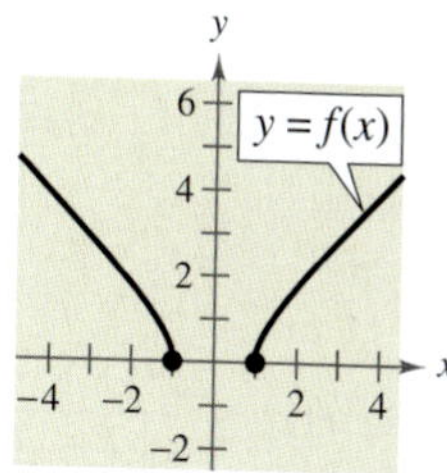

2.

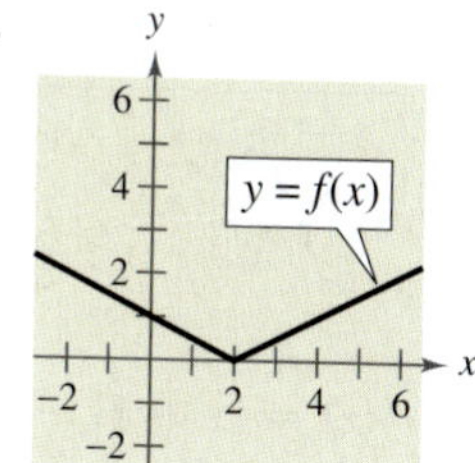

3.

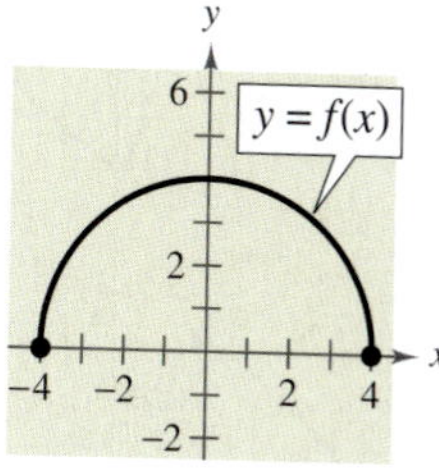

4.

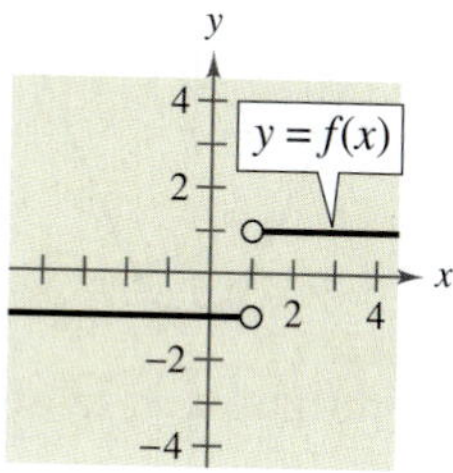

In Exercises 5–8, use the graph of the function to find the indicated function values.

5. (a) $f(-2)$ (b) $f(-1)$ (c) $f\left(\frac{1}{2}\right)$ (d) $f(1)$

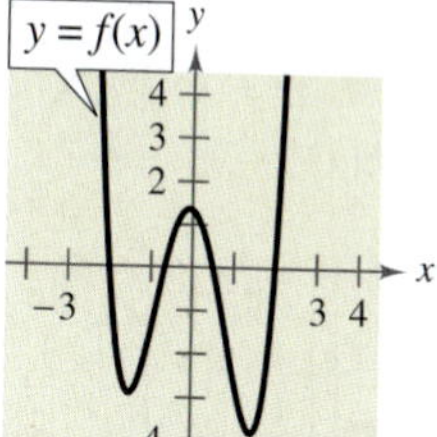

6. (a) $f(-1)$ (b) $f(2)$ (c) $f(0)$ (d) $f(1)$

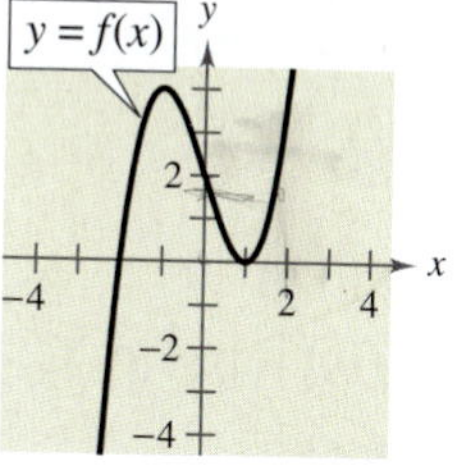

7. (a) $f(-2)$ (b) $f(1)$ (c) $f(0)$ (d) $f(2)$

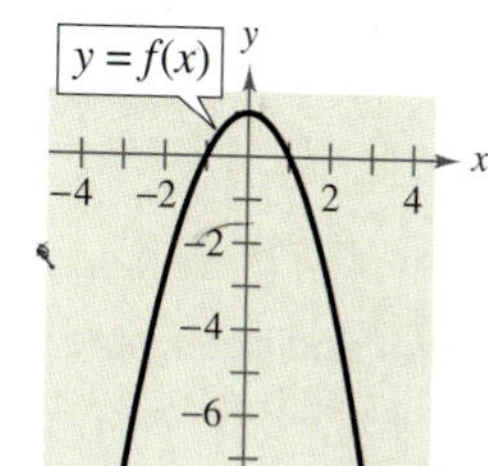

8. (a) $f(2)$ (b) $f(1)$ (c) $f(3)$ (d) $f(-1)$

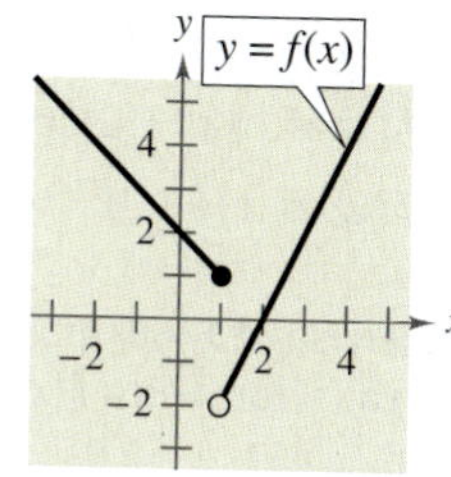

In Exercises 9–14, use the Vertical Line Test to determine whether *y* is a function of *x*. To print an enlarged copy of the graph, go to the website *www.mathgraphs.com*.

9. $y = \frac{1}{2}x^2$

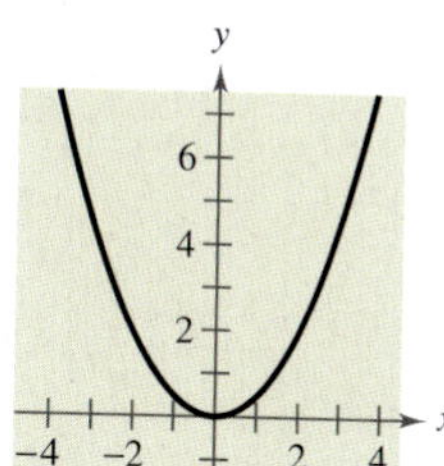

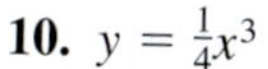

10. $y = \frac{1}{4}x^3$

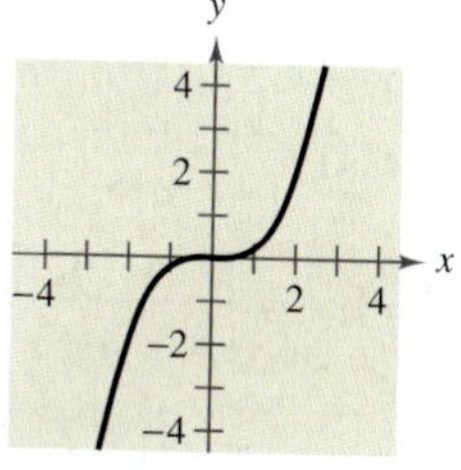

11. $x - y^2 = 1$

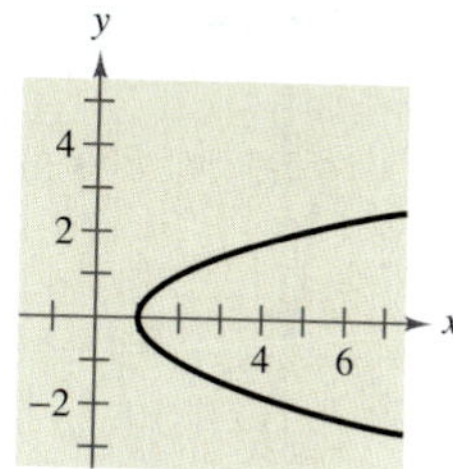

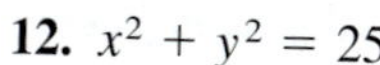

12. $x^2 + y^2 = 25$

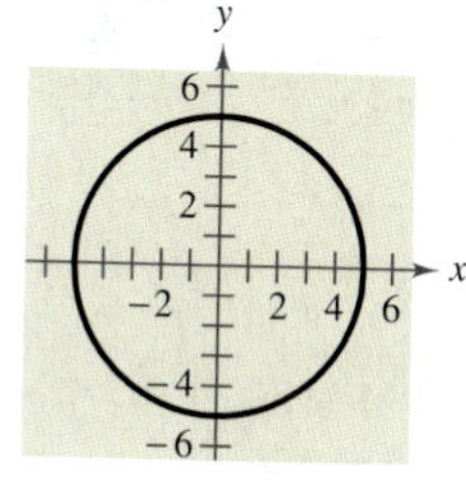

13. $x^2 = 2xy - 1$

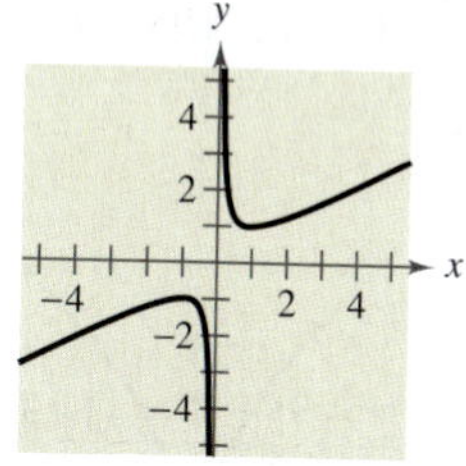

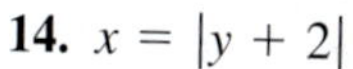

14. $x = |y + 2|$

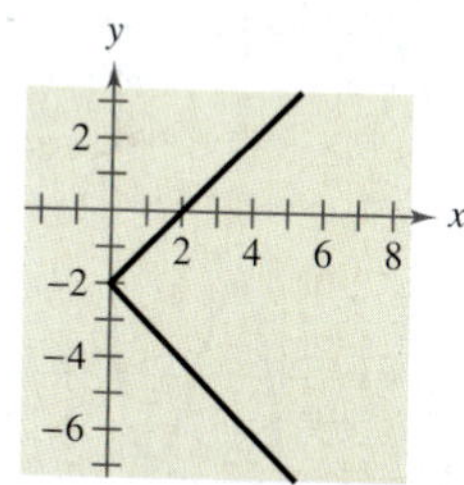

In Exercises 15–24, find the zeros of the function algebraically.

15. $f(x) = 2x^2 - 7x - 30$

16. $f(x) = 3x^2 + 22x - 16$

17. $f(x) = \dfrac{x}{9x^2 - 4}$

18. $f(x) = \dfrac{x^2 - 9x + 14}{4x}$

19. $f(x) = \frac{1}{2}x^3 - x$

20. $f(x) = x^3 - 4x^2 - 9x + 36$

21. $f(x) = 4x^3 - 24x^2 - x + 6$

22. $f(x) = 9x^4 - 25x^2$

23. $f(x) = \sqrt{2x} - 1$

24. $f(x) = \sqrt{3x + 2}$

In Exercises 25–30, use a graphing utility to graph the function and find the zeros of the function. Verify your results algebraically.

25. $f(x) = 3 + \frac{5}{x}$

26. $f(x) = x(x - 7)$

27. $f(x) = \sqrt{2x + 11}$

28. $f(x) = \sqrt{3x - 14} - 8$

29. $f(x) = \frac{3x - 1}{x - 6}$

30. $f(x) = \frac{2x^2 - 9}{3 - x}$

In Exercises 31–38, determine the intervals over which the function is increasing, decreasing, or constant.

31. $f(x) = \frac{3}{2}x$

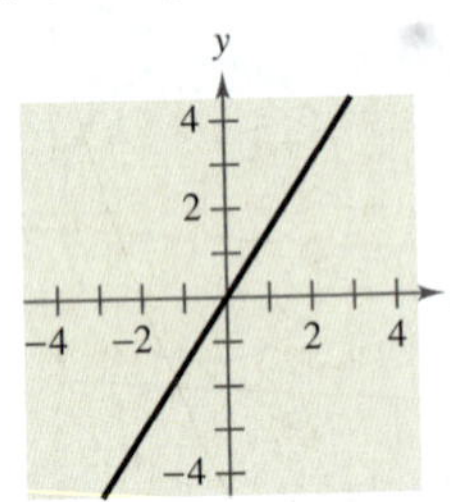

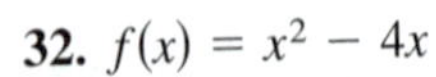

32. $f(x) = x^2 - 4x$

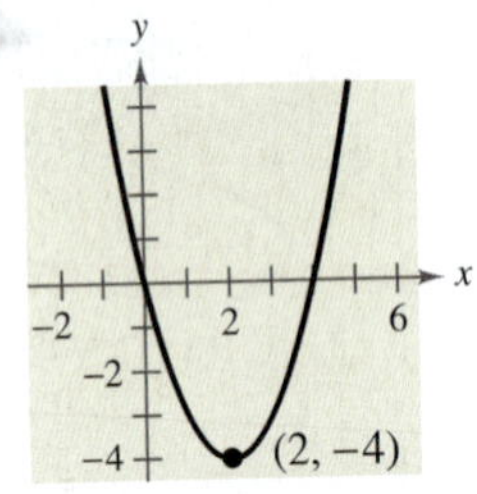

33. $f(x) = x^3 - 3x^2 + 2$

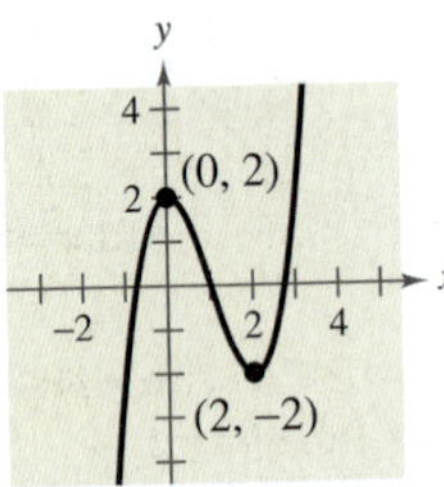

34. $f(x) = \sqrt{x^2 - 1}$

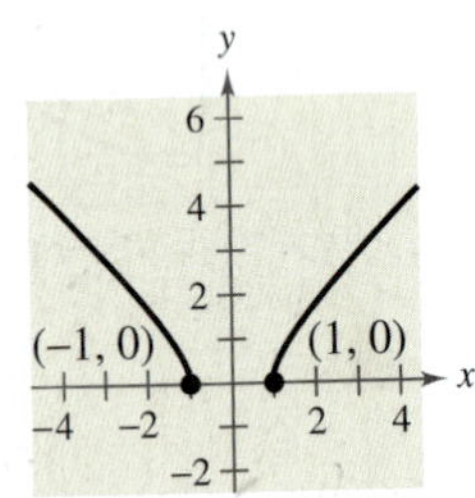

35. $f(x) = \begin{cases} x + 3, & x \le 0 \\ 3, & 0 < x \le 2 \\ 2x + 1, & x > 2 \end{cases}$

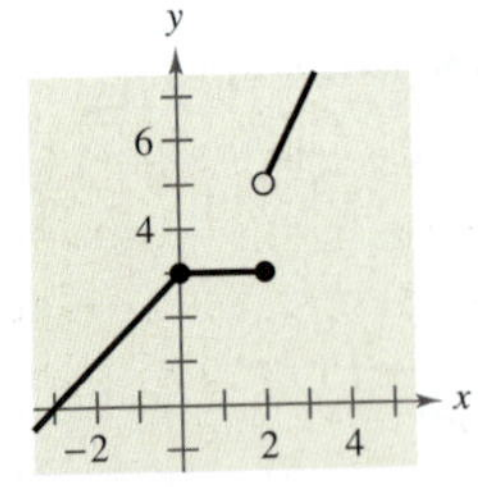

36. $f(x) = \begin{cases} 2x + 1, & x \le -1 \\ x^2 - 2, & x > -1 \end{cases}$

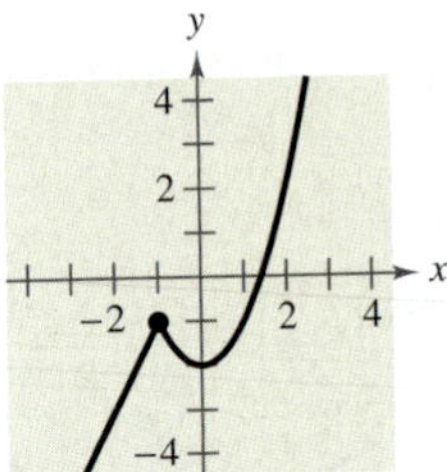

37. $f(x) = |x + 1| + |x - 1|$

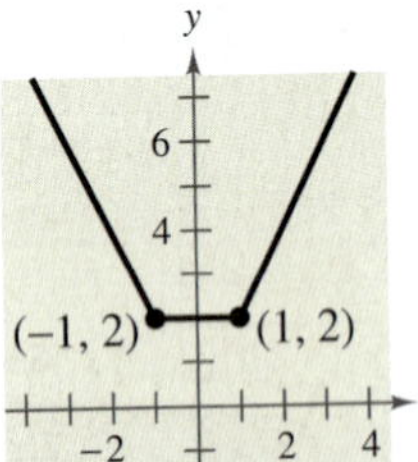

38. $f(x) = \frac{x^2 + x + 1}{x + 1}$

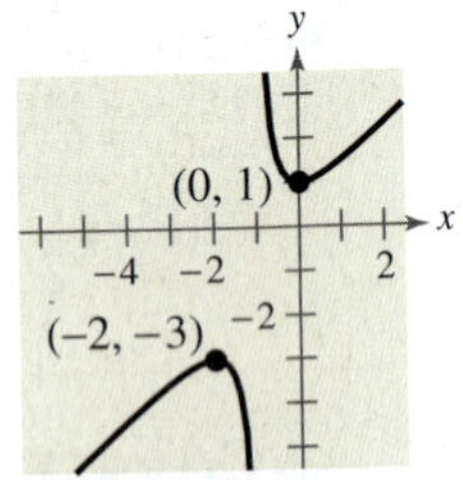

In Exercises 39–48, (a) use a graphing utility to graph the function and visually determine the intervals over which the function is increasing, decreasing, or constant, and (b) make a table of values to verify whether the function is increasing, decreasing, or constant over the intervals you identified in part (a).

39. $f(x) = 3$

40. $g(x) = x$

41. $g(s) = \frac{s^2}{4}$

42. $h(x) = x^2 - 4$

43. $f(t) = -t^4$

44. $f(x) = 3x^4 - 6x^2$

45. $f(x) = \sqrt{1 - x}$

46. $f(x) = x\sqrt{x + 3}$

47. $f(x) = x^{3/2}$

48. $f(x) = x^{2/3}$

In Exercises 49–52, use a graphing utility to approximate the relative minimum/relative maximum of each function.

49. $f(x) = (x - 4)(x + 2)$

50. $f(x) = 3x^2 - 2x - 5$

51. $f(x) = x(x - 2)(x + 3)$

52. $f(x) = x^3 - 3x^2 - x + 1$

In Exercises 53–60, graph the function and determine the interval(s) for which $f(x) \ge 0$.

53. $f(x) = 4 - x$

54. $f(x) = 4x + 2$

55. $f(x) = x^2 + x$

56. $f(x) = x^2 - 4x$

57. $f(x) = \sqrt{x - 1}$

58. $f(x) = \sqrt{x + 2}$

59. $f(x) = -(1 + |x|)$

60. $f(x) = \frac{1}{2}(2 + |x|)$

In Exercises 61–66, determine whether the function is even, odd, or neither.

61. $f(x) = x^6 - 2x^2 + 3$
62. $h(x) = x^3 - 5$
63. $g(x) = x^3 - 5x$
64. $f(x) = x\sqrt{1 - x^2}$
65. $f(t) = t^2 + 2t - 3$
66. $g(s) = 4s^{2/3}$

In Exercises 67–70, write the height h of the rectangle as a function of x.

67.
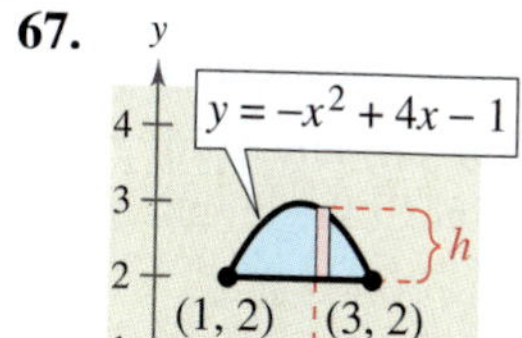

68.
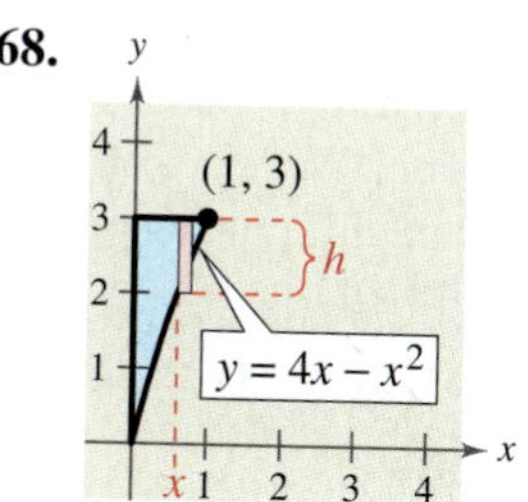

69.
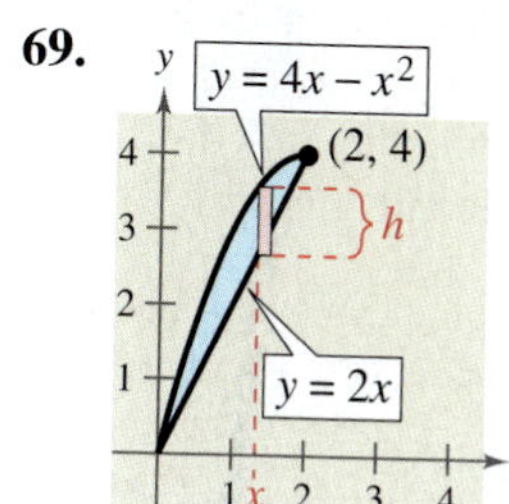

70.
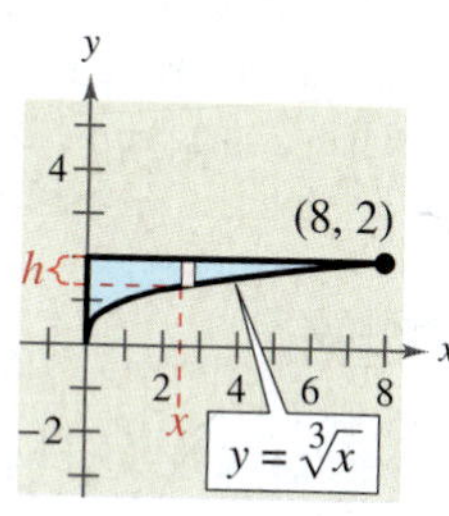

In Exercises 71–74, write the length L of the rectangle as a function of y.

71.
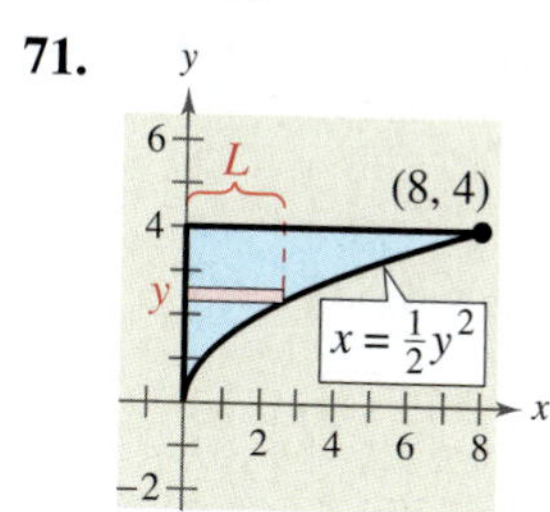

72.
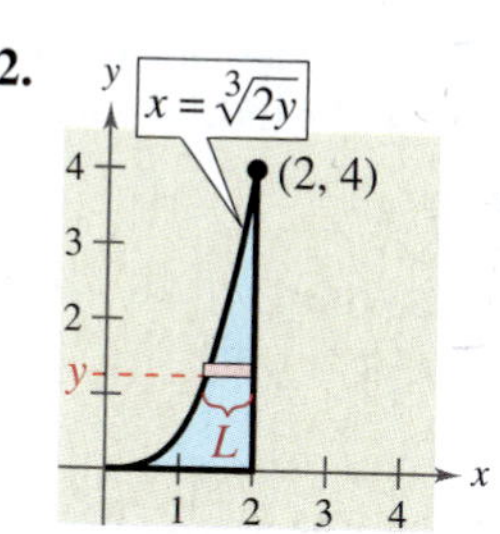

73.
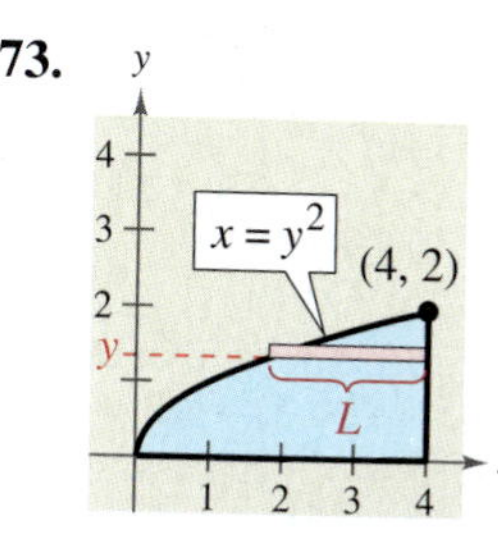

74.
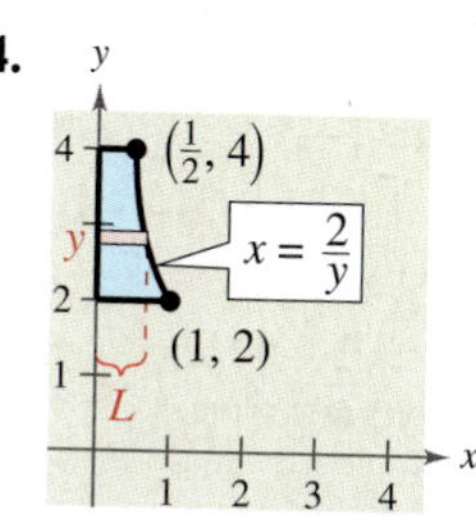

75. ***Electronics*** The number of lumens (time rate of flow of light) L from a fluorescent lamp can be approximated by the model

$$L = -0.294x^2 + 97.744x - 664.875, \quad 20 \le x \le 90$$

where x is the wattage of the lamp.

(a) Use a graphing utility to graph the function.

(b) Use the graph from part (a) to estimate the wattage necessary to obtain 2000 lumens.

▶ Model It

76. ***Data Analysis*** The table shows the amount y (in billions of dollars) of the merchandise trade balance of the United States for the years 1991 through 1999. The merchandise trade balance is the difference between the values of exports and imports. A negative merchandise trade balance indicates that imports exceeded exports. (Source: U.S. International Trade Administration and U.S. Foreign Trade Highlights)

Year, x	Trade balance, y
1991	−66.8
1992	−84.5
1993	−115.6
1994	−150.7
1995	−158.7
1996	−170.2
1997	−181.5
1998	−229.8
1999	−330.0

(a) Use a graphing utility to create a scatter plot of the data.

(b) Use the graph in part (a) to determine whether the data represents y as a function of x.

(c) Use the *regression* feature of a graphing utility to find a cubic model (a model of the form $y = ax^3 + bx^2 + cx + d$) for the data. Let x be the time (in years), with $x = 1$ corresponding to 1991.

(d) What is the domain of the model?

(e) Use a graphing utility to graph the model in the same viewing window you used in part (a).

(f) For which year does the model most accurately estimate the actual data? During which year is it least accurate?

77. ***Coordinate Axis Scale*** Each function models the specified data for the years 1995 through 2002, with $t = 5$ corresponding to 1995. Estimate a reasonable scale for the vertical axis (e.g., hundreds, thousands, millions, etc.) of the graph and justify your answer. (There are many correct answers.)

(a) $f(t)$ represents the average salary of college professors.

(b) $f(t)$ represents the U.S. population.

(c) $f(t)$ represents the percent of the civilian work force that is unemployed.

78. ***Geometry*** Corners of equal size are cut from a square with sides of length 8 meters (see figure).

(a) Write the area A of the resulting figure as a function of x. Determine the domain of the function.

(b) Use a graphing utility to graph the area function over its domain. Use the graph to find the range of the function.

(c) Identify the figure that would result if x were chosen to be the maximum value in the domain of the function. What would be the length of each side of the figure?

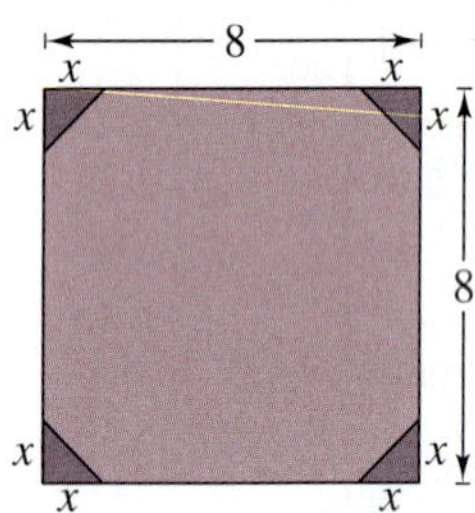

Synthesis

True or False? **In Exercises 79 and 80, determine whether the statement is true or false. Justify your answer.**

79. A function with a square root cannot have a domain that is the set of real numbers.

80. It is possible for an odd function to have the interval $[0, \infty)$ as its domain.

81. If f is an even function, determine whether g is even, odd, or neither. Explain.

(a) $g(x) = -f(x)$ (b) $g(x) = f(-x)$

(c) $g(x) = f(x) - 2$ (d) $g(x) = f(x - 2)$

82. ***Think About It*** Does the graph in Exercise 11 represent x as a function of y? Explain.

Think About It **In Exercises 83–86, find the coordinates of a second point on the graph of a function f if the given point is on the graph and the function is (a) even and (b) odd.**

83. $\left(-\frac{3}{2}, 4\right)$ **84.** $\left(-\frac{5}{3}, -7\right)$

85. $(4, 9)$ **86.** $(5, -1)$

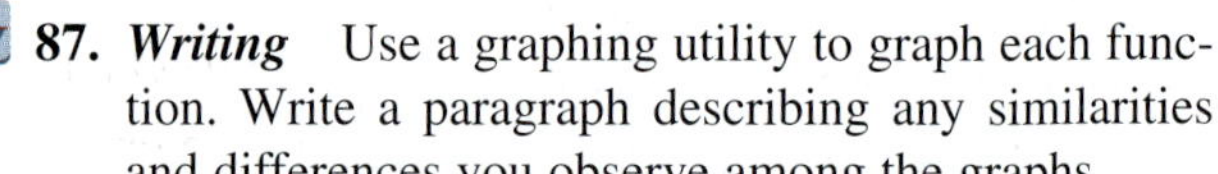

87. ***Writing*** Use a graphing utility to graph each function. Write a paragraph describing any similarities and differences you observe among the graphs.

(a) $y = x$ (b) $y = x^2$

(c) $y = x^3$ (d) $y = x^4$

(e) $y = x^5$ (f) $y = x^6$

88. ***Conjecture*** Use the results of Exercise 87 to make a conjecture about the graphs of $y = x^7$ and $y = x^8$. Use a graphing utility to graph the functions and compare the results with your conjecture.

Review

In Exercises 89–92, solve the equation.

89. $x^2 - 10x = 0$ **90.** $100 - (x - 5)^2 = 0$

91. $x^3 - x = 0$ **92.** $16x^2 - 40x + 25 = 0$

In Exercises 93–96, evaluate the function at each specified value of the independent variable and simplify.

93. $f(x) = 5x - 8$

(a) $f(9)$ (b) $f(-4)$ (c) $f(x - 7)$

94. $f(x) = x^2 - 10x$

(a) $f(4)$ (b) $f(-8)$ (c) $f(x - 4)$

95. $f(x) = \sqrt{x - 12} - 9$

(a) $f(12)$ (b) $f(40)$ (c) $f\left(-\sqrt{36}\right)$

96. $f(x) = x^4 - x - 5$

(a) $f(-1)$ (b) $f\left(\frac{1}{2}\right)$ (c) $f\left(2\sqrt{3}\right)$

In Exercises 97 and 98, find the difference quotient and simplify your answer.

97. $f(x) = x^2 - 2x + 9, \quad \dfrac{f(3 + h) - f(3)}{h}, \quad h \neq 0$

98. $f(x) = 5 + 6x - x^2, \quad \dfrac{f(6 + h) - f(6)}{h}, \quad h \neq 0$

2.4 A Library of Functions

▶ **What you should learn**

- How to identify and graph linear and squaring functions
- How to identify and graph cubic, square root, and reciprocal functions
- How to identify and graph step and other piecewise-defined functions
- How to recognize graphs of common functions

▶ **Why you should learn it**

Piecewise-defined functions can be used to model real-life situations. For instance, in Exercise 68 on page 218, you will use a piecewise-defined function to model the monthly revenue of a landscaping business.

Michael Newman/PhotoEdit

Linear and Squaring Functions

One of the goals of this text is to enable you to recognize the basic shapes of the graphs of different types of functions. For instance, you know that the graph of the **linear function** $f(x) = ax + b$ is a line with slope $m = a$ and y-intercept at $(0, b)$. The graph of the linear function has the following features.

- The domain of the function is the set of all real numbers.
- The range of the function is the set of all real numbers.
- The graph has one intercept, $(0, b)$.
- The graph is increasing if $m > 0$, decreasing if $m < 0$, and constant if $m = 0$.

Example 1 ▶ **Writing a Linear Function**

Write the linear function f for which $f(1) = 3$ and $f(4) = 0$.

Solution

To find the equation of the line that passes through $(x_1, y_1) = (1, 3)$ and $(x_2, y_2) = (4, 0)$, first find the slope of the line.

$$m = \frac{y_2 - y_1}{x_2 - x_1} = \frac{0 - 3}{4 - 1} = \frac{-3}{3} = -1$$

Next, use the point-slope form of the equation of a line.

$$y - y_1 = m(x - x_1) \quad \text{Point-slope form}$$

$$y - 3 = -1(x - 1) \quad \text{Substitute for } x_1, y_1, \text{ and } m.$$

$$y = -x + 4 \quad \text{Simplify.}$$

$$f(x) = -x + 4 \quad \text{Function notation}$$

The graph of this function is shown in Figure 2.36.

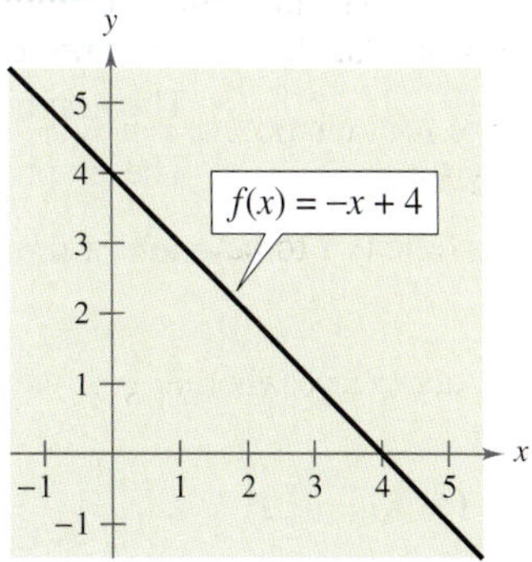

FIGURE 2.36

There are two special types of linear functions, the **constant function** and the **identity function.** A constant function has the form

$$f(x) = c$$

and has the domain of all real numbers with a range consisting of a single real number c. The graph of a constant function is a horizontal line, as shown in Figure 2.37. The identity function has the form

$$f(x) = x.$$

Its domain and range are the set of all real numbers. The identity function has a slope of $m = 1$ and a y-intercept $(0, 0)$. The graph of the identity function is a line for which each x-coordinate equals the corresponding y-coordinate. The graph is always increasing, as shown in Figure 2.38.

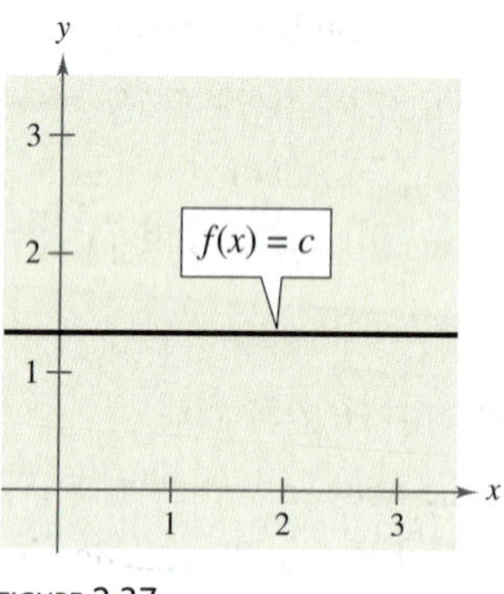

FIGURE 2.37

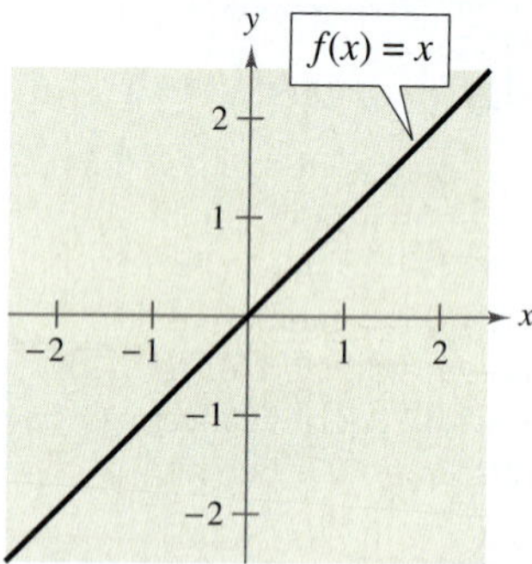

FIGURE 2.38

The graph of the **squaring function**

$$f(x) = x^2$$

is a U-shaped curve with the following features.

- The domain of the function is the set of all real numbers.
- The range of the function is the set of all nonnegative real numbers.
- The function is even.
- The graph has an intercept at $(0, 0)$.
- The graph is decreasing on the interval $(-\infty, 0)$ and increasing on the interval $(0, \infty)$.
- The graph is symmetric with respect to the y-axis.
- The graph has a relative minimum at $(0, 0)$.

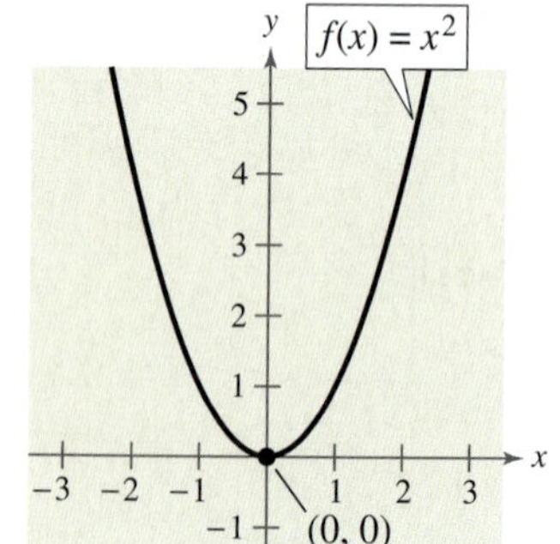

FIGURE 2.39

The graph of the squaring function is shown in Figure 2.39.

Cubic, Square Root, and Reciprocal Functions

Special features of the graphs of the **cubic, square root,** and **reciprocal functions** are summarized below.

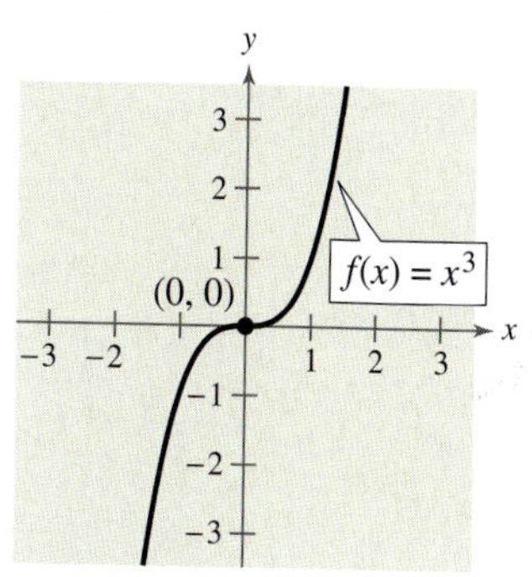

Cubic function
FIGURE 2.40

1. The graph of the *cubic* function

$$f(x) = x^3$$

has the following features.

- The domain of the function is the set of all real numbers.
- The range of the function is the set of all real numbers.
- The function is odd.
- The graph has an intercept at $(0, 0)$.
- The graph is increasing on the interval $(-\infty, \infty)$.
- The graph is symmetric with respect to the origin.

The graph of the cubic function is shown in Figure 2.40.

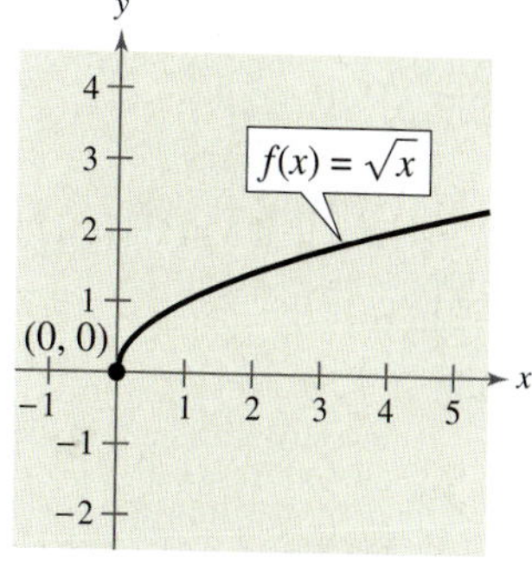

Square root function
FIGURE 2.41

2. The graph of the *square root* function

$$f(x) = \sqrt{x}$$

has the following features.

- The domain of the function is the set of all nonnegative real numbers.
- The range of the function is the set of all nonnegative real numbers.
- The graph has an intercept at $(0, 0)$.
- The graph is increasing on the interval $(0, \infty)$.

The graph of the square root function is shown in Figure 2.41.

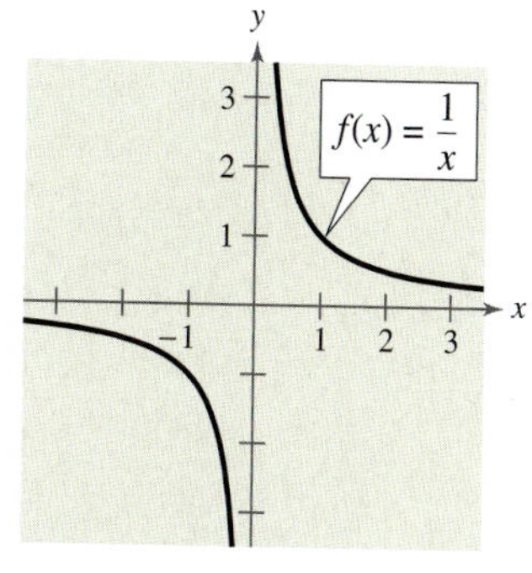

Reciprocal function
FIGURE 2.42

3. The graph of the *reciprocal* function

$$f(x) = \frac{1}{x}$$

has the following features.

- The domain of the function is $(-\infty, 0) \cup (0, \infty)$.
- The range of the function is $(-\infty, 0) \cup (0, \infty)$.
- The function is odd.
- The graph does not have any intercepts.
- The graph is decreasing on the intervals $(-\infty, 0)$ and $(0, \infty)$.
- The graph is symmetric with respect to the origin.

The graph of the reciprocal function is shown in Figure 2.42.

Step and Piecewise-Defined Functions

Functions whose graphs resemble sets of stairsteps are known as **step functions.** The most famous of the step functions is the **greatest integer function,** which is denoted by $[\![x]\!]$ and defined as

$$f(x) = [\![x]\!] = \textit{the greatest integer less than or equal to } x.$$

Some values of the greatest integer function are as follows.

$$[\![-1]\!] = (\text{greatest integer} \le -1) = -1$$

$$\left[\!\!\left[-\tfrac{1}{2}\right]\!\!\right] = \left(\text{greatest integer} \le -\tfrac{1}{2}\right) = -1$$

$$\left[\!\!\left[\tfrac{1}{10}\right]\!\!\right] = \left(\text{greatest integer} \le \tfrac{1}{10}\right) = 0$$

$$[\![1.5]\!] = (\text{greatest integer} \le 1.5) = 1$$

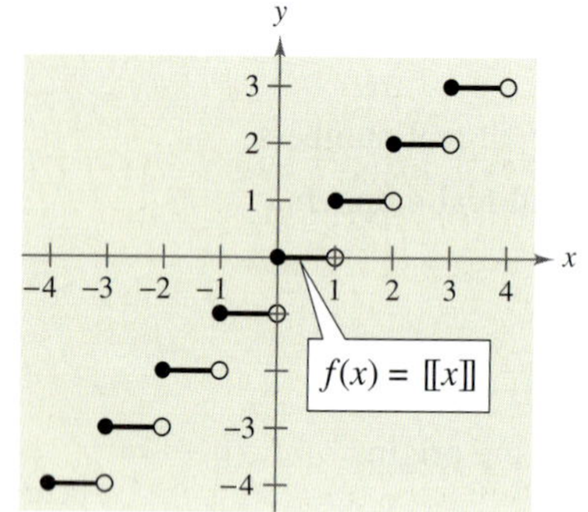

FIGURE 2.43

The graph of the greatest integer function

$$f(x) = [\![x]\!]$$

has the following features, as shown in Figure 2.43.

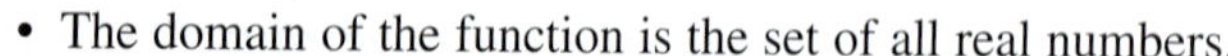

- The domain of the function is the set of all real numbers.
- The range of the function is the set of all integers.
- The graph has a y-intercept at $(0, 0)$ and x-intercepts in the interval $[0, 1)$.
- The graph is constant between each pair of consecutive integers.
- The graph jumps vertically one unit at each integer value.

Technology

When graphing a step function, you should set your graphing utility to *dot* mode.

Example 2 ▶ Evaluating a Step Function

Evaluate the function when $x = -1$, 2, and $\frac{3}{2}$.

$$f(x) = [\![x]\!] + 1$$

Solution

For $x = -1$, the greatest integer ≤ -1 is -1, so

$$f(-1) = [\![-1]\!] + 1 = -1 + 1 = 0.$$

For $x = 2$, the greatest integer ≤ 2 is 2, so

$$f(2) = [\![2]\!] + 1 = 2 + 1 = 3.$$

For $x = \frac{3}{2}$, the greatest integer $\le \frac{3}{2}$ is 1, so

$$f\left(\tfrac{3}{2}\right) = \left[\!\!\left[\tfrac{3}{2}\right]\!\!\right] + 1 = 1 + 1 = 2.$$

The graph of $f(x) = [\![x]\!] + 1$ is shown in Figure 2.44.

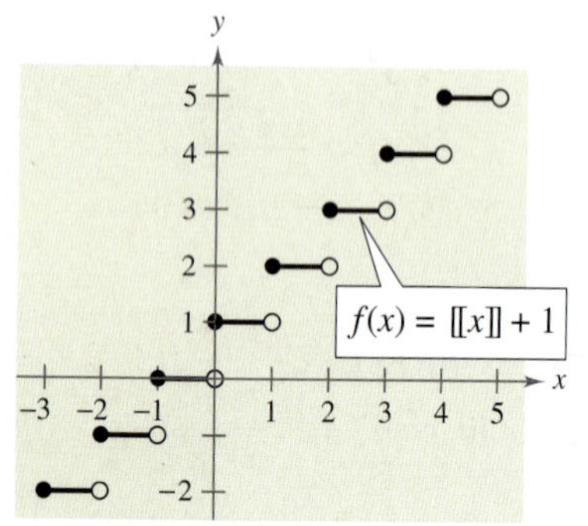

FIGURE 2.44

Recall from Section 2.2 that a piecewise-defined function is defined by two or more equations over a specified domain. To graph a piecewise-defined function, graph each equation separately over the specified domain, as shown in Example 3.

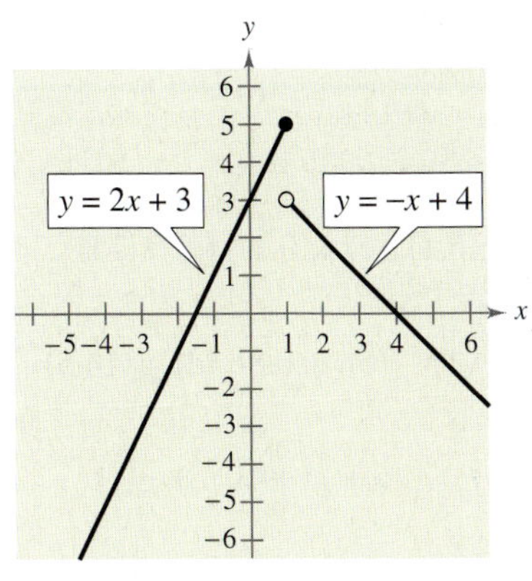

FIGURE 2.45

Example 3 ▶ Graphing a Piecewise-Defined Function

Sketch the graph of

$$f(x) = \begin{cases} 2x + 3, & x \le 1 \\ -x + 4, & x > 1 \end{cases}.$$

Solution

This piecewise-defined function is composed of two linear functions. At $x = 1$ and to the left of $x = 1$ the graph is the line $y = 2x + 3$, and to the right of $x = 1$ the graph is the line $y = -x + 4$, as shown in Figure 2.45.

Common Functions

The eight graphs shown in Figure 2.46 represent the most commonly used functions in algebra. Familiarity with the basic characteristics of these simple graphs will help you analyze the shapes of more complicated graphs—in particular, graphs obtained from these graphs by the rigid and nonrigid transformations studied in the next section.

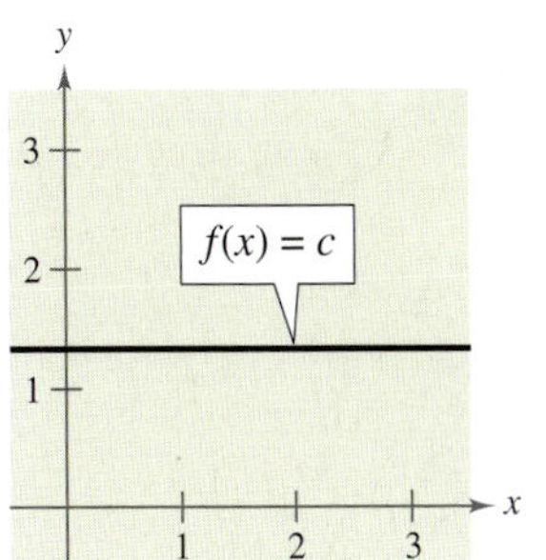

(a) Constant Function

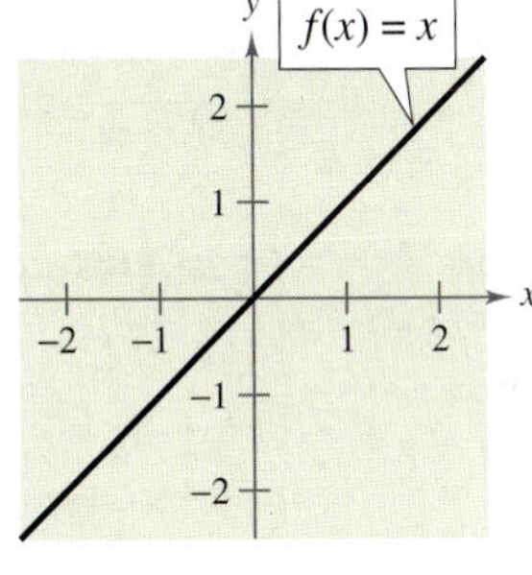

(b) Identity Function

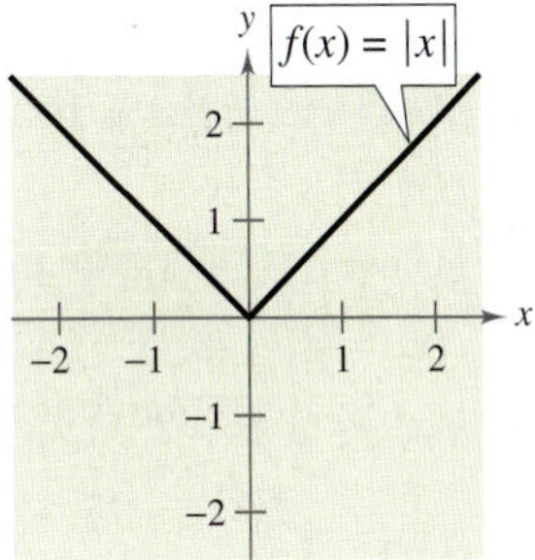

(c) Absolute Value Function

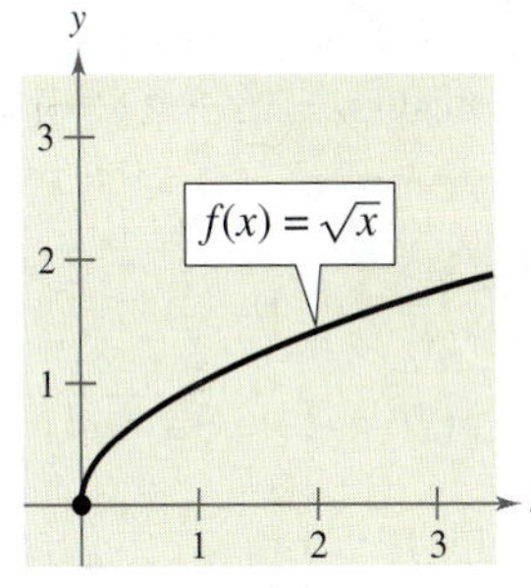

(d) Square Root Function

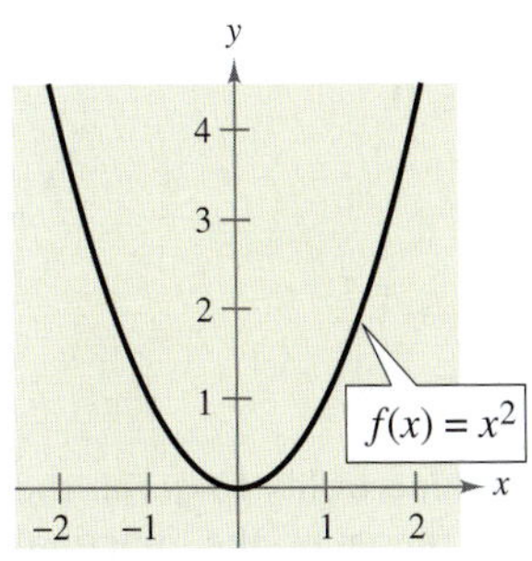

(e) Quadratic Function

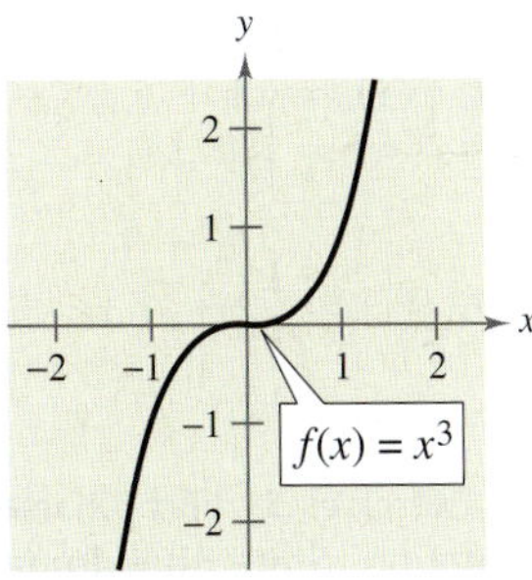

(f) Cubic Function

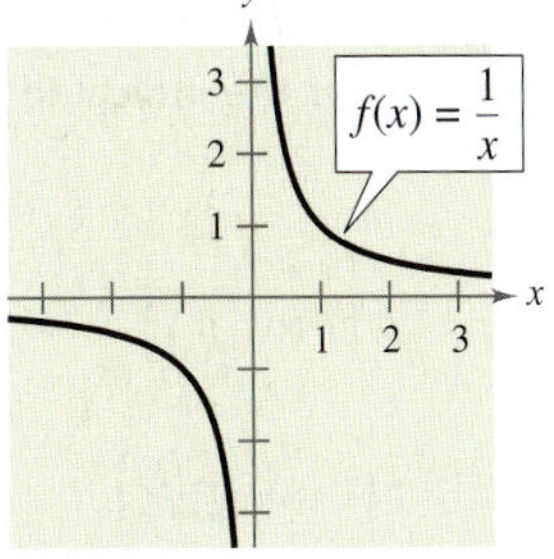

(g) Reciprocal Function

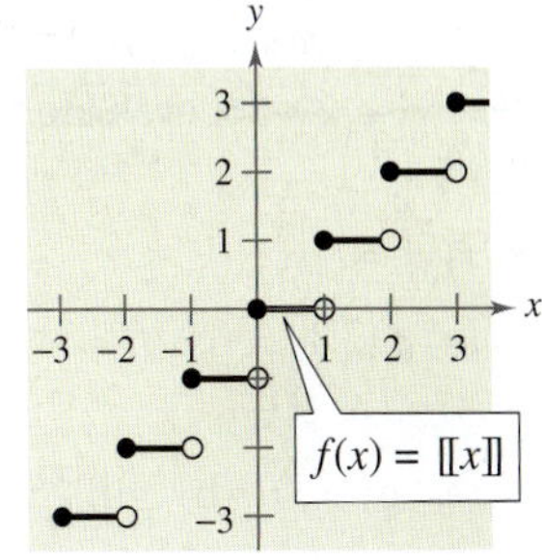

(h) Greatest Integer Function

FIGURE 2.46

2.4 Exercises

In Exercises 1–8, write the linear function that has the indicated function values. Then sketch the graph of the function.

1. $f(1) = 4, f(0) = 6$
2. $f(-3) = -8, f(1) = 2$
3. $f(5) = -4, f(-2) = 17$
4. $f(3) = 9, f(-1) = -11$
5. $f(-5) = -1, f(5) = -1$
6. $f(-10) = 12, f(16) = -1$
7. $f(\frac{1}{2}) = -6, f(4) = -3$
8. $f(\frac{2}{3}) = -\frac{15}{2}, f(-4) = -11$

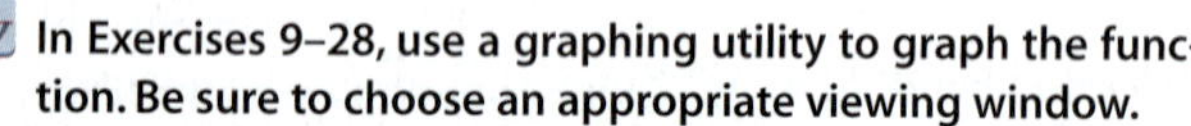

In Exercises 9–28, use a graphing utility to graph the function. Be sure to choose an appropriate viewing window.

9. $f(x) = -x - \frac{3}{4}$
10. $f(x) = 3x - \frac{5}{2}$
11. $f(x) = -\frac{1}{6}x - \frac{5}{2}$
12. $f(x) = \frac{5}{6} - \frac{2}{3}x$
13. $f(x) = x^2 - 2x$
14. $f(x) = -x^2 + 8x$
15. $h(x) = -x^2 + 4x + 12$
16. $g(x) = x^2 - 6x - 16$
17. $f(x) = x^3 - 1$
18. $f(x) = 8 - x^3$
19. $f(x) = (x - 1)^3 + 2$
20. $g(x) = 2(x + 3)^3 + 1$
21. $f(x) = 4\sqrt{x}$
22. $f(x) = 4 - 2\sqrt{x}$
23. $g(x) = 2 - \sqrt{x + 4}$
24. $h(x) = \sqrt{x + 2} + 3$
25. $f(x) = -\dfrac{1}{x}$
26. $f(x) = 4 + \dfrac{1}{x}$
27. $h(x) = \dfrac{1}{x + 2}$
28. $k(x) = \dfrac{1}{x - 3}$

In Exercise 29–36, evaluate the function for the indicated values.

29. $f(x) = [\![x]\!]$
    (a) $f(2.1)$ (b) $f(2.9)$ (c) $f(-3.1)$ (d) $f(\frac{7}{2})$
30. $g(x) = 2[\![x]\!]$
    (a) $g(-3)$ (b) $g(0.25)$ (c) $g(9.5)$ (d) $g(\frac{11}{3})$
31. $h(x) = [\![x + 3]\!]$
    (a) $h(-2)$ (b) $h(\frac{1}{2})$ (c) $h(4.2)$ (d) $h(-21.6)$
32. $f(x) = 4[\![x]\!] + 7$
    (a) $f(0)$ (b) $f(-1.5)$ (c) $f(6)$ (d) $f(\frac{5}{3})$
33. $h(x) = [\![3x - 1]\!]$
    (a) $h(2.5)$ (b) $h(-3.2)$ (c) $h(\frac{7}{3})$ (d) $h(-\frac{21}{3})$
34. $k(x) = [\![\frac{1}{2}x + 6]\!]$
    (a) $k(5)$ (b) $k(-6.1)$ (c) $k(0.1)$ (d) $k(15)$
35. $g(x) = 3[\![x - 2]\!] + 5$
    (a) $g(-2.7)$ (b) $g(-1)$ (c) $g(0.8)$ (d) $g(14.5)$
36. $g(x) = -7[\![x + 4]\!] + 6$
    (a) $g(\frac{1}{8})$ (b) $g(9)$ (c) $g(-4)$ (d) $g(\frac{3}{2})$

In Exercises 37–42, sketch the graph of the function.

37. $g(x) = -[\![x]\!]$
38. $g(x) = 4[\![x]\!]$
39. $g(x) = [\![x]\!] - 2$
40. $g(x) = [\![x]\!] - 1$
41. $g(x) = [\![x + 1]\!]$
42. $g(x) = [\![x - 3]\!]$

In Exercises 43–50, graph the function.

43. $f(x) = \begin{cases} 2x + 3, & x < 0 \\ 3 - x, & x \geq 0 \end{cases}$
44. $g(x) = \begin{cases} x + 6, & x \leq -4 \\ \frac{1}{2}x - 4, & x > -4 \end{cases}$
45. $f(x) = \begin{cases} \sqrt{4 + x}, & x < 0 \\ \sqrt{4 - x}, & x \geq 0 \end{cases}$
46. $f(x) = \begin{cases} 1 - (x - 1)^2, & x \leq 2 \\ \sqrt{x - 2}, & x > 2 \end{cases}$
47. $f(x) = \begin{cases} x^2 + 5, & x \leq 1 \\ -x^2 + 4x + 3, & x > 1 \end{cases}$
48. $h(x) = \begin{cases} 3 - x^2, & x < 0 \\ x^2 + 2, & x \geq 0 \end{cases}$
49. $h(x) = \begin{cases} 4 - x^2, & x < -2 \\ 3 + x, & -2 \leq x < 0 \\ x^2 + 1, & x \geq 0 \end{cases}$
50. $k(x) = \begin{cases} 2x + 1, & x \leq -1 \\ 2x^2 - 1, & -1 < x \leq 1 \\ 1 - x^2, & x > 1 \end{cases}$

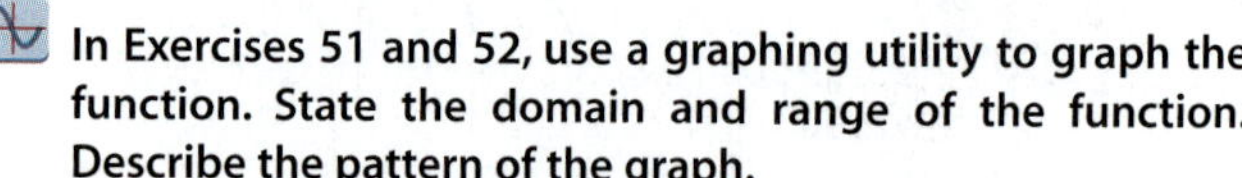

In Exercises 51 and 52, use a graphing utility to graph the function. State the domain and range of the function. Describe the pattern of the graph.

51. $s(x) = 2(\frac{1}{4}x - [\![\frac{1}{4}x]\!])$
52. $g(x) = 2(\frac{1}{4}x - [\![\frac{1}{4}x]\!])^2$

In Exercises 53–62, identify the common function and the transformed common function shown in the graph. Write an equation for the function shown in the graph. Then use a graphing utility to verify your answer.

53.

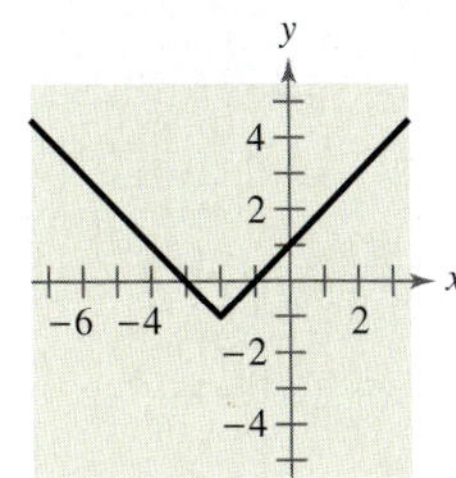

54.

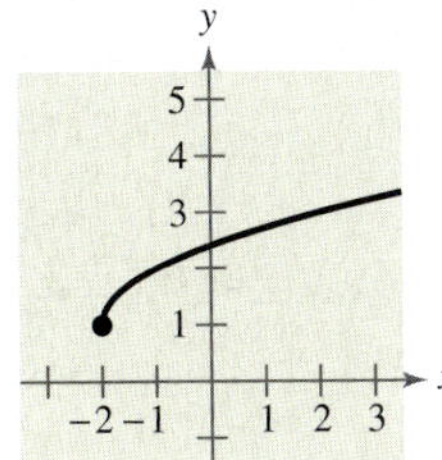

55.

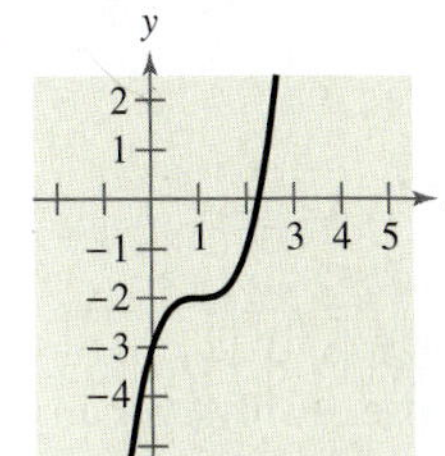

56.

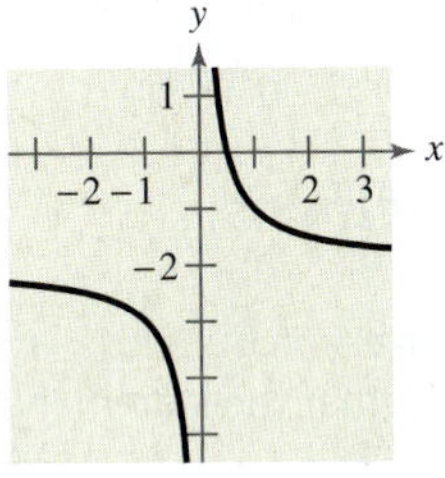

57.

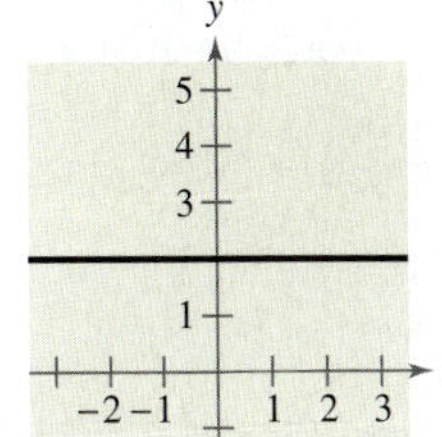

58.

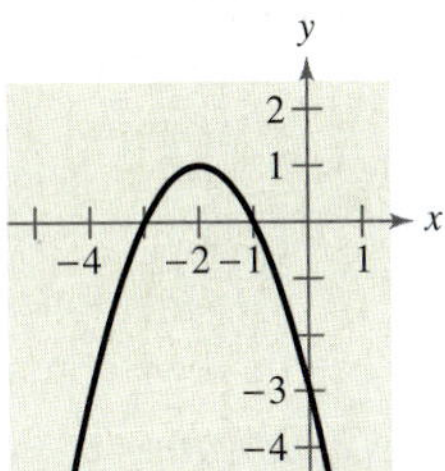

59.

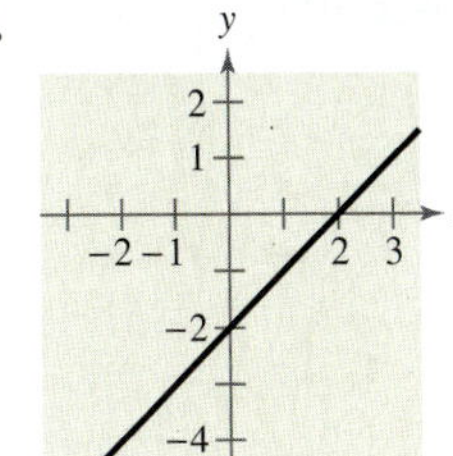

60.

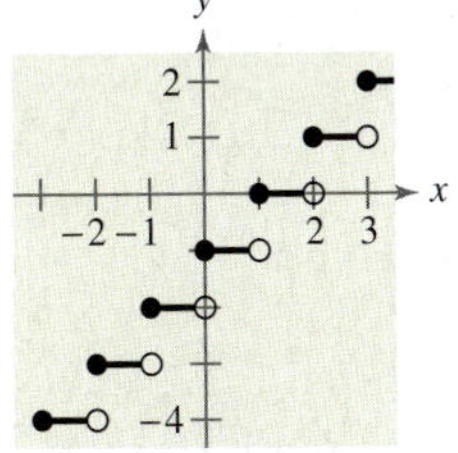

61.

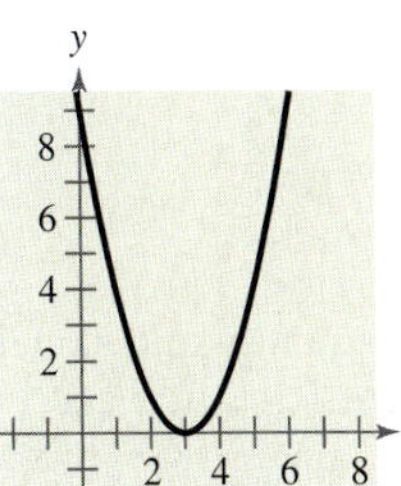

62.

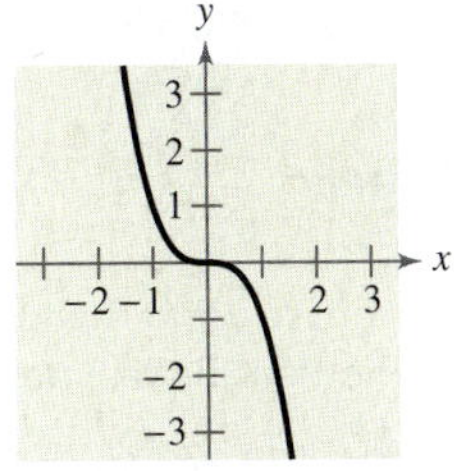

63. ***Communications*** The cost of a telephone call between Denver and Boise is \$0.60 for the first minute and \$0.42 for each additional minute or portion of a minute. A model for the total cost C (in dollars) of the phone call is

$$C = 0.60 - 0.42[\![1 - t]\!], \quad t > 0$$

where t is the length of the phone call in minutes.

(a) Sketch the graph of the model.

(b) Determine the cost of a call lasting 12 minutes and 30 seconds.

64. ***Communications*** The cost of using a telephone calling card is \$1.05 for the first minute and \$0.38 for each additional minute or portion of a minute.

(a) A customer needs a model for the cost C of using a calling card for a call lasting t minutes. Which of the following is the appropriate model? Explain.

$$C_1(t) = 1.05 + 0.38[\![t - 1]\!]$$

$$C_2(t) = 1.05 - 0.38[\![-(t - 1)]\!]$$

(b) Graph the appropriate model. Determine the cost of a call lasting 18 minutes and 45 seconds.

65. ***Delivery Charges*** The cost of sending an overnight package from Los Angeles to Miami is \$10.75 for a package weighing up to but not including 1 pound and \$3.95 for each additional pound or portion of a pound. A model for the total cost C (in dollars) of sending the package is

$$C = 10.75 + 3.95[\![x]\!], \quad x > 0$$

where x is the weight in pounds.

(a) Sketch a graph of the model.

(b) Determine the cost of sending a package that weighs 10.33 pounds.

66. ***Delivery Charges*** The cost of sending an overnight package from New York to Atlanta is \$9.80 for a package weighing up to but not including 1 pound and \$2.50 for each additional pound or portion of a pound.

(a) Use the greatest integer function to create a model for the cost C of overnight delivery of a package weighing x pounds, $x > 0$.

(b) Sketch the graph of the function.

67. ***Wages*** A mechanic is paid \$12.00 per hour for regular time and time-and-a-half for overtime. The weekly wage function is

$$W(h) = \begin{cases} 12h, & 0 < h \le 40 \\ 18(h - 40) + 480, & h > 40 \end{cases}$$

where h is the number of hours worked in a week.

(a) Evaluate $W(30)$, $W(40)$, $W(45)$, and $W(50)$.

(b) The company increased the regular work week to 45 hours. What is the new weekly wage function?

▶ Model It

68. ***Revenue*** The table shows the monthly revenue y (in thousands of dollars) of a landscaping business for the year 2002, with $x = 1$ representing January.

Month, x	Revenue, y
1	5.2
2	5.6
3	6.6
4	8.3
5	11.5
6	15.8
7	12.8
8	10.1
9	8.6
10	6.9
11	4.5
12	2.7

A mathematical model that represents this data is

$$f(x) = \begin{cases} -1.97x + 26.3 \\ 0.505x^2 - 1.47x + 6.3 \end{cases}.$$

(a) What is the domain of each part of the piecewise-defined function? How can you tell? Explain your reasoning.

(b) Sketch a graph of the model.

(c) Find $f(5)$ and $f(11)$, and interpret your results in the context of the problem.

(d) How do the values obtained from the model in part (b) compare with the actual data values?

69. ***Fluid Flow*** The intake pipe of a 100-gallon tank has a flow rate of 10 gallons per minute, and two drainpipes have flow rates of 5 gallons per minute each. The figure shows the volume V of fluid in the tank as a function of time t. Determine the combination of the input pipe and drain pipes in which the fluid is flowing in specific subintervals of the 1 hour of time shown on the graph. (There are many correct answers.)

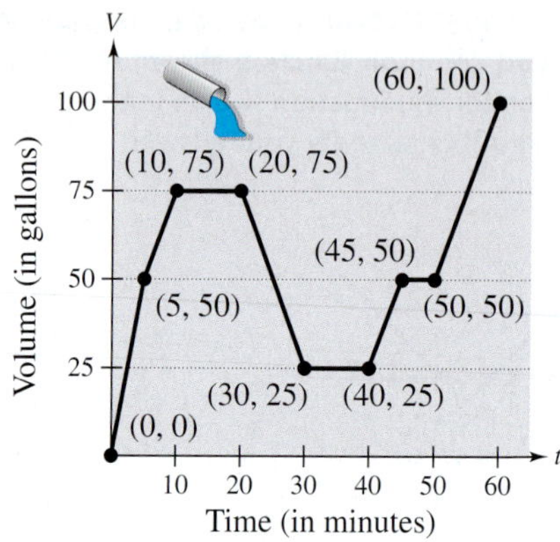

FIGURE FOR 69

Synthesis

True or False? **In Exercises 70 and 71, determine whether the statement is true or false. Justify your answer.**

70. A piecewise-defined function will always have at least one x-intercept or at least one y-intercept.

71. $f(x) = \begin{cases} 2, & 1 \le x < 2 \\ 4, & 2 \le x < 3 \\ 6, & 3 \le x < 4 \end{cases}$

can be rewritten as $f(x) = 2[\![x]\!], \quad 1 \le x < 4$.

72. ***Exploration*** Write equations for the piecewise-defined function shown in the graph.

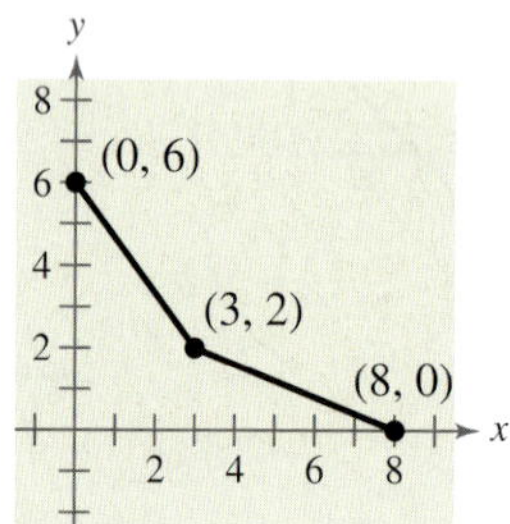

Review

In Exercise 73 and 74, solve the inequality and sketch the solution on the real number line.

73. $3x + 4 \le 12 - 5x$

74. $2x + 1 > 6x - 9$

In Exercises 75 and 76, determine whether the lines L_1 and L_2 passing through the pairs of points are parallel, perpendicular, or neither.

75. L_1: $(-2, -2), (2, 10)$
L_2: $(-1, 3), (3, 9)$

76. L_1: $(-1, -7), (4, 3)$
L_2: $(1, 5), (-2, -7)$

2.5 Shifting, Reflecting, and Stretching Graphs

▶ What you should learn

- How to use vertical and horizontal shifts to sketch graphs of functions
- How to use reflections to sketch graphs of functions
- How to use nonrigid transformations to sketch graphs of functions

▶ Why you should learn it

Knowing the graphs of common functions and knowing how to shift, reflect, and stretch graphs of functions can help you sketch a wide variety of simple functions by hand. This skill is useful in sketching graphs of functions that model real-life data, such as in Exercise 63 on page 227, where you are asked to sketch the graph of a function that models the amount of fuel used by trucks from 1980 through 1999.

Chuck Keeler/The Stock Market

Shifting Graphs 

Many functions have graphs that are simple transformations of the common graphs summarized in Section 2.4. For example, you can obtain the graph of

$$h(x) = x^2 + 2$$

by shifting the graph of $f(x) = x^2$ *up* two units, as shown in Figure 2.47. In function notation, h and f are related as follows.

$$h(x) = x^2 + 2 = f(x) + 2 \qquad \text{Upward shift of two units}$$

Similarly, you can obtain the graph of

$$g(x) = (x - 2)^2$$

by shifting the graph of $f(x) = x^2$ to the *right* two units, as shown in Figure 2.48. In this case, the functions g and f have the following relationship.

$$g(x) = (x - 2)^2 = f(x - 2) \qquad \text{Right shift of two units}$$

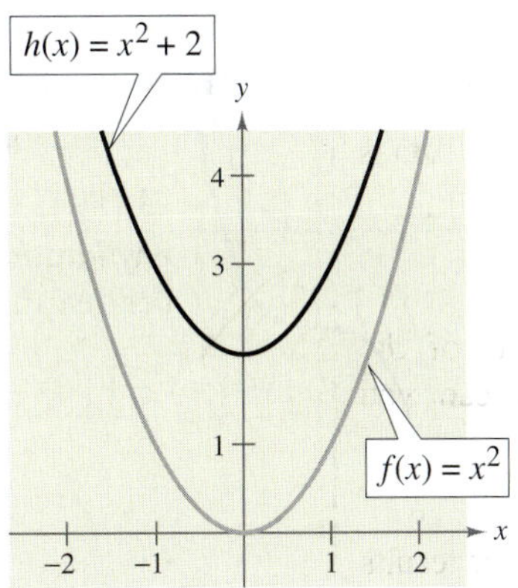

FIGURE 2.47

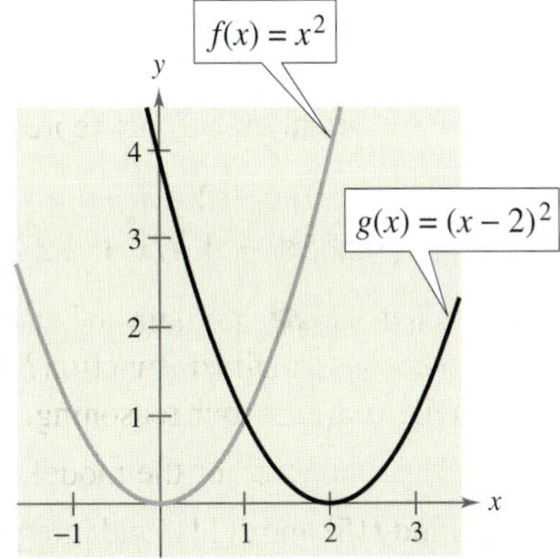

FIGURE 2.48

The following list summarizes this discussion about horizontal and vertical shifts.

Vertical and Horizontal Shifts

Let c be a positive real number. **Vertical and horizontal shifts** in the graph of $y = f(x)$ are represented as follows.

1. Vertical shift c units *upward:* $\quad h(x) = f(x) + c$
2. Vertical shift c units *downward:* $\quad h(x) = f(x) - c$
3. Horizontal shift c units to the *right:* $\quad h(x) = f(x - c)$
4. Horizontal shift c units to the *left:* $\quad h(x) = f(x + c)$

STUDY TIP

In items 3 and 4, be sure you see that $h(x) = f(x - c)$ corresponds to a *right* shift and $h(x) = f(x + c)$ corresponds to a *left* shift for $c > 0$.

Some graphs can be obtained from combinations of vertical and horizontal shifts, as demonstrated in Example 1(b). Vertical and horizontal shifts generate a *family of functions*, each with the same shape but at different locations in the plane.

Example 1 ▶ Shifts in the Graphs of a Function 

Use the graph of $f(x) = x^3$ to sketch the graph of each function.

a. $g(x) = x^3 - 1$

b. $h(x) = (x + 2)^3 + 1$

Solution

a. Relative to the graph of $f(x) = x^3$, the graph of $g(x) = x^3 - 1$ is a downward shift of one unit, as shown in Figure 2.49.

b. Relative to the graph of $f(x) = x^3$, the graph of $h(x) = (x + 2)^3 + 1$ involves a left shift of two units and an upward shift of one unit, as shown in Figure 2.50.

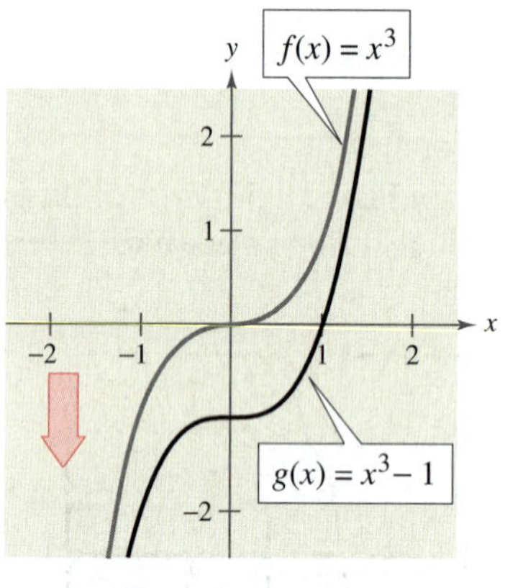

FIGURE 2.49

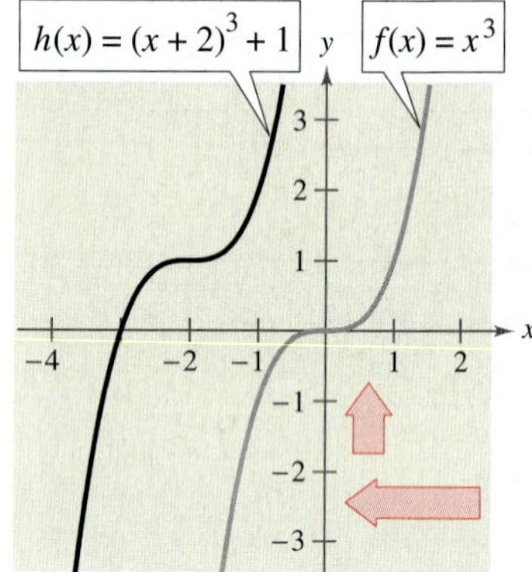

FIGURE 2.50

In Figure 2.50, notice that the same result is obtained if the vertical shift precedes the horizontal shift *or* if the horizontal shift precedes the vertical shift.

Exploration

Graphing utilities are ideal tools for exploring translations of functions. Graph f, g, and h in same viewing window. Before looking at the graphs, try to predict how the graphs of g and h relate to the graph of f.

a. $f(x) = x^2, \quad g(x) = (x - 4)^2, \quad h(x) = (x - 4)^2 + 3$

b. $f(x) = x^2, \quad g(x) = (x + 1)^2, \quad h(x) = (x + 1)^2 - 2$

c. $f(x) = x^2, \quad g(x) = (x + 4)^2, \quad h(x) = (x + 4)^2 + 2$

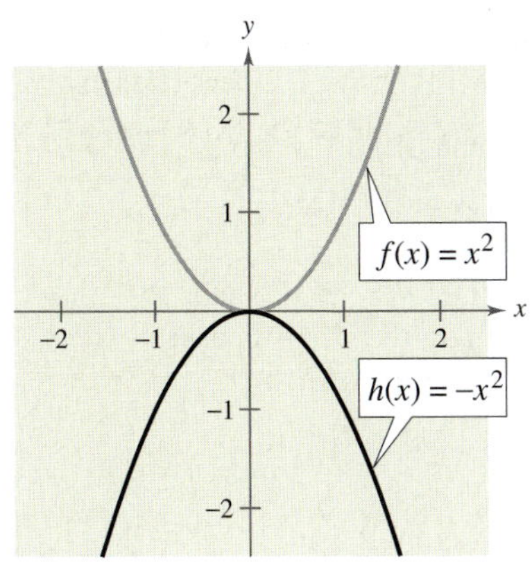

FIGURE 2.51

Reflecting Graphs 

The second common type of transformation is a **reflection.** For instance, if you consider the x-axis to be a mirror, the graph of

$$h(x) = -x^2$$

is the mirror image (or reflection) of the graph of $f(x) = x^2$, as shown in Figure 2.51.

Reflections in the Coordinate Axes

Reflections in the coordinate axes of the graph of $y = f(x)$ are represented as follows.

1. Reflection in the x-axis: $h(x) = -f(x)$
2. Reflection in the y-axis: $h(x) = f(-x)$

Example 2 ▶ Finding Equations from Graphs 

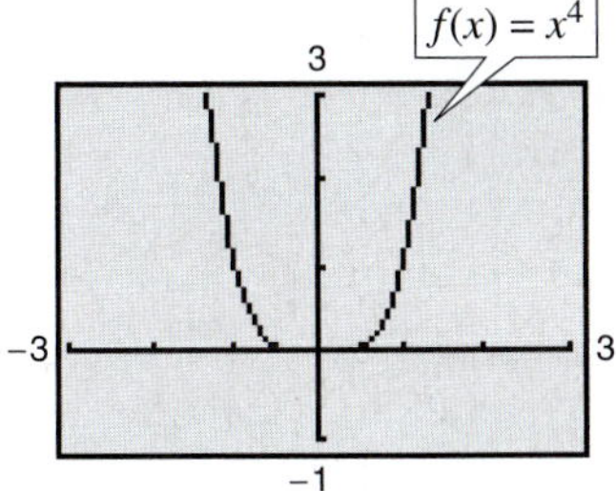

FIGURE 2.52

The graph of the function

$$f(x) = x^4$$

is shown in Figure 2.52. Each of the graphs in Figure 2.53 is a transformation of the graph of f. Find an equation for each of these functions.

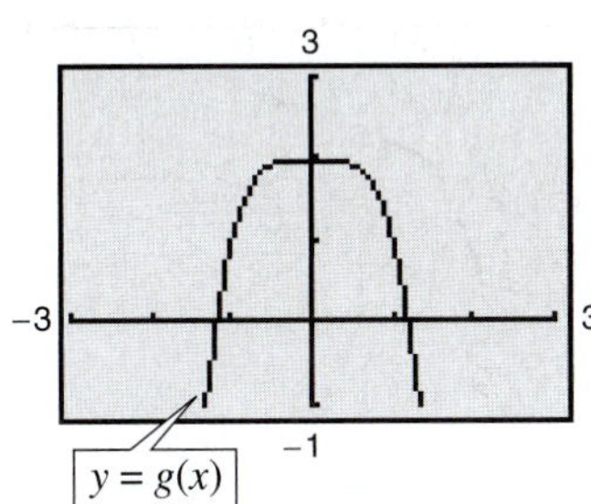

(a)

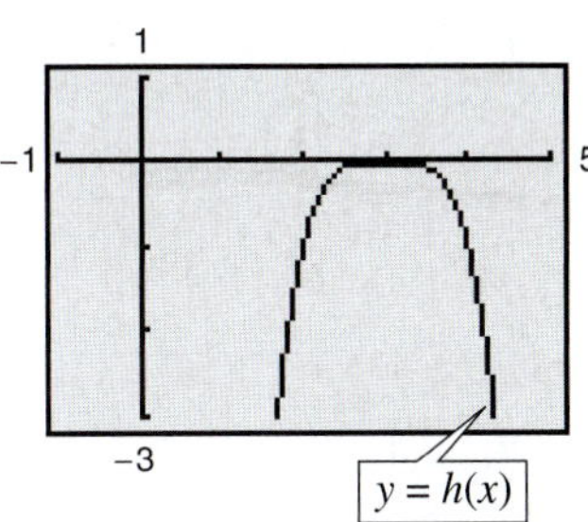

(b)

FIGURE 2.53

Solution

a. The graph of g is a reflection in the x-axis *followed by* an upward shift of two units of the graph of $f(x) = x^4$. So, the equation for g is

$$g(x) = -x^4 + 2.$$

b. The graph of h is a horizontal shift of three units to the right *followed by* a reflection in the x-axis of the graph of $f(x) = x^4$. So, the equation for h is

$$h(x) = -(x - 3)^4.$$

Exploration

Reverse the order of transformations in Example 2(a). Do you obtain the same graph? Do the same for Example 2(b). Do you obtain the same graph? Explain.

Example 3 ▶ Reflections and Shifts

Compare the graph of each function with the graph of $f(x) = \sqrt{x}$.

a. $g(x) = -\sqrt{x}$ **b.** $h(x) = \sqrt{-x}$ **c.** $k(x) = -\sqrt{x + 2}$

Solution

a. The graph of g is a reflection of the graph of f in the x-axis because

$$g(x) = -\sqrt{x}$$
$$= -f(x).$$

The graph of g compared with f is shown in Figure 2.54.

b. The graph of h is a reflection of the graph of f in the y-axis because

$$h(x) = \sqrt{-x}$$
$$= f(-x).$$

The graph of h compared with f is shown in Figure 2.55.

c. The graph of k is a left shift of two units, followed by a reflection in the x-axis because

$$k(x) = -\sqrt{x + 2}$$
$$= -f(x + 2).$$

The graph of k compared with f is shown in Figure 2.56.

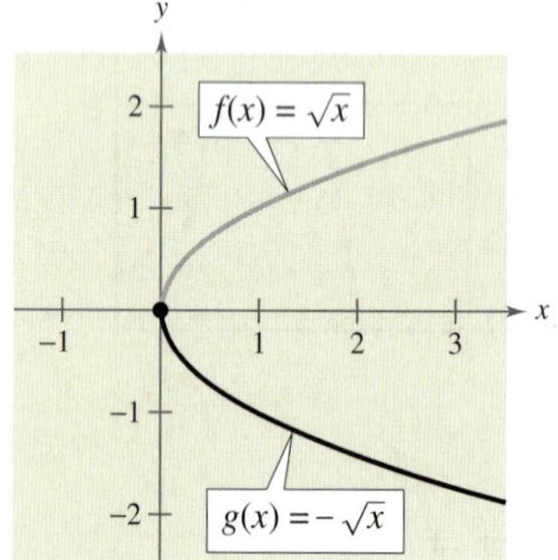

FIGURE 2.54

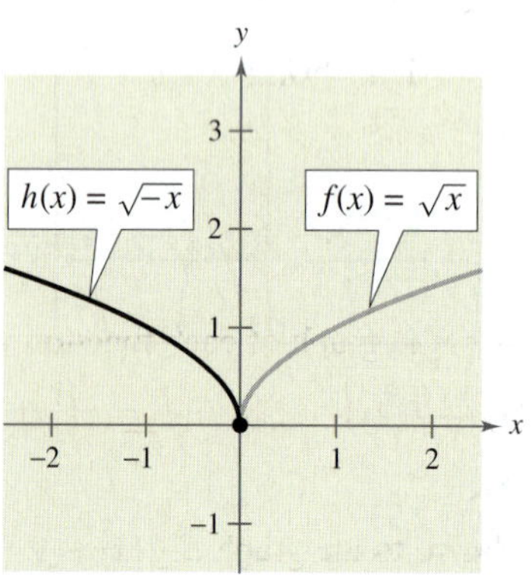

FIGURE 2.55

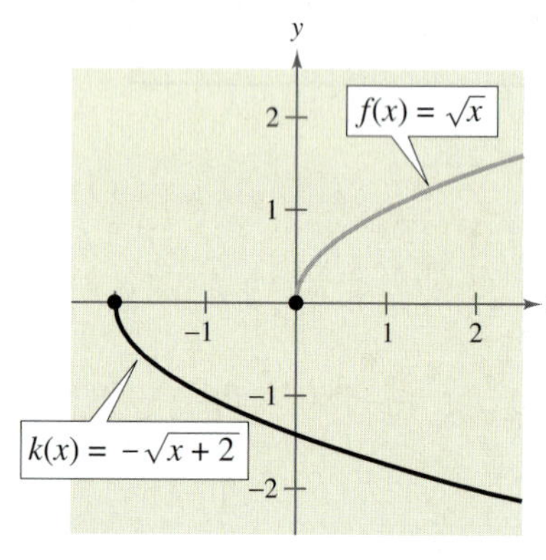

FIGURE 2.56

When sketching the graphs of functions involving square roots, remember that the domain must be restricted to exclude negative numbers inside the radical. For instance, here are the domains of the functions in Example 3.

Domain of $g(x) = -\sqrt{x}$: $x \geq 0$

Domain of $h(x) = \sqrt{-x}$: $x \leq 0$

Domain of $k(x) = -\sqrt{x + 2}$: $x \geq -2$

Nonrigid Transformations 

Horizontal shifts, vertical shifts, and reflections are **rigid transformations** because the basic shape of the graph is unchanged. These transformations change only the *position* of the graph in the xy-plane. **Nonrigid transformations** are those that cause a *distortion*—a change in the shape of the original graph. For instance, a nonrigid transformation of the graph of $y = f(x)$ is represented by $g(x) = cf(x)$, where the transformation is a **vertical stretch** if $c > 1$ and a **vertical shrink** if $0 < c < 1$. Another nonrigid transformation of the graph of $y = f(x)$ is represented by $h(x) = f(cx)$, where the transformation is a **horizontal shrink** if $c > 1$ and a **horizontal stretch** if $0 < c < 1$.

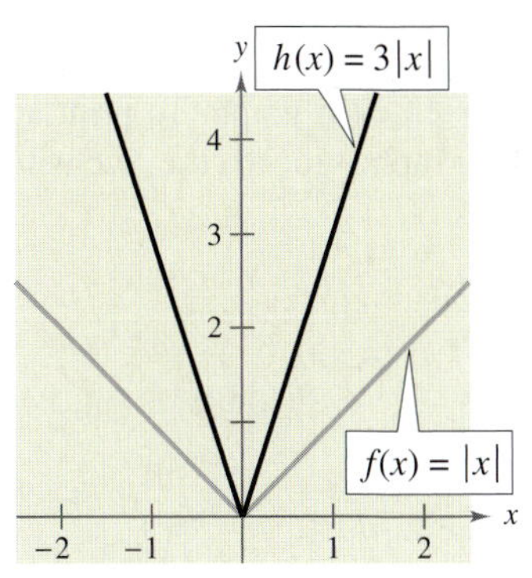

FIGURE 2.57

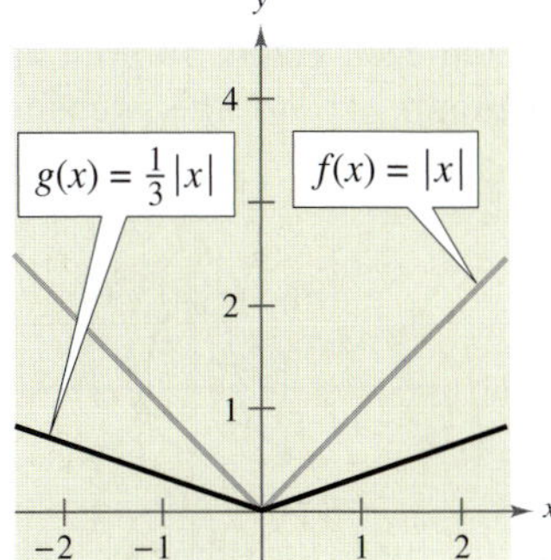

FIGURE 2.58

Example 4 ▶ Nonrigid Transformations 

Compare the graph of each function with the graph of $f(x) = |x|$.

a. $h(x) = 3|x|$ **b.** $g(x) = \frac{1}{3}|x|$

Solution

a. Relative to the graph of $f(x) = |x|$, the graph of

$$h(x) = 3|x| = 3f(x)$$

is a vertical stretch (each y-value is multiplied by 3) of the graph of f. (See Figure 2.57.)

b. Similarly, the graph of

$$g(x) = \tfrac{1}{3}|x| = \tfrac{1}{3}f(x)$$

is a vertical shrink (each y-value is multiplied by $\frac{1}{3}$) of the graph of f. (See Figure 2.58.)

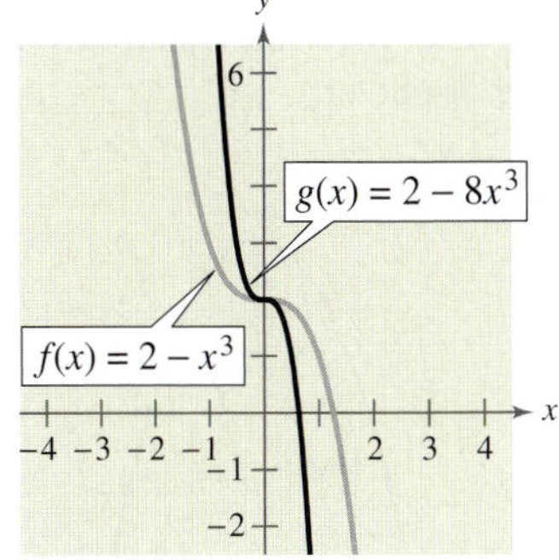

FIGURE 2.59

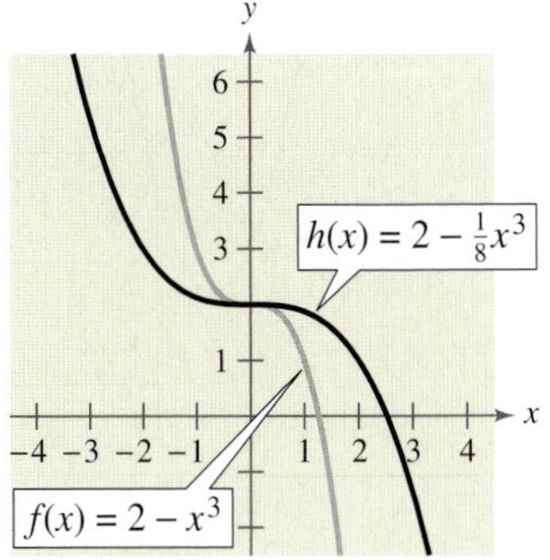

FIGURE 2.60

Example 5 ▶ Nonrigid Transformations

Compare the graph of each function with the graph of $f(x) = 2 - x^3$.

a. $g(x) = f(2x)$ **b.** $h(x) = f\left(\frac{1}{2}x\right)$

Solution

a. Relative to the graph of $f(x) = 2 - x^3$, the graph of

$$g(x) = f(2x) = 2 - (2x)^3 = 2 - 8x^3$$

is a horizontal shrink (each x-value is multiplied by $\frac{1}{2}$) of the graph of f. (See Figure 2.59.)

b. Similarly, the graph of

$$h(x) = f\left(\tfrac{1}{2}x\right) = 2 - \left(\tfrac{1}{2}x\right)^3 = 2 - \tfrac{1}{8}x^3$$

is a horizontal stretch (each x-value is multiplied by 2) of the graph of f. (See Figure 2.60.)

2.5 Exercises

1. For each function, sketch (on the same set of coordinate axes) a graph of each function for $c = -1, 1$, and 3.

(a) $f(x) = |x| + c$ (b) $f(x) = |x - c|$

(c) $f(x) = |x + 4| + c$

2. For each function, sketch (on the same set of coordinate axes) a graph of each function for $c = -3, -1, 1$, and 3.

(a) $f(x) = \sqrt{x} + c$ (b) $f(x) = \sqrt{x - c}$

(c) $f(x) = \sqrt{x - 3} + c$

3. For each function, sketch (on the same set of coordinate axes) a graph of each function for $c = -2, 0$, and 2.

(a) $f(x) = [\![x]\!] + c$ (b) $f(x) = [\![x + c]\!]$

(c) $f(x) = [\![x - 1]\!] + c$

4. For each function, sketch (on the same set of coordinate axes) a graph of each function for $c = -3, -1, 1$, and 3.

(a) $f(x) = \begin{cases} x^2 + c, & x < 0 \\ -x^2 + c, & x \ge 0 \end{cases}$

(b) $f(x) = \begin{cases} (x + c)^2, & x < 0 \\ -(x + c)^2, & x \ge 0 \end{cases}$

5. Use the graph of f to sketch each graph. To print an enlarged copy of the graph, go to the website *www.mathgraphs.com*.

(a) $y = f(x) + 2$

(b) $y = f(x - 2)$

(c) $y = 2f(x)$

(d) $y = -f(x)$

(e) $y = f(x + 3)$

(f) $y = f(-x)$

(g) $y = f\left(\frac{1}{2}x\right)$

6. Use the graph of f to sketch each graph. To print an enlarged copy of the graph, go to the website *www.mathgraphs.com*.

(a) $y = f(-x)$

(b) $y = f(x) + 4$

(c) $y = 2f(x)$

(d) $y = -f(x - 4)$

(e) $y = f(x) - 3$

(f) $y = -f(x) - 1$

(g) $y = f(2x)$

7. Use the graph of f to sketch each graph. To print an enlarged copy of the graph, go to the website *www.mathgraphs.com*.

(a) $y = f(x) - 1$

(b) $y = f(x - 1)$

(c) $y = f(-x)$

(d) $y = f(x + 1)$

(e) $y = -f(x - 2)$

(f) $y = \frac{1}{2}f(x)$

(g) $y = f(2x)$

8. Use the graph of f to sketch each graph. To print an enlarged copy of the graph, go to the website *www.mathgraphs.com*.

(a) $y = f(x - 5)$

(b) $y = -f(x) + 3$

(c) $y = \frac{1}{3}f(x)$

(d) $y = -f(x + 1)$

(e) $y = f(-x)$

(f) $y = f(x) - 10$

(g) $y = f\left(\frac{1}{3}x\right)$

9. Use the graph of $f(x) = x^2$ to write an equation for each function whose graph is shown.

(a)

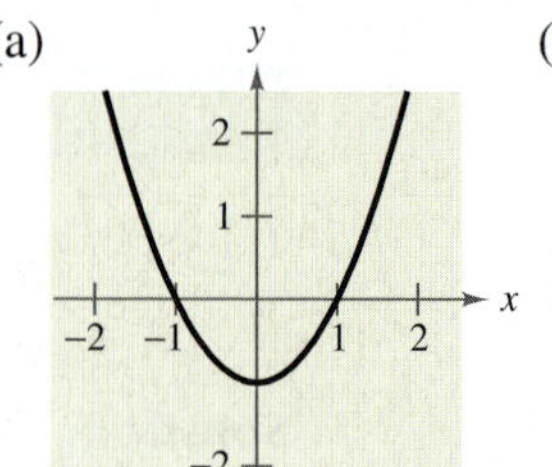

(b)

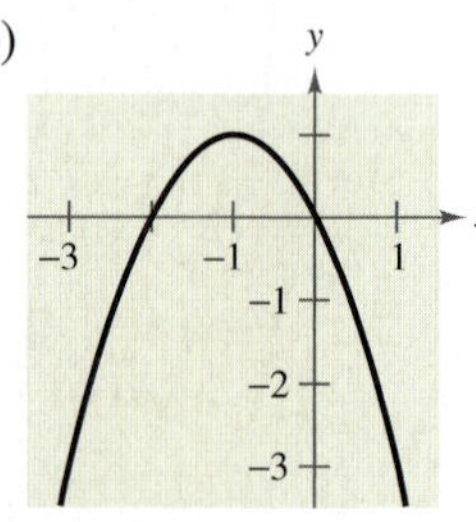

(c)

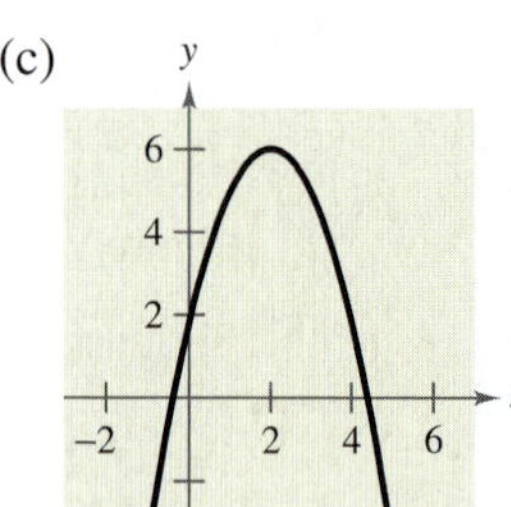

(d)

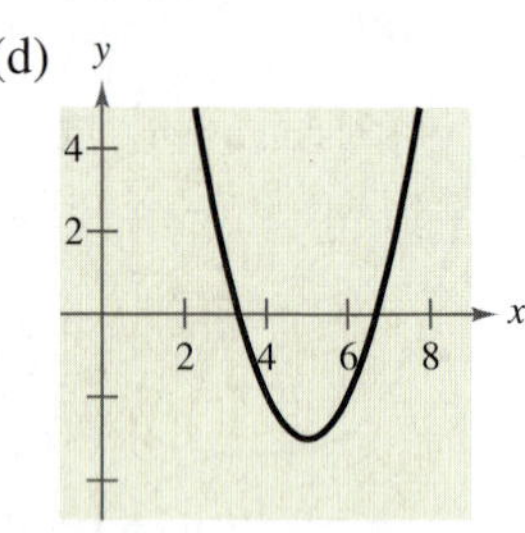

10. Use the graph of $f(x) = x^3$ to write an equation for each function whose graph is shown.

(a)

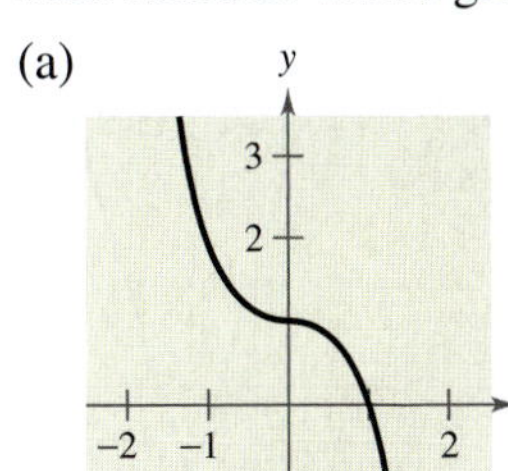

(b)

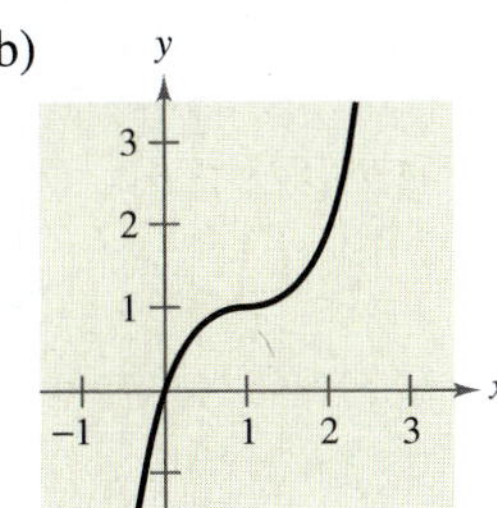

(c)

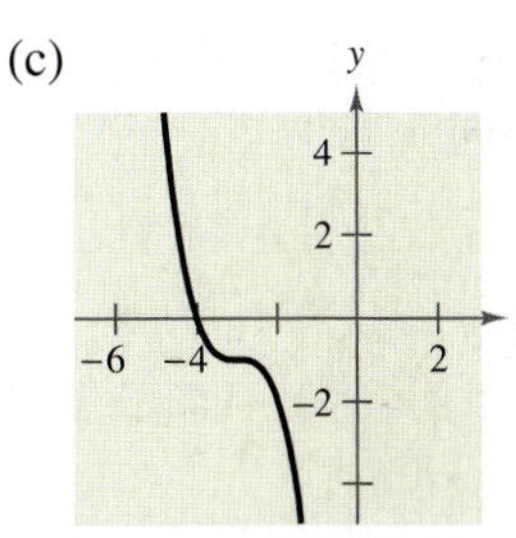

(d)

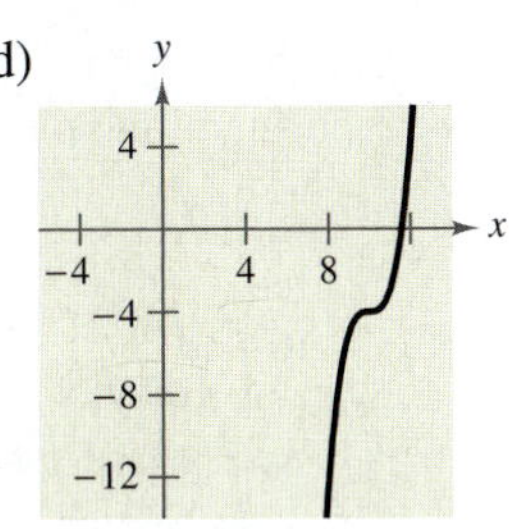

11. Use the graph of $f(x) = |x|$ to write an equation for each function whose graph is shown.

(a)

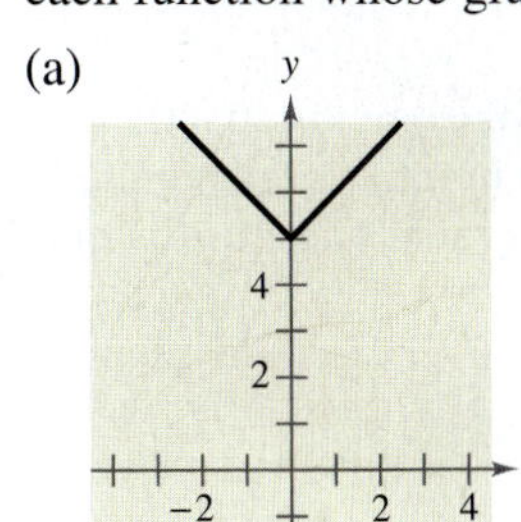

(b)

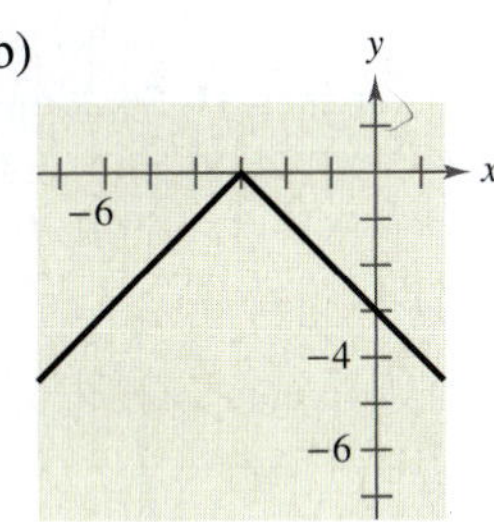

(c)

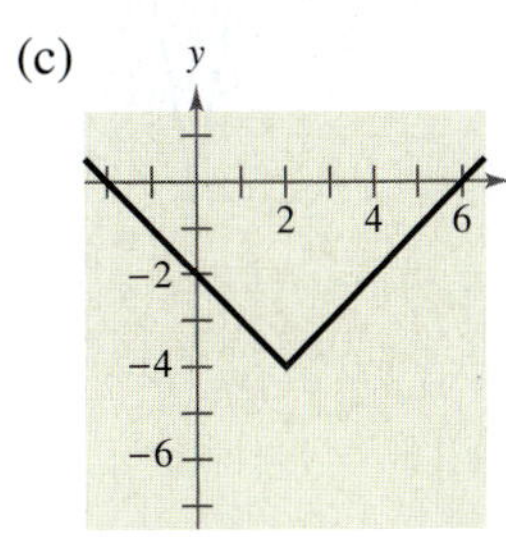

(d)

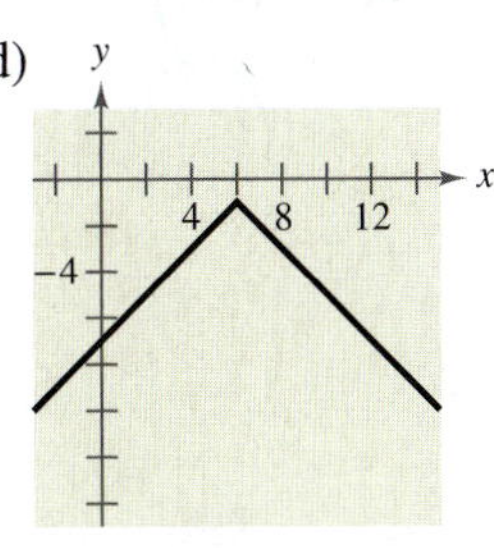

12. Use the graph of $f(x) = \sqrt{x}$ to write an equation for each function whose graph is shown.

(a)

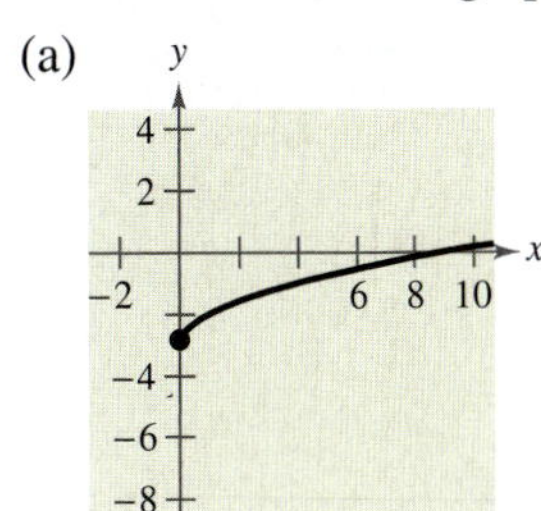

(b)

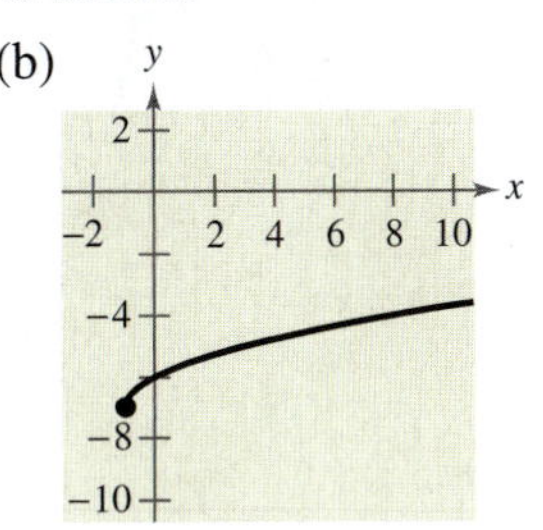

(c)

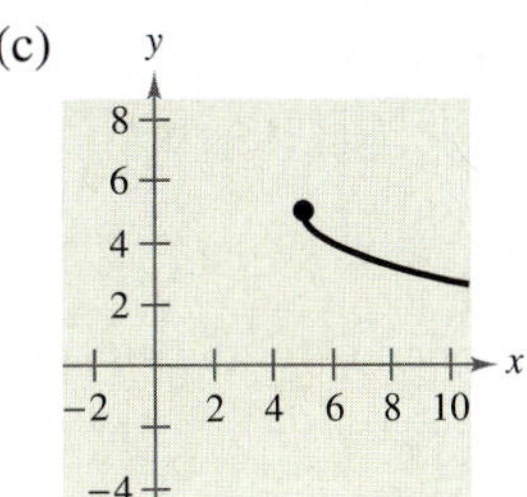

(d)

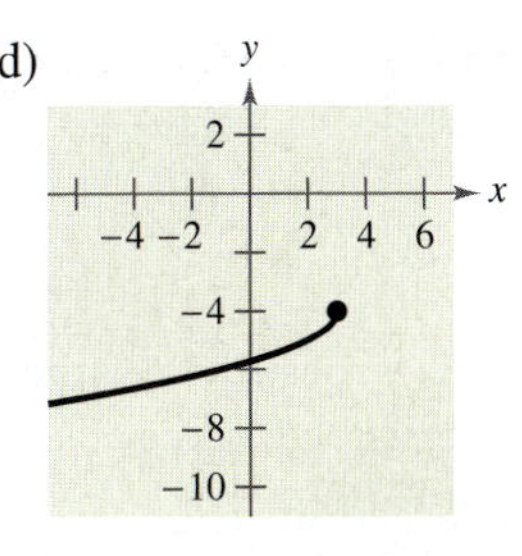

In Exercises 13–18, identify the common function and the transformation shown in the graph. Write an equation for the function shown in the graph.

13.

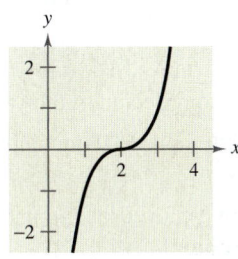

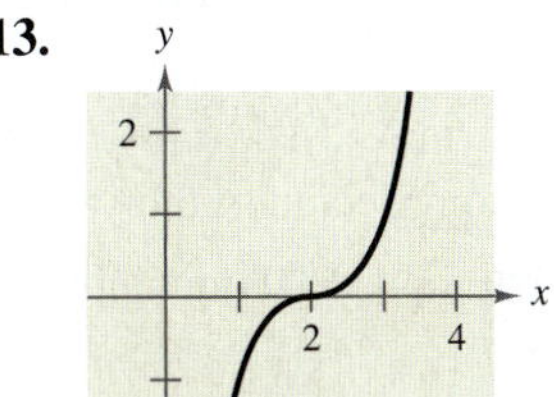

14.

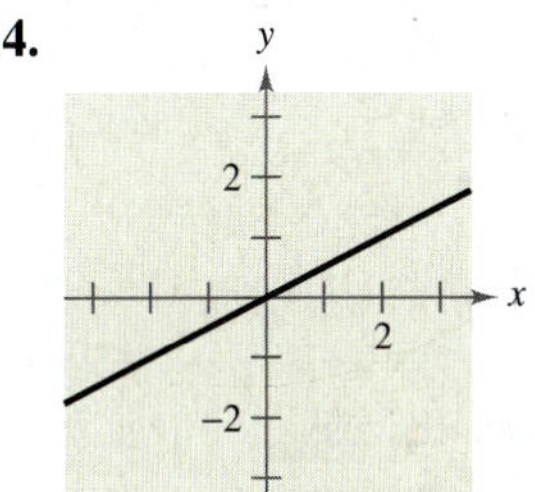

15.

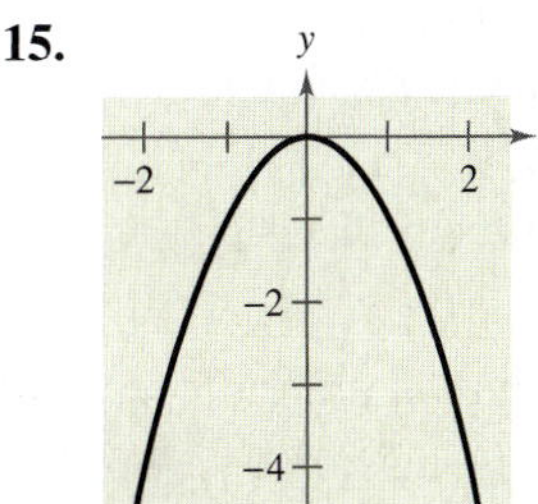

16.

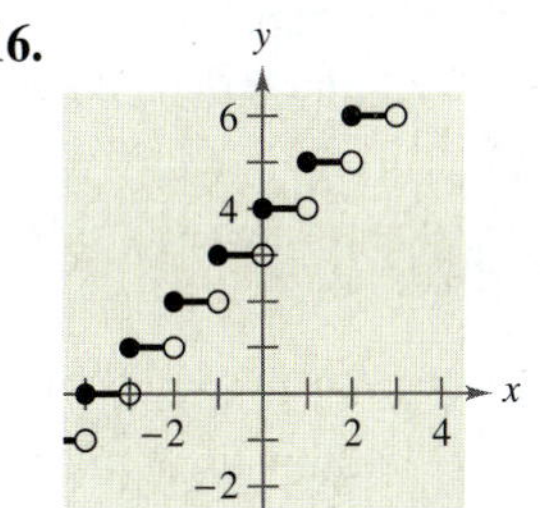

17.

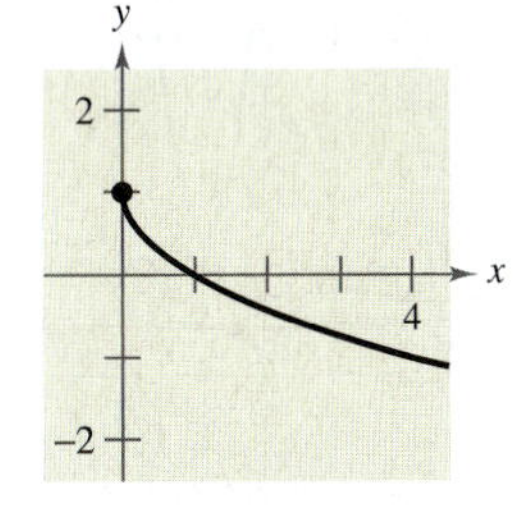

18.

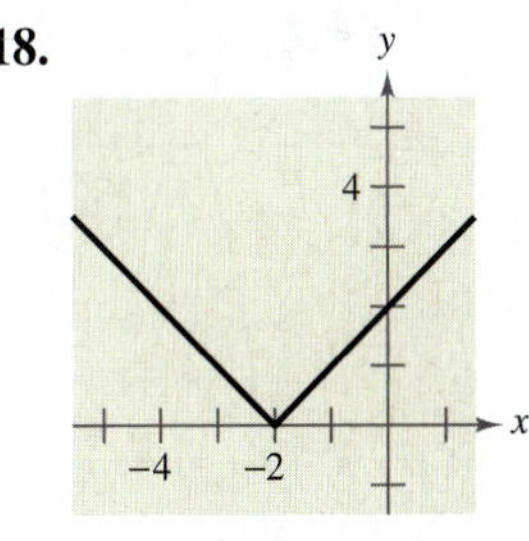

In Exercises 19–38, describe the transformation from a common function that occurs in the function. Then sketch its graph.

19. $f(x) = 12 - x^2$

20. $f(x) = (x - 8)^2$

21. $f(x) = x^3 + 7$

22. $f(x) = -x^3 - 1$

23. $f(x) = 2 - (x + 5)^2$

24. $f(x) = -(x + 10)^2 + 5$

25. $f(x) = (x - 1)^3 + 2$

26. $f(x) = (x + 3)^3 - 10$

27. $f(x) = -|x| - 2$

28. $f(x) = 6 - |x + 5|$

29. $f(x) = -|x + 4| + 8$

30. $f(x) = |-x + 3| + 9$

31. $f(x) = 3 - [\![x]\!]$

32. $f(x) = 2[\![x + 5]\!]$

33. $f(x) = \sqrt{x - 9}$

34. $f(x) = \sqrt{x + 4} + 8$

35. $f(x) = \sqrt{7 - x} - 2$

36. $f(x) = -\sqrt{x + 1} - 6$

37. $f(x) = \sqrt{\frac{1}{2}x} - 4$

38. $f(x) = \sqrt{3x} + 1$

In Exercises 39–46, write an equation for the function that is described by the given characteristics.

39. The shape of $f(x) = x^2$, but moved two units to the right and eight units downward

40. The shape of $f(x) = x^2$, but moved three units to the left, seven units upward, and reflected in the x-axis

41. The shape of $f(x) = x^3$, but moved 13 units to the right

42. The shape of $f(x) = x^3$, but moved six units to the left, six units downward, and reflected in the y-axis

43. The shape of $f(x) = |x|$, but moved 10 units upward and reflected in the x-axis

44. The shape of $f(x) = |x|$, but moved one unit to the left and seven units downward

45. The shape of $f(x) = \sqrt{x}$, but moved six units to the left and reflected in both the x-axis and the y-axis

46. The shape of $f(x) = \sqrt{x}$, but moved nine units downward and reflected in both the x-axis and the y-axis

47. Use the graph of $f(x) = x^2$ to write an equation for each function whose graph is shown.

(a)

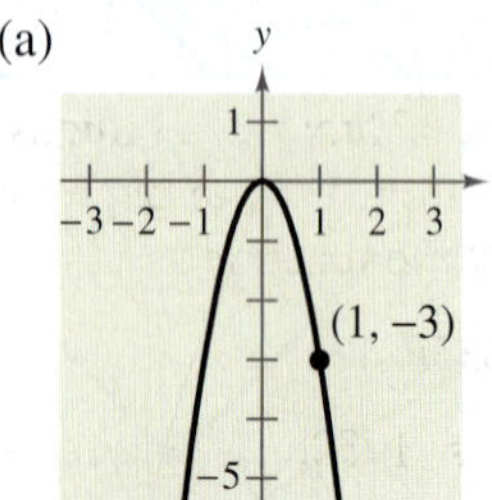

(b)

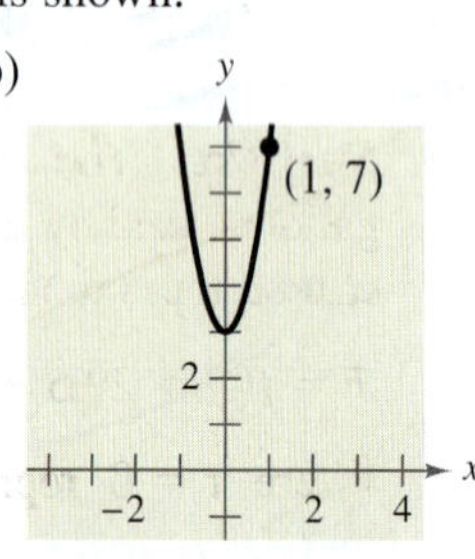

48. Use the graph of $f(x) = x^3$ to write an equation for each function whose graph is shown.

(a)

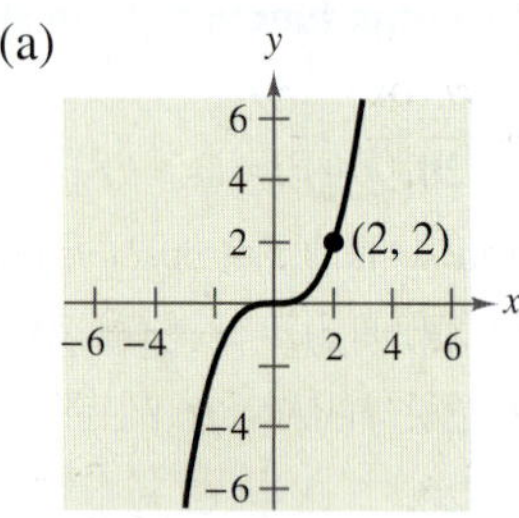

(b)

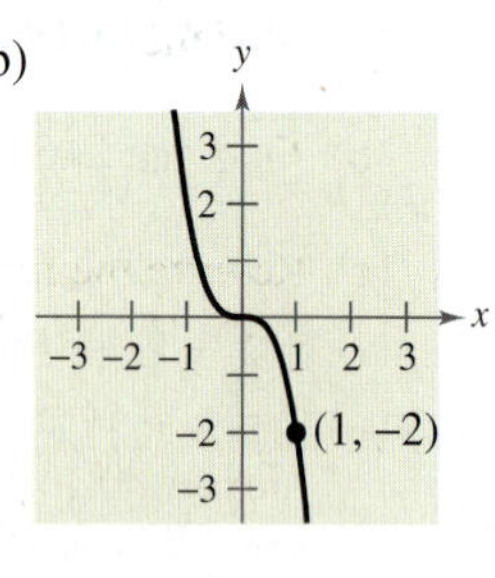

49. Use the graph of $f(x) = |x|$ to write an equation for each function whose graph is shown.

(a)

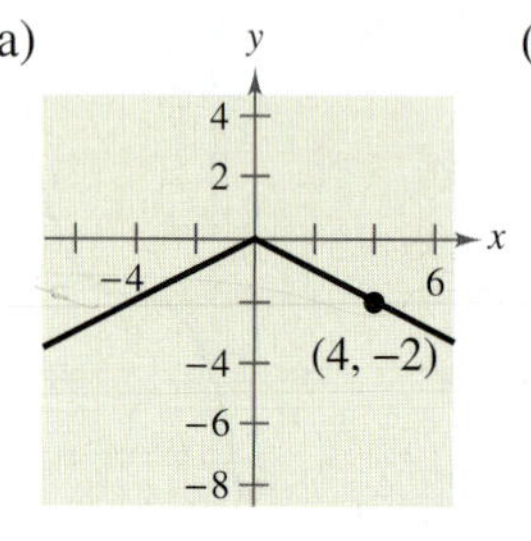

(b)

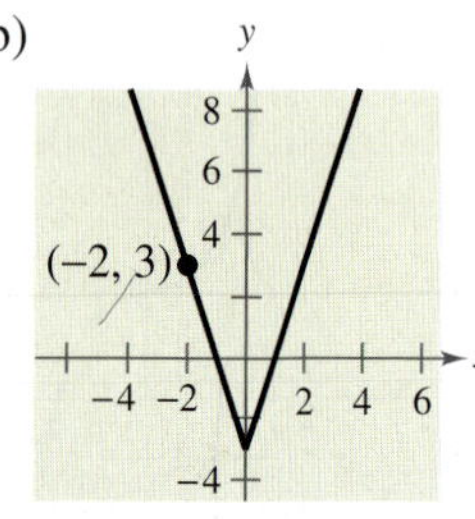

50. Use the graph of $f(x) = \sqrt{x}$ to write an equation for each function whose graph is shown.

(a)

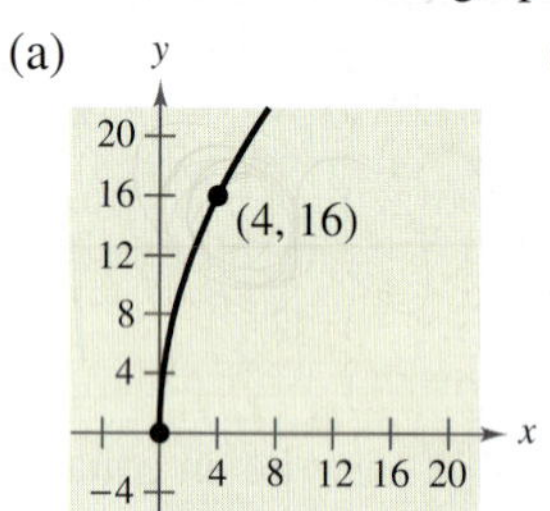

(b)

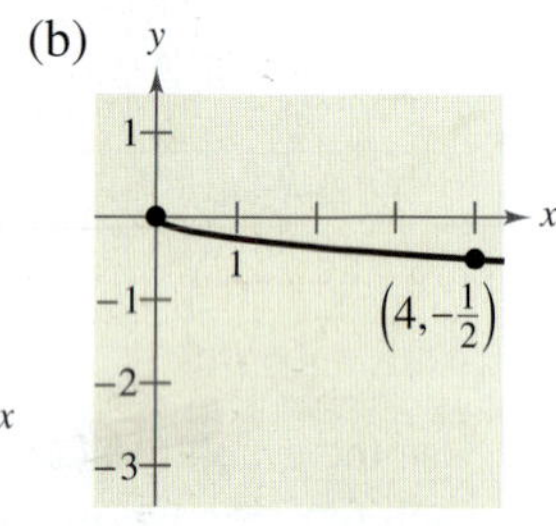

In Exercises 51–56, identify the common function and the transformation shown in the graph. Write an equation for the function shown in the graph. Then use a graphing utility to verify your answer.

51.

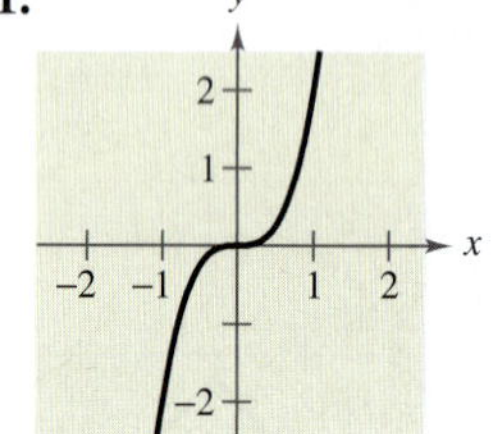

52.

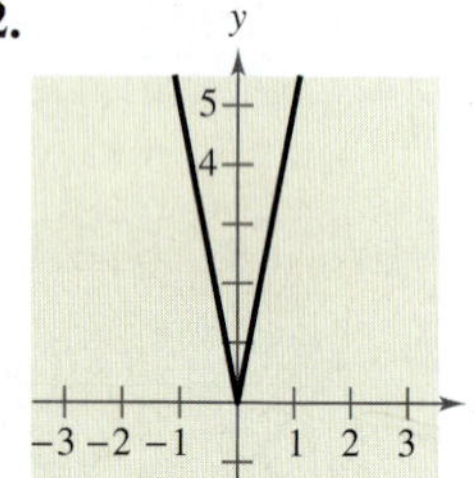

53.

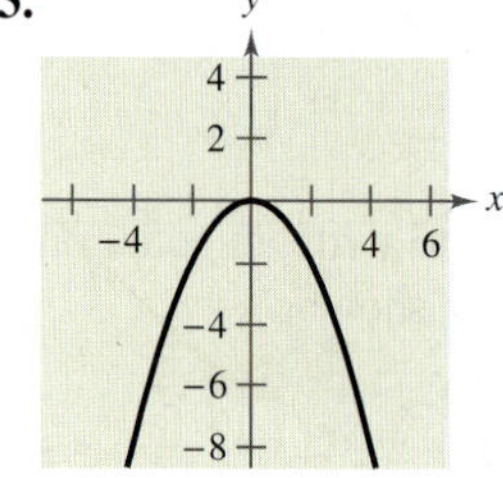

54.

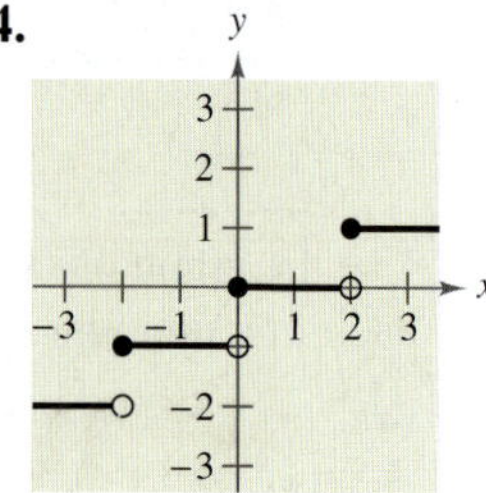

55.

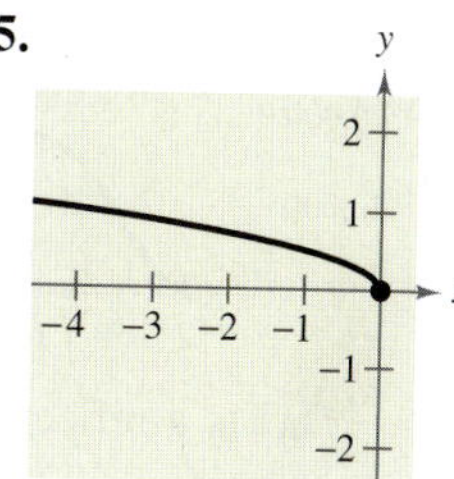

56.

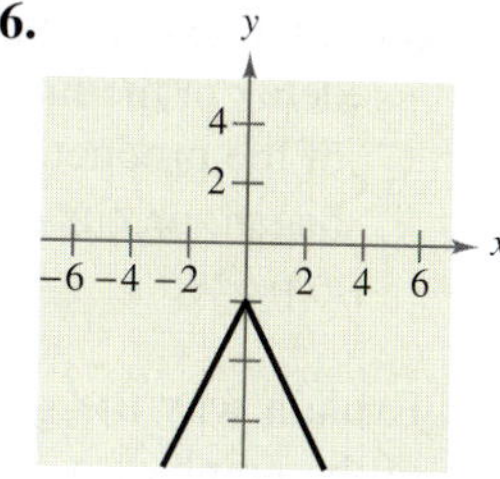

Graphical Analysis In Exercises 57–60, use the viewing window shown to write a possible equation for the transformation of the common function.

57.

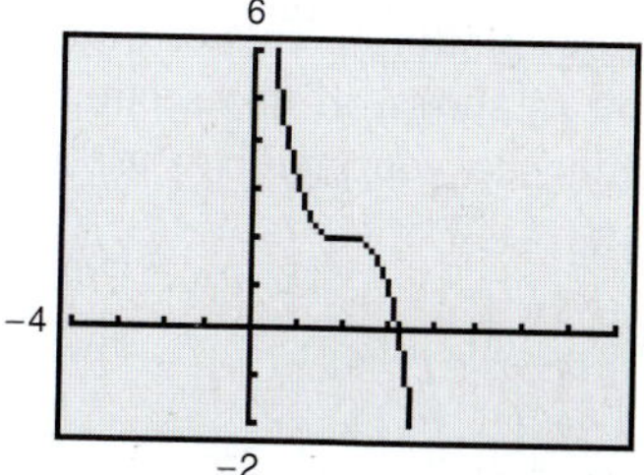

58.

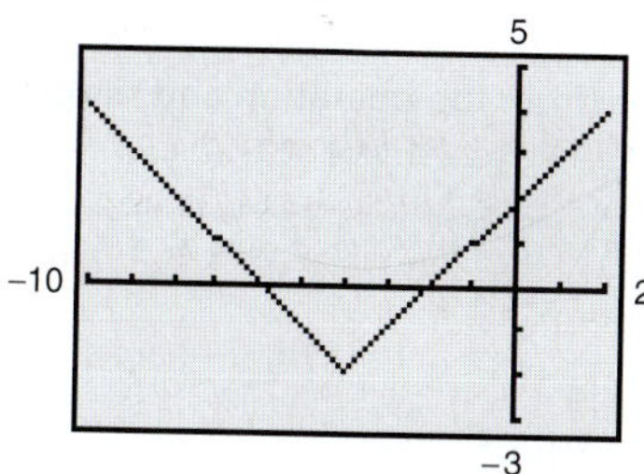

59.

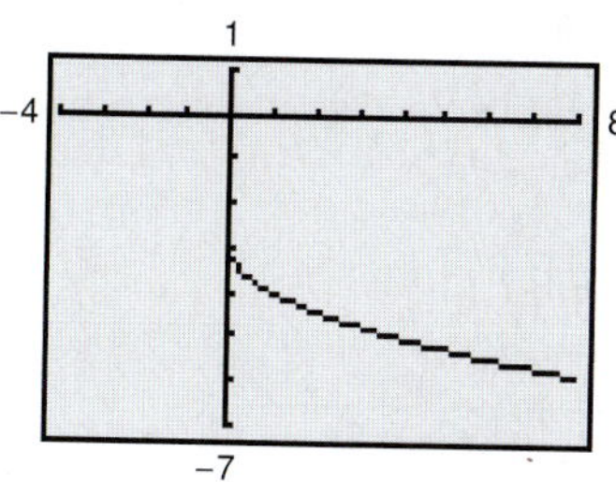

60.

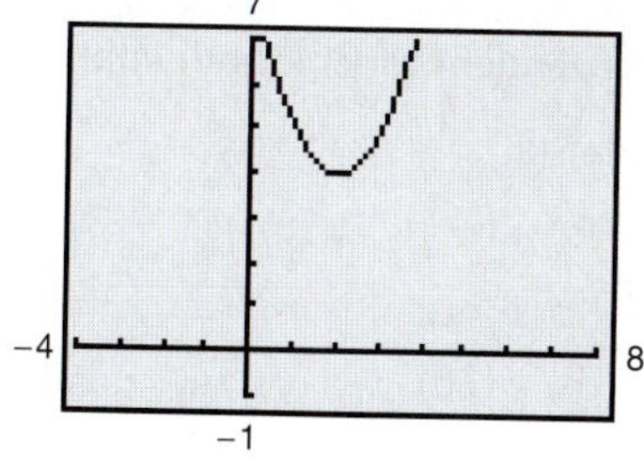

Graphical Reasoning In Exercises 61 and 62, use the graph of f to sketch the graph of g. To print an enlarged copy of the graph, go to the website *www.mathgraphs.com*.

61.

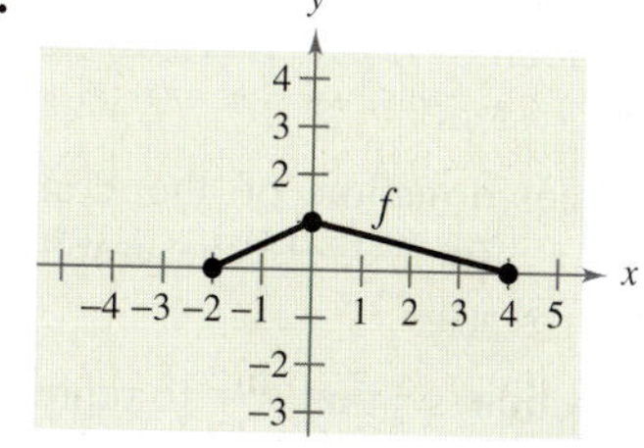

(a) $g(x) = f(x) + 2$ (b) $g(x) = f(x) - 1$

(c) $g(x) = f(-x)$ (d) $g(x) = -2f(x)$

(e) $g(x) = f(4x)$ (f) $g(x) = f\left(\frac{1}{2}x\right)$

62.

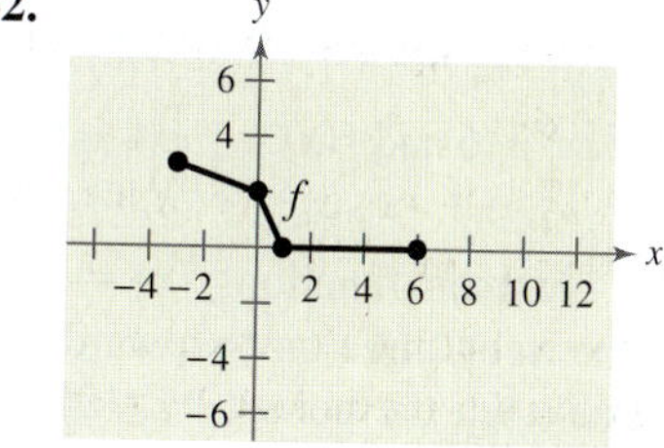

(a) $g(x) = f(x) - 5$ (b) $g(x) = f(x) + \frac{1}{2}$

(c) $g(x) = f(-x)$ (d) $g(x) = -4f(x)$

(e) $g(x) = f(2x) + 1$ (f) $g(x) = f\left(\frac{1}{4}x\right) - 2$

▶ Model It

63. *Fuel Use* The amount of fuel F (in billions of gallons) used by trucks from 1980 through 1999 can be approximated by the function

$$F = f(t) = 20.5 + 0.035t^2$$

where $t = 0$ represents 1980. (Source: U.S. Federal Highway Administration)

(a) Describe the transformation of the common function $f(x) = x^2$. Then sketch the graph over the interval $0 \le t \le 19$.

(b) Find and interpret $\dfrac{f(19) - f(0)}{19 - 0}$.

(c) Rewrite the function so that $t = 0$ represents 1990. Explain how you got your answer.

(d) Use the model from part (c) to predict the amount of fuel used by trucks in 2005. Does your answer seem reasonable? Explain.

64. ***Finance*** The amount M (in trillions of dollars) of mortgage debt outstanding in the United States from 1980 through 1999 can be approximated by the function $M = f(t) = 0.0037(t + 14.979)^2$, where $t = 0$ represents 1980. (Source: Board of Governors of the Federal Reserve System)

(a) Describe the transformation of the common function $f(x) = x^2$. Then sketch the graph over the interval $0 \le t \le 19$.

(b) Rewrite the function so that $t = 0$ represents 1990. Explain how you got your answer.

Synthesis

True or False? **In Exercises 65 and 66, determine whether the statement is true or false. Justify your answer.**

65. The graphs of $f(x) = |x| + 6$ and $f(x) = |-x| + 6$ are identical.

66. If the graph of the common function $f(x) = x^2$ is moved six units to the right, three units upward, and reflected in the x-axis, then the point $(-2, 19)$ will lie on the graph of the transformation.

67. ***Describing Profits*** Management originally predicted that the profits from the sales of a new product would be approximated by the graph of the function f shown. The actual profits are shown by the function g along with a verbal description. Use the concepts of transformations of graphs to write g in terms of f.

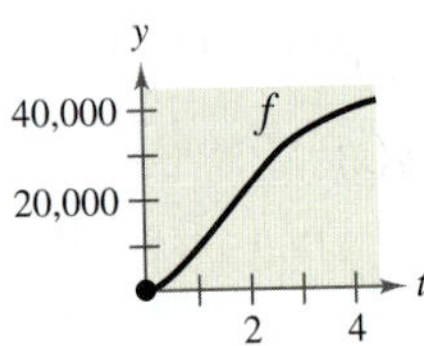

(a) The profits were only three-fourths as large as expected.

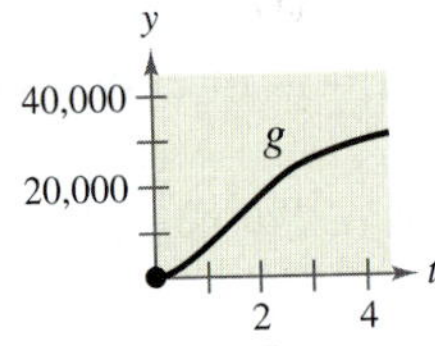

(b) The profits were consistently \$10,000 greater than predicted.

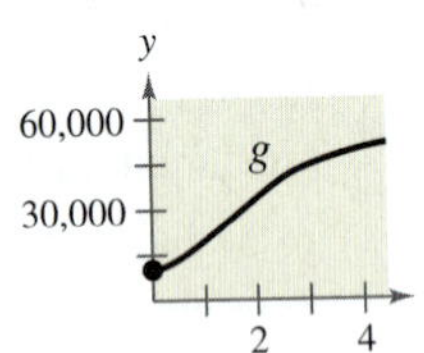

(c) There was a two-year delay in the introduction of the product. After sales began, profits grew as expected.

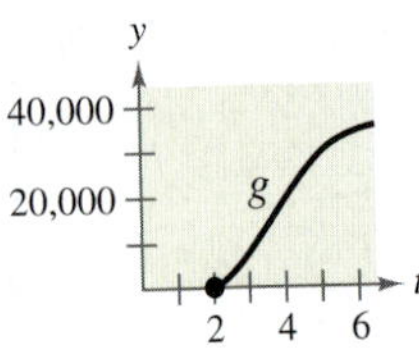

68. Explain why the graph of $y = -f(x)$ is a reflection of the graph of $y = f(x)$ about the x-axis.

69. The graph of $y = f(x)$ passes through the points $(0, 1)$, $(1, 2)$, and $(2, 3)$. Find the corresponding points on the graph of $y = f(x + 2) - 1$.

70. ***Think About It*** You can use either of two methods to graph a function: plotting points or translating a common function as shown in this section. Which method of graphing do you prefer to use for each function? Explain.

(a) $f(x) = 3x^2 - 4x + 1$

(b) $f(x) = 2(x - 1)^2 - 6$

Review

In Exercises 71–78, perform the operation and simplify.

71. $\dfrac{4}{x} + \dfrac{4}{1 - x}$

72. $\dfrac{2}{x + 5} - \dfrac{2}{x - 5}$

73. $\dfrac{3}{x - 1} - \dfrac{2}{x(x - 1)}$

74. $\dfrac{x}{x - 5} + \dfrac{1}{2}$

75. $(x - 4)\left(\dfrac{1}{\sqrt{x^2 - 4}}\right)$

76. $\left(\dfrac{x}{x^2 - 4}\right)\left(\dfrac{x^2 - x - 2}{x^2}\right)$

77. $(x^2 - 9) \div \left(\dfrac{x + 3}{5}\right)$

78. $\left(\dfrac{x}{x^2 - 3x - 28}\right) \div \left(\dfrac{x^2 + 3x}{x^2 + 5x + 4}\right)$

In Exercises 79 and 80, evaluate the function at the specified values of the independent variable and simplify.

79. $f(x) = x^2 - 6x + 11$

(a) $f(-3)$ (b) $f\left(-\frac{1}{2}\right)$ (c) $f(x - 3)$

80. $f(x) = \sqrt{x + 10} - 3$

(a) $f(-10)$ (b) $f(26)$ (c) $f(x - 10)$

In Exercises 81–84, find the domain of the function.

81. $f(x) = \dfrac{2}{11 - x}$

82. $f(x) = \dfrac{\sqrt{x - 3}}{x - 8}$

83. $f(x) = \sqrt{81 - x^2}$

84. $f(x) = \sqrt[3]{4 - x^2}$

2.6 Combinations of Functions

▶ **What you should learn**

- How to add, subtract, multiply, and divide functions
- How to find the composition of one function with another function
- How to use combinations of functions to model and solve real-life problems

▶ **Why you should learn it**

Combinations of functions can be used to model and solve real-life problems. For instance, in Exercise 33 on page 235, combinations of functions are used to analyze U.S. health expenditures.

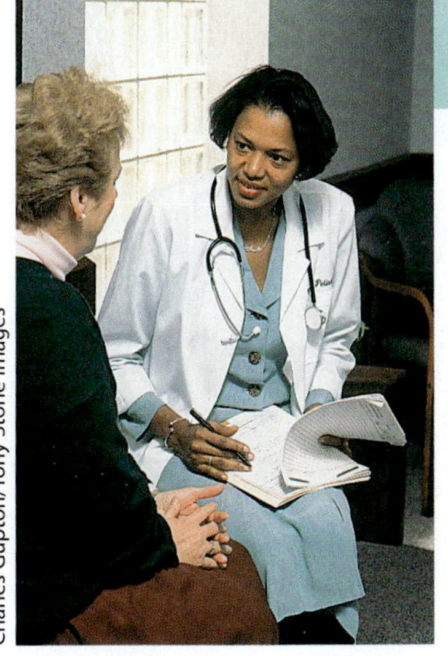

Charles Gupton/Tony Stone Images

Arithmetic Combinations of Functions 

Just as two real numbers can be combined by the operations of addition, subtraction, multiplication, and division to form other real numbers, two *functions* can be combined to create new functions. For example, the functions $f(x) = 2x - 3$ and $g(x) = x^2 - 1$ can be combined to form the sum, difference, product, and quotient of f and g.

$$f(x) + g(x) = (2x - 3) + (x^2 - 1)$$
$$= x^2 + 2x - 4 \quad \text{Sum}$$
$$f(x) - g(x) = (2x - 3) - (x^2 - 1)$$
$$= -x^2 + 2x - 2 \quad \text{Difference}$$
$$f(x)g(x) = (2x - 3)(x^2 - 1)$$
$$= 2x^3 - 3x^2 - 2x + 3 \quad \text{Product}$$
$$\frac{f(x)}{g(x)} = \frac{2x - 3}{x^2 - 1}, \quad x \neq \pm 1 \quad \text{Quotient}$$

The domain of an **arithmetic combination** of functions f and g consists of all real numbers that are common to the domains of f and g. In the case of the quotient $f(x)/g(x)$, there is the further restriction that $g(x) \neq 0$.

Sum, Difference, Product, and Quotient of Functions

Let f and g be two functions with overlapping domains. Then, for all x common to both domains, the *sum*, *difference*, *product*, and *quotient* of f and g are defined as follows.

1. *Sum:* $(f + g)(x) = f(x) + g(x)$
2. *Difference:* $(f - g)(x) = f(x) - g(x)$
3. *Product:* $(fg)(x) = f(x) \cdot g(x)$
4. *Quotient:* $\left(\frac{f}{g}\right)(x) = \frac{f(x)}{g(x)}, \quad g(x) \neq 0$

Example 1 ▶ Finding the Sum of Two Functions

Given $f(x) = 2x + 1$ and $g(x) = x^2 + 2x - 1$, find $(f + g)(x)$.

Solution

$$(f + g)(x) = f(x) + g(x) = (2x + 1) + (x^2 + 2x - 1) = x^2 + 4x$$

Example 2 ▶ Finding the Difference of Two Functions 

Given $f(x) = 2x + 1$ and $g(x) = x^2 + 2x - 1$, find $(f - g)(x)$. Then evaluate the difference when $x = 2$.

Solution

The difference of f and g is

$$\begin{aligned}(f - g)(x) &= f(x) - g(x)\\ &= (2x + 1) - (x^2 + 2x - 1)\\ &= -x^2 + 2.\end{aligned}$$

When $x = 2$, the value of this difference is

$$\begin{aligned}(f - g)(2) &= -(2)^2 + 2\\ &= -2.\end{aligned}$$

In Examples 1 and 2, both f and g have domains that consist of all real numbers. So, the domains of $(f + g)$ and $(f - g)$ are also the set of all real numbers. Remember that any restrictions on the domains of f and g must be considered when forming the sum, difference, product, or quotient of f and g.

Example 3 ▶ Finding the Domains of Quotients of Functions

Find the domains of $\left(\frac{f}{g}\right)(x)$ and $\left(\frac{g}{f}\right)(x)$ for the functions

$$f(x) = \sqrt{x} \quad \text{and} \quad g(x) = \sqrt{4 - x^2}.$$

Solution

The quotient of f and g is

$$\left(\frac{f}{g}\right)(x) = \frac{f(x)}{g(x)} = \frac{\sqrt{x}}{\sqrt{4 - x^2}}$$

and the quotient of g and f is

$$\left(\frac{g}{f}\right)(x) = \frac{g(x)}{f(x)} = \frac{\sqrt{4 - x^2}}{\sqrt{x}}.$$

The domain of f is $[0, \infty)$ and the domain of g is $[-2, 2]$. The intersection of these domains is $[0, 2]$. So, the domains of $\left(\frac{f}{g}\right)$ and $\left(\frac{g}{f}\right)$ are as follows.

$$\text{Domain of } \left(\frac{f}{g}\right): [0, 2) \qquad \text{Domain of } \left(\frac{g}{f}\right): (0, 2]$$

Can you see why these two domains differ slightly?

Composition of Functions 

Another way of combining two functions is to form the **composition** of one with the other. For instance, if $f(x) = x^2$ and $g(x) = x + 1$, the composition of f with g is

$$\begin{aligned} f(g(x)) &= f(x + 1) \\ &= (x + 1)^2. \end{aligned}$$

This composition is denoted as $(f \circ g)$.

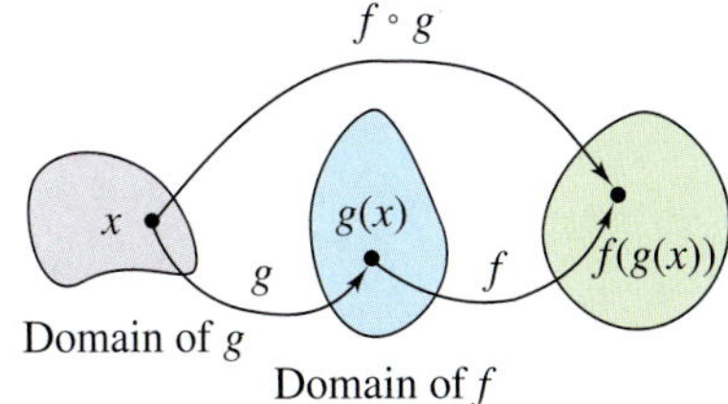

FIGURE 2.61

Definition of Composition of Two Functions

The **composition** of the function f with the function g is

$$(f \circ g)(x) = f(g(x)).$$

The domain of $(f \circ g)$ is the set of all x in the domain of g such that $g(x)$ is in the domain of f. (See Figure 2.61.)

STUDY TIP

The following tables of values help illustrate the composition $(f \circ g)(x)$ given in Example 4.

x	0	1	2	3
$g(x)$	4	3	0	-5

$g(x)$	4	3	0	-5
$f(g(x))$	6	5	2	-3

x	0	1	2	3
$f(g(x))$	6	5	2	-3

Note that the first two tables can be combined (or "composed") to produce the values given in the third table.

Example 4 ▶ Composition of Functions

Given $f(x) = x + 2$ and $g(x) = 4 - x^2$, find the following.

a. $(f \circ g)(x)$ **b.** $(g \circ f)(x)$ **c.** $(g \circ f)(-2)$

Solution

a. The composition of f with g is as follows.

$$\begin{aligned} (f \circ g)(x) &= f(g(x)) && \text{Definition of } f \circ g \\ &= f(4 - x^2) && \text{Definition of } g(x) \\ &= (4 - x^2) + 2 && \text{Definition of } f(x) \\ &= -x^2 + 6 && \text{Simplify.} \end{aligned}$$

b. The composition of g with f is as follows.

$$\begin{aligned} (g \circ f)(x) &= g(f(x)) && \text{Definition of } g \circ f \\ &= g(x + 2) && \text{Definition of } f(x) \\ &= 4 - (x + 2)^2 && \text{Definition of } g(x) \\ &= 4 - (x^2 + 4x + 4) && \text{Expand.} \\ &= -x^2 - 4x && \text{Simplify.} \end{aligned}$$

Note that, in this case, $(f \circ g)(x) \neq (g \circ f)(x)$.

c. Using the result of part (b), you can write the following.

$$\begin{aligned} (g \circ f)(-2) &= -(-2)^2 - 4(-2) && \text{Substitute.} \\ &= -4 + 8 && \text{Simplify.} \\ &= 4 && \text{Simplify.} \end{aligned}$$

Technology

You can use a graphing utility to determine the domain of a composition of functions. For the composition in Example 5, enter the function composition as

$$y = \left(\sqrt{9 - x^2}\right)^2 - 9.$$

You should obtain the graph shown below. Use the *trace* feature to determine that the x-coordinates of points on the graph extend from -3 to 3. So, the domain of $(f \circ g)(x)$ is $-3 \le x \le 3$.

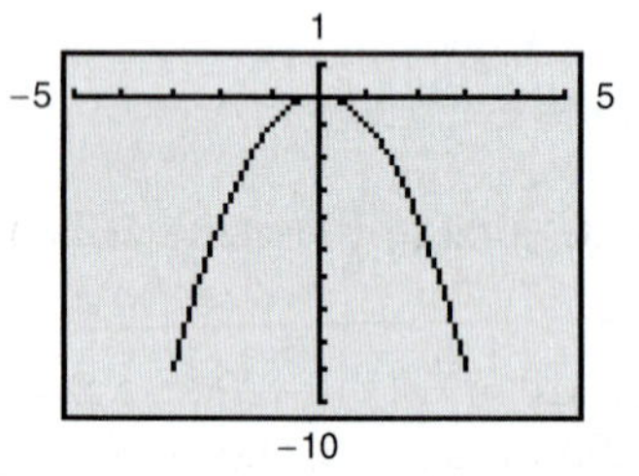

Example 5 ▶ Finding the Domain of a Composite Function

Find the composition $(f \circ g)(x)$ for the functions

$$f(x) = x^2 - 9 \quad \text{and} \quad g(x) = \sqrt{9 - x^2}.$$

Then find the domain of $(f \circ g)$.

Solution

$$\begin{aligned}(f \circ g)(x) &= f(g(x))\\ &= f\left(\sqrt{9 - x^2}\right)\\ &= \left(\sqrt{9 - x^2}\right)^2 - 9\\ &= 9 - x^2 - 9\\ &= -x^2\end{aligned}$$

From this, it might appear that the domain of the composition is the set of all real numbers. Because the domain of f is the set of all real numbers and the domain of g is $-3 \le x \le 3$, the domain of $(f \circ g)$ is $-3 \le x \le 3$.

In Examples 4 and 5, you formed the composition of two given functions. In calculus, it is also important to be able to identify two functions that make up a given composite function. For instance, the function h given by

$$h(x) = (3x - 5)^3$$

is the composition of f with g, where $f(x) = x^3$ and $g(x) = 3x - 5$. That is,

$$h(x) = (3x - 5)^3 = [g(x)]^3 = f(g(x)).$$

Basically, to "decompose" a composite function, look for an "inner" function and an "outer" function. In the function h above, $g(x) = 3x - 5$ is the inner function and $f(x) = x^3$ is the outer function.

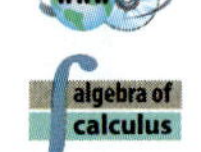

Example 6 ▶ Finding Components of Composite Functions

algebra of calculus

Express the function $h(x) = \dfrac{1}{(x - 2)^2}$ as a composition of two functions.

Solution

One way to write h as a composition of two functions is to take the inner function to be $g(x) = x - 2$ and the outer function to be

$$f(x) = \frac{1}{x^2} = x^{-2}.$$

Then you can write

$$h(x) = \frac{1}{(x - 2)^2} = (x - 2)^{-2} = f(x - 2) = f(g(x)).$$

Exploration

You are buying an automobile whose price is \$18,500. Which of the following options would you choose? Explain.

a. You are given a factory rebate of \$2000, followed by a dealer discount of 10%.

b. You are given a dealer discount of 10%, followed by a factory rebate of \$2000.

Let $f(x) = x - 2000$ and let $g(x) = 0.9x$. Which option is represented by the composite $f(g(x))$? Which is represented by the composite $g(f(x))$?

Application

Example 7 ▶ Bacteria Count 

The number N of bacteria in a refrigerated food is

$$N(T) = 20T^2 - 80T + 500, \qquad 2 \le T \le 14$$

where T is the temperature of the food in degrees Celsius. When the food is removed from refrigeration, the temperature is

$$T(t) = 4t + 2, \qquad 0 \le t \le 3$$

where t is the time in hours. (a) Find the composite $N(T(t))$ and interpret its meaning in context. (b) Find the time when the bacterial count reaches 2000.

Solution

a. $N(T(t)) = 20(4t + 2)^2 - 80(4t + 2) + 500$

$$= 20(16t^2 + 16t + 4) - 320t - 160 + 500$$

$$= 320t^2 + 320t + 80 - 320t - 160 + 500$$

$$= 320t^2 + 420$$

The composite function $N(T(t))$ represents the number of bacteria in the food as a function of time.

b. The bacterial count will reach 2000 when $320t^2 + 420 = 2000$. Solve this equation to find that the count will reach 2000 when $t \approx 2.2$ hours. When you solve this equation, note that the negative value is rejected because it is not in the domain of the composite function.

Writing ABOUT MATHEMATICS

Analyzing Arithmetic Combinations of Functions

a. Use the graphs of f and $(f + g)$ in Figure 2.62 to make a table showing the values of $g(x)$ when $x = 1, 2, 3, 4, 5,$ and 6. Explain your reasoning.

b. Use the graphs of f and $(f - h)$ in Figure 2.62 to make a table showing the values of $h(x)$ when $x = 1, 2, 3, 4, 5,$ and 6. Explain your reasoning.

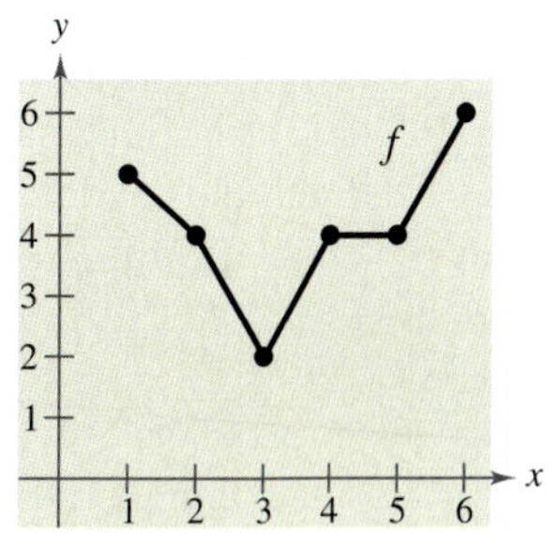

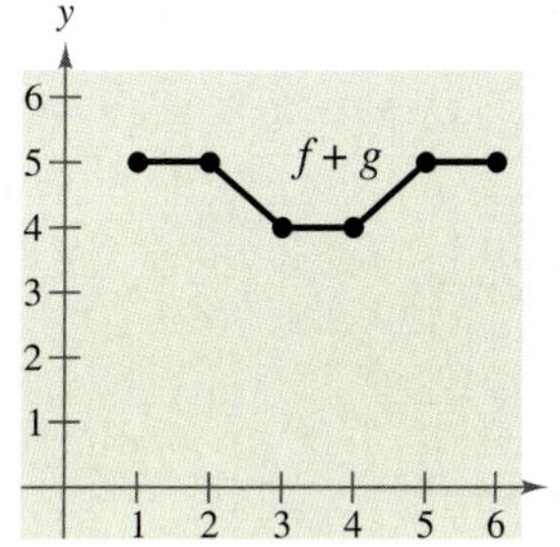

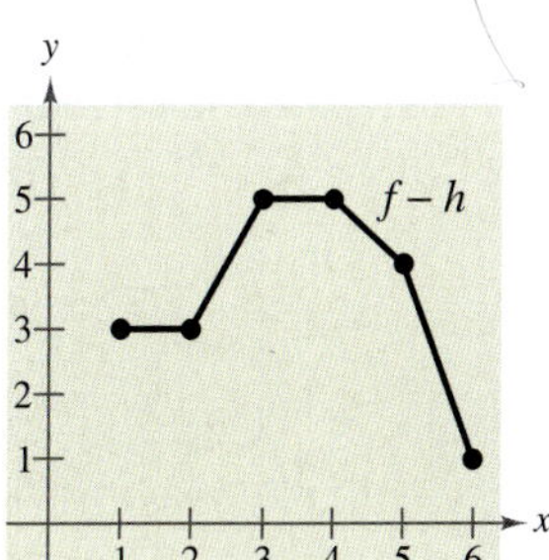

FIGURE 2.62

2.6 Exercises

In Exercises 1–4, use the graphs of f and g to graph $h(x) = (f + g)(x)$. To print an enlarged copy of the graph, go to the website *www.mathgraphs.com*.

1. 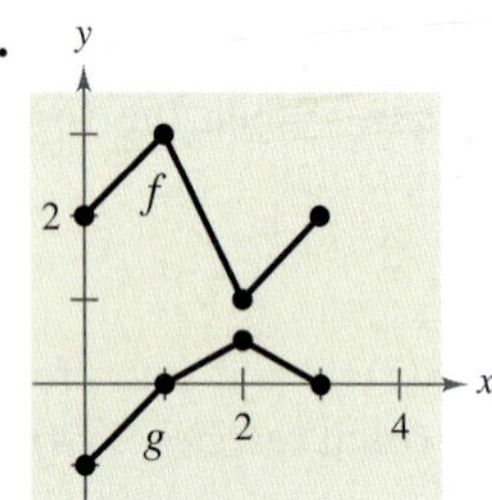

2. 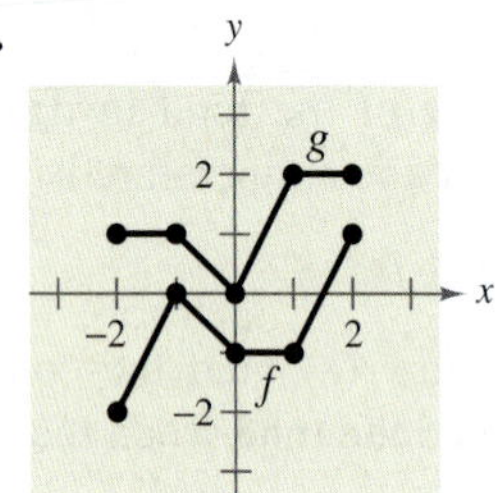

3. 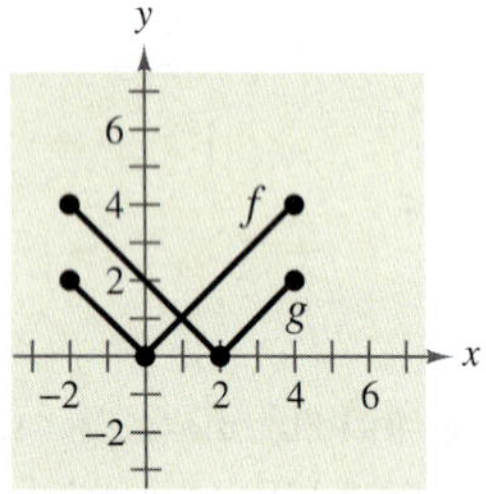

4. 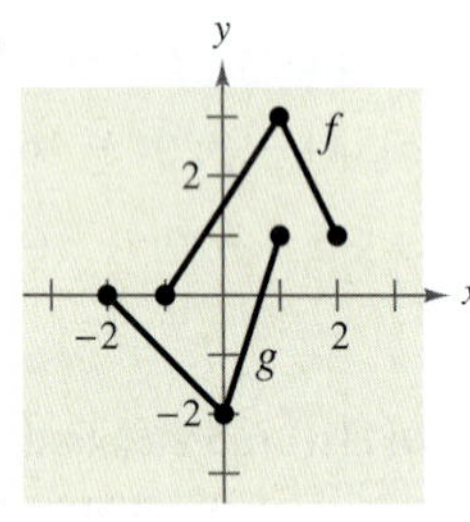

In Exercises 5–12, find (a) $(f + g)(x)$, (b) $(f - g)(x)$, (c) $(fg)(x)$, and (d) $(f/g)(x)$. What is the domain of f/g?

5. $f(x) = x + 2,\quad g(x) = x - 2$
6. $f(x) = 2x - 5,\quad g(x) = 2 - x$
7. $f(x) = x^2,\quad g(x) = 4x - 5$
8. $f(x) = 2x - 5,\quad g(x) = 4$
9. $f(x) = x^2 + 6,\quad g(x) = \sqrt{1 - x}$
10. $f(x) = \sqrt{x^2 - 4},\quad g(x) = \dfrac{x^2}{x^2 + 1}$
11. $f(x) = \dfrac{1}{x},\quad g(x) = \dfrac{1}{x^2}$
12. $f(x) = \dfrac{x}{x + 1},\quad g(x) = x^3$

In Exercises 13–24, evaluate the indicated function for $f(x) = x^2 + 1$ and $g(x) = x - 4$.

13. $(f + g)(2)$
14. $(f - g)(-1)$
15. $(f - g)(0)$
16. $(f + g)(1)$
17. $(f - g)(3t)$
18. $(f + g)(t - 2)$
19. $(fg)(6)$
20. $(fg)(-6)$
21. $\left(\dfrac{f}{g}\right)(5)$
22. $\left(\dfrac{f}{g}\right)(0)$
23. $\left(\dfrac{f}{g}\right)(-1) - g(3)$
24. $(fg)(5) + f(4)$

In Exercises 25–28, graph the functions f, g, and $f + g$ on the same set of coordinate axes.

25. $f(x) = \frac{1}{2}x,\quad g(x) = x - 1$
26. $f(x) = \frac{1}{3}x,\quad g(x) = -x + 4$
27. $f(x) = x^2,\quad g(x) = -2x$
28. $f(x) = 4 - x^2,\quad g(x) = x$

Graphical Reasoning **In Exercises 29 and 30, use a graphing utility to graph f, g, and $f + g$ in the same viewing window. Which function contributes most to the magnitude of the sum when $0 \le x \le 2$? Which function contributes most to the magnitude of the sum when $x > 6$?**

29. $f(x) = 3x,\quad g(x) = -\dfrac{x^3}{10}$
30. $f(x) = \dfrac{x}{2},\quad g(x) = \sqrt{x}$

31. ***Stopping Distance*** The research and development department of an automobile manufacturer has determined that when required to stop quickly to avoid an accident, the distance (in feet) a car travels during the driver's reaction time is given by $R(x) = \frac{3}{4}x$, where x is the speed of the car in miles per hour. The distance (in feet) traveled while the driver is braking is $B(x) = \frac{1}{15}x^2$. Find the function that represents the total stopping distance T. Graph the functions R, B, and T on the same set of coordinate axes for $0 \le x \le 60$.

32. ***Sales*** From 1997 to 2002, the sales R_1 (in thousands of dollars) for one of two restaurants owned by the same parent company can be modeled by

$$R_1 = 480 - 8t - 0.8t^2, \qquad t = 0, 1, 2, 3, 4, 5$$

where $t = 0$ represents 1997. During the same six-year period, the sales R_2 (in thousands of dollars) for the second restaurant can be modeled by

$$R_2 = 254 + 0.78t, \qquad t = 0, 1, 2, 3, 4, 5.$$

Write a function that represents the total sales of the two restaurants owned by the same parent company. Use a graphing utility to graph the total sales function.

▶ Model It

33. ***Health Care Costs*** The table shows the total amount (in billions of dollars) spent on health services and supplies in the United States (including Puerto Rico) for the years 1993 through 1999. The variables y_1, y_2, and y_3 represent out-of-pocket payments, insurance premiums, and other types of payments, respectively. (Source: Centers for Medicare and Medicaid Services)

Year	y_1	y_2	y_3
1993	148.9	295.7	39.1
1994	146.2	308.9	40.8
1995	149.2	322.3	44.8
1996	155.0	337.4	47.9
1997	165.5	355.6	52.0
1998	176.1	376.8	54.8
1999	186.5	401.2	58.9

(a) Use the *regression* feature of a graphing utility to find a quadratic model for y_1 and linear models for y_2 and y_3. Let $t = 3$ represent 1993.

(b) Find $y_1 + y_2 + y_3$. What does this sum represent?

(c) Use a graphing utility to graph y_1, y_2, y_3, and $y_1 + y_2 + y_3$ in the same viewing window.

(d) Use the model from part (b) to estimate the total amount spent on health services and supplies in the years 2003 and 2005.

34. ***Graphical Reasoning*** An electronically controlled thermostat in a home is programmed to lower the temperature automatically during the night. The temperature in the house T (in degrees Fahrenheit) is given in terms of t, the time in hours on a 24-hour clock (see figure).

(a) Explain why T is a function of t.

(b) Approximate $T(4)$ and $T(15)$.

(c) The thermostat is reprogrammed to produce a temperature H for which $H(t) = T(t - 1)$. How does this change the temperature?

(d) The thermostat is reprogrammed to produce a temperature H for which $H(t) = T(t) - 1$. How does this change the temperature?

(e) Write a piecewise-defined function that represents the graph.

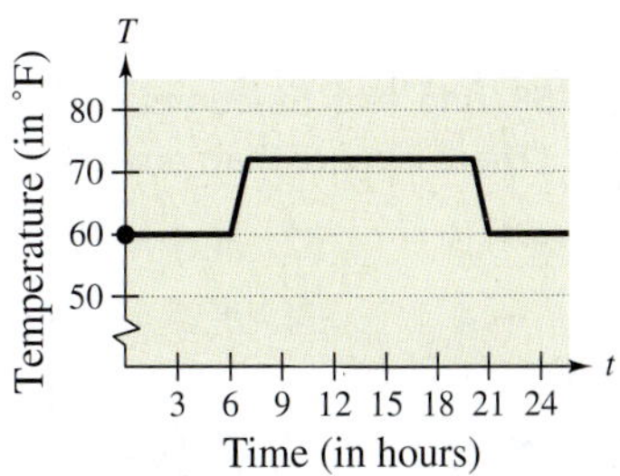

FIGURE FOR 34

In Exercises 35–38, find (a) $f \circ g$, (b) $g \circ f$, and (c) $f \circ f$.

35. $f(x) = x^2$, $\quad g(x) = x - 1$

36. $f(x) = 3x + 5$, $\quad g(x) = 5 - x$

37. $f(x) = \sqrt[3]{x - 1}$, $\quad g(x) = x^3 + 1$

38. $f(x) = x^3$, $\quad g(x) = \dfrac{1}{x}$

In Exercises 39–46, find (a) $f \circ g$ and (b) $g \circ f$. Find the domain of each function and each composite function.

39. $f(x) = \sqrt{x + 4}$, $\quad g(x) = x^2$

40. $f(x) = \sqrt[3]{x - 5}$, $\quad g(x) = x^3 + 1$

41. $f(x) = x^2 + 1$, $\quad g(x) = \sqrt{x}$

42. $f(x) = x^{2/3}$, $\quad g(x) = x^6$

43. $f(x) = |x|$, $\quad g(x) = x + 6$

44. $f(x) = |x - 4|$, $\quad g(x) = 3 - x$

45. $f(x) = \dfrac{1}{x}$, $\quad g(x) = x + 3$

46. $f(x) = \dfrac{3}{x^2 - 1}$, $\quad g(x) = x + 1$

In Exercises 47–50, use the graphs of f and g to evaluate the functions.

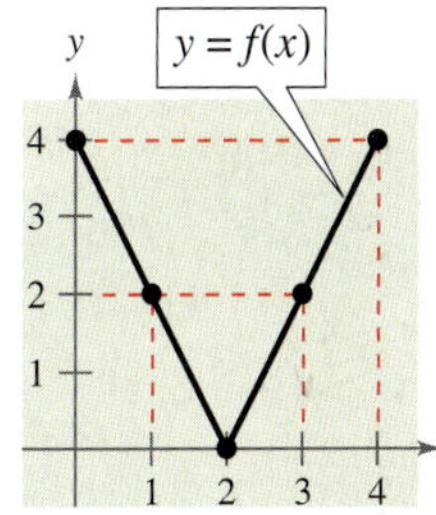

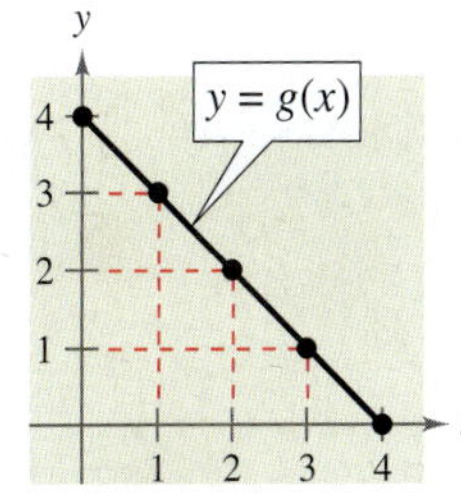

47. (a) $(f + g)(3)$ $\quad$ (b) $(f/g)(2)$

48. (a) $(f - g)(1)$ $\quad$ (b) $(fg)(4)$

49. (a) $(f \circ g)(2)$ (b) $(g \circ f)(2)$

50. (a) $(f \circ g)(1)$ (b) $(g \circ f)(3)$

In Exercises 51–58, find two functions f and g such that $(f \circ g)(x) = h(x)$. (There is more than one correct answer.)

51. $h(x) = (2x + 1)^2$

52. $h(x) = (1 - x)^3$

53. $h(x) = \sqrt[3]{x^2 - 4}$

54. $h(x) = \sqrt{9 - x}$

55. $h(x) = \dfrac{1}{x + 2}$

56. $h(x) = \dfrac{4}{(5x + 2)^2}$

57. $h(x) = \dfrac{-x^2 + 3}{4 - x^2}$

58. $h(x) = \dfrac{27x^3 + 6x}{10 - 27x^3}$

59. ***Geometry*** A square concrete foundation is prepared as a base for a cylindrical tank (see figure).

(a) Write the radius r of the tank as a function of the length x of the sides of the square.

(b) Write the area A of the circular base of the tank as a function of the radius r.

(c) Find and interpret $(A \circ r)(x)$.

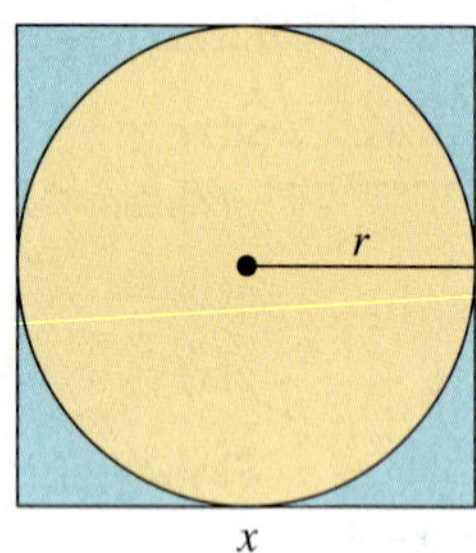

60. ***Physics*** A pebble is dropped into a calm pond, causing ripples in the form of concentric circles (see figure). The radius (in feet) of the outer ripple is $r(t) = 0.6t$, where t is the time in seconds after the pebble strikes the water. The area of the circle is given by the function $A(r) = \pi r^2$. Find and interpret $(A \circ r)(t)$.

Synthesis

True or False? **In Exercises 61 and 62, determine whether the statement is true or false. Justify your answer.**

61. If $f(x) = x + 1$ and $g(x) = 6x$, then $(f \circ g)(x) = (g \circ f)(x)$.

62. If you are given two functions $f(x)$ and $g(x)$, you can calculate $(f \circ g)(x)$ if and only if the range of g is a subset of the domain of f.

63. ***Think About It*** You are a sales representative for an automobile manufacturer. You are paid an annual salary, plus a bonus of 3% of your sales over \$500,000. Consider the two functions

$$f(x) = x - 500{,}000 \quad \text{and} \quad g(x) = 0.03x.$$

If x is greater than \$500,000, which of the following represents your bonus? Explain your reasoning.

(a) $f(g(x))$ (b) $g(f(x))$

64. ***Proof*** Prove that the product of two odd functions is an even function, and that the product of two even functions is an even function.

65. ***Conjecture*** Use examples to hypothesize whether the product of an odd function and an even function is even or odd. Then prove your hypothesis.

Review

66. Find the domain of the function.

$$f(x) = \frac{x}{5x + 7}$$

Average Rate of Change **In Exercises 67–70, find the difference quotient**

$$\frac{f(x + h) - f(x)}{h}$$

and simplify your answer.

67. $f(x) = 3x - 4$

68. $f(x) = 1 - x^2$

69. $f(x) = \dfrac{4}{x}$

70. $f(x) = \sqrt{2x + 1}$

In Exercises 71–74, find an equation of the line that passes through the given point and has the indicated slope. Sketch the line.

71. $(2, -4), m = 3$

72. $(-6, 3), m = -1$

73. $(8, -1), m = -\frac{3}{2}$

74. $(7, 0), m = \frac{5}{7}$

2.7 Inverse Functions

▶ **What you should learn**

- How to find inverse functions informally and verify that two functions are inverse functions of each other
- How to use graphs of functions to determine whether functions have inverse functions
- How to use the Horizontal Line Test to determine if functions are one-to-one
- How to find inverse functions algebraically

▶ **Why you should learn it**

Inverse functions can be used to model and solve real-life problems. For instance, in Exercise 79 on page 245, an inverse function can be used to determine the year in which there were a given number of households in the United States.

Michelle Bridwell/PhotoEdit

Inverse Functions

Recall from Section 2.2 that a function can be represented by a set of ordered pairs. For instance, the function $f(x) = x + 4$ from the set $A = \{1, 2, 3, 4\}$ to the set $B = \{5, 6, 7, 8\}$ can be written as follows.

$$f(x) = x + 4: \ \{(1, 5), (2, 6), (3, 7), (4, 8)\}$$

In this case, by interchanging the first and second coordinates of each of these ordered pairs, you can form the **inverse function** of f, which is denoted by f^{-1}. It is a function from the set B to the set A, and can be written as follows.

$$f^{-1}(x) = x - 4: \ \{(5, 1), (6, 2), (7, 3), (8, 4)\}$$

Note that the domain of f is equal to the range of f^{-1}, and vice versa, as shown in Figure 2.63. Also note that the functions f and f^{-1} have the effect of "undoing" each other. In other words, when you form the composition of f with f^{-1} or the composition of f^{-1} with f, you obtain the identity function.

$$f(f^{-1}(x)) = f(x - 4) = (x - 4) + 4 = x$$

$$f^{-1}(f(x)) = f^{-1}(x + 4) = (x + 4) - 4 = x$$

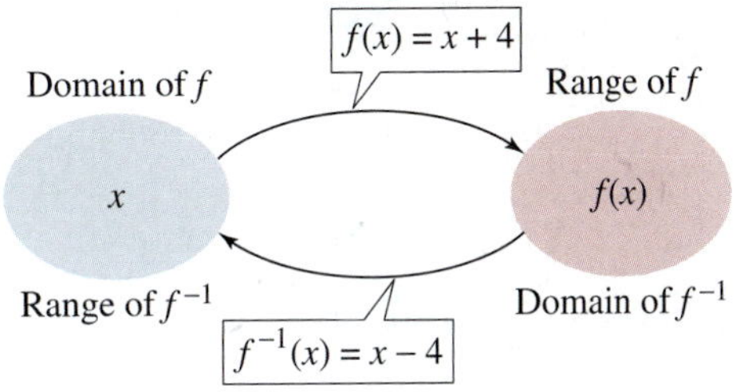

FIGURE 2.63

Example 1 ▶ Finding Inverse Functions Informally 

Find the inverse function of $f(x) = 4x$. Then verify that both $f(f^{-1}(x))$ and $f^{-1}(f(x))$ are equal to the identity function.

Solution

The function f *multiplies* each input by 4. To "undo" this function, you need to *divide* each input by 4. So, the inverse function of $f(x) = 4x$ is

$$f^{-1}(x) = \frac{x}{4}.$$

You can verify that both $f(f^{-1}(x))$ and $f^{-1}(f(x))$ are equal to the identity function as follows.

$$f(f^{-1}(x)) = f\left(\frac{x}{4}\right) = 4\left(\frac{x}{4}\right) = x \qquad f^{-1}(f(x)) = f^{-1}(4x) = \frac{4x}{4} = x$$

Exploration

Consider the functions

$$f(x) = x + 2$$

and

$$f^{-1}(x) = x - 2.$$

Evaluate $f(f^{-1}(x))$ and $f^{-1}(f(x))$ for the indicated values of x. What can you conclude about the functions?

x	-10	0	7	45
$f(f^{-1}(x))$				
$f^{-1}(f(x))$				

Definition of Inverse Function

Let f and g be two functions such that

$$f(g(x)) = x \quad \text{for every } x \text{ in the domain of } g$$

and

$$g(f(x)) = x \quad \text{for every } x \text{ in the domain of } f.$$

Under these conditions, the function g is the **inverse function** of the function f. The function g is denoted by f^{-1} (read "f-inverse"). So,

$$f(f^{-1}(x)) = x \quad \text{and} \quad f^{-1}(f(x)) = x.$$

The domain of f must be equal to the range of f^{-1}, and the range of f must be equal to the domain of f^{-1}.

Don't be confused by the use of -1 to denote the inverse function f^{-1}. In this text, whenever f^{-1} is written, it *always* refers to the inverse function of the function f and *not* to the reciprocal of $f(x)$.

If the function g is the inverse function of the function f, it must also be true that the function f is the inverse function of the function g. For this reason, you can say that the functions f and g are *inverse functions of each other*.

Example 2 ▶ Verifying Inverse Functions 

Which of the functions is the inverse function of $f(x) = \dfrac{5}{x-2}$?

$$g(x) = \frac{x-2}{5} \qquad h(x) = \frac{5}{x} + 2$$

Solution

By forming the composition of f with g, you have

$$f(g(x)) = f\left(\frac{x-2}{5}\right)$$

$$= \frac{5}{\left(\dfrac{x-2}{5}\right) - 2} \qquad \text{Substitute } \frac{x-2}{5} \text{ for } x.$$

$$= \frac{25}{x-12} \neq x.$$

Because this composition is not equal to the identity function x, it follows that g *is not* the inverse function of f. By forming the composition of f with h, you have

$$f(h(x)) = f\left(\frac{5}{x} + 2\right) = \frac{5}{\left(\dfrac{5}{x} + 2\right) - 2} = \frac{5}{\left(\dfrac{5}{x}\right)} = x.$$

So, it appears that h *is* the inverse function of f. You can confirm this by showing that the composition of h with f is also equal to the identity function.

The Graph of an Inverse Function 

The graphs of a function f and its inverse function f^{-1} are related to each other in the following way. If the point (a, b) lies on the graph of f, then the point (b, a) must lie on the graph of f^{-1}, and vice versa. This means that the graph of f^{-1} is a *reflection* of the graph of f in the line $y = x$, as shown in Figure 2.64.

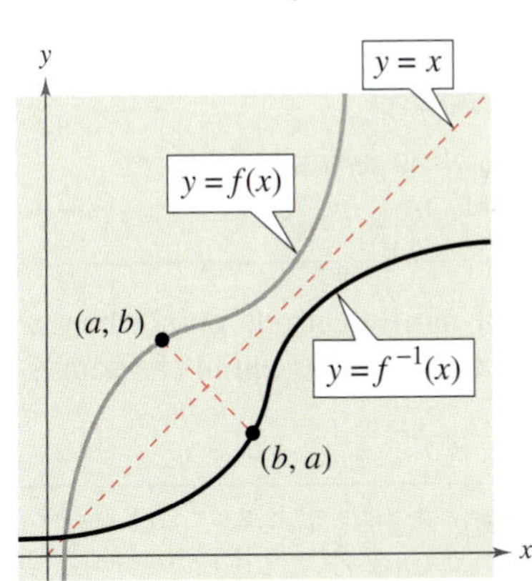

FIGURE 2.64

Example 3 ▶ The Graphs of f and f^{-1}

Sketch the graphs of the inverse functions $f(x) = 2x - 3$ and $f^{-1}(x) = \frac{1}{2}(x + 3)$ on the same rectangular coordinate system and show that the graphs are reflections of each other in the line $y = x$.

Solution

The graphs of f and f^{-1} are shown in Figure 2.65. It appears that the graphs are reflections of each other in the line $y = x$. You can further verify this reflective property by testing a few points on each graph. Note in the following list that if the point (a, b) is on the graph of f, the point (b, a) is on the graph of f^{-1}.

Graph of $f(x) = 2x - 3$	*Graph of* $f^{-1}(x) = \frac{1}{2}(x + 3)$
$(-1, -5)$	$(-5, -1)$
$(0, -3)$	$(-3, 0)$
$(1, -1)$	$(-1, 1)$
$(2, 1)$	$(1, 2)$
$(3, 3)$	$(3, 3)$

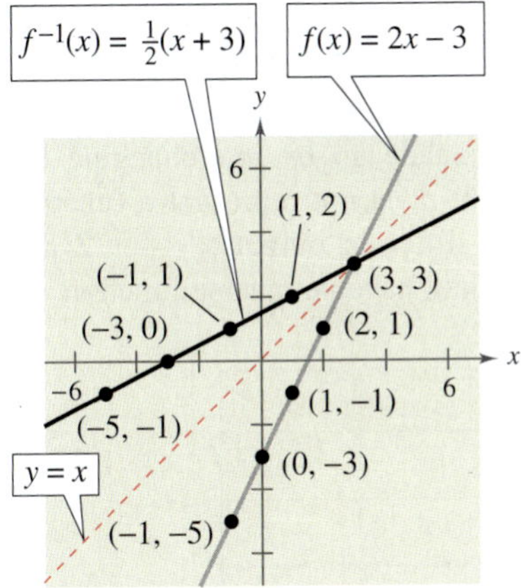

FIGURE 2.65

Example 4 ▶ Finding Inverse Functions Graphically

Sketch the graphs of the inverse functions $f(x) = x^2 (x \geq 0)$ and $f^{-1}(x) = \sqrt{x}$ on the same rectangular coordinate system and show that the graphs are reflections of each other in the line $y = x$.

Solution

The graphs of f and f^{-1} are shown in Figure 2.66. It appears that the graphs are reflections of each other in the line $y = x$. You can further verify this reflective property by testing a few points on each graph. Note in the following list that if the point (a, b) is on the graph of f, the point (b, a) is on the graph of f^{-1}.

Graph of $f(x) = x^2, \quad x \geq 0$	*Graph of* $f^{-1}(x) = \sqrt{x}$
$(0, 0)$	$(0, 0)$
$(1, 1)$	$(1, 1)$
$(2, 4)$	$(4, 2)$
$(3, 9)$	$(9, 3)$

Try showing that $f(f^{-1}(x)) = x$ and $f^{-1}(f(x)) = x$.

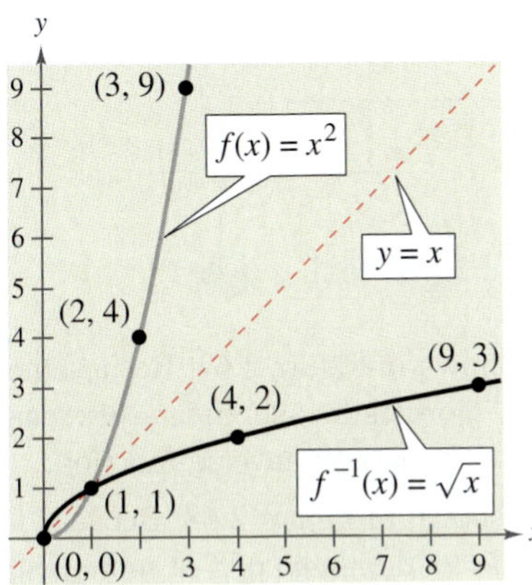

FIGURE 2.66

One-to-One Functions

The reflective property of the graphs of inverse functions gives you a nice *geometric* test for determining whether a function has an inverse function. This test is called the **Horizontal Line Test** for inverse functions.

> **Horizontal Line Test for Inverse Functions**
>
> A function f has an inverse function if and only if no *horizontal* line intersects the graph of f at more than one point.

If no horizontal line intersects the graph of f at more than one point, then no y-value is matched with more than one x-value. This is the essential characteristic of what are called **one-to-one** functions.

> **One-to-One Functions**
>
> A function f is **one-to-one** if each value of the dependent variable corresponds to exactly one value of the independent variable. A function f has an inverse function if and only if f is one-to-one.

Consider the function $f(x) = x^2$. The table on the left is a table of values for $f(x) = x^2$. The table of values on the right is made up by interchanging the columns of the first table. The table on the right does not represent a function because the input $x = 4$ is matched with two different outputs: $y = -2$ and $y = 2$. So, $f(x) = x^2$ is not one-to-one and does not have an inverse function.

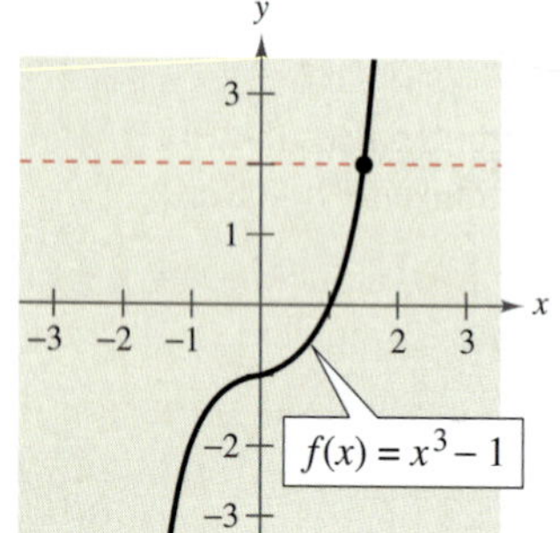

FIGURE 2.67

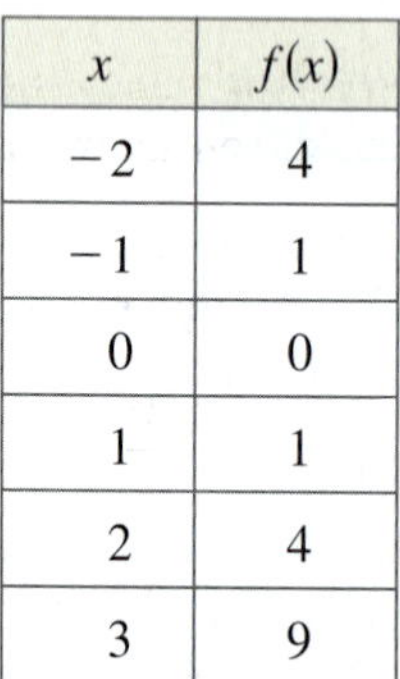

x	$f(x)$
-2	4
-1	1
0	0
1	1
2	4
3	9

x	y
4	-2
1	-1
0	0
1	1
4	2
9	3

FIGURE 2.68

Example 5 ▶ Applying the Horizontal Line Test 

a. The graph of the function $f(x) = x^3 - 1$ is shown in Figure 2.67. Because no horizontal line intersects the graph of f at more than one point, you can conclude that f *is* a one-to-one function and *does* have an inverse function.

b. The graph of the function $f(x) = x^2 - 1$ is shown in Figure 2.68. Because it is possible to find a horizontal line that intersects the graph of f at more than one point, you can conclude that f *is not* a one-to-one function and *does not* have an inverse function.

Finding Inverse Functions Algebraically 

For simple functions (such as the one in Example 1), you can find inverse functions by inspection. For more complicated functions, however, it is best to use the following guidelines. The key step in these guidelines is Step 3—interchanging the roles of x and y. This step corresponds to the fact that inverse functions have ordered pairs with the coordinates reversed.

Finding an Inverse Function

1. Use the Horizontal Line Test to decide whether f has an inverse function.
2. In the equation for $f(x)$, replace $f(x)$ by y.
3. Interchange the roles of x and y, and solve for y.
4. Replace y by $f^{-1}(x)$ in the new equation.
5. Verify that f and f^{-1} are inverse functions of each other by showing that the domain of f is equal to the range of f^{-1}, the range of f is equal to the domain of f^{-1}, and $f(f^{-1}(x)) = x = f^{-1}(f(x))$.

STUDY TIP

Note what happens when you try to find the inverse function of a function that is not one-to-one.

$f(x) = x^2 + 1$ Original function

$y = x^2 + 1$ Replace $f(x)$ by y.

$x = y^2 + 1$ Interchange x and y.

$x - 1 = y^2$ Isolate y-term.

$y = \pm\sqrt{x - 1}$ Solve for y.

You obtain two y-values for each x.

Example 6 ▶ Finding an Inverse Function Algebraically 

Find the inverse function of

$$f(x) = \frac{5 - 3x}{2}.$$

Solution

The graph of f is a line, as shown in Figure 2.69. This graph passes the Horizontal Line Test. So, you know that f is one-to-one and has an inverse function.

$f(x) = \dfrac{5 - 3x}{2}$ Write original function.

$y = \dfrac{5 - 3x}{2}$ Replace $f(x)$ by y.

$x = \dfrac{5 - 3y}{2}$ Interchange x and y.

$2x = 5 - 3y$ Multiply each side by 2.

$3y = 5 - 2x$ Isolate the y-term.

$y = \dfrac{5 - 2x}{3}$ Solve for y.

$f^{-1}(x) = \dfrac{5 - 2x}{3}$ Replace y by $f^{-1}(x)$.

Note that both f and f^{-1} have domains and ranges that consist of the entire set of real numbers. Check that $f(f^{-1}(x)) = x$ and $f^{-1}(f(x)) = x$.

Exploration

Restrict the domain of $f(x) = x^2 + 1$ to $x \geq 0$. Use a graphing utility to graph the function. Does the restricted function have an inverse function? Explain.

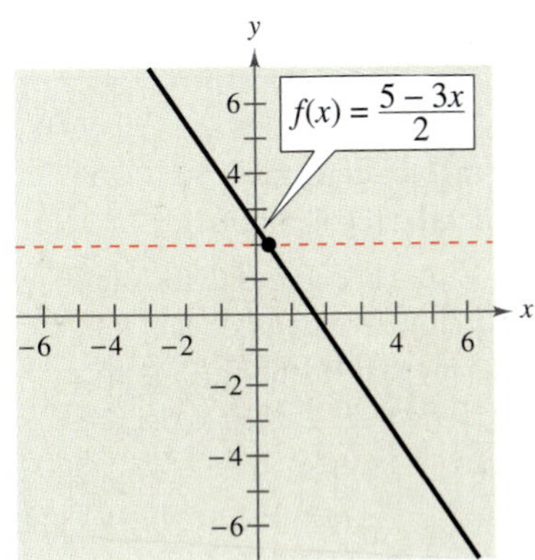

FIGURE 2.69

Example 7 ▶ Finding an Inverse Function 

Find the inverse function of

$$f(x) = \sqrt[3]{x+1}.$$

Solution

The graph of f is a curve, as shown in Figure 2.70. Because this graph passes the Horizontal Line Test, you know that f is one-to-one and has an inverse function.

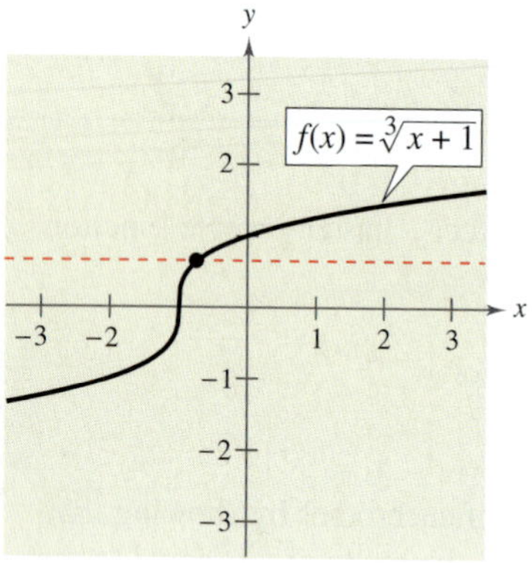

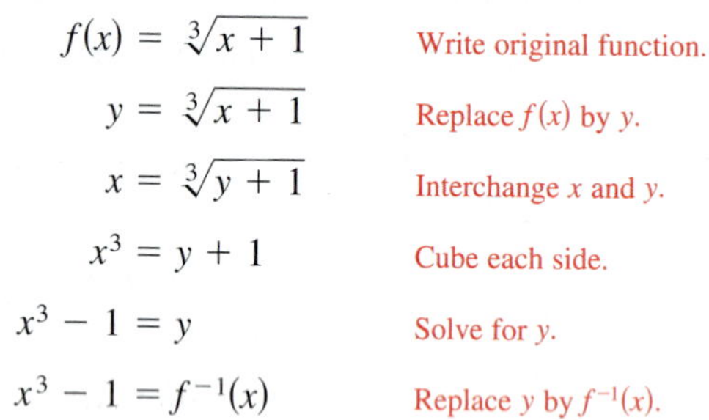

$$f(x) = \sqrt[3]{x+1} \quad \text{Write original function.}$$

$$y = \sqrt[3]{x+1} \quad \text{Replace } f(x) \text{ by } y.$$

$$x = \sqrt[3]{y+1} \quad \text{Interchange } x \text{ and } y.$$

$$x^3 = y + 1 \quad \text{Cube each side.}$$

$$x^3 - 1 = y \quad \text{Solve for } y.$$

$$x^3 - 1 = f^{-1}(x) \quad \text{Replace } y \text{ by } f^{-1}(x).$$

FIGURE 2.70

Both f and f^{-1} have domains and ranges that consist of the entire set of real numbers. You can verify this result numerically as shown in the tables below.

x	$f(x)$
-28	-3
-9	-2
-2	-1
-1	0
0	1
7	2
26	3

x	$f^{-1}(x)$
-3	-28
-2	-9
-1	-2
0	-1
1	0
2	7
3	26

Writing ABOUT MATHEMATICS

The Existence of an Inverse Function Write a short paragraph describing why the following functions do or do not have inverse functions.

a. Let x represent the retail price of an item (in dollars), and let $f(x)$ represent the sales tax on the item. Assume that the sales tax is 6% of the retail price *and* that the sales tax is rounded to the nearest cent. Does this function have an inverse function? (*Hint:* Can you undo this function? For instance, if you know that the sales tax is \$0.12, can you determine exactly what the retail price is?)

b. Let x represent the temperature in degrees Celsius, and let $f(x)$ represent the temperature in degrees Fahrenheit. Does this function have an inverse function? (*Hint:* The formula for converting from degrees Celsius to degrees Fahrenheit is $F = \frac{9}{5}C + 32$.)

2.7 Exercises

In Exercises 1–4, match the graph of the function with the graph of its inverse function. [The graphs of the inverse functions are labeled (a), (b), (c), and (d).]

(a)

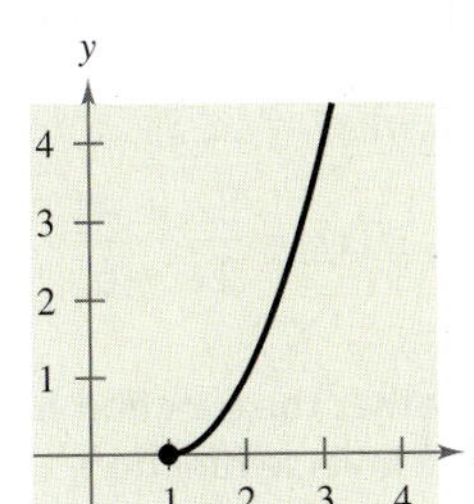

(b)

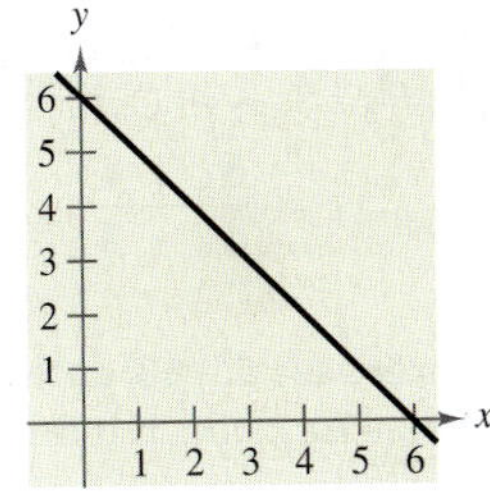

(c)

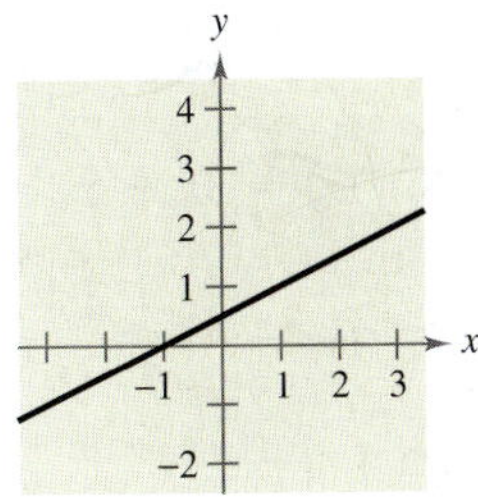

(d)

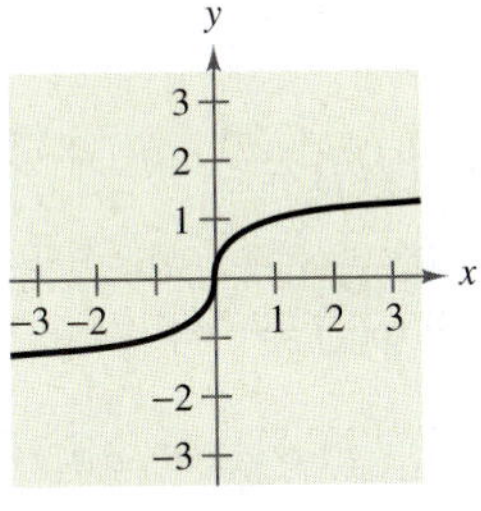

1.

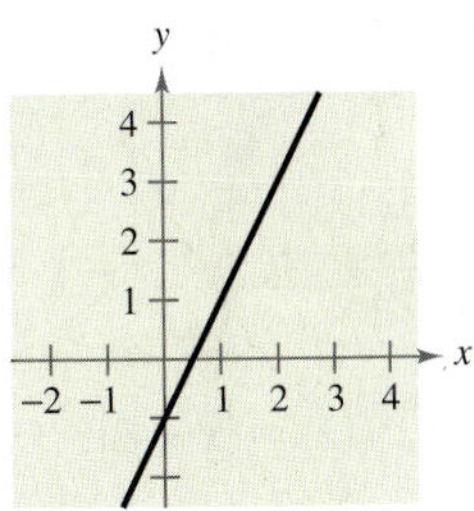

2.

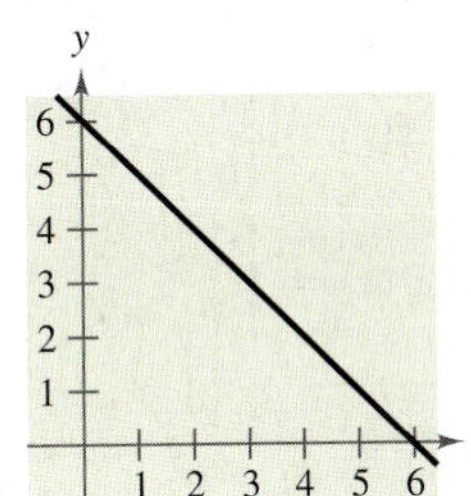

3.

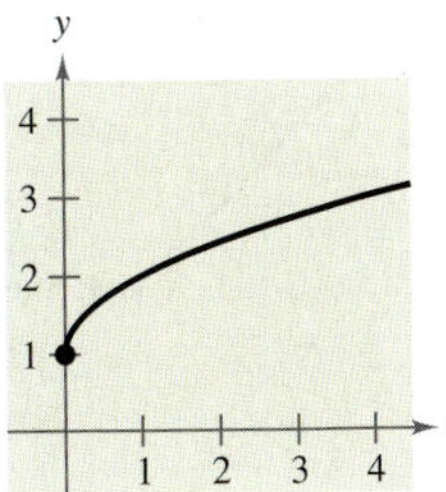

4.

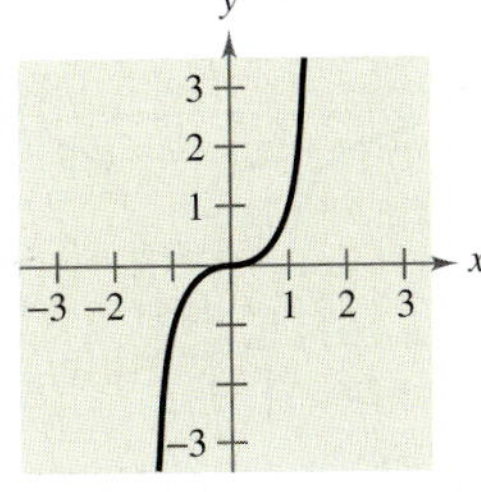

In Exercises 5–12, find the inverse function of *f* informally. Verify that $f(f^{-1}(x)) = x$ and $f^{-1}(f(x)) = x$.

5. $f(x) = 6x$

6. $f(x) = \frac{1}{3}x$

7. $f(x) = x + 9$

8. $f(x) = x - 4$

9. $f(x) = 3x + 1$

10. $f(x) = \dfrac{x - 1}{5}$

11. $f(x) = \sqrt[3]{x}$

12. $f(x) = x^5$

In Exercises 13–24, show that *f* and *g* are inverse functions (a) algebraically and (b) graphically.

13. $f(x) = 2x, \quad g(x) = \dfrac{x}{2}$

14. $f(x) = x - 5, \quad g(x) = x + 5$

15. $f(x) = 7x + 1, \quad g(x) = \dfrac{x - 1}{7}$

16. $f(x) = 3 - 4x, \quad g(x) = \dfrac{3 - x}{4}$

17. $f(x) = \dfrac{x^3}{8}, \quad g(x) = \sqrt[3]{8x}$

18. $f(x) = \dfrac{1}{x}, \quad g(x) = \dfrac{1}{x}$

19. $f(x) = \sqrt{x - 4}, \quad g(x) = x^2 + 4, \quad x \ge 0$

20. $f(x) = 1 - x^3, \quad g(x) = \sqrt[3]{1 - x}$

21. $f(x) = 9 - x^2, \quad x \ge 0, \quad g(x) = \sqrt{9 - x}, \quad x \le 9$

22. $f(x) = \dfrac{1}{1 + x}, \quad x \ge 0$

$g(x) = \dfrac{1 - x}{x}, \quad 0 < x \le 1$

23. $f(x) = \dfrac{x - 1}{x + 5}, \quad g(x) = -\dfrac{5x + 1}{x - 1}$

24. $f(x) = \dfrac{x + 3}{x - 2}, \quad g(x) = \dfrac{2x + 3}{x - 1}$

In Exercises 25 and 26, does the function have an inverse function?

25.

x	$f(x)$
−1	−2
0	1
1	2
2	1
3	−2
4	−6

26.

x	$f(x)$
−3	10
−2	6
−1	4
0	1
2	−3
3	−10

In Exercises 27 and 28, use the table of values for $y = f(x)$ to complete a table for $y = f^{-1}(x)$.

27.

x	-2	-1	0	1	2	3
$f(x)$	-2	0	2	4	6	8

28.

x	-3	-2	-1	0	1	2
$f(x)$	-10	-7	-4	-1	2	5

In Exercises 29–32, does the function have an inverse function?

29.

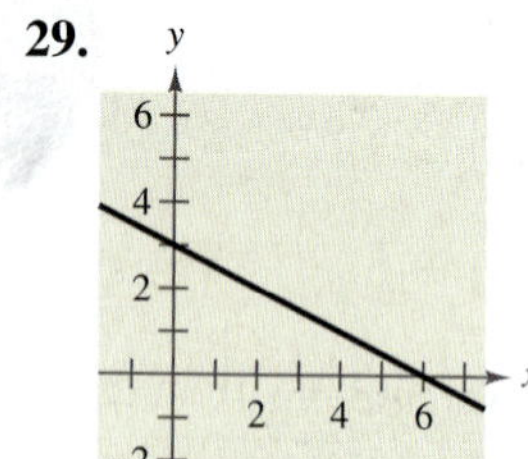

30.

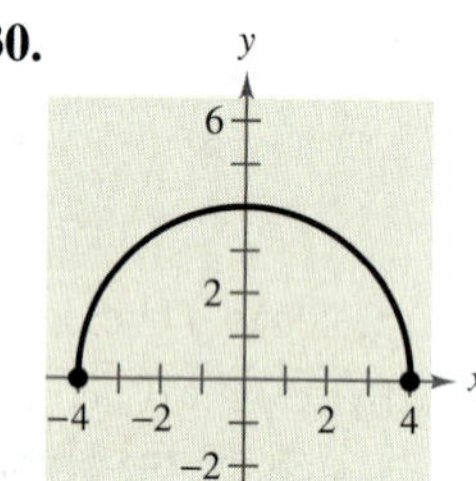

31.

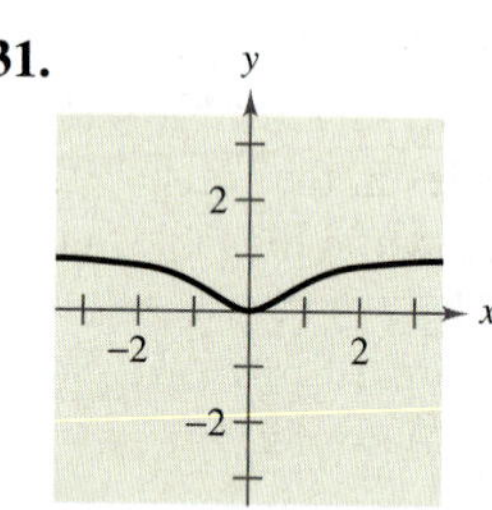

32.

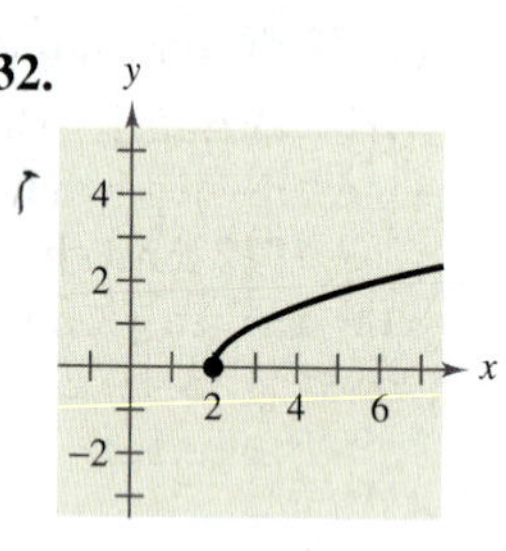

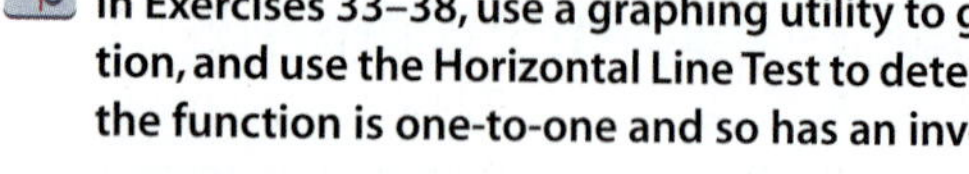

In Exercises 33–38, use a graphing utility to graph the function, and use the Horizontal Line Test to determine whether the function is one-to-one and so has an inverse function.

33. $g(x) = \dfrac{4 - x}{6}$

34. $f(x) = 10$

35. $h(x) = |x + 4| - |x - 4|$

36. $g(x) = (x + 5)^3$

37. $f(x) = -2x\sqrt{16 - x^2}$

38. $f(x) = \frac{1}{8}(x + 2)^2 - 1$

In Exercises 39–54, find the inverse function of *f*. Then graph both *f* and f^{-1} on the same set of coordinate axes.

39. $f(x) = 2x - 3$

40. $f(x) = 3x + 1$

41. $f(x) = x^5 - 2$

42. $f(x) = x^3 + 1$

43. $f(x) = \sqrt{x}$

44. $f(x) = x^2, \quad x \geq 0$

45. $f(x) = \sqrt{4 - x^2}, \quad 0 \leq x \leq 2$

46. $f(x) = x^2 - 2, \quad x \leq 0$

47. $f(x) = \dfrac{4}{x}$

48. $f(x) = -\dfrac{2}{x}$

49. $f(x) = \dfrac{x + 1}{x - 2}$

50. $f(x) = \dfrac{x - 3}{x + 2}$

51. $f(x) = \sqrt[3]{x - 1}$

52. $f(x) = x^{3/5}$

53. $f(x) = \dfrac{6x + 4}{4x + 5}$

54. $f(x) = \dfrac{8x - 4}{2x + 6}$

In Exercises 55–68, determine whether the function has an inverse function. If it does, find the inverse function.

55. $f(x) = x^4$

56. $f(x) = \dfrac{1}{x^2}$

57. $g(x) = \dfrac{x}{8}$

58. $f(x) = 3x + 5$

59. $p(x) = -4$

60. $f(x) = \dfrac{3x + 4}{5}$

61. $f(x) = (x + 3)^2, \quad x \geq -3$

62. $q(x) = (x - 5)^2$

63. $f(x) = \begin{cases} x + 3, & x < 0 \\ 6 - x, & x \geq 0 \end{cases}$

64. $f(x) = \begin{cases} -x, & x \leq 0 \\ x^2 - 3x, & x > 0 \end{cases}$

65. $h(x) = -\dfrac{4}{x^2}$

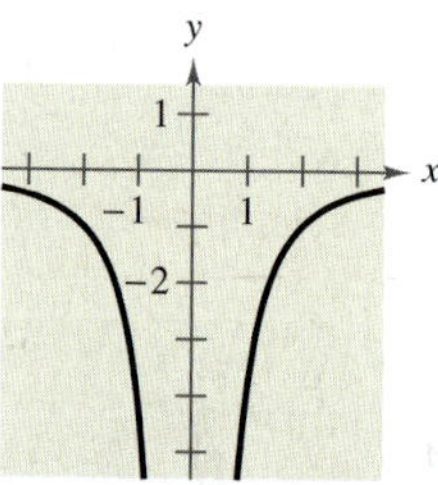

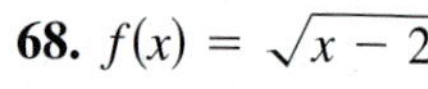

66. $f(x) = |x - 2|, \quad x \leq 2$

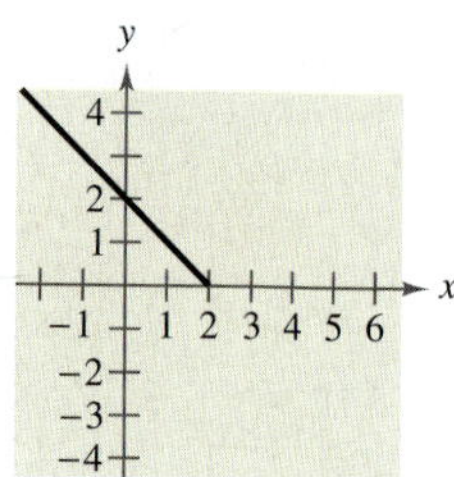

67. $f(x) = \sqrt{2x + 3}$

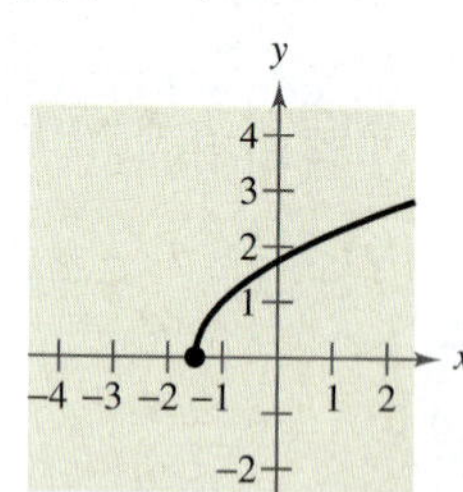

68. $f(x) = \sqrt{x - 2}$

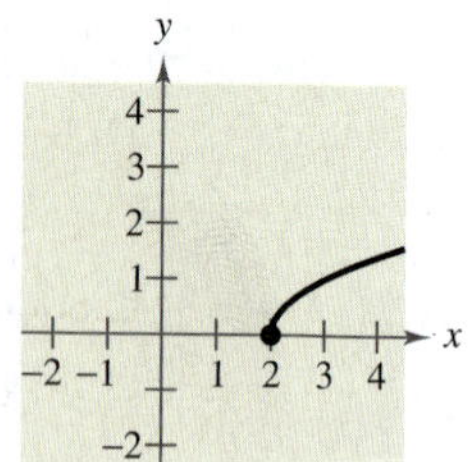

In Exercises 69–74, use the functions $f(x) = \frac{1}{8}x - 3$ and $g(x) = x^3$ to find the indicated value or function.

69. $(f^{-1} \circ g^{-1})(1)$

70. $(g^{-1} \circ f^{-1})(-3)$

71. $(f^{-1} \circ f^{-1})(6)$

72. $(g^{-1} \circ g^{-1})(-4)$

73. $(f \circ g)^{-1}$

74. $g^{-1} \circ f^{-1}$

In Exercises 75–78, use the functions $f(x) = x + 4$ and $g(x) = 2x - 5$ to find the specified function.

75. $g^{-1} \circ f^{-1}$

76. $f^{-1} \circ g^{-1}$

77. $(f \circ g)^{-1}$

78. $(g \circ f)^{-1}$

▶ Model It

79. ***U.S. Households*** The number of households f (in thousands) in the United States from 1994 to 2000 are shown in the table. The time (in years) is given by t, with $t = 4$ corresponding to 1994. (Source: U.S. Census Bureau)

Year, t	Households, $f(t)$
4	97,107
5	98,990
6	99,627
7	101,018
8	102,528
9	103,874
10	104,705

(a) Find $f^{-1}(103{,}874)$.

(b) What does f^{-1} mean in the context of the problem?

(c) Use the *regression* feature of a graphing utility to find a linear model for the data, $y = mx + b$. (Round m and b to two decimal places.)

(d) Algebraically find the inverse function of the linear model in part (c).

(e) Use the inverse function of the linear model you found in part (d) to approximate $f^{-1}(111, 254)$.

80. ***Bottled Water Consumption*** The per capita consumption f (in gallons) of bottled water in the United States from 1994 through 1999 is shown in the table. The time (in years) is given by t, with $t = 4$ corresponding to 1994. (Source: U.S. Department of Agriculture)

t	$f(t)$
4	10.7
5	11.6
6	12.5
7	13.1
8	16.0
9	18.1

(a) Does f^{-1} exist?

(b) If f^{-1} exists, what does it represent in the context of the problem?

(c) If f^{-1} exists, find $f^{-1}(16.0)$.

81. ***Miles Traveled*** The total number f (in billions) of miles traveled by motor vehicles in the United States from 1992 through 1999 is shown in the table below. The time (in years) is given by t, with $t = 2$ corresponding to 1992. (Source: U.S. Federal Highway Administration)

Year, t	Miles traveled, $f(t)$
2	2247
3	2296
4	2358
5	2423
6	2486
7	2562
8	2632
9	2691

(a) Does f^{-1} exist?

(b) If f^{-1} exists, what does it mean in the context of the problem?

(c) If f^{-1} exists, find $f^{-1}(2632)$.

(d) If the table was extended to 2000 and if the total number of miles traveled by motor vehicles for that year was 2423 billion, would f^{-1} exist? Explain.

82. ***Hourly Wage*** Your wage is \$8.00 per hour plus \$0.75 for each unit produced per hour. So, your hourly wage y in terms of the number of units produced is $y = 8 + 0.75x$.

(a) Find the inverse function.

(b) What does each variable represent in the inverse function?

(c) Determine the number of units produced when your hourly wage is \$22.25.

83. ***Diesel Mechanics*** The function

$$y = 0.03x^2 + 245.50, \qquad 0 < x < 100$$

approximates the exhaust temperature y in degrees Fahrenheit, where x is the percent load for a diesel engine.

(a) Find the inverse function. What does each variable represent in the inverse function?

(b) Use a graphing utility to graph the inverse function.

(c) The exhaust temperature of the engine must not exceed 500 degrees Fahrenheit. What is the percent load interval?

84. ***Cost*** You need a total of 50 pounds of two types of ground beef costing \$1.25 and \$1.60 per pound, respectively. A model for the total cost y of the two types of beef is

$$y = 1.25x + 1.60(50 - x)$$

where x is the number of pounds of the less expensive ground beef.

(a) Find the inverse function of the cost function. What does each variable represent in the inverse function?

(b) Use the context of the problem to determine the domain of the inverse function.

(c) Determine the number of pounds of the less expensive ground beef purchased when the total cost is \$73.

Synthesis

True or False? **In Exercises 85 and 86, determine whether the statement is true or false. Justify your answer.**

85. If f is an even function, f^{-1} exists.

86. If the inverse function of f exists and the graph of f has a y-intercept, the y-intercept of f is an x-intercept of f^{-1}.

In Exercises 87–90, use the graph of the function f to create a table of values for the given points. Then create a second table that can be used to find f^{-1}, and sketch the graph of f^{-1} if possible.

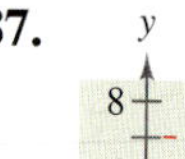

87.

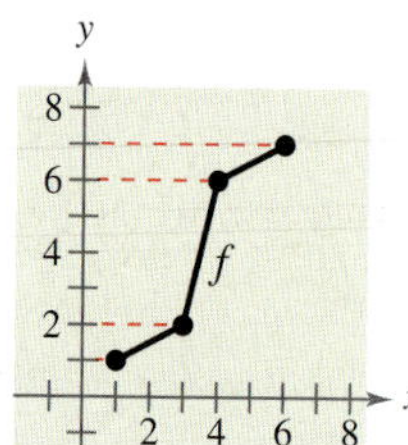

88.

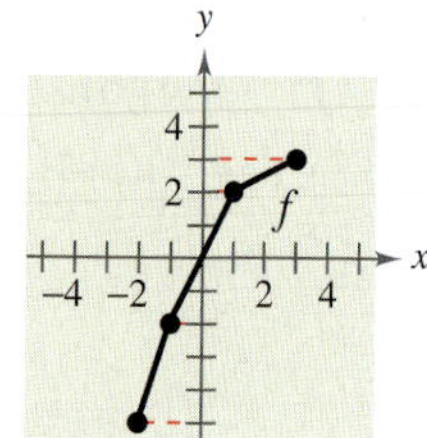

89.

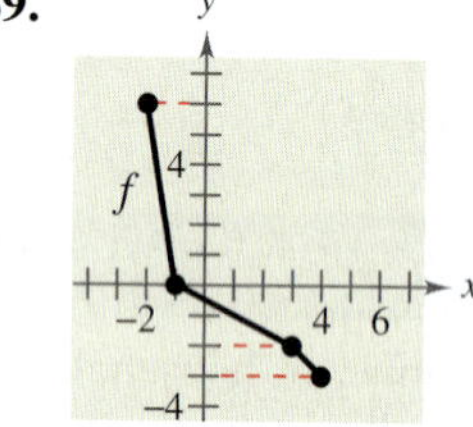

90.

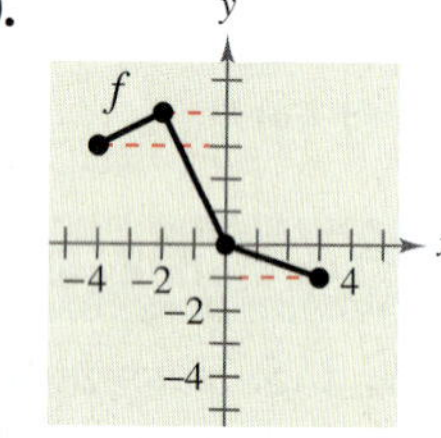

91. ***Think About It*** The function

$$f(x) = k(2 - x - x^3)$$

has an inverse function, and $f^{-1}(3) = -2$. Find k.

92. ***Think About It*** The function

$$f(x) = k(x^3 + 3x - 4)$$

has an inverse function, and $f^{-1}(-5) = 2$. Find k.

Review

In Exercises 93–100, solve the equation by any convenient method.

93. $x^2 = 64$

94. $(x - 5)^2 = 8$

95. $4x^2 - 12x + 9 = 0$

96. $9x^2 + 12x + 3 = 0$

97. $x^2 - 6x + 4 = 0$

98. $2x^2 - 4x - 6 = 0$

99. $50 + 5x = 3x^2$

100. $2x^2 + 4x - 9 = 2(x - 1)^2$

101. Find two consecutive positive even integers whose product is 288.

102. ***Geometry*** A triangular sign has a height that is twice its base. The area of the sign is 10 square feet. Find the base and height of the sign

Chapter Summary

▶ *What* did you learn?

Section 2.1	Review Exercises
☐ How to use slope to graph linear equations in two variables	1–14
☐ How to find slopes of lines	15–18
☐ How to write linear equations in two variables	19–26
☐ How to use slope to identify parallel and perpendicular lines	27, 28
☐ How to use linear equations in two variables to model and solve real-life problems	29–32
Section 2.2	
☐ How to determine whether relations between two variables are functions	33–38
☐ How to use function notation and evaluate functions	39–42
☐ How to find the domains of functions	43–48
☐ How to find the difference quotients	49, 50
☐ How to use functions to model and solve real-life problems	51–54
Section 2.3	
☐ How to use the Vertical Line Test for functions	55–58
☐ How to find the zeros of functions	59–62
☐ How to determine intervals on which functions are increasing or decreasing	63, 64
☐ How to identify even and odd functions	65–68
Section 2.4	
☐ How to identify and graph linear, squaring, cubic, square root, reciprocal, step, and other piecewise-defined functions	69–82
☐ How to recognize graphs of common functions	83, 84
Section 2.5	
☐ How to use vertical and horizontal shifts to sketch graphs of functions	85–88
☐ How to use reflections to sketch graphs of functions	89–94
☐ How to use nonrigid transformations to sketch graphs of functions	95–98
Section 2.6	
☐ How to add, subtract, multiply, and divide functions	99, 100
☐ How to find the composition of one function with another function	101–104
☐ How to use combinations of functions to model and solve real-life problems	105, 106
Section 2.7	
☐ How to find inverse functions informally and verify that two functions are inverse functions of each other	107, 108
☐ How to use graphs to determine whether functions have inverse functions	109, 110
☐ How to use the Horizontal Line Test to determine if functions are one-to-one	111–114
☐ How to find inverse functions algebraically	115–120

Review Exercises

2.1 In Exercises 1 and 2, identify the line that has each slope.

1. (a) $m = \frac{3}{2}$
(b) $m = 0$
(c) $m = -3$
(d) $m = -\frac{1}{5}$

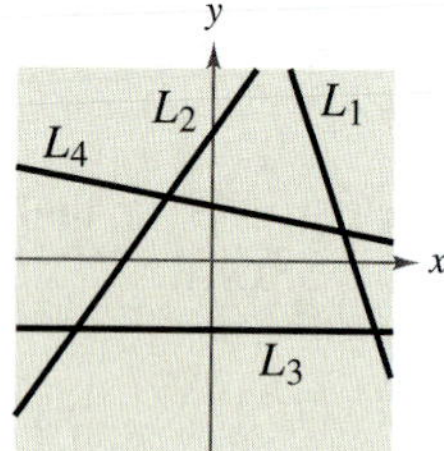

2. (a) m is undefined.
(b) $m = -1$
(c) $m = \frac{5}{2}$
(d) $m = \frac{1}{2}$

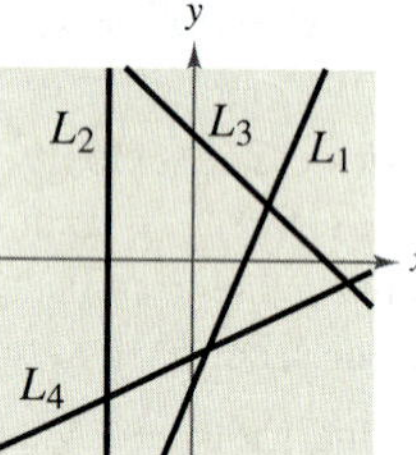

In Exercises 3–10, sketch the graph of the linear equation.

3. $y = -2x - 7$
4. $y = 4x - 3$
5. $y = 6$
6. $x = -3$
7. $y = 3x + 13$
8. $y = -10x + 9$
9. $y = -\frac{5}{2}x - 1$
10. $y = \frac{5}{6}x + 5$

In Exercises 11 and 12, use the concept of slope to find t such that the three points are on the same line.

11. $(-2, 5), (0, t), (1, 1)$
12. $(-6, 1), (1, t), (10, 5)$

In Exercises 13 and 14, use the point on the line and the slope of the line to find three additional points through which the line passes. (There are many correct answers.)

	Point	*Slope*
13.	$(2, -1)$	$m = \frac{1}{4}$
14.	$(-3, 5)$	$m = -\frac{3}{2}$

In Exercises 15–18, plot the points and find the slope of the line passing through the points.

15. $(3, -4), (-7, 1)$
16. $(-1, 8), (6, 5)$
17. $(-4.5, 6), (2.1, 3)$
18. $(-3, 2), (8, 2)$

In Exercises 19–22, find an equation of the line that passes through the points.

19. $(0, 0), (0, 10)$
20. $(2, 5), (-2, -1)$
21. $(-1, 4), (2, 0)$
22. $(11, -2), (6, -1)$

In Exercises 23–26, find an equation of the line that passes through the given point and has the specified slope. Sketch the line.

	Point	*Slope*
23.	$(0, -5)$	$m = \frac{3}{2}$
24.	$(-2, 6)$	$m = 0$
25.	$(10, -3)$	$m = -\frac{1}{2}$
26.	$(-8, 5)$	m is undefined.

In Exercises 27 and 28, write an equation of the line through the point (a) parallel to the given line and (b) perpendicular to the given line.

	Point	*Line*
27.	$(3, -2)$	$5x - 4y = 8$
28.	$(-8, 3)$	$2x + 3y = 5$

Rate of Change **In Exercises 29 and 30, you are given the dollar value of a product in the year 2004 *and* the rate at which the value of the item is expected to change during the next 5 years. Write a linear equation that gives the dollar value V of the product in terms of the year t. (Let $t = 4$ represent 2004.)**

	2004 Value	*Rate*
29.	\$12,500	\$850 increase per year
30.	\$72.95	\$5.15 increase per year

31. ***Sales*** During the second and third quarters of the year, a salvage yard had sales of \$160,000 and \$185,000, respectively. The growth of sales follows a linear pattern. Estimate sales during the fourth quarter.

32. ***Inflation*** The dollar value of a product in 2005 is \$85, and the product is expected to increase in value at a rate of \$3.75 per year.

(a) Write a linear equation that gives the dollar value V of the product in terms of the year t. (Let $t = 5$ represent 2005.)

(b) Use a graphing utility to graph the equation found in part (a).

(c) Move the cursor along the graph of the sales model to estimate the dollar value of the product in 2010.

2.2 In Exercises 33 and 34, determine which of the sets of ordered pairs represents a function from A to B. Give reasons for your answers.

33. $A = \{10, 20, 30, 40\}$ and $B = \{0, 2, 4, 6\}$

(a) $\{(20, 4), (40, 0), (20, 6), (30, 2)\}$

(b) $\{(10, 4), (20, 4), (30, 4), (40, 4)\}$

(c) $\{(40, 0), (30, 2), (20, 4), (10, 6)\}$

(d) $\{(20, 2), (10, 0), (40, 4)\}$

34. $A = \{u, v, w\}$ and $B = \{-2, -1, 0, 1, 2\}$

(a) $\{(v, -1), (u, 2), (w, 0), (u, -2)\}$

(b) $\{(u, -2), (v, 2), (w, 1)\}$

(c) $\{(u, 2), (v, 2), (w, 1), (w, 1)\}$

(d) $\{(w, -2), (v, 0), (w, 2)\}$

In Exercises 35–38, determine whether the equation represents y as a function of x.

35. $16x - y^4 = 0$ **36.** $2x - y - 3 = 0$

37. $y = \sqrt{1 - x}$ **38.** $|y| = x + 2$

In Exercises 39–42, evaluate the function as indicated. Simplify your answers.

39. $f(x) = x^2 + 1$

(a) $f(2)$ (b) $f(-4)$ (c) $f(t^2)$ (d) $f(t + 1)$

40. $g(x) = x^{4/3}$

(a) $g(8)$ (b) $g(t + 1)$ (c) $g(-27)$ (d) $g(-x)$

41. $h(x) = \begin{cases} 2x + 1, & x \leq -1 \\ x^2 + 2, & x > -1 \end{cases}$

(a) $h(-2)$ (b) $h(-1)$ (c) $h(0)$ (d) $h(2)$

42. $f(x) = \dfrac{4}{x^2 + 1}$

(a) $f(1)$ (b) $f(-5)$ (c) $f(-t)$ (d) $f(0)$

In Exercises 43–48, determine the domain of the function. Verify your result with a graph.

43. $f(x) = \sqrt{25 - x^2}$ **44.** $f(x) = 3x + 4$

45. $g(s) = \dfrac{5}{3s - 9}$ **46.** $f(x) = \sqrt{x^2 + 8x}$

47. $h(x) = \dfrac{x}{x^2 - x - 6}$ **48.** $h(t) = |t + 1|$

In Exercises 49 and 50, find the difference quotient and simplify your answer.

49. $f(x) = 2x^2 + 3x - 1,\quad \dfrac{f(x + h) - f(x)}{h},\quad h \neq 0$

50. $f(x) = x^3 - 5x^2 + x,\quad \dfrac{f(x + h) - f(x)}{h},\quad h \neq 0$

51. ***Physics*** The velocity of a ball thrown vertically upward from ground level is $v(t) = -32t + 48$, where t is the time in seconds and v is the velocity in feet per second.

(a) Find the velocity when $t = 1$.

(b) Find the time when the ball reaches its maximum height. [*Hint:* Find the time when $v(t) = 0$.]

(c) Find the velocity when $t = 2$.

52. ***Total Cost*** A hand tool manufacturer produces a product for which the variable cost is \$5.35 per unit and the fixed costs are \$16,000. The company sells the product for \$8.20 and can sell all that it produces.

(a) Find the total cost as a function of x, the number of units produced.

(b) Find the profit as a function of x.

53. ***Geometry*** A wire 24 inches long is to be cut into four pieces to form a rectangle with one side of length x.

(a) Write the area A of the rectangle as a function of x.

(b) Determine the domain of the function.

54. ***Mixture Problem*** From a full 50-liter container of a 40% concentration of acid, x liters is removed and replaced with 100% acid.

(a) Write the amount of acid in the final mixture as a function of x.

(b) Determine the domain and range of the function.

(c) Determine x if the final mixture is 50% acid.

2.3 **In Exercises 55–58, use the Vertical Line Test to determine whether y is a function of x. To print an enlarged copy of the graph, go to the website *www.mathgraphs.com*.**

55. $y = (x - 3)^2$

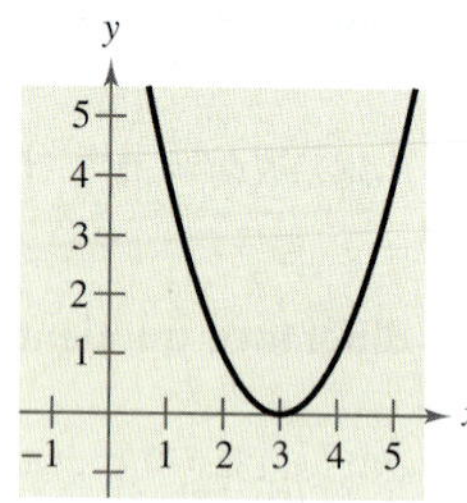

56. $y = -\frac{3}{5}x^3 - 2x + 1$

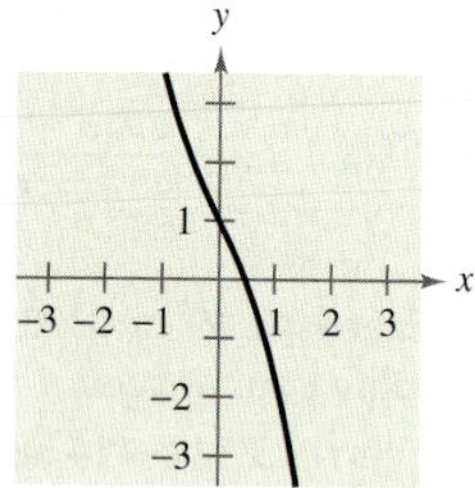

57. $x - 4 = y^2$

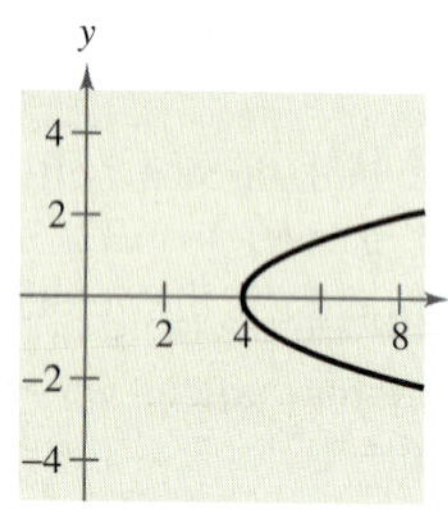

58. $x = -|4 - y|$

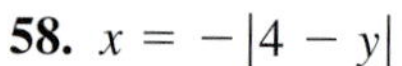

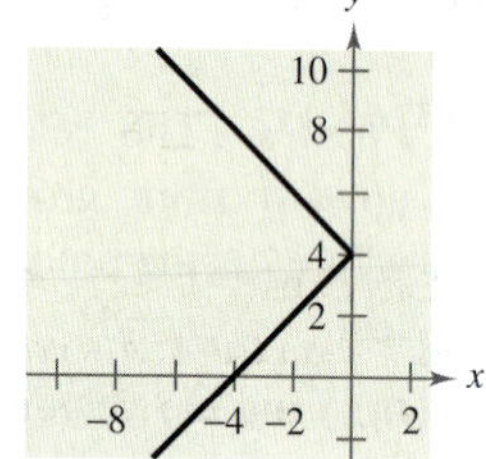

In Exercises 59–62, find the zeros of the function.

59. $f(x) = 3x^2 - 16x + 21$

60. $f(x) = 5x^2 + 4x - 1$

61. $f(x) = \dfrac{8x + 3}{11 - x}$

62. $f(x) = x^3 - x^2 - 25x + 25$

In Exercises 63 and 64, determine the intervals over which the function is increasing, decreasing, or constant.

63. $f(x) = |x| + |x + 1|$

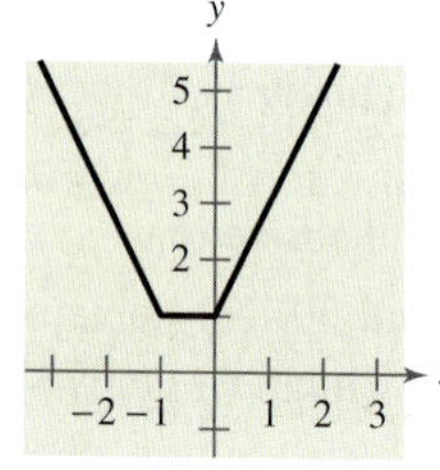

64. $f(x) = (x^2 - 4)^2$

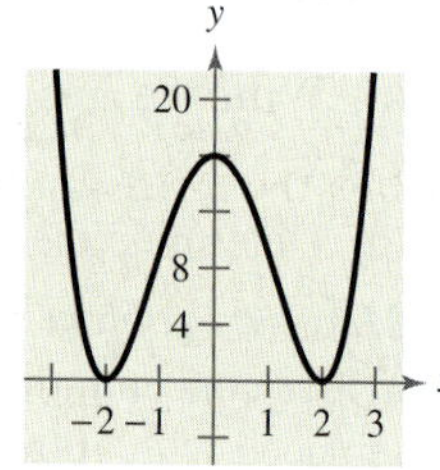

In Exercises 65–68, determine whether the function is even, odd, or neither.

65. $f(x) = x^5 + 4x - 7$

66. $f(x) = x^4 - 20x^2$

67. $f(x) = 2x\sqrt{x^2 + 3}$

68. $f(x) = \sqrt[5]{6x^2}$

2.4 **In Exercises 69–72, write the linear function f such that it has the indicated function values. Sketch a graph of the function.**

69. $f(2) = -6,\ f(-1) = 3$

70. $f(0) = -5,\ f(4) = -8$

71. $f\left(-\frac{4}{5}\right) = 2,\ f\left(\frac{11}{5}\right) = 7$

72. $f(3.3) = 5.6,\ f(-4.7) = -1.4$

In Exercises 73–82, graph the function.

73. $f(x) = 3 - x^2$

74. $h(x) = x^3 - 2$

75. $f(x) = -\sqrt{x}$

76. $f(x) = \sqrt{x + 1}$

77. $g(x) = \dfrac{3}{x}$

78. $g(x) = \dfrac{1}{x + 5}$

79. $f(x) = [\![x]\!] - 2$

80. $g(x) = [\![x + 4]\!]$

81. $f(x) = \begin{cases} 5x - 3, & x \geq -1 \\ -4x + 5, & x < -1 \end{cases}$

82. $f(x) = \begin{cases} x^2 - 2, & x < -2 \\ 5, & -2 \leq x \leq 0 \\ 8x - 5, & x > 0 \end{cases}$

In Exercises 83 and 84, identify the transformed common function shown in the graph.

83.

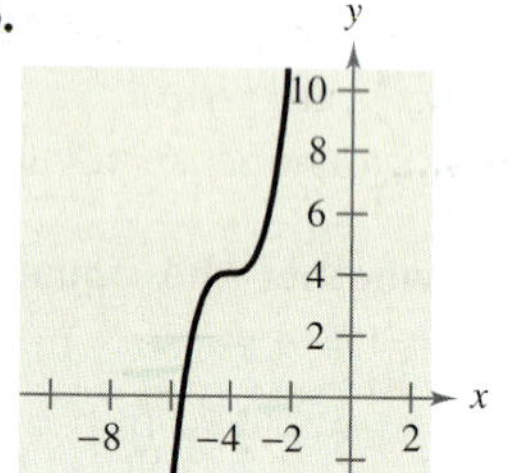

84.

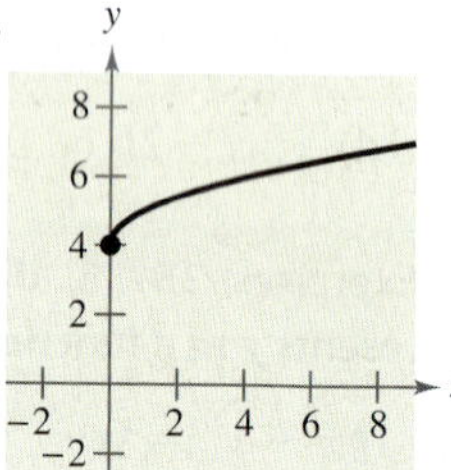

2.5 **In Exercises 85–98, identify the transformation of the graph of f and sketch the graph of h.**

85. $f(x) = x^2,\quad h(x) = x^2 - 9$

86. $f(x) = x^3,\quad h(x) = (x - 2)^3 + 2$

87. $f(x) = \sqrt{x},\quad h(x) = \sqrt{x - 7}$

88. $f(x) = |x|,\quad h(x) = |x + 3| - 5$

89. $f(x) = x^2,\quad h(x) = -(x + 3)^2 + 1$

90. $f(x) = x^3,\quad h(x) = -(x - 5)^3 - 5$

91. $f(x) = [\![x]\!],\quad h(x) = -[\![x]\!] + 6$

92. $f(x) = \sqrt{x},\quad h(x) = -\sqrt{x + 1} + 9$

93. $f(x) = |x|,\quad h(x) = -|-x + 4| + 6$

94. $f(x) = x^2,\quad h(x) = -(x + 1)^2 - 3$

95. $f(x) = [\![x]\!]$, $h(x) = 5[\![x - 9]\!]$

96. $f(x) = x^3$, $h(x) = -\frac{1}{3}x^3$

97. $f(x) = \sqrt{x}$, $h(x) = -2\sqrt{x - 4}$

98. $f(x) = |x|$, $h(x) = \frac{1}{2}|x| - 1$

2.6 In Exercises 99 and 100, find (a) $(f + g)(x)$, (b) $(f - g)(x)$, (c) $(fg)(x)$, and (d) $(f/g)(x)$. What is the domain of f/g?

99. $f(x) = x^2 + 3$, $g(x) = 2x - 1$

100. $f(x) = x^2 - 4$, $g(x) = \sqrt{3 - x}$

In Exercises 101 and 102, find (a) $f \circ g$ and (b) $g \circ f$. Find the domain of each function and each composite function.

101. $f(x) = \frac{1}{3}x - 3$, $g(x) = 3x + 1$

102. $f(x) = x^3 - 4$, $g(x) = \sqrt[3]{x + 7}$

In Exercise 103 and 104, find two functions f and g such that $(f \circ g)(x) = h(x)$. (There is more than one correct answer.)

103. $h(x) = (6x - 5)^3$

104. $h(x) = \sqrt[3]{x + 2}$

Data Analysis In Exercises 105 and 106, use the table, which shows the total values (in billions of dollars) of U.S. imports from Mexico and Canada for the years 1995 through 1999. The variables y_1 and y_2 represent the total values of imports from Mexico and Canada, respectively. (Source: U.S. Census Bureau)

Year	y_1	y_2
1995	62.1	144.4
1996	74.3	155.9
1997	85.9	168.2
1998	94.6	173.3
1999	109.7	198.3

105. Use a graphing utility to find quadratic models for y_1 and y_2. Let $t = 5$ represent 1995.

106. Use a graphing utility to graph y_1, y_2, and $y_1 + y_2$ in the same viewing window. Use the model to estimate the total value of U.S. imports from Canada and Mexico in 2005.

2.7 In Exercises 107 and 108, find the inverse function of f informally. Verify that $f(f^{-1}(x)) = x = f^{-1}(f(x))$.

107. $f(x) = x - 7$

108. $f(x) = x + 5$

In Exercises 109 and 110, determine whether the function has an inverse function.

109. **110.**

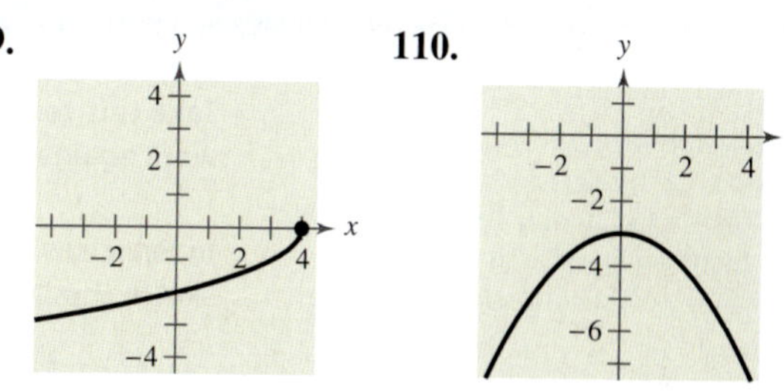

In Exercises 111–114, use the Horizontal Line Test to determine if the function is one-to-one and so has an inverse function.

111. $f(x) = 4 - \frac{1}{3}x$

112. $f(x) = (x - 1)^2$

113. $h(t) = \dfrac{2}{t - 3}$

114. $g(x) = \sqrt{x + 6}$

In Exercises 115–118, (a) find f^{-1}, (b) sketch the graphs of f and f^{-1} on the same coordinate system, and (c) verify that $f^{-1}(f(x)) = x = f(f^{-1}(x))$.

115. $f(x) = \frac{1}{2}x - 3$

116. $f(x) = 5x - 7$

117. $f(x) = \sqrt{x + 1}$

118. $f(x) = x^3 + 2$

In Exercises 119 and 120, restrict the domain of the function f to an interval over which the function is increasing and determine f^{-1} over that interval.

119. $f(x) = 2(x - 4)^2$

120. $f(x) = |x - 2|$

Synthesis

True or False? In Exercises 121 and 122, determine whether the statement is true or false. Justify your answer.

121. Relative to the graph of $f(x) = \sqrt{x}$, the function $h(x) = -\sqrt{x + 9} - 13$ is shifted 9 units to the left and 13 units downward, then reflected in the x-axis.

122. If f and g are two inverse functions, then the domain of g is equal to the range of f.

123. ***Writing*** Explain how to tell whether a relation between two variables is a function.

124. ***Writing*** Explain the difference between the Vertical Line Test and the Horizontal Line Test.

Chapter Test

The *Interactive* CD-ROM and *Internet* versions of this text offer Chapter Pre-Tests and Chapter Post-Tests, both of which have randomly generated exercises with diagnostic capabilities.

Take this test as you would take a test in class. When you are finished, check your work against the answers given in the back of the book.

In Exercises 1 and 2, find an equation of the line passing through the points. Then sketch the line.

1. $(2, -3), (-4, 9)$

2. $(3, 0.8), (7, -6)$

3. Find an equation of the line that passes through the point $(3, 8)$ and is (a) parallel to and (b) perpendicular to the line $-4x + 7y = -5$.

In Exercises 4 and 5, evaluate the function at each specified value.

4. $f(x) = |x + 2| - 15$

(a) $f(-8)$ (b) $f(14)$ (c) $f(x - 6)$

5. $f(x) = \dfrac{\sqrt{x + 9}}{x^2 - 81}$

(a) $f(7)$ (b) $f(-5)$ (c) $f(x - 9)$

In Exercises 6 and 7, determine the domain of the function.

6. $f(x) = \sqrt{100 - x^2}$

7. $f(x) = |-x + 6| + 2$

In Exercises 8–10, (a) use a graphing utility to graph the function, (b) approximate the intervals over which the function is increasing, decreasing, or constant, and (c) determine whether the function is even, odd, or neither.

8. $f(x) = 2x^6 + 5x^4 - x^2$

9. $f(x) = 4x\sqrt{3 - x}$

10. $f(x) = |x + 5|$

11. Sketch the graph of $f(x) = \begin{cases} 3x + 7, & x \le -3 \\ 4x^2 - 1, & x > -3 \end{cases}$.

In Exercises 12–14, identify the common function in the transformation. Then sketch a graph of the function.

12. $h(x) = -[\![x]\!]$

13. $h(x) = -\sqrt{x + 5} + 8$

14. $h(x) = \frac{1}{4}|x + 1| - 3$

In Exercises 15 and 16, find (a) $(f + g)(x)$, (b) $(f - g)(x)$, (c) $(fg)(x)$, (d) $(f/g)(x)$, (e) $(f \circ g)(x)$, and (f) $(g \circ f)(x)$.

15. $f(x) = 3x^2 - 7,\quad g(x) = -x^2 - 4x + 5$

16. $f(x) = \dfrac{1}{x},\quad g(x) = 2\sqrt{x}$

In Exercises 17–19, determine whether the function has an inverse function, and if so, find the inverse function.

17. $f(x) = x^3 + 8$

18. $f(x) = |x^2 - 3| + 6$

19. $f(x) = \dfrac{3x\sqrt{x}}{8}$

20. It costs a company \$58 to produce 6 units of a product and \$78 to produce 10 units. How much does it cost to produce 25 units, assuming that the cost function is linear?

Cumulative Test for Chapters P–2

Take this test to review the material from earlier chapters. When you are finished, check your work against the answers given in the back of the book.

In Exercises 1 and 2, simplify the expression.

1. $\dfrac{8x^2y^{-3}}{30x^{-1}y^2}$ **2.** $\sqrt{24x^4y^3}$

In Exercises 3–5, perform the operation and simplify the result.

3. $4x - [2x + 3(2 - x)]$ **4.** $(x - 2)(x^2 + x - 3)$ **5.** $\dfrac{2}{s + 3} - \dfrac{1}{s + 1}$

In Exercises 6–8, factor the expression completely.

6. $25 - (x - 2)^2$ **7.** $x - 5x^2 - 6x^3$ **8.** $54 - 16x^3$

In Exercises 9–11, graph the equation without using a graphing utility.

9. $x - 3y + 12 = 0$ **10.** $y = x^2 - 9$ **11.** $y = \sqrt{4 - x}$

In Exercises 12 and 13, write an expression for the area of the region.

12.

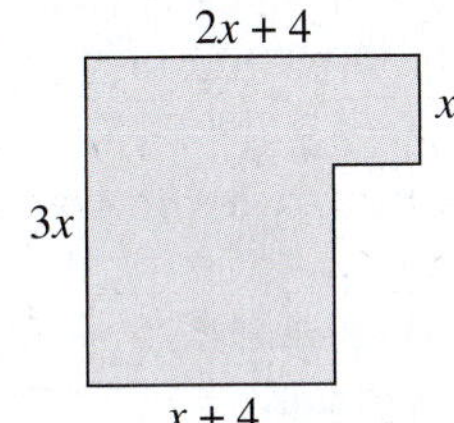

13.

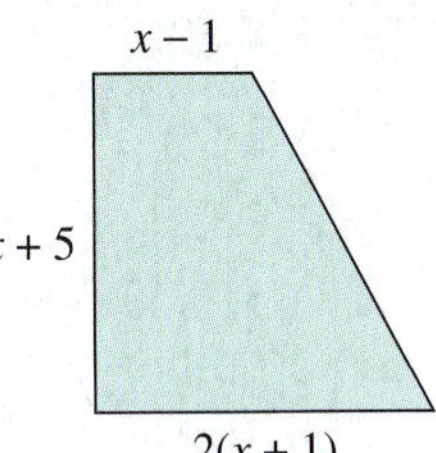

In Exercises 14–19, solve the equation by any convenient method. State the method you used.

14. $x^2 - 4x + 3 = 0$ **15.** $-2x^2 + 8x + 12 = 0$

16. $\frac{3}{4}x^2 = 12$ **17.** $3x^2 + 5x - 6 = 0$

18. $3x^2 + 9x + 1 = 0$ **19.** $\frac{1}{2}x^2 - 7 = 25$

In Exercises 20–25, solve the equation (if possible).

20. $x^4 + 12x^3 + 4x^2 + 48x = 0$

21. $8x^3 - 48x^2 + 72x = 0$

22. $x^{2/3} + 13 = 17$

23. $\sqrt{x + 10} = x - 2$

24. $|4(x - 2)| = 28$

25. $|x - 12| = -2$

In Exercises 26–28, determine whether each value of x is a solution of the inequality.

26. $4x + 2 > 7$

(a) $x = -1$ (b) $x = \frac{1}{2}$

(c) $x = \frac{3}{2}$ (d) $x = 2$

27. $3 - \frac{1}{2}x \leq -2$

(a) $x = -10$ (b) $x = 9$

(c) $x = 10$ (d) $x = 12$

28. $|5x - 1| < 4$

(a) $x = -1$ (b) $x = -\frac{1}{2}$

(c) $x = 1$ (d) $x = 2$

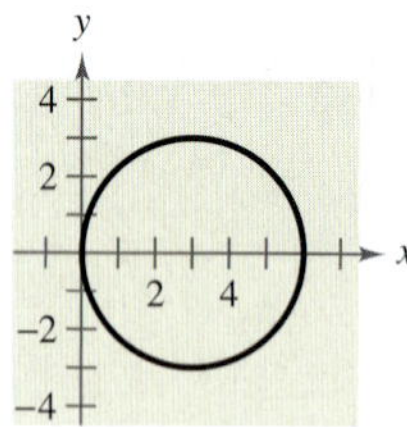

FIGURE FOR 34

In Exercises 29–32, solve the inequality and sketch the solution on the real number line.

29. $|x + 1| \leq 6$

30. $|7 + 8x| > 5$

31. $5x^2 + 12x + 7 \geq 0$

32. $-x^2 + x + 4 < 0$

33. Find an equation of the line passing through $\left(-\frac{1}{2}, 1\right)$ and $(3, 8)$.

34. Explain why the graph at the left does not represent y as a function of x.

35. Evaluate (if possible) the function $f(x) = \dfrac{x}{x - 2}$ for each value.

(a) $f(6)$ (b) $f(2)$ (c) $f(s + 2)$

36. Describe how the graph of each function would differ from the graph of $y = \sqrt[3]{x}$. (*Note:* It is not necessary to sketch the graphs.)

(a) $r(x) = \frac{1}{2}\sqrt[3]{x}$ (b) $h(x) = \sqrt[3]{x} + 2$ (c) $g(x) = \sqrt[3]{x + 2}$

In Exercises 37 and 38, find (a) $(f + g)(x)$, (b) $(f - g)(x)$, (c) $(fg)(x)$, and (d) $(f/g)(x)$. What is the domain of f/g?

37. $f(x) = x - 3, \quad g(x) = 4x + 1$

38. $f(x) = \sqrt{x - 1}, \quad g(x) = x^2 + 1$

In Exercises 39 and 40, find (a) $f \circ g$ and (b) $g \circ f$. Find the domain of each composite function.

39. $f(x) = 2x^2, \quad g(x) = \sqrt{x + 6}$

40. $f(x) = x - 2, \quad g(x) = |x|$

41. Determine whether $h(x) = 5x - 2$ has an inverse function. If so, find it.

42. A group of n people decide to buy a \$36,000 minibus. Each person will pay an equal share of the cost. If three additional people join the group, the cost per person will decrease by \$1000. Find n.

43. For groups of 80 or more, a charter bus company determines the rate per person according to the formula

$$\text{Rate} = \$8.00 - \$0.05(n - 80), \quad n \geq 80.$$

(a) Write the revenue R as a function of n.

(b) Use a graphing utility to graph the revenue function. Move the cursor along the function to estimate the number of passengers that will maximize the revenue.

Proofs in Mathematics

Biconditional Statements

Recall from the Proofs in Mathematics in Chapter 1 that a conditional statement is a statement of the form "if p, then q." A statement of the form "p if and only if q" is called a **biconditional statement.** A biconditional statement, denoted by

$p \leftrightarrow q$ Biconditional statement

is the conjunction of the conditional statement $p \rightarrow q$ and its converse $q \rightarrow p$.

A biconditional statement can be either true or false. To be true, *both* the conditional statement and its converse must be true.

Example 1 ▶ Analyzing a Biconditional Statement

Consider the statement $x = 3$ if and only if $x^2 = 9$.

a. Is the statement a biconditional statement? **b.** Is the statement true?

Solution

a. The statement is a biconditional statement because it is of the form "p if and only if q."

b. The statement can be rewritten as the following conditional statement and its converse.

Conditional statement: If $x = 3$, then $x^2 = 9$.
Converse: If $x^2 = 9$, then $x = 3$.

The first of these statements is true, but the second is false because x could also equal -3. So, the biconditional statement is false.

Knowing how to use biconditional statements is an important tool for reasoning in mathematics.

Example 2 ▶ Analyzing a Biconditional Statement

Determine whether the biconditional statement is true or false. If it is false, provide a counterexample.

A number is divisible by 5 if and only if it ends in 0.

Solution

The biconditional statement can be rewritten as the following conditional statement and its converse.

Conditional statement: If a number is divisible by 5, then it ends in 0.
Converse: If a number ends in 0, then it is divisible by 5.

The conditional statement is false. A counterexample is the number 15, which is divisible by 5 but does not end in 0.

P.S. Problem Solving

1. As a salesperson, you receive a monthly salary of \$2000, plus a commission of 7% of sales. You are offered a new job at \$2300 per month, plus a commission of 5% of sales.
   (a) Write a linear equation for your current monthly wage W_1 in terms of your monthly sales S.
   (b) Write a linear equation for the monthly wage W_2 of your new job offer in terms of the monthly sales S.
    (c) Use a graphing utility to graph both equations in the same viewing window. Find the point of intersection. What does it signify?
   (d) You think you can sell \$20,000 per month. Should you change jobs? Explain.
2. For the numbers 2 through 9 on a telephone keypad (see figure), create two relations: one mapping numbers onto letters, and the other mapping letters onto numbers. Are both relations functions? Explain.

3. What can be said about the sum and difference of each of the following?
   (a) Two even functions
   (b) Two odd functions
   (c) An odd function and an even function
4. The two functions

   $f(x) = x$ and $g(x) = -x$

   are their own inverse functions. Graph each function and explain why this is true. Graph other linear functions that are their own inverse functions. Find a general formula for a family of linear functions that are their own inverse functions.
5. Prove that a function of the following form is even.

   $y = a_{2n}x^{2n} + a_{2n-2}x^{2n-2} + \cdot \cdot \cdot + a_2x^2 + a_0$

6. A miniature golf professional is trying to make a hole-in-one on the miniature golf green shown. A coordinate plane is placed over the golf green. The golf ball is at the point (2.5, 2) and the hole is at the point (9.5, 2). The professional wants to bank the ball off the side wall of the green at the point (x, y). Find the coordinates of the point (x, y). Then write an equation for the path of the ball.

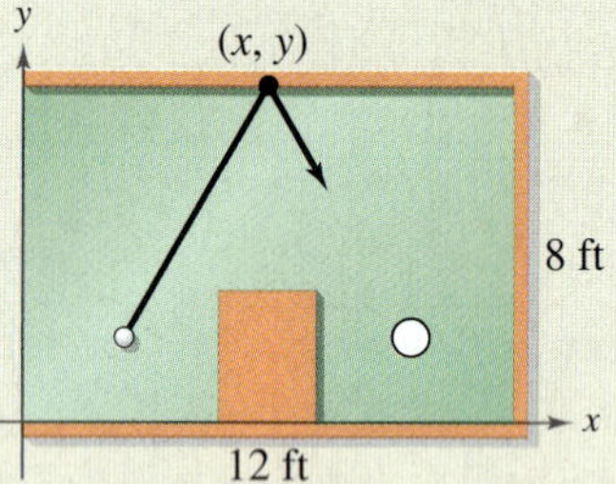

7. At 2:00 P.M. on April 11, 1912, the *Titanic* left Cobh, Ireland, on her voyage to New York City. At 11:40 P.M. on April 14, the *Titanic* struck an iceberg and sank, having covered only about 2100 miles of the approximately 3400-mile trip.
   (a) What was the total length of the *Titantic's* voyage in hours?
   (b) What was the *Titantic's* average speed in miles per hour?
   (c) Write a function relating the *Titantic's* distance from New York City and the number of hours traveled. Find the domain and range of the function.
   (d) Graph the function from part (c).
8. Consider the functions $f(x) = 4x$ and $g(x) = x + 6$.
   (a) Find $(f \circ g)(x)$.
   (b) Find $(f \circ g)^{-1}(x)$.
   (c) Find $f^{-1}(x)$ and $g^{-1}(x)$.
   (d) Find $(g^{-1} \circ f^{-1})(x)$ and compare the result with that of part (b).
   (e) Repeat parts (a) through (d) for $f(x) = x^3 + 1$ and $g(x) = 2x$.
   (f) Write two one-to-one functions f and g, and repeat parts (a) through (d) for these functions.
   (g) Make a conjecture about $(f \circ g)^{-1}(x)$ and $(g^{-1} \circ f^{-1})(x)$.

9. You are in a boat 2 miles from the nearest point on the coast. You are to travel to a point Q, 3 miles down the coast and 1 mile inland (see figure). You can row at 2 miles per hour and walk at 4 miles per hour.

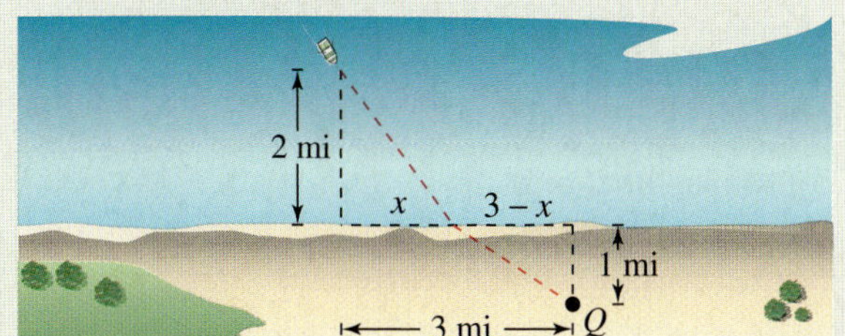

(a) Write the total time T of the trip as a function of x.

(b) Determine the domain of the function.

 (c) Use a graphing utility to graph the function. Be sure to choose an appropriate viewing window.

(d) Use the *zoom* and *trace* features to find the value of x that minimizes T.

(e) Write a brief paragraph interpreting these values.

10. The Heaviside function $H(x)$ is widely used in engineering applications. (See figure.) To print an enlarged copy of the graph, go to the website *www.mathgraphs.com*.

$$H(x) = \begin{cases} 1, & x \geq 0 \\ 0, & x < 0 \end{cases}$$

Sketch the graph of each function by hand.

(a) $H(x) - 2$ (b) $H(x - 2)$ (c) $-H(x)$

(d) $H(-x)$ (e) $\frac{1}{2}H(x)$ (f) $-H(x - 2) + 2$

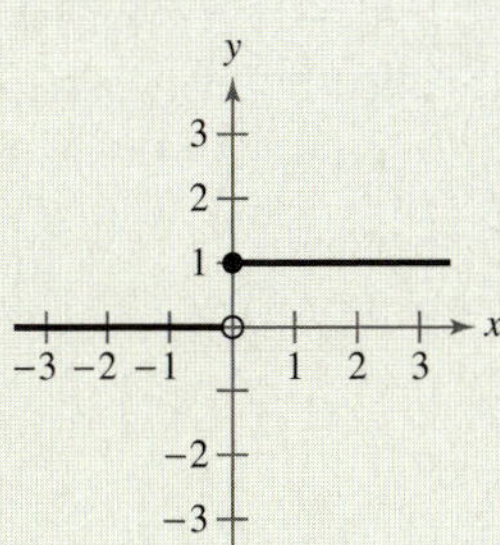

11. Let $f(x) = \dfrac{1}{1 - x}$.

(a) What are the domain and range of f?

(b) Find $f(f(x))$. What is the domain of this function?

(c) Find $f(f(f(x)))$. Is the graph a line? Why or why not?

12. Show that the Associative Property holds for compositions of functions—that is,

$$(f \circ (g \circ h))(x) = ((f \circ g) \circ h)(x).$$

13. Consider the graph of the function f shown in the figure. Use this graph to sketch the graph of each function. To print an enlarged copy of the graph, go to the website *www.mathgraphs.com*.

(a) $f(x + 1)$ (b) $f(x) + 1$ (c) $2f(x)$ (d) $f(-x)$

(e) $-f(x)$ (f) $|f(x)|$ (g) $f(|x|)$

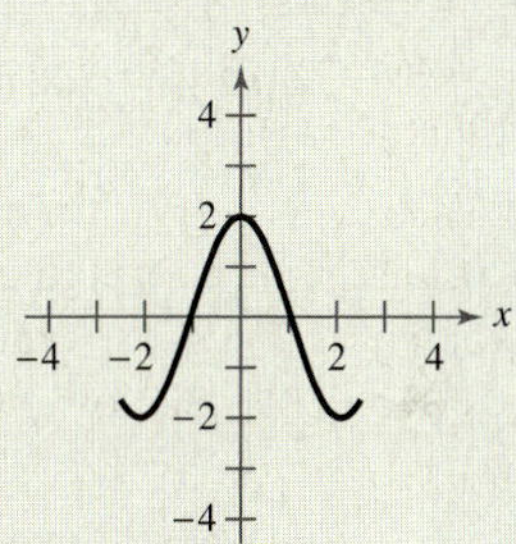

14. Use the graphs of f and f^{-1} to complete each table of function values.

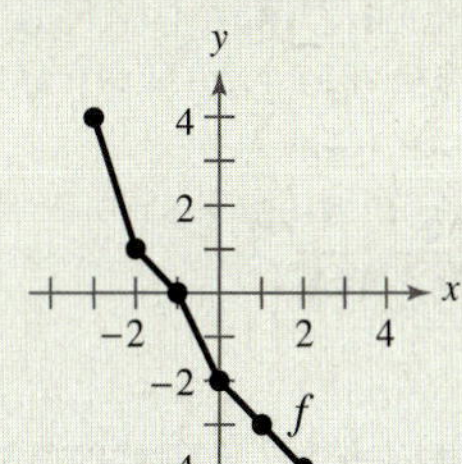

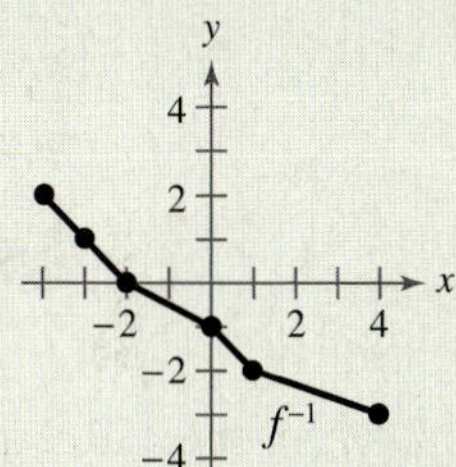

(a)

x	$f(f^{-1}(x))$
-4	
-2	
0	
4	

(b)

x	$(f + f^{-1})(x)$
-3	
-2	
0	
1	

(c)

x	$(f \cdot f^{-1})(x)$
-3	
-2	
0	
1	

(d)

x	$\lvert f^{-1}(x) \rvert$
-4	
-3	
0	
4	

How to study Chapter 3

What you should learn

In this chapter you will learn the following skills and concepts:

- How to sketch and analyze graphs of functions
- How to sketch and analyze graphs of polynomial functions
- How to use long division and synthetic division to divide polynomials by other polynomials
- How to determine the number of rational and real zeros of polynomial functions, and find the zeros
- How to write mathematical models for direct, inverse, and joint variation

Important Vocabulary

As you encounter each new vocabulary term in this chapter, add the term and its definition to your notebook glossary.

Polynomial function (p. 260)
Parabola (p. 260)
Axis (of a parabola) (p. 261)
Vertex (of a parabola) (p. 261)
Standard form of a quadratic function (p. 263)
Continuous (p. 271)
Leading Coefficient Test (p. 273)
Repeated zero (p. 275)
Multiplicity (p. 275)
Intermediate Value Theorem (p. 278)
Division Algorithm (p. 285)
Improper (rational expression) (p. 285)
Proper (rational expression) (p. 285)
Synthetic division (p. 287)
Rational Zero Test (p. 294)
Conjugates (p. 297)
Irreducible over the reals (p. 298)
Descartes's Rule of Signs (p. 300)
Variation in sign (p. 300)
Upper bound (p. 301)
Lower bound (p. 301)
Directly proportional (p. 309)
Constant of variation (p. 309)
Inversely proportional (p. 311)
Jointly proportional (p. 312)
Sum of square differences (p. 313)
Least squares regression line (p. 313)

Study Tools

Learning objectives in each section
Chapter Summary (p. 320)
Review Exercises (pp. 321–324)
Chapter Test (p. 325)

Additional Resources

Study and Solutions Guide
Interactive College Algebra
Videotapes/DVD for Chapter 3
College Algebra Website
Student Success Organizer

3

Polynomial Functions

3.1 Quadratic Functions

▶ **What you should learn**

- How to analyze graphs of quadratic functions
- How to write quadratic functions in standard form and use the results to sketch graphs of functions
- How to use quadratic functions to model and solve real-life problems

▶ **Why you should learn it**

Quadratic functions can be used to model data to analyze consumer behavior. For instance, in Exercise 86 on page 269, you will use a quadratic function to model the number of hairdressers and cosmetologists in the United States.

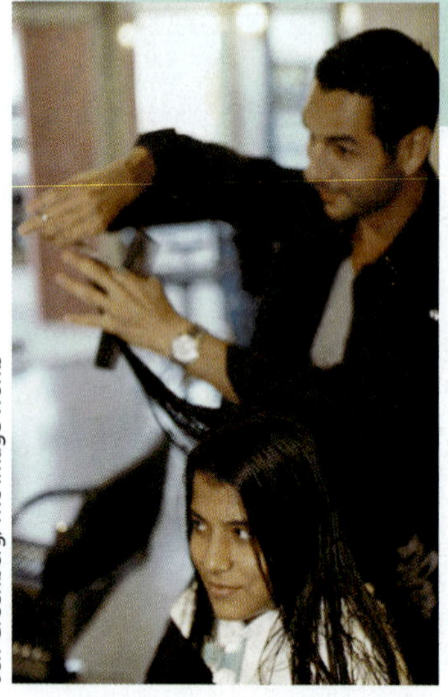

Jeff Greenberg/The Image Works

The Graph of a Quadratic Function

In this and the next section you will study the graphs of polynomial functions. In Section 2.4, you were introduced to the following basic functions.

$f(x) = ax + b$ Linear function

$f(x) = c$ Constant function

$f(x) = x^2$ Squaring function

These functions are examples of **polynomial functions.**

Definition of Polynomial Function

Let n be a nonnegative integer and let $a_n, a_{n-1}, \ldots, a_2, a_1, a_0$ be real numbers with $a_n \neq 0$. The function

$$f(x) = a_n x^n + a_{n-1}x^{n-1} + \cdots + a_2x^2 + a_1x + a_0$$

is called a **polynomial function of x with degree n.**

Polynomial functions are classified by degree. For instance, a constant function has degree 0 and a linear function has degree 1. In this section you will study second-degree polynomial functions, which are called **quadratic functions.**

For instance, each of the following functions is a quadratic function.

$$f(x) = x^2 + 6x + 2$$

$$g(x) = 2(x + 1)^2 - 3$$

$$h(x) = 9 + \tfrac{1}{4}x^2$$

$$k(x) = -3x^2 + 4$$

$$m(x) = (x - 2)(x + 1)$$

Note that the squaring function is a simple quadratic function that has degree 2.

Definition of Quadratic Function

Let a, b, and c be real numbers with $a \neq 0$. The function

$f(x) = ax^2 + bx + c$ Quadratic function

is called a **quadratic function.**

The graph of a quadratic function is a special type of "U"-shaped curve called a **parabola.** Parabolas occur in many real-life applications—especially those involving reflective properties of satellite dishes and flashlight reflectors. You will study these properties in Section 4.4.

All parabolas are symmetric with respect to a line called the **axis of symmetry,** or simply the **axis** of the parabola. The point where the axis intersects the parabola is the **vertex** of the parabola, as shown in Figure 3.1. If the leading coefficient is positive, the graph of $f(x) = ax^2 + bx + c$ is a parabola that opens upward. If the leading coefficient is negative, the graph of $f(x) = ax^2 + bx + c$ is a parabola that opens downward.

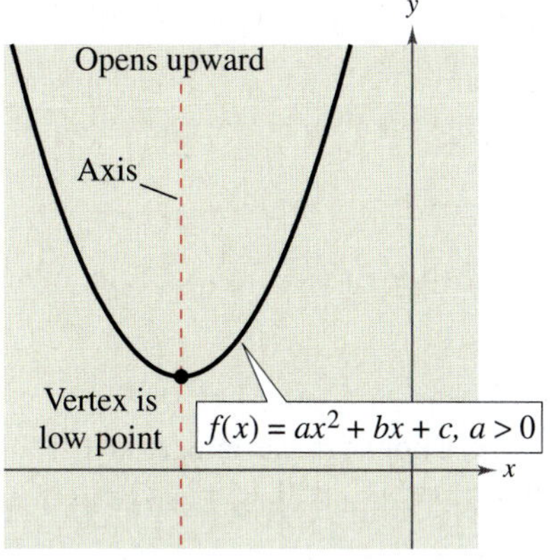

Leading coefficient is positive. *Leading coefficient is negative.*

FIGURE 3.1

The simplest type of quadratic function is

$$f(x) = ax^2.$$

Its graph is a parabola whose vertex is (0, 0). If $a > 0$, the vertex is the point with the *minimum* y-value on the graph, and if $a < 0$, the vertex is the point with the *maximum* y-value on the graph, as shown in Figure 3.2.

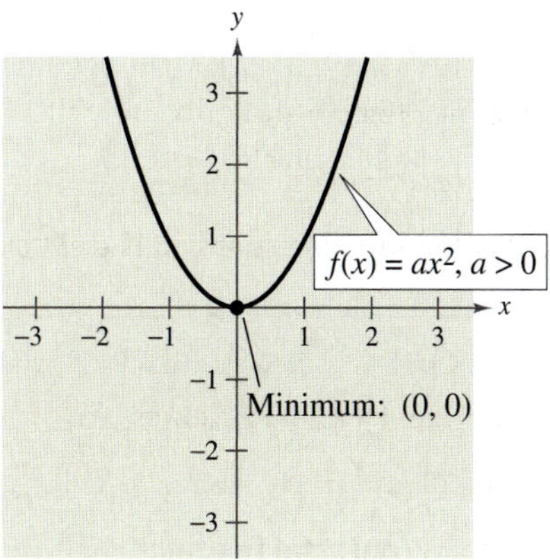

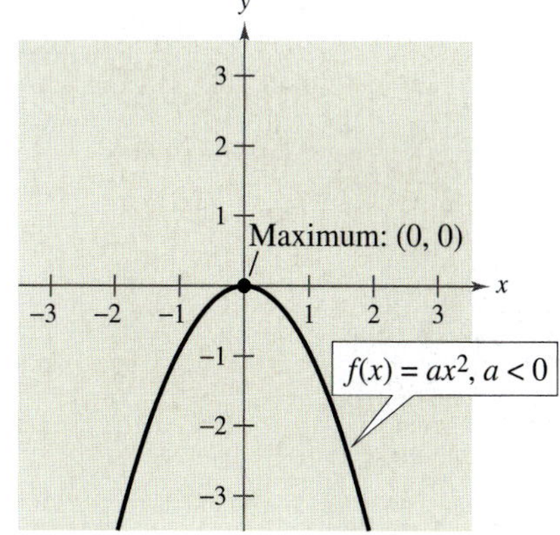

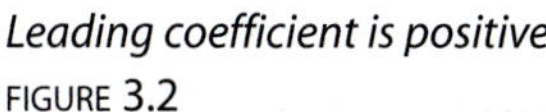

Leading coefficient is positive. *Leading coefficient is negative.*

FIGURE 3.2

When sketching the graph of $f(x) = ax^2$, it is helpful to use the graph of $y = x^2$ as a reference, as discussed in Section 2.5.

Exploration

Graph $y = ax^2$ for $a = -2, -1, -0.5, 0.5, 1$, and 2. How does changing the value of a affect the graph?

Graph $y = (x - h)^2$ for $h = -4, -2, 2$, and 4. How does changing the value of h affect the graph?

Graph $y = x^2 + k$ for $k = -4, -2, 2$, and 4. How does changing the value of k affect the graph?

The *Interactive* CD-ROM and *Internet* versions of this text offer a Try It for each example in the text.

Example 1 ▶ Sketching Graphs of Quadratic Functions

a. Compare the graphs of $y = x^2$ and $f(x) = \frac{1}{3}x^2$.

b. Compare the graphs of $y = x^2$ and $g(x) = 2x^2$.

Solution

a. Compared with $y = x^2$, each output of $f(x) = \frac{1}{3}x^2$ "shrinks" by a factor of $\frac{1}{3}$, creating the broader parabola shown in Figure 3.3.

b. Compared with $y = x^2$, each output of $g(x) = 2x^2$ "stretches" by a factor of 2, creating the narrower parabola shown in Figure 3.4.

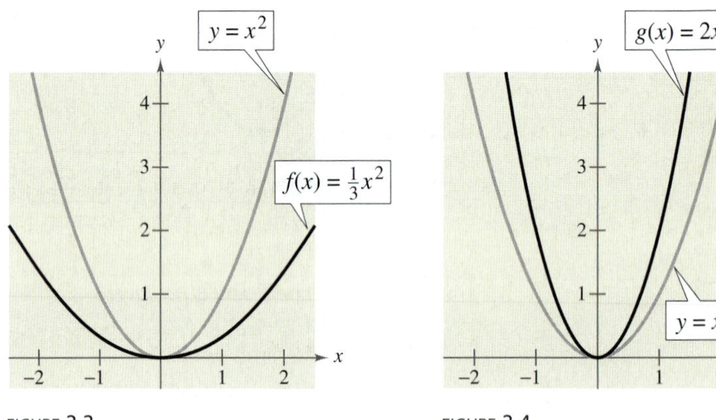

FIGURE 3.3

FIGURE 3.4

In Example 1, note that the coefficient a determines how widely the parabola given by $f(x) = ax^2$ opens. If $|a|$ is small, the parabola opens more widely than if $|a|$ is large.

Recall from Section 2.5 that the graphs of $y = f(x \pm c)$, $y = f(x) \pm c$, $y = f(-x)$, and $y = -f(x)$ are rigid transformations of the graph of $y = f(x)$. For instance, in Figure 3.5, notice how the graph of $y = x^2$ can be transformed to produce the graphs of $f(x) = -x^2 + 1$ and $g(x) = (x + 2)^2 - 3$.

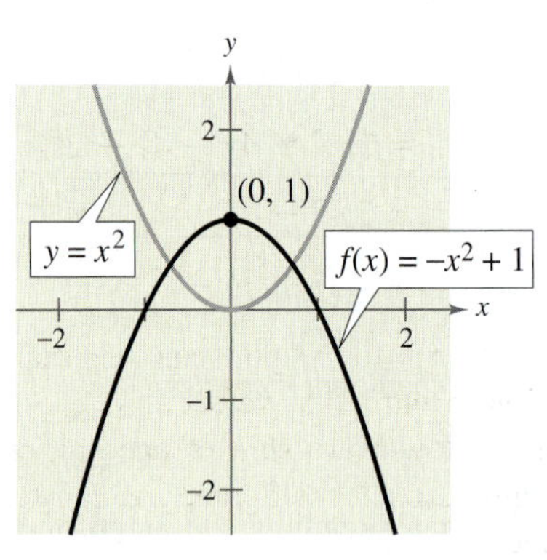

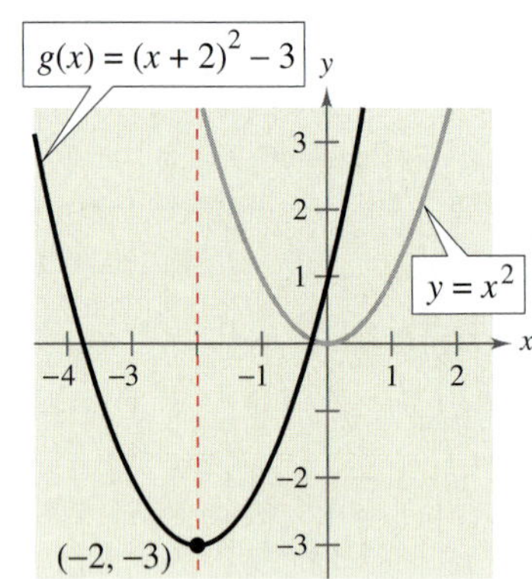

FIGURE 3.5

STUDY TIP

The standard form of a quadratic function identifies four basic transformations of the graph of $y = x^2$.

a. The factor $|a|$ produces a vertical stretch or shrink.

b. If $a < 0$, the graph is reflected in the x-axis.

c. The factor $(x - h)^2$ represents a horizontal shift of h units.

d. The term k represents a vertical shift of k units.

The Standard Form of a Quadratic Function

The **standard form** of a quadratic function is

$$f(x) = a(x - h)^2 + k.$$

This form is especially convenient for sketching a parabola because it identifies the vertex of the parabola.

Standard Form of a Quadratic Function

The quadratic function

$$f(x) = a(x - h)^2 + k, \qquad a \neq 0$$

is in **standard form.** The graph of f is a parabola whose axis is the vertical line $x = h$ and whose vertex is the point (h, k). If $a > 0$, the parabola opens upward, and if $a < 0$, the parabola opens downward.

To graph a parabola, it is helpful to begin by writing the quadratic function in standard form using the process of completing the square, as illustrated in Example 2.

Example 2 ▶ Graphing a Parabola in Standard Form 

Sketch the graph of

$$f(x) = 2x^2 + 8x + 7$$

and identify the vertex and the axis of the parabola.

Solution

Begin by writing the quadratic function in standard form. Notice that the first step in completing the square is to factor out any coefficient of x^2 that is not 1.

$$f(x) = 2x^2 + 8x + 7 \qquad \text{Write original function.}$$

$$= 2(x^2 + 4x) + 7 \qquad \text{Factor 2 out of } x\text{-terms.}$$

$$= 2(x^2 + 4x + 4 - 4) + 7 \qquad \text{Add and subtract 4 within parentheses.}$$

$(4/2)^2$

$$= 2(x^2 + 4x + 4) - 2(4) + 7 \qquad \text{Regroup terms.}$$

$$= 2(x^2 + 4x + 4) - 8 + 7 \qquad \text{Simplify.}$$

$$= 2(x + 2)^2 - 1 \qquad \text{Write in standard form.}$$

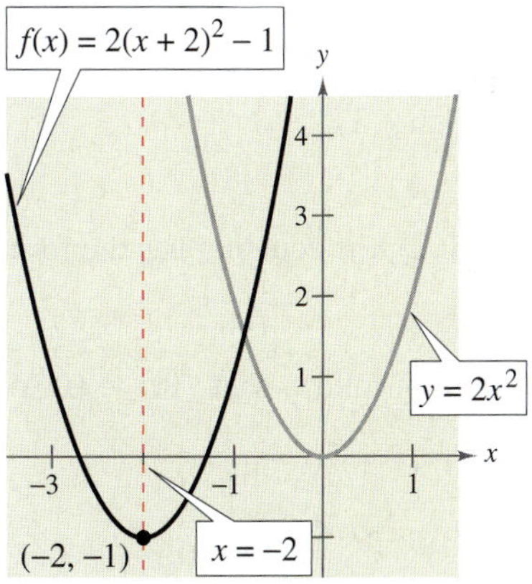

From this form, you can see that the graph of f is a parabola that opens upward and has its vertex at $(-2, -1)$. This corresponds to a left shift of two units and a downward shift of one unit relative to the graph of $y = 2x^2$, as shown in Figure 3.6. In the figure, you can see that the axis of the parabola is the vertical line through the vertex, $x = -2$.

FIGURE 3.6

The icon identifies examples and concepts related to features of the Learning Tools CD-ROM and the *Interactive* and *Internet* versions of this text. For more details see the chart on pages *xix–xxiii*.

To find the x-intercepts of the graph of $f(x) = ax^2 + bx + c$, you must solve the equation $ax^2 + bx + c = 0$. If $ax^2 + bx + c$ does not factor, you can use the Quadratic Formula to find the x-intercepts. Remember, however, that a parabola may have no x-intercepts.

The *Interactive* CD-ROM and *Internet* versions of this text offer a Quiz for every section of the text.

Example 3 ▶ Finding the Vertex and x-Intercepts of a Parabola

Sketch the graph of $f(x) = -x^2 + 6x - 8$ and identify the vertex and x-intercepts.

Solution

As in Example 2, begin by writing the quadratic function in standard form.

$$f(x) = -x^2 + 6x - 8$$ Write original function.

$$= -(x^2 - 6x) - 8$$ Factor -1 out of x-terms.

$$= -(x^2 - 6x + 9 - 9) - 8$$ Add and subtract 9 within parentheses.

$(-6/2)^2$

$$= -(x^2 - 6x + 9) - (-9) - 8$$ Regroup terms.

$$= -(x - 3)^2 + 1$$ Write in standard form.

FIGURE 3.7

From this form, you can see that the vertex is $(3, 1)$. To find the x-intercepts of the graph, solve the equation $-x^2 + 6x - 8 = 0$.

$$-(x^2 - 6x + 8) = 0$$ Factor out -1.

$$-(x - 2)(x - 4) = 0$$ Factor.

$$x - 2 = 0 \quad \Rightarrow \quad x = 2$$ Set 1st factor equal to 0.

$$x - 4 = 0 \quad \Rightarrow \quad x = 4$$ Set 2nd factor equal to 0.

The x-intercepts are $(2, 0)$ and $(4, 0)$. So, the graph of f is a parabola that opens downward, as shown in Figure 3.7.

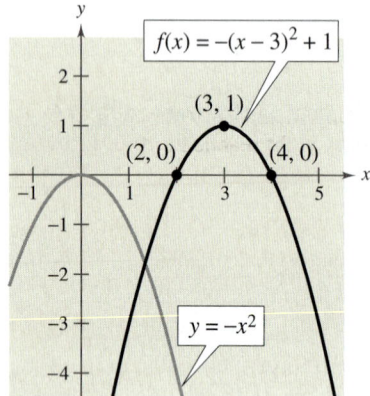

Example 4 ▶ Writing the Equation of a Parabola

Write the standard form of the equation of the parabola whose vertex is $(1, 2)$ and that passes through the point $(0, 0)$, as shown in Figure 3.8.

Solution

Because the vertex of the parabola is at $(h, k) = (1, 2)$, the equation has the form

$$f(x) = a(x - 1)^2 + 2.$$ Substitute for h and k in standard form.

Because the parabola passes through the point $(0, 0)$, it follows that $f(0) = 0$. So,

$$0 = a(0 - 1)^2 + 2 \quad \Rightarrow \quad a = -2$$ Substitute 0 for x; solve for a.

which implies that the equation is

$$f(x) = -2(x - 1)^2 + 2.$$ Substitute for a in standard form.

So, the equation of this parabola is $y = -2(x - 1)^2 + 2$.

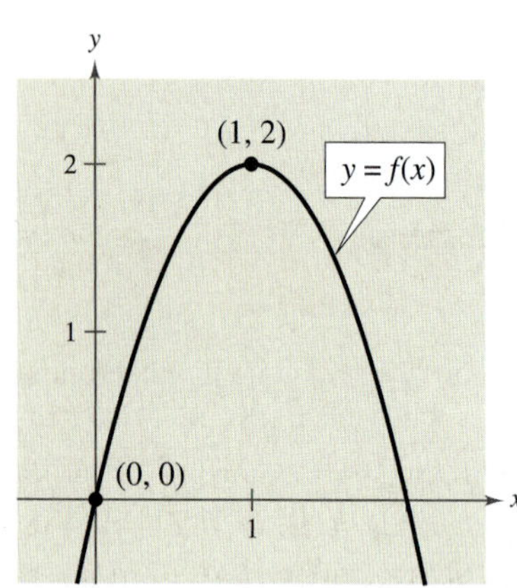

FIGURE 3.8

Application

Many applications involve finding the maximum or minimum value of a quadratic function. Some quadratic functions are not easily written in standard form. For such functions, it is useful to have an alternative method for finding the vertex. For a quadratic function in the form $f(x) = ax^2 + bx + c$, the vertex occurs when $x = -b/2a$. (You are asked to verify this in Exercise 91.)

Vertex of a Parabola

The vertex of the graph of $f(x) = ax^2 + bx + c$ is $\left(-\dfrac{b}{2a}, f\left(-\dfrac{b}{2a}\right)\right)$.

Example 5 ▶ The Maximum Height of a Baseball

A baseball is hit at a point 3 feet above the ground at a velocity of 100 feet per second and at an angle of 45° with respect to the ground. The path of the baseball is given by the function

$$f(x) = -0.0032x^2 + x + 3$$

where $f(x)$ is the height of the baseball (in feet) and x is the horizontal distance from home plate (in feet). What is the maximum height reached by the baseball?

Solution

For this quadratic function, you have

$$\begin{aligned} f(x) &= ax^2 + bx + c \\ &= -0.0032x^2 + x + 3. \end{aligned}$$

So, $a = -0.0032$ and $b = 1$. Because the function has a maximum at $x = -b/2a$, you can conclude that the baseball reaches its maximum height when it is x feet from home plate, where x is

$$\begin{aligned} x &= -\frac{b}{2a} \\ &= -\frac{1}{2(-0.0032)} \qquad \text{Substitute for } a \text{ and } b. \\ &= 156.25 \text{ feet.} \end{aligned}$$

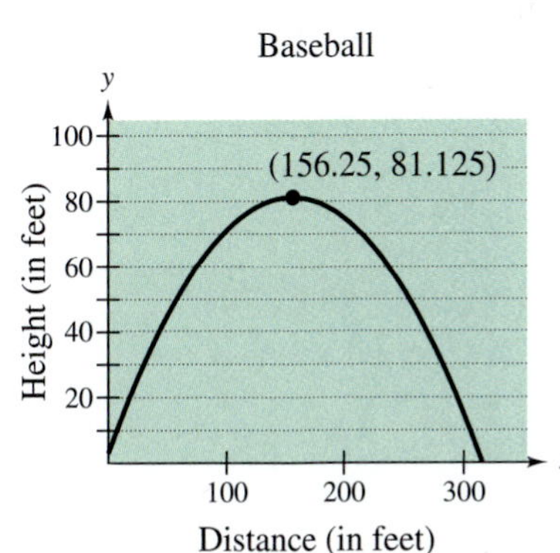

FIGURE 3.9

To find the maximum height, you must determine the value of the function when $x = 156.25$.

$$\begin{aligned} f(156.25) &= -0.0032(156.25)^2 + 156.25 + 3 \\ &= 81.125 \text{ feet.} \end{aligned}$$

The path of the baseball is shown in Figure 3.9. You can estimate from the graph in Figure 3.9 that the ball hits the ground at a distance of about 320 feet from home plate. The actual distance is the x-intercept of the graph of f, which you can find by solving the equation $-0.0032x^2 + x + 3 = 0$ and taking the positive solution, $x \approx 315.5$.

3.1 Exercises

The *Interactive* CD-ROM and *Internet* versions of this text contain step-by-step solutions to all odd-numbered exercises. They also provide Tutorial Exercises for additional help.

In Exercises 1–8, match the quadratic function with its graph. [The graphs are labeled (a), (b), (c), (d), (e), (f), (g), and (h).]

(a)

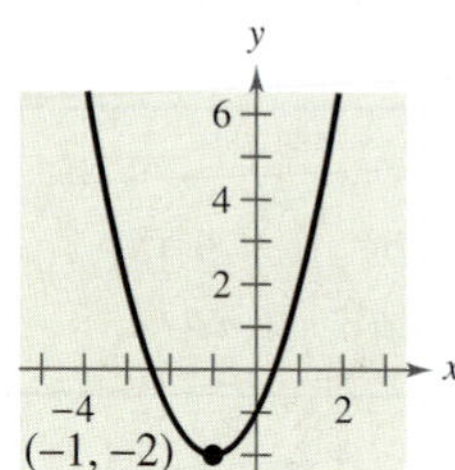

(b)

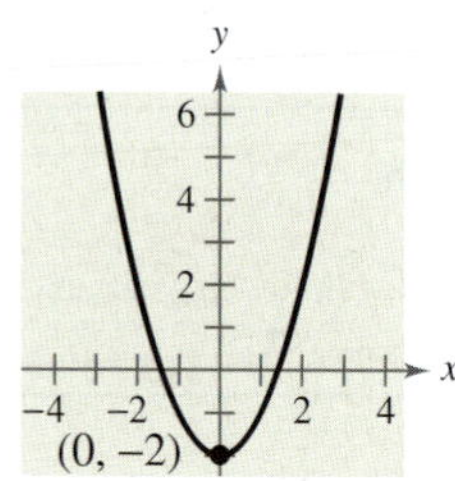

(c)

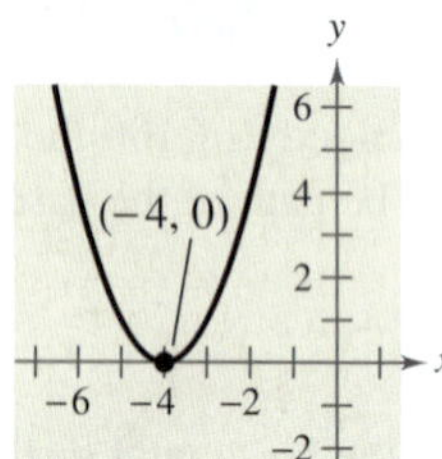

(d)

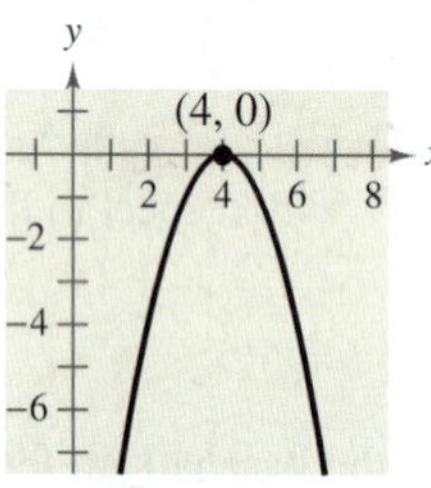

(e)

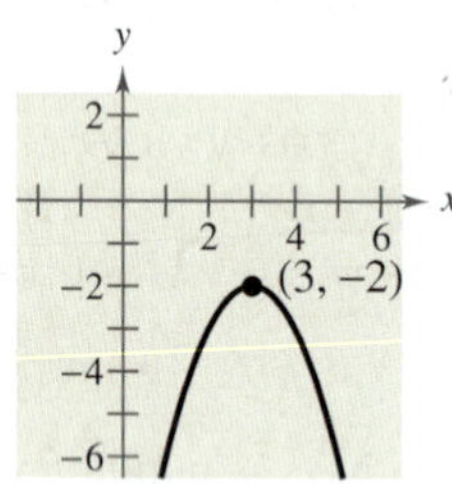

(f)

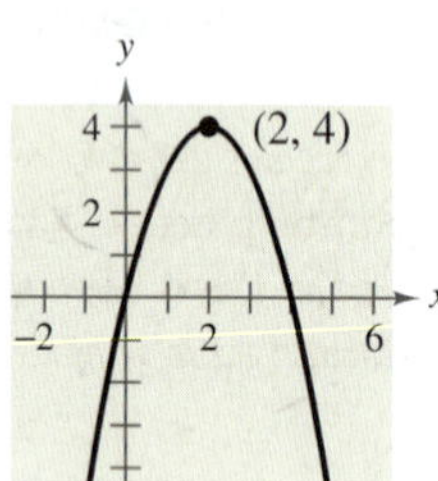

(g)

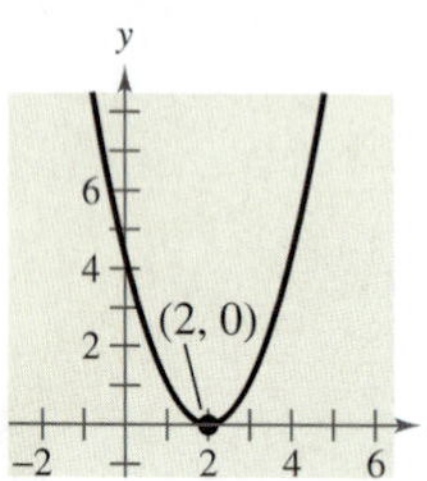

(h)

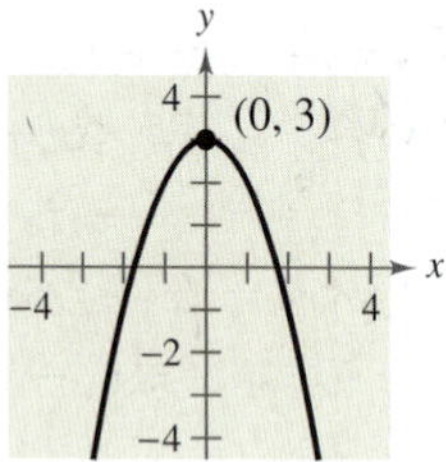

1. $f(x) = (x - 2)^2$

2. $f(x) = (x + 4)^2$

3. $f(x) = x^2 - 2$

4. $f(x) = 3 - x^2$

5. $f(x) = 4 - (x - 2)^2$

6. $f(x) = (x + 1)^2 - 2$

7. $f(x) = -(x - 3)^2 - 2$

8. $f(x) = -(x - 4)^2$

In Exercises 9–12, graph each function. Compare the graph of each function with the graph of $y = x^2$.

9. (a) $f(x) = \frac{1}{2}x^2$ (b) $g(x) = -\frac{1}{8}x^2$
(c) $h(x) = \frac{3}{2}x^2$ (d) $k(x) = -3x^2$

10. (a) $f(x) = x^2 + 1$ (b) $g(x) = x^2 - 1$
(c) $h(x) = x^2 + 3$ (d) $k(x) = x^2 - 3$

11. (a) $f(x) = (x - 1)^2$ (b) $g(x) = (3x)^2 + 1$
(c) $h(x) = \left(\frac{1}{3}x\right)^2 - 3$ (d) $k(x) = (x + 3)^2$

12. (a) $f(x) = -\frac{1}{2}(x - 2)^2 + 1$
(b) $g(x) = \left[\frac{1}{2}(x - 1)\right]^2 - 3$
(c) $h(x) = -\frac{1}{2}(x + 2)^2 - 1$
(d) $k(x) = [2(x + 1)]^2 + 4$

In Exercises 13–28, sketch the graph of the quadratic function without using a graphing utility. Identify the vertex and x-intercept(s).

13. $f(x) = x^2 - 5$

14. $h(x) = 25 - x^2$

15. $f(x) = \frac{1}{2}x^2 - 4$

16. $f(x) = 16 - \frac{1}{4}x^2$

17. $f(x) = (x + 5)^2 - 6$

18. $f(x) = (x - 6)^2 + 3$

19. $h(x) = x^2 - 8x + 16$

20. $g(x) = x^2 + 2x + 1$

21. $f(x) = x^2 - x + \frac{5}{4}$

22. $f(x) = x^2 + 3x + \frac{1}{4}$

23. $f(x) = -x^2 + 2x + 5$

24. $f(x) = -x^2 - 4x + 1$

25. $h(x) = 4x^2 - 4x + 21$

26. $f(x) = 2x^2 - x + 1$

27. $f(x) = \frac{1}{4}x^2 - 2x - 12$

28. $f(x) = -\frac{1}{3}x^2 + 3x - 6$

In Exercises 29–36, use a graphing utility to graph the quadratic function. Identify the vertex and x-intercepts. Then check your results algebraically by writing the quadratic function in standard form.

29. $f(x) = -(x^2 + 2x - 3)$

30. $f(x) = -(x^2 + x - 30)$

31. $g(x) = x^2 + 8x + 11$

32. $f(x) = x^2 + 10x + 14$

33. $f(x) = 2x^2 - 16x + 31$

34. $f(x) = -4x^2 + 24x - 41$

35. $g(x) = \frac{1}{2}(x^2 + 4x - 2)$

36. $f(x) = \frac{3}{5}(x^2 + 6x - 5)$

In Exercises 37–42, find the standard form of the quadratic function.

37.

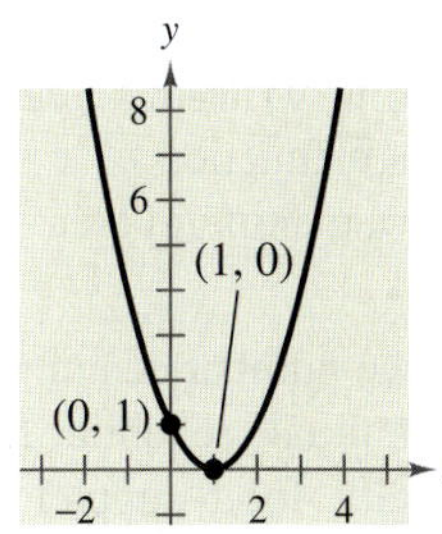

38.

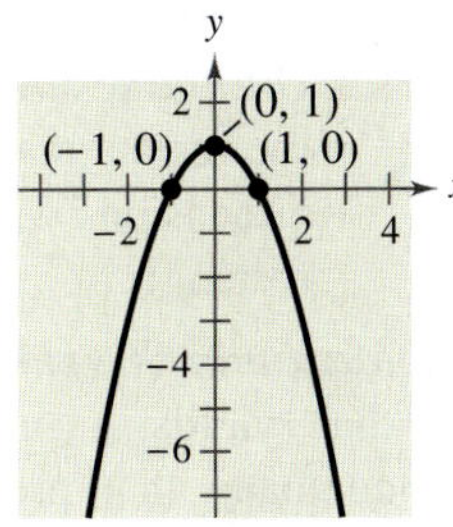

39.

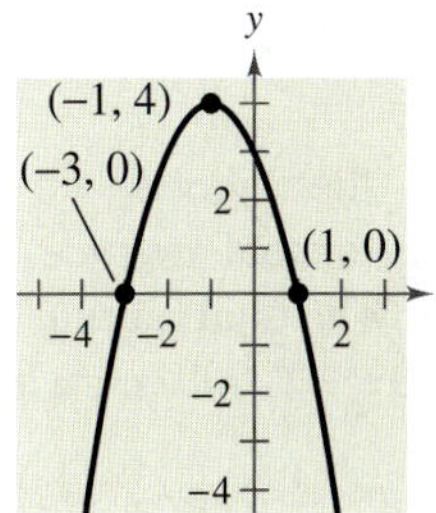

40.

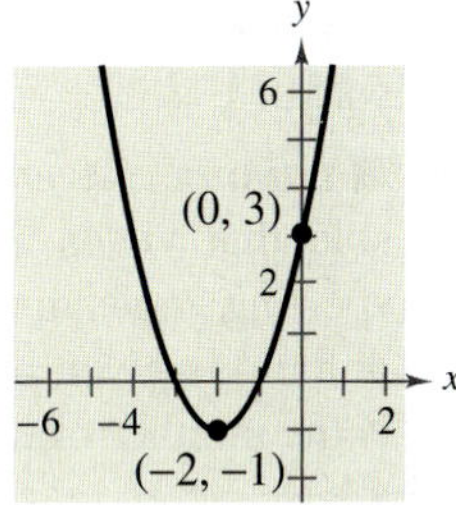

41.

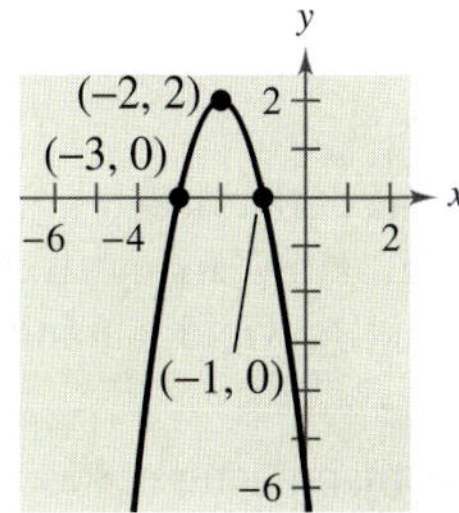

42.

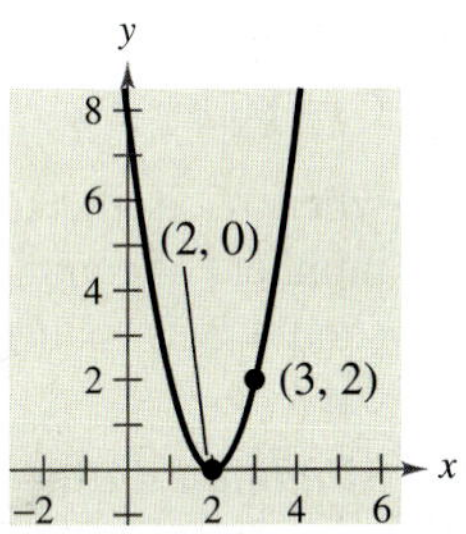

In Exercises 43–52, find the quadratic function that has the indicated vertex and whose graph passes through the given point.

43. Vertex: $(-2, 5)$; Point: $(0, 9)$

44. Vertex: $(4, -1)$; Point: $(2, 3)$

45. Vertex: $(3, 4)$; Point: $(1, 2)$

46. Vertex: $(2, 3)$; Point: $(0, 2)$

47. Vertex: $(5, 12)$; Point: $(7, 15)$

48. Vertex: $(-2, -2)$; Point: $(-1, 0)$

49. Vertex: $\left(-\frac{1}{4}, \frac{3}{2}\right)$; Point: $(-2, 0)$

50. Vertex: $\left(\frac{5}{2}, -\frac{3}{4}\right)$; Point: $(-2, 4)$

51. Vertex: $\left(-\frac{5}{2}, 0\right)$; Point: $\left(-\frac{7}{2}, -\frac{16}{3}\right)$

52. Vertex: $(6, 6)$; Point: $\left(\frac{61}{10}, \frac{3}{2}\right)$

Graphical Reasoning **In Exercises 53–56, determine the x-intercept(s) of the graph visually. Then find the x-intercepts algebraically to confirm your results.**

53. $y = x^2 - 16$

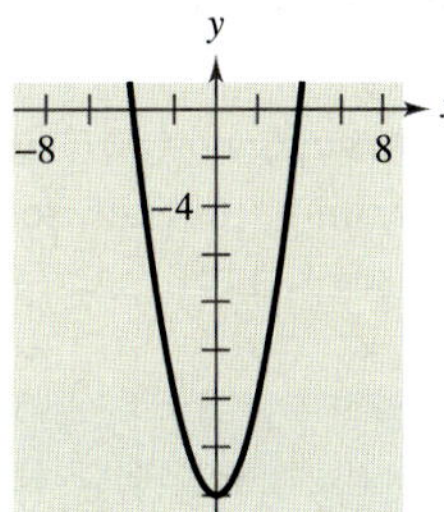

54. $y = x^2 - 6x + 9$

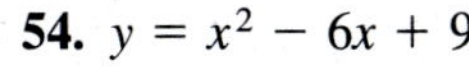

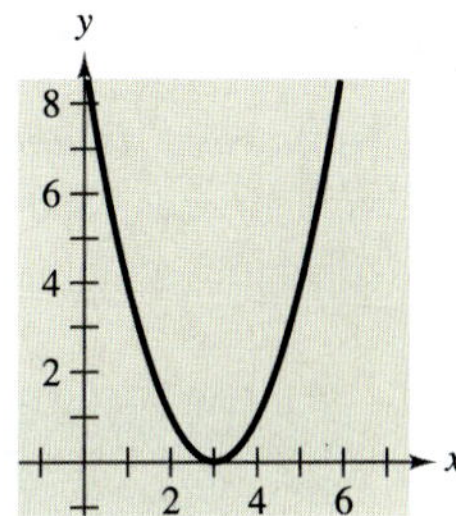

55. $y = x^2 - 4x - 5$

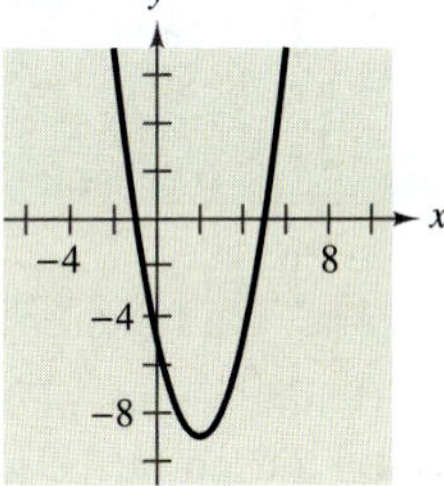

56. $y = 2x^2 + 5x - 3$

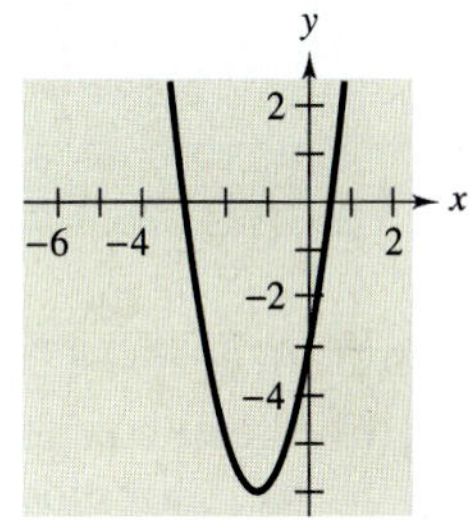

In Exercises 57–64, use a graphing utility to graph the quadratic function. Find the x-intercepts of the graph and compare them with the solutions of the corresponding quadratic equation when $y = 0$.

57. $f(x) = x^2 - 4x$

58. $f(x) = -2x^2 + 10x$

59. $f(x) = x^2 - 9x + 18$

60. $f(x) = x^2 - 8x - 20$

61. $f(x) = 2x^2 - 7x - 30$

62. $f(x) = 4x^2 + 25x - 21$

63. $f(x) = -\frac{1}{2}(x^2 - 6x - 7)$

64. $f(x) = \frac{7}{10}(x^2 + 12x - 45)$

In Exercises 65–70, find two quadratic functions, one that opens upward and one that opens downward, whose graphs have the given x-intercepts. (There are many correct answers.)

65. $(-1, 0), (3, 0)$

66. $(-5, 0), (5, 0)$

67. $(0, 0), (10, 0)$

68. $(4, 0), (8, 0)$

69. $(-3, 0), \left(-\frac{1}{2}, 0\right)$

70. $\left(-\frac{5}{2}, 0\right), (2, 0)$

In Exercises 71–74, find two positive real numbers whose product is a maximum.

71. The sum is 110.

72. The sum is S.

73. The sum of the first and twice the second is 24.

74. The sum of the first and three times the second is 42.

75. ***Numerical, Graphical, and Analytical Analysis*** A rancher has 200 feet of fencing to enclose two adjacent rectangular corrals (see figure).

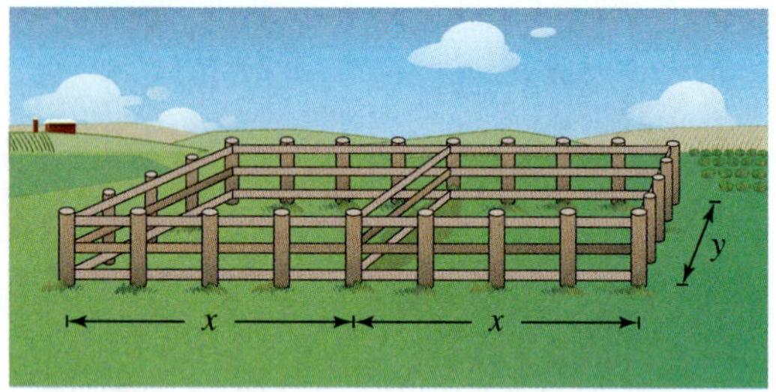

(a) Write the area A of the corral as a function of x.

(b) Create a table showing possible values of x and the corresponding area of the corral. Use the table to estimate the dimensions that will enclose the maximum area.

(c) Use a graphing utility to graph the area function. Use the graph to approximate the dimensions that will produce the maximum enclosed area.

(d) Write the area function in standard form to find analytically the dimensions that will produce the maximum area.

76. ***Geometry*** An indoor physical fitness room consists of a rectangular region with a semicircle on each end (see figure). The perimeter of the room is to be a 200-meter single-lane running track.

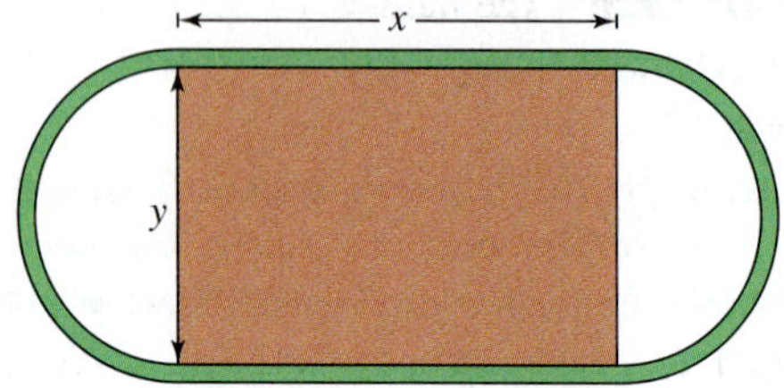

(a) Determine the radius of the semicircular ends of the room. Determine the distance, in terms of y, around the inside edge of the two semicircular parts of the track.

(b) Use the result of part (a) to write an equation, in terms of x and y, for the distance traveled in one lap around the track. Solve for y.

(c) Use the result of part (b) to write the area A of the rectangular region as a function of x. What dimensions will produce a maximum area of the rectangle?

77. ***Maximum Revenue*** Find the number of units sold that yields a maximum annual revenue for a company that produces health food supplements. The total revenue R (in dollars) is given by $R = 900x - 0.1x^2$, where x is the number of units sold.

78. ***Maximum Revenue*** Find the number of units sold that yields a maximum annual revenue for a sporting goods manufacturer. The total revenue R (in dollars) is given by $R = 100x - 0.0002x^2$, where x is the number of units sold.

79. ***Minimum Cost*** A manufacturer of lighting fixtures has daily production costs of

$$C = 800 - 10x + 0.25x^2$$

where C is the total cost (in dollars) and x is the number of units produced. How many fixtures should be produced each day to yield a minimum cost?

80. ***Minimum Cost*** A textile manufacturer has daily production costs of $C = 100{,}000 - 110x + 0.045x^2$, where C is the total cost (in dollars) and x is the number of units produced. How many units should be produced each day to yield a minimum cost?

81. ***Maximum Profit*** The profit P (in dollars) for a company that produces antivirus and system utilities software is

$$P = -0.0002x^2 + 140x - 250{,}000$$

where x is the number of units sold. What sales level will yield a maximum profit?

82. ***Maximum Profit*** The profit P (in hundreds of dollars) that a company makes depends on the amount x (in hundreds of dollars) the company spends on advertising according to the model

$$P = 230 + 20x - 0.5x^2.$$

What expenditure for advertising will yield a maximum profit?

83. ***Height of a Ball*** The height y (in feet) of a ball thrown by a child is

$$y = -\frac{1}{12}x^2 + 2x + 4$$

where x is the horizontal distance (in feet) from the point at which the ball is thrown (see figure).

(a) How high is the ball when it leaves the child's hand? (*Hint:* Find y when $x = 0$.)

(b) What is the maximum height of the ball?

(c) How far from the child does the ball strike the ground?

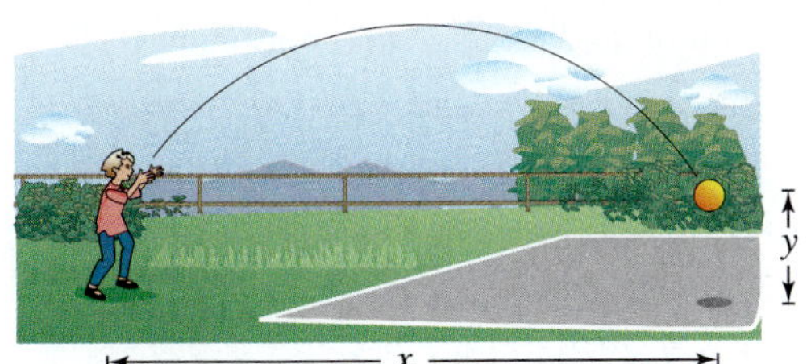

84. ***Path of a Diver*** The path of a diver is

$$y = -\frac{4}{9}x^2 + \frac{24}{9}x + 12$$

where y is the height (in feet) and x is the horizontal distance from the end of the diving board (in feet). What is the maximum height of the diver?

85. ***Graphical Analysis*** From 1960 to 2000, the per capita consumption C of cigarettes by Americans (age 18 and older) can be modeled by

$$C = 4258 + 6.5t - 1.62t^2, \qquad 0 \le t \le 40$$

where t is the year, with $t = 0$ corresponding to 1960. (Source: *Tobacco Situation and Outlook Yearbook*)

(a) Use a graphing utility to graph the model.

(b) Use the graph of the model to approximate the maximum average annual consumption. Beginning in 1966, all cigarette packages were required by law to carry a health warning. Do you think the warning had any effect? Explain.

(c) In 1990, the U.S. population (age 18 and over) was 185,105,441. Of those, about 63,423,167 were smokers. What was the average annual cigarette consumption *per smoker* in 1990? What was the average daily cigarette consumption *per smoker*?

▶ Model It

86. ***Data Analysis*** The numbers y (in thousands) of hairdressers and cosmetologists in the United States for the years 1995 through 2000 are shown in the table. (Source: U.S. Bureau of Labor Statistics)

Year	Number of hairdressers and cosmetologists, y
1995	750
1996	737
1997	748
1998	763
1999	784
2000	820

(a) Use a graphing utility to create a scatter plot of the data. Let x represent the year, with $x = 5$ corresponding to 1995.

(b) Use the *regression* feature of a graphing utility to find a quadratic model for the data.

(c) Use a graphing utility to graph the model in the same viewing window as the scatter plot. How well does the model fit the data?

(d) Use the *trace* feature of the graphing utility to approximate the year in which the number of hairdressers and cosmetologists was the least.

(e) Use the model to predict the number of hairdressers and cosmetologists in 2005.

87. ***Wind Drag*** The number of horsepower y required to overcome wind drag on an automobile is approximated by

$$y = 0.002s^2 + 0.005s - 0.029, \qquad 0 \le s \le 100$$

where s is the speed of the car (in miles per hour).

(a) Use a graphing utility to graph the function.

(b) Graphically estimate the maximum speed of the car if the power required to overcome wind drag is not to exceed 10 horsepower. Verify your estimate analytically.

88. ***Maximum Fuel Economy*** A study was done to compare the speed x (in miles per hour) with the mileage y (in miles per gallon) of an automobile. The results are shown in the table. (Source: Federal Highway Administration)

Speed, x	Mileage, y
15	22.3
20	25.5
25	27.5
30	29.0
35	28.8
40	30.0
45	29.9
50	30.2
55	30.4
60	28.8
65	27.4
70	25.3
75	23.3

(a) Use a graphing utility to create a scatter plot of the data.

(b) Use the *regression* feature of a graphing utility to find a quadratic model for the data.

(c) Use a graphing utility to graph the model in the same viewing window as the scatter plot.

(d) Estimate the speed for which the miles per gallon is greatest.

Synthesis

True or False? **In Exercises 89 and 90, determine whether the statement is true or false. Justify your answer.**

89. The function $f(x) = -12x^2 - 1$ has no x-intercepts.

90. The graphs of

$$f(x) = -4x^2 - 10x + 7$$

and

$$g(x) = 12x^2 + 30x + 1$$

have the same axis of symmetry.

91. Write the quadratic equation

$$f(x) = ax^2 + bx + c$$

in standard form to verify that the vertex occurs at

$$\left(-\frac{b}{2a}, f\left(-\frac{b}{2a}\right)\right).$$

92. ***Profit*** The profit P (in millions of dollars) for a recreational vehicle retailer is modeled by a quadratic function of the form

$$P = at^2 + bt + c$$

where t represents the year. If you were president of the company, which of the models below would you prefer? Explain your reasoning.

(a) a is positive and $-b/(2a) \le t$.

(b) a is positive and $t \le -b/(2a)$.

(c) a is negative and $-b/(2a) \le t$.

(d) a is negative and $t \le -b/(2a)$.

93. Is it possible for a quadratic equation to have only one x-intercept? Explain.

94. Assume that the function

$$f(x) = ax^2 + bx + c, \quad a \neq 0$$

has two real zeros. Show that the x-coordinate of the vertex of the graph is the average of the zeros of f. (*Hint:* Use the Quadratic Formula.)

Review

In Exercises 95–98, find the equation of the line in slope-intercept form that has the given characteristics.

95. Contains the points $(-4, 3)$ and $(2, 1)$

96. Contains the point $\left(\frac{7}{2}, 2\right)$ and has a slope of $\frac{3}{2}$

97. Contains the point $(0, 3)$ and is perpendicular to the line $4x + 5y = 10$

98. Contains the point $(-8, 4)$ and is parallel to the line $y = -3x + 2$

In Exercises 99–104, let $f(x) = 14x - 3$ and let $g(x) = 8x^2$. Find the indicated value.

99. $(f + g)(-3)$

100. $(g - f)(2)$

101. $(fg)\left(-\frac{4}{7}\right)$

102. $\left(\frac{f}{g}\right)(-1.5)$

103. $(f \circ g)(-1)$

104. $(g \circ f)(0)$

3.2 Polynomial Functions of Higher Degree

▶ **What you should learn**

- How to use transformations to sketch graphs of polynomial functions
- How to use the Leading Coefficient Test to determine the end behavior of graphs of polynomial functions
- How to use zeros of polynomial functions as sketching aids
- How to use the Intermediate Value Theorem to help locate zeros of polynomial functions

▶ **Why you should learn it**

You can use polynomial functions to model real-life processes, such as the growth of a red oak tree, as discussed in Exercise 91 on page 282.

Leonard Lee Rue/Earth Scenes

Graphs of Polynomial Functions

In this section you will study basic features of the graphs of polynomial functions. The first feature is that the graph of a polynomial function is **continuous.** Essentially, this means that the graph of a polynomial function has no breaks, holes, or gaps, as shown in Figure 3.10(a).

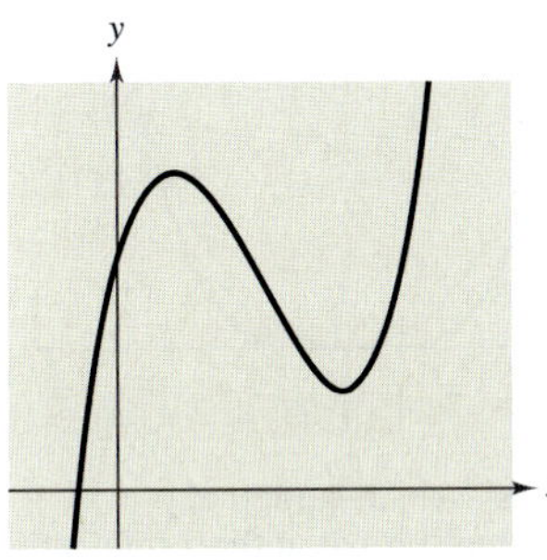

(a) Polynomial functions have continuous graphs.

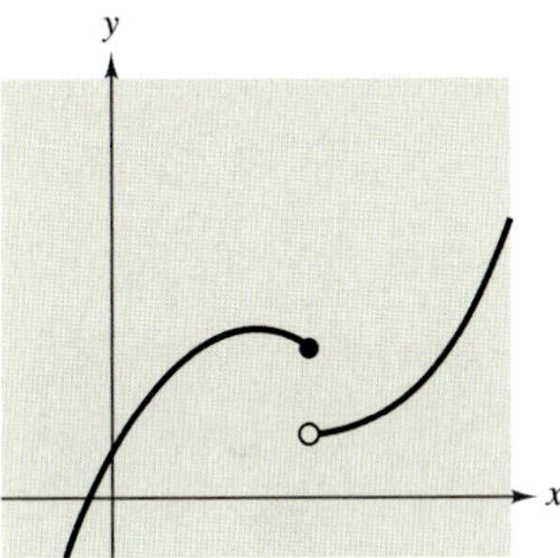

(b) Functions with graphs that are not continuous are not polynomial functions.

FIGURE 3.10

The second feature is that the graph of a polynomial function has only smooth, rounded turns, as shown in Figure 3.11. A polynomial function cannot have a sharp turn. For instance, the function $f(x) = |x|$, which has a sharp turn at the point $(0, 0)$, as shown in Figure 3.12, is not a polynomial function.

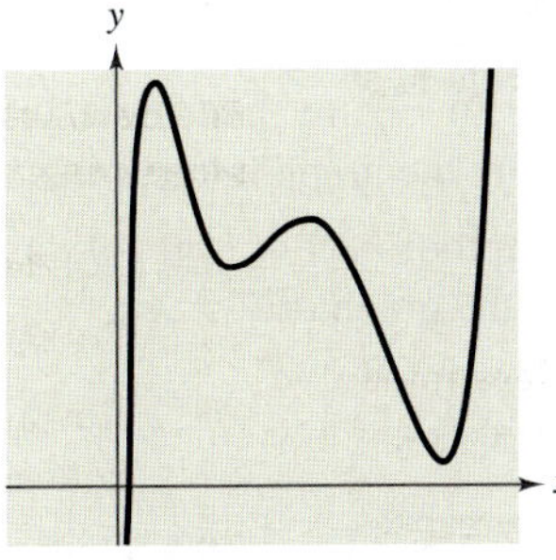

Polynomial functions have graphs with rounded turns.
FIGURE 3.11

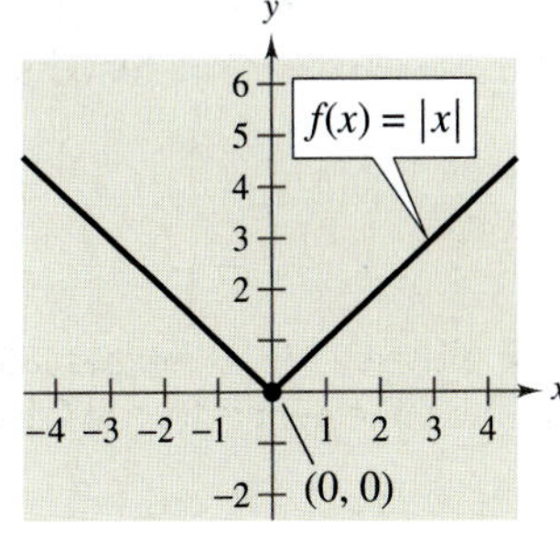

Functions whose graphs have sharp turns are not polynomial functions.
FIGURE 3.12

The graphs of polynomial functions of degree greater than 2 are more difficult to analyze than the graphs of polynomials of degree 0, 1, or 2. However, using the features presented in this section, together with point plotting, intercepts, and symmetry, you should be able to make reasonably accurate sketches *by hand.*

STUDY TIP

For power functions $f(x) = x^n$, if n is even, then the graph of the function is symmetric with respect to the y-axis, and if n is odd, then the graph of the function is symmetric with respect to the origin.

The polynomial functions that have the simplest graphs are monomials of the form $f(x) = x^n$, where n is an integer greater than zero. From Figure 3.13, you can see that when n is *even*, the graph is similar to the graph of $f(x) = x^2$, and when n is *odd*, the graph is similar to the graph of $f(x) = x^3$. Moreover, the greater the value of n, the flatter the graph near the origin. Polynomial functions of the form $f(x) = x^n$ are often referred to as **power functions.**

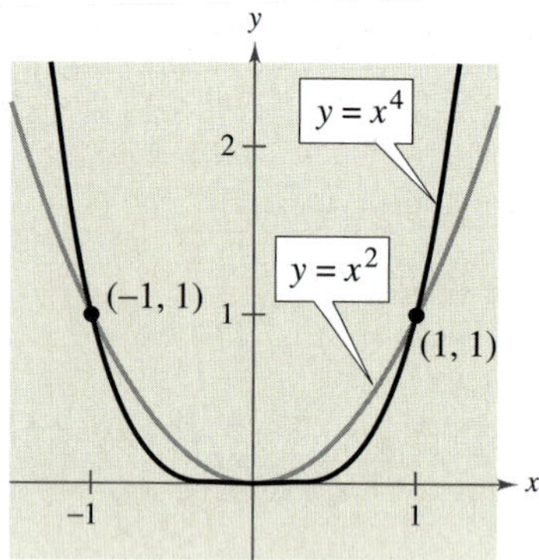

(a) If n is even, the graph of $y = x^n$ touches the axis at the x-intercept.

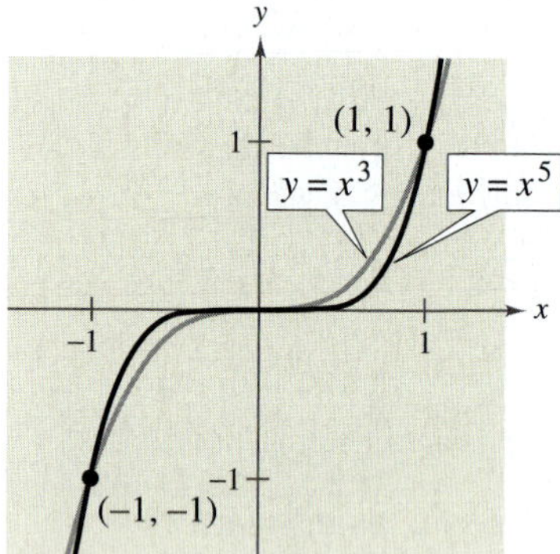

(b) If n is odd, the graph of $y = x^n$ crosses the axis at the x-intercept.

FIGURE 3.13

Example 1 ▶ Sketching Transformations of Monomial Functions

Sketch the graph of each function.

a. $f(x) = -x^5$

b. $h(x) = (x + 1)^4$

Solution

a. Because the degree of $f(x) = -x^5$ is odd, its graph is similar to the graph of $y = x^3$. In Figure 3.14, note that the negative coefficient has the effect of reflecting the graph in the x-axis.

b. The graph of $h(x) = (x + 1)^4$, as shown in Figure 3.15, is a left shift by one unit of the graph of $y = x^4$.

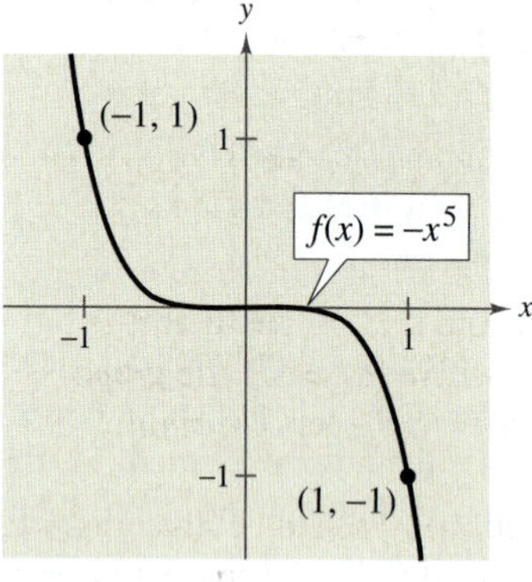

FIGURE 3.14

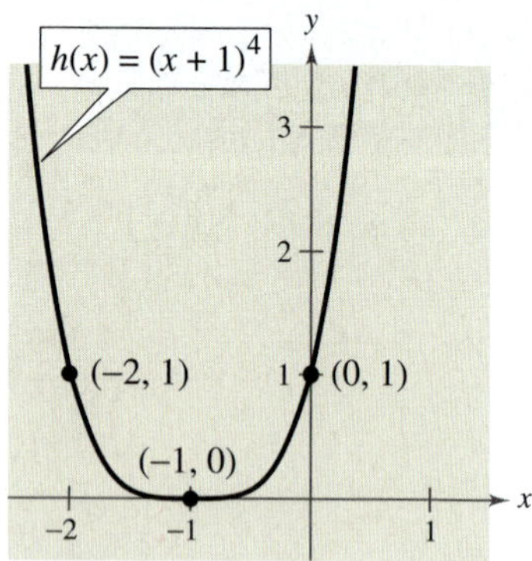

FIGURE 3.15

The Leading Coefficient Test

In Example 1, note that both graphs eventually rise or fall without bound as x moves to the right. Whether the graph of a polynomial function eventually rises or falls can be determined by the function's degree (even or odd) and by its leading coefficient, as indicated in the **Leading Coefficient Test.**

Leading Coefficient Test

As x moves without bound to the left or to the right, the graph of the polynomial function $f(x) = a_n x^n + \cdot \cdot \cdot + a_1 x + a_0$ eventually rises or falls in the following manner.

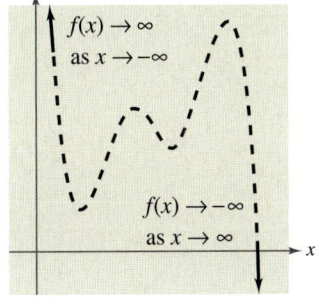

1. When n is *odd:*

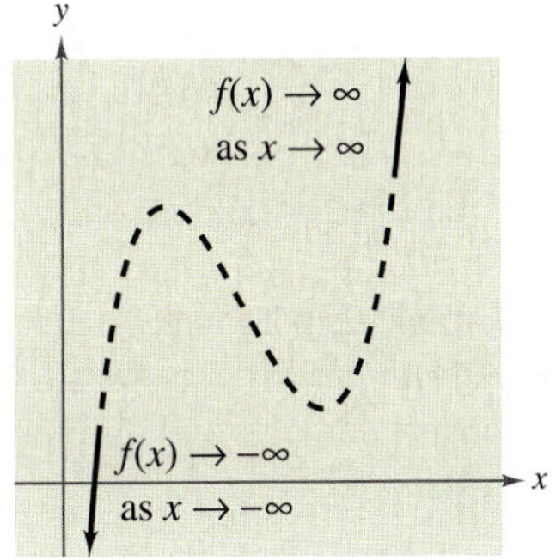

If the leading coefficient is positive ($a_n > 0$), the graph falls to the left and rises to the right.

y
f(x) → ∞
as x → −∞
f(x) → −∞
as x → ∞
x

If the leading coefficient is negative ($a_n < 0$), the graph rises to the left and falls to the right.

2. When n is *even:*

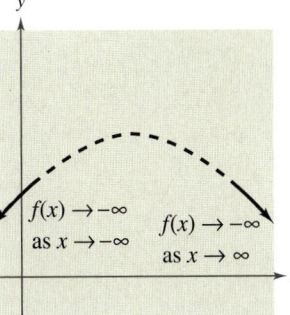

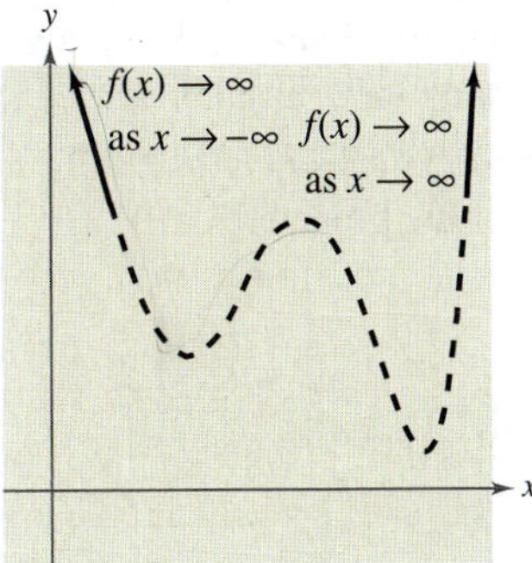

If the leading coefficient is positive ($a_n > 0$), the graph rises to the left and right.

y
f(x) → −∞
as x → −∞
f(x) → −∞
as x → ∞
x

If the leading coefficient is negative ($a_n < 0$), the graph falls to the left and right.

The dashed portions of the graphs indicate that the test determines *only* the right-hand and left-hand behavior of the graph.

Exploration

For each function, identify the degree of the function and whether the degree of the function is even or odd. Identify the leading coefficient and whether the leading coefficient is greater than 0 or less than 0. Use a graphing utility to graph each function. Describe the relationship between the degree and the sign of the leading coefficient of the function and the right- and left-hand behavior of the graph of the function.

a. $f(x) = x^3 - 2x^2 - x + 1$

b. $f(x) = 2x^5 + 2x^2 - 5x + 1$

c. $f(x) = -2x^5 - x^2 + 5x + 3$

d. $f(x) = -x^3 + 5x - 2$

e. $f(x) = 2x^2 + 3x - 4$

f. $f(x) = x^4 - 3x^2 + 2x - 1$

g. $f(x) = x^2 + 3x + 2$

STUDY TIP

The notation "$f(x) \to -\infty$ as $x \to -\infty$" indicates that the graph falls to the left. The notation "$f(x) \to \infty$ as $x \to \infty$" indicates that the graph rises to the right.

Example 2 ▶ Applying the Leading Coefficient Test

Describe the right-hand and left-hand behavior of the graph of $f(x) = -x^3 + 4x$.

Solution

Because the degree is odd and the leading coefficient is negative, the graph rises to the left and falls to the right, as shown in Figure 3.16.

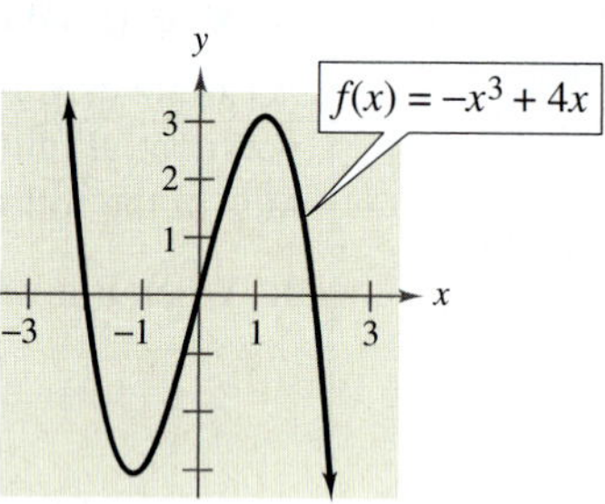

FIGURE 3.16

In Example 2, note that the Leading Coefficient Test tells you only whether the graph *eventually* rises or falls to the right or left. Other characteristics of the graph, such as intercepts and minimum and maximum points, must be determined by other tests.

Example 3 ▶ Applying the Leading Coefficient Test 

Describe the right-hand and left-hand behavior of the graph of each function.

a. $f(x) = x^4 - 5x^2 + 4$ **b.** $f(x) = x^5 - x$

Solution

a. Because the degree is even and the leading coefficient is positive, the graph rises to the left and right, as shown in Figure 3.17.

b. Because the degree is odd and the leading coefficient is positive, the graph falls to the left and rises to the right, as shown in Figure 3.18.

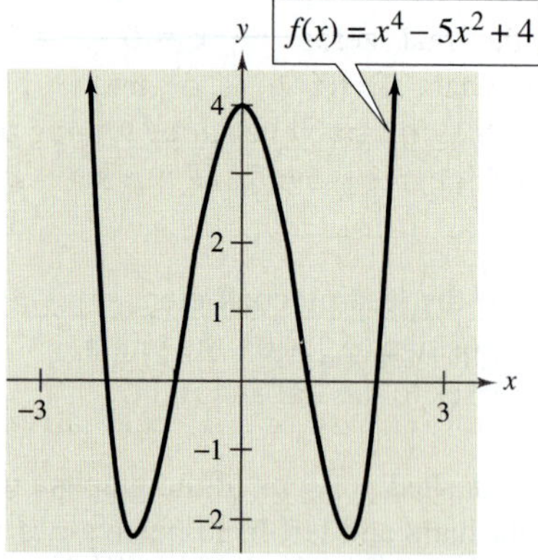

FIGURE 3.17

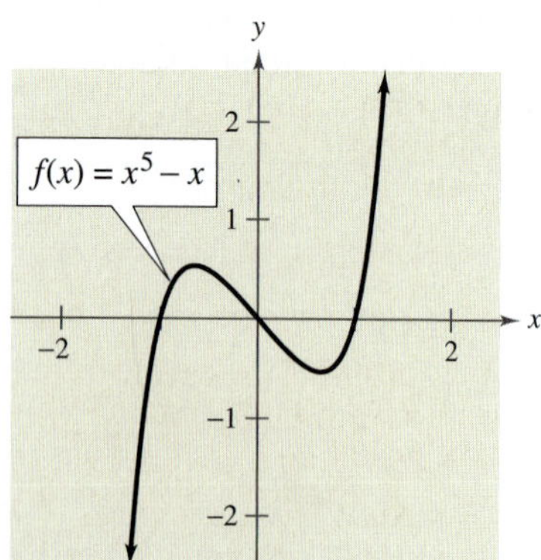

FIGURE 3.18

STUDY TIP

Remember that the *zeros* of a function of x are the x-values for which the function is zero.

Zeros of Polynomial Functions

It can be shown that for a polynomial function f of degree n, the following statements are true.

1. The graph of f has, at most, $n - 1$ turning points. (Turning points are points at which the graph changes from increasing to decreasing or vice versa.)
2. The function f has, at most, n real zeros. (You will study this result in detail in Section 3.4 on the Fundamental Theorem of Algebra.)

Finding the zeros of polynomial functions is one of the most important problems in algebra. There is a strong interplay between graphical and algebraic approaches to this problem. Sometimes you can use information about the graph of a function to help find its zeros, and in other cases you can use information about the zeros of a function to help sketch its graph.

Real Zeros of Polynomial Functions

If f is a polynomial function and a is a real number, the following statements are equivalent.

1. $x = a$ is a *zero* of the function f.
2. $x = a$ is a *solution* of the polynomial equation $f(x) = 0$.
3. $(x - a)$ is a *factor* of the polynomial $f(x)$.
4. $(a, 0)$ is an *x-intercept* of the graph of f.

Example 4 ▶ Finding the Zeros of a Polynomial Function 

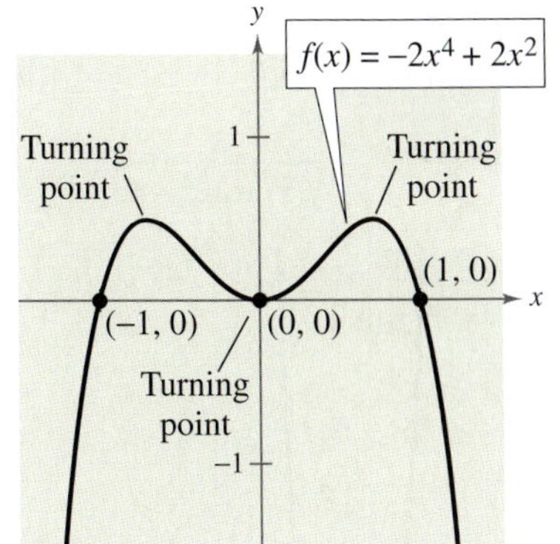

FIGURE 3.19

Find all real zeros of $f(x) = -2x^4 + 2x^2$. Use the graph in Figure 3.19 to determine the number of turning points of the graph of the function.

Solution

In this case, the polynomial factors as follows.

$$f(x) = -2x^2(x^2 - 1) \qquad \text{Remove common monomial factor.}$$

$$= -2x^2(x - 1)(x + 1) \qquad \text{Factor completely.}$$

So, the real zeros are $x = 0$, $x = 1$, and $x = -1$, and the corresponding x-intercepts are $(0, 0)$, $(1, 0)$, and $(-1, 0)$, as shown in Figure 3.19. Note in the figure that the graph has three turning points. This is consistent with the fact that a fourth-degree polynomial can have *at most* three turning points.

STUDY TIP

In Example 4, note that because k is even, the factor $-2x^2$ yields the repeated zero $x = 0$. The graph touches the x-axis at $x = 0$, as shown in Figure 3.19.

Repeated Zeros

A factor $(x - a)^k$, $k > 1$, yields a **repeated zero** $x = a$ of **multiplicity** k.

1. If k is odd, the graph *crosses* the x-axis at $x = a$.
2. If k is even, the graph *touches* the x-axis (but does not cross the x-axis) at $x = a$.

Technology

Example 5 uses an "algebraic approach" to describe the graph of the function. A graphing utility is a complement to this approach. Remember that an important aspect of using a graphing utility is to find a viewing window that shows all significant features of the graph. For instance, which of the graphs below shows all of the significant features of the function in Example 5?

a.

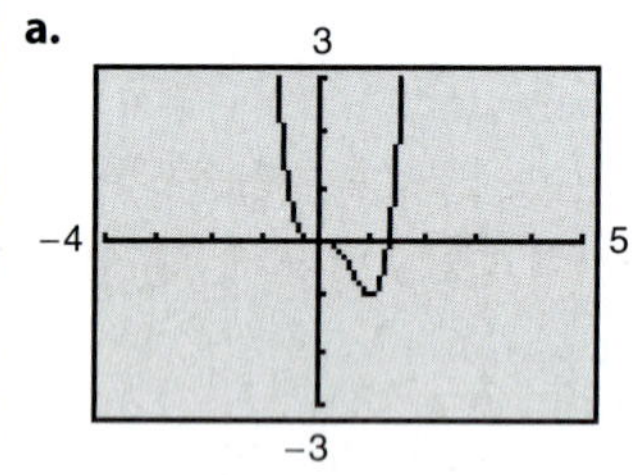

b.

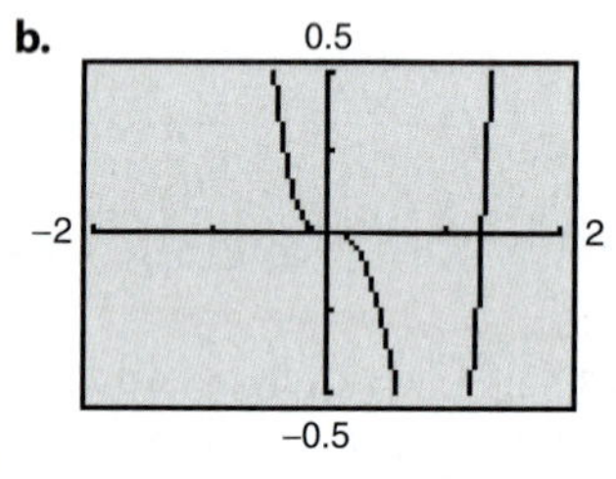

Example 5 ▶ Sketching the Graph of a Polynomial Function

Sketch the graph of $f(x) = 3x^4 - 4x^3$.

Solution

1. *Apply the Leading Coefficient Test.* Because the leading coefficient is positive and the degree is even, you know that the graph eventually rises to the left and to the right (see Figure 3.20).
2. *Find the Zeros of the Polynomial.* By factoring

$$f(x) = 3x^4 - 4x^3 = x^3(3x - 4) \qquad \text{Remove common factor.}$$

you can see that the zeros of f are $x = 0$ and $x = \frac{4}{3}$ (both of odd multiplicity). So, the x-intercepts occur at $(0, 0)$ and $\left(\frac{4}{3}, 0\right)$. Add these points to your graph, as shown in Figure 3.20.

3. *Plot a Few Additional Points.* To sketch the graph by hand, find a few additional points, as shown in the table. Then plot the points (see Figure 3.21).

x	$f(x)$
-1	7
0.5	-0.3125
1	-1
1.5	1.6875

4. *Draw the Graph.* Draw a continuous curve through the points, as shown in Figure 3.21. Because both zeros are of odd multiplicity, you know that the graph should cross the x-axis at $x = 0$ and $x = \frac{4}{3}$. If you are unsure of the shape of that portion of the graph, plot some additional points.

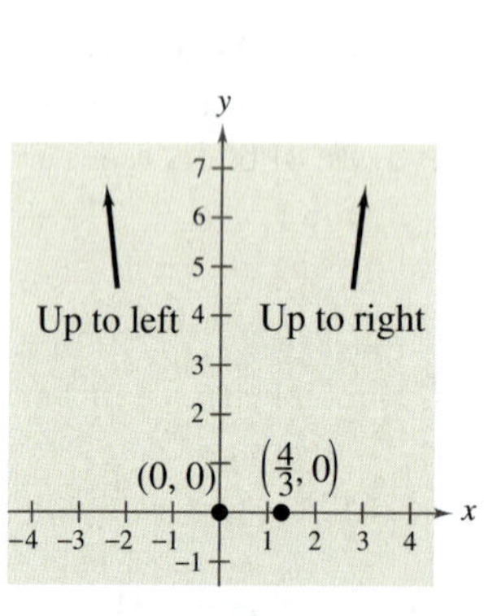

FIGURE 3.20

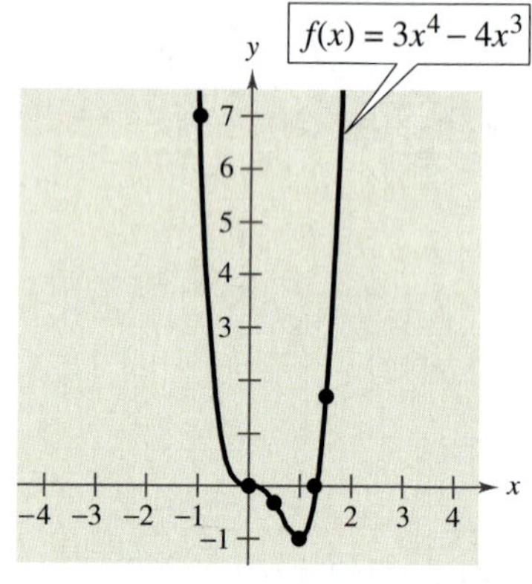

FIGURE 3.21

A polynomial function is written in **standard form** if its terms are written in descending order of exponents from left to right. Before applying the Leading Coefficient Test to a polynomial function, it is a good idea to check that the polynomial function is written in standard form.

Example 6 ▶ Sketching the Graph of a Polynomial Function

Sketch the graph of $f(x) = -2x^3 + 6x^2 - \frac{9}{2}x$.

Solution

1. *Apply the Leading Coefficient Test.* Because the leading coefficient is negative and the degree is odd, you know that the graph eventually rises to the left and falls to the right (see Figure 3.22).

2. *Find the Zeros of the Polynomial.* By factoring

$$\begin{aligned} f(x) &= -2x^3 + 6x^2 - \frac{9}{2}x \\ &= -\frac{1}{2}x(4x^2 - 12x + 9) \\ &= -\frac{1}{2}x(2x - 3)^2 \end{aligned}$$

you can see that the zeros of f are $x = 0$ (odd multiplicity) and $x = \frac{3}{2}$ (even multiplicity). So, the x-intercepts occur at $(0, 0)$ and $\left(\frac{3}{2}, 0\right)$. Add these points to your graph, as shown in Figure 3.22.

3. *Plot a Few Additional Points.* To sketch the graph by hand, find a few additional points, as shown in the table. Then plot the points (see Figure 3.23).

x	$f(x)$
-0.5	4
0.5	-1
1	-0.5
2	-1

4. *Draw the Graph.* Draw a continuous curve through the points, as shown in Figure 3.23. As indicated by the multiplicities of the zeros, the graph crosses the x-axis at $(0, 0)$ but does not cross the x-axis at $\left(\frac{3}{2}, 0\right)$.

STUDY TIP

Observe in Example 6 that the sign of $f(x)$ is positive to the left of and negative to the right of the zero $x = 0$. Similarly, the sign of $f(x)$ is negative to the left and to the right of the zero $x = \frac{3}{2}$. This suggests that if the zero of a polynomial function is of *odd* multiplicity, then the sign of $f(x)$ changes from one side to the other side of the zero. If the zero is of *even* multiplicity, then the sign of $f(x)$ does not change from one side of the zero to the other side. The following table helps to illustrate this result.

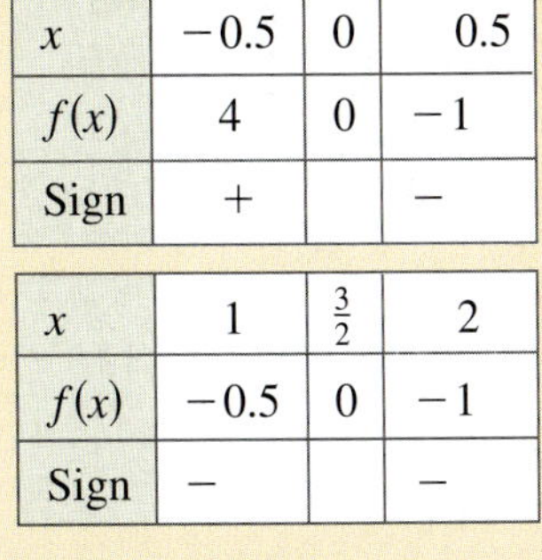

x	-0.5	0	0.5
$f(x)$	4	0	-1
Sign	+		−

x	1	$\frac{3}{2}$	2
$f(x)$	-0.5	0	-1
Sign	−		−

This sign analysis may be helpful in graphing polynomial functions.

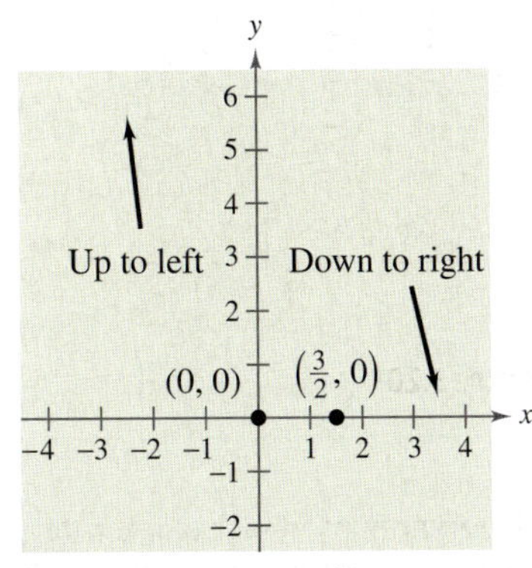

FIGURE 3.22

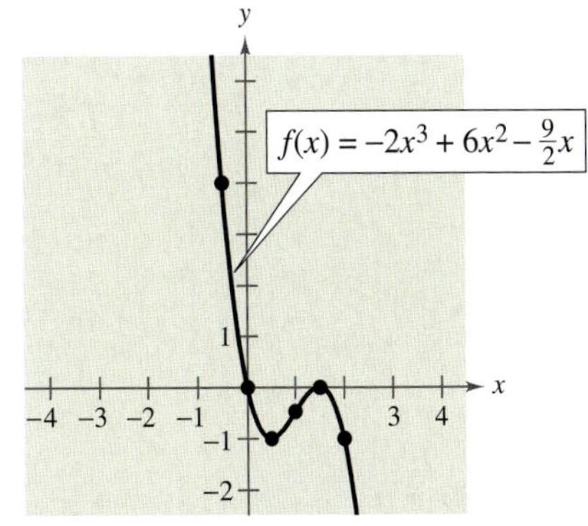

FIGURE 3.23

The Intermediate Value Theorem

The next theorem, called the **Intermediate Value Theorem,** tells you of the existence of real zeros of polynomial functions. This theorem implies that if $(a, f(a))$ and $(b, f(b))$ are two points on the graph of a polynomial function such that $f(a) \neq f(b)$, then for any number d between $f(a)$ and $f(b)$ there must be a number c between a and b such that $f(c) = d$. (See Figure 3.24.)

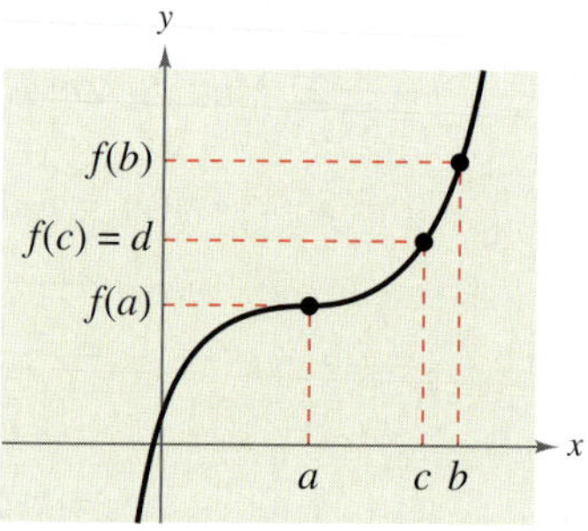

FIGURE 3.24

> **Intermediate Value Theorem**
>
> Let a and b be real numbers such that $a < b$. If f is a polynomial function such that $f(a) \neq f(b)$, then, in the interval $[a, b]$, f takes on every value between $f(a)$ and $f(b)$.

The Intermediate Value Theorem helps you locate the real zeros of a polynomial function in the following way. If you can find a value $x = a$ at which a polynomial function is positive, and another value $x = b$ at which it is negative, you can conclude that the function has at least one real zero between these two values. For example, the function $f(x) = x^3 + x^2 + 1$ is negative when $x = -2$ and positive when $x = -1$. Therefore, it follows from the Intermediate Value Theorem that f must have a real zero somewhere between -2 and -1, as shown in Figure 3.25.

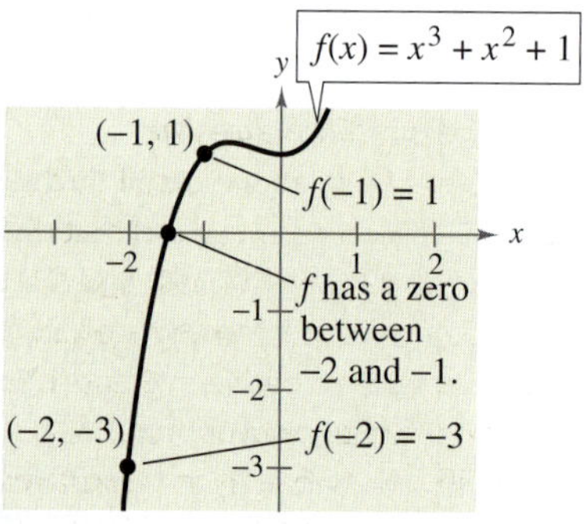

FIGURE 3.25

By continuing this line of reasoning, you can approximate any real zeros of a polynomial function to any desired accuracy. This concept is further demonstrated in Example 7.

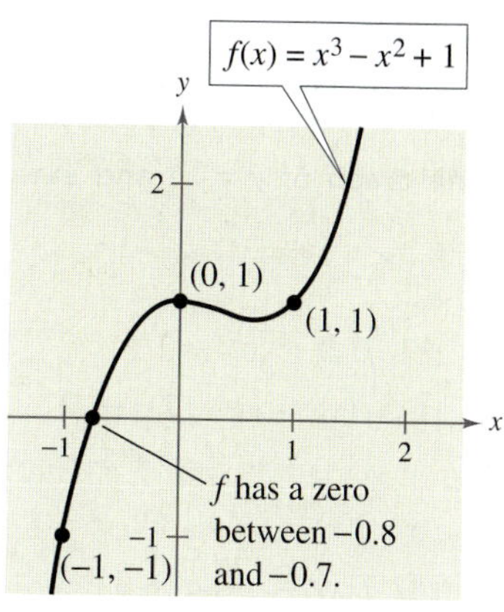

FIGURE 3.26

Example 7 ▶ Approximating a Zero of a Polynomial Function

Use the Intermediate Value Theorem to approximate the real zero of

$$f(x) = x^3 - x^2 + 1.$$

Solution

Begin by computing a few function values, as follows.

x	$f(x)$
-2	-11
-1	-1
0	1
1	1

Because $f(-1)$ is negative and $f(0)$ is positive, you can apply the Intermediate Value Theorem to conclude that the function has a zero between -1 and 0. To pinpoint this zero more closely, divide the interval $[-1, 0]$ into tenths and evaluate the function at each point. When you do this, you will find that

$$f(-0.8) = -0.152 \qquad \text{and} \qquad f(-0.7) = 0.167.$$

So, f must have a zero between -0.8 and -0.7, as shown in Figure 3.26. For a more accurate approximation, compute function values between $f(-0.8)$ and $f(-0.7)$ and apply the Intermediate Value Theorem again. By continuing this process, you can approximate this zero to any desired accuracy.

Technology

You can use the *table* feature of a graphing utility to approximate the zeros of a polynomial function. For instance, for the function

$$f(x) = -2x^3 - 3x^2 + 3$$

create a table that shows the function values for $-20 \le x \le 20$, as shown in the table above. Scroll through the table looking for consecutive function values that differ in sign. From the table above, you can see that $f(0)$ and $f(1)$ differ in sign. So, you can conclude from the Intermediate Value Theorem that the function has a zero between 0 and 1. You can adjust your table to show function values for $0 \le x \le 1$ using increments of 0.1, as shown below. By scrolling through the table on the right, you can see that $f(0.8)$ and $f(0.9)$ differ in sign. So, the function has a zero between 0.8 and 0.9. If you repeat this process several times, you should obtain $x \approx 0.806$ as the zero of the function. Use the *zero* or *root* feature of a graphing utility to confirm this result.

X	Y1
-2	7
-1	2
0	3
1	-2
2	-25
3	-78
4	-173

X=1

X	Y1
.4	2.392
.5	2
.6	1.488
.7	.844
.8	.056
.9	-.888
1	-2

X=.9

3.2 Exercises

In Exercises 1–8, match the polynomial function with its graph. [The graphs are labeled (a), (b), (c), (d), (e), (f), (g), and (h).]

(a)

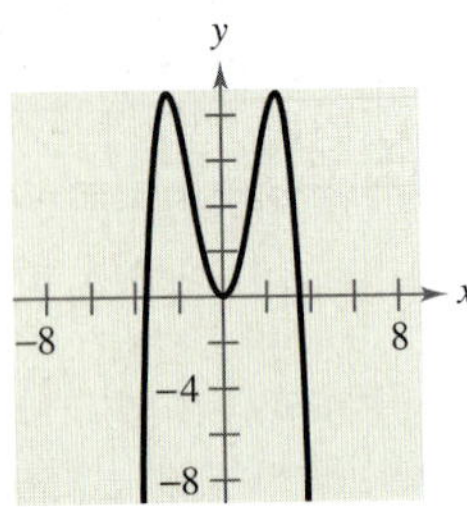

(b)

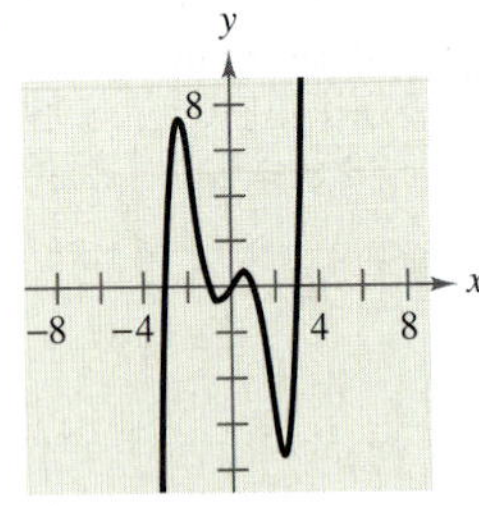

(c)

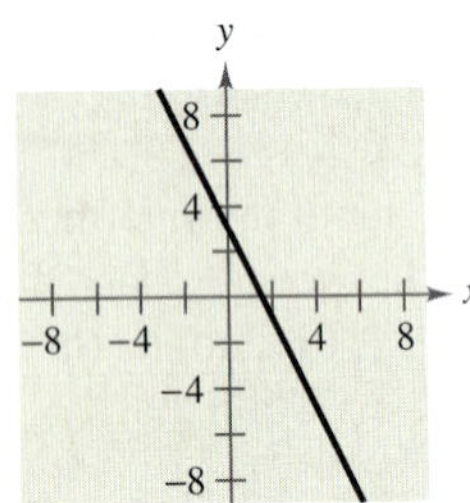

(d)

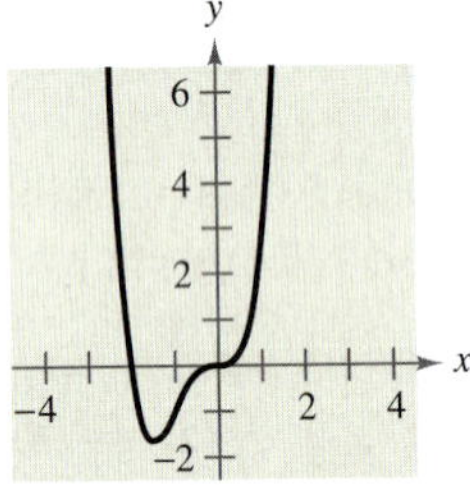

(e)

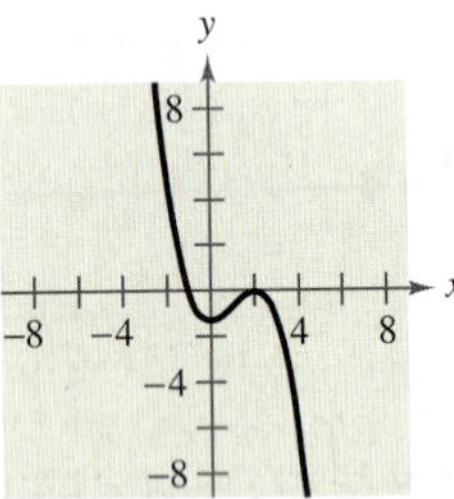

(f)

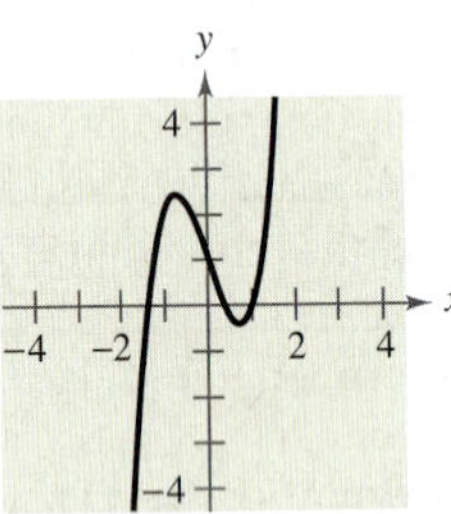

(g)

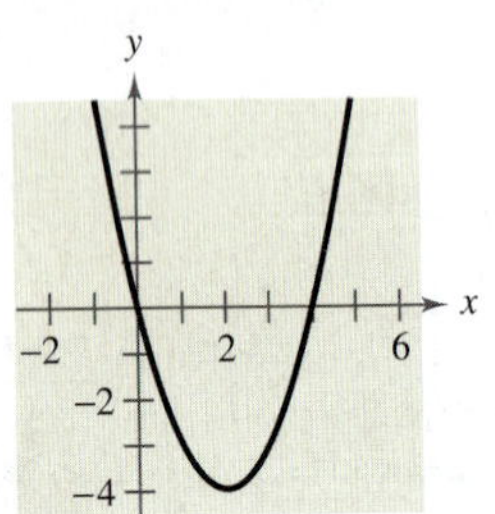

(h)

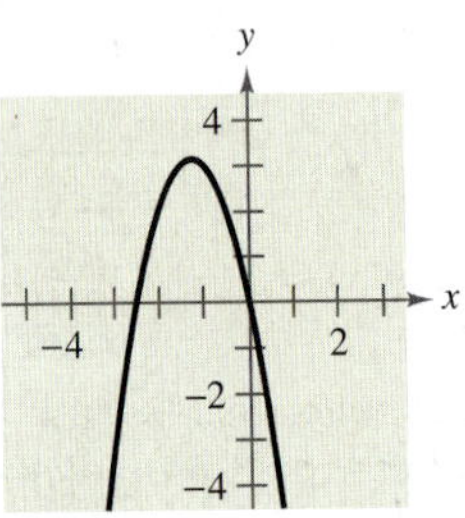

1. $f(x) = -2x + 3$

2. $f(x) = x^2 - 4x$

3. $f(x) = -2x^2 - 5x$

4. $f(x) = 2x^3 - 3x + 1$

5. $f(x) = -\frac{1}{4}x^4 + 3x^2$

6. $f(x) = -\frac{1}{3}x^3 + x^2 - \frac{4}{3}$

7. $f(x) = x^4 + 2x^3$

8. $f(x) = \frac{1}{5}x^5 - 2x^3 + \frac{9}{5}x$

In Exercises 9–12, sketch the graph of $y = x^n$ and each transformation.

9. $y = x^3$

(a) $f(x) = (x - 2)^3$ (b) $f(x) = x^3 - 2$
(c) $f(x) = -\frac{1}{2}x^3$ (d) $f(x) = (x - 2)^3 - 2$

10. $y = x^5$

(a) $f(x) = (x + 1)^5$ (b) $f(x) = x^5 + 1$
(c) $f(x) = 1 - \frac{1}{2}x^5$ (d) $f(x) = -\frac{1}{2}(x + 1)^5$

11. $y = x^4$

(a) $f(x) = (x + 3)^4$ (b) $f(x) = x^4 - 3$
(c) $f(x) = 4 - x^4$ (d) $f(x) = \frac{1}{2}(x - 1)^4$
(e) $f(x) = (2x)^4 + 1$ (f) $f(x) = \left(\frac{1}{2}x\right)^4 - 2$

12. $y = x^6$

(a) $f(x) = -\frac{1}{8}x^6$ (b) $f(x) = (x + 2)^6 - 4$
(c) $f(x) = x^6 - 4$ (d) $f(x) = -\frac{1}{4}x^6 + 1$
(e) $f(x) = \left(\frac{1}{4}x\right)^6 - 2$ (f) $f(x) = (2x)^6 - 1$

In Exercises 13–22, determine the right-hand and left-hand behavior of the graph of the polynomial function.

13. $f(x) = \frac{1}{3}x^3 + 5x$

14. $f(x) = 2x^2 - 3x + 1$

15. $g(x) = 5 - \frac{7}{2}x - 3x^2$

16. $h(x) = 1 - x^6$

17. $f(x) = -2.1x^5 + 4x^3 - 2$

18. $f(x) = 2x^5 - 5x + 7.5$

19. $f(x) = 6 - 2x + 4x^2 - 5x^3$

20. $f(x) = \dfrac{3x^4 - 2x + 5}{4}$

21. $h(t) = -\frac{2}{3}(t^2 - 5t + 3)$

22. $f(s) = -\frac{7}{8}(s^3 + 5s^2 - 7s + 1)$

Graphical Analysis **In Exercises 23–26, use a graphing utility to graph the functions f and g in the same viewing window. Zoom out sufficiently far to show that the right-hand and left-hand behaviors of f and g appear identical.**

23. $f(x) = 3x^3 - 9x + 1$, $g(x) = 3x^3$

24. $f(x) = -\frac{1}{3}(x^3 - 3x + 2)$, $g(x) = -\frac{1}{3}x^3$

25. $f(x) = -(x^4 - 4x^3 + 16x)$, $g(x) = -x^4$

26. $f(x) = 3x^4 - 6x^2$, $g(x) = 3x^4$

In Exercises 27–42, find all the real zeros of the polynomial function. Determine the multiplicity of each zero.

27. $f(x) = x^2 - 25$ **28.** $f(x) = 49 - x^2$

29. $h(t) = t^2 - 6t + 9$ **30.** $f(x) = x^2 + 10x + 25$

31. $f(x) = \frac{1}{3}x^2 + \frac{1}{3}x - \frac{2}{3}$ **32.** $f(x) = \frac{1}{2}x^2 + \frac{5}{2}x - \frac{3}{2}$

33. $f(x) = 3x^3 - 12x^2 + 3x$

34. $g(x) = 5x(x^2 - 2x - 1)$

35. $f(t) = t^3 - 4t^2 + 4t$

36. $f(x) = x^4 - x^3 - 20x^2$

37. $g(t) = t^5 - 6t^3 + 9t$

38. $f(x) = x^5 + x^3 - 6x$

39. $f(x) = 5x^4 + 15x^2 + 10$

40. $f(x) = 2x^4 - 2x^2 - 40$

41. $g(x) = x^3 + 3x^2 - 4x - 12$

42. $f(x) = x^3 - 4x^2 - 25x + 100$

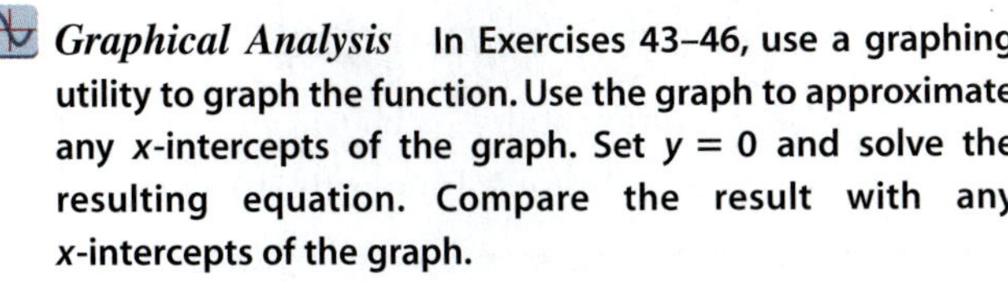

Graphical Analysis **In Exercises 43–46, use a graphing utility to graph the function. Use the graph to approximate any *x*-intercepts of the graph. Set $y = 0$ and solve the resulting equation. Compare the result with any *x*-intercepts of the graph.**

43. $y = 4x^3 - 20x^2 + 25x$

44. $y = 4x^3 + 4x^2 - 7x + 2$

45. $y = x^5 - 5x^3 + 4x$

46. $y = \frac{1}{4}x^3(x^2 - 9)$

In Exercises 47–56, find a polynomial function that has the given zeros. (There are many correct answers.)

47. 0, 10 **48.** $0, -3$

49. $2, -6$ **50.** $-4, 5$

51. $0, -2, -3$ **52.** 0, 2, 5

53. $4, -3, 3, 0$ **54.** $-2, -1, 0, 1, 2$

55. $1 + \sqrt{3}, 1 - \sqrt{3}$ **56.** $2, 4 + \sqrt{5}, 4 - \sqrt{5}$

In Exercises 57–66, find a polynomial of degree *n* that has the given zero(s). (There are many correct answers.)

57. Zero: $x = -2$ Degree: $n = 2$

58. Zeros: $x = -8, -4$ Degree: $n = 2$

59. Zeros: $x = -3, 0, 1$ Degree: $n = 3$

60. Zeros: $x = -2, 4, 7$ Degree: $n = 3$

61. Zeros: $x = 0, \sqrt{3}, -\sqrt{3}$ Degree: $n = 3$

62. Zero: $x = 9$ Degree: $n = 3$

63. Zeros: $x = -5, 1, 2$ Degree: $n = 4$

64. Zeros: $x = -4, -1, 3, 6$ Degree: $n = 4$

65. Zeros: $x = 0, -4$ Degree: $n = 5$

66. Zeros: $x = -3, 1, 5, 6$ Degree: $n = 5$

In Exercises 67–80, sketch the graph of the function by (a) applying the Leading Coefficient Test, (b) finding the zeros of the polynomial, (c) plotting sufficient solution points, and (d) drawing a continuous curve through the points.

67. $f(x) = x^3 - 9x$

68. $g(x) = x^4 - 4x^2$

69. $f(t) = \frac{1}{4}(t^2 - 2t + 15)$

70. $g(x) = -x^2 + 10x - 16$

71. $f(x) = x^3 - 3x^2$

72. $f(x) = 1 - x^3$

73. $f(x) = 3x^3 - 15x^2 + 18x$

74. $f(x) = -4x^3 + 4x^2 + 15x$

75. $f(x) = -5x^2 - x^3$

76. $f(x) = -48x^2 + 3x^4$

77. $f(x) = x^2(x - 4)$

78. $h(x) = \frac{1}{3}x^3(x - 4)^2$

79. $g(t) = -\frac{1}{4}(t - 2)^2(t + 2)^2$

80. $g(x) = \frac{1}{10}(x + 1)^2(x - 3)^3$

In Exercises 81–84, use a graphing utility to graph the function. Use the *zero* or *root* feature to approximate zeros of the function. Determine the multiplicity of each zero.

81. $f(x) = x^3 - 4x$

82. $f(x) = \frac{1}{4}x^4 - 2x^2$

83. $g(x) = \frac{1}{5}(x + 1)^2(x - 3)(2x - 9)$

84. $h(x) = \frac{1}{5}(x + 2)^2(3x - 5)^2$

In Exercises 85–88, use the Intermediate Value Theorem and the *table* feature of a graphing utility to find intervals one unit in length in which the polynomial function is guaranteed to have a zero. Adjust the table to approximate the zeros of the function. Use the *zero* or *root* feature of a graphing utility to verify your results.

85. $f(x) = x^3 - 3x^2 + 3$

86. $f(x) = 0.11x^3 - 2.07x^2 + 9.81x - 6.88$

87. $g(x) = 3x^4 + 4x^3 - 3$

88. $h(x) = x^4 - 10x^2 + 3$

89. ***Numerical and Graphical Analysis*** An open box is to be made from a square piece of material, 36 inches on a side, by cutting equal squares with sides of length x from the corners and turning up the sides (see figure).

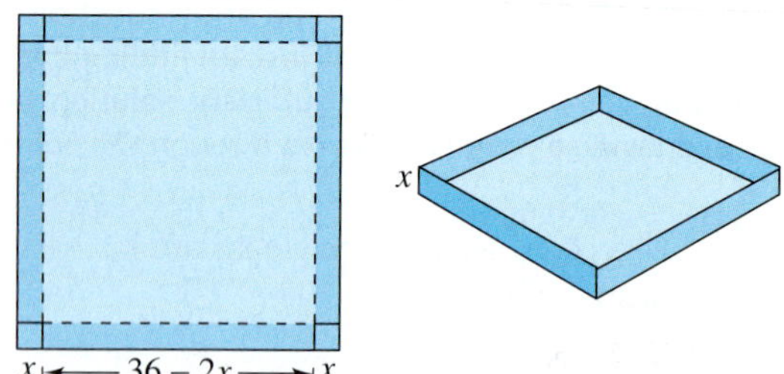

(a) Verify that the volume of the box is given by the function $V(x) = x(36 - 2x)^2$.

(b) Determine the domain of the function.

(c) Use a graphing utility to create a table that shows the box height x and the corresponding volumes V. Use the table to estimate the dimensions that will produce a maximum volume.

(d) Use a graphing utility to graph V and use the graph to estimate the value of x for which $V(x)$ is maximum. Compare your result with that of part (c).

90. ***Maximum Volume*** An open box with locking tabs is to be made from a square piece of material 24 inches on a side. This is to be done by cutting equal squares from the corners and folding along the dashed lines shown in the figure.

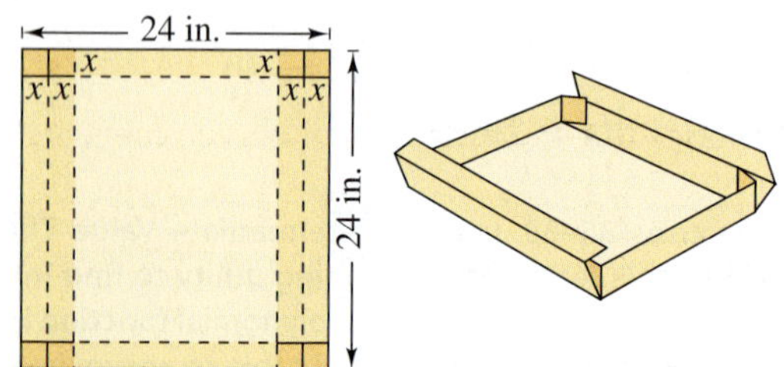

(a) Verify that the volume of the box is given by the function

$$V(x) = 8x(6 - x)(12 - x).$$

(b) Determine the domain of the function V.

(c) Sketch a graph of the function and estimate the value of x for which $V(x)$ is maximum.

▶ Model It

91. ***Tree Growth*** The growth of a red oak tree is approximated by the function

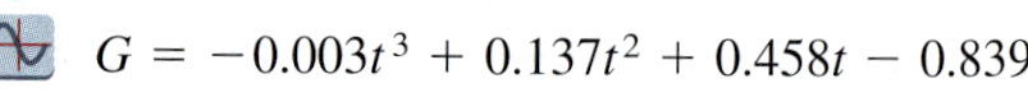

$$G = -0.003t^3 + 0.137t^2 + 0.458t - 0.839$$

where G is the height of the tree (in feet) and t $(2 \le t \le 34)$ is its age (in years).

(a) Use a graphing utility to graph the function. (*Hint:* Use a viewing window in which $-10 \le x \le 45$ and $-5 \le y \le 60$.)

(b) Estimate the age of the tree when it is growing most rapidly. This point is called the *point of diminishing returns* because the increase in size will be less with each additional year.

(c) Using calculus, the point of diminishing returns can also be found by finding the vertex of the parabola given by

$$y = -0.009t^2 + 0.274t + 0.458.$$

Find the vertex of this parabola.

(d) Compare your results from parts (b) and (c).

92. ***Revenue*** The total revenue R (in millions of dollars) for a company is related to its advertising expense by the function

$$R = \frac{1}{100,000}(-x^3 + 600x^2), \qquad 0 \le x \le 400$$

where x is the amount spent on advertising (in tens of thousands of dollars). Use the graph of this function, shown in the figure, to estimate the point on the graph at which the function is increasing most rapidly. This point is called the *point of diminishing returns* because any expense above this amount will yield less return per dollar invested in advertising.

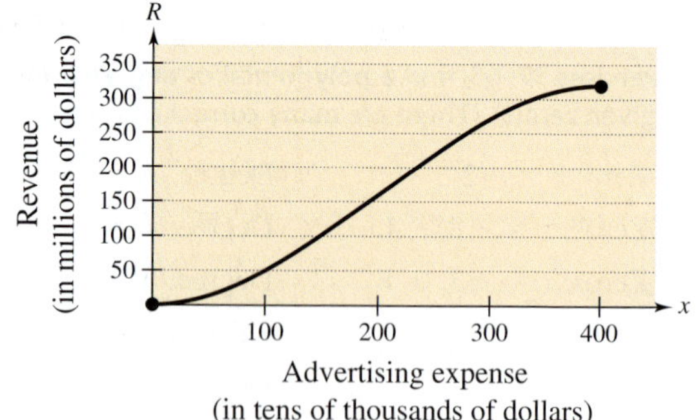

Synthesis

True or False? **In Exercises 93–95, determine whether the statement is true or false. Justify your answer.**

93. A fifth-degree polynomial can have five turning points in its graph.

94. It is possible for a sixth-degree polynomial to have only one solution.

95. The graph of the function

$$f(x) = 2 + x - x^2 + x^3 - x^4 + x^5 + x^6 - x^7$$

rises to the left and falls to the right.

96. ***Graphical Analysis*** Describe a polynomial function that could represent the graph. (Indicate the degree of the function and the sign of its leading coefficient.)

(a)

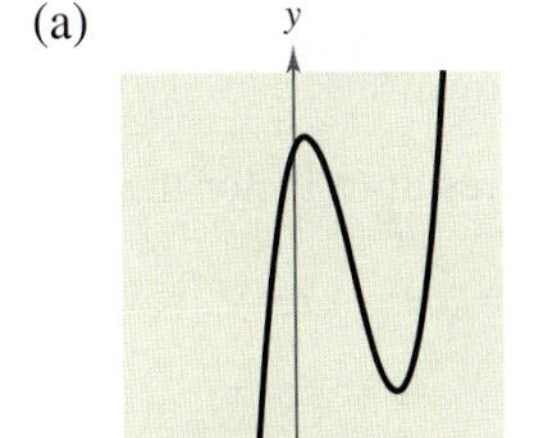

(b)

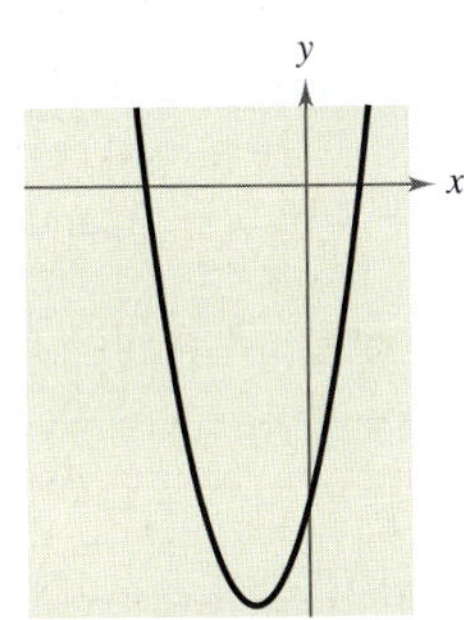

(c)

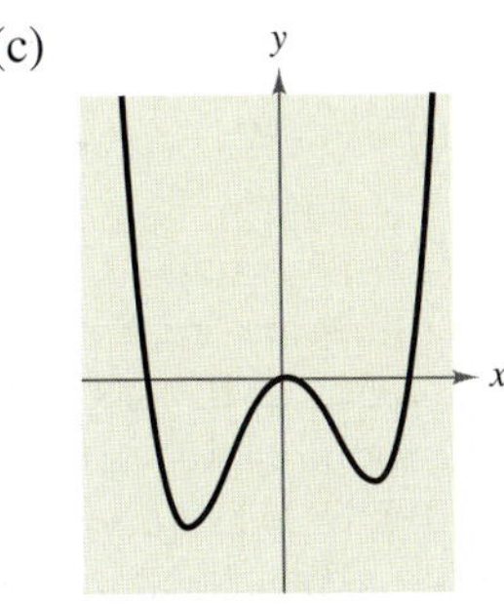

(d)

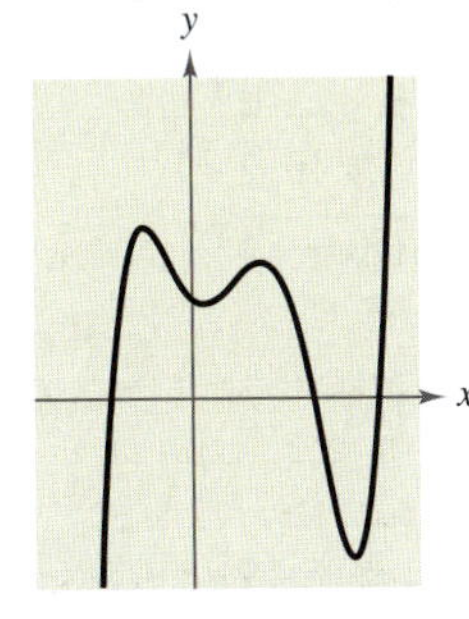

97. ***Graphical Reasoning*** Sketch a graph of the function $f(x) = x^4$. Explain how the graph of g differs (if it does) from the graph of f. Determine whether g is odd, even, or neither.

(a) $g(x) = f(x) + 2$ (b) $g(x) = f(x + 2)$

(c) $g(x) = f(-x)$ (d) $g(x) = -f(x)$

(e) $g(x) = f\left(\frac{1}{2}x\right)$ (f) $g(x) = \frac{1}{2}f(x)$

(g) $g(x) = f\left(x^{3/4}\right)$ (h) $g(x) = (f \circ f)(x)$

98. ***Exploration*** Explore the transformations of the form $g(x) = a(x - h)^5 + k$.

(a) Use a graphing utility to graph the functions

$$y_1 = -\frac{1}{3}(x - 2)^5 + 1$$

and

$$y_2 = \frac{3}{5}(x + 2)^5 - 3.$$

Determine whether the graphs are increasing or decreasing. Explain.

(b) Will the graph of g always be increasing or decreasing? If so, is this behavior determined by a, h, or k? Explain.

(c) Use a graphing utility to graph the function

$$H(x) = x^5 - 3x^3 + 2x + 1.$$

Use the graph and the result of part (b) to determine whether H can be written in the form $H(x) = a(x - h)^5 + k$. Explain.

Review

In Exercises 99–102, solve the equation by factoring.

99. $2x^2 - x - 28 = 0$ **100.** $3x^2 - 22x - 16 = 0$

101. $12x^2 + 11x - 5 = 0$ **102.** $x^2 + 24x + 144 = 0$

In Exercises 103–106, solve the equation by completing the square.

103. $x^2 - 2x - 21 = 0$ **104.** $x^2 - 8x + 2 = 0$

105. $2x^2 + 5x - 20 = 0$ **106.** $3x^2 + 4x - 9 = 0$

In Exercises 107–110, factor the expression completely.

107. $5x^2 + 7x - 24$ **108.** $6x^3 - 61x^2 + 10x$

109. $4x^4 - 7x^3 - 15x^2$ **110.** $y^3 + 216$

In Exercises 111–116, describe the transformation from a common function that occurs in the function. Then sketch its graph.

111. $f(x) = (x + 4)^2$ **112.** $f(x) = 3 - x^2$

113. $f(x) = \sqrt{x + 1} - 5$ **114.** $f(x) = 7 - \sqrt{x - 6}$

115. $f(x) = 2[\![x]\!] + 9$ **116.** $f(x) = 10 - \frac{1}{3}[\![x + 3]\!]$

3.3 Polynomial and Synthetic Division

▶ What you should learn

- How to use long division to divide polynomials by other polynomials
- How to use synthetic division to divide polynomials by binomials of the form $(x - k)$
- How to use the Remainder Theorem and the Factor Theorem

▶ Why you should learn it

Synthetic division can help you evaluate polynomial functions. For instance, in Exercise 75 on page 291, you will use synthetic division to determine the number of U.S. military personnel in 2005.

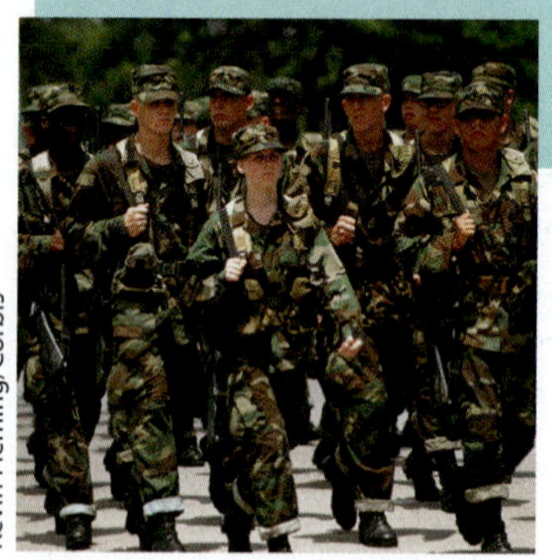

Kevin Fleming/Corbis

Long Division of Polynomials

In this section you will study two procedures for *dividing* polynomials. These procedures are especially valuable in factoring and finding the zeros of polynomial functions. To begin, suppose you are given the graph of

$$f(x) = 6x^3 - 19x^2 + 16x - 4.$$

Notice that a zero of f occurs at $x = 2$, as shown in Figure 3.27. Because $x = 2$ is a zero of f, you know that $(x - 2)$ is a factor of $f(x)$. This means that there exists a second-degree polynomial $q(x)$ such that

$$f(x) = (x - 2) \cdot q(x).$$

To find $q(x)$, you can use **long division,** as illustrated in Example 1.

Example 1 ▶ Long Division of Polynomials 

Divide $6x^3 - 19x^2 + 16x - 4$ by $x - 2$, and use the result to factor the polynomial completely.

Solution

Think $\frac{6x^3}{x} = 6x^2$.

Think $\frac{-7x^2}{x} = -7x$.

Think $\frac{2x}{x} = 2$.

$$
\begin{array}{rrl}
 & 6x^2 - 7x + 2 & \\
x - 2 \,\big) & 6x^3 - 19x^2 + 16x - 4 & \\
 & \underline{6x^3 - 12x^2} & \text{Multiply: } 6x^2(x - 2). \\
 & -7x^2 + 16x & \text{Subtract.} \\
 & \underline{-7x^2 + 14x} & \text{Multiply: } -7x(x - 2). \\
 & 2x - 4 & \text{Subtract.} \\
 & \underline{2x - 4} & \text{Multiply: } 2(x - 2). \\
 & 0 & \text{Subtract.}
\end{array}
$$

From this division, you can conclude that

$$6x^3 - 19x^2 + 16x - 4 = (x - 2)(6x^2 - 7x + 2)$$

and by factoring the quadratic $6x^2 - 7x + 2$, you have

$$6x^3 - 19x^2 + 16x - 4 = (x - 2)(2x - 1)(3x - 2).$$

Note that this factorization agrees with the graph shown in Figure 3.27 in that the three x-intercepts occur at $x = 2$, $x = \frac{1}{2}$, and $x = \frac{2}{3}$.

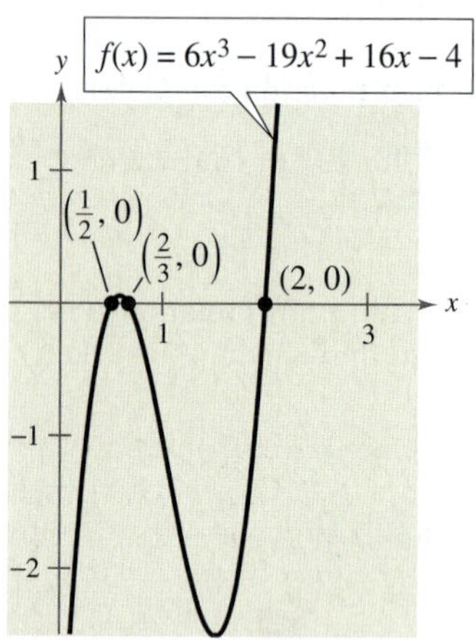

FIGURE 3.27

In Example 1, $x - 2$ is a factor of the polynomial $6x^3 - 19x^2 + 16x - 4$, and the long division process produces a remainder of zero. Often, long division will produce a nonzero remainder. For instance, if you divide $x^2 + 3x + 5$ by $x + 1$, you obtain the following.

$$\begin{array}{r} x + 2 \\ x + 1 \overline{) x^2 + 3x + 5} \\ \underline{x^2 + x \phantom{{}+5}} \\ 2x + 5 \\ \underline{2x + 2} \\ 3 \end{array}$$

Divisor → $x + 1$; Quotient ← $x + 2$; Dividend ← $x^2 + 3x + 5$; Remainder ← 3

In fractional form, you can write this result as follows.

$$\underbrace{\frac{x^2 + 3x + 5}{x + 1}}_{\text{Dividend / Divisor}} = \underbrace{x + 2}_{\text{Quotient}} + \frac{\overset{\text{Remainder}}{3}}{\underbrace{x + 1}_{\text{Divisor}}}$$

This implies that

$$x^2 + 3x + 5 = (x + 1)(x + 2) + 3$$ Multiply each side by $(x + 1)$.

which illustrates the following theorem, called the **Division Algorithm.**

The Division Algorithm

If $f(x)$ and $d(x)$ are polynomials such that $d(x) \neq 0$, and the degree of $d(x)$ is less than or equal to the degree of $f(x)$, there exist unique polynomials $q(x)$ and $r(x)$ such that

$$f(x) = d(x)q(x) + r(x)$$

Dividend: $f(x)$; Divisor: $d(x)$; Quotient: $q(x)$; Remainder: $r(x)$

where $r(x) = 0$ *or* the degree of $r(x)$ is less than the degree of $d(x)$. If the remainder $r(x)$ is zero, $d(x)$ *divides evenly* into $f(x)$.

The Division Algorithm can also be written as

$$\frac{f(x)}{d(x)} = q(x) + \frac{r(x)}{d(x)}.$$

In the Division Algorithm, the rational expression $f(x)/d(x)$ is **improper** because the degree of $f(x)$ is greater than or equal to the degree of $d(x)$. On the other hand, the rational expression $r(x)/d(x)$ is **proper** because the degree of $r(x)$ is less than the degree of $d(x)$.

Before you apply the Division Algorithm, follow these steps.

1. Write the dividend and divisor in descending powers of the variable.
2. Insert placeholders with zero coefficients for missing powers of the variable.

Example 2 ▶ Long Division of Polynomials 

Divide $x^3 - 1$ by $x - 1$.

Solution

Because there is no x^2-term or x-term in the dividend, you need to line up the subtraction by using zero coefficients (or leaving spaces) for the missing terms.

$$\begin{array}{r} x^2 + x + 1 \\ x - 1 \overline{) x^3 + 0x^2 + 0x - 1} \\ \underline{x^3 - x^2} \\ x^2 + 0x \\ \underline{x^2 - x} \\ x - 1 \\ \underline{x - 1} \\ 0 \end{array}$$

So, $x - 1$ divides evenly into $x^3 - 1$, and you can write

$$\frac{x^3 - 1}{x - 1} = x^2 + x + 1, \quad x \neq 1.$$

You can check the result of Example 2 by multiplying.

$$(x - 1)(x^2 + x + 1) = x^3 + x^2 + x - x^2 - x - 1 = x^3 - 1$$

Example 3 ▶ Long Division of Polynomials

Divide $2x^4 + 4x^3 - 5x^2 + 3x - 2$ by $x^2 + 2x - 3$.

Solution

$$\begin{array}{r} 2x^2 + 1 \\ x^2 + 2x - 3 \overline{) 2x^4 + 4x^3 - 5x^2 + 3x - 2} \\ \underline{2x^4 + 4x^3 - 6x^2} \\ x^2 + 3x - 2 \\ \underline{x^2 + 2x - 3} \\ x + 1 \end{array}$$

Note that the first subtraction eliminated two terms from the dividend. When this happens, the quotient skips a term. You can write the result as

$$\frac{2x^4 + 4x^3 - 5x^2 + 3x - 2}{x^2 + 2x - 3} = 2x^2 + 1 + \frac{x + 1}{x^2 + 2x - 3}.$$

Synthetic Division

There is a nice shortcut for long division of polynomials when dividing by divisors of the form $x - k$. This shortcut is called **synthetic division.** The pattern for synthetic division of a cubic polynomial is summarized as follows. (The pattern for higher-degree polynomials is similar.)

Synthetic Division (for a Cubic Polynomial)

To divide $ax^3 + bx^2 + cx + d$ by $x - k$, use the following pattern.

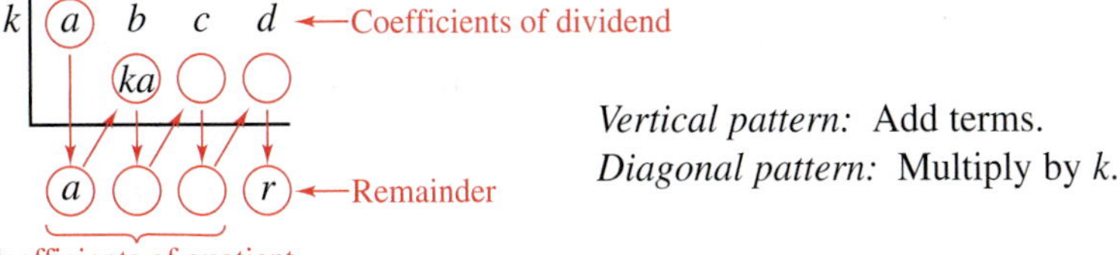

Synthetic division works only for divisors of the form $x - k$. [Remember that $x + k = x - (-k)$.] You cannot use synthetic division to divide a polynomial by a quadratic such as $x^2 - 3$.

Example 4 ▶ Using Synthetic Division

Use synthetic division to divide $x^4 - 10x^2 - 2x + 4$ by $x + 3$.

Solution

You should set up the array as follows. Note that a zero is included for the missing x^3-term in the dividend.

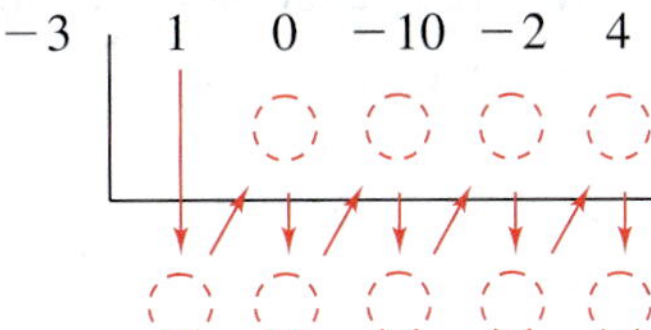

Then, use the synthetic division pattern by adding terms in columns and multiplying the results by -3.

Divisor: $x + 3$ Dividend: $x^4 - 10x^2 - 2x + 4$

$$
\begin{array}{r|rrrrr}
-3 & 1 & 0 & -10 & -2 & 4 \\
 & & -3 & 9 & 3 & -3 \\
\hline
 & 1 & -3 & -1 & 1 & 1
\end{array}
$$

Remainder: 1

Quotient: $x^3 - 3x^2 - x + 1$

So, you have

$$\frac{x^4 - 10x^2 - 2x + 4}{x + 3} = x^3 - 3x^2 - x + 1 + \frac{1}{x + 3}.$$

The Remainder and Factor Theorems

The remainder obtained in the synthetic division process has an important interpretation, as described in the **Remainder Theorem.**

The Remainder Theorem

If a polynomial $f(x)$ is divided by $x - k$, the remainder is

$$r = f(k).$$

For a proof of the Remainder Theorem, see Proofs in Mathematics on page 326.

The Remainder Theorem tells you that synthetic division can be used to evaluate a polynomial function. That is, to evaluate a polynomial function $f(x)$ when $x = k$, divide $f(x)$ by $x - k$. The remainder will be $f(k)$, as illustrated in Example 5.

Example 5 ▶ Using the Remainder Theorem

Use the Remainder Theorem to evaluate the following function at $x = -2$.

$$f(x) = 3x^3 + 8x^2 + 5x - 7$$

Solution

Using synthetic division, you obtain the following.

$$\begin{array}{r|rrrr} -2 & 3 & 8 & 5 & -7 \\ & & -6 & -4 & -2 \\ \hline & 3 & 2 & 1 & -9 \end{array}$$

Because the remainder is $r = -9$, you can conclude that

$$f(-2) = -9. \qquad r = f(x)$$

This means that $(-2, -9)$ is a point on the graph of f. You can check this by substituting $x = -2$ in the original function.

Check

$$f(-2) = 3(-2)^3 + 8(-2)^2 + 5(-2) - 7$$
$$= 3(-8) + 8(4) - 10 - 7 = -9$$

Another important theorem is the **Factor Theorem,** stated below. This theorem states that you can test to see whether a polynomial has $(x - k)$ as a factor by evaluating the polynomial at $x = k$. If the result is 0, $(x - k)$ is a factor.

The Factor Theorem

A polynomial $f(x)$ has a factor $(x - k)$ if and only if $f(k) = 0$.

For a proof of the Factor Theorem, see Proofs in Mathematics on page 326.

Example 6 ▶ Factoring a Polynomial: Repeated Division

Show that $(x - 2)$ and $(x + 3)$ are factors of

$$f(x) = 2x^4 + 7x^3 - 4x^2 - 27x - 18.$$

Then find the remaining factors of $f(x)$.

Solution

Using synthetic division with the factor $(x - 2)$, you obtain the following.

2	2	7	−4	−27	−18
		4	22	36	18
	2	11	18	9	0

0 remainder, so $f(2) = 0$ and $(x - 2)$ is a factor.

Take the result of this division and perform synthetic division again using the factor $(x + 3)$.

−3	2	11	18	9
		−6	−15	−9
	2	5	3	0

0 remainder, so $f(-3) = 0$ and $(x + 3)$ is a factor.

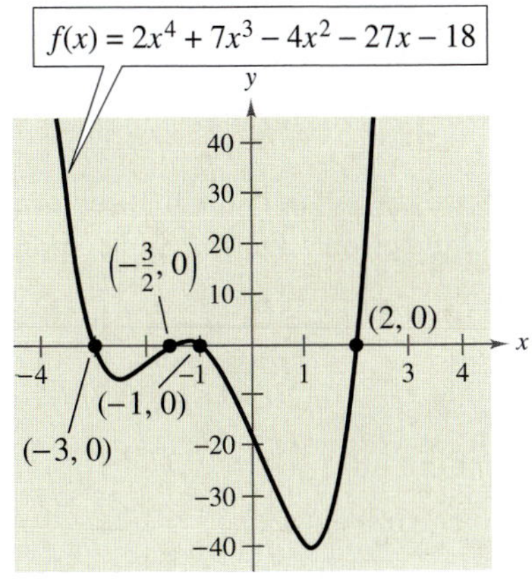

FIGURE 3.28

Because the resulting quadratic expression factors as

$$2x^2 + 5x + 3 = (2x + 3)(x + 1)$$

the complete factorization of $f(x)$ is

$$f(x) = (x - 2)(x + 3)(2x + 3)(x + 1).$$

Note that this factorization implies that f has four real zeros:

$$x = 2, x = -3, x = -\tfrac{3}{2}, \text{ and } x = -1.$$

This is confirmed by the graph of f, which is shown in Figure 3.28.

Uses of the Remainder in Synthetic Division

The remainder r, obtained in the synthetic division of $f(x)$ by $x - k$, provides the following information.

1. The remainder r gives the value of f at $x = k$. That is, $r = f(k)$.
2. If $r = 0$, $(x - k)$ is a factor of $f(x)$.
3. If $r = 0$, $(k, 0)$ is an x-intercept of the graph of f.

Throughout this text, the importance of developing several problem-solving strategies is emphasized. In the exercises for this section, try using more than one strategy to solve several of the exercises. For instance, if you find that $x - k$ divides evenly into $f(x)$ (with no remainder), try sketching the graph of f. You should find that $(k, 0)$ is an x-intercept of the graph.

3.3 Exercises

Analytical Analysis **In Exercises 1 and 2, use long division to verify that $y_1 = y_2$.**

1. $y_1 = \dfrac{x^2}{x+2}, \quad y_2 = x - 2 + \dfrac{4}{x+2}$

2. $y_1 = \dfrac{x^4 - 3x^2 - 1}{x^2 + 5}, \quad y_2 = x^2 - 8 + \dfrac{39}{x^2+5}$

Graphical Analysis **In Exercises 3 and 4, use a graphing utility to graph the two equations in the same viewing window. Use the graphs to verify that the expressions are equivalent. Use long division to verify the results algebraically.**

3. $y_1 = \dfrac{x^5 - 3x^3}{x^2+1}, \quad y_2 = x^3 - 4x + \dfrac{4x}{x^2+1}$

4. $y_1 = \dfrac{x^3 - 2x^2 + 5}{x^2 + x + 1}, \quad y_2 = x - 3 + \dfrac{2(x+4)}{x^2+x+1}$

In Exercises 5–18, use long division to divide.

5. $(2x^2 + 10x + 12) \div (x + 3)$
6. $(5x^2 - 17x - 12) \div (x - 4)$
7. $(4x^3 - 7x^2 - 11x + 5) \div (4x + 5)$
8. $(6x^3 - 16x^2 + 17x - 6) \div (3x - 2)$
9. $(x^4 + 5x^3 + 6x^2 - x - 2) \div (x + 2)$
10. $(x^3 + 4x^2 - 3x - 12) \div (x - 3)$
11. $(7x + 3) \div (x + 2)$
12. $(8x - 5) \div (2x + 1)$
13. $(6x^3 + 10x^2 + x + 8) \div (2x^2 + 1)$
14. $(x^3 - 9) \div (x^2 + 1)$
15. $\dfrac{x^4 + 3x^2 + 1}{x^2 - 2x + 3}$
16. $\dfrac{x^5 + 7}{x^3 - 1}$
17. $\dfrac{x^4}{(x-1)^3}$
18. $\dfrac{2x^3 - 4x^2 - 15x + 5}{(x-1)^2}$

In Exercises 19–36, use synthetic division to divide.

19. $(3x^3 - 17x^2 + 15x - 25) \div (x - 5)$
20. $(5x^3 + 18x^2 + 7x - 6) \div (x + 3)$
21. $(4x^3 - 9x + 8x^2 - 18) \div (x + 2)$
22. $(9x^3 - 16x - 18x^2 + 32) \div (x - 2)$
23. $(-x^3 + 75x - 250) \div (x + 10)$
24. $(3x^3 - 16x^2 - 72) \div (x - 6)$
25. $(5x^3 - 6x^2 + 8) \div (x - 4)$
26. $(5x^3 + 6x + 8) \div (x + 2)$
27. $\dfrac{10x^4 - 50x^3 - 800}{x - 6}$
28. $\dfrac{x^5 - 13x^4 - 120x + 80}{x + 3}$
29. $\dfrac{x^3 + 512}{x + 8}$
30. $\dfrac{x^3 - 729}{x - 9}$
31. $\dfrac{-3x^4}{x - 2}$
32. $\dfrac{-3x^4}{x + 2}$
33. $\dfrac{180x - x^4}{x - 6}$
34. $\dfrac{5 - 3x + 2x^2 - x^3}{x + 1}$
35. $\dfrac{4x^3 + 16x^2 - 23x - 15}{x + \frac{1}{2}}$
36. $\dfrac{3x^3 - 4x^2 + 5}{x - \frac{3}{2}}$

In Exercises 37–44, express the function in the form $f(x) = (x - k)q(x) + r$ for the given value of k, and demonstrate that $f(k) = r$.

	Function	*Value of k*
37.	$f(x) = x^3 - x^2 - 14x + 11$	$k = 4$
38.	$f(x) = x^3 - 5x^2 - 11x + 8$	$k = -2$
39.	$f(x) = 15x^4 + 10x^3 - 6x^2 + 14$	$k = -\frac{2}{3}$
40.	$f(x) = 10x^3 - 22x^2 - 3x + 4$	$k = \frac{1}{5}$
41.	$f(x) = x^3 + 3x^2 - 2x - 14$	$k = \sqrt{2}$
42.	$f(x) = x^3 + 2x^2 - 5x - 4$	$k = -\sqrt{5}$
43.	$f(x) = -4x^3 + 6x^2 + 12x + 4$	$k = 1 - \sqrt{3}$
44.	$f(x) = -3x^3 + 8x^2 + 10x - 8$	$k = 2 + \sqrt{2}$

In Exercises 45–48, use synthetic division to find each function value. Verify your answers using another method.

45. $f(x) = 4x^3 - 13x + 10$
    (a) $f(1)$ (b) $f(-2)$ (c) $f(\frac{1}{2})$ (d) $f(8)$
46. $g(x) = x^6 - 4x^4 + 3x^2 + 2$
    (a) $g(2)$ (b) $g(-4)$ (c) $g(3)$ (d) $g(-1)$
47. $h(x) = 3x^3 + 5x^2 - 10x + 1$
    (a) $h(3)$ (b) $h(\frac{1}{3})$ (c) $h(-2)$ (d) $h(-5)$
48. $f(x) = 0.4x^4 - 1.6x^3 + 0.7x^2 - 2$
    (a) $f(1)$ (b) $f(-2)$ (c) $f(5)$ (d) $f(-10)$

In Exercises 49–56, use synthetic division to show that x is a zero of the third-degree polynomial equation, and use the result to factor the polynomial completely. List all real zeros of the function.

Polynomial Equation	*Value of x*
49. $x^3 - 7x + 6 = 0$	$x = 2$
50. $x^3 - 28x - 48 = 0$	$x = -4$
51. $2x^3 - 15x^2 + 27x - 10 = 0$	$x = \frac{1}{2}$
52. $48x^3 - 80x^2 + 41x - 6 = 0$	$x = \frac{2}{3}$
53. $x^3 + 2x^2 - 3x - 6 = 0$	$x = \sqrt{3}$
54. $x^3 + 2x^2 - 2x - 4 = 0$	$x = \sqrt{2}$
55. $x^3 - 3x^2 + 2 = 0$	$x = 1 + \sqrt{3}$
56. $x^3 - x^2 - 13x - 3 = 0$	$x = 2 - \sqrt{5}$

In Exercises 57–64, (a) verify the given factors of the function f, (b) find the remaining factors of f, (c) use your results to write the complete factorization of f, (d) list all real zeros of f, and (e) confirm your results by using a graphing utility to graph the function.

Function	*Factors*
57. $f(x) = 2x^3 + x^2 - 5x + 2$	$(x + 2), (x - 1)$
58. $f(x) = 3x^3 + 2x^2 - 19x + 6$	$(x + 3), (x - 2)$
59. $f(x) = x^4 - 4x^3 - 15x^2 + 58x - 40$	$(x - 5), (x + 4)$
60. $f(x) = 8x^4 - 14x^3 - 71x^2 - 10x + 24$	$(x + 2), (x - 4)$
61. $f(x) = 6x^3 + 41x^2 - 9x - 14$	$(2x + 1), (3x - 2)$
62. $f(x) = 10x^3 - 11x^2 - 72x + 45$	$(2x + 5), (5x - 3)$
63. $f(x) = 2x^3 - x^2 - 10x + 5$	$(2x - 1), (x + \sqrt{5})$
64. $f(x) = x^3 + 3x^2 - 48x - 144$	$(x + 4\sqrt{3}), (x + 3)$

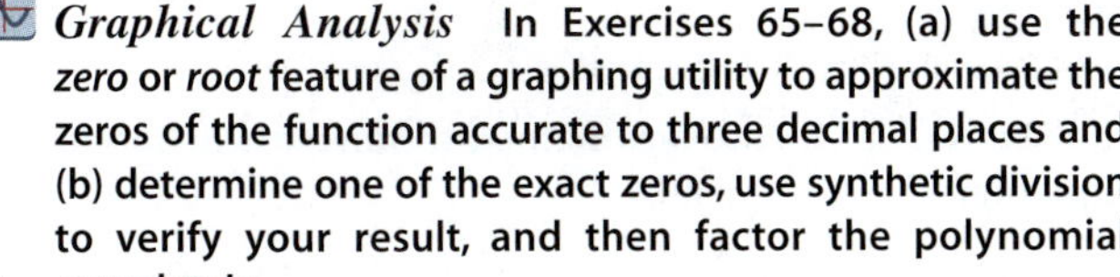

Graphical Analysis **In Exercises 65–68, (a) use the *zero* or *root* feature of a graphing utility to approximate the zeros of the function accurate to three decimal places and (b) determine one of the exact zeros, use synthetic division to verify your result, and then factor the polynomial completely.**

65. $f(x) = x^3 - 2x^2 - 5x + 10$

66. $g(x) = x^3 - 4x^2 - 2x + 8$

67. $h(t) = t^3 - 2t^2 - 7t + 2$

68. $f(s) = s^3 - 12s^2 + 40s - 24$

In Exercises 69–74, simplify the rational expression.

69. $\dfrac{4x^3 - 8x^2 + x + 3}{2x - 3}$

70. $\dfrac{x^3 + x^2 - 64x - 64}{x + 8}$

71. $\dfrac{x^3 + 3x^2 - x - 3}{x + 1}$

72. $\dfrac{2x^3 + 3x^2 - 3x - 2}{x - 1}$

73. $\dfrac{x^4 + 6x^3 + 11x^2 + 6x}{x^2 + 3x + 2}$

74. $\dfrac{x^4 + 9x^3 - 5x^2 - 36x + 4}{x^2 - 4}$

▶ Model It

75. ***Data Analysis*** The numbers M (in thousands) of United States military personnel on active duty for the years 1990 through 2000 are shown in the table, where t represents the time (in years), with $t = 0$ corresponding to 1990. (Source: U.S. Department of Defense)

Year, t	Military personnel, M
0	2044
1	1986
2	1807
3	1705
4	1610
5	1518
6	1472
7	1439
8	1407
9	1386
10	1384

(a) Use a graphing utility to create a scatter plot of the data.

(b) Use the *regression* feature of the graphing utility to find a cubic model for the data. Then graph the model in the same viewing window as the scatter plot.

(c) Use the model to create a table of estimated values of M. Compare the model with the data.

(d) Use synthetic division to evaluate the model for the year 2005. Even though the model is relatively accurate for estimating the given data, would you use this model to predict the number of military personnel in the future? Explain.

76. ***Data Analysis*** The average monthly basic rates R (in dollars) for cable television in the United States for the years 1990 through 1999 are shown in the table, where t represents the time (in years), with $t = 0$ corresponding to 1990. (Source: Paul Kagan Associates, Inc.)

Year, t	Basic rate, R
0	16.78
1	18.10
2	19.08
3	19.39
4	21.62
5	23.07
6	24.41
7	26.48
8	27.81
9	28.92

(a) Use a graphing utility to create a scatter plot of the data.

(b) Use the *regression* feature of the graphing utility to find a cubic model for the data. Then graph the model in the same viewing window as the scatter plot. Compare the model with the data.

(c) Use synthetic division to evaluate the model for the year 2005.

Synthesis

True or False? **In Exercises 77–79, determine whether the statement is true or false. Justify your answer.**

77. If $(7x + 4)$ is a factor of some polynomial function f, then $\frac{4}{7}$ is a zero of f.

78. $(2x - 1)$ is a factor of the polynomial $6x^6 + x^5 - 92x^4 + 45x^3 + 184x^2 + 4x - 48$.

79. The rational expression

$$\frac{x^3 + 2x^2 - 13x + 10}{x^2 - 4x - 12}$$

is improper.

80. ***Exploration*** Use the form

$$f(x) = (x - k)q(x) + r$$

to create a cubic function that (a) passes through the point $(2, 5)$ and rises to the right, and (b) passes through the point $(-3, 1)$ and falls to the right. (There are many correct answers.)

Think About It **In Exercises 81 and 82, perform the division by assuming that n is a positive integer.**

81. $\dfrac{x^{3n} + 9x^{2n} + 27x^n + 27}{x^n + 3}$

82. $\dfrac{x^{3n} - 3x^{2n} + 5x^n - 6}{x^n - 2}$

83. ***Writing*** Briefly explain what it means for a divisor to divide evenly into a dividend.

84. ***Writing*** Briefly explain how to check polynomial division, and justify your reasoning. Give an example.

Exploration **In Exercises 85 and 86, find the constant c such that the denominator will divide evenly into the numerator.**

85. $\dfrac{x^3 + 4x^2 - 3x + c}{x - 5}$

86. $\dfrac{x^5 - 2x^2 + x + c}{x + 2}$

Think About It **In Exercises 87 and 88, answer the questions about the division**

$$\frac{f(x)}{x - k}$$

where $f(x) = (x + 3)^2(x - 3)(x + 1)^3$.

87. What is the remainder when $k = -3$? Explain.

88. If it is necessary to find $f(2)$, is it easier to evaluate the function directly or to use synthetic division? Explain.

Review

In Exercises 89–94, use any method to solve the quadratic equation.

89. $9x^2 - 25 = 0$

90. $16x^2 - 21 = 0$

91. $5x^2 - 3x - 14 = 0$

92. $8x^2 - 22x + 15 = 0$

93. $2x^2 + 6x + 3 = 0$

94. $x^2 + 3x - 3 = 0$

In Exercises 95–98, find a polynomial function that has the given zeros. There are many correct answers.

95. $0, 3, 4$

96. $-6, 1$

97. $-3, 1 + \sqrt{2}, 1 - \sqrt{2}$

98. $1, -2, 2 + \sqrt{3}, 2 - \sqrt{3}$

3.4 Zeros of Polynomial Functions

▶ What you should learn

- How to use the Fundamental Theorem of Algebra to determine the number of zeros of polynomial functions
- How to find rational zeros of polynomial functions
- How to find conjugate pairs of complex zeros
- How to find zeros of polynomials by factoring
- How to use Descartes's Rule of Signs and the Upper and Lower Bound Rules to find zeros of polynomials

▶ Why you should learn it

Finding zeros of polynomial functions is an important part of solving real-life problems. For instance, in Exercise 107 on page 306, the zeros of a polynomial function can help you analyze the attendance at women's college basketball games.

Jonathan Daniel/Getty Images

The Fundamental Theorem of Algebra

You know that an nth-degree polynomial can have at most n real zeros. In the complex number system, this statement can be improved. That is, in the complex number system, every nth-degree polynomial function has *precisely* n zeros. This important result is derived from the **Fundamental Theorem of Algebra,** first proved by the German mathematician Carl Friedrich Gauss (1777–1855).

The Fundamental Theorem of Algebra

If $f(x)$ is a polynomial of degree n, where $n > 0$, then f has at least one zero in the complex number system.

Using the Fundamental Theorem of Algebra and the equivalence of zeros and factors, you obtain the **Linear Factorization Theorem.**

Linear Factorization Theorem

If $f(x)$ is a polynomial of degree n, where $n > 0$, then f has precisely n linear factors

$$f(x) = a_n(x - c_1)(x - c_2) \cdot \cdot \cdot (x - c_n)$$

where $c_1, c_2, \ldots, c_n$ are complex numbers.

For a proof of the Linear Factorization Theorem, see Proofs in Mathematics on page 327.

Note that the Fundamental Theorem of Algebra and the Linear Factorization Theorem tell you only that the zeros or factors of a polynomial exist, not how to find them. Such theorems are called *existence theorems.*

Example 1 ▶ Zeros of Polynomial Functions 

a. The first-degree polynomial $f(x) = x - 2$ has exactly *one* zero: $x = 2$.

b. Counting multiplicity, the second-degree polynomial function

$$f(x) = x^2 - 6x + 9 = (x - 3)(x - 3)$$

has exactly *two* zeros: $x = 3$ and $x = 3$. (This is called a *repeated zero.*)

c. The third-degree polynomial function

$$f(x) = x^3 + 4x = x(x^2 + 4) = x(x - 2i)(x + 2i)$$

has exactly *three* zeros: $x = 0$, $x = 2i$, and $x = -2i$.

d. The fourth-degree polynomial function

$$f(x) = x^4 - 1 = (x - 1)(x + 1)(x - i)(x + i)$$

has exactly *four* zeros: $x = 1$, $x = -1$, $x = i$, and $x = -i$.

The Rational Zero Test

The **Rational Zero Test** relates the possible rational zeros of a polynomial (having integer coefficients) to the leading coefficient and to the constant term of the polynomial.

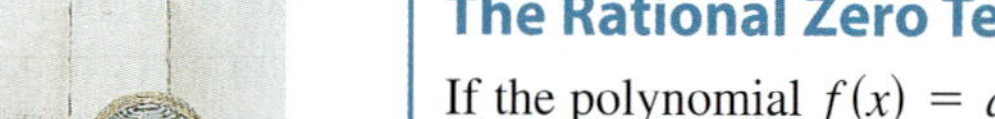

The Rational Zero Test

If the polynomial $f(x) = a_nx^n + a_{n-1}x^{n-1} + \cdot \cdot \cdot + a_2x^2 + a_1x + a_0$ has *integer* coefficients, every rational zero of f has the form

$$\text{Rational zero} = \frac{p}{q}$$

where p and q have no common factors other than 1, and

$p =$ a factor of the constant term a_0

$q =$ a factor of the leading coefficient a_n.

Historical Note
Although they were not contemporaries, Jean Le Rond d'Alembert (1717–1783) worked independently of Carl Gauss in trying to prove the Fundamental Theorem of Algebra. His efforts were such that, in France, the Fundamental Theorem of Algebra is frequently known as the Theorem of d'Alembert.

To use the Rational Zero Test, you should first list all rational numbers whose numerators are factors of the constant term and whose denominators are factors of the leading coefficient.

$$\text{Possible rational zeros} = \frac{\text{factors of constant term}}{\text{factors of leading coefficient}}$$

Having formed this list of *possible rational zeros*, use a trial-and-error method to determine which, if any, are actual zeros of the polynomial. Note that when the leading coefficient is 1, the possible rational zeros are simply the factors of the constant term.

Example 2 ▶ Rational Zero Test with Leading Coefficient of 1

Find the rational zeros of

$f(x) = x^3 + x + 1.$

Solution

Because the leading coefficient is 1, the possible rational zeros are ± 1, the factors of the constant term. By testing these possible zeros, you can see that neither works.

$$f(1) = (1)^3 + 1 + 1$$
$$= 3$$
$$f(-1) = (-1)^3 + (-1) + 1$$
$$= -1$$

So, you can conclude that the given polynomial has *no* rational zeros. Note from the graph of f in Figure 3.29 that f does have one real zero between -1 and 0. However, by the Rational Zero Test, you know that this real zero is *not* a rational number.

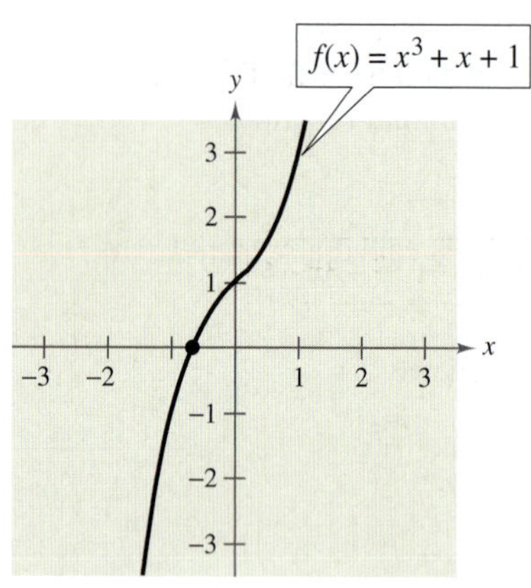

FIGURE 3.29

Example 3 ▶ Rational Zero Test with Leading Coefficient of 1

Find the rational zeros of $f(x) = x^4 - x^3 + x^2 - 3x - 6$.

Solution

Because the leading coefficient is 1, the possible rational zeros are the factors of the constant term.

Possible rational zeros: $\pm 1, \pm 2, \pm 3, \pm 6$

A test of these possible zeros shows that $x = -1$ and $x = 2$ are the only two rational zeros. Check the others to be sure.

If the leading coefficient of a polynomial is not 1, the list of possible rational zeros can increase dramatically. In such cases, the search can be shortened in several ways: (1) a programmable calculator can be used to speed up the calculations; (2) a graph, drawn either by hand or with a graphing utility, can give a good estimate of the locations of the zeros; (3) the Intermediate Value Theorem along with a table generated by a graphing utility can give approximations of zeros; and (4) synthetic division can be used to test the possible rational zeros.

To see how to use synthetic division to test the possible rational zeros, take another look at the function $f(x) = x^4 - x^3 + x^2 - 3x - 6$ from Example 3. To test that $x = -1$ and $x = 2$ are zeros of f, you can apply synthetic division successively, as follows.

$$\begin{array}{r|rrrrr} -1 & 1 & -1 & 1 & -3 & -6 \\ & & -1 & 2 & -3 & 6 \\ \hline & 1 & -2 & 3 & -6 & 0 \end{array}$$

$$\begin{array}{r|rrrr} 2 & 1 & -2 & 3 & -6 \\ & & 2 & 0 & 6 \\ \hline & 1 & 0 & 3 & 0 \end{array}$$

So, you have

$$f(x) = (x + 1)(x - 2)(x^2 + 3).$$

Because the factor $(x^2 + 3)$ produces no real zeros, you can conclude that $x = -1$ and $x = 2$ are the only *real* zeros of f, which is verified in Figure 3.30.

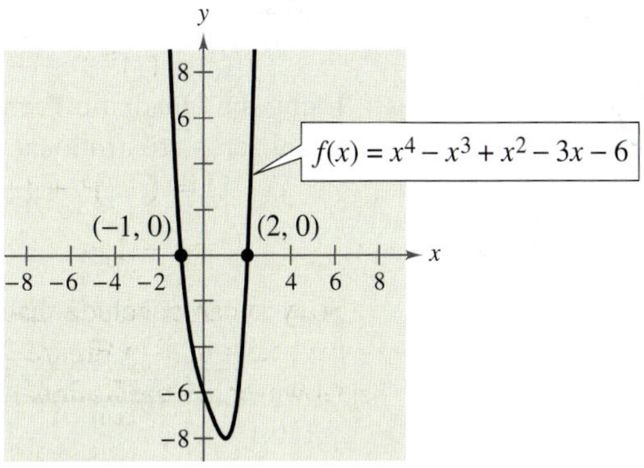

FIGURE 3.30

Finding the first zero is often the hardest part. After that, the search is simplified by using the lower-degree polynomial obtained in synthetic division.

Example 4 ▶ Using the Rational Zero Test

Find the rational zeros of $f(x) = 2x^3 + 3x^2 - 8x + 3$.

Solution

The leading coefficient is 2 and the constant term is 3.

Possible rational zeros: $\dfrac{\text{Factors of } 3}{\text{Factors of } 2} = \dfrac{\pm 1, \pm 3}{\pm 1, \pm 2} = \pm 1, \pm 3, \pm \dfrac{1}{2}, \pm \dfrac{3}{2}$

By synthetic division, you can determine that $x = 1$ is a rational zero.

$$\begin{array}{r|rrrr} 1 & 2 & 3 & -8 & 3 \\ & & 2 & 5 & -3 \\ \hline & 2 & 5 & -3 & 0 \end{array}$$

So, $f(x)$ factors as

$$\begin{aligned} f(x) &= (x-1)(2x^2+5x-3) \\ &= (x-1)(2x-1)(x+3) \end{aligned}$$

and you can conclude that the rational zeros of f are $x = 1$, $x = \frac{1}{2}$, and $x = -3$.

Example 5 ▶ Using the Rational Zero Test 

Find all the real zeros of $f(x) = -10x^3 + 15x^2 + 16x - 12$.

Solution

The leading coefficient is -10 and the constant term is -12.

Possible rational zeros: $\dfrac{\text{Factors of } -12}{\text{Factors of } -10} = \dfrac{\pm 1, \pm 2, \pm 3, \pm 4, \pm 6, \pm 12}{\pm 1, \pm 2, \pm 5, \pm 10}$

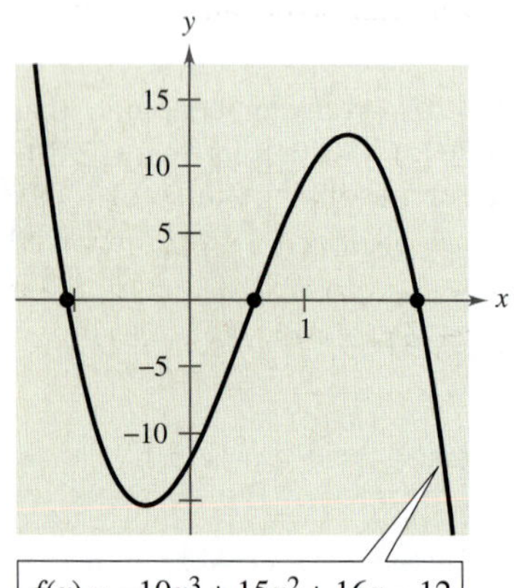

FIGURE 3.31

With so many possibilities (32, in fact), it is worth your time to stop and sketch a graph. From Figure 3.31, it looks like three reasonable choices would be $x = -\frac{6}{5}$, $x = \frac{1}{2}$, and $x = 2$. Testing these by synthetic division shows that only $x = 2$ is a zero. So, you have

$$f(x) = (x-2)(-10x^2 - 5x + 6).$$

Using the Quadratic Formula for the second factor, you find that the two additional zeros are irrational numbers.

$$x = \frac{-(-5) + \sqrt{265}}{-20} \approx -1.0639$$

and

$$x = \frac{-(-5) - \sqrt{265}}{-20} \approx 0.5639$$

Conjugate Pairs

In Example 1(c) and (d), note that the pairs of complex zeros are **conjugates.** That is, they are of the form $a + bi$ and $a - bi$.

Complex Zeros Occur in Conjugate Pairs

Let $f(x)$ be a polynomial function that has *real coefficients.* If $a + bi$, where $b \neq 0$, is a zero of the function, the conjugate $a - bi$ is also a zero of the function.

Be sure you see that this result is true only if the polynomial function has *real coefficients.* For instance, the result applies to the function $f(x) = x^2 + 1$ but not to the function $g(x) = x - i$.

Example 6 ▶ Finding a Polynomial with Given Zeros 

Find a fourth-degree polynomial function with real coefficients that has -1, -1, and $3i$ as zeros.

Solution

Because $3i$ is a zero *and* the polynomial is stated to have real coefficients, you know that the conjugate $-3i$ must also be a zero. So, from the Linear Factorization Theorem, $f(x)$ can be written as

$$f(x) = a(x + 1)(x + 1)(x - 3i)(x + 3i).$$

For simplicity, let $a = 1$ to obtain

$$f(x) = (x^2 + 2x + 1)(x^2 + 9)$$

$$= x^4 + 2x^3 + 10x^2 + 18x + 9.$$

Factoring a Polynomial

The Linear Factorization Theorem shows that you can write any nth-degree polynomial as the product of n linear factors.

$$f(x) = a_n(x - c_1)(x - c_2)(x - c_3) \cdot \cdot \cdot (x - c_n)$$

However, this result includes the possibility that some of the values of c_i are complex. The following theorem says that even if you do not want to get involved with "complex factors," you can still write $f(x)$ as the product of linear and/or quadratic factors. For a proof of this theorem, see Proofs in Mathematics on page 327.

Factors of a Polynomial

Every polynomial of degree $n > 0$ with real coefficients can be written as the product of linear and quadratic factors with real coefficients, where the quadratic factors have no real zeros.

A quadratic factor with no real zeros is said to be *prime* or **irreducible over the reals.** Be sure you see that this is not the same as being *irreducible over the rationals*. For example, the quadratic

$$x^2 + 1 = (x - i)(x + i)$$

is irreducible over the reals (and therefore over the rationals). On the other hand, the quadratic

$$x^2 - 2 = \left(x - \sqrt{2}\right)\left(x + \sqrt{2}\right)$$

is irreducible over the rationals but *reducible* over the reals.

Example 7 ▶ Finding the Zeros of a Polynomial Function 

Find all the zeros of

$$f(x) = x^4 - 3x^3 + 6x^2 + 2x - 60$$

given that $1 + 3i$ is a zero of f.

Solution

Because complex zeros occur in conjugate pairs, you know that $1 - 3i$ is also a zero of f. This means that both

$$[x - (1 + 3i)] \quad \text{and} \quad [x - (1 - 3i)]$$

are factors of f. Multiplying these two factors produces

$$\begin{aligned}[x - (1 + 3i)][x - (1 - 3i)] &= [(x - 1) - 3i][(x - 1) + 3i] \\ &= (x - 1)^2 - 9i^2 \\ &= x^2 - 2x + 1 - 9(-1) \\ &= x^2 - 2x + 10.\end{aligned}$$

Using long division, you can divide $x^2 - 2x + 10$ into f to obtain the following.

$$\begin{array}{r} x^2 - x - 6 \\ x^2 - 2x + 10 \,\big)\overline{\, x^4 - 3x^3 + 6x^2 + 2x - 60} \\ \underline{x^4 - 2x^3 + 10x^2} \\ -x^3 - 4x^2 + 2x \\ \underline{-x^3 + 2x^2 - 10x} \\ -6x^2 + 12x - 60 \\ \underline{-6x^2 + 12x - 60} \\ 0 \end{array}$$

So, you have

$$\begin{aligned} f(x) &= (x^2 - 2x + 10)(x^2 - x - 6) \\ &= (x^2 - 2x + 10)(x - 3)(x + 2) \end{aligned}$$

and you can conclude that the zeros of f are $x = 1 + 3i$, $x = 1 - 3i$, $x = 3$, and $x = -2$.

STUDY TIP

In Example 7, if you were not told that $1 + 3i$ is a zero of f, you could still find all zeros of the function by using synthetic division to find the real zeros -2 and 3. Then you could factor the polynomial as $(x + 2)(x - 3)(x^2 - 2x + 10)$. Finally, by using the Quadratic Formula, you could determine that the zeros are $x = -2$, $x = 3$, $x = 1 + 3i$, and $x = 1 - 3i$.

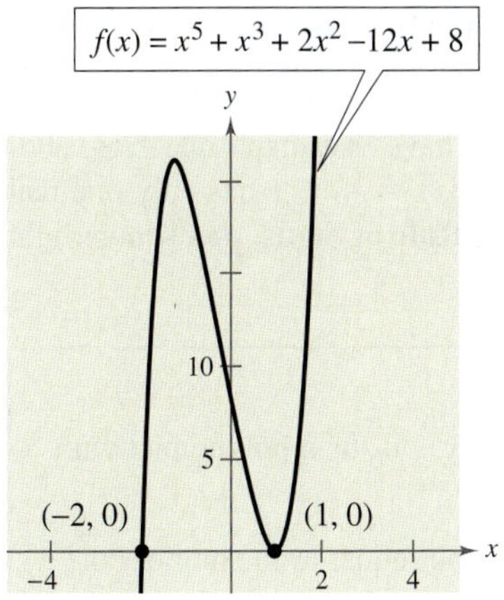

FIGURE 3.32

Example 8 shows how to find all the zeros of a polynomial function, including complex zeros.

Example 8 ▶ Finding the Zeros of a Polynomial Function

Write $f(x) = x^5 + x^3 + 2x^2 - 12x + 8$ as the product of linear factors, and list all of its zeros.

Solution

The possible rational zeros are ± 1, ± 2, ± 4, and ± 8. Synthetic division produces the following.

1	1	0	1	2	−12	8	
		1	1	2	4	−8	
	1	1	2	4	−8	0	⟶ 1 is a zero.

−2	1	1	2	4	−8	
		−2	2	−8	8	
	1	−1	4	−4	0	⟶ −2 is a zero.

So, you have

$$f(x) = x^5 + x^3 + 2x^2 - 12x + 8$$
$$= (x - 1)(x + 2)(x^3 - x^2 + 4x - 4).$$

You can factor $x^3 - x^2 + 4x - 4$ as $(x - 1)(x^2 + 4)$, and by factoring $x^2 + 4$ as

$$x^2 - (-4) = \left(x - \sqrt{-4}\right)\left(x + \sqrt{-4}\right)$$
$$= (x - 2i)(x + 2i)$$

you obtain

$$f(x) = (x - 1)(x - 1)(x + 2)(x - 2i)(x + 2i)$$

which gives the following five zeros of f.

$$x = 1, x = 1, x = -2, x = 2i, \quad \text{and} \quad x = -2i$$

From the graph of f shown in Figure 3.32, you can see that the *real* zeros are the only ones that appear as x-intercepts. Note that $x = 1$ is a repeated zero.

STUDY TIP

In Example 8, the fifth-degree polynomial function has three real zeros. In such cases, you can use the *zoom* and *trace* features or the *zero* or *root* feature of a graphing utility to approximate the real zeros. You can then use these real zeros to determine the complex zeros algebraically.

Technology

You can use the *table* feature of a graphing utility to help you determine which of the possible rational zeros are zeros of the polynomial in Example 8. The table should be set to *ask* mode. Then enter each of the possible rational zeros in the table. When you do this, you will see that there are two rational zeros, −2 and 1, as shown at the right.

X	Y1
-8	-33048
-4	-1000
-2	0
-1	20
1	0
2	32
4	1080

X=4

Other Tests for Zeros of Polynomials 

You know that an nth-degree polynomial function can have *at most n* real zeros. Of course, many nth-degree polynomials do not have that many real zeros. For instance, $f(x) = x^2 + 1$ has no real zeros, and $f(x) = x^3 + 1$ has only one real zero. The following theorem, called **Descartes's Rule of Signs,** sheds more light on the number of real zeros of a polynomial.

Descartes's Rule of Signs

Let $f(x) = a_n x^n + a_{n-1}x^{n-1} + \cdots + a_2x^2 + a_1x + a_0$ be a polynomial with real coefficients and $a_0 \neq 0$.

1. The number of *positive real zeros* of f is either equal to the number of variations in sign of $f(x)$ or less than that number by an even integer.
2. The number of *negative real zeros* of f is either equal to the number of variations in sign of $f(-x)$ or less than that number by an even integer.

A **variation in sign** means that two consecutive coefficients have opposite signs.

When using Descartes's Rule of Signs, a zero of multiplicity k should be counted as k zeros. For instance, the polynomial $x^3 - 3x + 2$ has two variations in sign, and so has either two positive or no positive real zeros. Because

$$x^3 - 3x + 2 = (x - 1)(x - 1)(x + 2)$$

you can see that the two positive real zeros are $x = 1$ of multiplicity 2.

Example 9 ▶ Using Descartes's Rule of Signs 

Describe the possible real zeros of

$$f(x) = 3x^3 - 5x^2 + 6x - 4.$$

Solution

The original polynomial has *three* variations in sign.

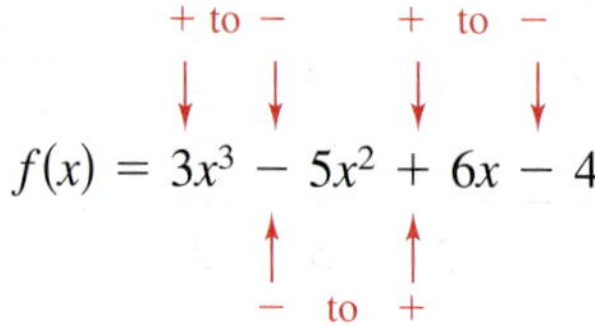

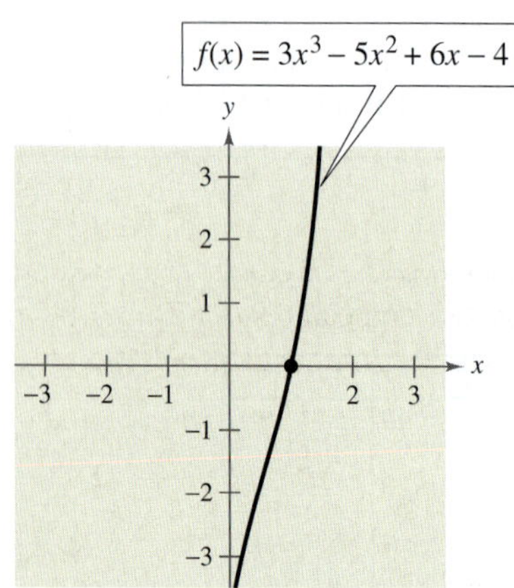

FIGURE 3.33

The polynomial

$$\begin{aligned} f(-x) &= 3(-x)^3 - 5(-x)^2 + 6(-x) - 4 \\ &= -3x^3 - 5x^2 - 6x - 4 \end{aligned}$$

has no variations in sign. So, from Descartes's Rule of Signs, the polynomial $f(x) = 3x^3 - 5x^2 + 6x - 4$ has either three positive real zeros or one positive real zero, and has no negative real zeros. From the graph in Figure 3.33, you can see that the function has only one real zero (it is a positive number, near $x = 1$).

Another test for zeros of a polynomial function is related to the sign pattern in the last row of the synthetic division array. This test can give you an upper or lower bound of the real zeros of f. A real number b is an **upper bound** for the real zeros of f if no zeros are greater than b. Similarly, b is a **lower bound** if no real zeros of f are less than b.

Upper and Lower Bound Rules

Let $f(x)$ be a polynomial with real coefficients and a positive leading coefficient. Suppose $f(x)$ is divided by $x - c$, using synthetic division.

1. If $c > 0$ and each number in the last row is either positive or zero, c is an *upper bound* for the real zeros of f.
2. If $c < 0$ and the numbers in the last row are alternately positive and negative (zero entries count as positive or negative), c is a *lower bound* for the real zeros of f.

Example 10 ▶ Finding the Zeros of a Polynomial Function 

Find the real zeros of

$$f(x) = 6x^3 - 4x^2 + 3x - 2.$$

Solution

The possible real zeros are as follows.

$$\frac{\text{Factors of } 2}{\text{Factors of } 6} = \frac{\pm 1, \pm 2}{\pm 1, \pm 2, \pm 3, \pm 6}$$

$$= \pm 1, \pm \frac{1}{2}, \pm \frac{1}{3}, \pm \frac{1}{6}, \pm \frac{2}{3}, \pm 2$$

Because $f(x)$ has three variations in sign and $f(-x)$ has none, you can apply Descartes's Rule of Signs to conclude that there are three positive real zeros or one positive real zero, and no negative zeros. Trying $x = 1$ produces the following.

$$\begin{array}{r|rrrr} 1 & 6 & -4 & 3 & -2 \\ & & 6 & 2 & 5 \\ \hline & 6 & 2 & 5 & 3 \end{array}$$

So, $x = 1$ is not a zero, but because the last row has all positive entries, you know that $x = 1$ is an upper bound for the real zeros. So, you can restrict the search to zeros between 0 and 1. By trial and error, you can determine that $x = \frac{2}{3}$ is a zero. So,

$$f(x) = \left(x - \frac{2}{3}\right)(6x^2 + 3).$$

Because $6x^2 + 3$ has no real zeros, it follows that $x = \frac{2}{3}$ is the only real zero.

Before concluding this section, here are two additional hints that can help you find the real zeros of a polynomial.

1. If the terms of $f(x)$ have a common monomial factor, it should be factored out before applying the tests in this section. For instance, by writing

$$f(x) = x^4 - 5x^3 + 3x^2 + x = x(x^3 - 5x^2 + 3x + 1)$$

you can see that $x = 0$ is a zero of f and that the remaining zeros can be obtained by analyzing the cubic factor.

2. If you are able to find all but two zeros of $f(x)$, you can always use the Quadratic Formula on the remaining quadratic factor. For instance, if you succeeded in writing

$$f(x) = x^4 - 5x^3 + 3x^2 + x = x(x - 1)(x^2 - 4x - 1)$$

you can apply the Quadratic Formula to $x^2 - 4x - 1$ to conclude that the two remaining zeros are $x = 2 + \sqrt{5}$ and $x = 2 - \sqrt{5}$.

Example 11 ▶ Using a Polynomial Model 

You are designing candle-making kits. Each kit contains 25 cubic inches of candle wax and a mold for making a pyramid-shaped candle. You want the height of the candle to be 2 inches less than the length of each side of the candle's square base. What should the dimensions of your candle mold be?

Solution

The volume of a pyramid is $V = \frac{1}{3}Bh$, where B is the area of the base and h is the height. The area of the base is x^2 and the height is $(x - 2)$. So, the volume of the pyramid is $V = \frac{1}{3}x^2(x - 2)$. Substituting 25 for the volume yields the following.

$$25 = \frac{1}{3}x^2(x - 2) \qquad \text{Substitute 25 for } V.$$

$$75 = x^3 - 2x^2 \qquad \text{Multiply each side by 3.}$$

$$0 = x^3 - 2x^2 - 75 \qquad \text{Write in general form.}$$

The possible rational zeros are $x = \pm1, \pm3, \pm5, \pm15, \pm25, \pm75$. Using synthetic division, you can determine that $x = 5$ is a solution. The other two solutions, which satisfy $x^2 + 3x + 15 = 0$, are imaginary and can be discarded. You can conclude that the base of the candle mold should be 5 inches by 5 inches and the height of the mold should be $5 - 2 = 3$ inches.

Writing ABOUT MATHEMATICS

Factoring Polynomials Compile a list of all the various techniques for factoring a polynomial that have been covered so far in the text. Give an example illustrating each technique, and write a paragraph discussing when the use of each technique is appropriate.

3.4 Exercises

In Exercises 1–6, find all the zeros of the function.

1. $f(x) = x(x - 6)^2$
2. $f(x) = x^2(x + 3)(x^2 - 1)$
3. $g(x) = (x - 2)(x + 4)^3$
4. $f(x) = (x + 5)(x - 8)^2$
5. $f(x) = (x + 6)(x + i)(x - i)$
6. $h(t) = (t - 3)(t - 2)(t - 3i)(t + 3i)$

In Exercises 7–10, use the Rational Zero Test to list all possible rational zeros of *f*. Verify that the zeros of *f* shown on the graph are contained in the list.

7. $f(x) = x^3 + 3x^2 - x - 3$

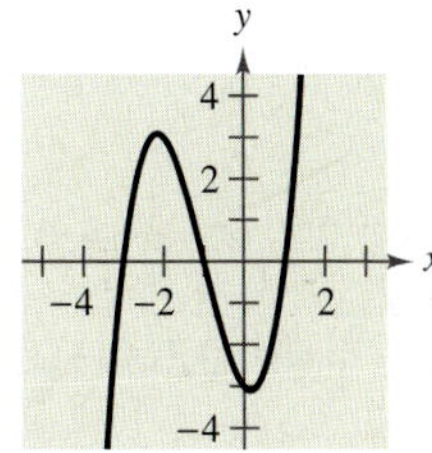

8. $f(x) = x^3 - 4x^2 - 4x + 16$

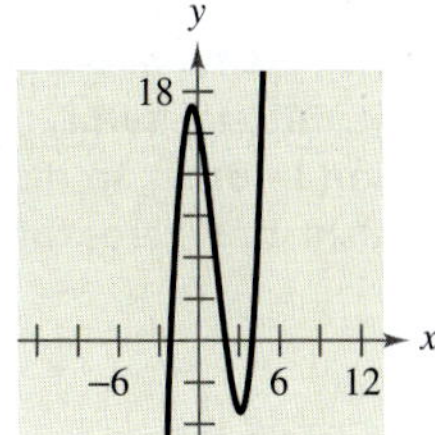

9. $f(x) = 2x^4 - 17x^3 + 35x^2 + 9x - 45$

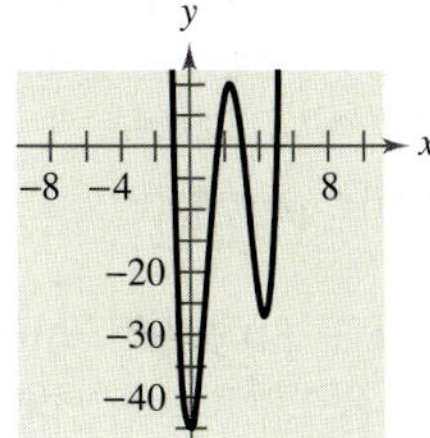

10. $f(x) = 4x^5 - 8x^4 - 5x^3 + 10x^2 + x - 2$

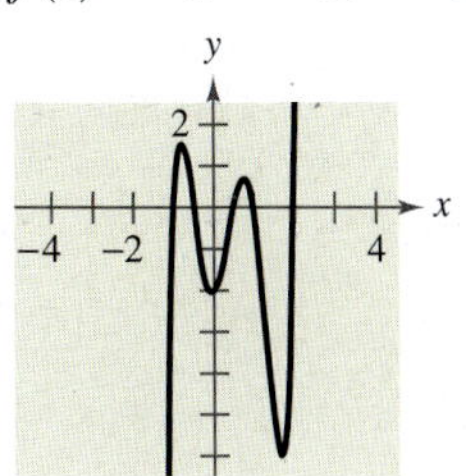

In Exercises 11–20, find all the real zeros of the function.

11. $f(x) = x^3 - 6x^2 + 11x - 6$
12. $f(x) = x^3 - 7x - 6$
13. $g(x) = x^3 - 4x^2 - x + 4$
14. $h(x) = x^3 - 9x^2 + 20x - 12$
15. $h(t) = t^3 + 12t^2 + 21t + 10$
16. $p(x) = x^3 - 9x^2 + 27x - 27$
17. $C(x) = 2x^3 + 3x^2 - 1$
18. $f(x) = 3x^3 - 19x^2 + 33x - 9$
19. $f(x) = 9x^4 - 9x^3 - 58x^2 + 4x + 24$
20. $f(x) = 2x^4 - 15x^3 + 23x^2 + 15x - 25$

In Exercises 21–24, find all real solutions of the polynomial equation.

21. $z^4 - z^3 - 2z - 4 = 0$
22. $x^4 - 13x^2 - 12x = 0$
23. $2y^4 + 7y^3 - 26y^2 + 23y - 6 = 0$
24. $x^5 - x^4 - 3x^3 + 5x^2 - 2x = 0$

In Exercises 25–28, (a) list the possible rational zeros of *f*, (b) sketch the graph of *f* so that some of the possible zeros in part (a) can be disregarded, and then (c) determine all real zeros of *f*.

25. $f(x) = x^3 + x^2 - 4x - 4$
26. $f(x) = -3x^3 + 20x^2 - 36x + 16$
27. $f(x) = -4x^3 + 15x^2 - 8x - 3$
28. $f(x) = 4x^3 - 12x^2 - x + 15$

In Exercises 29–32, (a) list the possible rational zeros of *f*, (b) use a graphing utility to graph *f* so that some of the possible zeros in part (a) can be disregarded, and then (c) determine all real zeros of *f*.

29. $f(x) = -2x^4 + 13x^3 - 21x^2 + 2x + 8$
30. $f(x) = 4x^4 - 17x^2 + 4$
31. $f(x) = 32x^3 - 52x^2 + 17x + 3$
32. $f(x) = 4x^3 + 7x^2 - 11x - 18$

Graphical Analysis **In Exercises 33–36, (a) use the *zero* or *root* feature of a graphing utility to approximate the zeros of the function accurate to three decimal places and (b) determine one of the exact zeros, use synthetic division to verify your result, and then factor the polynomial completely.**

33. $f(x) = x^4 - 3x^2 + 2$

34. $P(t) = t^4 - 7t^2 + 12$

35. $h(x) = x^5 - 7x^4 + 10x^3 + 14x^2 - 24x$

36. $g(x) = 6x^4 - 11x^3 - 51x^2 + 99x - 27$

In Exercises 37–42, find a polynomial function with integer coefficients that has the given zeros. (There are many correct answers.)

37. $1, 5i, -5i$

38. $4, 3i, -3i$

39. $6, -5 + 2i, -5 - 2i$

40. $2, 4 + i, 4 - i$

41. $\frac{2}{3}, -1, 3 + \sqrt{2}i$

42. $-5, -5, 1 + \sqrt{3}i$

In Exercises 43–46, write the polynomial (a) as the product of factors that are irreducible over the *rationals*, (b) as the product of linear and quadratic factors that are irreducible over the *reals*, and (c) in completely factored form.

43. $f(x) = x^4 + 6x^2 - 27$

44. $f(x) = x^4 - 2x^3 - 3x^2 + 12x - 18$
(*Hint:* One factor is $x^2 - 6$.)

45. $f(x) = x^4 - 4x^3 + 5x^2 - 2x - 6$
(*Hint:* One factor is $x^2 - 2x - 2$.)

46. $f(x) = x^4 - 3x^3 - x^2 - 12x - 20$
(*Hint:* One factor is $x^2 + 4$.)

In Exercises 47–54, use the given zero to find all the zeros of the function.

	Function	*Zero*
47.	$f(x) = 2x^3 + 3x^2 + 50x + 75$	$5i$
48.	$f(x) = x^3 + x^2 + 9x + 9$	$3i$
49.	$f(x) = 2x^4 - x^3 + 7x^2 - 4x - 4$	$2i$
50.	$g(x) = x^3 - 7x^2 - x + 87$	$5 + 2i$
51.	$g(x) = 4x^3 + 23x^2 + 34x - 10$	$-3 + i$
52.	$h(x) = 3x^3 - 4x^2 + 8x + 8$	$1 - \sqrt{3}i$
53.	$f(x) = x^4 + 3x^3 - 5x^2 - 21x + 22$	$-3 + \sqrt{2}i$
54.	$f(x) = x^3 + 4x^2 + 14x + 20$	$-1 - 3i$

In Exercises 55–72, find all the zeros of the function and write the polynomial as a product of linear factors.

55. $f(x) = x^2 + 25$

56. $f(x) = x^2 - x + 56$

57. $h(x) = x^2 - 4x + 1$

58. $g(x) = x^2 + 10x + 23$

59. $f(x) = x^4 - 81$

60. $f(y) = y^4 - 625$

61. $f(z) = z^2 - 2z + 2$

62. $h(x) = x^3 - 3x^2 + 4x - 2$

63. $g(x) = x^3 - 6x^2 + 13x - 10$

64. $f(x) = x^3 - 2x^2 - 11x + 52$

65. $h(x) = x^3 - x + 6$

66. $h(x) = x^3 + 9x^2 + 27x + 35$

67. $f(x) = 5x^3 - 9x^2 + 28x + 6$

68. $g(x) = 3x^3 - 4x^2 + 8x + 8$

69. $g(x) = x^4 - 4x^3 + 8x^2 - 16x + 16$

70. $h(x) = x^4 + 6x^3 + 10x^2 + 6x + 9$

71. $f(x) = x^4 + 10x^2 + 9$

72. $f(x) = x^4 + 29x^2 + 100$

In Exercises 73–78, find all the zeros of the function. When there is an extended list of possible rational zeros, use a graphing utility to graph the function in order to discard any rational zeros that are obviously not zeros of the function.

73. $f(x) = x^3 + 24x^2 + 214x + 740$

74. $f(s) = 2s^3 - 5s^2 + 12s - 5$

75. $f(x) = 16x^3 - 20x^2 - 4x + 15$

76. $f(x) = 9x^3 - 15x^2 + 11x - 5$

77. $f(x) = 2x^4 + 5x^3 + 4x^2 + 5x + 2$

78. $g(x) = x^5 - 8x^4 + 28x^3 - 56x^2 + 64x - 32$

In Exercises 79–86, use Descartes's Rule of Signs to determine the possible number of positive and negative zeros of the function.

79. $g(x) = 5x^5 + 10x$

80. $h(x) = 4x^2 - 8x + 3$

81. $h(x) = 3x^4 + 2x^2 + 1$

82. $h(x) = 2x^4 - 3x + 2$

83. $g(x) = 2x^3 - 3x^2 - 3$

84. $f(x) = 4x^3 - 3x^2 + 2x - 1$

85. $f(x) = -5x^3 + x^2 - x + 5$

86. $f(x) = 3x^3 + 2x^2 + x + 3$

In Exercises 87–90, use synthetic division to verify the upper and lower bounds of the real zeros of *f*.

87. $f(x) = x^4 - 4x^3 + 15$

(a) Upper: $x = 4$ (b) Lower: $x = -1$

88. $f(x) = 2x^3 - 3x^2 - 12x + 8$

(a) Upper: $x = 4$ (b) Lower: $x = -3$

89. $f(x) = x^4 - 4x^3 + 16x - 16$

(a) Upper: $x = 5$ (b) Lower: $x = -3$

90. $f(x) = 2x^4 - 8x + 3$

(a) Upper: $x = 3$ (b) Lower: $x = -4$

In Exercises 91–94, find all the real zeros of the function.

91. $f(x) = 4x^3 - 3x - 1$

92. $f(z) = 12z^3 - 4z^2 - 27z + 9$

93. $f(y) = 4y^3 + 3y^2 + 8y + 6$

94. $g(x) = 3x^3 - 2x^2 + 15x - 10$

In Exercises 95–98, find all the rational zeros of the polynomial function.

95. $P(x) = x^4 - \frac{25}{4}x^2 + 9 = \frac{1}{4}(4x^4 - 25x^2 + 36)$

96. $f(x) = x^3 - \frac{3}{2}x^2 - \frac{23}{2}x + 6 = \frac{1}{2}(2x^3 - 3x^2 - 23x + 12)$

97. $f(x) = x^3 - \frac{1}{4}x^2 - x + \frac{1}{4} = \frac{1}{4}(4x^3 - x^2 - 4x + 1)$

98. $f(z) = z^3 + \frac{11}{6}z^2 - \frac{1}{2}z - \frac{1}{3} = \frac{1}{6}(6z^3 + 11z^2 - 3z - 2)$

In Exercises 99–102, match the cubic function with the numbers of rational and irrational zeros.

(a) Rational zeros: 0; Irrational zeros: 1

(b) Rational zeros: 3; Irrational zeros: 0

(c) Rational zeros: 1; Irrational zeros: 2

(d) Rational zeros: 1; Irrational zeros: 0

99. $f(x) = x^3 - 1$

100. $f(x) = x^3 - 2$

101. $f(x) = x^3 - x$

102. $f(x) = x^3 - 2x$

103. ***Geometry*** An open box is to be made from a rectangular piece of material, 15 centimeters by 9 centimeters, by cutting equal squares from the corners and turning up the sides.

(a) Let x represent the length of the sides of the squares removed. Draw a diagram showing the squares removed from the original piece of material and the resulting dimensions of the open box.

(b) Use the diagram to write the volume V of the box as a function of x. Determine the domain of the function.

(c) Sketch the graph of the function and approximate the dimensions of the box that will yield a maximum volume.

(d) Find values of x such that $V = 56$. Which of these values is a physical impossibility in the construction of the box? Explain.

104. ***Geometry*** A rectangular package to be sent by a delivery service (see figure) can have a maximum combined length and girth (perimeter of a cross section) of 120 inches.

(a) Show that the volume of the package is

$$V(x) = 4x^2(30 - x).$$

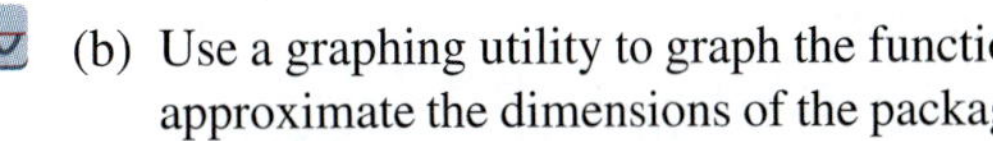

(b) Use a graphing utility to graph the function and approximate the dimensions of the package that will yield a maximum volume.

(c) Find values of x such that $V = 13{,}500$. Which of these values is a physical impossibility in the construction of the package? Explain.

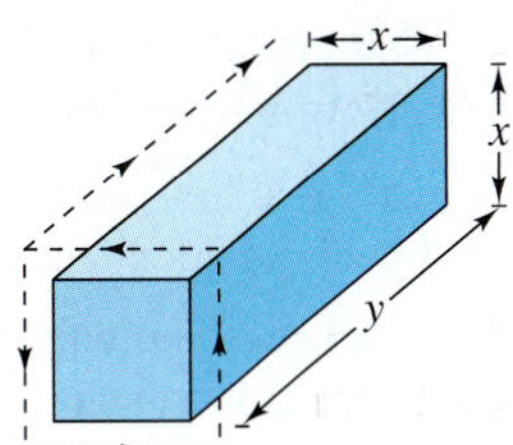

105. ***Advertising Cost*** A company that produces portable cassette players estimates that the profit P (in dollars) for selling a particular model is

$$P = -76x^3 + 4830x^2 - 320{,}000, \quad 0 \le x \le 60$$

where x is the advertising expense (in tens of thousands of dollars). Using this model, find the smaller of two advertising amounts that will yield a profit of \$2,500,000.

106. ***Advertising Cost*** A company that manufactures bicycles estimates that the profit P (in dollars) for selling a particular model is

$$P = -45x^3 + 2500x^2 - 275{,}000, \quad 0 \le x \le 50$$

where x is the advertising expense (in tens of thousands of dollars). Using this model, find the smaller of two advertising amounts that will yield a profit of \$800,000.

▶ Model It

107. ***Athletics*** The attendance A (in millions) at NCAA women's college basketball games for the years 1994 through 2000 is shown in the table, where t represents the year, with $t = 4$ corresponding to 1994. (Source: National Collegiate Athletic Association)

Year, t	Attendance, A
4	4.557
5	4.962
6	5.234
7	6.734
8	7.387
9	8.698
10	8.825

(a) Use the *regression* feature of a graphing utility to find a cubic model for the data.

(b) Use the graphing utility to create a scatter plot of the data. Then graph the model and the scatter plot in the same viewing window. How do they compare?

(c) According to the model found in part (a), in what year did attendance reach 5.5 million?

(d) According to the model found in part (a), in what year did attendance reach 8 million?

(e) According to the right-hand behavior of the model, will the attendance continue to increase? Explain.

108. ***Cost*** The ordering and transportation cost C (in thousands of dollars) for the components used in manufacturing a product is

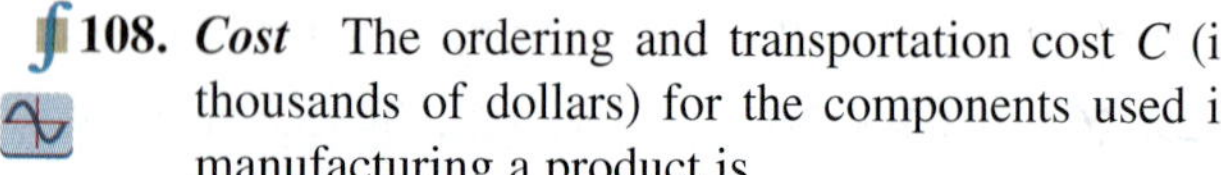

$$C = 100\left(\frac{200}{x^2} + \frac{x}{x + 30}\right), \quad x \geq 1$$

where x is the order size (in hundreds). In calculus, it can be shown that the cost is a minimum when

$$3x^3 - 40x^2 - 2400x - 36{,}000 = 0.$$

Use a calculator to approximate the optimal order size to the nearest hundred units.

109. ***Height of a Baseball*** A baseball is thrown upward from ground level with an initial velocity of 48 feet per second, and its height h (in feet) is

$$h(t) = -16t^2 + 48t, \quad 0 \leq t \leq 3$$

where t is the time (in seconds). You are told the ball reaches a height of 64 feet. Is this possible?

110. ***Profit*** The demand equation for a certain product is $p = 140 - 0.0001x$, where p is the unit price (in dollars) of the product and x is the number of units produced and sold. The cost equation for the product is $C = 80x + 150{,}000$, where C is the total cost (in dollars) and x is the number of units produced. The total profit obtained by producing and selling x units is

$$P = R - C = xp - C.$$

You are working in the marketing department of the company that produces this product, and you are asked to determine a price p that will yield a profit of 9 million dollars. Is this possible? Explain.

Synthesis

True or False? **In Exercises 111 and 112, decide whether the statement is true or false. Justify your answer.**

111. It is possible for a third-degree polynomial function with integer coefficients to have no real zeros.

112. If $x = -i$ is a zero of the function $f(x) = x^3 + ix^2 + ix - 1$, then $x = i$ must also be a zero of f.

Think About It **In Exercises 113–118, determine (if possible) the zeros of the function g if the function f has zeros at $x = r_1, x = r_2$, and $x = r_3$.**

113. $g(x) = -f(x)$

114. $g(x) = 3f(x)$

115. $g(x) = f(x - 5)$

116. $g(x) = f(2x)$

117. $g(x) = 3 + f(x)$

118. $g(x) = f(-x)$

119. ***Exploration*** Use a graphing utility to graph the function $f(x) = x^4 - 4x^2 + k$ for different values of k. Find values of k such that the zeros of f satisfy the specified characteristics. (Some parts do not have unique answers.)

(a) Four real zeros

(b) Two real zeros, each of multiplicity 2

(c) Two real zeros and two complex roots

(d) Four complex zeros

120. ***Think About It*** Will the answers to Exercise 119 change for the function g?

(a) $g(x) = f(x - 2)$ (b) $g(x) = f(2x)$

121. ***Think About It*** A third-degree polynomial function f has real zeros $-2, \frac{1}{2}$, and 3, and its leading coefficient is negative. Write an equation for f. Sketch the graph of f. How many different polynomial functions are possible for f?

122. ***Think About It*** Sketch the graph of a fifth-degree polynomial function whose leading coefficient is positive and that has one root at $x = 3$ of multiplicity 2.

123. Use the information in the table.

Interval	Value of $f(x)$
$(-\infty, -2)$	Positive
$(-2, 1)$	Negative
$(1, 4)$	Negative
$(4, \infty)$	Positive

(a) What are the three real zeros of the polynomial function f?

(b) What can be said about the behavior of the graph of f at $x = 1$?

(c) What is the least possible degree of f? Explain. Can the degree of f ever be odd? Explain.

(d) Is the leading coefficient of f positive or negative? Explain.

(e) Write an equation for f. There are many correct answers.

(f) Sketch a graph of the equation you wrote in part (e).

124. Use the information in the table.

Interval	Value of $f(x)$
$(-\infty, -2)$	Negative
$(-2, 0)$	Positive
$(0, 2)$	Positive
$(2, \infty)$	Negative

(a) What are the three real zeros of the polynomial function f?

(b) What can be said about the behavior of the graph of f at $x = 0$?

(c) What is the least possible degree of f? Explain. Can the degree of f ever be odd? Explain.

(d) Is the leading coefficient of f positive or negative? Explain.

(e) Write an equation for f. There are many correct answers.

(f) Sketch a graph of the equation you wrote in part (e).

125. (a) Find a quadratic function f (with integer coefficients) that has $\pm\sqrt{b}i$ as zeros. Assume that b is a positive integer.

(b) Find a quadratic function f (with integer coefficients) that has $a \pm bi$ as zeros. Assume that b is a positive integer.

126. ***Graphical Reasoning*** The graph of one of the following functions is shown below. Identify the function shown in the graph. Explain why each of the others is not the correct function. Use a graphing utility to verify your result.

(a) $f(x) = x^2(x + 2)(x - 3.5)$

(b) $g(x) = (x + 2)(x - 3.5)$

(c) $h(x) = (x + 2)(x - 3.5)(x^2 + 1)$

(d) $k(x) = (x + 1)(x + 2)(x - 3.5)$

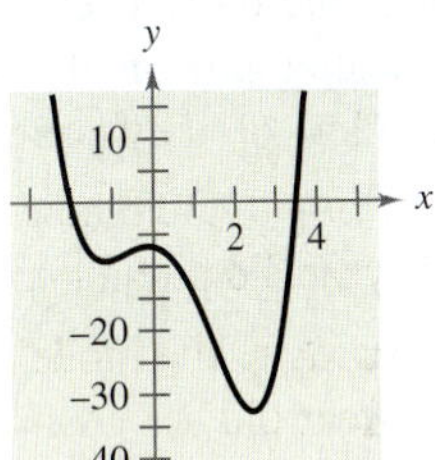

Review

In Exercises 127–130, perform the operation and simplify.

127. $(-3 + 6i) - (8 - 3i)$ **128.** $(12 - 5i) + 16i$

129. $(6 - 2i)(1 + 7i)$ **130.** $(9 - 5i)(9 + 5i)$

In Exercises 131–136, use the graph of f to sketch the graph of g. To print an enlarged copy of the graph, go to the website *www.mathgraphs.com*.

131. $g(x) = f(x - 2)$

132. $g(x) = f(x) - 2$

133. $g(x) = 2f(x)$

134. $g(x) = f(-x)$

135. $g(x) = f(2x)$

136. $g(x) = f\left(\frac{1}{2}x\right)$

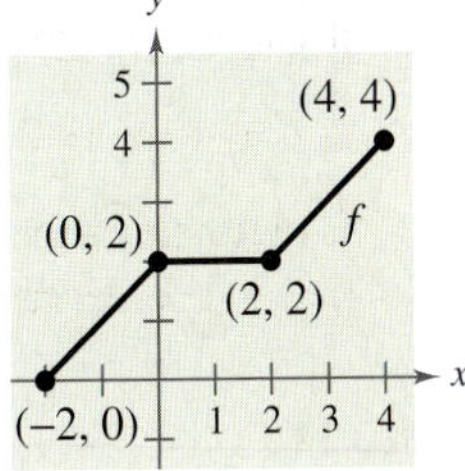

3.5 Mathematical Modeling

▶ **What you should learn**

- How to use mathematical models to approximate sets of data points
- How to write mathematical models for direct variation
- How to write mathematical models for direct variation as an nth power
- How to write mathematical models for inverse variation
- How to write mathematical models for joint variation
- How to use the *regression* feature of a graphing utility to find the equation of a least squares regression line

▶ **Why you should learn it**

You can use functions as models to represent a wide variety of real-life data sets. For instance, in Exercise 63 on page 317, a variation model can be used to model the water temperature of the ocean at various depths.

Introduction

You have already studied some techniques for fitting models to data. For instance, in Section 2.1, you learned how to find the equation of a line that passes through two points. In this section, you will study other techniques for fitting models to data: *direct and inverse variation* and *least squares regression*. The resulting models are either polynomial functions or rational functions. (Rational functions will be studied in Chapter 4.)

Example 1 ▶ A Mathematical Model

The numbers of insured commercial banks y (in thousands) in the United States for the years 1995 to 1999 are shown in the table. (Source: Federal Deposit Insurance Corporation)

Year	Insured commercial banks, y
1995	9.94
1996	9.53
1997	9.14
1998	8.77
1999	8.58

A linear model that approximates this data is $y = -0.348t + 11.63$ for $5 \le t \le 9$, where t is the year, with $t = 5$ corresponding to 1995. Plot the actual data *and* the model on the same graph. How closely does the model represent the data?

Solution

The actual data is plotted in Figure 3.34, along with the graph of the linear model. From the graph, it appears that the model is a "good fit" for the actual data. You can see how well the model fits by comparing the actual values of y with the values of y given by the model. The values given by the model are labeled y^* in the table below.

t	5	6	7	8	9
y	9.94	9.53	9.14	8.77	8.58
y^*	9.89	9.54	9.19	8.85	8.50

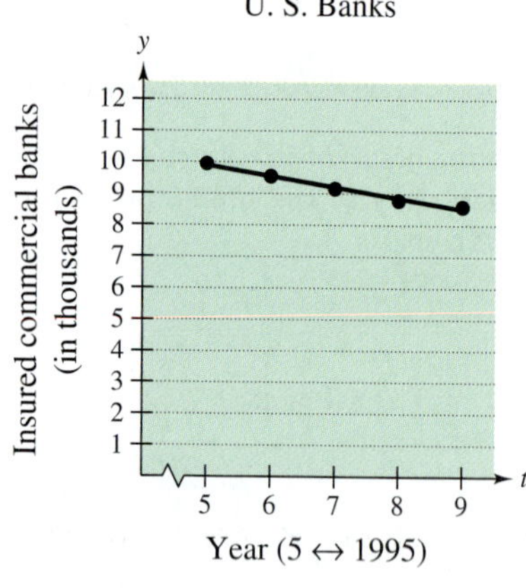

FIGURE 3.34

Note in Example 1 that you could have chosen any two points to find a line that fits the data. However, the given linear model was found using the *regression* feature of a graphing utility and is the line that *best* fits the data. This concept of a "best-fitting" line is discussed later in this section.

Direct Variation

There are two basic types of linear models. The more general model has a y-intercept that is nonzero.

$$y = mx + b, \quad b \neq 0$$

The simpler model

$$y = kx$$

has a y-intercept that is zero. In the simpler model, y is said to **vary directly** as x, or to be **directly proportional** to x.

Direct Variation

The following statements are equivalent.

1. y **varies directly** as x.
2. y is **directly proportional** to x.
3. $y = kx$ for some nonzero constant k.

k is the **constant of variation** or the **constant of proportionality.**

Example 2 ▶ Direct Variation  

In Pennsylvania, the state income tax is directly proportional to *gross income*. You are working in Pennsylvania and your state income tax deduction is \$42 for a gross monthly income of \$1500. Find a mathematical model that gives the Pennsylvania state income tax in terms of gross income.

Solution

Verbal Model: State income tax $=$ k $\cdot$ Gross income

Labels:
State income tax $= y$ (dollars)
Gross income $= x$ (dollars)
Income tax rate $= k$ (percent in decimal form)

Equation: $y = kx$

To solve for k, substitute the given information into the equation $y = kx$, and then solve for k.

$y = kx$ Write direct variation model.

$42 = k(1500)$ Substitute $y = 42$ and $x = 1500$.

$0.028 = k$ Simplify.

So, the equation (or model) for state income tax in Pennsylvania is

$$y = 0.028x.$$

In other words, Pennsylvania has a state income tax rate of 2.8% of gross income. The graph of this equation is shown in Figure 3.35.

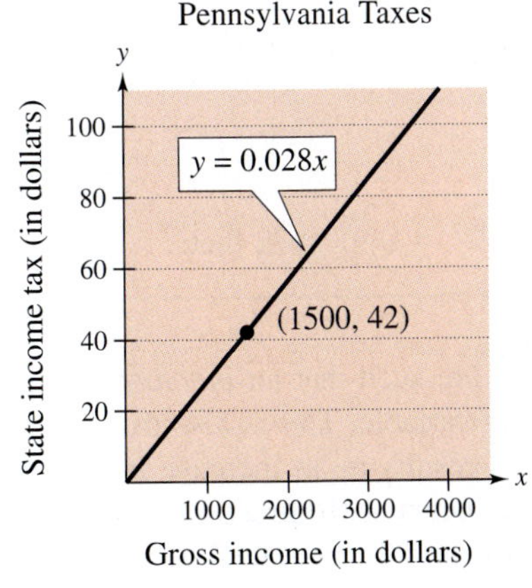

FIGURE 3.35

Direct Variation as an nth Power

Another type of direct variation relates one variable to a *power* of another variable. For example, in the formula for the area of a circle

$$A = \pi r^2$$

the area A is directly proportional to the square of the radius r. Note that for this formula, π is the constant of proportionality.

STUDY TIP

Note that the direct variation model $y = kx$ is a special case of $y = kx^n$ with $n = 1$.

Direct Variation as an nth Power

The following statements are equivalent.

1. y **varies directly as the nth power** of x.
2. y is **directly proportional to the nth power** of x.
3. $y = kx^n$ for some constant k.

Example 3 ▶ Direct Variation as nth Power  

The distance a ball rolls down an inclined plane is directly proportional to the square of the time it rolls. During the first second, the ball rolls 8 feet. (See Figure 3.36.)

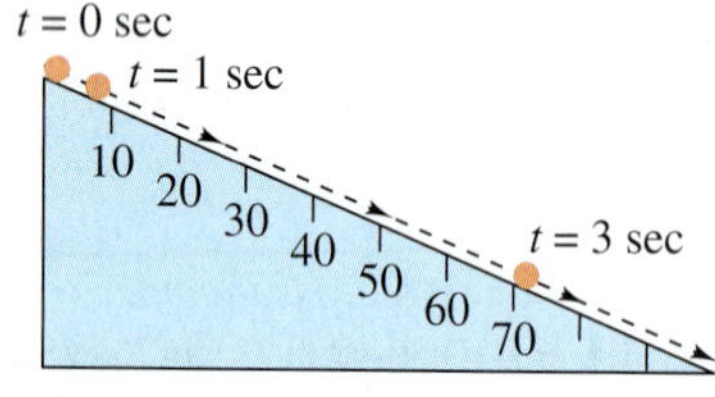

FIGURE 3.36

a. Write an equation relating the distance traveled to the time.

b. How far will the ball roll during the first 3 seconds?

Solution

a. Letting d be the distance (in feet) the ball rolls and letting t be the time (in seconds), you have

$$d = kt^2.$$

Now, because $d = 8$ when $t = 1$, you can see that $k = 8$, as follows.

$$d = kt^2$$

$$8 = k(1)^2$$

$$8 = k$$

So, the equation relating distance to time is

$$d = 8t^2.$$

b. When $t = 3$, the distance traveled is $d = 8(3)^2 = 8(9) = 72$ feet.

In Examples 2 and 3, the direct variations are such that an *increase* in one variable corresponds to an *increase* in the other variable. This is also true in the model $d = \frac{1}{5}F$, $F > 0$, where an increase in F results in an increase in d. You should not, however, assume that this always occurs with direct variation. For example, in the model $y = -3x$, an increase in x results in a *decrease* in y, and yet y is said to vary directly as x.

Inverse Variation

Inverse Variation

The following statements are equivalent.

1. y **varies inversely** as x.
2. y is **inversely proportional** to x.
3. $y = \frac{k}{x}$ for some constant k.

If x and y are related by an equation of the form $y = k/x^n$, then y varies inversely as the nth power of x (or y is inversely proportional to the nth power of x).

Example 4 ▶ Inverse Variation  

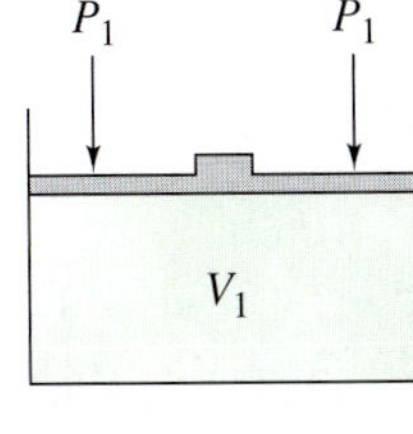

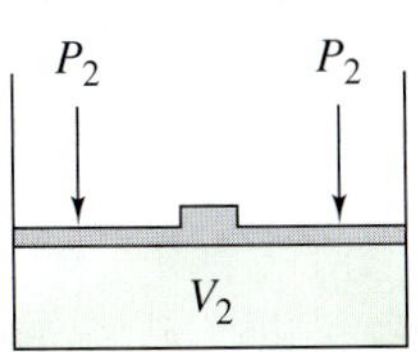

FIGURE 3.37 *If the temperature is held constant and pressure increases, volume decreases.*

A gas law states that the volume of an enclosed gas varies directly as the temperature *and* inversely as the pressure, as shown in Figure 3.37. The pressure of a gas is 0.75 kilogram per square centimeter when the temperature is 294 K and the volume is 8000 cubic centimeters.

a. Write an equation relating pressure, temperature, and volume.

b. Find the pressure when the temperature is 300 K and the volume is 7000 cubic centimeters.

Solution

a. Let V be volume (in cubic centimeters), let P be pressure (in kilograms per square centimeter), and let T be temperature (in Kelvin). Because V varies directly as T and inversely as P,

$$V = \frac{kT}{P}.$$

Now, because $P = 0.75$ when $T = 294$ and $V = 8000$,

$$8000 = \frac{k(294)}{0.75}$$

$$\frac{8000(0.75)}{294} = k$$

$$k = \frac{6000}{294} = \frac{1000}{49}.$$

So, the equation relating pressure, temperature, and volume is

$$V = \frac{1000}{49}\left(\frac{T}{P}\right).$$

b. When $T = 300$ and $V = 7000$, the pressure is

$$P = \frac{1000}{49}\left(\frac{300}{7000}\right) = \frac{300}{343} \approx 0.87 \text{ kilogram per square centimeter.}$$

Joint Variation

In Example 4, note that when a direct variation and an inverse variation occur in the same statement, they are coupled with the word "and." To describe two different *direct* variations in the same statement, the word **jointly** is used.

Joint Variation

The following statements are equivalent.

1. z **varies jointly** as x and y.
2. z is **jointly proportional** to x and y.
3. $z = kxy$ for some constant k.

If x, y, and z are related by an equation of the form

$$z = kx^n y^m$$

then z varies jointly as the nth power of x and the mth power of y.

Example 5 ▶ Joint Variation  

The *simple* interest for a certain savings account is jointly proportional to the time and the principal. After one quarter (3 months), the interest on a principal of \$5000 is \$43.75.

a. Write an equation relating the interest, principal, and time.

b. Find the interest after three quarters.

Solution

a. Let I = interest (in dollars), P = principal (in dollars), and t = time (in years). Because I is jointly proportional to P and t,

$$I = kPt.$$

For $I = 43.75$, $P = 5000$, and $t = \frac{1}{4}$,

$$43.75 = k(5000)\left(\frac{1}{4}\right)$$

which implies that $k = 4(43.75)/5000 = 0.035$. So, the equation relating interest, principal, and time is

$$I = 0.035Pt$$

which is the familiar equation for simple interest where the constant of proportionality, 0.035, represents an annual interest rate of 3.5%.

b. When $P = \$5000$ and $t = \frac{3}{4}$, the interest is

$$I = (0.035)(5000)\left(\frac{3}{4}\right)$$

$$= \$131.25.$$

Least Squares Regression and Graphing Utilities

So far in this text, you have worked with many different types of mathematical models that approximate real-life data. In some instances the model was given, whereas in other instances you were asked to find the model using simple algebraic techniques or a graphing utility.

To find a model that approximates the data most accurately, statisticians use a measure called the **sum of square differences,** which is the sum of the squares of the differences between actual data values and model values. The "best-fitting" linear model is the one with the least sum of square differences. This best-fitting linear model is called the **least squares regression line.** Recall that you can approximate this line visually by plotting the data points and drawing the line that appears to fit best—or you can enter the data points into a calculator or computer and use the calculator's or computer's linear regression program. When you run a linear regression program, the "r-value" or **correlation coefficient** gives a measure of how well the model fits the data. The closer the value of $|r|$ is to 1, the better the fit.

Example 6 ▶ Finding a Least Squares Regression Line  

The amounts p (in millions of dollars) of total annual prize money awarded at the Indianapolis 500 race from 1993 to 2001 are shown in the table. Construct a scatter plot that represents the data and find a linear model that approximates the data. (Source: Indy Racing League)

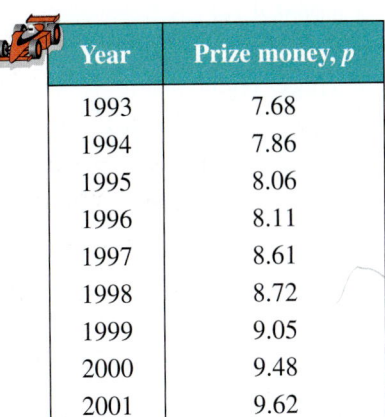

Year	Prize money, p
1993	7.68
1994	7.86
1995	8.06
1996	8.11
1997	8.61
1998	8.72
1999	9.05
2000	9.48
2001	9.62

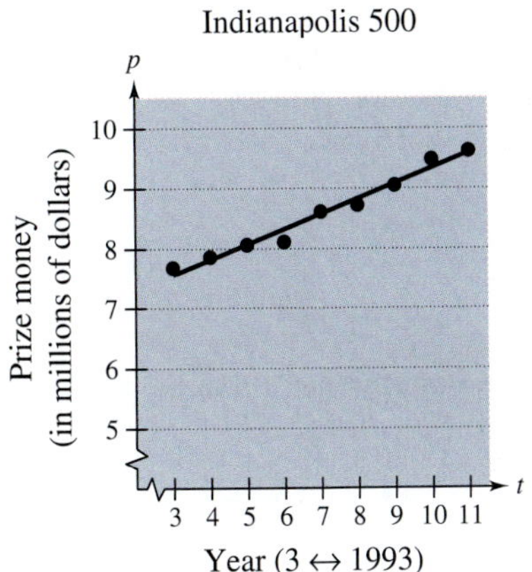

FIGURE 3.38

t	p	p^*
3	7.68	7.56
4	7.86	7.82
5	8.06	8.07
6	8.11	8.32
7	8.61	8.58
8	8.72	8.83
9	9.05	9.09
10	9.48	9.34
11	9.62	9.59

Solution

Let $t = 3$ represent 1993. The scatter plot for the points is shown in Figure 3.38. Using the *regression* feature of a graphing utility, you can determine that the equation of the least squares regression line is

$$p = 0.254t + 6.80.$$

To check this model, compare the actual p-values with the p-values given by the model, which are labeled $p*$ in the table at the left. The correlation coefficient for this model is $r \approx 0.988$, which implies that the model is a good fit.

3.5 Exercises

1. ***Employment*** The total numbers of employees (in thousands) in the United States from 1992 to 1999 are given by the following ordered pairs.

(1992, 128,105) (1996, 133,943)
(1993, 129,200) (1997, 136,297)
(1994, 131,056) (1998, 137,673)
(1995, 132,304) (1999, 139,368)

A linear model that approximates this data is $y = 124{,}420 + 1649.6t$, where y represents the number of employees (in thousands) and $t = 2$ represents 1992. Plot the actual data and the model on the same set of coordinate axes. How closely does the model represent the data? (Source: U.S. Bureau of Labor Statistics)

2. ***Sports*** The winning times (in minutes) in the women's 400-meter freestyle swimming event in the Olympics from 1948 to 2000 are given by the following ordered pairs.

(1948, 5.30) (1976, 4.16)
(1952, 5.20) (1980, 4.15)
(1956, 4.91) (1984, 4.12)
(1960, 4.84) (1988, 4.06)
(1964, 4.72) (1992, 4.12)
(1968, 4.53) (1996, 4.12)
(1972, 4.32) (2000, 4.10)

A linear model that approximates this data is $y = 5.06 - 0.024t$, where y represents the winning time in minutes and $t = 0$ represents 1950. Plot the actual data and the model on the same set of coordinate axes. How closely does the model represent the data? (Source: *The World Almanac and Book of Facts*)

Think About It **In Exercises 3 and 4, use the graph to determine whether y varies directly as some power of x or inversely as some power of x. Explain.**

3.

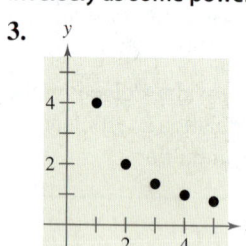

4.

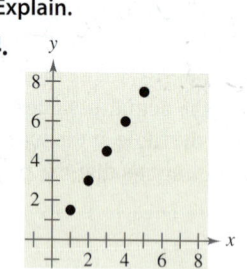

In Exercises 5–8, use the given value of k to complete the table for the direct variation model $y = kx^2$. Plot the points on a rectangular coordinate system.

x	2	4	6	8	10
$y = kx^2$					

5. $k = 1$
6. $k = 2$
7. $k = \frac{1}{2}$
8. $k = \frac{1}{4}$

In Exercises 9–12, use the given value of k to complete the table for the inverse variation model

$$y = \frac{k}{x^2}.$$

Plot the points on a rectangular coordinate system.

x	2	4	6	8	10
$y = \frac{k}{x^2}$					

9. $k = 2$
10. $k = 5$
11. $k = 10$
12. $k = 20$

In Exercises 13–16, determine whether the variation model is of the form $y = kx$ or $y = k/x$, and find k.

13.

x	y
5	1
10	$\frac{1}{2}$
15	$\frac{1}{3}$
20	$\frac{1}{4}$
25	$\frac{1}{5}$

14.

x	y
5	2
10	4
15	6
20	8
25	10

15.

x	y
5	-3.5
10	-7
15	-10.5
20	-14
25	-17.5

16.

x	y
5	24
10	12
15	8
20	6
25	$\frac{24}{5}$

Direct Variation **In Exercises 17–20, assume that y is directly proportional to x. Use the given x-value and y-value to find a linear model that relates y and x.**

	x-Value	*y-Value*		*x-Value*	*y-Value*
17.	$x = 5$	$y = 12$	**18.**	$x = 2$	$y = 14$
19.	$x = 10$	$y = 2050$	**20.**	$x = 6$	$y = 580$

21. ***Simple Interest*** The simple interest on an investment is directly proportional to the amount of the investment. By investing \$2500 in a certain bond issue, you obtained an interest payment of \$87.50 after 1 year. Find a mathematical model that gives the interest I for this bond issue after 1 year in terms of the amount invested P.

22. ***Simple Interest*** The simple interest on an investment is directly proportional to the amount of the investment. By investing \$5000 in a municipal bond, you obtained an interest payment of \$187.50 after 1 year. Find a mathematical model that gives the interest I for this municipal bond after 1 year in terms of the amount invested P.

23. ***Measurement*** On a yardstick with scales in inches and centimeters, you notice that 13 inches is approximately the same length as 33 centimeters. Use this information to find a mathematical model that relates centimeters to inches. Then use the model to find the number of centimeters in 10 inches and 20 inches.

24. ***Measurement*** When buying gasoline, you notice that 14 gallons of gasoline is approximately the same amount of gasoline as 53 liters. Use this information to find a linear model that relates gallons to liters. Use the model to find the number of liters in 5 gallons and 25 gallons.

25. ***Taxes*** Property tax is based on the assessed value of the property. A house that has an assessed value of \$150,000 has a property tax of \$5520. Find a mathematical model that gives the amount of property tax y in terms of the assessed value x of the property. Use the model to find the property tax on a house that has an assessed value of \$200,000.

26. ***Taxes*** State sales tax is based on retail price. An item that sells for \$145.99 has a sales tax of \$10.22. Find a mathematical model that gives the amount of sales tax y in terms of the retail price x. Use the model to find the sales tax on a \$540.50 purchase.

Hooke's Law **In Exercises 27–30, use Hooke's Law for springs, which states that the distance a spring is stretched (or compressed) varies directly as the force on the spring.**

27. A force of 265 newtons stretches a spring 0.15 meter (see figure).

(a) How far will a force of 90 newtons stretch the spring?

(b) What force is required to stretch the spring 0.1 meter?

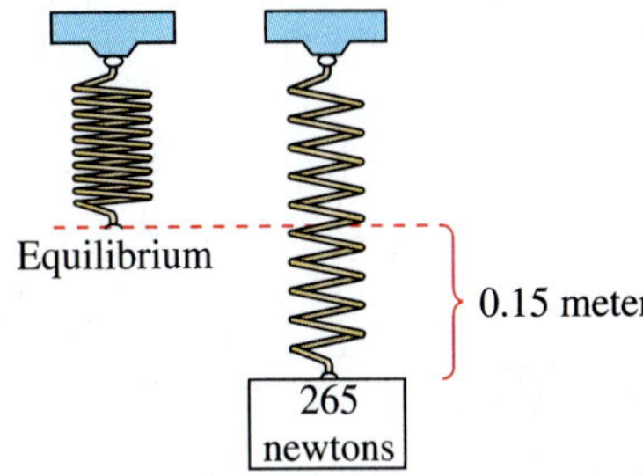

28. A force of 220 newtons stretches a spring 0.12 meter. What force is required to stretch the spring 0.16 meter?

29. The coiled spring of a toy supports the weight of a child. The spring is compressed a distance of 1.9 inches by the weight of a 25-pound child. The toy will not work properly if its spring is compressed more than 3 inches. What is the weight of the heaviest child who should be allowed to use the toy?

30. An overhead garage door has two springs, one on each side of the door (see figure). A force of 15 pounds is required to stretch each spring 1 foot. Because of a pulley system, the springs stretch only one-half the distance the door travels. The door moves a total of 8 feet, and the springs are at their natural length when the door is open. Find the combined lifting force applied to the door by the springs when the door is closed.

In Exercises 31–40, find a mathematical model for the verbal statement.

31. A varies directly as the square of r.

32. V varies directly as the cube of e.

33. y varies inversely as the square of x.

34. h varies inversely as the square root of s.

35. F varies directly as g and inversely as r^2.

36. z is jointly proportional to the square of x and y^3.

37. Boyle's Law: For a constant temperature, the pressure P of a gas is inversely proportional to the volume V of the gas.

38. Newton's Law of Cooling: The rate of change R of the temperature of an object is proportional to the difference between the temperature T of the object and the temperature T_e of the environment in which the object is placed.

39. Newton's Law of Universal Gravitation: The gravitational attraction F between two objects of masses m_1 and m_2 is proportional to the product of the masses and inversely proportional to the square of the distance r between the objects.

40. Logistic growth: The rate of growth R of a population is jointly proportional to the size S of the population and the difference between S and the maximum population size L that the environment can support.

In Exercises 41–46, write a sentence using the variation terminology of this section to describe the formula.

41. Area of a triangle: $A = \frac{1}{2}bh$

42. Surface area of a sphere: $S = 4\pi r^2$

43. Volume of a sphere: $V = \frac{4}{3}\pi r^3$

44. Volume of a right circular cylinder: $V = \pi r^2 h$

45. Average speed: $r = \dfrac{d}{t}$

46. Free vibrations: $\omega = \sqrt{\dfrac{kg}{W}}$

In Exercises 47–54, find a mathematical model representing the statement. (In each case, determine the constant of proportionality.)

47. A varies directly as r^2. ($A = 9\pi$ when $r = 3$.)

48. y varies inversely as x. ($y = 3$ when $x = 25$.)

49. y is inversely proportional to x. ($y = 7$ when $x = 4$.)

50. z varies jointly as x and y. ($z = 64$ when $x = 4$ and $y = 8$.)

51. F is jointly proportional to r and the third power of s. ($F = 4158$ when $r = 11$ and $s = 3$.)

52. P varies directly as x and inversely as the square of y. $\left(P = \frac{28}{3}\right.$ when $x = 42$ and $\left.y = 9.\right)$

53. z varies directly as the square of x and inversely as y. ($z = 6$ when $x = 6$ and $y = 4$.)

54. v varies jointly as p and q and inversely as the square of s. ($v = 1.5$ when $p = 4.1$, $q = 6.3$, and $s = 1.2$.)

Ecology **In Exercises 55 and 56, use the fact that the diameter of the largest particle that can be moved by a stream varies approximately directly as the square of the velocity of the stream.**

55. A stream with a velocity of $\frac{1}{4}$ mile per hour can move coarse sand particles about 0.02 inch in diameter. Approximate the velocity required to carry particles 0.12 inch in diameter.

56. A stream of velocity v can move particles of diameter d or less. By what factor does d increase when the velocity is doubled?

Resistance **In Exercises 57 and 58, use the fact that the resistance of a wire carrying an electrical current is directly proportional to its length and inversely proportional to its cross-sectional area.**

57. If #28 copper wire (which has a diameter of 0.0126 inch) has a resistance of 66.17 ohms per thousand feet, what length of #28 copper wire will produce a resistance of 33.5 ohms?

58. A 14-foot piece of copper wire produces a resistance of 0.05 ohm. Use the constant of proportionality from Exercise 57 to find the diameter of the wire.

59. ***Free Fall*** Neglecting air resistance, the distance s an object falls varies directly as the square of the duration t of the fall. An object falls a distance of 144 feet in 3 seconds. How far will it fall in 5 seconds?

60. ***Spending*** The prices of three sizes of pizza at a pizza shop are as follows.

9-inch: \$8.78, 12-inch: \$11.78, 15-inch: \$14.18

You would expect that the price of a certain size of pizza would be directly proportional to its surface area. Is that the case for this pizza shop? If not, which size of pizza is the best buy?

61. ***Fluid Flow*** The velocity v of a fluid flowing in a conduit is inversely proportional to the cross-sectional area of the conduit. (Assume that the volume of the flow per unit of time is held constant.) Determine the change in the velocity of water flowing from a hose when a person places a finger over the end of the hose to decrease its cross-sectional area by 25%.

62. ***Beam Load*** The maximum load that can be safely supported by a horizontal beam varies jointly as the width of the beam and the square of its depth, and inversely as the length of the beam. Determine the change in the maximum safe load under the following conditions.

(a) The width and length of the beam are doubled.

(b) The width and depth of the beam are doubled.

(c) All three of the dimensions are doubled.

(d) The depth of the beam is halved.

▶ Model It

63. ***Data Analysis*** An oceanographer took readings of the water temperature C (in degrees Celsius) at depth d (in meters). The data collected is shown in the table.

Depth, d	Temperature, C
1000	4.2°
2000	1.9°
3000	1.4°
4000	1.2°
5000	0.9°

(a) Sketch a scatter plot of the data.

(b) Does it appear that the data can be modeled by the inverse variation model $C = k/d$? If so, find k for each pair of coordinates.

(c) Determine the mean value of k from part (b) to find the inverse variation model $C = k/d$.

(d) Use a graphing utility to plot the data points and the inverse model in part (c).

(e) Use the model to approximate the depth at which the water temperature is 3°C.

64. ***Data Analysis*** An experiment in a physics lab requires a student to measure the compressed length y (in centimeters) of a spring when a force of F pounds is applied. The data is shown in the table.

Force, F	Length, y
0	0
2	1.15
4	2.3
6	3.45
8	4.6
10	5.75
12	6.9

(a) Sketch a scatter plot of the data.

(b) Does it appear that the data can be modeled by Hooke's Law? If so, estimate k. (See Exercises 27–30.)

(c) Use the model in part (b) to approximate the force required to compress the spring 9 centimeters.

65. ***Data Analysis*** A light probe is located x centimeters from a light source, and the intensity y (in microwatts per square centimeter) of the light is measured. The results are shown in the table.

x	y
30	0.1881
34	0.1543
38	0.1172
42	0.0998
46	0.0775
50	0.0645

A model for the data is $y = 262.76/x^{2.12}$.

(a) Use a graphing utility to plot the data points and the model in the same viewing window.

(b) Use the model to approximate the light intensity 25 centimeters from the light source.

66. ***Illumination*** The illumination from a light source varies inversely as the square of the distance from the light source. When the distance from a light source is doubled, how does the illumination change? Discuss this model in terms of the data given in Exercise 65. Give a possible explanation of the difference.

In Exercises 67–70, sketch the line that you think best approximates the data in the scatter plot. Then find an equation of the line. To print an enlarged copy of the graph, go to the website *www.mathgraphs.com*.

67.

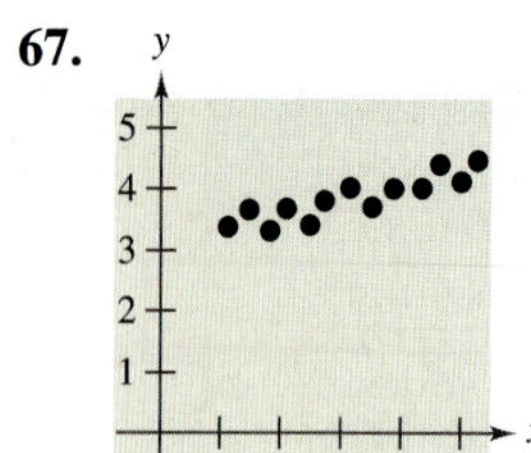

68.

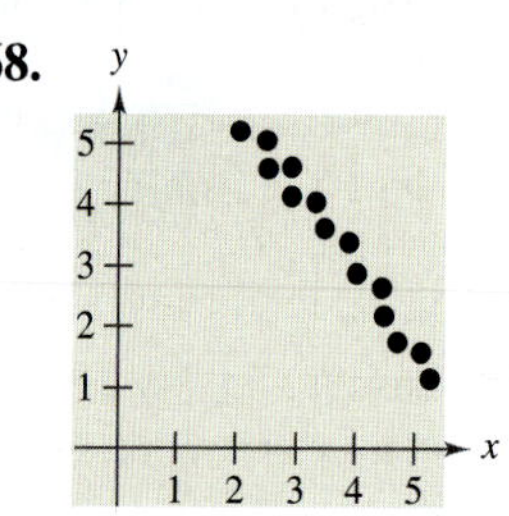

69.

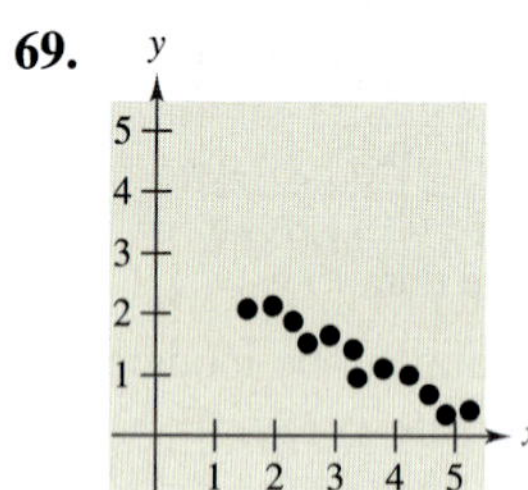

70.

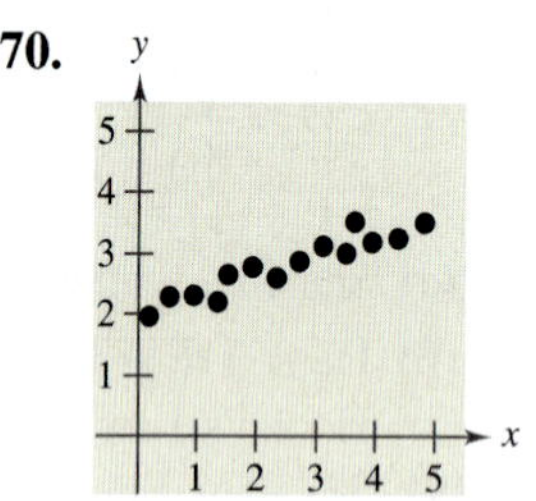

71. *Sports* The lengths (in feet) of the winning men's discus throws in the Olympics from 1908 to 2000 are listed below. (Source: *The World Almanac and Book of Facts*)

1908	134.2	1948	173.2	1976	221.4
1912	145.0	1952	180.5	1980	218.7
1920	146.6	1956	184.9	1984	218.5
1924	151.4	1960	194.2	1988	225.8
1928	155.2	1964	200.1	1992	213.7
1932	162.4	1968	212.5	1996	227.7
1936	165.6	1972	211.3	2000	227.3

(a) Sketch a scatter plot of the data. Let y represent the length of the winning discus throw (in feet) and let $t = 8$ represent 1908.

(b) Use a straightedge to sketch the best-fitting line through the points and find an equation of the line.

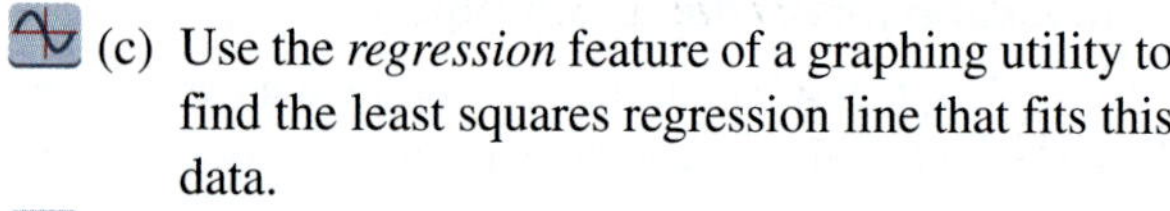

(c) Use the *regression* feature of a graphing utility to find the least squares regression line that fits this data.

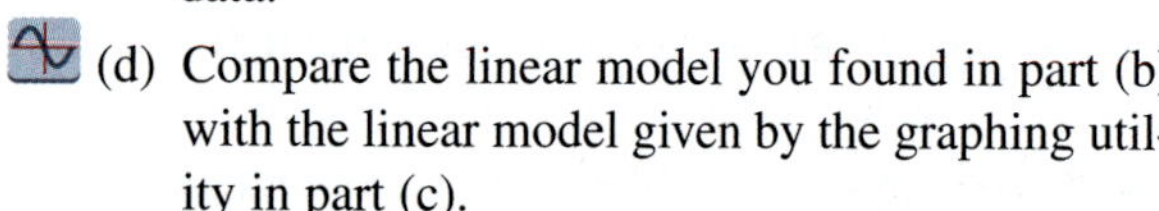

(d) Compare the linear model you found in part (b) with the linear model given by the graphing utility in part (c).

(e) Use the models from parts (b) and (c) to estimate the winning men's discus throw in the year 2004.

(f) Use your school's library, the Internet, or some other reference source to analyze the accuracy of the estimate in part (e).

72. *Sales* The total sales (in millions of dollars) for Barnes & Noble from 1992 to 2000 are listed below. (Source: Barnes & Noble, Inc.)

1992	1086.7	1995	1976.9	1998	3005.6
1993	1337.4	1996	2448.1	1999	3486.0
1994	1622.7	1997	2796.8	2000	4375.8

(a) Sketch a scatter plot of the data. Let y represent the total sales (in millions of dollars) and let $t = 2$ represent 1992.

(b) Use a straightedge to sketch the best-fitting line through the points and find an equation of the line.

(c) Use the *regression* feature of a graphing utility to find the least squares regression line that fits this data.

(d) Compare the linear model you found in part (b) with the linear model given by the graphing utility in part (c).

(e) Use the models from parts (b) and (c) to estimate the sales of Barnes & Noble in 2002.

(f) Use your school's library, the Internet, or some other reference source to analyze the accuracy of the estimate in part (e).

73. *Movie Theaters* The table shows the annual receipts R (in millions of dollars) for motion picture movie theaters in the United States from 1993 through 2001. (Source: Motion Picture Association of America)

Year	Receipts, R
1993	5154
1994	5396
1995	5494
1996	5912
1997	6366
1998	6949
1999	7448
2000	7661
2001	8413

(a) Use a graphing utility to create a scatter plot of the data. Let $t = 3$ represent 1993.

(b) Use the *regression* feature of a graphing utility to find the equation of the least squares regression line that fits this data.

(c) Use the graphing utility to graph the scatter plot you found in part (a) and the model you found in part (b) in the same viewing window.

(d) Use the model to estimate the annual receipts in 2000 and 2002.

(e) Interpret the meaning of the slope of the linear model in the context of the problem.

74. ***Data Analysis*** The table shows the number x (in millions) of households with cable television and the number y (in millions) of daily newspapers in circulation in the United States from 1993 through 1999. (Source: Nielsen Media Research and Editor & Publisher Co.)

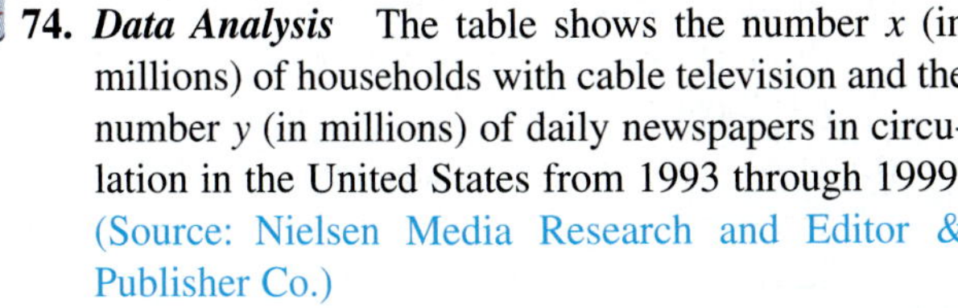

Households with cable, x	Daily newspapers, y
58.8	59.8
60.5	59.3
63.0	58.2
64.6	57.0
65.9	56.7
67.4	56.2
68.0	56.0

(a) Use the *regression* feature of a graphing utility to find the equation of the least squares regression line that fits this data.

(b) Use the graphing utility to create a scatter plot of the data. Then graph the model you found in part (a) and the scatter plot in the same viewing window.

(c) Use the model to estimate the number of daily newspapers in circulation if the number of households with cable television is 70 million.

(d) Interpret the meaning of the slope of the linear model in the context of the problem.

Synthesis

True or False? **In Exercises 75 and 76, decide whether the statement is true or false. Justify your answer.**

75. If y varies directly as x, then if x increases, y will increase as well.

76. In the equation for kinetic energy, $E = \frac{1}{2}mv^2$, the amount of kinetic energy E is directly proportional to the mass m of an object and the square of its velocity v.

77. ***Writing*** A linear mathematical model for predicting prize winnings at a race is based on data for 3 years. Write a paragraph discussing the potential accuracy or inaccuracy of such a model.

78. Discuss how well the data shown in each scatter plot can be approximated by a linear model.

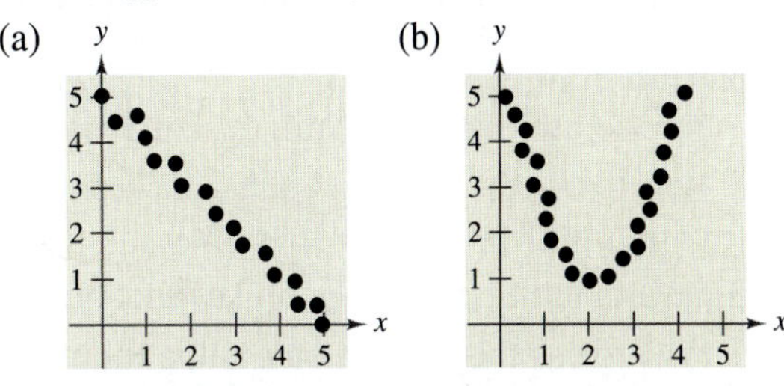

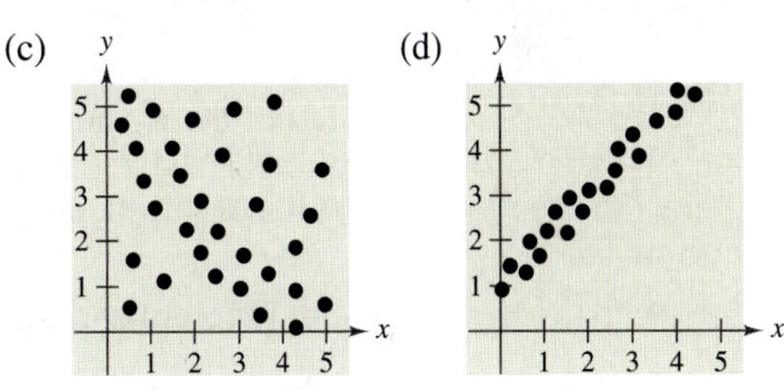

Review

In Exercises 79–82, solve the inequality and graph the solution on the real number line.

79. $(x - 5)^2 \geq 1$

80. $3(x + 1)(x - 3) < 0$

81. $6x^3 - 30x^2 > 0$

82. $x^4(x - 8) \geq 0$

In Exercises 83 and 84, evaluate the function at each value of the independent variable and simplify.

83. $f(x) = \dfrac{x^2 + 5}{x - 3}$

(a) $f(0)$ (b) $f(-3)$ (c) $f(4)$

84. $f(x) = \begin{cases} -x^2 + 10, & x \geq -2 \\ 6x^2 - 1, & x < -2 \end{cases}$

(a) $f(-2)$ (b) $f(1)$ (c) $f(-8)$

Chapter Summary

▶ *What* did you learn?

	Review Exercises
Section 3.1	
☐ How to analyze graphs of quadratic functions	1–6
☐ How to write quadratic functions in standard form and use the results to sketch graphs of functions	7–18
☐ How to use quadratic functions to model and solve real-life problems	19–24
Section 3.2	
☐ How to use transformations to sketch graphs of polynomial functions	25–30
☐ How to use the Leading Coefficient Test to determine the end behavior of graphs of polynomial functions	31–34
☐ How to use zeros of polynomial functions as sketching aids	35–44
☐ How to use the Intermediate Value Theorem to help locate zeros of polynomial functions	45–48
Section 3.3	
☐ How to use long division to divide polynomials by other polynomials	49–54
☐ How to use synthetic division to divide polynomials by binomials of the form $(x - k)$	55–62
☐ How to use the Remainder Theorem and the Factor Theorem	63–66
Section 3.4	
☐ How to use the Fundamental Theorem of Algebra to determine the number of zeros of polynomial functions	67–72
☐ How to find rational zeros of polynomial functions	73–80
☐ How to use conjugate pairs of complex zeros to find a polynomial with real coefficients	81, 82
☐ How to find zeros of polynomials by factoring	83–90
☐ How to use Descartes's Rule of Signs and the Upper and Lower Bound Rules to find zeros of polynomial functions	91–98
Section 3.5	
☐ How to use mathematical models to approximate sets of data points	99
☐ How to write mathematical models for direct variation	100
☐ How to write mathematical models for direct variation as an *n*th power	101, 102
☐ How to write mathematical models for inverse variation	103
☐ How to write mathematical models for joint variation	104
☐ How to use the *regression* feature of a graphing utility to find the equation of a least squares regression line	105

Review Exercises

3.1 In Exercises 1–4, find the quadratic function that has the indicated vertex and whose graph passes through the given point.

1. 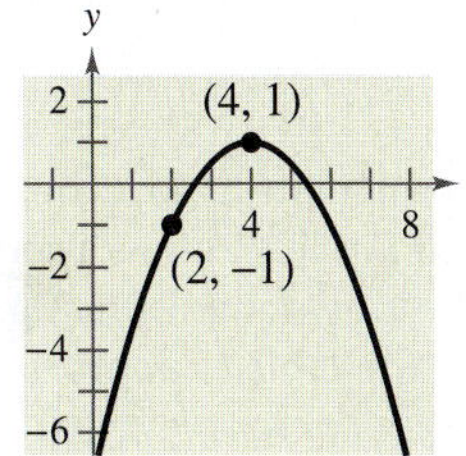

2. 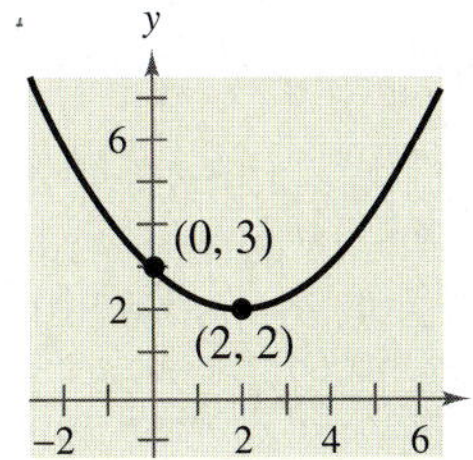

3. Vertex: $(1, -4)$; Point: $(2, -3)$

4. Vertex: $(2, 3)$; Point: $(-1, 6)$

In Exercises 5 and 6, graph each function. Compare the graph of each function with the graph of $y = x^2$.

5. (a) $f(x) = 2x^2$
(b) $g(x) = -2x^2$
(c) $h(x) = x^2 + 2$
(d) $k(x) = (x + 2)^2$

6. (a) $f(x) = x^2 - 4$
(b) $g(x) = 4 - x^2$
(c) $h(x) = (x - 3)^2$
(d) $k(x) = \frac{1}{2}x^2 - 1$

In Exercises 7–18, write the quadratic function in standard form and sketch its graph. Identify the vertex and x-intercepts.

7. $g(x) = x^2 - 2x$

8. $f(x) = 6x - x^2$

9. $f(x) = x^2 + 8x + 10$

10. $h(x) = 3 + 4x - x^2$

11. $f(t) = -2t^2 + 4t + 1$

12. $f(x) = x^2 - 8x + 12$

13. $h(x) = 4x^2 + 4x + 13$

14. $f(x) = x^2 - 6x + 1$

15. $h(x) = x^2 + 5x - 4$

16. $f(x) = 4x^2 + 4x + 5$

17. $f(x) = \frac{1}{3}(x^2 + 5x - 4)$

18. $f(x) = \frac{1}{2}(6x^2 - 24x + 22)$

19. ***Numerical, Graphical, and Analytical Analysis*** A rectangle is inscribed in the region bounded by the x-axis, the y-axis, and the graph of $x + 2y - 8 = 0$, as shown in the figure.

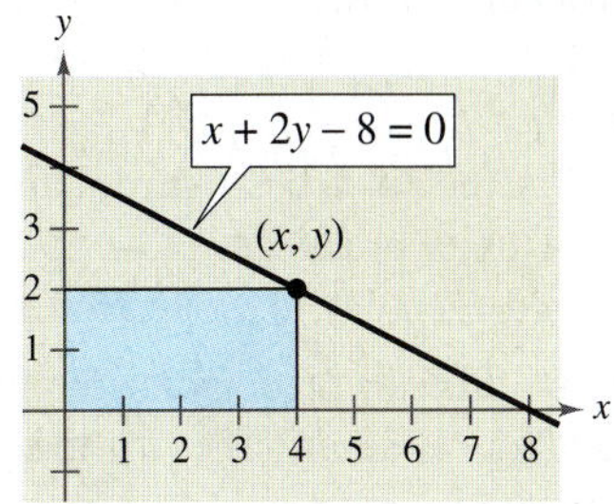

(a) Write the area A of the rectangle as a function of x.

(b) Determine the domain of the function in the context of the problem.

(c) Create a table showing possible values of x and the corresponding area of the rectangle.

(d) Use a graphing utility to graph the area function. Use the graph to approximate the dimensions that will produce the maximum area.

(e) Write the area function in standard form to find analytically the dimensions that will produce the maximum area.

20. ***Geometry*** The perimeter of a rectangle is 200 meters.

(a) Draw a rectangle that gives a visual representation of the problem. Label the length and width in terms of x and y, respectively.

(b) Write y as a function of x. Use the result to write the area as a function of x.

(c) Of all possible rectangles with perimeters of 200 meters, find the dimensions of the one with the maximum area.

21. ***Maximum Revenue*** Find the number of units that produces a maximum revenue for

$$R = 800x - 0.01x^2$$

where R is the total revenue (in dollars) for a cosmetics company and x is the number of units produced.

22. ***Maximum Profit*** A real estate office handles an apartment building that has 50 units. When the rent is \$540 per month, all units are occupied. However, for each \$30 increase in rent, one unit becomes vacant. Each occupied unit requires an average of \$18 per month for service and repairs. What rent should be charged to obtain the maximum profit?

23. ***Minimum Cost*** A soft-drink manufacturer has daily production costs of

$$C = 70{,}000 - 120x + 0.055x^2$$

where C is the total cost (in dollars) and x is the number of units produced. How many units should be produced each day to yield a minimum cost?

24. ***Sociology*** The average age of the groom at a first marriage for a given age of the bride can be approximated by the model

$$y = -0.107x^2 + 5.68x - 48.5, \quad 20 \le x \le 25$$

where y is the age of the groom and x is the age of the bride. For what age of the bride is the average age of the groom 26? (Source: U.S. Census Bureau)

3.2 In Exercises 25–30, sketch the graphs of $y = x^n$ and the transformation.

25. $y = x^3, \quad f(x) = -(x-4)^3$
26. $y = x^3, \quad f(x) = -4x^3$
27. $y = x^4, \quad f(x) = 2 - x^4$
28. $y = x^4, \quad f(x) = 2(x-2)^4$
29. $y = x^5, \quad f(x) = (x-3)^5$
30. $y = x^5, \quad f(x) = \frac{1}{2}x^5 + 3$

In Exercises 31–34, determine the right-hand and left- hand behavior of the graph of the polynomial function.

31. $f(x) = -x^2 + 6x + 9$
32. $f(x) = \frac{1}{2}x^3 + 2x$
33. $g(x) = \frac{3}{4}(x^4 + 3x^2 + 2)$
34. $h(x) = -x^5 - 7x^2 + 10x$

In Exercises 35–40, find all the real zeros of the polynomial function. Determine the multiplicity of each zero.

35. $f(x) = 2x^2 + 11x - 21$
36. $f(x) = x(x+3)^2$
37. $f(t) = t^3 - 3t$
38. $f(x) = x^3 - 8x^2$
39. $f(x) = -12x^3 + 20x^2$
40. $g(x) = x^4 - x^3 - 2x^2$

In Exercises 41–44, sketch the graph of the function by (a) applying the Leading Coefficient Test, (b) finding the zeros of the polynomial, (c) plotting sufficient solution points, and (d) drawing a continuous curve through the points.

41. $f(x) = -x^3 + x^2 - 2$
42. $g(x) = 2x^3 + 4x^2$
43. $f(x) = x(x^3 + x^2 - 5x + 3)$
44. $h(x) = 3x^2 - x^4$

In Exercises 45–48, use the Intermediate Value Theorem and the *table* feature of a graphing utility to find intervals one unit in length in which the polynomial function is guaranteed to have a zero. Adjust the table to approximate the zeros of the function. Use the *zero* or *root* feature of the graphing utility to verify your results.

45. $f(x) = 3x^3 - x^2 + 3$
46. $f(x) = 0.25x^3 - 3.65x + 6.12$
47. $f(x) = x^4 - 5x - 1$
48. $f(x) = 7x^4 + 3x^3 - 8x^2 + 2$

3.3 In Exercises 49–54, use long division to divide.

49. $\dfrac{24x^2 - x - 8}{3x - 2}$
50. $\dfrac{4x + 7}{3x - 2}$
51. $\dfrac{5x^3 - 13x^2 - x + 2}{x^2 - 3x + 1}$
52. $\dfrac{3x^4}{x^2 - 1}$
53. $\dfrac{x^4 - 3x^3 + 4x^2 - 6x + 3}{x^2 + 2}$
54. $\dfrac{6x^4 + 10x^3 + 13x^2 - 5x + 2}{2x^2 - 1}$

In Exercises 55–58, use synthetic division to divide.

55. $\dfrac{6x^4 - 4x^3 - 27x^2 + 18x}{x - 2}$
56. $\dfrac{0.1x^3 + 0.3x^2 - 0.5}{x - 5}$
57. $\dfrac{2x^3 - 19x^2 + 38x + 24}{x - 4}$
58. $\dfrac{3x^3 + 20x^2 + 29x - 12}{x + 3}$

In Exercises 59 and 60, use synthetic division to determine whether the given values of x are zeros of the function.

59. $f(x) = 20x^4 + 9x^3 - 14x^2 - 3x$
    (a) $x = -1$ (b) $x = \frac{3}{4}$ (c) $x = 0$ (d) $x = 1$

60. $f(x) = 3x^3 - 8x^2 - 20x + 16$
 (a) $x = 4$ (b) $x = -4$ (c) $x = \frac{2}{3}$ (d) $x = -1$

In Exercises 61 and 62, use synthetic division to find each specified value of the function.

61. $f(x) = x^4 + 10x^3 - 24x^2 + 20x + 44$
 (a) $f(-3)$ (b) $f(-1)$

62. $g(t) = 2t^5 - 5t^4 - 8t + 20$
 (a) $g(-4)$ (b) $g(\sqrt{2})$

In Exercises 63–66, (a) verify the given factor(s) of the function *f*, (b) find the remaining factors of *f*, (c) use your results to write the complete factorization of *f*, (d) list all real zeros of *f*, and (e) confirm your results by using a graphing utility to graph the function.

	Function	*Factor(s)*
63.	$f(x) = x^3 + 4x^2 - 25x - 28$	$(x - 4)$
64.	$f(x) = 2x^3 + 11x^2 - 21x - 90$	$(x + 6)$
65.	$f(x) = x^4 - 4x^3 - 7x^2 + 22x + 24$	$(x + 2)(x - 3)$
66.	$f(x) = x^4 - 11x^3 + 41x^2 - 61x + 30$	$(x - 2)(x - 5)$

3.4 In Exercises 67–72, find all the zeros of the function.

67. $f(x) = 3x(x - 2)^2$
68. $f(x) = (x - 4)(x + 9)^2$
69. $f(x) = x^2 - 9x + 8$
70. $f(x) = x^3 + 6x$
71. $f(x) = (x + 4)(x - 6)(x - 2i)(x + 2i)$
72. $f(x) = (x - 8)(x - 5)^2(x - 3 + i)(x - 3 - i)$

In Exercises 73 and 74, use the Rational Zero Test to list all possible rational zeros of *f*.

73. $f(x) = -4x^3 + 8x^2 - 3x + 15$
74. $f(x) = 3x^4 + 4x^3 - 5x^2 - 8$

In Exercises 75–80, find all the real zeros of the function.

75. $f(x) = x^3 - 2x^2 - 21x - 18$
76. $f(x) = 3x^3 - 20x^2 + 7x + 30$
77. $f(x) = x^3 - 10x^2 + 17x - 8$
78. $f(x) = x^3 + 9x^2 + 24x + 20$
79. $f(x) = x^4 + x^3 - 11x^2 + x - 12$
80. $f(x) = 25x^4 + 25x^3 - 154x^2 - 4x + 24$

In Exercises 81 and 82, find a polynomial with real coefficients that has the given zeros.

81. $\frac{2}{3}, 4, \sqrt{3}i$
82. $2, -3, 1 - 2i$

In Exercises 83–86, use the given zero to find all the zeros of the function.

	Function	*Zero*
83.	$f(x) = x^3 - 4x^2 + x - 4$	i
84.	$h(x) = -x^3 + 2x^2 - 16x + 32$	$-4i$
85.	$g(x) = 2x^4 - 3x^3 - 13x^2 + 37x - 15$	$2 + i$
86.	$f(x) = 4x^4 - 11x^3 + 14x^2 - 6x$	$1 - i$

In Exercises 87–90, find all the zeros of the function and write the polynomial as a product of linear factors.

87. $f(x) = x^3 + 4x^2 - 5x$
88. $g(x) = x^3 - 7x^2 + 36$
89. $g(x) = x^4 + 4x^3 - 3x^2 + 40x + 208$
90. $f(x) = x^4 + 8x^3 + 8x^2 - 72x - 153$

In Exercises 91–94, use a graphing utility to (a) graph the function, (b) determine the number of real zeros of the function, and (c) approximate the real zeros of the function to the nearest hundredth.

91. $f(x) = x^4 + 2x + 1$
92. $g(x) = x^3 - 3x^2 + 3x + 2$
93. $h(x) = x^3 - 6x^2 + 12x - 10$
94. $f(x) = x^5 + 2x^3 - 3x - 20$

In Exercises 95 and 96, use Descartes's Rule of Signs to determine the possible numbers of positive and negative zeros of the function.

95. $g(x) = 5x^3 + 3x^2 - 6x + 9$
96. $h(x) = -2x^5 + 4x^3 - 2x^2 + 5$

In Exercises 97 and 98, use synthetic division to verify the upper and lower bounds of the real zeros of *f*.

97. $f(x) = 4x^3 - 3x^2 + 4x - 3$
 (a) Upper: $x = 1$
 (b) Lower: $x = -\frac{1}{4}$

98. $f(x) = 2x^3 - 5x^2 - 14x + 8$
 (a) Upper: $x = 8$
 (b) Lower: $x = -4$

3.5 **99.** ***Data Analysis*** The federal minimum wage rates R (in dollars) in the United States for selected years from 1955 through 2000 are shown in the table. A linear model that approximates this data is

$$R = 0.099t - 0.08$$

where t represents the year, with $t = 5$ corresponding to 1955. (Source: U.S. Department of Labor)

Year	Wage rate, R
1955	0.75
1960	1.00
1965	1.25
1970	1.60
1975	2.10
1980	3.10
1985	3.35
1990	3.80
1995	4.25
2000	5.15

(a) Plot the actual data and the model on the same set of coordinate axes.

(b) How closely does the model represent the data?

100. ***Measurement*** You notice a billboard indicating that it is 2.5 miles or 4 kilometers to the next restaurant of a national fast-food chain. Use this information to find a linear model that relates miles to kilometers. Use the model to find the numbers of kilometers in 2 miles and 10 miles.

101. ***Energy*** The power P produced by a wind turbine is proportional to the cube of the wind speed S. A wind speed of 27 miles per hour produces a power output of 750 kilowatts. Find the output for a wind speed of 40 miles per hour.

102. ***Frictional Force*** The frictional force F between the tires and the road required to keep a car on a curved section of a highway is directly proportional to the square of the speed s of the car. If the speed of the car is doubled, the force will change by what factor?

In Exercises 103 and 104, find a mathematical model that represents the statement. (In each case, determine the constant of proportionality.)

103. y is inversely proportional to x. ($y = 9$ when $x = 5.5$.)

104. F is jointly proportional to x and the square root of y. ($F = 6$ when $x = 9$ and $y = 4$.)

 **105.** ***Recording Media*** The table shows the numbers y (in millions) of CDs shipped in the United States in the years 1990 through 1999. (Source: Recording Industry Association of America)

Year	CDs shipped, y
1990	286.5
1991	333.3
1992	407.5
1993	495.4
1994	662.1
1995	722.9
1996	778.9
1997	753.1
1998	847.0
1999	938.9

(a) Use a graphing utility to create a scatter plot of the data. Let t represent the year, with $t = 0$ corresponding to 1990.

(b) Use the *regression* feature of the graphing utility to find the equation of the least squares regression line that fits the data. Then graph the model and the scatter plot you found in part (a) in the same viewing window.

(c) Use the model to estimate the number of CDs that will be shipped in the year 2005.

(d) Interpret the meaning of the slope of the linear model in the context of the problem.

Synthesis

True or False? **In Exercises 106 and 107, determine whether the statement is true or false. Justify your answer.**

106. A fourth-degree polynomial can have -5, $-8i$, $4i$, and 5 as its zeros.

107. If y is directly proportional to x, then x is directly proportional to y.

Chapter Test

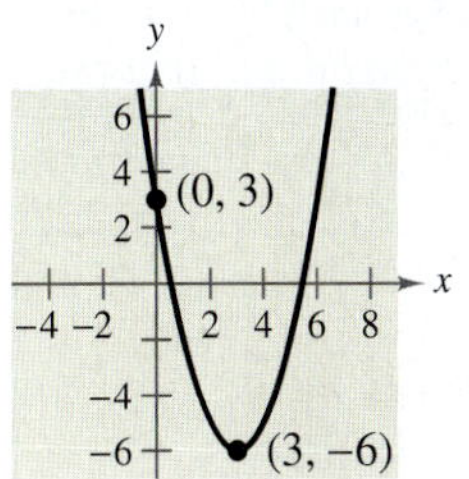

FIGURE FOR 3

The *Interactive* CD-ROM and *Internet* versions of this text offer Chapter Pre-Tests and Chapter Post-Tests, both of which have randomly generated exercises with diagnostic capabilities.

Take this test as you would take a test in class. When you are finished, check your work against the answers given in the back of the book.

1. Describe how the graph of g differs from the graph of $f(x) = x^2$.
   (a) $g(x) = 2 - x^2$ (b) $g(x) = \left(x - \frac{3}{2}\right)^2$
2. Identify the vertex and intercepts of the graph of $y = x^2 + 4x + 3$.
3. Find an equation of the parabola shown in the figure at the left.
4. The path of a ball is given by $y = -\frac{1}{20}x^2 + 3x + 5$, where y is the height (in feet) of the ball and x is the horizontal distance (in feet) from where the ball was thrown.
   (a) Find the maximum height of the ball.
   (b) Which number determines the height at which the ball was thrown? Does changing this constant change the coordinates of the maximum height of the ball? Explain.
5. Determine the right-hand and left-hand behavior of the graph of the function $h(t) = -\frac{3}{4}t^5 + 2t^2$. Then sketch its graph.
6. Divide by long division.
   $$\frac{3x^3 + 4x - 1}{x^2 + 1}$$
7. Divide by synthetic division.
   $$\frac{2x^4 - 5x^2 - 3}{x - 2}$$
8. Use synthetic division to show that $x = \sqrt{3}$ is a zero of the function
   $$f(x) = 4x^3 - x^2 - 12x + 3.$$
   Use the result to factor the polynomial function completely and list all the real zeros of the function.

In Exercises 9 and 10, find all the real zeros of the function.

9. $g(t) = 2t^4 - 3t^3 + 16t - 24$
10. $h(x) = 3x^5 + 2x^4 - 3x - 2$

In Exercises 11 and 12, find a polynomial function with integer coefficients that has the given zeros.

11. $0, 3, 3 + i, 3 - i$
12. $1 + \sqrt{3}i, 1 - \sqrt{3}i, 2, 2$

In Exercises 13 and 14, find all the zeros of the function.

13. $f(x) = x^3 + 2x^2 + 5x + 10$
14. $f(x) = x^4 - 9x^2 - 22x - 24$

In Exercises 15–17, find a mathematical model that represents the statement. (In each case, determine the constant of proportionality.)

15. v varies directly as the square root of s. ($v = 24$ when $s = 16$.)
16. A varies jointly as x and y. ($A = 500$ when $x = 15$ and $y = 8$.)
17. b varies inversely as a. ($b = 32$ when $a = 1.5$.)

Proofs in Mathematics

These two pages contain proofs of four important theorems and polynomial functions. The first two theorems are from Section 3.3, and the second two theorems are from Section 3.4.

The Remainder Theorem *(p. 288)*

If a polynomial $f(x)$ is divided by $x - k$, the remainder is

$$r = f(k).$$

Proof

From the Division Algorithm, you have

$$f(x) = (x - k)q(x) + r(x)$$

and because either $r(x) = 0$ or the degree of $r(x)$ is less than the degree of $x - k$, you know that $r(x)$ must be a constant. That is, $r(x) = r$. Now, by evaluating $f(x)$ at $x = k$, you have

$$\begin{aligned} f(k) &= (k - k)q(k) + r \\ &= (0)q(k) + r = r. \end{aligned}$$

To be successful in algebra, it is important that you understand the connection among *factors* of a polynomial, *zeros* of a polynomial function, and *solutions* or *roots* of a polynomial equation. The Factor Theorem is the basis for this connection.

The Factor Theorem *(p. 288)*

A polynomial $f(x)$ has a factor $(x - k)$ if and only if $f(k) = 0$.

Proof

Using the Division Algorithm with the factor $(x - k)$, you have

$$f(x) = (x - k)q(x) + r(x).$$

By the Remainder Theorem, $r(x) = r = f(k)$, and you have

$$f(x) = (x - k)q(x) + f(k)$$

where $q(x)$ is a polynomial of lesser degree than $f(x)$. If $f(k) = 0$, then

$$f(x) = (x - k)q(x)$$

and you see that $(x - k)$ is a factor of $f(x)$. Conversely, if $(x - k)$ is a factor of $f(x)$, division of $f(x)$ by $(x - k)$ yields a remainder of 0. So, by the Remainder Theorem, you have $f(k) = 0$.

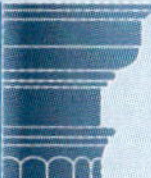

Linear Factorization Theorem (p. 293)

If $f(x)$ is a polynomial of degree n, where $n > 0$, then f has precisely n linear factors

$$f(x) = a_n(x - c_1)(x - c_2) \cdot \cdot \cdot (x - c_n)$$

where $c_1, c_2, \ldots, c_n$ are complex numbers.

Proof

Using the Fundamental Theorem of Algebra, you know that f must have at least one zero, c_1. Consequently, $(x - c_1)$ is a factor of $f(x)$, and you have

$$f(x) = (x - c_1)f_1(x).$$

If the degree of $f_1(x)$ is greater than zero, you again apply the Fundamental Theorem to conclude that f_1 must have a zero c_2, which implies that

$$f(x) = (x - c_1)(x - c_2)f_2(x).$$

It is clear that the degree of $f_1(x)$ is $n - 1$, that the degree of $f_2(x)$ is $n - 2$, and that you can repeatedly apply the Fundamental Theorem n times until you obtain

$$f(x) = a_n(x - c_1)(x - c_2) \cdot \cdot \cdot (x - c_n)$$

where a_n is the leading coefficient of the polynomial $f(x)$.

The Fundamental Theorem of Algebra

The Linear Factorization Theorem is closely related to the Fundamental Theorem of Algebra. The Fundamental Theorem of Algebra has a long and interesting history. In the early work with polynomial equations, The Fundamental Theorem of Algebra was thought to have been not true, because imaginary solutions were not considered. In fact, in the very early work by mathematicians such as Abu al-Khwarizmi (c. 800 A.D.), negative solutions were also not considered.

Once imaginary numbers were accepted, several mathematicians attempted to give a general proof of the Fundamental Theorem of Algebra. These included Gottfried von Leibniz (1702), Jean D`Alembert (1740), Leonhard Euler (1749), Joseph-Louis Lagrange (1772), and Pierre Simon Laplace (1795). The mathematician usually credited with the first correct proof of the Fundamental Theorem of Algebra is Carl Friedrich Gauss, who published the proof in his doctoral thesis in 1799.

Factors of a Polynomial (p. 297)

Every polynomial of degree $n > 0$ with real coefficients can be written as the product of linear and quadratic factors with real coefficients, where the quadratic factors have no real zeros.

Proof

To begin, you use the Linear Factorization Theorem to conclude that $f(x)$ can be *completely* factored in the form

$$f(x) = d(x - c_1)(x - c_2)(x - c_3) \cdot \cdot \cdot (x - c_n).$$

If each c_i is real, there is nothing more to prove. If any c_i is complex ($c_i = a + bi$, $b \neq 0$), then, because the coefficients of $f(x)$ are real, you know that the conjugate $c_j = a - bi$ is also a zero. By multiplying the corresponding factors, you obtain

$$\begin{aligned}(x - c_i)(x - c_j) &= [x - (a + bi)][x - (a - bi)] \\ &= x^2 - 2ax + (a^2 + b^2)\end{aligned}$$

where each coefficient is real.

P.S. Problem Solving

1. (a) Find the zeros of each quadratic function $g(x)$.

(i) $g(x) = x^2 - 4x - 12$

(ii) $g(x) = x^2 + 5x$

(iii) $g(x) = x^2 + 3x - 10$

(iv) $g(x) = x^2 - 4x + 4$

(v) $g(x) = x^2 - 2x - 6$

(vi) $g(x) = x^2 + 3x + 4$

(b) For each function in part (a), use a graphing utility to graph $f(x) = (x - 2) \cdot g(x)$. Verify that $(2, 0)$ is an x-intercept of the graph of $f(x)$. Describe any similarities or differences in the behavior of the six functions at this x-intercept.

(c) For each function in part (b), use the graph of $f(x)$ to approximate the other x-intercepts of the graph.

(d) Describe the connections that you find among the results of parts (a), (b), and (c).

2. Quonset huts were developed during World War II. They were temporary housing structures that could be assembled quickly and easily. A Quonset hut is shaped like a half cylinder. A manufacturer has 600 square feet of material with which to build a Quonset hut.

(a) The formula for the surface area of half a cylinder is $S = \pi r^2 + \pi rl$, where r is the radius and l is the length of the hut. Solve this equation for l when $S = 600$.

(b) The formula for the volume of the hut is $V = \frac{1}{2}\pi r^2 l$. Write the volume V of the Quonset hut as a polynomial function of r.

(c) Use the function you wrote in part (b) to find the maximum volume of a Quonset hut with a surface area of 600 square feet. What are the dimensions of the hut?

3. Show that if

$f(x) = ax^3 + bx^2 + cx + d$

then $f(k) = r$ where

$r = ak^3 + bk^2 + ck + d$

using long division. In other words, verify the Remainder Theorem for a third-degree polynomial function.

4. In 2000 B.C., the Babylonians solved polynomial equations by referring to tables of values. One such table gave the values of $y^3 + y^2$. To be able to use this table, the Babylonians sometimes had to manipulate the equation as shown below.

$$ax^3 + bx^2 = c \qquad \text{Original equation}$$

$$\frac{a^3x^3}{b^3} + \frac{a^2x^2}{b^2} = \frac{a^2c}{b^3} \qquad \text{Multiply each side by } \frac{a^2}{b^3}.$$

$$\left(\frac{ax}{b}\right)^3 + \left(\frac{ax}{b}\right)^2 = \frac{a^2c}{b^3} \qquad \text{Rewrite.}$$

Then they would find $(a^2c)/b^3$ in the $y^3 + y^2$ column of the table. Because they knew that the corresponding y-value was equal to $(ax)/b$, they could conclude that $x = (by)/a$.

(a) Calculate $y^3 + y^2$ for $y = 1, 2, 3, \ldots, 10$. Record the values in a table.

Use the table from part (a) and the method above to solve each equation.

(b) $x^3 + x^2 = 252$

(c) $x^3 + 2x^2 = 288$

(d) $3x^3 + x^2 = 90$

(e) $2x^3 + 5x^2 = 2500$

(f) $7x^3 + 6x^2 = 1728$

(g) $10x^3 + 3x^2 = 297$

Using the methods from this chapter, verify your solution to each equation.

5. At a glassware factory, molten cobalt glass is poured into molds to make paperweights. Each mold is a rectangular prism whose height is 3 inches greater than the length of each side of the square base. A machine pours 20 cubic inches of liquid glass into each mold. What are the dimensions of the mold?

6. (a) Complete the table.

Function	Zeros	Sum of zeros	Product of zeros
$f_1(x) = x^2 - 5x + 6$			
$f_2(x) = x^3 - 7x + 6$			
$f_3(x) = x^4 + 2x^3 + x^2 + 8x - 12$			
$f_4(x) = x^5 - 3x^4 - 9x^3 + 25x^2 - 6x$			

(b) Use the table to make a conjecture relating the sum of the zeros of a polynomial function with the coefficients of the polynomial function.

(c) Use the table to make a conjecture relating the product of the zeros of a polynomial function with the coefficients of the polynomial function.

7. The parabola shown in the figure has an equation of the form $y = ax^2 + bx + c$. Find the equation for this parabola by the following methods. (a) Find the equation analytically. (b) Use the *regression* feature of a graphing utility to find the equation.

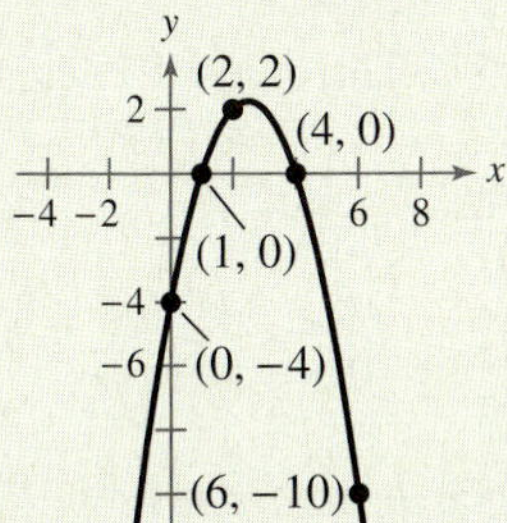

8. One of the fundamental themes of calculus is to find the slope of the tangent line to a curve at a point. To see how this can be done, consider the point (2, 4) on the graph of the quadratic function $f(x) = x^2$.

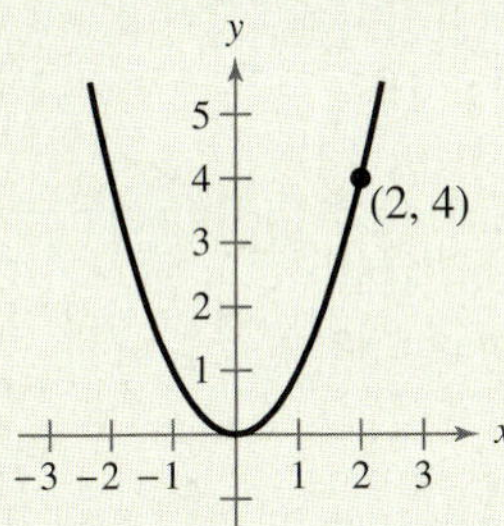

(a) Find the slope of the line joining (2, 4) and (3, 9). Is the slope of the tangent line at (2, 4) greater than or less than the slope of the line through (2, 4) and (3, 9)?

(b) Find the slope of the line joining (2, 4) and (1, 1). Is the slope of the tangent line at (2, 4) greater than or less than the slope of the line through (2, 4) and (1, 1)?

(c) Find the slope of the line joining (2, 4) and (2.1, 4.41). Is the slope of the tangent line at (2, 4) greater than or less than the slope of the line through (2, 4) and (2.1, 4.41)?

(d) Find the slope of the line joining (2, 4) and $(2 + h, f(2 + h))$ in terms of the nonzero number h.

(e) Evaluate the slope formula from part (d) for $h = -1$, 1, and 0.1. Compare these values with those in parts (a)–(c).

(f) What can you conclude the slope of the tangent line at (2, 4) to be? Explain your answer.

9. A rancher plans to fence a rectangular pasture adjacent to a river. The rancher has 100 meters of fence, and no fencing is needed along the river.

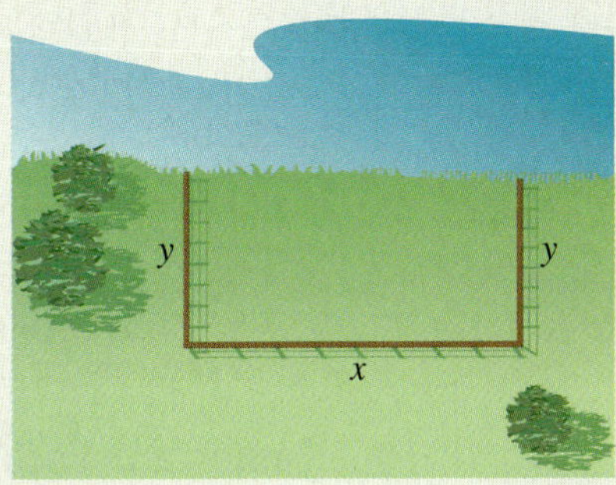

(a) Write the area as a function $A(x)$ of x, the length of the side of the pasture parallel to the river. What is the domain of $A(x)$?

(b) Graph the function $A(x)$ and estimate the dimensions that yield the maximum area of the pasture.

(c) Find the exact dimensions that yield the maximum area of the pasture by writing the quadratic function in standard form.

10. A wire 100 centimeters in length is cut into two pieces. One piece is bent to form a square and the other to form a circle. Let x equal the length of the wire used to form the square.

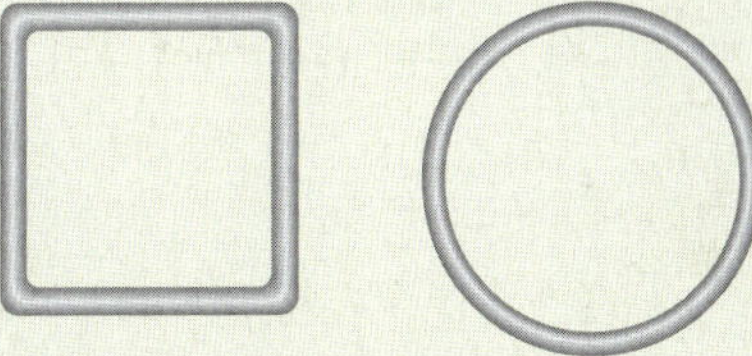

(a) Write the function that represents the combined area of the two figures.

(b) Determine the domain of the function.

(c) Find the value(s) of x that yield a maximum area and a minimum area.

(d) Explain your reasoning.

11. Find a formula for the polynomial division: $\dfrac{x^n - 1}{x - 1}$.

How to study Chapter 4

▶ What you should learn

In this chapter you will learn the following skills and concepts:

- How to determine the domains of rational functions and find asymptotes of rational functions.
- How to sketch the graphs of rational functions.
- How to recognize and find partial fraction decompositions of rational expressions.
- How to recognize, graph, and write equations of circles, ellipses, parabolas, and hyperbolas (vertex or center at origin).
- How to recognize, graph, and write equations of conics that have been shifted vertically or horizontally in the plane.

▶ Important Vocabulary

As you encounter each new vocabulary term in this chapter, add the term and its definition to your notebook glossary.

Rational function (p. 332)
Vertical asymptote (p. 333)
Horizontal asymptote (p. 333)
Slant (or oblique) asymptote (p. 344)
Partial fraction (p. 350)
Partial fraction decomposition (p. 350)
Basic equation (p. 351)
Conic section or conic (p. 358)
Degenerate conic (p. 358)
Parabola (p. 359)
Directrix (p. 359)
Focus (p. 359)
Standard form of the equation of a parabola (p. 359)
Ellipse (p. 361)
Foci (p. 361)
Vertices (p. 361)
Major axis (p. 361)
Center (p. 361)
Minor axis (p. 361)
Standard form of the equation of an ellipse (p. 361)
Hyperbola (p. 363)
Branches (p. 363)
Transverse axis (p. 363)
Standard form of the equation of a hyperbola (p. 363)
Conjugate axis (p. 364)
Asymptotes of a hyperbola (p. 365)

Study Tools

Learning objectives in each section
Chapter Summary (p. 379)
Review Exercises (pp. 380–383)
Chapter Test (p. 384)

Additional Resources

Study and Solutions Guide
Interactive College Algebra
Videotapes/DVD for Chapter 4
College Algebra Website
Student Success Organizer

NASA

4

Rational Functions and Conics

4.1 Rational Functions and Asymptotes

▶ **What you should learn**

- How to find the domains of rational functions
- How to find the horizontal and vertical asymptotes of graphs of rational functions
- How to use rational functions to model and solve real-life problems

▶ **Why you should learn it**

Rational functions can be used to model and solve real-life problems relating to business. For instance, in Exercise 41 on page 340, a rational function is used to model the number of business partnerships in the United States.

Stephen Derr/Getty Images

Introduction

A **rational function** can be written in the form

$$f(x) = \frac{N(x)}{D(x)}$$

where $N(x)$ and $D(x)$ are polynomials and $D(x)$ is not the zero polynomial. In this section it is assumed that $N(x)$ and $D(x)$ have no common factors.

In general, the *domain* of a rational function of x includes all real numbers except x-values that make the denominator zero. Much of the discussion of rational functions will focus on their graphical behavior near the x-values excluded from the domain.

Example 1 ▶ Finding the Domain of a Rational Function

Find the domain of $f(x) = 1/x$ and discuss the behavior of f near any excluded x-values.

Solution

Because the denominator is zero when $x = 0$, the domain of f is all real numbers except $x = 0$. To determine the behavior of f near this excluded value, evaluate $f(x)$ to the left and right of $x = 0$, as indicated in the following tables.

x	-1	-0.5	-0.1	-0.01	-0.001	$\rightarrow 0$
$f(x)$	-1	-2	-10	-100	-1000	$\rightarrow -\infty$

x	$0 \leftarrow$	0.001	0.01	0.1	0.5	1
$f(x)$	$\infty \leftarrow$	1000	100	10	2	1

Note that as x approaches 0 *from the left*, $f(x)$ decreases without bound. In contrast, as x approaches 0 *from the right*, $f(x)$ increases without bound. The graph of f is shown in Figure 4.1.

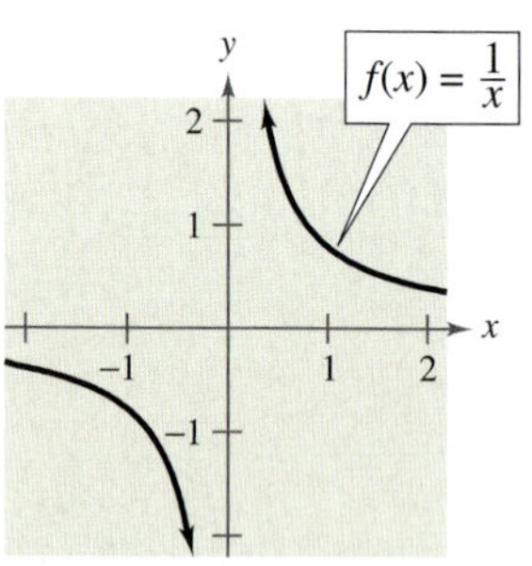

FIGURE 4.1

STUDY TIP

Note that the rational function

$$f(x) = \frac{1}{x}$$

is also referred to as the reciprocal function discussed in Section 2.4.

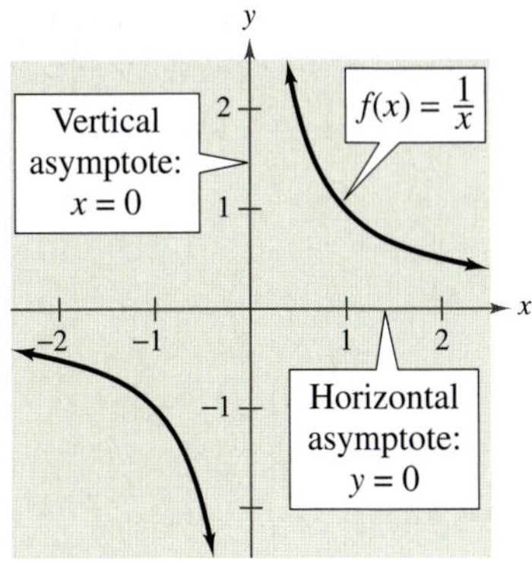

FIGURE 4.2

Horizontal and Vertical Asymptotes

In Example 1, the behavior of f near $x = 0$ is denoted as follows.

$$\underbrace{f(x) \longrightarrow -\infty \text{ as } x \longrightarrow 0^-}_{f(x) \text{ decreases without bound as } x \text{ approaches 0 from the left.}} \qquad \underbrace{f(x) \longrightarrow \infty \text{ as } x \longrightarrow 0^+}_{f(x) \text{ increases without bound as } x \text{ approaches 0 from the right.}}$$

The line $x = 0$ is a **vertical asymptote** of the graph of f, as shown in Figure 4.2. From this figure, you can see that the graph of f also has a **horizontal asymptote**—the line $y = 0$. This means that the values of $f(x) = 1/x$ approach zero as x increases or decreases without bound.

$$\underbrace{f(x) \longrightarrow 0 \text{ as } x \longrightarrow -\infty}_{f(x) \text{ approaches 0 as } x \text{ decreases without bound.}} \qquad \underbrace{f(x) \longrightarrow 0 \text{ as } x \longrightarrow \infty}_{f(x) \text{ approaches 0 as } x \text{ increases without bound.}}$$

Definition of Vertical and Horizontal Asymptotes

1. The line $x = a$ is a **vertical asymptote** of the graph of f if

$$f(x) \longrightarrow \infty \quad \text{or} \quad f(x) \longrightarrow -\infty$$

as $x \longrightarrow a$, either from the right or from the left.

2. The line $y = b$ is a **horizontal asymptote** of the graph of f if

$$f(x) \longrightarrow b$$

as $x \longrightarrow \infty$ or $x \longrightarrow -\infty$.

Eventually (as $x \longrightarrow \infty$ or $x \longrightarrow -\infty$), the distance between the horizontal asymptote and the points on the graph must approach zero. Figure 4.3 shows the horizontal and vertical asymptotes of the graphs of three rational functions.

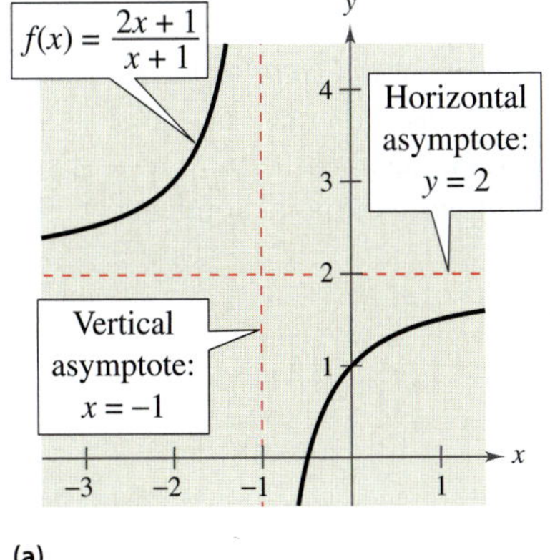

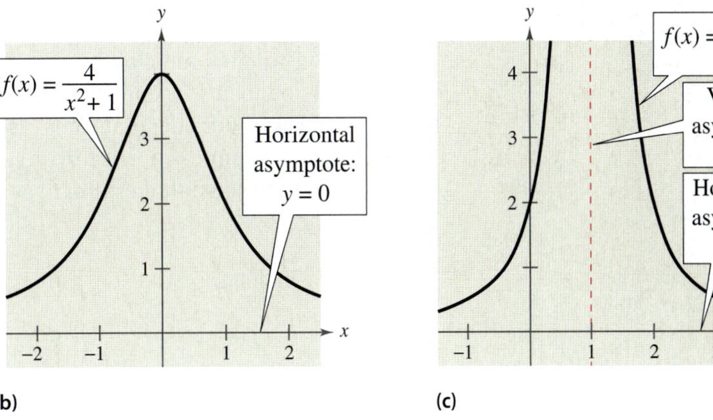

(a) (b) (c)

FIGURE 4.3

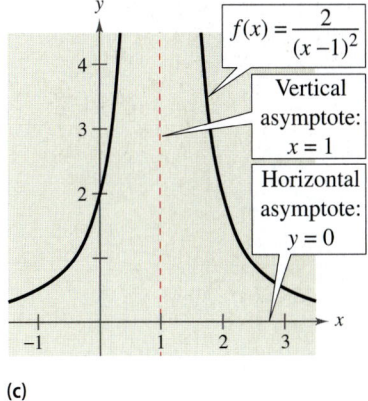

The graphs of $f(x) = 1/x$ in Figure 4.2 and $f(x) = (2x + 1)/(x + 1)$ in Figure 4.3(a) are **hyperbolas.** You will study hyperbolas in Sections 4.4 and 4.5.

The *Interactive* CD-ROM and *Internet* versions of this text offer a Try It for each example in the text.

Asymptotes of a Rational Function

Let f be the rational function given by

$$f(x) = \frac{N(x)}{D(x)} = \frac{a_n x^n + a_{n-1}x^{n-1} + \cdots + a_1 x + a_0}{b_m x^m + b_{m-1}x^{m-1} + \cdots + b_1 x + b_0}$$

where $N(x)$ and $D(x)$ have no common factors.

1. The graph of f has *vertical* asymptotes at the zeros of $D(x)$.
2. The graph of f has one or no *horizontal* asymptote determined by comparing the degrees of $N(x)$ and $D(x)$.
   a. If $n < m$, the graph of f has the line $y = 0$ (the x-axis) as a horizontal asymptote.
   b. If $n = m$, the graph of f has the line $y = a_n/b_m$ as a horizontal asymptote.
   c. If $n > m$, the graph of f has no horizontal asymptote.

Example 2 ▶ **Finding Horizontal and Vertical Asymptotes** 

Find all horizontal and vertical asymptotes of the graph of each rational function.

a. $f(x) = \dfrac{2x}{3x^2 + 1}$ **b.** $f(x) = \dfrac{2x^2}{x^2 - 1}$

Solution

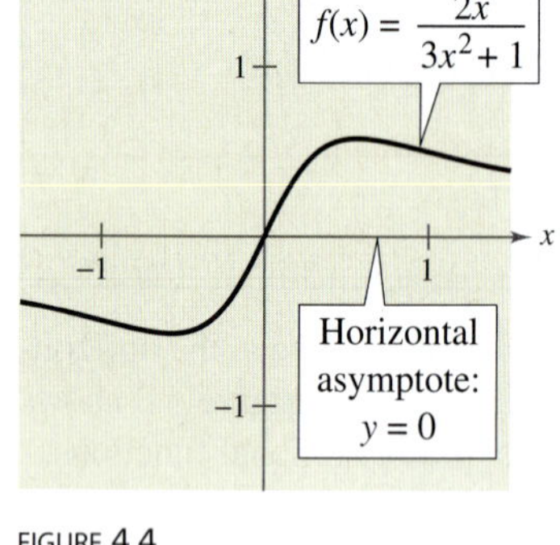

FIGURE 4.4

a. For this rational function, the degree of the numerator is *less than* the degree of the denominator, so the graph has the line $y = 0$ as a horizontal asymptote. To find any vertical asymptotes, set the denominator equal to zero and solve the resulting equation for x.

$$3x^2 + 1 = 0 \qquad \text{Set denominator equal to zero.}$$

Because this equation has no real solutions, you can conclude that the graph has no vertical asymptote. The graph of the function is shown in Figure 4.4.

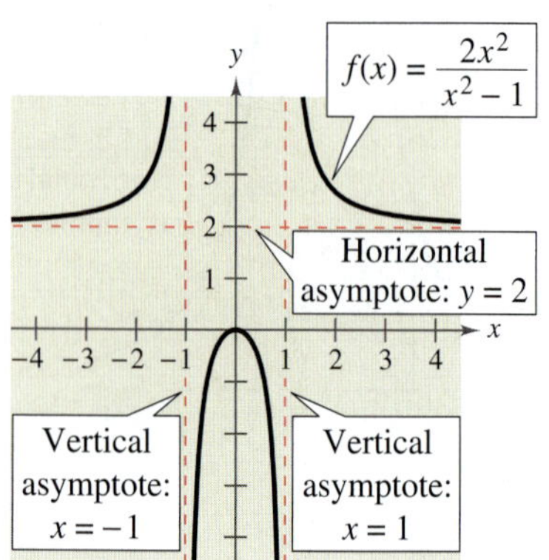

FIGURE 4.5

b. For this rational function, the degree of the numerator is *equal to* the degree of the denominator. The leading coefficient of the numerator is 2 and the leading coefficient of the denominator is 1, so the graph has the line $y = 2$ as a horizontal asymptote. To find any vertical asymptotes, set the denominator equal to zero and solve the resulting equation for x.

$$x^2 - 1 = 0 \qquad \text{Set denominator equal to zero.}$$

$$(x + 1)(x - 1) = 0 \qquad \text{Factor.}$$

$$x + 1 = 0 \quad \Rightarrow \quad x = -1 \qquad \text{Set 1st factor equal to 0.}$$

$$x - 1 = 0 \quad \Rightarrow \quad x = 1 \qquad \text{Set 2nd factor equal to 0.}$$

This equation has two real solutions $x = -1$ and $x = 1$, so the graph has the lines $x = -1$ and $x = 1$ as vertical asymptotes. The graph of the function is shown in Figure 4.5.

The icon identifies examples and concepts related to features of the Learning Tools CD-ROM and the *Interactive* and *Internet* versions of this text. For more details see the chart on pages *xix–xxiii*.

Applications

There are many examples of asymptotic behavior in real life. For instance, Example 3 shows how a vertical asymptote can be used to analyze the cost of removing pollutants from smokestack emissions.

Example 3 ▶ Cost-Benefit Model 

A utility company burns coal to generate electricity. The cost of removing a certain *percent* of the pollutants from smokestack emissions is typically not a linear function. That is, if it costs C dollars to remove 25% of the pollutants, it would cost more than $2C$ dollars to remove 50% of the pollutants. As the percent of removed pollutants approaches 100%, the cost tends to increase without bound, becoming prohibitive. The cost C (in dollars) of removing $p\%$ of the smokestack pollutants is $C = 80{,}000p/(100 - p)$ for $0 \le p < 100$. Sketch the graph of this function. You are a member of a state legislature considering a law that would require utility companies to remove 90% of the pollutants from their smokestack emissions. The current law requires 85% removal. How much additional cost would the utility company incur as a result of the new law?

Solution

The graph of this function is shown in Figure 4.6. Note that the graph has a vertical asymptote at $p = 100$. Because the current law requires 85% removal, the current cost to the utility company is

The *Interactive* CD-ROM and *Internet* versions of this text offer a Quiz for every section of the text.

$$C = \frac{80{,}000(85)}{100 - 85} \approx \$453{,}333. \qquad \text{Evaluate } C \text{ when } p = 85.$$

If the new law increases the percent removal to 90%, the cost to the utility company will be

$$C = \frac{80{,}000(90)}{100 - 90} = \$720{,}000. \qquad \text{Evaluate } C \text{ when } p = 90.$$

So, the new law would require the utility company to spend an additional

$$720{,}000 - 453{,}333 = \$266{,}667. \qquad \text{Subtract 85\% removal cost from 90\% removal cost.}$$

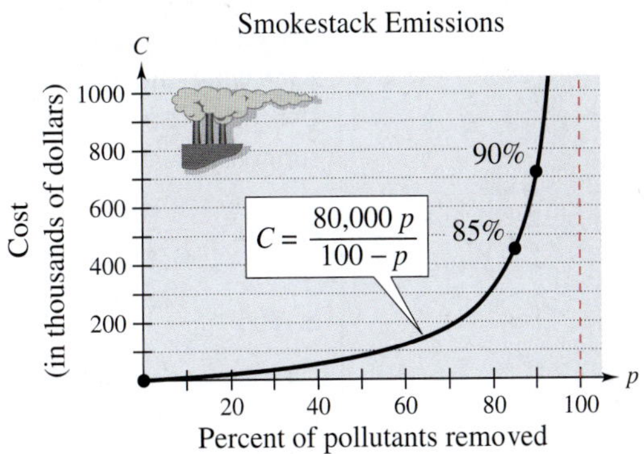

FIGURE 4.6

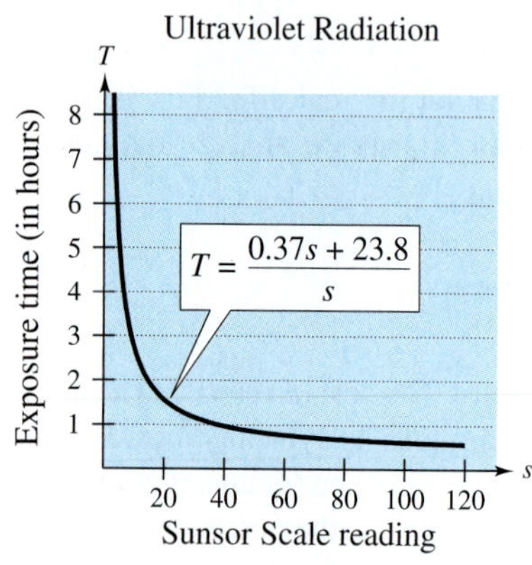

FIGURE 4.7

Example 4 ▶ Ultraviolet Radiation 

For a person with sensitive skin, the amount of time T (in hours) the person can be exposed to the sun with minimal burning can be modeled by

$$T = \frac{0.37s + 23.8}{s}, \qquad 0 < s \leq 120$$

where s is the Sunsor Scale reading. The Sunsor Scale is based on the level of intensity of UVB rays. (Source: Sunsor, Inc.)

a. Find the amount of time a person with sensitive skin can be exposed to the sun with minimal burning when $s = 10$, $s = 25$, and $s = 100$.

b. What is the horizontal asymptote of this function, and what does it represent?

Solution

a. When $s = 10$, $T = \dfrac{0.37(10) + 23.8}{10}$

$= 2.75$ hours.

When $s = 25$, $T = \dfrac{0.37(25) + 23.8}{25}$

≈ 1.32 hours.

When $s = 100$, $T = \dfrac{0.37(100) + 23.8}{100}$

≈ 0.61 hour.

b. As shown in Figure 4.7, the horizontal asymptote is the line $T = 0.37$. This line represents the shortest possible exposure time with minimal burning.

Writing ABOUT MATHEMATICS

Asymptotes of Graphs of Rational Functions Do you think it is possible for the graph of a rational function to cross its horizontal asymptote? If so, how can you determine when the graph of a rational function will cross its horizontal asymptote? Use the graphs of the following functions to investigate these questions. Write a summary of your conclusions. Explain your reasoning.

a. $f(x) = \dfrac{x}{x^2 + 1}$

b. $g(x) = \dfrac{x}{x^2 - 3}$

c. $h(x) = \dfrac{x^2}{2x^3 - x}$

4.1 Exercises

The *Interactive* CD-ROM and *Internet* versions of this text contain step-by-step solutions to all odd-numbered exercises. They also provide Tutorial Exercises for additional help.

In Exercises 1–4, (a) complete each table, (b) determine the vertical and horizontal asymptotes of the function, and (c) find the domain of the function.

x	$f(x)$
0.5	
0.9	
0.99	
0.999	

x	$f(x)$
1.5	
1.1	
1.01	
1.001	

x	$f(x)$
5	
10	
100	
1000	

1. $f(x) = \dfrac{1}{x - 1}$

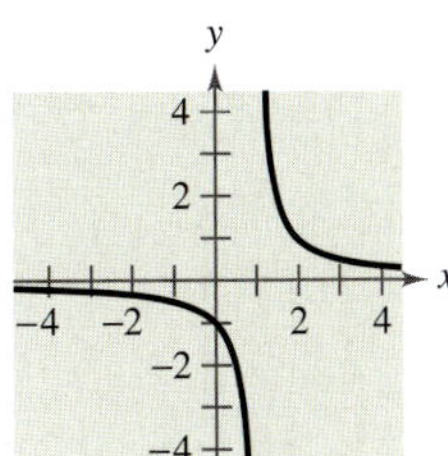

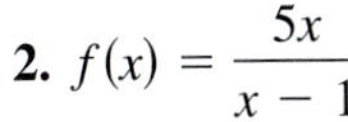

2. $f(x) = \dfrac{5x}{x - 1}$

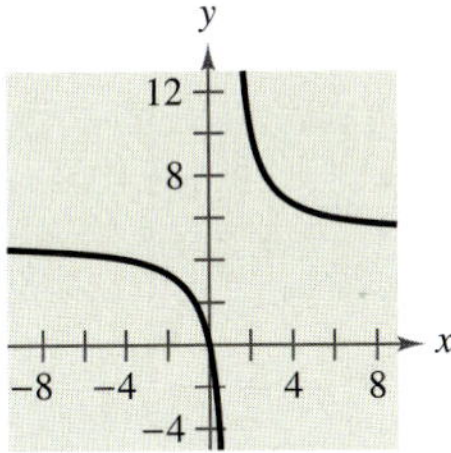

3. $f(x) = \dfrac{3x^2}{x^2 - 1}$

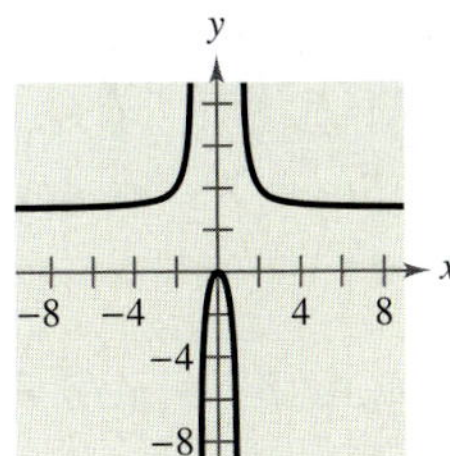

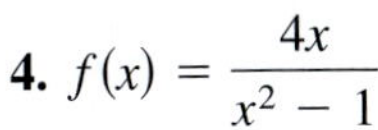

4. $f(x) = \dfrac{4x}{x^2 - 1}$

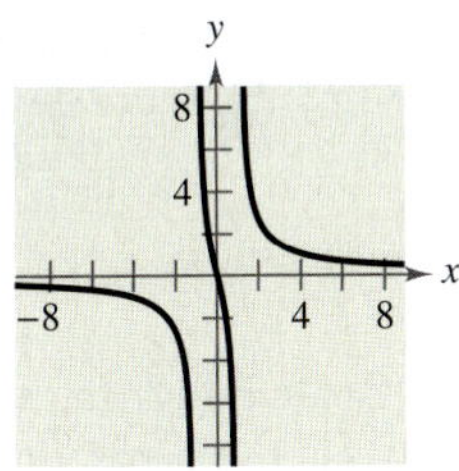

In Exercises 5–12, find the domain of the function and identify any horizontal and vertical asymptotes.

5. $f(x) = \dfrac{1}{x^2}$

6. $f(x) = \dfrac{4}{(x - 2)^3}$

7. $f(x) = \dfrac{2 + x}{2 - x}$

8. $f(x) = \dfrac{1 - 5x}{1 + 2x}$

9. $f(x) = \dfrac{x^3}{x^2 - 1}$

10. $f(x) = \dfrac{2x^2}{x + 1}$

11. $f(x) = \dfrac{3x^2 + 1}{x^2 + x + 9}$

12. $f(x) = \dfrac{3x^2 + x - 5}{x^2 + 1}$

In Exercises 13–16, match the rational function with its graph. [The graphs are labeled (a), (b), (c), and (d).]

(a)

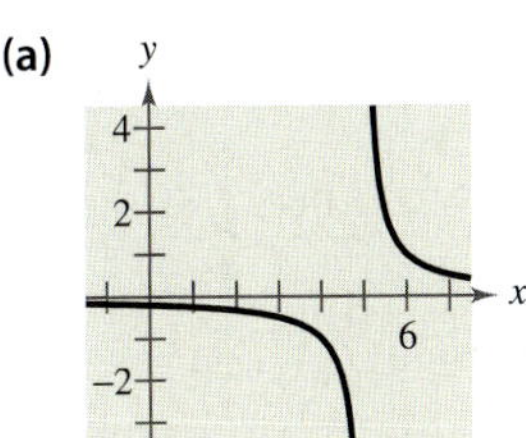

(b)

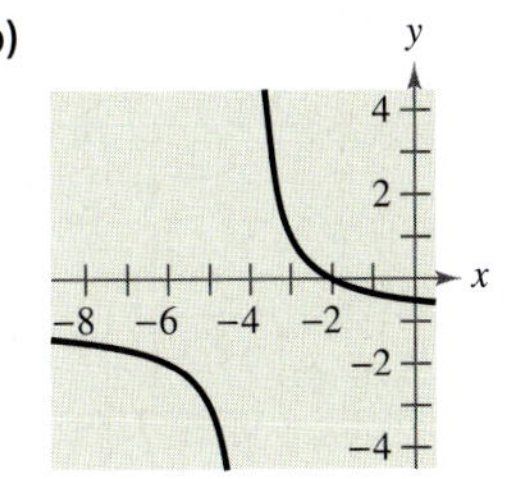

(c)

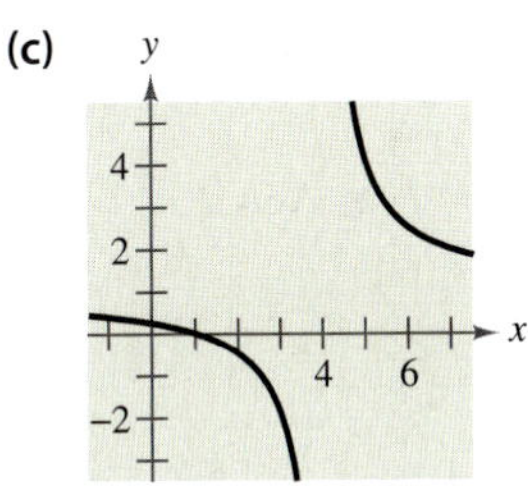

(d)

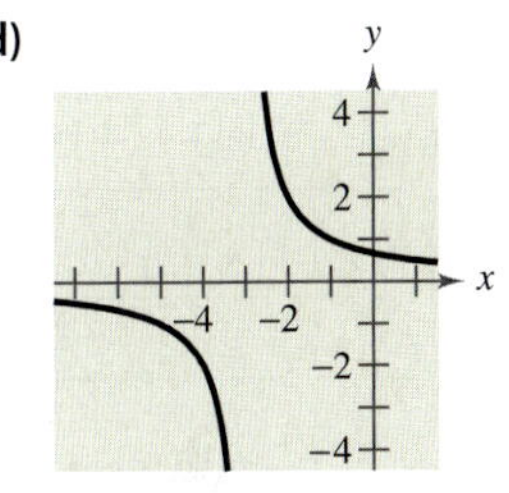

13. $f(x) = \dfrac{2}{x + 3}$

14. $f(x) = \dfrac{1}{x - 5}$

15. $f(x) = \dfrac{x - 1}{x - 4}$

16. $f(x) = -\dfrac{x + 2}{x + 4}$

In Exercises 17–24, find the zeros (if any) of the rational function.

17. $g(x) = \dfrac{x^2 - 1}{x + 1}$

18. $f(x) = \dfrac{x^2 - 2}{x - 3}$

19. $h(x) = 2 + \dfrac{5}{x^2 + 2}$

20. $f(x) = 1 + \dfrac{3}{x^2 - 4}$

21. $f(x) = 1 - \dfrac{3}{x - 3}$

22. $g(x) = 4 - \dfrac{2}{x + 5}$

23. $g(x) = \dfrac{x^3 - 8}{x^2 + 1}$

24. $f(x) = \dfrac{x^3 - 1}{x^2 + 6}$

Analytical and Numerical Analysis **In Exercises 25–28, (a) determine the domains of *f* and *g*, (b) simplify *f* and find any vertical asymptotes of *f*, (c) complete the table, and (d) explain how the two functions differ.**

25. $f(x) = \dfrac{x^2 - 4}{x + 2}, \quad g(x) = x - 2$

x	-4	-3	-2.5	-2	-1.5	-1	0
$f(x)$							
$g(x)$							

26. $f(x) = \dfrac{x^2(x + 3)}{x^2 + 3x}, \quad g(x) = x$

x	-3	-2	-1	0	1	2	3
$f(x)$							
$g(x)$							

27. $f(x) = \dfrac{2x - 1}{2x^2 - x}, \quad g(x) = \dfrac{1}{x}$

x	-1	-0.5	0	0.5	2	3	4
$f(x)$							
$g(x)$							

28. $f(x) = \dfrac{2x - 8}{x^2 - 9x + 20}, \quad g(x) = \dfrac{2}{x - 5}$

x	0	1	2	3	4	5	6
$f(x)$							
$g(x)$							

Exploration **In Exercises 29–32, (a) determine the value that the function *f* approaches as the magnitude of *x* increases. Is *f*(*x*) greater than or less than this functional value when (b) *x* is positive and large in magnitude and (c) *x* is negative and large in magnitude?**

29. $f(x) = 4 - \dfrac{1}{x}$

30. $f(x) = 2 + \dfrac{1}{x - 3}$

31. $f(x) = \dfrac{2x - 1}{x - 3}$

32. $f(x) = \dfrac{2x - 1}{x^2 + 1}$

Data Analysis **In Exercises 33 and 34, consider a physics laboratory experiment designed to determine an unknown mass. A flexible metal meter stick is clamped to a table with 50 centimeters overhanging the edge (see figure). Known masses *M* ranging from 200 grams to 2000 grams are attached to the end of the meter stick. For each mass, the meter stick is displaced vertically and then allowed to oscillate. The average time *t* of one oscillation (in seconds) for each mass is recorded in the table.**

Mass, M	Time, t
200	0.450
400	0.597
600	0.721
800	0.831
1000	0.906
1200	1.003
1400	1.008
1600	1.168
1800	1.218
2000	1.338

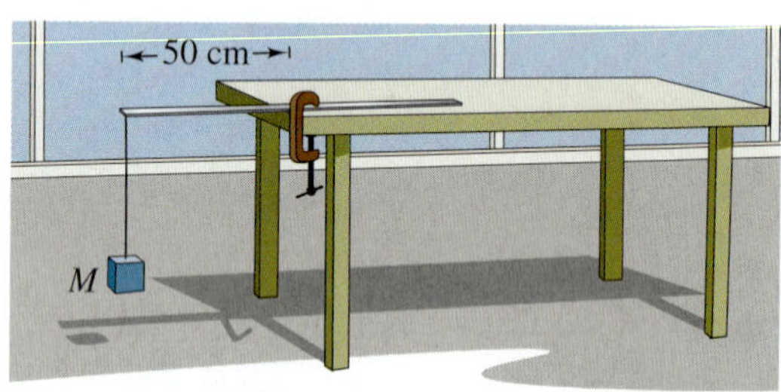

33. A model for the data that can be used to predict the time of one oscillation is

$$t = \frac{38M + 16{,}965}{10(M + 5000)}.$$

(a) Use this model to create a table showing the predicted time for each of the masses shown in the table.

(b) Compare the predicted times with the experimental times. What can you conclude?

34. Use the model in Exercise 33 to approximate the mass of an object for which $t = 1.056$ seconds.

35. *Pollution* The cost C (in millions of dollars) of removing $p\%$ of the industrial and municipal pollutants discharged into a river is

$$C = \frac{255p}{100 - p}, \qquad 0 \le p < 100.$$

(a) Find the cost of removing 10% of the pollutants.

(b) Find the cost of removing 40% of the pollutants.

(c) Find the cost of removing 75% of the pollutants.

(d) According to this model, would it be possible to remove 100% of the pollutants? Explain.

36. *Recycling* In a pilot project, a rural township is given recycling bins for separating and storing recyclable products. The cost C (in dollars) for supplying bins to $p\%$ of the population is

$$C = \frac{25{,}000p}{100 - p}, \qquad 0 \le p < 100.$$

(a) Find the cost of supplying bins to 15% of the population.

(b) Find the cost of supplying bins to 50% of the population.

(c) Find the cost of supplying bins to 90% of the population.

(d) According to this model, would it be possible to supply bins to 100% of the residents? Explain.

37. *Population Growth* The game commission introduces 100 deer into newly acquired state game lands. The population N of the herd is modeled by

$$N = \frac{20(5 + 3t)}{1 + 0.04t}, \qquad t \ge 0$$

where t is the time in years (see figure).

(a) Find the population when $t = 5$, $t = 10$, and $t = 25$.

(b) What is the limiting size of the herd as time increases?

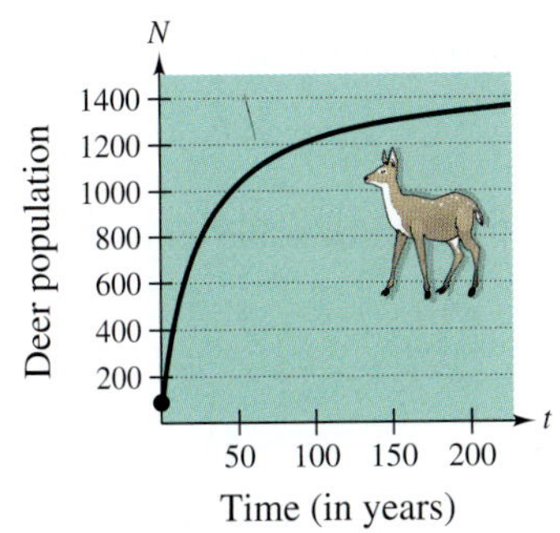

38. *Food Consumption* A biology class performs an experiment comparing the quantity of food consumed by a certain kind of moth with the quantity supplied. The model for the experimental data is

$$y = \frac{1.568x - 0.001}{6.360x + 1}, \qquad x > 0$$

where x is the quantity (in milligrams) of food supplied and y is the quantity (in milligrams) eaten (see figure). At what level of consumption will the moth become satiated?

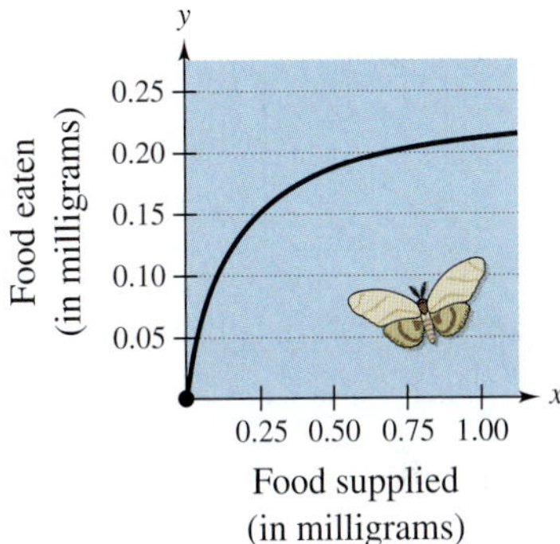

39. *Human Memory Model* Psychologists have developed mathematical models to predict performance as a function of the number of trials n of a certain task. Consider the learning curve

$$P = \frac{0.5 + 0.9(n - 1)}{1 + 0.9(n - 1)}, \qquad n > 0$$

where P is the fraction of correct responses after n trials.

(a) Complete the table for this model. What does it suggest?

n	1	2	3	4	5	6	7	8	9	10
P										

(b) According to this model, what is the limiting percent of correct responses as n increases?

40. *Human Memory Model* How would the limiting percent of correct responses change if the human memory model in Exercise 39 were changed to

$$P = \frac{0.5 + 0.6(n - 1)}{1 + 0.8(n - 1)}, \qquad n > 0?$$

▶ Model It

41. ***Business Partnerships*** The numbers P (in thousands) of business partnerships in the United States for the years 1995 through 2000 are shown in the table. (Source: U.S. Internal Revenue Service)

Year	Partnerships, P
1995	1581
1996	1654
1997	1759
1998	1826
1999	1966
2000	2048

A model for this data is

$$P = \frac{1402.79 + 9.482t^2}{1.0 + 0.0014t^2}$$

where t is the time (in years), with $t = 5$ corresponding to 1995.

(a) Use a graphing utility to plot the data and graph the model in the same viewing window.

(b) Use the model to estimate the number of partnerships in 2005.

(c) Would this model be useful for estimating the number of partnerships after 2005? Explain.

(d) Use the *regression* feature of a graphing utility to find a linear model for the data.

(e) Which model do you think is a better fit for this data? Explain.

42. ***Military*** The numbers M (in thousands) of United States military reserve personnel for the years 1996 through 2000 are shown in the table. (Source: U.S. Department of Defense)

Year	Reserve personnel, M
1996	1550
1997	1461
1998	1369
1999	1304
2000	1264

A model for this data is

$$M = \frac{2129.6 + 23.753t^2}{1.0 + 0.026t^2}$$

where t is the time (in years), with $t = 6$ corresponding to 1996.

(a) Use a graphing utility to plot the data and graph the model in the same viewing window.

(b) Use the model to estimate the number of military reserve personnel in 2005.

(c) Would this model be useful for estimating the number of military reserve personnel after 2005? Explain.

Synthesis

True or False? **In Exercises 43 and 44, determine whether the statement is true or false. Justify your answer.**

43. A polynomial can have infinitely many vertical asymptotes.

44. $f(x) = x^3 - 2x^2 - 5x + 6$ is a rational function.

Think About It **In Exercises 45 and 46, write a rational function f that has the specified characteristics. (There are many correct answers.)**

45. Vertical asymptote: None
Horizontal asymptote: $y = 2$

46. Vertical asymptotes: $x = 0, x = \frac{5}{2}$
Horizontal asymptote: $y = -3$

47. ***Think About It*** Give an example of a rational function whose domain is the set of all real numbers. Give an example of a rational function whose domain is the set of all real numbers except $x = 20$.

48. ***Writing*** Describe what is meant by an asymptote of a graph.

Review

In Exercises 49 and 50, find the inverse function of f. Then graph both f and f^{-1} in the same coordinate plane.

49. $f(x) = 8x - 7$ **50.** $f(x) = \frac{1}{9}x$

In Exercises 51–54, divide using long division.

51. $(x^2 + 5x + 6) \div (x - 4)$

52. $(x^2 - 10x + 15) \div (x - 3)$

53. $(2x^2 + x - 11) \div (x + 5)$

54. $(4x^2 + 3x - 10) \div (x + 6)$

4.2 Graphs of Rational Functions

▶ **What you should learn**

- How to analyze and sketch graphs of rational functions
- How to sketch graphs of rational functions that have slant asymptotes
- How to use graphs of rational functions to model and solve real-life problems

▶ **Why you should learn it**

You can use rational functions to model average speed over a distance. For example, see Exercise 73 on page 349.

Baron Wolman/Tony Stone Images

Analyzing Graphs of Rational Functions

Guidelines for Analyzing Graphs of Rational Functions

Let $f(x) = N(x)/D(x)$, where $N(x)$ and $D(x)$ are polynomials with no common factors.

1. Find and plot the y-intercept (if any) by evaluating $f(0)$.
2. Find the zeros of the numerator (if any) by solving the equation $N(x) = 0$. Then plot the corresponding x-intercepts.
3. Find the zeros of the denominator (if any) by solving the equation $D(x) = 0$. Then sketch the corresponding vertical asymptotes.
4. Find and sketch the horizontal asymptote (if any) by using the rule for finding the horizontal asymptote of a rational function.
5. Test for symmetry.
6. Plot at least one point *between* and one point *beyond* each x-intercept and vertical asymptote.
7. Use smooth curves to complete the graph between and beyond the vertical asymptotes.

Testing for symmetry can be useful, especially for simple rational functions. Recall from Section 2.4 that the graph of $f(x) = 1/x$ is symmetric with respect to the origin.

Technology

Some graphing utilities have difficulty graphing rational functions that have vertical asymptotes. Often, the utility will connect parts of the graph that are not supposed to be connected. For instance, the screen below on the left shows the graph of $f(x) = 1/(x - 2)$. Notice that the graph should consist of two *separated* portions—one to the left of $x = 2$ and the other to the right of $x = 2$. To eliminate this problem, you can try changing the *mode* of the graphing utility to *dot mode*. The problem with this is that the graph is then represented as a collection of dots (as shown in the screen on the right) rather than as a smooth curve.

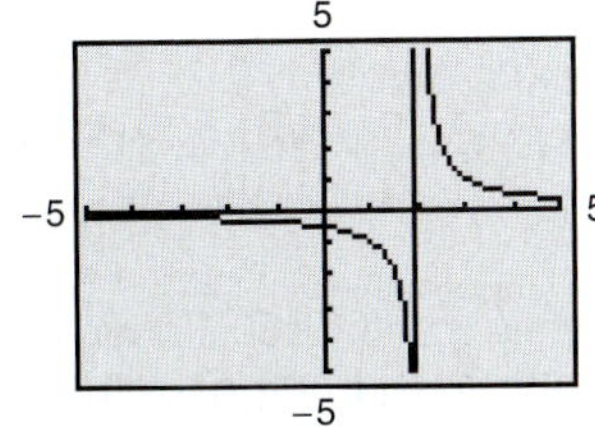

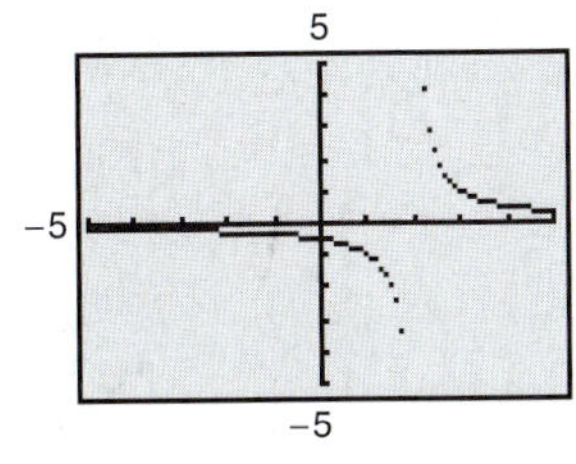

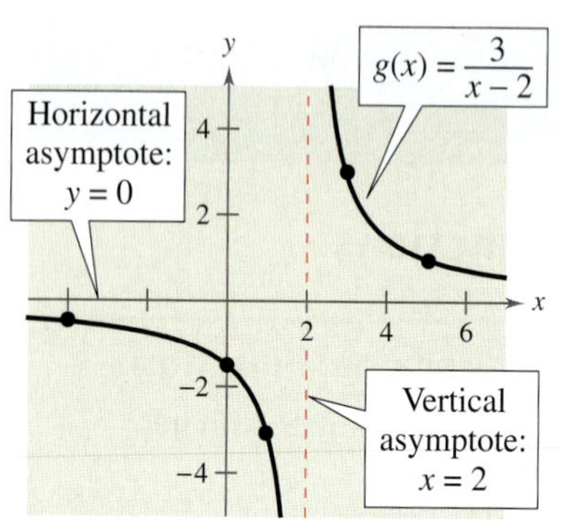

FIGURE 4.8

STUDY TIP

Note in the examples in this section that the vertical asymptotes are included in the table of additional points. This is done to emphasize numerically the behavior of the graph of the function.

Example 1 ▶ Sketching the Graph of a Rational Function

Sketch the graph of $g(x) = \dfrac{3}{x-2}$ and state its domain.

Solution

y-Intercept: $\left(0, -\frac{3}{2}\right)$, because $g(0) = -\frac{3}{2}$

x-Intercept: None, because $3 \neq 0$

Vertical asymptote: $x = 2$, zero of denominator

Horizontal asymptote: $y = 0$, because degree of $N(x) <$ degree of $D(x)$

Additional points:

x	$g(x)$
-4	-0.5
1	-3
2	Undefined
3	3
5	1

By plotting the intercepts, asymptotes, and a few additional points, you can obtain the graph shown in Figure 4.8. The domain of g is all real numbers except $x = 2$.

The graph of g in Example 1 is a vertical stretch and a right shift of the graph of $f(x) = 1/x$, because

$$g(x) = \frac{3}{x-2} = 3\left(\frac{1}{x-2}\right) = 3f(x-2).$$

Example 2 ▶ Sketching the Graph of a Rational Function

Sketch the graph of $f(x) = \dfrac{2x-1}{x}$ and state its domain.

Solution

y-Intercept: None, because $x = 0$ is not in the domain

x-Intercept: $\left(\frac{1}{2}, 0\right)$, from $2x - 1 = 0$

Vertical asymptote: $x = 0$, zero of denominator

Horizontal asymptote: $y = 2$, because degree of $N(x) =$ degree of $D(x)$

Additional points:

x	-4	-1	0	$\frac{1}{4}$	4
$f(x)$	2.25	3	Undefined	-2	1.75

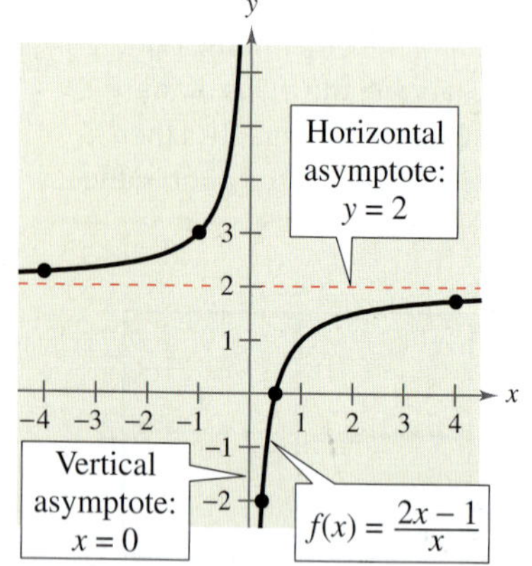

FIGURE 4.9

By plotting the intercepts, asymptotes, and a few additional points, you can obtain the graph shown in Figure 4.9. The domain of f is all real numbers except $x = 0$.

Example 3 ▶ Sketching the Graph of a Rational Function

Sketch the graph of $f(x) = \dfrac{x}{x^2 - x - 2}$.

Solution

Factor the denominator to determine more easily the zeros of the denominator.

$$f(x) = \frac{x}{x^2 - x - 2} = \frac{x}{(x + 1)(x - 2)}$$

y-Intercept: $(0, 0)$, because $f(0) = 0$

x-Intercept: $(0, 0)$

Vertical asymptotes: $x = -1$, $x = 2$, zeros of denominator

Horizontal asymptote: $y = 0$, because degree of $N(x) <$ degree of $D(x)$

Additional points:

x	-3	-1	-0.5	1	2	3
$f(x)$	-0.3	Undefined	0.4	-0.5	Undefined	0.75

The graph is shown in Figure 4.10.

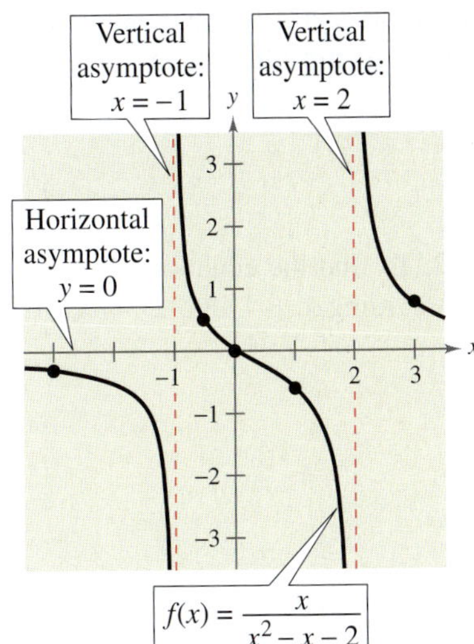

FIGURE 4.10

Example 4 ▶ Sketching the Graph of a Rational Function

Sketch the graph of $f(x) = \dfrac{2(x^2 - 9)}{x^2 - 4}$.

Solution

By factoring the numerator and denominator, you have

$$f(x) = \frac{2(x^2 - 9)}{x^2 - 4} = \frac{2(x - 3)(x + 3)}{(x - 2)(x + 2)}.$$

y-Intercept: $\left(0, \frac{9}{2}\right)$, because $f(0) = \frac{9}{2}$

x-Intercepts: $(-3, 0)$ and $(3, 0)$

Vertical asymptotes: $x = -2$, $x = 2$, zeros of denominator

Horizontal asymptote: $y = 2$, because degree of $N(x) =$ degree of $D(x)$

Symmetry: With respect to y-axis, because $f(-x) = f(x)$

Additional points:

x	-2	0.5	2	2.5	6
$f(x)$	Undefined	4.67	Undefined	-2.44	1.69

The graph is shown in Figure 4.11.

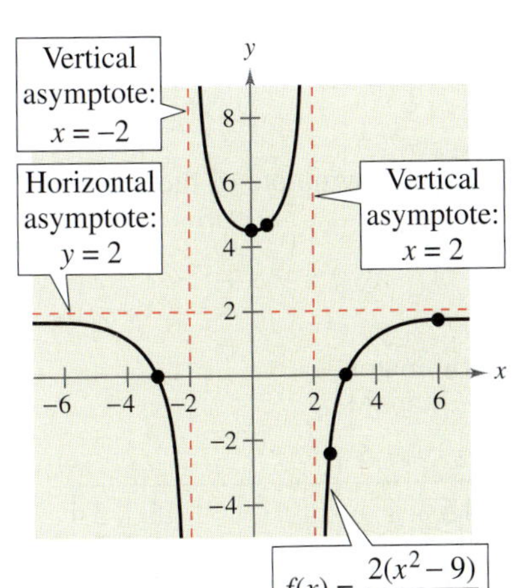

FIGURE 4.11

Slant Asymptotes

Consider a rational function whose denominator is of degree 1 or greater. If the degree of the numerator is exactly *one more* than the degree of the denominator, the graph of the function has a **slant** (or **oblique**) **asymptote.** For example, the graph of

$$f(x) = \frac{x^2 - x}{x + 1}$$

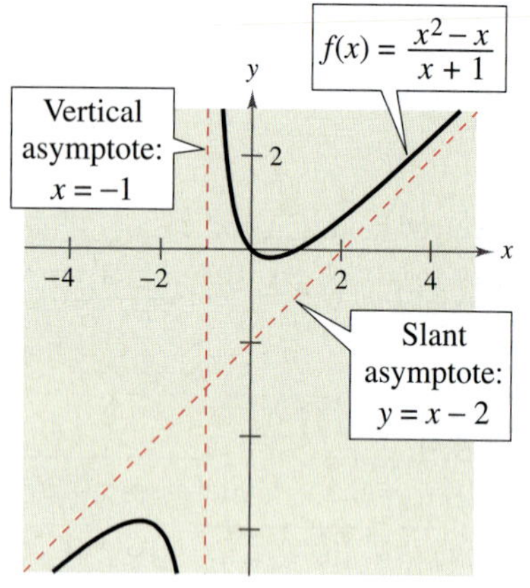

FIGURE 4.12

has a slant asymptote, as shown in Figure 4.12. To find the equation of a slant asymptote, use long division. For instance, by dividing $x + 1$ into $x^2 - x$, you obtain

$$f(x) = \frac{x^2 - x}{x + 1} = \underbrace{x - 2}_{\substack{\text{Slant asymptote} \\ (y = x - 2)}} + \frac{2}{x + 1}.$$

As x increases or decreases without bound, the remainder term $2/(x + 1)$ approaches 0, so the graph of f approaches the line $y = x - 2$, as shown in Figure 4.12.

Example 5 ▶ A Rational Function with a Slant Asymptote 

Sketch the graph of $f(x) = \dfrac{x^2 - x - 2}{x - 1}$.

Solution

First write $f(x)$ in two different ways. Factoring the numerator

$$f(x) = \frac{x^2 - x - 2}{x - 1} = \frac{(x - 2)(x + 1)}{x - 1}$$

allows you to recognize the x-intercepts. Long division

$$f(x) = \frac{x^2 - x - 2}{x - 1} = x - \frac{2}{x - 1}$$

allows you to recognize that the line $y = x$ is a slant asymptote of the graph.

y-Intercept: $(0, 2)$, because $f(0) = 2$

x-Intercepts: $(-1, 0)$ and $(2, 0)$

Vertical asymptote: $x = 1$, zero of denominator

Slant asymptote: $y = x$

Additional points:

x	-2	0.5	1	1.5	3
$f(x)$	-1.33	4.5	Undefined	-2.5	2

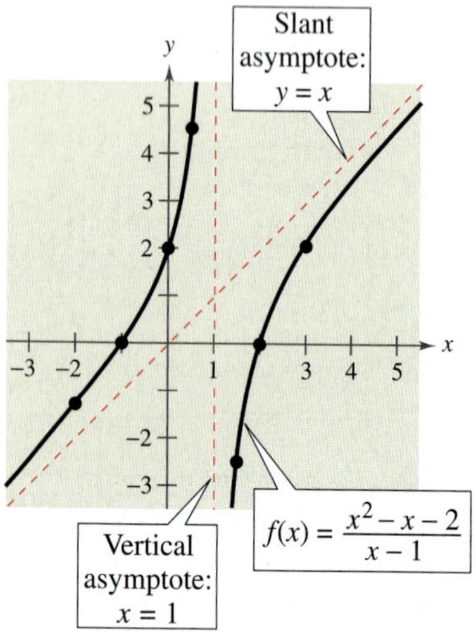

The graph is shown in Figure 4.13.

FIGURE 4.13

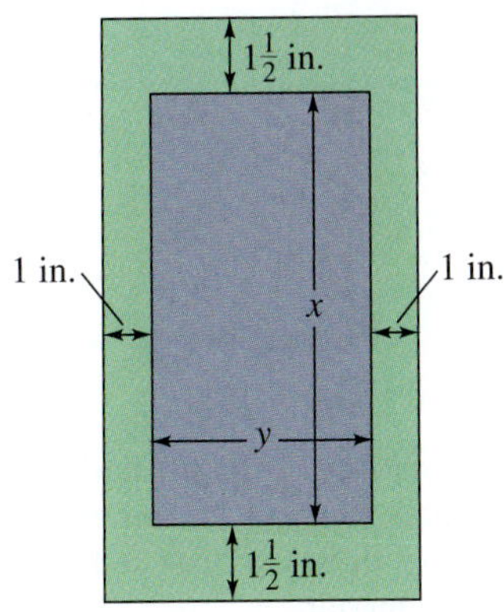

FIGURE 4.14

Application

Example 6 ▶ Finding a Minimum Area

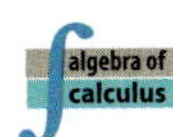

A rectangular page is designed to contain 48 square inches of print. The margins at the top and bottom of the page are each $1\frac{1}{2}$ inches deep. The margins on each side are 1 inch wide. What should the dimensions of the page be so that the least amount of paper is used?

Solution

Let A be the area to be minimized. From Figure 4.14, you can write

$$A = (x + 3)(y + 2).$$

The printed area inside the margins is modeled by $48 = xy$ or $y = 48/x$. To find the minimum area, rewrite the equation for A in terms of just one variable by substituting $48/x$ for y.

$$A = (x + 3)\left(\frac{48}{x} + 2\right)$$

$$= \frac{(x + 3)(48 + 2x)}{x}, \qquad x > 0$$

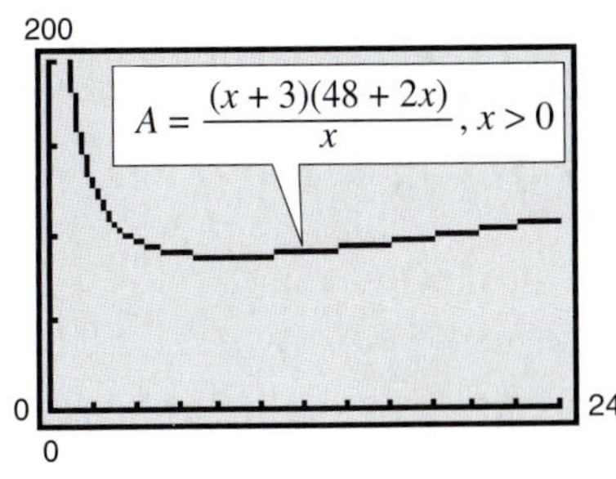

FIGURE 4.15

The graph of this rational function is shown in Figure 4.15. Because x represents the height of the printed area, you need consider only the portion of the graph for which x is positive. Using a graphing utility, you can approximate the minimum value of A to occur when $x \approx 8.5$ inches. The corresponding value of y is $48/8.5 \approx 5.6$ inches. So, the dimensions should be

$$x + 3 \approx 11.5 \text{ inches} \quad \text{by} \quad y + 2 \approx 7.6 \text{ inches}.$$

If you go on to take a course in calculus, you will learn an analytic technique for finding the exact value of x that produces a minimum area. In this case, that value is $x = 6\sqrt{2} \approx 8.485$.

Writing ABOUT MATHEMATICS

Common Factors in the Numerator and Denominator When sketching the graph of a rational function, be sure that the rational function has no factor that is common to its numerator and denominator. To see why, consider the function

$$f(x) = \frac{2x^2 + x - 1}{x + 1} = \frac{(x + 1)(2x - 1)}{x + 1}$$

which has a common factor of $x + 1$ in the numerator and denominator. Sketch the graph of this function. Does it have a vertical asymptote at $x = -1$?

Decide whether each function below has a vertical asymptote. Write a short paragraph to explain your reasoning. Include a graph of each function in your explanation.

a. $f(x) = \frac{x^2 - 4}{x + 2}$ **b.** $f(x) = \frac{x^2 - 4}{x}$ **c.** $f(x) = \frac{x^2 - 4}{2 - x}$

4.2 Exercises

In Exercises 1–4, use the graph of $f(x) = 2/x$ to sketch the graph of g.

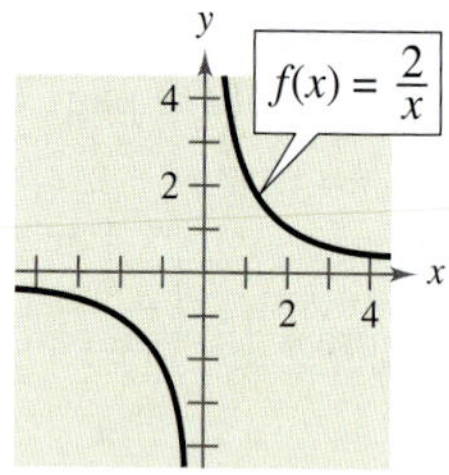

1. $g(x) = \frac{2}{x} + 3$
2. $g(x) = \frac{2}{x - 3}$
3. $g(x) = -\frac{2}{x}$
4. $g(x) = \frac{1}{x + 2}$

In Exercises 5–8, use the graph of $f(x) = 2/x^2$ to sketch the graph of g.

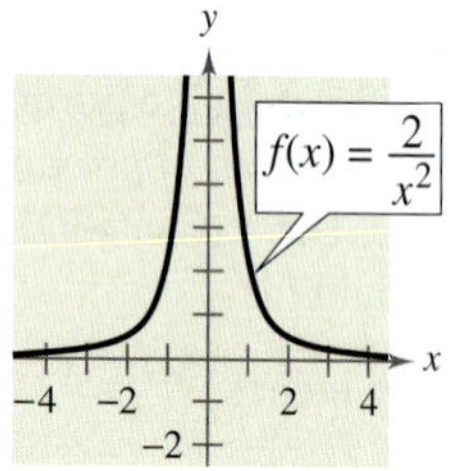

5. $g(x) = \frac{2}{x^2} - 1$
6. $g(x) = -\frac{2}{x^2}$
7. $g(x) = \frac{2}{(x - 1)^2}$
8. $g(x) = \frac{1}{2x^2}$

In Exercises 9–12, use the graph of $f(x) = 4/x^3$ to sketch the graph of g.

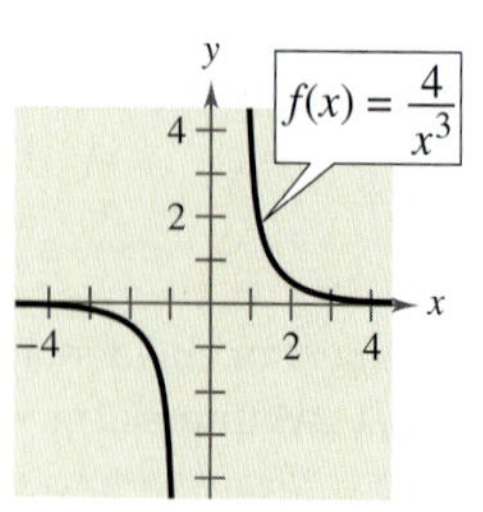

9. $g(x) = \frac{4}{(x + 3)^3}$
10. $g(x) = \frac{4}{x^3} + 3$
11. $g(x) = -\frac{4}{x^3}$
12. $g(x) = \frac{1}{x^3}$

In Exercises 13–34, (a) identify all intercepts, (b) find any vertical and horizontal asymptotes, (c) check for symmetry, and (d) plot additional solution points as needed and sketch the graph of the rational function.

13. $f(x) = \frac{1}{x + 2}$
14. $f(x) = \frac{1}{x - 3}$
15. $h(x) = \frac{-1}{x + 2}$
16. $g(x) = \frac{1}{3 - x}$
17. $C(x) = \frac{5 + 2x}{1 + x}$
18. $P(x) = \frac{1 - 3x}{1 - x}$
19. $g(x) = \frac{1}{x + 2} + 2$
20. $f(x) = 2 - \frac{3}{x^2}$
21. $f(x) = \frac{x^2}{x^2 + 9}$
22. $f(t) = \frac{1 - 2t}{t}$
23. $h(x) = \frac{x^2}{x^2 - 9}$
24. $g(x) = \frac{x}{x^2 - 9}$
25. $g(s) = \frac{s}{s^2 + 1}$
26. $f(x) = -\frac{1}{(x - 2)^2}$
27. $g(x) = \frac{4(x + 1)}{x(x - 4)}$
28. $h(x) = \frac{2}{x^2(x - 2)}$
29. $f(x) = \frac{3x}{x^2 - x - 2}$
30. $f(x) = \frac{2x}{x^2 + x - 2}$
31. $f(x) = \frac{6x}{x^2 - 5x - 14}$
32. $f(x) = \frac{3(x^2 + 1)}{x^2 + 2x - 15}$
33. $f(x) = \frac{2x^2 - 5x - 3}{x^3 - 2x^2 - x + 2}$
34. $f(x) = \frac{x^2 - x - 2}{x^3 - 2x^2 - 5x + 6}$

Analytical, Numerical, and Graphical Analysis **In Exercises 35–38, do the following.**

(a) Determine the domains of *f* and *g*.

(b) Simplify *f* and find any vertical asymptotes of *f*.

(c) Compare the functions by completing the table.

 **(d) Use a graphing utility to graph *f* and *g* in the same viewing window.**

 **(e) Explain why the graphing utility may not show the difference in the domains of *f* and *g*.**

35. $f(x) = \dfrac{x^2 - 1}{x + 1}, \quad g(x) = x - 1$

x	-3	-2	-1.5	-1	-0.5	0	1
$f(x)$							
$g(x)$							

36. $f(x) = \dfrac{x^2(x - 2)}{x^2 - 2x}, \quad g(x) = x$

x	-1	0	1	1.5	2	2.5	3
$f(x)$							
$g(x)$							

37. $f(x) = \dfrac{x - 2}{x^2 - 2x}, \quad g(x) = \dfrac{1}{x}$

x	-0.5	0	0.5	1	1.5	2	3
$f(x)$							
$g(x)$							

38. $f(x) = \dfrac{2x - 6}{x^2 - 7x + 12}, \quad g(x) = \dfrac{2}{x - 4}$

x	0	1	2	3	4	5	6
$f(x)$							
$g(x)$							

In Exercises 39–44, state the domain of the function and identify any vertical and slant asymptotes.

39. $h(x) = \dfrac{x^2 - 4}{x}$

40. $g(x) = \dfrac{x^2 + 5}{x}$

41. $f(t) = -\dfrac{t^2 + 1}{t + 5}$

42. $f(x) = \dfrac{x^2}{3x + 1}$

43. $f(x) = \dfrac{x^3 - 1}{x^2 - x}$

44. $f(x) = \dfrac{x^4 + x}{x^3}$

In Exercises 45–52, (a) identify all intercepts, (b) find any vertical and slant asymptotes, (c) check for symmetry, and (d) plot additional solution points as needed and sketch the graph of the rational function.

45. $f(x) = \dfrac{2x^2 + 1}{x}$

46. $f(x) = \dfrac{1 - x^2}{x}$

47. $g(x) = \dfrac{x^2 + 1}{x}$

48. $h(x) = \dfrac{x^2}{x - 1}$

49. $f(x) = \dfrac{x^3}{x^2 - 1}$

50. $g(x) = \dfrac{x^3}{2x^2 - 8}$

51. $f(x) = \dfrac{x^2 - x + 1}{x - 1}$

52. $f(x) = \dfrac{2x^2 - 5x + 5}{x - 2}$

 **In Exercises 53–56, use a graphing utility to graph the rational function. Give the domain of the function and identify any asymptotes. Then zoom out sufficiently far so that the graph appears as a line. Identify the line.**

53. $f(x) = \dfrac{x^2 + 5x + 8}{x + 3}$

54. $f(x) = \dfrac{2x^2 + x}{x + 1}$

55. $g(x) = \dfrac{1 + 3x^2 - x^3}{x^2}$

56. $h(x) = \dfrac{12 - 2x - x^2}{2(4 + x)}$

Graphical Reasoning **In Exercises 57–60, (a) use the graph to determine any *x*-intercepts of the rational function and (b) set $y = 0$ and solve the resulting equation to confirm your result in part (a).**

57. $y = \dfrac{x + 1}{x - 3}$

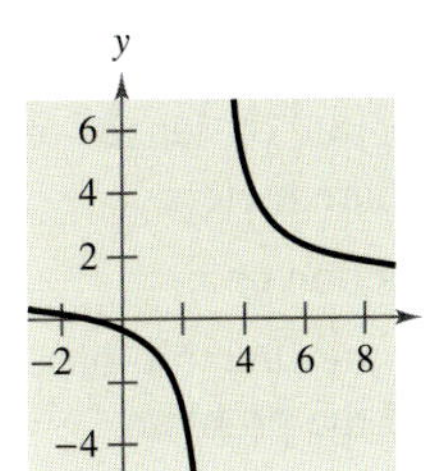

58. $y = \dfrac{2x}{x - 3}$

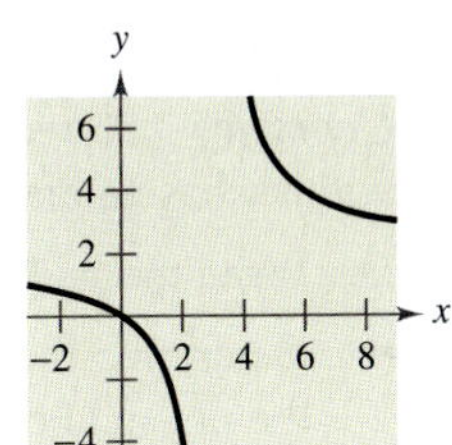

59. $y = \frac{1}{x} - x$

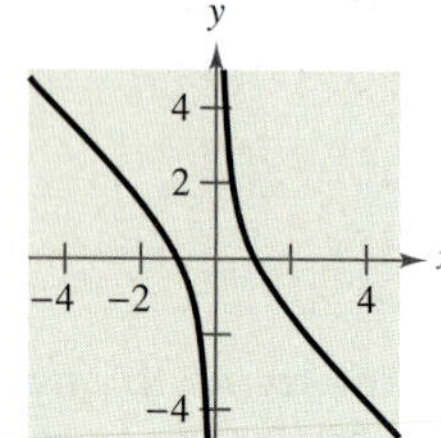

60. $y = x - 3 + \frac{2}{x}$

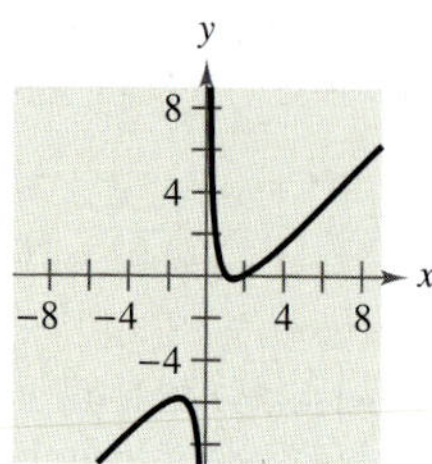

Graphical Reasoning **In Exercises 61–64, (a) use a graphing utility to graph the function and determine any *x*-intercepts and (b) set $y = 0$ and solve the resulting equation to confirm your result in part (a).**

61. $y = \frac{1}{x + 5} + \frac{4}{x}$

62. $y = 20\left(\frac{2}{x + 1} - \frac{3}{x}\right)$

63. $y = x - \frac{6}{x - 1}$

64. $y = x - \frac{9}{x}$

65. ***Concentration of a Mixture*** A 1000-liter tank contains 50 liters of a 25% brine solution. You add x liters of a 75% brine solution to the tank.

(a) Show that the concentration C, the proportion of brine to total solution, in the final mixture is

$$C = \frac{3x + 50}{4(x + 50)}.$$

(b) Determine the domain of the function based on the physical constraints of the problem.

(c) Sketch a graph of the concentration function.

(d) As the tank is filled, what happens to the rate at which the concentration of brine is increasing? What percent does the concentration of brine appear to approach?

66. ***Geometry*** A rectangular region of length x and width y has an area of 500 square meters.

(a) Write the width y as a function of x.

(b) Determine the domain of the function based on the physical constraints of the problem.

(c) Sketch a graph of the function and determine the width of the rectangle when $x = 30$ meters.

67. ***Minimum Area*** A page that is x inches wide and y inches high contains 30 square inches of print. The top and bottom margins are 1 inch deep and the margins on each side are 2 inches wide (see figure).

(a) Show that the total area A on the page is

$$A = \frac{2x(x + 11)}{x - 4}.$$

(b) Determine the domain of the function based on the physical constraints of the problem.

(c) Use a graphing utility to graph the area function and approximate the page size for which the least amount of paper will be used.

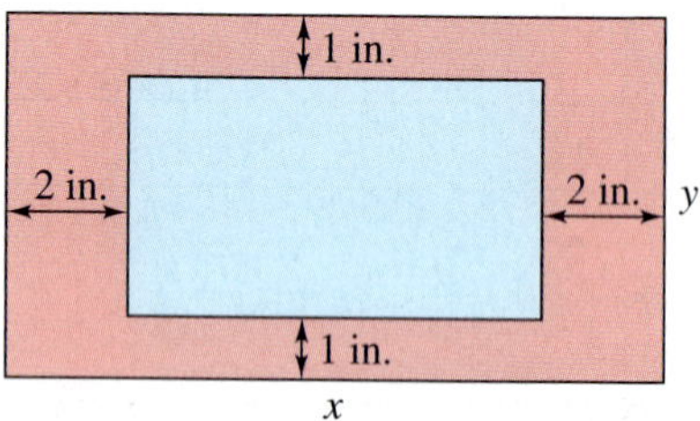

68. ***Minimum Area*** A rectangular page is designed to contain 64 square inches of print. The margins at the top and bottom of the page are each 1 inch deep. The margins on each side are $1\frac{1}{2}$ inches wide. What should the dimensions of the page be so that the least amount of paper is used?

In Exercises 69 and 70, use a graphing utility to graph the function and locate any relative maximum or minimum points on the graph.

69. $f(x) = \frac{3(x + 1)}{x^2 + x + 1}$

70. $C(x) = x + \frac{32}{x}$

71. ***Minimum Cost*** The ordering and transportation cost C (in thousands of dollars) for the components used in manufacturing a certain product is

$$C = 100\left(\frac{200}{x^2} + \frac{x}{x + 30}\right), \qquad x \geq 1$$

where x is the order size (in hundreds). Use a graphing utility to graph the cost function. From the graph, estimate the order size that minimizes cost.

72. ***Minimum Cost*** The cost C of producing x units is

$$C = 0.2x^2 + 10x + 5$$

and the average cost per unit is

$$\overline{C} = \frac{C}{x} = \frac{0.2x^2 + 10x + 5}{x}, \qquad x > 0.$$

Sketch the graph of the average cost function and estimate the number of units that should be produced to minimize the average cost per unit.

▶ Model It

73. ***Average Speed*** A driver averaged 50 miles per hour on the round trip between Akron, Ohio and Columbus, Ohio, 100 miles away. The average speeds for going and returning were x and y miles per hour, respectively.

(a) Show that $y = \dfrac{25x}{x - 25}$.

(b) Determine the vertical and horizontal asymptotes of the function.

(c) Use a graphing utility to graph the function.

(d) Complete the table.

x	30	35	40	45	50	55	60
y							

(e) Are the results in the table unexpected? Explain.

(f) Is it possible to average 20 miles per hour in one direction and still average 50 miles per hour on the round trip? Explain.

74. ***Medicine*** The concentration C of a chemical in the bloodstream t hours after injection into muscle tissue is

$$C = \frac{3t^2 + t}{t^3 + 50}, \qquad t > 0.$$

(a) Determine the horizontal asymptote of the function and interpret its meaning in the context of the problem.

(b) Use a graphing utility to graph the function and approximate the time when the bloodstream concentration is greatest.

Synthesis

True or False? **In Exercises 75–78, determine whether the statement is true or false. Justify your answer.**

75. If the graph of a rational function f has a vertical asymptote at $x = 5$, it is possible to sketch the graph without lifting your pencil from the paper.

76. The graph of a rational function can never cross one of its asymptotes.

77. The graph of

$$f(x) = \frac{2x^3}{x + 1}$$

has a slant asymptote.

78. Every rational function has a vertical asymptote.

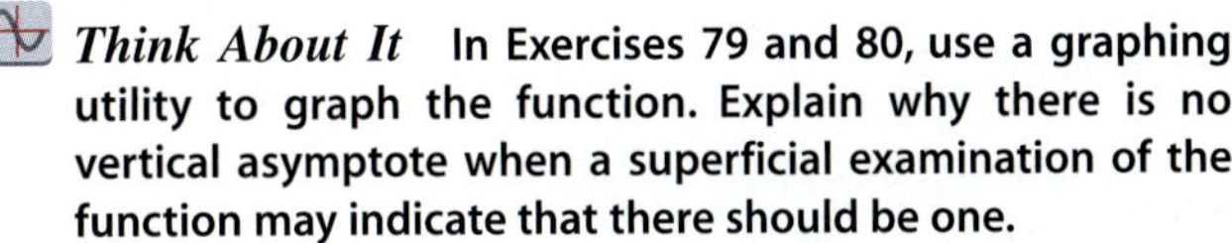

Think About It **In Exercises 79 and 80, use a graphing utility to graph the function. Explain why there is no vertical asymptote when a superficial examination of the function may indicate that there should be one.**

79. $h(x) = \dfrac{6 - 2x}{3 - x}$

80. $g(x) = \dfrac{x^2 + x - 2}{x - 1}$

81. ***Think About It*** Write a rational function satisfying the following criteria.

Vertical asymptote: $x = 2$

Slant asymptote: $y = x + 1$

Zero of the function: $x = -2$

82. ***Think About It*** Write a rational function satisfying the following criteria.

Vertical asymptote: $x = -1$

Slant asymptote: $y = x + 2$

Zero of the function: $x = 3$

Review

In Exercises 83–86, completely factor the expression.

83. $x^2 - 15x + 56$

84. $3x^2 + 23x - 36$

85. $x^3 - 5x^2 + 4x - 20$

86. $x^3 + 6x^2 - 2x - 12$

In Exercises 87–90, solve the inequality and graph the solution on the real number line.

87. $10 - 3x \le 0$

88. $5 - 2x > 5(x + 1)$

89. $|4(x - 2)| < 20$

90. $\frac{1}{2}|2x + 3| \ge 5$

4.3 Partial Fractions

▶ **What you should learn**

- How to recognize partial fraction decompositions of rational expressions
- How to find partial fraction decompositions of rational expressions

▶ **Why you should learn it**

Partial fractions can help you analyze the behavior of a rational function. For instance, in Exercise 57 on page 357 you can analyze the exhaust temperatures of a diesel engine using partial fractions.

Michael Rosenfeld/Tony Stone Images

Introduction

In this section, you will learn to write a rational expression as the sum of two or more simpler rational expressions. For example, the rational expression

$$\frac{x+7}{x^2-x-6}$$

can be written as the sum of two fractions with first-degree denominators. That is,

$$\frac{x+7}{x^2-x-6} = \underbrace{\overbrace{\underbrace{\frac{2}{x-3}}_{\text{Partial fraction}} + \underbrace{\frac{-1}{x+2}}_{\text{Partial fraction}}}^{\text{Partial fraction decomposition of } \frac{x+7}{x^2-x-6}}}_{}.$$

Each fraction on the right side of the equation is a **partial fraction,** and together they make up the **partial fraction decomposition** of the left side.

Decomposition of $N(x)/D(x)$ into Partial Fractions

1. *Divide if improper:* If $N(x)/D(x)$ is an improper fraction [degree of $N(x) \geq$ degree of $D(x)$], divide the denominator into the numerator to obtain

$$\frac{N(x)}{D(x)} = (\text{polynomial}) + \frac{N_1(x)}{D(x)}$$

and apply Steps 2, 3, and 4 below to the proper rational expression $N_1(x)/D(x)$. Note that $N_1(x)$ is the remainder from the division of $N(x)$ by $D(x)$.

2. *Factor the denominator:* Completely factor the denominator into factors of the form

$$(px+q)^m \quad \text{and} \quad (ax^2+bx+c)^n$$

where (ax^2+bx+c) is irreducible.

3. *Linear factors:* For *each* factor of the form $(px+q)^m$, the partial fraction decomposition must include the following sum of m fractions.

$$\frac{A_1}{(px+q)} + \frac{A_2}{(px+q)^2} + \cdots + \frac{A_m}{(px+q)^m}$$

4. *Quadratic factors:* For *each* factor of the form $(ax^2+bx+c)^n$, the partial fraction decomposition must include the following sum of n fractions.

$$\frac{B_1x+C_1}{ax^2+bx+c} + \frac{B_2x+C_2}{(ax^2+bx+c)^2} + \cdots + \frac{B_nx+C_n}{(ax^2+bx+c)^n}$$

STUDY TIP

In Section P.5, you learned how to combine expressions such as

$$\frac{1}{x-2} + \frac{-1}{x+3} = \frac{5}{(x-2)(x+3)}.$$

The method of partial fractions shows you how to reverse this process.

$$\frac{5}{(x-2)(x+3)} = \frac{?}{x-2} + \frac{?}{x+3}$$

Partial Fraction Decomposition

Algebraic techniques for determining the constants in the numerators of partial fractions are demonstrated in the examples that follow. Note that the techniques vary slightly, depending on the type of factors of the denominator: linear or quadratic, distinct or repeated.

Technology

You can use a graphing utility to check *graphically* the decomposition found in Example 1. To do this, graph

$$y_1 = \frac{x+7}{x^2-x-6}$$

and

$$y_2 = \frac{2}{x-3} + \frac{-1}{x+2}$$

in the same viewing window. The graphs should be identical, as shown below.

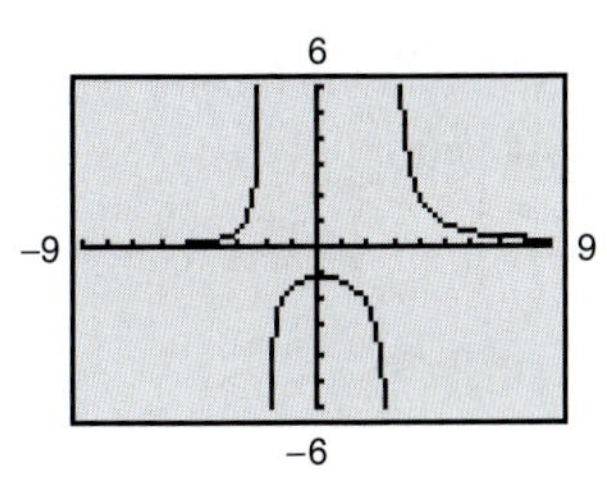

Example 1 ▶ Distinct Linear Factors 

Write the partial fraction decomposition of $\dfrac{x+7}{x^2-x-6}$.

Solution

The expression is not improper, so factor the denominator. Because $x^2 - x - 6 = (x-3)(x+2)$, you should include one partial fraction with a constant numerator for each linear factor of the denominator. Write the form of the decomposition as follows.

$$\frac{x+7}{x^2-x-6} = \frac{A}{x-3} + \frac{B}{x+2}$$ Write form of decomposition.

Multiplying each side of this equation by the least common denominator, $(x-3)(x+2)$, leads to the **basic equation**

$$x + 7 = A(x+2) + B(x-3).$$ Basic equation

Because this equation is true for all x, you can substitute any *convenient* values of x that will help determine the constants A and B. Values of x that are especially convenient are ones that make the factors $(x+2)$ and $(x-3)$ equal to zero. For instance, let $x = -2$. Then

$$-2 + 7 = A(-2+2) + B(-2-3)$$ Substitute -2 for x.

$$5 = A(0) + B(-5)$$

$$5 = -5B$$

$$-1 = B.$$

To solve for A, let $x = 3$ and obtain

$$3 + 7 = A(3+2) + B(3-3)$$ Substitute 3 for x.

$$10 = A(5) + B(0)$$

$$10 = 5A$$

$$2 = A.$$

So, the decomposition is

$$\frac{x+7}{x^2-x-6} = \frac{2}{x-3} + \frac{-1}{x+2}$$

as indicated at the beginning of this section. Check this result by combining the two partial fractions on the right side of the equation, or by using your graphing utility.

The next example shows how to find the partial fraction decomposition of a rational expression whose denominator has a repeated linear factor.

Example 2 ▶ Repeated Linear Factors 

Write the partial fraction decomposition of $\dfrac{x^4 + 2x^3 + 6x^2 + 20x + 6}{x^3 + 2x^2 + x}$.

Solution

This rational expression is improper, so you should begin by dividing the numerator by the denominator to obtain

$$x + \frac{5x^2 + 20x + 6}{x^3 + 2x^2 + x}.$$

Because the denominator of the remainder factors as

$$x^3 + 2x^2 + x = x(x^2 + 2x + 1) = x(x + 1)^2$$

you should include one partial fraction with a constant numerator for each power of x and $(x + 1)$ and write the form of the decomposition as follows.

$$\frac{5x^2 + 20x + 6}{x(x + 1)^2} = \frac{A}{x} + \frac{B}{x + 1} + \frac{C}{(x + 1)^2}$$

Multiplying by the LCD, $x(x + 1)^2$, leads to the basic equation

$$5x^2 + 20x + 6 = A(x + 1)^2 + Bx(x + 1) + Cx. \qquad \text{Basic equation}$$

Letting $x = -1$ eliminates the A- and B-terms and yields

$$5(-1)^2 + 20(-1) + 6 = A(-1 + 1)^2 + B(-1)(-1 + 1) + C(-1)$$
$$5 - 20 + 6 = 0 + 0 - C$$
$$C = 9.$$

Letting $x = 0$ eliminates the B- and C-terms and yields

$$5(0)^2 + 20(0) + 6 = A(0 + 1)^2 + B(0)(0 + 1) + C(0)$$
$$6 = A(1) + 0 + 0$$
$$6 = A.$$

At this point, you have exhausted the most convenient choices for x, so to find the value of B, use *any other value* for x along with the known values of A and C. So, using $x = 1$, $A = 6$, and $C = 9$,

$$5(1)^2 + 20(1) + 6 = 6(1 + 1)^2 + B(1)(1 + 1) + 9(1)$$
$$31 = 6(4) + 2B + 9$$
$$-2 = 2B$$
$$-1 = B.$$

Therefore, the partial fraction decomposition is

$$\frac{x^4 + 2x^3 + 6x^2 + 20x + 6}{x^3 + 2x^2 + x} = x + \frac{6}{x} + \frac{-1}{x + 1} + \frac{9}{(x + 1)^2}.$$

The procedure used to solve for the constants in Examples 1 and 2 works well when the factors of the denominator are linear. However, when the denominator contains irreducible quadratic factors, you should use a different procedure, which involves writing the right side of the basic equation in polynomial form and *equating the coefficients* of like terms.

Example 3 ▶ Distinct Linear and Quadratic Factors

The Granger Collection

Historical Note
John Bernoulli (1667–1748), a Swiss mathematician, introduced the method of partial fractions and was instrumental in the early development of calculus. Bernoulli was a professor at the University of Basel and taught many outstanding students, the most famous of whom was Leonhard Euler.

Write the partial fraction decomposition of

$$\frac{3x^2 + 4x + 4}{x^3 + 4x}.$$

Solution

This expression is not improper, so factor the denominator. Because the denominator factors as

$$x^3 + 4x = x(x^2 + 4)$$

you should include one partial fraction with a constant numerator and one partial fraction with a linear numerator and write the form of the decomposition as follows.

$$\frac{3x^2 + 4x + 4}{x^3 + 4x} = \frac{A}{x} + \frac{Bx + C}{x^2 + 4}$$

Multiplying by the LCD, $x(x^2 + 4)$, yields the basic equation

$$3x^2 + 4x + 4 = A(x^2 + 4) + (Bx + C)x. \qquad \text{Basic equation}$$

Expanding this basic equation and collecting like terms produces

$$\begin{aligned} 3x^2 + 4x + 4 &= Ax^2 + 4A + Bx^2 + Cx \\ &= (A + B)x^2 + Cx + 4A. \qquad \text{Polynomial form} \end{aligned}$$

Finally, because two polynomials are equal if and only if the coefficients of like terms are equal,

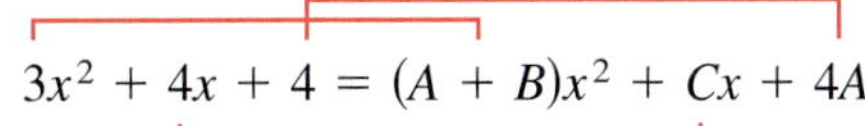

$$3x^2 + 4x + 4 = (A + B)x^2 + Cx + 4A \qquad \text{Equate coefficients of like terms.}$$

you obtain the equations

$$3 = A + B, \qquad 4 = C, \qquad \text{and} \qquad 4 = 4A.$$

So, $A = 1$ and $C = 4$. Moreover, substituting $A = 1$ in the equation $3 = A + B$ yields

$$3 = 1 + B$$

$$2 = B.$$

Therefore, the partial fraction decomposition is

$$\frac{3x^2 + 4x + 4}{x^3 + 4x} = \frac{1}{x} + \frac{2x + 4}{x^2 + 4}.$$

The next example shows how to find the partial fraction decomposition of a rational function whose denominator has a *repeated* quadratic factor.

Example 4 ▶ Repeated Quadratic Factors 

Write the partial fraction decomposition of

$$\frac{8x^3 + 13x}{(x^2 + 2)^2}.$$

Solution

You need to include one partial fraction with a linear numerator for each power of $(x^2 + 2)$.

$$\frac{8x^3 + 13x}{(x^2 + 2)^2} = \frac{Ax + B}{x^2 + 2} + \frac{Cx + D}{(x^2 + 2)^2}$$

Multiplying by the LCD, $(x^2 + 2)^2$, yields the basic equation

$$8x^3 + 13x = (Ax + B)(x^2 + 2) + Cx + D$$ Basic equation

$$= Ax^3 + 2Ax + Bx^2 + 2B + Cx + D$$

$$= Ax^3 + Bx^2 + (2A + C)x + (2B + D).$$ Polynomial form

Equating coefficients of like terms

$$8x^3 + 0x^2 + 13x + 0 = Ax^3 + Bx^2 + (2A + C)x + (2B + D)$$

produces

$$8 = A,\ 0 = B,\ 13 = 2A + C, \text{ and } 0 = 2B + D.$$ Equate coefficients.

Finally, use the values $A = 8$ and $B = 0$ to obtain the following.

$$13 = 2A + C$$

$$= 2(8) + C$$

$$-3 = C$$

$$0 = 2B + D$$

$$= 2(0) + D$$

$$0 = D$$

Therefore,

$$\frac{8x^3 + 13x}{(x^2 + 2)^2} = \frac{8x}{x^2 + 2} + \frac{-3x}{(x^2 + 2)^2}.$$

By equating coefficients of like terms in Examples 3 and 4, you obtained several equations involving A, B, C, and D, which were solved by *substitution*. In a later chapter you will study a more general method for solving such *systems of equations*.

Guidelines for Solving the Basic Equation

Linear Factors

1. Substitute the *zeros* of the distinct linear factors into the basic equation.
2. For repeated linear factors, use the coefficients determined in step 1 above to rewrite the basic equation. Then substitute *other* convenient values of x and solve for the remaining coefficients.

Quadratic Factors

1. Expand the basic equation.
2. Collect terms according to powers of x.
3. Equate the coefficients of like terms to obtain equations involving A, B, C, and so on.
4. Use substitution to solve for $A, B, C, \ldots$.

Keep in mind that for *improper* rational expressions such as

$$\frac{N(x)}{D(x)} = \frac{2x^3 + x^2 - 7x + 7}{x^2 + x - 2}$$

you must first divide before applying partial fraction decomposition.

Writing ABOUT MATHEMATICS

Error Analysis Suppose you are tutoring a student in algebra. In trying to find a partial fraction decomposition, your student writes the following.

$$\frac{x^2 + 1}{x(x-1)} = \frac{A}{x} + \frac{B}{x-1}$$

$$\frac{x^2 + 1}{x(x-1)} = \frac{A(x-1)}{x(x-1)} + \frac{Bx}{x(x-1)}$$

$$x^2 + 1 = A(x-1) + Bx \qquad \text{Basic equation}$$

By substituting $x = 0$ and $x = 1$ into the basic equation, your student concludes that $A = -1$ and $B = 2$. However, in checking this solution, your student obtains the following.

$$\frac{-1}{x} + \frac{2}{x-1} = \frac{(-1)(x-1) + 2(x)}{x(x-1)}$$

$$= \frac{x+1}{x(x-1)}$$

$$\neq \frac{x^2+1}{x(x-1)}$$

What has gone wrong?

4.3 Exercises

In Exercises 1–4, match the rational expression with the form of its decomposition. [The decompositions are labeled (a), (b), (c), and (d).]

(a) $\frac{A}{x} + \frac{B}{x+2} + \frac{C}{x-2}$

(b) $\frac{A}{x} + \frac{B}{x-4}$

(c) $\frac{A}{x} + \frac{B}{x^2} + \frac{C}{x-4}$

(d) $\frac{A}{x} + \frac{Bx+C}{x^2+4}$

1. $\frac{3x-1}{x(x-4)}$ **2.** $\frac{3x-1}{x^2(x-4)}$

3. $\frac{3x-1}{x(x^2+4)}$ **4.** $\frac{3x-1}{x(x^2-4)}$

In Exercises 5–14, write the form of the partial fraction decomposition of the rational expression. Do not solve for the constants.

5. $\frac{7}{x^2-14x}$ **6.** $\frac{x-2}{x^2+4x+3}$

7. $\frac{12}{x^3-10x^2}$ **8.** $\frac{x^2-3x+2}{4x^3+11x^2}$

9. $\frac{4x^2+3}{(x-5)^3}$ **10.** $\frac{6x+5}{(x+2)^4}$

11. $\frac{2x-3}{x^3+10x}$ **12.** $\frac{x-6}{2x^3+8x}$

13. $\frac{x-1}{x(x^2+1)^2}$ **14.** $\frac{x+4}{x^2(3x-1)^2}$

In Exercises 15–38, write the partial fraction decomposition of the rational expression. Check your result algebraically.

15. $\frac{1}{x^2-1}$ **16.** $\frac{1}{4x^2-9}$

17. $\frac{1}{x^2+x}$ **18.** $\frac{3}{x^2-3x}$

19. $\frac{1}{2x^2+x}$ **20.** $\frac{5}{x^2+x-6}$

21. $\frac{3}{x^2+x-2}$ **22.** $\frac{x+1}{x^2+4x+3}$

23. $\frac{x^2+12x+12}{x^3-4x}$ **24.** $\frac{x+2}{x(x-4)}$

25. $\frac{4x^2+2x-1}{x^2(x+1)}$ **26.** $\frac{2x-3}{(x-1)^2}$

27. $\frac{3x}{(x-3)^2}$ **28.** $\frac{6x^2+1}{x^2(x-1)^2}$

29. $\frac{x^2-1}{x(x^2+1)}$ **30.** $\frac{x}{(x-1)(x^2+x+1)}$

31. $\frac{x}{x^3-x^2-2x+2}$ **32.** $\frac{x+6}{x^3-3x^2-4x+12}$

33. $\frac{x^2}{x^4-2x^2-8}$ **34.** $\frac{2x^2+x+8}{(x^2+4)^2}$

35. $\frac{x}{16x^4-1}$ **36.** $\frac{x+1}{x^3+x}$

37. $\frac{x^2+5}{(x+1)(x^2-2x+3)}$ **38.** $\frac{x^2-4x+7}{(x+1)(x^2-2x+3)}$

In Exercises 39–44, write the partial fraction decomposition of the improper rational expression.

39. $\frac{x^2-x}{x^2+x+1}$ **40.** $\frac{x^2-4x}{x^2+x+6}$

41. $\frac{2x^3-x^2+x+5}{x^2+3x+2}$ **42.** $\frac{x^3+2x^2-x+1}{x^2+3x-4}$

43. $\frac{x^4}{(x-1)^3}$ **44.** $\frac{16x^4}{(2x-1)^3}$

In Exercises 45–52, write the partial fraction decomposition of the rational expression. Use a graphing utility to check your result graphically.

45. $\frac{5-x}{2x^2+x-1}$ **46.** $\frac{3x^2-7x-2}{x^3-x}$

47. $\frac{x-1}{x^3+x^2}$ **48.** $\frac{4x^2-1}{2x(x+1)^2}$

49. $\frac{x^2+x+2}{(x^2+2)^2}$

50. $\frac{x^3}{(x+2)^2(x-2)^2}$

51. $\frac{2x^3-4x^2-15x+5}{x^2-2x-8}$

52. $\frac{x^3-x+3}{x^2+x-2}$

Graphical Analysis **In Exercises 53–56, write the partial fraction decomposition of the rational function. Identify the graph of the rational function and the graph of each term of its decomposition. State any relationship between the vertical asymptotes of the rational function and the vertical asymptotes of the terms of the decomposition. To print an enlarged copy of the graph, go to the website *www.mathgraphs.com.***

53. $y = \dfrac{x - 12}{x(x - 4)}$

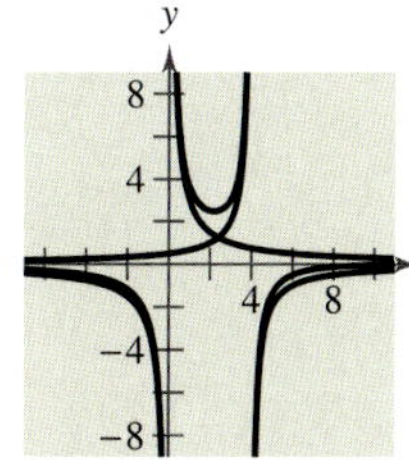

54. $y = \dfrac{2(x + 1)^2}{x(x^2 + 1)}$

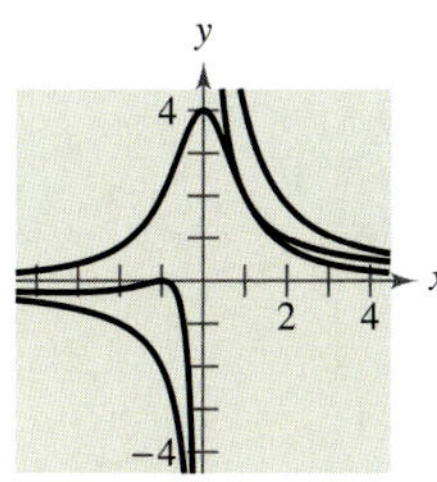

55. $y = \dfrac{2(4x - 3)}{x^2 - 9}$

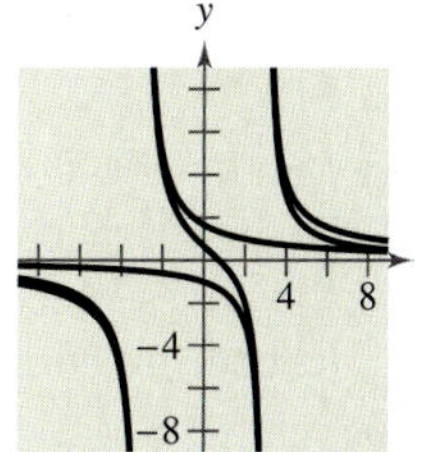

56. $y = \dfrac{2(4x^2 - 15x + 39)}{x^2(x^2 - 10x + 26)}$

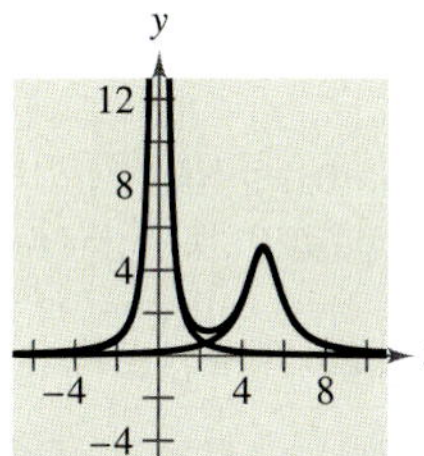

▶ Model It

57. ***Thermodynamics*** The magnitude of the range R of exhaust temperatures (in degrees Fahrenheit) in an experimental diesel engine is approximated by

$$R = \frac{2000(4 - 3x)}{(11 - 7x)(7 - 4x)}, \qquad 0 < x \le 1$$

where x is the relative load (in foot-pounds).

(a) Write the partial fraction decomposition of the equation.

(b) The decomposition in part (a) is the difference of two fractions. The absolute values of the terms give the expected maximum and minimum temperatures of the exhaust gases for different loads.

Ymax = |1st term| Ymin = |2nd term|

Write the equations for Ymax and Ymin.

▶ Model It *(continued)*

(c) Use a graphing utility to graph each equation from part (b) in the same viewing window.

(d) Determine the expected maximum and minimum temperatures for a relative load of 0.5.

Synthesis

58. ***Writing*** Describe two ways of solving for the constants in a partial fraction decomposition.

True or False? **In Exercises 59 and 60, determine whether the statement is true or false. Justify your answer.**

59. For the rational expression

$$\frac{x}{(x + 10)(x - 10)^2}$$

the partial fraction decomposition is of the form

$$\frac{A}{x + 10} + \frac{B}{(x - 10)^2}.$$

60. When writing the partial fraction decomposition of the expression

$$\frac{x^3 + x - 2}{x^2 - 5x + 14}$$

the first step is to factor the denominator.

In Exercises 61–64, write the partial fraction decomposition of the rational expression. Check your result algebraically. Then assign a value to the constant *a* to check the result graphically.

61. $\dfrac{1}{a^2 - x^2}$

62. $\dfrac{1}{x(x + a)}$

63. $\dfrac{1}{y(a - y)}$

64. $\dfrac{1}{(x + 1)(a - x)}$

Review

In Exercises 65–68, sketch the graph of the function.

65. $f(x) = x^2 - 9x + 18$

66. $f(x) = 2x^2 - 9x - 5$

67. $f(x) = -x^2(x - 3)$

68. $f(x) = \frac{1}{2}x^3 - 1$

In Exercises 69 and 70, sketch the graph of the rational function.

69. $f(x) = \dfrac{x^2 + x - 6}{x + 5}$

70. $f(x) = \dfrac{3x - 1}{x^2 + 4x - 12}$

4.4 Conics

▶ What you should learn

- How to recognize the four basic conics: circles, ellipses, parabolas, and hyperbolas
- How to recognize, graph, and write equations of parabolas (vertex at origin)
- How to recognize, graph, and write equations of ellipses (center at origin)
- How to recognize, graph, and write equations of hyperbolas (center at origin)

▶ Why you should learn it

Conics have been used for hundreds of years to model and solve engineering problems. For instance, in Exercise 33 on page 368, a parabola can be used to model the cables of the Golden Gate Bridge.

Cosmo Condina/Getty Images

Introduction 

Conic sections were discovered during the classical Greek period, 600 to 300 B.C. This early Greek study was largely concerned with the geometric properties of conics. It was not until the early 17th century that the broad applicability of conics became apparent and played a prominent role in the early development of calculus.

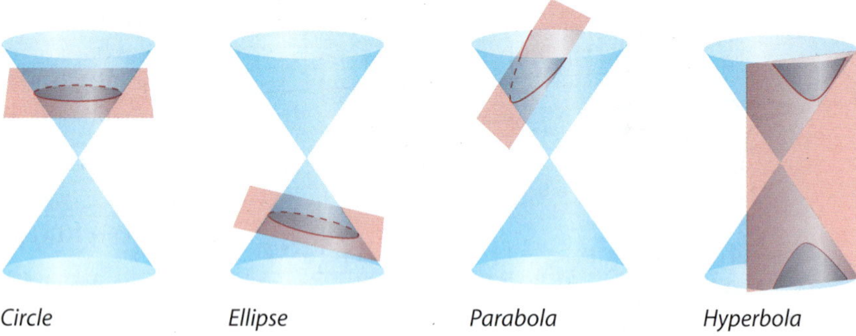

FIGURE 4.16 *Basic Conics*

A **conic section** (or simply **conic**) is the intersection of a plane and a double-napped cone. Notice in Figure 4.16 that in the formation of the four basic conics, the intersecting plane does not pass through the vertex of the cone. When the plane does pass through the vertex, the resulting figure is a **degenerate conic,** as shown in Figure 4.17.

FIGURE 4.17 *Degenerate Conics*

There are several ways to approach the study of conics. You could begin by defining conics in terms of the intersections of planes and cones, as the Greeks did, or you could define them algebraically, in terms of the general second-degree equation

$$Ax^2 + Bxy + Cy^2 + Dx + Ey + F = 0.$$

However, you will study a third approach, in which each of the conics is defined as a *locus* (collection) of points satisfying a certain geometric property. For example, in Section 1.1 you saw how the definition of a circle as *the collection of all points* (x, y) *that are equidistant from a fixed point* (h, k) led easily to the standard equation of a circle

$$(x - h)^2 + (y - k)^2 = r^2. \qquad \text{Equation of a circle}$$

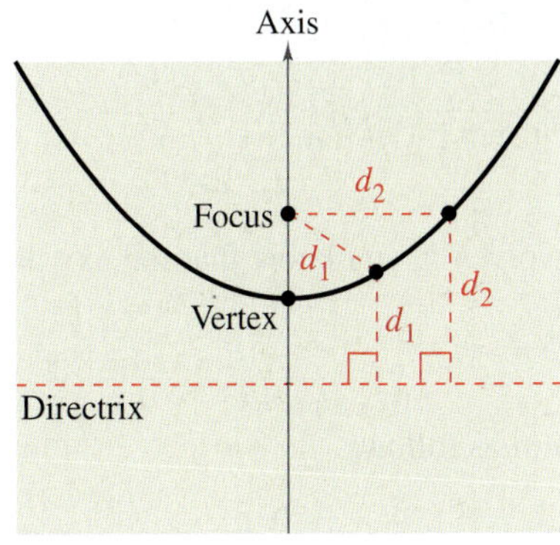

FIGURE 4.18 *Parabola*

Parabolas 

In Section 3.1, you learned that the graph of the quadratic function

$$f(x) = ax^2 + bx + c$$

is a parabola that opens upward or downward. The following definition of a parabola is more general in the sense that it is independent of the orientation of the parabola.

Definition of a Parabola

A **parabola** is the set of all points (x, y) in a plane that are equidistant from a fixed line, the **directrix,** and a fixed point, the **focus,** not on the line. (See Figure 4.18.) The **vertex** is the midpoint between the focus and the directrix. The **axis** of the parabola is the line passing through the focus and the vertex.

Standard Equation of a Parabola (Vertex at Origin)

The **standard form of the equation of a parabola** with vertex at $(0, 0)$ and directrix $y = -p$ is

$$x^2 = 4py, \qquad p \neq 0. \qquad \text{Vertical axis}$$

For directrix $x = -p$, the equation is

$$y^2 = 4px, \qquad p \neq 0. \qquad \text{Horizontal axis}$$

The focus is on the axis p units (directed distance) from the vertex.

For a proof of the standard form of the equation of a parabola, see Proofs in Mathematics on page 385.

Notice that a parabola can have a vertical or a horizontal axis. Examples of each are shown in Figure 4.19.

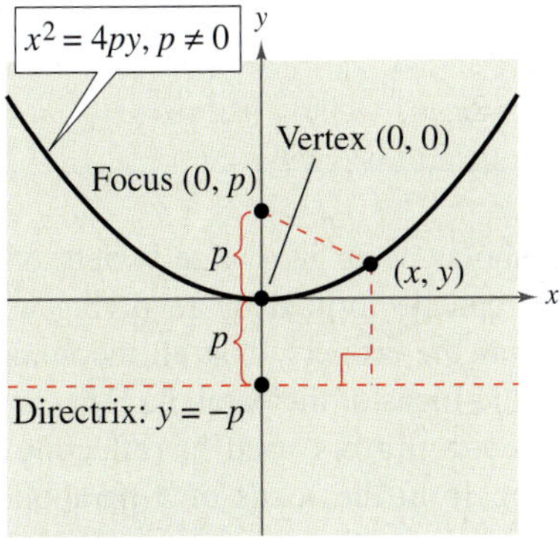

(a) Parabola with vertical axis

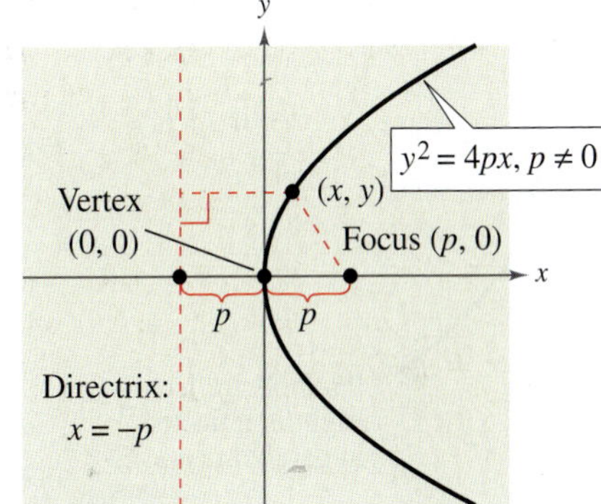

(b) Parabola with horizontal axis

FIGURE 4.19

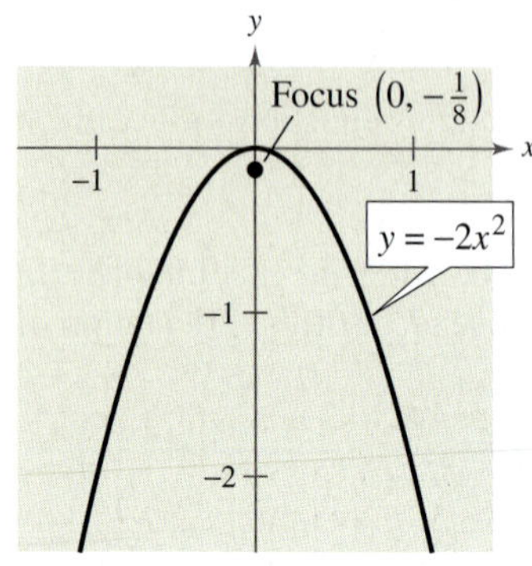

FIGURE 4.20

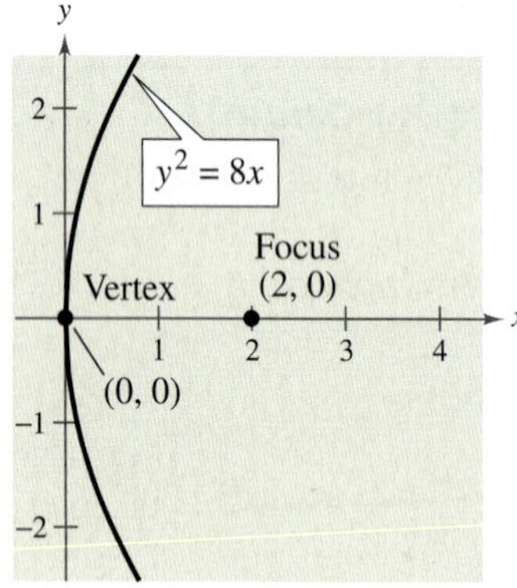

FIGURE 4.21

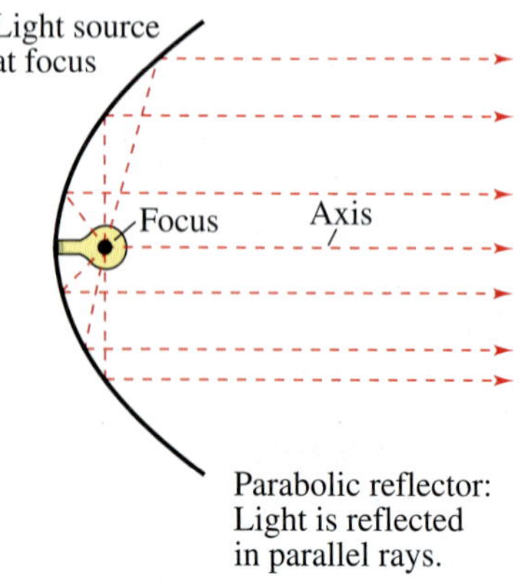

FIGURE 4.22

Example 1 ▶ Finding the Focus of a Parabola

Find the focus of the parabola whose equation is $y = -2x^2$.

Solution

Because the squared term in the equation involves x, you know that the axis is vertical, and the equation is of the form

$$x^2 = 4py.$$

You can write the original equation in this form as follows.

$$x^2 = -\frac{1}{2}y$$

$$x^2 = 4\left(-\frac{1}{8}\right)y \qquad \text{Write in standard form.}$$

So, $p = -\frac{1}{8}$. Because p is negative, the parabola opens downward (see Figure 4.20), and the focus of the parabola is

$$(0, p) = \left(0, -\frac{1}{8}\right). \qquad \text{Focus}$$

Example 2 ▶ A Parabola with a Horizontal Axis 

Write the standard form of the equation of the parabola with vertex at the origin and focus at (2, 0).

Solution

The axis of the parabola is horizontal, passing through (0, 0) and (2, 0), as shown in Figure 4.21. So, the standard form is

$$y^2 = 4px.$$

Because the focus is $p = 2$ units from the vertex, the equation is

$$y^2 = 4(2)x$$

$$y^2 = 8x.$$

Parabolas occur in a wide variety of applications. For instance, a parabolic reflector can be formed by revolving a parabola about its axis. The resulting surface has the property that all incoming rays parallel to the axis are reflected through the focus of the parabola. This is the principle behind the construction of the parabolic mirrors used in reflecting telescopes. Conversely, the light rays emanating from the focus of a parabolic reflector used in a flashlight are all parallel to one another, as shown in Figure 4.22.

Ellipses 

Definition of an Ellipse

An **ellipse** is the set of all points (x, y) in a plane the sum of whose distances from two distinct fixed points **(foci)** is constant. See Figure 4.23.

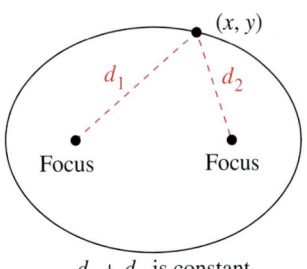

$d_1 + d_2$ is constant.

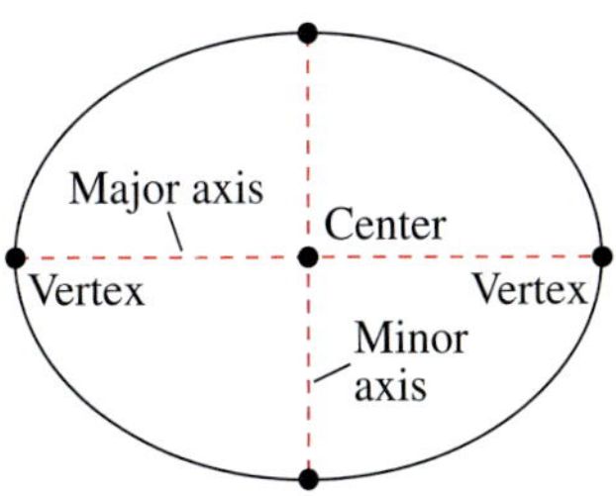

FIGURE 4.23

The line through the foci intersects the ellipse at two points **(vertices).** The chord joining the vertices is the **major axis,** and its midpoint is the **center** of the ellipse. The chord perpendicular to the major axis at the center is the **minor axis.**

You can visualize the definition of an ellipse by imagining two thumbtacks placed at the foci, as shown in Figure 4.24. If the ends of a fixed length of string are fastened to the thumbtacks and the string is *drawn taut* with a pencil, the path traced by the pencil will be an ellipse.

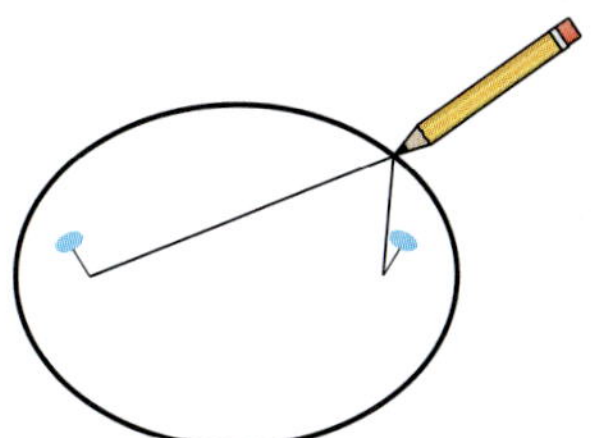

FIGURE 4.24

The standard form of the equation of an ellipse takes one of two forms, depending on whether the major axis is horizontal or vertical.

Standard Equation of an Ellipse (Center at Origin)

The **standard form of the equation of an ellipse** centered at the origin with major and minor axes of lengths $2a$ and $2b$ (where $0 < b < a$) is

$$\frac{x^2}{a^2} + \frac{y^2}{b^2} = 1 \qquad \text{or} \qquad \frac{x^2}{b^2} + \frac{y^2}{a^2} = 1.$$

The vertices and foci lie on the major axis, a and c units, respectively, from the center, as shown in Figure 4.25. Moreover, a, b, and c are related by the equation $c^2 = a^2 - b^2$.

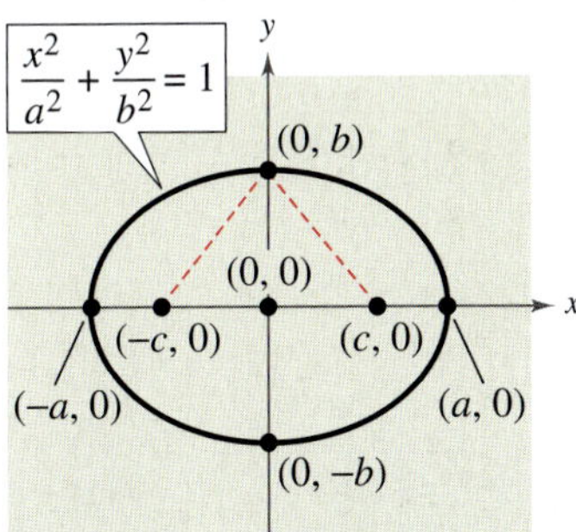

(a) **Major axis is horizontal; minor axis is vertical.**

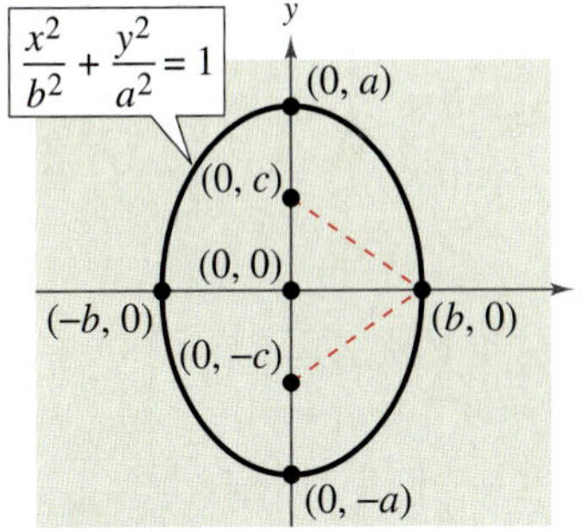

(b) **Major axis is vertical; minor axis is horizontal.**

FIGURE 4.25

In Figure 4.25(a), note that because the sum of the distances from a point on the ellipse to the two foci is constant,

$$\text{Sum of distances from } (0, b) \text{ to foci} = \text{sum of distances from } (a, 0) \text{ to foci}$$

$$2\sqrt{b^2 + c^2} = (a + c) + (a - c)$$

$$\sqrt{b^2 + c^2} = a$$

$$c^2 = a^2 - b^2.$$

Exploration

An ellipse can be drawn using two thumbtacks placed at the foci of the ellipse, a string of fixed length (greater than the distance between the tacks), and a pencil, as shown in Figure 4.24. Try doing this. Vary the length of the string and the distance between the thumbtacks. Explain how to obtain ellipses that are almost circular. Explain how to obtain ellipses that are long and narrow.

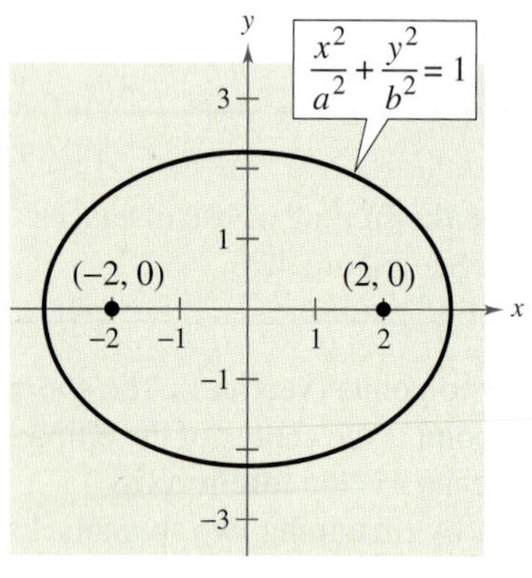

FIGURE 4.26

Example 3 ▶ Finding the Standard Equation of an Ellipse

Find the standard form of the equation of the ellipse that has a major axis of length 6 and foci at $(-2, 0)$ and $(2, 0)$, as shown in Figure 4.26.

Solution

Because the foci occur at $(-2, 0)$ and $(2, 0)$, the center of the ellipse is $(0, 0)$ and the major axis is horizontal. So, the ellipse has an equation of the form

$$\frac{x^2}{a^2} + \frac{y^2}{b^2} = 1. \quad \text{Standard form}$$

Because the length of the major axis is 6, $2a = 6$. This implies that $a = 3$. Moreover, the distance from the center to either focus is $c = 2$. Finally

$$b^2 = a^2 - c^2 = 3^2 - 2^2 = 9 - 4 = 5.$$

Substituting $a^2 = 3^2$ and $b^2 = \left(\sqrt{5}\right)^2$ yields the following equation in standard form.

$$\frac{x^2}{3^2} + \frac{y^2}{\left(\sqrt{5}\right)^2} = 1$$

This equation simplifies to

$$\frac{x^2}{9} + \frac{y^2}{5} = 1.$$

> **Technology**
>
> Conics can be graphed using a graphing calculator by first solving for y For example,
>
> $$\frac{(x-2)^2}{9} + \frac{(y-1)^2}{1} = 1$$
>
> can be graphed by graphing both
>
> $$y_1 = 1 - \sqrt{1 - \frac{(x-2)^2}{9}}$$
>
> and
>
> $$y_2 = 1 + \sqrt{1 - \frac{(x-2)^2}{9}}.$$

Example 4 ▶ Sketching an Ellipse

Sketch the ellipse given by $4x^2 + y^2 = 36$, and identify the vertices.

Solution

$$4x^2 + y^2 = 36 \quad \text{Write original equation.}$$

$$\frac{4x^2}{36} + \frac{y^2}{36} = \frac{36}{36} \quad \text{Divide each side by 36.}$$

$$\frac{x^2}{3^2} + \frac{y^2}{6^2} = 1 \quad \text{Write in standard form.}$$

$$\frac{x^2}{9} + \frac{y^2}{36} = 1 \quad \text{Simplify.}$$

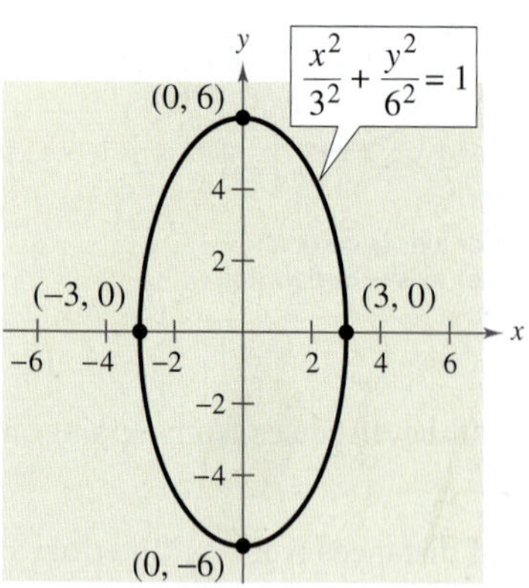

FIGURE 4.27

Because the denominator of the y^2-term is larger than the denominator of the x^2-term, you can conclude that the major axis is vertical. Moreover, because $a = 6$, the vertices are $(0, -6)$ and $(0, 6)$. Finally, because $b = 3$, the endpoints of the minor axis (or co-vertices) are $(-3, 0)$ and $(3, 0)$, as shown in Figure 4.27. Note that you can sketch the ellipse by locating the endpoints of the two axes. Because 3^2 is the denominator of the x^2-term, move three units to the *right and left* of the center to locate the endpoints of the horizontal axis. Similarly, because 6^2 is the denominator of the y^2-term, move six units *up and down* from the center to locate the endpoints of the vertical axis.

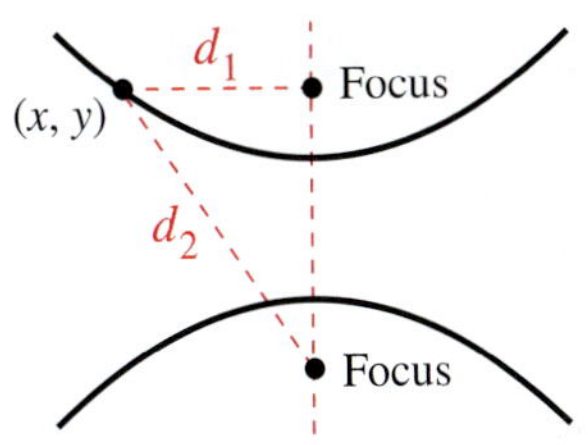

$d_2 - d_1$ is a positive constant.

(a)

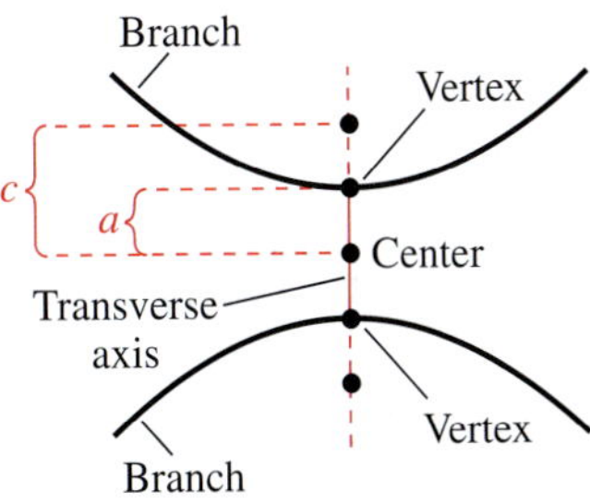

(b)

FIGURE 4.28

Hyperbolas

The definition of a **hyperbola** is similar to that of an ellipse. The difference is that for an ellipse the *sum* of the distances between the foci and a point on the ellipse is constant, whereas for a hyperbola the *difference* of the distances between the foci and a point on the hyperbola is constant.

Definition of a Hyperbola

A **hyperbola** is the set of all points (x, y) in a plane the difference of whose distances from two distinct fixed points **(foci)** is a positive constant. See Figure 4.28(a).

The graph of a hyperbola has two disconnected parts **(branches).** The line through the two foci intersects the hyperbola at two points **(vertices).** The line segment connecting the vertices is the **transverse axis,** and the midpoint of the transverse axis is the **center** of the hyperbola. See Figure 4.28(b).

Standard Equation of a Hyperbola (Center at Origin)

The **standard form of the equation of a hyperbola** with center at the origin (where $a \neq 0$ and $b \neq 0$) is

$$\frac{x^2}{a^2} - \frac{y^2}{b^2} = 1 \qquad \text{Transverse axis is horizontal.}$$

or

$$\frac{y^2}{a^2} - \frac{x^2}{b^2} = 1. \qquad \text{Transverse axis is vertical.}$$

The vertices and foci are, respectively, a and c units from the center. Moreover, a, b, and c are related by the equation $b^2 = c^2 - a^2$. See Figure 4.29.

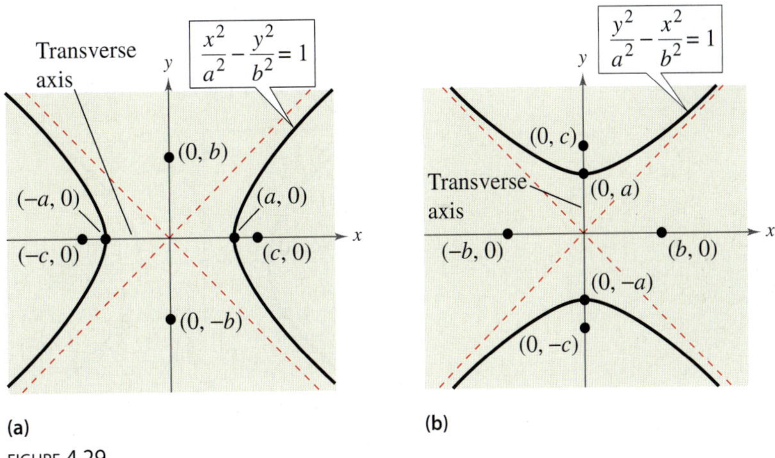

(a) (b)

FIGURE 4.29

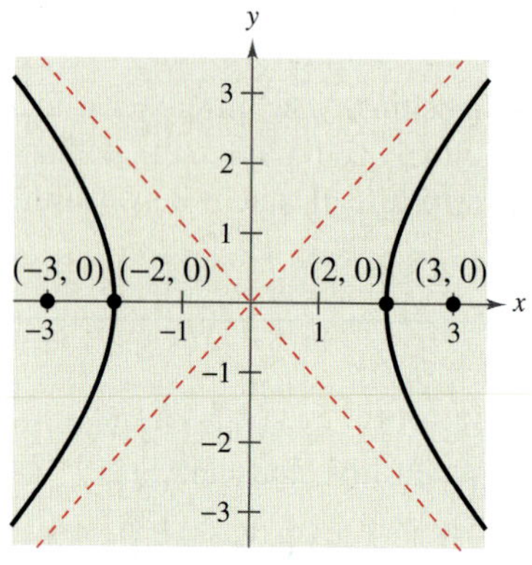

FIGURE 4.30

Example 5 ▶ Finding the Standard Equation of a Hyperbola

Find the standard form of the equation of the hyperbola with foci at $(-3, 0)$ and $(3, 0)$ and vertices at $(-2, 0)$ and $(2, 0)$, as shown in Figure 4.30.

Solution

From the graph, you can determine that $c = 3$, because the foci are three units from the center. Moreover, $a = 2$ because the vertices are two units from the center. So, it follows that

$$\begin{aligned} b^2 &= c^2 - a^2 \\ &= 3^2 - 2^2 \\ &= 9 - 4 \\ &= 5. \end{aligned}$$

Because the transverse axis is horizontal, the standard form of the equation is

$$\frac{x^2}{a^2} - \frac{y^2}{b^2} = 1.$$

Finally, substitute $a^2 = 2^2$ and $b^2 = \left(\sqrt{5}\right)^2$ to obtain

$$\frac{x^2}{2^2} - \frac{y^2}{\left(\sqrt{5}\right)^2} = 1 \qquad \text{Write in standard form.}$$

$$\frac{x^2}{4} - \frac{y^2}{5} = 1. \qquad \text{Simplify.}$$

An important aid in sketching the graph of a hyperbola is the determination of its *asymptotes*, as shown in Figure 4.31. Each hyperbola has two asymptotes that intersect at the center of the hyperbola. Furthermore, the asymptotes pass through the corners of a rectangle of dimensions $2a$ by $2b$. The line segment of length $2b$ joining $(0, b)$ and $(0, -b)$ [or $(-b, 0)$ and $(b, 0)$] is the **conjugate axis** of the hyperbola.

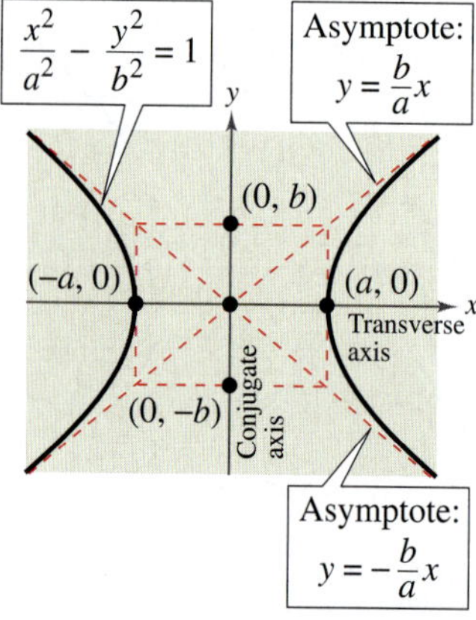

(a) Transverse axis is horizontal; conjugate axis is vertical.

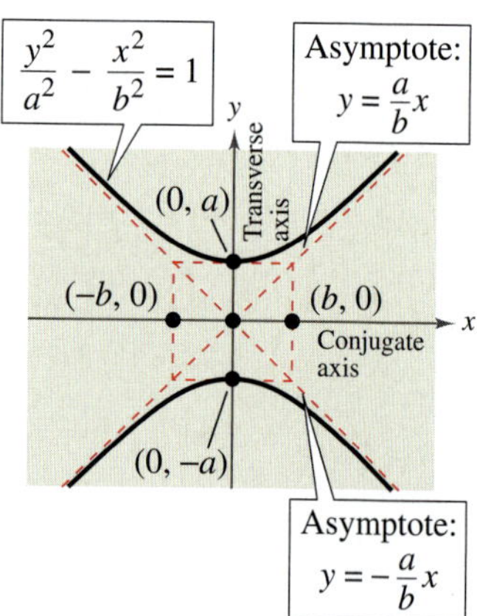

(b) Transverse axis is vertical; conjugate axis is horizontal.

FIGURE 4.31

Asymptotes of a Hyperbola (Center at Origin)

The **asymptotes of a hyperbola** with center at $(0, 0)$ are

$$y = \frac{b}{a}x \quad \text{and} \quad y = -\frac{b}{a}x \qquad \text{Transverse axis is horizontal.}$$

or

$$y = \frac{a}{b}x \quad \text{and} \quad y = -\frac{a}{b}x. \qquad \text{Transverse axis is vertical.}$$

Exploration

Use a graphing utility to graph the hyperbola in Example 6. Does your graph look like the one shown in Figure 4.33? If not, what must you do to obtain both the upper and lower portions of the hyperbola? Explain your reasoning.

Example 6 ▶ Sketching a Hyperbola 

Sketch the graph of the hyperbola whose equation is

$$4x^2 - y^2 = 16.$$

Solution

$$4x^2 - y^2 = 16 \qquad \text{Write original equation.}$$

$$\frac{4x^2}{16} - \frac{y^2}{16} = \frac{16}{16} \qquad \text{Divide each side by 16.}$$

$$\frac{x^2}{2^2} - \frac{y^2}{4^2} = 1 \qquad \text{Write in standard form.}$$

$$\frac{x^2}{4} - \frac{y^2}{16} = 1 \qquad \text{Simplify.}$$

Because the x^2-term is positive, you can conclude that the transverse axis is horizontal and the vertices occur at $(-2, 0)$ and $(2, 0)$. Moreover, the endpoints of the conjugate axis occur at $(0, -4)$ and $(0, 4)$, and you can sketch the rectangle shown in Figure 4.32. Finally, by drawing the asymptotes through the corners of this rectangle, you can complete the sketch shown in Figure 4.33.

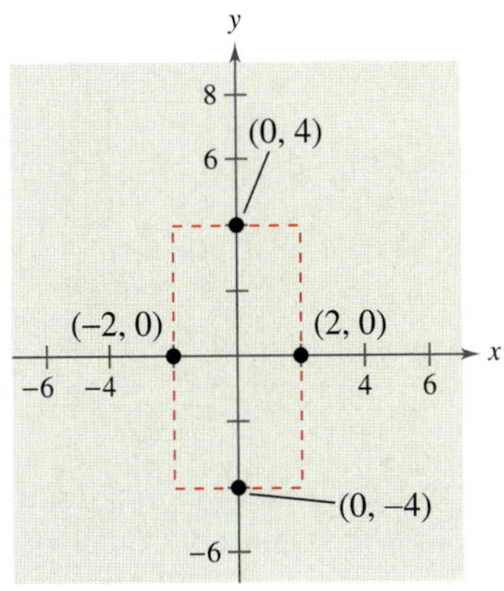

FIGURE 4.32

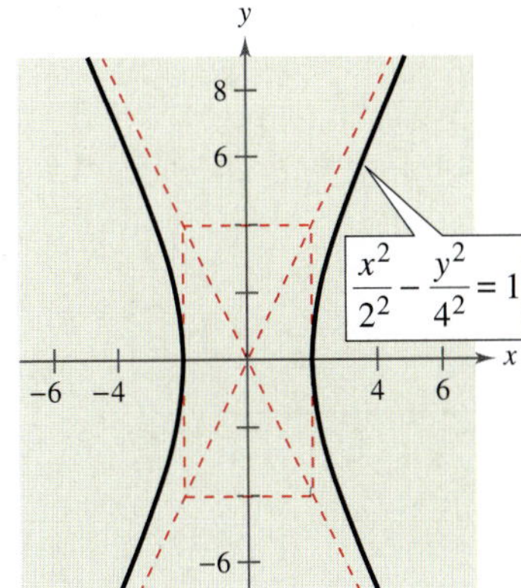

FIGURE 4.33

Example 7 ▶ Finding the Standard Equation of a Hyperbola

Find the standard form of the equation of the hyperbola that has vertices at $(0, -3)$ and $(0, 3)$ and asymptotes $y = -2x$ and $y = 2x$, as shown in Figure 4.34.

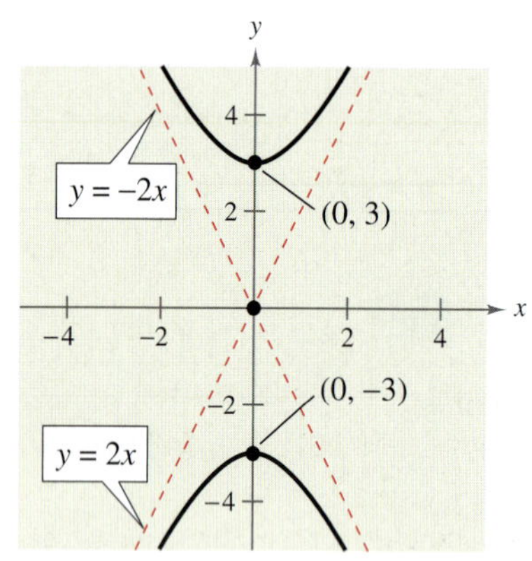

FIGURE 4.34

Solution

Because the transverse axis is vertical, the asymptotes are of the forms

$$y = \frac{a}{b}x \quad \text{and} \quad y = -\frac{a}{b}x.$$

Using the fact that $y = 2x$ and $y = -2x$, you can determine that

$$\frac{a}{b} = 2.$$

Because $a = 3$, you can determine that $b = \frac{3}{2}$. Finally, you can conclude that the hyperbola has the following equation.

$$\frac{y^2}{3^2} - \frac{x^2}{\left(\frac{3}{2}\right)^2} = 1 \qquad \text{Write in standard form.}$$

$$\frac{y^2}{9} - \frac{x^2}{\frac{9}{4}} = 1 \qquad \text{Simplify.}$$

Writing ABOUT MATHEMATICS

Hyperbolas in Application At the beginning of this section, you learned that each type of conic section can be formed by the intersection of a plane and a double-napped cone. The figure below shows three examples of how such intersections can occur in physical situations.

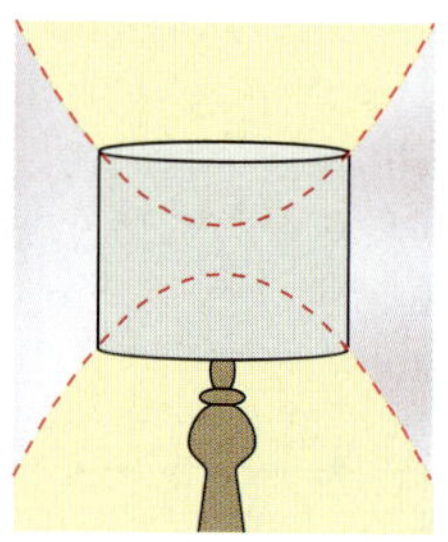

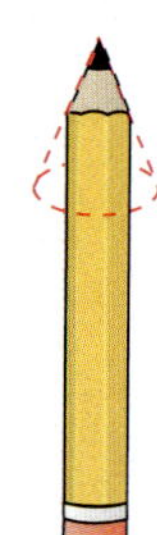

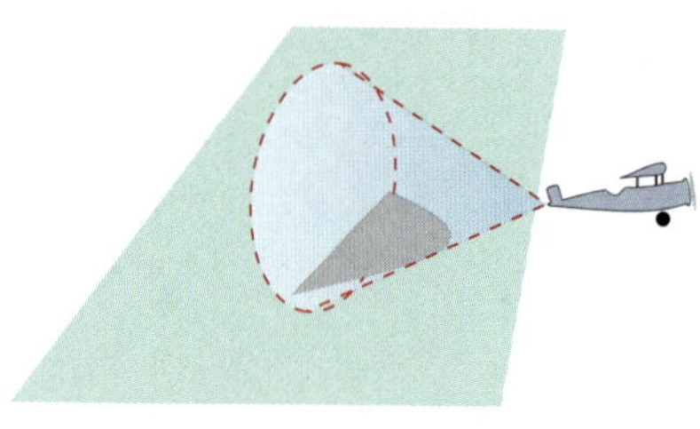

Identify the cone and hyperbola (or portion of a hyperbola) in each of the three situations. Write a short paragraph describing other examples of physical situations in which hyperbolas are formed.

4.4 Exercises

In Exercises 1–10, match the equation with its graph. If the graph of an equation is not shown, write "not shown." [The graphs are labeled (a), (b), (c), (d), (e), (f), (g), and (h).]

(a)

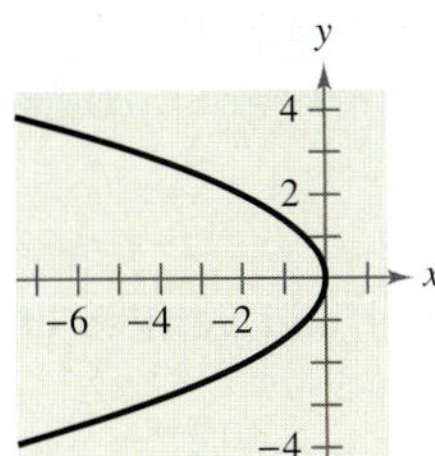

(b)

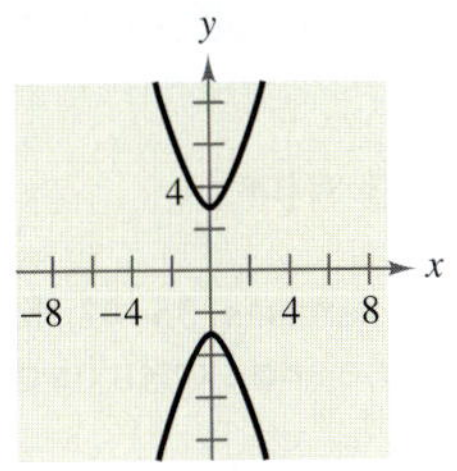

(c)

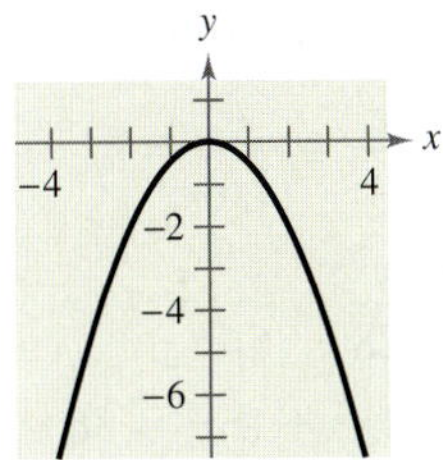

(d)

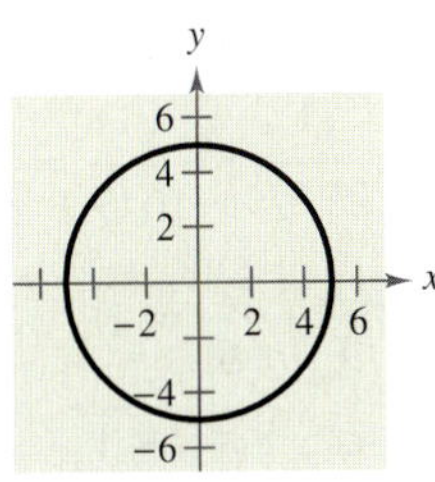

(e)

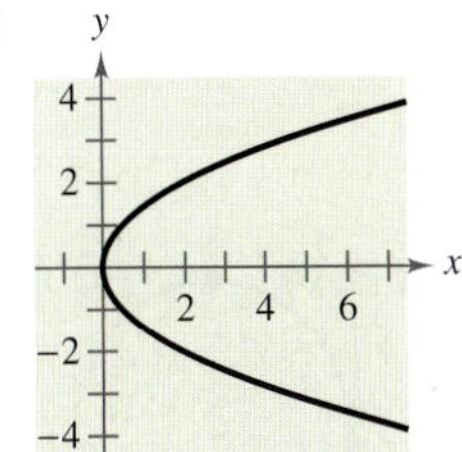

(f)

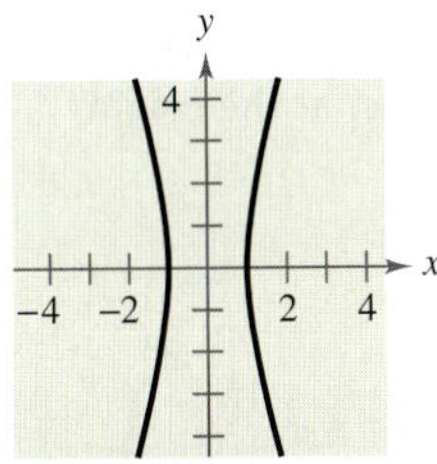

(g)

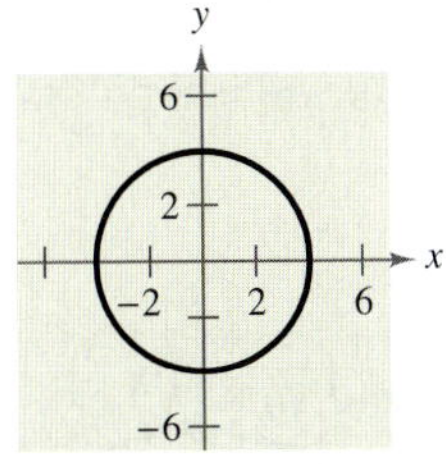

(h)

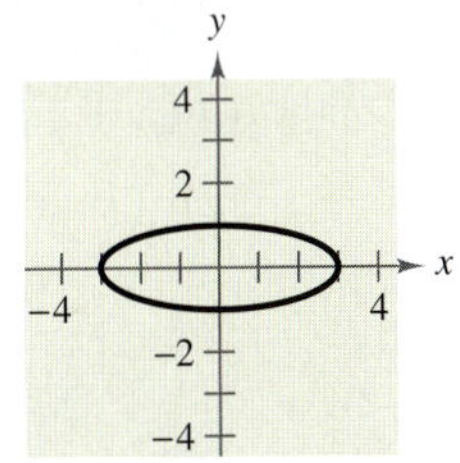

1. $x^2 = 2y$

2. $x^2 = -2y$

3. $y^2 = 2x$

4. $y^2 = -2x$

5. $9x^2 + y^2 = 9$

6. $x^2 + 9y^2 = 9$

7. $9x^2 - y^2 = 9$

8. $y^2 - 9x^2 = 9$

9. $x^2 + y^2 = 49$

10. $x^2 + y^2 = 16$

In Exercises 11–16, find the vertex and focus of the parabola and sketch its graph.

11. $y = \frac{1}{2}x^2$

12. $y = 2x^2$

13. $y^2 = -6x$

14. $y^2 = 3x$

15. $x^2 + 8y = 0$

16. $x + y^2 = 0$

In Exercises 17–26, find an equation of the parabola with vertex at the origin.

17. Focus: $(2, 0)$

18. Focus: $(-2, 0)$

19. Focus: $\left(0, -\frac{3}{2}\right)$

20. Focus: $(0, -2)$

21. Directrix: $y = -1$

22. Directrix: $y = 2$

23. Directrix: $x = 3$

24. Directrix: $x = -2$

25. Passes through the point $(4, 6)$; horizontal axis

26. Passes through the point $(-2, -2)$; vertical axis

In Exercises 27–30, find the standard form of the equation of the parabola and determine the coordinates of the focus.

27.

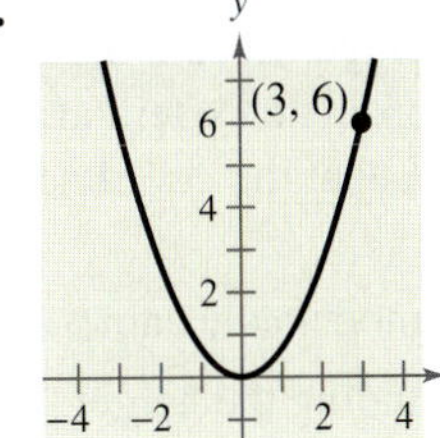

28.

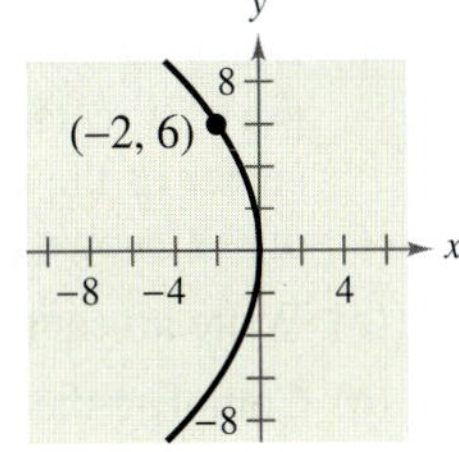

29.

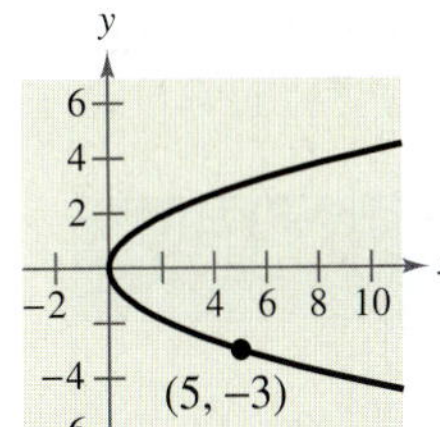

30.

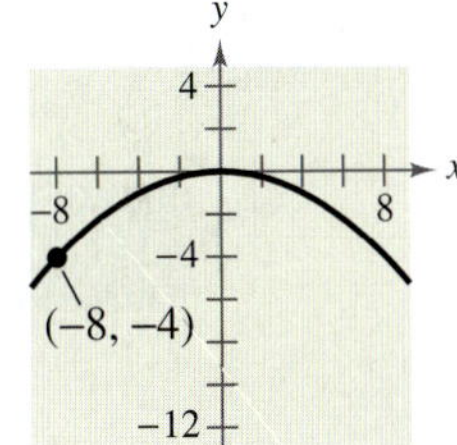

31. ***Flashlight*** The light bulb in a flashlight is at the focus of the parabolic reflector, 1.5 centimeters from the vertex of the reflector (see figure). Write an equation for a cross section of the flashlight's reflector with its focus on the positive x-axis and its vertex at the origin.

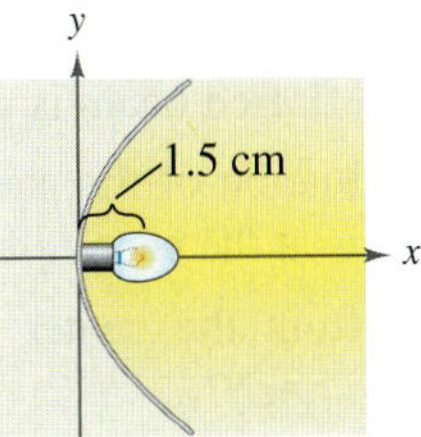

32. ***Satellite Antenna*** Write an equation for a cross section of the parabolic television dish antenna shown in the figure.

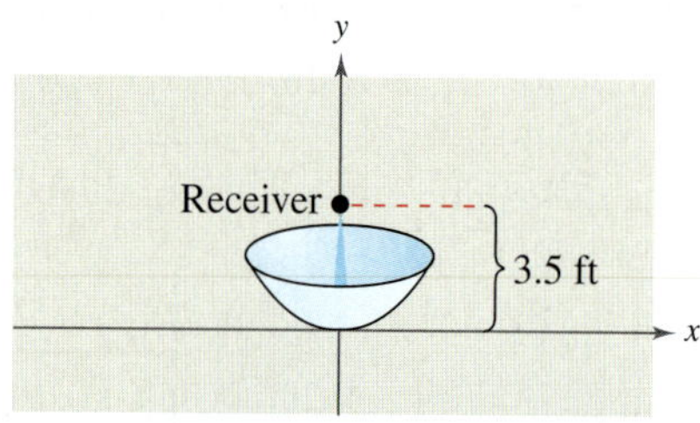

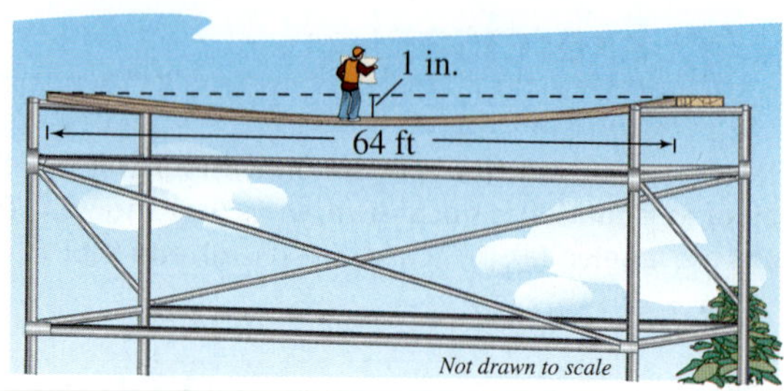

FIGURE FOR 34

▶ Model It

33. ***Suspension Bridge*** Each cable of the Golden Gate Bridge is suspended (in the shape of a parabola) between two towers that are 1280 meters apart. The top of each tower is 152 meters above the roadway. The cables touch the roadway at the midpoint between the towers.

(a) Draw a sketch of the bridge. Locate the origin of a rectangular coordinate system at the center of the roadway. Label the coordinates of the known points.

(b) Write an equation that models the cables.

(c) Complete the table by finding the height y of the suspension cables over the roadway at a distance of x meters from the center of the bridge.

Distance, x	Height, y
0	
200	
400	
500	
600	

34. ***Beam Deflection*** A simply supported beam (see figure) is 64 feet long and has a load at the center. The deflection of the beam at its center is 1 inch. The shape of the deflected beam is parabolic.

(a) Find an equation of the parabola. (Assume that the origin is at the center of the beam.)

(b) How far from the center of the beam is the deflection $\frac{1}{2}$ inch?

In Exercises 35–42, find the center and vertices of the ellipse and sketch its graph.

35. $\dfrac{x^2}{25} + \dfrac{y^2}{16} = 1$

36. $\dfrac{x^2}{144} + \dfrac{y^2}{169} = 1$

37. $\dfrac{x^2}{25/9} + \dfrac{y^2}{16/9} = 1$

38. $\dfrac{x^2}{4} + \dfrac{y^2}{1/4} = 1$

39. $\dfrac{x^2}{9} + \dfrac{y^2}{5} = 1$

40. $\dfrac{x^2}{28} + \dfrac{y^2}{64} = 1$

41. $4x^2 + y^2 = 1$

42. $4x^2 + 9y^2 = 36$

In Exercises 43–52, find an equation of the ellipse with center at the origin.

43.

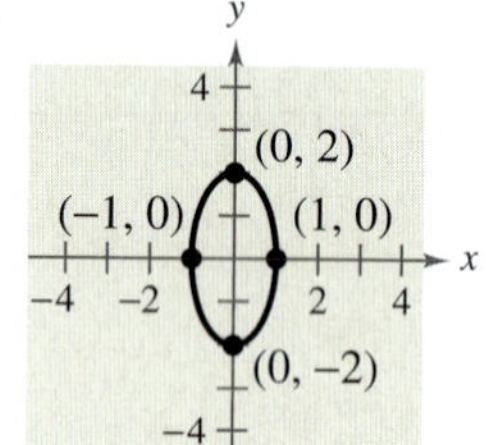

44.

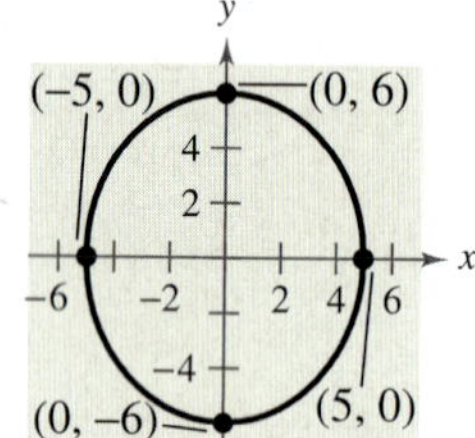

45.

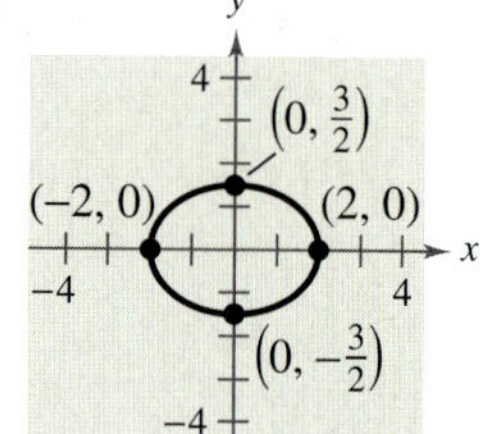

46.

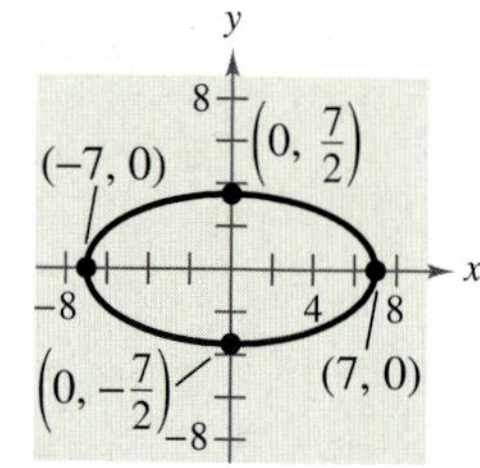

47. Vertices: $(\pm 5, 0)$; Foci: $(\pm 2, 0)$

48. Vertices: $(0, \pm 8)$; Foci: $(0, \pm 4)$

49. Foci: $(\pm 5, 0)$; Major axis of length 12

50. Foci: $(\pm 2, 0)$; Major axis of length 8

51. Vertices: $(0, \pm 5)$; Passes through the point $(4, 2)$

52. Major axis vertical; Passes through the points $(0, 4)$ and $(2, 0)$

53. ***Architecture*** A fireplace arch is to be constructed in the shape of a semiellipse. The opening is to have a height of 2 feet at the center and a width of 6 feet along the base (see figure). The contractor draws the outline of the ellipse on the wall by the method shown in Figure 4.24. Give the required positions of the tacks and the length of the string.

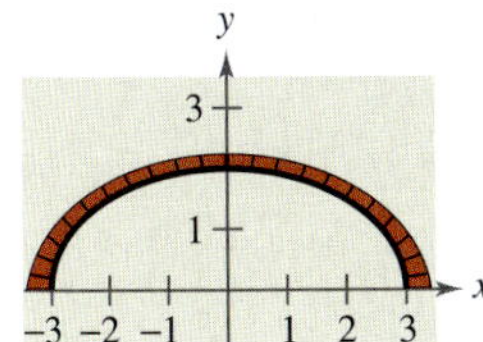

54. ***Architecture*** A semielliptical arch over a tunnel for a road through a mountain has a major axis of 100 feet and a height at the center of 30 feet.

(a) Draw a rectangular coordinate system on a sketch of the tunnel with the center of the road entering the tunnel at the origin. Identify the coordinates of the known points.

(b) Find an equation of the semielliptical arch over the tunnel.

(c) Determine the height of the arch 5 feet from each edge of the tunnel.

55. ***Architecture*** Repeat Exercise 54 for a semielliptical arch with a major axis of 110 feet and a height at the center of 40 feet.

56. ***Geometry*** A line segment through a focus of an ellipse with endpoints on the ellipse and perpendicular to the major axis is called a **latus rectum** of the ellipse. Therefore, an ellipse has two latera recta. Knowing the length of the latera recta is helpful in sketching an ellipse because it yields other points on the curve (see figure). Show that the length of each latus rectum is $2b^2/a$.

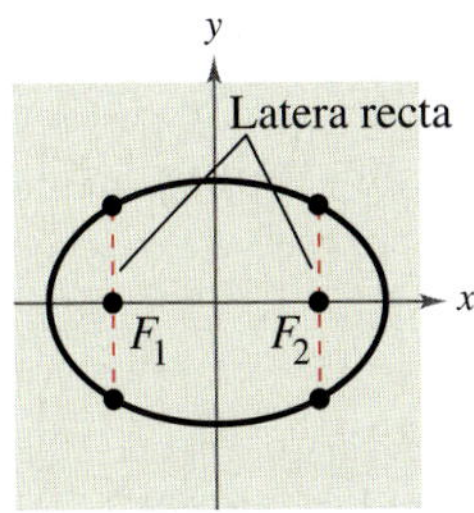

In Exercises 57–60, sketch the graph of the ellipse, using the latera recta (see Exercise 56).

57. $\dfrac{x^2}{4} + \dfrac{y^2}{1} = 1$ **58.** $\dfrac{x^2}{9} + \dfrac{y^2}{16} = 1$

59. $9x^2 + 4y^2 = 36$ **60.** $5x^2 + 3y^2 = 15$

In Exercises 61–68, find the center and vertices of the hyperbola and sketch its graph, using asymptotes as sketching aids.

61. $x^2 - y^2 = 1$ **62.** $\dfrac{x^2}{9} - \dfrac{y^2}{16} = 1$

63. $\dfrac{y^2}{1} - \dfrac{x^2}{4} = 1$ **64.** $\dfrac{y^2}{9} - \dfrac{x^2}{1} = 1$

65. $\dfrac{y^2}{25} - \dfrac{x^2}{144} = 1$ **66.** $\dfrac{x^2}{36} - \dfrac{y^2}{4} = 1$

67. $4y^2 - x^2 = 1$ **68.** $4y^2 - 9x^2 = 36$

In Exercises 69–76, find an equation of the hyperbola with center at the origin.

69. Vertices: $(0, \pm 2)$; Foci: $(0, \pm 4)$

70. Vertices: $(\pm 3, 0)$; Foci: $(\pm 5, 0)$

71. Vertices: $(\pm 1, 0)$; Asymptotes: $y = \pm 3x$

72. Vertices: $(0, \pm 3)$; Asymptotes: $y = \pm 3x$

73. Foci: $(0, \pm 8)$; Asymptotes: $y = \pm 4x$

74. Foci: $(\pm 10, 0)$; Asymptotes: $y = \pm \frac{3}{4}x$

75. **76.**

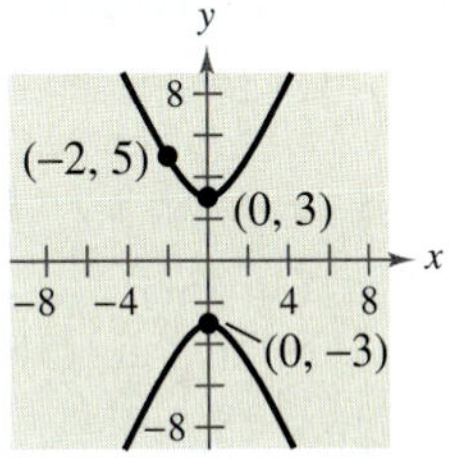

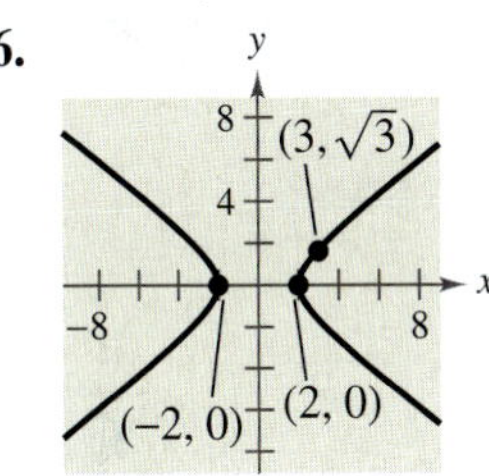

77. ***Optics*** A hyperbolic mirror (used in some telescopes) has the property that a light ray directed at the focus will be reflected to the other focus. The focus of a hyperbolic mirror (see figure) has coordinates (24, 0). Find the vertex of the mirror if its mount at the top edge of the mirror has coordinates (24, 24).

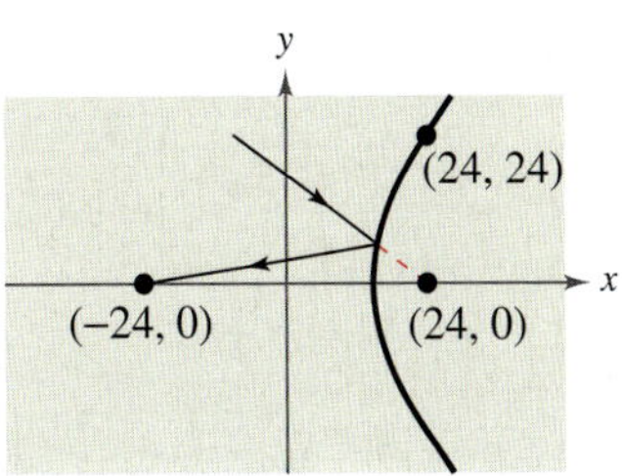

78. ***Navigation*** Long-distance radio navigation for aircraft and ships uses synchronized pulses transmitted by widely separated transmitting stations. These pulses travel at the speed of light (186,000 miles per second). The difference in the times of arrival of these pulses at an aircraft or ship is constant on a hyperbola having the transmitting stations as foci.

Assume that two stations 300 miles apart are positioned on a rectangular coordinate system at points with coordinates $(-150, 0)$ and $(150, 0)$ and that a ship is traveling on a path with coordinates $(x, 75)$, as shown in the figure. Find the x-coordinate of the position of the ship if the time difference between the pulses from the transmitting stations is 1000 microseconds (0.001 second).

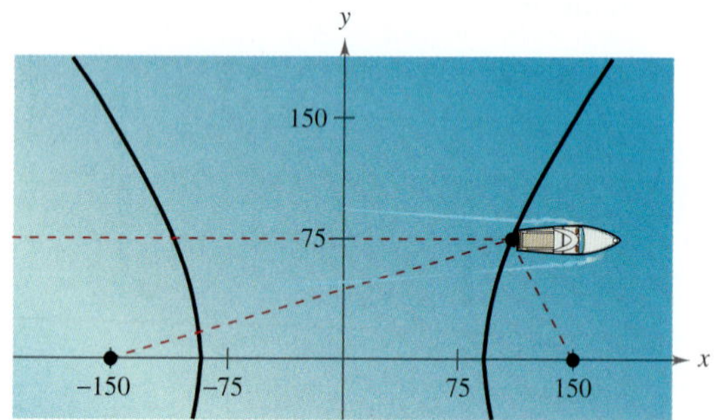

Synthesis

True or False? **In Exercises 79–81, determine whether the statement is true or false. Justify your answer.**

79. The equation $x^2 - y^2 = 144$ represents a circle.

80. The major axis of the ellipse $y^2 + 16x^2 = 64$ is vertical.

81. It is possible for a parabola to intersect its directrix.

82. ***Think About It*** How can you tell if an ellipse is a circle from the equation?

83. ***Exploration*** Consider the ellipse

$$\frac{x^2}{a^2} + \frac{y^2}{b^2} = 1, \qquad a + b = 20.$$

(a) The area of the ellipse is $A = \pi ab$. Write the area of the ellipse as a function of a.

(b) Find the equation of an ellipse with an area of 264 square centimeters.

(c) Complete the table using your equation from part (a), and make a conjecture about the shape of the ellipse with maximum area.

a	8	9	10	11	12	13
A						

(d) Use a graphing utility to graph the area function and use the graph to make a conjecture about the shape of the ellipse that yields a maximum area.

84. ***Think About It*** Is the graph of $x^2 + 4y^4 = 4$ an ellipse? Explain.

85. ***Think About It*** The graph of $x^2 - y^2 = 0$ is a degenerate conic. Sketch this graph and identify the degenerate conic.

86. ***Writing*** Write a paragraph discussing the changes in the shape and orientation of the graph of the ellipse

$$\frac{x^2}{a^2} + \frac{y^2}{4^2} = 1$$

as a increases from 1 to 8.

87. Use the definition of an ellipse to derive the standard form of the equation of an ellipse.

88. Use the definition of a hyperbola to derive the standard form of the equation of a hyperbola.

Review

In Exercises 89–92, sketch the graph of the quadratic function without using a graphing utility. Identify the vertex.

89. $f(x) = x^2 - 8$

90. $f(x) = 25 - x^2$

91. $f(x) = x^2 + 8x + 12$

92. $f(x) = x^2 - 4x - 21$

93. Find a polynomial with integer coefficients that has the zeros 3, $2 + i$, and $2 - i$.

94. Find all the zeros of $f(x) = 2x^3 - 3x^2 + 50x - 75$ if one of the zeros is $x = \frac{3}{2}$.

95. List the possible rational zeros of the function $g(x) = 6x^4 + 7x^3 - 29x^2 - 28x + 20$.

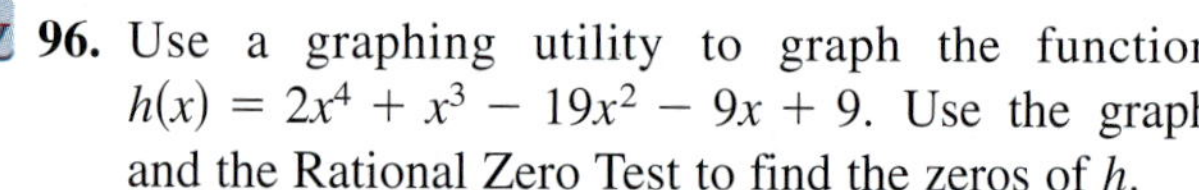

96. Use a graphing utility to graph the function $h(x) = 2x^4 + x^3 - 19x^2 - 9x + 9$. Use the graph and the Rational Zero Test to find the zeros of h.

In Exercises 97 and 98, write the partial fraction decomposition of the rational expression. Check your result algebraically.

97. $\dfrac{x + 8}{x^2 + 3x - 18}$

98. $\dfrac{x - 2}{x^2 - 6x - 7}$

4.5 Translations of Conics

▶ What you should learn

- How to recognize equations of conics that have been shifted vertically or horizontally in the plane
- How to write and graph equations of conics that have been shifted vertically or horizontally in the plane

▶ Why you should learn it

In some real-life applications, it is not convenient to use conics whose centers or vertices are at the origin. For instance, in Exercise 30 on page 376, a parabola can be used to model the maximum revenue of Manpower, Inc.

Vertical and Horizontal Shifts of Conics

In Section 4.4 you looked at conic sections whose graphs were in *standard position*. In this section you will study the equations of conic sections that have been shifted vertically or horizontally in the plane.

Standard Forms of Equations of Conics

Circle: Center $= (h, k)$; Radius $= r$

$$(x - h)^2 + (y - k)^2 = r^2$$

Ellipse: Center $= (h, k)$

Major axis length $= 2a$; Minor axis length $= 2b$

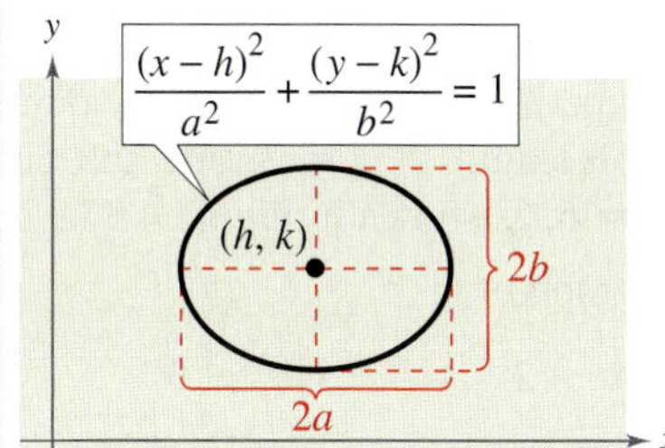

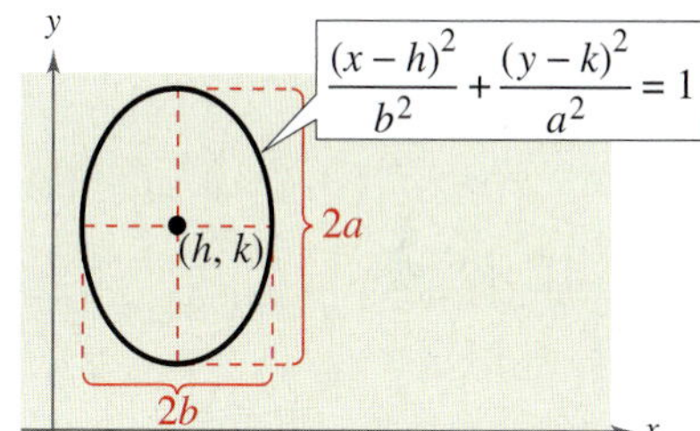

Hyperbola: Center $= (h, k)$

Transverse axis length $= 2a$; Conjugate axis length $= 2b$

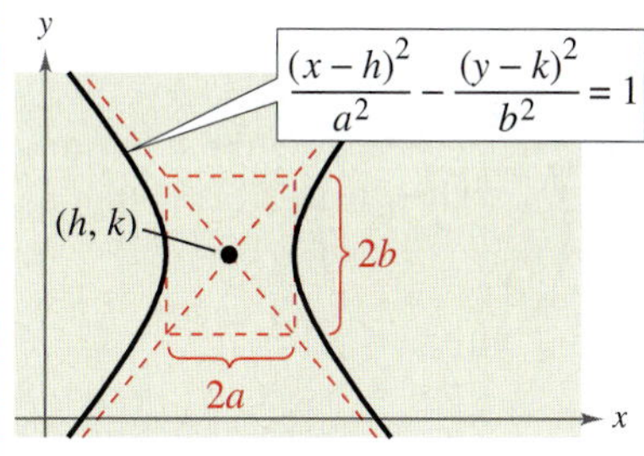

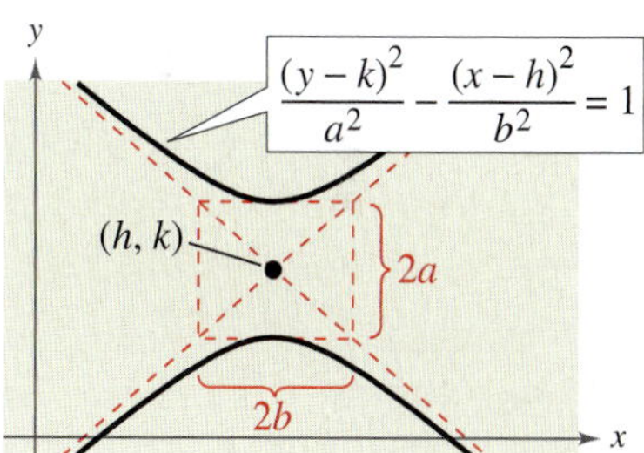

Parabola: Vertex $= (h, k)$

Directed distance from vertex to focus $= p$

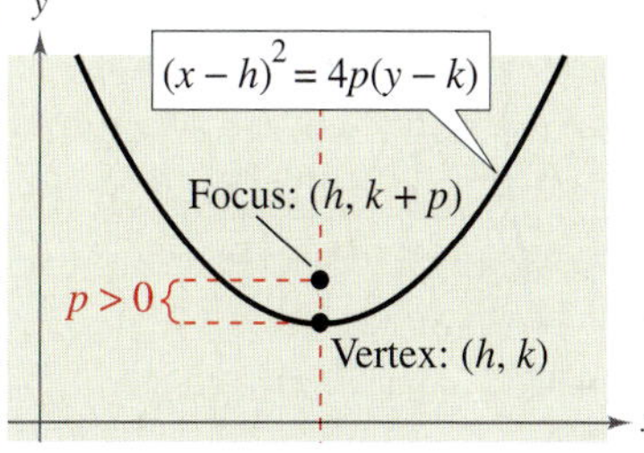

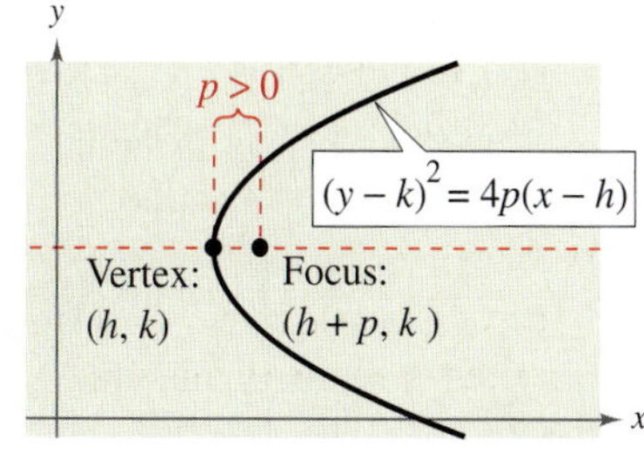

Example 1 ▶ Equations of Conic Sections 

a. The graph of

$$(x-1)^2+(y+2)^2=3^2$$

is a circle whose center is the point $(1, -2)$ and whose radius is 3, as shown in Figure 4.35.

b. The graph of

$$\frac{(x-2)^2}{3^2}+\frac{(y-1)^2}{2^2}=1$$

is an ellipse whose center is the point $(2, 1)$. The major axis of the ellipse is horizontal and of length $2(3)=6$, and the minor axis of the ellipse is vertical and of length $2(2)=4$, as shown in Figure 4.36.

c. The graph of

$$\frac{(x-3)^2}{1^2}-\frac{(y-2)^2}{3^2}=1$$

is a hyperbola whose center is the point $(3, 2)$. The transverse axis is horizontal and of length $2(1)=2$, and the conjugate axis is vertical and of length $2(3)=6$, as shown in Figure 4.37.

d. The graph of

$$(x-2)^2=4(-1)(y-3)$$

is a parabola whose vertex is the point $(2, 3)$. The axis of the parabola is vertical. The focus is one unit above or below the vertex. Moreover, because $p=-1$, it follows that the focus lies *below* the vertex, as shown in Figure 4.38.

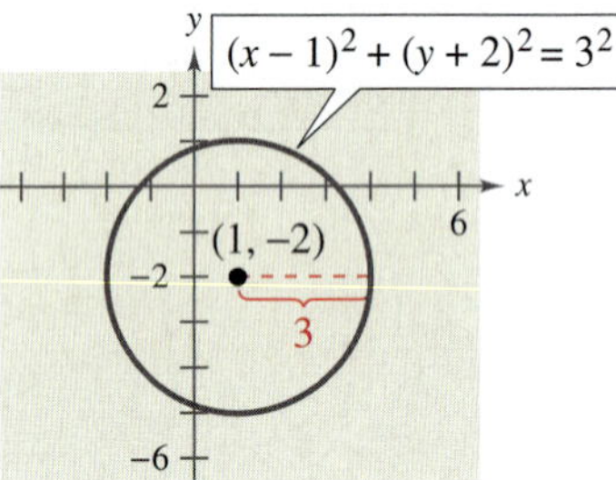

FIGURE 4.35

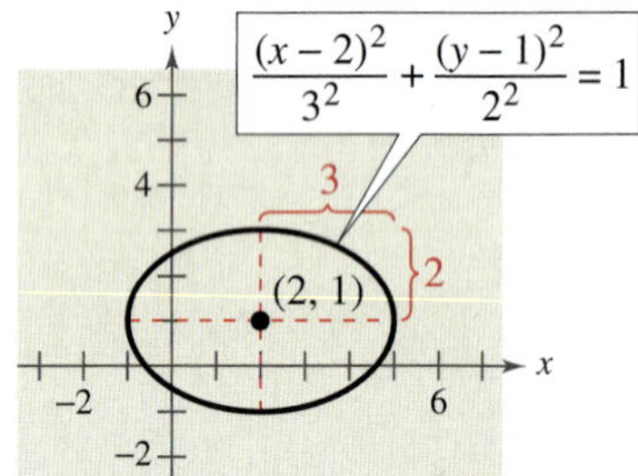

FIGURE 4.36

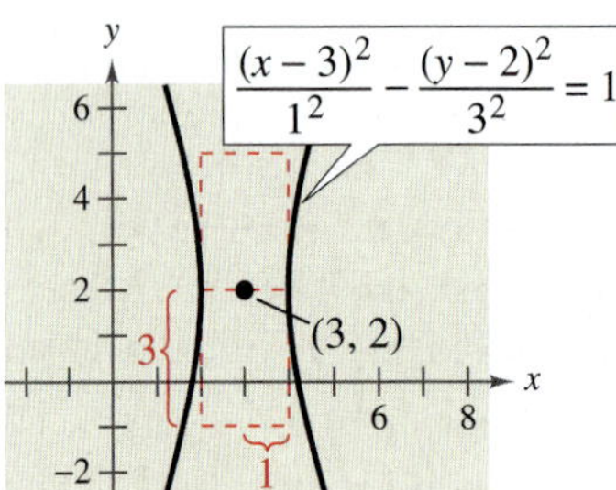

FIGURE 4.37

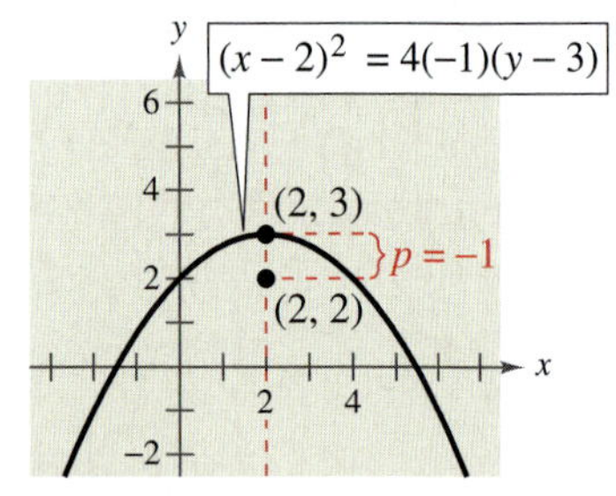

FIGURE 4.38

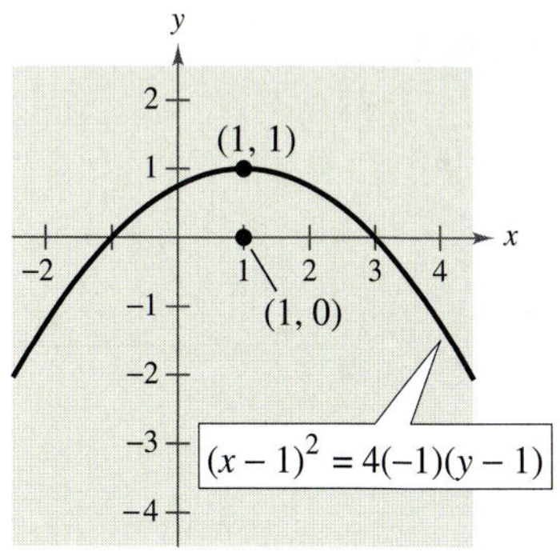

FIGURE 4.39

STUDY TIP

Note in Example 2 that p is the directed distance from the vertex to the focus. Because the axis of the parabola is vertical and $p = -1$, the focus is one unit below the vertex, and the parabola opens downward.

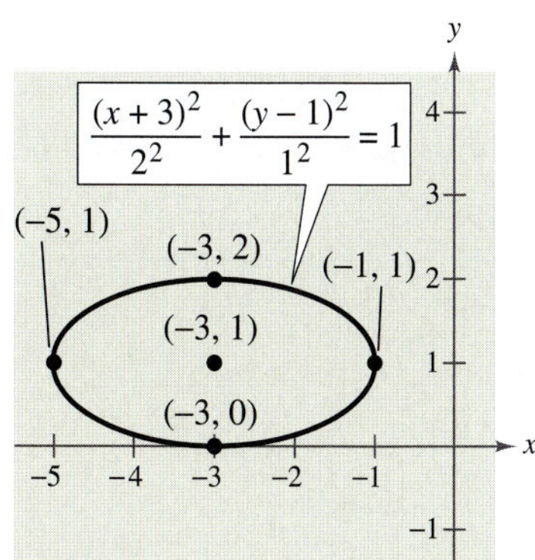

FIGURE 4.40

Equations of Conics in Standard Form 

Example 2 ▶ Finding the Standard Form of a Parabola 

Find the vertex and focus of the parabola $x^2 - 2x + 4y - 3 = 0$.

Solution

Complete the square to write the equation in standard form.

$$x^2 - 2x + 4y - 3 = 0 \qquad \text{Write original equation.}$$

$$x^2 - 2x = -4y + 3 \qquad \text{Group terms.}$$

$$x^2 - 2x + 1 = -4y + 3 + 1 \qquad \text{Add 1 to each side.}$$

$$(x-1)^2 = -4y + 4 \qquad \text{Write in completed square form.}$$

$$(x-1)^2 = 4(-1)(y-1) \qquad \text{Write in standard form, } (x-h)^2 = 4p(y-k).$$

From this standard form, it follows that $h = 1$, $k = 1$, and $p = -1$. Because the axis is vertical and p is negative, the parabola opens downward. The vertex is $(h, k) = (1, 1)$ and the focus is $(h, k + p) = (1, 0)$. (See Figure 4.39.)

Example 3 ▶ Sketching an Ellipse 

Sketch the graph of the ellipse $x^2 + 4y^2 + 6x - 8y + 9 = 0$.

Solution

Complete the square to write the equation in standard form.

$$x^2 + 4y^2 + 6x - 8y + 9 = 0 \qquad \text{Write original equation.}$$

$$(x^2 + 6x + \square) + (4y^2 - 8y + \square) = -9 \qquad \text{Group terms.}$$

$$(x^2 + 6x + \square) + 4(y^2 - 2y + \square) = -9 \qquad \text{Factor 4 out of } y\text{-terms.}$$

$$(x^2 + 6x + 9) + 4(y^2 - 2y + 1) = -9 + 9 + 4(1) \qquad \text{Add 9 and } 4(1) = 4 \text{ to each side.}$$

$$(x+3)^2 + 4(y-1)^2 = 4 \qquad \text{Write in completed square form.}$$

$$\frac{(x+3)^2}{4} + \frac{4(y-1)^2}{4} = 1 \qquad \text{Divide each side by 4.}$$

$$\frac{(x+3)^2}{2^2} + \frac{(y-1)^2}{1^2} = 1 \qquad \frac{(x-h)^2}{a^2} + \frac{(y-k)^2}{b^2} = 1$$

From this standard form, it follows that the center is $(h, k) = (-3, 1)$. Because the denominator of the x-term is $a^2 = 2^2$, the endpoints of the major axis lie two units to the right and left of the center. Similarly, because the denominator of the y-term is $b^2 = 1^2$, the endpoints of the minor axis lie one unit up and down from the center. The ellipse is shown in Figure 4.40.

Example 4 ▶ Sketching a Hyperbola 

Sketch the graph of the hyperbola given by the equation

$$y^2 - 4x^2 + 4y + 24x - 41 = 0.$$

Solution

Complete the square to write the equation in standard form.

$$y^2 - 4x^2 + 4y + 24x - 41 = 0 \qquad \text{Write original equation.}$$

$$(y^2 + 4y + \square) - (4x^2 - 24x + \square) = 41 \qquad \text{Group terms.}$$

$$(y^2 + 4y + \square) - 4(x^2 - 6x + \square) = 41 \qquad \text{Factor 4 out of } x\text{-terms.}$$

$$(y^2 + 4y + 4) - 4(x^2 - 6x + 9) = 41 + 4 - 4(9) \qquad \text{Add 4 and subtract } 4(9) = 36.$$

$$(y + 2)^2 - 4(x - 3)^2 = 9 \qquad \text{Write in completed square form.}$$

$$\frac{(y + 2)^2}{9} - \frac{4(x - 3)^2}{9} = 1 \qquad \text{Divide each side by 9.}$$

$$\frac{(y + 2)^2}{9} - \frac{(x - 3)^2}{\frac{9}{4}} = 1 \qquad \text{Change 4 to } \frac{1}{\frac{1}{4}}.$$

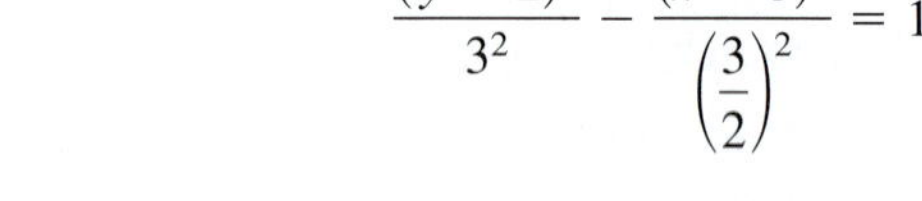

$$\frac{(y + 2)^2}{3^2} - \frac{(x - 3)^2}{\left(\frac{3}{2}\right)^2} = 1 \qquad \frac{(y - k)^2}{a^2} - \frac{(x - h)^2}{b^2} = 1$$

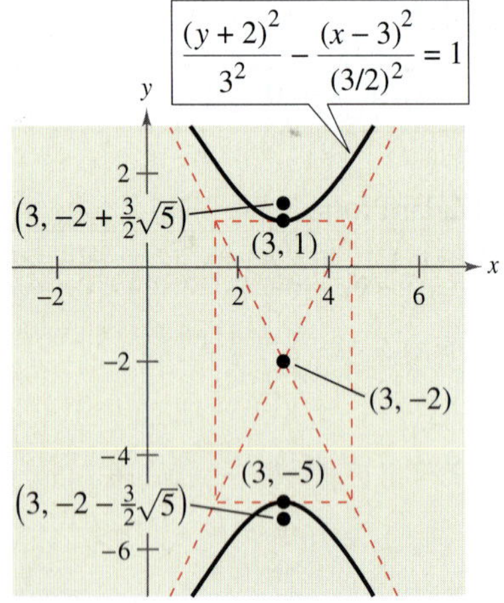

FIGURE 4.41

From this standard form, it follows that the transverse axis is vertical and the center lies at $(h, k) = (3, -2)$. Because the denominator of the y-term is $a^2 = 3^2$, you know that the vertices occur three units above and below the center.

$(3, -5)$ and $(3, 1)$ Vertices

To sketch the hyperbola, draw a rectangle whose top and bottom pass through the vertices. Because the denominator of the x-term is $b^2 = \left(\frac{3}{2}\right)^2$, locate the sides of the rectangle $\frac{3}{2}$ units to the right and left of the center, as shown in Figure 4.41. Finally, sketch the asymptotes by drawing lines through the opposite corners of the rectangle. Using these asymptotes, you can complete the graph of the hyperbola, as shown in Figure 4.41.

To find the foci in Example 4, first find c.

$$c^2 = a^2 + b^2$$

$$= 9 + \frac{9}{4} = \frac{45}{4} \quad \Rightarrow \quad c = \frac{3\sqrt{5}}{2}$$

Because the transverse axis is vertical, the foci lie c units above and below the center.

$\left(3, -2 + \frac{3}{2}\sqrt{5}\right)$ and $\left(3, -2 - \frac{3}{2}\sqrt{5}\right)$ Foci

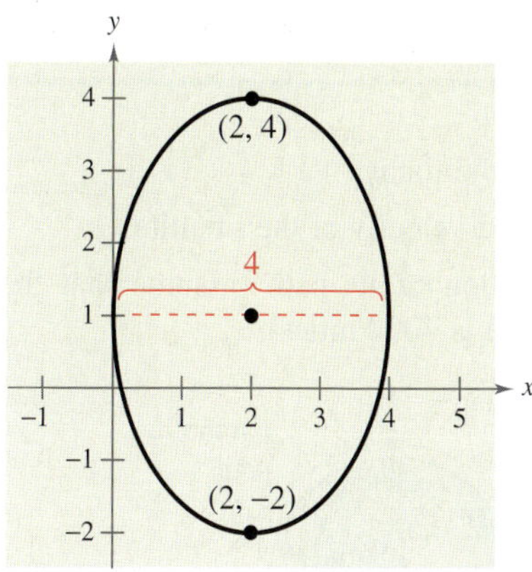

FIGURE 4.42

Example 5 ▶ Writing the Equation of an Ellipse 

Write the standard form of the equation of the ellipse whose vertices are $(2, -2)$ and $(2, 4)$. The length of the minor axis of the ellipse is 4, as shown in Figure 4.42.

Solution

The center of the ellipse lies at the midpoint of its vertices. So, the center is

$$(h, k) = (2, 1). \qquad \text{Center}$$

Because the vertices lie on a vertical line and are six units apart, it follows that the major axis is vertical and has a length of $2a = 6$. So, $a = 3$. Moreover, because the minor axis has a length of 4, it follows that $2b = 4$, which implies that $b = 2$. Therefore, the standard form of the ellipse is as follows.

$$\frac{(x-h)^2}{b^2} + \frac{(y-k)^2}{a^2} = 1 \qquad \text{Major axis is vertical.}$$

$$\frac{(x-2)^2}{2^2} + \frac{(y-1)^2}{3^2} = 1 \qquad \text{Write in standard form.}$$

An interesting application of conic sections involves the orbits of comets in our solar system. Of the 610 comets identified prior to 1970, 245 have elliptical orbits, 295 have parabolic orbits, and 70 have hyperbolic orbits. For example, Halley's comet has an elliptical orbit, and reappearance of this comet can be predicted every 76 years. The center of the sun is a focus of each of these orbits, and each orbit has a vertex at the point where the comet is closest to the sun, as shown in Figure 4.43.

If p is the distance between the vertex and the focus, and v is the speed of the comet at the vertex, then the orbit is:

an *ellipse* if $\quad v < \sqrt{\dfrac{2GM}{p}}$

a *parabola* if $\quad v = \sqrt{\dfrac{2GM}{p}}$

a *hyperbola* if $\quad v > \sqrt{\dfrac{2GM}{p}}$

where M is the mass of the sun and G is the universal gravitational constant, which is approximately equal to $6.67 \times 10^{-8}\ \text{cm}^3/(\text{kg} \cdot \text{sec}^2)$.

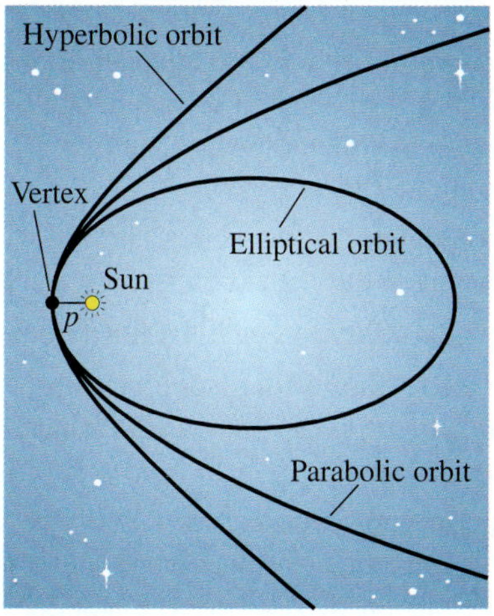

FIGURE 4.43

Writing ABOUT MATHEMATICS

Identifying Equations of Conics Use the Internet to research information about the orbits of comets in our solar system. What can you find about the orbits of comets that have been identified since 1970? Write a summary of your results. Identify your source. Does it seem reliable?

4.5 Exercises

In Exercises 1–6, identify the center and radius of the circle.

1. $x^2 + y^2 = 49$

2. $x^2 + y^2 = 1$

3. $(x + 3)^2 + (y - 8)^2 = 16$

4. $(x + 9)^2 + (y + 1)^2 = 36$

5. $(x - 1)^2 + y^2 = 10$

6. $x^2 + (y + 12)^2 = 24$

In Exercises 7–10, write the equation of the circle in standard form, and then identify its center and radius.

7. $x^2 + y^2 - 2x + 6y + 9 = 0$

8. $x^2 + y^2 - 10x - 6y + 25 = 0$

9. $4x^2 + 4y^2 + 12x - 24y + 41 = 0$

10. $9x^2 + 9y^2 + 54x - 36y + 17 = 0$

In Exercises 11–18, find the vertex, focus, and directrix of the parabola, and sketch its graph.

11. $(x - 1)^2 + 8(y + 2) = 0$

12. $(x + 3) + (y - 2)^2 = 0$

13. $\left(y + \frac{1}{2}\right)^2 = 2(x - 5)$

14. $\left(x + \frac{1}{2}\right)^2 = 4(y - 3)$

15. $y = \frac{1}{4}(x^2 - 2x + 5)$

16. $4x - y^2 - 2y - 33 = 0$

17. $y^2 + 6y + 8x + 25 = 0$

18. $y^2 - 4y - 4x = 0$

In Exercises 19–22, find the vertex, focus, and directrix of the parabola, and use a graphing utility to graph the parabola.

19. $y = -\frac{1}{6}(x^2 + 4x - 2)$

20. $x^2 - 2x + 8y + 9 = 0$

21. $y^2 + x + y = 0$

22. $y^2 - 4x - 4 = 0$

In Exercises 23–28, find an equation of the parabola.

23. Vertex: $(3, 2)$; Focus: $(1, 2)$

24. Vertex: $(-1, 2)$; Focus: $(-1, 0)$

25. Vertex: $(0, 4)$; Directrix: $y = 2$

26. Vertex: $(-2, 1)$; Directrix: $x = 1$

27. Focus: $(2, 2)$; Directrix: $x = -2$

28. Focus: $(0, 0)$; Directrix: $y = 4$

29. ***Satellite Orbit*** An Earth satellite in a 100-mile-high circular orbit around Earth has a velocity of approximately 17,500 miles per hour (see figure). If this velocity is multiplied by $\sqrt{2}$, the satellite will have the minimum velocity necessary to escape Earth's gravity and it will follow a parabolic path with the center of Earth as the focus.

(a) Find the escape velocity of the satellite.

(b) Find an equation of its path (assume that the radius of Earth is 4000 miles).

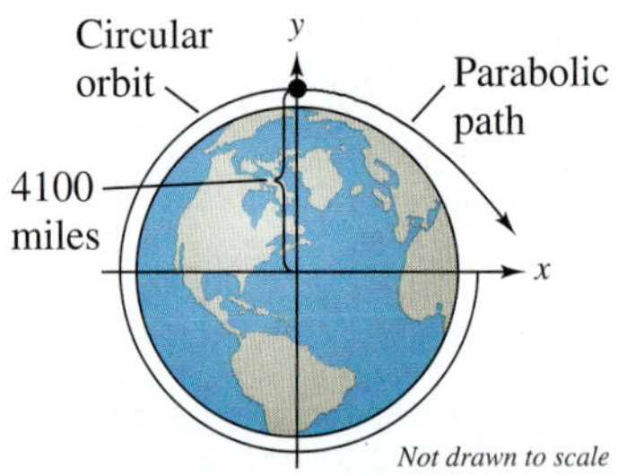

▶ Model It

30. ***Maximum Revenue*** The revenues R (in millions of dollars) for Manpower, Inc. for the years 1997 through 2001 are shown in the table. (Source: Manpower, Inc.)

Year	Revenue, R
1997	7,258.5
1998	8,814.3
1999	9,770.1
2000	10,842.8
2001	10,483.8

(a) Use a graphing utility to find an equation of the parabola $y = at^2 + bt + c$ that models the data. Let t represent the year, with $t = 7$ corresponding to 1997.

(b) Find the coordinates of the vertex and interpret its meaning in the context of the problem.

(c) Use a graphing utility to graph the function.

(d) Use the *trace* feature of the graphing utility to approximate *graphically* the year in which revenue was maximum.

(e) Use a table to approximate *numerically* the year in which revenue was maximum.

(f) Compare the results of parts (b), (d), and (e). What did you learn by using all three approaches?

31. ***Projectile Motion*** A bomber is flying at an altitude of 30,000 feet and a speed of 540 miles per hour (792 feet per second). How many feet will a bomb dropped from the plane travel horizontally before it hits the target if the path of the bomb is modeled by

$x^2 = -39{,}204(y - 30{,}000)$?

32. ***Path of a Projectile*** The path of a softball is modeled by $-12.5(y - 7.125) = (x - 6.25)^2$. The coordinates x and y are measured in feet, with $x = 0$ corresponding to the position from which the ball was thrown.

(a) Use a graphing utility to graph the trajectory of the softball.

(b) Use the *trace* feature of the graphing utility to approximate the highest point and the range of the trajectory.

In Exercises 33–40, find the center, foci, and vertices of the ellipse, and sketch its graph.

33. $\dfrac{(x-1)^2}{9} + \dfrac{(y-5)^2}{25} = 1$

34. $\dfrac{(x-6)^2}{4} + \dfrac{(y+7)^2}{16} = 1$

35. $(x+2)^2 + \dfrac{(y+4)^2}{\frac{1}{4}} = 1$

36. $\dfrac{(x-3)^2}{\frac{25}{9}} + (y-8)^2 = 1$

37. $9x^2 + 4y^2 + 36x - 24y + 36 = 0$

38. $9x^2 + 4y^2 - 36x + 8y + 31 = 0$

39. $16x^2 + 25y^2 - 32x + 50y + 16 = 0$

40. $9x^2 + 25y^2 - 36x - 50y + 61 = 0$

In Exercises 41–50, find an equation of the ellipse.

41. 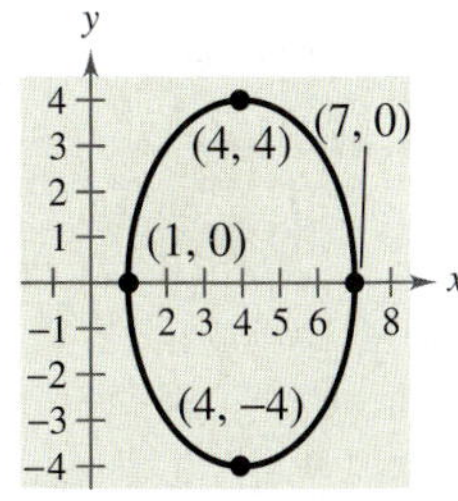

42. 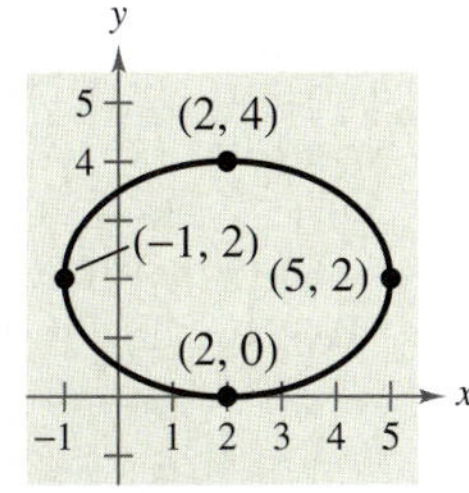

43. Vertices: (0, 2), (4, 2); Minor axis of length 2

44. Foci: (0, 0), (4, 0); Major axis of length 8

45. Foci: (0, 0), (0, 8); Major axis of length 16

46. Center: $(2, -1)$; Vertex: $\left(2, \frac{1}{2}\right)$;
Minor axis of length 2

47. Center: (0, 4); $a = 2c$; Vertices: $(-4, 4)$, (4, 4)

48. Center: (3, 2); $a = 3c$; Foci: (1, 2), (5, 2)

49. Vertices: (0, 2), (4, 2);
Endpoints of the minor axis: (2, 3), (2, 1)

50. Vertices: (5, 0), (5, 12);
Endpoints of the minor axis: (0, 6), (10, 6)

In Exercises 51 and 52, use the eccentricity of the ellipse, which is defined by $e = c/a$. The eccentricity measures the flatness of the ellipse.

51. Find an equation of the ellipse with vertices $(\pm 5, 0)$ and eccentricity $e = \frac{3}{5}$.

52. Find an equation of the ellipse with vertices $(0, \pm 8)$ and eccentricity $e = \frac{1}{2}$.

53. ***Distance from the Sun*** The planet Pluto moves in an elliptical orbit with the sun at one of the foci, as shown in the figure. The length of half of the major axis, a, is 3.666×10^9 miles, and the eccentricity is 0.248. Find the smallest distance (*perihelion*) and the greatest distance (*aphelion*) of Pluto from the center of the sun.

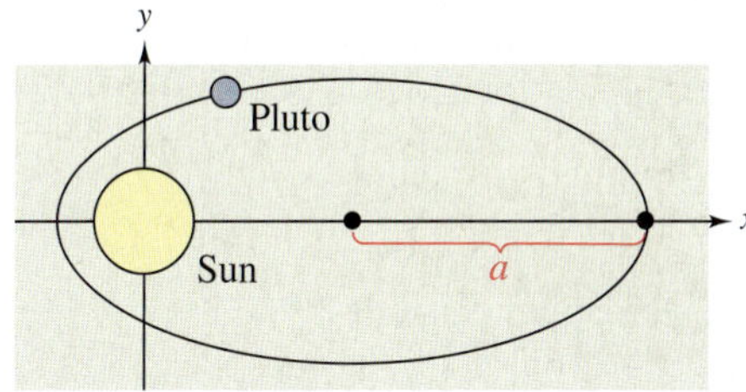

54. ***Australian Football*** In Australia, football by Australian Rules (or rugby) is played on elliptical fields. The field can be a maximum of 170 yards wide and a maximum of 200 yards long. Let the center of a field of maximum size be represented by the point (0, 85). Write an equation of the ellipse that represents this field. (Source: *Oxford Companion to World Sports and Games*)

In Exercises 55–64, find the center, foci, and vertices of the hyperbola, and sketch its graph. Sketch the asymptotes as an aid in obtaining the graph of the hyperbola.

55. $\dfrac{(x-1)^2}{4} - \dfrac{(y+2)^2}{1} = 1$

56. $\dfrac{(x+1)^2}{144} - \dfrac{(y-4)^2}{25} = 1$

57. $(y+6)^2 - (x-2)^2 = 1$

58. $\dfrac{(y-1)^2}{1/4} - \dfrac{(x+3)^2}{1/9} = 1$

59. $9x^2 - y^2 - 36x - 6y + 18 = 0$

60. $x^2 - 9y^2 + 36y - 72 = 0$

61. $x^2 - 9y^2 + 2x - 54y - 80 = 0$

62. $16y^2 - x^2 + 2x + 64y + 63 = 0$

63. $9y^2 - 4x^2 + 8x + 18y + 41 = 0$

64. $11y^2 - 3x^2 + 12x + 44y + 48 = 0$

In Exercises 65–74, find an equation of the hyperbola.

65.

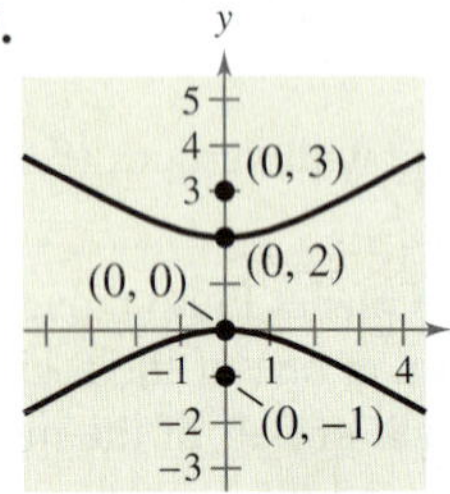

66.

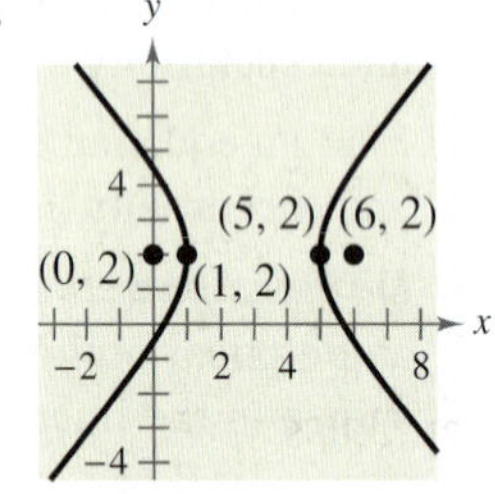

67. Vertices: $(2, 0), (6, 0)$; Foci: $(0, 0), (8, 0)$

68. Vertices: $(2, 3), (2, -3)$; Foci: $(2, 5), (2, -5)$

69. Vertices: $(4, 1), (4, 9)$; Foci: $(4, 0), (4, 10)$

70. Vertices: $(-2, 1), (2, 1)$; Foci: $(-3, 1), (3, 1)$

71. Vertices: $(2, 3), (2, -3)$;
Passes through the point $(0, 5)$

72. Vertices: $(-2, 1), (2, 1)$;
Passes through the point $(4, 3)$

73. Vertices: $(0, 2), (6, 2)$;
Asymptotes: $y = \frac{2}{3}x,\ y = 4 - \frac{2}{3}x$

74. Vertices: $(3, 0), (3, 4)$;
Asymptotes: $y = \frac{2}{3}x,\ y = 4 - \frac{2}{3}x$

In Exercises 75–82, identify the conic by writing its equation in standard form. Then sketch its graph.

75. $x^2 + y^2 - 6x + 4y + 9 = 0$

76. $x^2 + 4y^2 - 6x + 16y + 21 = 0$

77. $4x^2 - y^2 - 4x - 3 = 0$

78. $y^2 - 4y - 4x = 0$

79. $4x^2 + 3y^2 + 8x - 24y + 51 = 0$

80. $4y^2 - 2x^2 - 4y - 8x - 15 = 0$

81. $25x^2 - 10x - 200y - 119 = 0$

82. $4x^2 + 4y^2 - 16y + 15 = 0$

Synthesis

True or False? **In Exercises 83 and 84, determine whether the statement is true or false. Justify your answer.**

83. The conic represented by the equation $3x^2 + 2y^2 - 18x - 16y + 58 = 0$ is an ellipse.

84. The graphs of $x^2 + 10y - 10x + 5 = 0$ and $x^2 + 16y^2 + 10x - 32y - 23 = 0$ do not intersect.

Exploration **In Exercises 85 and 86, consider the ellipse**

$$\frac{x^2}{a^2} + \frac{y^2}{b^2} = 1.$$

85. Show that the equation of the ellipse can be written as

$$\frac{(x-h)^2}{a^2} + \frac{(y-k)^2}{a^2(1-e^2)} = 1$$

where e is the eccentricity.

86. Use a graphing utility to graph the ellipse

$$\frac{(x-2)^2}{4} + \frac{(y-3)^2}{4(1-e^2)} = 1$$

for $e = 0.95, 0.75, 0.5, 0.25$, and 0. Make a conjecture about the change in the shape of the ellipse as e approaches 0.

Review

In Exercises 87–90, identify the rule of algebra that is illustrated.

87. $(a+4) - (a+4) = 0$

88. $u + (v + 10) = (u + v) + 10$

89. $(x+3)(a+b) = x(a+b) + 3(a+b)$

90. $\dfrac{1}{z-3}(z-3) = 1, \quad z \neq 3$

In Exercises 91 and 92, determine whether f and g are inverse functions. If g is not the inverse function of f, find the correct inverse function.

91. $f(x) = 10 - 7x, \quad g(x) = \dfrac{x - 10}{7}$

92. $f(x) = \sqrt{x+8}, \quad g(x) = x^2 + 8$

Chapter Summary

▶ *What* did you learn?

	Review Exercises
Section 4.1	
☐ How to find the domains of rational functions	1–4
☐ How to find the horizontal and vertical asymptotes of graphs of rational functions	5–8
☐ How to use rational functions to model and solve real-life problems	9, 10
Section 4.2	
☐ How to analyze and sketch graphs of rational functions	11–22
☐ How to sketch graphs of rational functions that have slant asymptotes	23–26
☐ How to use graphs of rational functions to model and solve real-life problems	27–30
Section 4.3	
☐ How to recognize partial fraction decompositions of rational expressions	31–34
☐ How to find partial fraction decompositions of rational expressions	35–42
Section 4.4	
☐ How to recognize the four basic conics: circles, ellipses, parabolas and hyperbolas	43–50
☐ How to recognize, graph, and write equations of parabolas (vertex at origin)	51–56
☐ How to recognize, graph, and write equations of ellipses (center at origin)	57–64
☐ How to recognize, graph, and write equations of hyperbolas (center at origin)	65–68
Section 4.5	
☐ How to recognize equations of conics that have been shifted vertically or horizontally in the plane	69–76
☐ How to write and graph equations of conics that have been shifted vertically or horizontally in the plane	77–100

Review Exercises

4.1 In Exercises 1–4, find the domain of the rational function.

1. $f(x) = \dfrac{5x}{x + 12}$

2. $f(x) = \dfrac{3x^2}{1 + 3x}$

3. $f(x) = \dfrac{8}{x^2 - 10x + 24}$

4. $f(x) = \dfrac{x^2 + x - 2}{x^2 + 4}$

In Exercises 5–8, identify any horizontal or vertical asymptotes.

5. $f(x) = \dfrac{4}{x + 3}$

6. $f(x) = \dfrac{2x^2 + 5x - 3}{x^2 + 2}$

7. $g(x) = \dfrac{x^2}{x^2 - 4}$

8. $g(x) = \dfrac{1}{(x - 3)^2}$

9. ***Average Cost*** A business has a production cost of $C = 0.5x + 500$ for producing x units of a product. The average cost per unit is

$$\overline{C} = \frac{C}{x} = \frac{0.5x + 500}{x}, \qquad x > 0.$$

Determine the average cost per unit as x increases without bound. (Find the horizontal asymptote.)

10. ***Seizure of Illegal Drugs*** The cost C (in millions of dollars) for the federal government to seize $p\%$ of an illegal drug as it enters the country is

$$C = \frac{528p}{100 - p}, \qquad 0 \leq p < 100.$$

(a) Find the cost of seizing 25% of the drug.

(b) Find the cost of seizing 50% of the drug.

(c) Find the cost of seizing 75% of the drug.

(d) According to this model, would it be possible to seize 100% of the drug?

4.2 In Exercises 11–22, identify intercepts, check for symmetry, identify any vertical or horizontal asymptotes, and sketch the graph of the rational function.

11. $f(x) = \dfrac{-5}{x^2}$

12. $f(x) = \dfrac{4}{x}$

13. $g(x) = \dfrac{2 + x}{1 - x}$

14. $h(x) = \dfrac{x - 3}{x - 2}$

15. $p(x) = \dfrac{x^2}{x^2 + 1}$

16. $f(x) = \dfrac{2x}{x^2 + 4}$

17. $f(x) = \dfrac{x}{x^2 + 1}$

18. $h(x) = \dfrac{4}{(x - 1)^2}$

19. $f(x) = \dfrac{-6x^2}{x^2 + 1}$

20. $y = \dfrac{2x^2}{x^2 - 4}$

21. $y = \dfrac{x}{x^2 - 1}$

22. $g(x) = \dfrac{-2}{(x + 3)^2}$

In Exercises 23–26, state the domain of the function and identify any vertical and slant asymptotes. Then sketch the graph of the rational function.

23. $f(x) = \dfrac{2x^3}{x^2 + 1}$

24. $f(x) = \dfrac{x^2 + 1}{x + 1}$

25. $f(x) = \dfrac{x^2 + 3x - 10}{x + 2}$

26. $f(x) = \dfrac{x^3}{x^2 - 4}$

27. ***Average Cost*** The cost of producing x units of a product is C, and the average cost per unit is

$$\overline{C} = \frac{C}{x} = \frac{100{,}000 + 0.9x}{x}, \qquad x > 0.$$

(a) Graph the average cost function.

(b) Find the average cost of producing $x = 1000$, 10,000, and 100,000 units.

(c) By increasing the level of production, what is the smallest average cost per unit you can obtain? Explain your reasoning.

28. ***Minimum Area*** A page that is x inches wide and y inches high contains 30 square inches of print. The top and bottom margins are 2 inches deep and the margins on each side are 2 inches wide.

(a) Draw a diagram that gives a visual representation of the problem.

(b) Show that the total area A on the page is

$$A = \frac{2x(2x + 7)}{x - 4}.$$

(c) Determine the domain of the function based on the physical constraints of the problem.

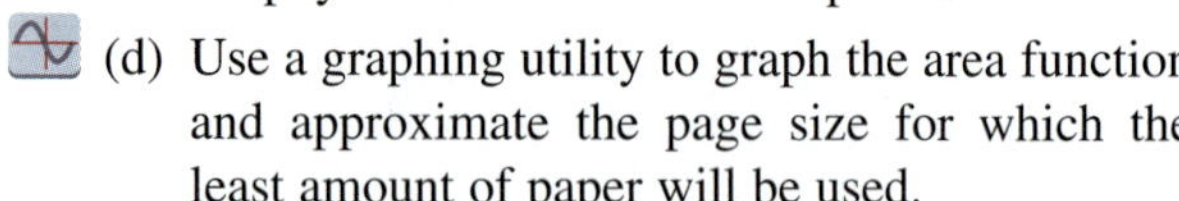

(d) Use a graphing utility to graph the area function and approximate the page size for which the least amount of paper will be used.

29. ***Photosynthesis*** The amount y of CO_2 uptake in milligrams per square decimeter per hour at optimal temperatures and with the natural supply of CO_2 is approximated by the model

$$y = \frac{18.47x - 2.96}{0.23x + 1}, \qquad x > 0$$

where x is the light intensity (in watts per square meter). Use a graphing utility to graph the function and determine the limiting amount of CO_2 uptake.

30. ***Medicine*** The concentration C of a medication in the bloodstream t hours after injection into muscle tissue is given by

$$C(t) = \frac{2t + 1}{t^2 + 4}, \qquad t > 0.$$

Use a graphing utility to graph the function and approximate the time when the bloodstream concentration is greatest.

4.3 In Exercises 31–34, write the form of the partial fraction decomposition for the rational expression. Do not solve for the constants.

31. $\dfrac{3}{x^2 + 20x}$ **32.** $\dfrac{x - 8}{x^2 - 3x - 28}$

33. $\dfrac{3x - 4}{x^3 - 5x^2}$ **34.** $\dfrac{x - 2}{x(x^2 + 2)^2}$

In Exercises 35–42, write the partial fraction decomposition for the rational expression.

35. $\dfrac{4 - x}{x^2 + 6x + 8}$ **36.** $\dfrac{-x}{x^2 + 3x + 2}$

37. $\dfrac{x^2}{x^2 + 2x - 15}$ **38.** $\dfrac{9}{x^2 - 9}$

39. $\dfrac{x^2 + 2x}{x^3 - x^2 + x - 1}$ **40.** $\dfrac{4x - 2}{3(x - 1)^2}$

41. $\dfrac{3x^3 + 4x}{(x^2 + 1)^2}$ **42.** $\dfrac{4x^2}{(x - 1)(x^2 + 1)}$

4.4 In Exercises 43–50, identify the conic.

43. $y^2 = -12x$ **44.** $16x^2 + y^2 = 16$

45. $\dfrac{x^2}{9} - \dfrac{y^2}{1} = 1$ **46.** $\dfrac{x^2}{1} + \dfrac{y^2}{9} = 1$

47. $x^2 + 20y = 0$ **48.** $x^2 + y^2 = 100$

49. $\dfrac{y^2}{49} - \dfrac{x^2}{144} = 1$ **50.** $\dfrac{x^2}{49} + \dfrac{y^2}{144} = 1$

In Exercises 51–54, write the equation of the specified parabola.

51.

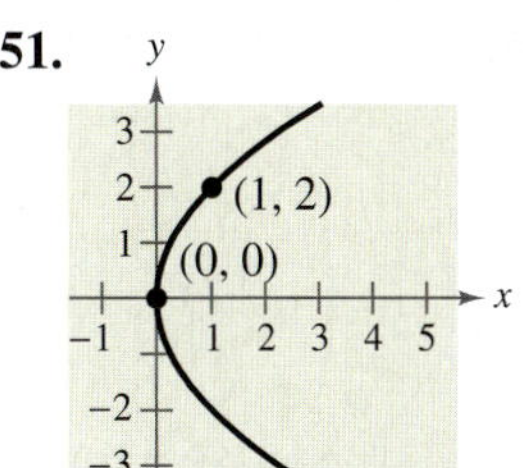

52.

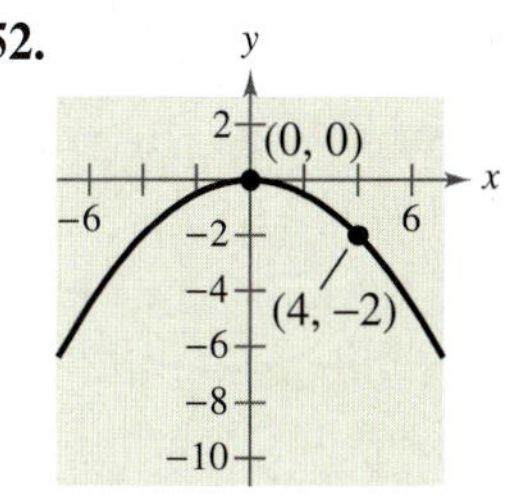

53. Vertex: $(0, 0)$; Focus: $(-6, 0)$

54. Vertex: $(0, 0)$; Focus: $(0, 3)$

55. ***Satellite Antenna*** A cross section of a large parabolic antenna (see figure) is modeled by

$$y = \frac{x^2}{200}, \qquad -100 \le x \le 100.$$

The receiving and transmitting equipment is positioned at the focus. Find the coordinates of the focus.

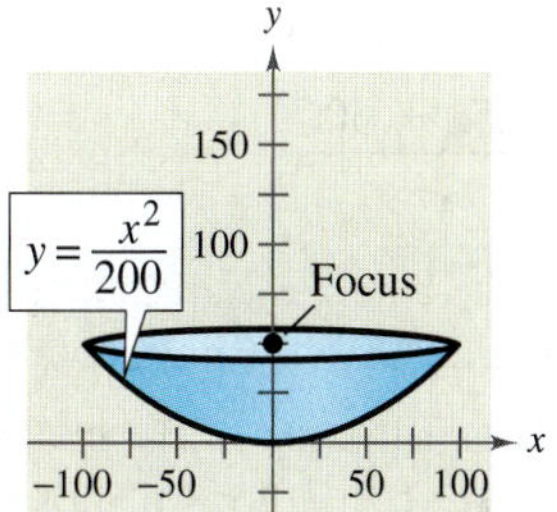

56. ***Suspension Bridge*** Each cable of a suspension bridge is suspended (in the shape of a parabola) between two towers (see figure). An equation that models the cables is given by

$$y = \frac{x^2}{180}$$

where x and y are measured in meters. Find the coordinates of the focus.

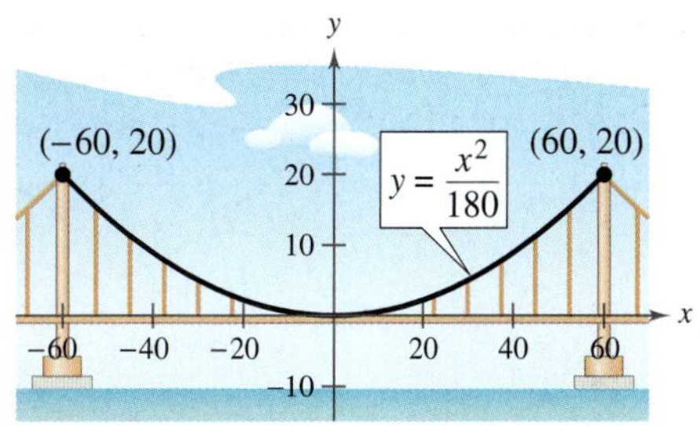

In Exercises 57–62, write the equation of the specified ellipse whose center is the origin.

57. **58.**

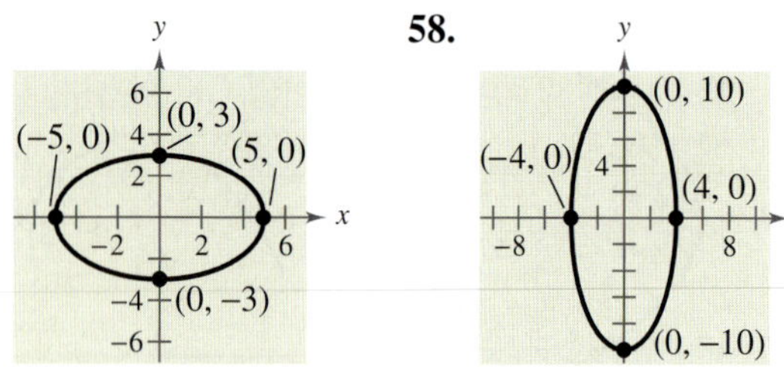

59. Vertices: $(0, \pm 6)$; Passes through the point $(2, 2)$

60. Vertices: $(\pm 7, 0)$; Foci: $(\pm 6, 0)$

61. Foci: $(\pm 6, 0)$; Minor axis of length 10

62. Foci: $(\pm 3, 0)$; Major axis of length 12

63. ***Architecture*** A semielliptical archway is to be formed over the entrance to an estate (see figure). The arch is to be set on pillars that are 10 feet apart and is to have a height (atop the pillars) of 4 feet. Where should the foci be placed in order to sketch the arch?

64. ***Wading Pool*** You are building a wading pool that is in the shape of an ellipse. Your plans give an equation for the elliptical shape of the pool measured in feet as

$$\frac{x^2}{324} + \frac{y^2}{196} = 1.$$

Find the longest distance across the pool, the shortest distance, and the distance between the foci.

In Exercises 65–68, write the equation of the specified hyperbola whose center is the origin.

65.

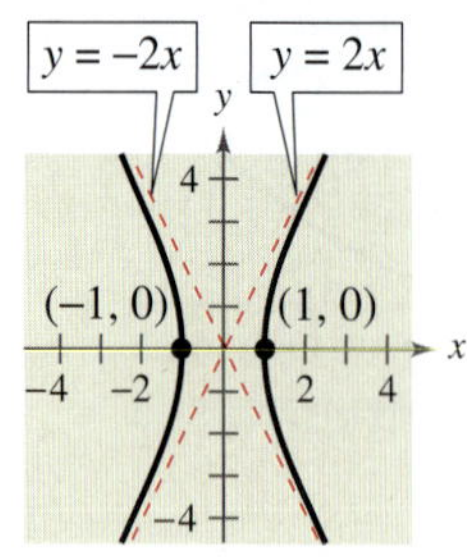

66.

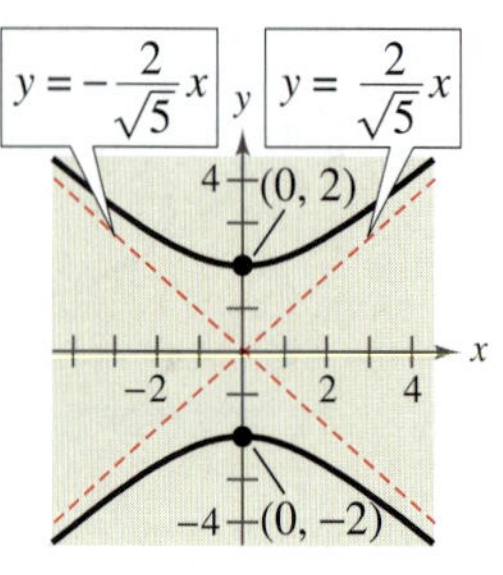

67. Vertices: $(0, \pm 1)$; Foci: $(0, \pm 3)$

68. Vertices: $(\pm 4, 0)$; Foci: $(\pm 6, 0)$

4.5 **In Exercises 69–76, identify the conic by writing its equation in standard form. Then sketch its graph.**

69. $x^2 - 6x + 2y + 9 = 0$

70. $y^2 - 12y - 8x + 20 = 0$

71. $x^2 + y^2 - 2x - 4y + 5 = 0$

72. $16x^2 + 16y^2 - 16x + 24y - 3 = 0$

73. $x^2 + 9y^2 + 10x - 18y + 25 = 0$

74. $4x^2 + y^2 - 16x + 15 = 0$

75. $9x^2 - y^2 - 72x + 8y + 119 = 0$

76. $x^2 - 9y^2 + 10x + 18y + 7 = 0$

In Exercises 77 and 78, identify the center and radius of the circle.

77. $(x - 2)^2 + (y + 6)^2 = 25$

78. $(x + 8)^2 + (y - 3)^2 = 64$

In Exercises 79–82, write the equation of the circle in standard form, and then identify its center and radius.

79. $x^2 + y^2 - 2x - 31 = 0$

80. $x^2 + y^2 - 6x + 8y + 9 = 0$

81. $4x^2 + 4y^2 + 40x + 16y + 80 = 0$

82. $2x^2 + 2y^2 - 4x - 28y + 66 = 0$

In Exercises 83–86, find an equation of the specified parabola.

83.

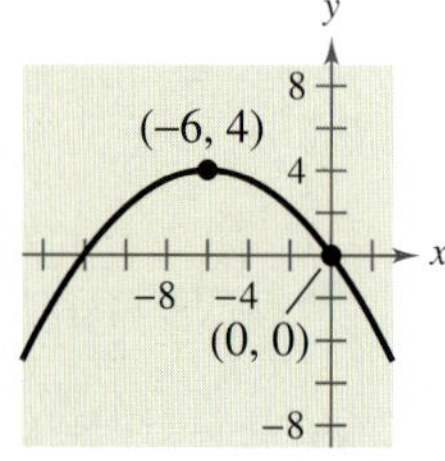

84.

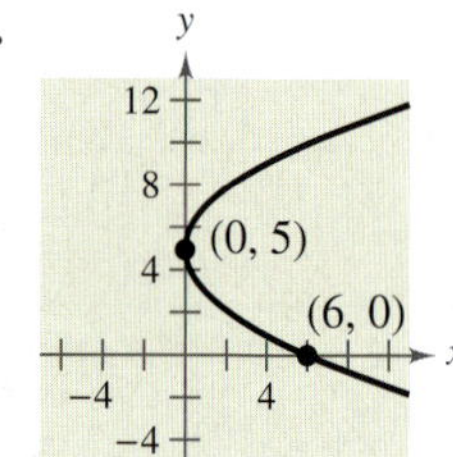

85. Vertex: $(4, 2)$; Focus: $(4, 0)$

86. Vertex: $(2, 0)$; Focus: $(0, 0)$

In Exercises 87–90, find an equation of the ellipse.

87.

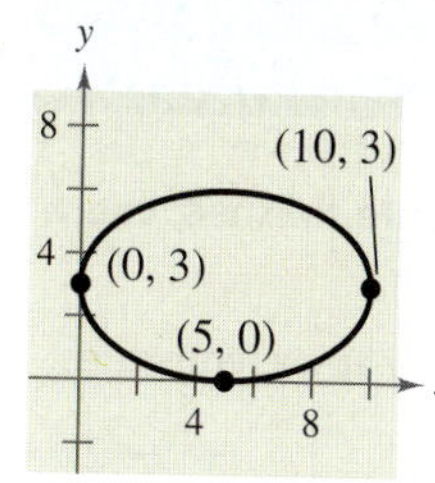

88.

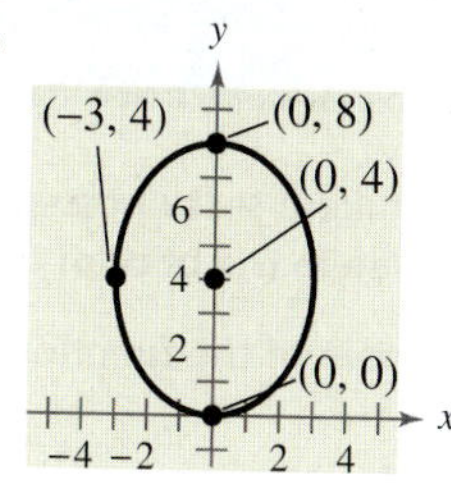

89. Vertices: $(-3, 0), (7, 0)$; Foci: $(0, 0), (4, 0)$

90. Vertices: $(2, 0), (2, 4)$; Foci: $(2, 1), (2, 3)$

In Exercises 91–96, find an equation of the hyperbola.

91.

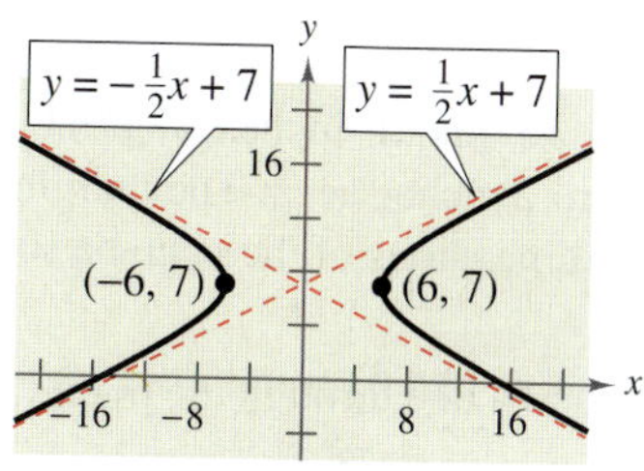

92.

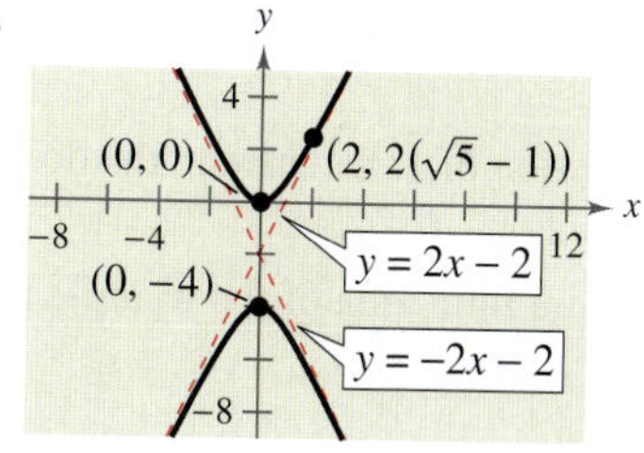

93. Vertices: $(-10, 3), (6, 3)$; Foci: $(-12, 3), (8, 3)$

94. Vertices: $(2, 2), (-2, 2)$; Foci: $(4, 2), (-4, 2)$

95. Foci: $(0, 0), (8, 0)$; Asymptotes: $y = \pm 2(x - 4)$

96. Foci: $(3, \pm 2)$; Asymptotes: $y = \pm 2(x - 3)$

97. ***Architecture*** A parabolic archway (see figure) is 12 meters high at the vertex. At a height of 10 meters, the width of the archway is 8 meters. How wide is the archway at ground level?

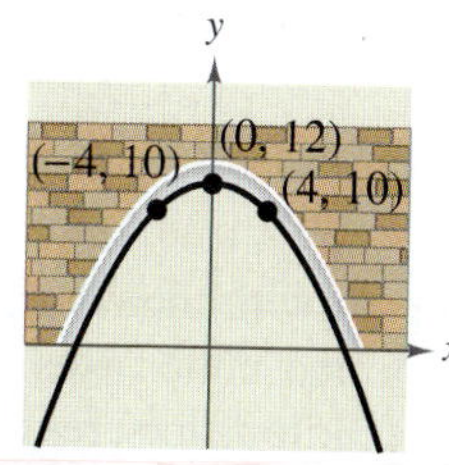

98. ***Architecture*** A church window (see figure) is bounded above by a parabola and below by the arc of a circle.

(a) Find equations for the parabola and the circle.

(b) Complete the table by filling in the vertical distance d between the circle and the parabola for each given value of x.

x	d
0	
1	
2	
3	
4	

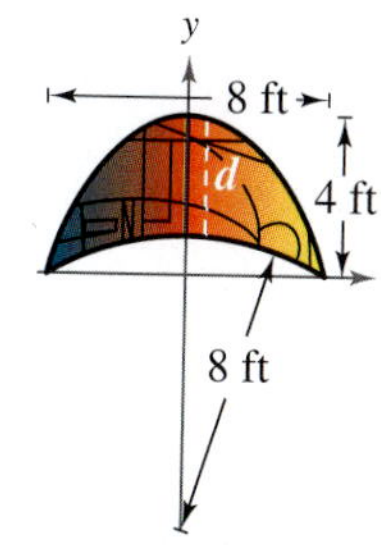

99. ***Heating and Plumbing*** Find the diameter d of the largest water pipe that can be placed in the corner behind the ventilation duct in the figure.

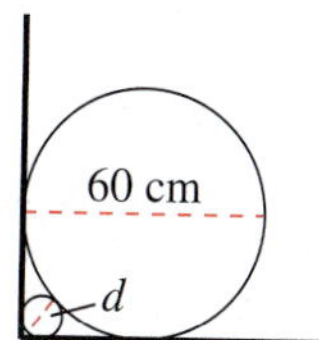

100. ***Eccentricity of Orbit*** Saturn moves in an elliptical orbit with the sun at one of the foci. The smallest distance and the greatest distance of the planet from the sun are 1.3495×10^9 and 1.5045×10^9 kilometers, respectively. Find the eccentricity of the orbit, defined by $e = c/a$. The eccentricity measures the flatness of an ellipse.

Synthesis

True or False? **In Exercises 101 and 102, determine whether the statement is true or false. Justify your answer.**

101. The domain of a rational function can never be the set of all real numbers.

102. The graph of the equation

$$Ax^2 + Bxy + Cy^2 + Dx + Ey + F = 0$$

can be a single point

Chapter Test

The *Interactive* CD-ROM and *Internet* versions of this text offer Chapter Pre-Tests and Chapter Post-Tests, both of which have randomly generated exercises with diagnostic capabilities.

Take this test as you would take a test in class. When you are finished, check your work against the answers given in the back of the book.

In Exercises 1–3, find the domain of the function and identify any asymptotes.

1. $y = \dfrac{2}{4 - x}$ **2.** $f(x) = \dfrac{3 - x^2}{3 + x^2}$ **3.** $g(x) = \dfrac{x^2 + 2x - 3}{x - 2}$

In Exercises 4–6, graph the function. Check for symmetry and identify any intercepts and asymptotes.

4. $h(x) = \dfrac{4}{x^2} - 1$ **5.** $g(x) = \dfrac{x^2 + 2}{x - 1}$ **6.** $f(x) = \dfrac{x + 1}{x^2 + x - 12}$

7. A rectangular page is designed to contain 36 square inches of print. The margins at the top and bottom of the page are 2 inches deep. The margins on each side are 1 inch wide. What should the dimensions of the page be so that the least amount of paper is used?

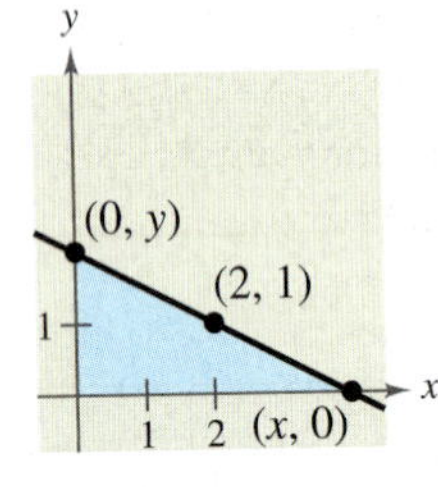

FIGURE FOR 8

8. A triangle is formed by the coordinate axes and a line through the point $(2, 1)$, as shown in the figure.

(a) Verify that $y = 1 + \dfrac{2}{x - 2}$.

(b) Write the area A of the triangle as a function of x. Determine the domain of the function in the context of the problem.

(c) Graph the area function. Estimate the minimum area of the triangle from the graph.

In Exercises 9–12, write the partial fraction decomposition for the rational expression.

9. $\dfrac{2x + 5}{x^2 - x - 2}$ **10.** $\dfrac{3x^2 - 2x + 4}{x^2(2 - x)}$ **11.** $\dfrac{x^2 + 5}{x^3 - x}$ **12.** $\dfrac{x^2 - 4}{x^3 + 2x}$

In Exercises 13–18, graph the conic and identify any centers, vertices, and foci.

13. $y^2 - 8x = 0$ **14.** $x^2 + y^2 - 10x + 4y + 4 = 0$

15. $x^2 - 10x - 2y + 19 = 0$ **16.** $x^2 - \dfrac{y^2}{4} = 1$

17. $x^2 - 4y^2 - 4x = 0$ **18.** $x^2 + 3y^2 - 2x + 36y + 100 = 0$

19. Find an equation of the ellipse with vertices $(0, 2)$ and $(8, 2)$ and minor axis of length 4.

20. Find an equation of the hyperbola with vertices $(0, \pm 3)$ and asymptotes $y = \pm \frac{3}{2}x$.

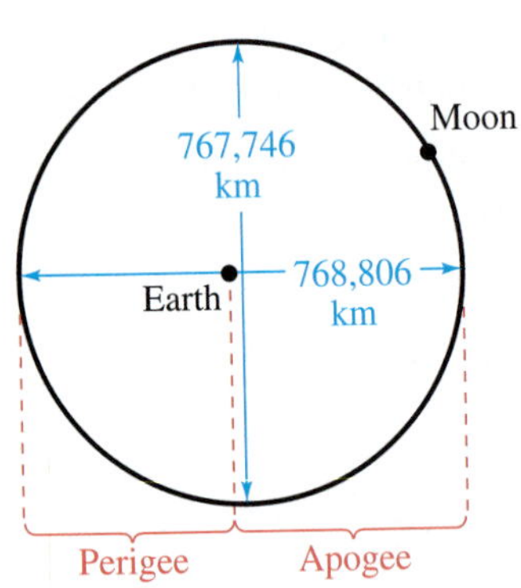

FIGURE FOR 21

21. The moon orbits Earth in an elliptical path with the center of Earth at one focus, as shown in the figure. The major and minor axes of the orbit have lengths of 768,806 kilometers and 767,746 kilometers, respectively. Find the smallest distance (perigee) and the greatest distance (apogee) from the center of the moon to the center of Earth.

Proofs in Mathematics

You can use the definition of a parabola to derive the standard form of the equation of a parabola whose directrix is parallel to the x-axis or to the y-axis.

Parabolic Patterns

There are many natural occurrences of parabolas in real life. For instance, the famous astronomer Galileo discovered in the 17th century that an object that is projected upward and obliquely to the pull of gravity travels in a parabolic path. Examples of this are the center of gravity of a jumping dolphin and the path of water molecules of a drinking water fountain.

Standard Equation of a Parabola (Vertex at Origin) *(p. 359)*

The standard form of the equation of a parabola with vertex at $(0, 0)$ and directrix $y = -p$ is

$$x^2 = 4py, \qquad p \neq 0. \qquad \text{Vertical axis}$$

For directrix $x = -p$, the equation is

$$y^2 = 4px, \qquad p \neq 0. \qquad \text{Horizontal axis}$$

The focus is on the axis p units (directed distance) from the vertex.

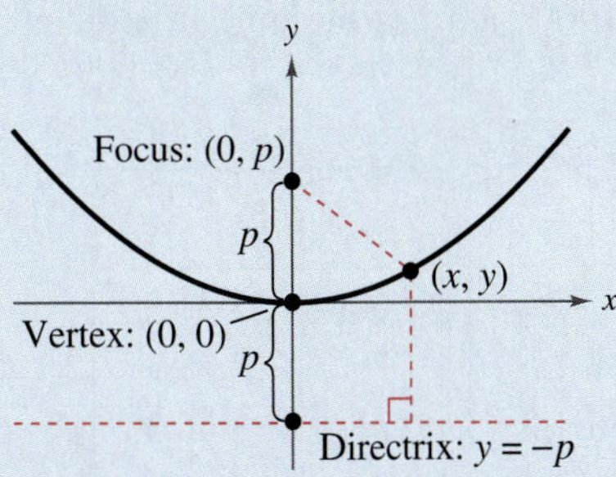

Parabola with vertical axis

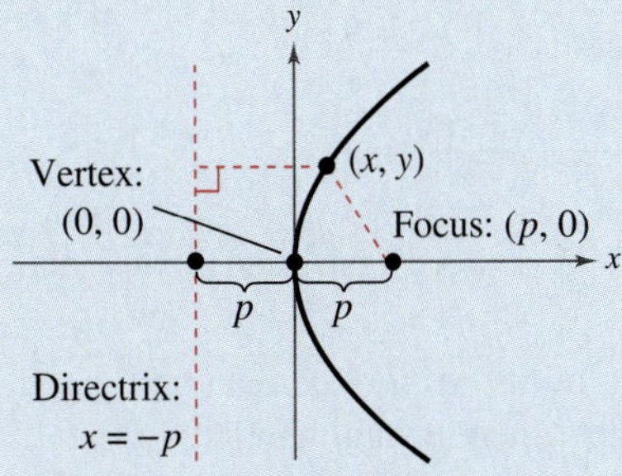

Parabola with horizontal axis

Proof

For the first case, suppose the directrix $(y = -p)$ is parallel to the x-axis. In the figure, you assume that $p > 0$, and because p is the directed distance from the vertex to the focus, the focus must lie above the vertex. Because the point (x, y) is equidistant from $(0, p)$ and $y = -p$, you can apply the Distance Formula to obtain

$$\sqrt{(x-0)^2 + (y-p)^2} = y + p \qquad \text{Distance Formula}$$

$$x^2 + (y-p)^2 = (y+p)^2 \qquad \text{Square each side.}$$

$$x^2 + y^2 - 2py + p^2 = y^2 + 2py + p^2 \qquad \text{Expand.}$$

$$x^2 = 4py. \qquad \text{Simplify.}$$

A proof of the second case is similar to the proof of the first case. Suppose the directrix $(x = -p)$ is parallel to the y-axis. In the figure, you assume that $p > 0$, and because p is the directed distance from the vertex to the focus, the focus must lie to the right of the vertex. Because the point (x, y) is equidistant from $(p, 0)$ and $x = -p$, you can apply the Distance Formula as follows.

$$\sqrt{(x-p)^2 + (y-0)^2} = x + p \qquad \text{Distance Formula}$$

$$(x-p)^2 + y^2 = (x+p)^2 \qquad \text{Square each side.}$$

$$x^2 - 2px + p^2 + y^2 = x^2 + 2px + p^2 \qquad \text{Expand.}$$

$$y^2 = 4px \qquad \text{Simplify.}$$

P.S. Problem Solving

1. Match the graph of the rational function

$$f(x) = \frac{ax + b}{cx + d}$$ with the given conditions.

(a)

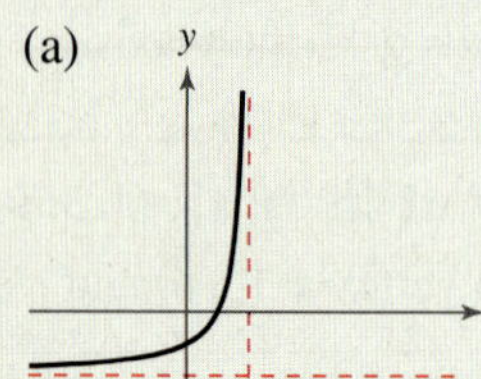

(b)

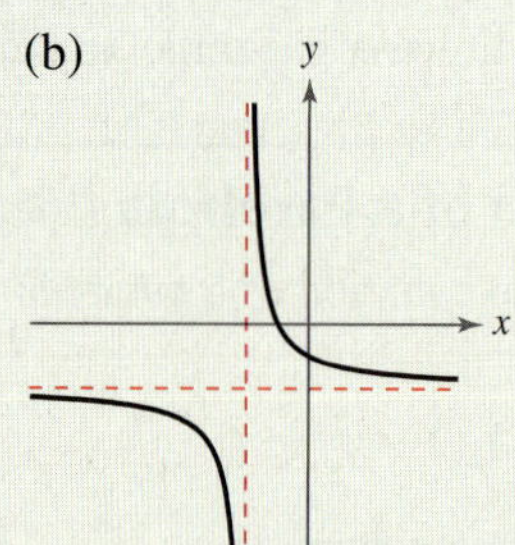

(c)

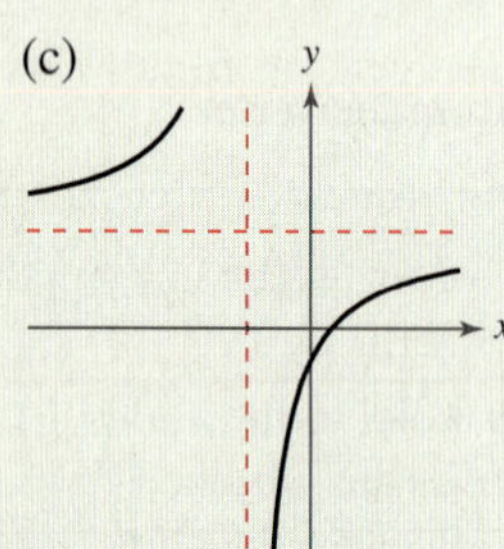

(d)

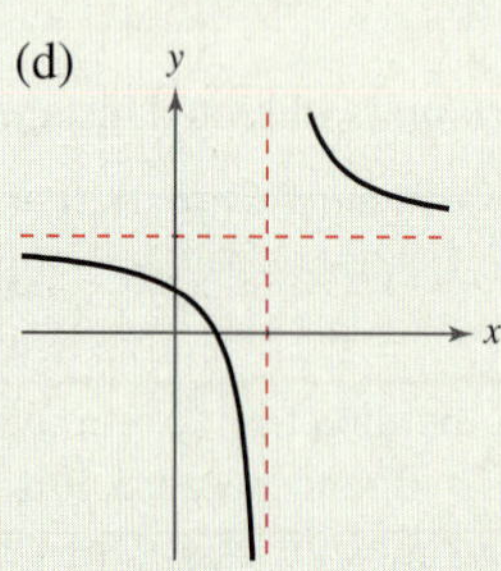

(i)	(ii)	(iii)	(iv)
$a > 0$	$a > 0$	$a < 0$	$a > 0$
$b < 0$	$b > 0$	$b > 0$	$b < 0$
$c > 0$	$c < 0$	$c > 0$	$c > 0$
$d < 0$	$d < 0$	$d < 0$	$d > 0$

2. Consider the function

$$f(x) = \frac{2x^2 + x - 1}{x + 1}.$$

 (a) Use a graphing utility to graph the function. Does the graph have a vertical asymptote at $x = -1$?

(b) Complete the table. What value is f approaching as x approaches -1 from the left and from the right?

x	-1.1	-1.01	-1.001	-1
$f(x)$				

x	-0.999	-0.99	-0.9
$f(x)$			

 (c) Use the *zoom* and *trace* features to determine the value of the graph near $x = -1$.

 (d) Use the following viewing window to adjust your graph in part (a).

Xmin: -9.4 Xmax: 9.4 Xscl: 1

Ymin: -6.2 Ymax: 6.2 Yscl: 1

What happens at the point $(-1, -3)$?

(e) Rewrite the function in simplified form. Evaluate the simplified function for $x = -1$.

(f) Sketch the graph of $g(x) = \dfrac{2x^2 - 5x - 12}{x - 4}$.

3. Consider the function

$$f(x) = \frac{ax}{(x - b)^2}.$$

(a) Determine the effect on the graph of f if $b \neq 0$ and a is varied. Consider cases in which a is positive and a is negative.

(b) Determine the effect on the graph of f if $a \neq 0$ and b is varied.

4. The table shows the world record times for running 1 mile, where y represents the year, with $y = 0$ corresponding to 1900, and t is the time in minutes and seconds.

y	t
23	4:10.4
33	4:07.6
45	4:01.3
54	3:59.4
58	3:54.5

y	t
66	3:51.3
79	3:48.9
85	3:46.3
99	3:43.1

A model for the data is

$$t = \frac{3.351y^2 + 42.461y - 543.730}{y^2}$$

where the seconds have been converted to decimal parts of a minute.

(a) Use a graphing utility to plot the data and graph the model in the same viewing window.

(b) Does there appear to be a limiting time for running 1 mile? Explain.

5. Statuary Hall is an elliptical room in the United States Capitol in Washington D.C. The room is also called the Whispering Gallery because a person standing at one focus of the room can hear even a whisper spoken by a person standing at the other focus. This occurs because any sound that is emitted from one focus of an ellipse will reflect off the side of the ellipse to the other focus. Statuary Hall is 46 feet wide and 97 feet long.
   (a) Find an equation that models the shape of the room.
   (b) How far apart are the two foci?
   (c) What is the area of the floor of the room? (The area of an ellipse is $A = \pi ab$.)

6. Use the figure to show that $|d_2 - d_1| = 2a$.

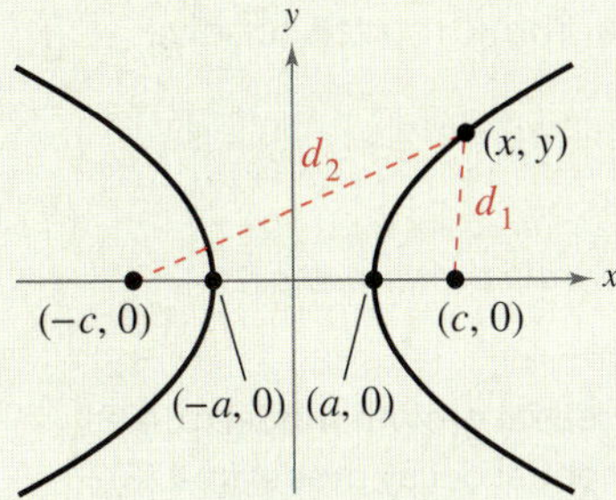

7. Find an equation of a hyperbola such that for any point on the hyperbola, the difference between its distances from the points $(2, 2)$ and $(10, 2)$ is 6.

8. The filament of a lightbulb is a thin wire that glows when electricity passes through it. The filament of a car headlight is at the focus of a parabolic reflector, which sends light out in a straight beam. Given that the filament is 1.5 inches from the vertex, find an equation for the cross section of the reflector. A reflector is 7 inches wide. How deep is it?

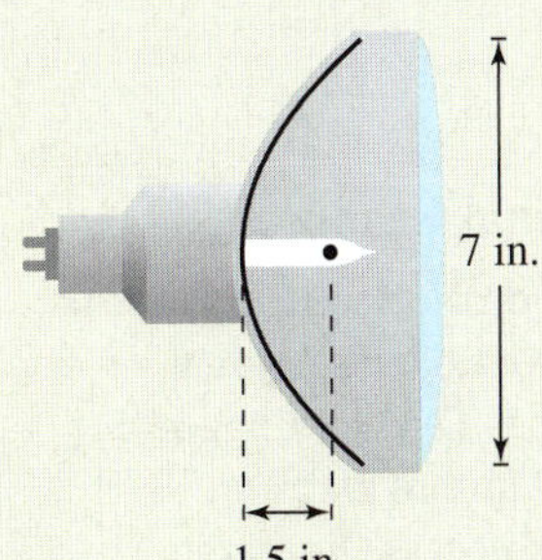

9. Consider the parabola $x^2 = 4py$.
   (a) Use a graphing utility to graph the parabola for $p = 1$, $p = 2$, $p = 3$, and $p = 4$. Describe the effect on the graph when p increases.
   (b) Locate the focus for each parabola in part (a).
   (c) For each parabola in part (a), find the length of the chord passing through the focus and parallel to the directrix. How can the length of this chord be determined directly from the standard form of the equation of the parabola?
   (d) Explain how the result of part (c) can be used as a sketching aid when graphing parabolas.

10. A tour boat travels between two islands that are 12 miles apart (see figure). For each trip between the islands, there is enough fuel for a 20-mile trip.
   (a) Explain why the region in which the boat can travel is bounded by an ellipse.
   (b) Let $(0, 0)$ represent the center of the ellipse. Find the coordinates of the center of each island.
   (c) The boat travels from one island, straight past the other island to the vertex of the ellipse, and back to the second island. How many miles does the boat travel? Use your answer to find the coordinates of the vertex.
   (d) Use the results of parts (b) and (c) to write an equation of the ellipse that bounds the region in which the boat can travel.

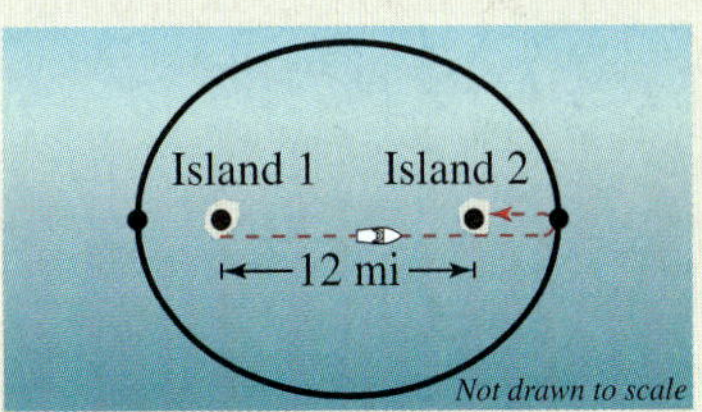

11. Prove that the graph of the equation

$$Ax^2 + Cy^2 + Dx + Ey + F = 0$$

is one of the following (except in degenerate cases).

Conic	*Condition*
(a) Circle	$A = C$
(b) Parabola	$A = 0$ or $C = 0$ (but not both)
(c) Ellipse	$AC > 0$
(d) Hyperbola	$AC < 0$

How to study Chapter 5

What you should learn

In this chapter you will learn the following skills and concepts:

- How to recognize and evaluate exponential and logarithmic functions
- How to graph exponential and logarithmic functions
- How to use the change-of-base formula to rewrite and evaluate logarithmic expressions
- How to use properties of logarithms to evaluate, rewrite, expand, or condense logarithmic expressions
- How to solve exponential and logarithmic equations
- How to use exponential growth models, exponential decay models, Gaussian models, logistic growth models, and logarithmic models to solve real-life problems

Important Vocabulary

As you encounter each new vocabulary term in this chapter, add the term and its definition to your notebook glossary.

Algebraic functions (p. 390)
Transcendental functions (p. 390)
Exponential function *f* with base *a* (p. 390)
Natural base *e* (p. 394)
Natural exponential function (p. 394)
Continuous compounding (p. 395)
Logarithmic function with base *a* (p. 401)
Common logarithmic function (p. 402)
Natural logarithmic function (p. 405)
Exponential growth model (p. 428)
Exponential decay model (p. 428)
Gaussian model (p. 428)
Logistic growth model (p. 428)
Logarithmic models (p. 428)
Bell-shaped curve (p. 432)
Logistic curve (p. 433)
Sigmoidal curve (p. 433)

Study Tools

Learning objectives in each section
Chapter Summary (p. 441)
Review Exercises (pp. 442–445)
Chapter Test (p. 446)
Cumulative Test for Chapters 3–5 (pp. 447, 448)

Additional Resources

Study and Solutions Guide
Interactive College Algebra
Videotapes/DVD for Chapter 5
College Algebra Website
Student Success Organizer

Bill Ross/Corbis

5

Exponential and Logarithmic Functions

5.1 Exponential Functions and Their Graphs

▶ **What you should learn**

- How to recognize and evaluate exponential functions with base *a*
- How to graph exponential functions
- How to recognize and evaluate exponential functions with base *e*
- How to use exponential functions to model and solve real-life applications

▶ **Why you should learn it**

Exponential functions can be used to model and solve real-life problems. For instance, in Exercise 59 on page 399, an exponential function is used to model the amount of defoliation caused by the gypsy moth.

Jenny Hager/The Image Works

Exponential Functions

So far, this book has dealt only with **algebraic functions,** which include polynomial functions and rational functions. In this chapter you will study two types of nonalgebraic functions—*exponential* functions and *logarithmic* functions. These functions are examples of **transcendental functions.**

> **Definition of Exponential Function**
>
> The **exponential function** f **with base** a is denoted by
>
> $$f(x) = a^x$$
>
> where $a > 0$, $a \neq 1$, and x is any real number.

The base $a = 1$ is excluded because it yields $f(x) = 1^x = 1$. This is a constant function, not an exponential function.

You already know how to evaluate a^x for integer and rational values of x. For example, you know that $4^3 = 64$ and $4^{1/2} = 2$. However, to evaluate 4^x for any real number x, you need to interpret forms with *irrational* exponents. For the purposes of this book, it is sufficient to think of

$$a^{\sqrt{2}} \quad (\text{where } \sqrt{2} \approx 1.41421356)$$

as the number that has the successively closer approximations

$$a^{1.4}, a^{1.41}, a^{1.414}, a^{1.4142}, a^{1.41421}, \ldots.$$

Example 1 shows how to use a calculator to evaluate exponential expressions.

Example 1 ▶ Evaluating Exponential Functions

Use a calculator to evaluate each function at the indicated value of x.

	Function	*Value*
a.	$f(x) = 2^x$	$x = -3.1$
b.	$f(x) = 2^{-x}$	$x = \pi$
c.	$f(x) = 12^x$	$x = \frac{5}{7}$
d.	$f(x) = 0.6^x$	$x = \frac{3}{2}$

Solution

	Function Value	*Graphing Calculator Keystrokes*	*Display*
a.	$f(-3.1) = 2^{-3.1}$	2 [^] [(−)] 3.1 [ENTER]	0.1166291
b.	$f(\pi) = 2^{-\pi}$	2 [^] [(−)] π [ENTER]	0.1133147
c.	$f\left(\frac{5}{7}\right) = 12^{5/7}$	12 [^] [(] 5 [÷] 7 [)] [ENTER]	5.8998877
d.	$f\left(\frac{3}{2}\right) = (0.6)^{3/2}$	.6 [^] [(] 3 [÷] 2 [)] [ENTER]	0.4647580

Graphs of Exponential Functions

The graphs of all exponential functions have similar characteristics, as shown in Examples 2, 3, and 4.

Example 2 ▶ Graphs of $y = a^x$

In the same coordinate plane, sketch the graph of each function.

a. $f(x) = 2^x$ **b.** $g(x) = 4^x$

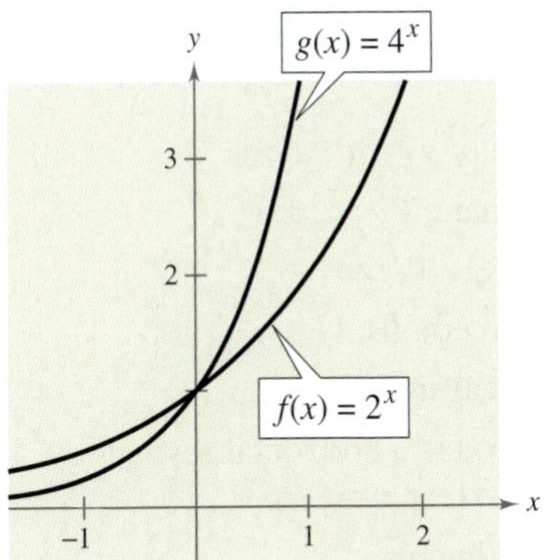

FIGURE 5.1

Solution

The table below lists some values for each function, and Figure 5.1 shows the graphs of the two functions. Note that both graphs are increasing. Moreover, the graph of $g(x) = 4^x$ is increasing more rapidly than the graph of $f(x) = 2^x$.

x	2^x	4^x
-2	$\frac{1}{4}$	$\frac{1}{16}$
-1	$\frac{1}{2}$	$\frac{1}{4}$
0	1	1
1	2	4
2	4	16
3	8	64

STUDY TIP

The table in Example 2 was evaluated by hand. You could, of course, use a graphing utility to construct tables with even more values.

Example 3 ▶ Graphs of $y = a^{-x}$

In the same coordinate plane, sketch the graph of each function.

a. $F(x) = 2^{-x}$ **b.** $G(x) = 4^{-x}$

Solution

The table below lists some values for each function, and Figure 5.2 shows the graphs of the two functions. Note that both graphs are decreasing. Moreover, the graph of $G(x) = 4^{-x}$ is decreasing more rapidly than the graph of $F(x) = 2^{-x}$.

x	-3	-2	-1	0	1	2
2^{-x}	8	4	2	1	$\frac{1}{2}$	$\frac{1}{4}$
4^{-x}	64	16	4	1	$\frac{1}{4}$	$\frac{1}{16}$

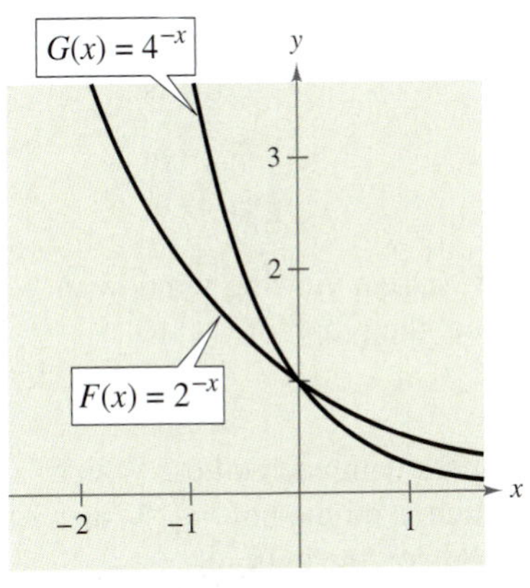

FIGURE 5.2

In Example 3, note that the functions $F(x) = 2^{-x}$ and $G(x) = 4^{-x}$ can be rewritten with positive exponents.

$$F(x) = 2^{-x} = \left(\frac{1}{2}\right)^x \quad \text{and} \quad G(x) = 4^{-x} = \left(\frac{1}{4}\right)^x$$

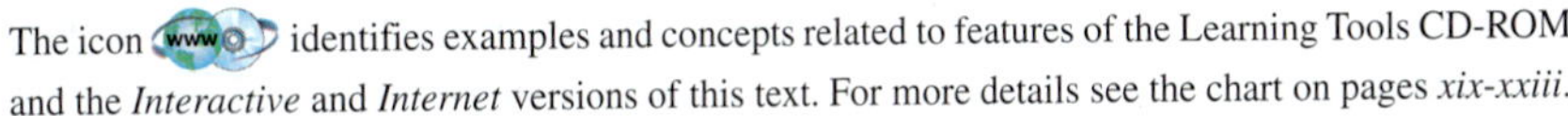
The icon identifies examples and concepts related to features of the Learning Tools CD-ROM and the *Interactive* and *Internet* versions of this text. For more details see the chart on pages *xix-xxiii*.

Comparing the functions in Examples 2 and 3, observe that

$$F(x) = 2^{-x} = f(-x) \qquad \text{and} \qquad G(x) = 4^{-x} = g(-x).$$

Consequently, the graph of F is a reflection (in the y-axis) of the graph of f. The graphs of G and g have the same relationship. The graphs in Figures 5.1 and 5.2 are typical of the exponential functions $y = a^x$ and $y = a^{-x}$. They have one y-intercept and one horizontal asymptote (the x-axis), and they are continuous. The basic characteristics of these exponential functions are summarized in Figures 5.3 and 5.4.

STUDY TIP

Notice that the range of an exponential function is $(0, \infty)$, which means that $a^x > 0$ for all values of x.

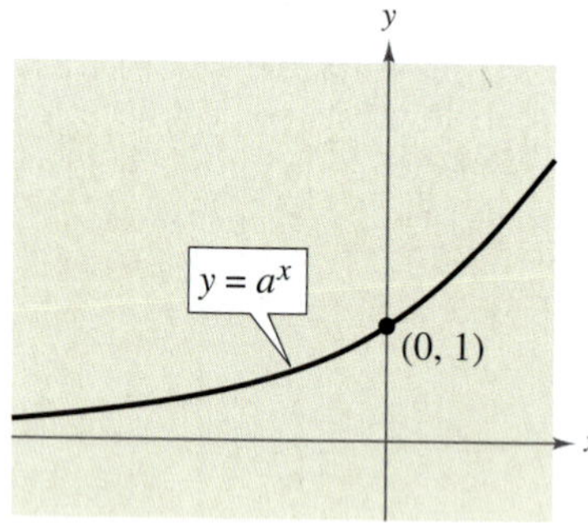

FIGURE 5.3

Graph of $y = a^x, a > 1$

- Domain: $(-\infty, \infty)$
- Range: $(0, \infty)$
- Intercept: $(0, 1)$
- Increasing
- x-Axis is a horizontal asymptote ($a^x \rightarrow 0$ as $x \rightarrow -\infty$)
- Continuous

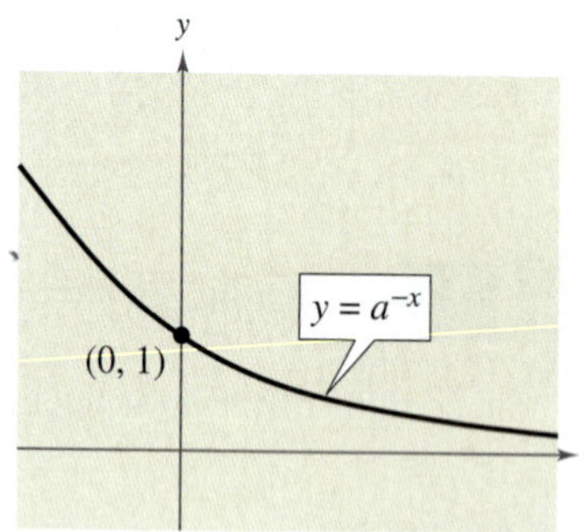

FIGURE 5.4

Graph of $y = a^{-x}, a > 1$

- Domain: $(-\infty, \infty)$
- Range: $(0, \infty)$
- Intercept: $(0, 1)$
- Decreasing
- x-Axis is a horizontal asymptote ($a^{-x} \rightarrow 0$ as $x \rightarrow \infty$)
- Continuous

Exploration

Use a graphing utility to graph

$$y = a^x$$

for $a = 3, 5,$ and 7 in the same viewing window. (Use a viewing window in which $-2 \le x \le 1$ and $0 \le y \le 2$.) For instance, the graph of

$$y = 3^x$$

is shown in Figure 5.5. How do the graphs compare with each other? Which graph is on the top in the interval $(-\infty, 0)$? Which is on the bottom? Which graph is on the top in the interval $(0, \infty)$? Which is on the bottom?

Repeat this experiment with the graphs of $y = b^x$ for $b = \frac{1}{3}, \frac{1}{5},$ and $\frac{1}{7}$. (Use a viewing window in which $-1 \le x \le 2$ and $0 \le y \le 2$.) What can you conclude about the shape of the graph of $y = b^x$ and the value of b?

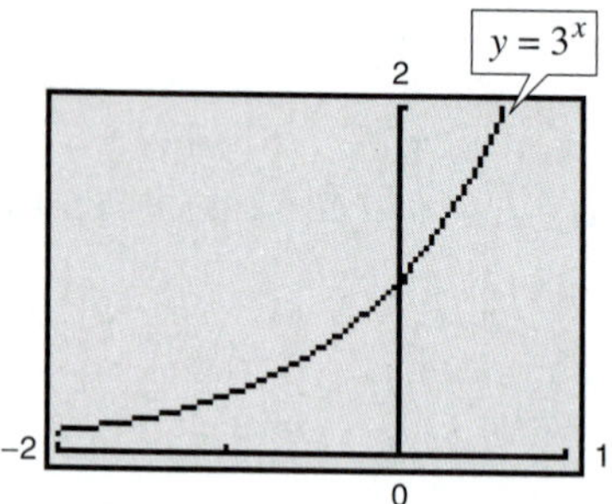

FIGURE 5.5

In the following example, notice how the graph of $y = a^x$ can be used to sketch the graphs of functions of the form $f(x) = b \pm a^{x+c}$.

Example 4 ▶ Transformations of Graphs of Exponential Functions

The *Interactive* CD-ROM and *Internet* versions of this text offer a Try It for each example in the text.

Each of the following graphs is a transformation of the graph of $f(x) = 3^x$.

a. Because $g(x) = 3^{x+1} = f(x + 1)$, the graph of g can be obtained by shifting the graph of f one unit to the *left*, as shown in Figure 5.6.

b. Because $h(x) = 3^x - 2 = f(x) - 2$, the graph of h can be obtained by shifting the graph of f *downward* two units, as shown in Figure 5.7.

c. Because $k(x) = -3^x = -f(x)$, the graph of k can be obtained by *reflecting* the graph of f in the x-axis, as shown in Figure 5.8.

d. Because $j(x) = 3^{-x} = f(-x)$, the graph of j can be obtained by *reflecting* the graph of f in the y-axis, as shown in Figure 5.9.

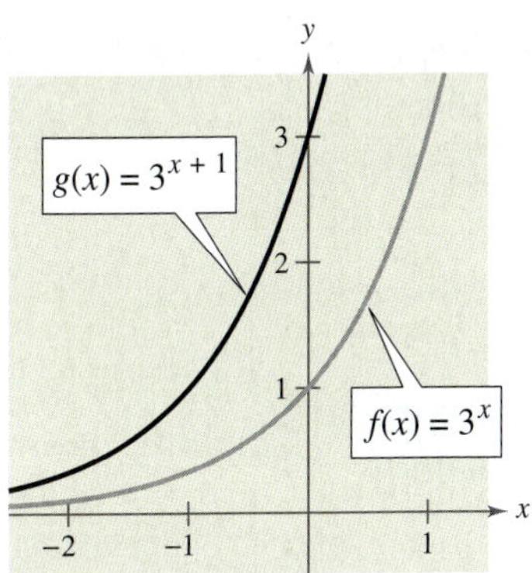

FIGURE 5.6 *Horizontal shift*

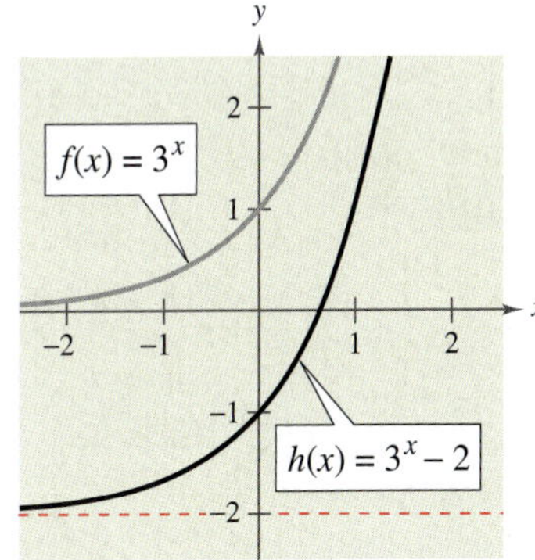

FIGURE 5.7 *Vertical shift*

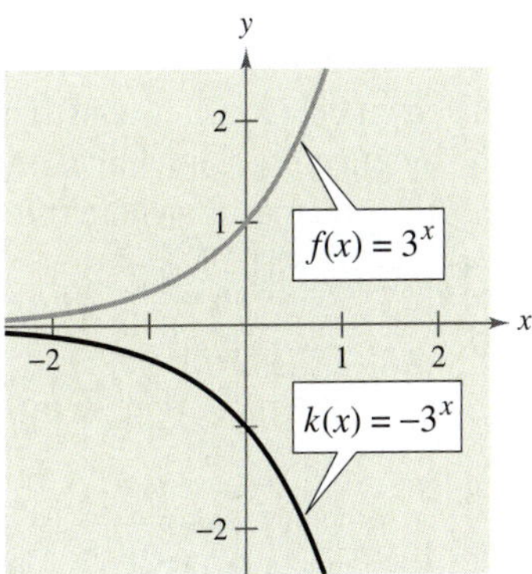

FIGURE 5.8 *Reflection in x-axis*

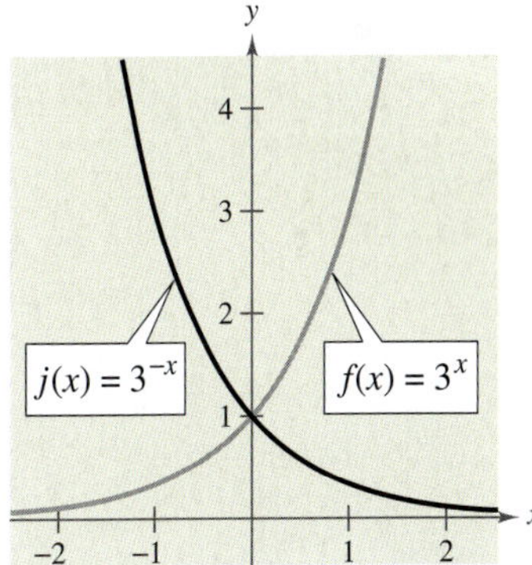

FIGURE 5.9 *Reflection in y-axis*

Notice that the transformations in Figures 5.6, 5.8, and 5.9 keep the x-axis as a horizontal asymptote, but the transformation in Figure 5.7 yields a new horizontal asymptote of $y = -2$. Also, be sure to note how the y-intercept is affected by each transformation.

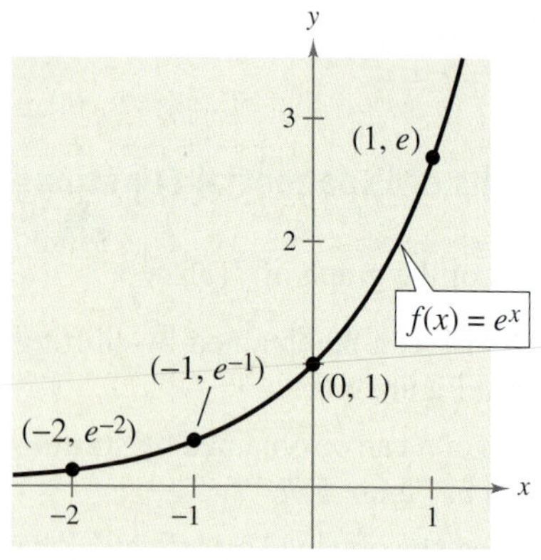

FIGURE 5.10

The Natural Base *e*

In many applications, the most convenient choice for a base is the irrational number

$$e \approx 2.718281828 \ldots.$$

This number is called the **natural base.** The function $f(x) = e^x$ is called the **natural exponential function.** Its graph is shown in Figure 5.10. Be sure you see that for the exponential function $f(x) = e^x$, e is the constant 2.718281828 . . . , whereas x is the variable.

Example 5 ▶ Evaluating the Natural Exponential Function 

Use a calculator to evaluate the function $f(x) = e^x$ at each indicated value of x.

a. $x = -2$ **b.** $x = -1$ **c.** $x = 0.25$ **d.** $x = -0.3$

Solution

Function Value	*Graphing Calculator Keystrokes*	*Display*
a. $f(-2) = e^{-2}$	[eˣ] [(−)] 2 [ENTER]	0.1353353
b. $f(-1) = e^{-1}$	[eˣ] [(−)] 1 [ENTER]	0.3678794
c. $f(0.25) = e^{0.25}$	[eˣ] 0.25 [ENTER]	1.2840254
d. $f(-0.3) = e^{-0.3}$	[eˣ] [(−)] 0.3 [ENTER]	0.7408182

FIGURE 5.11

Example 6 ▶ Graphing Natural Exponential Functions

Sketch the graph of each natural exponential function.

a. $f(x) = 2e^{0.24x}$ **b.** $g(x) = \frac{1}{2}e^{-0.58x}$

Solution

To sketch these two graphs, you can use a graphing utility to construct a table of values, as shown below. After constructing the table, plot the points and connect them with smooth curves, as shown in Figures 5.11 and 5.12. Note that the graph in Figure 5.11 is increasing whereas the graph in Figure 5.12 is decreasing.

x	$f(x)$	$g(x)$
−3	0.974	2.849
−2	1.238	1.595
−1	1.573	0.893
0	2.000	0.500
1	2.542	0.280
2	3.232	0.157
3	4.109	0.088

FIGURE 5.12

Exploration

Use the formula

$$A = P\left(1 + \frac{r}{n}\right)^{nt}$$

to calculate the amount in an account when $P = \$3000$, $r = 6\%$, $t = 10$ years, and compounding is done (1) by the day, (2) by the hour, (3) by the minute, and (4) by the second. Does increasing the number of compoundings per year result in unlimited growth of the amount in the account? Explain.

Applications

One of the most familiar examples of exponential growth is that of an investment earning *continuously compounded interest*. On page 136 in Section 1.6, you were introduced to the formula for the balance in an account that is compounded n times per year. Using exponential functions, you can now *develop* that formula and show how it leads to continuous compounding.

Suppose a principal P is invested at an annual interest rate r, compounded once a year. If the interest is added to the principal at the end of the year, the new balance P_1 is

$$\begin{aligned} P_1 &= P + Pr \\ &= P(1 + r). \end{aligned}$$

This pattern of multiplying the previous principal by $1 + r$ is then repeated each successive year, as shown below.

Year	*Balance After Each Compounding*
0	$P = P$
1	$P_1 = P(1 + r)$
2	$P_2 = P_1(1 + r) = P(1 + r)(1 + r) = P(1 + r)^2$
3	$P_3 = P_2(1 + r) = P(1 + r)^2(1 + r) = P(1 + r)^3$
	$\vdots$
t	$P_t = P(1 + r)^t$

To accommodate more frequent (quarterly, monthly, or daily) compounding of interest, let n be the number of compoundings per year and let t be the number of years. Then the rate per compounding is r/n and the account balance after t years is

$$A = P\left(1 + \frac{r}{n}\right)^{nt}. \qquad \text{Amount (balance) with } n \text{ compoundings per year}$$

If you let the number of compoundings n increase without bound, the process approaches what is called **continuous compounding.** In the formula for n compoundings per year, let $m = n/r$. This produces

$$\begin{aligned} A &= P\left(1 + \frac{r}{n}\right)^{nt} && \text{Amount with } n \text{ compoundings per year} \\ &= P\left(1 + \frac{r}{mr}\right)^{mrt} && \text{Substitute } mr \text{ for } n. \\ &= P\left(1 + \frac{1}{m}\right)^{mrt} && \text{Simplify.} \\ &= P\left[\left(1 + \frac{1}{m}\right)^{m}\right]^{rt}. && \text{Property of exponents} \end{aligned}$$

As m increases without bound, it can be shown that $[1 + (1/m)]^m$ approaches e. (Try the values $m = 10$, 10,000, and 10,000,000.) From this, you can conclude that the formula for continuous compounding is

$$A = Pe^{rt}. \qquad \text{Substitute } e \text{ for } (1 + 1/m)^m.$$

STUDY TIP

Be sure you see that the annual interest rate must be expressed in decimal form. For instance, 6% should be expressed as 0.06.

Formulas for Compound Interest

After t years, the balance A in an account with principal P and annual interest rate r (in decimal form) is given by the following formulas.

1. For n compoundings per year: $A = P\left(1 + \frac{r}{n}\right)^{nt}$
2. For continuous compounding: $A = Pe^{rt}$

Example 7 ▶ Compound Interest

A total of \$12,000 is invested at an annual interest rate of 9%. Find the balance after 5 years if it is compounded

a. quarterly.

b. monthly.

c. continuously.

Solution

a. For quarterly compoundings, you have $n = 4$. So, in 5 years at 9%, the balance is

$$A = P\left(1 + \frac{r}{n}\right)^{nt}$$ Formula for compound interest

$$= 12{,}000\left(1 + \frac{0.09}{4}\right)^{4(5)}$$ Substitute for P, r, n, and t.

$$\approx \$18{,}726.11.$$ Use a calculator.

b. For monthly compoundings, you have $n = 12$. So, in 5 years at 9%, the balance is

$$A = P\left(1 + \frac{r}{n}\right)^{nt}$$ Formula for compound interest

$$= 12{,}000\left(1 + \frac{0.09}{12}\right)^{12(5)}$$ Substitute for P, r, n, and t.

$$\approx \$18{,}788.17.$$ Use a calculator.

c. For continuous compounding, the balance is

$$A = Pe^{rt}$$ Formula for continuous compounding

$$= 12{,}000e^{0.09(5)}$$ Substitute for P, r, and t.

$$\approx \$18{,}819.75.$$ Use a calculator.

In Example 7, note that continuous compounding yields more than quarterly or monthly compounding. This is typical of the two types of compounding. That is, for a given principal, interest rate, and time, continuous compounding will always yield a larger balance than compounding n times a year.

Example 8 ▶ Radioactive Decay

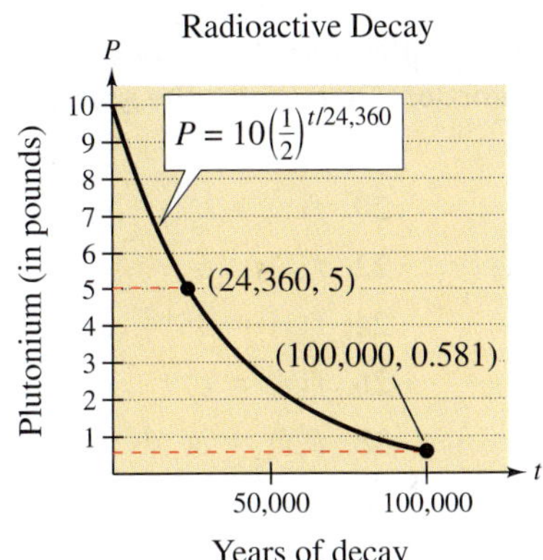

FIGURE 5.13

In 1986, a nuclear reactor accident occurred in Chernobyl in what was then the Soviet Union. The explosion spread highly toxic radioactive chemicals, such as plutonium, over hundreds of square miles, and the government evacuated the city and the surrounding area. To see why the city is now uninhabited, consider the model

$$P = 10\left(\frac{1}{2}\right)^{t/24,360}$$

which represents the amount of plutonium P that remains (from an initial amount of 10 pounds) after t years. Sketch the graph of this function over the interval from $t = 0$ to $t = 100,000$, where $t = 0$ represents 1986. How much of the 10 pounds will remain in the year 2005? How much of the 10 pounds will remain after 100,000 years?

Solution

The graph of this function is shown in Figure 5.13. Note from this graph that plutonium has a *half-life* of about 24,360 years. That is, after 24,360 years, *half* of the original amount will remain. After another 24,360 years, one-quarter of the original amount will remain, and so on. In the year 2005 ($t = 19$), there will still be

$$P = 10\left(\frac{1}{2}\right)^{19/24,360} \approx 10\left(\frac{1}{2}\right)^{0.0007800} \approx 9.995 \text{ pounds}$$

of plutonium remaining. After 100,000 years, there will still be

$$P = 10\left(\frac{1}{2}\right)^{100,000/24,360} \approx 10\left(\frac{1}{2}\right)^{4.105} \approx 0.581 \text{ pound}$$

of plutonium remaining.

The *Interactive* CD-ROM and *Internet* versions of this text offer a Quiz for every section of the text.

Writing ABOUT MATHEMATICS

Identifying Exponential Functions Which of the following functions generated the two tables below? Discuss how you were able to decide. What do these functions have in common? Are any of them the same? If so, explain why.

a. $f_1(x) = 2^{(x+3)}$ **b.** $f_2(x) = 8\left(\frac{1}{2}\right)^x$ **c.** $f_3(x) = \left(\frac{1}{2}\right)^{(x-3)}$

d. $f_4(x) = \left(\frac{1}{2}\right)^x + 7$ **e.** $f_5(x) = 7 + 2^x$ **f.** $f_6(x) = (8)2^x$

x	-1	0	1	2	3
$g(x)$	7.5	8	9	11	15

x	-2	-1	0	1	2
$h(x)$	32	16	8	4	2

Create two different exponential functions of the forms $y = a(b)^x$ and $y = c^x + d$ with y-intercepts of $(0, -3)$.

5.1 Exercises

The *Interactive* CD-ROM and *Internet* versions of this text contain step-by-step solutions to all odd-numbered exercises. They also provide Tutorial Exercises for additional help.

In Exercises 1–6, evaluate the function at the indicated value of x. Round your result to three decimal places.

Function	*Value*
1. $f(x) = 3.4^x$	$x = 5.6$
2. $f(x) = 2.3^x$	$x = \frac{3}{2}$
3. $f(x) = 5^x$	$x = -\pi$
4. $g(x) = 5000(2^x)$	$x = -1.5$
5. $h(x) = e^{-x}$	$x = \frac{3}{4}$
6. $f(x) = e^x$	$x = 3.2$

In Exercises 7–10, match the exponential function with its graph. [The graphs are labeled (a), (b), (c), and (d).]

(a)

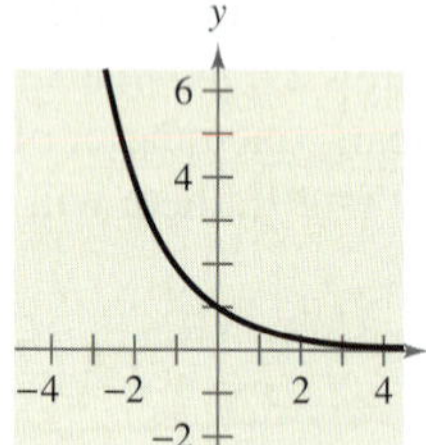

(b)

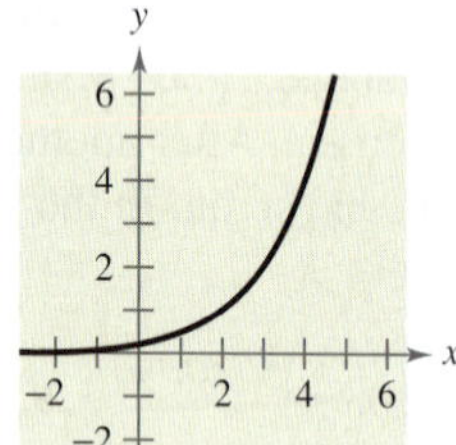

(c)

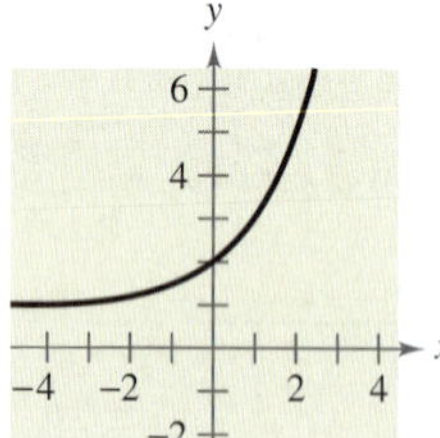

(d)

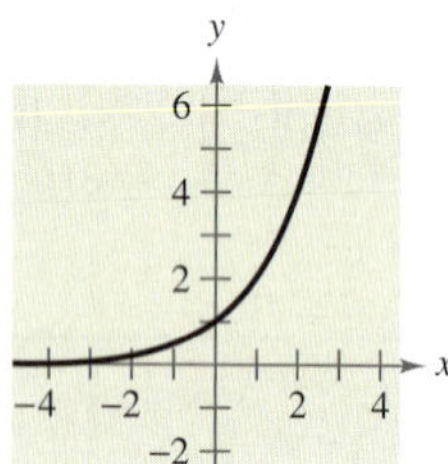

7. $f(x) = 2^x$

8. $f(x) = 2^x + 1$

9. $f(x) = 2^{-x}$

10. $f(x) = 2^{x-2}$

In Exercises 11–18, use the graph of f to describe the transformation that yields the graph of g.

11. $f(x) = 3^x$, $g(x) = 3^{x-4}$

12. $f(x) = 4^x$, $g(x) = 4^x + 1$

13. $f(x) = -2^x$, $g(x) = 5 - 2^x$

14. $f(x) = 10^x$, $g(x) = 10^{-x+3}$

15. $f(x) = \left(\frac{3}{5}\right)^x$, $g(x) = -\left(\frac{3}{5}\right)^{x+4}$

16. $f(x) = \left(\frac{7}{2}\right)^x$, $g(x) = -\left(\frac{7}{2}\right)^{-x+6}$

17. $f(x) = 0.3^x$, $g(x) = -0.3^x + 5$

18. $f(x) = 3.6^x$, $g(x) = -3.6^{-x} + 8$

In Exercises 19–32, use a graphing utility to construct a table of values for the function. Then sketch the graph of the function.

19. $f(x) = \left(\frac{1}{2}\right)^x$

20. $f(x) = \left(\frac{1}{2}\right)^{-x}$

21. $f(x) = 6^{-x}$

22. $f(x) = 6^x$

23. $f(x) = 2^{x-1}$

24. $f(x) = 3^{x+2}$

25. $f(x) = e^x$

26. $f(x) = e^{-x}$

27. $f(x) = 3e^{x+4}$

28. $f(x) = 2e^{-0.5x}$

29. $f(x) = 2e^{x-2} + 4$

30. $f(x) = 2 + e^{x-5}$

31. $f(x) = 4^{x-3} + 3$

32. $f(x) = -4^{x-3} - 3$

In Exercises 33–42, use a graphing utility to graph the exponential function.

33. $y = 2^{-x^2}$

34. $y = 3^{-|x|}$

35. $y = 3^{x-2} + 1$

36. $y = 4^{x+1} - 2$

37. $y = 1.08^{-5x}$

38. $y = 1.08^{5x}$

39. $s(t) = 2e^{0.12t}$

40. $s(t) = 3e^{-0.2t}$

41. $g(x) = 1 + e^{-x}$

42. $h(x) = e^{x-2}$

Compound Interest **In Exercises 43–46, complete the table to determine the balance A for P dollars invested at rate r for t years and compounded n times per year.**

n	1	2	4	12	365	Continuous
A						

43. $P = \$2500$, $r = 8\%$, $t = 10$ years

44. $P = \$1000$, $r = 6\%$, $t = 10$ years

45. $P = \$2500$, $r = 8\%$, $t = 20$ years

46. $P = \$1000$, $r = 6\%$, $t = 40$ years

Compound Interest **In Exercises 47–50, complete the table to determine the balance A for \$12,000 invested at rate r for t years, compounded continuously.**

t	10	20	30	40	50
A					

47. $r = 8\%$

48. $r = 6\%$

49. $r = 6.5\%$

50. $r = 7.5\%$

51. ***Trust Fund*** On the day of a child's birth, a deposit of \$25,000 is made in a trust fund that pays 8.75% interest, compounded continuously. Determine the balance in this account on the child's 25th birthday.

52. ***Trust Fund*** A deposit of \$5000 is made in a trust fund that pays 7.5% interest, compounded continuously. It is specified that the balance will be given to the college from which the donor graduated after the money has earned interest for 50 years. How much will the college receive?

53. ***Inflation*** If the annual rate of inflation averages 4% over the next 10 years, the approximate cost C of goods or services during any year in that decade will be modeled by $C(t) = P(1.04)^t$, where t is the time in years and P is the present cost. The price of an oil change for your car is presently \$23.95. Estimate the price 10 years from now.

54. ***Demand*** The demand equation for a product is

$$p = 5000\left(1 - \frac{4}{4 + e^{-0.002x}}\right).$$

(a) Use a graphing utility to graph the demand function for $x > 0$ and $p > 0$.

(b) Find the price p for a demand of $x = 500$ units.

(c) Use the graph in part (a) to approximate the greatest price that will still yield a demand of at least 600 units.

55. ***Population Growth*** A certain type of bacterium increases according to the model $P(t) = 100e^{0.2197t}$, where t is the time in hours. Find (a) $P(0)$, (b) $P(5)$, and (c) $P(10)$.

56. ***Population Growth*** The population of a town increases according to the model $P(t) = 2500e^{0.0293t}$, where t is the time in years, with $t = 0$ corresponding to 2000. Use the model to estimate the population in (a) 2010 and (b) 2020.

57. ***Radioactive Decay*** Let Q represent a mass of radioactive radium (^{226}Ra) (in grams), whose half-life is 1620 years. The quantity of radium present after t years is

$$Q = 25\left(\frac{1}{2}\right)^{t/1620}.$$

(a) Determine the initial quantity (when $t = 0$).

(b) Determine the quantity present after 1000 years.

(c) Use a graphing utility to graph the function over the interval $t = 0$ to $t = 5000$.

58. ***Radioactive Decay*** Let Q represent a mass of carbon 14 (^{14}C) (in grams), whose half-life is 5730 years. The quantity of carbon 14 present after t years is

$$Q = 10\left(\frac{1}{2}\right)^{t/5730}.$$

(a) Determine the initial quantity (when $t = 0$).

(b) Determine the quantity present after 2000 years.

(c) Sketch the graph of this function over the interval $t = 0$ to $t = 10{,}000$.

▶ Model It

59. ***Data Analysis*** To estimate the amount of defoliation caused by the gypsy moth during a given year, a forester counts the number x of egg masses on $\frac{1}{40}$ of an acre (circle of radius 18.6 feet) in the fall. The percent of defoliation y the next spring is shown in the table. (Source: USDA, Forest Service)

Egg masses, x	Percent of defoliation, y
0	12
25	44
50	81
75	96
100	99

A model for the data is

$$y = \frac{100}{1 + 7e^{-0.069x}}.$$

(a) Use a graphing utility to create a scatter plot of the data and graph the model in the same viewing window.

(b) Create a table that compares the model with the sample data.

(c) Estimate the percent of defoliation if 36 egg masses are counted on $\frac{1}{40}$ acre.

(d) You observe that $\frac{2}{3}$ of a forest is defoliated the following spring. Use the graph in part (a) to estimate the number of egg masses per $\frac{1}{40}$ acre.

60. ***Data Analysis*** A meteorologist measures the atmospheric pressure P (in pascals) at altitude h (in kilometers). The data is shown in the table.

Altitude, h	Pressure, P
0	101,293
5	54,735
10	23,294
15	12,157
20	5,069

A model for the data is given by

$P = 102{,}303e^{-0.137h}$.

(a) Sketch a scatter plot of the data and graph the model on the same set of axes.

(b) Estimate the atmospheric pressure at a height of 8 kilometers.

Synthesis

True or False? **In Exercises 61 and 62, determine whether the statement is true or false. Justify your answer.**

61. The line $y = -2$ is an asymptote for the graph of $f(x) = 10^x - 2$.

62. $e = \dfrac{271{,}801}{99{,}990}$.

Think About It **In Exercises 63–66, use properties of exponents to determine which functions (if any) are the same.**

63. $f(x) = 3^{x-2}$
$g(x) = 3^x - 9$
$h(x) = \frac{1}{9}(3^x)$

64. $f(x) = 4^x + 12$
$g(x) = 2^{2x+6}$
$h(x) = 64(4^x)$

65. $f(x) = 16(4^{-x})$
$g(x) = \left(\frac{1}{4}\right)^{x-2}$
$h(x) = 16(2^{-2x})$

66. $f(x) = 5^{-x} + 3$
$g(x) = 5^{3-x}$
$h(x) = -5^{x-3}$

67. Graph the functions $y = 3^x$ and $y = 4^x$ and use the graphs to solve the inequalities.

(a) $4^x < 3^x$ (b) $4^x > 3^x$

68. Graph the functions $y = \left(\frac{1}{2}\right)^x$ and $y = \left(\frac{1}{4}\right)^x$ and use the graphs to solve the inequalities.

(a) $\left(\frac{1}{4}\right)^x < \left(\frac{1}{2}\right)^x$ (b) $\left(\frac{1}{4}\right)^x > \left(\frac{1}{2}\right)^x$

69. Use a graphing utility to graph each function. Use the graph to find any asymptotes of the function.

(a) $f(x) = \dfrac{8}{1 + e^{-0.5x}}$ (b) $g(x) = \dfrac{8}{1 + e^{-0.5/x}}$

70. Use a graphing utility to graph each function. Use the graph to find where the function is increasing and decreasing, and approximate any relative maximum or minimum values.

(a) $f(x) = x^2e^{-x}$ (b) $g(x) = x2^{3-x}$

71. ***Graphical Analysis*** Use a graphing utility to graph

$$f(x) = \left(1 + \frac{0.5}{x}\right)^x \quad \text{and} \quad g(x) = e^{0.5}$$

in the same viewing window. What is the relationship between f and g as x increases and decreases without bound?

72. ***Conjecture*** Use the result of Exercise 71 to make a conjecture about the value of $[1 + (r/x)]^x$ as x increases without bound. Create a table that illustrates your conjecture for $r = 1$.

73. ***Think About It*** Which functions are exponential?

(a) $3x$ (b) $3x^2$

(c) 3^x (d) 2^{-x}

74. ***Writing*** Explain why $2^{\sqrt{2}}$ is greater than 2, but less than 4.

Review

In Exercises 75–78, solve for y.

75. $2x - 7y + 14 = 0$ **76.** $x^2 + 3y = 4$

77. $x^2 + y^2 = 25$ **78.** $x - |y| = 2$

In Exercises 79–82, sketch the graph of the rational function.

79. $f(x) = \dfrac{2}{9 + x}$ **80.** $f(x) = \dfrac{4x - 3}{x}$

81. $f(x) = \dfrac{6}{x^2 + 5x - 24}$ **82.** $f(x) = \dfrac{x^2 - 7x + 12}{x + 2}$

In Exercises 83–86, identify the conic and sketch its graph.

83. $16x^2 + 4y^2 = 64$ **84.** $x^2 + y^2 = 1$

85. $\dfrac{(x + 3)^2}{9} - \dfrac{(y - 2)^2}{36} = 1$

86. $\dfrac{(x - 1)^2}{25} + \dfrac{(y - 4)^2}{49} = 1$

5.2 Logarithmic Functions and Their Graphs

▶ What you should learn

- How to recognize and evaluate logarithmic functions with base a
- How to graph logarithmic functions
- How to recognize and evaluate natural logarithmic functions
- How to use logarithmic functions to model and solve real-life applications

▶ Why you should learn it

You can use logarithmic functions to model and solve real-life problems. For instance, in Exercise 57 on page 409, you can use a logarithmic function to approximate the length of a home mortgage.

Ryan McVay/Getty Images

Logarithmic Functions

In Section 2.7, you studied the concept of an inverse function. There, you learned that if a function is one-to-one—that is, if the function has the property that no horizontal line intersects the graph of the function more than once—the function must have an inverse function. By looking back at the graphs of the exponential functions introduced in Section 5.1, you will see that every function of the form

$$f(x) = a^x$$

passes the Horizontal Line Test and therefore must have an inverse function. This inverse function is called the **logarithmic function with base *a*.**

Definition of Logarithmic Function with Base *a*

For $x > 0$ and $0 < a \neq 1$,

$$y = \log_a x \text{ if and only if } x = a^y.$$

The function given by

$$f(x) = \log_a x$$

is called the **logarithmic function with base *a*.**

The equations

$$y = \log_a x \qquad \text{and} \qquad x = a^y$$

are equivalent. The first equation is in logarithmic form and the second is in exponential form.

When evaluating logarithms, remember that *a logarithm is an exponent*. This means that $\log_a x$ is the exponent to which a must be raised to obtain x. For instance, $\log_2 8 = 3$ because 2 must be raised to the third power to get 8.

Example 1 ▶ Evaluating Logarithms 

Use the definition of logarithmic function to evaluate each logarithm at the indicated value of x.

a. $f(x) = \log_2 x, \quad x = 32$ **b.** $f(x) = \log_3 x, \quad x = 1$

c. $f(x) = \log_4 x, \quad x = 2$ **d.** $f(x) = \log_{10} x, \quad x = \frac{1}{100}$

Solution

a. $f(32) = \log_2 32 = 5$ because $2^5 = 32.$

b. $f(1) = \log_3 1 = 0$ because $3^0 = 1.$

c. $f(2) = \log_4 2 = \frac{1}{2}$ because $4^{1/2} = \sqrt{4} = 2.$

d. $f\left(\frac{1}{100}\right) = \log_{10} \frac{1}{100} = -2$ because $10^{-2} = \frac{1}{10^2} = \frac{1}{100}.$

STUDY TIP

Remember that a logarithm is an exponent. So, to evaluate the logarithmic expression $\log_a x$, you need to ask the question, "To what power must a be raised to obtain x?"

Exploration

Complete the table for $f(x) = 10^x$.

x	-2	-1	0	1	2
$f(x)$					

Complete the table for $f(x) = \log_{10} x$.

x	$\frac{1}{100}$	$\frac{1}{10}$	1	10	100
$f(x)$					

Compare the two tables. What is the relationship between $f(x) = 10^x$ and $f(x) = \log_{10} x$?

The logarithmic function with base 10 is called the **common logarithmic function.** On most calculators, this function is denoted by LOG. Example 2 shows how to use a calculator to evaluate common logarithmic functions. You will learn how to use a calculator to calculate logarithms to any base in the next section.

Example 2 ▶ Evaluating Common Logarithms on a Calculator

Use a calculator to evaluate the function $f(x) = \log_{10} x$ at each value of x.

a. $x = 10$ **b.** $x = \frac{1}{3}$

c. $x = 2.5$ **d.** $x = -2$

Solution

	Function Value	*Graphing Calculator Keystrokes*	*Display*
a.	$f(10) = \log_{10} 10$	LOG 10 ENTER	1
b.	$f(\frac{1}{3}) = \log_{10} \frac{1}{3}$	LOG (1 ÷ 3) ENTER	-0.4771213
c.	$f(2.5) = \log_{10} 2.5$	LOG 2.5 ENTER	0.3979400
d.	$f(-2) = \log_{10}(-2)$	LOG (−) 2 ENTER	ERROR

Note that the calculator displays an error message (or a complex number) when you try to evaluate $\log_{10}(-2)$. The reason for this is that there is no real number power to which 10 can be raised to obtain -2.

The following properties follow directly from the definition of the logarithmic function with base a.

The logarithmic function can be one of the most difficult concepts for students to understand. Remind students that a logarithm is an exponent. Converting back and forth from logarithmic form to exponential form supports this concept.

Properties of Logarithms

1. $\log_a 1 = 0$ because $a^0 = 1$.
2. $\log_a a = 1$ because $a^1 = a$.
3. $\log_a a^x = x$ and $a^{\log_a x} = x$ — Inverse Properties
4. If $\log_a x = \log_a y$, then $x = y$. — One-to-One Property

Example 3 ▶ Using Properties of Logarithms

a. Solve for x: $\log_2 x = \log_2 3$ **b.** Solve for x: $\log_4 4 = x$

c. Simplify: $\log_5 5^x$ **d.** Simplify: $6^{\log_6 20}$

Solution

a. Using the One-to-One Property (Property 4), you can conclude that $x = 3$.

b. Using Property 2, you can conclude that $x = 1$.

c. Using the Inverse Property (Property 3), it follows that $\log_5 5^x = x$.

d. Using the Inverse Property (Property 3), it follows that $6^{\log_6 20} = 20$.

Graphs of Logarithmic Functions

To sketch the graph of $y = \log_a x$, you can use the fact that the graphs of inverse functions are reflections of each other in the line $y = x$.

Example 4 ▶ Graphs of Exponential and Logarithmic Functions

In the same coordinate plane, sketch the graph of each function.

a. $f(x) = 2^x$ **b.** $g(x) = \log_2 x$

Solution

a. For $f(x) = 2^x$, construct a table of values.

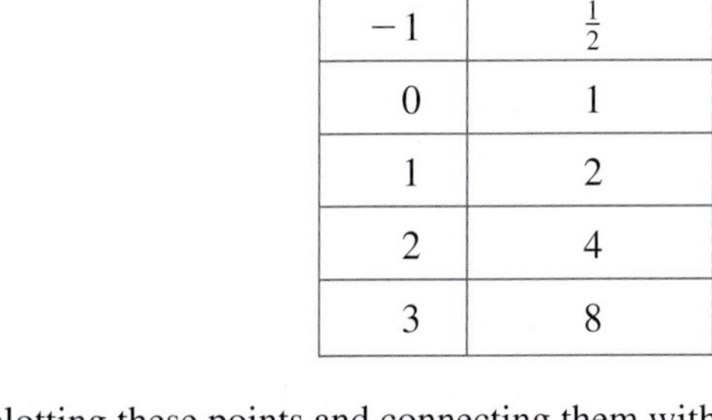

x	$f(x) = 2^x$
-2	$\frac{1}{4}$
-1	$\frac{1}{2}$
0	1
1	2
2	4
3	8

By plotting these points and connecting them with a smooth curve, you obtain the graph shown in Figure 5.14.

b. Because $g(x) = \log_2 x$ is the inverse of $f(x) = 2^x$, the graph of g is obtained by plotting the points $(f(x), x)$ and connecting them with a smooth curve. The graph of g is a reflection of the graph of f in the line $y = x$, as shown in Figure 5.14.

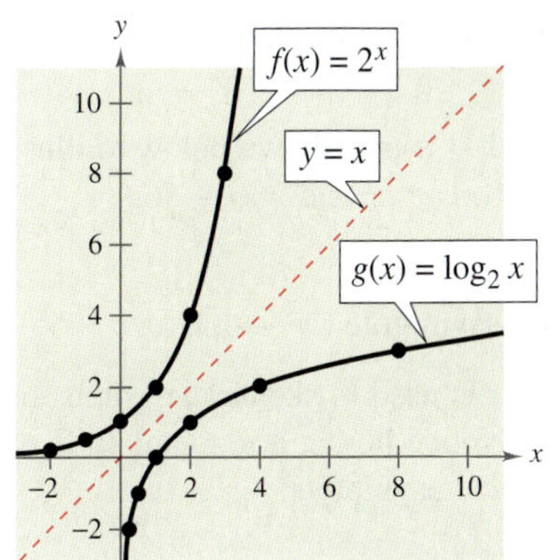

FIGURE 5.14

Example 5 ▶ Sketching the Graph of a Logarithmic Function

Sketch the graph of the common logarithmic function $f(x) = \log_{10} x$. Identify the x-intercept and the vertical asymptote.

Solution

Begin by constructing a table of values. Note that some of the values can be obtained without a calculator by using the Inverse Property of Logarithms. Others require a calculator. Next, plot the points and connect them with a smooth curve, as shown in Figure 5.15. The x-intercept of the graph is $(1, 0)$ and the vertical asymptote is $x = 0$ (y-axis).

	Without calculator				With calculator		
x	$\frac{1}{100}$	$\frac{1}{10}$	1	10	2	5	8
$\log_{10} x$	-2	-1	0	1	0.301	0.699	0.903

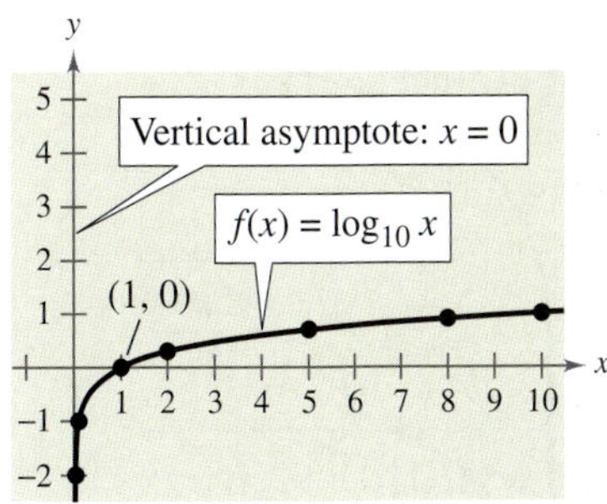

FIGURE 5.15

The nature of the graph in Figure 5.15 is typical of functions of the form $f(x) = \log_a x$, $a > 1$. They have one x-intercept and one vertical asymptote. Notice how slowly the graph rises for $x > 1$. The basic characteristics of logarithmic graphs are summarized in Figure 5.16.

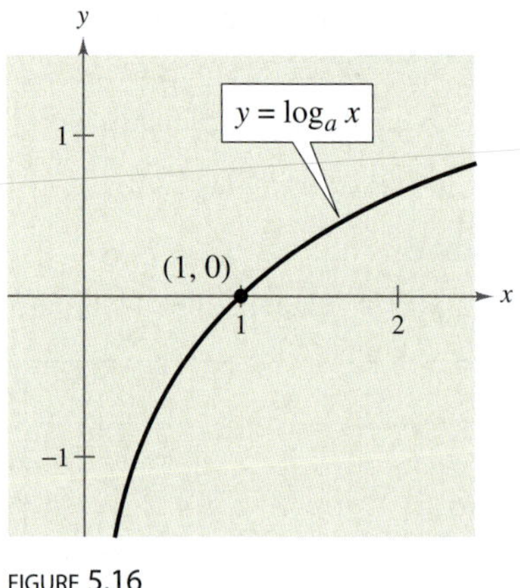

FIGURE 5.16

Graph of $y = \log_a x$, $a > 1$

- Domain: $(0, \infty)$
- Range: $(-\infty, \infty)$
- x-Intercept: $(1, 0)$
- Increasing
- One-to-one, therefore has an inverse function
- y-Axis is a vertical asymptote ($\log_a x \rightarrow -\infty$ as $x \rightarrow 0^+$).
- Continuous
- Reflection of graph of $y = a^x$ about the line $y = x$

The basic characteristics of the graph of $f(x) = a^x$ are shown below to illustrate the inverse relation between the functions $f(x) = a^x$ and $g(x) = \log_a x$.

- Domain: $(-\infty, \infty)$
- Range: $(0, \infty)$
- y-Intercept: $(0,1)$
- x-Axis is a horizontal asymptote ($a^x \rightarrow 0$ as $x \rightarrow -\infty$).

In the next example, the graph of $y = \log_a x$ is used to sketch the graphs of functions of the form $f(x) = b \pm \log_a(x + c)$. Notice how a horizontal shift of the graph results in a horizontal shift of the vertical asymptote.

STUDY TIP

You can use your understanding of transformations to identify vertical asymptotes of logarithmic functions. For instance, in Example 6(a) the graph of $g(x) = f(x - 1)$ shifts the graph of $f(x)$ one unit to the right. So the vertical asymptote of $g(x)$ is $x = 1$, one unit to the right of the asymptote of the graph of $f(x)$.

Example 6 ▶ Shifting Graphs of Logarithmic Functions 

The graph of each of the functions is similar to the graph of $f(x) = \log_{10} x$.

a. Because $g(x) = \log_{10}(x - 1) = f(x - 1)$, the graph of g can be obtained by shifting the graph of f one unit to the right, as shown in Figure 5.17.

b. Because $h(x) = 2 + \log_{10} x = 2 + f(x)$, the graph of h can be obtained by shifting the graph of f two units upward, as shown in Figure 5.18.

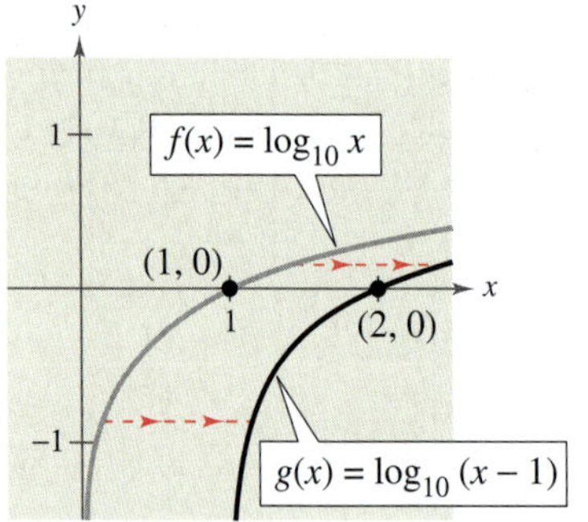

FIGURE 5.17

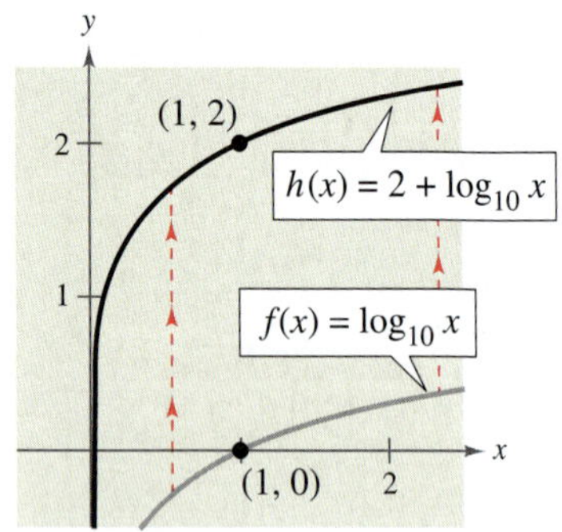

FIGURE 5.18

The Natural Logarithmic Function

By looking back at the graph of the natural exponential function introduced in Section 5.1, you will see that $f(x) = e^x$ is one-to-one and so has an inverse function. This inverse function is called the **natural logarithmic function** and is denoted by the special symbol ln x, read as "the natural log of x" or "el en of x."

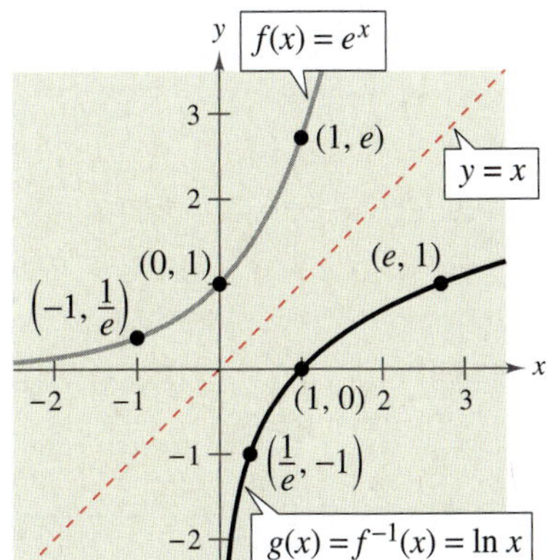

Reflection of graph of $f(x) = e^x$ about the line $y = x$
FIGURE 5.19

The Natural Logarithmic Function

The function defined by

$$f(x) = \log_e x = \ln x, \quad x > 0$$

is called the **natural logarithmic function.**

The definition above implies that the natural logarithmic function and the natural exponential function are inverse functions of each other. So, every logarithmic equation can be written in an equivalent exponential form and every exponential equation can be written in logarithmic form. That is, $y = \ln x$ and $x = e^y$ are equivalent equations.

Because the functions $f(x) = e^x$ and $g(x) = \ln x$ are inverse functions of each other, their graphs are reflections of each other in the line $y = x$. This reflective property is illustrated in Figure 5.19.

The four properties of logarithms listed on page 402 are also valid for natural logarithms.

STUDY TIP

Notice that as with every other logarithmic function, the domain of the natural logarithmic function is the set of *positive real numbers*—be sure you see that ln x is not defined for zero or for negative numbers.

Properties of Natural Logarithms

1. $\ln 1 = 0$ because $e^0 = 1$.
2. $\ln e = 1$ because $e^1 = e$.
3. $\ln e^x = x$ and $e^{\ln x} = x$ — Inverse Properties
4. If $\ln x = \ln y$, then $x = y$. — One-to-One Property

Example 7 ▶ Using Properties of Natural Logarithms 

Use the properties of natural logarithms to simplify each expression.

a. $\ln \frac{1}{e}$ **b.** $e^{\ln 5}$ **c.** $\frac{\ln 1}{3}$ **d.** $2 \ln e$

Solution

a. $\ln \frac{1}{e} = \ln e^{-1} = -1$ — Inverse Property

b. $e^{\ln 5} = 5$ — Inverse Property

c. $\frac{\ln 1}{3} = \frac{0}{3} = 0$ — Property 1

d. $2 \ln e = 2(1) = 2$ — Property 2

STUDY TIP

Some graphing utilities display a complex number instead of an ERROR message when evaluating an expression such as $\ln(-1)$.

On most calculators, the natural logarithm is denoted by [LN], as illustrated in Example 8.

Example 8 ▶ Evaluating the Natural Logarithmic Function 

Use a calculator to evaluate the function $f(x) = \ln x$ for each value of x.

a. $x = 2$ **b.** $x = 0.3$ **c.** $x = -1$ **d.** $x = 1 + \sqrt{2}$

Solution

	Function Value	*Graphing Calculator Keystrokes*	*Display*
a.	$f(2) = \ln 2$	[LN] 2 [ENTER]	0.6931472
b.	$f(0.3) = \ln 0.3$	[LN] .3 [ENTER]	–1.2039728
c.	$f(-1) = \ln(-1)$	[LN] [(–)] 1 [ENTER]	ERROR
d.	$f(1 + \sqrt{2}) = \ln(1 + \sqrt{2})$	[LN] [(] 1 [+] [√] 2 [)] [ENTER]	0.8813736

In Example 8, be sure you see that $\ln(-1)$ gives an error message on most calculators. This occurs because the domain of $\ln x$ is the set of positive real numbers (see Figure 5.19). So, $\ln(-1)$ is undefined.

Example 9 ▶ Finding the Domains of Logarithmic Functions

Find the domain of each function.

a. $f(x) = \ln(x - 2)$

b. $g(x) = \ln(2 - x)$

c. $h(x) = \ln x^2$

Solution

a. Because $\ln(x - 2)$ is defined only if $x - 2 > 0$, it follows that the domain of f is $(2, \infty)$. The graph of f is shown in Figure 5.20.

b. Because $\ln(2 - x)$ is defined only if $2 - x > 0$, it follows that the domain of g is $(-\infty, 2)$. The graph of g is shown in Figure 5.21.

c. Because $\ln x^2$ is defined only if $x^2 > 0$, it follows that the domain of h is all real numbers except $x = 0$. The graph of h is shown in Figure 5.22.

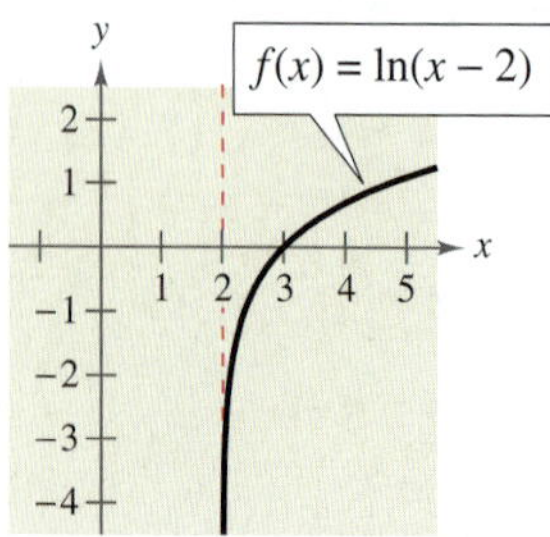

FIGURE 5.20

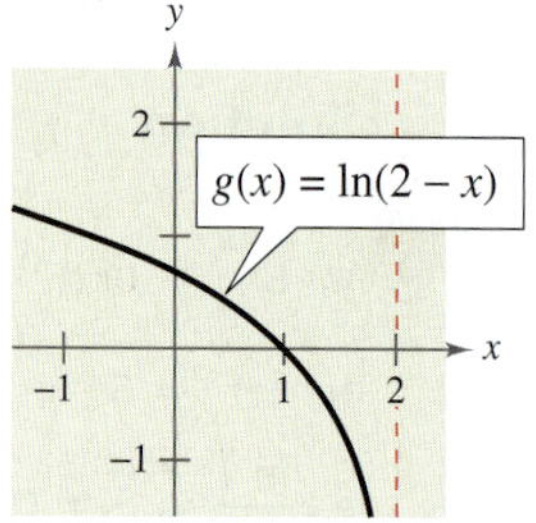

FIGURE 5.21

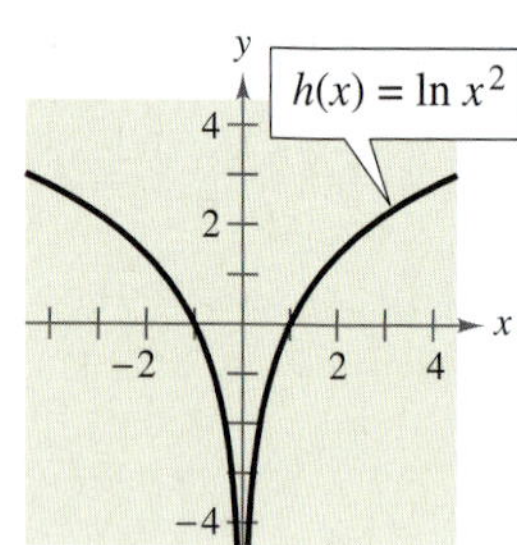

FIGURE 5.22

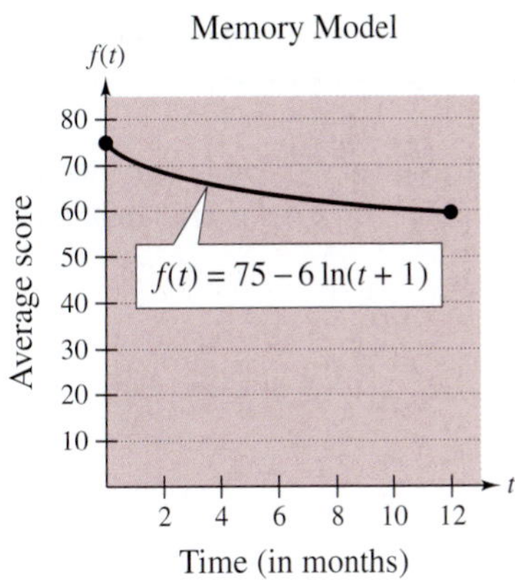

FIGURE 5.23

Application

Example 10 ▶ **Human Memory Model**  

Students participating in a psychology experiment attended several lectures on a subject and were given an exam. Every month for a year after the exam, the students were retested to see how much of the material they remembered. The average scores for the group are given by the *human memory model*

$$f(t) = 75 - 6\ln(t+1), \quad 0 \le t \le 12$$

where t is the time in months. The graph of f is shown in Figure 5.23.

a. What was the average score on the original ($t = 0$) exam?

b. What was the average score at the end of $t = 2$ months?

c. What was the average score at the end of $t = 6$ months?

Solution

a. The original average score was

$f(0) = 75 - 6\ln(0+1)$	Substitute 0 for t.
$= 75 - 6\ln 1$	Simplify.
$= 75 - 6(0)$	Property of natural logarithms
$= 75.$	Solution

b. After 2 months, the average score was

$f(2) = 75 - 6\ln(2+1)$	Substitute 2 for t.
$= 75 - 6\ln 3$	Simplify.
$\approx 75 - 6(1.0986)$	Use a calculator.
$\approx 68.4.$	Solution

c. After 6 months, the average score was

$f(6) = 75 - 6\ln(6+1)$	Substitute 6 for t.
$= 75 - 6\ln 7$	Simplify.
$\approx 75 - 6(1.9459)$	Use a calculator.
$\approx 63.3.$	Solution

Writing ABOUT MATHEMATICS

Analyzing a Human Memory Model Use a graphing utility to determine the time in months when the average score in Example 10 was 60. Explain your method of solving the problem. Describe another way that you can use a graphing utility to determine the answer.

5.2 Exercises

In Exercises 1–8, write the logarithmic equation in exponential form. For example, the exponential form of $\log_5 25 = 2$ is $5^2 = 25$.

1. $\log_4 64 = 3$ **2.** $\log_3 81 = 4$

3. $\log_7 \frac{1}{49} = -2$ **4.** $\log_{10} \frac{1}{1000} = -3$

5. $\log_{32} 4 = \frac{2}{5}$ **6.** $\log_{16} 8 = \frac{3}{4}$

7. $\ln \frac{1}{2} = -0.693 . . .$ **8.** $\ln 4 = 1.386 . . .$

In Exercises 9–18, write the exponential equation in logarithmic form. For example, the logarithmic form of $2^3 = 8$ is $\log_2 8 = 3$.

9. $5^3 = 125$ **10.** $8^2 = 64$

11. $81^{1/4} = 3$ **12.** $9^{3/2} = 27$

13. $6^{-2} = \frac{1}{36}$ **14.** $10^{-3} = 0.001$

15. $e^3 = 20.0855 . . .$ **16.** $e^{1/2} = 1.6487 . . .$

17. $e^x = 4$ **18.** $u^v = w$

In Exercises 19–26, evaluate the function at the indicated value of x without using a calculator.

	Function	*Value*
19.	$f(x) = \log_2 x$	$x = 16$
20.	$f(x) = \log_{16} x$	$x = 4$
21.	$f(x) = \log_7 x$	$x = 1$
22.	$f(x) = \log_{10} x$	$x = 10$
23.	$g(x) = \ln x$	$x = e^3$
24.	$g(x) = \ln x$	$x = e^{-2}$
25.	$g(x) = \log_a x$	$x = a^2$
26.	$g(x) = \log_b x$	$x = b^{-3}$

In Exercises 27–32, use a calculator to evaluate the function at the indicated value of x. Round your result to three decimal places.

	Function	*Value*
27.	$f(x) = \log_{10} x$	$x = \frac{4}{5}$
28.	$f(x) = \log_{10} x$	$x = 12.5$
29.	$f(x) = \ln x$	$x = 18.42$
30.	$f(x) = 3 \ln x$	$x = 0.32$
31.	$g(x) = 2 \ln x$	$x = 0.75$
32.	$g(x) = -\ln x$	$x = \frac{1}{2}$

In Exercises 33–38, use the graph of $y = \log_3 x$ to match the given function with its graph. [The graphs are labeled (a), (b), (c), (d), (e), and (f).]

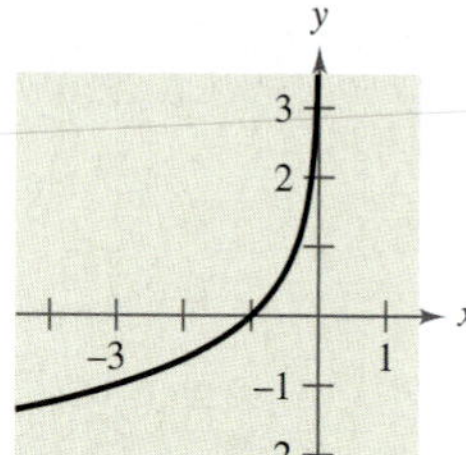

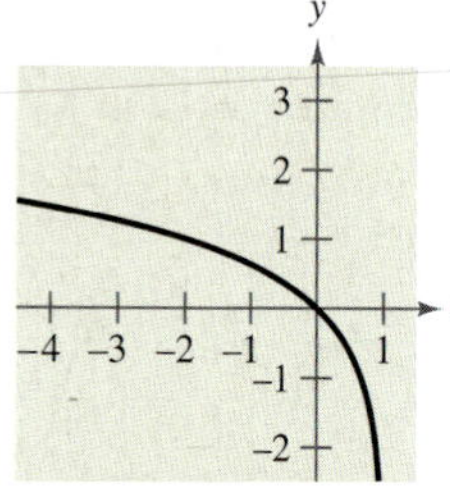

(c)

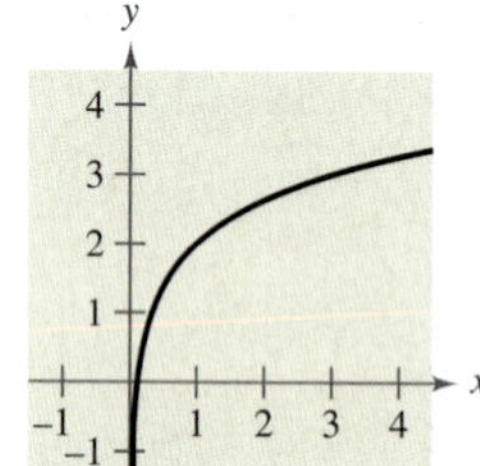

(d)

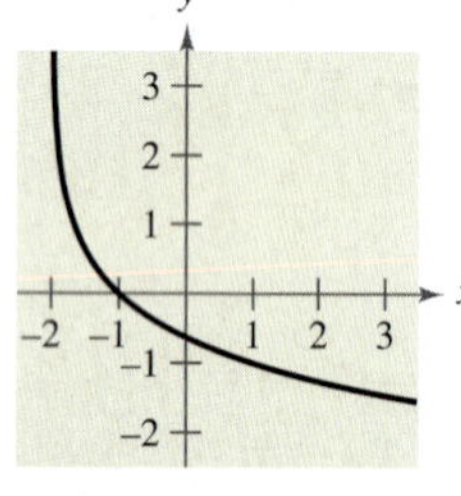

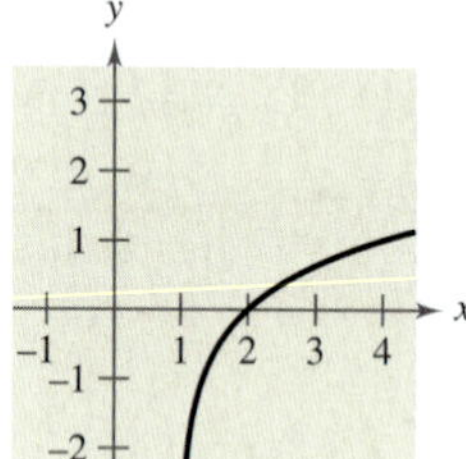

(f)

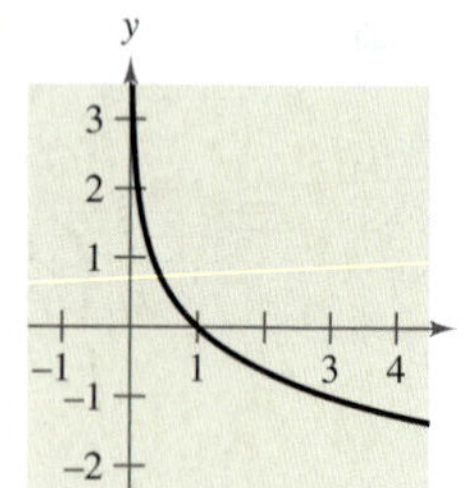

33. $f(x) = \log_3 x + 2$ **34.** $f(x) = -\log_3 x$

35. $f(x) = -\log_3(x + 2)$ **36.** $f(x) = \log_3(x - 1)$

37. $f(x) = \log_3(1 - x)$ **38.** $f(x) = -\log_3(-x)$

In Exercises 39–50, find the domain, x-intercept, and vertical asymptote of the logarithmic function and sketch its graph.

39. $f(x) = \log_4 x$ **40.** $g(x) = \log_6 x$

41. $y = -\log_3 x + 2$ **42.** $h(x) = \log_4(x - 3)$

43. $f(x) = -\log_6(x + 2)$ **44.** $y = \log_5(x - 1) + 4$

45. $y = \log_{10}\left(\frac{x}{5}\right)$ **46.** $y = \log_{10}(-x)$

47. $f(x) = \ln(x - 2)$ **48.** $h(x) = \ln(x + 1)$

49. $g(x) = \ln(-x)$ **50.** $f(x) = \ln(3 - x)$

In Exercises 51–56, use a graphing utility to graph the function. Be sure to use an appropriate viewing window.

51. $f(x) = \log_{10}(x + 1)$

52. $f(x) = \log_{10}(x - 1)$

53. $f(x) = \ln(x - 1)$

54. $f(x) = \ln(x + 2)$

55. $f(x) = \ln x + 2$

56. $f(x) = 3 \ln x - 1$

▶ Model It

57. ***Monthly Payment*** The model

$$t = 12.542 \ln\left(\frac{x}{x - 1000}\right), \qquad x > 1000$$

approximates the length of a home mortgage of \$150,000 at 8% in terms of the monthly payment. In the model, t is the length of the mortgage in years and x is the monthly payment in dollars (see figure).

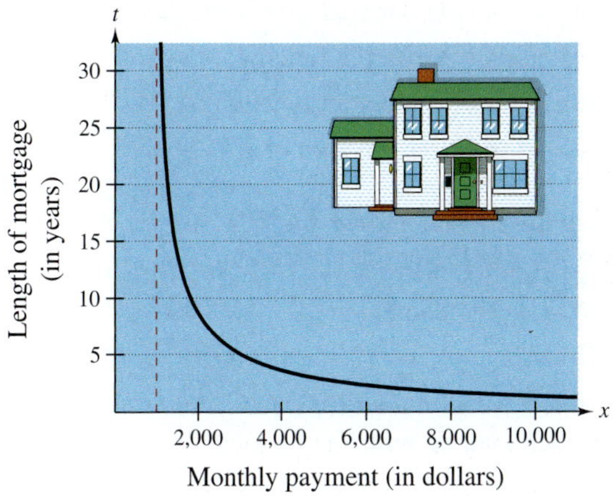

(a) Use the model to approximate the length of a \$150,000 mortgage at 8% when the monthly payment is \$1100.65 and when the monthly payment is \$1254.68.

(b) Approximate the total amount paid over the term of the mortgage with a monthly payment of \$1100.65 and with a monthly payment of \$1254.68.

(c) Approximate the total interest charge for a monthly payment of \$1100.65 and for a monthly payment of \$1254.68.

(d) What is the vertical asymptote for the model? Interpret its meaning in the context of the problem.

58. ***Compound Interest*** A principal P, invested at $9\frac{1}{2}\%$ and compounded continuously, increases to an amount K times the original principal after t years, where t is given by

$$t = \frac{\ln K}{0.095}.$$

(a) Complete the table and interpret your results.

K	1	2	4	6	8	10	12
t							

(b) Sketch a graph of the function.

59. ***Population*** The time t in years for the world population to double if it is increasing at a continuous rate of r is given by

$$t = \frac{\ln 2}{r}.$$

(a) Complete the table.

r	0.005	0.01	0.015	0.02	0.025	0.03
t						

(b) Sketch a graph of the function.

(c) Use a reference source to decide which value of r best approximates the actual rate of growth for the world population.

60. ***Sound Intensity*** The relationship between the number of decibels β and the intensity of a sound I in watts per square meter is

$$\beta = 10 \log_{10}\left(\frac{I}{10^{-12}}\right).$$

(a) Determine the number of decibels of a sound with an intensity of 1 watt per square meter.

(b) Determine the number of decibels of a sound with an intensity of 10^{-2} watt per square meter.

(c) The intensity of the sound in part (a) is 100 times as great as that in part (b). Is the number of decibels 100 times as great? Explain.

61. ***Human Memory Model*** Students in a mathematics class were given an exam and then retested monthly with an equivalent exam. The average scores for the class are given by the human memory model

$$f(t) = 80 - 17 \log_{10}(t + 1), \qquad 0 \le t \le 12$$

where t is the time in months.

(a) Use a graphing utility to graph the model over the specified domain.

(b) What was the average score on the original exam $(t = 0)$?

(c) What was the average score after 4 months?

(d) What was the average score after 10 months?

62. (a) Complete the table for the function

$$f(x) = \frac{\ln x}{x}.$$

x	1	5	10	10^2	10^4	10^6
$f(x)$						

(b) Use the table in part (a) to determine what value $f(x)$ approaches as x increases without bound.

(c) Use a graphing utility to confirm the result of part (b).

Synthesis

True or False? **In Exercises 63 and 64, determine whether the statement is true or false. Justify your answer.**

63. You can determine the graph of $f(x) = \log_6 x$ by graphing $g(x) = 6^x$ and reflecting it about the x-axis.

64. The graph of $f(x) = \log_3 x$ contains the point $(27, 3)$.

In Exercises 65–68, describe the relationship between the graphs of f and g. What is the relationship between the functions f and g?

65. $f(x) = 3^x$
$g(x) = \log_3 x$

66. $f(x) = 5^x$
$g(x) = \log_5 x$

67. $f(x) = e^x$
$g(x) = \ln x$

68. $f(x) = 10^x$
$g(x) = \log_{10} x$

69. ***Graphical Analysis*** Use a graphing utility to graph f and g in the same viewing window and determine which is increasing at the greater rate as x approaches $+\infty$. What can you conclude about the rate of growth of the natural logarithmic function?

(a) $f(x) = \ln x, \quad g(x) = \sqrt{x}$

(b) $f(x) = \ln x, \quad g(x) = \sqrt[4]{x}$

70. ***Think About It*** The table of values was obtained by evaluating a function. Determine which of the statements may be true and which must be false.

x	y
1	0
2	1
8	3

(a) y is an exponential function of x.

(b) y is a logarithmic function of x.

(c) x is an exponential function of y.

(d) y is a linear function of x.

In Exercises 71–73, answer the question for the function $f(x) = \log_{10} x$. Do not use a calculator.

71. What is the domain of f?

72. What is f^{-1}?

73. If x is a real number between 1000 and 10,000, in which interval will $f(x)$ be found?

74. ***Writing*** Explain why $\log_a x$ is defined only for $0 < a < 1$ and $a > 1$.

In Exercises 75 and 76, (a) use a graphing utility to graph the function, (b) use the graph to determine the intervals in which the function is increasing and decreasing, and (c) approximate any relative maximum or minimum values of the function.

75. $f(x) = |\ln x|$

76. $h(x) = \ln(x^2 + 1)$

Review

In Exercises 77 and 78, translate the statement into an algebraic expression.

77. The total cost for auto repairs if the cost of parts was \$83.95 and there were t hours of labor at \$37.50 per hour

78. The area of a rectangle if the length is 10 units more than the width w

5.3 Properties of Logarithms

▶ **What you should learn**

- How to use the change-of-base formula to rewrite and evaluate logarithmic expressions
- How to use properties of logarithms to evaluate or rewrite logarithmic expressions
- How to use properties of logarithms to expand or condense logarithmic expressions
- How to use logarithmic functions to model and solve real-life applications

▶ **Why you should learn it**

Logarithmic functions are often used to model scientific observations. For instance, in Exercise 81 on page 416, a logarithmic function is used to model human memory.

Gary Conner/PhotoEdit

Change of Base

Most calculators have only two types of log keys, one for common logarithms (base 10) and one for natural logarithms (base e). Although common logs and natural logs are the most frequently used, you may occasionally need to evaluate logarithms to other bases. To do this, you can use the following **change-of-base formula.**

Change-of-Base Formula

Let a, b, and x be positive real numbers such that $a \neq 1$ and $b \neq 1$. Then $\log_a x$ can be converted to a different base as follows.

Base b	*Base 10*	*Base e*
$\log_a x = \dfrac{\log_b x}{\log_b a}$	$\log_a x = \dfrac{\log_{10} x}{\log_{10} a}$	$\log_a x = \dfrac{\ln x}{\ln a}$

One way to look at the change-of-base formula is that logarithms to base a are simply *constant multiples* of logarithms to base b. The constant multiplier is $1/(\log_b a)$.

Example 1 ▶ Changing Bases Using Common Logarithms 

a. $\log_4 30 = \dfrac{\log_{10} 30}{\log_{10} 4}$ $\quad \log_a x = \dfrac{\log_{10} x}{\log_{10} a}$

$\approx \dfrac{1.47712}{0.60206}$ $\quad$ Use a calculator.

≈ 2.4534 $\quad$ Simplify.

b. $\log_2 14 = \dfrac{\log_{10} 14}{\log_{10} 2} \approx \dfrac{1.14613}{0.30103} \approx 3.8074$

Example 2 ▶ Changing Bases Using Natural Logarithms 

a. $\log_4 30 = \dfrac{\ln 30}{\ln 4}$ $\quad \log_a x = \dfrac{\ln x}{\ln a}$

$\approx \dfrac{3.40120}{1.38629}$ $\quad$ Use a calculator.

≈ 2.4535 $\quad$ Simplify.

b. $\log_2 14 = \dfrac{\ln 14}{\ln 2} \approx \dfrac{2.63906}{0.69315} \approx 3.8073$

Properties of Logarithms

You know from the preceding section that the logarithmic function with base a is the *inverse function* of the exponential function with base a. So, it makes sense that the properties of exponents should have corresponding properties involving logarithms. For instance, the exponential property $a^0 = 1$ has the corresponding logarithmic property $\log_a 1 = 0$.

STUDY TIP

There is no general property that can be used to rewrite $\log_a(u \pm v)$. Specifically, $\log_a(x + y)$ is not equal to $\log_a x + \log_a y$.

Properties of Logarithms

Let a be a positive number such that $a \neq 1$, and let n be a real number. If u and v are positive real numbers, the following properties are true.

Logarithm with Base a	*Natural Logarithm*
1. $\log_a(uv) = \log_a u + \log_a v$	**1.** $\ln(uv) = \ln u + \ln v$
2. $\log_a \frac{u}{v} = \log_a u - \log_a v$	**2.** $\ln \frac{u}{v} = \ln u - \ln v$
3. $\log_a u^n = n \log_a u$	**3.** $\ln u^n = n \ln u$

For a proof of the properties listed above, see Proofs in Mathematics on page 449.

Example 3 ▶ Using Properties of Logarithms

Write each logarithm in terms of ln 2 and ln 3.

a. $\ln 6$ **b.** $\ln \frac{2}{27}$

Solution

a. $\ln 6 = \ln(2 \cdot 3)$ Rewrite 6 as $2 \cdot 3$.

$= \ln 2 + \ln 3$ Property 1

b. $\ln \frac{2}{27} = \ln 2 - \ln 27$ Property 2

$= \ln 2 - \ln 3^3$ Rewrite 27 as 3^3.

$= \ln 2 - 3 \ln 3$ Property 3

The Granger Collection

Historical Note

John Napier, a Scottish mathematician, developed logarithms as a way to simplify some of the tedious calculations of his day. Beginning in 1594, Napier worked about 20 years on the invention of logarithms. Napier was only partially successful in his quest to simplify tedious calculations. Nonetheless, the development of logarithms was a step forward and received immediate recognition.

Example 4 ▶ Using Properties of Logarithms 

Use the properties of logarithms to verify that $-\log_{10} \frac{1}{100} = \log_{10} 100$.

Solution

$-\log_{10} \frac{1}{100} = -\log_{10}(100^{-1})$ Rewrite $\frac{1}{100}$ as 100^{-1}.

$= -(-1)\log_{10} 100$ Property 3

$= \log_{10} 100$ Simplify.

Try checking this result on your calculator.

Rewriting Logarithmic Expressions

The properties of logarithms are useful for rewriting logarithmic expressions in forms that simplify the operations of algebra. This is true because these properties convert complicated products, quotients, and exponential forms into simpler sums, differences, and products, respectively.

Example 5 ▶ Expanding Logarithmic Expressions 

Expand each logarithmic expression.

a. $\log_4 5x^3y$ **b.** $\ln \dfrac{\sqrt{3x-5}}{7}$

Exploration

Use a graphing utility to graph the functions

$$y_1 = \ln x - \ln(x-3)$$

and

$$y_2 = \ln \frac{x}{x-3}$$

in the same viewing window. Does the graphing utility show the functions with the same domain? If so, should it? Explain your reasoning.

Solution

a. $\log_4 5x^3y = \log_4 5 + \log_4 x^3 + \log_4 y$ Property 1

$= \log_4 5 + 3\log_4 x + \log_4 y$ Property 3

b. $\ln \dfrac{\sqrt{3x-5}}{7} = \ln \dfrac{(3x-5)^{1/2}}{7}$ Rewrite using rational exponent.

$= \ln(3x-5)^{1/2} - \ln 7$ Property 2

$= \dfrac{1}{2}\ln(3x-5) - \ln 7$ Property 3

In Example 5, the properties of logarithms were used to *expand* logarithmic expressions. In Example 6, this procedure is reversed and the properties of logarithms are used to *condense* logarithmic expressions.

Example 6 ▶ Condensing Logarithmic Expressions 

Condense each logarithmic expression.

a. $\frac{1}{2}\log_{10} x + 3\log_{10}(x+1)$ **b.** $2\ln(x+2) - \ln x$

c. $\frac{1}{3}[\log_2 x + \log_2(x+1)]$

Solution

a. $\frac{1}{2}\log_{10} x + 3\log_{10}(x+1) = \log_{10} x^{1/2} + \log_{10}(x+1)^3$ Property 3

$= \log_{10}\left[\sqrt{x}(x+1)^3\right]$ Property 1

b. $2\ln(x+2) - \ln x = \ln(x+2)^2 - \ln x$ Property 3

$= \ln \dfrac{(x+2)^2}{x}$ Property 2

c. $\frac{1}{3}[\log_2 x + \log_2(x+1)] = \frac{1}{3}\{\log_2[x(x+1)]\}$ Property 1

$= \log_2[x(x+1)]^{1/3}$ Property 3

$= \log_2 \sqrt[3]{x(x+1)}$ Rewrite with a radical.

Application

One method of determining how the x- and y-values for a set of nonlinear data are related begins by taking the natural log of each of the x- and y-values. If the points are graphed and fall on a straight line, then you can determine that the x- and y-values are related by the equation

$$\ln y = m \ln x$$

where m is the slope of the straight line.

Example 7 ▶ Finding a Mathematical Model 

The table shows the mean distance x and the period (the time it takes a planet to orbit the sun) y for each of the six planets that are closest to the sun. In the table, the mean distance is given in terms of astronomical units (where Earth's mean distance is defined as 1.0), and the period is given in terms of years. Find an equation that relates y and x.

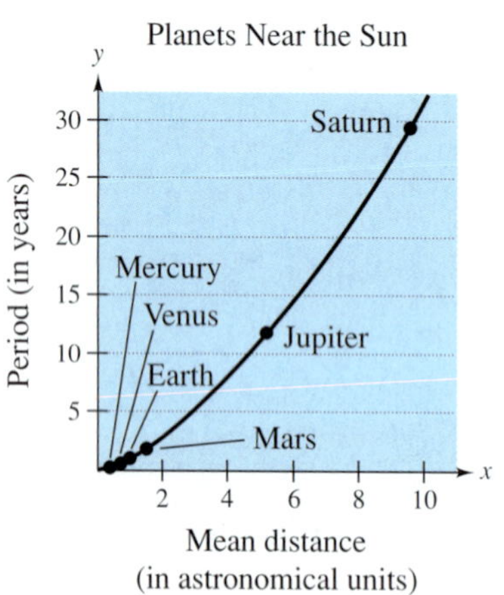

FIGURE 5.24

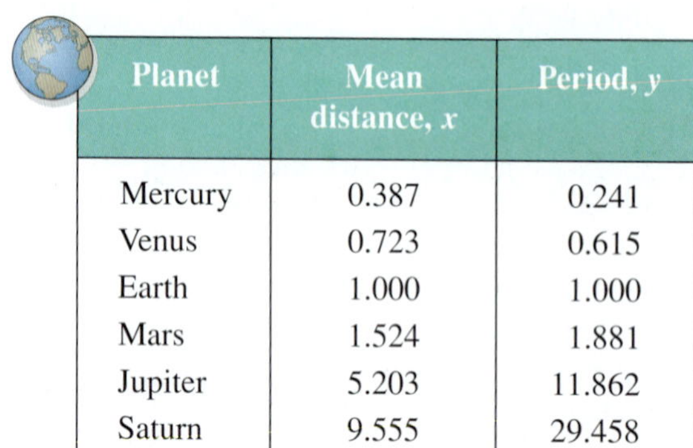

Planet	Mean distance, x	Period, y
Mercury	0.387	0.241
Venus	0.723	0.615
Earth	1.000	1.000
Mars	1.524	1.881
Jupiter	5.203	11.862
Saturn	9.555	29.458

Solution

The points in the table are plotted in Figure 5.24. From this figure it is not clear how to find an equation that relates y and x. To solve this problem, take the natural log of each of the x- and y-values in the table. This produces the following results.

Planet	$\ln x$	$\ln y$
Mercury	-0.949	-1.423
Venus	-0.324	-0.486
Earth	0.000	0.000
Mars	0.421	0.632
Jupiter	1.649	2.473
Saturn	2.257	3.383

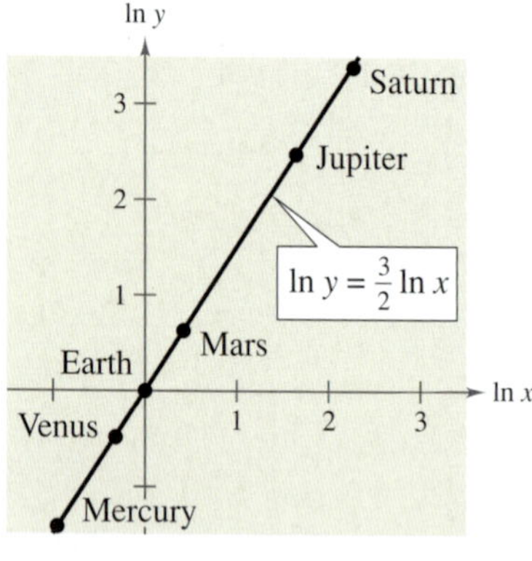

FIGURE 5.25

Now, by plotting the points in the second table, you can see that all six of the points appear to lie in a line (see Figure 5.25). You can use a graphical approach or the algebraic approach discussed in Section 3.5 to find that the slope of this line is $\frac{3}{2}$. You can therefore conclude that $\ln y = \frac{3}{2} \ln x$.

5.3 Exercises

In Exercises 1–8, rewrite the logarithm as a ratio of (a) common logarithms and (b) natural logarithms.

1. $\log_5 x$
2. $\log_3 x$
3. $\log_{1/5} x$
4. $\log_{1/3} x$
5. $\log_x \frac{3}{10}$
6. $\log_x \frac{3}{4}$
7. $\log_{2.6} x$
8. $\log_{7.1} x$

In Exercises 9–16, evaluate the logarithm using the change-of-base formula. Round your result to three decimal places.

9. $\log_3 7$
10. $\log_7 4$
11. $\log_{1/2} 4$
12. $\log_{1/4} 5$
13. $\log_9 0.4$
14. $\log_{20} 0.125$
15. $\log_{15} 1250$
16. $\log_3 0.015$

In Exercises 17–38, use the properties of logarithms to expand the expression as a sum, difference, and/or constant multiple of logarithms. (Assume all variables are positive.)

17. $\log_4 5x$
18. $\log_3 10z$
19. $\log_8 x^4$
20. $\log_{10} \frac{y}{2}$
21. $\log_5 \frac{5}{x}$
22. $\log_6 \frac{1}{z^3}$
23. $\ln \sqrt{z}$
24. $\ln \sqrt[3]{t}$
25. $\ln xyz^2$
26. $\log_{10} 4x^2 y$
27. $\ln z(z-1)^2,\ z > 1$
28. $\ln\left(\frac{x^2-1}{x^3}\right),\ x > 1$
29. $\log_2 \frac{\sqrt{a-1}}{9},\ a > 1$
30. $\ln \frac{6}{\sqrt{x^2+1}}$
31. $\ln \sqrt[3]{\frac{x}{y}}$
32. $\ln \sqrt{\frac{x^2}{y^3}}$
33. $\ln \frac{x^4\sqrt{y}}{z^5}$
34. $\log_2 \frac{\sqrt{x}\,y^4}{z^4}$
35. $\log_5 \frac{x^2}{y^2z^3}$
36. $\log_{10} \frac{xy^4}{z^5}$
37. $\ln \sqrt[4]{x^3(x^2+3)}$
38. $\ln \sqrt{x^2(x+2)}$

In Exercises 39–56, condense the expression to the logarithm of a single quantity.

39. $\ln x + \ln 3$
40. $\ln y + \ln t$
41. $\log_4 z - \log_4 y$
42. $\log_5 8 - \log_5 t$
43. $2 \log_2(x+4)$
44. $\frac{2}{3} \log_7(z-2)$
45. $\frac{1}{4} \log_3 5x$
46. $-4 \log_6 2x$
47. $\ln x - 3\ln(x+1)$
48. $2\ln 8 + 5\ln(z-4)$
49. $\log_{10} x - 2\log_{10} y + 3\log_{10} z$
50. $3\log_3 x + 4\log_3 y - 4\log_3 z$
51. $\ln x - 4[\ln(x+2) + \ln(x-2)]$
52. $4[\ln z + \ln(z+5)] - 2\ln(z-5)$
53. $\frac{1}{3}[2\ln(x+3) + \ln x - \ln(x^2-1)]$
54. $2[3\ln x - \ln(x+1) - \ln(x-1)]$
55. $\frac{1}{3}[\log_8 y + 2\log_8(y+4)] - \log_8(y-1)$
56. $\frac{1}{2}[\log_4(x+1) + 2\log_4(x-1)] + 6\log_4 x$

In Exercises 57 and 58, compare the logarithmic quantities. If two are equal, explain why.

57. $\frac{\log_2 32}{\log_2 4}$, $\log_2 \frac{32}{4}$, $\log_2 32 - \log_2 4$
58. $\log_7 \sqrt{70}$, $\log_7 35$, $\frac{1}{2} + \log_7 \sqrt{10}$

In Exercises 59–74, find the exact value of the logarithm without using a calculator. (If this is not possible, state the reason.)

59. $\log_3 9$
60. $\log_5 \frac{1}{125}$
61. $\log_2 \sqrt[4]{8}$
62. $\log_6 \sqrt[3]{6}$
63. $\log_4 16^{1.2}$
64. $\log_3 81^{-0.2}$
65. $\log_3(-9)$
66. $\log_2(-16)$
67. $\ln e^{4.5}$
68. $3\ln e^4$
69. $\ln \frac{1}{\sqrt{e}}$
70. $\ln \sqrt[4]{e^3}$
71. $\ln e^2 + \ln e^5$
72. $2\ln e^6 - \ln e^5$
73. $\log_5 75 - \log_5 3$
74. $\log_4 2 + \log_4 32$

In Exercises 75–80, use the properties of logarithms to rewrite and simplify the logarithmic expression.

75. $\log_4 8$

76. $\log_2(4^2 \cdot 3^4)$

77. $\log_5 \frac{1}{250}$

78. $\log_{10} \frac{9}{300}$

79. $\ln(5e^6)$

80. $\ln \dfrac{6}{e^2}$

▶ Model It

81. ***Human Memory Model*** Students participating in a psychology experiment attended several lectures and were given an exam. Every month for a year after the exam, the students were retested to see how much of the material they remembered. The average scores for the group can be modeled by the human memory model

$$f(t) = 90 - 15 \log_{10}(t + 1), \qquad 0 \le t \le 12$$

where t is the time in months.

(a) What was the average score on the original exam ($t = 0$)?

(b) What was the average score after 6 months?

(c) What was the average score after 12 months?

(d) When will the average score decrease to 75?

(e) Use the properties of logarithms to write the function in another form.

(f) Sketch the graph of the function over the specified domain.

82. ***Sound Intensity*** The relationship between the number of decibels β and the intensity of a sound I in watts per square meter is

$$\beta = 10 \log_{10}\left(\frac{I}{10^{-12}}\right).$$

Use the properties of logarithms to write the formula in simpler form, and determine the number of decibels of a sound with an intensity of 10^{-6} watt per square meter.

Synthesis

True or False? **In Exercises 83–88, determine whether the statement is true or false given that $f(x) = \ln x$. Justify your answer.**

83. $f(0) = 0$

84. $f(ax) = f(a) + f(x), \qquad a > 0, x > 0$

85. $f(x - 2) = f(x) - f(2), \qquad x > 2$

86. $\sqrt{f(x)} = \frac{1}{2}f(x)$

87. If $f(u) = 2f(v)$, then $v = u^2$.

88. If $f(x) < 0$, then $0 < x < 1$.

89. ***Proof*** Prove that $\log_b \dfrac{u}{v} = \log_b u - \log_b v$.

90. ***Proof*** Prove that $\log_b u^n = n \log_b u$.

In Exercises 91–96, use the change-of-base formula to rewrite the logarithm as a ratio of logarithms. Then use a graphing utility to graph both functions in the same viewing window to verify that the functions are equivalent.

91. $f(x) = \log_2 x$

92. $f(x) = \log_4 x$

93. $f(x) = \log_{1/2} x$

94. $f(x) = \log_{1/4} x$

95. $f(x) = \log_{11.8} x$

96. $f(x) = \log_{12.4} x$

97. ***Think About It*** Consider the functions below.

$$f(x) = \ln \frac{x}{2}, \quad g(x) = \frac{\ln x}{\ln 2}, \quad h(x) = \ln x - \ln 2$$

Which two functions should have identical graphs? Verify your answer by sketching the graphs of all three functions on the same set of coordinate axes.

98. ***Exploration*** For how many integers between 1 and 20 can the natural logarithms be approximated given that $\ln 2 \approx 0.6931$, $\ln 3 \approx 1.0986$, and $\ln 5 \approx 1.6094$? Approximate these logarithms (do not use a calculator).

Review

In Exercises 99–102, simplify the expression.

99. $\dfrac{24xy^{-2}}{16x^{-3}y}$

100. $\left(\dfrac{2x^2}{3y}\right)^{-3}$

101. $(18x^3y^4)^{-3}(18x^3y^4)^3$

102. $xy(x^{-1} + y^{-1})^{-1}$

In Exercises 103–106, solve the equation.

103. $3x^2 + 2x - 1 = 0$

104. $4x^2 - 5x + 1 = 0$

105. $\dfrac{2}{3x + 1} = \dfrac{x}{4}$

106. $\dfrac{5}{x - 1} = \dfrac{2x}{3}$

5.4 Exponential and Logarithmic Equations

▶ What you should learn

- How to solve simple exponential and logarithmic equations
- How to solve more complicated exponential equations
- How to solve more complicated logarithmic equations
- How to use exponential and logarithmic equations to model and solve real-life applications

▶ Why you should learn it

Applications of exponential and logarithmic equations are found in consumer safety testing. For instance, in Exercise 119, on page 426, a logarithmic function is used to model crumple zones for automobile crash tests.

David Woods/The Stock Market

Introduction

So far in this chapter, you have studied the definitions, graphs, and properties of exponential and logarithmic functions. In this section, you will study procedures for *solving equations* involving these exponential and logarithmic functions.

There are two basic strategies for solving exponential or logarithmic equations. The first is based on the One-to-One Properties and the second is based on the Inverse Properties. For $a > 0$ and $a \neq 1$, the following properties are true for all x and y for which $\log_a x$ and $\log_a y$ are defined.

One-to-One Properties

$a^x = a^y$ if and only if $x = y$.

$\log_a x = \log_a y$ if and only if $x = y$.

Inverse Properties

$a^{\log_a x} = x$

$\log_a a^x = x$

Example 1 ▶ Solving Simple Equations 

	Original Equation	*Rewritten Equation*	*Solution*	*Property*
a.	$2^x = 32$	$2^x = 2^5$	$x = 5$	One-to-One
b.	$\ln x - \ln 3 = 0$	$\ln x = \ln 3$	$x = 3$	One-to-One
c.	$\left(\frac{1}{3}\right)^x = 9$	$3^{-x} = 3^2$	$x = -2$	One-to-One
d.	$e^x = 7$	$\ln e^x = \ln 7$	$x = \ln 7$	Inverse
e.	$\ln x = -3$	$e^{\ln x} = e^{-3}$	$x = e^{-3}$	Inverse
f.	$\log_{10} x = -1$	$10^{\log_{10} x} = 10^{-1}$	$x = 10^{-1} = \frac{1}{10}$	Inverse

The strategies used in Example 1 are summarized as follows.

Strategies for Solving Exponential and Logarithmic Equations

1. Rewrite the original equation in a form that allows the use of the One-to-One Properties of exponential or logarithmic functions.
2. Rewrite an *exponential* equation in logarithmic form and apply the Inverse Property of logarithmic functions.
3. Rewrite a *logarithmic* equation in exponential form and apply the Inverse Property of exponential functions.

Solving Exponential Equations

Example 2 ▶ Solving Exponential Equations 

Solve each equation and approximate the result to three decimal places.

a. $4^x = 72$ **b.** $3(2^x) = 42$

Solution

a.

$4^x = 72$	Write original equation.
$\log_4 4^x = \log_4 72$	Take logarithm (base 4) of each side.
$x = \log_4 72$	Inverse Property
$x = \dfrac{\ln 72}{\ln 4}$	Change-of-base formula
$x \approx 3.085$	Use a calculator.

The solution is $x = \log_4 72 \approx 3.085$. Check this in the original equation.

b.

$3(2^x) = 42$	Write original equation.
$2^x = 14$	Divide each side by 3.
$\log_2 2^x = \log_2 14$	Take log (base 2) of each side.
$x = \log_2 14$	Inverse Property
$x = \dfrac{\ln 14}{\ln 2}$	Change-of-base formula
$x \approx 3.807$	Use a calculator.

The solution is $x = \log_2 14 \approx 3.807$. Check this in the original equation.

In Example 2(a), the exact solution is $x = \log_4 72$ and the approximate solution is $x \approx 3.085$. An exact answer is preferred when the solution is an intermediate step in a larger problem. For a final answer, an approximate solution is easier to comprehend.

Technology

When solving an exponential or logarithmic equation, remember that you can check your solution graphically by "graphing the left and right sides separately" and using the *intersect* feature of your graphing utility to determine the point of intersection. For instance, to check the solution of the equation in Example 2(a), you can graph

$$y_1 = 4^x \quad \text{and} \quad y_2 = 72$$

in the same viewing window, as shown below. Using the *intersect* feature of your graphing utility, you can determine that the graphs intersect when $x \approx 3.085$, which confirms the solution found in Example 2(a).

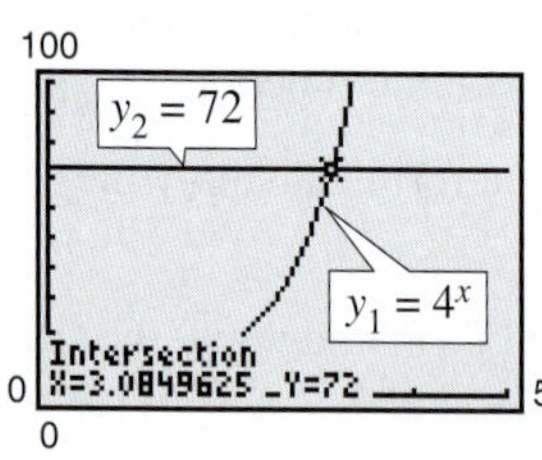

Example 3 ▶ Solving an Exponential Equation 

Solve $e^x + 5 = 60$ and approximate the result to three decimal places.

Solution

$e^x + 5 = 60$	Write original equation.
$e^x = 55$	Subtract 5 from each side.
$\ln e^x = \ln 55$	Take natural log of each side.
$x = \ln 55$	Inverse Property
$x \approx 4.007$	Use a calculator.

The solution is $x = \ln 55 \approx 4.007$. Check this in the original equation.

Example 4 ▶ Solving an Exponential Equation

Solve $2(3^{2t-5}) - 4 = 11$ and approximate the result to three decimal places.

Solution

$$2(3^{2t-5}) - 4 = 11 \quad \text{Write original equation.}$$

$$2(3^{2t-5}) = 15 \quad \text{Add 4 to each side.}$$

$$3^{2t-5} = \frac{15}{2} \quad \text{Divide each side by 2.}$$

$$\log_3 3^{2t-5} = \log_3 \frac{15}{2} \quad \text{Take log (base 3) of each side.}$$

$$2t - 5 = \log_3 \frac{15}{2} \quad \text{Inverse Property}$$

$$2t = 5 + \log_3 7.5 \quad \text{Add 5 to each side.}$$

$$t = \frac{5}{2} + \frac{1}{2}\log_3 7.5 \quad \text{Divide each side by 2.}$$

$$t \approx 3.417 \quad \text{Use a calculator.}$$

The solution is $t = \frac{5}{2} + \frac{1}{2}\log_3 7.5 \approx 3.417$. Check this in the original equation.

STUDY TIP

Remember that to evaluate a logarithm such as $\log_3 7.5$, you need to use the change-of-base formula.

$$\log_3 7.5 = \frac{\ln 7.5}{\ln 3} \approx 1.834$$

When an equation involves two or more exponential expressions, you can still use a procedure similar to that demonstrated in Examples 2, 3, and 4. However, the algebra is a bit more complicated.

Example 5 ▶ Solving an Exponential Equation of Quadratic Type

Solve $e^{2x} - 3e^x + 2 = 0$.

Solution

$$e^{2x} - 3e^x + 2 = 0 \quad \text{Write original equation.}$$

$$(e^x)^2 - 3e^x + 2 = 0 \quad \text{Write in quadratic form.}$$

$$(e^x - 2)(e^x - 1) = 0 \quad \text{Factor.}$$

$$e^x - 2 = 0 \quad \text{Set 1st factor equal to 0.}$$

$$x = \ln 2 \quad \text{Solution}$$

$$e^x - 1 = 0 \quad \text{Set 2nd factor equal to 0.}$$

$$x = 0 \quad \text{Solution}$$

The solutions are $x = \ln 2$ and $x = 0$. Check these in the original equation.

In Example 5, use a graphing utility to graph $y = e^{2x} - 3e^x + 2$. The graph should have two x-intercepts: one at $x = \ln 2 \approx 0.693$ and one at $x = 0$.

Solving Logarithmic Equations

To solve a logarithmic equation such as

$\ln x = 3$ Logarithmic form

write the equation in exponential form as follows.

$e^{\ln x} = e^3$ Exponentiate each side.

$x = e^3$ Exponential form

This procedure is called *exponentiating* each side of an equation.

Example 6 ▶ Solving a Logarithmic Equation 

a. Solve $\ln x = 2$.

b. Solve $\log_3(5x - 1) = \log_3(x + 7)$.

Solution

a. $\ln x = 2$ Write original equation.

$e^{\ln x} = e^2$ Exponentiate each side.

$x = e^2$ Inverse Property

The solution is $x = e^2$. Check this in the original equation.

b. $\log_3(5x - 1) = \log_3(x + 7)$ Write original equation.

$5x - 1 = x + 7$ One-to-One Property

$4x = 8$ Add $-x$ and 1 to each side.

$x = 2$ Divide each side by 4.

The solution is $x = 2$. Check this in the original equation.

Example 7 ▶ Solving a Logarithmic Equation 

Solve $5 + 2 \ln x = 4$ and approximate the result to three decimal places.

Solution

$5 + 2 \ln x = 4$ Write original equation.

$2 \ln x = -1$ Subtract 5 from each side.

$\ln x = -\frac{1}{2}$ Divide each side by 2.

$e^{\ln x} = e^{-1/2}$ Exponentiate each side.

$x = e^{-1/2}$ Inverse Property

$x \approx 0.607$ Use a calculator.

The solution is $x = e^{-1/2} \approx 0.607$. Check this in the original equation.

Example 8 ▶ Solving a Logarithmic Equation

Solve $2 \log_5 3x = 4$.

Solution

$$2 \log_5 3x = 4 \quad \text{Write original equation.}$$

$$\log_5 3x = 2 \quad \text{Divide each side by 2.}$$

$$5^{\log_5 3x} = 5^2 \quad \text{Exponentiate each side (base 5).}$$

$$3x = 25 \quad \text{Inverse Property}$$

$$x = \frac{25}{3} \quad \text{Divide each side by 3.}$$

The solution is $x = \frac{25}{3}$. Check this in the original equation.

STUDY TIP

Notice in Example 9 that the logarithmic part of the equation is condensed into a single logarithm before exponentiating each side of the equation.

Because the domain of a logarithmic function generally does not include all real numbers, you should be sure to check for extraneous solutions of logarithmic equations.

Technology

You can use a graphing utility to verify that the equation in Example 9 has $x = 5$ as its only solution. Graph

$$y_1 = \log_{10} 5x + \log_{10}(x - 1)$$

and

$$y_2 = 2$$

in the same viewing window. From the graph shown below, it appears that the graphs of the two equations intersect at one point. Use the *intersect* feature or the *zoom* and *trace* features to determine that $x = 5$ is an approximate solution. You can verify this algebraically by substituting $x = 5$ into the original equation.

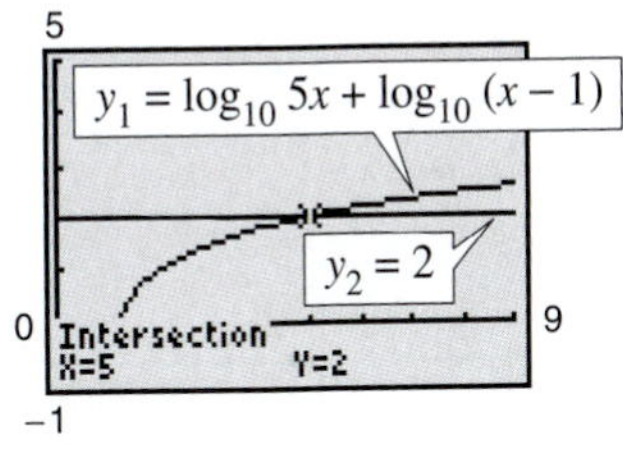

Example 9 ▶ Checking for Extraneous Solutions

Solve $\log_{10} 5x + \log_{10}(x - 1) = 2$.

Solution

$$\log_{10} 5x + \log_{10}(x - 1) = 2 \quad \text{Write original equation.}$$

$$\log_{10}[5x(x - 1)] = 2 \quad \text{Product Property of Logarithms}$$

$$10^{\log_{10}(5x^2 - 5x)} = 10^2 \quad \text{Exponentiate each side (base 10).}$$

$$5x^2 - 5x = 100 \quad \text{Inverse Property}$$

$$x^2 - x - 20 = 0 \quad \text{Write in general form.}$$

$$(x - 5)(x + 4) = 0 \quad \text{Factor.}$$

$$x - 5 = 0 \quad \text{Set 1st factor equal to 0.}$$

$$x = 5 \quad \text{Solution}$$

$$x + 4 = 0 \quad \text{Set 2nd factor equal to 0.}$$

$$x = -4 \quad \text{Solution}$$

The solutions appear to be $x = 5$ and $x = -4$. However, when you check these in the original equation, you can see that $x = 5$ is the only solution.

In Example 9, the domain of $\log_{10} 5x$ is $x > 0$ and the domain of $\log_{10}(x - 1)$ is $x > 1$, so the domain of the original equation is $x > 1$. Because the domain is all real numbers greater than 1, the solution $x = -4$ is extraneous.

Applications

Example 10 ▶ Doubling an Investment  

You have deposited \$500 in an account that pays 6.75% interest, compounded continuously. How long will it take your money to double?

Solution

Using the formula for continuous compounding, you can find that the balance in the account is

$$A = Pe^{rt}$$

$$A = 500e^{0.0675t}.$$

To find the time required for the balance to double, let $A = 1000$ and solve the resulting equation for t.

$$500e^{0.0675t} = 1000 \qquad \text{Let } A = 1000.$$

$$e^{0.0675t} = 2 \qquad \text{Divide each side by 500.}$$

$$\ln e^{0.0675t} = \ln 2 \qquad \text{Take natural log of each side.}$$

$$0.0675t = \ln 2 \qquad \text{Inverse Property}$$

$$t = \frac{\ln 2}{0.0675} \qquad \text{Divide each side by 0.0675.}$$

$$t \approx 10.27 \qquad \text{Use a calculator.}$$

The balance in the account will double after approximately 10.27 years. This result is demonstrated graphically in Figure 5.26.

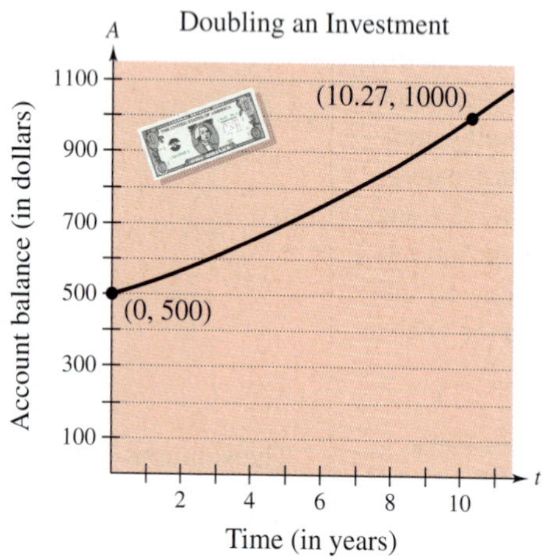

FIGURE 5.26

Exploration

The *effective yield* of a savings plan is the percent increase in the balance after 1 year. Find the effective yield for each savings plan when \$1000 is deposited in a savings account.

a. 7% annual interest rate, compounded annually

b. 7% annual interest rate, compounded continuously

c. 7% annual interest rate, compounded quarterly

d. 7.25% annual interest rate, compounded quarterly

Which savings plan has the greatest effective yield? Which savings plan will have the highest balance after 5 years?

In Example 10, an approximate answer of 10.27 years is given. Within the context of the problem, the exact solution, $(\ln 2)/0.0675$ years, does not make sense as an answer.

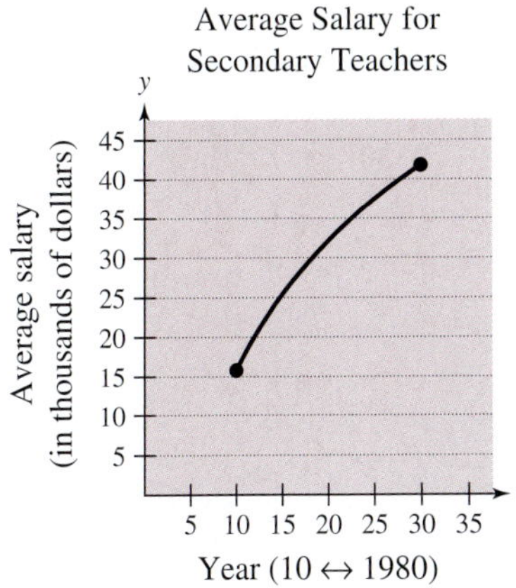

FIGURE 5.27

Example 11 ▶ Average Salary for Secondary Teachers

For selected years from 1980 to 2000, the average salary for secondary teachers y (in thousands of dollars) for the year t can be modeled by the equation

$$y = -38.8 + 23.7 \ln t, \qquad 10 \le t \le 30$$

where $t = 10$ represents 1980 (see Figure 5.27). During which year did the average salary for secondary teachers reach 2.5 times its 1980 level of \$16.5 thousand? (Source: National Education Association)

Solution

$-38.8 + 23.7 \ln t = y$	Write original equation.
$-38.8 + 23.7 \ln t = 41.25$	Let $y = (2.5)(16.5) = 41.25$.
$23.7 \ln t = 80.05$	Add 38.8 to each side.
$\ln t \approx 3.378$	Divide each side by 23.7.
$e^{\ln t} \approx e^{3.378}$	Exponentiate each side.
$t \approx e^{3.378}$	Inverse Property
$t \approx 29$	Use a calculator.

The solution is $t \approx 29$ years. Because $t = 10$ represents 1980, it follows that the average salary for secondary teachers reached 2.5 times its 1980 level in 1999.

Writing ABOUT MATHEMATICS

Comparing Mathematical Models The table shows the U.S. Postal Service rates y for sending an express mail package for selected years from 1985 through 2001, where $x = 5$ represents 1985. (Source: U.S. Postal Service)

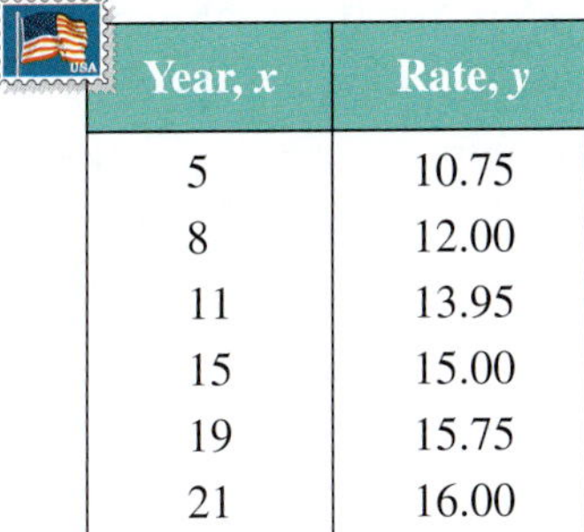

Year, x	Rate, y
5	10.75
8	12.00
11	13.95
15	15.00
19	15.75
21	16.00

a. Create a scatter plot of the data. Find a linear model for the data, and add its graph to your scatter plot. According to this model, when will the rate for sending an express mail package reach \$17.50?

b. Create a new table showing values for $\ln x$ and $\ln y$ and create a scatter plot of this transformed data. Use the method illustrated in Example 7 in Section 5.3 to find a model for the transformed data, and add its graph to your scatter plot. According to this model, when will the rate for sending an express mail package reach \$17.50?

c. Solve the model in part (b) for y, and add its graph to your scatter plot in part (a). Which model better fits the original data? Which model will better predict future shipments? Explain.

5.4 Exercises

In Exercises 1–6, determine whether each *x*-value is a solution (or an approximate solution) of the equation.

1. $4^{2x-7} = 64$
(a) $x = 5$
(b) $x = 2$

2. $2^{3x+1} = 32$
(a) $x = -1$
(b) $x = 2$

3. $3e^{x+2} = 75$
(a) $x = -2 + e^{25}$
(b) $x = -2 + \ln 25$
(c) $x \approx 1.219$

4. $5^{2x+3} = 812$
(a) $x = -\frac{3}{2} + \log_5 \sqrt{812}$
(b) $x \approx 0.581$
(c) $x = -1.5 + \dfrac{\ln 812}{\ln 5}$

5. $\log_4(3x) = 3$
(a) $x \approx 20.356$
(b) $x = -4$
(c) $x = \frac{64}{3}$

6. $\ln(x - 1) = 3.8$
(a) $x = 1 + e^{3.8}$
(b) $x \approx 45.701$
(c) $x = 1 + \ln 3.8$

In Exercises 7–26, solve for *x*.

7. $4^x = 16$ **8.** $3^x = 243$
9. $5^x = 625$ **10.** $3^x = 729$
11. $7^x = \frac{1}{49}$ **12.** $8^x = 4$
13. $\left(\frac{1}{2}\right)^x = 32$ **14.** $\left(\frac{1}{4}\right)^x = 64$
15. $\left(\frac{3}{4}\right)^x = \frac{27}{64}$ **16.** $\left(\frac{2}{3}\right)^x = \frac{4}{9}$
17. $\ln x - \ln 2 = 0$ **18.** $\ln x - \ln 5 = 0$
19. $e^x = 2$ **20.** $e^x = 4$
21. $\ln x = -1$ **22.** $\ln x = -7$
23. $\log_4 x = 3$ **24.** $\log_5 x = -3$
25. $\log_{10} x = -1$ **26.** $\log_{10} x - 2 = 0$

In Exercises 27–30, approximate the point of intersection of the graphs of *f* and *g*. Then solve the equation *f*(*x*) = *g*(*x*) algebraically.

27. $f(x) = 2^x$
$g(x) = 8$

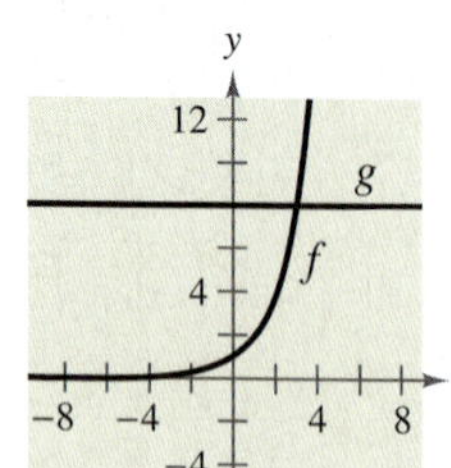

28. $f(x) = 27^x$
$g(x) = 9$

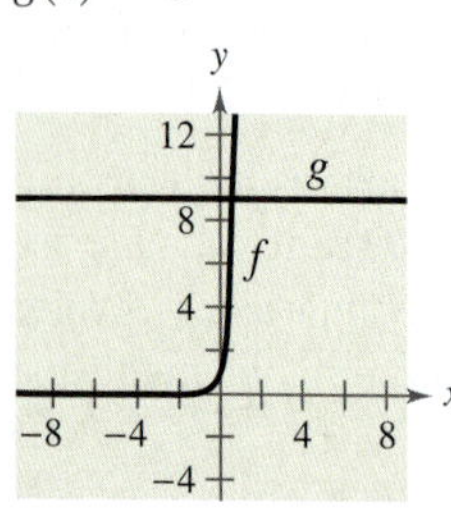

29. $f(x) = \log_3 x$
$g(x) = 2$

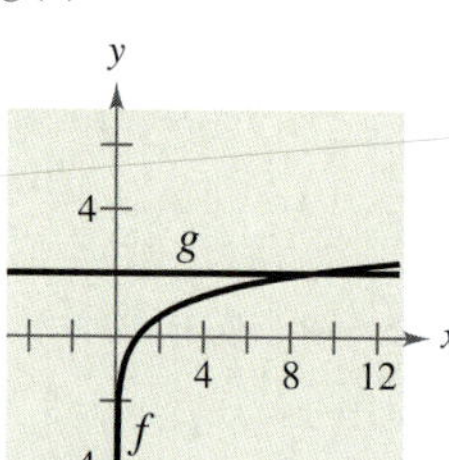

30. $f(x) = \ln(x - 4)$
$g(x) = 0$

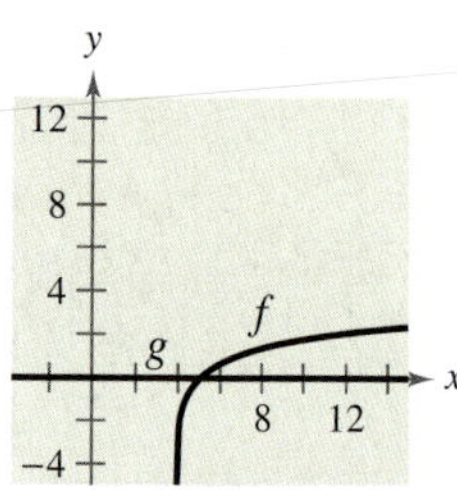

In Exercises 31–68, solve the exponential equation algebraically. Approximate the result to three decimal places.

31. $4(3^x) = 20$ **32.** $2(5^x) = 32$
33. $2e^x = 10$ **34.** $4e^x = 91$
35. $e^x - 9 = 19$ **36.** $6^x + 10 = 47$
37. $3^{2x} = 80$ **38.** $6^{5x} = 3000$
39. $5^{-t/2} = 0.20$ **40.** $4^{-3t} = 0.10$
41. $3^{x-1} = 27$ **42.** $2^{x-3} = 32$
43. $2^{3-x} = 565$ **44.** $8^{-2-x} = 431$
45. $8(10^{3x}) = 12$ **46.** $5(10^{x-6}) = 7$
47. $3(5^{x-1}) = 21$ **48.** $8(3^{6-x}) = 40$
49. $e^{3x} = 12$ **50.** $e^{2x} = 50$
51. $500e^{-x} = 300$ **52.** $1000e^{-4x} = 75$
53. $7 - 2e^x = 5$ **54.** $-14 + 3e^x = 11$
55. $6(2^{3x-1}) - 7 = 9$ **56.** $8(4^{6-2x}) + 13 = 41$
57. $e^{2x} - 4e^x - 5 = 0$ **58.** $e^{2x} - 5e^x + 6 = 0$
59. $e^{2x} - 3e^x - 4 = 0$ **60.** $e^{2x} + 9e^x + 36 = 0$
61. $\dfrac{500}{100 - e^{x/2}} = 20$ **62.** $\dfrac{400}{1 + e^{-x}} = 350$
63. $\dfrac{3000}{2 + e^{2x}} = 2$ **64.** $\dfrac{119}{e^{6x} - 14} = 7$
65. $\left(1 + \dfrac{0.065}{365}\right)^{365t} = 4$ **66.** $\left(4 - \dfrac{2.471}{40}\right)^{9t} = 21$
67. $\left(1 + \dfrac{0.10}{12}\right)^{12t} = 2$ **68.** $\left(16 - \dfrac{0.878}{26}\right)^{3t} = 30$

In Exercises 69–76, use a graphing utility to solve the equation. Approximate the result to three decimal places. Verify your result algebraically.

69. $6e^{1-x} = 25$

70. $-4e^{-x-1} + 15 = 0$

71. $3e^{3x/2} = 962$

72. $8e^{-2x/3} = 11$

73. $e^{0.09t} = 3$

74. $-e^{1.8x} + 7 = 0$

75. $e^{0.125t} - 8 = 0$

76. $e^{2.724x} = 29$

In Exercises 77–104, solve the logarithmic equation algebraically. Approximate the result to three decimal places.

77. $\ln x = -3$

78. $\ln x = 2$

79. $\ln 2x = 2.4$

80. $\ln 4x = 1$

81. $\log_{10} x = 6$

82. $\log_{10} 3z = 2$

83. $6 \log_3(0.5x) = 11$

84. $5 \log_{10}(x - 2) = 11$

85. $3 \ln 5x = 10$

86. $2 \ln x = 7$

87. $\ln\sqrt{x + 2} = 1$

88. $\ln\sqrt{x - 8} = 5$

89. $7 + 3 \ln x = 5$

90. $2 - 6 \ln x = 10$

91. $\ln x - \ln(x + 1) = 2$

92. $\ln x + \ln(x + 1) = 1$

93. $\ln x + \ln(x - 2) = 1$

94. $\ln x + \ln(x + 3) = 1$

95. $\ln(x + 5) = \ln(x - 1) - \ln(x + 1)$

96. $\ln(x + 1) - \ln(x - 2) = \ln x$

97. $\log_2(2x - 3) = \log_2(x + 4)$

98. $\log_{10}(x - 6) = \log_{10}(2x + 1)$

99. $\log_{10}(x + 4) - \log_{10} x = \log_{10}(x + 2)$

100. $\log_2 x + \log_2(x + 2) = \log_2(x + 6)$

101. $\log_4 x - \log_4(x - 1) = \frac{1}{2}$

102. $\log_3 x + \log_3(x - 8) = 2$

103. $\log_{10} 8x - \log_{10}\left(1 + \sqrt{x}\right) = 2$

104. $\log_{10} 4x - \log_{10}\left(12 + \sqrt{x}\right) = 2$

In Exercises 105–108, use a graphing utility to solve the equation. Approximate the result to three decimal places. Verify your result algebraically.

105. $7 = 2^x$

106. $500 = 1500e^{-x/2}$

107. $3 - \ln x = 0$

108. $10 - 4 \ln(x - 2) = 0$

Compound Interest **In Exercises 109 and 110, find the time required for a \$1000 investment to double at interest rate r, compounded continuously.**

109. $r = 0.085$

110. $r = 0.12$

Compound Interest **In Exercises 111 and 112, find the time required for a \$1000 investment to triple at interest rate r, compounded continuously.**

111. $r = 0.085$

112. $r = 0.12$

113. ***Demand*** The demand equation for a microwave oven is

$$p = 500 - 0.5(e^{0.004x}).$$

Find the demand x for a price of (a) $p = \$350$ and (b) $p = \$300$.

114. ***Demand*** The demand equation for a hand-held electronic organizer is

$$p = 5000\left(1 - \frac{4}{4 + e^{-0.002x}}\right).$$

Find the demand x for a price of (a) $p = \$600$ and (b) $p = \$400$.

115. ***Forest Yield*** The yield V (in millions of cubic feet per acre) for a forest at age t years is

$$V = 6.7e^{-48.1/t}.$$

(a) Use a graphing utility to graph the function.

(b) Determine the horizontal asymptote of the function. Interpret its meaning in the context of the problem.

(c) Find the time necessary to obtain a yield of 1.3 million cubic feet.

116. ***Trees per Acre*** The number of trees per acre N of a species is approximated by the model

$$N = 68(10^{-0.04x}), \quad 5 \le x \le 40$$

where x is the average diameter of the trees 3 feet above the ground. Use the model to approximate the average diameter of the trees in a test plot when $N = 21$.

117. ***Average Heights*** The percent of American males between the ages of 18 and 24 who are no more than x inches tall is

$$m(x) = \frac{100}{1 + e^{-0.6114(x-69.71)}}$$

and the percent of American females between the ages of 18 and 24 who are no more than x inches tall is

$$f(x) = \frac{100}{1 + e^{-0.66607(x-64.51)}}$$

where m and f are the percents and x is the height in inches. (Source: U.S. National Center for Health Statistics)

(a) Use the graph to determine any horizontal asymptotes of the functions. Interpret the meaning in the context of the problem.

(b) What is the average height of each sex?

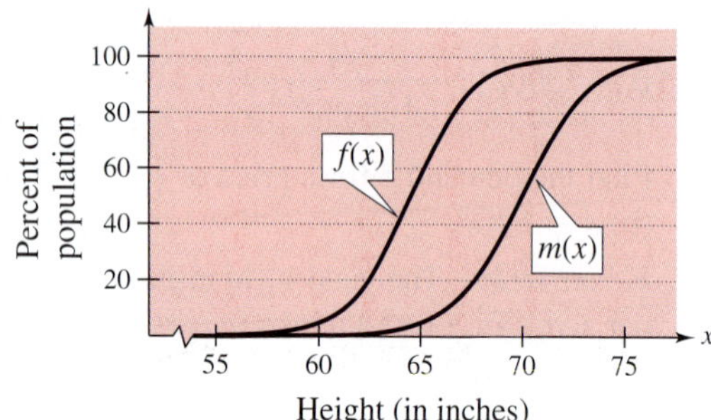

118. ***Learning Curve*** In a group project in learning theory, a mathematical model for the proportion P of correct responses after n trials was found to be

$$P = \frac{0.83}{1 + e^{-0.2n}}.$$

(a) Use a graphing utility to graph the function.

(b) Use the graph to determine any horizontal asymptotes of the function. Interpret the meaning of the upper asymptote in the context of this problem.

(c) After how many trials will 60% of the responses be correct?

▶ Model It

119. ***Automobiles*** Automobiles are designed with crumple zones that help protect their occupants in crashes. The crumple zones allow the occupants to move short distances when the automobiles come to abrupt stops. The greater the distance moved, the fewer g's the crash victims experience. (One g is equal to the acceleration due to gravity. For very short periods of time, humans have withstood as much as 40 g's.) In crash tests with vehicles moving at 90 kilometers per hour, analysts measured the numbers of g's experienced during deceleration by crash dummies that were permitted to move x meters during impact. The data is shown in the table.

x	g's
0.2	158
0.4	80
0.6	53
0.8	40
1.0	32

A model for this data is

$$y = -3.00 + 11.88 \ln x + \frac{36.94}{x}$$

where y is the number of g's.

(a) Complete the table using the model.

x	0.2	0.4	0.6	0.8	1.0
y					

(b) Use a graphing utility to graph the data points and the model in the same viewing window. How do they compare?

(c) Use the model to estimate the distance traveled during impact if the passenger deceleration must not exceed 30 g's.

(d) Do you think it is practical to lower the number of g's experienced during impact to fewer than 23? Explain your reasoning.

120. ***Data Analysis*** An object at a temperature of 160°C was removed from a furnace and placed in a room at 20°C. The temperature T of the object was measured each hour h and recorded in the table. A model for this data is $T = 20[1 + 7(2^{-h})]$. The graph of this model is shown in the figure.

Hour, h	Temperature, T
0	160°
1	90°
2	56°
3	38°
4	29°
5	24°

(a) Use the graph to identify the horizontal asymptote of the model and interpret the asymptote in the context of the problem.

(b) Use the model to approximate the time when the temperature of the object was 100°C.

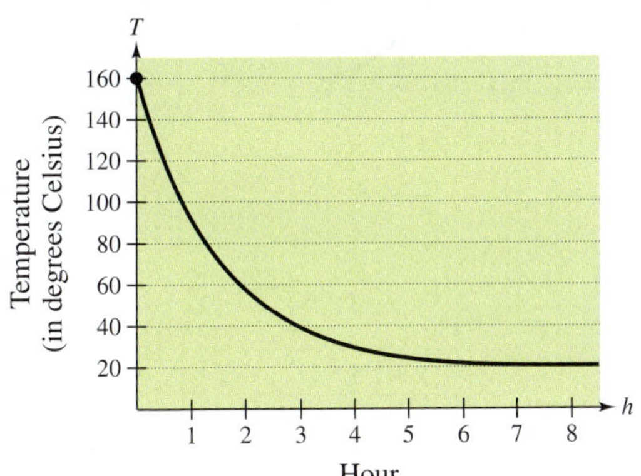

Synthesis

True or False? **In Exercises 121–124, rewrite each verbal statement as an equation. Then decide whether the statement is true or false. Justify your answer.**

121. The logarithm of the product of two numbers is equal to the sum of the logarithms of the numbers.

122. The logarithm of the sum of two numbers is equal to the product of the logarithms of the numbers.

123. The logarithm of the difference of two numbers is equal to the difference of the logarithms of the numbers.

124. The logarithm of the quotient of two numbers is equal to the difference of the logarithms of the numbers.

125. ***Think About It*** Is it possible for a logarithmic equation to have more than one extraneous solution? Explain.

126. ***Finance*** You are investing P dollars at an annual interest rate of r, compounded continuously, for t years. Which of the following would result in the highest value of the investment? Explain your reasoning.

(a) Double the amount you invest.

(b) Double your interest rate.

(c) Double the number of years.

127. ***Think About It*** Are the times required for the investments in Exercises 109 and 110 to quadruple twice as long as the times for them to double? Give a reason for your answer and verify your answer algebraically.

128. ***Writing*** Write two or three sentences stating the general guidelines that you follow when solving (a) exponential equations and (b) logarithmic equations.

Review

In Exercises 129–132, simplify the expression.

129. $\sqrt{48x^2y^5}$

130. $\sqrt{32} - 2\sqrt{25}$

131. $\sqrt[3]{25} \cdot \sqrt[3]{15}$

132. $\dfrac{3}{\sqrt{10} - 2}$

In Exercises 133–136, find a mathematical model for the verbal statement.

133. M varies directly as the cube of p.

134. t varies inversely as the cube of s.

135. d varies jointly as a and b.

136. x is inversely proportional to $b - 3$.

In Exercises 137–140, evaluate the logarithm using the change-of-base formula. Approximate your result to three decimal places.

137. $\log_6 9$

138. $\log_3 4$

139. $\log_{3/4} 5$

140. $\log_8 22$

5.5 Exponential and Logarithmic Models

▶ **What you should learn**

- How to recognize the five most common types of models involving exponential and logarithmic functions
- How to use exponential growth and decay functions to model and solve real-life problems
- How to use Gaussian functions to model and solve real-life problems
- How to use logistic growth functions to model and solve real-life problems
- How to use logarithmic functions to model and solve real-life problems

▶ **Why you should learn it**

Exponential growth and decay models are often used to model the population of a country. For instance, in Exercise 36 on page 436, you will use exponential growth and decay models to compare the populations of several countries.

Vittoriano Rastelli/Corbis

Introduction

The five most common types of mathematical models involving exponential functions and logarithmic functions are as follows.

1. **Exponential growth model:** $y = ae^{bx}, \quad b > 0$
2. **Exponential decay model:** $y = ae^{-bx}, \quad b > 0$
3. **Gaussian model:** $y = ae^{-(x-b)^2/c}$
4. **Logistic growth model:** $y = \dfrac{a}{1 + be^{-rx}}$
5. **Logarithmic models:** $y = a + b \ln x, \quad y = a + b \log_{10} x$

The graphs of the basic forms of these functions are shown in Figure 5.28.

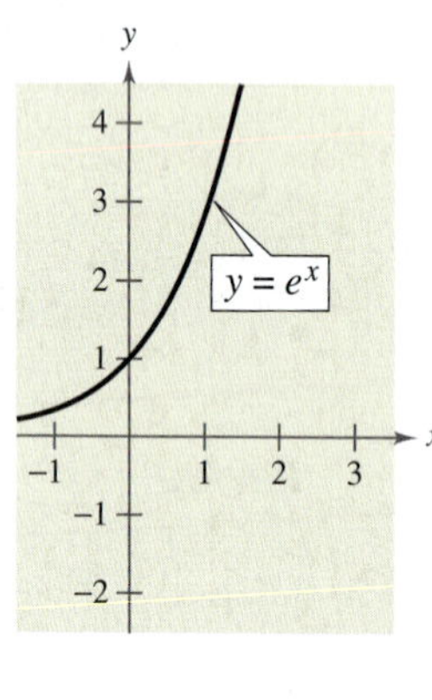

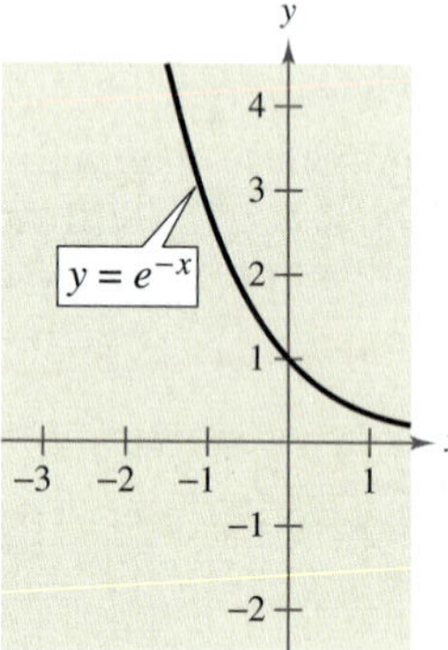

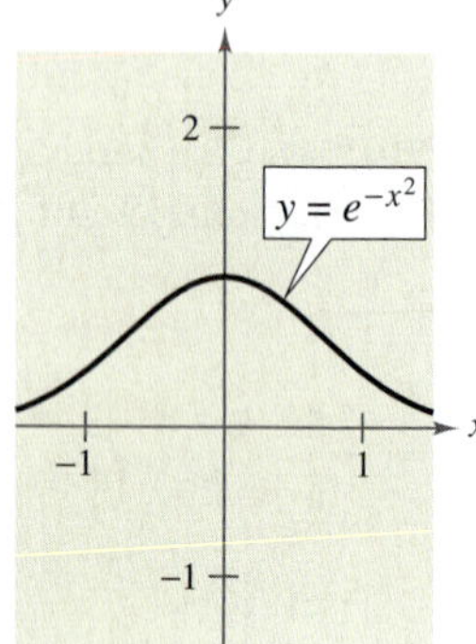

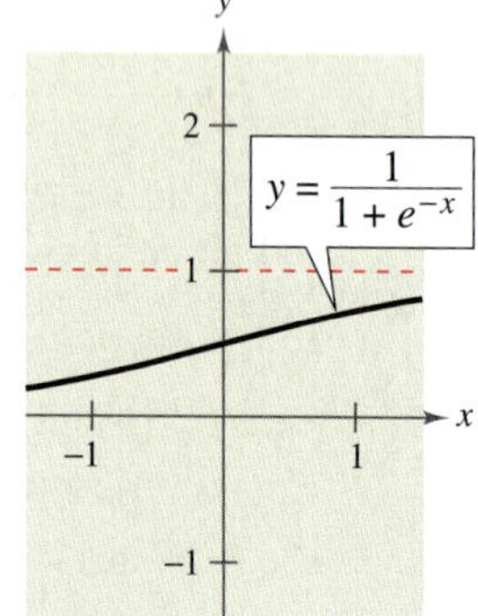

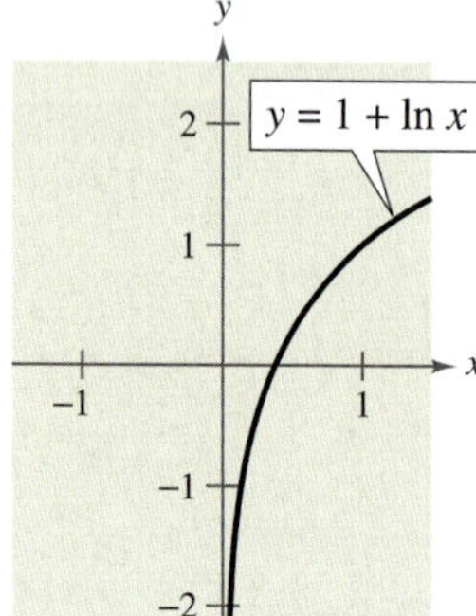

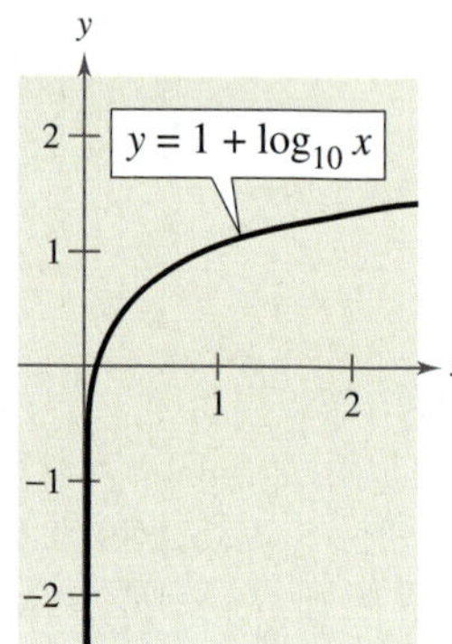

FIGURE 5.28

You can often gain quite a bit of insight into a situation modeled by an exponential or logarithmic function by identifying and interpreting the function's asymptotes. Use the graphs in Figure 5.28 to identify the asymptotes of each function.

Exponential Growth and Decay

Example 1 ▶ Population Increase  

Estimates of the world population (in millions) from 1995 through 2003 are shown in the table. The scatter plot of the data is shown in Figure 5.29. (Source: U.S. Census Bureau)

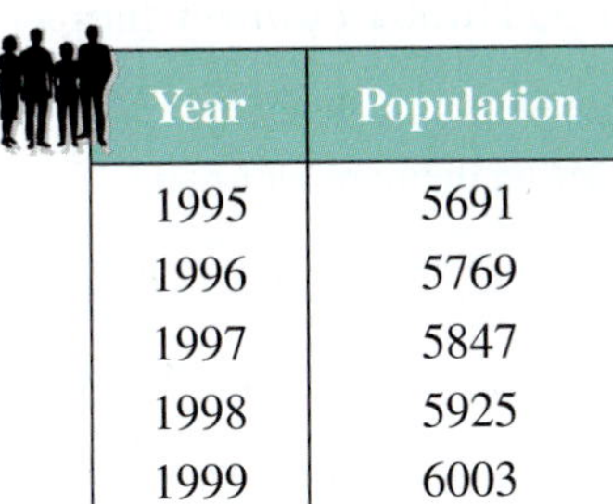

Year	Population
1995	5691
1996	5769
1997	5847
1998	5925
1999	6003

Year	Population
2000	6080
2001	6157
2002	6234
2003	6311

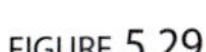

FIGURE 5.29

An exponential growth model that approximates this data is

$$P = 5340e^{0.012922t}, \quad 5 \le t \le 13$$

where P is the population (in millions) and $t = 5$ represents 1995. Compare the values given by the model with the estimates given by the U.S. Census Bureau. According to this model, when will the world population reach 6.8 billion?

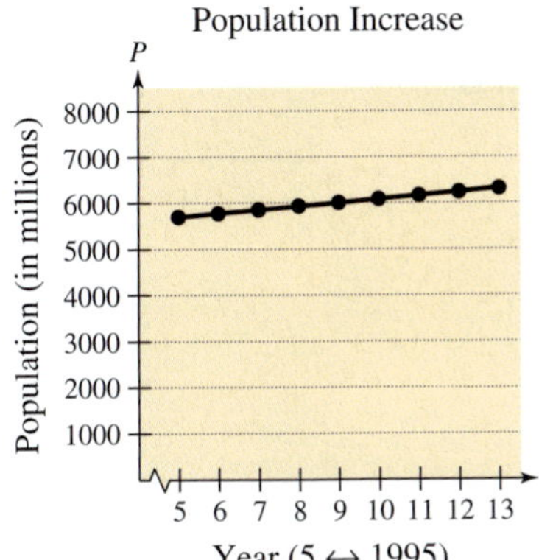

FIGURE 5.30

Solution

The following table compares the two sets of population figures. The graph of the model is shown in Figure 5.30.

Year	1995	1996	1997	1998	1999	2000	2001	2002	2003
Population	5691	5769	5847	5925	6003	6080	6157	6234	6311
Model	5696	5771	5846	5922	5999	6077	6156	6236	6317

To find when the world population will reach 6.8 billion, let $P = 6800$ in the model and solve for t.

$5340e^{0.012922t} = P$ — Write original model.

$5340e^{0.012922t} = 6800$ — Let $P = 6800$.

$e^{0.012922t} \approx 1.27341$ — Divide each side by 5340.

$\ln e^{0.012922t} \approx \ln 1.27341$ — Take natural log of each side.

$0.012922t \approx 0.241698$ — Inverse Property

$t \approx 18.7$ — Divide each side by 0.012922.

According to the model, the world population will reach 6.8 billion in 2008.

Technology

Some graphing utilities have curve-fitting capabilities that can be used to find models that represent data. If you have such a graphing utility, try using it to find a model for the data given in Example 1. How does your model compare with the model given in Example 1?

In Example 1, you were given the exponential growth model. But suppose this model were not given; how could you find such a model? One technique for doing this is demonstrated in Example 2.

Example 2 ▶ Modeling Population Growth  

In a research experiment, a population of fruit flies is increasing according to the law of exponential growth. After 2 days there are 100 flies, and after 4 days there are 300 flies. How many flies will there be after 5 days?

Solution

Let y be the number of flies at time t. From the given information, you know that $y = 100$ when $t = 2$ and $y = 300$ when $t = 4$. Substituting this information into the model $y = ae^{bt}$ produces

$$100 = ae^{2b} \qquad \text{and} \qquad 300 = ae^{4b}.$$

To solve for b, solve for a in the first equation.

$$100 = ae^{2b} \quad \Rightarrow \quad a = \frac{100}{e^{2b}} \qquad \text{Solve for } a \text{ in the first equation.}$$

Then substitute the result into the second equation.

$$300 = ae^{4b} \qquad \text{Write second equation.}$$

$$300 = \left(\frac{100}{e^{2b}}\right)e^{4b} \qquad \text{Substitute } 100/e^{2b} \text{ for } a.$$

$$\frac{300}{100} = e^{2b} \qquad \text{Divide each side by 100.}$$

$$\ln 3 = 2b \qquad \text{Take natural log of each side.}$$

$$\frac{1}{2}\ln 3 = b \qquad \text{Solve for } b.$$

Using $b = \frac{1}{2}\ln 3$ and the equation you found for a, you can determine that

$$a = \frac{100}{e^{2[(1/2)\ln 3]}} \qquad \text{Substitute } (1/2)\ln 3 \text{ for } b.$$

$$= \frac{100}{e^{\ln 3}} \qquad \text{Simplify.}$$

$$= \frac{100}{3} \qquad \text{Inverse Property}$$

$$\approx 33. \qquad \text{Simplify.}$$

So, with $a \approx 33$ and $b = \frac{1}{2}\ln 3 \approx 0.5493$, the exponential growth model is

$$y = 33e^{0.5493t}$$

as shown in Figure 5.31. This implies that, after 5 days, the population will be

$$y = 33e^{0.5493(5)} \approx 514 \text{ flies.}$$

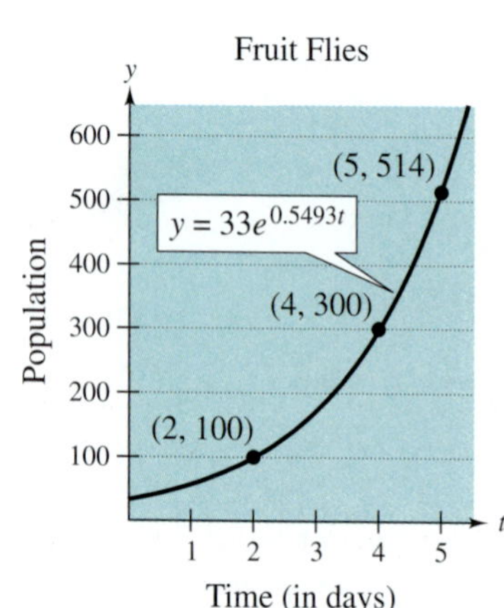

FIGURE 5.31

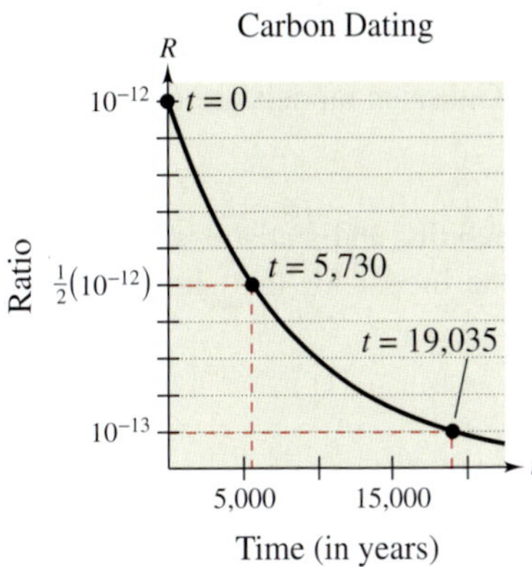

FIGURE 5.32

In living organic material, the ratio of the number of radioactive carbon isotopes (carbon 14) to the number of nonradioactive carbon isotopes (carbon 12) is about 1 to 10^{12}. When organic material dies, its carbon 12 content remains fixed, whereas its radioactive carbon 14 begins to decay with a half-life of 5730 years. To estimate the age of dead organic material, scientists use the following formula, which denotes the ratio of carbon 14 to carbon 12 present at any time t (in years).

$$R = \frac{1}{10^{12}}e^{-t/8267} \qquad \text{Carbon dating model}$$

The graph of R is shown in Figure 5.32. Note that R decreases as t increases.

Example 3 ▶ Carbon Dating

The ratio of carbon 14 to carbon 12 in a newly discovered fossil is

$$R = \frac{1}{10^{13}}.$$

Estimate the age of the fossil.

Solution

In the carbon dating model, substitute the given value of R to obtain the following.

$$\frac{1}{10^{12}}e^{-t/8267} = R \qquad \text{Write original model.}$$

$$\frac{e^{-t/8267}}{10^{12}} = \frac{1}{10^{13}} \qquad \text{Let } R = \frac{1}{10^{13}}.$$

$$e^{-t/8267} = \frac{1}{10} \qquad \text{Multiply each side by } 10^{12}.$$

$$\ln e^{-t/8267} = \ln\frac{1}{10} \qquad \text{Take natural log of each side.}$$

$$-\frac{t}{8267} \approx -2.3026 \qquad \text{Inverse Property}$$

$$t \approx 19{,}036 \qquad \text{Multiply each side by } -8267.$$

So, to the nearest thousand years, you can estimate the age of the fossil to be 19,000 years.

The carbon dating model in Example 3 assumed that the carbon 14/carbon 12 ratio was one part in 10,000,000,000,000. Suppose an error in measurement occurred and the actual ratio was only one part in 8,000,000,000,000. The fossil age corresponding to the actual ratio would then be approximately 17,000 years. Try checking this result.

STUDY TIP

An exponential model can be used to determine the *decay* of radioactive isotopes. For instance, to find how much of an initial 10 grams of ^{226}Ra isotope with a half-life of 1620 years is left after 500 years, you would use the exponential decay model.

$$y = ae^{-bt}$$

$$\frac{1}{2}(10) = 10e^{-b(1620)}$$

$$b = -\ln\left(\frac{1}{2}\right)/1620$$

Using the value of b found above and $a = 10$, the amount left is

$$y = 10e^{-[-\ln(1/2)/1620](500)}$$

$$y \approx 8.07 \text{ grams.}$$

Gaussian Models

As mentioned at the beginning of this section, Gaussian models are of the form

$$y = ae^{-(x-b)^2/c}.$$

This type of model is commonly used in probability and statistics to represent populations that are **normally distributed.** One model for this situation takes the form

$$y = \frac{1}{\sigma\sqrt{2\pi}}e^{-x^2/(2\sigma^2)}$$

where σ is the standard deviation (σ is the lowercase Greek letter sigma). The graph of a Gaussian model is called a **bell-shaped curve.**

The average value for a population can be found from the bell-shaped curve by observing where the maximum y-value of the function occurs. The x-value corresponding to the maximum y-value of the function represents the average value of the independent variable—in this case, x.

Example 4 ▶ SAT Scores 

In 2001, the Scholastic Aptitude Test (SAT) math scores for college-bound seniors roughly followed a normal distribution

$$y = 0.0035e^{-(x-514)^2/25{,}538}, \quad 200 \le x \le 800$$

where x is the SAT score for mathematics. Sketch the graph of this function. From the graph, estimate the average SAT score. (Source: College Board)

Solution

The graph of the function is shown in Figure 5.33. From the graph, you can see that the average mathematics score for college-bound seniors in 2001 was 514.

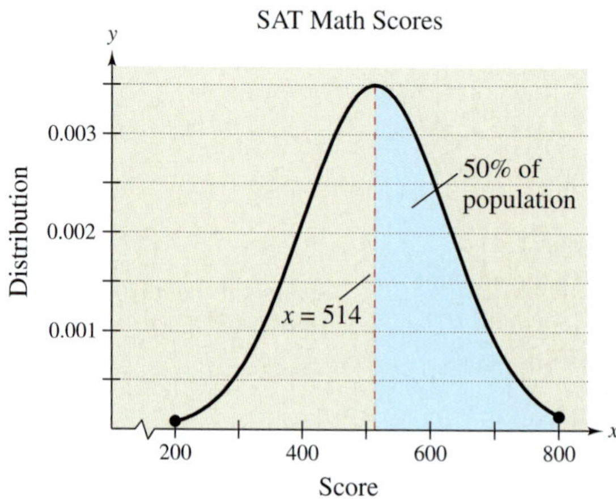

FIGURE 5.33

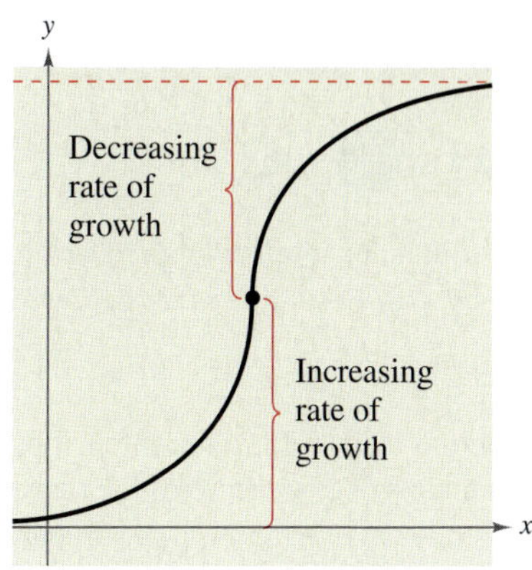

FIGURE 5.34

Logistic Growth Models

Some populations initially have rapid growth, followed by a declining rate of growth, as indicated by the graph in Figure 5.34. One model for describing this type of growth pattern is the **logistic curve** given by the function

$$y = \frac{a}{1 + be^{-rx}}$$

where y is the population size and x is the time. An example is a bacteria culture that is initially allowed to grow under ideal conditions, and then under less favorable conditions that inhibit growth. A logistic growth curve is also called a **sigmoidal curve.**

Example 5 ▶ Spread of a Virus  

On a college campus of 5000 students, one student returns from vacation with a contagious and long-lasting flu virus. The spread of the virus is modeled by

$$y = \frac{5000}{1 + 4999e^{-0.8t}}, \quad t \geq 0$$

where y is the total number of students infected after t days. The college will cancel classes when 40% or more of the students are infected.

a. How many students are infected after 5 days?

b. After how many days will the college cancel classes?

Solution

a. After 5 days, the number of students infected is

$$y = \frac{5000}{1 + 4999e^{-0.8(5)}} = \frac{5000}{1 + 4999e^{-4}} \approx 54.$$

b. Classes are canceled when the number infected is $(0.40)(5000) = 2000$.

$$2000 = \frac{5000}{1 + 4999e^{-0.8t}}$$

$$1 + 4999e^{-0.8t} = 2.5$$

$$e^{-0.8t} = \frac{1.5}{4999}$$

$$\ln e^{-0.8t} = \ln \frac{1.5}{4999}$$

$$-0.8t = \ln \frac{1.5}{4999}$$

$$t = -\frac{1}{0.8} \ln \frac{1.5}{4999}$$

$$t \approx 10.1$$

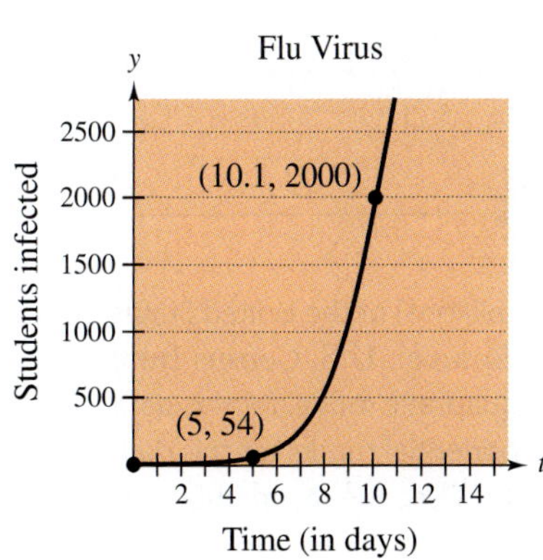

FIGURE 5.35

So, after 10 days, at least 40% of the students will be infected, and classes will be canceled. The graph of the function is shown in Figure 5.35.

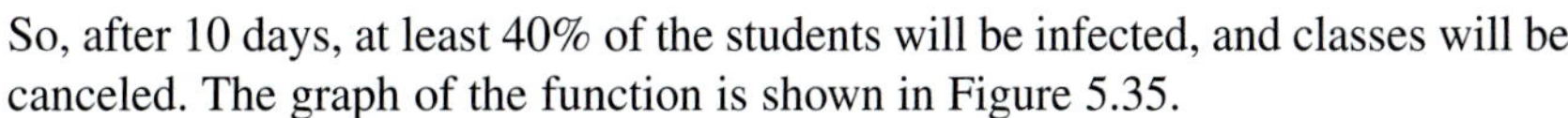

AFP/Corbis

On January 13, 2001 an earthquake of magnitude 7.7 in El Salvador caused 185 landslides and killed over 800 people.

Logarithmic Models

Example 6 ▶ Magnitude of Earthquakes

On the Richter scale, the magnitude R of an earthquake of intensity I is

$$R = \log_{10} \frac{I}{I_0}$$

where $I_0 = 1$ is the minimum intensity used for comparison. Find the intensities per unit of area for the following earthquakes. (Intensity is a measure of the wave energy of an earthquake.)

a. Tokyo and Yokohama, Japan in 1923: $R = 8.3$.

b. El Salvador in 2001: $R = 7.7$.

Solution

a. Because $I_0 = 1$ and $R = 8.3$, you have

$$8.3 = \log_{10} \frac{I}{1} \qquad \text{Substitute 1 for } I_0 \text{ and 8.3 for } R.$$

$$10^{8.3} = 10^{\log_{10} I} \qquad \text{Exponentiate each side.}$$

$$I = 10^{8.3} \approx 199{,}526{,}000. \qquad \text{Inverse property of exponents and logs}$$

b. For $R = 7.7$, you have

$$7.7 = \log_{10} \frac{I}{1} \qquad \text{Substitute 1 for } I_0 \text{ and 7.7 for } R.$$

$$10^{7.7} = 10^{\log_{10} I} \qquad \text{Exponentiate each side.}$$

$$I = 10^{7.7} \approx 50{,}119{,}000. \qquad \text{Inverse property of exponents and logs}$$

Note that an increase of 0.6 unit on the Richter scale (from 7.7 to 8.3) represents an increase in intensity by a factor of

$$\frac{199{,}526{,}000}{50{,}119{,}000} \approx 4.$$

In other words, the earthquake in 1923 had an intensity about 4 times greater than that of the 2001 earthquake.

t	Year	Population
1	1910	91.97
2	1920	105.71
3	1930	122.78
4	1940	131.67
5	1950	151.33
6	1960	179.32
7	1970	203.30
8	1980	226.54
9	1990	248.72
10	2000	281.42

Writing ABOUT MATHEMATICS

Comparing Population Models The population (in millions) of the United States from 1910 to 2000 is shown in the table at the left. (Source: U.S. Census Bureau) Least squares regression analysis gives the best quadratic model for this data as $P = 1.0317t^2 + 9.668t + 81.38$ and the best exponential model for this data as $P = 82.367e^{0.125t}$. Which model better fits the data? Describe the method you used to reach your conclusion.

5.5 Exercises

In Exercises 1–6, match the function with its graph. [The graphs are labeled (a), (b), (c), (d), (e), and (f).]

(a)

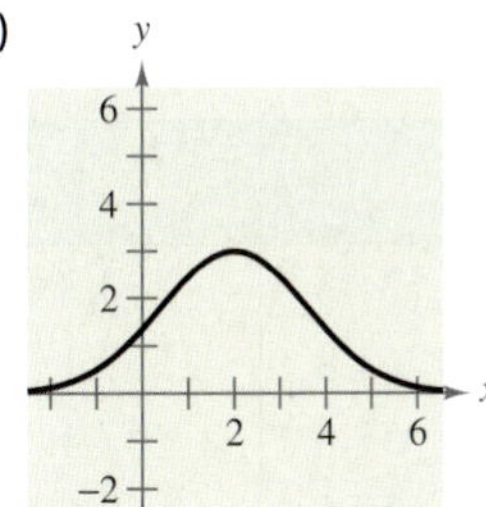

(b)

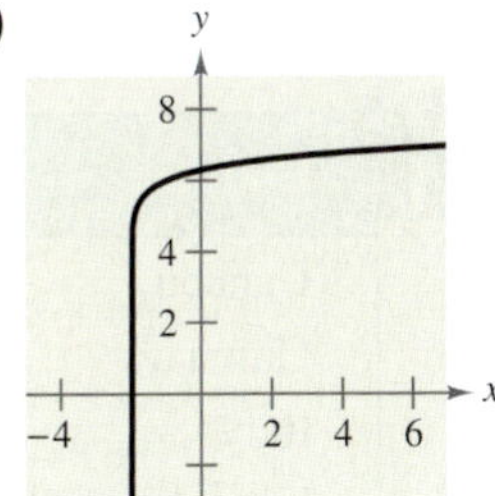

(c)

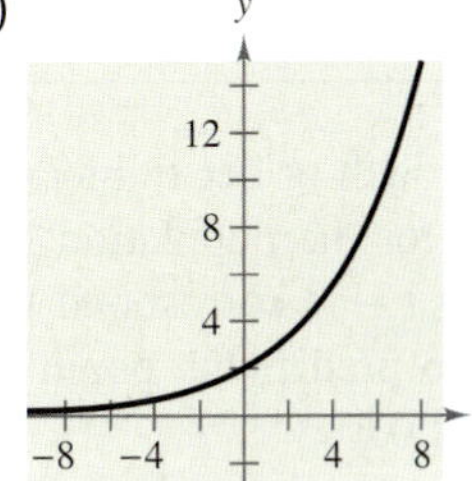

(d)

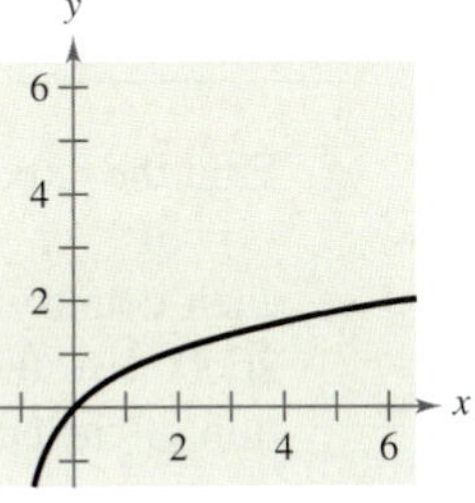

(e)

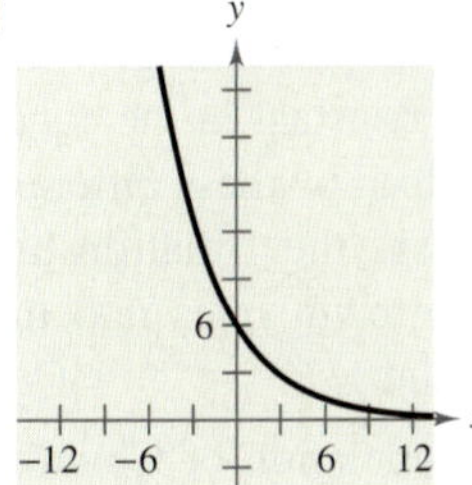

(f)

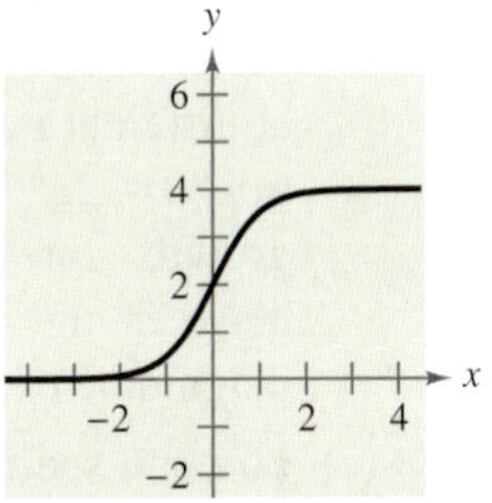

1. $y = 2e^{x/4}$

2. $y = 6e^{-x/4}$

3. $y = 6 + \log_{10}(x + 2)$

4. $y = 3e^{-(x-2)^2/5}$

5. $y = \ln(x + 1)$

6. $y = \dfrac{4}{1 + e^{-2x}}$

Compound Interest **In Exercises 7–14, complete the table for a savings account in which interest is compounded continuously.**

	Initial Investment	*Annual % Rate*	*Time to Double*	*Amount After 10 Years*
7.	\$1000	12%		
8.	\$20,000	$10\frac{1}{2}\%$		
9.	\$750		$7\frac{3}{4}$ yr	
10.	\$10,000		12 yr	
11.	\$500			\$1505.00
12.	\$600			\$19,205.00
13.		4.5%		\$10,000.00
14.		8%		\$20,000.00

Compound Interest **In Exercises 15 and 16, determine the principal P that must be invested at rate r, compounded monthly, so that \$500,000 will be available for retirement in t years.**

15. $r = 7\frac{1}{2}\%,\ t = 20$

16. $r = 12\%,\ t = 40$

Compound Interest **In Exercises 17 and 18, determine the time necessary for \$1000 to double if it is invested at interest rate r compounded (a) annually, (b) monthly, (c) daily, and (d) continuously.**

17. $r = 11\%$

18. $r = 10\frac{1}{2}\%$

19. ***Compound Interest*** Complete the table for the time t necessary for P dollars to triple if interest is compounded continuously at rate r.

r	2%	4%	6%	8%	10%	12%
t						

20. ***Modeling Data*** Draw a scatter plot of the data in Exercise 19. Use the *regression* feature of a graphing utility to find a model for the data.

21. ***Compound Interest*** Complete the table for the time t necessary for P dollars to triple if interest is compounded annually at rate r.

r	2%	4%	6%	8%	10%	12%
t						

22. ***Modeling Data*** Draw a scatter plot of the data in Exercise 21. Use the *regression* feature of a graphing utility to find a model for the data.

23. ***Comparing Models*** If \$1 is invested in an account over a 10-year period, the amount in the account, where t represents the time in years, is

$$A = 1 + 0.075[\![t]\!] \quad \text{or} \quad A = e^{0.07t}$$

depending on whether the account pays simple interest at $7\frac{1}{2}\%$ or continuous compound interest at 7%. Graph each function on the same set of axes. Which grows at the faster rate? (Remember that $[\![t]\!]$ is the greatest integer function discussed in Section 2.4.)

 **24.** ***Comparing Models*** If \$1 is invested in an account over a 10-year period, the amount in the account, where t represents the time in years, is

$$A = 1 + 0.06[\![t]\!] \quad \text{or} \quad A = \left(1 + \frac{0.055}{365}\right)^{[\![365t]\!]}$$

depending on whether the account pays simple interest at 6% or compound interest at $5\frac{1}{2}\%$ compounded daily. Use a graphing utility to graph each function in the same viewing window. Which grows at the faster rate?

Radioactive Decay **In Exercises 25–30, complete the table for the radioactive isotope.**

	Isotope	Half-life (years)	Initial Quantity	Amount After 1000 Years
25.	^{226}Ra	1620	10 g	
26.	^{226}Ra	1620		1.5 g
27.	^{14}C	5730		2 g
28.	^{14}C	5730	3 g	
29.	^{239}Pu	24,360		2.1 g
30.	^{239}Pu	24,360		0.4 g

In Exercises 31–34, find the exponential model $y = ae^{bx}$ that fits the points in the graph or table.

31.

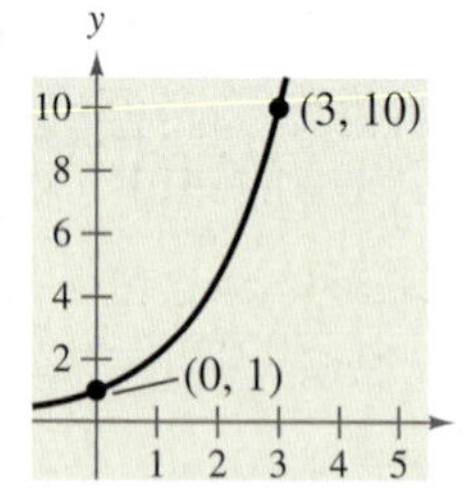

32.

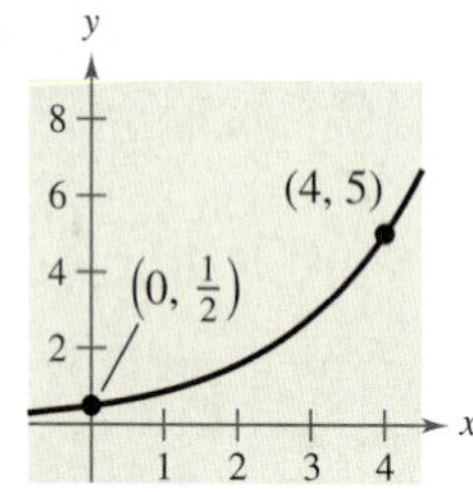

33.

x	0	4
y	5	1

34.

x	0	3
y	1	$\frac{1}{4}$

35. ***Population*** The population P of Texas (in thousands) from 1991 through 2000 can be modeled by

$$P = 16{,}968e^{0.019t}$$

where $t = 1$ represents the year 1991. According to this model, when will the population reach 22 million? (Source: U.S. Census Bureau)

▶ Model It

36. ***Population*** The table shows the populations (in millions) of five countries in 2000 and the projected populations (in millions) for the year 2010. (Source: U.S. Census Bureau)

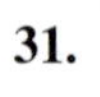

Country	2000	2010
Canada	31.3	34.3
China	1261.8	1359.1
Italy	57.6	57.4
United Kingdom	59.5	60.6
United States	275.6	300.1

(a) Find the exponential growth or decay model $y = ae^{bt}$ or $y = ae^{-bt}$ for the population in each country by letting $t = 0$ correspond to 2000. Use the model to predict the population of each country in 2030.

(b) You can see that the populations of the United States and the United Kingdom are growing at different rates. What constant in the equation $y = ae^{bt}$ is determined by these different growth rates? Discuss the relationship between the different growth rates and the magnitude of the constant.

(c) You can see that the population of China is increasing while the population of Italy is decreasing. What constant in the equation $y = ae^{bt}$ reflects this difference? Explain.

37. ***Population*** The population P of Charlotte, North Carolina (in thousands) is

$$P = 548e^{kt}$$

where $t = 0$ represents the year 2000. In 1970, the population was 241,000. Find the value of k, and use this result to predict the population in the year 2010. (Source: U.S. Census Bureau)

38. ***Population*** The population P of Lincoln, Nebraska (in thousands) is

$$P = 224e^{kt}$$

where $t = 0$ represents the year 2000. In 1980, the population was 172,000. Find the value of k, and use this result to predict the population in the year 2020. (Source: U.S. Census Bureau)

39. ***Bacteria Growth*** The number N of bacteria in a culture is modeled by

$$N = 100e^{kt}$$

where t is the time in hours. If $N = 300$ when $t = 5$, estimate the time required for the population to double in size.

40. ***Bacteria Growth*** The number N of bacteria in a culture is modeled by $N = 250e^{kt}$, where t is the time in hours. If $N = 280$ when $t = 10$, estimate the time required for the population to double in size.

41. ***Radioactive Decay*** The half-life of radioactive radium (^{226}Ra) is 1620 years. What percent of a present amount of radioactive radium will remain after 100 years?

42. ***Radioactive Decay*** Carbon 14 dating assumes that the carbon dioxide on Earth today has the same radioactive content as it did centuries ago. If this is true, the amount of ^{14}C absorbed by a tree that grew several centuries ago should be the same as the amount of ^{14}C absorbed by a tree growing today. A piece of ancient charcoal contains only 15% as much radioactive carbon as a piece of modern charcoal. How long ago was the tree burned to make the ancient charcoal if the half-life of ^{14}C is 5730 years?

43. ***Depreciation*** A car that cost \$22,000 new has a book value of \$13,000 after 2 years.

(a) Find the straight-line model $V = mt + b$.

(b) Find the exponential model $V = ae^{kt}$.

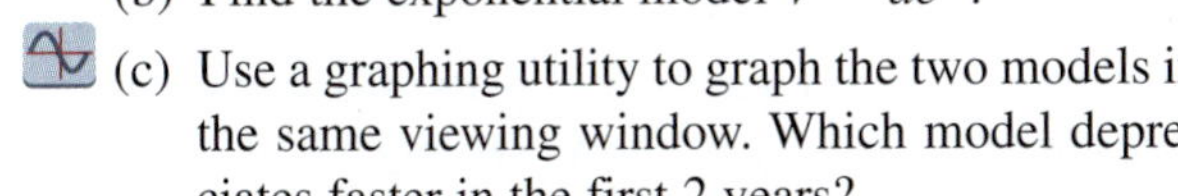

(c) Use a graphing utility to graph the two models in the same viewing window. Which model depreciates faster in the first 2 years?

(d) Find the book values of the car after 1 year and after 3 years using each model.

(e) Interpret the slope of the straight-line model.

44. ***Depreciation*** A computer that costs \$2000 new has a book value of \$500 after 2 years.

(a) Find the straight-line model $V = mt + b$.

(b) Find the exponential model $V = ae^{kt}$.

(c) Use a graphing utility to graph the two models in the same viewing window. Which model depreciates faster in the first 2 years?

(d) Find the book values of the computer after 1 year and after 3 years using each model.

(e) Interpret the slope of the straight-line model.

45. ***Sales*** The sales S (in thousands of units) of a new CD burner after it has been on the market t years are modeled by $S(t) = 100(1 - e^{kt})$. Fifteen thousand units of the new product were sold the first year.

(a) Complete the model by solving for k.

(b) Sketch the graph of the model.

(c) Use the model to estimate the number of units sold after 5 years.

46. ***Sales*** After discontinuing all advertising for a tool kit in 1998, the manufacturer noted that sales began to drop according to the model

$$S = \frac{500{,}000}{1 + 0.6e^{kt}}$$

where S represents the number of units sold and $t = 0$ represents 1998. In 2000, the company sold 300,000 units.

(a) Complete the model by solving for k.

(b) Estimate sales in 2005.

47. ***Sales*** The sales S (in thousands of units) of a cleaning solution after x hundred dollars is spent on advertising are modeled by $S = 10(1 - e^{kx})$. When \$500 is spent on advertising, 2500 units are sold.

(a) Complete the model by solving for k.

(b) Estimate the number of units that will be sold if advertising expenditures are raised to \$700.

48. ***Profit*** Because of a slump in the economy, a department store finds that its annual profits have dropped from \$742,000 in 2000 to \$632,000 in 2002. The profit follows an exponential pattern of decline. What is the expected profit for 2005? (Let $t = 0$ represent 2000.)

49. ***Learning Curve*** The management at a plastics factory has found that the maximum number of units a worker can produce in a day is 30. The learning curve for the number N of units produced per day after a new employee has worked t days is

$$N = 30(1 - e^{kt}).$$

After 20 days on the job, a new employee produces 19 units.

(a) Find the learning curve for this employee (first, find the value of k).

(b) How many days should pass before this employee is producing 25 units per day?

50. ***Population Growth*** A conservation organization releases 100 animals of an endangered species into a game preserve. The organization believes that the preserve has a carrying capacity of 1000 animals and that the growth of the herd will be modeled by the logistic curve

$$p(t) = \frac{1000}{1 + 9e^{-0.1656t}}$$

where t is measured in months (see figure).

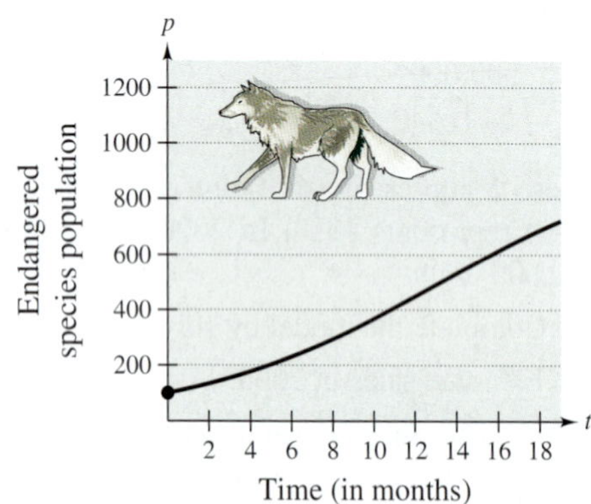

(a) Estimate the population after 5 months.

(b) After how many months will the population be 500?

 (c) Use a graphing utility to graph the function. Use the graph to determine the horizontal asymptotes, and interpret the meaning of the larger p-value in the context of the problem.

Geology **In Exercises 51 and 52, use the Richter scale for measuring the magnitudes of earthquakes.**

51. Find the magnitude R of an earthquake of intensity I (let $I_0 = 1$).

(a) $I = 80{,}500{,}000$

(b) $I = 48{,}275{,}000$

(c) $I = 251{,}200$

52. Find the intensity I of an earthquake measuring R on the Richter scale (let $I_0 = 1$).

(a) Chile in 1906, $R = 8.2$

(b) Los Angeles in 1971, $R = 6.7$

(c) India in 2001, $R = 7.7$

Intensity of Sound **In Exercises 53–56, use the following information for determining sound intensity. The level of sound β, in decibels, with an intensity of I is**

$$\beta = 10 \log_{10} \frac{I}{I_0}$$

where I_0 is an intensity of 10^{-12} watt per square meter, corresponding roughly to the faintest sound that can be heard by the human ear. In Exercises 53 and 54, find the level of sound, β.

53. (a) $I = 10^{-10}$ watt per m^2 (faint whisper)

(b) $I = 10^{-5}$ watt per m^2 (busy street corner)

(c) $I = 10^{-2.5}$ watt per m^2 (air hammer)

(d) $I = 10^{0}$ watt per m^2 (threshold of pain)

54. (a) $I = 10^{-9}$ watt per m^2 (whisper)

(b) $I = 10^{-3.5}$ watt per m^2 (jet 4 miles from takeoff)

(c) $I = 10^{-3}$ watt per m^2 (diesel truck at 25 feet)

(d) $I = 10^{-0.5}$ watt per m^2 (auto horn at 3 feet)

55. Due to the installation of noise suppression materials, the noise level in an auditorium was reduced from 93 to 80 decibels. Find the percent decrease in the intensity level of the noise as a result of the installation of these materials.

56. Due to the installation of a muffler, the noise level of an engine was reduced from 88 to 72 decibels. Find the percent decrease in the intensity level of the noise as a result of the installation of the muffler.

pH Levels **In Exercises 57–62, use the acidity model given by $\text{pH} = -\log_{10}[H^+]$, where acidity (pH) is a measure of the hydrogen ion concentration $[H^+]$ (measured in moles of hydrogen per liter) of a solution.**

57. Find the pH if $[H^+] = 2.3 \times 10^{-5}$.

58. Find the pH if $[H^+] = 11.3 \times 10^{-6}$.

59. Compute $[H^+]$ for a solution in which $\text{pH} = 5.8$.

60. Compute $[H^+]$ for a solution in which $\text{pH} = 3.2$.

61. A fruit has a pH of 2.5 and an antacid tablet has a pH of 9.5. The hydrogen ion concentration of the fruit is how many times the concentration of the tablet?

62. The pH of a solution is decreased by one unit. The hydrogen ion concentration is increased by what factor?

63. ***Forensics*** At 8:30 A.M., a coroner was called to the home of a person who had died during the night. In order to estimate the time of death, the coroner took the person's temperature twice. At 9:00 A.M. the temperature was 85.7°F, and at 11:00 A.M. the temperature was 82.8°F. From these two temperatures, the coroner was able to determine that the time elapsed since death and the body temperature were related by the formula

$$t = -10 \ln \frac{T - 70}{98.6 - 70}$$

where t is the time in hours elapsed since the person died and T is the temperature (in degrees Fahrenheit) of the person's body. Assume that the person had a normal body temperature of 98.6°F at death, and that the room temperature was a constant 70°F. (This formula is derived from a general cooling principle called Newton's Law of Cooling.) Use the formula to estimate the time of death of the person.

64. ***Home Mortgage*** A \$120,000 home mortgage for 35 years at $7\frac{1}{2}\%$ has a monthly payment of \$809.39. Part of the monthly payment goes for the interest charge on the unpaid balance, and the remainder of the payment is used to reduce the principal. The amount that goes toward the interest is

$$u = M - \left(M - \frac{Pr}{12}\right)\left(1 + \frac{r}{12}\right)^{12t}$$

and the amount that goes toward the reduction of the principal is

$$v = \left(M - \frac{Pr}{12}\right)\left(1 + \frac{r}{12}\right)^{12t}.$$

In these formulas, P is the size of the mortgage, r is the interest rate, M is the monthly payment, and t is the time in years.

(a) Use a graphing utility to graph each function in the same viewing window. (The viewing window should show all 35 years of mortgage payments.)

(b) In the early years of the mortgage, the larger part of the monthly payment goes for what purpose? Approximate the time when the monthly payment is evenly divided between interest and principal reduction.

(c) Repeat parts (a) and (b) for a repayment period of 20 years ($M = \$966.71$). What can you conclude?

65. ***Home Mortgage*** The total interest u paid on a home mortgage of P dollars at interest rate r for t years is

$$u = P\left[\frac{rt}{1 - \left(\frac{1}{1 + r/12}\right)^{12t}} - 1\right].$$

Consider a \$120,000 home mortgage at $7\frac{1}{2}\%$.

(a) Use a graphing utility to graph the total interest function.

(b) Approximate the length of the mortgage for which the total interest paid is the same as the size of the mortgage. Is it possible that some people are paying twice as much in interest charges as the size of the mortgage?

66. ***Data Analysis*** The table shows the time t (in seconds) required to attain a speed of s miles per hour from a standing start for a car.

Speed, s	Time, t
30	3.4
40	5.0
50	7.0
60	9.3
70	12.0
80	15.8
90	20.0

Two models for this data are as follows.

$$t_1 = 40.757 + 0.556s - 15.817 \ln s$$

$$t_2 = 1.2259 + 0.0023s^2$$

(a) Use a graphing utility to fit a linear model t_3 and an exponential model t_4 to the data.

(b) Use a graphing utility to graph the data points and each model in the same viewing window.

(c) Create a table comparing the data with estimates obtained from each model.

(d) Use the results of part (c) to find the sum of the absolute values of the differences between the data and estimated values given by each model. Based on the four sums, which model do you think better fits the data? Explain.

Synthesis

True or False? **In Exercises 67–70, determine whether the statement is true or false. Justify your answer.**

67. The domain of a logistic growth function cannot be the set of real numbers.

68. A logistic growth function will always have an x-intercept.

69. The graph of

$$f(x) = \frac{4}{1 + 6e^{-2x}} + 5$$

is the graph of

$$g(x) = \frac{4}{1 + 6e^{-2x}}$$

shifted to the right five units.

70. The graph of a Gaussian model will never have an x-intercept.

71. Identify each model as linear, logarithmic, exponential, logistic, or none of the above. Explain your reasoning.

(a)

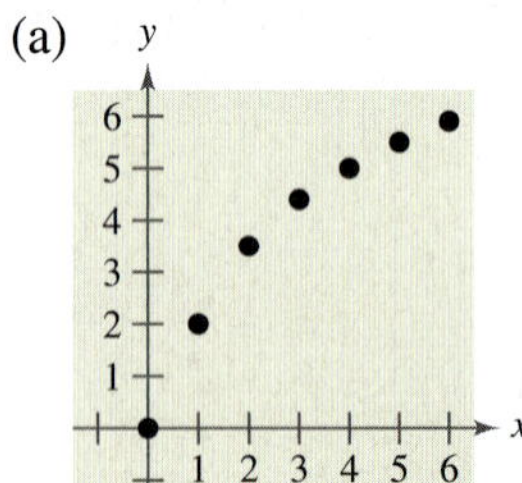

(b)

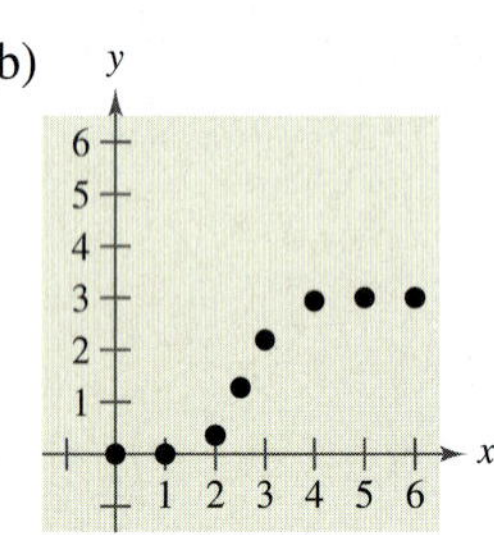

(c)

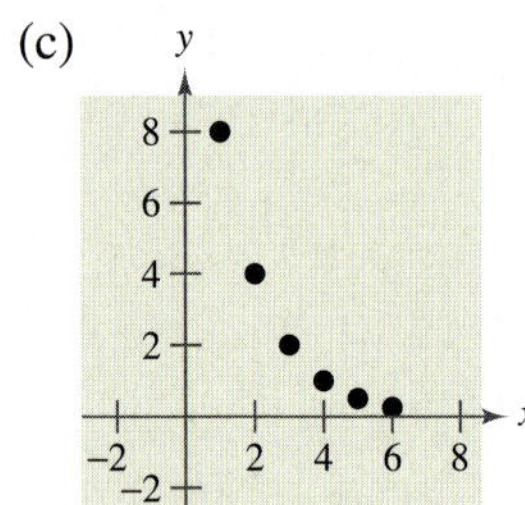

(d)

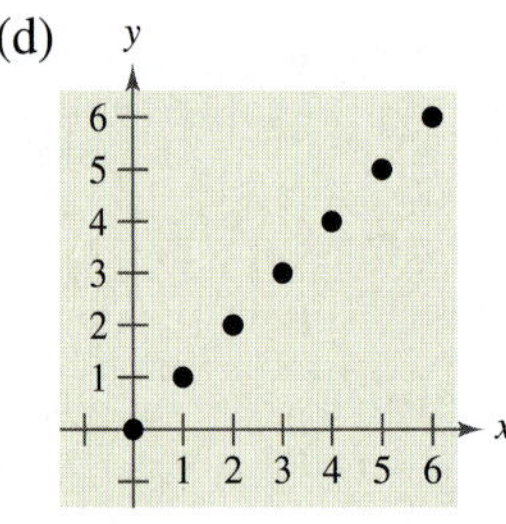

(e)

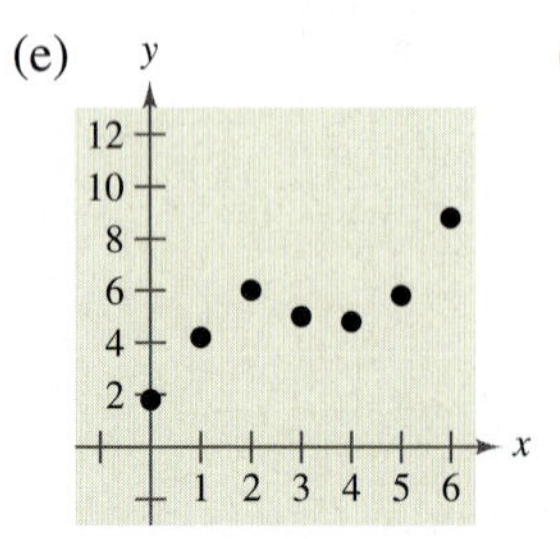

(f)

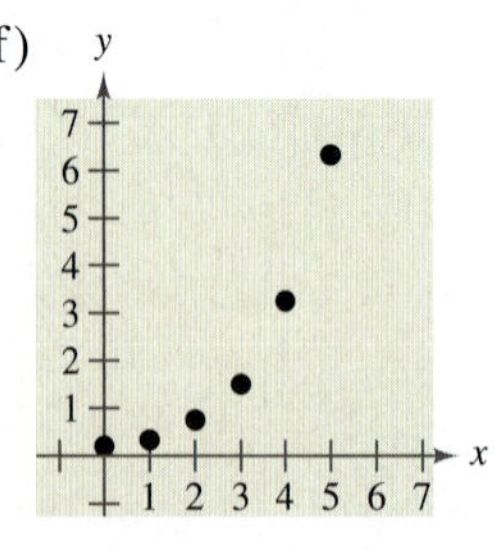

72. ***Writing*** Use your school's library, the Internet, or some other reference source to write a paper describing John Napier's work with logarithms.

Review

In Exercises 73–76, determine the right-hand and left-hand behavior of the polynomial function.

73. $f(x) = 2x^3 - 3x^2 + x - 1$

74. $f(x) = 5 - x^2 - 4x^4$

75. $g(x) = -1.6x^5 + 4x^2 - 2$

76. $g(x) = 7x^6 + 9.1x^5 - 3.2x^4 + 25x^3$

In Exercises 77–80, divide using synthetic division.

77. $\dfrac{4x^3 + 4x^2 - 39x + 36}{x + 4}$

78. $\dfrac{8x^3 - 36x^2 + 54x - 27}{x - \frac{3}{2}}$

79. $(2x^3 - 8x^2 + 3x - 9) \div (x - 4)$

80. $(x^4 - 3x + 1) \div (x + 5)$

In Exercises 81–90, sketch the graph of the equation.

81. $y = 10 - 3x$

82. $y = -4x - 1$

83. $y = -2x^2 - 3$

84. $y = 2x^2 - 7x - 30$

85. $3x^2 - 4y = 0$

86. $-x^2 - 8y = 0$

87. $y = \dfrac{4}{1 - 3x}$

88. $y = \dfrac{x^2}{-x - 2}$

89. $x^2 + (y - 8)^2 = 25$

90. $(x - 4)^2 + (y + 7) = 4$

In Exercises 91–94, graph the exponential function.

91. $f(x) = 2^{x-1} + 5$

92. $f(x) = -2^{-x-1} - 1$

93. $f(x) = 3^x - 4$

94. $f(x) = -3^x + 4$

Chapter Summary

▶ *What* did you learn?

Section 5.1	Review Exercises
☐ How to recognize and evaluate exponential functions with base *a*	1–10
☐ How to graph exponential functions	11–22, 27–30
☐ How to recognize and evaluate exponential functions with base *e*	23–26
☐ How to use exponential functions to model and solve real-life applications	31–36
Section 5.2	
☐ How to recognize and evaluate logarithmic functions with base *a*	37–42
☐ How to graph logarithmic functions	43–48, 55–58
☐ How to recognize and evaluate natural logarithmic functions	49–54
☐ How to use logarithmic functions to model and solve real-life applications	59, 60
Section 5.3	
☐ How to rewrite logarithmic functions with a different base	61–64
☐ How to use properties of logarithms to evaluate or rewrite and expand or condense logarithmic expressions	65–80
☐ How to use logarithmic functions to model and solve real-life applications	81, 82
Section 5.4	
☐ How to solve simple exponential and logarithmic equations	83–92
☐ How to solve more complicated exponential equations	93–106
☐ How to solve more complicated logarithmic equations	107–122
☐ How to use exponential and logarithmic equations to model and solve real-life applications	123, 124
Section 5.5	
☐ How to recognize the five most common types of models involving exponential and logarithmic functions	125–130
☐ How to use exponential growth and decay functions to model and solve real-life problems	131–136
☐ How to use Gaussian functions to model and solve real-life problems	137
☐ How to use logistic growth functions to model and solve real-life problems	138
☐ How to use logarithmic functions to model and solve real-life problems	139, 140

Review Exercises

5.1 **In Exercises 1–6, evaluate the expression. Approximate your result to three decimal places.**

1. $(6.1)^{2.4}$
2. $-14(5^{-0.8})$
3. $2^{-0.5\pi}$
4. $\sqrt[5]{1278}$
5. $60^{\sqrt{3}}$
6. $7^{-\sqrt{11}}$

In Exercises 7–10, match the function with its graph. [The graphs are labeled (a), (b), (c), and (d).]

(a)

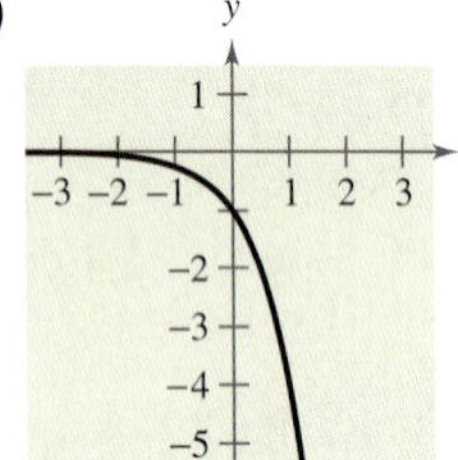

(b)

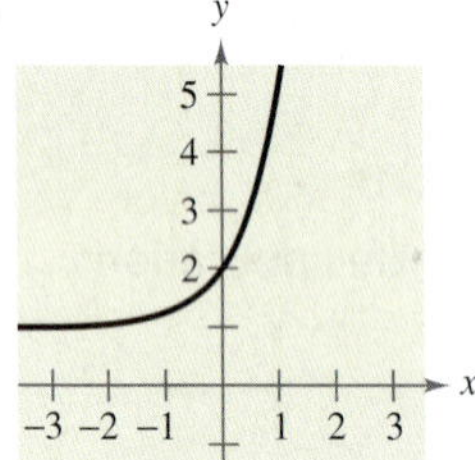

(c)

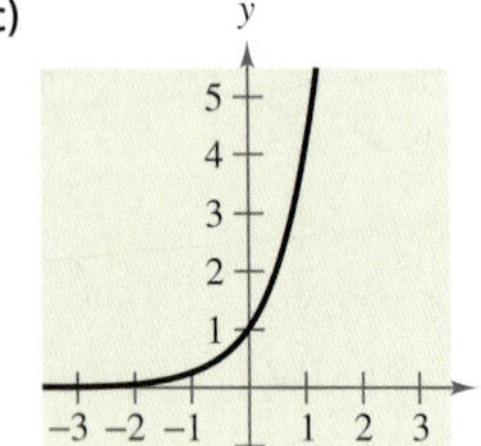

(d)

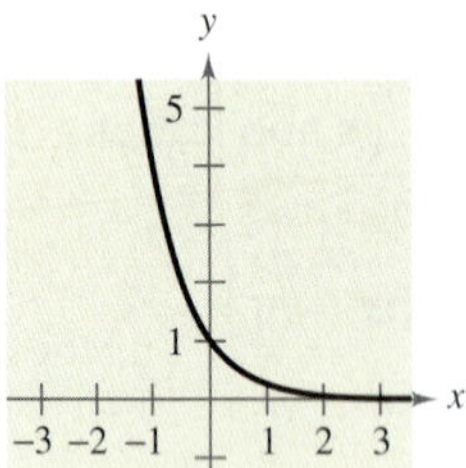

7. $f(x) = 4^x$
8. $f(x) = 4^{-x}$
9. $f(x) = -4^x$
10. $f(x) = 4^x + 1$

In Exercises 11–14, use the graph of f to describe the transformation that yields the graph of g.

11. $f(x) = 5^x, \quad g(x) = 5^{x-1}$
12. $f(x) = 4^x, \quad g(x) = 4^x - 3$
13. $f(x) = \left(\frac{1}{2}\right)^x, \quad g(x) = -\left(\frac{1}{2}\right)^{x+2}$
14. $f(x) = \left(\frac{2}{3}\right)^x, \quad g(x) = 8 - \left(\frac{2}{3}\right)^x$

In Exercises 15–22, use a graphing utility to construct a table of values. Then sketch the graph of the function.

15. $f(x) = 4^{-x} + 4$
16. $f(x) = -4^x - 3$
17. $f(x) = -2.65^{x+1}$
18. $f(x) = 2.65^{x-1}$
19. $f(x) = 5^{x-2} + 4$
20. $f(x) = 2^{x-6} - 5$
21. $f(x) = \left(\frac{1}{2}\right)^{-x} + 3$
22. $f(x) = \left(\frac{1}{8}\right)^{x+2} - 5$

In Exercises 23–26, evaluate the function $f(x) = e^x$ for the indicated value of x. Approximate your result to three decimal places.

23. $x = 8$
24. $x = \frac{5}{8}$
25. $x = -1.7$
26. $x = 0.278$

In Exercises 27–30, use a graphing utility to construct a table of values. Then sketch the graph of the function.

27. $h(x) = e^{-x/2}$
28. $h(x) = 2 - e^{-x/2}$
29. $f(x) = e^{x+2}$
30. $s(t) = 4e^{-2/t}, \quad t > 0$

Compound Interest **In Exercises 31 and 32, complete the table to determine the balance A for P dollars invested at rate r for t years and compounded n times per year.**

n	1	2	4	12	365	Continuous
A						

31. $P = \$3500, \; r = 6.5\%, \; t = 10$ years
32. $P = \$2000, \; r = 5\%, \; t = 30$ years

33. ***Waiting Times*** The average time between incoming calls at a switchboard is 3 minutes. The probability F of waiting less than t minutes until the next incoming call is approximated by the model $F(t) = 1 - e^{-t/3}$. A call has just come in. Find the probability that the next call will be within
   (a) $\frac{1}{2}$ minute. (b) 2 minutes. (c) 5 minutes.

34. ***Depreciation*** After t years, the value V of a car that cost \$14,000 is $V(t) = 14{,}000\left(\frac{3}{4}\right)^t$.
   (a) Use a graphing utility to graph the function.
   (b) Find the value of the car 2 years after it was purchased.
   (c) According to the model, when does the car depreciate most rapidly? Is this realistic? Explain.

35. ***Trust Fund*** On the day a person was born, a deposit of \$50,000 was made in a trust fund that pays 8.75% interest, compounded continuously.
   (a) Find the balance on the person's 35th birthday.
   (b) How much longer would the person have to wait to get twice as much?

36. ***Radioactive Decay*** Let Q represent a mass of plutonium 241 (^{241}Pu) (in grams), whose half-life is 13 years. The quantity of plutonium 241 present after t years is

$$Q = 100\left(\frac{1}{2}\right)^{t/13}.$$

(a) Determine the initial quantity (when $t = 0$).

(b) Determine the quantity present after 10 years.

(c) Sketch the graph of this function over the interval $t = 0$ to $t = 100$.

5.2 **In Exercises 37 and 38, write the exponential equation in logarithmic form.**

37. $4^3 = 64$ **38.** $25^{3/2} = 125$

In Exercises 39–42, evaluate the function at the indicated value of x without using a calculator.

	Function	*Value*
39.	$f(x) = \log_{10} x$	$x = 1000$
40.	$g(x) = \log_9 x$	$x = 3$
41.	$g(x) = \log_2 x$	$x = \frac{1}{8}$
42.	$f(x) = \log_4 x$	$x = \frac{1}{4}$

In Exercises 43–48, find the domain, x-intercept, and vertical asymptote of the logarithmic function and sketch its graph.

43. $g(x) = \log_7 x$ **44.** $g(x) = \log_5 x$

45. $f(x) = \log_{10}\left(\frac{x}{3}\right)$ **46.** $f(x) = 6 + \log_{10} x$

47. $f(x) = 4 - \log_{10}(x + 5)$

48. $f(x) = \log_{10}(x - 3) + 1$

In Exercises 49–54, use your calculator to evaluate the function $f(x) = \ln x$ for the indicated value of x. Approximate your result to three decimal places if necessary.

49. $x = 22.6$ **50.** $x = 0.98$

51. $x = e^{-12}$ **52.** $x = e^7$

53. $x = \sqrt{7} + 5$ **54.** $x = \frac{\sqrt{3}}{8}$

In Exercises 55–58, find the domain, x-intercept, and vertical asymptote of the logarithmic function and sketch its graph.

55. $f(x) = \ln x + 3$ **56.** $f(x) = \ln(x - 3)$

57. $h(x) = \ln(x^2)$ **58.** $f(x) = \frac{1}{4}\ln x$

59. ***Antler Spread*** The antler spread a (in inches) and shoulder height h (in inches) of an adult male American elk are related by the model

$$h = 116 \log_{10}(a + 40) - 176.$$

Approximate the shoulder height of a male American elk with an antler spread of 55 inches.

60. ***Snow Removal*** The number of miles s of roads cleared of snow is approximated by the model

$$s = 25 - \frac{13 \ln(h/12)}{\ln 3}, \qquad 2 \le h \le 15$$

where h is the depth of the snow in inches. Use this model to find s when $h = 10$ inches.

5.3 **In Exercises 61–64, evaluate the logarithm using the change-of-base formula. Do each problem twice, once with common logarithms and once with natural logarithms. Approximate the results to three decimal places.**

61. $\log_4 9$ **62.** $\log_{12} 200$

63. $\log_{1/2} 5$ **64.** $\log_3 0.28$

In Exercises 65–72, use the properties of logarithms to expand the expression as a sum, difference, and/or multiple of logarithms.

65. $\log_5 5x^2$ **66.** $\log_{10} 7x^4$

67. $\log_3 \frac{6}{\sqrt[3]{x}}$ **68.** $\log_7 \frac{\sqrt{x}}{4}$

69. $\ln x^2y^2z$ **70.** $\ln 3xy^2$

71. $\ln\left(\frac{x + 3}{xy}\right)$ **72.** $\ln\left(\frac{y - 1}{4}\right)^2, \quad y > 1$

In Exercises 73–80, condense the expression to the logarithm of a single quantity.

73. $\log_2 5 + \log_2 x$

74. $\log_6 y - 2 \log_6 z$

75. $\ln x - \frac{1}{4}\ln y$

76. $3 \ln x + 2 \ln(x + 1)$

77. $\frac{1}{3}\log_8(x + 4) + 7 \log_8 y$

78. $-2 \log_{10} x - 5 \log_{10}(x + 6)$

79. $\frac{1}{2}\ln(2x - 1) - 2 \ln(x + 1)$

80. $5 \ln(x - 2) - \ln(x + 2) - 3 \ln x$

81. ***Climb Rate*** The time t (in minutes) for a small plane to climb to an altitude of h feet is modeled by

$$t = 50 \log_{10} \frac{18{,}000}{18{,}000 - h}$$

where 18,000 feet is the plane's absolute ceiling.

(a) Determine the domain of the function appropriate for the context of the problem.

 (b) Use a graphing utility to graph the time function and identify any asymptotes.

(c) As the plane approaches its absolute ceiling, what can be said about the time required to increase its altitude further?

(d) Find the time for the plane to climb to an altitude of 4000 feet.

82. ***Human Memory Model*** Students in a sociology class were given an exam and then were retested monthly with an equivalent exam. The average score for the class was given by the human memory model

$$f(t) = 85 - 14 \log_{10}(t + 1), \quad 0 \le t \le 4$$

where t is the time in months. How did the average score change over the four-month period?

5.4 **In Exercises 83–92, solve for x.**

83. $8^x = 512$ **84.** $3^x = 729$

85. $6^x = \frac{1}{216}$ **86.** $5^x = \frac{1}{25}$

87. $e^x = 3$ **88.** $e^x = 6$

89. $\log_4 x = 2$ **90.** $\log_6 x = -1$

91. $\ln x = 4$ **92.** $\ln x = -3$

In Exercises 93–102, solve the exponential equation. Approximate your result to three decimal places.

93. $e^x = 12$ **94.** $e^{3x} = 25$

95. $3e^{-5x} = 132$ **96.** $14e^{3x+2} = 560$

97. $2^x + 13 = 35$ **98.** $6^x - 28 = -8$

99. $-4(5^x) = -68$ **100.** $2(12^x) = 190$

101. $e^{2x} - 7e^x + 10 = 0$ **102.** $e^{2x} - 6e^x + 8 = 0$

 **In Exercises 103–106, use a graphing utility to graph and solve the equation. Approximate the result to two decimal places.**

103. $2^{0.6x} - 3x = 0$ **104.** $4^{-0.2x} + x = 0$

105. $25e^{-0.3x} = 12$ **106.** $4e^{1.2x} = 9$

In Exercises 107–118, solve the logarithmic equation. Approximate the result to three decimal places.

107. $\ln 3x = 8.2$ **108.** $\ln 5x = 7.2$

109. $2 \ln 4x = 15$ **110.** $4 \ln 3x = 15$

111. $\ln x - \ln 3 = 2$ **112.** $\ln \sqrt{x + 8} = 3$

113. $\ln \sqrt{x + 1} = 2$ **114.** $\ln x - \ln 5 = 4$

115. $\log_{10}(x - 1) = \log_{10}(x - 2) - \log_{10}(x + 2)$

116. $\log_{10}(x + 2) - \log_{10} x = \log_{10}(x + 5)$

117. $\log_{10}(1 - x) = -1$

118. $\log_{10}(-x - 4) = 2$

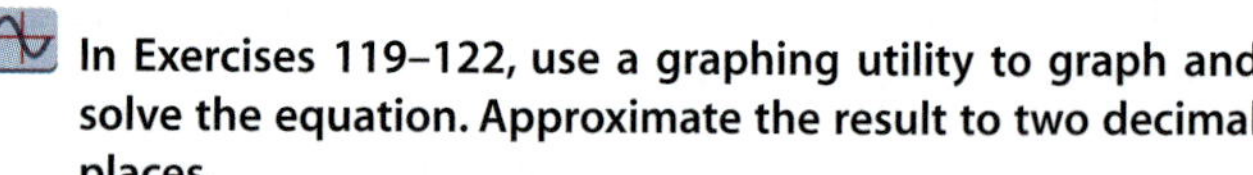

In Exercises 119–122, use a graphing utility to graph and solve the equation. Approximate the result to two decimal places.

119. $2 \ln(x + 3) + 3x = 8$

120. $6 \log_{10}(x^2 + 1) - x = 0$

121. $4 \ln(x + 5) - x = 10$

122. $x - 2 \log_{10}(x + 4) = 0$

123. ***Compound Interest*** $7550 is deposited in an account that pays 7.25% interest, compounded continuously. How long will it take the money to triple?

124. ***Demand*** The demand equation for a 32-inch television is modeled by $p = 500 - 0.5e^{0.004x}$. Find the demand x for a price of (a) $p = \$450$ and (b) $p = \$400$.

5.5 **In Exercises 125–130, match the function with its graph. [The graphs are labeled (a), (b), (c), (d), (e), and (f).]**

(a)

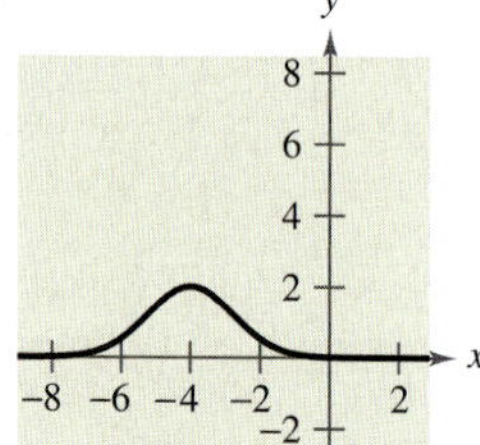

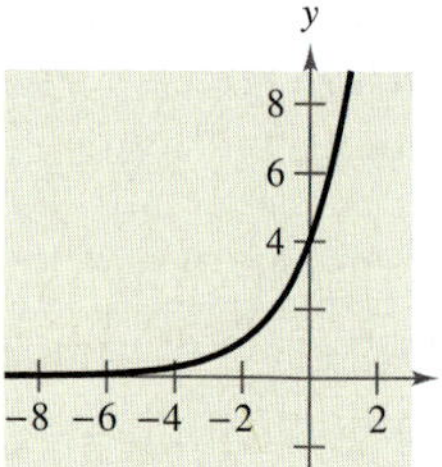
(b)

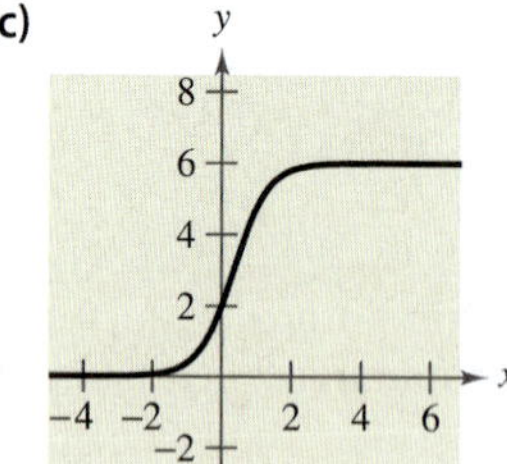
(c)

(d)

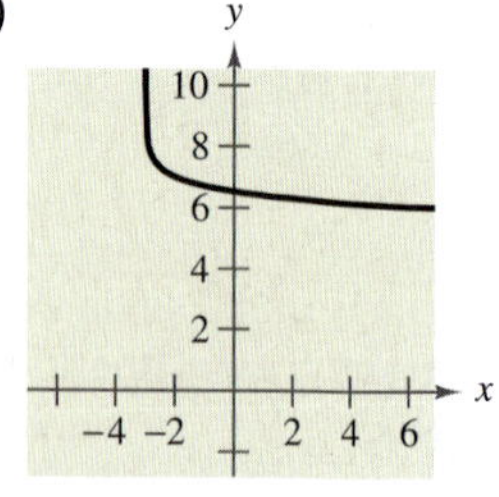

(e)

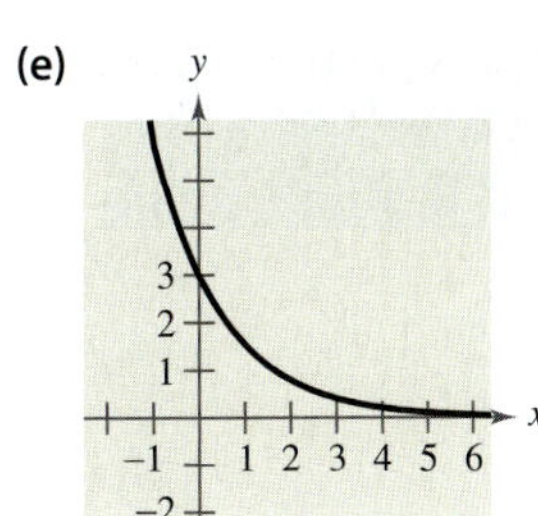

(f)

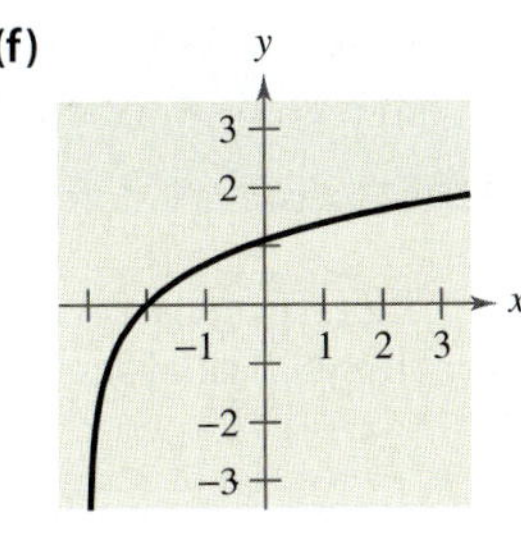

125. $y = 3e^{-2x/3}$

126. $y = 4e^{2x/3}$

127. $y = \ln(x + 3)$

128. $y = 7 - \log_{10}(x + 3)$

129. $y = 2e^{-(x+4)^2/3}$

130. $y = \dfrac{6}{1 + 2e^{-2x}}$

131. ***Population*** The population P of Phoenix, Arizona (in thousands) from 1970 through 2000 can be modeled by $P = 590e^{0.027t}$, where t represents the year, with $t = 0$ corresponding to 1970. According to this model, when will the population reach 1.5 million? (Source: U.S. Census Bureau)

132. ***Radioactive Decay*** The half-life of radioactive uranium II (^{234}U) is 250,000 years. What percent of a present amount of radioactive uranium II will remain after 5000 years?

133. ***Compound Interest*** A deposit of \$10,000 is made in a savings account for which the interest is compounded continuously. The balance will double in 5 years.

(a) What is the annual interest rate for this account?

(b) Find the balance after 1 year.

134. ***Bacteria Growth*** The number N of bacteria in a culture is given by the model $N = 200e^{kt}$, where t is the time in hours. If $N = 350$ when $t = 5$, estimate the time required for the population to triple in size.

In Exercises 135 and 136, find the exponential function $y = ae^{bx}$ that passes through the points.

135. $(0, 2), (4, 3)$

136. $\left(0, \frac{1}{2}\right), (5, 5)$

137. ***Test Scores*** The test scores for a biology test follow a normal distribution modeled by

$y = 0.0499e^{-(x-71)^2/128}, \quad 40 \le x \le 100$

where x is the test score.

(a) Use a graphing utility to graph the equation.

(b) From the graph, estimate the average test score.

138. ***Typing Speed*** In a typing class, the average number of words per minute typed after t weeks of lessons was found to be

$$N = \frac{157}{1 + 5.4e^{-0.12t}}.$$

Find the time necessary to type (a) 50 words per minute and (b) 75 words per minute.

139. ***Sound Intensity*** The relationship between the number of decibels β and the intensity of a sound I in watts per square centimeter is

$$\beta = 10 \log_{10}\left(\frac{I}{10^{-16}}\right).$$

Determine the intensity of a sound in watts per square centimeter if the decibel level is 125.

140. ***Geology*** On the Richter scale, the magnitude R of an earthquake of intensity I is

$$R = \log_{10} \frac{I}{I_0}$$

where $I_0 = 1$ is the minimum intensity used for comparison. Find the intensity per unit of area for each value of R.

(a) $R = 8.4$ (b) $R = 6.85$ (c) $R = 9.1$

Synthesis

True or False? **In Exercises 141 and 142, determine whether the equation or statement is true or false. Justify your answer.**

141. $\log_b b^{2x} = 2x$

142. $\ln(x + y) = \ln x + \ln y$

143. The graphs of $y = e^{kt}$ are shown for $k = a, b, c,$ and d. Use the graphs to order $a, b, c,$ and d. Which of the four values are negative? Which are positive?

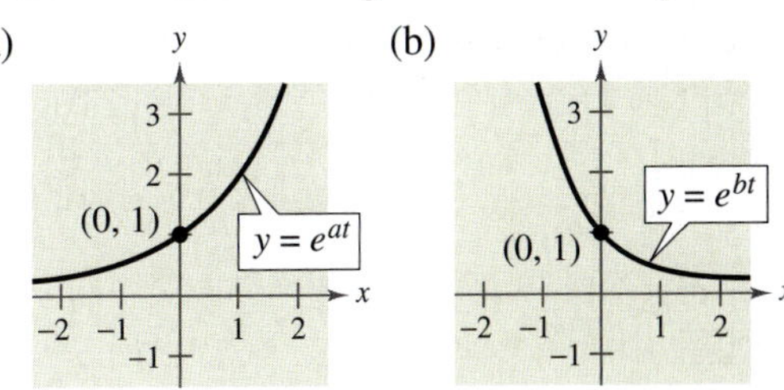

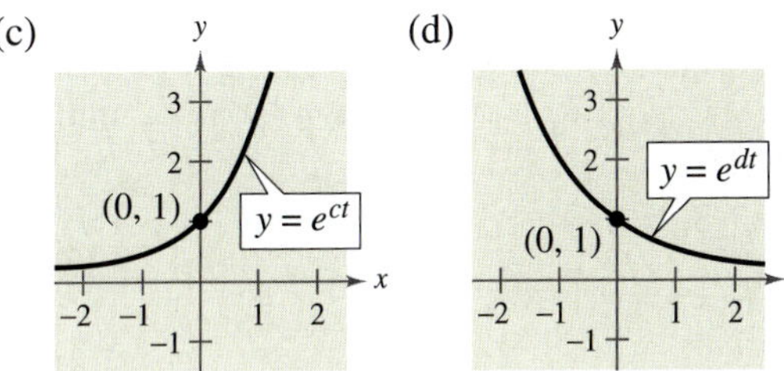

Chapter Test

The *Interactive* CD-ROM and *Internet* versions of this text offer Chapter Pre-Tests and Chapter Post-Tests, both of which have randomly generated exercises with diagnostic capabilities.

Take this test as you would take a test in class. When you are finished, check your work against the answers given in the back of the book.

In Exercises 1–4, evaluate the expression. Approximate your result to three decimal places.

1. $12.4^{2.79}$ **2.** $4^{3\pi/2}$ **3.** $e^{-7/10}$ **4.** $e^{3.1}$

In Exercises 5–7, construct a table of values. Then sketch the graph of the function.

5. $f(x) = 10^{-x}$ **6.** $f(x) = -6^{x-2}$ **7.** $f(x) = 1 - e^{2x}$

8. Evaluate (a) $\log_7 7^{-0.89}$ and (b) $4.6 \ln e^2$.

In Exercises 9–11, construct a table of values. Then sketch the graph of the function. Identify any asymptotes.

9. $f(x) = -\log_{10} x - 6$ **10.** $f(x) = \ln(x - 4)$ **11.** $f(x) = 1 + \ln(x + 6)$

In Exercises 12–14, evaluate the expression. Approximate your result to three decimal places.

12. $\log_7 44$ **13.** $\log_{2/5} 0.9$ **14.** $\log_{24} 68$

In Exercises 15 and 16, use the properties of logarithms to expand the expression as a sum, difference, and/or multiple of logarithms.

15. $\log_2 3a^4$ **16.** $\ln \dfrac{5\sqrt{x}}{6}$

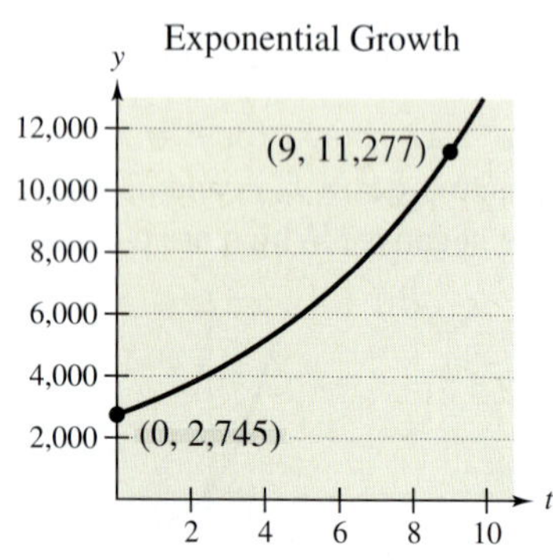

FIGURE FOR 21

In Exercises 17 and 18, condense the expression to the logarithm of a single quantity.

17. $\log_3 13 + \log_3 y$ **18.** $4 \ln x - 4 \ln y$

In Exercises 19 and 20, solve the equation algebraically. Approximate your result to three decimal places.

19. $\dfrac{1025}{8 + e^{4x}} = 5$ **20.** $\log_{10} x - \log_{10}(8 - 5x) = 2$

21. Find an exponential growth model for the graph shown in the figure.

22. The half-life of radioactive actinium (^{227}Ac) is 22 years. What percent of a present amount of radioactive actinium will remain after 19 years?

23. A model that can be used for predicting the height H (in centimeters) of a child based on his or her age is $H = 70.228 + 5.104x + 9.222 \ln x$, $\frac{1}{4} \le x \le 6$, where x is the age of the child in years. (Source: Snapshots of Applications in Mathematics)

(a) Construct a table of values. Then sketch the graph of the model.

(b) Use the graph from part (a) to estimate the height of a four-year-old child. Then calculate the actual height using the model.

Cumulative Test for Chapters 3–5

Take this test to review the material from earlier chapters. When you are finished, check your work against the answers given in the back of the book.

1. Find the quadratic function whose graph has a vertex at $(-8, 5)$ and passes through the point $(-4, -7)$.

In Exercises 2–4, sketch the graph of the function without the aid of a graphing utility.

2. $h(x) = -(x^2 + 4x)$ 3. $f(t) = \frac{1}{4}t(t - 2)^2$ 4. $g(s) = s^2 + 4s + 10$

In Exercises 5 and 6, find all the zeros of the function.

5. $f(x) = x^3 + 2x^2 + 4x + 8$ 6. $f(x) = x^4 + 4x^3 - 21x^2$

7. Divide: $\dfrac{6x^3 - 4x^2}{2x^2 + 1}$.

8. Use synthetic division to divide $2x^4 + 3x^3 - 6x + 5$ by $x + 2$.

9. Use a graphing utility to approximate the real zero of the function $g(x) = x^3 + 3x^2 - 6$ to the nearest hundredth.

10. Find a polynomial with integer coefficients that has -5, -2, and $2 + \sqrt{3}i$ as its zeros.

In Exercises 11–13, sketch the graph of the rational function by hand. Be sure to identify all intercepts and asymptotes.

11. $f(x) = \dfrac{2x}{x - 3}$ 12. $f(x) = \dfrac{4x^2}{x - 5}$ 13. $f(x) = \dfrac{2x}{x^2 - 9}$

In Exercises 14 and 15, write the partial fraction decomposition of the rational expression. Check your result algebraically.

14. $\dfrac{8}{x^2 - 4x - 21}$ 15. $\dfrac{5x}{(x - 4)^2}$

In Exercises 16 and 17, sketch a graph of the conic.

16. $\dfrac{(x + 3)^2}{16} - \dfrac{(y + 4)^2}{25} = 1$ 17. $\dfrac{(x - 2)^2}{4} + \dfrac{(y + 1)^2}{9} = 1$

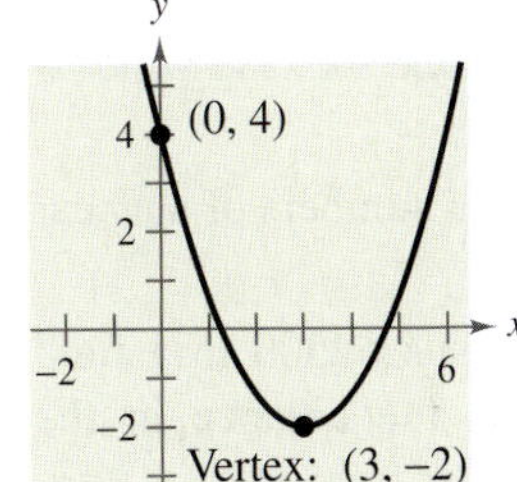

FIGURE FOR 18

18. Find an equation of the parabola shown in the figure.

19. Find an equation of the hyperbola with foci $(0, 0)$ and $(0, 4)$ and asymptotes $y = \pm\frac{1}{2}x + 2$.

In Exercises 20 and 21, use the graph of *f* to describe the transformation that yields the graph of *g*. Use a graphing utility to graph both equations in the same viewing window.

20. $f(x) = \left(\frac{2}{5}\right)^x$, $g(x) = -\left(\frac{2}{5}\right)^{-x+3}$ 21. $f(x) = 2.2^x$, $g(x) = -2.2^x + 4$

In Exercises 22–25, use a calculator to evaluate each expression. Approximate your result to three decimal places.

22. $\log_{10} 98$ **23.** $\log_{10}\left(\frac{6}{7}\right)$ **24.** $\ln\sqrt{31}$ **25.** $\ln\left(\sqrt{40} - 5\right)$

In Exercises 26–28, evaluate the logarithm using the change-of-base formula. Approximate your answer to three decimal places.

26. $\log_7 1.8$ **27.** $\log_3 0.149$ **28.** $\log_{1/2} 17$

29. Use the properties of logarithms to expand $\ln\left(\frac{x^2 - 16}{x^4}\right)$, where $x > 4$.

30. Write $2 \ln x - \frac{1}{2}\ln(x + 5)$ as a logarithm of a single quantity.

In Exercises 31–34, solve the equation.

31. $6e^{2x} = 72$ **32.** $4^{x-5} + 21 = 30$

33. $\log_2 x + \log_2 5 = 6$ **34.** $\ln 4x - \ln 2 = 8$

 **35.** Use a graphing utility to graph

$$f(x) = \frac{1000}{1 + 4e^{-0.2x}}$$

and determine the horizontal asymptotes.

36. Let x be the amount (in hundreds of dollars) that an online stock-trading company spends on advertising, and let P be the profit (in thousands of dollars), where $P = 230 + 20x - \frac{1}{2}x^2$. What amount of advertising will yield a maximum profit?

 **37.** The numbers D of new car dealerships in the United States from 1994 through 2000 are shown in the table. (Source: National Automobile Dealers Association)

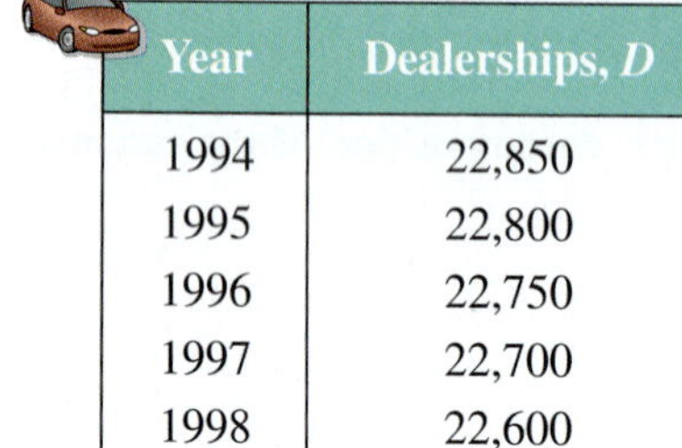

Year	Dealerships, D
1994	22,850
1995	22,800
1996	22,750
1997	22,700
1998	22,600
1999	22,400
2000	22,250

(a) Use a graphing utility to create a scatter plot of the data. Let t represent the year, with $t = 4$ corresponding to 1994.

(b) Use the *regression* feature of the graphing utility to find a quadratic model for the data.

(c) Use the graphing utility to graph the model in the same viewing window used for the scatter plot.

(d) Do you think this model could be used to predict the numbers of new car dealerships in the future? Explain.

38. On the day a grandchild is born, a grandparent deposits \$2500 in a fund earning 7.5%, compounded continuously. Determine the balance in the account at the time of the grandchild's 25th birthday.

39. The number N of bacteria in a culture is given by the model $N = 175e^{kt}$, where t is the time in hours. If $N = 420$ when $t = 8$, estimate the time required for the population to double in size.

40. The population P of Florida (in thousands) from 1991 through 2000 can be modeled by $P = 12{,}976e^{0.019t}$, where $t = 1$ represents the year 1991. According to this model, when will the population reach 17 million? (Source: U.S. Census Bureau)

Proofs in Mathematics

Each of the following three properties of logarithms can be proved by using properties of exponential functions.

Slide Rules

The slide rule was invented by William Oughtred (1574–1660) in 1625. The slide rule is a computational device with a sliding portion and a fixed portion. A slide rule enables you to perform multiplication by using Property 1 of the Properties of Logarithms. There are other slide rules that allow for the calculation of roots and trigonometric functions. Slide rules were used by mathematicians and engineers until the invention of the handheld calculator in 1972.

Properties of Logarithms *(p. 412)*

Let a be a positive number such that $a \neq 1$, and let n be a real number. If u and v are positive real numbers, the following properties are true.

Logarithm with Base a	*Natural Logarithm*
1. $\log_a(uv) = \log_a u + \log_a v$	**1.** $\ln(uv) = \ln u + \ln v$
2. $\log_a \dfrac{u}{v} = \log_a u - \log_a v$	**2.** $\ln \dfrac{u}{v} = \ln u - \ln v$
3. $\log_a u^n = n \log_a u$	**3.** $\ln u^n = n \ln u$

Proof

Let

$$x = \log_a u \quad \text{and} \quad y = \log_a v.$$

The corresponding exponential forms of these two equations are

$$a^x = u \quad \text{and} \quad a^y = v.$$

To prove Property 1, multiply u and v to obtain

$$uv = a^x a^y = a^{x+y}.$$

The corresponding logarithmic form of $uv = a^{x+y}$ is $\log_a(uv) = x + y$. So,

$$\log_a(uv) = \log_a u + \log_a v.$$

To prove Property 2, divide u by v to obtain

$$\frac{u}{v} = \frac{a^x}{a^y} = a^{x-y}.$$

The corresponding logarithmic form of $u/v = a^{x-y}$ is $\log_a(u/v) = x - y$. So,

$$\log_a(u/v) = \log_a u - \log_a v.$$

To prove Property 3, substitute a^x for u in the expression $\log_a u^n$, as follows.

$$\begin{aligned} \log_a u^n &= \log_a(a^x)^n && \text{Substitute } a^x \text{ for } u. \\ &= \log_a a^{nx} && \text{Property of exponents} \\ &= nx && \text{Inverse Property of Logarithms} \\ &= n \log_a u && \text{Substitute } \log_a u \text{ for } x. \end{aligned}$$

So, $\log_a u^n = n \log_a u$.

P.S. Problem Solving

1. Graph the exponential function $y = a^x$ for $a = 0.5$, 1.2, and 2.0. Which of these curves intersects the line $y = x$? Determine all positive numbers a for which the curve $y = a^x$ intersects the line $y = x$.

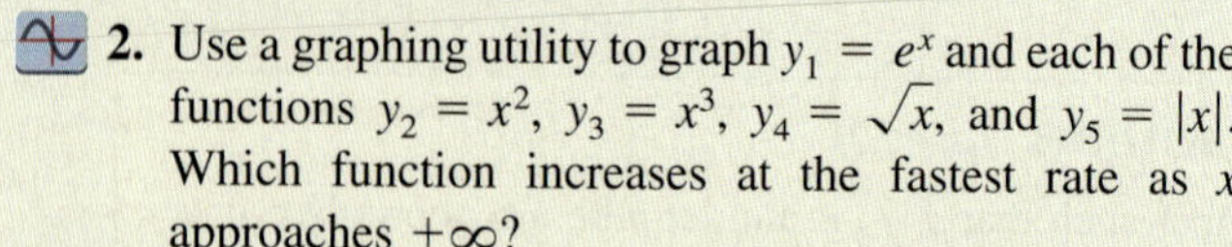

2. Use a graphing utility to graph $y_1 = e^x$ and each of the functions $y_2 = x^2$, $y_3 = x^3$, $y_4 = \sqrt{x}$, and $y_5 = |x|$. Which function increases at the fastest rate as x approaches $+\infty$?

3. Use the result of Exercise 2 to make a conjecture about the rate of growth of $y_1 = e^x$ and $y = x^n$, where n is a natural number and x approaches $+\infty$.

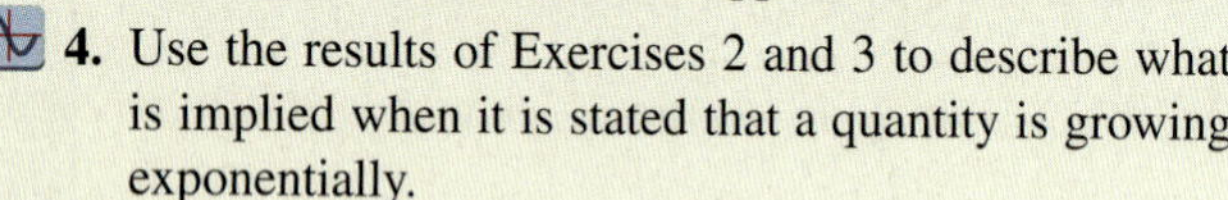

4. Use the results of Exercises 2 and 3 to describe what is implied when it is stated that a quantity is growing exponentially.

5. Given the exponential function

$$f(x) = a^x$$

show that

(a) $f(u + v) = f(u) \cdot f(v)$.

(b) $f(2x) = [f(x)]^2$.

6. Given that

$$f(x) = \frac{e^x + e^{-x}}{2} \text{ and } g(x) = \frac{e^x - e^{-x}}{2}$$

show that

$$[f(x)]^2 - [g(x)]^2 = 1.$$

7. Use a graphing utility to compare the graph of the function $y = e^x$ with the graph of each given function. [$n!$ (read "n factorial") is defined as $n! = 1 \cdot 2 \cdot 3 \cdot \cdot \cdot (n - 1) \cdot n$.]

(a) $y_1 = 1 + \dfrac{x}{1!}$

(b) $y_2 = 1 + \dfrac{x}{1!} + \dfrac{x^2}{2!}$

(c) $y_3 = 1 + \dfrac{x}{1!} + \dfrac{x^2}{2!} + \dfrac{x^3}{3!}$

8. Identify the pattern of successive polynomials given in Exercise 7. Extend the pattern one more term and compare the graph of the resulting polynomial function with the graph of $y = e^x$. What do you think this pattern implies?

9. Graph the function

$$f(x) = e^x - e^{-x}.$$

From the graph, the function appears to be one-to-one. Assuming that the function has an inverse function, find $f^{-1}(x)$.

10. Find a pattern for $f^{-1}(x)$ if

$$f(x) = \frac{a^x + 1}{a^x - 1}$$

where $a > 0$, $a \neq 1$.

11. By observation, identify the equation that corresponds to the graph. Explain your reasoning.

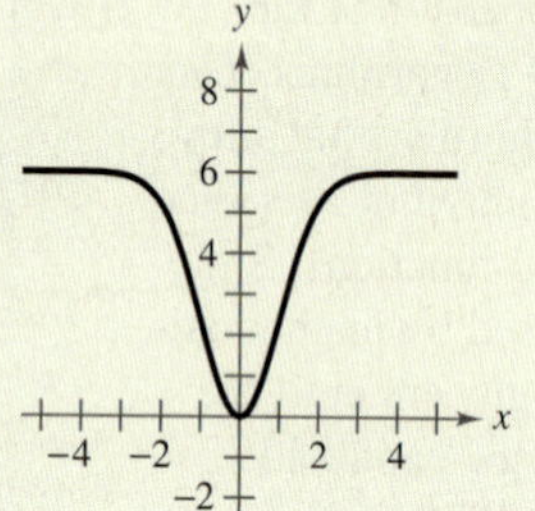

(a) $y = 6e^{-x^2/2}$ (b) $y = \dfrac{6}{1 + e^{-x/2}}$

(c) $y = 6(1 - e^{-x^2/2})$

12. There are two options for investing \$500. The first earns 7% compounded annually and the second earns 7% simple interest. The figure shows the growth of each investment over a 30-year period.

(a) Identify which graph represents each type of investment. Explain your reasoning.

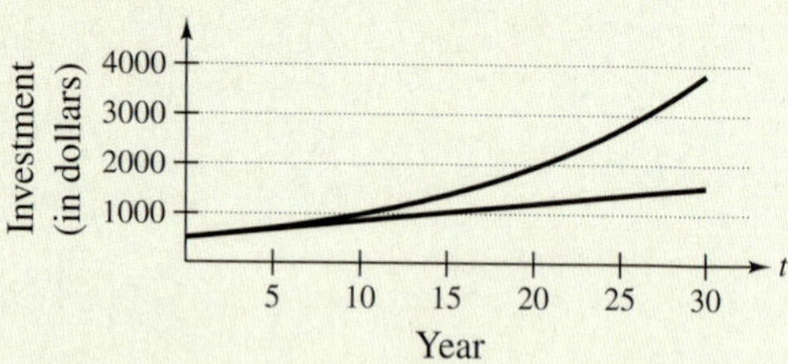

(b) Verify your answer in part (a) by finding the equations that model the investment growth and graphing the models.

13. Two different samples of radioactive isotopes are decaying. The isotopes have initial amounts of c_1 and c_2, as well as half-lives of k_1 and k_2, respectively. Find the time required for the samples to decay to equal amounts.

14. A lab culture initially contains 500 bacteria. Two hours later, the number of bacteria has decreased to 200. Find the exponential decay model of the form

$B = B_0 a^{kt}$

that can be used to approximate the number of bacteria after t hours.

15. The table shows the colonial population estimates of the American colonies from 1700 to 1780. (Source: U.S. Census Bureau)

Year	Population
1700	250,900
1710	331,700
1720	466,200
1730	629,400
1740	905,600
1750	1,170,800
1760	1,593,600
1770	2,148,100
1780	2,780,400

In each of the following, let y represent the population in the year t, with $t = 0$ corresponding to 1700.

(a) Use the *regression* feature of a graphing utility to find an exponential model for the data.

(b) Use the *regression* feature of the graphing utility to find a quadratic model for the data.

(c) Use the graphing utility to plot the data and the models from parts (a) and (b) in the same viewing window.

(d) Which model is a better fit for the data? Would you use this model to predict the population of the United States in 2010? Explain your reasoning.

16. Show that $\dfrac{\log_a x}{\log_{a/b} x} = 1 + \log_a \dfrac{1}{b}$.

17. Solve $(\ln x)^2 = \ln x^2$.

18. Use a graphing utility to compare the graph of the function $y = \ln x$ with the graph of each given function.

(a) $y_1 = x - 1$

(b) $y_2 = (x - 1) - \frac{1}{2}(x - 1)^2$

(c) $y_3 = (x - 1) - \frac{1}{2}(x - 1)^2 + \frac{1}{3}(x - 1)^3$

19. Identify the pattern of successive polynomials given in Exercise 18. Extend the pattern one more term and compare the graph of the resulting polynomial function with the graph of $y = \ln x$. What do you think the pattern implies?

20. Using

$y = ab^x$ and $y = ax^b$

take the natural logarithm of each side of each equation. What are the slope and y-intercept of the line relating x and $\ln y$ for $y = ab^x$? What are the slope and y-intercept of the line relating $\ln x$ and $\ln y$ for $y = ax^b$?

In Exercises 21 and 22, use the model

$y = 80.4 - 11 \ln x, \quad 100 \le x \le 1500$

which approximates the minimum required ventilation rate in terms of the air space per child in a public school classroom. In the model, x is the air space per child in cubic feet and y is the ventilation rate in cubic feet per minute.

21. Use a graphing utility to graph the function and approximate the required ventilation rate if there is 300 cubic feet of air space per child.

22. A classroom is designed for 30 students. The air conditioning system in the room has the capacity of moving 450 cubic feet of air per minute.

(a) Determine the ventilation rate per child, assuming that the room is filled to capacity.

(b) Estimate the air space required per child.

(c) Determine the minimum number of square feet of floor space required for the room if the ceiling height is 30 feet.

How to study Chapter 6

▶ What you should learn

In this chapter you will learn the following skills and concepts:

- How to describe an angle and to convert between degree and radian measures
- How to evaluate trigonometric functions of any angle
- How to use the fundamental trigonometric identities
- How to sketch the graphs of trigonometric functions and translations of graphs of sine and cosine functions
- How to evaluate the inverse trigonometric functions
- How to evaluate compositions of trigonometric functions and inverse trigonometric functions

▶ Important Vocabulary

As you encounter each new vocabulary term in this chapter, add the term and its definition to your notebook glossary.

Trigonometry (p. 454)
Angle (p. 454)
Initial side (of an angle) (p. 454)
Terminal side (of an angle) (p. 454)
Vertex (of an angle) (p. 454)
Standard position (p. 454)
Coterminal angles (p. 454)
Degree (p. 455)
Acute angles (p. 455)
Obtuse angles (p. 455)
Complementary angles (p. 456)
Supplementary angles (p. 456)
Central angle (p. 457)
Radian (p. 457)
Linear speed (p. 459)
Angular speed (p. 459)
Sine (p. 465)
Cosecant (p. 465)
Cosine (p. 465)
Secant (p. 465)
Tangent (p. 465)
Cotangent (p. 465)
Angle of elevation (p. 470)
Angle of depression (p. 470)
Reference angles (p. 478)
Period (pp. 483, 491)
Amplitude (p. 490)
Phase shift (p. 492)
Damping factor (p. 504)
Inverse sine function (p. 510)
Inverse cosine function (p. 512)
Inverse tangent function (p. 512)
Simple harmonic motion (pp. 523, 524)

Study Tools

Learning objectives in each section
Chapter Summary (p. 531)
Review Exercises (pp. 532–535)
Chapter Test (p. 536)

Additional Resources

Study and Solutions Guide
Interactive Algebra and Trigonometry
Videotapes/DVD for Chapter 6
Algebra and Trigonometry Website
Student Success Organizer

Raymond Forbes/SuperStock

6

Trigonometry

6.1 Angles and Their Measure

▶ **What you should learn**

- How to describe angles
- How to use degree measure
- How to use radian measure
- How to convert between degree and radian measures
- How to use angles to model and solve real-life problems

▶ **Why you should learn it**

You can use angles to model and solve real-life problems. For instance, in Exercise 97 on page 464, you can use angles to find the speed of a bicycle.

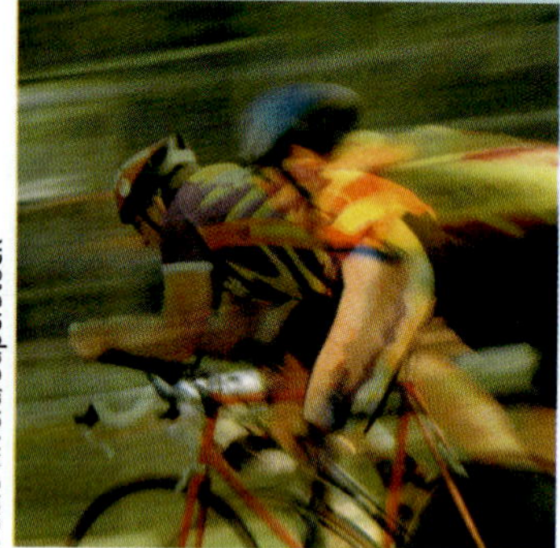

Pablo Rivera/SuperStock

Angles

As derived from the Greek language, the word **trigonometry** means "measurement of triangles." Initially, trigonometry dealt with relationships among the sides and angles of triangles and was used in the development of astronomy, navigation, and surveying. With the development of calculus and the physical sciences in the 17th century, a different perspective arose—one that viewed the classic trigonometric relationships as *functions* with the set of real numbers as their domains. Consequently, the applications of trigonometry expanded to include a vast number of physical phenomena involving rotations and vibrations. These phenomena include sound waves, light rays, planetary orbits, vibrating strings, pendulums, and orbits of atomic particles.

The approach in this text incorporates *both* perspectives, starting with angles and their measure.

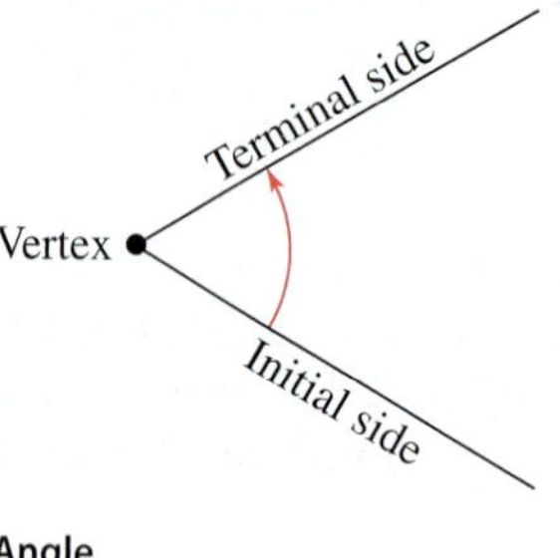

Angle
FIGURE 6.1

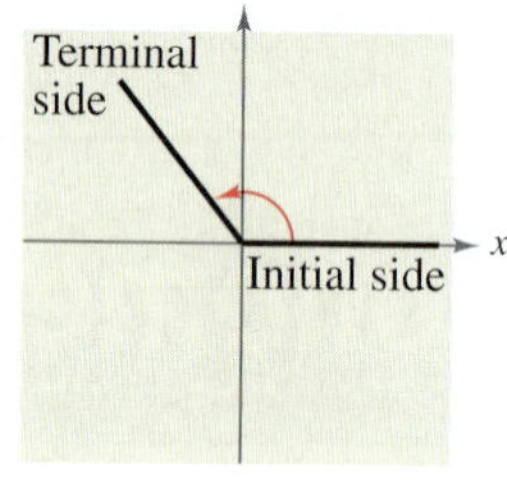

Angle in Standard Position
FIGURE 6.2

An **angle** is determined by rotating a ray (half-line) about its endpoint. The starting position of the ray is the **initial side** of the angle, and the position after rotation is the **terminal side,** as shown in Figure 6.1. The endpoint of the ray is the **vertex** of the angle. This perception of an angle fits a coordinate system in which the origin is the vertex and the initial side coincides with the positive x-axis. Such an angle is in **standard position,** as shown in Figure 6.2. **Positive angles** are generated by counterclockwise rotation, and **negative angles** by clockwise rotation, as shown in Figure 6.3. Angles are labeled with Greek letters α (alpha), β (beta), and θ (theta), as well as uppercase letters A, B, and C. In Figure 6.4, note that angles α and β have the same initial and terminal sides. Such angles are **coterminal.**

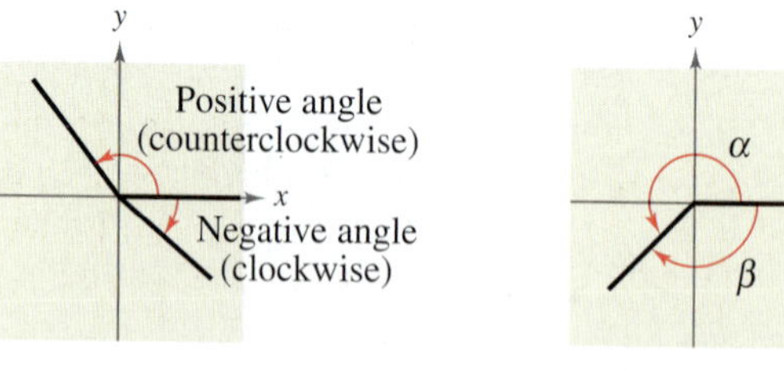

FIGURE 6.3

FIGURE 6.4

> **STUDY TIP**
>
> The phrase "the terminal side of θ lies in a quadrant" is often abbreviated by simply saying that "θ lies in a quadrant." The terminal sides of the "quadrant angles" 0°, 90°, 180°, and 270° do not lie within quadrants.

Degree Measure 

The measure of an angle is determined by the amount of rotation from the initial side to the terminal side. The most common unit of angle measure is the **degree,** denoted by the symbol °. A measure of one degree (1°) is equivalent to a rotation of $\frac{1}{360}$ of a complete revolution about the vertex. To measure angles, it is convenient to mark degrees on the circumference of a circle, as shown in Figure 6.5. So, a full revolution (counterclockwise) corresponds to 360°, a half revolution to 180°, a quarter revolution to 90°, and so on.

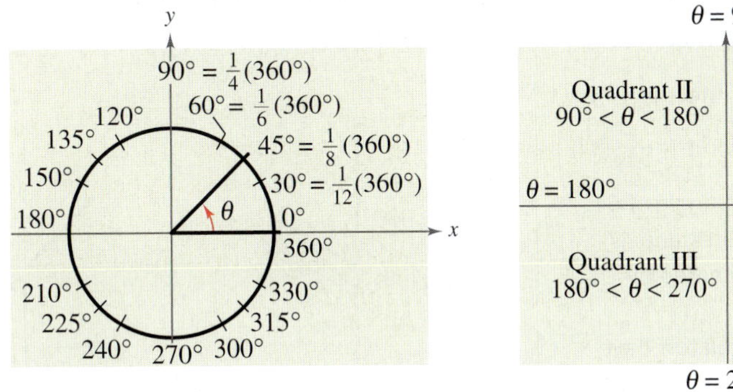

FIGURE 6.5

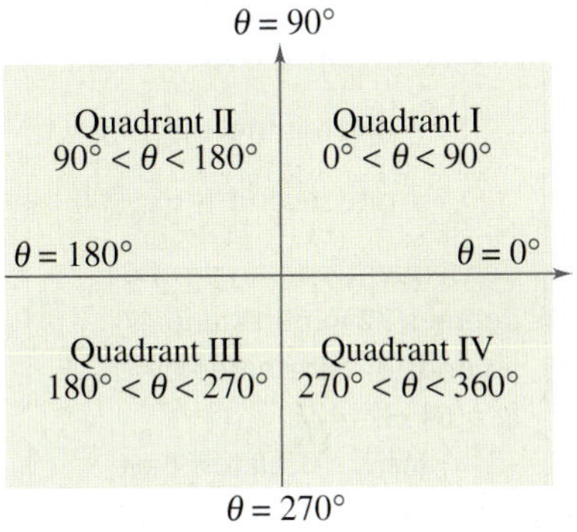

FIGURE 6.6

Recall that the four quadrants in a coordinate system are numbered I, II, III, and IV. Figure 6.6 shows which angles between 0° and 360° lie in each of the four quadrants. Figure 6.7 shows several common angles with their degree measures. Note that angles between 0° and 90° are **acute** and angles between 90° and 180° are **obtuse.**

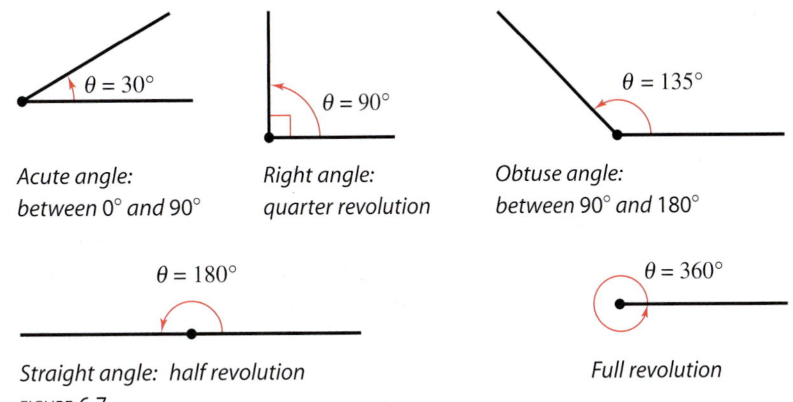

FIGURE 6.7

Two angles are coterminal if they have the same initial and terminal sides. For instance, the angles 0° and 360° are coterminal, as are the angles 30° and 390°. You can find an angle that is coterminal to a given angle θ by adding or subtracting 360° (one revolution), as demonstrated in Example 1. A given angle θ has infinitely many coterminal angles. For instance, $\theta = 30°$ is coterminal with $30° + n(360°)$ where n is an integer.

The icon identifies examples and concepts related to features of the Learning Tools CD-ROM and the *Interactive* and *Internet* versions of this text. For more details see the chart on pages *xix–xxiii*.

> **Technology**
>
> With calculators, it is convenient to use *decimal* degrees to denote fractional parts of degrees. Historically, however, fractional parts of degrees were expressed in *minutes* and *seconds*, using the prime (′) and double prime (″) notations, respectively. That is,
>
> $1' = \text{one minute} = \frac{1}{60}(1^\circ)$
>
> $1'' = \text{one second} = \frac{1}{3600}(1^\circ)$
>
> Consequently, an angle of 64 degrees, 32 minutes, and 47 seconds is represented by $\theta = 64^\circ\, 32'\, 47''$.
>
> Many calculators have special keys for converting an angle in degrees, minutes, and seconds (D° M′ S″) into decimal degree form, and vice versa.

Example 1 ▶ Sketching and Finding Coterminal Angles 

a. For the positive angle 390°, subtract 360° to obtain a coterminal angle.

$$390^\circ - 360^\circ = 30^\circ$$ See Figure 6.8.

b. For the positive angle 135°, subtract 360° to obtain a coterminal angle.

$$135^\circ - 360^\circ = -225^\circ$$ See Figure 6.9.

c. For the negative angle −120°, add 360° to obtain a coterminal angle.

$$-120^\circ + 360^\circ = 240^\circ$$ See Figure 6.10.

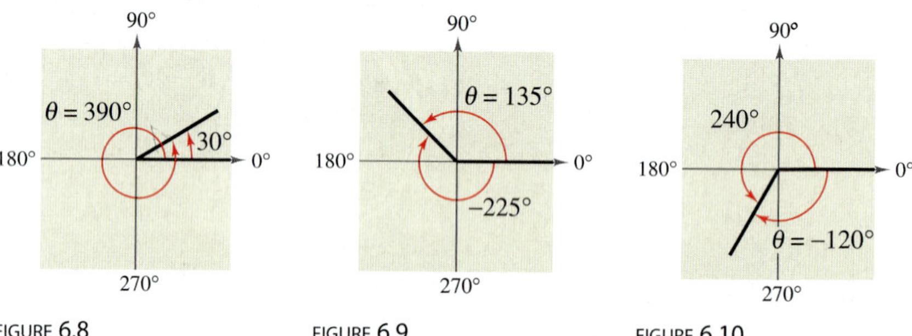

FIGURE 6.8 FIGURE 6.9 FIGURE 6.10

Two positive angles α and β are **complementary** (complements of each other) if their sum is 90°. Two positive angles are **supplementary** (supplements of each other) if their sum is 180°. See Figure 6.11.

The *Interactive* CD-ROM and *Internet* versions of this text offer a Try It for each example in the text.

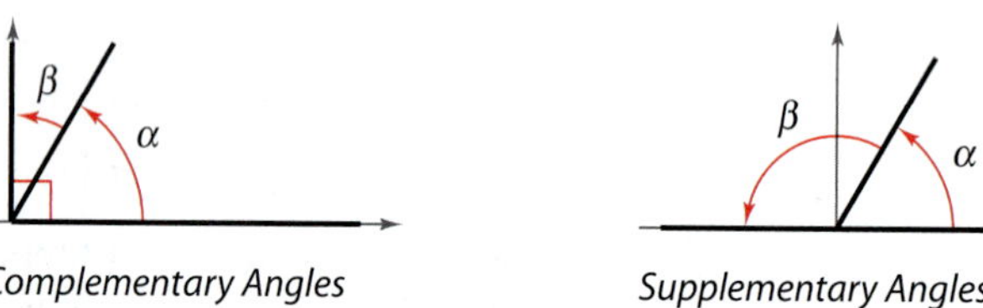

Complementary Angles *Supplementary Angles*

FIGURE 6.11

Example 2 ▶ Complementary and Supplementary Angles 

If possible, find the complement and the supplement of (a) 72° and (b) 148°.

Solution

a. The complement of $\theta = 72^\circ$ is

$$90^\circ - \theta = 90^\circ - 72^\circ = 18^\circ.$$

The supplement of $\theta = 72^\circ$ is

$$180^\circ - \theta = 180^\circ - 72^\circ = 108^\circ.$$

b. Because $\theta = 148^\circ$ is greater than 90°, it has no complement. (Remember to use only *positive* angles for complements.) The supplement is

$$180^\circ - \theta = 180^\circ - 148^\circ = 32^\circ.$$

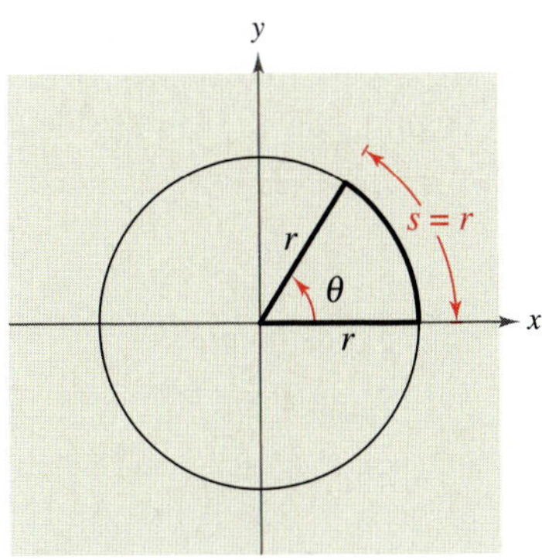

FIGURE 6.12 *Arc length = radius when θ = 1 radian*

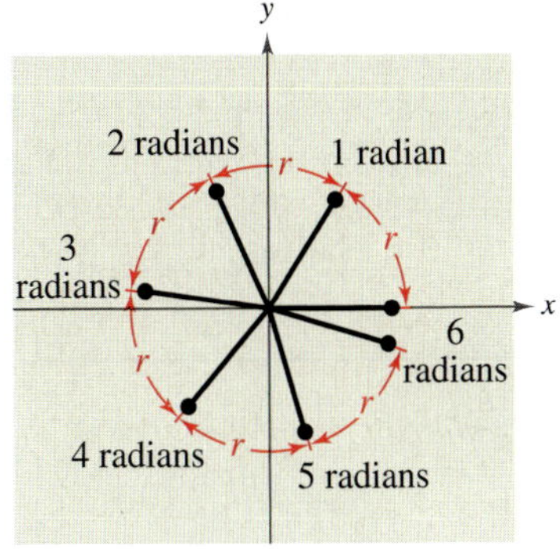

FIGURE 6.13

Radian Measure 

A second way to measure angles is in radians. This type of measure is especially useful in calculus. To define a radian, you can use a **central angle** of a circle, one whose vertex is the center of the circle, as shown in Figure 6.12.

Definition of Radian

One **radian** is the measure of a central angle θ that intercepts an arc s equal in length to the radius r of the circle. See Figure 6.12.

Because the circumference of a circle is $2\pi r$ units, it follows that a central angle of one full revolution (counterclockwise) corresponds to an arc length of $s = 2\pi r$. Moreover, because $2\pi \approx 6.28$, there are just over six radius lengths in a full circle, as shown in Figure 6.13. In general, the radian measure of a central angle θ is obtained by dividing the arc length s by r. That is, $s/r = \theta$, where θ *is measured in radians*. Because the units of measure for s and r are the same, this ratio has no units—it is simply a real number.

STUDY TIP

One revolution around a circle of radius r corresponds to an angle of 2π radians because

$$\theta = \frac{s}{r} = \frac{2\pi r}{r} = 2\pi \text{ radians.}$$

Example 3 ▶ Finding Angles 

Find each angle.

a. The complement of $\theta = \pi/12$

b. The supplement of $\theta = 5\pi/6$

c. A coterminal angle to $\theta = 17\pi/6$

Solution

a. In radian measure, the complement of an angle is found by subtracting the angle from $\pi/2$, which is equivalent to 90°. So, the complement of $\theta = \pi/12$ is

$$\pi/2 - \pi/12 = 6\pi/12 - \pi/12 = 5\pi/12.$$ See Figure 6.14.

b. In radian measure, the supplement of an angle is found by subtracting the angle from π, which is equivalent to 180°. So, the supplement of $\theta = 5\pi/6$ is

$$\pi - 5\pi/6 = 6\pi/6 - 5\pi/6 = \pi/6.$$ See Figure 6.15.

c. In radian measure, a coterminal angle is found by adding or subtracting 2π. For $\theta = 17\pi/6$, subtract 2π to obtain a coterminal angle.

$$17\pi/6 - 2\pi = 17\pi/6 - 12\pi/6 = 5\pi/6$$ See Figure 6.16.

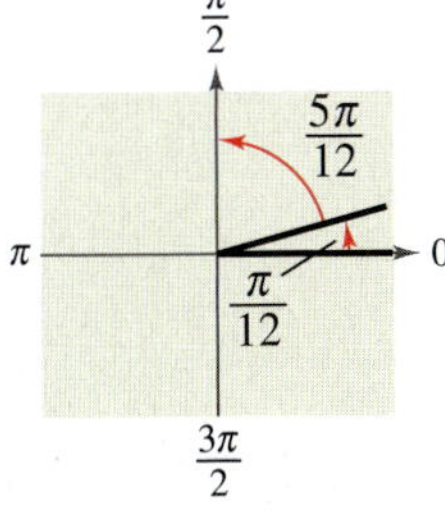

FIGURE 6.14

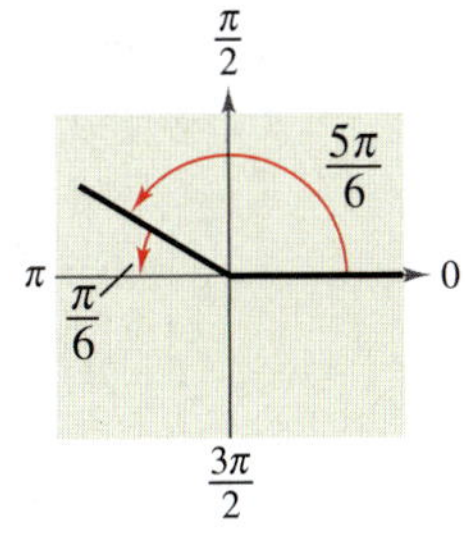

FIGURE 6.15

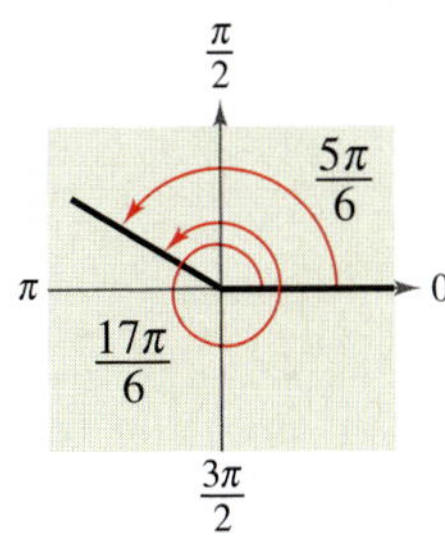

FIGURE 6.16

Conversion of Angle Measure

Because 2π radians corresponds to one complete revolution, degrees and radians are related by the equations

$$360^\circ = 2\pi \text{ rad} \qquad \text{and} \qquad 180^\circ = \pi \text{ rad}.$$

From the latter equation, you obtain

$$1^\circ = \frac{\pi}{180} \text{ rad} \qquad \text{and} \qquad 1 \text{ rad} = \left(\frac{180}{\pi}\right)^\circ$$

which lead to the following conversion rules.

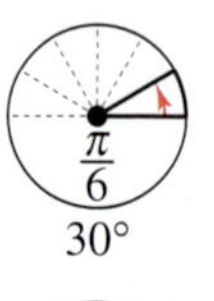

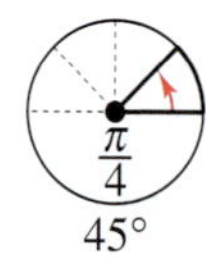

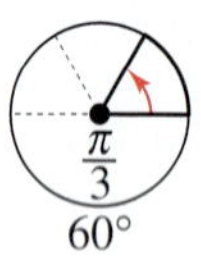

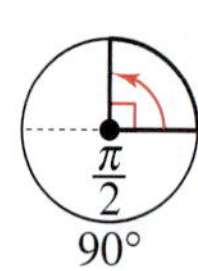

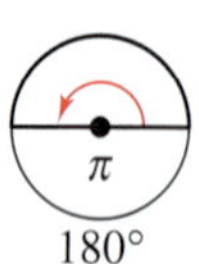

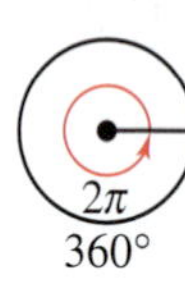

FIGURE 6.17

Conversions Between Degrees and Radians

1. To convert degrees to radians, multiply degrees by $\dfrac{\pi \text{ rad}}{180^\circ}$.

2. To convert radians to degrees, multiply radians by $\dfrac{180^\circ}{\pi \text{ rad}}$.

To apply these two conversion rules, use the basic relationship π rad $= 180^\circ$. (See Figure 6.17.)

When no units of angle measure are specified, *radian measure is implied*. For instance, if you write $\theta = \pi$ or $\theta = 2$, you should mean $\theta = \pi$ radians or $\theta = 2$ radians.

The *Interactive* CD-ROM and *Internet* versions of this text offer a Quiz for every section of the text.

Example 4 ▶ Converting from Degrees to Radians 

a. $135^\circ = (135 \text{ deg})\left(\dfrac{\pi \text{ rad}}{180 \text{ deg}}\right) = \dfrac{3\pi}{4}$ radians Multiply by $\pi/180$.

b. $540^\circ = (540 \text{ deg})\left(\dfrac{\pi \text{ rad}}{180 \text{ deg}}\right) = 3\pi$ radians Multiply by $\pi/180$.

c. $-270^\circ = (-270 \text{ deg})\left(\dfrac{\pi \text{ rad}}{180 \text{ deg}}\right) = -\dfrac{3\pi}{2}$ radians Multiply by $\pi/180$.

Example 5 ▶ Converting from Radians to Degrees 

a. $-\dfrac{\pi}{2} \text{ rad} = \left(-\dfrac{\pi}{2} \text{ rad}\right)\left(\dfrac{180 \text{ deg}}{\pi \text{ rad}}\right) = -90^\circ$ Multiply by $180/\pi$.

b. $\dfrac{9\pi}{2} \text{ rad} = \left(\dfrac{9\pi}{2} \text{ rad}\right)\left(\dfrac{180 \text{ deg}}{\pi \text{ rad}}\right) = 810^\circ$ Multiply by $180/\pi$.

c. $2 \text{ rad} = (2 \text{ rad})\left(\dfrac{180 \text{ deg}}{\pi \text{ rad}}\right) = \dfrac{360^\circ}{\pi} \approx 114.59^\circ$ Multiply by $180/\pi$.

If you have a calculator with a "radian-to-degree" conversion key, try using it to verify the result shown in part (c) of Example 5.

Applications

The *radian measure* formula, $\theta = s/r$, can be used to measure arc length along a circle. Specifically, for a circle of radius r, a central angle θ intercepts an arc of length s given by

$$s = r\theta \qquad \text{Length of circular arc}$$

where θ is measured in radians.

Example 6 ▶ Finding Arc Length 

A circle has a radius of 4 inches. Find the length of the arc intercepted by a central angle of 240°, as shown in Figure 6.18.

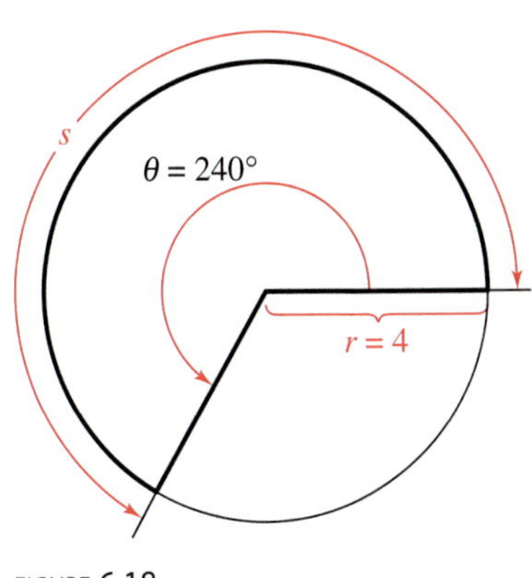

FIGURE 6.18

Solution

To use the formula $s = r\theta$, first convert 240° to radian measure.

$$240° = (240 \text{ deg})\left(\frac{\pi \text{ rad}}{180 \text{ deg}}\right) \qquad \text{Convert from degrees to radians.}$$

$$= \frac{4\pi}{3} \text{ radians} \qquad \text{Simplify.}$$

Then, using a radius of $r = 4$ inches, you can find the arc length to be

$$s = r\theta \qquad \text{Length of circular arc}$$

$$= 4\left(\frac{4\pi}{3}\right) \qquad \text{Substitute for } r \text{ and } \theta.$$

$$= \frac{16\pi}{3} \approx 16.76 \text{ inches} \qquad \text{Simplify.}$$

Note that the units for $r\theta$ are determined by the units for r, because θ is given in radian measure and so has no units.

The formula for the length of a circular arc can be used to analyze the motion of a particle moving at a *constant speed* along a circular path.

STUDY TIP

Linear speed measures how fast the particle moves, and angular speed measures how fast the angle changes.

Linear and Angular Speed

Consider a particle moving at a constant speed along a circular arc of radius r. If s is the length of the arc traveled in time t, then the **linear speed** of the particle is

$$\text{Linear speed} = \frac{\text{arc length}}{\text{time}} = \frac{s}{t}.$$

Moreover, if θ is the angle (in radian measure) corresponding to the arc length s, then the **angular speed** of the particle is

$$\text{Angular speed} = \frac{\text{central angle}}{\text{time}} = \frac{\theta}{t}.$$

Example 7 ▶ Finding Linear Speed 

FIGURE 6.19

The second hand of a clock is 10.2 centimeters long, as shown in Figure 6.19. Find the linear speed of the tip of this second hand as it passes around the clock face.

Solution

In one revolution, the arc length traveled is

$$s = 2\pi r$$

$$= 2\pi(10.2) \qquad \text{Substitute for } r.$$

$$= 20.4\pi \text{ centimeters.}$$

The time required for the second hand to travel this distance is

$$t = 1 \text{ minute} = 60 \text{ seconds.}$$

So, the linear speed of the tip of the second hand is

$$\text{Linear speed} = \frac{s}{t}$$

$$= \frac{20.4\pi \text{ centimeters}}{60 \text{ seconds}}$$

$$\approx 1.068 \text{ centimeters per second.}$$

Example 8 ▶ Finding Angular and Linear Speed  

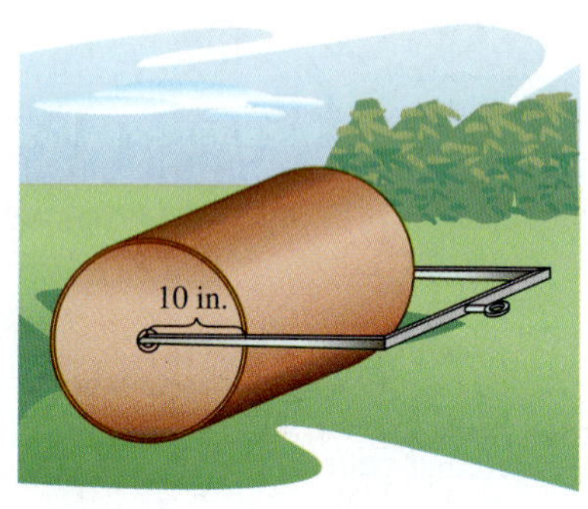

FIGURE 6.20

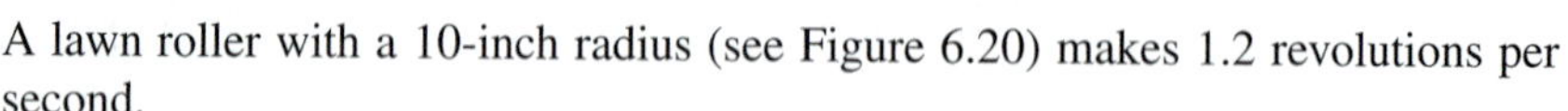

A lawn roller with a 10-inch radius (see Figure 6.20) makes 1.2 revolutions per second.

a. Find the angular speed of the roller in radians per second.

b. Find the speed of the tractor that is pulling the roller.

Solution

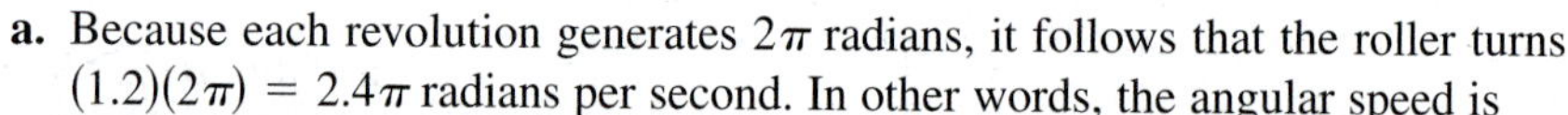

a. Because each revolution generates 2π radians, it follows that the roller turns $(1.2)(2\pi) = 2.4\pi$ radians per second. In other words, the angular speed is

$$\text{Angular speed} = \frac{\theta}{t}$$

$$= \frac{2.4\pi \text{ radians}}{1 \text{ second}} = 2.4\pi \text{ radians per second.}$$

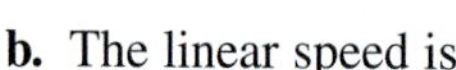

b. The linear speed is

$$\text{Linear speed} = \frac{s}{t}$$

$$= \frac{r\theta}{t}$$

$$= \frac{10(2.4\pi) \text{ inches}}{1 \text{ second}} \approx 75.4 \text{ inches per second.}$$

6.1 Exercises

The *Interactive* CD-ROM and *Internet* versions of this text contain step-by-step solutions to all odd-numbered exercises. They also provide Tutorial Exercises for additional help.

In Exercises 1–4, estimate the number of degrees in the angle.

1. 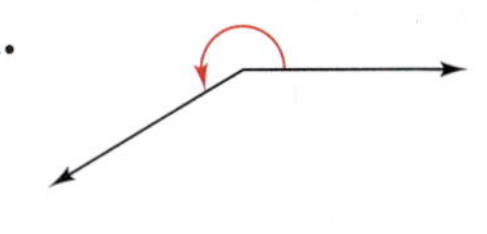

2. 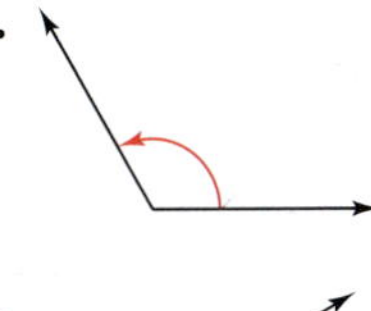

3. 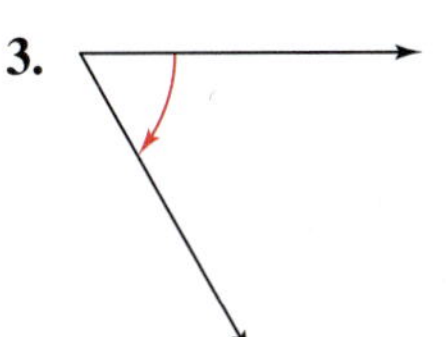

4. 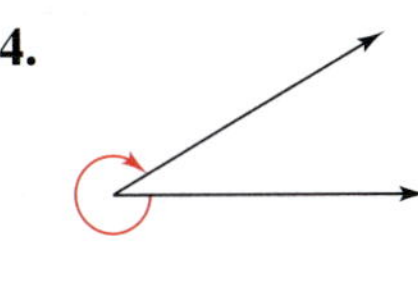

In Exercises 5–8, determine the quadrant in which each angle lies.

5. (a) $130°$ (b) $285°$

6. (a) $8.3°$ (b) $257° 30'$

7. (a) $-132° 50'$ (b) $-336°$

8. (a) $-260°$ (b) $-3.4°$

In Exercises 9–12, sketch each angle in standard position.

9. (a) $30°$ (b) $150°$

10. (a) $-270°$ (b) $-120°$

11. (a) $405°$ (b) $480°$

12. (a) $-750°$ (b) $-600°$

In Exercises 13–16, determine two coterminal angles (one positive and one negative) for each angle. Give your answers in degrees.

13. (a) 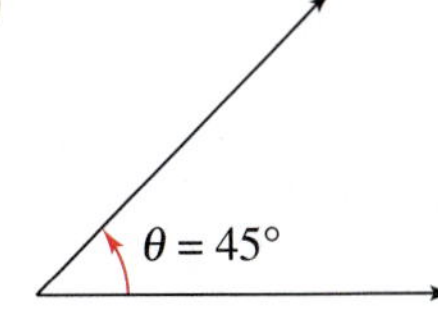

(b) 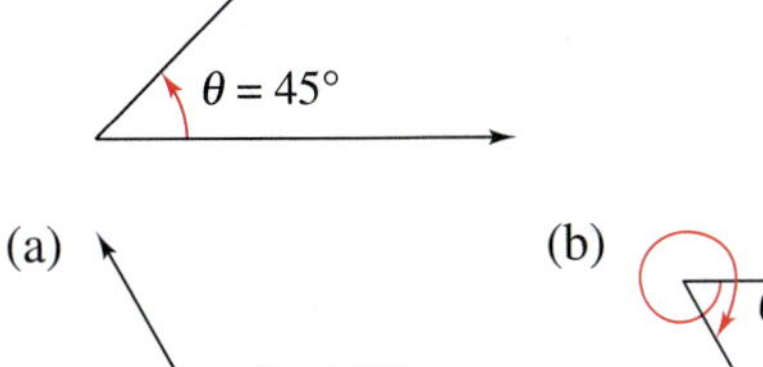

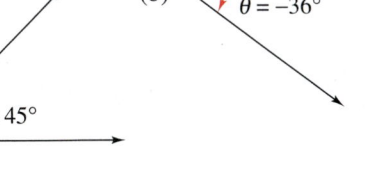

14. (a) (b) 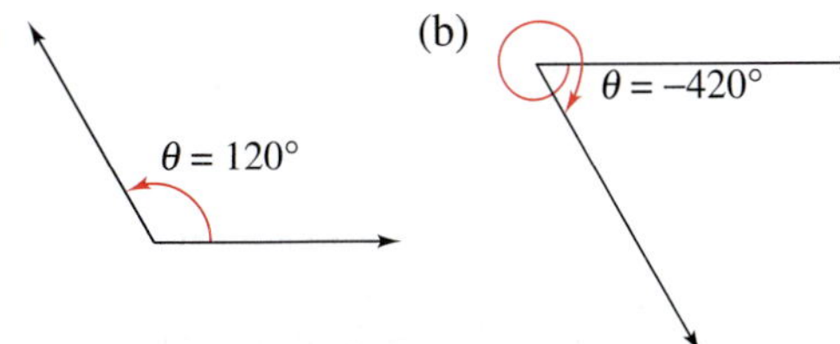

15. (a) $300°$ (b) $740°$

16. (a) $-520°$ (b) $230°$

In Exercises 17–20, convert each angle measure to decimal degree form.

17. (a) $54° 45'$ (b) $-128° 30'$

18. (a) $245° 10'$ (b) $2° 12'$

19. (a) $85° 18' 30''$ (b) $330° 25''$

20. (a) $-135° 36''$ (b) $-408° 16' 20''$

In Exercises 21–24, convert each angle measure to D° M′ S″ form.

21. (a) $240.6°$ (b) $-145.8°$

22. (a) $-345.12°$ (b) $0.45°$

23. (a) $2.5°$ (b) $-3.58°$

24. (a) $-0.355°$ (b) $0.7865°$

In Exercises 25–28, estimate the angle to the nearest one-half radian.

25. 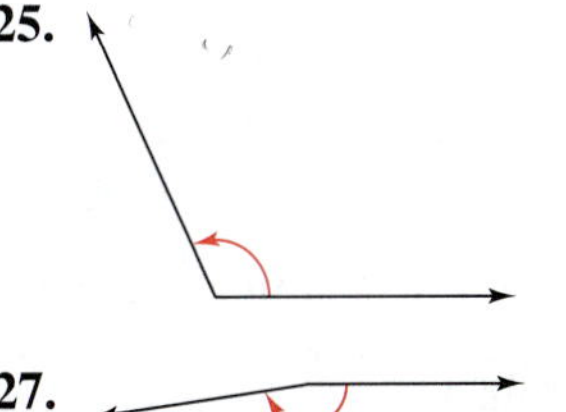

26. 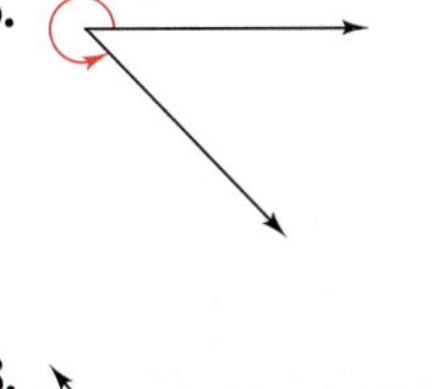

27.

28.

In Exercises 29–34, determine the quadrant in which each angle lies. (The angle measure is given in radians.)

29. (a) $\frac{\pi}{5}$ (b) $\frac{7\pi}{5}$

30. (a) $\frac{11\pi}{8}$ (b) $\frac{9\pi}{8}$

31. (a) $-\frac{\pi}{12}$ (b) $-\frac{11\pi}{9}$

32. (a) -1 (b) -2

33. (a) 3.5 (b) 2.25

34. (a) 6.02 (b) -4.25

In Exercises 35–38, sketch each angle in standard position.

35. (a) $\frac{5\pi}{4}$ (b) $-\frac{2\pi}{3}$ **36.** (a) $-\frac{7\pi}{4}$ (b) $\frac{5\pi}{2}$

37. (a) $\frac{11\pi}{6}$ (b) -3 **38.** (a) 4 (b) 7π

In Exercises 39–42, determine two coterminal angles (one positive and one negative) for each angle. Give your answers in radians.

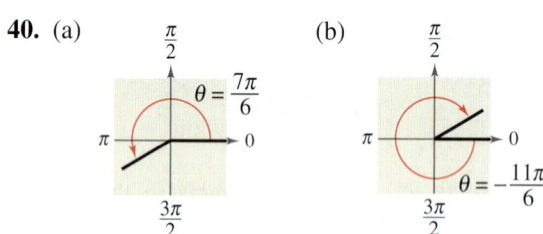

39. (a) $\theta = \frac{\pi}{6}$ (b) $\theta = \frac{5\pi}{6}$

40. (a) $\theta = \frac{7\pi}{6}$ (b) $\theta = -\frac{11\pi}{6}$

41. (a) $\theta = -\frac{9\pi}{4}$ (b) $\theta = -\frac{2\pi}{15}$

42. (a) $\theta = \frac{8\pi}{9}$ (b) $\theta = \frac{8\pi}{45}$

In Exercises 43–50, find (if possible) the complement and supplement of each angle.

43. (a) 18° (b) 115° **44.** (a) 79° (b) 150°

45. (a) 3° (b) 64° **46.** (a) 130° (b) 170°

47. (a) $\frac{\pi}{12}$ (b) $\frac{11\pi}{12}$ **48.** (a) $\frac{\pi}{3}$ (b) $\frac{3\pi}{4}$

49. (a) 3 (b) 1.5 **50.** (a) 1 (b) 2

In Exercises 51–54, express each angle in radian measure as a multiple of π. (Do not use a calculator.)

51. (a) 30° (b) 150° **52.** (a) 315° (b) 120°

53. (a) $-20°$ (b) $-240°$ **54.** (a) $-270°$ (b) 144°

In Exercises 55–58, express each angle in degree measure. (Do not use a calculator.)

55. (a) $\frac{3\pi}{2}$ (b) $\frac{7\pi}{6}$ **56.** (a) $-\frac{7\pi}{12}$ (b) $\frac{\pi}{9}$

57. (a) $\frac{7\pi}{3}$ (b) $-\frac{11\pi}{30}$ **58.** (a) $\frac{11\pi}{6}$ (b) $\frac{34\pi}{15}$

In Exercises 59–66, convert the measure from degrees to radians. Round to three decimal places.

59. 115° **60.** 87.4°

61. $-216.35°$ **62.** $-48.27°$

63. 532° **64.** 345°

65. $-0.83°$ **66.** 0.54°

In Exercises 67–74, convert the measure from radians to degrees. Round to three decimal places.

67. $\frac{\pi}{7}$ **68.** $\frac{5\pi}{11}$

69. $\frac{15\pi}{8}$ **70.** $\frac{13\pi}{2}$

71. -4.2π **72.** 4.8π

73. -2 **74.** -0.57

In Exercises 75–78, find the angle in radians.

75.

76.

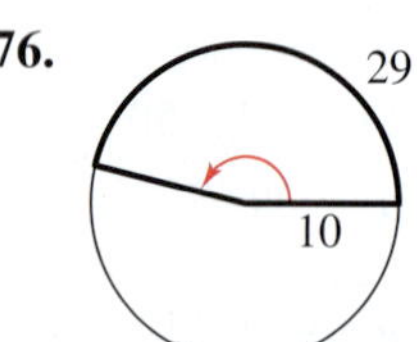

77.

78.

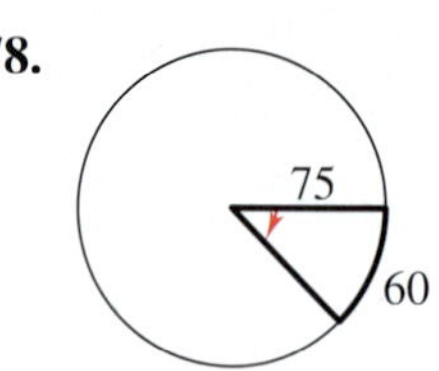

In Exercises 79–82, find the radian measure of the central angle of a circle of radius r that intercepts an arc of length s.

	Radius r	*Arc Length s*
79.	27 inches	6 inches
80.	14 feet	8 feet
81.	14.5 centimeters	25 centimeters
82.	80 kilometers	160 kilometers

In Exercises 83–86, find the length of the arc on a circle of radius r intercepted by a central angle θ.

	Radius r	*Central Angle θ*
83.	15 inches	180°
84.	9 feet	60°
85.	3 meters	1 radian
86.	20 centimeters	$\pi/4$ radian

Distance Between Cities **In Exercises 87–90, find the distance between the cities. Assume that Earth is a sphere of radius 4000 miles and that the cities are on the same longitude (one city is due north of the other).**

	City	*Latitude*
87.	Dallas, Texas	$32° 47' 9'' N$
	Omaha, Nebraska	$41° 15' 42'' N$
88.	San Francisco, California	$37° 46' 39'' N$
	Seattle, Washington	$47° 36' 32'' N$
89.	Miami, Florida	$25° 46' 37'' N$
	Erie, Pennsylvania	$42° 7' 15'' N$
90.	Johannesburg, South Africa	$26° 10' S$
	Jerusalem, Israel	$31° 47' N$

91. ***Difference in Latitudes*** Assuming that Earth is a sphere of radius 6378 kilometers, what is the difference in the latitudes of Syracuse, New York, and Annapolis, Maryland, where Syracuse is 450 kilometers due north of Annapolis?

92. ***Difference in Latitudes*** Assuming that Earth is a sphere of radius 6378 kilometers, what is the difference in the latitudes of Lynchburg, Virginia, and Myrtle Beach, South Carolina, where Lynchburg is 400 kilometers due north of Myrtle Beach?

93. ***Instrumentation*** The pointer on a voltmeter is 6 centimeters in length (see figure). Find the angle through which the pointer rotates when it moves 2.5 centimeters on the scale.

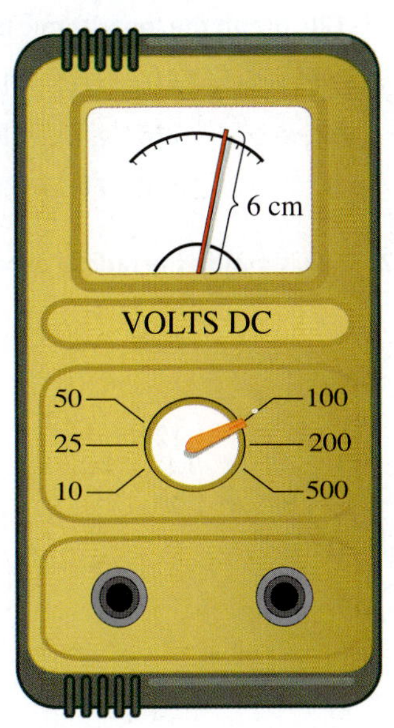

94. ***Electric Hoist*** An electric hoist is being used to lift a beam (see figure). The diameter of the drum on the hoist is 10 inches, and the beam must be raised 2 feet. Find the number of degrees through which the drum must rotate.

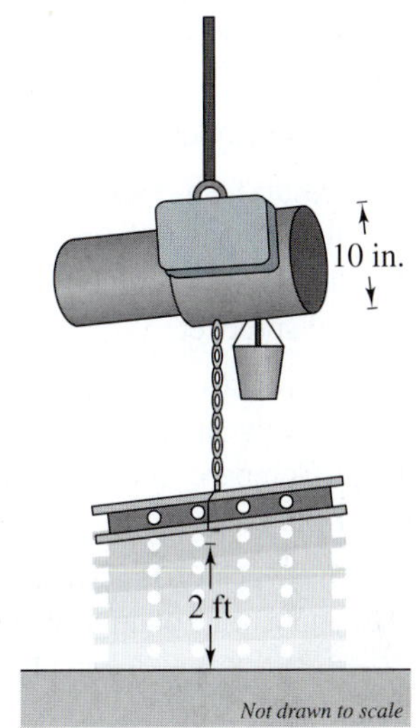

95. ***Angular Speed*** A car is moving at a rate of 65 miles per hour, and the diameter of its wheels is 2.5 feet.

(a) Find the number of revolutions per minute the wheels are rotating.

(b) Find the angular speed of the wheels in radians per minute.

96. ***Angular Speed*** A two-inch-diameter pulley on an electric motor that runs at 1700 revolutions per minute is connected by a belt to a four-inch-diameter pulley on a saw arbor.

(a) Find the angular speed (in radians per minute) of each pulley.

(b) Find the revolutions per minute of the saw.

▶ Model It

97. ***Speed of a Bicycle*** The radii of the pedal sprocket, the wheel sprocket, and the wheel of the bicycle in the figure are 4 inches, 2 inches, and 14 inches, respectively. The cyclist is pedaling at a rate of 1 revolution per second.

(a) Find the speed of the bicycle in feet per second and miles per hour.

(b) Use your result from part (a) to write a function for the distance d (in miles) a cyclist travels in terms of the number n of revolutions of the pedal sprocket.

(c) Write a function for the distance d (in miles) a cyclist travels in terms of the time t (in seconds). Compare this function with the function from part (b).

(d) Classify the types of functions you found in parts (b) and (c). Explain your reasoning.

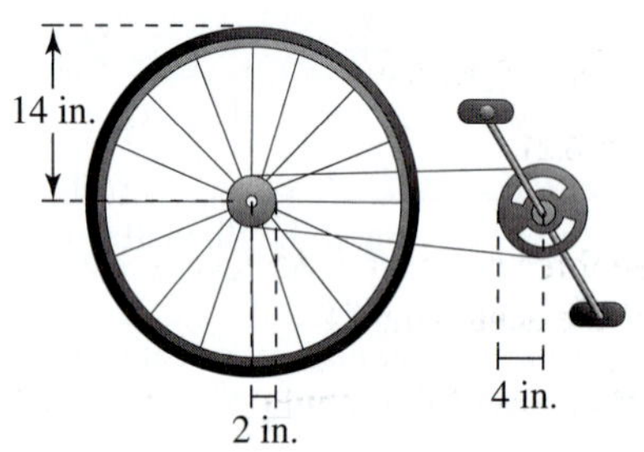

98. ***Floppy Disk*** The radius of the magnetic disk in a 3.5-inch diskette is 1.68 inches. Find the linear speed of a point on the circumference of the disk if it is rotating at a speed of 360 revolutions per minute.

Synthesis

True or False? **In Exercises 99–101, determine whether the statement is true or false. Justify your answer.**

99. A measurement of 4 radians corresponds to two complete revolutions from the initial side to the terminal side of an angle.

100. The difference between the measures of two coterminal angles is always a multiple of 360° if expressed in degrees and is always a multiple of 2π radians if expressed in radians.

101. An angle that measures $-1260°$ lies in Quadrant III.

102. ***Writing*** In your own words, explain the meanings of (a) an angle in standard position, (b) a negative angle, (c) coterminal angles, and (d) an obtuse angle.

103. A fan motor turns at a given angular speed. How does the speed of the tips of the blades change if a fan of greater diameter is installed on the motor? Explain.

104. ***Think About It*** Is a degree or a radian the larger unit of measure? Explain.

105. ***Writing*** If the radius of a circle is increasing and the magnitude of a central angle is held constant, how is the length of the intercepted arc changing? Explain your reasoning.

106. ***Proof*** Prove that the area of a circular sector of radius r with central angle θ is $A = \frac{1}{2}\theta r^2$, where θ is measured in radians.

Review

In Exercises 107–110, sketch the graphs of $y = x^5$ and the specified transformation.

107. $f(x) = (x - 2)^5$ **108.** $f(x) = x^5 - 4$

109. $f(x) = 2 - x^5$ **110.** $f(x) = -(x + 3)^5$

In Exercises 111–114, graph the exponential function.

111. $f(x) = 6^x$ **112.** $f(x) = 6^x - 2$

113. $f(x) = 6^{-x}$ **114.** $f(x) = 6^{x+1}$

In Exercises 115–120, graph the logarithmic function.

115. $f(x) = \log_4 x$ **116.** $f(x) = \log_4 x + 5$

117. $f(x) = -\log_4 x$ **118.** $f(x) = \log_4(x + 5)$

119. $f(x) = \log_4(-x)$ **120.** $f(x) = -\log_4(x + 5)$

In Exercises 121–128, simplify the radical expression.

121. $\dfrac{4}{4\sqrt{2}}$ **122.** $\dfrac{2}{\sqrt{3}}$

123. $\dfrac{2\sqrt{3}}{\sqrt{6}}$ **124.** $\dfrac{5\sqrt{5}}{2\sqrt{10}}$

125. $\sqrt{2^2 + 6^2}$

126. $\sqrt{18^2 + 12^2}$

127. $\sqrt{18^2 - 6^2}$

128. $\sqrt{17^2 - 9^2}$

6.2 Right Triangle Trigonometry

▶ What you should learn

- How to evaluate trigonometric functions of acute angles
- How to use the fundamental trigonometric identities
- How to use a calculator to evaluate trigonometric functions
- How to use trigonometric functions to model and solve real-life problems

▶ Why you should learn it

Trigonometric functions are often used to analyze real-life situations. For instance, in Exercise 61 on page 474, you can use trigonometric functions to find the height of a helium-filled balloon.

Joseph Sohm; ChromoSohm

The Six Trigonometric Functions 

Our first look at the trigonometric functions is from a *right triangle* perspective. Consider a right triangle, with one acute angle labeled θ, as shown in Figure 6.21. Relative to the angle θ, the three sides of the triangle are the **hypotenuse,** the **opposite side** (the side opposite the angle θ), and the **adjacent side** (the side adjacent to the angle θ).

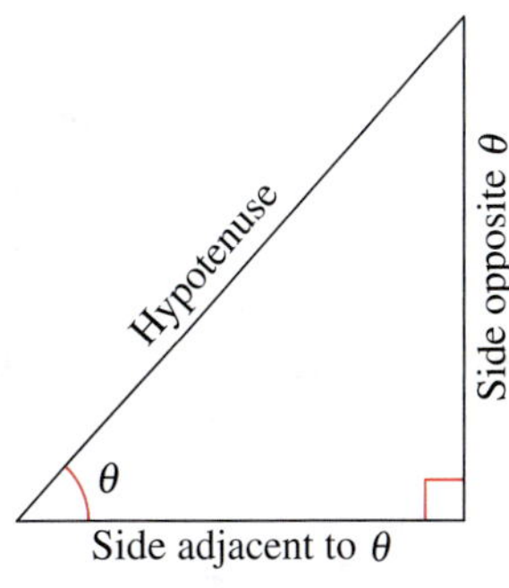

FIGURE 6.21

Using the lengths of these three sides, you can form six ratios that define the six trigonometric functions of the acute angle θ.

sine **cosecant** **cosine** **secant** **tangent** **cotangent**

These six functions are normally abbreviated as sin, csc, cos, sec, tan, and cot, respectively. In the following definition it is important to see that $0° < \theta < 90°$ and that for such angles the value of each trigonometric function is *positive*.

Right Triangle Definitions of Trigonometric Functions

Let θ be an *acute* angle of a right triangle. The six trigonometric functions of the angle θ are defined as follows. (Note that the functions in the second row are the *reciprocals* of the corresponding functions in the first row.)

$$\sin\theta = \frac{\text{opp}}{\text{hyp}} \qquad \cos\theta = \frac{\text{adj}}{\text{hyp}} \qquad \tan\theta = \frac{\text{opp}}{\text{adj}}$$

$$\csc\theta = \frac{\text{hyp}}{\text{opp}} \qquad \sec\theta = \frac{\text{hyp}}{\text{adj}} \qquad \cot\theta = \frac{\text{adj}}{\text{opp}}$$

The abbreviations opp, adj, and hyp represent the lengths of the three sides of a right triangle.

opp = the length of the side *opposite* θ

adj = the length of the side *adjacent to* θ

hyp = the length of the *hypotenuse*

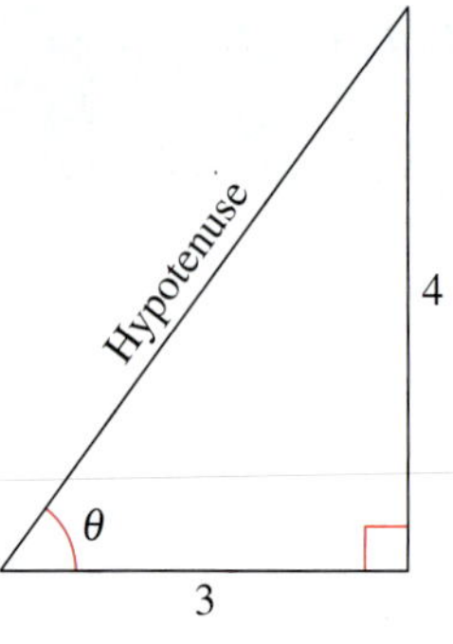

FIGURE 6.22

Example 1 ▶ Evaluating Trigonometric Functions

Use the triangle in Figure 6.22 to find the values of the six trigonometric functions of θ.

Solution

By the Pythagorean Theorem, $(\text{hyp})^2 = (\text{opp})^2 + (\text{adj})^2$, it follows that

$$\begin{aligned}\text{hyp} &= \sqrt{4^2 + 3^2}\\ &= \sqrt{25}\\ &= 5.\end{aligned}$$

So, the six trigonometric functions of θ are

$$\sin\theta = \frac{\text{opp}}{\text{hyp}} = \frac{4}{5} \qquad \csc\theta = \frac{\text{hyp}}{\text{opp}} = \frac{5}{4}$$

$$\cos\theta = \frac{\text{adj}}{\text{hyp}} = \frac{3}{5} \qquad \sec\theta = \frac{\text{hyp}}{\text{adj}} = \frac{5}{3}$$

$$\tan\theta = \frac{\text{opp}}{\text{adj}} = \frac{4}{3} \qquad \cot\theta = \frac{\text{adj}}{\text{opp}} = \frac{3}{4}.$$

Historical Note
Georg Joachim Rhaeticus (1514–1576) was the leading Teutonic mathematical astronomer of the 16th century. He was the first to define the trigonometric functions as ratios of the sides of a right triangle.

In Example 1, you were given the lengths of two sides of the right triangle, but not the angle θ. Often, you will be asked to find the trigonometric functions of a *given* acute angle θ. To do this, construct a right triangle having θ as one of its angles.

Example 2 ▶ Evaluating Trigonometric Functions of 45°

Find the values of $\sin 45°$, $\cos 45°$, and $\tan 45°$.

Solution

Construct a right triangle having 45° as one of its acute angles, as shown in Figure 6.23. Choose the length of the adjacent side to be 1. From geometry, you know that the other acute angle is also 45°. So, the triangle is isosceles and the length of the opposite side is also 1. Using the Pythagorean Theorem, you find the length of the hypotenuse to be $\sqrt{2}$.

$$\sin 45° = \frac{\text{opp}}{\text{hyp}} = \frac{1}{\sqrt{2}} = \frac{\sqrt{2}}{2}$$

$$\cos 45° = \frac{\text{adj}}{\text{hyp}} = \frac{1}{\sqrt{2}} = \frac{\sqrt{2}}{2}$$

$$\tan 45° = \frac{\text{opp}}{\text{adj}} = \frac{1}{1} = 1$$

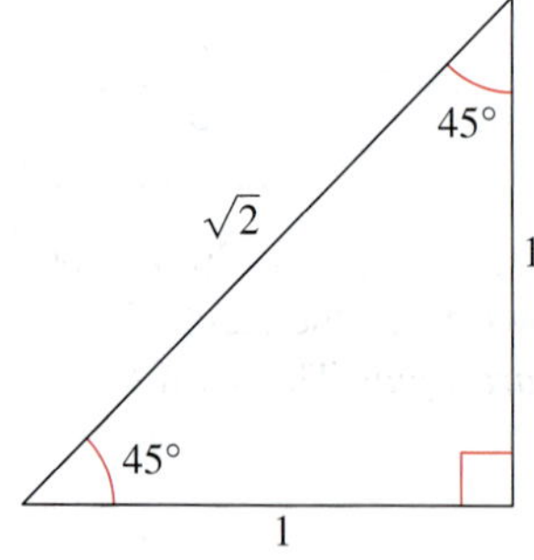

FIGURE 6.23

Example 3 ▶ Evaluating Trigonometric Functions of 30° and 60°

Use the equilateral triangle shown in Figure 6.24 to find the values of sin 60°, cos 60°, sin 30°, and cos 30°.

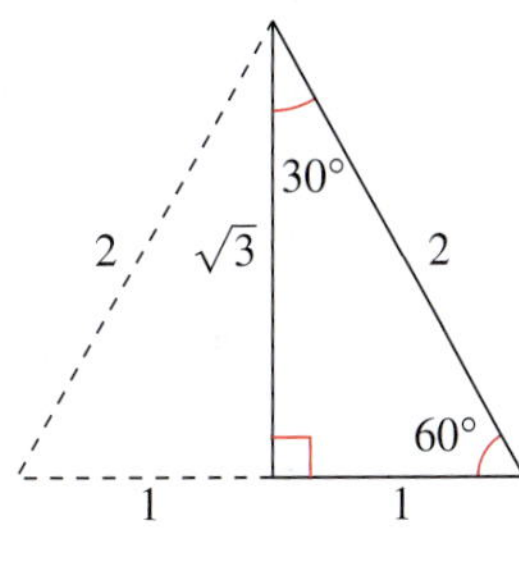

FIGURE 6.24

> **STUDY TIP**
>
> Because the angles 30°, 45°, and 60° ($\pi/6$, $\pi/4$, and $\pi/3$) occur frequently in trigonometry, you should learn to construct the triangles shown in Figures 6.23 and 6.24.

Solution

Use the Pythagorean Theorem and the equilateral triangle in Figure 6.24 to verify the lengths of the sides given in Figure 6.24. For $\theta = 60°$, you have adj $= 1$, opp $= \sqrt{3}$, and hyp $= 2$. So,

$$\sin 60° = \frac{\text{opp}}{\text{hyp}} = \frac{\sqrt{3}}{2} \quad \text{and} \quad \cos 60° = \frac{\text{adj}}{\text{hyp}} = \frac{1}{2}.$$

For $\theta = 30°$, adj $= \sqrt{3}$, opp $= 1$, and hyp $= 2$. So,

$$\sin 30° = \frac{\text{opp}}{\text{hyp}} = \frac{1}{2} \quad \text{and} \quad \cos 30° = \frac{\text{adj}}{\text{hyp}} = \frac{\sqrt{3}}{2}.$$

Sines, Cosines, and Tangents of Special Angles

$\sin 30° = \sin \frac{\pi}{6} = \frac{1}{2}$ $\cos 30° = \cos \frac{\pi}{6} = \frac{\sqrt{3}}{2}$ $\tan 30° = \tan \frac{\pi}{6} = \frac{\sqrt{3}}{3}$

$\sin 45° = \sin \frac{\pi}{4} = \frac{\sqrt{2}}{2}$ $\cos 45° = \cos \frac{\pi}{4} = \frac{\sqrt{2}}{2}$ $\tan 45° = \tan \frac{\pi}{4} = 1$

$\sin 60° = \sin \frac{\pi}{3} = \frac{\sqrt{3}}{2}$ $\cos 60° = \cos \frac{\pi}{3} = \frac{1}{2}$ $\tan 60° = \tan \frac{\pi}{3} = \sqrt{3}$

In the box, note that $\sin 30° = \frac{1}{2} = \cos 60°$. This occurs because 30° and 60° are complementary angles. In general, it can be shown from the right triangle definitions that *cofunctions of complementary angles are equal.* That is, if θ is an acute angle, the following relationships are true.

$\sin(90° - \theta) = \cos \theta$ $\cos(90° - \theta) = \sin \theta$

$\tan(90° - \theta) = \cot \theta$ $\cot(90° - \theta) = \tan \theta$

$\sec(90° - \theta) = \csc \theta$ $\csc(90° - \theta) = \sec \theta$

Trigonometric Identities 

In trigonometry, a great deal of time is spent studying relationships between trigonometric functions (identities).

Fundamental Trigonometric Identities

Reciprocal Identities

$$\sin\theta = \frac{1}{\csc\theta} \qquad \cos\theta = \frac{1}{\sec\theta} \qquad \tan\theta = \frac{1}{\cot\theta}$$

$$\csc\theta = \frac{1}{\sin\theta} \qquad \sec\theta = \frac{1}{\cos\theta} \qquad \cot\theta = \frac{1}{\tan\theta}$$

Quotient Identities

$$\tan\theta = \frac{\sin\theta}{\cos\theta} \qquad \cot\theta = \frac{\cos\theta}{\sin\theta}$$

Pythagorean Identities

$$\sin^2\theta + \cos^2\theta = 1 \qquad 1 + \tan^2\theta = \sec^2\theta$$

$$1 + \cot^2\theta = \csc^2\theta$$

Note that $\sin^2\theta$ represents $(\sin\theta)^2$, $\cos^2\theta$ represents $(\cos\theta)^2$, and so on.

Example 4 ▶ Applying Trigonometric Identities

Let θ be an acute angle such that $\sin\theta = 0.6$. Find the values of (a) $\cos\theta$ and (b) $\tan\theta$ using trigonometric identities.

Solution

a. To find the value of $\cos\theta$, use the Pythagorean identity

$$\sin^2\theta + \cos^2\theta = 1.$$

So, you have

$$(0.6)^2 + \cos^2\theta = 1 \qquad \text{Substitute 0.6 for } \sin\theta.$$

$$\cos^2\theta = 1 - (0.6)^2 = 0.64 \qquad \text{Subtract } (0.6)^2 \text{ from each side.}$$

$$\cos\theta = \sqrt{0.64} = 0.8. \qquad \text{Extract the positive square root.}$$

b. Now, knowing the sine and cosine of θ, you can find the tangent of θ to be

$$\tan\theta = \frac{\sin\theta}{\cos\theta} = \frac{0.6}{0.8} = 0.75.$$

Use the definitions of $\cos\theta$ and $\tan\theta$, and the triangle shown in Figure 6.25, to check these results.

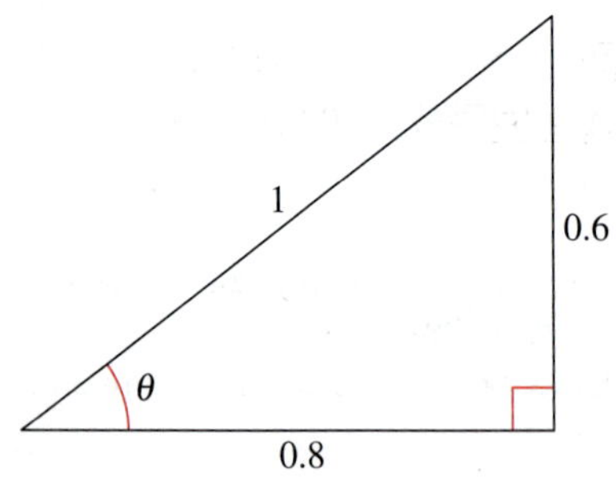

FIGURE 6.25

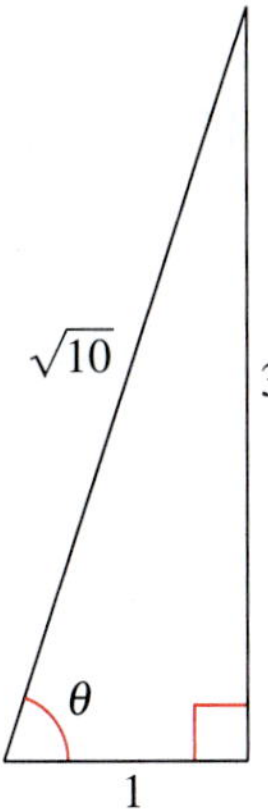

FIGURE 6.26

Example 5 ▶ Applying Trigonometric Identities 

Let θ be an acute angle such that $\tan\theta = 3$. Find the value of each trigonometric function using trigonometric identities.

a. $\cot\theta$ **b.** $\sec\theta$

Solution

a. $\cot\theta = \dfrac{1}{\tan\theta}$ Reciprocal identity

$\cot\theta = \dfrac{1}{3}$

b. $\sec^2\theta = 1 + \tan^2\theta$ Pythagorean identity

$\sec^2\theta = 1 + 3^2$

$\sec^2\theta = 10$

$\sec\theta = \sqrt{10}$

Use the definitions of $\cot\theta$ and $\sec\theta$, and the triangle shown in Figure 6.26, to check these results.

Technology

When evaluating trigonometric functions with a calculator, remember to enclose all fractional angle measures in parentheses. For instance, if you want to evaluate $\sin\theta$ for $\theta = \pi/6$, you should enter

SIN (π ÷ 6) ENTER

rather than

SIN π ÷ 6 ENTER.

The first set of keystrokes yields the correct value of 0.5. The second set yields the incorrect value of 0.

Evaluating Trigonometric Functions with a Calculator

When evaluating a trigonometric function with a calculator, you need to set the calculator to the desired *mode* of measurement (degrees or radians).

Most calculators do not have keys for the cosecant, secant, and cotangent functions. To evaluate these functions, you can use the x^{-1} key with their respective reciprocal functions sine, cosine, and tangent. For example, to evaluate $\csc(\pi/8)$, use the fact that

$$\csc\frac{\pi}{8} = \frac{1}{\sin(\pi/8)}$$

and enter the following keystroke sequence in radian mode.

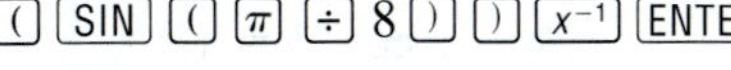
(SIN (π ÷ 8)) x^{-1} ENTER Display 2.6131259

Example 6 ▶ Using a Calculator 

	Function	*Mode*	*Calculator Keystrokes*	*Display*
a.	sin 76.4°	Degree	SIN 76.4 ENTER	0.9719610
b.	cot 1.5	Radian	(TAN 1.5) x^{-1} ENTER	0.0709148

You could also use the reciprocal identities for sine, cosine, and tangent to evaluate the cosecant, secant, and cotangent functions with a calculator. For instance, you could use the following keystroke sequence to evaluate the function in Example 6(b).

1 ÷ TAN (1.5) ENTER Display 0.0709148

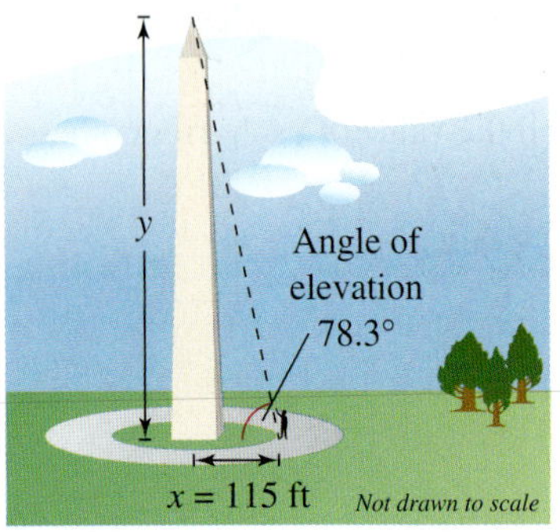

FIGURE 6.27

Applications Involving Right Triangles

Many applications of trigonometry involve a process called **solving right triangles.** In this type of application, you are usually given one side of a right triangle and one of the acute angles and asked to find one of the other sides, *or* you are given two sides and asked to find one of the acute angles.

Example 7 ▶ Using Trigonometry to Solve a Right Triangle

A surveyor is standing 115 feet from the base of the Washington Monument, as shown in Figure 6.27. The surveyor measures the angle of elevation to the top of the monument as 78.3°. How tall is the Washington Monument?

Solution

From Figure 6.27, you can see that

$$\tan 78.3° = \frac{\text{opp}}{\text{adj}} = \frac{y}{x}$$

where $x = 115$ and y is the height of the monument. So, the height of the Washington Monument is

$$y = x \tan 78.3° \approx 115(4.82882) \approx 555 \text{ feet.}$$

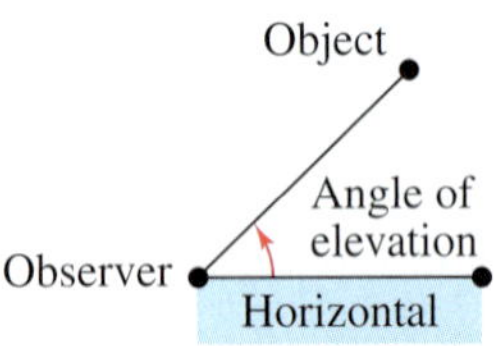

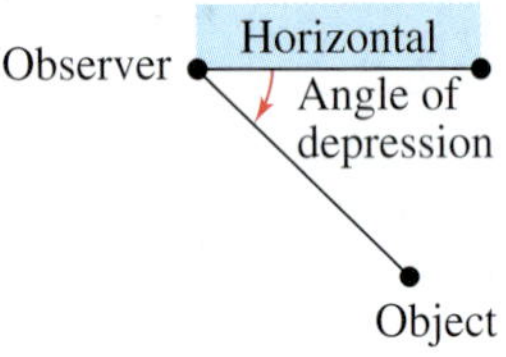

FIGURE 6.28

The term **angle of elevation** represents the angle from the horizontal upward to an object. For objects that lie below the horizontal, it is common to use the term **angle of depression,** as shown in Figure 6.28.

Example 8 ▶ Using Trigonometry to Solve a Right Triangle

You are 200 yards from a river. Rather than walking directly to the river, you walk 400 yards along a straight path to the river's edge. Find the acute angle θ between this path and the river's edge, as illustrated in Figure 6.29.

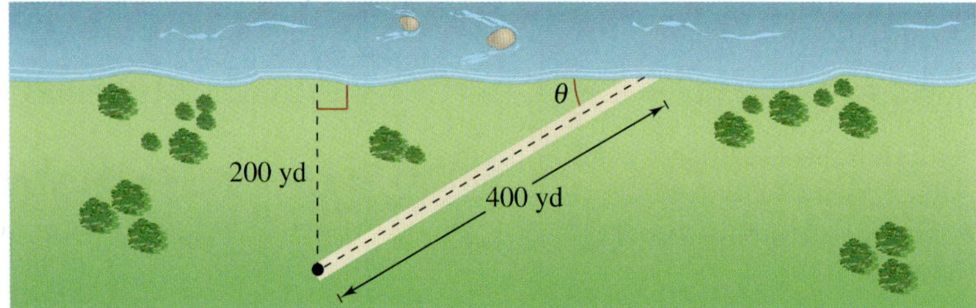

FIGURE 6.29

Solution

From Figure 6.29, you can see that the sine of the angle θ is

$$\sin \theta = \frac{\text{opp}}{\text{hyp}} = \frac{200}{400} = \frac{1}{2}.$$

So, $\theta = 30°$.

In Example 8, you were able to recognize that the acute angle that satisfies the equation $\sin \theta = \frac{1}{2}$ is $\theta = 30°$. Suppose, however, that you were given the equation $\sin \theta = 0.6$ and were asked to find the acute angle θ. Because

$$\sin 30° = \frac{1}{2}$$

$$= 0.5000$$

and

$$\sin 45° = \frac{1}{\sqrt{2}}$$

$$\approx 0.7071$$

you might guess that θ lies somewhere between 30° and 45°. A more precise value of θ can be found using the *inverse sine* key on a calculator. To do this, you can use the following keystroke sequence in *degree* mode.

[SIN⁻¹] .6 [ENTER] Display 36.8699

So, you can conclude that if $\sin \theta = 0.6$ and $0° < \theta < 90°$, then $\theta \approx 36.87°$.

Example 9 ▶ Solving a Right Triangle 

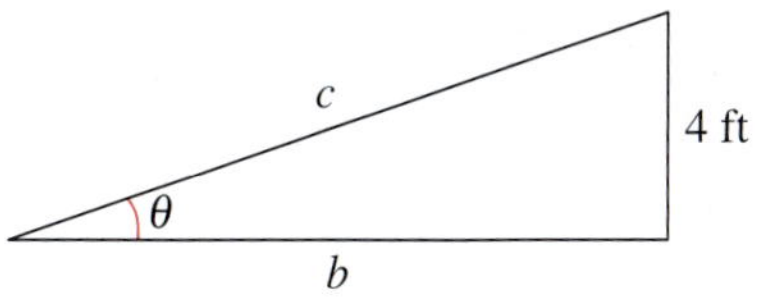

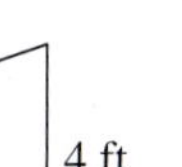

FIGURE 6.30

Specifications for a loading dock ramp require a rise of 1 foot for each 3 feet of horizontal length. In Figure 6.30, find the lengths of sides b and c and find the measure of θ.

Solution

From the given specifications, you can write

$$\frac{\text{rise}}{\text{run}} = \frac{1}{3} = \frac{4 \text{ ft}}{b \text{ ft}}$$

which implies that $b = 12$ feet. Using the Pythagorean Theorem, you can write

$c^2 = a^2 + b^2$ Pythagorean Theorem

$c^2 = 4^2 + 12^2$ Substitute for a and b.

$c^2 = 160$ Simplify.

$c = 4\sqrt{10}.$ Extract positive square root.

So, $c = 4\sqrt{10} \approx 12.65$ feet. To solve for θ, you can write

$$\tan \theta = \frac{4}{12}$$

$$= \frac{1}{3}.$$

Then, using the calculator keystrokes in *degree* mode

[TAN⁻¹] [(] 1 [÷] 3 [)] [ENTER]

you obtain $\theta \approx 18.43°$.

6.2 Exercises

In Exercises 1–4, find the exact values of the six trigonometric functions of the angle θ shown in the figure. (Use the Pythagorean Theorem to find the third side of the triangle.)

1.

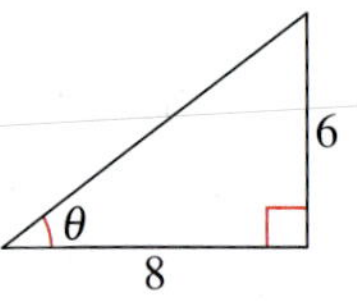

2.

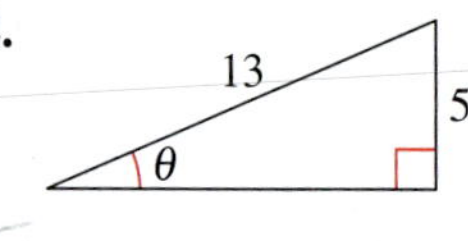

3.

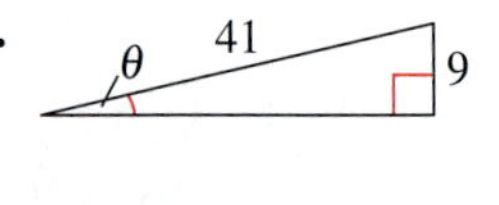

4.

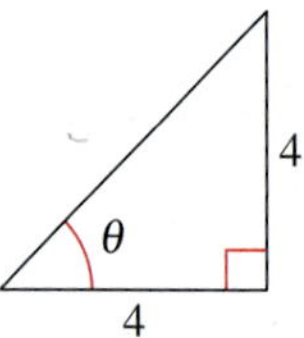

In Exercises 5–8, find the exact values of the six trigonometric functions of the angle θ for each of the two triangles. Explain why the function values are the same.

5.

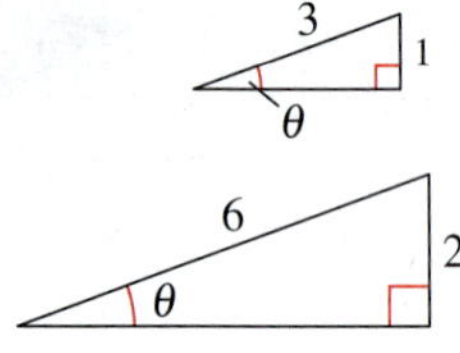

6.

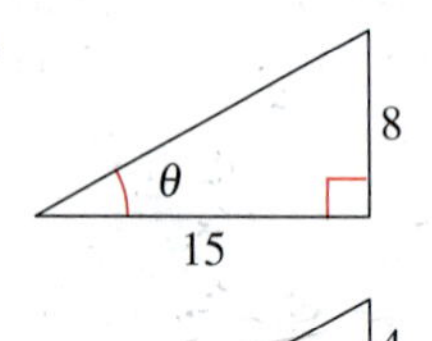

7.

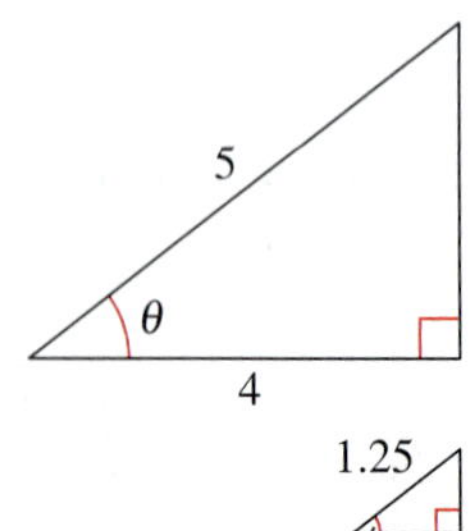

8.

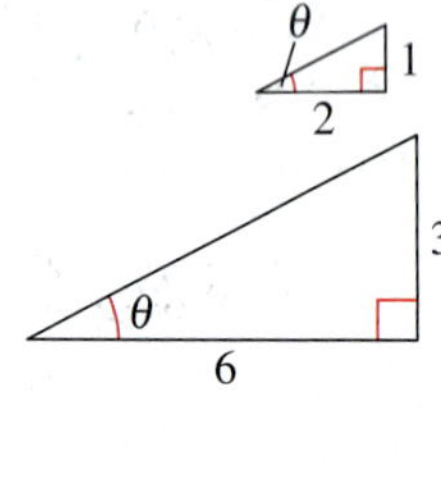

In Exercises 9–16, sketch a right triangle corresponding to the trigonometric function of the acute angle θ. Use the Pythagorean Theorem to determine the third side and then find the other five trigonometric functions of θ.

9. $\sin \theta = \frac{3}{4}$

10. $\cos \theta = \frac{5}{7}$

11. $\sec \theta = 2$

12. $\cot \theta = 5$

13. $\tan \theta = 3$

14. $\sec \theta = 6$

15. $\cot \theta = \frac{3}{2}$

16. $\csc \theta = \frac{17}{4}$

In Exercises 17–22, use the given function value(s), and trigonometric identities (including the cofunction identities), to find the indicated trigonometric functions.

17. $\sin 60° = \dfrac{\sqrt{3}}{2}, \quad \cos 60° = \dfrac{1}{2}$

(a) $\tan 60°$ (b) $\sin 30°$

(c) $\cos 30°$ (d) $\cot 60°$

18. $\sin 30° = \dfrac{1}{2}, \quad \tan 30° = \dfrac{\sqrt{3}}{3}$

(a) $\csc 30°$ (b) $\cot 60°$

(c) $\cos 30°$ (d) $\cot 30°$

19. $\csc \theta = \dfrac{\sqrt{13}}{2}, \quad \sec \theta = \dfrac{\sqrt{13}}{3}$

(a) $\sin \theta$ (b) $\cos \theta$

(c) $\tan \theta$ (d) $\sec(90° - \theta)$

20. $\sec \theta = 5, \quad \tan \theta = 2\sqrt{6}$

(a) $\cos \theta$ (b) $\cot \theta$

(c) $\cot(90° - \theta)$ (d) $\sin \theta$

21. $\cos \alpha = \frac{1}{3}$

(a) $\sec \alpha$ (b) $\sin \alpha$

(c) $\cot \alpha$ (d) $\sin(90° - \alpha)$

22. $\tan \beta = 5$

(a) $\cot \beta$ (b) $\cos \beta$

(c) $\tan(90° - \beta)$ (d) $\csc \beta$

In Exercises 23–26, evaluate the trigonometric function by memory or by constructing an appropriate triangle for the given special angle.

23. (a) $\cos 60°$ (b) $\csc 30°$ (c) $\tan 60°$

24. (a) $\cot 45°$ (b) $\cos 45°$ (c) $\csc 45°$

25. (a) $\sin \dfrac{\pi}{4}$ (b) $\cos \dfrac{\pi}{4}$ (c) $\tan \dfrac{\pi}{6}$

26. (a) $\sin \dfrac{\pi}{3}$ (b) $\tan \dfrac{\pi}{4}$ (c) $\sec \dfrac{\pi}{6}$

In Exercises 27–34, use a calculator to evaluate each function. Round your answers to four decimal places. (Be sure the calculator is in the correct angle mode.)

27. (a) $\tan 23.5°$ (b) $\cot 66.5°$

28. (a) $\sin 16.35°$ (b) $\csc 16.35°$

29. (a) $\cos 16° 18'$ (b) $\sin 73° 56'$

30. (a) $\sec 42° 12'$ (b) $\csc 48° 7'$

31. (a) $\cot \dfrac{\pi}{16}$ (b) $\tan \dfrac{\pi}{16}$

32. (a) $\sec 0.75$ (b) $\cos 0.75$

33. (a) $\csc 1$ (b) $\tan \frac{1}{2}$

34. (a) $\sec\left(\dfrac{\pi}{2} - 1\right)$ (b) $\cot\left(\dfrac{\pi}{2} - \dfrac{1}{2}\right)$

In Exercises 35–40, find the values of θ in degrees ($0° < \theta < 90°$) and radians ($0 < \theta < \pi/2$) without the aid of a calculator.

35. (a) $\sin \theta = \dfrac{1}{2}$ (b) $\csc \theta = 2$

36. (a) $\cos \theta = \dfrac{\sqrt{2}}{2}$ (b) $\tan \theta = 1$

37. (a) $\sec \theta = 2$ (b) $\cot \theta = 1$

38. (a) $\tan \theta = \sqrt{3}$ (b) $\cos \theta = \dfrac{1}{2}$

39. (a) $\csc \theta = \dfrac{2\sqrt{3}}{3}$ (b) $\sin \theta = \dfrac{\sqrt{2}}{2}$

40. (a) $\cot \theta = \dfrac{\sqrt{3}}{3}$ (b) $\sec \theta = \sqrt{2}$

In Exercises 41–44, find the values of θ in degrees ($0° < \theta < 90°$) and radians ($0 < \theta < \pi/2$) by using a calculator.

41. (a) $\sin \theta = 0.0145$ (b) $\sin \theta = 0.4565$

42. (a) $\cos \theta = 0.9848$ (b) $\cos \theta = 0.8746$

43. (a) $\tan \theta = 0.0125$ (b) $\tan \theta = 2.3545$

44. (a) $\sin \theta = 0.3746$ (b) $\cos \theta = 0.3746$

In Exercises 45–54, use trigonometric identities to transform the left side of the equation into the right side.

45. $\tan \theta \cot \theta = 1$

46. $\cos \theta \sec \theta = 1$

47. $\tan \alpha \cos \alpha = \sin \alpha$

48. $\cot \alpha \sin \alpha = \cos \alpha$

49. $(1 + \cos \theta)(1 - \cos \theta) = \sin^2 \theta$

50. $(1 + \sin \theta)(1 - \sin \theta) = \cos^2 \theta$

51. $(\sec \theta + \tan \theta)(\sec \theta - \tan \theta) = 1$

52. $\sin^2 \theta - \cos^2 \theta = 2 \sin^2 \theta - 1$

53. $\dfrac{\sin \theta}{\cos \theta} + \dfrac{\cos \theta}{\sin \theta} = \csc \theta \sec \theta$

54. $\dfrac{\tan \beta + \cot \beta}{\tan \beta} = \csc^2 \beta$

In Exercises 55–58, solve for *x*, *y*, or *r* as indicated.

55. Solve for x.

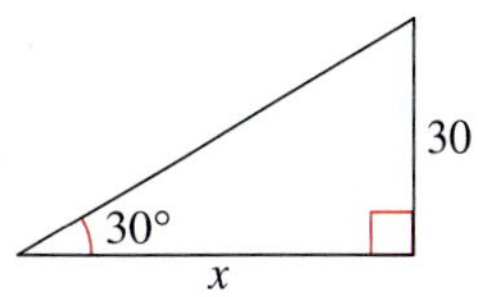

56. Solve for y.

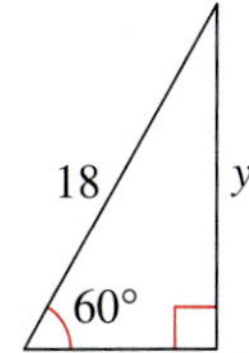

57. Solve for x.

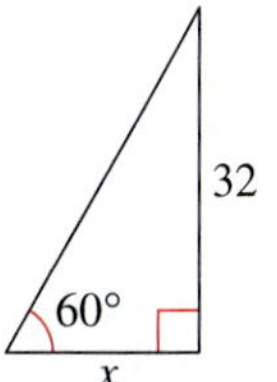

58. Solve for r.

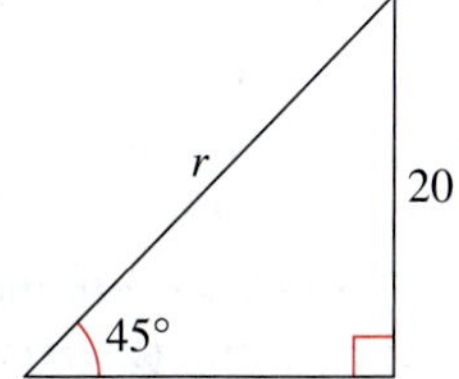

59. ***Height*** A six-foot person walks from the base of a broadcasting tower directly toward the tip of the shadow cast by the tower. When the person is 132 feet from the tower and 3 feet from the tip of the shadow, the person's shadow starts to appear beyond the tower's shadow.

(a) Draw a right triangle that gives a visual representation of the problem. Show the known quantities of the triangle and use a variable to indicate the height of the tower.

(b) Use a trigonometric function to write an equation involving the unknown quantity.

(c) What is the height of the tower?

60. ***Height*** A six-foot person standing 20 feet from a streetlight casts a 10-foot shadow (see figure). What is the height of the streetlight?

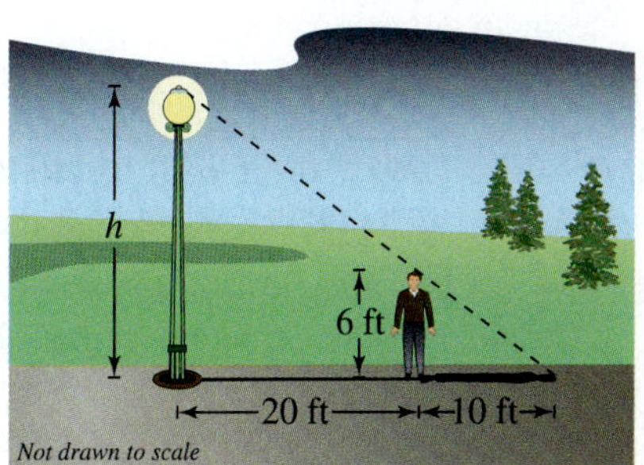

▶ Model It

61. ***Height*** A 20-meter line is used to tether a helium-filled balloon. Because of a breeze, the line makes an angle of approximately 85° with the ground.

(a) Draw a right triangle that gives a visual representation of the problem. Show the known quantities of the triangle and use a variable to indicate the height of the balloon.

(b) Use a trigonometric function to write an equation involving the unknown quantity.

(c) What is the height of the balloon?

(d) The breeze becomes stronger and the angle the balloon makes with the ground decreases. How does this affect the triangle you drew in part (a)?

(e) Complete the table, which shows the height (in meters) of the balloon for decreasing angle measures θ.

Angle, θ	Height
80°	
70°	
60°	
50°	
40°	
30°	
20°	
10°	

(f) As the angle the balloon makes with the ground approaches 0°, how does this affect the height of the balloon? Draw a right triangle to explain your reasoning.

62. ***Angle of Elevation*** A ramp 20 feet in length rises to a loading platform that is $3\frac{1}{3}$ feet off the ground.

(a) Draw a right triangle that gives a visual representation of the problem. Show the known quantities of the triangle and use a variable to indicate the angle of elevation of the ramp.

(b) Use a trigonometric function to write an equation involving the unknown quantity.

(c) What is the angle of elevation of the ramp?

63. ***Width of a River*** A biologist wants to know the width w of a river so that she can properly set instruments for studying the pollutants in the water. From point A, the biologist walks downstream 100 feet and sights to point C (see figure). From this sighting, it is determined that $\theta = 54°$. How wide is the river?

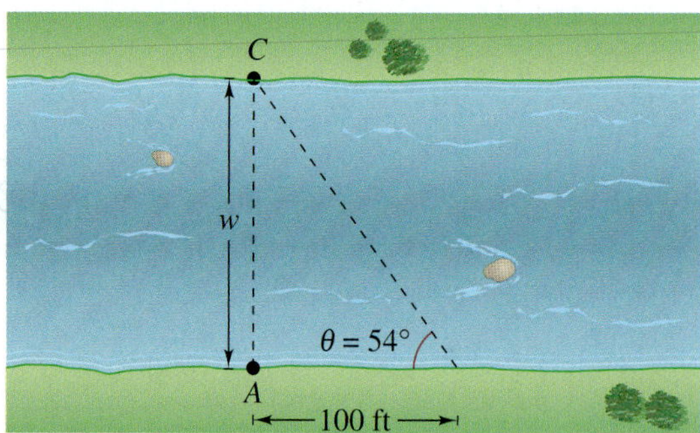

64. ***Height of a Mountain*** In traveling across flat land, you notice a mountain directly in front of you. Its angle of elevation (to the peak) is 3.5°. After you drive 13 miles closer to the mountain, the angle of elevation is 9°. Approximate the height of the mountain.

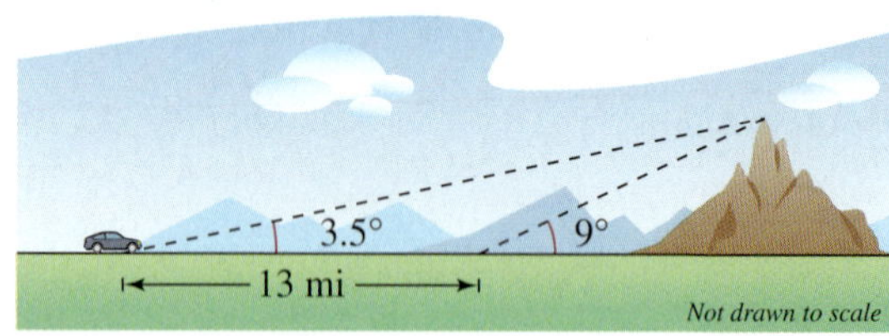

65. ***Machine Shop Calculations*** A steel plate has the form of one-fourth of a circle with a radius of 60 centimeters. Two two-centimeter holes are to be drilled in the plate positioned as shown in the figure. Find the coordinates of the center of each hole.

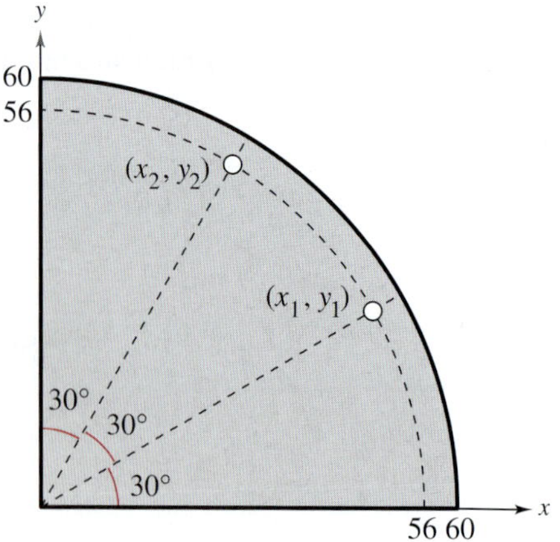

66. ***Machine Shop Calculations*** A tapered shaft has a diameter of 5 centimeters at the small end and is 15 centimeters long (see figure). The taper is 3°. Find the diameter d of the large end of the shaft.

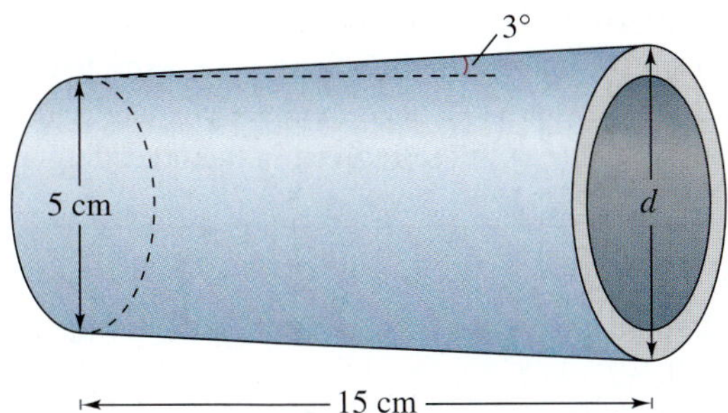

67. ***Geometry*** Use a compass to sketch a quarter of a circle of radius 10 centimeters. Using a protractor, construct an angle of 20° in standard position (see figure). Drop a perpendicular from the point of intersection of the terminal side of the angle and the arc of the circle. By actual measurement, calculate the coordinates (x, y) of the point of intersection and use these measurements to approximate the six trigonometric functions of a 20° angle.

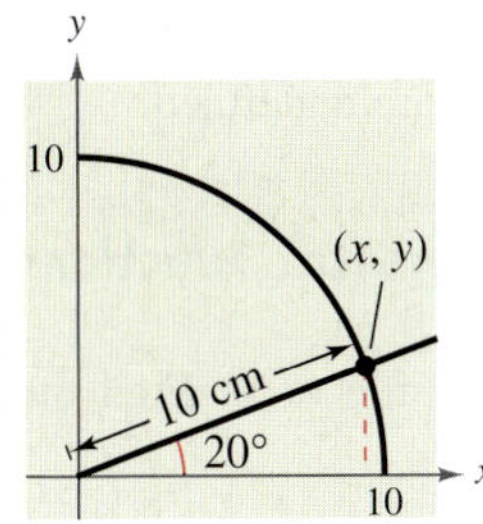

68. ***Geometry*** Repeat Exercise 67 using a 75° angle.

Synthesis

True or False? **In Exercises 69–74, determine whether the statement is true or false. Justify your answer.**

69. $\sin 60° \csc 60° = 1$

70. $\sec 30° = \csc 60°$

71. $\sin 45° + \cos 45° = 1$

72. $\cot^2 10° - \csc^2 10° = -1$

73. $\dfrac{\sin 60°}{\sin 30°} = \sin 2°$

74. $\tan[(0.8)^2] = \tan^2(0.8)$

75. ***Writing*** In right triangle trigonometry, explain why $\sin 30° = \frac{1}{2}$ regardless of the size of the triangle.

76. ***Think About It*** You are given only the value $\tan \theta$. Is it possible to find the value of $\sec \theta$ without finding the measure of θ? Explain.

77. ***Exploration***

(a) Complete the table.

θ	0.1	0.2	0.3	0.4	0.5
$\sin \theta$					

(b) Is θ or $\sin \theta$ greater for θ in the interval $(0, 0.5]$?

(c) As θ approaches 0, how do θ and $\sin \theta$ compare? Explain.

78. ***Exploration***

(a) Complete the table.

θ	0	0.3	0.6	0.9	1.2	1.5
$\sin \theta$						
$\cos \theta$						

(b) Discuss the behavior of the sine function for θ in the interval $[0, 1.5]$.

(c) Discuss the behavior of the cosine function for θ in the interval $[0, 1.5]$.

(d) Use the definitions of the sine and cosine functions to explain the results of parts (b) and (c).

Review

In Exercises 79–82, perform the operations and simplify.

79. $\dfrac{x^2 - 6x}{x^2 + 4x - 12} \cdot \dfrac{x^2 + 12x + 36}{x^2 - 36}$

80. $\dfrac{2t^2 + 5t - 12}{9 - 4t^2} \div \dfrac{t^2 - 16}{4t^2 + 12t + 9}$

81. $\dfrac{3}{x + 2} - \dfrac{2}{x - 2} + \dfrac{x}{x^2 + 4x + 4}$

82. $\dfrac{\left(\dfrac{3}{x} - \dfrac{1}{4}\right)}{\left(\dfrac{12}{x} - 1\right)}$

In Exercises 83 and 84, solve for x.

83. $\dfrac{2}{x + 3} + \dfrac{4}{x - 2} = \dfrac{12}{x^2 + x - 6}$

84. $\dfrac{3x + 2}{x^2 + x - 2} = \dfrac{4}{x + 2} - \dfrac{2}{1 - x}$

6.3 Trigonometric Functions of Any Angle

▶ **What you should learn**

- How to evaluate trigonometric functions of any angle
- How to use reference angles to evaluate trigonometric functions
- How to evaluate trigonometric functions of real numbers

▶ **Why you should learn it**

You can use trigonometric functions to model and solve real-life problems. For instance, in Exercise 103 on page 486, you can use trigonometric functions to model the monthly normal temperatures in New York City and Fairbanks, Alaska.

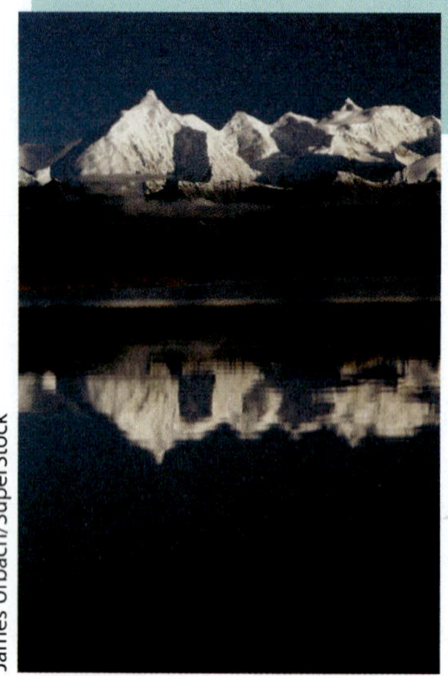

James Urbach/SuperStock

Introduction

In Section 6.2, the definitions of trigonometric functions were restricted to acute angles. In this section, the definitions are extended to cover *any* angle. If θ is an *acute* angle, these definitions coincide with those given in the preceding section.

Definitions of Trigonometric Functions of Any Angle

Let θ be an angle in standard position with (x, y) a point on the terminal side of θ and $r = \sqrt{x^2 + y^2} \neq 0$.

$\sin \theta = \dfrac{y}{r}$ $\qquad$ $\cos \theta = \dfrac{x}{r}$

$\tan \theta = \dfrac{y}{x}, \quad x \neq 0$ $\qquad$ $\cot \theta = \dfrac{x}{y}, \quad y \neq 0$

$\sec \theta = \dfrac{r}{x}, \quad x \neq 0$ $\qquad$ $\csc \theta = \dfrac{r}{y}, \quad y \neq 0$

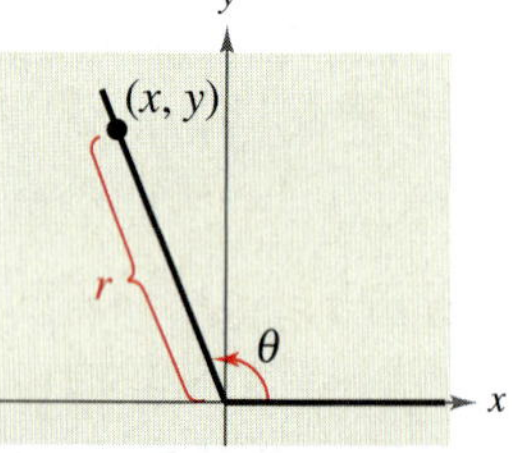

Because $r = \sqrt{x^2 + y^2}$ *cannot* be zero, it follows that the sine and cosine functions are defined for any real value of θ. However, if $x = 0$, the tangent and secant of θ are undefined. For example, the tangent of 90° is undefined. Similarly, if $y = 0$, the cotangent and cosecant of θ are undefined.

Example 1 ▶ **Evaluating Trigonometric Functions** 

Let $(-3, 4)$ be a point on the terminal side of θ. Find the sine, cosine, and tangent of θ.

Solution

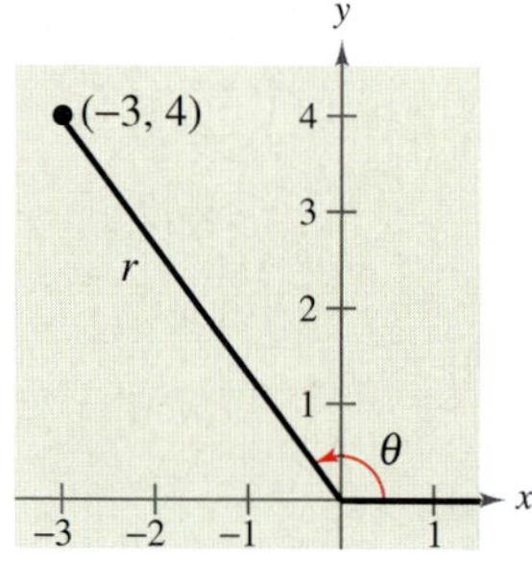

FIGURE 6.31

Referring to Figure 6.31, you can see that $x = -3$, $y = 4$, and

$$r = \sqrt{x^2 + y^2} = \sqrt{(-3)^2 + 4^2} = \sqrt{25} = 5.$$

So, you have the following.

$$\sin \theta = y/r = 4/5, \qquad \cos \theta = x/r = -3/5, \qquad \tan \theta = y/x = -4/3$$

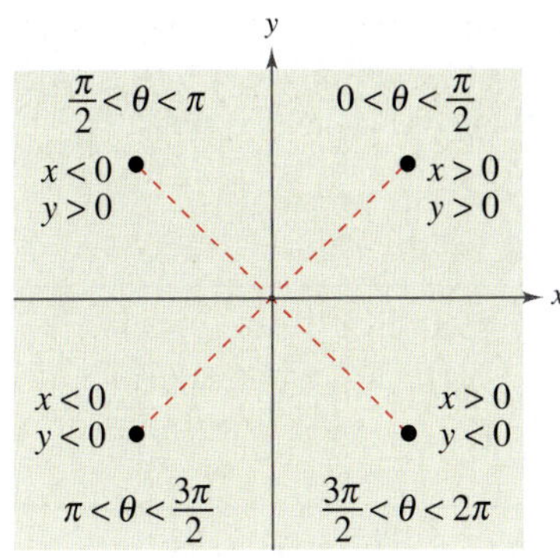

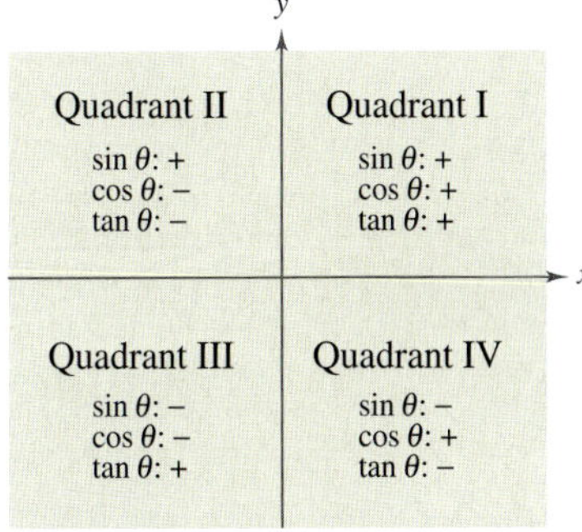

FIGURE 6.32

The *signs* of the trigonometric functions in the four quadrants can be determined easily from the definitions of the functions. For instance, because $\cos \theta = x/r$, it follows that $\cos \theta$ is positive wherever $x > 0$, which is in Quadrants I and IV. (Remember, r is always positive.) In a similar manner, you can verify the results shown in Figure 6.32.

Example 2 ▶ Evaluating Trigonometric Functions 

Given $\tan \theta = -\frac{5}{4}$ and $\cos \theta > 0$, find $\sin \theta$ and $\sec \theta$.

Solution

Note that θ lies in Quadrant IV because that is the only quadrant in which the tangent is negative and the cosine is positive. Moreover, using

$$\tan \theta = \frac{y}{x}$$

$$= -\frac{5}{4}$$

and the fact that y is negative in Quadrant IV, you can let $y = -5$ and $x = 4$. So, $r = \sqrt{16 + 25} = \sqrt{41}$ and you have

$$\sin \theta = \frac{y}{r} = \frac{-5}{\sqrt{41}}$$

$$\approx -0.7809$$

$$\sec \theta = \frac{r}{x} = \frac{\sqrt{41}}{4}$$

$$\approx 1.6008.$$

Example 3 ▶ Trigonometric Functions of Quadrant Angles

Evaluate the sine function at the four quadrant angles $0, \frac{\pi}{2}, \pi$, and $\frac{3\pi}{2}$.

Solution

To begin, choose a point on the terminal side of each angle, as shown in Figure 6.33. For each of the four points, $r = 1$, and you have

$$\sin 0 = \frac{y}{r} = \frac{0}{1} = 0 \qquad (x, y) = (1, 0)$$

$$\sin \frac{\pi}{2} = \frac{y}{r} = \frac{1}{1} = 1 \qquad (x, y) = (0, 1)$$

$$\sin \pi = \frac{y}{r} = \frac{0}{1} = 0 \qquad (x, y) = (-1, 0)$$

$$\sin \frac{3\pi}{2} = \frac{y}{r} = \frac{-1}{1} = -1. \qquad (x, y) = (0, -1)$$

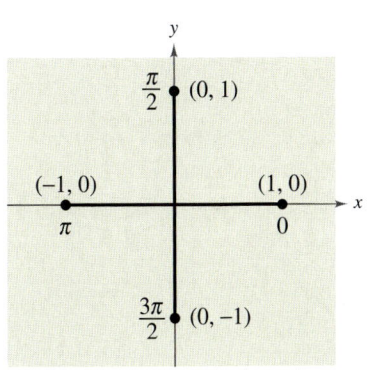

FIGURE 6.33

Reference Angles

The values of the trigonometric functions of angles greater than 90° (or less than 0°) can be determined from their values at corresponding acute angles called **reference angles.**

Definition of Reference Angle

Let θ be an angle in standard position. Its **reference angle** is the acute angle θ' formed by the terminal side of θ and the horizontal axis.

Figure 6.34 shows the reference angles for θ in Quadrants II, III, and IV.

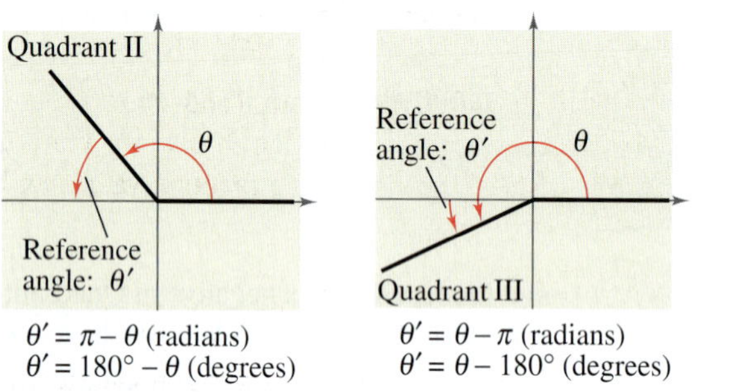

FIGURE 6.34

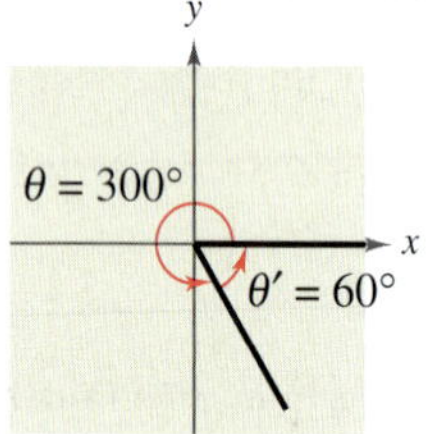

FIGURE 6.35

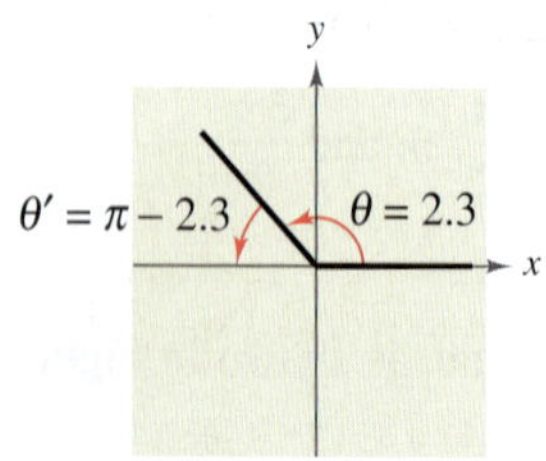

FIGURE 6.36

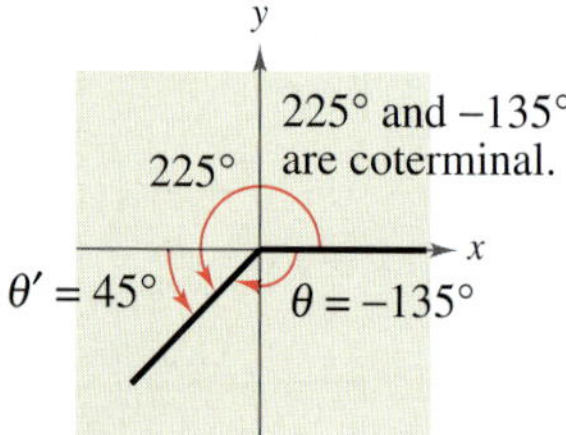

FIGURE 6.37

Example 4 ▶ Finding Reference Angles 

Find the reference angle θ'.

a. $\theta = 300°$ **b.** $\theta = 2.3$ **c.** $\theta = -135°$

Solution

a. Because 300° lies in Quadrant IV, the angle it makes with the x-axis is

$$\theta' = 360° - 300°$$

$$= 60°. \quad \text{Degrees}$$

Figure 6.35 shows the angle $\theta = 300°$ and its reference angle $\theta' = 60°$.

b. Because 2.3 lies between $\pi/2 \approx 1.5708$ and $\pi \approx 3.1416$, it follows that it is in Quadrant II and its reference angle is

$$\theta' = \pi - 2.3$$

$$\approx 0.8416. \quad \text{Radians}$$

Figure 6.36 shows the angle $\theta = 2.3$ and its reference angle $\theta' = \pi - 2.3$.

c. First, determine that $-135°$ is coterminal with 225°, which lies in Quadrant III. So, the reference angle is

$$\theta' = 225° - 180°$$

$$= 45°. \quad \text{Degrees}$$

Figure 6.37 shows the angle $\theta = -135°$ and its reference angle $\theta' = 45°$.

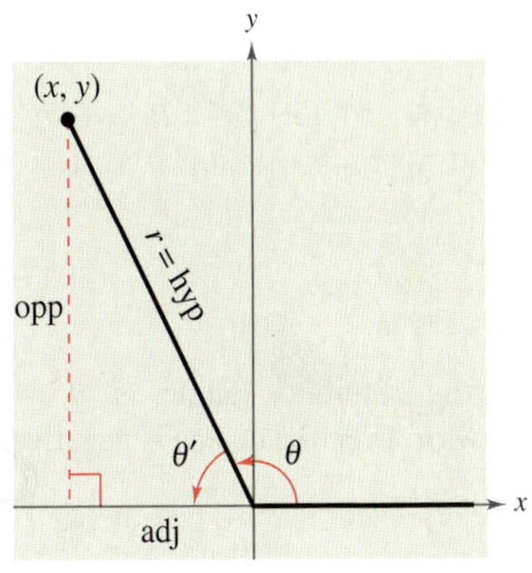

FIGURE 6.38

To see how a reference angle is used to evaluate a trigonometric function, consider the point (x, y) on the terminal side of θ, as shown in Figure 6.38. By definition, you know that

$$\sin \theta = \frac{y}{r} \qquad \text{and} \qquad \tan \theta = \frac{y}{x}.$$

For the right triangle with acute angle θ' and sides of lengths $|x|$ and $|y|$, you have

$$\sin \theta' = \frac{\text{opp}}{\text{hyp}} = \frac{|y|}{r}$$

and

$$\tan \theta' = \frac{\text{opp}}{\text{adj}} = \frac{|y|}{|x|}.$$

So, it follows that $\sin \theta$ and $\sin \theta'$ are equal, *except possibly in sign*. The same is true for $\tan \theta$ and $\tan \theta'$ *and* for the other four trigonometric functions. In all cases, the sign of the function value can be determined by the quadrant in which θ lies.

Evaluating Trigonometric Functions of Any Angle

To find the value of a trigonometric function of any angle θ,

1. determine the function value for the associated reference angle θ';
2. depending on the quadrant in which θ lies, affix the appropriate sign to the function value.

By using reference angles and the special angles discussed in the preceding section, you can greatly extend the scope of *exact* trigonometric values. For instance, knowing the function values of 30° means that you know the function values of all angles for which 30° is a reference angle. For convenience, the table below shows the exact values of the trigonometric functions of special angles and quadrant angles.

STUDY TIP

Learning the table of values at the right is worth the effort. Doing so will increase both your efficiency and your confidence. Here is a pattern for the sine function that may help you remember the values.

θ	0°	30°	45°	60°	90°
$\sin \theta$	$\frac{\sqrt{0}}{2}$	$\frac{\sqrt{1}}{2}$	$\frac{\sqrt{2}}{2}$	$\frac{\sqrt{3}}{2}$	$\frac{\sqrt{4}}{2}$

Reverse the order to get cosine values of the same angles.

Trigonometric Values of Common Angles

θ (degrees)	0°	30°	45°	60°	90°	180°	270°
θ (radians)	0	$\frac{\pi}{6}$	$\frac{\pi}{4}$	$\frac{\pi}{3}$	$\frac{\pi}{2}$	π	$\frac{3\pi}{2}$
$\sin \theta$	0	$\frac{1}{2}$	$\frac{\sqrt{2}}{2}$	$\frac{\sqrt{3}}{2}$	1	0	-1
$\cos \theta$	1	$\frac{\sqrt{3}}{2}$	$\frac{\sqrt{2}}{2}$	$\frac{1}{2}$	0	-1	0
$\tan \theta$	0	$\frac{\sqrt{3}}{3}$	1	$\sqrt{3}$	Undef.	0	Undef.

Example 5 ▶ Trigonometric Functions of Nonacute Angles 

Evaluate each trigonometric function.

a. $\cos \dfrac{4\pi}{3}$ **b.** $\tan(-210°)$ **c.** $\csc \dfrac{11\pi}{4}$

Solution

a. Because $\theta = 4\pi/3$ lies in Quadrant III, the reference angle is $\theta' = (4\pi/3) - \pi = \pi/3$, as shown in Figure 6.39. Moreover, the cosine is negative in Quadrant III, so that

$$\cos \frac{4\pi}{3} = (-) \cos \frac{\pi}{3}$$

$$= -\frac{1}{2}.$$

b. Because $-210° + 360° = 150°$, it follows that $-210°$ is coterminal with the second-quadrant angle $150°$. So, the reference angle is $\theta' = 180° - 150° = 30°$, as shown in Figure 6.40. Finally, because the tangent is negative in Quadrant II, you have

$$\tan(-210°) = (-) \tan 30°$$

$$= -\frac{\sqrt{3}}{3}.$$

c. Because $(11\pi/4) - 2\pi = 3\pi/4$, it follows that $11\pi/4$ is coterminal with the second-quadrant angle $3\pi/4$. So, the reference angle is $\theta' = \pi - (3\pi/4) = \pi/4$, as shown in Figure 6.41. Because the cosecant is positive in Quadrant II, you have

$$\csc \frac{11\pi}{4} = (+) \csc \frac{\pi}{4}$$

$$= \frac{1}{\sin(\pi/4)}$$

$$= \sqrt{2}.$$

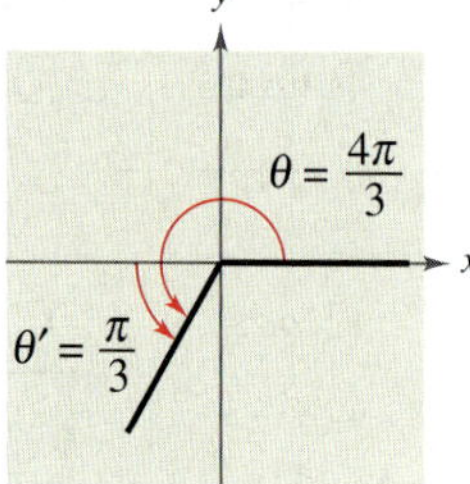

FIGURE 6.39

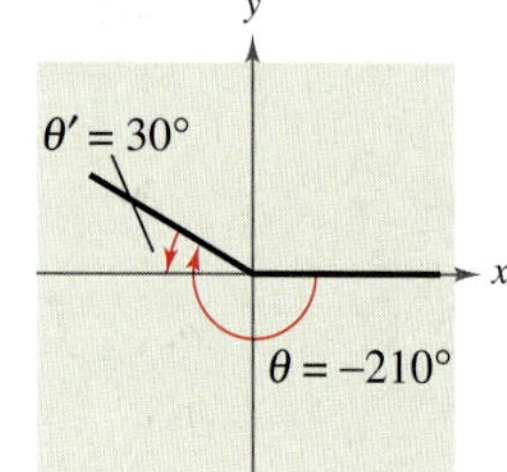

FIGURE 6.40

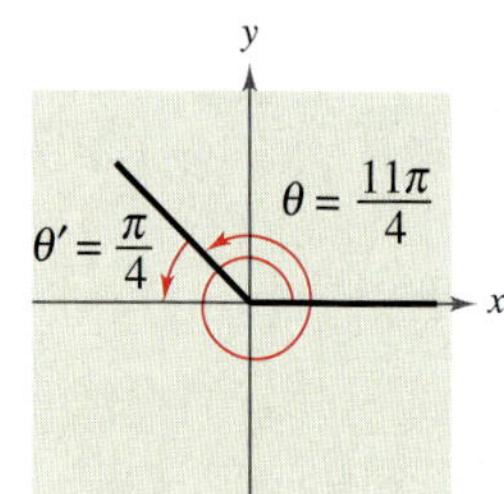

FIGURE 6.41

Example 6 ▶ Using Trigonometric Identities 

Let θ be an angle in Quadrant II such that $\sin \theta = \frac{1}{3}$. Find (a) $\cos \theta$ and (b) $\tan \theta$ by using trigonometric identities.

Solution

a. Using the Pythagorean identity $\sin^2 \theta + \cos^2 \theta = 1$, you obtain

$$\left(\frac{1}{3}\right)^2 + \cos^2 \theta = 1 \qquad \text{Substitute } \tfrac{1}{3} \text{ for } \sin \theta.$$

$$\cos^2 \theta = 1 - \frac{1}{9} = \frac{8}{9}.$$

Because $\cos \theta < 0$ in Quadrant II, you can use the negative root to obtain

$$\cos \theta = -\frac{\sqrt{8}}{\sqrt{9}}$$

$$= -\frac{2\sqrt{2}}{3}.$$

b. Using the trigonometric identity $\tan \theta = \sin \theta / \cos \theta$, you obtain

$$\tan \theta = \frac{1/3}{-2\sqrt{2}/3} \qquad \text{Substitute for } \sin \theta \text{ and } \cos \theta.$$

$$= -\frac{1}{2\sqrt{2}}$$

$$= -\frac{\sqrt{2}}{4}.$$

Example 7 ▶ Using a Calculator

a. Use a calculator to evaluate $\cot 410°$ and $\sin(-7)$.

b. Use a calculator to solve $\tan \theta = 4.812$, $0 \le \theta < 2\pi$.

Solution

	Function	*Mode*	*Calculator Keystrokes*	*Display*
a.	$\cot 410°$	Degree	(TAN 410) x^{-1} ENTER	0.8390996
	$\sin(-7)$	Radian	SIN ((−) 7) ENTER	−0.6569866

b. To solve the equation $\tan \theta = 4.812$, you can use the inverse tangent key, as follows.

Equation	*Mode*	*Calculator Keystrokes*	*Display*
$\tan \theta = 4.812$	Radian	TAN^{-1} 4.812 ENTER	1.365898912

The angle $\theta \approx 1.3659$ lies in Quadrant I. A second value of θ lies in Quadrant III (tangent is positive) and is

$$\theta \approx \pi + 1.3659 \approx 4.5075.$$

Exploration

With your graphing utility in *radian* and *parametric* modes, enter the equations X1T = cos T and Y1T = sin T and use the following settings.

Tmin = 0, Tmax = 6.3,
Tstep = 0.1
Xmin = -1.5, Xmax = 1.5,
Xscl = 1
Ymin = -1, Ymax = 1,
Yscl = 1

a. Graph the entered equations and describe the graph.

b. Use the *trace* feature to move the cursor around the graph. What do the t-values represent? What do the x- and y-values represent?

c. What are the least and greatest values of x and y?

Trigonometric Functions of Real Numbers

To define a trigonometric function of a real number (rather than an angle), let t represent any real number. Then imagine that the real number line is wrapped around a *unit circle*, as shown in Figure 6.42. Note that positive numbers correspond to a counterclockwise wrapping and negative numbers correspond to a clockwise wrapping.

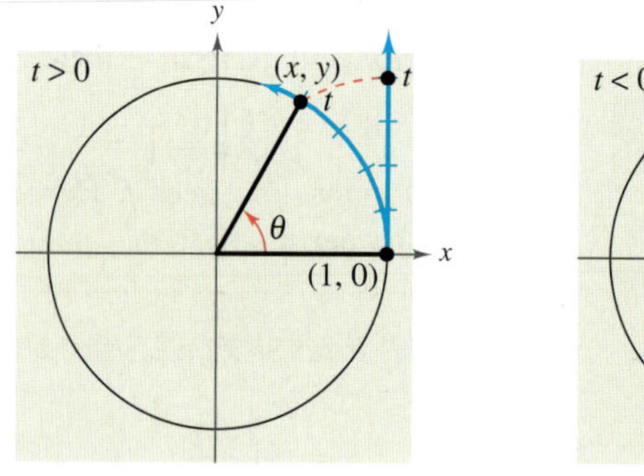

FIGURE 6.42

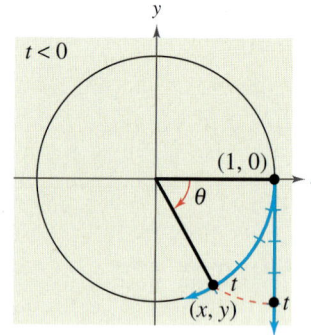

As the real line is wrapped around the unit circle, each real number t corresponds to a central angle θ. Moreover, because the circle has a radius of 1, the arc intercepted by the angle θ will have a length of t. The point is that *if θ is measured in radians*, then $t = \theta$. So, you can define $\sin t$ as

$$\sin t = \sin(t \text{ radians}).$$

Similarly, $\cos t = \cos(t \text{ radians})$, $\tan t = \tan(t \text{ radians})$, and so on.

Example 8 ▶ Evaluating Trigonometric Functions 

Evaluate $f(t) = \sin t$ for (a) $t = 1$ and (b) $t = 7\pi/2$.

Solution

a. $f(1) = \sin 1 \approx 0.841471$ Radian mode

b. $f\left(\dfrac{7\pi}{2}\right) = \sin \dfrac{7\pi}{2} = -1$ $\dfrac{7\pi}{2}$ and $\dfrac{3\pi}{2}$ are coterminal.

The *domain* of the sine and cosine functions is the set of all real numbers. To determine the *range* of these two functions, consider the unit circle shown in Figure 6.43. Because $r = 1$, it follows that $\sin t = y$ and $\cos t = x$. Moreover, because (x, y) is on the unit circle, you know that $-1 \le y \le 1$ and $-1 \le x \le 1$. So, the values of sine and cosine also range between -1 and 1.

$$-1 \le y \le 1 \qquad \text{and} \qquad -1 \le x \le 1$$
$$-1 \le \sin t \le 1 \qquad\qquad -1 \le \cos t \le 1$$

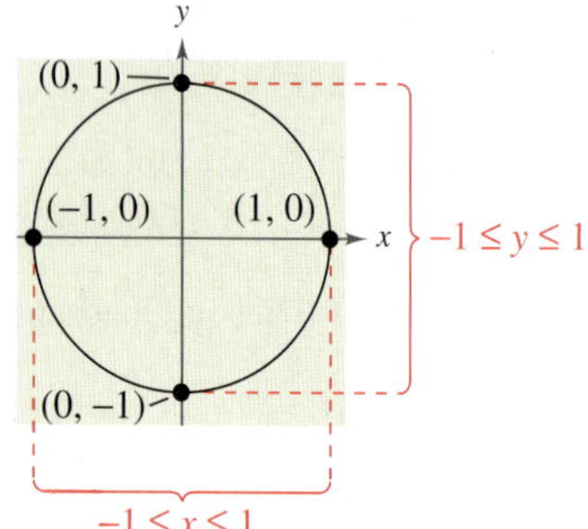

FIGURE 6.43

Adding 2π to each value of t in the interval $[0, 2\pi]$ completes a second revolution around the unit circle, as shown in Figure 6.44 on page 483. The values of $\sin(t + 2\pi)$ and $\cos(t + 2\pi)$ correspond to those of $\sin t$ and $\cos t$. Similar results

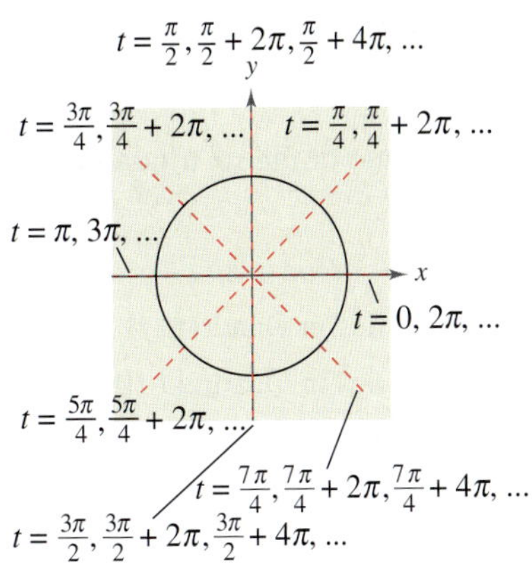

FIGURE 6.44

can be obtained for repeated revolutions (positive or negative) on the unit circle. This leads to the general result

$$\sin(t + 2\pi n) = \sin t \qquad \text{and} \qquad \cos(t + 2\pi n) = \cos t$$

for any integer n and real number t. Functions that behave in such a repetitive (or cyclic) manner are called **periodic.**

Definition of Periodic Function

A function f is **periodic** if there exists a positive real number c such that

$$f(t + c) = f(t)$$

for all t in the domain of f. The smallest number c for which f is periodic is called the **period** of f.

STUDY TIP

From this definition it follows that the sine and cosine functions are periodic and have a period of 2π. The other four trigonometric functions are also periodic, and more will be said about this in Section 6.5.

Recall from Section 2.3 that a function f is *even* if $f(-t) = f(t)$, and is *odd* if $f(-t) = -f(t)$.

Even and Odd Trigonometric Functions

The cosine and secant functions are *even.*

$$\cos(-t) = \cos t \qquad \sec(-t) = \sec t$$

The sine, cosecant, tangent, and cotangent functions are *odd.*

$$\sin(-t) = -\sin t \qquad \csc(-t) = -\csc t$$

$$\tan(-t) = -\tan t \qquad \cot(-t) = -\cot t$$

You have now defined the six trigonometric functions from a right triangle perspective and as functions of real numbers. In your remaining work with trigonometry you should continue to rely on both perspectives. A summary of basic trigonometry is included on the inside back cover of this text.

Writing ABOUT MATHEMATICS

Patterns in Trigonometric Functions Complete the table. Then identify and describe any inherent patterns in the trigonometric functions. What can you conclude?

Function	Domain	Range	Even/odd	Period	Zeros
Sine					
Cosine					
Tangent					
Cosecant					
Secant					
Cotangent					

6.3 Exercises

In Exercises 1–4, determine the exact values of the six trigonometric functions of the angle θ.

1. (a)

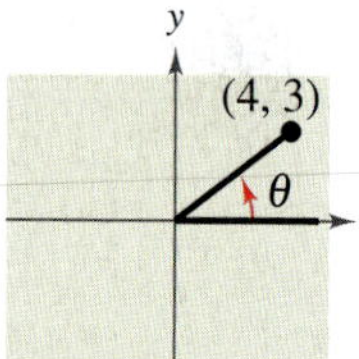

(b)

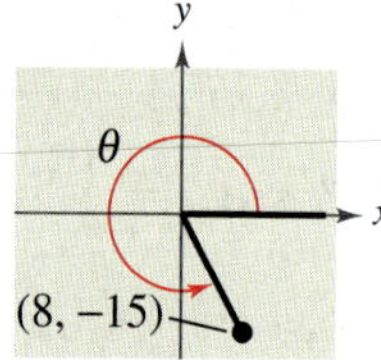

2. (a)

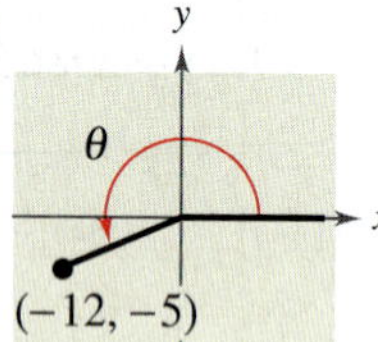

(b)

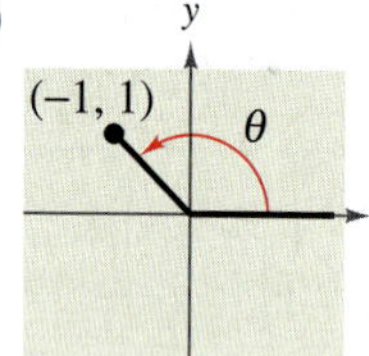

3. (a)

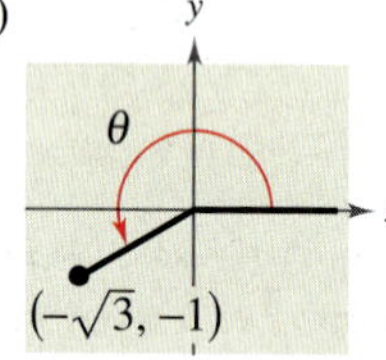

(b)

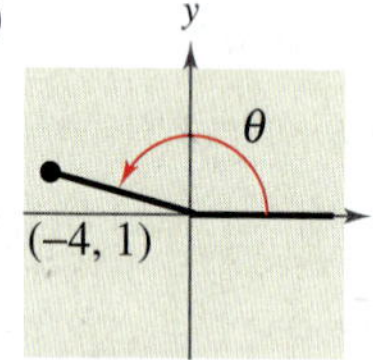

4. (a)

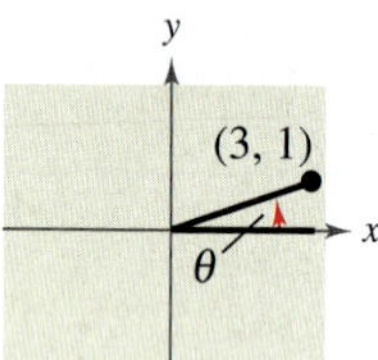

(b)

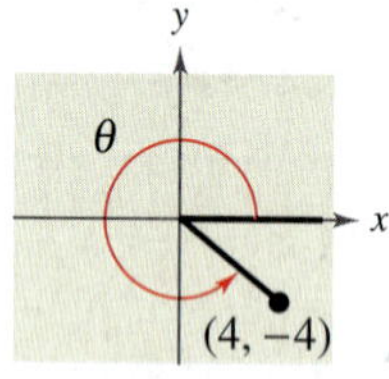

In Exercises 5–10, the point is on the terminal side of an angle in standard position. Determine the exact values of the six trigonometric functions of the angle.

5. $(7, 24)$
6. $(8, 15)$
7. $(-4, 10)$
8. $(-5, -2)$
9. $(-3.5, 6.8)$
10. $\left(3\frac{1}{2}, -7\frac{3}{4}\right)$

In Exercises 11–14, state the quadrant in which θ lies.

11. $\sin\theta < 0$ and $\cos\theta < 0$
12. $\sin\theta > 0$ and $\cos\theta > 0$
13. $\sin\theta > 0$ and $\tan\theta < 0$
14. $\sec\theta > 0$ and $\cot\theta < 0$

In Exercises 15–24, find the values of the six trigonometric functions of θ.

	Function Value	*Constraint*
15.	$\sin\theta = \frac{3}{5}$	θ lies in Quadrant II.
16.	$\cos\theta = -\frac{4}{5}$	θ lies in Quadrant III.
17.	$\tan\theta = -\frac{15}{8}$	$\sin\theta < 0$
18.	$\cos\theta = \frac{8}{17}$	$\tan\theta < 0$
19.	$\cot\theta = -3$	$\cos\theta > 0$
20.	$\csc\theta = 4$	$\cot\theta < 0$
21.	$\sec\theta = -2$	$\sin\theta > 0$
22.	$\sin\theta = 0$	$\sec\theta = -1$
23.	$\cot\theta$ is undefined.	$\pi/2 \le \theta \le 3\pi/2$
24.	$\tan\theta$ is undefined.	$\pi \le \theta \le 2\pi$

In Exercises 25–28, the terminal side of θ lies on the given line in the specified quadrant. Find the values of the six trigonometric functions of θ by finding a point on the line.

	Line	*Quadrant*
25.	$y = -x$	II
26.	$y = \frac{1}{3}x$	III
27.	$2x - y = 0$	III
28.	$4x + 3y = 0$	IV

In Exercises 29–36, evaluate the trigonometric function of the quadrant angle.

29. $\cos\pi$
30. $\cos\dfrac{3\pi}{2}$
31. $\sec\dfrac{3\pi}{2}$
32. $\sec\pi$
33. $\tan\dfrac{\pi}{2}$
34. $\tan\pi$
35. $\csc\pi$
36. $\cot\dfrac{\pi}{2}$

In Exercises 37–44, find the reference angle θ', and sketch θ and θ' in standard position.

37. $\theta = 203^\circ$
38. $\theta = 309^\circ$
39. $\theta = -245^\circ$
40. $\theta = -145^\circ$
41. $\theta = \dfrac{2\pi}{3}$
42. $\theta = \dfrac{7\pi}{4}$

43. $\theta = 3.5$

44. $\theta = \frac{11\pi}{3}$

In Exercises 45–58, evaluate the sine, cosine, and tangent of the angle without using a calculator.

45. $225°$

46. $300°$

47. $750°$

48. $-405°$

49. $-150°$

50. $-840°$

51. $\frac{4\pi}{3}$

52. $\frac{\pi}{4}$

53. $-\frac{\pi}{6}$

54. $-\frac{\pi}{2}$

55. $\frac{11\pi}{4}$

56. $\frac{10\pi}{3}$

57. $-\frac{3\pi}{2}$

58. $-\frac{25\pi}{4}$

In Exercises 59–68, use a calculator to evaluate the trigonometric function. Round your answers to four decimal places. (Be sure the calculator is set in the correct angle mode.)

59. $\sin 10°$

60. $\sec 225°$

61. $\cos(-110°)$

62. $\csc(-330°)$

63. $\tan 4.5$

64. $\cot 1.35$

65. $\tan \frac{\pi}{9}$

66. $\tan\left(-\frac{\pi}{9}\right)$

67. $\sin(-0.65)$

68. $\sin 0.65$

In Exercises 69–74, find two solutions of the equation. Give your answers in degrees ($0° \le \theta < 360°$) and radians ($0 \le \theta < 2\pi$). Do not use a calculator.

69. (a) $\sin \theta = \frac{1}{2}$ (b) $\sin \theta = -\frac{1}{2}$

70. (a) $\cos \theta = \frac{\sqrt{2}}{2}$ (b) $\cos \theta = -\frac{\sqrt{2}}{2}$

71. (a) $\csc \theta = \frac{2\sqrt{3}}{3}$ (b) $\cot \theta = -1$

72. (a) $\sec \theta = 2$ (b) $\sec \theta = -2$

73. (a) $\tan \theta = 1$ (b) $\cot \theta = -\sqrt{3}$

74. (a) $\sin \theta = \frac{\sqrt{3}}{2}$ (b) $\sin \theta = -\frac{\sqrt{3}}{2}$

In Exercises 75–78, use a calculator to approximate two values of θ ($0° \le \theta < 360°$) that satisfy the equation. Round the values to two decimal places.

75. $\sin \theta = 0.8191$

76. $\cos \theta = 0.8746$

77. $\cos \theta = -0.4367$

78. $\sin \theta = -0.6514$

In Exercises 79–84, use a calculator to approximate two values of θ ($0 \le \theta < 2\pi$) that satisfy the equation. Round the values to three decimal places.

79. $\cos \theta = 0.9848$

80. $\sin \theta = 0.0175$

81. $\tan \theta = 1.192$

82. $\cot \theta = 5.671$

83. $\sec \theta = -2.6667$

84. $\cos \theta = -0.3214$

In Exercises 85–90, find the indicated trigonometric value in the specified quadrant.

	Function	*Quadrant*	*Trigonometric Value*
85.	$\sin \theta = -\frac{3}{5}$	IV	$\cos \theta$
86.	$\cot \theta = -3$	II	$\sin \theta$
87.	$\tan \theta = \frac{3}{2}$	III	$\sec \theta$
88.	$\csc \theta = -2$	IV	$\cot \theta$
89.	$\cos \theta = \frac{5}{8}$	I	$\sec \theta$
90.	$\sec \theta = -\frac{9}{4}$	III	$\tan \theta$

In Exercises 91–98, find the point (x, y) on the unit circle that corresponds to the real number t. Use the result to evaluate $\sin t$, $\cos t$, and $\tan t$.

91. $t = \frac{\pi}{4}$

92. $t = \frac{\pi}{3}$

93. $t = \frac{5\pi}{6}$

94. $t = \frac{5\pi}{4}$

95. $t = \frac{4\pi}{3}$

96. $t = \frac{11\pi}{6}$

97. $t = \frac{3\pi}{2}$

98. $t = \pi$

Estimation **In Exercises 99 and 100, use the figure on the next page and a straightedge to approximate the value of each trigonometric function. To print an enlarged copy of the graph, go to the website *www.mathgraphs.com*.**

99. (a) $\sin 5$ (b) $\cos 2$

100. (a) $\sin 0.75$ (b) $\cos 2.5$

Estimation **In Exercises 101 and 102, use the figure on the next page to approximate the solution of each equation, where $0 \le t < 2\pi$. To print an enlarged copy of the graph, go to the website *www.mathgraphs.com*.**

101. (a) $\sin t = 0.25$ (b) $\cos t = -0.25$

102. (a) $\sin t = -0.75$ (b) $\cos t = 0.75$

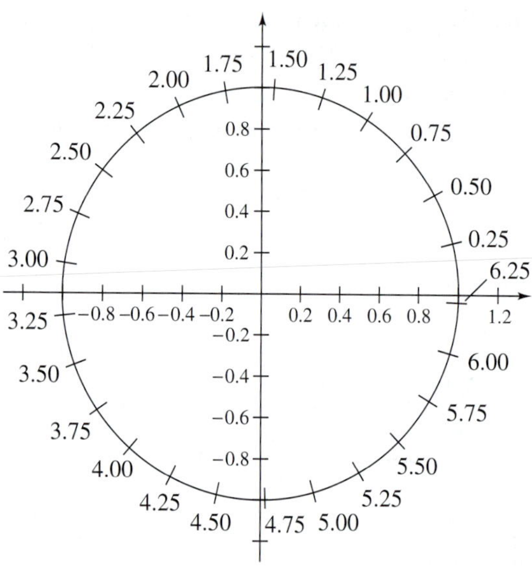

FIGURE FOR 99–102

▶ Model It

103. ***Data Analysis*** The table shows the monthly normal temperatures (in degrees Fahrenheit) for selected months for New York City (N) and Fairbanks, Alaska (F). (Source: National Climatic Data Center)

Month, t	New York City, N	Fairbanks, F
January	32	−10
April	53	31
July	77	63
October	58	25
December	37	−7

(a) Use the *regression* feature of a graphing utility to find a model of the form

$$y = a \sin(bt + c) + d$$

for each city. Let t represent the month, with $t = 1$ corresponding to January.

(b) Use the models from part (a) to find the monthly normal temperatures for the two cities in February, March, May, June, August, September, and November.

(c) Compare the models for the two cities.

104. ***Sales*** A company that produces snowboards, which are seasonal products, forecasts monthly sales for 2 years to be

$$S = 23.1 + 0.442t + 4.3 \cos \frac{\pi t}{6}$$

where S is measured in thousands of units and t is the time in months, with $t = 1$ representing January 2003. Predict sales for each of the following months.

(a) February 2003 (b) February 2004

(c) June 2003 (d) June 2004

105. ***Harmonic Motion*** The displacement from equilibrium of an oscillating weight suspended by a spring is

$$y(t) = 2 \cos 6t$$

where y is the displacement in centimeters and t is the time in seconds (see figure). Find the displacement when (a) $t = 0$, (b) $t = \frac{1}{4}$, and (c) $t = \frac{1}{2}$.

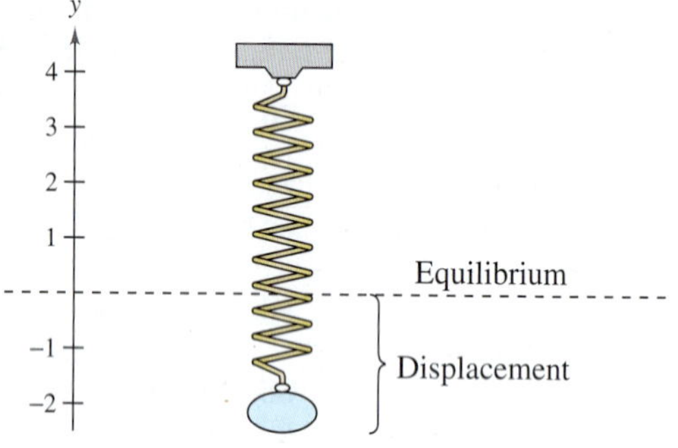

FIGURE FOR 105 AND 106

106. ***Harmonic Motion*** The displacement from equilibrium of an oscillating weight suspended by a spring and subject to the damping effect of friction is

$$y(t) = 2e^{-t} \cos 6t$$

where y is the displacement in centimeters and t is the time in seconds (see figure). Find the displacement when (a) $t = 0$, (b) $t = \frac{1}{4}$, and (c) $t = \frac{1}{2}$.

107. ***Electric Circuits*** The current I (in amperes) when 100 volts is applied to a circuit is

$$I = 5e^{-2t} \sin t$$

where t is the time in seconds after the voltage is applied. Approximate the current $t = 0.7$ second after the voltage is applied.

108. ***Distance*** An airplane, flying at an altitude of 6 miles, is on a flight path that passes directly over an observer (see figure). If θ is the angle of elevation from the observer to the plane, find the distance d from the observer to the plane when (a) $\theta = 30°$, (b) $\theta = 90°$, and (c) $\theta = 120°$.

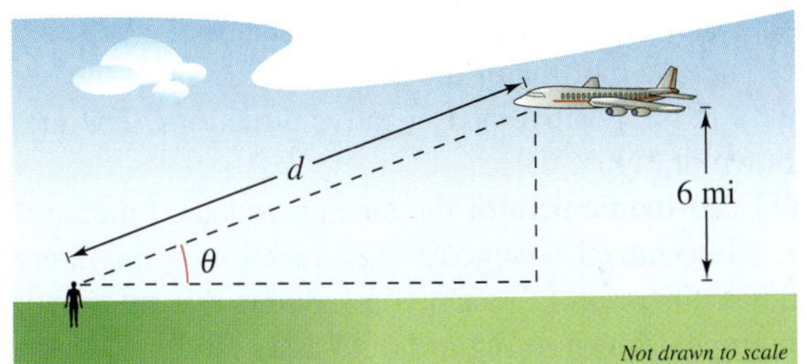

Synthesis

True or False? **In Exercises 109 and 110, determine whether the statement is true or false. Justify your answer.**

109. In each of the four quadrants, the signs of the secant function and sine function will be the same.

110. To find the reference angle for an angle θ (given in degrees), find the integer n such that $0 \le 360°n - \theta \le 360°$. The difference $360°n - \theta$ is the reference angle.

111. ***Think About It*** Because $f(t) = \sin t$ is an odd function and $g(t) = \cos t$ is an even function, what can be said about the function $h(t) = f(t)g(t)$?

112. ***Writing*** Consider an angle in standard position with $r = 12$ centimeters, as shown in the figure. Write a short paragraph describing the changes in the magnitudes of x, y, $\sin\theta$, $\cos\theta$, and $\tan\theta$ as θ increases continuously from 0° to 90°.

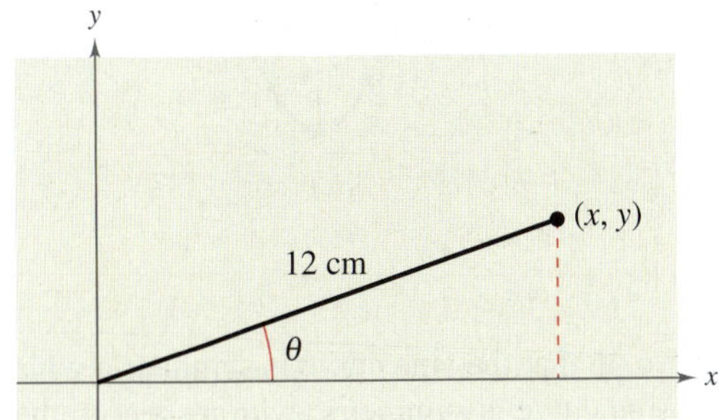

113. ***Writing*** Explain how reference angles are used to find the trigonometric functions of obtuse angles.

114. ***Conjecture***

(a) Use a graphing utility to complete the table.

θ	0	0.3	0.6	0.9	1.2	1.5
$\cos\left(\frac{3\pi}{2} - \theta\right)$						
$-\sin\theta$						

(b) Make a conjecture about the relationship between $\cos\left(\frac{3\pi}{2} - \theta\right)$ and $-\sin\theta$.

Review

In Exercises 115–118, graph the rational function. Identify any asymptotes.

115. $f(x) = \dfrac{2}{4 - x}$

116. $g(x) = -\dfrac{1}{x^2 - 4}$

117. $h(x) = \dfrac{2x^2 + 10x}{x - 6}$

118. $f(x) = \dfrac{x - 5}{x^2 + x - 12}$

In Exercises 119–122, graph the exponential function. Identify the intercepts, asymptotes, domain, and range of the function.

119. $y = 2^{x-1}$

120. $y = 3^{x+1} + 2$

121. $y = 3^{-x/2}$

122. $y = 3^{(x+1)/2}$

In Exercises 123–126, graph the logarithmic function. Identify the intercepts, asymptotes, domain, and range of the function.

123. $y = \ln(x - 1)$

124. $y = \ln x^4$

125. $y = \log_{10}(x + 2)$

126. $y = \log_{10}(-3x)$

6.4 Graphs of Sine and Cosine Functions

▶ **What you should learn**

- How to sketch the graphs of basic sine and cosine functions
- How to use amplitude and period to help sketch the graphs of sine and cosine functions
- How to sketch translations of the graphs of sine and cosine functions
- How to use sine and cosine functions to model real-life data

▶ **Why you should learn it**

Sine and cosine functions are often used in scientific calculations. For instance, in Exercise 74 on page 497, you can use a trigonometric function to model the percent of the moon's face that is illuminated for any given day in 2005.

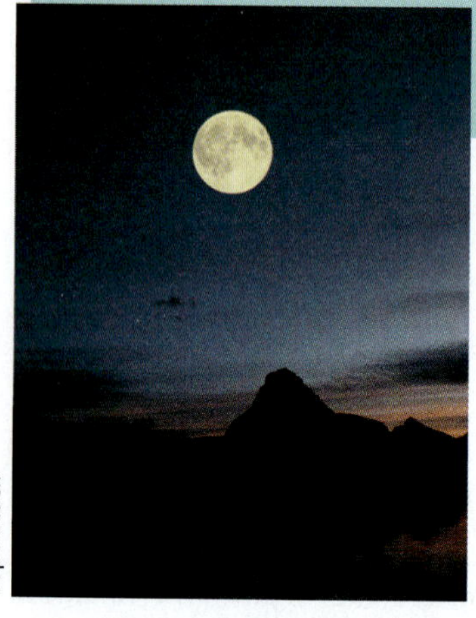

SuperStock

Basic Sine and Cosine Curves

In this section you will study techniques for sketching the graphs of the sine and cosine functions. The graph of the sine function is a **sine curve.** In Figure 6.45, the black portion of the graph represents one period of the function and is called **one cycle** of the sine curve. The gray portion of the graph indicates that the basic sine wave repeats indefinitely in the positive and negative directions. The graph of the cosine function is shown in Figure 6.46.

Recall from Section 6.3 that the domain of the sine and cosine functions is the set of all real numbers. Moreover, the range of each function is the interval $[-1, 1]$, and each function has a period of 2π. Do you see how this information is consistent with the basic graphs shown in Figures 6.45 and 6.46?

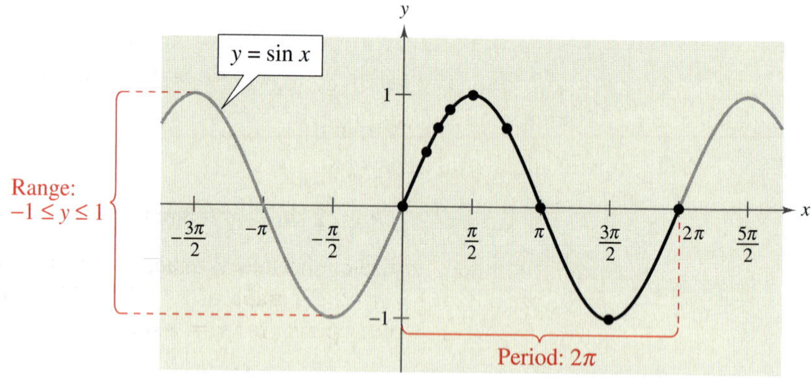

FIGURE 6.45

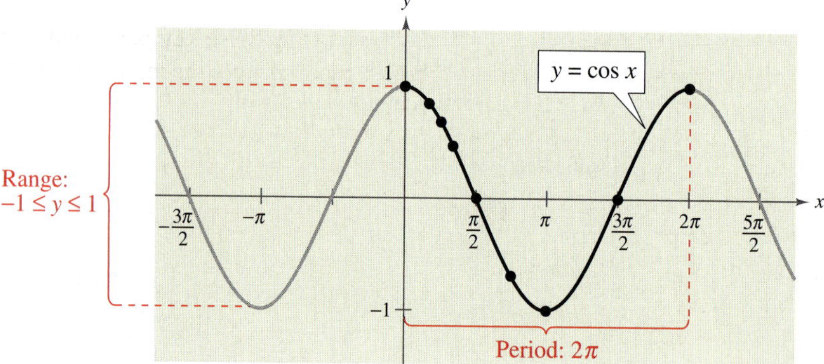

FIGURE 6.46

Note in Figures 6.45 and 6.46 that the sine curve is symmetric with respect to the *origin*, whereas the cosine curve is symmetric with respect to the *y-axis*. These properties of symmetry occur because the sine function is odd and the cosine function is even.

To sketch the graphs of the basic sine and cosine functions by hand, it helps to note five **key points** in one period of each graph: the *intercepts*, *maximum points*, and *minimum points* (see Figure 6.47).

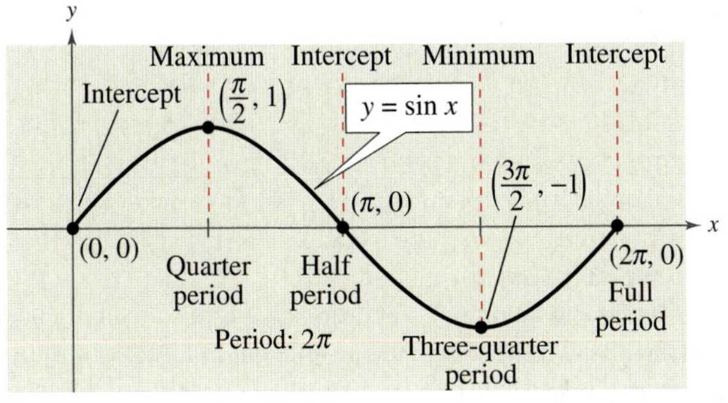

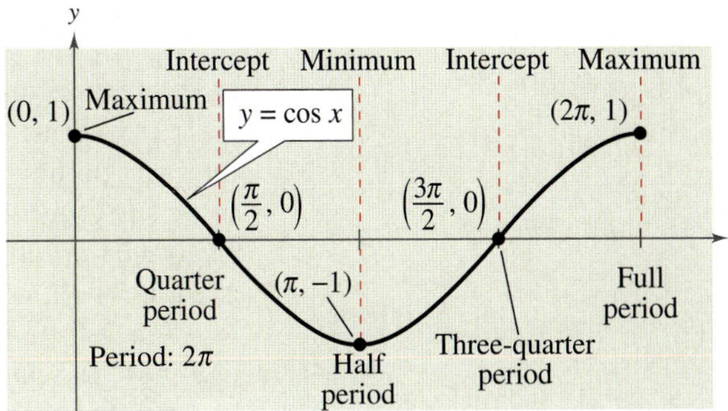

FIGURE 6.47

Example 1 ▶ Using Key Points to Sketch a Sine Curve

Sketch the graph of $y = 2 \sin x$ on the interval $[-\pi, 4\pi]$.

Solution

Note that

$$y = 2 \sin x = 2(\sin x)$$

indicates that the y-values for the key points will have twice the magnitude of those on the graph of $y = \sin x$. Divide the period 2π into four equal parts to get the key points for $y = 2 \sin x$.

Intercept	*Maximum*	*Intercept*	*Minimum*		*Intercept*
$(0, 0)$,	$\left(\frac{\pi}{2}, 2\right)$,	$(\pi, 0)$,	$\left(\frac{3\pi}{2}, -2\right)$,	and	$(2\pi, 0)$

By connecting these key points with a smooth curve and extending the curve in both directions over the interval $[-\pi, 4\pi]$, you obtain the graph shown in Figure 6.48.

Technology

When using a graphing utility to graph trigonometric functions, pay special attention to the viewing window you use. For instance, try graphing $y = [\sin(10x)]/10$ in the standard viewing window in *radian* mode. What do you observe? Use the *zoom* feature to find a viewing window that displays a good view of the graph.

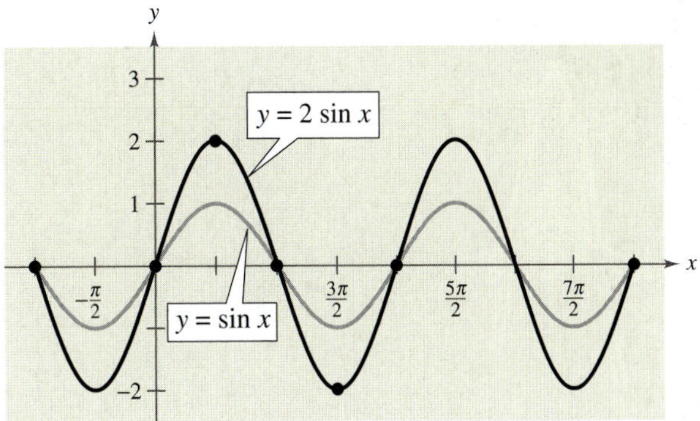

FIGURE 6.48

Amplitude and Period

In the remainder of this section you will study the graphic effect of each of the constants a, b, c, and d in equations of the forms

$$y = d + a \sin(bx - c)$$

and

$$y = d + a \cos(bx - c).$$

A quick review of the transformations you studied in Section 2.5 should help in this investigation.

The constant factor a in $y = a \sin x$ acts as a *scaling factor*—a *vertical stretch* or *vertical shrink* of the basic sine curve. If $|a| > 1$, the basic sine curve is stretched, and if $|a| < 1$, the basic sine curve is shrunk. The result is that the graph of $y = a \sin x$ ranges between $-a$ and a instead of between -1 and 1. The absolute value of a is the **amplitude** of the function $y = a \sin x$. The range of the function $y = a \sin x$ for $a > 0$ is $-a \le y \le a$.

Definition of Amplitude of Sine and Cosine Curves

The **amplitude** of $y = a \sin x$ and $y = a \cos x$ represents half the distance between the maximum and minimum values of the function and is given by

$$\text{Amplitude} = |a|.$$

Example 2 ▶ Scaling: Vertical Shrinking and Stretching 

On the same coordinate axes, sketch the graph of each function.

a. $y = \dfrac{1}{2} \cos x$

b. $y = 3 \cos x$

Solution

a. Because the amplitude of $y = \frac{1}{2} \cos x$ is $\frac{1}{2}$, the maximum value is $\frac{1}{2}$ and the minimum value is $-\frac{1}{2}$. Divide one cycle, $0 \le x \le 2\pi$, into four equal parts to get the key points

Maximum	*Intercept*	*Minimum*	*Intercept*		*Maximum*
$\left(0, \frac{1}{2}\right)$,	$\left(\frac{\pi}{2}, 0\right)$,	$\left(\pi, -\frac{1}{2}\right)$,	$\left(\frac{3\pi}{2}, 0\right)$,	and	$\left(2\pi, \frac{1}{2}\right)$.

b. A similar analysis shows that the amplitude of $y = 3 \cos x$ is 3, and the key points are

Maximum	*Intercept*	*Minimum*	*Intercept*		*Maximum*
$(0, 3)$,	$\left(\frac{\pi}{2}, 0\right)$,	$(\pi, -3)$,	$\left(\frac{3\pi}{2}, 0\right)$,	and	$(2\pi, 3)$.

The graphs of these two functions are shown in Figure 6.49.

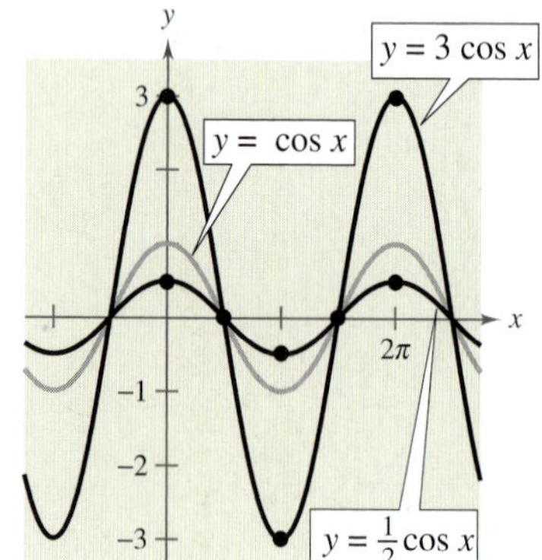

FIGURE 6.49

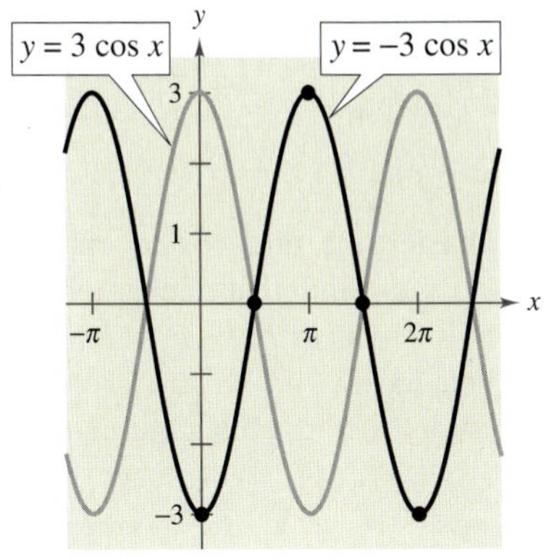

FIGURE 6.50

You know from Section 2.5 that the graph of $y = -f(x)$ is a **reflection** in the x-axis of the graph of $y = f(x)$. For instance, the graph of

$$y = -3 \cos x$$

is a reflection of the graph of

$$y = 3 \cos x$$

as shown in Figure 6.50.

Because $y = a \sin x$ completes one cycle from $x = 0$ to $x = 2\pi$, it follows that $y = a \sin bx$ completes one cycle from $x = 0$ to $x = 2\pi/b$.

Period of Sine and Cosine Functions

Let b be a positive real number. The **period** of $y = a \sin bx$ and $y = a \cos bx$ is $2\pi/b$.

Note that if $0 < b < 1$, the period of $y = a \sin bx$ is greater than 2π and represents a *horizontal stretching* of the graph of $y = a \sin x$. Similarly, if $b > 1$, the period of $y = a \sin bx$ is less than 2π and represents a *horizontal shrinking* of the graph of $y = a \sin x$. If b is negative, the identities $\sin(-x) = -\sin x$ and $\cos(-x) = \cos x$ are used to rewrite the function.

Exploration

Sketch the graph of $y = \cos bx$ for $b = \frac{1}{2}$, 2, and 3. How does the value of b affect the graph? How many complete cycles occur between 0 and 2π for each value of b?

Example 3 ▶ Scaling: Horizontal Stretching

Sketch the graph of $y = \sin \dfrac{x}{2}$.

Solution

The amplitude is 1. Moreover, because $b = \frac{1}{2}$, the period is

$$\frac{2\pi}{b} = \frac{2\pi}{\frac{1}{2}} = 4\pi. \qquad \text{Substitute for } b.$$

Now, divide the period-interval $[0, 4\pi]$ into four equal parts with the values π, 2π, and 3π to obtain the key points on the graph.

Intercept	*Maximum*	*Intercept*	*Minimum*		*Intercept*
$(0, 0)$,	$(\pi, 1)$,	$(2\pi, 0)$,	$(3\pi, -1)$,	and	$(4\pi, 0)$

The graph is shown in Figure 6.51.

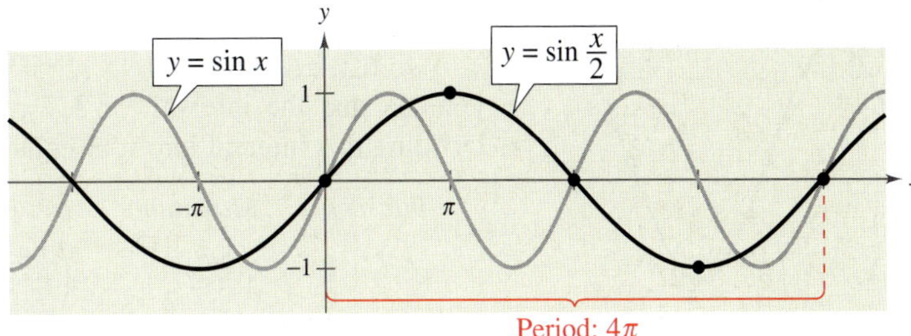

FIGURE 6.51

STUDY TIP

In general, to divide a period-interval into four equal parts, successively add "period/4," starting with the left endpoint of the interval. For instance, for the period-interval $[-\pi/6, \pi/2]$ of length $2\pi/3$, you would successively add

$$\frac{2\pi/3}{4} = \frac{\pi}{6}$$

to get $-\pi/6, 0, \pi/6, \pi/3$, and $\pi/2$.

Translations of Sine and Cosine Curves

The constant c in the general equations

$$y = a\sin(bx - c) \quad \text{and} \quad y = a\cos(bx - c)$$

creates a *horizontal translation* (shift) of the basic sine and cosine curves. Comparing $y = a\sin bx$ with $y = a\sin(bx - c)$, you find that the graph of $y = a\sin(bx - c)$ completes one cycle from $bx - c = 0$ to $bx - c = 2\pi$. By solving for x, you can find the interval for one cycle to be

$$\underbrace{\frac{c}{b}}_{\text{Left endpoint}} \le x \le \underbrace{\frac{c}{b} + \underbrace{\frac{2\pi}{b}}_{\text{Period}}}_{\text{Right endpoint}}.$$

This implies that the period of $y = a\sin(bx - c)$ is $2\pi/b$, and the graph of $y = a\sin bx$ is shifted by an amount c/b. The number c/b is the **phase shift**.

> **Exploration**
>
> Sketch the graph of
>
> $$y = \sin(x - c)$$
>
> where $c = -\pi/4$, 0, and $\pi/4$. How does the value of c affect the graph?

Graphs of Sine and Cosine Functions

The graphs of $y = a\sin(bx - c)$ and $y = a\cos(bx - c)$ have the following characteristics. (Assume $b > 0$.)

$$\text{Amplitude} = |a| \qquad \text{Period} = \frac{2\pi}{b}$$

The left and right endpoints of a one-cycle interval can be determined by solving the equations $bx - c = 0$ and $bx - c = 2\pi$.

Example 4 ▶ Horizontal Translation 

Sketch the graph of $y = \frac{1}{2}\sin(x - \pi/3)$.

Solution

The amplitude is $\frac{1}{2}$ and the period is 2π. By solving the equations

$$x - \frac{\pi}{3} = 0 \quad \Longrightarrow \quad x = \frac{\pi}{3}$$

and

$$x - \frac{\pi}{3} = 2\pi \quad \Longrightarrow \quad x = \frac{7\pi}{3}$$

you see that the interval $[\pi/3, 7\pi/3]$ corresponds to one cycle of the graph. Dividing this interval into four equal parts produces the key points

Intercept	*Maximum*	*Intercept*	*Minimum*		*Intercept*
$\left(\frac{\pi}{3}, 0\right)$,	$\left(\frac{5\pi}{6}, \frac{1}{2}\right)$,	$\left(\frac{4\pi}{3}, 0\right)$,	$\left(\frac{11\pi}{6}, -\frac{1}{2}\right)$,	and	$\left(\frac{7\pi}{3}, 0\right)$.

The graph is shown in Figure 6.52.

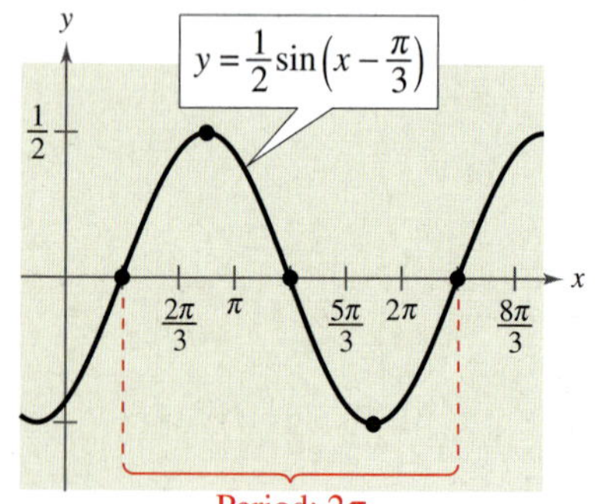

FIGURE 6.52

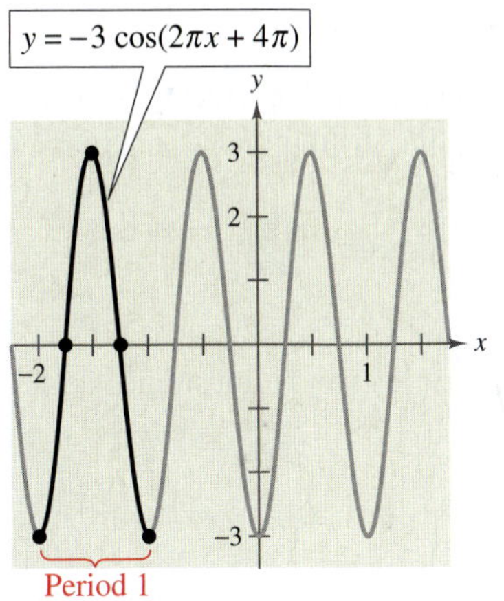

FIGURE 6.53

Example 5 ▶ Horizontal Translation 

Sketch the graph of

$$y = -3\cos(2\pi x + 4\pi).$$

Solution

The amplitude is 3 and the period is $2\pi/2\pi = 1$. By solving the equations

$$2\pi x + 4\pi = 0$$
$$2\pi x = -4\pi$$
$$x = -2$$

and

$$2\pi x + 4\pi = 2\pi$$
$$2\pi x = -2\pi$$
$$x = -1$$

you see that the interval $[-2, -1]$ corresponds to one cycle of the graph. Dividing this interval into four equal parts produces the key points

Minimum	*Intercept*	*Maximum*	*Intercept*		*Minimum*
$(-2, -3)$,	$\left(-\frac{7}{4}, 0\right)$,	$\left(-\frac{3}{2}, 3\right)$,	$\left(-\frac{5}{4}, 0\right)$,	and	$(-1, -3)$.

The graph is shown in Figure 6.53.

The final type of transformation is the *vertical translation* caused by the constant d in the equations

$$y = d + a\sin(bx - c)$$

and

$$y = d + a\cos(bx - c).$$

The shift is d units upward for $d > 0$ and downward for $d < 0$. In other words, the graph oscillates about the horizontal line $y = d$ instead of the x-axis.

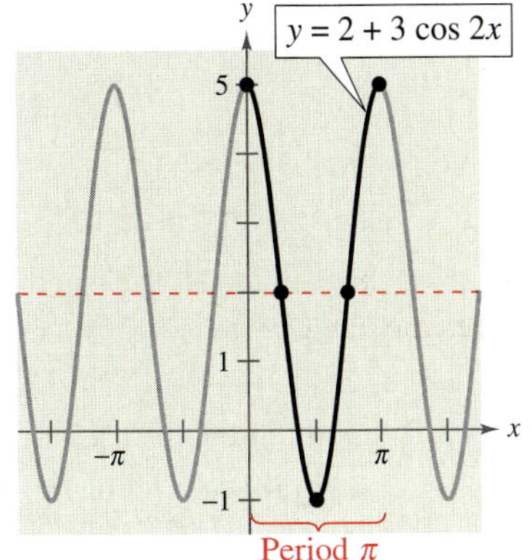

FIGURE 6.54

Example 6 ▶ Vertical Translation 

Sketch the graph of

$$y = 2 + 3\cos 2x.$$

Solution

The amplitude is 3 and the period is π. The key points over the interval $[0, \pi]$ are

$$(0, 5), \quad \left(\frac{\pi}{4}, 2\right), \quad \left(\frac{\pi}{2}, -1\right), \quad \left(\frac{3\pi}{4}, 2\right), \quad \text{and} \quad (\pi, 5).$$

The graph is shown in Figure 6.54.

Mathematical Modeling

Sine and cosine functions can be used to model many real-life situations, including electric currents, musical tones, radio waves, tides, and weather patterns.

Example 7 ▶ Finding a Trigonometric Model

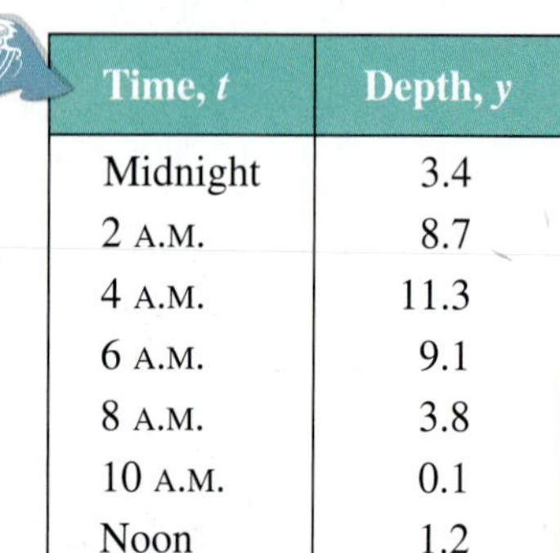

Time, t	Depth, y
Midnight	3.4
2 A.M.	8.7
4 A.M.	11.3
6 A.M.	9.1
8 A.M.	3.8
10 A.M.	0.1
Noon	1.2

Throughout the day, the depth of water at the end of a dock in Bar Harbor, Maine, varies with the tides. The table shows the depths (in feet) at various times during the morning. (Source: Nautical Software, Inc.)

a. Use a trigonometric function to model this data.

b. Find the depths at 9 A.M. and 3 P.M.

c. A boat needs at least 10 feet of water to moor at the dock. During what times in the afternoon can it safely dock?

Solution

a. Begin by graphing the data, as shown in Figure 6.55. You can use either a sine or cosine model. Suppose you use a cosine model of the form

$$y = a\cos(bt - c) + d.$$

The amplitude is given by

$$a = \frac{1}{2}[(\text{high}) - (\text{low})] = \frac{1}{2}(11.3 - 0.1) = 5.6.$$

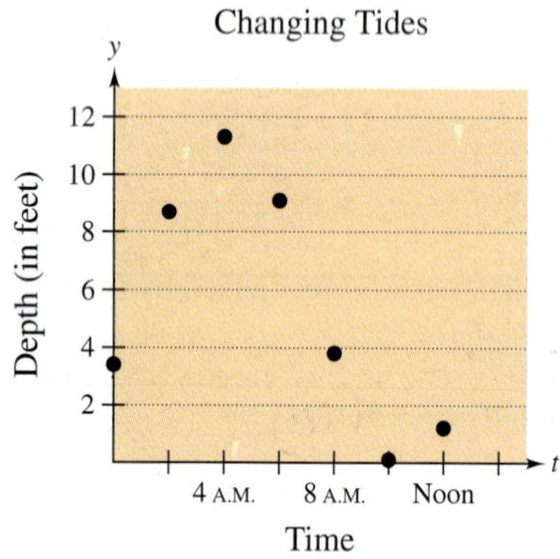

FIGURE 6.55

The period is

$$p = 2[(\text{low time}) - (\text{high time})] = 2(10 - 4) = 12$$

which implies that

$$b = \frac{2\pi}{p} \approx 0.524.$$

Because high tide occurs 4 hours after midnight, you can conclude that $c/b = 4$, so $c \approx 2.094$. Moreover, because the average depth is $\frac{1}{2}(11.3 + 0.1) = 5.7$, it follows that $d = 5.7$. So, you can model the depth with the function

$$y = 5.6\cos(0.524t - 2.094) + 5.7.$$

b. The depths at 9 A.M. and 3 P.M. are as follows.

$$y = 5.6\cos(0.524 \cdot 9 - 2.094) + 5.7$$

$$\approx 0.84 \text{ foot} \qquad \text{9 A.M.}$$

$$y = 5.6\cos(0.524 \cdot 15 - 2.094) + 5.7$$

$$\approx 10.57 \text{ feet} \qquad \text{3 P.M.}$$

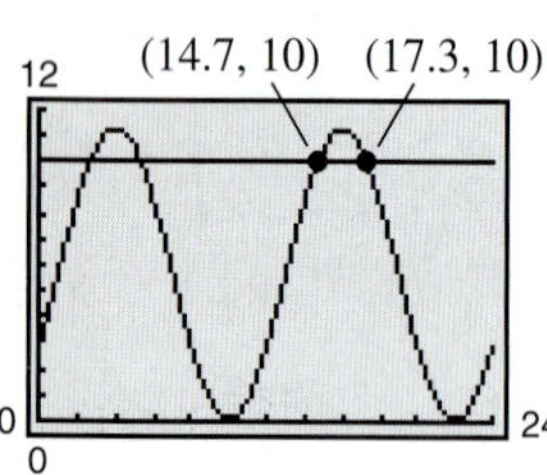

FIGURE 6.56

c. To find out when the depth y is at least 10 feet, you can graph the model with the line $y = 10$, as shown in Figure 6.56. From the graph, it follows that the depth is at least 10 feet between 2:42 P.M. ($t \approx 14.7$) and 5:18 P.M. ($t \approx 17.3$).

6.4 Exercises

In Exercises 1–14, find the period and amplitude.

1. $y = 3 \sin 2x$

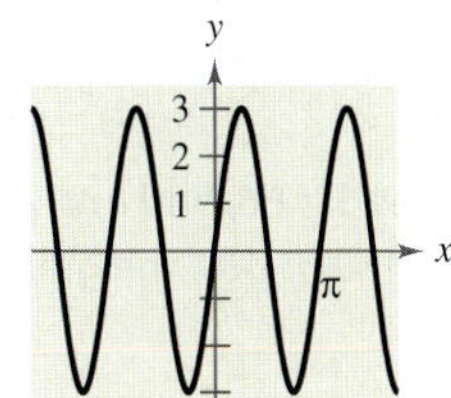

2. $y = 2 \cos 3x$

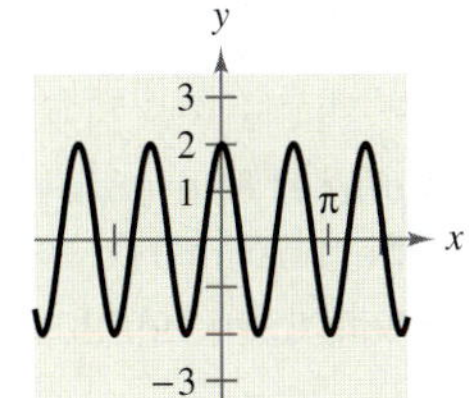

3. $y = \frac{5}{2} \cos \frac{x}{2}$

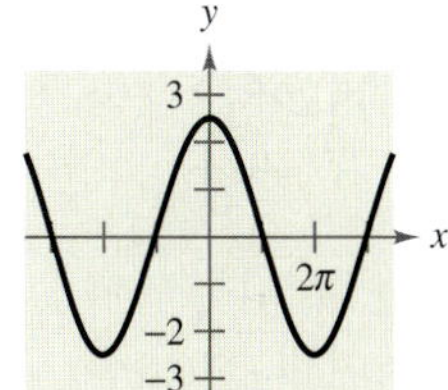

4. $y = -3 \sin \frac{x}{3}$

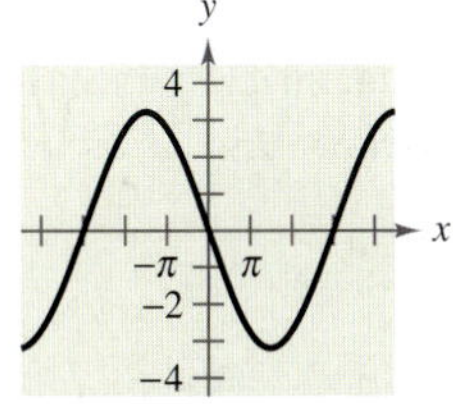

5. $y = \frac{1}{2} \sin \frac{\pi x}{3}$

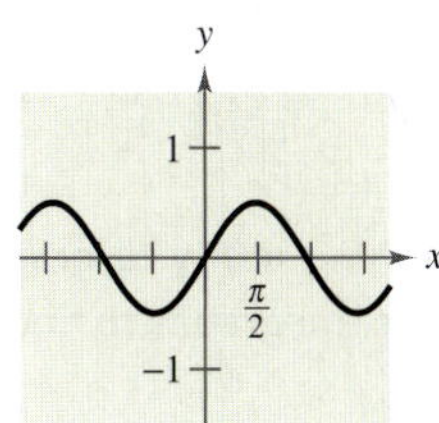

6. $y = \frac{3}{2} \cos \frac{\pi x}{2}$

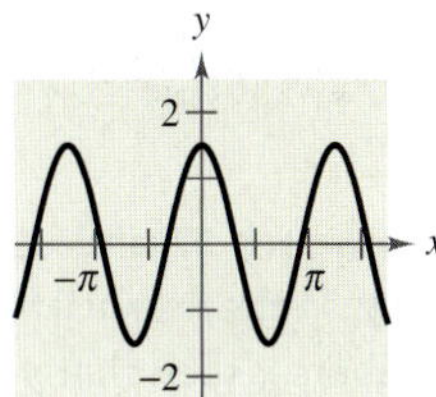

7. $y = -2 \sin x$
8. $y = -\cos \frac{2x}{3}$
9. $y = 3 \sin 10x$
10. $y = \frac{1}{3} \sin 8x$
11. $y = \frac{1}{2} \cos \frac{2x}{3}$
12. $y = \frac{5}{2} \cos \frac{x}{4}$
13. $y = \frac{1}{4} \sin 2\pi x$
14. $y = \frac{2}{3} \cos \frac{\pi x}{10}$

In Exercises 15–22, describe the relationship between the graphs of *f* and *g*. Consider amplitude, period, and shifts.

15. $f(x) = \sin x$
    $g(x) = \sin(x - \pi)$
16. $f(x) = \cos x$
    $g(x) = \cos(x + \pi)$
17. $f(x) = \cos 2x$
    $g(x) = -\cos 2x$
18. $f(x) = \sin 3x$
    $g(x) = \sin(-3x)$
19. $f(x) = \cos x$
    $g(x) = \cos 2x$
20. $f(x) = \sin x$
    $g(x) = \sin 3x$
21. $f(x) = \sin 2x$
    $g(x) = 3 + \sin 2x$
22. $f(x) = \cos 4x$
    $g(x) = -2 + \cos 4x$

In Exercises 23–26, describe the relationship between the graphs of *f* and *g*. Consider amplitude, period, and shifts.

23.

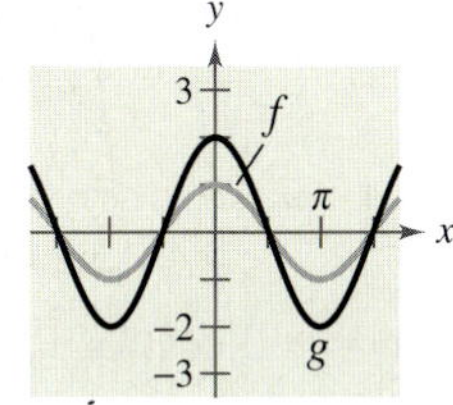

24.

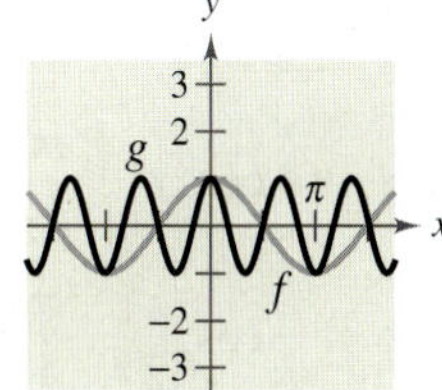

25.

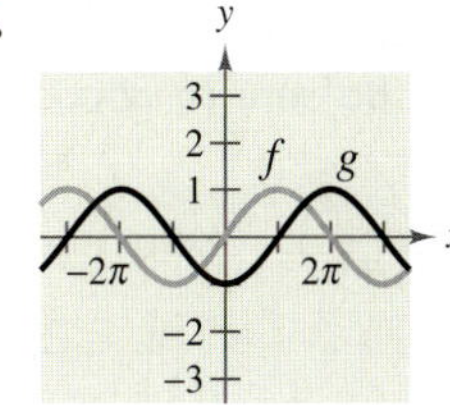

26.

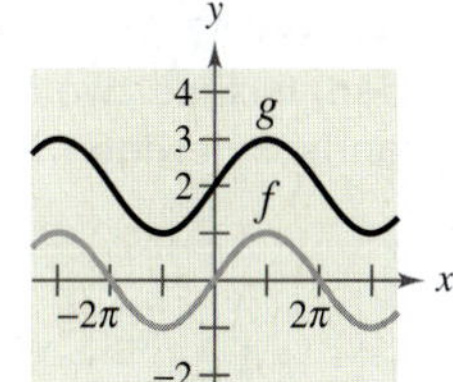

In Exercises 27–34, graph *f* and *g* on the same set of coordinate axes. (Include two full periods.)

27. $f(x) = -2 \sin x$
    $g(x) = 4 \sin x$
28. $f(x) = \sin x$
    $g(x) = \sin \frac{x}{3}$
29. $f(x) = \cos x$
    $g(x) = 1 + \cos x$
30. $f(x) = 2 \cos 2x$
    $g(x) = -\cos 4x$
31. $f(x) = -\frac{1}{2} \sin \frac{x}{2}$
    $g(x) = 3 - \frac{1}{2} \sin \frac{x}{2}$
32. $f(x) = 4 \sin \pi x$
    $g(x) = 4 \sin \pi x - 3$
33. $f(x) = 2 \cos x$
    $g(x) = 2 \cos(x + \pi)$
34. $f(x) = -\cos x$
    $g(x) = -\cos(x - \pi)$

In Exercises 35–52, sketch the graph of the function. (Include two full periods.)

35. $y = -2 \sin 6x$
36. $y = -3 \cos 4x$

37. $y = \cos 2\pi x$

38. $y = \sin \dfrac{\pi x}{4}$

39. $y = -\sin \dfrac{2\pi x}{3}$

40. $y = -10 \cos \dfrac{\pi x}{6}$

41. $y = \sin\left(x - \dfrac{\pi}{4}\right)$

42. $y = \sin(x - \pi)$

43. $y = 3 \cos(x + \pi)$

44. $y = 4 \cos\left(x + \dfrac{\pi}{4}\right)$

45. $y = 2 - \sin \dfrac{2\pi x}{3}$

46. $y = -3 + 5 \cos \dfrac{\pi t}{12}$

47. $y = 2 + \frac{1}{10} \cos 60\pi x$

48. $y = 2 \cos x - 3$

49. $y = 3 \cos(x + \pi) - 3$

50. $y = 4 \cos\left(x + \dfrac{\pi}{4}\right) + 4$

51. $y = \dfrac{2}{3} \cos\left(\dfrac{x}{2} - \dfrac{\pi}{4}\right)$

52. $y = -3 \cos(6x + \pi)$

In Exercises 53–58, use a graphing utility to graph the function. Include two full periods. Be sure to choose an appropriate viewing window.

53. $y = -2 \sin(4x + \pi)$

54. $y = -4 \sin\left(\dfrac{2}{3}x - \dfrac{\pi}{3}\right)$

55. $y = \cos\left(2\pi x - \dfrac{\pi}{2}\right) + 1$

56. $y = 3 \cos\left(\dfrac{\pi x}{2} + \dfrac{\pi}{2}\right) - 2$

57. $y = -0.1 \sin\left(\dfrac{\pi x}{10} + \pi\right)$

58. $y = \frac{1}{100} \sin 120\pi t$

Graphical Reasoning **In Exercises 59–62, find *a* and *d* for the function $f(x) = a \cos x + d$ such that the graph of *f* matches the figure.**

59.

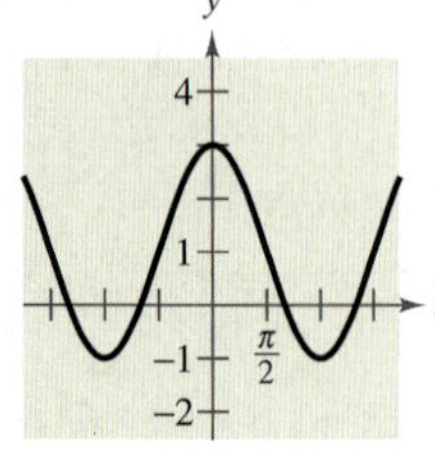

60.

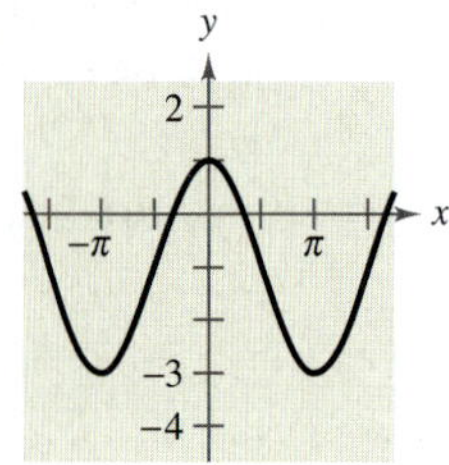

61.

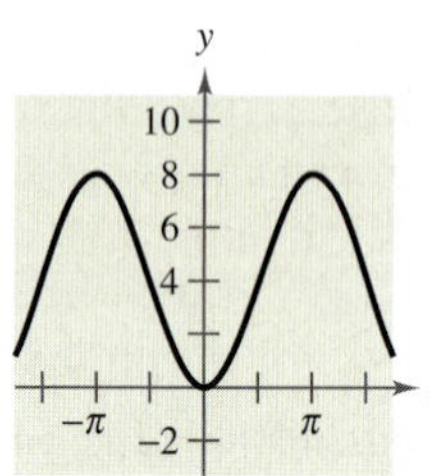

62.

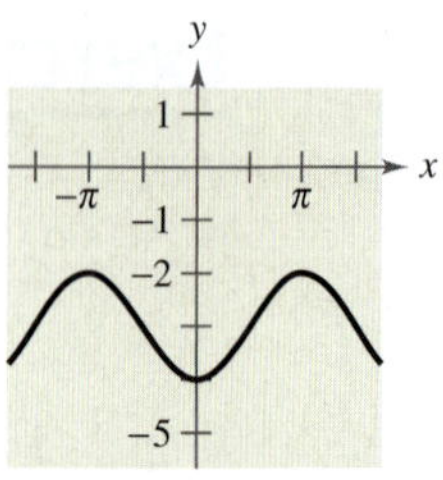

Graphical Reasoning **In Exercises 63–66, find *a*, *b*, and *c* for the function $f(x) = a \sin(bx - c)$ such that the graph of *f* matches the figure.**

63.

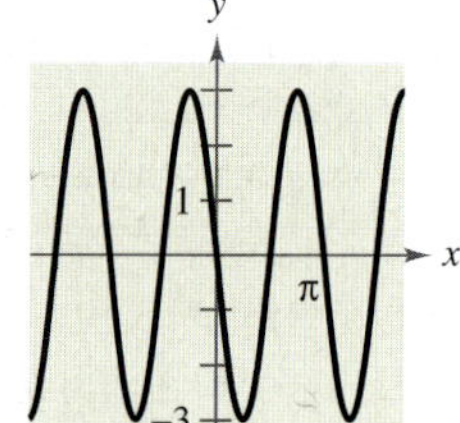

64.

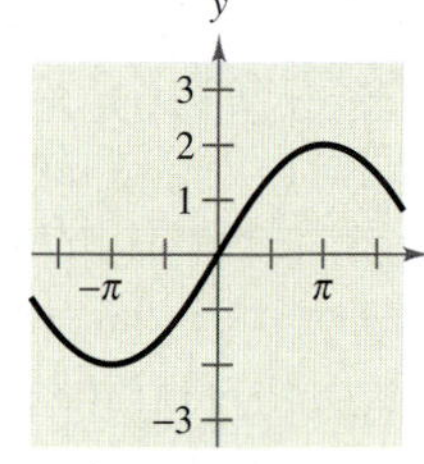

65.

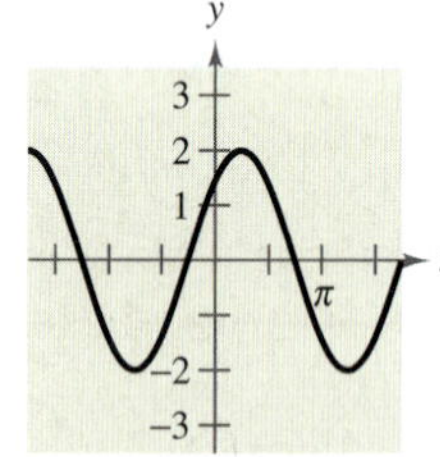

66.

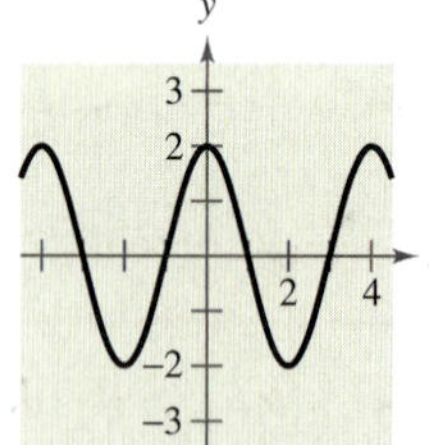

In Exercises 67 and 68, use a graphing utility to graph y_1 and y_2 in the interval $[-2\pi, 2\pi]$. Use the graphs to find real numbers x such that $y_1 = y_2$.

67. $y_1 = \sin x$
$y_2 = -\frac{1}{2}$

68. $y_1 = \cos x$
$y_2 = -1$

69. ***Respiratory Cycle*** For a person at rest, the velocity v (in liters per second) of air flow during a respiratory cycle (the time from the beginning of one breath to the beginning of the next) is

$$v = 0.85 \sin \frac{\pi t}{3}$$

where t is the time in seconds. (Inhalation occurs when $v > 0$, and exhalation occurs when $v < 0$.)

(a) Find the time for one full respiratory cycle.

(b) Find the number of cycles per minute.

(c) Sketch the graph of the velocity function.

70. *Respiratory Cycle* After exercising for a few minutes, a person has a respiratory cycle for which the velocity of air flow is approximated by

$$v = 1.75 \sin \frac{\pi t}{2}$$

where t is the time in seconds. (Inhalation occurs when $v > 0$, and exhalation occurs when $v < 0$.)

(a) Find the time for one full respiratory cycle.

(b) Find the number of cycles per minute.

(c) Sketch the graph of the velocity function.

71. *Piano Tuning* When tuning a piano, a technician strikes a tuning fork for the A above middle C and sets up a wave motion that can be approximated by $y = 0.001 \sin 880\pi t$, where t is the time in seconds.

(a) What is the period of the function?

(b) The frequency f is given by $f = 1/p$. What is the frequency of the note?

72. *Health* The function

$$P = 100 - 20 \cos \frac{5\pi t}{3}$$

approximates the blood pressure P in millimeters of mercury at time t in seconds for a person at rest.

(a) Find the period of the function.

(b) Find the number of heartbeats per minute.

73. *Data Analysis* The table shows the maximum daily high temperatures for Tallahassee T and Chicago C (in degrees Fahrenheit) for month t, with $t = 1$ corresponding to January. (Source: Southeast Regional Climate Center and National Climatic Data Center)

Month, t	Tallahassee, T	Chicago, C
1	63.9	29.0
2	67.1	33.5
3	73.1	45.8
4	79.9	58.6
5	86.5	70.1
6	90.5	79.6
7	91.3	83.7
8	91.0	81.8
9	88.1	74.8
10	80.8	63.3
11	72.2	48.4
12	65.5	34.0

(a) A model for the temperature in Tallahassee is

$$T(t) = 77.60 + 13.70 \cos\left(\frac{\pi t}{6} - 3.67\right).$$

Find a trigonometric model for Chicago.

(b) Use a graphing utility to graph the data points and the model for the temperatures in Tallahassee. How well does the model fit the data?

(c) Use a graphing utility to graph the data points and the model for the temperatures in Chicago. How well does the model fit the data?

(d) Use the models to estimate the average maximum temperature in each city. Which term of the models did you use? Explain.

(e) What is the period of each model? Are the periods what you expected? Explain.

(f) Which city has the greater variability in temperature throughout the year? Which factor of the models determines this variability? Explain.

▶ Model It

74. *Data Analysis* The percent y of the moon's face that is illuminated on day x of the year 2005, where $x = 70$ represents March 11, is shown in the table. (Source: U.S. Naval Observatory)

x	y
76	0.5
84	1.0
91	0.5
98	0.0
106	0.5
114	1.0

(a) Create a scatter plot of the data.

(b) Find a trigonometric model that fits the data.

(c) Add the graph of your model in part (b) to the scatter plot. How well does the model fit the data?

(d) What is the period of the model?

(e) Estimate the moon's percent illumination for May 8, 2005.

75. ***Fuel Consumption*** The daily consumption C (in gallons) of diesel fuel on a farm is modeled by

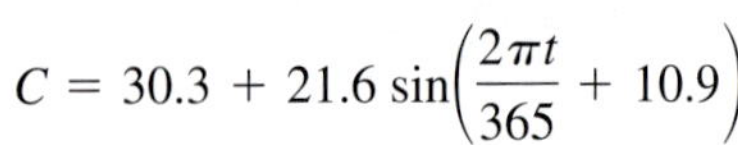

$$C = 30.3 + 21.6 \sin\left(\frac{2\pi t}{365} + 10.9\right)$$

where t is the time in days, with $t = 1$ corresponding to January 1.

(a) What is the period of the model? Is it what you expected? Explain.

(b) What is the average daily fuel consumption? Which term of the model did you use? Explain.

(c) Use a graphing utility to graph the model. Use the graph to approximate the time of the year when consumption exceeds 40 gallons per day.

Synthesis

True or False? **In Exercises 76–78, determine whether the statement is true or false. Justify your answer.**

76. The graph of the function $f(x) = \sin(x + 2\pi)$ translates the graph of $f(x) = \sin x$ exactly one period to the right so that the two graphs look identical.

77. The function $y = \frac{1}{2}\cos 2x$ has an amplitude that is twice that of the function $y = \cos x$.

78. The graph of $y = -\cos x$ is a reflection of the graph of $y = \sin(x + \pi/2)$ in the x-axis.

Conjecture **In Exercises 79 and 80, graph f and g on the same set of coordinate axes. Include two full periods. Make a conjecture about the functions.**

79. $f(x) = \sin x, \quad g(x) = \cos\left(x - \frac{\pi}{2}\right)$

80. $f(x) = \sin x, \quad g(x) = -\cos\left(x + \frac{\pi}{2}\right)$

81. ***Writing*** Use a graphing utility to graph the function $y = a \sin x$ for $a = \frac{1}{2}$, $a = \frac{3}{2}$, and $a = -3$. Write a paragraph describing the changes in the graph corresponding to the specified changes in a.

82. ***Writing*** Use a graphing utility to graph the function $y = d + \sin x$ for $d = 2$, $d = 3.5$, and $d = -2$. Write a paragraph describing the changes in the graph corresponding to the specified changes in d.

83. ***Writing*** Use a graphing utility to graph the function $y = \sin bx$ for $b = \frac{1}{2}$, $b = \frac{3}{2}$, and $b = 4$. Write a paragraph describing the changes in the graph corresponding to the specified changes in b.

84. ***Writing*** Use a graphing utility to graph the function $y = \sin(x - c)$ for $c = 1$, $c = 3$, and $c = -2$. Write a paragraph describing the changes in the graph corresponding to the specified changes in c.

85. ***Exploration*** Using calculus, it can be shown that the sine and cosine functions can be approximated by the polynomials

$$\sin x \approx x - \frac{x^3}{3!} + \frac{x^5}{5!} \text{ and } \cos x \approx 1 - \frac{x^2}{2!} + \frac{x^4}{4!}$$

where x is in radians.

(a) Use a graphing utility to graph the sine function and its polynomial approximation in the same viewing window. How do the graphs compare?

(b) Use a graphing utility to graph the cosine function and its polynomial approximation in the same viewing window. How do the graphs compare?

(c) Study the patterns in the polynomial approximations of the sine and cosine functions and guess the next term in each. Then repeat parts (a) and (b). How did the accuracy of the approximations change when additional terms were added?

86. ***Exploration*** Use the polynomial approximations for the sine and cosine functions from Exercise 85 to approximate the following functional values. Compare the results with those given by a calculator. Is the error in the approximation the same in each case? Explain.

(a) $\sin \frac{1}{2}$ (b) $\sin 1$ (c) $\sin \frac{\pi}{6}$

(d) $\cos(-0.5)$ (e) $\cos 1$ (f) $\cos \frac{\pi}{4}$

Review

In Exercises 87–90, use the properties of logarithms to write the expression as a sum, difference, and/or constant multiple of a logarithm.

87. $\log_{10} \sqrt{x - 2}$

88. $\log_2[x^2(x - 3)]$

89. $\ln \frac{t^3}{t - 1}$

90. $\ln \sqrt{\frac{z}{z^2 + 1}}$

In Exercises 91–94, write the expression as the logarithm of a single quantity.

91. $\frac{1}{2}(\log_{10} x + \log_{10} y)$

92. $2 \log_2 x + \log_2(xy)$

93. $\ln 3x - 4 \ln y$

94. $\frac{1}{2}(\ln 2x - 2 \ln x) + 3 \ln x$

6.5 Graphs of Other Trigonometric Functions

▶ **What you should learn**

- How to sketch the graphs of tangent functions
- How to sketch the graphs of cotangent functions
- How to sketch the graphs of secant and cosecant functions
- How to sketch the graphs of damped trigonometric functions

▶ **Why you should learn it**

Trigonometric functions can be used to model biological patterns. For instance, in Exercise 75 on page 508, trigonometric functions can be used to model predator-prey cycles.

Tom Brakefield/Corbis

Graph of the Tangent Function

Recall that the tangent function is odd. That is, $\tan(-x) = -\tan x$. Consequently, the graph of $y = \tan x$ is symmetric with respect to the origin. You also know from the identity $\tan x = \sin x/\cos x$ that the tangent is undefined for values at which $\cos x = 0$. Two such values are $x = \pm\pi/2 \approx \pm 1.5708$.

x	$-\frac{\pi}{2}$	-1.57	-1.5	$-\frac{\pi}{4}$	0	$\frac{\pi}{4}$	1.5	1.57	$\frac{\pi}{2}$
$\tan x$	Undef.	-1255.8	-14.1	-1	0	1	14.1	1255.8	Undef.

As indicated in the table, $\tan x$ increases without bound as x approaches $\pi/2$ from the left, and decreases without bound as x approaches $-\pi/2$ from the right. So, the graph of $y = \tan x$ has *vertical asymptotes* at $x = \pi/2$ and $x = -\pi/2$, as shown in Figure 6.57. Moreover, because the period of the tangent function is π, vertical asymptotes also occur when $x = \pi/2 + n\pi$, where n is an integer. The domain of the tangent function is the set of all real numbers other than $x = \pi/2 + n\pi$, and the range is the set of all real numbers.

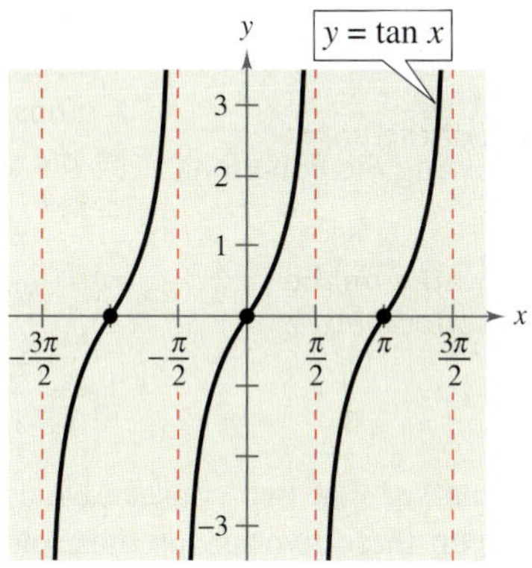

FIGURE 6.57

PERIOD: π
DOMAIN: ALL $x \neq \frac{\pi}{2} + n\pi$
RANGE: $(-\infty, \infty)$
VERTICAL ASYMPTOTES: $x = \frac{\pi}{2} + n\pi$

Sketching the graph of $y = a\tan(bx - c)$ is similar to sketching the graph of $y = a\sin(bx - c)$ in that you locate key points that identify the intercepts and asymptotes. Two consecutive asymptotes can be found by solving the equations

$$bx - c = -\frac{\pi}{2} \quad \text{and} \quad bx - c = \frac{\pi}{2}.$$

The midpoint between two consecutive asymptotes is an x-intercept of the graph. The period of the function $y = a\tan(bx - c)$ is the distance between two consecutive asymptotes. The amplitude of a tangent function is not defined. After plotting the asymptotes and the x-intercept, plot a few additional points between the two asymptotes and sketch one cycle. Finally, sketch one or two additional cycles to the left and right.

Example 1 ▶ Sketching the Graph of a Tangent Function

Sketch the graph of $y = \tan \frac{x}{2}$.

Solution

By solving the equations

$$\frac{x}{2} = -\frac{\pi}{2} \quad \text{and} \quad \frac{x}{2} = \frac{\pi}{2}$$

$$x = -\pi \qquad\qquad x = \pi$$

you can see that two consecutive asymptotes occur at $x = -\pi$ and $x = \pi$. Between these two asymptotes, plot a few points, including the x-intercept, as shown in the table. Three cycles of the graph are shown in Figure 6.58.

x	$-\frac{\pi}{2}$	0	$\frac{\pi}{2}$
$\tan \frac{x}{2}$	-1	0	1

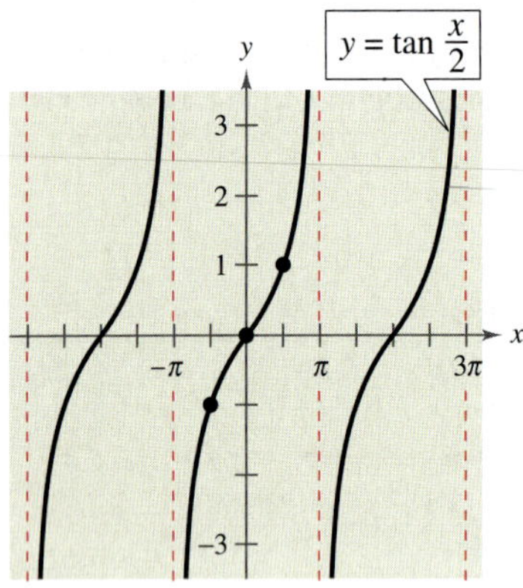

FIGURE 6.58

Example 2 ▶ Sketching the Graph of a Tangent Function

Sketch the graph of $y = -3 \tan 2x$.

Solution

By solving the equations

$$2x = -\frac{\pi}{2} \quad \text{and} \quad 2x = \frac{\pi}{2}$$

$$x = -\frac{\pi}{4} \qquad\qquad x = \frac{\pi}{4}$$

you can see that two consecutive asymptotes occur at $x = -\pi/4$ and $x = \pi/4$. Between these two asymptotes, plot a few points, including the x-intercept, as shown in the table. Three cycles of the graph are shown in Figure 6.59.

x	$-\frac{\pi}{8}$	0	$\frac{\pi}{8}$
$-3 \tan 2x$	3	0	-3

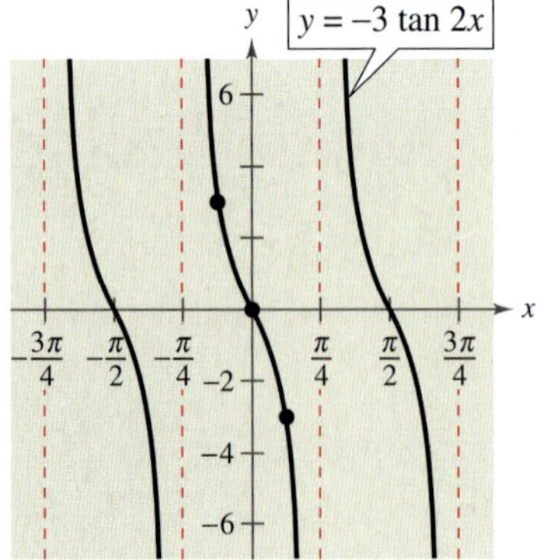

FIGURE 6.59

By comparing the graphs in Examples 1 and 2, you can see that the graph of $y = a \tan(bx - c)$ increases between consecutive vertical asymptotes when $a > 0$, and decreases between consecutive vertical asymptotes when $a < 0$. In other words, the graph for $a < 0$ is a reflection in the x-axis of the graph for $a > 0$.

Graph of the Cotangent Function

The graph of the cotangent function is similar to the graph of the tangent function. It also has a period of π. However, from the identity

$$y = \cot x = \frac{\cos x}{\sin x}$$

you can see that the cotangent function has vertical asymptotes when $\sin x$ is zero, which occurs at $x = n\pi$, where n is an integer. The graph of the cotangent function is shown in Figure 6.60.

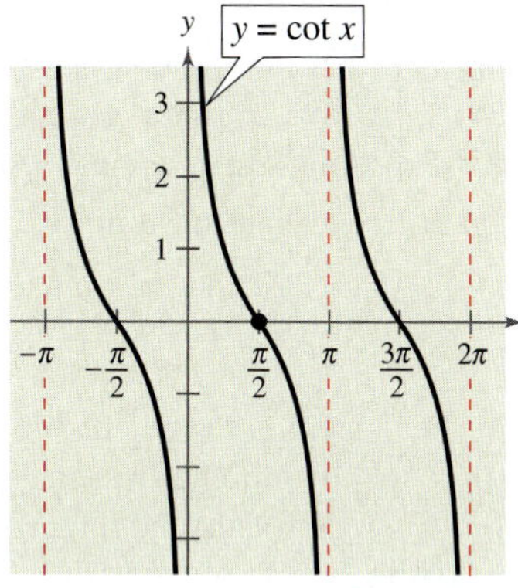

FIGURE 6.60

PERIOD: π
DOMAIN: ALL $x \neq n\pi$
RANGE: $(-\infty, \infty)$
VERTICAL ASYMPTOTES: $x = n\pi$

Example 3 ▶ Sketching the Graph of a Cotangent Function

Sketch the graph of $y = 2 \cot \frac{x}{3}$.

Solution

To locate two consecutive vertical asymptotes of the graph, solve the equations $x/3 = 0$ and $x/3 = \pi$, as follows.

$$\frac{x}{3} = 0 \qquad \text{and} \qquad \frac{x}{3} = \pi$$

$$x = 0 \qquad\qquad x = 3\pi$$

Then, between these two asymptotes, plot a few points, including the x-intercept, as shown in the table. Three cycles of the graph are shown in Figure 6.61. (Note that the period is 3π, the distance between consecutive asymptotes.)

x	$\frac{3\pi}{4}$	$\frac{3\pi}{2}$	$\frac{9\pi}{4}$
$2 \cot \frac{x}{3}$	2	0	-2

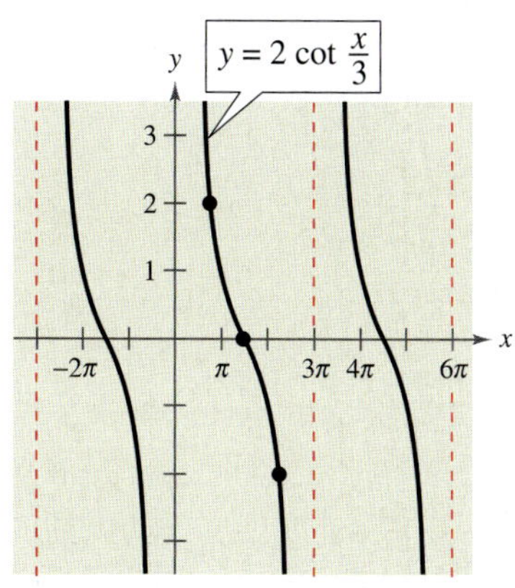

FIGURE 6.61

Graphs of the Reciprocal Functions 

The graphs of the two remaining trigonometric functions can be obtained from the graphs of the sine and cosine functions using the reciprocal identities

$$\csc x = \frac{1}{\sin x} \quad \text{and} \quad \sec x = \frac{1}{\cos x}.$$

For instance, at a given value of x, the y-coordinate of $\sec x$ is the reciprocal of the y-coordinate of $\cos x$. Of course, when $\cos x = 0$, the reciprocal does not exist. Near such values of x, the behavior of the secant function is similar to that of the tangent function. In other words, the graphs of

$$\tan x = \frac{\sin x}{\cos x} \quad \text{and} \quad \sec x = \frac{1}{\cos x}$$

have vertical asymptotes at $x = \pi/2 + n\pi$, where n is an integer, and the cosine is zero at these x-values. Similarly,

$$\cot x = \frac{\cos x}{\sin x} \quad \text{and} \quad \csc x = \frac{1}{\sin x}$$

have vertical asymptotes where $\sin x = 0$—that is, at $x = n\pi$.

To sketch the graph of a secant or cosecant function, you should first make a sketch of its reciprocal function. For instance, to sketch the graph of $y = \csc x$, first sketch the graph of $y = \sin x$. Then take reciprocals of the y-coordinates to obtain points on the graph of $y = \csc x$. This procedure is used to obtain the graphs shown in Figure 6.62.

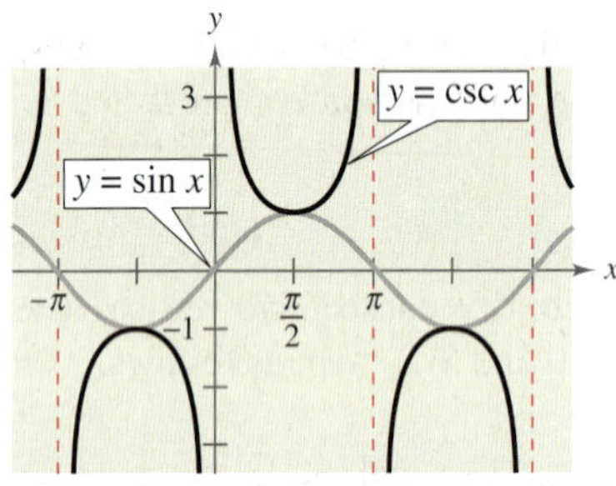

PERIOD: 2π
DOMAIN: ALL $x \neq n\pi$
RANGE: $(-\infty, -1]$ AND $[1, \infty)$
VERTICAL ASYMPTOTES: $x = n\pi$
SYMMETRY: ORIGIN

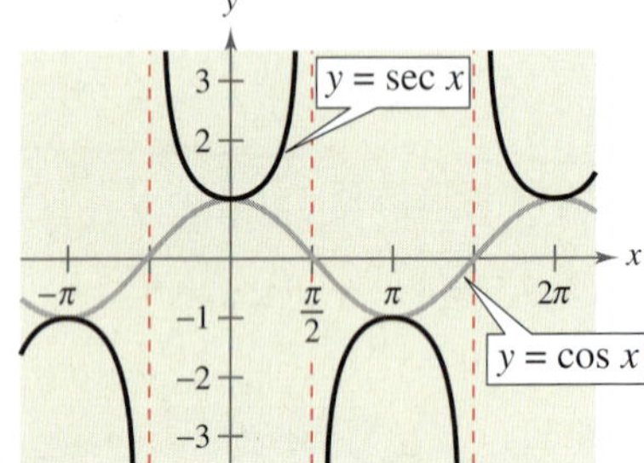

PERIOD: 2π
DOMAIN: ALL $x \neq \frac{\pi}{2} + n\pi$
RANGE: $(-\infty, -1]$ AND $[1, \infty)$
VERTICAL ASYMPTOTES: $x = \frac{\pi}{2} + n\pi$
SYMMETRY: y-AXIS

FIGURE 6.62

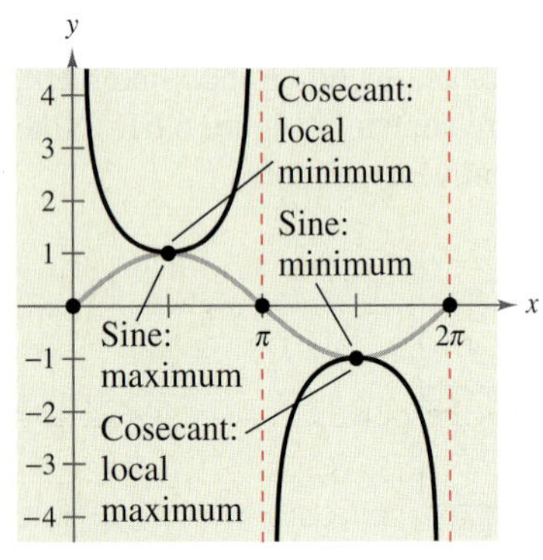

FIGURE 6.63

In comparing the graphs of the cosecant and secant functions with those of the sine and cosine functions, note that the "hills" and "valleys" are interchanged. For example, a hill (or maximum point) on the sine curve corresponds to a valley (a local minimum) on the cosecant curve, and a valley (or minimum point) on the sine curve corresponds to a hill (a local maximum) on the cosecant curve, as shown in Figure 6.63. Additionally, x-intercepts of the sine and cosine functions become vertical asymptotes of the cosecant and secant functions (see Figure 6.63).

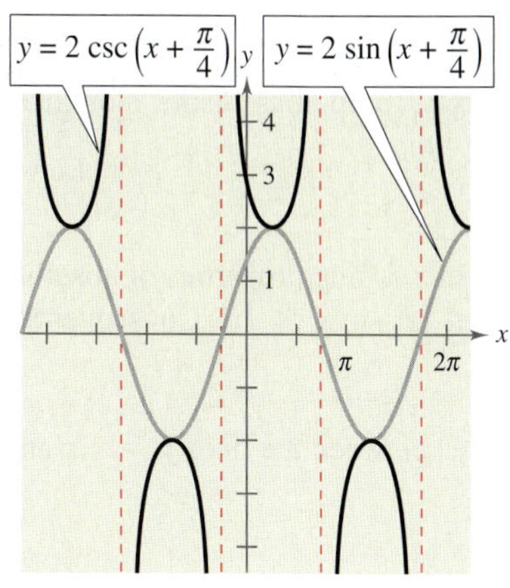

FIGURE 6.64

Example 4 ▶ Sketching the Graph of a Cosecant Function

Sketch the graph of $y = 2\csc\left(x + \dfrac{\pi}{4}\right)$.

Solution

Begin by sketching the graph of

$$y = 2\sin\left(x + \frac{\pi}{4}\right).$$

For this function, the amplitude is 2 and the period is 2π. By solving the equations

$$x + \frac{\pi}{4} = 0 \qquad \text{and} \qquad x + \frac{\pi}{4} = 2\pi$$

$$x = -\frac{\pi}{4} \qquad\qquad x = \frac{7\pi}{4}$$

you can see that one cycle of the sine function corresponds to the interval from $x = -\pi/4$ to $x = 7\pi/4$. The graph of this sine function is represented by the gray curve in Figure 6.64. Because the sine function is zero at the midpoint and endpoints of this interval, the corresponding cosecant function

$$y = 2\csc\left(x + \frac{\pi}{4}\right)$$

$$= 2\left(\frac{1}{\sin[x + (\pi/4)]}\right)$$

has vertical asymptotes at $x = -\pi/4$, $x = 3\pi/4$, $x = 7\pi/4$, etc. The graph of the cosecant function is represented by the black curve in Figure 6.64.

Example 5 ▶ Sketching the Graph of a Secant Function

Sketch the graph of $y = \sec 2x$.

Solution

Begin by sketching the graph of $y = \cos 2x$, as indicated by the gray curve in Figure 6.65. Then, form the graph of $y = \sec 2x$ as the black curve in the figure. Note that the x-intercepts of $y = \cos 2x$

$$\left(-\frac{\pi}{4}, 0\right), \quad \left(\frac{\pi}{4}, 0\right), \quad \left(\frac{3\pi}{4}, 0\right), \ldots$$

correspond to the vertical asymptotes

$$x = -\frac{\pi}{4}, \quad x = \frac{\pi}{4}, \quad x = \frac{3\pi}{4}, \ldots$$

of the graph of $y = \sec 2x$. Moreover, notice that the period of $y = \cos 2x$ and $y = \sec 2x$ is π.

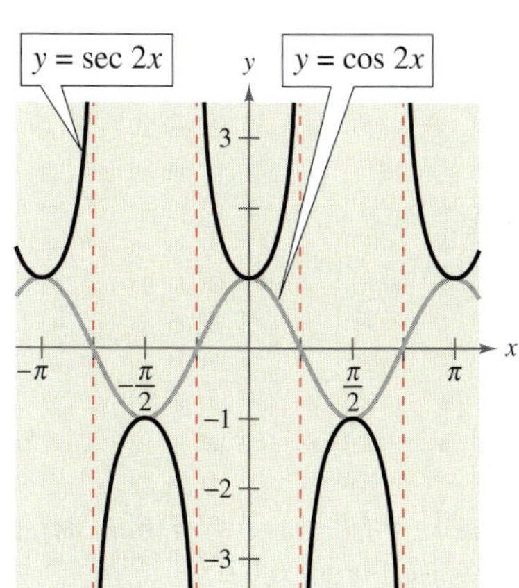

FIGURE 6.65

Damped Trigonometric Graphs

A *product* of two functions can be graphed using properties of the individual functions. For instance, consider the function

$$f(x) = x \sin x$$

as the product of the functions $y = x$ and $y = \sin x$. Using properties of absolute value and the fact that $|\sin x| \leq 1$, you have $0 \leq |x||\sin x| \leq |x|$. Consequently,

$$-|x| \leq x \sin x \leq |x|$$

which means that the graph of $f(x) = x \sin x$ lies between the lines $y = -x$ and $y = x$. Furthermore, because

$$f(x) = x \sin x = \pm x \qquad \text{at} \qquad x = \frac{\pi}{2} + n\pi$$

and

$$f(x) = x \sin x = 0 \qquad \text{at} \qquad x = n\pi$$

the graph of f touches the line $y = -x$ or the line $y = x$ at $x = \pi/2 + n\pi$ and has x-intercepts at $x = n\pi$. A sketch of f is shown in Figure 6.66. In the function $f(x) = x \sin x$, the factor x is called the **damping factor.**

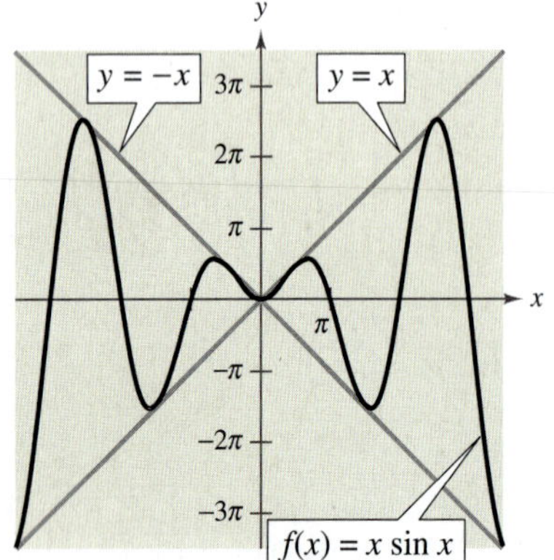

FIGURE 6.66

Example 6 ▶ Damped Sine Wave 

Sketch the graph of

$$f(x) = e^{-x} \sin 3x.$$

Solution

Consider $f(x)$ as the product of the two functions

$$y = e^{-x} \qquad \text{and} \qquad y = \sin 3x$$

each of which has the set of real numbers as its domain. For any real number x, you know that $e^{-x} \geq 0$ and $|\sin 3x| \leq 1$. So, $e^{-x}|\sin 3x| \leq e^{-x}$, which means that

$$-e^{-x} \leq e^{-x} \sin 3x \leq e^{-x}.$$

Furthermore, because

$$f(x) = e^{-x} \sin 3x = \pm e^{-x} \qquad \text{at} \qquad x = \frac{\pi}{6} + \frac{n\pi}{3}$$

and

$$f(x) = e^{-x} \sin 3x = 0 \qquad \text{at} \qquad x = \frac{n\pi}{3}$$

the graph of f touches the curves $y = -e^{-x}$ and $y = e^{-x}$ at $x = \pi/6 + n\pi/3$ and has intercepts at $x = n\pi/3$. A sketch is shown in Figure 6.67.

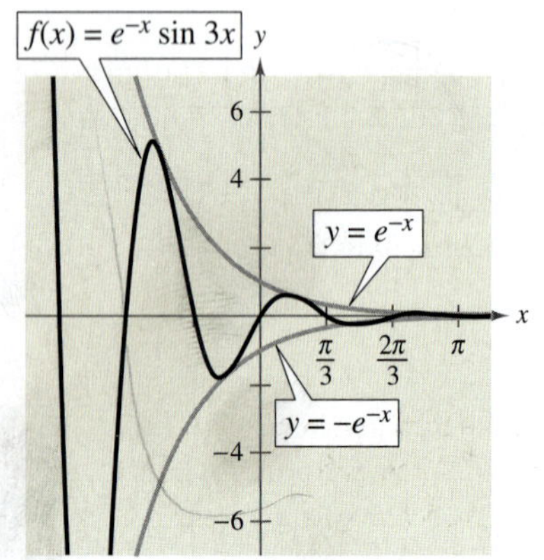

FIGURE 6.67

Figure 6.68 summarizes the six basic trigonometric functions.

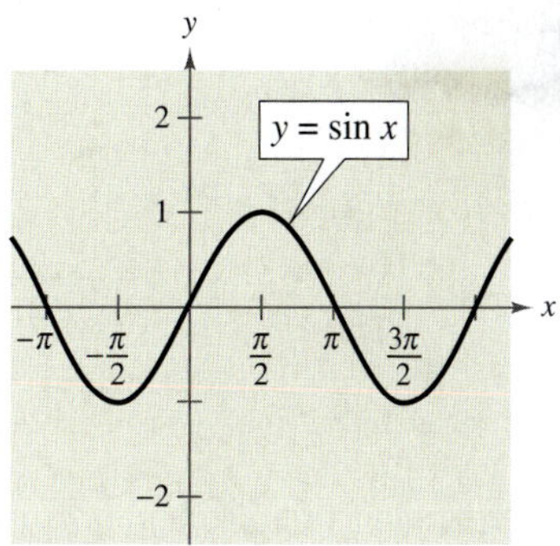

DOMAIN: ALL REALS
RANGE: $[-1, 1]$
PERIOD: 2π

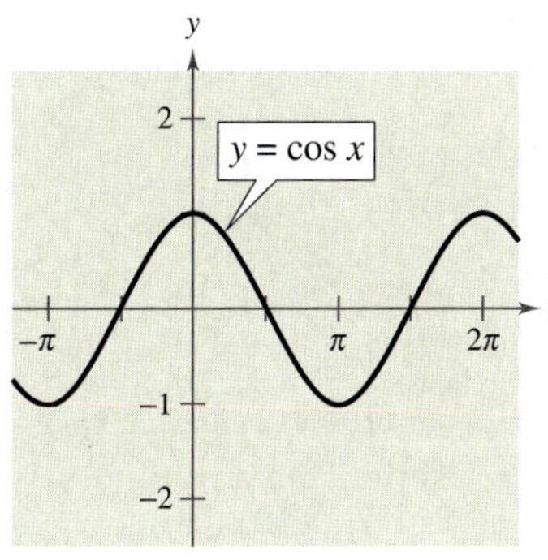

DOMAIN: ALL REALS
RANGE: $[-1, 1]$
PERIOD: 2π

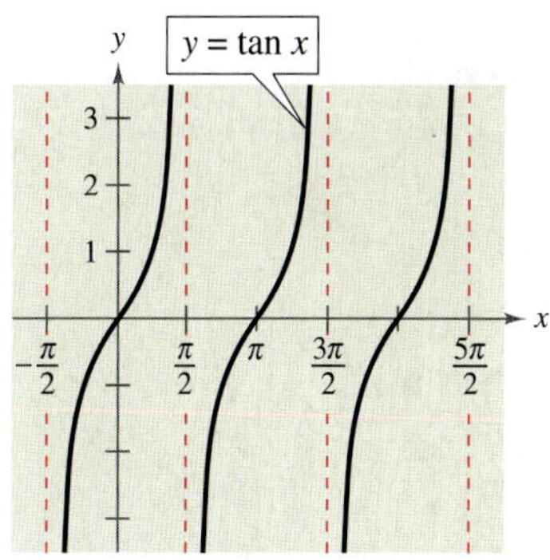

DOMAIN: ALL $x \neq \frac{\pi}{2} + n\pi$
RANGE: $(-\infty, \infty)$
PERIOD: π

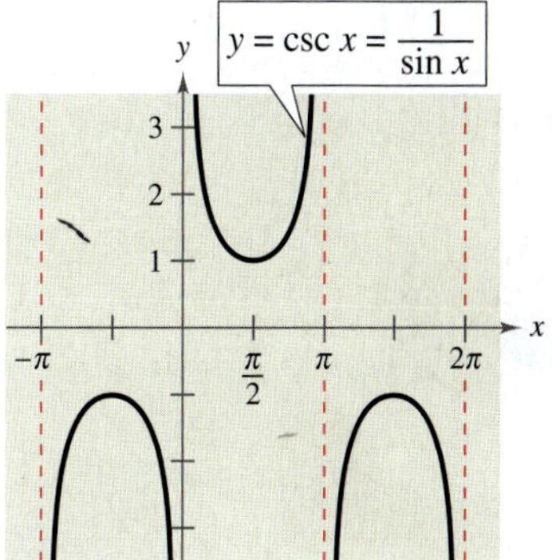

DOMAIN: ALL $x \neq n\pi$
RANGE: $(-\infty, -1]$ AND $[1, \infty)$
PERIOD: 2π

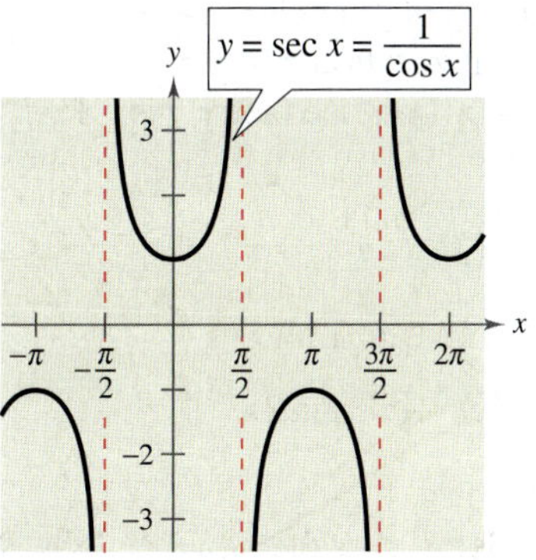

DOMAIN: ALL $x \neq \frac{\pi}{2} + n\pi$
RANGE: $(-\infty, -1]$ AND $[1, \infty)$
PERIOD: 2π

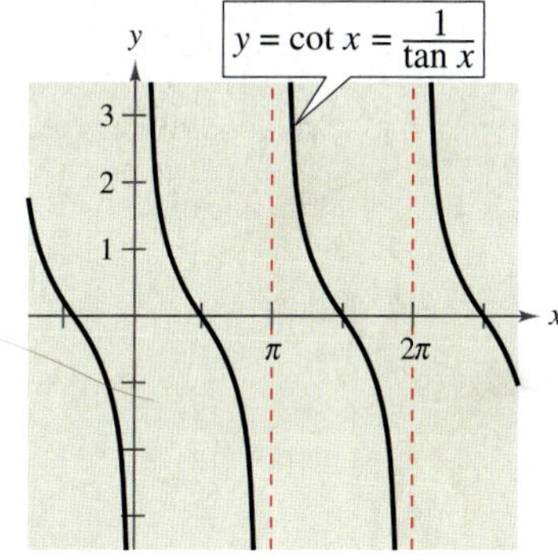

DOMAIN: ALL $x \neq n\pi$
RANGE: $(-\infty, \infty)$
PERIOD: π

FIGURE 6.68

Writing ABOUT MATHEMATICS

Combining Trigonometric Functions Recall from Section 2.6 that functions can be combined arithmetically. This also applies to trigonometric functions. For each of the functions

$$h(x) = x + \sin x \quad \text{and} \quad h(x) = \cos x - \sin 3x$$

(a) identify two simpler functions f and g that comprise the combination, (b) use a table to show how to obtain the numerical values of $h(x)$ from the numerical values of $f(x)$ and $g(x)$, and (c) use graphs of f and g to show how h may be formed.

Can you find functions

$$f(x) = d + a\sin(bx + c) \quad \text{and} \quad g(x) = d + a\cos(bx + c)$$

such that $f(x) + g(x) = 0$ for all x?

6.5 Exercises

In Exercises 1–6, match the function with its graph. State the period of the function. [The graphs are labeled (a), (b), (c), (d), (e), and (f).]

(a)

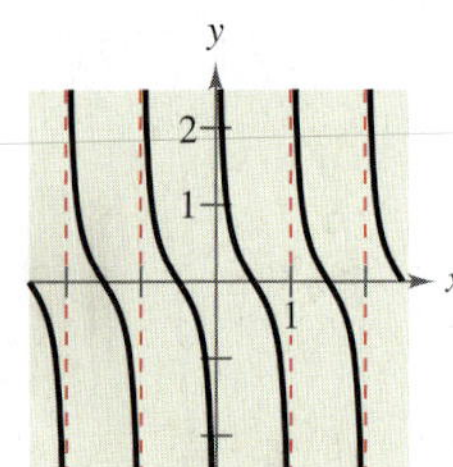

(b)

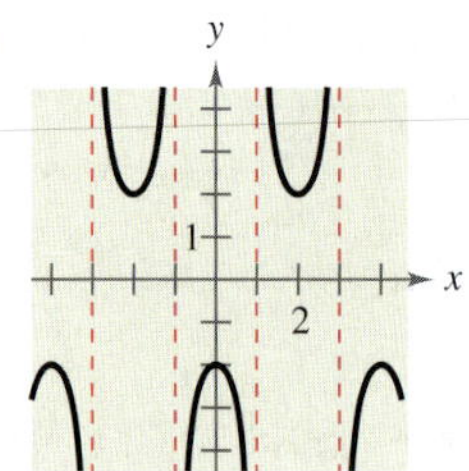

(c)

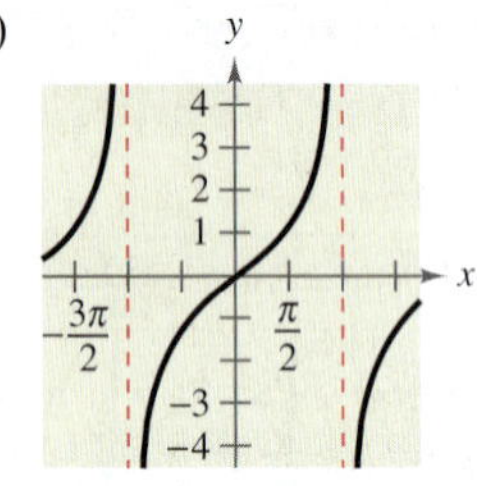

(d)

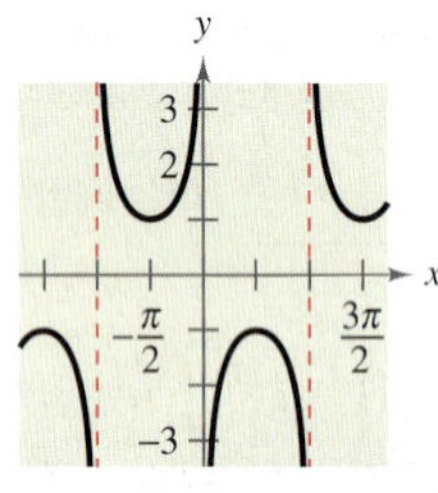

(e)

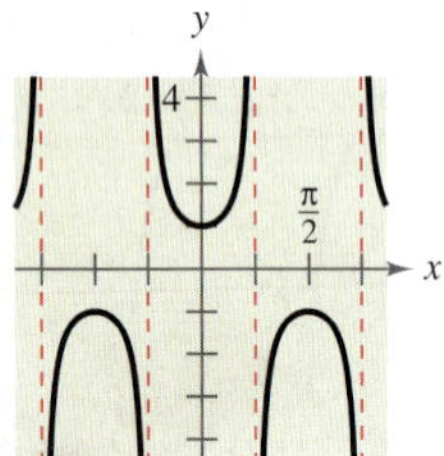

(f)

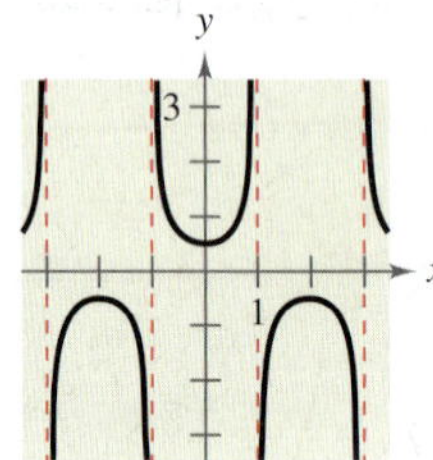

1. $y = \sec 2x$

2. $y = \tan \dfrac{x}{2}$

3. $y = \dfrac{1}{2} \cot \pi x$

4. $y = -\csc x$

5. $y = \dfrac{1}{2} \sec \dfrac{\pi x}{2}$

6. $y = -2 \sec \dfrac{\pi x}{2}$

In Exercises 7–28, sketch the graph of the function. Include two full periods.

7. $y = \frac{1}{3} \tan x$

8. $y = \frac{1}{4} \tan x$

9. $y = \tan 3x$

10. $y = -3 \tan \pi x$

11. $y = -\frac{1}{2} \sec x$

12. $y = \frac{1}{4} \sec x$

13. $y = \csc \pi x$

14. $y = 3 \csc 4x$

15. $y = \sec \pi x - 1$

16. $y = -2 \sec 4x + 2$

17. $y = \csc \dfrac{x}{2}$

18. $y = \csc \dfrac{x}{3}$

19. $y = \cot \dfrac{x}{2}$

20. $y = 3 \cot \dfrac{\pi x}{2}$

21. $y = \frac{1}{2} \sec 2x$

22. $y = -\frac{1}{2} \tan x$

23. $y = \tan \dfrac{\pi x}{4}$

24. $y = \tan(x + \pi)$

25. $y = \csc(\pi - x)$

26. $y = \sec(\pi - x)$

27. $y = \dfrac{1}{4} \csc\left(x + \dfrac{\pi}{4}\right)$

28. $y = 2 \cot\left(x + \dfrac{\pi}{2}\right)$

In Exercises 29–38, use a graphing utility to graph the function. Include two full periods.

29. $y = \tan \dfrac{x}{3}$

30. $y = -\tan 2x$

31. $y = -2 \sec 4x$

32. $y = \sec \pi x$

33. $y = \tan\left(x - \dfrac{\pi}{4}\right)$

34. $y = \dfrac{1}{4} \cot\left(x - \dfrac{\pi}{2}\right)$

35. $y = -\csc(4x - \pi)$

36. $y = 2 \sec(2x - \pi)$

37. $y = 0.1 \tan\left(\dfrac{\pi x}{4} + \dfrac{\pi}{4}\right)$

38. $y = \dfrac{1}{3} \sec\left(\dfrac{\pi x}{2} + \dfrac{\pi}{2}\right)$

In Exercises 39–46, use a graph to solve the equation on the interval $[-2\pi, 2\pi]$.

39. $\tan x = 1$

40. $\tan x = \sqrt{3}$

41. $\cot x = -\dfrac{\sqrt{3}}{3}$

42. $\cot x = 1$

43. $\sec x = -2$

44. $\sec x = 2$

45. $\csc x = \sqrt{2}$

46. $\csc x = -\dfrac{2\sqrt{3}}{3}$

In Exercises 47 and 48, use the graph of the function to determine whether the function is even, odd, or neither.

47. $f(x) = \sec x$

48. $f(x) = \tan x$

49. ***Graphical Reasoning*** Consider the functions

$$f(x) = 2 \sin x \quad \text{and} \quad g(x) = \frac{1}{2} \csc x$$

on the interval $(0, \pi)$.

(a) Graph f and g in the same coordinate plane.

(b) Approximate the interval in which $f > g$.

(c) Describe the behavior of each of the functions as x approaches π. How is the behavior of g related to the behavior of f as x approaches π?

50. *Graphical Reasoning* Consider the functions

$$f(x) = \tan \frac{\pi x}{2} \quad \text{and} \quad g(x) = \frac{1}{2} \sec \frac{\pi x}{2}$$

on the interval $(-1, 1)$.

(a) Use a graphing utility to graph f and g in the same viewing window.

(b) Approximate the interval in which $f < g$.

(c) Approximate the interval in which $2f < 2g$. How does the result compare with that of part (b)? Explain.

In Exercises 51–54, use a graphing utility to graph the two equations in the same viewing window. Determine analytically whether the expressions are equivalent.

51. $y_1 = \sin x \csc x, \quad y_2 = 1$

52. $y_1 = \sin x \sec x, \quad y_2 = \tan x$

53. $y_1 = \dfrac{\cos x}{\sin x}, \quad y_2 = \cot x$

54. $y_1 = \sec^2 x - 1, \quad y_2 = \tan^2 x$

In Exercises 55–58, match the function with its graph. Describe the behavior of the function as x approaches zero. [The graphs are labeled (a), (b), (c), and (d).]

(a)

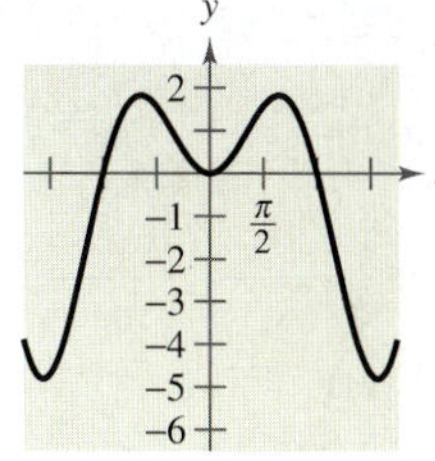

(b)

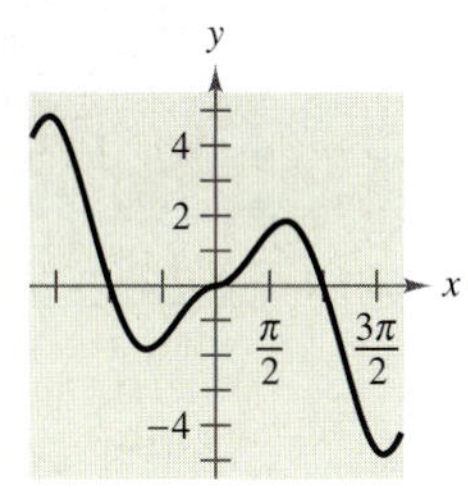

(c)

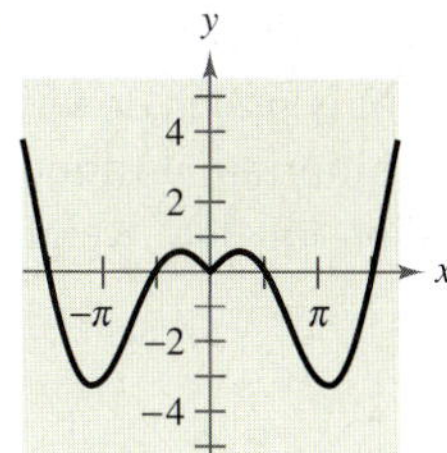

(d)

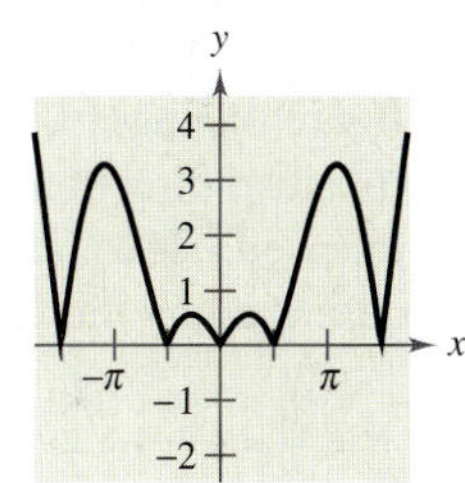

55. $f(x) = |x \cos x|$

56. $f(x) = x \sin x$

57. $g(x) = |x| \sin x$

58. $g(x) = |x| \cos x$

Conjecture **In Exercises 59–62, graph the functions f and g. Use the graphs to make a conjecture about the relationship between the functions.**

59. $f(x) = \sin x + \cos\left(x + \dfrac{\pi}{2}\right), \quad g(x) = 0$

60. $f(x) = \sin x - \cos\left(x + \dfrac{\pi}{2}\right), \quad g(x) = 2 \sin x$

61. $f(x) = \sin^2 x, \quad g(x) = \frac{1}{2}(1 - \cos 2x)$

62. $f(x) = \cos^2 \dfrac{\pi x}{2}, \quad g(x) = \dfrac{1}{2}(1 + \cos \pi x)$

In Exercises 63–66, use a graphing utility to graph the function and the damping factor of the function in the same viewing window. Describe the behavior of the function as x increases without bound.

63. $f(x) = 2^{-x/4} \cos \pi x$

64. $f(x) = e^{-x} \cos x$

65. $g(x) = e^{-x^2/2} \sin x$

66. $h(x) = 2^{-x^2/4} \sin x$

Exploration **In Exercises 67–72, use a graphing utility to graph the function. Describe the behavior of the function as x approaches zero.**

67. $y = \dfrac{6}{x} + \cos x, \quad x > 0$

68. $y = \dfrac{4}{x} + \sin 2x, \quad x > 0$

69. $g(x) = \dfrac{\sin x}{x}$

70. $f(x) = \dfrac{1 - \cos x}{x}$

71. $f(x) = \sin \dfrac{1}{x}$

72. $h(x) = x \sin \dfrac{1}{x}$

73. *Distance* A plane flying at an altitude of 7 miles above a radar antenna will pass directly over the radar antenna (see figure). Let d be the ground distance from the antenna to the point directly under the plane and let x be the angle of elevation to the plane from the antenna. (d is positive as the plane approaches the antenna.) Write d as a function of x and graph the function over the interval $0 < x < \pi$.

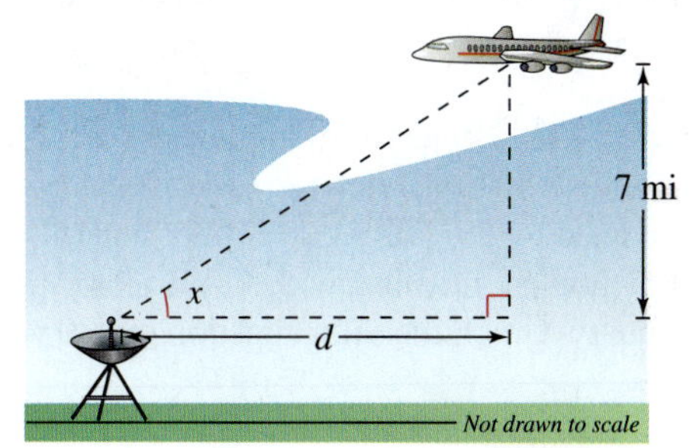

74. ***Television Coverage*** A television camera is on a reviewing platform 27 meters from the street on which a parade will be passing from left to right (see figure). Express the distance d from the camera to a particular unit in the parade as a function of the angle x, and graph the function over the interval $-\pi/2 < x < \pi/2$. (Consider x as negative when a unit in the parade approaches from the left.)

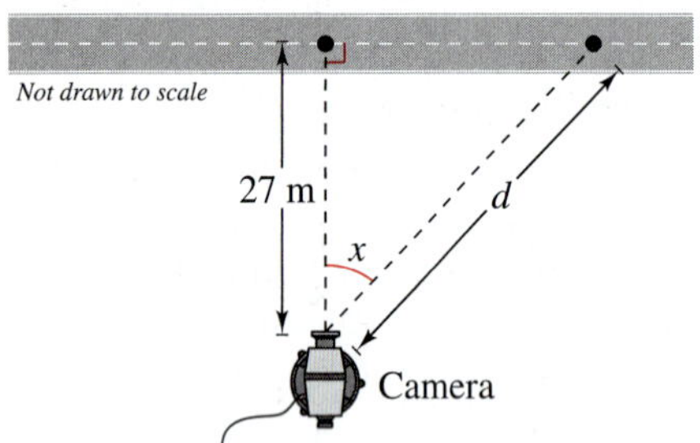

▶ Model It

75. ***Predator-Prey Model*** The population of coyotes (a predator) at time t (in months) in a region is estimated to be

$$C = 5000 + 2000 \sin \frac{\pi t}{12}$$

and the population of rabbits (its prey) is estimated to be

$$R = 25{,}000 + 15{,}000 \cos \frac{\pi t}{12}.$$

(a) Use a graphing utility to graph both models in the same viewing window. Use the window setting $0 \le t \le 100$.

(b) Use the graphs of the models in part (a) to explain the oscillations in the size of each population.

(c) The cycles of each population follow a periodic pattern. Find the period of each model and describe several factors that could be contributing to the cyclical patterns.

76. ***Sales*** The projected monthly sales S (in thousands of units) of lawn mowers (a seasonal product) are modeled by $S = 74 + 3t - 40\cos(\pi t/6)$, where t is the time in months, with $t = 1$ corresponding to January. Graph the sales function over 1 year.

77. ***Meterology*** The normal monthly high temperatures in degrees Fahrenheit for Erie, Pennsylvania are approximated by

$$H(t) = 54.33 - 20.38 \cos \frac{\pi t}{6} - 15.69 \sin \frac{\pi t}{6}$$

and the normal monthly low temperatures are approximated by

$$L(t) = 39.36 - 15.70 \cos \frac{\pi t}{6} - 14.16 \sin \frac{\pi t}{6}$$

where t is the time in months, with $t = 1$ corresponding to January (see figure). (Source: National Oceanic and Atmospheric Administration)

(a) What is the period of each function?

(b) During what part of the year is the difference between the normal high and low temperatures greatest? When is it smallest?

(c) The sun is northernmost in the sky around June 21, but the graph shows the warmest temperatures at a later date. Approximate the lag time of the temperatures relative to the position of the sun.

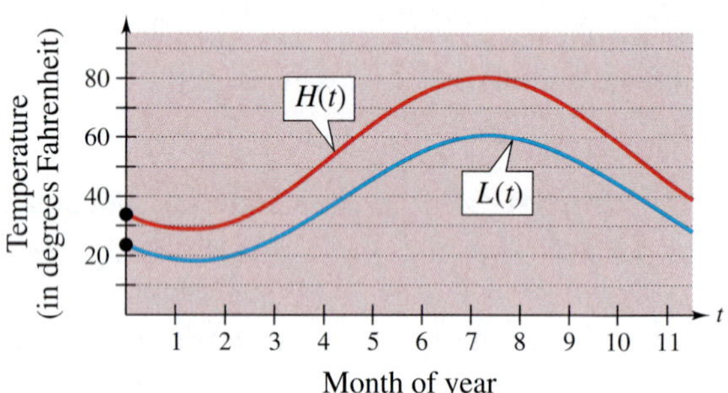

78. ***Harmonic Motion*** An object weighing W pounds is suspended from the ceiling by a steel spring (see figure on page 509). The weight is pulled downward (positive direction) from its equilibrium position and released. The resulting motion of the weight is described by the function

$$y = \frac{1}{2}e^{-t/4} \cos 4t, \qquad t > 0$$

where y is the distance in feet and t is the time in seconds.

(a) Use a graphing utility to graph the function.

(b) Describe the behavior of the displacement function for increasing values of time t.

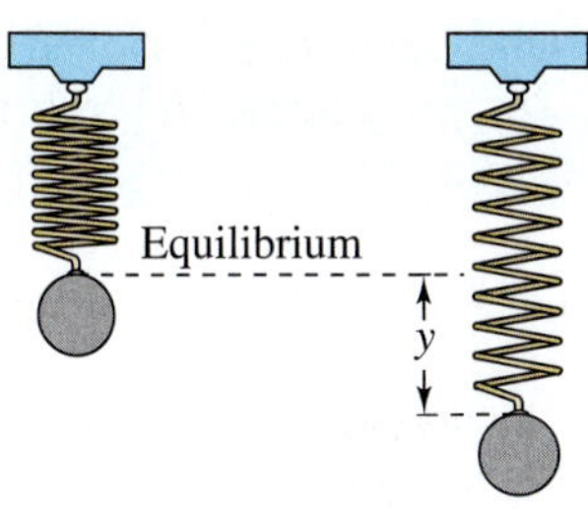

FIGURE FOR 78

Synthesis

True or False? **In Exercises 79 and 80, determine whether the statement is true or false. Justify your answer.**

79. The graph of $y = \csc x$ can be obtained on a calculator by graphing the reciprocal of $y = \sin x$.

80. The graph of $y = \sec x$ can be obtained on a calculator by graphing a translation of the reciprocal of $y = \sin x$.

81. ***Writing*** Describe the behavior of $f(x) = \tan x$ as x approaches $\pi/2$ from the left and from the right.

82. ***Writing*** Describe the behavior of $f(x) = \csc x$ as x approaches π from the left and from the right.

83. ***Exploration*** Consider the function

$$f(x) = x - \cos x.$$

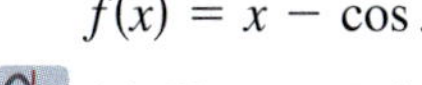

(a) Use a graphing utility to graph the function and verify that there exists a zero between 0 and 1. Use the graph to approximate the zero.

(b) Starting with $x_0 = 1$, generate a sequence $x_1, x_2, x_3, \ldots$ where $x_n = \cos(x_{n-1})$. For example,

$$x_0 = 1$$
$$x_1 = \cos(x_0)$$
$$x_2 = \cos(x_1)$$
$$x_3 = \cos(x_2)$$
$$\vdots$$

What value does the sequence approach?

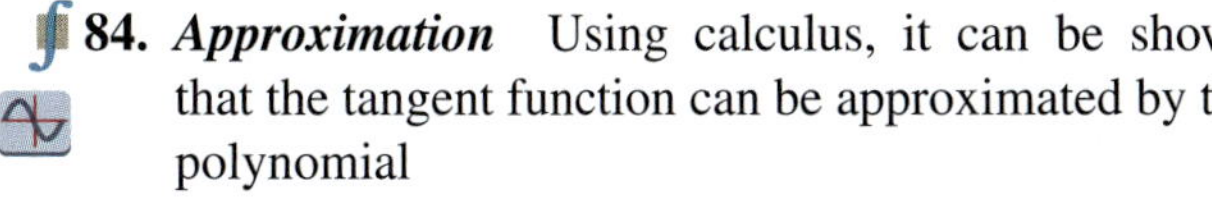

84. ***Approximation*** Using calculus, it can be shown that the tangent function can be approximated by the polynomial

$$\tan x \approx x + \frac{2x^3}{3!} + \frac{16x^5}{5!}$$

where x is in radians. Use a graphing utility to graph the tangent function and its polynomial approximation in the same viewing window. How do the graphs compare?

85. ***Approximation*** Using calculus, it can be shown that the secant function can be approximated by the polynomial

$$\sec x \approx 1 + \frac{x^2}{2!} + \frac{5x^4}{4!}$$

where x is in radians. Use a graphing utility to graph the secant function and its polynomial approximation in the same viewing window. How do the graphs compare?

86. ***Pattern Recognition***

(a) Use a graphing utility to graph each function.

$$y_1 = \frac{4}{\pi}\left(\sin \pi x + \frac{1}{3}\sin 3\pi x\right)$$

$$y_2 = \frac{4}{\pi}\left(\sin \pi x + \frac{1}{3}\sin 3\pi x + \frac{1}{5}\sin 5\pi x\right)$$

(b) Identify the pattern started in part (a) and find a function y_3 that continues the pattern one more term. Use a graphing utility to graph y_3.

(c) The graphs in parts (a) and (b) approximate the periodic function in the figure. Find a function y_4 that is a better approximation.

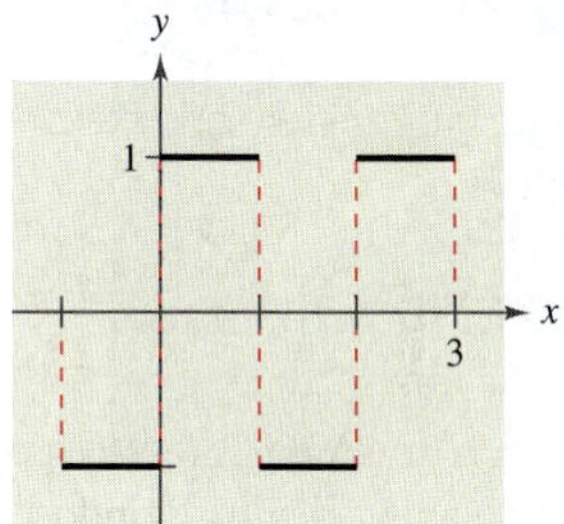

Review

In Exercises 87–90, solve the exponential equation. Round your answer to three decimal places.

87. $e^{2x} = 54$

88. $8^{3x} = 98$

89. $\dfrac{300}{1 + e^{-x}} = 100$

90. $\left(1 + \dfrac{0.15}{365}\right)^{365t} = 5$

In Exercises 91–96, solve the logarithmic equation. Round your answer to three decimal places.

91. $\ln(3x - 2) = 73$

92. $\ln(14 - 2x) = 68$

93. $\ln(x^2 + 1) = 3.2$

94. $\ln \sqrt{x + 4} = 5$

95. $\log_8 x + \log_8(x - 1) = \frac{1}{3}$

96. $\log_6 x + \log_6(x^2 - 1) = \log_6 64x$

6.6 Inverse Trigonometric Functions

▶ What you should learn

- How to evaluate the inverse sine function
- How to evaluate the other inverse trigonometric functions
- How to evaluate the compositions of trigonometric functions

▶ Why you should learn it

You can use inverse trigonometric functions to model and solve real-life problems. For instance, in Exercise 93 on page 518, an inverse trigonometric function can be used to model the angle subtended by a camera lens x feet from a painting.

John Foxx/Imagestate

Inverse Sine Function

Recall from Section 2.7 that, for a function to have an inverse function, it must be one-to-one—that is, it must pass the Horizontal Line Test. From Figure 6.69 you can see that $y = \sin x$ does not pass the test because different values of x yield the same y-value.

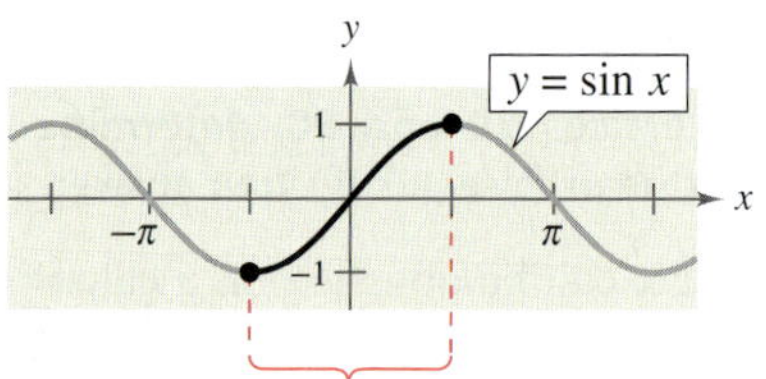

Sin x has an inverse function on this interval.

FIGURE 6.69

However, if you restrict the domain to the interval $-\pi/2 \leq x \leq \pi/2$ (corresponding to the black portion of the graph in Figure 6.69), the following properties hold.

1. On the interval $[-\pi/2, \pi/2]$, the function $y = \sin x$ is increasing.
2. On the interval $[-\pi/2, \pi/2]$, $y = \sin x$ takes on its full range of values, $-1 \leq \sin x \leq 1$.
3. On the interval $[-\pi/2, \pi/2]$, $y = \sin x$ is one-to-one.

So, on the restricted domain $-\pi/2 \leq x \leq \pi/2$, $y = \sin x$ has a unique inverse function called the **inverse sine function.** It is denoted by

$$y = \arcsin x \qquad \text{or} \qquad y = \sin^{-1} x.$$

The notation $\sin^{-1} x$ is consistent with the inverse function notation $f^{-1}(x)$. The $\arcsin x$ notation (read as "the arcsine of x") comes from the association of a central angle with its subtended *arc length* on a unit circle. So, $\arcsin x$ means the angle (or arc) whose sine is x. Both notations, $\arcsin x$ and $\sin^{-1} x$, are commonly used in mathematics, so remember that $\sin^{-1} x$ denotes the *inverse* sine function rather than $1/\sin x$. The values of $\arcsin x$ lie in the interval $-\pi/2 \leq \arcsin x \leq \pi/2$. The graph of $y = \arcsin x$ is shown in Figure 6.70 on page 511.

STUDY TIP

When evaluating the inverse sine function, it helps to remember the phrase "the arcsine of x is the angle (or number) whose sine is x."

Definition of Inverse Sine Function

The **inverse sine function** is defined by

$$y = \arcsin x \qquad \text{if and only if} \qquad \sin y = x$$

where $-1 \leq x \leq 1$ and $-\pi/2 \leq y \leq \pi/2$. The domain of $y = \arcsin x$ is $[-1, 1]$, and the range is $[-\pi/2, \pi/2]$.

STUDY TIP

As with the trigonometric functions, much of the work with the inverse trigonometric functions can be done by *exact* calculations rather than by calculator approximations. Exact calculations help to increase your understanding of the inverse functions by relating them to the right triangle definitions of the trigonometric functions.

Example 1 ▶ Evaluating the Inverse Sine Function

If possible, find the exact value.

a. $\arcsin\left(-\frac{1}{2}\right)$

b. $\sin^{-1}\frac{\sqrt{3}}{2}$

c. $\sin^{-1}2$

Solution

a. Because $\sin\left(-\frac{\pi}{6}\right) = -\frac{1}{2}$ for $-\frac{\pi}{2} \le y \le \frac{\pi}{2}$, it follows that

$$\arcsin\left(-\frac{1}{2}\right) = -\frac{\pi}{6}. \qquad \text{Angle whose sine is } -\tfrac{1}{2}$$

b. Because $\sin(\pi/3) = \sqrt{3}/2$ for $-\pi/2 \le y \le \pi/2$, it follows that

$$\sin^{-1}\frac{\sqrt{3}}{2} = \frac{\pi}{3}. \qquad \text{Angle whose sine is } \sqrt{3}/2$$

c. It is not possible to evaluate $y = \sin^{-1}x$ when $x = 2$ because there is no angle whose sine is 2. Remember that the domain of the inverse sine function is $[-1, 1]$.

Example 2 ▶ Graphing the Arcsine Function 

Sketch a graph of

$$y = \arcsin x.$$

Solution

By definition, the equations $y = \arcsin x$ and $\sin y = x$ are equivalent for $-\pi/2 \le y \le \pi/2$. So, their graphs are the same. From the interval $[-\pi/2, \pi/2]$, you can assign values to y in the second equation to make a table of values. Then plot the points and draw a smooth curve through the points.

y	$-\frac{\pi}{2}$	$-\frac{\pi}{4}$	$-\frac{\pi}{6}$	0	$\frac{\pi}{6}$	$\frac{\pi}{4}$	$\frac{\pi}{2}$
$x = \sin y$	-1	$-\frac{\sqrt{2}}{2}$	$-\frac{1}{2}$	0	$\frac{1}{2}$	$\frac{\sqrt{2}}{2}$	1

The resulting graph for $y = \arcsin x$ is shown in Figure 6.70. Note that it is the reflection (in the line $y = x$) of the black portion of the graph in Figure 6.69. Be sure you see that Figure 6.70 shows the *entire* graph of the inverse sine function. Remember that the range of $y = \arcsin x$ is the closed interval $[-\pi/2, \pi/2]$.

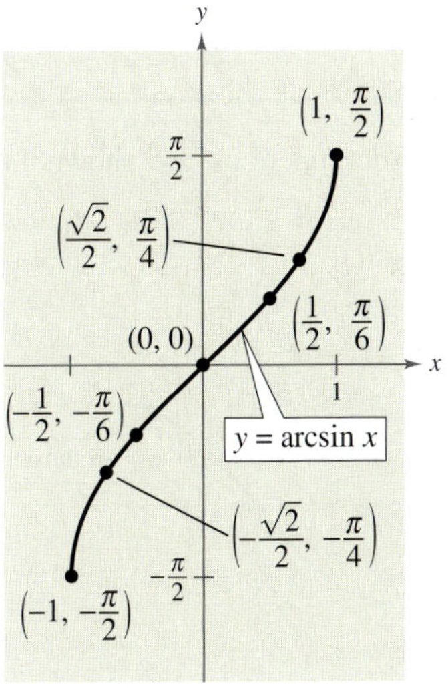

FIGURE 6.70

Other Inverse Trigonometric Functions 

The cosine function is decreasing and one-to-one on the interval $0 \le x \le \pi$, as shown in Figure 6.71.

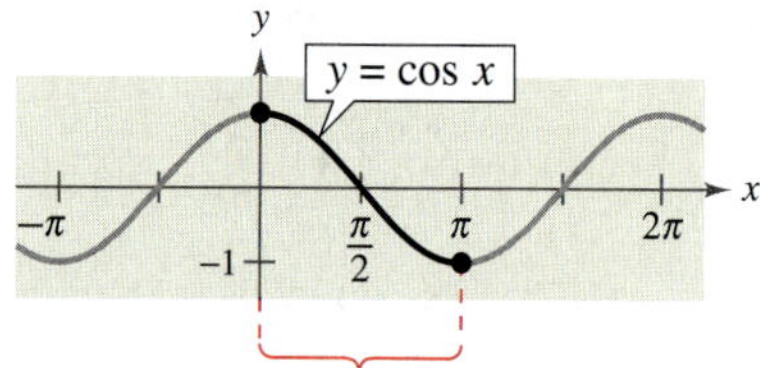

FIGURE 6.71

Consequently, on this interval the cosine function has an inverse function—the **inverse cosine function**—denoted by

$$y = \arccos x \quad \text{or} \quad y = \cos^{-1} x.$$

Similarly, you can define an **inverse tangent function** by restricting the domain of $y = \tan x$ to the interval $(-\pi/2, \pi/2)$. The following list summarizes the definitions of the three most common inverse trigonometric functions. The remaining three are defined in Exercises 100–102.

Definitions of the Inverse Trigonometric Functions

Function	*Domain*	*Range*
$y = \arcsin x$ if and only if $\sin y = x$	$-1 \le x \le 1$	$-\dfrac{\pi}{2} \le y \le \dfrac{\pi}{2}$
$y = \arccos x$ if and only if $\cos y = x$	$-1 \le x \le 1$	$0 \le y \le \pi$
$y = \arctan x$ if and only if $\tan y = x$	$-\infty < x < \infty$	$-\dfrac{\pi}{2} < y < \dfrac{\pi}{2}$

The graphs of these three inverse trigonometric functions are shown in Figure 6.72.

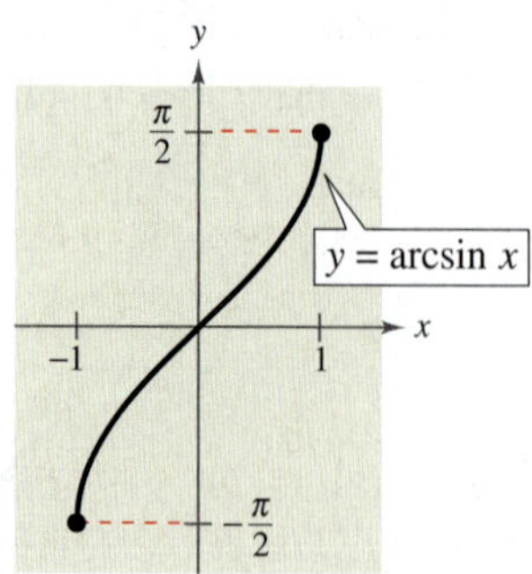

DOMAIN: $[-1, 1]$
RANGE: $\left[-\frac{\pi}{2}, \frac{\pi}{2}\right]$
FIGURE 6.72

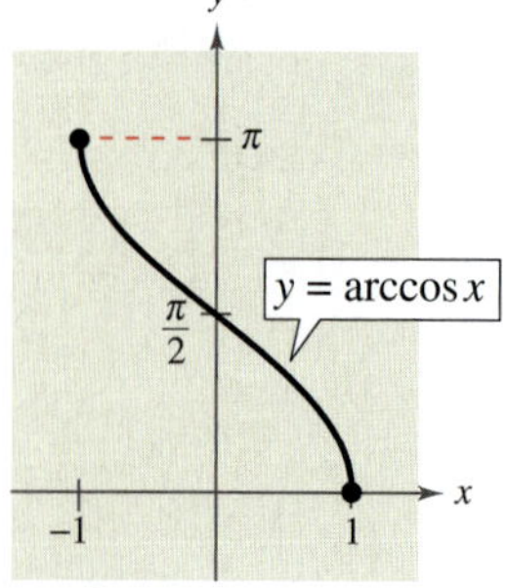

DOMAIN: $[-1, 1]$
RANGE: $[0, \pi]$

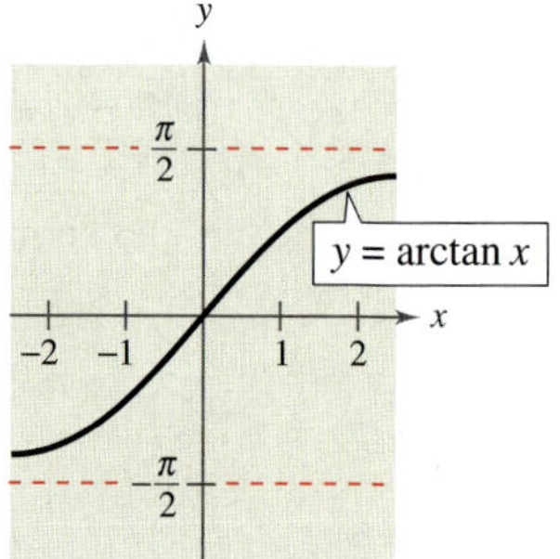

DOMAIN: $(-\infty, \infty)$
RANGE: $\left(-\frac{\pi}{2}, \frac{\pi}{2}\right)$

Example 3 ▶ **Evaluating Inverse Trigonometric Functions** 

Find the exact value.

a. $\arccos \dfrac{\sqrt{2}}{2}$ **b.** $\cos^{-1}(-1)$

c. $\arctan 0$ **d.** $\tan^{-1}(-1)$

Solution

a. Because $\cos(\pi/4) = \sqrt{2}/2$, and $\pi/4$ lies in $[0, \pi]$, it follows that

$$\arccos \frac{\sqrt{2}}{2} = \frac{\pi}{4}.$$ Angle whose cosine is $\sqrt{2}/2$

b. Because $\cos \pi = -1$, and π lies in $[0, \pi]$, it follows that

$$\cos^{-1}(-1) = \pi.$$ Angle whose cosine is -1

c. Because $\tan 0 = 0$, and 0 lies in $(-\pi/2, \pi/2)$, it follows that

$$\arctan 0 = 0.$$ Angle whose tangent is 0

d. Because $\tan(-\pi/4) = -1$, and $-\pi/4$ lies in $(-\pi/2, \pi/2)$, it follows that

$$\tan^{-1}(-1) = -\frac{\pi}{4}.$$ Angle whose tangent is -1

Example 4 ▶ **Calculators and Inverse Trigonometric Functions** 

Use a calculator to approximate the value (if possible).

a. $\arctan(-8.45)$

b. $\sin^{-1} 0.2447$

c. $\arccos 2$

Solution

	Function	*Mode*	*Calculator Keystrokes*
a.	$\arctan(-8.45)$	Radian	[TAN⁻¹] [(] [(−)] 8.45 [)] [ENTER]

From the display, it follows that $\arctan(-8.45) \approx -1.453001$.

b.	$\sin^{-1} 0.2447$	Radian	[SIN⁻¹] 0.2447 [ENTER]

From the display, it follows that $\arcsin 0.2447 \approx 0.2472103$.

c.	$\arccos 2$	Radian	[COS⁻¹] 2 [ENTER]

In real number mode, the calculator should display an *error message* because the domain of the inverse cosine function is $[-1, 1]$.

STUDY TIP

It is important to remember that the domain of the inverse sine function and the inverse cosine function is $[-1, 1]$, as indicated in Example 4(c).

In Example 4, if you had set the calculator to degree mode, the displays would have been in degrees rather than radians. This convention is peculiar to calculators. By definition, the values of inverse trigonometric functions are *always in radians*.

Compositions of Functions

Recall from Section 2.7 that for all x in the domains of f and f^{-1}, inverse functions have the properties

$$f(f^{-1}(x)) = x \quad \text{and} \quad f^{-1}(f(x)) = x.$$

Inverse Properties of Trigonometric Functions

If $-1 \le x \le 1$ and $-\pi/2 \le y \le \pi/2$, then

$$\sin(\arcsin x) = x \quad \text{and} \quad \arcsin(\sin y) = y.$$

If $-1 \le x \le 1$ and $0 \le y \le \pi$, then

$$\cos(\arccos x) = x \quad \text{and} \quad \arccos(\cos y) = y.$$

If x is a real number and $-\pi/2 < y < \pi/2$, then

$$\tan(\arctan x) = x \quad \text{and} \quad \arctan(\tan y) = y.$$

Keep in mind that these inverse properties do not apply for arbitrary values of x and y. For instance,

$$\arcsin\left(\sin \frac{3\pi}{2}\right) = \arcsin(-1) = -\frac{\pi}{2} \neq \frac{3\pi}{2}.$$

In other words, the property

$$\arcsin(\sin y) = y$$

is not valid for values of y outside the interval $[-\pi/2, \pi/2]$.

Example 5 ▶ Using Inverse Properties 

If possible, find the exact value.

a. $\tan[\arctan(-5)]$ **b.** $\arcsin\left(\sin \frac{5\pi}{3}\right)$ **c.** $\cos(\cos^{-1} \pi)$

Solution

a. Because -5 lies in the domain of the arctan function, the inverse property applies, and you have

$$\tan[\arctan(-5)] = -5.$$

b. In this case, $5\pi/3$ does not lie within the range of the arcsine function, $-\pi/2 \le y \le \pi/2$. However, $5\pi/3$ is coterminal with

$$\frac{5\pi}{3} - 2\pi = -\frac{\pi}{3}$$

which does lie in the range of the arcsine function, and you have

$$\arcsin\left(\sin \frac{5\pi}{3}\right) = \arcsin\left[\sin\left(-\frac{\pi}{3}\right)\right] = -\frac{\pi}{3}.$$

c. The expression $\cos(\cos^{-1} \pi)$ is not defined because $\cos^{-1} \pi$ is not defined. Remember that the domain of the inverse cosine function is $[-1, 1]$.

Example 6 shows how to use right triangles to find exact values of compositions of inverse functions. Then, Example 7 shows how to use triangles to convert a trigonometric expression into an algebraic expression. This conversion technique is used frequently in calculus.

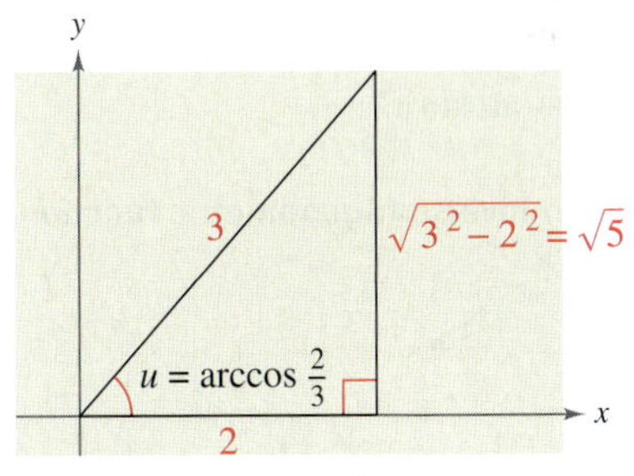

Angle whose cosine is $\frac{2}{3}$

FIGURE 6.73

Example 6 ▶ Evaluating Compositions of Functions

Find the exact value.

a. $\tan\left(\arccos \frac{2}{3}\right)$ **b.** $\cos\left[\arcsin\left(-\frac{3}{5}\right)\right]$

Solution

a. If you let $u = \arccos \frac{2}{3}$, then $\cos u = \frac{2}{3}$. Because $\cos u$ is positive, u is a *first*-quadrant angle. You can sketch and label angle u as shown in Figure 6.73. Consequently,

$$\tan\left(\arccos \frac{2}{3}\right) = \tan u = \frac{\text{opp}}{\text{adj}} = \frac{\sqrt{5}}{2}.$$

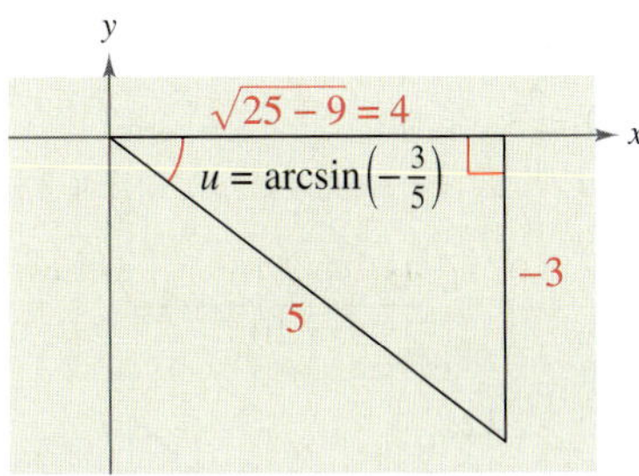

Angle whose sine is $-\frac{3}{5}$

FIGURE 6.74

b. If you let $u = \arcsin\left(-\frac{3}{5}\right)$, then $\sin u = -\frac{3}{5}$. Because $\sin u$ is negative, u is a *fourth*-quadrant angle. You can sketch and label angle u as shown in Figure 6.74. Consequently,

$$\cos\left[\arcsin\left(-\frac{3}{5}\right)\right] = \cos u = \frac{\text{adj}}{\text{hyp}} = \frac{4}{5}.$$

Example 7 ▶ Some Problems from Calculus

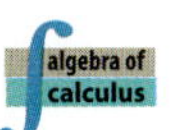

Write each of the following as an algebraic expression in x.

a. $\sin(\arccos 3x), \quad 0 \le x \le \frac{1}{3}$ **b.** $\cot(\arccos 3x), \quad 0 \le x < \frac{1}{3}$

Solution

If you let $u = \arccos 3x$, then $\cos u = 3x$. Because

$$\cos u = \frac{\text{adj}}{\text{hyp}} = \frac{3x}{1}$$

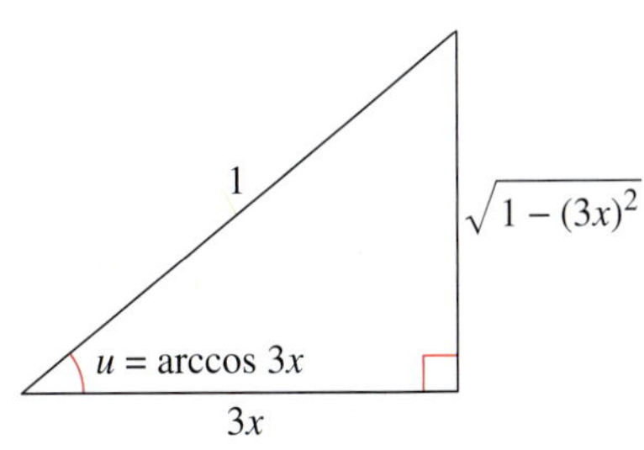

Angle whose cosine is $3x$

FIGURE 6.75

you can sketch a right triangle with acute angle u, as shown in Figure 6.75. From this triangle, you can easily convert each expression to algebraic form.

a. $\sin(\arccos 3x) = \sin u = \frac{\text{opp}}{\text{hyp}} = \sqrt{1 - 9x^2}, \quad 0 \le x \le \frac{1}{3}$

b. $\cot(\arccos 3x) = \cot u = \frac{\text{adj}}{\text{opp}} = \frac{3x}{\sqrt{1 - 9x^2}}, \quad 0 \le x < \frac{1}{3}$

In Example 7, similar arguments can be made for x-values lying in the interval $\left[-\frac{1}{3}, 0\right]$.

6.6 Exercises

In Exercises 1–16, evaluate the expression without the aid of a calculator.

1. $\arcsin \frac{1}{2}$
2. $\arcsin 0$
3. $\arccos \frac{1}{2}$
4. $\arccos 0$
5. $\arctan \frac{\sqrt{3}}{3}$
6. $\arctan(-1)$
7. $\cos^{-1}\left(-\frac{\sqrt{3}}{2}\right)$
8. $\sin^{-1}\left(-\frac{\sqrt{2}}{2}\right)$
9. $\arctan(-\sqrt{3})$
10. $\arctan \sqrt{3}$
11. $\arccos\left(-\frac{1}{2}\right)$
12. $\arcsin \frac{\sqrt{2}}{2}$
13. $\sin^{-1} \frac{\sqrt{3}}{2}$
14. $\tan^{-1}\left(-\frac{\sqrt{3}}{3}\right)$
15. $\tan^{-1} 0$
16. $\cos^{-1} 1$

In Exercises 17–32, use a calculator to approximate the expression. Round your result to two decimal places.

17. $\arccos 0.28$
18. $\arcsin 0.45$
19. $\arcsin(-0.75)$
20. $\arccos(-0.7)$
21. $\arctan(-3)$
22. $\arctan 15$
23. $\sin^{-1} 0.31$
24. $\cos^{-1} 0.26$
25. $\arccos(-0.41)$
26. $\arcsin(-0.125)$
27. $\arctan 0.92$
28. $\arctan 2.8$
29. $\arcsin \frac{3}{4}$
30. $\arccos\left(-\frac{1}{3}\right)$
31. $\tan^{-1} \frac{7}{2}$
32. $\tan^{-1}\left(-\frac{95}{7}\right)$

In Exercises 33 and 34, determine the missing coordinates of the points on the graph of the function.

33.

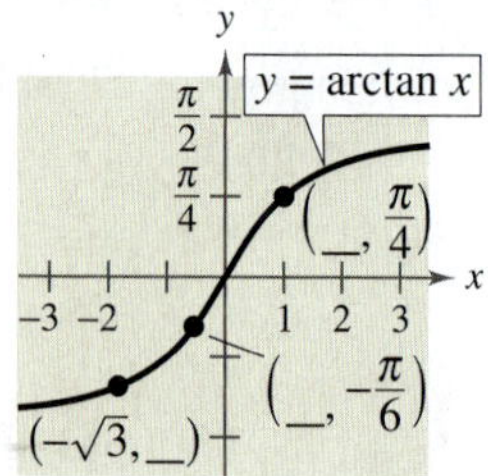

34.

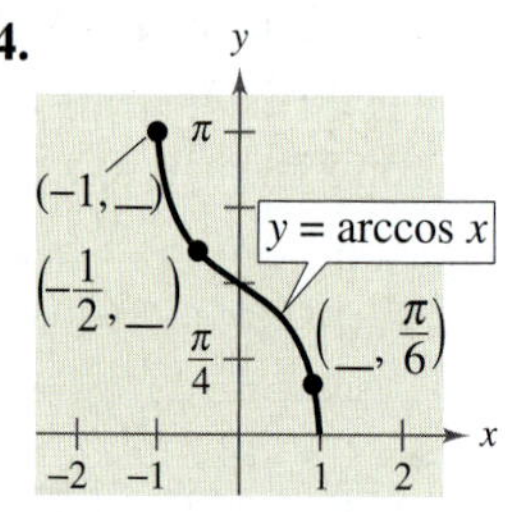

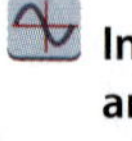

In Exercises 35 and 36, use a graphing utility to graph *f*, *g*, and *y* = *x* in the same viewing window to verify geometrically that *g* is the inverse function of *f*. (Be sure to restrict the domain of *f* properly.)

35. $f(x) = \tan x, \quad g(x) = \arctan x$
36. $f(x) = \sin x, \quad g(x) = \arcsin x$

In Exercises 37–42, use an inverse trigonometric function to write *θ* as a function of *x*.

37.

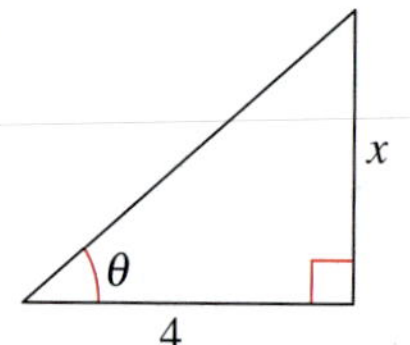

38.

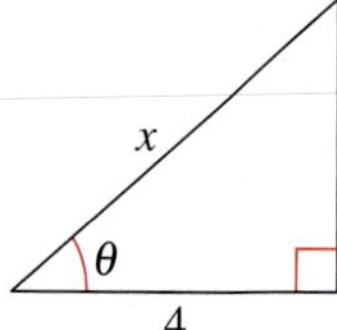

39.

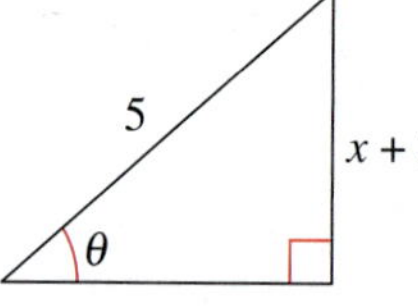

40.

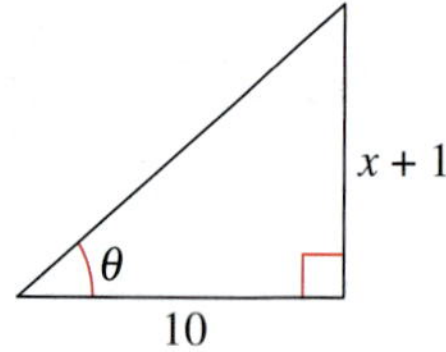

41.

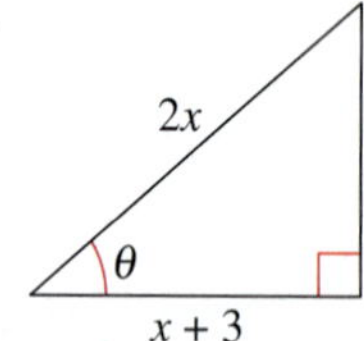

42.

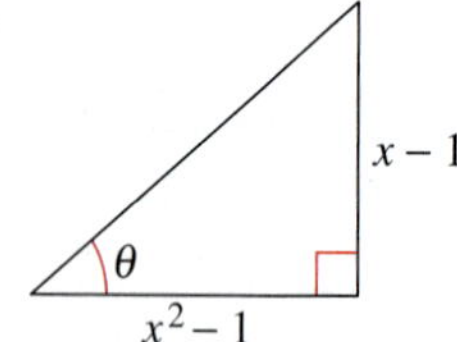

In Exercises 43–48, use the properties of inverse trigonometric functions to evaluate the expression.

43. $\sin(\arcsin 0.3)$
44. $\tan(\arctan 25)$
45. $\cos[\arccos(-0.1)]$
46. $\sin[\arcsin(-0.2)]$
47. $\arcsin(\sin 3\pi)$
48. $\arccos\left(\cos \frac{7\pi}{2}\right)$

In Exercises 49–58, find the exact value of the expression. (*Hint:* Make a sketch of a right triangle.)

49. $\sin\left(\arctan \frac{3}{4}\right)$
50. $\sec\left(\arcsin \frac{4}{5}\right)$
51. $\cos(\tan^{-1} 2)$
52. $\sin\left(\cos^{-1} \frac{\sqrt{5}}{5}\right)$
53. $\cos\left(\arcsin \frac{5}{13}\right)$
54. $\csc\left[\arctan\left(-\frac{5}{12}\right)\right]$
55. $\sec\left[\arctan\left(-\frac{3}{5}\right)\right]$
56. $\tan\left[\arcsin\left(-\frac{3}{4}\right)\right]$
57. $\sin\left[\arccos\left(-\frac{2}{3}\right)\right]$
58. $\cot\left(\arctan \frac{5}{8}\right)$

In Exercises 59–68, write an algebraic expression that is equivalent to the expression. (*Hint:* Sketch a right triangle, as demonstrated in Example 7.)

59. $\cot(\arctan x)$

60. $\sin(\arctan x)$

61. $\cos(\arcsin 2x)$

62. $\sec(\arctan 3x)$

63. $\sin(\arccos x)$

64. $\sec[\arcsin(x - 1)]$

65. $\tan\left(\arccos \dfrac{x}{3}\right)$

66. $\cot\left(\arctan \dfrac{1}{x}\right)$

67. $\csc\left(\arctan \dfrac{x}{\sqrt{2}}\right)$

68. $\cos\left(\arcsin \dfrac{x - h}{r}\right)$

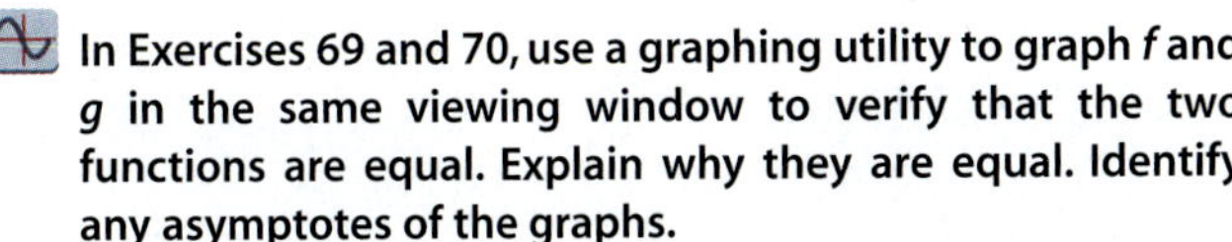

In Exercises 69 and 70, use a graphing utility to graph *f* and *g* in the same viewing window to verify that the two functions are equal. Explain why they are equal. Identify any asymptotes of the graphs.

69. $f(x) = \sin(\arctan 2x), \quad g(x) = \dfrac{2x}{\sqrt{1 + 4x^2}}$

70. $f(x) = \tan\left(\arccos \dfrac{x}{2}\right), \quad g(x) = \dfrac{\sqrt{4 - x^2}}{x}$

In Exercises 71–74, fill in the blank.

71. $\arctan \dfrac{9}{x} = \arcsin(\quad), \quad x \neq 0$

72. $\arcsin \dfrac{\sqrt{36 - x^2}}{6} = \arccos(\quad), \quad 0 \le x \le 6$

73. $\arccos \dfrac{3}{\sqrt{x^2 - 2x + 10}} = \arcsin(\quad)$

74. $\arccos \dfrac{x - 2}{2} = \arctan(\quad), \quad |x - 2| \le 2$

In Exercises 75–82, sketch a graph of the function.

75. $y = 2 \arccos x$

76. $y = \arcsin \dfrac{x}{2}$

77. $f(x) = \arcsin(x - 1)$

78. $g(t) = \arccos(t + 2)$

79. $f(x) = \arctan 2x$

80. $f(x) = \dfrac{\pi}{2} + \arctan x$

81. $h(v) = \tan(\arccos v)$

82. $f(x) = \arccos \dfrac{x}{4}$

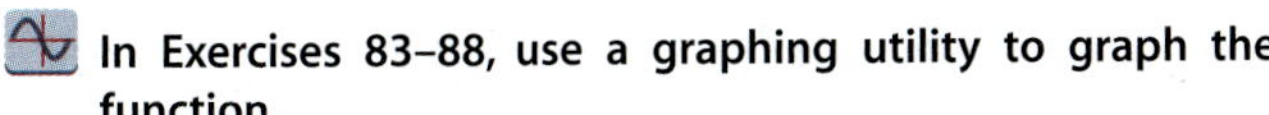

In Exercises 83–88, use a graphing utility to graph the function.

83. $f(x) = 2 \arccos(2x)$

84. $f(x) = \pi \arcsin(4x)$

85. $f(x) = \arctan(2x - 3)$

86. $f(x) = -3 + \arctan(\pi x)$

87. $f(x) = \pi - \sin^{-1}\left(\dfrac{2}{3}\right)$

88. $f(x) = \dfrac{\pi}{2} + \cos^{-1}\left(\dfrac{1}{\pi}\right)$

In Exercises 89 and 90, write the function in terms of the sine function by using the identity

$$A \cos \omega t + B \sin \omega t = \sqrt{A^2 + B^2} \sin\left(\omega t + \arctan \frac{A}{B}\right).$$

Use a graphing utility to graph both forms of the function. What does the graph imply?

89. $f(t) = 3 \cos 2t + 3 \sin 2t$

90. $f(t) = 4 \cos \pi t + 3 \sin \pi t$

91. ***Docking a Boat*** A boat is pulled in by means of a winch located on a dock 5 feet above the deck of the boat (see figure). Let θ be the angle of elevation from the boat to the winch and let s be the length of the rope from the winch to the boat.

(a) Write θ as a function of s.

(b) Find θ when $s = 40$ feet and $s = 20$ feet.

92. ***Photography*** A television camera at ground level is filming the lift-off of a space shuttle at a point 750 meters from the launch pad (see figure). Let θ be the angle of elevation to the shuttle and let s be the height of the shuttle.

(a) Write θ as a function of s.

(b) Find θ when $s = 300$ meters and $s = 1200$ meters.

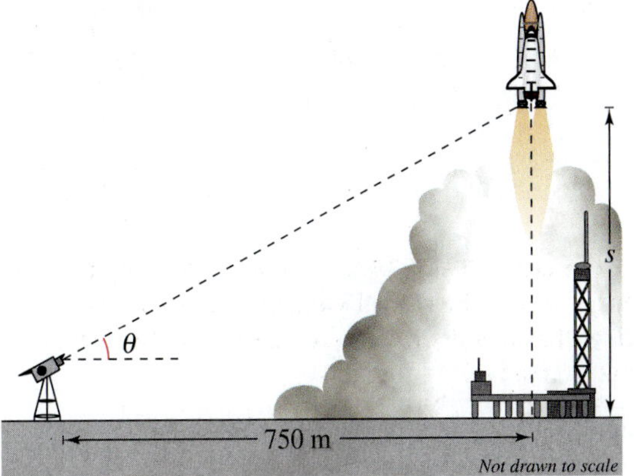

▶ Model It

93. ***Photography*** A photographer is taking a picture of a three-foot-tall painting hung in an art gallery. The camera lens is 1 foot below the lower edge of the painting (see figure). The angle β subtended by the camera lens x feet from the painting is

$$\beta = \arctan \frac{3x}{x^2 + 4}, \qquad x > 0.$$

(a) Use a graphing utility to graph β as a function of x.

(b) Move the cursor along the graph to approximate the distance from the picture when β is maximum.

(c) Identify the asymptote of the graph and discuss its meaning in the context of the problem.

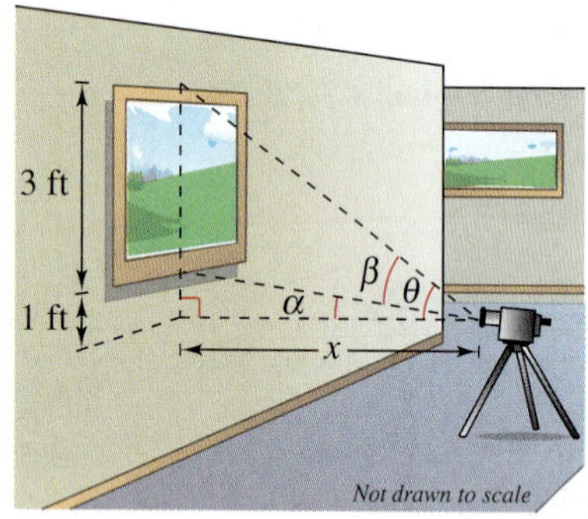

94. ***Granular Angle of Repose*** Different types of granular substances naturally settle at different angles when stored in cone-shaped piles. This angle θ is called the *angle of repose* (see figure). When rock salt is stored in a cone-shaped pile 11 feet high, the diameter of the pile's base is about 34 feet. (Source: Bulk-Store Structures, Inc.)

(a) Find the angle of repose for rock salt.

(b) How tall is a pile of rock salt that has a base diameter of 40 feet?

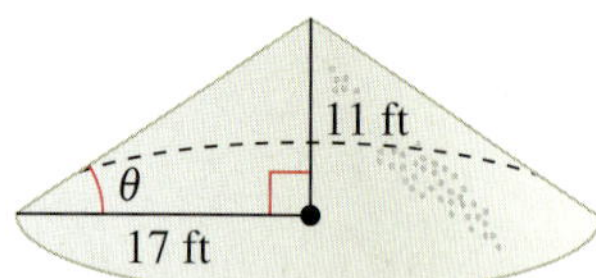

95. ***Granular Angle of Repose*** When whole corn is stored in a cone-shaped pile 20 feet high, the diameter of the pile's base is about 82 feet.

(a) Find the angle of repose for whole corn.

(b) How tall is a pile of corn that has a base diameter of 100 feet?

96. ***Angle of Elevation*** An airplane flies at an altitude of 6 miles toward a point directly over an observer. Consider θ and x as shown in the figure.

(a) Write θ as a function of x.

(b) Find θ when $x = 7$ miles and $x = 1$ mile.

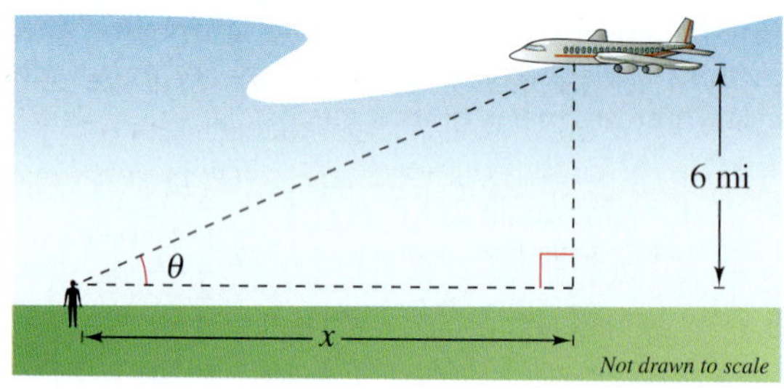

97. ***Security Patrol*** A security car with its spotlight on is parked 20 meters from a warehouse. Consider θ and x as shown in the figure.

(a) Write θ as a function of x.

(b) Find θ when $x = 5$ meters and $x = 12$ meters.

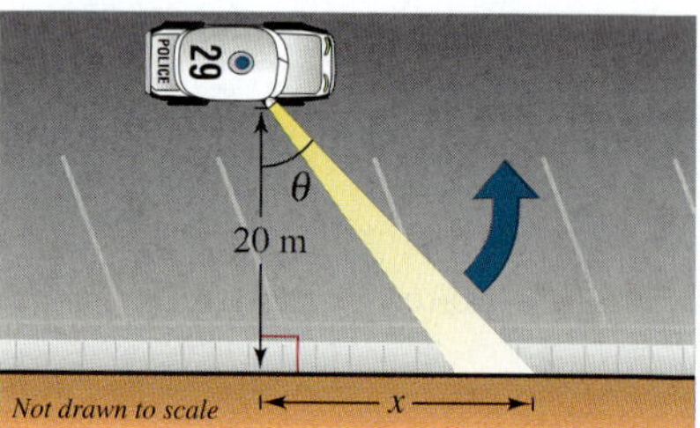

Synthesis

True or False? **In Exercises 98 and 99, determine whether the statement is true or false. Justify your answer.**

98. $\sin \frac{5\pi}{6} = \frac{1}{2}$ ⇨ $\arcsin \frac{1}{2} = \frac{5\pi}{6}$

99. $\tan \frac{5\pi}{4} = 1$ ⇨ $\arctan 1 = \frac{5\pi}{4}$

100. Define the inverse cotangent function by restricting the domain of the cotangent function to the interval $(0, \pi)$, and sketch its graph.

101. Define the inverse secant function by restricting the domain of the secant function to the intervals $[0, \pi/2)$ and $(\pi/2, \pi]$, and sketch its graph.

102. Define the inverse cosecant function by restricting the domain of the cosecant function to the intervals $[-\pi/2, 0)$ and $(0, \pi/2]$, and sketch its graph.

103. Use the results of Exercises 100–102 to evaluate each expression without using a calculator.

(a) $\operatorname{arcsec} \sqrt{2}$ (b) $\operatorname{arcsec} 1$

(c) $\operatorname{arccot}(-\sqrt{3})$ (d) $\operatorname{arccsc} 2$

104. ***Area*** In calculus, it is shown that the area of the region bounded by the graphs of $y = 0$, $y = 1/(x^2 + 1)$, $x = a$, and $x = b$ is given by

Area $= \arctan b - \arctan a$

(see figure). Find the area for the following values of a and b.

(a) $a = 0, b = 1$ (b) $a = -1, b = 1$

(c) $a = 0, b = 3$ (d) $a = -1, b = 3$

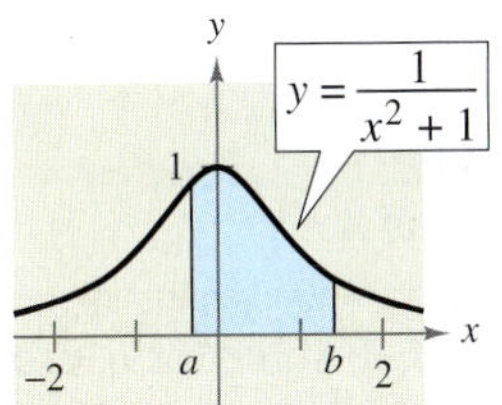

105. ***Think About It*** Use a graphing utility to graph the functions

$f(x) = \sqrt{x}$ and $g(x) = 6 \arctan x$.

For $x > 0$, it appears that $g > f$. Explain why you know that there exists a positive real number a such that $g < f$ for $x > a$. Approximate the number a.

106. ***Think About It*** Consider the functions

$f(x) = \sin x$ and $f^{-1}(x) = \arcsin x$.

(a) Use a graphing utility to graph the composite functions $f \circ f^{-1}$ and $f^{-1} \circ f$.

(b) Explain why the graphs in part (a) are not the graph of the line $y = x$. Why do the graphs of $f \circ f^{-1}$ and $f^{-1} \circ f$ differ?

Proof **In Exercises 107–112, prove the identity.**

107. $\arcsin(-x) = -\arcsin x$

108. $\arctan(-x) = -\arctan x$

109. $\arccos(-x) = \pi - \arccos x$

110. $\arctan x + \arctan \frac{1}{x} = \frac{\pi}{2}, \quad x > 0$

111. $\arcsin x + \arccos x = \frac{\pi}{2}$

112. $\arcsin x = \arctan \frac{x}{\sqrt{1 - x^2}}$

Review

In Exercises 113–116, sketch a right triangle corresponding to the trigonometric function of the acute angle θ. Use the Pythagorean Theorem to determine the third side.

113. $\sin \theta = \frac{3}{4}$

114. $\tan \theta = 2$

115. $\cos \theta = \frac{5}{6}$

116. $\sec \theta = 3$

In Exercises 117–120, evaluate the expression. Round your result to three decimal places.

117. $(8.2)^{3.4}$

118. $10(14)^{-2}$

119. $(1.1)^{50}$

120. $16^{-2\pi}$

121. ***Partnership Costs*** A group of people agree to share equally in the cost of a $250,000 endowment to a college. If they could find two more people to join the group, each person's share of the cost would decrease by $6250. How many people are presently in the group?

122. ***Speed*** A boat travels at a speed of 18 miles per hour in still water. It travels 35 miles upstream and then returns to the starting point in a total of 4 hours. Find the speed of the current.

6.7 Applications and Models

▶ What you should learn

- How to solve real-life problems involving right triangles
- How to solve real-life problems involving directional bearings
- How to solve real-life problems involving harmonic motion

▶ Why you should learn it

Trigonometric functions frequently model real-life problems involving cyclical patterns in business. For instance, in Exercise 63 on page 530, you can find a trigonometric model for the sales of an outerwear manufacturer.

Nancy Richmond/The Image Works

Applications Involving Right Triangles

In this section the three angles of a right triangle are denoted by the letters A, B, and C (where C is the right angle), and the lengths of the sides opposite these angles by the letters a, b, and c (where c is the hypotenuse).

Example 1 ▶ Solving a Right Triangle 

Solve the right triangle shown in Figure 6.76 for all unknown sides and angles.

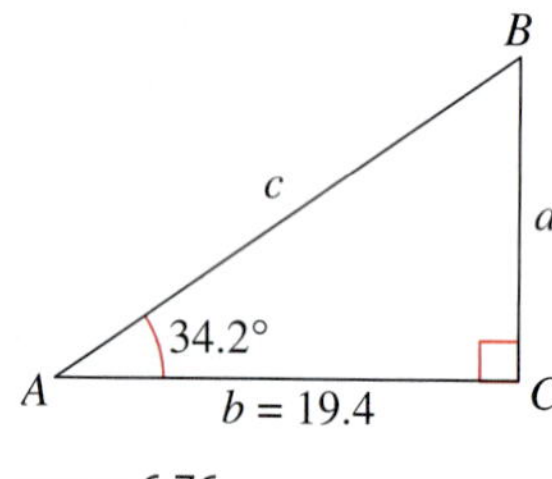

FIGURE 6.76

Solution

Because $C = 90°$, it follows that $A + B = 90°$ and $B = 90° - 34.2° = 55.8°$. To solve for a, use the fact that

$$\tan A = \frac{\text{opp}}{\text{adj}} = \frac{a}{b} \quad \Rightarrow \quad a = b \tan A.$$

So, $a = 19.4 \tan 34.2° \approx 13.18$. Similarly, to solve for c, use the fact that

$$\cos A = \frac{\text{adj}}{\text{hyp}} = \frac{b}{c} \quad \Rightarrow \quad c = \frac{b}{\cos A}.$$

So, $c = \dfrac{19.4}{\cos 34.2°} \approx 23.46$.

Example 2 ▶ Finding a Side of a Right Triangle 

A safety regulation states that the maximum angle of elevation for a rescue ladder is 72°. A fire department's longest ladder is 110 feet. What is the maximum safe rescue height?

Solution

A sketch is shown in Figure 6.77. From the equation $\sin A = a/c$, it follows that

$$a = c \sin A = 110 \sin 72° \approx 104.6.$$

So, the maximum safe rescue height is about 104.6 feet above the height of the fire truck.

FIGURE 6.77

Example 3 ▶ Finding a Side of a Right Triangle

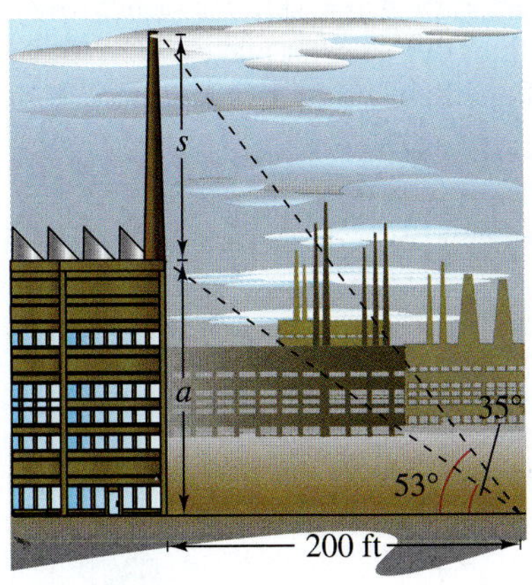

FIGURE 6.78

At a point 200 feet from the base of a building, the angle of elevation to the *bottom* of a smokestack is 35°, whereas the angle of elevation to the *top* is 53°, as shown in Figure 6.78. Find the height s of the smokestack alone.

Solution

Note from Figure 6.78 that this problem involves two right triangles. For the smaller right triangle, use the fact that

$$\tan 35^\circ = \frac{a}{200}$$

to conclude that the height of the building is

$$a = 200 \tan 35^\circ.$$

For the larger right triangle, use the equation

$$\tan 53^\circ = \frac{a + s}{200}$$

to conclude that $a + s = 200 \tan 53^\circ$. So, the height of the smokestack is

$$\begin{aligned} s &= 200 \tan 53^\circ - a \\ &= 200 \tan 53^\circ - 200 \tan 35^\circ \\ &\approx 125.4 \text{ feet.} \end{aligned}$$

Example 4 ▶ Finding an Acute Angle of a Right Triangle

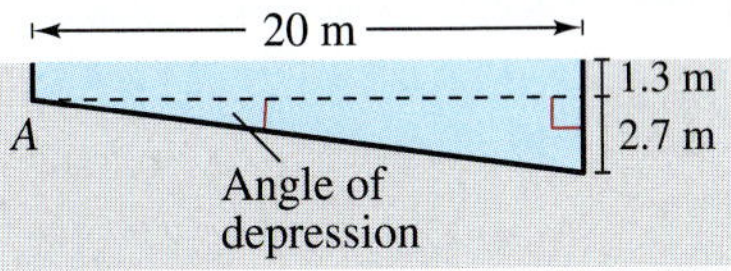

FIGURE 6.79

A swimming pool is 20 meters long and 12 meters wide. The bottom of the pool is slanted so that the water depth is 1.3 meters at the shallow end and 4 meters at the deep end, as shown in Figure 6.79. Find the angle of depression of the bottom of the pool.

Solution

Using the tangent function, you can see that

$$\begin{aligned} \tan A &= \frac{\text{opp}}{\text{adj}} \\ &= \frac{2.7}{20} \\ &= 0.135. \end{aligned}$$

So, the angle of depression is

$$\begin{aligned} A &= \arctan 0.135 \\ &\approx 0.13419 \text{ radian} \\ &\approx 7.69^\circ. \end{aligned}$$

Trigonometry and Bearings

In surveying and navigation, directions are generally given in terms of **bearings.** A bearing measures the acute angle that a path or line of sight makes with a fixed north-south line, as shown in Figure 6.80. For instance, the bearing S 35° E in Figure 6.80 means 35 degrees east of south.

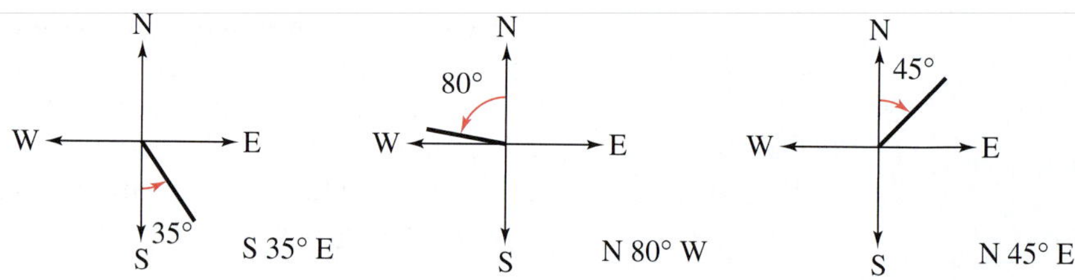

FIGURE 6.80

Example 5 ▶ Finding Directions in Terms of Bearings

A ship leaves port at noon and heads due west at 20 knots, or 20 nautical miles (nm) per hour. At 2 P.M. the ship changes course to N 54° W, as shown in Figure 6.81. Find the ship's bearing and distance from the port of departure at 3 P.M.

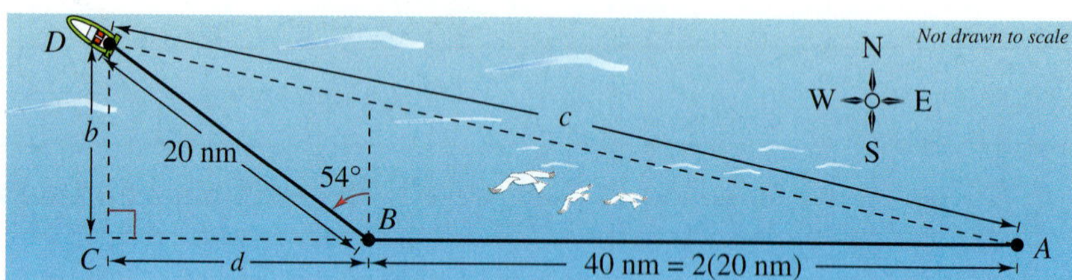

FIGURE 6.81

Solution

For triangle BCD, you have $B = 90° - 54° = 36°$. The two sides of this triangle can be determined to be

$$b = 20 \sin 36° \qquad \text{and} \qquad d = 20 \cos 36°.$$

For triangle ACD, you can find angle A as follows.

$$\tan A = \frac{b}{d + 40} = \frac{20 \sin 36°}{20 \cos 36° + 40} \approx 0.2092494$$

$$A \approx \arctan 0.2092494 \approx 0.2062732 \text{ radian} \approx 11.82°$$

The angle with the north-south line is $90° - 11.82° = 78.18°$. So, the bearing of the ship is N 78.18° W. Finally, from triangle ACD, you have $\sin A = b/c$, which yields

$$c = \frac{b}{\sin A} = \frac{20 \sin 36°}{\sin 11.82°}$$

$$\approx 57.4 \text{ nautical miles.}$$ Distance from port

STUDY TIP

In *air navigation*, bearings are measured in degrees *clockwise* from north. Examples of air navigation bearings are shown below.

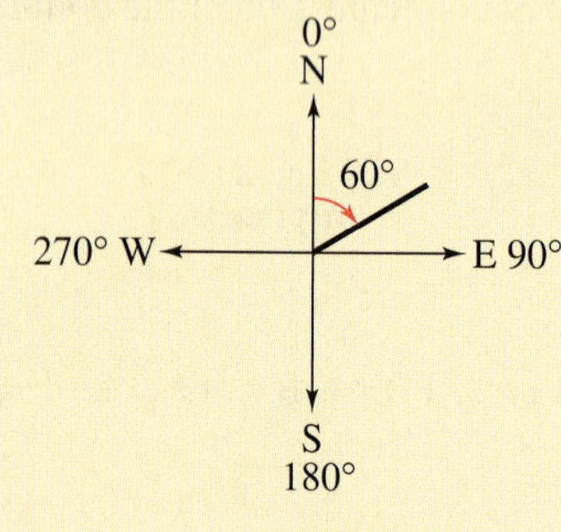

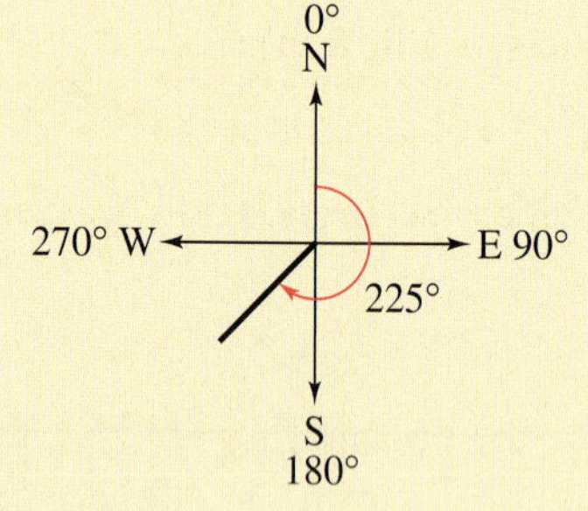

Harmonic Motion

The periodic nature of the trigonometric functions is useful for describing the motion of a point on an object that vibrates, oscillates, rotates, or is moved by wave motion.

For example, consider a ball that is bobbing up and down on the end of a spring, as shown in Figure 6.82. Suppose that 10 centimeters is the maximum distance the ball moves vertically upward or downward from its equilibrium (at rest) position. Suppose further that the time it takes for the ball to move from its maximum displacement above zero to its maximum displacement below zero and back again is $t = 4$ seconds. Assuming the ideal conditions of perfect elasticity and no friction or air resistance, the ball would continue to move up and down in a uniform and regular manner.

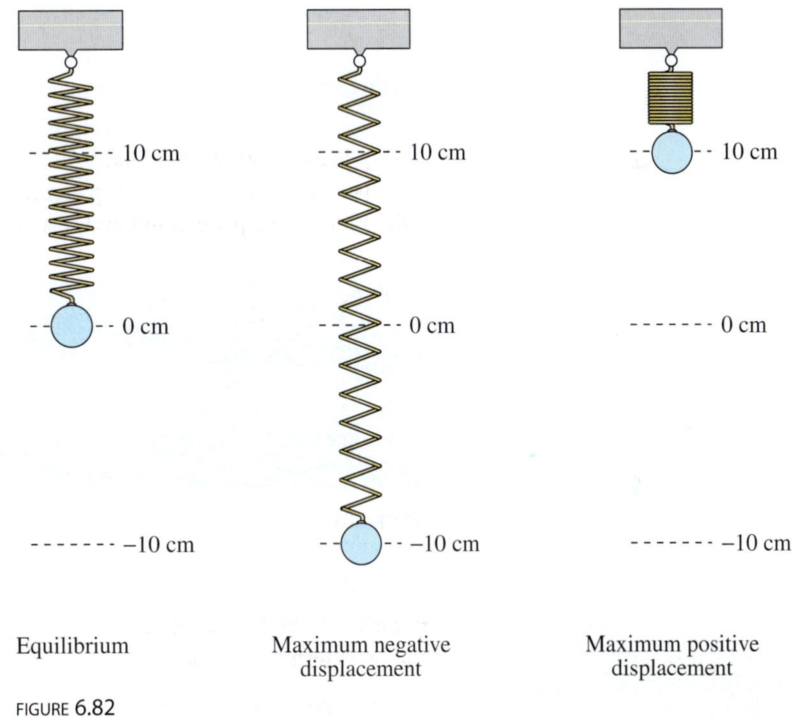

FIGURE 6.82

From this spring you can conclude that the period (time for one complete cycle) of the motion is

Period $=$ 4 seconds

and that its amplitude (maximum displacement from equilibrium) is

Amplitude $=$ 10 centimeters.

Motion of this nature can be described by a sine or cosine function, and is called **simple harmonic motion.**

> **Definition of Simple Harmonic Motion**
>
> A point that moves on a coordinate line is said to be in **simple harmonic motion** if its distance d from the origin at time t is given by either
>
> $$d = a \sin \omega t \qquad \text{or} \qquad d = a \cos \omega t$$
>
> where a and ω are real numbers such that $\omega > 0$. The motion has amplitude $|a|$, period $2\pi/\omega$, and frequency $\omega/(2\pi)$.

Example 6 ▶ Simple Harmonic Motion  

Write the equation for the simple harmonic motion of the ball described in Figure 6.82, where the period is 4 seconds. What is the frequency of this harmonic motion?

Solution

Because the spring is at equilibrium ($d = 0$) when $t = 0$, you use the equation

$$d = a \sin \omega t.$$

Moreover, because the maximum displacement from zero is 10 and the period is 4, you have

$$\text{Amplitude} = |a| = 10$$

$$\text{Period} = \frac{2\pi}{\omega} = 4 \quad \Longrightarrow \quad \omega = \frac{\pi}{2}.$$

Consequently, the equation of motion is

$$d = 10 \sin \frac{\pi}{2} t.$$

Note that the choice of $a = 10$ or $a = -10$ depends on whether the ball initially moves up or down. The frequency is

$$\text{Frequency} = \frac{\omega}{2\pi}$$

$$= \frac{\pi/2}{2\pi}$$

$$= \frac{1}{4} \text{ cycle per second.}$$

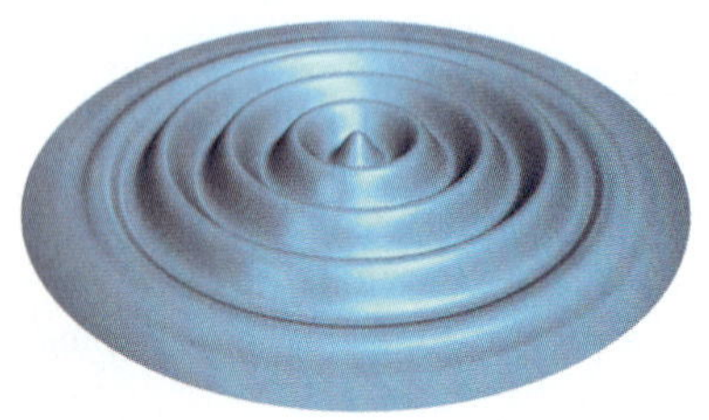

FIGURE 6.83

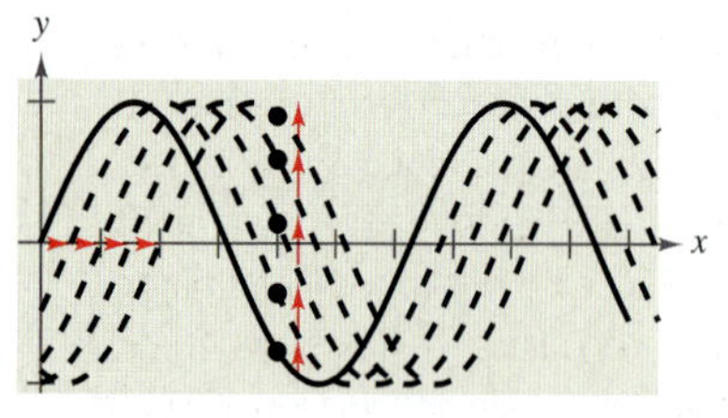

FIGURE 6.84

One illustration of the relationship between sine waves and harmonic motion can be seen in the wave motion resulting when a stone is dropped into a calm pool of water. The waves move outward in roughly the shape of sine (or cosine) waves, as shown in Figure 6.83. As an example, suppose you are fishing and your fishing bob is attached so that it does not move horizontally. As the waves move outward from the dropped stone, your fishing bob will move up and down in simple harmonic motion, as shown in Figure 6.84.

Example 7 ▶ Simple Harmonic Motion

Given the equation for simple harmonic motion

$$d = 6 \cos \frac{3\pi}{4} t$$

find (a) the maximum displacement, (b) the frequency, (c) the value of d when $t = 4$, and (d) the least positive value of t for which $d = 0$.

Solution

The given equation has the form $d = a \cos \omega t$, with $a = 6$ and $\omega = 3\pi/4$.

a. The maximum displacement (from the point of equilibrium) is given by the amplitude. So, the maximum displacement is 6.

b. $\text{Frequency} = \dfrac{\omega}{2\pi}$

$$= \frac{3\pi/4}{2\pi} = \frac{3}{8} \text{ cycle per unit of time}$$

c. $d = 6 \cos\left[\dfrac{3\pi}{4}(4)\right]$

$$= 6 \cos 3\pi$$

$$= 6(-1)$$

$$= -6$$

d. To find the least positive value of t for which $d = 0$, solve the equation

$$d = 6 \cos \frac{3\pi}{4} t = 0$$

to obtain

$$\frac{3\pi}{4} t = \frac{\pi}{2}, \frac{3\pi}{2}, \frac{5\pi}{2}, \ldots$$

$$t = \frac{2}{3}, 2, \frac{10}{3}, \ldots$$

So, the least positive value of t is $t = \frac{2}{3}$.

(a)

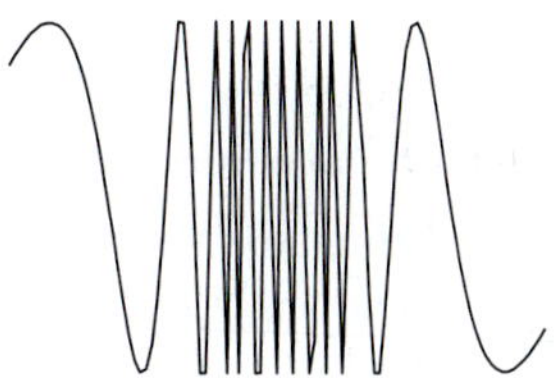

(b)

FIGURE 6.85

Writing ABOUT MATHEMATICS

Radio Waves Many different physical phenomena can be characterized by wave motion. These phenomena include electromagnetic waves such as radio waves, television waves, and microwaves. Radio waves transmit sound in two different ways. For an AM station, the *amplitude* of the wave is modified to carry sound. The letters AM stand for "amplitude modulation." An FM radio signal has its *frequency* modified in order to carry sound, hence the term "frequency modulation." The FM radio signal is preferred by listeners because of its low-noise and wide-bandwidth qualities. Of the two graphs in Figure 6.85, one shows an AM wave and the other shows an FM wave. Which is which? Explain your reasoning.

6.7 Exercises

In Exercises 1–10, solve the right triangle shown in the figure. Round your answer to two decimal places.

1. $A = 20°, \quad b = 10$
2. $B = 54°, \quad c = 15$
3. $B = 71°, \quad b = 24$
4. $A = 8.4°, \quad a = 40.5$
5. $a = 6, \quad b = 10$
6. $a = 25, \quad c = 35$
7. $b = 16, \quad c = 52$
8. $b = 1.32, \quad c = 9.45$
9. $A = 12°15', \quad c = 430.5$
10. $B = 65°12', \quad a = 14.2$

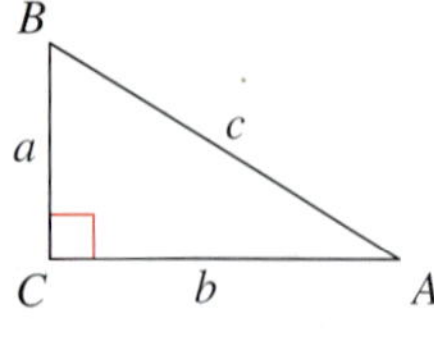

FIGURE FOR 1–10

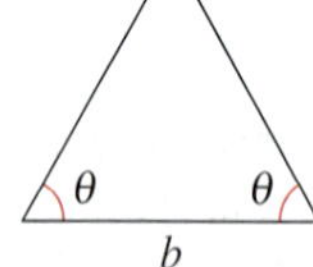

FIGURE FOR 11–14

In Exercises 11–14, find the altitude of the isosceles triangle shown in the figure. Round your answer to two decimal places.

11. $\theta = 52°, \; b = 4$ inches
12. $\theta = 18°, \; b = 10$ meters
13. $\theta = 41°, \; b = 46$ inches
14. $\theta = 27°, \; b = 11$ feet

15. ***Length*** The sun is 25° above the horizon. Find the length of a shadow cast by a silo that is 50 feet tall (see figure).

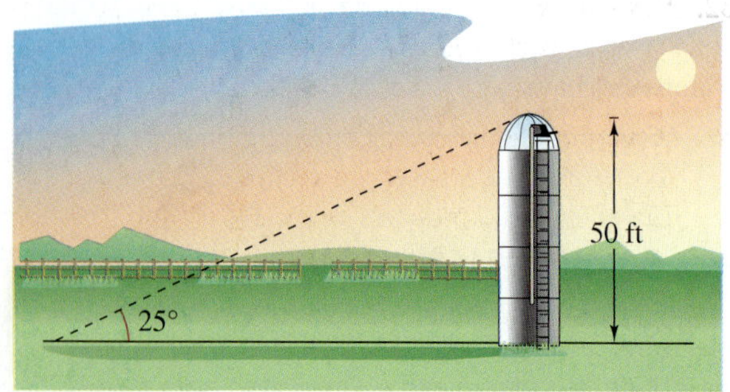

16. ***Length*** The sun is 20° above the horizon. Find the length of a shadow cast by a building that is 600 feet tall.

17. ***Height*** A ladder 20 feet long leans against the side of a house. Find the height h from the top of the ladder to the ground if the angle of elevation of the ladder is 80°.

18. ***Height*** The length of a shadow of a tree is 125 feet when the angle of elevation of the sun is 33°. Approximate the height h of the tree.

19. ***Height*** From a point 50 feet in front of a church, the angles of elevation to the base of the steeple and the top of the steeple are 35° and 47° 40′, respectively.
    (a) Draw right triangles that represent the problem. Label the known and unknown quantities.
    (b) Use a trigonometric function to write an equation involving the unknown height of the steeple.
    (c) Find the height of the steeple.

20. ***Height*** You are standing 100 feet from the base of a platform from which people are bungee jumping. The angle of elevation from your position to the top of the platform from which they jump is 51°. From what height are the people jumping?

21. ***Depth*** The sonar of a navy cruiser detects a submarine that is 4000 feet from the cruiser. The angle between the water line and the submarine is 34° (see figure). How deep is the submarine?

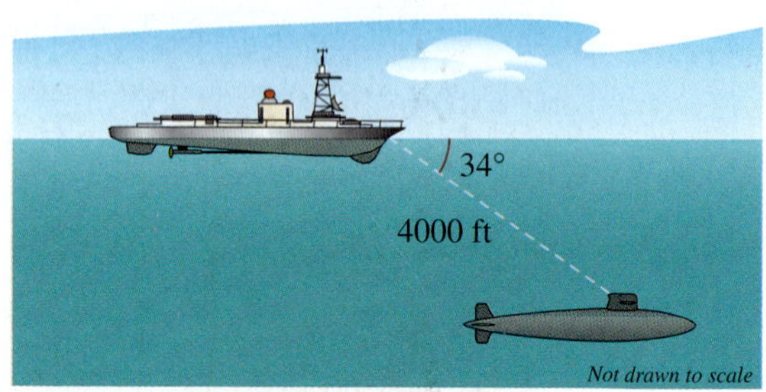

22. ***Angle of Elevation*** An amateur radio operator erects a 75-foot vertical tower for an antenna. Find the angle of elevation to the top of the tower at a point on level ground 50 feet from its base.

23. ***Angle of Elevation*** The height of an outdoor basketball backboard is $12\frac{1}{2}$ feet, and the backboard casts a shadow $17\frac{1}{3}$ feet long.
    (a) Draw a right triangle that represents the problem. Label the known and unknown quantities.
    (b) Use a trigonometric function to write an equation involving the unknown angle of elevation of the sun.
    (c) Find the angle of elevation of the sun.

24. ***Angle of Depression*** A Global Positioning System satellite orbits 12,500 miles above Earth's surface. Find the angle of depression from the satellite to the horizon. Assume the radius of Earth is 4000 miles.

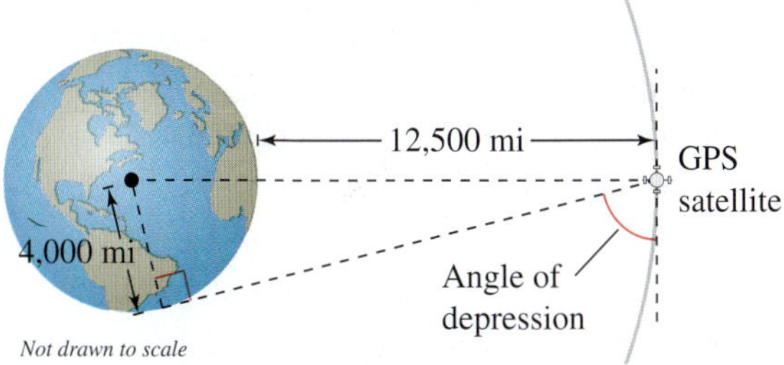

25. ***Angle of Depression*** A cellular telephone tower that is 150 feet tall is placed on top of a mountain that is 1200 feet above sea level. What is the angle of depression from the top of the tower to a cell phone user who is 5 horizontal miles away and 400 feet above sea level?

26. ***Airplane Ascent*** During takeoff, an airplane's angle of climb is 18° and its speed is 275 feet per second.

(a) Find the plane's altitude after 1 minute.

(b) How long will it take the plane to climb to an altitude of 10,000 feet?

27. ***Mountain Descent*** A sign on a roadway at the top of a mountain indicates that for the next 4 miles the grade is 10.5° (see figure). Find the change in elevation for a car descending the mountain.

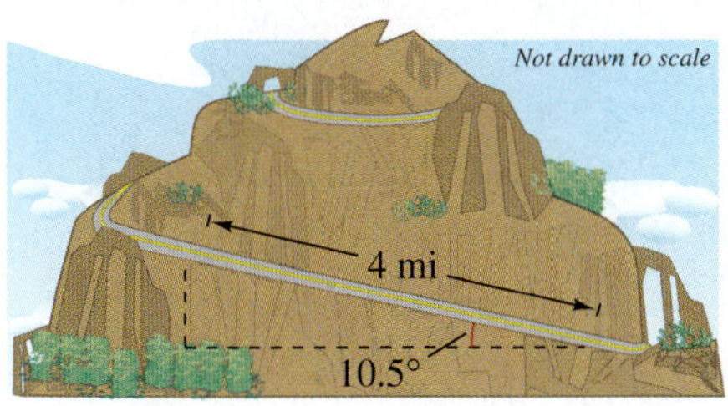

28. ***Mountain Descent*** A roadway sign at the top of a mountain indicates that for the next 4 miles the grade is 12%. Find the angle of the grade and the change in elevation for a car descending the mountain.

29. ***Navigation*** An airplane flying at 600 miles per hour has a bearing of 52°. After flying for 1.5 hours, how far north and how far east will the plane have traveled from its point of departure?

30. ***Surveying*** A surveyor wishes to find the distance across a swamp (see figure). The bearing from A to B is N 32° W. The surveyor walks 50 meters from A, and at the point C the bearing to B is N 68° W. Find (a) the bearing from A to C and (b) the distance from A to B.

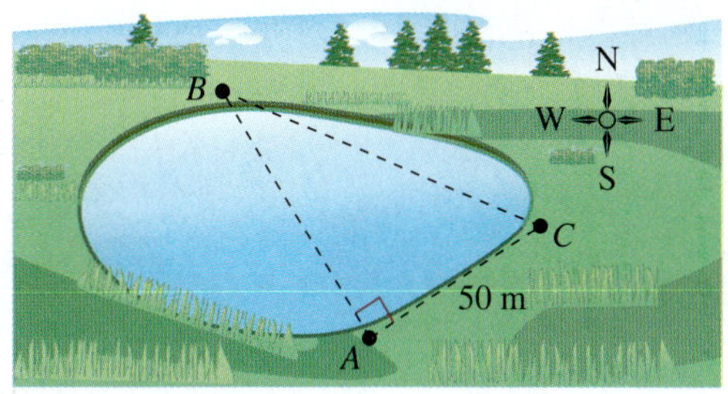

31. ***Location of a Fire*** Two fire towers are 30 kilometers apart, where tower A is due west of tower B. A fire is spotted from the towers, and the bearings from A and B are E 14° N and W 34° N, respectively (see figure). Find the distance d of the fire from the line segment AB.

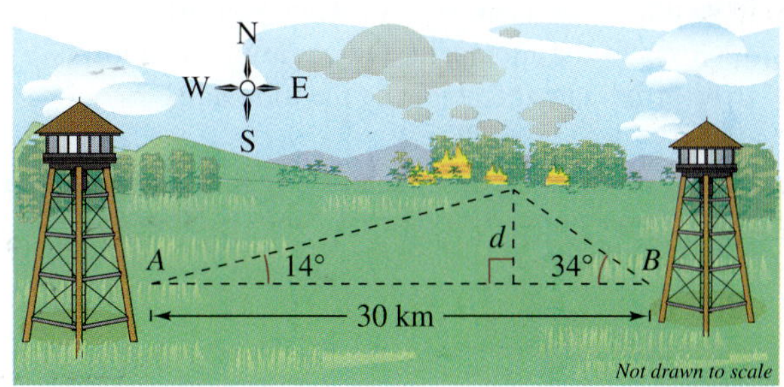

32. ***Navigation*** A ship is 45 miles east and 30 miles south of port. The captain wants to sail directly to port. What bearing should be taken?

33. ***Distance*** An observer in a lighthouse 350 feet above sea level observes two ships directly offshore. The angles of depression to the ships are 4° and 6.5° (see figure). How far apart are the ships?

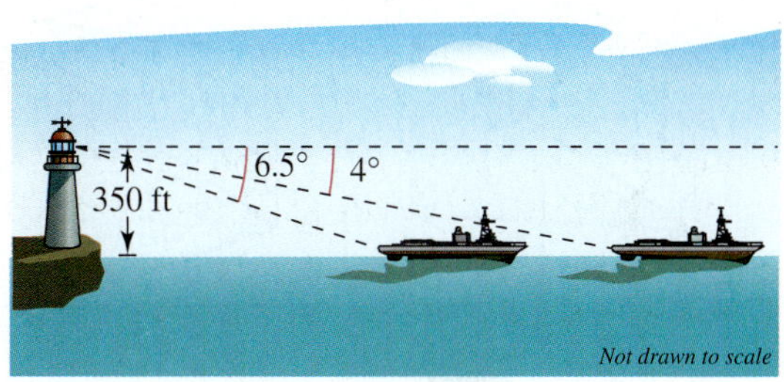

34. ***Distance*** A passenger in an airplane at an altitude of 10 kilometers sees two towns directly to the east of the plane. The angles of depression to the towns are 28° and 55° (see figure). How far apart are the towns?

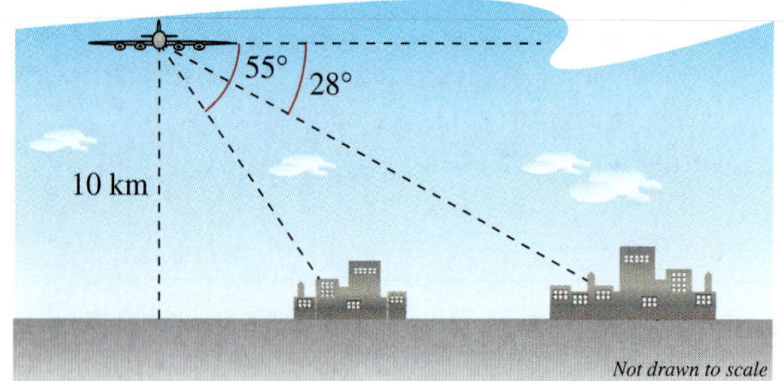

35. ***Altitude*** A plane is observed approaching your home and you assume that its speed is 550 miles per hour. The angle of elevation of the plane is 16° at one time and 57° one minute later. Approximate the altitude of the plane.

36. ***Height*** While traveling across flat land, you notice a mountain directly in front of you. The angle of elevation to the peak is 2.5°. After you drive 17 miles closer to the mountain, the angle of elevation is 9°. Approximate the height of the mountain.

Geometry **In Exercises 37 and 38, find the angle α between two nonvertical lines L_1 and L_2. The angle α satisfies the equation**

$$\tan \alpha = \left|\frac{m_2 - m_1}{1 + m_2 m_1}\right|$$

where m_1 and m_2 are the slopes of L_1 and L_2, respectively. (Assume that $m_1 m_2 \neq -1$.)

37. L_1: $3x - 2y = 5$
L_2: $x + y = 1$

38. L_1: $2x - y = 8$
L_2: $x - 5y = -4$

39. ***Geometry*** Determine the angle between the diagonal of a cube and the diagonal of its base, as shown in the figure.

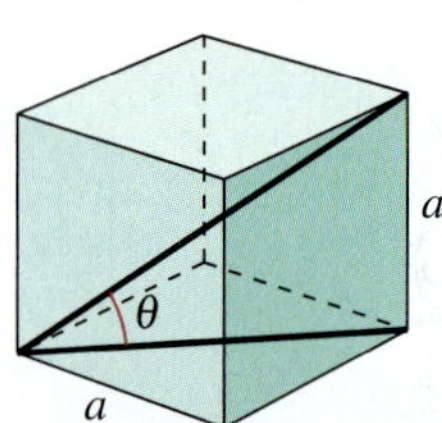

FIGURE FOR 39

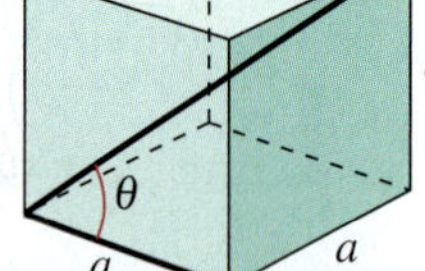

FIGURE FOR 40

40. ***Geometry*** Determine the angle between the diagonal of a cube and its edge, as shown in the figure.

41. ***Geometry*** Find the length of the sides of a regular pentagon inscribed in a circle of radius 25 inches.

42. ***Geometry*** Find the length of the sides of a regular hexagon inscribed in a circle of radius 25 inches.

43. ***Hardware*** Express the distance y across the flat sides of a hexagonal nut as a function of r, as shown in the figure.

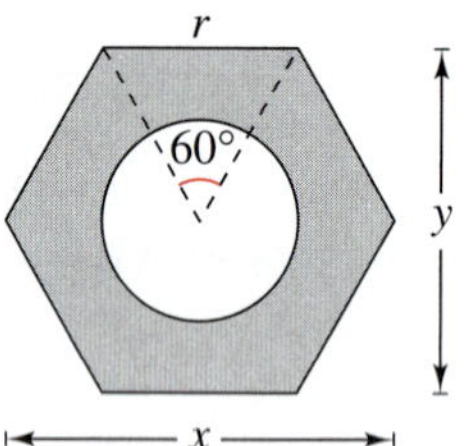

44. ***Bolt Holes*** The figure shows a circular piece of sheet metal that has a diameter of 40 centimeters and contains 12 equally spaced bolt holes. Determine the straight-line distance between the centers of consecutive bolt holes.

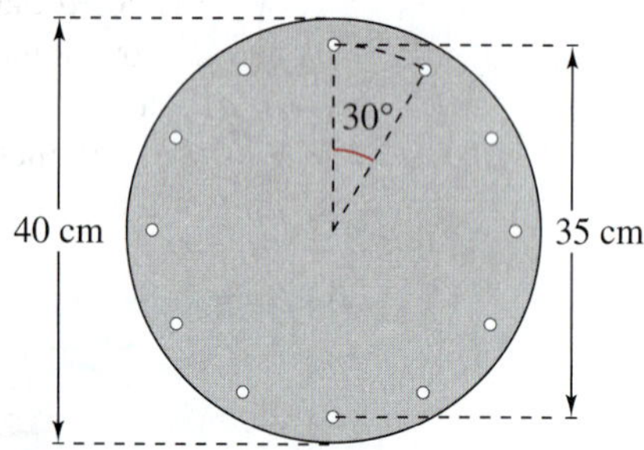

Trusses **In Exercises 45 and 46, find all the unknown lengths of the members of the truss.**

45.

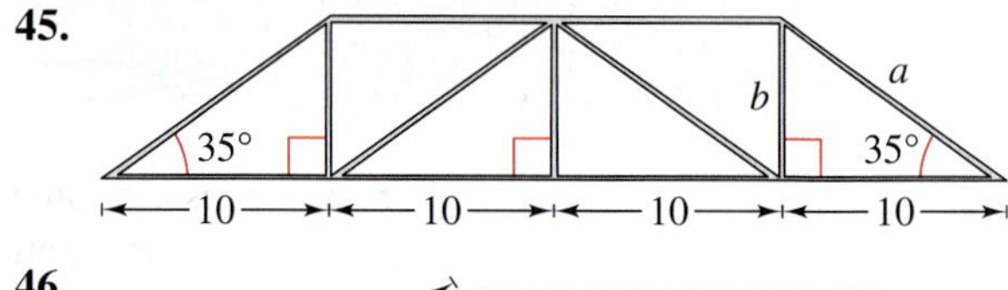

46.

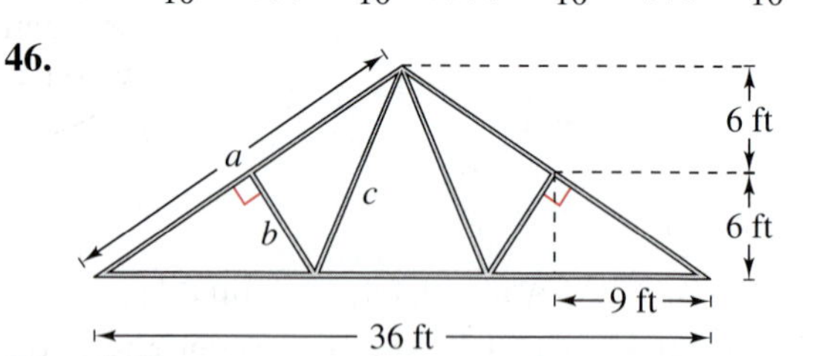

Harmonic Motion **In Exercises 47–50, for the simple harmonic motion described by the trigonometric function, find (a) the maximum displacement, (b) the frequency, and (c) the least positive value of *t* for which $d = 0$.**

47. $d = 4 \cos 8\pi t$

48. $d = \frac{1}{2} \cos 20\pi t$

49. $d = \frac{1}{16} \sin 120\pi t$

50. $d = \frac{1}{64} \sin 792\pi t$

Harmonic Motion **In Exercises 51–54, find a model for simple harmonic motion satisfying the specified conditions.**

	Displacement ($t = 0$)	*Amplitude*	*Period*
51.	0	4 centimeters	2 seconds
52.	0	3 meters	6 seconds
53.	3 inches	3 inches	1.5 seconds
54.	2 feet	2 feet	10 seconds

55. ***Tuning Fork*** A point on the end of a tuning fork moves in simple harmonic motion described by $d = a \sin \omega t$. Find ω given that the tuning fork for middle C has a frequency of 264 vibrations per second.

56. ***Wave Motion*** A buoy oscillates in simple harmonic motion as waves go past. It is noted that the buoy moves a total of 3.5 feet from its low point to its high point (see figure), and that it returns to its high point every 10 seconds. Write an equation that describes the motion of the buoy if its high point is at $t = 0$.

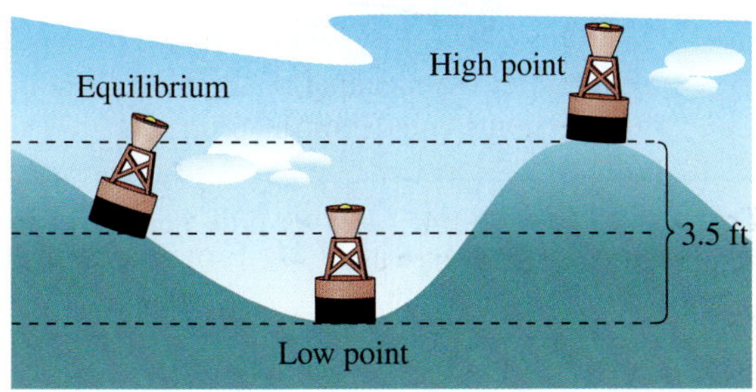

57. ***Oscillation of a Spring*** A ball that is bobbing up and down on the end of a spring has a maximum displacement of 3 inches. Its motion (in ideal conditions) is modeled by $y = \frac{1}{4} \cos 16t$ $(t > 0)$, where y is in feet and t is in seconds.

(a) Graph the function.

(b) What is the period of the oscillations?

(c) Determine the first time the weight passes the point of equilibrium $(y = 0)$.

Synthesis

True or False? **In Exercises 58 and 59, determine whether the statement is true or false. Justify your answer.**

58. The Leaning Tower of Pisa is not vertical, but if you know the exact angle of elevation θ to the 191-foot tower when you stand near it, then you can determine the exact distance to the tower d by using the formula

$$\tan \theta = \frac{191}{d}.$$

59. For the harmonic motion of a ball bobbing up and down on the end of a spring, one period can be described as the length of one coil of the spring.

60. ***Numerical and Graphical Analysis*** A two-meter-high fence is 3 meters from the side of a grain storage bin. A grain elevator must reach from ground level outside the fence to the storage bin (see figure). The objective is to determine the shortest elevator that meets the constraints.

(a) Complete four rows of the table.

θ	L_1	L_2	$L_1 + L_2$
0.1	$\frac{2}{\sin 0.1}$	$\frac{3}{\cos 0.1}$	23.0
0.2	$\frac{2}{\sin 0.2}$	$\frac{3}{\cos 0.2}$	13.1

 (b) Use a graphing utility to generate additional rows of the table. Use the table to estimate the minimum length of the elevator.

(c) Write the length $L_1 + L_2$ as a function of θ.

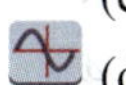 (d) Use a graphing utility to graph the function. Use the graph to estimate the minimum length. How does your estimate compare with that of part (b)?

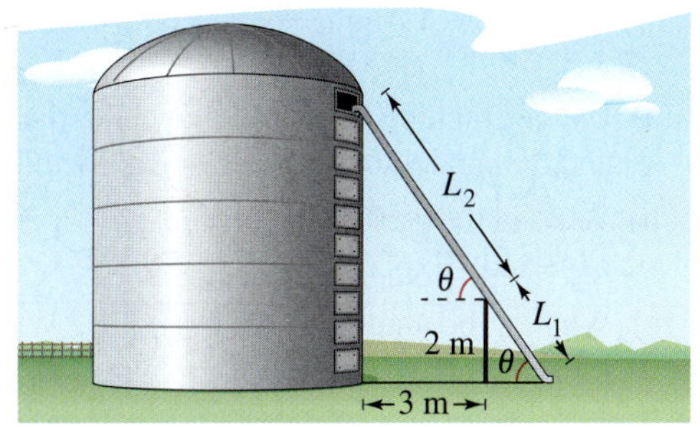

61. ***Numerical and Graphical Analysis*** The cross section of an irrigation canal is an isosceles trapezoid of which three of the sides are 8 feet long (see figure). The objective is to find the angle θ that maximizes the area of the cross section. [*Hint:* The area of a trapezoid is $(h/2)(b_1 + b_2)$.]

(a) Complete seven rows of the table.

Base 1	Base 2	Altitude	Area
8	$8 + 16 \cos 10°$	$8 \sin 10°$	22.1
8	$8 + 16 \cos 20°$	$8 \sin 20°$	42.5

(b) Use a graphing utility to generate additional rows of the table. Use the table to estimate the maximum cross-sectional area.

(c) Write the area A as a function of θ.

(d) Use a graphing utility to graph the function. Use the graph to estimate the maximum cross-sectional area. How does your estimate compare with that of part (b)?

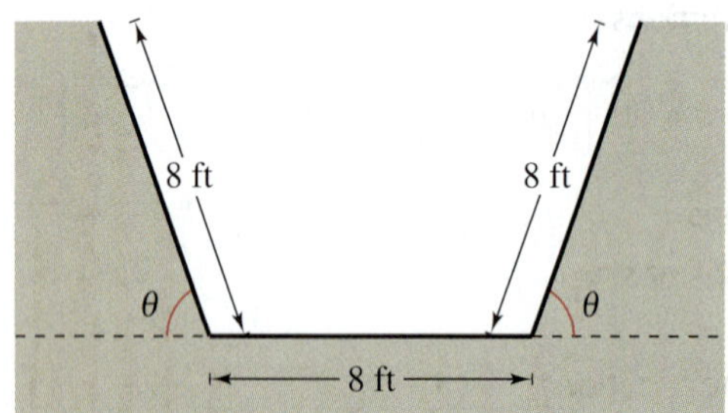

62. ***Data Analysis*** The times S of sunset (Greenwich Mean Time) at 40° north latitude on the 15th of each month are: 1(16:59), 2(17:35), 3(18:06), 4(18:38), 5(19:08), 6(19:30), 7(19:28), 8(18:57), 9(18:09), 10(17:21), 11(16:44), 12(16:36). The month is represented by t, with $t = 1$ corresponding to January. A model (in which minutes have been converted to the decimal parts of an hour) for this data is

$$S(t) = 18.09 + 1.41 \sin\left(\frac{\pi t}{6} + 4.60\right).$$

(a) Use a graphing utility to graph the data points and the model in the same viewing window.

(b) What is the period of the model? Is it what you expected? Explain.

(c) What is the amplitude of the model? What does it represent in the model? Explain.

▶ Model It

63. ***Data Analysis*** The table shows the average sales S (in millions of dollars) of an outerwear manufacturer for each month t, where $t = 1$ represents January.

Time, t	Sales, S
1	13.46
2	11.15
3	8.00
4	4.85
5	2.54
6	1.70
7	2.54
8	4.85
9	8.00
10	11.15
11	13.46
12	14.3

(a) Create a scatter plot of the data.

(b) Find a trigonometric model that fits the data. Graph the model on your scatter plot. How well does the model fit the data?

(c) What is the period of the model? Do you think it is reasonable given the context? Explain your reasoning.

(d) Interpret the meaning of the model's amplitude in the context of the problem.

64. ***Writing*** Is it true that N 24° E means 24 degrees north of east? Explain.

Review

In Exercises 65–72, sketch a graph of the equation.

65. $3x - 2y = 4$

66. $5y - 3x = 12$

67. $(y - 2)^2 = 8(x + 2)$

68. $(x + 3)^2 = 5y - 8$

69. $\dfrac{x^2}{4} + y^2 = 1$

70. $2x^2 + y^2 - 4 = 0$

71. $\dfrac{x^2}{4} + \dfrac{y^2}{4} = 1$

72. $(x - 2)^2 + y^2 = 25$

Chapter Summary

▶ *What* did you learn?

	Review Exercises
Section 6.1	
☐ How to describe angles	1–4
☐ How to use degree measure	5–8
☐ How to use radian measure	9–12
☐ How to convert between degree and radian measures	13–28
☐ How to use angles to model and solve real-life problems	29, 30
Section 6.2	
☐ How to evaluate trigonometric functions of acute angles	31–34
☐ How to use the fundamental trigonometric identities	35–38
☐ How to use a calculator to evaluate trigonometric functions	39–46
☐ How to use trigonometric functions to model and solve real-life problems	47, 48
Section 6.3	
☐ How to evaluate trigonometric functions of any angle	49–62
☐ How to use reference angles to evaluate trigonometric functions	63–72
☐ How to evaluate trigonometric functions of real numbers	73–84
Section 6.4	
☐ How to sketch the graphs of basic sine and cosine functions	85, 86
☐ How to use amplitude and period to help sketch the graphs of sine and cosine functions	87–90
☐ How to sketch translations of the graphs of sine and cosine functions	91–94
☐ How to use sine and cosine functions to model real-life data	95, 96
Section 6.5	
☐ How to sketch the graphs of tangent functions	97, 98
☐ How to sketch the graphs of cotangent functions	99, 100
☐ How to sketch the graphs of secant and cosecant functions	101–104
☐ How to sketch the graphs of damped trigonometric functions	105, 106
Section 6.6	
☐ How to evaluate the inverse sine function	107–112
☐ How to evaluate the other inverse trigonometric functions	113–124
☐ How to evaluate the compositions of trigonometric functions	125–132
Section 6.7	
☐ How to solve real-life problems involving right triangles	133, 134
☐ How to solve real-life problems involving directional bearings	135
☐ How to solve real-life problems involving harmonic motion	136

Review Exercises

6.1 In Exercises 1–4, estimate the number of degrees in the angle.

1. 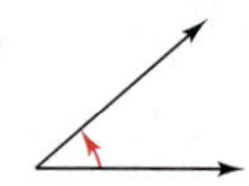

2. 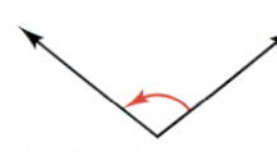

3. 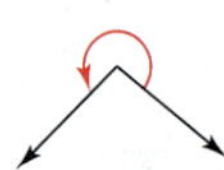

4. 

In Exercises 5–12, sketch the angle in standard position. List one positive and one negative coterminal angle.

5. $70°$

6. $280°$

7. $-110°$

8. $-405°$

9. $\frac{11\pi}{4}$

10. $\frac{2\pi}{9}$

11. $-\frac{4\pi}{3}$

12. $-\frac{23\pi}{3}$

In Exercises 13–20, convert the measure from radians to degrees. Round to two decimal places.

13. $\frac{5\pi}{7}$

14. $\frac{7\pi}{5}$

15. $-\frac{3\pi}{5}$

16. $-\frac{11\pi}{6}$

17. -3.5

18. -8.3

19. 1.75

20. 5.7

In Exercises 21–28, convert the measure from degrees to radians. Round to four decimal places.

21. $480°$

22. $120°$

23. $-16.5°$

24. $-127.5°$

25. $-33° 45'$

26. $-98° 25'$

27. $84° 15'$

28. $196° 77'$

29. ***Phonograph*** Compact discs have all but replaced phonograph records. Phonograph records are vinyl discs that rotate on a turntable. A typical record album is 12 inches in diameter and plays at $33\frac{1}{3}$ revolutions per minute.

(a) What is the angular speed of a record album?

(b) What is the linear speed of the outer edge of a record album?

30. ***Bicycle*** At what speed is a bicyclist traveling when his 27-inch-diameter tires are rotating at an angular speed of 5π radians per second?

6.2 In Exercises 31–34, find the exact values of the six trigonometric functions of the angle θ in the figure.

31. 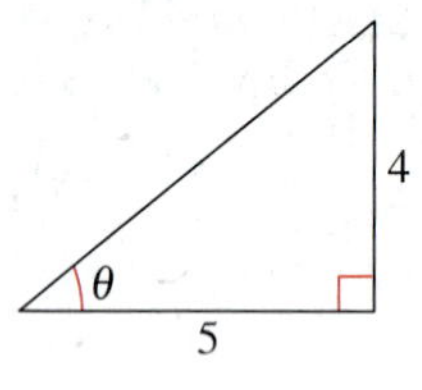

32. 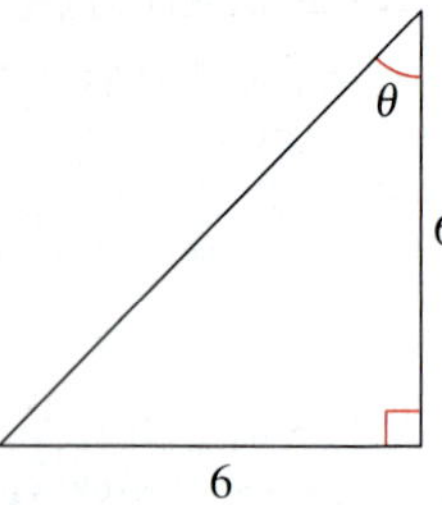

33. 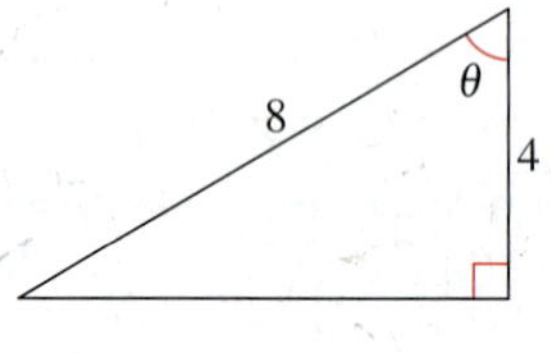

34. 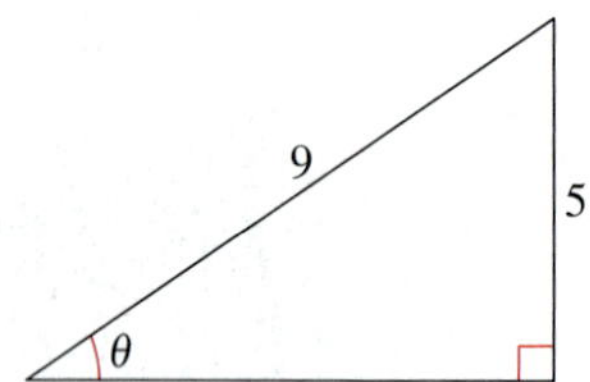

In Exercises 35–38, use the function value and trigonometric identities (including the cofunction identities) to find the indicated trigonometric functions.

35. $\sin \theta = \frac{1}{3}$ (a) $\csc \theta$ (b) $\cos \theta$ (c) $\sec \theta$ (d) $\tan \theta$

36. $\tan \theta = 4$ (a) $\cot \theta$ (b) $\sec \theta$ (c) $\cos \theta$ (d) $\csc \theta$

37. $\csc \theta = 4$ (a) $\sin \theta$ (b) $\cos \theta$ (c) $\sec \theta$ (d) $\tan \theta$

38. $\csc \theta = 5$ (a) $\sin \theta$ (b) $\cot \theta$ (c) $\tan \theta$ (d) $\sec(90° - \theta)$

In Exercises 39–46, use a calculator to evaluate the trigonometric function. Round your answer to two decimal places.

39. $\tan 33°$

40. $\csc 11°$

41. $\sin 34.2°$

42. $\sec 79.3°$

43. $\cot 15° 14'$

44. $\cos 78° 11' 58''$

45. $\cos \frac{\pi}{18}$

46. $\tan \frac{5\pi}{6}$

47. ***Railroad Grade*** A train travels 3.5 kilometers on a straight track with a grade of $1°\,10'$ (see figure). What is the vertical rise of the train in that distance?

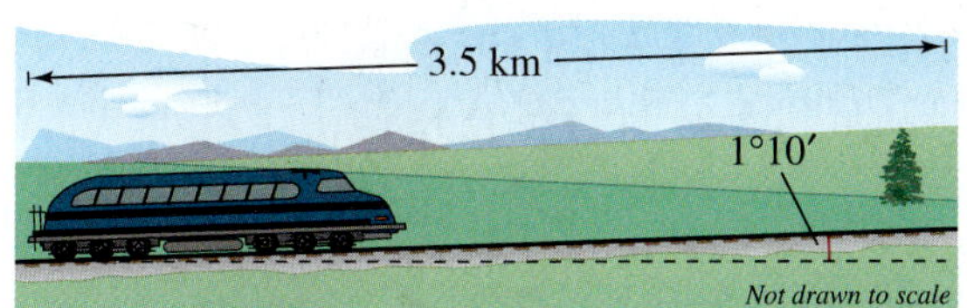

48. ***Guy Wire*** A guy wire runs from the ground to the top of a 25-foot telephone pole. The angle formed between the wire and the ground is 52°. How far from the base of the pole is the wire attached to the ground?

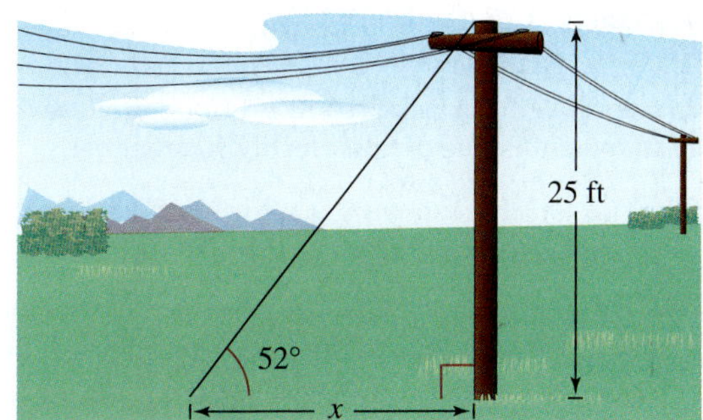

6.3 **In Exercises 49–56, find the exact values of the six trigonometric functions of the angle θ (in standard position) whose terminal side passes through the point.**

49. $(12, 16)$ **50.** $(3, -4)$

51. $\left(\frac{2}{3}, \frac{5}{2}\right)$ **52.** $\left(-\frac{10}{3}, -\frac{2}{3}\right)$

53. $(-0.5, 4.5)$ **54.** $(0.3, 0.4)$

55. $(x, 4x),\ x > 0$ **56.** $(-2x, -3x),\ x > 0$

In Exercises 57–62, find the remaining five trigonometric functions of θ satisfying the condition.

57. $\sec\theta = \frac{6}{5},\ \tan\theta < 0$ **58.** $\csc\theta = \frac{3}{2},\ \cos\theta < 0$

59. $\tan\theta = \frac{5}{4},\ \cos\theta < 0$ **60.** $\sin\theta = \frac{3}{8},\ \cos\theta < 0$

61. $\tan\theta = -\frac{12}{5},\ \sin\theta > 0$

62. $\cos\theta = -\frac{2}{5},\ \sin\theta > 0$

In Exercises 63–72, evaluate the sine, cosine, and tangent of the angle without using a calculator.

63. $\dfrac{\pi}{3}$ **64.** $\dfrac{\pi}{4}$

65. $\dfrac{5\pi}{6}$ **66.** $\dfrac{5\pi}{3}$

67. $-\dfrac{7\pi}{3}$ **68.** $-\dfrac{5\pi}{4}$

69. $495°$ **70.** $120°$

71. $-150°$ **72.** $-420°$

In Exercises 73–84, use a calculator to evaluate the trigonometric function of the real number. Round your answer to two decimal places.

73. $\sin 4$ **74.** $\tan 3$

75. $\sec 2.8$ **76.** $\cos 5.5$

77. $\tan(-3)$ **78.** $\csc(-1)$

79. $\sin 3\pi$ **80.** $\cot(1.5\pi)$

81. $\sec\dfrac{12\pi}{5}$ **82.** $\cos\dfrac{8\pi}{7}$

83. $\sin\left(-\dfrac{17\pi}{15}\right)$ **84.** $\tan\left(-\dfrac{25\pi}{7}\right)$

6.4 **In Exercises 85–94, sketch a graph of the function. Include two full periods.**

85. $y = \sin x$ **86.** $y = \cos x$

87. $y = 3\cos 2\pi x$ **88.** $y = -2\sin \pi x$

89. $f(x) = 5\sin\dfrac{2x}{5}$ **90.** $f(x) = 8\cos\left(-\dfrac{x}{4}\right)$

91. $y = 2 + \sin x$ **92.** $y = -4 - \cos \pi x$

93. $g(t) = \frac{5}{2}\sin(t - \pi)$ **94.** $g(t) = 3\cos(t + \pi)$

95. ***Sound Waves*** Sound waves can be modeled by sine functions of the form $y = a\sin bx$, where x is measured in seconds.

(a) Write an equation of a sound wave whose amplitude is 2 and whose period is $\frac{1}{264}$ second.

(b) What is the frequency of the sound wave described in part (a)?

96. ***Sound Waves*** Use the cosine function $y = a\cos bx$ to model the sound wave described in Exercise 95.

6.5 **In Exercises 97–106, sketch a graph of the function. Include two full periods.**

97. $f(x) = \tan x$ **98.** $f(t) = \tan\left(t - \dfrac{\pi}{4}\right)$

99. $f(x) = \cot x$ **100.** $g(t) = 2\cot 2t$

101. $f(x) = \sec x$ **102.** $h(t) = \sec\left(t - \dfrac{\pi}{4}\right)$

103. $f(x) = \csc x$

104. $f(t) = 3\csc\left(2t + \frac{\pi}{4}\right)$

105. $f(x) = x\cos x$

106. $g(x) = e^x \cos x$

6.6 **In Exercises 107–112, evaluate the expression. If necessary, round your answer to two decimal places.**

107. $\arcsin\left(-\frac{1}{2}\right)$

108. $\arcsin(-1)$

109. $\arcsin 0.4$

110. $\arcsin 0.213$

111. $\sin^{-1}(-0.44)$

112. $\sin^{-1} 0.89$

In Exercises 113–116, evaluate the expression without the aid of a calculator.

113. $\arccos \frac{\sqrt{3}}{2}$

114. $\arccos \frac{\sqrt{2}}{2}$

115. $\cos^{-1}(-1)$

116. $\cos^{-1} \frac{\sqrt{3}}{2}$

In Exercises 117–124, use a calculator to approximate the value of the expression. Round your answer to two decimal places.

117. $\arccos 0.324$

118. $\arccos(-0.888)$

119. $\arctan 0.123$

120. $\arctan 2.34$

121. $\arctan 5.783$

122. $\arctan 99.1$

123. $\tan^{-1}(-1.5)$

124. $\tan^{-1} 8.2$

In Exercises 125–132, find the exact value of the expression.

125. $\sin(\arcsin 0.72)$

126. $\cos(\arccos 0.25)$

127. $\arctan(\tan \pi)$

128. $\arccos[\cos(-5\pi)]$

129. $\cos\left(\arctan \frac{3}{4}\right)$

130. $\tan\left(\arccos \frac{3}{5}\right)$

131. $\sec\left(\arctan \frac{12}{5}\right)$

132. $\cot\left[\arcsin\left(-\frac{12}{13}\right)\right]$

6.7 **133.** ***Angle of Elevation*** The height of a radio transmission tower is 70 meters, and it casts a shadow of length 30 meters (see figure). Find the angle of elevation of the sun.

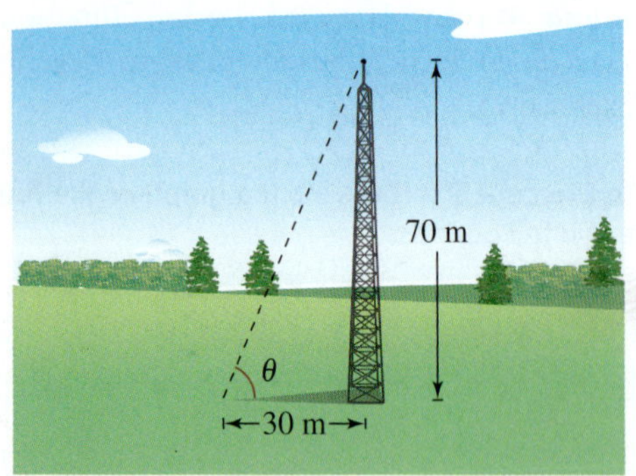

134. ***Height*** Your football has landed at the edge of the roof of your school building. When you are 25 feet from the base of the building, the angle of elevation to your football is 21°. How high off the ground is your football?

135. ***Distance*** From city A to city B, a plane flies 650 miles at a bearing of 48°. From city B to city C, the plane flies 810 miles at a bearing of 115°. Find the distance from A to C and the bearing from A to C.

136. ***Wave Motion*** Your fishing bobber oscillates in simple harmonic motion from the waves in the lake where you fish. Your bobber moves a total of 1.5 inches from its high point to its low point and returns to its high point every 3 seconds. Write an equation modeling the motion of your bobber if it is at its high point at time $t = 0$.

Synthesis

True or False? **In Exercises 137–140, determine whether the statement is true or false. Justify your answer.**

137. The tangent function is often useful for modeling simple harmonic motion.

138. The inverse sine function $y = \arcsin x$ cannot be defined as a function over any interval that is greater than the interval defined as $-\pi/2 \le y \le \pi/2$.

139. $y = \sin\theta$ is not a function because $\sin 30° = \sin 150°$.

140. Because $\tan 3\pi/4 = -1$, $\arctan(-1) = 3\pi/4$.

In Exercises 141–144, match the function $y = a \sin bx$ with its graph. Base your selection solely on your interpretation of the constants a and b. Explain your reasoning. [The graphs are labeled (a), (b), (c), and (d).]

(a)

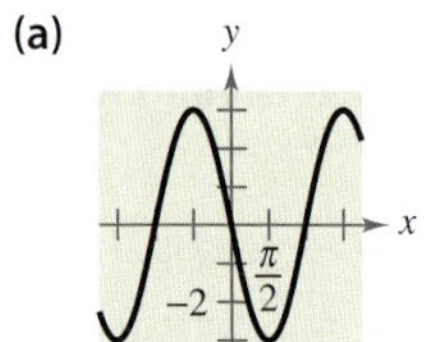

(b)

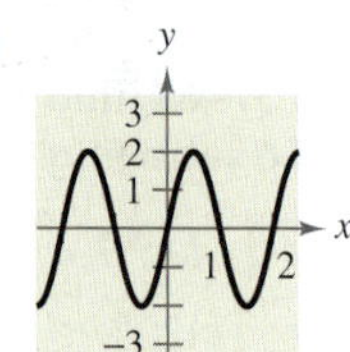

(c)

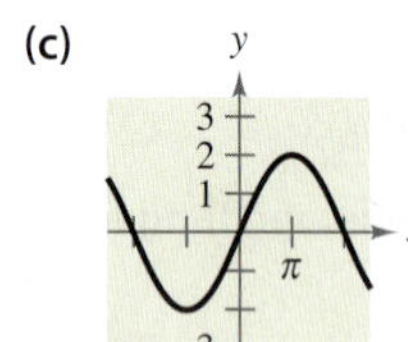

(d)

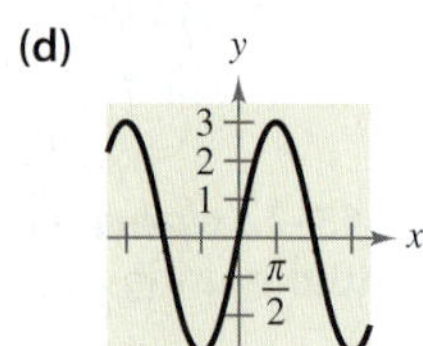

141. $y = 3 \sin x$

142. $y = -3 \sin x$

143. $y = 2 \sin \pi x$

144. $y = 2 \sin \dfrac{x}{2}$

145. ***Writing*** Describe the behavior of $f(\theta) = \sec \theta$ at the zeros of $g(\theta) = \cos \theta$. Explain your reasoning.

146. ***Conjecture***

(a) Use a graphing utility to complete the table.

θ	0.1	0.4	0.7	1.0	1.3
$\tan\left(\theta - \dfrac{\pi}{2}\right)$					
$-\cot \theta$					

(b) Make a conjecture about the relationship between $\tan\left(\theta - \dfrac{\pi}{2}\right)$ and $-\cot \theta$.

147. ***Writing*** When graphing the sine and cosine functions, determining the amplitude is part of the analysis. Explain why this is not true for the other four trigonometric functions.

148. ***Oscillation of a Spring*** A weight is suspended from a ceiling by a steel spring. The weight is lifted (positive direction) from the equilibrium position and released. The resulting motion of the weight is modeled by

$$y = Ae^{-kt} \cos bt = \frac{1}{5}e^{-t/10} \cos 6t$$

where y is the distance in feet from equilibrium and t is the time in seconds. The graph of the function is shown in the figure. For each of the following, describe the change in the system without graphing the resulting function.

(a) A is changed from $\frac{1}{5}$ to $\frac{1}{3}$.

(b) k is changed from $\frac{1}{10}$ to $\frac{1}{3}$.

(c) b is changed from 6 to 9.

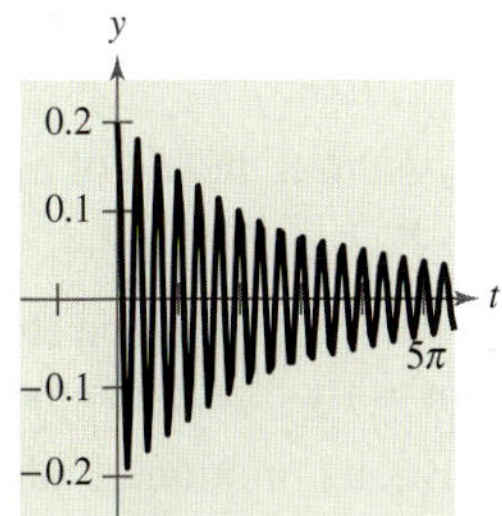

149. ***Exploration*** The base of the triangle shown in the figure is also the radius of a circular arc.

(a) Find the area A of the shaded region as a function of θ for $0 < \theta < \pi/2$.

(b) Use a graphing utility to graph the area function over the given domain. Interpret the graph in the context of the problem.

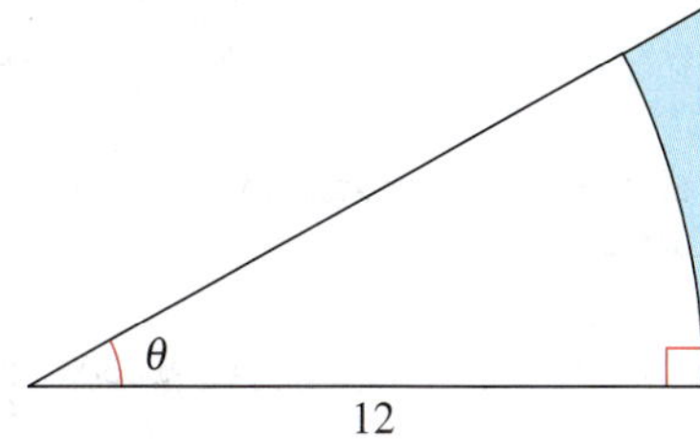

150. ***Graphical Reasoning*** The formulas for the area of a circular sector and arc length are $A = \frac{1}{2}r^2\theta$ and $s = r\theta$, respectively. (r is the radius and θ is the angle measured in radians.)

(a) For $\theta = 0.8$, write the area and arc length as functions of r. What is the domain of each function? Use a graphing utility to graph the functions. Use the graphs to determine which function changes more rapidly as r increases. Explain.

(b) For $r = 10$ centimeters, write the area and arc length as functions of θ. What is the domain of each function? Use a graphing utility to graph and identify the functions.

151. ***Writing*** Describe a real-life application that can be represented by a simple harmonic motion model and is different from any that you've seen in this chapter. Explain which function you would use to model your application and why. Explain how you would determine the amplitude, period, and frequency of the model for your application

Chapter Test

Take this test as you would take a test in class. When you are finished, check your work against the answers given in the back of the book.

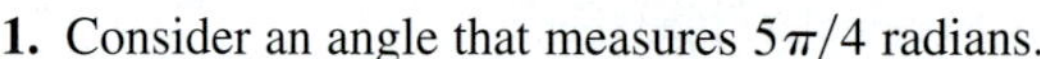

1. Consider an angle that measures $5\pi/4$ radians.
   (a) Sketch the angle in standard position.
   (b) Determine two coterminal angles (one positive and one negative).
   (c) Convert the angle to degree measure.
2. A truck is moving at a rate of 90 kilometers per hour, and the diameter of its wheels is 1 meter. Find the angular speed of the wheels in radians per minute.
3. Find the exact values of the six trigonometric functions of the angle θ shown in the figure.

(−2, 6)

FIGURE FOR 3

4. Given that $\tan \theta = \frac{3}{2}$, find the other five trigonometric functions of θ.
5. Determine the reference angle θ' of the angle $\theta = 290°$ and sketch θ and θ' in standard position.
6. Determine the quadrant in which θ lies if $\sec \theta < 0$ and $\tan \theta > 0$.
7. Find two values of θ in degrees $(0 \leq \theta < 360°)$ if $\cos \theta = -\sqrt{3}/2$. (Do not use a calculator.)
8. Use a calculator to approximate two values of θ in radians $(0 \leq \theta < 2\pi)$ if $\csc \theta = 1.030$. Round the result to two decimal places.

The *Interactive* CD-ROM and *Internet* versions of this text offer Chapter Pre-Tests and Chapter Post-Tests, both of which have randomly generated exercises with diagnostic capabilities.

In Exercises 9 and 10, find the remaining five trigonometric functions of θ satisfying the condition.

9. $\cos \theta = \frac{3}{5}$, $\tan \theta < 0$
10. $\sec \theta = -\frac{17}{8}$, $\sin \theta > 0$

In Exercises 11 and 12, graph the function through two full periods without the aid of a graphing utility.

11. $g(x) = -2\sin\left(x - \frac{\pi}{4}\right)$
12. $f(\alpha) = \frac{1}{2}\tan 2\alpha$

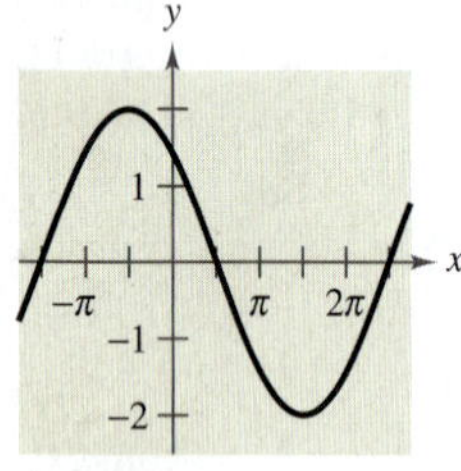

FIGURE FOR 15

In Exercises 13 and 14, use a graphing utility to graph the function. If the function is periodic, find its period.

13. $y = \sin 2\pi x + 2\cos \pi x$
14. $y = 6e^{-0.12t}\cos(0.25t)$, $0 \leq t \leq 32$
15. Find a, b, and c for the function $f(x) = a\sin(bx + c)$ such that the graph of f matches the figure.
16. Find the exact value of $\tan\left(\arccos \frac{2}{3}\right)$ without the aid of a calculator.
17. Graph the function $f(x) = 2\arcsin\left(\frac{1}{2}x\right)$.
18. A plane is 80 miles south and 95 miles east of Cleveland Hopkins International Airport. What bearing should be taken to fly directly to the airport?
19. Write the equation for the simple harmonic motion of a ball on a spring that starts at its lowest point of 6 inches below equilibrium, bounces to its maximum height of 6 inches above equilibrium, and returns to its lowest point in a total of 2 seconds.

Proofs in Mathematics

The Pythagorean Theorem

The Pythagorean Theorem is one of the most famous theorems in mathematics. More than 100 different proofs now exist. James A. Garfield, the twentieth president of the United States, developed a proof of the Pythagorean Theorem in 1876. His proof, shown below, involved the fact that a trapezoid can be formed from two congruent right triangles and an isosceles right triangle.

The Pythagorean Theorem

In a right triangle, the sum of the squares of the lengths of the legs is equal to the square of the length of the hypotenuse, where a and b are the legs and c is the hypotenuse.

$$a^2 + b^2 = c^2$$

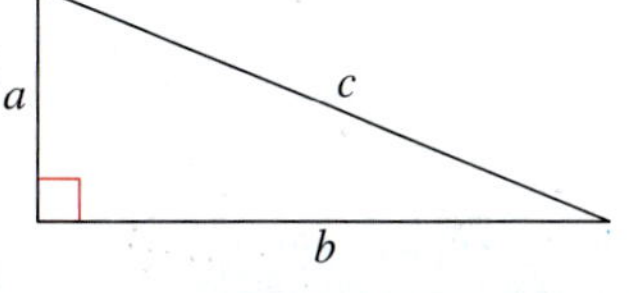

Proof

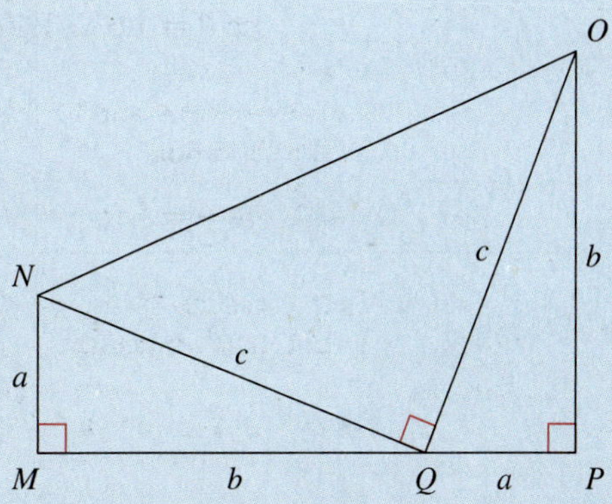

$$\text{Area of trapezoid } MNOP = \text{Area of } \triangle MNQ + \text{Area of } \triangle PQO + \text{Area of } \triangle NOQ$$

$$\frac{1}{2}(a + b)(a + b) = \frac{1}{2}ab + \frac{1}{2}ab + \frac{1}{2}c^2$$

$$\frac{1}{2}(a + b)(a + b) = ab + \frac{1}{2}c^2$$

$$(a + b)(a + b) = 2ab + c^2$$

$$a^2 + 2ab + b^2 = 2ab + c^2$$

$$a^2 + b^2 = c^2$$

P.S. Problem Solving

1. The restaurant at the top of the Space Needle in Seattle, Washington is circular and has a radius of 47.25 feet. The dining part of the restaurant revolves, making about one complete revolution every 48 minutes. A dinner party was seated at the edge of the revolving restaurant at 6:45 P.M. and was finished at 8:57 P.M.
   (a) Find the angle through which the dinner party rotated.
   (b) Find the distance the party traveled during dinner.

2. A bicycle's gear ratio is the number of times the freewheel turns for every one turn of the chainwheel (see figure). The table shows the numbers of teeth in the freewheel and chainwheel for the first five gears of an 18-speed touring bicycle. The chainwheel completes one rotation for each gear. Find the angle through which the freewheel turns for each gear. Give your answers in both degrees and radians.

Gear number	Number of teeth in freewheel	Number of teeth in chainwheel
1	32	24
2	26	24
3	22	24
4	32	40
5	19	24

3. A surveyor in a helicopter is trying to determine the width of an island, as shown in the figure.
   (a) What is the shortest distance d the helicopter would have to travel to land on the island?
   (b) What is the horizontal distance x that the helicopter would have to travel before it would be directly over the nearer end of the island?
   (c) Find the width w of the island. Explain how you obtained your answer.

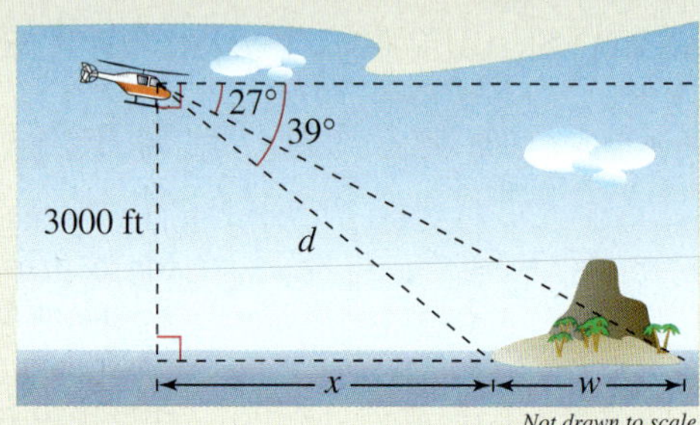

FIGURE FOR 3

4. Use the figure below.

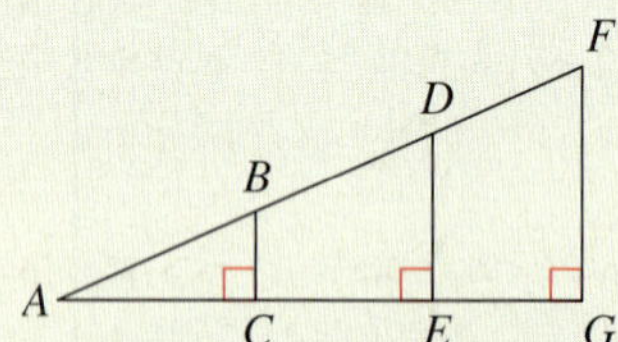

   (a) Explain why $\triangle ABC$, $\triangle ADE$, and $\triangle AFG$ are similar triangles.
   (b) What does similarity imply about the ratios
   $$\frac{BC}{AB}, \frac{DE}{AD}, \text{ and } \frac{FG}{AF}?$$
   (c) Does the value of $\sin A$ depend on which triangle from part (a) is used to calculate it? Would the value of $\sin A$ change if it were found using a different right triangle that was similar to the three given triangles?
   (d) Do your conclusions from part (c) apply to the other five trigonometric functions? Explain.

5. Use a graphing utility to graph h, and use the graph to decide whether h is even, odd, or neither.
   (a) $h(x) = \cos^2 x$ (b) $h(x) = \sin^2 x$

6. If f is an even function and g is an odd function, use the results of Exercise 5 to make a conjecture about h where
   (a) $h(x) = [f(x)]^2$ (b) $h(x) = [g(x)]^2$.

7. The model for the height h of a Ferris wheel car is
   $$h = 50 + 50 \sin 8\pi t$$
   where t is the time in minutes. (The Ferris wheel has a radius of 50 feet.) This model yields a height of 50 feet when $t = 0$. Alter the model so that the height of the car is 1 foot when $t = 0$.

8. A popular theory that attempts to explain the ups and downs of everyday life states that each of us has three cycles, called biorhythms, which begin at birth. These three cycles can be modeled by sine waves.

Physical (23 days): $P = \sin \dfrac{2\pi t}{23}, \quad t \geq 0$

Emotional (28 days): $E = \sin \dfrac{2\pi t}{28}, \quad t \geq 0$

Intellectual (33 days): $I = \sin \dfrac{2\pi t}{33}, \quad t \geq 0$

where t is the number of days since birth. Consider a person who was born on July 20, 1984.

(a) Use a graphing utility to graph the three models in the same viewing window for $7300 \leq t \leq 7380$

(b) Describe the person's biorhythms during the month of September 2004.

(c) Calculate the person's three energy levels on September 22, 2004.

9. (a) Use a graphing utility to graph the functions

$f(x) = 2\cos 2x + 3\sin 3x$

and

$g(x) = 2\cos 2x + 3\sin 4x.$

(b) Use the graphs from part (a) to find the period of each function.

(c) If α and β are positive integers, is the function

$h(x) = A\cos \alpha x + B\sin \beta x$

periodic? Explain your reasoning.

10. Two trigonometric functions f and g have periods of 2, and their graphs intersect at $x = 5.35$.

(a) Give one smaller and one larger positive value of x at which the functions have the same value.

(b) Determine one negative value of x at which the graphs intersect.

(c) Is it true that $f(13.35) = g(-4.65)$? Explain your reasoning.

11. The function f is periodic, with period c. So, $f(t + c) = f(t)$. Are the following equal? Explain.

(a) $f(t - 2c) = f(t)$ (b) $f\left(t + \frac{1}{2}c\right) = f\left(\frac{1}{2}t\right)$

(c) $f\left(\frac{1}{2}(t + c)\right) = f\left(\frac{1}{2}t\right)$

12. If you stand in shallow water and look at an object below the surface of the water, the object will look farther away from you than it really is. This is because when light rays pass between air and water, the water refracts, or bends, the light rays. The index of refraction for water is 1.333. This is the ratio of the sine of θ_1 and the sine of θ_2 (see figure).

(a) You are standing in water that is 2 feet deep and are looking at a rock at angle $\theta_1 = 60°$ (measured from a line perpendicular to the surface of the water). Find θ_2.

(b) Find the distances x and y.

(c) Find the distance d between where the rock is and where it appears to be.

(d) What happens to d as you move closer to the rock? Explain your reasoning.

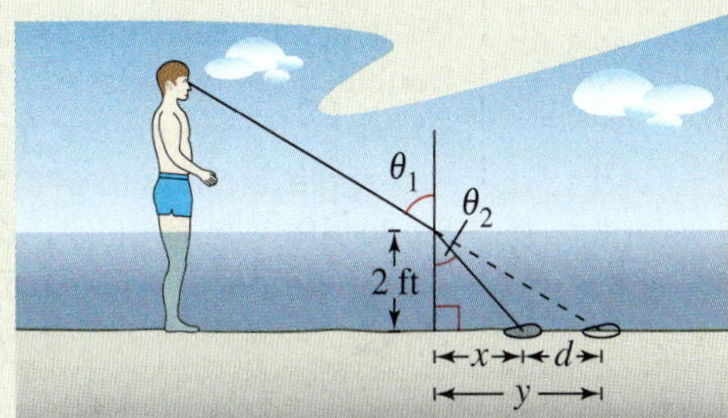

13. In calculus it can be shown that the arctangent function can be approximated by the polynomial

$$\arctan x \approx x - \frac{x^3}{3} + \frac{x^5}{5} - \frac{x^7}{7}$$

where x is in radians.

(a) Use a graphing utility to graph the arctangent function and its polynomial approximation in the same viewing window. How do the graphs compare?

(b) Study the pattern in the polynomial approximation of the arctangent function and guess the next term. Then repeat part (a). How does the accuracy of the approximation change when additional terms are added?

How to study Chapter 7

▶ What you should learn

In this chapter you will learn the following skills and concepts:

- How to use fundamental trigonometric identities to evaluate trigonometric functions and simplify trigonometric expressions
- How to verify trigonometric identities
- How to use standard algebraic techniques and inverse trigonometric functions to solve trigonometric equations
- How to use sum and difference formulas, multiple-angle formulas, power-reducing formulas, half-angle formulas, and product-to-sum formulas to rewrite and evaluate trigonometric functions

▶ Important Vocabulary

As you encounter each new vocabulary term in this chapter, add the term and its definition to your notebook glossary.

Sum and difference formulas (p. 568)
Reduction formulas (p. 570)
Double-angle formulas (p. 575)
Power-reducing formulas (p. 577)
Half-angle formulas (p. 578)
Product-to-sum formulas (p. 579)
Sum-to-product formulas (p. 580)

Study Tools

Learning objectives in each section
Chapter Summary (p. 586)
Review Exercises (pp. 587–589)
Chapter Test (p. 590)

Additional Resources

Study and Solutions Guide
Interactive Algebra and Trigonometry
Videotapes/DVD for Chapter 7
Algebra and Trigonometry Website
Student Success Organizer

Mike Powell/Allsport

7

Analytic Trigonometry

7.1 Using Fundamental Identities

▶ **What you should learn**

- How to recognize and write the fundamental trigonometric identities
- How to use the fundamental trigonometric identities to evaluate trigonometric functions, simplify trigonometric expressions, and rewrite trigonometric expressions

▶ **Why you should learn it**

Fundamental trigonometric identities can be used to simplify trigonometric expressions. For instance, in Exercise 99 on page 549, you can use trigonometric identities to simplify an expression for the coefficient of friction.

Introduction

In Chapter 6, you studied the basic definitions, properties, graphs, and applications of the individual trigonometric functions. In this chapter, you will learn how to use the fundamental identities to do the following.

1. Evaluate trigonometric functions.
2. Simplify trigonometric expressions.
3. Develop additional trigonometric identities.
4. Solve trigonometric equations.

Fundamental Trigonometric Identities

Reciprocal Identities

$$\sin u = \frac{1}{\csc u} \qquad \cos u = \frac{1}{\sec u} \qquad \tan u = \frac{1}{\cot u}$$

$$\csc u = \frac{1}{\sin u} \qquad \sec u = \frac{1}{\cos u} \qquad \cot u = \frac{1}{\tan u}$$

Quotient Identities

$$\tan u = \frac{\sin u}{\cos u} \qquad \cot u = \frac{\cos u}{\sin u}$$

Pythagorean Identities

$$\sin^2 u + \cos^2 u = 1 \qquad 1 + \tan^2 u = \sec^2 u \qquad 1 + \cot^2 u = \csc^2 u$$

Cofunction Identities

$$\sin\left(\frac{\pi}{2} - u\right) = \cos u \qquad \cos\left(\frac{\pi}{2} - u\right) = \sin u$$

$$\tan\left(\frac{\pi}{2} - u\right) = \cot u \qquad \cot\left(\frac{\pi}{2} - u\right) = \tan u$$

$$\sec\left(\frac{\pi}{2} - u\right) = \csc u \qquad \csc\left(\frac{\pi}{2} - u\right) = \sec u$$

Even/Odd Identities

$$\sin(-u) = -\sin u \qquad \cos(-u) = \cos u \qquad \tan(-u) = -\tan u$$

$$\csc(-u) = -\csc u \qquad \sec(-u) = \sec u \qquad \cot(-u) = -\cot u$$

Pythagorean identities are sometimes used in radical form such as

$$\sin u = \pm\sqrt{1 - \cos^2 u}$$

or

$$\tan u = \pm\sqrt{\sec^2 u - 1}$$

where the sign depends on the choice of u.

Using the Fundamental Identities

One common use of trigonometric identities is to use given values of trigonometric functions to evaluate other trigonometric functions.

STUDY TIP

You should learn the fundamental trigonometric identities well, because they are used frequently. They will also appear later in calculus. Also, u can be an angle, a real number, or a variable.

Example 1 ▶ Using Identities to Evaluate a Function 

Use the values $\sec u = -\frac{3}{2}$ and $\tan u > 0$ to find the values of all six trigonometric functions.

Solution

Using a reciprocal identity, you have

$$\cos u = \frac{1}{\sec u} = \frac{1}{-3/2} = -\frac{2}{3}.$$

Using a Pythagorean identity, you have

$$\sin^2 u = 1 - \cos^2 u$$
$$= 1 - \left(-\frac{2}{3}\right)^2$$
$$= 1 - \frac{4}{9} = \frac{5}{9}.$$

Because $\sec u < 0$ and $\tan u > 0$, it follows that u lies in Quadrant III. Moreover, because $\sin u$ is negative when u is in Quadrant III, you can choose the negative root and obtain $\sin u = -\sqrt{5}/3$. Now, knowing the values of the sine and cosine, you can find the values of all six trigonometric functions.

$$\sin u = -\frac{\sqrt{5}}{3} \qquad \csc u = \frac{1}{\sin u} = -\frac{3}{\sqrt{5}}$$

$$\cos u = -\frac{2}{3} \qquad \sec u = \frac{1}{\cos u} = -\frac{3}{2}$$

$$\tan u = \frac{\sin u}{\cos u} = \frac{-\sqrt{5}/3}{-2/3} = \frac{\sqrt{5}}{2} \qquad \cot u = \frac{1}{\tan u} = \frac{2}{\sqrt{5}}$$

Technology

You can use a graphing utility to check the result of Example 2. To do this, graph

$$y_1 = \sin x \cos^2 x - \sin x$$

and

$$y_2 = -\sin^3 x$$

in the same viewing window, as shown below. Because Example 2 shows the equivalence algebraically and the two graphs appear to coincide, you can conclude that the expressions are equivalent.

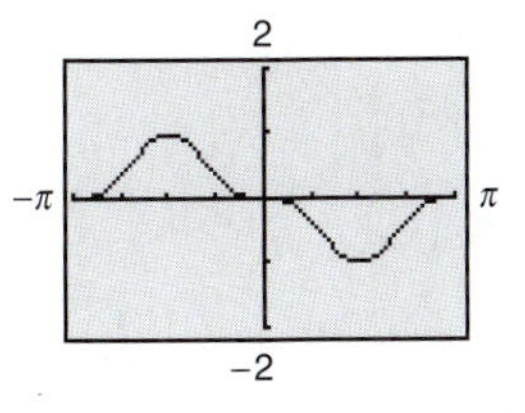

Example 2 ▶ Simplifying a Trigonometric Expression 

Simplify $\sin x \cos^2 x - \sin x$.

Solution

Factor the expression and then use a fundamental identity.

$$\sin x \cos^2 x - \sin x = \sin x(\cos^2 x - 1) \qquad \text{Monomial factor}$$
$$= -\sin x(1 - \cos^2 x) \qquad \text{Factor out } -1.$$
$$= -\sin x(\sin^2 x) \qquad \text{Pythagorean identity}$$
$$= -\sin^3 x \qquad \text{Multiply.}$$

The icon identifies examples and concepts related to features of the Learning Tools CD-ROM and the *Interactive* and *Internet* versions of this text. For more details see the chart on pages *xix–xxiii*.

Example 3 ▶ Factoring Trigonometric Expressions 

Factor each expression.

a. $\sec^2 \theta - 1$

b. $4 \tan^2 \theta + \tan \theta - 3$

Solution

a. Here you have the difference of two squares, which factors as

$$\sec^2 \theta - 1 = (\sec \theta - 1)(\sec \theta + 1).$$

b. This expression has the polynomial form $ax^2 + bx + c$, and it factors as

$$4 \tan^2 \theta + \tan \theta - 3 = (4 \tan \theta - 3)(\tan \theta + 1).$$

On occasion, factoring or simplifying can best be done by first rewriting the expression in terms of just *one* trigonometric function or in terms of *sine and cosine only*. These strategies are illustrated in Examples 4 and 5, respectively.

The *Interactive* CD-ROM and *Internet* versions of this text offer a Try It for each example in the text.

Example 4 ▶ Factoring a Trigonometric Expression 

Factor $\csc^2 x - \cot x - 3$.

Solution

You can use the identity $\csc^2 x = 1 + \cot^2 x$ to rewrite the expression in terms of the cotangent.

$$\begin{aligned} \csc^2 x - \cot x - 3 &= (1 + \cot^2 x) - \cot x - 3 && \text{Pythagorean identity} \\ &= \cot^2 x - \cot x - 2 && \text{Combine like terms.} \\ &= (\cot x - 2)(\cot x + 1) && \text{Factor.} \end{aligned}$$

Example 5 ▶ Simplifying a Trigonometric Expression 

Simplify $\sin t + \cot t \cos t$.

Solution

Begin by rewriting $\cot t$ in terms of sine and cosine.

$$\begin{aligned} \sin t + \cot t \cos t &= \sin t + \left(\frac{\cos t}{\sin t}\right) \cos t && \text{Quotient identity} \\ &= \frac{\sin^2 t + \cos^2 t}{\sin t} && \text{Add fractions.} \\ &= \frac{1}{\sin t} && \text{Pythagorean identity} \\ &= \csc t && \text{Reciprocal identity} \end{aligned}$$

Example 6 ▶ Adding Trigonometric Expressions 

Perform the addition and simplify.

$$\frac{\sin\theta}{1+\cos\theta}+\frac{\cos\theta}{\sin\theta}$$

Solution

$$\frac{\sin\theta}{1+\cos\theta}+\frac{\cos\theta}{\sin\theta}=\frac{(\sin\theta)(\sin\theta)+(\cos\theta)(1+\cos\theta)}{(1+\cos\theta)(\sin\theta)}$$

$$=\frac{\sin^2\theta+\cos^2\theta+\cos\theta}{(1+\cos\theta)(\sin\theta)}$$ Multiply.

$$=\frac{\cancel{1+\cos\theta}}{(\cancel{1+\cos\theta})(\sin\theta)}$$ Pythagorean identity

$$=\frac{1}{\sin\theta}$$ Divide out common factor.

$$=\csc\theta$$ Reciprocal identity

The *Interactive* CD-ROM and *Internet* versions of this text offer a Quiz for every section of the text.

The last two examples in this section involve techniques for rewriting expressions in forms that are used in calculus.

Example 7 ▶ Rewriting a Trigonometric Expression 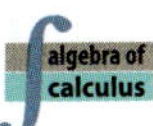

Rewrite $\dfrac{1}{1+\sin x}$ so that it is *not* in fractional form.

Solution

From the Pythagorean identity $\cos^2 x = 1 - \sin^2 x = (1 - \sin x)(1 + \sin x)$, you can see that by multiplying both the numerator and the denominator by $(1 - \sin x)$ you produce a monomial denominator.

$$\frac{1}{1+\sin x}=\frac{1}{1+\sin x}\cdot\frac{1-\sin x}{1-\sin x}$$ Multiply numerator and denominator by $(1 - \sin x)$.

$$=\frac{1-\sin x}{1-\sin^2 x}$$ Multiply.

$$=\frac{1-\sin x}{\cos^2 x}$$ Pythagorean identity

$$=\frac{1}{\cos^2 x}-\frac{\sin x}{\cos^2 x}$$ Separate fractions.

$$=\frac{1}{\cos^2 x}-\frac{\sin x}{\cos x}\cdot\frac{1}{\cos x}$$ Product of fractions

$$=\sec^2 x-\tan x\sec x$$ Identities

Example 8 ▶ Trigonometric Substitution

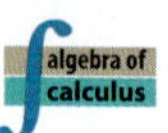

Use the substitution $x = 2 \tan \theta$, $0 < \theta < \pi/2$, to express

$$\sqrt{4 + x^2}$$

as a trigonometric function of θ.

Solution

Begin by letting $x = 2 \tan \theta$. Then, you can obtain

$$
\begin{aligned}
\sqrt{4 + x^2} &= \sqrt{4 + (2 \tan \theta)^2} && \text{Substitute } 2 \tan \theta \text{ for } x. \\
&= \sqrt{4 + 4 \tan^2 \theta} && \text{Rule of exponents} \\
&= \sqrt{4(1 + \tan^2 \theta)} && \text{Factor.} \\
&= \sqrt{4 \sec^2 \theta} && \text{Pythagorean identity} \\
&= 2 \sec \theta. && \sec \theta > 0 \text{ for } 0 < \theta < \pi/2
\end{aligned}
$$

Figure 7.1 shows the right triangle illustration of the trigonometric substitution in Example 8. You can use this triangle to check the solution of Example 8. For $0 < \theta < \pi/2$, you have

$$\text{opp} = x, \quad \text{adj} = 2, \quad \text{and} \quad \text{hyp} = \sqrt{4 + x^2}.$$

With these expressions, you can write the following.

$$
\begin{aligned}
\sec \theta &= \frac{\text{hyp}}{\text{adj}} \\
\sec \theta &= \frac{\sqrt{4 + x^2}}{2} \\
2 \sec \theta &= \sqrt{4 + x^2}
\end{aligned}
$$

So, the solution checks.

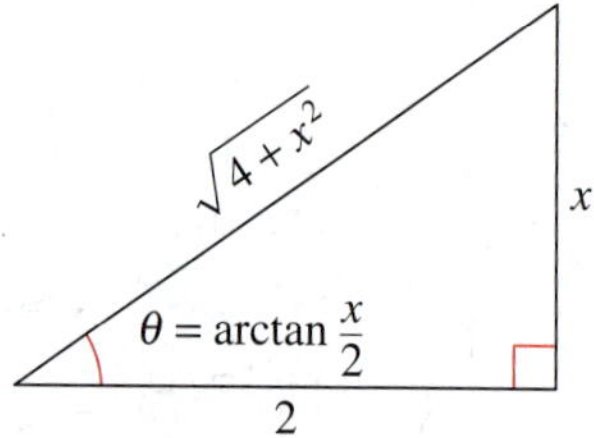

Angle whose tangent is $\frac{x}{2}$.

FIGURE 7.1

7.1 Exercises

The *Interactive* CD-ROM and *Internet* versions of this text contain step-by-step solutions to all odd-numbered exercises. They also provide Tutorial Exercises for additional help.

In Exercises 1–14, use the given values to evaluate (if possible) the remaining trigonometric functions.

1. $\sin x = \frac{\sqrt{3}}{2}, \quad \cos x = -\frac{1}{2}$
2. $\tan x = \frac{\sqrt{3}}{3}, \quad \cos x = -\frac{\sqrt{3}}{2}$
3. $\sec \theta = \sqrt{2}, \quad \sin \theta = -\frac{\sqrt{2}}{2}$
4. $\csc \theta = \frac{5}{3}, \quad \tan \theta = \frac{3}{4}$
5. $\tan x = \frac{5}{12}, \quad \sec x = -\frac{13}{12}$
6. $\cot \phi = -3, \quad \sin \phi = \frac{\sqrt{10}}{10}$
7. $\sec \phi = \frac{3}{2}, \quad \csc \phi = -\frac{3\sqrt{5}}{5}$
8. $\cos\left(\frac{\pi}{2} - x\right) = \frac{3}{5}, \quad \cos x = \frac{4}{5}$
9. $\sin(-x) = -\frac{1}{3}, \quad \tan x = -\frac{\sqrt{2}}{4}$
10. $\sec x = 4, \quad \sin x > 0$
11. $\tan \theta = 2, \quad \sin \theta < 0$
12. $\csc \theta = -5, \quad \cos \theta < 0$
13. $\sin \theta = -1, \quad \cot \theta = 0$
14. $\tan \theta$ is undefined, $\quad \sin \theta > 0$

In Exercises 15–20, match the trigonometric expression with one of the following.

(a) $\sec x$ (b) -1 (c) $\cot x$
(d) 1 (e) $-\tan x$ (f) $\sin x$

15. $\sec x \cos x$
16. $\tan x \csc x$
17. $\cot^2 x - \csc^2 x$
18. $(1 - \cos^2 x)(\csc x)$
19. $\frac{\sin(-x)}{\cos(-x)}$
20. $\frac{\sin[(\pi/2) - x]}{\cos[(\pi/2) - x]}$

In Exercises 21–26, match the trigonometric expression with one of the following.

(a) $\csc x$ (b) $\tan x$ (c) $\sin^2 x$
(d) $\sin x \tan x$ (e) $\sec^2 x$ (f) $\sec^2 x + \tan^2 x$

21. $\sin x \sec x$
22. $\cos^2 x(\sec^2 x - 1)$
23. $\sec^4 x - \tan^4 x$
24. $\cot x \sec x$
25. $\frac{\sec^2 x - 1}{\sin^2 x}$
26. $\frac{\cos^2[(\pi/2) - x]}{\cos x}$

In Exercises 27–44, use the fundamental identities to simplify the expression. There is more than one correct form of each answer.

27. $\cot \theta \sec \theta$
28. $\cos \beta \tan \beta$
29. $\sin \phi(\csc \phi - \sin \phi)$
30. $\sec^2 x(1 - \sin^2 x)$
31. $\frac{\cot x}{\csc x}$
32. $\frac{\csc \theta}{\sec \theta}$
33. $\frac{1 - \sin^2 x}{\csc^2 x - 1}$
34. $\frac{1}{\tan^2 x + 1}$
35. $\sec \alpha \cdot \frac{\sin \alpha}{\tan \alpha}$
36. $\frac{\tan^2 \theta}{\sec^2 \theta}$
37. $\cos\left(\frac{\pi}{2} - x\right)\sec x$
38. $\cot\left(\frac{\pi}{2} - x\right)\cos x$
39. $\frac{\cos^2 y}{1 - \sin y}$
40. $\cos t(1 + \tan^2 t)$
41. $\sin \beta \tan \beta + \cos \beta$
42. $\csc \phi \tan \phi + \sec \phi$
43. $\cot u \sin u + \tan u \cos u$
44. $\sin \theta \sec \theta + \cos \theta \csc \theta$

In Exercises 45–56, factor the expression and use the fundamental identities to simplify. There is more than one correct form of each answer.

45. $\tan^2 x - \tan^2 x \sin^2 x$
46. $\sin^2 x \csc^2 x - \sin^2 x$
47. $\sin^2 x \sec^2 x - \sin^2 x$
48. $\cos^2 x + \cos^2 x \tan^2 x$
49. $\frac{\sec^2 x - 1}{\sec x - 1}$
50. $\frac{\cos^2 x - 4}{\cos x - 2}$
51. $\tan^4 x + 2\tan^2 x + 1$
52. $1 - 2\cos^2 x + \cos^4 x$
53. $\sin^4 x - \cos^4 x$
54. $\sec^4 x - \tan^4 x$
55. $\csc^3 x - \csc^2 x - \csc x + 1$
56. $\sec^3 x - \sec^2 x - \sec x + 1$

In Exercises 57–60, perform the multiplication and use the fundamental identities to simplify. There is more than one correct form of each answer.

57. $(\sin x + \cos x)^2$

58. $(\cot x + \csc x)(\cot x - \csc x)$

59. $(2 \csc x + 2)(2 \csc x - 2)$

60. $(3 - 3 \sin x)(3 + 3 \sin x)$

In Exercises 61–64, perform the addition or subtraction and use the fundamental identities to simplify. There is more than one correct form of each answer.

61. $\dfrac{1}{1 + \cos x} + \dfrac{1}{1 - \cos x}$

62. $\dfrac{1}{\sec x + 1} - \dfrac{1}{\sec x - 1}$

63. $\dfrac{\cos x}{1 + \sin x} + \dfrac{1 + \sin x}{\cos x}$

64. $\tan x - \dfrac{\sec^2 x}{\tan x}$

In Exercises 65–68, rewrite the expression so that it is not in fractional form. There is more than one correct form of each answer.

65. $\dfrac{\sin^2 y}{1 - \cos y}$

66. $\dfrac{5}{\tan x + \sec x}$

67. $\dfrac{3}{\sec x - \tan x}$

68. $\dfrac{\tan^2 x}{\csc x + 1}$

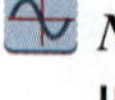

Numerical and Graphical Analysis **In Exercises 69–72, use a graphing utility to complete the table and graph the functions. Make a conjecture about y_1 and y_2.**

x	0.2	0.4	0.6	0.8	1.0	1.2	1.4
y_1							
y_2							

69. $y_1 = \cos\left(\dfrac{\pi}{2} - x\right), \quad y_2 = \sin x$

70. $y_1 = \sec x - \cos x, \quad y_2 = \sin x \tan x$

71. $y_1 = \dfrac{\cos x}{1 - \sin x}, \quad y_2 = \dfrac{1 + \sin x}{\cos x}$

72. $y_1 = \sec^4 x - \sec^2 x, \quad y_2 = \tan^2 x + \tan^4 x$

In Exercises 73–76, use a graphing utility to determine which of the six trigonometric functions is equal to the expression. Verify your answer algebraically.

73. $\cos x \cot x + \sin x$

74. $\sec x \csc x - \tan x$

75. $\dfrac{1}{\sin x}\left(\dfrac{1}{\cos x} - \cos x\right)$

76. $\dfrac{1}{2}\left(\dfrac{1 + \sin \theta}{\cos \theta} + \dfrac{\cos \theta}{1 + \sin \theta}\right)$

In Exercises 77–82, use the trigonometric substitution to write the algebraic expression as a trigonometric function of θ, where $0 < \theta < \pi/2$.

77. $\sqrt{9 - x^2}, \quad x = 3 \cos \theta$

78. $\sqrt{64 - 16x^2}, \quad x = 2 \cos \theta$

79. $\sqrt{x^2 - 9}, \quad x = 3 \sec \theta$

80. $\sqrt{x^2 - 4}, \quad x = 2 \sec \theta$

81. $\sqrt{x^2 + 25}, \quad x = 5 \tan \theta$

82. $\sqrt{x^2 + 100}, \quad x = 10 \tan \theta$

In Exercises 83–86, use the trigonometric substitution to write the algebraic equation as a trigonometric function of θ, where $-\pi/2 < \theta < \pi/2$. Then find $\sin \theta$ and $\cos \theta$.

83. $3 = \sqrt{9 - x^2}, \quad x = 3 \sin \theta$

84. $3 = \sqrt{36 - x^2}, \quad x = 6 \sin \theta$

85. $2\sqrt{2} = \sqrt{16 - 4x^2}, \quad x = 2 \cos \theta$

86. $-5\sqrt{3} = \sqrt{100 - x^2}, \quad x = 10 \cos \theta$

In Exercises 87–90, use a graphing utility to solve the equation for θ, where $0 \le \theta < 2\pi$.

87. $\sin \theta = \sqrt{1 - \cos^2 \theta}$

88. $\cos \theta = -\sqrt{1 - \sin^2 \theta}$

89. $\sec \theta = \sqrt{1 + \tan^2 \theta}$

90. $\csc \theta = \sqrt{1 + \cot^2 \theta}$

In Exercises 91–94, rewrite the expression as a single logarithm and simplify the result.

91. $\ln|\cos x| - \ln|\sin x|$

92. $\ln|\sec x| + \ln|\sin x|$

93. $\ln|\cot t| + \ln(1 + \tan^2 t)$

94. $\ln(\cos^2 t) + \ln(1 + \tan^2 t)$

In Exercises 95–98, use a calculator to demonstrate the identity for each value of θ.

95. $\csc^2 \theta - \cot^2 \theta = 1$

(a) $\theta = 132°$, (b) $\theta = \dfrac{2\pi}{7}$

96. $\tan^2 \theta + 1 = \sec^2 \theta$

(a) $\theta = 346°$, (b) $\theta = 3.1$

97. $\cos\left(\dfrac{\pi}{2} - \theta\right) = \sin \theta$

(a) $\theta = 80°$, (b) $\theta = 0.8$

98. $\sin(-\theta) = -\sin \theta$

(a) $\theta = 250°$, (b) $\theta = \frac{1}{2}$

99. ***Friction*** The forces acting on an object weighing W units on an inclined plane positioned at an angle of θ with the horizontal (see figure) are modeled by

$$\mu W \cos \theta = W \sin \theta$$

where μ is the coefficient of friction. Solve the equation for μ and simplify the result.

100. ***Rate of Change*** The rate of change of the function

$$f(x) = -\csc x - \sin x$$

is given by the expression

$$\csc x \cot x - \cos x.$$

Show that this expression can also be written as $\cos x \cot^2 x$.

Synthesis

True or False? **In Exercises 101 and 102, determine whether the statement is true or false. Justify your answer.**

101. The even and odd trigonometric identities are helpful for determining whether the value of a trigonometric function is positive or negative.

102. A cofunction identity can be used to transform a tangent function so that it can be represented by a cosecant function.

Calculus **In Exercises 103–106, fill in the blanks. (*Note:* The notation $x \to c^+$ indicates that x approaches c from the right and $x \to c^-$ indicates that x approaches c from the left.)**

103. As $x \to \dfrac{\pi}{2}^-$, $\sin x \to$ ☐ and $\csc x \to$ ☐.

104. As $x \to 0^+$, $\cos x \to$ ☐ and $\sec x \to$ ☐.

105. As $x \to \dfrac{\pi}{2}^-$, $\tan x \to$ ☐ and $\cot x \to$ ☐.

106. As $x \to \pi^+$, $\sin x \to$ ☐ and $\csc x \to$ ☐.

In Exercises 107–112, determine whether or not the equation is an identity, and give a reason for your answer.

107. $\cos \theta = \sqrt{1 - \sin^2 \theta}$

108. $\cot \theta = \sqrt{\csc^2 \theta + 1}$

109. $\dfrac{(\sin k\theta)}{(\cos k\theta)} = \tan \theta$, k is a constant.

110. $\dfrac{1}{(5 \cos \theta)} = 5 \sec \theta$

111. $\sin \theta \csc \theta = 1$

112. $\sin \theta \csc \theta = 1$

113. Use the definitions of sine and cosine to derive the Pythagorean identity $\sin^2 \theta + \cos^2 \theta = 1$.

114. ***Writing*** Use the Pythagorean identity

$$\sin^2 \theta + \cos^2 \theta = 1$$

to derive the other Pythagorean identities, $1 + \tan^2 \theta = \sec^2 \theta$ and $1 + \cot^2 \theta = \csc^2 \theta$. Discuss how to remember these identities and other fundamental identities.

Review

In Exercises 115 and 116, perform the operation and simplify.

115. $(\sqrt{x} + 5)(\sqrt{x} - 5)$

116. $(2\sqrt{z} + 3)^2$

In Exercises 117–120, perform the addition or subtraction and simplify.

117. $\dfrac{1}{x + 5} + \dfrac{x}{x - 8}$

118. $\dfrac{6x}{x - 4} - \dfrac{3}{4 - x}$

119. $\dfrac{2x}{x^2 - 4} - \dfrac{7}{x + 4}$

120. $\dfrac{x}{x^2 - 25} + \dfrac{x^2}{x - 5}$

7.2 Verifying Trigonometric Identities

▶ What you should learn

- How to plan a strategy for verifying trigonometric identities
- How to verify trigonometric identities

▶ Why you should learn it

You can use trigonometric identities to rewrite trigonometric equations that model real-life situations. For instance, in Exercise 58 on page 556, you can use trigonometric identities to simplify the equation that models the length of a shadow cast by a gnomon.

Robert Ginn/PhotoEdit

Introduction 

In this section, you will study techniques for verifying trigonometric identities. In the next section, you will study techniques for solving trigonometric equations. The key to verifying identities *and* solving equations is the ability to use the fundamental identities and the rules of algebra to rewrite trigonometric expressions.

Remember that a *conditional equation* is an equation that is true for only some of the values in its domain. For example, the conditional equation

$$\sin x = 0 \qquad \text{Conditional equation}$$

is true only for $x = n\pi$, where n is an integer. When you find these values, you are *solving* the equation.

On the other hand, an equation that is true for all real values in the domain of the variable is an *identity*. For example, the familiar equation

$$\sin^2 x = 1 - \cos^2 x \qquad \text{Identity}$$

is true for all real numbers x. So, it is an identity.

Although there are similarities, verifying that a trigonometric equation is an identity is quite different from solving an equation. There is no well-defined set of rules to follow in verifying trigonometric identities, and the process is best learned by practice.

For instance, to verify that the trigonometric equation $\tan\theta\cos\theta = \sin\theta$ is an identity, begin by working with the more complicated, left side of the equation.

$$\tan\theta\cos\theta = \left(\frac{\sin\theta}{\cos\theta}\right)\cos\theta \qquad \text{Rewrite } \tan\theta \text{ as } \frac{\sin\theta}{\cos\theta}.$$

$$= \left(\frac{\sin\theta}{\cancel{\cos\theta}}\right)\cancel{\cos\theta} \qquad \text{Divide out } \cos\theta.$$

$$= \sin\theta \qquad \text{Simplify.}$$

The result shows that the left side of the equation is equal to the right side. So, the identity has been verified.

Guidelines for Verifying Trigonometric Identities

1. Work with one side of the equation at a time. It is often better to work with the more complicated side first.
2. Look for opportunities to factor an expression, add fractions, square a binomial, or create a monomial denominator.
3. Look for opportunities to use the fundamental identities. Note which functions are in the final expression you want. Sines and cosines pair up well, as do secants and tangents, and cosecants and cotangents.
4. If the preceding guidelines do not help, try converting all terms to sines and cosines.
5. Always try *something*. Even paths that lead to dead ends provide insights.

Verifying Trigonometric Identities

Example 1 ▶ Verifying a Trigonometric Identity 

Verify the identity $\dfrac{\sec^2\theta - 1}{\sec^2\theta} = \sin^2\theta$.

Solution

Because the left side is more complicated, start with it.

$$\begin{aligned}
\frac{\sec^2\theta - 1}{\sec^2\theta} &= \frac{(\tan^2\theta + 1) - 1}{\sec^2\theta} && \text{Pythagorean identity}\\
&= \frac{\tan^2\theta}{\sec^2\theta} && \text{Simplify.}\\
&= \tan^2\theta(\cos^2\theta) && \text{Reciprocal identity}\\
&= \frac{\sin^2\theta}{\cancel{(\cos^2\theta)}}\cancel{(\cos^2\theta)} && \text{Quotient identity}\\
&= \sin^2\theta && \text{Simplify.}
\end{aligned}$$

Here is another way to verify the identity in Example 1.

$$\begin{aligned}
\frac{\sec^2\theta - 1}{\sec^2\theta} &= \frac{\sec^2\theta}{\sec^2\theta} - \frac{1}{\sec^2\theta} && \text{Rewrite as the difference of fractions.}\\
&= 1 - \cos^2\theta && \text{Reciprocal identity}\\
&= \sin^2\theta && \text{Pythagorean identity}
\end{aligned}$$

As you can see, there can be more than one way to verify an identity. Your method may differ from that used by your instructor or fellow students. Here is a good chance to be creative and establish your own style, but try to be as efficient as possible.

Example 2 ▶ Combining Fractions Before Using Identities

Verify the identity $\dfrac{1}{1 - \sin\alpha} + \dfrac{1}{1 + \sin\alpha} = 2\sec^2\alpha$.

Solution

$$\begin{aligned}
\frac{1}{1 - \sin\alpha} + \frac{1}{1 + \sin\alpha} &= \frac{1 + \sin\alpha + 1 - \sin\alpha}{(1 - \sin\alpha)(1 + \sin\alpha)} && \text{Add fractions.}\\
&= \frac{2}{1 - \sin^2\alpha} && \text{Simplify.}\\
&= \frac{2}{\cos^2\alpha} && \text{Pythagorean identity}\\
&= 2\sec^2\alpha && \text{Reciprocal identity}
\end{aligned}$$

Example 3 ▶ Verifying a Trigonometric Identity 

Verify the identity

$$(\tan^2 x + 1)(\cos^2 x - 1) = -\tan^2 x.$$

Solution

By applying identities before multiplying, you obtain the following.

$$\begin{aligned}(\tan^2 x + 1)(\cos^2 x - 1) &= (\sec^2 x)(-\sin^2 x) && \text{Pythagorean identities}\\ &= -\frac{\sin^2 x}{\cos^2 x} && \text{Reciprocal identity}\\ &= -\left(\frac{\sin x}{\cos x}\right)^2 && \text{Rule of exponents}\\ &= -\tan^2 x && \text{Quotient identity}\end{aligned}$$

Example 4 ▶ Converting to Sines and Cosines

Verify the identity

$$\tan x + \cot x = \sec x \csc x.$$

Solution

In this case there appear to be no fractions to add, no products to find, and no opportunities to use the Pythagorean identities. So, try converting the left side into sines and cosines to see what happens.

$$\begin{aligned}\tan x + \cot x &= \frac{\sin x}{\cos x} + \frac{\cos x}{\sin x} && \text{Quotient identities}\\ &= \frac{\sin^2 x + \cos^2 x}{\cos x \sin x} && \text{Add fractions.}\\ &= \frac{1}{\cos x \sin x} && \text{Pythagorean identity}\\ &= \frac{1}{\cos x} \cdot \frac{1}{\sin x} && \text{Product of fractions}\\ &= \sec x \csc x && \text{Reciprocal identities}\end{aligned}$$

Recall from algebra that *rationalizing the denominator* using conjugates is, on occasion, a powerful simplification technique. A related form of this technique works for simplifying trigonometric expressions as well. For instance, to simplify $1/(1 - \cos x)$, multiply the numerator and the denominator by $1 + \cos x$.

$$\begin{aligned}\frac{1}{1 - \cos x} &= \frac{1}{1 - \cos x}\left(\frac{1 + \cos x}{1 + \cos x}\right) = \frac{1 + \cos x}{1 - \cos^2 x} = \frac{1 + \cos x}{\sin^2 x}\\ &= \csc^2 x(1 + \cos x)\end{aligned}$$

This technique is demonstrated in the next example.

STUDY TIP

As shown at the right, $\csc^2 x(1 + \cos x)$ is considered a simplified form of $1/(1 - \cos x)$ because the expression does not contain any fractions.

Example 5 ▶ Verifying Trigonometric Identities 

Verify the identity $\sec y + \tan y = \dfrac{\cos y}{1 - \sin y}$.

Solution

Begin with the *right* side. Note that you can create a monomial denominator by multiplying the numerator and denominator by $1 + \sin y$.

$$\begin{aligned}
\frac{\cos y}{1 - \sin y} &= \frac{\cos y}{1 - \sin y}\left(\frac{1 + \sin y}{1 + \sin y}\right) && \text{Multiply numerator and denominator by } 1 + \sin y. \\
&= \frac{\cos y + \cos y \sin y}{1 - \sin^2 y} && \text{Multiply.} \\
&= \frac{\cos y + \cos y \sin y}{\cos^2 y} && \text{Pythagorean identity} \\
&= \frac{\cos y}{\cos^2 y} + \frac{\cos y \sin y}{\cos^2 y} && \text{Separate fractions.} \\
&= \frac{1}{\cos y} + \frac{\sin y}{\cos y} && \text{Simplify.} \\
&= \sec y + \tan y && \text{Identities}
\end{aligned}$$

In Examples 1 through 5, you have been verifying trigonometric identities by working with one side of the equation and converting to the form given on the other side. On occasion, it is practical to work with each side *separately*, to obtain one common form equivalent to both sides. This is illustrated in Example 6.

Example 6 ▶ Working with Each Side Separately

Verify the identity $\dfrac{\cot^2 \theta}{1 + \csc \theta} = \dfrac{1 - \sin \theta}{\sin \theta}$.

Solution

Working with the left side, you have

$$\begin{aligned}
\frac{\cot^2 \theta}{1 + \csc \theta} &= \frac{\csc^2 \theta - 1}{1 + \csc \theta} && \text{Pythagorean identity} \\
&= \frac{(\csc \theta - 1)\cancel{(\csc \theta + 1)}}{\cancel{1 + \csc \theta}} && \text{Factor.} \\
&= \csc \theta - 1. && \text{Simplify.}
\end{aligned}$$

Now, simplifying the right side, you have

$$\begin{aligned}
\frac{1 - \sin \theta}{\sin \theta} &= \frac{1}{\sin \theta} - \frac{\sin \theta}{\sin \theta} && \text{Separate fractions.} \\
&= \csc \theta - 1. && \text{Reciprocal identity}
\end{aligned}$$

The identity is verified because both sides are equal to $\csc \theta - 1$.

STUDY TIP

The technique of cross multiplication is not used in verifying trigonometric identities, because you do not know that the expressions are equal. Cross multiplication is used in solving an equation, when you know that the left side equals the right side.

In Example 7, powers of trigonometric functions are rewritten as more complicated sums of products of trigonometric functions. This is a common procedure used in calculus.

Example 7 ▶ Three Examples from Calculus

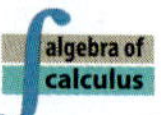

Verify each identity.

a. $\tan^4 x = \tan^2 x \sec^2 x - \tan^2 x$

b. $\sin^3 x \cos^4 x = (\cos^4 x - \cos^6 x) \sin x$

c. $\csc^4 x \cot x = \csc^2 x(\cot x + \cot^3 x)$

Solution

a. $\tan^4 x = (\tan^2 x)(\tan^2 x)$ — Separate factors.

$= \tan^2 x(\sec^2 x - 1)$ — Pythagorean identity

$= \tan^2 x \sec^2 x - \tan^2 x$ — Multiply.

b. $\sin^3 x \cos^4 x = \sin^2 x \cos^4 x \sin x$ — Separate factors.

$= (1 - \cos^2 x)\cos^4 x \sin x$ — Pythagorean identity

$= (\cos^4 x - \cos^6 x) \sin x$ — Multiply.

c. $\csc^4 x \cot x = \csc^2 x \csc^2 x \cot x$ — Separate factors.

$= \csc^2 x(1 + \cot^2 x) \cot x$ — Pythagorean identity

$= \csc^2 x(\cot x + \cot^3 x)$ — Multiply.

Writing ABOUT MATHEMATICS

Error Analysis You are tutoring a student in trigonometry. One of the homework problems your student encounters asks whether the following statement is an identity.

$$\tan^2 x \sin^2 x \stackrel{?}{=} \frac{5}{6}\tan^2 x$$

Your student does not attempt to verify the equivalence algebraically, but mistakenly uses only a graphical approach. Using range settings of

Xmin $= -3\pi$	Ymin $= -20$
Xmax $= 3\pi$	Ymax $= 20$
Xscl $= \pi/2$	Yscl $= 1$

your student graphs both sides of the expression on a graphing utility and concludes that the statement is an identity.

What is wrong with your student's reasoning? Explain. Discuss the limitations of verifying identities graphically.

7.2 Exercises

In Exercises 1–40, verify the identity.

1. $\sin t \csc t = 1$
2. $\sec y \cos y = 1$
3. $(1 + \sin\alpha)(1 - \sin\alpha) = \cos^2\alpha$
4. $\cot^2 y(\sec^2 y - 1) = 1$
5. $\cos^2\beta - \sin^2\beta = 1 - 2\sin^2\beta$
6. $\cos^2\beta - \sin^2\beta = 2\cos^2\beta - 1$
7. $\tan^2\theta + 4 = \sec^2\theta + 3$
8. $2 - \sec^2 z = 1 - \tan^2 z$
9. $\sin^2\alpha - \sin^4\alpha = \cos^2\alpha - \cos^4\alpha$
10. $\cos x + \sin x \tan x = \sec x$
11. $\dfrac{\csc^2\theta}{\cot\theta} = \csc\theta\sec\theta$
12. $\dfrac{\cot^3 t}{\csc t} = \cos t(\csc^2 t - 1)$
13. $\dfrac{\cot^2 t}{\csc t} = \csc t - \sin t$
14. $\dfrac{1}{\tan\beta} + \tan\beta = \dfrac{\sec^2\beta}{\tan\beta}$
15. $\sin^{1/2} x\cos x - \sin^{5/2} x\cos x = \cos^3 x\sqrt{\sin x}$
16. $\sec^6 x(\sec x\tan x) - \sec^4 x(\sec x\tan x) = \sec^5 x\tan^3 x$
17. $\dfrac{1}{\sec x\tan x} = \csc x - \sin x$
18. $\dfrac{\sec\theta - 1}{1 - \cos\theta} = \sec\theta$
19. $\cot\alpha + \tan\alpha = \csc\alpha\sec\alpha$
20. $\sec x - \cos x = \sin x\tan x$
21. $\dfrac{1}{\tan x} + \dfrac{1}{\cot x} = \tan x + \cot x$
22. $\dfrac{1}{\sin x} - \dfrac{1}{\csc x} = \csc x - \sin x$
23. $\dfrac{\cos\theta\cot\theta}{1 - \sin\theta} - 1 = \csc\theta$
24. $\dfrac{1 + \sin\theta}{\cos\theta} + \dfrac{\cos\theta}{1 + \sin\theta} = 2\sec\theta$
25. $\dfrac{1}{\sin x + 1} + \dfrac{1}{\csc x + 1} = 1$
26. $\cos x - \dfrac{\cos x}{1 - \tan x} = \dfrac{\sin x\cos x}{\sin x - \cos x}$
27. $\tan\left(\dfrac{\pi}{2} - \theta\right)\tan\theta = 1$
28. $\dfrac{\cos[(\pi/2) - x]}{\sin[(\pi/2) - x]} = \tan x$
29. $\dfrac{\csc(-x)}{\sec(-x)} = -\cot x$
30. $(1 + \sin y)[1 + \sin(-y)] = \cos^2 y$
31. $\dfrac{\sin x\cos y + \cos x\sin y}{\cos x\cos y - \sin x\sin y} = \dfrac{\tan x + \tan y}{1 - \tan x\tan y}$
32. $\dfrac{\tan x + \tan y}{1 - \tan x\tan y} = \dfrac{\cot x + \cot y}{\cot x\cot y - 1}$
33. $\dfrac{\tan x + \cot y}{\tan x\cot y} = \tan y + \cot x$
34. $\dfrac{\cos x - \cos y}{\sin x + \sin y} + \dfrac{\sin x - \sin y}{\cos x + \cos y} = 0$
35. $\sqrt{\dfrac{1 + \sin\theta}{1 - \sin\theta}} = \dfrac{1 + \sin\theta}{|\cos\theta|}$
36. $\sqrt{\dfrac{1 - \cos\theta}{1 + \cos\theta}} = \dfrac{1 - \cos\theta}{|\sin\theta|}$
37. $\cos^2\beta + \cos^2\left(\dfrac{\pi}{2} - \beta\right) = 1$
38. $\sec^2 y - \cot^2\left(\dfrac{\pi}{2} - y\right) = 1$
39. $\sin t\csc\left(\dfrac{\pi}{2} - t\right) = \tan t$
40. $\sec^2\left(\dfrac{\pi}{2} - x\right) - 1 = \cot^2 x$

In Exercises 41–52, use a graphing utility to determine whether the equation is an identity. If it is, confirm it algebraically.

41. $2\sec^2 x - 2\sec^2 x\sin^2 x - \sin^2 x - \cos^2 x = 1$
42. $\csc x(\csc x - \sin x) + \dfrac{\sin x - \cos x}{\sin x} + \cot x = \csc^2 x$
43. $2 + \cos^2 x - 3\cos^4 x = \sin^2 x(3 + 2\cos^2 x)$
44. $\tan^4 x + \tan^2 x - 3 = \sec^2 x(4\tan^2 x - 3)$
45. $\csc^4 x - 2\csc^2 x + 1 = \cot^4 x$
46. $(\sin^4\beta - 2\sin^2\beta + 1)\cos\beta = \cos^5\beta$
47. $\sec^4\theta - \tan^4\theta = 1 + 2\tan^2\theta$
48. $\csc^4\theta - \cot^4\theta = 2\csc^2\theta - 1$

49. $\dfrac{\cos x}{1 - \sin x} = \dfrac{1 - \sin x}{\cos x}$

50. $\dfrac{\cot \alpha}{\csc \alpha + 1} = \dfrac{\csc \alpha + 1}{\cot \alpha}$

51. $\dfrac{\tan^3 \alpha - 1}{\tan \alpha - 1} = \tan^2 \alpha + \tan \alpha + 1$

52. $\dfrac{\sin^3 \beta + \cos^3 \beta}{\sin \beta + \cos \beta} = 1 - \sin \beta \cos \beta$

In Exercises 53–56, use the cofunction identities to evaluate the expression without the aid of a calculator.

53. $\sin^2 25° + \sin^2 65°$

54. $\cos^2 55° + \cos^2 35°$

55. $\cos^2 20° + \cos^2 52° + \cos^2 38° + \cos^2 70°$

56. $\sin^2 12° + \sin^2 40° + \sin^2 50° + \sin^2 78°$

57. ***Rate of Change*** The rate of change of the function

$$f(x) = \sin x + \csc x$$

with respect to change in the variable x is given by the expression $\cos x - \csc x \cot x$. Show that the expression for the rate of change can also be $-\cos x \cot^2 x$.

▶ Model It

58. ***Shadow Length*** The length s of a shadow cast by a vertical gnomon (a device used to tell time) of height h when the angle of the sun above the horizon is θ (see figure) can be modeled by the equation

$$s = \frac{h \sin(90° - \theta)}{\sin \theta}.$$

(a) Verify that the equation above is equal to $h \cot \theta$.

(b) Use a graphing utility to complete the table. Let $h = 5$ feet.

θ	10°	20°	30°	40°	50°
s					

θ	60°	70°	80°	90°
s				

▶ Model It *(continued)*

(c) Use your table from part (b) to determine the angles of the sun for which the length of the shadow is the greatest and the least.

(d) Based on your results from part (c), what time of day do you think it is when the angle of the sun above the horizon is 90°?

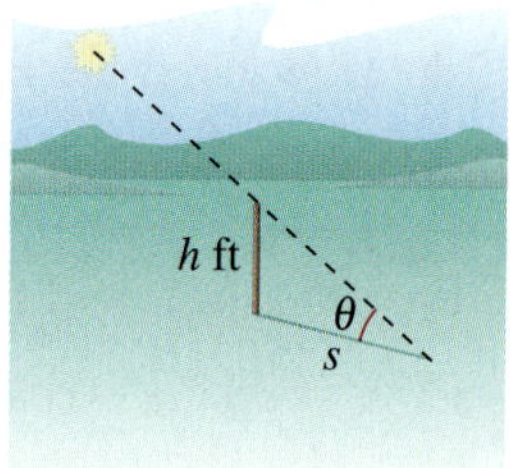

Synthesis

True or False? **In Exercises 59 and 60, determine whether the statement is true or false. Justify your answer.**

59. The equation

$$\sin^2 \theta + \cos^2 \theta = 1 + \tan^2 \theta$$

is an identity, because $\sin^2(0) + \cos^2(0) = 1$ and $1 + \tan^2(0) = 1$.

60. The equation $1 + \tan^2 \theta = 1 + \cot^2 \theta$ is *not* an identity, because it is true that $1 + \tan^2(\pi/6) = 1\frac{1}{3}$, and $1 + \cot^2(\pi/6) = 4$.

Think About It **In Exercises 61 and 62, explain why the equation is *not* an identity and find one value of the variable for which the equation is not true.**

61. $\sin \theta = \sqrt{1 - \cos^2 \theta}$

62. $\tan \theta = \sqrt{\sec^2 \theta - 1}$

Review

In Exercises 63–66, perform the operation and simplify.

63. $(2 + 3i) - \sqrt{-26}$

64. $(2 - 5i)^2$

65. $\sqrt{-16}\left(1 + \sqrt{-4}\right)$

66. $(3 + 2i)^3$

In Exercises 67–70, use the Quadratic Formula to solve the quadratic equation.

67. $x^2 - 6x + 12 = 0$

68. $x^2 + 5x + 7 = 0$

69. $3x^2 + 6x + 12 = 0$

70. $8x^2 - 4x + 3 = 0$

7.3 Solving Trigonometric Equations

▶ **What you should learn**

- How to use standard algebraic techniques to solve trigonometric equations
- How to solve trigonometric equations of quadratic type
- How to solve trigonometric equations involving multiple angles
- How to use inverse trigonometric functions to solve trigonometric equations

▶ **Why you should learn it**

You can use trigonometric equations to solve a variety of real-life problems. For instance, in Exercise 75 on page 567, you can solve a trigonometric equation to help answer questions about the unemployment rate in the United States.

M. Greenlar/The Image Works

Introduction

To solve a trigonometric equation, use standard algebraic techniques such as collecting like terms and factoring. Your preliminary goal in solving a trigonometric equation is to isolate the trigonometric function involved in the equation. For example, to solve the equation $2 \sin x = 1$, divide each side by 2 to obtain

$$\sin x = \frac{1}{2}.$$

To solve for x, note in Figure 7.2 that the equation $\sin x = \frac{1}{2}$ has solutions $x = \pi/6$ and $x = 5\pi/6$ in the interval $[0, 2\pi)$. Moreover, because $\sin x$ has a period of 2π, there are infinitely many other solutions, which can be written as

$$x = \frac{\pi}{6} + 2n\pi \quad \text{and} \quad x = \frac{5\pi}{6} + 2n\pi \qquad \text{General solution}$$

where n is an integer, as shown in Figure 7.2.

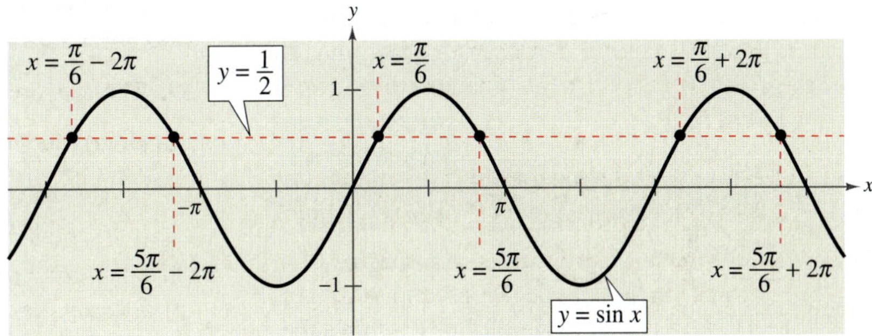

FIGURE 7.2

Another way to show that the equation $\sin x = \frac{1}{2}$ has infinitely many solutions is indicated in Figure 7.3. Any angles that are coterminal with $\pi/6$ or $5\pi/6$ will also be solutions of the equation.

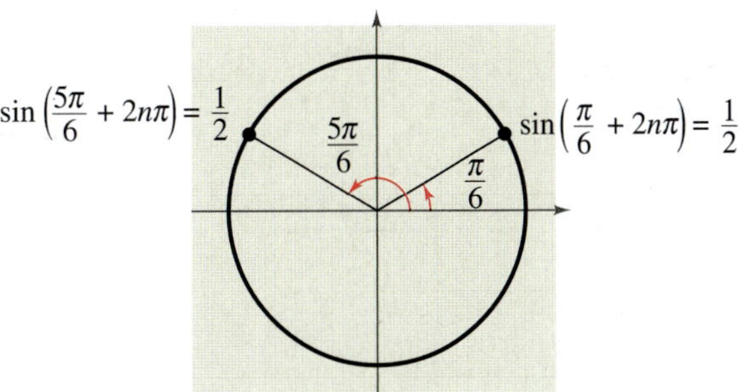

FIGURE 7.3

Example 1 ▶ Collecting Like Terms

Solve $\sin x + \sqrt{2} = -\sin x$.

Solution

Begin by rewriting the equation so that $\sin x$ is isolated on one side of the equation.

$$\sin x + \sqrt{2} = -\sin x \qquad \text{Write original equation.}$$

$$\sin x + \sin x + \sqrt{2} = 0 \qquad \text{Add } \sin x \text{ to each side.}$$

$$\sin x + \sin x = -\sqrt{2} \qquad \text{Subtract } \sqrt{2} \text{ from each side.}$$

$$2 \sin x = -\sqrt{2} \qquad \text{Combine like terms.}$$

$$\sin x = -\frac{\sqrt{2}}{2} \qquad \text{Divide each side by 2.}$$

Because $\sin x$ has a period of 2π, first find all solutions in the interval $[0, 2\pi)$. These solutions are $x = 5\pi/4$ and $x = 7\pi/4$. Finally, add $2n\pi$ to each of these solutions to get the general form

$$x = \frac{5\pi}{4} + 2n\pi \quad \text{and} \quad x = \frac{7\pi}{4} + 2n\pi \qquad \text{General solution}$$

where n is an integer.

Example 2 ▶ Extracting Square Roots

Solve $3 \tan^2 x - 1 = 0$.

Solution

Begin by rewriting the equation so that $\tan x$ is isolated on one side of the equation.

$$3 \tan^2 x - 1 = 0 \qquad \text{Write original equation.}$$

$$3 \tan^2 x = 1 \qquad \text{Add 1 to each side.}$$

$$\tan^2 x = \frac{1}{3} \qquad \text{Divide each side by 3.}$$

$$\tan x = \pm\frac{1}{\sqrt{3}} \qquad \text{Extract square roots.}$$

Because $\tan x$ has a period of π, first find all solutions in the interval $[0, \pi)$. These solutions are $x = \pi/6$ and $x = 5\pi/6$. Finally, add $n\pi$ to each of these solutions to get the general form

$$x = \frac{\pi}{6} + n\pi \quad \text{and} \quad x = \frac{5\pi}{6} + n\pi \qquad \text{General solution}$$

where n is an integer.

Technology

The solutions in Examples 1 and 2 are obtained analytically. You can use a graphing utility to confirm the solutions graphically. For instance, to confirm the solutions found in Example 2, graph

$$y = 3 \tan^2 x - 1$$

as shown below. Then use the *zero* or *root* feature or the *zoom* and *trace* features to approximate the x-intercepts.

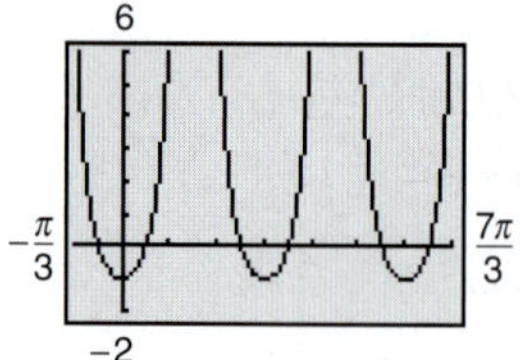

The equations in Examples 1 and 2 involved only one trigonometric function. When two or more functions occur in the same equation, collect all terms on one side and try to separate the functions by factoring or by using appropriate identities. This may produce factors that yield no solutions, as illustrated in Example 3.

Example 3 ▶ Factoring 

Solve $\cot x \cos^2 x = 2 \cot x$.

Solution

Begin by rewriting the equation so that all terms are collected on one side of the equation.

$$\cot x \cos^2 x = 2 \cot x$$ Write original equation.

$$\cot x \cos^2 x - 2 \cot x = 0$$ Subtract $2 \cot x$ from each side.

$$\cot x(\cos^2 x - 2) = 0$$ Factor.

By setting each of these factors equal to zero, you obtain

$$\cot x = 0 \quad \text{and} \quad \cos^2 x - 2 = 0$$

$$x = \frac{\pi}{2} \qquad\qquad \cos^2 x = 2$$

$$\cos x = \pm\sqrt{2}.$$

The equation $\cot x = 0$ has the solution $x = \pi/2$. No solution is obtained for $\cos x = \pm\sqrt{2}$ because $\pm\sqrt{2}$ are outside the range of the cosine function. Because $\cot x$ has a period of π, the general form of the solution is obtained by adding multiples of π to $x = \pi/2$, to get

$$x = \frac{\pi}{2} + n\pi$$ General solution

where n is an integer. You can confirm this graphically by sketching the graph of $y = \cot x \cos^2 x - 2 \cot x$, as shown in Figure 7.4.

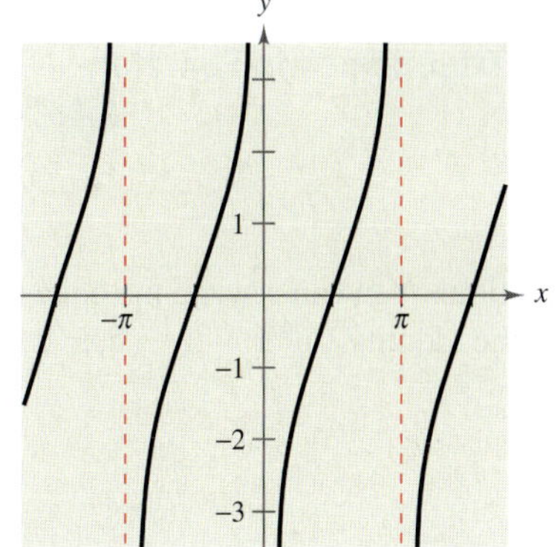

FIGURE 7.4

In Example 3, do not make the mistake of dividing each side of the equation by $\cot x$. If you do this, you lose the solutions. Can you see why?

Equations of Quadratic Type

Many trigonometric equations are of quadratic type. Here are a couple of examples.

Quadratic in sin x	*Quadratic in sec x*
$2 \sin^2 x - \sin x - 1 = 0$	$\sec^2 x - 3 \sec x - 2 = 0$
$2(\sin x)^2 - \sin x - 1 = 0$	$(\sec x)^2 - 3(\sec x) - 2 = 0$

To solve equations of this type, factor the quadratic or, if this is not possible, use the Quadratic Formula.

Example 4 ▶ Factoring an Equation of Quadratic Type

Find all solutions of $2\sin^2 x - \sin x - 1 = 0$ in the interval $[0, 2\pi)$.

Solution

Begin by treating the equation as a quadratic in $\sin x$ and factoring.

$$(2\sin x + 1)(\sin x - 1) = 0 \qquad \text{Factor.}$$

Setting each factor equal to zero, you can find the solutions in the interval $[0, 2\pi)$.

$$2\sin x + 1 = 0 \qquad \text{and} \qquad \sin x - 1 = 0$$

$$\sin x = -\frac{1}{2} \qquad\qquad \sin x = 1$$

$$x = \frac{7\pi}{6}, \frac{11\pi}{6} \qquad\qquad x = \frac{\pi}{2}$$

When working with an equation of quadratic type, be sure that the equation involves a *single* trigonometric function, as shown in the next example.

Example 5 ▶ Rewriting with a Single Trigonometric Function

Solve $2\sin^2 x + 3\cos x - 3 = 0$.

Solution

This equation contains both sine and cosine functions. You can rewrite the equation so that it has only cosine functions by using the identity $\sin^2 x = 1 - \cos^2 x$.

$$2\sin^2 x + 3\cos x - 3 = 0 \qquad \text{Write original equation.}$$

$$2(1 - \cos^2 x) + 3\cos x - 3 = 0 \qquad \text{Pythagorean identity}$$

$$2\cos^2 x - 3\cos x + 1 = 0 \qquad \text{Multiply each side by } -1.$$

$$(2\cos x - 1)(\cos x - 1) = 0 \qquad \text{Factor.}$$

By setting each factor equal to zero, you can find the solutions in the interval $[0, 2\pi)$.

$$2\cos x - 1 = 0 \qquad \text{and} \qquad \cos x - 1 = 0$$

$$\cos x = \frac{1}{2} \qquad\qquad \cos x = 1$$

$$x = 0$$

$$x = \frac{\pi}{3}, \frac{5\pi}{3}$$

Because $\cos x$ has a period of 2π, the general form of the solution is obtained by adding multiples of 2π to get

$$x = 2n\pi, \quad x = \frac{\pi}{3} + 2n\pi, \quad x = \frac{5\pi}{3} + 2n\pi \qquad \text{General solution}$$

where n is an integer.

Sometimes you must square each side of an equation to obtain a quadratic, as demonstrated in the next example. Because this procedure can introduce extraneous solutions, you should check any solutions in the original equation to see whether they are valid or extraneous.

Example 6 ▶ Squaring and Converting to Quadratic Type 

Find all solutions of $\cos x + 1 = \sin x$ in the interval $[0, 2\pi)$.

Solution

It is not clear how to rewrite this equation in terms of a single trigonometric function. See what happens when you square each side of the equation.

$$\cos x + 1 = \sin x$$ Write original equation.

$$\cos^2 x + 2\cos x + 1 = \sin^2 x$$ Square each side.

$$\cos^2 x + 2\cos x + 1 = 1 - \cos^2 x$$ Pythagorean identity

$$\cos^2 x + \cos^2 x + 2\cos x + 1 - 1 = 0$$ Rewrite equation.

$$2\cos^2 x + 2\cos x = 0$$ Combine like terms.

$$2\cos x(\cos x + 1) = 0$$ Factor.

Setting each factor equal to zero produces

$$2\cos x = 0 \qquad \text{and} \qquad \cos x + 1 = 0$$

$$\cos x = 0 \qquad\qquad \cos x = -1$$

$$x = \frac{\pi}{2}, \frac{3\pi}{2} \qquad\qquad x = \pi.$$

Because you squared the original equation, check for extraneous solutions.

Check for $x = \pi/2$

$$\cos\frac{\pi}{2} + 1 \stackrel{?}{=} \sin\frac{\pi}{2}$$ Substitute $\pi/2$ for x.

$$0 + 1 = 1$$ Solution checks. ✓

Check for $x = 3\pi/2$

$$\cos\frac{3\pi}{2} + 1 \stackrel{?}{=} \sin\frac{3\pi}{2}$$ Substitute $3\pi/2$ for x.

$$0 + 1 \neq -1$$ Solution does not check.

Check for $x = \pi$

$$\cos\pi + 1 \stackrel{?}{=} \sin\pi$$ Substitute π for x.

$$-1 + 1 = 0$$ Solution checks. ✓

Of the three possible solutions, $x = 3\pi/2$ is extraneous. So, in the interval $[0, 2\pi)$, the only two solutions are $x = \pi/2$ and $x = \pi$.

STUDY TIP

You square each side of the equation in Example 6 because the squares of the sine and cosine functions are related by a Pythagorean identity. The same is true for the squares of the secant and tangent functions and the cosecant and cotangent functions.

Exploration

Use a graphing utility to confirm the solutions found in Example 6 in two different ways. Do both methods produce the same x-values? Which method do you prefer? Why?

1. Graph both sides of the equation and find the x-coordinates of the points at which the graphs intersect.

   Left side: $y = \cos x + 1$

   Right side: $y = \sin x$

2. Graph the equation

   $$y = \cos x + 1 - \sin x$$

   and find the x-intercepts of the graph.

Functions Involving Multiple Angles

The next two examples involve trigonometric functions of multiple angles of the forms sin ku and cos ku. To solve equations of these forms, first solve the equation for ku, then divide your result by k.

Example 7 ▶ Functions of Multiple Angles

Find all solutions of $2 \cos 3t - 1 = 0$.

Solution

$$2 \cos 3t - 1 = 0 \qquad \text{Write original equation.}$$

$$2 \cos 3t = 1 \qquad \text{Add 1 to each side.}$$

$$\cos 3t = \frac{1}{2} \qquad \text{Divide each side by 2.}$$

In the interval $[0, 2\pi)$, you know that $3t = \pi/3$ and $3t = 5\pi/3$ are the only solutions, so, in general, you have

$$3t = \frac{\pi}{3} + 2n\pi \qquad \text{and} \qquad 3t = \frac{5\pi}{3} + 2n\pi.$$

Dividing these results by 3, you obtain the general solution

$$t = \frac{\pi}{9} + \frac{2n\pi}{3} \qquad \text{and} \qquad t = \frac{5\pi}{9} + \frac{2n\pi}{3} \qquad \text{General solution}$$

where n is an integer.

Example 8 ▶ Functions of Multiple Angles

Find all solutions of $3 \tan(x/2) + 3 = 0$.

Solution

$$3 \tan \frac{x}{2} + 3 = 0 \qquad \text{Write original equation.}$$

$$3 \tan \frac{x}{2} = -3 \qquad \text{Subtract 3 from each side.}$$

$$\tan \frac{x}{2} = -1 \qquad \text{Divide each side by 3.}$$

In the interval $[0, \pi)$, you know that $x/2 = 3\pi/4$ is the only solution, so, in general, you have

$$\frac{x}{2} = \frac{3\pi}{4} + n\pi.$$

Multiplying this result by 2, you obtain the general solution

$$x = \frac{3\pi}{2} + 2n\pi \qquad \text{General solution}$$

where n is an integer.

Using Inverse Functions

In the next example, you will see how inverse trigonometric functions can be used to solve an equation.

Example 9 ▶ Using Inverse Functions

Find all solutions of $\sec^2 x - 2\tan x = 4$.

Solution

$$\sec^2 x - 2\tan x = 4$$ Write original equation.

$$1 + \tan^2 x - 2\tan x - 4 = 0$$ Pythagorean identity

$$\tan^2 x - 2\tan x - 3 = 0$$ Combine like terms.

$$(\tan x - 3)(\tan x + 1) = 0$$ Factor.

Setting each factor equal to zero, you obtain two solutions in the interval $(-\pi/2, \pi/2)$. [Recall that the range of the inverse tangent function is $(-\pi/2, \pi/2)$.]

$$\tan x - 3 = 0 \qquad \text{and} \qquad \tan x + 1 = 0$$

$$\tan x = 3 \qquad\qquad \tan x = -1$$

$$x = \arctan 3 \qquad\qquad x = -\frac{\pi}{4}$$

Finally, because $\tan x$ has a period of π, you obtain the general solution by adding multiples of π

$$x = \arctan 3 + n\pi \qquad \text{and} \qquad x = -\frac{\pi}{4} + n\pi$$ General solution

where n is an integer. You can use a calculator to approximate the value of arctan 3.

Writing ABOUT MATHEMATICS

Equations with No Solutions One of the following equations has solutions and the other two do not. Which two equations do not have solutions?

a. $\sin^2 x - 5\sin x + 6 = 0$

b. $\sin^2 x - 4\sin x + 6 = 0$

c. $\sin^2 x - 5\sin x - 6 = 0$

Find conditions involving the constants b and c that will guarantee that the equation

$$\sin^2 x + b\sin x + c = 0$$

has at least one solution on some interval of length 2π.

7.3 Exercises

In Exercises 1–6, verify that the *x*-values are solutions of the equation.

1. $2 \cos x - 1 = 0$

(a) $x = \dfrac{\pi}{3}$ (b) $x = \dfrac{5\pi}{3}$

2. $\sec x - 2 = 0$

(a) $x = \dfrac{\pi}{3}$ (b) $x = \dfrac{5\pi}{3}$

3. $3 \tan^2 2x - 1 = 0$

(a) $x = \dfrac{\pi}{12}$ (b) $x = \dfrac{5\pi}{12}$

4. $2 \cos^2 4x - 1 = 0$

(a) $x = \dfrac{\pi}{16}$ (b) $x = \dfrac{3\pi}{16}$

5. $2 \sin^2 x - \sin x - 1 = 0$

(a) $x = \dfrac{\pi}{2}$ (b) $x = \dfrac{7\pi}{6}$

6. $\csc^4 x - 4 \csc^2 x = 0$

(a) $x = \dfrac{\pi}{6}$ (b) $x = \dfrac{5\pi}{6}$

In Exercises 7–20, solve the equation.

7. $2 \cos x + 1 = 0$
8. $2 \sin x + 1 = 0$
9. $\sqrt{3} \csc x - 2 = 0$
10. $\tan x + \sqrt{3} = 0$
11. $3 \sec^2 x - 4 = 0$
12. $3 \cot^2 x - 1 = 0$
13. $\sin x(\sin x + 1) = 0$
14. $(3 \tan^2 x - 1)(\tan^2 x - 3) = 0$
15. $4 \cos^2 x - 1 = 0$
16. $\sin^2 x = 3 \cos^2 x$
17. $2 \sin^2 2x = 1$
18. $\tan^2 3x = 3$
19. $\tan 3x(\tan x - 1) = 0$
20. $\cos 2x(2 \cos x + 1) = 0$

In Exercises 21–32, find all solutions of the equation in the interval $[0, 2\pi)$.

21. $\cos^3 x = \cos x$
22. $\sec^2 x - 1 = 0$
23. $3 \tan^3 x = \tan x$
24. $2 \sin^2 x = 2 + \cos x$
25. $\sec^2 x - \sec x = 2$
26. $\sec x \csc x = 2 \csc x$
27. $2 \sin x + \csc x = 0$
28. $\sec x + \tan x = 1$
29. $2 \cos^2 x + \cos x - 1 = 0$
30. $2 \sin^2 x + 3 \sin x + 1 = 0$
31. $2 \sec^2 x + \tan^2 x - 3 = 0$
32. $\cos x + \sin x \tan x = 2$

In Exercises 33–38, find all solutions of the equation.

33. $\cos 2x = \dfrac{1}{2}$
34. $\sin 2x = -\dfrac{\sqrt{3}}{2}$
35. $\tan 3x = 1$
36. $\sec 4x = 2$
37. $\cos \dfrac{x}{2} = \dfrac{\sqrt{2}}{2}$
38. $\sin \dfrac{x}{2} = -\dfrac{\sqrt{3}}{2}$

In Exercises 39–42, find the *x*-intercepts of the graph.

39. $y = \sin \dfrac{\pi x}{2} + 1$

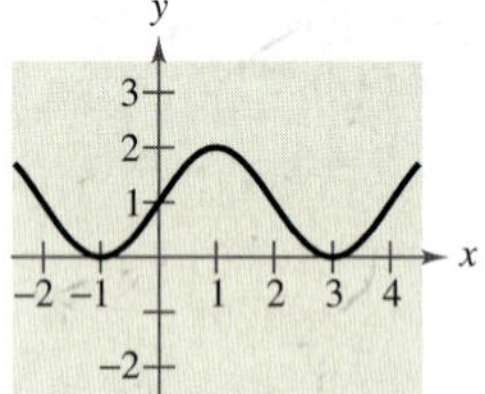

40. $y = \sin \pi x + \cos \pi x$

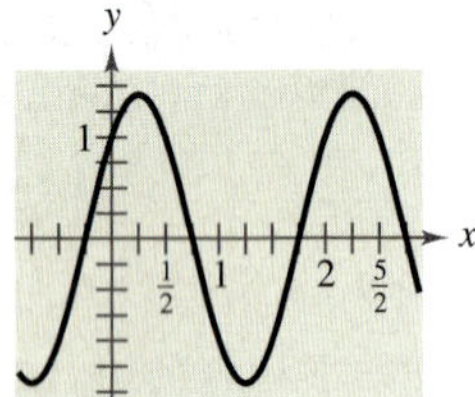

41. $y = \tan^2\left(\dfrac{\pi x}{6}\right) - 3$

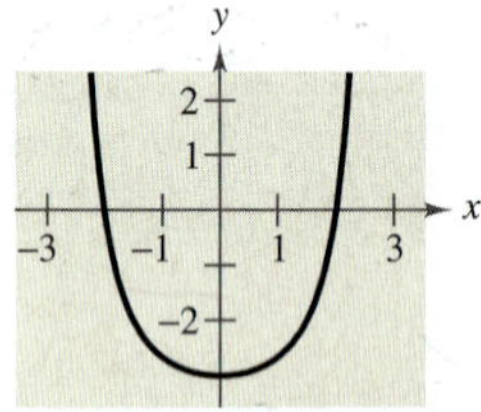

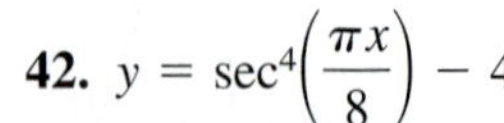

42. $y = \sec^4\left(\dfrac{\pi x}{8}\right) - 4$

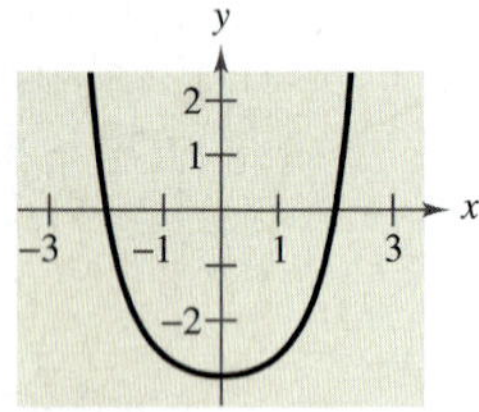

In Exercises 43 and 44, solve both equations. How do the solutions of the algebraic equation compare with the solutions of the trigonometric equation?

43. $6y^2 - 13y + 6 = 0$

$6 \cos^2 x - 13 \cos x + 6 = 0$

44. $y^2 + y - 20 = 0$

$\sin^2 x + \sin x - 20 = 0$

In Exercises 45–54, use a graphing utility to approximate the solutions of the equation in the interval $[0, 2\pi)$.

45. $2 \sin x + \cos x = 0$

46. $4 \sin^3 x + 2 \sin^2 x - 2 \sin x - 1 = 0$

47. $\dfrac{1 + \sin x}{\cos x} + \dfrac{\cos x}{1 + \sin x} = 4$

48. $\dfrac{\cos x \cot x}{1 - \sin x} = 3$

49. $x \tan x - 1 = 0$

50. $x \cos x - 1 = 0$

51. $\sec^2 x + 0.5 \tan x - 1 = 0$

52. $\csc^2 x + 0.5 \cot x - 5 = 0$

53. $2 \tan^2 x + 7 \tan x - 15 = 0$

54. $6 \sin^2 x - 7 \sin x + 2 = 0$

 **In Exercises 55–58, use the Quadratic Formula to solve the equation in the interval $[0, 2\pi)$. Then use a graphing utility to approximate the angle x.**

55. $12 \sin^2 x - 13 \sin x + 3 = 0$

56. $3 \tan^2 x + 4 \tan x - 4 = 0$

57. $\tan^2 x + 3 \tan x + 1 = 0$

58. $4 \cos^2 x - 4 \cos x - 1 = 0$

In Exercises 59–62, use inverse functions where needed to find all solutions of the equation in the interval $[0, 2\pi)$.

59. $\tan^2 x - 6 \tan x + 5 = 0$

60. $\sec^2 x + \tan x - 3 = 0$

61. $2 \cos^2 x - 5 \cos x + 2 = 0$

62. $2 \sin^2 x - 7 \sin x + 3 = 0$

 **In Exercises 63 and 64, (a) use a graphing utility to graph the function and approximate the maximum and minimum points on the graph in the interval $[0, 2\pi)$, and (b) solve the trigonometric equation and demonstrate that its solutions are the x-coordinates of the maximum and minimum points of f. (Calculus is required to find the trigonometric equation.)**

	Function	*Trigonometric Equation*
63.	$f(x) = \sin x + \cos x$	$\cos x - \sin x = 0$
64.	$f(x) = 2 \sin x + \cos 2x$	$2 \cos x - 4 \sin x \cos x = 0$

Fixed Point **In Exercises 65 and 66, find the smallest positive fixed point of the function f. [A fixed point of a function f is a real number c such that $f(c) = c$.]**

65. $f(x) = \tan \dfrac{\pi x}{4}$

66. $f(x) = \cos x$

67. ***Graphical Reasoning*** Consider the function

$$f(x) = \cos \frac{1}{x}$$

and its graph shown in the figure.

(a) What is the domain of the function?

(b) Identify any symmetry or asymptotes of the graph.

(c) Describe the behavior of the function as $x \to 0$.

(d) How many solutions does the equation

$$\cos \frac{1}{x} = 0$$

have in the interval $[-1, 1]$?

(e) Does the equation $\cos(1/x) = 0$ have a greatest solution? If so, approximate the solution. If not, explain why.

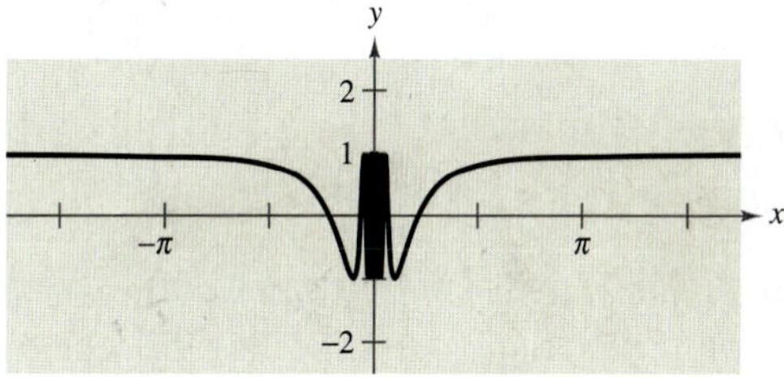

FIGURE FOR 67

68. ***Graphical Reasoning*** Consider the function

$$f(x) = \frac{\sin x}{x}$$

and its graph shown in the figure.

(a) What is the domain of the function?

(b) Identify any symmetry or asymptotes of the graph.

(c) Describe the behavior of the function as $x \to 0$.

(d) How many solutions does the equation

$$\frac{\sin x}{x} = 0$$

have in the interval $[-8, 8]$? Find the solutions.

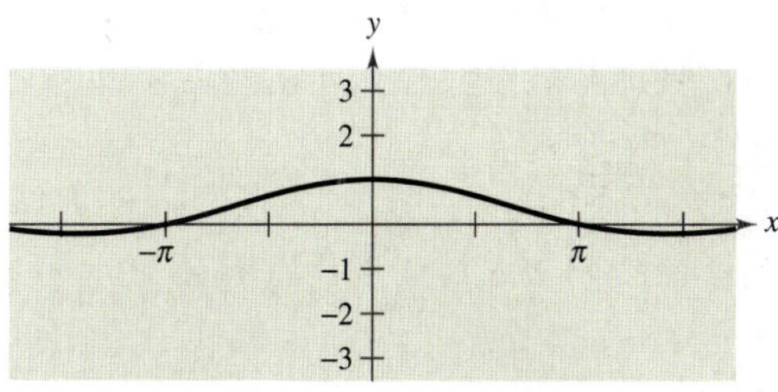

69. *Harmonic Motion* A weight is oscillating on the end of a spring (see figure). The position of the weight relative to the point of equilibrium is

$$y = \frac{1}{12}(\cos 8t - 3 \sin 8t)$$

where y is the displacement in meters and t is the time in seconds. Find the times when the weight is at the point of equilibrium ($y = 0$) for $0 \le t \le 1$.

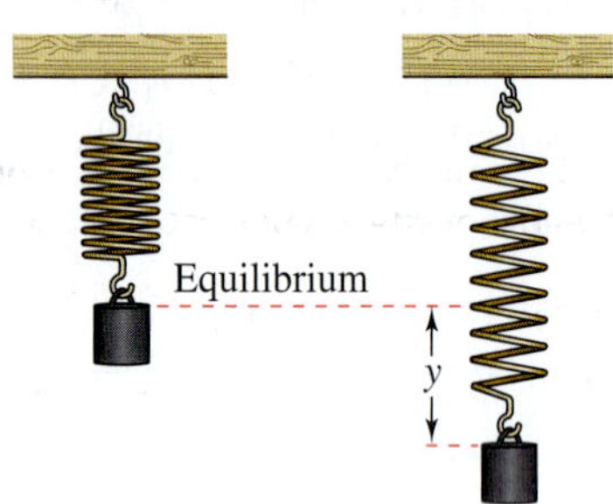

 **70. *Damped Harmonic Motion*** The displacement from equilibrium of a weight oscillating on the end of a spring is

$$y = 1.56e^{-0.22t} \cos 4.9t$$

where y is the displacement in feet and t is the time in seconds. Use a graphing utility to graph the displacement function for $0 \le t \le 10$. Find the time beyond which the displacement does not exceed 1 foot from equilibrium.

71. *Sales* The monthly sales (in thousands of units) of a seasonal product are approximated by

$$S = 74.50 + 43.75 \sin \frac{\pi t}{6}$$

where t is the time in months, with $t = 1$ corresponding to January. Determine the months when sales exceed 100,000 units.

72. *Projectile Motion* A batted baseball leaves the bat at an angle of θ with the horizontal and an initial velocity of $v_0 = 100$ feet per second. The ball is caught by an outfielder 300 feet from home plate (see figure). Find θ if the range r of a projectile is

$$r = \frac{1}{32}v_0^2 \sin 2\theta.$$

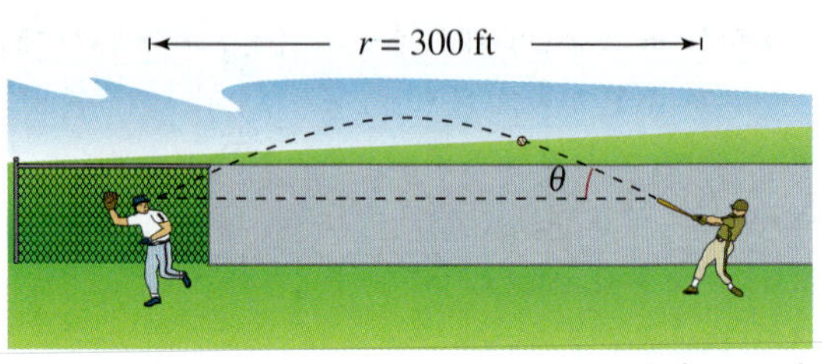

FIGURE FOR 72

73. *Projectile Motion* A sharpshooter intends to hit a target at a distance of 1000 yards with a gun that has a muzzle velocity of 1200 feet per second (see figure). Neglecting air resistance, determine the gun's minimum angle of elevation θ if the range r is

$$r = \frac{1}{32}v_0^2 \sin 2\theta.$$

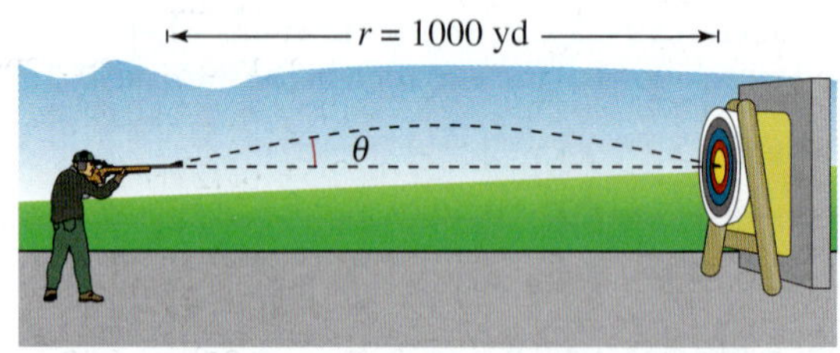

74. *Geometry* The area of a rectangle (see figure) inscribed in one arc of the graph of $y = \cos x$ is

$$A = 2x \cos x, \qquad 0 < x < \frac{\pi}{2}.$$

 (a) Use a graphing utility to graph the area function, and approximate the area of the largest inscribed rectangle.

(b) Determine the values of x for which $A \ge 1$.

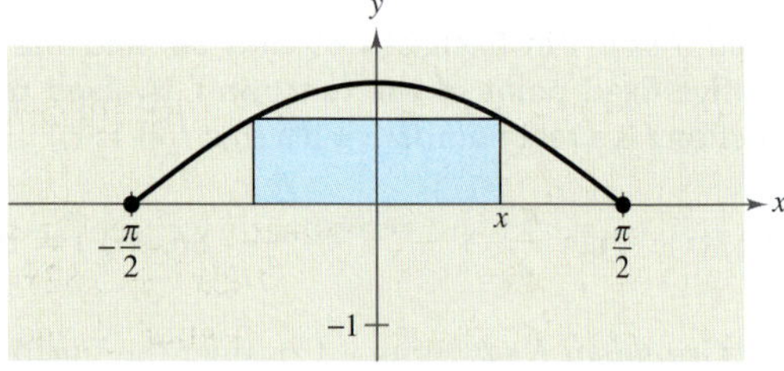

▶ Model It

75. ***Data Analysis*** The table shows the unemployment rate r for the years 1990 through 2001 in the United States. The time t is measured in years, with $t = 0$ corresponding to 1990. (Source: U.S. Bureau of Labor Statistics)

Time, t	Rate, r	Time, t	Rate, r
0	5.6	6	5.4
1	6.8	7	4.9
2	7.5	8	4.5
3	6.9	9	4.2
4	6.1	10	4.0
5	5.6	11	4.8

(a) Create a scatter plot of the data.

(b) Which of the following models best represents the data? Explain your reasoning.

(1) $r = 1.39 \sin(0.48t + 0.42) + 5.51$

(2) $r = 1.39 \sin(0.48t - 0.01) + 5.51$

(3) $r = \sin(0.10t + 5.61) + 4.80$

(4) $r = 896 \sin(0.57t - 2.05) + 6.48$

(c) What term in the model gives the average unemployment rate? What is the rate?

(d) Economists study the lengths of business cycles such as unemployment rates. Based on this short span of time, use the model to give the length of this cycle.

(e) Use the model to estimate the next time the unemployment rate will be 6.5% or more.

76. ***Quadratic Approximation*** Consider the function

$f(x) = 3 \sin(0.6x - 2)$.

(a) Approximate the zero of the function in the interval $[0, 6]$.

(b) A quadratic approximation agreeing with f at $x = 5$ is $g(x) = -0.45x^2 + 5.52x - 13.70$. Use a graphing utility to graph f and g in the same viewing window. Describe the result.

(c) Use the Quadratic Formula to find the zeros of g. Compare the zero in the interval $[0, 6]$ with the result of part (a).

Synthesis

True or False? **In Exercises 77 and 78, determine whether the statement is true or false. Justify your answer.**

77. The equation $2 \sin 4t - 1 = 0$ has four times the number of solutions in the interval $[0, 2\pi)$ as the equation $2 \sin t - 1 = 0$.

78. If you correctly solve a trigonometric equation down to the statement $\sin x = 3.4$, then you can finish solving the equation by using an inverse function.

In Exercises 79 and 80, use the graph to approximate the number of points of intersection of the graphs of y_1 and y_2.

79. $y_1 = 2 \sin x$
$y_2 = 3x + 1$

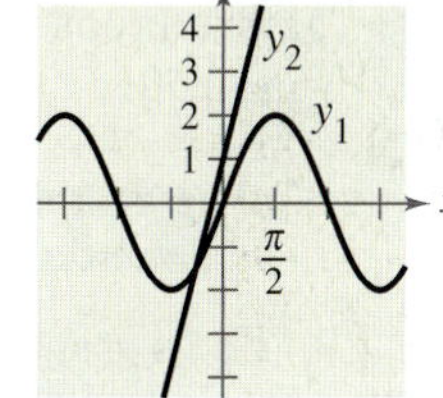

80. $y_1 = 2 \sin x$
$y_2 = \frac{1}{2}x + 1$

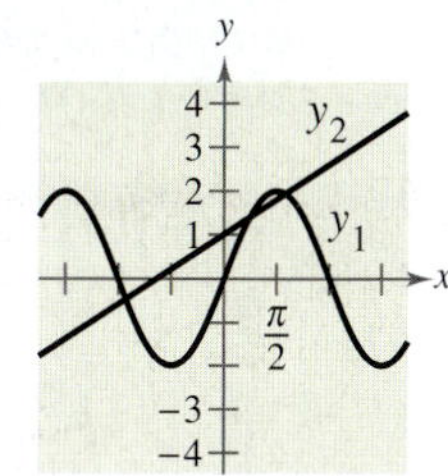

Review

In Exercises 81 and 82, solve triangle *ABC* by finding all missing angle measures and side lengths.

81. B, 22.3, 66°, A, C

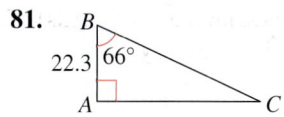

82.

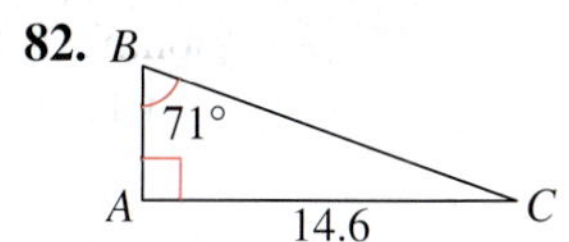

In Exercises 83–86, use reference angles to find the sine, cosine, and tangent of an angle with the given measure.

83. 390°

84. 570°

85. −1845°

86. −1410°

87. ***Angle of Depression*** Find the angle of depression from the top of a lighthouse 250 feet above water level to the water line of a ship 2 miles offshore.

88. ***Height*** From a point 100 feet in front of the public library, the angles of elevation to the base of the flagpole and the top of the pole are 28° and 39° 45′, respectively. The flagpole is mounted on the front of the library's roof. Find the height of the flagpole.

7.4 Sum and Difference Formulas

▶ What you should learn

- How to use sum and difference formulas to evaluate trigonometric functions, verify identities, and solve trigonometric equations

▶ Why you should learn it

You can use identities to rewrite trigonometric expressions. For instance, in Exercise 75 on page 573, you can use an identity to rewrite a trigonometric expression in a form that helps you analyze a harmonic motion equation.

Richard Megna/Fundamental Photographs

Using Sum and Difference Formulas 

In this and the following section, you will study the uses of several trigonometric identities and formulas.

Sum and Difference Formulas

$$\sin(u + v) = \sin u \cos v + \cos u \sin v$$

$$\sin(u - v) = \sin u \cos v - \cos u \sin v$$

$$\cos(u + v) = \cos u \cos v - \sin u \sin v$$

$$\cos(u - v) = \cos u \cos v + \sin u \sin v$$

$$\tan(u + v) = \frac{\tan u + \tan v}{1 - \tan u \tan v}$$

$$\tan(u - v) = \frac{\tan u - \tan v}{1 + \tan u \tan v}$$

For a proof of the sum and difference formulas, see Proofs in Mathematics on page 591.

Exploration

Use a graphing utility to graph $y_1 = \cos(x + 2)$ and $y_2 = \cos x + \cos 2$ in the same viewing window. What can you conclude about the graphs? Is it true that $\cos(x + 2) = \cos x + \cos 2$?

Use a graphing utility to graph $y_1 = \sin(x + 4)$ and $y_2 = \sin x + \sin 4$ in the same viewing window. What can you conclude about the graphs? Is it true that $\sin(x + 4) = \sin x + \sin 4$?

Examples 1 and 2 show how **sum and difference formulas** can be used to find exact values of trigonometric functions involving sums or differences of special angles.

Example 1 ▶ Evaluating a Trigonometric Function 

Find the exact value of $\cos 75°$.

Solution

To find the *exact* value of $\cos 75°$, use the fact that $75° = 30° + 45°$. Consequently, the formula for $\cos(u + v)$ yields

$$\begin{aligned}\cos 75° &= \cos(30° + 45°)\\ &= \cos 30° \cos 45° - \sin 30° \sin 45°\\ &= \frac{\sqrt{3}}{2}\left(\frac{\sqrt{2}}{2}\right) - \frac{1}{2}\left(\frac{\sqrt{2}}{2}\right) = \frac{\sqrt{6} - \sqrt{2}}{4}.\end{aligned}$$

Try checking this result on your calculator. You will find that $\cos 75° \approx 0.259$.

The Granger Collection, New York

Historical Note
Hipparchus, considered the most eminent of Greek astronomers, was born about 160 B.C. in Nicaea. He was credited with the invention of trigonometry. He also derived the sum and difference formulas for $\sin(A \pm B)$ and $\cos(A \pm B)$.

Example 2 ▶ Evaluating a Trigonometric Expression

Find the exact value of $\sin \dfrac{\pi}{12}$.

Solution

Using the fact that

$$\frac{\pi}{12} = \frac{\pi}{3} - \frac{\pi}{4}$$

together with the formula for $\sin(u - v)$, you obtain

$$\begin{aligned}\sin \frac{\pi}{12} &= \sin\left(\frac{\pi}{3} - \frac{\pi}{4}\right)\\ &= \sin \frac{\pi}{3} \cos \frac{\pi}{4} - \cos \frac{\pi}{3} \sin \frac{\pi}{4}\\ &= \frac{\sqrt{3}}{2}\left(\frac{\sqrt{2}}{2}\right) - \frac{1}{2}\left(\frac{\sqrt{2}}{2}\right)\\ &= \frac{\sqrt{6} - \sqrt{2}}{4}.\end{aligned}$$

Example 3 ▶ Evaluating a Trigonometric Expression

Find the exact value of $\sin 42° \cos 12° - \cos 42° \sin 12°$.

Solution

Recognizing that this expression fits the formula for $\sin(u - v)$, you can write

$$\begin{aligned}\sin 42° \cos 12° - \cos 42° \sin 12° &= \sin(42° - 12°)\\ &= \sin 30°\\ &= \frac{1}{2}.\end{aligned}$$

Example 4 ▶ An Application of a Sum Formula

Write $\cos(\arctan 1 + \arccos x)$ as an algebraic expression.

Solution

This expression fits the formula for $\cos(u + v)$. Angles $u = \arctan 1$ and $v = \arccos x$ are shown in Figure 7.5. So

$$\begin{aligned}\cos(u + v) &= \cos(\arctan 1) \cos(\arccos x) - \sin(\arctan 1) \sin(\arccos x)\\ &= \frac{1}{\sqrt{2}} \cdot x - \frac{1}{\sqrt{2}} \cdot \sqrt{1 - x^2}\\ &= \frac{x - \sqrt{1 - x^2}}{\sqrt{2}}.\end{aligned}$$

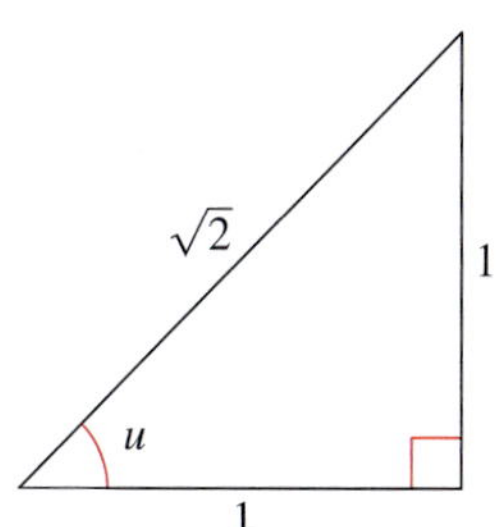

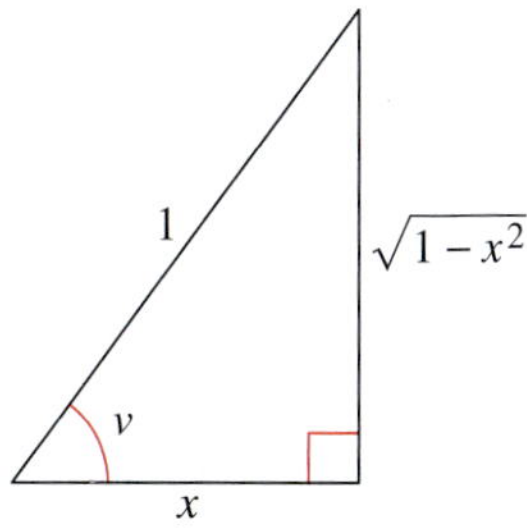

FIGURE 7.5

Example 5 shows how to use a difference formula to prove the cofunction identity

$$\cos\left(\frac{\pi}{2} - x\right) = \sin x.$$

Example 5 ▶ Proving a Cofunction Identity

Prove the cofunction identity $\cos\left(\frac{\pi}{2} - x\right) = \sin x.$

Solution

Using the formula for $\cos(u - v)$, you have

$$\cos\left(\frac{\pi}{2} - x\right) = \cos\frac{\pi}{2}\cos x + \sin\frac{\pi}{2}\sin x$$
$$= (0)(\cos x) + (1)(\sin x) = \sin x.$$

Sum and difference formulas can be used to derive **reduction formulas** involving expressions such as

$$\sin\left(\theta + \frac{n\pi}{2}\right) \quad \text{and} \quad \cos\left(\theta + \frac{n\pi}{2}\right), \quad \text{where } n \text{ is an integer.}$$

Example 6 ▶ Deriving Reduction Formulas

Simplify each expression.

a. $\cos\left(\theta - \frac{3\pi}{2}\right)$ **b.** $\tan(\theta + 3\pi)$

Solution

a. Using the formula for $\cos(u - v)$, you have

$$\cos\left(\theta - \frac{3\pi}{2}\right) = \cos\theta\cos\frac{3\pi}{2} + \sin\theta\sin\frac{3\pi}{2}$$
$$= (\cos\theta)(0) + (\sin\theta)(-1)$$
$$= -\sin\theta.$$

b. Using the formula for $\tan(u + v)$, you have

$$\tan(\theta + 3\pi) = \frac{\tan\theta + \tan 3\pi}{1 - \tan\theta\tan 3\pi}$$
$$= \frac{\tan\theta + 0}{1 - (\tan\theta)(0)}$$
$$= \tan\theta.$$

The next example was taken from calculus. It is used to derive the derivative of the sine function.

Example 7 ▶ An Application from Calculus

Verify that

$$\frac{\sin(x + h) - \sin x}{h} = (\cos x)\left(\frac{\sin h}{h}\right) - (\sin x)\left(\frac{1 - \cos h}{h}\right)$$

where $h \neq 0$.

Solution

Using the formula for $\sin(u + v)$, you have

$$\begin{aligned}\frac{\sin(x + h) - \sin x}{h} &= \frac{\sin x \cos h + \cos x \sin h - \sin x}{h} \\ &= \frac{\cos x \sin h - \sin x(1 - \cos h)}{h} \\ &= (\cos x)\left(\frac{\sin h}{h}\right) - (\sin x)\left(\frac{1 - \cos h}{h}\right).\end{aligned}$$

Example 8 ▶ Solving a Trigonometric Equation

Find all solutions of

$$\sin\left(x + \frac{\pi}{4}\right) + \sin\left(x - \frac{\pi}{4}\right) = -1$$

in the interval $[0, 2\pi)$.

Solution

Using sum and difference formulas, rewrite the equation as

$$\begin{aligned}\sin x \cos \frac{\pi}{4} + \cos x \sin \frac{\pi}{4} + \sin x \cos \frac{\pi}{4} - \cos x \sin \frac{\pi}{4} &= -1 \\ 2 \sin x \cos \frac{\pi}{4} &= -1 \\ 2(\sin x)\left(\frac{\sqrt{2}}{2}\right) &= -1 \\ \sin x &= -\frac{1}{\sqrt{2}} \\ \sin x &= -\frac{\sqrt{2}}{2}.\end{aligned}$$

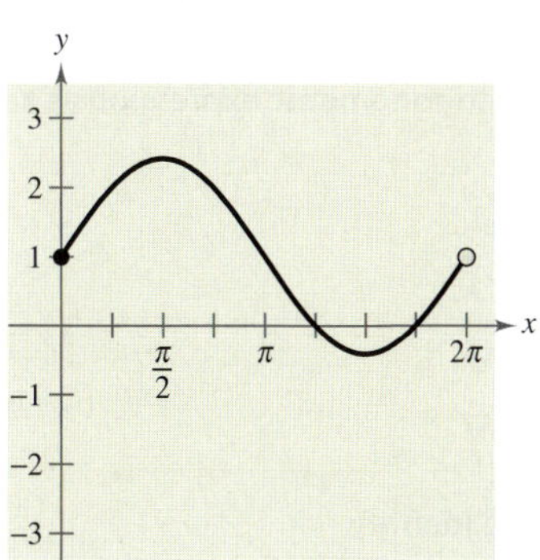

FIGURE 7.6

So, the only solutions in the interval $[0, 2\pi)$ are

$$x = \frac{5\pi}{4} \quad \text{and} \quad x = \frac{7\pi}{4}.$$

These solutions are checked graphically in Figure 7.6.

7.4 Exercises

In Exercises 1–6, find the exact value of each expression.

1. (a) $\cos\left(\frac{\pi}{4}+\frac{\pi}{3}\right)$ (b) $\cos\frac{\pi}{4}+\cos\frac{\pi}{3}$
2. (a) $\sin\left(\frac{3\pi}{4}+\frac{5\pi}{6}\right)$ (b) $\sin\frac{3\pi}{4}+\sin\frac{5\pi}{6}$
3. (a) $\sin\left(\frac{7\pi}{6}-\frac{\pi}{3}\right)$ (b) $\sin\frac{7\pi}{6}-\sin\frac{\pi}{3}$
4. (a) $\cos\left(\frac{2\pi}{3}-\frac{\pi}{6}\right)$ (b) $\cos\frac{2\pi}{3}+\cos\frac{\pi}{6}$
5. (a) $\cos(120° + 45°)$ (b) $\cos 120° + \cos 45°$
6. (a) $\sin(135° - 30°)$ (b) $\sin 135° - \cos 30°$

In Exercises 7–14, find the exact values of the sine, cosine, and tangent of the angle.

7. $105° = 60° + 45°$
8. $165° = 135° + 30°$
9. $195° = 225° - 30°$
10. $255° = 300° - 45°$
11. $\frac{11\pi}{12} = \frac{3\pi}{4} + \frac{\pi}{6}$
12. $\frac{7\pi}{12} = \frac{\pi}{3} + \frac{\pi}{4}$
13. $\frac{17\pi}{12} = \frac{9\pi}{4} - \frac{5\pi}{6}$
14. $-\frac{\pi}{12} = \frac{\pi}{6} - \frac{\pi}{4}$

In Exercises 15–22, find the exact values of the sine, cosine, and tangent of the angle.

15. $285°$
16. $-105°$
17. $-165°$
18. $15°$
19. $\frac{13\pi}{12}$
20. $-\frac{7\pi}{12}$
21. $-\frac{13\pi}{12}$
22. $\frac{5\pi}{12}$

In Exercises 23–30, write the expression as the sine, cosine, or tangent of an angle.

23. $\cos 25° \cos 15° - \sin 25° \sin 15°$
24. $\sin 140° \cos 50° + \cos 140° \sin 50°$
25. $\frac{\tan 325° - \tan 86°}{1 + \tan 325° \tan 86°}$
26. $\frac{\tan 140° - \tan 60°}{1 + \tan 140° \tan 60°}$
27. $\sin 3 \cos 1.2 - \cos 3 \sin 1.2$
28. $\cos\frac{\pi}{7}\cos\frac{\pi}{5} - \sin\frac{\pi}{7}\sin\frac{\pi}{5}$
29. $\frac{\tan 2x + \tan x}{1 - \tan 2x \tan x}$
30. $\cos 3x \cos 2y + \sin 3x \sin 2y$

In Exercises 31–36, find the exact value of the expression.

31. $\sin 330° \cos 30° - \cos 330° \sin 30°$
32. $\cos 15° \cos 60° + \sin 15° \sin 60°$
33. $\sin\frac{\pi}{12}\cos\frac{\pi}{4} + \cos\frac{\pi}{12}\sin\frac{\pi}{4}$
34. $\cos\frac{\pi}{16}\cos\frac{3\pi}{16} - \sin\frac{\pi}{16}\sin\frac{3\pi}{16}$
35. $\frac{\tan 25° + \tan 110°}{1 - \tan 25° \tan 110°}$
36. $\frac{\tan(5\pi/4) - \tan(\pi/12)}{1 + \tan(5\pi/4)\tan(\pi/12)}$

In Exercises 37–44, find the exact value of the trigonometric function given that $\sin u = \frac{5}{13}$ and $\cos v = -\frac{3}{5}$. (Both u and v are in Quadrant II.)

37. $\sin(u + v)$
38. $\cos(u - v)$
39. $\cos(u + v)$
40. $\sin(v - u)$
41. $\tan(u + v)$
42. $\csc(u - v)$
43. $\sec(v - u)$
44. $\cot(u + v)$

In Exercises 45–50, find the exact value of the trigonometric function given that $\sin u = -\frac{7}{25}$ and $\cos v = -\frac{4}{5}$. (Both u and v are in Quadrant III.)

45. $\cos(u + v)$
46. $\sin(u + v)$
47. $\tan(u - v)$
48. $\cot(v - u)$
49. $\sec(u + v)$
50. $\cos(u - v)$

In Exercises 51–54, write the trigonometric expression as an algebraic expression.

51. $\sin(\arcsin x + \arccos x)$
52. $\sin(\arctan 2x - \arccos x)$
53. $\cos(\arccos x + \arcsin x)$
54. $\cos(\arccos x - \arctan x)$

In Exercises 55–64, verify the identity.

55. $\sin(3\pi - x) = \sin x$
56. $\sin\left(\frac{\pi}{2} + x\right) = \cos x$

57. $\sin\left(\frac{\pi}{6} + x\right) = \frac{1}{2}(\cos x + \sqrt{3}\sin x)$

58. $\cos\left(\frac{5\pi}{4} - x\right) = -\frac{\sqrt{2}}{2}(\cos x + \sin x)$

59. $\cos(\pi - \theta) + \sin\left(\frac{\pi}{2} + \theta\right) = 0$

60. $\tan\left(\frac{\pi}{4} - \theta\right) = \frac{1 - \tan\theta}{1 + \tan\theta}$

61. $\cos(x + y)\cos(x - y) = \cos^2 x - \sin^2 y$

62. $\sin(x + y)\sin(x - y) = \sin^2 x - \sin^2 y$

63. $\sin(x + y) + \sin(x - y) = 2\sin x\cos y$

64. $\cos(x + y) + \cos(x - y) = 2\cos x\cos y$

In Exercises 65–68, simplify the expression algebraically and use a graphing utility to confirm your answer graphically.

65. $\cos\left(\frac{3\pi}{2} - x\right)$

66. $\cos(\pi + x)$

67. $\sin\left(\frac{3\pi}{2} + \theta\right)$

68. $\tan(\pi + \theta)$

In Exercises 69–72, find all solutions of the equation in the interval $[0, 2\pi)$.

69. $\sin\left(x + \frac{\pi}{3}\right) + \sin\left(x - \frac{\pi}{3}\right) = 1$

70. $\sin\left(x + \frac{\pi}{6}\right) - \sin\left(x - \frac{\pi}{6}\right) = \frac{1}{2}$

71. $\cos\left(x + \frac{\pi}{4}\right) - \cos\left(x - \frac{\pi}{4}\right) = 1$

72. $\tan(x + \pi) + 2\sin(x + \pi) = 0$

In Exercises 73 and 74, use a graphing utility to approximate the solutions in the interval $[0, 2\pi)$.

73. $\cos\left(x + \frac{\pi}{4}\right) + \cos\left(x - \frac{\pi}{4}\right) = 1$

74. $\tan(x + \pi) - \cos\left(x + \frac{\pi}{2}\right) = 0$

▶ Model It

75. ***Harmonic Motion*** A weight is attached to a spring suspended vertically from a ceiling. When a driving force is applied to the system, the weight moves vertically from its equilibrium position, and this motion is modeled by

▶ Model It *(continued)*

$$y = \frac{1}{3}\sin 2t + \frac{1}{4}\cos 2t$$

where y is the distance from equilibrium measured in feet and t is the time in seconds.

(a) Use the identity

$$a\sin B\theta + b\cos B\theta = \sqrt{a^2 + b^2}\sin(B\theta + C)$$

where $C = \arctan(b/a)$, $a > 0$, to write the model in the form

$$y = \sqrt{a^2 + b^2}\sin(Bt + C).$$

(b) Find the amplitude of the oscillations of the weight.

(c) Find the frequency of the oscillations of the weight.

76. ***Standing Waves*** The equation of a standing wave is obtained by adding the displacements of two waves traveling in opposite directions (see figure). Assume that each of the waves has amplitude A, period T, and wavelength λ. If the models for these waves are

$$y_1 = A\cos 2\pi\left(\frac{t}{T} - \frac{x}{\lambda}\right) \quad \text{and}$$

$$y_2 = A\cos 2\pi\left(\frac{t}{T} + \frac{x}{\lambda}\right)$$

show that

$$y_1 + y_2 = 2A\cos\frac{2\pi t}{T}\cos\frac{2\pi x}{\lambda}.$$

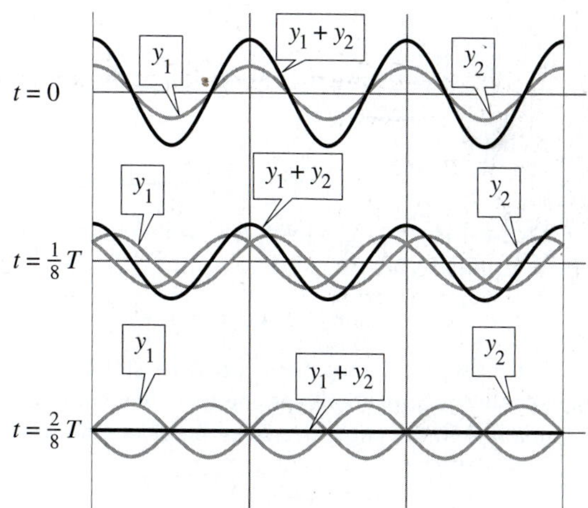

Synthesis

True or False? **In Exercises 77–80, determine whether the statement is true or false. Justify your answer.**

77. $\sin(u \pm v) = \sin u \pm \sin v$

78. $\cos(u \pm v) = \cos u \pm \cos v$

79. $\cos\left(x - \frac{\pi}{2}\right) = -\sin x$

80. $\sin\left(x - \frac{\pi}{2}\right) = -\cos x$

In Exercises 81–84, verify the identity.

81. $\cos(n\pi + \theta) = (-1)^n \cos \theta$, n is an integer

82. $\sin(n\pi + \theta) = (-1)^n \sin \theta$, n is an integer

83. $a \sin B\theta + b \cos B\theta = \sqrt{a^2 + b^2} \sin(B\theta + C)$, where $C = \arctan(b/a)$ and $a > 0$

84. $a \sin B\theta + b \cos B\theta = \sqrt{a^2 + b^2} \cos(B\theta - C)$, where $C = \arctan(a/b)$ and $b > 0$

In Exercises 85–88, use the formulas given in Exercises 83 and 84 to write the trigonometric expression in the following forms.

(a) $\sqrt{a^2 + b^2} \sin(B\theta + C)$

(b) $\sqrt{a^2 + b^2} \cos(B\theta - C)$

85. $\sin \theta + \cos \theta$

86. $3 \sin 2\theta + 4 \cos 2\theta$

87. $12 \sin 3\theta + 5 \cos 3\theta$

88. $\sin 2\theta - \cos 2\theta$

In Exercises 89 and 90, use the formulas given in Exercises 83 and 84 to write the trigonometric expression in the form $a \sin B\theta + b \cos B\theta$.

89. $2 \sin\left(\theta + \frac{\pi}{2}\right)$

90. $5 \cos\left(\theta + \frac{3\pi}{4}\right)$

In Exercises 91 and 92, use the figure, which shows two lines whose equations are

$y_1 = m_1x + b_1$ and $y_2 = m_2x + b_2$.

Assume that both lines have positive slopes. Derive a formula for the angle between the two lines. Then use your formula to find the angle between the given pair of lines.

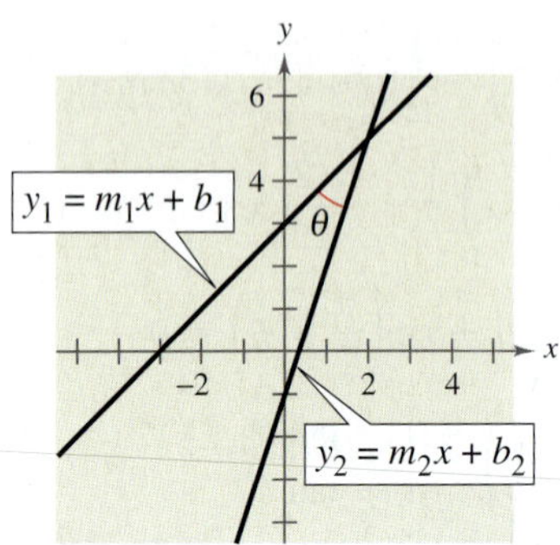

91. $y = x$ and $y = \sqrt{3}x$

92. $y = x$ and $y = \frac{1}{\sqrt{3}}x$

93. ***Conjecture*** Consider the function

$$f(\theta) = \sin^2\left(\theta + \frac{\pi}{4}\right) + \sin^2\left(\theta - \frac{\pi}{4}\right).$$

Use a graphing utility to graph the function and use the graph to create an identity. Prove your conjecture.

94. ***Conjecture*** Three squares of side s are placed side by side (see figure). Make a conjecture about the relationship between the sum $u + v$ and w. Prove your conjecture by using the identity for the tangent of the sum of two angles.

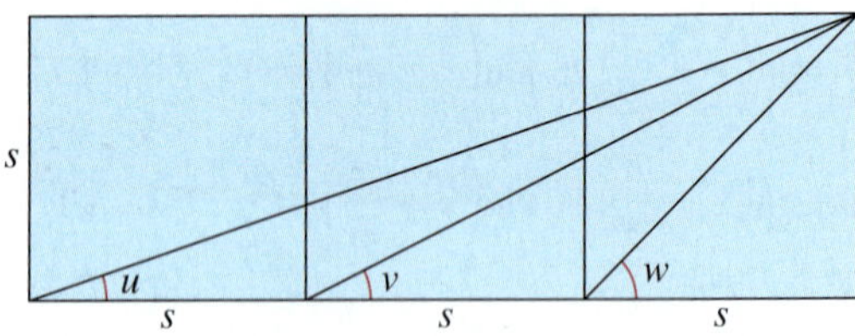

95. ***Proof*** Write a proof for the formula for $\sin(u + v)$.

96. ***Proof*** Write a proof for the formula for $\sin(u - v)$.

Review

In Exercises 97–100, find the inverse function of f. Verify that $f(f^{-1}(x)) = x$ and $f^{-1}(f(x)) = x$.

97. $f(x) = 5(x - 3)$

98. $f(x) = \frac{7 - x}{8}$

99. $f(x) = x^2 - 8$

100. $f(x) = \sqrt{x - 16}$

In Exercises 101–104, apply the inverse properties of $\ln x$ and e^x to simplify the expression.

101. $\log_3 3^{4x-3}$

102. $\log_8 8^{3x^2}$

103. $e^{\ln(6x-3)}$

104. $12x + e^{\ln x(x-2)}$

7.5 Multiple-Angle and Product-to-Sum Formulas

▶ **What you should learn**

- How to use multiple-angle formulas to rewrite and evaluate trigonometric functions
- How to use power-reducing formulas to rewrite and evaluate trigonometric functions
- How to use half-angle formulas to rewrite and evaluate trigonometric functions
- How to use product-to-sum and sum-to-product formulas to rewrite and evaluate trigonometric functions

▶ **Why you should learn it**

You can use a variety of trigonometric formulas to rewrite trigonometric functions in more convenient forms. For instance, in Exercise 121 on page 584, you can use a half-angle formula to determine the apex angle of a sound wave cone from the speed of an airplane.

NASA-Liaison Agency

Multiple-Angle Formulas 

In this section you will study four other categories of trigonometric identities.

1. The first category involves *functions of multiple angles* such as $\sin ku$ and $\cos ku$.
2. The second category involves *squares of trigonometric functions* such as $\sin^2 u$.
3. The third category involves *functions of half-angles* such as $\sin(u/2)$.
4. The fourth category involves *products of trigonometric functions* such as $\sin u \cos v$.

You should learn the **double-angle formulas** because they are used most often. For proofs of the formulas, see Proofs in Mathematics on page 592.

> **Double-Angle Formulas**
>
> $$\sin 2u = 2 \sin u \cos u \qquad \cos 2u = \cos^2 u - \sin^2 u$$
>
> $$\tan 2u = \frac{2 \tan u}{1 - \tan^2 u} \qquad = 2\cos^2 u - 1$$
>
> $$= 1 - 2\sin^2 u$$

Example 1 ▶ **Solving a Multiple-Angle Equation** 

Find all solutions of $2 \cos x + \sin 2x = 0$.

Solution

Begin by rewriting the equation so that it involves functions of x (rather than $2x$). Then factor and solve as usual.

$2 \cos x + \sin 2x = 0$ — Write original equation.

$2 \cos x + 2 \sin x \cos x = 0$ — Double-angle formula

$2 \cos x(1 + \sin x) = 0$ — Factor.

$2 \cos x = 0$ and $1 + \sin x = 0$ — Set factors equal to zero.

$x = \frac{\pi}{2}, \frac{3\pi}{2}$ $\qquad x = \frac{3\pi}{2}$ — Solutions in $[0, 2\pi)$

So, the general solution is

$$x = \frac{\pi}{2} + 2n\pi \quad \text{and} \quad x = \frac{3\pi}{2} + 2n\pi$$

where n is an integer. Try verifying these solutions graphically.

Example 2 ▶ Using Double-Angle Formulas to Analyze Graphs

Use a double-angle formula to rewrite the equation

$$y = 4\cos^2 x - 2.$$

Then sketch the graph of the equation over the interval $[0, 2\pi]$.

Solution

Using a double-angle formula, you can rewrite the original function as

$$\begin{aligned} y &= 4\cos^2 x - 2 \\ &= 2(2\cos^2 x - 1) \\ &= 2\cos 2x. \end{aligned}$$

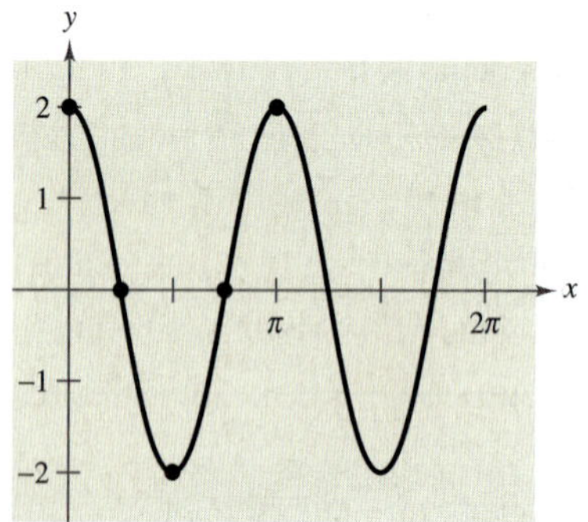

FIGURE 7.7

Using the techniques discussed in Section 6.4, you can recognize that the graph of this function has an amplitude of 2 and a period of π. The key points in the interval $[0, \pi]$ are as follows.

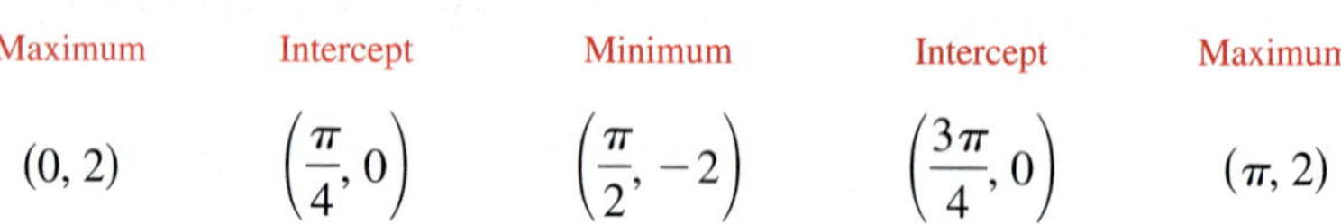

Maximum	Intercept	Minimum	Intercept	Maximum
$(0, 2)$	$\left(\frac{\pi}{4}, 0\right)$	$\left(\frac{\pi}{2}, -2\right)$	$\left(\frac{3\pi}{4}, 0\right)$	$(\pi, 2)$

Two cycles of the graph are shown in Figure 7.7.

Example 3 ▶ Evaluating Functions Involving Double Angles

Use the following to find $\sin 2\theta$, $\cos 2\theta$, and $\tan 2\theta$.

$$\cos\theta = \frac{5}{13}, \qquad \frac{3\pi}{2} < \theta < 2\pi$$

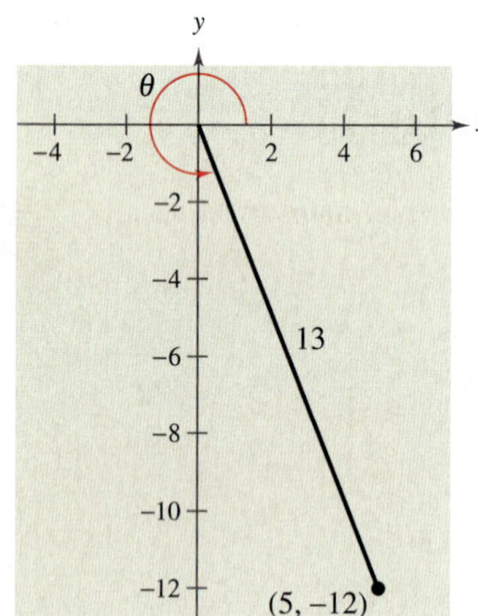

FIGURE 7.8

Solution

From Figure 7.8, you can see that $\sin\theta = y/r = -12/13$. Consequently, using each of the double-angle formulas, you can write

$$\sin 2\theta = 2\sin\theta\cos\theta = 2\left(-\frac{12}{13}\right)\left(\frac{5}{13}\right) = -\frac{120}{169}$$

$$\cos 2\theta = 2\cos^2\theta - 1 = 2\left(\frac{25}{169}\right) - 1 = -\frac{119}{169}$$

$$\tan 2\theta = \frac{\sin 2\theta}{\cos 2\theta} = \frac{120}{119}.$$

The double-angle formulas are not restricted to angles 2θ and θ. Other *double* combinations, such as 4θ and 2θ or 6θ and 3θ, are also valid. Here are two examples.

$$\sin 4\theta = 2\sin 2\theta\cos 2\theta \qquad \text{and} \qquad \cos 6\theta = \cos^2 3\theta - \sin^2 3\theta$$

By using double-angle formulas together with the sum formulas given in the preceding section, you can form other multiple-angle formulas.

Example 4 ▶ Deriving a Triple-Angle Formula

$$\begin{aligned}\sin 3x &= \sin(2x + x)\\ &= \sin 2x \cos x + \cos 2x \sin x\\ &= 2 \sin x \cos x \cos x + (1 - 2 \sin^2 x)\sin x\\ &= 2 \sin x \cos^2 x + \sin x - 2 \sin^3 x\\ &= 2 \sin x(1 - \sin^2 x) + \sin x - 2 \sin^3 x\\ &= 2 \sin x - 2 \sin^3 x + \sin x - 2 \sin^3 x\\ &= 3 \sin x - 4 \sin^3 x\end{aligned}$$

Power-Reducing Formulas 

The double-angle formulas can be used to obtain the following **power-reducing formulas.** Example 5 shows a typical power reduction that is used in calculus.

Power-Reducing Formulas

$$\sin^2 u = \frac{1 - \cos 2u}{2} \qquad \cos^2 u = \frac{1 + \cos 2u}{2} \qquad \tan^2 u = \frac{1 - \cos 2u}{1 + \cos 2u}$$

For a proof of the power-reducing formulas, see Proofs in Mathematics on page 592.

Example 5 ▶ Reducing a Power  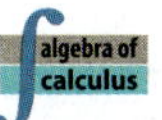

Rewrite $\sin^4 x$ as a sum of first powers of the cosines of multiple angles.

Solution

Note the repeated use of power-reducing formulas.

$$\begin{aligned}\sin^4 x &= (\sin^2 x)^2 && \text{Rewrite.}\\ &= \left(\frac{1 - \cos 2x}{2}\right)^2 && \text{Power-reducing formula}\\ &= \frac{1}{4}(1 - 2 \cos 2x + \cos^2 2x) && \text{Expand.}\\ &= \frac{1}{4}\left(1 - 2 \cos 2x + \frac{1 + \cos 4x}{2}\right) && \text{Power-reducing formula}\\ &= \frac{1}{4} - \frac{1}{2} \cos 2x + \frac{1}{8} + \frac{1}{8} \cos 4x && \text{Distributive Property}\\ &= \frac{1}{8}(3 - 4 \cos 2x + \cos 4x) && \text{Factor out common factor.}\end{aligned}$$

Half-Angle Formulas 

You can derive some useful alternative forms of the power-reducing formulas by replacing u with $u/2$. The results are called **half-angle formulas.**

Half-Angle Formulas

$$\sin\frac{u}{2} = \pm\sqrt{\frac{1-\cos u}{2}}$$

$$\cos\frac{u}{2} = \pm\sqrt{\frac{1+\cos u}{2}}$$

$$\tan\frac{u}{2} = \frac{1-\cos u}{\sin u} = \frac{\sin u}{1+\cos u}$$

The signs of $\sin\frac{u}{2}$ and $\cos\frac{u}{2}$ depend on the quadrant in which $\frac{u}{2}$ lies.

Example 6 ▶ Using a Half-Angle Formula 

Find the exact value of $\sin 105°$.

Solution

Begin by noting that 105° is half of 210°. Then, using the half-angle formula for $\sin(u/2)$ and the fact that 105° lies in Quadrant II, you have

$$\begin{aligned}\sin 105° &= \sqrt{\frac{1-\cos 210°}{2}}\\ &= \sqrt{\frac{1-(-\cos 30°)}{2}}\\ &= \sqrt{\frac{1+(\sqrt{3}/2)}{2}}\\ &= \frac{\sqrt{2+\sqrt{3}}}{2}.\end{aligned}$$

The positive square root is chosen because $\sin\theta$ is positive in Quadrant II.

Use your calculator to verify the result obtained in Example 6. That is, evaluate $\sin 105°$ and $\left(\sqrt{2+\sqrt{3}}\right)/2$.

$$\sin 105° \approx 0.9659258$$

$$\frac{\sqrt{2+\sqrt{3}}}{2} \approx 0.9659258$$

You can see that both values are approximately 0.9659258.

Example 7 ▶ Solving a Trigonometric Equation

Find all solutions of

$$2 - \sin^2 x = 2\cos^2\frac{x}{2}$$

in the interval $[0, 2\pi)$.

Solution

$$2 - \sin^2 x = 2\cos^2\frac{x}{2}$$ Write original equation.

$$2 - \sin^2 x = 2\left(\pm\sqrt{\frac{1+\cos x}{2}}\right)^2$$ Half-angle formula

$$2 - \sin^2 x = 2\left(\frac{1+\cos x}{2}\right)$$ Simplify.

$$2 - \sin^2 x = 1 + \cos x$$ Simplify.

$$2 - (1 - \cos^2 x) = 1 + \cos x$$ Pythagorean identity

$$\cos^2 x - \cos x = 0$$ Simplify.

$$\cos x(\cos x - 1) = 0$$ Factor.

By setting the factors $\cos x$ and $\cos x - 1$ equal to zero, you find that the solutions in the interval $[0, 2\pi)$ are

$$x = \frac{\pi}{2},\quad x = \frac{3\pi}{2},\quad \text{and}\quad x = 0.$$

STUDY TIP

A common error is to write

$$2\cos^2\frac{x}{2}$$

as $\cos^2 x$, rather than to use the correct identity

$$2\cos^2\frac{x}{2} = 2\left(\frac{1+\cos x}{2}\right).$$

Product-to-Sum Formulas

Each of the following **product-to-sum formulas** is easily verified using the sum and difference formulas discussed in the preceding section.

Product-to-Sum Formulas

$$\sin u \sin v = \frac{1}{2}[\cos(u - v) - \cos(u + v)]$$

$$\cos u \cos v = \frac{1}{2}[\cos(u - v) + \cos(u + v)]$$

$$\sin u \cos v = \frac{1}{2}[\sin(u + v) + \sin(u - v)]$$

$$\cos u \sin v = \frac{1}{2}[\sin(u + v) - \sin(u - v)]$$

Example 8 ▶ Writing Products as Sums

Rewrite the product $\cos 5x \sin 4x$ as a sum or difference.

Solution

Using the appropriate product-to-sum formula, you obtain

$$\cos 5x \sin 4x = \frac{1}{2}[\sin(5x + 4x) - \sin(5x - 4x)]$$

$$= \frac{1}{2}\sin 9x - \frac{1}{2}\sin x.$$

Occasionally, it is useful to reverse the procedure and write a sum of trigonometric functions as a product. This can be accomplished with the following **sum-to-product formulas.**

Sum-to-Product Formulas

$$\sin x + \sin y = 2\sin\left(\frac{x+y}{2}\right)\cos\left(\frac{x-y}{2}\right)$$

$$\sin x - \sin y = 2\cos\left(\frac{x+y}{2}\right)\sin\left(\frac{x-y}{2}\right)$$

$$\cos x + \cos y = 2\cos\left(\frac{x+y}{2}\right)\cos\left(\frac{x-y}{2}\right)$$

$$\cos x - \cos y = -2\sin\left(\frac{x+y}{2}\right)\sin\left(\frac{x-y}{2}\right)$$

For a proof of the sum-to-product formulas, see Proofs in Mathematics on page 593.

Example 9 ▶ Using a Sum-to-Product Formula 

Find the exact value of $\cos 195° + \cos 105°$.

Solution

Using the appropriate sum-to-product formula, you obtain

$$\cos 195° + \cos 105° = 2\cos\left(\frac{195° + 105°}{2}\right)\cos\left(\frac{195° - 105°}{2}\right)$$

$$= 2\cos 150° \cos 45°$$

$$= 2\left(-\frac{\sqrt{3}}{2}\right)\left(\frac{\sqrt{2}}{2}\right)$$

$$= -\frac{\sqrt{6}}{2}.$$

Example 10 ▶ Solving a Trigonometric Equation

Find all solutions of $\sin 5x + \sin 3x = 0$.

Solution

$$\sin 5x + \sin 3x = 0 \qquad \text{Write original equation.}$$

$$2 \sin\left(\frac{5x + 3x}{2}\right) \cos\left(\frac{5x - 3x}{2}\right) = 0 \qquad \text{Sum-to-product formula}$$

$$2 \sin 4x \cos x = 0 \qquad \text{Simplify.}$$

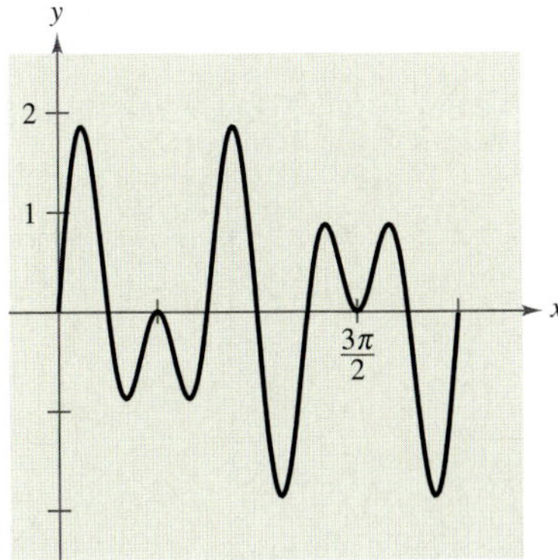

FIGURE 7.9

By setting the factor $2 \sin 4x$ equal to zero, you can find that the solutions in the interval $[0, 2\pi)$ are

$$x = 0, \frac{\pi}{4}, \frac{\pi}{2}, \frac{3\pi}{4}, \pi, \frac{5\pi}{4}, \frac{3\pi}{2}, \frac{7\pi}{4}.$$

The equation $\cos x = 0$ yields no additional solutions, and you can conclude that the solutions are of the form

$$x = \frac{n\pi}{4}$$

where n is an integer. These solutions are verified graphically in Figure 7.9.

Example 11 ▶ Verifying a Trigonometric Identity

Verify the identity

$$\frac{\sin t + \sin 3t}{\cos t + \cos 3t} = \tan 2t.$$

Solution

Using appropriate sum-to-product formulas, you have

$$\frac{\sin t + \sin 3t}{\cos t + \cos 3t} = \frac{2 \sin 2t \cos(-t)}{2 \cos 2t \cos(-t)}$$

$$= \frac{\sin 2t}{\cos 2t}$$

$$= \tan 2t.$$

Writing ABOUT MATHEMATICS

Deriving an Area Formula Describe how you can use a double-angle formula or a half-angle formula to derive a formula for the area of an isosceles triangle. Use a labeled sketch to illustrate your derivation. Then write two examples that show how your formula can be used.

7.5 Exercises

In Exercises 1–8, use the figure to find the exact value of the trigonometric function.

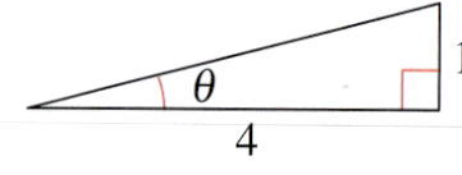

1. $\sin\theta$
2. $\tan\theta$
3. $\cos 2\theta$
4. $\sin 2\theta$
5. $\tan 2\theta$
6. $\sec 2\theta$
7. $\csc 2\theta$
8. $\cot 2\theta$

In Exercises 9–18, find the exact solutions of the equation in the interval $[0, 2\pi)$.

9. $\sin 2x - \sin x = 0$
10. $\sin 2x + \cos x = 0$
11. $4\sin x\cos x = 1$
12. $\sin 2x \sin x = \cos x$
13. $\cos 2x - \cos x = 0$
14. $\cos 2x + \sin x = 0$
15. $\tan 2x - \cot x = 0$
16. $\tan 2x - 2\cos x = 0$
17. $\sin 4x = -2\sin 2x$
18. $(\sin 2x + \cos 2x)^2 = 1$

In Exercises 19–22, use a double-angle formula to rewrite the expression.

19. $6\sin x\cos x$
20. $6\cos^2 x - 3$
21. $4 - 8\sin^2 x$
22. $(\cos x + \sin x)(\cos x - \sin x)$

In Exercises 23–28, find the exact values of sin 2*u*, cos 2*u*, and tan 2*u* using the double-angle formulas.

23. $\sin u = -\frac{4}{5}, \quad \pi < u < \frac{3\pi}{2}$
24. $\cos u = -\frac{2}{3}, \quad \frac{\pi}{2} < u < \pi$
25. $\tan u = \frac{3}{4}, \quad 0 < u < \frac{\pi}{2}$
26. $\cot u = -4, \quad \frac{3\pi}{2} < u < 2\pi$
27. $\sec u = -\frac{5}{2}, \quad \frac{\pi}{2} < u < \pi$
28. $\csc u = 3, \quad \frac{\pi}{2} < u < \pi$

In Exercises 29–34, use the power-reducing formulas to rewrite the expression in terms of the first power of the cosine.

29. $\cos^4 x$
30. $\sin^8 x$
31. $\sin^2 x\cos^2 x$
32. $\sin^4 x\cos^4 x$
33. $\sin^2 x\cos^4 x$
34. $\sin^4 x\cos^2 x$

In Exercises 35–40, use the figure to find the exact value of the trigonometric function.

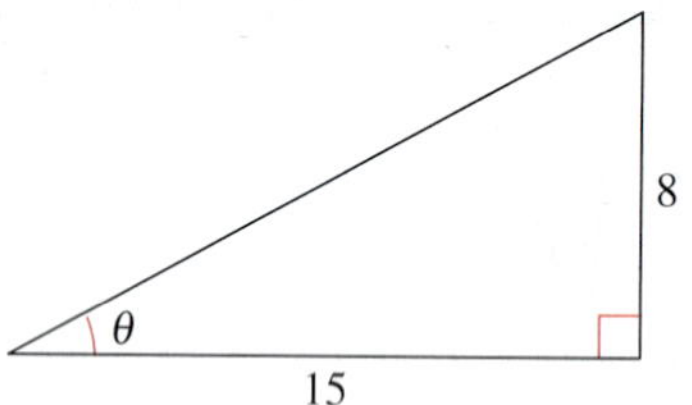

35. $\cos\frac{\theta}{2}$
36. $\sin\frac{\theta}{2}$
37. $\tan\frac{\theta}{2}$
38. $\sec\frac{\theta}{2}$
39. $\csc\frac{\theta}{2}$
40. $\cot\frac{\theta}{2}$

In Exercises 41–48, use the half-angle formulas to determine the exact values of the sine, cosine, and tangent of the angle.

41. 75°
42. 165°
43. $112^\circ\ 30'$
44. $67^\circ\ 30'$
45. $\frac{\pi}{8}$
46. $\frac{\pi}{12}$
47. $\frac{3\pi}{8}$
48. $\frac{7\pi}{12}$

In Exercises 49–54, find the exact values of sin(*u*/2), cos(*u*/2), and tan(*u*/2) using the half-angle formulas.

49. $\sin u = \frac{5}{13}, \quad \frac{\pi}{2} < u < \pi$
50. $\cos u = \frac{3}{5}, \quad 0 < u < \frac{\pi}{2}$

51. $\tan u = -\frac{5}{8}, \quad \frac{3\pi}{2} < u < 2\pi$

52. $\cot u = 3, \quad \pi < u < \frac{3\pi}{2}$

53. $\csc u = -\frac{5}{3}, \quad \pi < u < \frac{3\pi}{2}$

54. $\sec u = -\frac{7}{2}, \quad \frac{\pi}{2} < u < \pi$

In Exercises 55–58, use the half-angle formulas to simplify the expression.

55. $\sqrt{\frac{1 - \cos 6x}{2}}$ **56.** $\sqrt{\frac{1 + \cos 4x}{2}}$

57. $-\sqrt{\frac{1 - \cos 8x}{1 + \cos 8x}}$ **58.** $-\sqrt{\frac{1 - \cos(x - 1)}{2}}$

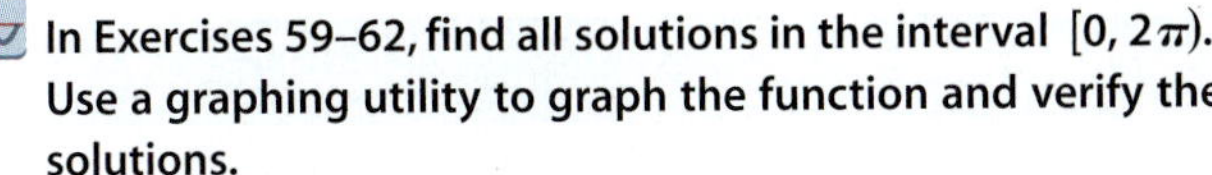

In Exercises 59–62, find all solutions in the interval $[0, 2\pi)$. Use a graphing utility to graph the function and verify the solutions.

59. $\sin \frac{x}{2} + \cos x = 0$

60. $\sin \frac{x}{2} + \cos x - 1 = 0$

61. $\cos \frac{x}{2} - \sin x = 0$

62. $\tan \frac{x}{2} - \sin x = 0$

In Exercises 63–74, use the product-to-sum formulas to write the product as a sum or difference.

63. $6 \sin \frac{\pi}{4} \cos \frac{\pi}{4}$ **64.** $4 \cos \frac{\pi}{3} \sin \frac{5\pi}{6}$

65. $\cos 4\theta \sin 6\theta$ **66.** $3 \sin 2\alpha \sin 3\alpha$

67. $5 \cos(-5\beta) \cos 3\beta$ **68.** $\cos 2\theta \cos 4\theta$

69. $\sin(x + y) \sin(x - y)$

70. $\sin(x + y) \cos(x - y)$

71. $\cos(\theta - \pi) \sin(\theta + \pi)$

72. $\sin(\theta + \pi) \sin(\theta - \pi)$

73. $10 \cos 75^\circ \cos 15^\circ$ **74.** $6 \sin 45^\circ \cos 15^\circ$

In Exercises 75–86, use the sum-to-product formulas to write the sum or difference as a product.

75. $\sin 60^\circ + \sin 30^\circ$ **76.** $\cos 120^\circ + \cos 30^\circ$

77. $\cos \frac{3\pi}{4} - \cos \frac{\pi}{4}$ **78.** $\sin \frac{5\pi}{4} - \sin \frac{3\pi}{4}$

79. $\sin 5\theta - \sin 3\theta$ **80.** $\sin 3\theta + \sin \theta$

81. $\cos 6x + \cos 2x$ **82.** $\sin x + \sin 5x$

83. $\sin(\alpha + \beta) - \sin(\alpha - \beta)$

84. $\cos(\phi + 2\pi) + \cos \phi$

85. $\cos\left(\theta + \frac{\pi}{2}\right) - \cos\left(\theta - \frac{\pi}{2}\right)$

86. $\sin\left(x + \frac{\pi}{2}\right) + \sin\left(x - \frac{\pi}{2}\right)$

In Exercises 87–90, find all solutions in the interval $[0, 2\pi)$. Use a graphing utility to graph the function and verify the solutions.

87. $\sin 6x + \sin 2x = 0$

88. $\cos 2x - \cos 6x = 0$

89. $\frac{\cos 2x}{\sin 3x - \sin x} - 1 = 0$

90. $\sin^2 3x - \sin^2 x = 0$

In Exercises 91–94, use the figure and trigonometric identities to find the exact value of the trigonometric function in two ways.

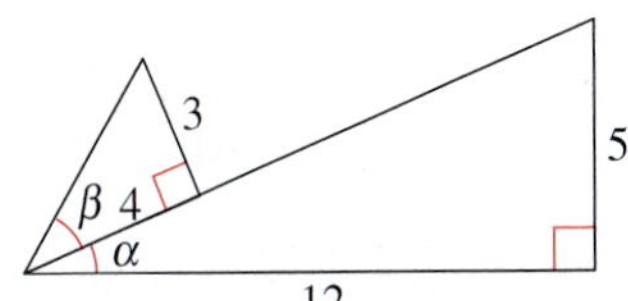

91. $\sin^2 \alpha$ **92.** $\cos^2 \alpha$

93. $\sin \alpha \cos \beta$ **94.** $\cos \alpha \sin \beta$

In Exercises 95–110, verify the identity.

95. $\csc 2\theta = \frac{\csc \theta}{2 \cos \theta}$

96. $\sec 2\theta = \frac{\sec^2 \theta}{2 - \sec^2 \theta}$

97. $\cos^2 2\alpha - \sin^2 2\alpha = \cos 4\alpha$

98. $\cos^4 x - \sin^4 x = \cos 2x$

99. $(\sin x + \cos x)^2 = 1 + \sin 2x$

100. $\sin \frac{\alpha}{3} \cos \frac{\alpha}{3} = \frac{1}{2} \sin \frac{2\alpha}{3}$

101. $1 + \cos 10y = 2 \cos^2 5y$

102. $\dfrac{\cos 3\beta}{\cos \beta} = 1 - 4\sin^2 \beta$

103. $\sec \dfrac{u}{2} = \pm\sqrt{\dfrac{2\tan u}{\tan u + \sin u}}$

104. $\tan \dfrac{u}{2} = \csc u - \cot u$

105. $\dfrac{\sin x \pm \sin y}{\cos x + \cos y} = \tan \dfrac{x \pm y}{2}$

106. $\dfrac{\sin x + \sin y}{\cos x - \cos y} = -\cot \dfrac{x - y}{2}$

107. $\dfrac{\cos 4x + \cos 2x}{\sin 4x + \sin 2x} = \cot 3x$

108. $\dfrac{\cos t + \cos 3t}{\sin 3t - \sin t} = \cot t$

109. $\sin\left(\dfrac{\pi}{6} + x\right) + \sin\left(\dfrac{\pi}{6} - x\right) = \cos x$

110. $\cos\left(\dfrac{\pi}{3} + x\right) + \cos\left(\dfrac{\pi}{3} - x\right) = \cos x$

 **In Exercises 111–114, use a graphing utility to verify the identity. Confirm that it is an identity algebraically.**

111. $\cos 3\beta = \cos^3 \beta - 3\sin^2 \beta \cos \beta$

112. $\sin 4\beta = 4\sin \beta \cos \beta(1 - 2\sin^2 \beta)$

113. $(\cos 4x - \cos 2x)/(2\sin 3x) = -\sin x$

114. $(\cos 3x - \cos x)/(\sin 3x - \sin x) = -\tan 2x$

In Exercises 115 and 116, graph the function by hand in the interval $[0, 2\pi)$ by using the power-reducing formulas.

115. $f(x) = \sin^2 x$

116. $f(x) = \cos^2 x$

In Exercises 117 and 118, write the trigonometric expression as an algebraic expression.

117. $\sin(2 \arcsin x)$

118. $\cos(2 \arccos x)$

119. ***Geometry*** The length of each of the two equal sides of an isosceles triangle is 10 meters (see figure). The angle between the two sides is θ.

(a) Write the area of the triangle as a function of $\theta/2$.

(b) Write the area of the triangle as a function of θ. Determine the value of θ such that the area is a maximum.

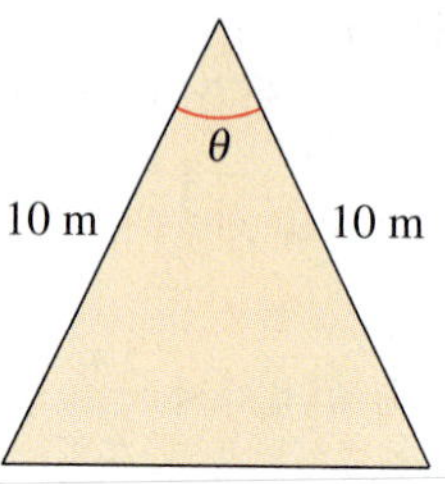

FIGURE FOR 119

120. ***Projectile Motion*** The range of a projectile fired at an angle θ with the horizontal and with an initial velocity of v_0 feet per second is

$$r = \frac{1}{32} v_0^2 \sin 2\theta$$

where r is measured in feet. Determine the expression for the range in terms of θ.

▶ Model It

121. ***Mach Number*** The mach number M of an airplane is the ratio of its speed to the speed of sound. When an airplane travels faster than the speed of sound, the sound waves form a cone behind the airplane. The mach number is related to the apex angle θ of the cone by

$$\sin \frac{\theta}{2} = \frac{1}{M}.$$

(a) Find the angle θ that corresponds to a mach number of 1.

(b) Find the angle θ that corresponds to a mach number of 4.5.

(c) The speed of sound is about 760 miles per hour. Determine the speed of an object with the mach numbers from parts (a) and (b).

(d) Rewrite the equation in terms of θ.

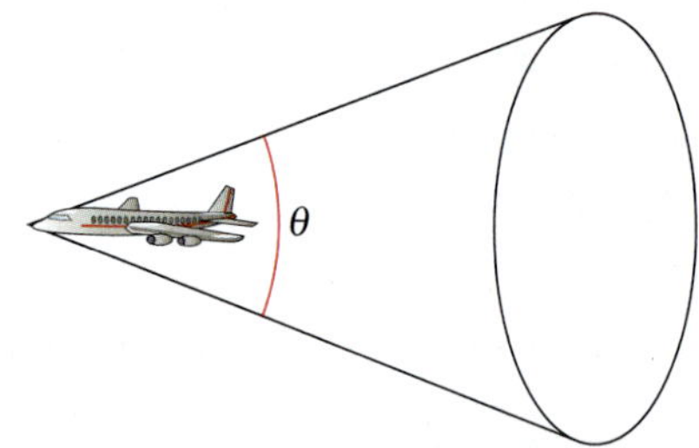

122. ***Railroad Track*** When two railroad tracks merge, the overlapping portions of the tracks are in the shapes of circular arcs (see figure). The radius of each arc r (in feet) and the angle θ are related by

$$\frac{x}{2} = 2r \sin^2 \frac{\theta}{2}.$$

Write a formula for x in terms of $\cos \theta$.

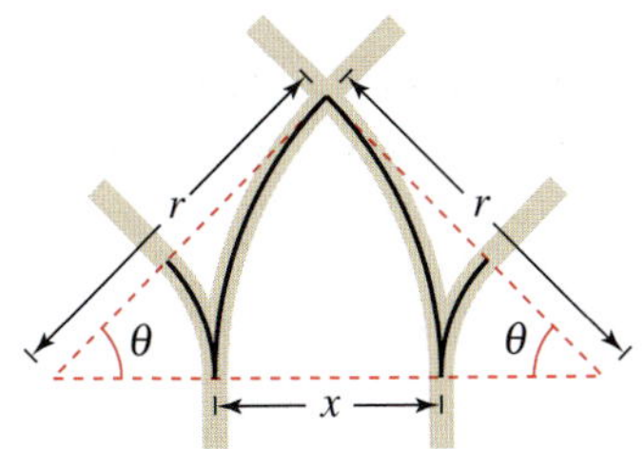

Synthesis

True or False? **In Exercises 123 and 124, determine whether the statement is true or false. Justify your answer.**

123. Because the sine function is an odd function, for a negative number u, $\sin 2u = -2 \sin u \cos u$.

124. $\sin \frac{u}{2} = -\sqrt{\frac{1 - \cos u}{2}}$ when u is in the second quadrant.

In Exercises 125 and 126, (a) use a graphing utility to graph the function and approximate the maximum and minimum points on the graph in the interval $[0, 2\pi)$ and (b) solve the trigonometric equation and verify that its solutions are the x-coordinates of the maximum and minimum points of f. (Calculus is required to find the trigonometric equation.)

	Function	*Trigonometric Equation*
125.	$f(x) = 4 \sin \frac{x}{2} + \cos x$	$2 \cos \frac{x}{2} - \sin x = 0$
126.	$f(x) = \cos 2x - 2 \sin x$	$-2 \cos x(2 \sin x + 1) = 0$

127. ***Exploration*** Consider the function

$$f(x) = \sin^4 x + \cos^4 x.$$

(a) Use the power-reducing formulas to write the function in terms of cosine to the first power.

(b) Determine another way of rewriting the function. Use a graphing utility to rule out incorrectly rewritten functions.

(c) Add a trigonometric term to the function so that it becomes a perfect square trinomial. Rewrite the function as a perfect square trinomial minus the term that you added. Use a graphing utility to rule out incorrectly rewritten functions.

(d) Rewrite the result of part (c) in terms of the sine of a double angle. Use a graphing utility to rule out incorrectly rewritten functions.

(e) When you rewrite a trigonometric expression, the result may not be the same as a friend's. Does this mean that one of you is wrong? Explain.

128. ***Conjecture*** Consider the function

$$f(x) = 2 \sin x\left(2 \cos^2 \frac{x}{2} - 1\right).$$

(a) Use a graphing utility to graph the function.

(b) Make a conjecture about the function that is an identity with f.

(c) Verify your conjecture analytically.

Review

In Exercises 129–132, write the equation of the parabola in standard form and identify its vertex.

129. $y = x^2 - 6x + 13$

130. $y = x^2 + 10x + 23$

131. $y = 2x^2 - 4x + 3$

132. $y = \frac{1}{2}x^2 + 4x + 2$

In Exercises 133 and 134, find the vertex of the parabola and sketch its graph.

133. $(x - 5)^2 + y + 8 = 0$

134. $x^2 - 2x + 12y + 13 = 0$

135. ***Profit*** The total profit for a car manufacturer in October was 16% higher than it was in September. The total profit for the 2 months was $507,600. Find the profit for each month.

136. ***Mixture Problem*** A 55-gallon barrel contains a mixture with a concentration of 30%. How much of this mixture must be withdrawn and replaced by 100% concentrate to bring the mixture up to 50% concentration?

137. ***Distance*** A baseball diamond has the shape of a square in which the distance between each of the consecutive bases is 90 feet. Approximate the distance from home plate to second base.

Chapter Summary

▶ *What* did you learn?

	Review Exercises
Section 7.1	
☐ How to recognize and write the fundamental trigonometric identities	1–6
☐ How to use the fundamental trigonometric identities to evaluate trigonometric functions, simplify trigonometric expressions, and rewrite trigonometric expressions	7–24
Section 7.2	
☐ How to plan a strategy for verifying trigonometric identities	25–32
☐ How to verify trigonometric identities	25–32
Section 7.3	
☐ How to use standard algebraic techniques to solve trigonometric equations	33–38
☐ How to solve trigonometric equations of quadratic type	39–42
☐ How to solve trigonometric equations involving multiple angles	43–46
☐ How to use inverse trigonometric functions to solve trigonometric equations	47–50
Section 7.4	
☐ How to use sum and difference formulas to evaluate trigonometric functions, verify identities, and solve trigonometric equations	51–68
Section 7.5	
☐ How to use multiple-angle formulas to rewrite and evaluate trigonometric functions	69–74
☐ How to use power-reducing formulas to rewrite and evaluate trigonometric functions	75–78
☐ How to use half-angle formulas to rewrite and evaluate trigonometric functions	79–86
☐ How to use product-to-sum and sum-to-product formulas to rewrite and evaluate trigonometric functions	87–95

Review Exercises

7.1 In Exercises 1–6, name the trigonometric function that is equivalent to the expression.

1. $\dfrac{1}{\cos x}$
2. $\dfrac{1}{\sin x}$
3. $\dfrac{1}{\sec x}$
4. $\dfrac{1}{\tan x}$
5. $\dfrac{\cos x}{\sin x}$
6. $\sqrt{1 + \tan^2 x}$

In Exercises 7–10, use the given values and trigonometric identities to evaluate (if possible) the other trigonometric functions of the angle.

7. $\sin x = \frac{3}{5}, \quad \cos x = \frac{4}{5}$
8. $\tan \theta = \dfrac{2}{3}, \quad \sec \theta = \dfrac{\sqrt{13}}{3}$
9. $\sin\left(\dfrac{\pi}{2} - x\right) = \dfrac{\sqrt{2}}{2}, \quad \sin x = -\dfrac{\sqrt{2}}{2}$
10. $\csc\left(\dfrac{\pi}{2} - \theta\right) = 9, \quad \sin \theta = \dfrac{4\sqrt{5}}{9}$

In Exercises 11–22, use the fundamental trigonometric identities to simplify the trigonometric expression.

11. $\dfrac{1}{\cot^2 x + 1}$
12. $\dfrac{\tan \theta}{1 - \cos^2 \theta}$
13. $\tan^2 x(\csc^2 x - 1)$
14. $\cot^2 x(\sin^2 x)$
15. $\dfrac{\sin\left(\dfrac{\pi}{2} - \theta\right)}{\sin \theta}$
16. $\dfrac{\cot\left(\dfrac{\pi}{2} - u\right)}{\cos u}$
17. $\cos^2 x + \cos^2 x \cot^2 x$
18. $\tan^2 \theta \csc^2 \theta - \tan^2 \theta$
19. $(\tan x + 1)^2 \cos x$
20. $(\sec x - \tan x)^2$
21. $\dfrac{1}{\csc \theta + 1} - \dfrac{1}{\csc \theta - 1}$
22. $\dfrac{\cos^2 x}{1 - \sin x}$

23. ***Rate of Change*** The rate of change of the function $f(x) = \csc x - \cot x$ is the expression $\csc^2 x - \csc x \cot x$. Show that this expression can also be written as

$$\frac{(1 - \cos x)}{\sin^2 x}.$$

24. ***Rate of Change*** The rate of change of the function $f(x) = 2\sqrt{\sin x}$ is the expression $\sin^{-1/2} x \cos x$. Show that this expression can also be written as $\cot x\sqrt{\sin x}$.

7.2 In Exercises 25–32, verify the identity.

25. $\cos x(\tan^2 x + 1) = \sec x$
26. $\sec^2 x \cot x - \cot x = \tan x$
27. $\cos\left(x + \dfrac{\pi}{2}\right) = -\sin x$
28. $\cot\left(\dfrac{\pi}{2} - x\right) = \tan x$
29. $\dfrac{1}{\tan \theta \csc \theta} = \cos \theta$
30. $\dfrac{1}{\tan x \csc x \sin x} = \cot x$
31. $\sin^5 x \cos^2 x = (\cos^2 x - 2\cos^4 x + \cos^6 x) \sin x$
32. $\cos^3 x \sin^2 x = (\sin^2 x - \sin^4 x) \cos x$

7.3 In Exercises 33–38, solve the equation.

33. $\sin x = \sqrt{3} - \sin x$
34. $4\cos \theta = 1 + 2\cos \theta$
35. $3\sqrt{3} \tan u = 3$
36. $\frac{1}{2} \sec x - 1 = 0$
37. $3\csc^2 x = 4$
38. $4\tan^2 u - 1 = \tan^2 u$

In Exercises 39–46, find all solutions of the equation in the interval $[0, 2\pi)$.

39. $2\cos^2 x - \cos x = 1$
40. $2\sin^2 x - 3\sin x = -1$
41. $\cos^2 x + \sin x = 1$
42. $\sin^2 x + 2\cos x = 2$
43. $2\sin 2x - \sqrt{2} = 0$
44. $\sqrt{3} \tan 3x = 0$
45. $\cos 4x(\cos x - 1) = 0$
46. $3\csc^2 5x = -4$

In Exercises 47–50, use inverse functions where needed to find all solutions of the equation in the interval $[0, 2\pi)$.

47. $\sin^2 x - 2\sin x = 0$
48. $2\cos^2 x + 3\cos x = 0$
49. $\tan^2 \theta + \tan \theta - 12 = 0$
50. $\sec^2 x + 6\tan x + 4 = 0$

7.4 **In Exercises 51–54, find the exact values of the sine, cosine, and tangent of the angle by using a sum or difference formula.**

51. $285° = 315° - 30°$

52. $345° = 300° + 45°$

53. $\frac{25\pi}{12} = \frac{11\pi}{6} + \frac{\pi}{4}$

54. $\frac{19\pi}{12} = \frac{11\pi}{6} - \frac{\pi}{4}$

In Exercises 55–58, write the expression as the sine, cosine, or tangent of an angle.

55. $\sin 60° \cos 45° - \cos 60° \sin 45°$

56. $\cos 45° \cos 120° - \sin 45° \sin 120°$

57. $\frac{\tan 25° + \tan 10°}{1 - \tan 25° \tan 10°}$

58. $\frac{\tan 68° - \tan 115°}{1 + \tan 68° \tan 115°}$

In Exercises 59–64, find the exact value of the trigonometric function given that $\sin u = \frac{3}{4}$, $\cos v = -\frac{5}{13}$, and u and v are in Quadrant II.

59. $\sin(u + v)$

60. $\tan(u + v)$

61. $\cos(u - v)$

62. $\sin(u - v)$

63. $\cos(u + v)$

64. $\tan(u - v)$

In Exercises 65–68, find all solutions of the equation in the interval $[0, 2\pi)$.

65. $\sin\left(x + \frac{\pi}{4}\right) - \sin\left(x - \frac{\pi}{4}\right) = 1$

66. $\cos\left(x + \frac{\pi}{6}\right) - \cos\left(x - \frac{\pi}{6}\right) = 1$

67. $\sin\left(x + \frac{\pi}{2}\right) - \sin\left(x - \frac{\pi}{2}\right) = \sqrt{3}$

68. $\cos\left(x + \frac{3\pi}{4}\right) - \cos\left(x - \frac{3\pi}{4}\right) = 0$

7.5 **In Exercises 69 and 70, use double-angle formulas to verify the identity algebraically and use a graphing utility to confirm it graphically.**

69. $\sin 4x = 8 \cos^3 x \sin x - 4 \cos x \sin x$

70. $\tan^2 x = \frac{1 - \cos 2x}{1 + \cos 2x}$

In Exercises 71 and 72, find the exact values of $\sin 2u$, $\cos 2u$, and $\tan 2u$ using the double-angle formulas.

71. $\sin u = -\frac{4}{5}, \quad \pi < u < \frac{3\pi}{2}$

72. $\cos u = -\frac{2}{\sqrt{5}}, \quad \frac{\pi}{2} < u < \pi$

73. ***Projectile Motion*** A baseball leaves the hand of the person at first base at an angle of θ with the horizontal and at an initial velocity of $v_0 = 80$ feet per second. The ball is caught by the person at second base 100 feet away. Find θ if the range r of a projectile is

$$r = \frac{1}{32} v_0^2 \sin 2\theta.$$

74. ***Projectile Motion*** Use the equation in Exercise 73 to find θ when a golf ball is hit at an initial velocity of 50 feet per second and lands 77 feet away.

In Exercises 75–78, use the power-reducing formulas to rewrite the expression in terms of the first power of the cosine.

75. $\tan^2 2x$

76. $\cos^2 3x$

77. $\sin^2 x \tan^2 x$

78. $\cos^2 x \tan^2 x$

In Exercises 79–82, use the half-angle formulas to determine the exact values of the sine, cosine, and tangent of the angle.

79. $-75°$

80. $15°$

81. $\frac{19\pi}{12}$

82. $-\frac{17\pi}{12}$

In Exercises 83 and 84, use the half-angle formulas to simplify the expression.

83. $-\sqrt{\frac{1 + \cos 10x}{2}}$

84. $\frac{\sin 6x}{1 + \cos 6x}$

85. Find the exact values of $\sin (u/2)$, $\cos (u/2)$, and $\tan (u/2)$ for $\sin u = \frac{3}{5}$, $0 < u < \frac{\pi}{2}$.

86. ***Geometry*** A trough for feeding cattle is 4 meters long and its cross sections are isosceles triangles with the two equal sides being $\frac{1}{2}$ meter (see figure on page 589). The angle between the two sides is θ.

(a) Write the trough's volume as a function of $\frac{\theta}{2}$.

(b) Write the volume of the trough as a function of θ and determine the value of θ such that the volume is maximum.

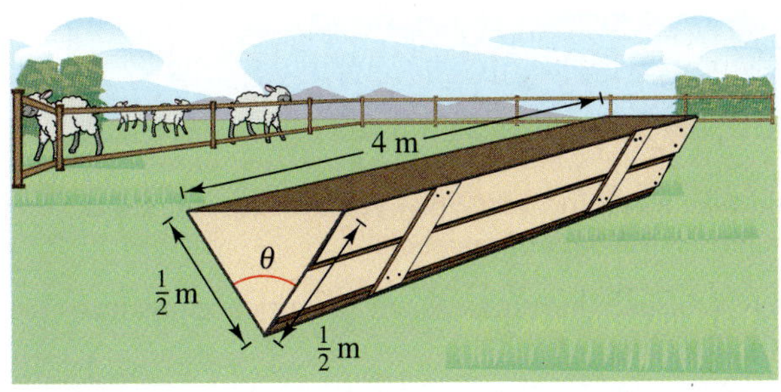

FIGURE FOR 86

In Exercises 87–90, use the product-to-sum formulas to write the product as a sum or difference.

87. $\cos \dfrac{\pi}{6} \sin \dfrac{\pi}{6}$

88. $6 \sin 15° \sin 45°$

89. $\cos 5\theta \cos 3\theta$

90. $4 \sin 3\alpha \cos 2\alpha$

In Exercises 91–94, use the sum-to-product formulas to write the sum or difference as a product.

91. $\sin 60° + \sin 90°$

92. $\cos 3\theta + \cos 2\theta$

93. $\cos\left(x + \dfrac{\pi}{6}\right) - \cos\left(x - \dfrac{\pi}{6}\right)$

94. $\sin\left(x + \dfrac{\pi}{4}\right) - \sin\left(x - \dfrac{\pi}{4}\right)$

95. ***Harmonic Motion*** A weight is attached to a spring suspended vertically from a ceiling. When a driving force is applied to the system, the weight moves vertically from its equilibrium position, and this motion is described by the model

$y = 1.5 \sin 8t - 0.5 \cos 8t$

where y is the distance from equilibrium measured in feet and t is the time in seconds.

(a) Write the model in the form

$y = \sqrt{a^2 + b^2} \sin(Bt + C)$.

(b) Find the amplitude of the oscillations of the weight.

(c) Find the frequency of the oscillations of the weight.

Synthesis

True or False? **In Exercises 96–99, determine whether the statement is true or false. Justify your answer.**

96. If $\dfrac{\pi}{2} < \theta < \pi$, then $\cos \dfrac{\theta}{2} < 0$.

97. $\sin(x + y) = \sin x + \sin y$

98. $4 \sin(-x) \cos(-x) = -2 \sin 2x$

99. $4 \sin 45° \cos 15° = 1 + \sqrt{3}$

100. List the reciprocal identities, quotient identities, and Pythagorean identities from memory.

101. ***Think About It*** If a trigonometric equation has an infinite number of solutions, is it true that the equation is an identity? Explain.

102. ***Think About It*** Explain why you know from observation that the equation $a \sin x - b = 0$ has no solution if $|a| < |b|$.

In Exercises 103 and 104, use the graphs of y_1 and y_2 to determine how to change one function to form the identity $y_1 = y_2$.

103. $y_1 = \sec^2\left(\dfrac{\pi}{2} - x\right)$

$y_2 = \cot^2 x$

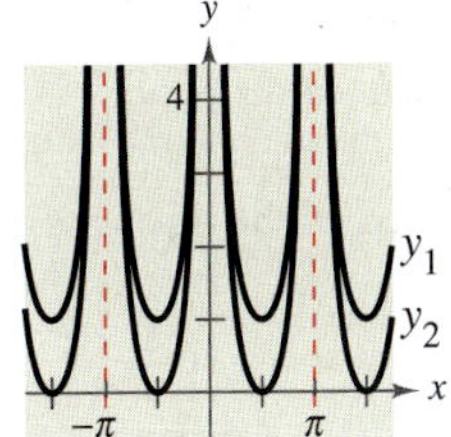

104. $y_1 = \dfrac{\cos 3x}{\cos x}$

$y_2 = (2 \sin x)^2$

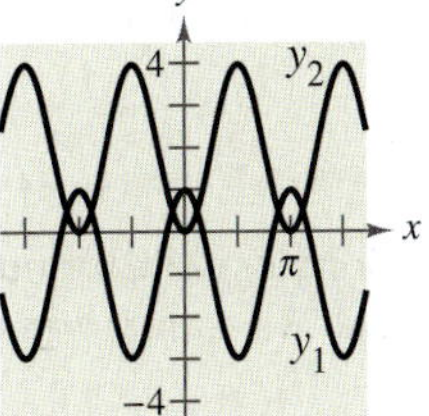

In Exercises 105 and 106, use the *zero* or *root* feature of a graphing utility to approximate the zeros of the function.

105. $y = \sqrt{x + 3} + 4 \cos x$

106. $y = 2 - \dfrac{1}{2}x^2 + 3 \sin \dfrac{\pi x}{2}$

Chapter Test

The *Interactive* CD-ROM and *Internet* versions of this text offer Chapter Pre-Tests and Chapter Post-Tests, both of which have randomly generated exercises with diagnostic capabilities.

Take this test as you would take a test in class. When you are finished, check your work against the answers given in the back of the book.

1. If $\tan \theta = \frac{3}{2}$ and $\cos \theta < 0$, use the fundamental identities to evaluate the other five trigonometric functions of θ.
2. Use the fundamental identities to simplify $\csc^2 \beta(1 - \cos^2 \beta)$.
3. Factor and simplify $\dfrac{\sec^4 x - \tan^4 x}{\sec^2 x + \tan^2 x}$.
4. Add and simplify $\dfrac{\cos \theta}{\sin \theta} + \dfrac{\sin \theta}{\cos \theta}$.
5. Determine the values of θ, $0 \le \theta < 2\pi$, for which $\tan \theta = -\sqrt{\sec^2 \theta - 1}$ is true.

6. Use a graphing utility to graph the functions $y_1 = \cos x + \sin x \tan x$ and $y_2 = \sec x$. Make a conjecture about y_1 and y_2. Verify the result analytically.

In Exercises 7–12, verify the identity.

7. $\sin \theta \sec \theta = \tan \theta$
8. $\sec^2 x \tan^2 x + \sec^2 x = \sec^4 x$
9. $\dfrac{\csc \alpha + \sec \alpha}{\sin \alpha + \cos \alpha} = \cot \alpha + \tan \alpha$
10. $\cos\left(x + \dfrac{\pi}{2}\right) = -\sin x$
11. $\sin(n\pi + \theta) = (-1)^n \sin \theta$, n is an integer.
12. $(\sin x + \cos x)^2 = 1 + \sin 2x$
13. Rewrite $\sin^4 x \tan^2 x$ in terms of the first power of the cosine.
14. Use a half-angle formula to simplify the expression $\dfrac{\sin 4\theta}{1 + \cos 4\theta}$.
15. Write $4 \cos 2\theta \sin 4\theta$ as a sum or difference.
16. Write $\sin 3\theta - \sin 4\theta$ as a product.

In Exercises 17–20, find all solutions of the equation in the interval $[0, 2\pi)$.

17. $\tan^2 x + \tan x = 0$
18. $\sin 2\alpha - \cos \alpha = 0$
19. $4 \cos^2 x - 3 = 0$
20. $\csc^2 x - \csc x - 2 = 0$
21. Use a graphing utility to approximate the solutions of the equation $3 \cos x - x = 0$ accurate to three decimal places.
22. Find the exact value of $\cos 105°$ using the fact that $105° = 135° - 30°$.
23. Use the figure to find the exact values of $\sin 2u$ and $\tan 2u$.

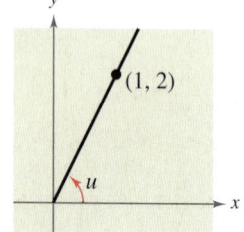

FIGURE FOR 23

24. Cheyenne, Wyoming has a latitude of 41°N. At this latitude, the position of the sun at sunrise can be modeled by

$$D = 31 \sin\left(\frac{2\pi}{365}t - 1.4\right)$$

where t is the time (in days) and $t = 1$ represents January 1. In this model, D represents the number of degrees north or south of due east that the sun rises. Use a graphing utility to determine the days on which the sun is more than 20° north of due east at sunrise.

Proofs in Mathematics

Sum and Difference Formulas *(p. 568)*

$$\sin(u + v) = \sin u \cos v + \cos u \sin v$$

$$\sin(u - v) = \sin u \cos v - \cos u \sin v$$

$$\cos(u + v) = \cos u \cos v - \sin u \sin v$$

$$\cos(u - v) = \cos u \cos v + \sin u \sin v$$

$$\tan(u + v) = \frac{\tan u + \tan v}{1 - \tan u \tan v}$$

$$\tan(u - v) = \frac{\tan u - \tan v}{1 + \tan u \tan v}$$

Proof

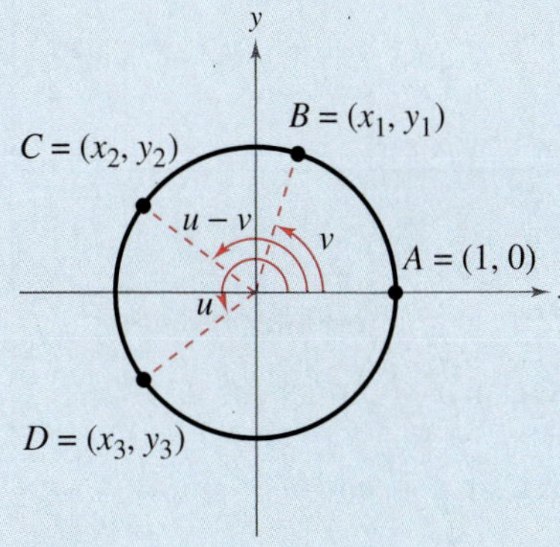

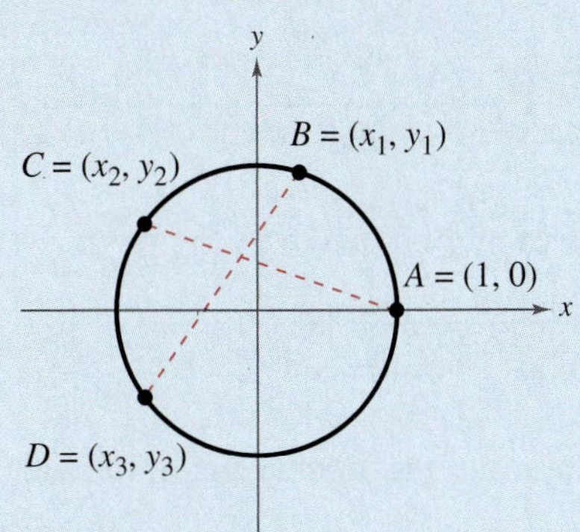

You can use the figures at the left for the proofs of the formulas for $\cos(u \pm v)$. In the top figure, let A be the point $(1, 0)$ and then use u and v to locate the points $B = (x_1, y_1)$, $C = (x_2, y_2)$, and $D = (x_3, y_3)$ on the unit circle. So, $x_i^2 + y_i^2 = 1$ for $i = 1, 2,$ and 3. For convenience, assume that $0 < v < u < 2\pi$. In the bottom figure, note that arcs AC and BD have the same length. So, line segments AC and BD are also equal in length, which implies that

$$\sqrt{(x_2 - 1)^2 + (y_2 - 0)^2} = \sqrt{(x_3 - x_1)^2 + (y_3 - y_1)^2}$$

$$x_2^2 - 2x_2 + 1 + y_2^2 = x_3^2 - 2x_1x_3 + x_1^2 + y_3^2 - 2y_1y_3 + y_1^2$$

$$(x_2^2 + y_2^2) + 1 - 2x_2 = (x_3^2 + y_3^2) + (x_1^2 + y_1^2) - 2x_1x_3 - 2y_1y_3$$

$$1 + 1 - 2x_2 = 1 + 1 - 2x_1x_3 - 2y_1y_3$$

$$x_2 = x_3x_1 + y_3y_1.$$

Finally, by substituting the values $x_2 = \cos(u - v)$, $x_3 = \cos u$, $x_1 = \cos v$, $y_3 = \sin u$, and $y_1 = \sin v$, you obtain $\cos(u - v) = \cos u \cos v + \sin u \sin v$. The formula for $\cos(u + v)$ can be established by considering $u + v = u - (-v)$ and using the formula just derived to obtain

$$\cos(u + v) = \cos[u - (-v)] = \cos u \cos(-v) + \sin u \sin(-v)$$

$$= \cos u \cos v - \sin u \sin v.$$

You can use the sum and difference formulas for sine and cosine to prove the formulas for $\tan(u \pm v)$.

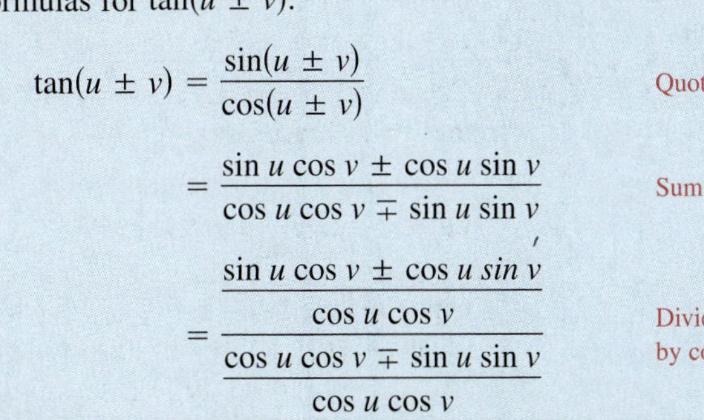

$$\tan(u \pm v) = \frac{\sin(u \pm v)}{\cos(u \pm v)}$$ Quotient identity

$$= \frac{\sin u \cos v \pm \cos u \sin v}{\cos u \cos v \mp \sin u \sin v}$$ Sum and difference formulas

$$= \frac{\dfrac{\sin u \cos v \pm \cos u \sin v}{\cos u \cos v}}{\dfrac{\cos u \cos v \mp \sin u \sin v}{\cos u \cos v}}$$ Divide numerator and denominator by $\cos u \cos v$.

$$= \frac{\dfrac{\sin u \cos v}{\cos u \cos v} \pm \dfrac{\cos u \sin v}{\cos u \cos v}}{\dfrac{\cos u \cos v}{\cos u \cos v} \mp \dfrac{\sin u \sin v}{\cos u \cos v}}$$ Separate fractions.

$$= \frac{\dfrac{\sin u}{\cos u} \pm \dfrac{\sin v}{\cos v}}{1 \mp \dfrac{\sin u}{\cos u} \cdot \dfrac{\sin v}{\cos v}}$$ Product of fractions.

$$= \frac{\tan u \pm \tan v}{1 \mp \tan u \tan v}$$ Quotient identity

Trigonometry and Astronomy

Trigonometry was used by early astronomers to calculate measurements in the universe. Trigonometry was used to calculate the circumference of Earth and the distance from Earth to the moon. Another major accomplishment in astronomy using trigonometry was computing distances to stars.

Double-Angle Formulas *(p. 575)*

$$\sin 2u = 2 \sin u \cos u \qquad \cos 2u = \cos^2 u - \sin^2 u$$

$$\tan 2u = \frac{2 \tan u}{1 - \tan^2 u} \qquad = 2\cos^2 u - 1 = 1 - 2\sin^2 u$$

Proof

To prove all three formulas, let $v = u$ in the corresponding sum formulas.

$$\sin 2u = \sin(u + u) = \sin u \cos u + \cos u \sin u = 2 \sin u \cos u$$

$$\cos 2u = \cos(u + u) = \cos u \cos u - \sin u \sin u = \cos^2 u - \sin^2 u$$

$$\tan 2u = \tan(u + u) = \frac{\tan u + \tan u}{1 - \tan u \tan u} = \frac{2 \tan u}{1 - \tan^2 u}$$

Power-Reducing Formulas *(p. 577)*

$$\sin^2 u = \frac{1 - \cos 2u}{2} \qquad \cos^2 u = \frac{1 + \cos 2u}{2} \qquad \tan^2 u = \frac{1 - \cos 2u}{1 + \cos 2u}$$

Proof

To prove the first formula, solve for $\sin^2 u$ in the double-angle formula $\cos 2u = 1 - 2\sin^2 u$, as follows.

$$\cos 2u = 1 - 2\sin^2 u$$ Write double-angle formula.

$$2\sin^2 u = 1 - \cos 2u$$ Subtract $\cos 2u$ from and add $2 \sin^2 u$ to each side.

$$\sin^2 u = \frac{1 - \cos 2u}{2}$$ Divide each side by 2.

In a similar way you can prove the second formula, by solving for $\cos^2 u$ in the double-angle formula

$$\cos 2u = 2\cos^2 u - 1.$$

To prove the third formula, use a quotient identity, as follows.

$$\tan^2 u = \frac{\sin^2 u}{\cos^2 u}$$

$$= \frac{\dfrac{1 - \cos 2u}{2}}{\dfrac{1 + \cos 2u}{2}}$$

$$= \frac{1 - \cos 2u}{1 + \cos 2u}$$

Sum-to-Product Formulas *(p. 580)*

$$\sin x + \sin y = 2\sin\left(\frac{x+y}{2}\right)\cos\left(\frac{x-y}{2}\right)$$

$$\sin x - \sin y = 2\cos\left(\frac{x+y}{2}\right)\sin\left(\frac{x-y}{2}\right)$$

$$\cos x + \cos y = 2\cos\left(\frac{x+y}{2}\right)\cos\left(\frac{x-y}{2}\right)$$

$$\cos x - \cos y = -2\sin\left(\frac{x+y}{2}\right)\sin\left(\frac{x-y}{2}\right)$$

Proof

To prove the first formula, let $x = u + v$ and $y = u - v$. Then substitute $u = (x + y)/2$ and $v = (x - y)/2$ in the product-to-sum formula.

$$\sin u \cos v = \frac{1}{2}[\sin(u + v) + \sin(u - v)]$$

$$\sin\left(\frac{x+y}{2}\right)\cos\left(\frac{x-y}{2}\right) = \frac{1}{2}(\sin x + \sin y)$$

$$2\sin\left(\frac{x+y}{2}\right)\cos\left(\frac{x-y}{2}\right) = \sin x + \sin y$$

The other sum-to-product formulas can be proved in a similar manner.

P.S. Problem Solving

1. (a) Write each of the other trigonometric functions of θ in terms of $\sin\theta$.

   (b) Write each of the other trigonometric functions of θ in terms of $\cos\theta$.

2. Verify that for all integers n,

$$\cos\left[\frac{(2n+1)\pi}{2}\right] = 0.$$

3. Verify that for all integers n,

$$\sin\left[\frac{(12n+1)\pi}{6}\right] = \frac{1}{2}.$$

4. A particular sound wave is modeled by

$$p(t) = \frac{1}{4\pi}(p_1(t) + 30p_2(t) + p_3(t) + p_5(t) + 30p_6(t))$$

where $p_n(t) = \frac{1}{n}\sin(524n\pi t)$, and t is the time in seconds.

   (a) Find the sine components $p_n(t)$ and use a graphing utility to graph each component. Then verify the graph of p that is shown.

   (b) Find the period of each sine component of p. Is p periodic? If so, what is its period?

   (c) Use the *zero* or *root* feature or the *zoom* and *trace* features of a graphing utility to find the t-intercepts of the graph of p over one cycle.

   (d) Use the *maximum* and *minimum* features of a graphing utility to approximate the absolute maximum and absolute minimum values of p over one cycle.

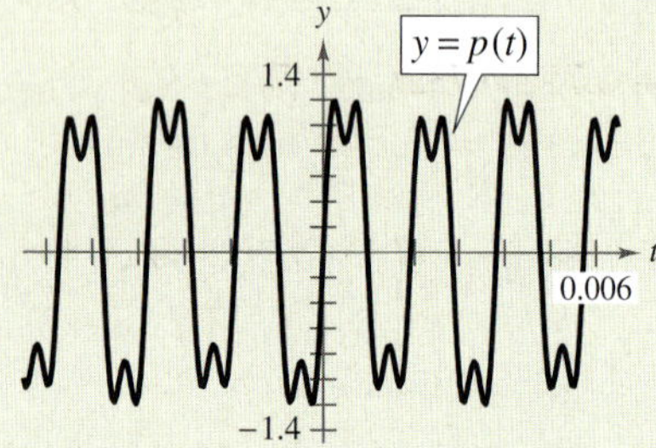

5. The path traveled by an object (neglecting air resistance) that is projected at an initial height of h_0 feet, an initial velocity of v feet per second, and an initial angle θ is given by

$$y = -\frac{16}{v^2\cos^2\theta}x^2 + (\tan\theta)x + h_0$$

where x and y are measured in feet. Find a formula for the maximum height of an object projected from ground level at velocity v and angle θ. To do this, find half of the horizontal distance

$$\frac{1}{32}v^2\sin 2\theta$$

and then substitute it for x in the general model for the path of a projectile (where $h_0 = 0$).

6. Verify the following identity used in calculus.

$$\frac{\cos(x+h) - \cos x}{h} = \cos x\left(\frac{\cos h - 1}{h}\right) - \sin x\left(\frac{\sin h}{h}\right)$$

7. Use the figure to derive the formulas for

$$\sin\frac{\theta}{2},\ \cos\frac{\theta}{2},\ \text{and}\ \tan\frac{\theta}{2}$$

when θ is an acute angle.

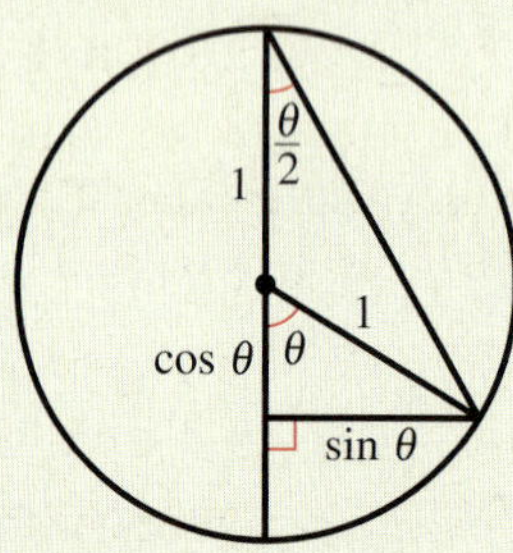

8. The force F (in pounds) on a person's back when he or she bends over at an angle θ is modeled by

$$F = \frac{0.6W\sin(\theta + 90°)}{\sin 12°}$$

where W is the person's weight (in pounds).

   (a) Simplify the model.

   (b) Use a graphing utility to graph the model, where $W = 185$ and $0° < \theta < 90°$.

   (c) At what angle is the force a maximum? At what angle is the force a minimum?

9. The number of hours of daylight that occur at any location on Earth depends on the time of year and the latitude of the location. The following equations model the number of hours of daylight in Seward, Alaska (60° latitude) and New Orleans, Louisiana (30° latitude).

$$D = 12.2 - 6.4 \cos\left[\frac{\pi(t + 0.2)}{182.6}\right] \qquad \text{Seward}$$

$$D = 12.2 - 1.9 \cos\left[\frac{\pi(t + 0.2)}{182.6}\right] \qquad \text{New Orleans}$$

In these models, D represents the number of hours of daylight and t represents the day, with $t = 0$ corresponding to January 1.

(a) Use a graphing utility to graph both models in the same viewing window. Use a viewing window of $0 \le t \le 365$.

(b) Find the days of the year on which both cities receive the same amount of daylight. What are these days called?

(c) Which city has the greater variation in the number of daylight hours? Which constant in each model would you use to determine the difference between the greatest and least numbers of hours of daylight?

(d) Determine the period of each model.

10. The tide, or depth of the ocean near the shore, changes throughout the day. The water depth d (in feet) of a bay can be modeled by

$$d = 35 - 28 \cos \frac{\pi}{6.2} t$$

where t is the time in hours, with $t = 0$ corresponding to 12:00 A.M.

(a) Algebraically find the times at which the high and low tides occur.

(b) Algebraically find the time(s) at which the water depth is 3.5 feet.

(c) Use a graphing utility to verify your results from parts (a) and (b).

11. Find the solution of each inequality in the interval $[0, 2\pi]$.

(a) $\sin x \ge 0.5$ (b) $\cos x \le -0.5$

(c) $\tan x < \sin x$ (d) $\cos x \ge \sin x$

12. The index of refraction n of a transparent material is the ratio of the speed of light in a vacuum to the speed of light in the material. Some common materials and their indices are air (1.00), water (1.33), and glass (1.50). Triangular prisms are often used to measure the index of refraction based on the formula

$$n = \frac{\sin\left(\frac{\theta}{2} + \frac{\alpha}{2}\right)}{\sin \frac{\theta}{2}}.$$

For the prism shown in the figure, $\alpha = 60°$.

(a) Write the index of refraction as a function of $\cot(\theta/2)$.

(b) Find θ for a prism made of glass.

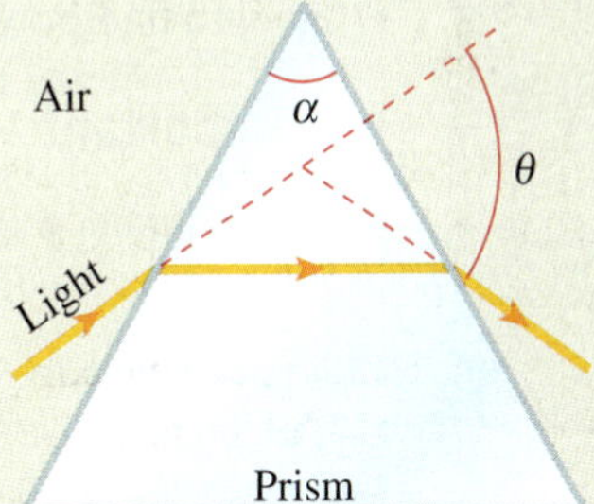

13. (a) Write a sum formula for $\sin(u + v + w)$.

(b) Write a sum formula for $\tan(u + v + w)$.

14. (a) Derive a formula for $\cos 3\theta$.

(b) Derive a formula for $\cos 4\theta$.

15. The heights h (in inches) of pistons 1 and 2 in an automobile engine can be modeled by

$$h_1 = 3.75 \sin 733t + 7.5$$

and

$$h_2 = 3.75 \sin 733\left(t + \frac{4\pi}{3}\right) + 7.5$$

where t is measured in seconds.

(a) Use a graphing utility to graph the heights of these two pistons in the same viewing window for $0 \le t \le 1$.

(b) How often are the pistons at the same height?

How to study Chapter 8

▶ What you should learn

In this chapter you will learn the following skills and concepts:

- How to use the Law of Sines and the Law of Cosines to solve oblique triangles
- How to find the areas of oblique triangles
- How to write the component forms of vectors and perform basic vector operations
- How to find the direction angles of vectors and the angle between two vectors
- How to multiply and divide complex numbers written in trigonometric form
- How to find powers and *n*th roots of complex numbers

▶ Important Vocabulary

As you encounter each new vocabulary term in this chapter, add the term and its definition to your notebook glossary.

Oblique triangle (p. 598)
Law of Sines (p. 598)
Law of Cosines (p. 607)
Directed line segment (p. 615)
Initial point (p. 615)
Terminal point (p. 615)
Magnitude of directed line segment (p. 615)
Vector **v** in the plane (p. 615)
Standard position (p. 616)
Component form of a vector **v** (p. 616)
Zero vector (p. 616)
Magnitude of **v** (p. 616)
Unit vector (p. 616)
Parallelogram law (p. 617)
Resultant (p. 617)
Standard unit vectors (p. 620)
Linear combination of vectors (p. 620)
Direction angle (p. 621)
Dot product (p. 628)
Angle between two nonzero vectors (p. 629)
Orthogonal vectors (p. 630)
Work (p. 634)
Complex plane (p. 637)
Real axis (p. 637)
Imaginary axis (p. 637)
Absolute value of a complex number (p. 637)
Trigonometric form of a complex number (p. 638)
*n*th root of a complex number (p. 642)
*n*th roots of unity (p. 644)

Study Tools

Learning objectives in each section
Chapter Summary (p. 648)
Review Exercises (pp. 649–652)
Chapter Test (p. 653)
Cumulative Test for Chapters 6–8 (pp. 654 and 655)

Additional Resources

Study and Solutions Guide
Interactive Algebra and Trigonometry
Videotapes/DVD for Chapter 8
Algebra and Trigonometry Website
Student Success Organizer

Michael Rutherford/SuperStock

8

Additional Topics in Trigonometry

8.1 Law of Sines

▶ What you should learn

- How to use the Law of Sines to solve oblique triangles (AAS or ASA)
- How to use the Law of Sines to solve oblique triangles (SSA)
- How to find the areas of oblique triangles
- How to use the Law of Sines to model and solve real-life problems

▶ Why you should learn it

You can use the Law of Sines to solve real-life problems involving oblique triangles. For instance, in Exercise 44 on page 606, you can use the Law of Sines to determine the length of the shadow of the Leaning Tower of Pisa.

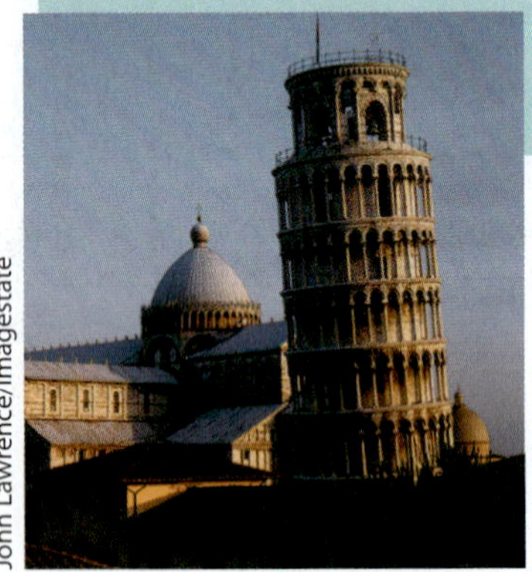

John Lawrence/Imagestate

Introduction

In Chapter 6, you studied techniques for solving right triangles. In this section and the next, you will solve **oblique triangles**—triangles that have no right angles. As standard notation, the angles of a triangle are labeled A, B, and C, and their opposite sides are labeled a, b, and c, as shown in Figure 8.1.

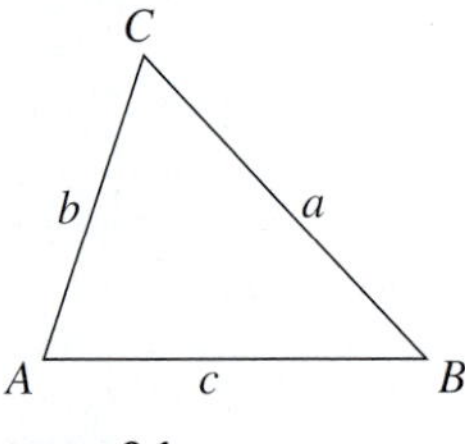

FIGURE 8.1

To solve an oblique triangle, you need to know the measure of at least one side and any two other parts of the triangle—either two sides, two angles, or one angle and one side. This breaks down into the following four cases.

1. Two angles and any side (AAS or ASA)
2. Two sides and an angle opposite one of them (SSA)
3. Three sides (SSS)
4. Two sides and their included angle (SAS)

The first two cases can be solved using the **Law of Sines,** whereas the last two cases require the Law of Cosines (see Section 8.2).

Law of Sines

If ABC is a triangle with sides a, b, and c, then

$$\frac{a}{\sin A} = \frac{b}{\sin B} = \frac{c}{\sin C}.$$

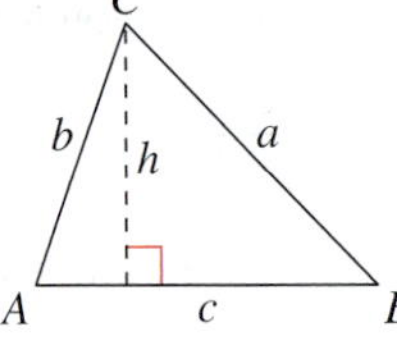

A is acute.

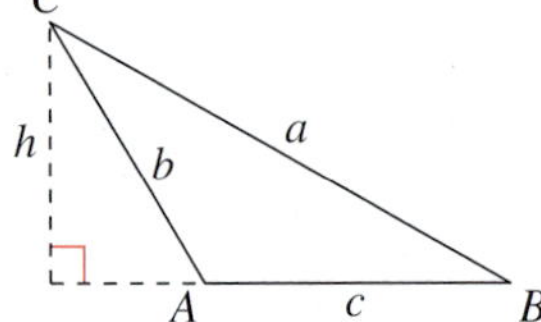

A is obtuse.

The Law of Sines can also be written in the reciprocal form

$$\frac{\sin A}{a} = \frac{\sin B}{b} = \frac{\sin C}{c}.$$

For a proof of the Law of Sines, see Proofs in Mathematics on page 656.

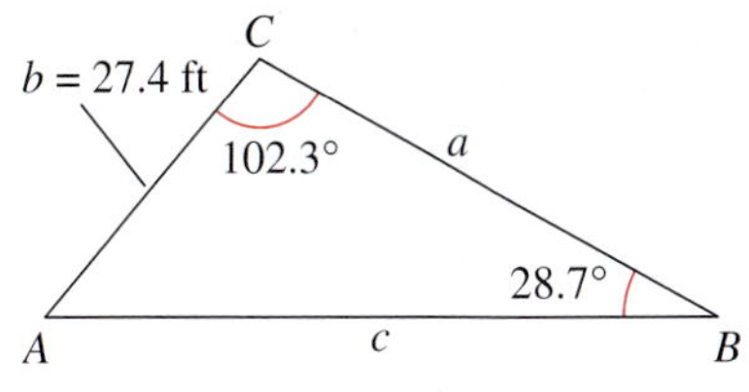

FIGURE 8.2

Example 1 ▶ Given Two Angles and One Side—AAS 

For the triangle in Figure 8.2, $C = 102.3°$, $B = 28.7°$, and $b = 27.4$ feet. Find the remaining angle and sides.

Solution

The third angle of the triangle is

$$\begin{aligned} A &= 180° - B - C \\ &= 180° - 28.7° - 102.3° \\ &= 49.0°. \end{aligned}$$

By the Law of Sines, you have

$$\frac{a}{\sin A} = \frac{b}{\sin B} = \frac{c}{\sin C}.$$

Using $b = 27.4$ produces

$$a = \frac{b}{\sin B}(\sin A) = \frac{27.4}{\sin 28.7°}(\sin 49.0°) \approx 43.06 \text{ feet}$$

and

$$c = \frac{b}{\sin B}(\sin C) = \frac{27.4}{\sin 28.7°}(\sin 102.3°) \approx 55.75 \text{ feet.}$$

STUDY TIP

When solving triangles, a careful sketch is useful as a quick test for the feasibility of an answer. Remember that the longest side lies opposite the largest angle, and the shortest side lies opposite the smallest angle.

Example 2 ▶ Given Two Angles and One Side—ASA 

A pole tilts *toward* the sun at an 8° angle from the vertical, and it casts a 22-foot shadow. The angle of elevation from the tip of the shadow to the top of the pole is 43°. How tall is the pole?

Solution

From Figure 8.3, note that $A = 43°$ and $B = 90° + 8° = 98°$. So, the third angle is

$$\begin{aligned} C &= 180° - A - B \\ &= 180° - 43° - 98° \\ &= 39°. \end{aligned}$$

By the Law of Sines, you have

$$\frac{a}{\sin A} = \frac{c}{\sin C}.$$

Because $c = 22$ feet, the length of the pole is

$$a = \frac{c}{\sin C}(\sin A) = \frac{22}{\sin 39°}(\sin 43°) \approx 23.84 \text{ feet.}$$

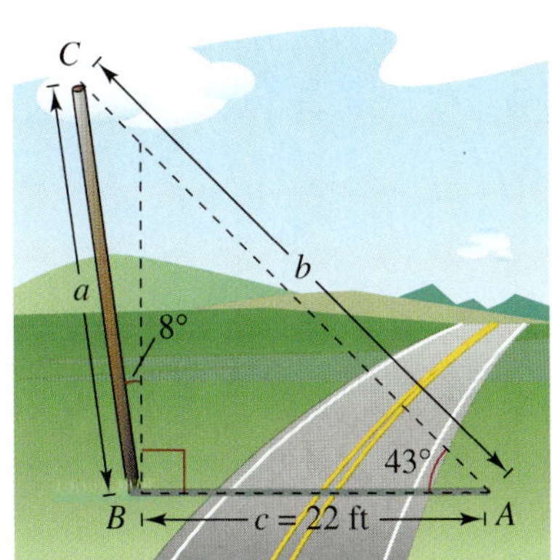

FIGURE 8.3

For practice, try reworking Example 2 for a pole that tilts *away from* the sun under the same conditions.

The icon identifies examples and concepts related to features of the Learning Tools CD-ROM and the *Interactive* and *Internet* versions of this text. For more details see the chart on pages *xix–xxiii*.

The Ambiguous Case (SSA)

In Examples 1 and 2 you saw that two angles and one side determine a unique triangle. However, if two sides and one opposite angle are given, three possible situations can occur: (1) no such triangle exists, (2) one such triangle exists, or (3) two distinct triangles may satisfy the conditions.

The Ambiguous Case (SSA)

Consider a triangle in which you are given a, b, and A. $(h = b \sin A)$

	A is acute.	A is acute.	A is acute.	A is acute.	A is obtuse.	A is obtuse.
Sketch	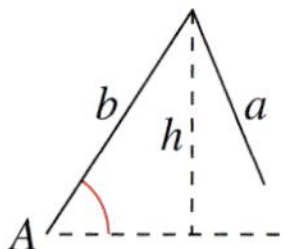	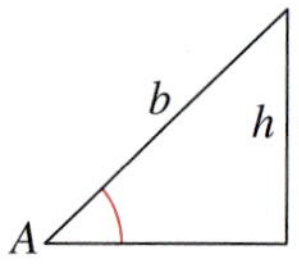	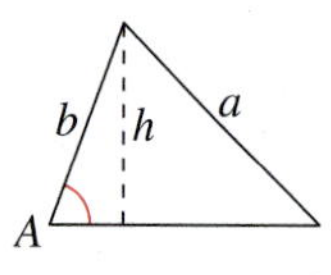	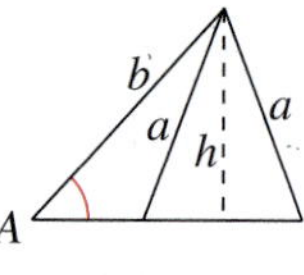	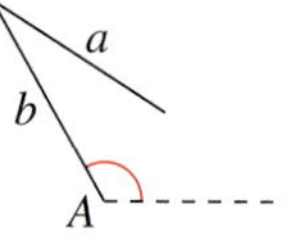	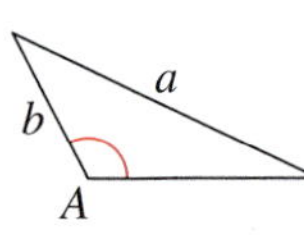
Necessary condition	$a < h$	$a = h$	$a > b$	$h < a < b$	$a \le b$	$a > b$
Triangles possible	None	One	One	Two	None	One

Example 3 ▶ **Single-Solution Case—SSA** 

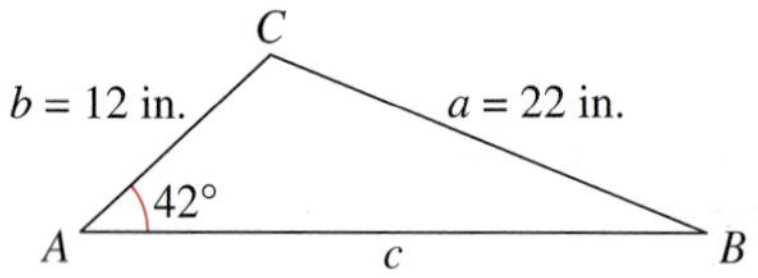

For the triangle in Figure 8.4, $a = 22$ inches, $b = 12$ inches, and $A = 42°$. Find the remaining side and angles.

One solution: $a > b$

FIGURE 8.4

Solution

By the Law of Sines, you have

$$\frac{\sin B}{b} = \frac{\sin A}{a}$$ Reciprocal form

$$\sin B = b\left(\frac{\sin A}{a}\right)$$ Multiply each side by b.

$$\sin B = 12\left(\frac{\sin 42°}{22}\right)$$ Substitute for A, a, and b.

$$B \approx 21.41°.$$ B is acute.

The *Interactive* CD-ROM and *Internet* versions of this text offer a Try It for each example in the text.

Now, you can determine that

$$C \approx 180° - 42° - 21.41° = 116.59°.$$

Then, the remaining side is

$$\frac{c}{\sin C} = \frac{a}{\sin A}$$

$$c = \frac{a}{\sin A}(\sin C) = \frac{22}{\sin 42°}(\sin 116.59°) \approx 29.40 \text{ inches.}$$

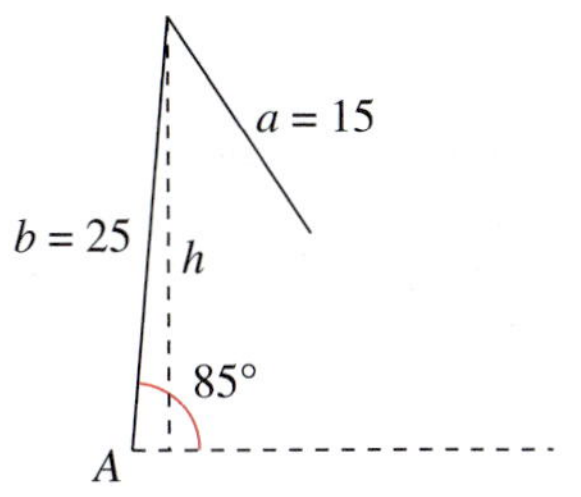

No solution: $a < h$

FIGURE 8.5

Example 4 ▶ No-Solution Case—SSA

Show that there is no triangle for which $a = 15$, $b = 25$, and $A = 85°$.

Solution

Begin by making the sketch shown in Figure 8.5. From this figure it appears that no triangle is formed. You can verify this using the Law of Sines.

$$\frac{\sin B}{b} = \frac{\sin A}{a}$$ Reciprocal form

$$\sin B = b\left(\frac{\sin A}{a}\right)$$ Multiply each side by b.

$$\sin B = 25\left(\frac{\sin 85°}{15}\right) \approx 1.660 > 1$$

This contradicts the fact that $|\sin B| \le 1$. So, no triangle can be formed having sides $a = 15$ and $b = 25$ and an angle of $A = 85°$.

Example 5 ▶ Two-Solution Case—SSA 

Find two triangles for which $a = 12$ meters, $b = 31$ meters, and $A = 20.5°$.

Solution

By the Law of Sines, you have

$$\frac{\sin B}{b} = \frac{\sin A}{a}$$ Reciprocal form

$$\sin B = b\left(\frac{\sin A}{a}\right) = 31\left(\frac{\sin 20.5°}{12}\right) \approx 0.9047.$$

There are two angles $B_1 \approx 64.8°$ and $B_2 \approx 180° - 64.8° = 115.2°$ between 0° and 180° whose sine is 0.9047. For $B_1 \approx 64.8°$, you obtain

$$C \approx 180° - 20.5° - 64.8° = 94.7°$$

$$c = \frac{a}{\sin A}(\sin C) = \frac{12}{\sin 20.5°}(\sin 94.7°) \approx 34.15 \text{ meters.}$$

For $B_2 \approx 115.2°$, you obtain

$$C \approx 180° - 20.5° - 115.2° = 44.3°$$

$$c = \frac{a}{\sin A}(\sin C) = \frac{12}{\sin 20.5°}(\sin 44.3°) \approx 23.93 \text{ meters.}$$

The resulting triangles are shown in Figure 8.6.

The *Interactive* CD-ROM and *Internet* versions of this text offer a Quiz for every section of the text.

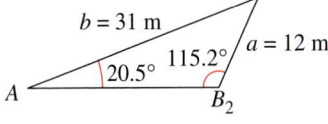

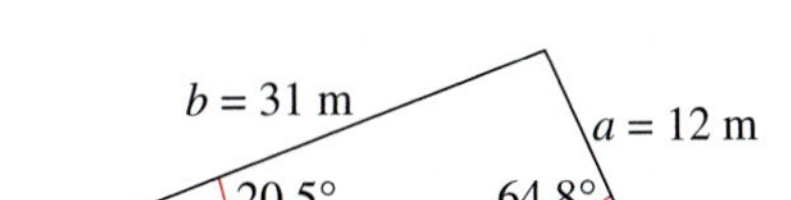

FIGURE 8.6

STUDY TIP

To see how to obtain the height of the obtuse triangle in Figure 8.7, notice the use of the reference angle $180° - A$ and the difference formula for sine, as follows.

$$h = b\sin(180° - A)$$
$$= b(\sin 180° \cos A - \cos 180° \sin A)$$
$$= b[0 \cdot \cos A - (-1) \cdot \sin A]$$
$$= b\sin A$$

Area of an Oblique Triangle

The procedure used to prove the Law of Sines leads to a simple formula for the area of an oblique triangle. Referring to Figure 8.7, note that each triangle has a height of $h = b\sin A$. Consequently, the area of each triangle is

$$\text{Area} = \frac{1}{2}(\text{base})(\text{height}) = \frac{1}{2}(c)(b\sin A) = \frac{1}{2}bc\sin A.$$

By similar arguments, you can develop the formulas

$$\text{Area} = \frac{1}{2}ab\sin C = \frac{1}{2}ac\sin B.$$

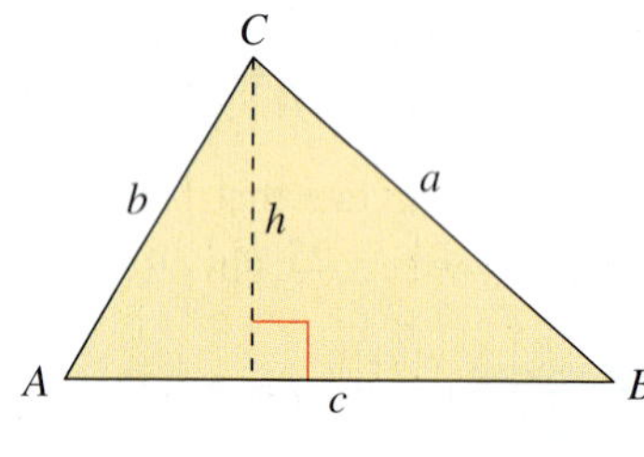

A is acute

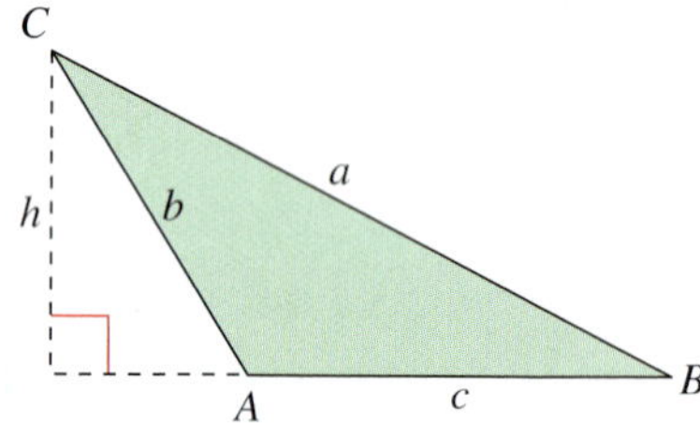

A is obtuse

FIGURE 8.7

Area of an Oblique Triangle

The area of any triangle is one-half the product of the lengths of two sides times the sine of their included angle. That is,

$$\text{Area} = \frac{1}{2}bc\sin A = \frac{1}{2}ab\sin C = \frac{1}{2}ac\sin B.$$

Note that if angle A is $90°$, the formula gives the area for a right triangle:

$$\text{Area} = \frac{1}{2}bc(\sin 90°) = \frac{1}{2}bc = \frac{1}{2}(\text{base})(\text{height}).$$

$\sin 90° = 1$

Similar results are obtained for angles C and B equal to $90°$.

Example 6 ▶ Finding the Area of a Triangular Lot  

Find the area of a triangular lot having two sides of lengths 90 meters and 52 meters and an included angle of $102°$.

Solution

Consider $a = 90$ meters, $b = 52$ meters, and angle $C = 102°$, as shown in Figure 8.8. Then, the area of the triangle is

$$\text{Area} = \frac{1}{2}ab\sin C = \frac{1}{2}(90)(52)(\sin 102°) \approx 2289 \text{ square meters.}$$

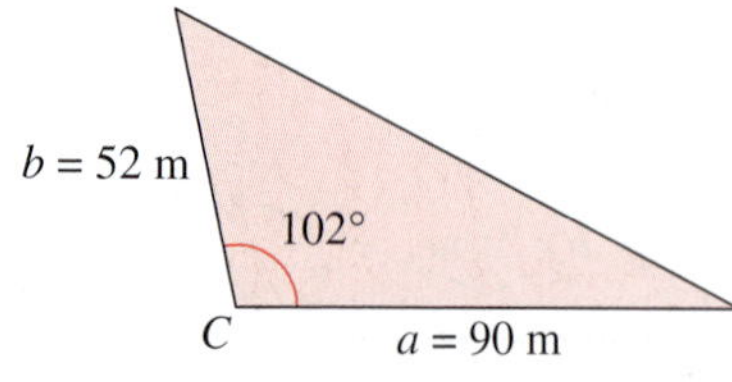

FIGURE 8.8

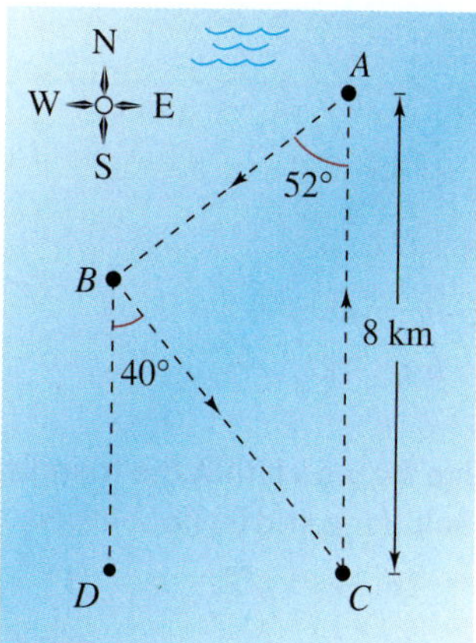

FIGURE 8.9

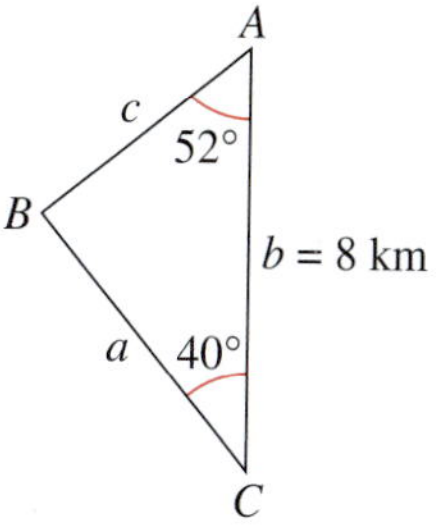

FIGURE 8.10

Application

Example 7 ▶ An Application of the Law of Sines

The course for a boat race starts at point A and proceeds in the direction S 52° W to point B, then in the direction S 40° E to point C, and finally back to A, as shown in Figure 8.9. Point C lies 8 kilometers directly south of point A. Approximate the total distance of the race course.

Solution

Because lines BD and AC are parallel, it follows that $\angle BCA \cong \angle DBC$. Consequently, triangle ABC has the measures shown in Figure 8.10. For angle B, you have $B = 180° - 52° - 40° = 88°$. Using the Law of Sines

$$\frac{a}{\sin 52°} = \frac{b}{\sin 88°} = \frac{c}{\sin 40°}$$

you can let $b = 8$ and obtain

$$a = \frac{8}{\sin 88°}(\sin 52°) \approx 6.308$$

and

$$c = \frac{8}{\sin 88°}(\sin 40°) \approx 5.145.$$

The total length of the course is approximately

$$\begin{aligned} \text{Length} &\approx 8 + 6.308 + 5.145 \\ &= 19.453 \text{ kilometers.} \end{aligned}$$

Writing ABOUT MATHEMATICS

Using the Law of Sines In this section, you have been using the Law of Sines to solve *oblique* triangles. Can the Law of Sines also be used to solve a right triangle? If so, write a short paragraph explaining how to use the Law of Sines to solve each triangle. Is there an easier way to solve these triangles?

a. (AAS)

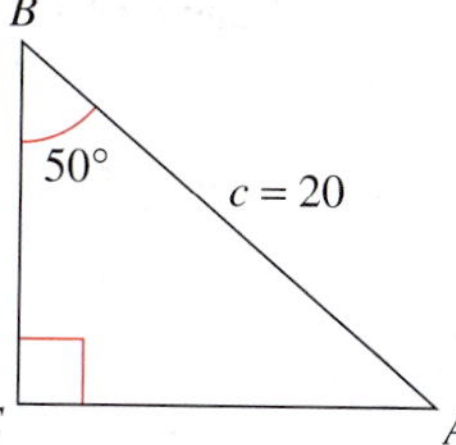

b. (ASA)

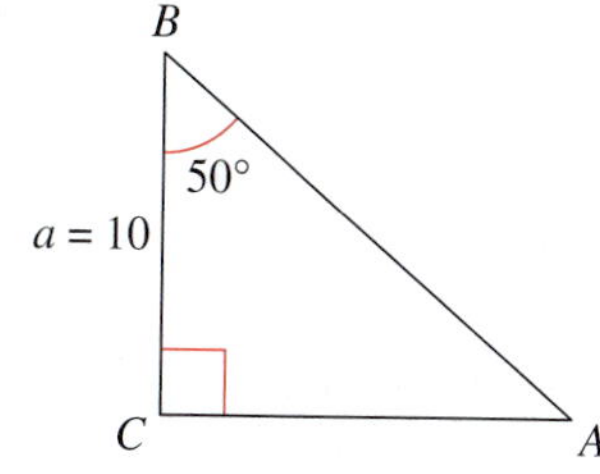

8.1 Exercises

The *Interactive* CD-ROM and *Internet* versions of this text contain step-by-step solutions to all odd-numbered exercises. They also provide Tutorial Exercises for additional help.

In Exercises 1–18, use the information to solve the triangle.

1.

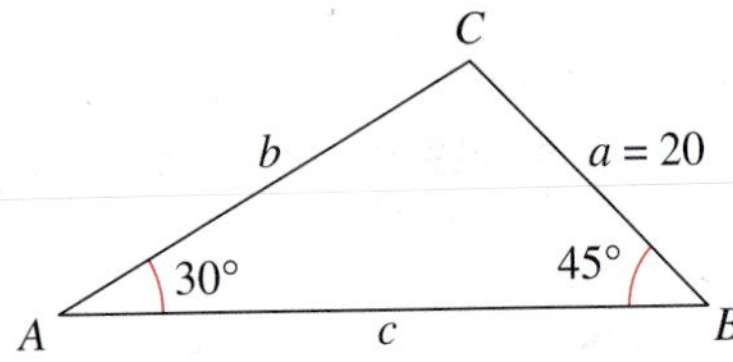

2.

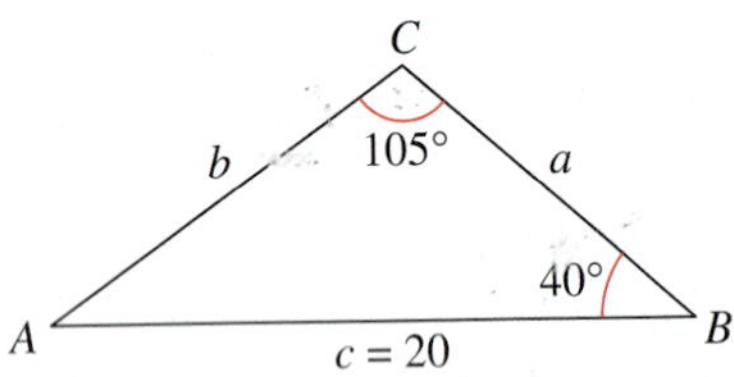

3.

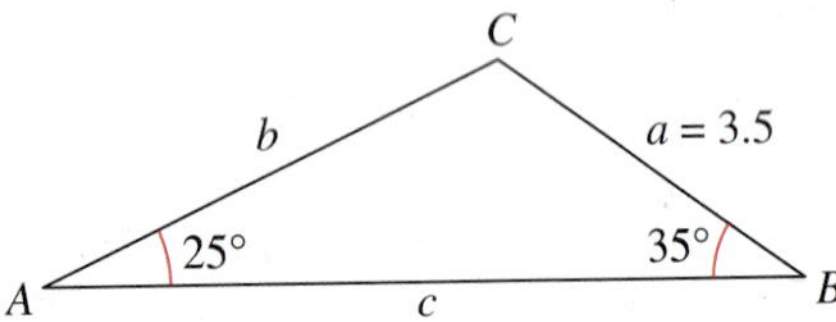

4.

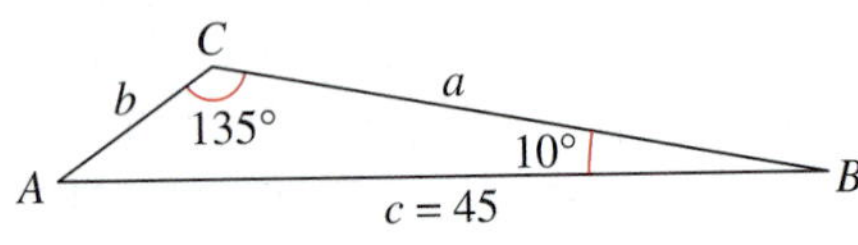

5. $A = 36°, \quad a = 8, \quad b = 5$

6. $A = 60°, \quad a = 9, \quad c = 10$

7. $A = 102.4°, \quad C = 16.7°, \quad a = 21.6$

8. $A = 24.3°, \quad C = 54.6°, \quad c = 2.68$

9. $A = 83° \, 20', \quad C = 54.6°, \quad c = 18.1$

10. $A = 5° \, 40', \quad B = 8° \, 15', \quad b = 4.8$

11. $B = 15° \, 30', \quad a = 4.5, \quad b = 6.8$

12. $B = 2° \, 45', \quad b = 6.2, \quad c = 5.8$

13. $C = 145°, \quad b = 4, \quad c = 14$

14. $A = 100°, \quad a = 125, \quad c = 10$

15. $A = 110° \, 15', \quad a = 48, \quad b = 16$

16. $C = 85° \, 20', \quad a = 35, \quad c = 50$

17. $A = 55°, \quad B = 42°, \quad c = \frac{3}{4}$

18. $B = 28°, \quad C = 104°, \quad a = 3\frac{5}{8}$

In Exercises 19–24, use the information to solve (if possible) the triangle. If two solutions exist, find both.

19. $A = 58°, \quad a = 4.5, \quad b = 12.8$

20. $A = 58°, \quad a = 11.4, \quad b = 12.8$

21. $A = 76°, \quad a = 18, \quad b = 20$

22. $A = 76°, \quad a = 34, \quad b = 21$

23. $A = 110°, \quad a = 125, \quad b = 200$

24. $A = 110°, \quad a = 125, \quad b = 100$

In Exercises 25–28, find a value for b such that the triangle has (a) one solution, (b) two solutions, and (c) no solution.

25. $A = 36°, \quad a = 5$

26. $A = 60°, \quad a = 10$

27. $A = 10°, \quad a = 10.8$

28. $A = 88°, \quad a = 315.6$

In Exercises 29–34, find the area of the triangle having the indicated sides and angle.

29. $C = 120°, \quad a = 4, \quad b = 6$

30. $B = 130°, \quad a = 62, \quad c = 20$

31. $A = 43° \, 45', \quad b = 57, \quad c = 85$

32. $A = 5° \, 15', \quad b = 4.5, \quad c = 22$

33. $B = 72° \, 30', \quad a = 105, \quad c = 64$

34. $C = 84° \, 30', \quad a = 16, \quad b = 20$

35. ***Height*** Because of prevailing winds, a tree grew so that it was leaning 4° from the vertical. At a point 35 meters from the tree, the angle of elevation to the top of the tree is 23° (see figure). Find the height h of the tree.

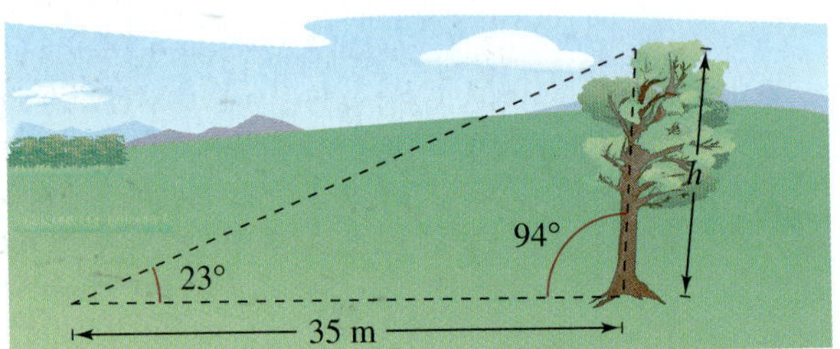

36. ***Height*** A flagpole at a right angle to the horizontal is located on a slope that makes an angle of 12° with the horizontal. The flagpole's shadow is 16 meters long and points directly up the slope. The angle of elevation from the tip of the shadow to the sun is 20°.

(a) Draw a triangle that represents the problem. Show the known quantities on the triangle and use a variable to indicate the height of the flagpole.

(b) Write an equation involving the unknown quantity.

(c) Find the height of the flagpole.

37. *Angle of Elevation* A 10-meter telephone pole casts a 17-meter shadow directly down a slope when the angle of elevation of the sun is 42° (see figure). Find θ, the angle of elevation of the ground.

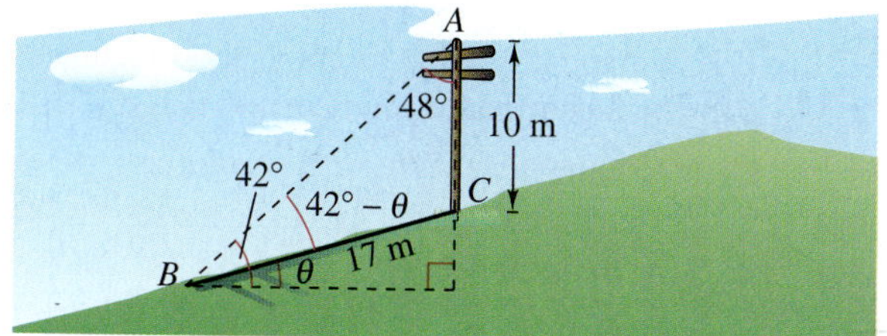

38. *Flight Path* A plane flies 500 kilometers with a bearing of 316° from Naples to Elgin (see figure). The plane then flies 720 kilometers from Elgin to Canton. Find the bearing of the flight from Elgin to Canton.

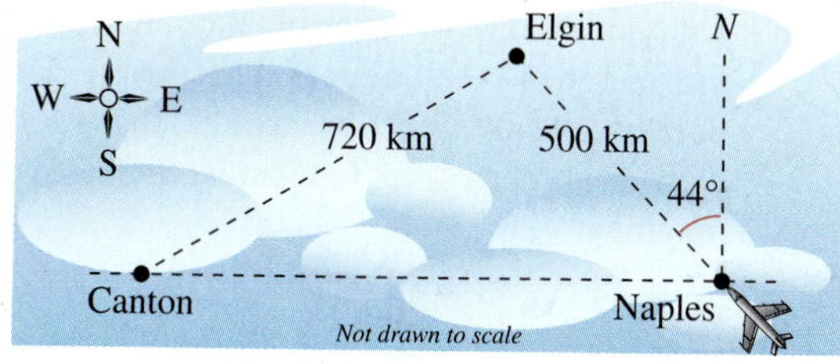

39. *Bridge Design* A bridge is to be built across a small lake from a gazebo to a dock (see figure). The bearing from the gazebo to the dock is S 41° W. From a tree 100 meters from the gazebo, the bearings to the gazebo and the dock are S 74° E and S 28° E, respectively. Find the distance from the gazebo to the dock.

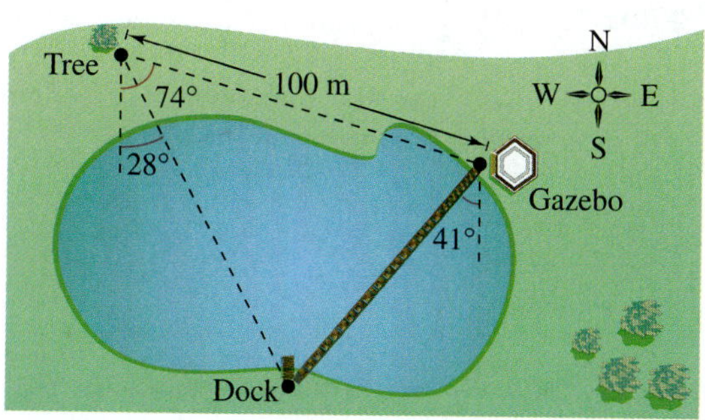

40. *Railroad Track Design* The circular arc of a railroad curve has a chord of length 3000 feet and a central angle of 40°.

(a) Draw a figure that visually represents the problem. Show the known quantities on the figure and use variables r and s to represent the radius of the arc and the length of the arc, respectively.

(b) Find the radius r of the circular arc.

(c) Find the length s of the circular arc.

41. *Glide Path* A pilot has just started on the glide path for landing at an airport where the length of the runway is 9000 feet. The angles of depression from the plane to the ends of the runway are 17.5° and 18.8°.

(a) Draw a figure that visually represents the problem.

(b) Find the air distance the plane must travel until touching down on the near end of the runway.

(c) Find the ground distance the plane must travel until touching down.

(d) Find the altitude of the plane when the pilot begins the descent.

42. *Locating a Fire* The bearing from the Pine Knob fire tower to the Colt Station fire tower is N 65° E, and the two towers are 30 kilometers apart. A fire spotted by rangers in each tower has a bearing of N 80° E from Pine Knob and S 70° E from Colt Station. Find the distance of the fire from each tower.

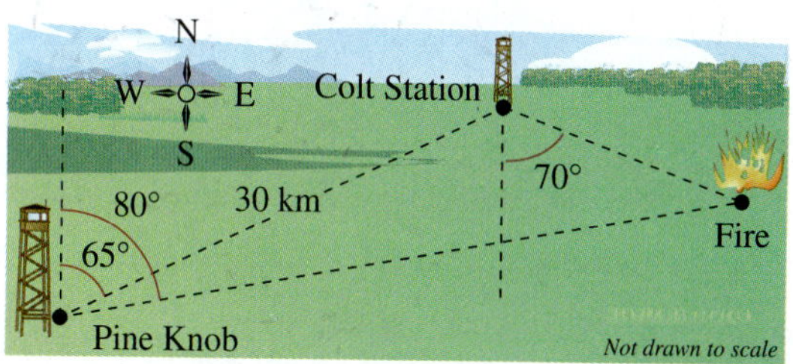

43. *Distance* A boat is sailing due east parallel to the shoreline at a speed of 10 miles per hour. At a given time, the bearing to the lighthouse is S 70° E, and 15 minutes later the bearing is S 63° E (see figure). The lighthouse is located at the shoreline. What is the distance from the boat to the shoreline?

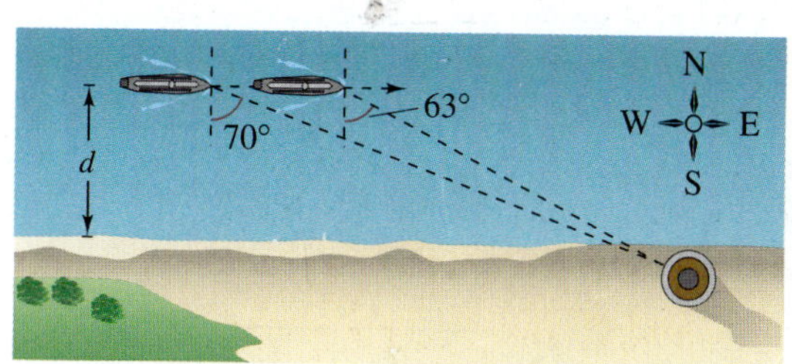

▶ Model It

44. ***Shadow Length*** The Leaning Tower of Pisa in Italy is characterized by its tilt. The tower leans because it was built on a layer of unstable soil—clay, sand, and water. The tower is approximately 58.36 meters tall from its foundation (see figure). The top of the tower leans about 5.45 meters off center.

(a) Find the angle of lean α of the tower.

(b) Write β as a function of d and θ, where θ is the angle of elevation to the sun.

(c) Use the Law of Sines to write an equation for the length d of the shadow cast by the tower.

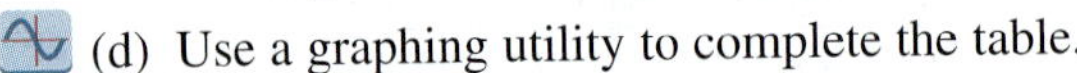

(d) Use a graphing utility to complete the table.

θ	10°	20°	30°	40°	50°	60°
d						

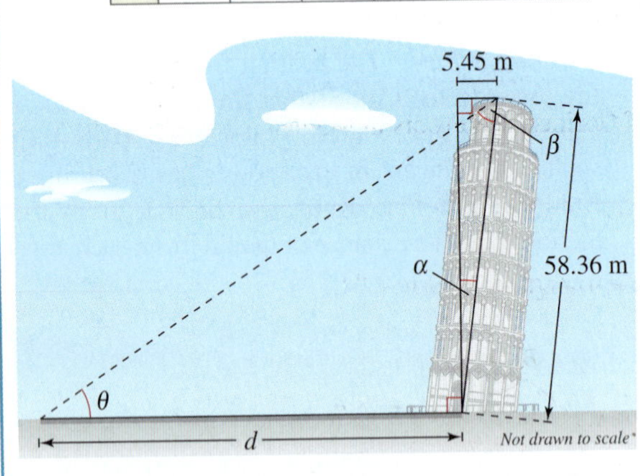

Synthesis

True or False? **In Exercises 45 and 46, determine whether the statement is true or false. Justify your answer.**

45. If a triangle contains an obtuse angle, then it must be oblique.

46. Two angles and one side of a triangle do not necessarily determine a unique triangle.

47. ***Graphical and Numerical Analysis*** In the figure, α and β are positive angles.

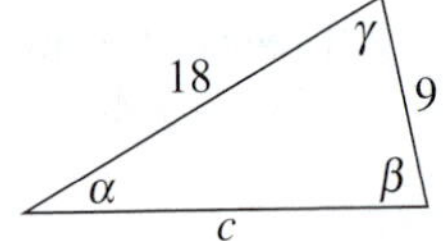

(a) Write α as a function of β.

(b) Use a graphing utility to graph the function. Determine its domain and range.

(c) Use the result of part (a) to write c as a function of β.

(d) Use a graphing utility to graph the function in part (c). Determine its domain and range.

(e) Complete the table. What can you infer?

β	0.4	0.8	1.2	1.6	2.0	2.4	2.8
α							
c							

48. ***Graphical Analysis***

(a) Write the area A of the shaded region in the figure as a function of θ.

(b) Use a graphing utility to graph the area function.

(c) Determine the domain of the area function. Explain how the area of the region and the domain of the function would change if the eight-centimeter line segment were decreased in length.

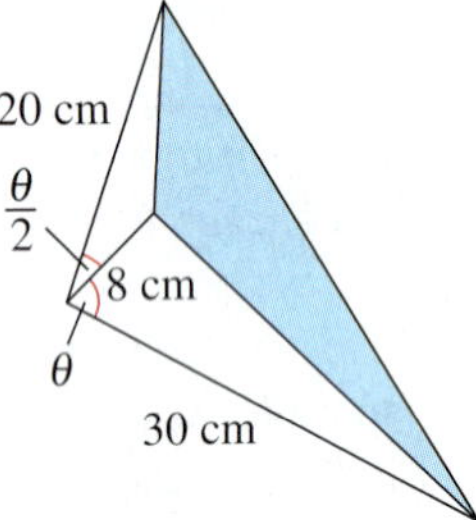

Review

In Exercises 49–52, use the fundamental trigonometric identities to simplify the expression.

49. $\sin x \cot x$

50. $\tan x \cos x \sec x$

51. $1 - \sin^2\left(\frac{\pi}{2} - x\right)$

52. $1 + \cot^2\left(\frac{\pi}{2} - x\right)$

In Exercises 53 and 54, write the product as a sum or difference.

53. $6 \sin 8\theta \cos 3\theta$

54. $2 \cos 5\theta \sin 2\theta$

8.2 Law of Cosines

▶ **What you should learn**

- How to use the Law of Cosines to solve oblique triangles (SSS or SAS)
- How to use the Law of Cosines to model and solve real-life problems
- How to use Heron's Area Formula to find the area of a triangle

▶ **Why you should learn it**

You can use the Law of Cosines to solve real-life problems involving oblique triangles. For instance, in Exercise 44 on page 613, you can use the Law of Cosines to determine the maximum distance a piston moves in one cycle.

Rosenthal/SuperStock

Introduction

Two cases remain in the list of conditions needed to solve an oblique triangle—SSS and SAS. If you are given three sides (SSS), or two sides and their included angle (SAS), none of the ratios in the Law of Sines would be complete. In such cases, you can use the **Law of Cosines.**

Law of Cosines

Standard Form	*Alternative Form*
$a^2 = b^2 + c^2 - 2bc\cos A$	$\cos A = \dfrac{b^2 + c^2 - a^2}{2bc}$
$b^2 = a^2 + c^2 - 2ac\cos B$	$\cos B = \dfrac{a^2 + c^2 - b^2}{2ac}$
$c^2 = a^2 + b^2 - 2ab\cos C$	$\cos C = \dfrac{a^2 + b^2 - c^2}{2ab}$

For a proof of the Law of Cosines, see Proofs in Mathematics on page 657.

Example 1 ▶ **Three Sides of a Triangle—SSS** 

Find the three angles of the triangle in Figure 8.11.

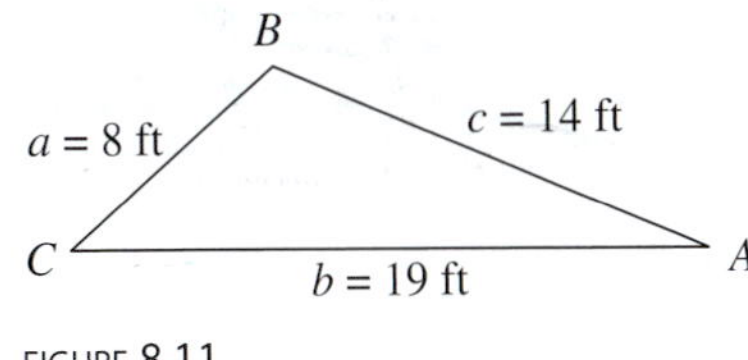

FIGURE 8.11

Solution

It is a good idea first to find the angle opposite the longest side—side b in this case (see Figure 8.11). Using the Law of Cosines, you find that

$$\cos B = \frac{a^2 + c^2 - b^2}{2ac} = \frac{8^2 + 14^2 - 19^2}{2(8)(14)} \approx -0.45089.$$

Because $\cos B$ is negative, you know that B is an *obtuse* angle given by $B \approx 116.80°$. At this point, it is simpler to use the Law of Sines to determine A.

$$\sin A = a\left(\frac{\sin B}{b}\right) \approx 8\left(\frac{\sin 116.80°}{19}\right) \approx 0.37583$$

Because B is obtuse, you know that A must be acute, because a triangle can have, at most, one obtuse angle. So, $A \approx 22.08°$ and $C \approx 180° - 22.08° - 116.80° = 41.12°$.

Exploration

What familiar formula do you obtain when you use the third form of the Law of Cosines

$$c^2 = a^2 + b^2 - 2ab\cos C$$

and you let $C = 90°$? What is the relationship between the Law of Cosines and this formula?

Do you see why it was wise to find the largest angle *first* in Example 1? Knowing the cosine of an angle, you can determine whether the angle is acute or obtuse. That is,

$$\cos\theta > 0 \quad \text{for} \quad 0° < \theta < 90° \qquad \text{Acute}$$

$$\cos\theta < 0 \quad \text{for} \quad 90° < \theta < 180°. \qquad \text{Obtuse}$$

So, in Example 1, once you found that angle B was obtuse, you knew that angles A and C were both acute. If the largest angle is acute, the remaining two angles are acute also.

Example 2 ▶ Two Sides and the Included Angle—SAS 

Find the remaining angles and side of the triangle in Figure 8.12.

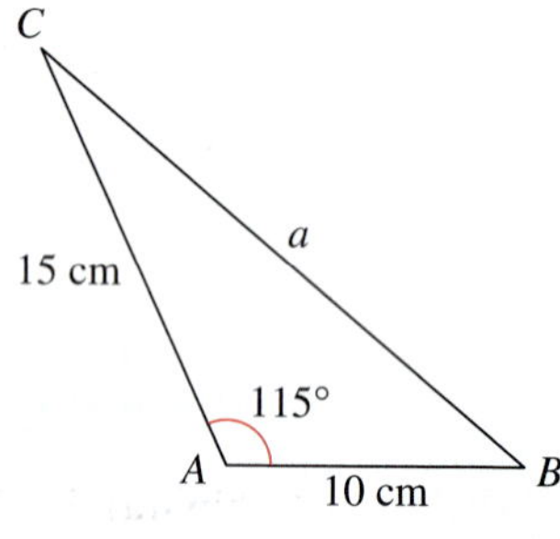

FIGURE 8.12

Solution

Use the Law of Cosines to find the unknown side a in the figure.

$$a^2 = b^2 + c^2 - 2bc\cos A$$

$$a^2 = 15^2 + 10^2 - 2(15)(10)\cos 115°$$

$$a^2 \approx 451.79$$

$$a \approx 21.26$$

Because $a \approx 21.26$ centimeters, you now know the ratio $\sin A/a$ and you can use the Law of Sines

$$\frac{\sin B}{b} = \frac{\sin A}{a}$$

to solve for B.

$$\sin B = b\left(\frac{\sin A}{a}\right)$$

$$= 15\left(\frac{\sin 115°}{21.26}\right)$$

$$\approx 0.63945$$

So, $B = \arcsin 0.63945 \approx 39.75°$ and $C \approx 180° - 115° - 39.75° = 25.25°$.

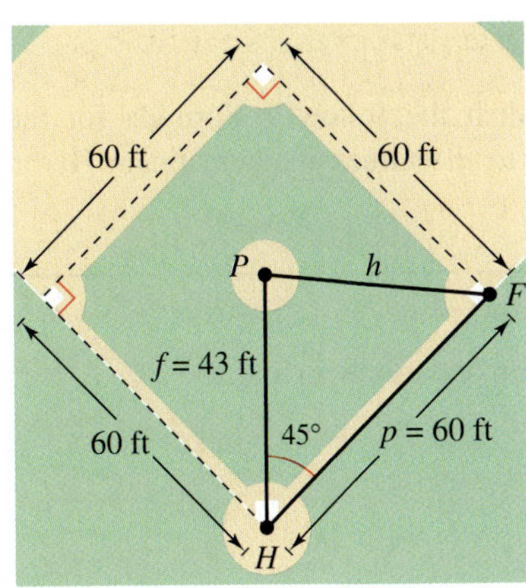

FIGURE 8.13

Applications

Example 3 ▶ **An Application of the Law of Cosines**

The pitcher's mound on a women's softball field is 43 feet from home plate and the distance between the bases is 60 feet, as shown in Figure 8.13. (The pitcher's mound is not halfway between home plate and second base.) How far is the pitcher's mound from first base?

Solution

In triangle HPF, $H = 45°$ (line HP bisects the right angle at H), $f = 43$, and $p = 60$. Using the Law of Cosines for this SAS case, you have

$$h^2 = f^2 + p^2 - 2fp \cos H$$

$$= 43^2 + 60^2 - 2(43)(60) \cos 45°$$

$$\approx 1800.3$$

So, the approximate distance from the pitcher's mound to first base is

$$h \approx \sqrt{1800.3} \approx 42.43 \text{ feet.}$$

Example 4 ▶ **An Application of the Law of Cosines**

A ship travels 60 miles due east, then adjusts its course northward, as shown in Figure 8.14. After traveling 80 miles in that direction, the ship is 139 miles from its point of departure. Describe the bearing from point B to point C.

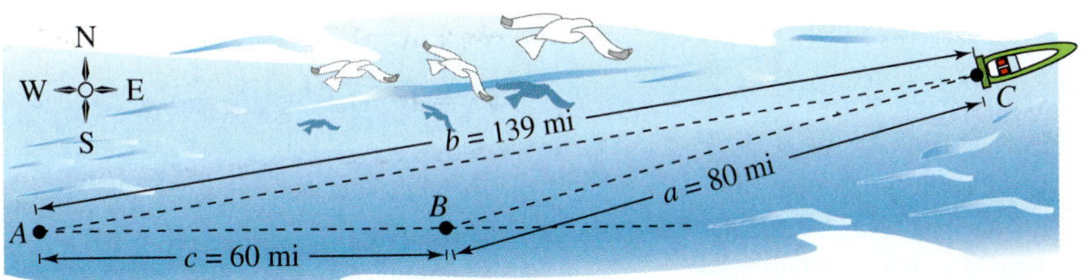

FIGURE 8.14

Solution

You have $a = 80$, $b = 139$, and $c = 60$; so, using the alternative form of the Law of Cosines, you have

$$\cos B = \frac{a^2 + c^2 - b^2}{2ac}$$

$$= \frac{80^2 + 60^2 - 139^2}{2(80)(60)}$$

$$\approx -0.97094.$$

So, $B \approx \arccos(-0.97094) \approx 166.15°$, and thus the bearing measured from due north from point B to point C is $166.15° - 90° = 76.15°$, or N 76.15° E.

Heron's Area Formula

The Law of Cosines can be used to establish the following formula for the area of a triangle. This formula is credited to the Greek mathematician Heron (c. 100 B.C.).

Heron's Area Formula

Given any triangle with sides of lengths a, b, and c, the area of the triangle is

$$\text{Area} = \sqrt{s(s-a)(s-b)(s-c)}$$

where $s = (a + b + c)/2$.

For a proof of Heron's Area Formula, see Proofs in Mathematics on page 658.

Example 5 ▶ Using Heron's Area Formula 

Find the area of a triangle having sides of lengths $a = 43$ meters, $b = 53$ meters, and $c = 72$ meters.

Solution

Because $s = (a + b + c)/2 = 168/2 = 84$, Heron's Area Formula yields

$$\begin{aligned}\text{Area} &= \sqrt{s(s-a)(s-b)(s-c)} \\ &= \sqrt{84(41)(31)(12)} \approx 1131.89 \text{ square meters.}\end{aligned}$$

Historical Note

Heron of Alexandria (c. 100 B.C.) was a Greek geometer and inventor. His works describe how to find the areas of triangles, quadrilaterals, regular polygons having 3 to 12 sides, and circles as well as the surface areas and volumes of three-dimensional objects.

Writing ABOUT MATHEMATICS

The Area of a Triangle You have now studied three different formulas for the area of a triangle. Use the most appropriate formula to find the area of each triangle below. Show your work and give your reasons for choosing each formula.

Standard Formula $\text{Area} = \frac{1}{2}bh$

Oblique Triangle $\text{Area} = \frac{1}{2}bc \sin A = \frac{1}{2}ab \sin C = \frac{1}{2}ac \sin B$

Heron's Area Formula $\text{Area} = \sqrt{s(s-a)(s-b)(s-c)}$

a. 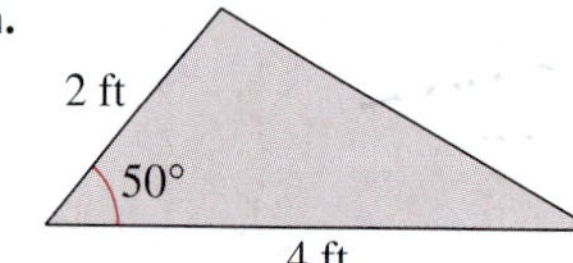

b. 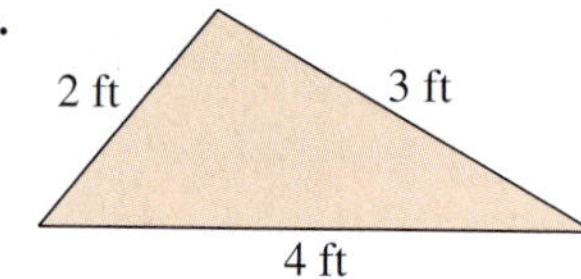

c. 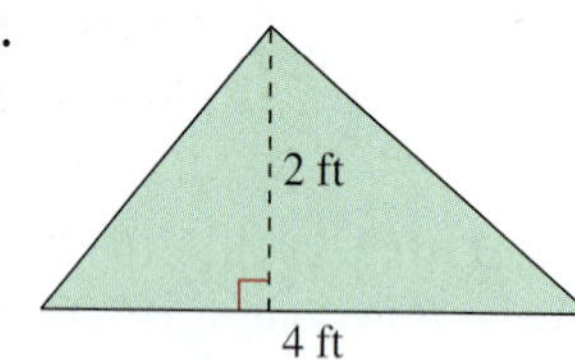

d. 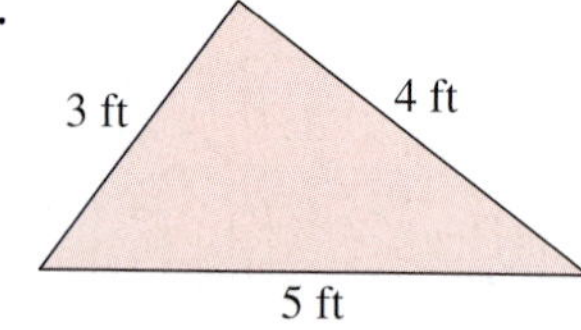

8.2 Exercises

In Exercises 1–16, use the Law of Cosines to solve the triangle.

1.

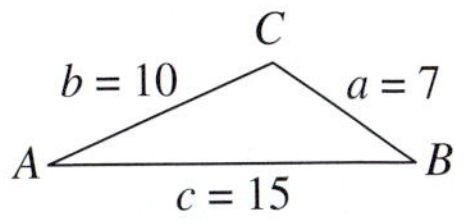

2.

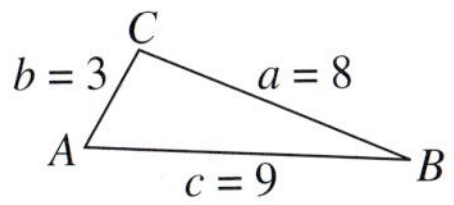

3.

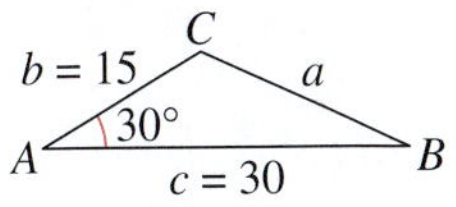

4.

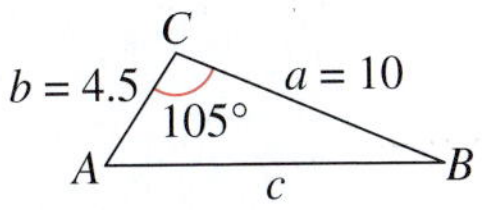

5. $a = 11, \quad b = 14, \quad c = 20$

6. $a = 55, \quad b = 25, \quad c = 72$

7. $a = 75.4, \quad b = 52, \quad c = 52$

8. $a = 1.42, \quad b = 0.75, \quad c = 1.25$

9. $A = 135°, \quad b = 4, \quad c = 9$

10. $A = 55°, \quad b = 3, \quad c = 10$

11. $B = 10° \, 35', \quad a = 40, \quad c = 30$

12. $B = 75° \, 20', \quad a = 6.2, \quad c = 9.5$

13. $B = 125° \, 40', \quad a = 32, \quad c = 32$

14. $C = 15° \, 15', \quad a = 6.25, \quad b = 2.15$

15. $C = 43°, \quad a = \frac{4}{9}, \quad b = \frac{7}{9}$

16. $C = 103°, \quad a = \frac{3}{8}, \quad b = \frac{3}{4}$

In Exercises 17–22, complete the table by solving the parallelogram shown in the figure. (The lengths of the diagonals are given by *c* and *d*.)

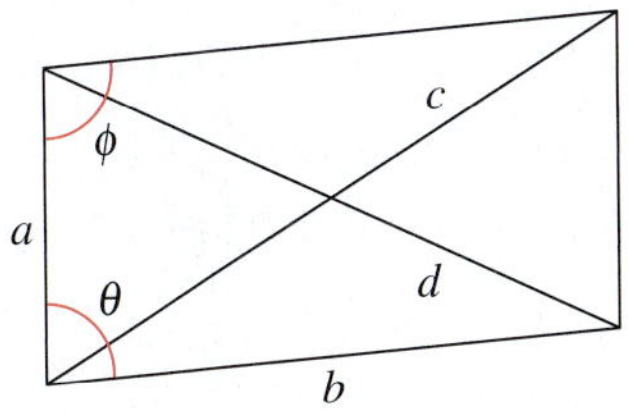

	a	b	c	d	θ	ϕ
17.	5	8			45°	
18.	25	35				120°
19.	10	14	20			
20.	40	60		80		
21.	15		25	20		
22.		25	50	35		

In Exercises 23–28, use Heron's Area Formula to find the area of the triangle.

23. $a = 5, \quad b = 7, \quad c = 10$

24. $a = 12, \quad b = 15, \quad c = 9$

25. $a = 2.5, \quad b = 10.2, \quad c = 9$

26. $a = 75.4, \quad b = 52, \quad c = 52$

27. $a = 12.32, \quad b = 8.46, \quad c = 15.05$

28. $a = 3.05, \quad b = 0.75, \quad c = 2.45$

29. ***Navigation*** A boat race runs along a triangular course marked by buoys A, B, and C. The race starts with the boats headed west for 3700 meters. The other two sides of the course lie to the north of the first side, and their lengths are 1700 meters and 3000 meters. Draw a figure that gives a visual representation of the problem, and find the bearings for the last two legs of the race.

30. ***Navigation*** A plane flies 810 miles from Niagara to Cuyahoga with a bearing of 75°. Then it flies 648 miles from Cuyahoga to Rosemount with a bearing of 32°. Draw a figure that visually represents the problem, and find the straight-line distance and bearing from Niagara to Rosemount.

31. ***Surveying*** To approximate the length of a marsh, a surveyor walks 250 meters from point A to point B, then turns 75° and walks 220 meters to point C (see figure). Approximate the length AC of the marsh.

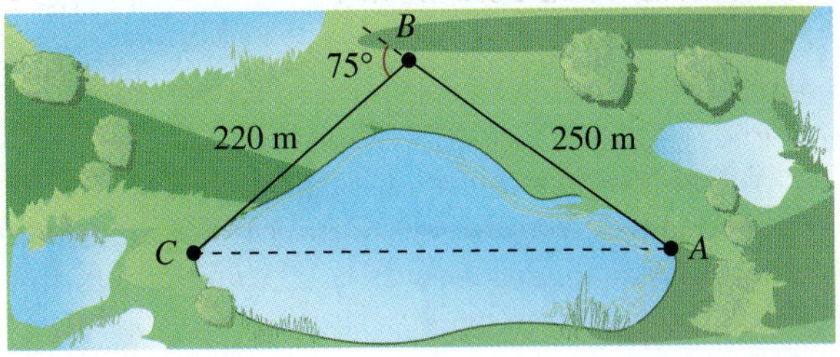

32. ***Surveying*** A triangular parcel of land has 115 meters of frontage, and the other boundaries have lengths of 76 meters and 92 meters. What angles does the frontage make with the two other boundaries?

33. ***Surveying*** A triangular parcel of ground has sides of lengths 725 feet, 650 feet, and 575 feet. Find the measure of the largest angle.

34. ***Streetlight Design*** Determine the angle θ in the design of the streetlight shown in the figure.

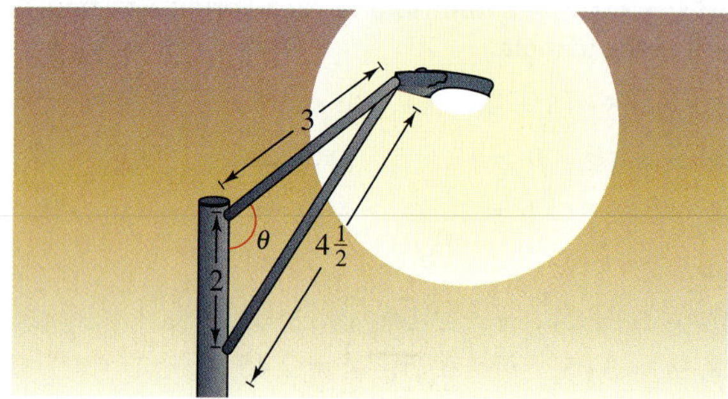

35. ***Distance*** Two ships leave a port at 9 A.M. One travels at a bearing of N 53° W at 12 miles per hour, and the other travels at a bearing of S 67° W at 16 miles per hour. Approximate how far apart they are at noon that day.

36. ***Length*** A 100-foot vertical tower is to be erected on the side of a hill that makes a 6° angle with the horizontal (see figure). Find the length of each of the two guy wires that will be anchored 75 feet uphill and downhill from the base of the tower.

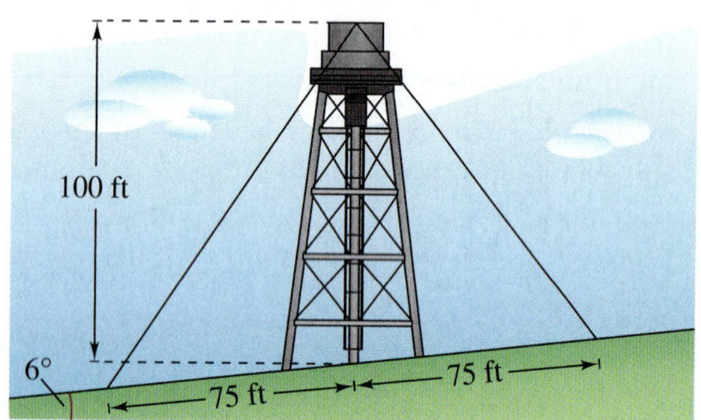

37. ***Navigation*** On a map, Orlando is 178 millimeters due south of Niagara Falls, Denver is 273 millimeters from Orlando, and Denver is 235 millimeters from Niagara Falls (see figure).

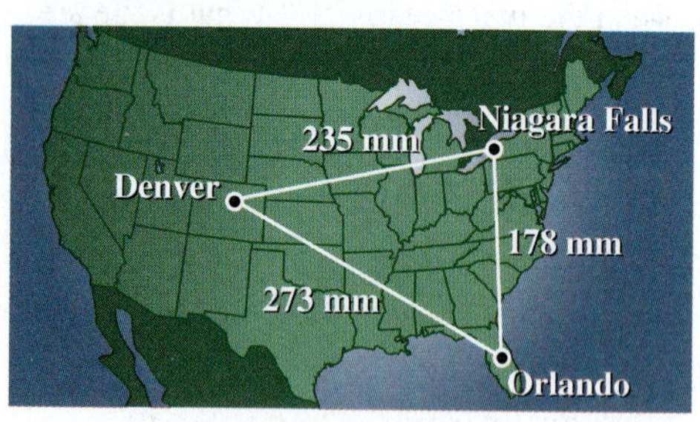

(a) Find the bearing of Denver from Orlando.

(b) Find the bearing of Denver from Niagara Falls.

38. ***Navigation*** On a map, Minneapolis is 165 millimeters due west of Albany, Phoenix is 216 millimeters from Minneapolis, and Phoenix is 368 millimeters from Albany (see figure).

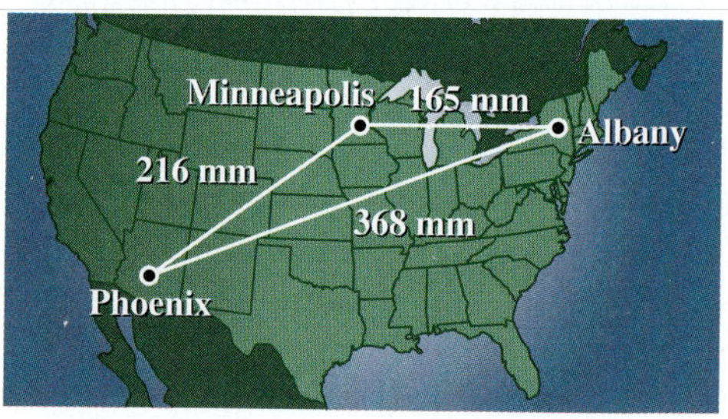

(a) Find the bearing of Minneapolis from Phoenix.

(b) Find the bearing of Albany from Phoenix.

39. ***Baseball*** On a baseball diamond with 90-foot sides, the pitcher's mound is 60.5 feet from home plate. How far is it from the pitcher's mound to third base?

40. ***Baseball*** The baseball player in center field is playing approximately 330 feet from the television camera that is behind home plate. A batter hits a fly ball that goes to the wall 420 feet from the camera (see figure). The camera turns 8° to follow the play. Approximately how far does the center fielder have to run to make the catch?

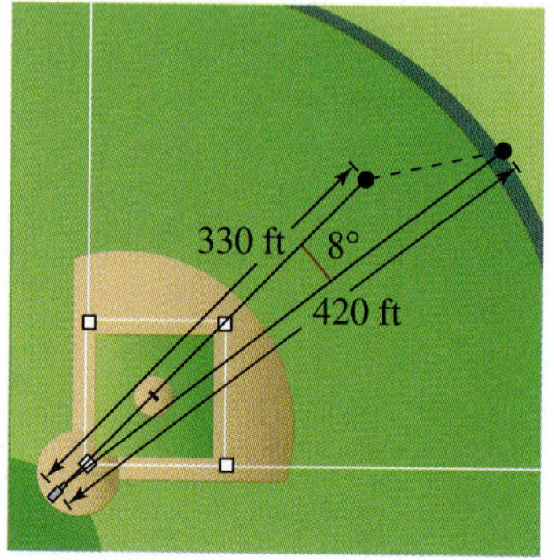

41. ***Aircraft Tracking*** To determine the distance between two aircraft, a tracking station continuously determines the distance to each aircraft and the angle A between them (see figure). Determine the distance a between the planes when $A = 42°$, $b = 35$ miles, and $c = 20$ miles.

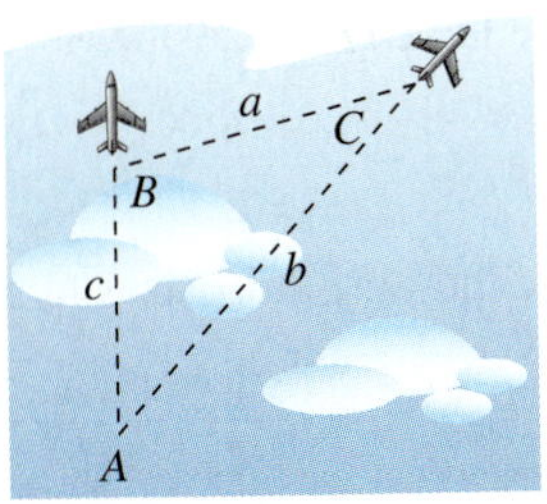

FIGURE FOR 41

42. *Aircraft Tracking* Use the figure for Exercise 41 to determine the distance a between the planes when $A = 11°$, $b = 20$ miles, and $c = 20$ miles.

43. *Trusses* Q is the midpoint of the line segment $\overline{PR}$ in the truss rafter shown in the figure. What are the lengths of the line segments $\overline{PQ}$, $\overline{QS}$, and $\overline{RS}$?

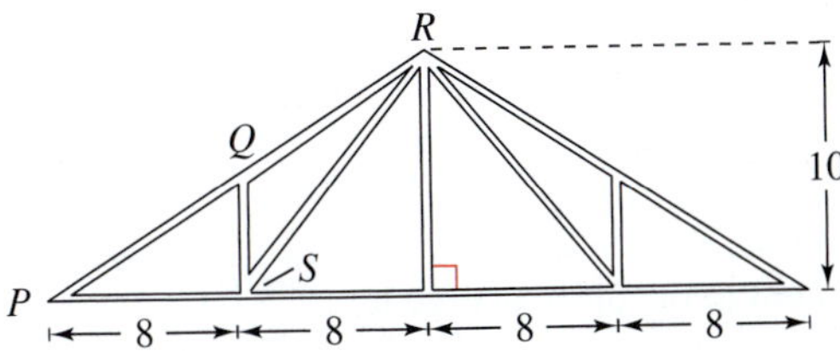

▶ Model It

44. *Engine Design* An engine has a seven-inch connecting rod fastened to a crank (see figure).

(a) Use the Law of Cosines to write an equation giving the relationship between x and θ.

(b) Write x as a function of θ. (Select the sign that yields positive values of x.)

(c) Use a graphing utility to graph the function in part (b).

(d) Use the graph in part (c) to determine the maximum distance the piston moves in one cycle.

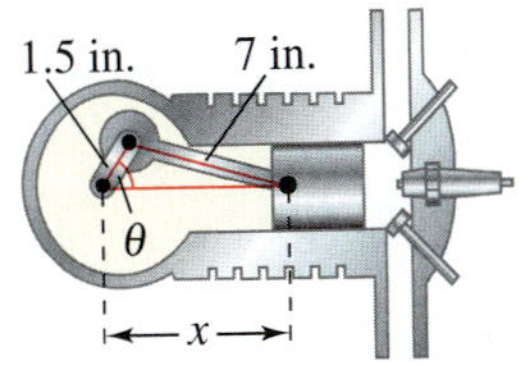

45. *Paper Manufacturing* In a process with continuous paper, the paper passes across three rollers of radii 3 inches, 4 inches, and 6 inches (see figure). The centers of the three-inch and six-inch rollers are d inches apart, and the length of the arc in contact with the paper on the four-inch roller is s inches. Complete the table.

d (inches)	9	10	12	13	14	15	16
θ (degrees)							
s (inches)							

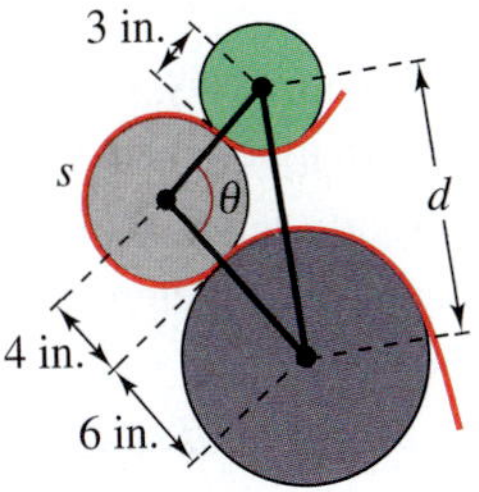

FIGURE FOR 45

46. *Awning Design* A retractable awning above a patio door lowers at an angle of 50° from the exterior wall at a height of 10 feet above the ground (see figure). No direct sunlight is to enter the door when the angle of elevation of the sun is greater than 70°. What is the length x of the awning?

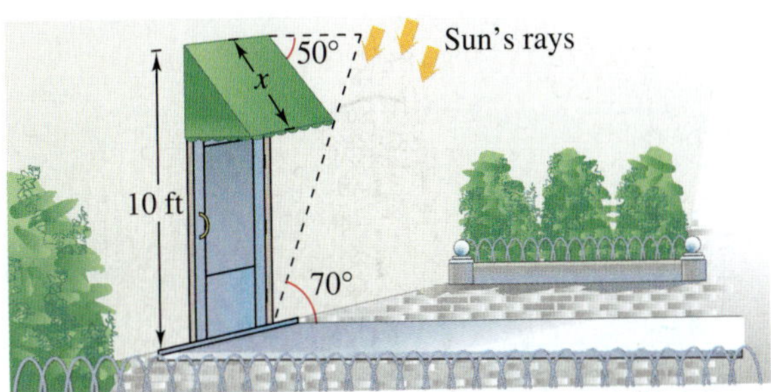

47. *Geometry* The lengths of the sides of a triangular parcel of land are approximately 200 feet, 500 feet, and 600 feet. Approximate the area of the parcel.

48. *Geometry* A parking lot has the shape of a parallelogram (see figure). The lengths of two adjacent sides are 70 meters and 100 meters. The angle between the two sides is 70°. What is the area of the parking lot?

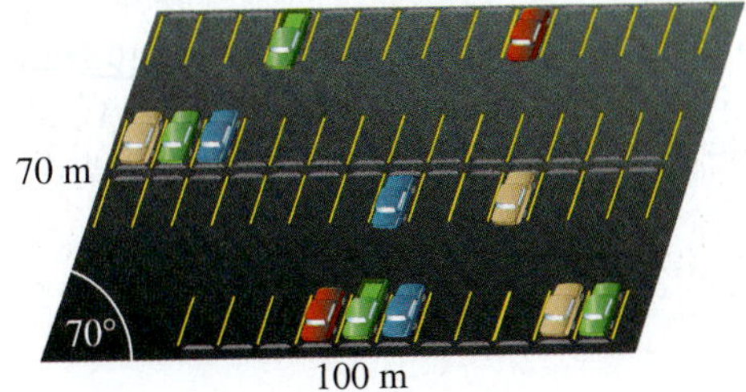

Synthesis

True or False? **In Exercises 49–51, determine whether the statement is true or false. Justify your answer.**

49. In Heron's Area Formula

$$\text{Area} = \sqrt{s(s-a)(s-b)(s-c)}$$

s is the average of the lengths of the three sides of the triangle.

50. In addition to SSS and SAS, the Law of Cosines can be used to solve triangles with SSA conditions.

51. A triangle with side lengths of 10 centimeters, 16 centimeters, and 5 centimeters can be solved using the Law of Cosines.

52. ***Circumscribed and Inscribed Circles*** Let R and r be the radii of the circumscribed and inscribed circles of a triangle ABC, respectively (see figure), and let

$$s = \frac{a+b+c}{2}.$$

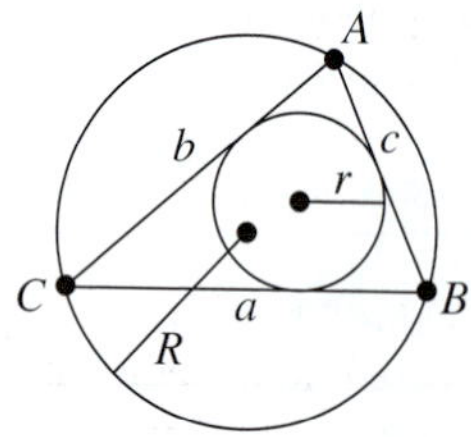

(a) Prove that $2R = \dfrac{a}{\sin A} = \dfrac{b}{\sin B} = \dfrac{c}{\sin C}$.

(b) Prove that $r = \sqrt{\dfrac{(s-a)(s-b)(s-c)}{s}}$.

Circumscribed and Inscribed Circles **In Exercises 53 and 54, use the results of Exercise 52.**

53. Given a triangle with

$$a = 25,\ b = 55,\ \text{and } c = 72$$

find the areas of (a) the triangle, (b) the circumscribed circle, and (c) the inscribed circle.

54. Find the length of the largest circular running track that can be built on a triangular piece of property with sides of lengths 200 feet, 250 feet, and 325 feet.

55. ***Proof*** Use the Law of Cosines to prove that

$$\frac{1}{2}bc(1+\cos A) = \frac{a+b+c}{2}\cdot\frac{-a+b+c}{2}.$$

56. ***Proof*** Use the Law of Cosines to prove that

$$\frac{1}{2}bc(1-\cos A) = \frac{a-b+c}{2}\cdot\frac{a+b-c}{2}.$$

Review

In Exercises 57–62, evaluate the expression without the aid of a calculator.

57. $\arcsin(-1)$

58. $\arccos 0$

59. $\arctan \sqrt{3}$

60. $\arctan -\sqrt{3}$

61. $\arcsin\left(-\dfrac{\sqrt{3}}{2}\right)$

62. $\arccos\left(-\dfrac{\sqrt{3}}{2}\right)$

In Exercises 63–66, write an algebraic expression that is equivalent to the given expression.

63. $\sec(\arcsin 2x)$

64. $\tan(\arccos 3x)$

65. $\cot[\arctan(x-2)]$

66. $\cos\left(\arcsin \dfrac{x-1}{2}\right)$

In Exercises 67–70, use trigonometric substitution to write the algebraic equation as a trigonometric function of θ, where $-\pi/2 < \theta < \pi/2$. Then find $\sec\theta$ and $\csc\theta$.

67. $5 = \sqrt{25-x^2},\quad x = 5\sin\theta$

68. $-\sqrt{2} = \sqrt{4-x^2},\quad x = 2\cos\theta$

69. $-\sqrt{3} = \sqrt{x^2-9},\quad x = 3\sec\theta$

70. $12 = \sqrt{36+x^2},\quad x = 6\tan\theta$

In Exercises 71 and 72, write the sum or difference as a product.

71. $\cos\dfrac{5\pi}{6} - \cos\dfrac{\pi}{3}$

72. $\sin\left(x-\dfrac{\pi}{2}\right) - \sin\left(x+\dfrac{\pi}{2}\right)$

8.3 Vectors in the Plane

▶ What you should learn

- How to represent vectors as directed line segments
- How to write the component forms of vectors
- How to perform basic vector operations and represent them graphically
- How to write vectors as linear combinations of unit vectors
- How to find the direction angles of vectors
- How to use vectors to model and solve real-life problems

▶ Why you should learn it

You can use vectors to model and solve real-life problems involving magnitude and direction. For instance, in Exercise 80 on page 626, you can use vectors to determine the true direction of a commercial jet.

Introduction

Quantities such as force and velocity involve both *magnitude* and *direction* and cannot be completely characterized by a single real number. To represent such a quantity, you can use a **directed line segment,** as shown in Figure 8.15. The directed line segment $\overrightarrow{PQ}$ has **initial point** P and **terminal point** Q. Its **magnitude** (or length) is denoted by $\|\overrightarrow{PQ}\|$ and can be found using the Distance Formula.

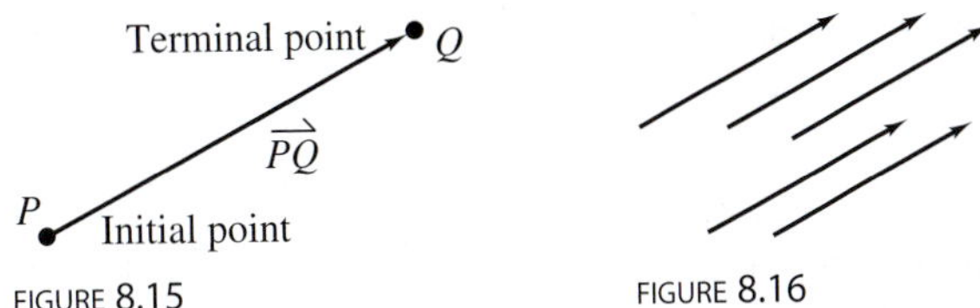

FIGURE 8.15 FIGURE 8.16

Two directed line segments that have the same magnitude and direction are equivalent. For example, the directed line segments in Figure 8.16 are all equivalent. The set of all directed line segments that are equivalent to the directed line segment $\overrightarrow{PQ}$ is a **vector v in the plane,** written $\mathbf{v} = \overrightarrow{PQ}$. Vectors are denoted by lowercase, boldface letters such as **u**, **v**, and **w**.

Example 1 ▶ Vector Representation by Directed Line Segments

Let **u** be represented by the directed line segment from $P = (0, 0)$ to $Q = (3, 2)$, and let **v** be represented by the directed line segment from $R = (1, 2)$ to $S = (4, 4)$, as shown in Figure 8.17. Show that $\mathbf{u} = \mathbf{v}$.

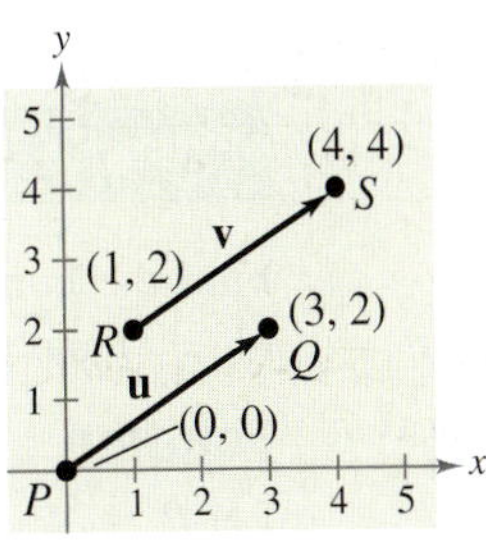

FIGURE 8.17

Solution

From the Distance Formula, it follows that $\overrightarrow{PQ}$ and $\overrightarrow{RS}$ have the *same magnitude.*

$$\|\overrightarrow{PQ}\| = \sqrt{(3-0)^2 + (2-0)^2} = \sqrt{13}$$

$$\|\overrightarrow{RS}\| = \sqrt{(4-1)^2 + (4-2)^2} = \sqrt{13}$$

Moreover, both line segments have the *same direction* because they are both directed toward the upper right on lines having a slope of $\frac{2}{3}$. So, $\overrightarrow{PQ}$ and $\overrightarrow{RS}$ have the same magnitude and direction, and it follows that $\mathbf{u} = \mathbf{v}$.

Component Form of a Vector

The directed line segment whose initial point is the origin is often the most convenient representative of a set of equivalent directed line segments. This representative of the vector **v** is in **standard position.**

A vector whose initial point is at the origin (0, 0) can be uniquely represented by the coordinates of its terminal point (v_1, v_2). This is the **component form of a vector v,** written

$$\mathbf{v} = \langle v_1, v_2 \rangle.$$

The coordinates v_1 and v_2 are the *components* of **v**. If both the initial point and the terminal point lie at the origin, **v** is the **zero vector** and is denoted by $\mathbf{0} = \langle 0, 0 \rangle$.

Component Form of a Vector

The component form of the vector with initial point $P = (p_1, p_2)$ and terminal point $Q = (q_1, q_2)$ is

$$\overrightarrow{PQ} = \langle q_1 - p_1, q_2 - p_2 \rangle = \langle v_1, v_2 \rangle = \mathbf{v}.$$

The **magnitude** (or length) of **v** is

$$\|\mathbf{v}\| = \sqrt{(q_1 - p_1)^2 + (q_2 - p_2)^2} = \sqrt{v_1^2 + v_2^2}.$$

If $\|\mathbf{v}\| = 1$, **v** is a **unit vector.** Moreover, $\|\mathbf{v}\| = 0$ if and only if **v** is the zero vector **0**.

Two vectors $\mathbf{u} = \langle u_1, u_2 \rangle$ and $\mathbf{v} = \langle v_1, v_2 \rangle$ are *equal* if and only if $u_1 = v_1$ and $u_2 = v_2$. For instance, in Example 1, the vector **u** from $P = (0, 0)$ to $Q = (3, 2)$ is

$$\mathbf{u} = \overrightarrow{PQ} = \langle 3 - 0, 2 - 0 \rangle = \langle 3, 2 \rangle$$

and the vector **v** from $R = (1, 2)$ to $S = (4, 4)$ is

$$\mathbf{v} = \overrightarrow{RS} = \langle 4 - 1, 4 - 2 \rangle = \langle 3, 2 \rangle.$$

Example 2 ▶ Finding the Component Form of a Vector 

Find the component form and magnitude of the vector **v** that has initial point $(4, -7)$ and terminal point $(-1, 5)$.

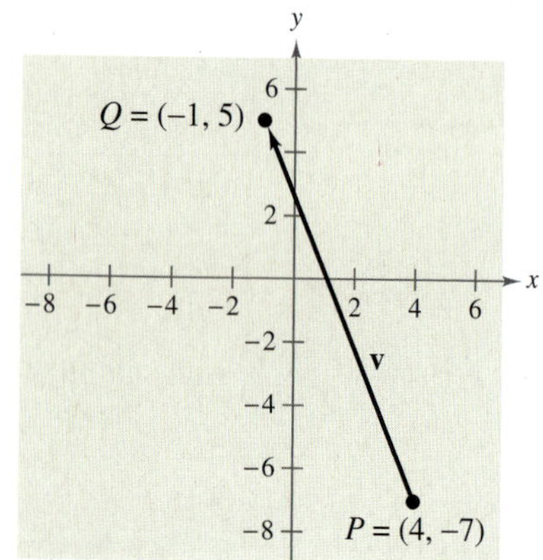

FIGURE 8.18

Solution

Let $P = (4, -7) = (p_1, p_2)$ and let $Q = (-1, 5) = (q_1, q_2)$, as shown in Figure 8.18. Then, the components of $\mathbf{v} = \langle v_1, v_2 \rangle$ are

$$v_1 = q_1 - p_1 = -1 - 4 = -5$$

$$v_2 = q_2 - p_2 = 5 - (-7) = 12.$$

So, $\mathbf{v} = \langle -5, 12 \rangle$ and the magnitude of **v** is

$$\|\mathbf{v}\| = \sqrt{(-5)^2 + 12^2}$$

$$= \sqrt{169} = 13.$$

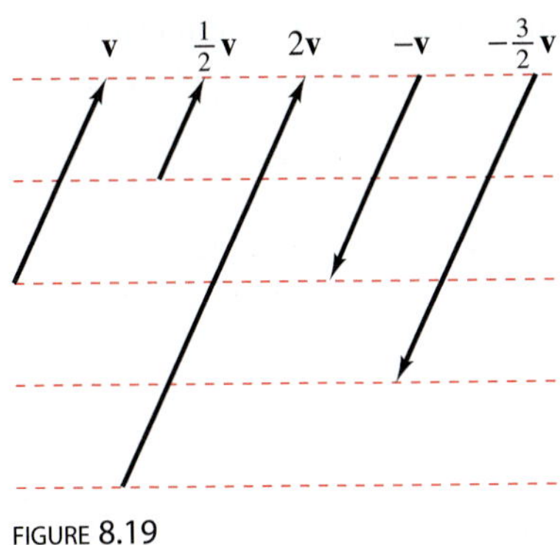

FIGURE 8.19

Vector Operations 

The two basic vector operations are **scalar multiplication** and **vector addition.** In operations with vectors, numbers are usually referred to as **scalars.** In this text, scalars will always be real numbers. Geometrically, the product of a vector **v** and a scalar k is the vector that is $|k|$ times as long as **v**. If k is positive, $k\mathbf{v}$ has the same direction as **v**, and if k is negative, $k\mathbf{v}$ has the direction opposite that of **v**, as shown in Figure 8.19.

To add two vectors geometrically, position them (without changing length or direction) so that the initial point of one coincides with the terminal point of the other. The sum $\mathbf{u} + \mathbf{v}$ is formed by joining the initial point of the second vector **v** with the terminal point of the first vector **u**, as shown in Figure 8.20. This technique is called the **parallelogram law** for vector addition because the vector $\mathbf{u} + \mathbf{v}$, often called the **resultant** of vector addition, is the diagonal of a parallelogram having **u** and **v** as its adjacent sides.

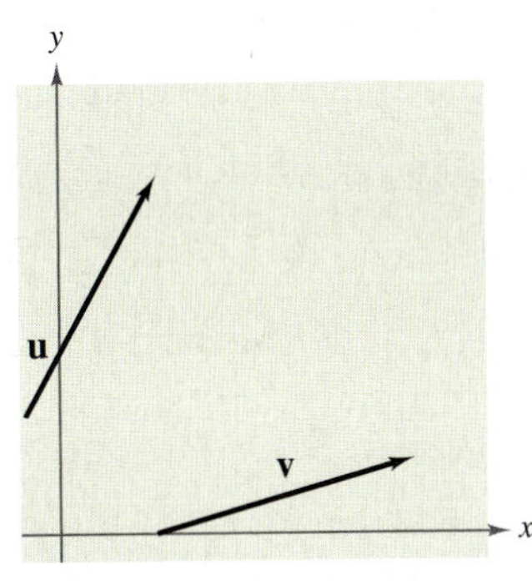

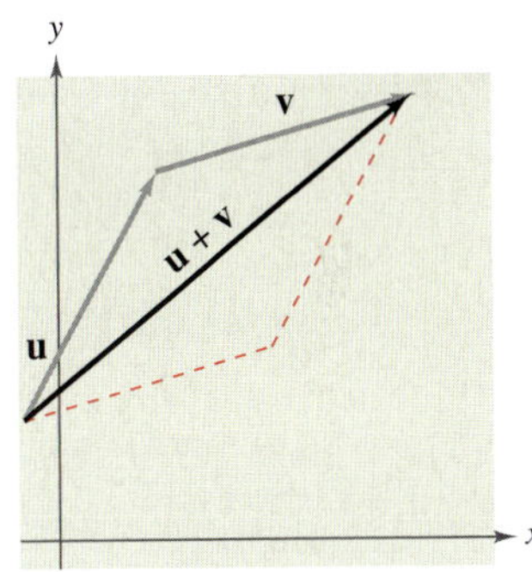

FIGURE 8.20

Definitions of Vector Addition and Scalar Multiplication

Let $\mathbf{u} = \langle u_1, u_2 \rangle$ and $\mathbf{v} = \langle v_1, v_2 \rangle$ be vectors and let k be a scalar (a real number). Then the *sum* of **u** and **v** is the vector

$$\mathbf{u} + \mathbf{v} = \langle u_1 + v_1, u_2 + v_2 \rangle \qquad \text{Sum}$$

and the *scalar multiple* of k times **u** is the vector

$$k\mathbf{u} = k\langle u_1, u_2 \rangle = \langle ku_1, ku_2 \rangle. \qquad \text{Scalar multiple}$$

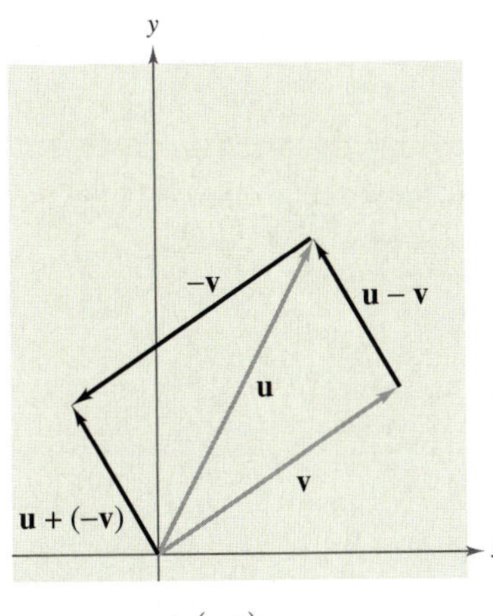

$\mathbf{u} - \mathbf{v} = \mathbf{u} + (-\mathbf{v})$

FIGURE 8.21

The *negative* of $\mathbf{v} = \langle v_1, v_2 \rangle$ is

$$-\mathbf{v} = (-1)\mathbf{v}$$

$$= \langle -v_1, -v_2 \rangle \qquad \text{Negative}$$

and the *difference* of **u** and **v** is

$$\mathbf{u} - \mathbf{v} = \mathbf{u} + (-\mathbf{v}) \qquad \text{Add } (-\mathbf{v}). \text{ See Figure 8.21.}$$

$$= \langle u_1 - v_1, u_2 - v_2 \rangle. \qquad \text{Difference}$$

To represent $\mathbf{u} - \mathbf{v}$ geometrically, you can use directed line segments with the *same* initial point. The difference $\mathbf{u} - \mathbf{v}$ is the vector from the terminal point of **v** to the terminal point of **u**, which is equal to $\mathbf{u} + (-\mathbf{v})$, as shown in Figure 8.21.

The component definitions of vector addition and scalar multiplication are illustrated in Example 3. In this example, notice that each of the vector operations can be interpreted geometrically.

Example 3 ▶ Vector Operations 

Let $\mathbf{v} = \langle -2, 5 \rangle$ and $\mathbf{w} = \langle 3, 4 \rangle$, and find each of the following vectors.

a. $2\mathbf{v}$ **b.** $\mathbf{w} - \mathbf{v}$ **c.** $\mathbf{v} + 2\mathbf{w}$

Solution

a. Because $\mathbf{v} = \langle -2, 5 \rangle$, you have

$$\begin{aligned} 2\mathbf{v} &= 2\langle -2, 5 \rangle \\ &= \langle 2(-2), 2(5) \rangle \\ &= \langle -4, 10 \rangle. \end{aligned}$$

A sketch of $2\mathbf{v}$ is shown in Figure 8.22.

b. The difference of $\mathbf{w}$ and $\mathbf{v}$ is

$$\begin{aligned} \mathbf{w} - \mathbf{v} &= \langle 3 - (-2), 4 - 5 \rangle \\ &= \langle 5, -1 \rangle. \end{aligned}$$

A sketch of $\mathbf{w} - \mathbf{v}$ is shown in Figure 8.23.

c. The sum of $\mathbf{v}$ and $2\mathbf{w}$ is

$$\begin{aligned} \mathbf{v} + 2\mathbf{w} &= \langle -2, 5 \rangle + 2\langle 3, 4 \rangle \\ &= \langle -2, 5 \rangle + \langle 2(3), 2(4) \rangle \\ &= \langle -2, 5 \rangle + \langle 6, 8 \rangle \\ &= \langle -2 + 6, 5 + 8 \rangle \\ &= \langle 4, 13 \rangle. \end{aligned}$$

A sketch of $\mathbf{v} + 2\mathbf{w}$ is shown in Figure 8.24.

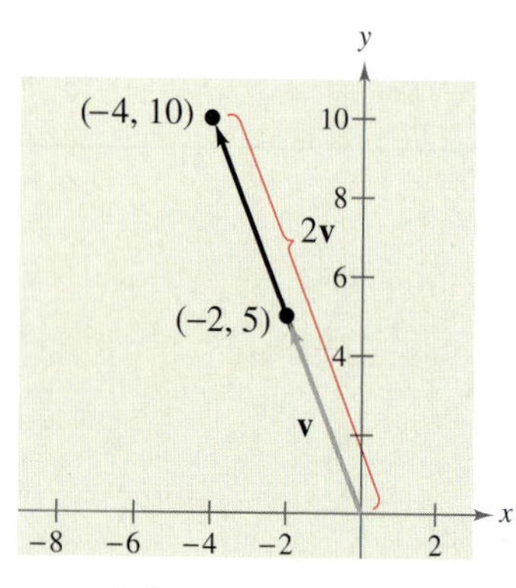

FIGURE 8.22

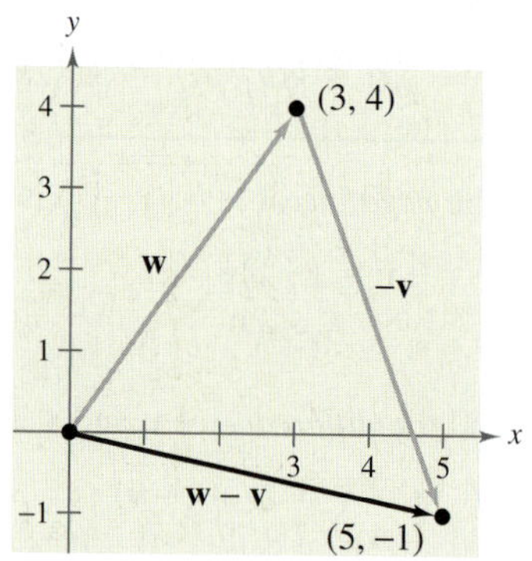

FIGURE 8.23

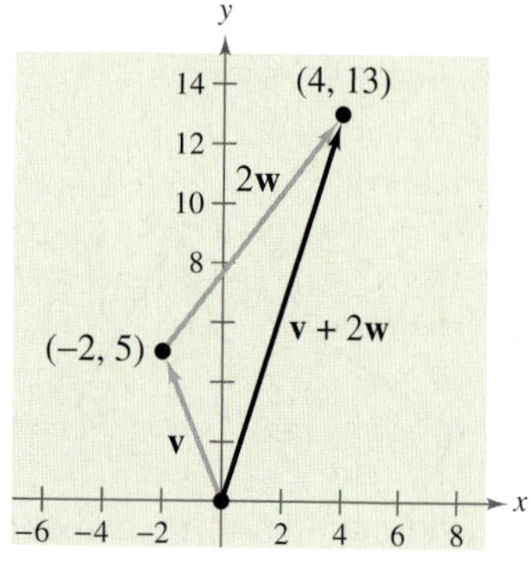

FIGURE 8.24

Vector addition and scalar multiplication share many of the properties of ordinary arithmetic.

Properties of Vector Addition and Scalar Multiplication

Let **u**, **v**, and **w** be vectors and let c and d be scalars. Then the following properties are true.

1. $\mathbf{u} + \mathbf{v} = \mathbf{v} + \mathbf{u}$

2. $(\mathbf{u} + \mathbf{v}) + \mathbf{w} = \mathbf{u} + (\mathbf{v} + \mathbf{w})$

3. $\mathbf{u} + \mathbf{0} = \mathbf{u}$

4. $\mathbf{u} + (-\mathbf{u}) = \mathbf{0}$

5. $c(d\mathbf{u}) = (cd)\mathbf{u}$

6. $(c + d)\mathbf{u} = c\mathbf{u} + d\mathbf{u}$

7. $c(\mathbf{u} + \mathbf{v}) = c\mathbf{u} + c\mathbf{v}$

8. $1(\mathbf{u}) = \mathbf{u}, 0(\mathbf{u}) = \mathbf{0}$

9. $\|c\mathbf{v}\| = |c|\,\|\mathbf{v}\|$

Property 9 can be stated as follows: the magnitude of the vector $c\mathbf{v}$ is the absolute value of c times the magnitude of **v**.

The Granger Collection

Historical Note

William Rowan Hamilton (1805–1865), an Irish mathematician, did some of the earliest work with vectors. Hamilton spent many years developing a system of vector-like quantities called quaternions. Although Hamilton was convinced of the benefits of quaternions, the operations he defined did not produce good models for physical phenomena. It wasn't until the latter half of the nineteenth century that the Scottish physicist James Maxwell (1831–1879) restructured Hamilton's quaternions in a form useful for representing physical quantities such as force, velocity, and acceleration.

Unit Vectors 

In many applications of vectors, it is useful to find a unit vector that has the same direction as a given nonzero vector **v**. To do this, you can divide **v** by its magnitude to obtain

$$\mathbf{u} = \text{unit vector} = \frac{\mathbf{v}}{\|\mathbf{v}\|} = \left(\frac{1}{\|\mathbf{v}\|}\right)\mathbf{v}.$$

Unit vector in direction of **v**

Note that **u** is a scalar multiple of **v**. The vector **u** has magnitude of 1 and the same direction as **v**. The vector **u** is called a **unit vector in the direction of v**.

Example 4 ▶ Finding a Unit Vector

Find a unit vector in the direction of $\mathbf{v} = \langle -2, 5 \rangle$ and verify that the result has a magnitude of 1.

Solution

The unit vector in the direction of **v** is

$$\frac{\mathbf{v}}{\|\mathbf{v}\|} = \frac{\langle -2, 5 \rangle}{\sqrt{(-2)^2 + (5)^2}}$$

$$= \frac{1}{\sqrt{29}}\langle -2, 5 \rangle$$

$$= \left\langle \frac{-2}{\sqrt{29}}, \frac{5}{\sqrt{29}} \right\rangle.$$

This vector has a magnitude of 1 because

$$\sqrt{\left(\frac{-2}{\sqrt{29}}\right)^2 + \left(\frac{5}{\sqrt{29}}\right)^2} = \sqrt{\frac{4}{29} + \frac{25}{29}} = \sqrt{\frac{29}{29}} = 1.$$

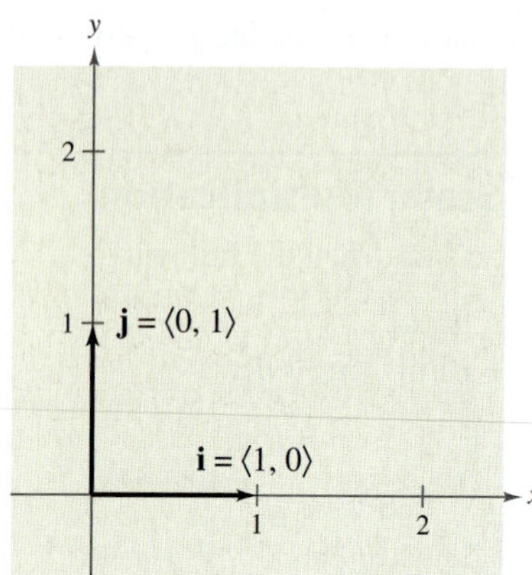

FIGURE 8.25

The unit vectors $\langle 1, 0\rangle$ and $\langle 0, 1\rangle$ are called the **standard unit vectors** and are denoted by

$$\mathbf{i} = \langle 1, 0\rangle \qquad \text{and} \qquad \mathbf{j} = \langle 0, 1\rangle$$

as shown in Figure 8.25. (Note that the lowercase letter **i** is written in boldface to distinguish it from the imaginary number $i = \sqrt{-1}$.) These vectors can be used to represent any vector $\mathbf{v} = \langle v_1, v_2\rangle$, as follows.

$$\begin{aligned}\mathbf{v} &= \langle v_1, v_2\rangle \\ &= v_1\langle 1, 0\rangle + v_2\langle 0, 1\rangle \\ &= v_1\mathbf{i} + v_2\mathbf{j}\end{aligned}$$

The scalars v_1 and v_2 are called the **horizontal** and **vertical components of v,** respectively. The vector sum

$$v_1\mathbf{i} + v_2\mathbf{j}$$

is called a **linear combination** of the vectors **i** and **j**. Any vector in the plane can be expressed as a linear combination of the standard unit vectors **i** and **j**.

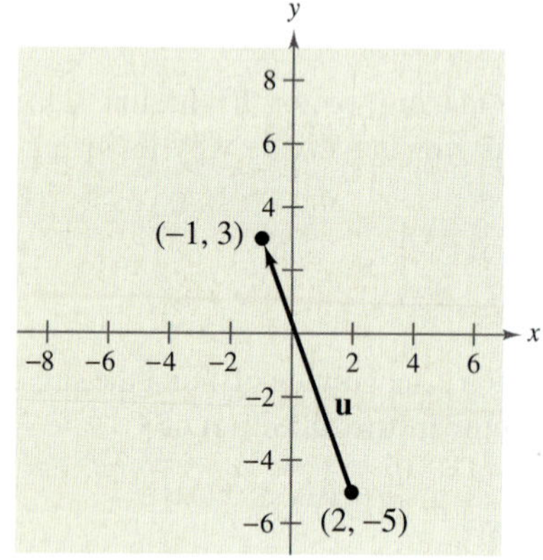

FIGURE 8.26

Example 5 ▶ Writing a Linear Combination of Unit Vectors

Let **u** be the vector with initial point $(2, -5)$ and terminal point $(-1, 3)$. Write **u** as a linear combination of the standard unit vectors **i** and **j**.

Solution

Begin by writing the component form of the vector **u**.

$$\begin{aligned}\mathbf{u} &= \langle -1 - 2, 3 + 5\rangle \\ &= \langle -3, 8\rangle \\ &= -3\mathbf{i} + 8\mathbf{j}\end{aligned}$$

This result is shown graphically in Figure 8.26.

Example 6 ▶ Vector Operations 

Let $\mathbf{u} = -3\mathbf{i} + 8\mathbf{j}$ and let $\mathbf{v} = 2\mathbf{i} - \mathbf{j}$. Find $2\mathbf{u} - 3\mathbf{v}$.

Solution

You could solve this problem by converting **u** and **v** to component form. This, however, is not necessary. It is just as easy to perform the operations in unit vector form.

$$\begin{aligned}2\mathbf{u} - 3\mathbf{v} &= 2(-3\mathbf{i} + 8\mathbf{j}) - 3(2\mathbf{i} - \mathbf{j}) \\ &= -6\mathbf{i} + 16\mathbf{j} - 6\mathbf{i} + 3\mathbf{j} \\ &= -12\mathbf{i} + 19\mathbf{j}\end{aligned}$$

Direction Angles 

If $\mathbf{u}$ is a *unit vector* such that θ is the angle (measured counterclockwise) from the positive x-axis to $\mathbf{u}$, the terminal point of $\mathbf{u}$ lies on the unit circle and you have

$$\mathbf{u} = \langle x, y \rangle = \langle \cos\theta, \sin\theta \rangle = (\cos\theta)\mathbf{i} + (\sin\theta)\mathbf{j}$$

as shown in Figure 8.27. The angle θ is the **direction angle** of the vector $\mathbf{u}$.

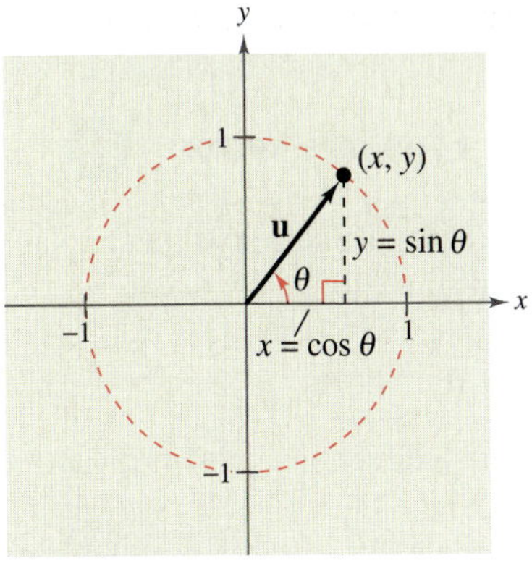

FIGURE 8.27 $\|\mathbf{u}\| = 1$

Suppose that $\mathbf{u}$ is a unit vector with direction angle θ. If $\mathbf{v}$ is any vector that makes an angle θ with the positive x-axis, it has the same direction as $\mathbf{u}$ and you can write

$$\begin{aligned}\mathbf{v} &= \|\mathbf{v}\| \langle \cos\theta, \sin\theta \rangle \\ &= \|\mathbf{v}\| (\cos\theta)\mathbf{i} + \|\mathbf{v}\| (\sin\theta)\mathbf{j}.\end{aligned}$$

Because $\mathbf{v} = a\mathbf{i} + b\mathbf{j} = \|\mathbf{v}\| (\cos\theta)\,\mathbf{i} + \|\mathbf{v}\| (\sin\theta)\,\mathbf{j}$, it follows that the direction angle θ for $\mathbf{v}$ is determined from

$$\tan\theta = \frac{\sin\theta}{\cos\theta} \qquad \text{Quotient identity}$$

$$= \frac{\|\mathbf{v}\|\sin\theta}{\|\mathbf{v}\|\cos\theta} \qquad \text{Multiply numerator and denominator by } \|\mathbf{v}\|.$$

$$= \frac{b}{a}. \qquad \text{Simplify.}$$

Example 7 ▶ Finding Direction Angles of Vectors 

Find the direction angle of each vector.

a. $\mathbf{u} = 3\mathbf{i} + 3\mathbf{j}$

b. $\mathbf{v} = 3\mathbf{i} - 4\mathbf{j}$

Solution

a. The direction angle is

$$\tan\theta = \frac{b}{a} = \frac{3}{3} = 1.$$

So, $\theta = 45°$, as shown in Figure 8.28.

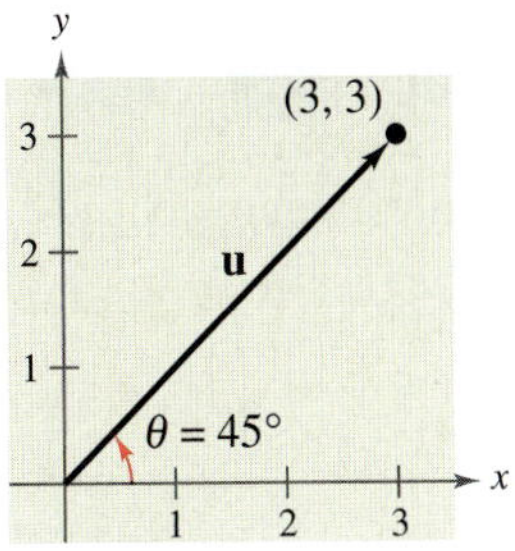

FIGURE 8.28

b. The direction angle is

$$\tan\theta = \frac{b}{a} = \frac{-4}{3}.$$

Moreover, because $\mathbf{v} = 3\mathbf{i} - 4\mathbf{j}$ lies in Quadrant IV, θ lies in Quadrant IV and its reference angle is

$$\theta = \left|\arctan\left(-\frac{4}{3}\right)\right| \approx |-53.13°| = 53.13°.$$

So, it follows that $\theta \approx 360° - 53.13° = 306.87°$, as shown in Figure 8.29.

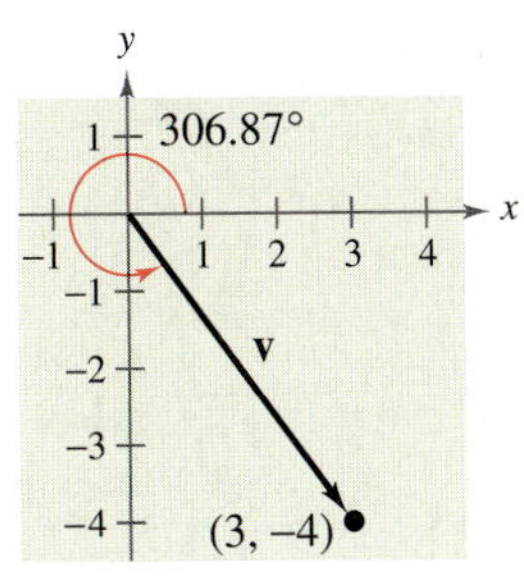

FIGURE 8.29

Applications of Vectors

Example 8 ▶ Finding the Component Form of a Vector 

Find the component form of the vector that represents the velocity of an airplane descending at a speed of 100 miles per hour at an angle 30° below the horizontal, as shown in Figure 8.30.

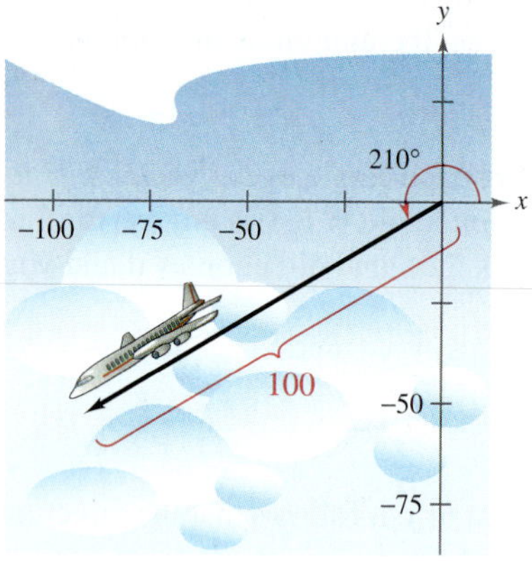

FIGURE 8.30

Solution

The velocity vector **v** has a magnitude of 100 and a direction angle of $\theta = 210°$.

$$\begin{aligned}\mathbf{v} &= \|\mathbf{v}\|(\cos\theta)\mathbf{i} + \|\mathbf{v}\|(\sin\theta)\mathbf{j}\\ &= 100(\cos 210°)\mathbf{i} + 100(\sin 210°)\mathbf{j}\\ &= 100\left(\frac{-\sqrt{3}}{2}\right)\mathbf{i} + 100\left(\frac{-1}{2}\right)\mathbf{j}\\ &= -50\sqrt{3}\,\mathbf{i} - 50\mathbf{j}\\ &= \langle -50\sqrt{3}, -50\rangle\end{aligned}$$

You can check that **v** has a magnitude of 100, as follows.

$$\begin{aligned}\|\mathbf{v}\| &= \sqrt{\left(-50\sqrt{3}\right)^2 + (50)^2}\\ &= \sqrt{7500 + 2500}\\ &= \sqrt{10{,}000}\\ &= 100\end{aligned}$$

Example 9 ▶ Using Vectors to Determine Weight  

A force of 600 pounds is required to pull a boat and trailer up a ramp inclined at 15° from the horizontal. Find the combined weight of the boat and trailer.

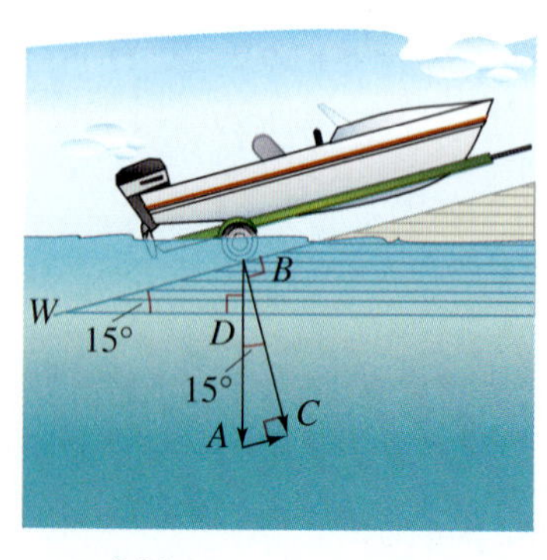

FIGURE 8.31

Solution

Based on Figure 8.31, you can make the following observations.

$\|\overrightarrow{BA}\|$ = force of gravity = combined weight of boat and trailer

$\|\overrightarrow{BC}\|$ = force against ramp

$\|\overrightarrow{AC}\|$ = force required to move boat up ramp = 600 pounds

By construction, triangles *BWD* and *ABC* are similar. So, angle *ABC* is 15°, and so in triangle *ABC* you have

$$\sin 15° = \frac{\|\overrightarrow{AC}\|}{\|\overrightarrow{BA}\|} = \frac{600}{\|\overrightarrow{BA}\|}$$

$$\|BA\| = \frac{600}{\sin 15°} \approx 2318.$$

Consequently, the combined weight is approximately 2318 pounds. (In Figure 8.31, note that $\overrightarrow{AC}$ is parallel to the ramp.)

Example 10 ▶ Using Vectors to Find Speed and Direction

An airplane is traveling at a speed of 500 miles per hour on a bearing of 330° at a fixed altitude with a negligible wind velocity as shown in Figure 8.32(a). When the airplane reaches a certain point, it encounters a wind with a velocity of 70 miles per hour in the direction N 45° E, as shown in Figure 8.32(b). What are the resultant speed and direction of the airplane?

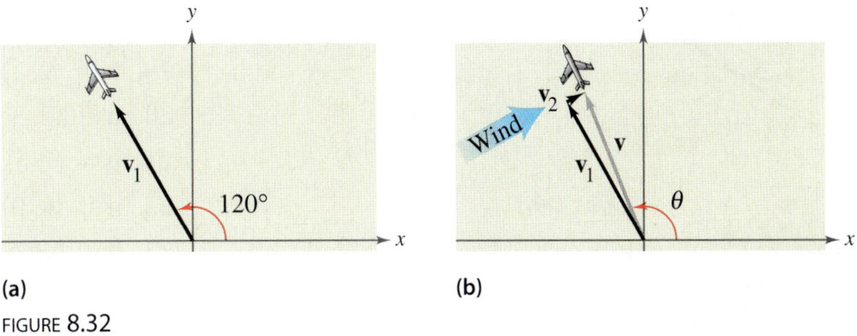

FIGURE 8.32

Solution

Using Figure 8.32, the velocity of the airplane (alone) is

$$\begin{aligned}\mathbf{v}_1 &= 500\langle\cos 120°, \sin 120°\rangle\\ &= \langle -250, 250\sqrt{3}\rangle\end{aligned}$$

and the velocity of the wind is

$$\begin{aligned}\mathbf{v}_2 &= 70\langle\cos 45°, \sin 45°\rangle\\ &= \langle 35\sqrt{2}, 35\sqrt{2}\rangle.\end{aligned}$$

So, the velocity of the airplane (in the wind) is

$$\begin{aligned}\mathbf{v} &= \mathbf{v}_1 + \mathbf{v}_2\\ &= \langle -250 + 35\sqrt{2}, 250\sqrt{3} + 35\sqrt{2}\rangle\\ &\approx \langle -200.5, 482.5\rangle\end{aligned}$$

and the resultant speed of the airplane is

$$\begin{aligned}\|\mathbf{v}\| &= \sqrt{(-200.5)^2 + (482.5)^2}\\ &\approx 522.5 \text{ miles per hour.}\end{aligned}$$

Finally, if θ is the direction angle of the flight path, you have

$$\begin{aligned}\tan\theta &= \frac{482.5}{-200.5}\\ &\approx -2.4065\end{aligned}$$

which implies that

$$\theta \approx 180° + \arctan(-2.4065) \approx 180° - 67.4° = 112.6°.$$

So, the true direction of the airplane is 337.4°.

8.3 Exercises

In Exercises 1–12, find the component form and the magnitude of the vector v.

1.

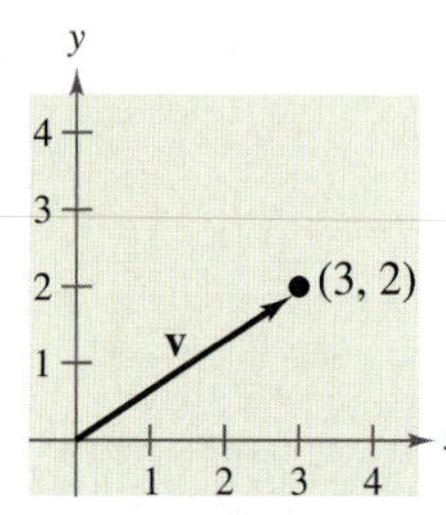

2.

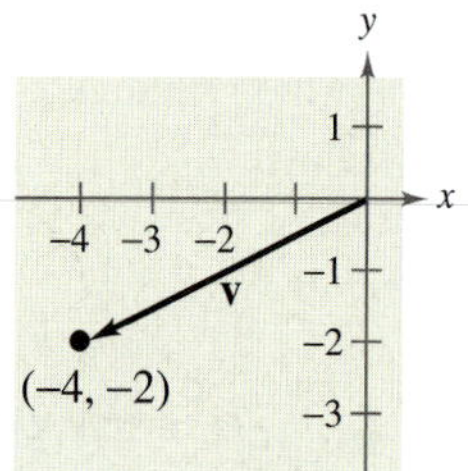

3.

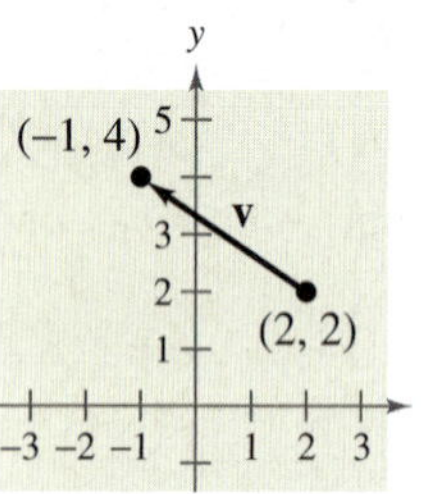

4.

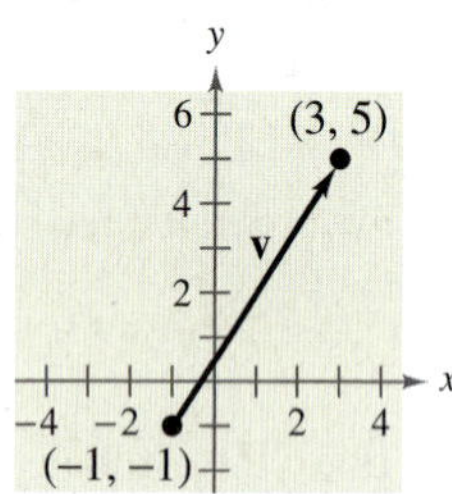

5.

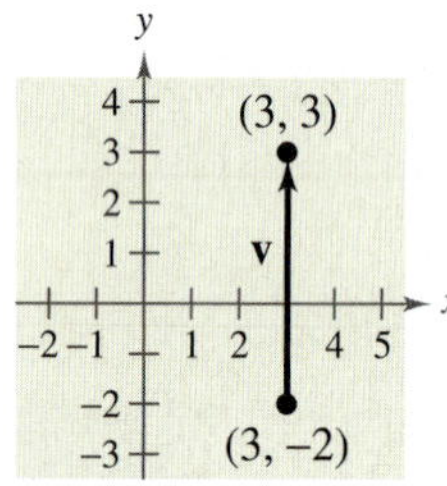

6.

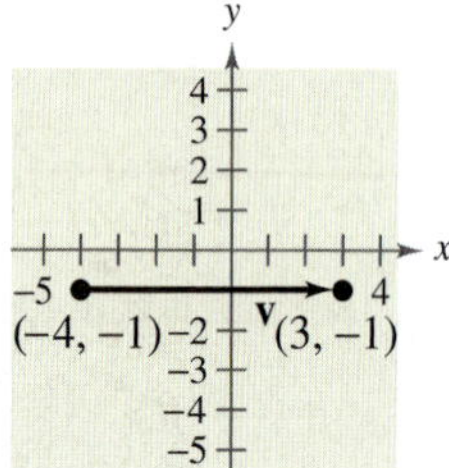

	Initial Point	*Terminal Point*
7.	$(-1, 5)$	$(15, 12)$
8.	$(1, 11)$	$(9, 3)$
9.	$(-3, -5)$	$(5, 1)$
10.	$(-3, 11)$	$(9, 40)$
11.	$(1, 3)$	$(-8, -9)$
12.	$(-2, 7)$	$(5, -17)$

In Exercises 13–18, use the figure to sketch a graph of the specified vector. To print an enlarged copy of the graph, go to the website, *www.mathgraphs.com*.

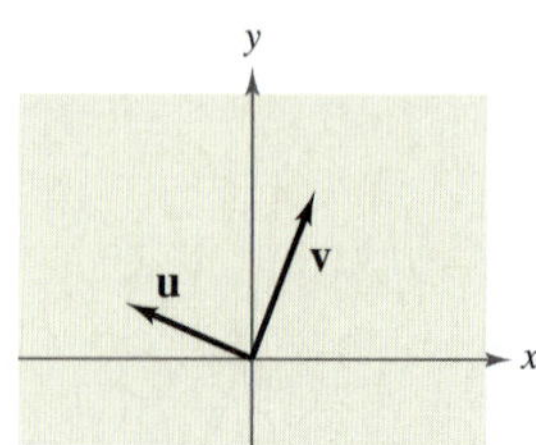

13. $-\mathbf{v}$ **14.** $5\mathbf{v}$

15. $\mathbf{u} + \mathbf{v}$ **16.** $\mathbf{u} - \mathbf{v}$

17. $\mathbf{u} + 2\mathbf{v}$ **18.** $\mathbf{v} - \frac{1}{2}\mathbf{u}$

In Exercises 19–26, find (a) u + v, (b) u − v, and (c) 2u − 3v. Then sketch the resultant vector.

19. $\mathbf{u} = \langle 2, 1 \rangle, \quad \mathbf{v} = \langle 1, 3 \rangle$

20. $\mathbf{u} = \langle 2, 3 \rangle, \quad \mathbf{v} = \langle 4, 0 \rangle$

21. $\mathbf{u} = \langle -5, 3 \rangle, \quad \mathbf{v} = \langle 0, 0 \rangle$

22. $\mathbf{u} = \langle 0, 0 \rangle, \quad \mathbf{v} = \langle 2, 1 \rangle$

23. $\mathbf{u} = \mathbf{i} + \mathbf{j}, \quad \mathbf{v} = 2\mathbf{i} - 3\mathbf{j}$

24. $\mathbf{u} = -2\mathbf{i} + \mathbf{j}, \quad \mathbf{v} = -\mathbf{i} + 2\mathbf{j}$

25. $\mathbf{u} = 2\mathbf{i}, \quad \mathbf{v} = \mathbf{j}$

26. $\mathbf{u} = 3\mathbf{j}, \quad \mathbf{v} = 2\mathbf{i}$

In Exercises 27–36, find a unit vector in the direction of the given vector.

27. $\mathbf{u} = \langle 3, 0 \rangle$ **28.** $\mathbf{u} = \langle 0, -2 \rangle$

29. $\mathbf{v} = \langle -2, 2 \rangle$ **30.** $\mathbf{v} = \langle 5, -12 \rangle$

31. $\mathbf{v} = 6\mathbf{i} - 2\mathbf{j}$ **32.** $\mathbf{v} = \mathbf{i} + \mathbf{j}$

33. $\mathbf{w} = 4\mathbf{j}$ **34.** $\mathbf{w} = -6\mathbf{i}$

35. $\mathbf{w} = \mathbf{i} - 2\mathbf{j}$ **36.** $\mathbf{w} = 7\mathbf{j} - 3\mathbf{i}$

In Exercises 37–40, find the vector v with the given magnitude and the same direction as u.

	Magnitude	*Direction*
37.	$\\|\mathbf{v}\\| = 5$	$\mathbf{u} = \langle 3, 3 \rangle$
38.	$\\|\mathbf{v}\\| = 6$	$\mathbf{u} = \langle -3, 3 \rangle$
39.	$\\|\mathbf{v}\\| = 9$	$\mathbf{u} = \langle 2, 5 \rangle$
40.	$\\|\mathbf{v}\\| = 10$	$\mathbf{u} = \langle -10, 0 \rangle$

In Exercises 41–46, find the component form of v and sketch the specified vector operations geometrically, where u = 2i − j and w = i + 2j.

41. $\mathbf{v} = \frac{3}{2}\mathbf{u}$ **42.** $\mathbf{v} = \frac{3}{4}\mathbf{w}$

43. $\mathbf{v} = \mathbf{u} + 2\mathbf{w}$ **44.** $\mathbf{v} = -\mathbf{u} + \mathbf{w}$

45. $\mathbf{v} = \frac{1}{2}(3\mathbf{u} + \mathbf{w})$ **46.** $\mathbf{v} = \mathbf{u} - 2\mathbf{w}$

In Exercises 47–50, find the magnitude and direction angle of the vector v.

47. $\mathbf{v} = 3(\cos 60^\circ\mathbf{i} + \sin 60^\circ\mathbf{j})$

48. $\mathbf{v} = 8(\cos 135°\mathbf{i} + \sin 135°\mathbf{j})$

49. $\mathbf{v} = 6\mathbf{i} - 6\mathbf{j}$ **50.** $\mathbf{v} = -5\mathbf{i} + 4\mathbf{j}$

In Exercises 51–58, find the component form of v given its magnitude and the angle it makes with the positive *x*-axis. Sketch v.

	Magnitude	*Angle*
51.	$\|\mathbf{v}\| = 3$	$\theta = 0°$
52.	$\|\mathbf{v}\| = 1$	$\theta = 45°$
53.	$\|\mathbf{v}\| = \frac{7}{2}$	$\theta = 150°$
54.	$\|\mathbf{v}\| = \frac{5}{2}$	$\theta = 45°$
55.	$\|\mathbf{v}\| = 3\sqrt{2}$	$\theta = 150°$
56.	$\|\mathbf{v}\| = 4\sqrt{3}$	$\theta = 90°$
57.	$\|\mathbf{v}\| = 2$	$\mathbf{v}$ in the direction $\mathbf{i} + 3\mathbf{j}$
58.	$\|\mathbf{v}\| = 3$	$\mathbf{v}$ in the direction $3\mathbf{i} + 4\mathbf{j}$

In Exercises 59–62, find the component form of the sum of u and v with direction angles $\theta_\mathbf{u}$ and $\theta_\mathbf{v}$.

	Magnitude	*Angle*
59.	$\|\mathbf{u}\| = 5$	$\theta_\mathbf{u} = 0°$
	$\|\mathbf{v}\| = 5$	$\theta_\mathbf{v} = 90°$
60.	$\|\mathbf{u}\| = 4$	$\theta_\mathbf{u} = 60°$
	$\|\mathbf{v}\| = 4$	$\theta_\mathbf{v} = 90°$
61.	$\|\mathbf{u}\| = 20$	$\theta_\mathbf{u} = 45°$
	$\|\mathbf{v}\| = 50$	$\theta_\mathbf{v} = 180°$
62.	$\|\mathbf{u}\| = 50$	$\theta_\mathbf{u} = 30°$
	$\|\mathbf{v}\| = 30$	$\theta_\mathbf{v} = 110°$

In Exercises 63–66, use the Law of Cosines to find the angle α between the vectors. (Assume $0° \le \alpha \le 180°$.)

63. $\mathbf{v} = \mathbf{i} + \mathbf{j}, \quad \mathbf{w} = 2\mathbf{i} - 2\mathbf{j}$

64. $\mathbf{v} = 3\mathbf{i} - 2\mathbf{j}, \quad \mathbf{w} = 2\mathbf{i} + 2\mathbf{j}$

65. $\mathbf{v} = \mathbf{i} + \mathbf{j}, \quad \mathbf{w} = 3\mathbf{i} - \mathbf{j}$

66. $\mathbf{v} = \mathbf{i} + 2\mathbf{j}, \quad \mathbf{w} = 2\mathbf{i} - \mathbf{j}$

Resultant Force **In Exercises 67 and 68, find the angle between the forces given the magnitude of their resultant. (*Hint:* Write force 1 as a vector in the direction of the positive *x*-axis and force 2 as a vector at an angle θ with the positive *x*-axis.)**

	Force 1	*Force 2*	*Resultant Force*
67.	45 pounds	60 pounds	90 pounds
68.	3000 pounds	1000 pounds	3750 pounds

69. ***Resultant Force*** Forces with magnitudes of 125 newtons and 300 newtons act on a hook (see figure). The angle between the two forces is 45°. Find the direction and magnitude of the resultant of these forces.

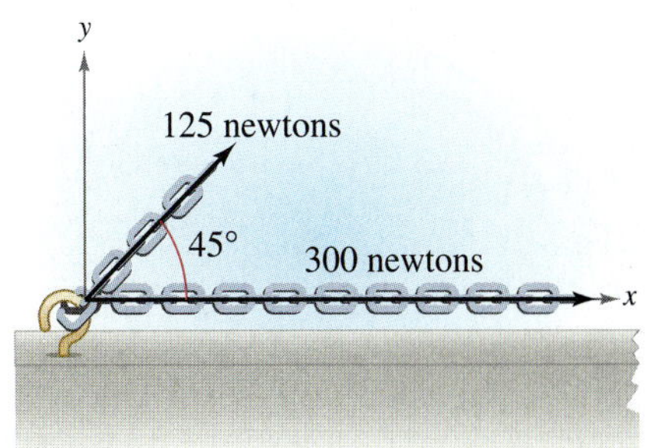

70. ***Resultant Force*** Forces with magnitudes of 2000 newtons and 900 newtons act on a machine part at angles of 30° and −45°, respectively, with the *x*-axis (see figure). Find the direction and magnitude of the resultant of these forces.

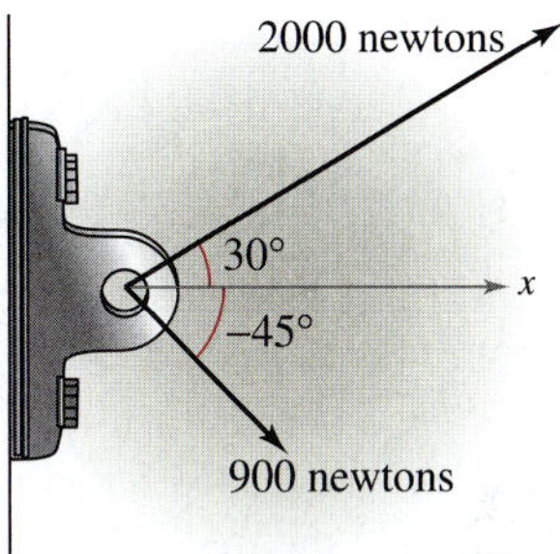

71. ***Resultant Force*** Three forces with magnitudes of 75 pounds, 100 pounds, and 125 pounds act on an object at angles of 30°, 45°, and 120°, respectively, with the positive *x*-axis. Find the direction and magnitude of the resultant of these forces.

72. ***Resultant Force*** Three forces with magnitudes of 70 pounds, 40 pounds, and 60 pounds act on an object at angles of −30°, 45°, and 135°, respectively, with the positive *x*-axis. Find the direction and magnitude of the resultant of these forces.

73. ***Velocity*** A ball is thrown with an initial velocity of 70 feet per second, at an angle of 35° with the horizontal (see figure). Find the vertical and horizontal components of the velocity.

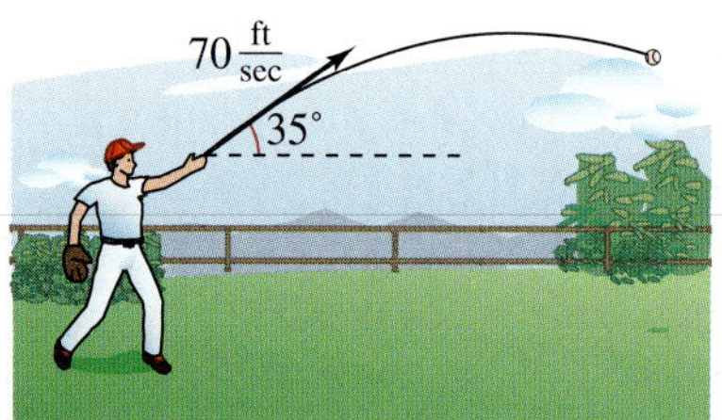

74. ***Velocity*** A gun with a muzzle velocity of 1200 feet per second is fired at an angle of 6° with the horizontal. Find the vertical and horizontal components of the velocity.

Cable Tension **In Exercises 75 and 76, use the figure to determine the tension in each cable supporting the load.**

75. **76.**

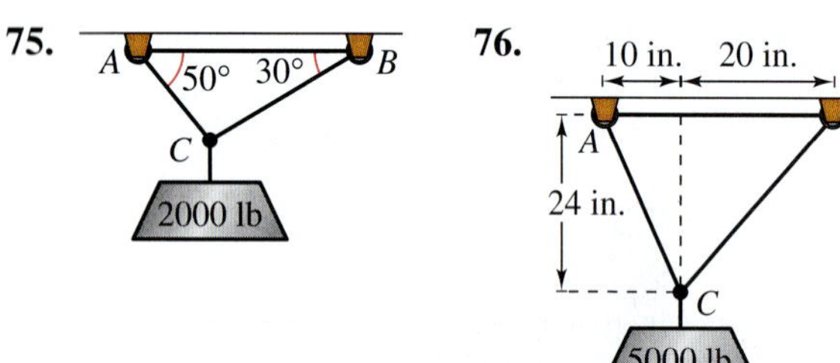

77. ***Tow Line Tension*** A loaded barge is being towed by two tugboats, and the magnitude of the resultant is 6000 pounds directed along the axis of the barge (see figure). Find the tension in the tow lines if they each make an 18° angle with the axis of the barge.

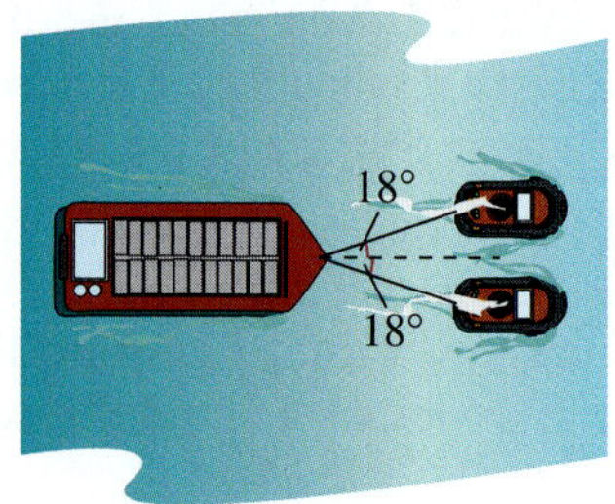

78. ***Rope Tension*** To carry a 100-pound cylindrical weight, two people lift on the ends of short ropes that are tied to an eyelet on the top center of the cylinder. Each rope makes a 20° angle with the vertical. Draw a figure that gives a visual representation of the problem, and find the tension in the ropes.

79. ***Navigation*** An airplane is flying in the direction of 148°, with an airspeed of 875 kilometers per hour. Because of the wind, its groundspeed and direction are 800 kilometers per hour and 140°, respectively (see figure). Find the direction and speed of the wind.

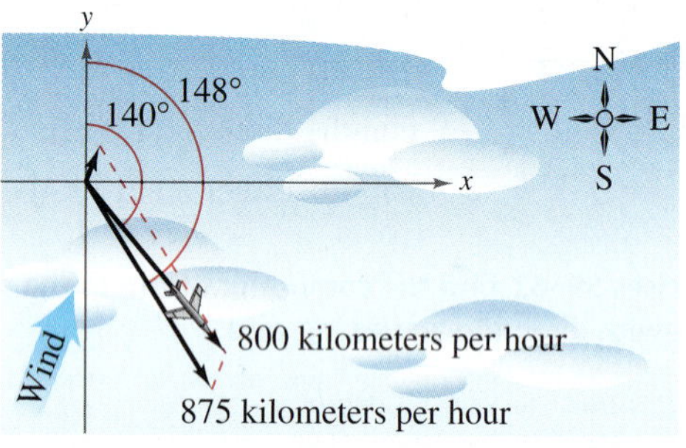

▶ Model It

80. ***Navigation*** A commercial jet is flying from Miami to Seattle. The jet's velocity with respect to the air is 580 miles per hour, and its bearing is 332°. The wind, at the altitude of the plane, is blowing from the southwest with a velocity of 60 miles per hour.

(a) Draw a figure that gives a visual representation of the problem.

(b) Write the velocity of the wind as a vector in component form.

(c) Write the velocity of the jet relative to the air in component form.

(d) What is the speed of the jet with respect to the ground?

(e) What is the true direction of the jet?

81. ***Work*** A heavy implement is pulled 30 feet across a floor, using a force of 100 pounds. The force is 50° above the horizontal (see figure). Find the work done. (Use the formula for work, $W = FD$, where F is the component of the force in the direction of motion and D is the distance.)

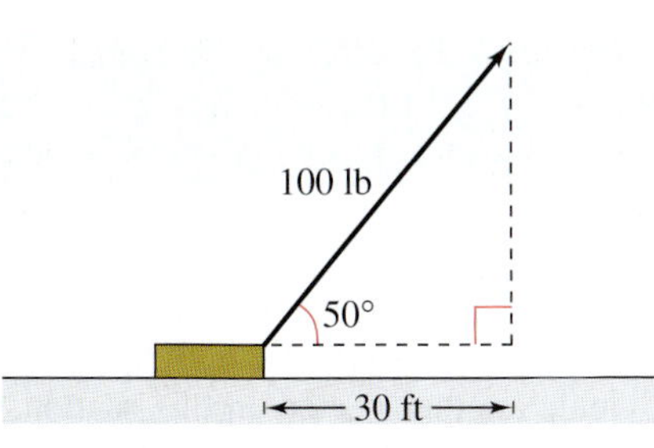

FIGURE FOR 81

82. ***Rope Tension*** A tetherball weighing 1 pound is pulled outward from the pole by a horizontal force **u** until the rope makes a 45° angle with the pole (see figure). Determine the resulting tension in the rope and the magnitude of **u**.

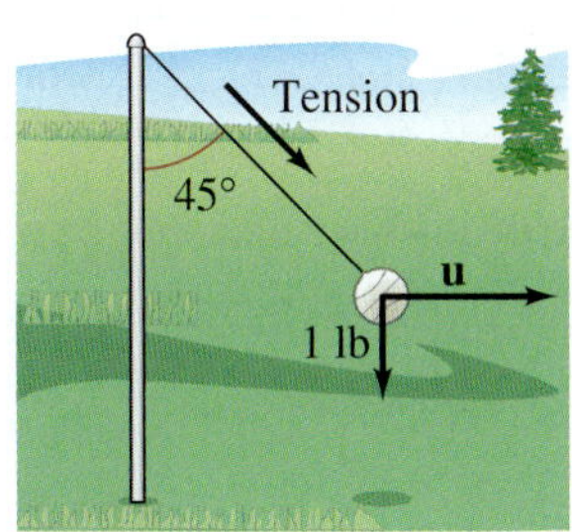

Synthesis

True or False? **In Exercises 83 and 84, decide whether the statement is true or false. Justify your answer.**

83. If **u** and **v** have the same magnitude and direction, then **u** = **v**.

84. If **u** = $a\mathbf{i} + b\mathbf{j}$ is a unit vector, then $a^2 + b^2 = 1$.

85. ***Think About It*** Consider two forces of equal magnitude acting on a point.
   (a) If the magnitude of the resultant is the sum of the magnitudes of the two forces, make a conjecture about the angle between the forces.
   (b) If the resultant of the forces is **0**, make a conjecture about the angle between the forces.
   (c) Can the magnitude of the resultant be greater than the sum of the magnitudes of the two forces? Explain.

86. ***Graphical Reasoning*** Consider two forces

$\mathbf{F}_1 = \langle 10, 0\rangle$ and $\mathbf{F}_2 = 5\langle \cos\theta, \sin\theta\rangle$.

   (a) Find $\|\mathbf{F}_1 + \mathbf{F}_2\|$ as a function of θ.
    (b) Use a graphing utility to graph the function in part (a) for $0 \le \theta < 2\pi$.
    (c) Use the graph in part (b) to determine the range of the function. What is its maximum, and for what value of θ does it occur? What is its minimum, and for what value of θ does it occur?
   (d) Explain why the magnitude of the resultant is never 0.

87. ***Proof*** Prove that $(\cos\theta)\mathbf{i} + (\sin\theta)\mathbf{j}$ is a unit vector for any value of θ.

 88. ***Technology*** Write a program for your graphing utility that graphs two vectors and their difference given the vectors in component form.

In Exercises 89 and 90, use the program in Exercise 88 to find the difference of the vectors shown in the figure.

89.

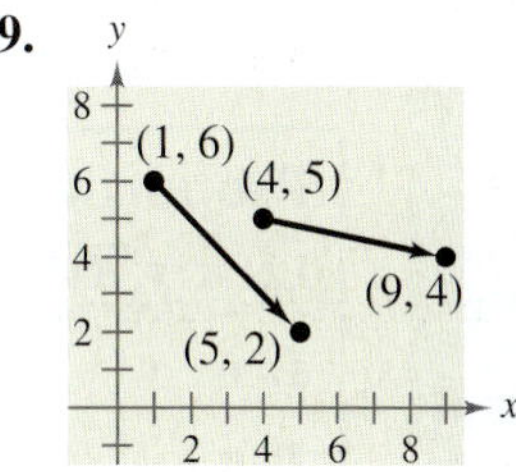

90.

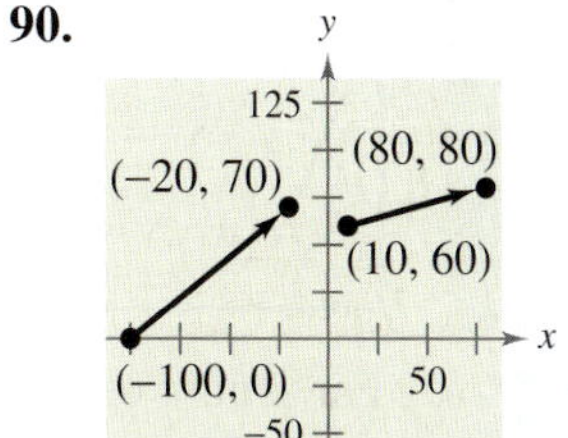

Review

In Exercises 91–94, use the specified trigonometric substitution to write the algebraic expression as a trigonometric function of θ, where $0 < \theta < \pi/2$.

91. $\sqrt{x^2 - 64}$, $x = 8 \sec\theta$

92. $\sqrt{64 - x^2}$, $x = 8 \sin\theta$

93. $\sqrt{x^2 + 36}$, $x = 6 \tan\theta$

94. $\sqrt{(x^2 - 25)^3}$, $x = 5 \sec\theta$

In Exercises 95–98, solve the equation.

95. $\cos x(\cos x + 1) = 0$

96. $\sin x\left(2 \sin x + \sqrt{2}\right) = 0$

97. $3 \sec x \sin x - 2\sqrt{3} \sin x = 0$

98. $\cos x \csc x + \cos x\sqrt{2} = 0$

8.4 Vectors and Dot Products

▶ **What you should learn**

- How to find the dot product of two vectors and use the Properties of the Dot Product
- How to find the angle between two vectors and determine whether two vectors are orthogonal
- How to write a vector as the sum of two vector components
- How to use vectors to find the work done by a force

▶ **Why you should learn it**

You can use the dot product of two vectors to solve real-life problems involving two vector quantities. For instance, in Exercise 48 on page 636, you can use the dot product to find the force necessary to keep a truck from rolling down a hill.

Alan Thornton/Tony Stone Images

The Dot Product of Two Vectors

So far you have studied two vector operations—vector addition and multiplication by a scalar—each of which yields another vector. In this section you will study a third vector operation, the **dot product.** This product yields a scalar, rather than a vector.

Definition of Dot Product

The **dot product** of $\mathbf{u} = \langle u_1, u_2 \rangle$ and $\mathbf{v} = \langle v_1, v_2 \rangle$ is

$$\mathbf{u} \cdot \mathbf{v} = u_1 v_1 + u_2 v_2.$$

Properties of the Dot Product

Let **u**, **v**, and **w** be vectors in the plane or in space and let c be a scalar.

1. $\mathbf{u} \cdot \mathbf{v} = \mathbf{v} \cdot \mathbf{u}$
2. $\mathbf{0} \cdot \mathbf{v} = 0$
3. $\mathbf{u} \cdot (\mathbf{v} + \mathbf{w}) = \mathbf{u} \cdot \mathbf{v} + \mathbf{u} \cdot \mathbf{w}$
4. $\mathbf{v} \cdot \mathbf{v} = \|\mathbf{v}\|^2$
5. $c(\mathbf{u} \cdot \mathbf{v}) = c\mathbf{u} \cdot \mathbf{v} = \mathbf{u} \cdot c\mathbf{v}$

For proofs of the Properties of the Dot Product, see Proofs in Mathematics on page 659.

Example 1 ▶ Finding Dot Products 

Find each dot product.

a. $\langle 4, 5 \rangle \cdot \langle 2, 3 \rangle$ **b.** $\langle 2, -1 \rangle \cdot \langle 1, 2 \rangle$ **c.** $\langle 0, 3 \rangle \cdot \langle 4, -2 \rangle$

Solution

a. $\langle 4, 5 \rangle \cdot \langle 2, 3 \rangle = 4(2) + 5(3)$
$= 8 + 15$
$= 23$

b. $\langle 2, -1 \rangle \cdot \langle 1, 2 \rangle = 2(1) + (-1)(2) = 2 - 2 = 0$

c. $\langle 0, 3 \rangle \cdot \langle 4, -2 \rangle = 0(4) + 3(-2) = 0 - 6 = -6$

In Example 1, be sure you see that the dot product of two vectors is a scalar (a real number), not a vector. Moreover, notice that the dot product can be positive, zero, or negative.

Example 2 ▶ Using Properties of Dot Products 

Let $\mathbf{u} = \langle -1, 3 \rangle$, $\mathbf{v} = \langle 2, -4 \rangle$, and $\mathbf{w} = \langle 1, -2 \rangle$. Find each dot product.

a. $(\mathbf{u} \cdot \mathbf{v})\mathbf{w}$ **b.** $\mathbf{u} \cdot 2\mathbf{v}$

Solution

Begin by finding the dot product of **u** and **v**.

$$\begin{aligned}\mathbf{u} \cdot \mathbf{v} &= \langle -1, 3 \rangle \cdot \langle 2, -4 \rangle \\ &= (-1)(2) + 3(-4) \\ &= -14\end{aligned}$$

a. $$\begin{aligned}(\mathbf{u} \cdot \mathbf{v})\mathbf{w} &= -14\langle 1, -2 \rangle \\ &= \langle -14, 28 \rangle\end{aligned}$$

b. $$\begin{aligned}\mathbf{u} \cdot 2\mathbf{v} &= 2(\mathbf{u} \cdot \mathbf{v}) \\ &= 2(-14) \\ &= -28\end{aligned}$$

Notice that the first product is a vector, whereas the second is a scalar. Can you see why?

Example 3 ▶ Dot Product and Magnitude

The dot product of **u** with itself is 5. What is the magnitude of **u**?

Solution

Because $\|\mathbf{u}\|^2 = \mathbf{u} \cdot \mathbf{u}$ and $\mathbf{u} \cdot \mathbf{u} = 5$, it follows that

$$\begin{aligned}\|\mathbf{u}\| &= \sqrt{\mathbf{u} \cdot \mathbf{u}} \\ &= \sqrt{5}.\end{aligned}$$

The Angle Between Two Vectors

The **angle between two nonzero vectors** is the angle θ, $0 \le \theta \le \pi$, between their respective standard position vectors, as shown in Figure 8.33 on page 630. This angle can be found using the dot product. (Note that the angle between the zero vector and another vector is not defined.)

Angle Between Two Vectors

If θ is the angle between two nonzero vectors **u** and **v**, then

$$\cos \theta = \frac{\mathbf{u} \cdot \mathbf{v}}{\|\mathbf{u}\| \, \|\mathbf{v}\|}.$$

For a proof of the angle between two vectors, see Proofs in Mathematics on page 659.

Example 4 Finding the Angle Between Two Vectors

Find the angle between $\mathbf{u} = \langle 4, 3 \rangle$ and $\mathbf{v} = \langle 3, 5 \rangle$.

Solution

$$\cos \theta = \frac{\mathbf{u} \cdot \mathbf{v}}{\|\mathbf{u}\|\ \|\mathbf{v}\|}$$

$$= \frac{\langle 4, 3 \rangle \cdot \langle 3, 5 \rangle}{\|\langle 4, 3 \rangle\|\ \|\langle 3, 5 \rangle\|}$$

$$= \frac{27}{5\sqrt{34}}$$

This implies that the angle between the two vectors is

$$\theta = \arccos \frac{27}{5\sqrt{34}} \approx 22.2°$$

as shown in Figure 8.33.

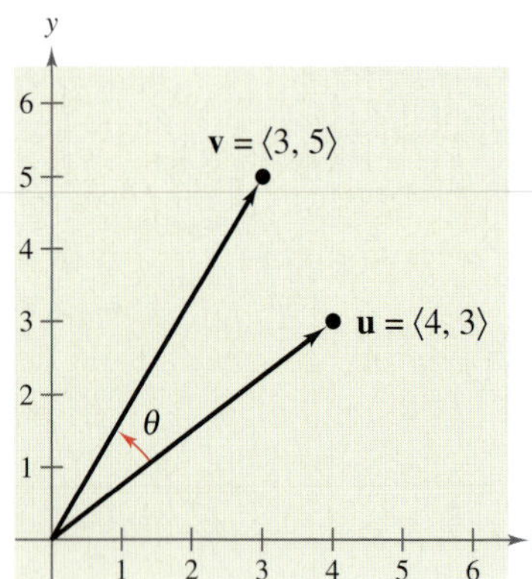

FIGURE 8.33

Rewriting the expression for the angle between two vectors in the form

$$\mathbf{u} \cdot \mathbf{v} = \|\mathbf{u}\|\ \|\mathbf{v}\| \cos \theta \qquad \text{Alternative form of dot product}$$

produces an alternative way to calculate the dot product. From this form, you can see that because $\|\mathbf{u}\|$ and $\|\mathbf{v}\|$ are always positive, $\mathbf{u} \cdot \mathbf{v}$ and $\cos \theta$ will always have the same sign. Figure 8.34 shows the five possible orientations of two vectors.

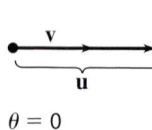

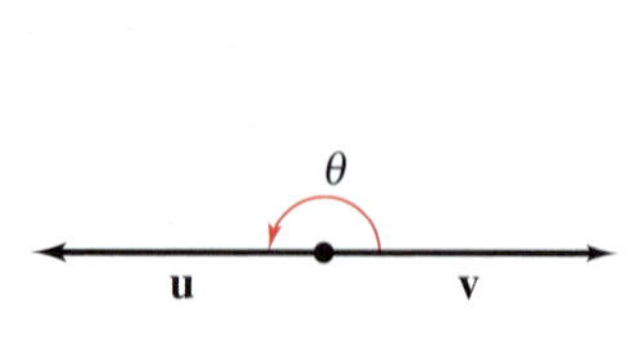

$\theta = \pi$

$\cos \theta = -1$

Opposite Direction

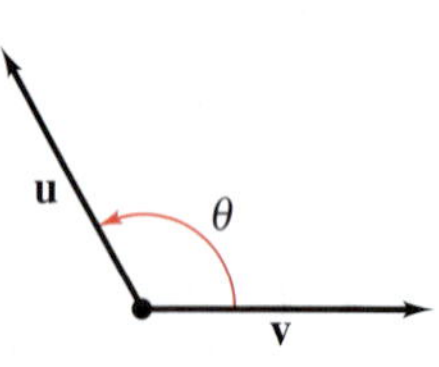

$\frac{\pi}{2} < \theta < \pi$

$-1 < \cos \theta < 0$

Obtuse Angle

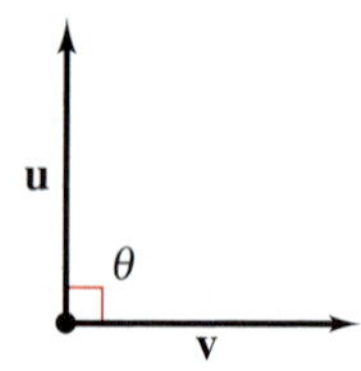

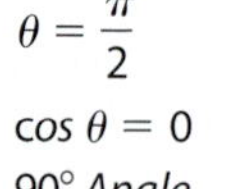

$\theta = \frac{\pi}{2}$

$\cos \theta = 0$

90° Angle

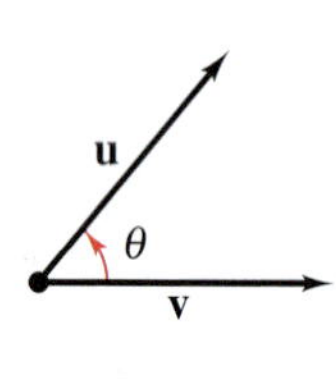

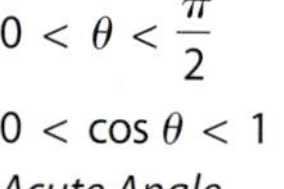

$0 < \theta < \frac{\pi}{2}$

$0 < \cos \theta < 1$

Acute Angle

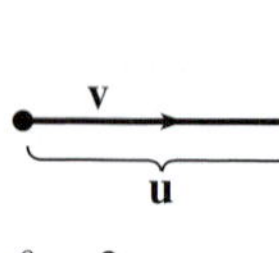

$\theta = 0$

$\cos \theta = 1$

Same Direction

FIGURE 8.34

> **Definition of Orthogonal Vectors**
>
> The vectors $\mathbf{u}$ and $\mathbf{v}$ are **orthogonal** if $\mathbf{u} \cdot \mathbf{v} = 0$.

The terms "orthogonal" and "perpendicular" mean essentially the same thing—meeting at right angles. Even though the angle between the zero vector and another vector is not defined, it is convenient to extend the definition of orthogonality to include the zero vector. In other words, the zero vector is orthogonal to every vector $\mathbf{u}$, because $\mathbf{0} \cdot \mathbf{u} = 0$.

Example 5 ▶ Determining Orthogonal Vectors

Are the vectors $\mathbf{u} = \langle 2, -3 \rangle$ and $\mathbf{v} = \langle 6, 4 \rangle$ orthogonal?

Solution

Begin by finding the dot product of the two vectors.

$$\mathbf{u} \cdot \mathbf{v} = \langle 2, -3 \rangle \cdot \langle 6, 4 \rangle = 2(6) + (-3)(4) = 0$$

Because the dot product is 0, the two vectors are orthogonal (see Figure 8.35).

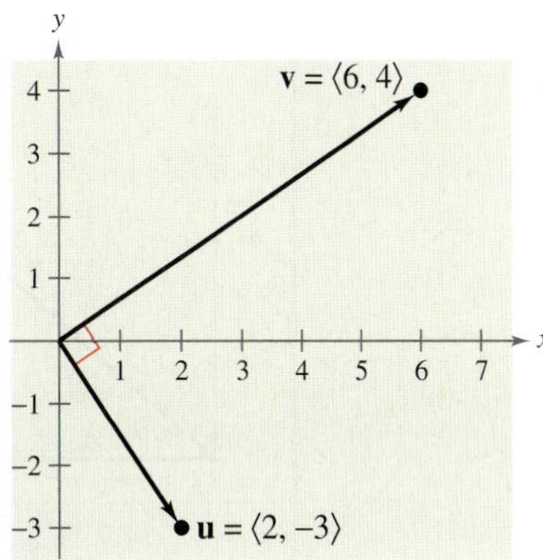

FIGURE 8.35

Finding Vector Components 

You have already seen applications in which two vectors are added to produce a resultant vector. Many applications in physics and engineering pose the reverse problem—decomposing a given vector into the sum of two **vector components.**

Consider a boat on an inclined ramp, as shown in Figure 8.36. The force $\mathbf{F}$ due to gravity pulls the boat *down* the ramp and *against* the ramp. These two orthogonal forces, $\mathbf{w}_1$ and $\mathbf{w}_2$, are vector components of $\mathbf{F}$. That is,

$$\mathbf{F} = \mathbf{w}_1 + \mathbf{w}_2. \qquad \text{Vector components of } \mathbf{F}$$

The negative of component $\mathbf{w}_1$ represents the force needed to keep the boat from rolling down the ramp, whereas $\mathbf{w}_2$ represents the force that the tires must withstand against the ramp. A procedure for finding $\mathbf{w}_1$ and $\mathbf{w}_2$ is shown on the following page.

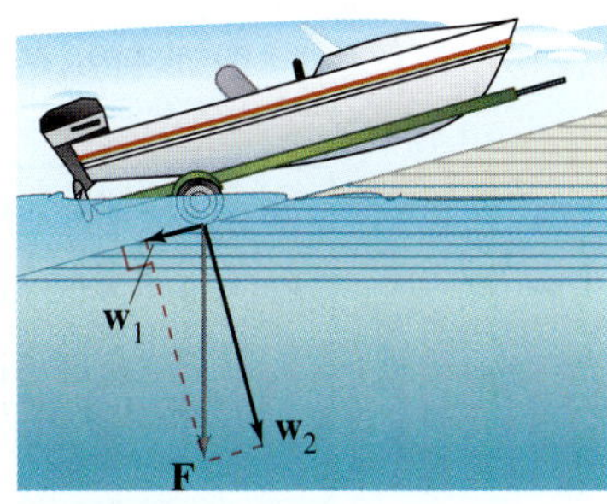

FIGURE 8.36

Definition of Vector Components

Let **u** and **v** be nonzero vectors such that

$$\mathbf{u} = \mathbf{w}_1 + \mathbf{w}_2$$

where $\mathbf{w}_1$ and $\mathbf{w}_2$ are orthogonal and $\mathbf{w}_1$ is parallel to (or a scalar multiple of) **v**, as shown in Figure 8.37. The vectors $\mathbf{w}_1$ and $\mathbf{w}_2$ are called vector components of **u**. The vector $\mathbf{w}_1$ is the **projection** of **u** onto **v** and is denoted by

$$\mathbf{w}_1 = \text{proj}_{\mathbf{v}}\mathbf{u}.$$

The vector $\mathbf{w}_2$ is given by $\mathbf{w}_2 = \mathbf{u} - \mathbf{w}_1$.

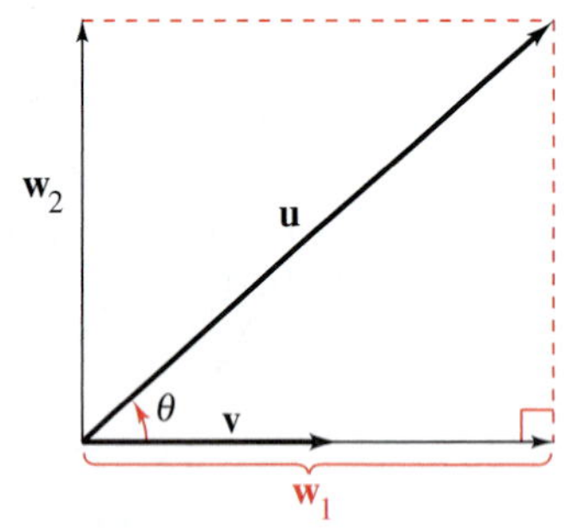

θ is acute.

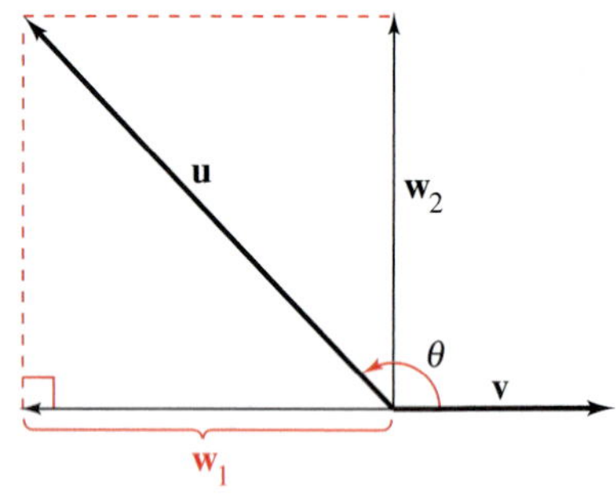

θ is obtuse.

FIGURE 8.37

From the definition of vector components, you can see that it is easy to find the component $\mathbf{w}_2$ once you have found the projection of **u** onto **v**. To find the projection, you can use the dot product, as follows.

$$\mathbf{u} = \mathbf{w}_1 + \mathbf{w}_2 = c\mathbf{v} + \mathbf{w}_2 \qquad \text{$\mathbf{w}_1$ is a scalar multiple of $\mathbf{v}$.}$$

$$\mathbf{u} \cdot \mathbf{v} = (c\mathbf{v} + \mathbf{w}_2) \cdot \mathbf{v} \qquad \text{Take dot product of each side with $\mathbf{v}$.}$$

$$= c\mathbf{v} \cdot \mathbf{v} + \mathbf{w}_2 \cdot \mathbf{v}$$

$$= c\|\mathbf{v}\|^2 + 0 \qquad \text{$\mathbf{w}_2$ and $\mathbf{v}$ are orthogonal.}$$

So,

$$c = \frac{\mathbf{u} \cdot \mathbf{v}}{\|\mathbf{v}\|^2}$$

and

$$\mathbf{w}_1 = \text{proj}_{\mathbf{v}}\,\mathbf{u} = c\mathbf{v} = \frac{\mathbf{u} \cdot \mathbf{v}}{\|\mathbf{v}\|^2}\mathbf{v}.$$

Projection of u onto v

Let **u** and **v** be nonzero vectors. The projection of **u** onto **v** is

$$\text{proj}_{\mathbf{v}}\mathbf{u} = \left(\frac{\mathbf{u} \cdot \mathbf{v}}{\|\mathbf{v}\|^2}\right)\mathbf{v}.$$

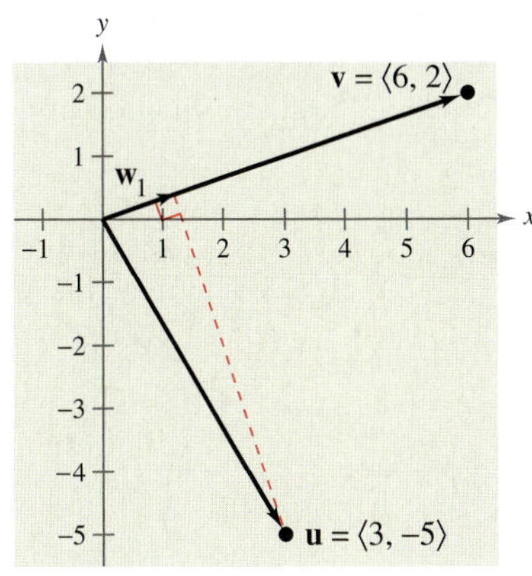

FIGURE 8.38

Example 6 ▶ Decomposing a Vector into Components 

Find the projection of $\mathbf{u} = \langle 3, -5\rangle$ onto $\mathbf{v} = \langle 6, 2\rangle$. Then write $\mathbf{u}$ as the sum of two orthogonal vectors, one of which is $\text{proj}_{\mathbf{v}}\mathbf{u}$.

Solution

The projection of $\mathbf{u}$ onto $\mathbf{v}$ is

$$\mathbf{w}_1 = \text{proj}_{\mathbf{v}}\mathbf{u} = \left(\frac{\mathbf{u}\cdot\mathbf{v}}{\|\mathbf{v}\|^2}\right)\mathbf{v} = \left(\frac{8}{40}\right)\langle 6, 2\rangle = \left\langle \frac{6}{5}, \frac{2}{5}\right\rangle$$

as shown in Figure 8.38. The other component, $\mathbf{w}_2$, is

$$\mathbf{w}_2 = \mathbf{u} - \mathbf{w}_1 = \langle 3, -5\rangle - \left\langle \frac{6}{5}, \frac{2}{5}\right\rangle = \left\langle \frac{9}{5}, -\frac{27}{5}\right\rangle.$$

So,

$$\mathbf{u} = \mathbf{w}_1 + \mathbf{w}_2 = \left\langle \frac{6}{5}, \frac{2}{5}\right\rangle + \left\langle \frac{9}{5}, -\frac{27}{5}\right\rangle = \langle 3, -5\rangle.$$

Example 7 ▶ Finding a Force 

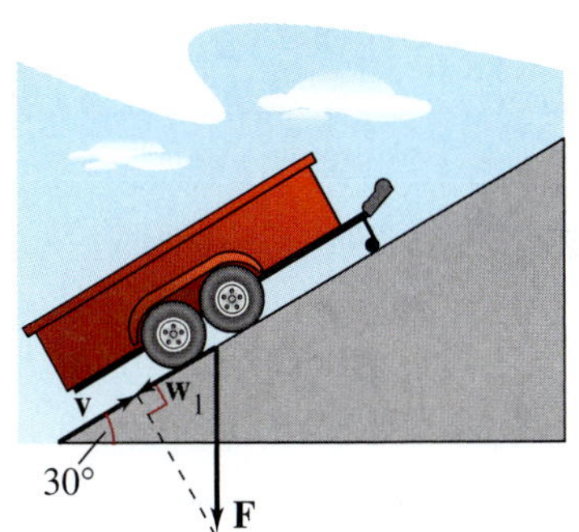

FIGURE 8.39

A 200-pound cart sits on a ramp inclined at 30°, as shown in Figure 8.39. What force is required to keep the cart from rolling down the ramp?

Solution

Because the force due to gravity is vertical and downward, you can represent the gravitational force by the vector

$$\mathbf{F} = -200\mathbf{j}. \qquad \text{Force due to gravity}$$

To find the force required to keep the cart from rolling down the ramp, project $\mathbf{F}$ onto a unit vector $\mathbf{v}$ in the direction of the ramp, as follows.

$$\mathbf{v} = (\cos 30°)\mathbf{i} + (\sin 30°)\mathbf{j} = \frac{\sqrt{3}}{2}\mathbf{i} + \frac{1}{2}\mathbf{j} \qquad \text{Unit vector along ramp}$$

Therefore, the projection of $\mathbf{F}$ onto $\mathbf{v}$ is

$$\begin{aligned}\mathbf{w}_1 &= \text{proj}_{\mathbf{v}}\mathbf{F}\\ &= \left(\frac{\mathbf{F}\cdot\mathbf{v}}{\|\mathbf{v}\|^2}\right)\mathbf{v}\\ &= (\mathbf{F}\cdot\mathbf{v})\mathbf{v}\\ &= (-200)\left(\frac{1}{2}\right)\mathbf{v}\\ &= -100\left(\frac{\sqrt{3}}{2}\mathbf{i} + \frac{1}{2}\mathbf{j}\right).\end{aligned}$$

The magnitude of this force is 100, and so a force of 100 pounds is required to keep the cart from rolling down the ramp.

Work

The work W done by a *constant* force $\mathbf{F}$ acting along the line of motion of an object is

$$W = (\text{magnitude of force})(\text{distance}) = \|\mathbf{F}\|\,\|\overrightarrow{PQ}\|$$

as shown in Figure 8.40. If the constant force $\mathbf{F}$ is not directed along the line of motion, as shown in Figure 8.41, the work W done by the force is

$$W = \|\text{proj}_{\overrightarrow{PQ}}\mathbf{F}\|\,\|\overrightarrow{PQ}\|$$

$$= (\cos\theta)\|\mathbf{F}\|\,\|\overrightarrow{PQ}\| \qquad \|\text{proj}_{\overrightarrow{PQ}}\mathbf{F}\| = (\cos\theta)\|\mathbf{F}\|$$

$$= \mathbf{F}\cdot\overrightarrow{PQ}. \qquad \cos\theta = (\mathbf{F}\cdot\overrightarrow{PQ})/(\|\mathbf{F}\|\,\|\overrightarrow{PQ}\|)$$

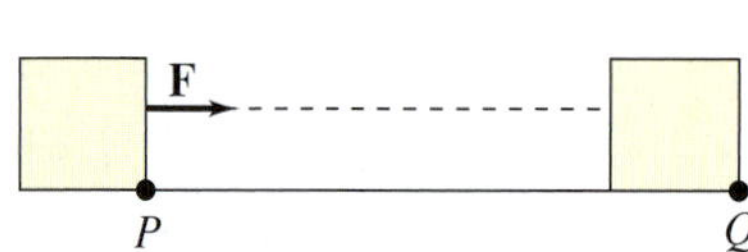

Force acts along the line of motion.
FIGURE 8.40

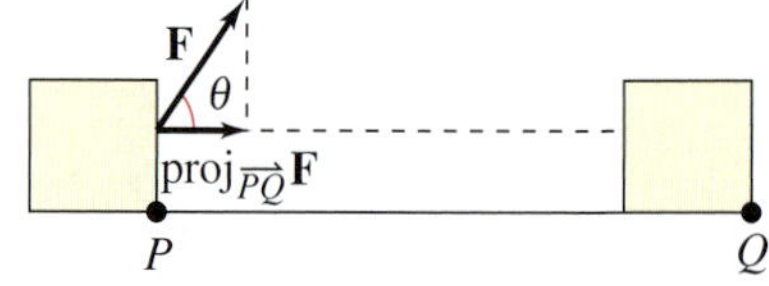

Force acts at angle θ with the line of motion.
FIGURE 8.41

This notion of work is summarized in the following definition.

Definition of Work

The **work** W done by a constant force $\mathbf{F}$ as its point of application moves along the vector $\overrightarrow{PQ}$ is given by either of the following.

1. $W = \|\text{proj}_{\overrightarrow{PQ}}\mathbf{F}\|\,\|\overrightarrow{PQ}\|$ Projection form
2. $W = \mathbf{F}\cdot\overrightarrow{PQ}$ Dot product form

Example 8 ▶ Finding Work  

To close a sliding door, a person pulls on a rope with a constant force of 50 pounds at a constant angle of 60°, as shown in Figure 8.42. Find the work done in moving the door 12 feet to its closed position.

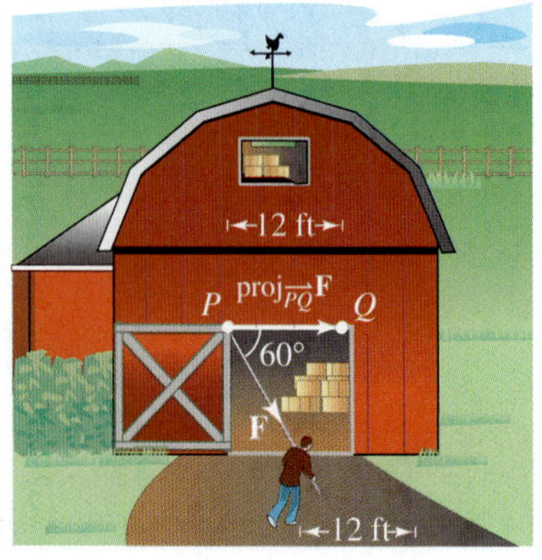

FIGURE 8.42

Solution

Using a projection, you can calculate the work as follows.

$$W = \|\text{proj}_{\overrightarrow{PQ}}\mathbf{F}\|\,\|\overrightarrow{PQ}\| \qquad \text{Projection form for work}$$

$$= (\cos 60^\circ)\|\mathbf{F}\|\,\|\overrightarrow{PQ}\|$$

$$= \frac{1}{2}(50)(12) = 300 \text{ foot-pounds}$$

So, the work done is 300 foot-pounds. You can verify this result by finding the vectors $\mathbf{F}$ and $\overrightarrow{PQ}$ and calculating their dot product.

8.4 Exercises

In Exercises 1–4, find the dot product of u and v.

1. $\mathbf{u} = \langle 6, 1 \rangle$
$\mathbf{v} = \langle -2, 3 \rangle$

2. $\mathbf{u} = \langle 5, 12 \rangle$
$\mathbf{v} = \langle -3, 2 \rangle$

3. $\mathbf{u} = 4\mathbf{i} - 2\mathbf{j}$
$\mathbf{v} = \mathbf{i} - \mathbf{j}$

4. $\mathbf{u} = 3\mathbf{i} + 4\mathbf{j}$
$\mathbf{v} = 7\mathbf{i} - 2\mathbf{j}$

In Exercises 5–8, use the vectors $\mathbf{u} = \langle 2, 2 \rangle$ and $\mathbf{v} = \langle -3, 4 \rangle$ to find the indicated quantity. State whether the result is a vector or a scalar.

5. $\mathbf{u} \cdot \mathbf{u}$

6. $\|\mathbf{v}\| + 3$

7. $(\mathbf{u} \cdot \mathbf{v})\mathbf{v}$

8. $3\mathbf{u} \cdot \mathbf{v}$

In Exercises 9–14, use the dot product to find the magnitude of u.

9. $\mathbf{u} = \langle -5, 12 \rangle$

10. $\mathbf{u} = \langle 2, -4 \rangle$

11. $\mathbf{u} = 20\mathbf{i} + 25\mathbf{j}$

12. $\mathbf{u} = 12\mathbf{i} - 16\mathbf{j}$

13. $\mathbf{u} = 6\mathbf{j}$

14. $\mathbf{u} = -21\mathbf{i}$

In Exercises 15–24, find the angle θ between the vectors.

15. $\mathbf{u} = \langle 1, 0 \rangle$
$\mathbf{v} = \langle 0, -2 \rangle$

16. $\mathbf{u} = \langle 3, 2 \rangle$
$\mathbf{v} = \langle 4, 0 \rangle$

17. $\mathbf{u} = 3\mathbf{i} + 4\mathbf{j}$
$\mathbf{v} = -2\mathbf{j}$

18. $\mathbf{u} = 2\mathbf{i} - 3\mathbf{j}$
$\mathbf{v} = \mathbf{i} - 2\mathbf{j}$

19. $\mathbf{u} = 2\mathbf{i} - \mathbf{j}$
$\mathbf{v} = 6\mathbf{i} + 4\mathbf{j}$

20. $\mathbf{u} = -6\mathbf{i} - 3\mathbf{j}$
$\mathbf{v} = -8\mathbf{i} + 4\mathbf{j}$

21. $\mathbf{u} = 5\mathbf{i} + 5\mathbf{j}$
$\mathbf{v} = -6\mathbf{i} + 6\mathbf{j}$

22. $\mathbf{u} = 2\mathbf{i} - 3\mathbf{j}$
$\mathbf{v} = 4\mathbf{i} + 3\mathbf{j}$

23. $\mathbf{u} = \cos\left(\frac{\pi}{3}\right)\mathbf{i} + \sin\left(\frac{\pi}{3}\right)\mathbf{j}$
$\mathbf{v} = \cos\left(\frac{3\pi}{4}\right)\mathbf{i} + \sin\left(\frac{3\pi}{4}\right)\mathbf{j}$

24. $\mathbf{u} = \cos\left(\frac{\pi}{4}\right)\mathbf{i} + \sin\left(\frac{\pi}{4}\right)\mathbf{j}$
$\mathbf{v} = \cos\left(\frac{\pi}{2}\right)\mathbf{i} + \sin\left(\frac{\pi}{2}\right)\mathbf{j}$

In Exercises 25–28, use vectors to find the interior angles of the triangle with the given vertices.

25. $(1, 2), (3, 4), (2, 5)$

26. $(-3, -4), (1, 7), (8, 2)$

27. $(-3, 0), (2, 2), (0, 6)$

28. $(-3, 5), (-1, 9), (7, 9)$

In Exercises 29 and 30, find $\mathbf{u} \cdot \mathbf{v}$, where θ is the angle between u and v.

29. $\|\mathbf{u}\| = 4, \|\mathbf{v}\| = 10, \theta = \frac{2\pi}{3}$

30. $\|\mathbf{u}\| = 100, \ \|\mathbf{v}\| = 250, \ \theta = \frac{\pi}{6}$

In Exercises 31–36, determine whether u and v are orthogonal, parallel, or neither.

31. $\mathbf{u} = \langle -12, 30 \rangle$
$\mathbf{v} = \left\langle \frac{1}{2}, -\frac{5}{4} \right\rangle$

32. $\mathbf{u} = \langle 3, 15 \rangle$
$\mathbf{v} = \langle -1, 5 \rangle$

33. $\mathbf{u} = \frac{1}{4}(3\mathbf{i} - \mathbf{j})$
$\mathbf{v} = 5\mathbf{i} + 6\mathbf{j}$

34. $\mathbf{u} = \mathbf{i}$
$\mathbf{v} = -2\mathbf{i} + 2\mathbf{j}$

35. $\mathbf{u} = 2\mathbf{i} - 2\mathbf{j}$
$\mathbf{v} = -\mathbf{i} - \mathbf{j}$

36. $\mathbf{u} = \langle \cos \theta, \sin \theta \rangle$
$\mathbf{v} = \langle \sin \theta, -\cos \theta \rangle$

In Exercises 37–40, find the projection of u onto v and the vector component of u orthogonal to v.

37. $\mathbf{u} = \langle 2, 2 \rangle$
$\mathbf{v} = \langle 6, 1 \rangle$

38. $\mathbf{u} = \langle 4, 2 \rangle$
$\mathbf{v} = \langle 1, -2 \rangle$

39. $\mathbf{u} = \langle 0, 3 \rangle$
$\mathbf{v} = \langle 2, 15 \rangle$

40. $\mathbf{u} = \langle -3, -2 \rangle$
$\mathbf{v} = \langle -4, -1 \rangle$

In Exercises 41–44, find two vectors in opposite directions that are orthogonal to the vector u. (The answers are not unique.)

41. $\mathbf{u} = \langle 3, 5 \rangle$

42. $\mathbf{u} = \langle -8, 3 \rangle$

43. $\mathbf{u} = \frac{1}{2}\mathbf{i} - \frac{2}{3}\mathbf{j}$

44. $\mathbf{u} = -\frac{5}{2}\mathbf{i} - 3\mathbf{j}$

Work **In Exercises 45 and 46, find the work done in moving a particle from P to Q if the magnitude and direction of the force are given by v.**

45. $P = (0, 0), \quad Q = (4, 7), \quad \mathbf{v} = \langle 1, 4 \rangle$

46. $P = (1, 3), \quad Q = (-3, 5), \quad \mathbf{v} = -2\mathbf{i} + 3\mathbf{j}$

47. ***Revenue*** The vector $\mathbf{u} = \langle 1650, 3200 \rangle$ gives the numbers of units of two types of baking pans produced by a company. The vector $\mathbf{v} = \langle 15.25, 10.50 \rangle$ gives the price (in dollars) of each pan, respectively.
(a) Find the dot product $\mathbf{u} \cdot \mathbf{v}$ and explain what information it gives.
(b) Identify the vector operation used to increase the prices by 5%.

▶ Model It

48. ***Braking Load*** A truck with a gross weight of 30,000 pounds is parked on a slope of $d°$ (see figure). Assume that the only force to overcome is the force of gravity.

(a) Find the force required to keep the truck from rolling down the hill in terms of the slope d.

(b) Use a graphing utility to complete the table.

d	0°	1°	2°	3°	4°	5°
Force						

d	6°	7°	8°	9°	10°
Force					

(c) Find the force perpendicular to the hill when $d = 5°$.

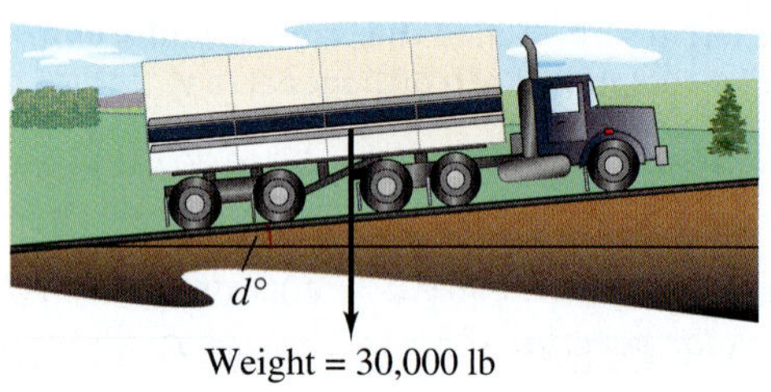

49. ***Work*** A 25-kilogram (245-newton) bag of sugar is lifted 3 meters. Determine the work done.

50. ***Work*** Determine the work done by a crane lifting a 2400-pound car 5 feet.

51. ***Work*** A force of 45 pounds in the direction of 30° above the horizontal is required to slide a table across a floor (see figure). The table is dragged 20 feet. Determine the work done.

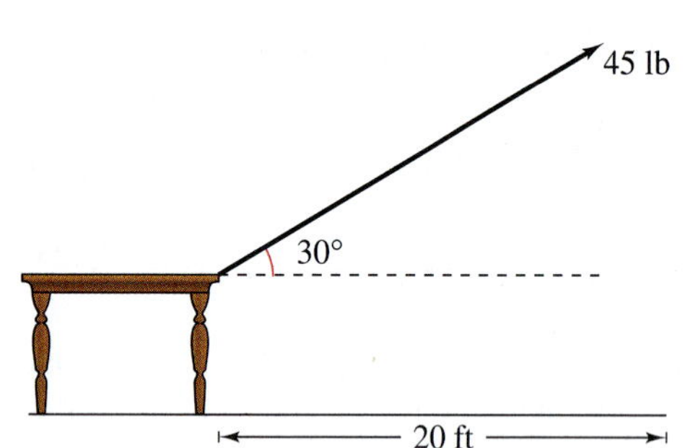

52. ***Work*** A tractor pulls a log 800 meters, and the tension in the cable connecting the tractor and log is approximately 1600 kilograms (15,691 newtons). The direction of the force is 35° above the horizontal. Approximate the work done.

Synthesis

True or False? **In Exercises 53 and 54, determine whether the statement is true or false. Justify your answer.**

53. The work W done by a constant force $\mathbf{F}$ acting along the line of motion of an object is represented by a vector.

54. A sliding door moves along the line of vector $\overrightarrow{PQ}$. If a force is applied to the door along a vector that is orthogonal to $\overrightarrow{PQ}$, then no work is done.

55. ***Think About It*** What is known about θ, the angle between two nonzero vectors $\mathbf{u}$ and $\mathbf{v}$, under each condition?

(a) $\mathbf{u} \cdot \mathbf{v} = 0$ (b) $\mathbf{u} \cdot \mathbf{v} > 0$ (c) $\mathbf{u} \cdot \mathbf{v} < 0$

56. ***Think About It*** What can be said about the vectors $\mathbf{u}$ and $\mathbf{v}$ under each condition?

(a) The projection of $\mathbf{u}$ onto $\mathbf{v}$ equals $\mathbf{u}$.

(b) The projection of $\mathbf{u}$ onto $\mathbf{v}$ equals $\mathbf{0}$.

57. ***Proof*** Use vectors to prove that the diagonals of a rhombus are perpendicular.

58. ***Proof*** Prove the following.

$$\|\mathbf{u} - \mathbf{v}\|^2 = \|\mathbf{u}\|^2 + \|\mathbf{v}\|^2 - 2\mathbf{u} \cdot \mathbf{v}$$

Review

In Exercises 59–62, perform the operation and write the result in standard form.

59. $\sqrt{42} \cdot \sqrt{24}$ **60.** $\sqrt{18} \cdot \sqrt{112}$

61. $\sqrt{-3} \cdot \sqrt{-8}$ **62.** $\sqrt{-12} \cdot \sqrt{-96}$

In Exercises 63–66, find the exact solutions of the equation in the interval $[0, 2\pi)$.

63. $\sin 2x - \sqrt{3} \sin x = 0$ **64.** $\sin 2x + \sqrt{2} \cos x = 0$

65. $2 \tan x = \tan 2x$ **66.** $\cos 2x - 3 \sin x = 2$

In Exercises 67–70, find the exact value of the trigonometric function given that $\sin u = -\frac{12}{13}$ and $\cos v = \frac{24}{25}$. (Both u and v are in Quadrant IV.)

67. $\sin(u - v)$ **68.** $\sin(u + v)$

69. $\cos(v - u)$ **70.** $\tan(u - v)$

8.5 Trigonometric Form of a Complex Number

▶ What you should learn

- How to plot complex numbers in the complex plane
- How to write the trigonometric forms of complex numbers
- How to multiply and divide complex numbers written in trigonometric form
- How to use DeMoivre's Theorem to find powers of complex numbers
- How to find nth roots of complex numbers

▶ Why you should learn it

You can use the trigonometric form of a complex number to perform operations with complex numbers. For instance, in Exercises 107–114 on page 647, you can use the trigonometric forms of complex numbers to help you solve polynomial equations.

The Complex Plane

Just as real numbers can be represented by points on the real number line, you can represent a complex number

$$z = a + bi$$

as the point (a, b) in a coordinate plane (the **complex plane**). The horizontal axis is called the **real axis** and the vertical axis is called the **imaginary axis,** as shown in Figure 8.43.

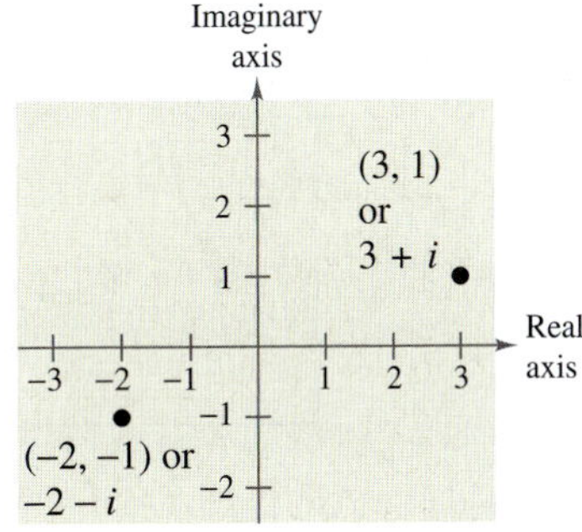

FIGURE 8.43

The **absolute value** of the complex number $a + bi$ is defined as the distance between the origin $(0, 0)$ and the point (a, b).

Definition of the Absolute Value of a Complex Number

The **absolute value** of the complex number $z = a + bi$ is

$$|a + bi| = \sqrt{a^2 + b^2}.$$

If the complex number $a + bi$ is a real number (that is, if $b = 0$), then this definition agrees with that given for the absolute value of a real number

$$|a + 0i| = \sqrt{a^2 + 0^2} = |a|.$$

Example 1 ▶ Finding the Absolute Value of a Complex Number

Plot $z = -2 + 5i$ and find its absolute value.

Solution

The number is plotted in Figure 8.44. It has an absolute value of

$$|z| = \sqrt{(-2)^2 + 5^2}$$
$$= \sqrt{29}.$$

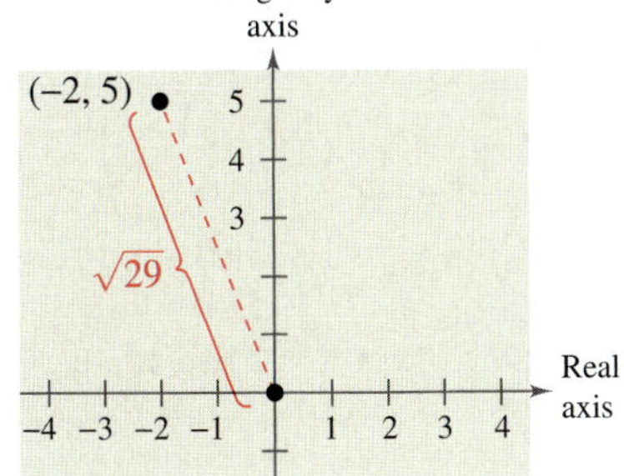

FIGURE 8.44

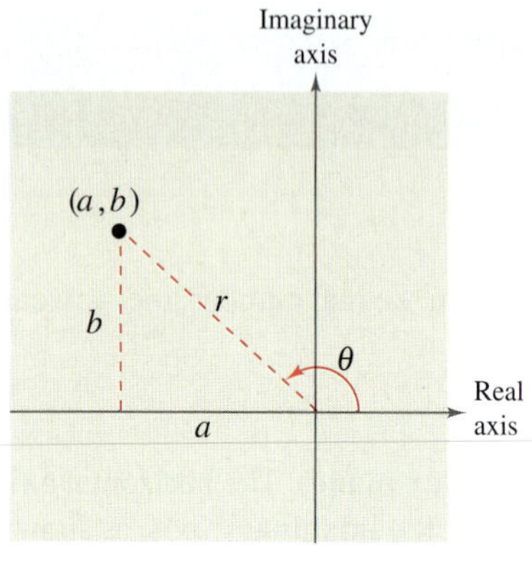

FIGURE 8.45

Trigonometric Form of a Complex Number

In Section 1.5 you learned how to add, subtract, multiply, and divide complex numbers. To work effectively with *powers* and *roots* of complex numbers, it is helpful to write complex numbers in trigonometric form. In Figure 8.45, consider the nonzero complex number $a + bi$. By letting θ be the angle from the positive real axis (measured counterclockwise) to the line segment connecting the origin and the point (a, b), you can write

$$a = r\cos\theta \qquad \text{and} \qquad b = r\sin\theta$$

where $r = \sqrt{a^2 + b^2}$. Consequently, you have

$$a + bi = (r\cos\theta) + (r\sin\theta)i$$

from which you can obtain the **trigonometric form of a complex number.**

Trigonometric Form of a Complex Number

The **trigonometric form** of the complex number $z = a + bi$ is

$$z = r(\cos\theta + i\sin\theta)$$

where $a = r\cos\theta$, $b = r\sin\theta$, $r = \sqrt{a^2 + b^2}$, and $\tan\theta = b/a$. The number r is the **modulus** of z, and θ is called an **argument** of z.

The trigonometric form of a complex number is also called the *polar form*. Because there are infinitely many choices for θ, the trigonometric form of a complex number is not unique. Normally, θ is restricted to the interval $0 \le \theta < 2\pi$, although on occasion it is convenient to use $\theta < 0$.

Example 2 ▶ Writing a Complex Number in Trigonometric Form

Write the complex number $z = -2 - 2\sqrt{3}i$ in trigonometric form.

Solution

The absolute value of z is

$$r = \left|-2 - 2\sqrt{3}i\right| = \sqrt{(-2)^2 + \left(-2\sqrt{3}\right)^2} = \sqrt{16} = 4$$

and the angle θ is

$$\tan\theta = \frac{b}{a} = \frac{-2\sqrt{3}}{-2} = \sqrt{3}.$$

Because $\tan(\pi/3) = \sqrt{3}$ and $z = -2 - 2\sqrt{3}i$ lies in Quadrant III, you choose θ to be $\theta = \pi + \pi/3 = 4\pi/3$. So, the trigonometric form is

$$z = r(\cos\theta + i\sin\theta)$$
$$= 4\left(\cos\frac{4\pi}{3} + i\sin\frac{4\pi}{3}\right).$$

See Figure 8.46.

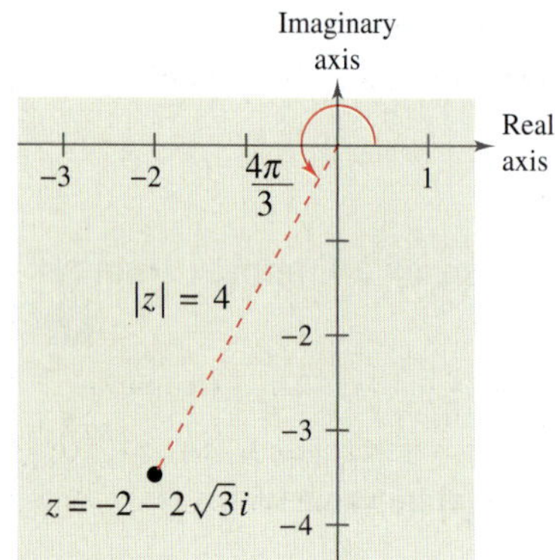

FIGURE 8.46

Example 3 ▶ **Writing a Complex Number in Standard Form** 

Write the complex number in standard form $a + bi$.

$$z = \sqrt{8}\left[\cos\left(-\frac{\pi}{3}\right) + i\sin\left(-\frac{\pi}{3}\right)\right]$$

Solution

Because $\cos(-\pi/3) = \frac{1}{2}$ and $\sin(-\pi/3) = -\sqrt{3}/2$, you can write

$$z = \sqrt{8}\left[\cos\left(-\frac{\pi}{3}\right) + i\sin\left(-\frac{\pi}{3}\right)\right]$$

$$= 2\sqrt{2}\left(\frac{1}{2} - \frac{\sqrt{3}}{2}i\right)$$

$$= \sqrt{2} - \sqrt{6}i.$$

Technology

A graphing utility can be used to convert a complex number in trigonometric (or polar) form to rectangular form, and vice versa. For specific keystrokes, see the user's manual for your graphing utility.

Multiplication and Division of Complex Numbers

The trigonometric form adapts nicely to multiplication and division of complex numbers. Suppose you are given two complex numbers

$$z_1 = r_1(\cos\theta_1 + i\sin\theta_1) \quad \text{and} \quad z_2 = r_2(\cos\theta_2 + i\sin\theta_2).$$

The product of z_1 and z_2 is

$$z_1z_2 = r_1r_2(\cos\theta_1 + i\sin\theta_1)(\cos\theta_2 + i\sin\theta_2)$$

$$= r_1r_2[(\cos\theta_1\cos\theta_2 - \sin\theta_1\sin\theta_2) + i(\sin\theta_1\cos\theta_2 + \cos\theta_1\sin\theta_2)].$$

Using the sum and difference formulas for cosine and sine, you can rewrite this equation as

$$z_1z_2 = r_1r_2[\cos(\theta_1 + \theta_2) + i\sin(\theta_1 + \theta_2)].$$

This establishes the first part of the following rule. The second part is left to you (see Exercise 119).

Product and Quotient of Two Complex Numbers

Let $z_1 = r_1(\cos\theta_1 + i\sin\theta_1)$ and $z_2 = r_2(\cos\theta_2 + i\sin\theta_2)$ be complex numbers.

$$z_1z_2 = r_1r_2[\cos(\theta_1 + \theta_2) + i\sin(\theta_1 + \theta_2)] \qquad \text{Product}$$

$$\frac{z_1}{z_2} = \frac{r_1}{r_2}[\cos(\theta_1 - \theta_2) + i\sin(\theta_1 - \theta_2)], \quad z_2 \neq 0 \qquad \text{Quotient}$$

Note that this rule says that to *multiply* two complex numbers you multiply moduli and add arguments, whereas to *divide* two complex numbers you divide moduli and subtract arguments.

Example 4 ▶ Multiplying Complex Numbers 

Find the product of the complex numbers.

$$z_1 = 2\left(\cos\frac{2\pi}{3} + i\sin\frac{2\pi}{3}\right) \qquad z_2 = 8\left(\cos\frac{11\pi}{6} + i\sin\frac{11\pi}{6}\right)$$

Solution

$$\begin{aligned}
z_1 z_2 &= 2\left(\cos\frac{2\pi}{3} + i\sin\frac{2\pi}{3}\right) \times 8\left(\cos\frac{11\pi}{6} + i\sin\frac{11\pi}{6}\right)\\
&= 16\left[\cos\left(\frac{2\pi}{3} + \frac{11\pi}{6}\right) + i\sin\left(\frac{2\pi}{3} + \frac{11\pi}{6}\right)\right]\\
&= 16\left(\cos\frac{5\pi}{2} + i\sin\frac{5\pi}{2}\right)\\
&= 16\left(\cos\frac{\pi}{2} + i\sin\frac{\pi}{2}\right)\\
&= 16[0 + i(1)]\\
&= 16i
\end{aligned}$$

You can check this result by first converting the complex numbers to the standard forms $z_1 = -1 + \sqrt{3}i$ and $z_2 = 4\sqrt{3} - 4i$ and then multiplying algebraically, as in Section 1.5.

$$\begin{aligned}
z_1 z_2 &= \left(-1 + \sqrt{3}i\right)\left(4\sqrt{3} - 4i\right)\\
&= -4\sqrt{3} + 4i + 12i + 4\sqrt{3}\\
&= 16i
\end{aligned}$$

Example 5 ▶ Dividing Complex Numbers 

Find the quotient, z_1/z_2, of the complex numbers.

$$z_1 = 24(\cos 300° + i\sin 300°) \qquad z_2 = 8(\cos 75° + i\sin 75°)$$

Solution

$$\begin{aligned}
\frac{z_1}{z_2} &= \frac{24(\cos 300° + i\sin 300°)}{8(\cos 75° + i\sin 75°)}\\
&= \frac{24}{8}[\cos(300° - 75°) + i\sin(300° - 75°)]\\
&= 3(\cos 225° + i\sin 225°)\\
&= 3\left[\left(-\frac{\sqrt{2}}{2}\right) + i\left(-\frac{\sqrt{2}}{2}\right)\right]\\
&= -\frac{3\sqrt{2}}{2} - \frac{3\sqrt{2}}{2}i
\end{aligned}$$

Powers of Complex Numbers

The trigonometric form of a complex number is used to raise a complex number to a power. To accomplish this, consider repeated use of the multiplication rule.

$$\begin{aligned}
z &= r(\cos\theta + i\sin\theta)\\
z^2 &= r(\cos\theta + i\sin\theta)r(\cos\theta + i\sin\theta) = r^2(\cos 2\theta + i\sin 2\theta)\\
z^3 &= r^2(\cos 2\theta + i\sin 2\theta)r(\cos\theta + i\sin\theta) = r^3(\cos 3\theta + i\sin 3\theta)\\
z^4 &= r^4(\cos 4\theta + i\sin 4\theta)\\
z^5 &= r^5(\cos 5\theta + i\sin 5\theta)\\
&\ \vdots
\end{aligned}$$

This pattern leads to the following important theorem, which is named after the French mathematician Abraham DeMoivre (1667–1754).

The Granger Collection

Historical Note

Abraham DeMoivre (1667–1754) is remembered for his work in probability theory and DeMoivre's Theorem. His book *The Doctrine of Chances* (published in 1718) includes the theory of recurring series and the theory of partial fractions.

DeMoivre's Theorem

If $z = r(\cos\theta + i\sin\theta)$ is a complex number and n is a positive integer, then

$$\begin{aligned}
z^n &= [r(\cos\theta + i\sin\theta)]^n\\
&= r^n(\cos n\theta + i\sin n\theta).
\end{aligned}$$

Example 6 ▶ Finding Powers of a Complex Number

Use DeMoivre's Theorem to find $\left(-1 + \sqrt{3}i\right)^{12}$.

Solution

First convert the complex number to trigonometric form using

$$r = \sqrt{(-1)^2 + \left(\sqrt{3}\right)^2} = 2 \text{ and } \theta = \arctan\frac{\sqrt{3}}{-1} = \frac{2\pi}{3}.$$

So, the trigonometric form is

$$z = -1 + \sqrt{3}i = 2\left(\cos\frac{2\pi}{3} + i\sin\frac{2\pi}{3}\right).$$

Then, by DeMoivre's Theorem, you have

$$\begin{aligned}
(-1 + \sqrt{3}i)^{12} &= \left[2\left(\cos\frac{2\pi}{3} + i\sin\frac{2\pi}{3}\right)\right]^{12}\\
&= 2^{12}\left[\cos(12)\frac{2\pi}{3} + i\sin(12)\frac{2\pi}{3}\right]\\
&= 4096(\cos 8\pi + i\sin 8\pi)\\
&= 4096(1 + 0)\\
&= 4096.
\end{aligned}$$

Roots of Complex Numbers

Recall that a consequence of the Fundamental Theorem of Algebra is that a polynomial equation of degree n has n solutions in the complex number system. So, the equation $x^6 = 1$ has six solutions, and in this particular case you can find the six solutions by factoring and using the Quadratic Formula.

$$x^6 - 1 = (x^3 - 1)(x^3 + 1)$$
$$= (x - 1)(x^2 + x + 1)(x + 1)(x^2 - x + 1) = 0$$

Consequently, the solutions are

$$x = \pm 1, \qquad x = \frac{-1 \pm \sqrt{3}i}{2}, \qquad \text{and} \qquad x = \frac{1 \pm \sqrt{3}i}{2}.$$

Each of these numbers is a sixth root of 1. In general, the ***n*th root** of a complex number is defined as follows.

Definition of *n*th Root of a Complex Number

The complex number $u = a + bi$ is an ***n*th root** of the complex number z if

$$z = u^n = (a + bi)^n.$$

Exploration

The nth roots of a complex number are useful for solving some polynomial equations. For instance, explain how you can use DeMoivre's Theorem to solve the polynomial equation

$$x^4 + 16 = 0.$$

[*Hint*: Write -16 as $16(\cos \pi + i \sin \pi)$.]

To find a formula for an nth root of a complex number, let u be an nth root of z, where

$$u = s(\cos \beta + i \sin \beta)$$

and

$$z = r(\cos \theta + i \sin \theta).$$

By DeMoivre's Theorem and the fact that $u^n = z$, you have

$$s^n(\cos n\beta + i \sin n\beta) = r(\cos \theta + i \sin \theta).$$

Taking the absolute value of each side of this equation, it follows that $s^n = r$. Substituting back into the previous equation and dividing by r, you get

$$\cos n\beta + i \sin n\beta = \cos \theta + i \sin \theta.$$

So, it follows that

$$\cos n\beta = \cos \theta \qquad \text{and} \qquad \sin n\beta = \sin \theta.$$

Because both sine and cosine have a period of 2π, these last two equations have solutions if and only if the angles differ by a multiple of 2π. Consequently, there must exist an integer k such that

$$n\beta = \theta + 2\pi k$$
$$\beta = \frac{\theta + 2\pi k}{n}.$$

By substituting this value of β into the trigonometric form of u, you get the result stated on the following page.

*n*th Roots of a Complex Number

For a positive integer n, the complex number $z = r(\cos\theta + i\sin\theta)$ has exactly n distinct nth roots given by

$$\sqrt[n]{r}\left(\cos\frac{\theta + 2\pi k}{n} + i\sin\frac{\theta + 2\pi k}{n}\right)$$

where $k = 0, 1, 2, \ldots, n - 1$.

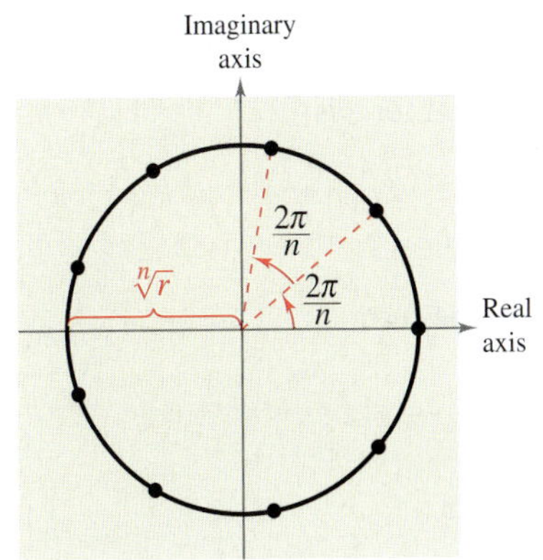

FIGURE 8.47

When k exceeds $n - 1$, the roots begin to repeat. For instance, if $k = n$, the angle

$$\frac{\theta + 2\pi n}{n} = \frac{\theta}{n} + 2\pi$$

is coterminal with θ/n, which is also obtained when $k = 0$.

The formula for the nth roots of a complex number z has a nice geometrical interpretation, as shown in Figure 8.47. Note that because the nth roots of z all have the same magnitude $\sqrt[n]{r}$, they all lie on a circle of radius $\sqrt[n]{r}$ with center at the origin. Furthermore, because successive nth roots have arguments that differ by $2\pi/n$, the n roots are equally spaced around the circle.

You have already found the sixth roots of 1 by factoring and by using the Quadratic Formula. Example 7 shows how you can solve the same problem with the formula for nth roots.

Example 7 ▶ Finding the *n*th Roots of a Real Number

Find all the sixth roots of 1.

Solution

First write 1 in the trigonometric form $1 = 1(\cos 0 + i\sin 0)$. Then, by the nth root formula, with $n = 6$ and $r = 1$, the roots have the form

$$\sqrt[6]{1}\left(\cos\frac{0 + 2\pi k}{6} + i\sin\frac{0 + 2\pi k}{6}\right)$$

or simply $\cos(\pi k/3) + i\sin(\pi k/3)$. So, for $k = 0, 1, 2, 3, 4,$ and 5, the sixth roots are as follows. (See Figure 8.48.)

$$\cos 0 + i\sin 0 = 1$$

$$\cos\frac{\pi}{3} + i\sin\frac{\pi}{3} = \frac{1}{2} + \frac{\sqrt{3}}{2}i \qquad \text{Increment by } \frac{2\pi}{n} = \frac{2\pi}{6} = \frac{\pi}{3}$$

$$\cos\frac{2\pi}{3} + i\sin\frac{2\pi}{3} = -\frac{1}{2} + \frac{\sqrt{3}}{2}i$$

$$\cos\pi + i\sin\pi = -1$$

$$\cos\frac{4\pi}{3} + i\sin\frac{4\pi}{3} = -\frac{1}{2} - \frac{\sqrt{3}}{2}i$$

$$\cos\frac{5\pi}{3} + i\sin\frac{5\pi}{3} = \frac{1}{2} - \frac{\sqrt{3}}{2}i$$

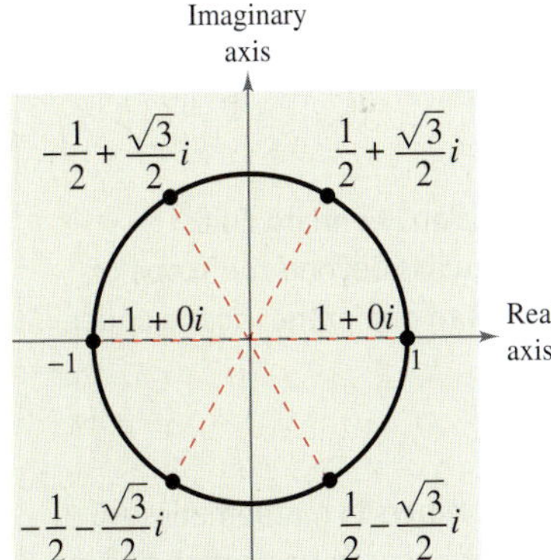

FIGURE 8.48

In Figure 8.48, notice that the roots obtained in Example 7 all have a magnitude of 1 and are equally spaced around the unit circle. Also notice that the complex roots occur in conjugate pairs, as discussed in Section 3.4. The n distinct nth roots of 1 are called the ***n*th roots of unity.**

Example 8 ▶ Finding the *n*th Roots of a Complex Number 

Find the three cube roots of

$$z = -2 + 2i.$$

Solution

Because z lies in Quadrant II, the trigonometric form for z is

$$z = -2 + 2i$$
$$= \sqrt{8}\,(\cos 135° + i \sin 135°). \qquad \theta = \arctan(2/-2) = 135°$$

By the formula for nth roots, the cube roots have the form

$$\sqrt[6]{8}\left(\cos \frac{135° + 360°k}{3} + i \sin \frac{135° + 360°k}{3}\right).$$

Finally, for $k = 0, 1,$ and 2, you obtain the roots

$$\sqrt[6]{8}\left(\cos \frac{135° + 360°(0)}{3} + i \sin \frac{135° + 360°(0)}{3}\right) = \sqrt{2}(\cos 45° + i \sin 45°)$$
$$= 1 + i$$

$$\sqrt[6]{8}\left(\cos \frac{135° + 360°(1)}{3} + i \sin \frac{135° + 360°(1)}{3}\right) = \sqrt{2}(\cos 165° + i \sin 165°)$$
$$\approx -1.3660 + 0.3660i$$

$$\sqrt[6]{8}\left(\cos \frac{135° + 360°(2)}{3} + i \sin \frac{135° + 360°(2)}{3}\right) = \sqrt{2}(\cos 285° + i \sin 285°)$$
$$\approx 0.3660 - 1.3660i.$$

Writing ABOUT MATHEMATICS

A Famous Mathematical Formula The famous formula

$$e^{a+bi} = e^a(\cos b + i \sin b)$$

is called Euler's Formula, after the German mathematician Leonhard Euler (1707–1783). Although the interpretation of this formula is beyond the scope of this text, we decided to include it because it gives rise to one of the most wonderful equations in mathematics.

$$e^{\pi i} + 1 = 0$$

This elegant equation relates the five most famous numbers in mathematics—0, 1, π, e, and i—in a single equation. Show how Euler's Formula can be used to derive this equation.

8.5 Exercises

In Exercises 1–6, plot the complex number and find its absolute value.

1. $-7i$

2. -7

3. $-4 + 4i$

4. $5 - 12i$

5. $6 - 7i$

6. $-8 + 3i$

In Exercises 7–10, write the complex number in trigonometric form.

7.

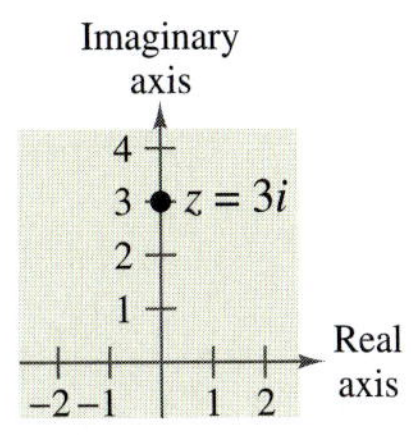

8.

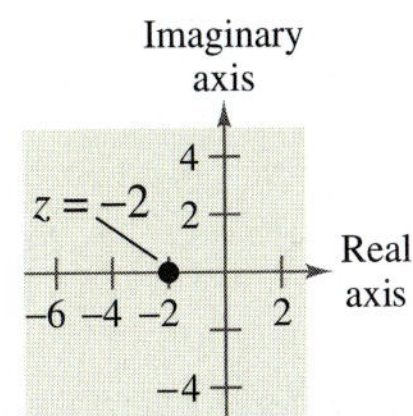

9.

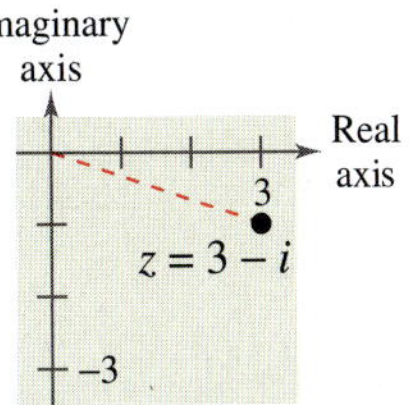

10.

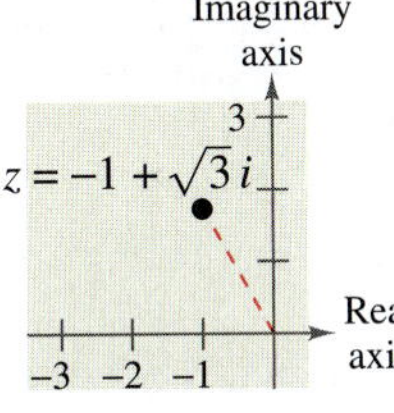

In Exercises 11–26, represent the complex number graphically, and find the trigonometric form of the number.

11. $3 - 3i$

12. $2 + 2i$

13. $\sqrt{3} + i$

14. $4 - 4\sqrt{3}i$

15. $-2(1 + \sqrt{3}i)$

16. $\frac{5}{2}(\sqrt{3} - i)$

17. $-5i$

18. $4i$

19. $-7 + 4i$

20. $3 - i$

21. 7

22. 4

23. $3 + \sqrt{3}i$

24. $2\sqrt{2} - i$

25. $-3 - i$

26. $1 + 3i$

In Exercises 27–34, use a graphing utility to represent the complex number in trigonometric form.

27. $5 + 2i$

28. $8 + 3i$

29. $-3 + i$

30. $-5 - i$

31. $3\sqrt{2} - 7i$

32. $4\sqrt{5} - 4i$

33. $-8 - 5\sqrt{3}i$

34. $-9 - 2\sqrt{10}i$

In Exercises 35–44, represent the complex number graphically, and find the standard form of the number.

35. $3(\cos 120° + i \sin 120°)$

36. $5(\cos 135° + i \sin 135°)$

37. $\frac{3}{2}(\cos 300° + i \sin 300°)$

38. $\frac{1}{4}(\cos 225° + i \sin 225°)$

39. $3.75\left(\cos \frac{3\pi}{4} + i \sin \frac{3\pi}{4}\right)$

40. $6\left(\cos \frac{5\pi}{12} + i \sin \frac{5\pi}{12}\right)$

41. $8\left(\cos \frac{\pi}{2} + i \sin \frac{\pi}{2}\right)$

42. $7(\cos 0 + i \sin 0)$

43. $3[\cos(18° 45') + i \sin(18° 45')]$

44. $6[\cos(230° 30') + i \sin(230° 30')]$

In Exercises 45–48, use a graphing utility to represent the complex number in standard form.

45. $5\left(\cos \frac{\pi}{9} + i \sin \frac{\pi}{9}\right)$

46. $10\left(\cos \frac{2\pi}{5} + i \sin \frac{2\pi}{5}\right)$

47. $3(\cos 165.5° + i \sin 165.5°)$

48. $9(\cos 58° + i \sin 58°)$

In Exercises 49–60, perform the operation and leave the result in trigonometric form.

49. $\left[2\left(\cos \frac{\pi}{4} + i \sin \frac{\pi}{4}\right)\right]\left[6\left(\cos \frac{\pi}{12} + i \sin \frac{\pi}{12}\right)\right]$

50. $\left[\frac{3}{4}\left(\cos \frac{\pi}{3} + i \sin \frac{\pi}{3}\right)\right]\left[4\left(\cos \frac{3\pi}{4} + i \sin \frac{3\pi}{4}\right)\right]$

51. $\left[\frac{5}{3}(\cos 140° + i \sin 140°)\right]\left[\frac{2}{3}(\cos 60° + i \sin 60°)\right]$

52. $[0.5(\cos 100° + i \sin 100°)] \times [0.8(\cos 300° + i \sin 300°)]$

53. $[0.45(\cos 310° + i \sin 310°)] \times [0.60(\cos 200° + i \sin 200°)]$

54. $(\cos 5° + i \sin 5°)(\cos 20° + i \sin 20°)$

55. $\dfrac{\cos 50° + i \sin 50°}{\cos 20° + i \sin 20°}$

56. $\dfrac{2(\cos 120° + i \sin 120°)}{4(\cos 40° + i \sin 40°)}$

57. $\dfrac{\cos(5\pi/3) + i \sin(5\pi/3)}{\cos \pi + i \sin \pi}$

58. $\dfrac{5(\cos 4.3 + i \sin 4.3)}{4(\cos 2.1 + i \sin 2.1)}$

59. $\dfrac{12(\cos 52° + i \sin 52°)}{3(\cos 110° + i \sin 110°)}$

60. $\dfrac{6(\cos 40° + i \sin 40°)}{7(\cos 100° + i \sin 100°)}$

In Exercises 61–68, (a) give the trigonometric forms of the complex numbers, (b) perform the indicated operation using the trigonometric forms, and (c) perform the indicated operation using the standard forms, and check your result with that of part (b).

61. $(2 + 2i)(1 - i)$

62. $(\sqrt{3} + i)(1 + i)$

63. $-2i(1 + i)$

64. $4(1 - \sqrt{3}i)$

65. $\dfrac{3 + 4i}{1 - \sqrt{3}i}$

66. $\dfrac{1 + \sqrt{3}i}{6 - 3i}$

67. $\dfrac{5}{2 + 3i}$

68. $\dfrac{4i}{-4 + 2i}$

In Exercises 69–72, sketch the graphs of all complex numbers *z* satisfying the given condition.

69. $|z| = 2$

70. $|z| = 3$

71. $\theta = \dfrac{\pi}{6}$

72. $\theta = \dfrac{5\pi}{4}$

In Exercises 73–90, use DeMoivre's Theorem to find the indicated power of the complex number. Write the result in standard form.

73. $(1 + i)^5$

74. $(2 + 2i)^6$

75. $(-1 + i)^{10}$

76. $(3 - 2i)^8$

77. $2(\sqrt{3} + i)^7$

78. $4(1 - \sqrt{3}i)^3$

79. $[5(\cos 20° + i \sin 20°)]^3$

80. $[3(\cos 150° + i \sin 150°)]^4$

81. $\left(\cos \dfrac{\pi}{4} + i \sin \dfrac{\pi}{4}\right)^{12}$

82. $\left[2\left(\cos \dfrac{\pi}{2} + i \sin \dfrac{\pi}{2}\right)\right]^8$

83. $[5(\cos 3.2 + i \sin 3.2)]^4$

84. $(\cos 0 + i \sin 0)^{20}$

85. $(3 - 2i)^5$

86. $(\sqrt{5} - 4i)^3$

87. $[3(\cos 15° + i \sin 15°)]^4$

88. $[2(\cos 10° + i \sin 10°)]^8$

89. $\left[2\left(\cos \dfrac{\pi}{10} + i \sin \dfrac{\pi}{10}\right)\right]^5$

90. $\left[2\left(\cos \dfrac{\pi}{8} + i \sin \dfrac{\pi}{8}\right)\right]^6$

In Exercises 91–106, (a) use the formula on page 643 to find the indicated roots of the complex number, (b) represent each of the roots graphically, and (c) express each of the roots in standard form.

91. Square roots of $5(\cos 120° + i \sin 120°)$

92. Square roots of $16(\cos 60° + i \sin 60°)$

93. Cube roots of $8\left(\cos \dfrac{2\pi}{3} + i \sin \dfrac{2\pi}{3}\right)$

94. Fifth roots of $32\left(\cos \dfrac{5\pi}{6} + i \sin \dfrac{5\pi}{6}\right)$

95. Square roots of $-25i$

96. Fourth roots of $625i$

97. Cube roots of $-\frac{125}{2}(1 + \sqrt{3}i)$

98. Cube roots of $-4\sqrt{2}(1 - i)$

99. Fourth roots of 16

100. Fourth roots of i

101. Fifth roots of 1

102. Cube roots of 1000

103. Cube roots of -125

104. Fourth roots of -4

105. Fifth roots of $128(-1 + i)$

106. Sixth roots of $64i$

In Exercises 107–114, use the formula on page 643 to find all the solutions of the equation and represent the solutions graphically.

107. $x^4 + i = 0$

108. $x^3 + 1 = 0$

109. $x^5 + 243 = 0$

110. $x^3 - 27 = 0$

111. $x^4 + 16i = 0$

112. $x^6 + 64i = 0$

113. $x^3 - (1 - i) = 0$

114. $x^4 + (1 + i) = 0$

Synthesis

True or False? **In Exercises 115–118, determine whether the statement is true or false. Justify your answer.**

115. Although the square of the complex number bi is given by $(bi)^2 = -b^2$, the absolute value of the complex number $z = a + bi$ is defined as

$$|a + bi| = \sqrt{a^2 + b^2}.$$

116. Geometrically, the nth roots of any complex number z are all equally spaced around the unit circle centered at the origin.

117. The product of two complex numbers

$$z_1 = r_1(\cos \theta_1 + i \sin \theta_1)$$

and

$$z_2 = r_2(\cos \theta_2 + i \sin \theta_2)$$

is zero only when $r_1 = 0$ and/or $r_2 = 0$.

118. By DeMoivre's Theorem,

$$(4 + \sqrt{6}i)^8 = \cos(32) + i \sin(8\sqrt{6}).$$

119. Given two complex numbers $z_1 = r_1(\cos \theta_1 + i \sin \theta_1)$ and $z_2 = r_2(\cos \theta_2 + i \sin \theta_2)$, $z_2 \neq 0$, show that

$$\frac{z_1}{z_2} = \frac{r_1}{r_2}[\cos(\theta_1 - \theta_2) + i \sin(\theta_1 - \theta_2)].$$

120. Show that $\bar{z} = r[\cos(-\theta) + i \sin(-\theta)]$ is the complex conjugate of $z = r(\cos \theta + i \sin \theta)$.

121. Use the trigonometric forms of z and $\bar{z}$ in Exercise 120 to find (a) $z\bar{z}$ and (b) $z/\bar{z}$, $\bar{z} \neq 0$.

122. Show that the negative of $z = r(\cos \theta + i \sin \theta)$ is $-z = r[\cos(\theta + \pi) + i \sin(\theta + \pi)]$.

123. Show that $-\frac{1}{2}(1 + \sqrt{3}i)$ is a sixth root of 1.

124. Show that $2^{-1/4}(1 - i)$ is a fourth root of -2.

Graphical Reasoning **In Exercises 125 and 126, use the graph of the roots of a complex number.**

(a) Write each of the roots in trigonometric form.

(b) Identify the complex number whose roots are given.

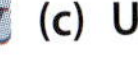 **(c) Use a graphing utility to verify the results of part (b).**

125.

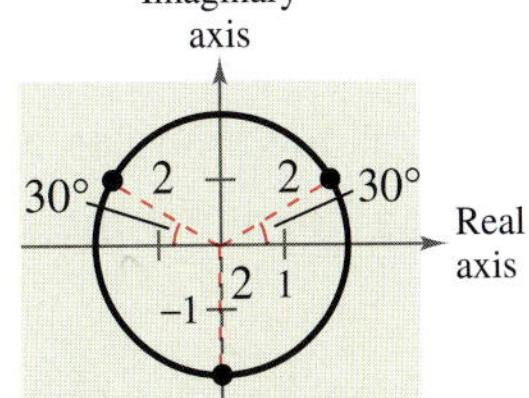

126.

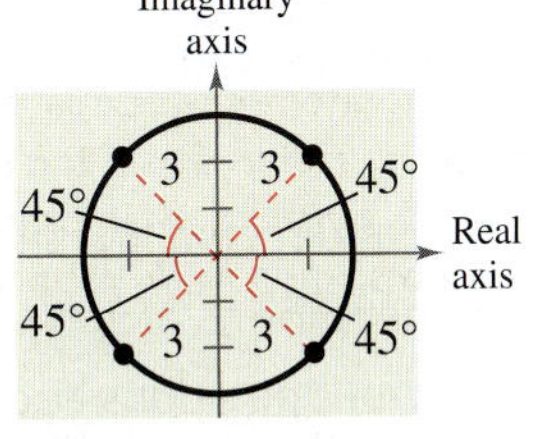

Review

In Exercises 127–132, solve the right triangle shown in the figure. Approximate the result to two decimal places.

127. $A = 22°, \quad a = 8$

128. $B = 66°, \quad a = 33.5$

129. $A = 30°, \quad b = 112.6$

130. $B = 6°, \quad b = 211.2$

131. $A = 42° 15', \quad c = 11.2$

132. $B = 81° 30', \quad c = 6.8$

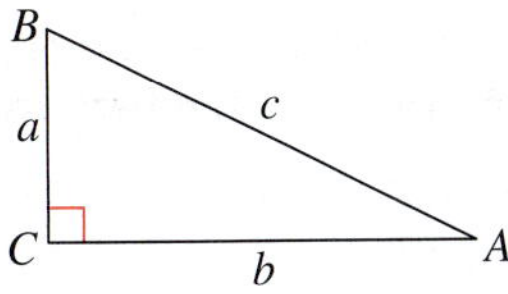

Harmonic Motion **In Exercises 133–136, for the simple harmonic motion described by the trigonometric function, find the maximum displacement and the lowest possible value of t for which $d = 0$.**

133. $d = 16 \cos \dfrac{\pi}{4}t$

134. $d = \dfrac{1}{8} \cos 12\pi t$

135. $d = \frac{1}{16} \sin \frac{5}{4}\pi t$

136. $d = \frac{1}{12} \sin 60\pi t$

Chapter Summary

▶ *What* did you learn?

	Review Exercises
Section 8.1	
☐ How to use the Law of Sines to solve oblique triangles (AAS, ASA, or SSA)	1–12
☐ How to find the areas of oblique triangles	13–16
☐ How to use the Law of Sines to model and solve real-life problems	17–20
Section 8.2	
☐ How to use the Law of Cosines to solve oblique triangles (SSS or SAS)	21–28
☐ How to use the Law of Cosines to model and solve real-life problems	29, 30
☐ How to use Heron's Area Formula to find the area of a triangle	31–34
Section 8.3	
☐ How to represent vectors as directed line segments	35–38
☐ How to write the component forms of vectors	39–44
☐ How to perform basic vector operations and represent them graphically	45–48
☐ How to write vectors as linear combinations of unit vectors	49–54
☐ How to find the direction angles of vectors	55–60
☐ How to use vectors to model and solve real-life problems	61–64
Section 8.4	
☐ How to find the dot product of two vectors and use the Properties of the Dot Product	65–72
☐ How to find the angle between two vectors and determine whether two vectors are orthogonal	73–80
☐ How to write a vector as the sum of two vector components	81–84
☐ How to use vectors to find the work done by a force	85, 86
Section 8.5	
☐ How to plot complex numbers in the complex plane	87–90
☐ How to write the trigonometric forms of complex numbers	91–94
☐ How to multiply and divide complex numbers written in trigonometric form	95, 96
☐ How to use DeMoivre's Theorem to find powers of complex numbers	97–100
☐ How to find *n*th roots of complex numbers	101–108

Review Exercises

8.1 **In Exercises 1–12, use the Law of Sines to solve (if possible) the triangle. If two solutions exist, list both. Round your answers to two decimal places.**

1.

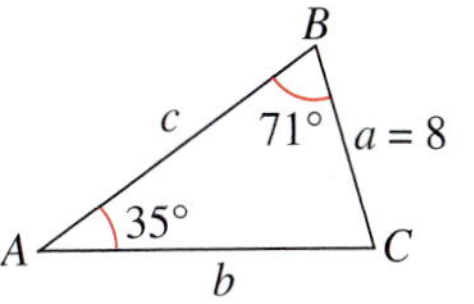

2.

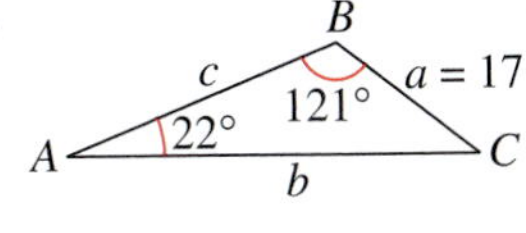

3. $B = 72°,\ C = 82°,\ b = 54$

4. $B = 10°,\ C = 20°,\ c = 33$

5. $A = 16°,\ B = 98°,\ c = 8.4$

6. $A = 95°,\ B = 45°,\ c = 104.8$

7. $A = 24°,\ C = 48°,\ b = 27.5$

8. $B = 64°,\ C = 36°,\ a = 367$

9. $B = 150°,\ b = 30,\ c = 10$

10. $B = 150°,\ a = 10,\ b = 3$

11. $A = 75°,\ a = 51.2,\ b = 33.7$

12. $B = 25°,\ a = 6.2,\ b = 4$

In Exercises 13–16, use the information to find the area of the triangle.

13. $A = 27°,\ b = 5,\ c = 7$

14. $B = 80°,\ a = 4,\ c = 8$

15. $C = 123°,\ a = 16,\ b = 5$

16. $A = 11°,\ b = 22,\ c = 21$

17. ***Height*** From a certain distance, the angle of elevation to the top of a building is 17°. At a point 50 meters closer to the building, the angle of elevation is 31°. Approximate the height of the building.

18. ***Geometry*** Find the length of the side w of the parallelogram.

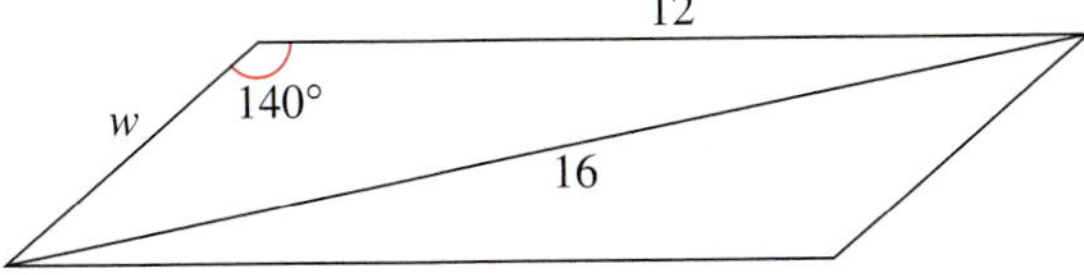

19. ***Height*** A tree stands on a hillside of slope 28°(from the horizontal). From a point 75 feet down the hill, the angle of elevation to the top of the tree is 45° (see figure). What is the height of the tree?

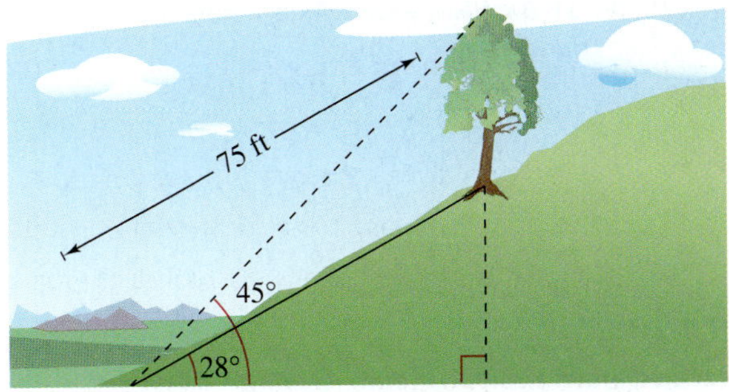

20. ***River Width*** A surveyor finds that a tree on the opposite bank of a river flowing due east has a bearing of N 22° 30′ E from a certain point and a bearing of N 15° W from a point 400 feet downstream. What is the width of the river?

8.2 **In Exercises 21–28, use the Law of Cosines to solve the triangle. Round your answers to two decimal places.**

21. $a = 5,\ b = 8,\ c = 10$

22. $a = 80,\ b = 60,\ c = 100$

23. $a = 2.5,\ b = 5.0,\ c = 4.5$

24. $a = 16.4,\ b = 8.8,\ c = 12.2$

25. $B = 110°,\ a = 4,\ c = 4$

26. $B = 150°,\ a = 10,\ c = 20$

27. $C = 43°,\ a = 22.5,\ b = 31.4$

28. $A = 62°,\ b = 11.34,\ c = 19.52$

29. ***Surveying*** To approximate the length of a marsh, a surveyor walks 425 meters from point A to point B. Then the surveyor turns 65° and walks 300 meters to point C (see figure). Approximate the length AC of the marsh.

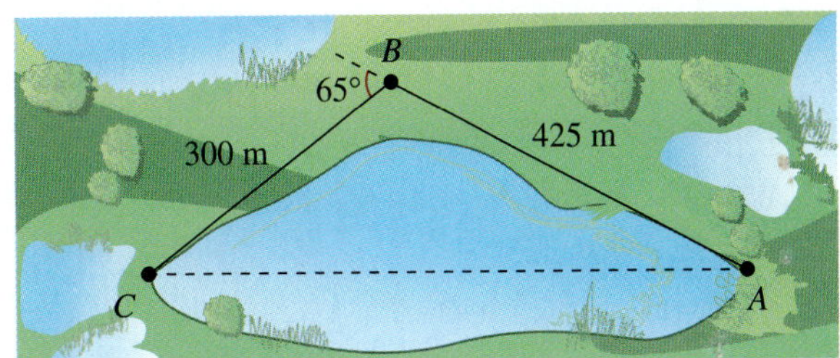

30. ***Navigation*** Two planes leave Raleigh-Durham Airport at approximately the same time. One is flying 425 miles per hour at a bearing of 355°, and the other is flying 530 miles per hour at a bearing of 67°. Draw a figure that gives a visual representation of the problem and determine the distance between the planes after they have flown for 2 hours.

In Exercises 31–34, use Heron's Area Formula to find the area of the triangle.

31. $a = 4,\ b = 5,\ c = 7$

32. $a = 15,\ b = 8,\ c = 10$

33. $a = 12.3,\ b = 15.8,\ c = 3.7$

34. $a = 38.1,\ b = 26.7,\ c = 19.4$

8.3 In Exercises 35–38, graph the vector with the specified initial point and terminal point.

	Initial Point	*Terminal Point*
35.	$(0, 0)$	$(8, 7)$
36.	$(3, 4)$	$(-5, -7)$
37.	$(-3, 9)$	$(8, -4)$
38.	$(-6, -8)$	$(8, 3)$

In Exercises 39–44, find the component form of the vector v satisfying the conditions.

39. 

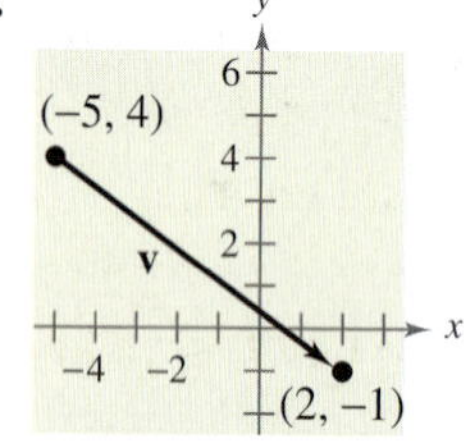

40.

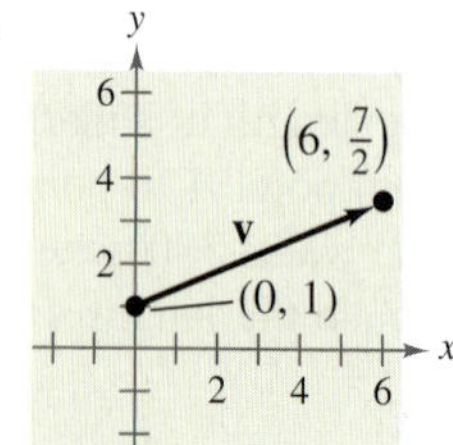

41. Initial point: $(0, 10)$; Terminal point: $(7, 3)$

42. Initial point: $(1, 5)$; Terminal point: $(15, 9)$

43. $\|\mathbf{v}\| = 8, \quad \theta = 120°$

44. $\|\mathbf{v}\| = \frac{1}{2}, \quad \theta = 225°$

In Exercises 45–48, find the component form of the vector given that u = 6i − 5j and v = 10i + 3j. Then sketch your result.

45. $2\mathbf{u} + \mathbf{v}$

46. $4\mathbf{u} - 5\mathbf{v}$

47. $3\mathbf{v}$

48. $\frac{1}{2}\mathbf{v}$

In Exercises 49–52, write vector u as a linear combination of the standard unit vectors i and j.

49. $\mathbf{u} = \langle -3, 4 \rangle$

50. $\mathbf{u} = \langle -6, -8 \rangle$

51. $\mathbf{u}$ has initial point $(3, 4)$ and terminal point $(9, 8)$.

52. $\mathbf{u}$ has initial point $(-2, 7)$ and terminal point $(5, -9)$.

In Exercises 53 and 54, write the vector v in the form $\|\mathbf{v}\|(\cos\theta\mathbf{i} + \sin\theta\mathbf{j})$.

53. $\mathbf{v} = -10\mathbf{i} + 10\mathbf{j}$

54. $\mathbf{v} = 4\mathbf{i} - \mathbf{j}$

In Exercises 55–60, find the magnitude and the direction angle of the vector v.

55. $\mathbf{v} = 7(\cos 60°\mathbf{i} + \sin 60°\mathbf{j})$

56. $\mathbf{v} = 3(\cos 150°\mathbf{i} + \sin 150°\mathbf{j})$

57. $\mathbf{v} = 5\mathbf{i} + 4\mathbf{j}$

58. $\mathbf{v} = -4\mathbf{i} + 7\mathbf{j}$

59. $\mathbf{v} = -3\mathbf{i} - 3\mathbf{j}$

60. $\mathbf{v} = 8\mathbf{i} - \mathbf{j}$

61. ***Resultant Force*** Forces of 85 pounds and 50 pounds act on a single point. The angle between the forces is 15°. Describe the resultant force.

62. ***Rope Tension*** A 180-pound weight is supported by two ropes, as shown in the figure. Find the tension exerted on each rope.

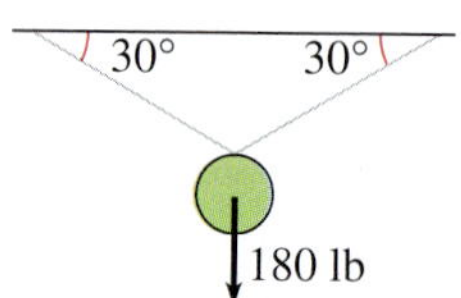

63. ***Navigation*** An airplane has an airspeed of 430 miles per hour at a bearing of 135°. The wind velocity is 35 miles per hour in the direction of N 30° E. What are the groundspeed and the direction of the plane?

64. ***Navigation*** An airplane has an airspeed of 724 kilometers per hour at a bearing of 30°. The wind velocity is 32 kilometers per hour from the west. What are the groundspeed and the direction of the plane?

8.4 In Exercises 65–68, find the dot product of **u** and **v**.

65. $\mathbf{u} = \langle 6, 7 \rangle$, $\mathbf{v} = \langle -3, 9 \rangle$

66. $\mathbf{u} = \langle -7, 12 \rangle$, $\mathbf{v} = \langle -4, -14 \rangle$

67. $\mathbf{u} = 3\mathbf{i} + 7\mathbf{j}$, $\mathbf{v} = 11\mathbf{i} - 5\mathbf{j}$

68. $\mathbf{u} = -7\mathbf{i} + 2\mathbf{j}$, $\mathbf{v} = 16\mathbf{i} - 12\mathbf{j}$

In Exercises 69–72, use the vectors $\mathbf{u} = \langle -3, 4 \rangle$ and $\mathbf{v} = \langle 2, 1 \rangle$ to find the quantity. State whether the result is a vector or a scalar.

69. $2\mathbf{u} \cdot \mathbf{u}$

70. $\|\mathbf{v}\|^2$

71. $\mathbf{u}(\mathbf{u} \cdot \mathbf{v})$

72. $3\mathbf{u} \cdot \mathbf{v}$

In Exercises 73–76, find the angle between **u** and **v**.

73. $\mathbf{u} = \cos \frac{7\pi}{4}\mathbf{i} + \sin \frac{7\pi}{4}\mathbf{j}$

$\mathbf{v} = \cos \frac{5\pi}{6}\mathbf{i} + \sin \frac{5\pi}{6}\mathbf{j}$

74. $\mathbf{u} = \cos 45°\mathbf{i} + \sin 45°\mathbf{j}$

$\mathbf{v} = \cos 300°\mathbf{i} + \sin 300°\mathbf{j}$

75. $\mathbf{u} = \langle 2\sqrt{2}, -4 \rangle$, $\mathbf{v} = \langle -\sqrt{2}, 1 \rangle$

76. $\mathbf{u} = \langle 3, \sqrt{3} \rangle$, $\mathbf{v} = \langle 4, 3\sqrt{3} \rangle$

In Exercises 77–80, determine whether **u** and **v** are orthogonal, parallel, or neither.

77. $\mathbf{u} = \langle -3, 8 \rangle$, $\mathbf{v} = \langle 8, 3 \rangle$

78. $\mathbf{u} = \langle \frac{1}{4}, -\frac{1}{2} \rangle$, $\mathbf{v} = \langle -2, 4 \rangle$

79. $\mathbf{u} = -\mathbf{i}$, $\mathbf{v} = \mathbf{i} + 2\mathbf{j}$

80. $\mathbf{u} = -2\mathbf{i} + \mathbf{j}$, $\mathbf{v} = 3\mathbf{i} + 6\mathbf{j}$

In Exercises 81–84, find $\text{proj}_{\mathbf{v}}\mathbf{u}$ and the vector component of **u** orthogonal to **v**.

81. $\mathbf{u} = \langle -4, 3 \rangle$, $\mathbf{v} = \langle -8, -2 \rangle$

82. $\mathbf{u} = \langle 5, 6 \rangle$, $\mathbf{v} = \langle 10, 0 \rangle$

83. $\mathbf{u} = \langle 2, 7 \rangle$, $\mathbf{v} = \langle 1, -1 \rangle$

84. $\mathbf{u} = \langle -3, 5 \rangle$, $\mathbf{v} = \langle -5, 2 \rangle$

Work In Exercises 85 and 86, find the work done in moving a particle from P to Q if the magnitude and direction of the force are given by **v**.

85. $P = (5, 3)$, $Q = (8, 9)$, $\mathbf{v} = \langle 2, 7 \rangle$

86. $P = (-2, -9)$, $Q = (-12, 8)$, $\mathbf{v} = 3\mathbf{i} - 6\mathbf{j}$

8.5 In Exercises 87–90, plot the complex number and find its absolute value.

87. $7i$

88. $-6i$

89. $5 + 3i$

90. $-10 - 4i$

In Exercises 91–94, write the trigonometric form of the complex number.

91. $5 - 5i$

92. $5 + 12i$

93. $-3\sqrt{3} + 3i$

94. -7

In Exercises 95 and 96, (a) write the two complex numbers in trigonometric form, and (b) use the trigonometric form to find $z_1 z_2$ and z_1/z_2.

95. $z_1 = 2\sqrt{3} - 2i$, $z_2 = -10i$

96. $z_1 = -3(1 + i)$, $z_2 = 2(\sqrt{3} + i)$

In Exercises 97–100, use DeMoivre's Theorem to find the indicated power of the complex number. Write the result in standard form.

97. $\left[5\left(\cos \frac{\pi}{12} + i \sin \frac{\pi}{12}\right)\right]^4$

98. $\left[2\left(\cos \frac{4\pi}{15} + i \sin \frac{4\pi}{15}\right)\right]^5$

99. $(2 + 3i)^6$

100. $(1 - i)^8$

Graphical Reasoning In Exercises 101 and 102, use the graph of the roots of a complex number.

(a) Write each of the roots in trigonometric form.

(b) Identify the complex number whose roots are given.

(c) Use a graphing utility to verify the results of part (b).

101.

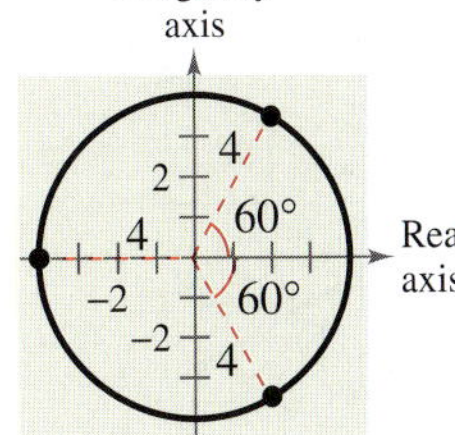

102.

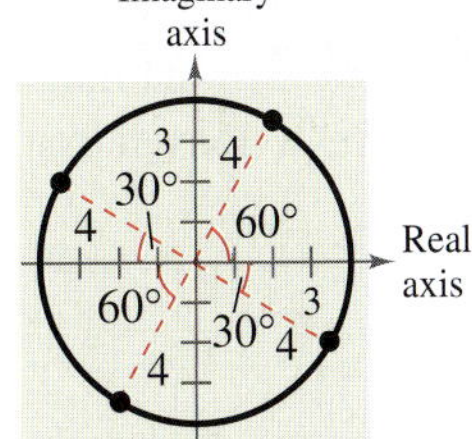

In Exercises 103 and 104, use the formula on page 643 to find the roots of the complex number.

103. Sixth roots of $-729i$

104. Fourth roots of 256

In Exercises 105–108, find all solutions of the equation and represent the solutions graphically.

105. $x^4 + 81 = 0$

106. $x^5 - 32 = 0$

107. $x^3 + 8i = 0$

108. $(x^3 - 1)(x^2 + 1) = 0$

Synthesis

True or False? **In Exercises 109–113, determine whether the statement is true or false. Justify your answer.**

109. The Law of Sines is true if one of the angles in the triangle is a right angle.

110. When the Law of Sines is used, the solution is always unique.

111. If $\mathbf{u}$ is a unit vector in the direction of $\mathbf{v}$, then $\mathbf{v} = \|\mathbf{v}\|\,\mathbf{u}$.

112. If $\mathbf{v} = a\mathbf{i} + b\mathbf{j} = \mathbf{0}$, then $a = -b$.

113. $x = \sqrt{3} + i$ is a solution of the equation $x^2 - 8i = 0$.

114. State the Law of Sines from memory.

115. State the Law of Cosines from memory.

116. What characterizes a vector in the plane?

117. Which vectors in the figure appear to be equivalent?

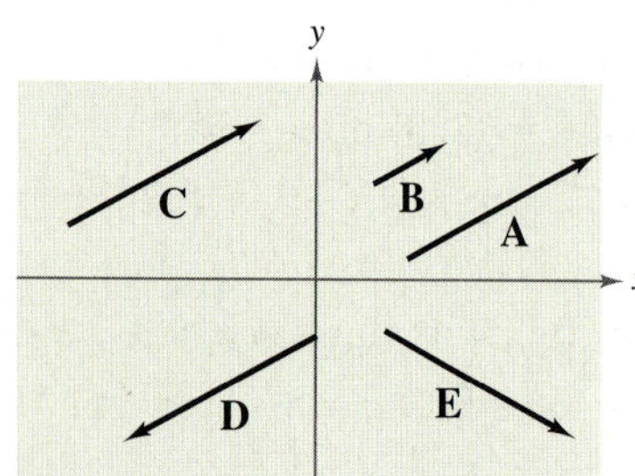

118. The vectors $\mathbf{u}$ and $\mathbf{v}$ have the same magnitudes in the two figures. In which figure will the magnitude of the sum be greater? Give a reason for your answer.

(a)

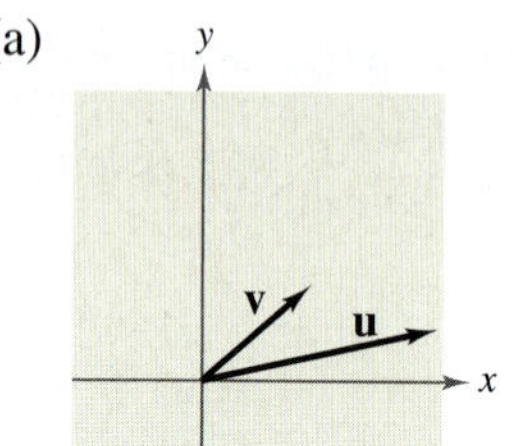

(b)

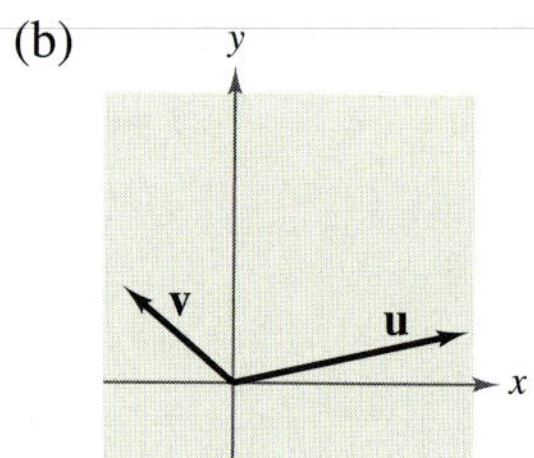

119. Give a geometric description of the scalar multiple $k\mathbf{u}$ of the vector $\mathbf{u}$, for $k > 0$ and for $k < 0$.

120. Give a geometric description of the sum of the vectors $\mathbf{u}$ and $\mathbf{v}$.

121. The figure shows z_1 and z_2. Describe z_1z_2 and z_1/z_2.

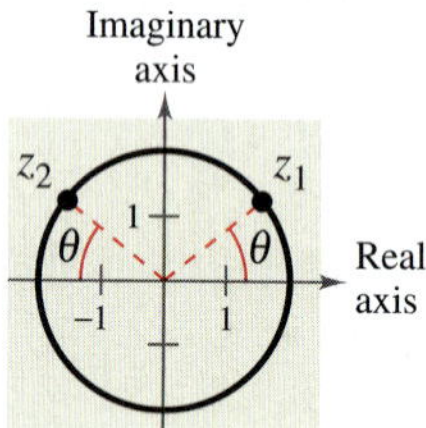

122. One of the fourth roots of a complex number z is shown in the figure.

(a) How many roots are not shown?

(b) Describe the other roots.

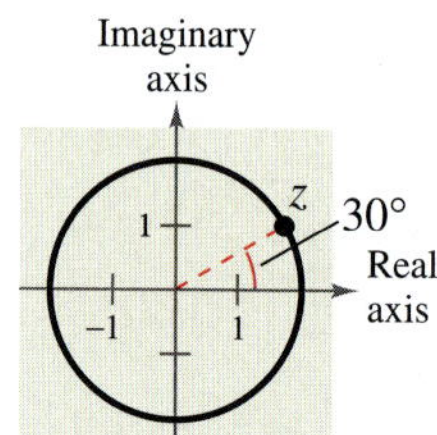

Chapter Test

The *Interactive* CD-ROM and *Internet* versions of this text offer Chapter Pre-Tests and Chapter Post-Tests, both of which have randomly generated exercises with diagnostic capabilities.

Take this test as you would take a test in class. When you are finished, check your work against the answers given in the back of the book.

In Exercises 1–6, use the information to solve the triangle. If two solutions exist, find both.

1. $A = 24°,\ B = 68°,\ a = 12.2$

2. $B = 104°,\ C = 33°,\ a = 18.1$

3. $A = 24°,\ a = 11.2,\ b = 13.4$

4. $a = 4.0,\ b = 7.3,\ c = 12.4$

5. $B = 100°,\ a = 15,\ b = 23$

6. $C = 123°,\ a = 41,\ b = 57$

7. A triangular parcel of land has borders of lengths 60 meters, 70 meters, and 82 meters. Find the area of the parcel of land.

8. An airplane flies 370 miles from point A to point B with a bearing of 24°. It then flies 240 miles from point B to point C with a bearing of 37° (see figure). Find the distance and bearing from point A to point C.

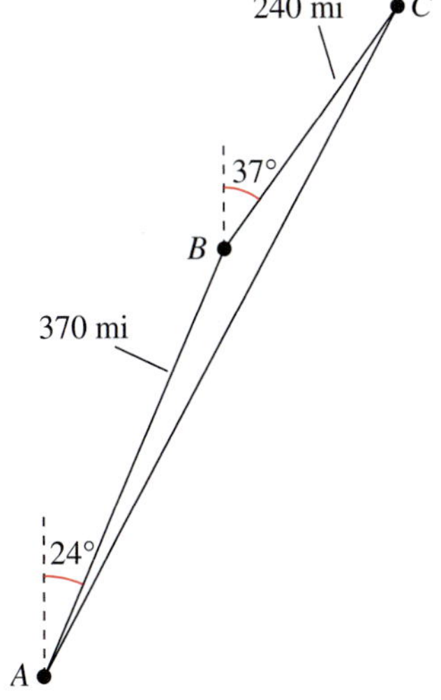

FIGURE FOR 8

In Exercises 9 and 10, find the component form of the vector v with the given components.

9. Initial point of $\mathbf{v}$: $(-3, 7)$; Terminal point of $\mathbf{v}$: $(11, -16)$

10. Magnitude of $\mathbf{v}$: $\|\mathbf{v}\| = 12$; Direction of $\mathbf{v}$: $\mathbf{u} = \langle 3, -5 \rangle$

In Exercises 11–13, u = ⟨3, 5⟩ and v = ⟨−7, 1⟩. Find the resultant vector and sketch its graph.

11. $\mathbf{u} + \mathbf{v}$

12. $\mathbf{u} - \mathbf{v}$

13. $5\mathbf{u} - 3\mathbf{v}$

14. Find a unit vector in the direction of $\mathbf{u} = \langle 4, -3 \rangle$.

15. Forces with magnitudes of 250 pounds and 130 pounds act on an object at angles of 45° and −60°, respectively, with the x-axis. Find the direction and magnitude of the resultant of these forces.

16. Find the angle between the vectors $\mathbf{u} = \langle -1, 5 \rangle$ and $\mathbf{v} = \langle 3, -2 \rangle$.

17. Are the vectors $\mathbf{u} = \langle 6, 10 \rangle$ and $\mathbf{v} = \langle 2, 3 \rangle$ orthogonal?

18. Find the projection of $\mathbf{u} = \langle 6, 7 \rangle$ onto $\mathbf{v} = \langle -5, -1 \rangle$ and the vector component of $\mathbf{u}$ orthogonal to $\mathbf{v}$.

19. Write the complex number $z = 5 - 5i$ in trigonometric form.

20. Write the complex number $z = 6(\cos 120° + i \sin 120°)$ in standard form.

In Exercises 21 and 22, use DeMoivre's Theorem to find the indicated power of the complex number.

21. $\left[3\left(\cos \frac{7\pi}{6} + i \sin \frac{7\pi}{6}\right)\right]^8$

22. $(3 - 3i)^6$

23. Find the fourth roots of $256(1 + \sqrt{3}i)$.

24. Find all solutions of the equation $x^3 - 27i = 0$ and represent the solutions graphically.

Cumulative Test for Chapters 6–8

Take this test to review the material from earlier chapters. When you are finished, check your work against the answers given in the back of the book.

1. Consider the angle $\theta = -120°$.
   (a) Sketch the angle in standard position.
   (b) Determine a coterminal angle in the interval $[0°, 360°)$.
   (c) Convert the angle to radian measure.
   (d) Find the reference angle θ'.
   (e) Find the exact values of the six trigonometric functions of θ.
2. Convert the angle of measure 2.35 radians to degrees. Round the answer to one decimal place.
3. Find $\cos\theta$ if $\tan\theta = -\frac{4}{3}$ and $\sin\theta < 0$.

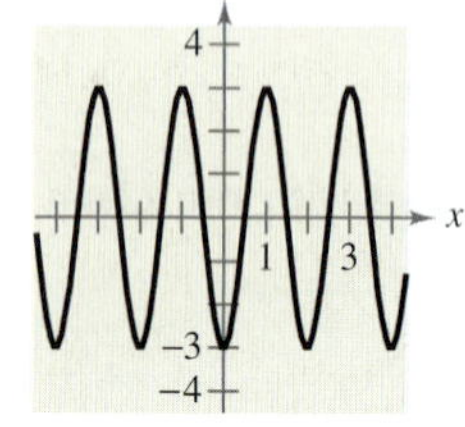

FIGURE FOR 6

In Exercises 4 and 5, find the period and amplitude, and sketch the graph of the trigonometric function.

4. $f(x) = 3 - 2\sin\pi x$
5. $g(x) = \frac{1}{2}\tan\left(x - \frac{\pi}{2}\right)$
6. Find a, b, and c such that the graph of the function $h(x) = a\cos(bx + c)$ matches the graph in the figure.
7. Sketch the graph of the function $f(x) = \frac{1}{2}x\sin x$ over the interval $-3\pi \le x \le 3\pi$.

In Exercises 8 and 9, find the exact value of the expression without the aid of a calculator.

8. $\tan(\arctan 6.7)$
9. $\tan\left(\arcsin\frac{3}{5}\right)$
10. Write an algebraic expression equivalent to $\sin(\arccos 2x)$.
11. Use the fundamental identities to simplify: $\cos\left(\frac{\pi}{2} - x\right)\csc x$.
12. Subtract and simplify: $\dfrac{\sin\theta - 1}{\cos\theta} - \dfrac{\cos\theta}{\sin\theta - 1}$.

In Exercises 13–15, prove the identity.

13. $\cot^2\alpha(\sec^2\alpha - 1) = 1$
14. $\sin(x + y)\sin(x - y) = \sin^2 x - \sin^2 y$
15. $\sin^2 x\cos^2 x = \frac{1}{8}(1 - \cos 4x)$

In Exercises 16 and 17, find all solutions of the equation in the interval $[0, 2\pi)$.

16. $2\cos^2\beta - \cos\beta = 0$
17. $3\tan\theta - \cot\theta = 0$
18. Use the Quadratic Formula to solve the equation in the interval $[0, 2\pi)$: $\sin^2 x + 2\sin x + 1 = 0$.

19. Given that $\sin u = \frac{12}{13}$, $\cos v = \frac{3}{5}$, and angles u and v are both in Quadrant I, find $\tan(u - v)$.

20. If $\tan \theta = \frac{1}{2}$, find the exact value of $\tan(2\theta)$.

21. If $\tan \theta = \dfrac{4}{3}$, find the exact value of $\sin \dfrac{\theta}{2}$.

22. Write the product $5 \sin \dfrac{3\pi}{4} \cdot \cos \dfrac{7\pi}{4}$ as a sum or difference.

In Exercises 23–26, use the information to solve the triangle shown in the figure.

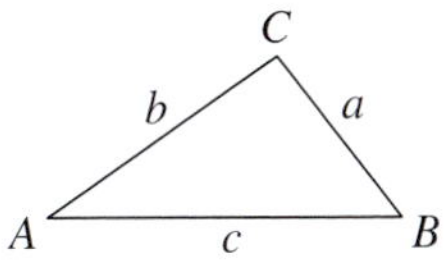

FIGURE FOR 23–26

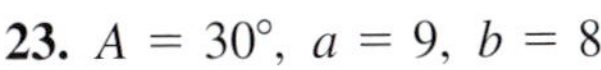

23. $A = 30°,\ a = 9,\ b = 8$

24. $A = 30°, b = 8,\ c = 10$

25. $A = 30°,\ C = 90°,\ b = 10$

26. $a = 4,\ b = 8,\ c = 9$

27. Two sides of a triangle have lengths 7 inches and 12 inches. Their included angle measures 60°. Find the area of the triangle.

28. Find the area of a triangle with sides of lengths 11 inches, 16 inches, and 17 inches.

29. Write vector $\mathbf{u} = \langle 3, 5 \rangle$ as a linear combination of the standard unit vectors $\mathbf{i}$ and $\mathbf{j}$.

30. Find $\mathbf{u} \cdot \mathbf{v}$ for $\mathbf{u} = 3\mathbf{i} + 4\mathbf{j}$ and $\mathbf{v} = \mathbf{i} - 2\mathbf{j}$.

31. Find the projection of $\mathbf{u} = \langle 8, -2 \rangle$ onto $\mathbf{v} = \langle 1, 5 \rangle$ and the vector component of $\mathbf{u}$ orthogonal to $\mathbf{v}$.

32. Find the trigonometric form of the complex number $-2 + 2i$.

33. Find the product of $[4(\cos 30° + i \sin 30°)][6(\cos 120° + i \sin 120°)]$. Write the answer in standard form.

34. Find the three cube roots of 1.

35. Write all the solutions of the equation $x^4 - 256i = 0$.

36. From a point 200 feet from a flagpole, the angles of elevation to the bottom and top of the flag are $16° 45'$ and 18°, respectively. Approximate the height of the flag to the nearest foot.

37. A compact disc can have an angular speed up to 3142 radians per minute. At this angular speed, how many revolutions per minute would the CD make? How long would it take the CD to make 10,000 revolutions?

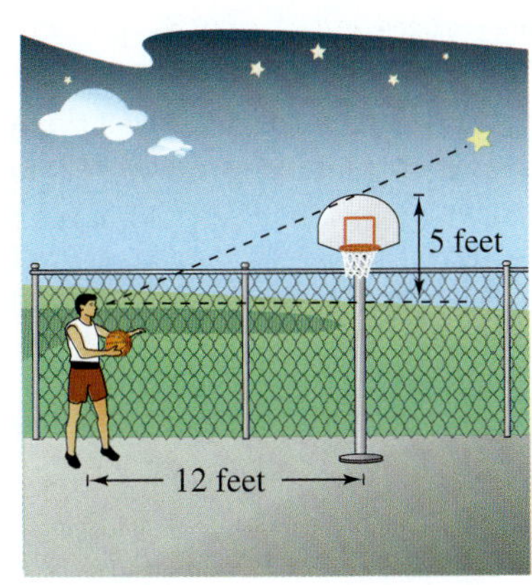

FIGURE FOR 38

38. To determine the angle of elevation of a star in the sky, you get the star in your line of vision with the backboard of a basketball hoop that is 5 feet higher than your eyes (see figure). Your horizontal distance from the backboard is 12 feet. What is the angle of elevation of the star?

39. Write a model for a particle in simple harmonic motion with a displacement of 4 inches and a period of 8 seconds.

40. An airplane's velocity with respect to the air is 500 kilometers per hour, with a bearing of 30°. The wind at the altitude of the plane has a velocity of 50 kilometers per hour with a bearing of N 60° E. What is the true direction of the plane, and what is its speed relative to the ground?

Proofs in Mathematics

Law of Tangents

Besides the Law of Sines and the Law of Cosines in trigonometry there is also a Law of Tangents developed by Francois Viète (1540–1603). The Law of Tangents follows from the Law of Sines and the sum-to-product formulas for sine and is defined as follows.

$$\frac{a+b}{a-b} = \frac{\tan(A+B)/2}{\tan(A-B)/2}$$

The Law of Tangents can be used to solve a triangle when two sides and the included angle are given (SAS). Before calculators were invented, the Law of Tangents was used to solve the SAS case instead of the Law of Cosines, because computation with a table of tangent values was easier.

Law of Sines (p. 598)

If ABC is a triangle with sides a, b, and c, then

$$\frac{a}{\sin A} = \frac{b}{\sin B} = \frac{c}{\sin C}.$$

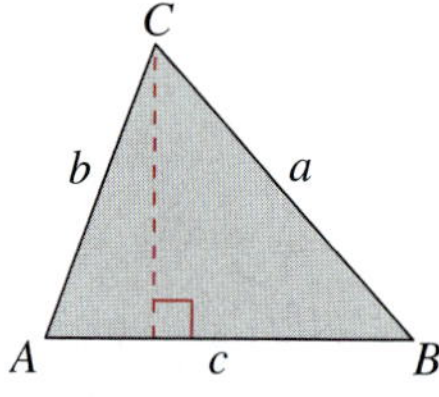

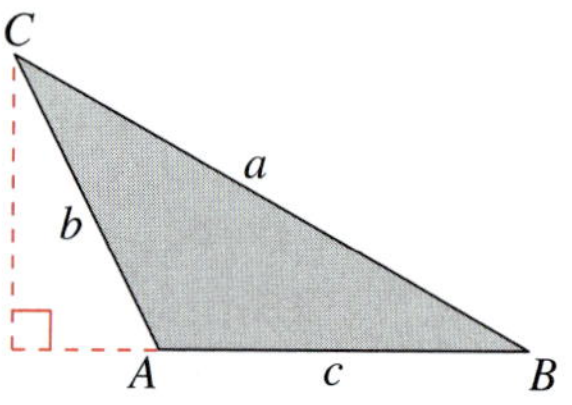

Proof

Let h be the altitude of either triangle found in the figure above. Then you have

$$\sin A = \frac{h}{b} \quad \text{or} \quad h = b \sin A$$

$$\sin B = \frac{h}{a} \quad \text{or} \quad h = a \sin B.$$

Equating these two values of h, you have

$$a \sin B = b \sin A \quad \text{or} \quad \frac{a}{\sin A} = \frac{b}{\sin B}.$$

Note that $\sin A \neq 0$ and $\sin B \neq 0$ because no angle of a triangle can have a measure of 0° or 180°. In a similar manner, construct an altitude from vertex B to side AC (extended in the obtuse triangle), as shown at the left. Then you have

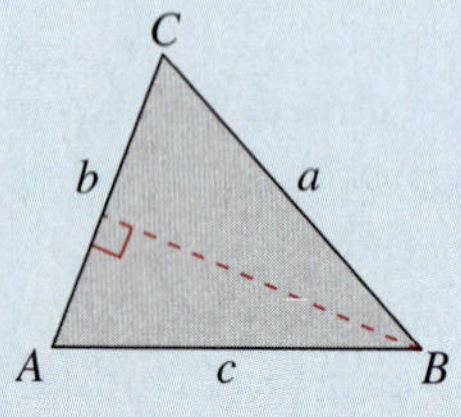

A is acute

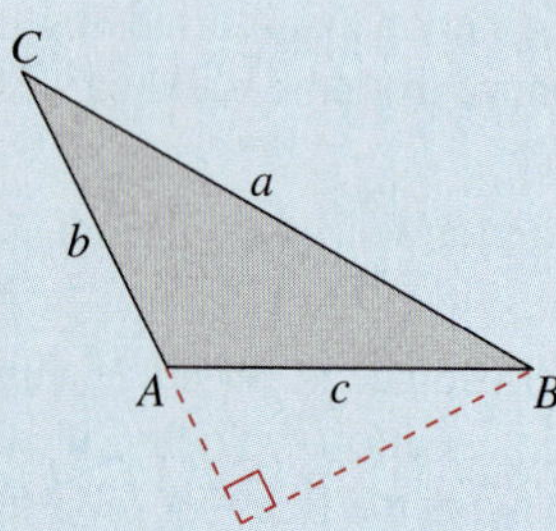

A is obtuse

$$\sin A = \frac{h}{c} \quad \text{or} \quad h = c \sin A$$

$$\sin C = \frac{h}{a} \quad \text{or} \quad h = a \sin C.$$

Equating these two values of h, you have

$$a \sin C = c \sin A \quad \text{or} \quad \frac{a}{\sin A} = \frac{c}{\sin C}.$$

By the Transitive Property of Equality you know that

$$\frac{a}{\sin A} = \frac{b}{\sin B} = \frac{c}{\sin C}.$$

So, the Law of Sines is established.

Law of Cosines *(p. 607)*

Standard Form	*Alternative Form*
$a^2 = b^2 + c^2 - 2bc \cos A$	$\cos A = \dfrac{b^2 + c^2 - a^2}{2bc}$
$b^2 = a^2 + c^2 - 2ac \cos B$	$\cos B = \dfrac{a^2 + c^2 - b^2}{2ac}$
$c^2 = a^2 + b^2 - 2ab \cos C$	$\cos C = \dfrac{a^2 + b^2 - c^2}{2ab}$

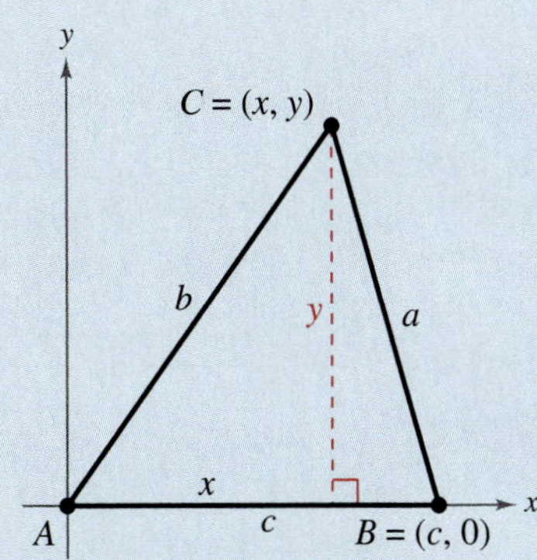

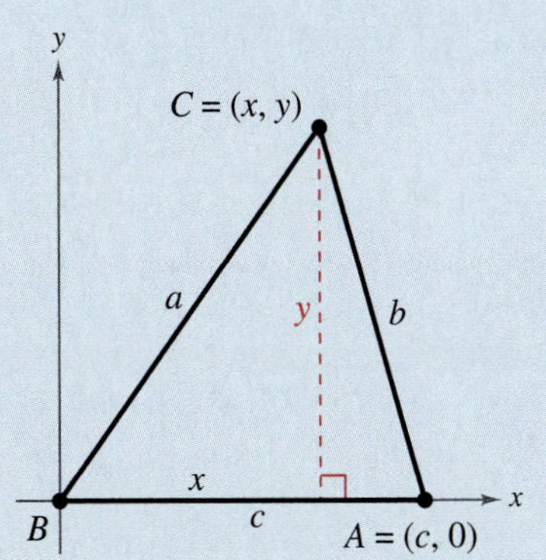

Proof

To prove the first formula, consider the top triangle at the left, which has three acute angles. Note that vertex B has coordinates $(c, 0)$. Furthermore, C has coordinates (x, y), where $x = b \cos A$ and $y = b \sin A$. Because a is the distance from vertex C to vertex B, it follows that

$$a = \sqrt{(x - c)^2 + (y - 0)^2}$$ Distance Formula

$$a^2 = (x - c)^2 + (y - 0)^2$$ Square each side.

$$a^2 = (b \cos A - c)^2 + (b \sin A)^2$$ Substitute for x and y.

$$a^2 = b^2 \cos^2 A - 2bc \cos A + c^2 + b^2 \sin^2 A$$ Expand.

$$a^2 = b^2(\sin^2 A + \cos^2 A) + c^2 - 2ab \cos A$$ Factor out b^2.

$$a^2 = b^2 + c^2 - 2bc \cos A.$$ $\sin^2 A + \cos^2 A = 1$

To prove the second formula, consider the bottom triangle at the left, which also has three acute angles. Note that vertex A has coordinates $(c, 0)$. Furthermore, C has coordinates (x, y), where $x = a \cos B$ and $y = a \sin B$. Because b is the distance from vertex C to vertex A, it follows that

$$b = \sqrt{(x - c)^2 + (y - 0)^2}$$ Distance Formula

$$b^2 = (x - c)^2 + (y - 0)^2$$ Square each side.

$$b^2 = (a \cos B - c)^2 + (a \sin B)^2$$ Substitute for x and y.

$$b^2 = a^2 \cos^2 B - 2ac \cos B + c^2 + a^2 + \sin^2 B$$ Expand.

$$b^2 = a^2(\sin^2 B + \cos^2 B) + c^2 - 2ac \cos B$$ Factor out a^2.

$$b^2 = a^2 + c^2 - 2ac \cos B.$$ $\sin^2 B + \cos^2 B = 1$

Try using a similar argument to establish the third formula.

Heron's Area Formula *(p. 610)*

Given any triangle with sides of lengths a, b, and c, the area of the triangle is

$$\text{Area} = \sqrt{s(s-a)(s-b)(s-c)}$$

where $s = \dfrac{(a+b+c)}{2}$.

Proof

From Section 8.1, you know that

$$\text{Area} = \frac{1}{2}bc \sin A \qquad \text{Formula for the area of an oblique triangle.}$$

$$= \frac{1}{4}b^2c^2 \sin^2 A \qquad \text{Square the right side.}$$

$$= \sqrt{\frac{1}{4}b^2c^2 \sin^2 A} \qquad \text{Take the square root of the right side.}$$

$$= \sqrt{\frac{1}{4}b^2c^2(1-\cos^2 A)} \qquad \text{Pythagorean Identity}$$

$$= \sqrt{\left[\frac{1}{2}bc(1+\cos A)\right]\left[\frac{1}{2}bc(1-\cos A)\right]}. \qquad \text{Factor.}$$

Using the Law of Cosines, you can show that

$$\frac{1}{2}bc(1+\cos A) = \frac{a+b+c}{2} \cdot \frac{-a+b+c}{2}$$

and

$$\frac{1}{2}bc(1-\cos A) = \frac{a-b+c}{2} \cdot \frac{a+b-c}{2}.$$

Letting $s = (a+b+c)/2$, these two equations can be rewritten as

$$\frac{1}{2}bc(1+\cos A) = s(s-a)$$

and

$$\frac{1}{2}bc(1-\cos A) = (s-b)(s-c).$$

By substituting into the last formula for area, you can conclude that

$$\text{Area} = \sqrt{s(s-a)(s-b)(s-c)}.$$

Properties of the Dot Product (p. 628)

Let **u**, **v**, and **w** be vectors in the place or in space and let c be a scalar.

1. $\mathbf{u} \cdot \mathbf{v} = \mathbf{v} \cdot \mathbf{u}$
2. $\mathbf{0} \cdot \mathbf{v} = 0$
3. $\mathbf{u} \cdot (\mathbf{v} + \mathbf{w}) = \mathbf{u} \cdot \mathbf{v} + \mathbf{u} \cdot \mathbf{w}$
4. $\mathbf{v} \cdot \mathbf{v} = \|\mathbf{v}\|^2$
5. $c(\mathbf{u} \cdot \mathbf{v}) = c\mathbf{u} \cdot \mathbf{v} = \mathbf{u} \cdot c\mathbf{v}$

Proof

Let $\mathbf{u} = \langle u_1, u_2 \rangle$, $\mathbf{v} = \langle v_1, v_2 \rangle$, $\mathbf{w} = \langle w_1, w_2 \rangle$, $\mathbf{0} = \langle 0, 0 \rangle$, and let c be a scalar.

1. $\mathbf{u} \cdot \mathbf{v} = u_1v_1 + u_2v_2 = v_1u_1 + v_2u_2 = \mathbf{v} \cdot \mathbf{u}$
2. $\mathbf{0} \cdot \mathbf{v} = 0 \cdot v_1 + 0 \cdot v_2 = 0$
3. $$\begin{aligned} \mathbf{u} \cdot (\mathbf{v} + \mathbf{w}) &= \mathbf{u} \cdot \langle v_1 + w_1, v_2 + w_2 \rangle \\ &= u_1(v_1 + w_1) + u_2(v_2 + w_2) \\ &= u_1v_1 + u_1w_1 + u_2v_2 + u_2w_2 \\ &= (u_1v_1 + u_2v_2) + (u_1w_1 + u_2w_2) = \mathbf{u} \cdot \mathbf{v} + \mathbf{u} \cdot \mathbf{w} \end{aligned}$$
4. $\mathbf{v} \cdot \mathbf{v} = v_1^2 + v_2^2 = \left(\sqrt{v_1^2 + v_2^2}\right)^2 = \|\mathbf{v}\|^2$
5. $$\begin{aligned} c(\mathbf{u} \cdot \mathbf{v}) &= c(\langle u_1, u_2 \rangle \cdot \langle v_1, v_2 \rangle) \\ &= c(u_1v_1 + u_2v_2) = (cu_1)v_1 + (cu_2)v_2 \\ &= \langle cu_1, cu_2 \rangle \cdot \langle v_1, v_2 \rangle = c\mathbf{u} \cdot \mathbf{v} \end{aligned}$$

Angle Between Two Vectors (p. 629)

If θ is the angle between two nonzero vectors **u** and **v**, then $\cos\theta = \dfrac{\mathbf{u} \cdot \mathbf{v}}{\|\mathbf{u}\|\,\|\mathbf{v}\|}$.

Proof

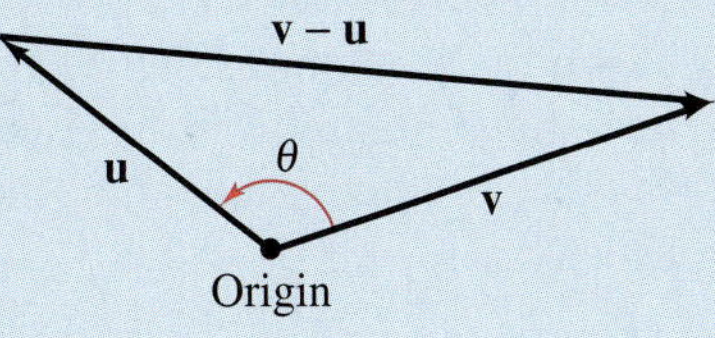

Consider the triangle determined by vectors **u**, **v**, and **v** − **u**, as shown in the figure. By the Law of Cosines, you can write

$$\begin{aligned} \|\mathbf{v} - \mathbf{u}\|^2 &= \|\mathbf{u}\|^2 + \|\mathbf{v}\|^2 - 2\|\mathbf{u}\|\,\|\mathbf{v}\|\cos\theta \\ (\mathbf{v} - \mathbf{u}) \cdot (\mathbf{v} - \mathbf{u}) &= \|\mathbf{u}\|^2 + \|\mathbf{v}\|^2 - 2\|\mathbf{u}\|\,\|\mathbf{v}\|\cos\theta \\ (\mathbf{v} - \mathbf{u}) \cdot \mathbf{v} - (\mathbf{v} - \mathbf{u}) \cdot \mathbf{u} &= \|\mathbf{u}\|^2 + \|\mathbf{v}\|^2 - 2\|\mathbf{u}\|\,\|\mathbf{v}\|\cos\theta \\ \mathbf{v} \cdot \mathbf{v} - \mathbf{u} \cdot \mathbf{v} - \mathbf{v} \cdot \mathbf{u} + \mathbf{u} \cdot \mathbf{u} &= \|\mathbf{u}\|^2 + \|\mathbf{v}\|^2 - 2\|\mathbf{u}\|\,\|\mathbf{v}\|\cos\theta \\ \|\mathbf{v}\|^2 - 2\mathbf{u} \cdot \mathbf{v} + \|\mathbf{u}\|^2 &= \|\mathbf{u}\|^2 + \|\mathbf{v}\|^2 - 2\|\mathbf{u}\|\,\|\mathbf{v}\|\cos\theta \\ \cos\theta &= \frac{\mathbf{u} \cdot \mathbf{v}}{\|\mathbf{u}\|\,\|\mathbf{v}\|}. \end{aligned}$$

P.S. Problem Solving

1. In the figure, a beam of light is directed at the blue mirror, reflected to the red mirror, and then reflected back to the blue mirror. Find the distance PT that the light travels from the red mirror back to the blue mirror.

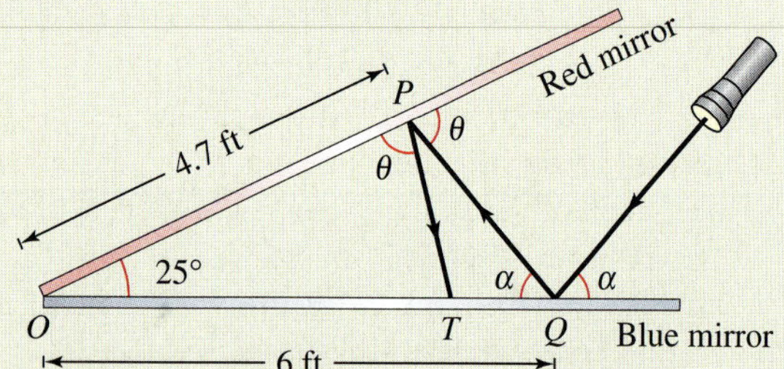

2. A triathlete sets a course to swim S 25° E from a point on shore to a buoy $\frac{3}{4}$ mile away. After swimming 300 yards through a strong current, the triathlete is off course at a bearing of S 35° E. Find the bearing and distance the triathlete needs to swim to correct her course.

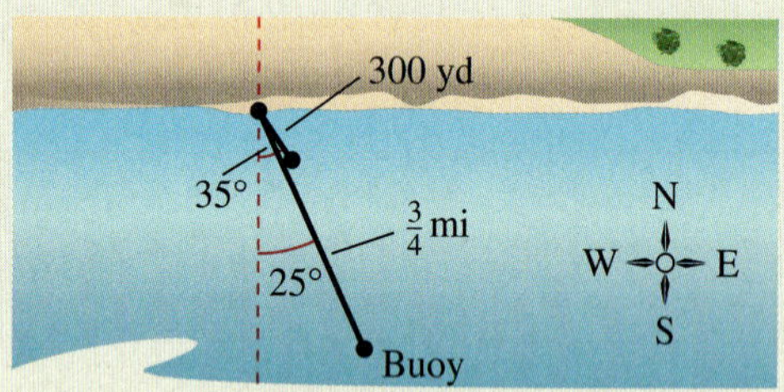

3. A hiking party is lost in a national park. Two ranger stations have received an emergency SOS signal from the party. Station B is 75 miles due east of station A. The bearing from station A to the signal is S 60° E and the bearing from station B to the signal is S 75° W.

(a) Find the distance from each station to the SOS signal.

(b) A rescue party is in the park 20 miles from station A at a bearing of S 80° E. Find the distance and the bearing the rescue party must travel to reach the lost hiking party.

4. You are seeding a triangular courtyard. One side of the courtyard is 52 feet long and the other side is 46 feet long. The angle opposite the 52-foot side is 65°.

(a) Draw a diagram that gives a visual representation of the problem.

(b) How long is the third side of the courtyard?

(c) One bag of grass covers an area of 50 square feet. How many bags of grass will you need to cover the courtyard?

5. For each pair of vectors, find the following.

(i) $\|\mathbf{u}\|$ (ii) $\|\mathbf{v}\|$ (iii) $\|\mathbf{u}+\mathbf{v}\|$

(iv) $\left\|\frac{\mathbf{u}}{\|\mathbf{u}\|}\right\|$ (v) $\left\|\frac{\mathbf{v}}{\|\mathbf{v}\|}\right\|$ (vi) $\left\|\frac{\mathbf{u}+\mathbf{v}}{\|\mathbf{u}+\mathbf{v}\|}\right\|$

(a) $\mathbf{u} = \langle 1, -1\rangle$
$\mathbf{v} = \langle -1, 2\rangle$

(b) $\mathbf{u} = \langle 0, 1\rangle$
$\mathbf{v} = \langle 3, -3\rangle$

(c) $\mathbf{u} = \left\langle 1, \frac{1}{2}\right\rangle$
$\mathbf{v} = \langle 2, 3\rangle$

(d) $\mathbf{u} = \langle 2, -4\rangle$
$\mathbf{v} = \langle 5, 5\rangle$

6. A skydiver is falling at a constant downward velocity of 120 miles per hour. In the figure, vector **u** represents the skydiver's velocity. A steady breeze pushes the skydiver to the east at 40 miles per hour. Vector **v** represents the wind velocity.

(a) Write the vectors **u** and **v** in component form.

(b) Let $\mathbf{s} = \mathbf{u} + \mathbf{v}$. Use the figure to sketch **s**. To print an enlarged copy of the graph, go to the website, *www.mathgraphs.com*.

(c) Find the magnitude of **s**. What information does the magnitude give you about the skydiver's fall?

(d) If there were no wind, the skydiver would fall in a path perpendicular to Earth. At what angle to the ground is the path of the skydiver when the skydiver is affected by the 40 mile per hour wind from due west?

(e) The skydiver is blown to the west at 30 miles per hour. Draw a new figure that gives a visual representation of the problem and find the skydiver's new velocity.

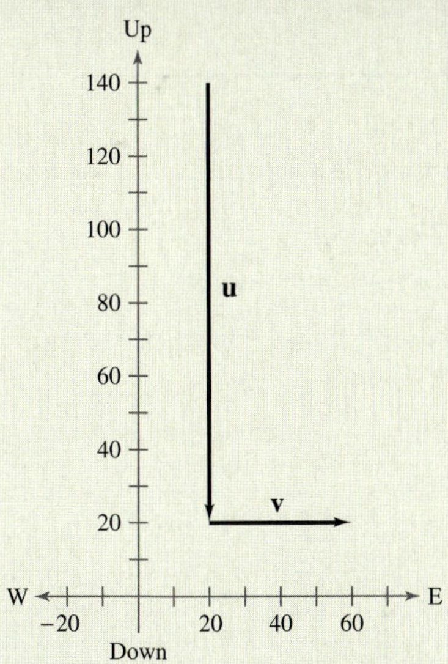

FIGURE FOR 6

7. Write the vector $\mathbf{w}$ in terms of $\mathbf{u}$ and $\mathbf{v}$, given that the terminal point of $\mathbf{w}$ bisects the line segment.

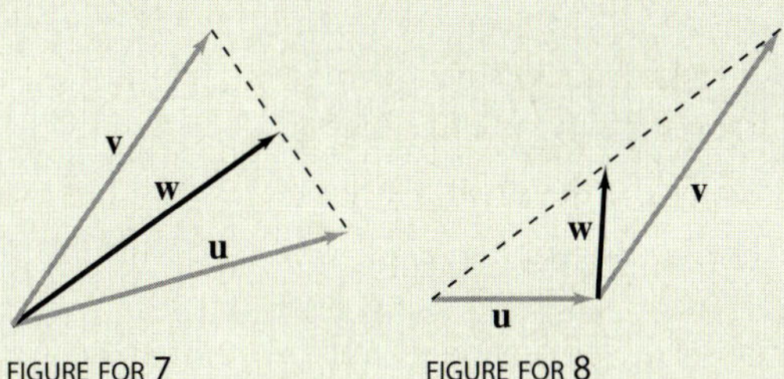

FIGURE FOR 7 FIGURE FOR 8

8. Prove that if $\mathbf{u}$ is orthogonal to $\mathbf{v}$ and $\mathbf{w}$, then $\mathbf{u}$ is orthogonal to

$c\mathbf{v} + d\mathbf{w}$

for any scalars c and d.

9. Two forces of the same magnitude $\mathbf{F}_1$ and $\mathbf{F}_2$ act at angles θ_1 and θ_2, respectively. Use a diagram to compare the work done by $\mathbf{F}_1$ with the work done by $\mathbf{F}_2$ in moving along the vector PQ if

(a) $\theta_1 = -\theta_2$

(b) $\theta_1 = 60°$ and $\theta_2 = 30°$.

10. Four basic forces are in action during flight: weight, lift, thrust, and drag. To fly through the air, an object must overcome its own *weight*. To do this, it must create an upward force called *lift*. To generate lift, a forward motion called *thrust* is needed. The thrust must be great enough to overcome air resistance, which is called *drag*.

For a commercial jet aircraft, a quick climb is important to maximize efficiency, because the performance of an aircraft at high altitudes is enhanced. In addition, it is necessary to clear obstacles such as buildings and mountains and reduce noise in residential areas. In the diagram, the angle θ is called the climb angle. The velocity of the plane can be represented by a vector $\mathbf{v}$ with a vertical component $\|\mathbf{v}\| \sin \theta$ (called climb speed) and a horizontal component $\|\mathbf{v}\| \cos \theta$, where $\|\mathbf{v}\|$ is the speed of the plane. When taking off, a pilot must decide how much of the thrust to apply to each component. The more the thrust is applied to the horizontal component, the faster the airplane will gain speed. The more the thrust is applied to the vertical component, the quicker the airplane will climb.

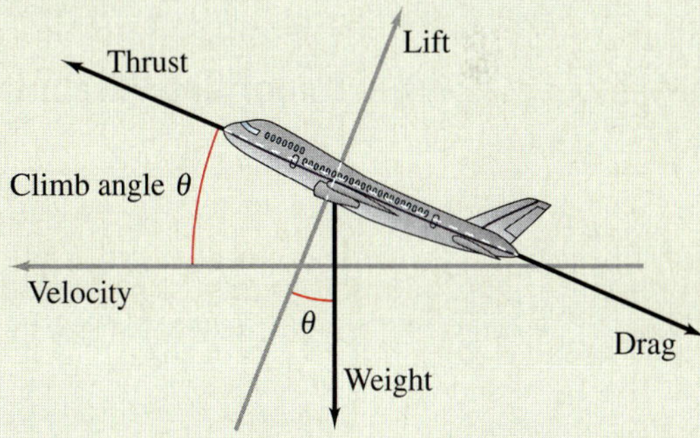

(a) Complete the table for an airplane that has a speed of $\|\mathbf{v}\| = 100$ miles per hour.

θ	0.5°	1.0°	1.5°	2.0°	2.5°	3.0°
$\|\mathbf{v}\| \sin \theta$						
$\|\mathbf{v}\| \cos \theta$						

(b) Does an airplane's speed equal the sum of the vertical and horizontal components of its velocity? If not, how could you find the speed of an airplane whose velocity components were known?

(c) Use the result of part (b) to find the speed of an airplane with the given velocity components.

(i) $\|\mathbf{v}\| \sin \theta = 5.235$ miles per hour
$\|\mathbf{v}\| \cos \theta = 149.909$ miles per hour

(ii) $\|\mathbf{v}\| \sin \theta = 10.463$ miles per hour
$\|\mathbf{v}\| \cos \theta = 149.634$ miles per hour

How to study Chapter 9

What you should learn

In this chapter you will learn the following skills and concepts:

- How to solve systems of equations by substitution, by elimination, by Gaussian elimination, and by graphing
- How to recognize linear systems in row-echelon form and to use back-substitution to solve the systems
- How to solve nonsquare systems of equations
- How to sketch the graphs of inequalities in two variables and to solve systems of inequalities
- How to solve linear programming problems
- How to use systems of equations and inequalities to model and solve real-life problems

Important Vocabulary

As you encounter each new vocabulary term in this chapter, add the term and its definition to your notebook glossary.

System of equations (p. 664)
Solution of a system of equations (p. 664)
Solving a system of equations (p. 664)
Method of substitution (p. 664)
Graphical method (p. 668)
Points of intersection (p. 668)
Break-even point (p. 669)
Method of elimination (p. 675)
Equivalent systems (p. 676)
Consistent system (p. 678)
Inconsistent system (p. 678)
Row-echelon form (p. 687)
Ordered triple (p. 687)
Row operations (p. 688)
Gaussian elimination (p. 688)
Nonsquare system of equations (p. 692)
Position equation (p. 693)
Solution of an inequality (p. 700)
Graph of an inequality (p. 700)
Linear inequalities (p. 701)
Solution of a system of inequalities (p. 702)
Consumer surplus (p. 705)
Producer surplus (p. 705)
Optimization (p. 711)
Linear programming (p. 711)
Objective function (p. 711)
Constraints (p. 711)
Feasible solutions (p. 711)

Study Tools

Learning objectives in each section
Chapter Summary (p. 721)
Review Exercises (pp. 722–725)
Chapter Test (p. 726)

Additional Resources

Study and Solutions Guide
Interactive College Algebra
Videotapes/DVD for Chapter 9
Algebra and Trigonometry Website
Student Success Organizer

Gavin Hellier/Getty Images

9

Systems of Equations and Inequalities

9.1 Solving Systems of Equations

▶ What you should learn

- How to use the method of substitution to solve systems of equations in two variables
- How to use a graphical approach to solve systems of equations in two variables
- How to use systems of equations to model and solve real-life problems

▶ Why you should learn it

Graphs of systems of equations help you solve real-life problems. For instance, in Exercise 75 on page 673, you can use the graph of a system of equations to approximate the points of intersection of two equations that model the number of visits to hospital emergency departments.

Spencer Grant/PhotoEdit

The Method of Substitution

Up to this point in the book, most problems have involved either a function of one variable or a single equation in two variables. However, many problems in science, business, and engineering involve two or more equations in two or more variables. To solve such problems, you need to find solutions of a **system of equations.** Here is an example of a system of two equations in two unknowns.

$$\begin{cases} 2x + y = 5 & \text{Equation 1} \\ 3x - 2y = 4 & \text{Equation 2} \end{cases}$$

A **solution** of this system is an ordered pair that satisfies each equation in the system. Finding the set of all solutions is called **solving the system of equations.** For instance, the ordered pair $(2, 1)$ is a solution of this system. To check this, you can substitute 2 for x and 1 for y in *each* equation.

Check (2, 1) in Equation 1 and Equation 2:

$2x + y = 5$ — Write Equation 1.

$2(2) + 1 \stackrel{?}{=} 5$ — Substitute 2 for x and 1 for y.

$4 + 1 = 5$ — Solution checks in Equation 1. ✓

$3x - 2y = 4$ — Write Equation 2.

$3(2) - 2(1) \stackrel{?}{=} 4$ — Substitute 2 for x and 1 for y.

$6 - 2 = 4$ — Solution checks in Equation 2. ✓

In this chapter you will study four ways to solve equations, beginning with the **method of substitution.**

Method	*Section*	*Type of System*
1. Substitution	9.1	Linear or nonlinear, two variables
2. Graphical method	9.1	Linear or nonlinear, two variables
3. Elimination	9.2	Linear, two variables
4. Gaussian elimination	9.3	Linear, three or more variables

Method of Substitution

1. *Solve* one of the equations for one variable in terms of the other.
2. *Substitute* the expression found in Step 1 into the other equation to obtain an equation in one variable.
3. *Solve* the equation obtained in Step 2.
4. *Back-substitute* the value obtained in Step 3 into the expression obtained in Step 1 to find the value of the other variable.
5. *Check* that the solution satisfies *each* of the original equations.

Exploration

Use a graphing utility to graph $y_1 = 4 - x$ and $y_2 = x - 2$ in the same viewing window. Use the *zoom* and *trace* features to find the coordinates of the point of intersection. Are the coordinates the same as the solution found in Example 1? Explain.

Example 1 ▶ Solving a System of Equations by Substitution

Solve the system of equations.

$$\begin{cases} x + y = 4 & \text{Equation 1} \\ x - y = 2 & \text{Equation 2} \end{cases}$$

Solution

Begin by solving for y in Equation 1.

$y = 4 - x$ — Solve for y in Equation 1.

Next, substitute this expression for y into Equation 2 and solve the resulting single-variable equation for x.

$x - y = 2$ — Write Equation 2.

$x - (4 - x) = 2$ — Substitute $4 - x$ for y.

$x - 4 + x = 2$ — Distributive Property

$2x = 6$ — Combine like terms.

$x = 3$ — Divide each side by 2.

Finally, you can solve for y by *back-substituting* $x = 3$ into the equation $y = 4 - x$, to obtain

$y = 4 - x$ — Write revised Equation 1.

$y = 4 - 3$ — Substitute 3 for x.

$y = 1.$ — Solve for y.

The solution is the ordered pair $(3, 1)$. You can check this solution as follows.

Check

Substitute $(3, 1)$ into Equation 1:

$x + y = 4$ — Write Equation 1.

$3 + 1 \stackrel{?}{=} 4$ — Substitute for x and y.

$4 = 4$ — Solution checks in Equation 1. ✓

Substitute $(3, 1)$ into Equation 2:

$x - y = 2$ — Write Equation 2.

$3 - 1 \stackrel{?}{=} 2$ — Substitute for x and y.

$2 = 2$ — Solution checks in Equation 2. ✓

Because $(3, 1)$ satisfies both equations in the system, it is a solution of the system of equations.

STUDY TIP

Because many steps are required to solve a system of equations, it is very easy to make errors in arithmetic. So, you should always check your solution by substituting it into *each* equation in the original system.

The term *back-substitution* implies that you work *backwards*. First you solve for one of the variables, and then you substitute that value *back* into one of the equations in the system to find the value of the other variable.

The icon identifies examples and concepts related to features of the Learning Tools CD-ROM and the *Interactive* and *Internet* versions of this text. For more details see the chart on pages *xix-xxiii*.

Example 2 ▶ Solving a System by Substitution 

A total of \$12,000 is invested in two funds paying 9% and 11% simple interest. The yearly interest is \$1180. How much is invested at each rate?

Solution

Verbal Model:

9% fund	+	11% fund	=	Total investment
9% interest	+	11% interest	=	Total interest

Labels:

Amount in 9% fund $= x$ (dollars)
Interest for 9% fund $= 0.09x$ (dollars)
Amount in 11% fund $= y$ (dollars)
Interest for 11% fund $= 0.11y$ (dollars)
Total investment $= \$12{,}000$ (dollars)
Total interest $= \$1{,}180$ (dollars)

System:

$$\begin{cases} x + y = 12{,}000 & \text{Equation 1} \\ 0.09x + 0.11y = 1{,}180 & \text{Equation 2} \end{cases}$$

To begin, it is convenient to multiply each side of Equation 2 by 100. This eliminates the need to work with decimals.

$$100(0.09x + 0.11y) = 100(1180) \quad \text{Multiply each side by 100.}$$

$$9x + 11y = 118{,}000 \quad \text{Revised Equation 2}$$

To solve this system, you can solve for x in Equation 1.

$$x = 12{,}000 - y \quad \text{Revised Equation 1}$$

Then, substitute this expression for x into revised Equation 2 and solve the resulting equation for y.

$$9x + 11y = 118{,}000 \quad \text{Write revised Equation 2.}$$

$$9(12{,}000 - y) + 11y = 118{,}000 \quad \text{Substitute } 12{,}000 - y \text{ for } x.$$

$$108{,}000 - 9y + 11y = 118{,}000 \quad \text{Distributive Property}$$

$$2y = 10{,}000 \quad \text{Combine like terms.}$$

$$y = 5{,}000 \quad \text{Divide each side by 2.}$$

Next, back-substitute the value $y = 5000$ to solve for x.

$$x = 12{,}000 - y \quad \text{Write revised Equation 1.}$$

$$x = 12{,}000 - 5000 \quad \text{Substitute 5000 for } y.$$

$$x = 7000 \quad \text{Simplify.}$$

The solution is (7000, 5000). So, \$7000 is invested at 9% and \$5000 is invested at 11%. Check this in the original problem.

Technology

One way to check the answers you obtain in this section is to use a graphing utility. For instance, enter the two equations in Example 2

$$y_1 = 12{,}000 - x$$

$$y_2 = \frac{1180 - 0.09x}{0.11}$$

and find an appropriate viewing window that shows where the two lines intersect. Then use the *intersect* feature or the *zoom* and *trace* features to find the point of intersection. Does this point agree with the solution obtained at the right?

The *Interactive* CD-ROM and *Internet* versions of this text offer a Try It for each example in the text.

Exploration

Use a graphing utility to graph the two equations in Example 3

$y_1 = x^2 + 4x - 7$

$y_2 = 2x + 1$

in the same viewing window. How many solutions do you think this system has? Repeat this experiment for the equations in Example 4. How many solutions does this system have? Explain your reasoning.

The equations in Examples 1 and 2 are linear. Substitution can also be used to solve systems in which one or both of the equations are nonlinear.

Example 3 ▶ Substitution: Two-Solution Case 

Solve the system of equations.

$$\begin{cases} x^2 + 4x - y = 7 & \text{Equation 1} \\ 2x - y = -1 & \text{Equation 2} \end{cases}$$

Solution

Begin by solving for y in Equation 2 to obtain $y = 2x + 1$. Next, substitute this expression for y into Equation 1 and solve for x.

$x^2 + 4x - y = 7$ Write Equation 1.

$x^2 + 4x - (2x + 1) = 7$ Substitute $2x + 1$ for y.

$x^2 + 2x - 1 = 7$ Simplify.

$x^2 + 2x - 8 = 0$ General form

$(x + 4)(x - 2) = 0$ Factor.

$x = -4, 2$ Solve for x.

Back-substituting these values of x to solve for the corresponding values of y produces the solutions $(-4, -7)$ and $(2, 5)$. Check these in the original system.

Example 4 ▶ Substitution: No-Real-Solution Case 

Solve the system of equations.

$$\begin{cases} -x + y = 4 & \text{Equation 1} \\ x^2 + y = 3 & \text{Equation 2} \end{cases}$$

Solution

Begin by solving for y in Equation 1 to obtain $y = x + 4$. Next, substitute this expression for y into Equation 2 and solve for x.

$x^2 + y = 3$ Write Equation 2.

$x^2 + (x + 4) = 3$ Substitute $x + 4$ for y.

$x^2 + x + 1 = 0$ Simplify.

$x = \dfrac{-1 \pm \sqrt{1^2 - 4(1)(1)}}{2}$ Use the Quadratic Formula.

$x = \dfrac{-1 \pm \sqrt{-3}}{2}$ Simplify.

Because the discriminant is negative, the equation $x^2 + x + 1 = 0$ has no (real) solution. So, this system has no (real) solution.

Technology

Most graphing calculators have built-in features that approximate the point(s) of intersection of two graphs. Typically, you must enter the equations of the graphs and visually locate a point of intersection before using the *intersect* feature.

Use this feature to find the points of intersection for the graphs in Figures 9.1–9.3. Be sure to adjust your viewing window so that you see all the points of intersection.

Graphical Approach to Finding Solutions

From Examples 2, 3, and 4, you can see that a system of two equations in two unknowns can have exactly one solution, more than one solution, or no solution. By using a **graphical method,** you can gain insight about the number of solutions and the location(s) of the solution(s) of a system of equations by graphing each of the equations in the same coordinate plane. The solutions of the system correspond to the **points of intersection** of the graphs. For instance, the two equations in Figure 9.1 graph as two lines with a *single point* of intersection; the two equations in Figure 9.2 graph as a parabola and a line with *two points* of intersection; and the two equations in Figure 9.3 graph as a line and a parabola that have *no points* of intersection.

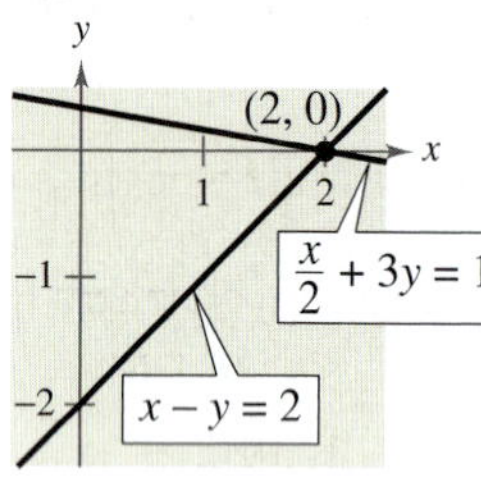

One intersection point
FIGURE 9.1

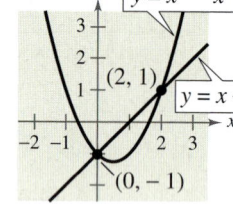

Two intersection points
FIGURE 9.2

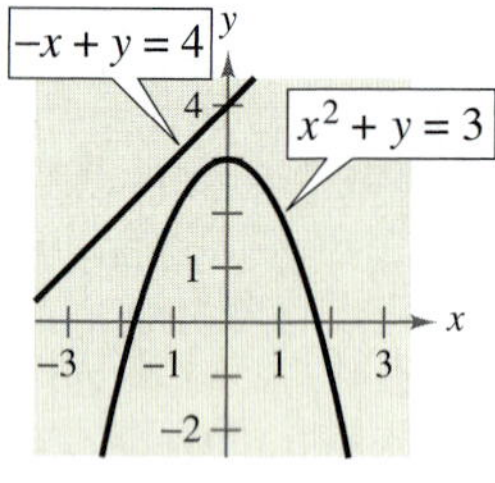

No intersection points
FIGURE 9.3

Example 5 ▶ Solving a System of Equations Graphically

Solve the system of equations.

$$\begin{cases} y = \ln x & \text{Equation 1} \\ x + y = 1 & \text{Equation 2} \end{cases}$$

Solution

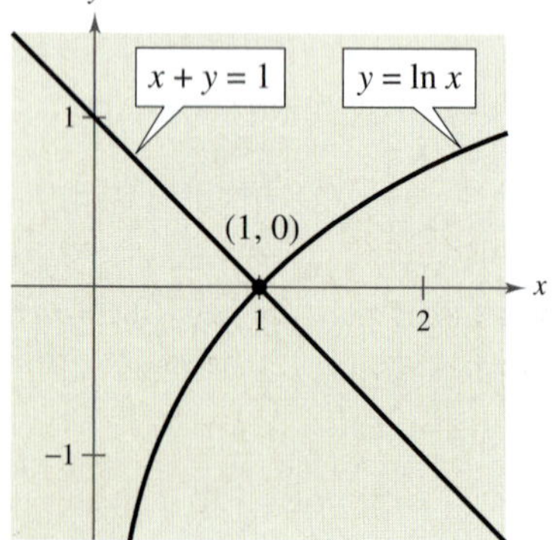

FIGURE 9.4

Sketch the graphs of the two equations, as shown in Figure 9.4. From the graphs, it is clear that there is only one point of intersection and that $(1, 0)$ is the solution point. You can confirm this by substituting 1 for x and 0 for y in *both* equations.

Check $(1, 0)$ in Equation 1:

$y = \ln x$ Write Equation 1.

$0 = \ln 1$ Equation 1 checks. ✓

Check $(1, 0)$ in Equation 2:

$x + y = 1$ Write Equation 2.

$1 + 0 = 1$ Equation 2 checks. ✓

Example 5 shows the value of a graphical approach to solving systems of equations in two variables. Notice what would happen if you tried only the substitution method in Example 5. You would obtain the equation $x + \ln x = 1$. It would be difficult to solve this equation for x using standard algebraic techniques.

Applications

The total cost C of producing x units of a product typically has two components—the initial cost and the cost per unit. When enough units have been sold so that the total revenue R equals the total cost C, the sales are said to have reached the **break-even point.** You will find that the break-even point corresponds to the point of intersection of the cost and revenue curves.

Example 6 ▶ Break-Even Analysis

A small business invests $10,000 in equipment to produce a product. Each unit of the product costs $0.65 to produce and is sold for $1.20. How many items must be sold before the business breaks even?

Solution

The total cost of producing x units is

$$\boxed{\text{Total cost}} = \boxed{\text{Cost per unit}} \cdot \boxed{\text{Number of units}} + \boxed{\text{Initial cost}}$$

$$C = 0.65x + 10{,}000. \qquad \text{Equation 1}$$

The revenue obtained by selling x units is

$$\boxed{\text{Total revenue}} = \boxed{\text{Price per unit}} \cdot \boxed{\text{Number of units}}$$

$$R = 1.20x. \qquad \text{Equation 2}$$

Because the break-even point occurs when $R = C$, you have $C = 1.20x$, and the system of equations to solve is

$$\begin{cases} C = 0.65x + 10{,}000 \\ C = 1.20x \end{cases}.$$

Now you can solve by substitution.

$1.20x = 0.65x + 10{,}000$ — Substitute $1.20x$ for C in Equation 1.

$0.55x = 10{,}000$ — Subtract $0.65x$ from each side.

$x \approx 18{,}182$ units — Divide each side by 0.55.

Note in Figure 9.5 that revenue less than the break-even point corresponds to an overall loss, whereas revenue greater than the break-even point corresponds to a profit.

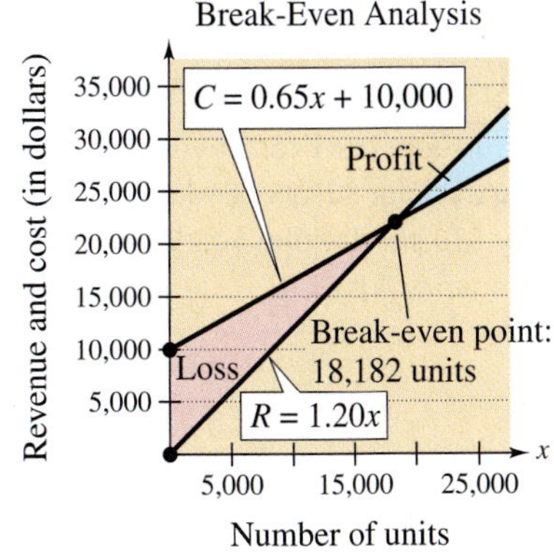

FIGURE 9.5

Another way to view the solution in Example 6 is to consider the profit function

$$P = R - C.$$

The break-even point occurs when the profit is 0, which is the same as saying that $R = C$.

The *Interactive* CD-ROM and *Internet* versions of this text offer a Quiz for every section of the text.

Example 7 ▶ State Population  

From 1990 to 2000, the population of Tennessee was increasing at a faster rate than the population of Missouri. Models that approximate the two populations P (in thousands) are

$$\begin{cases} P = 4866 + 74.5t & \text{Tennessee} \\ P = 5109 + 43.5t & \text{Missouri} \end{cases}$$

where $t = 0$ represents 1990 (see Figure 9.6). According to these two models, when would you expect the population of Tennessee to have exceeded the population of Missouri? (Source: U.S. Census Bureau)

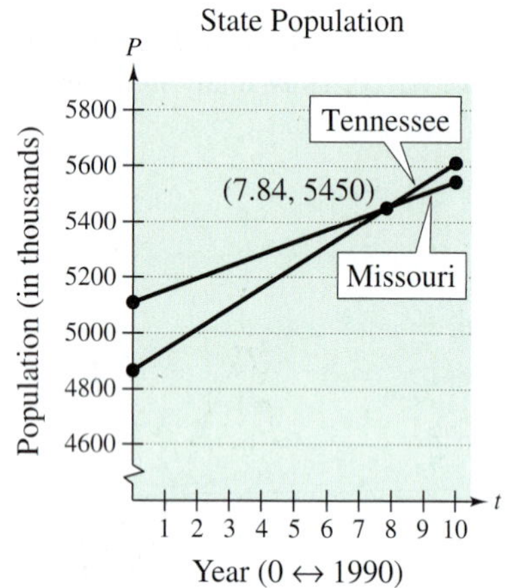

FIGURE 9.6

Solution

Because the first equation has already been solved for P in terms of t, substitute this value into the second equation and solve for t, as follows.

$$4866 + 74.5t = 5109 + 43.5t \quad \text{Substitute for } P \text{ in Equation 2.}$$

$$74.5t - 43.5t = 5109 - 4866 \quad \text{Subtract } 43.5t \text{ and 4866 from each side.}$$

$$31t = 243 \quad \text{Combine like terms.}$$

$$t \approx 7.84 \quad \text{Divide each side by 31.}$$

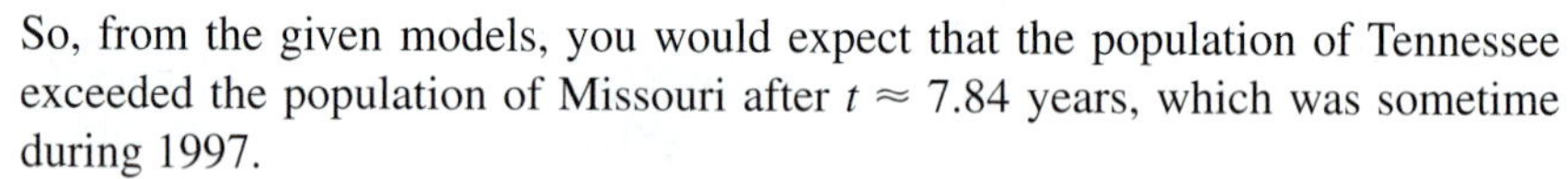

So, from the given models, you would expect that the population of Tennessee exceeded the population of Missouri after $t \approx 7.84$ years, which was sometime during 1997.

Writing ABOUT MATHEMATICS

Interpreting Points of Intersection You plan to rent a 14-foot truck for a two-day local move. At truck rental agency A, you can rent a truck for \$29.95 per day plus \$0.49 per mile. At agency B, you can rent a truck for \$50 per day plus \$0.25 per mile. The total cost y (in dollars) for the truck from agency A is

$$y = (\$29.95 \text{ per day})(2 \text{ days}) + 0.49x$$

$$= 59.90 + 0.49x$$

where x is the total number of miles the truck is driven.

a. Write a total cost equation in terms of x and y for the total cost of the truck from agency B.

b. Use a graphing utility to graph the two equations in the same viewing window and find the point of intersection. Interpret the meaning of the point of intersection in the context of the problem.

c. Which agency should you choose if you plan to travel a total of 100 miles during the two-day move? Why?

d. How does the situation change if you plan to drive 200 miles during the two-day move?

9.1 Exercises

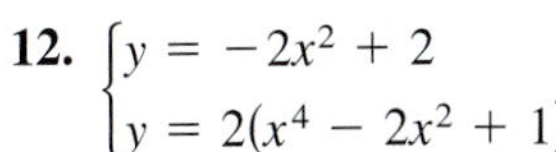
The *Interactive* CD-ROM and *Internet* versions of this text contain step-by-step solutions to all odd-numbered exercises. They also provide Tutorial Exercises for additional help.

In Exercises 1–4, determine which ordered pairs are solutions of the system of equations.

1. $\begin{cases} 4x - y = 1 \\ 6x + y = -6 \end{cases}$ (a) $(0, -3)$ (b) $(-1, -4)$ (c) $\left(-\frac{3}{2}, -2\right)$ (d) $\left(-\frac{1}{2}, -3\right)$

2. $\begin{cases} 4x^2 + y = 3 \\ -x - y = 11 \end{cases}$ (a) $(2, -13)$ (b) $(2, -9)$ (c) $\left(-\frac{3}{2}, -\frac{31}{3}\right)$ (d) $\left(-\frac{7}{4}, -\frac{37}{4}\right)$

3. $\begin{cases} y = -2e^x \\ 3x - y = 2 \end{cases}$ (a) $(-2, 0)$ (b) $(0, -2)$ (c) $(0, -3)$ (d) $(-1, 2)$

4. $\begin{cases} -\log x + 3 = y \\ \frac{1}{9}x + y = \frac{28}{9} \end{cases}$ (a) $\left(9, \frac{37}{9}\right)$ (b) $(10, 2)$ (c) $(1, 3)$ (d) $(2, 4)$

In Exercises 5–14, solve the system by the method of substitution. Check your solution graphically.

5. $\begin{cases} 2x + y = 6 \\ -x + y = 0 \end{cases}$

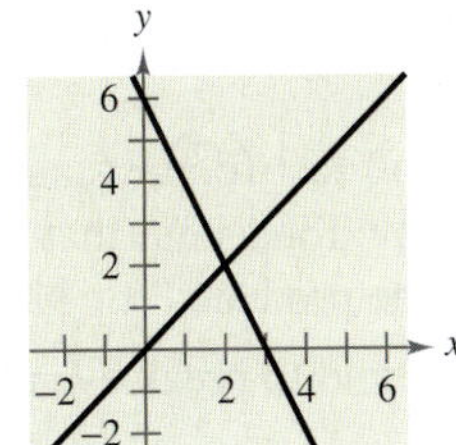

6. $\begin{cases} x - y = -4 \\ x + 2y = 5 \end{cases}$

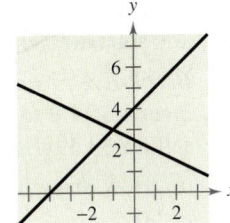

7. $\begin{cases} x - y = -4 \\ x^2 - y = -2 \end{cases}$

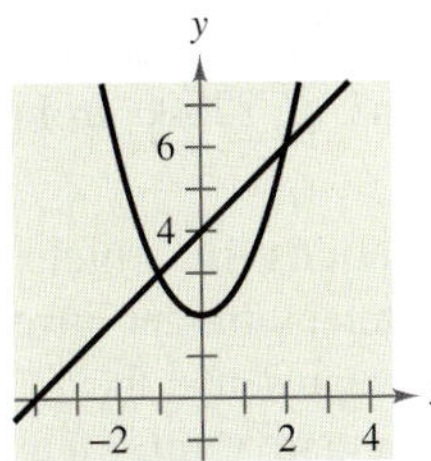

8. $\begin{cases} 3x + y = 2 \\ x^3 - 2 + y = 0 \end{cases}$

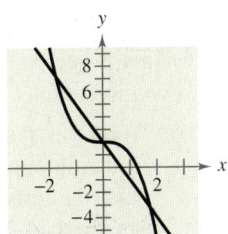

9. $\begin{cases} -2x + y = -5 \\ x^2 + y^2 = 25 \end{cases}$

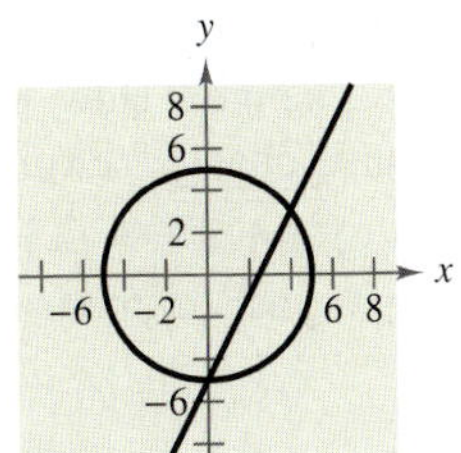

10. $\begin{cases} x + y = 0 \\ x^3 - 5x - y = 0 \end{cases}$

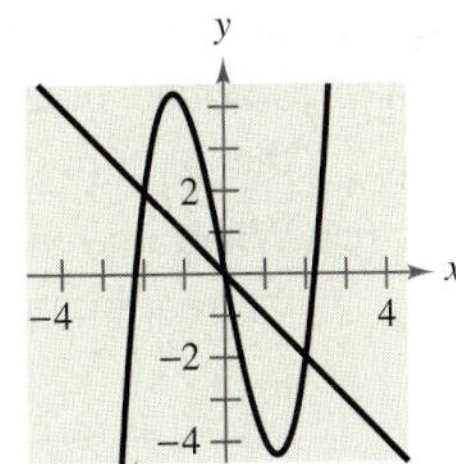

11. $\begin{cases} x^2 + y = 0 \\ x^2 - 4x - y = 0 \end{cases}$

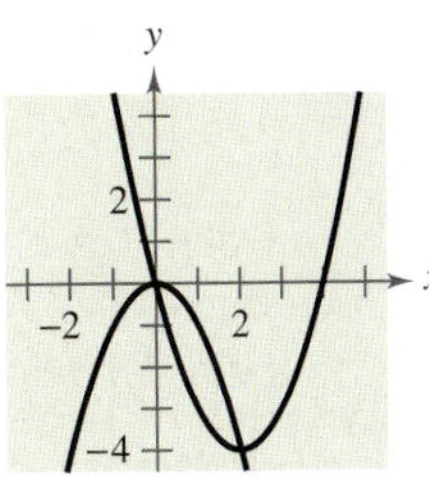

12. $\begin{cases} y = -2x^2 + 2 \\ y = 2(x^4 - 2x^2 + 1) \end{cases}$

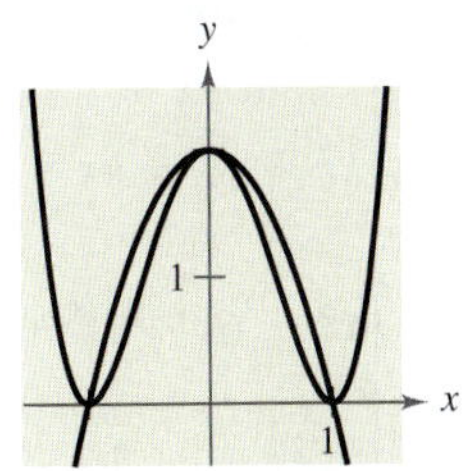

13. $\begin{cases} y = x^3 - 3x^2 + 1 \\ y = x^2 - 3x + 1 \end{cases}$

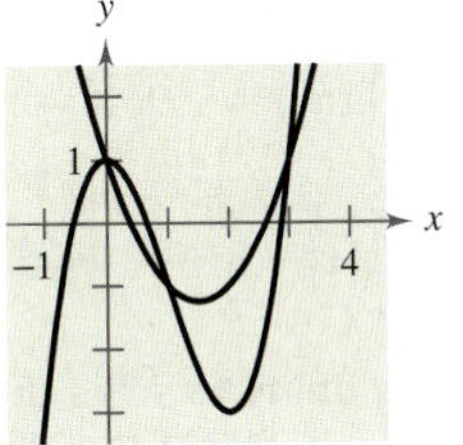

14. $\begin{cases} y = x^3 - 3x^2 + 4 \\ y = -2x + 4 \end{cases}$

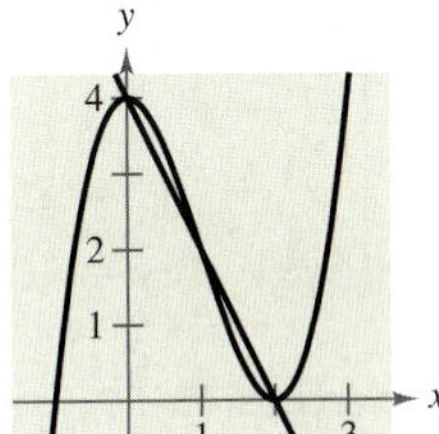

In Exercises 15–28, solve the system by the method of substitution.

15. $\begin{cases} x - y = 0 \\ 5x - 3y = 10 \end{cases}$

16. $\begin{cases} x + 2y = 1 \\ 5x - 4y = -23 \end{cases}$

17. $\begin{cases} 2x - y + 2 = 0 \\ 4x + y - 5 = 0 \end{cases}$

18. $\begin{cases} 6x - 3y - 4 = 0 \\ x + 2y - 4 = 0 \end{cases}$

19. $\begin{cases} 1.5x + 0.8y = 2.3 \\ 0.3x - 0.2y = 0.1 \end{cases}$

20. $\begin{cases} 0.5x + 3.2y = 9.0 \\ 0.2x - 1.6y = -3.6 \end{cases}$

21. $\begin{cases} \frac{1}{5}x + \frac{1}{2}y = 8 \\ x + y = 20 \end{cases}$

22. $\begin{cases} \frac{1}{2}x + \frac{3}{4}y = 10 \\ \frac{3}{4}x - y = 4 \end{cases}$

23. $\begin{cases} 6x + 5y = -3 \\ -x - \frac{5}{6}y = -7 \end{cases}$

24. $\begin{cases} -\frac{2}{3}x + y = 2 \\ 2x - 3y = 6 \end{cases}$

25. $\begin{cases} x^2 - y = 0 \\ 2x + y = 0 \end{cases}$

26. $\begin{cases} x - 2y = 0 \\ 3x - y^2 = 0 \end{cases}$

27. $\begin{cases} x^3 - y = 0 \\ x - y = 0 \end{cases}$

28. $\begin{cases} y = -x \\ y = x^3 + 3x^2 + 2x \end{cases}$

In Exercises 29–42, solve the system graphically.

29. $\begin{cases} -x + 2y = 2 \\ 3x + y = 15 \end{cases}$

30. $\begin{cases} x + y = 0 \\ 3x - 2y = 10 \end{cases}$

31. $\begin{cases} x - 3y = -2 \\ 5x + 3y = 17 \end{cases}$

32. $\begin{cases} -x + 2y = 1 \\ x - y = 2 \end{cases}$

33. $\begin{cases} x + y = 4 \\ x^2 + y^2 - 4x = 0 \end{cases}$

34. $\begin{cases} -x + y = 3 \\ x^2 - 6x - 27 + y^2 = 0 \end{cases}$

35. $\begin{cases} x - y + 3 = 0 \\ x^2 - 4x + 7 = y \end{cases}$

36. $\begin{cases} y^2 - 4x + 11 = 0 \\ -\frac{1}{2}x + y = -\frac{1}{2} \end{cases}$

37. $\begin{cases} 7x + 8y = 24 \\ x - 8y = 8 \end{cases}$

38. $\begin{cases} x - y = 0 \\ 5x - 2y = 6 \end{cases}$

39. $\begin{cases} 3x - 2y = 0 \\ x^2 - y^2 = 4 \end{cases}$

40. $\begin{cases} 2x - y + 3 = 0 \\ x^2 + y^2 - 4x = 0 \end{cases}$

41. $\begin{cases} x^2 + y^2 = 25 \\ 3x^2 - 16y = 0 \end{cases}$

42. $\begin{cases} x^2 + y^2 = 25 \\ (x - 8)^2 + y^2 = 41 \end{cases}$

In Exercises 43–50, use a graphing utility to solve the system of equations. Find the solution accurate to two decimal places.

43. $\begin{cases} y = e^x \\ x - y + 1 = 0 \end{cases}$

44. $\begin{cases} y = -4e^{-x} \\ y + 3x + 8 = 0 \end{cases}$

45. $\begin{cases} x + 2y = 8 \\ y = \log_2 x \end{cases}$

46. $\begin{cases} y = -2 + \ln(x - 1) \\ 3y + 2x = 9 \end{cases}$

47. $\begin{cases} y = \sqrt{x} \\ y = x \end{cases}$

48. $\begin{cases} x - y = 3 \\ x - y^2 = 1 \end{cases}$

49. $\begin{cases} x^2 + y^2 = 169 \\ x^2 - 8y = 104 \end{cases}$

50. $\begin{cases} x^2 + y^2 = 4 \\ 2x^2 - y = 2 \end{cases}$

In Exercises 51–62, solve the system graphically or algebraically. Explain your choice of method.

51. $\begin{cases} y = 2x \\ y = x^2 + 1 \end{cases}$

52. $\begin{cases} x + y = 4 \\ x^2 + y = 2 \end{cases}$

53. $\begin{cases} 3x - 7y + 6 = 0 \\ x^2 - y^2 = 4 \end{cases}$

54. $\begin{cases} x^2 + y^2 = 25 \\ 2x + y = 10 \end{cases}$

55. $\begin{cases} x - 2y = 4 \\ x^2 - y = 0 \end{cases}$

56. $\begin{cases} y = (x + 1)^3 \\ y = \sqrt{x - 1} \end{cases}$

57. $\begin{cases} y - e^{-x} = 1 \\ y - \ln x = 3 \end{cases}$

58. $\begin{cases} x^2 + y = 4 \\ e^x - y = 0 \end{cases}$

59. $\begin{cases} y = x^4 - 2x^2 + 1 \\ y = 1 - x^2 \end{cases}$

60. $\begin{cases} y = x^3 - 2x^2 + x - 1 \\ y = -x^2 + 3x - 1 \end{cases}$

61. $\begin{cases} xy - 1 = 0 \\ 2x - 4y + 7 = 0 \end{cases}$

62. $\begin{cases} x - 2y = 1 \\ y = \sqrt{x - 1} \end{cases}$

Break-Even Analysis **In Exercises 63–66, find the sales necessary to break even ($R = C$) for the cost C of producing x units and the revenue R obtained by selling x units. (Round to the nearest whole unit.)**

63. $C = 8650x + 250{,}000, \quad R = 9950x$

64. $C = 2.65x + 350{,}000, \quad R = 4.15x$

65. $C = 5.5\sqrt{x} + 10{,}000, \quad R = 3.29x$

66. $C = 7.8\sqrt{x} + 18{,}500, \quad R = 12.84x$

67. ***Break-Even Analysis*** A small software company invests \$16,000 to produce a software package that will sell for \$55.95. Each unit can be produced for \$35.45.

(a) How many units must be sold to break even?

(b) How many units must be sold to make a profit of \$60,000?

68. ***Break-Even Analysis*** A small fast-food restaurant invests \$5000 to produce a new food item that will sell for \$3.49. Each item can be produced for \$2.16.

(a) How many items must be sold to break even?

(b) How many items must be sold to make a profit of \$8500?

69. ***Investment Portfolio*** A total of \$25,000 is invested in two funds paying 6% and 8.5% simple interest. The 6% investment has a lower risk. The investor wants a yearly interest income of \$2000 from the two investments.

(a) Write a system of equations in which one equation represents the total amount invested and the other equation represents the \$2000 required in interest. Let x and y represent the amounts invested at 6% and 8.5%, respectively.

(b) Use a graphing utility to graph the two equations in the same viewing window. As the amount invested at 6% increases, how does the amount invested at 8.5% change? How does the amount of interest income change? Explain.

(c) What amount should be invested at 6% to meet the requirement of \$2000 per year in interest?

70. ***Investment Portfolio*** A total of \$20,000 is invested in two funds paying 6.5% and 8.5% simple interest. The 6.5% investment has a lower risk. The investor wants a yearly interest check of \$1600 from the two investments.

(a) Write a system of equations in which one equation represents the total amount invested and the other equation represents the \$1600 required in interest. Let x and y represent the amounts invested at 6.5% and 8.5%, respectively.

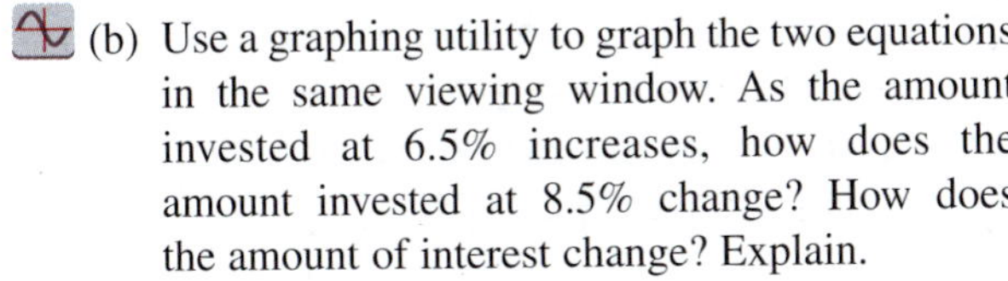

(b) Use a graphing utility to graph the two equations in the same viewing window. As the amount invested at 6.5% increases, how does the amount invested at 8.5% change? How does the amount of interest change? Explain.

(c) What amount should be invested at 6.5% to meet the requirement of \$1600 per year in interest?

71. ***Choice of Two Jobs*** You are offered two jobs selling dental supplies. One company offers a straight commission of 6% of sales. The other company offers a salary of \$350 per week plus 3% of sales. How much would you have to sell in a week in order to make the straight commission offer better?

72. ***Choice of Two Jobs*** You are offered two different jobs selling college textbooks. One company offers an annual salary of \$25,000 plus a year-end bonus of 2% of your total sales. The other company offers an annual salary of \$20,000 plus a year-end bonus of 3% of your total sales. Determine the annual sales required to make the second offer better.

73. ***Log Volume*** You are offered two different rules for estimating the number of board feet in a 16-foot log. (A board foot is a unit of measure for lumber equal to a board 1 foot square and 1 inch thick.) The first rule is the *Doyle Log Rule* and is modeled by

$$V_1 = (D - 4)^2, \qquad 5 \le D \le 40$$

and the other is the *Scribner Log Rule* and is modeled by

$$V_2 = 0.79D^2 - 2D - 4, \qquad 5 \le D \le 40$$

where D is the diameter (in inches) of the log and V is its volume in board feet.

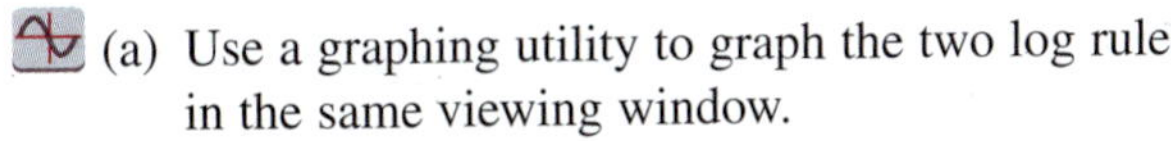

(a) Use a graphing utility to graph the two log rules in the same viewing window.

(b) For what diameter do the two scales agree?

(c) You are selling large logs by the board foot. Which scale would you use?

74. ***Supply and Demand*** The supply and demand curves for a business dealing with wheat are

Supply: $p = 1.45 + 0.00014x^2$

Demand: $p = (2.388 - 0.007x)^2$

where p is the price in dollars per bushel and x is the quantity in bushels per day. Use a graphing utility to graph the supply and demand equations and find the market equilibrium. (The market equilibrium is the point of intersection of the graphs for $x > 0$.)

▶ Model It

75. ***Data Analysis*** The table shows the numbers y (in thousands) of visits to hospital emergency departments in the United States for the years 1996 to 1999. (Source: U.S. National Center for Health Statistics)

Year, t	Number of visits, y
1996	90,347
1997	94,936
1998	100,384
1999	102,765

(a) Use the *regression* feature of a graphing utility to find a linear model f and a quadratic model g for the data. Let t represent the year, with $t = 6$ corresponding to 1996.

(b) Use a graphing utility to graph the data and the two models in the same viewing window.

(c) Use your graph from part (b) to approximate the points of intersection of the graphs of the models.

(d) Approximate the points of intersection of the graphs of the models algebraically. Compare this result with the one obtained in part (c).

(e) Use the models to estimate the number of visits to hospital emergency departments in 2005.

(f) Which model do you think gives the more accurate estimate? Explain.

76. ***Data Analysis*** The table shows the average hourly earnings y of production workers in manufacturing industries in the United States for the years 1993 to 2000. (Source: U.S. Bureau of Labor Statistics)

Year, t	Average hourly earnings, y
1993	\$11.74
1994	\$12.07
1995	\$12.37
1996	\$12.77
1997	\$13.17
1998	\$13.49
1999	\$13.91
2000	\$14.38

A linear model that represents the data is given by

$f(t) = 0.374t + 10.55.$

A quadratic model that represents the data is given by

$g(t) = 0.0092t^2 + 0.255t + 10.89.$

For both models, t represents the year, with $t = 3$ corresponding to 1993.

(a) Use a graphing utility to graph the data and the two models in the same viewing window.

(b) Approximate the points of intersection of the graphs of the models.

Geometry **In Exercises 77–80, find the dimensions of the rectangle meeting the specified conditions.**

77. The perimeter is 30 meters and the length is 3 meters greater than the width.

78. The perimeter is 280 centimeters and the width is 20 centimeters less than the length.

79. The perimeter is 42 inches and the width is three-fourths the length.

80. The perimeter is 210 feet and the length is $1\frac{1}{2}$ times the width.

81. ***Geometry*** What are the dimensions of a rectangular tract of land if its perimeter is 40 kilometers and its area is 96 square kilometers?

82. ***Geometry*** What are the dimensions of an isosceles right triangle with a 2-inch hypotenuse and an area of 1 square inch?

Synthesis

True or False? **In Exercises 83 and 84, determine whether the statement is true or false. Justify your answer.**

83. In order to solve a system of equations by substitution, you must always solve for y in one of the two equations and then back-substitute.

84. If a system consists of a parabola and a circle, then the system can have at most two solutions.

85. ***Writing*** List and explain the steps used to solve a system of equations by substitution.

86. ***Think About It*** When solving a system of equations by substitution, how do you recognize that the system has no solution?

87. ***Exploration*** Find an equation of a line whose graph intersects the graph of the parabola $y = x^2$ at (a) two points, (b) one point, and (c) no points. (There is more than one correct answer.)

88. ***Conjecture*** Consider the system of equations

$$\begin{cases} y = b^x \\ y = x^b \end{cases}.$$

(a) Use a graphing utility to graph the system for $b = 1, 2, 3$, and 4.

(b) For a fixed even value of $b > 1$, make a conjecture about the number of points of intersection of the graphs in part (a).

Review

In Exercises 89–94, find the general form of the equation of the line through the two points.

89. $(-2, 7), (5, 5)$

90. $(3.5, 4), (10, 6)$

91. $(6, 3), (10, 3)$

92. $(4, -2), (4, 5)$

93. $\left(\frac{3}{5}, 0\right), (4, 6)$

94. $\left(-\frac{7}{3}, 8\right), \left(\frac{5}{2}, \frac{1}{2}\right)$

In Exercises 95–98, find the domain of the function and identify any horizontal or vertical asymptotes.

95. $f(x) = \dfrac{5}{x - 6}$

96. $f(x) = \dfrac{2x - 7}{3x + 2}$

97. $f(x) = \dfrac{x^2 + 2}{x^2 - 16}$

98. $f(x) = 3 - \dfrac{2}{x^2}$

9.2 Two-Variable Linear Systems

▶ **What you should learn**

- How to use the method of elimination to solve systems of linear equations in two variables
- How to interpret graphically the numbers of solutions of systems of linear equations in two variables
- How to use systems of equations in two variables to model and solve real-life problems

▶ **Why you should learn it**

You can use systems of equations in two variables to model and solve real-life problems. For instance, in Exercise 61 on page 685, you will solve a system of equations to find a linear model that represents the average room rate for a hotel room in the United States.

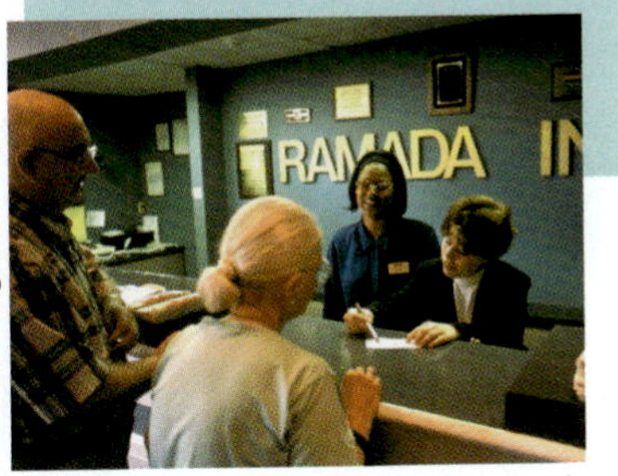

Jeff Greenberg/The Image Works

The Method of Elimination

In Section 9.1, you studied two methods for solving a system of equations: substitution and graphing. Now you will study the **method of elimination.** The key step in this method is to obtain, for one of the variables, coefficients that differ only in sign so that *adding* the equations eliminates the variable.

$$3x + 5y = 7 \qquad \text{Equation 1}$$
$$-3x - 2y = -1 \qquad \text{Equation 2}$$
$$3y = 6 \qquad \text{Add equations.}$$

Note that by adding the two equations, you eliminate the x-terms and obtain a single equation in y. Solving this equation for y produces $y = 2$, which you can then back-substitute into one of the original equations to solve for x.

Example 1 ▶ **Solving a System of Equations by Elimination**

Solve the system of linear equations.

$$\begin{cases} 3x + 2y = 4 & \text{Equation 1} \\ 5x - 2y = 8 & \text{Equation 2} \end{cases}$$

Solution

Because the coefficients of y differ only in sign, you can eliminate the y-terms by adding the two equations.

$$3x + 2y = 4 \qquad \text{Write Equation 1.}$$
$$5x - 2y = 8 \qquad \text{Write Equation 2.}$$
$$8x \qquad = 12 \qquad \text{Add equations.}$$

So, $x = \frac{3}{2}$. By back-substituting this value into Equation 1, you can solve for y.

$$3x + 2y = 4 \qquad \text{Write Equation 1.}$$
$$3\left(\frac{3}{2}\right) + 2y = 4 \qquad \text{Substitute } \tfrac{3}{2} \text{ for } x.$$
$$\frac{9}{2} + 2y = 4 \qquad \text{Simplify.}$$
$$y = -\frac{1}{4} \qquad \text{Solve for } y.$$

The solution is $\left(\frac{3}{2}, -\frac{1}{4}\right)$. Check this in the original system.

Try using the method of substitution to solve the system given in Example 1. Which method do you think is easier? Many people find that the method of elimination is more efficient.

Example 2 ▶ Solving a System of Equations by Elimination

Solve the system of linear equations.

$$\begin{cases} 2x - 3y = -7 & \text{Equation 1} \\ 3x + y = -5 & \text{Equation 2} \end{cases}$$

Solution

For this system, you can obtain coefficients that differ only in sign by multiplying Equation 2 by 3.

$$\begin{array}{lll} 2x - 3y = -7 & \Rightarrow \quad 2x - 3y = -7 & \text{Write Equation 1.} \\ 3x + y = -5 & \Rightarrow \quad 9x + 3y = -15 & \text{Multiply Equation 2 by 3.} \\ & \qquad 11x = -22 & \text{Add equations.} \end{array}$$

> **STUDY TIP**
>
> To obtain coefficients (for one of the variables) that differ only in sign, you often need to multiply one or both of the equations by suitably chosen constants.

So, you can see that $x = -2$. By back-substituting this value of x into Equation 1, you can solve for y.

$$\begin{aligned} 2x - 3y &= -7 && \text{Write Equation 1.} \\ 2(-2) - 3y &= -7 && \text{Substitute } -2 \text{ for } x. \\ -3y &= -3 && \text{Combine like terms.} \\ y &= 1 && \text{Solve for } y. \end{aligned}$$

The solution is $(-2, 1)$. Check this in the original system, as follows.

Check

$$\begin{aligned} 2x - 3y &= -7 && \text{Write original Equation 1.} \\ 2(-2) - 3(1) &\stackrel{?}{=} -7 && \text{Substitute into Equation 1.} \\ -4 - 3 &= -7 && \text{Equation 1 checks. } \checkmark \\ 3x + y &= -5 && \text{Write original Equation 2.} \\ 3(-2) + 1 &\stackrel{?}{=} -5 && \text{Substitute into Equation 2.} \\ -6 + 1 &= -5 && \text{Equation 2 checks. } \checkmark \end{aligned}$$

In Example 2, the two systems of linear equations

$$\begin{cases} 2x - 3y = -7 \\ 3x + y = -5 \end{cases}$$

and

$$\begin{cases} 2x - 3y = -7 \\ 9x + 3y = -15 \end{cases}$$

are called **equivalent systems** because they have precisely the same solution set. The operations that can be performed on a system of linear equations to produce an equivalent system are (1) interchanging any two equations, (2) multiplying an equation by a nonzero constant, and (3) adding a multiple of one equation to any other equation in the system.

STUDY TIP

Keep in mind that the terminology and methods discussed in this section and the following section apply only to systems of *linear* equations.

Method of Elimination

1. *Obtain coefficients* for x (or y) that differ only in sign by multiplying all terms of one or both equations by suitably chosen constants.
2. *Add* the equations to eliminate one variable, and solve the resulting equation.
3. *Back-substitute* the value obtained in Step 2 into either of the original equations and solve for the other variable.
4. *Check* your solution in both of the original equations.

Exploration

Rewrite each system of equations in slope-intercept form and sketch the graph of each system. What is the relationship between the slopes of the two lines and the number of points of intersection?

a. $\begin{cases} 5x - y = -1 \\ -x + y = -5 \end{cases}$

b. $\begin{cases} 4x - 3y = 1 \\ -8x + 6y = -2 \end{cases}$

c. $\begin{cases} x + 2y = 3 \\ x + 2y = -8 \end{cases}$

Example 3 ▶ Solving a System of Equations by Elimination

Solve the system of linear equations.

$$\begin{cases} 5x + 3y = 9 & \text{Equation 1} \\ 2x - 4y = 14 & \text{Equation 2} \end{cases}$$

Solution

You can obtain coefficients that differ only in sign by multiplying Equation 1 by 4 and multiplying Equation 2 by 3.

$5x + 3y = 9 \quad \Rightarrow \quad 20x + 12y = 36$ Multiply Equation 1 by 4.

$2x - 4y = 14 \quad \Rightarrow \quad 6x - 12y = 42$ Multiply Equation 2 by 3.

$26x = 78$ Add equations.

From this equation, you can see that $x = 3$. By back-substituting this value of x into Equation 2, you can solve for y.

$2x - 4y = 14$ Write Equation 2.

$2(3) - 4y = 14$ Substitute 3 for x.

$-4y = 8$ Collect like terms.

$y = -2$ Solve for y.

The solution is $(3, -2)$. Check this in the original system.

Technology

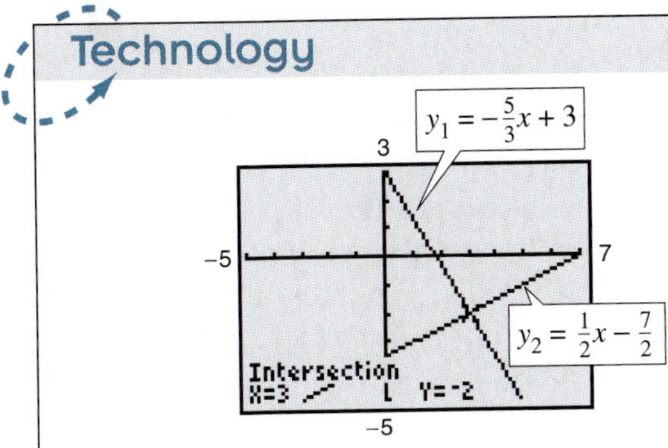

Remember that you can check the solution of a system of equations graphically. For instance, to check the solution found in Example 3, solve each equation for y. Then use a graphing utility to graph $y_1 = -\frac{5}{3}x + 3$ and $y_2 = \frac{1}{2}x - \frac{7}{2}$ in the same viewing window, as shown in the figure at the left. Use the *intersect* feature or the *zoom* and *trace* features to approximate the point of intersection of the graphs. From the graph, the point of intersection is $(3, -2)$.

Graphical Interpretation of Solutions

It is possible for a *general* system of equations to have exactly one solution, two or more solutions, or no solution. If a system of *linear* equations has two different solutions, it must have an *infinite* number of solutions. To see why this is true, consider the following graphical interpretations of a system of two linear equations in two variables.

Graphical Interpretations of Solutions

For a system of two linear equations in two variables, the number of solutions is one of the following.

Number of Solutions	*Graphical Interpretation*	*Slopes of Lines*
1. Exactly one solution	The two lines intersect at one point.	The slopes of the two lines are not equal.
2. Infinitely many solutions	The two lines are coincident (identical).	The slopes of the two lines are equal.
3. No solution	The two lines are parallel.	The slopes of the two lines are equal.

A system of linear equations is **consistent** if it has at least one solution. It is **inconsistent** if it has no solution.

Example 4 ▶ Recognizing Graphs of Linear Systems 

Match the system of linear equations with its graph in Figure 9.7. State whether the system is consistent or inconsistent and describe the number of solutions.

a. $\begin{cases} 2x - 3y = 3 \\ -4x + 6y = 6 \end{cases}$ **b.** $\begin{cases} 2x - 3y = 3 \\ x + 2y = 5 \end{cases}$ **c.** $\begin{cases} 2x - 3y = 3 \\ -4x + 6y = -6 \end{cases}$

i.

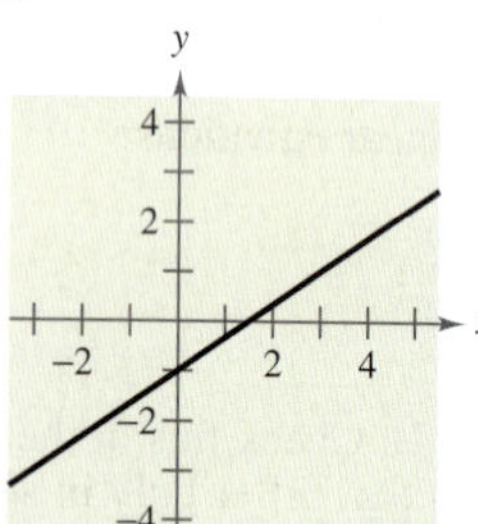

ii.

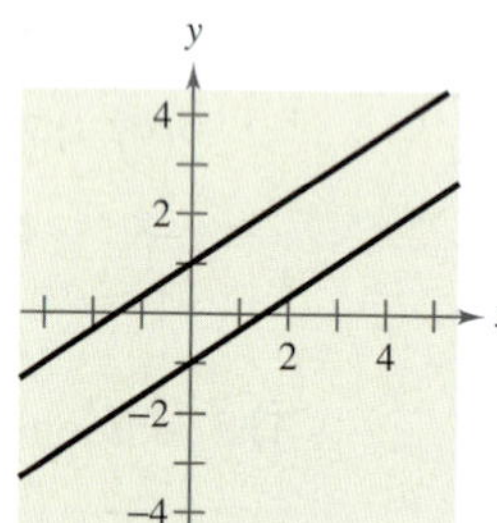

iii.

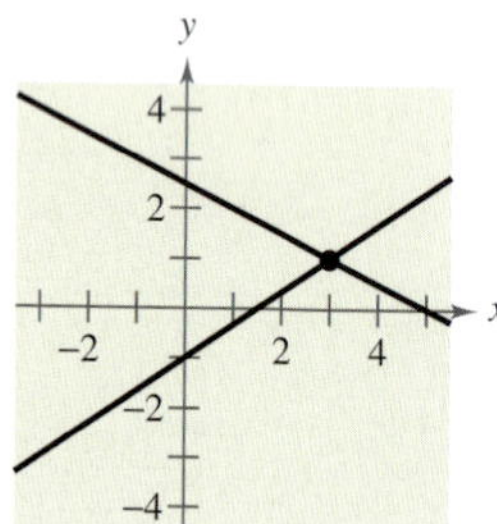

FIGURE 9.7

Solution

a. The graph of system (a) is a pair of parallel lines (ii). The lines have no point of intersection, so the system has no solution. The system is inconsistent.

b. The graph of system (b) is a pair of intersecting lines (iii). The lines have one point of intersection, so the system has exactly one solution. The system is consistent.

c. The graph of system (c) is a pair of lines that coincide (i). The lines have infinitely many points of intersection, so the system has infinitely many solutions. The system is consistent.

STUDY TIP

A comparison of the slopes of two lines gives useful information about the number of solutions of the corresponding system of equations. To solve a system of equations graphically, it helps to begin by writing the equations in slope-intercept form. Try doing this for the systems in Example 4.

In Examples 5 and 6, note how you can use the method of elimination to determine that a system of linear equations has no solution or infinitely many solutions.

Example 5 ▶ No-Solution Case: Method of Elimination 

Solve the system of linear equations.

$$\begin{cases} x - 2y = 3 & \text{Equation 1} \\ -2x + 4y = 1 & \text{Equation 2} \end{cases}$$

Solution

To obtain coefficients that differ only in sign, multiply Equation 1 by 2.

$$\begin{array}{rcrl} x - 2y = 3 & \Rightarrow & 2x - 4y = 6 & \text{Multiply Equation 1 by 2.} \\ \underline{-2x + 4y = 1} & \Rightarrow & \underline{-2x + 4y = 1} & \text{Write Equation 2.} \\ & & 0 = 7 & \text{False statement} \end{array}$$

Because there are no values of x and y for which $0 = 7$, you can conclude that the system is inconsistent and has no solution. The lines corresponding to the two equations in this system are shown in Figure 9.8. Note that the two lines are parallel and therefore have no point of intersection.

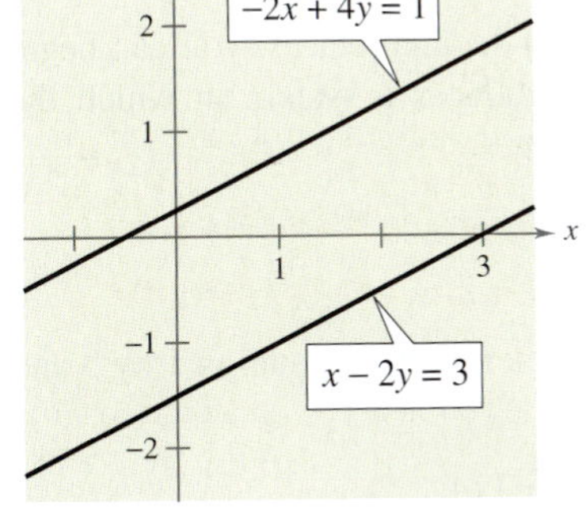

FIGURE 9.8

In Example 5, note that the occurrence of a false statement, such as $0 = 7$, indicates that the system has no solution. In the next example, note that the occurrence of a statement that is true for all values of the variables, such as $0 = 0$, indicates that the system has infinitely many solutions.

Example 6 ▶ Many-Solution Case: Method of Elimination 

Solve the system of linear equations.

$$\begin{cases} 2x - y = 1 & \text{Equation 1} \\ 4x - 2y = 2 & \text{Equation 2} \end{cases}$$

Solution

To obtain coefficients that differ only in sign, multiply Equation 2 by $-\frac{1}{2}$.

$$\begin{array}{rcrl} 2x - y = 1 & \Rightarrow & 2x - y = 1 & \text{Write Equation 1.} \\ \underline{4x - 2y = 2} & \Rightarrow & \underline{-2x + y = -1} & \text{Multiply Equation 2 by } -\tfrac{1}{2}. \\ & & 0 = 0 & \text{Add equations.} \end{array}$$

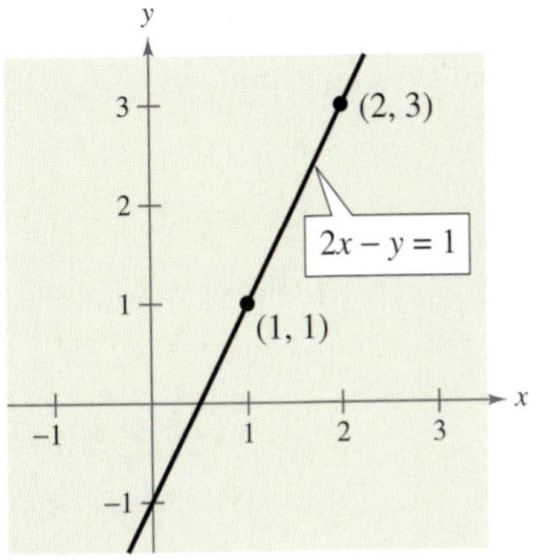

Because the two equations turn out to be equivalent (have the same solution set), you can conclude that the system has infinitely many solutions. The solution set consists of all points (x, y) lying on the line $2x - y = 1$, as shown in Figure 9.9. Letting $x = a$, where a is any real number, you can see that the solutions to the system are $(a, 2a - 1)$.

FIGURE 9.9

Technology

The general solution of the linear system

$$\begin{cases} ax + by = c \\ dx + ey = f \end{cases}$$

is $x = (ce - bf)/(ae - db)$ and $y = (af - cd)/(ae - db)$. If $ae - db = 0$, the system does not have a unique solution. Graphing utility programs for solving such a system can be found at our website *college.hmco.com*. Try using the program for your graphing utility to solve the system in Example 7.

Example 7 illustrates a strategy for solving a system of linear equations that has decimal coefficients.

Example 7 ▶ A Linear System Having Decimal Coefficients

Solve the system of linear equations.

$$\begin{cases} 0.02x - 0.05y = -0.38 & \text{Equation 1} \\ 0.03x + 0.04y = 1.04 & \text{Equation 2} \end{cases}$$

Solution

Because the coefficients in this system have two decimal places, you can begin by multiplying each equation by 100. (This produces a system in which the coefficients are all integers.)

$$\begin{cases} 2x - 5y = -38 & \text{Revised Equation 1} \\ 3x + 4y = 104 & \text{Revised Equation 2} \end{cases}$$

Now, to obtain coefficients that differ only in sign, multiply Equation 1 by 3 and multiply Equation 2 by -2.

$$\begin{aligned} 2x - 5y = -38 \quad &\Rightarrow \quad 6x - 15y = -114 && \text{Multiply Equation 1 by 3.} \\ 3x + 4y = 104 \quad &\Rightarrow \quad \underline{-6x - 8y = -208} && \text{Multiply Equation 2 by } -2. \\ & \qquad\quad -23y = -322 && \text{Add equations.} \end{aligned}$$

So, you can conclude that

$$\begin{aligned} y &= \frac{-322}{-23} \\ &= 14. \end{aligned}$$

Back-substituting this value into Equation 2 produces the following.

$$\begin{aligned} 3x + 4y &= 104 && \text{Write revised Equation 2.} \\ 3x + 4(14) &= 104 && \text{Substitute 14 for } y. \\ 3x &= 48 && \text{Combine like terms.} \\ x &= 16 && \text{Solve for } x. \end{aligned}$$

The solution is $(16, 14)$. Check this in the original system, as follows.

Check

$$\begin{aligned} 0.02x - 0.05y &= -0.38 && \text{Write original Equation 1.} \\ 0.02(16) - 0.05(14) &\stackrel{?}{=} -0.38 && \text{Substitute into Equation 1.} \\ 0.32 - 0.70 &= -0.38 && \text{Equation 1 checks. ✓} \\ 0.03x + 0.04y &= 1.04 && \text{Write original Equation 2.} \\ 0.03(16) + 0.04(14) &\stackrel{?}{=} 1.04 && \text{Substitute into Equation 2.} \\ 0.48 + 0.56 &= 1.04 && \text{Equation 2 checks. ✓} \end{aligned}$$

Applications 

At this point, you may be asking the question "How can I tell which application problems can be solved using a system of linear equations?" The answer comes from the following considerations.

1. Does the problem involve more than one unknown quantity?
2. Are there two (or more) equations or conditions to be satisfied?

If one or both of these situations occur, the appropriate mathematical model for the problem may be a system of linear equations. Example 8 shows how to construct such a model.

Example 8 ▶ An Application of a Linear System 

An airplane flying into a headwind travels the 2000-mile flying distance between Chicopee, Massachusetts and Salt Lake City, Utah in 4 hours and 24 minutes. On the return flight, the same distance is traveled in 4 hours. Find the airspeed of the plane and the speed of the wind, assuming that both remain constant.

Original flight

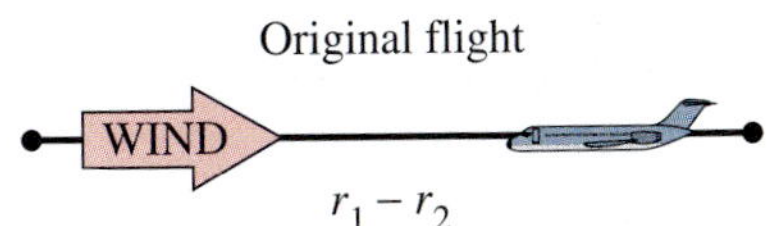

$r_1 - r_2$

Return flight

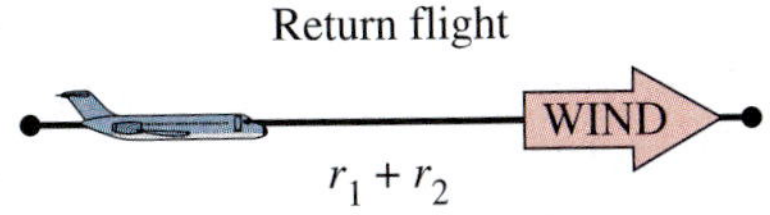

$r_1 + r_2$

FIGURE 9.10

Solution

The two unknown quantities are the speeds of the wind and the plane. If r_1 is the speed of the plane and r_2 is the speed of the wind, then

$$r_1 - r_2 = \text{speed of the plane against the wind}$$

$$r_1 + r_2 = \text{speed of the plane with the wind}$$

as shown in Figure 9.10. Using the formula distance = (rate)(time) for these two speeds, you obtain the following equations.

$$2000 = (r_1 - r_2)\left(4 + \frac{24}{60}\right)$$

$$2000 = (r_1 + r_2)(4)$$

These two equations simplify as follows.

$$\begin{cases} 5000 = 11r_1 - 11r_2 & \text{Equation 1} \\ 500 = r_1 + r_2 & \text{Equation 2} \end{cases}$$

To solve this system by elimination, multiply Equation 2 by 11.

$$5000 = 11r_1 - 11r_2 \quad \Longrightarrow \quad 5000 = 11r_1 - 11r_2 \qquad \text{Write Equation 1.}$$

$$500 = r_1 + r_2 \quad \Longrightarrow \quad \underline{5500 = 11r_1 + 11r_2} \qquad \text{Multiply Equation 2 by 11.}$$

$$10{,}500 = 22r_1 \qquad \text{Add equations.}$$

So,

$$r_1 = \frac{10{,}500}{22} = \frac{5250}{11} \approx 477.27 \text{ miles per hour} \qquad \text{Speed of plane}$$

$$r_2 = 500 - \frac{5250}{11} = \frac{250}{11} \approx 22.73 \text{ miles per hour.} \qquad \text{Speed of wind}$$

Check this solution in the original statement of the problem.

In a free market, the demands for many products are related to the prices of the products. As the prices decrease, the demands by consumers increase and the amounts that producers are able or willing to supply decrease.

Example 9 ▶ Finding the Equilibrium Point  

The demand and supply functions for a new type of calculator are

$$\begin{cases} p = 150 - 0.00001x & \text{Demand equation} \\ p = 60 + 0.00002x & \text{Supply equation} \end{cases}$$

where p is the price in dollars and x represents the number of units. Find the equilibrium point for this market. The equilibrium point is the price p and number of units x that satisfy both the demand and supply equations.

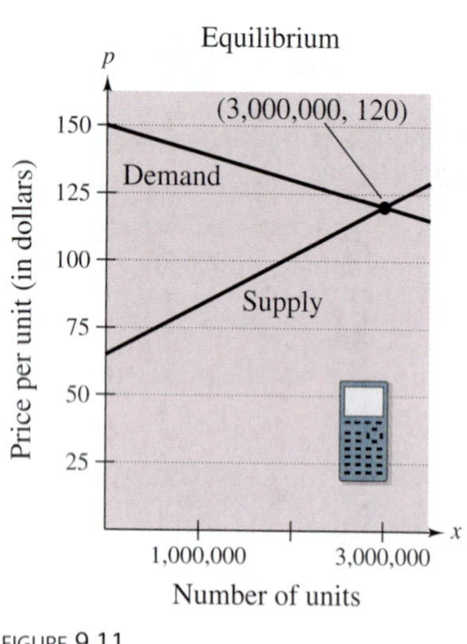

FIGURE 9.11

Solution

Begin by substituting the value of p given in the supply equation into the demand equation.

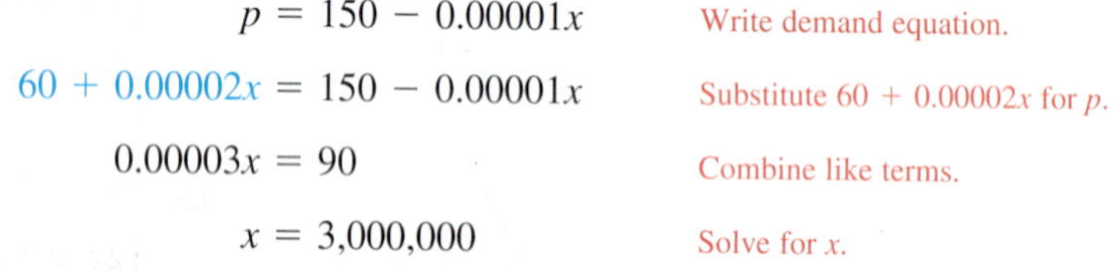

$$p = 150 - 0.00001x \qquad \text{Write demand equation.}$$

$$60 + 0.00002x = 150 - 0.00001x \qquad \text{Substitute } 60 + 0.00002x \text{ for } p.$$

$$0.00003x = 90 \qquad \text{Combine like terms.}$$

$$x = 3{,}000{,}000 \qquad \text{Solve for } x.$$

So, the equilibrium point occurs when the demand and supply are each 3 million units. (See Figure 9.11.) The price that corresponds to this x-value is obtained by back-substituting $x = 3{,}000{,}000$ into either of the original equations. For instance, back-substituting into the demand equation produces

$$\begin{aligned} p &= 150 - 0.00001(3{,}000{,}000) \\ &= 150 - 30 \\ &= \$120. \end{aligned}$$

The solution is (3,000,000, 120). You can check this as follows.

Check

Substitute (3,000,000, 120) into the demand equation.

$$p = 150 - 0.00001x \qquad \text{Write demand equation.}$$

$$120 \stackrel{?}{=} 150 - 0.00001(3{,}000{,}000) \qquad \text{Substitute 120 for } p \text{ and 3,000,000 for } x.$$

$$120 = 120 \qquad \text{Solution checks in demand equation. ✓}$$

Substitute (3,000,000, 120) into the supply equation.

$$p = 60 + 0.00002x \qquad \text{Write supply equation.}$$

$$120 \stackrel{?}{=} 60 + 0.00002(3{,}000{,}000) \qquad \text{Substitute 120 for } p \text{ and 3,000,000 for } x.$$

$$120 = 120 \qquad \text{Solution checks in supply equation. ✓}$$

9.2 Exercises

In Exercises 1–10, solve by elimination. Label each line with its equation. To print an enlarged copy of the graph, go to the website *www.mathgraphs.com*.

1. $\begin{cases} 2x + y = 5 \\ x - y = 1 \end{cases}$

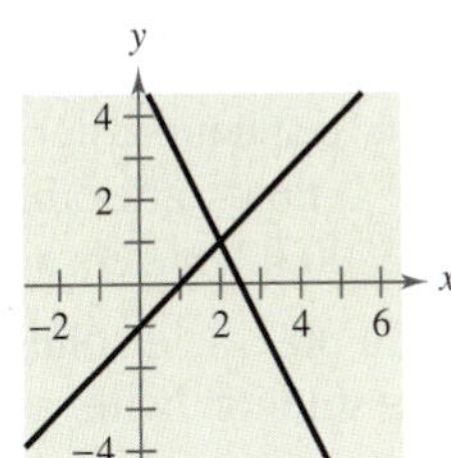

2. $\begin{cases} x + 3y = 1 \\ -x + 2y = 4 \end{cases}$

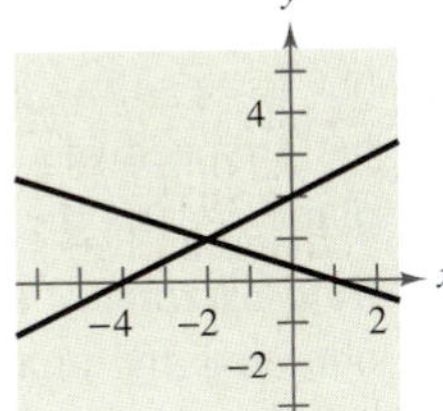

3. $\begin{cases} x + y = 0 \\ 3x + 2y = 1 \end{cases}$

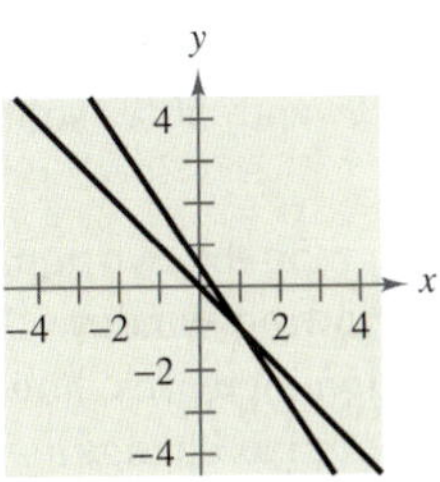

4. $\begin{cases} 2x - y = 3 \\ 4x + 3y = 21 \end{cases}$

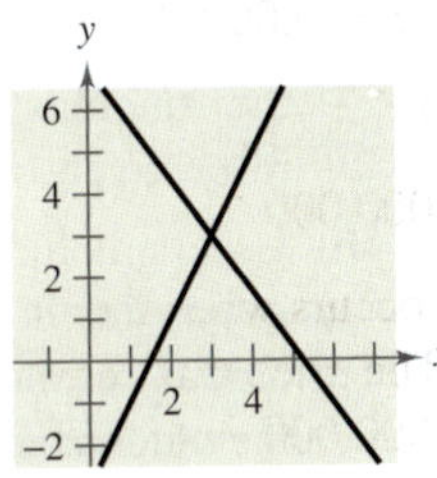

5. $\begin{cases} x - y = 2 \\ -2x + 2y = 5 \end{cases}$

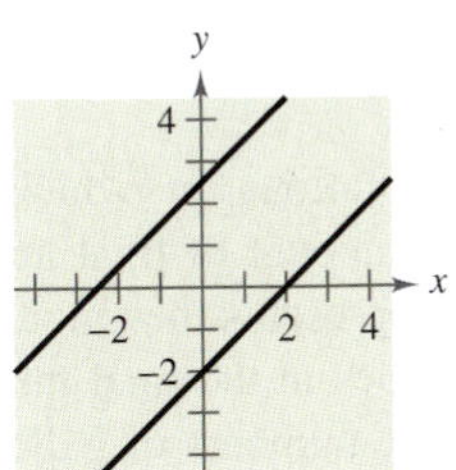

6. $\begin{cases} 3x + 2y = 3 \\ 6x + 4y = 14 \end{cases}$

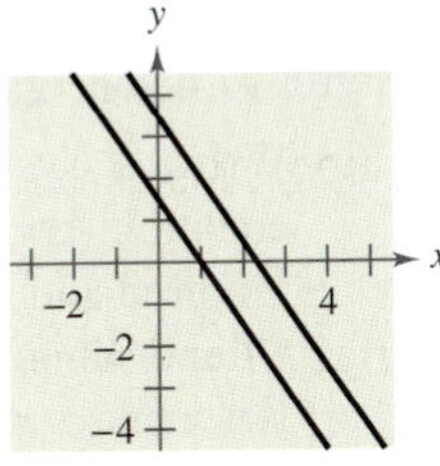

7. $\begin{cases} 3x - 2y = 5 \\ -6x + 4y = -10 \end{cases}$

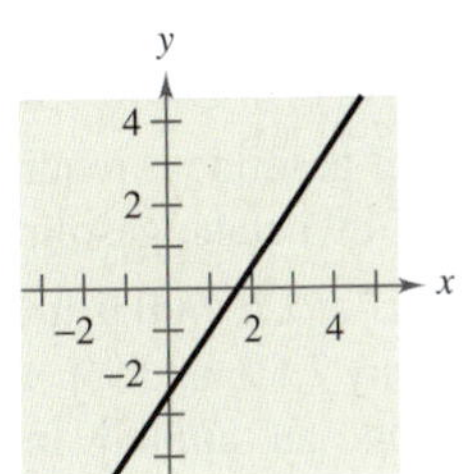

8. $\begin{cases} 9x - 3y = -15 \\ -3x + y = 5 \end{cases}$

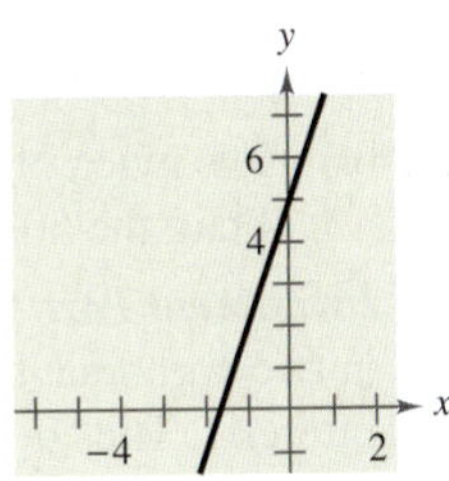

9. $\begin{cases} 9x + 3y = 1 \\ 3x - 6y = 5 \end{cases}$

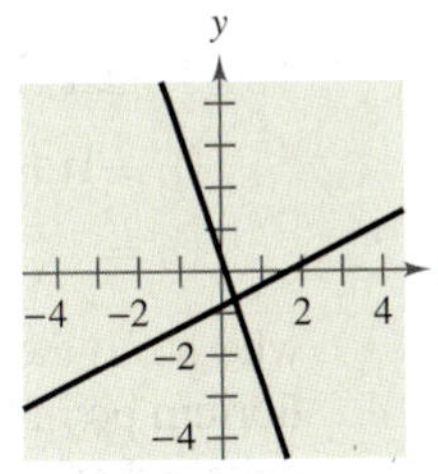

10. $\begin{cases} 5x + 3y = -18 \\ 2x - 6y = 1 \end{cases}$

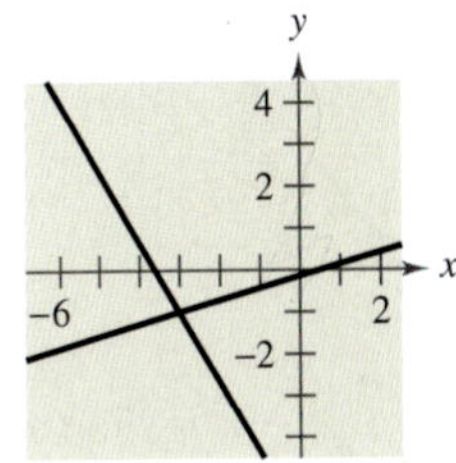

In Exercises 11–30, solve the system by elimination and check any solutions algebraically.

11. $\begin{cases} x + 2y = 4 \\ x - 2y = 1 \end{cases}$

12. $\begin{cases} 3x - 5y = 2 \\ 2x + 5y = 13 \end{cases}$

13. $\begin{cases} 2x + 3y = 18 \\ 5x - y = 11 \end{cases}$

14. $\begin{cases} x + 7y = 12 \\ 3x - 5y = 10 \end{cases}$

15. $\begin{cases} 3x + 2y = 10 \\ 2x + 5y = 3 \end{cases}$

16. $\begin{cases} 2r + 4s = 5 \\ 16r + 50s = 55 \end{cases}$

17. $\begin{cases} 5u + 6v = 24 \\ 3u + 5v = 18 \end{cases}$

18. $\begin{cases} 3x + 11y = 4 \\ -2x - 5y = 9 \end{cases}$

19. $\begin{cases} 1.8x + 1.2y = 4 \\ 9x + 6y = 3 \end{cases}$

20. $\begin{cases} 3.1x - 2.9y = -10.2 \\ 15.5x - 14.5y = 21 \end{cases}$

21. $\begin{cases} \dfrac{x}{4} + \dfrac{y}{6} = 1 \\ x - y = 3 \end{cases}$

22. $\begin{cases} \dfrac{2}{3}x + \dfrac{1}{6}y = \dfrac{2}{3} \\ 4x + y = 4 \end{cases}$

23. $\begin{cases} 2.5x - 3y = 1.5 \\ 2x - 2.4y = 1.2 \end{cases}$

24. $\begin{cases} 6.3x + 7.2y = 5.4 \\ 5.6x + 6.4y = 4.8 \end{cases}$

25. $\begin{cases} 0.05x - 0.03y = 0.21 \\ 0.07x + 0.02y = 0.16 \end{cases}$

26. $\begin{cases} 0.2x - 0.5y = -27.8 \\ 0.3x + 0.4y = 68.7 \end{cases}$

27. $\begin{cases} 4b + 3m = 3 \\ 3b + 11m = 13 \end{cases}$

28. $\begin{cases} 2x + 5y = 8 \\ 5x + 8y = 10 \end{cases}$

29. $\begin{cases} \dfrac{x+3}{4} + \dfrac{y-1}{3} = 1 \\ 2x - y = 12 \end{cases}$

30. $\begin{cases} \dfrac{x-1}{2} + \dfrac{y+2}{3} = 4 \\ x - 2y = 5 \end{cases}$

In Exercises 31–34, match the system of linear equations with its graph. [The graphs are labeled (a), (b), (c) and (d).]

(a)

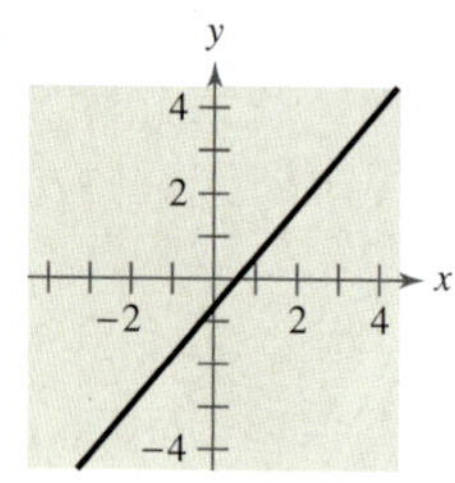

(b)

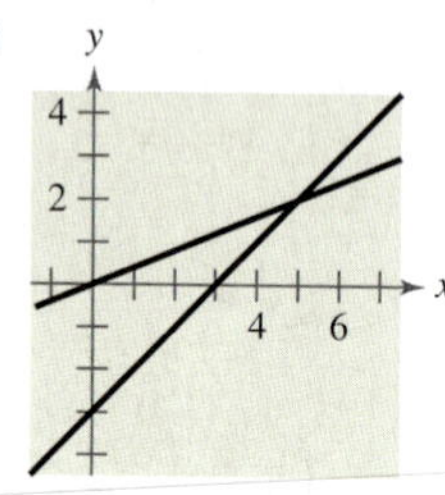

(c)

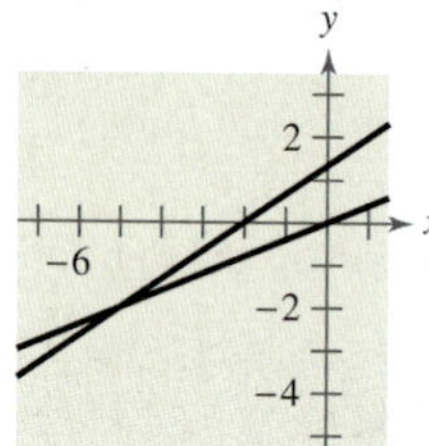

(d)

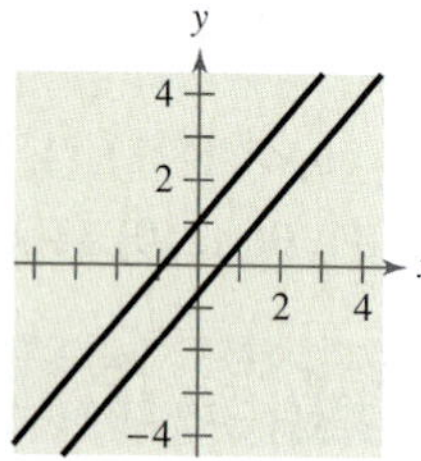

31. $\begin{cases} 2x - 5y = 0 \\ x - y = 3 \end{cases}$

32. $\begin{cases} -7x + 6y = -4 \\ 14x - 12y = 8 \end{cases}$

33. $\begin{cases} 2x - 5y = 0 \\ 2x - 3y = -4 \end{cases}$

34. $\begin{cases} 7x - 6y = -6 \\ -7x + 6y = -4 \end{cases}$

In Exercises 35–42, use any method to solve the system.

35. $\begin{cases} 3x - 5y = 7 \\ 2x + y = 9 \end{cases}$

36. $\begin{cases} -x + 3y = 17 \\ 4x + 3y = 7 \end{cases}$

37. $\begin{cases} y = 2x - 5 \\ y = 5x - 11 \end{cases}$

38. $\begin{cases} 7x + 3y = 16 \\ y = x + 2 \end{cases}$

39. $\begin{cases} x - 5y = 21 \\ 6x + 5y = 21 \end{cases}$

40. $\begin{cases} y = -3x - 8 \\ y = 15 - 2x \end{cases}$

41. $\begin{cases} -2x + 8y = 19 \\ y = x - 3 \end{cases}$

42. $\begin{cases} 4x - 3y = 6 \\ -5x + 7y = -1 \end{cases}$

Supply and Demand **In Exercises 43–46, find the equilibrium point of the demand and supply equations.**

	Demand	*Supply*
43.	$p = 50 - 0.5x$	$p = 0.125x$
44.	$p = 100 - 0.05x$	$p = 25 + 0.1x$
45.	$p = 140 - 0.00002x$	$p = 80 + 0.00001x$
46.	$p = 400 - 0.0002x$	$p = 225 + 0.0005x$

47. ***Airplane Speed*** An airplane flying into a headwind travels the 1800-mile flying distance between Pittsburgh, Pennsylvania and Phoenix, Arizona in 3 hours and 36 minutes. On the return flight, the distance is traveled in 3 hours. Find the airspeed of the plane and the speed of the wind, assuming that both remain constant.

48. ***Airplane Speed*** Two planes start from Los Angeles International Airport and fly in opposite directions. The second plane starts $\frac{1}{2}$ hour after the first plane, but its speed is 80 kilometers per hour faster. Find the airspeed of each plane if 2 hours after the first plane departs the planes are 3200 kilometers apart.

49. ***Acid Mixture*** Ten liters of a 30% acid solution is obtained by mixing a 20% solution with a 50% solution.

(a) Write a system of equations in which one equation represents the amount of final mixture required and the other represents the percent of acid in the final mixture. Let x and y represent the amounts of the 20% and 50% solutions, respectively.

 (b) Use a graphing utility to graph the two equations in part (a) in the same viewing window. As the amount of the 20% solution increases, how does the amount of the 50% solution change?

(c) How much of each solution is required to obtain the specified concentration of the final mixture?

50. ***Fuel Mixture*** Five hundred gallons of 89 octane gasoline is obtained by mixing 87 octane gasoline with 92 octane gasoline.

(a) Write a system of equations in which one equation represents the amount of final mixture required and the other represents the amounts of 87 and 92 octane gasolines in the final mixture. Let x and y represent the numbers of gallons of 87 octane and 92 octane gasolines, respectively.

 (b) Use a graphing utility to graph the two equations in part (a) in the same viewing window. As the amount of 87 octane gasoline increases, how does the amount of 92 octane gasoline change?

(c) How much of each type of gasoline is required to obtain the 500 gallons of 89 octane gasoline?

51. ***Investment Portfolio*** A total of \$12,000 is invested in two corporate bonds that pay 7.5% and 9% simple interest. The investor wants an annual interest income of \$990 from the investments. What amount should be invested in the 7.5% bond?

52. ***Investment Portfolio*** A total of \$32,000 is invested in two municipal bonds that pay 5.75% and 6.25% simple interest. The investor wants an annual interest income of \$1900 from the investments. What amount should be invested in the 5.75% bond?

53. ***Ticket Sales*** At a local high school city championship basketball game, 1435 tickets were sold. A student admission ticket cost \$1.50 and an adult admission ticket cost \$5.00. The total ticket receipts for the basketball game were \$3552.50. How many of each type of ticket were sold?

54. ***Consumer Awareness*** A department store held a sale to sell all of the 214 winter jackets that remained after the season ended. Until noon, each jacket in the store was priced at \$31.95. At noon, the price of the jackets was further reduced to \$18.95. After the last jacket was sold, total receipts for the clearance sale were \$5108.30. How many jackets were sold before noon and how many were sold after noon?

Fitting a Line to Data **In Exercises 55–60, find the least squares regression line $y = ax + b$ for the points**

$$(x_1, y_1), (x_2, y_2), \ldots, (x_n, y_n)$$

by solving the system for a and b.

$$nb + \left(\sum_{i=1}^{n} x_i\right)a = \sum_{i=1}^{n} y_i$$

$$\left(\sum_{i=1}^{n} x_i\right)b + \left(\sum_{i=1}^{n} x_i^2\right)a = \sum_{i=1}^{n} x_i y_i$$

Then use a graphing utility to confirm the result. (If you are unfamiliar with summation notation, look at the discussion in Section 11.1 or in Appendix A at the website for this text at *college.hmco.com*.)

55. $\begin{cases} 5b + 10a = 20.2 \\ 10b + 30a = 50.1 \end{cases}$

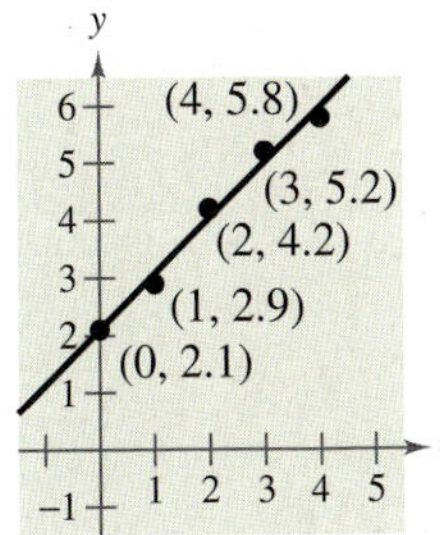

56. $\begin{cases} 5b + 10a = 11.7 \\ 10b + 30a = 25.6 \end{cases}$

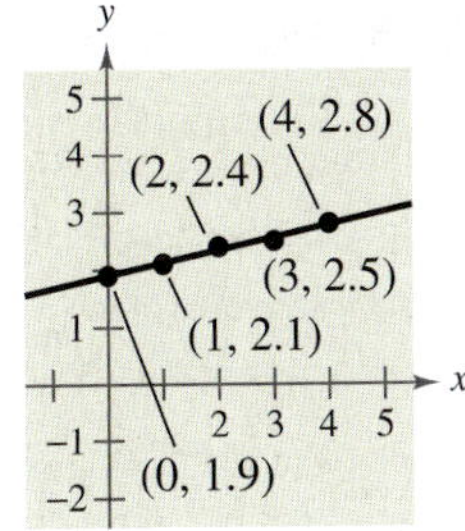

57. $\begin{cases} 7b + 21a = 35.1 \\ 21b + 91a = 114.2 \end{cases}$

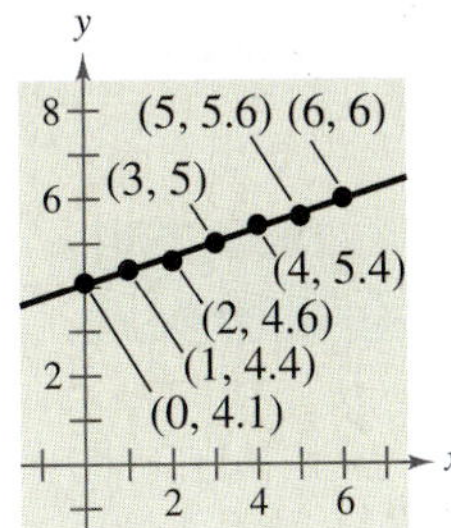

58. $\begin{cases} 6b + 15a = 23.6 \\ 15b + 55a = 48.8 \end{cases}$

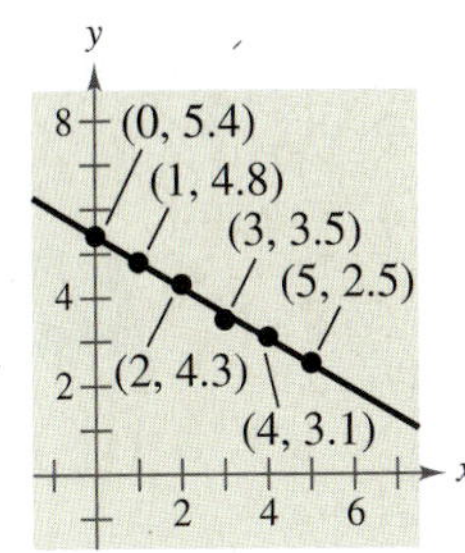

59. (0, 4), (1, 3), (1, 1), (2, 0)

60. (1, 0), (2, 0), (3, 0), (3, 1), (4, 1), (4, 2), (5, 2), (6, 2)

▶ Model It

61. ***Data Analysis*** The table shows the average room rates y for a hotel room in the United States for the years 1995 through 1999. (Source: American Hotel & Motel Association)

Year, t	Average room rate, y
1995	\$66.65
1996	\$70.93
1997	\$75.31
1998	\$78.62
1999	\$81.33

(a) Use the technique demonstrated in Exercises 55–60 to find the least squares regression line $y = at + b$. Let t represent the year, with $t = 5$ corresponding to 1995.

(b) Use the *regression* feature of a graphing utility to find a linear model for the data. How does this model compare with the model obtained in part (a)?

(c) Use the linear model to create a table of estimated values of y. Compare the estimated values with the actual data.

(d) Use the linear model to predict the average room rate in 2005.

(e) Use the linear model to predict when the average room rate will be \$100.00.

62. ***Data Analysis*** A farmer used four test plots to determine the relationship between wheat yield in bushels per acre and the amount of fertilizer in hundreds of pounds per acre. The results are shown in the table.

Fertilizer, x	Yield, y
1.0	32
1.5	41
2.0	48
2.5	53

(a) Use the technique demonstrated in Exercises 55–60 to find the least squares regression line $y = ax + b$.

(b) Use the linear model to predict the yield for a fertilizer application of 160 pounds per acre.

Synthesis

True or False? **In Exercises 63 and 64, determine whether the statement is true or false. Justify your answer.**

63. If two lines do not have exactly one point of intersection, then they must be parallel.

64. Solving a system of equations graphically will always give an exact solution.

Think About It **In Exercises 65 and 66, the graphs of the two equations appear to be parallel. Yet, when the system is solved algebraically, you find that the system does have a solution. Find the solution and explain why it does not appear on the portion of the graph that is shown.**

65. $\begin{cases} 100y - x = 200 \\ 99y - x = -198 \end{cases}$

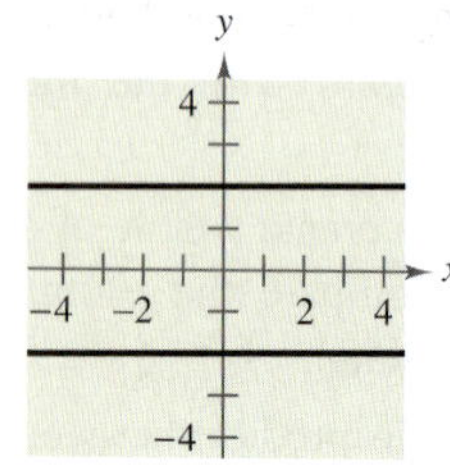

66. $\begin{cases} 21x - 20y = 0 \\ 13x - 12y = 120 \end{cases}$

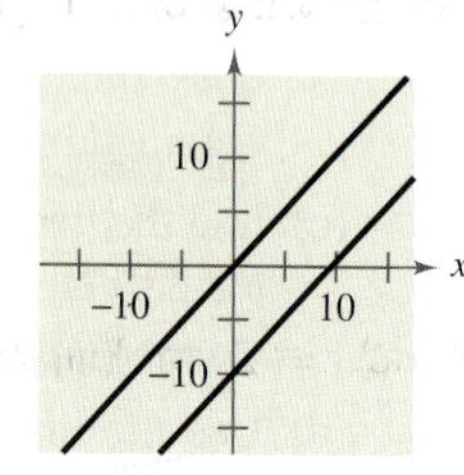

67. ***Writing*** Briefly explain whether or not it is possible for a consistent system of linear equations to have exactly two solutions.

68. ***Think About It*** Give examples of (a) a system of linear equations that has no solution and (b) a system that has an infinite number of solutions.

In Exercises 69 and 70, find the value of k such that the system of linear equations is inconsistent.

69. $\begin{cases} 4x - 8y = -3 \\ 2x + ky = 16 \end{cases}$

70. $\begin{cases} 15x + 3y = 6 \\ -10x + ky = 9 \end{cases}$

Review

In Exercises 71–78, solve the inequality and graph the solution on the real number line.

71. $-11 - 6x \geq 33$

72. $2(x - 3) > -5x + 1$

73. $8x - 15 \leq -4(2x - 1)$

74. $-6 \leq 3x - 10 < 6$

75. $|x - 8| < 10$

76. $|x + 10| \geq -3$

77. $2x^2 + 3x - 35 < 0$

78. $3x^2 + 12x > 0$

In Exercises 79 and 80, write the partial fraction decomposition for the rational expression.

79. $\dfrac{x - 1}{x^2 + 11x + 30}$

80. $\dfrac{3}{x(x^2 - 1)}$

In Exercises 81–84, write the expression as the logarithm of a single quantity.

81. $\ln x + \ln 6$

82. $\ln x - 5\ln(x + 3)$

83. $\log_9 12 - \log_9 x$

84. $\frac{1}{4}\log_6 3x$

In Exercises 85 and 86, solve the system by the method of substitution.

85. $\begin{cases} 2x - y = 4 \\ -4x + 2y = -12 \end{cases}$

86. $\begin{cases} 30x - 40y - 33 = 0 \\ 10x + 20y - 21 = 0 \end{cases}$

9.3 Multivariable Linear Systems

▶ **What you should learn**

- How to use back-substitution to solve linear systems in row-echelon form
- How to use Gaussian elimination to solve systems of linear equations
- How to solve nonsquare systems of linear equations
- How to use systems of linear equations in three or more variables to model and solve application problems

▶ **Why you should learn it**

Systems of linear equations in three or more variables can be used to model and solve real-life problems. For instance, in Exercise 71 on page 698, a system of linear equations can be used to analyze the reproduction rates of deer in a wildlife preserve.

Jeanne Drake/Tony Stone Images

Row-Echelon Form and Back-Substitution

The method of elimination can be applied to a system of linear equations in more than two variables. In fact, this method easily adapts to computer use for solving linear systems with dozens of variables.

When elimination is used to solve a system of linear equations, the goal is to rewrite the system in a form to which back-substitution can be applied. To see how this works, consider the following two systems of linear equations.

System of Three Linear Equations in Three Variables: (See Example 3.)

$$\begin{cases} x - 2y + 3z = 9 \\ -x + 3y = -4 \\ 2x - 5y + 5z = 17 \end{cases}$$

Equivalent System in Row-Echelon Form: (See Example 1.)

$$\begin{cases} x - 2y + 3z = 9 \\ y + 3z = 5 \\ z = 2 \end{cases}$$

The second system is said to be in **row-echelon form,** which means that it has a "stair-step" pattern with leading coefficients of 1. After comparing the two systems, it should be clear that it is easier to solve the system in row-echelon form, using back-substitution.

Example 1 ▶ Using Back-Substitution in Row-Echelon Form

Solve the system of linear equations.

$$\begin{cases} x - 2y + 3z = 9 & \text{Equation 1} \\ y + 3z = 5 & \text{Equation 2} \\ z = 2 & \text{Equation 3} \end{cases}$$

Solution

From Equation 3, you know the value of z. To solve for y, substitute $z = 2$ into Equation 2 to obtain

$$y + 3(2) = 5 \qquad \text{Substitute 2 for } z.$$

$$y = -1. \qquad \text{Solve for } y.$$

Finally, substitute $y = -1$ and $z = 2$ into Equation 1 to obtain

$$x - 2(-1) + 3(2) = 9 \qquad \text{Substitute } -1 \text{ for } y \text{ and 2 for } z.$$

$$x = 1. \qquad \text{Solve for } x.$$

The solution is $x = 1$, $y = -1$, and $z = 2$, which can be written as the **ordered triple** $(1, -1, 2)$. Check this in the original system of equations.

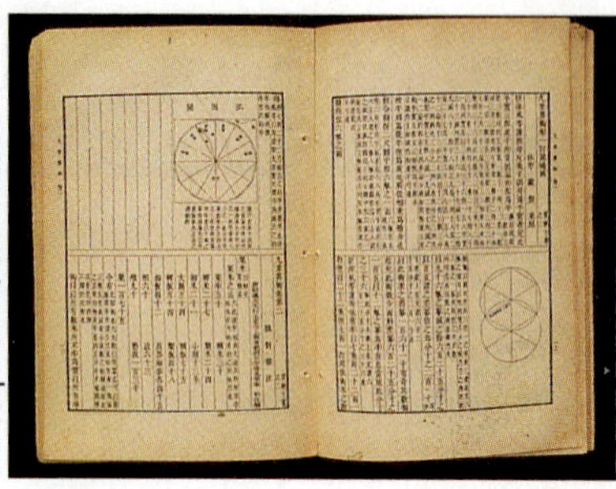
Christopher Lui/China Stock

Historical Note

One of the most influential Chinese mathematics books was the *Chui-chang suan-shu* or *Nine Chapters on the Mathematical Art* (written in approximately 250 B.C.). Chapter Eight of the *Nine Chapters* contained solutions of systems of linear equations using positive and negative numbers. One such system was as follows.

$$\begin{cases} 3x + 2y + z = 39 \\ 2x + 3y + z = 34 \\ x + 2y + 3z = 26 \end{cases}$$

This system was solved using column operations on a matrix. Matrices (plural for matrix) will be discussed in the next chapter.

STUDY TIP

As demonstrated in the first step in the solution of Example 2, interchanging rows is an easy way of obtaining a leading coefficient of 1.

Gaussian Elimination

Two systems of equations are *equivalent* if they have the same solution set. To solve a system that is not in row-echelon form, first convert it to an *equivalent* system that is in row-echelon form by using the following operations.

Operations That Produce Equivalent Systems

Each of the following **row operations** on a system of linear equations produces an *equivalent* system of linear equations.

1. Interchange two equations.
2. Multiply one of the equations by a nonzero constant.
3. Add a multiple of one of the equations to another equation to replace the latter equation.

To see how this is done, take another look at the method of elimination, as applied to a system of two linear equations.

Example 2 ▶ Using Gaussian Elimination to Solve a System

Solve the system of linear equations.

$$\begin{cases} 3x - 2y = -1 \\ x - y = 0 \end{cases}$$

Solution

There are two strategies that seem reasonable: eliminate the variable x or eliminate the variable y. The following steps show how to use the first strategy.

$$\begin{cases} x - y = 0 \\ 3x - 2y = -1 \end{cases}$$ Interchange two equations in the system.

$$-3x + 3y = 0$$ Multiply the first equation by -3.

$$\begin{array}{r} -3x + 3y = 0 \\ 3x - 2y = -1 \\ \hline y = -1 \end{array}$$ Add the multiple of the first equation to the second equation to obtain a new equation.

$$\begin{cases} x - y = 0 \\ y = -1 \end{cases}$$ New system in row-echelon form

Now, using back-substitution, you can determine that the solution is $y = -1$ and $x = -1$, which can be written as the ordered pair $(-1, -1)$. Check this solution in the original system of equations.

As shown in Example 2, rewriting a system of linear equations in row-echelon form usually involves a chain of equivalent systems, each of which is obtained by using one of the three basic row operations listed above. This process is called **Gaussian elimination,** after the German mathematician Carl Friedrich Gauss (1777–1855).

Example 3 ▶ Using Gaussian Elimination to Solve a System

Solve the system of linear equations.

$$\begin{cases} x - 2y + 3z = 9 & \text{Equation 1} \\ -x + 3y = -4 & \text{Equation 2} \\ 2x - 5y + 5z = 17 & \text{Equation 3} \end{cases}$$

Solution

Because the leading coefficient of the first equation is 1, you can begin by saving the x at the upper left and eliminating the other x-terms from the first column.

$$x - 2y + 3z = 9 \qquad \text{Write Equation 1.}$$

$$-x + 3y = -4 \qquad \text{Write Equation 2.}$$

$$y + 3z = 5 \qquad \text{Add Equation 1 to Equation 2.}$$

$$\begin{cases} x - 2y + 3z = 9 \\ y + 3z = 5 \\ 2x - 5y + 5z = 17 \end{cases}$$

Adding the first equation to the second equation produces a new second equation.

$$-2x + 4y - 6z = -18 \qquad \text{Multiply Equation 1 by } -2.$$

$$2x - 5y + 5z = 17 \qquad \text{Write Equation 3.}$$

$$-y - z = -1 \qquad \text{Add revised Equation 1 to Equation 3.}$$

$$\begin{cases} x - 2y + 3z = 9 \\ y + 3z = 5 \\ -y - z = -1 \end{cases}$$

Adding -2 times the first equation to the third equation produces a new third equation.

Now that all but the first x have been eliminated from the first column, go to work on the second column. (You need to eliminate y from the third equation.)

$$\begin{cases} x - 2y + 3z = 9 \\ y + 3z = 5 \\ 2z = 4 \end{cases}$$

Adding the second equation to the third equation produces a new third equation.

Finally, you need a coefficient of 1 for z in the third equation.

$$\begin{cases} x - 2y + 3z = 9 \\ y + 3z = 5 \\ z = 2 \end{cases}$$

Multiplying the third equation by $\frac{1}{2}$ produces a new third equation.

This is the same system that was solved in Example 1, and, as in that example, you can conclude that the solution is

$$x = 1, \qquad y = -1, \qquad \text{and} \qquad z = 2.$$

In Example 3, you can check the solution by substituting $x = 1$, $y = -1$, and $z = 2$ into each original equation, as follows.

Equation 1: $1 - 2(-1) + 3(2) = 9$ ✓

Equation 2: $-1 + 3(-1) = -4$ ✓

Equation 3: $2(1) - 5(-1) + 5(2) = 17$ ✓

The next example involves an inconsistent system—one that has no solution. The key to recognizing an inconsistent system is that at some stage in the elimination process you obtain a false statement such as $0 = -2$.

Example 4 ▶ An Inconsistent System 

Solve the system of linear equations.

$$\begin{cases} x - 3y + z = 1 & \text{Equation 1} \\ 2x - y - 2z = 2 & \text{Equation 2} \\ x + 2y - 3z = -1 & \text{Equation 3} \end{cases}$$

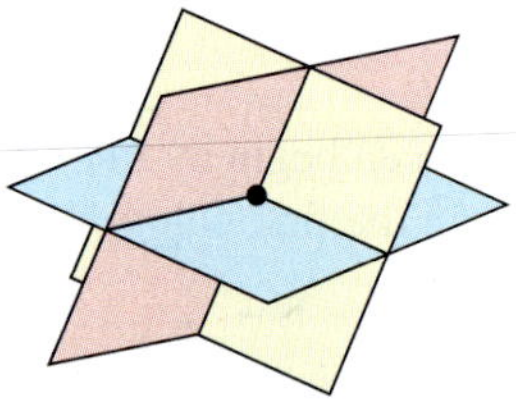

Solution

$$\begin{cases} x - 3y + z = 1 \\ 5y - 4z = 0 \\ x + 2y - 3z = -1 \end{cases}$$

Adding -2 times the first equation to the second equation produces a new second equation.

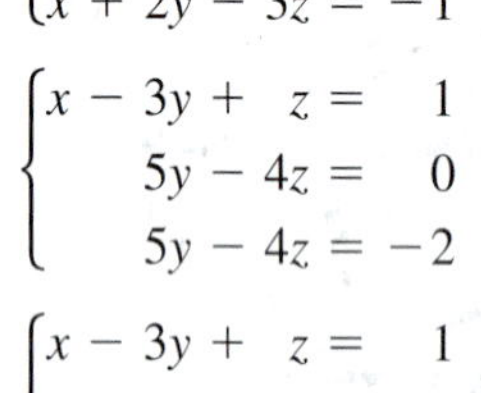

$$\begin{cases} x - 3y + z = 1 \\ 5y - 4z = 0 \\ 5y - 4z = -2 \end{cases}$$

Adding -1 times the first equation to the third equation produces a new third equation.

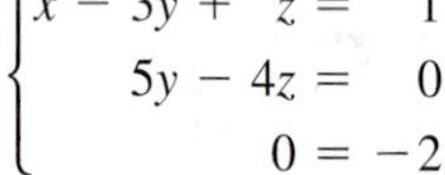

$$\begin{cases} x - 3y + z = 1 \\ 5y - 4z = 0 \\ 0 = -2 \end{cases}$$

Adding -1 times the second equation to the third equation produces a new third equation.

Because the third "equation" is impossible, you can conclude that this system is inconsistent and so has no solution. Moreover, because this system is equivalent to the original system, you can conclude that the original system also has no solution.

FIGURE 9.12 *Solution: one point*

FIGURE 9.13 *Solution: one line*

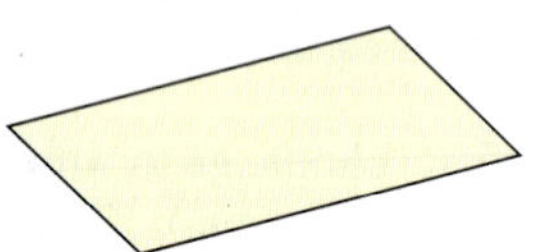

FIGURE 9.14 *Solution: one plane*

As with a system of linear equations in two variables, the solution(s) of a system of linear equations in more than two variables must fall into one of three categories.

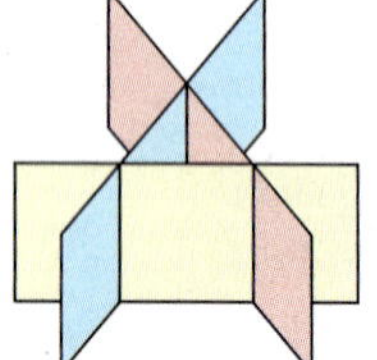

FIGURE 9.15 *Solution: none*

The Number of Solutions of a Linear System

For a system of linear equations, exactly one of the following is true.

1. There is exactly one solution.
2. There are infinitely many solutions.
3. There is no solution.

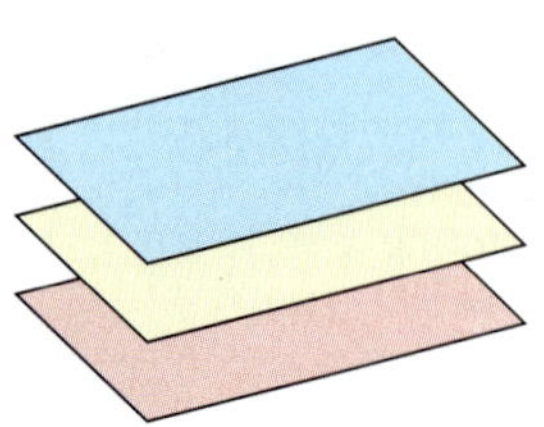

FIGURE 9.16 *Solution: none*

In Section 9.2, you learned that a system of two linear equations in two variables can be represented graphically as a pair of lines that are intersecting, coincident, or parallel. A system of three linear equations in three variables has a similar graphical representation—it can be represented as three planes in space that intersect in one point (exactly one solution) [see Figure 9.12], intersect in a line or a plane (infinitely many solutions) [see Figures 9.13 and 9.14], or have no points common to all three planes (no solution) [see Figures 9.15 and 9.16].

Example 5 ▶ A System with Infinitely Many Solutions

Solve the system of linear equations.

$$\begin{cases} x + y - 3z = -1 & \text{Equation 1} \\ y - z = 0 & \text{Equation 2} \\ -x + 2y = 1 & \text{Equation 3} \end{cases}$$

Solution

$$\begin{cases} x + y - 3z = -1 \\ y - z = 0 \\ 3y - 3z = 0 \end{cases}$$

Adding the first equation to the third equation produces a new third equation.

$$\begin{cases} x + y - 3z = -1 \\ y - z = 0 \\ 0 = 0 \end{cases}$$

Adding -3 times the second equation to the third equation produces a new third equation.

This means that Equation 3 depends on Equations 1 and 2 in the sense that it gives us no additional information about the variables. So, the original system is equivalent to the system

$$\begin{cases} x + y - 3z = -1 \\ y - z = 0. \end{cases}$$

In this last equation, solve for y in terms of z to obtain $y = z$. Back-substituting for y in the previous equation produces $x = 2z - 1$. Finally, letting $z = a$, where a is a real number, you can see that the solutions to the given system are all of the form

$$x = 2a - 1, \quad y = a, \quad \text{and} \quad z = a.$$

So, every ordered triple of the form

$$(2a - 1, a, a), \quad a \text{ is a real number}$$

is a solution of the system.

STUDY TIP

In Example 5, x and y are solved in terms of the third variable z. To write a solution to the system that does not use any of the three variables of the system, let a represent any real number and let $z = a$. Then solve for x and y. The solution can then be written in terms of a, which is not one of the variables of the system.

In Example 5, there are other ways to write the same infinite set of solutions. For instance, letting $x = b$, the solutions could have been written as

$$\left(b, \tfrac{1}{2}(b + 1), \tfrac{1}{2}(b + 1)\right), \quad b \text{ is a real number.}$$

To convince yourself that this description produces the same set of solutions, consider the following.

Substitution	*Solution*	
$a = 0$	$(2(0) - 1, 0, 0) = (-1, 0, 0)$	Same solution
$b = -1$	$\left(-1, \tfrac{1}{2}(-1 + 1), \tfrac{1}{2}(-1 + 1)\right) = (-1, 0, 0)$	
$a = 1$	$(2(1) - 1, 1, 1) = (1, 1, 1)$	Same solution
$b = 1$	$\left(1, \tfrac{1}{2}(1 + 1), \tfrac{1}{2}(1 + 1)\right) = (1, 1, 1)$	
$a = 2$	$(2(2) - 1, 2, 2) = (3, 2, 2)$	Same solution
$b = 3$	$\left(3, \tfrac{1}{2}(3 + 1), \tfrac{1}{2}(3 + 1)\right) = (3, 2, 2)$	

STUDY TIP

When comparing descriptions of an infinite solution set, keep in mind that there is more than one way to describe the set.

Nonsquare Systems

So far, each system of linear equations you have looked at has been *square*, which means that the number of equations is equal to the number of variables. In a **nonsquare** system, the number of equations differs from the number of variables. A system of linear equations cannot have a unique solution unless there are at least as many equations as there are variables in the system.

Example 6 ▶ A System with Fewer Equations than Variables

Solve the system of linear equations.

$$\begin{cases} x - 2y + z = 2 & \text{Equation 1} \\ 2x - y - z = 1 & \text{Equation 2} \end{cases}$$

Solution

Begin by rewriting the system in row-echelon form.

$$\begin{cases} x - 2y + z = 2 \\ 3y - 3z = -3 \end{cases}$$

Adding -2 times the first equation to the second equation produces a new second equation.

$$\begin{cases} x - 2y + z = 2 \\ y - z = -1 \end{cases}$$

Multiplying the second equation by $\frac{1}{3}$ produces a new second equation.

Solve for y in terms of z, to obtain

$$y = z - 1.$$

By back-substituting into Equation 1, you can solve for x, as follows.

$$x - 2y + z = 2 \qquad \text{Write Equation 1.}$$

$$x - 2(z - 1) + z = 2 \qquad \text{Substitute for } y \text{ in Equation 1.}$$

$$x - 2z + 2 + z = 2 \qquad \text{Distributive Property}$$

$$x = z \qquad \text{Solve for } x.$$

Finally, by letting $z = a$, where a is a real number, you have the solution

$$x = a, \quad y = a - 1, \quad \text{and} \quad z = a.$$

So, every ordered triple of the form

$$(a, a - 1, a), \quad a \text{ is a real number}$$

is a solution of the system. Because there were originally three variables and only two equations, the system cannot have a unique solution.

In Example 6, try choosing some values of a to obtain different solutions of the system, such as $(1, 0, 1)$, $(2, 1, 2)$, and $(3, 2, 3)$. Then check each of the solutions in the original system.

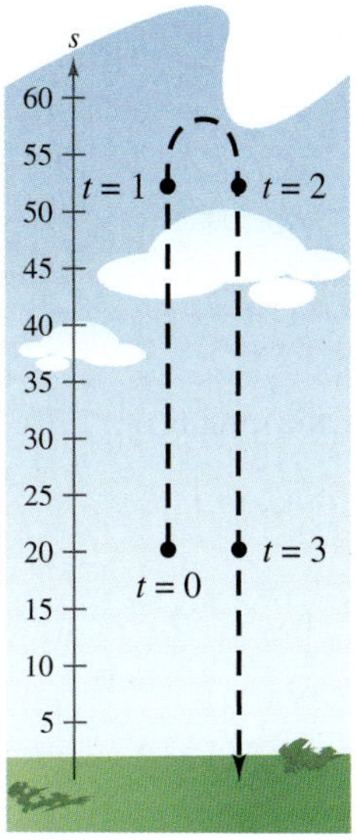

Applications 

Example 7 ▶ Vertical Motion  

The height at time t of an object that is moving in a (vertical) line with constant acceleration a is given by the **position equation**

$$s = \frac{1}{2}at^2 + v_0t + s_0.$$

The height s is measured in feet, the acceleration a is measured in feet per second squared, t is measured in seconds, v_0 is the initial velocity (at $t = 0$), and s_0 is the initial height. Find the values of a, v_0, and s_0 if $s = 52$ at $t = 1$, $s = 52$ at $t = 2$, and $s = 20$ at $t = 3$. (See Figure 9.17.)

FIGURE 9.17

Solution

By substituting the three values of t and s into the position equation, you can obtain three linear equations in a, v_0, and s_0.

When $t = 1$: $\frac{1}{2}a(1)^2 + v_0(1) + s_0 = 52 \quad \Rightarrow \quad a + 2v_0 + 2s_0 = 104$

When $t = 2$: $\frac{1}{2}a(2)^2 + v_0(2) + s_0 = 52 \quad \Rightarrow \quad 2a + 2v_0 + s_0 = 52$

When $t = 3$: $\frac{1}{2}a(3)^2 + v_0(3) + s_0 = 20 \quad \Rightarrow \quad 9a + 6v_0 + 2s_0 = 40$

Solving this system yields $a = -32$, $v_0 = 48$, and $s_0 = 20$. This solution results in a position equation of $s = -16t^2 + 48t + 20$ and implies that the object was thrown upward at a velocity of 48 feet per second from a height of 20 feet.

Example 8 ▶ Partial Fractions 

Write the partial fraction decomposition of $\dfrac{3x + 4}{x^3 - 2x - 4}$.

Solution

Because $x^3 - 2x - 4 = (x - 2)(x^2 + 2x + 2)$, you can write

$$\frac{3x + 4}{x^3 - 2x - 4} = \frac{A}{x - 2} + \frac{Bx + C}{x^2 + 2x + 2}$$

$$3x + 4 = A(x^2 + 2x + 2) + (Bx + C)(x - 2)$$

$$3x + 4 = (A + B)x^2 + (2A - 2B + C)x + (2A - 2C).$$

By equating coefficients of like powers on each side of the expanded equation, you obtain the following system in A, B, and C.

$$\begin{cases} A + B = 0 \\ 2A - 2B + C = 3 \\ 2A - 2C = 4 \end{cases}$$

You can solve this system to find that $A = 1$, $B = -1$, and $C = -1$. So, the partial fraction decomposition is

$$\frac{3x + 4}{x^3 - 2x - 4} = \frac{1}{x - 2} + \frac{-x - 1}{x^2 + 2x + 2} = \frac{1}{x - 2} - \frac{x + 1}{x^2 + 2x + 2}.$$

Example 9 ▶ Data Analysis: Curve-Fitting 

Find a quadratic equation,

$$y = ax^2 + bx + c$$

whose graph passes through the points $(-1, 3)$, $(1, 1)$, and $(2, 6)$.

Solution

Because the graph of $y = ax^2 + bx + c$ passes through the points $(-1, 3)$, $(1, 1)$, and $(2, 6)$, you can write the following.

When $x = -1, y = 3$: $\quad a(-1)^2 + b(-1) + c = 3$

When $x = 1, y = 1$: $\quad a(1)^2 + b(1) + c = 1$

When $x = 2, y = 6$: $\quad a(2)^2 + b(2) + c = 6$

This produces the following system of linear equations.

$$\begin{cases} a - b + c = 3 & \text{Equation 1} \\ a + b + c = 1 & \text{Equation 2} \\ 4a + 2b + c = 6 & \text{Equation 3} \end{cases}$$

The solution of this system is $a = 2$, $b = -1$, and $c = 0$. So, the equation of the parabola is $y = 2x^2 - x$, as shown in Figure 9.18.

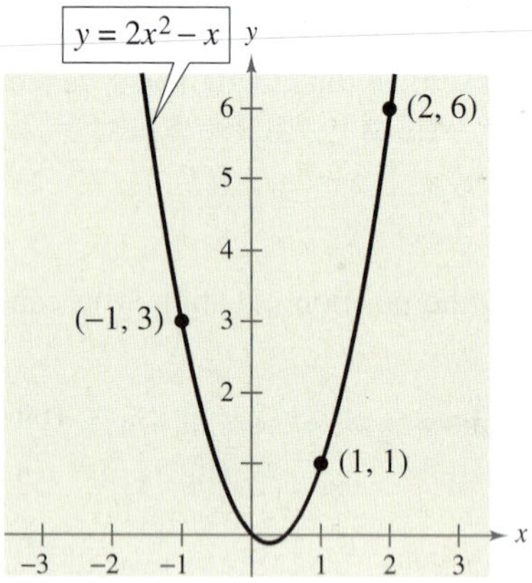

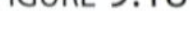

FIGURE 9.18

Example 10 ▶ Investment Analysis  

An inheritance of \$12,000 was invested among three funds: a money-market fund that paid 5% annually, municipal bonds that paid 6% annually, and mutual funds that paid 12% annually. The amount invested in mutual funds was \$4000 more than the amount invested in municipal bonds. The total interest earned during the first year was \$1120. How much was invested in each type of fund?

Solution

Let x, y, and z represent the amounts invested in the money-market fund, municipal bonds, and mutual funds, respectively. From the given information, you can write the following equations.

$$\begin{cases} x + y + z = 12{,}000 & \text{Equation 1} \\ z = y + 4000 & \text{Equation 2} \\ 0.05x + 0.06y + 0.12z = 1120 & \text{Equation 3} \end{cases}$$

Rewriting this system in standard form without decimals produces the following.

$$\begin{cases} x + y + z = 12{,}000 & \text{Equation 1} \\ -y + z = 4{,}000 & \text{Equation 2} \\ 5x + 6y + 12z = 112{,}000 & \text{Equation 3} \end{cases}$$

Using Gaussian elimination to solve this system yields $x = 2000$, $y = 3000$, and $z = 7000$. So, \$2000 was invested in the money-market fund, \$3000 was invested in municipal bonds, and \$7000 was invested in mutual funds.

9.3 Exercises

In Exercises 1–4, determine which ordered triples are solutions of the system of equations.

1. $\begin{cases} 3x - y + z = 1 \\ 2x - 3z = -14 \\ 5y + 2z = 8 \end{cases}$

(a) $(2, 0, -3)$ (b) $(-2, 0, 8)$
(c) $(0, -1, 3)$ (d) $(-1, 0, 4)$

2. $\begin{cases} 3x + 4y - z = 17 \\ 5x - y + 2z = -2 \\ 2x - 3y + 7z = -21 \end{cases}$

(a) $(3, -1, 2)$ (b) $(1, 3, -2)$
(c) $(4, 1, -3)$ (d) $(1, -2, 2)$

3. $\begin{cases} 4x + y - z = 0 \\ -8x - 6y + z = -\frac{7}{4} \\ 3x - y = -\frac{9}{4} \end{cases}$

(a) $\left(\frac{1}{2}, -\frac{3}{4}, -\frac{7}{4}\right)$ (b) $\left(-\frac{3}{2}, \frac{5}{4}, -\frac{5}{4}\right)$
(c) $\left(-\frac{1}{2}, \frac{3}{4}, -\frac{5}{4}\right)$ (d) $\left(-\frac{1}{2}, \frac{1}{6}, -\frac{3}{4}\right)$

4. $\begin{cases} -4x - y - 8z = -6 \\ y + z = 0 \\ 4x - 7y = 6 \end{cases}$

(a) $(-2, -2, 2)$ (b) $\left(-\frac{33}{2}, -10, 10\right)$
(c) $\left(\frac{1}{8}, -\frac{1}{2}, \frac{1}{2}\right)$ (d) $\left(-\frac{11}{2}, -4, 4\right)$

In Exercises 5–10, use back-substitution to solve the system of linear equations.

5. $\begin{cases} 2x - y + 5z = 24 \\ y + 2z = 6 \\ z = 4 \end{cases}$

6. $\begin{cases} 4x - 3y - 2z = 21 \\ 6y - 5z = -8 \\ z = -2 \end{cases}$

7. $\begin{cases} 2x + y - 3z = 10 \\ y + z = 12 \\ z = 2 \end{cases}$

8. $\begin{cases} x - y + 2z = 22 \\ 3y - 8z = -9 \\ z = -3 \end{cases}$

9. $\begin{cases} 4x - 2y + z = 8 \\ -y + z = 4 \\ z = 2 \end{cases}$

10. $\begin{cases} 5x - 8z = 22 \\ 3y - 5z = 10 \\ z = -4 \end{cases}$

In Exercises 11 and 12, perform the row operation and write the equivalent system.

11. Add Equation 1 to Equation 2.

$\begin{cases} x - 2y + 3z = 5 & \text{Equation 1} \\ -x + 3y - 5z = 4 & \text{Equation 2} \\ 2x - 3z = 0 & \text{Equation 3} \end{cases}$

What did this operation accomplish?

12. Add -2 times Equation 1 to Equation 3.

$\begin{cases} x - 2y + 3z = 5 & \text{Equation 1} \\ -x + 3y - 5z = 4 & \text{Equation 2} \\ 2x - 3z = 0 & \text{Equation 3} \end{cases}$

What did this operation accomplish?

In Exercises 13–38, solve the system of linear equations and check any solution algebraically.

13. $\begin{cases} x + y + z = 6 \\ 2x - y + z = 3 \\ 3x - z = 0 \end{cases}$

14. $\begin{cases} x + y + z = 3 \\ x - 2y + 4z = 5 \\ 3y + 4z = 5 \end{cases}$

15. $\begin{cases} 2x + 2z = 2 \\ 5x + 3y = 4 \\ 3y - 4z = 4 \end{cases}$

16. $\begin{cases} 2x + 4y + z = 1 \\ x - 2y - 3z = 2 \\ x + y - z = -1 \end{cases}$

17. $\begin{cases} 6y + 4z = -12 \\ 3x + 3y = 9 \\ 2x - 3z = 10 \end{cases}$

18. $\begin{cases} 2x + 4y - z = 7 \\ 2x - 4y + 2z = -6 \\ x + 4y + z = 0 \end{cases}$

19. $\begin{cases} 2x + y - z = 7 \\ x - 2y + 2z = -9 \\ 3x - y + z = 5 \end{cases}$

20. $\begin{cases} 5x - 3y + 2z = 3 \\ 2x + 4y - z = 7 \\ x - 11y + 4z = 3 \end{cases}$

21. $\begin{cases} 3x - 5y + 5z = 1 \\ 5x - 2y + 3z = 0 \\ 7x - y + 3z = 0 \end{cases}$

22. $\begin{cases} 2x + y + 3z = 1 \\ 2x + 6y + 8z = 3 \\ 6x + 8y + 18z = 5 \end{cases}$

23. $\begin{cases} x + 2y - 7z = -4 \\ 2x + y + z = 13 \\ 3x + 9y - 36z = -33 \end{cases}$

24. $\begin{cases} 2x + y - 3z = 4 \\ 4x + 2z = 10 \\ -2x + 3y - 13z = -8 \end{cases}$

25. $\begin{cases} 3x - 3y + 6z = 6 \\ x + 2y - z = 5 \\ 5x - 8y + 13z = 7 \end{cases}$

26. $\begin{cases} x + 2z = 5 \\ 3x - y - z = 1 \\ 6x - y + 5z = 16 \end{cases}$

27. $\begin{cases} x - 2y + 5z = 2 \\ 4x \qquad\quad - z = 0 \end{cases}$

28. $\begin{cases} x - 3y + 2z = 18 \\ 5x - 13y + 12z = 80 \end{cases}$

29. $\begin{cases} 2x - 3y + z = -2 \\ -4x + 9y \qquad = 7 \end{cases}$

30. $\begin{cases} 2x + 3y + 3z = 7 \\ 4x + 18y + 15z = 44 \end{cases}$

31. $\begin{cases} x \qquad\qquad\quad + 3w = 4 \\ \quad 2y - z - w = 0 \\ \quad 3y \qquad - 2w = 1 \\ 2x - y + 4z \qquad = 5 \end{cases}$

32. $\begin{cases} x + y + z + w = 6 \\ 2x + 3y \qquad - w = 0 \\ -3x + 4y + z + 2w = 4 \\ x + 2y - z + w = 0 \end{cases}$

33. $\begin{cases} x \qquad + 4z = 1 \\ x + y + 10z = 10 \\ 2x - y + 2z = -5 \end{cases}$

34. $\begin{cases} 2x - 2y - 6z = -4 \\ -3x + 2y + 6z = 1 \\ x - y - 5z = -3 \end{cases}$

35. $\begin{cases} 2x + 3y \qquad = 0 \\ 4x + 3y - z = 0 \\ 8x + 3y + 3z = 0 \end{cases}$

36. $\begin{cases} 4x + 3y + 17z = 0 \\ 5x + 4y + 22z = 0 \\ 4x + 2y + 19z = 0 \end{cases}$

37. $\begin{cases} 12x + 5y + z = 0 \\ 23x + 4y - z = 0 \end{cases}$

38. $\begin{cases} 2x - y - z = 0 \\ -2x + 6y + 4z = 2 \end{cases}$

In Exercises 39–42, find the equation of the parabola

$$y = ax^2 + bx + c$$

that passes through the points. To verify your result, use a graphing utility to plot the points and graph the parabola.

39. $(0, 0), (2, -2), (4, 0)$

40. $(0, 3), (1, 4), (2, 3)$

41. $(2, 0), (3, -1), (4, 0)$

42. $(1, 3), (2, 2), (3, -3)$

In Exercises 43–46, find the equation of the circle

$$x^2 + y^2 + Dx + Ey + F = 0$$

that passes through the points. To verify your result, use a graphing utility to plot the points and graph the circle.

43. $(0, 0), (2, 2), (4, 0)$

44. $(0, 0), (0, 6), (3, 3)$

45. $(-3, -1), (2, 4), (-6, 8)$

46. $(0, 0), (0, -2), (3, 0)$

Vertical Motion **In Exercises 47–50, an object moving vertically is at the given heights at the specified times. Find the position equation $s = \frac{1}{2}at^2 + v_0t + s_0$ for the object.**

47. At $t = 1$ second, $s = 128$ feet
At $t = 2$ seconds, $s = 80$ feet
At $t = 3$ seconds, $s = 0$ feet

48. At $t = 1$ second, $s = 48$ feet
At $t = 2$ seconds, $s = 64$ feet
At $t = 3$ seconds, $s = 48$ feet

49. At $t = 1$ second, $s = 452$ feet
At $t = 2$ seconds, $s = 372$ feet
At $t = 3$ seconds, $s = 260$ feet

50. At $t = 1$ second, $s = 132$ feet
At $t = 2$ seconds, $s = 100$ feet
At $t = 3$ seconds, $s = 36$ feet

51. ***Sports*** The University of Michigan and the University of Tennessee scored a total of 62 points during the 2002 Florida Citrus Bowl. The points came from a total of 18 different scoring plays, which were a combination of touchdowns, extra-point kicks, and field goals, worth 6, 1, and 3 points, respectively. The same number of extra points and touchdowns were scored. How many touchdowns, extra-point kicks, and field goals were scored? (Source: National Collegiate Athletic Association)

52. ***Sports*** During the second game of the 2002 Western Conference finals, the Los Angeles Lakers scored a total of 90 points, resulting from a combination of three-point baskets, two-point baskets, and one-point free-throws. There were 11 times as many two-point baskets as three-point baskets and five times as many free-throws as three-point baskets. What combination of scoring accounted for the Lakers' 90 points? (Source: National Basketball Association)

53. ***Finance*** A small corporation borrowed $775,000 to expand its clothing line. Some of the money was borrowed at 8%, some at 9%, and some at 10%. How much was borrowed at each rate if the annual interest owed was $67,500 and the amount borrowed at 8% was four times the amount borrowed at 10%?

54. ***Finance*** A small corporation borrowed $800,000 to expand its line of toys. Some of the money was borrowed at 8%, some at 9%, and some at 10%. How much was borrowed at each rate if the annual interest owed was $67,000 and the amount borrowed at 8% was five times the amount borrowed at 10%?

Investment Portfolio **In Exercises 55 and 56, consider an investor with a portfolio totaling $500,000 that is invested in certificates of deposit, municipal bonds, blue-chip stocks, and growth or speculative stocks. How much is invested in each type of investment?**

55. The certificates of deposit pay 10% annually, and the municipal bonds pay 8% annually. Over a five-year period, the investor expects the blue-chip stocks to return 12% annually and the growth stocks to return 13% annually. The investor wants a combined annual return of 10% and also wants to have only one-fourth of the portfolio invested in stocks.

56. The certificates of deposit pay 9% annually, and the municipal bonds pay 5% annually. Over a five-year period, the investor expects the blue-chip stocks to return 12% annually and the growth stocks to return 14% annually. The investor wants a combined annual return of 10% and also wants to have only one-fourth of the portfolio invested in stocks.

57. ***Truck Scheduling*** A small company that manufactures two models of exercise machines has an order for 15 units of the standard model and 16 units of the deluxe model. The company has trucks of three different sizes that can haul the products, as shown in the table.

Truck	Standard	Deluxe
Large	6	3
Medium	4	4
Small	0	3

How many trucks of each size are needed to deliver the order? Give two possible solutions.

58. ***Agriculture*** A mixture of 12 liters of chemical A, 16 liters of chemical B, and 26 liters of chemical C is required to kill a destructive crop insect. Commercial spray X contains 1, 2, and 2 parts, respectively, of these chemicals. Commercial spray Y contains only chemical C. Commercial spray Z contains only chemicals A and B in equal amounts. How much of each type of commercial spray is needed to get the desired mixture?

59. ***Acid Mixture*** A chemist needs 10 liters of a 25% acid solution. The solution is to be mixed from three solutions whose concentrations are 10%, 20%, and 50%. How many liters of each solution should the chemist use so that as little as possible of the 50% solution is used?

60. ***Electrical Network*** Applying Kirchhoff's Laws to the electrical network in the figure, the currents I_1, I_2, and I_3 are the solution of the system

$$\begin{cases} I_1 - I_2 + I_3 = 0 \\ 3I_1 + 2I_2 = 7 \\ 2I_2 + 4I_3 = 8. \end{cases}$$

Find the currents.

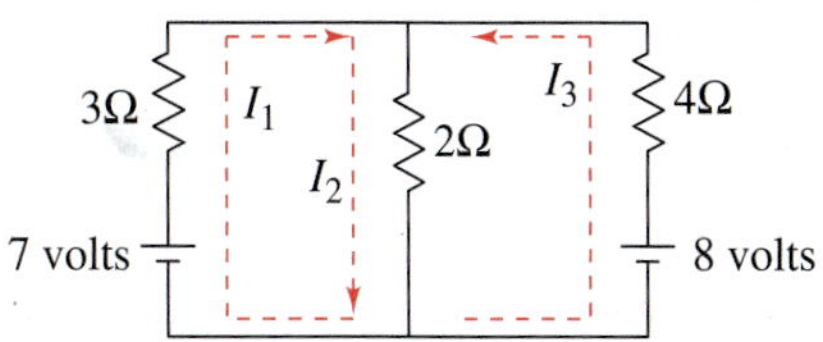

61. ***Pulley System*** A system of pulleys is loaded with 128-pound and 32-pound weights (see figure). The tensions t_1 and t_2 in the ropes and the acceleration a of the 32-pound weight are found by solving the system of equations

$$\begin{cases} t_1 - 2t_2 = 0 \\ t_1 - 2a = 128 \\ t_2 + a = 32 \end{cases}$$

where t_1 and t_2 are measured in pounds and a is measured in feet per second squared. Solve this system.

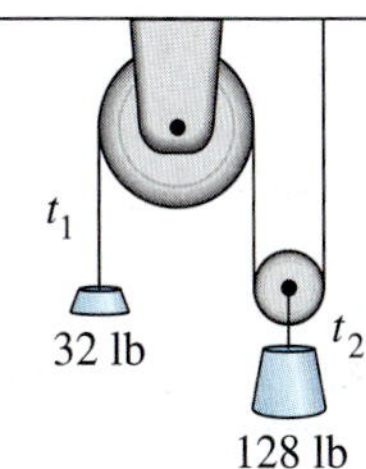

62. ***Pulley System*** The 32-pound weight in the pulley system in Exercise 61 is replaced by a 64-pound weight. The new pulley system will be modeled by the following system of equations.

$$\begin{cases} t_1 - 2t_2 = 0 \\ t_1 - 2a = 128 \\ t_2 + 2a = 64 \end{cases}$$

Solve this system and use your answer for the acceleration to describe what (if anything) is happening in the pulley system.

Partial Fraction Decomposition **In Exercises 63–66, write the partial fraction decomposition of the rational expression.**

63. $\dfrac{1}{x^3 - x} = \dfrac{A}{x} + \dfrac{B}{x - 1} + \dfrac{C}{x + 1}$

64. $\dfrac{3}{x^2 + x - 2} = \dfrac{A}{x - 1} + \dfrac{B}{x + 2}$

65. $\dfrac{x^2 - 3x - 3}{x(x - 2)(x + 3)} = \dfrac{A}{x} + \dfrac{B}{x - 2} + \dfrac{C}{x + 3}$

66. $\dfrac{12}{x(x - 2)(x + 3)} = \dfrac{A}{x} + \dfrac{B}{x - 2} + \dfrac{C}{x + 3}$

Fitting a Parabola **In Exercises 67–70, find the least squares regression parabola $y = ax^2 + bx + c$ for the points $(x_1, y_1), (x_2, y_2), \ldots, (x_n, y_n)$ by solving the following system of linear equations for a, b, and c. Then use the *regression* feature of a graphing utility to confirm the result. (If you are unfamiliar with summation notation, look at the discussion in Section 11.1 or in Appendix A at the website for this text at *college.hmco.com*.)**

$$nc + \left(\sum_{i=1}^{n} x_i\right)b + \left(\sum_{i=1}^{n} x_i^2\right)a = \sum_{i=1}^{n} y_i$$

$$\left(\sum_{i=1}^{n} x_i\right)c + \left(\sum_{i=1}^{n} x_i^2\right)b + \left(\sum_{i=1}^{n} x_i^3\right)a = \sum_{i=1}^{n} x_i y_i$$

$$\left(\sum_{i=1}^{n} x_i^2\right)c + \left(\sum_{i=1}^{n} x_i^3\right)b + \left(\sum_{i=1}^{n} x_i^4\right)a = \sum_{i=1}^{n} x_i^2 y_i$$

67.

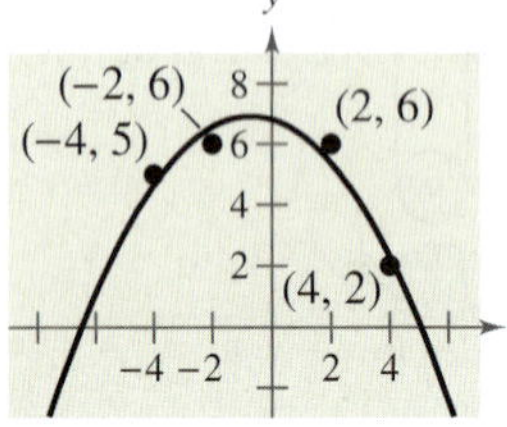

68.

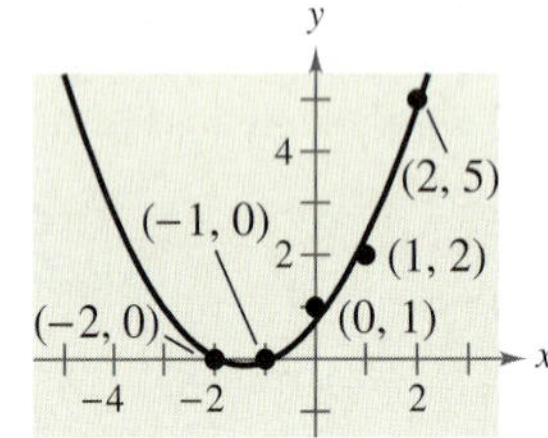

69.

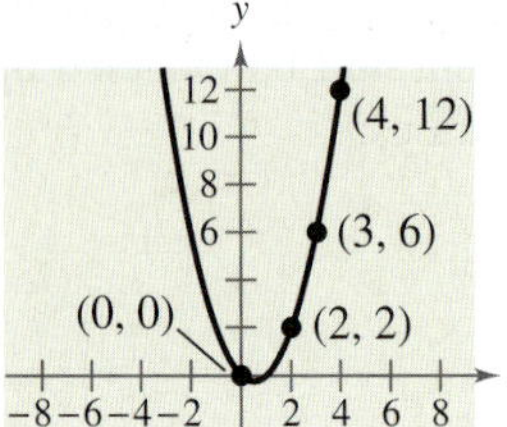

70.

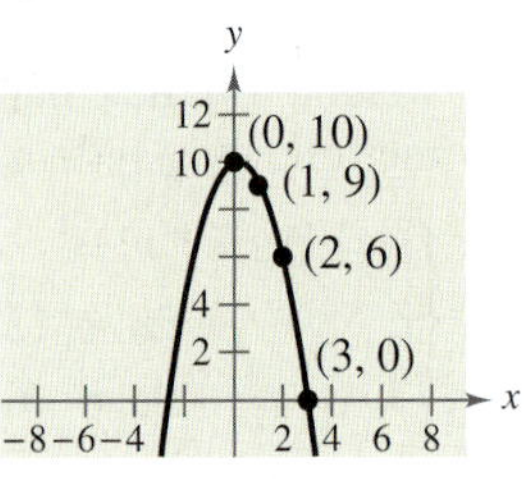

▶ Model It

71. ***Data Analysis*** A wildlife management team studied the reproduction rates of deer in three tracts of a wildlife preserve. Each tract contained 5 acres. In each tract, the number of females, and the percent of females that had offspring the following year, were recorded. The results are shown in the table.

Number, x	Percent, y
100	75
120	68
140	55

(a) Use the technique demonstrated in Exercises 67–70 to find a least squares regression parabola that models the data.

(b) Use a graphing utility to graph the parabola and the data in the same viewing window.

(c) Use the model to create a table of estimated values of y. Compare the estimated values with the actual data.

(d) Use the model to estimate the percent of females that had offspring when there were 170 females.

(e) Use the model to estimate the number of females when 40% of the females had offspring.

72. ***Data Analysis*** In testing a new automobile braking system, the speed in miles per hour and the stopping distance in feet were recorded in the table.

Speed, x	Stopping distance, y
30	55
40	105
50	188

(a) Use the technique demonstrated in Exercises 67–70 to find a least squares regression parabola that models the data.

(b) Graph the parabola and the data on the same set of axes.

(c) Use the model to estimate the stopping distance when the speed is 70 miles per hour.

Advanced Applications **In Exercises 73–76, find x, y, and λ satisfying the system. These systems arise in certain optimization problems in calculus, and λ is called a Lagrange multiplier.**

73. $\begin{cases} y + \lambda = 0 \\ x + \lambda = 0 \\ x + y - 10 = 0 \end{cases}$

74. $\begin{cases} 2x + \lambda = 0 \\ 2y + \lambda = 0 \\ x + y - 4 = 0 \end{cases}$

75. $\begin{cases} 2x - 2x\lambda = 0 \\ -2y + \lambda = 0 \\ y - x^2 = 0 \end{cases}$

76. $\begin{cases} 2 + 2y + 2\lambda = 0 \\ 2x + 1 + \lambda = 0 \\ 2x + y - 100 = 0 \end{cases}$

Synthesis

True or False? **In Exercises 77 and 78, determine whether the statement is true or false. Justify your answer.**

77. The system

$$\begin{cases} x + 3y - 6z = -16 \\ 2y - z = -1 \\ z = 3 \end{cases}$$

is in row-echelon form.

78. If a system of three linear equations is inconsistent, then its graph has no points common to all three equations.

79. ***Think About It*** Are the following two systems of equations equivalent? Give reasons for your answer.

$$\begin{cases} x + 3y - z = 6 \\ 2x - y + 2z = 1 \\ 3x + 2y - z = 2 \end{cases} \qquad \begin{cases} x + 3y - z = 6 \\ -7y + 4z = 1 \\ -7y - 4z = -16 \end{cases}$$

80. ***Writing*** When using Gaussian elimination to solve a system of linear equations, explain how you can recognize that the system has no solution. Give an example that illustrates your answer.

In Exercises 81–84, find two systems of linear equations that have the ordered triple as a solution. (The answers are not unique.)

81. $(4, -1, 2)$

82. $(-5, -2, 1)$

83. $\left(3, -\frac{1}{2}, \frac{7}{4}\right)$

84. $\left(-\frac{3}{2}, 4, -7\right)$

Review

In Exercises 85–88, solve the percent problem.

85. What is $7\frac{1}{2}\%$ of 85?

86. 225 is what percent of 150?

87. 0.5% of what number is 400?

88. 48% of what number is 132?

In Exercises 89–94, perform the operation and write the result in standard form.

89. $(7 - i) + (4 + 2i)$

90. $(-6 + 3i) - (1 + 6i)$

91. $(4 - i)(5 + 2i)$

92. $(1 + 2i)(3 - 4i)$

93. $\dfrac{i}{1 + i} + \dfrac{6}{1 - i}$

94. $\dfrac{i}{4 + i} - \dfrac{2i}{8 - 3i}$

In Exercises 95–98, (a) determine the real zeros of f and (b) sketch the graph of f.

95. $f(x) = x^3 + x^2 - 12x$

96. $f(x) = -8x^4 + 32x^2$

97. $f(x) = 2x^3 + 5x^2 - 21x - 36$

98. $f(x) = 6x^3 - 29x^2 - 6x + 5$

In Exercises 99–102, use a graphing utility to construct a table of values for the equation. Then sketch the graph of the equation by hand.

99. $y = 4^{x-4} - 5$

100. $y = \left(\frac{5}{2}\right)^{-x+1} - 4$

101. $y = 1.9^{-0.8x} + 3$

102. $y = 3.5^{-x+2} + 6$

In Exercises 103 and 104, solve the system by elimination.

103. $\begin{cases} 2x + y = 120 \\ x + 2y = 120 \end{cases}$

104. $\begin{cases} 6x - 5y = 3 \\ 10x - 12y = 5 \end{cases}$

9.4 Systems of Inequalities

▶ What you should learn

- How to sketch the graphs of inequalities in two variables
- How to solve systems of inequalities
- How to use systems of inequalities in two variables to model and solve real-life problems

▶ Why you should learn it

You can use systems of inequalities in two variables to model and solve real-life problems. For instance, in Exercise 73 on page 709, you will use a system of inequalities to analyze the number of electric-powered vehicles in the United States.

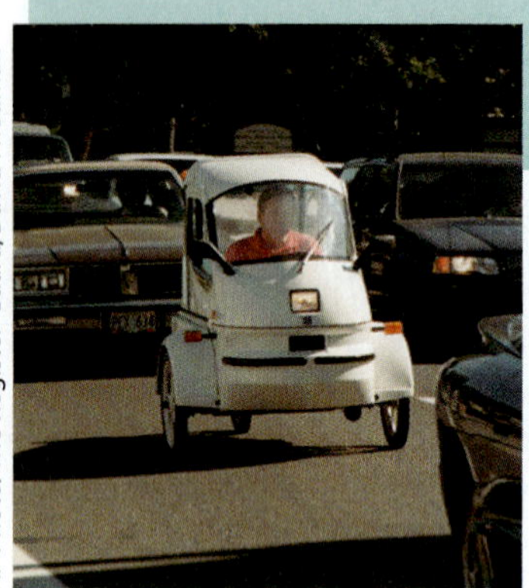

AP Photo/The Register-Guard, David Friedman

The Graph of an Inequality 

The statements $3x - 2y < 6$ and $2x^2 + 3y^2 \geq 6$ are inequalities in two variables. An ordered pair (a, b) is a **solution of an inequality** in x and y if the inequality is true when a and b are substituted for x and y, respectively. The **graph of an inequality** is the collection of all solutions of the inequality. To sketch the graph of an inequality, begin by sketching the graph of the *corresponding equation*. The graph of the equation will normally separate the plane into two or more regions. In each such region, one of the following must be true.

1. *All* points in the region are solutions of the inequality.
2. *No* point in the region is a solution of the inequality.

So, you can determine whether the points in an entire region satisfy the inequality by simply testing *one* point in the region.

Sketching the Graph of an Inequality in Two Variables

1. Replace the inequality sign by an equal sign, and sketch the graph of the resulting equation. (Use a dashed line for $<$ or $>$ and a solid line for $\leq$ or $\geq$.)
2. Test one point in each of the regions formed by the graph in Step 1. If the point satisfies the inequality, shade the entire region to denote that every point in the region satisfies the inequality.

Example 1 ▶ Sketching the Graph of an Inequality 

To sketch the graph of $y \geq x^2 - 1$, begin by graphing the corresponding *equation* $y = x^2 - 1$, which is a parabola, as shown in Figure 9.19. By testing a point *above* the parabola $(0, 0)$ and a point *below* the parabola $(0, -2)$, you can see that the points that satisfy the inequality are those lying above (or on) the parabola.

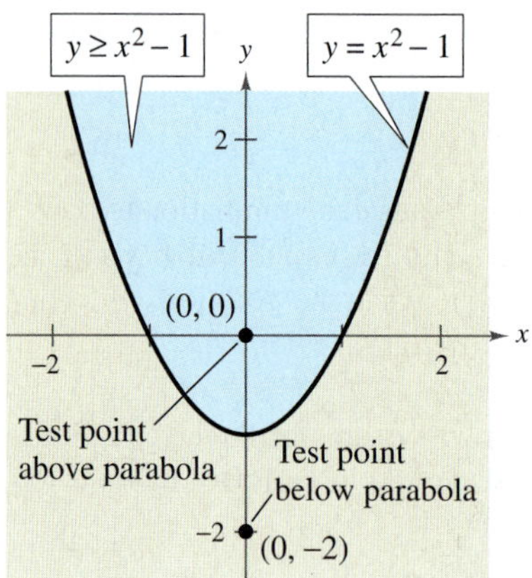

FIGURE 9.19

The inequality in Example 1 is a nonlinear inequality in two variables. Most of the following examples involve **linear inequalities** such as $ax + by < c$ (a and b are not both zero). The graph of a linear inequality is a half-plane lying on one side of the line $ax + by = c$.

Example 2 ▶ Sketching the Graph of a Linear Inequality

Sketch the graph of each linear inequality.

a. $x > -2$ **b.** $y \leq 3$

Solution

a. The graph of the corresponding equation $x = -2$ is a vertical line. The points that satisfy the inequality $x > -2$ are those lying to the right of this line, as shown in Figure 9.20.

b. The graph of the corresponding equation $y = 3$ is a horizontal line. The points that satisfy the inequality $y \leq 3$ are those lying below (or on) this line, as shown in Figure 9.21.

Technology

A graphing utility can be used to graph an inequality or a system of inequalities. Consult the user's guide for your graphing utility for the required keystrokes. The graph of the inequality in Example 3 is shown below.

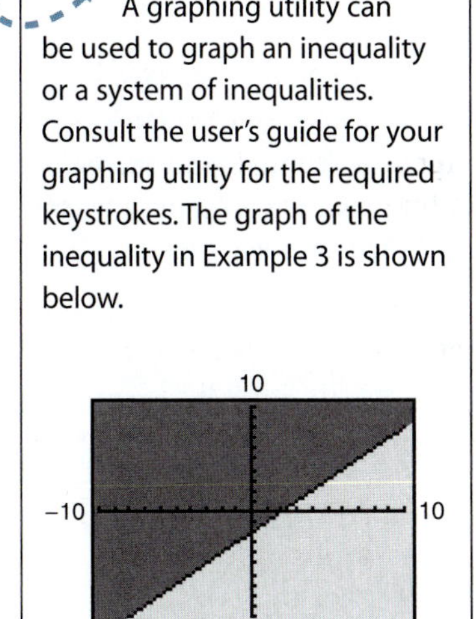

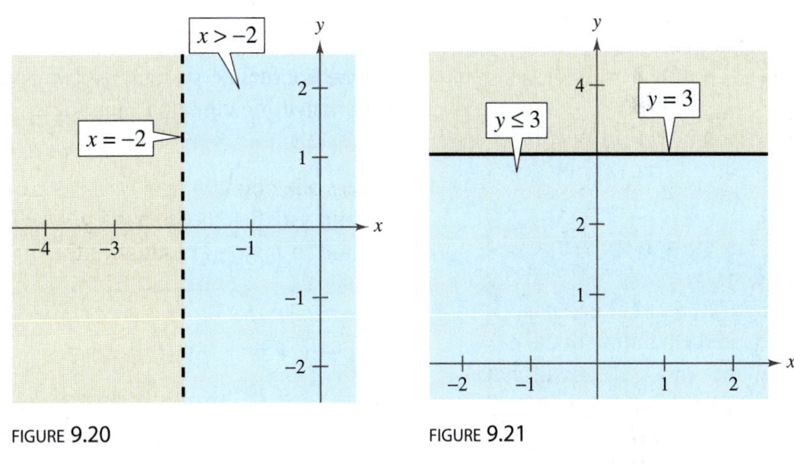

FIGURE 9.20

FIGURE 9.21

Example 3 ▶ Sketching the Graph of a Linear Inequality

Sketch the graph of $x - y < 2$.

Solution

The graph of the corresponding equation $x - y = 2$ is a line, as shown in Figure 9.22. Because the origin (0, 0) satisfies the inequality, the graph consists of the half-plane lying above the line. (Try checking a point below the line. Regardless of which point you choose, you will see that it does not satisfy the inequality.)

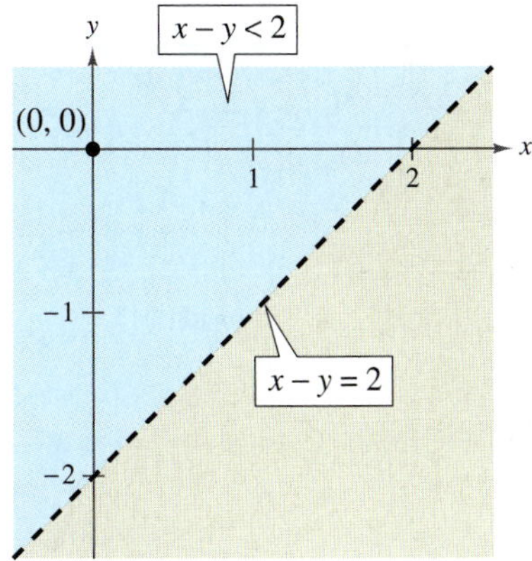

FIGURE 9.22

To graph a linear inequality, it can help to write the inequality in slope-intercept form. For instance, by writing $x - y < 2$ in the form

$$y > x - 2$$

you can see that the solution points lie *above* the line $x - y = 2$ (or $y = x - 2$), as shown in Figure 9.22.

Systems of Inequalities

Many practical problems in business, science, and engineering involve systems of linear inequalities. A **solution** of a system of inequalities in x and y is a point (x, y) that satisfies each inequality in the system.

To sketch the graph of a system of inequalities in two variables, first sketch the graph of each individual inequality (on the same coordinate system) and then find the region that is *common* to every graph in the system. For systems of *linear* inequalities, it is helpful to find the vertices of the solution region.

Example 4 ▶ Solving a System of Inequalities 

Sketch the graph (and label the vertices) of the solution set of the system.

$$\begin{cases} x - y < 2 & \text{Inequality 1} \\ x > -2 & \text{Inequality 2} \\ y \le 3 & \text{Inequality 3} \end{cases}$$

Solution

The graphs of these inequalities are shown in Figures 9.20 to 9.22. The triangular region common to all three graphs can be found by superimposing the graphs on the same coordinate system, as shown in Figure 9.23. To find the vertices of the region, solve the three systems of corresponding equations obtained by taking *pairs* of equations representing the boundaries of the individual regions.

Vertex A: $(-2, -4)$

$$\begin{cases} x - y = 2 & \text{Boundary of Inequality 1} \\ x = -2 & \text{Boundary of Inequality 2} \end{cases}$$

Vertex B: $(5, 3)$

$$\begin{cases} x - y = 2 & \text{Boundary of Inequality 1} \\ y = 3 & \text{Boundary of Inequality 3} \end{cases}$$

Vertex C: $(-2, 3)$

$$\begin{cases} x = -2 & \text{Boundary of Inequality 2} \\ y = 3 & \text{Boundary of Inequality 3} \end{cases}$$

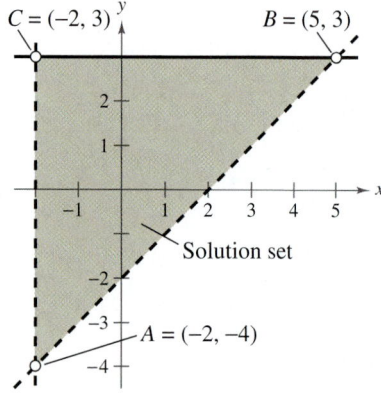

STUDY TIP

Using different colored pencils to shade the solution of each inequality in a system will make identifying the solution of the system of inequalities easier.

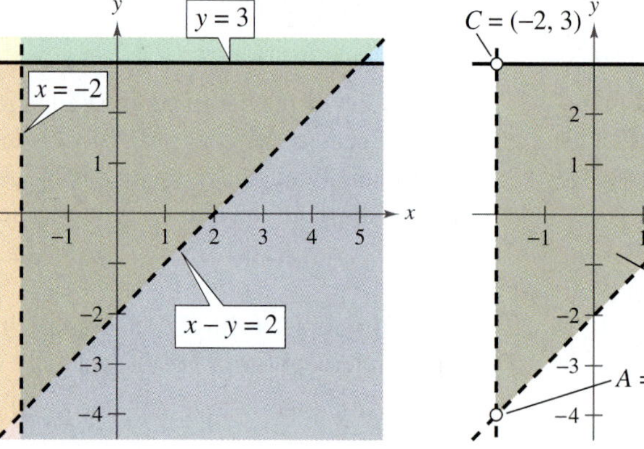

FIGURE 9.23

For the triangular region shown in Figure 9.23, each point of intersection of a pair of boundary lines corresponds to a vertex. With more complicated regions, two border lines can sometimes intersect at a point that is not a vertex of the region, as shown in Figure 9.24. To keep track of which points of intersection are actually vertices of the region, you should sketch the region and refer to your sketch as you find each point of intersection.

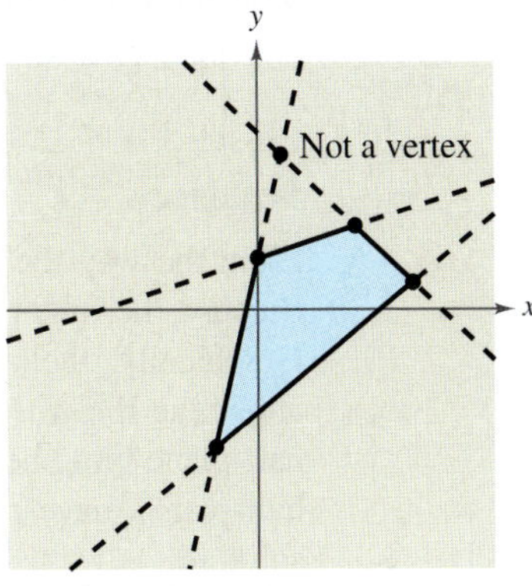

FIGURE 9.24

Example 5 ▶ Solving a System of Inequalities 

Sketch the region containing all points that satisfy the system of inequalities.

$$\begin{cases} x^2 - y \le 1 & \text{Inequality 1} \\ -x + y \le 1 & \text{Inequality 2} \end{cases}$$

Solution

As shown in Figure 9.25, the points that satisfy the inequality

$$x^2 - y \le 1 \qquad \text{Inequality 1}$$

are the points lying above (or on) the parabola given by

$$y = x^2 - 1. \qquad \text{Parabola}$$

The points satisfying the inequality

$$-x + y \le 1 \qquad \text{Inequality 2}$$

are the points lying below (or on) the line given by

$$y = x + 1. \qquad \text{Line}$$

To find the points of intersection of the parabola and the line, solve the system of corresponding equations.

$$\begin{cases} x^2 - y = 1 \\ -x + y = 1 \end{cases}$$

Using the method of substitution, you can find the solutions to be $(-1, 0)$ and $(2, 3)$, as shown in Figure 9.25.

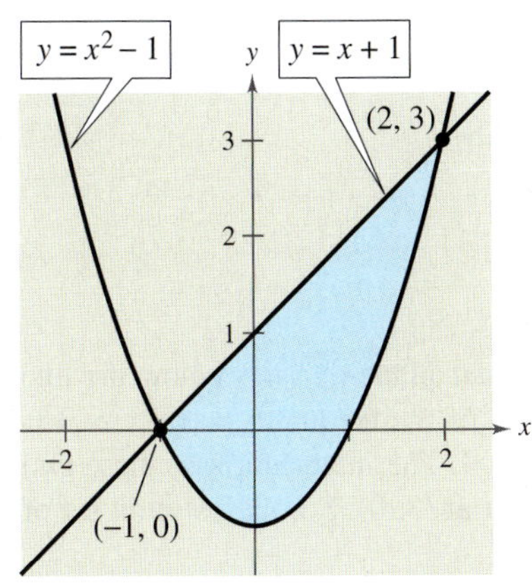

FIGURE 9.25

When solving a system of inequalities, you should be aware that the system might have no solution *or* it might be represented by an unbounded region in the plane. These two possibilities are shown in Examples 6 and 7.

Example 6 ▶ A System with No Solution

Sketch the solution set of the system of inequalities.

$$\begin{cases} x + y > 3 & \text{Inequality 1} \\ x + y < -1 & \text{Inequality 2} \end{cases}$$

Solution

From the way the system is written, it is clear that the system has no solution, because the quantity $(x + y)$ cannot be both less than -1 and greater than 3. Graphically, the inequality $x + y > 3$ is represented by the half-plane lying above the line $x + y = 3$, and the inequality $x + y < -1$ is represented by the half-plane lying below the line $x + y = -1$, as shown in Figure 9.26. These two half-planes have no points in common. So, the system of inequalities has no solution.

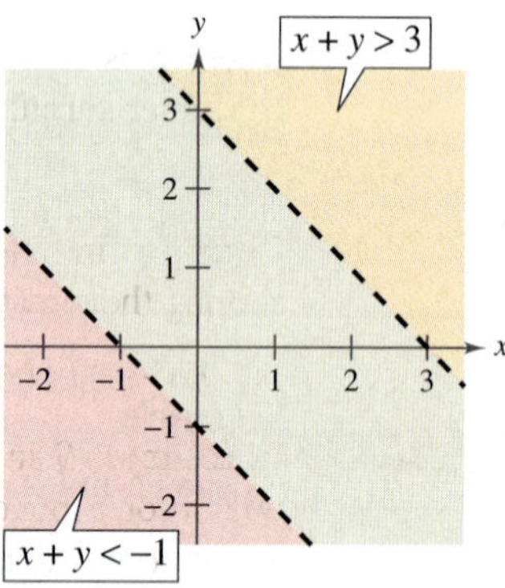

FIGURE 9.26

Example 7 ▶ An Unbounded Solution Set

Sketch the solution set of the system of inequalities.

$$\begin{cases} x + y < 3 & \text{Inequality 1} \\ x + 2y > 3 & \text{Inequality 2} \end{cases}$$

Solution

The graph of the inequality $x + y < 3$ is the half-plane that lies below the line $x + y = 3$, as shown in Figure 9.27. The graph of the inequality $x + 2y > 3$ is the half-plane that lies above the line $x + 2y = 3$. The intersection of these two half-planes is an *infinite wedge* that has a vertex at $(3, 0)$. So, the solution set of the system of inequalities is unbounded.

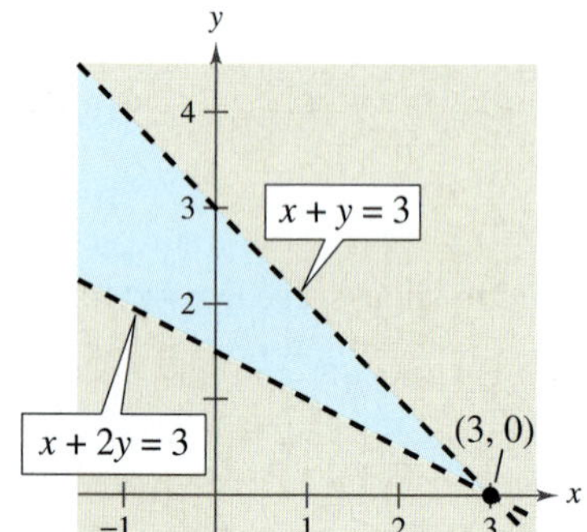

FIGURE 9.27

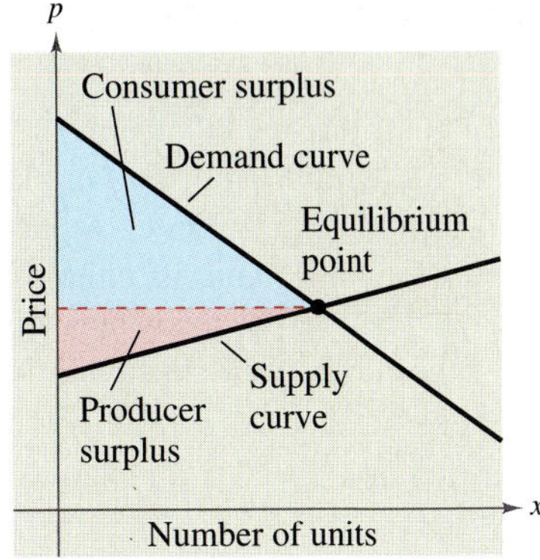

FIGURE 9.28

Applications

Example 9 in Section 9.2 discussed the *equilibrium point* for a system of demand and supply functions. The next example discusses two related concepts that economists call **consumer surplus** and **producer surplus.** As shown in Figure 9.28, the consumer surplus is defined as the area of the region that lies *below* the demand curve, *above* the horizontal line passing through the equilibrium point, and to the right of the p-axis. Similarly, the producer surplus is defined as the area of the region that lies *above* the supply curve, *below* the horizontal line passing through the equilibrium point, and to the right of the p-axis. The consumer surplus is a measure of the amount that consumers would have been willing to pay *above what they actually paid*, whereas the producer surplus is a measure of the amount that producers would have been willing to receive *below what they actually received.*

Example 8 ▶ Consumer Surplus and Producer Surplus 

The demand and supply functions for a new type of calculator are given by

$$\begin{cases} p = 150 - 0.00001x & \text{Demand equation} \\ p = 60 + 0.00002x & \text{Supply equation} \end{cases}$$

where p is the price in dollars and x represents the number of units. Find the consumer surplus and producer surplus for these two equations.

Solution

Begin by finding the equilibrium point (when supply and demand are equal) by solving the equation

$$60 + 0.00002x = 150 - 0.00001x.$$

In Example 9 in Section 9.2, you saw that the solution is $x = 3{,}000{,}000$ units, which corresponds to an equilibrium price of $p = \$120$. So, the consumer surplus and producer surplus are the areas of the following triangular regions.

Consumer Surplus

$$\begin{cases} p \le 150 - 0.00001x \\ p \ge 120 \\ x \ge 0 \end{cases}$$

Producer Surplus

$$\begin{cases} p \ge 60 + 0.00002x \\ p \le 120 \\ x \ge 0 \end{cases}$$

In Figure 9.29, you can see that the consumer and producer surpluses are defined as the areas of the shaded triangles.

FIGURE 9.29

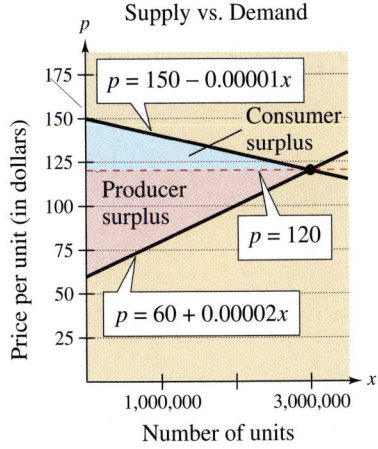

$$\text{Consumer surplus} = \frac{1}{2}(\text{base})(\text{height})$$

$$= \frac{1}{2}(3{,}000{,}000)(30) = \$45{,}000{,}000$$

$$\text{Producer surplus} = \frac{1}{2}(\text{base})(\text{height})$$

$$= \frac{1}{2}(3{,}000{,}000)(60) = \$90{,}000{,}000$$

Example 9 ▶ Nutrition  

The minimum daily requirements from the liquid portion of a diet are 300 calories, 36 units of vitamin A, and 90 units of vitamin C. A cup of dietary drink X provides 60 calories, 12 units of vitamin A, and 10 units of vitamin C. A cup of dietary drink Y provides 60 calories, 6 units of vitamin A, and 30 units of vitamin C. Set up a system of linear inequalities that describes how many cups of each drink should be consumed each day to meet the minimum daily requirements for calories and vitamins.

Solution

Begin by letting x and y represent the following.

$x =$ number of cups of dietary drink X

$y =$ number of cups of dietary drink Y

To meet the minimum daily requirements, the following inequalities must be satisfied.

$$\begin{cases} 60x + 60y \geq 300 & \text{Calories} \\ 12x + 6y \geq 36 & \text{Vitamin A} \\ 10x + 30y \geq 90 & \text{Vitamin C} \\ x \geq 0 & \\ y \geq 0 & \end{cases}$$

The last two inequalities are included because x and y cannot be negative. The graph of this system of inequalities is shown in Figure 9.30. (More is said about this application in Example 6 in Section 9.5.)

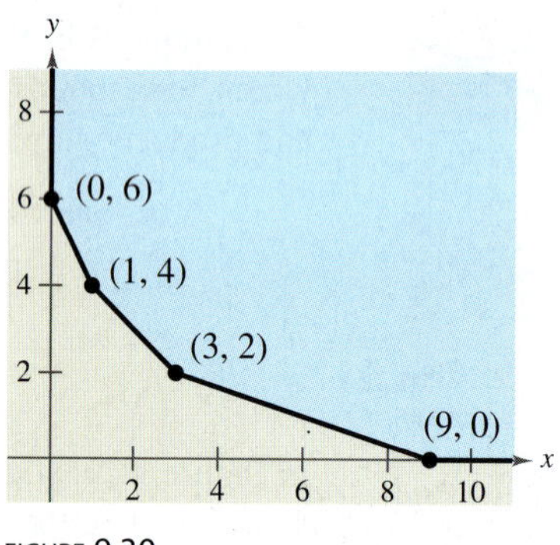

FIGURE 9.30

Writing ABOUT MATHEMATICS

Creating a System of Inequalities Plot the points (0, 0), (4, 0), (3, 2), and (0, 2) in a coordinate plane. Draw the quadrilateral that has these four points as its vertices. Write a system of linear inequalities that has the quadrilateral as its solution. Explain how you found the system of inequalities.

9.4 Exercises

In Exercises 1–12, sketch the graph of the inequality.

1. $x \geq 2$
2. $x \leq 4$
3. $y \geq -1$
4. $y \leq 3$
5. $y < 2 - x$
6. $y > 2x - 4$
7. $2y - x \geq 4$
8. $5x + 3y \geq -15$
9. $(x + 1)^2 + (y - 2)^2 < 9$
10. $y^2 - x < 0$
11. $y \leq \dfrac{1}{1 + x^2}$
12. $y > \dfrac{-15}{x^2 + x + 4}$

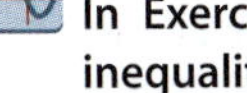 In Exercises 13–24, use a graphing utility to graph the inequality. Shade the region representing the solution.

13. $y < \ln x$
14. $y \geq 6 - \ln(x + 5)$
15. $y < 3^{-x-4}$
16. $y \leq 2^{2x-0.5} - 7$
17. $y \geq \frac{2}{3}x - 1$
18. $y \leq 6 - \frac{3}{2}x$
19. $y < -3.8x + 1.1$
20. $y \geq -20.74 + 2.66x$
21. $x^2 + 5y - 10 \leq 0$
22. $2x^2 - y - 3 > 0$
23. $\frac{5}{2}y - 3x^2 - 6 \geq 0$
24. $-\frac{1}{10}x^2 - \frac{3}{8}y < -\frac{1}{4}$

In Exercises 25–28, write an inequality for the shaded region shown in the figure.

25.

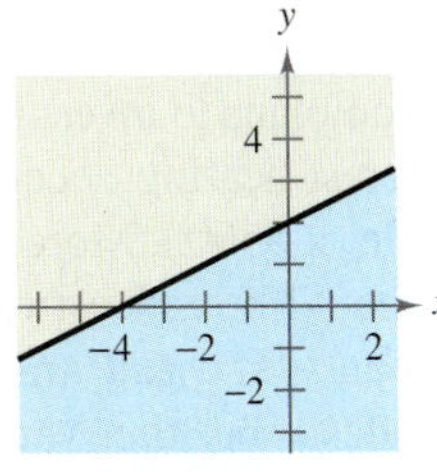

26.

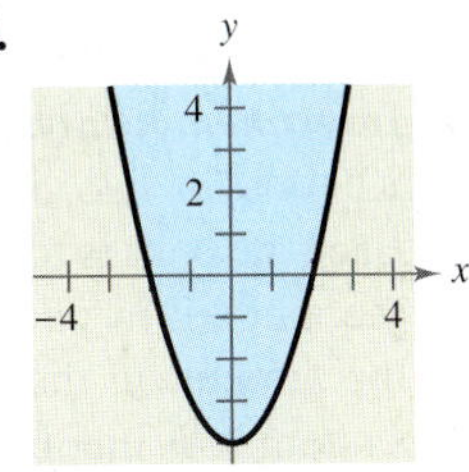

27.

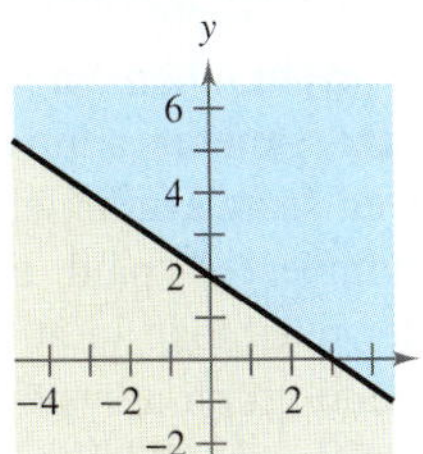

28.

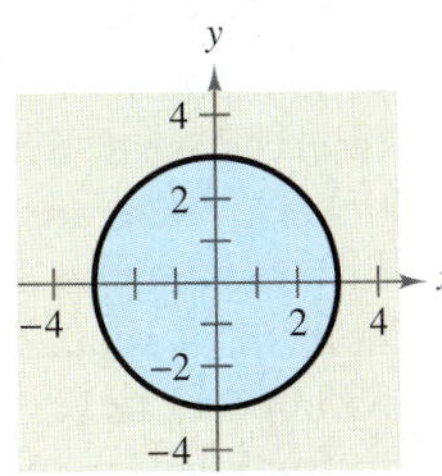

In Exercises 29–32, determine which ordered pairs are solutions of the system of linear inequalities.

29. $\begin{cases} x \geq -4 \\ y > -3 \\ y \leq -8x - 3 \end{cases}$

(a) $(0, 0)$ (b) $(-1, -3)$
(c) $(-4, 0)$ (d) $(-3, 11)$

30. $\begin{cases} -2x + 5y \geq 3 \\ y < 4 \\ -4x + 2y < 7 \end{cases}$

(a) $(0, 2)$ (b) $(-6, 4)$
(c) $(-8, -2)$ (d) $(-3, 2)$

31. $\begin{cases} 3x + y > 1 \\ -y - \frac{1}{2}x^2 \leq -4 \\ -15x + 4y > 0 \end{cases}$

(a) $(0, 10)$ (b) $(0, -1)$
(c) $(2, 9)$ (d) $(-1, 6)$

32. $\begin{cases} x^2 + y^2 \geq 36 \\ -3x + y \leq 10 \\ \frac{2}{3}x - y \geq 5 \end{cases}$

(a) $(-1, 7)$ (b) $(-5, 1)$
(c) $(6, 0)$ (d) $(4, -8)$

In Exercises 33–46, sketch the graph and label the vertices of the solution of the system of inequalities.

33. $\begin{cases} x + y \leq 1 \\ -x + y \leq 1 \\ y \geq 0 \end{cases}$

34. $\begin{cases} 3x + 2y < 6 \\ x > 0 \\ y > 0 \end{cases}$

35. $\begin{cases} x^2 + y \leq 5 \\ x \geq -1 \\ y \geq 0 \end{cases}$

36. $\begin{cases} 2x^2 + y \geq 2 \\ x \leq 2 \\ y \leq 1 \end{cases}$

37. $\begin{cases} -3x + 2y < 6 \\ x - 4y > -2 \\ 2x + y < 3 \end{cases}$

38. $\begin{cases} x - 7y > -36 \\ 5x + 2y > 5 \\ 6x - 5y > 6 \end{cases}$

39. $\begin{cases} 2x + y > 2 \\ 6x + 3y < 2 \end{cases}$

40. $\begin{cases} x - 2y < -6 \\ 5x - 3y > -9 \end{cases}$

41. $\begin{cases} x > y^2 \\ x < y + 2 \end{cases}$

42. $\begin{cases} x - y^2 > 0 \\ x - y > 2 \end{cases}$

43. $\begin{cases} x^2 + y^2 \leq 9 \\ x^2 + y^2 \geq 1 \end{cases}$

44. $\begin{cases} x^2 + y^2 \leq 25 \\ 4x - 3y \leq 0 \end{cases}$

45. $\begin{cases} 3x + 4 \geq y^2 \\ x - y < 0 \end{cases}$

46. $\begin{cases} x < 2y - y^2 \\ 0 < x + y \end{cases}$

In Exercises 47–52, use a graphing utility to graph the inequalities. Shade the region representing the solution of the system.

47. $\begin{cases} y \le \sqrt{3x} + 1 \\ y \ge x^2 + 1 \end{cases}$

48. $\begin{cases} y < -x^2 + 2x + 3 \\ y > \quad x^2 - 4x + 3 \end{cases}$

49. $\begin{cases} y < x^3 - 2x + 1 \\ y > -2x \\ x \le 1 \end{cases}$

50. $\begin{cases} y \ge x^4 - 2x^2 + 1 \\ y \le 1 - x^2 \end{cases}$

51. $\begin{cases} x^2y \ge 1 \\ 0 < x \le 4 \\ y \le 4 \end{cases}$

52. $\begin{cases} y \le e^{-x^2/2} \\ y \ge 0 \\ -2 \le x \le 2 \end{cases}$

In Exercises 53–62, derive a set of inequalities to describe the region.

53. 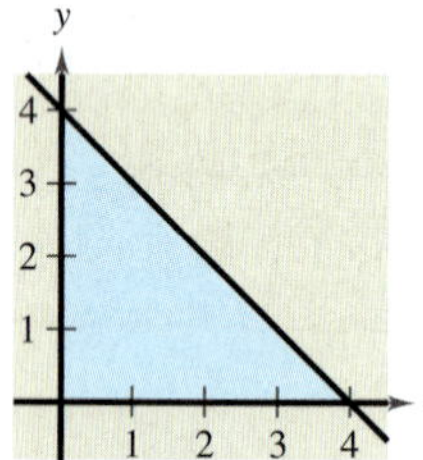

54. 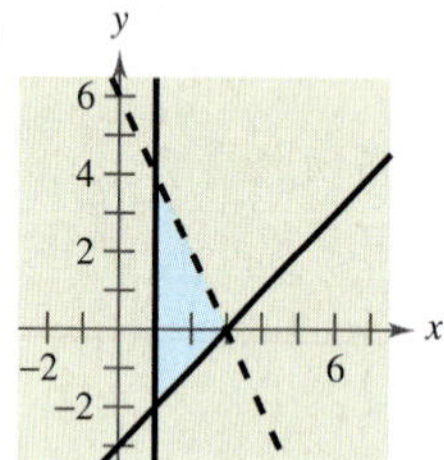

55. 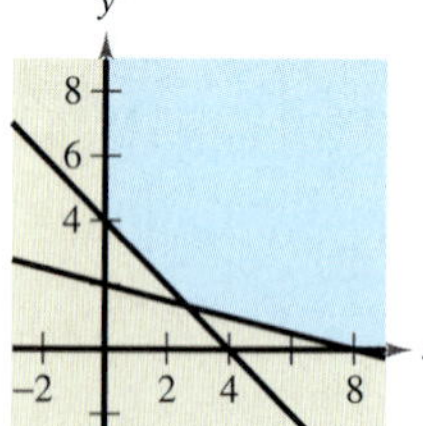

56. 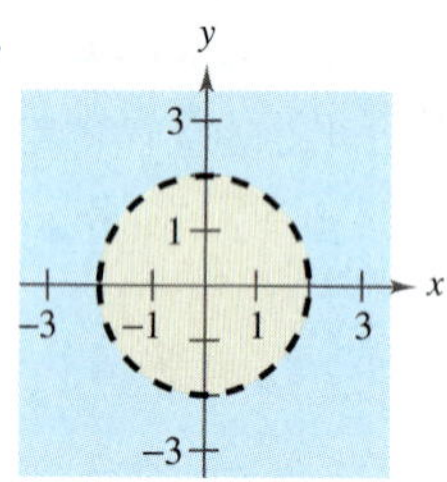

57. 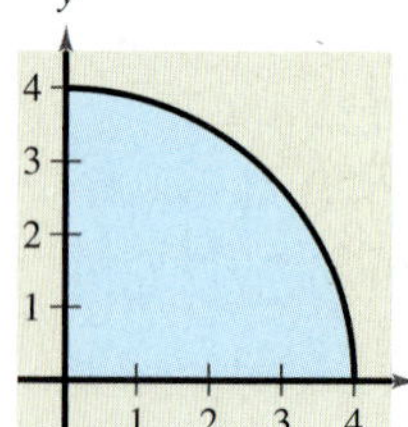

58. 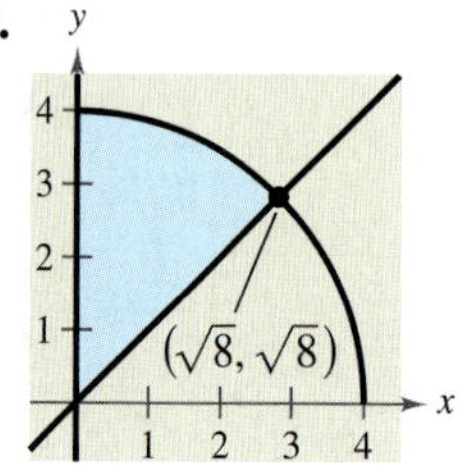

59. Rectangle: Vertices at (2, 1), (5, 1), (5, 7), (2, 7)

60. Parallelogram: Vertices at (0, 0), (4, 0), (1, 4), (5, 4)

61. Triangle: Vertices at (0, 0), (5, 0), (2, 3)

62. Triangle: Vertices at (−1, 0), (1, 0), (0, 1)

Supply and Demand **In Exercises 63–66, find the consumer surplus and producer surplus for the demand and supply equations.**

	Demand	*Supply*
63.	$p = 50 - 0.5x$	$p = 0.125x$
64.	$p = 100 - 0.05x$	$p = 25 + 0.1x$
65.	$p = 140 - 0.00002x$	$p = 80 + 0.00001x$
66.	$p = 400 - 0.0002x$	$p = 225 + 0.0005x$

67. ***Production*** A furniture company can sell all the tables and chairs it produces. Each table requires 1 hour in the assembly center and $1\frac{1}{3}$ hours in the finishing center. Each chair requires $1\frac{1}{2}$ hours in the assembly center and $1\frac{1}{2}$ hours in the finishing center. The company's assembly center is available 12 hours per day, and its finishing center is available 15 hours per day. Find and graph a system of inequalities describing all possible production levels.

68. ***Inventory*** A store sells two models of computers. Because of the demand, the store stocks at least twice as many units of model A as of model B. The costs to the store for the two models are \$800 and \$1200, respectively. The management does not want more than \$20,000 in computer inventory at any one time, and it wants at least four model A computers and two model B computers in inventory at all times. Find and graph a system of inequalities describing all possible inventory levels.

69. ***Investment Analysis*** A person plans to invest up to \$20,000 in two different interest-bearing accounts. Each account is to contain at least \$5000. Moreover, the amount in one account should be at least twice the amount in the other account. Find and graph a system of inequalities to describe the various amounts that can be deposited in each account.

70. ***Ticket Sales*** For a concert event, there are \$30 reserved seat tickets and \$20 general admission tickets. There are 2000 reserved seats available, and fire regulations limit the number of paid ticket holders to 3000. The promoter must take in at least \$75,000 in ticket sales. Find and graph a system of inequalities describing the different numbers of tickets that can be sold.

71. ***Shipping*** A warehouse supervisor is told to ship at least 50 packages of gravel that weigh 55 pounds each and at least 40 bags of stone that weigh 70 pounds each. The maximum weight capacity in the truck he is loading is 7500 pounds. Find and graph a system of inequalities describing the numbers of bags of stone and gravel that he can send.

72. ***Nutrition*** A dietitian is asked to design a special dietary supplement using two different foods. Each ounce of food X contains 20 units of calcium, 15 units of iron, and 10 units of vitamin B. Each ounce of food Y contains 10 units of calcium, 10 units of iron, and 20 units of vitamin B. The minimum daily requirements of the diet are 300 units of calcium, 150 units of iron, and 200 units of vitamin B. Find and graph a system of inequalities describing the different amounts of food X and food Y that can be used.

▶ Model It

73. ***Data Analysis*** The table shows the numbers y of electric-powered vehicles in the United States for the years 1996 through 2000. (Source: U.S. Energy Information Administration)

Year, t	Number of vehicles, y
1996	3280
1997	4040
1998	5243
1999	6417
2000	7590

(a) Use the *regression* feature of a graphing utility to find a linear model for the data. Let t represent the year, with $t = 6$ corresponding to 1996.

(b) The total number of electric-powered vehicles in the United States during this five-year period can be approximated by finding the area of the trapezoid bounded by the linear model you found in part (a) and the lines $y = 0$, $t = 5.5$, and $t = 10.5$. Use a graphing utility to graph this region.

(c) Use the formula for the area of a trapezoid to approximate the total number of electric-powered vehicles.

74. ***Physical Fitness Facility*** An indoor running track is to be constructed with a space for body-building equipment inside the track (see figure). The track must be at least 125 meters long, and the body-building space must have an area of at least 500 square meters.

(a) Find a system of inequalities describing the requirements of the facility.

(b) Graph the system from part (a).

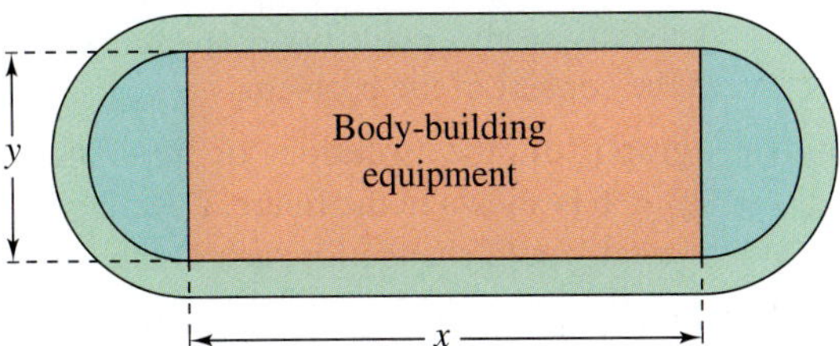

Synthesis

True or False? **In Exercises 75 and 76, determine whether the statement is true or false. Justify your answer.**

75. The area of the figure defined by the system

$$\begin{cases} x \geq -3 \\ x \leq 6 \\ y \leq 5 \\ y \geq -6 \end{cases}$$

is 99 square units.

76. The graph below shows the solution of the system

$$\begin{cases} y \leq 6 \\ -4x - 9y > 6. \\ 3x + y^2 \geq 2 \end{cases}$$

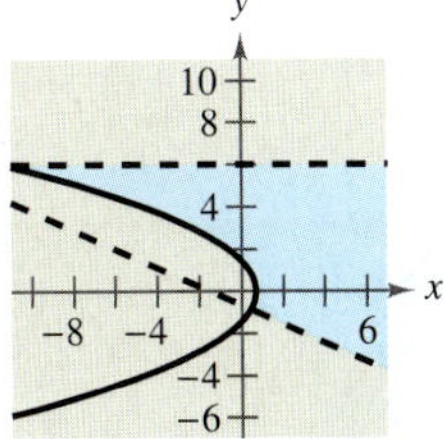

77. ***Writing*** Explain the difference between the graphs of the inequality $x \leq 4$ on the real number line and on the rectangular coordinate system.

78. ***Think About It*** After graphing the boundary of an inequality in x and y, how do you decide on which side of the boundary the solution set of the inequality lies?

79. ***Graphical Reasoning*** Two concentric circles have radii x and y, where $y > x$. The area between the circles must be at least 10 square units.

(a) Find a system of inequalities describing the constraints on the circles.

(b) Use a graphing utility to graph the system of inequalities in part (a). Graph the line $y = x$ in the same viewing window.

(c) Identify the graph of the line in relation to the boundary of the inequality. Explain its meaning in the context of the problem.

80. The graph of the solution of the inequality $x + 2y < 6$ is shown in the figure. Describe how the solution set would change for each of the following.

(a) $x + 2y \le 6$ (b) $x + 2y > 6$

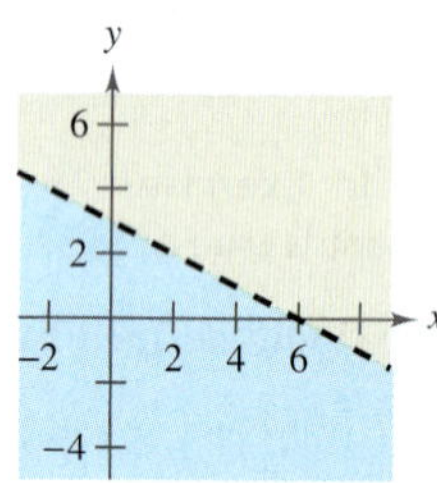

In Exercises 81–84, match the system of inequalities with the graph of its solution. [The graphs are labeled (a), (b), (c), and (d).]

(a)

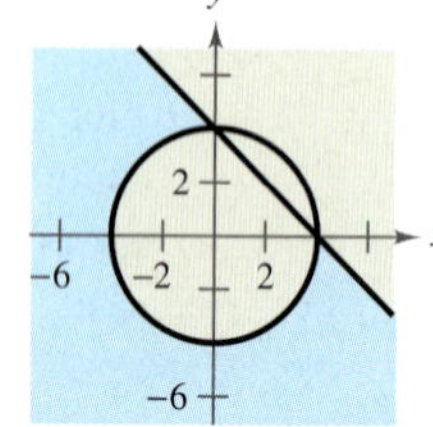

(b)

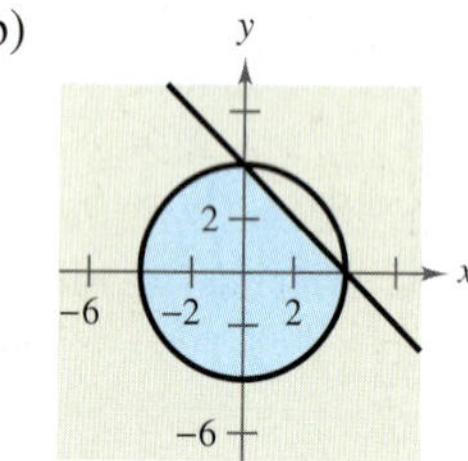

(c)

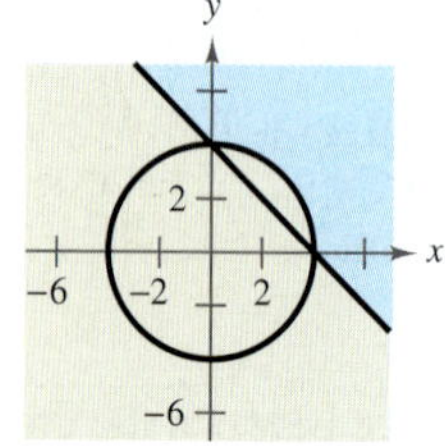

(d)

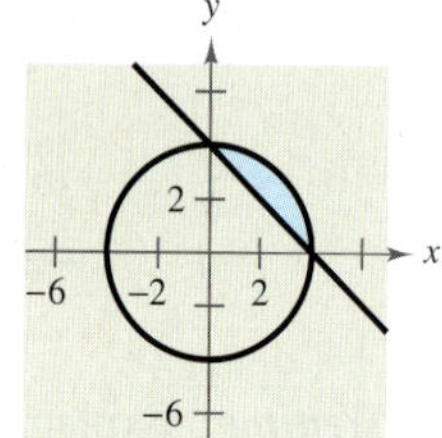

81. $\begin{cases} x^2 + y^2 \le 16 \\ x + y \ge 4 \end{cases}$

82. $\begin{cases} x^2 + y^2 \le 16 \\ x + y \le 4 \end{cases}$

83. $\begin{cases} x^2 + y^2 \ge 16 \\ x + y \ge 4 \end{cases}$

84. $\begin{cases} x^2 + y^2 \ge 16 \\ x + y \le 4 \end{cases}$

Review

In Exercises 85–90, find the equation of the line passing through the two points.

85. $(-2, 6), (4, -4)$

86. $(-8, 0), (3, -1)$

87. $\left(\frac{3}{4}, -2\right), \left(-\frac{7}{2}, 5\right)$

88. $\left(-\frac{1}{2}, 0\right), \left(\frac{11}{2}, 12\right)$

89. $(3.4, -5.2), (-2.6, 0.8)$

90. $(-4.1, -3.8), (2.9, 8.2)$

91. ***Data Analysis*** The table shows the amount y (in trillions of dollars) of personal expenditures for medical care in the United States for the years 1994 through 1999. (Source: U.S. Bureau of Economic Analysis)

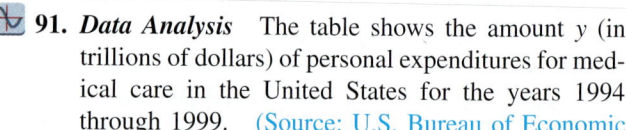

Year, t	Expenditures, y
1994	833.7
1995	888.6
1996	912.8
1997	977.6
1998	1040.9
1999	1102.6

(a) Use the *regression* feature of a graphing utility to find a linear model and a quadratic model for the data. Let t represent the year, with $t = 4$ corresponding to 1994.

(b) Use a graphing utility to plot the data and the models in the same viewing window.

(c) How closely do the models represent the data?

92. ***Compound Interest*** Determine the amount after 5 years if \$4000 is invested in an account earning 6% interest compounded monthly.

In Exercises 93–96, evaluate the function at the indicated value of x. Round your result to three decimal places.

	Function	*Value*
93.	$f(x) = 2.7^x$	$x = 3.99$
94.	$f(x) = 1.5^{-x}$	$x = 3\pi$
95.	$f(x) = e^x$	$x = -\frac{11}{4}$
96.	$f(x) = e^{-x}$	$x = \sqrt{13}$

9.5 Linear Programming

▶ **What you should learn**

- How to solve linear programming problems
- How to use linear programming to model and solve real-life problems

▶ **Why you should learn it**

Linear programming is often useful in making real-life economic decisions. For example, Exercise 35 on page 718 shows how a merchant can use linear programming to analyze the profitability of two models of compact disc players.

Jeff Greenberg/Unicorn Stock Photos

Linear Programming: A Graphical Approach

Many applications in business and economics involve a process called **optimization,** in which you are asked to find the minimum or maximum of a quantity. In this section you will study an optimization strategy called **linear programming.**

A two-dimensional linear programming problem consists of a linear **objective function** and a system of linear inequalities called **constraints.** The objective function gives the quantity that is to be maximized (or minimized), and the constraints determine the set of **feasible solutions.** For example, suppose you are asked to maximize the value of

$$z = ax + by \qquad \text{Objective function}$$

subject to a set of constraints that determines the shaded region in Figure 9.31.

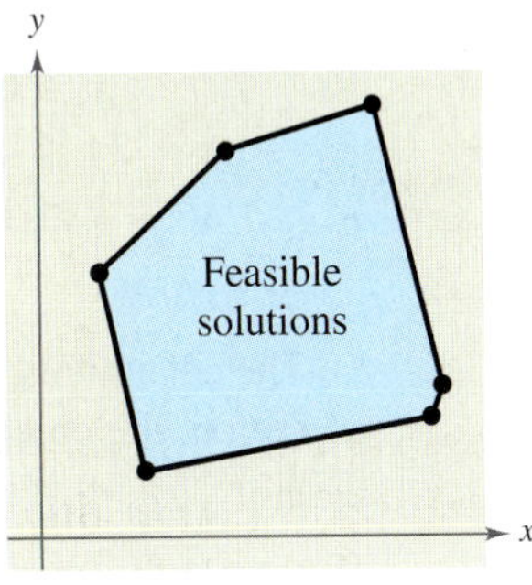

FIGURE 9.31

Because every point in the shaded region satisfies each constraint, it is not clear how you should find the point that yields a maximum value of z. Fortunately, it can be shown that if there is an optimal solution, it must occur at one of the vertices. This means that *you can find the maximum value of z by testing z at each of the vertices.*

Optimal Solution of a Linear Programming Problem

If a linear programming problem has a solution, it must occur at a vertex of the set of feasible solutions. If there is more than one solution, at least one of them must occur at such a vertex. In either case, the value of the objective function is unique.

Some guidelines for solving a linear programming problem in two variables are listed at the top of the next page.

Solving a Linear Programming Problem

1. Sketch the region corresponding to the system of constraints. (The points inside or on the boundary of the region are *feasible solutions.*)
2. Find the vertices of the region.
3. Test the objective function at each of the vertices and select the values of the variables that optimize the objective function. For a bounded region, both a minimum and a maximum value will exist. (For an unbounded region, *if* an optimal solution exists, it will occur at a vertex.)

Example 1 ▶ Solving a Linear Programming Problem 

Find the maximum value of

$$z = 3x + 2y \qquad \text{Objective function}$$

subject to the following constraints.

$$\left.\begin{aligned} x &\geq 0 \\ y &\geq 0 \\ x + 2y &\leq 4 \\ x - y &\leq 1 \end{aligned}\right\} \qquad \text{Constraints}$$

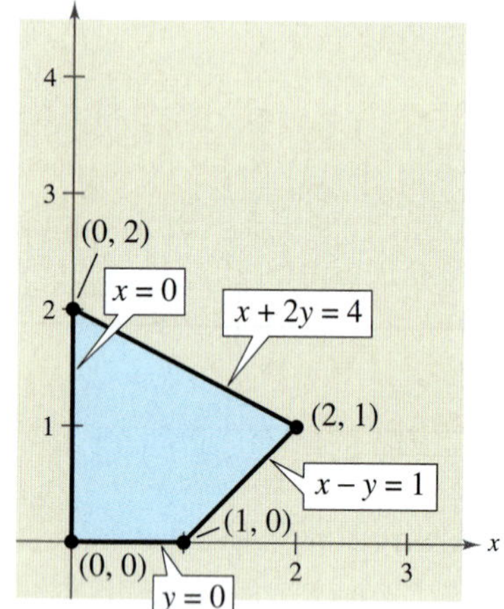

FIGURE 9.32

Solution

The constraints form the region shown in Figure 9.32. At the four vertices of this region, the objective function has the following values.

At (0, 0): $z = 3(0) + 2(0) = 0$

At (1, 0): $z = 3(1) + 2(0) = 3$

At (2, 1): $z = 3(2) + 2(1) = 8$ Maximum value of z

At (0, 2): $z = 3(0) + 2(2) = 4$

So, the maximum value of z is 8, and this occurs when $x = 2$ and $y = 1$.

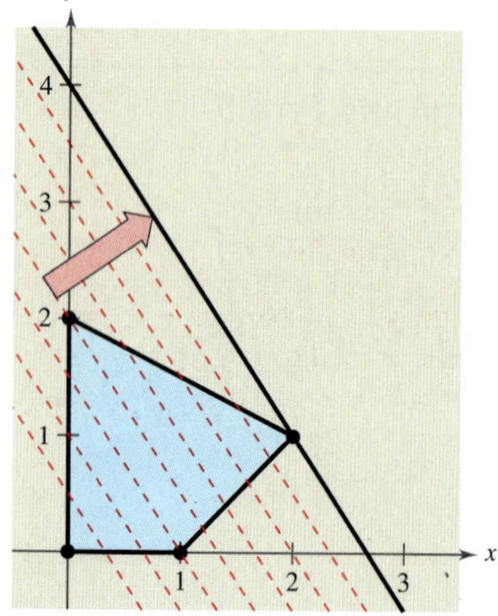

FIGURE 9.33

In Example 1, try testing some of the *interior* points in the region. You will see that the corresponding values of z are less than 8. Here are some examples.

At (1, 1): $z = 3(1) + 2(1) = 5$ At $\left(\frac{1}{2}, \frac{3}{2}\right)$: $z = 3\left(\frac{1}{2}\right) + 2\left(\frac{3}{2}\right) = \frac{9}{2}$

To see why the maximum value of the objective function in Example 1 must occur at a vertex, consider writing the objective function in slope-intercept form

$$y = -\frac{3}{2}x + \frac{z}{2} \qquad \text{Family of lines}$$

where $z/2$ is the y-intercept of the objective function. This equation represents a family of lines, each of slope $-\frac{3}{2}$. Of these infinitely many lines, you want the one that has the largest z-value while still intersecting the region determined by the constraints. In other words, of all the lines whose slope is $-\frac{3}{2}$, you want the one that has the largest y-intercept *and* intersects the given region, as shown in Figure 9.33. From the graph you can see that such a line will pass through one (or more) of the vertices of the region.

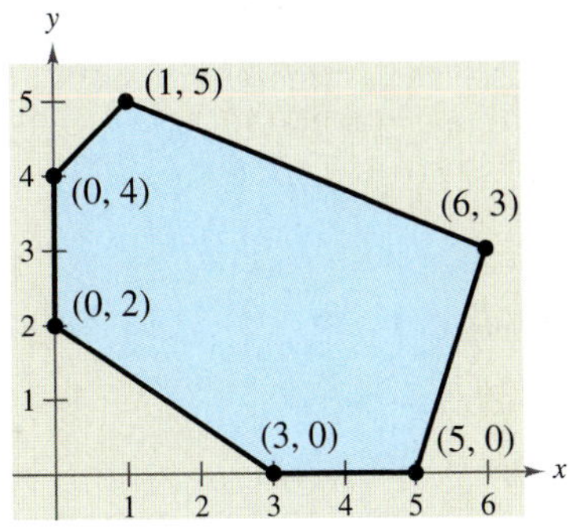

FIGURE 9.34

The next example shows that the same basic procedure can be used to solve a problem in which the objective function is to be *minimized.*

Example 2 ▶ Minimizing an Objective Function

Find the minimum value of

$$z = 5x + 7y \qquad \text{Objective function}$$

where $x \geq 0$ and $y \geq 0$, subject to the following constraints.

$$\left.\begin{aligned} 2x + 3y &\geq 6 \\ 3x - y &\leq 15 \\ -x + y &\leq 4 \\ 2x + 5y &\leq 27 \end{aligned}\right\} \qquad \text{Constraints}$$

Solution

The region bounded by the constraints is shown in Figure 9.34. By testing the objective function at each vertex, you obtain the following.

At $(0, 2)$: $z = 5(0) + 7(2) = 14$ Minimum value of z

At $(0, 4)$: $z = 5(0) + 7(4) = 28$

At $(1, 5)$: $z = 5(1) + 7(5) = 40$

At $(6, 3)$: $z = 5(6) + 7(3) = 51$

At $(5, 0)$: $z = 5(5) + 7(0) = 25$

At $(3, 0)$: $z = 5(3) + 7(0) = 15$

So, the minimum value of z is 14, and this occurs when $x = 0$ and $y = 2$.

Example 3 ▶ Maximizing an Objective Function

Find the maximum value of

$$z = 5x + 7y \qquad \text{Objective function}$$

where $x \geq 0$ and $y \geq 0$, subject to the following constraints.

$$\left.\begin{aligned} 2x + 3y &\geq 6 \\ 3x - y &\leq 15 \\ -x + y &\leq 4 \\ 2x + 5y &\leq 27 \end{aligned}\right\} \qquad \text{Constraints}$$

Solution

This linear programming problem is identical to that given in Example 2 above, *except* that the objective function is maximized instead of minimized. Using the values of z at the vertices shown above, you can conclude that the maximum value of

$$z = 5(6) + 7(3) = 51$$

occurs when $x = 6$ and $y = 3$.

Edward W. Souza/News Service/Stanford University

Historical Note

George Dantzig (1914–) was the first to propose the simplex method, or linear programming, in 1947. This technique defined the steps needed to find the optimal solution to a complex multivariable problem.

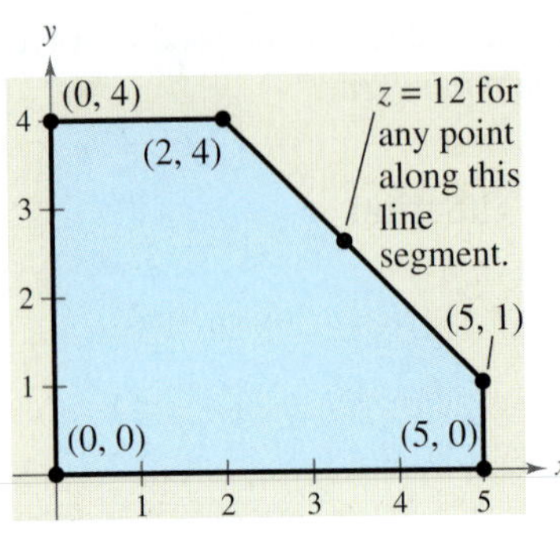

FIGURE 9.35

It is possible for the maximum (or minimum) value in a linear programming problem to occur at *two* different vertices. For instance, at the vertices of the region shown in Figure 9.35, the objective function

$$z = 2x + 2y \qquad \text{Objective function}$$

has the following values.

At (0, 0): $z = 2(0) + 2(0) = 0$

At (0, 4): $z = 2(0) + 2(4) = 8$

At (2, 4): $z = 2(2) + 2(4) = 12$ — Maximum value of z

At (5, 1): $z = 2(5) + 2(1) = 12$ — Maximum value of z

At (5, 0): $z = 2(5) + 2(0) = 10$

In this case, you can conclude that the objective function has a maximum value not only at the vertices (2, 4) and (5, 1); it also has a maximum value (of 12) at *any point on the line segment connecting these two vertices.* Note that the objective function in slope-intercept form $y = -x + \frac{1}{2}z$ has the same slope as the line through the vertices (2, 4) and (5, 1).

Some linear programming problems have no optimal solutions. This can occur if the region determined by the constraints is *unbounded.* Example 4 illustrates such a problem.

Example 4 ▶ An Unbounded Region 

Find the maximum value of

$$z = 4x + 2y \qquad \text{Objective function}$$

where $x \ge 0$ and $y \ge 0$, subject to the following constraints.

$$\left.\begin{aligned} x + 2y &\ge 4 \\ 3x + y &\ge 7 \\ -x + 2y &\le 7 \end{aligned}\right\} \qquad \text{Constraints}$$

Solution

The region determined by the constraints is shown in Figure 9.36. For this unbounded region, there is no maximum value of z. To see this, note that the point $(x, 0)$ lies in the region for all values of $x \ge 4$. Substituting this point into the objective function, you get

$$z = 4(x) + 2(0) = 4x.$$

By choosing x to be large, you can obtain values of z that are as large as you want. So, there is no maximum value of z.

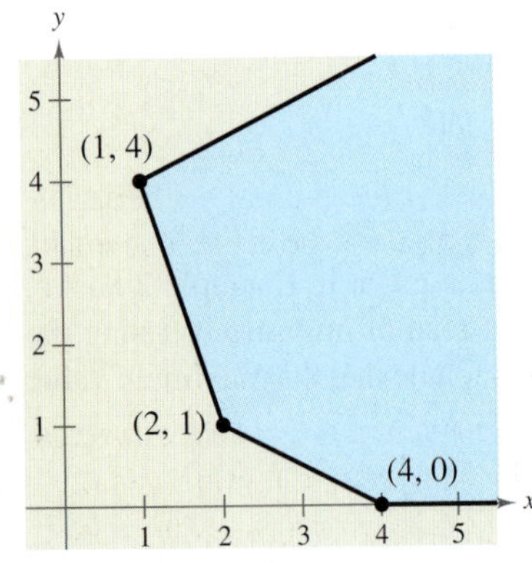

FIGURE 9.36

Although the objective function in Example 4 has no maximum, there *is* a minimum value of z.

At (1, 4): $z = 4(1) + 2(4) = 12$

At (2, 1): $z = 4(2) + 2(1) = 10$ — Minimum value of z

At (4, 0): $z = 4(4) + 2(0) = 16$

So, the minimum value of z is 10, and this occurs when $x = 2$ and $y = 1$.

Applications

Example 5 shows how linear programming can be used to find the maximum profit in a business application.

Example 5 ▶ Maximum Profit  

A candy manufacturer wants to maximize the profit for two types of boxed chocolates. A box of chocolate covered creams yields a profit of \$1.50 per box, and a box of chocolate covered nuts yields a profit of \$2.00 per box. Market tests and available resources have indicated the following constraints.

1. The combined production level should not exceed 1200 boxes per month.
2. The demand for a box of chocolate covered nuts is no more than half the demand for a box of chocolate covered creams.
3. The production level of a box of chocolate covered creams should be less than or equal to 600 boxes plus three times the production level of a box of chocolate covered nuts.

Solution

Let x be the number of boxes of chocolate covered creams and let y be the number of boxes of chocolate covered nuts. So, the objective function (for the combined profit) is given by

$$P = 1.5x + 2y.$$ Objective function

The three constraints translate into the following linear inequalities.

1. $x + y \le 1200$ ⇨ $x + y \le 1200$
2. $y \le \frac{1}{2}x$ ⇨ $-x + 2y \le 0$
3. $x \le 600 + 3y$ ⇨ $x - 3y \le 600$

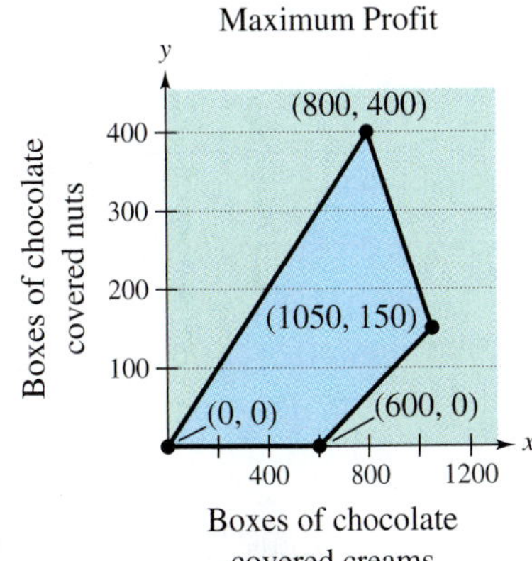

FIGURE 9.37

Because neither x nor y can be negative, you also have the two additional constraints of $x \ge 0$ and $y \ge 0$. Figure 9.37 shows the region determined by the constraints. To find the maximum profit, test the values of P at the vertices of the region.

At $(0, 0)$: $P = 1.5(0) + 2(0) = 0$

At $(800, 400)$: $P = 1.5(800) + 2(400) = 2000$ Maximum profit

At $(1050, 150)$: $P = 1.5(1050) + 2(150) = 1875$

At $(600, 0)$: $P = 1.5(600) + 2(0) = 900$

So, the maximum profit is \$2000, and it occurs when the monthly production consists of 800 boxes of chocolate covered creams and 400 boxes of chocolate covered nuts.

In Example 5, the manufacturer improved the production of chocolate covered creams so that it yielded a profit of \$2.50 per unit. The maximum profit can then be found using the objective function $P = 2.5x + 2y$. By testing the values of P at the vertices of the region, you find the maximum profit is \$2925 and that it now occurs when $x = 1050$ and $y = 150$.

Example 6 ▶ Minimum Cost

The minimum daily requirements from the liquid portion of a diet are 300 calories, 36 units of vitamin A, and 90 units of vitamin C. A cup of dietary drink X costs \$0.12 and provides 60 calories, 12 units of vitamin A, and 10 units of vitamin C. A cup of dietary drink Y costs \$0.15 and provides 60 calories, 6 units of vitamin A, and 30 units of vitamin C. How many cups of each drink should be consumed each day to minimize the cost and still meet the daily requirements?

Solution

As in Example 9 on page 706, let x be the number of cups of dietary drink X and let y be the number of cups of dietary drink Y.

$$\left.\begin{array}{lr} \text{For calories:} & 60x + 60y \geq 300 \\ \text{For vitamin A:} & 12x + 6y \geq 36 \\ \text{For vitamin C:} & 10x + 30y \geq 90 \\ & x \geq 0 \\ & y \geq 0 \end{array}\right\} \quad \text{Constraints}$$

The cost C is given by $C = 0.12x + 0.15y$. Objective function

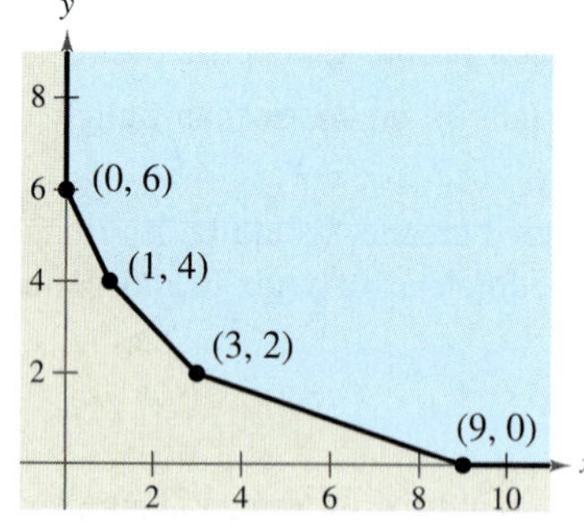

FIGURE 9.38

The graph of the region corresponding to the constraints is shown in Figure 9.38. To determine the minimum cost, test C at each vertex of the region.

At $(0, 6)$: $C = 0.12(0) + 0.15(6) = 0.90$

At $(1, 4)$: $C = 0.12(1) + 0.15(4) = 0.72$

At $(3, 2)$: $C = 0.12(3) + 0.15(2) = 0.66$ Minimum value of C

At $(9, 0)$: $C = 0.12(9) + 0.15(0) = 1.08$

So, the minimum cost is \$0.66 per day, and this occurs when three cups of drink X and two cups of drink Y are consumed each day.

Writing ABOUT MATHEMATICS

Creating a Linear Programming Problem Sketch the region determined by the following constraints.

$$\left.\begin{array}{r} x + 2y \leq 8 \\ x + y \leq 5 \\ x \geq 0 \\ y \geq 0 \end{array}\right\} \quad \text{Constraints}$$

Find, if possible, an objective function of the form $z = ax + by$ that has a maximum at the indicated vertex of the region.

a. $(0, 4)$ **b.** $(2, 3)$

c. $(5, 0)$ **d.** $(0, 0)$

Explain how you found each objective function.

9.5 Exercises

In Exercises 1–12, find the minimum and maximum values of the objective function and where they occur, subject to the indicated constraints. (For each exercise, the graph of the region determined by the constraints is provided.)

1. Objective function:

$z = 4x + 3y$

Constraints:

$x \geq 0$

$y \geq 0$

$x + y \leq 5$

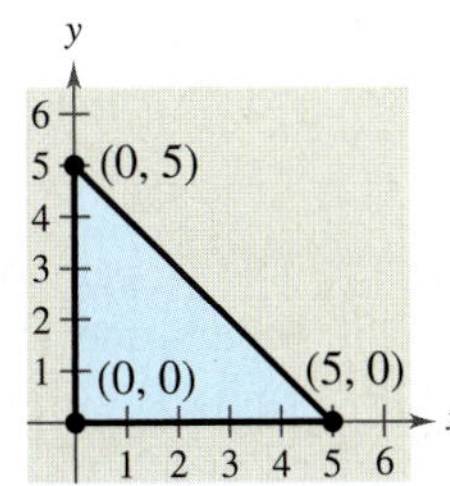

2. Objective function:

$z = 2x + 8y$

Constraints:

$x \geq 0$

$y \geq 0$

$2x + y \leq 4$

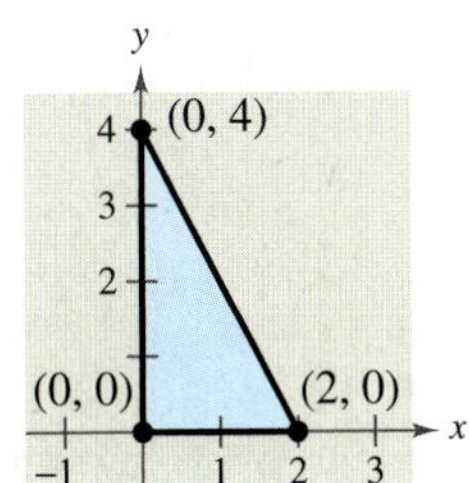

3. Objective function:

$z = 3x + 8y$

Constraints:
(See Exercise 1.)

4. Objective function:

$z = 7x + 3y$

Constraints:
(See Exercise 2.)

5. Objective function:

$z = 3x + 2y$

Constraints:

$x \geq 0$

$y \geq 0$

$x + 3y \leq 15$

$4x + y \leq 16$

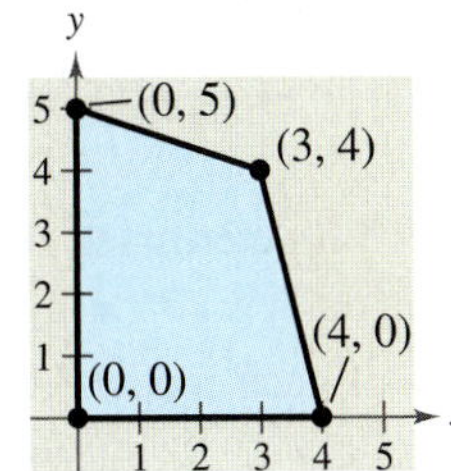

6. Objective function:

$z = 4x + 5y$

Constraints:

$x \geq 0$

$2x + 3y \geq 6$

$3x - y \leq 9$

$x + 4y \leq 16$

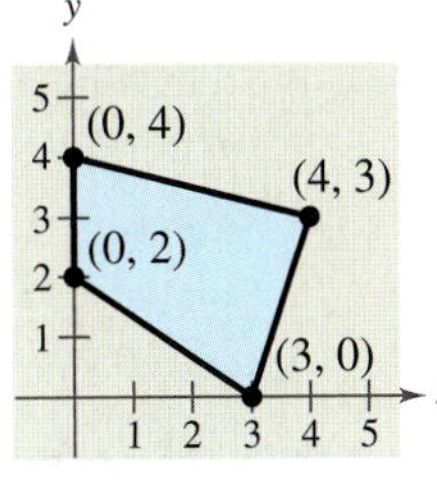

7. Objective function:

$z = 5x + 0.5y$

Constraints:
(See Exercise 5.)

8. Objective function:

$z = 2x + y$

Constraints:
(See Exercise 6.)

9. Objective function:

$z = 10x + 7y$

Constraints:

$0 \leq x \leq 60$

$0 \leq y \leq 45$

$5x + 6y \leq 420$

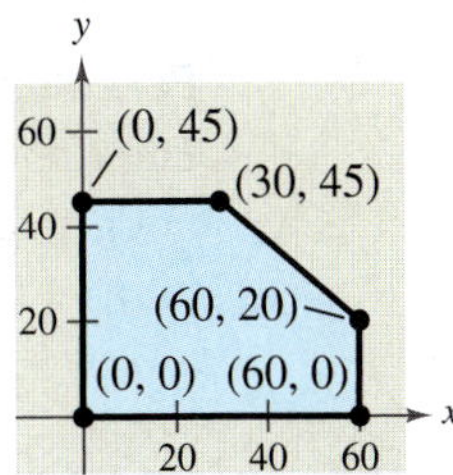

10. Objective function:

$z = 25x + 35y$

Constraints:

$x \geq 0$

$y \geq 0$

$8x + 9y \leq 7200$

$8x + 9y \geq 3600$

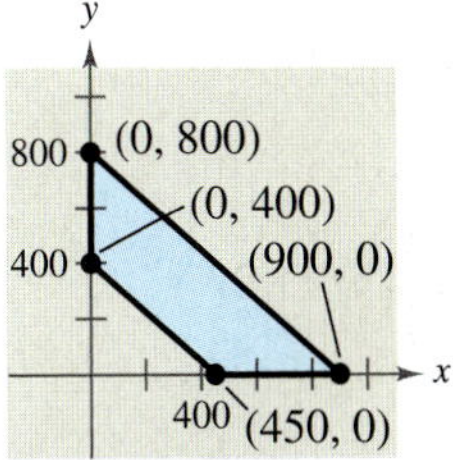

11. Objective function:

$z = 25x + 30y$

Constraints:
(See Exercise 9.)

12. Objective function:

$z = 15x + 20y$

Constraints:
(See Exercise 10.)

In Exercises 13–20, sketch the region determined by the constraints. Then find the minimum and maximum values of the objective function and where they occur, subject to the indicated constraints.

13. Objective function:

$z = 6x + 10y$

Constraints:

$x \geq 0$

$y \geq 0$

$2x + 5y \leq 10$

14. Objective function:

$z = 7x + 8y$

Constraints:

$x \geq 0$

$y \geq 0$

$x + \frac{1}{2}y \leq 4$

15. Objective function:

$z = 9x + 24y$

Constraints:
(See Exercise 13.)

16. Objective function:

$z = 7x + 2y$

Constraints:
(See Exercise 14.)

17. Objective function:

$z = 4x + 5y$

Constraints:

$x \geq 0$

$y \geq 0$

$x + y \geq 8$

$3x + 5y \geq 30$

18. Objective function:

$z = 4x + 5y$

Constraints:

$x \geq 0$

$y \geq 0$

$2x + 2y \leq 10$

$x + 2y \leq 6$

19. Objective function:

$z = 2x + 7y$

Constraints:

(See Exercise 17.)

20. Objective function:

$z = 2x - y$

Constraints:

(See Exercise 18.)

In Exercises 21–24, use a graphing utility to graph the region determined by the constraints. Then find the minimum and maximum values of the objective function and where they occur, subject to the constraints.

21. Objective function:

$z = 4x + y$

Constraints:

$$x \geq 0$$
$$y \geq 0$$
$$x + 2y \leq 40$$
$$2x + 3y \geq 72$$

22. Objective function:

$z = x$

Constraints:

$$x \geq 0$$
$$y \geq 0$$
$$2x + 3y \leq 60$$
$$2x + y \leq 28$$
$$4x + y \leq 48$$

23. Objective function:

$z = x + 4y$

Constraints:

(See Exercise 21.)

24. Objective function:

$z = y$

Constraints:

(See Exercise 22.)

In Exercises 25–28, find the maximum value of the objective function and where it occurs, subject to the constraints $x \geq 0$, $y \geq 0$, $3x + y \leq 15$, and $4x + 3y \leq 30$.

25. $z = 2x + y$ **26.** $z = 5x + y$

27. $z = x + y$ **28.** $z = 3x + y$

In Exercises 29–32, find the maximum value of the objective function and where it occurs, subject to the constraints $x \geq 0$, $y \geq 0$, $x + 4y \leq 20$, $x + y \leq 18$, and $2x + 2y \leq 21$.

29. $z = x + 5y$ **30.** $z = 2x + 4y$

31. $z = 4x + 5y$ **32.** $z = 4x + y$

33. ***Maximum Profit*** A manufacturer produces two models of bicycles. The times (in hours) required for assembling, painting, and packaging each model are shown in the table.

Process	Hours, model A	Hours, model B
Assembling	2	2.5
Painting	4	1
Packaging	1	0.75

The total times available for assembling, painting, and packaging are 4000 hours, 4800 hours, and 1500 hours, respectively. The profits per unit are $45 for model A and $50 for model B. How many of each type should be produced to maximize profit? What is the maximum profit?

34. ***Maximum Profit*** A manufacturer produces two models of bicycles. The times (in hours) required for assembling, painting, and packaging each model are shown in the table.

Process	Hours, model A	Hours, model B
Assembling	2.5	3
Painting	2	1
Packaging	0.75	1.25

The total times available for assembling, painting, and packaging are 4000 hours, 2500 hours, and 1500 hours, respectively. The profits per unit are $50 for model A and $52 for model B. How many of each type should be produced to maximize profit? What is the maximum profit?

▶ Model It

35. ***Maximum Profit*** A merchant plans to sell two models of compact disc players at costs of $150 and $200. The $150 model yields a profit of $25 per unit and the $200 model yields a profit of $40 per unit. The merchant estimates that the total monthly demand will not exceed 250 units. The merchant does not want to invest more than $40,000 in inventory for these products.

(a) Write an objective function that models the profit for the compact disc players.

(b) Determine the constraints for the objective function.

(c) Sketch a graph of the region determined by the constraints from part (b).

(d) Find the number of units of each model that should be stocked in order to maximize profit.

(e) What is the maximum profit?

36. ***Maximum Profit*** A fruit grower has 150 acres of land available to raise two crops, A and B. It takes 1 day to trim an acre of crop A and 2 days to trim an acre of crop B, and there are 240 days per year available for trimming. It takes 0.3 day to pick an acre of crop A and 0.1 day to pick an acre of crop B, and there are 30 days available for picking. The profit is \$140 per acre for crop A and \$235 per acre for crop B. Find the number of acres of each fruit that should be planted to maximize profit. What is the maximum profit?

37. ***Minimum Cost*** A farming cooperative mixes two brands of cattle feed. Brand X costs \$25 per bag and contains two units of nutritional element A, two units of element B, and two units of element C. Brand Y costs \$20 per bag and contains one unit of nutritional element A, nine units of element B, and three units of element C. The minimum requirements of nutrients A, B, and C are 12 units, 36 units, and 24 units, respectively. Find the number of bags of each brand that should be mixed to produce a mixture having a minimum cost. What is the minimum cost?

38. ***Minimum Cost*** Two gasolines, type A and type B, have octane ratings of 87 and 93, respectively. Type A costs \$1.13 per gallon and type B costs \$1.28 per gallon. Determine the blend of minimum cost with an octane rating of at least 89. What is the minimum cost? Is the cost cheaper than the advertised cost of \$1.20 per gallon for gasoline with an octane rating of 89? (*Hint:* Let x be the fraction of each gallon that is type A and let y be the fraction that is type B.)

39. ***Maximum Revenue*** An accounting firm has 800 hours of staff time and 96 hours of reviewing time available each week. The firm charges \$2000 for an audit and \$300 for a tax return. Each audit requires 100 hours of staff time and 8 hours of review time. Each tax return requires 12.5 hours of staff time and 2 hours of review time. What numbers of audits and tax returns will yield the maximum revenue? What is the maximum revenue?

40. ***Maximum Revenue*** The accounting firm in Exercise 39 lowers its charge for an audit to \$1000. What numbers of audits and tax returns will yield the maximum revenue? What is the maximum revenue?

41. ***Investment Portfolio*** An investor has up to \$250,000 to invest in two types of investments. Type A pays 8% annually and type B pays 10% annually. To have a well-balanced portfolio, the investor imposes the following conditions. At least one-fourth of the total portfolio is to be allocated to type A investments and at least one-fourth of the portfolio is to be allocated to type B investments. How much should be allocated to each type of investment to obtain a maximum return?

42. ***Investment Portfolio*** An investor has up to \$450,000 to invest in two types of investments. Type A pays 6% annually and type B pays 10% annually. To have a well-balanced portfolio, the investor imposes the following conditions. At least one-half of the total portfolio is to be allocated to type A investments and at least one-fourth of the portfolio is to be allocated to type B investments. How much should be allocated to each type of investment to obtain a maximum return?

In Exercises 43–48, the linear programming problem has an unusual characteristic. Sketch a graph of the solution region for the problem and describe the unusual characteristic. Find the maximum value of the objective function and where it occurs.

43. Objective function:

$z = 2.5x + y$

Constraints:

$x \geq 0$

$y \geq 0$

$3x + 5y \leq 15$

$5x + 2y \leq 10$

44. Objective function:

$z = x + y$

Constraints:

$x \geq 0$

$y \geq 0$

$-x + y \leq 1$

$-x + 2y \leq 4$

45. Objective function:

$z = -x + 2y$

Constraints:

$x \geq 0$

$y \geq 0$

$x \leq 10$

$x + y \leq 7$

46. Objective function:

$z = x + y$

Constraints:

$x \geq 0$

$y \geq 0$

$-x + y \leq 0$

$-3x + y \geq 3$

47. Objective function:

$z = 3x + 4y$

Constraints:

$x \geq 0$

$y \geq 0$

$x + y \leq 1$

$2x + y \leq 4$

48. Objective function:

$z = x + 2y$

Constraints:

$x \geq 0$

$y \geq 0$

$x + 2y \leq 4$

$2x + y \leq 4$

Synthesis

True or False? **In Exercises 49 and 50, determine whether the statement is true or false. Justify your answer.**

49. If an objective function has a maximum value at the vertices $(4, 7)$ and $(8, 3)$, you can conclude that it also has a maximum value at the points $(4.5, 6.5)$ and $(7.8, 3.2)$.

50. When solving a linear programming problem, if the objective function has a maximum value at more than one vertex, you can assume that there are an infinite number of points that will produce the maximum value.

In Exercises 51 and 52, determine values of *t* such that the objective function has maximum values at the indicated vertices.

51. Objective function:

$z = 3x + ty$

Constraints:

$x \geq 0$

$y \geq 0$

$x + 3y \leq 15$

$4x + y \leq 16$

(a) $(0, 5)$

(b) $(3, 4)$

52. Objective function:

$z = 3x + ty$

Constraints:

$x \geq 0$

$y \geq 0$

$x + 2y \leq 4$

$x - y \leq 1$

(a) $(2, 1)$

(b) $(0, 2)$

Think About It **In Exercises 53–56, find an objective function that has a maximum or minimum value at the indicated vertex of the constraint region shown below. (There are many correct answers.)**

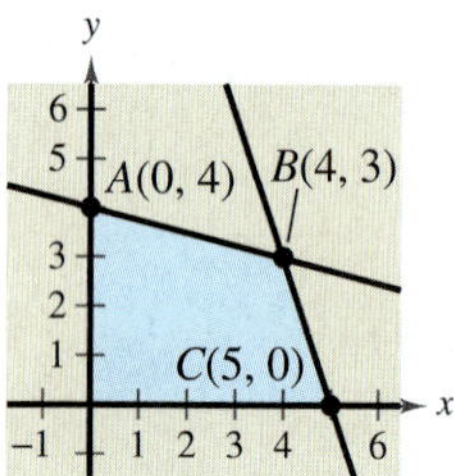

53. The maximum occurs at vertex A.

54. The maximum occurs at vertex B.

55. The maximum occurs at vertex C.

56. The minimum occurs at vertex C.

Review

In Exercises 57–60, simplify the compound fraction.

57. $\dfrac{\left(\dfrac{9}{x}\right)}{\left(\dfrac{6}{x} + 2\right)}$

58. $\dfrac{\left(1 + \dfrac{2}{x}\right)}{\left(x - \dfrac{4}{x}\right)}$

59. $\dfrac{\left(\dfrac{4}{x^2 - 9} + \dfrac{2}{x - 2}\right)}{\left(\dfrac{1}{x + 3} + \dfrac{1}{x - 3}\right)}$

60. $\dfrac{\left(\dfrac{1}{x + 1} + \dfrac{1}{2}\right)}{\left(\dfrac{3}{2x^2 + 4x + 2}\right)}$

In Exercises 61–66, identify the conic and sketch its graph.

61. $y^2 = 6x$

62. $3x^2 - 4 = y$

63. $\dfrac{x^2}{9} + \dfrac{y^2}{49} = 1$

64. $\dfrac{y^2}{16} - \dfrac{x^2}{169} = 1$

65. $\dfrac{(x + 5)^2}{4} - \dfrac{y^2}{36} = 1$

66. $\dfrac{(x - 3)^2}{25} + \dfrac{(y - 6)^2}{36} = 1$

In Exercises 67–72, solve the equation algebraically. Round the result to three decimal places.

67. $e^{2x} + 2e^x - 15 = 0$

68. $e^{2x} - 10e^x + 24 = 0$

69. $8(62 - e^{x/4}) = 192$

70. $\dfrac{150}{e^{-x} - 4} = 75$

71. $7 \ln 3x = 12$

72. $\ln(x + 9)^2 = 2$

In Exercises 73 and 74, solve the system of linear equations and check any solution algebraically.

73. $\begin{cases} -x - 2y + 3z = -23 \\ 2x + 6y - z = 17 \\ 5y + z = 8 \end{cases}$

74. $\begin{cases} 7x - 3y + 5z = -28 \\ 4x + 4z = -16 \\ 7x + 2y - z = 0 \end{cases}$

Chapter Summary

▶ *What* did you learn?

	Review Exercises
Section 9.1	
☐ How to use the method of substitution to solve systems of equations in two variables	1–4
☐ How to use a graphical approach to solve systems of equations in two variables	5–10
☐ How to use systems of equations to model and solve real-life problems	11–14
Section 9.2	
☐ How to use the method of elimination to solve systems of linear equations in two variables	15–22
☐ How to interpret graphically the numbers of solutions of systems of linear equations in two variables	23–26
☐ How to use systems of equations in two variables to model and solve real-life problems	27, 28
Section 9.3	
☐ How to recognize linear systems in row-echelon form and use back-substitution to solve the systems	29, 30
☐ How to use Gaussian elimination to solve systems of linear equations	31–34
☐ How to solve nonsquare systems of linear equations	35, 36
☐ How to use systems of linear equations in three or more variables to model and solve application problems	37–48
Section 9.4	
☐ How to sketch the graphs of inequalities in two variables	49–52
☐ How to solve systems of inequalities	53–60
☐ How to use systems of inequalities in two variables to model and solve real-life problems	61–64
Section 9.5	
☐ How to solve linear programming problems	65–70
☐ How to use linear programming to model and solve real-life problems	71–74

Review Exercises

9.1 In Exercises 1–4, solve the system by the method of substitution.

1. $\begin{cases} x^2 - y^2 = 9 \\ x - y = 1 \end{cases}$

2. $\begin{cases} x^2 + y^2 = 169 \\ 3x + 2y = 39 \end{cases}$

3. $\begin{cases} y = 2x^2 \\ y = x^4 - 2x^2 \end{cases}$

4. $\begin{cases} x = y + 3 \\ x = y^2 + 1 \end{cases}$

In Exercises 5–8, solve the system graphically.

5. $\begin{cases} 2x - y = 10 \\ x + 5y = -6 \end{cases}$

6. $\begin{cases} y^2 - 2y + x = 0 \\ x + y = 0 \end{cases}$

7. $\begin{cases} y = 2x^2 - 4x + 1 \\ y = x^2 - 4x + 3 \end{cases}$

8. $\begin{cases} y = 2(6 - x) \\ y = 2^{x-2} \end{cases}$

In Exercises 9 and 10, use a graphing utility to solve the system of equations. Find the solution accurate to two decimal places.

9. $\begin{cases} y = -2e^{-x} \\ 2e^x + y = 0 \end{cases}$

10. $\begin{cases} y = \ln(x - 1) - 3 \\ y = 4 - \frac{1}{2}x \end{cases}$

11. ***Break-Even Analysis*** You set up a business and make an initial investment of \$50,000. The unit cost of the product is \$2.15 and the selling price is \$6.95. How many units must you sell to break even?

12. ***Choice of Two Jobs*** You are offered two sales jobs. One company offers an annual salary of \$22,500 plus a year-end bonus of 1.5% of your total sales. The other company offers an annual salary of \$20,000 plus a year-end bonus of 2% of your total sales. What amount of sales will make the second offer better? Explain.

13. ***Geometry*** The perimeter of a rectangle is 480 meters and its length is 150% of its width. Find the dimensions of the rectangle.

14. ***Geometry*** The perimeter of a rectangle is 68 feet and its width is $\frac{8}{9}$ times its length. Find the dimensions of the rectangle.

9.2 In Exercises 15–22, solve the system by elimination.

15. $\begin{cases} 2x - y = 2 \\ 6x + 8y = 39 \end{cases}$

16. $\begin{cases} 40x + 30y = 24 \\ 20x - 50y = -14 \end{cases}$

17. $\begin{cases} 0.2x + 0.3y = 0.14 \\ 0.4x + 0.5y = 0.20 \end{cases}$

18. $\begin{cases} 12x + 42y = -17 \\ 30x - 18y = 19 \end{cases}$

19. $\begin{cases} 3x - 2y = 0 \\ 3x + 2(y + 5) = 10 \end{cases}$

20. $\begin{cases} 7x + 12y = 63 \\ 2x + 3(y + 2) = 21 \end{cases}$

21. $\begin{cases} 1.25x - 2y = 3.5 \\ 5x - 8y = 14 \end{cases}$

22. $\begin{cases} 1.5x + 2.5y = 8.5 \\ 6x + 10y = 24 \end{cases}$

In Exercises 23–26, match the system of linear equations with its graph. [The graphs are labeled (a), (b), (c), and (d).]

(a)

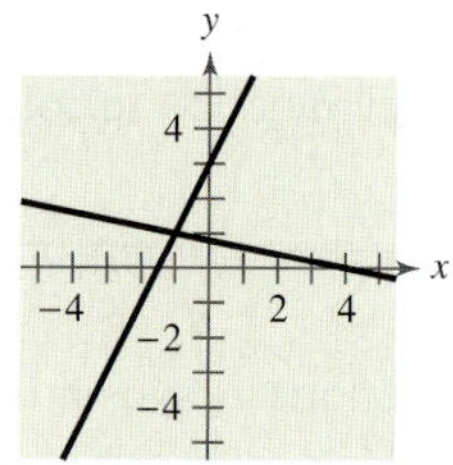

(b)

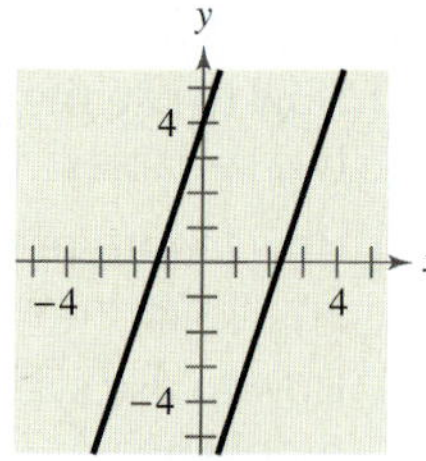

(c)

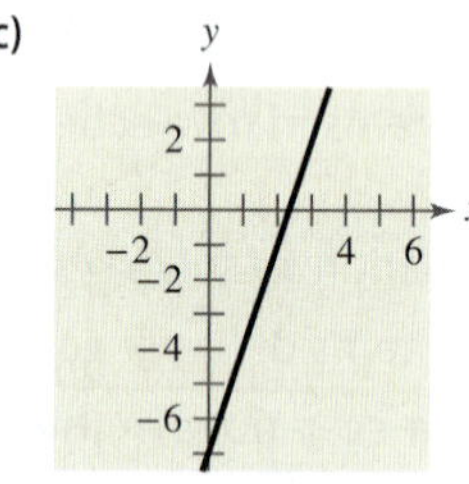

(d)

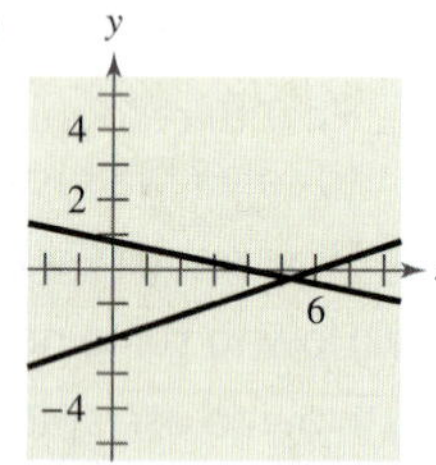

23. $\begin{cases} x + 5y = 4 \\ x - 3y = 6 \end{cases}$

24. $\begin{cases} -3x + y = -7 \\ 9x - 3y = 21 \end{cases}$

25. $\begin{cases} 3x - y = 7 \\ -6x + 2y = 8 \end{cases}$

26. $\begin{cases} 2x - y = -3 \\ x + 5y = 4 \end{cases}$

Supply and Demand **In Exercises 27 and 28, find the equilibrium point.**

	Demand Function	*Supply Function*
27.	$p = 37 - 0.0002x$	$p = 22 + 0.00001x$
28.	$p = 120 - 0.0001x$	$p = 45 + 0.0002x$

9.3 In Exercises 29 and 30, use back-substitution to solve the system.

29. $\begin{cases} x - 4y + 3z = 3 \\ -y + z = -1 \\ z = -5 \end{cases}$

30. $\begin{cases} x - 7y + 8z = 85 \\ y - 9z = -35 \\ z = 3 \end{cases}$

In Exercises 31–34, use Gaussian elimination to solve the system of equations.

31. $\begin{cases} x + 2y + 6z = 4 \\ -3x + 2y - z = -4 \\ 4x + 2z = 16 \end{cases}$

32. $\begin{cases} x + 3y - z = 13 \\ 2x - 5z = 23 \\ 4x - y - 2z = 14 \end{cases}$

33. $\begin{cases} x - 2y + z = -6 \\ 2x - 3y = -7 \\ -x + 3y - 3z = 11 \end{cases}$

34. $\begin{cases} 2x + 6z = -9 \\ 3x - 2y + 11z = -16 \\ 3x - y + 7z = -11 \end{cases}$

In Exercises 35 and 36, solve the nonsquare system of equations.

35. $\begin{cases} 5x - 12y + 7z = 16 \\ 3x - 7y + 4z = 9 \end{cases}$

36. $\begin{cases} 2x + 5y - 19z = 34 \\ 3x + 8y - 31z = 54 \end{cases}$

In Exercises 37 and 38, find the equation of the parabola $y = ax^2 + bx + c$ that passes through the points. Use a graphing utility to verify your result.

37.

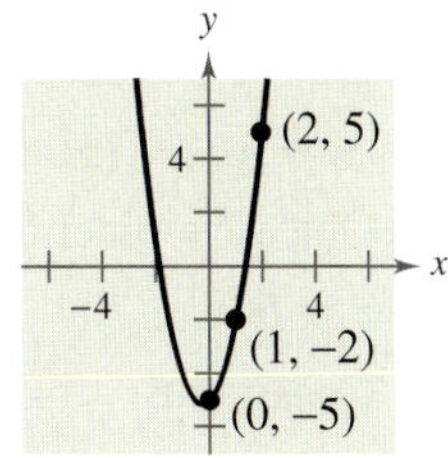

38.

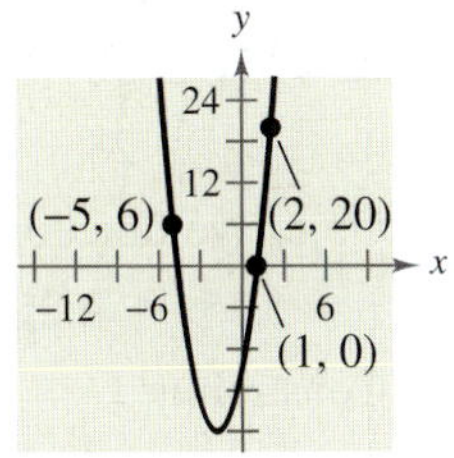

In Exercises 39 and 40, find the equation of the circle $x^2 + y^2 + Dx + Ey + F = 0$ that passes through the points. Use a graphing utility to verify your result.

39.

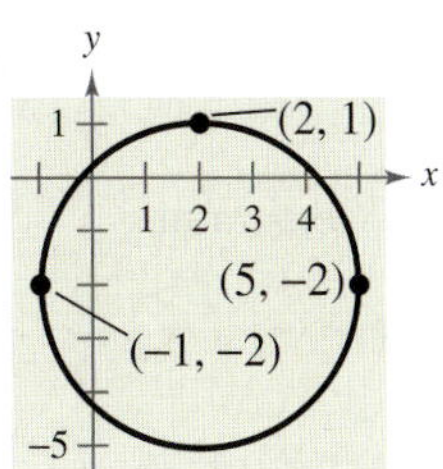

40.

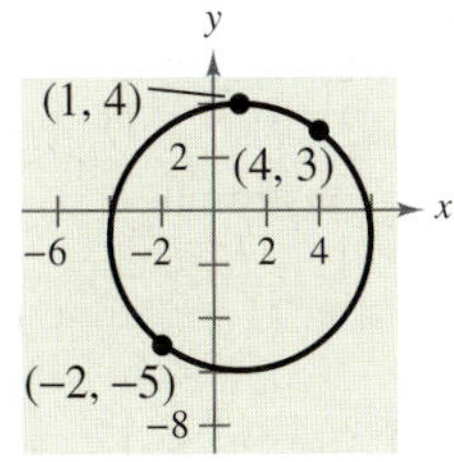

In Exercises 41–44, write the partial fraction decomposition of the rational expression.

41. $\dfrac{4 - x}{x^2 + 6x + 8} = \dfrac{A}{x + 4} + \dfrac{B}{x + 2}$

42. $\dfrac{9}{x^2 - 9} = \dfrac{A}{x + 3} + \dfrac{B}{x - 3}$

43. $\dfrac{x}{x^2 + 3x + 2} = \dfrac{A}{x + 1} + \dfrac{B}{x + 2}$

44. $\dfrac{x^2}{x^2 + 2x - 15} = \dfrac{A}{x + 5} + \dfrac{B}{x - 3}$

45. ***Data Analysis*** The table shows the numbers y (in thousands) of DVD players sold in the United States for the years 1998 through 2000. (Source: Electronics Industries Association)

Year, x	DVD players sold, y
1998	1079
1999	4072
2000	8258

(a) Use the technique demonstrated in Exercises 67–70 in Section 9.3 to find a least squares regression parabola that models the data. Let x represent the year, with $x = 8$ corresponding to 1998.

(b) Use a graphing utility to graph the parabola and the data in the same viewing window. How well does the model fit the data?

(c) Use the model to estimate the number of DVD players sold in 2005. Does your answer seem reasonable?

46. ***Agriculture*** A mixture of 6 gallons of chemical A, 8 gallons of chemical B, and 13 gallons of chemical C is required to kill a destructive crop insect. Commercial spray X contains 1, 2, and 2 parts, respectively, of these chemicals. Commercial spray Y contains only chemical C. Commercial spray Z contains chemicals A, B, and C in equal amounts. How much of each type of commercial spray is needed to get the desired mixture?

47. ***Investment Analysis*** An inheritance of \$40,000 was divided among three investments yielding \$3500 in interest per year. The interest rates for the three investments were 7%, 9%, and 11%. Find the amount placed in each investment if the second and third were \$3000 and \$5000 less than the first, respectively.

48. ***Vertical Motion*** An object moving vertically is at the given heights at the specified times. Find the position equation $s = \frac{1}{2}at^2 + v_0t + s_0$ for the object.

(a) At $t = 1$ second, $s = 134$ feet

At $t = 2$ seconds, $s = 86$ feet

At $t = 3$ seconds, $s = 6$ feet

(b) At $t = 1$ second, $s = 184$ feet

At $t = 2$ seconds, $s = 116$ feet

At $t = 3$ seconds, $s = 16$ feet

9.4 **In Exercises 49–52, sketch the graph of the inequality.**

49. $y \le 5 - \frac{1}{2}x$

50. $3y - x \ge 7$

51. $y - 4x^2 > -1$

52. $y \le \dfrac{3}{x^2 + 2}$

In Exercises 53–60, sketch a graph and label the vertices of the solution set of the system of inequalities.

53. $\begin{cases} x + 2y \le 160 \\ 3x + y \le 180 \\ x \ge 0 \\ y \ge 0 \end{cases}$

54. $\begin{cases} 2x + 3y \le 24 \\ 2x + y \le 16 \\ x \ge 0 \\ y \ge 0 \end{cases}$

55. $\begin{cases} 3x + 2y \ge 24 \\ x + 2y \ge 12 \\ 2 \le x \le 15 \\ y \le 15 \end{cases}$

56. $\begin{cases} 2x + y \ge 16 \\ x + 3y \ge 18 \\ 0 \le x \le 25 \\ 0 \le y \le 25 \end{cases}$

57. $\begin{cases} y < x + 1 \\ y > x^2 - 1 \end{cases}$

58. $\begin{cases} y \le 6 - 2x - x^2 \\ y \ge x + 6 \end{cases}$

59. $\begin{cases} 2x - 3y \ge 0 \\ 2x - y \le 8 \\ y \ge 0 \end{cases}$

60. $\begin{cases} x^2 + y^2 \le 9 \\ (x - 3)^2 + y^2 \le 9 \end{cases}$

In Exercises 61 and 62, determine a system of inequalities that models the description. Use a graphing utility to graph and shade the solution of the system.

61. ***Fruit Distribution*** A Pennsylvania fruit grower has 1500 bushels of apples that are to be divided between markets in Harrisburg and Philadelphia. These two markets need at least 400 bushels and 600 bushels, respectively.

62. ***Inventory Costs*** A warehouse operator has 24,000 square feet of floor space in which to store two products. Each unit of product I requires 20 square feet of floor space and costs \$12 per day to store. Each unit of product II requires 30 square feet of floor space and costs \$8 per day to store. The total storage cost per day cannot exceed \$12,400.

Supply and Demand **In Exercises 63 and 64, find the consumer surplus and producer surplus for the demand and supply equations. Sketch the graph of the equations and shade the regions representing the consumer surplus and producer surplus.**

	Demand	*Supply*
63.	$p = 160 - 0.0001x$	$p = 70 + 0.0002x$
64.	$p = 130 - 0.0002x$	$p = 30 + 0.0003x$

9.5 **In Exercises 65–70, sketch the region determined by the constraints. Then find the minimum and maximum values of the objective function and where they occur, subject to the indicated restraints.**

65. Objective function:

$z = 3x + 4y$

Constraints:

$x \ge 0$

$y \ge 0$

$2x + 5y \le 50$

$4x + y \le 28$

66. Objective function:

$z = 10x + 7y$

Constraints:

$x \ge 0$

$y \ge 0$

$2x + y \ge 100$

$x + y \ge 75$

67. Objective function:

$z = 1.75x + 2.25y$

Constraints:

$x \ge 0$

$y \ge 0$

$2x + y \ge 25$

$3x + 2y \ge 45$

68. Objective function:

$z = 50x + 70y$

Constraints:

$x \ge 0$

$y \ge 0$

$x + 2y \le 1500$

$5x + 2y \le 3500$

69. Objective function:

$z = 5x + 11y$

Constraints:

$x \ge 0$

$y \ge 0$

$x + 3y \le 12$

$3x + 2y \le 15$

70. Objective function:

$z = -2x + y$

Constraints:

$x \ge 0$

$y \ge 0$

$x + y \ge 7$

$5x + 2y \ge 20$

71. ***Maximum Revenue*** A student is working part time as a hairdresser to pay college expenses. The student may work no more than 24 hours per week. Haircuts cost \$25 and require an average of 20 minutes, and permanents cost \$70 and require an average of 1 hour and 10 minutes. What combination of haircuts and/or permanents will yield a maximum revenue? What is the maximum revenue?

72. ***Maximum Profit*** A shoe manufacturer produces a walking shoe and a running shoe yielding profits of \$18 and \$24, respectively. Each shoe must go through three processes, for which the required times per unit are shown in the table.

	Process I	Process II	Process III
Hours for Walking Shoe	4	1	1
Hours for Running Shoe	2	2	1
Hours Available per Day	24	9	8

Find the daily production level for each shoe to maximize the profit. What is the maximum profit?

73. ***Minimum Cost*** A pet supply company mixes two brands of dry dog food. Brand X costs \$15 per bag and contains eight units of nutritional element A, one unit of nutritional element B, and two units of nutritional element C. Brand Y costs \$30 per bag and contains two units of nutritional element A, one unit of nutritional element B, and seven units of nutritional element C. Each bag of mixed dog food must contain at least 16 units, 5 units, and 20 units of nutritional elements A, B, and C, respectively. Find the numbers of bags of brands X and Y that should be mixed to produce a mixture meeting the minimum nutritional requirements and having a minimum cost. What is the minimum cost?

74. ***Minimum Cost*** Two gasolines, type A and type B, have octane ratings of 87 and 92, respectively. Type A costs \$1.25 per gallon and type B costs \$1.55 per gallon. Determine the blend of minimum cost with an octane rating of at least 89. What is the minimum cost? (*Hint:* Let x be the fraction of each gallon that is type A and let y be the fraction that is type B.)

Synthesis

True or False? **In Exercises 75 and 76, determine whether the statement is true or false. Justify your answer.**

75. The system

$$\begin{cases} y \le 5 \\ y \ge -2 \\ y \ge \frac{7}{2}x - 9 \\ y \ge -\frac{7}{2}x + 26 \end{cases}$$

represents the region covered by an isosceles trapezoid.

76. It is possible for an objective function of a linear programming problem to have exactly 10 maximum value points.

In Exercises 77–80, find a system of linear equations having the ordered pair as a solution. (There is more than one correct answer.)

77. $(-6, 8)$

78. $(5, -4)$

79. $\left(\frac{4}{3}, 3\right)$

80. $\left(-1, \frac{9}{4}\right)$

In Exercises 81–84, find a system of linear equations having the ordered triple as a solution. (There is more than one correct answer.)

81. $(4, -1, 3)$

82. $(-3, 5, 6)$

83. $\left(5, \frac{3}{2}, 2\right)$

84. $\left(\frac{3}{4}, -2, 8\right)$

85. ***Writing*** Explain what is meant by an inconsistent system of linear equations.

86. How can you tell graphically that a system of linear equations in two variables has no solution? Give an example.

87. ***Writing*** Write a brief paragraph describing any advantages of substitution over the graphical method of solving a system of equations.

Chapter Test

The *Interactive* CD-ROM and *Internet* versions of this text offer Chapter Pre-Tests and Chapter Post-Tests, both of which have randomly generated exercises with diagnostic capabilities.

Take this test as you would take a test in class. When you are finished, check your work against the answers given in the back of the book.

In Exercises 1–3, solve the system by the method of substitution.

1. $\begin{cases} x - y = -7 \\ 4x + 5y = 8 \end{cases}$ **2.** $\begin{cases} y = x - 1 \\ y = (x-1)^3 \end{cases}$ **3.** $\begin{cases} 2x - y^2 = 0 \\ x - y = 4 \end{cases}$

In Exercises 4–6, solve the system graphically.

4. $\begin{cases} 2x - 3y = 0 \\ 2x + 3y = 12 \end{cases}$ **5.** $\begin{cases} y = 9 - x^2 \\ y = x + 3 \end{cases}$ **6.** $\begin{cases} y - \ln x = 12 \\ 7x - 2y + 11 = -6 \end{cases}$

In Exercises 7–10, solve the linear system by elimination.

7. $\begin{cases} 2x + 3y = 17 \\ 5x - 4y = -15 \end{cases}$ **8.** $\begin{cases} 2.5x - y = 6 \\ 3x + 4y = 2 \end{cases}$

9. $\begin{cases} x - 2y + 3z = 11 \\ 2x - z = 3 \\ 3y + z = -8 \end{cases}$ **10.** $\begin{cases} 3x + 2y + z = 17 \\ -x + y + z = 4 \\ x - y - z = 3 \end{cases}$

In Exercises 11–13, sketch the graph and label the vertices of the solution of the system of inequalities.

11. $\begin{cases} 2x + y \le 4 \\ 2x - y \ge 0 \\ x \ge 0 \end{cases}$ **12.** $\begin{cases} y < -x^2 + x + 4 \\ y > 4x \end{cases}$ **13.** $\begin{cases} x^2 + y^2 \le 16 \\ x \ge 1 \\ y \ge -3 \end{cases}$

14. Find the maximum value of the objective function $z = 20x + 12y$ and where it occurs, subject to the following constraints.

$$\left.\begin{aligned} x &\ge 0 \\ y &\ge 0 \\ x + 4y &\le 32 \\ 3x + 2y &\le 36 \end{aligned}\right\} \quad \text{Constraints}$$

15. A total of \$50,000 is invested in two funds paying 8% and 8.5% simple interest. The yearly interest is \$4150. How much is invested at each rate?

16. Find the equation of the parabola $y = ax^2 + bx + c$ passing through the points $(0, 6)$, $(-2, 2)$, and $\left(3, \frac{9}{2}\right)$.

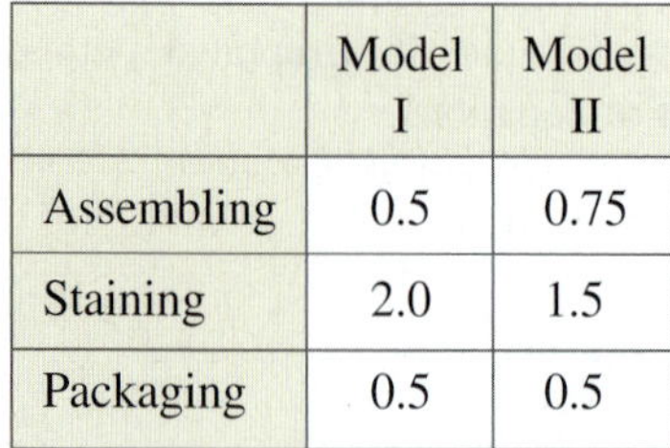

	Model I	Model II
Assembling	0.5	0.75
Staining	2.0	1.5
Packaging	0.5	0.5

17. A manufacturer produces two types of television stands. The amounts of time for assembling, staining, and packaging the two models are shown in the table at the left. The total amounts of time available for assembling, staining, and packaging are 4000, 8950, and 2650 hours, respectively. The profits per unit are \$30 (model I) and \$40 (model II). How many of each model should be produced to maximize the profit? What is the maximum profit?

Proofs in Mathematics

An **indirect proof** can be useful in proving statements of the form "p implies q." Recall that the conditional statement $p \to q$ is false only when p is true and q is false. To prove a conditional statement indirectly, assume that p is true and q is false. If this assumption leads to an impossibility, then you have proved that the conditional statement is true. An indirect proof is also called a **proof by contradiction.**

You can use an indirect proof to prove the conditional statement, "If a is a positive integer and a^2 is divisible by 2, then a is divisible by 2," as follows. Assume that p, "a is a positive integer and a^2 is divisible by 2," is true and q, "a is divisible by 2," is false. This means that a is not divisible by 2. If so, a is odd and can be written as $a = 2n + 1$, where n is an integer.

$$a = 2n + 1 \qquad \text{Definition of an odd integer}$$

$$a^2 = 4n^2 + 4n + 1 \qquad \text{Square each side.}$$

$$a^2 = 2(2n^2 + 2n) + 1 \qquad \text{Distributive Property}$$

So, by the definition of an odd integer, a^2 is odd. This contradicts the assumption, and you can conclude that a is divisible by 2.

Example ▶ Using an Indirect Proof

Use an indirect proof to prove that $\sqrt{2}$ is an irrational number.

Solution

Begin by assuming that $\sqrt{2}$ is *not* an irrational number. Then $\sqrt{2}$ can be written as the quotient of two integers a and $b(b \neq 0)$ that have no common factors.

$$\sqrt{2} = \frac{a}{b} \qquad \text{Assume that } \sqrt{2} \text{ is a rational number.}$$

$$2 = \frac{a^2}{b^2} \qquad \text{Square each side.}$$

$$2b^2 = a^2 \qquad \text{Multiply each side by } b^2.$$

This implies that 2 is a factor of a^2. So, 2 is also a factor of a, and a can be written as $2c$, where c is an integer.

$$2b^2 = (2c)^2 \qquad \text{Substitute } 2c \text{ for } a.$$

$$2b^2 = 4c^2 \qquad \text{Simplify.}$$

$$b^2 = 2c^2 \qquad \text{Divide each side by 2.}$$

This implies that 2 is a factor of b^2 and also a factor of b. So, 2 is a factor of both a and b. This contradicts the assumption that a and b have no common factors. So, you can conclude that $\sqrt{2}$ is an irrational number.

P.S. Problem Solving

1. A theorem from geometry states that if a triangle is inscribed in a circle such that one side of the triangle is a diameter of the circle, then the triangle is a right triangle. Show that this theorem is true for the circle

$$x^2 + y^2 = 100$$

and the triangle formed by the lines

$$y = 0,\ y = \tfrac{1}{2}x + 5, \text{ and } y = -2x + 20.$$

2. Find k_1 and k_2 such that the system of equations has an infinite number of solutions.

$$\begin{cases} 3x - 5y = 8 \\ 2x + k_1 y = k_2 \end{cases}$$

3. Consider the following system of linear equations in x and y.

$$\begin{cases} ax + by = e \\ cx + dy = f \end{cases}$$

Under what conditions will the system have exactly one solution?

4. Graph the lines determined by each system of linear equations. Then use Gaussian elimination to solve each system. At each step of the elimination process, graph the corresponding lines. What do you observe?

(a) $\begin{cases} x - 4y = -3 \\ 5x - 6y = 13 \end{cases}$ (b) $\begin{cases} 2x - 3y = 7 \\ -4x + 6y = -14 \end{cases}$

5. A system of two equations in two unknowns is solved and has a finite number of solutions. Determine the maximum number of solutions of the system satisfying each condition.

(a) Both equations are linear.

(b) One equation is linear and the other is quadratic.

(c) Both equations are quadratic.

6. In the 2000 presidential election, approximately 104.779 million voters divided their votes among four presidential candidates. George W. Bush received 537,000 votes less than Al Gore. Ralph Nader and Pat Buchanan together received 3.18% of the votes. Write and solve a system of equations to find the total number of votes cast for each candidate. Let B represent the total votes cast for Bush, G the total votes cast for Gore, and T the total votes cast for Nader and Buchanan together. (Source: Congressional Quarterly)

7. The Vietnam Veterans Memorial (or "The Wall") in Washington, D.C. was designed by Maya Ying Lin when she was a student at Yale University. This monument has two vertical, triangular sections of black granite with a common side (see figure). The top of each section is level with the ground. The bottoms of the two sections can be approximately modeled by the equations $2x + 50y = -505$ and $2x - 50y = 505$ when the x-axis is superimposed on the top of the wall. Each unit in the coordinate system represents 1 foot. How deep is the memorial at the point where the two sections meet? How long is each section?

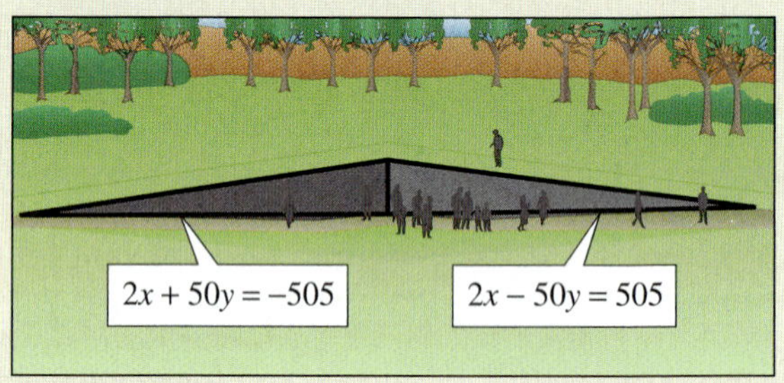

8. Weights of atoms and molecules are measured in atomic mass units (u). A molecule of C_2H_6 (ethane) is made up of two carbon atoms and six hydrogen atoms and weighs 30.07 u. A molecule of C_3H_8 (propane) is made up of three carbon atoms and eight hydrogen atoms and weighs 44.097 u. Find the weights of a carbon atom and a hydrogen atom.

9. To connect a VCR to a television set, a cable with special connectors is required at both ends. You buy a six-foot cable for \$15.50 and a three-foot cable for \$10.25. Assuming that the cost of a cable is the sum of the cost of the two connectors and the cost of the cable itself, what is the cost of a four-foot cable? Explain your reasoning.

10. A hotel 35 miles from the airport runs a shuttle service to and from the airport. The 9:00 A.M. bus leaves for the airport traveling at 30 miles per hour. The 9:15 A.M. bus leaves for the airport traveling at 40 miles per hour. Write a system of linear equations that represents distance as a function of time for each bus. Graph and solve the system. How far from the airport will the 9:15 A.M. bus catch up to the 9:00 A.M. bus?

11. Solve each system of equations by letting $X = 1/x$, $Y = 1/y$, and $Z = 1/z$.

(a) $$\begin{cases} \dfrac{12}{x} - \dfrac{12}{y} = 7 \\ \dfrac{3}{x} + \dfrac{4}{y} = 0 \end{cases}$$

(b) $$\begin{cases} \dfrac{2}{x} + \dfrac{1}{y} - \dfrac{3}{z} = 4 \\ \dfrac{4}{x} + \dfrac{2}{z} = 10 \\ -\dfrac{2}{x} + \dfrac{3}{y} - \dfrac{13}{z} = -8 \end{cases}$$

12. What values should be given to a, b, and c so that the linear system shown has $(-1, 2, -3)$ as its only solution?

$$\begin{cases} x + 2y - 3z = a \\ -x - y + z = b \\ 2x + 3y - 2z = c \end{cases}$$

13. The following system has one solution: $x = 1$, $y = -1$, and $z = 2$.

$$\begin{cases} 4x - 2y + 5z = 16 & \text{Equation 1} \\ x + y = 0 & \text{Equation 2} \\ -x - 3y + 2z = 6 & \text{Equation 3} \end{cases}$$

Solve the system given by (a) Equation 1 and Equation 2, (b) Equation 1 and Equation 3, and (c) Equation 2 and Equation 3. (d) How many solutions does each of these systems have?

14. Each day, an average adult moose can process about 32 kilograms of terrestrial vegetation (twigs and leaves) and aquatic vegetation. From this food, it needs to obtain about 1.9 grams of sodium and 11,000 calories of energy. Aquatic vegetation has about 0.15 gram of sodium per kilogram and about 193 calories of energy per kilogram, whereas terrestrial vegetation has minimal sodium and about four times more energy than aquatic vegetation. Write and graph a system of inequalities that describes the amounts t and a of terrestrial and aquatic vegetation, respectively, for the daily diet of an average adult moose. (Source: *Biology by Numbers*)

15. For a healthy person who is 4 feet 10 inches tall, the recommended minimum weight is about 91 pounds and increases by about 3.7 pounds for each additional inch of height. The recommended maximum weight is about 119 pounds and increases by about 4.9 pounds for each additional inch of height. (Source: Dietary Guidelines Advisory Committee)

(a) Let x be the number of inches by which a person's height exceeds 4 feet 10 inches and let y be the person's weight in pounds. Write a system of inequalities that describes the possible values of x and y for a healthy person.

(b) Use a graphing utility to graph the system of inequalities from part (a).

(c) What is the recommended weight range for someone 6 feet tall?

16. The cholesterol in human blood is necessary, but too much cholesterol can lead to health problems. A blood cholesterol test gives three readings: LDL "bad" cholesterol, HDL "good" cholesterol, and total cholesterol (LDL + HDL). It is recommended that your LDL cholesterol level be less than 130 milligrams per deciliter, your HDL cholesterol level be at least 35 milligrams per deciliter, and your total cholesterol level be no more than 200 milligrams per deciliter. (Source: WebMD, Inc.)

(a) Write a system of linear inequalities for the recommended cholesterol levels. Let x represent HDL cholesterol and let y represent LDL cholesterol.

(b) Graph the system of inequalities from part (a). Label any vertices of the solution region.

(c) Are the following cholesterol levels within recommendations? Explain your reasoning.

LDL: 120 milligrams per deciliter
HDL: 90 milligrams per deciliter
Total: 210 milligrams per deciliter

(d) Give an example of cholesterol levels in which the LDL cholesterol level is too high but the HDL and total cholesterol levels are acceptable.

(e) Another recommendation is that the ratio of total cholesterol to HDL cholesterol be less than 4. Find a point in your solution region from part (b) that meets this recommendation, and show that it does.

How to study Chapter 10

▶ What you should learn

In this chapter you will learn the following skills and concepts:

- How to use matrices, Gaussian elimination, and Gauss-Jordan elimination to solve systems of linear equations
- How to add and subtract matrices, multiply matrices by scalars, and multiply two matrices
- How to find the inverses of matrices and use inverse matrices to solve systems of linear equations
- How to find minors, cofactors, and determinants of square matrices
- How to use Cramer's Rule to solve systems of linear equations
- How to use determinants and matrices to model and solve problems

▶ Important Vocabulary

As you encounter each new vocabulary term in this chapter, add the term and its definition to your notebook glossary.

Matrix (p. 732)
Entry of a matrix (p. 732)
Order of a matrix (p. 732)
Square matrix (p. 732)
Main diagonal (p. 732)
Row matrix (p. 732)
Column matrix (p. 732)
Augmented matrix (p. 733)
Coefficient matrix (p. 733)
Elementary row operations (p. 734)
Row-equivalent matrices (p. 734)
Row-echelon form (p. 736)
Reduced row-echelon form (p. 736)
Gauss-Jordan elimination (p. 739)
Scalars (p. 748)
Scalar multiple (p. 748)
Zero matrix (p. 751)
Matrix multiplication (p. 752)
Identity matrix of order *n* (p. 754)
Inverse of a matrix (p. 761)
Determinant (pp. 765, 770)
Minors (p. 772)
Cofactors (p. 772)
Cramer's Rule (p. 778)

Study Tools

Learning objectives in each section
Chapter Summary (p. 790)
Review Exercises (pp. 791–795)
Chapter Test (p. 796)

Additional Resources

Study and Solutions Guide
Interactive College Algebra
Videotapes/DVD for Chapter 10
Algebra and Trigonometry Website
Student Success Organizer

Layne Kennedy/Corbis

10

Matrices and Determinants

10.1 Matrices and Systems of Equations

▶ **What you should learn**

- How to write a matrix and identify its order
- How to perform elementary row operations on matrices
- How to use matrices and Gaussian elimination to solve systems of linear equations
- How to use matrices and Gauss-Jordan elimination to solve systems of linear equations

▶ **Why you should learn it**

You can use matrices to solve systems of linear equations in two or more variables. For instance, in Exercise 90 on page 745, you will use a matrix to find a model for the retail sales for drug stores in the United States.

Larry Mulvehill/The Image Works

Matrices

In this section you will study a streamlined technique for solving systems of linear equations. This technique involves the use of a rectangular array of real numbers called a **matrix.** The plural of matrix is *matrices*.

Definition of Matrix

If m and n are positive integers, an $m \times n$ (read "m by n") matrix is a rectangular array

$$\begin{array}{l} \\ \text{Row 1} \\ \text{Row 2} \\ \text{Row 3} \\ \vdots \\ \text{Row } m \end{array} \begin{array}{c} \begin{array}{ccccc} \text{Column 1} & \text{Column 2} & \text{Column 3} & \cdots & \text{Column } n \end{array} \\ \begin{bmatrix} a_{11} & a_{12} & a_{13} & \cdots & a_{1n} \\ a_{21} & a_{22} & a_{23} & \cdots & a_{2n} \\ a_{31} & a_{32} & a_{33} & \cdots & a_{3n} \\ \vdots & \vdots & \vdots & & \vdots \\ a_{m1} & a_{m2} & a_{m3} & \cdots & a_{mn} \end{bmatrix} \end{array}$$

in which each **entry,** a_{ij}, of the matrix is a number. An $m \times n$ matrix has m rows and n columns.

The entry in the ith row and jth column is denoted by the *double subscript* notation a_{ij}. For instance, a_{23} refers to the entry in the second row, third column. A matrix having m rows and n columns is said to be of **order** $m \times n$. If $m = n$, the matrix is **square** of order n. For a square matrix, the entries $a_{11}, a_{22}, a_{33}, \ldots$ are the **main diagonal** entries.

Example 1 ▶ **Order of Matrices** 

Determine the order of each matrix.

a. $[2]$ **b.** $\begin{bmatrix} 1 & -3 & 0 & \frac{1}{2} \end{bmatrix}$

c. $\begin{bmatrix} 0 & 0 \\ 0 & 0 \end{bmatrix}$ **d.** $\begin{bmatrix} 5 & 0 \\ 2 & -2 \\ -7 & 4 \end{bmatrix}$

Solution

a. This matrix has *one* row and *one* column. The order of the matrix is 1×1.

b. This matrix has *one* row and *four* columns. The order of the matrix is 1×4.

c. This matrix has *two* rows and *two* columns. The order of the matrix is 2×2.

d. This matrix has *three* rows and *two* columns. The order of the matrix is 3×2.

A matrix that has only one row is called a **row matrix,** and a matrix that has only one column is called a **column matrix.**

The icon (www) identifies examples and concepts related to features of the Learning Tools CD-ROM and the *Interactive* and *Internet* versions of this text. For more details see the chart on pages *xix-xxiii*.

A matrix derived from a system of linear equations (each written in standard form with the constant term on the right) is the **augmented matrix** of the system. Moreover, the matrix derived from the coefficients of the system (but not including the constant terms) is the **coefficient matrix** of the system.

STUDY TIP

The vertical dots in an augmented matrix separate the coefficients of the linear system from the constant terms.

System
$$\begin{cases} x - 4y + 3z = 5 \\ -x + 3y - z = -3 \\ 2x - 4z = 6 \end{cases}$$

Augmented Matrix
$$\begin{bmatrix} 1 & -4 & 3 & \vdots & 5 \\ -1 & 3 & -1 & \vdots & -3 \\ 2 & 0 & -4 & \vdots & 6 \end{bmatrix}$$

Coefficient Matrix
$$\begin{bmatrix} 1 & -4 & 3 \\ -1 & 3 & -1 \\ 2 & 0 & -4 \end{bmatrix}$$

Note the use of 0 for the missing y-variable in the third equation, and also note the fourth column of constant terms in the augmented matrix.

When forming either the coefficient matrix or the augmented matrix of a system, you should begin by vertically aligning the variables in the equations and using zeros for the missing variables.

Example 2 ▶ Writing an Augmented Matrix 

Write the augmented matrix for the system of linear equations.

$$\begin{cases} x + 3y - w = 9 \\ -y + 4z + 2w = -2 \\ x - 5z - 6w = 0 \\ 2x + 4y - 3z = 4 \end{cases}$$

What is the order of the augmented matrix?

Solution

The *Interactive* CD-ROM and *Internet* versions of this text offer a Try It for each example in the text.

Begin by rewriting the linear system and aligning the variables.

$$\begin{cases} x + 3y \qquad\quad - w = 9 \\ \quad -y + 4z + 2w = -2 \\ x \qquad - 5z - 6w = 0 \\ 2x + 4y - 3z \qquad = 4 \end{cases}$$

Next, use the coefficients as the matrix entries. Include zeros for the missing coefficients.

$$\begin{matrix} R_1 \\ R_2 \\ R_3 \\ R_4 \end{matrix}\begin{bmatrix} 1 & 3 & 0 & -1 & \vdots & 9 \\ 0 & -1 & 4 & 2 & \vdots & -2 \\ 1 & 0 & -5 & -6 & \vdots & 0 \\ 2 & 4 & -3 & 0 & \vdots & 4 \end{bmatrix}$$

The augmented matrix has four rows and five columns, so it is a 4×5 matrix. The notation R_n is used to designate each row in the matrix. For example, Row 1 is represented by R_1.

Elementary Row Operations

In Section 9.3, you studied three operations that can be used on a system of linear equations to produce an equivalent system.

1. Interchange two equations.
2. Multiply an equation by a nonzero constant.
3. Add a multiple of an equation to another equation.

In matrix terminology, these three operations correspond to **elementary row operations.** An elementary row operation on an augmented matrix of a given system of linear equations produces a new augmented matrix corresponding to a new (but equivalent) system of linear equations. Two matrices are **row-equivalent** if one can be obtained from the other by a sequence of elementary row operations.

Elementary Row Operations

1. Interchange two rows.
2. Multiply a row by a nonzero constant.
3. Add a multiple of a row to another row.

Although elementary row operations are simple to perform, they involve a lot of arithmetic. Because it is easy to make a mistake, you should get in the habit of noting the elementary row operations performed in each step so that you can go back and check your work.

Example 3 ▶ Elementary Row Operations

a. Interchange the first and second rows.

Original Matrix

$$\begin{bmatrix} 0 & 1 & 3 & 4 \\ -1 & 2 & 0 & 3 \\ 2 & -3 & 4 & 1 \end{bmatrix}$$

New Row-Equivalent Matrix

$$\begin{matrix} R_2 \\ R_1 \\ \end{matrix}\begin{bmatrix} -1 & 2 & 0 & 3 \\ 0 & 1 & 3 & 4 \\ 2 & -3 & 4 & 1 \end{bmatrix}$$

b. Multiply the first row by $\frac{1}{2}$.

Original Matrix

$$\begin{bmatrix} 2 & -4 & 6 & -2 \\ 1 & 3 & -3 & 0 \\ 5 & -2 & 1 & 2 \end{bmatrix}$$

New Row-Equivalent Matrix

$$\begin{matrix} \frac{1}{2}R_1 \rightarrow \\ \\ \end{matrix}\begin{bmatrix} 1 & -2 & 3 & -1 \\ 1 & 3 & -3 & 0 \\ 5 & -2 & 1 & 2 \end{bmatrix}$$

c. Add -2 times the first row to the third row.

Original Matrix

$$\begin{bmatrix} 1 & 2 & -4 & 3 \\ 0 & 3 & -2 & -1 \\ 2 & 1 & 5 & -2 \end{bmatrix}$$

New Row-Equivalent Matrix

$$\begin{matrix} \\ \\ -2R_1 + R_3 \rightarrow \end{matrix}\begin{bmatrix} 1 & 2 & -4 & 3 \\ 0 & 3 & -2 & -1 \\ 0 & -3 & 13 & -8 \end{bmatrix}$$

Note that the elementary row operation is written beside the row that is *changed.*

Technology

Most graphing utilities can perform elementary row operations on matrices. Consult the user's guide for your graphing utility for the required keystrokes.

After performing a row operation, the new row-equivalent matrix that is displayed on your graphing utility is stored in the *answer* variable. You should use the *answer* variable and not the original matrix for subsequent row operations.

In Example 3 in Section 9.3, you used Gaussian elimination with back-substitution to solve a system of linear equations. The next example demonstrates the matrix version of Gaussian elimination. The two methods are essentially the same. The basic difference is that with matrices you do not need to keep writing the variables.

Example 4 ▶ Comparing Linear Systems and Matrix Operations

Linear System

$$\begin{cases} x - 2y + 3z = 9 \\ -x + 3y = -4 \\ 2x - 5y + 5z = 17 \end{cases}$$

Associated Augmented Matrix

$$\begin{bmatrix} 1 & -2 & 3 & \vdots & 9 \\ -1 & 3 & 0 & \vdots & -4 \\ 2 & -5 & 5 & \vdots & 17 \end{bmatrix}$$

Add the first equation to the second equation.

$$\begin{cases} x - 2y + 3z = 9 \\ y + 3z = 5 \\ 2x - 5y + 5z = 17 \end{cases}$$

Add the first row to the second row.

$$R_1 + R_2 \rightarrow \begin{bmatrix} 1 & -2 & 3 & \vdots & 9 \\ 0 & 1 & 3 & \vdots & 5 \\ 2 & -5 & 5 & \vdots & 17 \end{bmatrix}$$

Add -2 times the first equation to the third equation.

$$\begin{cases} x - 2y + 3z = 9 \\ y + 3z = 5 \\ -y - z = -1 \end{cases}$$

Add -2 times the first row to the third row.

$$-2R_1 + R_3 \rightarrow \begin{bmatrix} 1 & -2 & 3 & \vdots & 9 \\ 0 & 1 & 3 & \vdots & 5 \\ 0 & -1 & -1 & \vdots & -1 \end{bmatrix}$$

Add the second equation to the third equation.

$$\begin{cases} x - 2y + 3z = 9 \\ y + 3z = 5 \\ 2z = 4 \end{cases}$$

Add the second row to the third row.

$$R_2 + R_3 \rightarrow \begin{bmatrix} 1 & -2 & 3 & \vdots & 9 \\ 0 & 1 & 3 & \vdots & 5 \\ 0 & 0 & 2 & \vdots & 4 \end{bmatrix}$$

Multiply the third equation by $\frac{1}{2}$.

$$\begin{cases} x - 2y + 3z = 9 \\ y + 3z = 5 \\ z = 2 \end{cases}$$

Multiply the third row by $\frac{1}{2}$.

$$\tfrac{1}{2}R_3 \rightarrow \begin{bmatrix} 1 & -2 & 3 & \vdots & 9 \\ 0 & 1 & 3 & \vdots & 5 \\ 0 & 0 & 1 & \vdots & 2 \end{bmatrix}$$

At this point, you can use back-substitution to find x and y.

$$y + 3(2) = 5 \qquad \text{Substitute 2 for } z.$$

$$y = -1 \qquad \text{Solve for}$$

$$x - 2(-1) + 3(2) = 9 \qquad \text{Substitute } -1 \text{ for } y \text{ and 2 for } z.$$

$$x = 1 \qquad \text{Solve for } x.$$

The solution is $x = 1$, $y = -1$, and $z = 2$.

Remember that you can check a solution by substituting the values of x, y, and z into each equation in the original system.

The last matrix in Example 4 is said to be in **row-echelon form.** The term *echelon* refers to the stair-step pattern formed by the nonzero elements of the matrix. To be in this form, a matrix must have the following properties.

Row-Echelon Form and Reduced Row-Echelon Form

A matrix in **row-echelon form** has the following properties.

1. All rows consisting entirely of zeros occur at the bottom of the matrix.
2. For each row that does not consist entirely of zeros, the first nonzero entry is 1 (called a leading 1).
3. For two successive (nonzero) rows, the leading 1 in the higher row is farther to the left than the leading 1 in the lower row.

A matrix in *row-echelon form* is in **reduced row-echelon form** if every column that has a leading 1 has zeros in every position above and below its leading 1.

The *Interactive* CD-ROM and *Internet* versions of this text offer a Quiz for every section of the text.

Example 5 ▶ Row-Echelon Form 

The following matrices are in row-echelon form.

a. $\begin{bmatrix} 1 & 2 & -1 & 4 \\ 0 & 1 & 0 & 3 \\ 0 & 0 & 1 & -2 \end{bmatrix}$ **b.** $\begin{bmatrix} 0 & 1 & 0 & 5 \\ 0 & 0 & 1 & 3 \\ 0 & 0 & 0 & 0 \end{bmatrix}$

c. $\begin{bmatrix} 1 & -5 & 2 & -1 & 3 \\ 0 & 0 & 1 & 3 & -2 \\ 0 & 0 & 0 & 1 & 4 \\ 0 & 0 & 0 & 0 & 1 \end{bmatrix}$ **d.** $\begin{bmatrix} 1 & 0 & 0 & -1 \\ 0 & 1 & 0 & 2 \\ 0 & 0 & 1 & 3 \\ 0 & 0 & 0 & 0 \end{bmatrix}$

The matrices in (b) and (d) also happen to be in *reduced* row-echelon form. The following matrices are not in row-echelon form.

e. $\begin{bmatrix} 1 & 2 & -3 & 4 \\ 0 & 2 & 1 & -1 \\ 0 & 0 & 1 & -3 \end{bmatrix}$ **f.** $\begin{bmatrix} 1 & 2 & -1 & 2 \\ 0 & 0 & 0 & 0 \\ 0 & 1 & 2 & -4 \end{bmatrix}$

For the matrix in (e), the first nonzero row entry in Row 2 is not 1. For the matrix in (f), the row that consists entirely of zeros is not at the bottom of the matrix.

Every matrix is row-equivalent to a matrix in row-echelon form. For instance, in Example 5, you can change the matrix in part (e) to row-echelon form by multiplying its second row by $\frac{1}{2}$.

$$\frac{1}{2}R_2 \rightarrow \begin{bmatrix} 1 & 2 & -3 & 4 \\ 0 & 1 & \frac{1}{2} & -\frac{1}{2} \\ 0 & 0 & 1 & -3 \end{bmatrix}$$

What elementary row operation could you perform on the matrix in part (f) so that it would be in row-echelon form?

Gaussian Elimination with Back-Substitution

Gaussian elimination with back-substitution works well for solving systems of linear equations by hand or with a computer. For this algorithm, the order in which the elementary row operations are performed is important. You should operate from left to right by columns, using elementary row operations to obtain zeros in all entries directly below the leading 1's.

Example 6 ▶ Gaussian Elimination with Back-Substitution

Solve the system $\begin{cases} y + z - 2w = -3 \\ x + 2y - z = 2 \\ 2x + 4y + z - 3w = -2 \\ x - 4y - 7z - w = -19 \end{cases}$.

Solution

$$\begin{bmatrix} 0 & 1 & 1 & -2 & \vdots & -3 \\ 1 & 2 & -1 & 0 & \vdots & 2 \\ 2 & 4 & 1 & -3 & \vdots & -2 \\ 1 & -4 & -7 & -1 & \vdots & -19 \end{bmatrix}$$ Write augmented matrix.

$$\begin{matrix} R_2 \\ R_1 \\ \\ \\ \end{matrix}\begin{bmatrix} 1 & 2 & -1 & 0 & \vdots & 2 \\ 0 & 1 & 1 & -2 & \vdots & -3 \\ 2 & 4 & 1 & -3 & \vdots & -2 \\ 1 & -4 & -7 & -1 & \vdots & -19 \end{bmatrix}$$ Interchange R_1 and R_2 so first column has leading 1 in upper left corner.

$$\begin{matrix} \\ \\ -2R_1 + R_3 \rightarrow \\ -R_1 + R_4 \rightarrow \end{matrix}\begin{bmatrix} 1 & 2 & -1 & 0 & \vdots & 2 \\ 0 & 1 & 1 & -2 & \vdots & -3 \\ 0 & 0 & 3 & -3 & \vdots & -6 \\ 0 & -6 & -6 & -1 & \vdots & -21 \end{bmatrix}$$ Perform operations on R_3 and R_4 so first column has zeros below its leading 1.

$$\begin{matrix} \\ \\ \\ 6R_2 + R_4 \rightarrow \end{matrix}\begin{bmatrix} 1 & 2 & -1 & 0 & \vdots & 2 \\ 0 & 1 & 1 & -2 & \vdots & -3 \\ 0 & 0 & 3 & -3 & \vdots & -6 \\ 0 & 0 & 0 & -13 & \vdots & -39 \end{bmatrix}$$ Perform operations on R_4 so second column has zeros below its leading 1.

$$\begin{matrix} \\ \\ \frac{1}{3}R_3 \rightarrow \\ -\frac{1}{13}R_4 \rightarrow \end{matrix}\begin{bmatrix} 1 & 2 & -1 & 0 & \vdots & 2 \\ 0 & 1 & 1 & -2 & \vdots & -3 \\ 0 & 0 & 1 & -1 & \vdots & -2 \\ 0 & 0 & 0 & 1 & \vdots & 3 \end{bmatrix}$$ Perform operations on R_3 and R_4 so third and fourth columns have leading 1's.

The matrix is now in row-echelon form, and the corresponding system is

$$\begin{cases} x + 2y - z = 2 \\ y + z - 2w = -3 \\ z - w = -2 \\ w = 3. \end{cases}$$

Using back-substitution, the solution is $x = -1$, $y = 2$, $z = 1$, and $w = 3$.

The procedure for using Gaussian elimination with back-substitution is summarized below.

Gaussian Elimination with Back-Substitution

1. Write the augmented matrix of the system of linear equations.
2. Use elementary row operations to rewrite the augmented matrix in row-echelon form.
3. Write the system of linear equations corresponding to the matrix in row-echelon form, and use back-substitution to find the solution.

When solving a system of linear equations, remember that it is possible for the system to have no solution. If, in the elimination process, you obtain a row with zeros except for the last entry, it is unnecessary to continue the elimination process. You can simply conclude that the system has no solution, or is *inconsistent*.

Example 7 ▶ A System with No Solution 

Solve the system $\begin{cases} x - y + 2z = 4 \\ x + z = 6 \\ 2x - 3y + 5z = 4 \\ 3x + 2y - z = 1 \end{cases}$.

Solution

$$\begin{bmatrix} 1 & -1 & 2 & \vdots & 4 \\ 1 & 0 & 1 & \vdots & 6 \\ 2 & -3 & 5 & \vdots & 4 \\ 3 & 2 & -1 & \vdots & 1 \end{bmatrix}$$ Write augmented matrix.

$$\begin{matrix} \\ -R_1 + R_2 \rightarrow \\ -2R_1 + R_3 \rightarrow \\ -3R_1 + R_4 \rightarrow \end{matrix} \begin{bmatrix} 1 & -1 & 2 & \vdots & 4 \\ 0 & 1 & -1 & \vdots & 2 \\ 0 & -1 & 1 & \vdots & -4 \\ 0 & 5 & -7 & \vdots & -11 \end{bmatrix}$$ Perform row operations.

$$\begin{matrix} \\ \\ R_2 + R_3 \rightarrow \\ \\ \end{matrix} \begin{bmatrix} 1 & -1 & 2 & \vdots & 4 \\ 0 & 1 & -1 & \vdots & 2 \\ 0 & 0 & 0 & \vdots & -2 \\ 0 & 5 & -7 & \vdots & -11 \end{bmatrix}$$ Perform row operations.

Note that the third row of this matrix consists of zeros except for the last entry. This means that the original system of linear equations is inconsistent. You can see why this is true by converting back to a system of linear equations.

$$\begin{cases} x - y + 2z = 4 \\ y - z = 2 \\ 0 = -2 \\ 5y - 7z = -11 \end{cases}$$

Because the third equation is not possible, the system has no solution.

Gauss-Jordan Elimination

With Gaussian elimination, elementary row operations are applied to a matrix to obtain a (row-equivalent) row-echelon form of the matrix. A second method of elimination, called **Gauss-Jordan elimination,** after Carl Friedrich Gauss and Wilhelm Jordan (1842–1899), continues the reduction process until a *reduced* row-echelon form is obtained. This procedure is demonstrated in Example 8.

Example 8 ▶ Gauss-Jordan Elimination 

Use Gauss-Jordan elimination to solve the system $\begin{cases} x - 2y + 3z = 9 \\ -x + 3y = -4. \\ 2x - 5y + 5z = 17 \end{cases}$

Solution

In Example 4, Gaussian elimination was used to obtain the row-echelon form of the linear system above.

$$\begin{bmatrix} 1 & -2 & 3 & \vdots & 9 \\ 0 & 1 & 3 & \vdots & 5 \\ 0 & 0 & 1 & \vdots & 2 \end{bmatrix}$$

Now, apply elementary row operations until you obtain zeros above each of the leading 1's, as follows.

$$2R_2 + R_1 \rightarrow \begin{bmatrix} 1 & 0 & 9 & \vdots & 19 \\ 0 & 1 & 3 & \vdots & 5 \\ 0 & 0 & 1 & \vdots & 2 \end{bmatrix}$$

Perform operations on R_1 so second column has a zero above its leading 1.

$$\begin{matrix} -9R_3 + R_1 \rightarrow \\ -3R_3 + R_2 \rightarrow \\ \end{matrix} \begin{bmatrix} 1 & 0 & 0 & \vdots & 1 \\ 0 & 1 & 0 & \vdots & -1 \\ 0 & 0 & 1 & \vdots & 2 \end{bmatrix}$$

Perform operations on R_1 and R_2 so third column has zeros above its leading 1.

The matrix is now in reduced row-echelon form. Converting back to a system of linear equations, you have

$$\begin{cases} x = 1 \\ y = -1. \\ z = 2 \end{cases}$$

Now you can simply read the solution.

Technology

For a demonstration of a graphical approach to Gauss-Jordan elimination on a 2×3 matrix, see the graphing calculator program available for several models of graphing calculators at our website *college.hmco.com*.

STUDY TIP

The advantage of using Gauss-Jordan elimination to solve a system of linear equations is that the solution of the system is easily found without using back-substitution, as illustrated in Example 8.

The elimination procedures described in this section sometimes result in fractional coefficients. For instance, in the elimination procedure for the system

$$\begin{cases} 2x - 5y + 5z = 17 \\ 3x - 2y + 3z = 11 \\ -3x + 3y = -6 \end{cases}$$

you may be inclined to multiply the first row by $\frac{1}{2}$ to produce a leading 1, which will result in working with fractional coefficients. You can sometimes avoid fractions by judiciously choosing the order in which you apply elementary row operations.

Example 9 ▶ A System with an Infinite Number of Solutions

Solve the system.

$$\begin{cases} 2x + 4y - 2z = 0 \\ 3x + 5y \qquad\; = 1 \end{cases}$$

Solution

$$\begin{bmatrix} 2 & 4 & -2 & \vdots & 0 \\ 3 & 5 & 0 & \vdots & 1 \end{bmatrix}$$

$$\tfrac{1}{2}R_1 \rightarrow \begin{bmatrix} 1 & 2 & -1 & \vdots & 0 \\ 3 & 5 & 0 & \vdots & 1 \end{bmatrix}$$

$$-3R_1 + R_2 \rightarrow \begin{bmatrix} 1 & 2 & -1 & \vdots & 0 \\ 0 & -1 & 3 & \vdots & 1 \end{bmatrix}$$

$$-R_2 \rightarrow \begin{bmatrix} 1 & 2 & -1 & \vdots & 0 \\ 0 & 1 & -3 & \vdots & -1 \end{bmatrix}$$

$$-2R_2 + R_1 \rightarrow \begin{bmatrix} 1 & 0 & 5 & \vdots & 2 \\ 0 & 1 & -3 & \vdots & -1 \end{bmatrix}$$

The corresponding system of equations is

$$\begin{cases} x + 5z = 2 \\ y - 3z = -1 \end{cases}$$

Solving for x and y in terms of z, you have

$$x = -5z + 2$$

and

$$y = 3z - 1.$$

To write a solution to the system that does not use any of the three variables of the system, let a represent any real number and let

$$z = a.$$

Now substitute a for z in the equations for x and y.

$$x = -5z + 2 = -5a + 2$$

$$y = 3z - 1 = 3a - 1$$

So, the solution set has the form

$$(-5a + 2, 3a - 1, a)$$

where a is any real number. Try substituting values for a to obtain a few solutions. Then check each solution in the original equation.

STUDY TIP

In Example 9, x and y are solved for in terms of the third variable z. To write a solution to the system that does not use any of the three variables of the system, let a represent any real number and let $z = a$. Then solve for x and y. The solution can then be written in terms of a, which is not one of the variables of the system.

It is worth noting that the row-echelon form of a matrix is not unique. That is, two different sequences of elementary row operations may yield different row-echelon forms. This is demonstrated in Example 10.

Example 10 ▶ Comparing Row-Echelon Forms

Compare the following row-echelon form with the one found in Example 4. Is it the same? Does it yield the same solution?

$$\begin{cases} x - 2y + 3z = 9 \\ -x + 3y = -4 \\ 2x - 5y + 5z = 17 \end{cases}$$

$$\begin{bmatrix} 1 & -2 & 3 & \vdots & 9 \\ -1 & 3 & 0 & \vdots & -4 \\ 2 & -5 & 5 & \vdots & 17 \end{bmatrix}$$

$$\begin{matrix} R_2 \\ R_1 \\ \\ \end{matrix} \begin{bmatrix} -1 & 3 & 0 & \vdots & -4 \\ 1 & -2 & 3 & \vdots & 9 \\ 2 & -5 & 5 & \vdots & 17 \end{bmatrix}$$

$$\begin{matrix} -R_1 \rightarrow \\ \\ \\ \end{matrix} \begin{bmatrix} 1 & -3 & 0 & \vdots & 4 \\ 1 & -2 & 3 & \vdots & 9 \\ 2 & -5 & 5 & \vdots & 17 \end{bmatrix}$$

$$\begin{matrix} \\ -R_1 + R_2 \rightarrow \\ -2R_1 + R_3 \rightarrow \end{matrix} \begin{bmatrix} 1 & -3 & 0 & \vdots & 4 \\ 0 & 1 & 3 & \vdots & 5 \\ 0 & 1 & 5 & \vdots & 9 \end{bmatrix}$$

$$\begin{matrix} \\ \\ -R_2 + R_3 \rightarrow \end{matrix} \begin{bmatrix} 1 & -3 & 0 & \vdots & 4 \\ 0 & 1 & 3 & \vdots & 5 \\ 0 & 0 & 2 & \vdots & 4 \end{bmatrix}$$

$$\begin{matrix} \\ \\ \frac{1}{2}R_3 \rightarrow \end{matrix} \begin{bmatrix} 1 & -3 & 0 & \vdots & 4 \\ 0 & 1 & 3 & \vdots & 5 \\ 0 & 0 & 1 & \vdots & 2 \end{bmatrix}$$

Solution

This row-echelon form is different from that obtained in Example 4. The corresponding system of linear equations for this row-echelon matrix is

$$\begin{cases} x - 3y = 4 \\ y + 3z = 5. \\ z = 2 \end{cases}$$

Using back-substitution on this system, you obtain the solution

$$x = 1,\ y = -1, \text{ and } z = 2$$

which is the same solution that was obtained in Example 4.

You have seen that the row-echelon form of a given matrix *is not* unique; however, the *reduced* row-echelon form of a given matrix *is* unique. Try applying Gauss-Jordan elimination to the row-echelon matrix in Example 10 to see that you obtain the same reduced row-echelon form as in Example 8.

10.1 Exercises

The *Interactive* CD-ROM and *Internet* versions of this text contain step-by-step solutions to all odd-numbered exercises. They also provide Tutorial Exercises for additional help.

In Exercises 1–6, determine the order of the matrix.

1. $\begin{bmatrix} 7 & 0 \end{bmatrix}$

2. $\begin{bmatrix} 5 & -3 & 8 & 7 \end{bmatrix}$

3. $\begin{bmatrix} 2 \\ 36 \\ 3 \end{bmatrix}$

4. $\begin{bmatrix} -3 & 7 & 15 & 0 \\ 0 & 0 & 3 & 3 \\ 1 & 1 & 6 & 7 \end{bmatrix}$

5. $\begin{bmatrix} 33 & 45 \\ -9 & 20 \end{bmatrix}$

6. $\begin{bmatrix} -7 & 6 & 4 \\ 0 & -5 & 1 \end{bmatrix}$

In Exercises 7–12, write the augmented matrix for the system of linear equations.

7. $\begin{cases} 4x - 3y = -5 \\ -x + 3y = 12 \end{cases}$

8. $\begin{cases} 7x + 4y = 22 \\ 5x - 9y = 15 \end{cases}$

9. $\begin{cases} x + 10y - 2z = 2 \\ 5x - 3y + 4z = 0 \\ 2x + y = 6 \end{cases}$

10. $\begin{cases} -x - 8y + 5z = 8 \\ -7x - 15z = -38 \\ 3x - y + 8z = 20 \end{cases}$

11. $\begin{cases} 7x - 5y + z = 13 \\ 19x - 8z = 10 \end{cases}$

12. $\begin{cases} 9x + 2y - 3z = 20 \\ -25y + 11z = -5 \end{cases}$

In Exercises 13–18, write the system of linear equations represented by the augmented matrix. (Use variables *x*, *y*, *z*, and *w*.)

13. $\begin{bmatrix} 1 & 2 & \vdots & 7 \\ 2 & -3 & \vdots & 4 \end{bmatrix}$

14. $\begin{bmatrix} 7 & -5 & \vdots & 0 \\ 8 & 3 & \vdots & -2 \end{bmatrix}$

15. $\begin{bmatrix} 2 & 0 & 5 & \vdots & -12 \\ 0 & 1 & -2 & \vdots & 7 \\ 6 & 3 & 0 & \vdots & 2 \end{bmatrix}$

16. $\begin{bmatrix} 4 & -5 & -1 & \vdots & 18 \\ -11 & 0 & 6 & \vdots & 25 \\ 3 & 8 & 0 & \vdots & -29 \end{bmatrix}$

17. $\begin{bmatrix} 9 & 12 & 3 & 0 & \vdots & 0 \\ -2 & 18 & 5 & 2 & \vdots & 10 \\ 1 & 7 & -8 & 0 & \vdots & -4 \\ 3 & 0 & 2 & 0 & \vdots & -10 \end{bmatrix}$

18. $\begin{bmatrix} 6 & 2 & -1 & -5 & \vdots & -25 \\ -1 & 0 & 7 & 3 & \vdots & 7 \\ 4 & -1 & -10 & 6 & \vdots & 23 \\ 0 & 8 & 1 & -11 & \vdots & -21 \end{bmatrix}$

In Exercises 19–22, fill in the blank(s) using elementary row operations to form a row-equivalent matrix.

19. $\begin{bmatrix} 1 & 4 & 3 \\ 2 & 10 & 5 \end{bmatrix}$

$\begin{bmatrix} 1 & 4 & 3 \\ 0 & \square & -1 \end{bmatrix}$

20. $\begin{bmatrix} 3 & 6 & 8 \\ 4 & -3 & 6 \end{bmatrix}$

$\begin{bmatrix} 1 & \square & \frac{8}{3} \\ 4 & -3 & 6 \end{bmatrix}$

21. $\begin{bmatrix} 1 & 1 & 4 & -1 \\ 3 & 8 & 10 & 3 \\ -2 & 1 & 12 & 6 \end{bmatrix}$

$\begin{bmatrix} 1 & 1 & 4 & -1 \\ 0 & 5 & \square & \square \\ 0 & 3 & \square & \square \end{bmatrix}$

$\begin{bmatrix} 1 & 1 & 4 & -1 \\ 0 & 1 & -\frac{2}{5} & \frac{6}{5} \\ 0 & 3 & \square & \square \end{bmatrix}$

22. $\begin{bmatrix} 2 & 4 & 8 & 3 \\ 1 & -1 & -3 & 2 \\ 2 & 6 & 4 & 9 \end{bmatrix}$

$\begin{bmatrix} 1 & \square & \square & \square \\ 1 & -1 & -3 & 2 \\ 2 & 6 & 4 & 9 \end{bmatrix}$

$\begin{bmatrix} 1 & 2 & 4 & \frac{3}{2} \\ 0 & \square & -7 & \frac{1}{2} \\ 0 & 2 & \square & \square \end{bmatrix}$

In Exercises 23–26, identify the elementary row operation(s) being performed to obtain the new row-equivalent matrix.

23. *Original Matrix* $\begin{bmatrix} -2 & 5 & 1 \\ 3 & -1 & -8 \end{bmatrix}$ *New Row-Equivalent Matrix* $\begin{bmatrix} 13 & 0 & -39 \\ 3 & -1 & -8 \end{bmatrix}$

24. *Original Matrix* $\begin{bmatrix} 3 & -1 & -4 \\ -4 & 3 & 7 \end{bmatrix}$ *New Row-Equivalent Matrix* $\begin{bmatrix} 3 & -1 & -4 \\ 5 & 0 & -5 \end{bmatrix}$

25. *Original Matrix* $\begin{bmatrix} 0 & -1 & -5 & 5 \\ -1 & 3 & -7 & 6 \\ 4 & -5 & 1 & 3 \end{bmatrix}$ *New Row-Equivalent Matrix* $\begin{bmatrix} -1 & 3 & -7 & 6 \\ 0 & -1 & -5 & 5 \\ 0 & 7 & -27 & 27 \end{bmatrix}$

26. *Original Matrix* $\begin{bmatrix} -1 & -2 & 3 & -2 \\ 2 & -5 & 1 & -7 \\ 5 & 4 & -7 & 6 \end{bmatrix}$ *New Row-Equivalent Matrix* $\begin{bmatrix} -1 & -2 & 3 & -2 \\ 0 & -9 & 7 & -11 \\ 0 & -6 & 8 & -4 \end{bmatrix}$

In Exercises 27–30, determine whether the matrix is in row-echelon form. If it is, determine if it is also in reduced row-echelon form.

27. $\begin{bmatrix} 1 & 0 & 0 & 0 \\ 0 & 1 & 1 & 5 \\ 0 & 0 & 0 & 0 \end{bmatrix}$

28. $\begin{bmatrix} 1 & 3 & 0 & 0 \\ 0 & 0 & 1 & 8 \\ 0 & 0 & 0 & 0 \end{bmatrix}$

29. $\begin{bmatrix} 2 & 0 & 4 & 0 \\ 0 & -1 & 3 & 6 \\ 0 & 0 & 1 & 5 \end{bmatrix}$

30. $\begin{bmatrix} 1 & 0 & 2 & 1 \\ 0 & 1 & -3 & 10 \\ 0 & 0 & 1 & 0 \end{bmatrix}$

31. Perform the sequence of row operations on the matrix. What did the operations accomplish?

$$\begin{bmatrix} 1 & 2 & 3 \\ 2 & -1 & -4 \\ 3 & 1 & -1 \end{bmatrix}$$

(a) Add -2 times R_1 to R_2.

(b) Add -3 times R_1 to R_3.

(c) Add -1 times R_2 to R_3.

(d) Multiply R_2 by $-\frac{1}{5}$.

(e) Add -2 times R_2 to R_1.

32. Perform the sequence of row operations on the matrix. What did the operations accomplish?

$$\begin{bmatrix} 7 & 1 \\ 0 & 2 \\ -3 & 4 \\ 4 & 1 \end{bmatrix}$$

(a) Add R_3 to R_4.

(b) Interchange R_1 and R_4.

(c) Add 3 times R_1 to R_3.

(d) Add -7 times R_1 to R_4.

(e) Multiply R_2 by $\frac{1}{2}$.

(f) Add the appropriate multiples of R_2 to R_1, R_3, and R_4.

In Exercises 33–36, write the matrix in row-echelon form. Remember that the row-echelon form of a matrix is not unique.

33. $\begin{bmatrix} 1 & 1 & 0 & 5 \\ -2 & -1 & 2 & -10 \\ 3 & 6 & 7 & 14 \end{bmatrix}$

34. $\begin{bmatrix} 1 & 2 & -1 & 3 \\ 3 & 7 & -5 & 14 \\ -2 & -1 & -3 & 8 \end{bmatrix}$

35. $\begin{bmatrix} 1 & -1 & -1 & 1 \\ 5 & -4 & 1 & 8 \\ -6 & 8 & 18 & 0 \end{bmatrix}$

36. $\begin{bmatrix} 1 & -3 & 0 & -7 \\ -3 & 10 & 1 & 23 \\ 4 & -10 & 2 & -24 \end{bmatrix}$

In Exercises 37–42, use the matrix capabilities of a graphing utility to write the matrix in *reduced* row-echelon form.

37. $\begin{bmatrix} 3 & 3 & 3 \\ -1 & 0 & -4 \\ 2 & 4 & -2 \end{bmatrix}$ **38.** $\begin{bmatrix} 1 & 3 & 2 \\ 5 & 15 & 9 \\ 2 & 6 & 10 \end{bmatrix}$

39. $\begin{bmatrix} 1 & 2 & 3 & -5 \\ 1 & 2 & 4 & -9 \\ -2 & -4 & -4 & 3 \\ 4 & 8 & 11 & -14 \end{bmatrix}$

40. $\begin{bmatrix} -2 & 3 & -1 & -2 \\ 4 & -2 & 5 & 8 \\ 1 & 5 & -2 & 0 \\ 3 & 8 & -10 & -30 \end{bmatrix}$

41. $\begin{bmatrix} -3 & 5 & 1 & 12 \\ 1 & -1 & 1 & 4 \end{bmatrix}$

42. $\begin{bmatrix} 5 & 1 & 2 & 4 \\ -1 & 5 & 10 & -32 \end{bmatrix}$

In Exercises 43–46, write the system of linear equations represented by the augmented matrix. Then use back-substitution to solve. (Use variables *x*, *y*, and *z*.)

43. $\left[\begin{array}{cc|c} 1 & -2 & 4 \\ 0 & 1 & -3 \end{array}\right]$ **44.** $\left[\begin{array}{cc|c} 1 & 5 & 0 \\ 0 & 1 & -1 \end{array}\right]$

45. $\left[\begin{array}{ccc|c} 1 & -1 & 2 & 4 \\ 0 & 1 & -1 & 2 \\ 0 & 0 & 1 & -2 \end{array}\right]$

46. $\left[\begin{array}{ccc|c} 1 & 2 & -2 & -1 \\ 0 & 1 & 1 & 9 \\ 0 & 0 & 1 & -3 \end{array}\right]$

In Exercises 47–50, an augmented matrix that represents a system of linear equations (in variables *x*, *y*, and *z*) has been reduced using Gauss-Jordan elimination. Write the solution represented by the augmented matrix.

47. $\left[\begin{array}{cc|c} 1 & 0 & 3 \\ 0 & 1 & -4 \end{array}\right]$ **48.** $\left[\begin{array}{cc|c} 1 & 0 & -6 \\ 0 & 1 & 10 \end{array}\right]$

49. $\left[\begin{array}{ccc|c} 1 & 0 & 0 & -4 \\ 0 & 1 & 0 & -10 \\ 0 & 0 & 1 & 4 \end{array}\right]$

50. $\left[\begin{array}{ccc|c} 1 & 0 & 0 & 5 \\ 0 & 1 & 0 & -3 \\ 0 & 0 & 1 & 0 \end{array}\right]$

In Exercises 51–70, use matrices to solve the system of equations (if possible). Use Gaussian elimination with back-substitution or Gauss-Jordan elimination.

51. $\begin{cases} x + 2y = 7 \\ 2x + y = 8 \end{cases}$ **52.** $\begin{cases} 2x + 6y = 16 \\ 2x + 3y = 7 \end{cases}$

53. $\begin{cases} 3x - 2y = -27 \\ x + 3y = 13 \end{cases}$ **54.** $\begin{cases} -x + y = 4 \\ 2x - 4y = -34 \end{cases}$

55. $\begin{cases} -2x + 6y = -22 \\ x + 2y = -9 \end{cases}$

56. $\begin{cases} 5x - 5y = -5 \\ -2x - 3y = 7 \end{cases}$

57. $\begin{cases} -x + 2y = 1.5 \\ 2x - 4y = 3 \end{cases}$

58. $\begin{cases} x - 3y = 5 \\ -2x + 6y = -10 \end{cases}$

59. $\begin{cases} x - 3z = -2 \\ 3x + y - 2z = 5 \\ 2x + 2y + z = 4 \end{cases}$

60. $\begin{cases} 2x - y + 3z = 24 \\ 2y - z = 14 \\ 7x - 5y = 6 \end{cases}$

61. $\begin{cases} -x + y - z = -14 \\ 2x - y + z = 21 \\ 3x + 2y + z = 19 \end{cases}$

62. $\begin{cases} 2x + 2y - z = 2 \\ x - 3y + z = -28 \\ -x + y = 14 \end{cases}$

63. $\begin{cases} x + 2y - 3z = -28 \\ 4y + 2z = 0 \\ -x + y - z = -5 \end{cases}$

64. $\begin{cases} 3x - 2y + z = 15 \\ -x + y + 2z = -10 \\ x - y - 4z = 14 \end{cases}$

65. $\begin{cases} x + y - 5z = 3 \\ x - 2z = 1 \\ 2x - y - z = 0 \end{cases}$

66. $\begin{cases} 2x + 3z = 3 \\ 4x - 3y + 7z = 5 \\ 8x - 9y + 15z = 9 \end{cases}$

67. $\begin{cases} x + 2y + z + 2w = 8 \\ 3x + 7y + 6z + 9w = 26 \end{cases}$

68. $\begin{cases} 4x + 12y - 7z - 20w = 22 \\ 3x + 9y - 5z - 28w = 30 \end{cases}$

69. $\begin{cases} -x + y = -22 \\ 3x + 4y = 4 \\ 4x - 8y = 32 \end{cases}$

70. $\begin{cases} x + 2y = 0 \\ x + y = 6 \\ 3x - 2y = 8 \end{cases}$

In Exercises 71–76, use the matrix capabilities of a graphing utility to reduce the augmented matrix corresponding to the system of equations, and solve the system.

71. $\begin{cases} 3x + 3y + 12z = 6 \\ x + y + 4z = 2 \\ 2x + 5y + 20z = 10 \\ -x + 2y + 8z = 4 \end{cases}$

72. $\begin{cases} 2x + 10y + 2z = 6 \\ x + 5y + 2z = 6 \\ x + 5y + z = 3 \\ -3x - 15y - 3z = -9 \end{cases}$

73. $\begin{cases} 2x + y - z + 2w = -6 \\ 3x + 4y + w = 1 \\ x + 5y + 2z + 6w = -3 \\ 5x + 2y - z - w = 3 \end{cases}$

74. $\begin{cases} x + 2y + 2z + 4w = 11 \\ 3x + 6y + 5z + 12w = 30 \\ x + 3y - 3z + 2w = -5 \\ 6x - y - z + w = -9 \end{cases}$

75. $\begin{cases} x + y + z + w = 0 \\ 2x + 3y + z - 2w = 0 \\ 3x + 5y + z = 0 \end{cases}$

76. $\begin{cases} x + 2y + z + 3w = 0 \\ x - y + w = 0 \\ y - z + 2w = 0 \end{cases}$

In Exercises 77–80, determine whether the two systems of linear equations yield the same solutions. If so, find the solutions using matrices.

77. (a) $\begin{cases} x - 2y + z = -6 \\ y - 5z = 16 \\ z = -3 \end{cases}$ (b) $\begin{cases} x + y - 2z = 6 \\ y + 3z = -8 \\ z = -3 \end{cases}$

78. (a) $\begin{cases} x - 3y + 4z = -11 \\ y - z = -4 \\ z = 2 \end{cases}$

(b) $\begin{cases} x + 4y = -11 \\ y + 3z = 4 \\ z = 2 \end{cases}$

79. (a) $\begin{cases} x - 4y + 5z = 27 \\ y - 7z = -54 \\ z = 8 \end{cases}$ (b) $\begin{cases} x - 6y + z = 15 \\ y + 5z = 42 \\ z = 8 \end{cases}$

80. (a) $\begin{cases} x + 3y - z = 19 \\ y + 6z = -18 \\ z = -4 \end{cases}$

(b) $\begin{cases} x - y + 3z = -15 \\ y - 2z = 14 \\ z = -4 \end{cases}$

81. Use the system $\begin{cases} x + 3y + z = 3 \\ x + 5y + 5z = 1 \\ 2x + 6y + 3z = 8 \end{cases}$ to write two different matrices in row-echelon form that yield the same solutions.

82. ***Electrical Network*** The currents in an electrical network are given by the solution of the system

$$\begin{cases} I_1 - I_2 + I_3 = 0 \\ 3I_1 + 4I_2 = 18 \\ I_2 + 3I_3 = 6 \end{cases}$$

where I_1, I_2, and I_3 are measured in amperes. Solve the system of equations using matrices.

83. ***Partial Fractions*** Use a system of equations to write the partial fraction decomposition of the rational expression. Solve the system using matrices.

$$\frac{4x^2}{(x+1)^2(x-1)} = \frac{A}{x-1} + \frac{B}{x+1} + \frac{C}{(x+1)^2}$$

84. ***Partial Fractions*** Use a system of equations to write the partial fraction decomposition of the rational expression. Solve the system using matrices.

$$\frac{8x^2}{(x-1)^2(x+1)} = \frac{A}{x+1} + \frac{B}{x-1} + \frac{C}{(x-1)^2}$$

85. ***Finance*** A small shoe corporation borrowed \$1,500,000 to expand its line of shoes. Some of the money was borrowed at 7%, some at 8%, and some at 10%. Use a system of equations to determine how much was borrowed at each rate if the annual interest was \$130,500 and the amount borrowed at 10% was 4 times the amount borrowed at 7%. Solve the system using matrices.

86. ***Finance*** A small software corporation borrowed \$500,000 to expand its software line. Some of the money was borrowed at 9%, some at 10%, and some at 12%. Use a system of equations to determine how much was borrowed at each rate if the annual interest was \$52,000 and the amount borrowed at 10% was $2\frac{1}{2}$ times the amount borrowed at 9%. Solve the system using matrices.

In Exercises 87 and 88, use a system of equations to find the specified equation that passes through the points. Solve the system using matrices. Use a graphing utility to verify your results.

87. Parabola: $y = ax^2 + bx + c$

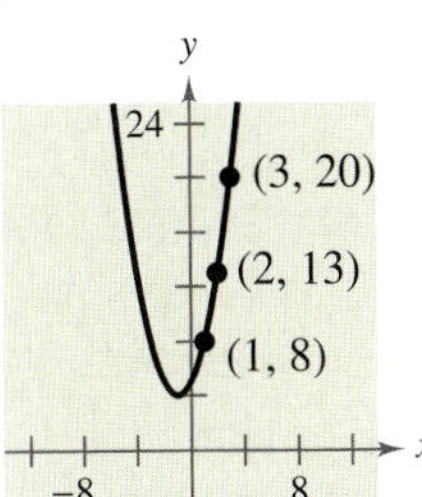

88. Parabola: $y = ax^2 + bx + c$

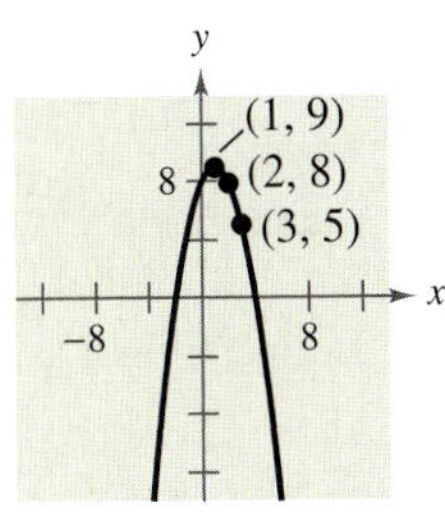

89. ***Mathematical Modeling*** A videotape of the path of a ball thrown by a baseball player was analyzed with a grid covering the TV screen. The tape was paused three times, and the position of the ball was measured each time. The coordinates obtained are shown in the table. (x and y are measured in feet.)

Horizontal distance, x	Height, y
0	5.0
15	9.6
30	12.4

(a) Use a system of equations to find the equation of the parabola $y = ax^2 + bx + c$ that passes through the three points. Solve the system using matrices.

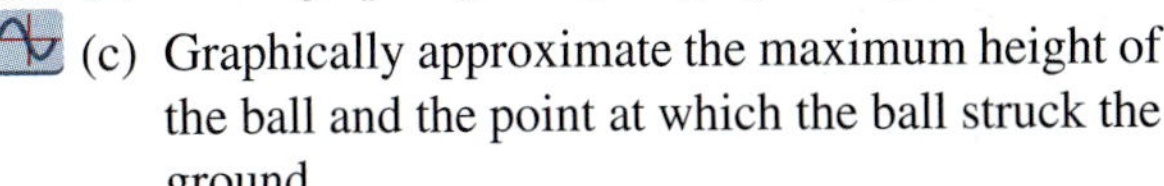

(b) Use a graphing utility to graph the parabola.

(c) Graphically approximate the maximum height of the ball and the point at which the ball struck the ground.

(d) Analytically find the maximum height of the ball and the point at which the ball struck the ground.

(e) Compare your results from parts (c) and (d).

▶ Model It

90. ***Data Analysis*** The table shows the retail sales y (in billions of dollars) for drug stores in the United States for the years 1998 through 2000. (Source: U.S. Department of Commerce and National Association of Chain Drug Stores Economics Department)

Year, t	Retail sales, y
1998	106.7
1999	122.6
2000	134.4

(a) Use a system of equations to find the equation of the parabola $y = at^2 + bt + c$ that passes through the points. Let $t = 8$ represent 1998. Solve the system using matrices.

(b) Use a graphing utility to graph the parabola.

(c) Use the equation in part (a) to estimate the retail sales in 2001. How does this value compare with the actual 2001 sales of \$147.4 billion?

(d) Use the equation in part (a) to estimate y in the year 2004. Is the estimate reasonable? Explain.

Network Analysis **In Exercises 91 and 92, answer the questions about the specified network. (In a network it is assumed that the total flow into each junction is equal to the total flow out of each junction.)**

91. Water flowing through a network of pipes (in thousands of cubic meters per hour) is shown in the figure.

(a) Solve this system using matrices for the water flow represented by x_i, $i = 1, 2, \ldots, 7$.

(b) Find the network flow pattern when $x_6 = x_7 = 0$.

(c) Find the network flow pattern when $x_5 = 1000$ and $x_6 = 0$.

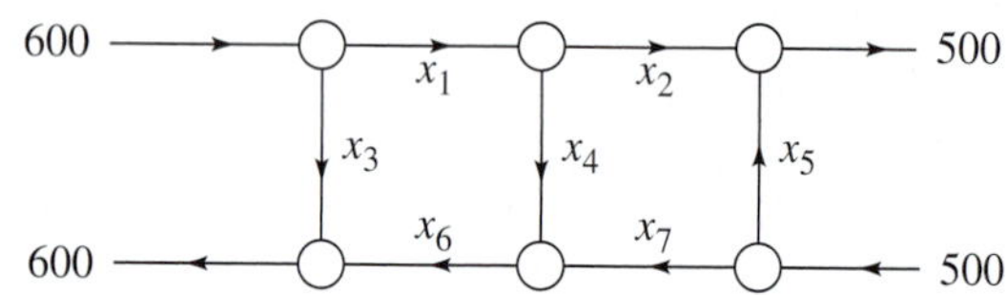

92. The flow of traffic (in vehicles per hour) through a network of streets is shown in the figure.

(a) Solve this system using matrices for the traffic flow represented by x_i, $i = 1, 2, \ldots, 5$.

(b) Find the traffic flow when $x_2 = 200$ and $x_3 = 50$.

(c) Find the traffic flow when $x_2 = 150$ and $x_3 = 0$.

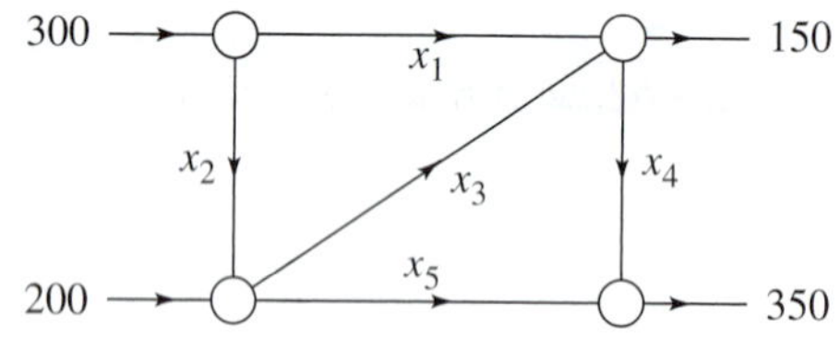

Synthesis

True or False? **In Exercises 93–95, determine whether the statement is true or false. Justify your answer.**

93. $\begin{bmatrix} 5 & 0 & -2 & 7 \\ -1 & 3 & -6 & 0 \end{bmatrix}$ is a 4×2 matrix.

94. The matrix $\begin{bmatrix} 0 & 0 & 0 & 0 \\ 0 & 0 & 1 & -4 \\ 0 & 1 & 0 & 2 \\ 1 & 0 & 0 & 5 \end{bmatrix}$ is in reduced row-echelon form.

95. Gaussian elimination reduces a matrix until a reduced row-echelon form is obtained.

96. ***Think About It*** The augmented matrix represents a system of linear equations (in variables x, y, and z) that has been reduced using Gauss-Jordan elimination. Write a system of equations with nonzero coefficients that is represented by the reduced matrix. (The answer is not unique.)

$$\left[\begin{array}{ccc|c} 1 & 0 & 3 & -2 \\ 0 & 1 & 4 & 1 \\ 0 & 0 & 0 & 0 \end{array}\right]$$

97. ***Think About It***

(a) Describe the row-echelon form of an augmented matrix that corresponds to a system of linear equations that is inconsistent.

(b) Describe the row-echelon form of an augmented matrix that corresponds to a system of linear equations that has an infinite number of solutions.

98. Describe the three elementary row operations that can be performed on an augmented matrix.

99. What is the relationship between the three elementary row operations performed on an augmented matrix and the operations that lead to equivalent systems of equations?

100. ***Writing*** In your own words, describe the difference between a matrix in row-echelon form and a matrix in reduced row-echelon form.

Review

In Exercises 101–106, identify the equation as a line, a circle, a parabola, an ellipse, or a hyperbola.

101. $x^2 + y^2 = 100$

102. $y = -x^2 + 8$

103. $5x - 2y = 8$

104. $\dfrac{x^2}{9} - \dfrac{(y-3)^2}{36} = 1$

105. $\dfrac{(x+4)^2}{9} + \dfrac{(y-6)^2}{16} = 1$

106. $(x+5)^2 + (y-7)^2 = 15$

In Exercises 107–110, sketch the graph of the function. Do not use a graphing utility.

107. $f(x) = 2^{x-1}$

108. $g(x) = 3^{-x+2}$

109. $h(x) = \ln(x - 1)$

110. $f(x) = 3 + \ln x$

10.2 Operations with Matrices

▶ **What you should learn**

- How to decide whether two matrices are equal
- How to add and subtract matrices and multiply matrices by real numbers
- How to multiply two matrices
- How to use matrix operations to model and solve real-life problems

▶ **Why you should learn it**

Matrix operations can be used to model and solve real-life problems. For instance, in Exercise 61 on page 759, matrix multiplication is used to analyze the profit a fruit farmer makes on two fruit crops.

Kelly-Mooney Photography/Corbis

Equality of Matrices

In Section 10.1, you used matrices to solve systems of linear equations. Matrices, however, can do much more than this. There is a rich mathematical theory of matrices, and its applications are numerous. This section and the next two introduce some fundamentals of matrix theory. It is standard mathematical convention to represent matrices in any of the following three ways.

1. A matrix can be denoted by an uppercase letter such as A, B, or C.
2. A matrix can be denoted by a representative element enclosed in brackets, such as $[a_{ij}]$, $[b_{ij}]$, or $[c_{ij}]$.
3. A matrix can be denoted by a rectangular array of numbers such as

$$A = [a_{ij}] = \begin{bmatrix} a_{11} & a_{12} & a_{13} & \cdots & a_{1n} \\ a_{21} & a_{22} & a_{23} & \cdots & a_{2n} \\ a_{31} & a_{32} & a_{33} & \cdots & a_{3n} \\ \vdots & \vdots & \vdots & & \vdots \\ a_{m1} & a_{m2} & a_{m3} & \cdots & a_{mn} \end{bmatrix}.$$

Two matrices $A = [a_{ij}]$ and $B = [b_{ij}]$ are equal if they have the same order $(m \times n)$ and $a_{ij} = b_{ij}$ for $1 \le i \le m$ and $1 \le j \le n$. In other words, two matrices are equal if their corresponding entries are equal.

Example 1 ▶ **Equality of Matrices** 

Solve for a_{11}, a_{12}, a_{21}, and a_{22} in the following matrix equation.

$$\begin{bmatrix} a_{11} & a_{12} \\ a_{21} & a_{22} \end{bmatrix} = \begin{bmatrix} 2 & -1 \\ -3 & 0 \end{bmatrix}$$

Solution

Because two matrices are equal only if their corresponding entries are equal, you can conclude that

$$a_{11} = 2, \quad a_{12} = -1, \quad a_{21} = -3, \quad \text{and} \quad a_{22} = 0.$$

Be sure you see that for two matrices to be equal, they must have the same order *and* their corresponding entries must be equal. For instance,

$$\begin{bmatrix} 2 & -1 \\ \sqrt{4} & \frac{1}{2} \end{bmatrix} = \begin{bmatrix} 2 & -1 \\ 2 & 0.5 \end{bmatrix}$$

but

$$\begin{bmatrix} 2 & -1 \\ 3 & 4 \\ 0 & 0 \end{bmatrix} \neq \begin{bmatrix} 2 & -1 \\ 3 & 4 \end{bmatrix}.$$

The Granger Collection

Historical Note

Arthur Cayley (1821–1895), a British mathematician, invented matrices around 1858. Cayley was a Cambridge University graduate and a lawyer by profession. His groundbreaking work on matrices was begun as he studied the theory of transformations. Cayley also was instrumental in the development of determinants. Cayley and two American mathematicians, Benjamin Peirce (1809–1880) and his son Charles S. Peirce (1839–1914), are credited with developing "matrix algebra."

Matrix Addition and Scalar Multiplication

In this section, three basic matrix operations will be covered. The first two are matrix addition and scalar multiplication. With matrix addition, you can add two matrices (of the same order) by adding their corresponding entries.

Definition of Matrix Addition

If $A = [a_{ij}]$ and $B = [b_{ij}]$ are matrices of order $m \times n$, their sum is the $m \times n$ matrix given by

$$A + B = [a_{ij} + b_{ij}].$$

The sum of two matrices of different orders is undefined.

Example 2 ▶ Addition of Matrices

a. $\begin{bmatrix} -1 & 2 \\ 0 & 1 \end{bmatrix} + \begin{bmatrix} 1 & 3 \\ -1 & 2 \end{bmatrix} = \begin{bmatrix} -1+1 & 2+3 \\ 0-1 & 1+2 \end{bmatrix}$

$= \begin{bmatrix} 0 & 5 \\ -1 & 3 \end{bmatrix}$

b. $\begin{bmatrix} 0 & 1 & -2 \\ 1 & 2 & 3 \end{bmatrix} + \begin{bmatrix} 0 & 0 & 0 \\ 0 & 0 & 0 \end{bmatrix} = \begin{bmatrix} 0 & 1 & -2 \\ 1 & 2 & 3 \end{bmatrix}$

c. $\begin{bmatrix} 1 \\ -3 \\ -2 \end{bmatrix} + \begin{bmatrix} -1 \\ 3 \\ 2 \end{bmatrix} = \begin{bmatrix} 0 \\ 0 \\ 0 \end{bmatrix}$

d. The sum of

$$A = \begin{bmatrix} 2 & 1 & 0 \\ 4 & 0 & -1 \\ 3 & -2 & 2 \end{bmatrix} \quad \text{and}$$

$$B = \begin{bmatrix} 0 & 1 \\ -1 & 3 \\ 2 & 4 \end{bmatrix}$$

is undefined because A and B have different orders.

In operations with matrices, numbers are usually referred to as **scalars.** In this text, scalars will always be real numbers. You can multiply a matrix A by a scalar c by multiplying each entry in A by c.

Definition of Scalar Multiplication

If $A = [a_{ij}]$ is an $m \times n$ matrix and c is a scalar, the **scalar multiple** of A by c is the $m \times n$ matrix given by

$$cA = [ca_{ij}].$$

The symbol $-A$ represents the negation of A, or the scalar product $(-1)A$. Moreover, if A and B are of the same order, then $A - B$ represents the sum of A and $(-1)B$. That is,

$$A - B = A + (-1)B.$$ Subtraction of matrices

Example 3 ▶ Scalar Multiplication and Matrix Subtraction 

For the following matrices, find (a) $3A$, (b) $-B$, and (c) $3A - B$.

$$A = \begin{bmatrix} 2 & 2 & 4 \\ -3 & 0 & -1 \\ 2 & 1 & 2 \end{bmatrix} \quad \text{and} \quad B = \begin{bmatrix} 2 & 0 & 0 \\ 1 & -4 & 3 \\ -1 & 3 & 2 \end{bmatrix}$$

Solution

a. $$3A = 3\begin{bmatrix} 2 & 2 & 4 \\ -3 & 0 & -1 \\ 2 & 1 & 2 \end{bmatrix}$$ Scalar multiplication

$$= \begin{bmatrix} 3(2) & 3(2) & 3(4) \\ 3(-3) & 3(0) & 3(-1) \\ 3(2) & 3(1) & 3(2) \end{bmatrix}$$ Multiply each entry by 3.

$$= \begin{bmatrix} 6 & 6 & 12 \\ -9 & 0 & -3 \\ 6 & 3 & 6 \end{bmatrix}$$ Simplify.

b. $$-B = (-1)\begin{bmatrix} 2 & 0 & 0 \\ 1 & -4 & 3 \\ -1 & 3 & 2 \end{bmatrix}$$ Definition of negation

$$= \begin{bmatrix} -2 & 0 & 0 \\ -1 & 4 & -3 \\ 1 & -3 & -2 \end{bmatrix}$$ Multiply each entry by -1.

c. $$3A - B = \begin{bmatrix} 6 & 6 & 12 \\ -9 & 0 & -3 \\ 6 & 3 & 6 \end{bmatrix} - \begin{bmatrix} 2 & 0 & 0 \\ 1 & -4 & 3 \\ -1 & 3 & 2 \end{bmatrix}$$ Matrix subtraction

$$= \begin{bmatrix} 4 & 6 & 12 \\ -10 & 4 & -6 \\ 7 & 0 & 4 \end{bmatrix}$$ Subtract corresponding entries.

STUDY TIP

The order of operations for matrix expressions is similar to that for real numbers. In particular, you perform scalar multiplication before matrix addition and subtraction, as shown in Example 3(c).

It is often convenient to rewrite the scalar multiple cA by factoring c out of every entry in the matrix. For instance, in the following example, the scalar $\frac{1}{2}$ has been factored out of the matrix.

$$\begin{bmatrix} \frac{1}{2} & -\frac{3}{2} \\ \frac{5}{2} & \frac{1}{2} \end{bmatrix} = \begin{bmatrix} \frac{1}{2}(1) & \frac{1}{2}(-3) \\ \frac{1}{2}(5) & \frac{1}{2}(1) \end{bmatrix}$$

$$= \frac{1}{2}\begin{bmatrix} 1 & -3 \\ 5 & 1 \end{bmatrix}$$

The properties of matrix addition and scalar multiplication are similar to those of addition and multiplication of real numbers.

Properties of Matrix Addition and Scalar Multiplication

Let A, B, and C be $m \times n$ matrices and let c and d be scalars.

1. $A + B = B + A$ — Commutative Property of Matrix Addition
2. $A + (B + C) = (A + B) + C$ — Associative Property of Matrix Addition
3. $(cd)A = c(dA)$ — Associative Property of Scalar Multiplication
4. $1A = A$ — Scalar Identity Property
5. $c(A + B) = cA + cB$ — Distributive Property
6. $(c + d)A = cA + dA$ — Distributive Property

Note that the Associative Property of Matrix Addition allows you to write expressions such as $A + B + C$ without ambiguity because the same sum occurs no matter how the matrices are grouped. In other words, you obtain the same sum whether you group $A + B + C$ as $(A + B) + C$ or as $A + (B + C)$. This same reasoning applies to sums of four or more matrices.

Example 4 ▶ Addition of More than Two Matrices

By adding corresponding entries, you obtain the following sum of four matrices.

$$\begin{bmatrix} 1 \\ 2 \\ -3 \end{bmatrix} + \begin{bmatrix} -1 \\ -1 \\ 2 \end{bmatrix} + \begin{bmatrix} 0 \\ 1 \\ 4 \end{bmatrix} + \begin{bmatrix} 2 \\ -3 \\ -2 \end{bmatrix} = \begin{bmatrix} 2 \\ -1 \\ 1 \end{bmatrix}$$

Example 5 ▶ Using the Distributive Property

$$3\left(\begin{bmatrix} -2 & 0 \\ 4 & 1 \end{bmatrix} + \begin{bmatrix} 4 & -2 \\ 3 & 7 \end{bmatrix}\right) = 3\begin{bmatrix} -2 & 0 \\ 4 & 1 \end{bmatrix} + 3\begin{bmatrix} 4 & -2 \\ 3 & 7 \end{bmatrix}$$

$$= \begin{bmatrix} -6 & 0 \\ 12 & 3 \end{bmatrix} + \begin{bmatrix} 12 & -6 \\ 9 & 21 \end{bmatrix}$$

$$= \begin{bmatrix} 6 & -6 \\ 21 & 24 \end{bmatrix}$$

Technology

Most graphing utilities can add and subtract matrices and multiply matrices by scalars. Try using a graphing utility to find the sum of the matrices

$$A = \begin{bmatrix} 2 & -3 \\ -1 & 0 \end{bmatrix}$$

and

$$B = \begin{bmatrix} -1 & 4 \\ 2 & -5 \end{bmatrix}.$$

In Example 5, you could add the two matrices first and then multiply the matrix by 3, as follows.

$$3\left(\begin{bmatrix} -2 & 0 \\ 4 & 1 \end{bmatrix} + \begin{bmatrix} 4 & -2 \\ 3 & 7 \end{bmatrix}\right) = 3\begin{bmatrix} 2 & -2 \\ 7 & 8 \end{bmatrix}$$

$$= \begin{bmatrix} 6 & -6 \\ 21 & 24 \end{bmatrix}$$

Note that the result is the same.

One important property of addition of real numbers is that the number 0 is the additive identity. That is, $c + 0 = c$ for any real number c. For matrices, a similar property holds. That is, if A is an $m \times n$ matrix and O is the $m \times n$ **zero matrix** consisting entirely of zeros, then $A + O = A$.

In other words, O is the **additive identity** for the set of all $m \times n$ matrices. For example, the following matrices are the additive identities for the set of all 2×3 and 2×2 matrices.

$$O = \underbrace{\begin{bmatrix} 0 & 0 & 0 \\ 0 & 0 & 0 \end{bmatrix}}_{2 \times 3 \text{ zero matrix}} \quad \text{and} \quad O = \underbrace{\begin{bmatrix} 0 & 0 \\ 0 & 0 \end{bmatrix}}_{2 \times 2 \text{ zero matrix}}$$

The algebra of real numbers and the algebra of matrices have many similarities. For example, compare the following solutions.

Real Numbers *(Solve for x.)*	*$m \times n$ Matrices* *(Solve for X.)*
$x + a = b$	$X + A = B$
$x + a + (-a) = b + (-a)$	$X + A + (-A) = B + (-A)$
$x + 0 = b - a$	$X + O = B - A$
$x = b - a$	$X = B - A$

The algebra of real numbers and the algebra of matrices also have important differences, which will be discussed later.

Example 6 ▶ Solving a Matrix Equation

Solve for X in the equation $3X + A = B$, where

$$A = \begin{bmatrix} 1 & -2 \\ 0 & 3 \end{bmatrix} \quad \text{and} \quad B = \begin{bmatrix} -3 & 4 \\ 2 & 1 \end{bmatrix}.$$

Solution

Begin by solving the equation for X to obtain

$$3X = B - A$$

$$X = \frac{1}{3}(B - A).$$

Now, using the matrices A and B, you have

$$X = \frac{1}{3}\left(\begin{bmatrix} -3 & 4 \\ 2 & 1 \end{bmatrix} - \begin{bmatrix} 1 & -2 \\ 0 & 3 \end{bmatrix}\right) \qquad \text{Substitute the matrices.}$$

$$= \frac{1}{3}\begin{bmatrix} -4 & 6 \\ 2 & -2 \end{bmatrix} \qquad \text{Subtract matrix } A \text{ from matrix } B.$$

$$= \begin{bmatrix} -\frac{4}{3} & 2 \\ \frac{2}{3} & -\frac{2}{3} \end{bmatrix}. \qquad \text{Multiply the matrix by } \tfrac{1}{3}.$$

Matrix Multiplication

The third basic matrix operation is **matrix multiplication.** At first glance, the definition may seem unusual. You will see later, however, that this definition of the product of two matrices has many practical applications.

Definition of Matrix Multiplication

If $A = [a_{ij}]$ is an $m \times n$ matrix and $B = [b_{ij}]$ is an $n \times p$ matrix, the product AB is an $m \times p$ matrix

$$AB = [c_{ij}]$$

where $c_{ij} = a_{i1}b_{1j} + a_{i2}b_{2j} + a_{i3}b_{3j} + \cdot \cdot \cdot + a_{in}b_{nj}$.

The definition of matrix multiplication indicates a *row-by-column* multiplication, where the entry in the ith row and jth column of the product AB is obtained by multiplying the entries in the ith row of A by the corresponding entries in the jth column of B and then adding the results. The general pattern for matrix multiplication is as follows.

$$\begin{bmatrix} a_{11} & a_{12} & a_{13} & \cdots & a_{1n} \\ a_{21} & a_{22} & a_{23} & \cdots & a_{2n} \\ a_{31} & a_{32} & a_{33} & \cdots & a_{3n} \\ \vdots & \vdots & \vdots & & \vdots \\ a_{i1} & a_{i2} & a_{i3} & \cdots & a_{in} \\ \vdots & \vdots & \vdots & & \vdots \\ a_{m1} & a_{m2} & a_{m3} & \cdots & a_{mn} \end{bmatrix} \begin{bmatrix} b_{11} & b_{12} & \cdots & b_{1j} & \cdots & b_{1p} \\ b_{21} & b_{22} & \cdots & b_{2j} & \cdots & b_{2p} \\ b_{31} & b_{32} & \cdots & b_{3j} & \cdots & b_{3p} \\ \vdots & \vdots & & \vdots & & \vdots \\ b_{n1} & b_{n2} & \cdots & b_{nj} & \cdots & b_{np} \end{bmatrix} = \begin{bmatrix} c_{11} & c_{12} & \cdots & c_{1j} & \cdots & c_{1p} \\ c_{21} & c_{22} & \cdots & c_{2j} & \cdots & c_{2p} \\ \vdots & \vdots & & \vdots & & \vdots \\ c_{i1} & c_{i2} & \cdots & c_{ij} & \cdots & c_{ip} \\ \vdots & \vdots & & \vdots & & \vdots \\ c_{m1} & c_{m2} & \cdots & c_{mj} & \cdots & c_{mp} \end{bmatrix}$$

$$a_{i1}b_{1j} + a_{i2}b_{2j} + a_{i3}b_{3j} + \cdot \cdot \cdot + a_{in}b_{nj} = c_{ij}$$

Technology

Some graphing utilities are able to add, subtract, and multiply matrices. If you have such a graphing utility, enter the matrices

$$A = \begin{bmatrix} 1 & 2 & 3 \\ 2 & -5 & 1 \end{bmatrix} \text{ and }$$

$$B = \begin{bmatrix} -3 & 2 & 1 \\ 4 & -2 & 0 \\ 1 & 2 & 3 \end{bmatrix}$$

and find their product *AB*. You should obtain the following.

$$\begin{bmatrix} 8 & 4 & 10 \\ -25 & 16 & 5 \end{bmatrix}$$

Example 7 ▶ Finding the Product of Two Matrices 

First, note that the product AB is defined because the number of columns of A is equal to the number of rows of B. Moreover, the product AB has order 3×2. To find the entries of the product, multiply each row of A by each column of B, as follows.

$$AB = \begin{bmatrix} -1 & 3 \\ 4 & -2 \\ 5 & 0 \end{bmatrix} \begin{bmatrix} -3 & 2 \\ -4 & 1 \end{bmatrix}$$

$$= \begin{bmatrix} (-1)(-3) + (3)(-4) & (-1)(2) + (3)(1) \\ (4)(-3) + (-2)(-4) & (4)(2) + (-2)(1) \\ (5)(-3) + (0)(-4) & (5)(2) + (0)(1) \end{bmatrix}$$

$$= \begin{bmatrix} -9 & 1 \\ -4 & 6 \\ -15 & 10 \end{bmatrix}$$

Be sure you understand that for the product of two matrices to be defined, the number of *columns* of the first matrix must equal the number of *rows* of the second matrix. That is, the middle two indices must be the same. The outside two indices give the order of the product, as shown below.

$$\underset{m \times n}{A} \quad \times \quad \underset{n \times p}{B} \quad = \quad \underset{m \times p}{AB}$$

Equal (the middle indices n and n)

Order of AB (the outside indices m and p)

Example 8 ▶ Finding the Product of Two Matrices 

Find the product AB where

$$A = \begin{bmatrix} 1 & 0 & 3 \\ 2 & -1 & -2 \end{bmatrix} \quad \text{and} \quad B = \begin{bmatrix} -2 & 4 \\ 1 & 0 \\ -1 & 1 \end{bmatrix}.$$

Solution

Note that the order of A is 2×3 and the order of B is 3×2. So, the product AB has order 2×2.

$$AB = \begin{bmatrix} 1 & 0 & 3 \\ 2 & -1 & -2 \end{bmatrix}\begin{bmatrix} -2 & 4 \\ 1 & 0 \\ -1 & 1 \end{bmatrix}$$

$$= \begin{bmatrix} 1(-2) + 0(1) + 3(-1) & 1(4) + 0(0) + 3(1) \\ 2(-2) + (-1)(1) + (-2)(-1) & 2(4) + (-1)(0) + (-2)(1) \end{bmatrix}$$

$$= \begin{bmatrix} -5 & 7 \\ -3 & 6 \end{bmatrix}$$

Example 9 ▶ Patterns in Matrix Multiplication

a. $\underset{2 \times 2}{\begin{bmatrix} 3 & 4 \\ -2 & 5 \end{bmatrix}}\underset{2 \times 2}{\begin{bmatrix} 1 & 0 \\ 0 & 1 \end{bmatrix}} = \underset{2 \times 2}{\begin{bmatrix} 3 & 4 \\ -2 & 5 \end{bmatrix}}$

b. $\underset{3 \times 3}{\begin{bmatrix} 6 & 2 & 0 \\ 3 & -1 & 2 \\ 1 & 4 & 6 \end{bmatrix}}\underset{3 \times 1}{\begin{bmatrix} 1 \\ 2 \\ -3 \end{bmatrix}} = \underset{3 \times 1}{\begin{bmatrix} 10 \\ -5 \\ -9 \end{bmatrix}}$

c. The product AB for the following matrices is not defined.

$$A = \underset{3 \times 2}{\begin{bmatrix} -2 & 1 \\ 1 & -3 \\ 1 & 4 \end{bmatrix}} \quad \text{and} \quad B = \underset{3 \times 4}{\begin{bmatrix} -2 & 3 & 1 & 4 \\ 0 & 1 & -1 & 2 \\ 2 & -1 & 0 & 1 \end{bmatrix}}$$

Exploration

Use a graphing utility to multiply the matrices

$$A = \begin{bmatrix} 1 & 2 \\ 3 & 4 \end{bmatrix} \quad \text{and}$$

$$B = \begin{bmatrix} 0 & 1 \\ 2 & 3 \end{bmatrix}.$$

Do you obtain the same result for the product AB as for the product BA? What does this tell you about matrix multiplication and commutativity?

Example 10 ▶ Patterns in Matrix Multiplication

a. $\begin{bmatrix} 1 & -2 & -3 \end{bmatrix}\begin{bmatrix} 2 \\ -1 \\ 1 \end{bmatrix} = \begin{bmatrix} 1 \end{bmatrix}$

$1 \times 3 \qquad 3 \times 1 \qquad 1 \times 1$

b. $\begin{bmatrix} 2 \\ -1 \\ 1 \end{bmatrix}\begin{bmatrix} 1 & -2 & -3 \end{bmatrix} = \begin{bmatrix} 2 & -4 & -6 \\ -1 & 2 & 3 \\ 1 & -2 & -3 \end{bmatrix}$

$3 \times 1 \qquad 1 \times 3 \qquad 3 \times 3$

In Example 10, note that the two products are different. Even if AB and BA are defined, matrix multiplication is not, in general, commutative. That is, for most matrices, $AB \neq BA$.

Properties of Matrix Multiplication

Let A, B, and C be matrices and let c be a scalar.

1. $A(BC) = (AB)C$ — Associative Property of Multiplication

2. $A(B + C) = AB + AC$ — Distributive Property

3. $(A + B)C = AC + BC$ — Distributive Property

4. $c(AB) = (cA)B = A(cB)$ — Associative Property of Scalar Multiplication

Definition of Identity Matrix

The $n \times n$ matrix that consists of 1's on its main diagonal and 0's elsewhere is called the **identity matrix of order n** and is denoted by

$$I_n = \begin{bmatrix} 1 & 0 & 0 & \cdots & 0 \\ 0 & 1 & 0 & \cdots & 0 \\ 0 & 0 & 1 & \cdots & 0 \\ \vdots & \vdots & \vdots & & \vdots \\ 0 & 0 & 0 & \cdots & 1 \end{bmatrix}. \qquad \text{Identity matrix}$$

Note that an identity matrix must be *square*. When the order is understood to be n, you can denote I_n simply by I.

If A is an $n \times n$ matrix, the identity matrix has the property that $AI_n = A$ and $I_nA = A$. For example,

$$\begin{bmatrix} 3 & -2 & 5 \\ 1 & 0 & 4 \\ -1 & 2 & -3 \end{bmatrix}\begin{bmatrix} 1 & 0 & 0 \\ 0 & 1 & 0 \\ 0 & 0 & 1 \end{bmatrix} = \begin{bmatrix} 3 & -2 & 5 \\ 1 & 0 & 4 \\ -1 & 2 & -3 \end{bmatrix}$$

and

$$\begin{bmatrix} 1 & 0 & 0 \\ 0 & 1 & 0 \\ 0 & 0 & 1 \end{bmatrix}\begin{bmatrix} 3 & -2 & 5 \\ 1 & 0 & 4 \\ -1 & 2 & -3 \end{bmatrix} = \begin{bmatrix} 3 & -2 & 5 \\ 1 & 0 & 4 \\ -1 & 2 & -3 \end{bmatrix}.$$

Applications

Matrix multiplication can be used to represent a system of linear equations. Note how the system

$$\begin{cases} a_{11}x_1 + a_{12}x_2 + a_{13}x_3 = b_1 \\ a_{21}x_1 + a_{22}x_2 + a_{23}x_3 = b_2 \\ a_{31}x_1 + a_{32}x_2 + a_{33}x_3 = b_3 \end{cases}$$

can be written as the matrix equation $AX = B$, where A is the *coefficient matrix* of the system, and X and B are column matrices.

$$\begin{bmatrix} a_{11} & a_{12} & a_{13} \\ a_{21} & a_{22} & a_{23} \\ a_{31} & a_{32} & a_{33} \end{bmatrix} \begin{bmatrix} x_1 \\ x_2 \\ x_3 \end{bmatrix} = \begin{bmatrix} b_1 \\ b_2 \\ b_3 \end{bmatrix}$$

$$A \qquad\qquad \times \; X \; = \; B$$

Example 11 ▶ Solving a System of Linear Equations 

Consider the following system of linear equations.

$$\begin{cases} x_1 - 2x_2 + x_3 = -4 \\ x_2 + 2x_3 = 4 \\ 2x_1 + 3x_2 - 2x_3 = 2 \end{cases}$$

a. Write this system as a matrix equation, $AX = B$.

b. Use Gauss-Jordan elimination on the augmented matrix $[A \vdots B]$ to solve for the matrix X.

> **STUDY TIP**
>
> The notation $[A \vdots B]$ represents the augmented matrix formed when matrix B is adjoined to matrix A. The notation $[I \vdots X]$ represents the reduced row-echelon form of the augmented matrix that yields the *solution* to the system.

Solution

a. In matrix form, $AX = B$, the system can be written as follows.

$$\begin{bmatrix} 1 & -2 & 1 \\ 0 & 1 & 2 \\ 2 & 3 & -2 \end{bmatrix} \begin{bmatrix} x_1 \\ x_2 \\ x_3 \end{bmatrix} = \begin{bmatrix} -4 \\ 4 \\ 2 \end{bmatrix}$$

b. The augmented matrix is formed by adjoining matrix B to matrix A.

$$[A \vdots B] = \begin{bmatrix} 1 & -2 & 1 & \vdots & -4 \\ 0 & 1 & 2 & \vdots & 4 \\ 2 & 3 & -2 & \vdots & 2 \end{bmatrix}$$

Using Gauss-Jordan elimination, you can rewrite this equation as

$$[I \vdots X] = \begin{bmatrix} 1 & 0 & 0 & \vdots & -1 \\ 0 & 1 & 0 & \vdots & 2 \\ 0 & 0 & 1 & \vdots & 1 \end{bmatrix}.$$

So, the solution of the system of linear equations is $x_1 = -1$, $x_2 = 2$, and $x_3 = 1$, and the solution of the matrix equation is

$$X = \begin{bmatrix} x_1 \\ x_2 \\ x_3 \end{bmatrix} = \begin{bmatrix} -1 \\ 2 \\ 1 \end{bmatrix}.$$

Example 12 ▶ Softball Team Expenses 

Two softball teams submit equipment lists to their sponsors.

	Women's Team	*Men's Team*
Bats	12	15
Balls	45	38
Gloves	15	17

Each bat costs \$80, each ball costs \$6, and each glove costs \$60. Use matrices to find the total cost of equipment for each team.

Solution

The equipment lists E and the costs per item C can be written in matrix form as

$$E = \begin{bmatrix} 12 & 15 \\ 45 & 38 \\ 15 & 17 \end{bmatrix}$$

and

$$C = [80 \quad 6 \quad 60].$$

The total cost of equipment for each team is given by the product

$$\begin{aligned} CE &= [80 \quad 6 \quad 60]\begin{bmatrix} 12 & 15 \\ 45 & 38 \\ 15 & 17 \end{bmatrix} \\ &= [80(12) + 6(45) + 60(15) \quad 80(15) + 6(38) + 60(17)] \\ &= [2130 \quad 2448]. \end{aligned}$$

So, the total cost of equipment for the women's team is \$2130 and the total cost of equipment for the men's team is \$2448.

Writing ABOUT MATHEMATICS

Problem Posing Write a matrix multiplication application problem that uses the matrix

$$A = \begin{bmatrix} 20 & 42 & 33 \\ 17 & 30 & 50 \end{bmatrix}.$$

Exchange problems with another student in your class. Form the matrices that represent the problem, and solve the problem. Interpret your solution in the context of the problem. Check with the creator of the problem to see if you are correct. Discuss other ways to represent and/or approach the problem.

10.2 Exercises

In Exercises 1–4, find *x* and *y*.

1. $\begin{bmatrix} x & -2 \\ 7 & y \end{bmatrix} = \begin{bmatrix} -4 & -2 \\ 7 & 22 \end{bmatrix}$

2. $\begin{bmatrix} -5 & x \\ y & 8 \end{bmatrix} = \begin{bmatrix} -5 & 13 \\ 12 & 8 \end{bmatrix}$

3. $\begin{bmatrix} 16 & 4 & 5 & 4 \\ -3 & 13 & 15 & 6 \\ 0 & 2 & 4 & 0 \end{bmatrix} = \begin{bmatrix} 16 & 4 & 2x+1 & 4 \\ -3 & 13 & 15 & 3x \\ 0 & 2 & 3y-5 & 0 \end{bmatrix}$

4. $\begin{bmatrix} x+2 & 8 & -3 \\ 1 & 2y & 2x \\ 7 & -2 & y+2 \end{bmatrix} = \begin{bmatrix} 2x+6 & 8 & -3 \\ 1 & 18 & -8 \\ 7 & -2 & 11 \end{bmatrix}$

In Exercises 5–12, if possible, find (a) $A + B$, (b) $A - B$, (c) $3A$, and (d) $3A - 2B$.

5. $A = \begin{bmatrix} 1 & -1 \\ 2 & -1 \end{bmatrix}, \quad B = \begin{bmatrix} 2 & -1 \\ -1 & 8 \end{bmatrix}$

6. $A = \begin{bmatrix} 1 & 2 \\ 2 & 1 \end{bmatrix}, \quad B = \begin{bmatrix} -3 & -2 \\ 4 & 2 \end{bmatrix}$

7. $A = \begin{bmatrix} 6 & -1 \\ 2 & 4 \\ -3 & 5 \end{bmatrix}, \quad B = \begin{bmatrix} 1 & 4 \\ -1 & 5 \\ 1 & 10 \end{bmatrix}$

8. $A = \begin{bmatrix} 2 & 1 & 1 \\ -1 & -1 & 4 \end{bmatrix}, \quad B = \begin{bmatrix} 2 & -3 & 4 \\ -3 & 1 & -2 \end{bmatrix}$

9. $A = \begin{bmatrix} 2 & 2 & -1 & 0 & 1 \\ 1 & 1 & -2 & 0 & -1 \end{bmatrix},$

$B = \begin{bmatrix} 1 & 1 & -1 & 1 & 0 \\ -3 & 4 & 9 & -6 & -7 \end{bmatrix}$

10. $A = \begin{bmatrix} -1 & 4 & 0 \\ 3 & -2 & 2 \\ 5 & 4 & -1 \\ 0 & 8 & -6 \\ -4 & -1 & 0 \end{bmatrix}, \quad B = \begin{bmatrix} -3 & 5 & 1 \\ 2 & -4 & -7 \\ 10 & -9 & -1 \\ 3 & 2 & -4 \\ 0 & 1 & -2 \end{bmatrix}$

11. $A = \begin{bmatrix} 6 & 0 & 3 \\ -1 & -4 & 0 \end{bmatrix}, \quad B = \begin{bmatrix} 8 & -1 \\ 4 & -3 \end{bmatrix}$

12. $A = \begin{bmatrix} 3 \\ 2 \\ -1 \end{bmatrix}, \quad B = \begin{bmatrix} -4 & 6 & 2 \end{bmatrix}$

In Exercises 13–18, evaluate the expression.

13. $\begin{bmatrix} -5 & 0 \\ 3 & -6 \end{bmatrix} + \begin{bmatrix} 7 & 1 \\ -2 & -1 \end{bmatrix} + \begin{bmatrix} -10 & -8 \\ 14 & 6 \end{bmatrix}$

14. $\begin{bmatrix} 6 & 8 \\ -1 & 0 \end{bmatrix} + \begin{bmatrix} 0 & 5 \\ -3 & -1 \end{bmatrix} + \begin{bmatrix} -11 & -7 \\ 2 & -1 \end{bmatrix}$

15. $4\left(\begin{bmatrix} -4 & 0 & 1 \\ 0 & 2 & 3 \end{bmatrix} - \begin{bmatrix} 2 & 1 & -2 \\ 3 & -6 & 0 \end{bmatrix}\right)$

16. $\frac{1}{2}(\begin{bmatrix} 5 & -2 & 4 & 0 \end{bmatrix} + \begin{bmatrix} 14 & 6 & -18 & 9 \end{bmatrix})$

17. $-3\left(\begin{bmatrix} 0 & -3 \\ 7 & 2 \end{bmatrix} + \begin{bmatrix} -6 & 3 \\ 8 & 1 \end{bmatrix}\right) - 2\begin{bmatrix} 4 & -4 \\ 7 & -9 \end{bmatrix}$

18. $-1\begin{bmatrix} 4 & 11 \\ -2 & -1 \\ 9 & 3 \end{bmatrix} + \frac{1}{6}\left(\begin{bmatrix} -5 & -1 \\ 3 & 4 \\ 0 & 13 \end{bmatrix} + \begin{bmatrix} 7 & 5 \\ -9 & -1 \\ 6 & -1 \end{bmatrix}\right)$

In Exercises 19–22, use the matrix capabilities of a graphing utility to evaluate each expression. Round your results to three decimal places, if necessary.

19. $\frac{3}{7}\begin{bmatrix} 2 & 5 \\ -1 & -4 \end{bmatrix} + 6\begin{bmatrix} -3 & 0 \\ 2 & 2 \end{bmatrix}$

20. $55\left(\begin{bmatrix} 14 & -11 \\ -22 & 19 \end{bmatrix} + \begin{bmatrix} -22 & 20 \\ 13 & 6 \end{bmatrix}\right)$

21. $-\begin{bmatrix} 3.211 & 6.829 \\ -1.004 & 4.914 \\ 0.055 & -3.889 \end{bmatrix} - \begin{bmatrix} -1.630 & -3.090 \\ 5.256 & 8.335 \\ -9.768 & 4.251 \end{bmatrix}$

22. $-12\left(\begin{bmatrix} 6 & 20 \\ 1 & -9 \\ -2 & 5 \end{bmatrix} + \begin{bmatrix} 14 & -15 \\ -8 & -6 \\ 7 & 0 \end{bmatrix} + \begin{bmatrix} -31 & -19 \\ 16 & 10 \\ 24 & -10 \end{bmatrix}\right)$

In Exercises 23–26, solve for X when

$$A = \begin{bmatrix} -2 & -1 \\ 1 & 0 \\ 3 & -4 \end{bmatrix} \quad \text{and} \quad B = \begin{bmatrix} 0 & 3 \\ 2 & 0 \\ -4 & -1 \end{bmatrix}.$$

23. $X = 3A - 2B$

24. $2X = 2A - B$

25. $2X + 3A = B$

26. $2A + 4B = -2X$

In Exercises 27–34, find *AB*, if possible.

27. $A = \begin{bmatrix} 2 & 1 \\ -3 & 4 \\ 1 & 6 \end{bmatrix}, \quad B = \begin{bmatrix} 0 & -1 & 0 \\ 4 & 0 & 2 \\ 8 & -1 & 7 \end{bmatrix}$

28. $A = \begin{bmatrix} 1 & 0 & 3 & -2 \\ 6 & 13 & 8 & -17 \end{bmatrix}, \quad B = \begin{bmatrix} 1 & 6 \\ 4 & 2 \end{bmatrix}$

29. $A = \begin{bmatrix} 0 & -1 & 0 \\ 4 & 0 & 2 \\ 8 & -1 & 7 \end{bmatrix}, \quad B = \begin{bmatrix} 2 & 1 \\ -3 & 4 \\ 1 & 6 \end{bmatrix}$

30. $A = \begin{bmatrix} -1 & 3 \\ 4 & -5 \\ 0 & 2 \end{bmatrix}, \quad B = \begin{bmatrix} 1 & 2 \\ 0 & 7 \end{bmatrix}$

31. $A = \begin{bmatrix} 1 & 0 & 0 \\ 0 & 4 & 0 \\ 0 & 0 & -2 \end{bmatrix}, \quad B = \begin{bmatrix} 3 & 0 & 0 \\ 0 & -1 & 0 \\ 0 & 0 & 5 \end{bmatrix}$

32. $A = \begin{bmatrix} 5 & 0 & 0 \\ 0 & -8 & 0 \\ 0 & 0 & 7 \end{bmatrix}, \quad B = \begin{bmatrix} \frac{1}{5} & 0 & 0 \\ 0 & -\frac{1}{8} & 0 \\ 0 & 0 & \frac{1}{2} \end{bmatrix}$

33. $A = \begin{bmatrix} 0 & 0 & 5 \\ 0 & 0 & -3 \\ 0 & 0 & 4 \end{bmatrix}, \quad B = \begin{bmatrix} 6 & -11 & 4 \\ 8 & 16 & 4 \\ 0 & 0 & 0 \end{bmatrix}$

34. $A = \begin{bmatrix} 10 \\ 12 \end{bmatrix}, \quad B = \begin{bmatrix} 6 & -2 & 1 & 6 \end{bmatrix}$

In Exercises 35–40, use the matrix capabilities of a graphing utility to find *AB*.

35. $A = \begin{bmatrix} 5 & 6 & -3 \\ -2 & 5 & 1 \\ 10 & -5 & 5 \end{bmatrix}, \quad B = \begin{bmatrix} 1 & -1 & 2 \\ 8 & 1 & 4 \\ 4 & -2 & 9 \end{bmatrix}$

36. $A = \begin{bmatrix} 11 & -12 & 4 \\ 14 & 10 & 12 \\ 6 & -2 & 9 \end{bmatrix}, \quad B = \begin{bmatrix} 12 & 10 \\ -5 & 12 \\ 15 & 16 \end{bmatrix}$

37. $A = \begin{bmatrix} -3 & 8 & -6 & 8 \\ -12 & 15 & 9 & 6 \\ 5 & -1 & 1 & 5 \end{bmatrix}, \quad B = \begin{bmatrix} 3 & 1 & 6 \\ 24 & 15 & 14 \\ 16 & 10 & 21 \\ 8 & -4 & 10 \end{bmatrix}$

38. $A = \begin{bmatrix} -2 & 4 & 8 \\ 21 & 5 & 6 \\ 13 & 2 & 6 \end{bmatrix}, \quad B = \begin{bmatrix} 2 & 0 \\ -7 & 15 \\ 32 & 14 \\ 0.5 & 1.6 \end{bmatrix}$

39. $A = \begin{bmatrix} 9 & 10 & -38 & 18 \\ 100 & -50 & 250 & 75 \end{bmatrix},$

$B = \begin{bmatrix} 52 & -85 & 27 & 45 \\ 40 & -35 & 60 & 82 \end{bmatrix}$

40. $A = \begin{bmatrix} 15 & -18 \\ -4 & 12 \\ -8 & 22 \end{bmatrix}, \quad B = \begin{bmatrix} -7 & 22 & 1 \\ 8 & 16 & 24 \end{bmatrix}$

In Exercises 41–46, find (a) *AB*, (b) *BA*, and, if possible, (c) A^2. (*Note:* $A^2 = AA$.)

41. $A = \begin{bmatrix} 1 & 2 \\ 4 & 2 \end{bmatrix}, \quad B = \begin{bmatrix} 2 & -1 \\ -1 & 8 \end{bmatrix}$

42. $A = \begin{bmatrix} 2 & -1 \\ 1 & 4 \end{bmatrix}, \quad B = \begin{bmatrix} 0 & 0 \\ 3 & -3 \end{bmatrix}$

43. $A = \begin{bmatrix} 3 & -1 \\ 1 & 3 \end{bmatrix}, \quad B = \begin{bmatrix} 1 & -3 \\ 3 & 1 \end{bmatrix}$

44. $A = \begin{bmatrix} 1 & -1 \\ 1 & 1 \end{bmatrix}, \quad B = \begin{bmatrix} 1 & 3 \\ -3 & 1 \end{bmatrix}$

45. $A = \begin{bmatrix} 7 \\ 8 \\ -1 \end{bmatrix}, \quad B = \begin{bmatrix} 1 & 1 & 2 \end{bmatrix}$

46. $A = \begin{bmatrix} 3 & 2 & 1 \end{bmatrix}, \quad B = \begin{bmatrix} 2 \\ 3 \\ 0 \end{bmatrix}$

In Exercises 47–50, evaluate the expression. Use the matrix capabilities of a graphing utility to verify your answer.

47. $\begin{bmatrix} 3 & 1 \\ 0 & -2 \end{bmatrix} \begin{bmatrix} 1 & 0 \\ -2 & 2 \end{bmatrix} \begin{bmatrix} 1 & 0 \\ 2 & 4 \end{bmatrix}$

48. $-3\left(\begin{bmatrix} 6 & 5 & -1 \\ 1 & -2 & 0 \end{bmatrix} \begin{bmatrix} 0 & 3 \\ -1 & -3 \\ 4 & 1 \end{bmatrix}\right)$

49. $\begin{bmatrix} 0 & 2 & -2 \\ 4 & 1 & 2 \end{bmatrix}\left(\begin{bmatrix} 4 & 0 \\ 0 & -1 \\ -1 & 2 \end{bmatrix} + \begin{bmatrix} -2 & 3 \\ -3 & 5 \\ 0 & -3 \end{bmatrix}\right)$

50. $\begin{bmatrix} 3 \\ -1 \\ 5 \\ 7 \end{bmatrix}\left(\begin{bmatrix} 5 & -6 \end{bmatrix} + \begin{bmatrix} 7 & -1 \end{bmatrix} + \begin{bmatrix} -8 & 9 \end{bmatrix}\right)$

In Exercises 51–58, (a) write each system of linear equations as a matrix equation, $AX = B$, and (b) use Gauss-Jordan elimination on the augmented matrix $[A \vdots B]$ to solve for the matrix X.

51. $\begin{cases} -x_1 + x_2 = 4 \\ -2x_1 + x_2 = 0 \end{cases}$

52. $\begin{cases} 2x_1 + 3x_2 = 5 \\ x_1 + 4x_2 = 10 \end{cases}$

53. $\begin{cases} -2x_1 - 3x_2 = -4 \\ 6x_1 + x_2 = -36 \end{cases}$

54. $\begin{cases} -4x_1 + 9x_2 = -13 \\ x_1 - 3x_2 = 12 \end{cases}$

55. $\begin{cases} x_1 - 2x_2 + 3x_3 = 9 \\ -x_1 + 3x_2 - x_3 = -6 \\ 2x_1 - 5x_2 + 5x_3 = 17 \end{cases}$

56. $\begin{cases} x_1 + x_2 - 3x_3 = 9 \\ -x_1 + 2x_2 = 6 \\ x_1 - x_2 + x_3 = -5 \end{cases}$

57. $\begin{cases} x_1 - 5x_2 + 2x_3 = -20 \\ -3x_1 + x_2 - x_3 = 8 \\ -2x_2 + 5x_3 = -16 \end{cases}$

58. $\begin{cases} x_1 - x_2 + 4x_3 = 17 \\ x_1 + 3x_2 = -11 \\ -6x_2 + 5x_3 = 40 \end{cases}$

59. ***Manufacturing*** A corporation has three factories, each of which manufactures acoustic guitars and electric guitars. The number of units of guitars produced at factory j in one day is represented by a_{ij} in the matrix

$$A = \begin{bmatrix} 70 & 50 & 25 \\ 35 & 100 & 70 \end{bmatrix}.$$

Find the production levels if production is increased by 20%.

60. ***Manufacturing*** A corporation has four factories, each of which manufactures sport utility vehicles and pickup trucks. The number of units of vehicle i produced at factory j in one day is represented by a_{ij} in the matrix

$$A = \begin{bmatrix} 100 & 90 & 70 & 30 \\ 40 & 20 & 60 & 60 \end{bmatrix}.$$

Find the production levels if production is increased by 10%.

▶ Model It

61. ***Agriculture*** A fruit grower raises two crops, apples and peaches. Each of these crops is sent to three different outlets for sale. These outlets are The Farmer's Market, The Fruit Stand, and The Fruit Farm. The numbers of bushels of apples sent to the three outlets are 125, 100, and 75, respectively. The numbers of bushels of peaches sent to the three outlets are 100, 175, and 125, respectively. The profit per bushel for apples is $3.50 and the profit per bushel for peaches is $6.00.

(a) Write a matrix A that represents the number of bushels of each crop i that are shipped to each outlet j. State what each entry a_{ij} of the matrix represents.

(b) Write a matrix B that represents the profit per bushel of each fruit. State what each entry b_{ij} of the matrix represents.

(c) Find the product BA and state what each entry of the matrix represents.

62. ***Revenue*** A manufacturer of electronics produces three models of portable CD players, which are shipped to two warehouses. The number of units of model i that are shipped to warehouse j is represented by a_{ij} in the matrix

$$A = \begin{bmatrix} 5{,}000 & 4{,}000 \\ 6{,}000 & 10{,}000 \\ 8{,}000 & 5{,}000 \end{bmatrix}.$$

The price per unit is represented by the matrix

$$B = [\$39.50 \quad \$44.50 \quad \$56.50].$$

Compute BA and interpret the result.

63. ***Inventory*** A company sells five models of computers through three retail outlets. The inventories are represented by S.

Model (columns A, B, C, D, E); Outlet (rows 1, 2, 3)

$$S = \begin{bmatrix} 3 & 2 & 2 & 3 & 0 \\ 0 & 2 & 3 & 4 & 3 \\ 4 & 2 & 1 & 3 & 2 \end{bmatrix}$$

The wholesale and retail prices are represented by T.

Price (columns Wholesale, Retail); Model (rows A, B, C, D, E)

$$T = \begin{bmatrix} \$840 & \$1100 \\ \$1200 & \$1350 \\ \$1450 & \$1650 \\ \$2650 & \$3000 \\ \$3050 & \$3200 \end{bmatrix}$$

Compute ST and interpret the result.

64. ***Voting Preferences*** The matrix

From (columns R, D, I); To (rows R, D, I)

$$P = \begin{bmatrix} 0.6 & 0.1 & 0.1 \\ 0.2 & 0.7 & 0.1 \\ 0.2 & 0.2 & 0.8 \end{bmatrix}$$

is called a *stochastic matrix*. Each entry p_{ij} $(i \neq j)$ represents the proportion of the voting population that changes from party i to party j, and p_{ii} represents the proportion that remains loyal to the party from one election to the next. Compute and interpret P^2.

65. ***Voting Preferences*** Use a graphing utility to find P^3, P^4, P^5, P^6, P^7, and P^8 for the matrix given in Exercise 64. Can you detect a pattern as P is raised to higher powers?

66. ***Labor/Wage Requirements*** A company that manufactures boats has the following labor-hour and wage requirements.

Labor per boat

$$S = \begin{bmatrix} 1.0 \text{ hr} & 0.5 \text{ hr} & 0.2 \text{ hr} \\ 1.6 \text{ hr} & 1.0 \text{ hr} & 0.2 \text{ hr} \\ 2.5 \text{ hr} & 2.0 \text{ hr} & 1.4 \text{ hr} \end{bmatrix}$$

Department: columns Cutting, Assembly, Packaging; Boat size: rows Small, Medium, Large

Wages per hour

$$T = \begin{bmatrix} \$12 & \$10 \\ \$9 & \$8 \\ \$8 & \$7 \end{bmatrix}$$

Plant: columns A, B; Department: rows Cutting, Assembly, Packaging

Compute ST and interpret the result.

Synthesis

True or False? **In Exercises 67 and 68, determine whether the statement is true or false. Justify your answer.**

67. Two matrices can be added only if they have the same order.

68. $\begin{bmatrix} -6 & -2 \\ 2 & -6 \end{bmatrix}\begin{bmatrix} 4 & 0 \\ 0 & -1 \end{bmatrix} = \begin{bmatrix} 4 & 0 \\ 0 & -1 \end{bmatrix}\begin{bmatrix} -6 & -2 \\ 2 & -6 \end{bmatrix}$

Think About It **In Exercises 69–76, let matrices A, B, C, and D be of orders 2×3, 2×3, 3×2, and 2×2, respectively. Determine whether the matrices are of proper order to perform the operation(s). If so, give the order of the answer.**

69. $A + 2C$

70. $B - 3C$

71. AB

72. BC

73. $BC - D$

74. $CB - D$

75. $D(A - 3B)$

76. $(BC - D)A$

77. ***Think About It*** If a, b, and c are real numbers such that $c \neq 0$ and $ac = bc$, then $a = b$. However, if A, B, and C are nonzero matrices such that $AC = BC$, then A is *not* necessarily equal to B. Illustrate this using the following matrices.

$$A = \begin{bmatrix} 0 & 1 \\ 0 & 1 \end{bmatrix}, \quad B = \begin{bmatrix} 1 & 0 \\ 1 & 0 \end{bmatrix}, \quad C = \begin{bmatrix} 2 & 3 \\ 2 & 3 \end{bmatrix}$$

78. ***Think About It*** If a and b are real numbers such that $ab = 0$, then $a = 0$ or $b = 0$. However, if A and B are matrices such that $AB = O$, it is *not* necessarily true that $A = O$ or $B = O$. Illustrate this using the following matrices.

$$A = \begin{bmatrix} 3 & 3 \\ 4 & 4 \end{bmatrix}, \quad B = \begin{bmatrix} 1 & -1 \\ -1 & 1 \end{bmatrix}$$

79. ***Exploration*** Let A and B be unequal diagonal matrices of the same order. (A diagonal matrix is a square matrix in which each entry not on the main diagonal is zero.) Determine the products AB for several pairs of such matrices. Make a conjecture about a quick rule for such products.

80. ***Exploration*** Let $i = \sqrt{-1}$ and let

$$A = \begin{bmatrix} i & 0 \\ 0 & i \end{bmatrix} \quad \text{and} \quad B = \begin{bmatrix} 0 & -i \\ i & 0 \end{bmatrix}.$$

(a) Find A^2, A^3, and A^4. Identify any similarities with i^2, i^3, and i^4.

(b) Find and identify B^2.

Review

In Exercises 81–86, solve the equation.

81. $3x^2 + 20x - 32 = 0$

82. $8x^2 - 10x - 3 = 0$

83. $4x^3 + 10x^2 - 3x = 0$

84. $3x^3 + 22x^2 - 45x = 0$

85. $3x^3 - 12x^2 + 5x - 20 = 0$

86. $2x^3 - 5x^2 - 12x + 30 = 0$

In Exercises 87–90, solve the system of linear equations both graphically and algebraically.

87. $\begin{cases} -x + 4y = -9 \\ 5x - 8y = 39 \end{cases}$

88. $\begin{cases} 8x - 3y = -17 \\ -6x + 7y = 27 \end{cases}$

89. $\begin{cases} -x + 2y = -5 \\ -3x - y = -8 \end{cases}$

90. $\begin{cases} 6x - 13y = 11 \\ 9x + 5y = 41 \end{cases}$

10.3 The Inverse of a Square Matrix

▶ **What you should learn**

- How to verify that two matrices are inverses of each other
- How to use Gauss-Jordan elimination to find the inverses of matrices
- How to use a formula to find the inverses of 2×2 matrices
- How to use inverse matrices to solve systems of linear equations

▶ **Why you should learn it**

You can use inverse matrices to model and solve real-life problems. For instance, in Exercise 72 on page 769, an inverse matrix is used to find a linear model for the number of vehicle registrations in the United States.

Jon Love/Getty Images

The Inverse of a Matrix

This section further develops the algebra of matrices. To begin, consider the real number equation $ax = b$. To solve this equation for x, multiply each side of the equation by a^{-1} (provided that $a \neq 0$).

$$ax = b$$
$$(a^{-1}a)x = a^{-1}b$$
$$(1)x = a^{-1}b$$
$$x = a^{-1}b$$

The number a^{-1} is called the *multiplicative inverse of a* because $a^{-1}a = 1$. The definition of the multiplicative **inverse of a matrix** is similar.

Definition of the Inverse of a Square Matrix

Let A be an $n \times n$ matrix and let I_n be the $n \times n$ identity matrix. If there exists matrix A^{-1} such that

$$AA^{-1} = I_n = A^{-1}A$$

then A^{-1} is called the **inverse** of A. The symbol A^{-1} is read "A inverse."

Example 1 ▶ The Inverse of a Matrix 

Show that B is the inverse of A, where

$$A = \begin{bmatrix} -1 & 2 \\ -1 & 1 \end{bmatrix}$$

and

$$B = \begin{bmatrix} 1 & -2 \\ 1 & -1 \end{bmatrix}.$$

Solution

To show that B is the inverse of A, show that $AB = I = BA$, as follows.

$$AB = \begin{bmatrix} -1 & 2 \\ -1 & 1 \end{bmatrix}\begin{bmatrix} 1 & -2 \\ 1 & -1 \end{bmatrix} = \begin{bmatrix} -1+2 & 2-2 \\ -1+1 & 2-1 \end{bmatrix} = \begin{bmatrix} 1 & 0 \\ 0 & 1 \end{bmatrix}$$

$$BA = \begin{bmatrix} 1 & -2 \\ 1 & -1 \end{bmatrix}\begin{bmatrix} -1 & 2 \\ -1 & 1 \end{bmatrix} = \begin{bmatrix} -1+2 & 2-2 \\ -1+1 & 2-1 \end{bmatrix} = \begin{bmatrix} 1 & 0 \\ 0 & 1 \end{bmatrix}$$

Recall that it is not always true that $AB = BA$, even if both products are defined. However, if A and B are both square matrices and $AB = I_n$, it can be shown that $BA = I_n$. So, in Example 1, you need only to check that $AB = I_2$.

Finding Inverse Matrices

If a matrix A has an inverse, A is called **invertible** (or **nonsingular**); otherwise, A is called **singular.** A nonsquare matrix cannot have an inverse. To see this, note that if A is of order $m \times n$ and B is of order $n \times m$ (where $m \neq n$), the products AB and BA are of different orders and so cannot be equal to each other. Not all square matrices have inverses (see the matrix at the bottom of page 764). If, however, a matrix does have an inverse, that inverse is unique. Example 2 shows how to use a system of equations to find the inverse of a matrix.

Example 2 ▶ Finding the Inverse of a Matrix

Find the inverse of

$$A = \begin{bmatrix} 1 & 4 \\ -1 & -3 \end{bmatrix}.$$

Solution

To find the inverse of A, try to solve the matrix equation $AX = I$ for X.

$$\underset{A}{\begin{bmatrix} 1 & 4 \\ -1 & -3 \end{bmatrix}} \underset{X}{\begin{bmatrix} x_{11} & x_{12} \\ x_{21} & x_{22} \end{bmatrix}} = \underset{I}{\begin{bmatrix} 1 & 0 \\ 0 & 1 \end{bmatrix}}$$

$$\begin{bmatrix} x_{11} + 4x_{21} & x_{12} + 4x_{22} \\ -x_{11} - 3x_{21} & -x_{12} - 3x_{22} \end{bmatrix} = \begin{bmatrix} 1 & 0 \\ 0 & 1 \end{bmatrix}$$

Equating corresponding entries, you obtain two systems of linear equations.

$$\begin{cases} x_{11} + 4x_{21} = 1 \\ -x_{11} - 3x_{21} = 0 \end{cases}$$

Linear system with two variables, x_{11} and x_{21}.

$$\begin{cases} x_{12} + 4x_{22} = 0 \\ -x_{12} - 3x_{22} = 1 \end{cases}$$

Linear system with two variables, x_{12} and x_{22}.

From the first system you can determine that $x_{11} = -3$ and $x_{21} = 1$, and from the second system you can determine that $x_{12} = -4$ and $x_{22} = 1$. Therefore, the inverse of A is

$$X = A^{-1}$$

$$= \begin{bmatrix} -3 & -4 \\ 1 & 1 \end{bmatrix}.$$

You can use matrix multiplication to check this result.

Check

$$AA^{-1} = \begin{bmatrix} 1 & 4 \\ -1 & -3 \end{bmatrix} \begin{bmatrix} -3 & -4 \\ 1 & 1 \end{bmatrix} = \begin{bmatrix} 1 & 0 \\ 0 & 1 \end{bmatrix} \checkmark$$

$$A^{-1}A = \begin{bmatrix} -3 & -4 \\ 1 & 1 \end{bmatrix} \begin{bmatrix} 1 & 4 \\ -1 & -3 \end{bmatrix} = \begin{bmatrix} 1 & 0 \\ 0 & 1 \end{bmatrix} \checkmark$$

In Example 2, note that the two systems of linear equations have the *same coefficient matrix* A. Rather than solve the two systems represented by

$$\begin{bmatrix} 1 & 4 & \vdots & 1 \\ -1 & -3 & \vdots & 0 \end{bmatrix}$$

and

$$\begin{bmatrix} 1 & 4 & \vdots & 0 \\ -1 & -3 & \vdots & 1 \end{bmatrix}$$

separately, you can solve them *simultaneously* by *adjoining* the identity matrix to the coefficient matrix to obtain

$$\overset{A}{}\qquad\qquad \overset{I}{}$$
$$\begin{bmatrix} 1 & 4 & \vdots & 1 & 0 \\ -1 & -3 & \vdots & 0 & 1 \end{bmatrix}.$$

This "doubly augmented" matrix can be represented as $[A \,\vdots\, I]$. By applying Gauss-Jordan elimination to this matrix, you can solve *both* systems with a single elimination process.

$$\begin{bmatrix} 1 & 4 & \vdots & 1 & 0 \\ -1 & -3 & \vdots & 0 & 1 \end{bmatrix}$$

$$R_1 + R_2 \rightarrow \begin{bmatrix} 1 & 4 & \vdots & 1 & 0 \\ 0 & 1 & \vdots & 1 & 1 \end{bmatrix}$$

$$-4R_2 + R_1 \rightarrow \begin{bmatrix} 1 & 0 & \vdots & -3 & -4 \\ 0 & 1 & \vdots & 1 & 1 \end{bmatrix}$$

So, from the "doubly augmented" matrix $[A \,\vdots\, I]$, you obtained the matrix $[I \,\vdots\, A^{-1}]$.

$$\underbrace{\begin{bmatrix} 1 & 4 \\ -1 & -3 \end{bmatrix}}_{A} \vdots \underbrace{\begin{bmatrix} 1 & 0 \\ 0 & 1 \end{bmatrix}}_{I} \Rightarrow \underbrace{\begin{bmatrix} 1 & 0 \\ 0 & 1 \end{bmatrix}}_{I} \vdots \underbrace{\begin{bmatrix} -3 & -4 \\ 1 & 1 \end{bmatrix}}_{A^{-1}}$$

This procedure (or algorithm) works for an arbitrary square matrix that has an inverse.

Technology

You can find the inverse of a matrix with a graphing utility. Enter the matrix in the graphing utility, press the inverse key $\boxed{x^{-1}}$, and press $\boxed{\text{ENTER}}$. The inverse of the matrix will be displayed on the screen.

Finding an Inverse Matrix

Let A be a square matrix of order n.

1. Write the $n \times 2n$ matrix that consists of the given matrix A on the left and the $n \times n$ identity matrix I on the right to obtain $[A \,\vdots\, I]$.
2. If possible, row reduce A to I using elementary row operations on the *entire* matrix $[A \,\vdots\, I]$. The result will be the matrix $[I \,\vdots\, A^{-1}]$. If this is not possible, A is not invertible.
3. Check your work by multiplying to see that $AA^{-1} = I = A^{-1}A$.

Example 3 ▶ Finding the Inverse of a Matrix 

Find the inverse of

$$A = \begin{bmatrix} 1 & -1 & 0 \\ 1 & 0 & -1 \\ 6 & -2 & -3 \end{bmatrix}.$$

Solution

Begin by adjoining the identity matrix to A to form the matrix

$$[A \vdots I] = \left[\begin{array}{ccc|ccc} 1 & -1 & 0 & 1 & 0 & 0 \\ 1 & 0 & -1 & 0 & 1 & 0 \\ 6 & -2 & -3 & 0 & 0 & 1 \end{array}\right].$$

Use elementary row operations to obtain the form $[I \ \vdots \ A^{-1}]$, as follows.

$$\begin{array}{r} \\ -R_1 + R_2 \rightarrow \\ -6R_1 + R_3 \rightarrow \end{array} \left[\begin{array}{ccc|ccc} 1 & -1 & 0 & 1 & 0 & 0 \\ 0 & 1 & -1 & -1 & 1 & 0 \\ 0 & 4 & -3 & -6 & 0 & 1 \end{array}\right]$$

$$\begin{array}{r} R_2 + R_1 \rightarrow \\ \\ -4R_2 + R_3 \rightarrow \end{array} \left[\begin{array}{ccc|ccc} 1 & 0 & -1 & 0 & 1 & 0 \\ 0 & 1 & -1 & -1 & 1 & 0 \\ 0 & 0 & 1 & -2 & -4 & 1 \end{array}\right]$$

$$\begin{array}{r} R_3 + R_1 \rightarrow \\ R_3 + R_2 \rightarrow \\ \\ \end{array} \left[\begin{array}{ccc|ccc} 1 & 0 & 0 & -2 & -3 & 1 \\ 0 & 1 & 0 & -3 & -3 & 1 \\ 0 & 0 & 1 & -2 & -4 & 1 \end{array}\right] = [I \vdots A^{-1}]$$

So, the matrix A is invertible and its inverse is

$$A^{-1} = \begin{bmatrix} -2 & -3 & 1 \\ -3 & -3 & 1 \\ -2 & -4 & 1 \end{bmatrix}.$$

Confirm this result by multiplying A and A^{-1} to obtain I, as follows.

Check

$$AA^{-1} = \begin{bmatrix} 1 & -1 & 0 \\ 1 & 0 & -1 \\ 6 & -2 & -3 \end{bmatrix}\begin{bmatrix} -2 & -3 & 1 \\ -3 & -3 & 1 \\ -2 & -4 & 1 \end{bmatrix} = \begin{bmatrix} 1 & 0 & 0 \\ 0 & 1 & 0 \\ 0 & 0 & 1 \end{bmatrix} = I$$

The process shown in Example 3 applies to any $n \times n$ matrix A. If A has an inverse, this process will find it. On the other hand, if A does not have an inverse (if A is *singular*), the process will tell you so. That is, matrix A will not reduce to the identity matrix. For instance, the following matrix has no inverse.

$$A = \begin{bmatrix} 1 & 2 & 0 \\ 3 & -1 & 2 \\ -2 & 3 & -2 \end{bmatrix}$$

Explain how the elimination process shows that this matrix is singular.

Exploration

Use a graphing utility with matrix capabilities to find the inverse of the matrix

$$A = \begin{bmatrix} 1 & -3 \\ -2 & 6 \end{bmatrix}.$$

What message appears on the screen? Why does the graphing utility display this message?

The Inverse of a 2 × 2 Matrix

Using Gauss-Jordan elimination to find the inverse of a matrix works well (even as a computer technique) for matrices of order 3×3 or greater. For 2×2 matrices, however, many people prefer to use a formula for the inverse rather than Gauss-Jordan elimination. This simple formula, which works *only* for 2×2 matrices, is explained as follows. If A is a 2×2 matrix given by

$$A = \begin{bmatrix} a & b \\ c & d \end{bmatrix}$$

then A is invertible if and only if $ad - bc \neq 0$. Moreover, if $ad - bc \neq 0$, the inverse is given by

$$A^{-1} = \frac{1}{ad - bc}\begin{bmatrix} d & -b \\ -c & a \end{bmatrix}.$$ Formula for inverse of matrix A

The denominator $ad - bc$ is called the **determinant** of the 2×2 matrix A. You will study determinants in the next section.

Example 4 ▶ Finding the Inverse of a 2 × 2 Matrix 

If possible, find the inverse of each matrix.

a. $A = \begin{bmatrix} 3 & -1 \\ -2 & 2 \end{bmatrix}$

b. $B = \begin{bmatrix} 3 & -1 \\ -6 & 2 \end{bmatrix}$

Solution

a. For the matrix A, apply the formula for the inverse of a 2×2 matrix to obtain

$$ad - bc = (3)(2) - (-1)(-2)$$
$$= 4.$$

Because this quantity is not zero, the inverse is formed by interchanging the entries on the main diagonal, changing the signs of the other two entries, and multiplying by the scalar $\frac{1}{4}$, as follows.

$$A^{-1} = \frac{1}{4}\begin{bmatrix} 2 & 1 \\ 2 & 3 \end{bmatrix}$$ Substitute for a, b, c, d, and the determinant.

$$= \begin{bmatrix} \frac{1}{2} & \frac{1}{4} \\ \frac{1}{2} & \frac{3}{4} \end{bmatrix}$$ Multiply by the scalar $\frac{1}{4}$.

b. For the matrix B, you have

$$ad - bc = (3)(2) - (-1)(-6)$$
$$= 0$$

which means that B is not invertible.

Systems of Linear Equations

You know that a system of linear equations can have exactly one solution, infinitely many solutions, or no solution. If the coefficient matrix A of a *square* system (a system that has the same number of equations as variables) is invertible, the system has a unique solution, which is defined as follows.

A System of Equations with a Unique Solution

If A is an invertible matrix, the system of linear equations represented by $AX = B$ has a unique solution given by

$$X = A^{-1}B.$$

Technology

To solve a system of equations with a graphing utility, enter the matrices A and B in the matrix editor. Then, using the inverse key, solve for X.

A [x^{-1}] B [ENTER]

The screen will display the solution, matrix X.

Example 5 ▶ Solving a System Using an Inverse  

You are going to invest \$10,000 in AAA-rated bonds, AA-rated bonds, and B-rated bonds and want an annual return of \$730. The average yields are 6% on AAA bonds, 7.5% on AA bonds, and 9.5% on B bonds. You will invest twice as much in AAA bonds as in B bonds. Your investment can be represented as

$$\begin{cases} x + y + z = 10{,}000 \\ 0.06x + 0.075y + 0.095z = 730 \\ x - 2z = 0 \end{cases}$$

where x, y, and z represent the amounts invested in AAA, AA, and B bonds, respectively. Use an inverse matrix to solve the system.

Solution

Begin by writing the system in the matrix form $AX = B$.

$$\begin{bmatrix} 1 & 1 & 1 \\ 0.06 & 0.075 & 0.095 \\ 1 & 0 & -2 \end{bmatrix}\begin{bmatrix} x \\ y \\ z \end{bmatrix} = \begin{bmatrix} 10{,}000 \\ 730 \\ 0 \end{bmatrix}$$

Then, use Gauss-Jordan elimination to find A^{-1}.

$$A^{-1} = \begin{bmatrix} 15 & -200 & -2 \\ -21.5 & 300 & 3.5 \\ 7.5 & -100 & -1.5 \end{bmatrix}$$

Finally, multiply B by A^{-1} on the left to obtain the solution.

$$X = A^{-1}B$$

$$= \begin{bmatrix} 15 & -200 & -2 \\ -21.5 & 300 & 3.5 \\ 7.5 & -100 & -1.5 \end{bmatrix}\begin{bmatrix} 10{,}000 \\ 730 \\ 0 \end{bmatrix} = \begin{bmatrix} 4000 \\ 4000 \\ 2000 \end{bmatrix}$$

The solution to the system is $x = 4000$, $y = 4000$, and $z = 2000$. So, you will invest \$4000 in AAA bonds, \$4000 in AA bonds, and \$2000 in B bonds.

10.3 Exercises

In Exercises 1–10, show that B is the inverse of A.

1. $A = \begin{bmatrix} 2 & 1 \\ 5 & 3 \end{bmatrix},\quad B = \begin{bmatrix} 3 & -1 \\ -5 & 2 \end{bmatrix}$

2. $A = \begin{bmatrix} 1 & -1 \\ -1 & 2 \end{bmatrix},\quad B = \begin{bmatrix} 2 & 1 \\ 1 & 1 \end{bmatrix}$

3. $A = \begin{bmatrix} 1 & 2 \\ 3 & 4 \end{bmatrix},\quad B = \begin{bmatrix} -2 & 1 \\ \frac{3}{2} & -\frac{1}{2} \end{bmatrix}$

4. $A = \begin{bmatrix} 1 & -1 \\ 2 & 3 \end{bmatrix},\quad B = \begin{bmatrix} \frac{3}{5} & \frac{1}{5} \\ -\frac{2}{5} & \frac{1}{5} \end{bmatrix}$

5. $A = \begin{bmatrix} 2 & -17 & 11 \\ -1 & 11 & -7 \\ 0 & 3 & -2 \end{bmatrix},\quad B = \begin{bmatrix} 1 & 1 & 2 \\ 2 & 4 & -3 \\ 3 & 6 & -5 \end{bmatrix}$

6. $A = \begin{bmatrix} -4 & 1 & 5 \\ -1 & 2 & 4 \\ 0 & -1 & -1 \end{bmatrix},\quad B = \begin{bmatrix} -\frac{1}{2} & 1 & \frac{3}{2} \\ \frac{1}{4} & -1 & -\frac{11}{4} \\ -\frac{1}{4} & 1 & \frac{7}{4} \end{bmatrix}$

7. $A = \begin{bmatrix} 2 & 0 & 1 & 1 \\ 3 & 0 & 0 & 1 \\ -1 & 1 & -2 & 1 \\ 4 & -1 & 1 & 0 \end{bmatrix},$

$B = \begin{bmatrix} -1 & 2 & -1 & -1 \\ -4 & 9 & -5 & -6 \\ 0 & 1 & -1 & -1 \\ 3 & -5 & 3 & 3 \end{bmatrix}$

8. $A = \begin{bmatrix} -2 & 0 & 1 & 0 \\ 1 & -1 & -3 & 0 \\ -2 & -1 & 0 & -2 \\ 0 & 1 & 3 & -1 \end{bmatrix},$

$B = \begin{bmatrix} -3 & -3 & 1 & -2 \\ 12 & 14 & -5 & 10 \\ -5 & -6 & 2 & -4 \\ -3 & -4 & 1 & -3 \end{bmatrix}$

9. $A = \begin{bmatrix} -2 & 2 & 3 \\ 1 & -1 & 0 \\ 0 & 1 & 4 \end{bmatrix},\quad B = \frac{1}{3}\begin{bmatrix} -4 & -5 & 3 \\ -4 & -8 & 3 \\ 1 & 2 & 0 \end{bmatrix}$

10. $A = \begin{bmatrix} -1 & 1 & 0 & -1 \\ 1 & -1 & 1 & 0 \\ -1 & 1 & 2 & 0 \\ 0 & -1 & 1 & 1 \end{bmatrix},$

$B = \frac{1}{3}\begin{bmatrix} -3 & 1 & 1 & -3 \\ -3 & -1 & 2 & -3 \\ 0 & 1 & 1 & 0 \\ -3 & -2 & 1 & 0 \end{bmatrix}$

In Exercises 11–26, find the inverse of the matrix (if it exists).

11. $\begin{bmatrix} 2 & 0 \\ 0 & 3 \end{bmatrix}$

12. $\begin{bmatrix} 1 & 2 \\ 3 & 7 \end{bmatrix}$

13. $\begin{bmatrix} 1 & -2 \\ 2 & -3 \end{bmatrix}$

14. $\begin{bmatrix} -7 & 33 \\ 4 & -19 \end{bmatrix}$

15. $\begin{bmatrix} -1 & 1 \\ -2 & 1 \end{bmatrix}$

16. $\begin{bmatrix} 11 & 1 \\ -1 & 0 \end{bmatrix}$

17. $\begin{bmatrix} 2 & 4 \\ 4 & 8 \end{bmatrix}$

18. $\begin{bmatrix} 2 & 3 \\ 1 & 4 \end{bmatrix}$

19. $\begin{bmatrix} 2 & 7 & 1 \\ -3 & -9 & 2 \end{bmatrix}$

20. $\begin{bmatrix} -2 & 5 \\ 6 & -15 \\ 0 & 1 \end{bmatrix}$

21. $\begin{bmatrix} 1 & 1 & 1 \\ 3 & 5 & 4 \\ 3 & 6 & 5 \end{bmatrix}$

22. $\begin{bmatrix} 1 & 2 & 2 \\ 3 & 7 & 9 \\ -1 & -4 & -7 \end{bmatrix}$

23. $\begin{bmatrix} 1 & 0 & 0 \\ 3 & 4 & 0 \\ 2 & 5 & 5 \end{bmatrix}$

24. $\begin{bmatrix} 1 & 0 & 0 \\ 3 & 0 & 0 \\ 2 & 5 & 5 \end{bmatrix}$

25. $\begin{bmatrix} -8 & 0 & 0 & 0 \\ 0 & 1 & 0 & 0 \\ 0 & 0 & 4 & 0 \\ 0 & 0 & 0 & -5 \end{bmatrix}$

26. $\begin{bmatrix} 1 & 3 & -2 & 0 \\ 0 & 2 & 4 & 6 \\ 0 & 0 & -2 & 1 \\ 0 & 0 & 0 & 5 \end{bmatrix}$

In Exercises 27–38, use the matrix capabilities of a graphing utility to find the inverse of the matrix (if it exists).

27. $\begin{bmatrix} 1 & 2 & -1 \\ 3 & 7 & -10 \\ -5 & -7 & -15 \end{bmatrix}$

28. $\begin{bmatrix} 10 & 5 & -7 \\ -5 & 1 & 4 \\ 3 & 2 & -2 \end{bmatrix}$

29. $\begin{bmatrix} 1 & 1 & 2 \\ 3 & 1 & 0 \\ -2 & 0 & 3 \end{bmatrix}$

30. $\begin{bmatrix} 3 & 2 & 2 \\ 2 & 2 & 2 \\ -4 & 4 & 3 \end{bmatrix}$

31. $\begin{bmatrix} -\frac{1}{2} & \frac{3}{4} & \frac{1}{4} \\ 1 & 0 & -\frac{3}{2} \\ 0 & -1 & \frac{1}{2} \end{bmatrix}$

32. $\begin{bmatrix} -\frac{5}{6} & \frac{1}{3} & \frac{11}{6} \\ 0 & \frac{2}{3} & 2 \\ 1 & -\frac{1}{2} & -\frac{5}{2} \end{bmatrix}$

33. $\begin{bmatrix} 0.1 & 0.2 & 0.3 \\ -0.3 & 0.2 & 0.2 \\ 0.5 & 0.4 & 0.4 \end{bmatrix}$

34. $\begin{bmatrix} 0.6 & 0 & -0.3 \\ 0.7 & -1 & 0.2 \\ 1 & 0 & -0.9 \end{bmatrix}$

35. $\begin{bmatrix} 1 & 0 & 3 & 0 \\ 0 & 2 & 0 & 4 \\ 1 & 0 & 3 & 0 \\ 0 & 2 & 0 & 4 \end{bmatrix}$

36. $\begin{bmatrix} 4 & 8 & -7 & 14 \\ 2 & 5 & -4 & 6 \\ 0 & 2 & 1 & -7 \\ 3 & 6 & -5 & 10 \end{bmatrix}$

37. $\begin{bmatrix} -1 & 0 & 1 & 0 \\ 0 & 2 & 0 & -1 \\ 2 & 0 & -1 & 0 \\ 0 & -1 & 0 & 1 \end{bmatrix}$

38. $\begin{bmatrix} 1 & -2 & -1 & -2 \\ 3 & -5 & -2 & -3 \\ 2 & -5 & -2 & -5 \\ -1 & 4 & 4 & 11 \end{bmatrix}$

In Exercises 39–44, use the formula on page 765 to find the inverse of the matrix.

39. $\begin{bmatrix} 5 & -2 \\ 2 & 3 \end{bmatrix}$

40. $\begin{bmatrix} 7 & 12 \\ -8 & -5 \end{bmatrix}$

41. $\begin{bmatrix} -4 & -6 \\ 2 & 3 \end{bmatrix}$

42. $\begin{bmatrix} -12 & 3 \\ 5 & -2 \end{bmatrix}$

43. $\begin{bmatrix} \frac{7}{2} & -\frac{3}{4} \\ \frac{1}{5} & \frac{4}{5} \end{bmatrix}$

44. $\begin{bmatrix} -\frac{1}{4} & \frac{9}{4} \\ \frac{5}{3} & \frac{8}{9} \end{bmatrix}$

In Exercises 45–48, use the inverse matrix found in Exercise 13 to solve the system of linear equations.

45. $\begin{cases} x - 2y = 5 \\ 2x - 3y = 10 \end{cases}$

46. $\begin{cases} x - 2y = 0 \\ 2x - 3y = 3 \end{cases}$

47. $\begin{cases} x - 2y = 4 \\ 2x - 3y = 2 \end{cases}$

48. $\begin{cases} x - 2y = 1 \\ 2x - 3y = -2 \end{cases}$

In Exercises 49 and 50, use the inverse matrix found in Exercise 21 to solve the system of linear equations.

49. $\begin{cases} x + y + z = 0 \\ 3x + 5y + 4z = 5 \\ 3x + 6y + 5z = 2 \end{cases}$

50. $\begin{cases} x + y + z = -1 \\ 3x + 5y + 4z = 2 \\ 3x + 6y + 5z = 0 \end{cases}$

In Exercises 51 and 52, use the inverse matrix found in Exercise 38 to solve the system of linear equations.

51. $\begin{cases} x_1 - 2x_2 - x_3 - 2x_4 = 0 \\ 3x_1 - 5x_2 - 2x_3 - 3x_4 = 1 \\ 2x_1 - 5x_2 - 2x_3 - 5x_4 = -1 \\ -x_1 + 4x_2 + 4x_3 + 11x_4 = 2 \end{cases}$

52. $\begin{cases} x_1 - 2x_2 - x_3 - 2x_4 = 1 \\ 3x_1 - 5x_2 - 2x_3 - 3x_4 = -2 \\ 2x_1 - 5x_2 - 2x_3 - 5x_4 = 0 \\ -x_1 + 4x_2 + 4x_3 + 11x_4 = -3 \end{cases}$

In Exercises 53–60, use an inverse matrix to solve (if possible) the system of linear equations.

53. $\begin{cases} 3x + 4y = -2 \\ 5x + 3y = 4 \end{cases}$

54. $\begin{cases} 18x + 12y = 13 \\ 30x + 24y = 23 \end{cases}$

55. $\begin{cases} -0.4x + 0.8y = 1.6 \\ 2x - 4y = 5 \end{cases}$

56. $\begin{cases} 0.2x - 0.6y = 2.4 \\ -x + 1.4y = -8.8 \end{cases}$

57. $\begin{cases} -\frac{1}{4}x + \frac{3}{8}y = -2 \\ \frac{3}{2}x + \frac{3}{4}y = -12 \end{cases}$

58. $\begin{cases} \frac{5}{6}x - y = -20 \\ \frac{4}{3}x - \frac{7}{2}y = -51 \end{cases}$

59. $\begin{cases} 4x - y + z = -5 \\ 2x + 2y + 3z = 10 \\ 5x - 2y + 6z = 1 \end{cases}$

60. $\begin{cases} 4x - 2y + 3z = -2 \\ 2x + 2y + 5z = 16 \\ 8x - 5y - 2z = 4 \end{cases}$

In Exercises 61–66, use the matrix capabilities of a graphing utility to solve (if possible) the system of linear equations.

61. $\begin{cases} 5x - 3y + 2z = 2 \\ 2x + 2y - 3z = 3 \\ x - 7y + 8z = -4 \end{cases}$

62. $\begin{cases} 2x + 3y + 5z = 4 \\ 3x + 5y + 9z = 7 \\ 5x + 9y + 17z = 13 \end{cases}$

63. $\begin{cases} 3x - 2y + z = -29 \\ -4x + y - 3z = 37 \\ x - 5y + z = -24 \end{cases}$

64. $\begin{cases} -8x + 7y - 10z = -151 \\ 12x + 3y - 5z = 86 \\ 15x - 9y + 2z = 187 \end{cases}$

65. $\begin{cases} 7x - 3y + 2w = 41 \\ -2x + y - w = -13 \\ 4x + z - 2w = 12 \\ -x + y - w = -8 \end{cases}$

66. $\begin{cases} 2x + 5y + w = 11 \\ x + 4y + 2z - 2w = -7 \\ 2x - 2y + 5z + w = 3 \\ x - 3w = -1 \end{cases}$

Investment Portfolio **In Exercises 67–70, consider a person who invests in AAA-rated bonds, A-rated bonds, and B-rated bonds. The average yields are 6.5% on AAA bonds, 7% on A bonds, and 9% on B bonds. The person invests twice as much in B bonds as in A bonds. Let *x*, *y*, and *z* represent the amounts invested in AAA, A, and B bonds, respectively.**

$$\begin{cases} x + y + z = \text{(total investment)} \\ 0.065x + 0.07y + 0.09z = \text{(annual return)} \\ 2y - z = 0 \end{cases}$$

Use the inverse of the coefficient matrix of this system to find the amount invested in each type of bond.

	Total Investment	*Annual Return*
67.	\$10,000	\$705
68.	\$10,000	\$760
69.	\$12,000	\$835
70.	\$500,000	\$38,000

71. ***Circuit Analysis*** Consider the circuit in the figure. The currents I_1, I_2, and I_3, in amperes, are the solution of the system of linear equations

$$\begin{cases} 2I_1 \qquad + 4I_3 = E_1 \\ \qquad I_2 + 4I_3 = E_2 \\ I_1 + I_2 - \ I_3 = 0 \end{cases}$$

where E_1 and E_2 are voltages. Use the inverse of the coefficient matrix of this system to find the unknown currents for the voltages.

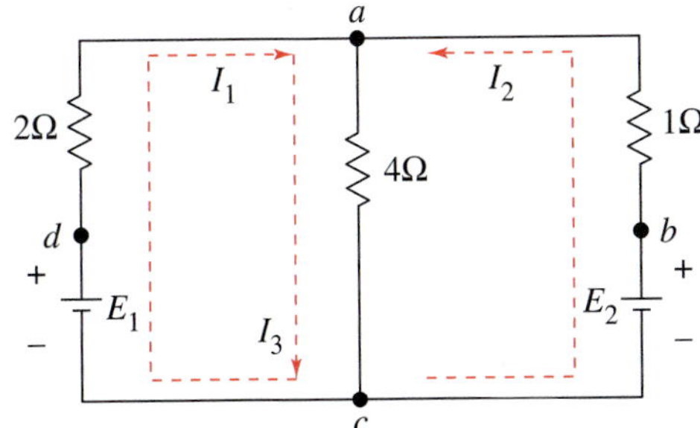

(a) $E_1 = 14$ volts, $E_2 = 28$ volts

(b) $E_1 = 24$ volts, $E_2 = 23$ volts

▶ Model It

72. ***Data Analysis*** The table shows the numbers y (in millions) of motor vehicle registrations in the United States for the years 1997 through 1999. (Source: U.S. Federal Highway Administration)

Year, t	Registrations, y
1997	207.8
1998	211.6
1999	216.3

(a) Use the technique demonstrated in Exercises 55–60 in Section 9.2 to create a system of linear equations for the data. Let t represent the year, with $t = 7$ corresponding to 1997.

(b) Use the matrix capabilities of a graphing utility to find an inverse matrix to solve the system from part (a) and find the least squares regression line $y = at + b$.

(c) Use the result of part (b) to estimate the number of motor vehicle registrations in 2000.

▶ Model It *(continued)*

(d) The actual number of motor vehicle registrations in 2000 was 221.5 million. How does this value compare with your estimate from part (c)?

(e) Use the result of part (b) to estimate when the number of vehicle registrations will reach 240 million.

Synthesis

True or False? **In Exercises 73 and 74, determine whether the statement is true or false. Justify your answer.**

73. Multiplication of an invertible matrix and its inverse is commutative.

74. If you multiply two square matrices and obtain the identity matrix, you can assume that the matrices are inverses of one another.

75. If A is a 2×2 matrix $A = \begin{bmatrix} a & b \\ c & d \end{bmatrix}$, then A is invertible if and only if $ad - bc \neq 0$. If $ad - bc \neq 0$, verify that the inverse is

$$A^{-1} = \frac{1}{ad - bc}\begin{bmatrix} d & -b \\ -c & a \end{bmatrix}.$$

76. ***Exploration*** Consider matrices of the form

$$A = \begin{bmatrix} a_{11} & 0 & 0 & 0 & \cdots & 0 \\ 0 & a_{22} & 0 & 0 & \cdots & 0 \\ 0 & 0 & a_{33} & 0 & \cdots & 0 \\ \vdots & \vdots & \vdots & \vdots & \cdots & \vdots \\ 0 & 0 & 0 & 0 & \cdots & a_{nn} \end{bmatrix}.$$

(a) Write a 2×2 matrix and a 3×3 matrix in the form of A. Find the inverse of each.

(b) Use the result of part (a) to make a conjecture about the inverses of matrices in the form of A.

Review

In Exercises 77 and 78, solve the inequality and sketch the solution on the real number line.

77. $|x + 7| \geq 2$

78. $|2x - 1| < 3$

In Exercises 79–82, solve the equation.

79. $3^{x/2} = 315$

80. $2000e^{-x/5} = 400$

81. $\log_2 x - 2 = 4.5$

82. $\ln x + \ln(x - 1) = 0$

10.4 The Determinant of a Square Matrix

▶ **What you should learn**

- How to find the determinants of 2×2 matrices
- How to find minors and cofactors of square matrices
- How to find the determinants of square matrices

▶ **Why you should learn it**

Determinants are often used in other branches of mathematics. For instance, Exercises 79–84 on page 777 show some types of determinants that are useful when changes in variables are made in calculus.

The Determinant of a 2 × 2 Matrix 

Every *square* matrix can be associated with a real number called its **determinant.** Determinants have many uses, and several will be discussed in this and the next section. Historically, the use of determinants arose from special number patterns that occur when systems of linear equations are solved. For instance, the system

$$\begin{cases} a_1x + b_1y = c_1 \\ a_2x + b_2y = c_2 \end{cases}$$

has a solution

$$x = \frac{c_1b_2 - c_2b_1}{a_1b_2 - a_2b_1} \quad \text{and} \quad y = \frac{a_1c_2 - a_2c_1}{a_1b_2 - a_2b_1}$$

provided that $a_1b_2 - a_2b_1 \neq 0$. Note that the denominators of the two fractions are the same. This denominator is called the *determinant* of the coefficient matrix of the system.

Coefficient Matrix   *Determinant*

$$A = \begin{bmatrix} a_1 & b_1 \\ a_2 & b_2 \end{bmatrix} \qquad \det(A) = a_1b_2 - a_2b_1$$

The determinant of the matrix A can also be denoted by vertical bars on both sides of the matrix, as indicated in the following definition.

Definition of the Determinant of a 2 × 2 Matrix

The **determinant** of the matrix

$$A = \begin{bmatrix} a_1 & b_1 \\ a_2 & b_2 \end{bmatrix}$$

is given by

$$\det(A) = |A| = \begin{vmatrix} a_1 & b_1 \\ a_2 & b_2 \end{vmatrix} = a_1b_2 - a_2b_1.$$

In this book, $\det(A)$ and $|A|$ are used interchangeably to represent the determinant of A. Although vertical bars are also used to denote the absolute value of a real number, the context will show which use is intended.

A convenient method for remembering the formula for the determinant of a 2×2 matrix is shown in the following diagram.

$$\det(A) = \begin{vmatrix} a_1 & b_1 \\ a_2 & b_2 \end{vmatrix} = a_1b_2 - a_2b_1$$

Note that the determinant is the difference of the products of the two diagonals of the matrix.

Example 1 ▶ The Determinant of a 2 × 2 Matrix 

Find the determinant of each matrix.

a. $A = \begin{bmatrix} 2 & -3 \\ 1 & 2 \end{bmatrix}$

b. $B = \begin{bmatrix} 2 & 1 \\ 4 & 2 \end{bmatrix}$

c. $C = \begin{bmatrix} 0 & \frac{3}{2} \\ 2 & 4 \end{bmatrix}$

Solution

a. $\det(A) = \begin{vmatrix} 2 & -3 \\ 1 & 2 \end{vmatrix}$

$= 2(2) - 1(-3)$

$= 4 + 3 = 7$

b. $\det(B) = \begin{vmatrix} 2 & 1 \\ 4 & 2 \end{vmatrix}$

$= 2(2) - 4(1)$

$= 4 - 4 = 0$

c. $\det(C) = \begin{vmatrix} 0 & \frac{3}{2} \\ 2 & 4 \end{vmatrix}$

$= 0(4) - 2(\frac{3}{2})$

$= 0 - 3 = -3$

Exploration

Use a graphing utility to find the determinant of the following matrix.

$$A = \begin{bmatrix} 1 & 2 \\ -1 & 0 \\ 3 & -2 \end{bmatrix}$$

What message appears on the screen? Why does the graphing utility display this message?

Notice in Example 1 that the determinant of a matrix can be positive, zero, or negative.

The determinant of a matrix of order 1×1 is defined simply as the entry of the matrix. For instance, if $A = [-2]$, then $\det(A) = -2$.

Technology

Most graphing utilities can evaluate the determinant of a matrix. For instance, you can evaluate the determinant of

$$A = \begin{bmatrix} 2 & -3 \\ 1 & 2 \end{bmatrix}$$

by entering the matrix as $[A]$ and then choosing the *determinant* feature. The result should be 7, as in Example 1(a). Try evaluating the determinants of other matrices.

Minors and Cofactors

To define the determinant of a square matrix of order 3×3 or higher, it is convenient to introduce the concepts of **minors** and **cofactors.**

Sign Pattern for Cofactors

$$\begin{bmatrix} + & - & + \\ - & + & - \\ + & - & + \end{bmatrix}$$

3 × 3 matrix

$$\begin{bmatrix} + & - & + & - \\ - & + & - & + \\ + & - & + & - \\ - & + & - & + \end{bmatrix}$$

4 × 4 matrix

$$\begin{bmatrix} + & - & + & - & + & \cdots \\ - & + & - & + & - & \cdots \\ + & - & + & - & + & \cdots \\ - & + & - & + & - & \cdots \\ + & - & + & - & + & \cdots \\ \vdots & \vdots & \vdots & \vdots & \vdots & \end{bmatrix}$$

$n \times n$ matrix

Minors and Cofactors of a Square Matrix

If A is a square matrix, the **minor** M_{ij} of the entry a_{ij} is the determinant of the matrix obtained by deleting the ith row and jth column of A. The **cofactor** C_{ij} of the entry a_{ij} is

$$C_{ij} = (-1)^{i+j}M_{ij}.$$

In the sign pattern for cofactors at the left, notice that *odd* positions (where $i + j$ is odd) have negative signs and *even* positions (where $i + j$ is even) have positive signs.

Example 2 ▶ Finding the Minors and Cofactors of a Matrix

Find all the minors and cofactors of

$$A = \begin{bmatrix} 0 & 2 & 1 \\ 3 & -1 & 2 \\ 4 & 0 & 1 \end{bmatrix}.$$

Solution

To find the minor M_{11}, delete the first row and first column of A and evaluate the determinant of the resulting matrix.

$$\begin{bmatrix} 0 & 2 & 1 \\ 3 & -1 & 2 \\ 4 & 0 & 1 \end{bmatrix}, \quad M_{11} = \begin{vmatrix} -1 & 2 \\ 0 & 1 \end{vmatrix} = -1(1) - 0(2) = -1$$

Similarly, to find M_{12}, delete the first row and second column.

$$\begin{bmatrix} 0 & 2 & 1 \\ 3 & -1 & 2 \\ 4 & 0 & 1 \end{bmatrix}, \quad M_{12} = \begin{vmatrix} 3 & 2 \\ 4 & 1 \end{vmatrix} = 3(1) - 4(2) = -5$$

Continuing this pattern, you obtain the minors.

$$\begin{array}{lll} M_{11} = -1 & M_{12} = -5 & M_{13} = 4 \\ M_{21} = 2 & M_{22} = -4 & M_{23} = -8 \\ M_{31} = 5 & M_{32} = -3 & M_{33} = -6 \end{array}$$

Now, to find the cofactors, combine the checkerboard pattern of signs for a 3×3 matrix (at left above) with these minors.

$$\begin{array}{lll} C_{11} = -1 & C_{12} = 5 & C_{13} = 4 \\ C_{21} = -2 & C_{22} = -4 & C_{23} = 8 \\ C_{31} = 5 & C_{32} = 3 & C_{33} = -6 \end{array}$$

The Determinant of a Square Matrix

The definition below is called *inductive* because it uses determinants of matrices of order $n - 1$ to define determinants of matrices of order n.

Determinant of a Square Matrix

If A is a square matrix (of order 2×2 or greater), the determinant of A is the sum of the entries in any row (or column) of A multiplied by their respective cofactors. For instance, expanding along the first row yields

$$|A| = a_{11}C_{11} + a_{12}C_{12} + \cdot \cdot \cdot + a_{1n}C_{1n}.$$

Applying this definition to find a determinant is called expanding by *cofactors.*

Try checking that for a 2×2 matrix

$$A = \begin{bmatrix} a_1 & b_1 \\ a_2 & b_2 \end{bmatrix}$$

this definition of the determinant yields $|A| = a_1b_2 - a_2b_1$, as previously defined.

Example 3 ▶ The Determinant of a Matrix of Order 3 × 3

Find the determinant of

$$A = \begin{bmatrix} 0 & 2 & 1 \\ 3 & -1 & 2 \\ 4 & 0 & 1 \end{bmatrix}.$$

Solution

Note that this is the same matrix that was in Example 2. There you found the cofactors of the entries in the first row to be

$$C_{11} = -1, \quad C_{12} = 5, \quad \text{and} \quad C_{13} = 4.$$

So, by the definition of a determinant, you have

$$\begin{aligned} |A| &= a_{11}C_{11} + a_{12}C_{12} + a_{13}C_{13} && \text{First-row expansion} \\ &= 0(-1) + 2(5) + 1(4) \\ &= 14. \end{aligned}$$

In Example 3, the determinant was found by expanding by the cofactors in the first row. You could have used any row or column. For instance, you could have expanded along the second row to obtain

$$\begin{aligned} |A| &= a_{21}C_{21} + a_{22}C_{22} + a_{23}C_{23} && \text{Second-row expansion} \\ &= 3(-2) + (-1)(-4) + 2(8) \\ &= 14. \end{aligned}$$

When expanding by cofactors, you do not need to find cofactors of zero entries, because zero times its cofactor is zero.

$$a_{ij}C_{ij} = (0)C_{ij} = 0$$

So, the row (or column) containing the most zeros is usually the best choice for expansion by cofactors. This is demonstrated in the next example.

Example 4 ▶ The Determinant of a Matrix of Order 4 × 4 

Find the determinant of

$$A = \begin{bmatrix} 1 & -2 & 3 & 0 \\ -1 & 1 & 0 & 2 \\ 0 & 2 & 0 & 3 \\ 3 & 4 & 0 & 2 \end{bmatrix}.$$

Solution

After inspecting this matrix, you can see that three of the entries in the third column are zeros. So, you can eliminate some of the work in the expansion by using the third column.

$$|A| = 3(C_{13}) + 0(C_{23}) + 0(C_{33}) + 0(C_{43})$$

Because C_{23}, C_{33}, and C_{43} have zero coefficients, you need only find the cofactor C_{13}. To do this, delete the first row and third column of A and evaluate the determinant of the resulting matrix.

$$C_{13} = (-1)^{1+3}\begin{vmatrix} -1 & 1 & 2 \\ 0 & 2 & 3 \\ 3 & 4 & 2 \end{vmatrix} \qquad \text{Delete 1st row and 3rd column.}$$

$$= \begin{vmatrix} -1 & 1 & 2 \\ 0 & 2 & 3 \\ 3 & 4 & 2 \end{vmatrix} \qquad \text{Simplify.}$$

Expanding by cofactors in the second row yields

$$C_{13} = 0(-1)^3\begin{vmatrix} 1 & 2 \\ 4 & 2 \end{vmatrix} + 2(-1)^4\begin{vmatrix} -1 & 2 \\ 3 & 2 \end{vmatrix} + 3(-1)^5\begin{vmatrix} -1 & 1 \\ 3 & 4 \end{vmatrix}$$

$$= 0 + 2(1)(-8) + 3(-1)(-7)$$

$$= 5.$$

So, you obtain

$$|A| = 3C_{13}$$

$$= 3(5)$$

$$= 15.$$

Try using a graphing utility to confirm the result of Example 4.

10.4 Exercises

In Exercises 1–16, find the determinant of the matrix.

1. $[5]$

2. $[-8]$

3. $\begin{bmatrix} 2 & 1 \\ 3 & 4 \end{bmatrix}$

4. $\begin{bmatrix} -3 & 1 \\ 5 & 2 \end{bmatrix}$

5. $\begin{bmatrix} 5 & 2 \\ -6 & 3 \end{bmatrix}$

6. $\begin{bmatrix} 2 & -2 \\ 4 & 3 \end{bmatrix}$

7. $\begin{bmatrix} -7 & 0 \\ 3 & 0 \end{bmatrix}$

8. $\begin{bmatrix} 4 & -3 \\ 0 & 0 \end{bmatrix}$

9. $\begin{bmatrix} 2 & 6 \\ 0 & 3 \end{bmatrix}$

10. $\begin{bmatrix} 2 & -3 \\ -6 & 9 \end{bmatrix}$

11. $\begin{bmatrix} -3 & -2 \\ -6 & -1 \end{bmatrix}$

12. $\begin{bmatrix} 4 & 7 \\ -2 & 5 \end{bmatrix}$

13. $\begin{bmatrix} 9 & 0 \\ 7 & 8 \end{bmatrix}$

14. $\begin{bmatrix} 0 & 6 \\ -3 & 2 \end{bmatrix}$

15. $\begin{bmatrix} -\frac{1}{2} & \frac{1}{3} \\ -6 & \frac{1}{3} \end{bmatrix}$

16. $\begin{bmatrix} \frac{2}{3} & \frac{4}{3} \\ -1 & -\frac{1}{3} \end{bmatrix}$

In Exercises 17–22, use the matrix capabilities of a graphing utility to find the determinant of the matrix.

17. $\begin{bmatrix} 0.3 & 0.2 & 0.2 \\ 0.2 & 0.2 & 0.2 \\ -0.4 & 0.4 & 0.3 \end{bmatrix}$

18. $\begin{bmatrix} 0.1 & 0.2 & 0.3 \\ -0.3 & 0.2 & 0.2 \\ 0.5 & 0.4 & 0.4 \end{bmatrix}$

19. $\begin{bmatrix} 0.9 & 0.7 & 0 \\ -0.1 & 0.3 & 1.3 \\ -2.2 & 4.2 & 6.1 \end{bmatrix}$

20. $\begin{bmatrix} 0.1 & 0.1 & -4.3 \\ 7.5 & 6.2 & 0.7 \\ 0.3 & 0.6 & -1.2 \end{bmatrix}$

21. $\begin{bmatrix} 1 & 4 & -2 \\ 3 & 6 & -6 \\ -2 & 1 & 4 \end{bmatrix}$

22. $\begin{bmatrix} 2 & 3 & 1 \\ 0 & 5 & -2 \\ 0 & 0 & -2 \end{bmatrix}$

In Exercises 23–30, find all (a) minors and (b) cofactors of the matrix.

23. $\begin{bmatrix} 3 & 4 \\ 2 & -5 \end{bmatrix}$

24. $\begin{bmatrix} 11 & 0 \\ -3 & 2 \end{bmatrix}$

25. $\begin{bmatrix} 3 & 1 \\ -2 & -4 \end{bmatrix}$

26. $\begin{bmatrix} -6 & 5 \\ 7 & -2 \end{bmatrix}$

27. $\begin{bmatrix} 4 & 0 & 2 \\ -3 & 2 & 1 \\ 1 & -1 & 1 \end{bmatrix}$

28. $\begin{bmatrix} 1 & -1 & 0 \\ 3 & 2 & 5 \\ 4 & -6 & 4 \end{bmatrix}$

29. $\begin{bmatrix} 3 & -2 & 8 \\ 3 & 2 & -6 \\ -1 & 3 & 6 \end{bmatrix}$

30. $\begin{bmatrix} -2 & 9 & 4 \\ 7 & -6 & 0 \\ 6 & 7 & -6 \end{bmatrix}$

In Exercises 31–36, find the determinant of the matrix by the method of expansion by cofactors. Expand using the indicated row or column.

31. $\begin{bmatrix} -3 & 2 & 1 \\ 4 & 5 & 6 \\ 2 & -3 & 1 \end{bmatrix}$

(a) Row 1

(b) Column 2

32. $\begin{bmatrix} -3 & 4 & 2 \\ 6 & 3 & 1 \\ 4 & -7 & -8 \end{bmatrix}$

(a) Row 2

(b) Column 3

33. $\begin{bmatrix} 5 & 0 & -3 \\ 0 & 12 & 4 \\ 1 & 6 & 3 \end{bmatrix}$

(a) Row 2

(b) Column 2

34. $\begin{bmatrix} 10 & -5 & 5 \\ 30 & 0 & 10 \\ 0 & 10 & 1 \end{bmatrix}$

(a) Row 3

(b) Column 1

35. $\begin{bmatrix} 6 & 0 & -3 & 5 \\ 4 & 13 & 6 & -8 \\ -1 & 0 & 7 & 4 \\ 8 & 6 & 0 & 2 \end{bmatrix}$

(a) Row 2

(b) Column 2

36. $\begin{bmatrix} 10 & 8 & 3 & -7 \\ 4 & 0 & 5 & -6 \\ 0 & 3 & 2 & 7 \\ 1 & 0 & -3 & 2 \end{bmatrix}$

(a) Row 3

(b) Column 1

In Exercises 37–52, find the determinant of the matrix. Expand by cofactors on the row or column that appears to make the computations easiest.

37. $\begin{bmatrix} 2 & -1 & 0 \\ 4 & 2 & 1 \\ 4 & 2 & 1 \end{bmatrix}$

38. $\begin{bmatrix} -2 & 2 & 3 \\ 1 & -1 & 0 \\ 0 & 1 & 4 \end{bmatrix}$

39. $\begin{bmatrix} 6 & 3 & -7 \\ 0 & 0 & 0 \\ 4 & -6 & 3 \end{bmatrix}$

40. $\begin{bmatrix} 1 & 1 & 2 \\ 3 & 1 & 0 \\ -2 & 0 & 3 \end{bmatrix}$

41. $\begin{bmatrix} -1 & 2 & -5 \\ 0 & 3 & 4 \\ 0 & 0 & 3 \end{bmatrix}$

42. $\begin{bmatrix} 1 & 0 & 0 \\ -4 & -1 & 0 \\ 5 & 1 & 5 \end{bmatrix}$

43. $\begin{bmatrix} 1 & 4 & -2 \\ 3 & 2 & 0 \\ -1 & 4 & 3 \end{bmatrix}$

44. $\begin{bmatrix} 2 & -1 & 3 \\ 1 & 4 & 4 \\ 1 & 0 & 2 \end{bmatrix}$

45. $\begin{bmatrix} 2 & 4 & 6 \\ 0 & 3 & 1 \\ 0 & 0 & -5 \end{bmatrix}$

46. $\begin{bmatrix} -3 & 0 & 0 \\ 7 & 11 & 0 \\ 1 & 2 & 2 \end{bmatrix}$

47. $\begin{bmatrix} 2 & 6 & 6 & 2 \\ 2 & 7 & 3 & 6 \\ 1 & 5 & 0 & 1 \\ 3 & 7 & 0 & 7 \end{bmatrix}$

48. $\begin{bmatrix} 3 & 6 & -5 & 4 \\ -2 & 0 & 6 & 0 \\ 1 & 1 & 2 & 2 \\ 0 & 3 & -1 & -1 \end{bmatrix}$

49. $\begin{bmatrix} 5 & 3 & 0 & 6 \\ 4 & 6 & 4 & 12 \\ 0 & 2 & -3 & 4 \\ 0 & 1 & -2 & 2 \end{bmatrix}$

50. $\begin{bmatrix} 1 & 4 & 3 & 2 \\ -5 & 6 & 2 & 1 \\ 0 & 0 & 0 & 0 \\ 3 & -2 & 1 & 5 \end{bmatrix}$

51. $\begin{bmatrix} 3 & 2 & 4 & -1 & 5 \\ -2 & 0 & 1 & 3 & 2 \\ 1 & 0 & 0 & 4 & 0 \\ 6 & 0 & 2 & -1 & 0 \\ 3 & 0 & 5 & 1 & 0 \end{bmatrix}$

52. $\begin{bmatrix} 5 & 2 & 0 & 0 & -2 \\ 0 & 1 & 4 & 3 & 2 \\ 0 & 0 & 2 & 6 & 3 \\ 0 & 0 & 3 & 4 & 1 \\ 0 & 0 & 0 & 0 & 2 \end{bmatrix}$

In Exercises 53–60, use the matrix capabilities of a graphing utility to evaluate the determinant.

53. $\begin{vmatrix} 3 & 8 & -7 \\ 0 & -5 & 4 \\ 8 & 1 & 6 \end{vmatrix}$

54. $\begin{vmatrix} 5 & -8 & 0 \\ 9 & 7 & 4 \\ -8 & 7 & 1 \end{vmatrix}$

55. $\begin{vmatrix} 7 & 0 & -14 \\ -2 & 5 & 4 \\ -6 & 2 & 12 \end{vmatrix}$

56. $\begin{vmatrix} 3 & 0 & 0 \\ -2 & 5 & 0 \\ 12 & 5 & 7 \end{vmatrix}$

57. $\begin{vmatrix} 1 & -1 & 8 & 4 \\ 2 & 6 & 0 & -4 \\ 2 & 0 & 2 & 6 \\ 0 & 2 & 8 & 0 \end{vmatrix}$

58. $\begin{vmatrix} 0 & -3 & 8 & 2 \\ 8 & 1 & -1 & 6 \\ -4 & 6 & 0 & 9 \\ -7 & 0 & 0 & 14 \end{vmatrix}$

59. $\begin{vmatrix} 3 & -2 & 4 & 3 & 1 \\ -1 & 0 & 2 & 1 & 0 \\ 5 & -1 & 0 & 3 & 2 \\ 4 & 7 & -8 & 0 & 0 \\ 1 & 2 & 3 & 0 & 2 \end{vmatrix}$

60. $\begin{vmatrix} -2 & 0 & 0 & 0 & 0 \\ 0 & 3 & 0 & 0 & 0 \\ 0 & 0 & -1 & 0 & 0 \\ 0 & 0 & 0 & 2 & 0 \\ 0 & 0 & 0 & 0 & -4 \end{vmatrix}$

In Exercises 61–68, find (a) $|A|$, (b) $|B|$, (c) AB, and (d) $|AB|$.

61. $A = \begin{bmatrix} -1 & 0 \\ 0 & 3 \end{bmatrix}, \quad B = \begin{bmatrix} 2 & 0 \\ 0 & -1 \end{bmatrix}$

62. $A = \begin{bmatrix} -2 & 1 \\ 4 & -2 \end{bmatrix}, \quad B = \begin{bmatrix} 1 & 2 \\ 0 & -1 \end{bmatrix}$

63. $A = \begin{bmatrix} 4 & 0 \\ 3 & -2 \end{bmatrix}, \quad B = \begin{bmatrix} -1 & 1 \\ -2 & 2 \end{bmatrix}$

64. $A = \begin{bmatrix} 5 & 4 \\ 3 & -1 \end{bmatrix}, \quad B = \begin{bmatrix} 0 & 6 \\ 1 & -2 \end{bmatrix}$

65. $A = \begin{bmatrix} 0 & 1 & 2 \\ -3 & -2 & 1 \\ 0 & 4 & 1 \end{bmatrix}, \quad B = \begin{bmatrix} 3 & -2 & 0 \\ 1 & -1 & 2 \\ 3 & 1 & 1 \end{bmatrix}$

66. $A = \begin{bmatrix} 3 & 2 & 0 \\ -1 & -3 & 4 \\ -2 & 0 & 1 \end{bmatrix}, \quad B = \begin{bmatrix} -3 & 0 & 1 \\ 0 & 2 & -1 \\ -2 & -1 & 1 \end{bmatrix}$

67. $A = \begin{bmatrix} -1 & 2 & 1 \\ 1 & 0 & 1 \\ 0 & 1 & 0 \end{bmatrix}, \quad B = \begin{bmatrix} -1 & 0 & 0 \\ 0 & 2 & 0 \\ 0 & 0 & 3 \end{bmatrix}$

68. $A = \begin{bmatrix} 2 & 0 & 1 \\ 1 & -1 & 2 \\ 3 & 1 & 0 \end{bmatrix}, \quad B = \begin{bmatrix} 2 & -1 & 4 \\ 0 & 1 & 3 \\ 3 & -2 & 1 \end{bmatrix}$

In Exercises 69–74, evaluate the determinant(s) to verify the equation.

69. $\begin{vmatrix} w & x \\ y & z \end{vmatrix} = -\begin{vmatrix} y & z \\ w & x \end{vmatrix}$

70. $\begin{vmatrix} w & cx \\ y & cz \end{vmatrix} = c\begin{vmatrix} w & x \\ y & z \end{vmatrix}$

71. $\begin{vmatrix} w & x \\ y & z \end{vmatrix} = \begin{vmatrix} w & x + cw \\ y & z + cy \end{vmatrix}$

72. $\begin{vmatrix} w & x \\ cw & cx \end{vmatrix} = 0$

73. $\begin{vmatrix} 1 & x & x^2 \\ 1 & y & y^2 \\ 1 & z & z^2 \end{vmatrix} = (y - x)(z - x)(z - y)$

74. $\begin{vmatrix} a + b & a & a \\ a & a + b & a \\ a & a & a + b \end{vmatrix} = b^2(3a + b)$

In Exercises 75–78, solve for x.

75. $\begin{vmatrix} x - 1 & 2 \\ 3 & x - 2 \end{vmatrix} = 0$

76. $\begin{vmatrix} x - 2 & -1 \\ -3 & x \end{vmatrix} = 0$

77. $\begin{vmatrix} x + 3 & 2 \\ 1 & x + 2 \end{vmatrix} = 0$

78. $\begin{vmatrix} x + 4 & -2 \\ 7 & x - 5 \end{vmatrix} = 0$

In Exercises 79–84, evaluate the determinant in which the entries are functions. Determinants of this type occur when changes in variables are made in calculus.

79. $\begin{vmatrix} 4u & -1 \\ -1 & 2v \end{vmatrix}$

80. $\begin{vmatrix} 3x^2 & -3y^2 \\ 1 & 1 \end{vmatrix}$

81. $\begin{vmatrix} e^{2x} & e^{3x} \\ 2e^{2x} & 3e^{3x} \end{vmatrix}$

82. $\begin{vmatrix} e^{-x} & xe^{-x} \\ -e^{-x} & (1-x)e^{-x} \end{vmatrix}$

83. $\begin{vmatrix} x & \ln x \\ 1 & 1/x \end{vmatrix}$

84. $\begin{vmatrix} x & x \ln x \\ 1 & 1 + \ln x \end{vmatrix}$

Synthesis

True or False? **In Exercises 85 and 86, determine whether the statement is true or false. Justify your answer.**

85. If a square matrix has an entire row of zeros, the determinant will always be zero.

86. If two columns of a square matrix are the same, the determinant of the matrix will be zero.

87. ***Exploration*** Find square matrices A and B to demonstrate that

$|A + B| \neq |A| + |B|$.

88. ***Exploration*** Consider square matrices in which the entries are consecutive integers. An example of such a matrix is

$$\begin{bmatrix} 4 & 5 & 6 \\ 7 & 8 & 9 \\ 10 & 11 & 12 \end{bmatrix}.$$

(a) Use a graphing utility to evaluate the determinants of four matrices of this type. Make a conjecture based on the results.

(b) Verify your conjecture.

89. ***Writing*** Write a brief paragraph explaining the difference between a square matrix and its determinant.

90. ***Think About It*** If A is a matrix of order 3×3 such that $|A| = 5$, is it possible to find $|2A|$? Explain.

Properties of Determinants **In Exercises 91 and 92, a property of determinants is given. State how the property has been applied to the given determinants and use a graphing utility to verify the results.**

91. If A and B are square matrices and B is obtained from A by interchanging two rows of A or interchanging two columns of A, then $|B| = -|A|$.

(a) $\begin{vmatrix} 1 & 3 & 4 \\ -7 & 2 & -5 \\ 6 & 1 & 2 \end{vmatrix} = -\begin{vmatrix} 1 & 4 & 3 \\ -7 & -5 & 2 \\ 6 & 2 & 1 \end{vmatrix}$

(b) $\begin{vmatrix} 1 & 3 & 4 \\ -2 & 2 & 0 \\ 1 & 6 & 2 \end{vmatrix} = -\begin{vmatrix} 1 & 6 & 2 \\ -2 & 2 & 0 \\ 1 & 3 & 4 \end{vmatrix}$

92. If A and B are square matrices and B is obtained from A by adding a multiple of a row of A to another row of A or by adding a multiple of a column of A to another column of A, then $|B| = |A|$.

(a) $\begin{vmatrix} 1 & -3 \\ 5 & 2 \end{vmatrix} = \begin{vmatrix} 1 & -3 \\ 0 & 17 \end{vmatrix}$

(b) $\begin{vmatrix} 5 & 4 & 2 \\ 2 & -3 & 4 \\ 7 & 6 & 3 \end{vmatrix} = \begin{vmatrix} 1 & 10 & -6 \\ 2 & -3 & 4 \\ 7 & 6 & 3 \end{vmatrix}$

Review

In Exercises 93–98, find the domain of the function.

93. $f(x) = x^3 - 2x$

94. $g(x) = \sqrt[3]{x}$

95. $h(x) = \sqrt{16 - x^2}$

96. $A(x) = \dfrac{3}{36 - x^2}$

97. $g(t) = \ln(t - 1)$

98. $f(s) = 625e^{-0.5s}$

In Exercises 99–102, find the equation of the conic satisfying the conditions.

99. Parabola: Vertex: $(0, 3)$; Focus: $(2, 3)$

100. Ellipse: Vertices: $(0, \pm 4)$; Foci: $(0, \pm 3)$

101. Ellipse: Vertices: $(\pm 8, 0)$; Foci: $(\pm 6, 0)$

102. Hyperbola: Vertices: $(\pm 5, 0)$; Foci: $(\pm 6, 0)$

In Exercises 103 and 104, sketch the graph of the system of inequalities.

103. $\begin{cases} x + y \leq 8 \\ x \geq -3 \\ 2x - y < 5 \end{cases}$

104. $\begin{cases} -x - y > 4 \\ y \leq 1 \\ 7x + 4y \leq -10 \end{cases}$

In Exercises 105–108, find the inverse of the matrix (if it exists).

105. $\begin{bmatrix} -4 & 1 \\ 8 & -1 \end{bmatrix}$

106. $\begin{bmatrix} -5 & -8 \\ 3 & 6 \end{bmatrix}$

107. $\begin{bmatrix} -7 & 2 & 9 \\ 2 & -4 & -6 \\ 3 & 5 & 2 \end{bmatrix}$

108. $\begin{bmatrix} -6 & 2 & 0 \\ 1 & 3 & -2 \\ -2 & 0 & 1 \end{bmatrix}$

10.5 Applications of Matrices and Determinants

▶ **What you should learn**

- How to use Cramer's Rule to solve systems of linear equations
- How to use determinants to find the areas of triangles
- How to use a determinant to test for collinear points and find an equation of a line passing through two points
- How to use matrices to code and decode messages

▶ **Why you should learn it**

You can use Cramer's Rule to solve real-life problems. For instance, in Exercise 56 on page 789, Cramer's Rule is used to find a quadratic model for the number of U.S. Supreme Court cases waiting to be tried.

Lester Lefkowitz/Corbis

Cramer's Rule

So far, you have studied three methods for solving a system of linear equations: substitution, elimination with equations, and elimination with matrices. In this section, you will study one more method, **Cramer's Rule,** named after Gabriel Cramer (1704–1752). This rule uses determinants to write the solution of a system of linear equations. To see how Cramer's Rule works, take another look at the solution described at the beginning of Section 10.4. There, it was pointed out that the system

$$\begin{cases} a_1x + b_1y = c_1 \\ a_2x + b_2y = c_2 \end{cases}$$

has a solution

$$x = \frac{c_1b_2 - c_2b_1}{a_1b_2 - a_2b_1}$$

and

$$y = \frac{a_1c_2 - a_2c_1}{a_1b_2 - a_2b_1}$$

provided that $a_1b_2 - a_2b_1 \neq 0$. Each numerator and denominator in this solution can be expressed as a determinant, as follows.

$$x = \frac{c_1b_2 - c_2b_1}{a_1b_2 - a_2b_1} = \frac{\begin{vmatrix} c_1 & b_1 \\ c_2 & b_2 \end{vmatrix}}{\begin{vmatrix} a_1 & b_1 \\ a_2 & b_2 \end{vmatrix}}$$

$$y = \frac{a_1c_2 - a_2c_1}{a_1b_2 - a_2b_1} = \frac{\begin{vmatrix} a_1 & c_1 \\ a_2 & c_2 \end{vmatrix}}{\begin{vmatrix} a_1 & b_1 \\ a_2 & b_2 \end{vmatrix}}$$

Relative to the original system, the denominator for x and y is simply the determinant of the *coefficient* matrix of the system. This determinant is denoted by D. The numerators for x and y are denoted by D_x and D_y, respectively. They are formed by using the column of constants as replacements for the coefficients of x and y, as follows.

Coefficient Matrix	D	D_x	D_y
$\begin{bmatrix} a_1 & b_1 \\ a_2 & b_2 \end{bmatrix}$	$\begin{vmatrix} a_1 & b_1 \\ a_2 & b_2 \end{vmatrix}$	$\begin{vmatrix} c_1 & b_1 \\ c_2 & b_2 \end{vmatrix}$	$\begin{vmatrix} a_1 & c_1 \\ a_2 & c_2 \end{vmatrix}$

Example 1 ▶ Using Cramer's Rule for a 2 × 2 System 

Use Cramer's Rule to solve the system of linear equations.

$$\begin{cases} 4x - 2y = 10 \\ 3x - 5y = 11 \end{cases}$$

Solution

To begin, find the determinant of the coefficient matrix.

$$D = \begin{vmatrix} 4 & -2 \\ 3 & -5 \end{vmatrix} = -20 - (-6) = -14$$

Because this determinant is not zero, you can apply Cramer's Rule to find the solution, as follows.

$$x = \frac{D_x}{D} = \frac{\begin{vmatrix} 10 & -2 \\ 11 & -5 \end{vmatrix}}{-14} = \frac{-50 - (-22)}{-14} = \frac{-28}{-14} = 2$$

$$y = \frac{D_y}{D} = \frac{\begin{vmatrix} 4 & 10 \\ 3 & 11 \end{vmatrix}}{-14} = \frac{44 - 30}{-14} = \frac{14}{-14} = -1$$

So, the solution is $x = 2$ and $y = -1$. Check this in the original system.

Cramer's Rule generalizes easily to systems of n equations in n variables. The value of each variable is given as the quotient of two determinants. The denominator is the determinant of the coefficient matrix, and the numerator is the determinant of the matrix formed by replacing the column corresponding to the variable (being solved for) with the column representing the constants. For instance, the solution for x_3 in the system

$$\begin{cases} a_{11}x_1 + a_{12}x_2 + a_{13}x_3 = b_1 \\ a_{21}x_1 + a_{22}x_2 + a_{23}x_3 = b_2 \\ a_{31}x_1 + a_{32}x_2 + a_{33}x_3 = b_3 \end{cases}$$

is given by

$$x_3 = \frac{|A_3|}{|A|} = \frac{\begin{vmatrix} a_{11} & a_{12} & b_1 \\ a_{21} & a_{22} & b_2 \\ a_{31} & a_{32} & b_3 \end{vmatrix}}{\begin{vmatrix} a_{11} & a_{12} & a_{13} \\ a_{21} & a_{22} & a_{23} \\ a_{31} & a_{32} & a_{33} \end{vmatrix}}.$$

STUDY TIP

When using Cramer's Rule, remember that this method does not apply if the determinant of the coefficient matrix is zero.

Cramer's Rule

If a system of n linear equations in n variables has a coefficient matrix A with a nonzero determinant $|A|$, the solution of the system is

$$x_1 = \frac{|A_1|}{|A|}, \quad x_2 = \frac{|A_2|}{|A|}, \quad \ldots, \quad x_n = \frac{|A_n|}{|A|}$$

where the ith column of A_i is the column of constants in the system of equations. If the determinant of the coefficient matrix is zero, the system has either no solution or infinitely many solutions.

Example 2 ▶ Using Cramer's Rule for a 3 × 3 System 

Use Cramer's Rule to solve the system of linear equations.

$$\begin{cases} -x + 2y - 3z = 1 \\ 2x \qquad\quad + z = 0 \\ 3x - 4y + 4z = 2 \end{cases}$$

Solution

The coefficient matrix

$$\begin{bmatrix} -1 & 2 & -3 \\ 2 & 0 & 1 \\ 3 & -4 & 4 \end{bmatrix}$$

can be expanded along the second row, as follows.

$$\begin{aligned} D &= 2(-1)^3\begin{vmatrix} 2 & -3 \\ -4 & 4 \end{vmatrix} + 0(-1)^4\begin{vmatrix} -1 & -3 \\ 3 & 4 \end{vmatrix} + 1(-1)^5\begin{vmatrix} -1 & 2 \\ 3 & -4 \end{vmatrix} \\ &= -2(-4) + 0 - 1(-2) = 10 \end{aligned}$$

Because this determinant is not zero, you can apply Cramer's Rule to find the solution, as follows.

$$x = \frac{D_x}{D} = \frac{\begin{vmatrix} 1 & 2 & -3 \\ 0 & 0 & 1 \\ 2 & -4 & 4 \end{vmatrix}}{10} = \frac{8}{10} = \frac{4}{5}$$

$$y = \frac{D_y}{D} = \frac{\begin{vmatrix} -1 & 1 & -3 \\ 2 & 0 & 1 \\ 3 & 2 & 4 \end{vmatrix}}{10} = \frac{-15}{10} = -\frac{3}{2}$$

$$z = \frac{D_z}{D} = \frac{\begin{vmatrix} -1 & 2 & 1 \\ 2 & 0 & 0 \\ 3 & -4 & 2 \end{vmatrix}}{10} = \frac{-16}{10} = -\frac{8}{5}$$

The solution is $\left(\frac{4}{5}, -\frac{3}{2}, -\frac{8}{5}\right)$. Check this in the original system.

Area of a Triangle

Another application of matrices and determinants is finding the area of a triangle whose vertices are given as points in a coordinate plane.

> **Area of a Triangle**
>
> The area of a triangle with vertices (x_1, y_1), (x_2, y_2), and (x_3, y_3) is
>
> $$\text{Area} = \pm\frac{1}{2}\begin{vmatrix} x_1 & y_1 & 1 \\ x_2 & y_2 & 1 \\ x_3 & y_3 & 1 \end{vmatrix}$$
>
> where the symbol $\pm$ indicates that the appropriate sign should be chosen to yield a positive area.

Example 3 ▶ **Finding the Area of a Triangle** 

Find the area of a triangle whose vertices are $(1, 0)$, $(2, 2)$, and $(4, 3)$, as shown in Figure 10.1.

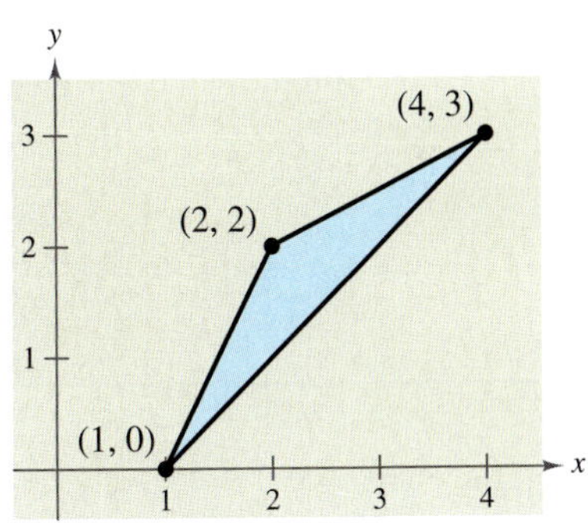

FIGURE 10.1

Solution

Let $(x_1, y_1) = (1, 0)$, $(x_2, y_2) = (2, 2)$, and $(x_3, y_3) = (4, 3)$. Then, to find the area of the triangle, evaluate the determinant.

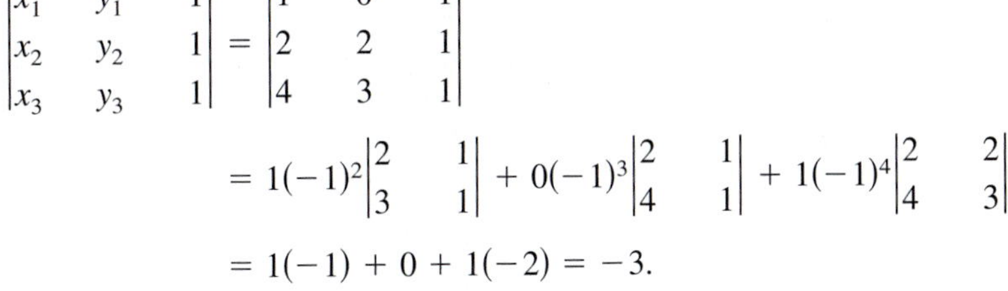

$$\begin{vmatrix} x_1 & y_1 & 1 \\ x_2 & y_2 & 1 \\ x_3 & y_3 & 1 \end{vmatrix} = \begin{vmatrix} 1 & 0 & 1 \\ 2 & 2 & 1 \\ 4 & 3 & 1 \end{vmatrix}$$

$$= 1(-1)^2\begin{vmatrix} 2 & 1 \\ 3 & 1 \end{vmatrix} + 0(-1)^3\begin{vmatrix} 2 & 1 \\ 4 & 1 \end{vmatrix} + 1(-1)^4\begin{vmatrix} 2 & 2 \\ 4 & 3 \end{vmatrix}$$

$$= 1(-1) + 0 + 1(-2) = -3.$$

Using this value, you can conclude that the area of the triangle is

$$\text{Area} = -\frac{1}{2}\begin{vmatrix} 1 & 0 & 1 \\ 2 & 2 & 1 \\ 4 & 3 & 1 \end{vmatrix} \qquad \text{Choose } (-) \text{ so that the area is positive.}$$

$$= -\frac{1}{2}(-3)$$

$$= \frac{3}{2} \text{ square units.}$$

Try using determinants to find the area of a triangle with vertices $(3, -1)$, $(7, -1)$, and $(7, 5)$. Confirm your answer by plotting the points in a coordinate plane and using the formula

$$\text{Area} = \frac{1}{2}(\text{base})(\text{height}).$$

The test for collinear points can be adapted to another use. That is, if you are given two points on a rectangular coordinate system, you can find an equation of the line passing through the two points, as follows.

Two-Point Form of the Equation of a Line

An equation of the line passing through the distinct points (x_1, y_1) and (x_2, y_2) is given by

$$\begin{vmatrix} x & y & 1 \\ x_1 & y_1 & 1 \\ x_2 & y_2 & 1 \end{vmatrix} = 0.$$

Example 5 ▶ **Finding an Equation of a Line** 

Find an equation of the line passing through the two points $(2, 4)$ and $(-1, 3)$, as shown in Figure 10.4.

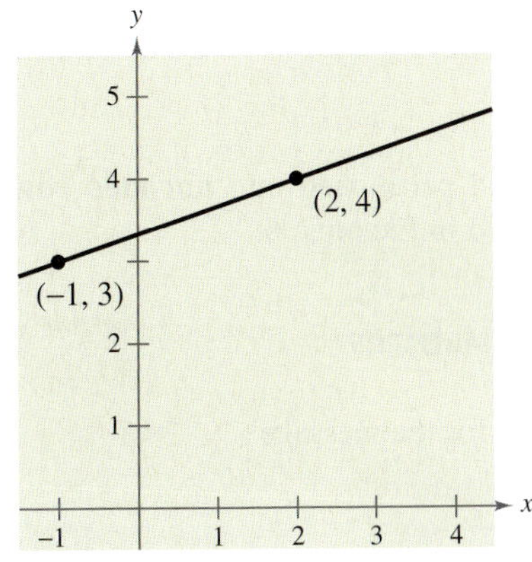

FIGURE 10.4

Solution

Applying the determinant formula for the equation of a line produces

$$\begin{vmatrix} x & y & 1 \\ 2 & 4 & 1 \\ -1 & 3 & 1 \end{vmatrix} = 0.$$

To evaluate this determinant, you can expand by cofactors along the first row to obtain the following.

$$x(-1)^2\begin{vmatrix} 4 & 1 \\ 3 & 1 \end{vmatrix} + y(-1)^3\begin{vmatrix} 2 & 1 \\ -1 & 1 \end{vmatrix} + 1(-1)^4\begin{vmatrix} 2 & 4 \\ -1 & 3 \end{vmatrix} = 0$$

$$x(1)(1) + y(-1)(3) + (1)(1)(10) = 0$$

$$x - 3y + 10 = 0$$

So, an equation of the line is

$$x - 3y + 10 = 0.$$

Note that this method of finding the equation of a line works for all lines, including horizontal and vertical lines. For instance, the equation of the vertical line through $(2, 0)$ and $(2, 2)$ is

$$\begin{vmatrix} x & y & 1 \\ 2 & 0 & 1 \\ 2 & 2 & 1 \end{vmatrix} = 0$$

$$4 - 2x = 0$$

$$x = 2.$$

Cryptography

A **cryptogram** is a message written according to a secret code. (The Greek word *kryptos* means "hidden.") Matrix multiplication can be used to encode and decode messages. To begin, you need to assign a number to each letter in the alphabet (with 0 assigned to a blank space), as follows.

0 = _	9 = I	18 = R
1 = A	10 = J	19 = S
2 = B	11 = K	20 = T
3 = C	12 = L	21 = U
4 = D	13 = M	22 = V
5 = E	14 = N	23 = W
6 = F	15 = O	24 = X
7 = G	16 = P	25 = Y
8 = H	17 = Q	26 = Z

Then the message is converted to numbers and partitioned into **uncoded row matrices,** each having n entries, as demonstrated in Example 6.

Example 6 ▶ Forming Uncoded Row Matrices

Write the uncoded row matrices of order 1×3 for the message

MEET ME MONDAY.

Solution

Partitioning the message (including blank spaces, but ignoring punctuation) into groups of three produces the following uncoded row matrices.

$$\begin{matrix}[13 & 5 & 5] & [20 & 0 & 13] & [5 & 0 & 13] & [15 & 14 & 4] & [1 & 25 & 0] \\ \text{M} & \text{E} & \text{E} & \text{T} & & \text{M} & \text{E} & & \text{M} & \text{O} & \text{N} & \text{D} & \text{A} & \text{Y} & \end{matrix}$$

Note that a blank space is used to fill out the last uncoded row matrix.

To encode a message, choose an $n \times n$ invertible matrix such as

$$A = \begin{bmatrix} 1 & -2 & 2 \\ -1 & 1 & 3 \\ 1 & -1 & -4 \end{bmatrix}$$

and multiply the uncoded row matrices by A (on the right) to obtain **coded row matrices.** Here is an example.

Uncoded Matrix   *Encoding Matrix A*   *Coded Matrix*

$$[13 \quad 5 \quad 5]\begin{bmatrix} 1 & -2 & 2 \\ -1 & 1 & 3 \\ 1 & -1 & -4 \end{bmatrix} = [13 \quad -26 \quad 21]$$

Example 7 ▶ Encoding a Message 

Use the following invertible matrix to encode the message MEET ME MONDAY.

$$A = \begin{bmatrix} 1 & -2 & 2 \\ -1 & 1 & 3 \\ 1 & -1 & -4 \end{bmatrix}$$

Solution

The coded row matrices are obtained by multiplying each of the uncoded row matrices found in Example 6 by the matrix A, as follows.

Uncoded Matrix *Encoding Matrix A* *Coded Matrix*

$$[13 \quad 5 \quad 5] \begin{bmatrix} 1 & -2 & 2 \\ -1 & 1 & 3 \\ 1 & -1 & -4 \end{bmatrix} = [13 \quad -26 \quad 21]$$

$$[20 \quad 0 \quad 13] \begin{bmatrix} 1 & -2 & 2 \\ -1 & 1 & 3 \\ 1 & -1 & -4 \end{bmatrix} = [33 \quad -53 \quad -12]$$

$$[5 \quad 0 \quad 13] \begin{bmatrix} 1 & -2 & 2 \\ -1 & 1 & 3 \\ 1 & -1 & -4 \end{bmatrix} = [18 \quad -23 \quad -42]$$

$$[15 \quad 14 \quad 4] \begin{bmatrix} 1 & -2 & 2 \\ -1 & 1 & 3 \\ 1 & -1 & -4 \end{bmatrix} = [5 \quad -20 \quad 56]$$

$$[1 \quad 25 \quad 0] \begin{bmatrix} 1 & -2 & 2 \\ -1 & 1 & 3 \\ 1 & -1 & -4 \end{bmatrix} = [-24 \quad 23 \quad 77]$$

So, the sequence of coded row matrices is

$$[13 \ -26 \ 21]\,[33 \ -53 \ -12]\,[18 \ -23 \ -42]\,[5 \ -20 \ 56]\,[-24 \ 23 \ 77].$$

Finally, removing the matrix notation produces the following cryptogram.

$$13 \ -26 \ 21 \ 33 \ -53 \ -12 \ 18 \ -23 \ -42 \ 5 \ -20 \ 56 \ -24 \ 23 \ 77$$

For those who do not know the encoding matrix A, decoding the cryptogram found in Example 7 is difficult. But for an authorized receiver who knows the encoding matrix A, decoding is simple. The receiver need only multiply the coded row matrices by A^{-1} (on the right) to retrieve the uncoded row matrices. Here is an example.

$$\underbrace{[13 \ -26 \quad 21]}_{\text{Coded}} A^{-1} = \underbrace{[13 \quad 5 \quad 5]}_{\text{Uncoded}}$$

Example 8 ▶ Decoding a Message 

Use the inverse of the matrix

$$A = \begin{bmatrix} 1 & -2 & 2 \\ -1 & 1 & 3 \\ 1 & -1 & -4 \end{bmatrix}$$

to decode the cryptogram

13 −26 21 33 −53 −12 18 −23 −42 5 −20 56 −24 23 77.

Solution

First find A^{-1} by using the techniques demonstrated in Section 10.3. A^{-1} is the decoding matrix. Then partition the message into groups of three to form the coded row matrices. Finally, multiply each coded row matrix by A^{-1} (on the right).

Coded Matrix Decoding Matrix A^{-1} *Decoded Matrix*

$$[13 \quad -26 \quad 21] \begin{bmatrix} -1 & -10 & -8 \\ -1 & -6 & -5 \\ 0 & -1 & -1 \end{bmatrix} = [13 \quad 5 \quad 5]$$

$$[33 \quad -53 \quad -12] \begin{bmatrix} -1 & -10 & -8 \\ -1 & -6 & -5 \\ 0 & -1 & -1 \end{bmatrix} = [20 \quad 0 \quad 13]$$

$$[18 \quad -23 \quad -42] \begin{bmatrix} -1 & -10 & -8 \\ -1 & -6 & -5 \\ 0 & -1 & -1 \end{bmatrix} = [5 \quad 0 \quad 13]$$

$$[5 \quad -20 \quad 56] \begin{bmatrix} -1 & -10 & -8 \\ -1 & -6 & -5 \\ 0 & -1 & -1 \end{bmatrix} = [15 \quad 14 \quad 4]$$

$$[-24 \quad 23 \quad 77] \begin{bmatrix} -1 & -10 & -8 \\ -1 & -6 & -5 \\ 0 & -1 & -1 \end{bmatrix} = [1 \quad 25 \quad 0]$$

So, the message is as follows.

[13	5	5]	[20	0	13]	[5	0	13]	[15	14	4]	[1	25	0]
M	E	E	T		M	E		M	O	N	D	A	Y	

Writing ABOUT MATHEMATICS

Cryptography Use your school's library, the Internet, or some other reference source to research information about another type of cryptography. Write a short paragraph describing how mathematics is used to code and decode messages.

10.5 Exercises

In Exercises 1–8, use Cramer's Rule to solve (if possible) the system of equations.

1. $\begin{cases} 3x + 4y = -2 \\ 5x + 3y = 4 \end{cases}$

2. $\begin{cases} -4x - 7y = 47 \\ -x + 6y = -27 \end{cases}$

3. $\begin{cases} -0.4x + 0.8y = 1.6 \\ 0.2x + 0.3y = 2.2 \end{cases}$

4. $\begin{cases} 2.4x - 1.3y = 14.63 \\ -4.6x + 0.5y = -11.51 \end{cases}$

5. $\begin{cases} 4x - y + z = -5 \\ 2x + 2y + 3z = 10 \\ 5x - 2y + 6z = 1 \end{cases}$

6. $\begin{cases} 4x - 2y + 3z = -2 \\ 2x + 2y + 5z = 16 \\ 8x - 5y - 2z = 4 \end{cases}$

7. $\begin{cases} x + 2y + 3z = -3 \\ -2x + y - z = 6 \\ 3x - 3y + 2z = -11 \end{cases}$

8. $\begin{cases} 5x - 4y + z = -14 \\ -x + 2y - 2z = 10 \\ 3x + y + z = 1 \end{cases}$

In Exercises 9–12, use a graphing utility and Cramer's Rule to solve (if possible) the system of equations.

9. $\begin{cases} 3x + 3y + 5z = 1 \\ 3x + 5y + 9z = 2 \\ 5x + 9y + 17z = 4 \end{cases}$

10. $\begin{cases} x + 2y - z = -7 \\ 2x - 2y - 2z = -8 \\ -x + 3y + 4z = 8 \end{cases}$

11. $\begin{cases} 2x + y + 2z = 6 \\ -x + 2y - 3z = 0 \\ 3x + 2y - z = 6 \end{cases}$

12. $\begin{cases} 2x + 3y + 5z = 4 \\ 3x + 5y + 9z = 7 \\ 5x + 9y + 17z = 13 \end{cases}$

In Exercises 13–22, use a determinant and the given vertices of a triangle to find the area of the triangle.

13.

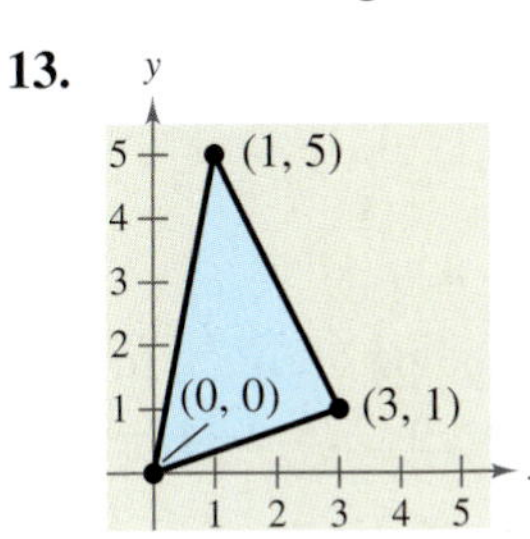

14.

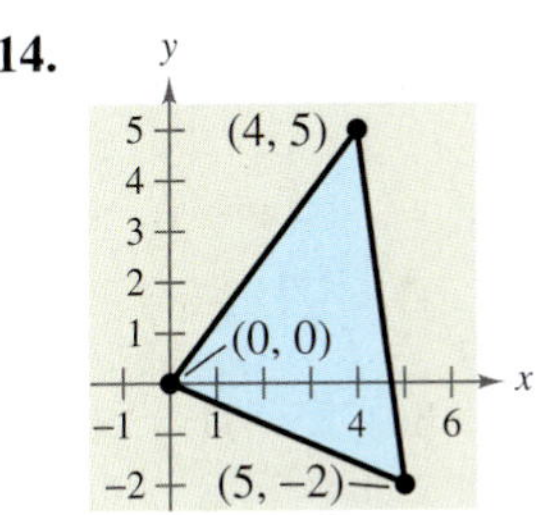

15.

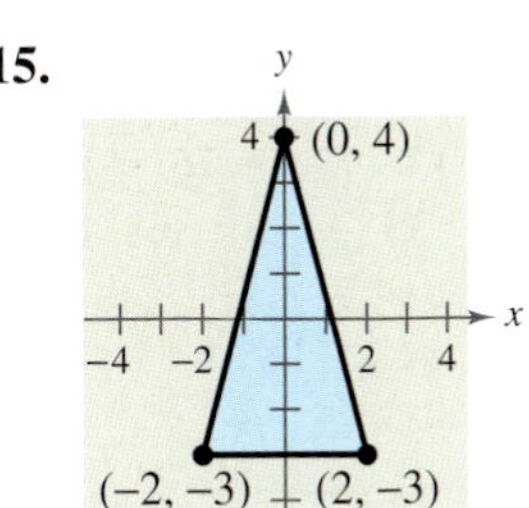

16.

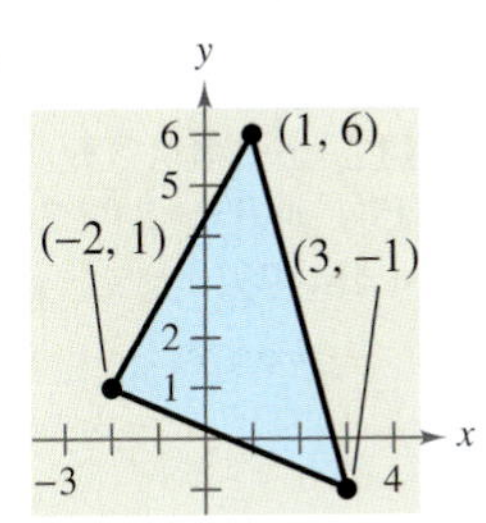

17.

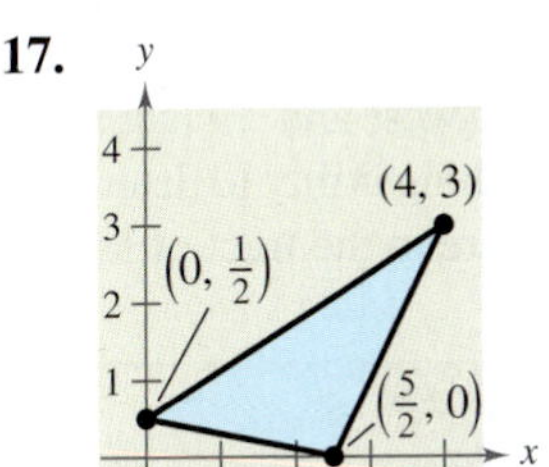

18.

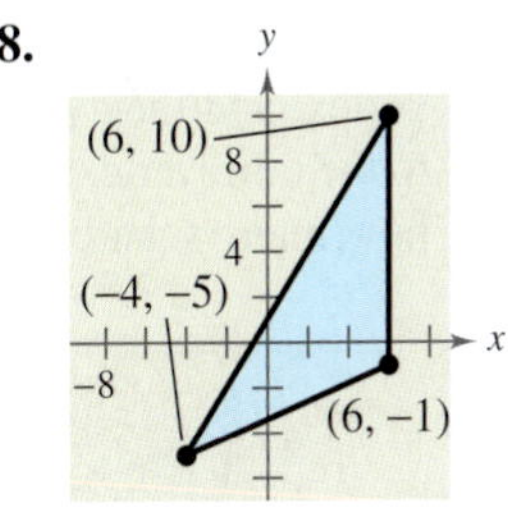

19. $(-2, 4), (2, 3), (-1, 5)$

20. $(0, -2), (-1, 4), (3, 5)$

21. $(-3, 5), (2, 6), (3, -5)$

22. $(-2, 4), (1, 5), (3, -2)$

In Exercises 23 and 24, find a value of x such that the triangle with the given vertices has an area of 4 square units.

23. $(-5, 1), (0, 2), (-2, x)$

24. $(-4, 2), (-3, 5), (-1, x)$

In Exercises 25 and 26, find a value of x such that the triangle with the given vertices has an area of 6 square units.

25. $(-2, -3),\ (1, -1),\ (-8, x)$

26. $(1, 0),\ (5, -3),\ (-3, x)$

27. ***Area of a Region*** A large region of forest has been infested with gypsy moths. The region is roughly triangular, as shown in the figure. From the northernmost vertex A of the region, the distances to the other vertices are 25 miles south and 10 miles east (for vertex B), and 20 miles south and 28 miles east (for vertex C). Use a graphing utility to approximate the number of square miles in this region.

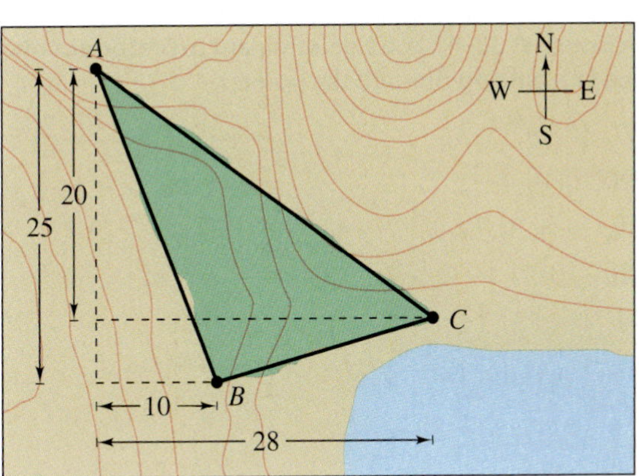

28. ***Area of a Region*** You own a triangular tract of land, as shown in the figure. To estimate the number of square feet in the tract, you start at one vertex, walk 65 feet east and 50 feet north to the second vertex, and then walk 85 feet west and 30 feet north to the third vertex. Use a graphing utility to determine how many square feet there are in the tract of land.

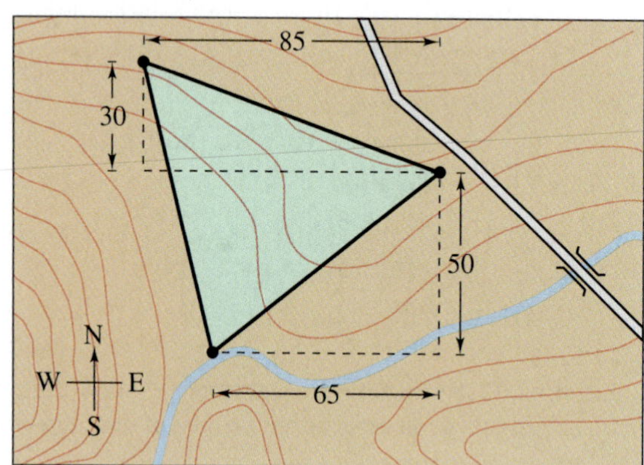

In Exercises 29–34, use a determinant to determine whether the points are collinear.

29. $(3, -1), (0, -3), (12, 5)$
30. $(-3, -5), (6, 1), (10, 2)$
31. $\left(2, -\frac{1}{2}\right), (-4, 4), (6, -3)$
32. $(0, 1), (4, -2), \left(-2, \frac{5}{2}\right)$
33. $(0, 2), (1, 2.4), (-1, 1.6)$
34. $(2, 3), (3, 3.5), (-1, 2)$

In Exercises 35 and 36, find x such that the points are collinear.

35. $(2, -5), (4, x), (5, -2)$
36. $(-6, 2), (-5, x), (-3, 5)$

In Exercises 37–42, use a determinant to find an equation of the line passing through the points.

37. $(0, 0), (5, 3)$
38. $(0, 0), (-2, 2)$
39. $(-4, 3), (2, 1)$
40. $(10, 7), (-2, -7)$
41. $\left(-\frac{1}{2}, 3\right), \left(\frac{5}{2}, 1\right)$
42. $\left(\frac{2}{3}, 4\right), (6, 12)$

In Exercises 43 and 44, find the uncoded 1×3 row matrices for the message. Then encode the message using the encoding matrix.

	Message	*Encoding Matrix*
43.	TROUBLE IN RIVER CITY	$\begin{bmatrix} 1 & -1 & 0 \\ 1 & 0 & -1 \\ -6 & 2 & 3 \end{bmatrix}$
44.	PLEASE SEND MONEY	$\begin{bmatrix} 4 & 2 & 1 \\ -3 & -3 & -1 \\ 3 & 2 & 1 \end{bmatrix}$

In Exercises 45–48, write a cryptogram for the message using the matrix A.

$$A = \begin{bmatrix} 1 & 2 & 2 \\ 3 & 7 & 9 \\ -1 & -4 & -7 \end{bmatrix}.$$

45. CALL AT NOON
46. ICEBERG DEAD AHEAD
47. HAPPY BIRTHDAY
48. OPERATION OVERLOAD

In Exercises 49–52, use A^{-1} to decode the cryptogram.

49. $A = \begin{bmatrix} 1 & 2 \\ 3 & 5 \end{bmatrix}$

11 21 64 112 25 50 29 53 23 46
40 75 55 92

50. $A = \begin{bmatrix} -5 & 2 \\ -7 & 3 \end{bmatrix}$

-136 58 -173 72 -120 51 -95 38
-178 73 -70 28 -242 101 -115 47
-90 36 -115 49 -199 82

51. $A = \begin{bmatrix} 1 & -1 & 0 \\ 1 & 0 & -1 \\ -6 & 2 & 3 \end{bmatrix}$

9 -1 -9 38 -19 -19 28 -9 -19
-80 25 41 -64 21 31 9 -5 -4

52. $A = \begin{bmatrix} 3 & -4 & 2 \\ 0 & 2 & 1 \\ 4 & -5 & 3 \end{bmatrix}$

112 -140 83 19 -25 13 72 -76 61
95 -118 71 20 21 38 35 -23 36 42
-48 32

In Exercises 53 and 54, decode the cryptogram by using the inverse of the matrix A.

$$A = \begin{bmatrix} 1 & 2 & 2 \\ 3 & 7 & 9 \\ -1 & -4 & -7 \end{bmatrix}$$

53. 20 17 −15 −12 −56 −104 1 −25 −65 62 143 181

54. 13 −9 −59 61 112 106 −17 −73 −131 11 24 29 65 144 172

55. The following cryptogram was encoded with a 2×2 matrix.

8 21 −15 −10 −13 −13 5 10 5 25 5 19 −1 6 20 40 −18 −18 1 16

The last word of the message is _RON. What is the message?

▶ Model It

56. ***Data Analysis*** The table shows the numbers y of U.S. Supreme Court cases waiting to be tried for the years 1997 through 1999. (Source: Office of the Clerk, Supreme Court of the United States)

Year, t	Number of cases, y
1997	7692
1998	8083
1999	8445

(a) Use the technique demonstrated in Exercises 67–70 in Section 9.3 to create a system of linear equations for the data. Let t represent the year, with $t = 7$ corresponding to 1997.

(b) Use Cramer's Rule to solve the system from part (a) and find the least squares regression parabola $y = at^2 + bt + c$.

(c) Use a graphing utility to graph the parabola from part (b).

(d) Use the graph from part (c) to estimate when the number of U.S. Supreme Court cases waiting to be tried will reach 9800.

Synthesis

True or False? **In Exercises 57–59, determine whether the statement is true or false. Justify your answer.**

57. In Cramer's Rule, the numerator is the determinant of the coefficient matrix.

58. You cannot use Cramer's Rule when solving a system of linear equations if the determinant of the coefficient matrix is zero.

59. In a system of linear equations, if the determinant of the coefficient matrix is zero, the system has no solution.

60. ***Writing*** At this point in the book, you have learned several methods for solving systems of linear equations. Briefly describe which method(s) you find easiest to use and which method(s) you find most difficult to use.

Review

In Exercises 61–64, use any method to solve the system of equations.

61. $\begin{cases} -x - 7y = -22 \\ 5x + y = -26 \end{cases}$

62. $\begin{cases} 3x + 8y = 11 \\ -2x + 12y = -16 \end{cases}$

63. $\begin{cases} -x - 3y + 5z = -14 \\ 4x + 2y - z = -1 \\ 5x - 3y + 2z = -11 \end{cases}$

64. $\begin{cases} 5x - y - z = 7 \\ -2x + 3y + z = -5 \\ 4x + 10y - 5z = -37 \end{cases}$

In Exercises 65 and 66, sketch the constraint region. Then find the minimum and maximum values of the objective function and where they occur, subject to the constraints.

65. Objective function:

$z = 6x + 4y$

Constraints:

$x \geq 0$

$y \geq 0$

$x + 6y \leq 30$

$6x + y \leq 40$

66. Objective function:

$z = 6x + 7y$

Constraints:

$x \geq 0$

$y \geq 0$

$4x + 3y \geq 24$

$x + 3y \geq 15$

Chapter Summary

▶ *What* did you learn?

Section 10.1	Review Exercises
☐ How to write a matrix and identify its order	1–8
☐ How to perform elementary row operations on matrices	9, 10
☐ How to use matrices and Gaussian elimination to solve systems of linear equations	11–24
☐ How to use matrices and Gauss-Jordan elimination to solve systems of linear equations	25–30
Section 10.2	
☐ How to decide whether two matrices are equal	31–34
☐ How to add and subtract matrices and multiply matrices by real numbers	35–48
☐ How to multiply two matrices	49–62
☐ How to use matrix operations to model and solve real-life problems	63, 64
Section 10.3	
☐ How to verify that two matrices are inverses of each other	65–68
☐ How to use Gauss-Jordan elimination to find the inverses of matrices	69–76
☐ How to use a formula to find the inverses of 2×2 matrices	77–80
☐ How to use inverse matrices to solve systems of linear equations	81–92
Section 10.4	
☐ How to find the determinants of 2×2 matrices	93–96
☐ How to find minors and cofactors of square matrices	97–100
☐ How to find the determinants of square matrices	101–104
Section 10.5	
☐ How to use Cramer's Rule to solve systems of linear equations	105–108
☐ How to use determinants to find the areas of triangles	109–112
☐ How to use a determinant to test for collinear points and find an equation of a line passing through two points	113–118
☐ How to use matrices to code and decode messages	119–122

Review Exercises

10.1 In Exercises 1–4, determine the order of the matrix.

1. $\begin{bmatrix} -4 \\ 0 \\ 5 \end{bmatrix}$

2. $\begin{bmatrix} 3 & -1 & 0 & 6 \\ -2 & 7 & 1 & 4 \end{bmatrix}$

3. $[3]$

4. $[6 \quad 2 \quad -5 \quad 8 \quad 0]$

In Exercises 5 and 6, form the augmented matrix for the system of linear equations.

5. $\begin{cases} 3x - 10y = 15 \\ 5x + 4y = 22 \end{cases}$

6. $\begin{cases} 8x - 7y + 4z = 12 \\ 3x - 5y + 2z = 20 \\ 5x + 3y - 3z = 26 \end{cases}$

In Exercises 7 and 8, write the system of linear equations represented by the augmented matrix. (Use variables *x, y, z,* and *w*.)

7. $\begin{bmatrix} 5 & 1 & 7 & \vdots & -9 \\ 4 & 2 & 0 & \vdots & 10 \\ 9 & 4 & 2 & \vdots & 3 \end{bmatrix}$

8. $\begin{bmatrix} 13 & 16 & 7 & 3 & \vdots & 2 \\ 1 & 21 & 8 & 5 & \vdots & 12 \\ 4 & 10 & -4 & 3 & \vdots & -1 \end{bmatrix}$

In Exercises 9 and 10, write the matrix in row-echelon form. Remember that the row-echelon form of a matrix is not unique.

9. $\begin{bmatrix} 0 & 1 & 1 \\ 1 & 2 & 3 \\ 2 & 2 & 2 \end{bmatrix}$

10. $\begin{bmatrix} 4 & 8 & 16 \\ 3 & -1 & 2 \\ -2 & 10 & 12 \end{bmatrix}$

In Exercises 11–14, write the system of linear equations represented by the augmented matrix. Then use back-substitution to solve. (Use variables *x, y,* and *z*.)

11. $\begin{bmatrix} 1 & 2 & 3 & \vdots & 9 \\ 0 & 1 & -2 & \vdots & 2 \\ 0 & 0 & 1 & \vdots & 0 \end{bmatrix}$

12. $\begin{bmatrix} 1 & 3 & -9 & \vdots & 4 \\ 0 & 1 & -1 & \vdots & 10 \\ 0 & 0 & 1 & \vdots & -2 \end{bmatrix}$

13. $\begin{bmatrix} 1 & -5 & 4 & \vdots & 1 \\ 0 & 1 & 2 & \vdots & 3 \\ 0 & 0 & 1 & \vdots & 4 \end{bmatrix}$

14. $\begin{bmatrix} 1 & -8 & 0 & \vdots & -2 \\ 0 & 1 & -1 & \vdots & -7 \\ 0 & 0 & 1 & \vdots & 1 \end{bmatrix}$

In Exercises 15–24, use matrices to solve the system of equations (if possible). Use Gaussian elimination with back-substitution.

15. $\begin{cases} 5x + 4y = 2 \\ -x + y = -22 \end{cases}$

16. $\begin{cases} 2x - 5y = 2 \\ 3x - 7y = 1 \end{cases}$

17. $\begin{cases} 0.3x - 0.1y = -0.13 \\ 0.2x - 0.3y = -0.25 \end{cases}$

18. $\begin{cases} 0.2x - 0.1y = 0.07 \\ 0.4x - 0.5y = -0.01 \end{cases}$

19. $\begin{cases} 2x + 3y + z = 10 \\ 2x - 3y - 3z = 22 \\ 4x - 2y + 3z = -2 \end{cases}$

20. $\begin{cases} 2x + 3y + 3z = 3 \\ 6x + 6y + 12z = 13 \\ 12x + 9y - z = 2 \end{cases}$

21. $\begin{cases} 2x + y + 2z = 4 \\ 2x + 2y = 5 \\ 2x - y + 6z = 2 \end{cases}$

22. $\begin{cases} x + 2y + 6z = 1 \\ 2x + 5y + 15z = 4 \\ 3x + y + 3z = -6 \end{cases}$

23. $\begin{cases} 2x + y + z = 6 \\ -2y + 3z - w = 9 \\ 3x + 3y - 2z - 2w = -11 \\ x + z + 3w = 14 \end{cases}$

24. $\begin{cases} x + 2y + w = 3 \\ -3y + 3z = 0 \\ 4x + 4y + z + 2w = 0 \\ 2x + z = 3 \end{cases}$

In Exercises 25–28, use matrices to solve the system of equations. Use Gauss-Jordan elimination.

25. $\begin{cases} -x + y + 2z = 1 \\ 2x + 3y + z = -2 \\ 5x + 4y + 2z = 4 \end{cases}$

26. $\begin{cases} 4x + 4y + 4z = 5 \\ 4x - 2y - 8z = 1 \\ 5x + 3y + 8z = 6 \end{cases}$

27. $\begin{cases} 2x - y + 9z = -8 \\ -x - 3y + 4z = -15 \\ 5x + 2y - z = 17 \end{cases}$

28. $\begin{cases} -3x + y + 7z = -20 \\ 5x - 2y - z = 34 \\ -x + y + 4z = -8 \end{cases}$

In Exercises 29 and 30, use the matrix capabilities of a graphing utility to reduce the augmented matrix corresponding to the system of equations, and solve the system.

29. $\begin{cases} 3x - y + 5z - 2w = -44 \\ x + 6y + 4z - w = 1 \\ 5x - y + z + 3w = -15 \\ 4y - z - 8w = 58 \end{cases}$

30. $\begin{cases} 4x + 12y + 2z = 20 \\ x + 6y + 4z = 12 \\ x + 6y + z = 8 \\ -2x - 10y - 2z = -10 \end{cases}$

10.2 **In Exercises 31–34, find *x* and *y*.**

31. $\begin{bmatrix} -1 & x \\ y & 9 \end{bmatrix} = \begin{bmatrix} -1 & 12 \\ -7 & 9 \end{bmatrix}$

32. $\begin{bmatrix} -1 & 0 \\ x & 5 \\ -4 & y \end{bmatrix} = \begin{bmatrix} -1 & 0 \\ 8 & 5 \\ -4 & 0 \end{bmatrix}$

33. $\begin{bmatrix} x+3 & -4 & 4y \\ 0 & -3 & 2 \\ -2 & y+5 & 6x \end{bmatrix} = \begin{bmatrix} 5x-1 & -4 & 44 \\ 0 & -3 & 2 \\ -2 & 16 & 6 \end{bmatrix}$

34. $\begin{bmatrix} -9 & 4 & 2 & -5 \\ 0 & -3 & 7 & -4 \\ 6 & -1 & 1 & 0 \end{bmatrix} = \begin{bmatrix} -9 & 4 & x-10 & -5 \\ 0 & -3 & 7 & 2y \\ \frac{1}{2}x & -1 & 1 & 0 \end{bmatrix}$

In Exercises 35–38, if possible, find (a) $A + B$, (b) $A - B$, (c) $4A$, and (d) $A + 3B$.

35. $A = \begin{bmatrix} 2 & -2 \\ 3 & 5 \end{bmatrix}, \quad B = \begin{bmatrix} -3 & 10 \\ 12 & 8 \end{bmatrix}$

36. $A = \begin{bmatrix} 5 & 4 \\ -7 & 2 \\ 11 & 2 \end{bmatrix}, \quad B = \begin{bmatrix} 4 & 12 \\ 20 & 40 \\ 15 & 30 \end{bmatrix}$

37. $A = \begin{bmatrix} 5 & 4 \\ -7 & 2 \\ 11 & 2 \end{bmatrix}, \quad B = \begin{bmatrix} 0 & 3 \\ 4 & 12 \\ 20 & 40 \end{bmatrix}$

38. $A = \begin{bmatrix} 6 & -5 & 7 \end{bmatrix}, \quad B = \begin{bmatrix} -1 \\ 4 \\ 8 \end{bmatrix}$

In Exercises 39–42, perform the matrix operations. If it is not possible, explain why.

39. $\begin{bmatrix} 7 & 3 \\ -1 & 5 \end{bmatrix} + \begin{bmatrix} 10 & -20 \\ 14 & -3 \end{bmatrix}$

40. $\begin{bmatrix} -11 & 16 & 19 \\ -7 & -2 & 1 \end{bmatrix} - \begin{bmatrix} 6 & 0 \\ 8 & -4 \\ -2 & 10 \end{bmatrix}$

41. $-2\begin{bmatrix} 1 & 2 \\ 5 & -4 \\ 6 & 0 \end{bmatrix} + 8\begin{bmatrix} 7 & 1 \\ 1 & 2 \\ 1 & 4 \end{bmatrix}$

42. $-\begin{bmatrix} 8 & -1 & 8 \\ -2 & 4 & 12 \\ 0 & -6 & 0 \end{bmatrix} - 5\begin{bmatrix} -2 & 0 & -4 \\ 3 & -1 & 1 \\ 6 & 12 & -8 \end{bmatrix}$

In Exercises 43 and 44, use a graphing utility to perform the matrix operations.

43. $3\begin{bmatrix} 8 & -2 & 5 \\ 1 & 3 & -1 \end{bmatrix} + 6\begin{bmatrix} 4 & -2 & -3 \\ 2 & 7 & 6 \end{bmatrix}$

44. $-5\begin{bmatrix} 2 & 0 \\ 7 & -2 \\ 8 & 2 \end{bmatrix} + 4\begin{bmatrix} 4 & -2 \\ 6 & 11 \\ -1 & 3 \end{bmatrix}$

In Exercises 45–48, solve for X when

$$A = \begin{bmatrix} -4 & 0 \\ 1 & -5 \\ -3 & 2 \end{bmatrix} \quad \text{and} \quad B = \begin{bmatrix} 1 & 2 \\ -2 & 1 \\ 4 & 4 \end{bmatrix}.$$

45. $X = 3A - 2B$

46. $6X = 4A + 3B$

47. $3X + 2A = B$

48. $2A - 5B = 3X$

In Exercises 49–52, find *AB*, if possible.

49. $A = \begin{bmatrix} 2 & -2 \\ 3 & 5 \end{bmatrix}, \quad B = \begin{bmatrix} -3 & 10 \\ 12 & 8 \end{bmatrix}$

50. $A = \begin{bmatrix} 5 & 4 \\ -7 & 2 \\ 11 & 2 \end{bmatrix}, \quad B = \begin{bmatrix} 4 & 12 \\ 20 & 40 \\ 15 & 30 \end{bmatrix}$

51. $A = \begin{bmatrix} 5 & 4 \\ -7 & 2 \\ 11 & 2 \end{bmatrix}, \quad B = \begin{bmatrix} 4 & 12 \\ 20 & 40 \end{bmatrix}$

52. $A = \begin{bmatrix} 6 & -5 & 7 \end{bmatrix}, \quad B = \begin{bmatrix} -1 \\ 4 \\ 8 \end{bmatrix}$

In Exercises 53–60, perform the matrix operations. If it is not possible, explain why.

53. $\begin{bmatrix} 1 & 2 \\ 5 & -4 \\ 6 & 0 \end{bmatrix}\begin{bmatrix} 6 & -2 & 8 \\ 4 & 0 & 0 \end{bmatrix}$

54. $\begin{bmatrix} 1 & 5 & 6 \\ 2 & -4 & 0 \end{bmatrix}\begin{bmatrix} 6 & -2 & 8 \\ 4 & 0 & 0 \end{bmatrix}$

55. $\begin{bmatrix} 1 & 5 & 6 \\ 2 & -4 & 0 \end{bmatrix}\begin{bmatrix} 6 & 4 \\ -2 & 0 \\ 8 & 0 \end{bmatrix}$

56. $\begin{bmatrix} 1 & 3 & 2 \\ 0 & 2 & -4 \\ 0 & 0 & 3 \end{bmatrix}\begin{bmatrix} 4 & -3 & 2 \\ 0 & 3 & -1 \\ 0 & 0 & 2 \end{bmatrix}$

57. $\begin{bmatrix} 4 \\ 6 \end{bmatrix}\begin{bmatrix} 6 & -2 \end{bmatrix}$

58. $\begin{bmatrix} 4 & -2 & 6 \end{bmatrix}\begin{bmatrix} -2 & 1 \\ 0 & -3 \\ 2 & 0 \end{bmatrix}$

59. $\begin{bmatrix} 2 & 1 \\ 6 & 0 \end{bmatrix}\left(\begin{bmatrix} 4 & 2 \\ -3 & 1 \end{bmatrix} + \begin{bmatrix} -2 & 4 \\ 0 & 4 \end{bmatrix}\right)$

60. $-3\begin{bmatrix} 1 & -1 \\ 4 & 2 \end{bmatrix}\left(\begin{bmatrix} 0 & 3 \\ 1 & 2 \end{bmatrix}\begin{bmatrix} 1 & 0 \\ 5 & -3 \end{bmatrix}\right)$

In Exercises 61 and 62, use a graphing utility to perform the matrix operations.

61. $\begin{bmatrix} 4 & 1 \\ 11 & -7 \\ 12 & 3 \end{bmatrix}\begin{bmatrix} 3 & -5 & 6 \\ 2 & -2 & -2 \end{bmatrix}$

62. $\begin{bmatrix} -2 & 3 & 10 \\ 4 & -2 & 2 \end{bmatrix}\begin{bmatrix} 1 & 1 \\ -5 & 2 \\ 3 & 2 \end{bmatrix}$

63. ***Manufacturing*** A corporation has four factories, each of which manufactures three types of cordless power tools. The number of units of cordless power tools produced at factory j in one day is represented by a_{ij} in the matrix

$$A = \begin{bmatrix} 80 & 70 & 90 & 40 \\ 50 & 30 & 80 & 20 \\ 90 & 60 & 100 & 50 \end{bmatrix}.$$

Find the production levels if production is increased by 20%.

64. ***Manufacturing*** A manufacturing company produces three kinds of computer games that are shipped to two warehouses. The number of units of game i that are shipped to warehouse j is represented by a_{ij} in the matrix

$$A = \begin{bmatrix} 8200 & 7400 \\ 6500 & 9800 \\ 5400 & 4800 \end{bmatrix}.$$

The price per unit is represented by the matrix

$B = [\$10.25 \quad \$14.50 \quad \$17.75]$.

Compute BA and interpret the result.

10.3 In Exercises 65–68, show that B is the inverse of A.

65. $A = \begin{bmatrix} -4 & -1 \\ 7 & 2 \end{bmatrix}, \quad B = \begin{bmatrix} -2 & -1 \\ 7 & 4 \end{bmatrix}$

66. $A = \begin{bmatrix} 5 & -1 \\ 11 & -2 \end{bmatrix}, \quad B = \begin{bmatrix} -2 & 1 \\ -11 & 5 \end{bmatrix}$

67. $A = \begin{bmatrix} 1 & 1 & 0 \\ 1 & 0 & 1 \\ 6 & 2 & 3 \end{bmatrix}, \quad B = \begin{bmatrix} -2 & -3 & 1 \\ 3 & 3 & -1 \\ 2 & 4 & -1 \end{bmatrix}$

68. $A = \begin{bmatrix} 1 & -1 & 0 \\ -1 & 0 & -1 \\ 8 & -4 & 2 \end{bmatrix}, \quad B = \begin{bmatrix} -2 & 1 & \frac{1}{2} \\ -3 & 1 & \frac{1}{2} \\ 2 & -2 & -\frac{1}{2} \end{bmatrix}$

In Exercises 69–72, find the inverse of the matrix (if it exists).

69. $\begin{bmatrix} -6 & 5 \\ -5 & 4 \end{bmatrix}$

70. $\begin{bmatrix} -3 & -5 \\ 2 & 3 \end{bmatrix}$

71. $\begin{bmatrix} -1 & -2 & -2 \\ 3 & 7 & 9 \\ 1 & 4 & 7 \end{bmatrix}$

72. $\begin{bmatrix} 0 & -2 & 1 \\ -5 & -2 & -3 \\ 7 & 3 & 4 \end{bmatrix}$

In Exercises 73–76, use a graphing utility to find the inverse of the matrix (if it exists).

73. $\begin{bmatrix} 2 & 0 & 3 \\ -1 & 1 & 1 \\ 2 & -2 & 1 \end{bmatrix}$

74. $\begin{bmatrix} 1 & 4 & 6 \\ 2 & -3 & 1 \\ -1 & 18 & 16 \end{bmatrix}$

75. $\begin{bmatrix} 1 & 3 & 1 & 6 \\ 4 & 4 & 2 & 6 \\ 3 & 4 & 1 & 2 \\ -1 & 2 & -1 & -2 \end{bmatrix}$

76. $\begin{bmatrix} 8 & 0 & 2 & 8 \\ 4 & -2 & 0 & -2 \\ 1 & 2 & 1 & 4 \\ -1 & 4 & 1 & 1 \end{bmatrix}$

In Exercises 77–80, let

$$A = \begin{bmatrix} a & b \\ c & d \end{bmatrix}.$$

Find the inverse of the matrix above

$$A^{-1} = \frac{1}{ad - bc}\begin{bmatrix} d & -b \\ -c & a \end{bmatrix}$$

(if it exists).

77. $\begin{bmatrix} -7 & 2 \\ -8 & 2 \end{bmatrix}$ **78.** $\begin{bmatrix} 10 & 4 \\ 7 & 3 \end{bmatrix}$

79. $\begin{bmatrix} -\frac{1}{2} & 20 \\ \frac{3}{10} & -6 \end{bmatrix}$ **80.** $\begin{bmatrix} -\frac{3}{4} & \frac{5}{2} \\ -\frac{4}{5} & -\frac{8}{3} \end{bmatrix}$

In Exercises 81–88, use an inverse matrix to solve (if possible) the system of linear equations.

81. $\begin{cases} -x + 4y = 8 \\ 2x - 7y = -5 \end{cases}$ **82.** $\begin{cases} 5x - y = 13 \\ -9x + 2y = -24 \end{cases}$

83. $\begin{cases} -3x + 10y = 8 \\ 5x - 17y = -13 \end{cases}$ **84.** $\begin{cases} 4x - 2y = -10 \\ -19x + 9y = 47 \end{cases}$

85. $\begin{cases} 3x + 2y - z = 6 \\ x - y + 2z = -1 \\ 5x + y + z = 7 \end{cases}$

86. $\begin{cases} -x + 4y - 2z = 12 \\ 2x - 9y + 5z = -25 \\ -x + 5y - 4z = 10 \end{cases}$

87. $\begin{cases} -2x + y + 2z = -13 \\ -x - 4y + z = -11 \\ -y - z = 0 \end{cases}$

88. $\begin{cases} 3x - y + 5z = -14 \\ -x + y + 6z = 8 \\ -8x + 4y - z = 44 \end{cases}$

In Exercises 89–92, use a graphing utility to solve (if possible) the system of linear equations using the inverse of the coefficient matrix.

89. $\begin{cases} x + 2y = -1 \\ 3x + 4y = -5 \end{cases}$ **90.** $\begin{cases} x + 3y = 23 \\ -6x + 2y = -18 \end{cases}$

91. $\begin{cases} -3x - 3y - 4z = 2 \\ y + z = -1 \\ 4x + 3y + 4z = -1 \end{cases}$

92. $\begin{cases} x - 3y - 2z = 8 \\ -2x + 7y + 3z = -19 \\ x - y - 3z = 3 \end{cases}$

10.4 In Exercises 93–96, find the determinant of the matrix.

93. $\begin{bmatrix} 8 & 5 \\ 2 & -4 \end{bmatrix}$ **94.** $\begin{bmatrix} -9 & 11 \\ 7 & -4 \end{bmatrix}$

95. $\begin{bmatrix} 50 & -30 \\ 10 & 5 \end{bmatrix}$ **96.** $\begin{bmatrix} 14 & -24 \\ 12 & -15 \end{bmatrix}$

In Exercises 97–100, find all (a) minors and (b) cofactors of the matrix.

97. $\begin{bmatrix} 2 & -1 \\ 7 & 4 \end{bmatrix}$ **98.** $\begin{bmatrix} 3 & 6 \\ 5 & -4 \end{bmatrix}$

99. $\begin{bmatrix} 3 & 2 & -1 \\ -2 & 5 & 0 \\ 1 & 8 & 6 \end{bmatrix}$ **100.** $\begin{bmatrix} 8 & 3 & 4 \\ 6 & 5 & -9 \\ -4 & 1 & 2 \end{bmatrix}$

In Exercises 101–104, find the determinant of the matrix. Expand by cofactors on the row or column that appears to make the computations easiest.

101. $\begin{bmatrix} -2 & 4 & 1 \\ -6 & 0 & 2 \\ 5 & 3 & 4 \end{bmatrix}$ **102.** $\begin{bmatrix} 4 & 7 & -1 \\ 2 & -3 & 4 \\ -5 & 1 & -1 \end{bmatrix}$

103. $\begin{bmatrix} 3 & 0 & -4 & 0 \\ 0 & 8 & 1 & 2 \\ 6 & 1 & 8 & 2 \\ 0 & 3 & -4 & 1 \end{bmatrix}$

104. $\begin{bmatrix} -5 & 6 & 0 & 0 \\ 0 & 1 & -1 & 2 \\ -3 & 4 & -5 & 1 \\ 1 & 6 & 0 & 3 \end{bmatrix}$

10.5 In Exercises 105–108, use Cramer's Rule to solve (if possible) the system of equations.

105. $\begin{cases} 5x - 2y = 6 \\ -11x + 3y = -23 \end{cases}$

106. $\begin{cases} 3x + 8y = -7 \\ 9x - 5y = 37 \end{cases}$

107. $\begin{cases} -2x + 3y - 5z = -11 \\ 4x - y + z = -3 \\ -x - 4y + 6z = 15 \end{cases}$

108. $\begin{cases} 5x - 2y + z = 15 \\ 3x - 3y - z = -7 \\ 2x - y - 7z = -3 \end{cases}$

In Exercises 109–112, use a determinant and the given vertices of a triangle to find the area of the triangle.

109.

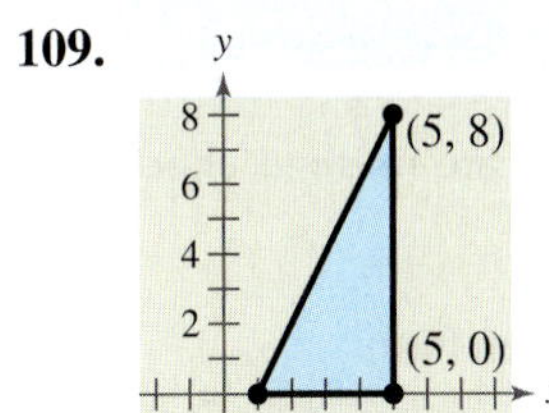

110.

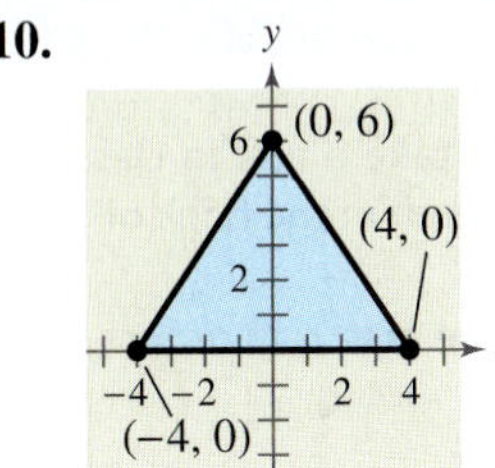

111.

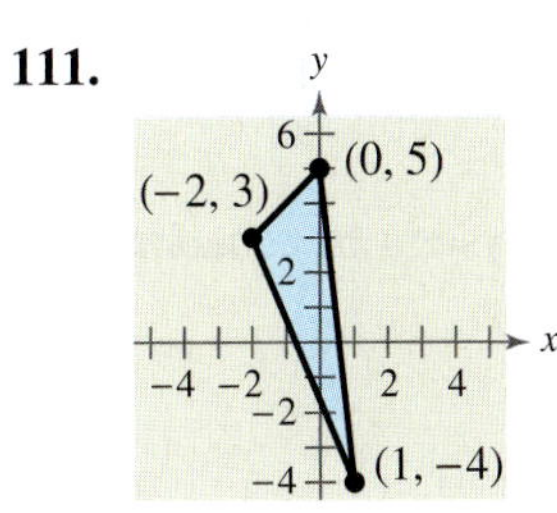

112.

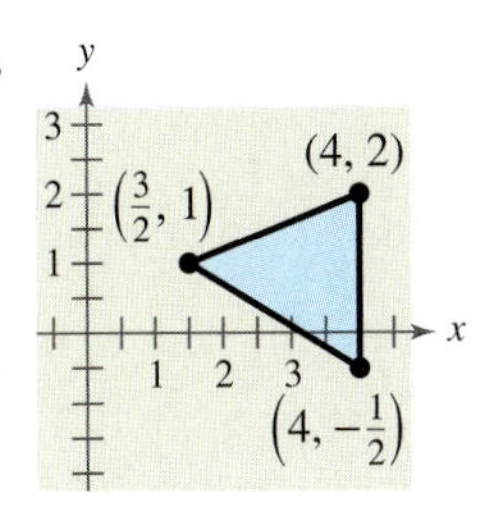

In Exercises 113 and 114, use a determinant to decide whether the points are collinear.

113. $(-1, 7), (3, -9), (-3, 15)$

114. $(0, -5), (-2, -6), (8, -1)$

In Exercises 115–118, use a determinant to find an equation of the line passing through the points.

115. $(-4, 0), (4, 4)$ **116.** $(2, 5), (6, -1)$

117. $\left(-\frac{5}{2}, 3\right), \left(\frac{7}{2}, 1\right)$ **118.** $(-0.8, 0.2), (0.7, 3.2)$

In Exercises 119 and 120, find the uncoded 1 × 3 row matrices for the message. Then encode the message using the encoding matrix.

	Message	*Encoding Matrix*
119.	LOOK OUT BELOW	$\begin{bmatrix} 2 & -2 & 0 \\ 3 & 0 & -3 \\ -6 & 2 & 3 \end{bmatrix}$
120.	RETURN TO BASE	$\begin{bmatrix} 2 & 1 & 0 \\ -6 & -6 & -2 \\ 3 & 2 & 1 \end{bmatrix}$

In Exercises 121 and 122, decode the cryptogram by using the inverse of the matrix

$$A = \begin{bmatrix} -5 & 4 & -3 \\ 10 & -7 & 6 \\ 8 & -6 & 5 \end{bmatrix}.$$

121. −5 11 −2 370 −265 225 −57 48 −33 32 −15 20 245 −171 147

122. 145 −105 92 264 −188 160 23 −16 15 129 −84 78 −9 8 −5 159 −118 100 219 −152 133 370 −265 225 −105 84 −63

Synthesis

True or False? In Exercises 123 and 124, determine whether the statement is true or false. Justify your answer.

123. It is possible to find the determinant of a 4 × 5 matrix.

124. $$\begin{vmatrix} a_{11} & a_{12} & a_{13} \\ a_{21} & a_{22} & a_{23} \\ a_{31}+c_1 & a_{32}+c_2 & a_{33}+c_3 \end{vmatrix} = \begin{vmatrix} a_{11} & a_{12} & a_{13} \\ a_{21} & a_{22} & a_{23} \\ a_{31} & a_{32} & a_{33} \end{vmatrix} + \begin{vmatrix} a_{11} & a_{12} & a_{13} \\ a_{21} & a_{22} & a_{23} \\ c_1 & c_2 & c_3 \end{vmatrix}$$

125. Under what conditions does a matrix have an inverse?

126. ***Writing*** What is meant by the cofactor of an entry of a matrix? How are cofactors used to find the determinant of the matrix?

127. Three people were asked to solve a system of equations using an augmented matrix. Each person reduced the matrix to row-echelon form. The reduced matrices were

$$\begin{bmatrix} 1 & 2 & \vdots & 3 \\ 0 & 1 & \vdots & 1 \end{bmatrix},$$

$$\begin{bmatrix} 1 & 0 & \vdots & 1 \\ 0 & 1 & \vdots & 1 \end{bmatrix},$$

and

$$\begin{bmatrix} 1 & 2 & \vdots & 3 \\ 0 & 0 & \vdots & 0 \end{bmatrix}.$$

Can all three be right? Explain.

128. ***Think About It*** Describe the row-echelon form of an augmented matrix that corresponds to a system of linear equations that has a unique solution.

129. Solve the equation

$$\begin{vmatrix} 2-\lambda & 5 \\ 3 & -8-\lambda \end{vmatrix} = 0.$$

Chapter Test

The *Interactive* CD-ROM and *Internet* versions of this text offer Chapter Pre-Tests and Chapter Post-Tests, both of which have randomly generated exercises with diagnostic capabilities.

Take this test as you would take a test in class. When you are finished, check your work against the answers given in the back of the book.

In Exercises 1 and 2, write the matrix in reduced row-echelon form.

1. $\begin{bmatrix} 1 & -1 & 5 \\ 6 & 2 & 3 \\ 5 & 3 & -3 \end{bmatrix}$

2. $\begin{bmatrix} 1 & 0 & -1 & 2 \\ -1 & 1 & 1 & -3 \\ 1 & 1 & -1 & 1 \\ 3 & 2 & -3 & 4 \end{bmatrix}$

3. Write the augmented matrix corresponding to the system of equations and solve the system.

$$\begin{cases} 4x + 3y - 2z = 14 \\ -x - y + 2z = -5 \\ 3x + y - 4z = 8 \end{cases}$$

4. Find (a) $A - B$, (b) $3A$, (c) $3A - 2B$, and (d) AB (if possible).

$$A = \begin{bmatrix} 5 & 4 \\ -4 & -4 \end{bmatrix}, \quad B = \begin{bmatrix} 4 & -1 \\ -4 & 0 \end{bmatrix}$$

In Exercises 5 and 6, find the inverse of the matrix (if it exists).

5. $\begin{bmatrix} -6 & 4 \\ 10 & -5 \end{bmatrix}$

6. $\begin{bmatrix} -2 & 4 & -6 \\ 2 & 1 & 0 \\ 4 & -2 & 5 \end{bmatrix}$

7. Use the result of Exercise 5 to solve the system.

$$\begin{cases} -6x + 4y = 10 \\ 10x - 5y = 20 \end{cases}$$

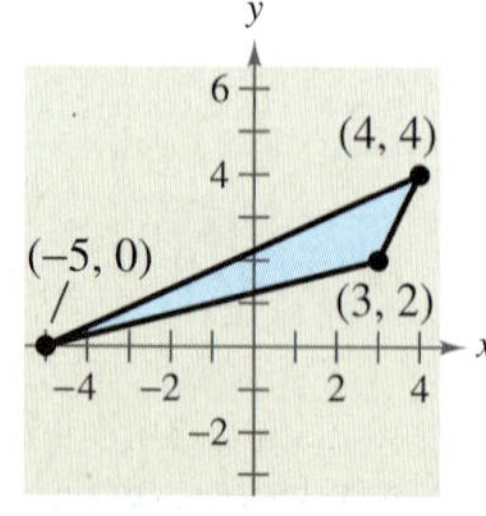

FIGURE FOR 13

$$A = \begin{bmatrix} 1 & -1 & 0 \\ 1 & 0 & -1 \\ 6 & -2 & -3 \end{bmatrix}$$

MATRIX FOR 14

In Exercises 8–10, evaluate the determinant of the matrix.

8. $\begin{bmatrix} -9 & 4 \\ 13 & 16 \end{bmatrix}$

9. $\begin{bmatrix} \frac{5}{2} & \frac{13}{4} \\ -8 & \frac{6}{5} \end{bmatrix}$

10. $\begin{bmatrix} 6 & -7 & 2 \\ 3 & -2 & 0 \\ 1 & 5 & 1 \end{bmatrix}$

In Exercises 11 and 12, use Cramer's Rule to solve (if possible) the system of equations.

11. $\begin{cases} 7x + 6y = 9 \\ -2x - 11y = -49 \end{cases}$

12. $\begin{cases} 6x - y + 2z = -4 \\ -2x + 3y - z = 10 \\ 4x - 4y + z = -18 \end{cases}$

13. Use a determinant to find the area of the triangle in the figure.

14. Find the uncoded 1×3 row matrices for the message KNOCK ON WOOD. Then encode the message using the matrix A at the left.

15. One hundred liters of a 50% solution is obtained by mixing a 60% solution with a 20% solution. How many liters of each solution must be used to obtain the desired mixture?

Proofs in Mathematics

Proofs without words are pictures or diagrams that give a visual understanding of why a theorem or statement is true. They can also provide a starting point for writing a formal proof. The following proof shows that a 2×2 determinant is the area of a parallelogram.

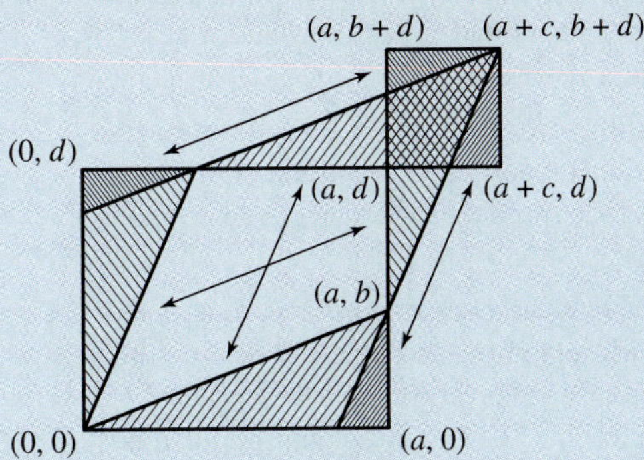

$$\begin{vmatrix} a & b \\ c & d \end{vmatrix} = ad - bc = \|\text{▭}\| - \|\text{□}\| = \|\text{▱}\|$$

The following is a color-coded version of the proof along with a brief explanation of why this proof works.

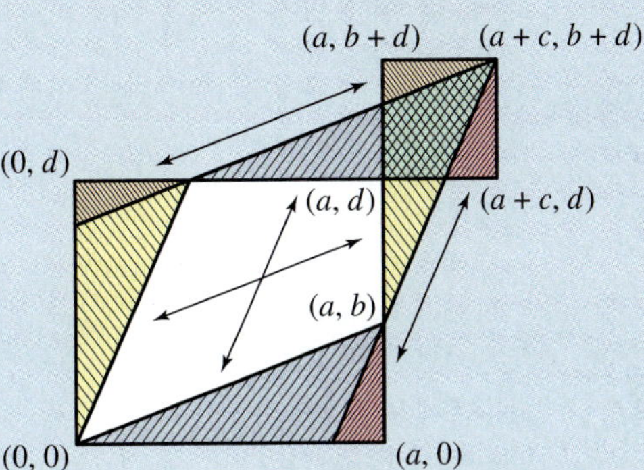

$$\begin{vmatrix} a & b \\ c & d \end{vmatrix} = ad - bc = \|\text{▭}\| - \|\text{□}\| = \|\text{▱}\|$$

Area of ▭ = Area of orange △ + Area of yellow △ + Area of blue △ + Area of pink △ + Area of white quadrilateral

Area of □ = Area of orange △ + Area of pink △ + Area of green quadrilateral

Area of ▱ = Area of white quadrilateral + Area of blue △ + Area of yellow △ − Area of green quadrilateral
= Area of ▭ − Area of □

Exercise taken from "Proof Without Words" by Solomon W. Golomb, *Mathematics Magazine*, March 1985. Used by permission of the author.

P.S. Problem Solving

1. The columns of matrix T show the coordinates of the vertices of a triangle. Matrix A is a transformation matrix.

$$A = \begin{bmatrix} 0 & -1 \\ 1 & 0 \end{bmatrix} \quad T = \begin{bmatrix} 1 & 2 & 3 \\ 1 & 4 & 2 \end{bmatrix}$$

   (a) Find AT and AAT. Then sketch the original triangle and the two transformed triangles. What transformation does A represent?

   (b) Given the triangle determined by AAT, describe the transformation process that produces the triangle determined by AT and then the triangle determined by T.

2. The matrices show the number of people (in thousands) who lived in each region of the United States in 2000 and the number of people (in thousands) projected to live in each region in 2015. The regional populations are separated into three age categories. (Source: U.S. Census Bureau)

2000

	0–17	18–64	65 +
Northeast	13,049	33,175	7,372
Midwest	16,646	39,386	8,263
South	25,569	62,235	12,437
Mountain	4,935	11,210	2,031
Pacific	12,098	28,036	4,893

2015

	0–17	18–64	65 +
Northeast	12,589	34,081	8,165
Midwest	15,886	41,038	10,101
South	25,916	68,998	17,470
Mountain	5,226	12,626	3,270
Pacific	14,906	33,296	6,565

   (a) The total population in 2000 was 281,335,000 and the projected total population in 2015 is 310,133,000. Rewrite the matrices to give the information as percents of the total population.

   (b) Write a matrix that gives the projected change in the percent of the population in each region and age group from 2000 to 2015.

   (c) Based on the result of part (b), which region(s) and age group(s) are projected to show relative growth from 2000 to 2015?

3. Determine whether the matrix is idempotent. A square matrix is **idempotent** if $A^2 = A$.

   (a) $\begin{bmatrix} 1 & 0 \\ 0 & 0 \end{bmatrix}$ (b) $\begin{bmatrix} 0 & 1 \\ 1 & 0 \end{bmatrix}$

   (c) $\begin{bmatrix} 2 & 3 \\ -1 & -2 \end{bmatrix}$ (d) $\begin{bmatrix} 2 & 3 \\ 1 & 2 \end{bmatrix}$

4. Let $A = \begin{bmatrix} 1 & 2 \\ -2 & 1 \end{bmatrix}$.

   (a) Show that $A^2 - 2A + 5I = 0$, where I is the identity matrix of order 2.

   (b) Show that $A^{-1} = \frac{1}{5}(2I - A)$.

   (c) Show in general that for any square matrix satisfying

   $$A^2 - 2A + 5I = 0$$

   the inverse of A is given by

   $$A^{-1} = \tfrac{1}{5}(2I - A).$$

5. Find x such that the matrix is equal to its own inverse.

$$A = \begin{bmatrix} 3 & x \\ -2 & -3 \end{bmatrix}$$

6. Find x such that the matrix is singular.

$$A = \begin{bmatrix} 4 & x \\ -2 & -3 \end{bmatrix}$$

7. Find an example of a singular 2×2 matrix satisfying $A^2 = A$.

8. Verify the following equation.

$$\begin{vmatrix} 1 & 1 & 1 \\ a & b & c \\ a^2 & b^2 & c^2 \end{vmatrix} = (a - b)(b - c)(c - a)$$

9. Verify the following equation.

$$\begin{vmatrix} 1 & 1 & 1 \\ a & b & c \\ a^3 & b^3 & c^3 \end{vmatrix} = (a - b)(b - c)(c - a)(a + b + c)$$

10. Verify the following equation.

$$\begin{vmatrix} x & 0 & c \\ -1 & x & b \\ 0 & -1 & a \end{vmatrix} = ax^2 + bx + c$$

11. Use the equation given in Exercise 10 as a model to find a determinant that is equal to $ax^3 + bx^2 + cx + d$.

12. The atomic masses of three compounds are shown in the table. Use a linear system and Cramer's Rule to find the atomic masses of sulfur (S), nitrogen (N), and fluorine (F).

Compound	Formula	Atomic mass
Tetrasulphur tetranitride	S_4N_4	184
Sulfur hexafluoride	SF_6	146
Dinitrogen tetrafluoride	N_2F_4	104

13. A walkway lighting package includes a transformer, a certain length of wire, and a certain number of lights on the wire. The price of each lighting package depends on the length of wire and the number of lights on the wire. Use the following information to find the cost of a transformer, the cost per foot of wire, and the cost of a light. Assume that the cost of each item is the same in each lighting package.

- A package that contains a transformer, 25 feet of wire, and 5 lights costs $20.
- A package that contains a transformer, 50 feet of wire, and 15 lights costs $35.
- A package that contains a transformer, 100 feet of wire, and 20 lights costs $50.

14. Use the inverse of matrix A to decode the cryptogram.

$$A = \begin{bmatrix} 1 & -2 & 2 \\ 1 & 1 & -3 \\ 1 & -1 & 4 \end{bmatrix}$$

23 13 −34 31 −34 63 25 −17 61
24 14 −37 41 −17 −8 20 −29 40
38 −56 116 13 −11 1 22 −3 −6
41 −53 85 28 −32 16

15. The **transpose** of a matrix, denoted A^T, is formed by writing its columns as rows. Find the transpose of each matrix and verify that $(AB)^T = B^TA^T$.

$$A = \begin{bmatrix} -1 & 1 & -2 \\ 2 & 0 & 1 \end{bmatrix} \text{ and } B = \begin{bmatrix} -3 & 0 \\ 1 & 2 \\ 1 & -1 \end{bmatrix}$$

16. A code breaker intercepted the encoded message below.

45 −35 38 −30 18 −18 35 −30 81 −60
42 −28 75 −55 2 −2 22 −21 15
−10

Let

$$A^{-1} = \begin{bmatrix} w & x \\ y & z \end{bmatrix}.$$

(a) You know that $[45 \quad -35]A^{-1} = [10 \quad 15]$ and that $[38 \quad -30]A^{-1} = [8 \quad 14]$, where A^{-1} is the inverse of the encoding matrix A. Write and solve two systems of equations to find w, x, y, and z.

(b) Decode the message.

17. Let

$$A = \begin{bmatrix} 6 & 4 & 1 \\ 0 & 2 & 3 \\ 1 & 1 & 2 \end{bmatrix}.$$

Use a graphing utility to find A^{-1}. Compare $|A^{-1}|$ with $|A|$. Make a conjecture about the determinant of the inverse of a matrix.

18. Let A be an $n \times n$ matrix each of whose rows adds up to zero. Find $|A|$.

19. Consider matrices of the form

$$A = \begin{bmatrix} 0 & a_{12} & a_{13} & a_{14} & \cdots & a_{1n} \\ 0 & 0 & a_{23} & a_{24} & \cdots & a_{2n} \\ 0 & 0 & 0 & a_{34} & \cdots & a_{3n} \\ \vdots & \vdots & \vdots & \vdots & \cdots & \vdots \\ 0 & 0 & 0 & 0 & \cdots & a_{(n-1)n} \\ 0 & 0 & 0 & 0 & \cdots & 0 \end{bmatrix}$$

(a) Write a 2×2 matrix and a 3×3 matrix in the form of A.

(b) Use a graphing utility to raise each of the matrices to higher powers. Describe the result.

(c) Use the result of part (b) to make a conjecture about powers of A if A is a 4×4 matrix. Use a graphing utility to test your conjecture.

(d) Use the results of parts (b) and (c) to make a conjecture about powers of A if A is an $n \times n$ matrix.

How to study Chapter 11

▶ What you should learn

In this chapter you will learn the following skills and concepts:

- How to use sequence, factorial, and summation notation to write the terms and sum of a sequence
- How to recognize, write, and manipulate arithmetic sequences and geometric sequences
- How to use mathematical induction to prove a statement involving a positive integer n
- How to use the Binomial Theorem and Pascal's Triangle to calculate binomial coefficients and binomial expansions
- How to solve counting problems using the Fundamental Counting Principle, permutations, and combinations
- How to find the probabilities of events and their complements

▶ Important Vocabulary

As you encounter each new vocabulary term in this chapter, add the term and its definition to your notebook glossary.

Infinite sequence (p. 802)
Terms of a sequence (p. 802)
Finite sequence (p. 802)
Factorial (p. 804)
Summation or sigma notation (p. 806)
Finite series (p. 807)
Infinite series (p. 807)
Arithmetic sequence (p. 813)
Common difference (p. 813)
Geometric sequence (p. 822)
Common ratio (p. 822)
Geometric series (p. 826)
Mathematical induction (p. 832)
Binomial coefficients (p. 842)
Binomial Theorem (p. 842)
Fundamental Counting Principle (p. 851)
Permutation (p. 852)
Distinguishable permutations (p. 854)
Combination (p. 855)
Experiment (p. 860)
Outcomes (p. 860)
Sample space (p. 860)
Event (p. 860)
Probability (p. 861)
Mutually exclusive (p. 864)
Independent events (p. 866)
Complement of an event (p. 867)

Study Tools

Learning objectives in each section
Chapter Summary (p. 873)
Review Exercises (pp. 874–877)
Chapter Test (p. 878)
Cumulative Test for Chapters 9–11 (pp. 879, 880)

Additional Resources

Study and Solutions Guide
Interactive College Algebra
Videotapes/DVD for Chapter 11
Algebra and Trigonometry Website
Student Success Organizer

Gavin Hellier/Getty Images

11

Sequences, Series, and Probability

11.1 Sequences and Series

▶ What you should learn

- How to use sequence notation to write the terms of a sequence
- How to use factorial notation
- How to use summation notation to write sums
- How to find the sum of an infinite series
- How to use sequences and series to model and solve real-life problems

▶ Why you should learn it

Sequences and series can be used to model real-life problems. For instance, in Exercise 107 on page 811, sequences are used to model the number of Circuit City Stores from 1991 through 2000.

Bob Daemmrich/The Image Works

Sequences

In mathematics, the word *sequence* is used in much the same way as in ordinary English. Saying that a collection is listed in *sequence* means that it is ordered so that it has a first member, a second member, a third member, and so on.

Mathematically, you can think of a sequence as a *function* whose domain is the set of positive integers.

$$f(1) = a_1,\ f(2) = a_2,\ f(3) = a_3,\ f(4) = a_4, \ldots,\ f(n) = a_n, \ldots$$

Rather than using function notation, however, sequences are usually written using subscript notation, as indicated in the following definition.

Definition of Sequence

An **infinite sequence** is a function whose domain is the set of positive integers. The function values

$$a_1, a_2, a_3, a_4, \ldots, a_n, \ldots$$

are the **terms** of the sequence. If the domain of the function consists of the first n positive integers only, the sequence is a **finite sequence.**

On occasion it is convenient to begin subscripting a sequence with 0 instead of 1 so that the terms of the sequence become $a_0, a_1, a_2, a_3, \ldots$.

Example 1 ▶ Finding Terms of a Sequence

Find the first four terms of the sequences given by

a. $a_n = 3n - 2$ **b.** $a_n = 3 + (-1)^n$.

Solution

a. The first four terms of the sequence given by $a_n = 3n - 2$ are

$a_1 = 3(1) - 2 = 1$ 1st term

$a_2 = 3(2) - 2 = 4$ 2nd term

$a_3 = 3(3) - 2 = 7$ 3rd term

$a_4 = 3(4) - 2 = 10.$ 4th term

b. The first four terms of the sequence given by $a_n = 3 + (-1)^n$ are

$a_1 = 3 + (-1)^1 = 3 - 1 = 2$ 1st term

$a_2 = 3 + (-1)^2 = 3 + 1 = 4$ 2nd term

$a_3 = 3 + (-1)^3 = 3 - 1 = 2$ 3rd term

$a_4 = 3 + (-1)^4 = 3 + 1 = 4.$ 4th term

Exploration

Write out the first five terms of the sequence whose nth term is

$$a_n = \frac{(-1)^{n+1}}{2n-1}.$$

Are they the same as the first five terms of the sequence in Example 2? If not, how do they differ?

Example 2 ▶ A Sequence Whose Terms Alternate in Sign

The first five terms of the sequence given by $a_n = \dfrac{(-1)^n}{2n-1}$ are as follows.

$$a_1 = \frac{(-1)^1}{2(1)-1} = \frac{-1}{2-1} = -1 \qquad \text{1st term}$$

$$a_2 = \frac{(-1)^2}{2(2)-1} = \frac{1}{4-1} = \frac{1}{3} \qquad \text{2nd term}$$

$$a_3 = \frac{(-1)^3}{2(3)-1} = \frac{-1}{6-1} = -\frac{1}{5} \qquad \text{3rd term}$$

$$a_4 = \frac{(-1)^4}{2(4)-1} = \frac{1}{8-1} = \frac{1}{7} \qquad \text{4th term}$$

$$a_5 = \frac{(-1)^5}{2(5)-1} = \frac{-1}{10-1} = -\frac{1}{9} \qquad \text{5th term}$$

Simply listing the first few terms is not sufficient to define a unique sequence—the nth term *must be given.* To see this, consider the following sequences, both of which have the same first three terms.

$$\frac{1}{2}, \frac{1}{4}, \frac{1}{8}, \frac{1}{16}, \ldots, \frac{1}{2^n}, \ldots$$

$$\frac{1}{2}, \frac{1}{4}, \frac{1}{8}, \frac{1}{15}, \ldots, \frac{6}{(n+1)(n^2-n+6)}, \ldots$$

Example 3 ▶ Finding the nth Term of a Sequence 

Write an expression for the apparent nth term (a_n) of each sequence.

a. 1, 3, 5, 7, . . .   **b.** 2, −5, 10, −17, . . .

Solution

a. *n:* 1 2 3 4 . . . n

Terms: 1 3 5 7 . . . a_n

Apparent pattern: Each term is 1 less than twice n, which implies that

$$a_n = 2n - 1.$$

b. *n:* 1 2 3 4 . . . n

Terms: 2 −5 10 −17 . . . a_n

Apparent pattern: The terms have alternating signs with those in the even positions being negative. Each term is 1 more than the square of n, which implies that

$$a_n = (-1)^{n+1}(n^2 + 1)$$

Technology

To graph a sequence using a graphing utility, set the mode to *seq* and *dot* and enter the sequence. The graph of the sequence in Example 3(a) is shown below. You can use the *trace* feature to identify its terms.

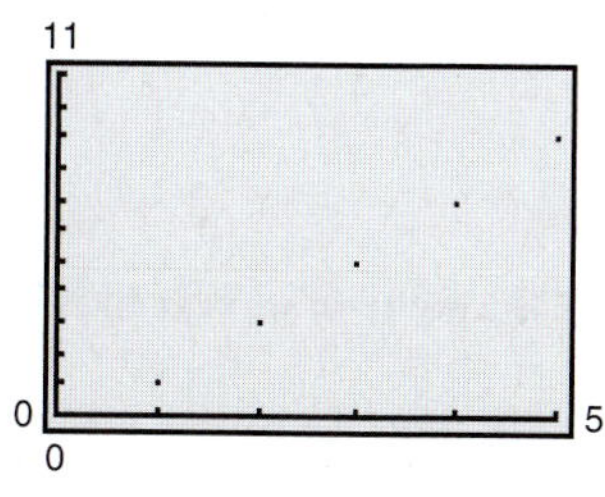

The icon (www) identifies examples and concepts related to features of the Learning Tools CD-ROM and the *Interactive* and *Internet* versions of this text. For more details see the chart on pages *xix-xxiii*.

Some sequences are defined **recursively.** To define a sequence recursively, you need to be given one or more of the first few terms. All other terms of the sequence are then defined using previous terms. A well-known example is the Fibonacci sequence shown in Example 4.

Example 4 ▶ The Fibonacci Sequence: A Recursive Sequence

The Fibonacci sequence is defined recursively, as follows.

$$a_0 = 1,\ a_1 = 1,\ a_k = a_{k-2} + a_{k-1},\ \text{where } k \geq 2$$

Write the first six terms of this sequence.

Solution

$a_0 = 1$ — 0th term is given.

$a_1 = 1$ — 1st term is given.

$a_2 = a_{2-2} + a_{2-1} = a_0 + a_1 = 1 + 1 = 2$ — Use recursion formula.

$a_3 = a_{3-2} + a_{3-1} = a_1 + a_2 = 1 + 2 = 3$ — Use recursion formula.

$a_4 = a_{4-2} + a_{4-1} = a_2 + a_3 = 2 + 3 = 5$ — Use recursion formula.

$a_5 = a_{5-2} + a_{5-1} = a_3 + a_4 = 3 + 5 = 8$ — Use recursion formula.

STUDY TIP

Any variable can be used as a subscript. The most commonly used variable subscripts in sequence and series notation are i, j, k, and n.

Factorial Notation

Some very important sequences in mathematics involve terms that are defined with special types of products called **factorials.**

The *Interactive* CD-ROM and *Internet* versions of this text offer a Try It for each example in the text.

Definition of Factorial

If n is a positive integer, ***n*** **factorial** is defined as

$$n! = 1 \cdot 2 \cdot 3 \cdot 4 \cdot \cdot \cdot (n-1) \cdot n.$$

As a special case, zero factorial is defined as $0! = 1$.

Here are some values of $n!$ for the first several nonnegative integers. Notice that $0!$ is 1 by definition.

$0! = 1$

$1! = 1$

$2! = 1 \cdot 2 = 2$

$3! = 1 \cdot 2 \cdot 3 = 6$

$4! = 1 \cdot 2 \cdot 3 \cdot 4 = 24$

$5! = 1 \cdot 2 \cdot 3 \cdot 4 \cdot 5 = 120$

The value of n does not have to be very large before the value of $n!$ becomes huge. For instance, $10! = 3{,}628{,}800$.

Factorials follow the same conventions for order of operations as do exponents. For instance,

$$2n! = 2(n!) = 2(1 \cdot 2 \cdot 3 \cdot 4 \cdot \cdot \cdot n)$$

whereas $(2n)! = 1 \cdot 2 \cdot 3 \cdot 4 \cdot \cdot \cdot 2n$.

Example 5 ▶ Finding Terms of a Sequence Involving Factorials

List the first five terms of the sequence given by

$$a_n = \frac{2^n}{n!}.$$

Begin with $n = 0$. Then plot the points on a set of coordinate axes.

Solution

$$a_0 = \frac{2^0}{0!} = \frac{1}{1} = 1 \qquad \text{0th term}$$

$$a_1 = \frac{2^1}{1!} = \frac{2}{1} = 2 \qquad \text{1st term}$$

$$a_2 = \frac{2^2}{2!} = \frac{4}{2} = 2 \qquad \text{2nd term}$$

$$a_3 = \frac{2^3}{3!} = \frac{8}{6} = \frac{4}{3} \qquad \text{3rd term}$$

$$a_4 = \frac{2^4}{4!} = \frac{16}{24} = \frac{2}{3} \qquad \text{4th term}$$

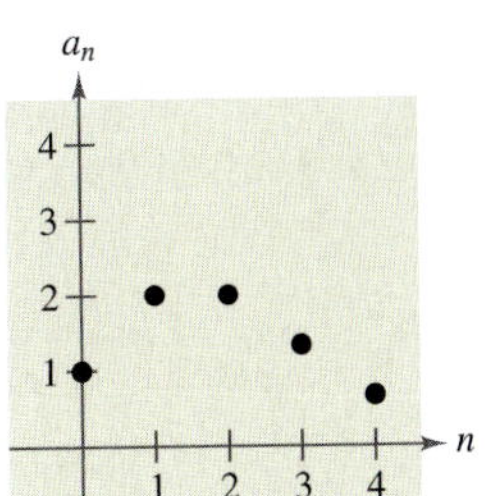

FIGURE 11.1

Figure 11.1 shows the first five terms of the sequence.

When working with fractions involving factorials, you will often find that the fractions can be reduced.

Example 6 ▶ Evaluating Factorial Expressions 

Evaluate each factorial expression.

a. $\dfrac{8!}{2! \cdot 6!}$ **b.** $\dfrac{2! \cdot 6!}{3! \cdot 5!}$ **c.** $\dfrac{n!}{(n-1)!}$

Solution

a. $\dfrac{8!}{2! \cdot 6!} = \dfrac{\cancel{1 \cdot 2 \cdot 3 \cdot 4 \cdot 5 \cdot 6} \cdot 7 \cdot 8}{1 \cdot 2 \cdot \cancel{1 \cdot 2 \cdot 3 \cdot 4 \cdot 5 \cdot 6}} = \dfrac{7 \cdot 8}{2} = 28$

b. $\dfrac{2! \cdot 6!}{3! \cdot 5!} = \dfrac{\cancel{1 \cdot 2} \cdot \cancel{1 \cdot 2 \cdot 3 \cdot 4 \cdot 5} \cdot 6}{\cancel{1 \cdot 2} \cdot 3 \cdot \cancel{1 \cdot 2 \cdot 3 \cdot 4 \cdot 5}} = \dfrac{6}{3} = 2$

c. $\dfrac{n!}{(n-1)!} = \dfrac{\cancel{1 \cdot 2 \cdot 3 \cdots (n-1)} \cdot n}{\cancel{1 \cdot 2 \cdot 3 \cdots (n-1)}} = n$

Summation Notation 

Technology Most graphing utilities are able to sum the first n terms of a sequence. Check your user's guide for a *sum sequence* feature or a *series* feature.

There is a convenient notation for the sum of the terms of a finite sequence. It is called **summation notation** or **sigma notation** because it involves the use of the uppercase Greek letter sigma, written as Σ.

Definition of Summation Notation

The sum of the first n terms of a sequence is represented by

$$\sum_{i=1}^{n} a_i = a_1 + a_2 + a_3 + a_4 + \cdot \cdot \cdot + a_n$$

where i is called the **index of summation,** n is the **upper limit of summation,** and 1 is the **lower limit of summation.**

Example 7 ▶ Summation Notation for Sums 

STUDY TIP

Summation notation is an instruction to add the terms of a sequence. From the definition at the right, the upper limit of summation tells you where to end the sum. Summation notation helps you generate the appropriate terms of the sequence prior to finding the actual sum, which may not be clear.

Find each sum.

a. $\displaystyle\sum_{i=1}^{5} 3i$ **b.** $\displaystyle\sum_{k=3}^{6} (1 + k^2)$ **c.** $\displaystyle\sum_{i=0}^{8} \frac{1}{i!}$

Solution

a.
$$\begin{aligned}\sum_{i=1}^{5} 3i &= 3(1) + 3(2) + 3(3) + 3(4) + 3(5)\\ &= 3(1 + 2 + 3 + 4 + 5)\\ &= 3(15)\\ &= 45\end{aligned}$$

b.
$$\begin{aligned}\sum_{k=3}^{6} (1 + k^2) &= (1 + 3^2) + (1 + 4^2) + (1 + 5^2) + (1 + 6^2)\\ &= 10 + 17 + 26 + 37\\ &= 90\end{aligned}$$

c.
$$\begin{aligned}\sum_{i=0}^{8} \frac{1}{i!} &= \frac{1}{0!} + \frac{1}{1!} + \frac{1}{2!} + \frac{1}{3!} + \frac{1}{4!} + \frac{1}{5!} + \frac{1}{6!} + \frac{1}{7!} + \frac{1}{8!}\\ &= 1 + 1 + \frac{1}{2} + \frac{1}{6} + \frac{1}{24} + \frac{1}{120} + \frac{1}{720} + \frac{1}{5040} + \frac{1}{40{,}320}\\ &\approx 2.71828\end{aligned}$$

For this summation, note that the sum is very close to the irrational number $e \approx 2.718281828$. It can be shown that as more terms of the sequence whose nth term is $1/n!$ are added, the sum becomes closer and closer to e.

The *Interactive* CD-ROM and *Internet* versions of this text offer a Quiz for every section of the text.

In Example 7, note that the lower limit of a summation does not have to be 1. Also note that the index of summation does not have to be the letter i. For instance, in part (b), the letter k is the index of summation.

STUDY TIP

Variations in the upper and lower limits of summation can produce quite different-looking summation notations for *the same sum*. For example, the following two sums have the same terms.

$$\sum_{i=1}^{3} 3(2^i) = 3(2^1 + 2^2 + 2^3)$$

$$\sum_{i=0}^{2} 3(2^{i+1}) = 3(2^1 + 2^2 + 2^3)$$

Properties of Sums

1. $\sum_{i=1}^{n} c = cn$, c is a constant.
2. $\sum_{i=1}^{n} ca_i = c\sum_{i=1}^{n} a_i$, c is a constant.
3. $\sum_{i=1}^{n} (a_i + b_i) = \sum_{i=1}^{n} a_i + \sum_{i=1}^{n} b_i$
4. $\sum_{i=1}^{n} (a_i - b_i) = \sum_{i=1}^{n} a_i - \sum_{i=1}^{n} b_i$

For proofs of these properties, see Proofs in Mathematics on page 881.

Series

Many applications involve the sum of the terms of a finite or infinite sequence. Such a sum is called a **series.**

Definition of Series

Consider the infinite sequence $a_1, a_2, a_3, \ldots, a_i, \ldots$.

1. The sum of the first n terms of the sequence is called a **finite series** or the ***n*th partial sum** of the sequence and is denoted by

$$a_1 + a_2 + a_3 + \cdots + a_n = \sum_{i=1}^{n} a_i.$$

2. The sum of all terms of the infinite sequence is called an **infinite series** and is denoted by

$$a_1 + a_2 + a_3 + \cdots + a_i + \cdots = \sum_{i=1}^{\infty} a_i.$$

Example 8 ▶ **Finding the Sum of a Series**

For the series $\sum_{i=1}^{\infty} \frac{3}{10^i}$, find (a) the third partial sum and (b) the sum.

Solution

a. The third partial sum is

$$\sum_{i=1}^{3} \frac{3}{10^i} = \frac{3}{10^1} + \frac{3}{10^2} + \frac{3}{10^3} = 0.3 + 0.03 + 0.003 = 0.333.$$

b. The sum of the series is

$$\sum_{i=1}^{\infty} \frac{3}{10^i} = \frac{3}{10^1} + \frac{3}{10^2} + \frac{3}{10^3} + \frac{3}{10^4} + \frac{3}{10^5} + \cdots$$

$$= 0.3 + 0.03 + 0.003 + 0.0003 + 0.00003 + \cdots$$

$$= 0.33333\ldots = \frac{1}{3}.$$

Application

Sequences have many applications in business and science. One such application is illustrated in Example 9.

Example 9 ▶ Population of the United States 

For the years 1970 to 2000, the resident population of the United States can be approximated by the model

$$a_n = 204.8 + 2.09n + 0.009n^2, \qquad n = 0, 1, \ldots, 30$$

where a_n is the population in millions and n represents the calendar year, with $n = 0$ corresponding to 1970. Find the last five terms of this finite sequence, which represent the U.S. population for the years 1996 to 2000. (Source: U.S. Census Bureau)

Solution

The last five terms of this finite sequence are as follows.

$$a_{26} = 204.8 + 2.09(26) + 0.009(26)^2 \approx 265.2 \qquad \text{1996 population}$$

$$a_{27} = 204.8 + 2.09(27) + 0.009(27)^2 \approx 267.8 \qquad \text{1997 population}$$

$$a_{28} = 204.8 + 2.09(28) + 0.009(28)^2 \approx 270.4 \qquad \text{1998 population}$$

$$a_{29} = 204.8 + 2.09(29) + 0.009(29)^2 \approx 273.0 \qquad \text{1999 population}$$

$$a_{30} = 204.8 + 2.09(30) + 0.009(30)^2 \approx 275.6 \qquad \text{2000 population}$$

The bar graph in Figure 11.2 graphically represents the population given by this sequence for the entire 31-year period from 1970 to 2000.

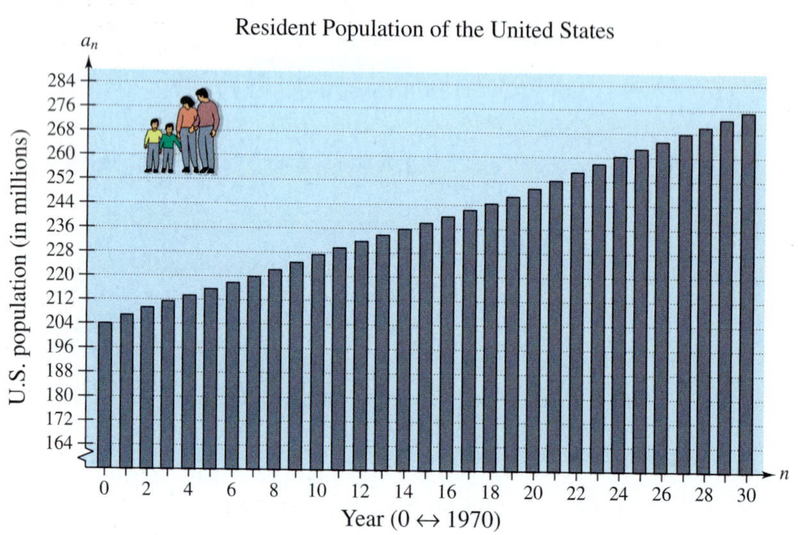

FIGURE 11.2

11.1 Exercises

The *Interactive* CD-ROM and *Internet* versions of this text contain step-by-step solutions to all odd-numbered exercises. They also provide Tutorial Exercises for additional help.

In Exercises 1–24, write the first five terms of the sequence. (Assume that *n* begins with 1.)

1. $a_n = 3n + 1$ **2.** $a_n = 5n - 3$

3. $a_n = 2^n$ **4.** $a_n = \left(\frac{1}{2}\right)^n$

5. $a_n = (-2)^n$ **6.** $a_n = \left(-\frac{1}{2}\right)^n$

7. $a_n = \dfrac{n+2}{n}$ **8.** $a_n = \dfrac{n}{n+2}$

9. $a_n = \dfrac{6n}{3n^2 - 1}$ **10.** $a_n = \dfrac{3n^2 - n + 4}{2n^2 + 1}$

11. $a_n = \dfrac{1 + (-1)^n}{n}$ **12.** $a_n = 1 + (-1)^n$

13. $a_n = 2 - \dfrac{1}{3^n}$ **14.** $a_n = \dfrac{2^n}{3^n}$

15. $a_n = \dfrac{1}{n^{3/2}}$ **16.** $a_n = \dfrac{10}{n^{2/3}}$

17. $a_n = \dfrac{3^n}{n!}$ **18.** $a_n = \dfrac{n!}{n}$

19. $a_n = \dfrac{(-1)^n}{n^2}$ **20.** $a_n = (-1)^n\left(\dfrac{n}{n+1}\right)$

21. $a_n = \frac{2}{3}$ **22.** $a_n = 0.3$

23. $a_n = n(n-1)(n-2)$ **24.** $a_n = n(n^2 - 6)$

In Exercises 25–30, find the indicated term of the sequence.

25. $a_n = (-1)^n(3n - 2)$

$a_{25} =$

26. $a_n = (-1)^{n-1}[n(n-1)]$

$a_{16} =$

27. $a_n = \dfrac{2^n}{n!}$ **28.** $a_n = \dfrac{n!}{2n}$

$a_{10} =$ $a_8 =$

29. $a_n = \dfrac{4n}{2n^2 - 3}$ **30.** $a_n = \dfrac{4n^2 - n + 3}{n(n-1)(n+2)}$

$a_{11} =$ $a_{13} =$

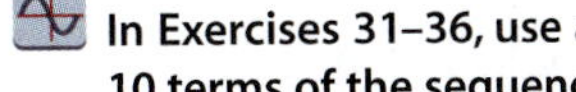

In Exercises 31–36, use a graphing utility to graph the first 10 terms of the sequence.

31. $a_n = \dfrac{3}{4}n$ **32.** $a_n = 2 - \dfrac{4}{n}$

33. $a_n = 16(-0.5)^{n-1}$ **34.** $a_n = 8(0.75)^{n-1}$

35. $a_n = \dfrac{2n}{n+1}$ **36.** $a_n = \dfrac{n^2}{n^2 + 2}$

In Exercises 37–40, match the sequence with the graph of its first 10 terms. [The graphs are labeled (a), (b), (c), and (d).]

(a)

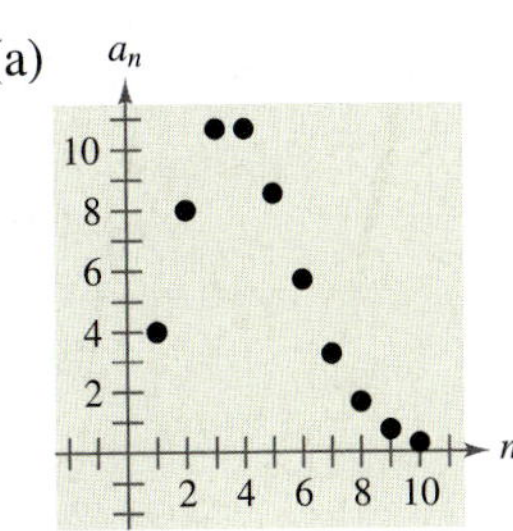

(b)

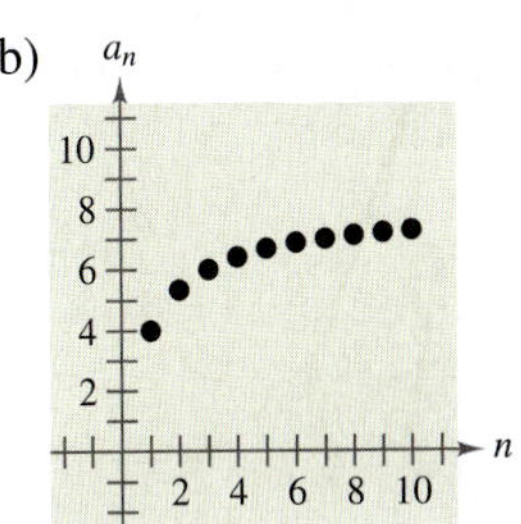

(c)

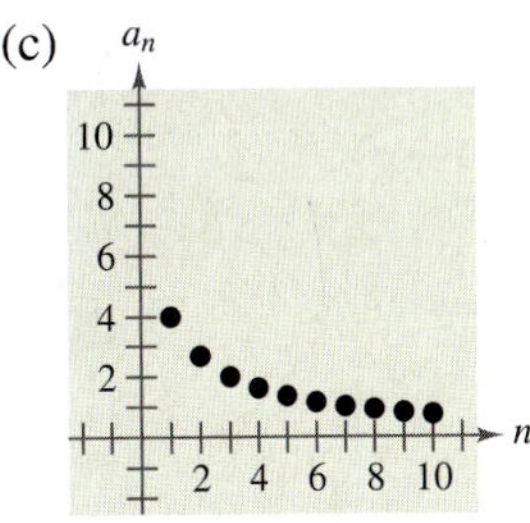

(d)

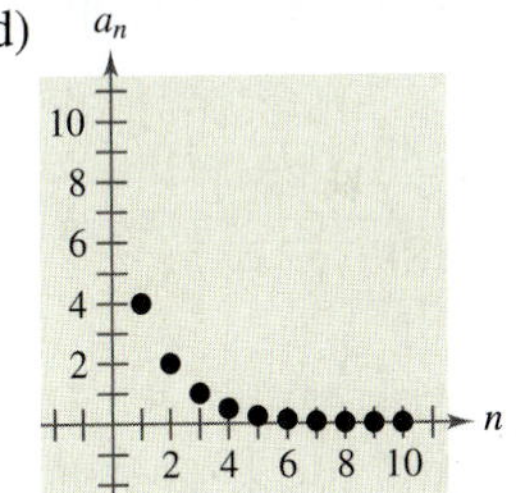

37. $a_n = \dfrac{8}{n+1}$ **38.** $a_n = \dfrac{8n}{n+1}$

39. $a_n = 4(0.5)^{n-1}$ **40.** $a_n = \dfrac{4^n}{n!}$

In Exercises 41–54, write an expression for the *apparent* *n*th term of the sequence. (Assume that *n* begins with 1.)

41. $1, 4, 7, 10, 13, . . .$ **42.** $3, 7, 11, 15, 19, . . .$

43. $0, 3, 8, 15, 24, . . .$ **44.** $2, -4, 6, -8, 10, . . .$

45. $\frac{-2}{3}, \frac{3}{4}, \frac{-4}{5}, \frac{5}{6}, \frac{-6}{7}, . . .$

46. $\frac{1}{2}, \frac{-1}{4}, \frac{1}{8}, \frac{-1}{16}, . . .$ **47.** $\frac{2}{1}, \frac{3}{3}, \frac{4}{5}, \frac{5}{7}, \frac{6}{9}, . . .$

48. $\frac{1}{3}, \frac{2}{9}, \frac{4}{27}, \frac{8}{81}, . . .$ **49.** $1, \frac{1}{4}, \frac{1}{9}, \frac{1}{16}, \frac{1}{25}, . . .$

50. $1, \frac{1}{2}, \frac{1}{6}, \frac{1}{24}, \frac{1}{120}, . . .$ **51.** $1, -1, 1, -1, 1, . . .$

52. $1, 2, \dfrac{2^2}{2}, \dfrac{2^3}{6}, \dfrac{2^4}{24}, \dfrac{2^5}{120}, . . .$

53. $1 + \frac{1}{1}, 1 + \frac{1}{2}, 1 + \frac{1}{3}, 1 + \frac{1}{4}, 1 + \frac{1}{5}, . . .$

54. $1 + \frac{1}{2}, 1 + \frac{3}{4}, 1 + \frac{7}{8}, 1 + \frac{15}{16}, 1 + \frac{31}{32}, . . .$

In Exercises 55–58, write the first five terms of the sequence defined recursively.

55. $a_1 = 28, \quad a_{k+1} = a_k - 4$

56. $a_1 = 15, \quad a_{k+1} = a_k + 3$

57. $a_1 = 3, \quad a_{k+1} = 2(a_k - 1)$

58. $a_1 = 32, \quad a_{k+1} = \frac{1}{2}a_k$

In Exercises 59–62, write the first five terms of the sequence defined recursively. Use the pattern to write the *n*th term of the sequence as a function of *n*. (Assume that *n* begins with 1.)

59. $a_1 = 6, \quad a_{k+1} = a_k + 2$

60. $a_1 = 25, \quad a_{k+1} = a_k - 5$

61. $a_1 = 81, \quad a_{k+1} = \frac{1}{3}a_k$

62. $a_1 = 14, \quad a_{k+1} = (-2)a_k$

In Exercises 63–70, simplify the factorial expression.

63. $\frac{4!}{6!}$

64. $\frac{5!}{8!}$

65. $\frac{10!}{8!}$

66. $\frac{25!}{23!}$

67. $\frac{(n+1)!}{n!}$

68. $\frac{(n+2)!}{n!}$

69. $\frac{(2n-1)!}{(2n+1)!}$

70. $\frac{(3n+1)!}{(3n)!}$

In Exercises 71–82, find the sum.

71. $\sum_{i=1}^{5} (2i + 1)$

72. $\sum_{i=1}^{6} (3i - 1)$

73. $\sum_{k=1}^{4} 10$

74. $\sum_{k=1}^{5} 5$

75. $\sum_{i=0}^{4} i^2$

76. $\sum_{i=0}^{5} 2i^2$

77. $\sum_{k=0}^{3} \frac{1}{k^2 + 1}$

78. $\sum_{j=3}^{5} \frac{1}{j^2 - 3}$

79. $\sum_{k=2}^{5} (k + 1)^2(k - 3)$

80. $\sum_{i=1}^{4} [(i - 1)^2 + (i + 1)^3]$

81. $\sum_{i=1}^{4} 2^i$

82. $\sum_{j=0}^{4} (-2)^j$

In Exercises 83–86, use a calculator to find the sum.

83. $\sum_{j=1}^{6} (24 - 3j)$

84. $\sum_{j=1}^{10} \frac{3}{j + 1}$

85. $\sum_{k=0}^{4} \frac{(-1)^k}{k + 1}$

86. $\sum_{k=0}^{4} \frac{(-1)^k}{k!}$

In Exercises 87–96, use sigma notation to write the sum.

87. $\frac{1}{3(1)} + \frac{1}{3(2)} + \frac{1}{3(3)} + \cdots + \frac{1}{3(9)}$

88. $\frac{5}{1+1} + \frac{5}{1+2} + \frac{5}{1+3} + \cdots + \frac{5}{1+15}$

89. $\left[2\left(\frac{1}{8}\right) + 3\right] + \left[2\left(\frac{2}{8}\right) + 3\right] + \cdots + \left[2\left(\frac{8}{8}\right) + 3\right]$

90. $\left[1 - \left(\frac{1}{6}\right)^2\right] + \left[1 - \left(\frac{2}{6}\right)^2\right] + \cdots + \left[1 - \left(\frac{6}{6}\right)^2\right]$

91. $3 - 9 + 27 - 81 + 243 - 729$

92. $1 - \frac{1}{2} + \frac{1}{4} - \frac{1}{8} + \cdots - \frac{1}{128}$

93. $\frac{1}{1^2} - \frac{1}{2^2} + \frac{1}{3^2} - \frac{1}{4^2} + \cdots - \frac{1}{20^2}$

94. $\frac{1}{1 \cdot 3} + \frac{1}{2 \cdot 4} + \frac{1}{3 \cdot 5} + \cdots + \frac{1}{10 \cdot 12}$

95. $\frac{1}{4} + \frac{3}{8} + \frac{7}{16} + \frac{15}{32} + \frac{31}{64}$

96. $\frac{1}{2} + \frac{2}{4} + \frac{6}{8} + \frac{24}{16} + \frac{120}{32} + \frac{720}{64}$

In Exercises 97–100, find the indicated partial sum of the series.

97. $\sum_{i=1}^{\infty} 5\left(\frac{1}{2}\right)^i$, Fourth partial sum

98. $\sum_{i=1}^{\infty} 2\left(\frac{1}{3}\right)^i$, Fifth partial sum

99. $\sum_{n=1}^{\infty} 4\left(-\frac{1}{2}\right)^n$, Third partial sum

100. $\sum_{n=1}^{\infty} 8\left(-\frac{1}{4}\right)^n$, Fourth partial sum

In Exercises 101–104, find the sum of the infinite series.

101. $\sum_{i=1}^{\infty} 6\left(\frac{1}{10}\right)^i$

102. $\sum_{k=1}^{\infty} \left(\frac{1}{10}\right)^k$

103. $\sum_{k=1}^{\infty} 7\left(\frac{1}{10}\right)^k$

104. $\sum_{i=1}^{\infty} 2\left(\frac{1}{10}\right)^i$

105. ***Compound Interest*** A deposit of \$5000 is made in an account that earns 8% interest compounded quarterly. The balance in the account after n quarters is

$$A_n = 5000\left(1 + \frac{0.08}{4}\right)^n, \quad n = 1, 2, 3, \ldots.$$

(a) Compute the first eight terms of this sequence.

(b) Find the balance in this account after 10 years by computing the 40th term of the sequence.

106. ***Compound Interest*** A deposit of \$100 is made each month in an account that earns 12% interest compounded monthly. The balance in the account after n months is

$$A_n = 100(101)[(1.01)^n - 1], n = 1, 2, 3, \ldots .$$

(a) Compute the first six terms of this sequence.

(b) Find the balance in this account after 5 years by computing the 60th term of the sequence.

(c) Find the balance in this account after 20 years by computing the 240th term of the sequence.

▶ Model It

107. ***Data Analysis*** The table shows the numbers a_n of Circuit City stores for the years 1991 to 2000. (Source: Circuit City Stores, Inc.)

Year, n	Number of stores, a_n
1991	228
1992	260
1993	294
1994	352
1995	419
1996	493
1997	556
1998	587
1999	616
2000	629

(a) Use the *regression* feature of a graphing utility to find a linear sequence that models the data. Let n represent the year, with $n = 1$ corresponding to 1991.

(b) Use the *regression* feature of a graphing utility to find a quadratic sequence that models the data.

(c) Evaluate the sequences from parts (a) and (b) for $n = 1, 2, \ldots, 10$. Compare these values with those shown in the table. Which model is a better fit for the data? Explain.

(d) Which model do you think would better predict the number of Circuit City stores in the future? Use the model you chose to predict the number of Circuit City stores in 2005.

108. ***Medicine*** The numbers a_n (in millions) of AIDS cases reported from 1994 to 1999 can be approximated by the model

$$a_n = 0.3944n^3 - 7.738n^2 + 41.48n + 10.2,$$
$$n = 4, 5, \ldots, 9$$

where n is the year, with $n = 4$ corresponding to 1994. Find the terms of this finite sequence. Use a graphing utility to construct a bar graph that represents the sequence. (Source: U.S. Centers for Disease Control and Prevention)

109. ***Federal Debt*** From 1990 to 2000, the federal debt of the United States rose from just over \$3 trillion to more than \$5 trillion. The federal debt a_n (in billions of dollars) from 1990 to 2000 is approximated by the model

$$a_n = -20.613n^2 + 452.92n + 3183.1,$$
$$n = 0, 1, \ldots, 10$$

where n is the year, with $n = 0$ corresponding to 1990. Find the terms of this finite sequence. Use a graphing utility to construct a bar graph that represents the sequence. (Source: U.S. Office of Management and Budget)

110. ***Net Profit*** The net profits a_n (in millions of dollars) of Walgreen for the years 1991 through 2001 are shown in the figure. These profits can be approximated by the model

$$a_n = 5.547n^2 + 184.7, \qquad n = 1, 2, \ldots, 11$$

where $n = 1$ represents 1991. Use this model to approximate the total net income from 1991 through 2001. Compare this sum with the result of adding the incomes shown in the figure. (Source: Walgreen Company)

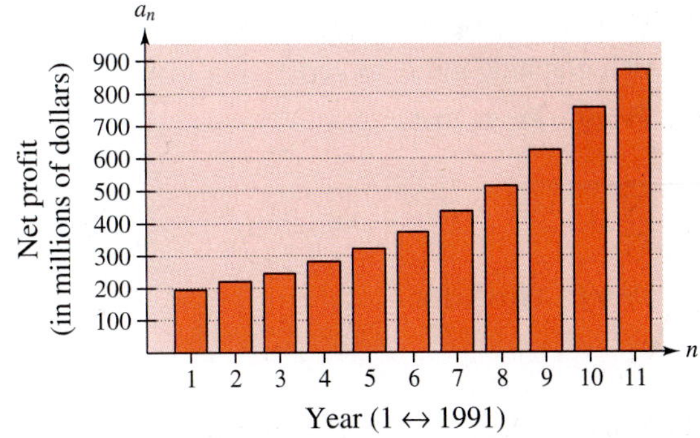

Synthesis

True or False? In Exercises 111 and 112, determine whether the statement is true or false. Justify your answer.

111. $\sum_{i=1}^{4}(i^2 + 2i) = \sum_{i=1}^{4} i^2 + 2\sum_{i=1}^{4} i$

112. $\sum_{j=1}^{4} 2^j = \sum_{j=3}^{6} 2^{j-2}$

Fibonacci Sequence In Exercises 113 and 114, use the Fibonacci sequence. (See Example 4.)

113. Write the first 12 terms of the Fibonacci sequence a_n and the first 10 terms of the sequence given by

$$b_n = \frac{a_{n+1}}{a_n}, \qquad n \geq 1.$$

114. Using the definition for b_n in Exercise 113, show that b_n can be defined recursively by

$$b_n = 1 + \frac{1}{b_{n-1}}.$$

Arithmetic Mean In Exercises 115–118, use the following definition of the arithmetic mean $\bar{x}$ of a set of n measurements $x_1, x_2, x_3, \ldots, x_n$.

$$\bar{x} = \frac{1}{n}\sum_{i=1}^{n} x_i$$

115. Find the arithmetic mean of the six checking account balances \$327.15, \$785.69, \$433.04, \$265.38, \$604.12, and \$590.30. Use the statistical capabilities of a graphing utility to verify your result.

116. Find the arithmetic mean of the following prices per gallon for regular unleaded gasoline at five gasoline stations in a city: \$1.279, \$1.259, \$1.289, \$1.329, and \$1.349. Use the statistical capabilities of a graphing utility to verify your result.

117. ***Proof*** Prove that

$$\sum_{i=1}^{n}(x_i - \bar{x}) = 0.$$

118. ***Proof*** Prove that

$$\sum_{i=1}^{n}(x_i - \bar{x})^2 = \sum_{i=1}^{n} x_i^2 - \frac{1}{n}\left(\sum_{i=1}^{n} x_i\right)^2.$$

In Exercises 119–122, find the first five terms of the sequence.

119. $a_n = \dfrac{x^n}{n!}$

120. $a_n = \dfrac{(-1)^n x^{2n+1}}{2n+1}$

121. $a_n = \dfrac{(-1)^n x^{2n}}{(2n)!}$

122. $a_n = \dfrac{(-1)^n x^{2n+1}}{(2n+1)!}$

Review

In Exercises 123–126, determine whether the function has an inverse function. If it does, find its inverse function.

123. $f(x) = 4x - 3$

124. $g(x) = \dfrac{3}{x}$

125. $h(x) = \sqrt{5x + 1}$

126. $f(x) = (x - 1)^2$

In Exercises 127–130, find (a) $A - B$, (b) $4B - 3A$, (c) AB, and (d) BA.

127. $A = \begin{bmatrix} 6 & 5 \\ 3 & 4 \end{bmatrix}, \quad B = \begin{bmatrix} -2 & 4 \\ 6 & -3 \end{bmatrix}$

128. $A = \begin{bmatrix} 10 & 7 \\ -4 & 6 \end{bmatrix}, \quad B = \begin{bmatrix} 0 & -12 \\ 8 & 11 \end{bmatrix}$

129. $A = \begin{bmatrix} -2 & -3 & 6 \\ 4 & 5 & 7 \\ 1 & 7 & 4 \end{bmatrix}, \quad B = \begin{bmatrix} 1 & 4 & 2 \\ 0 & 1 & 6 \\ 0 & 3 & 1 \end{bmatrix}$

130. $A = \begin{bmatrix} -1 & 4 & 0 \\ 5 & 1 & 2 \\ 0 & -1 & 3 \end{bmatrix}, \quad B = \begin{bmatrix} 0 & 4 & 0 \\ 3 & 1 & -2 \\ -1 & 0 & 2 \end{bmatrix}$

In Exercises 131–134, find the determinant of the matrix.

131. $A = \begin{bmatrix} 3 & 5 \\ -1 & 7 \end{bmatrix}$

132. $A = \begin{bmatrix} -2 & 8 \\ 12 & 15 \end{bmatrix}$

133. $A = \begin{bmatrix} 3 & 4 & 5 \\ 0 & 7 & 3 \\ 4 & 9 & -1 \end{bmatrix}$

134. $A = \begin{bmatrix} 16 & 11 & 10 & 2 \\ 9 & 8 & 3 & 7 \\ -2 & -1 & 12 & 3 \\ -4 & 6 & 2 & 1 \end{bmatrix}$

11.2 Arithmetic Sequences and Partial Sums

▶ **What you should learn**

- How to recognize and write arithmetic sequences
- How to find an nth partial sum of an arithmetic sequence
- How to use arithmetic sequences to model and solve real-life problems

▶ **Why you should learn it**

Arithmetic sequences have practical real-life applications. For instance, in Exercise 85 on page 821, an arithmetic sequence is used to model the per capita personal income in the United States from 1990 through 2000.

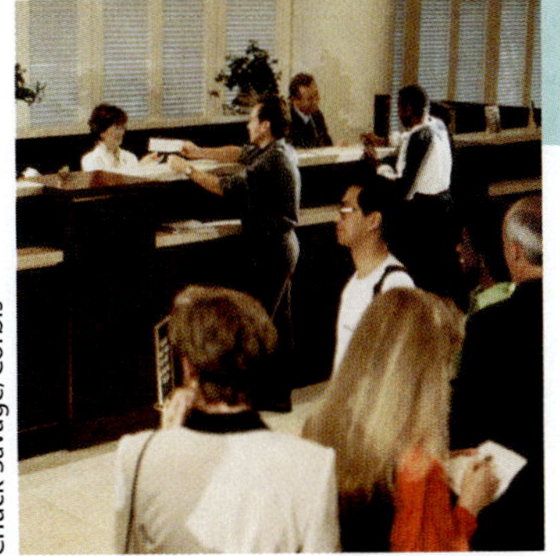

Chuck Savage/Corbis

Arithmetic Sequences

A sequence whose consecutive terms have a common difference is called an **arithmetic sequence.**

Definition of Arithmetic Sequence

A sequence is **arithmetic** if the differences between consecutive terms are the same. So, the sequence

$$a_1, a_2, a_3, a_4, \ldots, a_n, \ldots$$

is arithmetic if there is a number d such that

$$a_2 - a_1 = a_3 - a_2 = a_4 - a_3 = \cdots = d$$

and so on. The number d is the **common difference** of the arithmetic sequence.

Example 1 ▶ Examples of Arithmetic Sequences

a. The sequence whose nth term is $4n + 3$ is arithmetic. For this sequence, the common difference between consecutive terms is 4.

$$7, 11, 15, 19, \ldots, 4n + 3, \ldots \qquad \text{Begin with } n = 1.$$

$11 - 7 = 4$

b. The sequence whose nth term is $7 - 5n$ is arithmetic. For this sequence, the common difference between consecutive terms is -5.

$$2, -3, -8, -13, \ldots, 7 - 5n, \ldots \qquad \text{Begin with } n = 1.$$

$-3 - 2 = -5$

c. The sequence whose nth term is $\frac{1}{4}(n + 3)$ is arithmetic. For this sequence, the common difference between consecutive terms is $\frac{1}{4}$.

$$1, \frac{5}{4}, \frac{3}{2}, \frac{7}{4}, \ldots, \frac{n + 3}{4}, \ldots \qquad \text{Begin with } n = 1.$$

$\frac{5}{4} - 1 = \frac{1}{4}$

The sequence 1, 4, 9, 16, . . . , whose nth term is n^2, is *not* arithmetic. The difference between the first two terms is

$$a_2 - a_1 = 4 - 1 = 3$$

but the difference between the second and third terms is

$$a_3 - a_2 = 9 - 4 = 5.$$

In Example 1, notice that each of the arithmetic sequences has an nth term that is of the form $dn + c$, where the common difference of the sequence is d. An arithmetic sequence may be thought of as a linear function whose domain is the set of natural numbers.

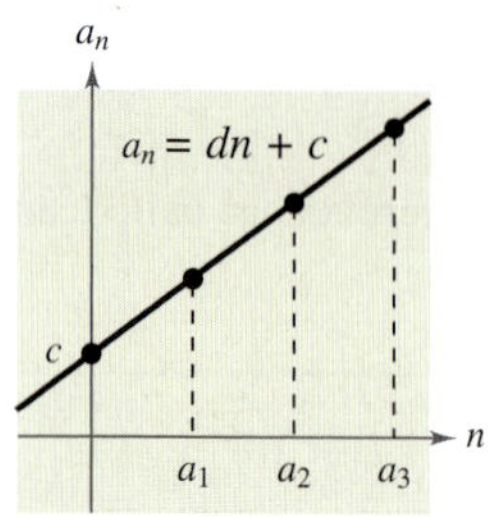

FIGURE 11.3

The nth Term of an Arithmetic Sequence

The nth term of an arithmetic sequence has the form

$$a_n = dn + c$$

where d is the common difference between consecutive terms of the sequence and $c = a_1 - d$. A graphical representation of this definition is shown in Figure 11.3.

STUDY TIP

An alternative form of the nth term of an arithmetic sequence can be derived from the pattern below.

$a_1 = a_1$ — 1st term

$a_2 = a_1 + d$ — 2nd term

$a_3 = a_1 + 2d$ — 3rd term

$a_4 = a_1 + 3d$ — 4th term

$a_5 = a_1 + 4d$ — 5th term

1 less

⋮

$a_n = a_1 + (n - 1)d$ — nth term

1 less

Example 2 ▶ Finding the nth Term of an Arithmetic Sequence

Find a formula for the nth term of the arithmetic sequence whose common difference is 3 and whose first term is 2.

Solution

Because the sequence is arithmetic, you know that the formula for the nth term is of the form $a_n = dn + c$. Moreover, because the common difference is $d = 3$, the formula must have the form

$$a_n = 3n + c. \qquad \text{Substitute 3 for } d.$$

Because $a_1 = 2$, it follows that

$$\begin{aligned} c &= a_1 - d \\ &= 2 - 3 \qquad \text{Substitute 2 for } a_1 \text{ and 3 for } d. \\ &= -1. \end{aligned}$$

So, the formula for the nth term is

$$a_n = 3n - 1.$$

The sequence therefore has the following form.

$$2, 5, 8, 11, 14, \ldots, 3n - 1, \ldots$$

Another way to find a formula for the nth term of the sequence in Example 2 is to begin by writing the terms of the sequence.

a_1	a_2	a_3	a_4	a_5	a_6	a_7	. . .
2	2 + 3	5 + 3	8 + 3	11 + 3	14 + 3	17 + 3	. . .
2	5	8	11	14	17	20	. . .

From these terms, you can reason that the nth term is of the form

$$\begin{aligned} a_n &= dn + c \\ &= 3n - 1. \end{aligned}$$

STUDY TIP

You can find a_1 in Example 3 by using the alternative form of the nth term of an arithmetic sequence, as follows.

$$a_n = a_1 + (n - 1)d$$

$$a_4 = a_1 + (n - 1)d$$

$$20 = a_1 + (4 - 1)5$$

$$20 = a_1 + 15$$

$$5 = a_1$$

Example 3 ▶ Writing the Terms of an Arithmetic Sequence

The fourth term of an arithmetic sequence is 20, and the 13th term is 65. Write the first several terms of this sequence.

Solution

The fourth and 13th terms of the sequence are related by

$$a_{13} = a_4 + 9d.$$

Using $a_4 = 20$ and $a_{13} = 65$, you can conclude that $d = 5$, which implies that the sequence is as follows.

a_1	a_2	a_3	a_4	a_5	a_6	a_7	a_8	a_9	a_{10}	a_{11} . . .
5	10	15	20	25	30	35	40	45	50	55 . . .

If you know the nth term of an arithmetic sequence *and* you know the common difference of the sequence, you can find the $(n + 1)$th term by using the *recursion formula*

$$a_{n+1} = a_n + d. \qquad \text{Recursion formula}$$

With this formula, you can find any term of an arithmetic sequence, *provided* that you know the preceding term. For instance, if you know the first term, you can find the second term. Then, knowing the second term, you can find the third term, and so on.

If you substitute $a_1 - d$ for c in the formula $a_n = dn + c$, the nth term of an arithmetic sequence has the alternative recursion formula

$$a_n = a_1 + (n - 1)d. \qquad \text{Alternative recursion formula}$$

Use this formula to solve Example 4. You should get the same answer.

Example 4 ▶ Using a Recursion Formula

Find the ninth term of the arithmetic sequence that begins with 2 and 9.

Solution

For this sequence, the common difference is $d = 9 - 2 = 7$. There are two ways to find the ninth term. One way is simply to write out the first nine terms (by repeatedly adding 7).

2, 9, 16, 23, 30, 37, 44, 51, 58

Another way to find the ninth term is first to find a formula for the nth term. Because the first term is 2, it follows that

$$c = a_1 - d = 2 - 7 = -5.$$

Therefore, a formula for the nth term is

$$a_n = 7n - 5$$

which implies that the ninth term is

$$a_9 = 7(9) - 5 = 58.$$

The Sum of a Finite Arithmetic Sequence

There is a simple formula for the *sum* of a finite arithmetic sequence.

STUDY TIP

Be sure you see that this formula works only for *arithmetic* sequences.

The Sum of a Finite Arithmetic Sequence

The sum of a finite arithmetic sequence with n terms is

$$S_n = \frac{n}{2}(a_1 + a_n).$$

For a proof of the sum of a finite arithmetic sequence, see Proofs in Mathematics on page 882.

Example 5 ▶ Finding the Sum of a Finite Arithmetic Sequence

Find the sum: $1 + 3 + 5 + 7 + 9 + 11 + 13 + 15 + 17 + 19$.

Solution

To begin, notice that the sequence is arithmetic (with a common difference of 2). Moreover, the sequence has 10 terms. So, the sum of the sequence is

$$S_n = \frac{n}{2}(a_1 + a_n) \qquad \text{Formula for sum of a sequence}$$

$$= \frac{10}{2}(1 + 19) \qquad \text{Substitute 10 for } n, \text{ 1 for } a_1, \text{ 19 for } a_n.$$

$$= 5(20) = 100. \qquad \text{Simplify.}$$

The Granger Collection

Historical Note

A teacher of Carl Friedrich Gauss (1777–1855) asked him to add all the integers from 1 to 100. When Gauss returned with the correct answer after only a few moments, the teacher could only look at him in astounded silence. This is what Gauss did:

$$\begin{array}{r} 1 + \quad 2 + \quad 3 + \cdots + 100 \\ \underline{100 + \quad 99 + \quad 98 + \cdots + \quad 1} \\ 101 + 101 + 101 + \cdots + 101 \end{array}$$

$$\frac{100 \times 101}{2} = 5050$$

Example 6 ▶ Finding the Sum of a Finite Arithmetic Sequence

Find the sum of the integers (a) from 1 to 100 and (b) from 1 to N.

Solution

The integers from 1 to 100 form an arithmetic sequence that has 100 terms. So, you can use the formula for the sum of an arithmetic sequence, as follows.

a. $S_n = 1 + 2 + 3 + 4 + 5 + 6 + \cdots + 99 + 100$

$$= \frac{n}{2}(a_1 + a_n) \qquad \text{Formula for sum of a sequence}$$

$$= \frac{100}{2}(1 + 100) \qquad \text{Substitute 100 for } n, \text{ 1 for } a_1, \text{ 100 for } a_n.$$

$$= 50(101) = 5050 \qquad \text{Simplify.}$$

b. $S_n = 1 + 2 + 3 + 4 + \cdots + N$

$$= \frac{n}{2}(a_1 + a_n) \qquad \text{Formula for sum of a sequence}$$

$$= \frac{N}{2}(1 + N) \qquad \text{Substitute } N \text{ for } n, \text{ 1 for } a_1, N \text{ for } a_n.$$

The sum of the first n terms of an infinite sequence is the *nth partial sum*. The nth partial sum can be found by using the formula for the sum of a finite arithmetic sequence.

Example 7 ▶ Finding a Partial Sum of an Arithmetic Sequence

Find the 150th partial sum of the arithmetic sequence

$$5, 16, 27, 38, 49, \ldots.$$

Solution

For this arithmetic sequence, $a_1 = 5$ and $d = 16 - 5 = 11$. So,

$$c = a_1 - d = 5 - 11 = -6$$

and the nth term is $a_n = 11n - 6$. Therefore, $a_{150} = 11(150) - 6 = 1644$, and the sum of the first 150 terms is

$$S_n = \frac{n}{2}(a_1 + a_n) \qquad n\text{th partial sum formula}$$

$$= \frac{150}{2}(5 + 1644) \qquad \text{Substitute for } n, a_1, \text{ and } a_{150}.$$

$$= 75(1649) \qquad \text{Simplify.}$$

$$= 123{,}675. \qquad n\text{th partial sum}$$

Applications

Example 8 ▶ Seating Capacity

An auditorium has 20 rows of seats. There are 20 seats in the first row, 21 seats in the second row, 22 seats in the third row, and so on. How many seats are there in all 20 rows?

Solution

The numbers of seats in the 20 rows form an arithmetic sequence in which the common difference is $d = 1$. Because

$$c = a_1 - d = 20 - 1 = 19$$

you can determine that the formula for the nth term of the sequence is $a_n = n + 19$. Therefore, the 20th term in the sequence is $a_{20} = 20 + 19 = 39$ (see Figure 11.4), and the total number of seats is

$$S_n = 20 + 21 + 22 + \cdots + 39$$

$$= \frac{n}{2}(a_1 + a_{20}) \qquad n\text{th partial sum formula}$$

$$= \frac{20}{2}(20 + 39) \qquad \text{Substitute for } n, a_1, \text{ and } a_{20}.$$

$$= 10(59) = 590. \qquad \text{Simplify.}$$

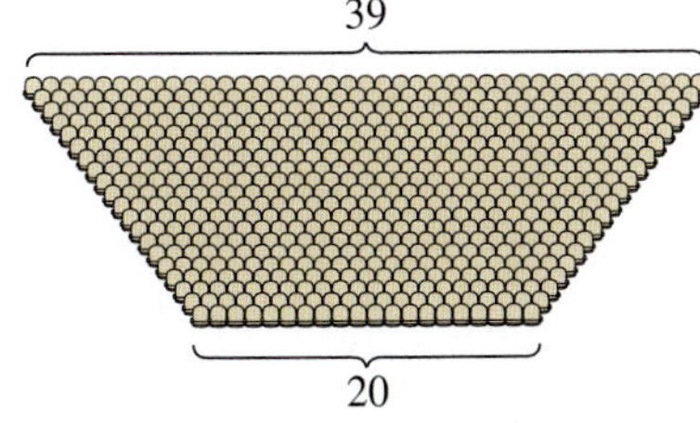

FIGURE 11.4

Example 9 ▶ Total Sales 

A small business sells \$10,000 worth of skin care products during its first year. The owner of the business has set a goal of increasing annual sales by \$7500 each year for 9 years. Assuming that this goal is met, find the total sales during the first 10 years this business is in operation.

Solution

The annual sales form an arithmetic sequence in which $a_1 = 10{,}000$ and $d = 7500$. So,

$$\begin{aligned} c &= a_1 - d \\ &= 10{,}000 - 7500 \\ &= 2500 \end{aligned}$$

and the nth term of the sequence is

$$a_n = 7500n + 2500.$$

This implies that the 10th term of the sequence is

$$a_{10} = 77{,}500. \qquad \text{See Figure 11.5.}$$

The sum of the first 10 terms of the sequence is

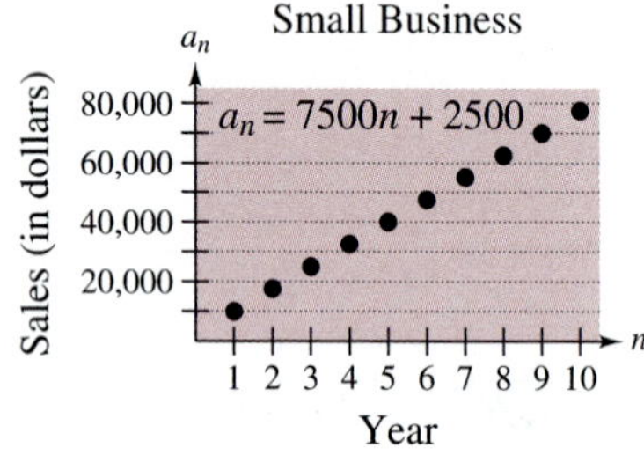

FIGURE 11.5

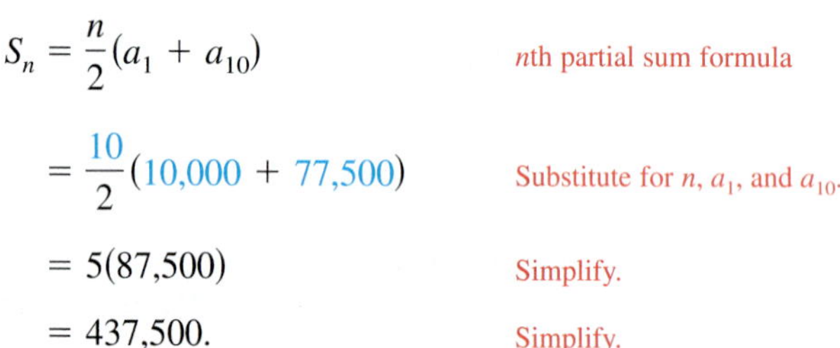

$$\begin{aligned} S_n &= \frac{n}{2}(a_1 + a_{10}) && n\text{th partial sum formula} \\ &= \frac{10}{2}(10{,}000 + 77{,}500) && \text{Substitute for } n, a_1, \text{ and } a_{10}. \\ &= 5(87{,}500) && \text{Simplify.} \\ &= 437{,}500. && \text{Simplify.} \end{aligned}$$

So, the total sales for the first 10 years will be \$437,500.

Writing ABOUT MATHEMATICS

Numerical Relationships Decide whether it is possible to fill in the blanks in each of the sequences such that the resulting sequence is arithmetic. If so, find a recursion formula for the sequence.

a. −7, ___, ___, ___, ___, ___, 11

b. 17, ___, ___, ___, ___, ___, ___, ___, ___, 71

c. 2, 6, ___, ___, 162

d. 4, 7.5, ___, ___, ___, ___, ___, ___, ___, ___, 39

e. 8, 12, ___, ___, ___, 60.75

11.2 Exercises

In Exercises 1–10, determine whether the sequence is arithmetic. If it is, find the common difference.

1. $10, 8, 6, 4, 2, \ldots$

2. $4, 7, 10, 13, 16, \ldots$

3. $1, 2, 4, 8, 16, \ldots$

4. $80, 40, 20, 10, 5, \ldots$

5. $\frac{9}{4}, 2, \frac{7}{4}, \frac{3}{2}, \frac{5}{4}, \ldots$

6. $3, \frac{5}{2}, 2, \frac{3}{2}, 1, \ldots$

7. $\frac{1}{3}, \frac{2}{3}, 1, \frac{4}{3}, \frac{5}{6}, \ldots$

8. $5.3, 5.7, 6.1, 6.5, 6.9, \ldots$

9. $\ln 1, \ln 2, \ln 3, \ln 4, \ln 5, \ldots$

10. $1^2, 2^2, 3^2, 4^2, 5^2, \ldots$

In Exercises 11–18, write the first five terms of the sequence. Determine whether the sequence is arithmetic, and if it is, find the common difference. (Assume that *n* begins with 1.)

11. $a_n = 5 + 3n$

12. $a_n = 100 - 3n$

13. $a_n = 3 - 4(n - 2)$

14. $a_n = 1 + (n - 1)4$

15. $a_n = (-1)^n$

16. $a_n = 2^{n-1}$

17. $a_n = \dfrac{(-1)^n 3}{n}$

18. $a_n = (2^n)n$

In Exercises 19–24, write the first five terms of the arithmetic sequence. Find the common difference and write the *n*th term of the sequence as a function of *n*.

19. $a_1 = 15, \quad a_{k+1} = a_k + 4$

20. $a_1 = 6, \quad a_{k+1} = a_k + 5$

21. $a_1 = 200, \quad a_{k+1} = a_k - 10$

22. $a_1 = 72, \quad a_{k+1} = a_k - 6$

23. $a_1 = \frac{5}{8}, \quad a_{k+1} = a_k - \frac{1}{8}$

24. $a_1 = 0.375, \quad a_{k+1} = a_k + 0.25$

In Exercises 25–32, write the first five terms of the arithmetic sequence.

25. $a_1 = 5, d = 6$

26. $a_1 = 5, d = -\frac{3}{4}$

27. $a_1 = -2.6, d = -0.4$

28. $a_1 = 16.5, d = 0.25$

29. $a_1 = 2, a_{12} = 46$

30. $a_4 = 16, a_{10} = 46$

31. $a_8 = 26, a_{12} = 42$

32. $a_3 = 19, a_{15} = -1.7$

In Exercises 33–44, find a formula for a_n for the arithmetic sequence.

33. $a_1 = 1, d = 3$

34. $a_1 = 15, d = 4$

35. $a_1 = 100, d = -8$

36. $a_1 = 0, d = -\frac{2}{3}$

37. $a_1 = x, d = 2x$

38. $a_1 = -y, d = 5y$

39. $4, \frac{3}{2}, -1, -\frac{7}{2}, \ldots$

40. $10, 5, 0, -5, -10, \ldots$

41. $a_1 = 5, a_4 = 15$

42. $a_1 = -4, a_5 = 16$

43. $a_3 = 94, a_6 = 85$

44. $a_5 = 190, a_{10} = 115$

In Exercises 45–48, match the arithmetic sequence with its graph. [The graphs are labeled (a), (b), (c), and (d).]

(a)

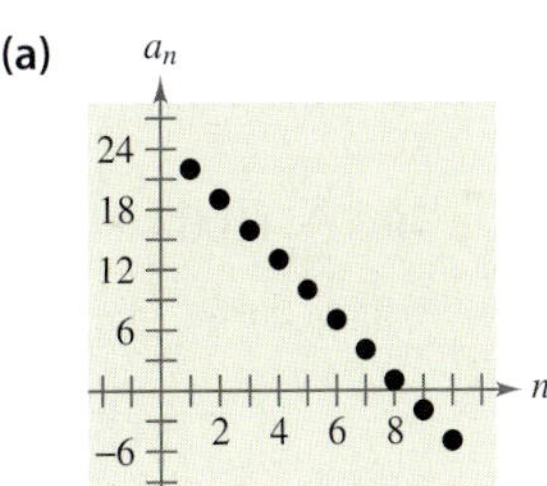

(b)

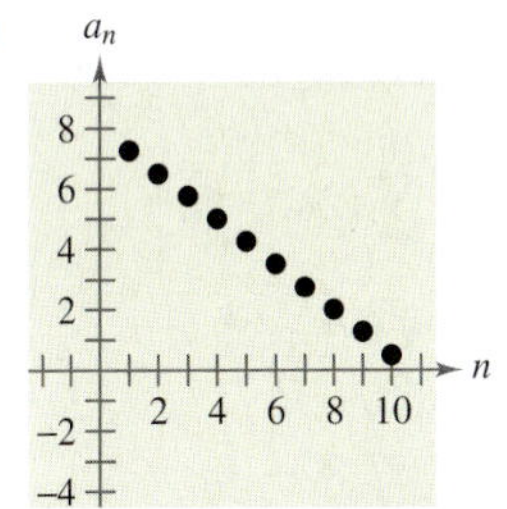

(c)

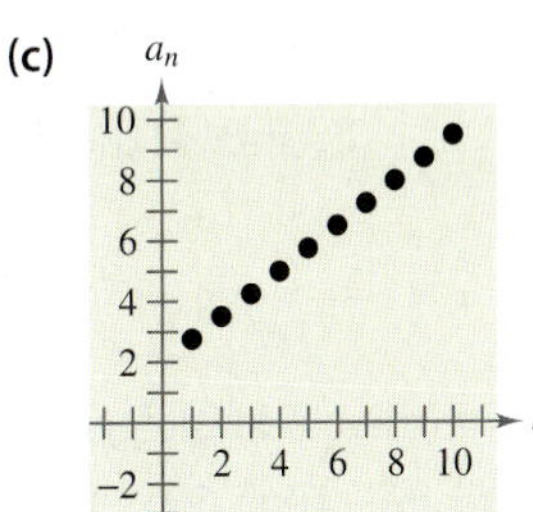

(d)

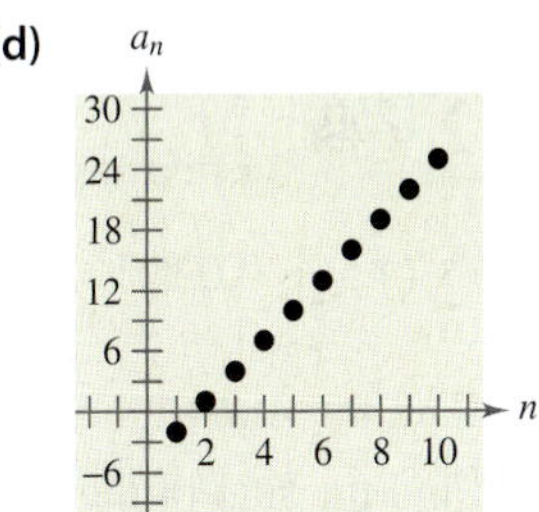

45. $a_n = -\frac{3}{4}n + 8$

46. $a_n = 3n - 5$

47. $a_n = 2 + \frac{3}{4}n$

48. $a_n = 25 - 3n$

In Exercises 49–52, use a graphing utility to graph the first 10 terms of the sequence.

49. $a_n = 15 - \frac{3}{2}n$

50. $a_n = -5 + 2n$

51. $a_n = 0.2n + 3$

52. $a_n = -0.3n + 8$

In Exercises 53–60, find the indicated *n*th partial sum of the arithmetic sequence.

53. $8, 20, 32, 44, \ldots, \quad n = 10$

54. $2, 8, 14, 20, \ldots, \quad n = 25$

55. $4.2, 3.7, 3.2, 2.7, \ldots, \quad n = 12$

56. $0.5, 0.9, 1.3, 1.7, \ldots, \quad n = 10$

57. $40, 37, 34, 31, \ldots, \quad n = 10$

58. $75, 70, 65, 60, \ldots, \quad n = 25$

59. $a_1 = 100, \quad a_{25} = 220, \quad n = 25$

60. $a_1 = 15, \quad a_{100} = 307, \quad n = 100$

In Exercises 61–68, find the partial sum.

61. $\sum_{n=1}^{50} n$

62. $\sum_{n=1}^{100} 2n$

63. $\sum_{n=10}^{100} 6n$

64. $\sum_{n=51}^{100} 7n$

65. $\sum_{n=11}^{30} n - \sum_{n=1}^{10} n$

66. $\sum_{n=51}^{100} n - \sum_{n=1}^{50} n$

67. $\sum_{n=1}^{400} (2n - 1)$

68. $\sum_{n=1}^{250} (1000 - n)$

In Exercises 69–74, use a calculator to find the partial sum.

69. $\sum_{n=1}^{20} (2n + 5)$

70. $\sum_{n=0}^{50} (1000 - 5n)$

71. $\sum_{n=1}^{100} \frac{n + 4}{2}$

72. $\sum_{n=0}^{100} \frac{8 - 3n}{16}$

73. $\sum_{i=1}^{60} \left(250 - \frac{8}{3}i\right)$

74. $\sum_{j=1}^{200} (4.5 + 0.025j)$

75. Find the sum of the first 100 positive odd integers.

76. Find the sum of the integers from -10 to 50.

Job Offer **In Exercises 77 and 78, consider a job offer with the given starting salary and the given annual raise.**

(a) Determine the salary during the sixth year of employment.

(b) Determine the total compensation from the company through six full years of employment.

	Starting Salary	*Annual Raise*
77.	\$32,500	\$1500
78.	\$36,800	\$1750

79. ***Seating Capacity*** Determine the seating capacity of an auditorium with 30 rows of seats if there are 20 seats in the first row, 24 seats in the second row, 28 seats in the third row, and so on.

80. ***Seating Capacity*** Determine the seating capacity of an auditorium with 36 rows of seats if there are 15 seats in the first row, 18 seats in the second row, 21 seats in the third row, and so on.

81. ***Brick Pattern*** A brick patio has the approximate shape of a trapezoid (see figure). The patio has 18 rows of bricks. The first row has 14 bricks and the 18th row has 31 bricks. How many bricks are in the patio?

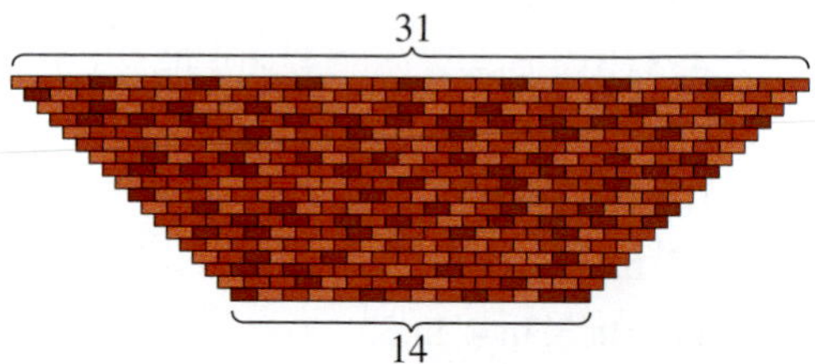

82. ***Brick Pattern*** A triangular brick wall is made by cutting some bricks in half to use in the first column of every other row. The wall has 28 rows. The top row is one-half brick wide and the bottom row is 14 bricks wide. How many bricks are used in the finished wall?

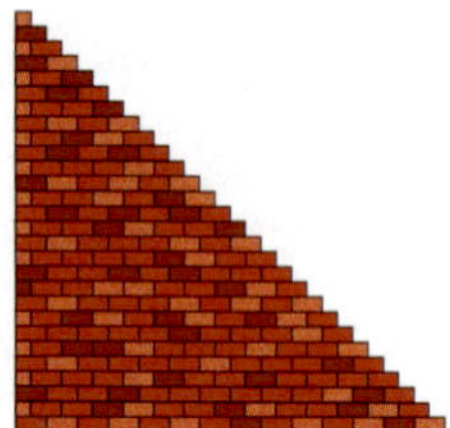

83. ***Falling Object*** An object with negligible air resistance is dropped from a plane. During the first second of fall, the object falls 4.9 meters; during the second second, it falls 14.7 meters; during the third second, it falls 24.5 meters; during the fourth second, it falls 34.3 meters. If this arithmetic pattern continues, how many meters will the object fall in 10 seconds?

84. ***Falling Object*** An object with negligible air resistance is dropped from the top of the Petronas Towers in Malaysia at a height of 1483 feet. During the first second of fall, the object falls 16 feet; during the second second, it falls 48 feet; during the third second, it falls 80 feet; during the fourth second, it falls 112 feet. If this arithmetic pattern continues, how many feet will the object fall in 7 seconds?

▶ Model It

85. ***Data Analysis*** The table shows the per capita personal income in the United States from 1990 to 2000. (Source: U.S. Department of Commerce, Bureau of Economic Analysis)

Year, n	Per capita personal income, a_n
1990	\$19,188
1991	\$19,652
1992	\$20,576
1993	\$21,231
1994	\$22,086
1995	\$23,562
1996	\$24,651
1997	\$25,924
1998	\$27,203
1999	\$28,546
2000	\$29,676

(a) Find an arithmetic sequence that models the data. Let a_n represent the per capita personal income and let n represent the year, with $n = 0$ corresponding to 1990.

(b) Use the *regression* feature of a graphing utility to find a linear model for the data. How does this model compare with the arithmetic sequence you found in part (a)?

(c) Use a graphing utility to graph the terms of the finite sequence you found in part (a).

(d) Use the sequence from part (a) to estimate the per capita personal income in 2001 and 2003.

(e) Use your school's library, the Internet, or some other reference source to find the actual per capita personal income in 2001 and 2003, and compare these values with the estimates from part (d).

Synthesis

86. ***Writing*** Explain how to use the first two terms of an arithmetic sequence to find the nth term.

True or False? **In Exercises 87 and 88, determine whether the statement is true or false. Justify your answer.**

87. Given an arithmetic sequence for which only the first two terms are known, it is possible to find the nth term.

88. If the only known information about a finite arithmetic sequence is its first term and its last term, then it is possible to find the sum of the sequence.

89. ***Exploration***

(a) Graph the first 10 terms of the arithmetic sequence $a_n = 2 + 3n$.

(b) Graph the equation of the line $y = 3x + 2$.

(c) Discuss any differences between the graph of $a_n = 2 + 3n$ and the graph of $y = 3x + 2$.

(d) Compare the slope of the line in part (b) with the common difference of the sequence in part (a). What can you conclude about the slope of a line and the common difference of an arithmetic sequence?

90. ***Pattern Recognition***

(a) Compute the following sums of positive odd integers.

$1 + 3 = \square$

$1 + 3 + 5 = \square$

$1 + 3 + 5 + 7 = \square$

$1 + 3 + 5 + 7 + 9 = \square$

$1 + 3 + 5 + 7 + 9 + 11 = \square$

(b) Use the sums in part (a) to make a conjecture about the sums of positive odd integers. Check your conjecture for the sum

$1 + 3 + 5 + 7 + 9 + 11 + 13 = \square$.

(c) Verify your conjecture analytically.

91. ***Think About It*** The sum of the first 20 terms of an arithmetic sequence with a common difference of 3 is 650. Find the first term.

92. ***Think About It*** The sum of the first n terms of an arithmetic sequence with first term a_1 and common difference d is S_n. Determine the sum if each term is increased by 5. Explain.

Review

In Exercises 93–96, find the slope and y-intercept (if possible) of the equation of the line. Sketch the line.

93. $2x - 4y = 3$

94. $9x + y = -8$

95. $x - 7 = 0$

96. $y + 11 = 0$

11.3 Geometric Sequences and Series

▶ **What you should learn**

- How to recognize and write geometric sequences
- How to find the sum of a geometric sequence
- How to find the sum of an infinite geometric series
- How to use geometric sequences to model and solve real-life problems

▶ **Why you should learn it**

Geometric sequences can be used to model and solve real-life problems. For instance, in Exercise 97 on page 829, you will use a geometric sequence to model the population of China.

Bob Krist/Corbis

Geometric Sequences

In Section 11.2, you learned that a sequence whose consecutive terms have a common *difference* is an arithmetic sequence. In this section, you will study another important type of sequence called a **geometric sequence.** Consecutive terms of a geometric sequence have a common *ratio*.

Definition of Geometric Sequence

A sequence is **geometric** if the ratios of consecutive terms are the same. So, the sequence $a_1, a_2, a_3, a_4, \ldots, a_n \ldots$ is geometric if there is a number r such that

$$\frac{a_2}{a_1}, \quad \frac{a_3}{a_2}, \quad \frac{a_4}{a_3} = \ldots = r, \qquad r \neq 0$$

and so on. The number r is the **common ratio** of the sequence.

Example 1 ▶ Examples of Geometric Sequences 

a. The sequence whose nth term is 2^n is geometric. For this sequence, the common ratio of consecutive terms is 2.

$$2, 4, 8, 16, \ldots, 2^n, \ldots \qquad \text{Begin with } n = 1.$$

$$\tfrac{4}{2} = 2$$

b. The sequence whose nth term is $4(3^n)$ is geometric. For this sequence, the common ratio of consecutive terms is 3.

$$12, 36, 108, 324, \ldots, 4(3^n), \ldots \qquad \text{Begin with } n = 1.$$

$$\tfrac{36}{12} = 3$$

c. The sequence whose nth term is $\left(-\frac{1}{3}\right)^n$ is geometric. For this sequence, the common ratio of consecutive terms is $-\frac{1}{3}$.

$$-\frac{1}{3}, \frac{1}{9}, -\frac{1}{27}, \frac{1}{81}, \ldots, \left(-\frac{1}{3}\right)^n, \ldots \qquad \text{Begin with } n = 1.$$

$$\frac{1/9}{-1/3} = -\frac{1}{3}$$

The sequence 1, 4, 9, 16, . . . , whose nth term is n^2, is *not* geometric. The ratio of the second term to the first term is

$$\frac{a_2}{a_1} = \frac{4}{1} = 4$$

but the ratio of the third term to the second term is $\dfrac{a_3}{a_2} = \dfrac{9}{4}$.

In Example 1, notice that each of the geometric sequences has an nth term that is of the form ar^n, where the common ratio of the sequence is r. A geometric sequence may be thought of as an exponential function whose domain is the set of natural numbers.

The nth Term of a Geometric Sequence

The nth term of a geometric sequence has the form

$$a_n = a_1 r^{n-1}$$

where r is the common ratio of consecutive terms of the sequence. So, every geometric sequence can be written in the following form.

$$a_1, \ a_2, \ a_3, \qquad a_5, \ldots\ldots, \quad a_n, \ldots\ldots$$

$$\downarrow \quad \downarrow \quad \downarrow \quad \downarrow \quad \downarrow \qquad\qquad \downarrow$$

$$a_1, a_1r, a_1r^2, a_1r^3, a_1r^4, \ldots, a_1r^{n-1}, \ldots$$

If you know the nth term of a geometric sequence, you can find the $(n + 1)$th term by multiplying by r. That is, $a_{n+1} = ra_n$.

Example 2 ▶ Finding the Terms of a Geometric Sequence

Write the first five terms of the geometric sequence whose first term is $a_1 = 3$ and whose common ratio is $r = 2$. Then plot the points on a set of coordinate axes.

Solution

Starting with 3, repeatedly multiply by 2 to obtain the following.

$a_1 = 3$ — 1st term

$a_2 = 3(2^1) = 6$ — 2nd term

$a_3 = 3(2^2) = 12$ — 3rd term

$a_4 = 3(2^3) = 24$ — 4th term

$a_5 = 3(2^4) = 48$ — 5th term

Figure 11.6 shows the first five terms of this geometric sequence.

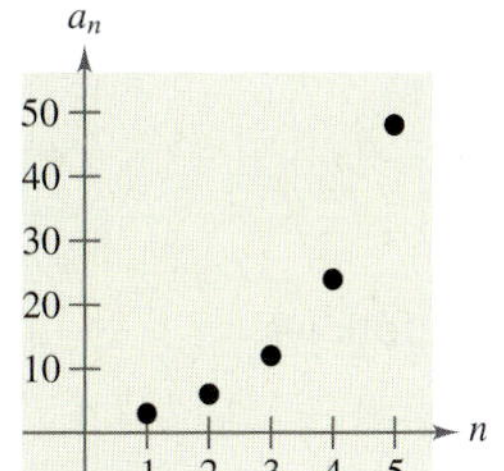

FIGURE 11.6

Example 3 ▶ Finding a Term of a Geometric Sequence

Find the 15th term of the geometric sequence whose first term is 20 and whose common ratio is 1.05.

Solution

$a_{15} = a_1 r^{n-1}$ — Formula for geometric sequence

$= 20(1.05)^{15-1}$ — Substitute for a_1, r, and n.

≈ 39.599 — Use a calculator.

Example 4 ▶ Finding a Term of a Geometric Sequence 

Find the 12th term of the geometric sequence

$$5, 15, 45, \ldots.$$

Solution

The common ratio of this sequence is

$$r = \frac{15}{5} = 3.$$

Because the first term is $a_1 = 5$, you can determine the 12th term $(n = 12)$ to be

$$a_n = a_1 r^{n-1} \quad \text{Formula for geometric sequence}$$

$$a_{12} = 5(3)^{12-1} \quad \text{Substitute for } a_1, r, \text{ and } n.$$

$$= 5(177{,}147) \quad \text{Use a calculator.}$$

$$= 885{,}735. \quad \text{Simplify.}$$

If you know any two terms of a geometric sequence, you can use that information to find a formula for the nth term of the sequence.

STUDY TIP

Remember that r is the common ratio of consecutive terms of a sequence. So in Example 5,

$$a_{10} = a_1 r^9$$

$$= a_1 \cdot r \cdot r \cdot r \cdot r^6$$

$$= a_1 \cdot \frac{a_2}{a_1} \cdot \frac{a_3}{a_2} \cdot \frac{a_4}{a_3} \cdot r^6$$

$$= a_4 r^6.$$

Example 5 ▶ Finding a Term of a Geometric Sequence

The fourth term of a geometric sequence is 125, and the 10th term is 125/64. Find the 14th term. (Assume that the terms of the sequence are positive.)

Solution

The 10th term is related to the fourth term by the equation

$$a_{10} = a_4 r^6. \quad \text{Multiply 4th term by } r^{10-4}.$$

Because $a_{10} = 125/64$ and $a_4 = 125$, you can solve for r as follows.

$$\frac{125}{64} = 125r^6 \quad \text{Substitute } \tfrac{125}{64} \text{ for } a_{10} \text{ and 125 for } a_4.$$

$$\frac{1}{64} = r^6 \quad \text{Divide each side by 125.}$$

$$\frac{1}{2} = r \quad \text{Take the sixth root of each side.}$$

You can obtain the 14th term by multiplying the 10th term by r^4.

$$a_{14} = a_{10} r^4 \quad \text{Multiply the 10th term by } r^{14-10}.$$

$$= \frac{125}{64}\left(\frac{1}{2}\right)^4 \quad \text{Substitute } \tfrac{125}{64} \text{ for } a_{10} \text{ and } \tfrac{1}{2} \text{ for } r.$$

$$= \frac{125}{1024} \quad \text{Simplify.}$$

The Sum of a Finite Geometric Sequence

The formula for the sum of a *finite* geometric sequence is as follows.

The Sum of a Finite Geometric Sequence

The sum of the finite geometric sequence

$$a_1,\ a_1r,\ a_1r^2,\ a_1r^3,\ a_1r^4, \ldots, a_1r^{n-1}$$

with common ratio $r \neq 1$ is given by

$$S_n = \sum_{i=1}^{n} a_1 r^{i-1} = a_1\left(\frac{1 - r^n}{1 - r}\right).$$

For a proof of the sum of a finite geometric sequence, see Proofs in Mathematics on page 882.

Example 6 ▶ **Finding the Sum of a Finite Geometric Sequence**

Find the sum $\sum_{n=1}^{12} 4(0.3)^n$.

Solution

By writing out a few terms, you have

$$\sum_{n=1}^{12} 4(0.3)^n = 4(0.3)^1 + 4(0.3)^2 + 4(0.3)^3 + \cdots + 4(0.3)^{12}.$$

Now, because $a_1 = 4(0.3)$, $r = 0.3$, and $n = 12$, you can apply the formula for the sum of a finite geometric sequence to obtain

$$S_n = a_1\left(\frac{1 - r^n}{1 - r}\right) \qquad \text{Formula for sum of a sequence}$$

$$= 4(0.3)\left[\frac{1 - (0.3)^{12}}{1 - 0.3}\right] \qquad \text{Substitute for } a_1, r, \text{ and } n.$$

$$\approx 1.714. \qquad \text{Use a calculator.}$$

When using the formula for the sum of a geometric sequence, be careful to check that the index begins at $i = 1$. If the index begins at $i = 0$, you must adjust the formula for the nth partial sum. For instance, if the index in Example 6 had begun with $n = 0$, the sum would have been

$$\sum_{n=0}^{12} 4(0.3)^n = 4(0.3)^0 + \sum_{n=1}^{12} 4(0.3)^n$$

$$= 4 + \sum_{n=1}^{12} 4(0.3)^n$$

$$\approx 4 + 1.714$$

$$= 5.714.$$

Exploration

Use a graphing utility to graph

$$y = \left(\frac{1 - r^x}{1 - r}\right)$$

for $r = \frac{1}{2}, \frac{2}{3}$, and $\frac{4}{5}$. What happens as $x \to \infty$?

Use a graphing utility to graph

$$y = \left(\frac{1 - r^x}{1 - r}\right)$$

for $r = 1.5, 2$, and 3. What happens as $x \to \infty$?

Geometric Series 

The summation of the terms of an infinite geometric *sequence* is called an **infinite geometric series** or simply a **geometric series.**

The formula for the sum of a *finite* geometric *sequence* can, depending on the value of r, be extended to produce a formula for the sum of an *infinite* geometric *series*. Specifically, if the common ratio r has the property that $|r| < 1$, it can be shown that r^n becomes arbitrarily close to zero as n increases without bound. Consequently,

$$a_1\left(\frac{1 - r^n}{1 - r}\right) \longrightarrow a_1\left(\frac{1 - 0}{1 - r}\right) \quad \text{as} \quad n \longrightarrow \infty.$$

This result is summarized as follows.

The Sum of an Infinite Geometric Series

If $|r| < 1$, the infinite geometric series

$$a_1 + a_1r + a_1r^2 + a_1r^3 + \cdot \cdot \cdot + a_1r^{n-1} + \cdot \cdot \cdot$$

has the sum

$$S = \sum_{i=0}^{\infty} a_1r^i = \frac{a_1}{1 - r}.$$

Note that if $|r| \geq 1$, the series does not have a sum.

Example 7 ▶ Finding the Sum of an Infinite Geometric Series

Find each sum.

a. $\sum_{n=1}^{\infty} 4(0.6)^{n-1}$

b. $3 + 0.3 + 0.03 + 0.003 + \cdot \cdot \cdot$

Solution

a. $$\sum_{n=1}^{\infty} 4(0.6)^{n-1} = 4 + 4(0.6) + 4(0.6)^2 + 4(0.6)^3 + \cdot \cdot \cdot + 4(0.6)^{n-1} + \cdot \cdot \cdot$$

$$= \frac{4}{1 - (0.6)} \qquad \frac{a_1}{1 - r}$$

$$= 10$$

b. $$3 + 0.3 + 0.03 + 0.003 + \cdot \cdot \cdot = 3 + 3(0.1) + 3(0.1)^2 + 3(0.1)^3 + \cdot \cdot \cdot$$

$$= \frac{3}{1 - (0.1)} \qquad \frac{a_1}{1 - r}$$

$$= \frac{10}{3}$$

$$\approx 3.33$$

Application

Example 8 ▶ Increasing Annuity  

A deposit of \$50 is made on the first day of each month in a savings account that pays 6% compounded monthly. What is the balance at the end of 2 years? (This type of savings plan is called an **increasing annuity.**)

Solution

The first deposit will gain interest for 24 months, and its balance will be

$$A_{24} = 50\left(1 + \frac{0.06}{12}\right)^{24}$$
$$= 50(1.005)^{24}.$$

The second deposit will gain interest for 23 months, and its balance will be

$$A_{23} = 50\left(1 + \frac{0.06}{12}\right)^{23}$$
$$= 50(1.005)^{23}.$$

The last deposit will gain interest for only 1 month, and its balance will be

$$A_1 = 50\left(1 + \frac{0.06}{12}\right)^{1}$$
$$= 50(1.005).$$

The total balance in the annuity will be the sum of the balances of the 24 deposits. Using the formula for the sum of a finite geometric sequence, with $A_1 = 50(1.005)$ and $r = 1.005$, you have

$$S_{24} = 50(1.005)\left[\frac{1 - (1.005)^{24}}{1 - 1.005}\right]$$ Substitute for A_1 r, and n.

$$= \$1277.96.$$

Writing ABOUT MATHEMATICS

An Experiment You will need a piece of string or yarn, a pair of scissors, and a tape measure. Measure out any length of string at least 5 feet long. Double over the string and cut it in half. Take one of the resulting halves, double it over, and cut it in half. Continue this process until you are no longer able to cut a length of string in half. How many cuts were you able to make? Construct a sequence of the resulting string lengths after each cut, starting with the original length of the string. Find a formula for the *n*th term of this sequence. How many cuts could you theoretically make? Discuss why you were not able to make that many cuts.

11.3 Exercises

In Exercises 1–10, determine whether the sequence is geometric. If it is, find the common ratio.

1. $5, 15, 45, 135, \ldots$

2. $3, 12, 48, 192, \ldots$

3. $3, 12, 21, 30, \ldots$

4. $36, 27, 18, 9, \ldots$

5. $1, -\frac{1}{2}, \frac{1}{4}, -\frac{1}{8}, \ldots$

6. $5, 1, 0.2, 0.04, \ldots$

7. $\frac{1}{8}, \frac{1}{4}, \frac{1}{2}, 1, \ldots$

8. $9, -6, 4, -\frac{8}{3}, \ldots$

9. $1, \frac{1}{2}, \frac{1}{3}, \frac{1}{4}, \ldots$

10. $\frac{1}{5}, \frac{2}{7}, \frac{3}{9}, \frac{4}{11}, \ldots$

In Exercises 11–20, write the first five terms of the geometric sequence.

11. $a_1 = 2, r = 3$

12. $a_1 = 6, r = 2$

13. $a_1 = 1, r = \frac{1}{2}$

14. $a_1 = 1, r = \frac{1}{3}$

15. $a_1 = 5, r = -\frac{1}{10}$

16. $a_1 = 6, r = -\frac{1}{4}$

17. $a_1 = 1, r = e$

18. $a_1 = 3, r = \sqrt{5}$

19. $a_1 = 2, r = \dfrac{x}{4}$

20. $a_1 = 5, r = 2x$

In Exercises 21–26, write the first five terms of the geometric sequence. Determine the common ratio and write the *n*th term of the sequence as a function of *n*.

21. $a_1 = 64, \quad a_{k+1} = \frac{1}{2}a_k$

22. $a_1 = 81, \quad a_{k+1} = \frac{1}{3}a_k$

23. $a_1 = 7, \quad a_{k+1} = 2a_k$

24. $a_1 = 5, \quad a_{k+1} = -2a_k$

25. $a_1 = 6, \quad a_{k+1} = -\frac{3}{2}a_k$

26. $a_1 = 48, \quad a_{k+1} = -\frac{1}{2}a_k$

In Exercises 27–38, write an expression for the *n*th term of the geometric sequence.

27. $a_1 = 4, r = \frac{1}{2}, n = 10$

28. $a_1 = 5, r = \frac{3}{2}, n = 8$

29. $a_1 = 6, r = -\frac{1}{3}, n = 12$

30. $a_1 = 64, r = -\frac{1}{4}, n = 10$

31. $a_1 = 100, r = e^x, n = 9$

32. $a_1 = 1, r = \sqrt{3}, n = 8$

33. $a_1 = 500, r = 1.02, n = 40$

34. $a_1 = 1000, r = 1.005, n = 60$

35. $a_1 = 16, a_4 = \frac{27}{4}, n = 3$

36. $a_2 = 3, a_5 = \frac{3}{64}, n = 1$

37. $a_4 = -18, a_7 = \frac{2}{3}, n = 6$

38. $a_3 = \frac{16}{3}, a_5 = \frac{64}{27}, n = 7$

In Exercises 39–42, match the geometric sequence with its graph. [The graphs are labeled (a), (b), (c), and (d).]

(a)

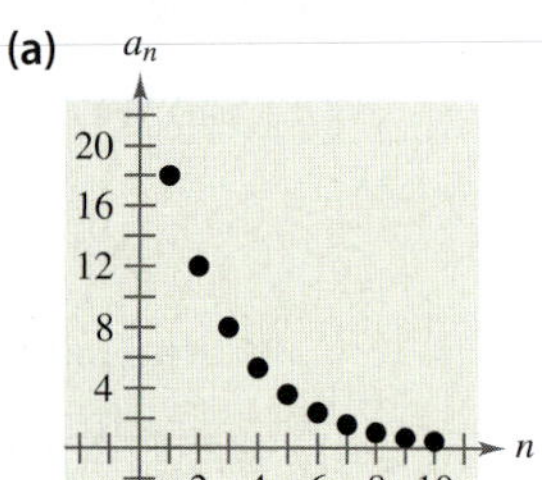

(b)

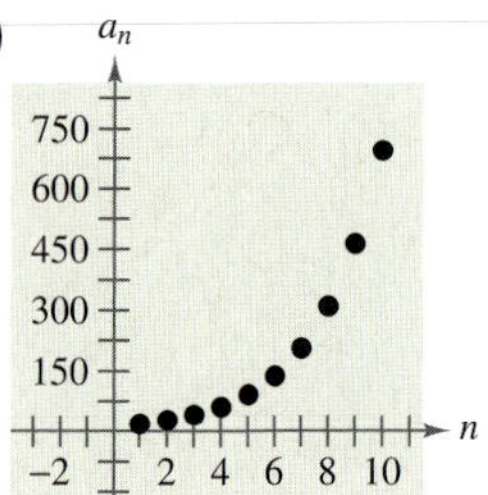

(c)

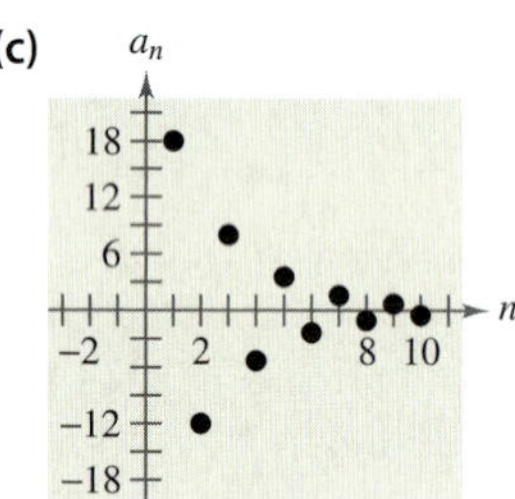

(d)

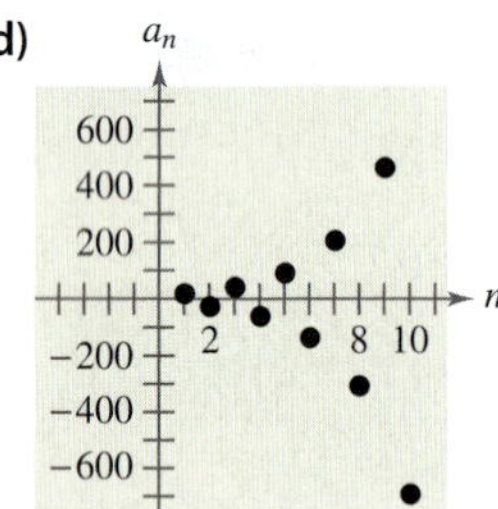

39. $a_n = 18\left(\frac{2}{3}\right)^{n-1}$

40. $a_n = 18\left(-\frac{2}{3}\right)^{n-1}$

41. $a_n = 18\left(\frac{3}{2}\right)^{n-1}$

42. $a_n = 18\left(-\frac{3}{2}\right)^{n-1}$

In Exercises 43–50, use a graphing utility to graph the first 10 terms of the sequence.

43. $a_n = 12(-0.75)^{n-1}$

44. $a_n = 10(1.5)^{n-1}$

45. $a_n = 12(-0.4)^{n-1}$

46. $a_n = 20(-1.25)^{n-1}$

47. $a_n = 2(1.3)^{n-1}$

48. $a_n = 10(1.2)^{n-1}$

49. $a_n = 2(-1.4)^{n-1}$

50. $a_n = 12(-1.2)^{n-1}$

In Exercises 51–70, find the sum of the finite geometric sequence.

51. $\displaystyle\sum_{n=1}^{9} 2^{n-1}$

52. $\displaystyle\sum_{n=1}^{10} \left(\frac{5}{2}\right)^{n-1}$

53. $\displaystyle\sum_{n=1}^{9} (-2)^{n-1}$

54. $\displaystyle\sum_{n=1}^{8} 5\left(-\frac{3}{2}\right)^{n-1}$

55. $\displaystyle\sum_{i=1}^{7} 64\left(-\frac{1}{2}\right)^{i-1}$

56. $\displaystyle\sum_{i=1}^{10} 2\left(\frac{1}{4}\right)^{i-1}$

57. $\displaystyle\sum_{i=1}^{6} 32\left(\frac{1}{4}\right)^{i-1}$

58. $\displaystyle\sum_{i=1}^{12} 16\left(\frac{1}{2}\right)^{i-1}$

59. $\sum_{n=0}^{20} 3\left(\frac{3}{2}\right)^n$

60. $\sum_{n=0}^{40} 5\left(\frac{3}{5}\right)^n$

61. $\sum_{n=0}^{15} 2\left(\frac{4}{3}\right)^n$

62. $\sum_{n=0}^{20} 10\left(\frac{1}{5}\right)^n$

63. $\sum_{n=0}^{5} 300(1.06)^n$

64. $\sum_{n=0}^{6} 500(1.04)^n$

65. $\sum_{n=0}^{40} 2\left(-\frac{1}{4}\right)^n$

66. $\sum_{n=0}^{50} 10\left(\frac{2}{3}\right)^{n-1}$

67. $\sum_{i=1}^{10} 8\left(-\frac{1}{4}\right)^{i-1}$

68. $\sum_{i=0}^{25} 8\left(-\frac{1}{2}\right)^i$

69. $\sum_{i=1}^{10} 5\left(-\frac{1}{3}\right)^{i-1}$

70. $\sum_{i=1}^{100} 15\left(\frac{2}{3}\right)^{i-1}$

In Exercises 71–76, use summation notation to express the sum.

71. $5 + 15 + 45 + \cdots + 3645$

72. $7 + 14 + 28 + \cdots + 896$

73. $2 - \frac{1}{2} + \frac{1}{8} - \cdots + \frac{1}{2048}$

74. $15 - 3 + \frac{3}{5} - \cdots - \frac{3}{625}$

75. $0.1 + 0.4 + 1.6 + \cdots + 102.4$

76. $32 + 24 + 18 + \cdots + 10.125$

In Exercises 77–90, find the sum of the infinite geometric series.

77. $\sum_{n=0}^{\infty} \left(\frac{1}{2}\right)^n$

78. $\sum_{n=0}^{\infty} 2\left(\frac{2}{3}\right)^n$

79. $\sum_{n=0}^{\infty} \left(-\frac{1}{2}\right)^n$

80. $\sum_{n=0}^{\infty} 2\left(-\frac{2}{3}\right)^n$

81. $\sum_{n=0}^{\infty} 4\left(\frac{1}{4}\right)^n$

82. $\sum_{n=0}^{\infty} \left(\frac{1}{10}\right)^n$

83. $\sum_{n=0}^{\infty} (0.4)^n$

84. $\sum_{n=0}^{\infty} 4(0.2)^n$

85. $\sum_{n=0}^{\infty} -3(0.9)^n$

86. $\sum_{n=0}^{\infty} -10(0.2)^n$

87. $8 + 6 + \frac{9}{2} + \frac{27}{8} + \cdots$

88. $9 + 6 + 4 + \frac{8}{3} + \cdots$

89. $\frac{1}{9} - \frac{1}{3} + 1 - 3 + \cdots$

90. $-\frac{125}{36} + \frac{25}{6} - 5 + 6 - \cdots$

In Exercises 91–94, find the rational number representation of the repeating decimal.

91. $0.\overline{36}$

92. $0.\overline{297}$

93. $0.3\overline{18}$

94. $1.3\overline{8}$

Graphical Reasoning **In Exercises 95 and 96, use a graphing utility to graph the function. Identify the horizontal asymptote of the graph and determine its relationship to the sum.**

95. $f(x) = 6\left[\frac{1 - (0.5)^x}{1 - (0.5)}\right], \quad \sum_{n=0}^{\infty} 6\left(\frac{1}{2}\right)^n$

96. $f(x) = 2\left[\frac{1 - (0.8)^x}{1 - (0.8)}\right], \quad \sum_{n=0}^{\infty} 2\left(\frac{4}{5}\right)^n$

▶ Model It

97. ***Data Analysis*** The table shows the population of China (in millions) from 1995 through 2001. (Source: U.S. Census Bureau)

Year, n	Population, a_n
1995	1203.1
1996	1210.0
1997	1221.6
1998	1236.9
1999	1246.9
2000	1261.8
2001	1273.1

(a) Use the *regression* feature of a graphing utility to find a geometric (exponential) sequence that models the data. Let a_n represent the population and let n represent the year, with $n = 5$ corresponding to 1995.

(b) Use the sequence from part (a) to describe the rate at which the population of China is growing.

(c) Use the sequence from part (a) to predict the population of China in 2010.

(d) Use the sequence from part (a) to determine when the population of China will reach 1.3 billion.

98. ***Compound Interest*** A principal of \$1000 is invested at 6% interest. Find the amount after 10 years if the interest is compounded (a) annually, (b) semiannually, (c) quarterly, (d) monthly, and (e) daily.

99. ***Compound Interest*** A principal of \$2500 is invested at 8% interest. Find the amount after 20 years if the interest is compounded (a) annually, (b) semiannually, (c) quarterly, (d) monthly, and (e) daily.

100. ***Depreciation*** A tool and die company buys a machine for \$135,000 and it depreciates at a rate of 30% per year. (In other words, at the end of each year the depreciated value is 70% of what it was at the beginning of the year.) Find the depreciated value of the machine after 5 full years.

101. ***Annuities*** A deposit of \$100 is made at the beginning of each month in an account that pays 6%, compounded monthly. The balance A in the account at the end of 5 years is

$$A = 100\left(1 + \frac{0.06}{12}\right)^1 + \cdots + 100\left(1 + \frac{0.06}{12}\right)^{60}.$$

Find A.

102. ***Annuities*** A deposit of \$50 is made at the beginning of each month in an account that pays 8%, compounded monthly. The balance A in the account at the end of 5 years is

$$A = 50\left(1 + \frac{0.08}{12}\right)^1 + \cdots + 50\left(1 + \frac{0.08}{12}\right)^{60}.$$

Find A.

103. ***Annuities*** A deposit of P dollars is made at the beginning of each month in an account earning an annual interest rate r, compounded monthly. The balance A after t years is

$$A = P\left(1 + \frac{r}{12}\right) + P\left(1 + \frac{r}{12}\right)^2 + \cdots + P\left(1 + \frac{r}{12}\right)^{12t}.$$

Show that the balance is

$$A = P\left[\left(1 + \frac{r}{12}\right)^{12t} - 1\right]\left(1 + \frac{12}{r}\right).$$

104. ***Annuities*** A deposit of P dollars is made at the beginning of each month in an account earning an annual interest rate r, compounded continuously. The balance A after t years is

$$A = Pe^{r/12} + Pe^{2r/12} + \cdots + Pe^{12tr/12}.$$

Show that the balance is

$$A = \frac{Pe^{r/12}(e^{rt} - 1)}{e^{r/12} - 1}.$$

Annuities **In Exercises 105–108, consider making monthly deposits of P dollars in a savings account earning an annual interest rate r. Use the results of Exercises 103 and 104 to find the balance A after t years if the interest is compounded (a) monthly and (b) continuously.**

105. $P = \$50,\ r = 7\%,\ t = 20$ years

106. $P = \$75,\ r = 9\%,\ t = 25$ years

107. $P = \$100,\ r = 10\%,\ t = 40$ years

108. $P = \$20,\ r = 6\%,\ t = 50$ years

109. ***Annuities*** Consider an initial deposit of P dollars in an account earning an annual interest rate r, compounded monthly. At the end of each month, a withdrawal of W dollars will occur and the account will be depleted in t years. The amount of the initial deposit required is

$$P = W\left(1 + \frac{r}{12}\right)^{-1} + W\left(1 + \frac{r}{12}\right)^{-2} + \cdots + W\left(1 + \frac{r}{12}\right)^{-12t}.$$

Show that the initial deposit is

$$P = W\left(\frac{12}{r}\right)\left[1 - \left(1 + \frac{r}{12}\right)^{-12t}\right].$$

110. ***Annuities*** Determine the amount required in a retirement account for an individual who retires at age 65 and wants an income of \$2000 from the account each month for 20 years. Use the result of Exercise 109 and assume that the account earns 9% compounded monthly.

111. ***Geometry*** The sides of a square are 16 inches in length. A new square is formed by connecting the midpoints of the sides of the original square, and two of the resulting triangles are shaded (see figure). If this process is repeated five more times, determine the total area of the shaded region.

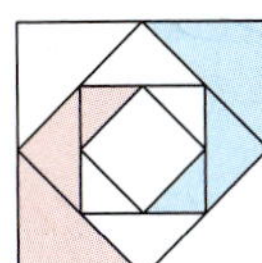
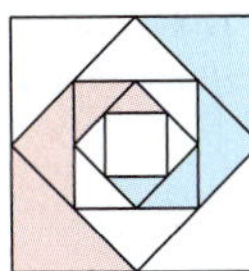

112. ***Sales*** The annual sales a_n (in millions of dollars) for Merck & Co., Inc. for 1991 through 2000 can be approximated by the model

$$a_n = 6942.2e^{0.173n}, \qquad n = 1, 2, \ldots, 10$$

where $n = 1$ represents 1991. Use this model and the formula for the sum of a finite geometric sequence to approximate the total sales earned during this 10-year period. (Source: Merck & Co., Inc.)

113. ***Salary*** An investment firm has a job opening with a salary of $30,000 for the first year. Suppose that during the next 39 years, there is a 5% raise each year. Find the total compensation over the 40-year period.

114. ***Distance*** A ball is dropped from a height of 16 feet. Each time it drops h feet, it rebounds $0.81h$ feet.

(a) Find the total vertical distance traveled by the ball.

(b) The ball takes the following times for each fall.

$$\begin{aligned}
s_1 &= -16t^2 + 16, & s_1 &= 0 \text{ if } t = 1\\
s_2 &= -16t^2 + 16(0.81), & s_2 &= 0 \text{ if } t = 0.9\\
s_3 &= -16t^2 + 16(0.81)^2, & s_3 &= 0 \text{ if } t = (0.9)^2\\
s_4 &= -16t^2 + 16(0.81)^3, & s_4 &= 0 \text{ if } t = (0.9)^3\\
&\vdots & &\vdots\\
s_n &= -16t^2 + 16(0.81)^{n-1}, & s_n &= 0 \text{ if } t = (0.9)^{n-1}
\end{aligned}$$

Beginning with s_2, the ball takes the same amount of time to bounce up as it does to fall, and so the total time elapsed before it comes to rest is

$$t = 1 + 2\sum_{n=1}^{\infty} (0.9)^n.$$

Find this total.

Synthesis

True or False? **In Exercises 115 and 116, determine whether the statement is true or false. Justify your answer.**

115. A sequence is geometric if the ratios of consecutive differences of consecutive terms are the same.

116. You can find the nth term of a geometric sequence by multiplying its common ratio by the first term of the sequence raised to the $(n - 1)$th power.

117. ***Writing*** Write a brief paragraph explaining why the terms of a geometric sequence decrease in magnitude when $-1 < r < 1$.

118. Find two different geometric series with sums of 4.

Review

In Exercises 119–122, evaluate the function for $f(x) = 3x + 1$ and $g(x) = x^2 - 1$.

119. $g(x + 1)$ **120.** $f(x + 1)$

121. $f(g(x + 1))$ **122.** $g(f(x + 1))$

In Exercises 123–126, completely factor the expression over the rational numbers.

123. $9x^3 - 64x$

124. $x^2 + 4x - 63$

125. $6x^2 - 13x - 5$

126. $16x^2 - 4x^4$

In Exercises 127–132, perform the indicated operation(s) and simplify.

127. $\dfrac{3}{x + 3} \cdot \dfrac{x(x + 3)}{x - 3}$

128. $\dfrac{x - 2}{x + 7} \cdot \dfrac{2x(x + 7)}{6x(x - 2)}$

129. $\dfrac{x}{3} \div \dfrac{3x}{6x + 3}$

130. $\dfrac{x - 5}{x - 3} \div \dfrac{10 - 2x}{2(3 - x)}$

131. $5 + \dfrac{7}{x + 2} + \dfrac{2}{x - 2}$

132. $8 - \dfrac{x - 1}{x + 4} - \dfrac{4}{x - 1} - \dfrac{x + 4}{(x - 1)(x + 4)}$

11.4 Mathematical Induction

▶ What you should learn

- How to use mathematical induction to prove a statement
- How to find the sums of powers of integers
- How to recognize patterns and write the *n*th term of a sequence
- How to find finite differences of a sequence

▶ Why you should learn it

Finite differences can be used to determine what type of model can be used to represent a sequence. For instance, in Exercise 59 on page 841, you will use finite differences to find a model that represents the total prize money awarded at professional rodeos.

Gary Brettnacher/Getty Images

Introduction 

In this section you will study a form of mathematical proof called **mathematical induction.** It is important that you see clearly the logical need for it, so take a closer look at the problem discussed in Example 5 on page 816.

$$S_1 = 1 = 1^2$$

$$S_2 = 1 + 3 = 2^2$$

$$S_3 = 1 + 3 + 5 = 3^2$$

$$S_4 = 1 + 3 + 5 + 7 = 4^2$$

$$S_5 = 1 + 3 + 5 + 7 + 9 = 5^2$$

Judging from the pattern formed by these first five sums, it appears that the sum of the first n odd integers is

$$S_n = 1 + 3 + 5 + 7 + 9 + \cdot \cdot \cdot + (2n - 1) = n^2.$$

Although this particular formula *is* valid, it is important for you to see that recognizing a pattern and then simply *jumping to the conclusion* that the pattern must be true for all values of n is *not* a logically valid method of proof. There are many examples in which a pattern appears to be developing for small values of n and then at some point the pattern fails. One of the most famous cases of this was the conjecture by the French mathematician Pierre de Fermat (1601–1665), who speculated that all numbers of the form

$$F_n = 2^{2^n} + 1, \quad n = 0, 1, 2, . . .$$

are prime. For $n = 0, 1, 2, 3,$ and 4, the conjecture is true.

$$F_0 = 3$$

$$F_1 = 5$$

$$F_2 = 17$$

$$F_3 = 257$$

$$F_4 = 65{,}537$$

The size of the next Fermat number ($F_5 = 4{,}294{,}967{,}297$) is so great that it was difficult for Fermat to determine whether it was prime or not. However, another well-known mathematician, Leonhard Euler (1707–1783), later found the factorization

$$F_5 = 4{,}294{,}967{,}297$$

$$= 641(6{,}700{,}417)$$

which proved that F_5 is not prime and therefore Fermat's conjecture was false.

Just because a rule, pattern, or formula seems to work for several values of n, you cannot simply decide that it is valid for all values of n without going through a *legitimate proof.* Mathematical induction is one method of proof.

STUDY TIP

It is important to recognize that both parts of the Principle of Mathematical Induction are necessary.

The Principle of Mathematical Induction

Let P_n be a statement involving the positive integer n. If

1. P_1 is true, and

2. the truth of P_k implies the truth of P_{k+1} for every positive k,

then P_n must be true for all positive integers n.

To apply the Principle of Mathematical Induction, you need to be able to determine the statement P_{k+1} for a given statement P_k.

Example 1 ▶ A Preliminary Example 

Find P_{k+1} for each P_k.

a. $P_k : S_k = \dfrac{k^2(k+1)^2}{4}$

b. $P_k : S_k = 1 + 5 + 9 + \cdot \cdot \cdot + [4(k-1) - 3] + (4k - 3)$

c. $P_k : k + 3 < 5k^2$

d. $P_k : 3^k \geq 2k + 1$

Solution

a. $P_{k+1} : S_{k+1} = \dfrac{(k+1)^2(k+1+1)^2}{4}$ Replace k by $k + 1$.

$= \dfrac{(k+1)^2(k+2)^2}{4}$ Simplify.

b. $P_{k+1} : S_{k+1} = 1 + 5 + 9 + \cdot \cdot \cdot + \{4[(k+1) - 1] - 3\} + [4(k+1) - 3]$

$= 1 + 5 + 9 + \cdot \cdot \cdot + (4k - 3) + (4k + 1)$

c. P_{k+1}: $(k+1) + 3 < 5(k+1)^2$

$k + 4 < 5(k^2 + 2k + 1)$

d. $P_{k+1} : 3^{k+1} \geq 2(k+1) + 1$

$3^{k+1} \geq 2k + 3$

FIGURE 11.7

A well-known illustration used to explain why the Principle of Mathematical Induction works is the unending line of dominoes shown in Figure 11.7. If the line actually contains infinitely many dominoes, it is clear that you could not knock the entire line down by knocking down only *one domino* at a time. However, suppose it were true that each domino would knock down the next one as it fell. Then you could knock them all down simply by pushing the first one and starting a chain reaction. Mathematical induction works in the same way. If the truth of P_k implies the truth of P_{k+1} and if P_1 is true, the chain reaction proceeds as follows: P_1 implies P_2, P_2 implies P_3, P_3 implies P_4, and so on.

When using mathematical induction to prove a *summation* formula (such as the one in Example 2), it is helpful to think of S_{k+1} as

$$S_{k+1} = S_k + a_{k+1}$$

where a_{k+1} is the $(k + 1)$th term of the original sum.

Example 2 ▶ Using Mathematical Induction

Use mathematical induction to prove the following formula.

$$\begin{aligned} S_n &= 1 + 3 + 5 + 7 + \cdot \cdot \cdot + (2n - 1) \\ &= n^2 \end{aligned}$$

Solution

Mathematical induction consists of two distinct parts. First, you must show that the formula is true when $n = 1$.

1. When $n = 1$, the formula is valid, because

$$S_1 = 1 = 1^2.$$

The second part of mathematical induction has two steps. The first step is to assume that the formula is valid for *some* integer k. The second step is to use this assumption to prove that the formula is valid for the next integer, $k + 1$.

2. Assuming that the formula

$$\begin{aligned} S_k &= 1 + 3 + 5 + 7 + \cdot \cdot \cdot + (2k - 1) \\ &= k^2 \end{aligned}$$

is true, you must show that the formula $S_{k+1} = (k + 1)^2$ is true.

$$\begin{aligned} S_{k+1} &= 1 + 3 + 5 + 7 + \cdot \cdot \cdot + (2k - 1) + [2(k + 1) - 1] \\ &= [1 + 3 + 5 + 7 + \cdot \cdot \cdot + (2k - 1)] + (2k + 2 - 1) \\ &= S_k + (2k + 1) && \text{Group terms to form } S_k. \\ &= k^2 + 2k + 1 && \text{Replace } S_k \text{ by } k^2. \\ &= (k + 1)^2 \end{aligned}$$

Combining the results of parts (1) and (2), you can conclude by mathematical induction that the formula is valid for all positive integer values of n.

It occasionally happens that a statement involving natural numbers is not true for the first $k - 1$ positive integers but is true for all values of $n \geq k$. In these instances, you use a slight variation of the Principle of Mathematical Induction in which you verify P_k rather than P_1. This variation is called the *extended principle of mathematical induction*. To see the validity of this, note from Figure 11.7 that all but the first $k - 1$ dominoes can be knocked down by knocking over the kth domino. This suggests that you can prove a statement P_n to be true for $n \geq k$ by showing that P_k is true and that P_k implies P_{k+1}. In Exercises 35–40 of this section, you are asked to apply this extension of mathematical induction.

Example 3 ▶ Using Mathematical Induction

Use mathematical induction to prove the formula

$$S_n = 1^2 + 2^2 + 3^2 + 4^2 + \cdots + n^2$$
$$= \frac{n(n+1)(2n+1)}{6}.$$

Solution

1. When $n = 1$, the formula is valid, because

$$S_1 = 1^2 = \frac{1(2)(3)}{6}.$$

2. Assuming that

$$S_k = 1^2 + 2^2 + 3^2 + 4^2 + \cdots + k^2$$
$$= \frac{k(k+1)(2k+1)}{6}$$

you must show that

$$S_{k+1} = \frac{(k+1)(k+1+1)[2(k+1)+1]}{6}$$
$$= \frac{(k+1)(k+2)(2k+3)}{6}.$$

To do this, write the following.

$$S_{k+1} = S_k + a_{k+1}$$
$$= (1^2 + 2^2 + 3^2 + 4^2 + \cdots + k^2) + (k+1)^2$$
$$= \frac{k(k+1)(2k+1)}{6} + (k+1)^2 \qquad \text{By assumption}$$
$$= \frac{k(k+1)(2k+1) + 6(k+1)^2}{6}$$
$$= \frac{(k+1)[k(2k+1) + 6(k+1)]}{6}$$
$$= \frac{(k+1)(2k^2 + 7k + 6)}{6}$$
$$= \frac{(k+1)(k+2)(2k+3)}{6}$$

Combining the results of parts (1) and (2), you can conclude by mathematical induction that the formula is valid for *all* integers $n \geq 1$.

When proving a formula using mathematical induction, the only statement that you *need* to verify is P_1. As a check, however, it is good to try verifying other statements. For instance, in Example 3, try verifying P_2 and P_3.

Sums of Powers of Integers

The formula in Example 3 is one of a collection of useful summation formulas. This and other formulas dealing with the sums of various powers of the first n positive integers are as follows.

Sums of Powers of Integers

1. $1 + 2 + 3 + 4 + \cdots + n = \dfrac{n(n+1)}{2}$

2. $1^2 + 2^2 + 3^2 + 4^2 + \cdots + n^2 = \dfrac{n(n+1)(2n+1)}{6}$

3. $1^3 + 2^3 + 3^3 + 4^3 + \cdots + n^3 = \dfrac{n^2(n+1)^2}{4}$

4. $1^4 + 2^4 + 3^4 + 4^4 + \cdots + n^4 = \dfrac{n(n+1)(2n+1)(3n^2+3n-1)}{30}$

5. $1^5 + 2^5 + 3^5 + 4^5 + \cdots + n^5 = \dfrac{n^2(n+1)^2(2n^2+2n-1)}{12}$

Example 4 ▶ Finding a Sum of Powers of Integers 

Find each sum.

a. $\displaystyle\sum_{i=1}^{7} i^3 = 1^3 + 2^3 + 3^3 + 4^3 + 5^3 + 6^3 + 7^3$ **b.** $\displaystyle\sum_{i=1}^{4} (6i - 4i^2)$

Solution

a. Using the formula for the sum of the cubes of the first n positive integers, you obtain

$$\sum_{i=1}^{7} i^3 = 1^3 + 2^3 + 3^3 + 4^3 + 5^3 + 6^3 + 7^3$$

$$= \frac{7^2(7+1)^2}{4} = \frac{49(64)}{4} = 784.$$

b. $$\sum_{i=1}^{4} (6i - 4i^2) = \sum_{i=1}^{4} 6i - \sum_{i=1}^{4} 4i^2$$

$$= 6\sum_{i=1}^{4} i - 4\sum_{i=1}^{4} i^2$$

$$= 6\left[\frac{4(4+1)}{2}\right] - 4\left[\frac{4(4+1)(8+1)}{6}\right]$$

$$= 6(10) - 4(30)$$

$$= 60 - 120 = -60$$

Example 5 ▶ Proving an Inequality by Mathematical Induction

Prove that $n < 2^n$ for all positive integers n.

Solution

1. For $n = 1$ or 2, the statement is true, because

$$1 < 2^1 \quad \text{and} \quad 2 < 2^2.$$

2. Assuming that

$$k < 2^k$$

you need to show that $k + 1 < 2^{k+1}$. For $n = k$, you have

$$2^{k+1} = 2(2^k) > 2(k) = 2k. \qquad \text{By assumption}$$

Because $2k = k + k > k + 1$ for all $k > 1$, it follows that

$$2^{k+1} > 2k > k + 1$$

or

$$k + 1 < 2^{k+1}.$$

So, $n < 2^n$ for all integers $n \geq 1$.

To check a result that you have proved by mathematical induction, it helps to list the statement for several values of n. For instance, in Example 5, you could list

$$1 < 2^1 = 2, \qquad 2 < 2^2 = 4, \qquad 3 < 2^3 = 8,$$

$$4 < 2^4 = 16, \qquad 5 < 2^5 = 32, \qquad 6 < 2^6 = 64.$$

From this list, your intuition confirms that the statement $n < 2^n$ is reasonable.

Pattern Recognition

Although choosing a formula on the basis of a few observations does *not* guarantee the validity of the formula, pattern recognition *is* important. Once you have a pattern or formula that you think works, you can try using mathematical induction to prove your formula.

Finding a Formula for the *n*th Term of a Sequence

To find a formula for the nth term of a sequence, consider these guidelines.

1. Calculate the first several terms of the sequence. It is often a good idea to write the terms in both simplified and factored forms.

2. Try to find a recognizable pattern for the terms and write a formula for the nth term of the sequence. This is your *hypothesis* or *conjecture*. You might try computing one or two more terms in the sequence to test your hypothesis.

3. Use mathematical induction to prove your hypothesis.

Example 6 ▶ Finding a Formula for a Finite Sum

Find a formula for the finite sum and prove its validity.

$$\frac{1}{1 \cdot 2} + \frac{1}{2 \cdot 3} + \frac{1}{3 \cdot 4} + \frac{1}{4 \cdot 5} + \cdots + \frac{1}{n(n+1)}$$

Solution

Begin by writing out the first few sums.

$$S_1 = \frac{1}{1 \cdot 2} = \frac{1}{2} = \frac{1}{1+1}$$

$$S_2 = \frac{1}{1 \cdot 2} + \frac{1}{2 \cdot 3} = \frac{4}{6} = \frac{2}{3} = \frac{2}{2+1}$$

$$S_3 = \frac{1}{1 \cdot 2} + \frac{1}{2 \cdot 3} + \frac{1}{3 \cdot 4} = \frac{9}{12} = \frac{3}{4} = \frac{3}{3+1}$$

$$S_4 = \frac{1}{1 \cdot 2} + \frac{1}{2 \cdot 3} + \frac{1}{3 \cdot 4} + \frac{1}{4 \cdot 5} = \frac{48}{60} = \frac{4}{5} = \frac{4}{4+1}$$

From this sequence, it appears that the formula for the kth sum is

$$S_k = \frac{1}{1 \cdot 2} + \frac{1}{2 \cdot 3} + \frac{1}{3 \cdot 4} + \frac{1}{4 \cdot 5} + \cdots + \frac{1}{k(k+1)} = \frac{k}{k+1}.$$

To prove the validity of this hypothesis, use mathematical induction, as follows. Note that you have already verified the formula for $n = 1$, so you can begin by assuming that the formula is valid for $n = k$ and trying to show that it is valid for $n = k + 1$.

$$S_{k+1} = \left[\frac{1}{1 \cdot 2} + \frac{1}{2 \cdot 3} + \frac{1}{3 \cdot 4} + \frac{1}{4 \cdot 5} + \cdots + \frac{1}{k(k+1)}\right] + \frac{1}{(k+1)(k+2)}$$

$$= \frac{k}{k+1} + \frac{1}{(k+1)(k+2)} \qquad \text{By assumption}$$

$$= \frac{k(k+2)+1}{(k+1)(k+2)}$$

$$= \frac{k^2+2k+1}{(k+1)(k+2)}$$

$$= \frac{(k+1)^2}{(k+1)(k+2)}$$

$$= \frac{k+1}{k+2}$$

So, the hypothesis is valid.

STUDY TIP

For a linear model, the *first* differences should be the same nonzero number. For a quadratic model, the *second* differences are the same nonzero number.

Finite Differences

The **first differences** of a sequence are found by subtracting consecutive terms. The **second differences** are found by subtracting consecutive first differences. The first and second differences of the sequence 3, 5, 8, 12, 17, 23, . . . are as follows.

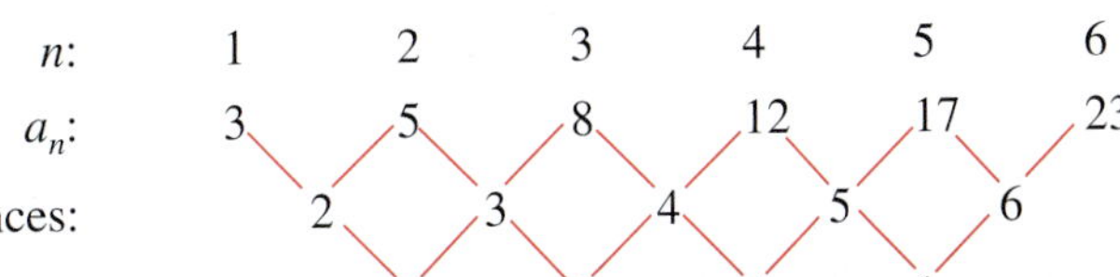

For this sequence, the second differences are all the same. When this happens, the sequence has a perfect *quadratic* model. If the first differences are all the same, the sequence has a *linear* model. That is, it is arithmetic.

Example 7 ▶ Finding a Quadratic Model

Find the quadratic model for the sequence

3, 5, 8, 12, 17, 23, ...

Solution

You know from the second differences shown above that the model is quadratic and has the form

$$a_n = an^2 + bn + c.$$

By substituting 1, 2, and 3 for n, you can obtain a system of three linear equations in three variables.

$a_1 = a(1)^2 + b(1) + c = 3$ Substitute 1 for n.

$a_2 = a(2)^2 + b(2) + c = 5$ Substitute 2 for n.

$a_3 = a(3)^2 + b(3) + c = 8$ Substitute 3 for n.

You now have a system of three equations in a, b, and c.

$$\begin{cases} a + b + c = 3 & \text{Equation 1} \\ 4a + 2b + c = 5 & \text{Equation 2} \\ 9a + 3b + c = 8 & \text{Equation 3} \end{cases}$$

Using the techniques discussed in Chapter 9, you can find the solution to be $a = \frac{1}{2}$, $b = \frac{1}{2}$, and $c = 2$. So, the quadratic model is

$$a_n = \frac{1}{2}n^2 + \frac{1}{2}n + 2.$$

Try checking the values of a_1, a_2, and a_3.

11.4 Exercises

In Exercises 1–4, find P_{k+1} for the given P_k.

1. $P_k = \dfrac{5}{k(k+1)}$
2. $P_k = \dfrac{1}{2(k+2)}$
3. $P_k = \dfrac{k^2(k+1)^2}{4}$
4. $P_k = \dfrac{k}{3}(2k+1)$

In Exercises 5–18, use mathematical induction to prove the formula for every positive integer *n*.

5. $2 + 4 + 6 + 8 + \cdots + 2n = n(n+1)$
6. $3 + 7 + 11 + 15 + \cdots + (4n-1) = n(2n+1)$
7. $2 + 7 + 12 + 17 + \cdots + (5n-3) = \dfrac{n}{2}(5n-1)$
8. $1 + 4 + 7 + 10 + \cdots + (3n-2) = \dfrac{n}{2}(3n-1)$
9. $1 + 2 + 2^2 + 2^3 + \cdots + 2^{n-1} = 2^n - 1$
10. $2(1 + 3 + 3^2 + 3^3 + \cdots + 3^{n-1}) = 3^n - 1$
11. $1 + 2 + 3 + 4 + \cdots + n = \dfrac{n(n+1)}{2}$
12. $1^2 + 2^2 + 3^2 + 4^2 + \cdots + n^2 = \dfrac{n(n+1)(2n+1)}{6}$
13. $1^3 + 2^3 + 3^3 + 4^3 + \cdots + n^3 = \dfrac{n^2(n+1)^2}{4}$
14. $\left(1 + \dfrac{1}{1}\right)\left(1 + \dfrac{1}{2}\right)\left(1 + \dfrac{1}{3}\right) \cdots \left(1 + \dfrac{1}{n}\right) = n + 1$
15. $\displaystyle\sum_{i=1}^{n} i^5 = \frac{n^2(n+1)^2(2n^2+2n-1)}{12}$
16. $\displaystyle\sum_{i=1}^{n} i^4 = \frac{n(n+1)(2n+1)(3n^2+3n-1)}{30}$
17. $\displaystyle\sum_{i=1}^{n} i(i+1) = \frac{n(n+1)(n+2)}{3}$
18. $\displaystyle\sum_{i=1}^{n} \frac{1}{(2i-1)(2i+1)} = \frac{n}{2n+1}$

In Exercises 19–28, find the sum using the formulas for the sums of powers of integers.

19. $\displaystyle\sum_{n=1}^{15} n$
20. $\displaystyle\sum_{n=1}^{30} n$
21. $\displaystyle\sum_{n=1}^{6} n^2$
22. $\displaystyle\sum_{n=1}^{10} n^3$
23. $\displaystyle\sum_{n=1}^{5} n^4$
24. $\displaystyle\sum_{n=1}^{8} n^5$
25. $\displaystyle\sum_{n=1}^{6} (n^2 - n)$
26. $\displaystyle\sum_{n=1}^{20} (n^3 - n)$
27. $\displaystyle\sum_{i=1}^{6} (6i - 8i^3)$
28. $\displaystyle\sum_{j=1}^{10} \left(3 - \tfrac{1}{2}j + \tfrac{1}{2}j^2\right)$

In Exercises 29–34, find a formula for the sum of the first *n* terms of the sequence.

29. $1, 5, 9, 13, \ldots$
30. $25, 22, 19, 16, \ldots$
31. $1, \frac{9}{10}, \frac{81}{100}, \frac{729}{1000}, \ldots$
32. $3, -\frac{9}{2}, \frac{27}{4}, -\frac{81}{8}, \ldots$
33. $\dfrac{1}{4}, \dfrac{1}{12}, \dfrac{1}{24}, \dfrac{1}{40}, \ldots, \dfrac{1}{2n(n+1)}, \ldots$
34. $\dfrac{1}{2 \cdot 3}, \dfrac{1}{3 \cdot 4}, \dfrac{1}{4 \cdot 5}, \dfrac{1}{5 \cdot 6}, \ldots, \dfrac{1}{(n+1)(n+2)}, \ldots$

In Exercises 35–40, prove the inequality for the indicated integer values of *n*.

35. $n! > 2^n, \quad n \geq 4$
36. $\left(\frac{4}{3}\right)^n > n, \quad n \geq 7$
37. $\dfrac{1}{\sqrt{1}} + \dfrac{1}{\sqrt{2}} + \dfrac{1}{\sqrt{3}} + \cdots + \dfrac{1}{\sqrt{n}} > \sqrt{n}, \quad n \geq 2$
38. $\left(\dfrac{x}{y}\right)^{n+1} < \left(\dfrac{x}{y}\right)^n, \quad n \geq 1 \text{ and } 0 < x < y$
39. $(1 + a)^n \geq na, \quad n \geq 1 \text{ and } a > 0$
40. $2n^2 > (n+1)^2, \quad n \geq 3$

In Exercises 41–48, use mathematical induction to prove the property for all positive integers *n*.

41. $(ab)^n = a^n b^n$
42. $\left(\dfrac{a}{b}\right)^n = \dfrac{a^n}{b^n}$
43. If $x_1 \neq 0, x_2 \neq 0, \ldots, x_n \neq 0$, then
$$(x_1 x_2 x_3 \cdots x_n)^{-1} = x_1^{-1} x_2^{-1} x_3^{-1} \cdots x_n^{-1}.$$
44. If $x_1 > 0, x_2 > 0, \ldots, x_n > 0$, then
$$\ln(x_1 x_2 \cdots x_n) = \ln x_1 + \ln x_2 + \cdots + \ln x_n.$$
45. Generalized Distributive Law:
$$x(y_1 + y_2 + \cdots + y_n) = xy_1 + xy_2 + \cdots + xy_n$$
46. $(a + bi)^n$ and $(a - bi)^n$ are complex conjugates for all $n \geq 1$.

47. A factor of $(n^3 + 3n^2 + 2n)$ is 3.

48. A factor of $(2^{2n-1} + 3^{2n-1})$ is 5.

In Exercises 49–54, write the first six terms of the sequence. Then calculate the first and second differences of the sequence. Does the sequence have a linear model, a quadratic model, or neither?

49. $a_1 = 0$
$a_n = a_{n-1} + 3$

50. $a_1 = 2$
$a_n = a_{n-1} + 2$

51. $a_1 = 3$
$a_n = a_{n-1} - n$

52. $a_2 = -3$
$a_n = -2a_{n-1}$

53. $a_0 = 2$
$a_n = (a_{n-1})^2$

54. $a_0 = 0$
$a_n = a_{n-1} + n$

In Exercises 55–58, find a quadratic model for the sequence with the indicated terms.

55. $a_0 = 3,\ a_1 = 3,\ a_4 = 15$

56. $a_0 = 7,\ a_1 = 6,\ a_3 = 10$

57. $a_0 = -3,\ a_2 = 1,\ a_4 = 9$

58. $a_0 = 3,\ a_2 = 0,\ a_6 = 36$

▶ Model It

59. ***Data Analysis*** The table shows the total prize money awarded (in millions of dollars) at professional rodeos from 1995 through 1999. (Source: Professional Rodeo Cowboys Association)

Year, n	Prize money, a_n
1995	24.5
1996	26.4
1997	28.0
1998	29.9
1999	31.1

(a) Find the first differences of the data shown in the table.

(b) Use your results from part (a) to determine whether a linear model can be used to approximate the data. If so, find a model. Let a_n represent the total prize money awarded and let n represent the year, with $n = 5$ corresponding to 1995.

(c) Use the model to estimate the total prize money awarded in 2004.

Synthesis

60. ***Writing*** In your own words, explain what is meant by a proof by mathematical induction.

True or False? **In Exercises 61–64, determine whether the statement is true or false. Justify your answer.**

61. If the statement P_1 is true but the true statement P_6 does *not* imply that the statement P_7 is true, then P_n is not necessarily true for all positive integers n.

62. If the statement P_k is true and P_k implies P_{k+1}, then P_1 is also true.

63. If the second differences of a sequence are all zero, then the sequence is arithmetic.

64. A sequence with n terms has $n - 1$ second differences.

Review

In Exercises 65–68, find the product.

65. $(2x^2 - 1)^2$

66. $(2x - y)^2$

67. $(5 - 4x)^3$

68. $(2x - 4y)^3$

In Exercises 69–72, use synthetic division to divide.

69. $(x^3 + x^2 - 10x + 8) \div (x - 1)$

70. $(x^3 - 4x^2 - 29x - 24) \div (x - 8)$

71. $(4x^3 + 11x^2 - 43x + 10) \div (x + 5)$

72. $(6x^3 - 35x^2 - 8x + 12) \div (x - 6)$

In Exercises 73–76, (a) identify all intercepts, (b) find any vertical and horizontal asymptotes, (c) check for symmetry, and (d) plot additional solution points as needed and sketch the graph of the rational function.

73. $f(x) = \dfrac{x}{x + 3}$

74. $g(x) = \dfrac{x^2}{x^2 - 4}$

75. $h(t) = \dfrac{t - 7}{t}$

76. $f(x) = \dfrac{5 + x}{1 - x}$

11.5 The Binomial Theorem

▶ What you should learn

- How to use the Binomial Theorem to calculate binomial coefficients
- How to use Pascal's Triangle to calculate binomial coefficients
- How to use binomial coefficients to write binomial expansions

▶ Why you should learn it

You can use binomial coefficients to model and solve real-life problems. For instance, in Exercise 79 on page 848, you will use binomial coefficients to write the expansion of a model that represents the per capita consumption of bottled water.

Rob Gage/Getty Images

Binomial Coefficients

Recall that a *binomial* is a polynomial that has two terms. In this section, you will study a formula that gives a quick method of raising a binomial to a power. To begin, look at the expansion of $(x + y)^n$ for several values of n.

$$(x + y)^0 = 1$$

$$(x + y)^1 = x + y$$

$$(x + y)^2 = x^2 + 2xy + y^2$$

$$(x + y)^3 = x^3 + 3x^2y + 3xy^2 + y^3$$

$$(x + y)^4 = x^4 + 4x^3y + 6x^2y^2 + 4xy^3 + y^4$$

$$(x + y)^5 = x^5 + 5x^4y + 10x^3y^2 + 10x^2y^3 + 5xy^4 + y^5$$

There are several observations you can make about these expansions.

1. In each expansion, there are $n + 1$ terms.
2. In each expansion, x and y have symmetrical roles. The powers of x decrease by 1 in successive terms, whereas the powers of y increase by 1.
3. The sum of the powers of each term is n. For instance, in the expansion of $(x + y)^5$, the sum of the powers of each term is 5.

$$(x + y)^5 = x^5 + 5\overbrace{x^4y^1}^{4+1=5} + 10\overbrace{x^3y^2}^{3+2=5} + 10x^2y^3 + 5x^1y^4 + y^5$$

4. The coefficients increase and then decrease in a symmetric pattern.

The coefficients of a binomial expansion are called **binomial coefficients.** To find them, you can use the **Binomial Theorem.**

The Binomial Theorem

In the expansion of $(x + y)^n$

$$(x + y)^n = x^n + nx^{n-1}y + \cdots + {}_nC_r\, x^{n-r}y^r + \cdots + nxy^{n-1} + y^n$$

the coefficient of $x^{n-r}y^r$ is

$${}_nC_r = \frac{n!}{(n-r)!r!}.$$

The symbol $\binom{n}{r}$ is often used in place of ${}_nC_r$ to denote binomial coefficients.

For a proof of the Binomial Theorem, see Proofs in Mathematics on page 883.

Technology

Most graphing calculators are programmed to evaluate ${}_nC_r$. Consult your user's manual and then evaluate ${}_8C_5$. You should get an answer of 56.

Example 1 ▶ Finding Binomial Coefficients

Find each binomial coefficient.

a. ${}_8C_2$ **b.** $\binom{10}{3}$ **c.** ${}_7C_0$ **d.** $\binom{8}{8}$

Solution

a. $${}_8C_2 = \frac{8!}{6! \cdot 2!} = \frac{(8 \cdot 7) \cdot \cancel{6!}}{\cancel{6!} \cdot 2!} = \frac{8 \cdot 7}{2 \cdot 1} = 28$$

b. $$\binom{10}{3} = \frac{10!}{7! \cdot 3!} = \frac{(10 \cdot 9 \cdot 8) \cdot \cancel{7!}}{\cancel{7!} \cdot 3!} = \frac{10 \cdot 9 \cdot 8}{3 \cdot 2 \cdot 1} = 120$$

c. $${}_7C_0 = \frac{\cancel{7!}}{\cancel{7!} \cdot 0!} = 1$$

d. $$\binom{8}{8} = \frac{\cancel{8!}}{0! \cdot \cancel{8!}} = 1$$

When $r \neq 0$ and $r \neq n$, as in parts (a) and (b) above, there is a simple pattern for evaluating binomial coefficients.

$${}_8C_2 = \frac{\overbrace{8 \cdot 7}^{\text{2 factors}}}{\underbrace{2 \cdot 1}_{\text{2 factors}}} \quad \text{and} \quad \binom{10}{3} = \frac{\overbrace{10 \cdot 9 \cdot 8}^{\text{3 factors}}}{\underbrace{3 \cdot 2 \cdot 1}_{\text{3 factors}}}$$

Example 2 ▶ Finding Binomial Coefficients 

Find each binomial coefficient.

a. ${}_7C_3$ **b.** $\binom{7}{4}$ **c.** ${}_{12}C_1$ **d.** $\binom{12}{11}$

Solution

a. $${}_7C_3 = \frac{7 \cdot \cancel{6} \cdot 5}{\cancel{3} \cdot \cancel{2} \cdot 1} = 35$$

b. $$\binom{7}{4} = \frac{7 \cdot \cancel{6} \cdot 5 \cdot \cancel{4}}{\cancel{4} \cdot \cancel{3} \cdot \cancel{2} \cdot 1} = 35$$

c. $${}_{12}C_1 = \frac{12}{1} = 12$$

d. $$\binom{12}{11} = \frac{12!}{1! \cdot 11!} = \frac{(12) \cdot \cancel{11!}}{1! \cdot \cancel{11!}} = \frac{12}{1} = 12$$

It is not a coincidence that the results in parts (a) and (b) of Example 2 are the same and that the results in parts (c) and (d) are the same. In general, it is true that

$${}_nC_r = {}_nC_{n-r}.$$

This shows the symmetric property of binomial coefficients that was identified earlier.

Pascal's Triangle

There is a convenient way to remember the pattern for binomial coefficients. By arranging the coefficients in a triangular pattern, you obtain **Pascal's Triangle.** This triangle is named after the famous French mathematician Blaise Pascal (1623–1662).

Exploration

Complete the table and describe the result.

n	r	${}_nC_r$	${}_nC_{n-r}$
9	5		
7	1		
12	4		
6	0		
10	7		

What characteristic of Pascal's Triangle is illustrated by this table?

```
                    1
                  1   1
                1   2   1
              1   3   3   1
            1   4   6   4   1
          1   5   10  10  5   1          4 + 6 = 10
        1   6   15  20  15  6   1
      1   7   21  35  35  21  7   1      15 + 6 = 21
```

The first and last numbers in each row of Pascal's Triangle are 1. Every other number in each row is formed by adding the two numbers immediately above the number. Pascal noticed that numbers in this triangle are precisely the same numbers that are the coefficients of binomial expansions, as follows.

$$(x + y)^0 = 1 \qquad \text{0th row}$$

$$(x + y)^1 = 1x + 1y \qquad \text{1st row}$$

$$(x + y)^2 = 1x^2 + 2xy + 1y^2 \qquad \text{2nd row}$$

$$(x + y)^3 = 1x^3 + 3x^2y + 3xy^2 + 1y^3 \qquad \text{3rd row}$$

$$(x + y)^4 = 1x^4 + 4x^3y + 6x^2y^2 + 4xy^3 + 1y^4 \qquad \vdots$$

$$(x + y)^5 = 1x^5 + 5x^4y + 10x^3y^2 + 10x^2y^3 + 5xy^4 + 1y^5$$

$$(x + y)^6 = 1x^6 + 6x^5y + 15x^4y^2 + 20x^3y^3 + 15x^2y^4 + 6xy^5 + 1y^6$$

$$(x + y)^7 = 1x^7 + 7x^6y + 21x^5y^2 + 35x^4y^3 + 35x^3y^4 + 21x^2y^5 + 7xy^6 + 1y^7$$

The top row in Pascal's Triangle is called the *zeroth row* because it corresponds to the binomial expansion $(x + y)^0 = 1$. Similarly, the next row is called the *first row* because it corresponds to the binomial expansion $(x + y)^1 = 1(x) + 1(y)$. In general, the *nth row* in Pascal's Triangle gives the coefficients of $(x + y)^n$.

Example 3 ▶ Using Pascal's Triangle

Use the seventh row of Pascal's Triangle to find the binomial coefficients.

${}_8C_0, {}_8C_1, {}_8C_2, {}_8C_3, {}_8C_4, {}_8C_5, {}_8C_6, {}_8C_7, {}_8C_8$

Solution

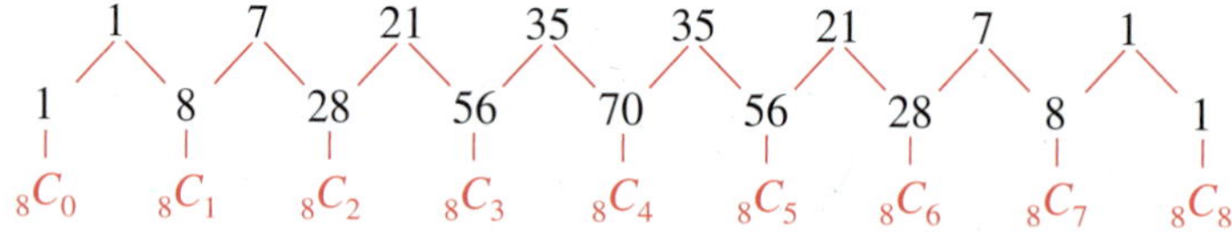

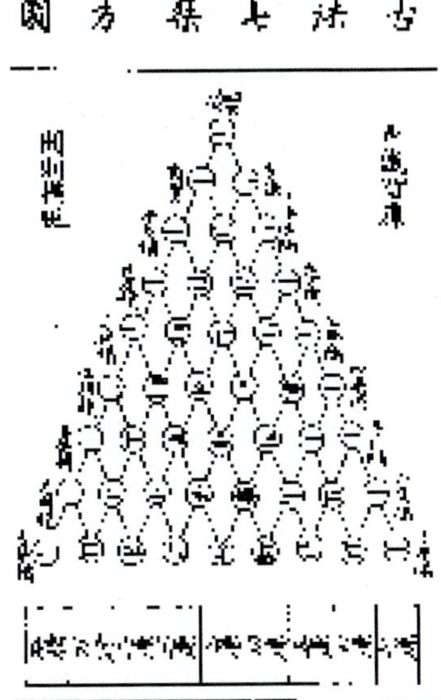

Historical Note

Precious Mirror "Pascal's" Triangle and forms of the Binomial Theorem were known in Eastern cultures prior to the Western "discovery" of the theorem. A Chinese text entitled *Precious Mirror* contains a triangle of binomial expansions through the eighth power.

Binomial Expansions

As mentioned at the beginning of this section, when you write out the coefficients for a binomial that is raised to a power, you are **expanding a binomial.** The formulas for binomial coefficients give you an easy way to expand binomials, as demonstrated in the next four examples.

Example 4 ▶ Expanding a Binomial 

Write the expansion for the expression

$(x + 1)^3$.

Solution

The binomial coefficients from the third row of Pascal's Triangle are

1, 3, 3, 1.

So, the expansion is as follows.

$$(x+1)^3 = (1)x^3 + (3)x^2(1) + (3)x(1^2) + (1)(1^3)$$
$$= x^3 + 3x^2 + 3x + 1$$

To expand binomials representing *differences* rather than sums, you alternate signs. Here are two examples.

$$(x-1)^3 = x^3 - 3x^2 + 3x - 1$$
$$(x-1)^4 = x^4 - 4x^3 + 6x^2 - 4x + 1$$

Example 5 ▶ Expanding a Binomial 

Write the expansion for each expression.

a. $(2x - 3)^4$ **b.** $(x - 2y)^4$

Solution

a. The binomial coefficients from the fourth row of Pascal's Triangle are

1, 4, 6, 4, 1.

So, the expansion is as follows.

$$(2x-3)^4 = (1)(2x)^4 - (4)(2x)^3(3) + (6)(2x)^2(3^2) - (4)(2x)(3^3) + (1)(3^4)$$
$$= 16x^4 - 96x^3 + 216x^2 - 216x + 81$$

b. The binomial coefficients from the fourth row of Pascal's Triangle are

1, 4, 6, 4, 1.

So, the expansion is as follows.

$$(x-2y)^4 = (1)x^4 - (4)x^3(2y) + (6)x^2(2y)^2 - (4)x(2y)^3 + (1)(2y)^4$$
$$= x^4 - 8x^3y + 24x^2y^2 - 32xy^3 + 16y^4$$

Example 6 ▶ Expanding a Binomial

Write the expansion for $(x^2 + 4)^3$.

Solution

Use the third row of Pascal's Triangle, as follows.

$$(x^2 + 4)^3 = (1)(x^2)^3 + (3)(x^2)^2(4) + (3)x^2(4^2) + (1)(4^3)$$
$$= x^6 + 12x^4 + 48x^2 + 64$$

Example 7 ▶ Finding a Term in a Binomial Expansion

a. Find the sixth term of $(a + 2b)^8$.

b. Find the coefficient of the term a^6b^5 in the expansion of $(3a - 2b)^{11}$.

Solution

a. For the first term of the binomial expansion, you would use $n = 8$ and $r = 0$ to get ${}_8C_0 a^{8-0}(2b)^0$. For the second term of the binomial expansion, you would use $n = 8$ and $r = 1$ to get ${}_8C_1 a^{8-1}(2b)^1$. So, for the sixth term of this binomial expansion, use $n = 8$ and $r = 5$ to obtain

$${}_8C_5 a^{8-5}(2b)^5 = 56 \cdot a^3 \cdot (2b)^5$$
$$= 56(2^5)a^3b^5$$
$$= 1792a^3b^5.$$

b. From the Binomial Theorem, you can see that the $(r + 1)$th term is ${}_nC_r x^{n-r} y^r$. So in this case, $n = 11$, $r = 5$, $x = 3a$, and $y = -2b$. Substitute these values to obtain

$${}_nC_r x^{n-r} y^r = {}_{11}C_5(3a)^6(-2b)^5$$
$$= (462)(729a^6)(-32b^5)$$
$$= -10{,}777{,}536a^6b^5.$$

So, the coefficient is $-10{,}777{,}536$.

Writing ABOUT MATHEMATICS

Error Analysis You are a math instructor and receive the following solutions from one of your students on a quiz. Find the error(s) in each solution. Discuss ways that your student could avoid the error(s) in the future.

a. Find the second term in the expansion of $(2x - 3y)^5$.

$5(2x)^4(3y)^2 = 720x^4y^2$ (crossed out)

b. Find the fourth term in the expansion of $\left(\frac{1}{2}x + 7y\right)^6$.

${}_6C_4\left(\frac{1}{2}x\right)^2(7y)^4 = 9003.75x^2y^4$ (crossed out)

11.5 Exercises

In Exercises 1–10, find the binomial coefficient.

1. ${}_5C_3$
2. ${}_8C_6$
3. ${}_{12}C_0$
4. ${}_{20}C_{20}$
5. ${}_{20}C_{15}$
6. ${}_{12}C_5$
7. $\binom{10}{4}$
8. $\binom{10}{6}$
9. $\binom{100}{98}$
10. $\binom{100}{2}$

In Exercises 11–14, evaluate using Pascal's Triangle.

11. $\binom{8}{5}$
12. $\binom{8}{7}$
13. ${}_7C_4$
14. ${}_6C_3$

In Exercises 15–34, use the Binomial Theorem to expand and simplify the expression.

15. $(x + 1)^4$
16. $(x + 1)^6$
17. $(a + 6)^4$
18. $(a + 5)^5$
19. $(y - 4)^3$
20. $(y - 2)^5$
21. $(x + y)^5$
22. $(c + d)^3$
23. $(r + 3s)^6$
24. $(x + 2y)^4$
25. $(3a - b)^5$
26. $(2x - y)^5$
27. $(1 - 2x)^3$
28. $(5 - 3y)^3$
29. $(x^2 + 5)^4$
30. $(x^2 + y^2)^6$
31. $\left(\frac{1}{x} + y\right)^5$
32. $\left(\frac{1}{x} + 2y\right)^6$
33. $2(x - 3)^4 + 5(x - 3)^2$
34. $3(x + 1)^5 - 4(x + 1)^3$

In Exercises 35–38, expand the binomial using Pascal's Triangle to determine the coefficients.

35. $(2t - s)^5$
36. $(3 - 2z)^4$
37. $(x + 2y)^5$
38. $(2v + 3)^6$

In Exercises 39–46, find the specified nth term in the expansion of the binomial.

39. $(x + y)^{10}$, $n = 4$
40. $(x - y)^6$, $n = 7$
41. $(x - 6y)^5$, $n = 3$
42. $(x - 10z)^7$, $n = 4$
43. $(4x + 3y)^9$, $n = 8$
44. $(5a + 6b)^5$, $n = 5$
45. $(10x - 3y)^{12}$, $n = 9$
46. $(7x + 2y)^{15}$, $n = 7$

In Exercises 47–54, find the coefficient a of the term in the expansion of the binomial.

	Binomial	*Term*
47.	$(x + 3)^{12}$	ax^5
48.	$(x^2 + 3)^{12}$	ax^8
49.	$(x - 2y)^{10}$	ax^8y^2
50.	$(4x - y)^{10}$	ax^2y^8
51.	$(3x - 2y)^9$	ax^4y^5
52.	$(2x - 3y)^8$	ax^6y^2
53.	$(x^2 + y)^{10}$	ax^8y^6
54.	$(z^2 - t)^{10}$	az^4t^8

In Exercises 55–58, use the Binomial Theorem to expand and simplify the expression.

55. $\left(\sqrt{x} + 3\right)^4$
56. $\left(2\sqrt{t} - 1\right)^3$
57. $(x^{2/3} - y^{1/3})^3$
58. $(u^{3/5} + 2)^5$

In Exercises 59–62, expand the expression in the difference quotient and simplify.

$$\frac{f(x + h) - f(x)}{h} \quad \text{Difference quotient}$$

59. $f(x) = x^3$
60. $f(x) = x^4$
61. $f(x) = \sqrt{x}$
62. $f(x) = \frac{1}{x}$

In Exercises 63–68, use the Binomial Theorem to expand the complex number. Simplify your result.

63. $(1 + i)^4$
64. $(2 - i)^5$
65. $(2 - 3i)^6$
66. $\left(5 + \sqrt{-9}\right)^3$
67. $\left(-\frac{1}{2} + \frac{\sqrt{3}}{2}i\right)^3$
68. $\left(5 - \sqrt{3}i\right)^4$

Approximation **In Exercises 69–72, use the Binomial Theorem to approximate the quantity accurate to three decimal places. For example, in Exercise 69, use the expansion**

$$(1.02)^8 = (1 + 0.02)^8 = 1 + 8(0.02) + 28(0.02)^2 + \cdots.$$

69. $(1.02)^8$
70. $(2.005)^{10}$
71. $(2.99)^{12}$
72. $(1.98)^9$

Graphical Reasoning **In Exercises 73 and 74, use a graphing utility to graph f and g in the same viewing window. What is the relationship between the two graphs? Use the Binomial Theorem to write the polynomial function g in standard form.**

73. $f(x) = x^3 - 4x, \quad g(x) = f(x + 4)$

74. $f(x) = -x^4 + 4x^2 - 1, \quad g(x) = f(x - 3)$

Probability **In Exercises 75–78, consider n independent trials of an experiment in which each trial has two possible outcomes: "success" or "failure." The probability of a success on each trial is p, and the probability of a failure is $q = 1 - p$. In this context, the term ${}_nC_k\, p^k q^{n-k}$ in the expansion of $(p + q)^n$ gives the probability of k successes in the n trials of the experiment.**

75. A fair coin is tossed seven times. To find the probability of obtaining four heads, evaluate the term

$${}_7C_4\left(\frac{1}{2}\right)^4\left(\frac{1}{2}\right)^3$$

in the expansion of $\left(\frac{1}{2} + \frac{1}{2}\right)^7$.

76. The probability of a baseball player getting a hit during any given time at bat is $\frac{1}{4}$. To find the probability that the player gets three hits during the next 10 times at bat, evaluate the term

$${}_{10}C_3\left(\frac{1}{4}\right)^3\left(\frac{3}{4}\right)^7$$

in the expansion of $\left(\frac{1}{4} + \frac{3}{4}\right)^{10}$.

77. The probability of a sales representative making a sale with any one customer is $\frac{1}{3}$. The sales representative makes eight contacts a day. To find the probability of making four sales, evaluate the term

$${}_8C_4\left(\frac{1}{3}\right)^4\left(\frac{2}{3}\right)^4$$

in the expansion of $\left(\frac{1}{3} + \frac{2}{3}\right)^8$.

78. To find the probability that the sales representative in Exercise 77 makes four sales if the probability of a sale with any one customer is $\frac{1}{2}$, evaluate the term

$${}_8C_4\left(\frac{1}{2}\right)^4\left(\frac{1}{2}\right)^4$$

in the expansion of $\left(\frac{1}{2} + \frac{1}{2}\right)^8$.

▶ Model It

79. ***Data Analysis*** The table shows the per capita consumption of bottled water $f(t)$ (in gallons) in the United States from 1985 through 1999. (Source: U.S. Department of Agriculture)

Year, t	Consumption, $f(t)$
1985	4.5
1986	5.0
1987	5.7
1988	6.5
1989	7.3
1990	8.0
1991	8.0
1992	8.2
1993	9.4
1994	10.7
1995	11.6
1996	12.5
1997	13.1
1998	16.0
1999	18.1

(a) Use the *regression* feature of a graphing utility to find a cubic model for the data. Let t represent the year, with $t = 5$ corresponding to 1985.

(b) Use a graphing utility to plot the data and the model in the same viewing window.

(c) You want to adjust the model so that $t = 0$ corresponds to 1990 rather than 1980. To do this, you shift the graph of f 10 units *to the left* to obtain $g(t) = f(t + 10)$. Write $g(t)$ in standard form.

(d) Use a graphing utility to graph g in the same viewing window as f.

(e) Use both models to estimate the per capita consumption of bottled water in 2004. Do you obtain the same answer?

(f) Describe the overall trend in the data. What factors do you think may have contributed to the increase in the per capita consumption of bottled water?

80. ***Life Insurance*** The average amount of life insurance per household $f(t)$ (in thousands of dollars) from 1980 through 1999 can be approximated by

$$f(t) = 0.020t^2 + 5.16t + 41.2, \quad 0 \le t \le 19$$

where $t = 0$ represents 1980 (see figure). (Source: American Council of Life Insurance)

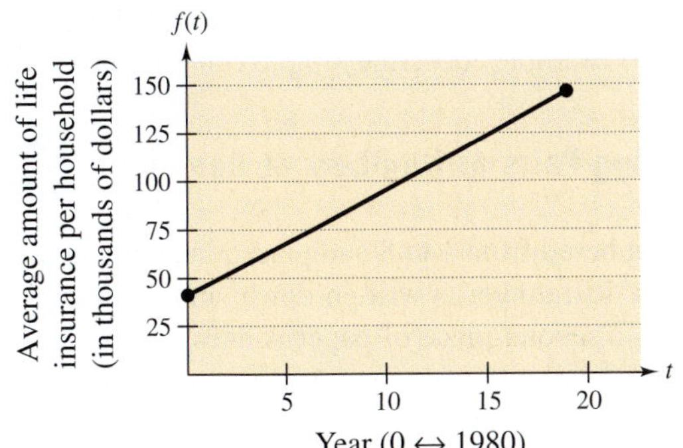

You want to adjust this model so that $t = 0$ corresponds to 1990 rather than 1980. To do this, you shift the graph of f 10 units *to the left* and obtain $g(t) = f(t + 10)$.

(a) Write $g(t)$ in standard form.

(b) Use a graphing utility to graph f and g in the same viewing window.

Synthesis

True or False? **In Exercises 81–83, determine whether the statement is true or false. Justify your answer.**

81. The Binomial Theorem could be used to produce each row of Pascal's Triangle.

82. A binomial that represents a difference cannot always be accurately expanded using the Binomial Theorem.

83. The x^{10}-term and the x^{14}-term of the expansion of $(x^2 + 3)^{12}$ have identical coefficients.

84. ***Writing*** In your own words, explain how to form the rows of Pascal's Triangle.

85. Form rows 8–10 of Pascal's Triangle.

86. ***Think About It*** How many terms are in the expansion of $(x + y)^n$?

87. ***Think About It*** How do the expansions of $(x + y)^n$ and $(x - y)^n$ differ?

88. ***Graphical Reasoning*** Which two functions have identical graphs, and why? Use a graphing utility to graph the functions in the given order and in the same viewing window. Compare the graphs.

(a) $f(x) = (1 - x)^3$ (b) $g(x) = 1 - x^3$

(c) $h(x) = 1 + 3x + 3x^2 + x^3$

(d) $k(x) = 1 - 3x + 3x^2 - x^3$

(e) $p(x) = 1 + 3x - 3x^2 + x^3$

Proof **In Exercises 89–92, prove the property for all integers r and n where $0 \le r \le n$.**

89. ${}_nC_r = {}_nC_{n-r}$

90. ${}_nC_0 - {}_nC_1 + {}_nC_2 - \cdots \pm {}_nC_n = 0$

91. ${}_{n+1}C_r = {}_nC_r + {}_nC_{r-1}$

92. The sum of the numbers in the nth row of Pascal's Triangle is 2^n.

Review

In Exercises 93–96, the graph of $y = g(x)$ is shown. Graph f and use the graph to write an equation for the graph of g.

93. $f(x) = x^2$

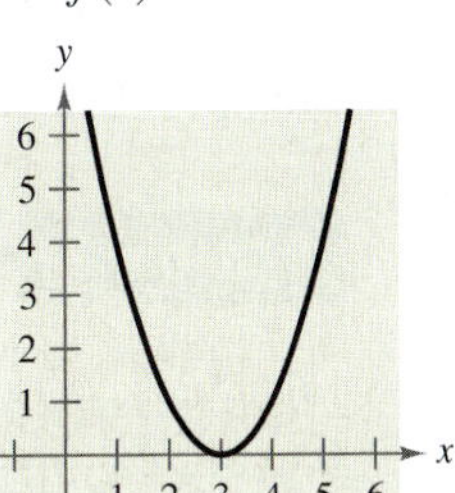

94. $f(x) = x^2$

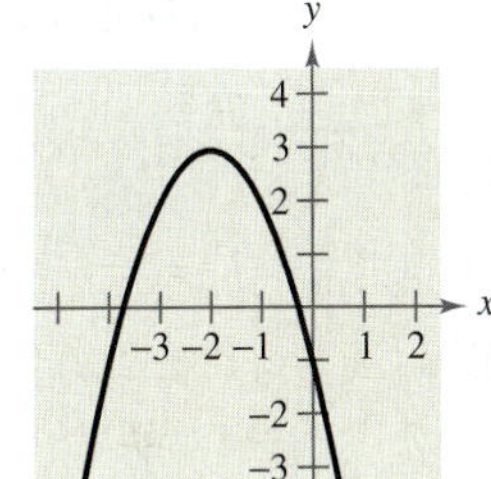

95. $f(x) = \sqrt{x}$

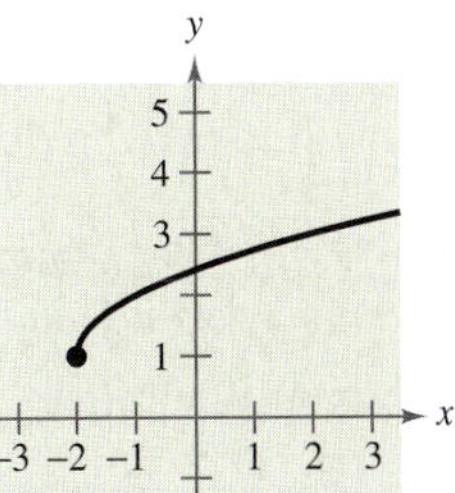

96. $f(x) = \sqrt{x}$

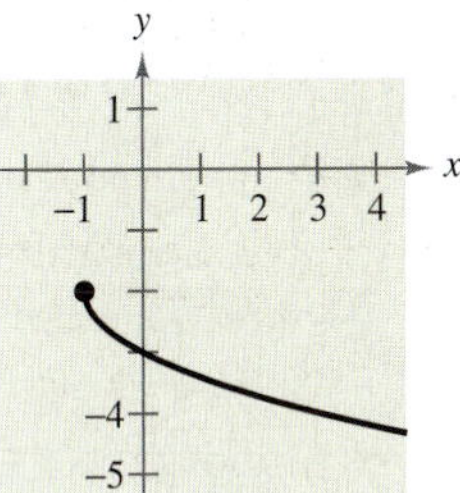

In Exercises 97 and 98, find the inverse of the matrix.

97. $\begin{bmatrix} -6 & 5 \\ -5 & 4 \end{bmatrix}$

98. $\begin{bmatrix} 1.2 & -2.3 \\ -2 & 4 \end{bmatrix}$

11.6 Counting Principles

▶ **What you should learn**

- How to solve simple counting problems
- How to use the Fundamental Counting Principle to solve counting problems
- How to use permutations to solve counting problems
- How to use combinations to solve counting problems

▶ **Why you should learn it**

You can use counting principles to solve counting problems that occur in real life. For instance, in Exercise 67 on page 859, you are asked to use counting principles to determine the number of possible ways of selecting the winning numbers in the Powerball lottery.

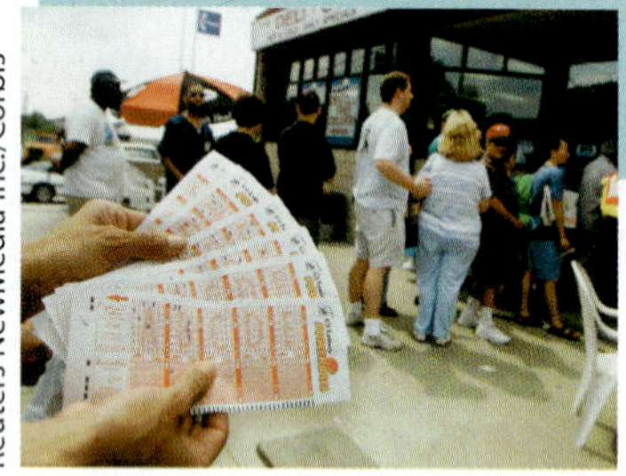

Reuters NewMedia Inc./Corbis

Simple Counting Problems

This section and Section 11.7 present a brief introduction to some of the basic counting principles and their application to probability. In Section 11.7 you will see that much of probability has to do with counting the number of ways an event can occur. The following two examples describe simple counting problems.

Example 1 ▶ Selecting Pairs of Numbers at Random

Eight pieces of paper are numbered from 1 to 8 and placed in a box. One piece of paper is drawn from the box, its number is written down, and the piece of paper is *replaced in the box*. Then, a second piece of paper is drawn from the box, and its number is written down. Finally, the two numbers are added together. How many different ways can a sum of 12 be obtained?

Solution

To solve this problem, count the different ways that a sum of 12 can be obtained using two numbers from 1 to 8.

First number	4	5	6	7	8
Second number	8	7	6	5	4

From this list, you can see that a sum of 12 can occur in five different ways.

Example 2 ▶ Selecting Pairs of Numbers at Random

Eight pieces of paper are numbered from 1 to 8 and placed in a box. Two pieces of paper are drawn from the box *at the same time*, and the numbers on the pieces of paper are written down and totaled. How many different ways can a sum of 12 be obtained?

Solution

To solve this problem, count the different ways that a sum of 12 can be obtained *using two different numbers* from 1 to 8.

First number	4	5	7	8
Second number	8	7	5	4

So, a sum of 12 can be obtained in four different ways.

The difference between the counting problems in Examples 1 and 2 can be expressed by saying that the random selection in Example 1 occurs **with replacement,** whereas the random selection in Example 2 occurs **without replacement,** which eliminates the possibility of choosing two 6's.

The Fundamental Counting Principle

Examples 1 and 2 describe simple counting problems in which you can *list* each possible way that an event can occur. When it is possible, this is always the best way to solve a counting problem. However, some events can occur in so many different ways that it is not feasible to write out the entire list. In such cases, you must rely on formulas and counting principles. The most important of these is the **Fundamental Counting Principle.**

> ### Fundamental Counting Principle
> Let E_1 and E_2 be two events. The first event E_1 can occur in m_1 different ways. After E_1 has occurred, E_2 can occur in m_2 different ways. The number of ways that the two events can occur is $m_1 \cdot m_2$.

The Fundamental Counting Principle can be extended to three or more events. For instance, the number of ways that three events E_1, E_2, and E_3 can occur is $m_1 \cdot m_2 \cdot m_3$.

Example 3 ▶ Using the Fundamental Counting Principle

How many different pairs of letters from the English alphabet are possible?

Solution

There are two events in this situation. The first event is the choice of the first letter, and the second event is the choice of the second letter. Because the English alphabet contains 26 letters, it follows that the number of two-letter pairs is $26 \cdot 26 = 676$.

Example 4 ▶ Using the Fundamental Counting Principle

Telephone numbers in the United States currently have 10 digits. The first three are the *area code* and the next seven are the *local telephone number*. How many different telephone numbers are possible within each area code? (Note that at this time, a local telephone number cannot begin with 0 or 1.)

Solution

Because the first digit cannot be 0 or 1, there are only eight choices for the first digit. For each of the other six digits, there are 10 choices.

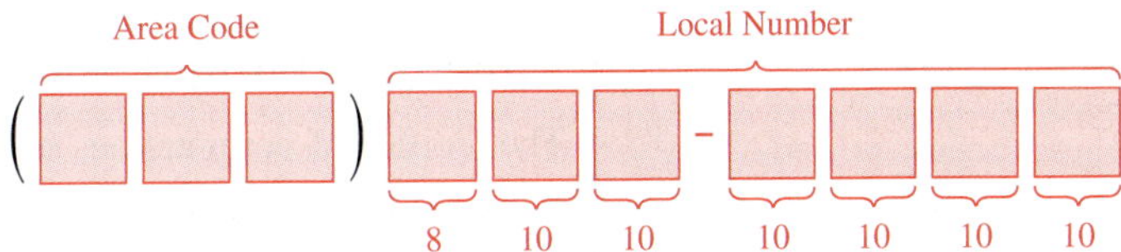

So, the number of local telephone numbers that are possible *within* each area code is $8 \cdot 10 \cdot 10 \cdot 10 \cdot 10 \cdot 10 \cdot 10 = 8{,}000{,}000$.

Permutations

One important application of the Fundamental Counting Principle is in determining the number of ways that n elements can be arranged (in order). An ordering of n elements is called a **permutation** of the elements.

Definition of Permutation

A **permutation** of n different elements is an ordering of the elements such that one element is first, one is second, one is third, and so on.

Example 5 ▶ Finding the Number of Permutations of n Elements

How many permutations are possible for the letters A, B, C, D, E, and F?

Solution

Consider the following reasoning.

First position: Any of the *six* letters

Second position: Any of the remaining *five* letters

Third position: Any of the remaining *four* letters

Fourth position: Any of the remaining *three* letters

Fifth position: Any of the remaining *two* letters

Sixth position: The *one* remaining letter

So, the numbers of choices for the six positions are as follows.

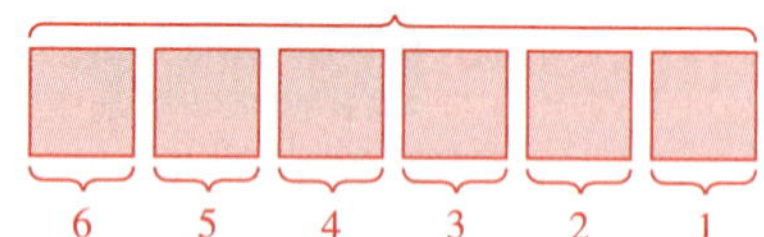

The total number of permutations of the six letters is

$$6! = 6 \cdot 5 \cdot 4 \cdot 3 \cdot 2 \cdot 1$$
$$= 720.$$

Number of Permutations of n Elements

The number of permutations of n elements is

$$n \cdot (n - 1) \cdot \cdot \cdot 4 \cdot 3 \cdot 2 \cdot 1 = n!.$$

In other words, there are $n!$ different ways that n elements can be ordered.

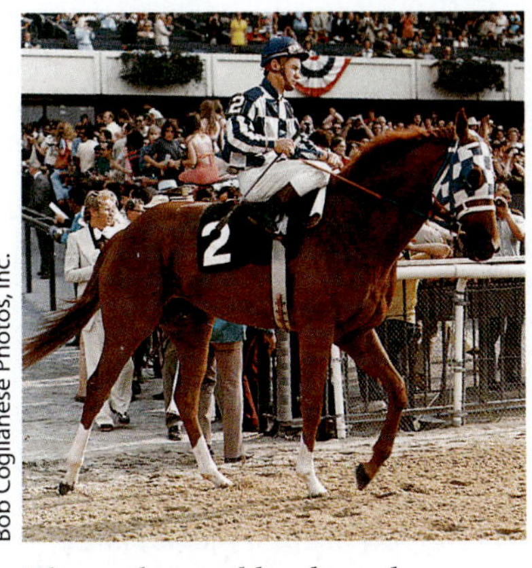

Bob Coglianese Photos, Inc.

Eleven thoroughbred racehorses hold the title of Triple Crown winner for winning the Kentucky Derby, the Preakness, and the Belmont Stakes in the same year. Forty-six horses have won two out of the three races.

It is useful, on occasion, to order a *subset* of a collection of elements rather than the entire collection. For example, you might want to choose and order r elements out of a collection of n elements. Such an ordering is called a **permutation of n elements taken r at a time.**

Example 6 ▶ Counting Horse Race Finishes  

Eight horses are running in a race. In how many different ways can these horses come in first, second, and third? (Assume that there are no ties.)

Solution

Here are the different possibilities.

Win (first position): *Eight* choices

Place (second position): *Seven* choices

Show (third position): *Six* choices

Using the Fundamental Counting Principle, multiply these three numbers together to obtain the following.

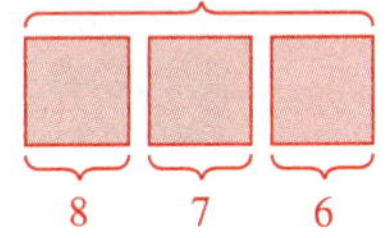

So, there are $8 \cdot 7 \cdot 6 = 336$ different orders.

Technology

Most graphing calculators are programmed to evaluate ${}_nP_r$. Consult your user's manual and then evaluate ${}_8P_5$. You should get an answer of 6720.

Permutations of n Elements Taken r at a Time

The number of permutations of n elements taken r at a time is

$$\begin{aligned} {}_nP_r &= \frac{n!}{(n-r)!} \\ &= n(n-1)(n-2) \cdot \cdot \cdot (n-r+1). \end{aligned}$$

Using this formula, you can rework Example 6 to find that the number of permutations of eight horses taken three at a time is

$$\begin{aligned} {}_8P_3 &= \frac{8!}{(8-3)!} \\ &= \frac{8!}{5!} \\ &= \frac{8 \cdot 7 \cdot 6 \cdot \cancel{5!}}{\cancel{5!}} \\ &= 336 \end{aligned}$$

which is the same answer obtained in the example.

Remember that for permutations, order is important. So, if you are looking at the possible permutations of the letters A, B, C, and D taken three at a time, the permutations (A, B, D) and (B, A, D) are counted as different because the *order* of the elements is different.

Suppose, however, that you are asked to find the possible permutations of the letters A, A, B, and C. The total number of permutations of the four letters would be ${}_4P_4 = 4!$. However, not all of these arrangements would be *distinguishable* because there are two A's in the list. To find the number of distinguishable permutations, you can use the following formula.

Distinguishable Permutations

Suppose a set of n objects has n_1 of one kind of object, n_2 of a second kind, n_3 of a third kind, and so on, with

$$n = n_1 + n_2 + n_3 + \cdot \cdot \cdot + n_k.$$

Then the number of **distinguishable permutations** of the n objects is

$$\frac{n!}{n_1! \cdot n_2! \cdot n_3! \cdot \cdot \cdot \cdot \cdot n_k!}.$$

Example 7 ▶ Distinguishable Permutations 

In how many distinguishable ways can the letters in BANANA be written?

Solution

This word has six letters, of which three are A's, two are N's, and one is a B. So, the number of distinguishable ways the letters can be written is

$$\frac{n!}{n_1! \cdot n_2! \cdot n_3!} = \frac{6!}{3! \cdot 2! \cdot 1!}$$

$$= \frac{6 \cdot 5 \cdot 4 \cdot \cancel{3!}}{\cancel{3!} \cdot 2!}$$

$$= 60.$$

The 60 different distinguishable permutations are as follows.

AAABNN	AAANBN	AAANNB	AABANN	AABNAN	AABNNA
AANABN	AANANB	AANBAN	AANBNA	AANNAB	AANNBA
ABAANN	ABANAN	ABANNA	ABNAAN	ABNANA	ABNNAA
ANAABN	ANAANB	ANABAN	ANABNA	ANANAB	ANANBA
ANBAAN	ANBANA	ANBNAA	ANNAAB	ANNABA	ANNBAA
BAAANN	BAANAN	BAANNA	BANAAN	BANANA	BANNAA
BNAAAN	BNAANA	BNANAA	BNNAAA	NAAABN	NAAANB
NAABAN	NAABNA	NAANAB	NAANBA	NABAAN	NABANA
NABNAA	NANAAB	NANABA	NANBAA	NBAAAN	NBAANA
NBANAA	NBNAAA	NNAAAB	NNAABA	NNABAA	NNBAAA

Combinations

When you count the number of possible permutations of a set of elements, *order* is important. As a final topic in this section, you will look at a method of selecting subsets of a larger set in which order is *not* important. Such subsets are called **combinations of n elements taken r at a time.** For instance, the combinations

$$\{A, B, C\} \qquad \text{and} \qquad \{B, A, C\}$$

are equivalent because both sets contain the same three elements, and the order in which the elements are listed is not important. So, you would count only one of the two sets. A common example of how a combination occurs is a card game in which the player is free to reorder the cards after they have been dealt.

Example 8 ▶ Combinations of n Elements Taken r at a Time

In how many different ways can three letters be chosen from the letters A, B, C, D, and E? (The order of the three letters is not important.)

Solution

The following subsets represent the different combinations of three letters that can be chosen from the five letters.

$\{A, B, C\}$ $\{A, B, D\}$
$\{A, B, E\}$ $\{A, C, D\}$
$\{A, C, E\}$ $\{A, D, E\}$
$\{B, C, D\}$ $\{B, C, E\}$
$\{B, D, E\}$ $\{C, D, E\}$

From this list, you can conclude that there are 10 different ways that three letters can be chosen from five letters.

STUDY TIP

When solving problems involving counting principles, you need to be able to distinguish among the various counting principles in order to determine which is necessary to solve the problem correctly. To do this, ask yourself the following questions.

1. Is the order of the elements important? *Permutation*
2. Are the chosen elements a subset of a larger set in which order is not important? *Combination*
3. Does the problem involve two or more separate events? *Fundamental Counting Principle*

Combinations of n Elements Taken r at a Time

The number of combinations of n elements taken r at a time is

$${}_nC_r = \frac{n!}{(n-r)!r!}.$$

Note that the formula for ${}_nC_r$ is the same one given for binomial coefficients. To see how this formula is used, solve the counting problem in Example 8. In that problem, you are asked to find the number of combinations of five elements taken three at a time. So, $n = 5$, $r = 3$, and the number of combinations is

$${}_5C_3 = \frac{5!}{2!3!} = \frac{5 \cdot \overset{2}{\cancel{4}} \cdot \cancel{3!}}{\cancel{2} \cdot 1 \cdot \cancel{3!}} = 10$$

which is the same answer obtained in the example.

Example 9 ▶ Counting Card Hands  

A standard poker hand consists of five cards dealt from a deck of 52. How many different poker hands are possible? (After the cards are dealt, the player may reorder them, and so order is not important.)

Solution

You can find the number of different poker hands by using the formula for the number of combinations of 52 elements taken five at a time, as follows.

$$\begin{aligned}{}_{52}C_5 &= \frac{52!}{(52-5)!5!}\\ &= \frac{52!}{47!5!}\\ &= \frac{52 \cdot 51 \cdot 50 \cdot 49 \cdot 48 \cdot 47!}{5 \cdot 4 \cdot 3 \cdot 2 \cdot 1 \cdot 47!}\\ &= 2{,}598{,}960\end{aligned}$$

Example 10 ▶ The Number of Subsets of a Set

Find the total number of subsets of a set that has 10 elements.

Solution

Begin by considering the number of subsets with 0 elements, the number with 1 element, the number with 2 elements, and so on.

Number of subsets	=	Subsets with 0 elements	+	Subsets with 1 element	+	Subsets with 2 elements	+ . . . +	Subsets with 10 elements

$$= \binom{10}{0} + \binom{10}{1} + \binom{10}{2} + \cdots + \binom{10}{10}$$

By comparing this expression with the binomial expansion of $(1 + 1)^{10}$, you see that they are the same.

$$\begin{aligned}(1+1)^{10} &= \binom{10}{0}1^{10}1^0 + \binom{10}{1}1^9 1^1 + \binom{10}{2}1^8 1^2 + \cdots + \binom{10}{10}1^0 1^{10}\\ &= \binom{10}{0} + \binom{10}{1} + \binom{10}{2} + \cdots + \binom{10}{10}\end{aligned}$$

This implies that the total number of subsets of a set of 10 elements is

$$\begin{aligned}(1+1)^{10} &= 2^{10}\\ &= 1024.\end{aligned}$$

The result of Example 10 can be generalized to conclude that the total number of subsets of a set of n elements is 2^n.

11.6 Exercises

Random Selection **In Exercises 1–8, determine the number of ways a computer can randomly generate one or more such integers from 1 through 12.**

1. An odd integer
2. An even integer
3. A prime integer
4. An integer that is greater than 9
5. An integer that is divisible by 4
6. An integer that is divisible by 3
7. Two integers whose sum is 8
8. Two *distinct* integers whose sum is 8

9. ***Entertainment Systems*** A customer can choose one of three amplifiers, one of two compact disc players, and one of five speaker models for an entertainment system. Determine the number of possible system configurations.

10. ***Computer Systems*** A customer in a computer store can choose one of four monitors, one of three keyboards, and one of five computers. If all the choices are compatible, determine the number of possible system configurations.

11. ***Job Applicants*** A college needs two additional faculty members: a chemist and a statistician. In how many ways can these positions be filled if there are five applicants for the chemistry position and three applicants for the statistics position?

12. ***Course Schedule*** A college student is preparing a course schedule for the next semester. The student may select one of two mathematics courses, one of three science courses, and one of five courses from the social sciences and humanities. How many schedules are possible?

13. ***True-False Exam*** In how many ways can a six-question true-false exam be answered? (Assume that no questions are omitted.)

14. ***True-False Exam*** In how many ways can a 12-question true-false exam be answered? (Assume that no questions are omitted.)

15. ***Toboggan Ride*** Three people are lining up for a ride on a toboggan, but only two of the three are willing to take the first position. With that constraint, in how many ways can the three people be seated on the toboggan?

16. ***Aircraft Boarding*** Eight people are boarding an aircraft. Two have tickets for first class and board before those in the economy class. In how many ways can the eight people board the aircraft?

17. ***License Plate Numbers*** In the state of Pennsylvania, each automobile license plate number consists of three letters followed by a four-digit number. How many distinct license plate numbers can be formed?

18. ***License Plate Numbers*** In a certain state, each automobile license plate number consists of two letters followed by a four-digit number. To avoid confusion between "O" and "zero" and between "I" and "one," the letters "O" and "I" are not used. How many distinct license plate numbers can be formed?

19. ***Three-Digit Numbers*** How many three-digit numbers can be formed under each condition?
    (a) The leading digit cannot be zero.
    (b) The leading digit cannot be zero and no repetition of digits is allowed.
    (c) The leading digit cannot be zero and the number must be a multiple of 5.
    (d) The number is at least 400.

20. ***Four-Digit Numbers*** How many four-digit numbers can be formed under each condition?
    (a) The leading digit cannot be zero.
    (b) The leading digit cannot be zero and no repetition of digits is allowed.
    (c) The leading digit cannot be zero and the number must be less than 5000.
    (d) The leading digit cannot be zero and the number must be even.

21. ***Combination Lock*** A combination lock will open when the right choice of three numbers (from 1 to 40, inclusive) is selected. How many different lock combinations are possible?

22. ***Combination Lock*** A combination lock will open when the right choice of three numbers (from 1 to 50, inclusive) is selected. How many different lock combinations are possible?

23. ***Concert Seats*** Four couples have reserved seats in a row for a concert. In how many different ways can they be seated if

(a) there are no seating restrictions?

(b) the two members of each couple wish to sit together?

24. ***Single File*** In how many orders can four girls and four boys walk through a doorway single file if

(a) there are no restrictions?

(b) the girls walk through before the boys?

In Exercises 25–30, evaluate $_nP_r$.

25. $_4P_4$

26. $_5P_5$

27. $_8P_3$

28. $_{20}P_2$

29. $_5P_4$

30. $_7P_4$

In Exercises 31 and 32, solve for *n*.

31. $14 \cdot {_nP_3} = {_{n+2}P_4}$

32. ${_nP_5} = 18 \cdot {_{n-2}P_4}$

In Exercises 33–38, evaluate using a calculator.

33. $_{20}P_5$

34. $_{100}P_5$

35. $_{100}P_3$

36. $_{10}P_8$

37. $_{20}C_5$

38. $_{10}C_7$

In Exercises 39–42, find the number of distinguishable permutations of the group of letters.

39. A, A, G, E, E, E, M

40. B, B, B, T, T, T, T, T

41. A, L, G, E, B, R, A

42. M, I, S, S, I, S, S, I, P, P, I

43. Write all permutations of the letters A, B, C, and D.

44. Write all permutations of the letters A, B, C, and D if the letters B and C must remain between the letters A and D.

45. Write all possible selections of two letters that can be formed from the letters A, B, C, D, E, and F. (The order of the two letters is not important.)

46. Write all possible selections of three letters that can be formed from the letters A, B, C, D, E, and F. (The order of the three letters is not important.)

47. ***Posing for a Photograph*** In how many ways can five children line up in a row?

48. ***Riding in a Car*** In how many ways can six people sit in a six-passenger car?

49. ***Choosing Officers*** From a pool of 12 candidates, the offices of president, vice-president, secretary, and treasurer will be filled. In how many different ways can the offices be filled?

50. ***Assembly Line Production*** There are four processes involved in assembling a product, and these processes can be performed in any order. The management wants to test each order to determine which is the least time-consuming. How many different orders will have to be tested?

51. ***Forming an Experimental Group*** In order to conduct an experiment, five students are randomly selected from a class of 20. How many different groups of five students are possible?

52. ***Test Questions*** You can answer any 10 questions from a total of 12 questions on an exam. In how many different ways can you select the questions?

53. ***Lottery Choices*** There are 35 numbers in the Massachusetts Mass Cash game. In how many ways can a player select five of the numbers?

54. ***Lottery Choices*** There are 40 numbers in the Louisiana Lotto game. In how many ways can a player select six of the numbers?

55. ***Number of Subsets*** How many subsets of four elements can be formed from a set of 100 elements?

56. ***Number of Subsets*** How many subsets of five elements can be formed from a set of 80 elements?

57. ***Geometry*** Three points that are not on a line determine three lines. How many lines are determined by seven points, no three of which are on a line?

58. ***Defective Units*** A shipment of 10 microwave ovens contains three defective units. In how many ways can a vending company purchase four of these units and receive (a) all good units, (b) two good units, and (c) at least two good units?

59. ***Job Applicants*** A toy manufacturer interviews eight people for four openings in the research and development department of the company. Three of the eight people are women. If all eight are qualified, in how many ways can the employer fill the four positions if (a) the selection is random and (b) exactly two selections are women?

60. ***Poker Hand*** You are dealt five cards from an ordinary deck of 52 playing cards. In how many ways can you get a full house? (A full house consists of three of one kind and two of another. For example, A-A-A-5-5 and K-K-K-10-10 are full houses.)

61. ***Forming a Committee*** Four people are to be selected at random from a group of four couples. In how many ways can this be done under the following conditions?
(a) There are no restrictions.
(b) The group must have at least one couple.
(c) Each couple must be represented in the group.

62. ***Interpersonal Relationships*** The complexity of the interpersonal relationships increases dramatically as the size of a group increases. Determine the number of different two-person relationships in a group of people of size (a) 3, (b) 8, (c) 12, and (d) 20.

Geometry **In Exercises 63–66, find the number of diagonals of the polygon. (A line segment connecting any two nonadjacent vertices is called a diagonal of the polygon.)**

63. Pentagon
64. Hexagon
65. Octagon
66. Decagon (10 sides)

▶ Model It

67. ***Lottery*** Powerball is a lottery game that is operated by the Multi-State Lottery Association and its 24 current state members. The game is played by drawing five white balls out of a drum of 53 white balls (numbered 1–53) and one red powerball out of a drum of 42 red balls (numbered 1–42). The jackpot is won by matching all five white balls in any order and the red powerball.
(a) Find the possible number of winning Powerball numbers.
(b) Find the possible number of winning Powerball numbers if the jackpot is won by matching all five white balls in order and the red powerball.
(c) Compare the results of part (a) with a state lottery in which a jackpot is won by matching six balls from a drum of 53 balls.

Synthesis

True or False? **In Exercises 68–70, determine whether the statement is true or false. Justify your answer.**

68. The number of letter pairs that can be formed from any of the first 13 letters in the alphabet (A–M) is an example of a permutation.

69. The number of permutations of n elements can be determined by using the Fundamental Counting Principle.

70. The value of ${}_nP_r$ is always greater than the value of ${}_nC_r$.

71. What is the relationship between ${}_nC_r$ and ${}_nC_{n-r}$?

72. Without calculating the numbers, determine which of the following is greater. Explain.
(a) The combinations of 10 elements taken six at a time
(b) The permutations of 10 elements taken six at a time

Proof **In Exercises 73–76, prove the identity.**

73. ${}_nP_{n-1} = {}_nP_n$
74. ${}_nC_n = {}_nC_0$
75. ${}_nC_{n-1} = {}_nC_1$
76. ${}_nC_r = \dfrac{{}_nP_r}{r!}$

77. ***Think About It*** Can your calculator evaluate ${}_{100}P_{80}$? If not, explain why.

78. ***Writing*** Explain in words the meaning of ${}_nP_r$.

Review

In Exercises 79–82, evaluate the function at the specified values of the independent variable.

79. $f(x) = 3x^2 + 8$
(a) $f(3)$ (b) $f(0)$ (c) $f(-5)$

80. $g(x) = \sqrt{x-3} + 2$
(a) $g(3)$ (b) $g(7)$ (c) $g(x + 3)$

81. $f(x) = -|x - 5| + 6$
(a) $f(-5)$ (b) $f(-1)$ (c) $f(11)$

82. $f(x) = \begin{cases} x^2 - 2x + 5, & x \le -4 \\ -x^2 - 2, & x > -4 \end{cases}$
(a) $f(-4)$ (b) $f(-1)$ (c) $f(-20)$

In Exercises 83–86, solve the equation. Round your answer to two decimal places, if necessary.

83. $\sqrt{x-3} = x - 6$
84. $\dfrac{4}{t} + \dfrac{3}{2t} = 1$
85. $\log_2(x - 3) = 5$
86. $e^{x/3} = 16$

In Exercises 87–90, use the Binomial Theorem to expand and simplify the expression.

87. $(x + 1)^5$
88. $(y - 2)^6$
89. $(x^2 + 2y)^5$
90. $(x^2 - y^2)^4$

11.7 Probability

▶ **What you should learn**

- How to find the probabilities of events
- How to find the probabilities of mutually exclusive events
- How to find the probabilities of independent events
- How to find the probability of the complement of an event

▶ **Why you should learn it**

You can use probability to solve a variety of problems that occur in real life. For instance, in Exercise 35 on page 869, you are asked to use probability to help analyze the voting-age population distribution in the United States.

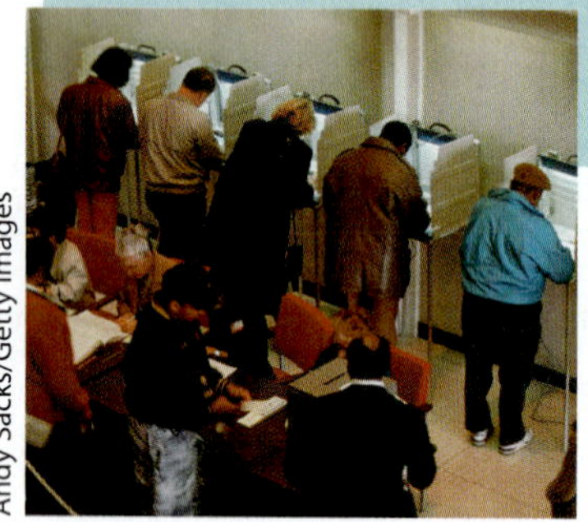
Andy Sacks/Getty Images

The Probability of an Event

Any happening for which the result is uncertain is called an **experiment.** The possible results of the experiment are **outcomes,** the set of all possible outcomes of the experiment is the **sample space** of the experiment, and any subcollection of a sample space is an **event.**

For instance, when a six-sided die is tossed, the sample space can be represented by the numbers 1 through 6. For this experiment, each of the outcomes is *equally likely.*

To describe sample spaces in such a way that each outcome is equally likely, you must sometimes distinguish between or among various outcomes in ways that appear artificial. Example 1 illustrates such a situation.

Example 1 ▶ Finding the Sample Space

Find the sample space for each of the following.

a. One coin is tossed.

b. Two coins are tossed.

c. Three coins are tossed.

Solution

a. Because the coin will land either heads up (denoted by H) or tails up (denoted by T), the sample space is

$$S = \{H, T\}.$$

b. Because either coin can land heads up or tails up, the possible outcomes are as follows.

HH = heads up on both coins

HT = heads up on first coin and tails up on second coin

TH = tails up on first coin and heads up on second coin

TT = tails up on both coins

So, the sample space is

$$S = \{HH, HT, TH, TT\}.$$

Note that this list distinguishes between the two cases HT and TH, even though these two outcomes appear to be similar.

c. Following the notation of part (b), the sample space is

$$S = \{HHH, HHT, HTH, HTT, THH, THT, TTH, TTT\}.$$

Note that this list distinguishes among the cases HHT, HTH, and THH, and among the cases HTT, THT, and TTH.

To calculate the probability of an event, count the number of outcomes in the event and in the sample space. The *number of outcomes* in event E is denoted by $n(E)$, and the number of outcomes in the sample space S is denoted by $n(S)$. The probability that event E will occur is given by $n(E)/n(S)$.

The Probability of an Event

If an event E has $n(E)$ equally likely outcomes and its sample space S has $n(S)$ equally likely outcomes, the **probability** of event E is

$$P(E) = \frac{n(E)}{n(S)}.$$

Because the number of outcomes in an event must be less than or equal to the number of outcomes in the sample space, the probability of an event must be a number between 0 and 1. That is,

$$0 \leq P(E) \leq 1.$$

If $P(E) = 0$, event E *cannot occur*, and E is called an impossible event. If $P(E) = 1$, event E *must occur*, and E is called a certain event.

Example 2 ▶ Finding the Probability of an Event

a. Two coins are tossed. What is the probability that both land heads up?

b. A card is drawn from a standard deck of playing cards. What is the probability that it is an ace?

Solution

a. Following the procedure in Example 1(b), let

$$E = \{HH\}$$

and

$$S = \{HH, HT, TH, TT\}.$$

The probability of getting two heads is

$$P(E) = \frac{n(E)}{n(S)} = \frac{1}{4}.$$

b. Because there are 52 cards in a standard deck of playing cards and there are four aces (one in each suit), the probability of drawing an ace is

$$P(E) = \frac{n(E)}{n(S)}$$

$$= \frac{4}{52}$$

$$= \frac{1}{13}.$$

STUDY TIP

You can express a probability as a fraction, decimal, or percent. For instance, in Example 2(a), the probability of getting two heads can be written as $\frac{1}{4}$, 0.25, or 25%.

FIGURE 11.8

Example 3 ▶ Finding the Probability of an Event  

Two six-sided dice are tossed. What is the probability that the total of the two dice is 7? (See Figure 11.8.)

Solution

Because there are six possible outcomes on each die, you can use the Fundamental Counting Principle to conclude that there are $6 \cdot 6$ or 36 different outcomes when two dice are tossed. To find the probability of rolling a total of 7, you must first count the number of ways in which this can occur.

First die	Second die
1	6
2	5
3	4
4	3
5	2
6	1

So, a total of 7 can be rolled in six ways, which means that the probability of rolling a 7 is

$$P(E) = \frac{n(E)}{n(S)} = \frac{6}{36} = \frac{1}{6}.$$

STUDY TIP

You could have written out each sample space in Examples 2 and 3 and simply counted the outcomes in the desired events. For larger sample spaces, however, you should use the counting principles discussed in Section 11.6.

Example 4 ▶ Finding the Probability of an Event 

Twelve-sided dice, as shown in Figure 11.9, can be constructed (in the shape of regular dodecahedrons) such that each of the numbers from 1 to 6 appears twice on each die. Prove that these dice can be used in any game requiring ordinary six-sided dice without changing the probabilities of different outcomes.

Solution

For an ordinary six-sided die, each of the numbers 1, 2, 3, 4, 5, and 6 occurs only once, so the probability of any particular number coming up is

$$P(E) = \frac{n(E)}{n(S)} = \frac{1}{6}.$$

For one of the 12-sided dice, each number occurs twice, so the probability of any particular number coming up is

$$P(E) = \frac{n(E)}{n(S)} = \frac{2}{12} = \frac{1}{6}.$$

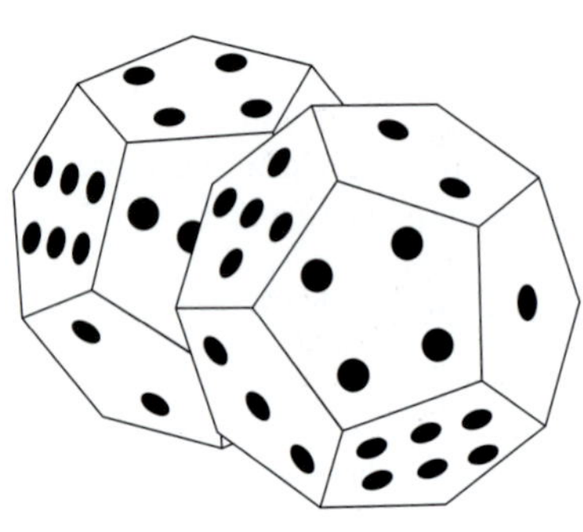

FIGURE 11.9

Example 5 ▶ The Probability of Winning a Lottery  

In the Arizona state lottery, a player chooses six different numbers from 1 to 41. If these six numbers match the six numbers drawn by the lottery commission, the player wins (or shares) the top prize. What is the probability of winning?

Solution

To find the number of elements in the sample space, use the formula for the number of combinations of 41 elements taken six at a time.

$$n(S) = {}_{41}C_6$$

$$= \frac{41 \cdot 40 \cdot 39 \cdot 38 \cdot 37 \cdot 36}{6 \cdot 5 \cdot 4 \cdot 3 \cdot 2 \cdot 1}$$

$$= 4{,}496{,}388$$

If a person buys only one ticket, the probability of winning is

$$P(E) = \frac{n(E)}{n(S)}$$

$$= \frac{1}{4{,}496{,}388}.$$

Example 6 ▶ Random Selection 

The numbers of colleges and universities in various regions of the United States in 2001 are shown in Figure 11.10. One institution is selected at random. What is the probability that the institution is in one of the three southern regions? (Source: U.S. National Center for Education Statistics)

Solution

From the figure, the total number of colleges and universities is 4178. Because there are $687 + 274 + 383 = 1344$ colleges and universities in the three southern regions, the probability that the institution is in one of these regions is

$$P(E) = \frac{n(E)}{n(S)} = \frac{1344}{4178} \approx 0.322.$$

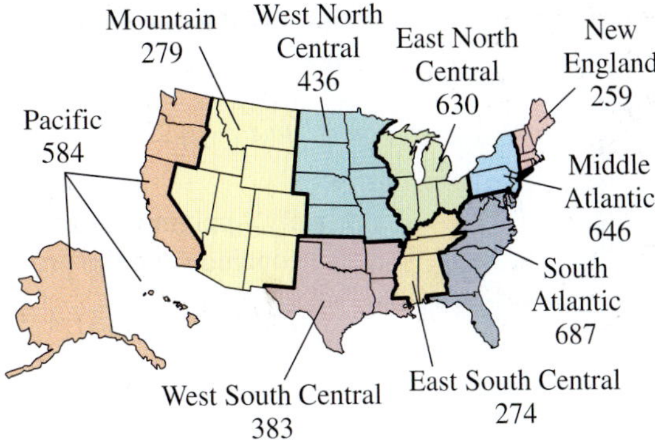

FIGURE 11.10

Mutually Exclusive Events

Two events A and B (from the same sample space) are **mutually exclusive** if A and B have no outcomes in common. In the terminology of sets, the intersection of A and B is the empty set, which is expressed as

$$P(A \cap B) = 0.$$

For instance, if two dice are tossed, the event A of rolling a total of 6 and the event B of rolling a total of 9 are mutually exclusive. To find the probability that one or the other of two mutually exclusive events will occur, you can *add* their individual probabilities.

Probability of the Union of Two Events

If A and B are events in the same sample space, the probability of A or B occurring is given by

$$P(A \cup B) = P(A) + P(B) - P(A \cap B).$$

If A and B are mutually exclusive, then

$$P(A \cup B) = P(A) + P(B).$$

Example 7 ▶ The Probability of a Union of Events 

One card is selected from a standard deck of 52 playing cards. What is the probability that the card is either a heart or a face card?

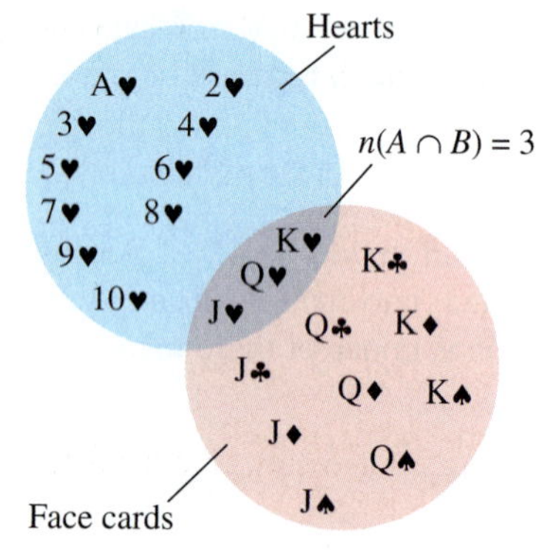

FIGURE 11.11

Solution

Because the deck has 13 hearts, the probability of selecting a heart (event A) is

$$P(A) = \frac{13}{52}.$$

Similarly, because the deck has 12 face cards, the probability of selecting a face card (event B) is

$$P(B) = \frac{12}{52}.$$

Because three of the cards are hearts *and* face cards (see Figure 11.11), it follows that

$$P(A \cap B) = \frac{3}{52}.$$

Finally, applying the formula for the probability of the union of two events, you can conclude that the probability of selecting a heart or a face card is

$$P(A \cup B) = P(A) + P(B) - P(A \cap B)$$

$$= \frac{13}{52} + \frac{12}{52} - \frac{3}{52} = \frac{22}{52} \approx 0.423.$$

Example 8 ▶ Probability of Mutually Exclusive Events

The personnel department of a company has compiled data on the numbers of employees who have been with the company for various periods of time. The results are shown in the table.

Years of service	Number of Employees
0–4	157
5–9	89
10–14	74
15–19	63
20–24	42
25–29	38
30–34	37
35–39	21
40–44	8

If an employee is chosen at random, what is the probability that the employee has 9 or fewer years of service?

Solution

To begin, add the number of employees and find that the total is 529. Next, let event A represent choosing an employee with 0 to 4 years of service and let event B represent choosing an employee with 5 to 9 years of service. Then

$$P(A) = \frac{157}{529}$$

and

$$P(B) = \frac{89}{529}.$$

Because A and B have no outcomes in common, you can conclude that these two events are mutually exclusive and that

$$\begin{aligned} P(A \cup B) &= P(A) + P(B) \\ &= \frac{157}{529} + \frac{89}{529} \\ &= \frac{246}{529} \\ &\approx 0.465. \end{aligned}$$

So, the probability of choosing an employee who has 9 or fewer years of service is about 0.465.

Independent Events

Two events are **independent** if the occurrence of one has no effect on the occurrence of the other. For instance, rolling a total of 12 with two six-sided dice has no effect on the outcome of future rolls of the dice. To find the probability that two independent events will occur, *multiply* the probabilities of each.

> **Probability of Independent Events**
>
> If A and B are independent events, the probability that both A and B will occur is
>
> $$P(A \text{ and } B) = P(A) \cdot P(B).$$

Example 9 ▶ Probability of Independent Events

A random number generator on a computer selects three integers from 1 to 20. What is the probability that all three numbers are less than or equal to 5?

Solution

The probability of selecting a number from 1 to 5 is

$$P(A) = \frac{5}{20}$$
$$= \frac{1}{4}.$$

So, the probability that all three numbers are less than or equal to 5 is

$$P(A) \cdot P(A) \cdot P(A) = \left(\frac{1}{4}\right)\left(\frac{1}{4}\right)\left(\frac{1}{4}\right)$$
$$= \frac{1}{64}.$$

Example 10 ▶ Probability of Independent Events

In 2000, approximately 65% of the population of the United States was 25 years old or older. In a survey, 10 people were chosen at random from the population. What is the probability that all 10 were 25 years old or older? (Source: U.S. Census Bureau)

Solution

Let A represent choosing a person who was 25 years old or older. Because the probability of choosing a person who was 25 years old or older was 0.65, you can conclude that the probability that all 10 people were 25 years old or older is

$$[P(A)]^{10} = (0.65)^{10}$$
$$\approx 0.0135.$$

Exploration

You are in a class with 22 other people. What is the probability that at least two out of the 23 people will have a birthday on the same day of the year?

The complement of the probability that at least two people have the same birthday is the probability that all 23 birthdays are different. So, first find the probability that all 23 people have different birthdays and then find the complement.

Now, determine the probability that in a room with 50 people at least two people have the same birthday.

The Complement of an Event

The **complement of an event** A is the collection of all outcomes in the sample space that are *not* in A. The complement of event A is denoted by A'. Because $P(A \text{ or } A') = 1$ and because A and A' are mutually exclusive, it follows that $P(A) + P(A') = 1$. So, the probability of A' is

$$P(A') = 1 - P(A).$$

For instance, if the probability of *winning* a certain game is

$$P(A) = \frac{1}{4}$$

the probability of *losing* the game is

$$P(A') = 1 - \frac{1}{4} = \frac{3}{4}.$$

Probability of a Complement

Let A be an event and let A' be its complement. If the probability of A is $P(A)$, the probability of the complement is

$$P(A') = 1 - P(A).$$

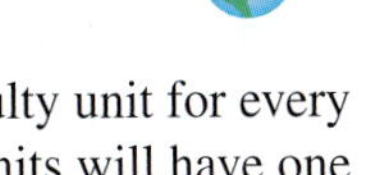

Example 11 ▶ Finding the Probability of a Complement

A manufacturer has determined that a machine averages one faulty unit for every 1000 it produces. What is the probability that an order of 200 units will have one or more faulty units?

Solution

To solve this problem as stated, you would need to find the probabilities of having exactly one faulty unit, exactly two faulty units, exactly three faulty units, and so on. However, using complements, you can simply find the probability that all units are perfect and then subtract this value from 1. Because the probability that any given unit is perfect is 999/1000, the probability that all 200 units are perfect is

$$P(A) = \left(\frac{999}{1000}\right)^{200} \approx 0.8186.$$

So, the probability that at least one unit is faulty is

$$P(A') = 1 - P(A) \approx 0.1814.$$

11.7 Exercises

In Exercises 1–6, determine the sample space for the experiment.

1. A coin and a six-sided die are tossed.
2. A six-sided die is tossed twice and the sum of the points is recorded.
3. A taste tester has to rank three varieties of yogurt, A, B, and C, according to preference.
4. Two marbles are selected from a sack containing two red marbles, two blue marbles, and one black marble. The color of each marble is recorded.
5. Two county supervisors are selected from five supervisors, A, B, C, D, and E, to study a recycling plan.
6. A sales representative makes presentations of a product in three homes per day. In each home, there may be a sale (denote by S) or there may be no sale (denote by F).

Heads or Tails **In Exercises 7–10, find the probability for the experiment of tossing a coin three times. Use the sample space $S = \{HHH, HHT, HTH, HTT, THH, THT, TTH, TTT\}$.**

7. The probability of getting exactly one tail
8. The probability of getting a head on the first toss
9. The probability of getting at least one head
10. The probability of getting at least two heads

Drawing a Card **In Exercises 11–14, find the probability for the experiment of selecting one card from a standard deck of 52 playing cards.**

11. The card is a face card.
12. The card is not a face card.
13. The card is a red face card.
14. The card is a 6 or less.

Tossing a Die **In Exercises 15–20, find the probability for the experiment of tossing a six-sided die twice.**

15. The sum is 4.
16. The sum is at least 7.
17. The sum is less than 11.
18. The sum is 2, 3, or 12.
19. The sum is odd and no more than 7.
20. The sum is odd or prime.

Drawing Marbles **In Exercises 21–24, find the probability for the experiment of drawing two marbles (without replacement) from a bag containing one green, two yellow, and three red marbles.**

21. Both marbles are red.
22. Both marbles are yellow.
23. Neither marble is yellow.
24. The marbles are of different colors.

In Exercises 25–28, you are given the probability that an event *will* happen. Find the probability that the event *will not* happen.

25. $p = 0.7$
26. $p = 0.36$
27. $p = \frac{1}{3}$
28. $p = \frac{5}{6}$

In Exercises 29–32, you are given the probability that an event *will not* happen. Find the probability that the event *will* happen.

29. $p = 0.15$
30. $p = 0.84$
31. $p = \frac{13}{20}$
32. $p = \frac{87}{100}$

33. ***Data Analysis*** A study of the effectiveness of a flu vaccine was conducted with a sample of 500 people. Some in the study were given no vaccine, some were given one injection, and others were given two injections. The results of the study are listed in the table.

	No vaccine	One injection	Two injections	Total
Flu	7	2	13	22
No flu	149	52	277	478
Total	156	54	290	500

A person is selected at random from the sample. Find the specified probability.

(a) The person had two injections.

(b) The person did not get the flu.

(c) The person got the flu and had one injection.

34. ***Data Analysis*** One hundred college students were interviewed to determine their political party affiliations and whether they favored a balanced-budget amendment to the Constitution. The results of the study are listed in the table, where D represents Democrat and R represents Republican.

	Favor	Not Favor	Unsure	Total
D	23	25	7	55
R	32	9	4	45
Total	55	34	11	100

A person is selected at random from the sample. Find the probability that the described person is selected.

(a) A person who doesn't favor the amendment

(b) A Republican

(c) A Democrat who favors the amendment

▶ Model It

35. ***Graphical Reasoning*** In 2000, the voting-age population of the United States was approximately 203 million. The figure shows the regions in which these people lived. (Source: U.S. Census Bureau)

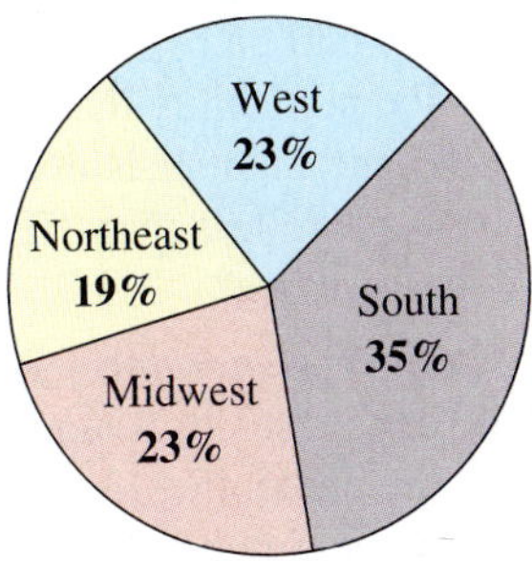

(a) Estimate the voting-age population of the Northeast.

(b) A person is selected at random from the voting-age population. What is the probability that the person lives in the West?

(c) A person is selected at random from the voting-age population. What is the probability that the person lives in the Midwest or the South?

36. ***Graphical Reasoning*** In 2000, there were approximately 131.4 million employees in nonfarm establishments in the United States. The figure shows the types of occupations of these employees. (Source: U.S. Bureau of Labor Statistics)

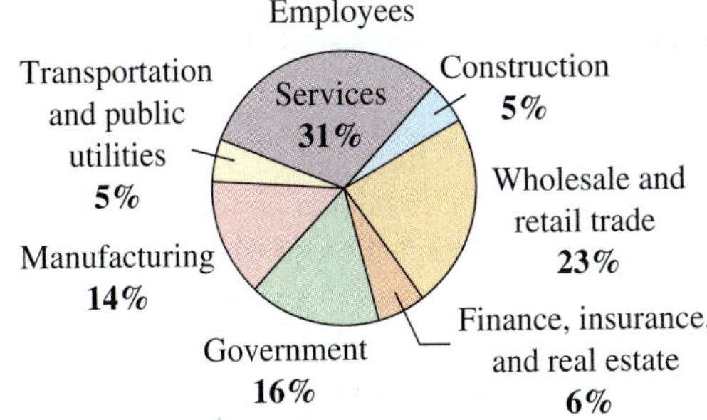

(a) Estimate the number of employees in the manufacturing industry.

(b) What is the probability that a person selected at random from the population of employees in nonfarm establishments works in construction?

(c) What is the probability that a person selected at random from the population of employees in nonfarm establishments works in the service industry or works for the government?

37. ***Alumni Association*** A college sends a survey to selected members of the class of 2002. Of the 1254 people who graduated that year, 672 are women, of whom 124 went on to graduate school. Of the 582 male graduates, 198 went on to graduate school. An alumnus member is selected at random. What is the probability that the person is (a) female, (b) male, and (c) female and did not attend graduate school?

38. ***Education*** In a high school graduating class of 72 students, 28 are on the honor roll. Of these, 18 are going on to college, and of the other 44 students, 12 are going on to college. A student is selected at random from the class. What is the probability that the person chosen is (a) going to college, (b) not going to college, and (c) on the honor roll, but not going to college?

39. ***Winning an Election*** Taylor, Moore, and Jenkins are candidates for public office. It is estimated that Moore and Jenkins have about the same probability of winning, and Taylor is believed to be twice as likely to win as either of the others. Find the probability of each candidate winning the election.

40. ***Winning an Election*** Three people have been nominated for president of a class. From a poll, it is estimated that the first has a 37% chance of winning and the second has a 44% chance of winning. What is the probability that the third candidate will win?

In Exercises 41–52, the sample spaces are large and you should use the counting principles discussed in Section 11.6.

41. ***Preparing for a Test*** A class is given a list of 20 study problems, from which 10 will be part of an upcoming exam. A student knows how to solve 15 of the problems. Find the probability that the student will be able to answer (a) all 10 questions on the exam, (b) exactly eight questions on the exam, and (c) at least nine questions on the exam.

42. ***Preparing for a Test*** A class is given a list of eight study problems, from which five will be part of an upcoming exam. A student knows how to solve six of the problems. Find the probability that the student will be able to answer (a) all five questions on the exam, (b) exactly four questions on the exam, and (c) at least four questions on the exam.

43. ***Letter Mix-Up*** Four letters and envelopes are addressed to four different people. The letters are randomly inserted into the envelopes. What is the probability that (a) exactly one will be inserted in the correct envelope and (b) at least one will be inserted in the correct envelope?

44. ***Payroll Mix-Up*** Five paychecks and envelopes are addressed to five different people. The paychecks are randomly inserted into the envelopes. What is the probability that (a) exactly one will be inserted in the correct envelope and (b) at least one will be inserted in the correct envelope?

45. ***Game Show*** On a game show, you are given five digits to arrange in the proper order to form the price of a car. If you are correct, you win the car. What is the probability of winning, given the following conditions?
    (a) You guess the position of each digit.
    (b) You know the first digit and guess the positions of the others.

46. ***Game Show*** On a game show, you are given five digits in the price of a car. The first digit is 1, and you are given the other four digits to arrange in the correct order to win the car. What is your probability of winning given the following conditions?
    (a) You guess the position of each digit.
    (b) You know the second digit but guess the others.

47. ***Drawing Cards from a Deck*** Two cards are selected at random from an ordinary deck of 52 playing cards. Find the probability that two aces are selected, given the following conditions.
    (a) The cards are drawn in sequence, with the first card being replaced and the deck reshuffled prior to the second drawing.
    (b) The two cards are drawn consecutively, without replacement.

48. ***Poker Hand*** Five cards are drawn from an ordinary deck of 52 playing cards. What is the probability that the hand drawn is a full house? (A full house is a hand that consists of two of one kind and three of another kind.)

49. ***Defective Units*** A shipment of 12 microwave ovens contains three defective units. A vending company has ordered four of these units, and because each is identically packaged, the selection will be random. What is the probability that (a) all four units are good, (b) exactly two units are good, and (c) at least two units are good?

50. ***Defective Units*** A shipment of 20 compact disc players contains four defective units. A retail outlet has ordered five of these units. What is the probability that (a) all five units are good, (b) exactly four units are good, and (c) at least one unit is defective?

51. ***Random Number Generator*** Two integers from 1 through 30 are chosen by a random number generator. What is the probability that (a) the numbers are both even, (b) one number is even and one is odd, (c) both numbers are less than 10, and (d) the same number is chosen twice?

52. ***Random Number Generator*** Two integers from 1 through 40 are chosen by a random number generator. What is the probability that (a) the numbers are both even, (b) one number is even and one is odd, (c) both numbers are less than 30, and (d) the same number is chosen twice?

53. ***Backup System*** A space vehicle has an independent backup system for one of its communication networks. The probability that either system will function satisfactorily during a flight is 0.985. What is the probability that during a given flight (a) both systems function satisfactorily, (b) at least one system functions satisfactorily, and (c) both systems fail?

54. ***Backup Vehicle*** A fire company keeps two rescue vehicles. Because of the demand on the vehicles and the chance of mechanical failure, the probability that a specific vehicle is available when needed is 90%. The availability of one vehicle is *independent* of the availability of the other. Find the probability that (a) both vehicles are available at a given time, (b) neither vehicle is available at a given time, and (c) at least one vehicle is available at a given time.

55. ***Making a Sale*** A sales representative makes a sale on approximately one-fourth of all calls. On a given day, the representative contacts five potential clients. What is the probability that a sale will be made with (a) each of the five contacts, (b) none of the contacts, and (c) at least one contact?

56. ***A Boy or a Girl?*** Assume that the probability of the birth of a child of a particular sex is 50%. In a family with four children, what is the probability that (a) all the children are boys, (b) all the children are the same sex, and (c) there is at least one boy?

57. ***Flexible Work Hours*** In a survey, people were asked if they would prefer to work flexible hours—even if it meant slower career advancement—so they could spend more time with their families. The results of the survey are shown in the figure. Three people from the survey were chosen at random. What is the probability that all three people would prefer flexible work hours?

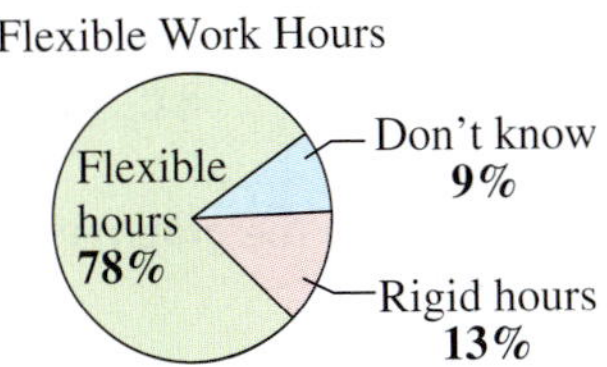

58. ***Consumer Awareness*** Suppose that the methods used by shoppers to pay for merchandise are as shown in the circle graph. Two shoppers are chosen at random. What is the probability that both shoppers paid for their purchases only in cash?

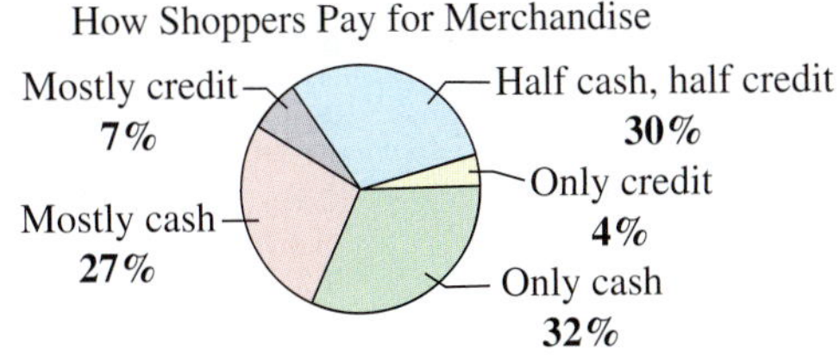

59. ***Geometry*** You and a friend agree to meet at your favorite fast-food restaurant between 5:00 and 6:00 P.M. The one who arrives first will wait 15 minutes for the other, and then will leave (see figure). What is the probability that the two of you will actually meet, assuming that your arrival times are random within the hour?

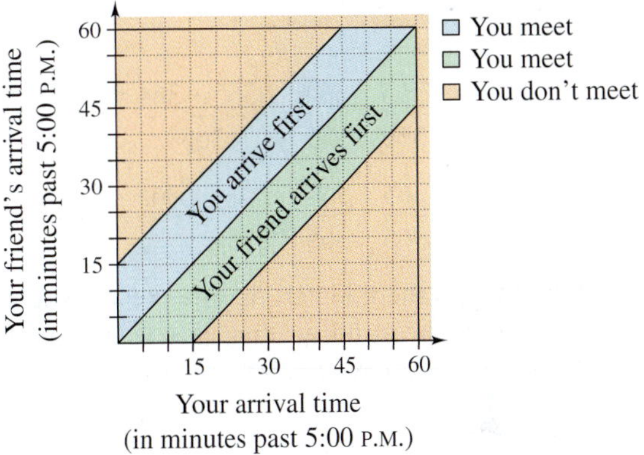

60. ***Estimating π*** A coin of diameter d is dropped onto a paper that contains a grid of squares d units on a side (see figure).

(a) Find the probability that the coin covers a vertex of one of the squares on the grid.

(b) Perform the experiment 100 times and use the results to approximate π.

Synthesis

True or False? **In Exercises 61 and 62, determine whether the statement is true or false. Justify your answer.**

61. If A and B are independent events with nonzero probabilities, then A can occur when B occurs.

62. Rolling a number less than 3 on a normal six-sided die has a probability of $\frac{1}{3}$. The complement of this event is to roll a number greater than 3, and its probability is $\frac{1}{2}$.

63. ***Pattern Recognition and Exploration*** Consider a group of n people.

(a) Explain why the following pattern gives the probabilities that the n people have distinct birthdays.

$$n = 2: \quad \frac{365}{365} \cdot \frac{364}{365} = \frac{365 \cdot 364}{365^2}$$

$$n = 3: \quad \frac{365}{365} \cdot \frac{364}{365} \cdot \frac{363}{365} = \frac{365 \cdot 364 \cdot 363}{365^3}$$

(b) Use the pattern in part (a) to write an expression for the probability that $n = 4$ people have distinct birthdays.

(c) Let P_n be the probability that the n people have distinct birthdays. Verify that this probability can be obtained recursively by

$$P_1 = 1 \text{ and } P_n = \frac{365 - (n - 1)}{365} P_{n-1}.$$

(d) Explain why $Q_n = 1 - P_n$ gives the probability that at least two people in a group of n people have the same birthday.

(e) Use the results of parts (c) and (d) to complete the table.

n	10	15	20	23	30	40	50
P_n							
Q_n							

(f) How many people must be in a group so that the probability of at least two of them having the same birthday is greater than $\frac{1}{2}$? Explain.

64. ***Think About It*** A weather forecast indicates that the probability of rain is 40%. What does this mean?

Review

In Exercises 65–68, find all real solutions of the polynomial equation.

65. $6x^2 + 8 = 0$

66. $4x^2 + 6x - 12 = 0$

67. $x^3 - x^2 - 3x = 0$

68. $x^5 + x^3 - 2x = 0$

In Exercises 69–74, find all real solutions of the rational equation.

69. $\dfrac{12}{x} = -3$

70. $\dfrac{32}{x} = 2x$

71. $\dfrac{2}{x - 5} = 4$

72. $\dfrac{3}{2x + 3} - 4 = \dfrac{-1}{2x + 3}$

73. $\dfrac{3}{x - 2} + \dfrac{x}{x + 2} = 1$

74. $\dfrac{2}{x} - \dfrac{5}{x - 2} = \dfrac{-13}{x^2 - 2x}$

In Exercises 75–80, find all real solutions of the exponential equation.

75. $e^x = 27$

76. $e^x + 7 = 35$

77. $e^{2x} - 4e^x + 3 = 0$

78. $e^{2x} - 7e^x + 12 = 0$

79. $200e^{-x} = 75$

80. $800e^{-x} = 250$

In Exercises 81–84, find all real solutions of the logarithmic equation.

81. $\ln x = 8$

82. $3 - 4 \ln x = 6$

83. $4 \ln 6x = 16$

84. $5 \ln 2x - 4 = 11$

In Exercises 85–88, sketch the graph of the solution of the system of inequalities.

85. $\begin{cases} y \geq -3 \\ x \geq -1 \\ -x - y \geq -8 \end{cases}$

86. $\begin{cases} x \leq 3 \\ y \leq 6 \\ 5x + 2y \geq 10 \end{cases}$

87. $\begin{cases} x^2 + y \geq -2 \\ y \geq x - 4 \end{cases}$

88. $\begin{cases} x^2 + y^2 \leq 4 \\ x + y \geq -2 \end{cases}$

In Exercises 89–92, evaluate ${}_nC_r$.

89. ${}_6C_2$

90. ${}_9C_5$

91. ${}_{11}C_8$

92. ${}_{16}C_3$

Chapter Summary

▶ *What* did you learn?

Section 11.1	Review Exercises
☐ How to use sequence notation to write the terms of a sequence	1–8
☐ How to use factorial notation	9–12
☐ How to use summation notation to write sums	13–18
☐ How to find the sum of an infinite series	19–24
☐ How to use sequences and series to model and solve real-life problems	25, 26
Section 11.2	
☐ How to recognize and write arithmetic sequences	27–40
☐ How to find an *n*th partial sum of an arithmetic sequence	41–46
☐ How to use arithmetic sequences to model and solve real-life problems	47, 48
Section 11.3	
☐ How to recognize and write geometric sequences	49–60
☐ How to find the sum of a geometric sequence	61–70
☐ How to find the sum of an infinite geometric series	71–76
☐ How to use geometric sequences to model and solve real-life problems	77, 78
Section 11.4	
☐ How to use mathematical induction to prove a statement	79–82
☐ How to find the sums of powers of integers	83–86
☐ How to recognize patterns and write the *n*th term of a sequence	87–90
☐ How to find finite differences of a sequence	91–94
Section 11.5	
☐ How to use the Binomial Theorem to calculate binomial coefficients	95–98
☐ How to use Pascal's Triangle to calculate binomial coefficients	99–102
☐ How to use binomial coefficients to write binomial expansions	103–108
Section 11.6	
☐ How to solve simple counting problems	109, 110
☐ How to use the Fundamental Counting Principle to solve counting problems	111, 112
☐ How to use permutations to solve counting problems	113, 114
☐ How to use combinations to solve counting problems	115, 116
Section 11.7	
☐ How to find the probabilities of events	117, 118
☐ How to find the probabilities of mutually exclusive events	119, 120
☐ How to find the probabilities of independent events	121, 122
☐ How to find the probability of the complement of an event	123, 124

Review Exercises

11.1 In Exercises 1–4, write the first five terms of the sequence. (Assume that *n* begins with 1.)

1. $a_n = 2 + \frac{6}{n}$
2. $a_n = \frac{(-1)^n 5n}{2n - 1}$
3. $a_n = \frac{72}{n!}$
4. $a_n = n(n - 1)$

In Exercises 5–8, write an expression for the *apparent* *n*th term of the sequence. (Assume that *n* begins with 1.)

5. $-2, 2, -2, 2, -2, \ldots$
6. $-1, 2, 7, 14, 23, \ldots$
7. $4, 2, \frac{4}{3}, 1, \frac{4}{5}, \ldots$
8. $1, -\frac{1}{2}, \frac{1}{3}, -\frac{1}{4}, \frac{1}{5}, \ldots$

In Exercises 9–12, evaluate the factorial expression.

9. $5!$
10. $3! \cdot 2!$
11. $\frac{3! \cdot 5!}{6!}$
12. $\frac{7! \cdot 6!}{6! \cdot 8!}$

In Exercises 13–18, find the sum.

13. $\sum_{i=1}^{6} 5$
14. $\sum_{k=2}^{5} 4k$
15. $\sum_{j=1}^{4} \frac{6}{j^2}$
16. $\sum_{i=1}^{8} \frac{i}{i + 1}$
17. $\sum_{k=1}^{10} 2k^3$
18. $\sum_{j=0}^{4} (j^2 + 1)$

In Exercises 19 and 20, use sigma notation to write the sum.

19. $\frac{1}{2(1)} + \frac{1}{2(2)} + \frac{1}{2(3)} + \cdots + \frac{1}{2(20)}$
20. $\frac{1}{2} + \frac{2}{3} + \frac{3}{4} + \cdots + \frac{9}{10}$

In Exercises 21–24, find the sum of the infinite series.

21. $\sum_{i=1}^{\infty} \frac{5}{10^i}$
22. $\sum_{i=1}^{\infty} \frac{3}{10^i}$
23. $\sum_{k=1}^{\infty} \frac{2}{100^k}$
24. $\sum_{k=2}^{\infty} \frac{9}{10^k}$

25. ***Job Offer*** The starting salary for an accountant is \$34,000 with a guaranteed salary increase of \$2250 per year. Determine (a) the salary during the fifth year and (b) the total compensation through 5 full years of employment.

26. ***Baling Hay*** In the first two trips baling hay around a large field, a farmer obtains 123 bales and 112 bales, respectively. Because each round gets shorter, the farmer estimates that the same pattern will continue. Estimate the total number of bales made if there are another six trips around the field.

11.2 In Exercises 27–30, determine whether the sequence is arithmetic. If it is, find the common difference.

27. $5, 3, 1, -1, -3, \ldots$
28. $0, 1, 3, 6, 10, \ldots$
29. $\frac{1}{2}, 1, \frac{3}{2}, 2, \frac{5}{2}, \ldots$
30. $\frac{9}{9}, \frac{8}{9}, \frac{7}{9}, \frac{6}{9}, \frac{5}{9}, \ldots$

In Exercises 31–34, write the first five terms of the arithmetic sequence.

31. $a_1 = 4, d = 3$
32. $a_1 = 6, d = -2$
33. $a_1 = 25, a_{k+1} = a_k + 3$
34. $a_1 = 4.2, a_{k+1} = a_k + 0.4$

In Exercises 35–40, find a formula for a_n for the arithmetic sequence.

35. $a_1 = 7, d = 12$
36. $a_1 = 25, d = -3$
37. $a_1 = y, d = 3y$
38. $a_1 = -2x, d = x$
39. $a_2 = 93, a_6 = 65$
40. $a_7 = 8, a_{13} = 6$

In Exercises 41–44, find the partial sum.

41. $\sum_{j=1}^{10} (2j - 3)$
42. $\sum_{j=1}^{8} (20 - 3j)$
43. $\sum_{k=1}^{11} \left(\frac{2}{3}k + 4\right)$
44. $\sum_{k=1}^{25} \left(\frac{3k + 1}{4}\right)$

45. Find the sum of the first 100 positive multiples of 5.

46. Find the sum of the integers from 20 to 80 (inclusive).

47. ***Running*** The first time you run a five-mile distance course, it takes you 49 minutes. You run the same course 30 seconds faster each week. How fast can you run the five-mile course after 12 weeks?

48. ***Running*** On the first day of a new training schedule, you run 2 miles. You increase your distance by one-half mile every day. How many total miles will you run in 14 days?

11.3 **In Exercises 49–52, determine whether the sequence is geometric. If it is, find the common ratio.**

49. $5, 10, 20, 40, . . .$

50. $54, -18, 6, -2, . . .$

51. $\frac{1}{3}, -\frac{2}{3}, \frac{4}{3}, -\frac{8}{3}, . . .$

52. $\frac{1}{4}, \frac{2}{5}, \frac{3}{6}, \frac{4}{7}, . . .$

In Exercises 53–56, write the first five terms of the geometric sequence.

53. $a_1 = 4,\ r = -\frac{1}{4}$

54. $a_1 = 2,\ r = 2$

55. $a_1 = 9,\ a_3 = 4$

56. $a_1 = 2,\ a_3 = 12$

In Exercises 57–60, write an expression for the *n*th term of the geometric sequence and find the sum of the first 20 terms of the sequence.

57. $a_1 = 16, a_2 = -8$

58. $a_3 = 6, a_4 = 1$

59. $a_1 = 100, r = 1.05$

60. $a_1 = 5, r = 0.2$

In Exercises 61–66, find the sum.

61. $\sum_{i=1}^{7} 2^{i-1}$

62. $\sum_{i=1}^{5} 3^{i-1}$

63. $\sum_{i=1}^{4} \left(\frac{1}{2}\right)^i$

64. $\sum_{i=1}^{6} \left(\frac{1}{3}\right)^{i-1}$

65. $\sum_{i=1}^{5} (2)^{i-1}$

66. $\sum_{i=1}^{4} 6(3)^i$

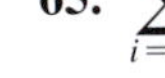 **In Exercises 67–70, use a graphing utility to find the sum.**

67. $\sum_{i=1}^{10} 10\left(\frac{3}{5}\right)^{i-1}$

68. $\sum_{i=1}^{15} 20(0.2)^{i-1}$

69. $\sum_{i=1}^{25} 100(1.06)^{i-1}$

70. $\sum_{i=1}^{20} 8\left(\frac{6}{5}\right)^{i-1}$

In Exercises 71–76, find the sum of the infinite geometric series.

71. $\sum_{i=1}^{\infty} \left(\frac{7}{8}\right)^{i-1}$

72. $\sum_{i=1}^{\infty} \left(\frac{1}{3}\right)^{i-1}$

73. $\sum_{i=1}^{\infty} (0.1)^{i-1}$

74. $\sum_{i=1}^{\infty} (0.5)^{i-1}$

75. $\sum_{k=1}^{\infty} 4\left(\frac{2}{3}\right)^{k-1}$

76. $\sum_{k=1}^{\infty} 1.3\left(\frac{1}{10}\right)^{k-1}$

77. ***Depreciation*** A paper manufacturer buys a machine for \$120,000. During the next 5 years, it will depreciate at a rate of 30% per year. (That is, at the end of each year the depreciated value will be 70% of what it was at the beginning of the year.)

(a) Find the formula for the *n*th term of a geometric sequence that gives the value of the machine *t* full years after it was purchased.

(b) Find the depreciated value of the machine at the end of 5 full years.

78. ***Total Compensation*** A computer programming position pays a salary of \$32,000 the first year. During the next 39 years, there is a 5.5% raise each year. What would be the total amount earned over the 40-year period?

11.4 **In Exercises 79–82, use mathematical induction to prove the formula for every positive integer *n*.**

79. $3 + 5 + 7 + \cdot \cdot \cdot + (2n + 1) = n(n + 2)$

80. $1 + \frac{3}{2} + 2 + \frac{5}{2} + \cdot \cdot \cdot + \frac{1}{2}(n + 1) = \frac{n}{4}(n + 3)$

81. $\sum_{i=0}^{n-1} ar^i = \frac{a(1 - r^n)}{1 - r}$

82. $\sum_{k=0}^{n-1} (a + kd) = \frac{n}{2}[2a + (n - 1)d]$

In Exercises 83–86, find the sum using the formulas for the sums of powers of integers.

83. $\sum_{n=1}^{30} n$

84. $\sum_{n=1}^{10} n^2$

85. $\sum_{n=1}^{7} (n^4 - n)$

86. $\sum_{n=1}^{6} (n^5 - n^2)$

In Exercises 87–90, find a formula for the sum of the first *n* terms of the sequence.

87. $9, 13, 17, 21, . . .$

88. $68, 60, 52, 44, . . .$

89. $1, \frac{3}{5}, \frac{9}{25}, \frac{27}{125}, . . .$

90. $12, -1, \frac{1}{12}, -\frac{1}{144}, . . .$

In Exercises 91–94, find the first five terms of the sequence. Then calculate the first and second differences of the sequence. Does the sequence have a linear model, a quadratic model, or neither?

91. $a_1 = 5$
$a_n = a_{n-1} + 5$

92. $a_1 = -3$
$a_n = a_{n-1} - 2n$

93. $a_1 = 16$
$a_n = a_{n-1} - 1$

94. $a_0 = 0$
$a_n = n - a_{n-1}$?

11.5 **In Exercises 95–98, use the Binomial Theorem to calculate the binomial coefficient.**

95. ${}_6C_4$

96. ${}_{10}C_7$

97. ${}_8C_5$

98. ${}_{12}C_3$

In Exercises 99–102, use Pascal's Triangle to calculate the binomial coefficient.

99. $\binom{7}{3}$

100. $\binom{9}{4}$

101. $\binom{8}{6}$

102. $\binom{5}{3}$

In Exercises 103–108, use the Binomial Theorem to expand the binomial. Simplify your answer. (Remember that $i = \sqrt{-1}$.)

103. $\left(\frac{x}{2} + y\right)^4$

104. $\left(\frac{2}{x} - 3x\right)^6$

105. $(a - 3b)^5$

106. $(3x + y^2)^7$

107. $(5 + 2i)^4$

108. $(4 - 5i)^3$

11.6 **109.** ***Dice*** In how many different ways can a pair of dice be rolled to obtain a total of 10?

110. ***Numbers in a Hat*** Slips of paper numbered 1 through 14 are placed in a hat. In how many ways can you draw two numbers with replacement that total 12?

111. ***Telephone Numbers*** The same three-digit prefix is used for all of the telephone numbers in a small town. How many different telephone numbers are possible by changing only the last four digits?

112. ***Telephone Numbers*** A telephone number in another town can use any one of five different three-digit prefixes. How many different telephone numbers are possible in this town?

113. ***Bike Race*** There are 10 bicyclists entered in a race. In how many different orders could these 10 bicyclists finish?

114. ***Bike Race*** There are 10 bicyclists entered in a race. In how many different ways could the top three places be decided?

115. ***Apparel*** You have eight different suits to choose from to take on a trip. How many combinations of three suits could you take?

116. ***Apparel*** You have 20 different neckties in your wardrobe. How many combinations of three ties could you choose?

11.7 **117.** ***Apparel*** A man has five pairs of socks, of which no two pairs are the same color. He randomly selects two socks from a drawer. What is the probability that he gets a matched pair?

118. ***Bookshelf Order*** A child returns a five-volume set of books to a bookshelf. The child is not able to read, and so cannot distinguish one volume from another. What is the probability that the books are shelved in the correct order?

119. ***Students by Class*** At a particular high school, the numbers of students in the four classes are broken down by percents, as shown in the table.

Class	Percent
Freshman	31
Sophomores	26
Juniors	25
Seniors	18

A single student is picked randomly by lottery for a cash scholarship. What is the probability that the scholarship winner is

(a) a junior or senior?

(b) a freshman, sophomore, or junior?

120. ***Data Analysis*** A sample of college students, faculty, and administration were asked whether they favored a proposed increase in the annual activity fee to enhance student life on campus. The results of the study are listed in the table.

	Students	Faculty	Admin.	Total
Favor	237	37	18	292
Oppose	163	38	7	208
Total	400	75	25	500

A person is selected at random from the sample. Find the specified probability.

(a) The person is not in favor of the proposal.

(b) The person is a student.

(c) The person is a faculty member and is in favor of the proposal.

121. ***Tossing a Die*** A six-sided die is rolled three times. What is the probability of a 6 on each roll?

122. ***Tossing a Die*** A six-sided die is rolled six times. What is the probability that each side appears exactly once?

123. ***Drawing a Card*** You randomly select a card from a 52-card deck. What is the probability that the card is *not* a club?

124. ***Tossing a Coin*** Find the probability of obtaining at least one tail when a coin is tossed five times.

Synthesis

True or False? **In Exercises 125–128, determine whether the statement is true or false. Justify your answer.**

125. $\dfrac{(n+2)!}{n!} = (n+2)(n+1)$

126. $\displaystyle\sum_{i=1}^{5} (i^3 + 2i) = \sum_{i=1}^{5} i^3 + \sum_{i=1}^{5} 2i$

127. $\displaystyle\sum_{k=1}^{8} 3k = 3\sum_{k=1}^{8} k$

128. $\displaystyle\sum_{j=1}^{6} 2^j = \sum_{j=3}^{8} 2^{j-2}$

129. ***Think About It*** An infinite sequence is a function. What is the domain of the function?

130. ***Think About It*** How do the two sequences differ?

(a) $a_n = \dfrac{(-1)^n}{n}$

(b) $a_n = \dfrac{(-1)^{n+1}}{n}$

131. ***Writing*** In your own words, explain what makes a sequence (a) arithmetic and (b) geometric.

132. ***Graphical Reasoning*** The graphs of two sequences are shown below. Identify each sequence as arithmetic or geometric. Explain your reasoning.

(a)

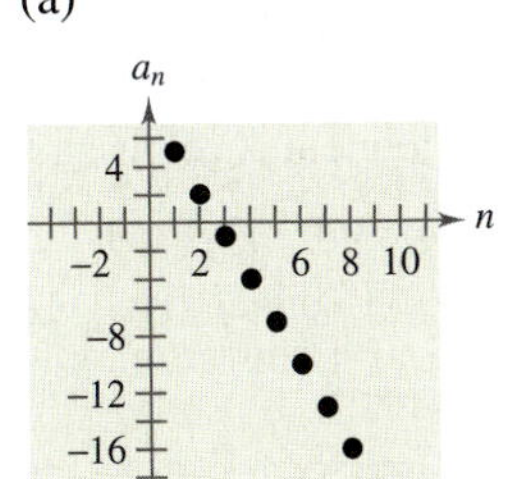

(b)

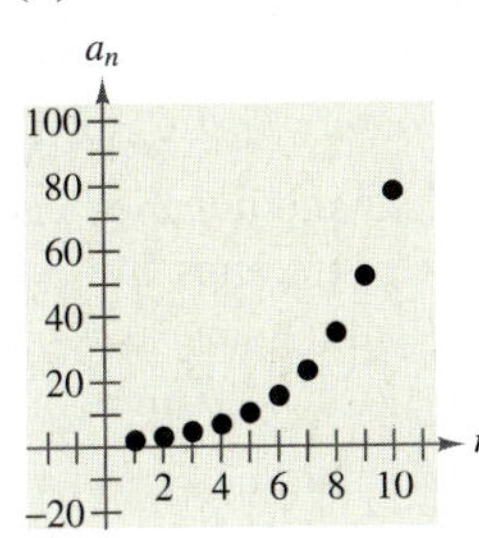

133. ***Writing*** Explain what a recursion formula is.

134. ***Writing*** Explain why the terms of a geometric sequence decrease when $0 < r < 1$.

Graphical Reasoning **In Exercises 135–138, match the sequence or sum of a sequence with its graph without doing any calculations. Explain your reasoning. [The graphs are labeled (a), (b), (c), and (d).]**

(a)

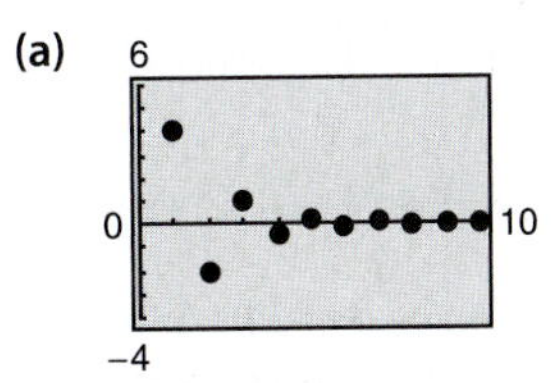

(b)

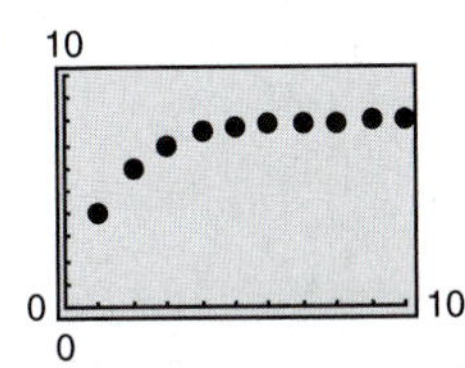

(c)

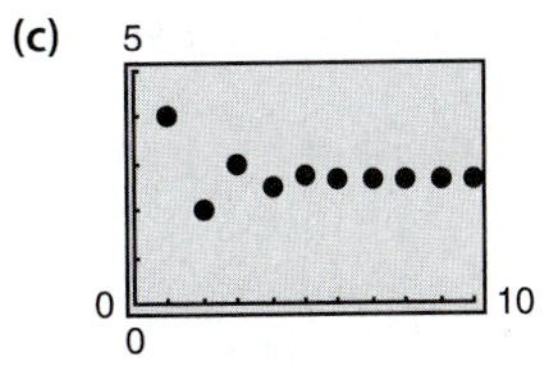

(d)

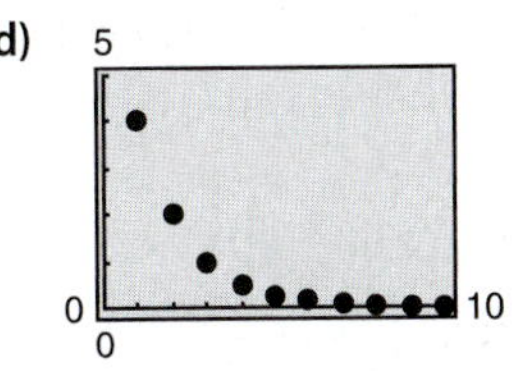

135. $a_n = 4\left(\frac{1}{2}\right)^{n-1}$

136. $a_n = 4\left(-\frac{1}{2}\right)^{n-1}$

137. $a_n = \displaystyle\sum_{k=1}^{n} 4\left(\frac{1}{2}\right)^{k-1}$

138. $a_n = \displaystyle\sum_{k=1}^{n} 4\left(-\frac{1}{2}\right)^{k-1}$

139. Consider an idealized population with the characteristic that each member of the population produces one offspring at the end of every time period. If each member has a life span of three time periods and the population begins with 10 newborn members, then the following table shows the population during the first five time periods.

	Time Period				
Age Bracket	1	2	3	4	5
0–1	10	10	20	40	70
1–2		10	10	20	40
2–3			10	10	20
Total	10	20	40	70	130

The sequence for the total population has the property that

$$S_n = S_{n-1} + S_{n-2} + S_{n-3}, \quad n > 3.$$

Find the total population during the next five time periods.

140. The probability of an event must be a real number in what interval? Is the interval open or closed?

Chapter Test

The *Interactive* CD-ROM and *Internet* versions of this text offer Chapter Pre-Tests and Chapter Post-Tests, both of which have randomly generated exercises with diagnostic capabilities.

Take this test as you would take a test in class. When you are finished, check your work against the answers given in the back of the book.

1. Write the first five terms of the sequence $a_n = \dfrac{(-1)^n}{3n + 2}$.

2. Write an expression for the nth term of the sequence.

$\frac{3}{1!}, \frac{4}{2!}, \frac{5}{3!}, \frac{6}{4!}, \frac{7}{5!}, \ldots$

3. Find the next three terms of the series. Then find the fifth partial sum of the series.

$6 + 17 + 28 + 39 + \cdots$

4. The fifth term of an arithmetic series is 5.4, and the 12th term is 11.0. Find the nth term.

5. Write the first five terms of the sequence $a_n = 5(2)^{n-1}$. (Assume that n begins with 1.)

6. Find the sum of the finite series $\displaystyle\sum_{i=1}^{50} (2i^2 + 5)$.

7. Find the sum of the infinite series $\displaystyle\sum_{i=1}^{\infty} 4\left(\tfrac{1}{2}\right)^i$.

8. Use mathematical induction to prove the formula.

$$5 + 10 + 15 + \cdots + 5n = \frac{5n(n + 1)}{2}$$

9. Use the Binomial Theorem to expand the expression $(x + 2y)^4$.

10. To dress for a party, you can choose from six different pairs of slacks, 10 different shirts, and three different pairs of shoes. How many possible outfit combinations do you have?

In Exercises 11 and 12, evaluate each expression.

11. (a) ${}_9P_2$ (b) ${}_{70}P_3$ **12.** (a) ${}_{11}C_4$ (b) ${}_{66}C_4$

13. Eight people are going for a ride in a boat that seats eight people. The owner of the boat will drive, and only three of the remaining people are willing to ride in the two bow seats. How many seating arrangements are possible?

14. You attend a karaoke night and hope to hear your favorite song. The karaoke song book has 300 different songs. Assuming that the singers are equally likely to pick any song and no song is repeated, what is the probability that your favorite song is one of the 20 that you hear?

15. You are with seven of your friends at a party. Names of all of the 60 guests are placed in a hat and drawn randomly to award eight door prizes. Each guest is limited to one prize. What is the probability that you and your friends win all eight of the prizes?

16. The weather report calls for a 75% chance of rain. According to this report, what is the probability that it will not rain?

Cumulative Test for Chapters 9–11

Take this test to review the material from earlier chapters. When you are finished, check your work against the answers given in the back of the book.

In Exercises 1–4, solve the system by the specified method.

1. Substitution

$$\begin{cases} y = 3 - x^2 \\ 2(y - 2) = x - 1 \end{cases}$$

2. Elimination

$$\begin{cases} x + 3y = -1 \\ 2x + 4y = 0 \end{cases}$$

3. Elimination

$$\begin{cases} -2x + 4y - z = 3 \\ x - 2y + 2z = -6 \\ x - 3y - z = 1 \end{cases}$$

4. Gauss-Jordan Elimination

$$\begin{cases} x + 3y - 2z = -7 \\ -2x + y - z = -5 \\ 4x + y + z = 3 \end{cases}$$

In Exercises 5 and 6, sketch the graph of the solution set of the system of inequalities.

5. $\begin{cases} 2x + y \geq -3 \\ x - 3y \leq 2 \end{cases}$

6. $\begin{cases} x - y > 6 \\ 5x + 2y < 10 \end{cases}$

7. Sketch a graph of the solution of the constraints and maximize the objective function $z = 3x + 2y$ subject to the constraints.

$$\begin{aligned} x + 4y &\leq 20 \\ 2x + y &\leq 12 \\ x &\geq 0 \\ y &\geq 0 \end{aligned}$$

8. A custom-blend bird seed is to be mixed from seed mixtures costing \$0.75 per pound and \$1.25 per pound. How many pounds of each seed mixture are used to make 200 pounds of custom-blend bird seed costing \$0.95 per pound?

9. Find the equation of the parabola $y = ax^2 + bx + c$ passing through the points $(0, 4)$, $(3, 1)$, and $(6, 4)$.

$$\begin{cases} -x + 2y - z = 9 \\ 2x - y + 2z = -9 \\ 3x + 3y - 4z = 7 \end{cases}$$

SYSTEM FOR 10 AND 11

In Exercises 10 and 11, use the system of equations at the left.

10. Write the augmented matrix corresponding to the system of equations.

11. Solve the system using the matrix and Gauss-Jordan elimination.

In Exercises 12–15, use the following matrices.

$$A = \begin{bmatrix} 4 & 0 \\ -1 & 2 \end{bmatrix}, \quad B = \begin{bmatrix} -1 & 3 \\ 1 & 0 \end{bmatrix}$$

12. Find $A - B$.

13. Find $-2B$.

14. Find $A - 2B$.

15. Find AB, if possible.

$$\begin{bmatrix} 8 & 0 & -5 \\ 1 & 3 & -1 \\ -2 & 6 & 4 \end{bmatrix}$$

MATRIX FOR 16

16. Find the determinant of the matrix at the left.

17. Find the inverse (if it exists): $\begin{bmatrix} 1 & 2 & -1 \\ 3 & 7 & -10 \\ -5 & -7 & -15 \end{bmatrix}$.

		Gym shoes	Jogging shoes	Walking shoes
Age group	14–17	0.13	0.14	0.03
	18–24	0.06	0.11	0.04
	25–34	0.10	0.17	0.09

MATRIX FOR 18

18. The percents (by age group) of the total amounts spent on three types of footwear in 1999 are shown in the matrix. The total amounts (in millions) spent by each age group on the three types of footwear were \$554.93 (14–17 age group), \$405.34 (18–24 age group), and \$727.85 (25–34 age group). How many dollars worth of gym shoes, jogging shoes, and walking shoes were sold in 1999? (Source: National Sporting Goods Association)

In Exercises 19 and 20, use Cramer's Rule to solve the system of equations.

19. $\begin{cases} 8x - 3y = -52 \\ 3x + 5y = 5 \end{cases}$

20. $\begin{cases} 5x + 4y + 3z = 7 \\ -3x - 8y + 7z = -9 \\ 7x - 5y - 6z = -53 \end{cases}$

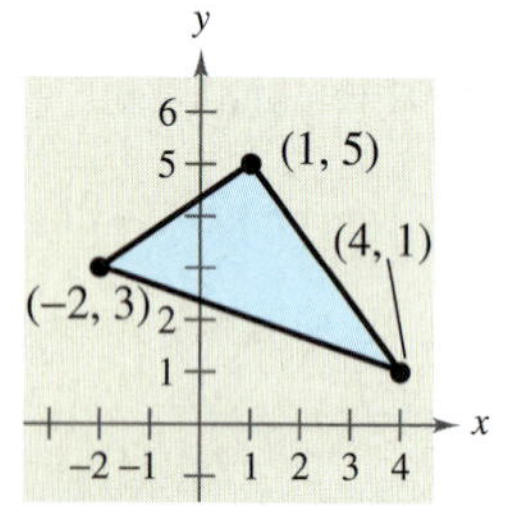

FIGURE FOR 21

21. Find the area of the triangle in the figure.

22. Write the first five terms of the sequence $a_n = \frac{(-1)^{n+1}}{2n + 3}$ (assume that n begins with 1).

23. Write an expression for the nth term of the sequence.

$\frac{2!}{4}, \frac{3!}{5}, \frac{4!}{6}, \frac{5!}{7}, \frac{6!}{8}, \ldots$

24. Sum the first 20 terms of the arithmetic sequence 8, 12, 16, 20, ...

25. The sixth term of an arithmetic sequence is 20.6, and the ninth term is 30.2.
    (a) Find the 20th term.
    (b) Find the nth term

26. Write the first five terms of the sequence $a_n = 3(2)^{n-1}$ (assume that n begins with 1).

27. Find the sum: $\sum_{i=0}^{\infty} 1.3\left(\frac{1}{10}\right)^{i-1}$.

28. Use mathematical induction to prove the formula

$$3 + 7 + 11 + 15 + \cdots + (4n - 1) = n(2n + 1).$$

29. Use the Binomial Theorem to expand and simplify $(z - 3)^4$.

In Exercises 30–33, evaluate the expression.

30. ${}_7P_3$ 31. ${}_{25}P_2$ 32. $\binom{8}{4}$ 33. ${}_{10}C_3$

34. A personnel manager at a department store has 10 applicants to fill three different sales positions. In how many ways can this be done, assuming that all the applicants are qualified for any of the three positions?

35. On a game show, the digits 3, 4, and 5 must be arranged in the proper order to form the price of an appliance. If the digits are arranged correctly, the contestant wins the appliance. What is the probability of winning if the contestant knows that the price is at least \$400?

Proofs in Mathematics

Properties of Sums (p. 807)

1. $\sum_{i=1}^{n} c = nc$, c is a constant.
2. $\sum_{i=1}^{n} ca_i = c\sum_{i=1}^{n} a_i$, c is a constant.
3. $\sum_{i=1}^{n} (a_i + b_i) = \sum_{i=1}^{n} a_i + \sum_{i=1}^{n} b_i$
4. $\sum_{i=1}^{n} (a_i - b_i) = \sum_{i=1}^{n} a_i - \sum_{i=1}^{n} b_i$

Proof

Each of these properties follows directly from the properties of real numbers.

1. $\sum_{i=1}^{n} c = c + c + c + \cdot \cdot \cdot + c = nc$ $\quad$ n terms

The Distributive Property is used in the proof of Property 2.

2.
$$\sum_{i=1}^{n} ca_i = ca_1 + ca_2 + ca_3 + \cdot \cdot \cdot + ca_n$$
$$= c(a_1 + a_2 + a_3 + \cdot \cdot \cdot + a_n) \quad = c\sum_{i=1}^{n} a_i$$

The proof of Property 3 uses the Commutative and Associative Properties of Addition.

3.
$$\sum_{i=1}^{n} (a_i + b_i) = (a_1 + b_1) + (a_2 + b_2) + (a_3 + b_3) + \cdot \cdot \cdot + (a_n + b_n)$$
$$= (a_1 + a_2 + a_3 + \cdot \cdot \cdot + a_n) + (b_1 + b_2 + b_3 + \cdot \cdot \cdot + b_n)$$
$$= \sum_{i=1}^{n} a_i + \sum_{i=1}^{n} b_i$$

The proof of Property 4 uses the Commutative and Associative Properties of Addition and the Distributive Property.

4.
$$\sum_{i=1}^{n} (a_i - b_i) = (a_1 - b_1) + (a_2 - b_2) + (a_3 - b_3) + \cdot \cdot \cdot + (a_n - b_n)$$
$$= (a_1 + a_2 + a_3 + \cdot \cdot \cdot + a_n) + (-b_1 - b_2 - b_3 - \cdot \cdot \cdot - b_n)$$
$$= (a_1 + a_2 + a_3 + \cdot \cdot \cdot + a_n) - (b_1 + b_2 + b_3 + \cdot \cdot \cdot + b_n)$$
$$= \sum_{i=1}^{n} a_i - \sum_{i=1}^{n} b_i$$

Infinite Series

The study of infinite series was considered a novelty in the fourteenth century. Logician Richard Suiseth, whose nickname was Calculator, solved this problem.

If throughout the first half of a given time interval a variation continues at a certain intensity; throughout the next quarter of the interval at double the intensity; throughout the following eighth at triple the intensity and so ad infinitum; The average intensity for the whole interval will be the intensity of the variation during the second subinterval (or double the intensity).

This is the same as saying that the sum of the infinite series

$$\frac{1}{2} + \frac{2}{4} + \frac{3}{8} + \cdot \cdot \cdot + \frac{n}{2^n} + \cdot \cdot \cdot$$

is 2.

The Sum of a Finite Arithmetic Sequence (p. 816)

The sum of a finite arithmetic sequence with n terms is

$$S_n = \frac{n}{2}(a_1 + a_n).$$

Proof

Begin by generating the terms of the arithmetic sequence in two ways. In the first way, repeatedly add d to the first term to obtain

$$\begin{aligned} S_n &= a_1 + a_2 + a_3 + \cdot \cdot \cdot + a_{n-2} + a_{n-1} + a_n \\ &= a_1 + [a_1 + d] + [a_1 + 2d] + \cdot \cdot \cdot + [a_1 + (n-1)d]. \end{aligned}$$

In the second way, repeatedly subtract d from the nth term to obtain

$$\begin{aligned} S_n &= a_n + a_{n-1} + a_{n-2} + \cdot \cdot \cdot + a_3 + a_2 + a_1 \\ &= a_n + [a_n - d] + [a_n - 2d] + \cdot \cdot \cdot + [a_n - (n-1)d]. \end{aligned}$$

If you add these two versions of S_n, the multiples of d subtract out and you obtain

$$2S_n = (a_1 + a_n) + (a_1 + a_n) + (a_1 + a_n) + \cdot \cdot \cdot + (a_1 + a_n) \quad n \text{ terms}$$

$$2S_n = n(a_1 + a_n)$$

$$S_n = \frac{n}{2}(a_1 + a_n).$$

The Sum of a Finite Geometric Sequence (p. 825)

The sum of the finite geometric sequence

$$a_1,\ a_1r,\ a_1r^2,\ a_1r^3,\ a_1r^4,\ . . . ,\ a_1r^{n-1}$$

with common ratio $r \neq 1$ is given by $S_n = \sum_{i=1}^{n} a_1 r^{i-1} = a_1\left(\frac{1 - r^n}{1 - r}\right)$.

Proof

$$S_n = a_1 + a_1r + a_1r^2 + \cdot \cdot \cdot + a_1r^{n-2} + a_1r^{n-1}$$

$$rS_n = a_1r + a_1r^2 + a_1r^3 + \cdot \cdot \cdot + a_1r^{n-1} + a_1r^n \quad \text{Multiply by } r.$$

Subtracting the second equation from the first yields

$$S_n - rS_n = a_1 - a_1r^n.$$

So, $S_n(1 - r) = a_1(1 - r^n)$, and, because $r \neq 1$, you have $S_n = a_1\left(\frac{1 - r^n}{1 - r}\right)$.

The Binomial Theorem (p. 842)

In the expansion of $(x + y)^n$

$$(x + y)^n = x^n + nx^{n-1}y + \cdot \cdot \cdot + {}_nC_r\, x^{n-r}y^r + \cdot \cdot \cdot + nxy^{n-1} + y^n$$

the coefficient of $x^{n-r}y^r$ is

$${}_nC_r = \frac{n!}{(n-r)!r!}.$$

Proof

The Binomial Theorem can be proved quite nicely using mathematical induction. The steps are straightforward but look a little messy, so only an outline of the proof is presented.

1. If $n = 1$, you have $(x + y)^1 = x^1 + y^1 = {}_1C_0x + {}_1C_1y$, and the formula is valid.
2. Assuming that the formula is true for $n = k$, the coefficient of $x^{k-r}y^r$ is

$${}_kC_r = \frac{k!}{(k-r)!r!} = \frac{k(k-1)(k-2) \cdot \cdot \cdot (k-r+1)}{r!}.$$

To show that the formula is true for $n = k + 1$, look at the coefficient of $x^{k+1-r}y^r$ in the expansion of

$$(x + y)^{k+1} = (x + y)^k(x + y).$$

From the right-hand side, you can determine that the term involving $x^{k+1-r}y^r$ is the sum of two products.

$$\begin{aligned}
&({}_kC_r x^{k-r}y^r)(x) + ({}_kC_{r-1}x^{k+1-r}y^{r-1})(y) \\
&= \left[\frac{k!}{(k-r)!r!} + \frac{k!}{(k+1-r)!(r-1)!}\right]x^{k+1-r}y^r \\
&= \left[\frac{(k+1-r)k!}{(k+1-r)!r!} + \frac{k!r}{(k+1-r)!r!}\right]x^{k+1-r}y^r \\
&= \left[\frac{k!(k+1-r+r)}{(k+1-r)!r!}\right]x^{k+1-r}y^r \\
&= \left[\frac{(k+1)!}{(k+1-r)!r!}\right]x^{k+1-r}y^r \\
&= {}_{k+1}C_r x^{k+1-r}y^r
\end{aligned}$$

So, by mathematical induction, the Binomial Theorem is valid for all positive integers n.

P.S. Problem Solving

 **1.** Let $x_0 = 1$ and consider the sequence x_n given by

$$x_n = \frac{1}{2}x_{n-1} + \frac{1}{x_{n-1}}, \quad n = 1, 2, . . .$$

Use a graphing utility to compute the first 10 terms of the sequence and make a conjecture about the value of x_n as n approaches infinity.

2. Consider the sequence

$$a_n = \frac{n+1}{n^2+1}.$$

 (a) Use a graphing utility to graph the first 10 terms of the sequence.

(b) Use the graph from part (a) to estimate the value of a_n as n approaches infinity.

(c) Complete the table.

n	1	10	100	1000	10,000
a_n					

(d) Use the table from part (c) to determine (if possible) the value of a_n as n approaches infinity.

3. Consider the sequence

$$a_n = 3 + (-1)^n.$$

 (a) Use a graphing utility to graph the first 10 terms of the sequence.

(b) Use the graph from part (a) to describe the behavior of the graph of the sequence.

(c) Complete the table.

n	1	10	101	1000	10,001
a_n					

(d) Use the table from part (c) to determine (if possible) the value of a_n as n approaches infinity.

4. The following operations are performed on each term of an arithmetic sequence. Determine if the resulting sequence is arithmetic, and if so, state the common difference.

(a) A constant C is added to each term.

(b) Each term is multiplied by a nonzero constant C.

(c) Each term is squared.

5. The following sequence of perfect squares is not arithmetic.

1, 4, 9, 16, 25, 36, 49, 64, 81, . . .

However, you can form a related sequence that is arithmetic by finding the differences of consecutive terms.

(a) Write the first eight terms of the related arithmetic sequence described above. What is the nth term of this sequence?

(b) Describe how you can find an arithmetic sequence that is related to the following sequence of perfect cubes.

1, 8, 27, 64, 125, 216, 343, 512, 729, . . .

(c) Write the first seven terms of the related sequence in part (b) and find the nth term of the sequence.

(d) Describe how you can find the arithmetic sequence that is related to the following sequence of perfect fourth powers.

1, 16, 81, 256, 625, 1296, 2401, 4096, 6561, . . .

(e) Write the first six terms of the related sequence in part (d) and find the nth term of the sequence.

6. Can the Greek hero Achilles, running at 20 feet per second, ever catch a tortoise, starting 20 feet ahead of Achilles and running at 10 feet per second? The Greek mathematician Zeno said no. When Achilles runs 20 feet, the tortoise will be 10 feet ahead. Then, when Achilles runs 10 feet, the tortoise will be 5 feet ahead. Achilles will keep cutting the distance in half but will never catch the tortoise. The table shows Zeno's reasoning. From the table you can see that both the distances and the times required to achieve them form infinite geometric series. Using the table, show that both series have finite sums. What do these sums represent?

Distance (in feet)	Time (in seconds)
20	1
10	0.5
5	0.25
2.5	0.125
1.25	0.0625
0.625	0.03125

7. Recall that a *fractal* is a geometric figure that consists of a pattern that is repeated infinitely on a smaller and smaller scale. A well-known fractal is called the *Sierpinski Triangle*. In the first stage, the midpoints of the three sides are used to create the vertices of a new triangle, which is then removed, leaving three triangles. The first three stages are shown below. Note that each remaining triangle is similar to the original triangle. Assume that the length of each side of the original triangle is one unit. Write a formula that describes the side length of the triangles that will be generated in the nth stage. Write a formula for the area of the triangles that will be generated in the nth stage.

8. You can define a sequence using a piecewise formula. The following is an example of a piecewise-defined sequence.

$$a_1 = 7,\ a_n = \begin{cases} \dfrac{a_{n-1}}{2}, & \text{if } a_{n-1} \text{ is even} \\ 3a_{n-1} + 1, & \text{if } a_{n-1} \text{ is odd} \end{cases}$$

   (a) Write the first 10 terms of the sequence.

   (b) Choose three different values for a_1 (other than $a_1 = 7$). For each value of a_1, find the first 10 terms of the sequence. What conclusions can you make about the behavior of this sequence?

9. The numbers 1, 5, 12, 22, 35, 51, . . . are called pentagonal numbers because they represent the numbers of dots used to make pentagons, as shown below. Use mathematical induction to prove that the nth pentagonal number P_n is given by

$$P_n = \frac{n(3n - 1)}{2}.$$

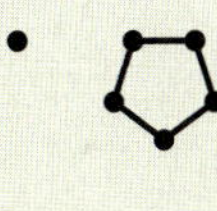
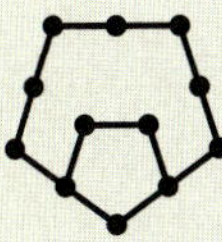
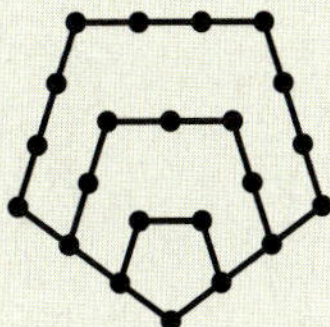

10. What conclusion can be drawn from the information about the sequence of statements P_n?

   (a) P_3 is true and P_k implies P_{k+1}.

   (b) $P_1, P_2, P_3, \ldots, P_{50}$ are all true.

   (c) P_1, P_2, and P_3 are all true, but the truth of P_k does not imply that P_{k+1} is true.

   (d) P_2 is true and P_{2k} implies P_{2k+2}.

11. Let $f_1, f_2, \ldots, f_n, \ldots$ be the Fibonacci sequence.

   (a) Use mathematical induction to prove that $f_1 + f_2 + \cdots + f_n = f_{n+2} - 1$.

   (b) Find the sum of the first 20 terms of the Fibonacci sequence.

12. The odds in favor of an event occurring are the ratio of the probability that the event will occur to the probability that the event will not occur. The reciprocal of this ratio represents the odds against the event occurring.

   (a) Six marbles in a bag are red. The odds against choosing a red marble are "4 to 1." How many marbles are in the bag?

   (b) A bag contains three blue marbles and seven yellow marbles. What are the odds in favor of choosing a blue marble? What are the odds against choosing a blue marble?

   (c) Write a formula for converting the odds in favor of an event to the probability of the event.

   (d) Write a formula for converting the probability of an event to the odds in favor of the event.

13. You are taking a test that contains only multiple choice questions. You are on the last question and you know that the answer is not B or D, but you are not sure about answers A, C, and E. What is the probability that you will get the right answer if you take a guess?

14. A dart is thrown at the circular target shown. The dart is equally likely to hit any point inside the target. What is the probability that it hits the region outside the triangle?

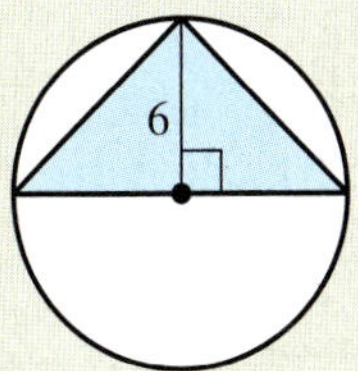

Answers to Odd-Numbered Exercises and Tests

Chapter P

Section P.1 *(page 9)*

1. (a) $5, 1, 2$ (b) $-9, 5, 0, 1, -4, 2, -11$
(c) $-9, -\frac{7}{2}, 5, \frac{2}{3}, 0, 1, -4, 2, -11$ (d) $\sqrt{2}$

3. (a) 1 (b) $-13, 1, -6$
(c) $2.01, 0.666\ldots, -13, 1, -6$ (d) $0.010110111\ldots$

5. (a) $\frac{6}{3}, 8$ (b) $\frac{6}{3}, -1, 8, -22$
(c) $-\frac{1}{3}, \frac{6}{3}, -7.5, -1, 8, -22$ (d) $-\pi, \frac{1}{2}\sqrt{2}$

7. 0.625 **9.** $0.\overline{123}$ **11.** $-1 < 2.5$

13. $-4 > -8$

15. $\frac{3}{2} < 7$

17. $\frac{5}{6} > \frac{2}{3}$

19. $x \le 5$ denotes the set of all real numbers less than or equal to 5. Unbounded

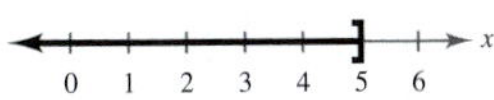

21. $x < 0$ denotes the set of all negative real numbers. Unbounded

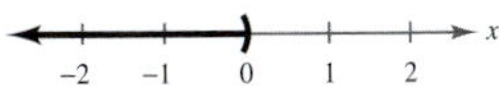

23. $x \ge 4$ denotes the set of all real numbers greater than or equal to 4. Unbounded

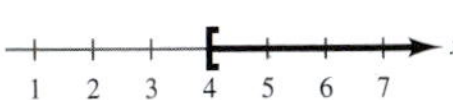

25. $-2 < x < 2$ denotes the set of all real numbers greater than -2 and less than 2. Bounded

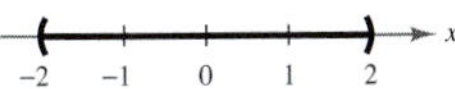

27. $-1 \le x < 0$ denotes the set of all negative real numbers greater than or equal to -1. Bounded

29. $-2 < x \le 4$ **31.** $y \ge 0$

33. $10 \le t \le 22$ **35.** $W > 65$

37. This interval consists of all real numbers greater than or equal to 0 and less than 8.

39. This interval consists of all real numbers greater than -6.

41. 10 **43.** 5 **45.** -1 **47.** -1 **49.** -1

51. $|-3| > -|-3|$ **53.** $-5 = -|5|$

55. $-|-2| = -|2|$ **57.** 4 **59.** 51 **61.** $\frac{5}{2}$ **63.** $\frac{128}{75}$

65. $|\$113{,}356 - \$112{,}700| = \$656 > \$500$
$0.05(\$112{,}700) = \$5635$

Because the actual expenses differ from the budget by more than \$500, there is failure to meet the "budget variance test."

67. $|\$37{,}335 - \$37{,}640| = \$305 < \$500$
$0.05(\$37{,}640) = \$1882$

Because the difference between the actual expenses and the budget is less than \$500 and less than 5% of the budgeted amount, there is compliance with the "budget variance test."

69. (a)

Year	Expenditures (in billions)	Surplus or deficit (in billions)
1960	\$92.2	\$0.3 (s)
1970	\$195.6	\$2.8 (d)
1980	\$590.9	\$73.8 (d)
1990	\$1253.2	\$221.2 (d)
2000	\$1788.8	\$236.4 (s)

(b)

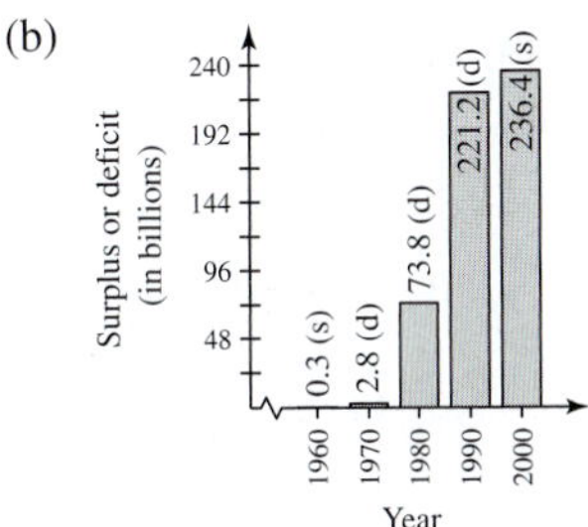

71. $|57 - 236| = 179$ miles **73.** $|60 - 23| = 37°$

75. $|x - 5| \le 3$ **77.** $|y| \ge 6$

79. $7x$ and 4 are the terms; 7 is the coefficient.

81. $\sqrt{3}x^2, -8x$, and -11 are the terms; $\sqrt{3}$ and -8 are the coefficients.

83. $4x^3, x/2$, and -5 are the terms; 4 and $\frac{1}{2}$ are the coefficients.

85. (a) -10 (b) -6 **87.** (a) 14 (b) 2

89. (a) Division by 0 is undefined. (b) 0

91. Commutative Property of Addition

93. Multiplicative Inverse Property

95. Distributive Property

97. Multiplicative Identity Property

99. Associative and Commutative Properties of Multiplication

101. $\frac{1}{2}$ **103.** $\frac{3}{8}$ **105.** 48 **107.** $\frac{5x}{12}$

109. (a)

n	1	0.5	0.01	0.0001	0.000001
$5/n$	5	10	500	50,000	5,000,000

(b) The value of $5/n$ approaches infinity as n approaches 0.

111. False. If $a < b$, then $\frac{1}{a} > \frac{1}{b}$, where $a \neq b \neq 0$.

113. (a) No. If one variable is negative and the other is positive, the expressions are unequal.

(b) $|u + v| \leq |u| + |v|$

The expressions are equal when u and v have the same sign. If u and v differ in sign, $|u + v|$ is less than $|u| + |v|$.

115. The only even prime number is 2, because its only factors are itself and 1.

117. (a) Negative (b) Negative

119. Yes. $|a| = -a$ if $a < 0$.

Section P.2 *(page 21)*

1. $8 \times 8 \times 8 \times 8 \times 8$

3. $-(0.4 \times 0.4 \times 0.4 \times 0.4 \times 0.4 \times 0.4)$ **5.** 4.9^6

7. $(-10)^5$ **9.** (a) 27 (b) 81

11. (a) 729 (b) -9 **13.** (a) $\frac{243}{64}$ (b) -1

15. (a) $\frac{5}{6}$ (b) 4 **17.** -1600 **19.** 2.125

21. -24 **23.** 6 **25.** -54 **27.** 1

29. (a) $-125z^3$ (b) $5x^6$ **31.** (a) $24y^{10}$ (b) $3x^2$

33. (a) $\frac{7}{x}$ (b) $\frac{4}{3}(x + y)^2$ **35.** (a) 1 (b) $\frac{1}{4x^4}$

37. (a) $-2x^3$ (b) $\frac{10}{x}$ **39.** (a) 3^{3n} (b) $\frac{b^5}{a^5}$

41. 5.73×10^7 square miles

43. 8.99×10^{-5} gram per cubic centimeter

45. 4,568,000,000 servings

47. 0.0000000000000000001602 coulomb

49. (a) 50,000 (b) 200,000

51. (a) 954.448 (b) 3.077×10^{10}

53. (a) 67,082.039 (b) 39.791 **55.** $9^{1/2}$

57. $\sqrt[5]{32}$ **59.** $\sqrt{196}$ **61.** $(-216)^{1/3}$

63. $\sqrt[3]{27^2}$ **65.** $81^{3/4}$ **67.** (a) 3 (b) 2

69. (a) -125 (b) 3 **71.** (a) $\frac{1}{8}$ (b) $\frac{27}{8}$

73. (a) -4 (b) 2 **75.** (a) 7.550 (b) -7.225

77. (a) -0.011 (b) 0.005 **79.** (a) $2\sqrt{2}$ (b) $2\sqrt[3]{3}$

81. (a) $6x\sqrt{2x}$ (b) $\frac{18}{z\sqrt{z}}$

83. (a) $2x\sqrt[3]{2x^2}$ (b) $\frac{5|x|\sqrt{3}}{y^2}$ **85.** $\frac{2}{x}$

87. $\frac{1}{x^3}$, $x > 0$ **89.** $\frac{\sqrt{3}}{3}$ **91.** $\frac{5 + \sqrt{3}}{11}$

93. $\frac{2}{\sqrt{2}}$ **95.** $\frac{2}{3(\sqrt{5} - \sqrt{3})}$

97. (a) $\sqrt{3}$ (b) $\sqrt[3]{(x + 1)^2}$

99. (a) $2\sqrt[4]{2}$ (b) $\sqrt[8]{2x}$

101. (a) $34\sqrt{2}$ (b) $22\sqrt{2}$ **103.** (a) $2\sqrt{x}$ (b) $4\sqrt{y}$

105. (a) $13\sqrt{x + 1}$ (b) $18\sqrt{5x}$

107. $\sqrt{5} + \sqrt{3} > \sqrt{5 + 3}$

109. $5 > \sqrt{3^2 + 2^2}$ **111.** $\frac{\pi}{2} \approx 1.57$ seconds

113. (a)

h	0	1	2	3	4	5	6
t	0	2.93	5.48	7.67	9.53	11.08	12.32

h	7	8	9	10	11	12
t	13.29	14.00	14.50	14.80	14.93	14.96

(b) Yes. $t = 8.64\sqrt{3} \approx 14.96$

115. True. When dividing variables, you subtract exponents.

117. $a^0 = 1, a \neq 0$, using the property $\frac{a^m}{a^n} = a^{m-n}$: $\frac{a^m}{a^m} = a^{m-m} = a^0 = 1$.

119. When any positive integer is squared, the units digit is 0, 1, 4, 5, 6, or 9. Therefore, $\sqrt{5233}$ is not an integer.

Section P.3 *(page 29)*

1. d **2.** e **3.** b **4.** a **5.** f **6.** c

7. $-2x^3 + 4x^2 - 3x + 20$ (Answers will vary.)

9. $-15x^4 + 1$ (Answers will vary.)

11. Degree: 2; Leading coefficient: 2

13. Degree: 5; Leading coefficient: 1

15. Degree: 5; Leading coefficient: -4

17. Degree: 5; Leading coefficient: 1

19. Polynomial: $-3x^3 + 2x + 8$

21. Not a polynomial because of the operation of division

23. Polynomial: $-y^4 + y^3 + y^2$ **25.** $-2x - 10$

27. $3x^3 - 2x + 2$ **29.** $8.3x^3 + 29.7x^2 + 11$

31. $12z + 8$ **33.** $3x^3 - 6x^2 + 3x$ **35.** $-15z^2 + 5z$

37. $-4x^4 + 4x$ **39.** $7.5x^3 + 9x$ **41.** $-\frac{1}{2}x^2 - 12x$

43. $4x^3 - 2x^2 + 4$ **45.** $5x^2 - 4x + 11$

47. $-30x^3 + 57x^2 + 25x - 12$

49. $x^4 - x^3 + 5x^2 - 9x - 36$ **51.** $x^4 + x^2 + 1$

53. $x^2 + 7x + 12$ **55.** $6x^2 - 7x - 5$

57. $4x^2 + 12x + 9$ **59.** $4x^2 - 20xy + 25y^2$

61. $x^2 - 100$ **63.** $x^2 - 4y^2$ **65.** $m^2 - n^2 - 6m + 9$

67. $x^2 + 2xy + y^2 - 6x - 6y + 9$ **69.** $4r^4 - 25$

71. $x^3 + 3x^2 + 3x + 1$ **73.** $8x^3 - 12x^2y + 6xy^2 - y^3$

75. $16x^6 - 24x^3 + 9$ **77.** $\frac{1}{4}x^2 - 3x + 9$

79. $\frac{1}{9}x^2 - 4$ **81.** $1.44x^2 + 7.2x + 9$

83. $2.25x^2 - 16$ **85.** $2x^2 + 2x$ **87.** $u^4 - 16$

89. $x - y$ **91.** $x^2 - 2\sqrt{5}x + 5$ **93.** $85,000

95. (a) $500r^2 + 1000r + 500$

(b)

r	$2\frac{1}{2}\%$	3%	4%
$500(1 + r)^2$	$525.31	$530.45	$540.80

r	$4\frac{1}{2}\%$	5%
$500(1 + r)^2$	$546.01	$551.25

(c) The amount increases with increasing r.

97. $V = x(26 - 2x)(18 - 2x)$
$= 4x(x - 13)(x - 9)$

x (cm)	1	2	3
V (cm^3)	384	616	720

99. (a) $3x^2 + 8x$ (b) $30x^2$
(c) $x^2 + \frac{7}{2}x$ (d) $\frac{3}{2}x^2 + 14x + 30$

101. $44x + 308$

103. (a) Estimates will vary.
(b) The difference in safe load decreases in magnitude.

105. $(x + 1)(x + 4) = x(x + 4) + 1(x + 4)$
Distributive Property

107. False. $(4x^2 + 1)(3x + 1) = 12x^3 + 4x^2 + 3x + 1$

109. $m + n$

111. The student omitted the middle term when squaring the binomial. $(x - 3)^2 = x^2 - 6x + 9 \neq x^2 + 9$

113. No. $(x^2 + 1) + (-x^2 + 3) = 4$, which is not a second-degree polynomial. (Examples will vary.)

115. $(3 + 4)^2 = 49 \neq 25 = 3^2 + 4^2$.
If either x or y is zero, then $(x + y)^2 = x^2 + y^2$.

Section P.4 *(page 38)*

1. 30 **3.** $6x^2y$ **5.** $3(x + 2)$ **7.** $2x(x^2 - 3)$

9. $(x - 1)(x + 6)$ **11.** $(x + 3)(x - 1)$ **13.** $\frac{1}{2}(x + 8)$

15. $\frac{1}{2}x(x^2 + 4x - 10)$ **17.** $\frac{2}{3}(x - 6)(x - 3)$

19. $(x + 6)(x - 6)$ **21.** $(4y + 3)(4y - 3)$

23. $\left(4x + \frac{1}{3}\right)\left(4x - \frac{1}{3}\right)$ **25.** $(x + 1)(x - 3)$

27. $(3u + 2v)(3u - 2v)$ **29.** $(x - 2)^2$ **31.** $(2t + 1)^2$

33. $(5y - 1)^2$ **35.** $(3u + 4v)^2$ **37.** $\left(x - \frac{2}{3}\right)^2$

39. $(x - 2)(x^2 + 2x + 4)$ **41.** $(y + 4)(y^2 - 4y + 16)$

43. $(2t - 1)(4t^2 + 2t + 1)$

45. $(u + 3v)(u^2 - 3uv + 9v^2)$

47. $-5(x^2 - 5)$ **49.** $-2(t^3 - 2t - 3)$

51. $(x + 2)(x - 1)$ **53.** $(s - 3)(s - 2)$

55. $-(y + 5)(y - 4)$ **57.** $(x - 20)(x - 10)$

59. $(3x - 2)(x - 1)$ **61.** $(5x + 1)(x + 5)$

63. $-(3z - 2)(3z + 1)$ **65.** $(x - 1)(x^2 + 2)$

67. $(2x - 1)(x^2 - 3)$ **69.** $(3 + x)(2 - x^3)$

71. $(3x^2 - 1)(2x + 1)$ **73.** $(x + 2)(3x + 4)$

75. $(2x - 1)(3x + 2)$ **77.** $(3x - 1)(5x - 2)$

79. $6(x + 3)(x - 3)$ **81.** $x^2(x - 4)$ **83.** $(x - 1)^2$

85. $(1 - 2x)^2$ **87.** $-2x(x + 1)(x - 2)$

89. $(9x + 1)(x + 1)$ **91.** $\frac{1}{81}(x + 36)(x - 18)$

93. $(3x + 1)(x^2 + 5)$ **95.** $x(x - 4)(x^2 + 1)$

97. $\frac{1}{4}(x^2 + 3)(x + 12)$ **99.** $(t + 6)(t - 8)$

101. $(x + 2)(x + 4)(x - 2)(x - 4)$

103. $5(x + 2)(x^2 - 2x + 4)$ **105.** $(3 - 4x)(23 - 60x)$

107. $5(1 - x)^2(3x + 2)(4x + 3)$

109. $(x - 2)^2(x + 1)^3(7x - 5)$

111. $3(x^2 + 1)^4(x^4 - x^2 + 1)^4(3x + 2)^2(33x^6 + 20x^5 + 3)$

113. b **114.** c **115.** a **116.** d

117.

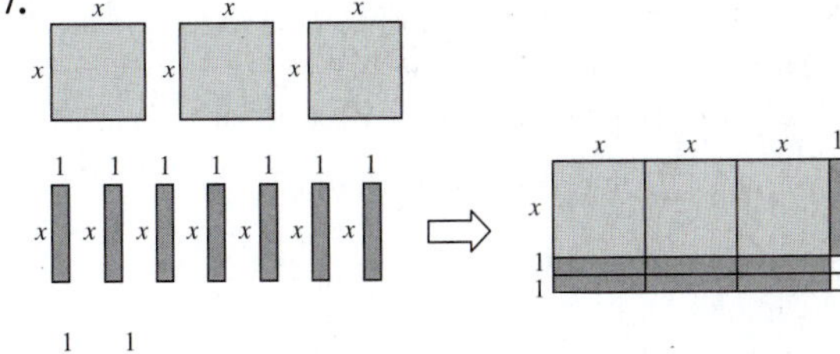

119.

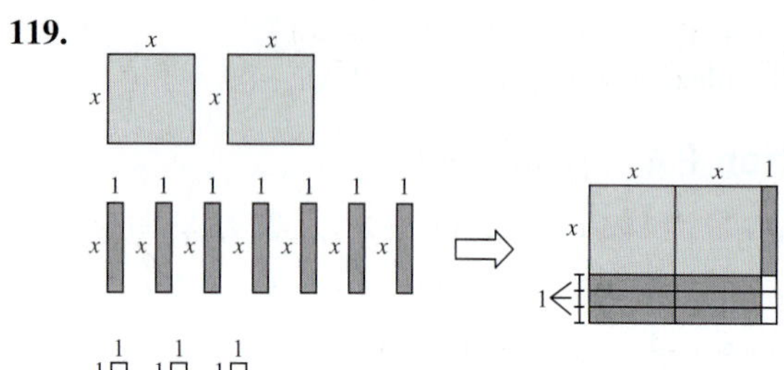

121. $4\pi(r + 1)$ **123.** $4(6 - x)(6 + x)$

125. $-14, 14, -2, 2$ **127.** $-11, 11, -4, 4, -1, 1$

129. Two possible answers: $2, -12$

131. Two possible answers: $-2, -4$

133. A 3 was not factored out of the second binomial.

135. $kx(Q - x)$

137. True. $a^2 - b^2 = (a + b)(a - b)$

139. $(x^n + y^n)(x^n - y^n)$

141. $x^{3n} - y^{2n}$ is completely factored.

Section P.5 *(page 48)*

1. All real numbers **3.** All nonnegative real numbers

5. All real numbers x such that $x \neq 2$

7. All real numbers x such that $x \geq -1$

9. $3x, \; x \neq 0$ **11.** $\frac{3x}{2}, \; x \neq 0$ **13.** $\frac{3y}{y + 1}, \; x \neq 0$

15. $\frac{-4y}{5}, \; y \neq \frac{1}{2}$ **17.** $-\frac{1}{2}, \; x \neq 5$

19. $y - 4, \; y \neq -4$ **21.** $\frac{x(x + 3)}{x - 2}, \; x \neq -2$

23. $\frac{y - 4}{y + 6}, \; y \neq 3$ **25.** $\frac{-(x^2 + 1)}{(x + 2)}, \; x \neq 2$

27. $z - 2$

29.

x	0	1	2	3	4	5	6
$\frac{x^2 - 2x - 3}{x - 3}$	1	2	3	Undef.	5	6	7
$x + 1$	1	2	3	4	5	6	7

The expressions are equivalent except at $x = 3$.

31. The expression cannot be simplified.

33. $\frac{\pi}{4}, \; r \neq 0$ **35.** $\frac{1}{5(x - 2)}, \; x \neq 1$

37. $\frac{r + 1}{r}, \; r \neq 1$ **39.** $\frac{t - 3}{(t + 3)(t - 2)}, \; t \neq -2$

41. $\frac{(x + 6)(x + 1)}{x^2}, \; x \neq 6$ **43.** $\frac{x + 5}{x - 1}$ **45.** $\frac{6x + 13}{x + 3}$

47. $-\frac{2}{x - 2}$ **49.** $-\frac{x^2 + 3}{(x + 1)(x - 2)(x - 3)}$

51. $\frac{2 - x}{x^2 + 1}, \; x \neq 0$ **53.** $\frac{x^7 - 2}{x^2}$ **55.** $\frac{-1}{(x^2 + 1)^5}$

57. $\frac{2x^3 - 2x^2 - 5}{(x - 1)^{1/2}}$

59. The error was incorrect subtraction in the numerator.

61. $\frac{1}{2}, \; x \neq 2$ **63.** $x(x + 1), \; x \neq -1, 0$

65. $-\frac{2x + h}{x^2(x + h)^2}, \; h \neq 0$ **67.** $\frac{2x - 1}{2x}, \; x > 0$

69. $\frac{3x - 1}{3}, \; x \neq 0$ **71.** $\frac{1}{\sqrt{x + 2} + \sqrt{x}}$

73. $\frac{x}{2(2x + 1)}, \; x \neq 0$

75. (a) $\frac{1}{16}$ minute (b) $\frac{x}{16}$ minute(s) (c) $\frac{60}{16} = \frac{15}{4}$ minutes

77. (a)

t	0	2	4	6	8	10
T	75	55.9	48.3	45	43.3	42.3

t	12	14	16	18	20	22
T	41.7	41.3	41.1	40.9	40.7	40.6

(b) The model is approaching a T-value of 40.

79. False. In order for the simplified expression to be equivalent to the original expression, the domain of the simplified expression needs to be restricted. If n is even, $x \neq -1, 1$. If n is odd, $x \neq 1$.

81. Completely factor each polynomial in the numerator and in the denominator. Then conclude that there are no common factors.

Section P.6 *(page 56)*

1. Change all signs when distributing the minus sign.

$2x - (3y + 4) = 2x - 3y - 4$

3. Change all signs when distributing the minus sign.

$$\frac{4}{16x - (2x + 1)} = \frac{4}{14x - 1}$$

5. z occurs twice as a factor.

$(5z)(6z) = 30z^2$

7. The fraction as a whole is multiplied by a, not the numerator and denominator separately.

$a\left(\frac{x}{y}\right) = \frac{ax}{y}$

9. $\sqrt{x+9}$ cannot be simplified.

11. Divide out common factors, not common terms.

$\frac{2x^2+1}{5x}$ cannot be simplified.

13. To get rid of negative exponents:

$\frac{1}{a^{-1}+b^{-1}} = \frac{1}{a^{-1}+b^{-1}} \cdot \frac{ab}{ab} = \frac{ab}{b+a}.$

15. Factor within grouping symbols before applying exponent to each factor.

$(x^2+5x)^{1/2} = [x(x+5)]^{1/2} = x^{1/2}(x+5)^{1/2}$

17. To add fractions, first find a common denominator.

$\frac{3}{x} + \frac{4}{y} = \frac{3y+4x}{xy}$

19. $3x+2$ **21.** $2x^2+x+15$ **23.** $\frac{1}{3}$ **25.** 2

27. $\frac{1}{2x^2}$ **29.** $\frac{25}{9}, \frac{49}{16}$ **31.** 1, 2 **33.** $1-5x$

35. $1-7x$ **37.** $3x-1$ **39.** $\frac{16}{x}-5-x$

41. $4x^{8/3}-7x^{5/3}+\frac{1}{x^{1/3}}$ **43.** $\frac{3}{\sqrt{x}}-5x^{3/2}-x^{7/2}$

45. $\frac{-7x^2-4x+9}{(x^2-3)^3(x+1)^4}$ **47.** $\frac{27x^2-24x+2}{(6x+1)^4}$

49. $\frac{-1}{(x+3)^{2/3}(x+2)^{7/4}}$ **51.** $\frac{4x-3}{(3x-1)^{4/3}}$ **53.** $\frac{x}{x^2+4}$

55. $\frac{(3x-2)^{1/2}(15x^2-4x+45)}{2(x^2+5)^{1/2}}$

57. (a)

x	0.5	1.0	1.5	2.0
t	1.70	1.72	1.78	1.89

x	2.5	3.0	3.5	4.0
t	2.02	2.18	2.36	2.57

(b) $x = 0.5$ mile

(c) $\frac{3x\sqrt{x^2-8x+20}+(x-4)\sqrt{x^2+4}}{6\sqrt{x^2+4}\sqrt{x^2-8x+20}}$

59. True. $x^{-1}+y^{-2} = \frac{1}{x}+\frac{1}{y^2}$

$= \frac{y^2+x}{xy^2}$

61. True. $\frac{1}{\sqrt{x}+4} = \frac{1}{\sqrt{x}+4} \cdot \frac{\sqrt{x}-4}{\sqrt{x}-4}$

$= \frac{\sqrt{x}-4}{x-16}$

63. Add exponents when multiplying powers with like bases.

$x^n \cdot x^{3n} = x^{4n}$

65. When a binomial is squared, there is also a middle term.

$(x^n+y^n)^2 = x^{2n}+2x^ny^n+y^{2n} \neq x^{2n}+y^{2n}$

67. The two answers are equivalent and can be obtained by factoring.

$\frac{1}{10}(2x-1)^{5/2}+\frac{1}{6}(2x-1)^{3/2}$

$= \frac{1}{60}(2x-1)^{3/2}[6(2x-1)+10]$

$= \frac{1}{60}(2x-1)^{3/2}(12x+4)$

$= \frac{4}{60}(2x-1)^{3/2}(3x+1)$

$= \frac{1}{15}(2x-1)^{3/2}(3x+1)$

Section P.7 *(page 64)*

1. A: (2, 6), B: (−6, −2), C: (4, −4), D: (−3, 2)

3. (−3, 4) **5.** (−5, −5)

7. Quadrant IV **9.** Quadrant II

11. Quadrant III or IV **13.** Quadrant III

15. Quadrant I or III **17.** (0, 1), (4, 2), (1, 4)

19. (−3, 6), (2, 10), (2, 4), (−3, 4)

21.

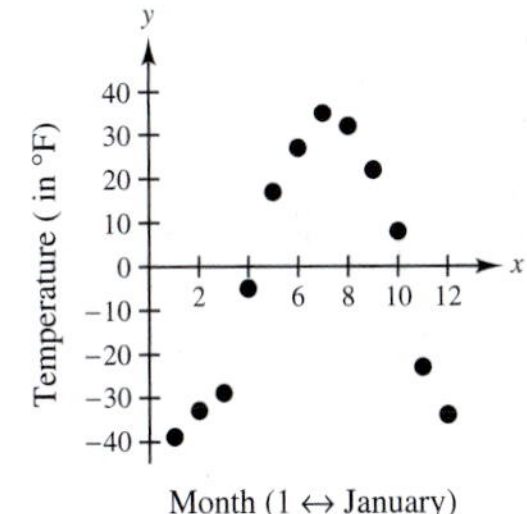

23. \$3.20 per pound; 1998 **25.** 242.9%

27. (a) 1990s (b) 11.8%; 21.2% (c) \$6.24 (d) Yes

29. 8 **31.** 5 **33.** (a) 4, 3, 5 (b) $4^2+3^2=5^2$

35. (a) 10, 3, $\sqrt{109}$ (b) $10^2+3^2=\left(\sqrt{109}\right)^2$

37. (a)

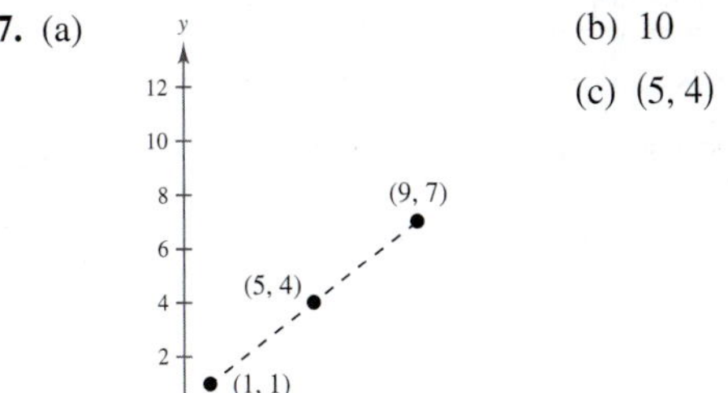

(b) 10

(c) (5, 4)

39. (a)

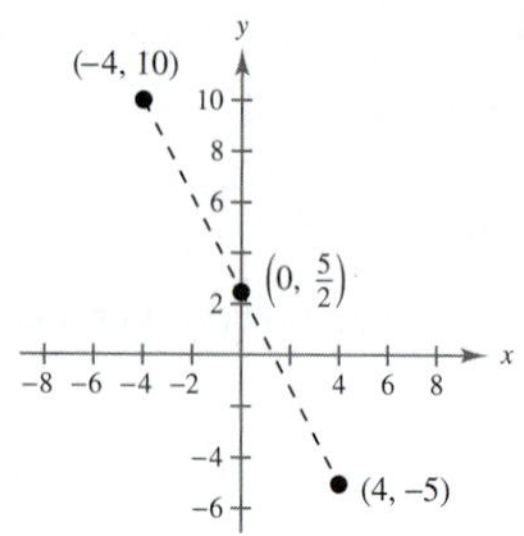

(b) 17

(c) $\left(0, \frac{5}{2}\right)$

41. (a)

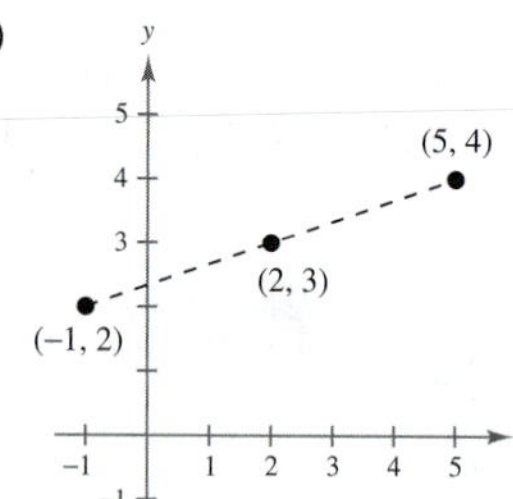

(b) $2\sqrt{10}$

(c) (2, 3)

43. (a)

(b) $\dfrac{\sqrt{82}}{3}$

(c) $\left(-1, \frac{7}{6}\right)$

45. (a)

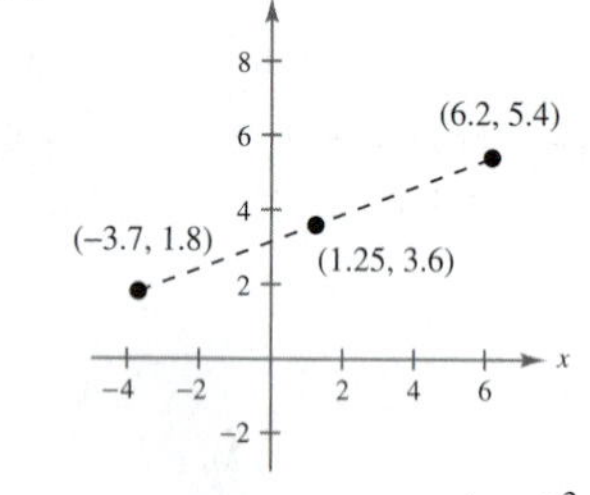

(b) $\sqrt{110.97}$

(c) (1.25, 3.6)

47. \$31,137 million **49.** $(\sqrt{5})^2 + (\sqrt{45})^2 = (\sqrt{50})^2$

51. $(2x_m - x_1, 2y_m - y_1)$

53. $\left(\dfrac{3x_1 + x_2}{4}, \dfrac{3y_1 + y_2}{4}\right), \left(\dfrac{x_1 + x_2}{2}, \dfrac{y_1 + y_2}{2}\right),$

$\left(\dfrac{x_1 + 3x_2}{4}, \dfrac{y_1 + 3y_2}{4}\right)$

55. $5\sqrt{74} \approx 43$ yards

57.

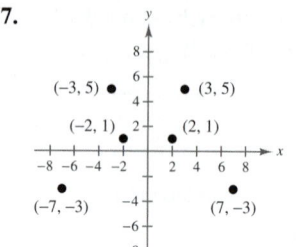

(a) The point is reflected through the y-axis.

(b) The point is reflected through the x-axis.

(c) The point is reflected through the origin.

59. \$1969.45 million

61. False. The Midpoint Formula would be used 15 times.

63. Point on x-axis: $y = 0$; Point on y-axis: $x = 0$

65. b **66.** c **67.** d **68.** a

69. Use the Midpoint Formula to prove that the diagonals of the parallelogram bisect each other.

$\left(\dfrac{b + a}{2}, \dfrac{c + 0}{2}\right) = \left(\dfrac{a + b}{2}, \dfrac{c}{2}\right)$

$\left(\dfrac{a + b + 0}{2}, \dfrac{c + 0}{2}\right) = \left(\dfrac{a + b}{2}, \dfrac{c}{2}\right)$

Review Exercises *(page 69)*

1. (a) 11 (b) 11, −14

(c) 11, −14, $-\frac{8}{9}$, $\frac{5}{2}$, 0.4 (d) $\sqrt{6}$

3. (a) $0.8\overline{3}$ (b) 0.875

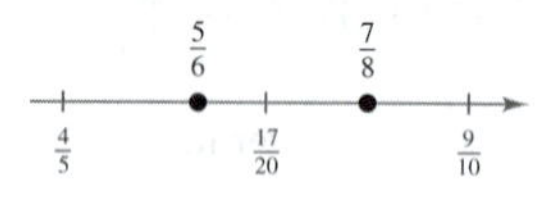

$\frac{5}{6} < \frac{7}{8}$

5. The set consists of all real numbers less than or equal to 7.

7. 155 **9.** $|x - 7| \geq 4$ **11.** $|y + 30| < 5$

13. (a) −7 (b) −19 **15.** (a) −1 (b) −3

17. Associative Property of Addition

19. Additive Identity Property **21.** −11 **23.** $\frac{1}{12}$

25. −144 **27.** (a) $192x^{11}$ (b) $\dfrac{y^5}{2}$, $y \neq 0$

29. (a) $\dfrac{3u^5}{v^4}$ (b) $\dfrac{1}{m^2}$ **31.** 1.8809×10^9

33. 483,600,000 **35.** (a) 9 (b) 343

37. (a) 216 (b) 32 **39.** (a) $2\sqrt{2}$ (b) $26\sqrt{2}$

41. Radicals cannot be combined by addition or subtraction unless the index and the radicand are the same.

43. $2 + \sqrt{3}$ **45.** $\frac{3}{\sqrt{7} + 1}$ **47.** 64 **49.** $6x^{9/10}$

51. $-11x^2 + 3$; Degree: 2; Leading coefficient: -11

53. $-12x^2 - 4$; Degree: 2; Leading coefficient: -12

55. $-3x^2 - 7x + 1$ **57.** $2x^3 - 10x^2 + 12x$

59. $15x^2 - 27x - 6$ **61.** $4x^2 - 12x + 9$ **63.** 41

65. (a)

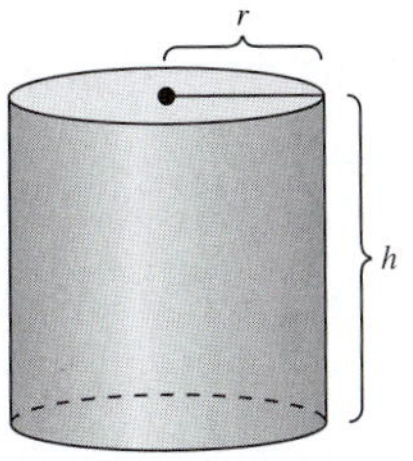

Explanations will vary.

(b) $168\pi \approx 527.79$ square inches

67. $x(x + 1)(x - 1)$ **69.** $(5x + 7)(5x - 7)$

71. $(x - 4)(x^2 + 4x + 16)$ **73.** $(x + 10)(2x + 1)$

75. $(x - 1)(x^2 + 2)$ **77.** All real numbers $x \neq -6$

79. $\frac{x - 8}{15}, x \neq -8$ **81.** $\frac{1}{x^2}, x \neq \pm 2$

83. $\frac{3x}{(x - 1)(x^2 + x + 1)}$ **85.** $\frac{3ax^2}{(a^2 - x)(a - x)}$

87. The multiplication in parentheses comes first.
$10(4 \cdot 7) = 10(28) = 280$

89. Multiply exponents when raising a power to a power.
$(3^4)^4 = 3^{16}$

91. Add the numbers in parentheses before squaring.
$(5 + 8)^2 = 13^2 \neq 5^2 + 8^2$

93. $16x^2 - 9x + 20$ **95.** $-5x^2 + 2x + 15$

97. $x^2 + 5x + \frac{7}{x}$ **99.** $\frac{2(x + 1)}{(x + 2)^2}$

101. Quadrant IV **103.** (2, 5), (4, 5), (2, 0), (4, 0)

105. (a) (b) 5

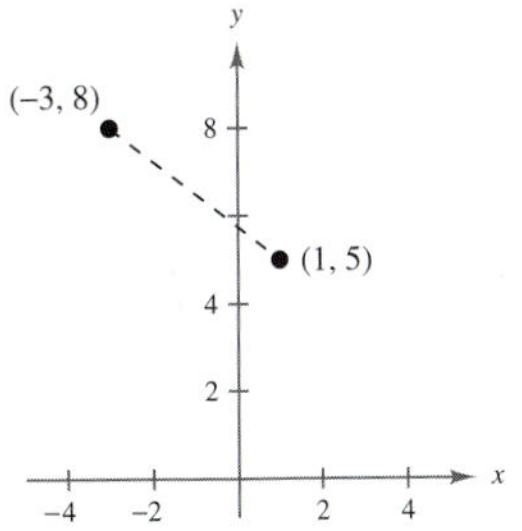

107. (a) (b) $\left(1, \frac{3}{2}\right)$

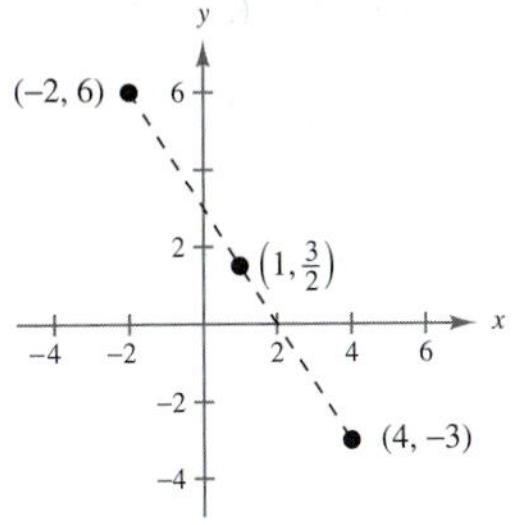

109. \$443.3 million

111. False. There is also a cross-product term when a binomial sum is squared.
$(x + a)^2 = x^2 + 2ax + a^2$

Chapter Test *(page 72)*

1. $-\frac{10}{3} > -|-4|$ **2.** 9.15

3. Additive Identity Property

4. (a) -18 (b) $\frac{4}{27}$ (c) $-\frac{27}{125}$ (d) $\frac{8}{729}$

5. (a) 25 (b) 6 (c) 1.8×10^5 (d) 2.7×10^{13}

6. (a) $12z^8$ (b) $(u - 2)^{-7}$ (c) $\frac{3x^2}{y^2}$

7. (a) $15z\sqrt{2z}$ (b) $4x^{14/15}$ (c) $\frac{2\sqrt[3]{2v}}{v^2}$

8. $4x^4 - 3x^2 + x - 5$; Degree: 4; Leading coefficient: 4

9. $2x^2 - 3x - 5$ **10.** $x^2 - 5$ **11.** $8, x \neq 3$

12. $\frac{x - 1}{2x}, x \neq \pm 1$

13. (a) $x^2(2x + 1)(x - 2)$ (b) $(x - 2)(x + 2)^2$

14. (a) $4\sqrt[3]{4}$ (b) $-3(1 + \sqrt{3})$

15. All real numbers $x \neq 4$ **16.** $\frac{4}{y + 4}, y \neq 2$

17. \$545

18.

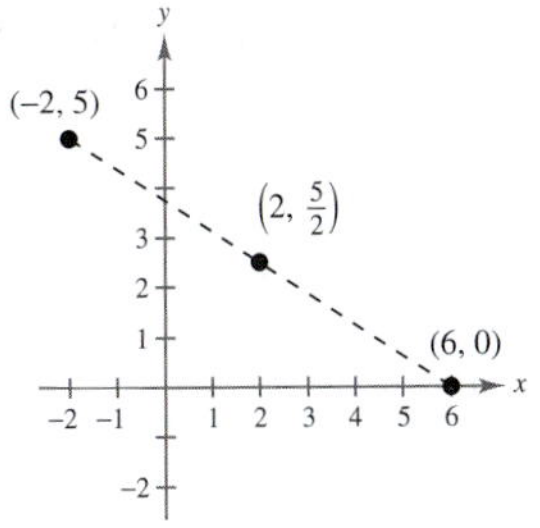

Midpoint: $\left(2, \frac{5}{2}\right)$; Distance: $\sqrt{89}$

19. $\frac{5}{6}\sqrt{3}\,x^2$

Problem Solving (page 74)

1. (a) Men's: 1,150,347 cubic millimeters;
   696,910 cubic millimeters
   Women's: 696,910 cubic millimeters;
   448,921 cubic millimeters
   (b) Men's: 1.04×10^{-5} kilograms per cubic millimeter;
   6.31×10^{-6} kilograms per cubic millimeter
   Women's: 8.91×10^{-6} kilograms per cubic millimeter;
   5.74×10^{-6} kilograms per cubic millimeter
   (c) No. Iron has a greater density than cork.

3. 1.62 ounces 5. Answers will vary.

7. $r \approx 0.28$ 9. 9.57 square feet

11. $y_1(0) = 0, y_2(0) = 2$

$$y_2 = \frac{x(2 - 3x^2)}{\sqrt{1 - x^2}}$$

13. (a) $(2, -1), (3, 0)$ (b) $\left(-\frac{4}{3}, -2\right), \left(-\frac{2}{3}, -1\right)$

Chapter 1

Section 1.1 (page 85)

1. (a) Yes (b) Yes 3. (a) No (b) Yes

5.

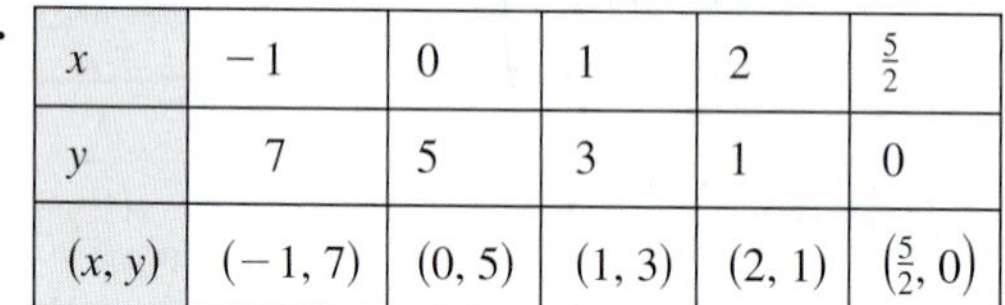

x	-1	0	1	2	$\frac{5}{2}$
y	7	5	3	1	0
(x, y)	$(-1, 7)$	$(0, 5)$	$(1, 3)$	$(2, 1)$	$\left(\frac{5}{2}, 0\right)$

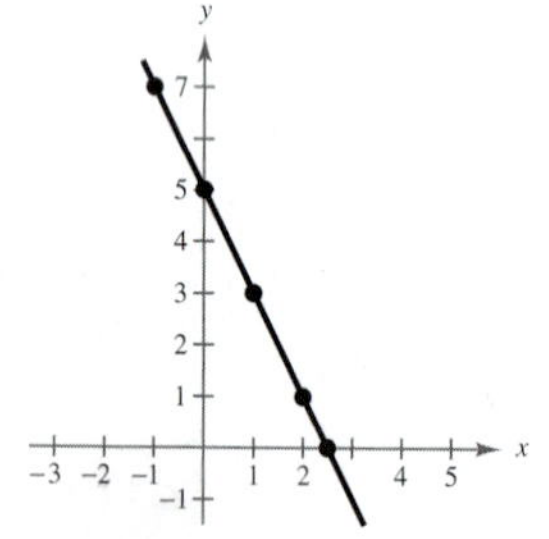

7.

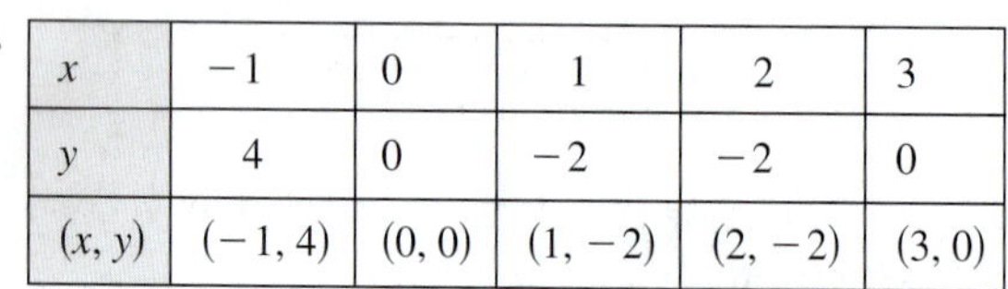

x	-1	0	1	2	3
y	4	0	-2	-2	0
(x, y)	$(-1, 4)$	$(0, 0)$	$(1, -2)$	$(2, -2)$	$(3, 0)$

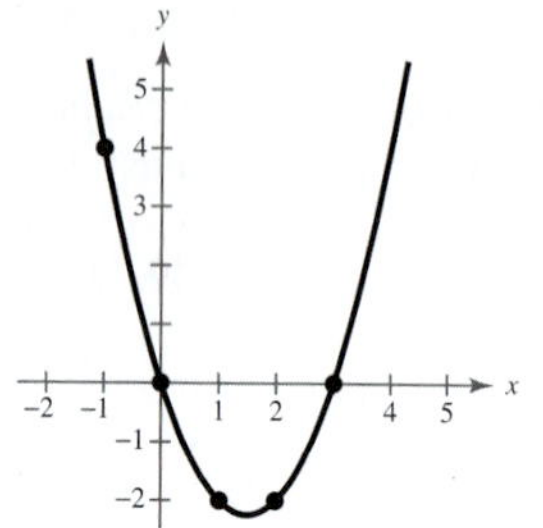

9. x-intercepts: $(\pm 2, 0)$
   y-intercept: $(0, 16)$

11. x-intercepts: $(0, 0), (2, 0)$
    y-intercept: $(0, 0)$

13. y-axis symmetry 15. Origin symmetry

17. Origin symmetry 19. x-axis symmetry

21.

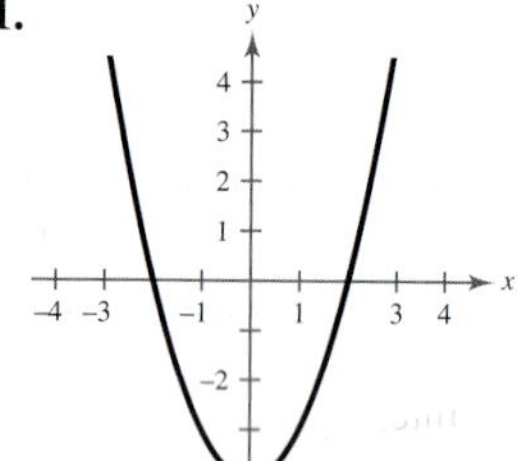

23.

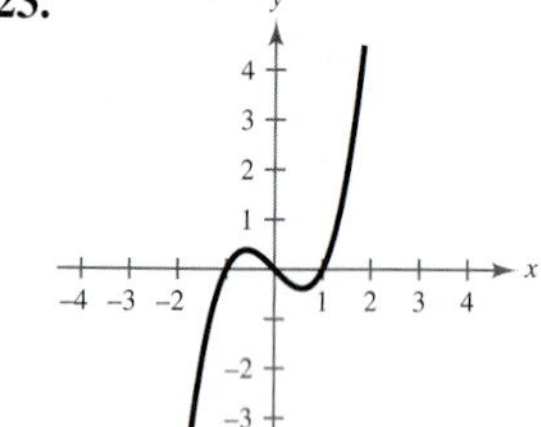

25.

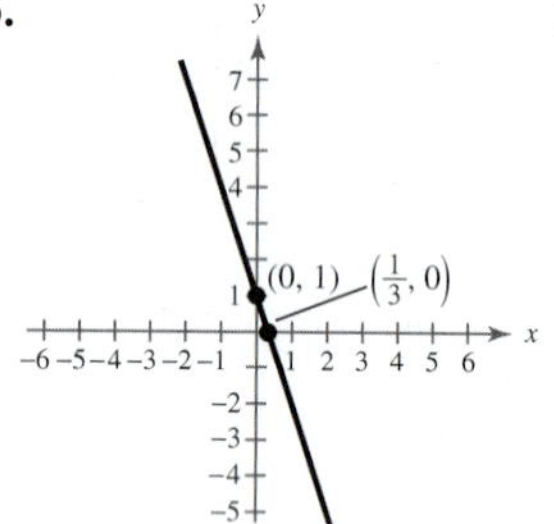

27.

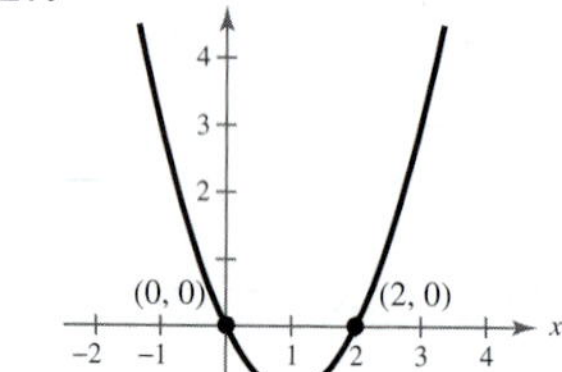

29.

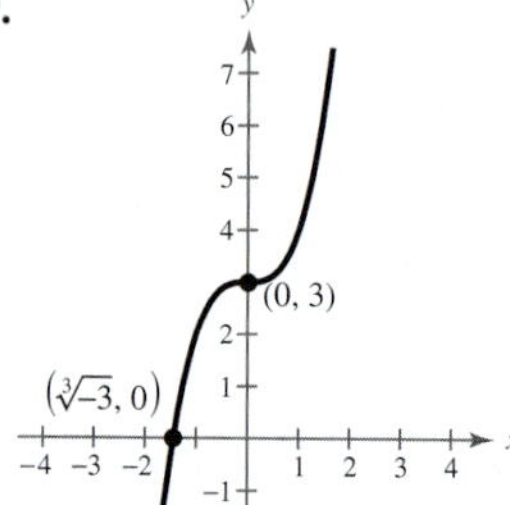

31.

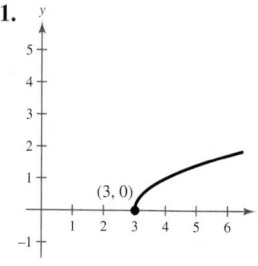

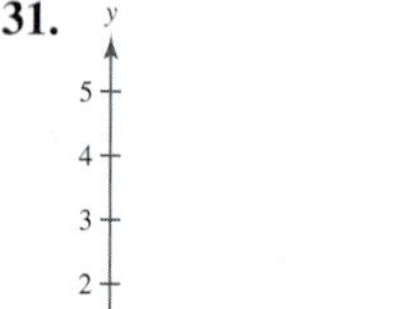

33.

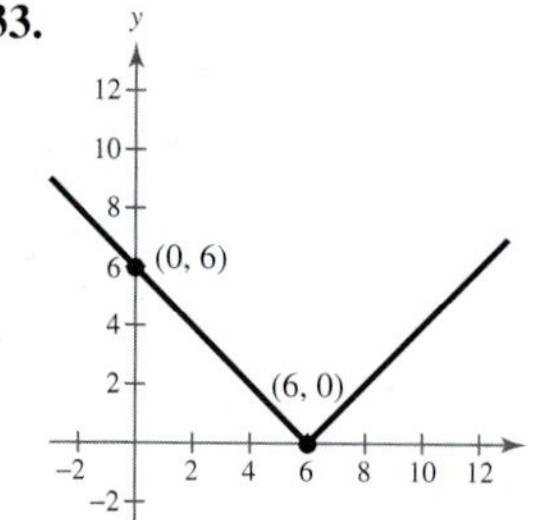

35.

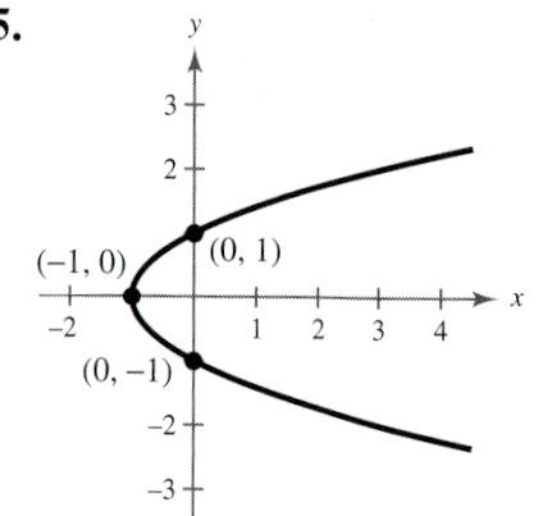

37.

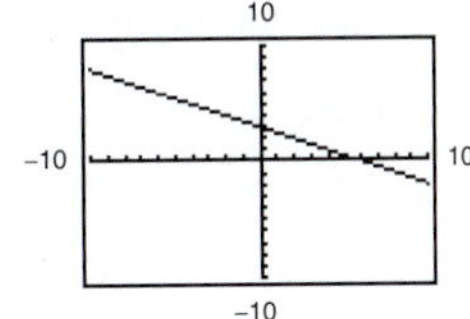

Intercepts: $(6, 0)$, $(0, 3)$

39.

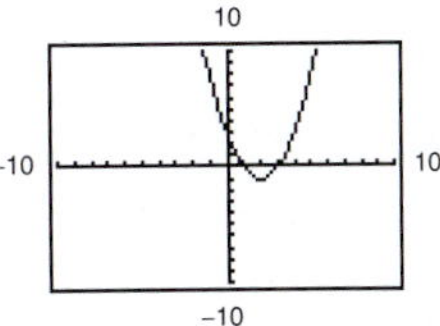

Intercepts: $(3, 0)$, $(1, 0)$, $(0, 3)$

41.

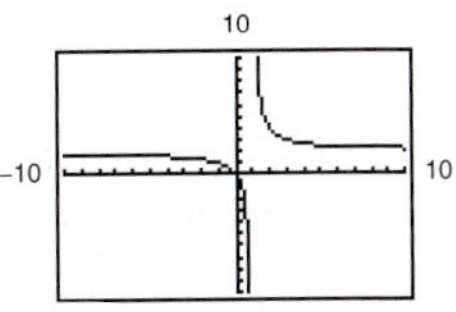

Intercept: $(0, 0)$

43.

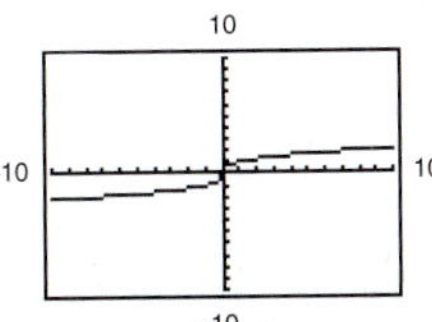

Intercept: $(0, 0)$

45.

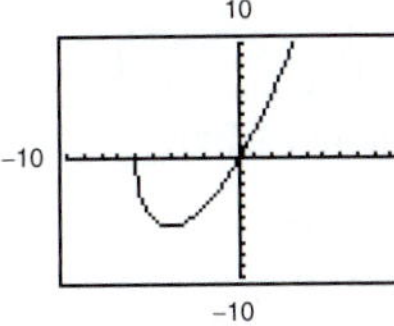

Intercepts: $(0, 0)$, $(-6, 0)$

47.

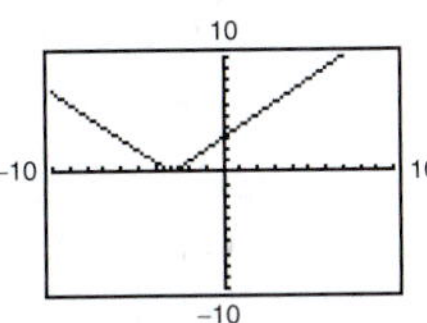

Intercepts: $(-3, 0)$, $(0, 3)$

49. $x^2 + y^2 = 16$

51. $(x - 2)^2 + (y + 1)^2 = 16$

53. $(x + 1)^2 + (y - 2)^2 = 5$

55. $(x - 3)^2 + (y - 4)^2 = 25$

57. Center: $(0, 0)$; Radius: 5

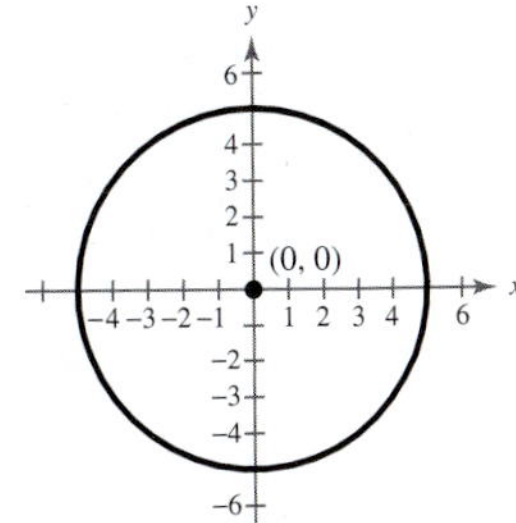

59. Center: $(1, -3)$; Radius: 3

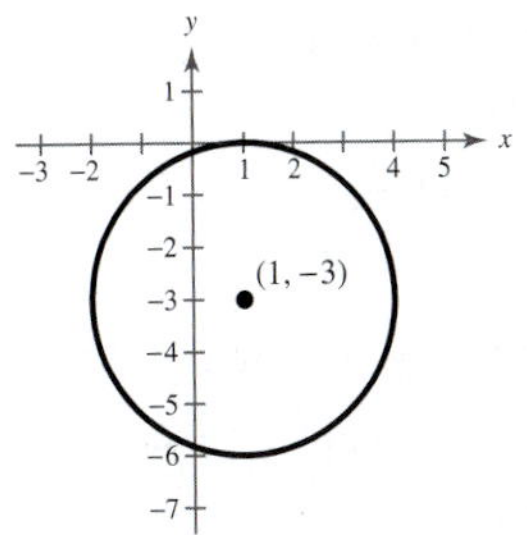

61. Center: $\left(\frac{1}{2}, \frac{1}{2}\right)$; Radius: $\frac{3}{2}$

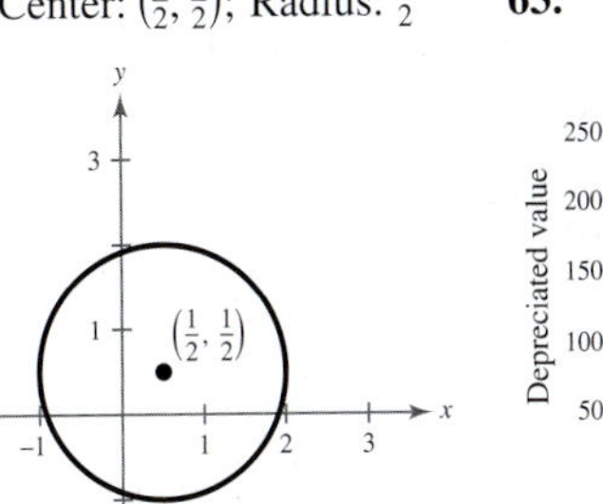

63.

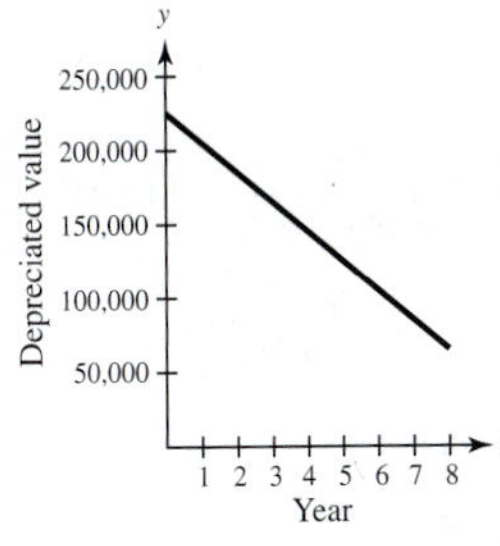

65. (a)

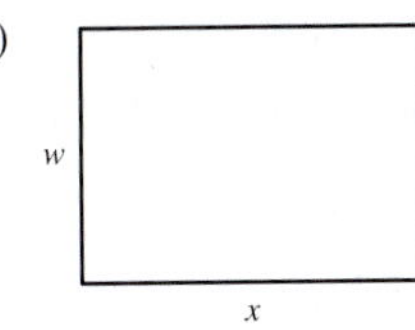

(b) Answers will vary.

(c)

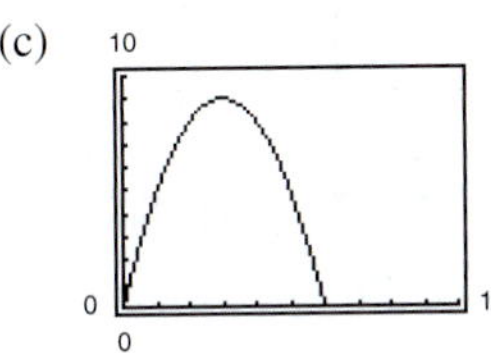

(d) $x = 3$, $w = 3$

67. (a) and (b)

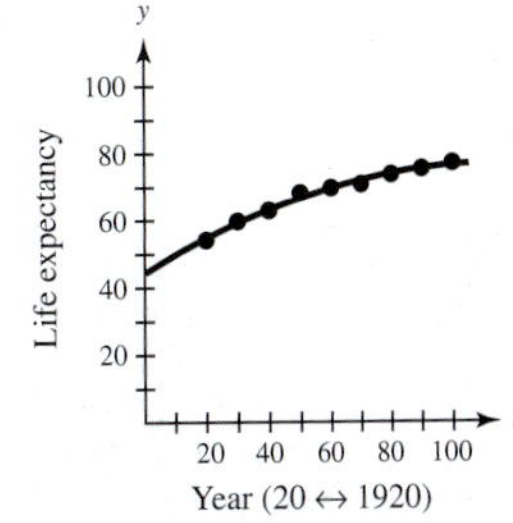

The curve seems to be a good fit for the data.

(c) 2005: 76.8 years; 2010: 77.0 years

(d) Answers will vary.

69. 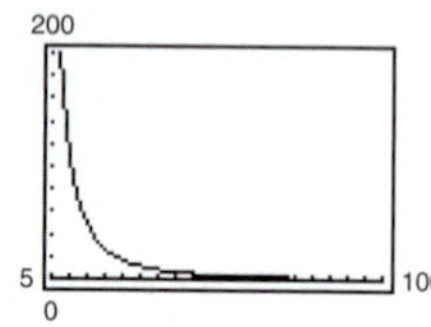

3.9 ohms

71. True. All linear equations of the form $y = mx + b$, which excludes vertical lines, cross the y-axis one time.

73. (a) $a = 1, b = 0$ (b) $a = 0, b = 1$

75. $9x^5$, $4x^3$, -7 **77.** $2\sqrt{2x}$ **79.** $\dfrac{10\sqrt{7x}}{x}$

81. $\sqrt[3]{|t|}$

Section 1.2 *(page 93)*

1. (a) No (b) No (c) Yes (d) No

3. (a) Yes (b) Yes (c) No (d) No

5. (a) Yes (b) No (c) No (d) No

7. (a) No (b) No (c) No (d) No

9. (a) Yes (b) No (c) Yes (d) No

11. Identity **13.** Conditional equation **15.** Identity

17. Identity **19.** Conditional equation

21. Original equation
Subtract 32 from each side.
Simplify.
Divide each side by 4.
Simplify.

23. 4 **25.** -9 **27.** 5 **29.** 9 **31.** No solution

33. -4 **35.** $-\frac{6}{5}$ **37.** 9 **39.** $\frac{7}{3}$ **41.** 13

43. 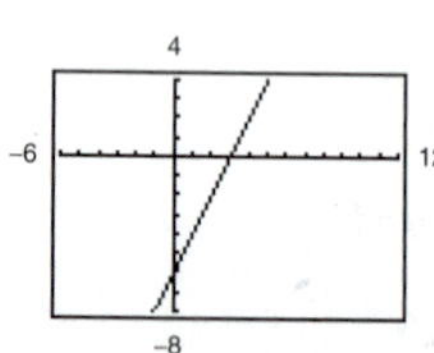

$x = 3$

45. 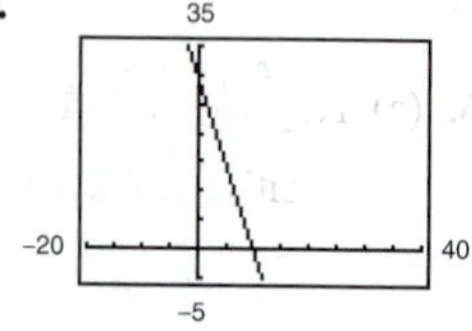

$x = 10$

47. 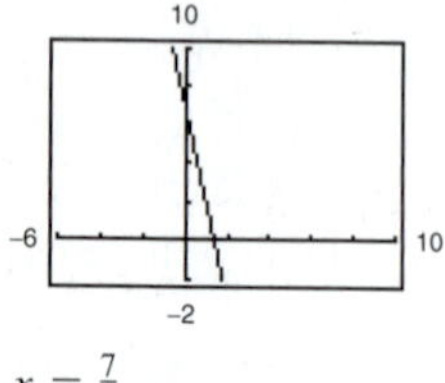

$x = \frac{7}{5}$

49. x-intercept: $\left(\frac{12}{5}, 0\right)$
y-intercept: $(0, 12)$

51. x-intercept: $\left(-\frac{1}{2}, 0\right)$
y-intercept: $(0, -3)$

53. x-intercept: $(5, 0)$
y-intercept: $\left(0, \frac{10}{3}\right)$

55. x-intercept: $(-20, 0)$
y-intercept: $\left(0, \frac{8}{3}\right)$

57. x-intercept: $(1.6, 0)$
y-intercept: $(0, -0.3)$

59. No solution. The x-terms sum to zero. **61.** 10

63. 4 **65.** 3 **67.** 0

69. No solution. The variable is divided out.

71. No solution. The solution is extraneous.

73. 5 **75.** No solution. The solution is extraneous.

77. 0 **79.** All real numbers

81. $\dfrac{1}{3-a}$, $a \neq 3$ **83.** $\dfrac{5}{4+a}$, $a \neq -4$

85. $\dfrac{18}{36+a}$, $a \neq -36$ **87.** $\dfrac{-17}{10-2a}$, $a \neq 5$

89. 138.889 **91.** 19.993 **93.** $x = 0$ feet

95. (a) 61.2 inches

(b) Yes. The estimated height of a male with a 19-inch femur is 69.4 inches.

(c)

Height, x	Female femur length	Male femur length
60	15.48	14.79
70	19.80	19.28
80	24.12	23.77
90	28.44	28.26
100	32.76	32.75
110	37.08	37.24

100 inches

(d) $x \approx 100.59$; There would not be a problem because it is not likely for either a male or a female to be 100 inches tall (or 8 feet 4 inches tall).

97. 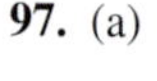 (a)

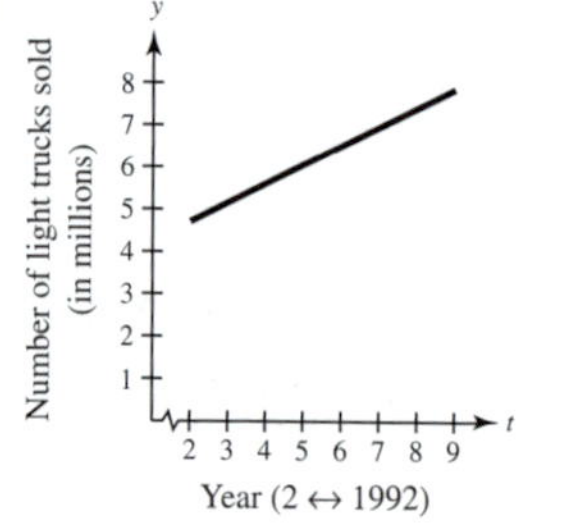

(b) 1994

99. 23,437.5 miles

101. False. $x(3 - x) = 10$

$3x - x^2 = 10$

The equation cannot be written in the form $ax + b = 0$.

103. Equivalent equations have the same solution set, and one is derived from the other by steps for generating equivalent equations.

$2x = 5,\ 2x + 3 = 8$

105. (a)

x	-1	0	1	2	3	4
$3.2x - 5.8$	-9	-5.8	-2.6	0.6	3.8	7

(b) $1 < x < 2$. The expression changes from negative to positive in this interval.

(c)

x	1.5	1.6	1.7	1.8	1.9	2
$3.2x - 5.8$	-1	-0.68	-0.36	-0.04	0.28	0.6

(d) $1.8 < x < 1.9$. To improve accuracy, evaluate the expression in this interval and determine where the sign changes.

107. $\dfrac{x - 4}{2x - 1},\ x \neq -9$

109.

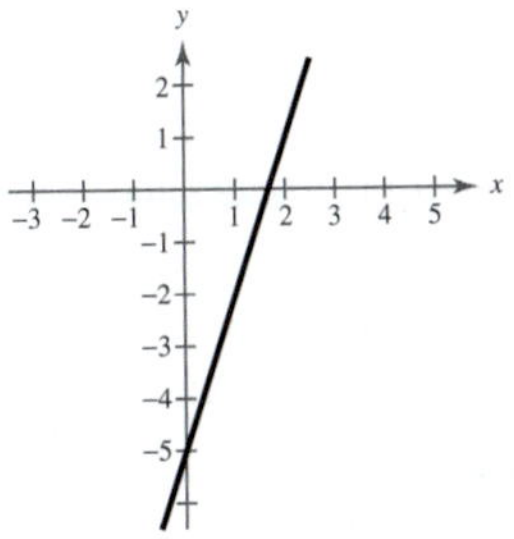

111.

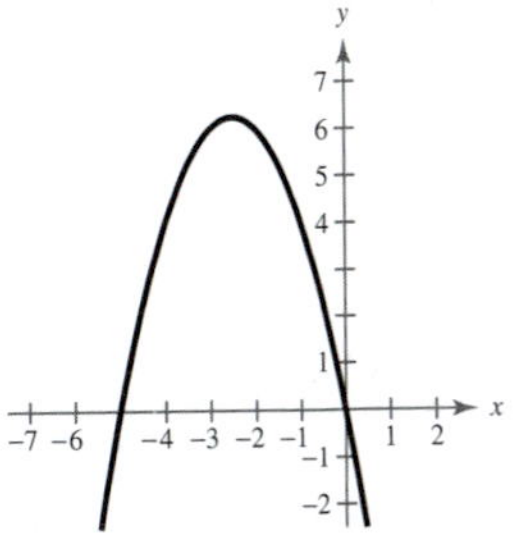

Section 1.3 *(page 104)*

1. A number increased by 4 **3.** A number divided by 5

5. A number decreased by 4 is divided by 5.

7. Negative 3 is multiplied by a number increased by 2.

9. 12 is multiplied by a number and that product is multiplied by the number decreased by 5.

11. $n + (n + 1) = 2n + 1$

13. $(2n - 1)(2n + 1) = 4n^2 - 1$

15. $50t$ **17.** $0.20x$ **19.** $6x$ **21.** $25x + 1200$

23. $0.30L$ **25.** $N = p(500)$ **27.** $4x + 8x = 12x$

29. 262, 263 **31.** 37, 185 **33.** $-5, -4$

35. 13.5 **37.** 27% **39.** 2400 **41.** \$22,316.98

43. \$1450.00 **45.** 14.6% increase **47.** 54.3% decrease

49. (a)

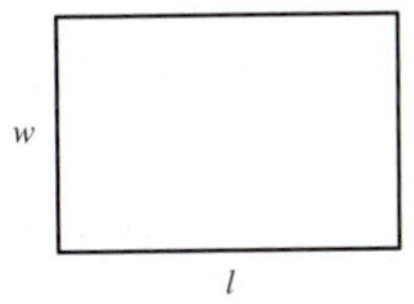

(b) $l = 1.5w;\ p = 5w$

(c) 7.5 meters × 5 meters

51. 97 **53.** 3 hours

55. 3 hours at 58 miles per hour; 2 hours and 45 minutes at 52 miles per hour

57. (a) 3.8 hours, 3.2 hours (b) 1.1 hours

(c) 25.6 miles

59. $66\frac{2}{3}$ kilometers per hour **61.** 1.29 seconds

63. 1044 feet **65.** 4.36 feet **67.** \$16,666.67

69. Compact: \$450,000; Midsize: \$150,000

71. ≈ 32.1 gallons **73.** 50 pounds of each kind

75. 8064 units **77.** $x = 6$ feet from the 50-pound child

79. $\dfrac{2A}{b}$ **81.** $\dfrac{S}{1 + R}$ **83.** $\dfrac{3V}{4\pi a^2}$ **85.** $\dfrac{2(h - v_0 t)}{t^2}$

87. $\dfrac{CC_2}{C_2 - C}$ **89.** $\dfrac{L - a + d}{d}$

91. $\sqrt[3]{\dfrac{4.47}{\pi}} \approx 1.12$ inches

93. False. The expression should be $\dfrac{z^3 - 8}{z^2 - 9}$.

95. (a) Negative; explanations will vary.

(b) Positive; explanations will vary.

97. $\dfrac{x^2}{5},\ x \neq 0$ **99.** $\dfrac{1}{x - 5}$ **101.** $\dfrac{10\sqrt{3}}{21}$

103. $-(\sqrt{6} - \sqrt{11})$

Section 1.4 *(page 118)*

1. $2x^2 + 8x - 3 = 0$ **3.** $x^2 - 6x + 6 = 0$

5. $3x^2 - 90x - 10 = 0$ **7.** $0, -\frac{1}{2}$ **9.** $4, -2$

11. -5 **13.** $3, -\frac{1}{2}$ **15.** $2, -6$ **17.** $-\frac{20}{3}, -4$

19. $-a$ **21.** $\pm 7; \pm 7.00$ **23.** $\pm\sqrt{11}; \pm 3.32$

25. $\pm 3\sqrt{3}; \pm 5.20$ **27.** 8, 16; 8.00, 16.00

29. $-2 \pm \sqrt{14}$; 1.74, -5.74 **31.** $\frac{1 \pm 3\sqrt{2}}{2}$; 2.62, -1.62

33. 2; 2.00 **35.** 0, 2 **37.** 4, -8 **39.** $-3 \pm \sqrt{7}$

41. $1 \pm \frac{\sqrt{6}}{3}$ **43.** $2 \pm 2\sqrt{3}$ **45.** $\frac{1}{(x+1)^2 + 4}$

47. $\frac{4}{(x+2)^2 - 7}$ **49.** $\frac{1}{\sqrt{9 - (x-3)^2}}$

51.

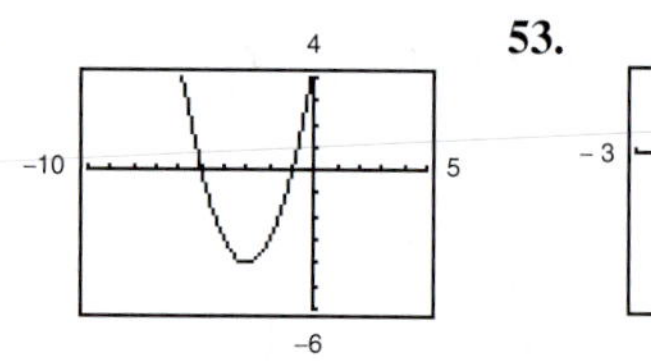

$x = -1, -5$

53.

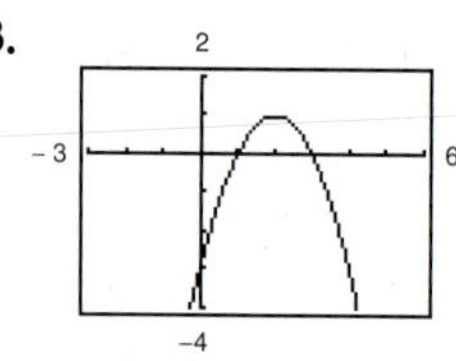

$x = 3, 1$

55.

5, -5, 7, -3

$x = -\frac{1}{2}, \frac{3}{2}$

57.

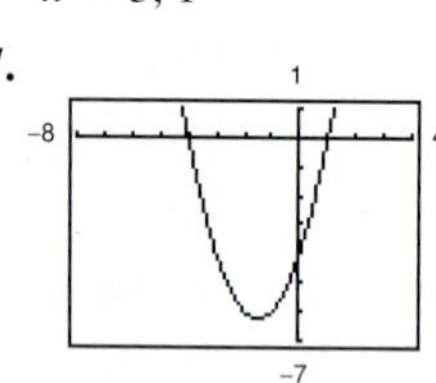

$x = 1, -4$

59. No real solution **61.** Two real solutions

63. No real solution **65.** Two real solutions

67. $\frac{1}{2}, -1$ **69.** $\frac{1}{4}, -\frac{3}{4}$ **71.** $1 \pm \sqrt{3}$

73. $-7 \pm \sqrt{5}$ **75.** $-4 \pm 2\sqrt{5}$ **77.** $\frac{2}{3} \pm \frac{\sqrt{7}}{3}$

79. $-\frac{4}{3}$ **81.** $-\frac{1}{2} \pm \sqrt{2}$ **83.** $\frac{2}{7}$

85. $2 \pm \frac{\sqrt{6}}{2}$ **87.** $6 \pm \sqrt{11}$ **89.** $-\frac{3}{8} \pm \frac{\sqrt{265}}{8}$

91. 0.976, -0.643 **93.** 1.355, -14.071

95. 1.687, -0.488 **97.** $-0.290, -2.200$

99. $1 \pm \sqrt{2}$ **101.** 6, -12 **103.** $\frac{1}{2} \pm \sqrt{3}$

105. $-\frac{1}{2}$ **107.** $\frac{3}{4} \pm \frac{\sqrt{97}}{4}$

109. (a) w; $w + 14$

(b) $w(w + 14) = 1632$

(c) $w = 34$ feet

$l = 48$ feet

111. 6 inches × 6 inches × 2 inches

113. 19.098 feet; 9.5 trips

115. (a) $20\sqrt{5} \approx 44.72$ seconds

(b) The bomb will travel approximately 7.5 miles horizontally.

117. (a) $s = -16t^2 + 45t + 5.5$ (b) 24 feet

(c) ≈ 2.8 seconds

119. $\frac{5\sqrt{2}}{2} \approx 3.54$ centimeters

121. ≈ 550 miles per hour and 600 miles per hour

123. 50,000 units **125.** 258 units

127. 653 units **129.** 1990. Yes.

131. (a) 12, 2, 0, 9

(b) No. Using the model, sales will never reach 12 billion dollars.

133. False. $b^2 - 4ac < 0$, so the quadratic equation has no real solution.

135. Yes. The student should have subtracted $15x$ from both sides to make the right side of the equation equal to zero. Factoring out an x shows that there are two solutions, $x = 0$ and $x = 6$.

137. (a) and (b) $x = -5, -\frac{10}{3}$

(c) The method used in part (a) reduces the number of algebraic steps.

139. Answers will vary. Sample answer: $x^2 - 3x - 18 = 0$

141. $x^2 - 22x + 112 = 0$

143. Associative Property of Multiplication

145. Additive Inverse Property **147.** $x^2 - 3x - 18$

149. $x^3 + 3x^2 - 2x + 8$

Section 1.5 *(page 128)*

1. $a = -10,\ b = 6$ **3.** $a = 6,\ b = 5$ **5.** $4 + 3i$

7. $2 - 3\sqrt{3}i$ **9.** $5\sqrt{3}i$ **11.** 8 **13.** $-1 - 6i$

15. $0.3i$ **17.** $11 - i$ **19.** 4 **21.** $3 - 3\sqrt{2}i$

23. $-14 + 20i$ **25.** $\frac{1}{6} + \frac{7}{6}i$ **27.** $-2\sqrt{3}$ **29.** -10

31. $5 + i$ **33.** $12 + 30i$ **35.** 24 **37.** $-9 + 40i$

39. -10 **41.** $6 - 3i$, 45 **43.** $-1 + \sqrt{5}i$, 6

45. $-2\sqrt{5}i$, 20 **47.** $\sqrt{8}$, 8 **49.** $-5i$

51. $\frac{8}{41} + \frac{10}{41}i$ **53.** $\frac{4}{5} + \frac{3}{5}i$ **55.** $-5 - 6i$

57. $-\frac{120}{1681} - \frac{27}{1681}i$ **59.** $-\frac{1}{2} - \frac{5}{2}i$ **61.** $\frac{62}{949} + \frac{297}{949}i$

63. $1 \pm i$ **65.** $-2 \pm \frac{1}{2}i$ **67.** $-\frac{3}{2}, -\frac{5}{2}$

69. $2 \pm \sqrt{2}i$ **71.** $\frac{5}{7} \pm \frac{5\sqrt{15}}{7}$ **73.** $-1 + 6i$

75. $-5i$ **77.** $-375\sqrt{3}i$ **79.** i

81. (a) 8 (b) 8 (c) 8

83. (a) 1 (b) i (c) -1 (d) $-i$

85. False. If the complex number is real, the number equals its conjugate.

87. False.
$i^{44} + i^{150} - i^{74} - i^{109} + i^{61} = 1 - 1 + 1 - i + i = 1$

89. Answers will vary. **91.** $-x^2 - 3x + 12$

93. $3x^2 + \frac{23}{2}x - 2$ **95.** -31 **97.** $\frac{27}{2}$

99. $a = \dfrac{\sqrt{3V\pi b}}{2\pi b}$ **101.** 1 liter

Section 1.6 *(page 137)*

1. $0, \pm\dfrac{3\sqrt{2}}{2}$ **3.** $\pm 3, \pm 3i$ **5.** $-6, 3 \pm 3\sqrt{3}i$

7. $-3, 0$ **9.** $3, 1, -1$ **11.** $\pm 1, \dfrac{1}{2} \pm \dfrac{\sqrt{3}}{2}i$

13. $\pm\sqrt{3}, \pm 1$ **15.** $\pm\frac{1}{2}, \pm 4$

17. $1, -2, 1 \pm \sqrt{3}i, -\dfrac{1}{2} \pm \dfrac{\sqrt{3}i}{2}$

19. $-\frac{1}{5}, -\frac{1}{3}$ **21.** $\frac{1}{4}$ **23.** $1, -\frac{125}{8}$

25. (a)

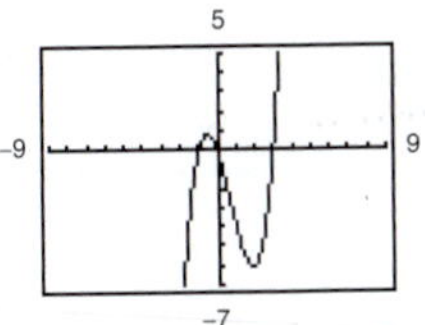

(b) $(0, 0), (3, 0), (-1, 0)$

(c) $x = 0, 3, -1$

(d) The x-intercepts and the solutions are the same.

27. (a)

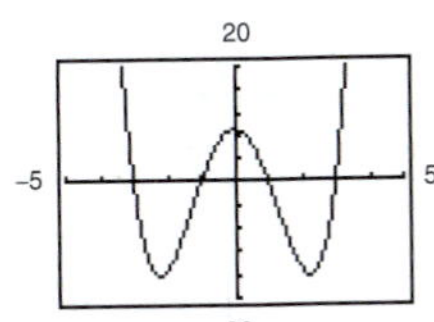

(b) $(\pm 3, 0), (\pm 1, 0)$

(c) $x = \pm 3, \pm 1$

(d) The x-intercepts and the solutions are the same.

29. 50 **31.** 26 **33.** -16 **35.** $2, -5$

37. 0 **39.** 9 **41.** $\frac{101}{4}$ **43.** 14 **45.** 9

47. $-3 \pm 16\sqrt{2}$ **49.** $\pm\sqrt{14}$ **51.** 1

53. (a)

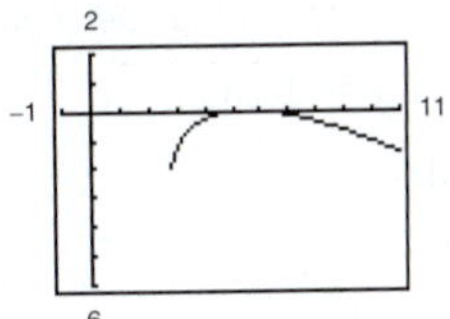

(b) $(5, 0), (6, 0)$

(c) $x = 5, 6$

(d) The x-intercepts and the solutions are the same.

55. (a)

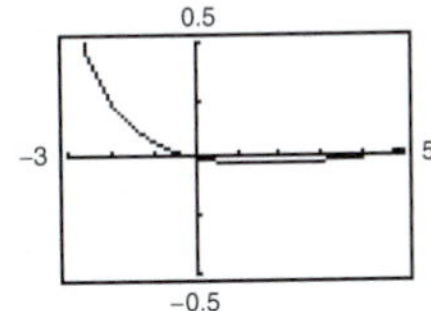

(b) $(0, 0), (4, 0)$

(c) $x = 0, 4$

(d) The x-intercepts and the solutions are the same.

57. $2, -\dfrac{3}{2}$ **59.** $\dfrac{-3 \pm \sqrt{21}}{6}$ **61.** $4, -5$

63. $\dfrac{1 \pm \sqrt{31}}{3}$ **65.** $3, -2$ **67.** $\sqrt{3}, -3$

69. $3, \dfrac{-1 - \sqrt{17}}{2}$

71. (a)

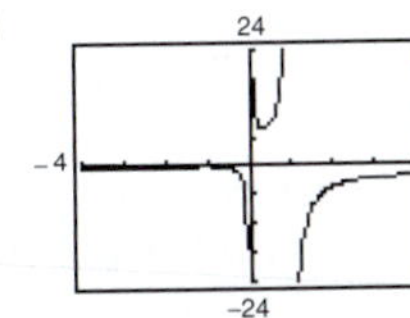

(b) $(-1, 0)$

(c) $x = -1$

(d) The x-intercept and the solution are the same.

73. (a)

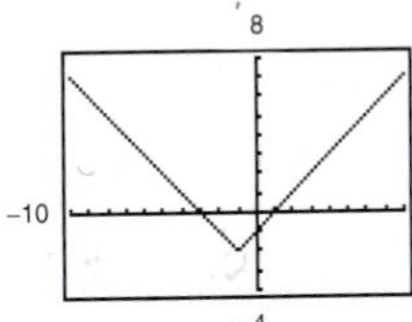

(b) $(1, 0), (-3, 0)$

(c) $x = 1, -3$

(d) The x-intercepts and the solutions are the same.

75. ± 1.038 **77.** 16.756 **79.** $x^2 - 3x - 10 = 0$

81. $21x^2 + 31x - 42 = 0$ **83.** $x^3 - 4x^2 - 3x + 12 = 0$

85. $x^4 - 1 = 0$ **87.** 34 students

89. 191.5 miles per hour

91. 4%

93. (a)

x	T
5	162.56
10	192.31
15	212.68
20	228.20
25	240.62
30	250.83
35	259.38
40	266.60

(b) ≈ 15 pounds per square inch

(c) $x \approx 14.81$

(d)

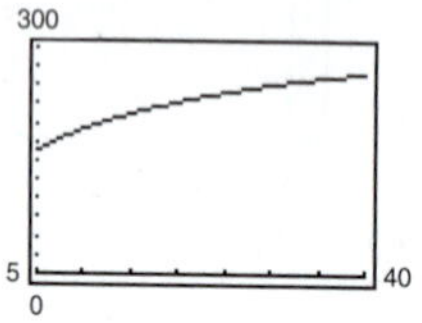

95. 500 units **97.** 90 feet

99. (a)

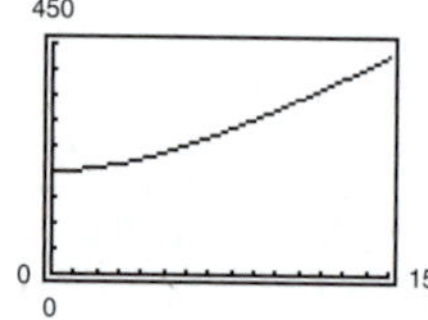

The height $h \approx 11.4$ when $S = 350$.

(b)

h	8	9	10	11	12	13
S	284.3	302.6	321.9	341.8	362.5	383.6

The height h is between 11 and 12 inches when $S = 350$.

(c) $h = 11.4$ when $S = 350$.

(d) Solving graphically or numerically yields an approximate solution. An exact solution is obtained algebraically.

101. $\dfrac{21 + \sqrt{585}}{2} \approx 23$ hours

$\dfrac{27 + \sqrt{585}}{2} \approx 26$ hours

103. $g = \dfrac{\mu s v^2}{R}$ **105.** False. See Example 7 on page 134.

107. $6, -4$ **109.** ± 15 **111.** $a = 9, b = 9$

113. $a = 4, b = 24$ **115.** $\dfrac{25}{6x}$

117. $\dfrac{-3z^2 - 2z + 4}{z(z + 2)}$ **119.** 11

Section 1.7 *(page 147)*

1. $-1 \le x \le 5$. Bounded **3.** $x > 11$. Unbounded

5. $x < -2$. Unbounded

7. b **8.** f **9.** d **10.** c **11.** e **12.** a

13. (a) Yes (b) No (c) Yes (d) No

15. (a) Yes (b) No (c) No (d) Yes

17. (a) Yes (b) Yes (c) Yes (d) No

19. $x < 3$

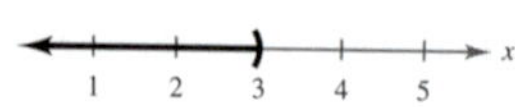

21. $x < \frac{3}{2}$

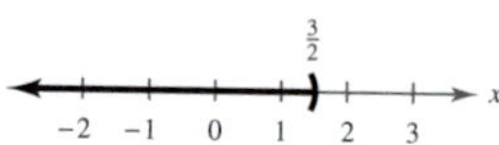

23. $x \ge 12$

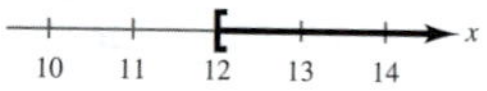

25. $x > 2$

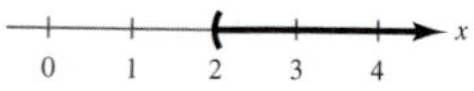

27. $x \ge \frac{2}{7}$

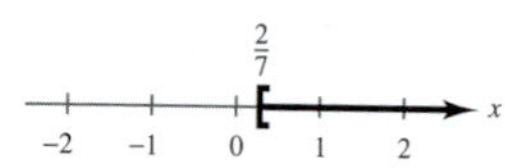

29. $x < 5$

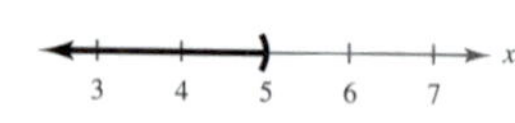

31. $x \ge 4$

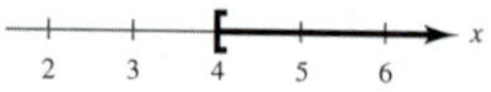

33. $x \ge 2$

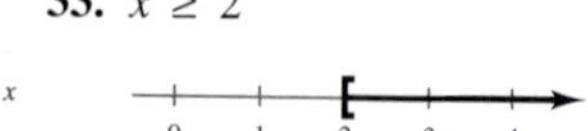

35. $x \ge -4$

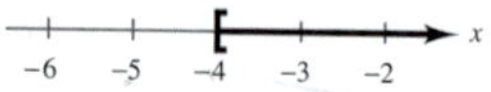

37. $-1 < x < 3$

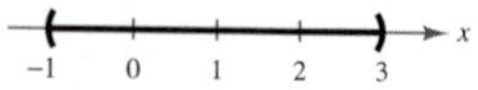

39. $-\frac{9}{2} < x < \frac{15}{2}$

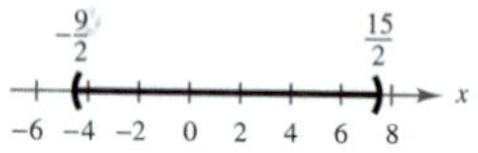

41. $-\frac{3}{4} < x < -\frac{1}{4}$

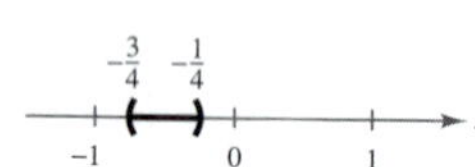

43. $10.5 \le x \le 13.5$

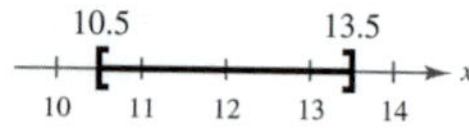

45. $-6 < x < 6$

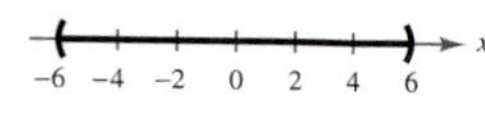

47. $x < -2, x > 2$

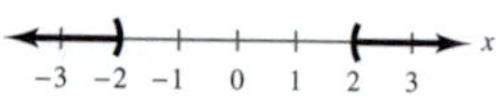

49. No solution

51. $14 \le x \le 26$

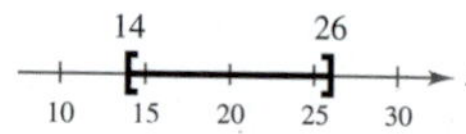

53. $x \le -\frac{3}{2}, x \ge 3$

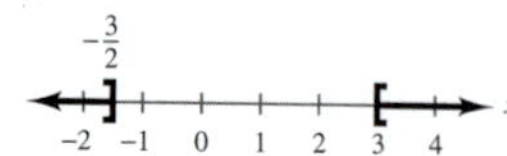

55. $x \le -5,\ x \ge 11$

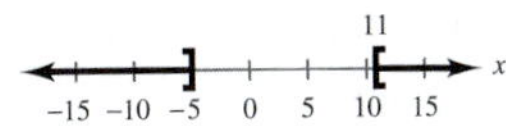

57. $4 < x < 5$

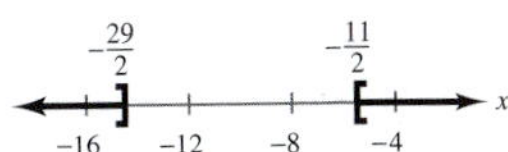

59. $x \le -\frac{29}{2},\ x \ge -\frac{11}{2}$

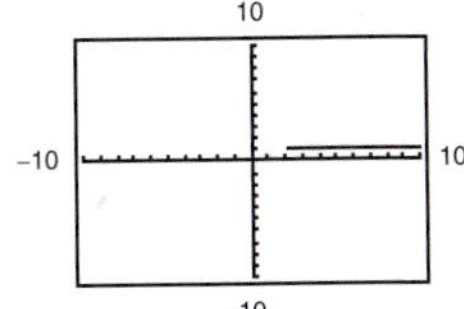

61.

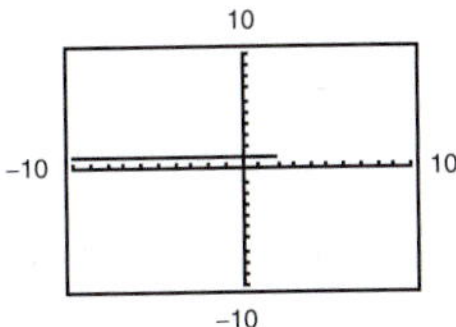

$x > 2$

63.

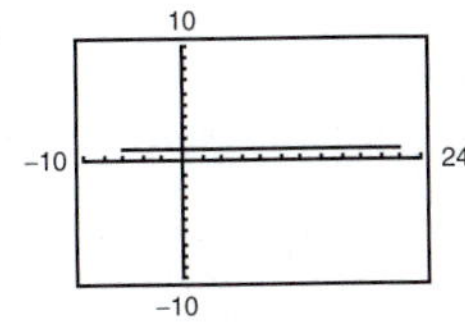

$x \le 2$

65.

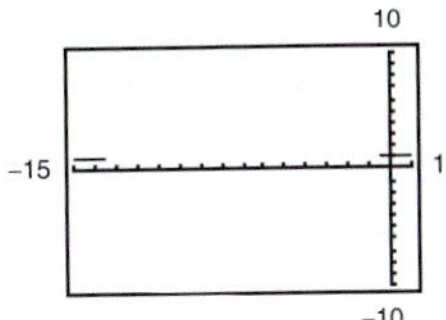

$-6 \le x \le 22$

67.

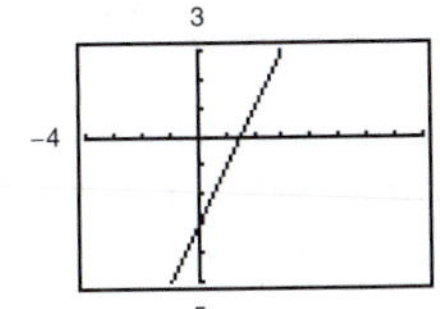

$x \le -\frac{27}{2}, x \ge -\frac{1}{2}$

69.

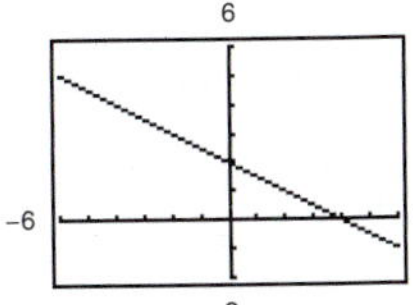

(a) $x \ge 2$

(b) $x \le \frac{3}{2}$

71.

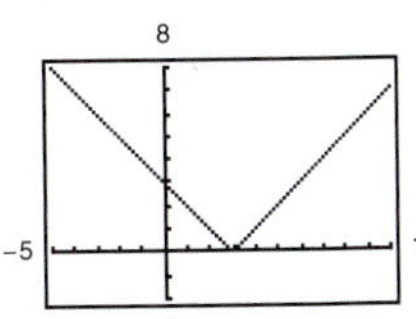

(a) $4 \ge x \ge -2$

(b) $x \le 4$

73.

(a) $1 \le x \le 5$

(b) $x \le -1,\ x \ge 7$

75. $[5, \infty)$ **77.** $[-3, \infty)$ **79.** $\left(-\infty, \frac{7}{2}\right]$

81. All real numbers within 8 units of 10

83. $|x| \le 3$ **85.** $|x - 7| \ge 3$ **87.** $|x - 12| < 10$

89. $|x + 3| > 5$ **91.** More than 400 miles

93. $r > 3.125\%$ **95.** $x \ge 36$

97. (a)

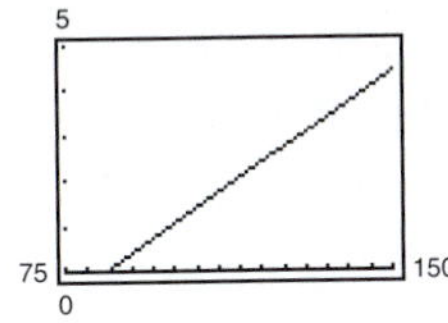

(b) $x \ge 129$

99. $t \ge 21.2$, or during 2001

101. 106.864 square inches $\le$ area $\le$ 109.464 square inches

103. 6¢

105. $13.7 < t < 17.5$

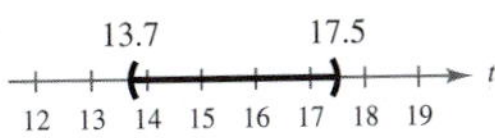

107. $20 \le h \le 80$

109. False. c has to be greater than zero. **111.** b

113. $5\sqrt{5}; \left(-\frac{3}{2}, 7\right)$ **115.** $2\sqrt{65}; (-1, -1)$ **117.** 11

119. 10 **121.** $-\frac{1}{2}, 10$ **123.** $\frac{1}{7}, -\frac{1}{2}$ **125.** $(-3, 10)$

Section 1.8 *(page 158)*

1. (a) No (b) Yes (c) Yes (d) No

3. (a) Yes (b) No (c) No (d) Yes

5. $2, -\frac{3}{2}$ **7.** $\frac{7}{2}, 5$

9. $[-3, 3]$

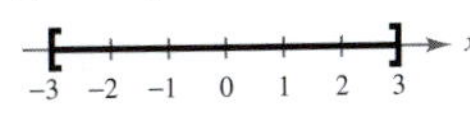

11. $(-7, 3)$

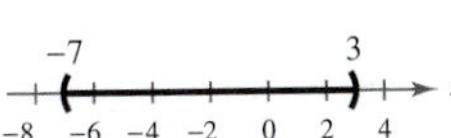

13. $(-\infty, -5] \cup [1, \infty)$

15. $(-3, 2)$

17. $(-3, 1)$

19. $\left(-\infty, -4 - \sqrt{21}\right] \cup \left[-4 + \sqrt{21}, \infty\right)$

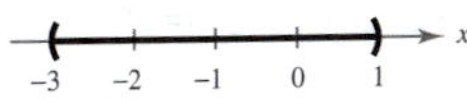

21. $(-1, 1) \cup (3, \infty)$

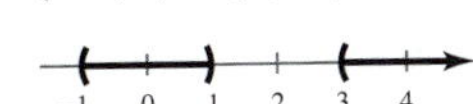

23. $[-3, 2] \cup [3, \infty)$

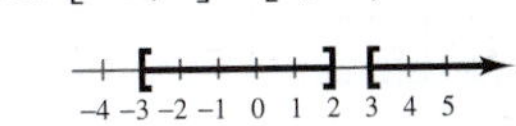

25. $(-\infty, 0) \cup \left(0, \frac{3}{2}\right)$ **27.** $[-2, 0] \cup [2, \infty)$

29. $[-2, \infty)$

31.

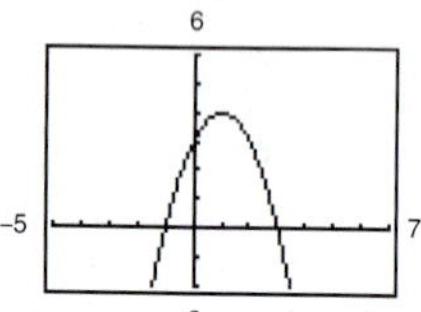

(a) $x \leq -1,\ x \geq 3$

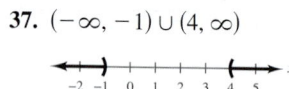

(b) $0 \leq x \leq 2$

33.

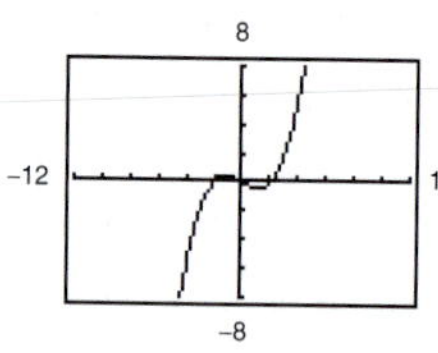

(a) $-2 \leq x \leq 0$,

$2 \leq x < \infty$

(b) $x \leq 4$

35. $(-\infty, -1) \cup (0, 1)$

37. $(-\infty, -1) \cup (4, \infty)$

39. $(5, 15)$

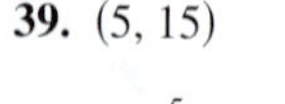

41. $\left(-5, -\frac{3}{2}\right) \cup (-1, \infty)$

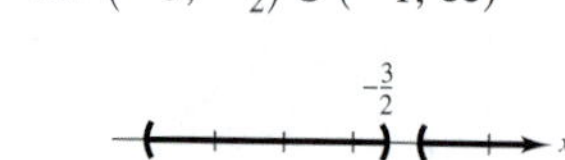

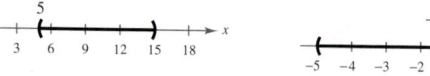

43. $\left(-\frac{3}{4}, 3\right) \cup [6, \infty)$

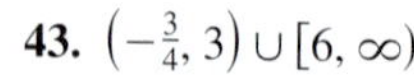

45. $(-3, -2] \cup [0, 3)$

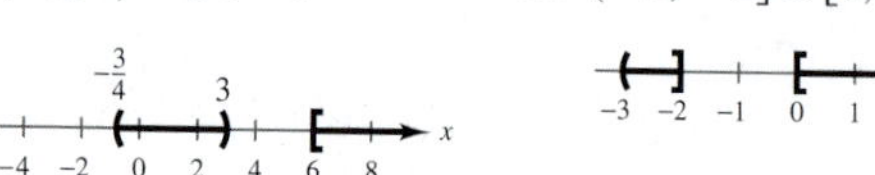

47. $(-\infty, -1) \cup \left(-\frac{2}{3}, 1\right) \cup (3, \infty)$

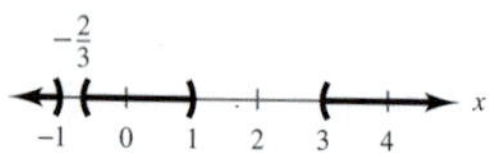

49.

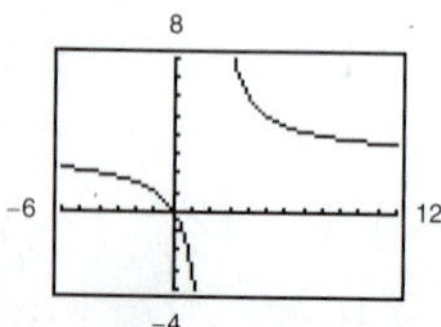

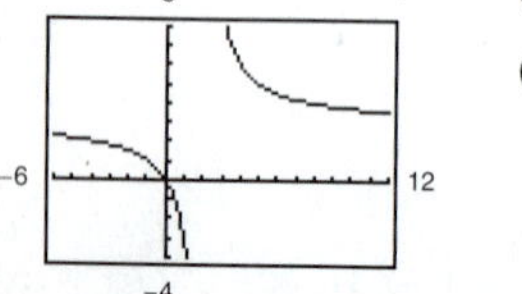

(a) $0 \leq x < 2$

(b) $2 < x \leq 4$

51.

(a) $|x| \geq 2$

(b) $-\infty < x < \infty$

53. $[-2, 2]$ **55.** $(-\infty, 3] \cup [4, \infty)$

57. $(-5, 0] \cup (7, \infty)$ **59.** $(-3.51, 3.51)$

61. $(-0.13, 25.13)$ **63.** $(2.26, 2.39)$

65. (a) $t = 10$ seconds (b) 4 seconds $< t <$ 6 seconds

67. 13.8 meters $\leq L \leq$ 36.2 meters **69.** $r > 4.88\%$

71. (a)

(b)

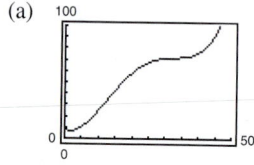

t	30	32	34	36	38	40
C	71.00	71.41	71.68	72.20	73.45	76.00

2006

(c) $t \approx 36$

(d)

t	40	41	42	43
C	76.00	77.95	80.48	83.65

t	44	45	46	47
C	87.59	92.38	98.12	104.94

2006 to 2016

(e) $36 < t < 46$

(f) Answers will vary. Sample answer: The model is derived from the data for 1970 to 2002. Therefore it is not necessarily accurate for years prior to 1970 or after 2002.

73. $R_1 \geq 2$ ohms

75. True. The test intervals are $(-\infty, -3)$, $(-3, 1)$, $(1, 4)$, and $(4, \infty)$.

77. $(-\infty, -4] \cup [4, \infty)$

79. $\left(-\infty, -2\sqrt{30}\right] \cup \left[2\sqrt{30}, \infty\right)$

81. (a) If $a > 0$ and $c \leq 0$, b can be any real number. If $a > 0$ and $c > 0$, $b < -2\sqrt{ac}$ or $b > 2\sqrt{ac}$.

(b) 0

83. $(2x + 5)^2$ **85.** $(x + 3)(x + 2)(x - 2)$

87. $2x^2 + x$

Review Exercises *(page 162)*

1.

x	-2	-1	0	1	2
y	-11	-8	-5	-2	1

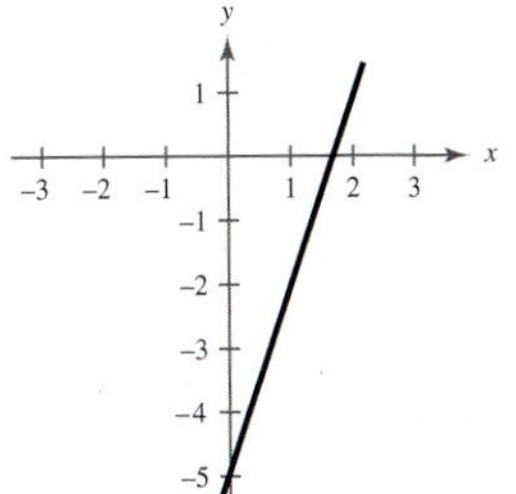

3.

x	-1	0	1	2	3	4
y	4	0	-2	-2	0	4

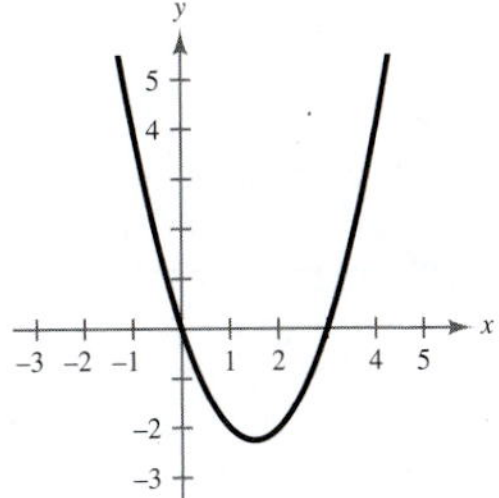

5.

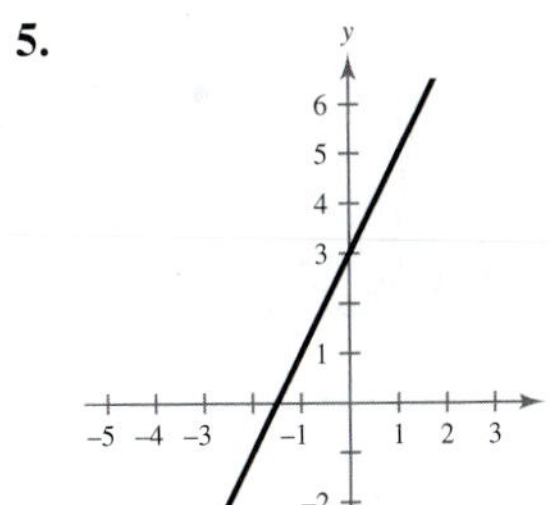

7.

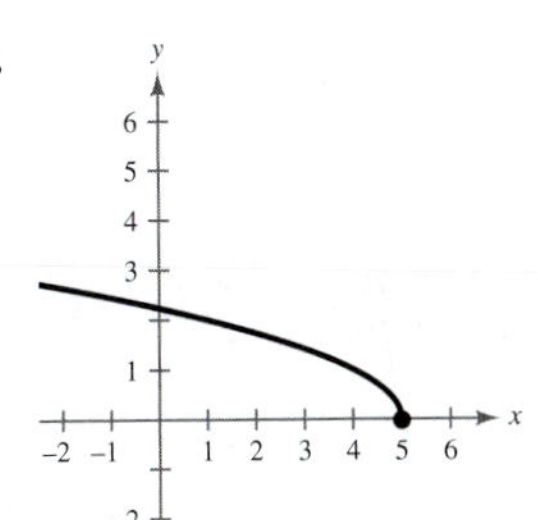

9.

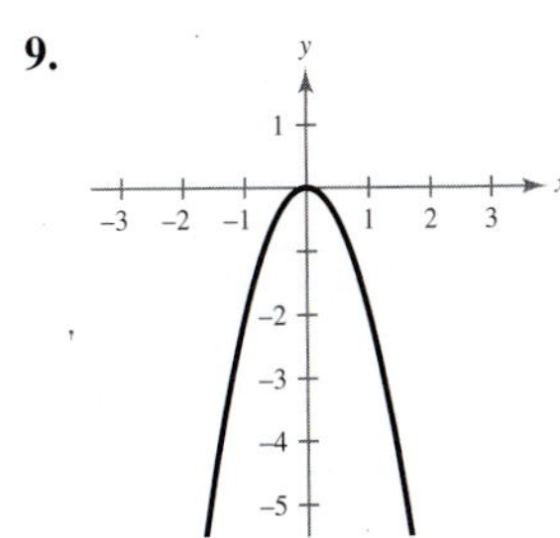

11. x-intercepts: $(1, 0)$, $(5, 0)$
y-intercept: $(0, 5)$

13. No symmetry

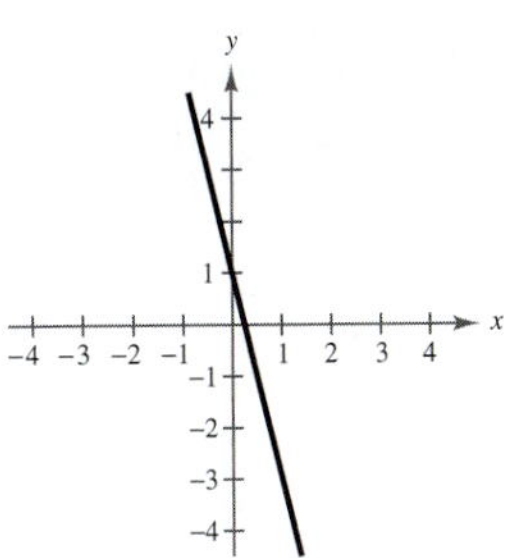

15. y-axis symmetry

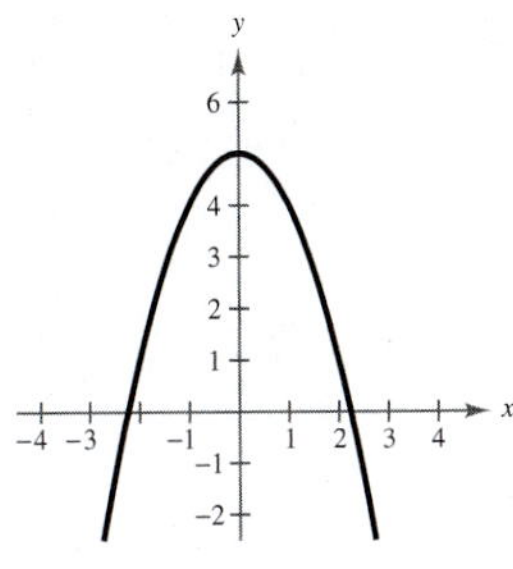

17. No symmetry

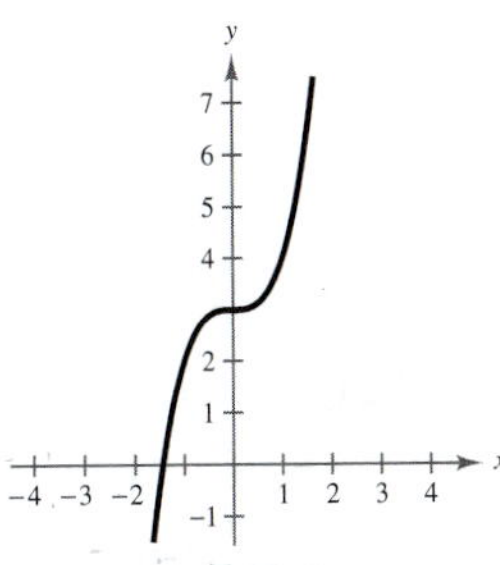

19. No symmetry

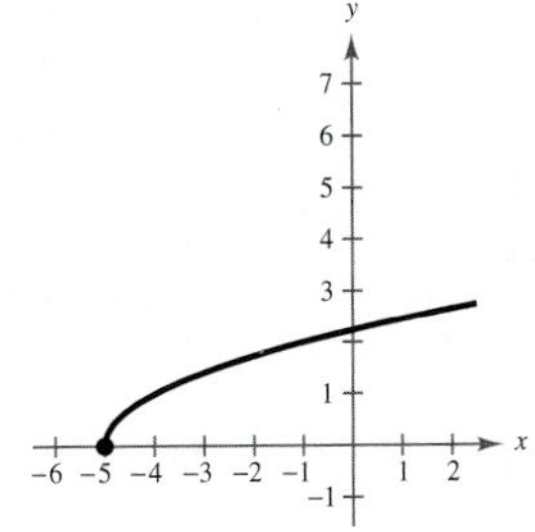

21. Center: $(0, 0)$; Radius: 3

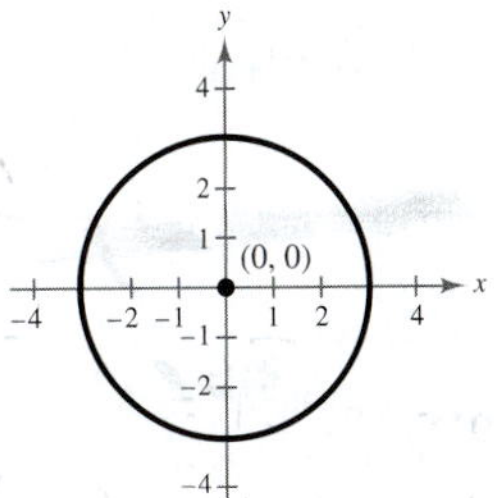

23. Center: $(-2, 0)$; Radius: 4

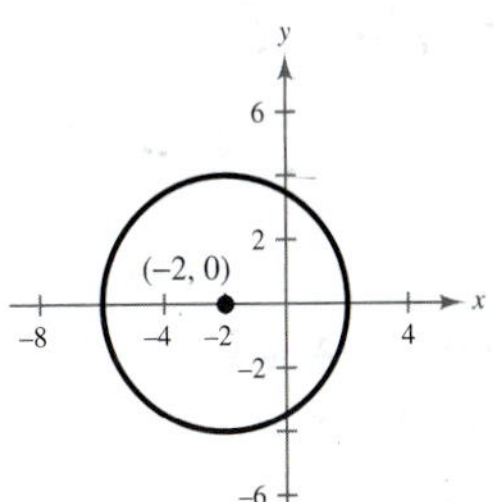

25. Center: $\left(\frac{1}{2}, -1\right)$; Radius: 6

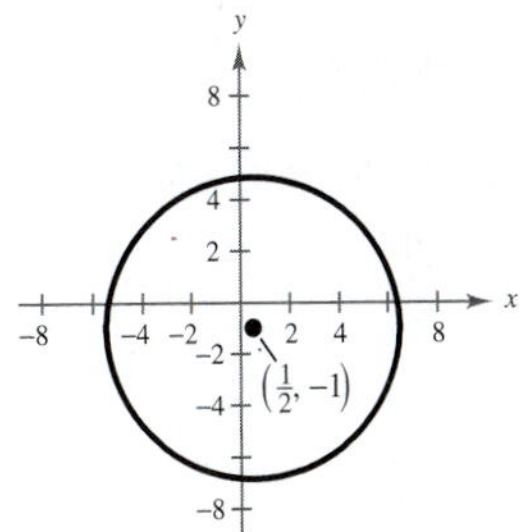

27. $(x - 2)^2 + (y + 3)^2 = 13$

29. (a)

x	0	4	8	12	16	20
F	0	5	10	15	20	25

(b)

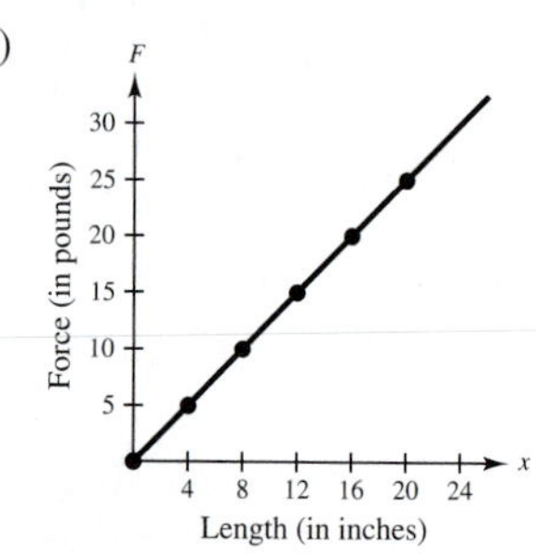

(c) 12.5 pounds

31. Identity **33.** Identity

35. 20 **37.** $-\frac{1}{2}$ **39.** -5 **41.** 9

43. x-intercept: $\left(\frac{1}{3}, 0\right)$
y-intercept: $(0, -1)$

45. x-intercept: $(4, 0)$
y-intercept: $(0, -8)$

47. x-intercept: $\left(\frac{4}{3}, 0\right)$
y-intercept: $\left(0, \frac{2}{3}\right)$

49. x-intercept: $(2, 0)$
y-intercept: $\left(0, -\frac{5}{19}\right)$

51. $h = 10$ inches

53. September: \$325,000; October: \$364,000

55. 24 feet **57.** Nine **59.** $2\frac{6}{7}$ liters **61.** $h = \dfrac{3V}{\pi r^2}$

63. $\frac{2}{3}$ hour **65.** $-\frac{5}{2}, 3$ **67.** $\pm\sqrt{2}$

69. $-4 \pm 3\sqrt{2}$ **71.** $6 \pm \sqrt{6}$ **73.** $-\dfrac{5}{4} \pm \dfrac{\sqrt{241}}{4}$

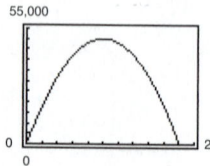

75. (a) $x = 0, 20$

(b) 55,000 0 0 22

(c) $x = 10$

77. $6 + 2i$ **79.** $-1 + 3i$ **81.** $3 + 7i$

83. $40 + 65i$ **85.** $-4 - 46i$ **87.** $\frac{23}{17} + \frac{10}{17}i$

89. $\frac{21}{13} - \frac{1}{13}i$ **91.** $\pm\dfrac{\sqrt{3}}{3}i$ **93.** $1 \pm 3i$ **95.** $0, \frac{12}{5}$

97. $\pm\sqrt{2}, \pm\sqrt{3}$ **99.** 5 **101.** No solution

103. $-124, 126$ **105.** $-2 \pm \dfrac{\sqrt{95}}{5}, -4$ **107.** $\pm\sqrt{10}$

109. $-5, 15$ **111.** 1, 3 **113.** 143,203 units

115. $-7 < x \le 2$. Bounded

117. $-\infty < x \le -10$. Unbounded **119.** $(-\infty, 12]$

121. $\left[\frac{32}{15}, \infty\right)$ **123.** $\left(-\frac{2}{3}, 17\right]$ **125.** $[-4, 4]$

127. $(-\infty, -1) \cup (7, \infty)$

129. 353.44 square centimeters $\le$ area $\le$ 392.04 square centimeters

131. $(-3, 9)$ **133.** $\left(-\frac{4}{3}, \frac{1}{2}\right)$ **135.** $[-5, -1) \cup (1, \infty)$

137. $[-4, -3] \cup (0, \infty)$ **139.** 4.9%

141. False. $\sqrt{-18}\sqrt{-2} = \left(3\sqrt{2}i\right)\left(\sqrt{2}i\right) = 6i^2 = -6$ and $\sqrt{(-18)(-2)} = \sqrt{36} = 6$

143. Some solutions to certain types of equations may be extraneous solutions, which do not satisfy the original equations. So, checking is crucial.

Chapter Test *(page 166)*

1. No symmetry

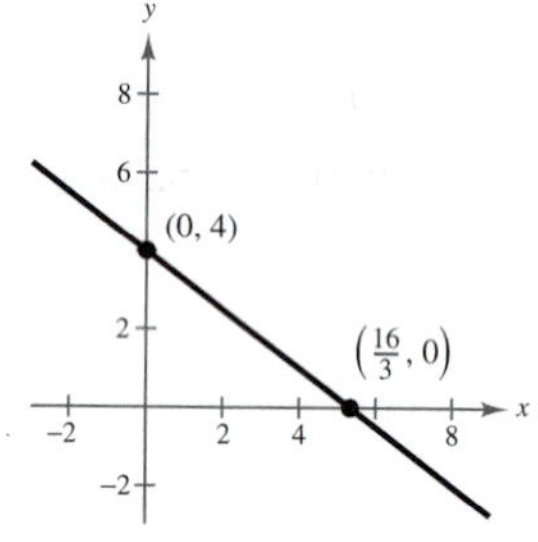

2. y-axis symmetry

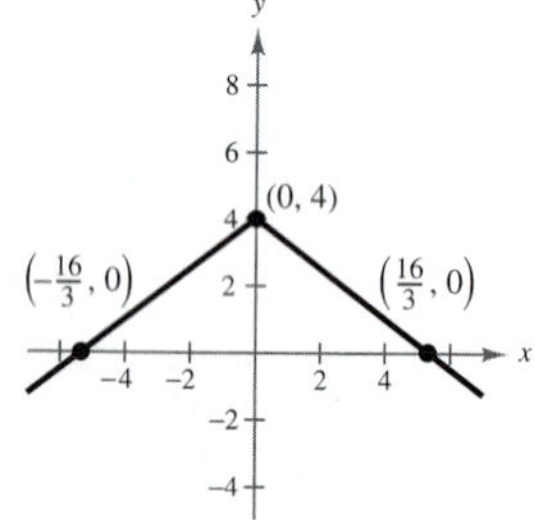

3. No symmetry

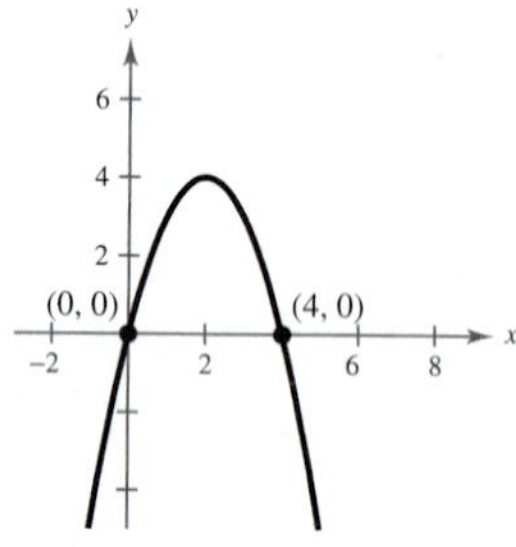

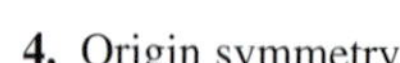

4. Origin symmetry

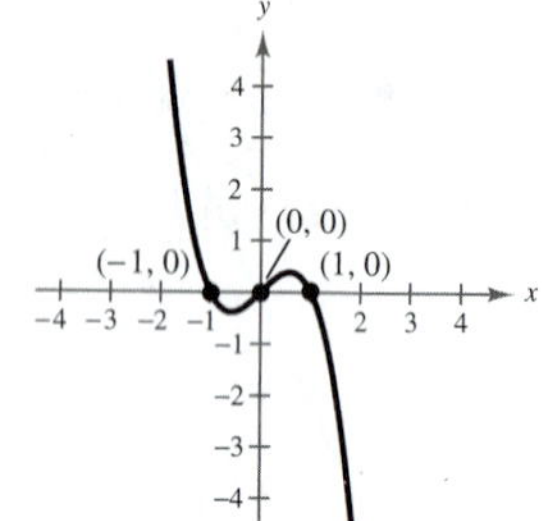

5. No symmetry

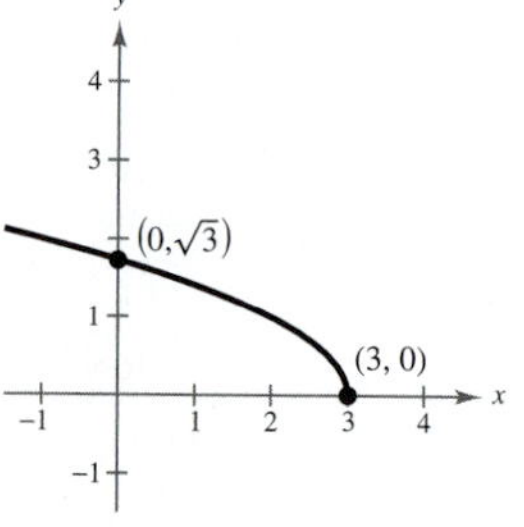

6. x-axis symmetry

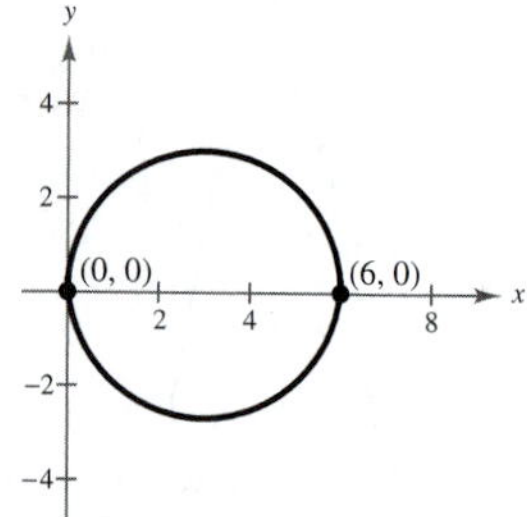

7. $\frac{128}{11}$ **8.** $-4, 5$ **9.** No solution

10. $\pm\sqrt{2}, \pm\sqrt{3}i$ **11.** 4 **12.** $-2, \frac{8}{3}$

13. $-\frac{11}{2} \le x < 3$

14. $x < -6$ or $0 < x < 4$

15. $x < -4$ or $x > \frac{3}{2}$

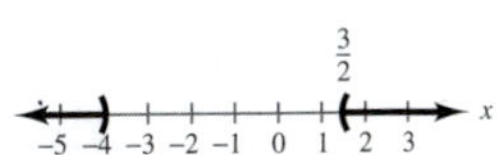

16. $x \le 10$ or $x \ge 20$

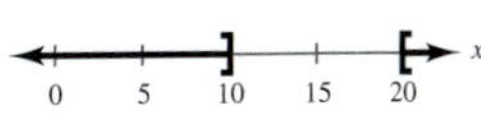

17. (a) $-3 + 5i$ (b) 7 **18.** $2 - i$

19. (a)

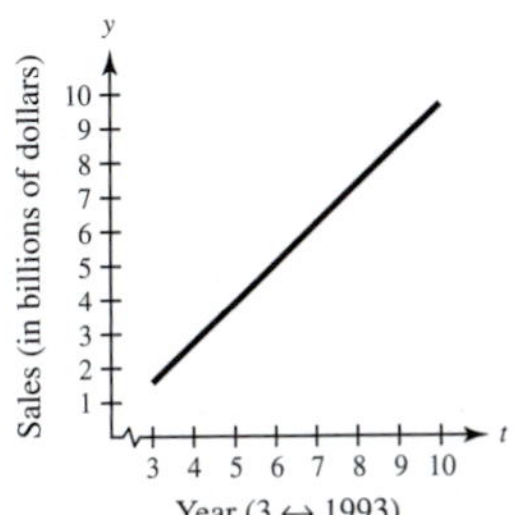

(b) \$14.3 billion

(c) \$14.34 billion

20. $93\frac{3}{4}$ kilometers per hour **21.** $a = 80, b = 20$

Problem Solving *(page 168)*

1.

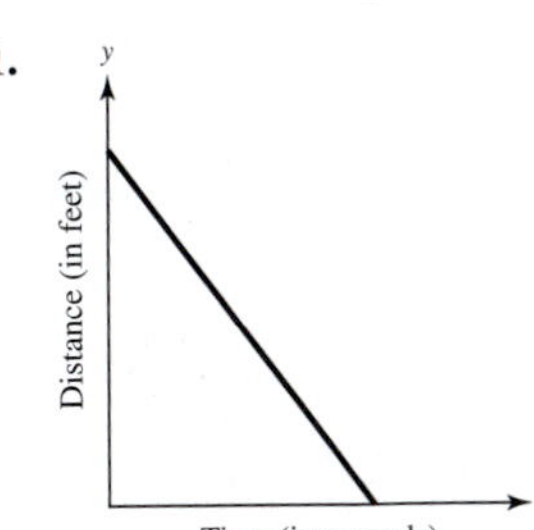

3. (a) Answers will vary.

Sample answer:

$A = \pi ab$

$b = 20 - a$, since $a + b = 20$

$A = \pi a(20 - a)$

(b)

a	4	7	10	13	16
A	64π	91π	100π	91π	64π

(c) $\dfrac{10\pi + 10\sqrt{\pi(\pi - 3)}}{\pi} \approx 12.12$ or $\dfrac{10\pi - 10\sqrt{\pi(\pi - 3)}}{\pi} \approx 7.88$

(d)

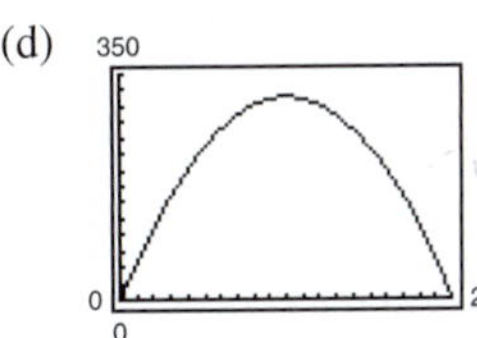

(e) (0, 0), (20, 0)

They represent the minimum and maximum values of a.

(f) 100π; $a = b = 10$

5. (a) $\approx$ 60.6 seconds (b) $\approx$ 146.2 seconds

(c) The speed at which water drains decreases as the amount of water in the bathtub decreases.

7. (a) 5, 12, 13; 8, 15, 17. Answers will vary. (b) Yes

(c) The product of the three numbers in a Pythagorean Triple is divisible by 60.

9. $x_1 + x_2 = -\dfrac{b}{a}; x_1 \cdot x_2 = \dfrac{c}{a}$

11. (a) $\frac{1}{2} - \frac{1}{2}i$ (b) $\frac{3}{10} + \frac{1}{10}i$ (c) $-\frac{1}{34} - \frac{2}{17}i$

13. (a) Yes (b) No (c) Yes

15. $(-\infty\ -2) \cup (-1, 2) \cup (2, \infty)$

Chapter 2

Section 2.1 *(page 181)*

1. (a) L_2 (b) L_3 (c) L_1

3.

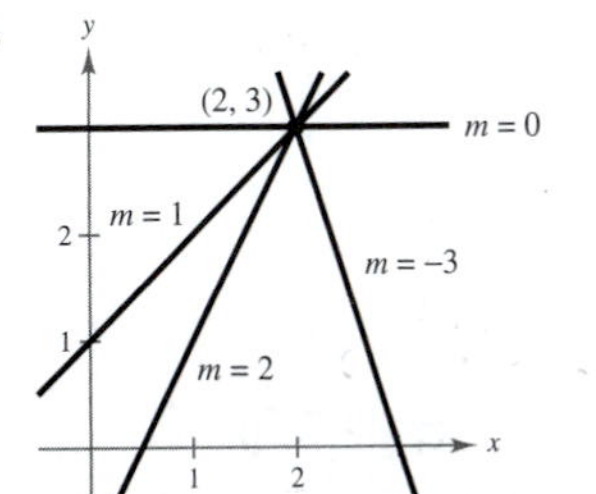

5. $\frac{8}{5}$ **7.** -4

9. $m = 5$; y-intercept: (0, 3) **11.** $m = -\frac{1}{2}$; y-intercept: (0, 4)

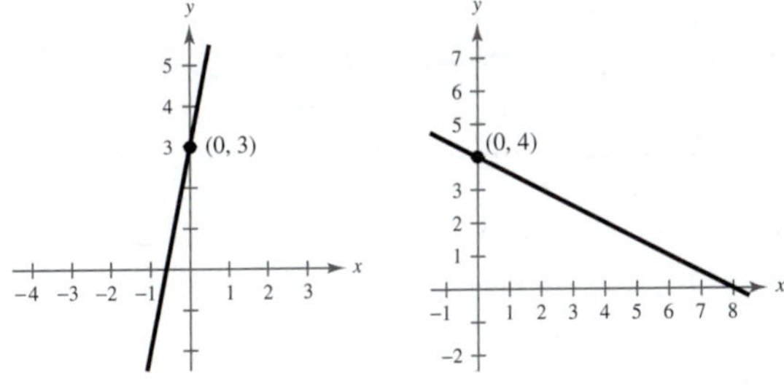

13. m is undefined. There is no y-intercept.

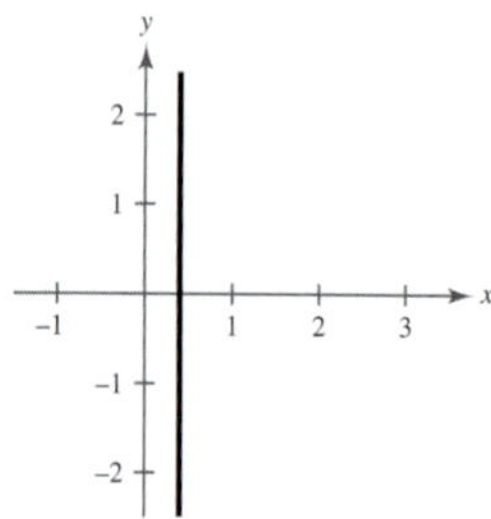

15. $m = -\frac{7}{6}$; y-intercept: $(0, 5)$

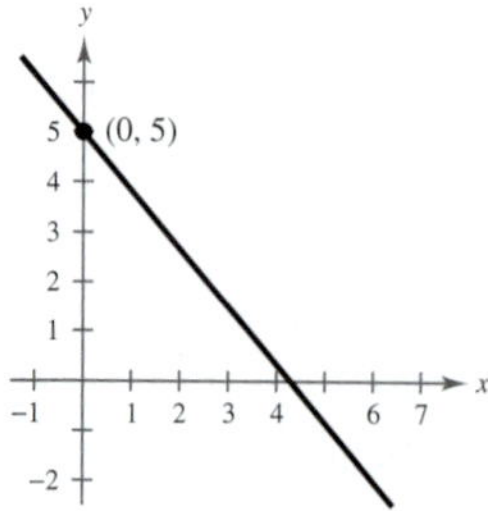

17. $m = 0$; y-intercept: $(0, 3)$

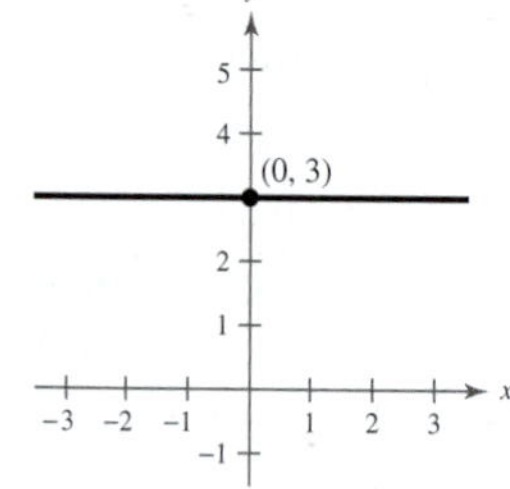

19. m is undefined. There is no y-intercept.

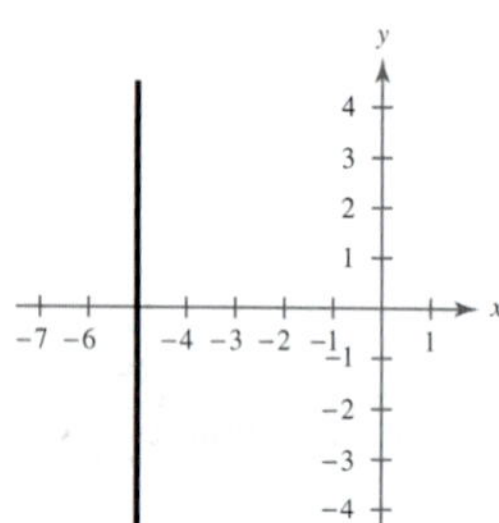

21.

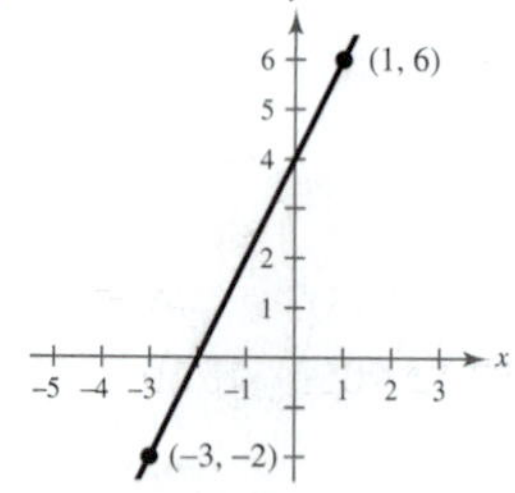

$m = 2$

23.

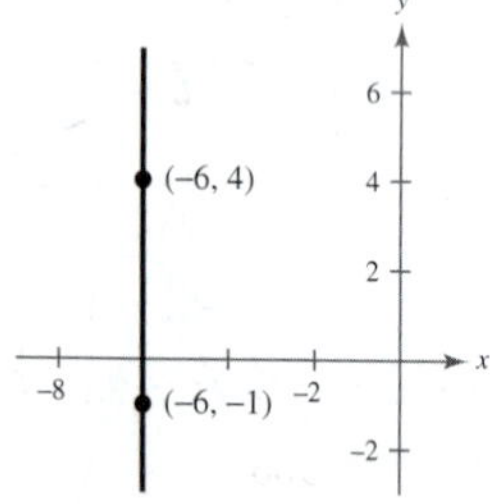

m is undefined.

25.

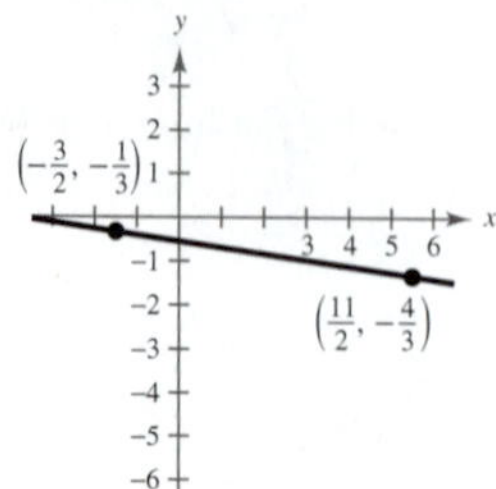

$m = -\frac{1}{7}$

27.

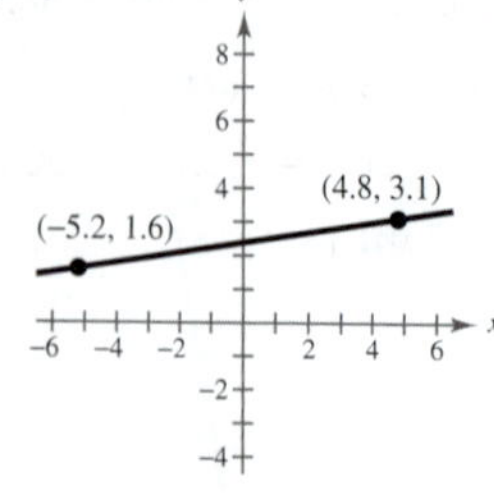

$m = 0.15$

29. Answers will vary. Sample answer: $(0, 1)$, $(3, 1)$, $(-1, 1)$

31. Answers will vary. Sample answer: $(6, -5)$, $(7, -4)$, $(8, -3)$

33. Answers will vary. Sample answer: $(-8, 0)$, $(-8, 2)$, $(-8, 3)$

35. Answers will vary. Sample answer: $(-4, 6)$, $(-3, 8)$, $(-2, 10)$

37. Answers will vary. Sample answer: $(9, -1)$, $(11, 0)$, $(13, 1)$

39. Perpendicular **41.** Parallel

43. (a) Sales increasing 135 units per year
(b) No change in sales
(c) Sales decreasing 40 units per year

45. (a) Greatest increase: 2000-2001
Smallest increase: 1991-1992
(b) 0.205
(c) Each year, the earnings per share increase by \$0.205.

47. (a) and (b)

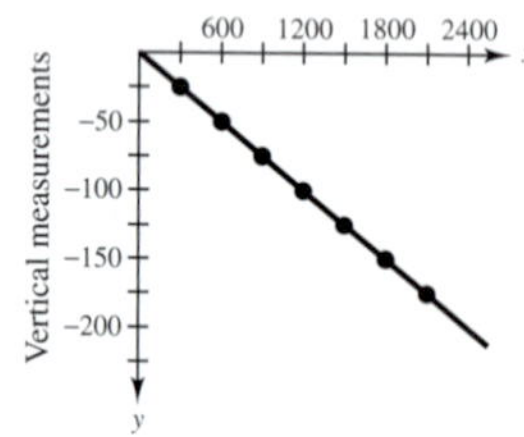

(c) $y = -\frac{1}{12}x$
(d) For every 12 horizontal measurements, the vertical measurement decreases by 1.
(e) "8.3% grade"

49. $V = 125t + 2165$

51. b; The slope is -20, which represents the decrease in the amount of the loan each week.

52. c; The slope is 2, which represents the hourly wage per unit produced.

53. a; The slope is 0.32, which represents the increase in travel cost for each mile driven.

54. d; The slope is -100, which represents the decrease in the value of the word processor each year.

55. $y = 3x - 2$

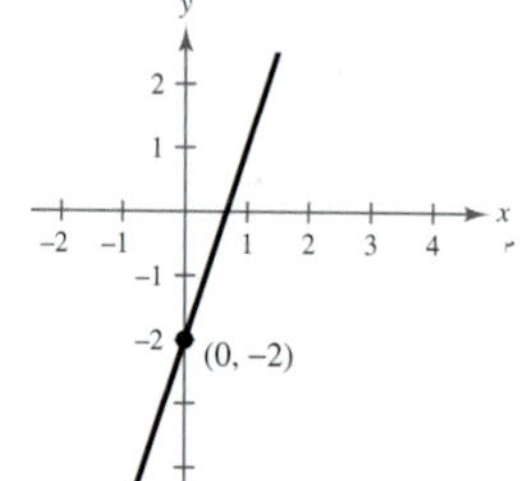

57. $y = -2x$

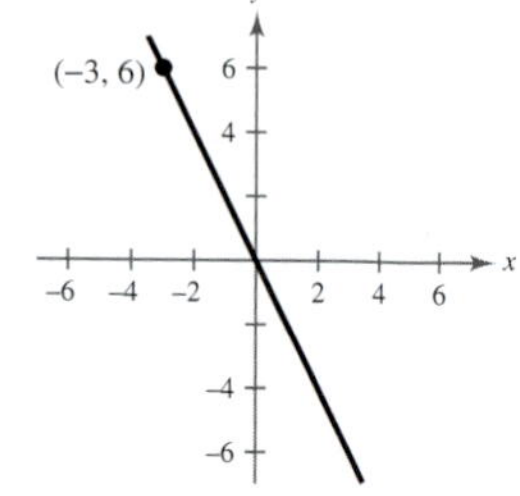

59. $y = -\frac{1}{3}x + \frac{4}{3}$

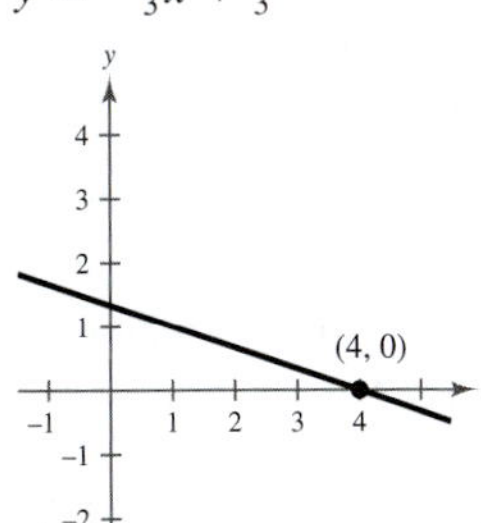

61. $x = 6$

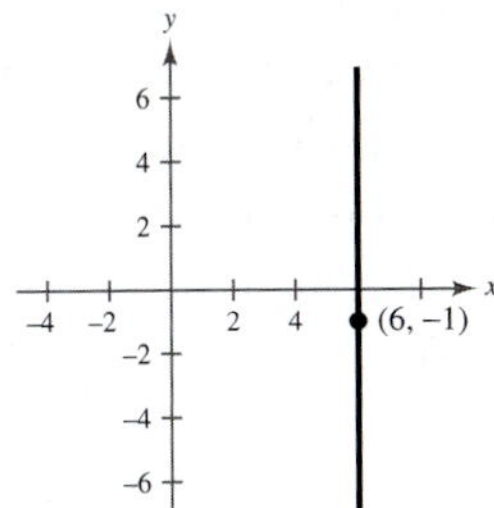

63. $y = \frac{5}{2}$

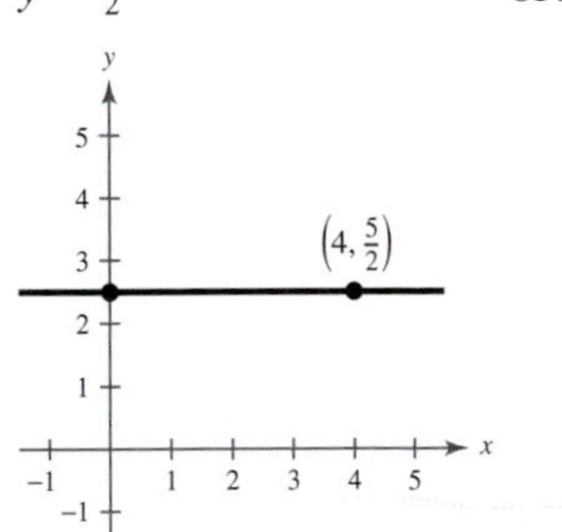

65. $y = 5x + 27.3$

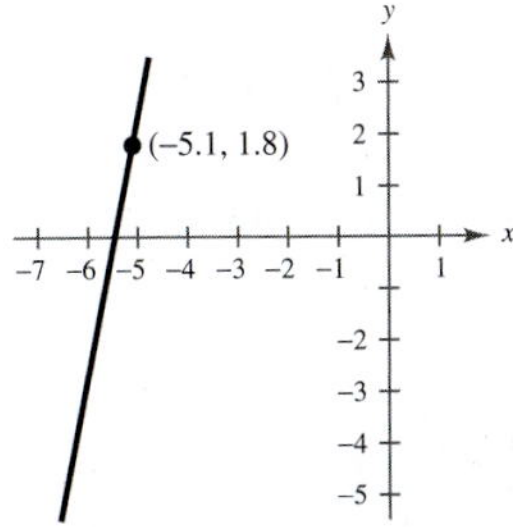

67. $y = -\frac{3}{5}x + 2$

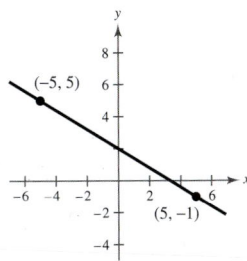

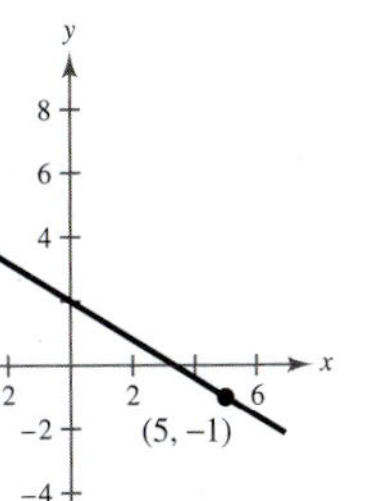

69. $x = -8$

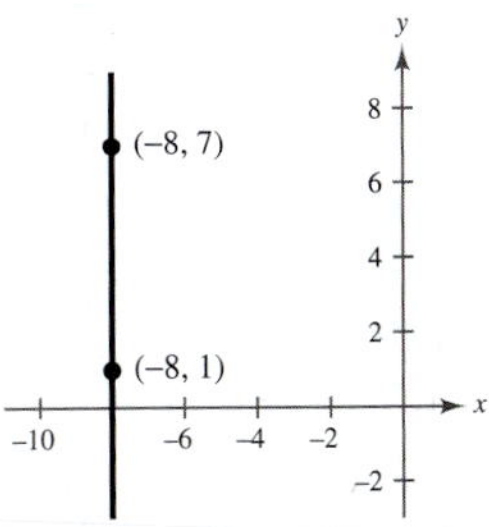

71. $y = -\frac{1}{2}x + \frac{3}{2}$

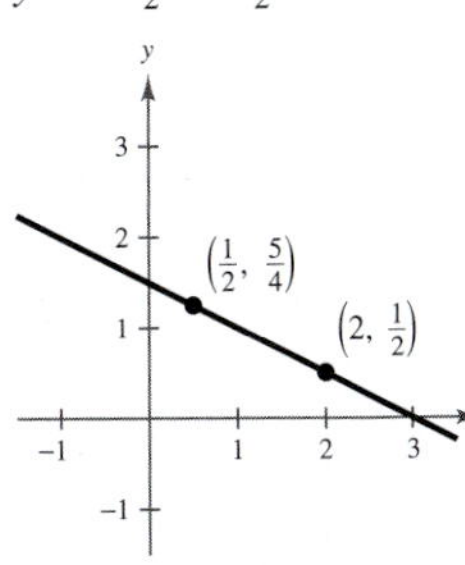

73. $y = -\frac{6}{5}x - \frac{18}{25}$

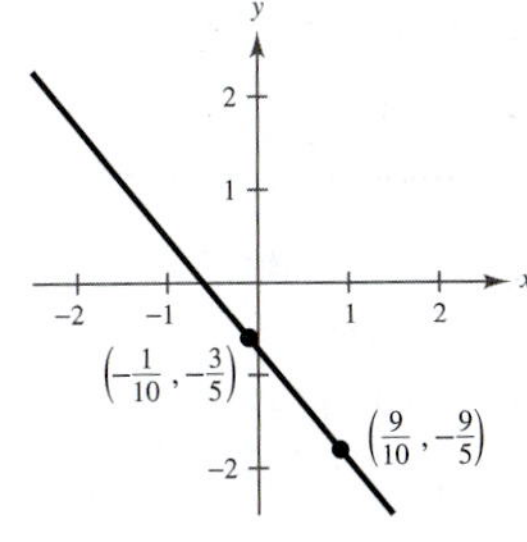

75. $y = 0.4x + 0.2$

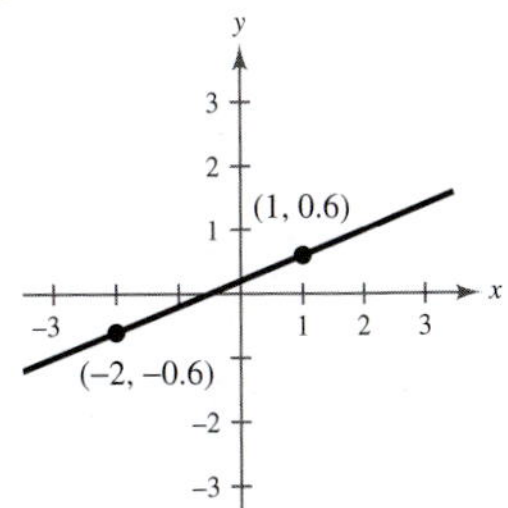

77. $y = -1$

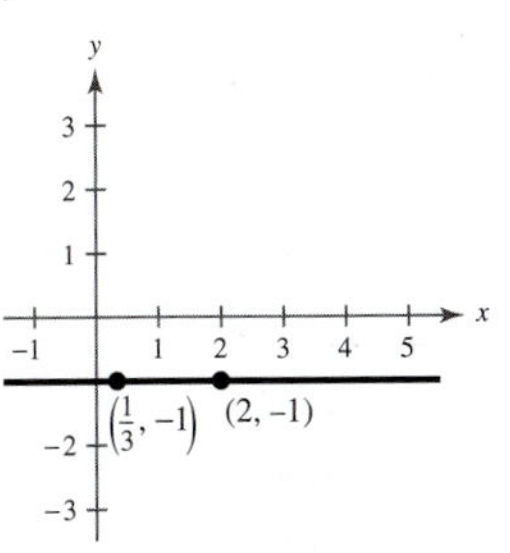

79. $x = \frac{7}{3}$

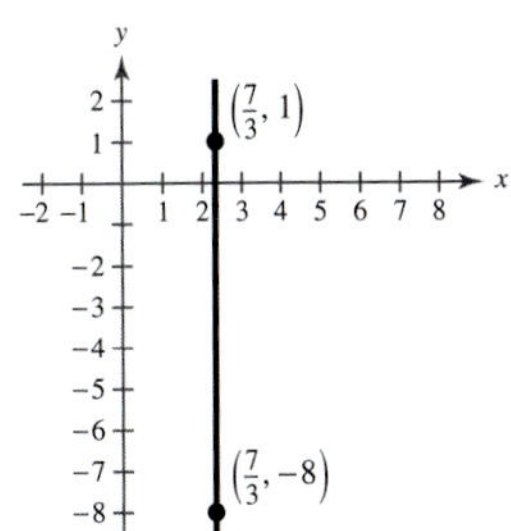

81. $3x + 2y - 6 = 0$ **83.** $12x + 3y + 2 = 0$

85. $x + y - 3 = 0$

87. (a) $y = 2x - 3$ (b) $y = -\frac{1}{2}x + 2$

89. (a) $y = -\frac{3}{4}x + \frac{3}{8}$ (b) $y = \frac{4}{3}x + \frac{127}{72}$

91. (a) $y = 0$ (b) $x = -1$

93. (a) $x = 2$ (b) $y = 5$

95. (a) $y = x + 4.3$ (b) $y = -x + 9.3$

97. Line (b) is perpendicular to line (c).

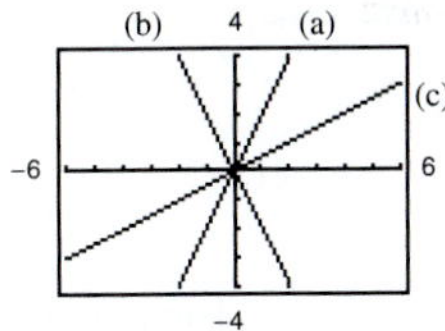

99. Line (a) is parallel to line (b).

Line (c) is perpendicular to line (a) and line (b).

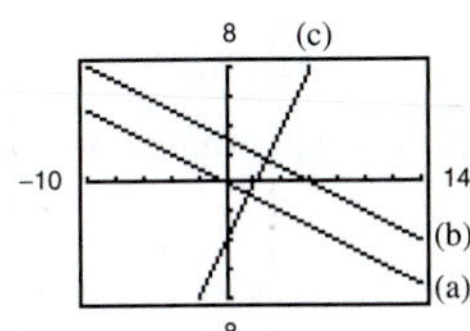

101. $3x - 2y - 1 = 0$ **103.** $80x + 12y + 139 = 0$

105. $y = 0.694t + 0.18$; 2005: \$7.12, 2010: \$10.59

107. \$43,900

109. $V = -175t + 875$ **111.** $S = 0.85L$

113. (a) $C = 16.75t + 36{,}500$

(b) $R = 27t$

(c) $P = 10.25t - 36{,}500$

(d) $t \approx 3561$ hours

115. (a)

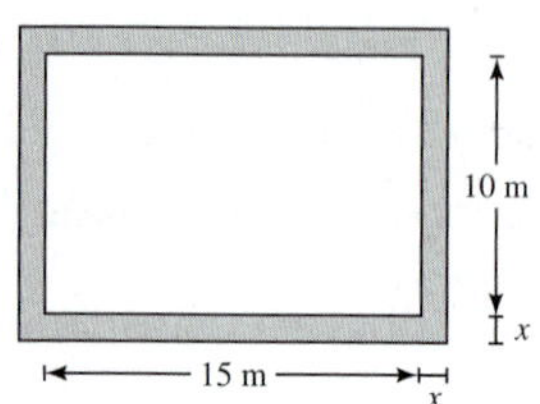

(b) $y = 8x + 50$

(c)

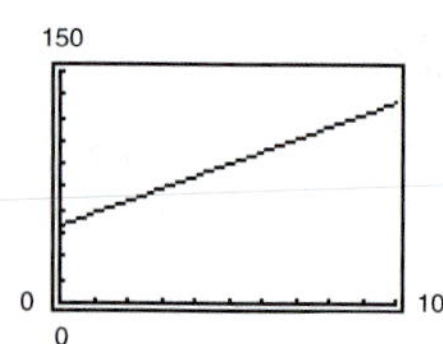

(d) $m = 8$, 8 meters

117. $C = 0.35x + 120$

119. (a) and (b)

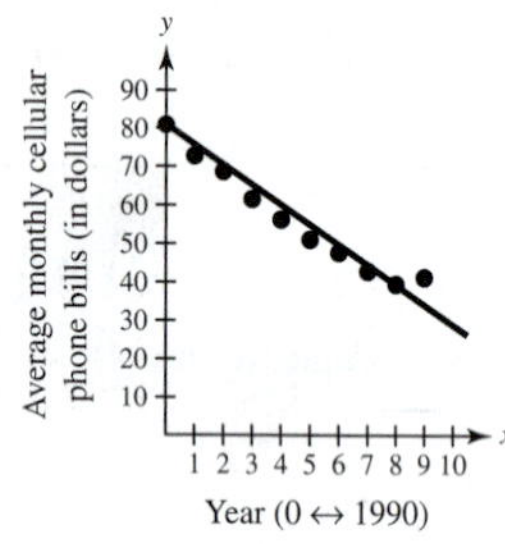

(c) Answers will vary. Sample answer: $y = -5.2x + 80.9$

(d) Answers will vary. Sample answer: The initial cost is \$80.90. Each year, the cellular phone bill decreases an additional \$5.20.

(e) The model is accurate.

(f) Answers will vary. Sample answer: \$2.90

121. False. The slope with the greatest magnitude corresponds to the steepest line.

123. Find the distance between each two points and use the Pythagorean Theorem.

125. No. The slope cannot be determined without knowing the scale on the y-axis. The slopes could be the same.

127. V-intercept: initial cost; Slope: annual depreciation

129. d **130.** c **131.** a **132.** b

133. -1 **135.** $\frac{7}{2}, 7$ **137.** No solution

Section 2.2 *(page 195)*

1. Yes **3.** No

5. Yes, each input value has exactly one output value.

7. No, the input values of 7 and 10 each have two different output values.

9. (a) Function

(b) Not a function, because the element 1 in A corresponds to two elements, -2 and 1, in B.

(c) Function

(d) Not a function, because not every element in A is matched with an element in B.

11. Each is a function. For each year there corresponds one and only one circulation.

13. Not a function **15.** Function **17.** Function

19. Not a function **21.** Function

23. (a) -1 (b) -9 (c) $2x - 5$

25. (a) 36π (b) $\frac{9}{2}\pi$ (c) $\frac{32}{3}\pi r^3$

27. (a) 1 (b) 2.5 (c) $3 - 2|x|$

29. (a) $-\frac{1}{9}$ (b) Undefined (c) $\frac{1}{y^2 + 6y}$

31. (a) 1 (b) -1 (c) $\frac{|x-1|}{x-1}$

33. (a) -1 (b) 2 (c) 6

35. (a) -7 (b) 4 (c) 9

37.

x	-2	-1	0	1	2
$f(x)$	1	-2	-3	-2	1

39.

t	-5	-4	-3	-2	-1
$h(t)$	1	$\frac{1}{2}$	0	$\frac{1}{2}$	1

41.

x	-2	-1	0	1	2
$f(x)$	5	$\frac{9}{2}$	4	1	0

43. 5 **45.** $\frac{4}{3}$ **47.** ± 3 **49.** $0, \pm 1$

51. $2, -1$ **53.** 3, 0 **55.** All real numbers

57. All real numbers $t \neq 0$

59. $y \geq 10$ **61.** $-1 \leq x \leq 1$

63. All real numbers $x \neq 0, -2$ **65.** $s \geq 1,\ s \neq 4$

67. $x > 0$

69. $\{(-2, 4), (-1, 1), (0, 0), (1, 1), (2, 4)\}$

71. $\{(-2, 4), (-1, 3), (0, 2), (1, 3), (2, 4)\}$

73. $g(x) = cx^2;\ c = -2$ **75.** $r(x) = \frac{c}{x};\ c = 32$

77. $3 + h,\ h \neq 0$

79. $3x^2 + 3xh + h^2 + 3,\ h \neq 0$

81. $-\frac{x+3}{9x^2},\ x \neq 3$ **83.** $\frac{\sqrt{5x} - 5}{x - 5}$

85. $A = \frac{P^2}{16}$

87. (a) The maximum volume is 1024 cubic centimeters.

(b)

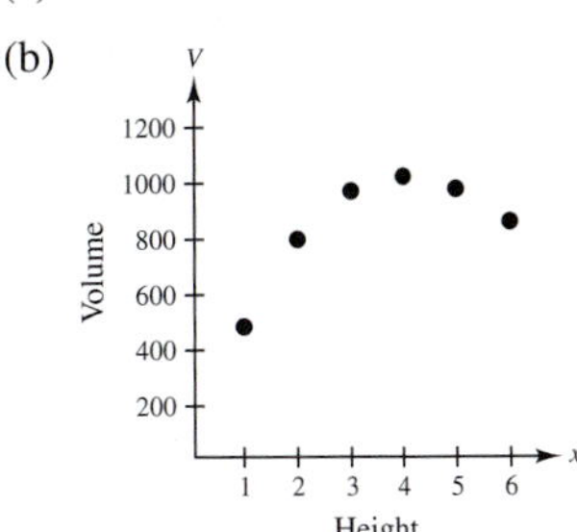

Yes, V is a function of x.

(c) $V = x(24 - 2x)^2, 0 < x < 12$

89. $A = \dfrac{x^2}{2(x-2)}, x > 2$

91. 1990: \$27,800; 1994: \$33,488; 1996: \$38,440; 1999: \$44,110

93. (a) $C = 12.30x + 98{,}000$

(b) $R = 17.98x$

(c) $P = 5.68x - 98{,}000$

95. (a) $R = \dfrac{240n - n^2}{20}, n \geq 80$

(b)

n	90	100	110	120	130	140	150
$R(n)$	\$675	\$700	\$715	\$720	\$715	\$700	\$675

The revenue is maximum when 120 people take the trip.

97. (a)

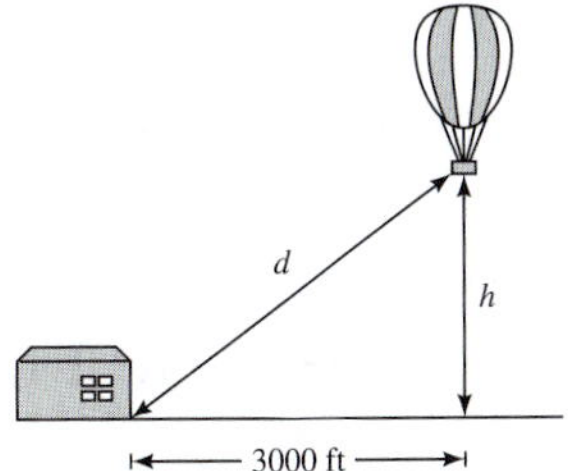

(b) $h = \sqrt{d^2 - 3000^2}, d \geq 3000$

99. (a) The result is 1.8 which is the average increase per year of the number of threatened and endangered fish.

(b) $\begin{cases} 2x + 104 & 6 \leq x \leq 7 \\ 2x + 103 & 8 \leq x \leq 11 \end{cases}$

(c)

x	6	7	8	9	10	11
N	116	118	119	121	123	125

(d) The results are the same.

(e) $y = 1.8x + 105$

The answer in part (b) is more accurate.

101. True. Each x-value corresponds to one y-value.

103. $\frac{15}{8}$ **105.** $-\frac{1}{5}$ **107.** $2x - 3y - 11 = 0$

109. $10x + 9y + 15 = 0$

Section 2.3 *(page 207)*

1. Domain: $(-\infty, -1], [1, \infty)$
Range: $[0, \infty)$

3. Domain: $[-4, 4]$
Range: $[0, 4]$

5. (a) 0 (b) -1 (c) 0 (d) -2

7. (a) -3 (b) 0 (c) 1 (d) -3

9. Function **11.** Not a function **13.** Function

15. $-\frac{5}{2}, 6$ **17.** 0 **19.** $0, \pm\sqrt{2}$ **21.** $\pm\frac{1}{2}, 6$ **23.** $\frac{1}{2}$

25. $-\frac{5}{3}$

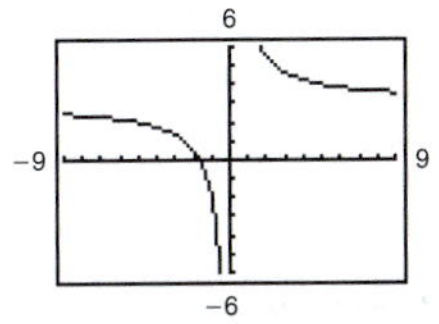

27. $-\frac{11}{2}$

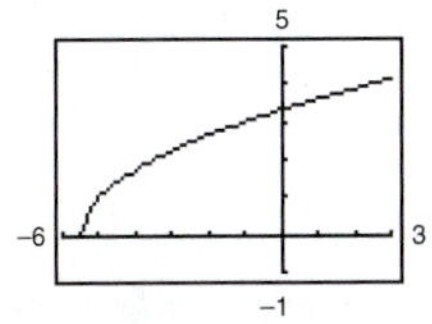

29.

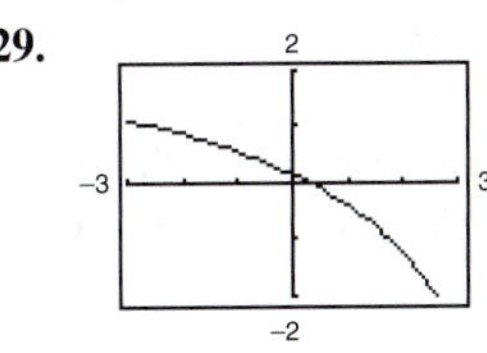

$\frac{1}{3}$

31. Increasing on $(-\infty, \infty)$

33. Increasing on $(-\infty, 0)$ and $(2, \infty)$

Decreasing on $(0, 2)$

35. Increasing on $(-\infty, 0)$ and $(2, \infty)$

Constant on $(0, 2)$

37. Increasing on $(1, \infty)$

Decreasing on $(-\infty, -1)$

Constant on $(-1, 1)$

39. (a)

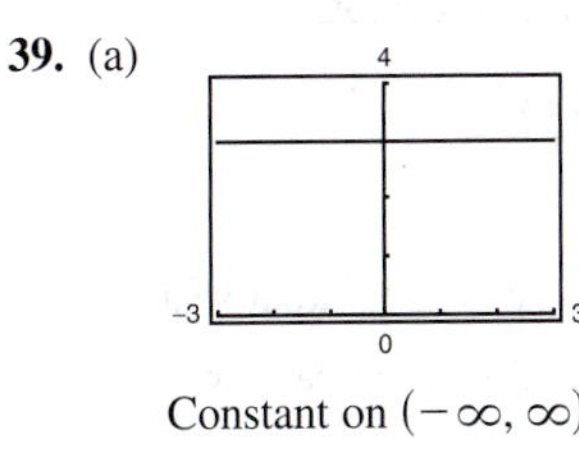

Constant on $(-\infty, \infty)$

(b)

x	-2	-1	0	1	2
$f(x)$	3	3	3	3	3

41. (a)

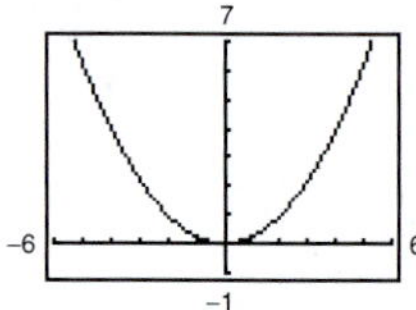

Decreasing on $(-\infty, 0)$; Increasing on $(0, \infty)$

(b)

s	-4	-2	0	2	4
$g(s)$	4	1	0	1	4

43. (a)

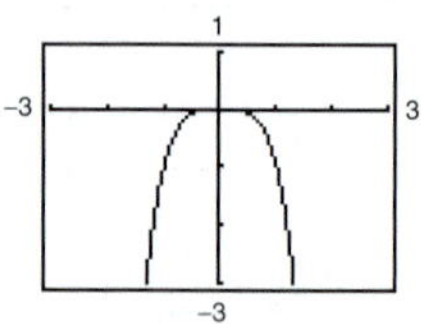

Increasing on $(-\infty, 0)$; Decreasing on $(0, \infty)$

(b)

t	-2	-1	0	1	2
$f(t)$	-16	-1	0	-1	-16

45. (a)

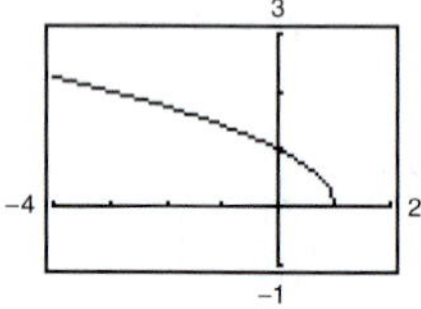

Decreasing on $(-\infty, 1)$

(b)

x	-3	-2	-1	0	1
$f(x)$	2	$\sqrt{3}$	$\sqrt{2}$	1	0

47. (a)

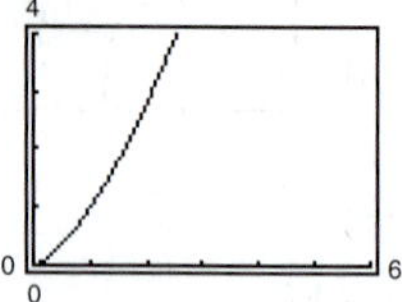

Increasing on $(0, \infty)$

(b)

x	0	1	2	3	4
$f(x)$	0	1	2.8	5.2	8

49.

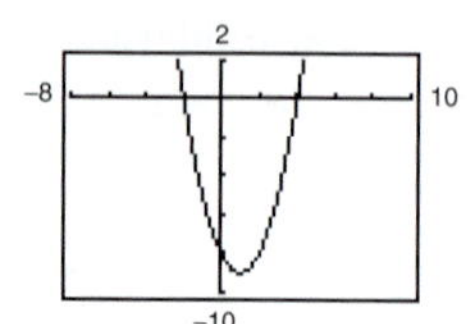

Relative minimum: $(1, -9)$

51.

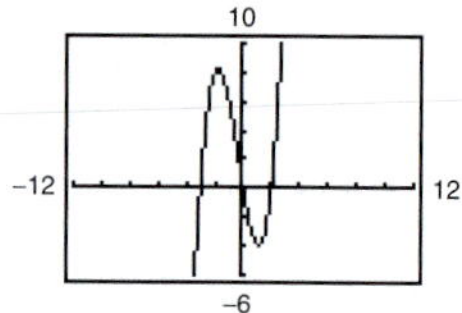

Relative maximum: $(-1.79, 8.21)$

Relative minimum: $(1.12, -4.06)$

53.

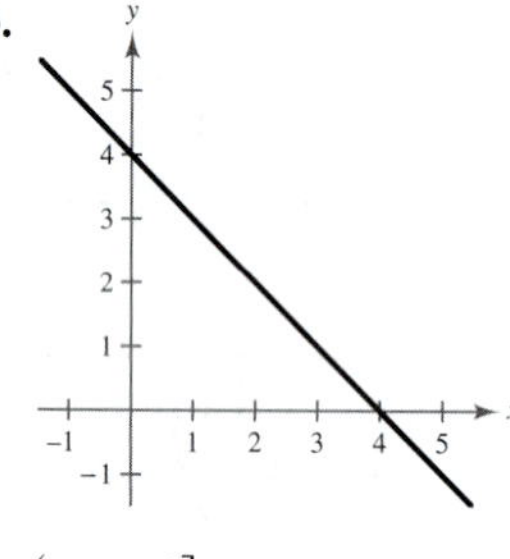

$(-\infty, 4]$

55.

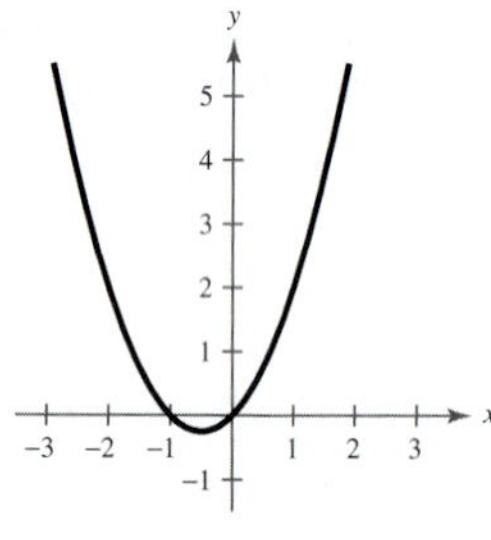

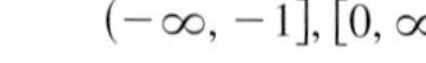

$(-\infty, -1], [0, \infty)$

57.

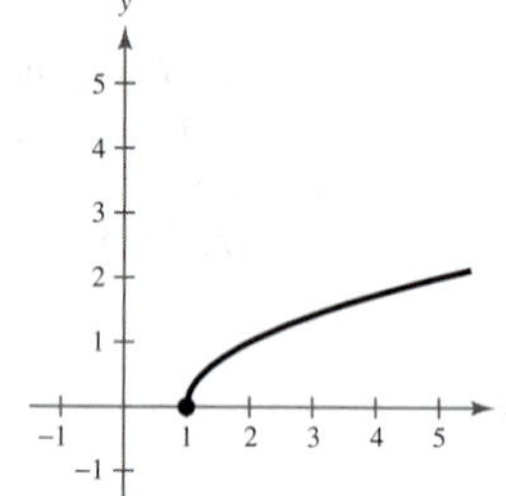

$[1, \infty)$

59.

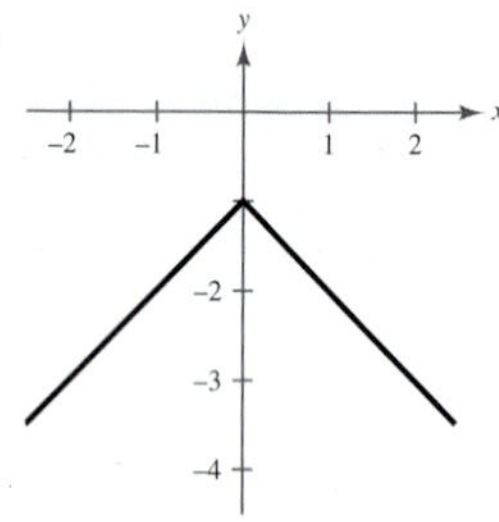

$f(x) < 0$ for all x

61. Even **63.** Odd **65.** Neither even nor odd

67. $h = -x^2 + 4x - 3$ **69.** $h = 2x - x^2$

71. $L = \frac{1}{2}y^2$ **73.** $L = 4 - y^2$

75. (a)

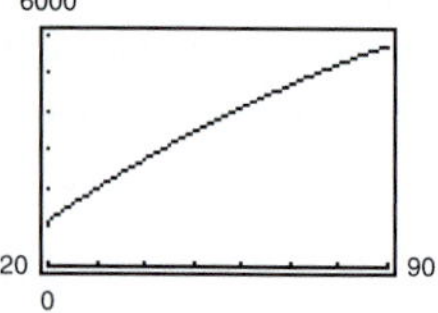

(b) 30 watts

77. (a) ten thousands (b) ten millions (c) percents

79. False. The function $f(x) = \sqrt{x^2 + 1}$ has a domain of all real numbers.

81. (a) Even. The graph is a reflection in the x-axis.

(b) Even. The graph is a reflection in the y-axis.

(c) Even. The graph is a vertical translation of f.

(d) Neither. The graph is a horizontal translation of f.

83. (a) $\left(\frac{3}{2}, 4\right)$ (b) $\left(\frac{3}{2}, -4\right)$

85. (a) $(-4, 9)$ (b) $(-4, -9)$

87. (a) (b)

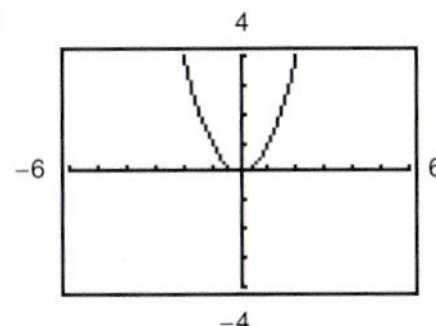

(c) (d)

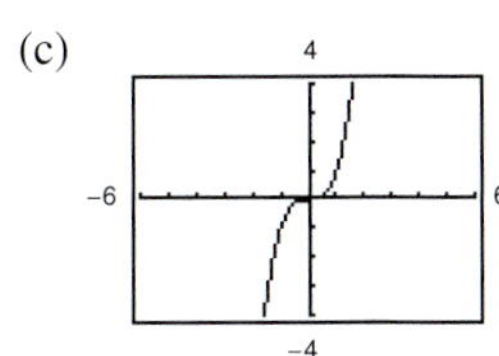

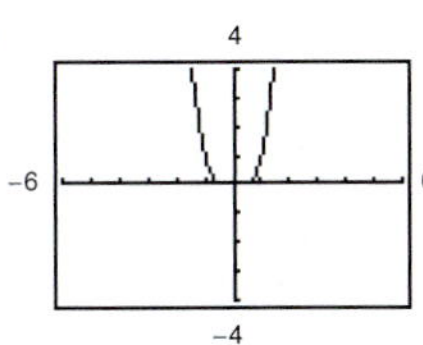

(e) (f)

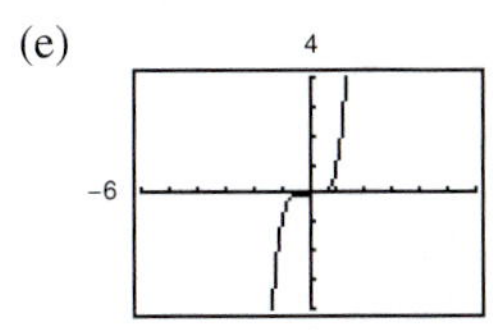

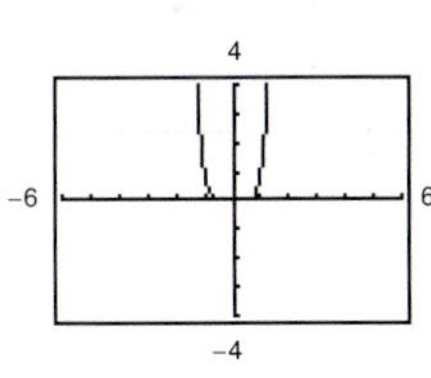

All the graphs pass through the origin. The graphs of the odd powers of x are symmetric with respect to the origin, and the graphs of the even powers are symmetric with respect to the y-axis. As the powers increase, the graphs become flatter in the interval $-1 < x < 1$.

89. 0, 10 **91.** $0, \pm 1$

93. (a) 37 (b) -28 (c) $5x - 43$

95. (a) -9 (b) $2\sqrt{7} - 9$

(c) The given value is not in the domain of the function.

97. $h + 4,\ h \neq 0$

Section 2.4 *(page 216)*

1. $f(x) = -2x + 6$ **3.** $f(x) = -3x + 11$

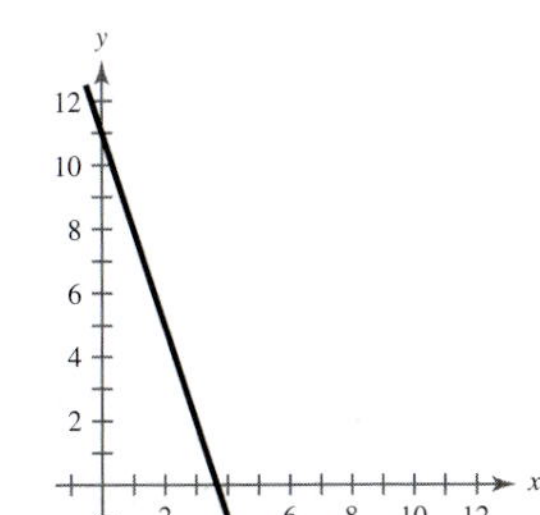

5. $f(x) = -1$

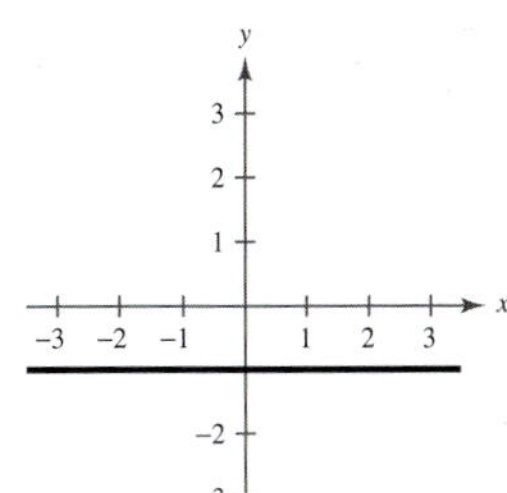

7. $f(x) = \frac{6}{7}x - \frac{45}{7}$

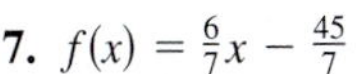

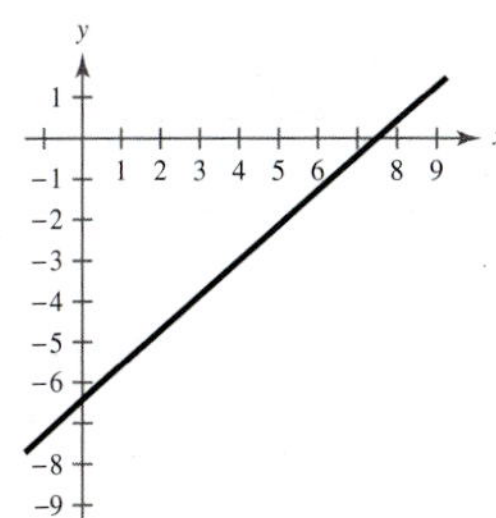

9.

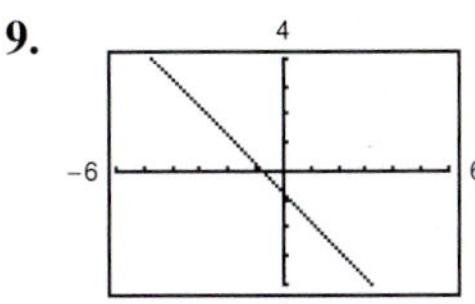

11.

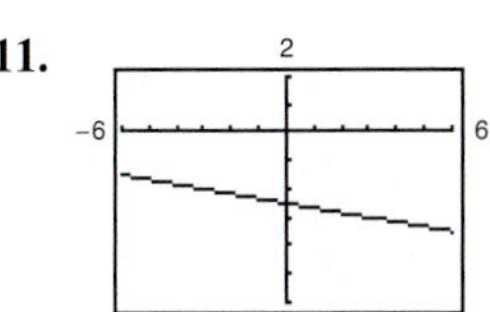

13.

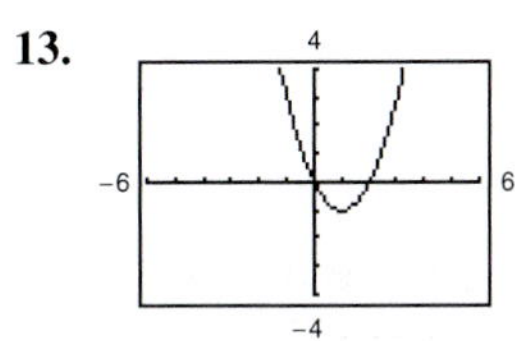

15.

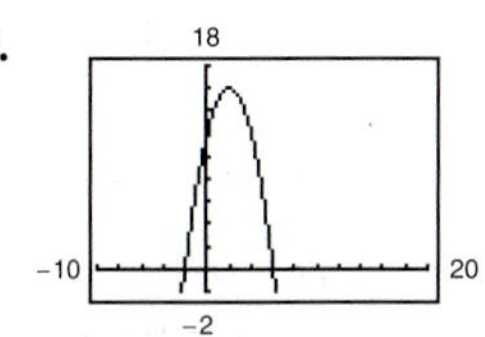

17.

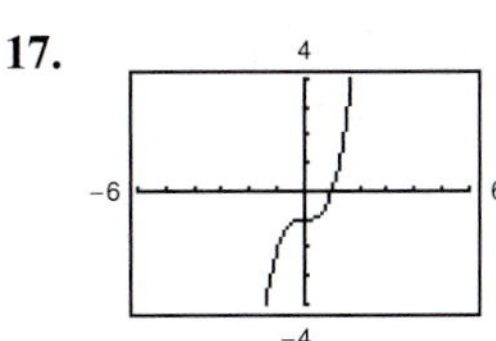

19.

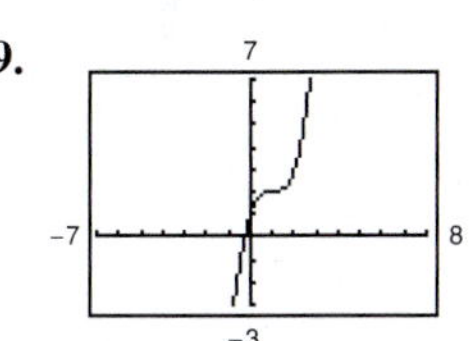

21.

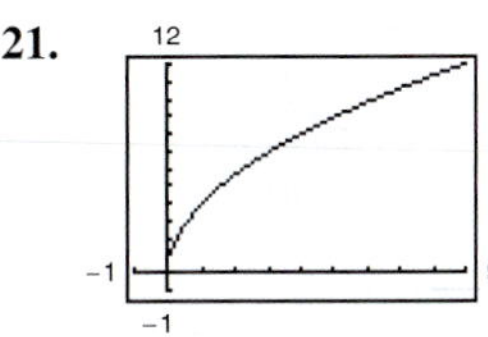

23.

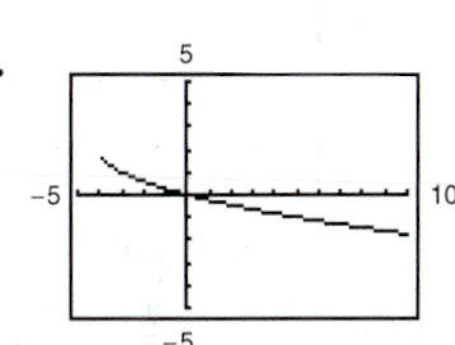

25.

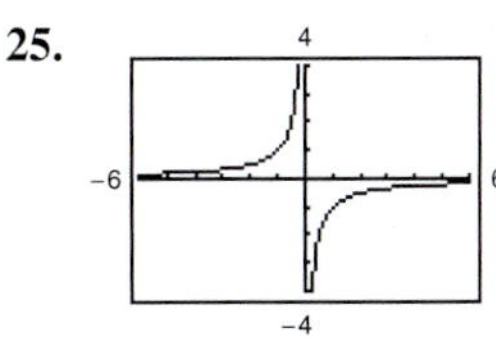

27.

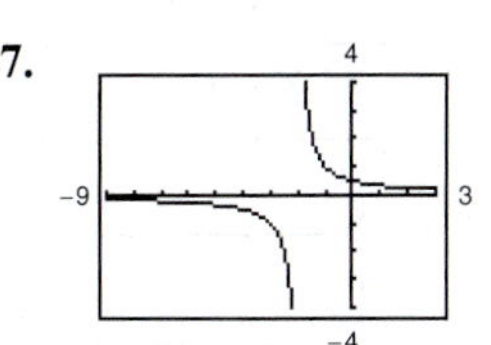

29. (a) 2 (b) 2 (c) -4 (d) 3

31. (a) 1 (b) 3 (c) 7 (d) -19

33. (a) 6 (b) -11 (c) 6 (d) -22

35. (a) -10 (b) -4 (c) -1 (d) 41

37.

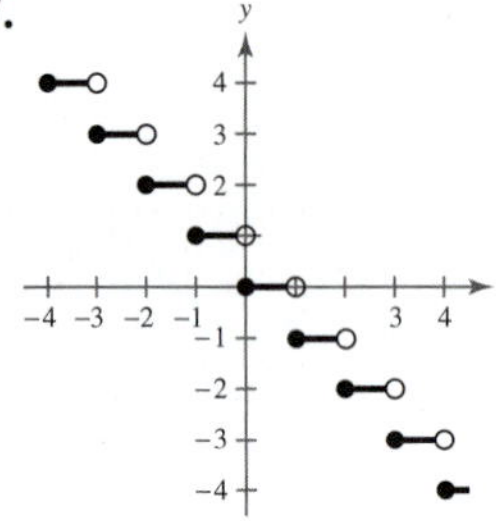

39.

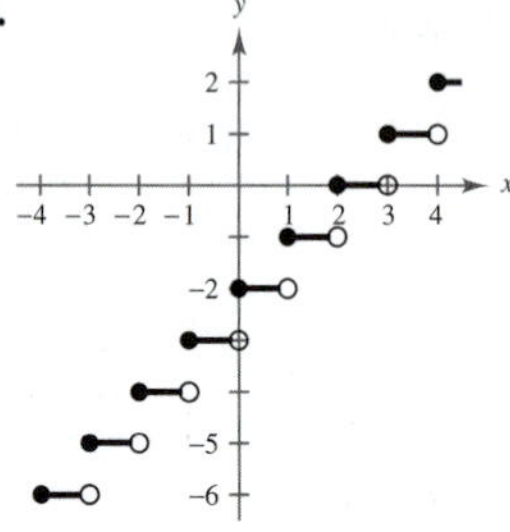

41.

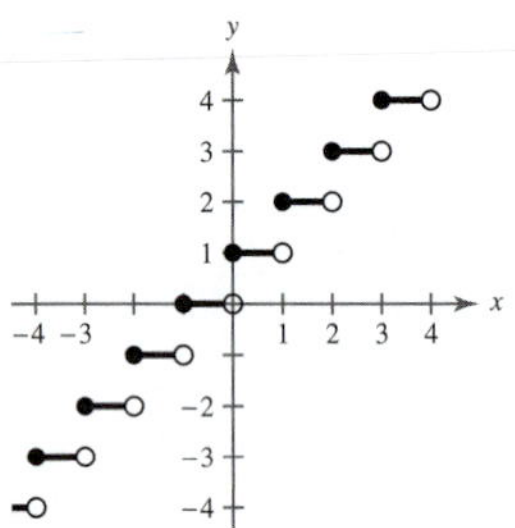

43.

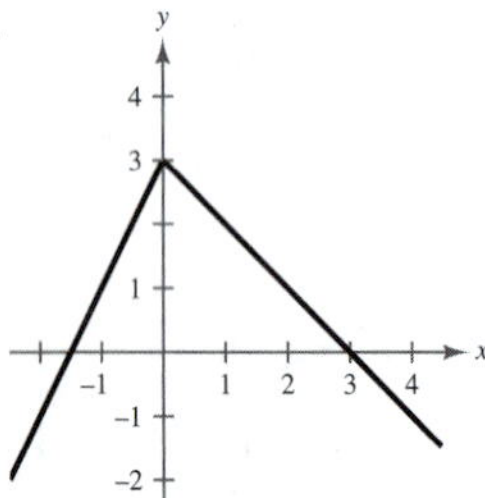

45.

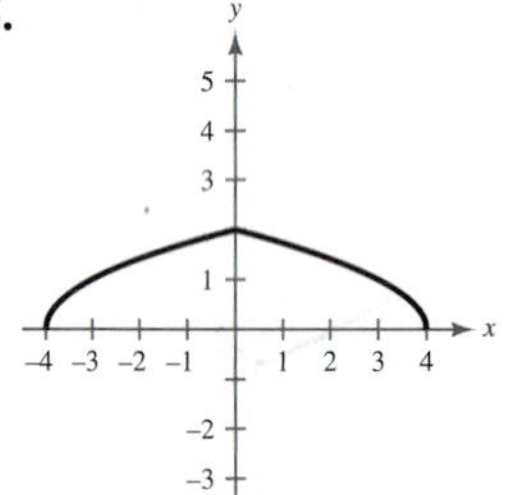

47.

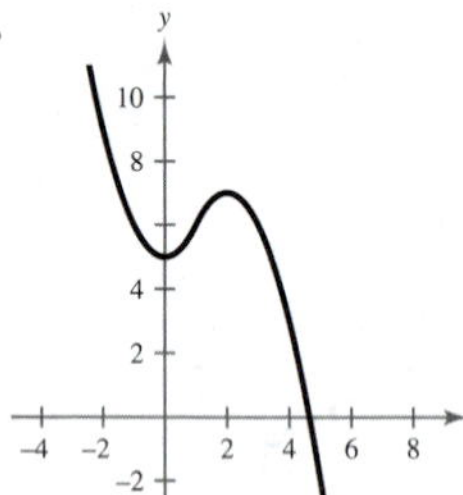

49.

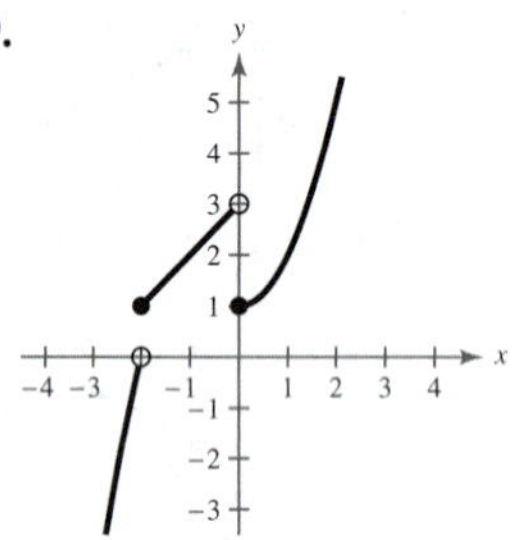

51.

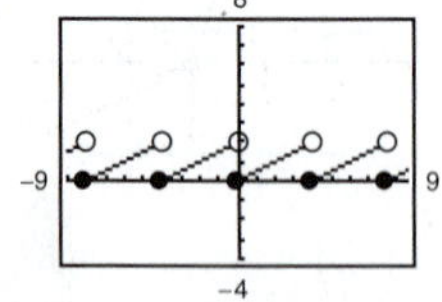

Domain: $(-\infty, \infty)$

Range: $[0, 2)$

Sawtooth pattern

53. $f(x) = |x|;\ g(x) = |x + 2| - 1$

55. $f(x) = x^3;\ g(x) = (x - 1)^3 - 2$

57. $f(x) = 2;\ g(x) = 2$

59. $f(x) = x;\ g(x) = x - 2$

61. $f(x) = x^2;\ g(x) = (x - 3)^2$

63. (a)

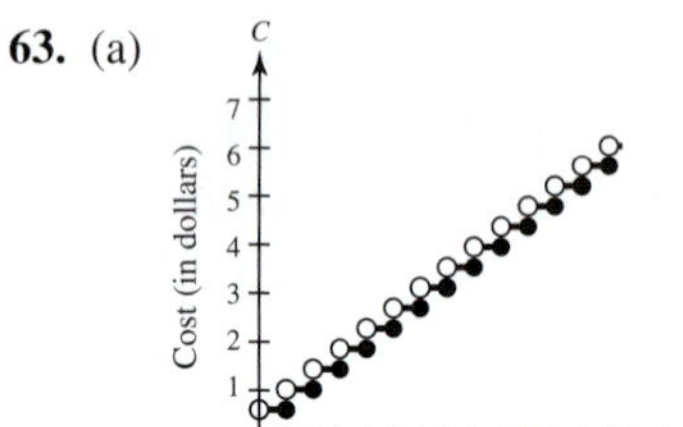

(b) \$5.64

65. (a)

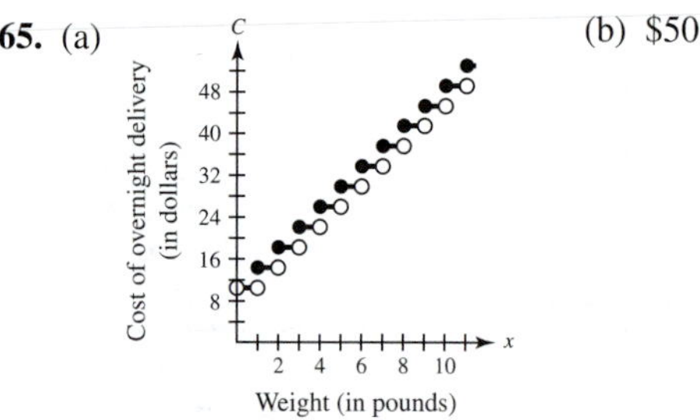

(b) \$50.25

67. (a) $W(30) = 360;\ W(40) = 480;$
$W(45) = 570;\ W(50) = 660$

(b) $W(h) = \begin{cases} 12h, & 0 < h \le 45 \\ 18(h - 45) + 540, & h > 45 \end{cases}$

69.

Interval	*Input Pipe*	*Drainpipe 1*	*Drainpipe 2*
[0, 5]	Open	Closed	Closed
[5, 10]	Open	Open	Closed
[10, 20]	Closed	Closed	Closed
[20, 30]	Closed	Closed	Open
[30, 40]	Open	Open	Open
[40, 45]	Open	Closed	Open
[45, 50]	Open	Open	Open
[50, 60]	Open	Open	Closed

71. True. The solution sets are the same.

73. $x \le 1$

−3 −2 −1 0 1 2 3 x

75. Neither

Section 2.5 *(page 224)*

1. (a)

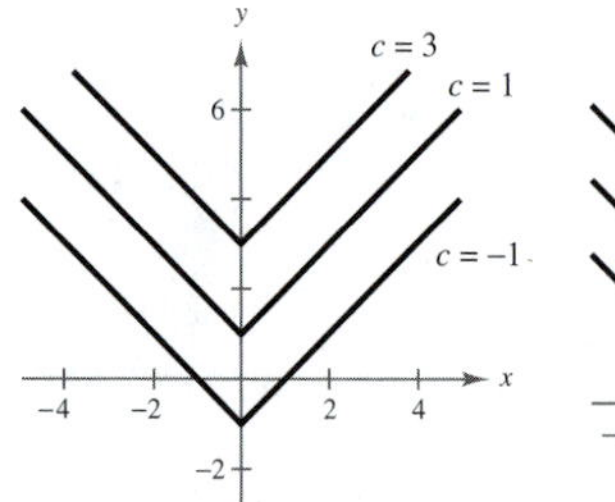

(b)

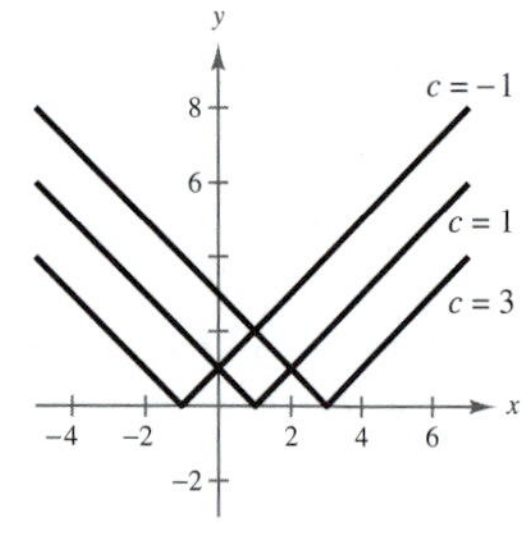

(c)

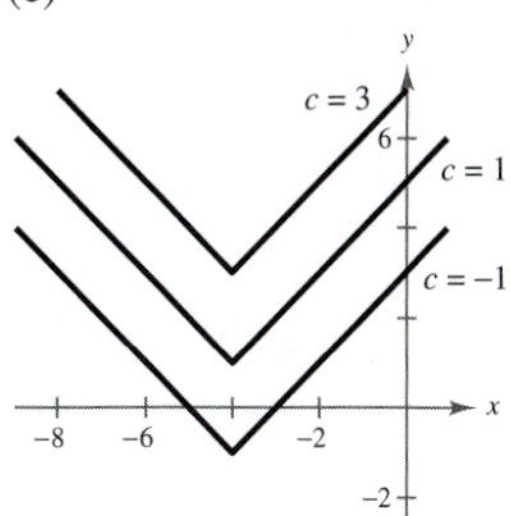

3. (a)

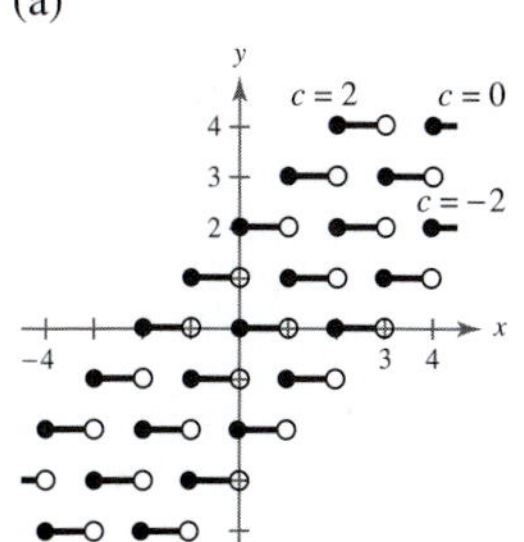

(b)

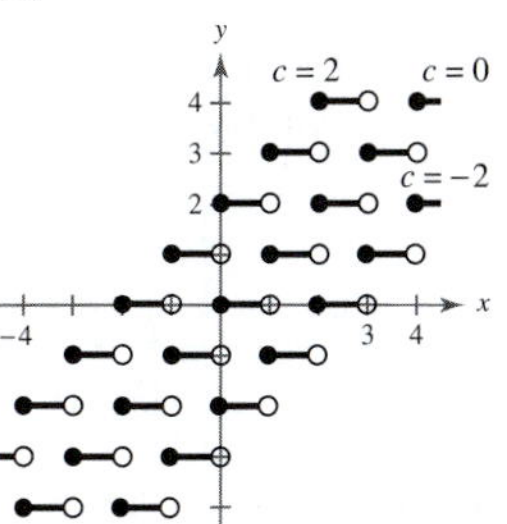

(c)

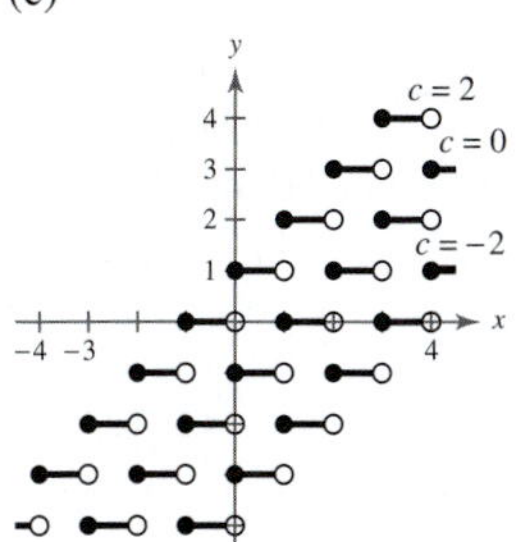

5. (a)

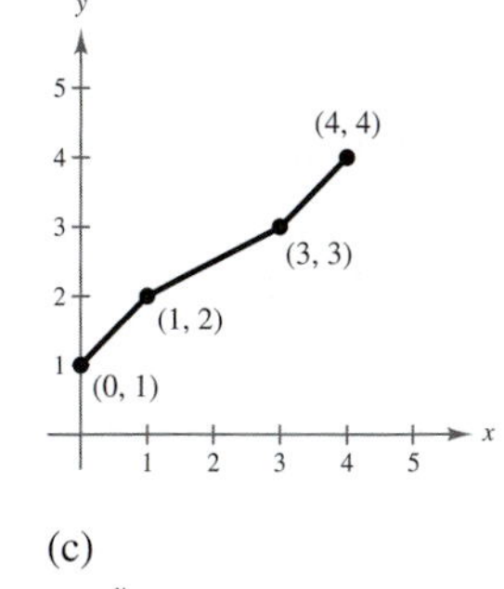

(b)

(c)

(d)

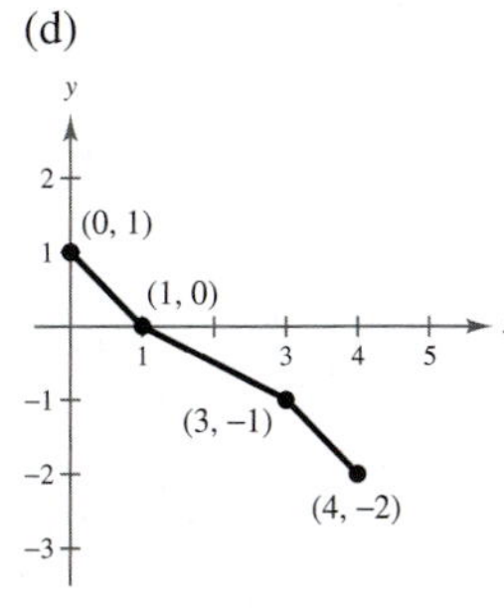

(e)

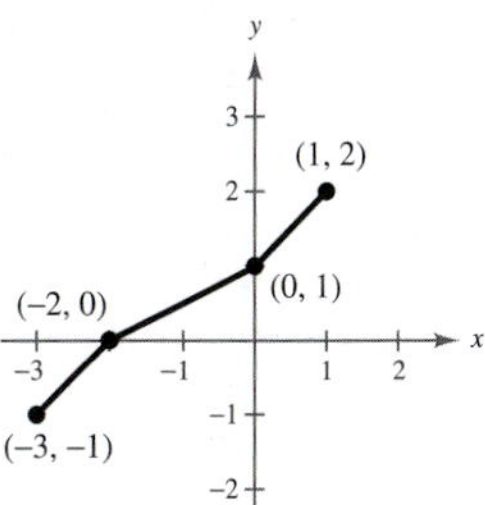

(f)

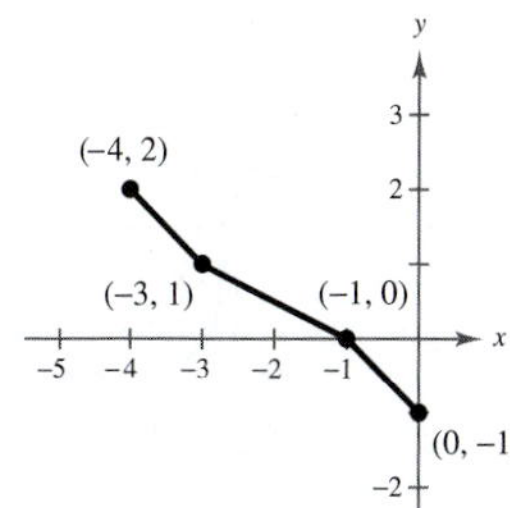

(g)

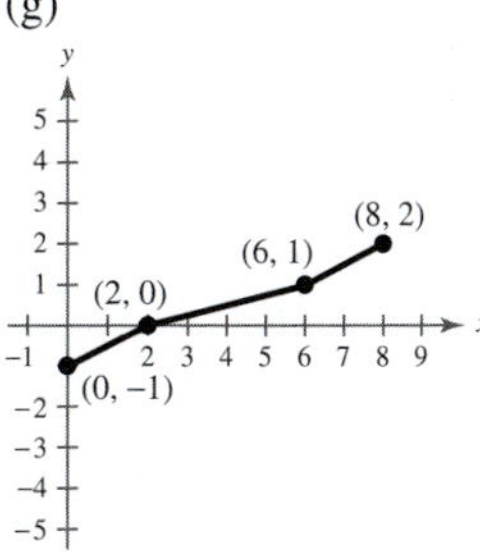

7. (a)

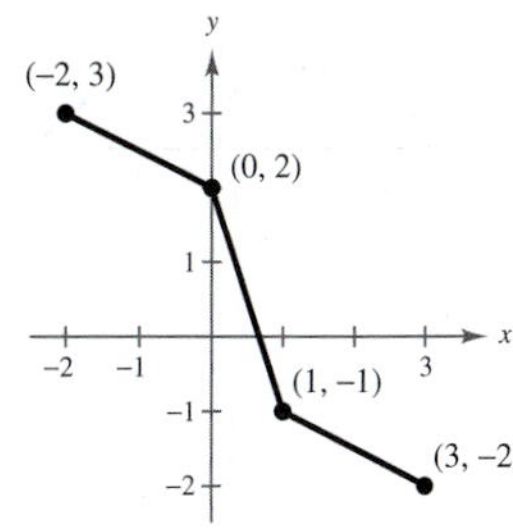

(b)

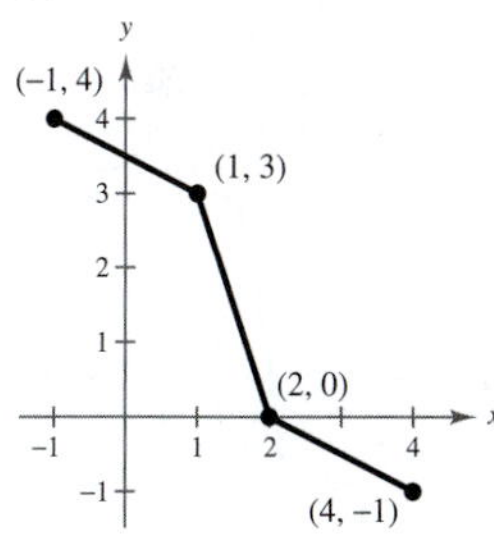

(c)

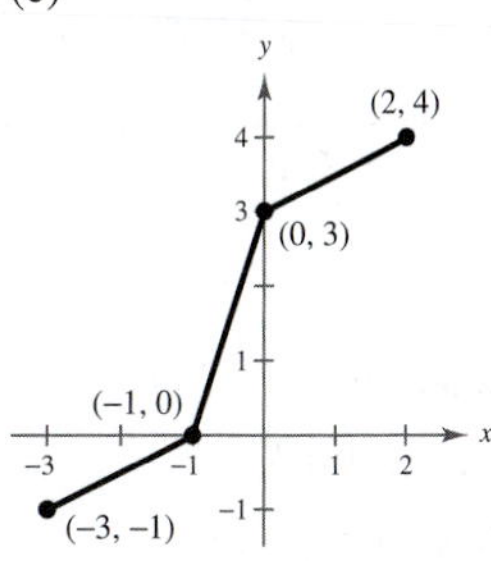

(d)

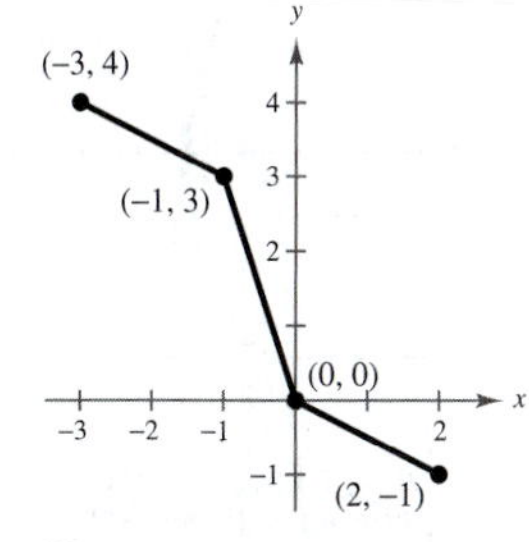

(e)

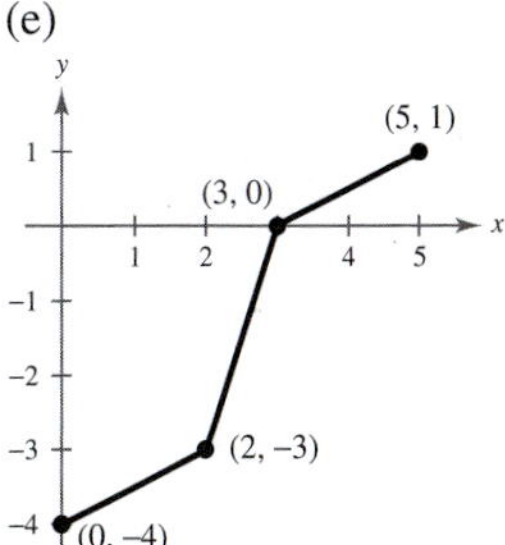

(f)

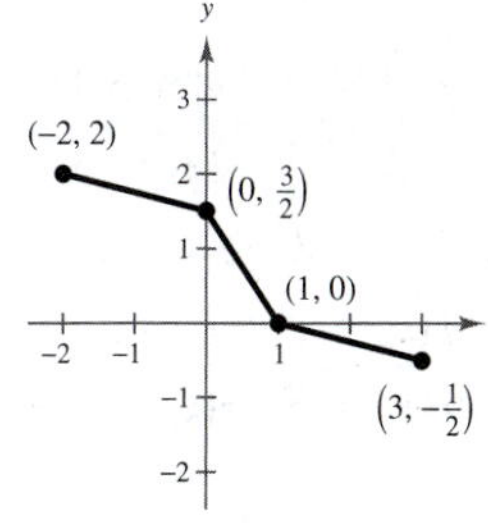

(g)

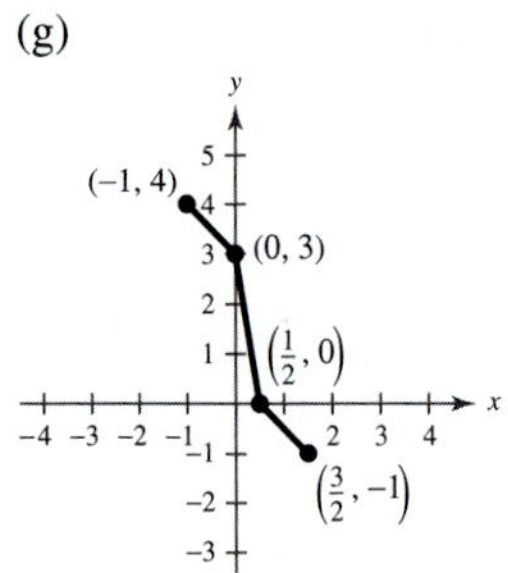

9. (a) $y = x^2 - 1$ (b) $y = 1 - (x + 1)^2$

(c) $y = -(x - 2)^2 + 6$ (d) $y = (x - 5)^2 - 3$

11. (a) $y = |x| + 5$ (b) $y = -|x + 3|$

(c) $y = |x - 2| - 4$ (d) $y = -|x - 6| - 1$

13. Horizontal shift of $y = x^3$; $y = (x - 2)^3$

15. Reflection in the x-axis of $y = x^2$; $y = -x^2$

17. Reflection in the x-axis and vertical shift of $y = \sqrt{x}$; $y = 1 - \sqrt{x}$

19. Reflection in the x-axis, and vertical shift 12 units upward, of $f(x) = x^2$

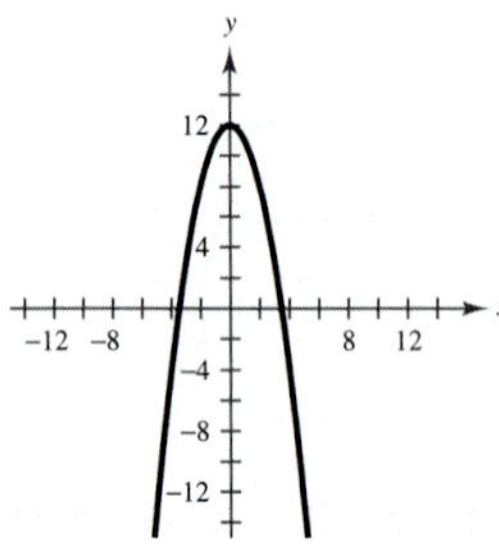

21. Vertical shift seven units upward of $f(x) = x^3$

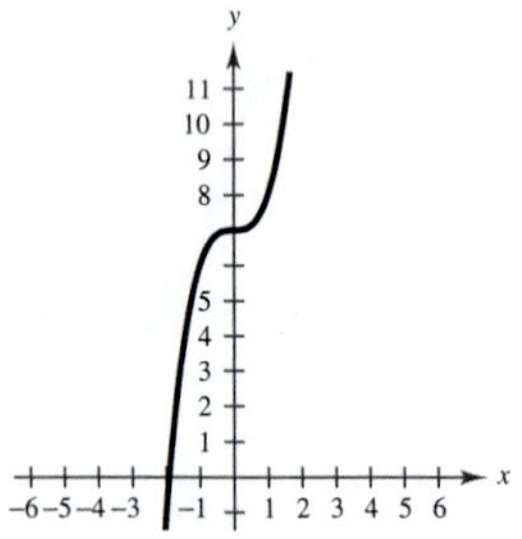

23. Reflection in the x-axis, vertical shift two units upward, and horizontal shift five units to the left, of $f(x) = x^2$

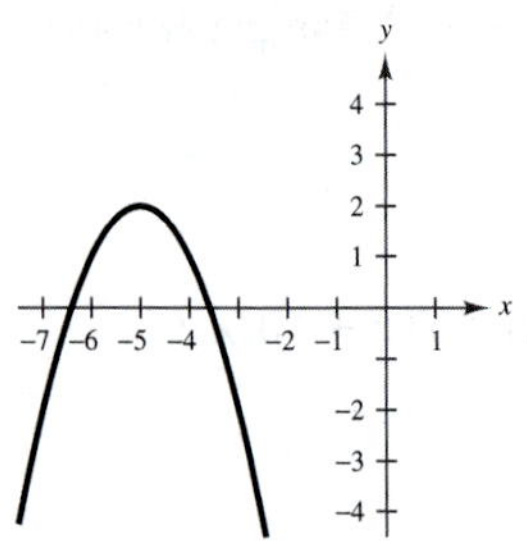

25. Vertical shift two units upward, and horizontal shift one unit to the right, of $f(x) = x^3$

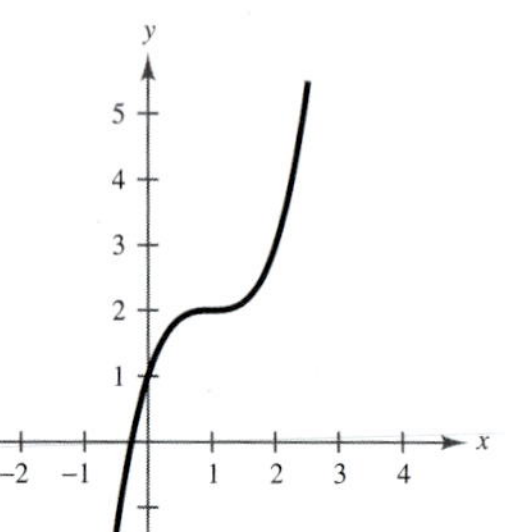

27. Reflection in the x-axis, and vertical shift two units downward, of $f(x) = |x|$

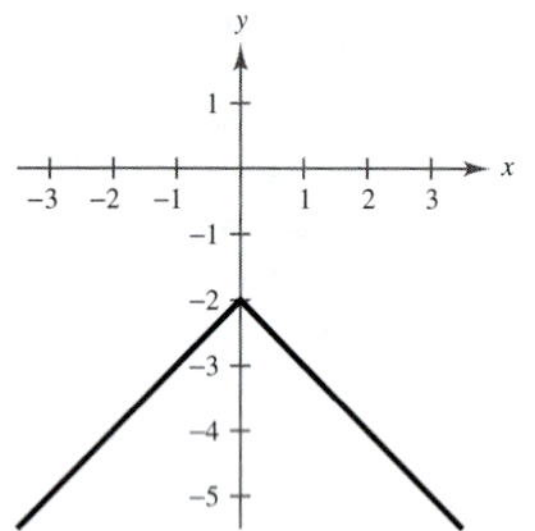

29. Reflection in the x-axis, vertical shift eight units upward, and horizontal shift four units to the left, of $f(x) = |x|$

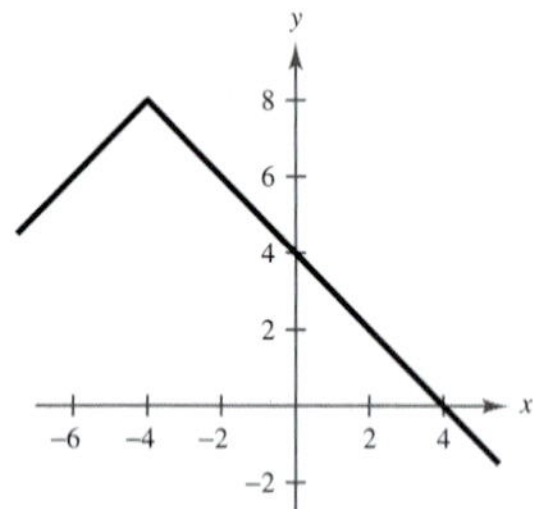

31. Reflection in the x-axis, and vertical shift three units upward, of $f(x) = [\![x]\!]$

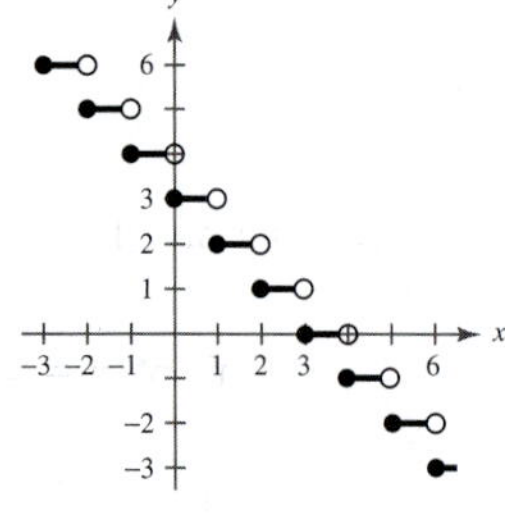

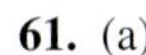

33. Horizontal shift nine units to the right of $f(x) = \sqrt{x}$

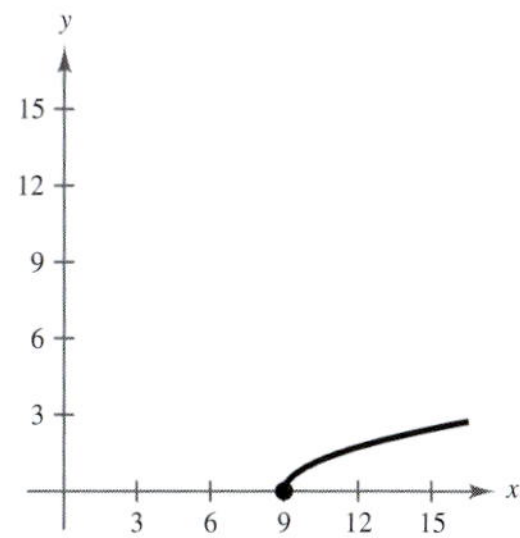

35. Reflection in the y-axis, vertical shift two units downward, and horizontal shift seven units to the right, of $f(x) = \sqrt{x}$

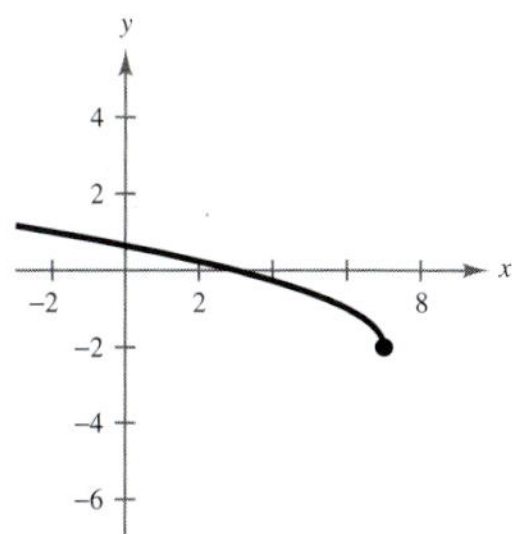

37. Vertical shift four units downward, and horizontal stretch of two, of $f(x) = \sqrt{x}$

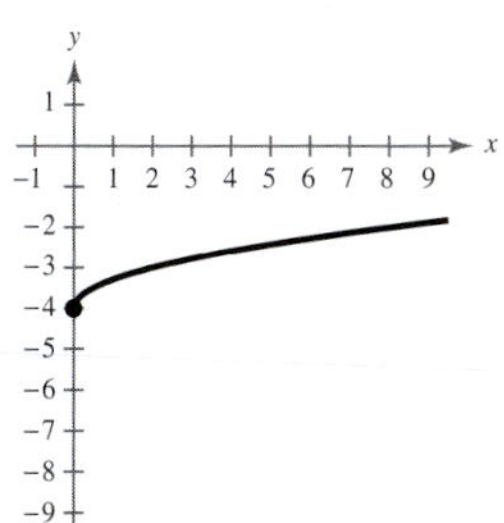

39. $f(x) = (x - 2)^2 - 8$ **41.** $f(x) = (x - 13)^3$

43. $f(x) = -|x| - 10$ **45.** $f(x) = -\sqrt{-x + 6}$

47. (a) $y = -3x^2$ (b) $y = 4x^2 + 3$

49. (a) $y = -\frac{1}{2}|x|$ (b) $y = 3|x| - 3$

51. Vertical stretch of $y = x^3$; $y = 2x^3$

53. Reflection in the x-axis and vertical shrink of $y = x^2$; $y = -\frac{1}{2}x^2$

55. Reflection in the y-axis and vertical shrink of $y = \sqrt{x}$; $y = \frac{1}{2}\sqrt{-x}$

57. $y = -(x - 2)^3 + 2$ **59.** $y = -\sqrt{x} - 3$

61. (a) (b)

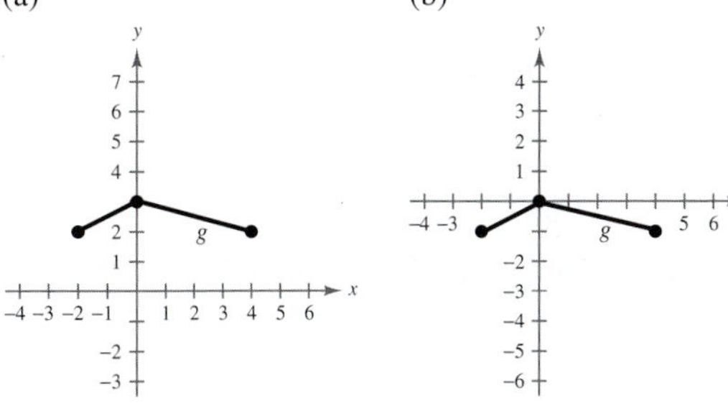

(c) (d)

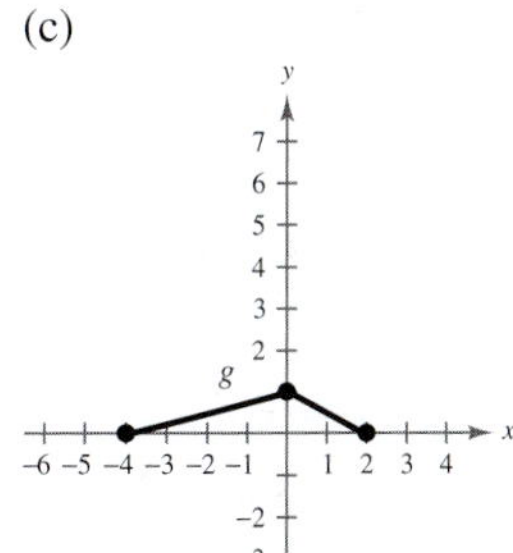

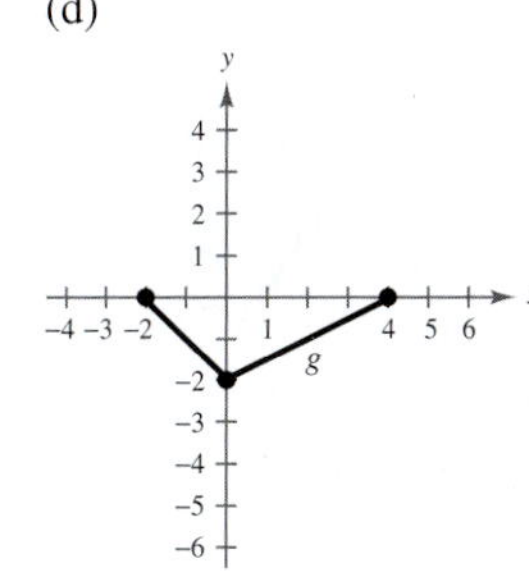

(e) (f)

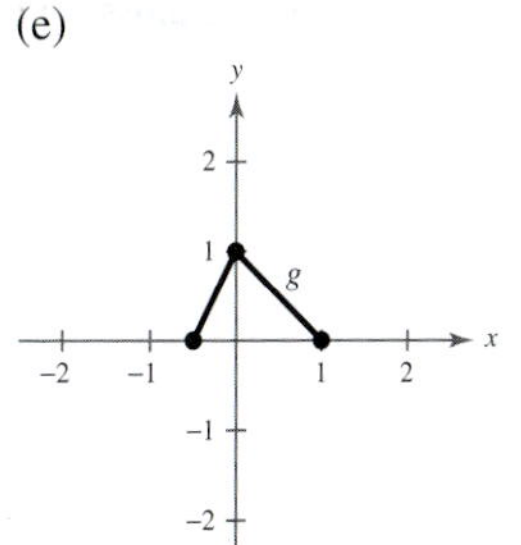

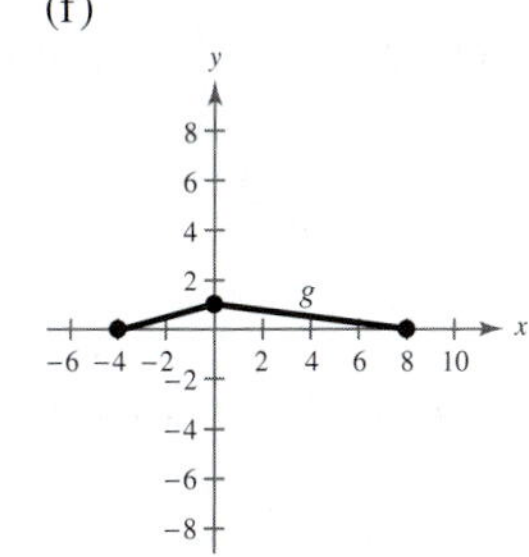

63. (a) Vertical shift of 20.5 units upward and vertical shrink of 0.035

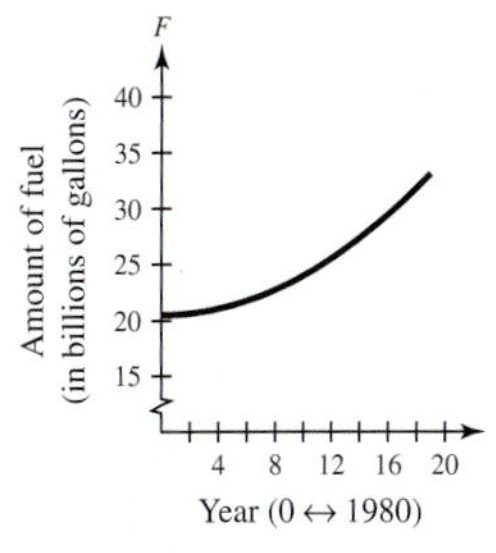

(b) 0.665-billion-gallon increase in fuel usage by trucks each year

(c) $f(t) = 20.5 + 0.035(t + 10)^2$. The graph was shifted 10 units to the left.

(d) 42.375 billion gallons. Yes.

65. True. $|-x| = |x|$

67. (a) $g(t) = \frac{3}{4}f(t)$ (b) $g(t) = f(t) + 10{,}000$

(c) $g(t) = f(t - 2)$

69. $(-2, 0), (-1, 1), (0, 2)$

71. $\dfrac{4}{x(1-x)}$ **73.** $\dfrac{3x-2}{x(x-1)}$ **75.** $\dfrac{(x-4)\sqrt{x^2-4}}{x^2-4}$

77. $5(x-3),\ x \neq -3$

79. (a) 38 (b) $\frac{57}{4}$ (c) $x^2 - 12x + 38$

81. All real numbers $x \neq 11$ **83.** $-9 \leq x \leq 9$

Section 2.6 *(page 234)*

1.

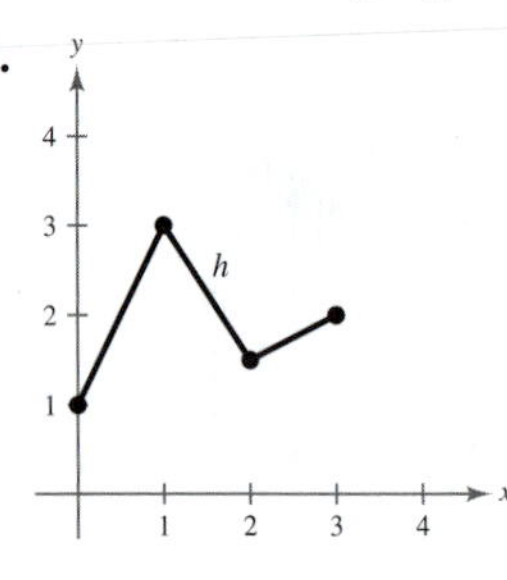

3.

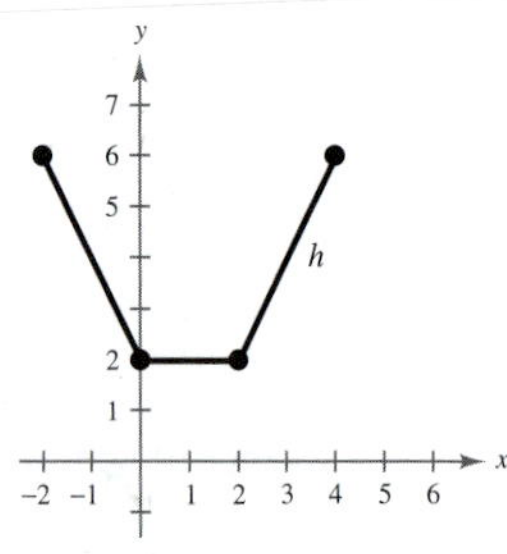

5. (a) $2x$ (b) 4 (c) $x^2 - 4$ (d) $\dfrac{x+2}{x-2};\ x \neq 2$

7. (a) $x^2 + 4x - 5$ (b) $x^2 - 4x + 5$

(c) $4x^3 - 5x^2$ (d) $\dfrac{x^2}{4x-5};\ x \neq \dfrac{5}{4}$

9. (a) $x^2 + 6 + \sqrt{1-x}$ (b) $x^2 + 6 - \sqrt{1-x}$

(c) $(x^2+6)\sqrt{1-x}$ (d) $\dfrac{(x^2+6)\sqrt{1-x}}{1-x};\ x < 1$

11. (a) $\dfrac{x+1}{x^2}$ (b) $\dfrac{x-1}{x^2}$ (c) $\dfrac{1}{x^3}$ (d) $x;\ x \neq 0$

13. 3 **15.** 5 **17.** $9t^2 - 3t + 5$ **19.** 74

21. 26 **23.** $\frac{3}{5}$

25.

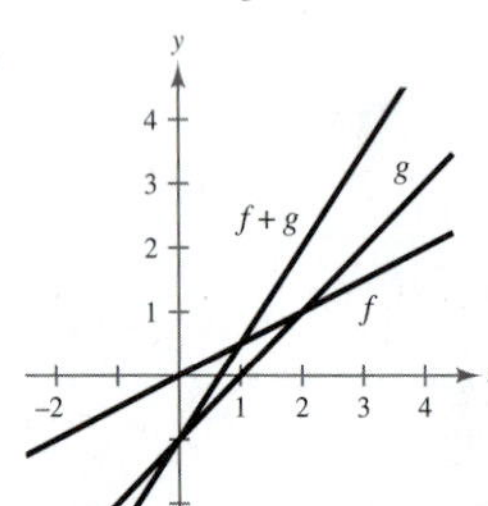

27.

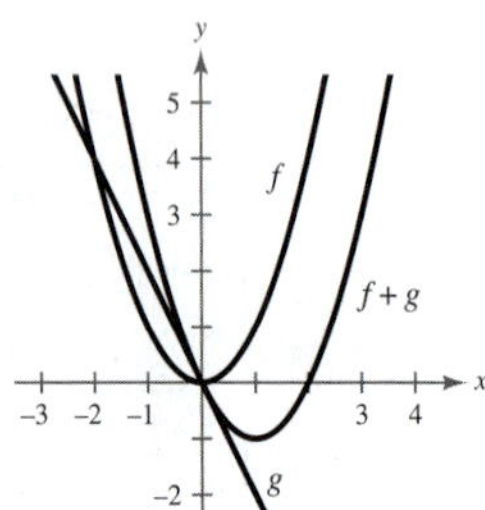

29.

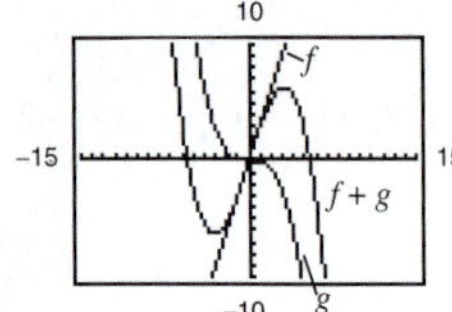

$f(x), g(x)$

31. $T = \frac{3}{4}x + \frac{1}{15}x^2$

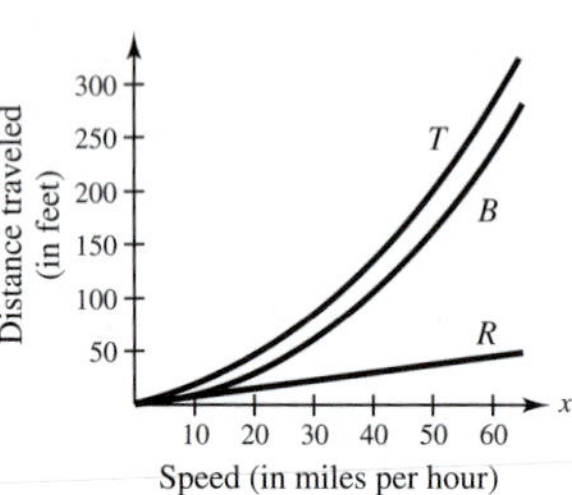

33. (a) $y_1 = 1.344t^2 - 9.38t + 163.6$

$y_2 = 17.34t + 238.5$

$y_3 = 3.38t + 28.1$

(b) $y_1 + y_2 + y_3 = 1.344t^2 + 11.34t + 430.2$; the total amount spent on health services and supplies

(c)

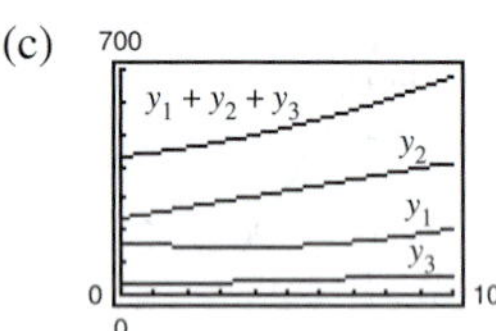

(d) 2003: $804.76 billion; 2005: $902.70 billion

35. (a) $(x-1)^2$ (b) $x^2 - 1$ (c) x^4

37. (a) x (b) x (c) $\sqrt[3]{\sqrt[3]{x-1} - 1}$

39. (a) $\sqrt{x^2+4}$ (b) $x + 4$

Domain of f and $g \circ f$: $x \geq -4$;

Domain of g and $f \circ g$: all real numbers

41. (a) $x + 1$ (b) $\sqrt{x^2+1}$

Domain of f and $g \circ f$: all real numbers

Domain of g and $f \circ g$: $x \geq 0$

43. (a) $|x + 6|$ (b) $|x| + 6$

Domain of $f, g, f \circ g$, and $g \circ f$: all real numbers

45. (a) $\dfrac{1}{x+3}$ (b) $\dfrac{1}{x} + 3$

Domain of f and $g \circ f$: all real numbers $x \neq 0$

Domain of g: all real numbers

Domain of $f \circ g$: all real numbers $x \neq -3$

47. (a) 3 (b) 0 **49.** (a) 0 (b) 4

51. Answers will vary. Sample answer:
$f(x) = x^2,\ g(x) = 2x + 1$

53. Answers will vary. Sample answer:
$f(x) = \sqrt[3]{x},\ g(x) = x^2 - 4$

55. Answers will vary. Sample answer:

$f(x) = \dfrac{1}{x},\ g(x) = x + 2$

57. Answers will vary. Sample answer:
$f(x) = \dfrac{x+3}{4+x}$, $g(x) = -x^2$

59. (a) $r(x) = \dfrac{x}{2}$ (b) $A(r) = \pi r^2$

(c) $(A \circ r)(x) = \pi\left(\dfrac{x}{2}\right)^2$; $(A \circ r)(x)$ represents the area of the circular base of the tank on the square foundation with side length x.

61. False. $(f \circ g)(x) = 6x + 1$ and $(g \circ f)(x) = 6x + 6$

63. $g(f(x))$ represents 3 percent of an amount over \$500,000.

65. Odd **67.** 3 **69.** $\dfrac{-4}{x(x+h)}$

71. $3x - y - 10 = 0$ **73.** $3x + 2y - 22 = 0$

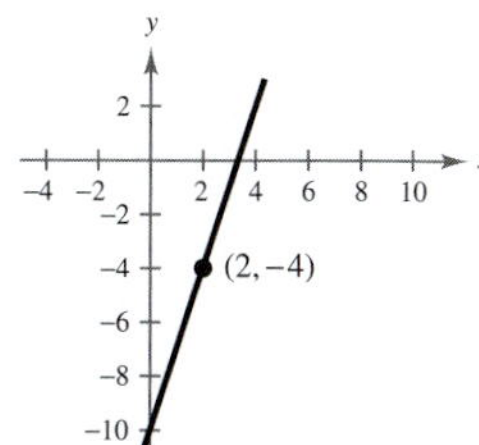

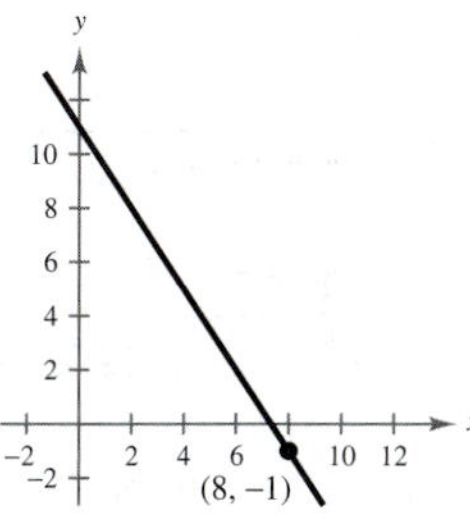

Section 2.7 *(page 243)*

1. c **2.** b **3.** a **4.** d **5.** $f^{-1}(x) = \frac{1}{6}x$

7. $f^{-1}(x) = x - 9$ **9.** $f^{-1}(x) = \dfrac{x-1}{3}$

11. $f^{-1}(x) = x^3$

13. (a) $f(g(x)) = f\left(\dfrac{x}{2}\right) = 2\left(\dfrac{x}{2}\right) = x$

$g(f(x)) = g(2x) = \dfrac{(2x)}{2} = x$

(b)

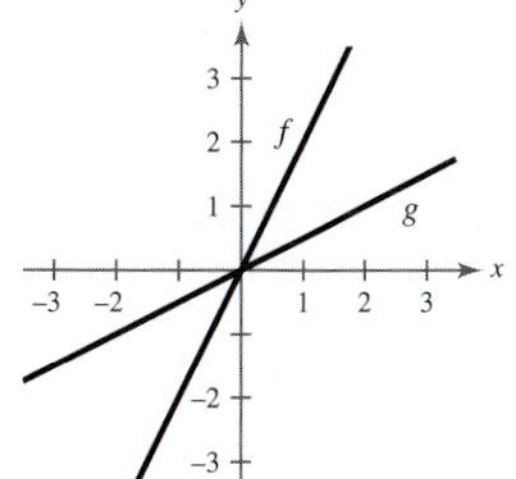

15. (a) $f(g(x)) = f\left(\dfrac{x-1}{7}\right) = 7\left(\dfrac{x-1}{7}\right) + 1 = x$

$g(f(x)) = g(7x+1) = \dfrac{(7x+1)-1}{7} = x$

(b)

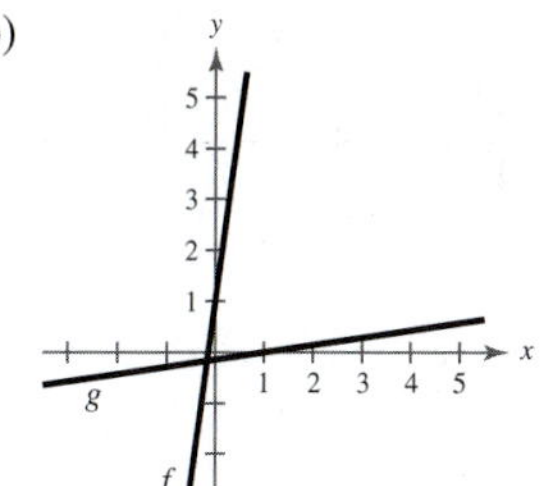

17. (a) $f(g(x)) = f\left(\sqrt[3]{8x}\right) = \dfrac{\left(\sqrt[3]{8x}\right)^3}{8} = x$

$g(f(x)) = g\left(\dfrac{x^3}{8}\right) = \sqrt[3]{8\left(\dfrac{x^3}{8}\right)} = x$

(b)

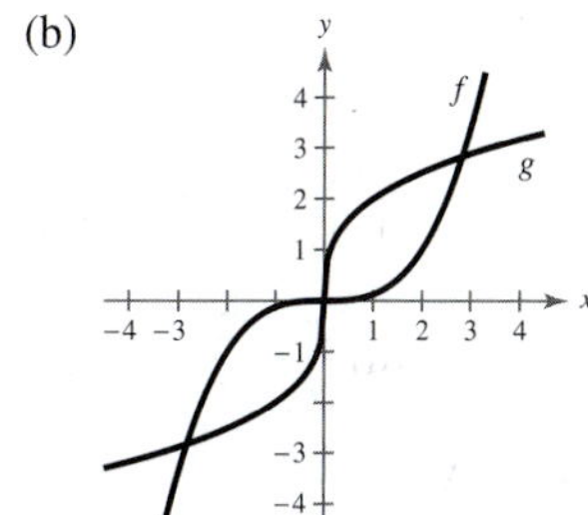

19. (a) $f(g(x)) = f(x^2 + 4),\ x \ge 0$

$= \sqrt{(x^2+4) - 4} = x$

$g(f(x)) = g\left(\sqrt{x-4}\right)$

$= \left(\sqrt{x-4}\right)^2 + 4 = x$

(b)

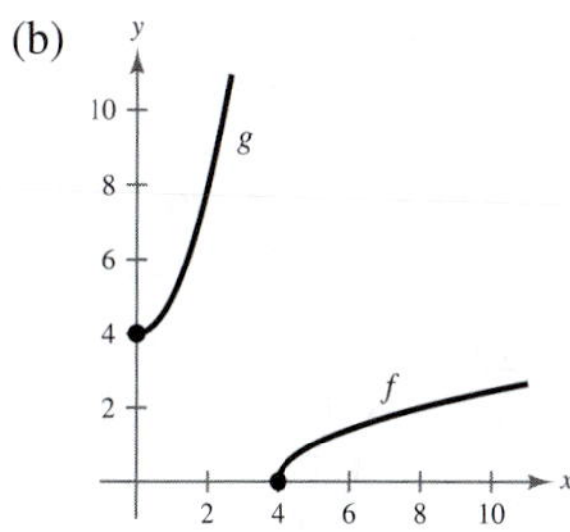

21. (a) $f(g(x)) = f\left(\sqrt{9-x}\right),\ x \le 9$

$= 9 - \left(\sqrt{9-x}\right)^2 = x$

$g(f(x)) = g(9 - x^2),\ x \ge 0$

$= \sqrt{9 - (9 - x^2)} = x$

(b)

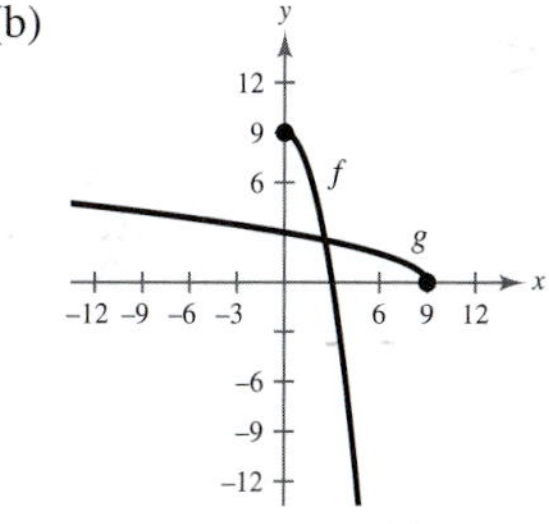

23. (a) $f(g(x)) = f\left(-\dfrac{5x+1}{x-1}\right) = \dfrac{-\left(\dfrac{5x+1}{x-1}\right) - 1}{-\left(\dfrac{5x+1}{x-1}\right) + 5}$

$$= \frac{-5x - 1 - x + 1}{-5x - 1 + 5x - 5} = x$$

$g(f(x)) = g\left(\dfrac{x-1}{x+5}\right) = \dfrac{-5\left(\dfrac{x-1}{x+5}\right) - 1}{\dfrac{x-1}{x+5} - 1}$

$$= \frac{-5x + 5 - x - 5}{x - 1 - x - 5} = x$$

(b)

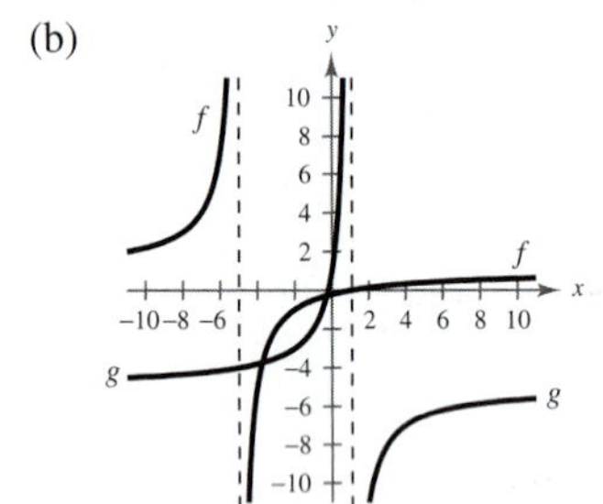

25. No

27.

x	-2	0	2	4	6	8
$f^{-1}(x)$	-2	-1	0	1	2	3

29. Yes **31.** No

33.

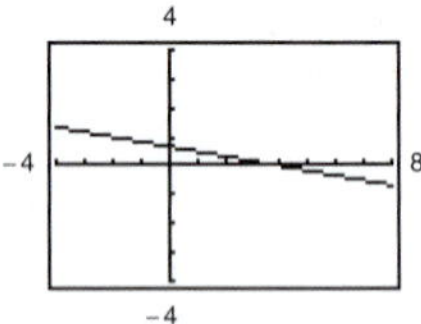

The function has an inverse.

35.

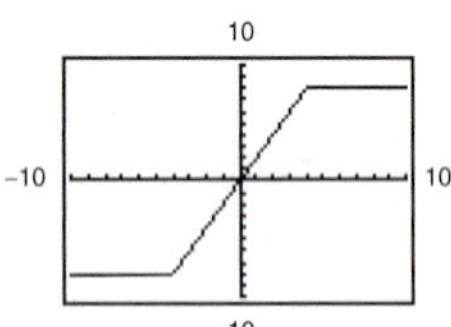

The function does not have an inverse.

37.

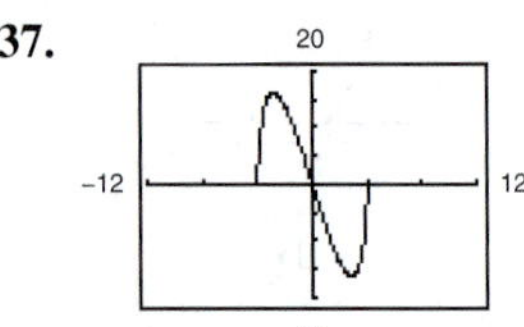

The function does not have an inverse.

39. $f^{-1}(x) = \dfrac{x+3}{2}$

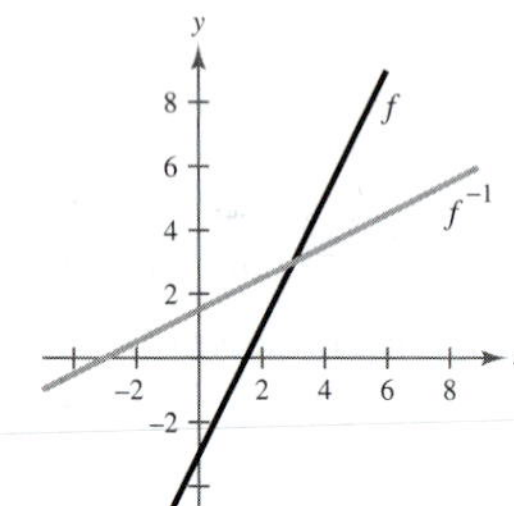

41. 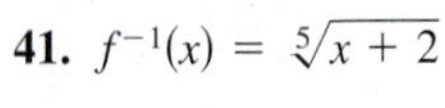 $f^{-1}(x) = \sqrt[5]{x+2}$

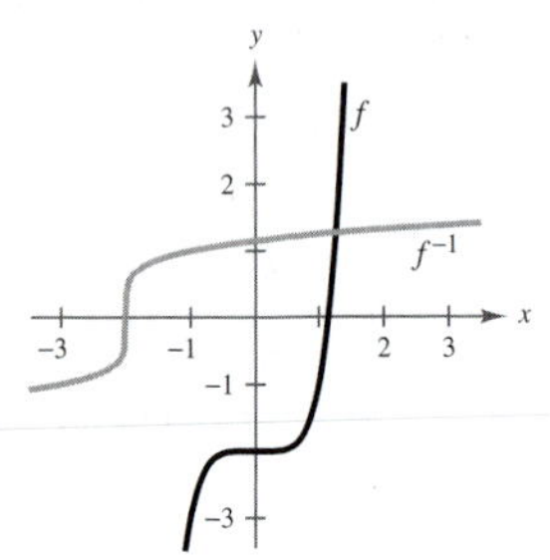

43. $f^{-1}(x) = x^2,\ x \geq 0$

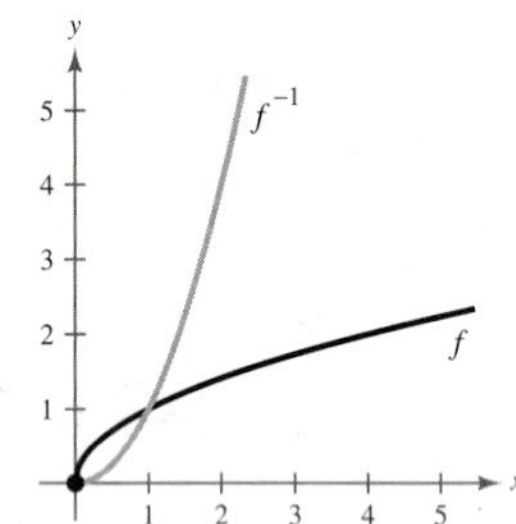

45. 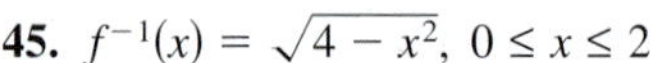 $f^{-1}(x) = \sqrt{4 - x^2},\ 0 \leq x \leq 2$

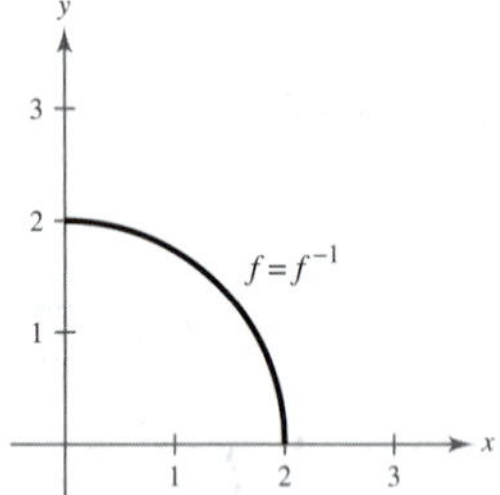

47. 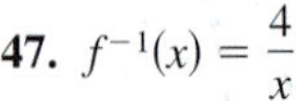 $f^{-1}(x) = \dfrac{4}{x}$

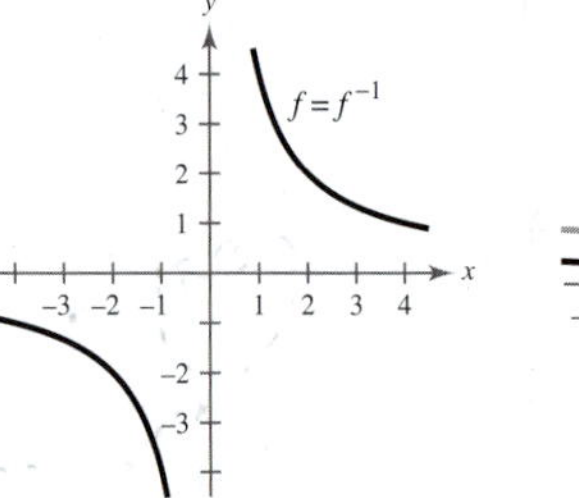

49. $f^{-1}(x) = \dfrac{2x+1}{x-1}$

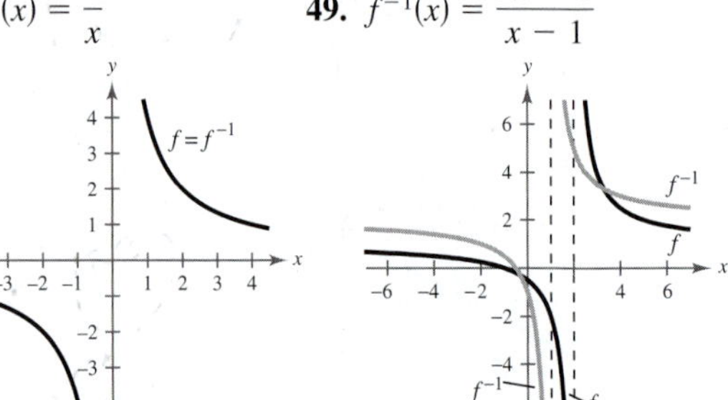

51. $f^{-1}(x) = x^3 + 1$

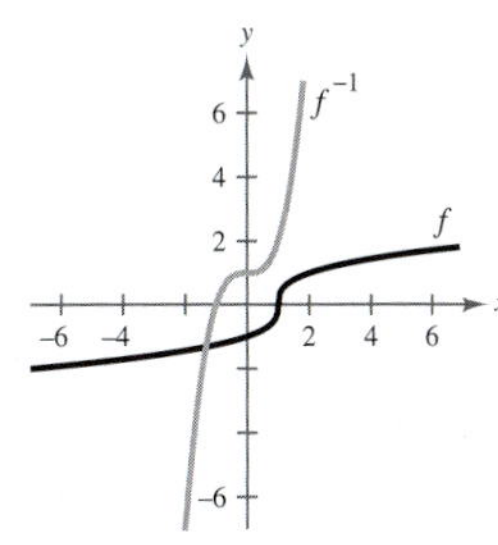

53. $f^{-1}(x) = \dfrac{5x - 4}{6 - 4x}$

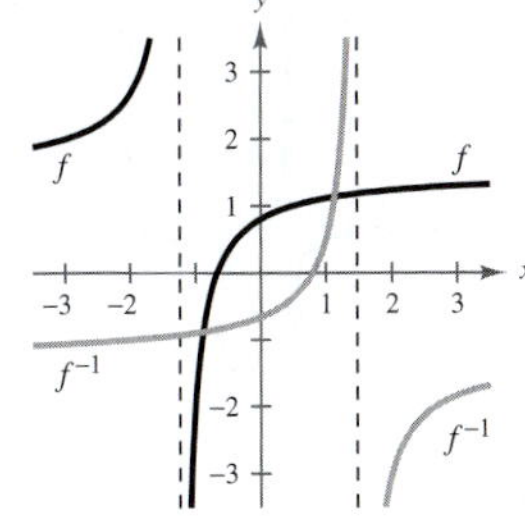

55. No inverse **57.** $g^{-1}(x) = 8x$ **59.** No inverse

61. $f^{-1}(x) = \sqrt{x} - 3$ **63.** No inverse

65. No inverse **67.** $f^{-1}(x) = \dfrac{x^2 - 3}{2},\ x \geq 0$ **69.** 32

71. 600 **73.** $2\sqrt[3]{x + 3}$ **75.** $\dfrac{x + 1}{2}$ **77.** $\dfrac{x + 1}{2}$

79. (a) 9

(b) f^{-1} yields the year for a given number of households.

(c) $y = 1266.54x + 92{,}255.54$

(d) $f^{-1} = \dfrac{x - 92{,}255.54}{1266.54}$ (e) 15

81. (a) Yes

(b) f^{-1} yields the year for a given number of miles traveled by motor vehicles.

(c) 8

(d) No. $f(t)$ would not pass the Horizontal Line Test.

83. (a) $y = \sqrt{\dfrac{x - 245.50}{0.03}},\ 245.5 < x < 545.5$

$x =$ degrees Fahrenheit; $y =$ % load

(b) 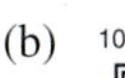

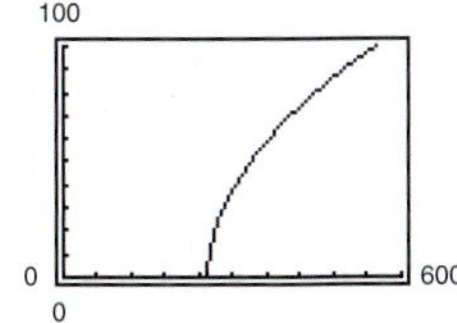

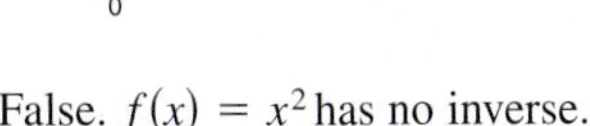

(c) $0 < x < 92.11$

85. False. $f(x) = x^2$ has no inverse.

87.

x	1	3	4	6
y	1	2	6	7

x	1	2	6	7
$f^{-1}(x)$	1	3	4	6

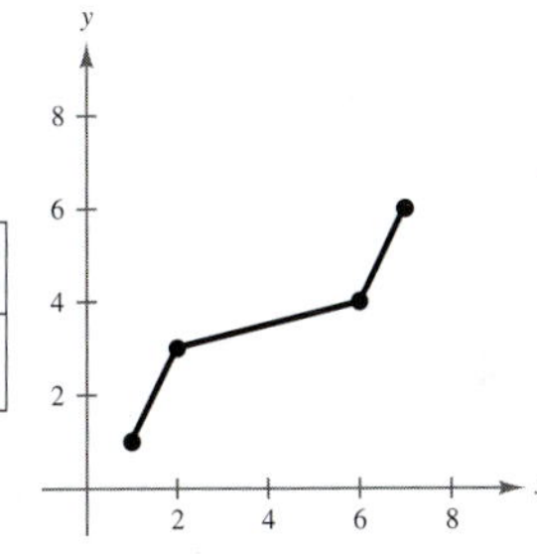

89.

x	-2	-1	3	4
y	6	0	-2	-3

x	-3	-2	0	6
$f^{-1}(x)$	4	3	-1	-2

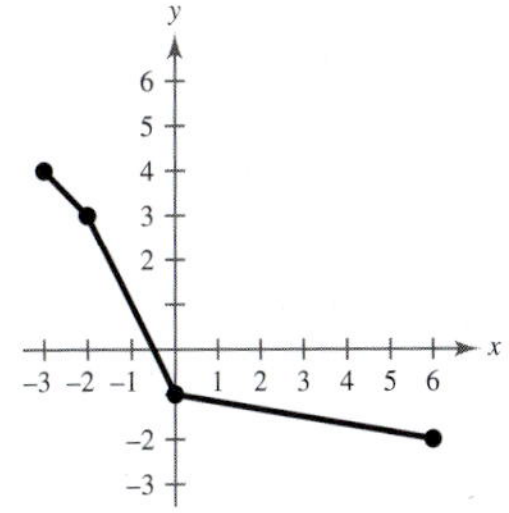

91. $k = \frac{1}{4}$

93. ± 8 **95.** $\frac{3}{2}$ **97.** $3 \pm \sqrt{5}$ **99.** $5, -\frac{10}{3}$

101. 16, 18

Review Exercises *(page 248)*

1. (a) L_2 (b) L_3 (c) L_1 (d) L_4

3.

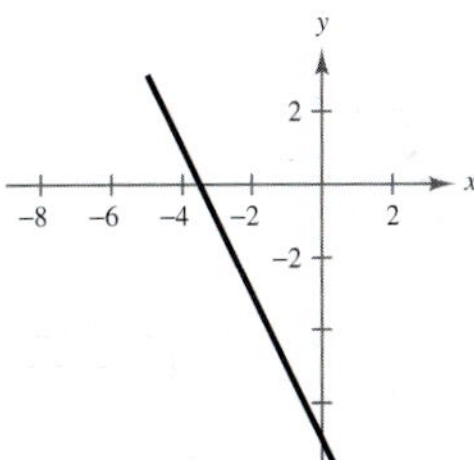

5.

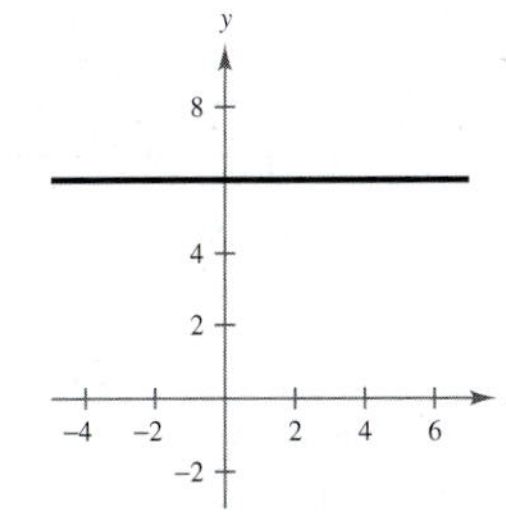

7.

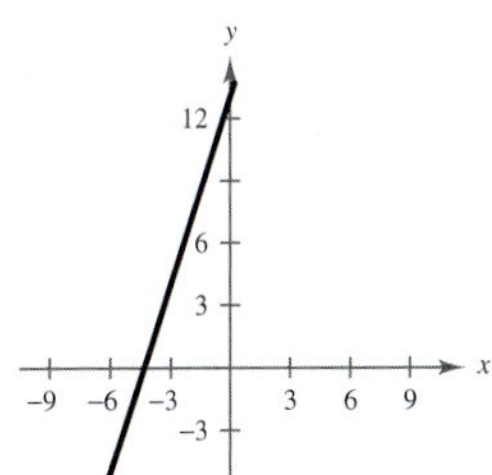

9.

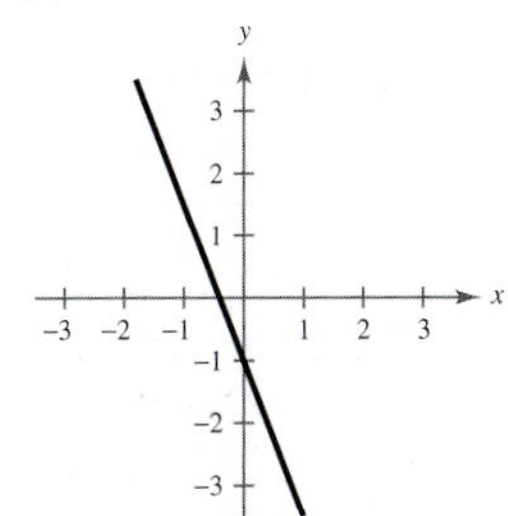

11. $t = \frac{7}{3}$ **13.** $(6, 0), (10, 1), (-2, -2)$

15.

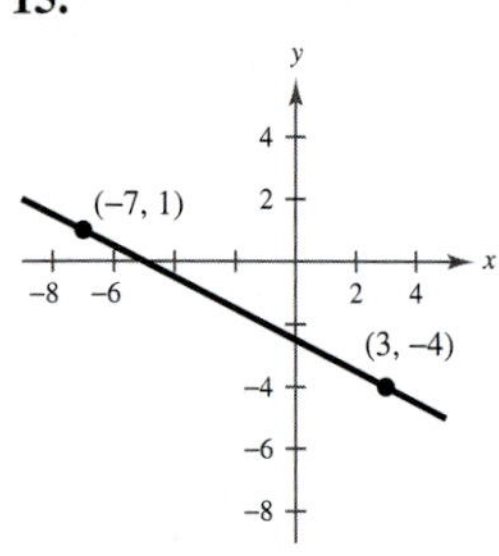

$m = -\frac{1}{2}$

17.

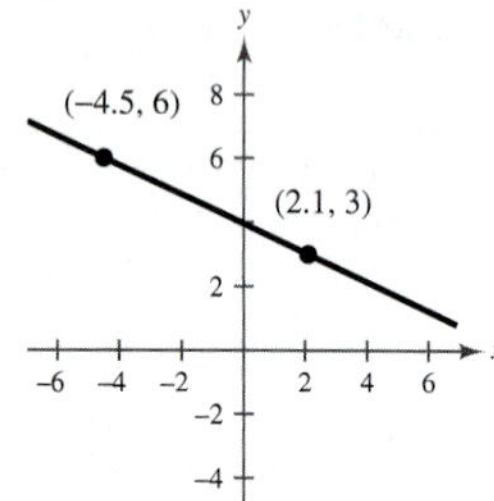

$m = -\frac{5}{11}$

19. $x = 0$ **21.** $4x + 3y - 8 = 0$

23. $3x - 2y - 10 = 0$ **25.** $x + 2y - 4 = 0$

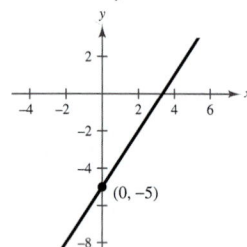

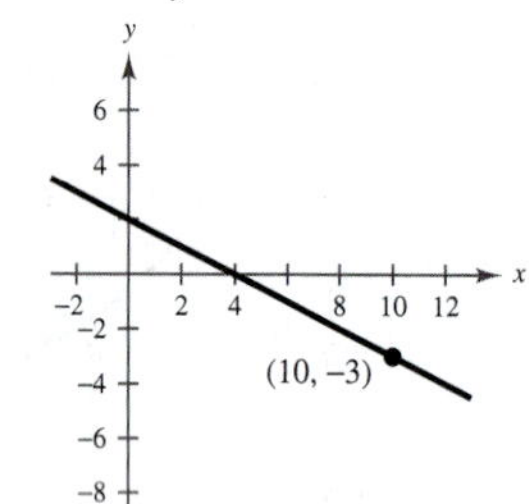

27. (a) $5x - 4y - 23 = 0$ (b) $4x + 5y - 2 = 0$

29. $V = 850t + 9100, 4 \le t \le 9$

31. $210,000

33. (a) Not a function, because 20 in the domain corresponds to two values in the range

(b) A function, because each input value has exactly one output value

(c) A function, because each input value has exactly one output value

(d) Not a function, because 30 in A is not matched with any element in B

35. No **37.** Yes

39. (a) 5 (b) 17 (c) $t^4 + 1$ (d) $t^2 + 2t + 2$

41. (a) -3 (b) -1 (c) 2 (d) 6

43. $-5 \le x \le 5$

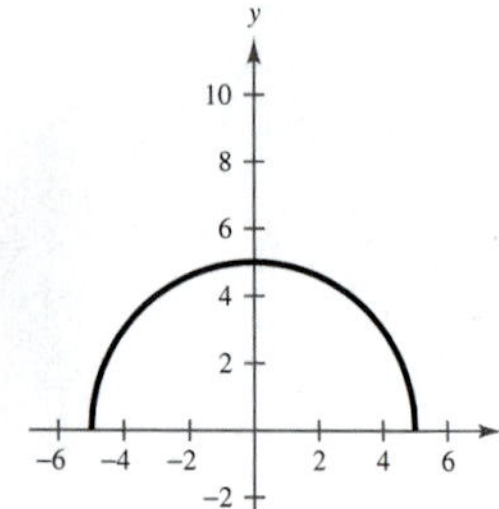

45. All real numbers $s \ne 3$

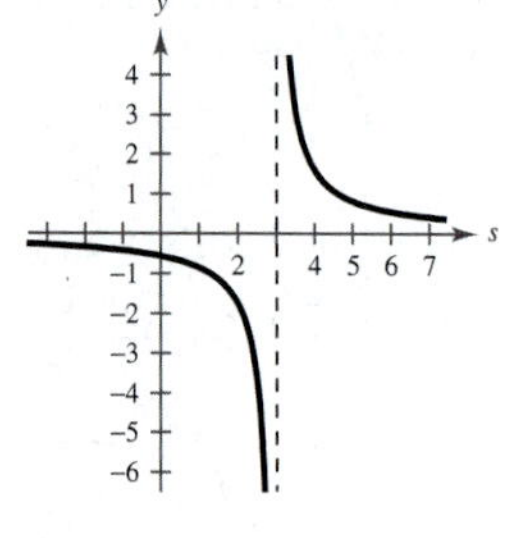

47. All real numbers $x \ne 3, -2$

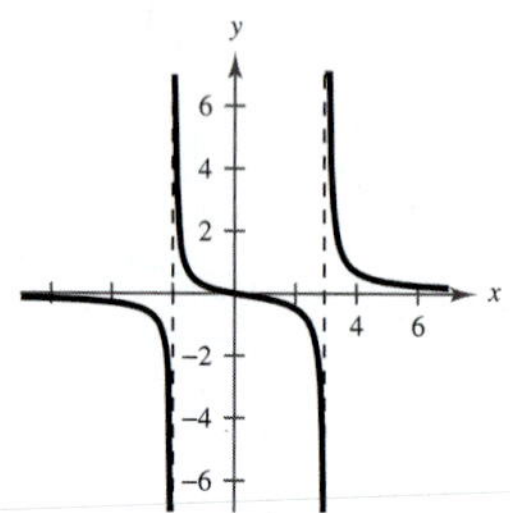

49. $4x + 2h + 3, \quad h \ne 0$

51. (a) 16 feet per second (b) 1.5 seconds

(c) -16 feet per second

53. (a) $A = x(12 - x)$ (b) $0 < x < 12$

55. Function **57.** Not a function

59. $\frac{7}{3}, 3$ **61.** $-\frac{3}{8}$

63. Increasing on $(0, \infty)$

Decreasing on $(-\infty, -1)$

Constant on $(0, -1)$

65. Neither even nor odd **67.** Odd

69. $f(x) = -3x$

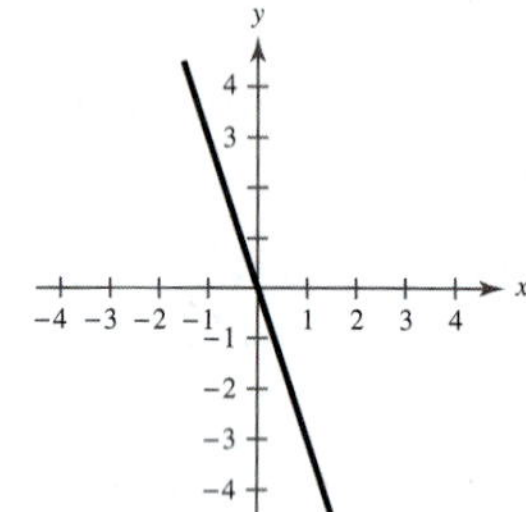

71. $f(x) = \frac{5}{3}x + \frac{10}{3}$

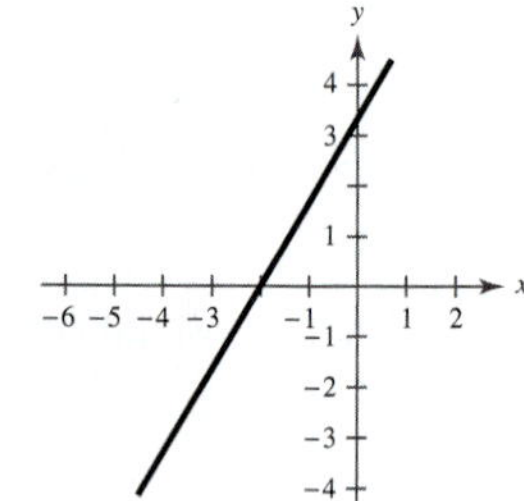

73.

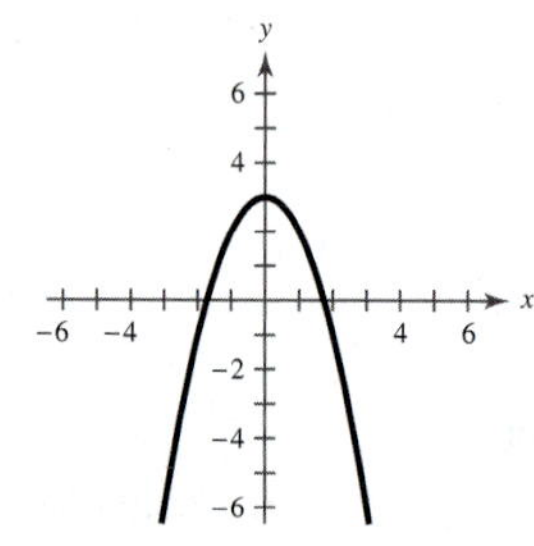

75.

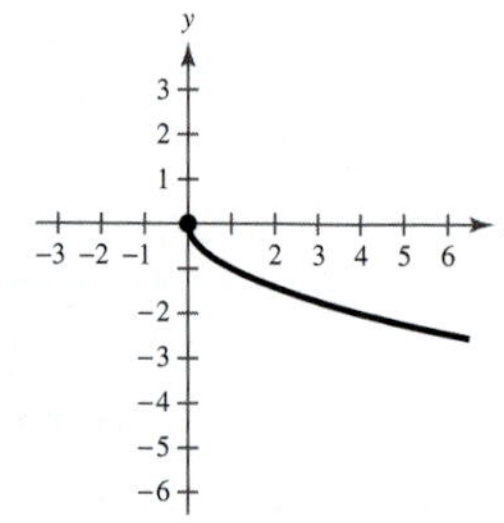

77.

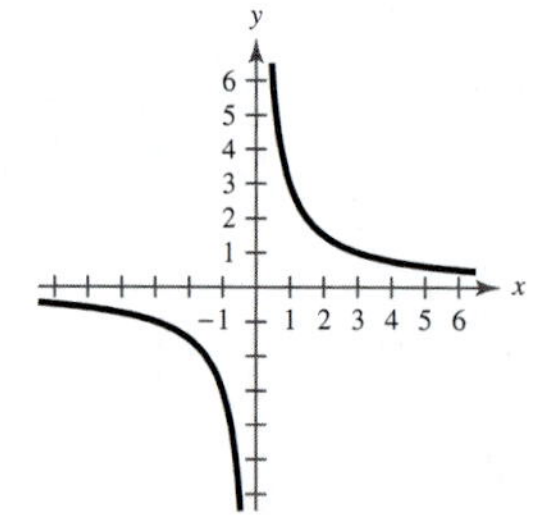

79.

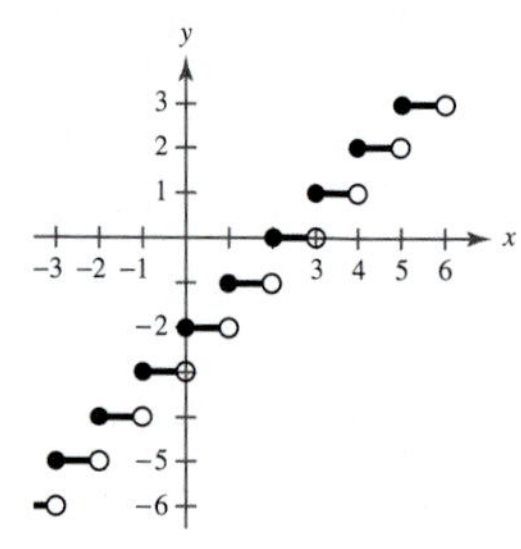

81.

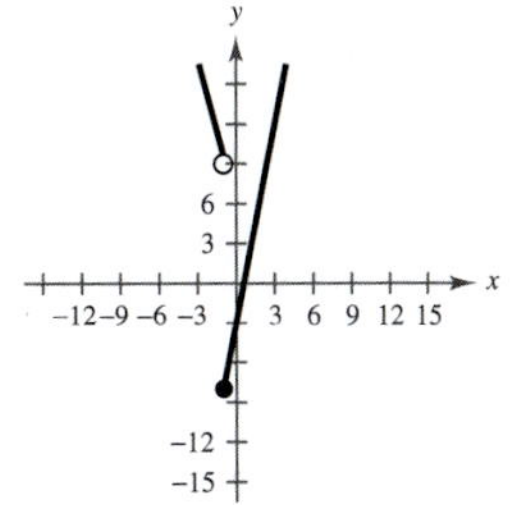

83. $y = x^3$

85. Vertical shift of nine units downward

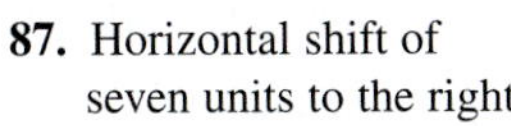

87. Horizontal shift of seven units to the right

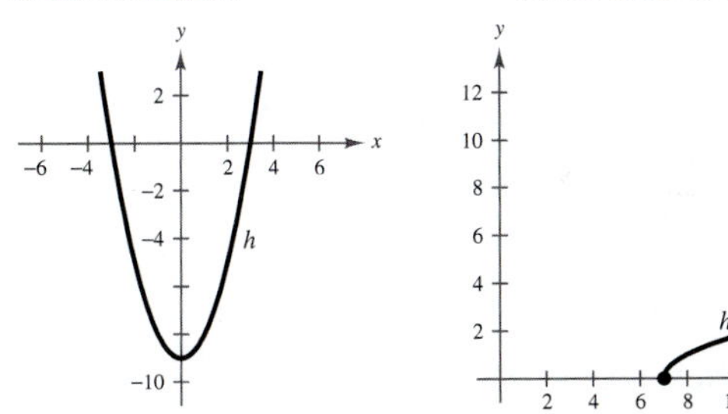

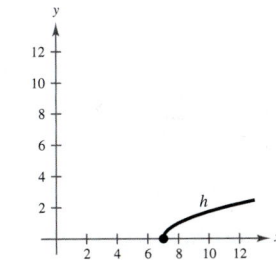

89. Reflection in the x-axis, horizontal shift of three units to the left, and vertical shift of one unit upward

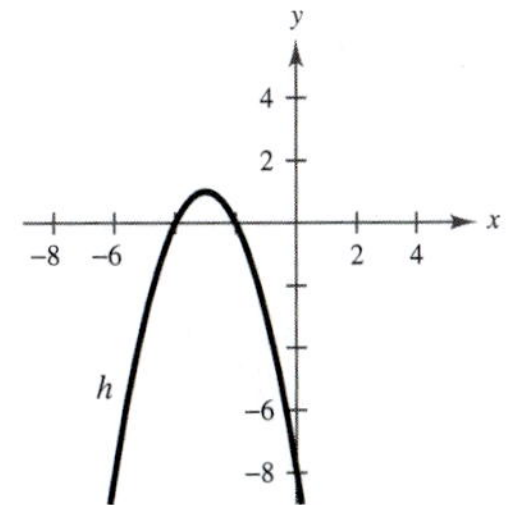

91. Reflection in the x-axis and vertical shift of six units upward

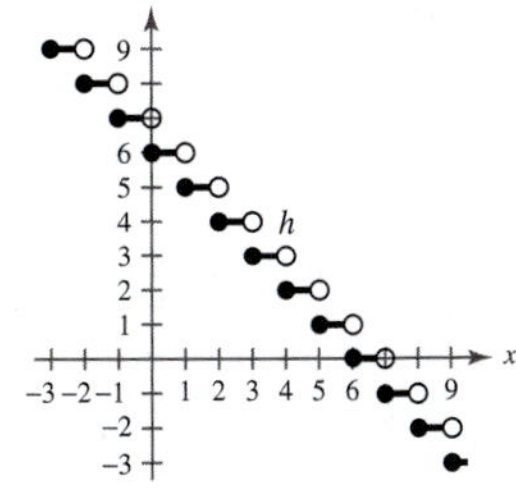

93. Reflections in the x-axis and the y-axis, horizontal shift of four units to the right, and vertical shift of six units upward

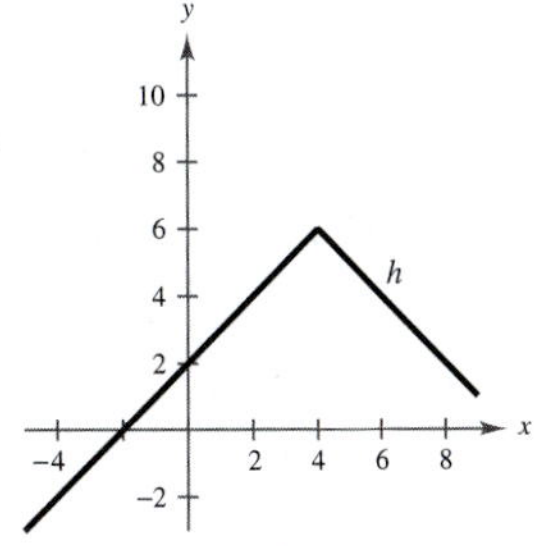

95. Horizontal shift of nine units to the right and vertical stretch

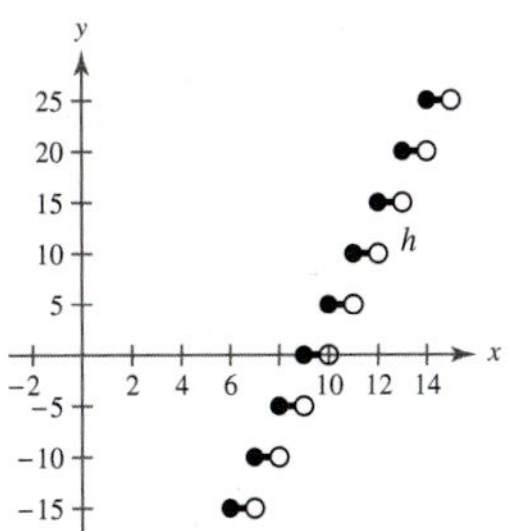

97. Reflection in the x-axis, vertical stretch, and horizontal shift of four units to the right

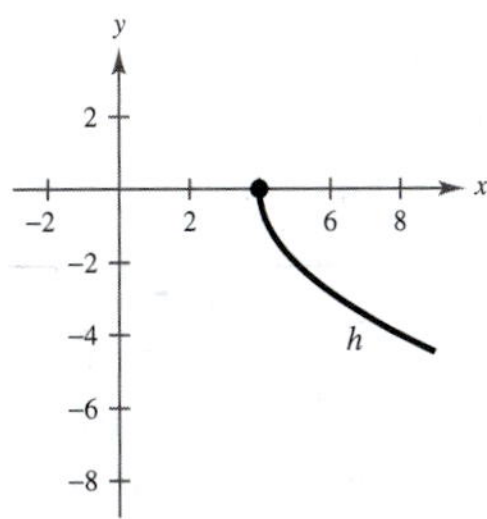

99. (a) $x^2 + 2x + 2$ (b) $x^2 - 2x + 4$

(c) $2x^3 - x^2 + 6x - 3$ (d) $\dfrac{x^2 + 3}{2x - 1}$; $x \neq \dfrac{1}{2}$

101. (a) $x - \frac{8}{3}$ (b) $x - 8$

Domain of f, g, $f \circ g$, and $g \circ f$: all real numbers

103. $f(x) = x^3$, $g(x) = 6x - 5$

105. $y_1 = 0.207t^2 + 8.65t + 14.2$

$y_2 = 1.414t^2 - 7.28t + 146.9$

107. $f^{-1}(x) = x + 7$

$f(f^{-1}(x)) = x + 7 - 7 = x$

$f^{-1}(f(x)) = x - 7 + 7 = x$

109. The function has an inverse.

111. The function has an inverse.

113. The function has an inverse.

115. (a) $f^{-1}(x) = 2x + 6$

(b)

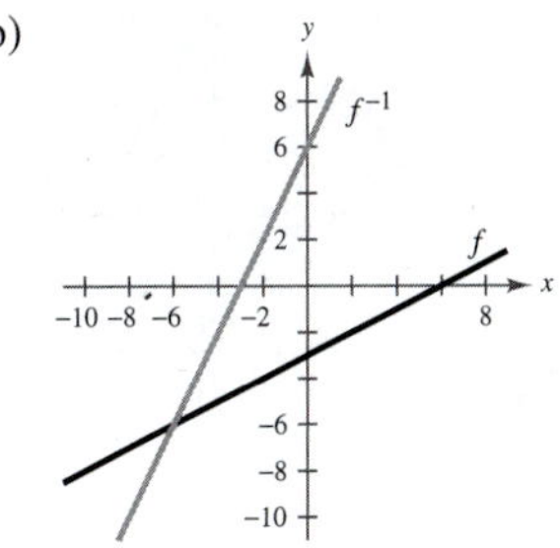

(c) $f^{-1}(f(x)) = 2\left(\frac{1}{2}x - 3\right) + 6 = x - 6 + 6 = x$

$f(f^{-1}(x)) = \frac{1}{2}(2x + 6) - 3 = x + 3 - 3 = x$

117. (a) $f^{-1}(x) = x^2 - 1, \ x \geq 0$

(b)

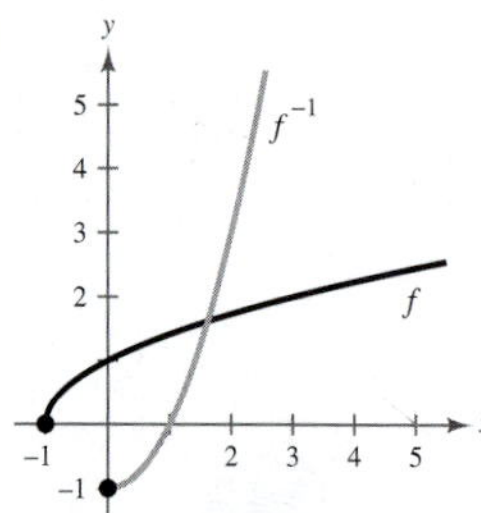

(c) $f^{-1}(f(x)) = f^{-1}(\sqrt{x+1})$
$= (x + 1) - 1$
$= x$
$f(f^{-1}(x)) = f(x^2 - 1), \ x \geq 0$
$= \sqrt{x^2 - 1 + 1}$
$= x$

119. $x \geq 4; \ f^{-1}(x) = \sqrt{\dfrac{x}{2}} + 4$

121. False. The graph is reflected in the x-axis, shifted 9 units to the left, and then shifted 13 units downward.

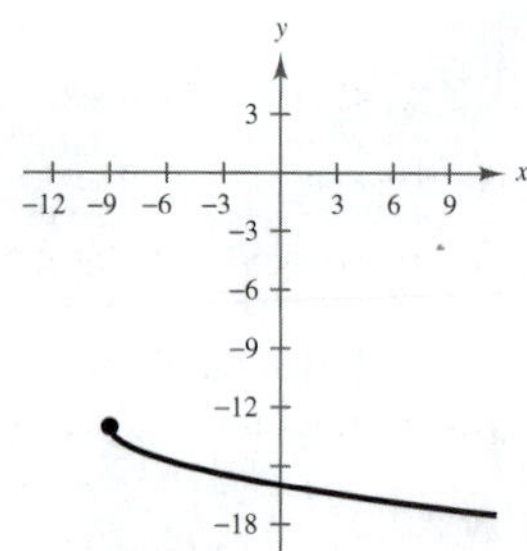

123. A function from a set A to a set B is a relation that assigns to each element x in the set A exactly one element y in the set B.

Chapter Test *(page 252)*

1. $2x + y - 1 = 0$

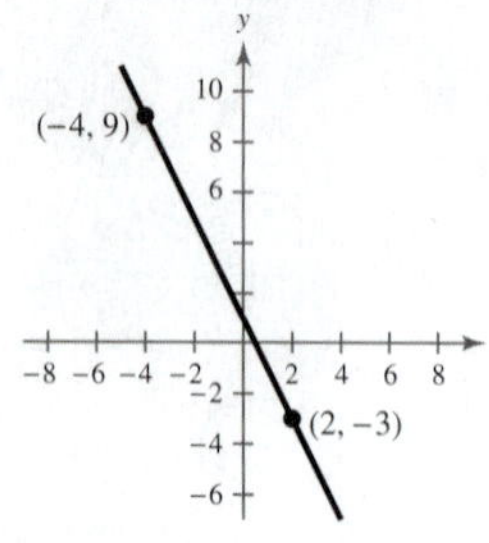

2. $17x + 10y - 59 = 0$

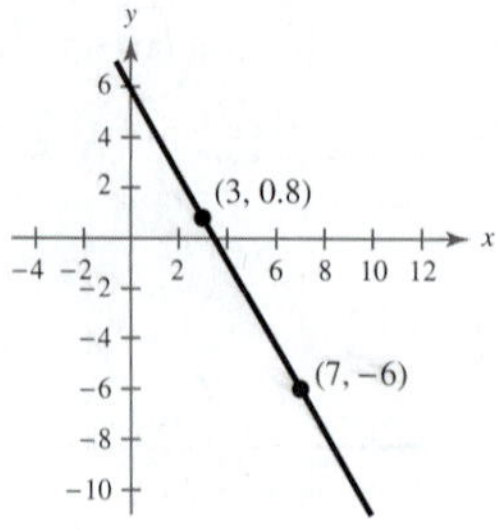

3. (a) $4x - 7y + 44 = 0$ (b) $7x + 4y - 53 = 0$

4. (a) -9 (b) 1 (c) $|x - 4| - 15$

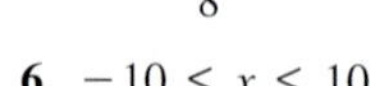

5. (a) $-\dfrac{1}{8}$ (b) $-\dfrac{1}{28}$ (c) $\dfrac{\sqrt{x}}{x^2 - 18x}$

6. $-10 \leq x \leq 10$ **7.** All real numbers

8. (a)

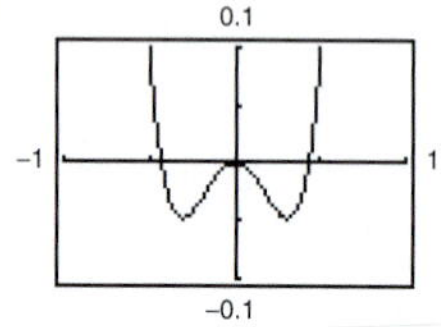

(b) Increasing on $(-0.31, 0), (0.31, \infty)$
Decreasing on $(-\infty, -0.31), (0, 0.31)$

(c) Even

9. (a)

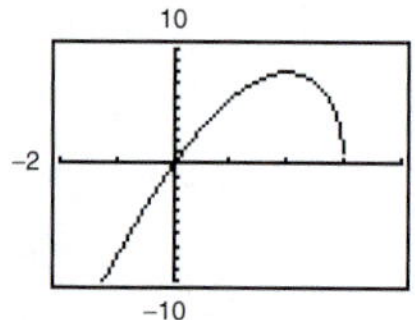

(b) Increasing on $(-\infty, 2)$
Decreasing on $(2, 3)$

(c) Neither even nor odd

10. (a)

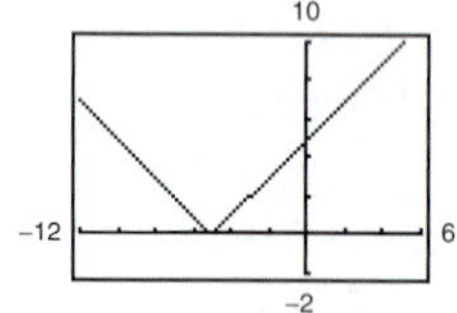

(b) Increasing on $(-5, \infty)$
Decreasing on $(-\infty, -5)$

(c) Neither even nor odd

11.

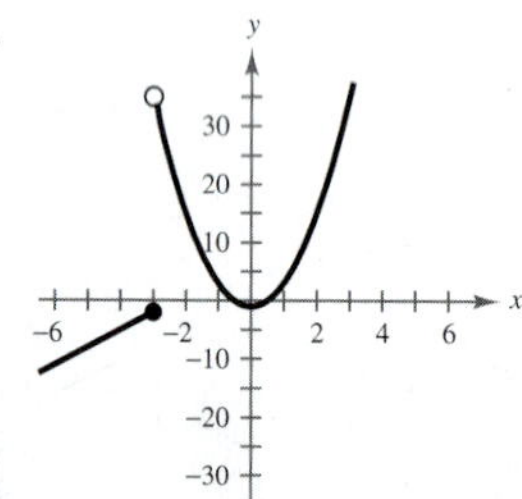

12. Reflection in the x-axis of $y = [\![x]\!]$

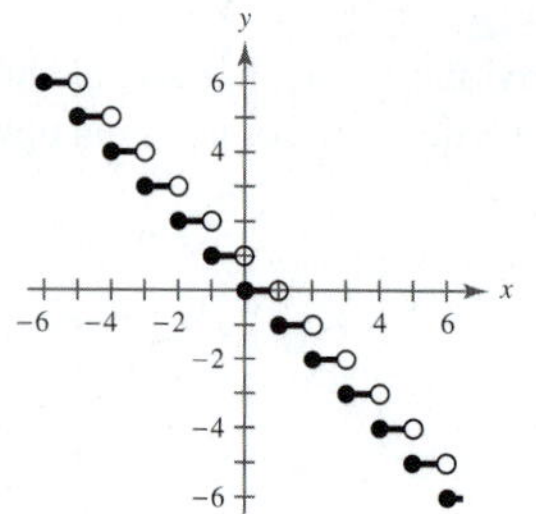

13. Reflection in the x-axis, horizontal shift, and vertical shift of $y = \sqrt{x}$

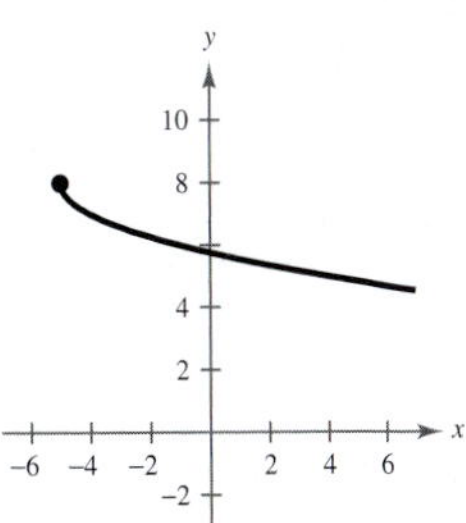

14. Vertical shrink, horizontal shift, and vertical shift of $y = |x|$

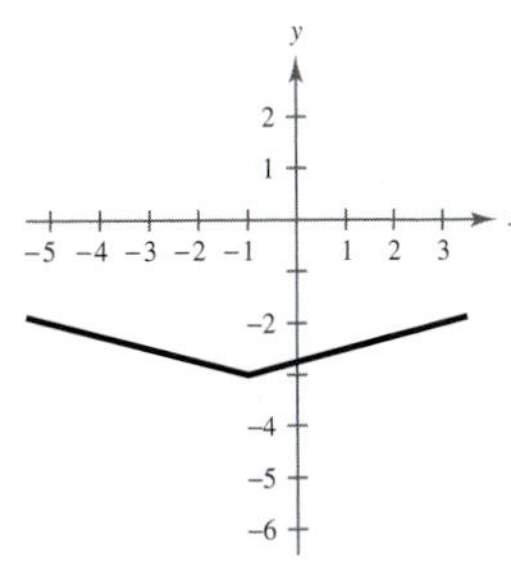

15. (a) $2x^2 - 4x - 2$ (b) $4x^2 + 4x - 12$

(c) $-3x^4 - 12x^3 + 22x^2 + 28x - 35$

(d) $\dfrac{3x^2 - 7}{-x^2 - 4x + 5}$, $x \neq 1, -5$

(e) $3x^4 + 24x^3 + 18x^2 - 120x + 68$

(f) $-9x^4 + 30x^2 - 16$

16. (a) $\dfrac{1 + 2x^{3/2}}{x}$, $x > 0$ (b) $\dfrac{1 - 2x^{3/2}}{x}$, $x > 0$

(c) $\dfrac{2\sqrt{x}}{x}$, $x > 0$ (d) $\dfrac{1}{2x^{3/2}}$, $x > 0$

(e) $\dfrac{\sqrt{x}}{2x}$, $x > 0$ (f) $\dfrac{2\sqrt{x}}{x}$, $x > 0$

17. $f^{-1}(x) = \sqrt[3]{x - 8}$ **18.** No inverse

19. $f^{-1}(x) = \left(\frac{8}{3}x\right)^{2/3}$, $x \geq 0$ **20.** \$153

Cumulative Test for Chapters P–2
(page 253)

1. $\dfrac{4x^3}{15y^5}$, $x \neq 0$ **2.** $2x^2y\sqrt{6y}$ **3.** $5x - 6$

4. $x^3 - x^2 - 5x + 6$ **5.** $\dfrac{s - 1}{(s + 1)(s + 3)}$

6. $(x + 3)(7 - x)$ **7.** $x(x + 1)(1 - 6x)$

8. $2(3 - 2x)(9 + 6x + 4x^2)$

9.

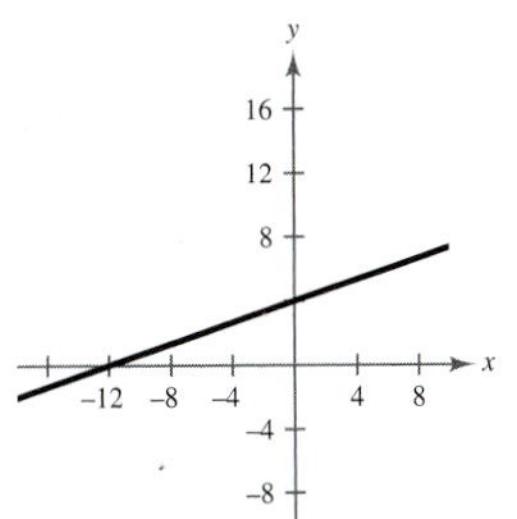

10.

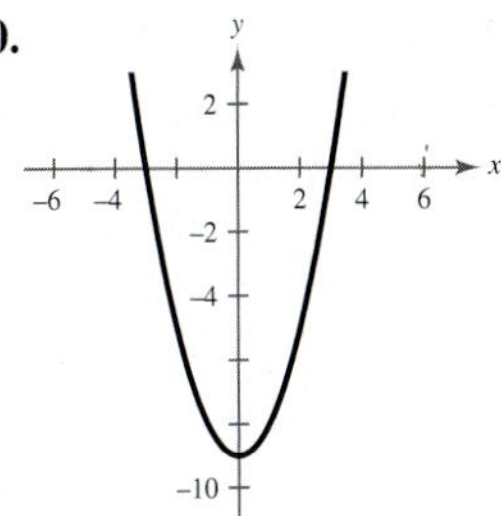

11.

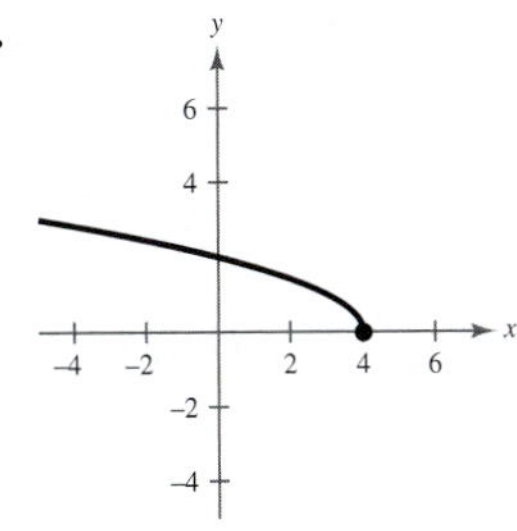

12. $4x^2 + 12x$ **13.** $\frac{3}{2}x^2 + 8x + \frac{5}{2}$ **14.** 1, 3

15. $2 \pm \sqrt{10}$ **16.** ± 4

17. $\dfrac{-5 \pm \sqrt{97}}{6}$ **18.** $-\dfrac{3}{2} \pm \dfrac{\sqrt{69}}{6}$

19. ± 8 **20.** $0, -12, \pm 2i$ **21.** 0, 3 **22.** ± 8

23. 6 **24.** $-5, 9$ **25.** No solution

26. (a) Not a solution (b) Not a solution (c) Solution (d) Solution

27. (a) Not a solution (b) Not a solution (c) Solution (d) Solution

28. (a) Not a solution (b) Solution (c) Not a solution (d) Not a solution

29. $-7 \leq x \leq 5$

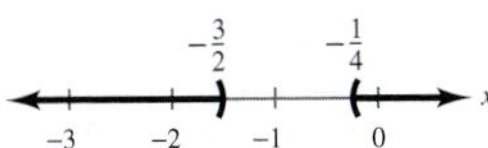

30. $x < -\frac{3}{2}$, $x > -\frac{1}{4}$

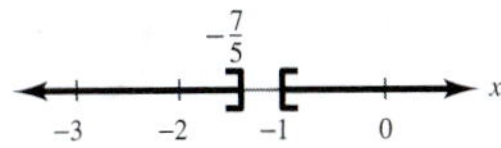

31. $x \leq -\frac{7}{5}$, $x \geq -1$

32. $x < \dfrac{1 - \sqrt{17}}{2}$, $x > \dfrac{1 + \sqrt{17}}{2}$

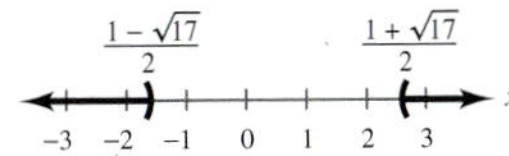

33. $2x - y + 2 = 0$

34. For some values of x there correspond two values of y.

35. (a) $\frac{3}{2}$ (b) Division by 0 is undefined. (c) $\frac{s+2}{s}$

36. (a) Vertical shrink by $\frac{1}{2}$

(b) Vertical shift of two units upward

(c) Horizontal shift of two units to the left

37. (a) $5x - 2$ (b) $-3x - 4$ (c) $4x^2 - 11x - 3$

(d) $\frac{x-3}{4x+1}$; Domain: all real numbers except $x = -\frac{1}{4}$

38. (a) $\sqrt{x-1} + x^2 + 1$ (b) $\sqrt{x-1} - x^2 - 1$

(c) $x^2\sqrt{x-1} + \sqrt{x-1}$ (d) $\frac{\sqrt{x-1}}{x^2+1}$; Domain: $x \geq 1$

39. (a) $2x + 12$ (b) $\sqrt{2x^2 + 6}$

Domain of $f \circ g$: $x > -6$

Domain of $g \circ f$: all real numbers

40. (a) $|x| - 2$ (b) $|x - 2|$

Domain of $f \circ g$ and $g \circ f$: all real numbers

41. $h^{-1}(x) = \frac{1}{5}(x + 2)$ **42.** $n = 9$

43. (a) $R(n) = n[8 - 0.05(n - 80)],\ n \geq 80$

(b)

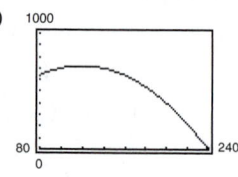

120 passengers

Problem Solving *(page 256)*

1. (a) $W_1 = 2000 + 0.07S$ (b) $W_2 = 2300 + 0.05S$

(c)

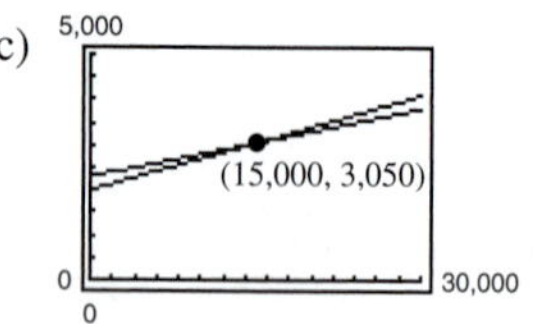

Both jobs pay the same monthly salary if sales equal \$15,000.

(d) No. Job 1 would pay \$3400 and job 2 would pay \$3300.

3. (a) The function will be even.

(b) The function will be odd if the two functions are not equal.

(c) The function will be neither even nor odd.

5. $f(x) = a_{2n}x^{2n} + a_{2n-2}x^{2n-2} + \cdots + a_2x^2 + a_0$

$f(-x) = a_{2n}(-x)^{2n} + a_{2n-2}(-x)^{2n-2} + \cdots + a_2(-x)^2 + a_0$

$= f(x)$

7. (a) $81\frac{2}{3}$ hours (b) $25\frac{5}{7}$ miles per hour

(c) $y = \frac{-180}{7}x + 3400$

Domain: $0 \leq x \leq \frac{1190}{9}$

Range: $0 \leq y \leq 3400$

(d)

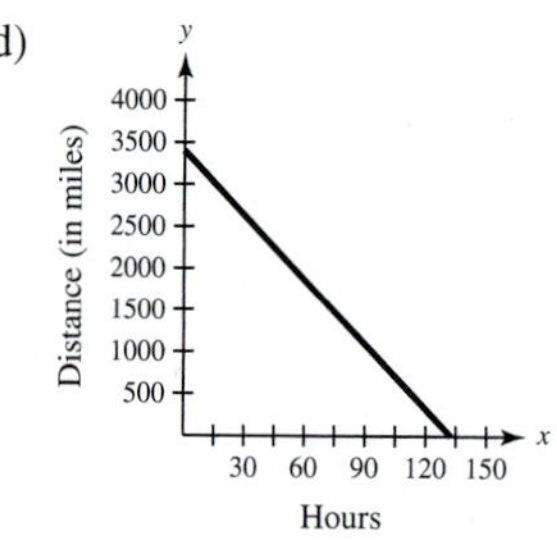

9. (a) $T = \frac{1}{2}\sqrt{4 + x^2} + \frac{1}{4}\sqrt{x^2 - 6x + 10}$

(b) $0 \leq x \leq 3$

(c)

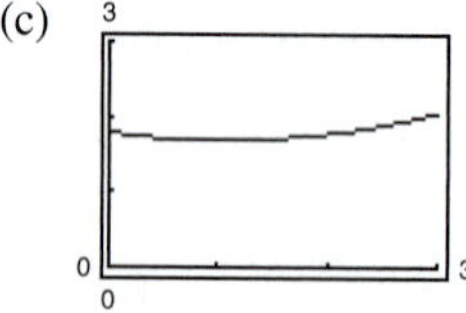

(d) $x = 1$

(e) The distance $x = 1$ yields a time of 1.68 hours.

11. (a) Domain: all real numbers $x \neq 1$

Range: all real numbers

(b) $f(f(x)) = \frac{x-1}{x}$

Domain: all real numbers $x \neq 0, 1$

(c) $f(f(f(x))) = x$

The graph is not a line because there are holes at $x = 0$ and $x = 1$.

13. (a) (b)

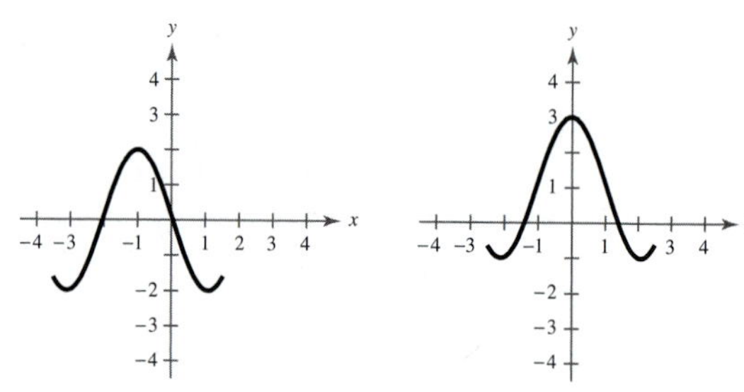

(c)

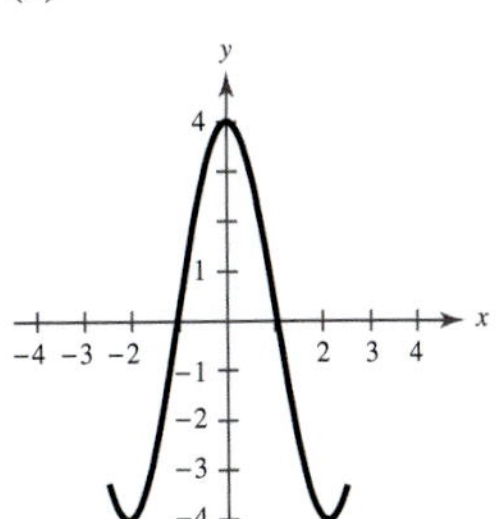

(d)

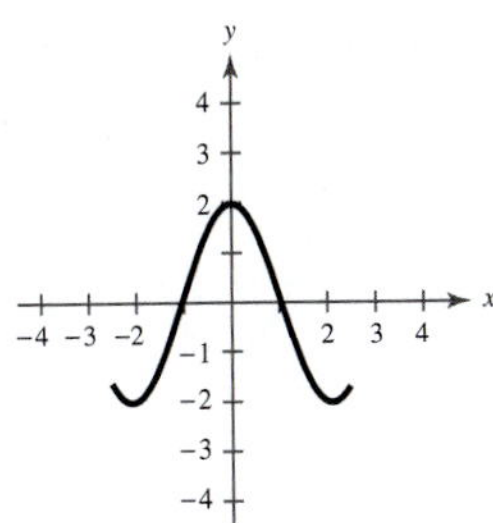

(e)

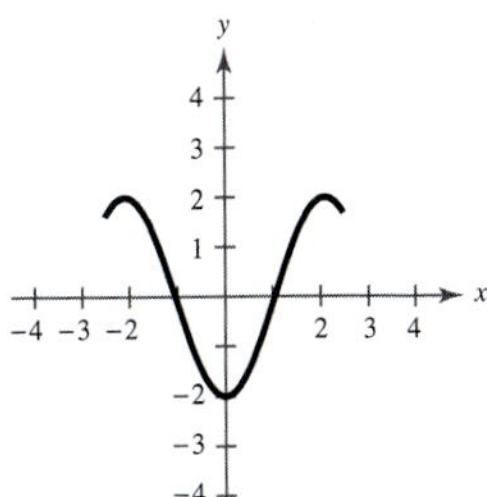

(f)

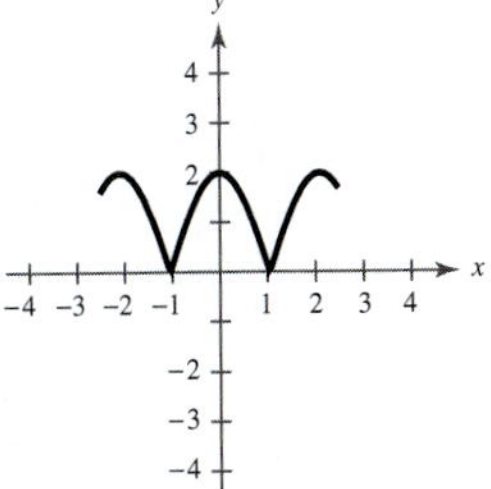

(g)

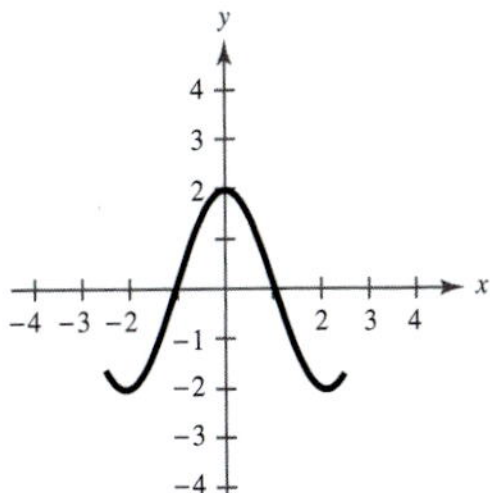

Chapter 3

Section 3.1 *(page 266)*

1. g **2.** c **3.** b **4.** h

5. f **6.** a **7.** e **8.** d

9. (a)

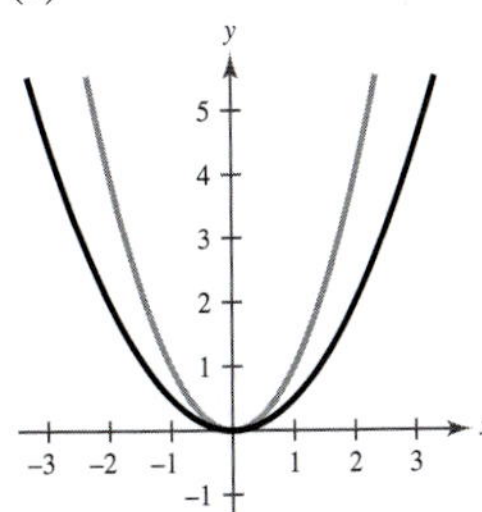

Vertical shrink

(b)

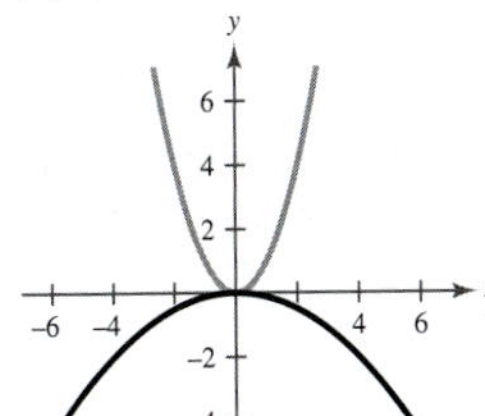

Vertical shrink and reflection in the x-axis

(c)

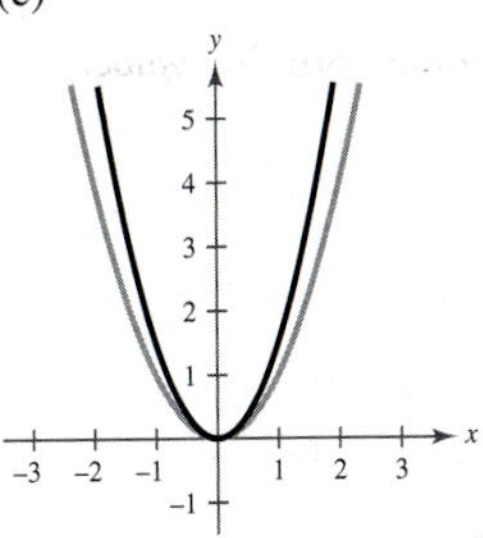

Vertical stretch

(d)

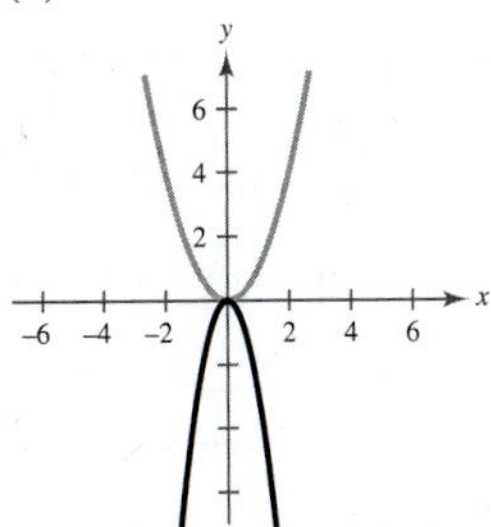

Vertical stretch and reflection in the x-axis

11. (a)

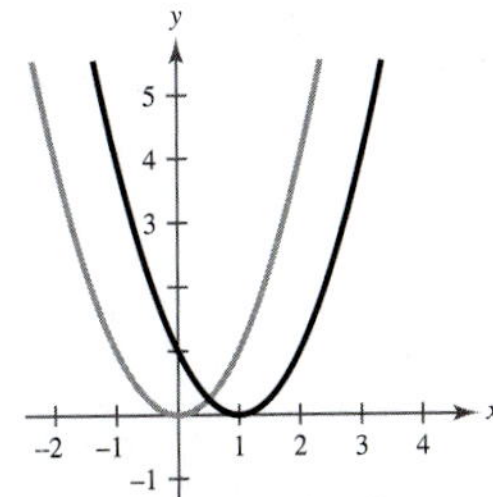

Horizontal translation

(b)

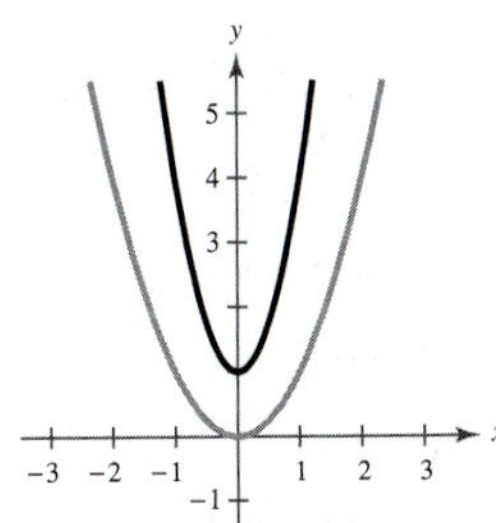

Horizontal shrink and vertical translation

(c)

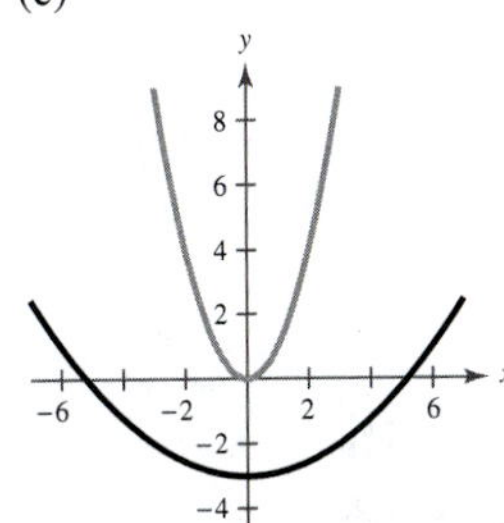

Horizontal stretch and vertical translation

(d)

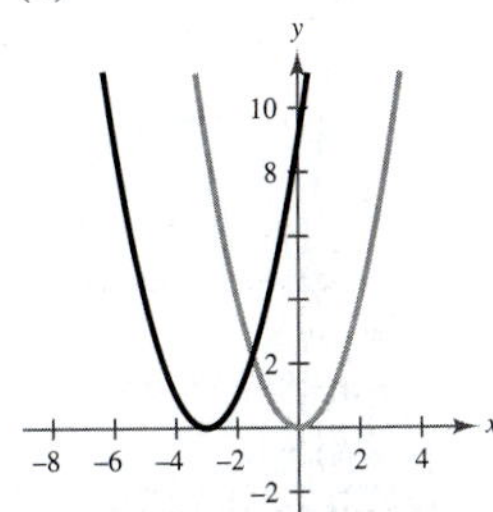

Horizontal translation

13.

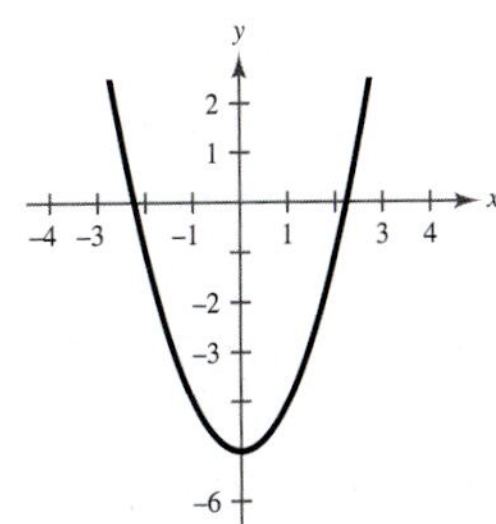

Vertex: $(0, -5)$

x-intercepts: $(\pm\sqrt{5}, 0)$

15.

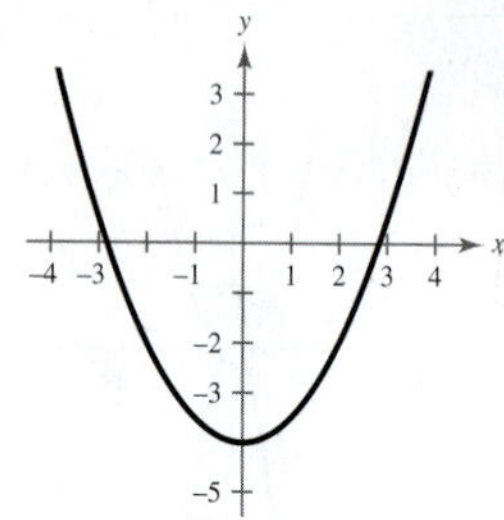

Vertex: $(0, -4)$

x-intercepts: $(\pm 2\sqrt{2}, 0)$

17.

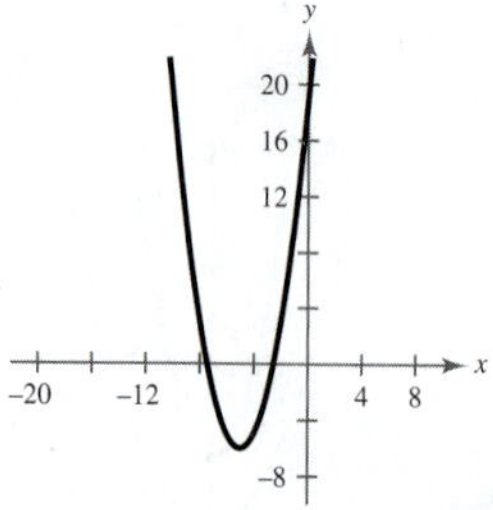

Vertex: $(-5, -6)$

x-intercepts: $(-5 \pm \sqrt{6}, 0)$

19.

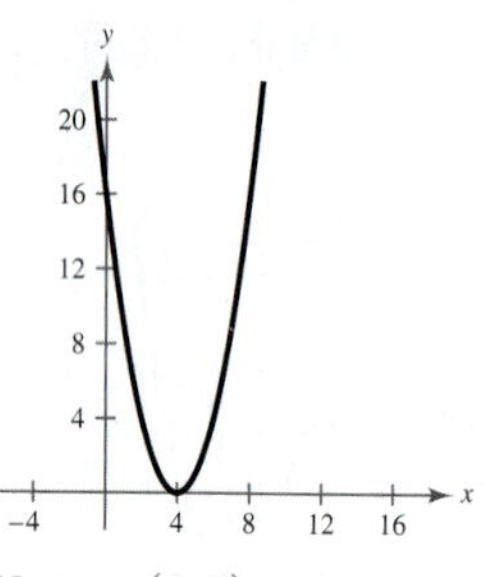

Vertex: $(4, 0)$

x-intercept: $(4, 0)$

21.

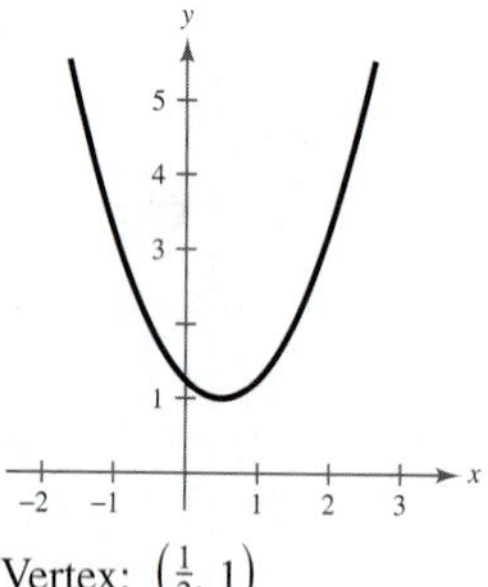

Vertex: $\left(\frac{1}{2}, 1\right)$

No x-intercept

23.

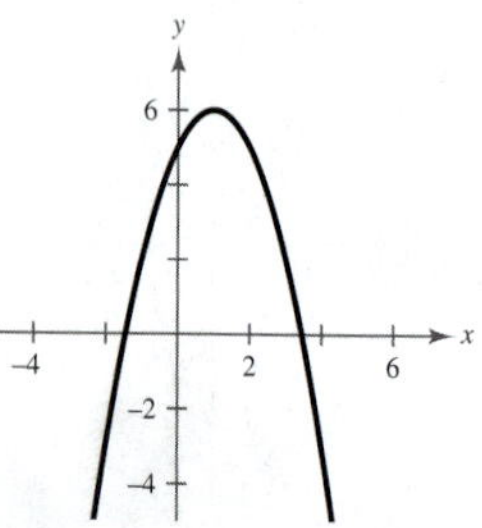

Vertex: $(1, 6)$

x-intercepts: $(1 \pm \sqrt{6}, 0)$

25.

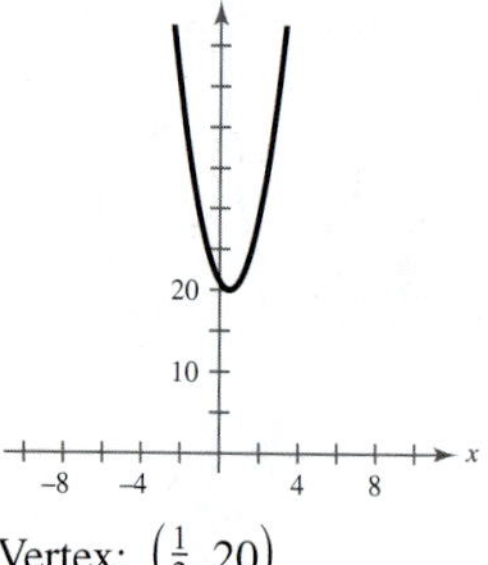

Vertex: $\left(\frac{1}{2}, 20\right)$

No x-intercept

27.

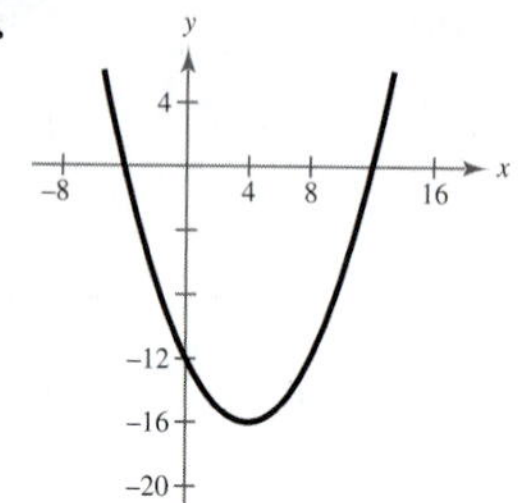

Vertex: $(4, -16)$

x-intercepts: $(-4, 0), (12, 0)$

29.

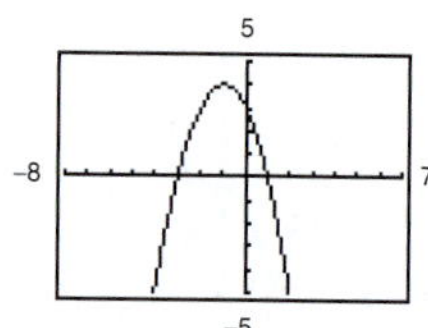

Vertex: $(-1, 4)$

x-intercepts: $(1, 0), (-3, 0)$

31.

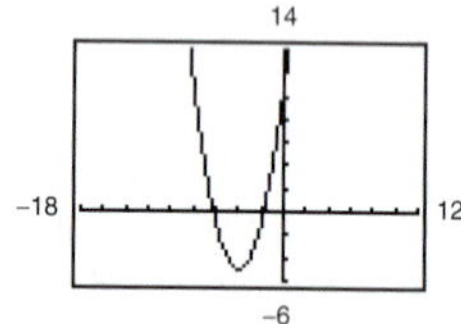

Vertex: $(-4, -5)$

x-intercepts: $(-4 \pm \sqrt{5}, 0)$

33.

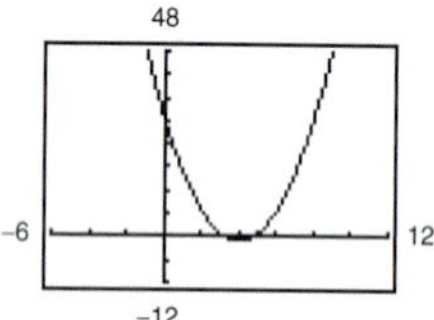

Vertex: $(4, -1)$

x-intercepts: $\left(4 \pm \frac{1}{2}\sqrt{2}, 0\right)$

35.

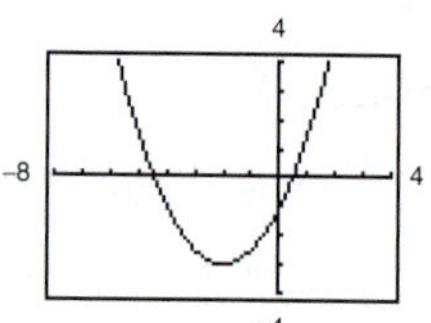

Vertex: $(-2, -3)$

x-intercepts: $(-2 \pm \sqrt{6}, 0)$

37. $y = (x - 1)^2$ **39.** $y = -(x + 1)^2 + 4$

41. $y = -2(x + 2)^2 + 2$ **43.** $f(x) = (x + 2)^2 + 5$

45. $f(x) = -\frac{1}{2}(x - 3)^2 + 4$ **47.** $f(x) = \frac{3}{4}(x - 5)^2 + 12$

49. $f(x) = -\frac{24}{49}\left(x + \frac{1}{4}\right)^2 + \frac{3}{2}$ **51.** $f(x) = -\frac{16}{3}\left(x + \frac{5}{2}\right)^2$

53. $(\pm 4, 0)$ **55.** $(5, 0), (-1, 0)$

57.

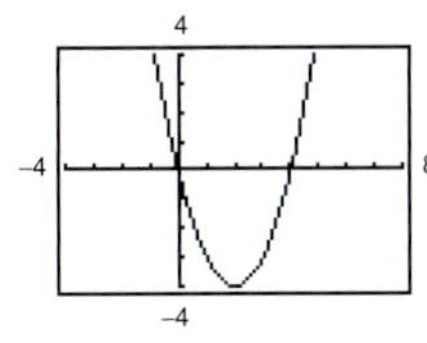

(0, 0), (4, 0)

59.

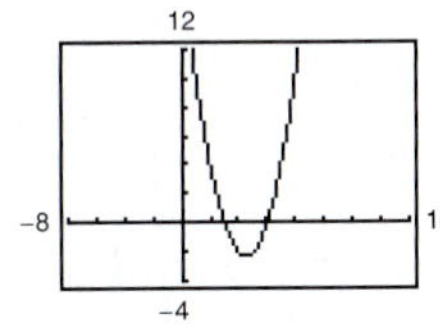

(3, 0), (6, 0)

61.

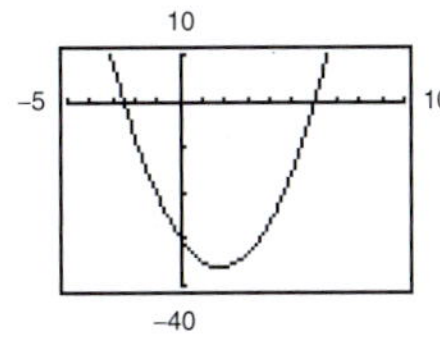

$\left(-\frac{5}{2}, 0\right)$, (6, 0)

63.

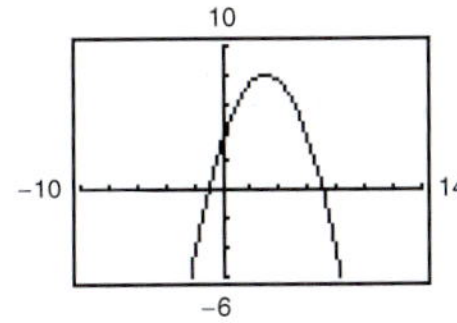

(7, 0), (−1, 0)

65. $f(x) = x^2 - 2x - 3$
$g(x) = -x^2 + 2x + 3$

67. $f(x) = x^2 - 10x$
$g(x) = -x^2 + 10x$

69. $f(x) = 2x^2 + 7x + 3$
$g(x) = -2x^2 - 7x - 3$

71. 55, 55 **73.** 12, 6

75. (a) $A = \dfrac{8x(50 - x)}{3}$

(b)

x	5	10	15	20	25	30
A	600	1067	1400	1600	1667	1600

$x = 25$ feet, $y = 33\frac{1}{3}$ feet

(c)

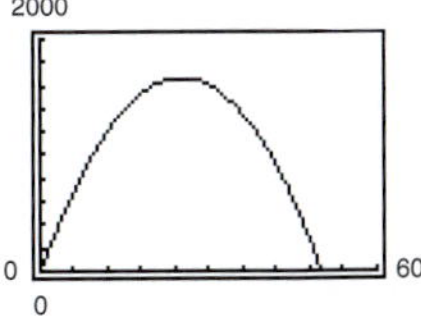

$x = 25$ feet, $y = 33\frac{1}{3}$ feet

(d) $A = -\frac{8}{3}(x - 25)^2 + \frac{5000}{3}$

77. 4500 units **79.** 20 fixtures **81.** 350,000 units

83. (a) 4 feet (b) 16 feet (c) 25.86 feet

85. (a)

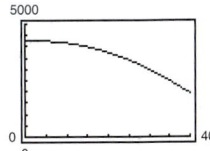

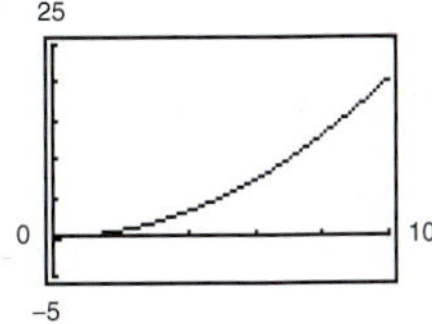

(b) 4265; Yes

(c) 8741 annually; 24 daily

87. (a) [graph: 25, 0, 100, −5]

(b) 69.6 miles per hour

89. True. The equation has no real solution, so the graph has no x-intercepts.

91. $f(x) = a\left(x + \dfrac{b}{2a}\right)^2 + \dfrac{4ac - b^2}{4a}$

93. Yes. A graph of a quadratic equation whose vertex is on the x-axis has only one x-intercept.

95. $y = -\frac{1}{3}x + \frac{5}{3}$ **97.** $y = \frac{5}{4}x + 3$ **99.** 27

101. $-\frac{1408}{49}$ **103.** 109

Section 3.2 *(page 280)*

1. c **2.** g **3.** h **4.** f

5. a **6.** e **7.** d **8.** b

9. (a)

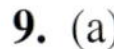

(b)

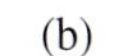

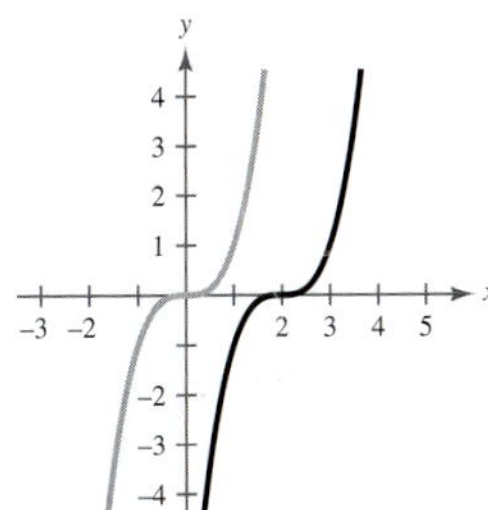

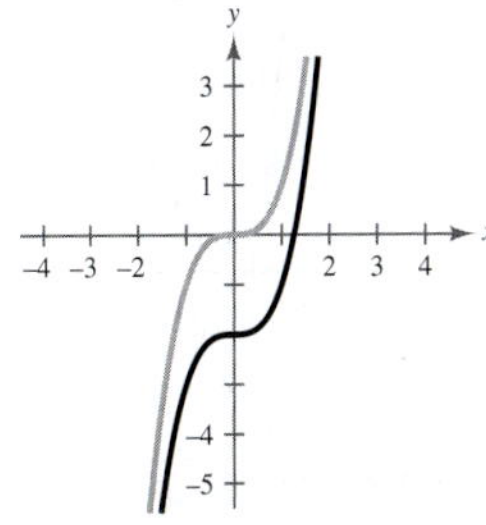

(c)

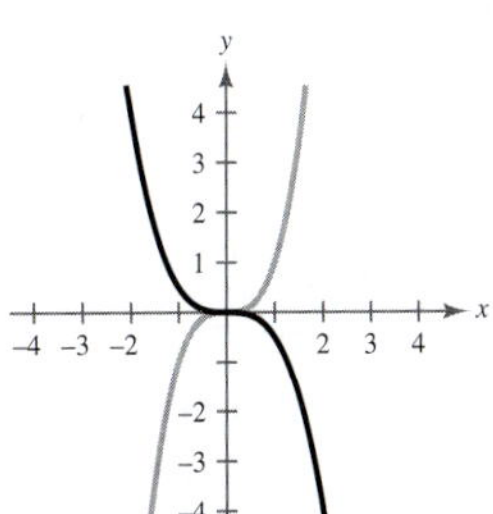

(d)

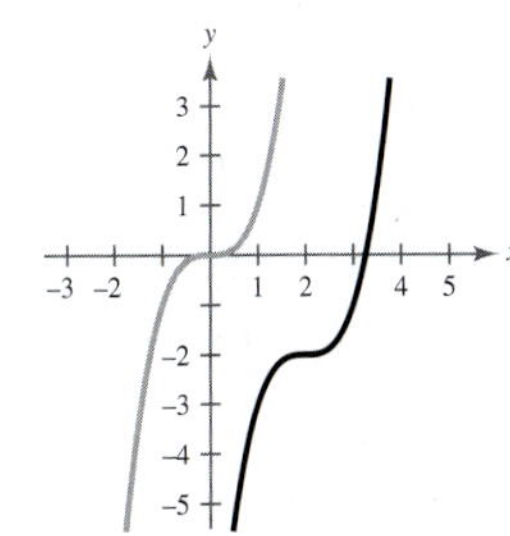

11. (a)

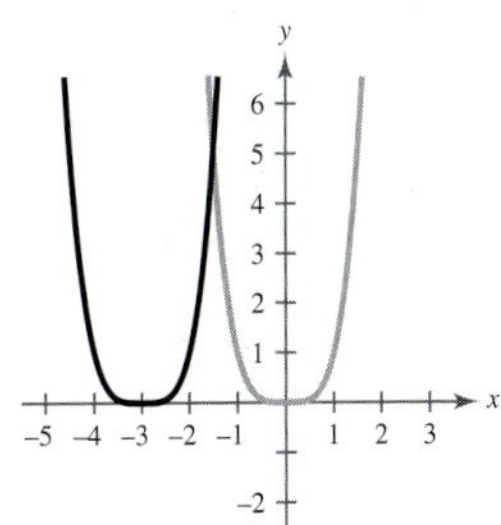

(b)

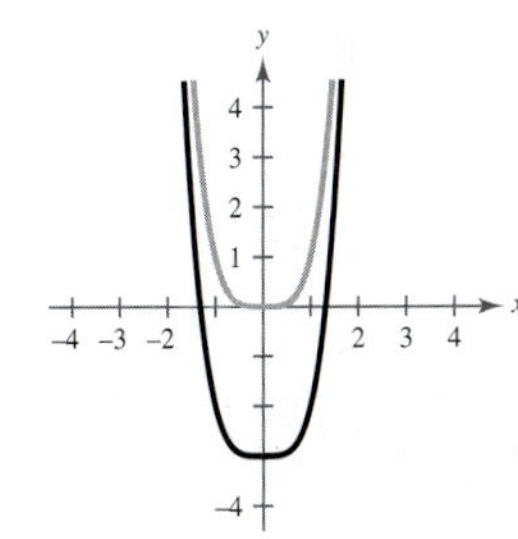

(c)

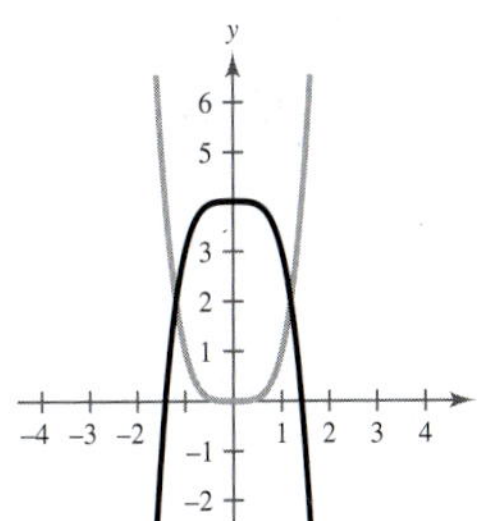

(d)

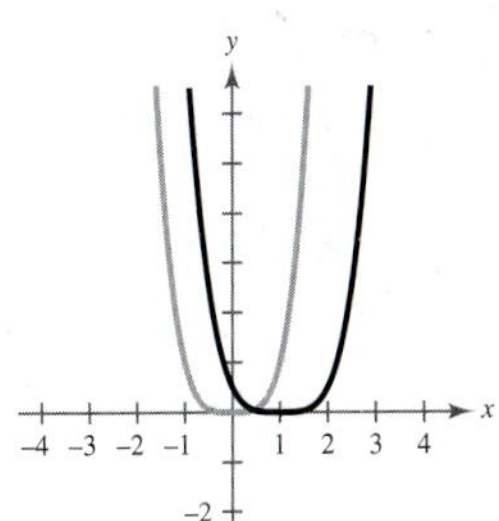

(e)

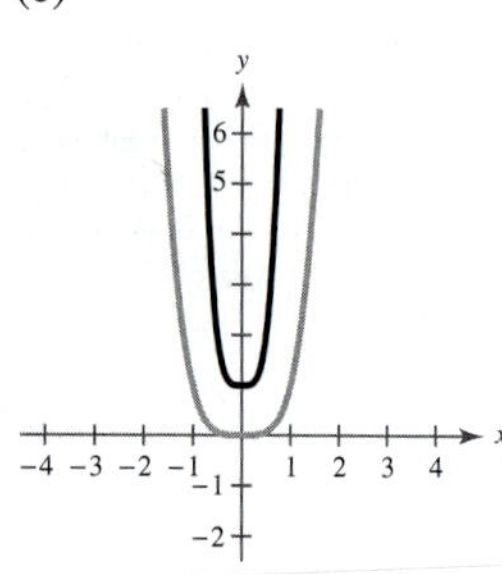

(f)

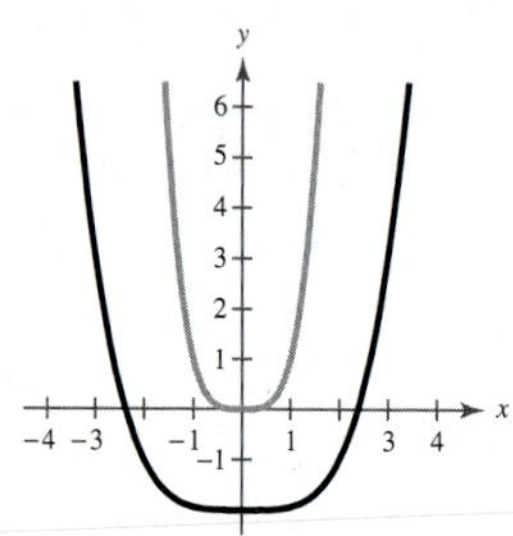

13. Falls to the left, rises to the right

15. Falls to the left, falls to the right

17. Rises to the left, falls to the right

19. Rises to the left, falls to the right

21. Falls to the left, falls to the right

23.

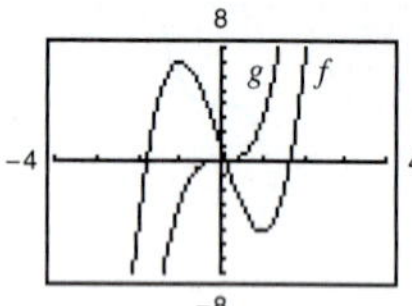

25.

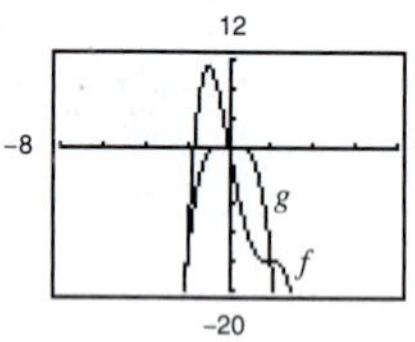

27. ± 5, odd multiplicity **29.** 3, even multiplicity

31. $-2, 1$, odd multiplicity

33. $0, 2 \pm \sqrt{3}$, odd multiplicity

35. 0, odd multiplicity; 2, even multiplicity

37. 0, odd multiplicity; $\pm\sqrt{3}$, even multiplicity

39. No real zeros **41.** $\pm 2, -3$, odd multiplicity

43.

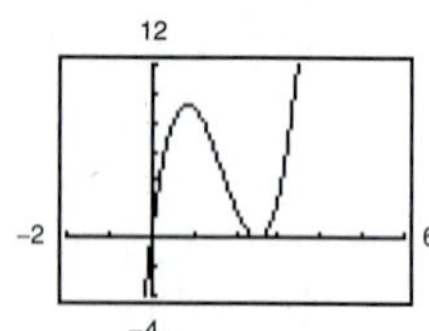

$(0, 0), \left(\frac{5}{2}, 0\right)$

45.

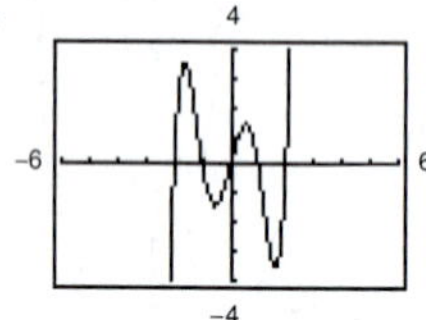

$(0, 0), (\pm 1, 0), (\pm 2, 0)$

47. $f(x) = x^2 - 10x$ **49.** $f(x) = x^2 + 4x - 12$

51. $f(x) = x^3 + 5x^2 + 6x$

53. $f(x) = x^4 - 4x^3 - 9x^2 + 36x$

55. $f(x) = x^2 - 2x - 2$ **57.** $f(x) = x^2 + 4x + 4$

59. $f(x) = x^3 + 2x^2 - 3x$ **61.** $f(x) = x^3 - 3x$

63. $f(x) = x^4 + x^3 - 15x^2 + 23x - 10$

65. $f(x) = x^5 + 16x^4 + 96x^3 + 256x^2 + 256x$

67. (a) Falls to the left, rises to the right

(b) $0, \pm 3$ (c) Answers will vary.

(d)

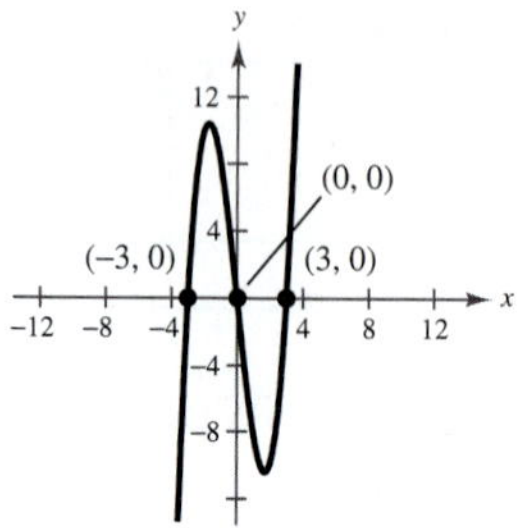

69. (a) Rises to the left, rises to the right

(b) No zeros (c) Answers will vary.

(d)

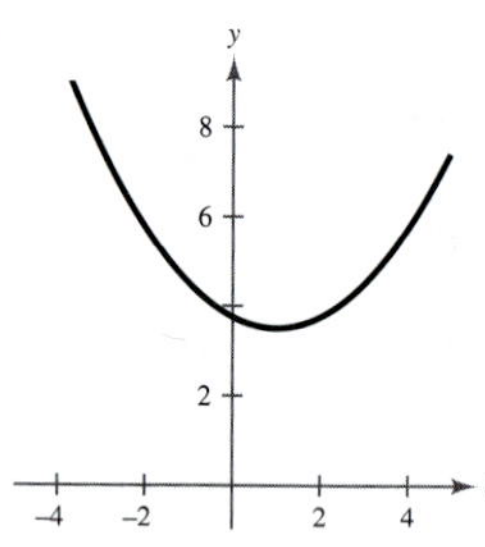

71. (a) Falls to the left, rises to the right

(b) 0, 3 (c) Answers will vary.

(d)

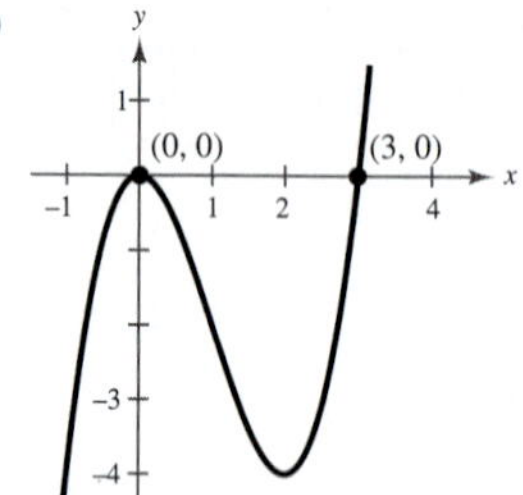

73. (a) Falls to the left, rises to the right

(b) 0, 2, 3 (c) Answers will vary.

(d)

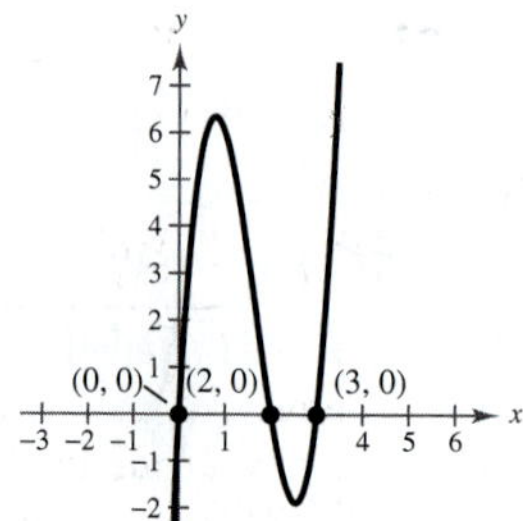

75. (a) Rises to the left, falls to the right

(b) $-5, 0$ (c) Answers will vary.

(d)

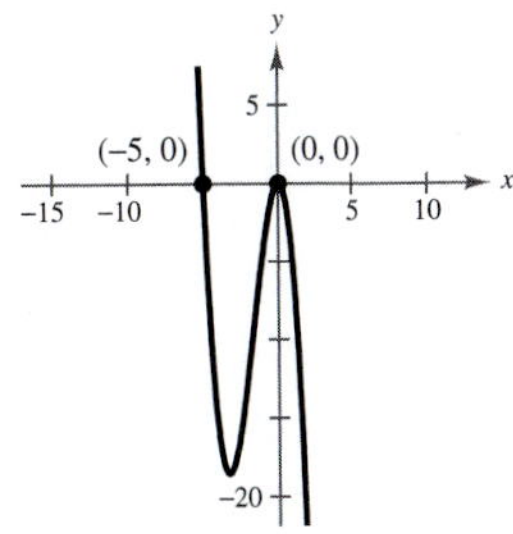

77. (a) Falls to the left, rises to the right

(b) 0, 4 (c) Answers will vary.

(d)

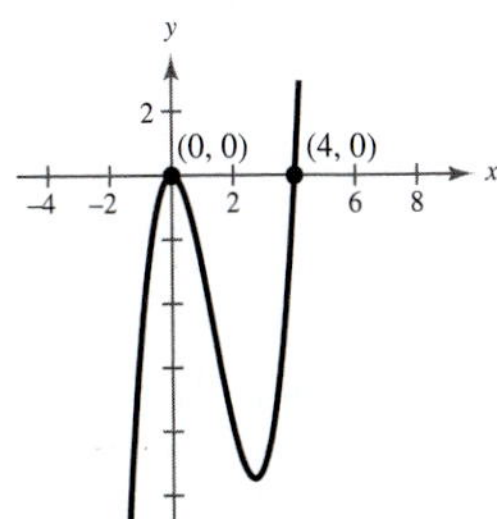

79. (a) Falls to the left, falls to the right

(b) ± 2 (c) Answers will vary.

(d)

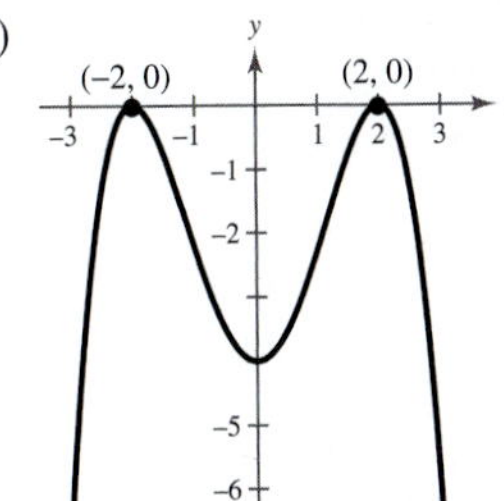

81.

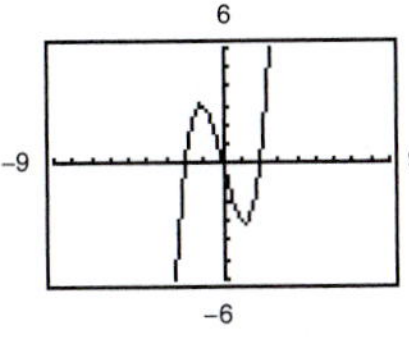

Zeros: $0, \pm 2$, odd multiplicity

83.

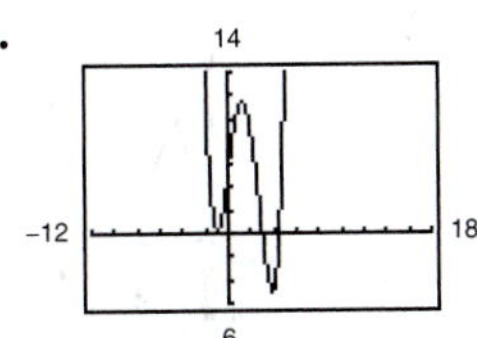

Zeros: -1, even multiplicity; $3, \frac{9}{2}$, odd multiplicity

85. $[-1, 0], [1, 2], [2, 3]$; $\approx -0.879, 1.347, 2.532$

87. $[-2, -1], [0, 1]$; $\approx -1.585, 0.779$

89. (a) $V = l \times w \times h$

$= (36 - 2x)(36 - 2x)x$

$= x(36 - 2x)^2$

(b) Domain: $0 < x < 18$

(c)

x	1	2	3	4	5	6	7
V	1156	2048	2700	3136	3380	3456	3388

6 inches $\times$ 24 inches $\times$ 24 inches

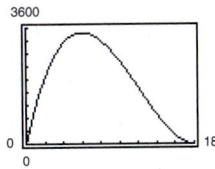

(d) 3600 0 18 0

$x = 6$

91. (a)

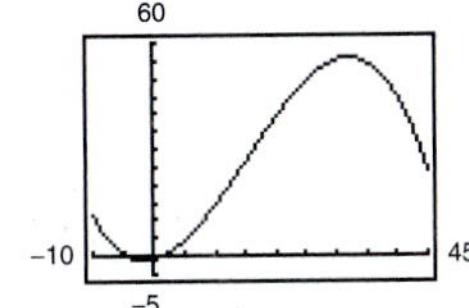

(b) $t \approx 15$

(c) Vertex: (15.22, 2.54)

(d) The results are approximately equal.

93. False. A fifth-degree polynomial can have at most four turning points.

95. True. The degree of the function is odd and its leading coefficient is negative, so the graph rises to the left and falls to the right.

97.

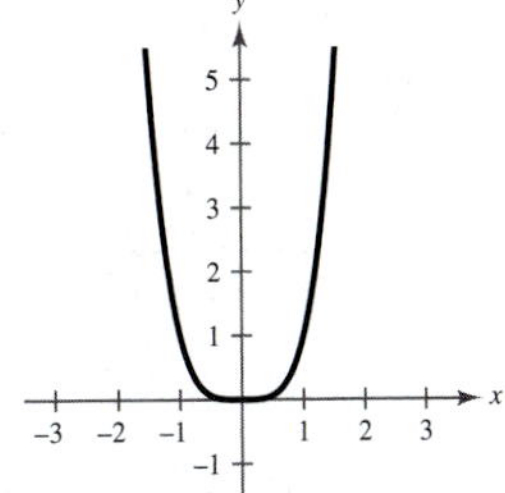

(a) Vertical shift of 2 units; Even

(b) Horizontal shift of 2 units; Neither even nor odd

(c) Reflection in the y-axis; Even

(d) Reflection in the x-axis; Even

(e) Horizontal stretch; Even

(f) Vertical shrink; Even

(g) $g(x) = x^3$; Odd

(h) $g(x) = x^{16}$; Even

99. $-\frac{7}{2}, 4$ **101.** $-\frac{5}{4}, \frac{1}{3}$ **103.** $1 \pm \sqrt{22}$

105. $\dfrac{-5 \pm \sqrt{185}}{4}$ **107.** $(5x - 8)(x + 3)$

109. $x^2(4x + 5)(x - 3)$

111. Horizontal translation four units to the left of $y = x^2$.

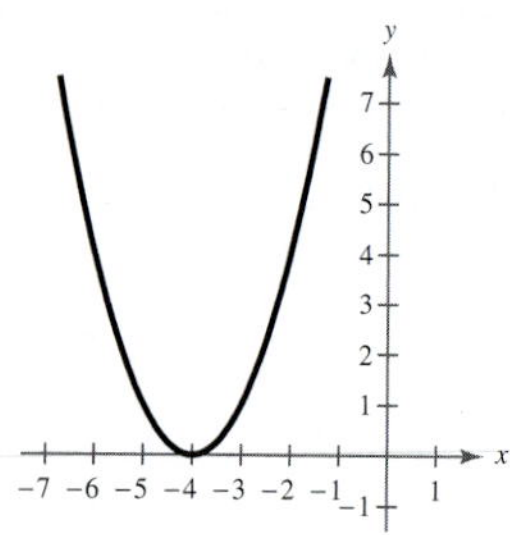

113. Horizontal translation one unit left and vertical translation five units down of $y = \sqrt{x}$

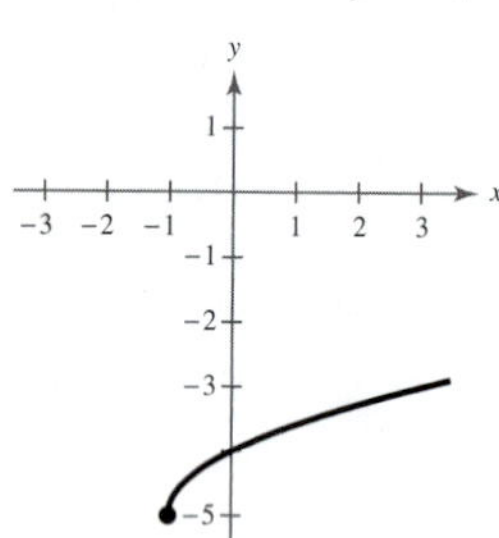

115. Vertical stretch of a factor of 2 and vertical translation nine units up of $y = [\![x]\!]$

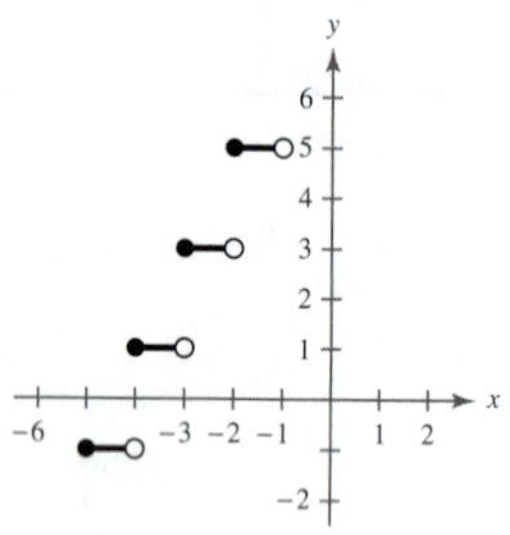

Section 3.3 *(page 290)*

1. Answers will vary.

3.

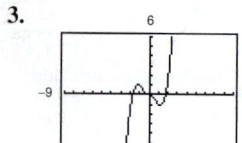

5. $2x + 4$ **7.** $x^2 - 3x + 1$ **9.** $x^3 + 3x^2 - 1$

11. $7 - \dfrac{11}{x + 2}$ **13.** $3x + 5 - \dfrac{2x - 3}{2x^2 + 1}$

15. $x^2 + 2x + 4 + \dfrac{2x - 11}{x^2 - 2x + 3}$

17. $x + 3 + \dfrac{6x^2 - 8x + 3}{(x - 1)^3}$ **19.** $3x^2 - 2x + 5$

21. $4x^2 - 9$ **23.** $-x^2 + 10x - 25$

25. $5x^2 + 14x + 56 + \dfrac{232}{x - 4}$

27. $10x^3 + 10x^2 + 60x + 360 + \dfrac{1360}{x - 6}$

29. $x^2 - 8x + 64$

31. $-3x^3 - 6x^2 - 12x - 24 - \dfrac{48}{x - 2}$

33. $-x^3 - 6x^2 - 36x - 36 - \dfrac{216}{x - 6}$

35. $4x^2 + 14x - 30$

37. $f(x) = (x - 4)(x^2 + 3x - 2) + 3, \ f(4) = 3$

39. $f(x) = \left(x + \frac{2}{3}\right)(15x^3 - 6x + 4) + \frac{34}{3}$,
$f\left(-\frac{2}{3}\right) = \frac{34}{3}$

41. $f(x) = \left(x - \sqrt{2}\right)\left[x^2 + \left(3 + \sqrt{2}\right)x + 3\sqrt{2}\right] - 8$,
$f\left(\sqrt{2}\right) = -8$

43. $f(x) = \left(x - 1 + \sqrt{3}\right)\left[-4x^2 + \left(2 + 4\sqrt{3}\right)x + \left(2 + 2\sqrt{3}\right)\right]$,
$f\left(1 - \sqrt{3}\right) = 0$

45. (a) 1 (b) 4 (c) 4 (d) 1954

47. (a) 97 (b) $-\frac{5}{3}$ (c) 17 (d) -199

49. $(x - 2)(x + 3)(x - 1)$; Zeros: $2, -3, 1$

51. $(2x - 1)(x - 5)(x - 2)$; Zeros: $\frac{1}{2}, 5, 2$

53. $\left(x + \sqrt{3}\right)\left(x - \sqrt{3}\right)(x + 2)$; Zeros: $-\sqrt{3}, \sqrt{3}, -2$

55. $(x - 1)\left(x - 1 - \sqrt{3}\right)\left(x - 1 + \sqrt{3}\right)$;
Zeros: $1,\ 1 + \sqrt{3},\ 1 - \sqrt{3}$

57. (a) Answers will vary. (b) $2x - 1$
(c) $f(x) = (2x - 1)(x + 2)(x - 1)$ (d) $\frac{1}{2}, -2, 1$
(e)

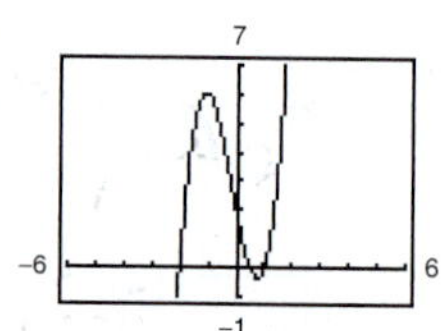

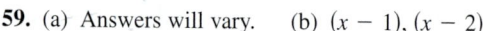

59. (a) Answers will vary. (b) $(x - 1), (x - 2)$
(c) $f(x) = (x - 1)(x - 2)(x - 5)(x + 4)$
(d) $1, 2, 5, -4$
(e)

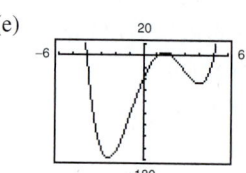

61. (a) Answers will vary. (b) $x + 7$
(c) $f(x) = (x + 7)(2x + 1)(3x - 2)$ (d) $-7, -\frac{1}{2}, \frac{2}{3}$

(e)

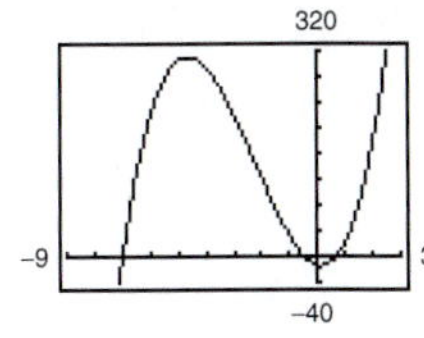

63. (a) Answers will vary. (b) $(x - \sqrt{5})$
(c) $f(x) = (x - \sqrt{5})(x + \sqrt{5})(2x - 1)$ (d) $\pm\sqrt{5}, \frac{1}{2}$
(e)

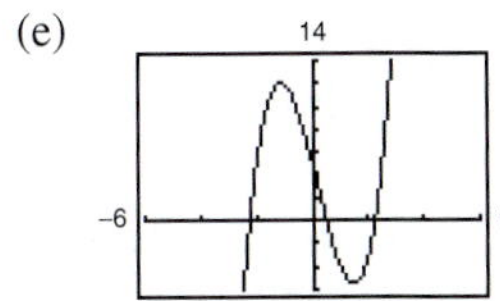

65. (a) Zeros are 2 and $\approx \pm 2.236$.
(b) $f(x) = (x - 2)(x - \sqrt{5})(x + \sqrt{5})$

67. (a) Zeros are -2, ≈ 0.268, and ≈ 3.732.
(b) $h(t) = (t + 2)[t - (2 + \sqrt{3})][t - (2 - \sqrt{3})]$

69. $2x^2 - x - 1,\ x \neq \frac{3}{2}$ **71.** $x^2 + 2x - 3,\ x \neq -1$

73. $x^2 + 3x,\ x \neq -2, -1$

75. (a) and (b)

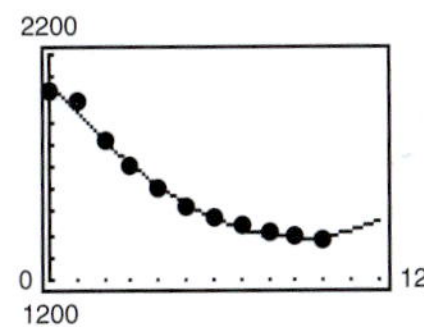

$M = 0.126t^3 + 5.86t^2 - 139.2t + 2070$

(c)

t	0	1	2	3	4	5
M	2070	1937	1816	1709	1615	1536

t	6	7	8	9	10
M	1473	1426	1396	1384	1390

(d) 1726 thousand. No, because the model will approach infinity quickly.

77. False. $-\frac{4}{7}$ is a zero of f.

79. True. The degree of the numerator is greater than the degree of the denominator.

81. $x^{2n} + 6x^n + 9$ **83.** The remainder is 0.

85. $c = -210$ **87.** 0; $x + 3$ is a factor of f.

89. $\pm\dfrac{5}{3}$ **91.** $-\dfrac{7}{5}, 2$ **93.** $\dfrac{-3 \pm \sqrt{3}}{2}$

95. $f(x) = x^3 - 7x^2 + 12x$

97. $f(x) = x^3 + x^2 - 7x - 3$

Section 3.4 *(page 303)*

1. 0, 6 **3.** 2, -4 **5.** $-6, \pm i$ **7.** $\pm 1, \pm 3$

9. $\pm 1, \pm 3, \pm 5, \pm 9, \pm 15, \pm 45, \pm\frac{1}{2}, \pm\frac{3}{2}, \pm\frac{5}{2}, \pm\frac{9}{2}, \pm\frac{15}{2}, \pm\frac{45}{2}$

11. 1, 2, 3 **13.** 1, -1, 4 **15.** $-1, -10$

17. $\frac{1}{2}, -1$ **19.** $-2, 3, \pm\frac{2}{3}$ **21.** $-1, 2$ **23.** $-6, \frac{1}{2}, 1$

25. (a) $\pm 1, \pm 2, \pm 4$
(b)

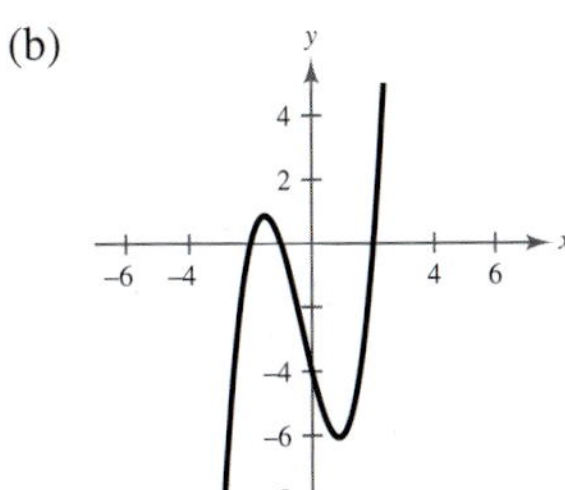

(c) $-2, -1, 2$

27. (a) $\pm 1, \pm 3, \pm\frac{1}{2}, \pm\frac{3}{2}, \pm\frac{1}{4}, \pm\frac{3}{4}$
(b)

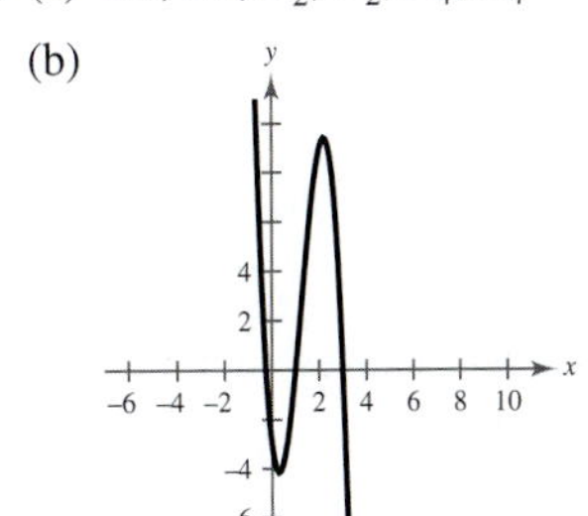

(c) $-\frac{1}{4}, 1, 3$

29. (a) $\pm 1, \pm 2, \pm 4, \pm 8, \pm\frac{1}{2}$
(b)

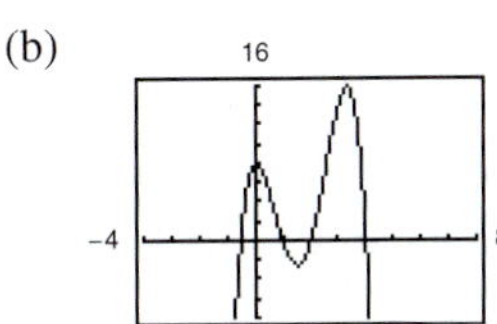

(c) $-\frac{1}{2}, 1, 2, 4$

31. (a) $\pm 1, \pm 3, \pm\frac{1}{2}, \pm\frac{3}{2}, \pm\frac{1}{4}, \pm\frac{3}{4}, \pm\frac{1}{8}, \pm\frac{3}{8}, \pm\frac{1}{16}, \pm\frac{3}{16}, \pm\frac{1}{32}, \pm\frac{3}{32}$
(b)

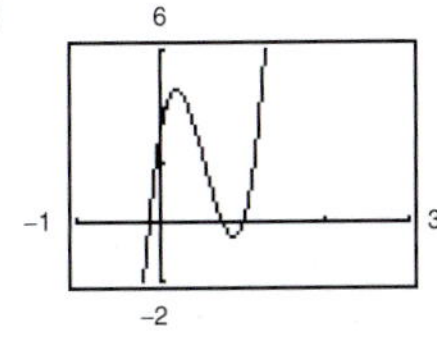

(c) $1, \frac{3}{4}, -\frac{1}{8}$

33. (a) $\pm 1, \approx \pm 1.414$
(b) $f(x) = (x + 1)(x - 1)(x + \sqrt{2})(x - \sqrt{2})$

35. (a) 0, 3, 4, $\approx \pm 1.414$
(b) $h(x) = x(x - 3)(x - 4)(x + \sqrt{2})(x - \sqrt{2})$

37. $x^3 - x^2 + 25x - 25$ **39.** $x^3 + 4x^2 - 31x - 174$

41. $3x^4 - 17x^3 + 25x^2 + 23x - 22$

43. (a) $(x^2 + 9)(x^2 - 3)$
(b) $(x^2 + 9)(x + \sqrt{3})(x - \sqrt{3})$
(c) $(x + 3i)(x - 3i)(x + \sqrt{3})(x - \sqrt{3})$

45. (a) $(x^2 - 2x - 2)(x^2 - 2x + 3)$
(b) $(x - 1 + \sqrt{3})(x - 1 - \sqrt{3})(x^2 - 2x + 3)$
(c) $(x - 1 + \sqrt{3})(x - 1 - \sqrt{3})(x - 1 + \sqrt{2}i)$
$(x - 1 - \sqrt{2}i)$

47. $-\frac{3}{2}, \pm 5i$ **49.** $\pm 2i, 1, -\frac{1}{2}$ **51.** $-3 \pm i, \frac{1}{4}$

53. $2, -3 \pm \sqrt{2}i, 1$ **55.** $\pm 5i$; $(x + 5i)(x - 5i)$

57. $2 \pm \sqrt{3}$; $(x - 2 - \sqrt{3})(x - 2 + \sqrt{3})$

59. $\pm 3, \pm 3i$; $(x + 3)(x - 3)(x + 3i)(x - 3i)$

61. $1 \pm i$; $(z - 1 + i)(z - 1 - i)$

63. $2, 2 \pm i$; $(x - 2)(x - 2 + i)(x - 2 - i)$

65. $-2, 1 \pm \sqrt{2}i$; $(x + 2)(x - 1 + \sqrt{2}i)(x - 1 - \sqrt{2}i)$

67. $-\frac{1}{5}, 1 \pm \sqrt{5}i$; $(5x + 1)(x - 1 + \sqrt{5}i)(x - 1 - \sqrt{5}i)$

69. $2, \pm 2i$; $(x - 2)^2(x + 2i)(x - 2i)$

71. $\pm i, \pm 3i$; $(x + i)(x - i)(x + 3i)(x - 3i)$

73. $-10, -7 \pm 5i$ **75.** $-\frac{3}{4}, 1 \pm \frac{1}{2}i$ **77.** $-2, -\frac{1}{2}, \pm i$

79. No real zeros **81.** No real zeros

83. One positive zero **85.** One or three positive zeros

87. Answers will vary. **89.** Answers will vary.

91. $1, -\frac{1}{2}$ **93.** $-\frac{3}{4}$ **95.** $\pm 2, \pm\frac{3}{2}$ **97.** $\pm 1, \frac{1}{4}$

99. d **100.** a **101.** b **102.** c

103. (a)

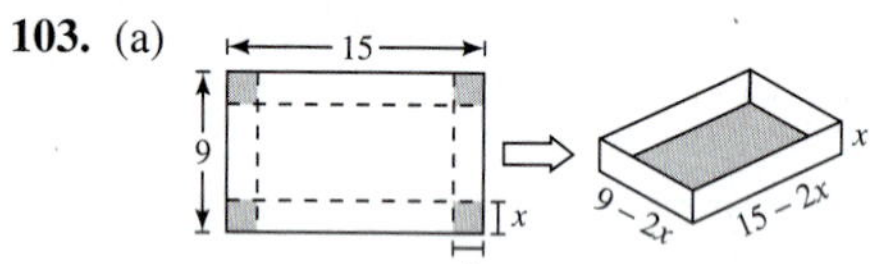

(b) $V = x(9 - 2x)(15 - 2x)$
Domain: $0 < x < \frac{9}{2}$

(c)

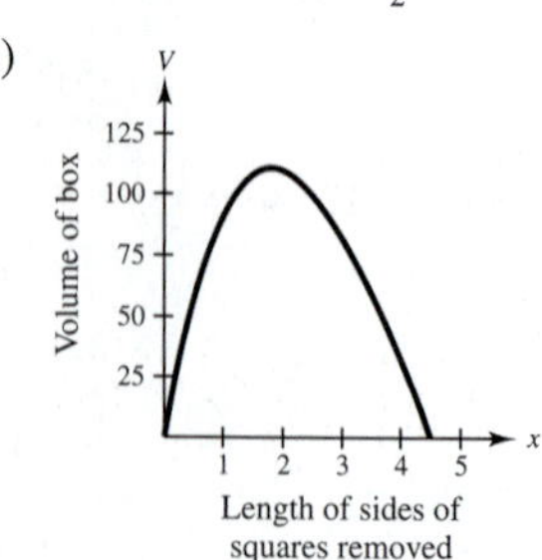

1.82 centimeters × 5.36 centimeters × 11.36 centimeters

(d) $\frac{1}{2}, \frac{7}{2}, 8$; 8 is not in the domain of V.

105. $x \approx 38.4$, or \$384,000

107. (a) $A = -0.045028t^3 + 0.97071t^2 - 5.8547t + 15.390$
(b)

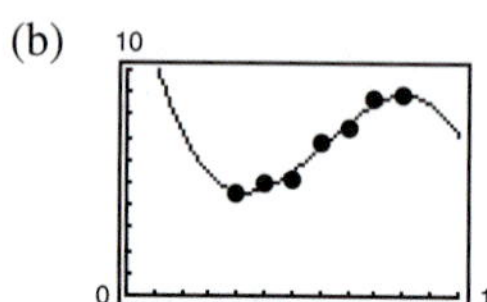

(c) 1996 (d) 1998

(e) The attendance will increase until 2000. It will then decrease quickly.

109. No. Setting $h = 64$ and solving the resulting equation yields imaginary roots.

111. False. The most complex zeros it can have is two, and the Linear Factorization Theorem guarantees that there are three linear factors, so one zero must be real.

113. r_1, r_2, r_3 **115.** $5 + r_1, 5 + r_2, 5 + r_3$

117. The zeros cannot be determined.

119. (a) $0 < k < 4$ (b) $k = 4$ (c) $k < 0$ (d) $k > 4$

121. $f(x) = -2x^3 + 3x^2 + 11x - 6$

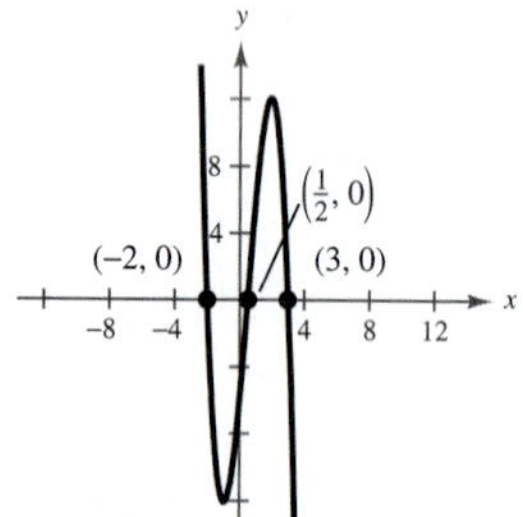

(Equations and graphs will vary.) There are infinitely many possible functions for f.

123. (a) $-2, 1, 4$
(b) The graph touches the x-axis at $x = 1$.
(c) The least possible degree of the function is 4, because there are at least four real zeros (1 is repeated) and a function can have at most the number of real zeros equal to the degree of the function. The degree cannot be odd by the definition of multiplicity.
(d) Positive. From the information in the table, it can be concluded that the graph will eventually rise to the left and rise to the right.
(e) $f(x) = x^4 - 4x^3 - 3x^2 + 14x - 8$
(f)

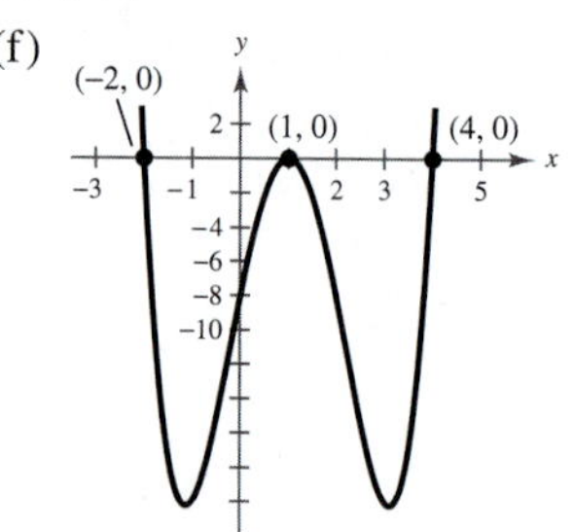

125. (a) $x^2 + b$ (b) $x^2 - 2ax + a^2 + b^2$

127. $-11 + 9i$ **129.** $20 + 40i$

131.

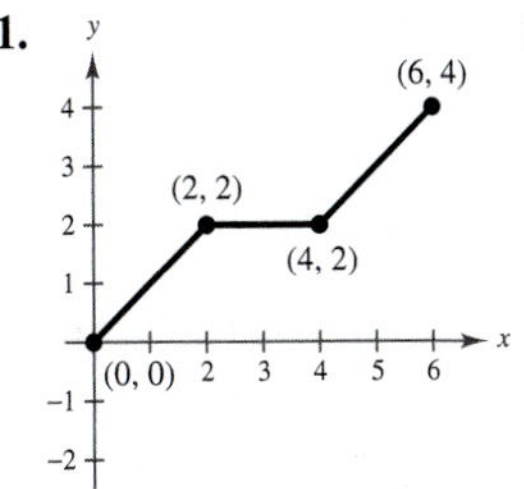

133.

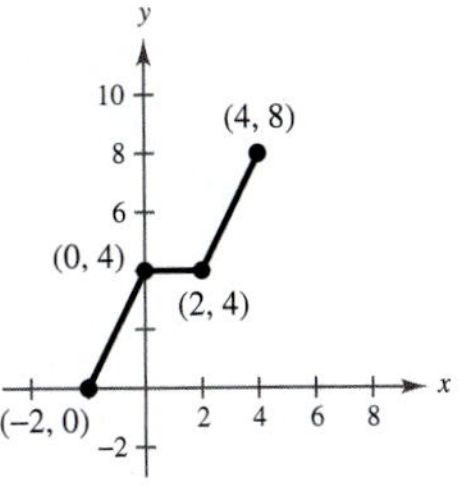

135.

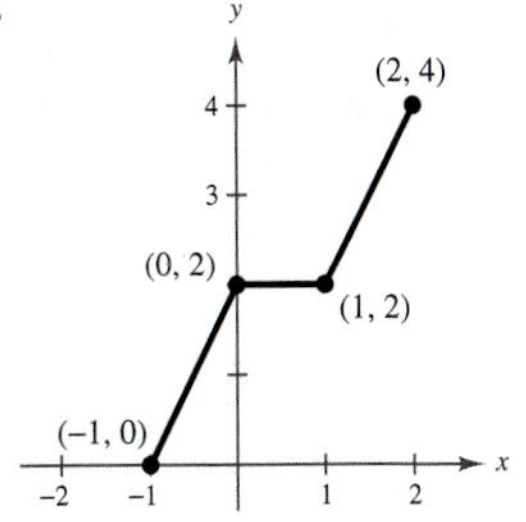

Section 3.5 *(page 314)*

1.

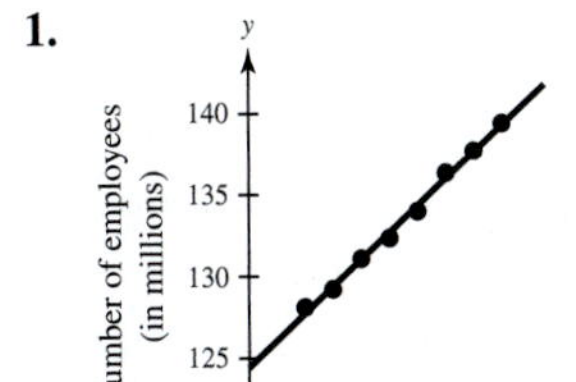

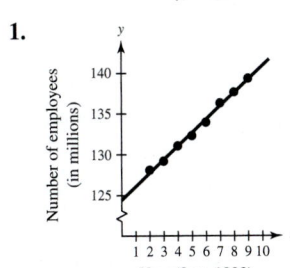

The model is a "good fit" for the actual data.

3. Inversely

5.

x	2	4	6	8	10
$y = kx^2$	4	16	36	64	100

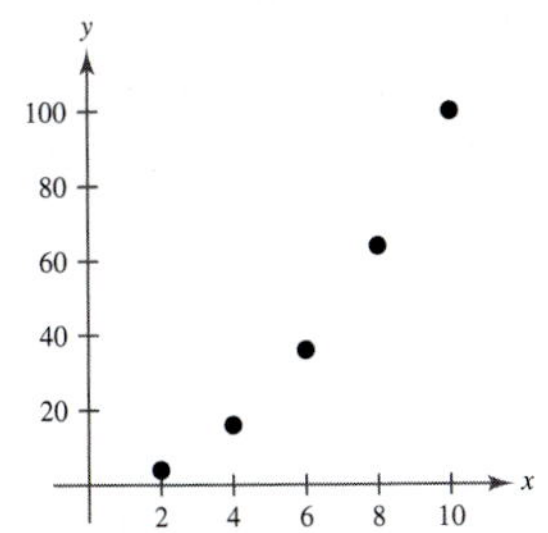

7.

x	2	4	6	8	10
$y = kx^2$	2	8	18	32	50

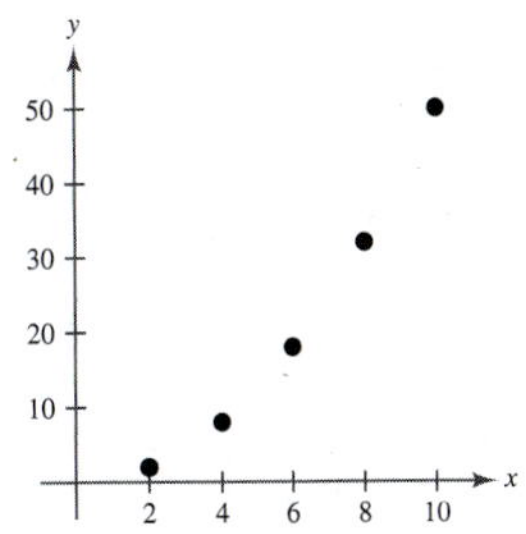

9.

x	2	4	6	8	10
$y = k/x^2$	$\frac{1}{2}$	$\frac{1}{8}$	$\frac{1}{18}$	$\frac{1}{32}$	$\frac{1}{50}$

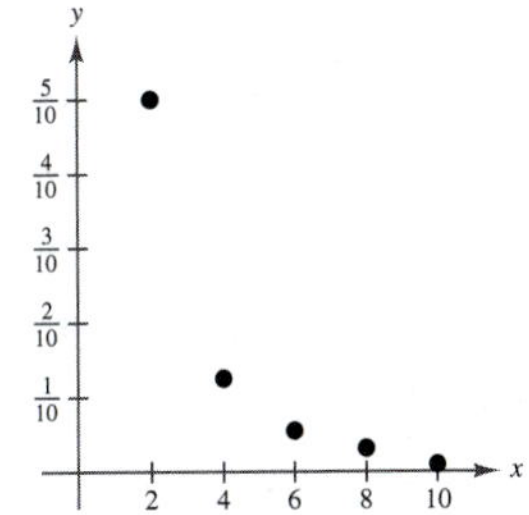

11.

x	2	4	6	8	10
$y = k/x^2$	$\frac{5}{2}$	$\frac{5}{8}$	$\frac{5}{18}$	$\frac{5}{32}$	$\frac{1}{10}$

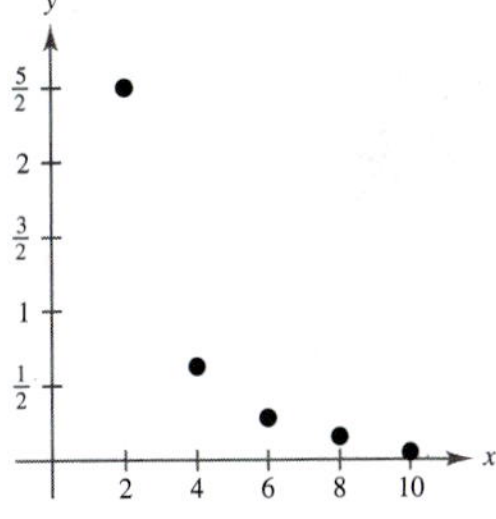

13. $y = \dfrac{5}{x}$ **15.** $y = -\dfrac{7}{10}x$ **17.** $y = \dfrac{12}{5}x$

19. $y = 205x$ **21.** $I = 0.035P$

23. Model: $y = \frac{33}{13}x$; 25.4 centimeters, 50.8 centimeters

25. $y = 0.0368x$; \$7360

27. (a) 0.05 meter (b) $176\frac{2}{3}$ newtons **29.** 39.47 pounds

31. $A = kr^2$ **33.** $y = \dfrac{k}{x^2}$ **35.** $F = \dfrac{kg}{r^2}$

37. $P = \dfrac{k}{V}$ **39.** $F = \dfrac{km_1m_2}{r^2}$

41. The area of a triangle is jointly proportional to its base and height.

43. The volume of a sphere varies directly as the cube of its radius.

45. Average speed is directly proportional to the distance and inversely proportional to the time.

47. $A = \pi r^2$ **49.** $y = \dfrac{28}{x}$ **51.** $F = 14rs^3$

53. $z = \dfrac{2x^2}{3y}$ **55.** ≈ 0.61 mile per hour **57.** 506 feet

59. 400 feet **61.** The velocity is increased by one-third.

63. (a)

(b) Yes. $k_1 = 4200$, $k_2 = 3800$, $k_3 = 4200$, $k_4 = 4800$, $k_5 = 4500$

(c) $C = \dfrac{4300}{d}$

(d) (e) $\approx$ 1433 meters

65. (a) (b) 0.2857 microwatt per square centimeter

67. **69.**

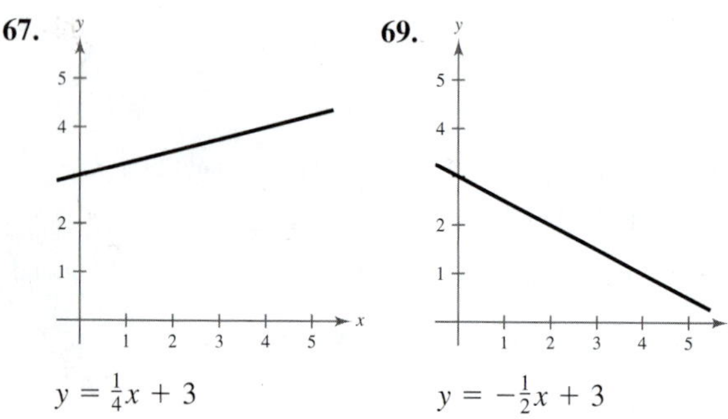

$y = \frac{1}{4}x + 3$ $y = -\frac{1}{2}x + 3$

71. (a) and (b)

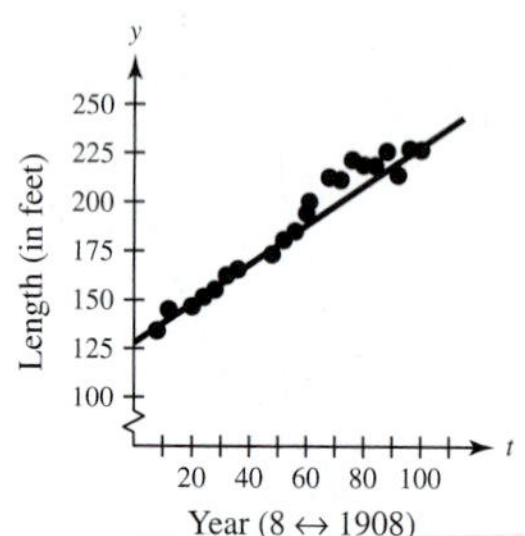

$y = t + 128$

(c) $y = 1.08t + 127.7$ (d) The models are similar.

(e) Part (b): 232 feet; Part (c): 240.02 feet

(f) Analyses will vary.

73. (a) and (c)

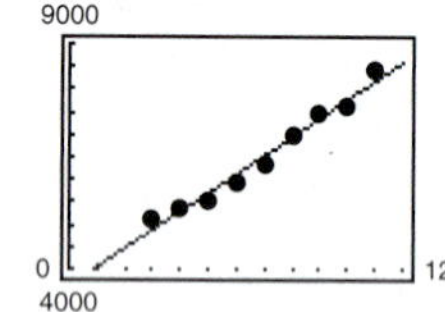

(b) $R = 412.9t + 3642$

(d) 2000: \$7771.0 million; 2002: \$8596.8 million

(e) Each year, the annual receipts for motion picture movie theaters increases by \$412.9 million.

75. False. y will increase if k is positive and y will decrease if k is negative.

77. The accuracy is questionable when based on such limited data.

79. $x \le 4$, $x \ge 6$ **81.** $x > 5$

83. (a) $-\frac{5}{3}$ (b) $-\frac{7}{3}$ (c) 21

Review Exercises *(page 321)*

1. $f(x) = -\frac{1}{2}(x - 4)^2 + 1$ **3.** $f(x) = (x - 1)^2 - 4$

5. (a)

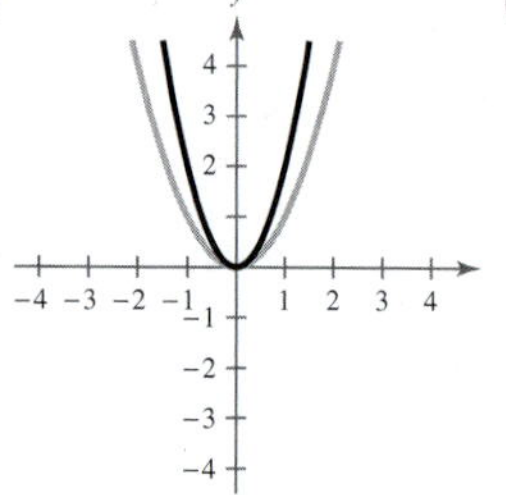

Vertical stretch

(b)

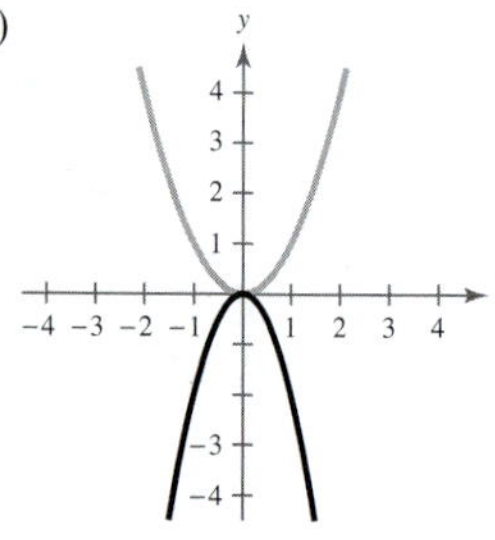

Vertical stretch and reflection in the x-axis

(c)

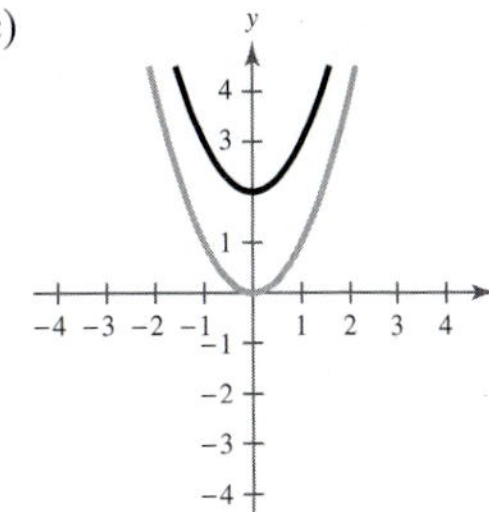

Vertical translation

(d)

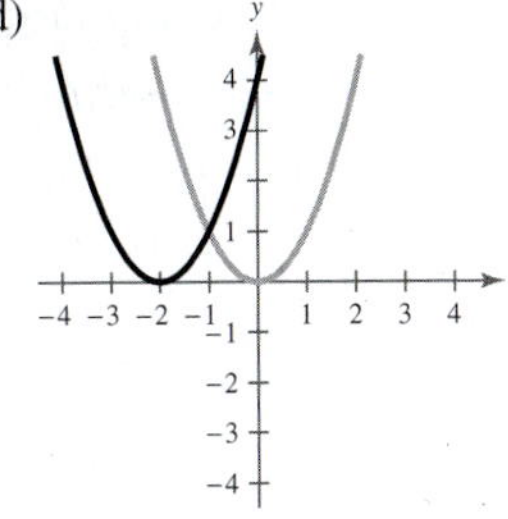

Horizontal translation

7. $g(x) = (x - 1)^2 - 1$

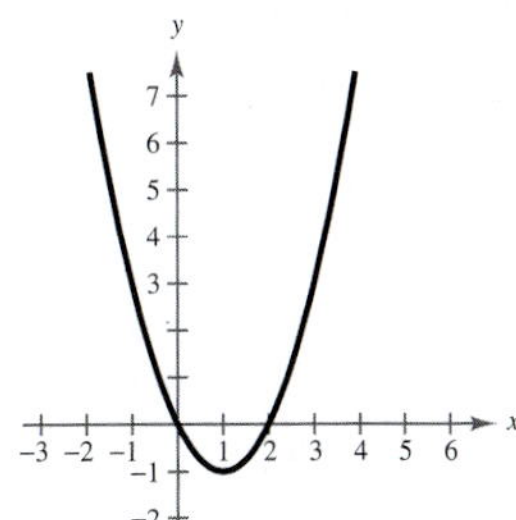

Vertex: $(1, -1)$

x-intercepts: $(0, 0)$, $(2, 0)$

9. $f(x) = (x + 4)^2 - 6$

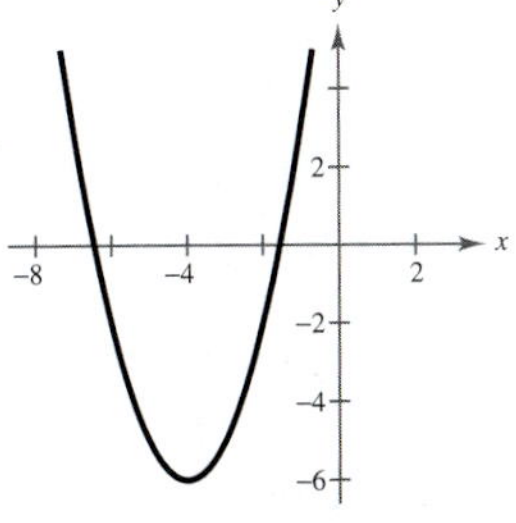

Vertex: $(-4, -6)$

x-intercepts: $(-4 \pm \sqrt{6}, 0)$

11. $f(t) = -2(t - 1)^2 + 3$

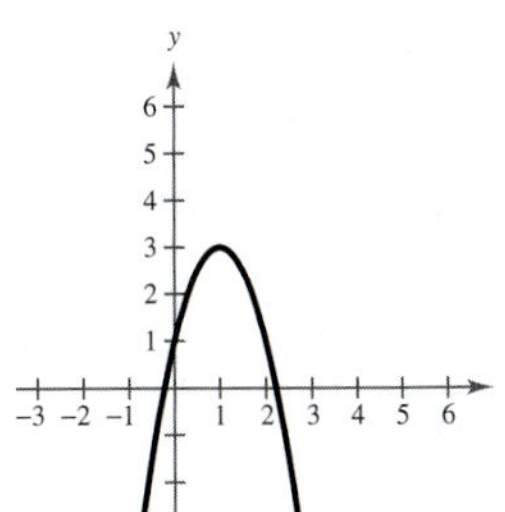

Vertex: $(1, 3)$

x-intercepts: $\left(1 \pm \dfrac{\sqrt{6}}{2}, 0\right)$

13. $h(x) = 4\left(x + \frac{1}{2}\right)^2 + 12$

Vertex: $\left(-\frac{1}{2}, 12\right)$

No x-intercept

15. $h(x) = \left(x + \frac{5}{2}\right)^2 - \frac{41}{4}$

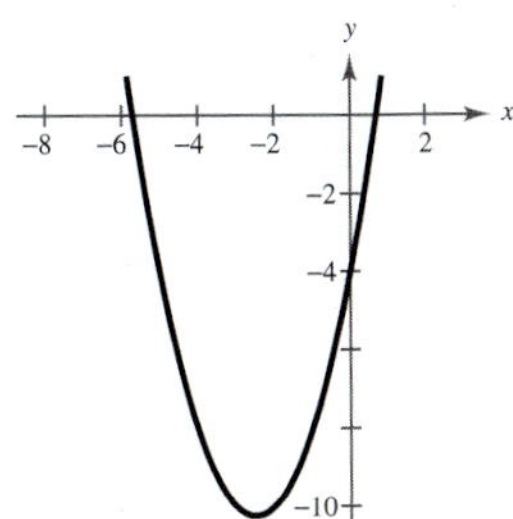

Vertex: $\left(-\frac{5}{2}, -\frac{41}{4}\right)$

x-intercepts: $\left(\dfrac{\pm\sqrt{41} - 5}{2}, 0\right)$

17. $f(x) = \frac{1}{3}\left(x + \frac{5}{2}\right)^2 - \frac{41}{12}$

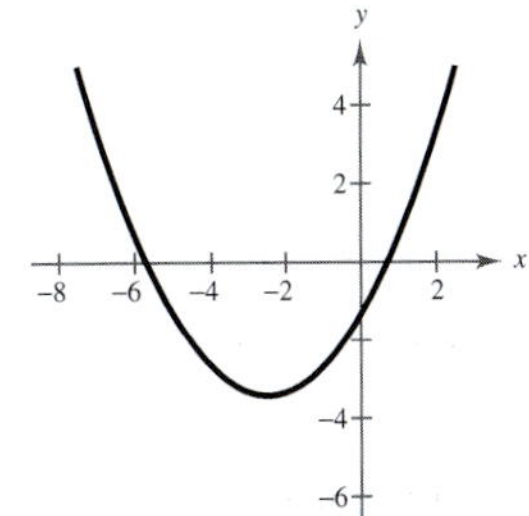

Vertex: $\left(-\frac{5}{2}, -\frac{41}{12}\right)$

x-intercepts: $\left(\dfrac{\pm\sqrt{41} - 5}{2}, 0\right)$

19. (a) $A = x\left(\dfrac{8 - x}{2}\right)$

(b) $0 < x < 8$

(c)

x	1	2	3	4	5	6
A	$\frac{7}{2}$	6	$\frac{15}{2}$	8	$\frac{15}{2}$	6

(d)

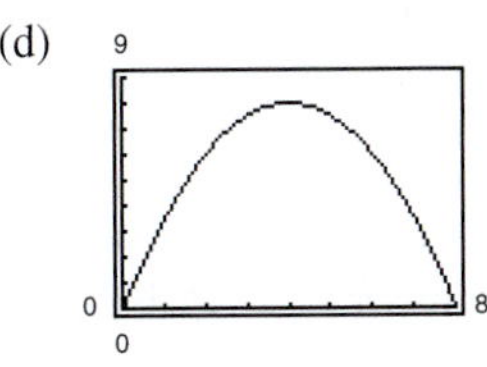

$x = 4, y = 2$

(e) $A = -\frac{1}{2}(x - 4)^2 + 8$; $x = 4, y = 2$

21. 40,000 units **23.** 1091 units

25.

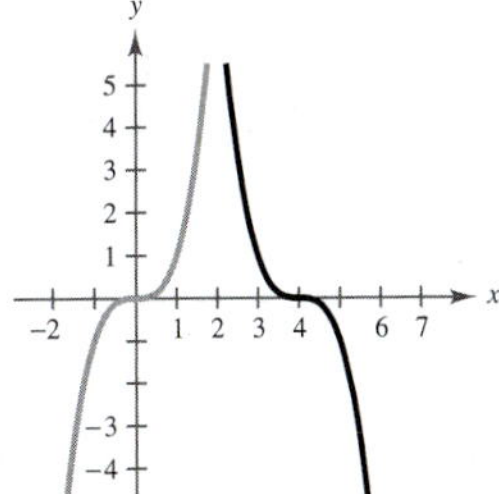

27.

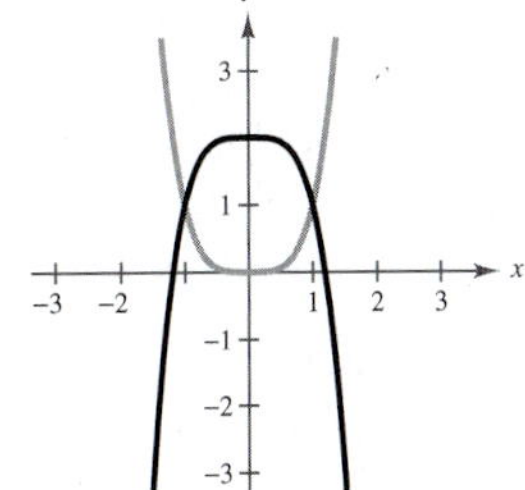

29.

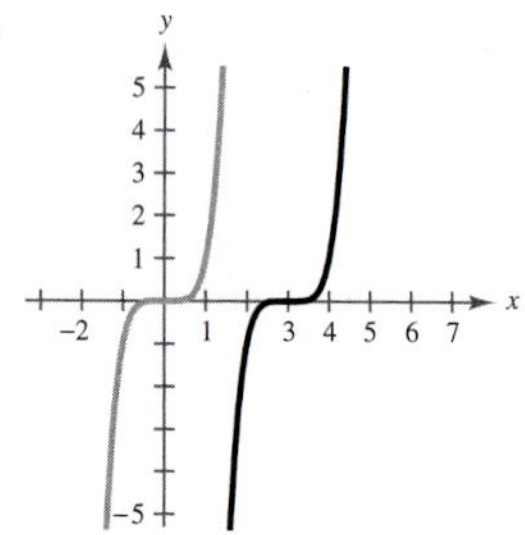

31. Falls to the left, falls to the right

33. Rises to the left, rises to the right

35. $-7, \frac{3}{2}$, odd multiplicity **37.** $0, \pm\sqrt{3}$, odd multiplicity

39. 0, even multiplicity; $\frac{5}{3}$, odd multiplicity

41. (a) Rises to the left, falls to the right (b) -1

(c) Answers will vary.

(d)

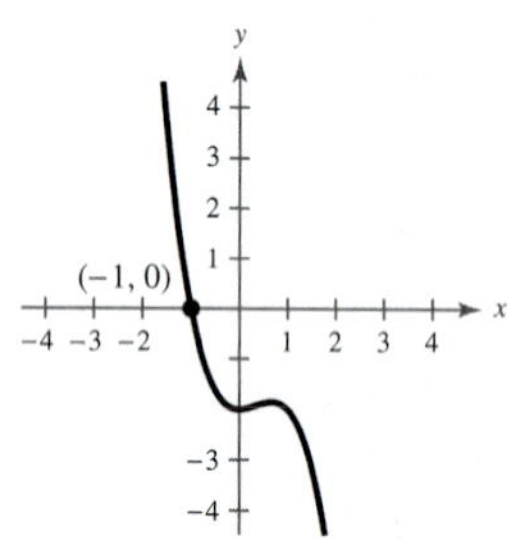

43. (a) Rises to the right, rises to the left (b) $-3, 0, 1$

(c) Answers will vary.

(d)

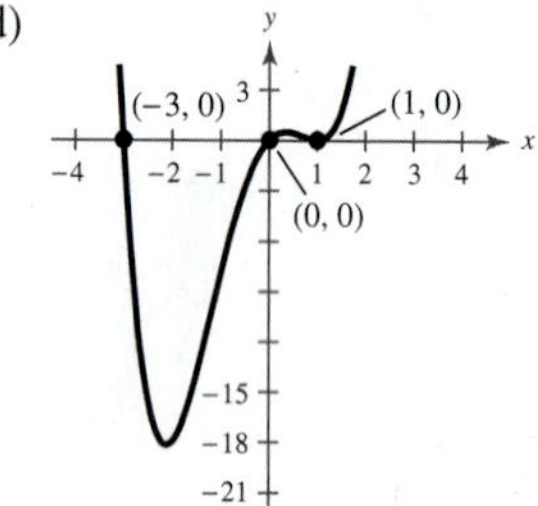

45. $[-1, 0]; \approx -0.900$

47. $[-1, 0], [1, 2]; \approx -0.200, \approx 1.772$

49. $8x + 5 + \dfrac{2}{3x - 2}$ **51.** $5x + 2$

53. $x^2 - 3x + 2 - \dfrac{1}{x^2 + 2}$

55. $6x^3 + 8x^2 - 11x - 4 - \dfrac{8}{x - 2}$ **57.** $2x^2 - 11x - 6$

59. (a) Yes (b) Yes (c) Yes (d) No

61. (a) -421 (b) -9

63. (a) Answers will vary.

(b) $(x + 7), (x + 1)$

(c) $f(x) = (x + 7)(x + 1)(x - 4)$

(d) $-7, -1, 4$

(e)

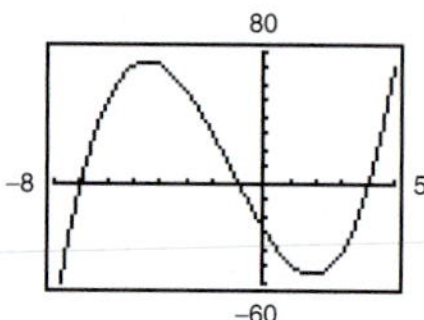

65. (a) Answers will vary.

(b) $(x + 1), (x - 4)$

(c) $f(x) = (x + 1)(x - 4)(x + 2)(x - 3)$

(d) $-2, -1, 3, 4$

(e)

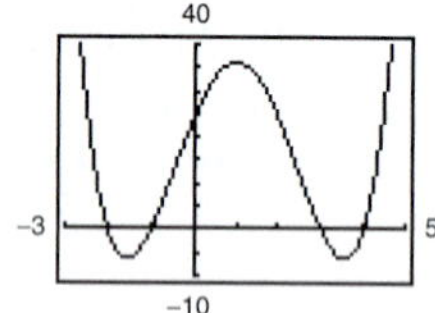

67. 0, 2 **69.** 8, 1 **71.** $-4, 6, \pm 2i$

73. $\pm 1, \pm 3, \pm 5, \pm 15, \pm\frac{1}{2}, \pm\frac{3}{2}, \pm\frac{5}{2}, \pm\frac{15}{2}, \pm\frac{1}{4}, \pm\frac{3}{4}, \pm\frac{5}{4}, \pm\frac{15}{4}$

75. $-1, -3, 6$ **77.** 1, 8 **79.** $-4, 3$

81. $3x^4 - 14x^3 + 17x^2 - 42x + 24$ **83.** $4, \pm i$

85. $-3, \frac{1}{2}, 2 \pm i$ **87.** $0, 1, -5; f(x) = x(x - 1)(x + 5)$

89. $-4, 2 \pm 3i; g(x) = (x + 4)^2(x - 2 - 3i)(x - 2 + 3i)$

91. (a)

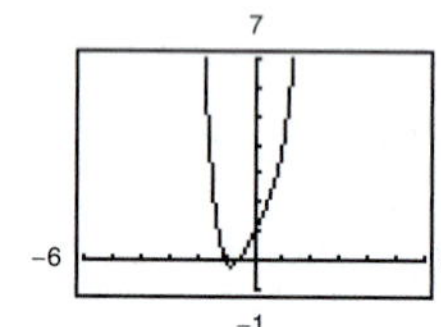

(b) Two zeros

(c) $-1, -0.54$

93. (a)

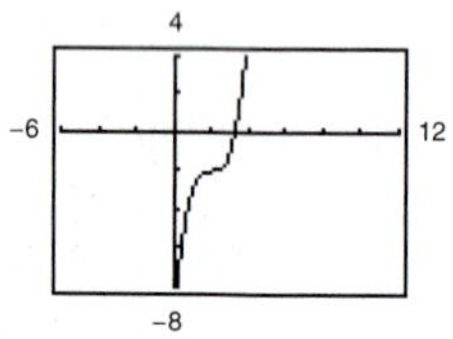

(b) One zero

(c) 3.26

95. Two or no positive real zeros, one negative real zero

97. Answers will vary.

99. (a)

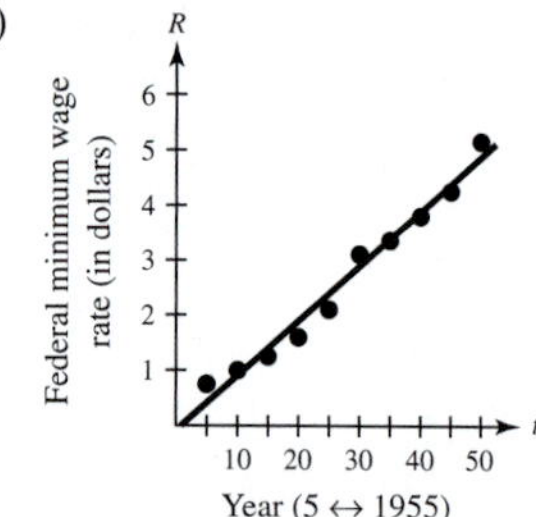

(b) The model is a fairly "good fit" for the actual data.

101. 2438.7 kilowatts **103.** $y = \dfrac{49.5}{x}$

105. (a) and (b)

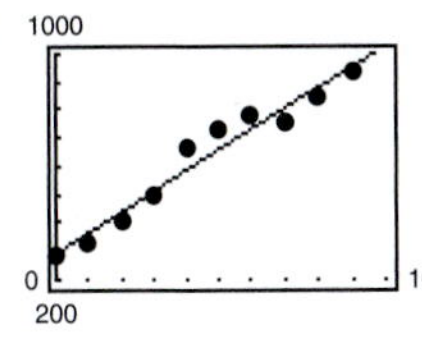

$y = 73.37t + 292.4$

(c) 1392.95 million

(d) Each year, the number of CDs shipped in the United States increases by 73.37 million.

107. True. If y is directly proportional to x, then $y = kx$, so $x = (1/k)y$. Therefore, x is directly proportional to y.

Chapter Test *(page 325)*

1. (a) Reflection in the x-axis followed by a vertical translation

(b) Horizontal translation

2. Vertex: $(-2, -1)$;

Intercepts: $(0, 3)$, $(-3, 0)$, $(-1, 0)$

3. $y = (x - 3)^2 - 6$

4. (a) 50 feet

(b) 5. Yes, changing the constant term results in a vertical translation of the graph and therefore changes the maximum height.

5. Rises to the left, falls to the right

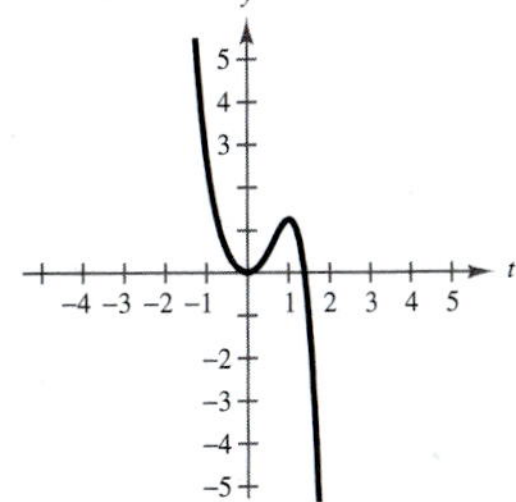

6. $3x + \dfrac{x - 1}{x^2 + 1}$ **7.** $2x^3 + 4x^2 + 3x + 6 + \dfrac{9}{x - 2}$

8. $(4x - 1)(x - \sqrt{3})(x + \sqrt{3})$;

Real zeros: solutions: $\frac{1}{4}, \pm\sqrt{3}$

9. $-2, \frac{3}{2}$ **10.** $\pm 1, -\frac{2}{3}$

11. $f(x) = x^4 - 9x^3 + 28x^2 - 30x$

12. $f(x) = x^4 - 6x^3 + 16x^2 - 24x + 16$

13. $-2, \pm\sqrt{5}i$ **14.** $-2, 4, -1 \pm \sqrt{2}i$

15. $v = 6\sqrt{s}$ **16.** $A = \dfrac{25}{6}xy$ **17.** $b = \dfrac{48}{a}$

Problem Solving *(page 328)*

1. (a) (i) $6, -2$ (iv) 2

(ii) $0, -5$ (v) $1 \pm \sqrt{7}$

(iii) $-5, 2$ (vi) $\dfrac{-3 \pm \sqrt{7}i}{2}$

(b) (i)

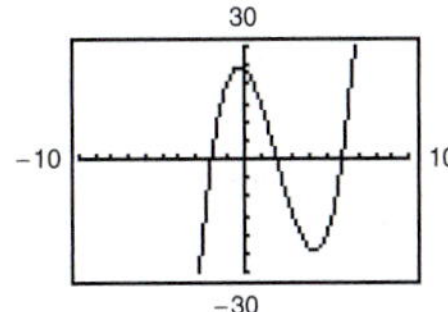

(ii)

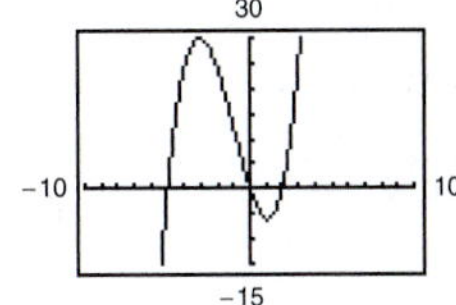

(iii)

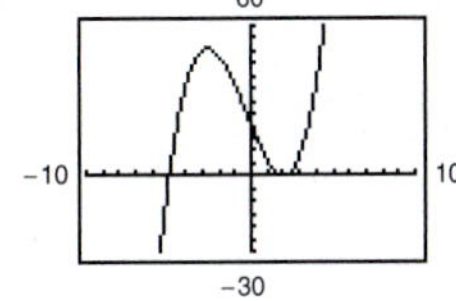

(iv)

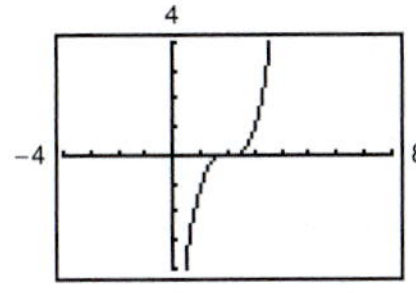

(v)

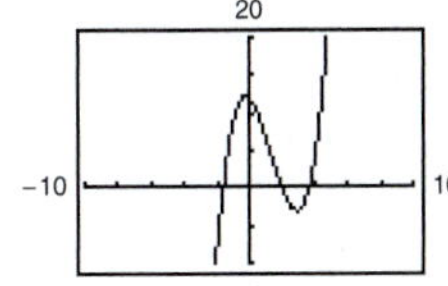

(vi)

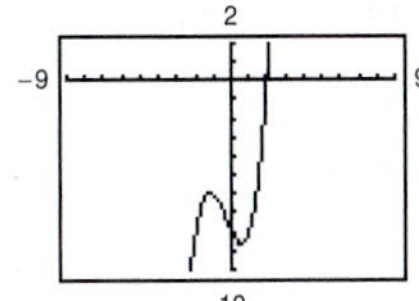

Graph (iii) touches the x-axis at $(2, 0)$, and all the other graphs pass through the x-axis at $(2, 0)$.

(c) (i) $(6, 0), (-2, 0)$ (iv) Other x-intercepts

(ii) $(0, 0), (-5, 0)$ (v) $(-1.6, 0), (3.6, 0)$

(iii) $(-5, 0)$ (vi) No other x-intercepts

(d) When the function has two real zeros, the results are the same. When the function has one real zero, the graph touches the x-axis at the zero. When there are no real zeros, there is no x-intercept.

3. Answers will vary. **5.** 2 inches × 2 inches × 5 inches

7. (a) and (b) $y = -x^2 + 5x - 4$

9. (a) $A(x) = x\left(\frac{100 - x}{2}\right)$; Domain: $0 < x < 100$

(b)

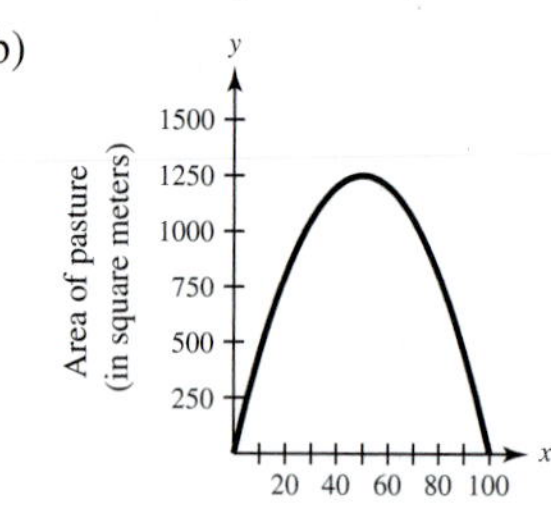

$x = 50, y = 25$

(c) $A(x) = -\frac{1}{2}(x - 50)^2 + 1250;$

$x = 50, y = 25$

11. $\frac{x^n - 1}{x - 1} = x^{n-1} + x^{n-2} + x^{n-3} + \cdot \cdot \cdot + 1$

Chapter 4

Section 4.1 *(page 337)*

1. (a)

x	$f(x)$
0.5	-2
0.9	-10
0.99	-100
0.999	-1000

x	$f(x)$
1.5	2
1.1	10
1.01	100
1.001	1000

x	$f(x)$
5	0.25
10	$0.\overline{1}$
100	$0.\overline{01}$
1000	$0.\overline{001}$

(b) Vertical asymptote: $x = 1$

Horizontal asymptote: $y = 0$

(c) Domain: all real numbers $x \neq 1$

3. (a)

x	$f(x)$
0.5	-1
0.9	-12.79
0.99	-147.8
0.999	-1498

x	$f(x)$
1.5	5.4
1.1	17.29
1.01	152.3
1.001	1502

x	$f(x)$
5	3.125
10	$3.\overline{03}$
100	$3.\overline{0003}$
1000	3

(b) Vertical asymptotes: $x = \pm 1$

Horizontal asymptote: $y = 3$

(c) Domain: all real numbers $x \neq \pm 1$

5. Domain: all real numbers $x \neq 0$

Vertical asymptote: $x = 0$

Horizontal asymptote: $y = 0$

7. Domain: all real numbers $x \neq 2$

Vertical asymptote: $x = 2$

Horizontal asymptote: $y = -1$

9. Domain: all real numbers $x \neq \pm 1$

Vertical asymptotes: $x = \pm 1$

11. Domain: all real numbers

Horizontal asymptote: $y = 3$

13. d **14.** a **15.** c **16.** b **17.** 1

19. None **21.** 6 **23.** 2

25. (a) Domain of f: all real numbers $x \neq -2$

Domain of g: all real numbers

(b) $x - 2$; Vertical asymptote: none

(c)

x	-4	-3	-2.5	-2	-1.5	-1	0
$f(x)$	-6	-5	-4.5	Undef.	-3.5	-3	-2
$g(x)$	-6	-5	-4.5	-4	-3.5	-3	-2

(d) The functions differ only at $x - 2$ where f is undefined.

27. (a) Domain of f: all real numbers $x \neq 0, \frac{1}{2}$

Domain of g: all real numbers $x \neq 0$

(b) $\frac{1}{x}$; Vertical asymptote: $x = 0$

(c)

x	-1	-0.5	0	0.5	2	3	4
$f(x)$	-1	-2	Undef.	Undef.	$\frac{1}{2}$	$\frac{1}{3}$	$\frac{1}{4}$
$g(x)$	-1	-2	Undef.	2	$\frac{1}{2}$	$\frac{1}{3}$	$\frac{1}{4}$

(d) The functions differ only at $x = 0.5$ where f is undefined.

29. (a) 4 (b) Less than (c) Greater than

31. (a) 2 (b) Greater than (c) Less than

33. (a)

M	200	400	600	800	1000
t	0.472	0.596	0.710	0.817	0.916

M	1200	1400	1600	1800	2000
t	1.009	1.096	1.178	1.255	1.328

(b) The greater the mass, the more time required per oscillation.

35. (a) \$28.33 million

(b) \$170 million

(c) \$765 million

(d) No. The function is undefined at $p = 100$.

37. (a) 333 deer, 500 deer, 800 deer (b) 1500 deer

39. (a)

n	1	2	3	4	5	6
P	0.50	0.74	0.82	0.86	0.89	0.91

n	7	8	9	10
P	0.92	0.93	0.94	0.95

P approaches 1 as n increases.

(b) 100%

41. (a)

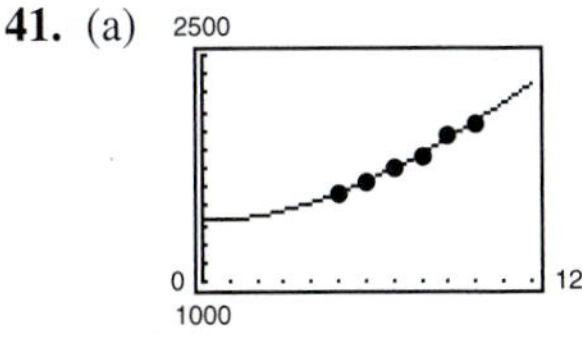

(b) 2689 thousand

(c) Possible, if the present trend continues.

(d) $P = 95.4t + 1090$

(e) Answers will vary.

43. False. Polynomials do not have vertical asymptotes.

45. Answers will vary. Sample answer: $f(x) = \dfrac{2x^2}{x^2 + 1}$

47. Answers will vary. Sample answer:

$f(x) = \dfrac{1}{x^2 + 2}; f(x) = \dfrac{1}{x - 20}$

49. $f^{-1}(x) = \dfrac{x + 7}{8}$

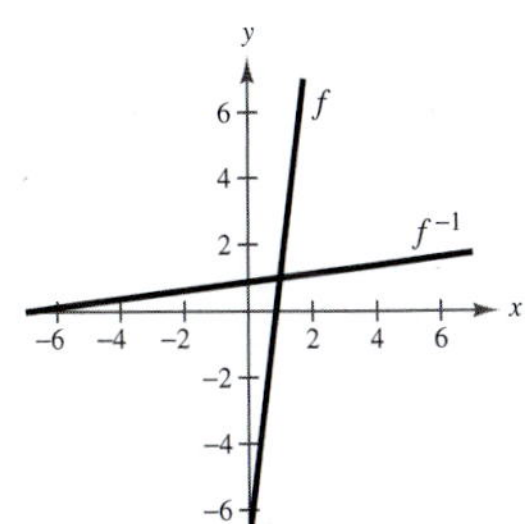

51. $x + 9 + \dfrac{42}{x - 4}$ **53.** $2x - 9 + \dfrac{34}{x + 5}$

Section 4.2 *(page 346)*

1.

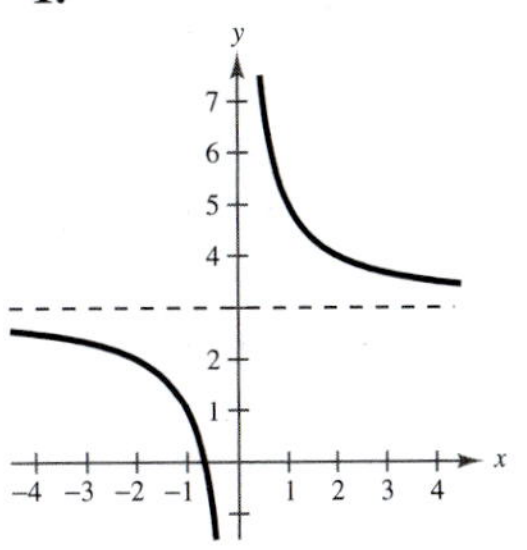

3.

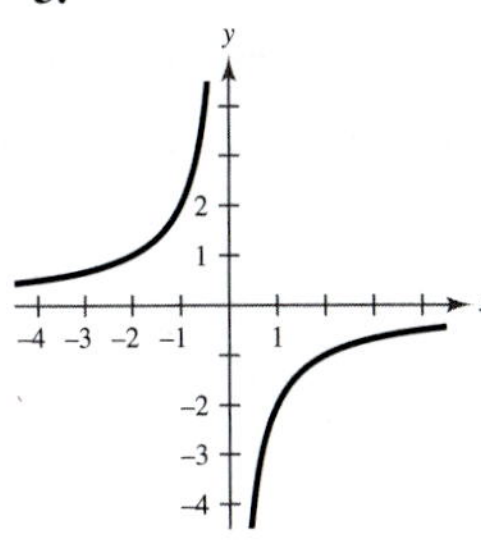

5.

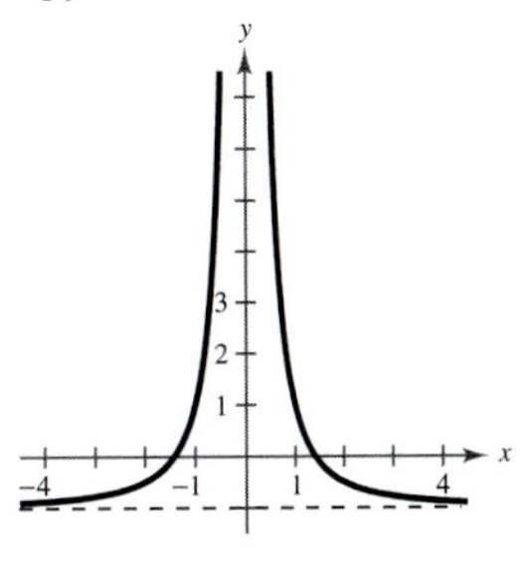

7.

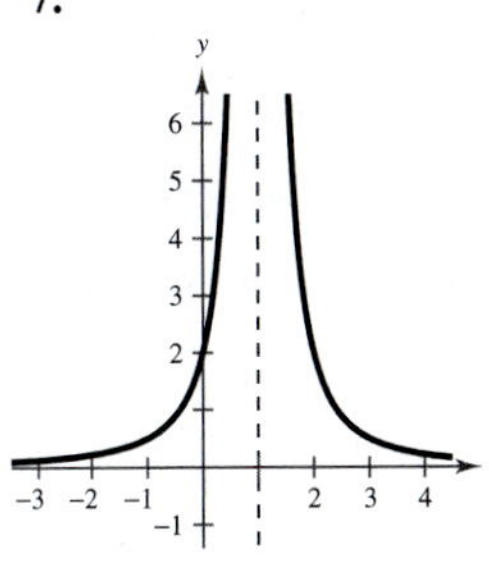

9.

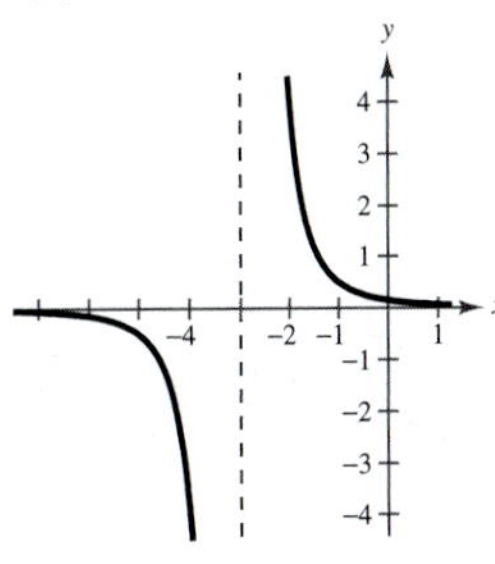

11.

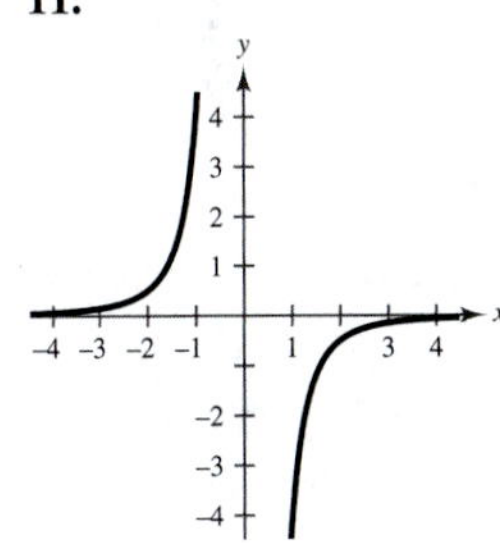

13. (a) y-intercept: $(0, \frac{1}{2})$

(b) Vertical asymptote: $x = -2$

Horizontal asymptote: $y = 0$

(c) No axis or origin symmetry

(d)

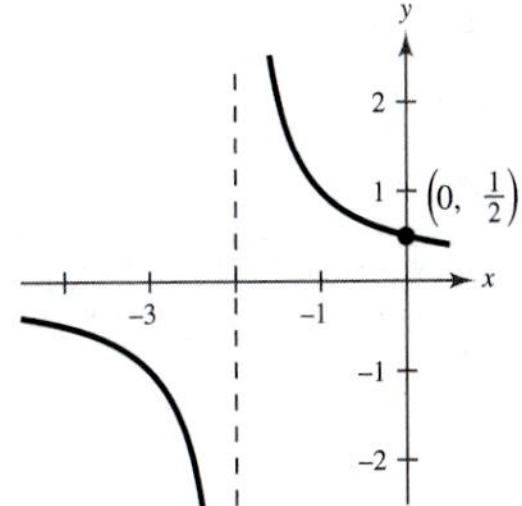

15. (a) y-intercept: $(0, -\frac{1}{2})$

(b) Vertical asymptote: $x = -2$

Horizontal asymptote: $y = 0$

(c) No axis or origin symmetry

(d)

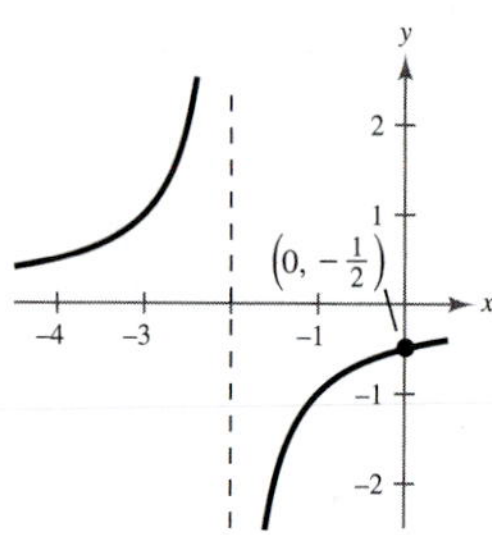

17. (a) x-intercept: $\left(-\frac{5}{2}, 0\right)$

y-intercept: $(0, 5)$

(b) Vertical asymptote: $x = -1$

Horizontal asymptote: $y = 2$

(c) No axis or origin symmetry

(d)

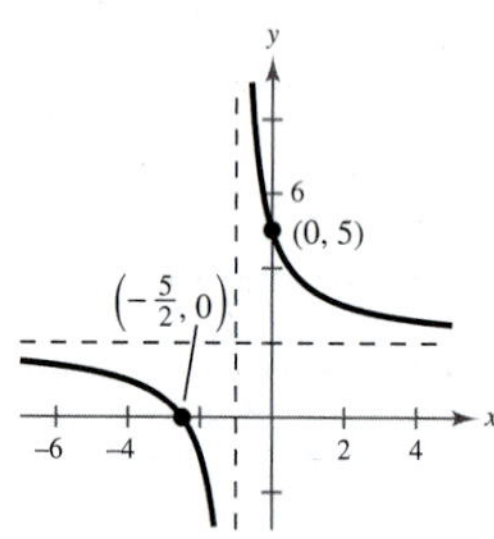

19. (a) x-intercept: $\left(-\frac{5}{2}, 0\right)$

y-intercept: $\left(0, \frac{5}{2}\right)$

(b) Vertical asymptote: $x = -2$

Horizontal asymptote: $y = 2$

(c) No axis or origin symmetry

(d)

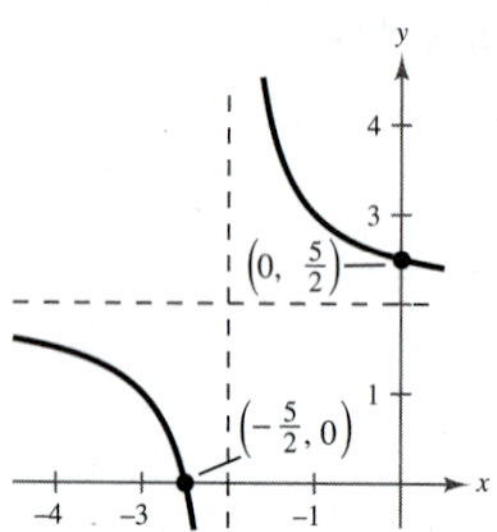

21. (a) Intercept: $(0, 0)$

(b) Horizontal asymptote: $y = 1$

(c) y-axis symmetry

(d)

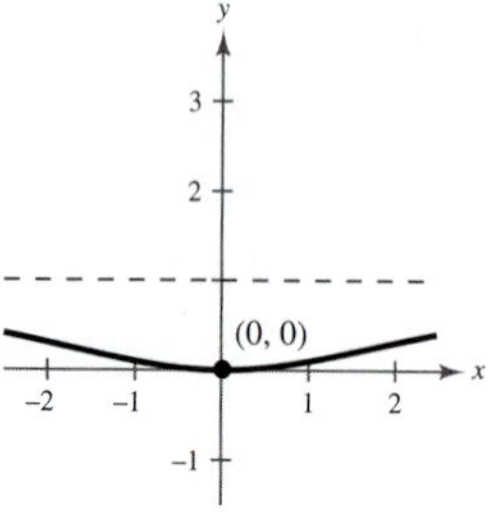

23. (a) Intercept: $(0, 0)$

(b) Vertical asymptotes: $x = 3, x = -3$

Horizontal asymptote: $y = 1$

(c) y-axis symmetry

(d)

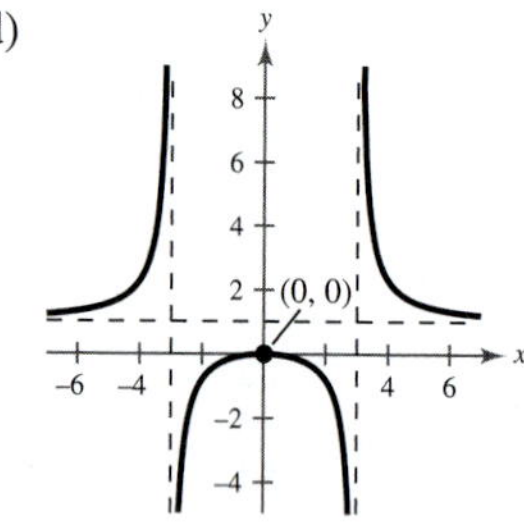

25. (a) Intercept: $(0, 0)$

(b) Horizontal asymptote: $y = 0$

(c) Origin symmetry

(d)

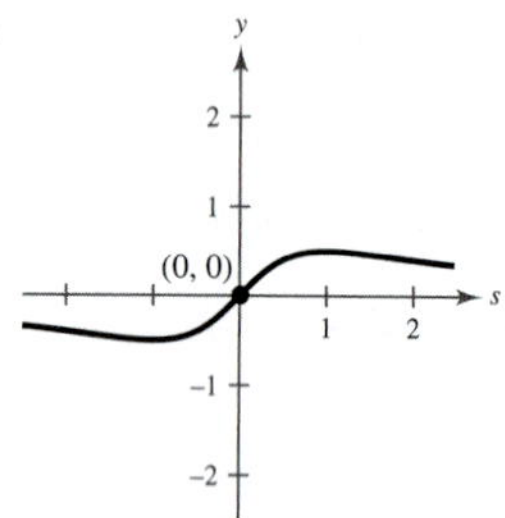

27. (a) x-intercept: $(-1, 0)$

(b) Vertical asymptotes: $x = 0,\ x = 4$

Horizontal asymptote: $y = 0$

(c) No axis or origin symmetry

(d)

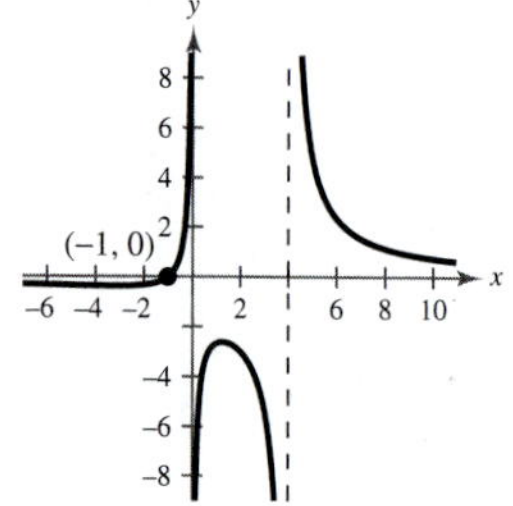

29. (a) Intercept: $(0, 0)$

(b) Vertical asymptotes: $x = -1,\ x = 2$

Horizontal asymptote: $y = 0$

(c) No axis or origin symmetry

(d)

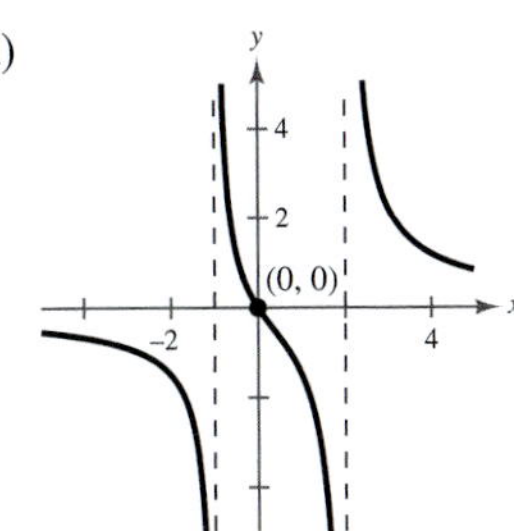

31. (a) Intercept: $(0, 0)$

(b) Vertical asymptotes: $x = -2,\ x = 7$

Horizontal asymptote: $y = 0$

(c) No axis or origin symmetry

(d)

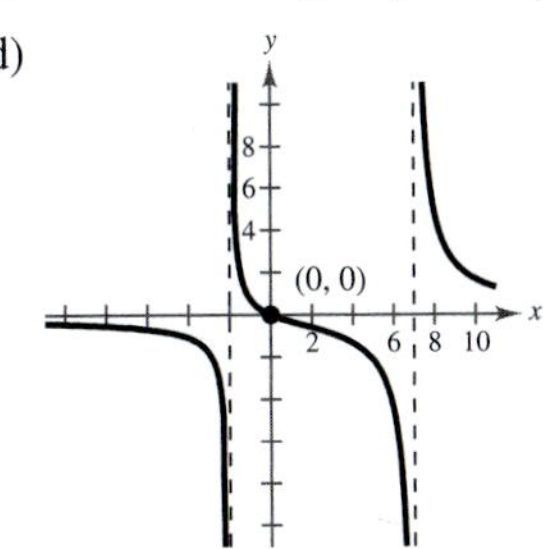

33. (a) x-intercepts: $(3, 0), \left(-\frac{1}{2}, 0\right)$

y-intercept: $\left(0, -\frac{3}{2}\right)$

(b) Vertical asymptotes: $x = 2, x = 1, x = -1$

Horizontal asymptote: $y = 0$

(c) No axis or origin symmetry

(d)

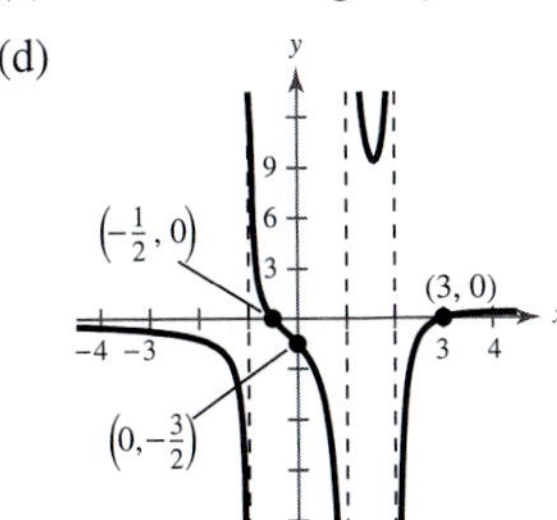

35. (a) Domain of f: all real numbers $x \neq -1$

Domain of g: all real numbers

(b) $x - 1$; Vertical asymptote: none

(c)

x	-3	-2	-1.5	-1	-0.5	0	1
$f(x)$	-4	-3	-2.5	Undef.	-1.5	-1	0
$g(x)$	-4	-3	-2.5	-2	-1.5	-1	0

(d)

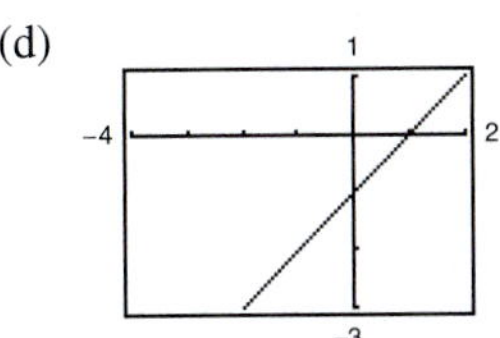

(e) Because there are only a finite number of pixels, the utility may not attempt to evaluate the function where it does not exist.

37. (a) Domain of f: all real numbers $x \neq 0, 2$

Domain of g: all real numbers $x \neq 0$

(b) $\frac{1}{x}$; Vertical asymptote: $x = 0$

(c)

x	-0.5	0	0.5	1	1.5	2	3
$f(x)$	-2	Undef.	2	1	$\frac{2}{3}$	Undef.	$\frac{1}{3}$
$g(x)$	-2	Undef.	2	1	$\frac{2}{3}$	$\frac{1}{2}$	$\frac{1}{3}$

(d)

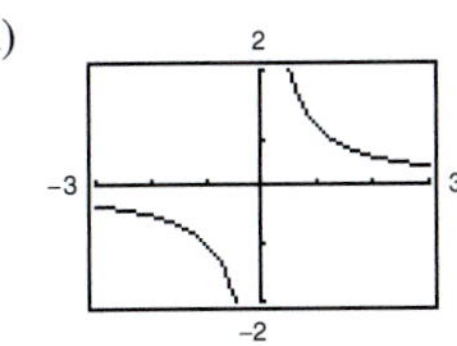

(e) Because there are only a finite number of pixels, the utility may not attempt to evaluate the function where it does not exist.

39. Domain: all real numbers $x \neq 0$

Vertical asymptote: $x = 0$

Slant asymptote: $y = x$

41. Domain: all real numbers $t \neq -5$

Vertical asymptote: $t = -5$

Slant asymptote: $y = -t + 5$

43. Domain: all real numbers $x \neq 0, 1$

Vertical asymptote: $x = 0$

Slant asymptote: $y = x + 1$

45. (a) No intercepts

(b) Vertical asymptote: $x = 0$

Slant asymptote: $y = 2x$

(c) Origin symmetry

(d) 

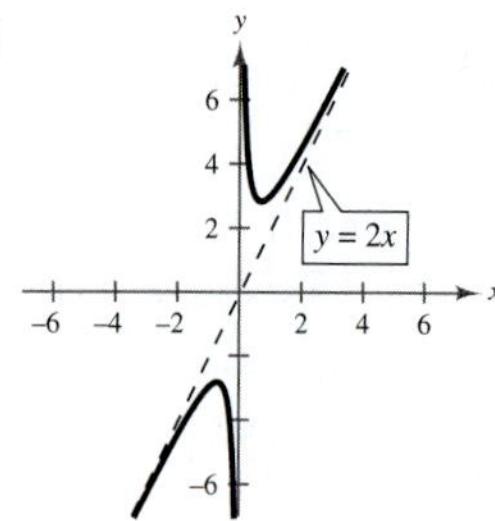

47. (a) No intercepts

(b) Vertical asymptote: $x = 0$

Slant asymptote: $y = x$

(c) Origin symmetry

(d) 

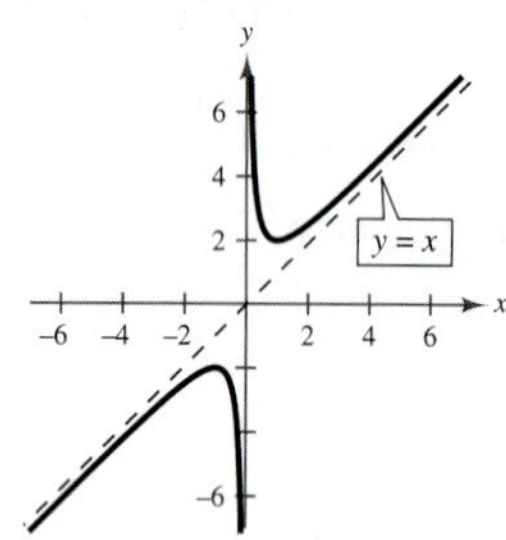

49. (a) Intercept: $(0, 0)$

(b) Vertical asymptotes: $x = \pm 1$

Slant asymptote: $y = x$

(c) Origin symmetry

(d) 

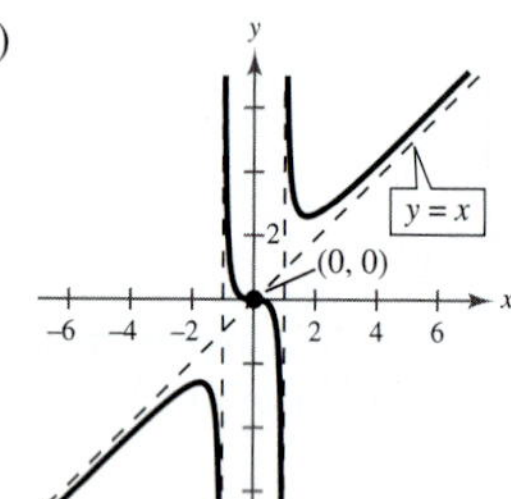

51. (a) y-intercept: $(0, -1)$

(b) Vertical asymptote: $x = 1$

Slant asymptote: $y = x$

(c) No axis or origin symmetry

(d)

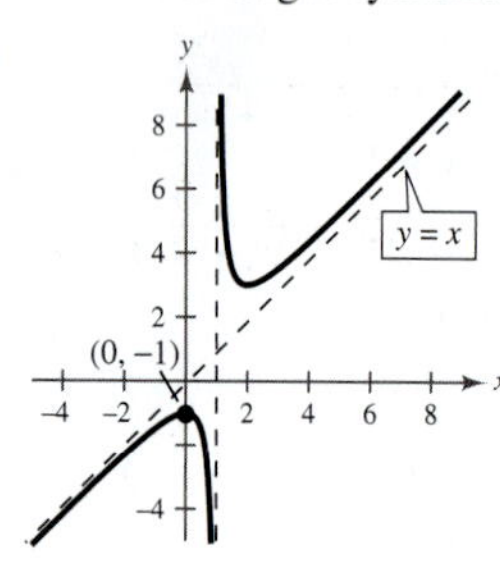

53.

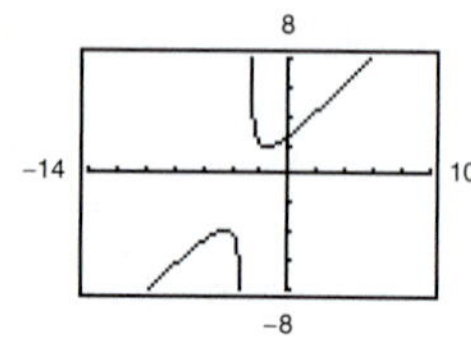

Domain: all real numbers $x \neq -3$

Vertical asymptote: $x = -3$

Slant asymptote: $y = x + 2$

$y = x + 2$

55.

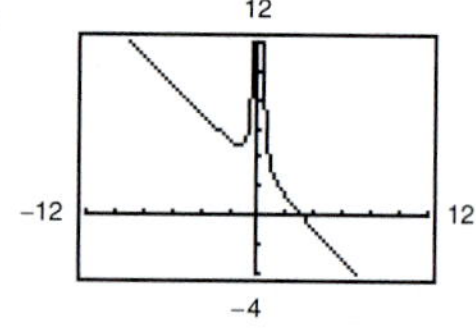

Domain: all real numbers $x \neq 0$

Vertical asymptote: $x = 0$

Slant asymptote: $y = -x + 3$

$y = -x + 3$

57. (a) $(-1, 0)$ (b) -1

59. (a) $(1, 0)$, $(-1, 0)$ (b) ± 1

61. (a) $(-4, 0)$

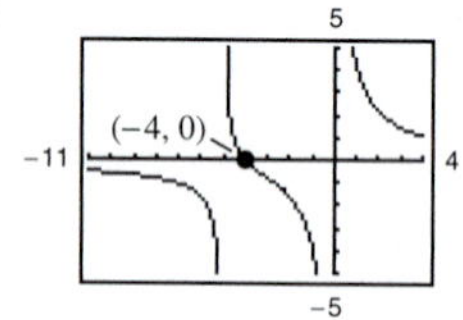

(b) -4

63. (a) $(3, 0)$, $(-2, 0)$

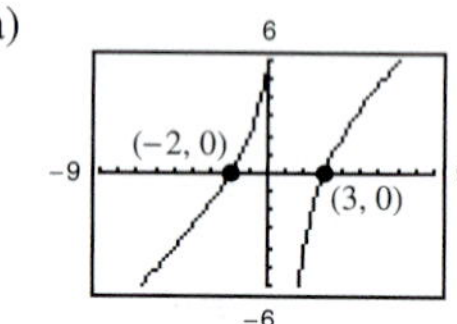

(b) $3, -2$

65. (a) Answers will vary. (b) $[0, 950]$

(c)

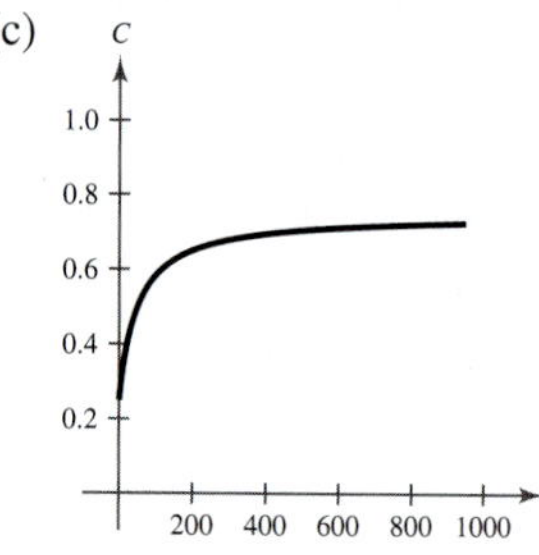

(d) Increases more slowly; 0.75

67. (a) Answers will vary. (b) $(4, \infty)$

(c)

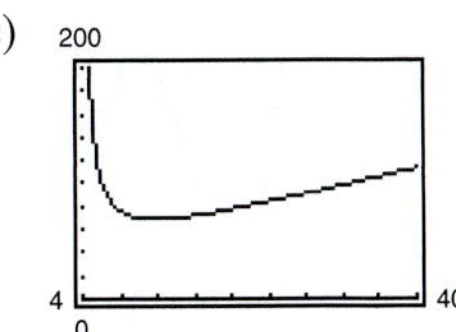

11.75 inches × 5.87 inches

69.

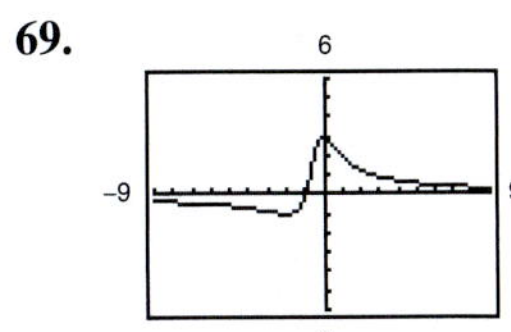

Minimum: $(-2, -1)$

Maximum: $(0, 3)$

71.

$x \approx 40.45$, or 4045 components.

73. (a) Answers will vary.

(b) Vertical asymptote: $x = 25$

Horizontal asymptote: $y = 25$

(c)

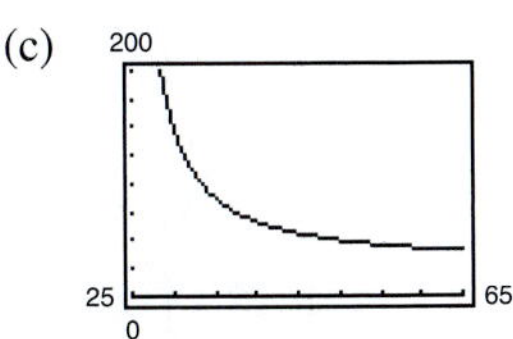

(d)

x	30	35	40	45	50	55	60
y	150	87.5	66.7	56.3	50	45.8	42.9

(e) Yes. You would expect the average speed for the round trip to be the average of the average speeds for the two parts of the trip.

(f) No. At 20 miles per hour you would use more time in one direction than is required for the round trip at an average speed of 50 miles per hour.

75. False. There are two distinct branches of the graph.

77. False. The degree of the numerator is 2 more than the degree of the denominator.

79.

The fraction is not reduced.

81. $f(x) = \dfrac{x^2 - x - 6}{x - 2}$

83. $(x - 7)(x - 8)$

85. $(x - 5)(x + 2i)(x - 2i)$

87. $x \geq \frac{10}{3}$

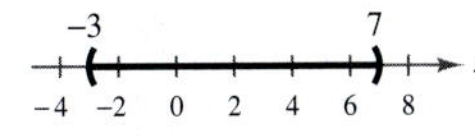

89. $-3 < x < 7$

Section 4.3 *(page 356)*

1. b **2.** c **3.** d **4.** a

5. $\dfrac{A}{x} + \dfrac{B}{x - 14}$

7. $\dfrac{A}{x} + \dfrac{B}{x^2} + \dfrac{C}{x - 10}$

9. $\dfrac{A}{x - 5} + \dfrac{B}{(x - 5)^2} + \dfrac{C}{(x - 5)^3}$

11. $\dfrac{A}{x} + \dfrac{Bx + C}{x^2 + 10}$

13. $\dfrac{A}{x} + \dfrac{Bx + C}{x^2 + 1} + \dfrac{Dx + E}{(x^2 + 1)^2}$

15. $\dfrac{1}{2}\left(\dfrac{1}{x - 1} - \dfrac{1}{x + 1}\right)$

17. $\dfrac{1}{x} - \dfrac{1}{x + 1}$

19. $\dfrac{1}{x} - \dfrac{2}{2x + 1}$

21. $\dfrac{1}{x - 1} - \dfrac{1}{x + 2}$

23. $-\dfrac{3}{x} - \dfrac{1}{x + 2} + \dfrac{5}{x - 2}$

25. $\dfrac{3}{x} - \dfrac{1}{x^2} + \dfrac{1}{x + 1}$

27. $\dfrac{3}{x - 3} + \dfrac{9}{(x - 3)^2}$

29. $-\dfrac{1}{x} + \dfrac{2x}{x^2 + 1}$

31. $-\dfrac{1}{x - 1} + \dfrac{x + 2}{x^2 - 2}$

33. $\dfrac{1}{6}\left(\dfrac{2}{x^2 + 2} - \dfrac{1}{x + 2} + \dfrac{1}{x - 2}\right)$

35. $\dfrac{1}{8}\left(\dfrac{1}{2x + 1} + \dfrac{1}{2x - 1} - \dfrac{4x}{4x^2 + 1}\right)$

37. $\dfrac{1}{x + 1} + \dfrac{2}{x^2 - 2x + 3}$

39. $1 - \dfrac{2x + 1}{x^2 + x + 1}$

41. $2x - 7 + \dfrac{17}{x + 2} + \dfrac{1}{x + 1}$

43. $x + 3 + \dfrac{6}{x - 1} + \dfrac{4}{(x - 1)^2} + \dfrac{1}{(x - 1)^3}$

45. $\dfrac{3}{2x - 1} - \dfrac{2}{x + 1}$

47. $\dfrac{2}{x} - \dfrac{1}{x^2} - \dfrac{2}{x + 1}$

49. $\dfrac{1}{x^2 + 2} + \dfrac{x}{(x^2 + 2)^2}$

51. $2x + \dfrac{1}{2}\left(\dfrac{3}{x - 4} - \dfrac{1}{x + 2}\right)$

53. $\dfrac{3}{x} - \dfrac{2}{x-4}$

$y = \dfrac{x-12}{x(x-4)}$ $\qquad y = \dfrac{3}{x}, y = -\dfrac{2}{x-4}$

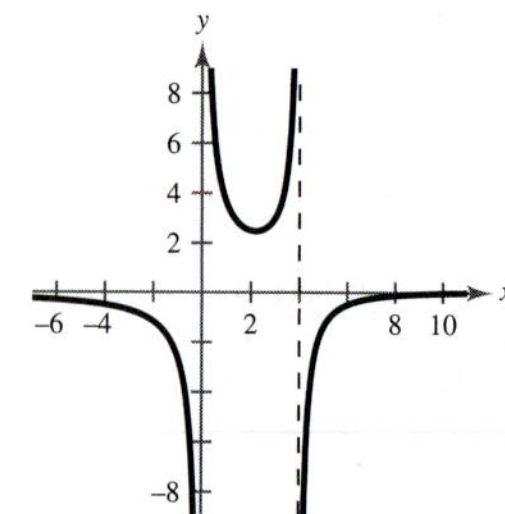

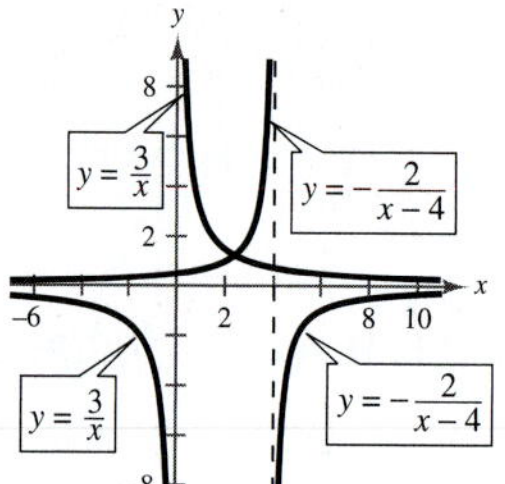

The vertical asymptotes are the same.

55. $\dfrac{3}{x-3} + \dfrac{5}{x+3}$

$y = \dfrac{2(4x-3)}{x^2-9}$ $\qquad y = \dfrac{3}{x-3}, y = \dfrac{5}{x+3}$

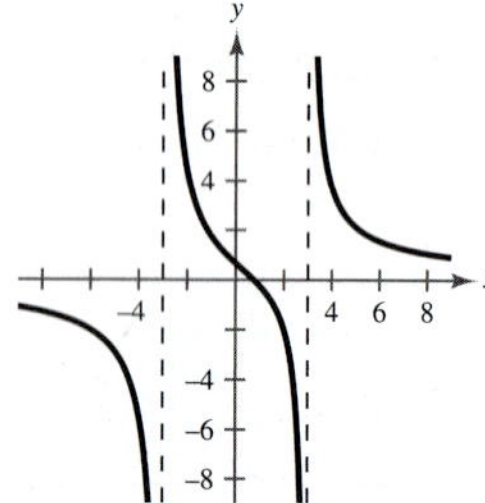

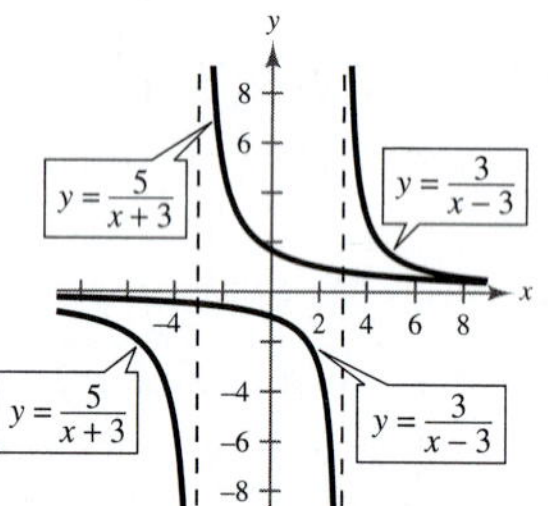

The vertical asymptotes are the same.

57. (a) $\dfrac{2000}{7-4x} - \dfrac{2000}{11-7x}, \quad 0 < x \le 1$

(b) $\text{Ymax} = \left|\dfrac{2000}{7-4x}\right|$

$\text{Ymin} = \left|\dfrac{-2000}{11-7x}\right|$

(c) 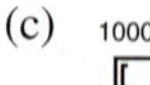

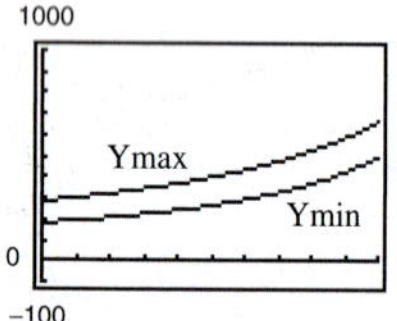

(d) Maximum: 400°F

Minimum: 266.7°F

59. False. The partial fraction decomposition is

$$\frac{A}{x+10} + \frac{B}{x-10} + \frac{C}{(x-10)^2}.$$

61. $\dfrac{1}{2a}\left(\dfrac{1}{a+x} + \dfrac{1}{a-x}\right)$ **63.** $\dfrac{1}{a}\left(\dfrac{1}{y} + \dfrac{1}{a-y}\right)$

65.

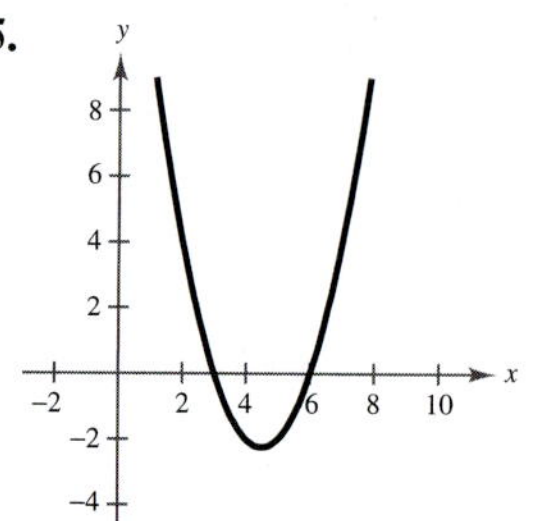

67.

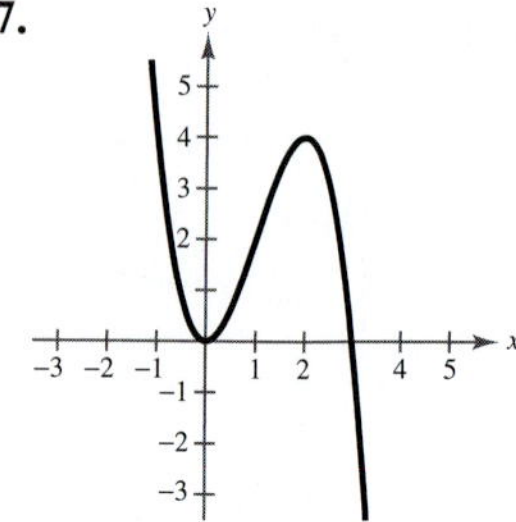

69.

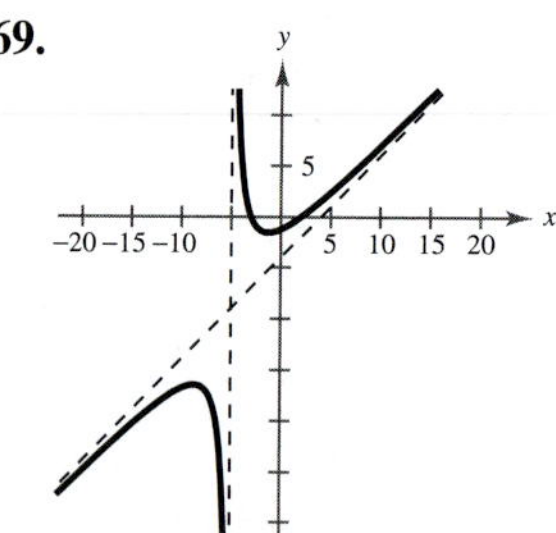

Section 4.4 *(page 367)*

1. Not shown **2.** c **3.** e **4.** a **5.** Not shown

6. h **7.** f **8.** b **9.** Not shown **10.** g

11. Vertex: $(0, 0)$

Focus: $\left(0, \frac{1}{2}\right)$

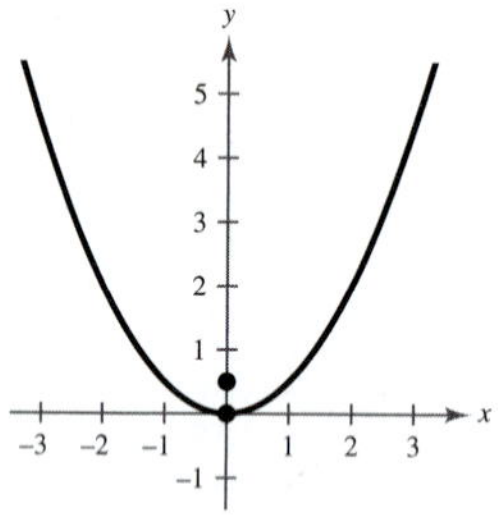

13. Vertex: $(0, 0)$

Focus: $\left(-\frac{3}{2}, 0\right)$

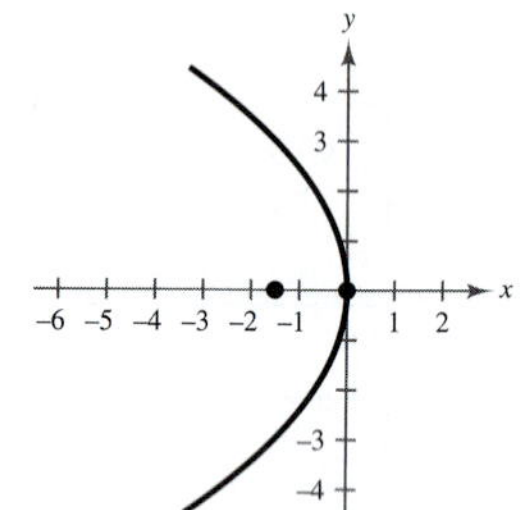

15. Vertex: $(0, 0)$

Focus: $(0, -2)$

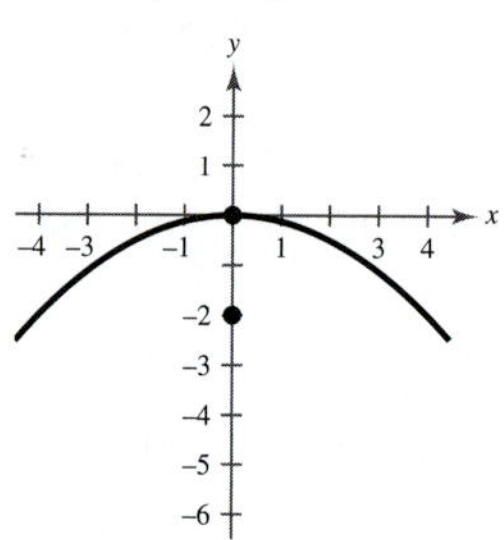

17. $y^2 = 8x$ **19.** $x^2 = -6y$ **21.** $x^2 = 4y$

23. $y^2 = -12x$ **25.** $y^2 = 9x$

27. $x^2 = \frac{3}{2}y$; Focus: $\left(0, \frac{3}{8}\right)$ **29.** $y^2 = \frac{9}{5}x$; Focus: $\left(\frac{9}{20}, 0\right)$

31. $y^2 = 6x$

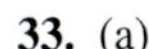

33. (a)

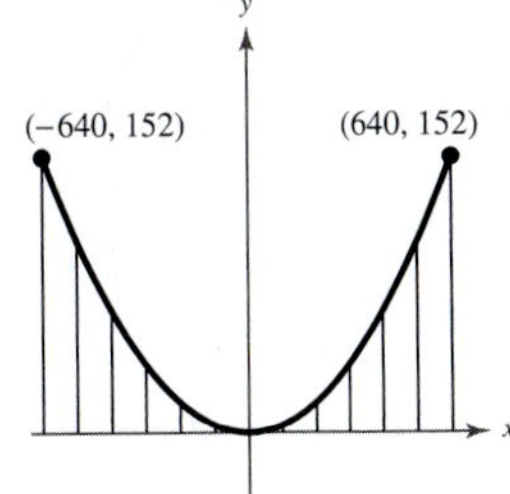

(b) $y = \dfrac{19x^2}{51{,}200}$

(c)

Distance, x	0	200	400	500	600
Height, y	0	$14\frac{27}{32}$	$59\frac{3}{8}$	$92\frac{99}{128}$	$133\frac{19}{32}$

35. Center: $(0, 0)$

Vertices: $(\pm 5, 0)$

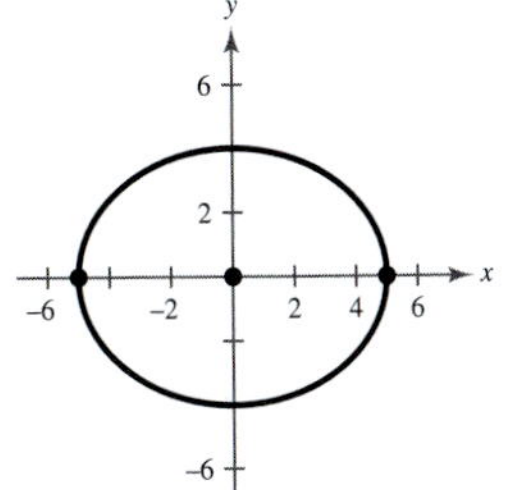

37. Center: $(0, 0)$

Vertices: $\left(\pm\frac{5}{3}, 0\right)$

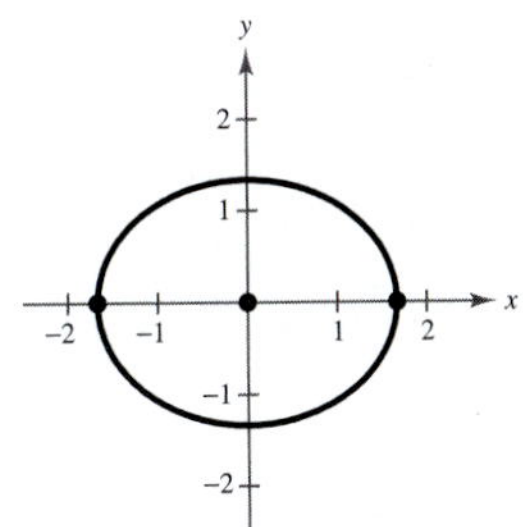

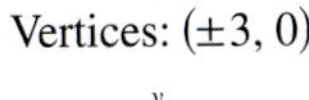

39. Center: $(0, 0)$

Vertices: $(\pm 3, 0)$

41. Center: $(0, 0)$

Vertices: $(0, \pm 1)$

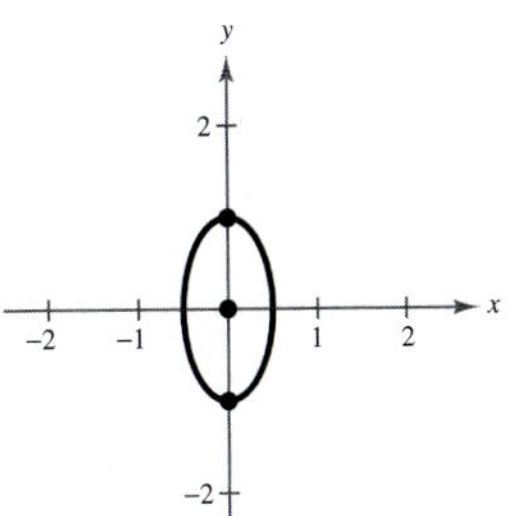

43. $\dfrac{x^2}{1} + \dfrac{y^2}{4} = 1$

45. $\dfrac{x^2}{4} + \dfrac{y^2}{9/4} = 1$

47. $\dfrac{x^2}{25} + \dfrac{y^2}{21} = 1$

49. $\dfrac{x^2}{36} + \dfrac{y^2}{11} = 1$

51. $\dfrac{21x^2}{400} + \dfrac{y^2}{25} = 1$

53. $(\pm\sqrt{5}, 0)$; Length of string: 6 feet

55. (a)

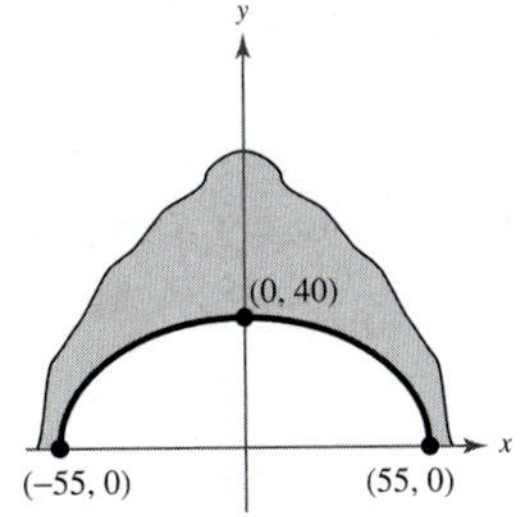

(b) $y = \frac{8}{11}\sqrt{3025 - x^2}$ (c) 16.66 feet

57.

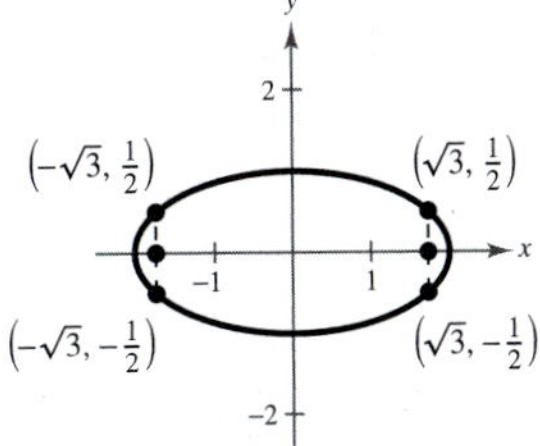

59.

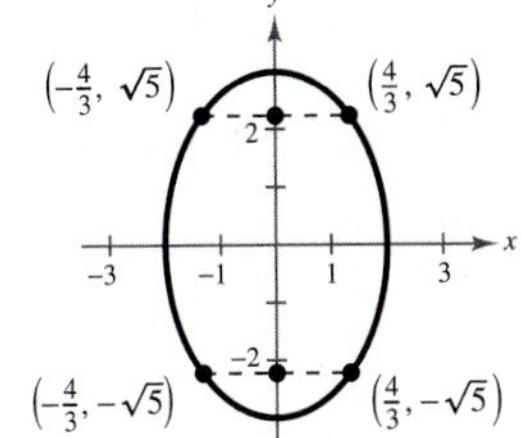

61. Center: $(0, 0)$

Vertices: $(\pm 1, 0)$

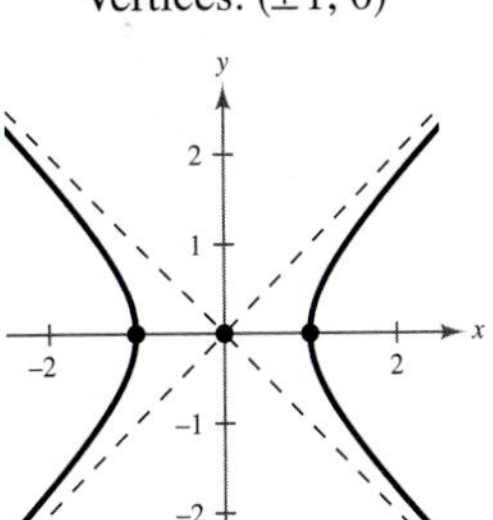

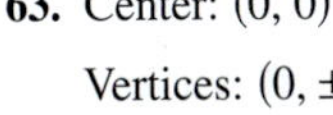

63. Center: $(0, 0)$

Vertices: $(0, \pm 1)$

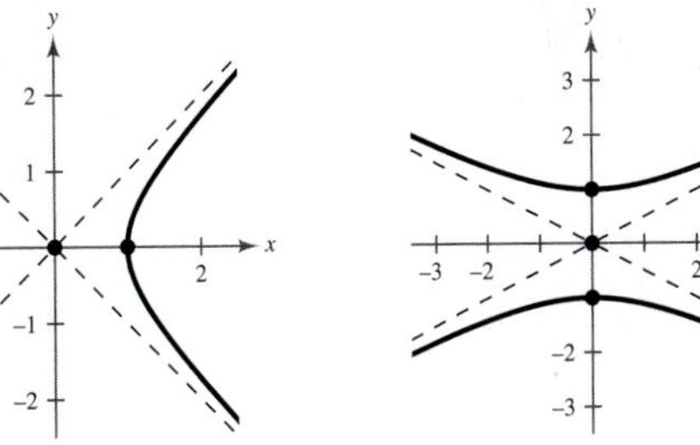

65. Center: $(0, 0)$

Vertices: $(0, \pm 5)$

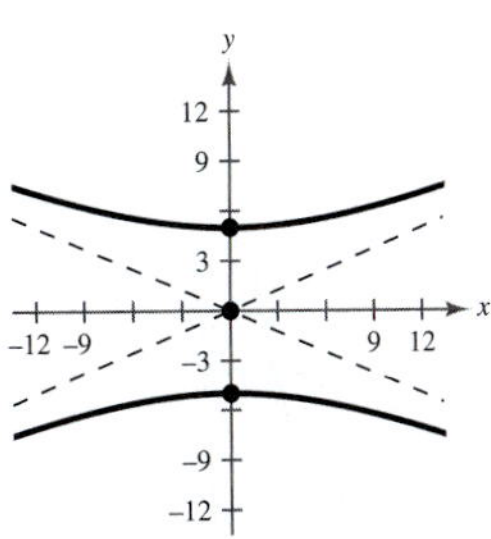

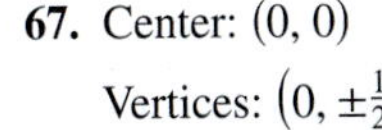

67. Center: $(0, 0)$

Vertices: $\left(0, \pm\frac{1}{2}\right)$

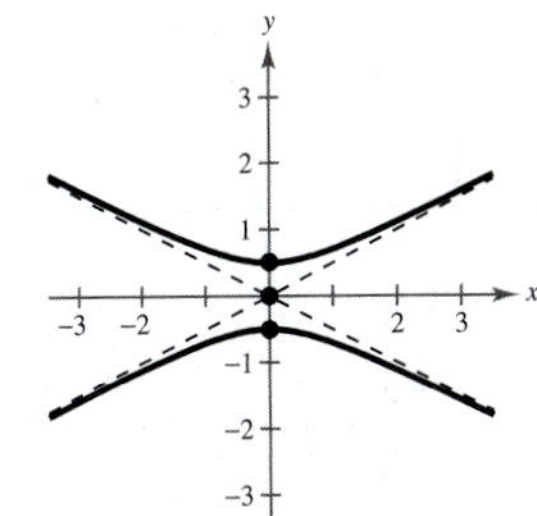

69. $\dfrac{y^2}{4} - \dfrac{x^2}{12} = 1$

71. $\dfrac{x^2}{1} - \dfrac{y^2}{9} = 1$

73. $\dfrac{17y^2}{1024} - \dfrac{17x^2}{64} = 1$

75. $\dfrac{y^2}{9} - \dfrac{x^2}{9/4} = 1$

77. $\left(12(\sqrt{5} - 1), 0\right) \approx (14.83, 0)$

79. False. The equation represents a hyperbola:
$$\frac{x^2}{144} - \frac{y^2}{144} = 1.$$

81. False. If the graph crossed the directrix, there would exist points nearer the directrix than the focus.

83. (a) $A = \pi a(20 - a)$

(b) $\dfrac{x^2}{196} + \dfrac{y^2}{36} = 1$

(c)

a	8	9	10	11	12	13
A	301.6	311.0	314.2	311.0	301.6	285.9

$a = 10$, circle

(d) 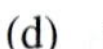

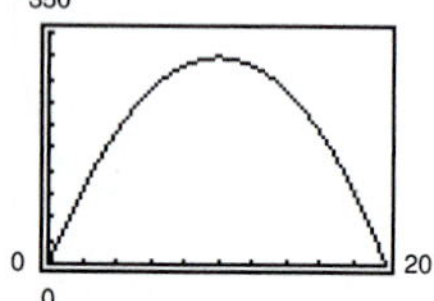

$a = 10$, circle

85.

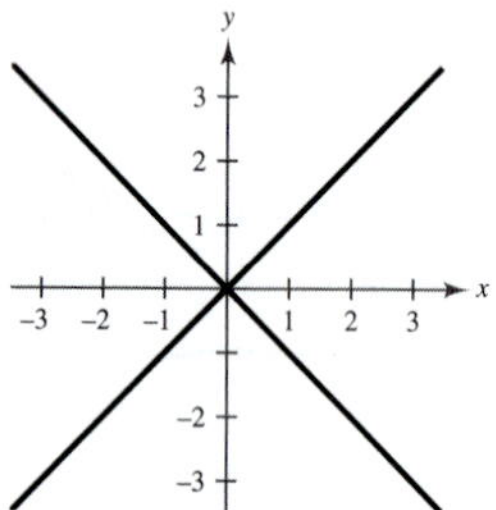

Two intersecting lines

87. Answers will vary.

89.

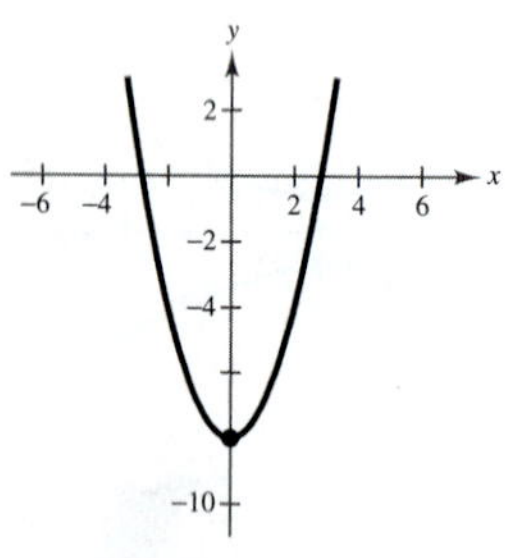

Vertex: $(0, -8)$

91.

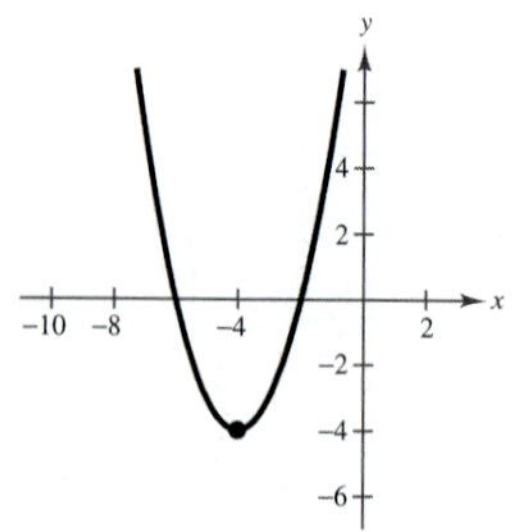

Vertex: $(-4, -4)$

93. $x^3 - 7x^2 + 17x - 15$

95. $\pm 1, \pm 2, \pm 4, \pm 5, \pm 10, \pm 20, \pm\frac{1}{2}, \pm\frac{5}{2}, \pm\frac{1}{3}, \pm\frac{2}{3}, \pm\frac{4}{3}, \pm\frac{5}{3}, \pm\frac{10}{3}, \pm\frac{20}{3}, \pm\frac{1}{6}, \pm\frac{5}{6}$

97. $\dfrac{1}{9}\left(-\dfrac{2}{x + 6} + \dfrac{11}{x - 3}\right)$

Section 4.5 *(page 376)*

1. Center: $(0, 0)$
Radius: 7

3. Center: $(-3, 8)$
Radius: 4

5. Center: $(1, 0)$
Radius: $\sqrt{10}$

7. $(x - 1)^2 + (y + 3)^2 = 1$
Center: $(1, -3)$
Radius: 1

9. $\left(x + \frac{3}{2}\right)^2 + (y - 3)^2 = 1$
Center: $\left(-\frac{3}{2}, 3\right)$
Radius: 1

11. Vertex: $(1, -2)$
Focus: $(1, -4)$
Directrix: $y = 0$

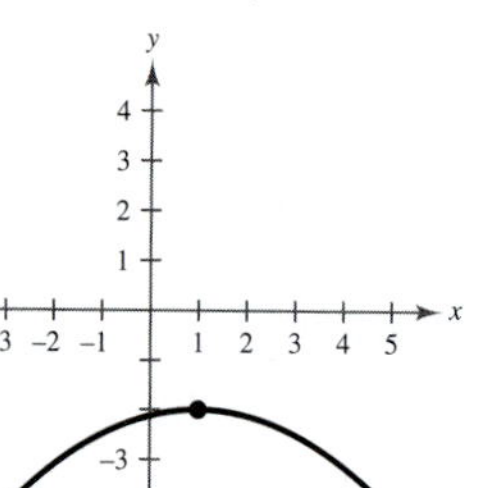

13. Vertex: $\left(5, -\frac{1}{2}\right)$
Focus: $\left(\frac{11}{2}, -\frac{1}{2}\right)$
Directrix: $x = \frac{9}{2}$

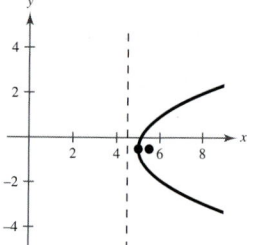

15. Vertex: $(1, 1)$
Focus: $(1, 2)$
Directrix: $y = 0$

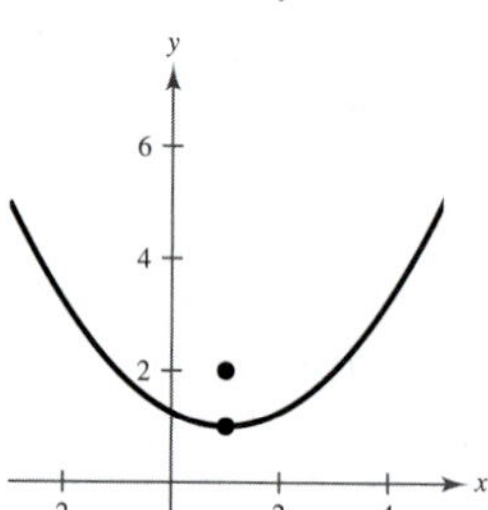

17. Vertex: $(-2, -3)$
Focus: $(-4, -3)$
Directrix: $x = 0$

19. Vertex: $(-2, 1)$
Focus: $\left(-2, -\frac{1}{2}\right)$
Directrix: $y = \frac{5}{2}$

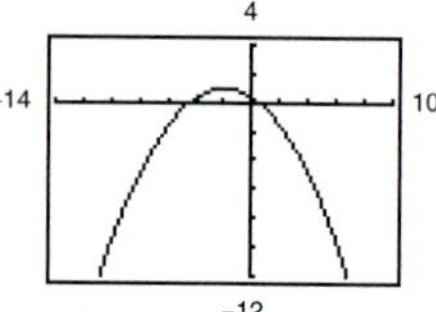

21. Vertex: $\left(\frac{1}{4}, -\frac{1}{2}\right)$
Focus: $\left(0, -\frac{1}{2}\right)$
Directrix: $x = \frac{1}{2}$

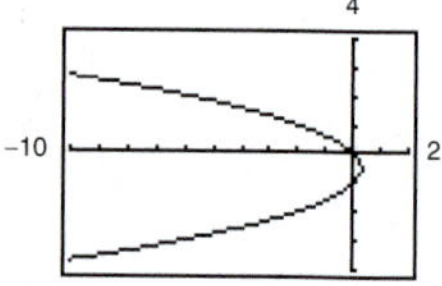

23. $(y - 2)^2 = -8(x - 3)$

25. $x^2 = 8(y - 4)$

27. $(y - 2)^2 = 8x$

29. (a) $17{,}500\sqrt{2}$ miles per hour
(b) $x^2 = -16{,}400(y - 4100)$

31. 34,295 feet

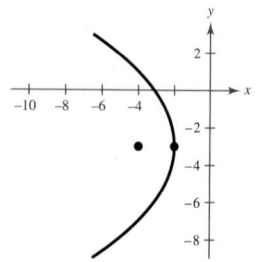

33. Center: $(1, 5)$
Foci: $(1, 9), (1, 1)$
Vertices: $(1, 10), (1, 0)$

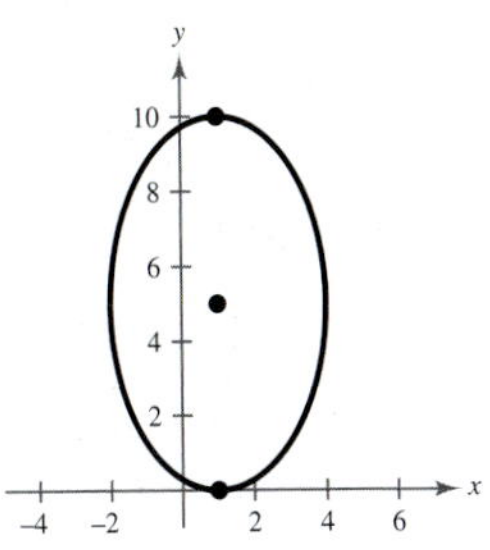

35. Center: $(-2, -4)$
Foci: $\left(\frac{-4 \pm \sqrt{3}}{2}, -4\right)$
Vertices: $(-3, -4), (-1, -4)$

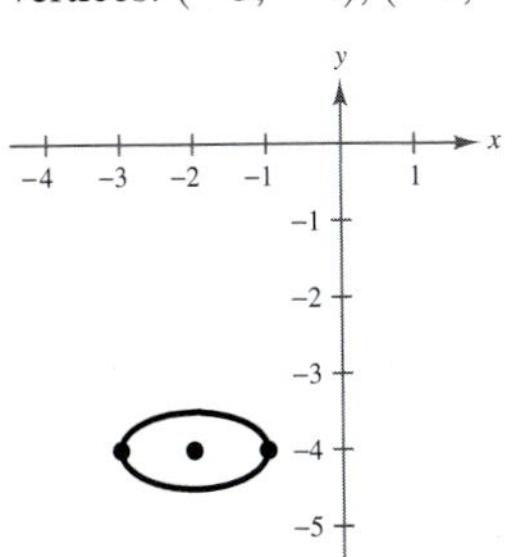

37. Center: $(-2, 3)$
Foci: $(-2, 3 \pm \sqrt{5})$
Vertices: $(-2, 6), (-2, 0)$

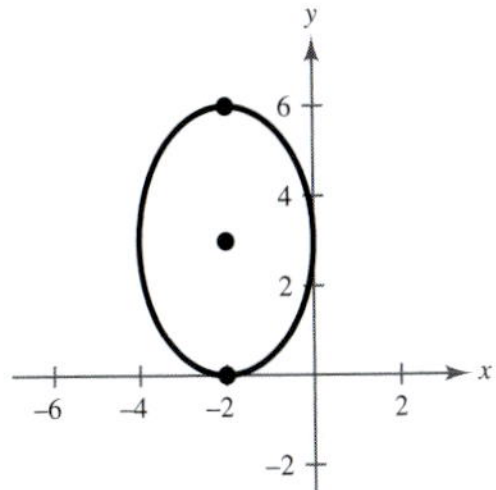

39. Center: $(1, -1)$
Foci: $\left(\frac{7}{4}, -1\right), \left(\frac{1}{4}, -1\right)$
Vertices: $\left(\frac{9}{4}, -1\right), \left(-\frac{1}{4}, -1\right)$

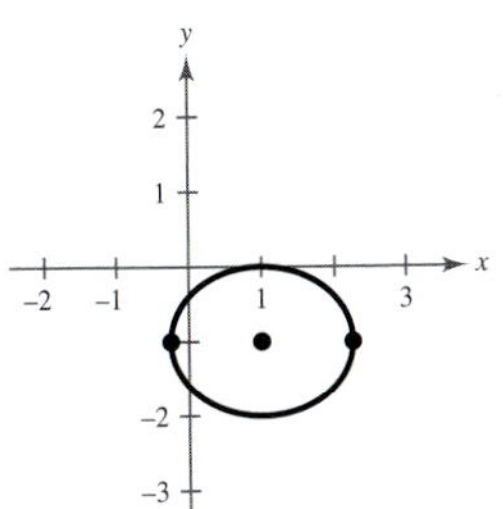

41. $\dfrac{(x-4)^2}{9} + \dfrac{y^2}{16} = 1$ **43.** $\dfrac{(x-2)^2}{4} + \dfrac{(y-2)^2}{1} = 1$

45. $\dfrac{x^2}{48} + \dfrac{(y-4)^2}{64} = 1$ **47.** $\dfrac{x^2}{16} + \dfrac{(y-4)^2}{12} = 1$

49. $\dfrac{(x-2)^2}{4} + \dfrac{(y-2)^2}{1} = 1$ **51.** $\dfrac{x^2}{25} + \dfrac{y^2}{16} = 1$

53. 2,756,832,000 miles; 4,575,168,000 miles

55. Center: $(1, -2)$
Foci: $(1 \pm \sqrt{5}, -2)$
Vertices: $(3, -2), (-1, -2)$

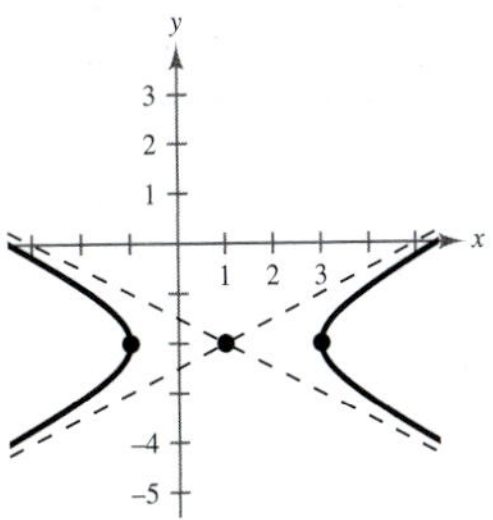

57. Center: $(2, -6)$
Foci: $(2, -6 \pm \sqrt{2})$
Vertices: $(2, -5), (2, -7)$

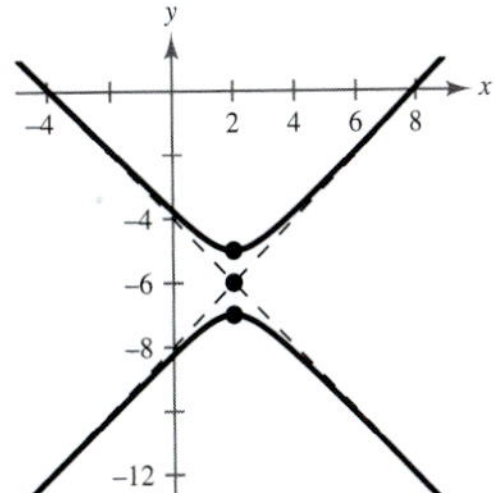

59. Center: $(2, -3)$
Foci: $(2 \pm \sqrt{10}, -3)$
Vertices: $(3, -3), (1, -3)$

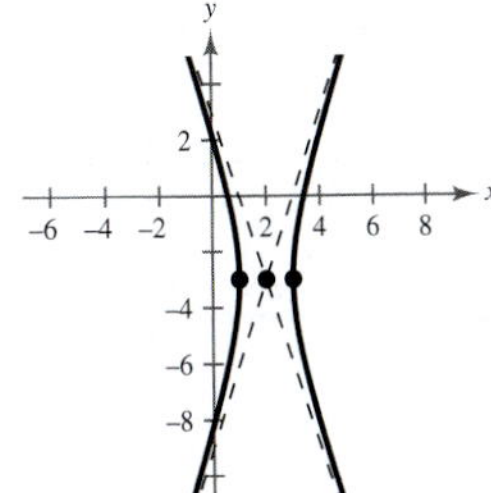

61. The graph of this equation is two lines intersecting at $(-1, -3)$.

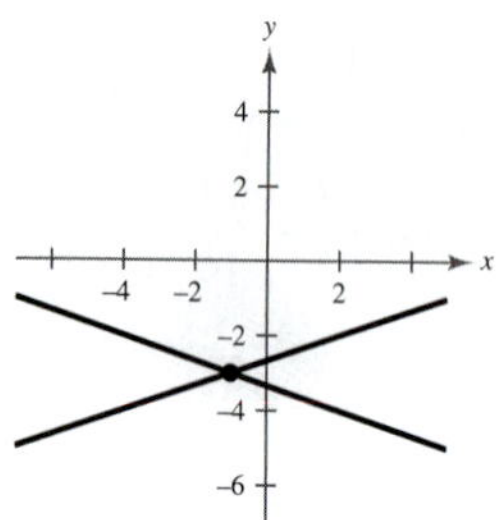

63. Center: $(1, -1)$

Foci: $\left(1 \pm \sqrt{13}, -1\right)$

Vertices: $(-2, -1), (4, -1)$

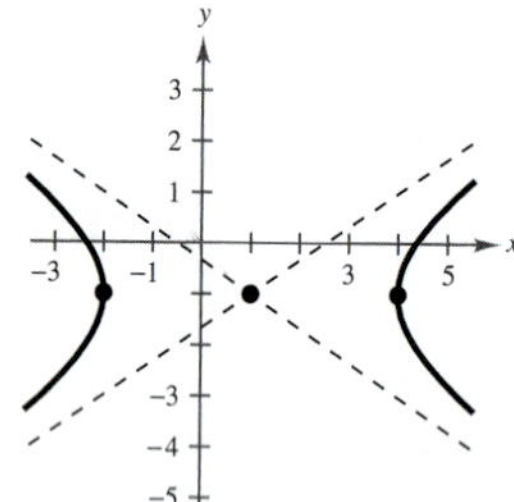

65. $(y-1)^2 - \frac{x^2}{3} = 1$

67. $\frac{(x-4)^2}{4} - \frac{y^2}{12} = 1$

69. $\frac{(y-5)^2}{16} - \frac{(x-4)^2}{9} = 1$

71. $\frac{y^2}{9} - \frac{4(x-2)^2}{9} = 1$

73. $\frac{(x-3)^2}{9} - \frac{(y-2)^2}{4} = 1$

75. $(x-3)^2 + (y+2)^2 = 4$

Circle

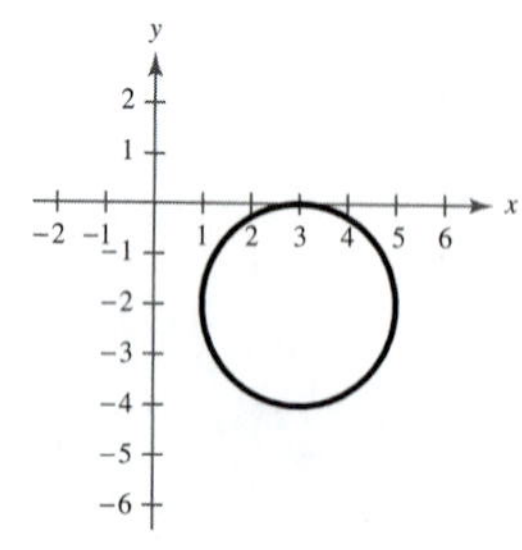

77. $\left(x - \frac{1}{2}\right)^2 - \frac{y^2}{4} = 1$

Hyperbola

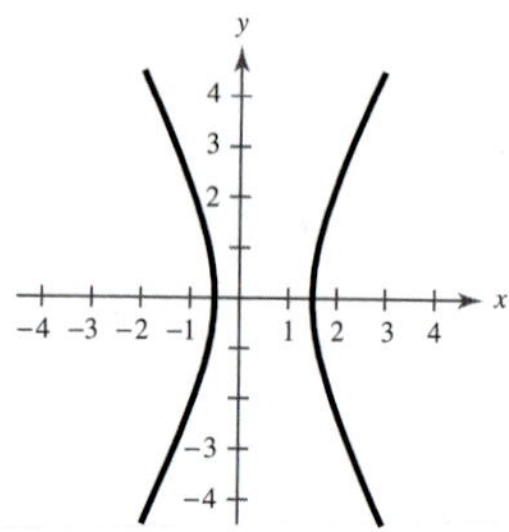

79. $\frac{(x+1)^2}{1/4} + \frac{(y-4)^2}{1/3} = 1$

Ellipse

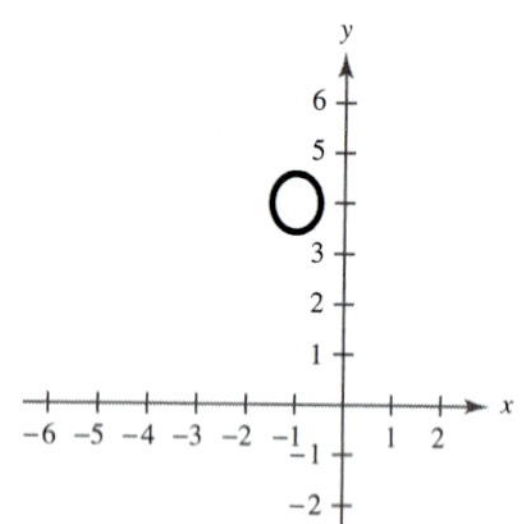

81. $\left(x - \frac{1}{5}\right)^2 = 8\left(y + \frac{3}{5}\right)$

Parabola

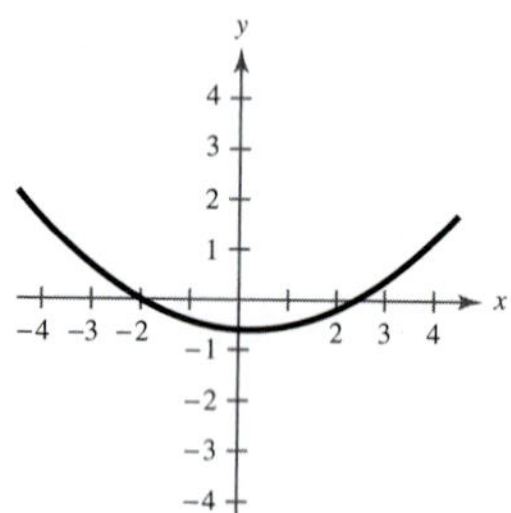

83. True. The equation in standard form is

$$\frac{(x-3)^2}{1/3} + \frac{(y-4)^2}{1/2} = 1.$$

85. $e = \frac{c}{a} \Rightarrow e^2a^2 = c^2$

For an ellipse:

$$a^2 = b^2 + c^2$$
$$a^2 = b^2 + e^2a^2$$
$$a^2(1 - e^2) = b^2$$
$$\frac{(x-h)^2}{a^2} + \frac{(y-k)^2}{a^2(1-e^2)} = 1$$

87. Additive Inverse Property

89. Distributive Property

91. $f^{-1}(x) = -\frac{x-10}{7}$

Review Exercises (page 380)

1. Domain: all real numbers $x \neq -12$
3. Domain: all real numbers $x \neq 6, 4$
5. Vertical asymptote: $x = -3$
   Horizontal asymptote: $y = 0$
7. Vertical asymptotes: $x = -2, x = 2$
   Horizontal asymptote: $y = 1$
9. \$0.50 is the horizontal asymptote of the function.
11. No intercepts
    y-axis symmetry
    Vertical asymptote: $x = 0$
    Horizontal asymptote: $y = 0$

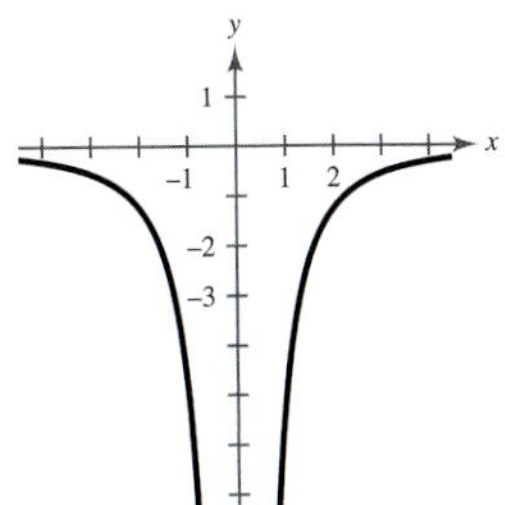

13. x-intercept: $(-2, 0)$
    y-intercept: $(0, 2)$
    No axis or origin symmetry
    Vertical asymptote: $x = 1$
    Horizontal asymptote: $y = -1$

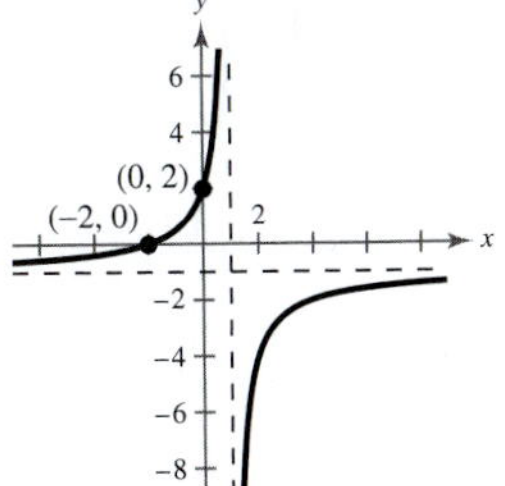

15. Intercept: $(0, 0)$
    y-axis symmetry
    Horizontal asymptote: $y = 1$

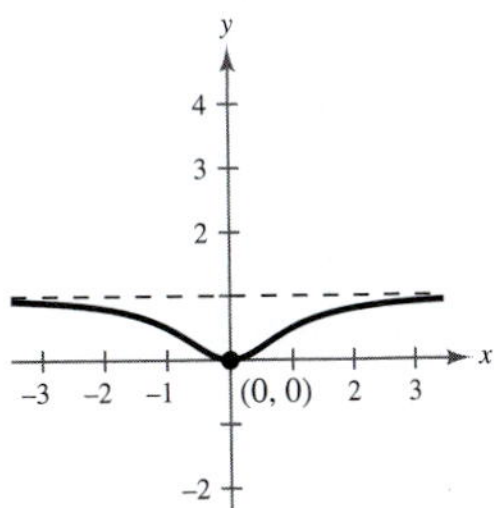

17. Intercept: $(0, 0)$
    Origin symmetry
    Horizontal asymptote: $y = 0$

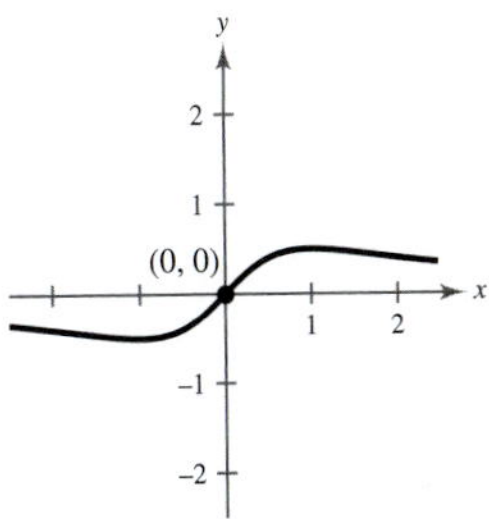

19. Intercept: $(0, 0)$
    y-axis symmetry
    Horizontal asymptote: $y = -6$

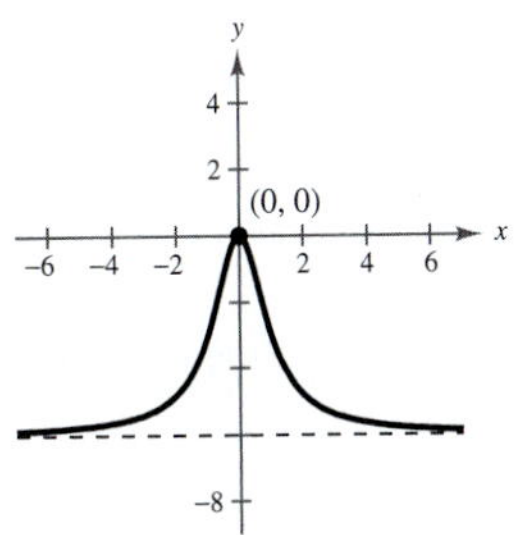

21. Intercept: $(0, 0)$
    Origin symmetry
    Vertical asymptotes: $x = \pm 1$
    Horizontal asymptote: $y = 0$

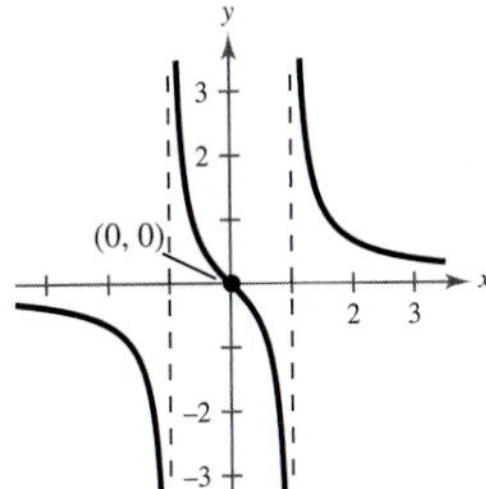

23. Domain: all real numbers
    Slant asymptote: $y = 2x$

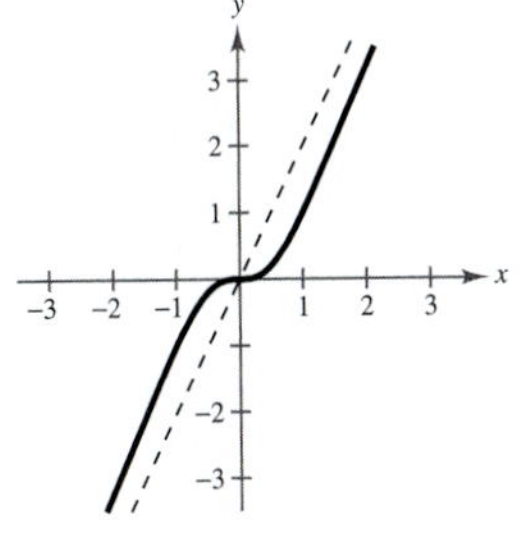

25. Domain: all real numbers $x \neq -2$

Vertical asymptote: $x = -2$

Slant asymptote: $y = x + 1$

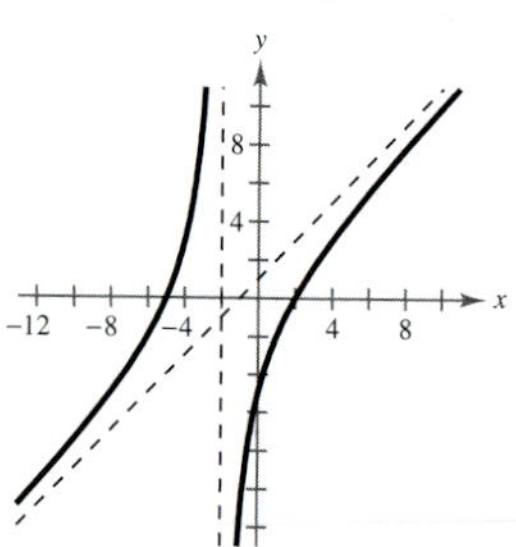

27. (a)

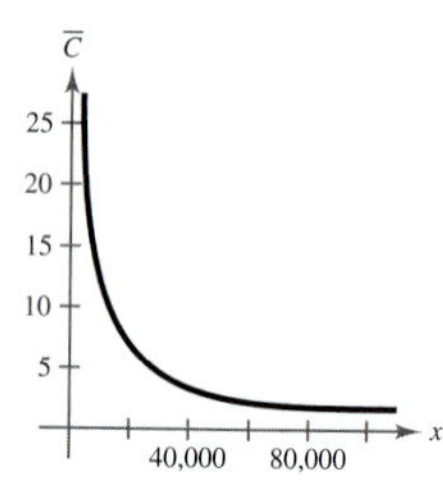

(b) \$100.90, \$10.90, \$1.90

(c) \$0.90 is the horizontal asymptote of the function.

29.

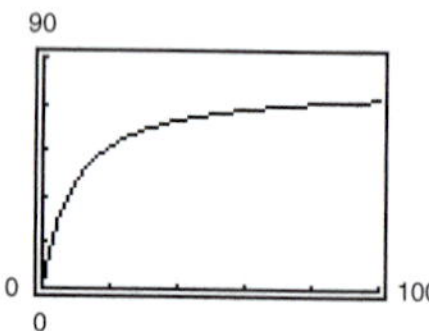

80.3 milligrams per square decimeter per hour

31. $\dfrac{A}{x} + \dfrac{B}{x + 20}$ **33.** $\dfrac{A}{x} + \dfrac{B}{x^2} + \dfrac{C}{x - 5}$

35. $\dfrac{3}{x + 2} - \dfrac{4}{x + 4}$ **37.** $1 - \dfrac{25}{8(x + 5)} + \dfrac{9}{8(x - 3)}$

39. $\dfrac{1}{2}\left(\dfrac{3}{x - 1} - \dfrac{x - 3}{x^2 + 1}\right)$ **41.** $\dfrac{3x}{x^2 + 1} + \dfrac{x}{(x^2 + 1)^2}$

43. Parabola **45.** Hyperbola **47.** Parabola

49. Hyperbola **51.** $y^2 = 4x$ **53.** $y^2 = -24x$

55. (0, 50) **57.** $\dfrac{x^2}{25} + \dfrac{y^2}{9} = 1$

59. $\dfrac{2x^2}{9} + \dfrac{y^2}{36} = 1$ **61.** $\dfrac{x^2}{61} + \dfrac{y^2}{25} = 1$

63. The foci should be placed 3 feet on either side of the center and have the same height as the pillars.

65. $\dfrac{x^2}{1} - \dfrac{y^2}{4} = 1$ **67.** $\dfrac{y^2}{1} - \dfrac{x^2}{8} = 1$

69. $(x - 3)^2 = -2y$

Parabola

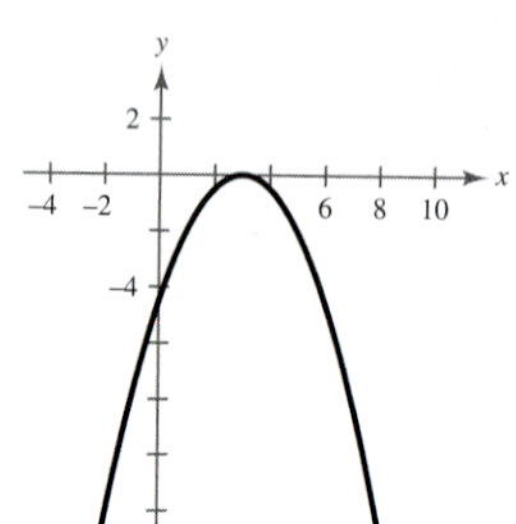

71. (1, 2)

Degenerate conic (a point)

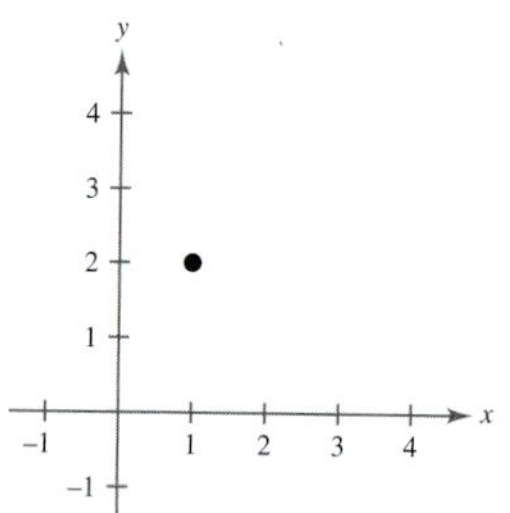

73. $\dfrac{(x + 5)^2}{9} + (y - 1)^2 = 1$ **75.** $(x - 4)^2 - \dfrac{(y - 4)^2}{9} = 1$

Ellipse Hyperbola

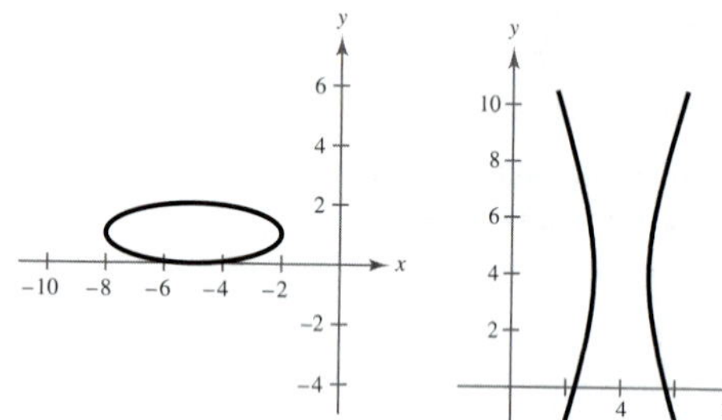

77. Center: (2, −6)

Radius: 5

79. $(x - 1)^2 + y^2 = 32$

Center: (1, 0)

Radius: $4\sqrt{2}$

81. $(x + 5)^2 + (y + 2)^2 = 9$

Center: (−5, −2)

Radius: 3

83. $(x + 6)^2 = -9(y - 4)$ **85.** $(x - 4)^2 = -8(y - 2)$

87. $\dfrac{(x - 5)^2}{25} + \dfrac{(y - 3)^2}{9} = 1$ **89.** $\dfrac{(x - 2)^2}{25} + \dfrac{y^2}{21} = 1$

91. $\dfrac{x^2}{36} - \dfrac{(y - 7)^2}{9} = 1$ **93.** $\dfrac{(x + 2)^2}{64} - \dfrac{(y - 3)^2}{36} = 1$

95. $\dfrac{5(x - 4)^2}{16} - \dfrac{5y^2}{64} = 1$ **97.** $8\sqrt{6}$ meters

99. 10.29 centimeters

101. False. The domain of $f(x) = \dfrac{1}{x^2 + 1}$ is the set of all real numbers.

Chapter Test (page 384)

1. Domain: all real numbers $x \neq 4$

Vertical asymptote: $x = 4$

Horizontal asymptote: $y = 0$

2. Domain: all real numbers

Vertical asymptote: none

Horizontal asymptote: $y = -1$

3. Domain: all real numbers $x \neq 2$

Vertical asymptote: $x = 2$

Slant asymptote: $y = x + 4$

4.

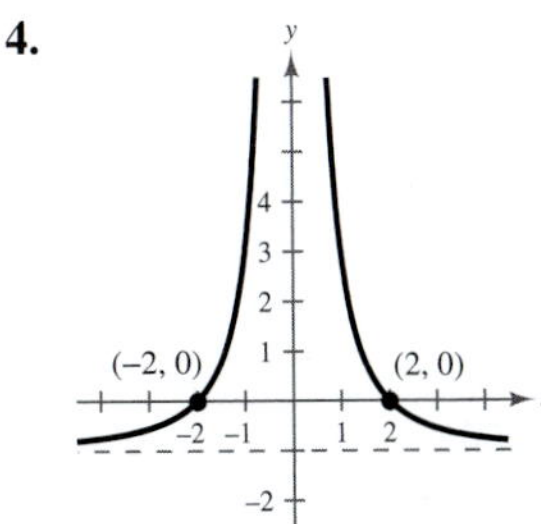

y-axis symmetry

x-intercepts: $(-2, 0), (2, 0)$

no y-intercept

Vertical asymptote: $x = 0$

Horizontal asymptote: $y = -1$

5.

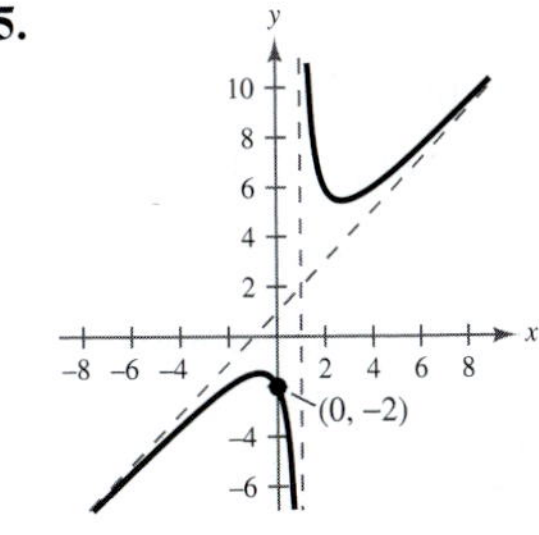

No axis or origin symmetry

no x-intercepts

y-intercept: $(0, -2)$

Vertical asymptote: $x = 1$

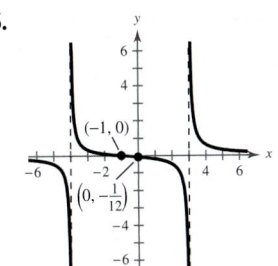

Slant asymptote: $y = x + 1$

6.

No axis or origin symmetry

x-intercept: $(-1, 0)$

y-intercept: $\left(0, -\frac{1}{12}\right)$

Vertical asymptotes: $x = 3, x = -4$

Horizontal asymptote: $y = 0$

7. 6.24 inches $\times$ 12.49 inches

8. (a) Answers will vary.

(b) $A = \dfrac{x^2}{2(x - 2)}, \; x > 2$

(c)

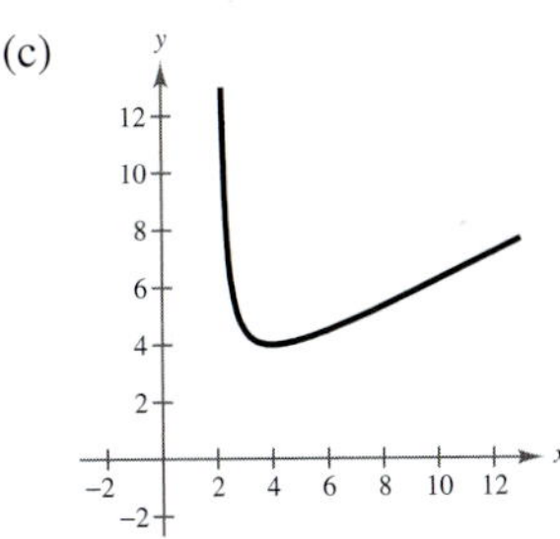

$A = 4$

9. $\dfrac{3}{x - 2} - \dfrac{1}{x + 1}$ **10.** $\dfrac{2}{x^2} - \dfrac{3}{x - 2}$

11. $-\dfrac{5}{x} + \dfrac{3}{x - 1} + \dfrac{3}{x + 1}$ **12.** $-\dfrac{2}{x} + \dfrac{3x}{x^2 + 2}$

13. Vertex: $(0, 0)$

Focus: $(2, 0)$

14. Center: $(5, -2)$

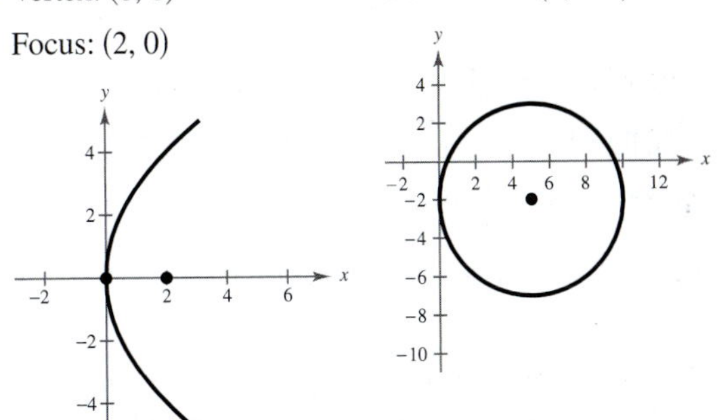

15. Vertex: $(5, -3)$

Focus: $\left(5, -\frac{5}{2}\right)$

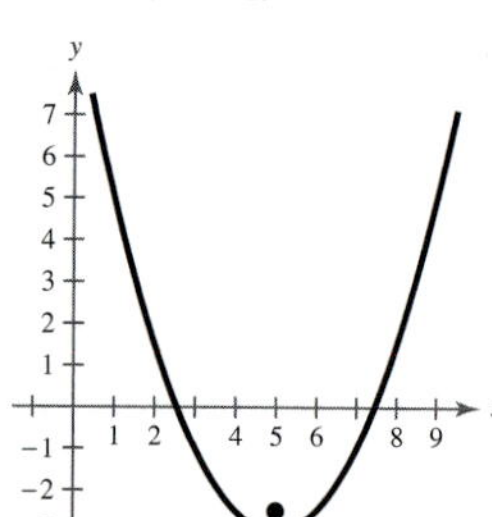

16. Vertices: $(\pm 1, 0)$

Foci: $\left(\pm\sqrt{5}, 0\right)$

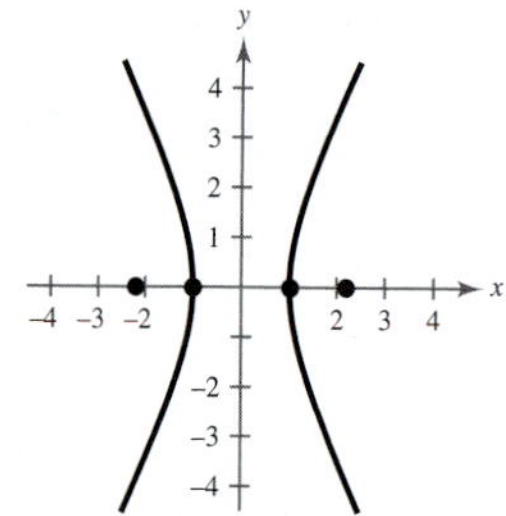

17. Vertices: $(0, 0)$, $(4, 0)$

Foci: $\left(2 \pm \sqrt{5}, 0\right)$

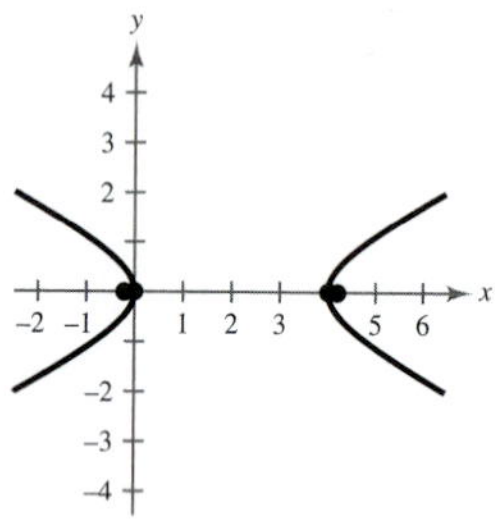

18. Center: $(1, -6)$

Vertices: $(4, -6)$, $(-2, -6)$

Foci: $\left(1 \pm \sqrt{6}, -6\right)$

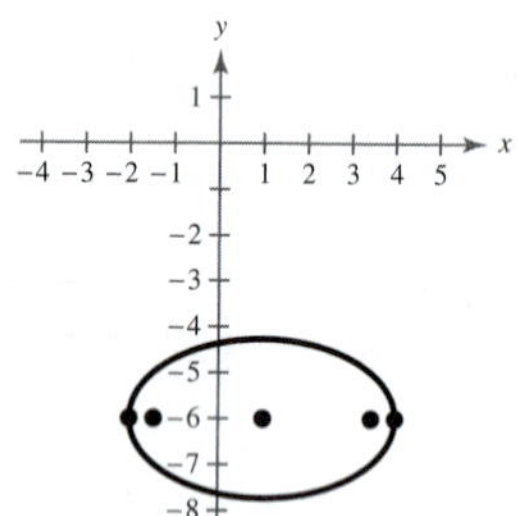

19. $\dfrac{(x-4)^2}{16} + \dfrac{(y-2)^2}{4} = 1$ **20.** $\dfrac{y^2}{9} - \dfrac{x^2}{4} = 1$

21. Smallest distance: $\approx$ 364,224 kilometers

Greatest distance: $\approx$ 404,582 kilometers

Problem Solving (page 386)

1. (i) d (ii) b (iii) a (iv) c

3. (a) As $|a|$ increases, the graph becomes wider. For $a < 0$, the graph is reflected in the x-axis.

(b) As $|b|$ increases, the graph becomes wider. For $b > 0$, the graph is translated to the right. For $b < 0$, the graph is reflected in the x-axis and is translated to the left.

5. (a) $\dfrac{x^2}{2352.25} + \dfrac{y^2}{529} = 1$ (b) $\approx$ 85.4 feet

(c) 1115.5π square feet

7. $\dfrac{(x-6)^2}{9} - \dfrac{(y-2)^2}{7} = 1$

9. (a)

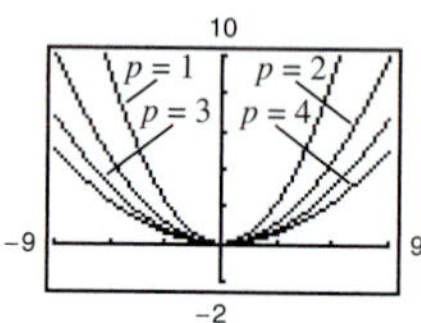

As p increases, the graph becomes wider.

(b) $p = 1$: focus $(0, 1)$

$p = 2$: focus $(0, 2)$

$p = 3$: focus $(0, 3)$

$p = 4$: focus $(0, 4)$

(c) $p = 1$: 4 units

$p = 2$: 8 units

$p = 3$: 12 units

$p = 4$: 16 units

Length of chord $= 4|p|$ units

(d) Answers will vary.

11. Answers will vary.

Chapter 5

Section 5.1 (page 398)

1. 946.852 **3.** 0.006 **5.** 0.472

7. d **8.** c **9.** a **10.** b

11. Shift the graph of four units to the right.

13. Shift the graph of five units upward.

15. Reflect f in the x-axis and shift four units to the left.

17. Reflect f in the x-axis and shift five units upward.

19.

x	-2	-1	0	1	2
$f(x)$	4	2	1	0.5	0.25

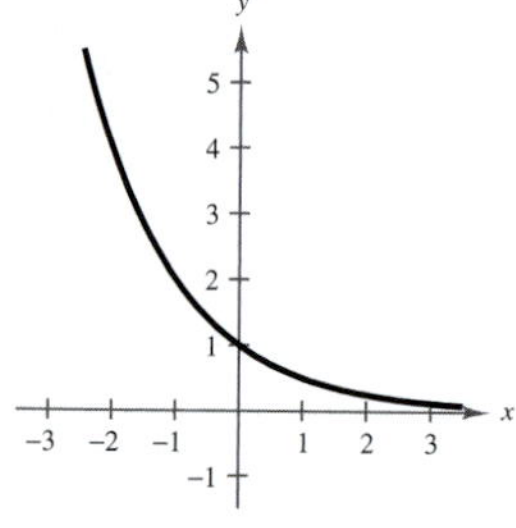

21.

x	-2	-1	0	1	2
$f(x)$	36	6	1	0.167	0.028

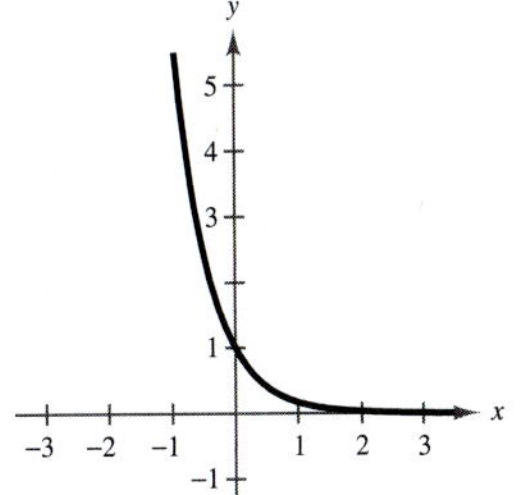

23.

x	-2	-1	0	1	2
$f(x)$	0.125	0.25	0.5	1	2

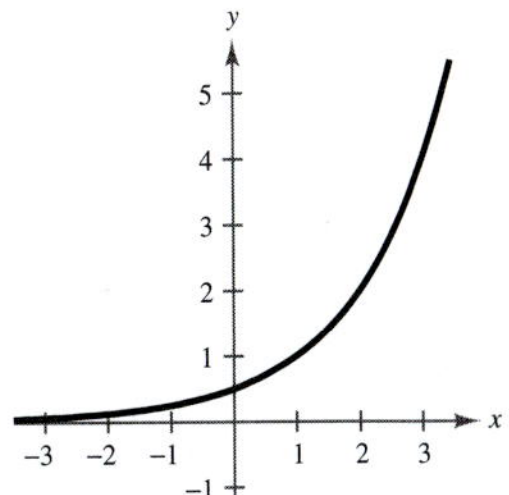

25.

x	-2	-1	0	1	2
$f(x)$	0.135	0.368	1	2.718	7.389

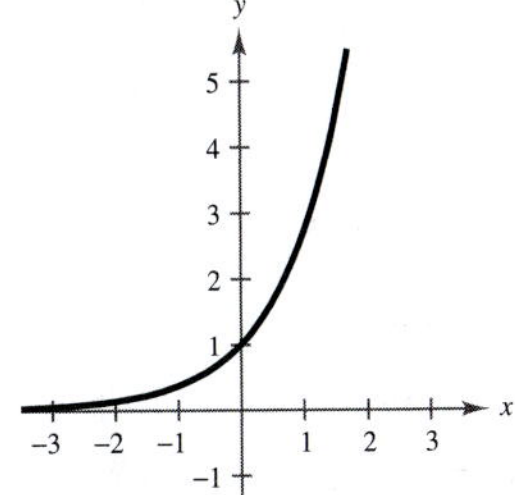

27.

x	-8	-7	-6	-5	-4
$f(x)$	0.055	0.149	0.406	1.104	3

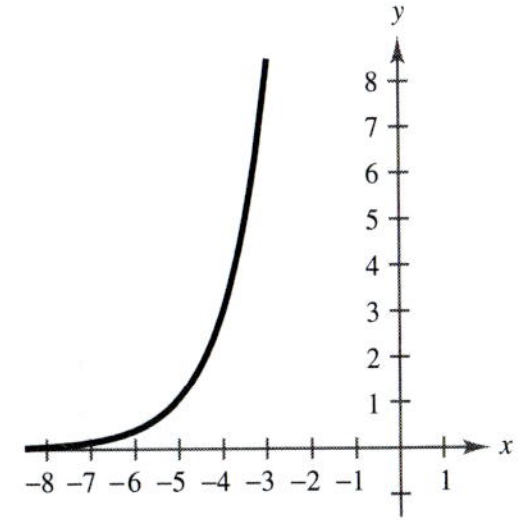

29.

x	-2	-1	0	1	2
$f(x)$	4.037	4.100	4.271	4.736	6

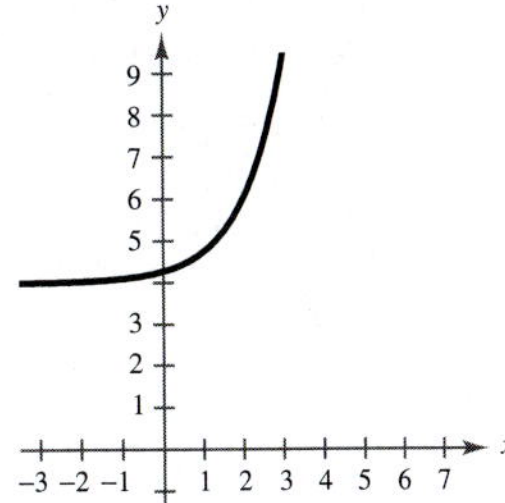

31.

x	-1	0	1	2	3
$f(x)$	3.004	3.016	3.063	3.25	4

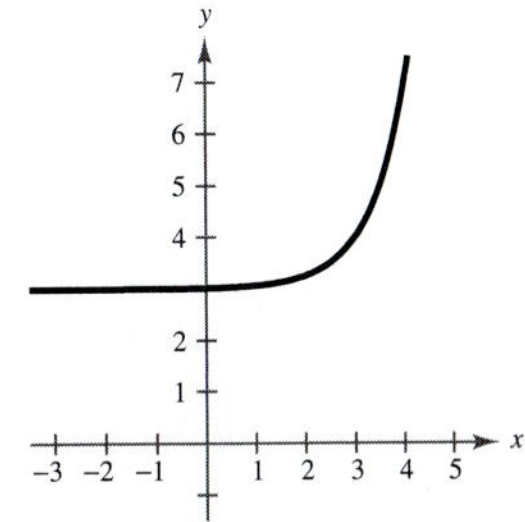

33.

35.

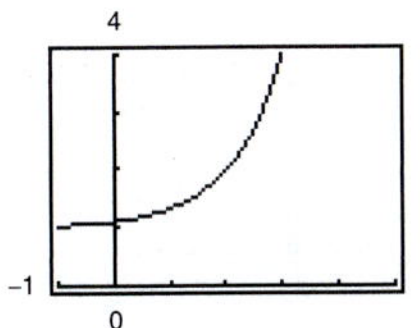

37.

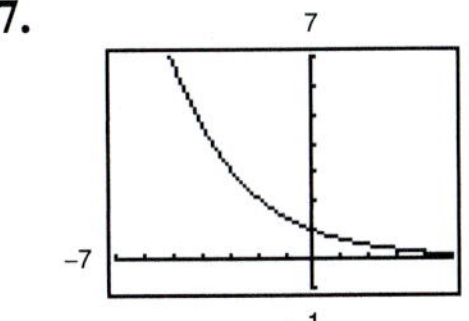

39.

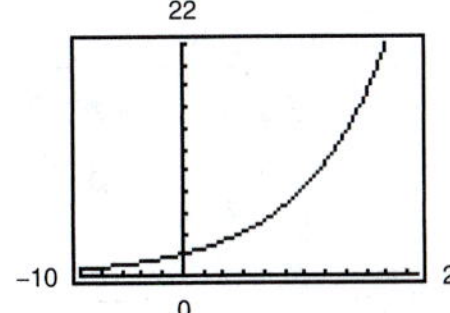

41.

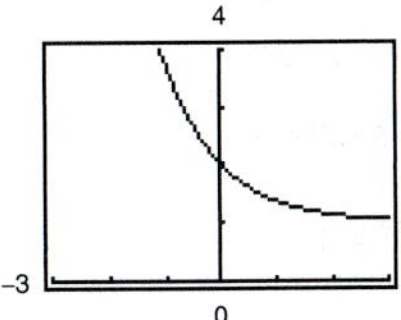

43.

n	1	2	4
A	\$5397.31	\$5477.81	\$5520.10

n	12	365	Continuous
A	\$5549.10	\$5563.36	\$5563.85

45.

n	1	2	4
A	\$11,652.39	\$12,002.55	\$12,188.60

n	12	365	Continuous
A	\$12,317.01	\$12,380.41	\$12,382.58

47.

t	10	20
A	\$26,706.49	\$59,436.39

t	30	40	50
A	\$132,278.12	\$294,390.36	\$655,177.80

49.

t	10	20
A	\$22,986.49	\$44,031.56

t	30	40	50
A	\$84,344.25	\$161,564.86	\$309,484.08

51. \$222,822.57

53. \$35.45 **55.** (a) 100 (b) 300 (c) 900

57. (a) 25 grams (b) 16.30 grams

(c)

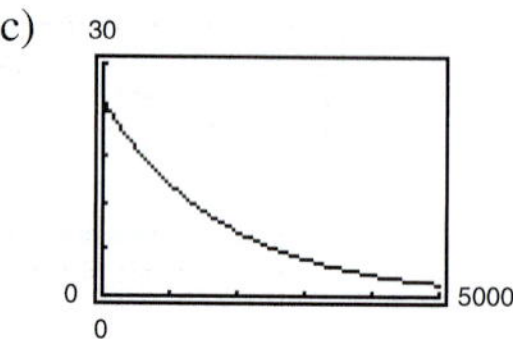

59. (a)

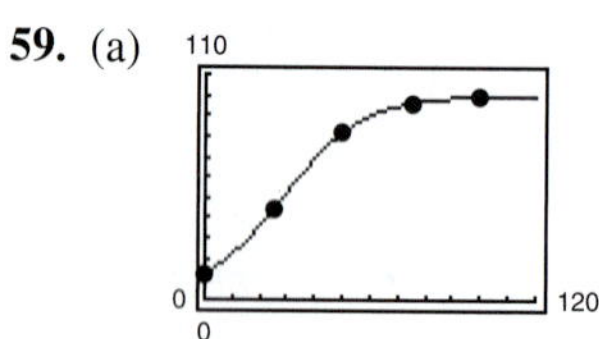

(b)

x	0	25	50	75	100
y	13	45	82	96	99

(c) 63.1% (d) 38.2

61. True. As $x \to -\infty$, $f(x) \to -2$ but never reaches -2.

63. $f(x) = h(x)$ **65.** $f(x) = g(x) = h(x)$

67. (a) $x < 0$ (b) $x > 0$

69. (a)

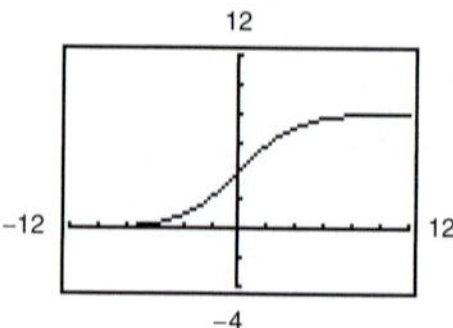

Horizontal asymptotes: $y = 0, y = 8$

(b)

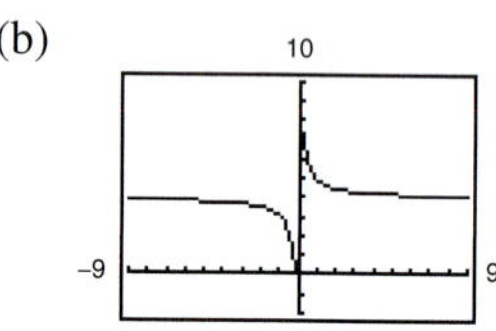

Horizontal asymptote: $y = 4$

Vertical asymptote: $x = 0$

71.

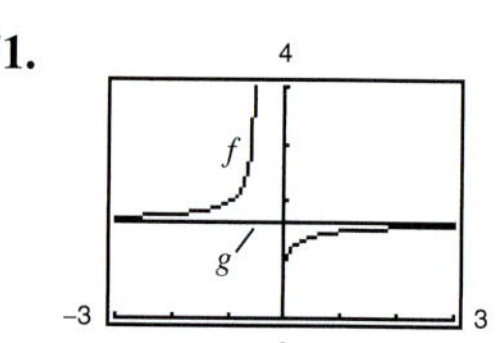

As $x \to \infty$, $f(x) \to g(x)$.

As $x \to -\infty$, $f(x) \to g(x)$.

73. c, d **75.** $y = \frac{1}{7}(2x + 14)$ **77.** $y = \pm\sqrt{25 - x^2}$

79.

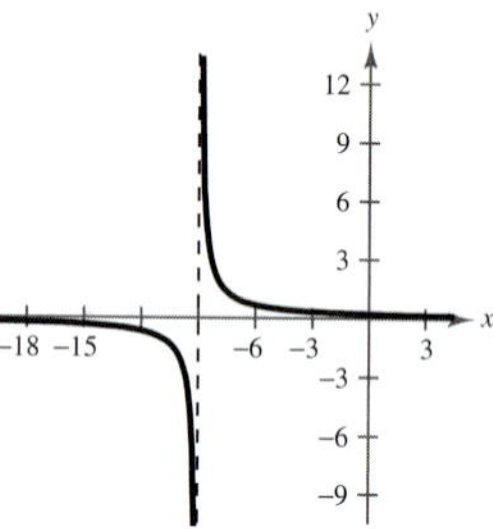

81.

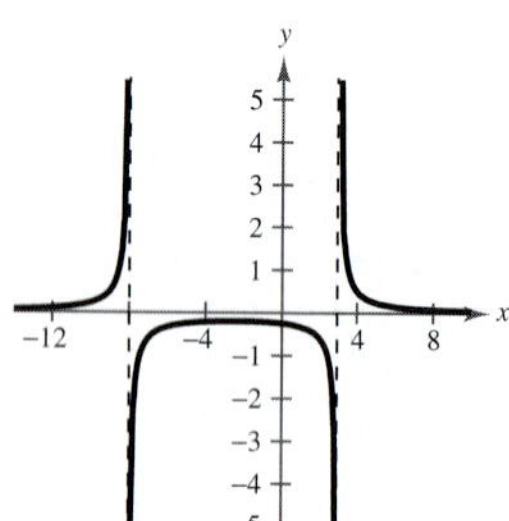

83. Ellipse

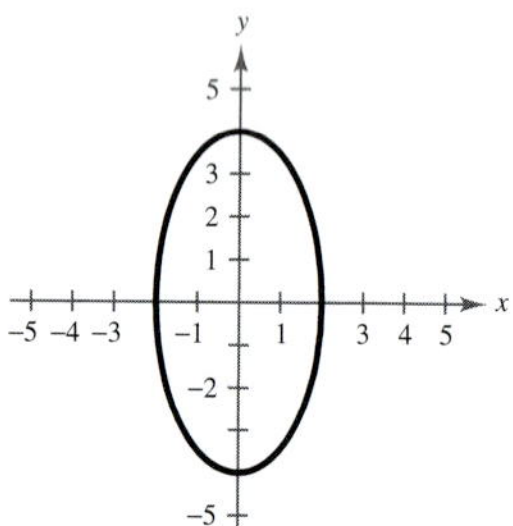

85. Hyperbola

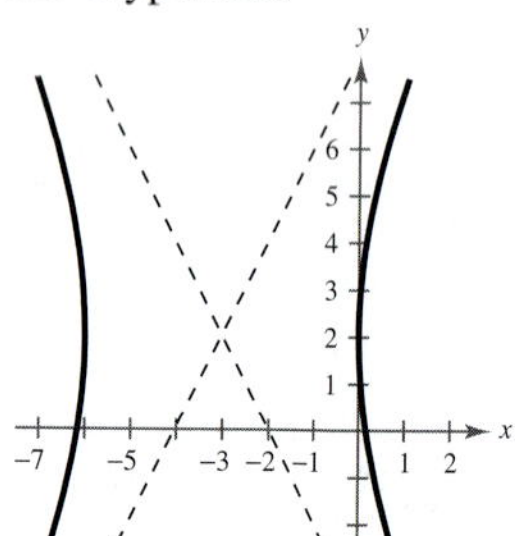

Section 5.2 *(page 408)*

1. $4^3 = 64$ **3.** $7^{-2} = \frac{1}{49}$ **5.** $32^{2/5} = 4$

7. $e^{-0.693\ldots} = \frac{1}{2}$ **9.** $\log_5 125 = 3$ **11.** $\log_{81} 3 = \frac{1}{4}$

13. $\log_6 \frac{1}{36} = -2$ **15.** $\ln 20.0855\ldots = 3$

17. $\ln 4 = x$ **19.** 4 **21.** 0 **23.** 3 **25.** 2

27. -0.097 **29.** 2.913 **31.** -0.575 **33.** c

34. f **35.** d **36.** e **37.** b **38.** a

39. Domain: $(0, \infty)$

x-intercept: $(1, 0)$

Vertical asymptote: $x = 0$

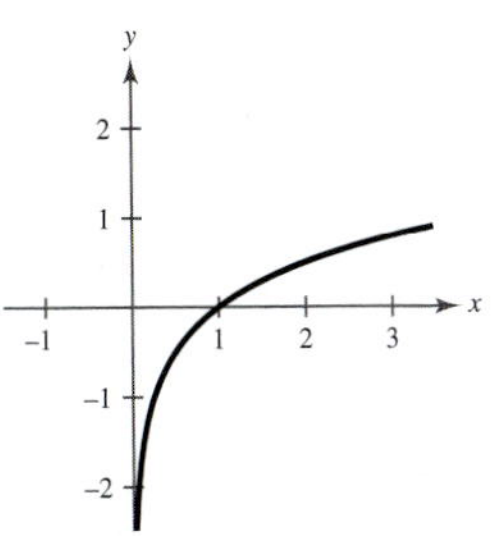

41. Domain: $(0, \infty)$

x-intercept: $(9, 0)$

Vertical asymptote: $x = 0$

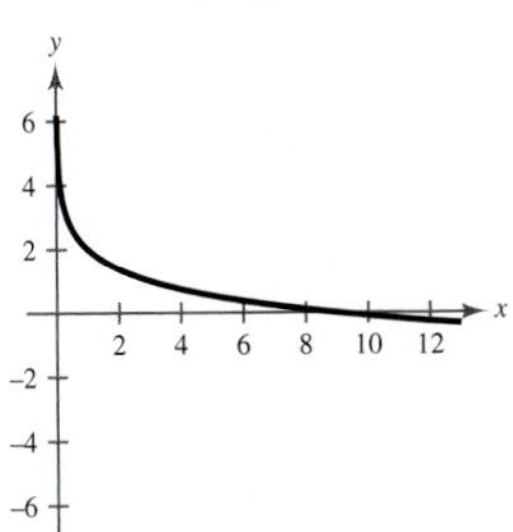

43. Domain: $(-2, \infty)$

x-intercept: $(-1, 0)$

Vertical asymptote: $x = -2$

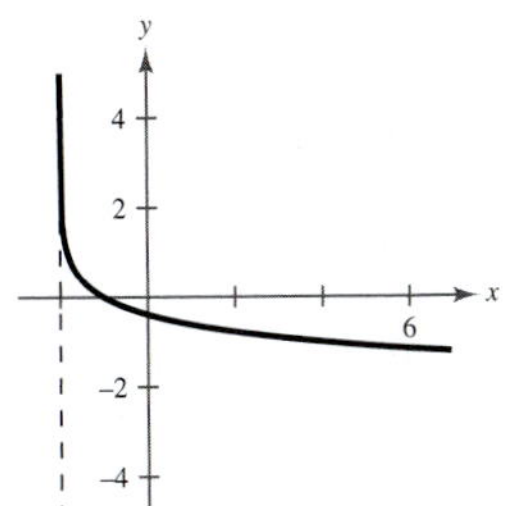

45. Domain: $(0, \infty)$

x-intercept: $(5, 0)$

Vertical asymptote: $x = 0$

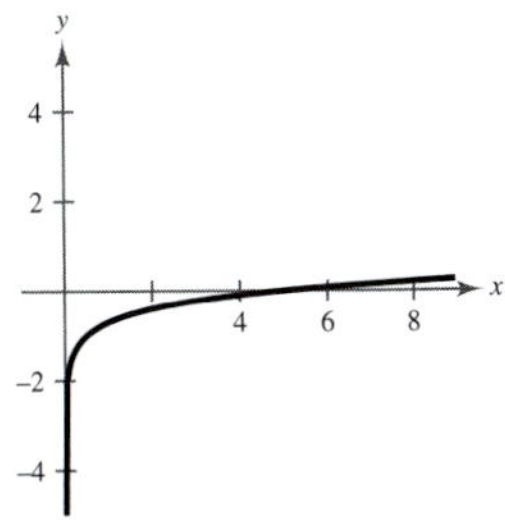

47. Domain: $(2, \infty)$

x-intercept: $(3, 0)$

Vertical asymptote: $x = 2$

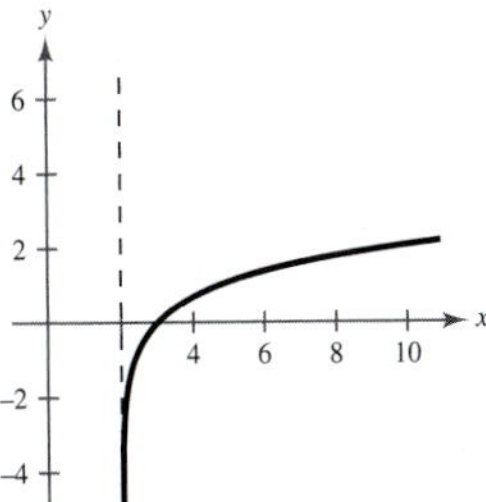

49. Domain: $(-\infty, 0)$

x-intercept: $(-1, 0)$

Vertical asymptote: $x = 0$

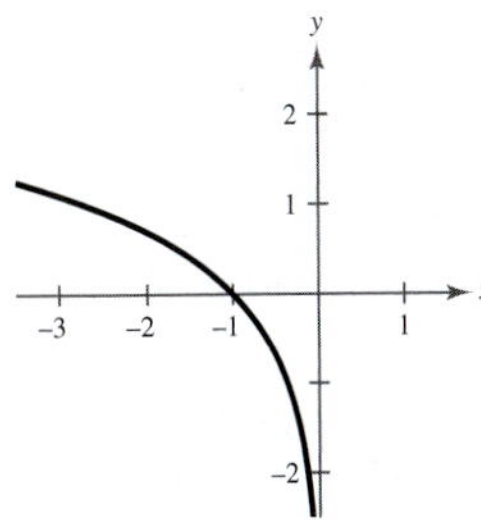

51.

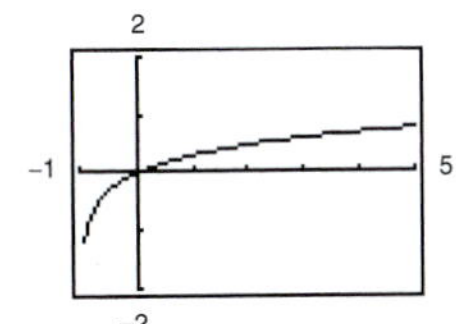

53.

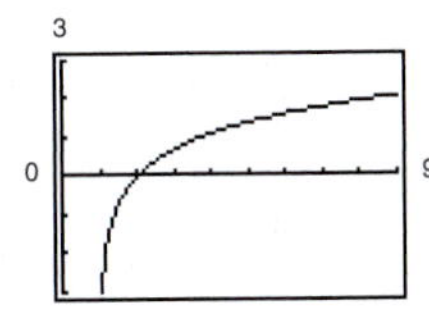

55.

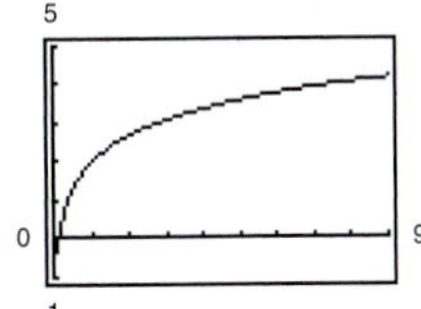

57. (a) 30 years; 20 years (b) \$396,234; \$301,123.20

(c) \$246,234; \$151,123.20

(d) $x = 1000$; The monthly payment must be greater than \$1000.

59. (a)

r	0.005	0.01	0.015	0.02	0.025	0.03
t	138.6	69.3	46.2	34.7	27.7	23.1

(b)

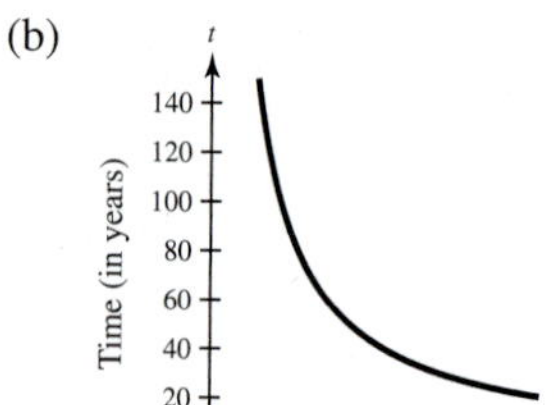

(c) Answers will vary.

61. (a)

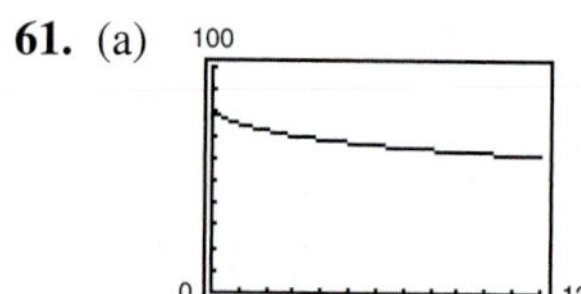

(b) 80 (c) 68.1 (d) 62.3

63. False. Reflecting $g(x)$ about the line $y = x$ will determine the graph of $f(x)$.

65.

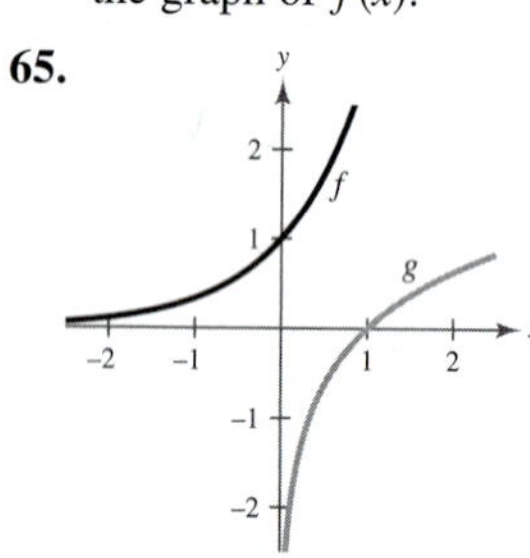

$g = f^{-1}$

67.

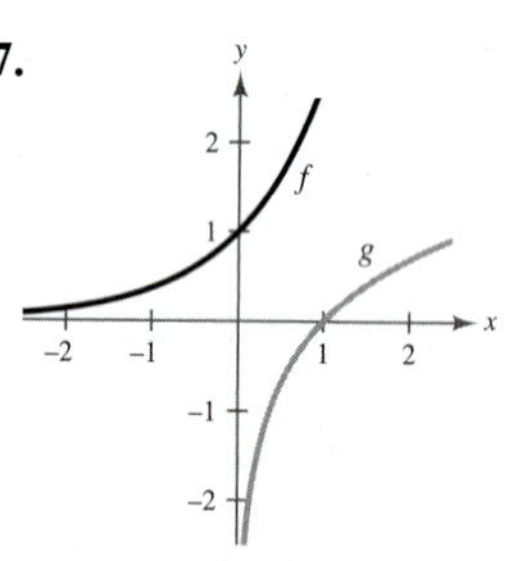

$g = f^{-1}$

69. (a)

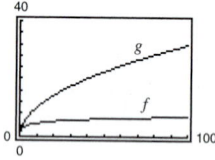

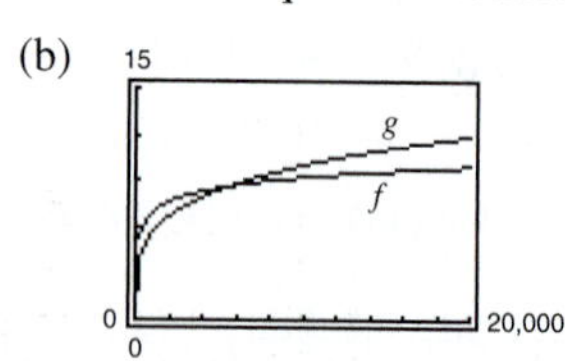

$g(x)$; The natural log function grows at a slower rate than the square root function.

(b) 15

g

f

0 20,000

0

$g(x)$; The natural log function grows at a slower rate than the fourth root function.

71. $(0, \infty)$ **73.** $3 < x < 4$

75. (a)

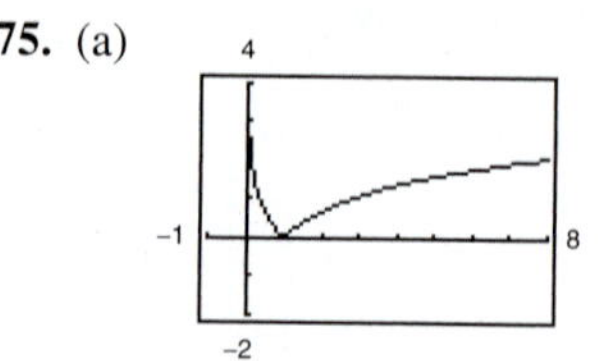

(b) Increasing: $(1, \infty)$

Decreasing: $(0, 1)$

(c) Relative minimum: $(1, 0)$

77. $83.95 + 37.50t$

Section 5.3 *(page 415)*

1. (a) $\dfrac{\log_{10} x}{\log_{10} 5}$ (b) $\dfrac{\ln x}{\ln 5}$ **3.** (a) $\dfrac{\log_{10} x}{\log_{10} \frac{1}{5}}$ (b) $\dfrac{\ln x}{\ln \frac{1}{5}}$

5. (a) $\dfrac{\log_{10} \frac{3}{10}}{\log_{10} x}$ (b) $\dfrac{\ln \frac{3}{10}}{\ln x}$ **7.** (a) $\dfrac{\log_{10} x}{\log_{10} 2.6}$ (b) $\dfrac{\ln x}{\ln 2.6}$

9. 1.771 **11.** -2.000 **13.** -0.417 **15.** 2.633

17. $\log_4 5 + \log_4 x$ **19.** $4 \log_8 x$ **21.** $1 - \log_5 x$

23. $\frac{1}{2} \ln z$ **25.** $\ln x + \ln y + 2 \ln z$

27. $\ln z + 2 \ln(z - 1)$ **29.** $\frac{1}{2} \log_2(a - 1) - \log_2 9$

31. $\frac{1}{3} \ln x - \frac{1}{3} \ln y$ **33.** $4 \ln x + \frac{1}{2} \ln y - 5 \ln z$

35. $2 \log_5 x - 2 \log_5 y - 3 \log_5 z$

37. $\frac{3}{4} \ln x + \frac{1}{4} \ln(x^2 + 3)$ **39.** $\ln 3x$ **41.** $\log_4 \dfrac{z}{y}$

43. $\log_2(x + 4)^2$ **45.** $\log_3 \sqrt[4]{5x}$

47. $\ln \dfrac{x}{(x + 1)^3}$ **49.** $\log_{10} \dfrac{xz^3}{y^2}$ **51.** $\ln \dfrac{x}{(x^2 - 4)^4}$

53. $\ln \sqrt[3]{\dfrac{x(x + 3)^2}{x^2 - 1}}$ **55.** $\log_8 \dfrac{\sqrt[3]{y(y + 4)^2}}{y - 1}$

57. $\log_2 \frac{32}{4} = \log_2 32 - \log_2 4$; Property 2 **59.** 2

61. $\frac{3}{4}$ **63.** 2.4 **65.** -9 is not in the domain of $\log_3 x$.

67. 4.5 **69.** $-\frac{1}{2}$ **71.** 7 **73.** 2

75. $\frac{3}{2}$ **77.** $-3 - \log_5 2$ **79.** $6 + \ln 5$

81. (a) 90 (b) 77 (c) 73 (d) 9 months

(e) $90 - \log_{10}(t + 1)^{15}$

(f)

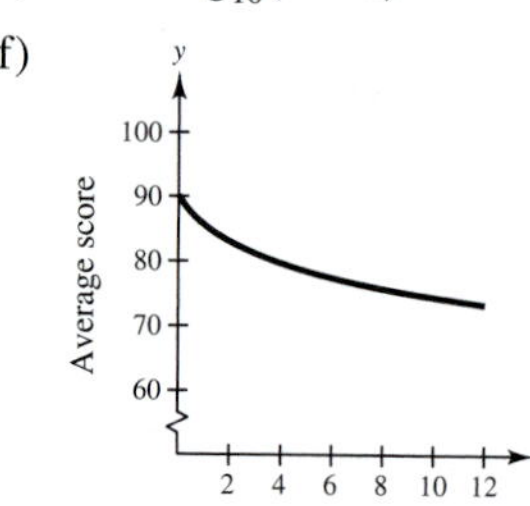

83. False. $\ln 1 = 0$ **85.** False. $\ln(x-2) \neq \ln x - \ln 2$

87. False. $u = v^2$ **89.** Answers will vary.

91. $f(x) = \dfrac{\log_{10} x}{\log_{10} 2} = \dfrac{\ln x}{\ln 2}$ **93.** $f(x) = \dfrac{\log_{10} x}{\log_{10} \frac{1}{2}} = \dfrac{\ln x}{\ln \frac{1}{2}}$

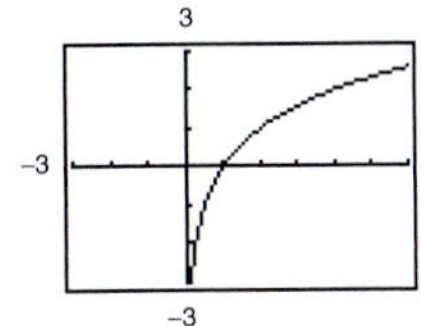

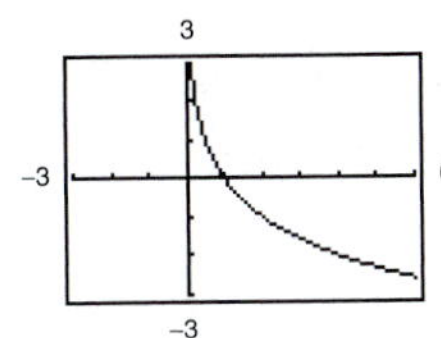

95. $f(x) = \dfrac{\log_{10} x}{\log_{10} 11.8} = \dfrac{\ln x}{\ln 11.8}$

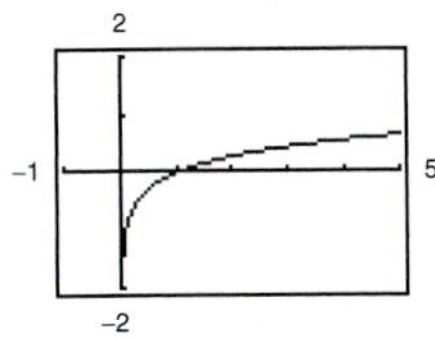

97. $f(x) = h(x)$; Property 2

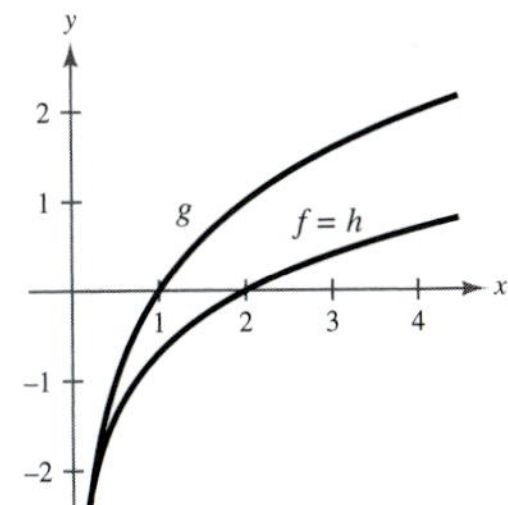

99. $\dfrac{3x^4}{2y^3}, \ x \neq 0$ **101.** $1, x \neq 0, y \neq 0$

103. $-1, \dfrac{1}{3}$ **105.** $\dfrac{-1 \pm \sqrt{97}}{6}$

Section 5.4 *(page 424)*

1. (a) Yes (b) No

3. (a) No (b) Yes (c) Yes, approximate

5. (a) Yes, approximate (b) No (c) Yes **7.** 2

9. 4 **11.** -2 **13.** -5 **15.** 3 **17.** 2

19. $\ln 2 \approx 0.693$ **21.** $e^{-1} \approx 0.368$ **23.** 64

25. $\frac{1}{10}$ **27.** (3, 8) **29.** (9, 2)

31. $\dfrac{\ln 5}{\ln 3} \approx 1.465$ **33.** $\ln 5 \approx 1.609$

35. $\ln 28 \approx 3.332$ **37.** $\dfrac{\ln 80}{2 \ln 3} \approx 1.994$ **39.** 2

41. 4 **43.** $3 - \dfrac{\ln 565}{\ln 2} \approx -6.142$

45. $\dfrac{1}{3} \log_{10}\left(\dfrac{3}{2}\right) \approx 0.059$ **47.** $1 + \dfrac{\ln 7}{\ln 5} \approx 2.209$

49. $\dfrac{\ln 12}{3} \approx 0.828$ **51.** $-\ln \dfrac{3}{5} \approx 0.511$ **53.** 0

55. $\dfrac{\ln \frac{8}{3}}{3 \ln 2} + \dfrac{1}{3} \approx 0.805$ **57.** $\ln 5 \approx 1.609$

59. $\ln 4 \approx 1.386$ **61.** $2 \ln 75 \approx 8.635$

63. $\frac{1}{2} \ln 1498 \approx 3.656$ **65.** $\dfrac{\ln 4}{365 \ln\left(1 + \frac{0.065}{365}\right)} \approx 21.330$

67. $\dfrac{\ln 2}{12 \ln\left(1 + \frac{0.10}{12}\right)} \approx 6.960$

69.

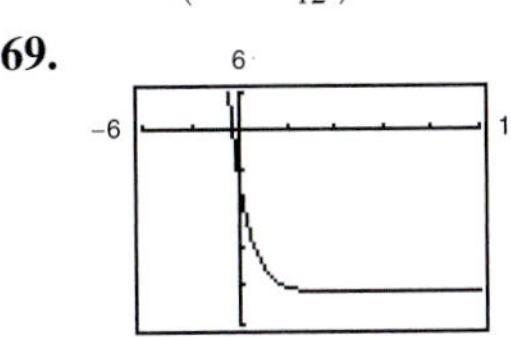

-0.427

71.

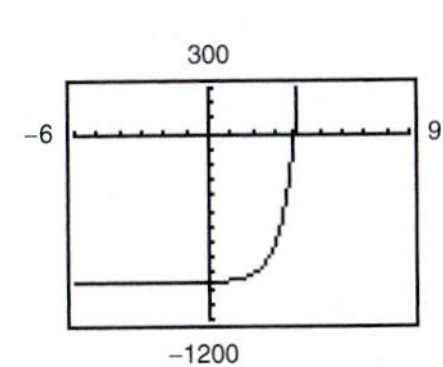

3.847

73.

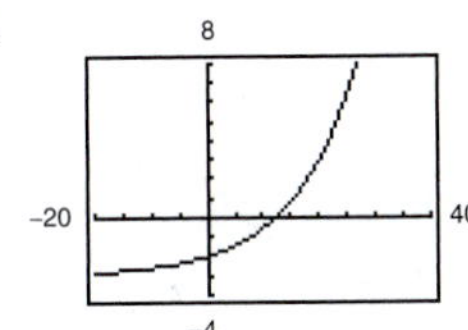

12.207

75.

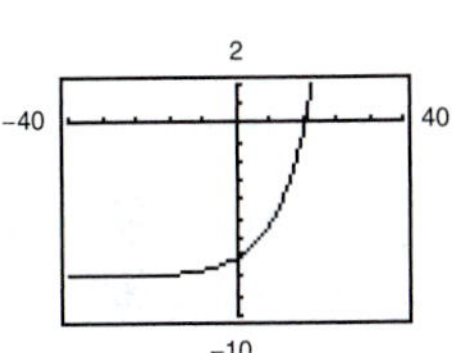

16.636

77. $e^{-3} \approx 0.050$ **79.** $\dfrac{e^{2.4}}{2} \approx 5.512$ **81.** 1,000,000

83. $2(3^{11/6}) \approx 14.988$ **85.** $\dfrac{e^{10/3}}{5} \approx 5.606$

87. $e^2 - 2 \approx 5.389$ **89.** $e^{-2/3} \approx 0.513$

91. No solution **93.** $1 + \sqrt{1 + e} \approx 2.928$

95. No solution **97.** 7 **99.** $\dfrac{-1 + \sqrt{17}}{2} \approx 1.562$

101. 2 **103.** $\dfrac{725 + 125\sqrt{33}}{8} \approx 180.384$

105.

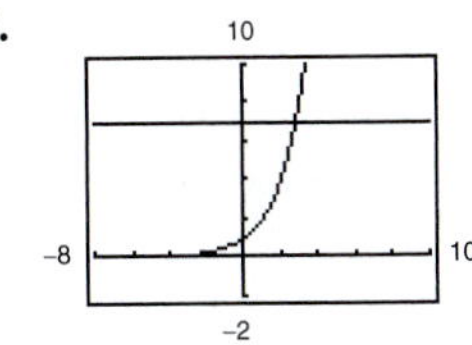

2.807

107.

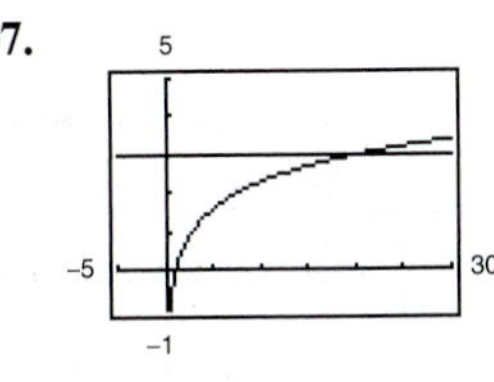

20.086

109. 8.2 years **111.** 12.9 years

113. (a) 1426 units (b) 1498 units

115. (a)

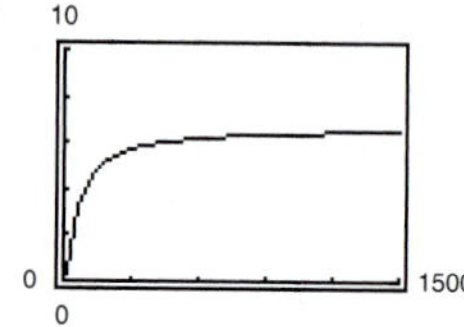

(b) $V = 6.7$; The yield will approach 6.7 million cubic feet per acre.

(c) 29.3 years

117. (a) $y = 100$ and $y = 0$; The range falls between 0% and 100%.

(b) Males: 69.71 inches Females: 64.51 inches

119. (a)

x	0.2	0.4	0.6	0.8	1.0
y	162.6	78.5	52.5	40.5	33.9

(b)

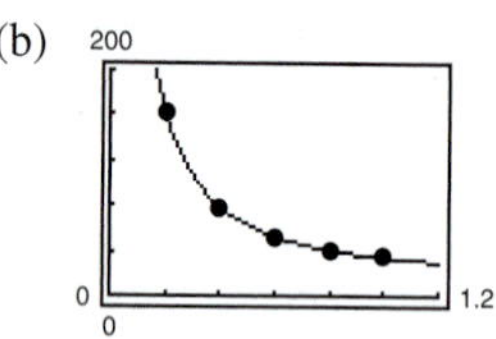

The model appears to fit the data well.

(c) 1.2 meters

(d) No. According to the model, when the number of g's is less than 23, x is between 2.276 meters and 4.404 meters, which isn't realistic in most vehicles.

121. $\log_b uv = \log_b u + \log_b v$

True by Property 1 in Section 5.3.

123. $\log_b(u - v) = \log_b u - \log_b v$

False.

$1.95 \approx \log_{10}(100 - 10) \neq \log_{10} 100 - \log_{10} 10 = 1$

125. Yes. See Exercise 95.

127. Yes. Time to double: $t = \dfrac{\ln 2}{r}$;

Time to quadruple: $t = \dfrac{\ln 4}{r} = 2\left(\dfrac{\ln 2}{r}\right)$

129. $4|x|y^2\sqrt{3y}$ **131.** $5\sqrt[3]{3}$ **133.** $M = kp^3$

135. $d = kab$ **137.** 1.226 **139.** -5.595

Section 5.5 *(page 435)*

1. c **2.** e **3.** b

4. a **5.** d **6.** f

	Initial Investment	*Annual % Rate*	*Time to Double*	*Amount After 10 years*
7.	\$1000	12%	5.78 yr	\$3320.12
9.	\$750	8.9438%	7.75 yr	\$1834.37
11.	\$500	11.0%	6.3 yr	\$1505.00
13.	\$6376.28	4.5%	15.4 yr	\$10,000.00

15. \$112,087.09

17. (a) 6.642 years (b) 6.330 years

(c) 6.302 years (d) 6.301 years

19.

r	2%	4%	6%	8%	10%	12%
t	54.93	27.47	18.31	13.73	10.99	9.16

21.

r	2%	4%	6%	8%	10%	12%
t	55.48	28.01	18.85	14.27	11.53	9.69

23.

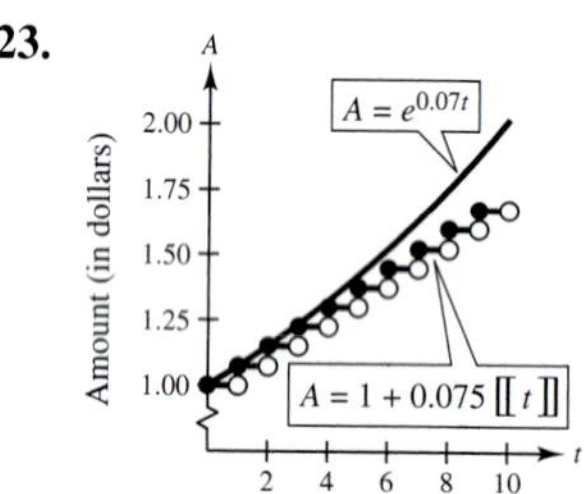

Continuous compounding

	Isotope	*Half-life (years)*	*Initial Quantity*	*Amount After 1000 Years*
25.	^{226}Ra	1620	10 g	6.52 g
27.	^{14}C	5730	2.26 g	2 g
29.	^{239}Pu	24,360	2.16 g	2.1 g

31. $y = e^{0.7675x}$ **33.** $y = 5e^{-0.4024x}$ **35.** 2003

37. $k = 0.0274$; 720,738 **39.** 3.15 hours **41.** 95.8%

43. (a) $V = -4500t + 22{,}000$ (b) $V = 22{,}000e^{-0.263t}$

(c) 25,000 0 5 0 (b) (a)

Exponential

(d) 1 year: Straight-line, \$17,500;

Exponential, \$16,912

3 years: Straight-line, \$8500;

Exponential, \$9995

(e) The value decreases \$4500 per year.

45. (a) $S(t) = 100(1 - e^{-0.1625t})$

(b)

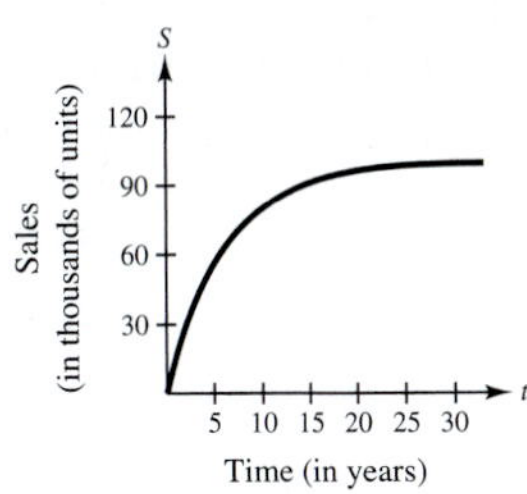

(c) 55,625

47. (a) $S = 10(1 - e^{-0.0575x})$ (b) 3314 units

49. (a) $N = 30(1 - e^{-0.050t})$ (b) 36 days

51. (a) 7.91 (b) 7.68 (c) 5.40

53. (a) 20 decibels (b) 70 decibels

(c) 95 decibels (d) 120 decibels

55. 95% **57.** 4.64 **59.** 1.58×10^{-6} moles per liter

61. 10^7 **63.** 3:00 A.M.

65. (a)

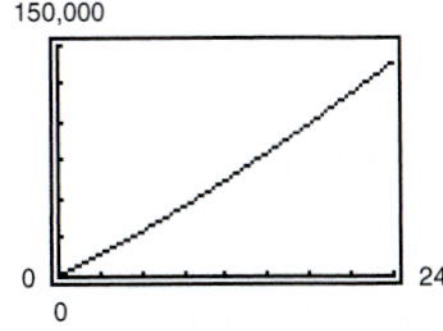

(b) ≈ 21 years; Yes

67. False. The domain can be the set of real numbers for a logistic growth function.

69. False. The graph of $f(x)$ is the graph of $g(x)$ shifted upward five units.

71. (a) Logarithmic (b) Logistic (c) Exponential (d) Linear (e) None of the above (f) Exponential

73. Rises to the right.

Falls to the left.

75. Rises to the left.

Falls to the right.

77. $4x^2 - 12x + 9$ **79.** $2x^2 + 3 + \dfrac{3}{x - 4}$

81.

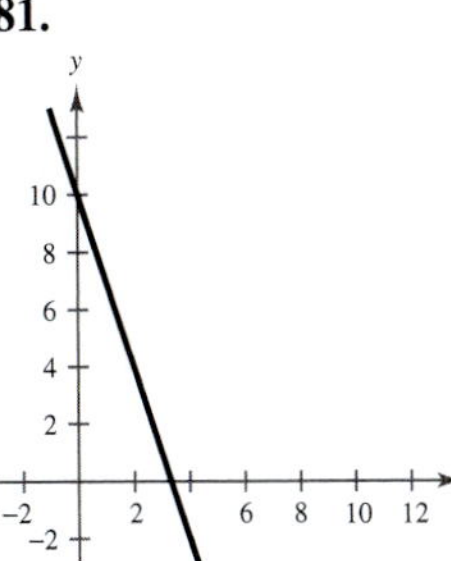

83.

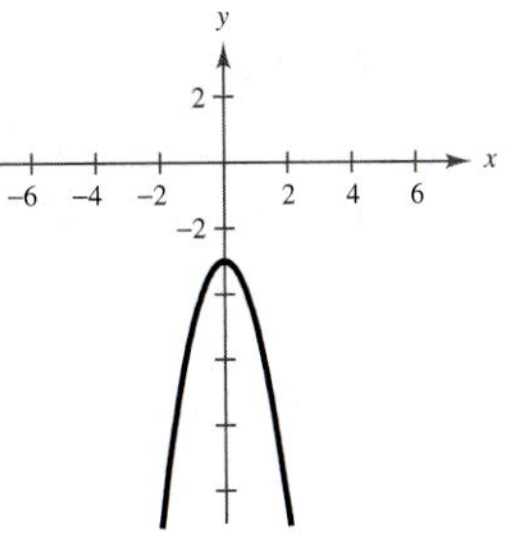

85.

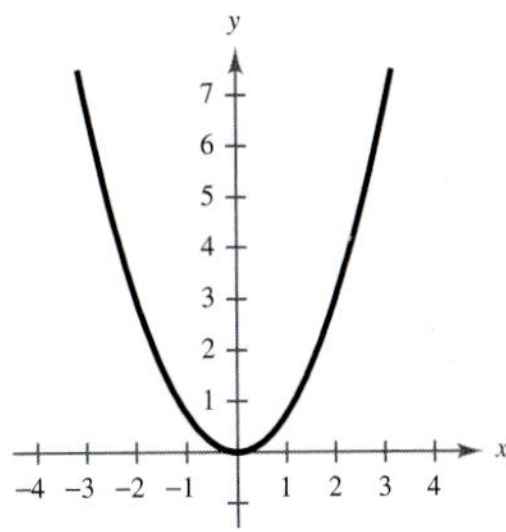

87.

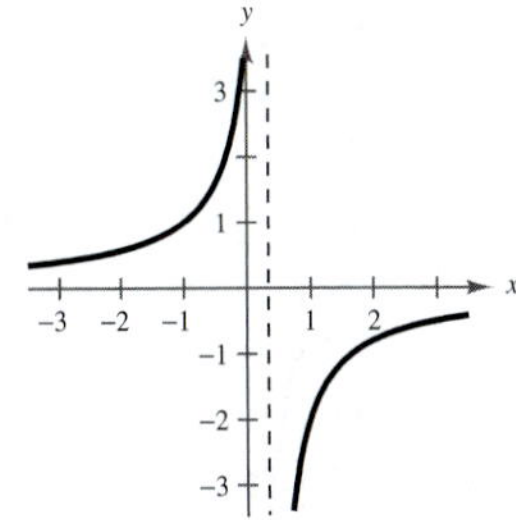

89.

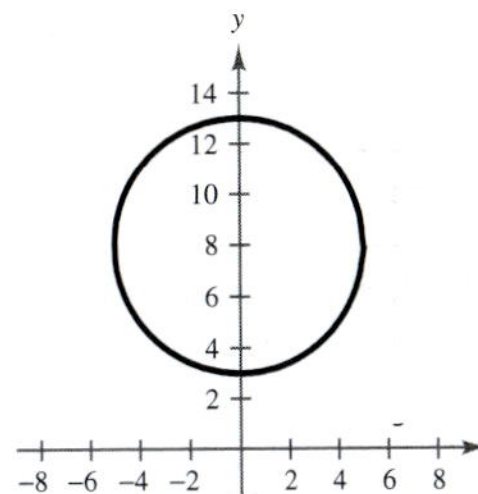

91.

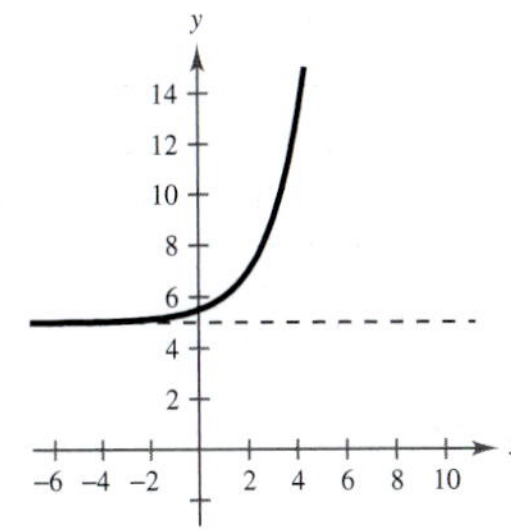

93.

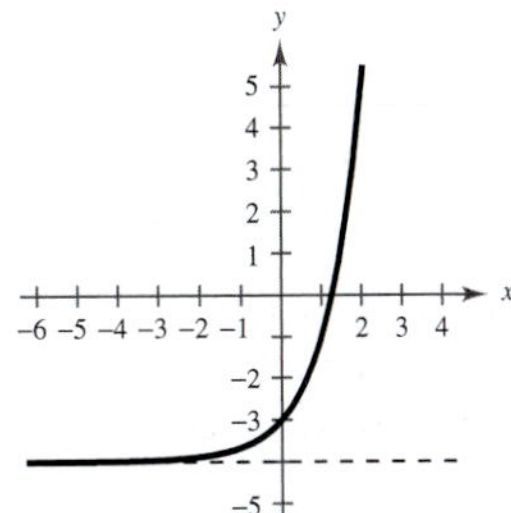

Review Exercises *(page 442)*

1. 76.699 **3.** 0.337 **5.** 1201.845 **7.** c

8. d **9.** a **10.** b

11. Shift the graph of f one unit to the right.

13. Reflect f in the x-axis and shift two units to the left.

15.

x	-1	0	1	2	3
$f(x)$	8	5	4.25	4.063	4.016

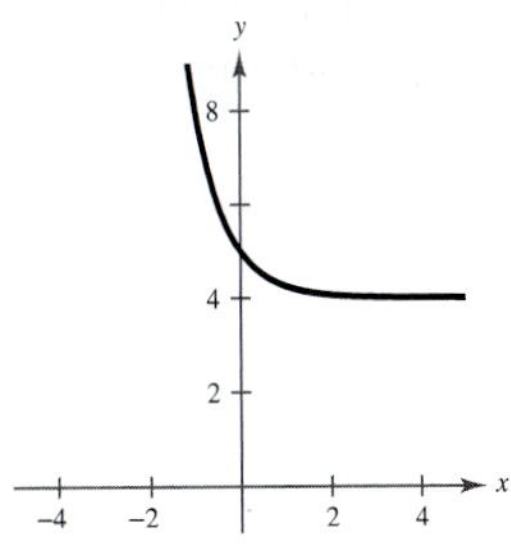

17.

x	-2	-1	0	1	2
$f(x)$	-0.377	-1	-2.65	-7.023	-18.61

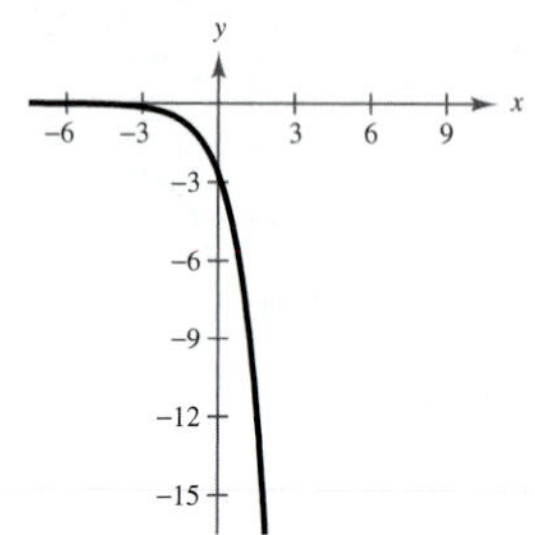

19.

x	-1	0	1	2	3
$f(x)$	4.008	4.04	4.2	5	9

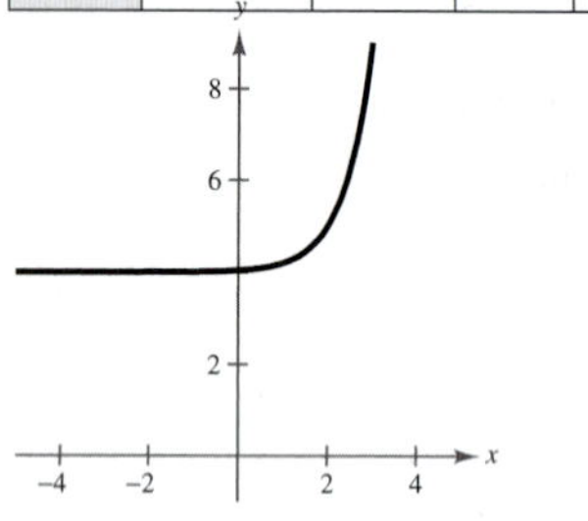

21.

x	-2	-1	0	1	2
$f(x)$	3.25	3.5	4	5	7

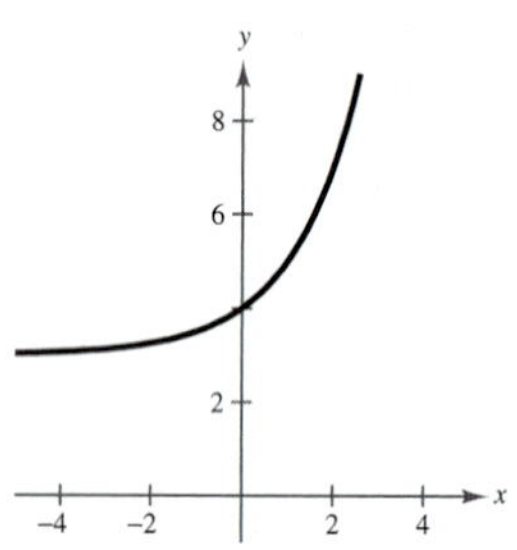

23. 2980.958 **25.** 0.183

27.

x	-2	-1	0	1	2
$h(x)$	2.72	1.65	1	0.61	0.37

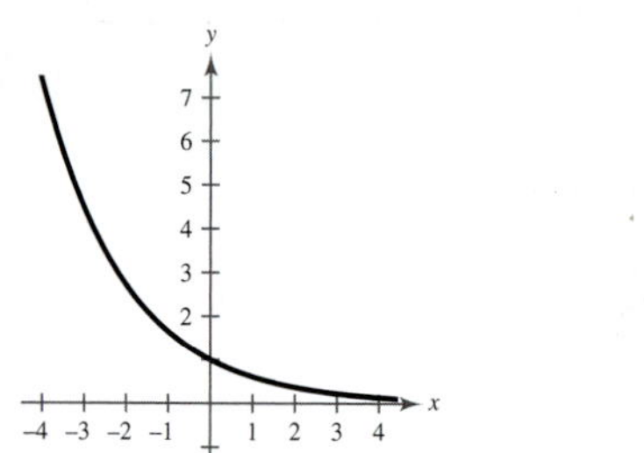

29.

x	-3	-2	-1	0	1
$f(x)$	0.37	1	2.72	7.39	20.09

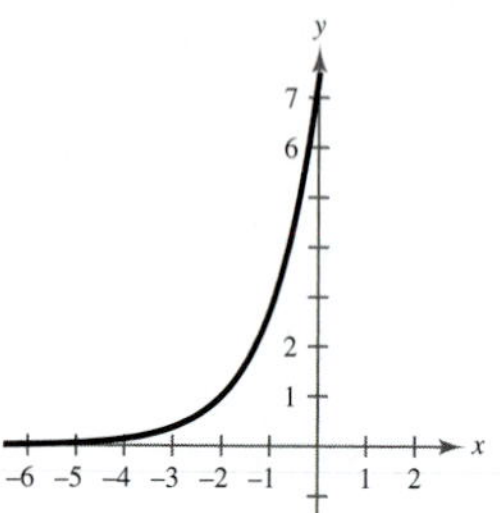

31.

n	1	2	4	12
A	\$6569.98	\$6635.43	\$6669.46	\$6692.64

n	365	Continuous
A	\$6704.00	\$6704.39

33. (a) 0.154 (b) 0.487 (c) 0.811

35. (a) \$1,069,047.14 (b) 7.9 years

37. $\log_4 64 = 3$ **39.** 3 **41.** -3

43. Domain: $(0, \infty)$
x-intercept: $(1, 0)$
Vertical asymptote: $x = 0$

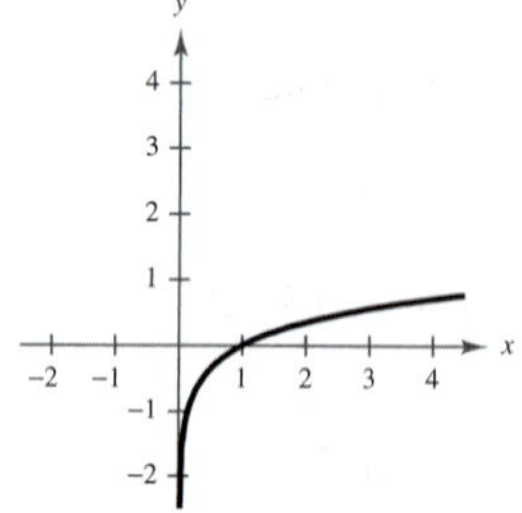

45. Domain: $(0, \infty)$
x-intercept: $(3, 0)$
Vertical asymptote: $x = 0$

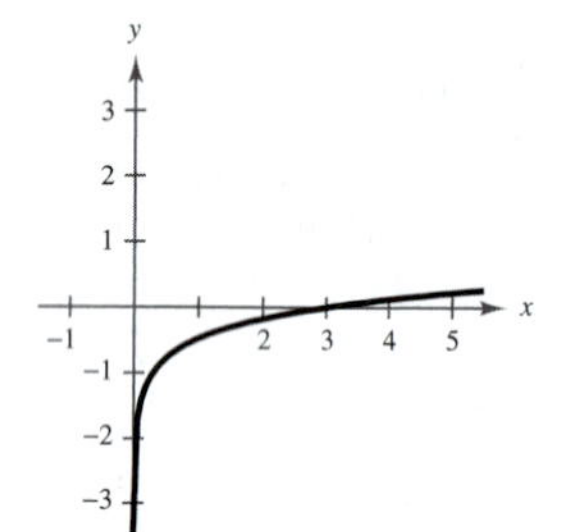

47. Domain: $(-5, \infty)$
x-intercept: $(9995, 0)$
Vertical asymptote: $x = -5$

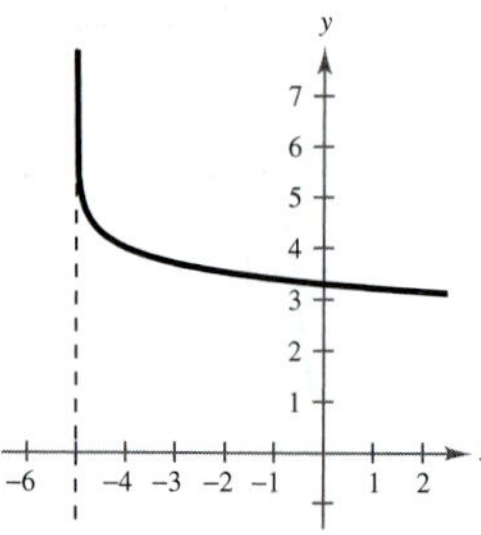

49. 3.118 **51.** -12 **53.** 2.034

55. Domain: $(0, \infty)$
x-intercept: $(e^{-3}, 0)$
Vertical asymptote: $x = 0$

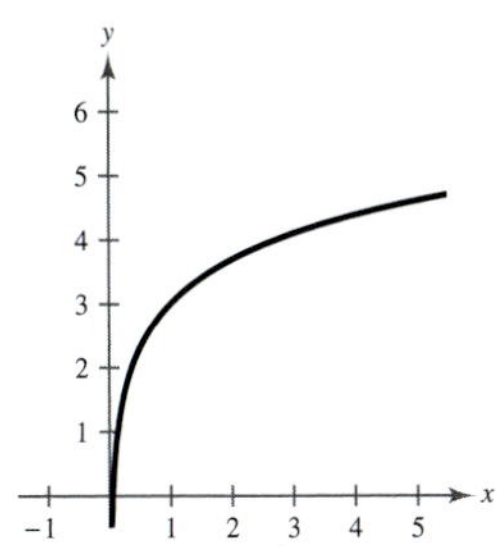

57. Domain: $(-\infty, 0), (0, \infty)$
x-intercept: $(\pm 1, 0)$
Vertical asymptote: $x = 0$

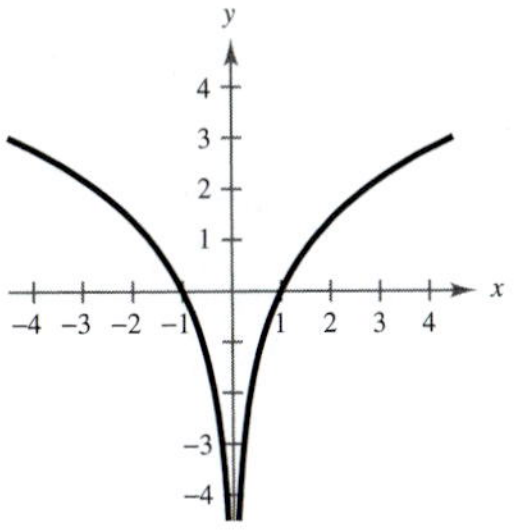

59. 53.4 inches **61.** 1.585 **63.** -2.322

65. $1 + 2 \log_5 |x|$ **67.** $1 + \log_3 2 - \frac{1}{3} \log_3 x$

69. $2 \ln x + 2 \ln y + \ln z$ **71.** $\ln(x + 3) - \ln x - \ln y$

73. $\log_2 5x$ **75.** $\ln \dfrac{x}{\sqrt[4]{y}}$ **77.** $\log_8 y^7 \sqrt[3]{x + 4}$

79. $\ln \dfrac{\sqrt{|2x - 1|}}{(x + 1)^2}$

81. (a) $0 \le h < 18{,}000$
(b)

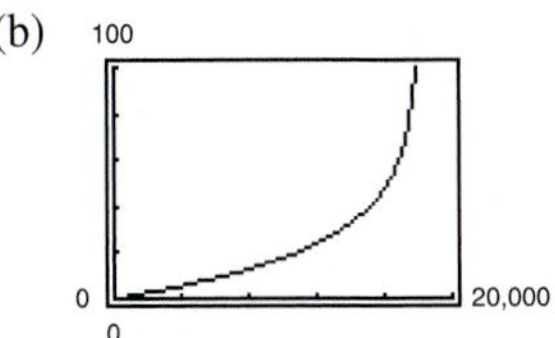

Vertical asymptote: $h = 18{,}000$
(c) The plane is climbing at a slower rate, so the time required increases.
(d) 5.46 minutes

83. 3 **85.** -3 **87.** $\ln 3 \approx 1.099$ **89.** 16

91. $e^4 \approx 54.598$

93. $\ln 12 \approx 2.485$ **95.** $-\dfrac{\ln 44}{5} \approx -0.757$

97. $\dfrac{\ln 22}{\ln 2} \approx 4.459$ **99.** $\dfrac{\ln 17}{\ln 5} \approx 1.760$

101. $\ln 2 \approx 0.693$, $\ln 5 \approx 1.609$

103.

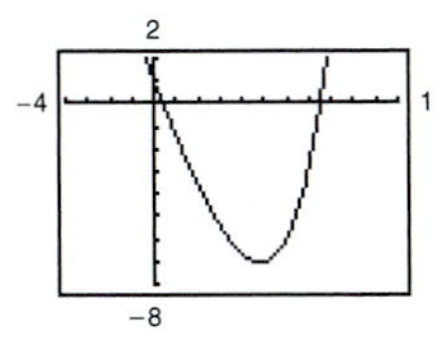

0.39, 7.48

105.

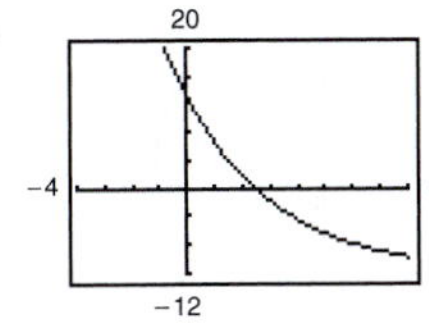

2.45

107. $\frac{1}{3}e^{8.2} \approx 1213.650$ **109.** $\frac{1}{4}e^{7.5} \approx 452.011$

111. $3e^2 \approx 22.167$ **113.** $e^4 - 1 \approx 53.598$

115. No solution **117.** 0.900

119.

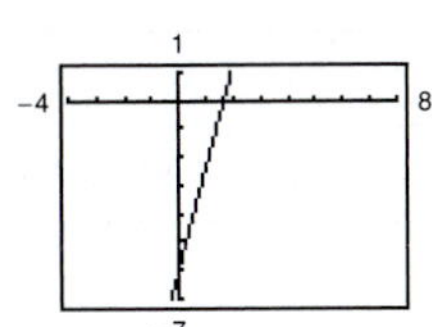

1.64

121.

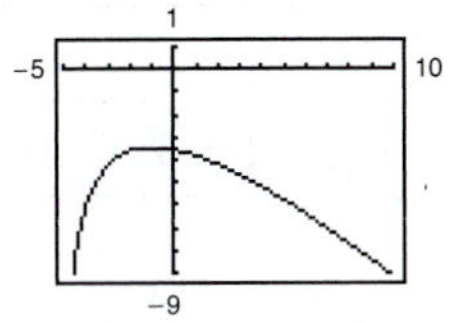

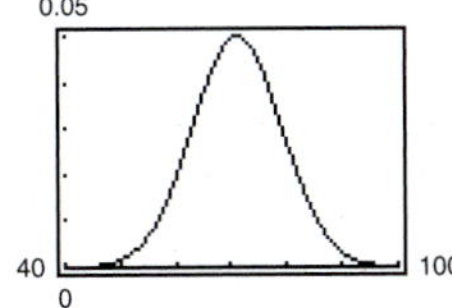

No solution

123. 15.2 years **125.** e **126.** b **127.** f

128. d **129.** a **130.** c **131.** 2004

133. (a) 13.8629% (b) \$11,486.98 **135.** $y = 2e^{0.1014x}$

137. (a)

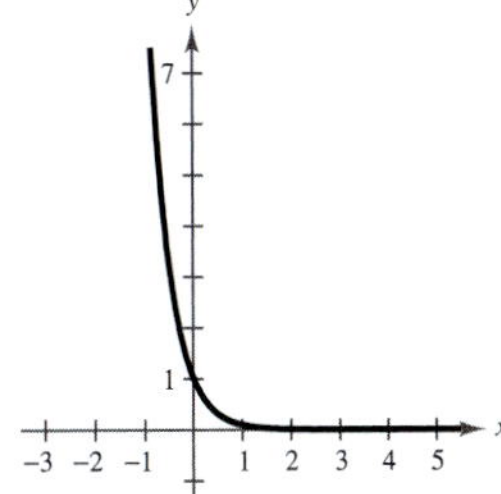

(b) 71

139. $10^{-3.5}$ watt per square centimeter

141. True by the inverse properties

143. $b < d < a < c$
b and d are negative.
a and c are positive.

Chapter Test *(page 446)*

1. 1123.690 **2.** 687.291 **3.** 0.497 **4.** 22.198

5.

x	-1	$-\frac{1}{2}$	0	$\frac{1}{2}$	1
$f(x)$	10	3.162	1	0.316	0.1

6.

x	-1	0	1	2	3
$f(x)$	-0.005	-0.028	-0.167	-1	-6

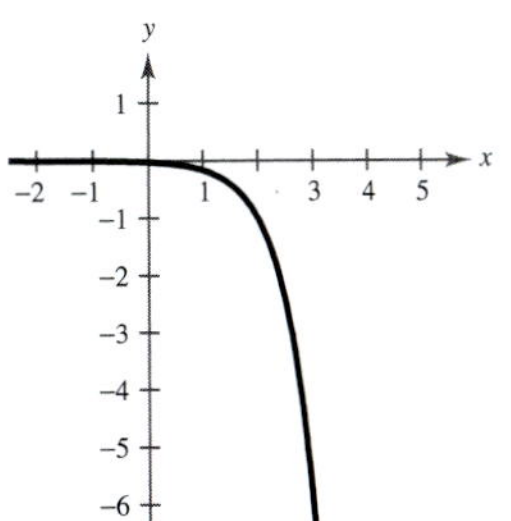

7.

x	-1	$-\frac{1}{2}$	0	$\frac{1}{2}$	1
$f(x)$	0.865	0.632	0	-1.718	-6.389

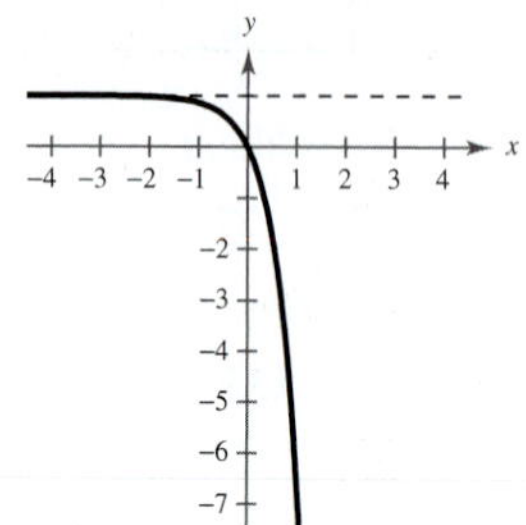

8. (a) -0.89 (b) 9.2

9.

x	$\frac{1}{2}$	1	$\frac{3}{2}$	2	4
$f(x)$	-5.699	-6	-6.176	-6.301	-6.602

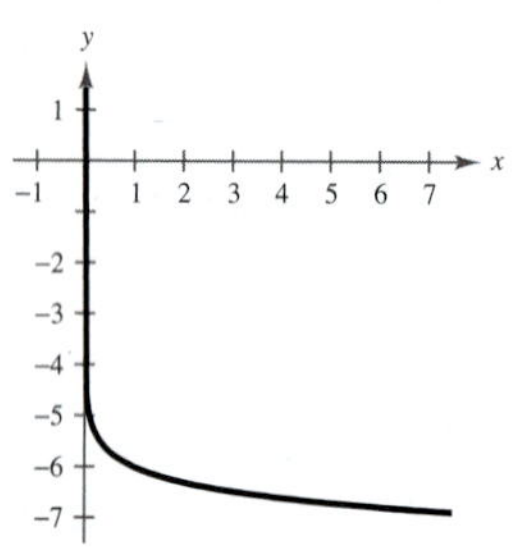

Vertical asymptote: $x = 0$

10.

x	5	7	9	11	13
$f(x)$	0	1.099	1.609	1.946	2.197

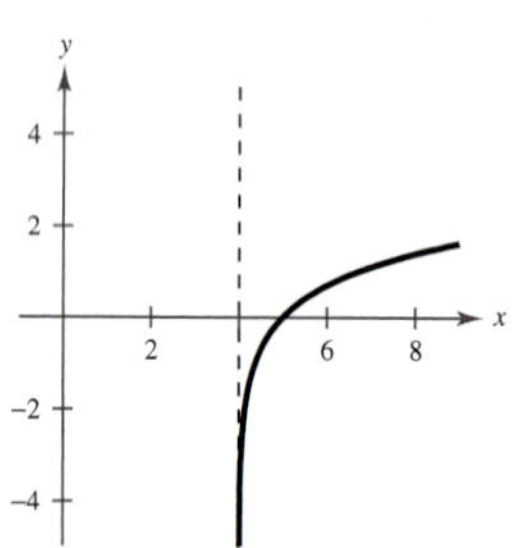

Vertical asymptote: $x = 4$

11.

x	-5	-3	-1	0	1
$f(x)$	1	2.099	2.609	2.792	2.946

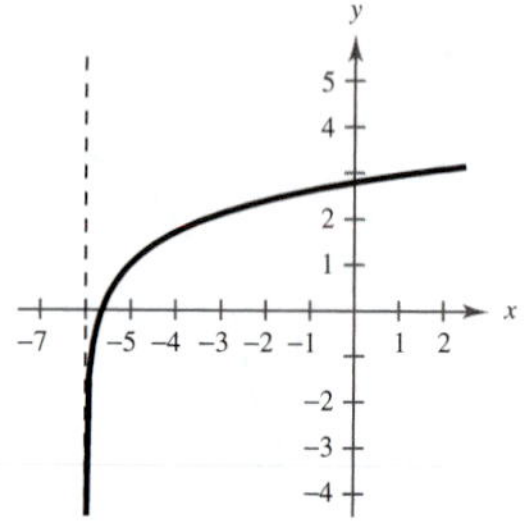

Vertical asymptote: $x = -6$

12. 1.945 **13.** 0.115 **14.** 1.328

15. $\log_2 3 + 4 \log_2 |a|$ **16.** $\ln 5 + \frac{1}{2}\ln x - \ln 6$

17. $\log_3 13y$ **18.** $\ln \dfrac{x^4}{y^4}$ **19.** $\dfrac{\ln 197}{4} \approx 1.321$

20. $\frac{800}{501} \approx 1.597$ **21.** $y = 2745e^{0.1570x}$ **22.** 55%

23. (a)

x	$\frac{1}{4}$	1	2	4	5	6
H	58.720	75.332	86.828	103.43	110.59	117.38

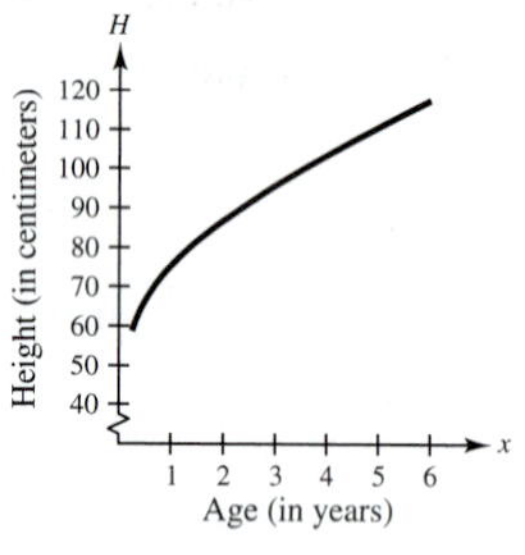

(b) 103 centimeters; 103.43 centimeters

Cumulative Test for Chapters 3–5 *(page 447)*

1. $y = -\frac{3}{4}(x + 8)^2 + 5$

2.

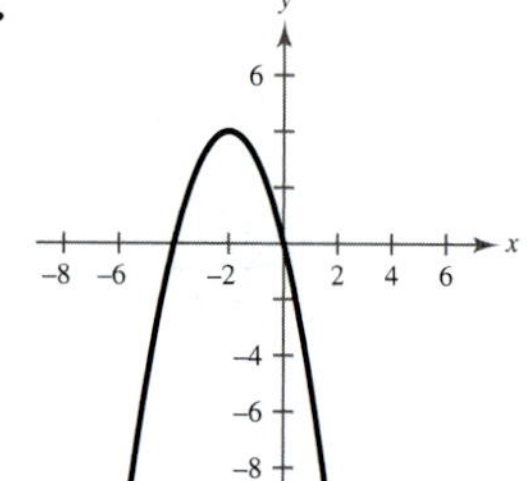

3.

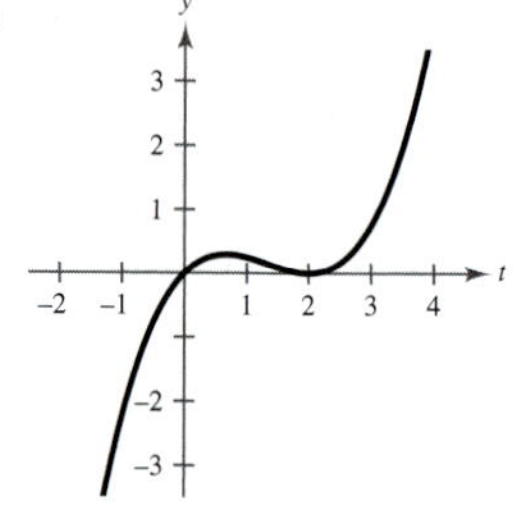

4.

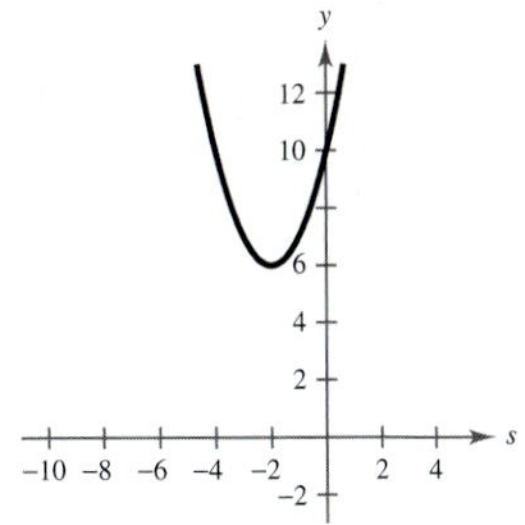

5. $-2, \pm 2i$ **6.** $-7, 0, 3$ **7.** $3x - 2 - \dfrac{3x - 2}{2x^2 + 1}$

8. $2x^3 - x^2 + 2x - 10 + \dfrac{25}{x + 2}$ **9.** 1.20

10. $x^4 + 3x^3 - 11x^2 + 9x + 70$

11.

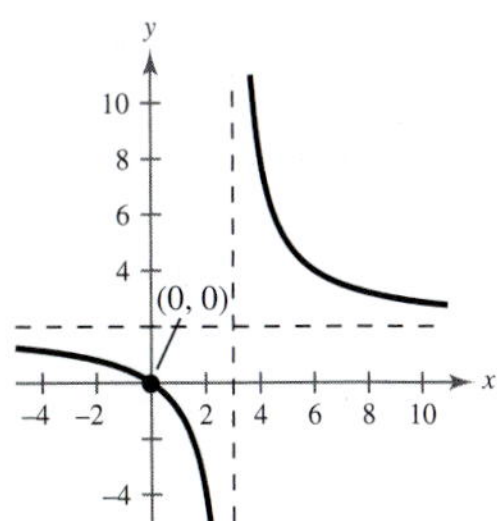

Vertical asymptote: $x = 3$

Horizontal asymptote: $y = 2$

12.

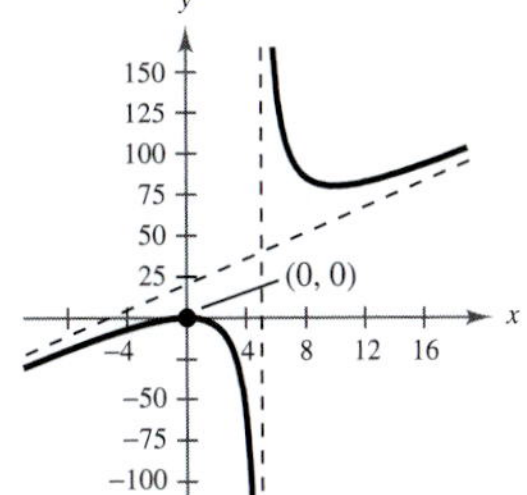

Vertical asymptote: $x = 5$

Slant asymptote:
$y = 4x + 20$

13.

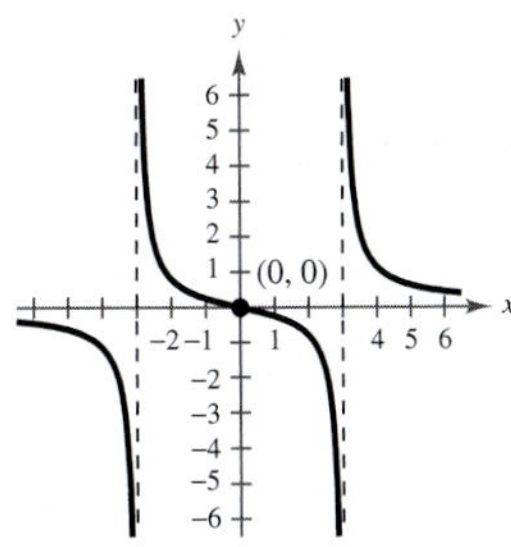

Vertical asymptotes: $x = \pm 3$

Horizontal asymptote:
$y = 0$

14. $\dfrac{1}{5}\left(\dfrac{4}{x - 7} - \dfrac{4}{x + 3}\right)$ **15.** $\dfrac{5}{x - 4} + \dfrac{20}{(x - 4)^2}$

16.

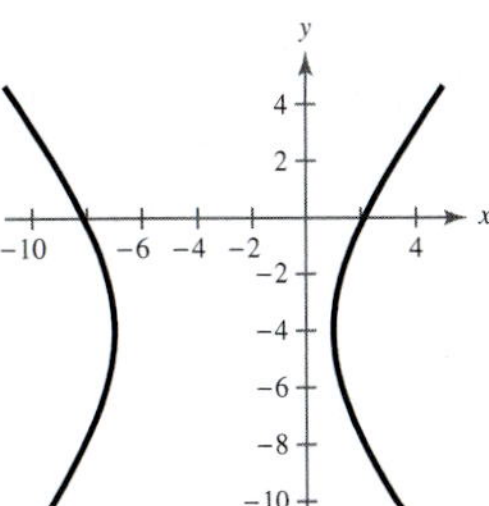

17.

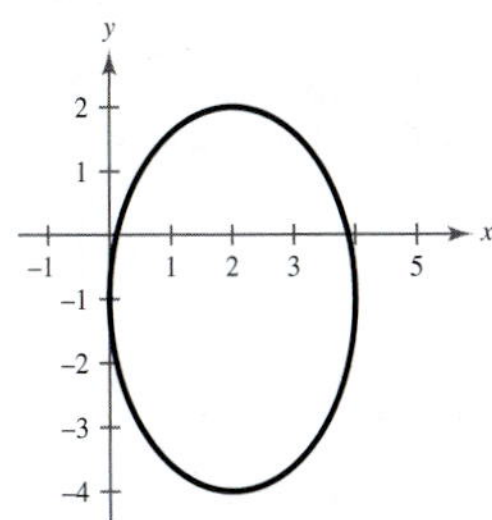

18. $(x - 3)^2 = \dfrac{3}{2}(y + 2)$ **19.** $\dfrac{(y - 2)^2}{\frac{4}{5}} - \dfrac{x^2}{\frac{16}{5}} = 1$

20. Reflect f in the x-axis and y-axis, and shift three units to the right.

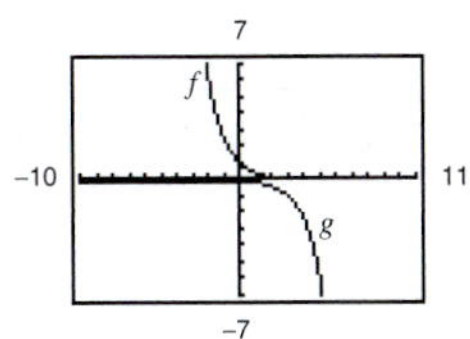

21. Reflect f in the x-axis, and shift four units upward.

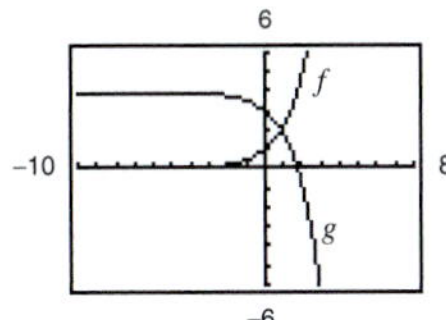

22. 1.991 **23.** -0.067 **24.** 1.717 **25.** 0.281

26. 0.302 **27.** -1.733 **28.** -4.087

29. $\ln(x + 4) + \ln(x - 4) - 4 \ln x,\ x > 4$

30. $\ln \dfrac{x^2}{\sqrt{x + 5}},\ x > 0$ **31.** $\dfrac{\ln 12}{2} \approx 1.242$

32. $\dfrac{\ln 9}{\ln 4} + 5 \approx 6.585$ **33.** $\dfrac{64}{5} = 12.8$

34. $\frac{1}{2}e^8 \approx 1490.479$

35.

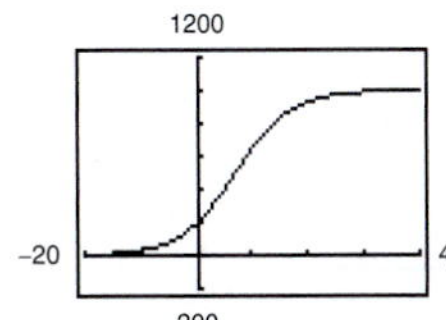

Horizontal asymptotes: $y = 0,\ y = 1000$

36. \$2000

37. (a) and (c)

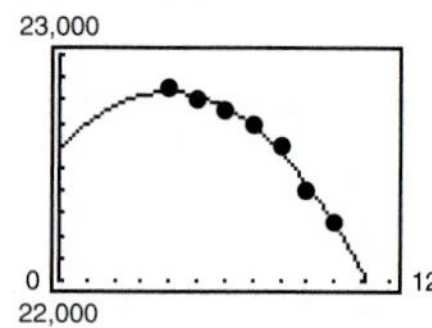

(b) $D = -16.07x^2 + 126.8x + 22{,}586$

(d) No, because the model will eventually become negative.

38. \$16,302.05 **39.** 6.3 hours **40.** 2004

Problem Solving (page 450)

1.

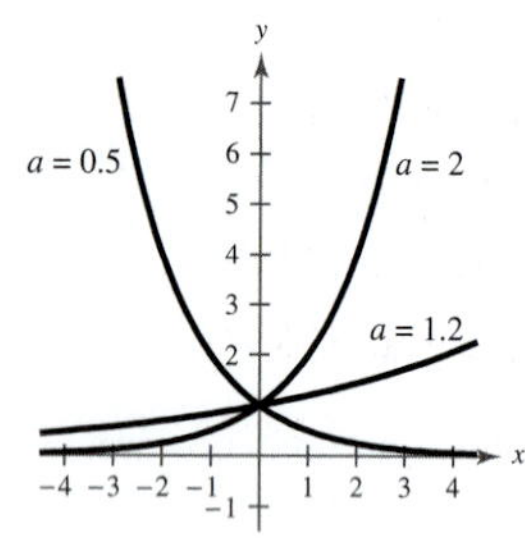

$y = 0.5^x$ and $y = 1.2^x$.

$0 \leq a \leq 1.44$.

3. As $x \to \infty$, the graph of e^x increases at a greater rate than the graph of x^n.

5. Answers will vary.

7. (a)

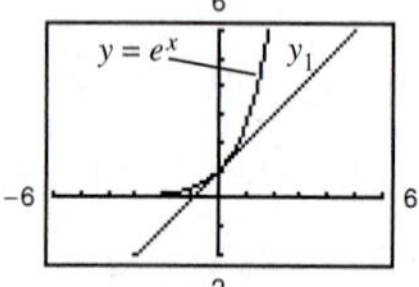

(b)

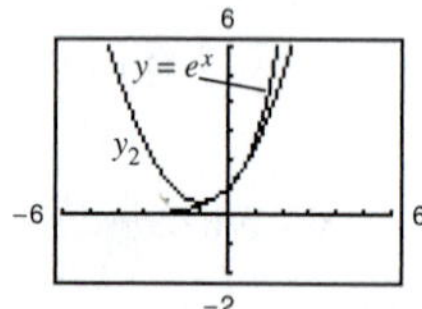

(c)

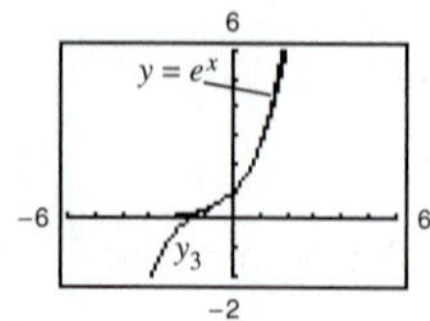

9.

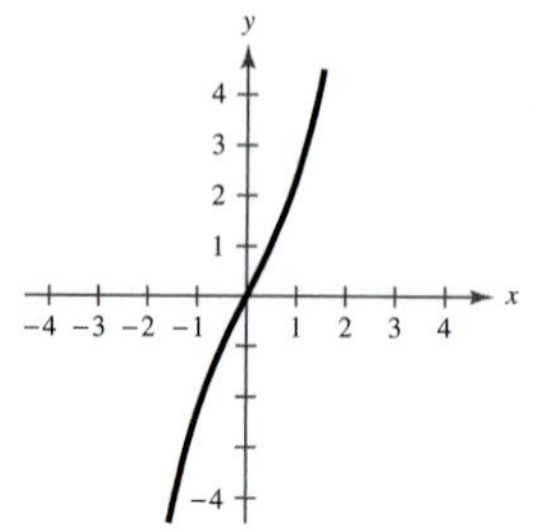

$$f^{-1}(x) = \ln\left(\frac{x + \sqrt{x^2 + 4}}{2}\right)$$

11. c **13.** $t = \dfrac{\ln c_1 - \ln c_2}{\left(\dfrac{1}{k_2} - \dfrac{1}{k_1}\right)\ln \dfrac{1}{2}}$

15. (a) $y_1 = 252{,}606(1.0310)^t$

(b) $y_2 = 400.88t^2 - 1464.6t + 291{,}782$

(c)

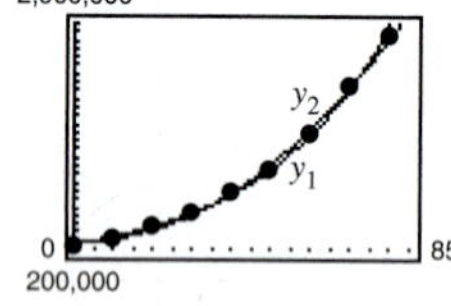

(d) The exponential model is a better fit. No, because the model is rapidly approaching infinity.

17. $1, e^2$

19. $y_4 = (x - 1) - \frac{1}{2}(x - 1)^2 + \frac{1}{3}(x - 1)^3 - \frac{1}{4}(x - 1)^4$

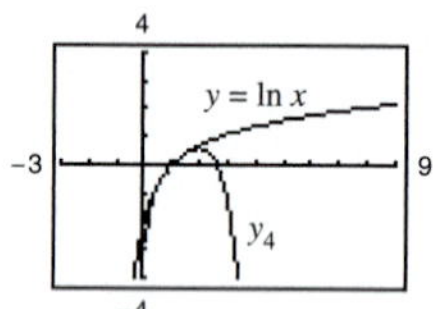

The pattern implies that

$\ln x = (x - 1) - \frac{1}{2}(x - 1)^2 + \frac{1}{3}(x - 1)^3 - \cdots$

21.

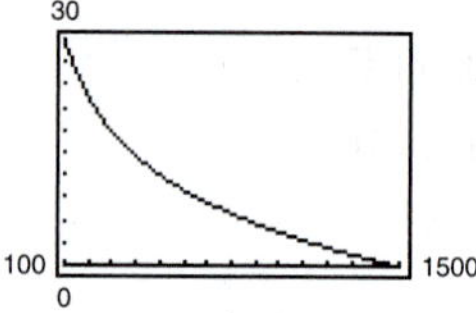

17.7 cubic feet per minute

Chapter 6

Section 6.1 *(page 461)*

1. 210° **3.** $-60°$

5. (a) Quadrant II (b) Quadrant IV

7. (a) Quadrant III (b) Quadrant I

9. (a) (b)

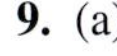

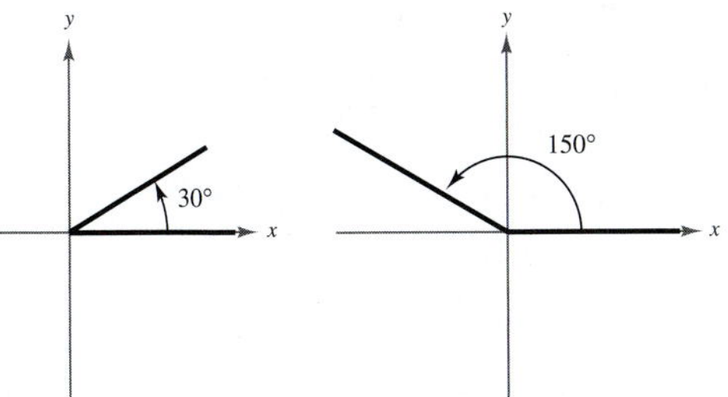

11. (a) (b)

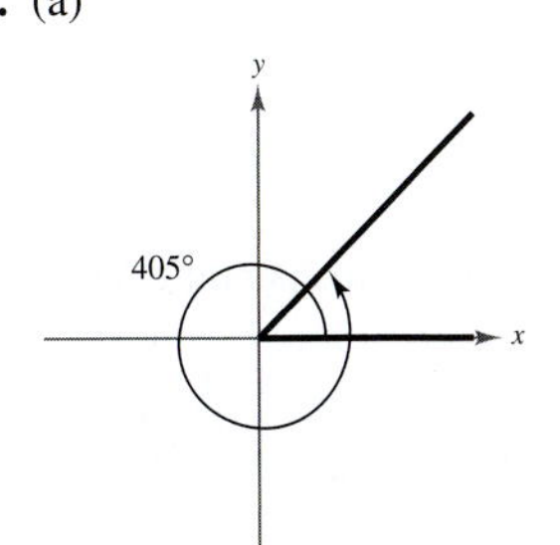

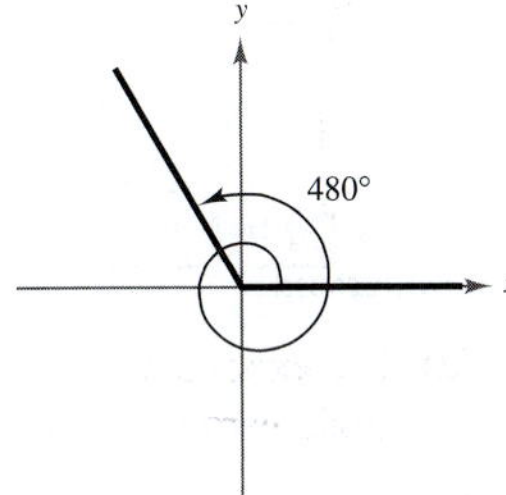

13. (a) 405°, $-315°$ (b) 324°, $-396°$

15. (a) 660°, $-60°$ (b) 20°, $-340°$

17. (a) 54.75° (b) $-128.5°$

19. (a) 85.308° (b) 330.007°

21. (a) 240° 36′ (b) $-145°48'$

23. (a) 2° 30′ (b) $-3° 34' 48''$

25. 2 radians **27.** -3 radians

29. (a) Quadrant I (b) Quadrant III

31. (a) Quadrant IV (b) Quadrant II

33. (a) Quadrant III (b) Quadrant II

35. (a) (b)

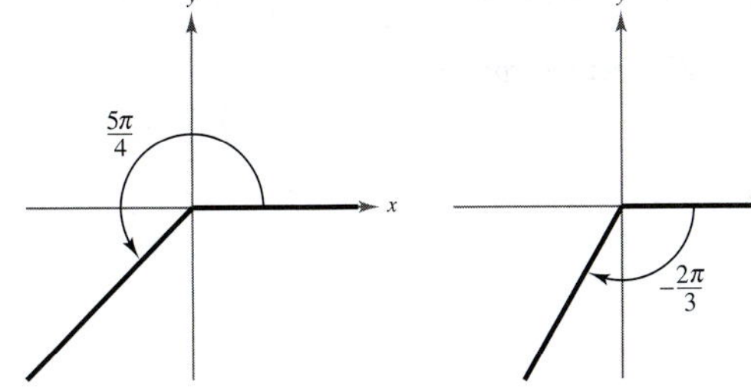

37. (a) (b)

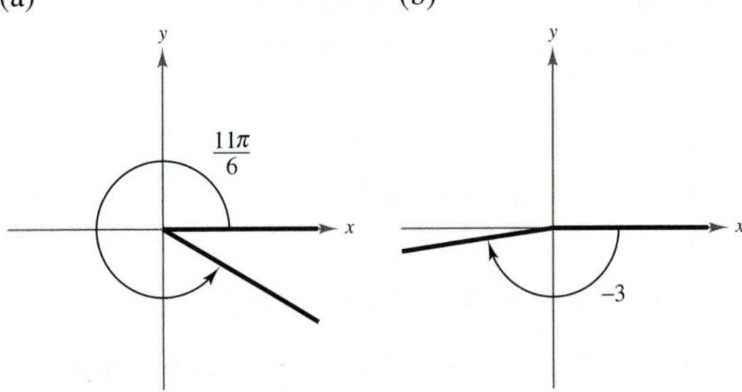

39. (a) $\frac{13\pi}{6}, -\frac{11\pi}{6}$ (b) $\frac{17\pi}{6}, -\frac{7\pi}{6}$

41. (a) $\frac{7\pi}{4}, -\frac{\pi}{4}$ (b) $\frac{28\pi}{15}, -\frac{32\pi}{15}$

43. (a) Complement: 72°; Supplement: 162°

(b) Complement: None; Supplement: 65°

45. (a) Complement: 87°; Supplement: 177°

(b) Complement: 26°; Supplement: 116°

47. (a) Complement: $\frac{5\pi}{12}$; Supplement: $\frac{11\pi}{12}$

(b) Complement: None; Supplement: $\frac{\pi}{12}$

49. (a) Complement: None; Supplement: $\pi - 3 \approx 0.14$

(b) Complement: $\frac{\pi}{2} - 1.5 \approx 0.07$;

Supplement: $\pi - 1.5 \approx 1.64$

51. (a) $\frac{\pi}{6}$ (b) $\frac{5\pi}{6}$ **53.** (a) $-\frac{\pi}{9}$ (b) $-\frac{4\pi}{3}$

55. (a) 270° (b) 210° **57.** (a) 420° (b) $-66°$

59. 2.007 **61.** -3.776 **63.** 9.285 **65.** -0.014

67. 25.714° **69.** 337.500° **71.** $-756.000°$

73. $-114.592°$ **75.** $\frac{6}{5}$ radians **77.** $\frac{32}{7}$ radians

79. $\frac{2}{9}$ radian **81.** $\frac{50}{29}$ radians

83. 15π inches ≈ 47.12 inches **85.** 3 meters

87. 591.7 miles **89.** 1141.0 miles

91. 0.071 radian $\approx 4.04°$ **93.** $\frac{5}{12}$ radian

95. (a) 728.3 revolutions per minute

(b) 4576 radians per minute

97. (a) $\frac{14\pi}{3}$ feet per second; ≈ 10 miles per hour

(b) $d = \frac{7\pi}{7920}n$ (c) $d = \frac{7\pi}{7920}t$

(d) The functions are both linear.

99. False. A measurement of 4π radians corresponds to two complete revolutions from the initial to the terminal side of an angle.

101. False. The terminal side of the angle lies on the x-axis.

103. The speed increases. The linear velocity is proportional to the radius.

105. If θ is constant, the length of the arc is proportional to the radius ($s = r\theta$).

107.

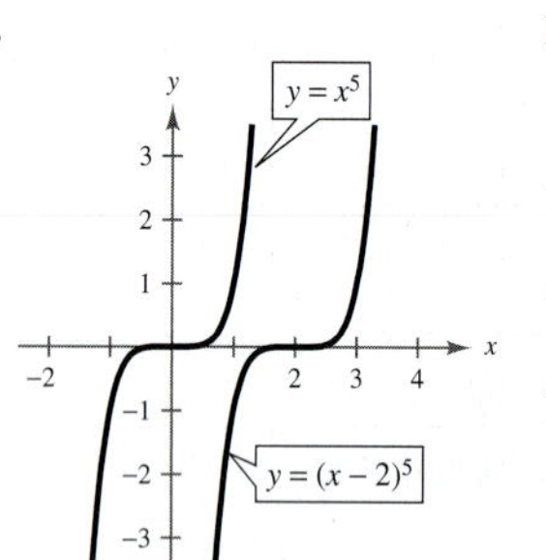

109.

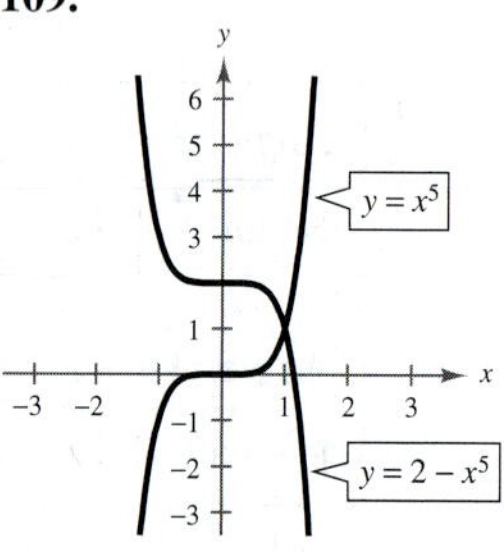

111.

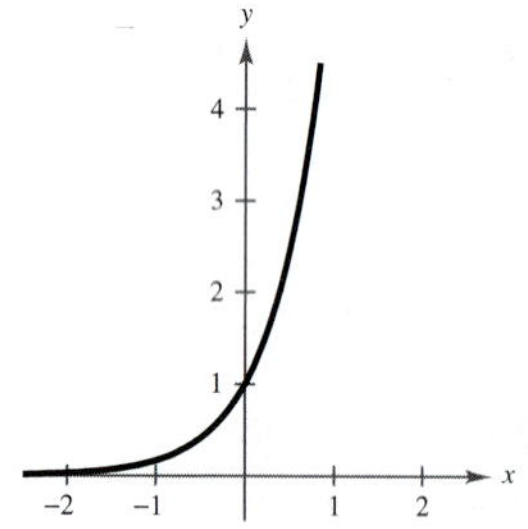

113.

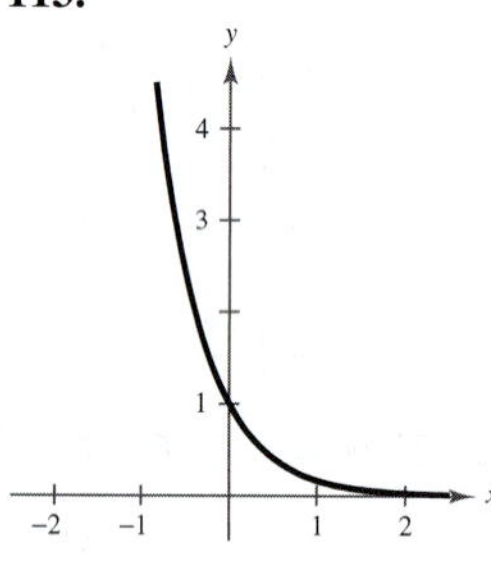

115.

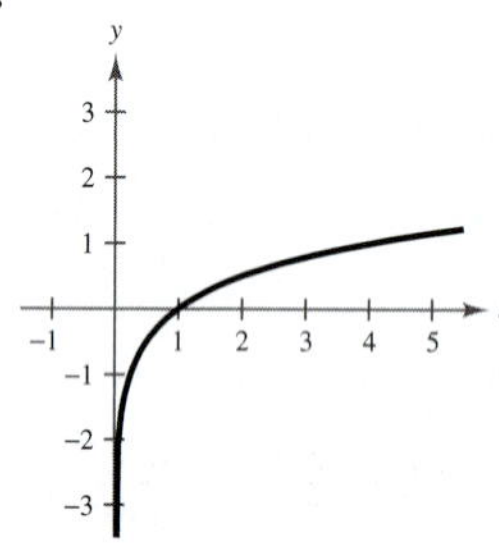

117.

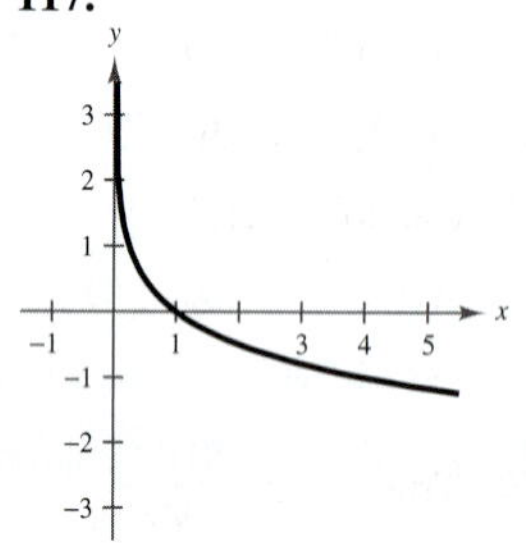

119.

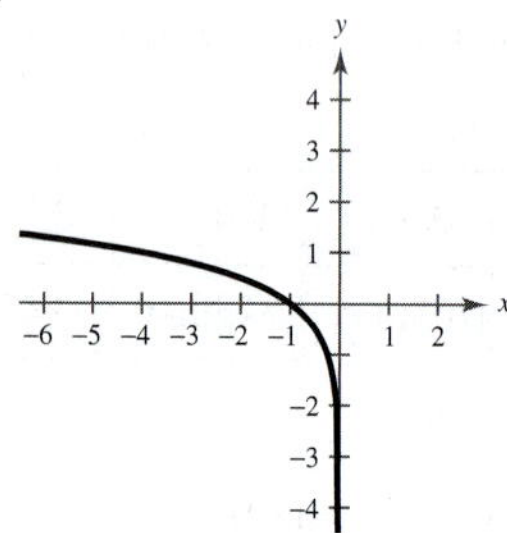

121. $\frac{\sqrt{2}}{2}$ **123.** $\sqrt{2}$ **125.** $2\sqrt{10}$ **127.** $12\sqrt{2}$

Section 6.2 *(page 472)*

1. $\sin\theta = \frac{3}{5}$ $\csc\theta = \frac{5}{3}$
$\cos\theta = \frac{4}{5}$ $\sec\theta = \frac{5}{4}$
$\tan\theta = \frac{3}{4}$ $\cot\theta = \frac{4}{3}$

3. $\sin\theta = \frac{9}{41}$ $\csc\theta = \frac{41}{9}$
$\cos\theta = \frac{40}{41}$ $\sec\theta = \frac{41}{40}$
$\tan\theta = \frac{9}{40}$ $\cot\theta = \frac{40}{9}$

5. $\sin\theta = \dfrac{1}{3}$ $\csc\theta = 3$
$\cos\theta = \dfrac{2\sqrt{2}}{3}$ $\sec\theta = \dfrac{3\sqrt{2}}{4}$
$\tan\theta = \dfrac{\sqrt{2}}{4}$ $\cot\theta = 2\sqrt{2}$
The triangles are similar, and corresponding sides are proportional.

7. $\sin\theta = \frac{3}{5}$ $\csc\theta = \frac{5}{3}$
$\cos\theta = \frac{4}{5}$ $\sec\theta = \frac{5}{4}$
$\tan\theta = \frac{3}{4}$ $\cot\theta = \frac{4}{3}$
The triangles are similar, and corresponding sides are proportional.

9.

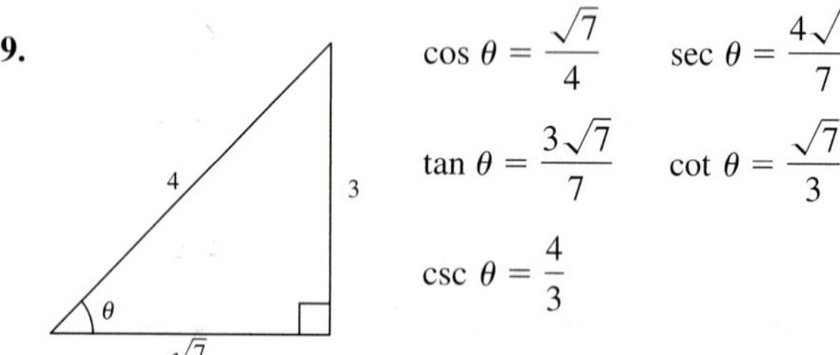

$\cos\theta = \dfrac{\sqrt{7}}{4}$ $\sec\theta = \dfrac{4\sqrt{7}}{7}$
$\tan\theta = \dfrac{3\sqrt{7}}{7}$ $\cot\theta = \dfrac{\sqrt{7}}{3}$
$\csc\theta = \dfrac{4}{3}$

11.

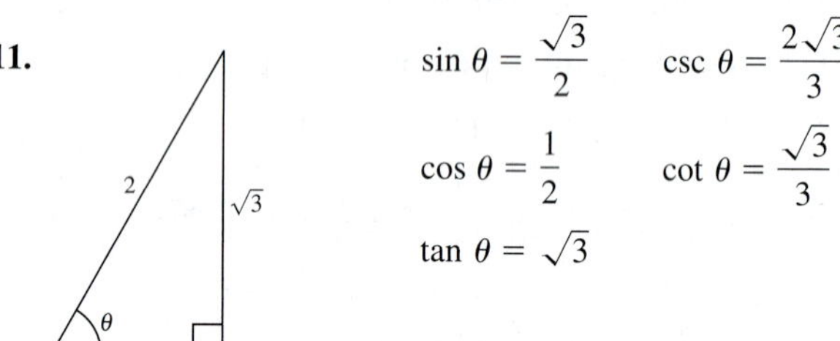

$\sin\theta = \dfrac{\sqrt{3}}{2}$ $\csc\theta = \dfrac{2\sqrt{3}}{3}$
$\cos\theta = \dfrac{1}{2}$ $\cot\theta = \dfrac{\sqrt{3}}{3}$
$\tan\theta = \sqrt{3}$

13.

$\sin\theta = \dfrac{3\sqrt{10}}{10}$ $\sec\theta = \sqrt{10}$
$\cos\theta = \dfrac{\sqrt{10}}{10}$ $\cot\theta = \dfrac{1}{3}$
$\csc\theta = \dfrac{\sqrt{10}}{3}$

15.

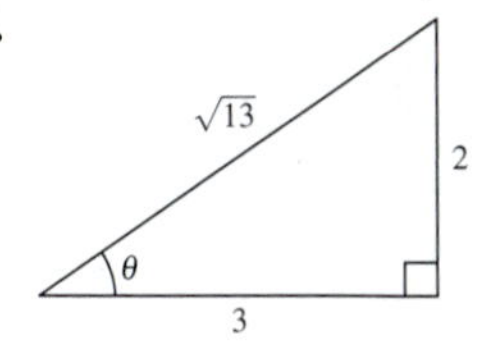

$\sin\theta = \frac{2\sqrt{13}}{13}$

$\cos\theta = \frac{3\sqrt{13}}{13}$

$\tan\theta = \frac{2}{3}$

$\csc\theta = \frac{\sqrt{13}}{2}$

$\sec\theta = \frac{\sqrt{13}}{3}$

17. (a) $\sqrt{3}$ (b) $\frac{1}{2}$ (c) $\frac{\sqrt{3}}{2}$ (d) $\frac{\sqrt{3}}{3}$

19. (a) $\frac{2\sqrt{13}}{13}$ (b) $\frac{3\sqrt{13}}{13}$ (c) $\frac{2}{3}$ (d) $\frac{\sqrt{13}}{2}$

21. (a) 3 (b) $\frac{2\sqrt{2}}{3}$ (c) $\frac{\sqrt{2}}{4}$ (d) $\frac{1}{3}$

23. (a) $\frac{1}{2}$ (b) 2 (c) $\sqrt{3}$

25. (a) $\frac{\sqrt{2}}{2}$ (b) $\frac{\sqrt{2}}{2}$ (c) $\frac{\sqrt{3}}{3}$

27. (a) 0.4348 (b) 0.4348 **29.** (a) 0.9598 (b) 0.9609

31. (a) 5.0273 (b) 0.1989

33. (a) 1.1884 (b) 0.5463

35. (a) $30° = \frac{\pi}{6}$ (b) $30° = \frac{\pi}{6}$

37. (a) $60° = \frac{\pi}{3}$ (b) $45° = \frac{\pi}{4}$

39. (a) $60° = \frac{\pi}{3}$ (b) $45° = \frac{\pi}{4}$

41. (a) $0.83° \approx 0.015$ (b) $27° \approx 0.474$

43. (a) $0.72° \approx 0.012$ (b) $67° \approx 1.169$

45–53. Answers will vary. **55.** $30\sqrt{3}$ **57.** $\frac{32\sqrt{3}}{3}$

59. (a)

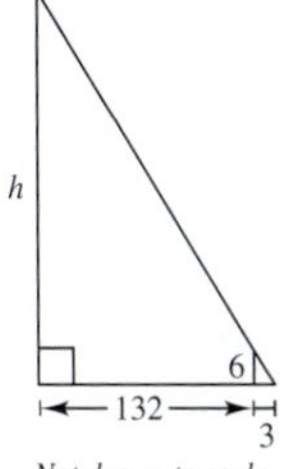

Not drawn to scale

(b) $\tan\theta = \frac{6}{3} = \frac{h}{135}$

(c) 270 feet

61. (a)

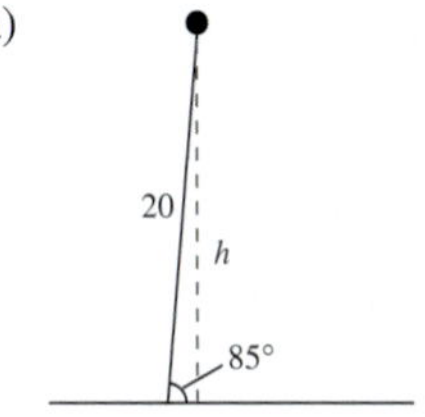

(b) $\sin 85° = \frac{h}{20}$ (c) 19.9 meters

(d) The side of the triangle labeled h will become shorter.

(e)

Angle, θ	80°	70°	60°	50°
Height	19.7	18.8	17.3	15.3

Angle, θ	40°	30°	20°	10°
Height	12.9	10.0	6.8	3.5

(f) The height of the balloon decreases.

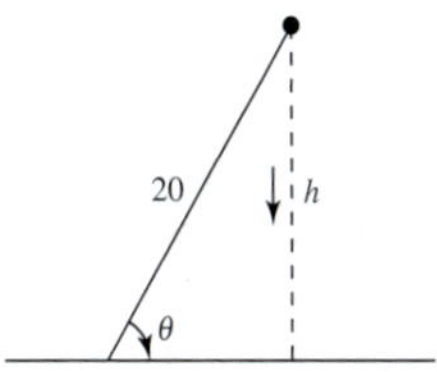

63. 137.6 feet **65.** $(x_1, y_1) = (28\sqrt{3}, 28)$
$(x_2, y_2) = (28, 28\sqrt{3})$

67. $\sin 20° \approx 0.34$
$\cos 20° \approx 0.94$
$\tan 20° \approx 0.36$
$\csc 20° \approx 2.92$
$\sec 20° \approx 1.06$
$\cot 20° \approx 2.75$

69. True, $\csc x = \frac{1}{\sin x}$. **71.** False, $\frac{\sqrt{2}}{2} + \frac{\sqrt{2}}{2} \neq 1$.

73. False, $1.7321 \neq 0.0349$.

75. Corresponding sides of similar triangles are proportional.

77. (a)

θ	0.1	0.2	0.3	0.4	0.5
$\sin\theta$	0.0998	0.1987	0.2955	0.3894	0.4794

(b) θ is greater.

(c) As θ approaches 0, $\sin\theta$ approaches θ.

79. $\frac{x}{x-2}$, $x \neq \pm 6$ **81.** $\frac{2(x^2 - 5x - 10)}{(x-2)(x+2)^2}$

83. $x = \frac{2}{3}$

Section 6.3 *(page 484)*

1. (a) $\sin\theta = \frac{3}{5}$ (b) $\sin\theta = -\frac{15}{17}$
$\cos\theta = \frac{4}{5}$ $\cos\theta = \frac{8}{17}$
$\tan\theta = \frac{3}{4}$ $\tan\theta = -\frac{15}{8}$
$\csc\theta = \frac{5}{3}$ $\csc\theta = -\frac{17}{15}$
$\sec\theta = \frac{5}{4}$ $\sec\theta = \frac{17}{8}$
$\cot\theta = \frac{4}{3}$ $\cot\theta = -\frac{8}{15}$

3. (a) $\sin\theta = -\frac{1}{2}$ (b) $\sin\theta = \frac{\sqrt{17}}{17}$
$\cos\theta = -\frac{\sqrt{3}}{2}$ $\cos\theta = -\frac{4\sqrt{17}}{17}$
$\tan\theta = \frac{\sqrt{3}}{3}$ $\tan\theta = -\frac{1}{4}$
$\csc\theta = -2$ $\csc\theta = \sqrt{17}$
$\sec\theta = -\frac{2\sqrt{3}}{3}$ $\sec\theta = -\frac{\sqrt{17}}{4}$
$\cot\theta = \sqrt{3}$ $\cot\theta = -4$

5. $\sin\theta = \frac{24}{25}$ $\csc\theta = \frac{25}{24}$
$\cos\theta = \frac{7}{25}$ $\sec\theta = \frac{25}{7}$
$\tan\theta = \frac{24}{7}$ $\cot\theta = \frac{7}{24}$

7. $\sin\theta = \frac{5\sqrt{29}}{29}$ $\csc\theta = \frac{\sqrt{29}}{5}$
$\cos\theta = -\frac{2\sqrt{29}}{29}$ $\sec\theta = -\frac{\sqrt{29}}{2}$
$\tan\theta = -\frac{5}{2}$ $\cot\theta = -\frac{2}{5}$

9. $\sin\theta \approx 0.9$ $\csc\theta \approx 1.1$
$\cos\theta \approx -0.5$ $\sec\theta \approx -2.2$
$\tan\theta \approx -1.9$ $\cot\theta \approx -0.5$

11. Quadrant III **13.** Quadrant II

15. $\sin\theta = \frac{3}{5}$ $\csc\theta = \frac{5}{3}$
$\cos\theta = -\frac{4}{5}$ $\sec\theta = -\frac{5}{4}$
$\tan\theta = -\frac{3}{4}$ $\cot\theta = -\frac{4}{3}$

17. $\sin\theta = -\frac{15}{17}$ $\csc\theta = -\frac{17}{15}$
$\cos\theta = \frac{8}{17}$ $\sec\theta = \frac{17}{8}$
$\tan\theta = -\frac{15}{8}$ $\cot\theta = -\frac{8}{15}$

19. $\sin\theta = -\frac{\sqrt{10}}{10}$ $\csc\theta = -\sqrt{10}$
$\cos\theta = \frac{3\sqrt{10}}{10}$ $\sec\theta = \frac{\sqrt{10}}{3}$
$\tan\theta = -\frac{1}{3}$ $\cot\theta = -3$

21. $\sin\theta = \frac{\sqrt{3}}{2}$ $\csc\theta = \frac{2\sqrt{3}}{3}$
$\cos\theta = -\frac{1}{2}$ $\sec\theta = -2$
$\tan\theta = -\sqrt{3}$ $\cot\theta = -\frac{\sqrt{3}}{3}$

23. $\sin\theta = 0$ $\csc\theta$ is undefined.
$\cos\theta = -1$ $\sec\theta = -1$
$\tan\theta = 0$ $\cot\theta$ is undefined.

25. $\sin\theta = \frac{\sqrt{2}}{2}$ $\csc\theta = \sqrt{2}$
$\cos\theta = -\frac{\sqrt{2}}{2}$ $\sec\theta = -\sqrt{2}$
$\tan\theta = -1$ $\cot\theta = -1$

27. $\sin\theta = -\frac{2\sqrt{5}}{5}$ $\csc\theta = -\frac{\sqrt{5}}{2}$
$\cos\theta = -\frac{\sqrt{5}}{5}$ $\sec\theta = -\sqrt{5}$
$\tan\theta = 2$ $\cot\theta = \frac{1}{2}$

29. -1 **31.** Undefined

33. Undefined **35.** Undefined

37. $\theta' = 23°$

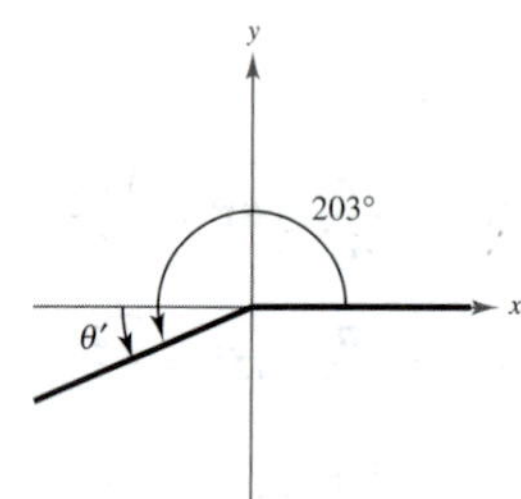

39. $\theta' = 65°$

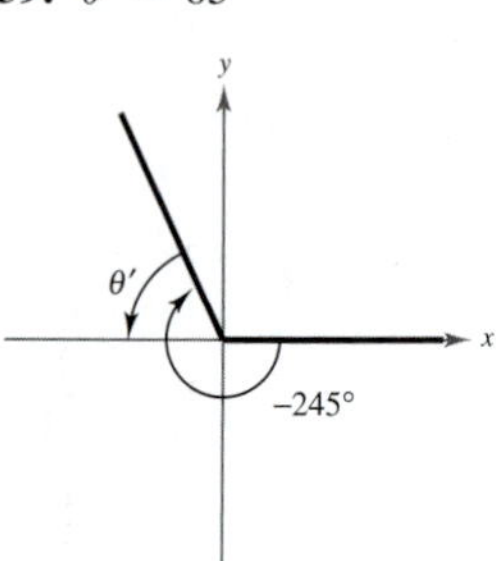

41. $\theta' = \frac{\pi}{3}$

y
$\frac{2\pi}{3}$
θ′
x

43. $\theta' = 3.5 - \pi$

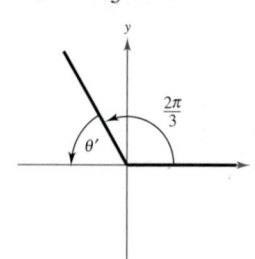

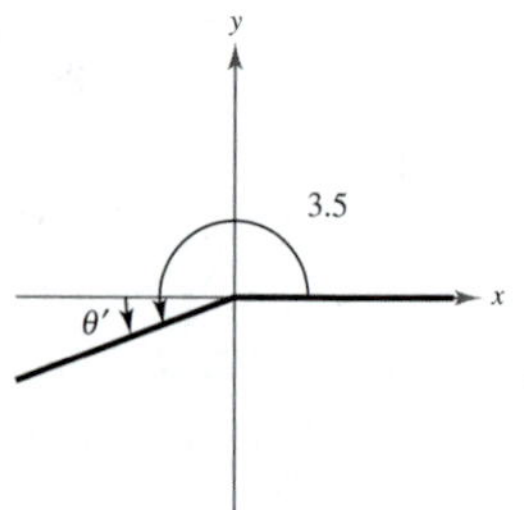

45. $\sin 225° = -\frac{\sqrt{2}}{2}$

$\cos 225° = -\frac{\sqrt{2}}{2}$

$\tan 225° = 1$

47. $\sin 750° = \frac{1}{2}$

$\cos 750° = \frac{\sqrt{3}}{2}$

$\tan 750° = \frac{\sqrt{3}}{3}$

49. $\sin(-150°) = -\frac{1}{2}$

$\cos(-150°) = -\frac{\sqrt{3}}{2}$

$\tan(-150°) = \frac{\sqrt{3}}{3}$

51. $\sin \frac{4\pi}{3} = -\frac{\sqrt{3}}{2}$

$\cos \frac{4\pi}{3} = -\frac{1}{2}$

$\tan \frac{4\pi}{3} = \sqrt{3}$

53. $\sin\left(-\frac{\pi}{6}\right) = -\frac{1}{2}$

$\cos\left(-\frac{\pi}{6}\right) = \frac{\sqrt{3}}{2}$

$\tan\left(-\frac{\pi}{6}\right) = -\frac{\sqrt{3}}{3}$

55. $\sin \frac{11\pi}{4} = \frac{\sqrt{2}}{2}$

$\cos \frac{11\pi}{4} = -\frac{\sqrt{2}}{2}$

$\tan \frac{11\pi}{4} = -1$

57. $\sin\left(-\frac{3\pi}{2}\right) = 1$

$\cos\left(-\frac{3\pi}{2}\right) = 0$

$\tan\left(-\frac{3\pi}{2}\right)$ is undefined.

59. 0.1736 **61.** -0.3420

63. 4.6373 **65.** 0.3640 **67.** -0.6052

69. (a) $30° = \frac{\pi}{6}$, $150° = \frac{5\pi}{6}$ (b) $210° = \frac{7\pi}{6}$, $330° = \frac{11\pi}{6}$

71. (a) $60° = \frac{\pi}{3}$, $120° = \frac{2\pi}{3}$ (b) $135° = \frac{3\pi}{4}$, $315° = \frac{7\pi}{4}$

73. (a) $45° = \frac{\pi}{4}$, $225° = \frac{5\pi}{4}$ (b) $150° = \frac{5\pi}{6}$, $330° = \frac{11\pi}{6}$

75. 54.99°, 125.01° **77.** 115.89°, 244.11°

79. 0.175, 6.109 **81.** 0.873, 4.014 **83.** 1.955, 4.328

85. $\frac{4}{5}$ **87.** $-\frac{\sqrt{13}}{2}$ **89.** $\frac{8}{5}$

91. $\left(\frac{\sqrt{2}}{2}, \frac{\sqrt{2}}{2}\right)$

$\sin \frac{\pi}{4} = \frac{\sqrt{2}}{2}$

$\cos \frac{\pi}{4} = \frac{\sqrt{2}}{2}$

$\tan \frac{\pi}{4} = 1$

93. $\left(-\frac{\sqrt{3}}{2}, \frac{1}{2}\right)$

$\sin \frac{5\pi}{6} = \frac{1}{2}$

$\cos \frac{5\pi}{6} = -\frac{\sqrt{3}}{2}$

$\tan \frac{5\pi}{6} = -\frac{\sqrt{3}}{3}$

95. $\left(-\frac{1}{2}, -\frac{\sqrt{3}}{2}\right)$

$\sin \frac{4\pi}{3} = -\frac{\sqrt{3}}{2}$

$\cos \frac{4\pi}{3} = -\frac{1}{2}$

$\tan \frac{4\pi}{3} = \sqrt{3}$

97. $(0, -1)$

$\sin \frac{3\pi}{2} = -1$

$\cos \frac{3\pi}{2} = 0$

$\tan \frac{3\pi}{2}$ is undefined.

99. (a) -1 (b) -0.4

101. (a) 0.25 or 2.89 (b) 1.82 or 4.46

103. (a) $N = 22.66 \sin(0.51t - 2.12) + 54.58$
$F = 37.18 \sin(0.51t - 1.91) + 26.17$
(b) February: $N = 34°$, $F = -3°$
March: $N = 42°$, $F = 12°$
May: $N = 64°$, $F = 48°$
June: $N = 73°$, $F = 60°$
August: $N = 76°$, $F = 57°$
September: $N = 69°$, $F = 43°$
November: $N = 47°$, $F = 6°$
(c) Answers will vary.

105. (a) 2 centimeters
(b) 0.14 centimeter
(c) -1.98 centimeters

107. 0.79 ampere

109. False. In each of the four quadrants, the signs of the secant function and cosine function will be the same, because these functions are reciprocals of each other.

111. $h(t)$ is an odd function.

113. Determine the trigonometric function of the reference angle and prefix the appropriate sign.

115.

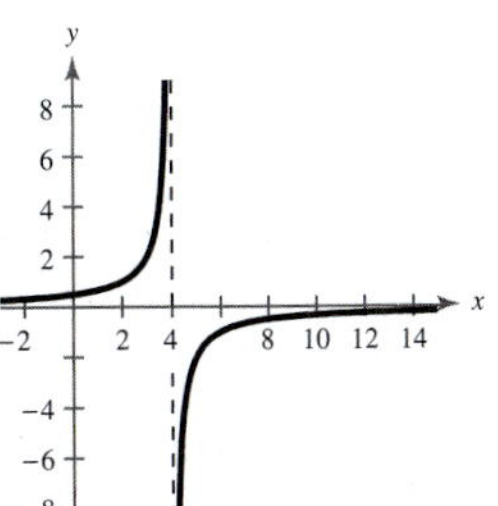

Vertical asymptote: $x = 4$
Horizontal asymptote: $y = 0$

117.

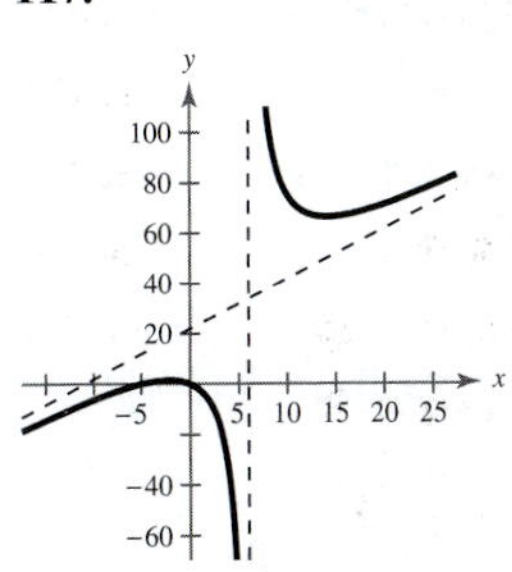

Vertical asymptote: $x = 6$
Slant asymptote: $y = 2x + 22$

119. 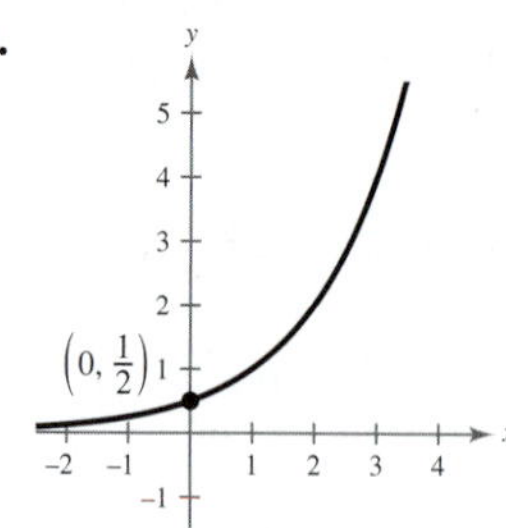

y-intercept: $(0, \frac{1}{2})$
Horizontal asymptote: $y = 0$
Domain: All real numbers
Range: $y > 0$

121. 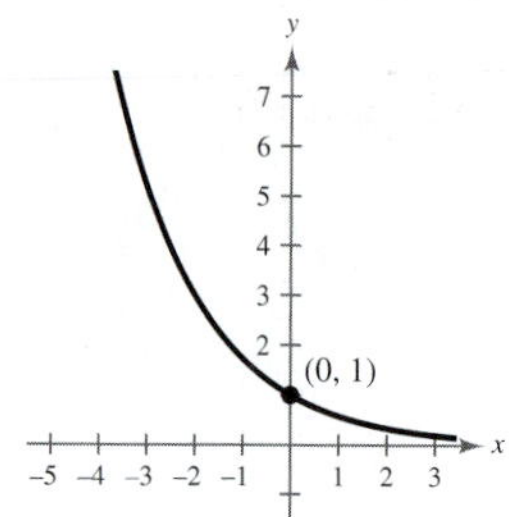

y-intercept: $(0, 1)$
Horizontal asymptote: $y = 0$
Domain: All real numbers
Range: $y > 0$

123. 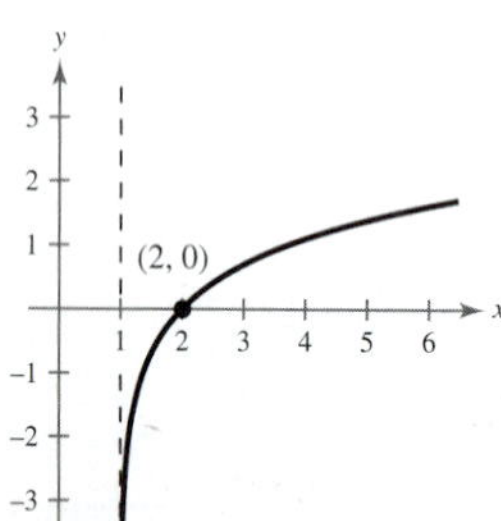

x-intercept: $(2, 0)$
Vertical asymptote: $x = 1$
Domain: $x > 1$
Range: All real numbers

125. 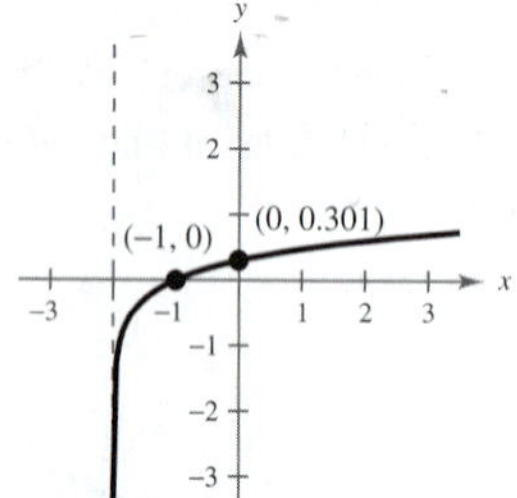

x-intercept: $(-1, 0)$,
y-intercept: $(0, 0.301)$
Vertical asymptote: $x = -2$
Domain: $x > -2$
Range: All real numbers

Section 6.4 *(page 495)*

1. Period: π
Amplitude: 3

3. Period: 4π
Amplitude: $\frac{5}{2}$

5. Period: 6
Amplitude: $\frac{1}{2}$

7. Period: 2π
Amplitude: 2

9. Period: $\frac{\pi}{5}$
Amplitude: 3

11. Period: 3π
Amplitude: $\frac{1}{2}$

13. Period: 1
Amplitude: $\frac{1}{4}$

15. g is a shift of f π units to the right.

17. g is a reflection of f in the x-axis.

19. The period of f is twice the period of g.

21. g is a shift of f three units upward.

23. The graph of g has twice the amplitude of the graph of f.

25. The graph of g is a horizontal shift of the graph of f π units to the right.

27. 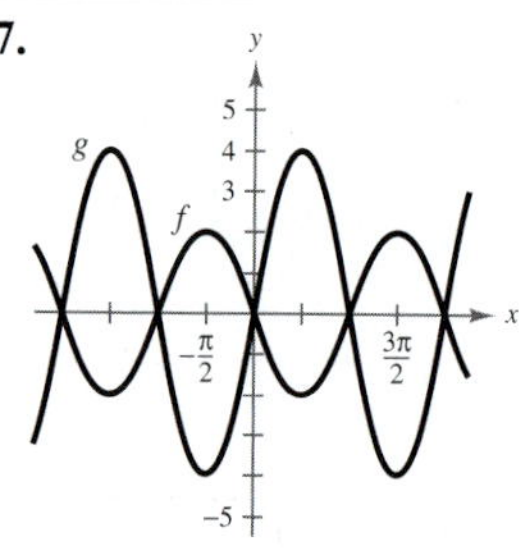

29. 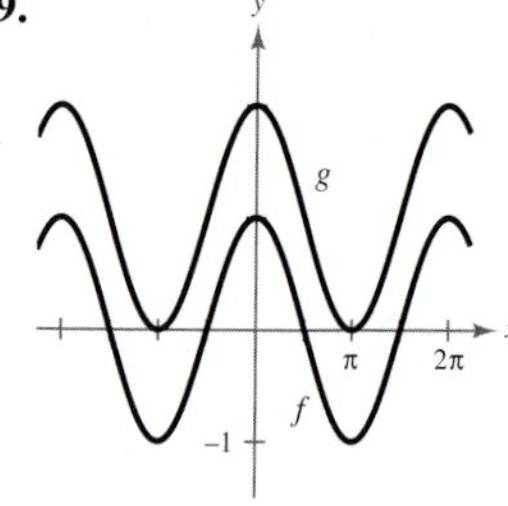

31. 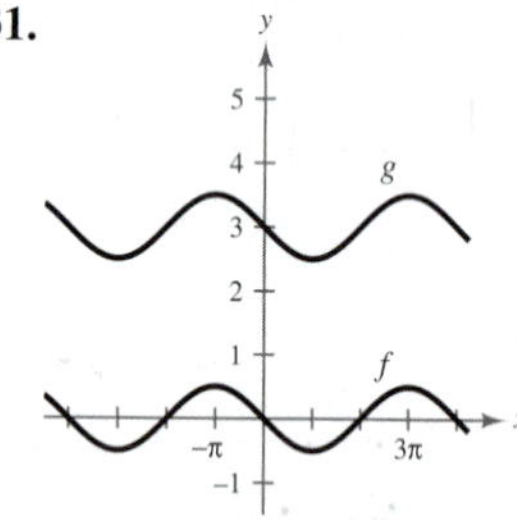

33. 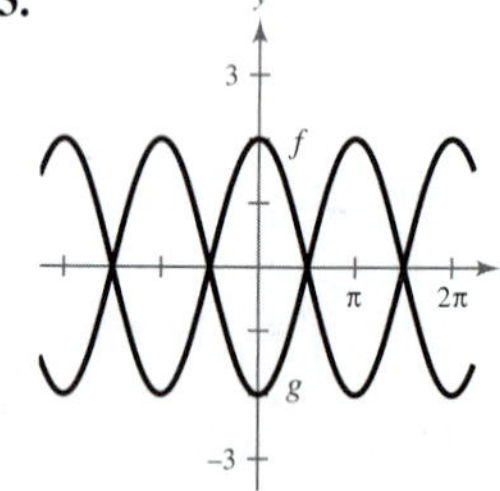

35. 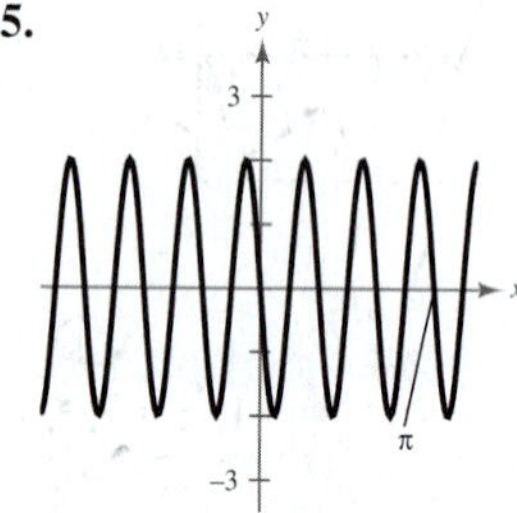

37. 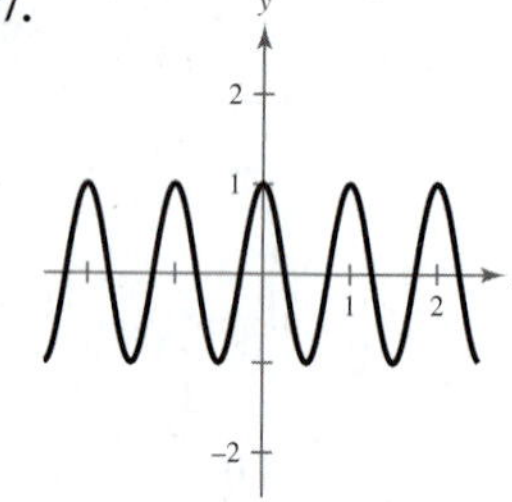

39. 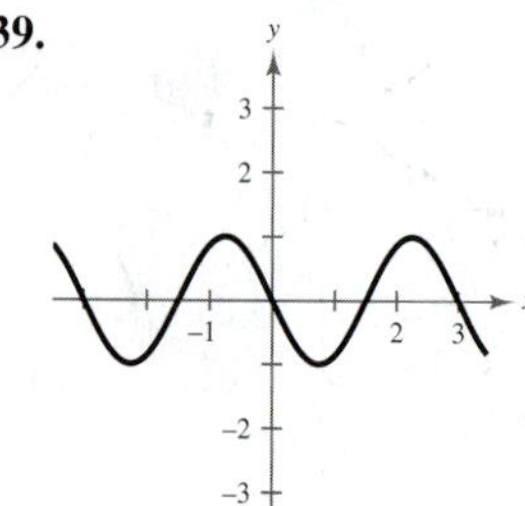

41. 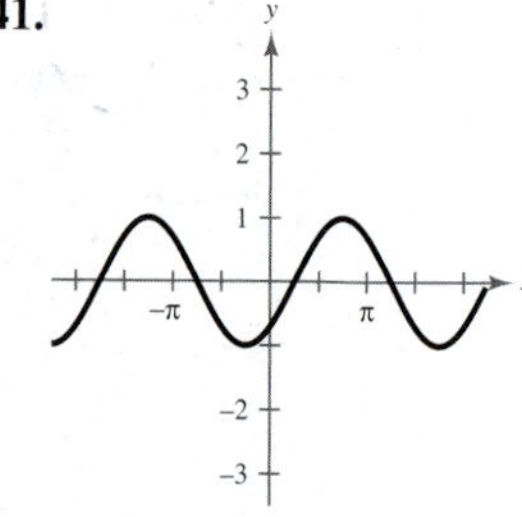

43.

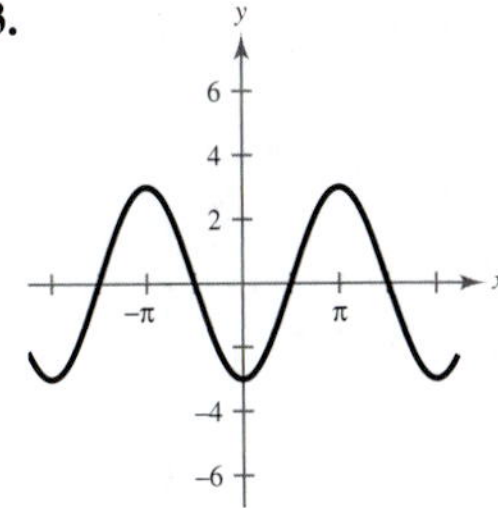

45.

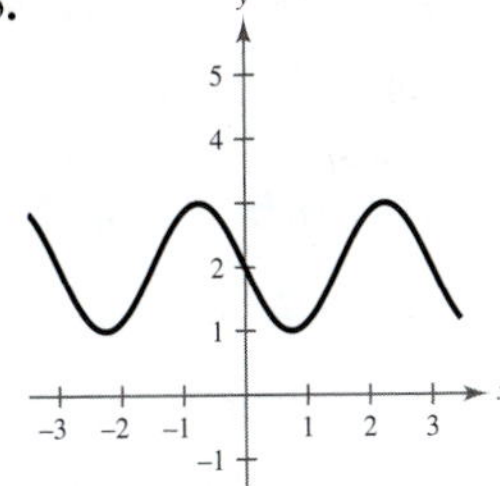

47.

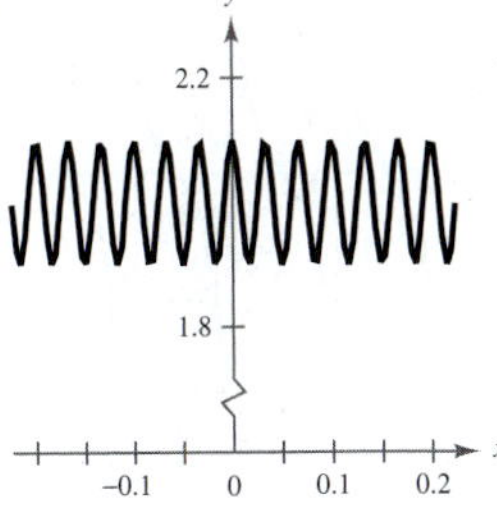

49.

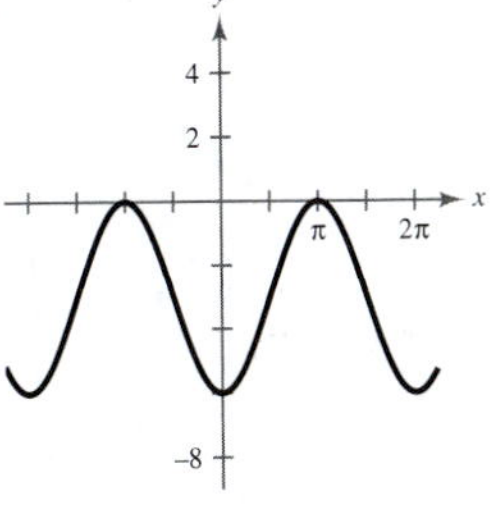

51.

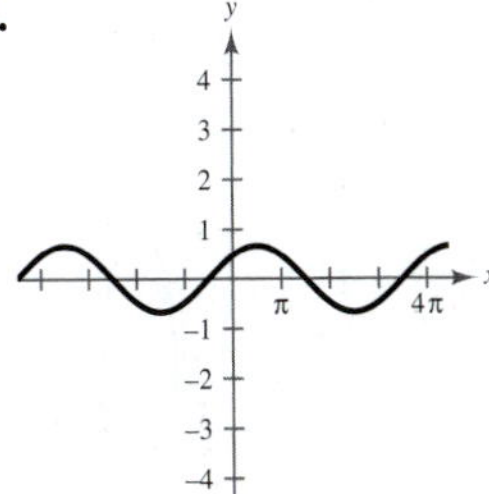

53.

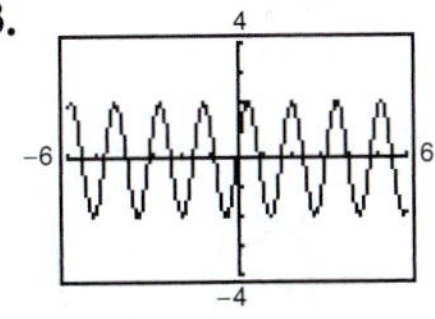

55.

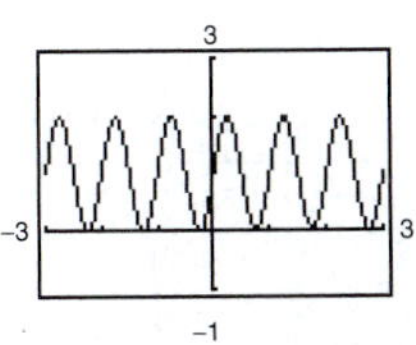

57.

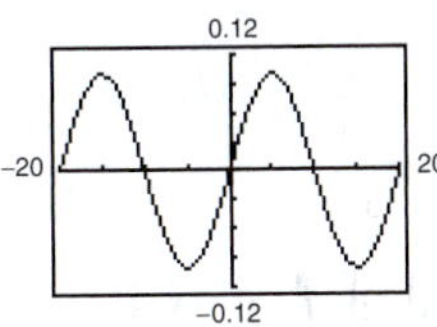

59. $a = 2, d = 1$ **61.** $a = -4, d = 4$

63. $a = -3, b = 2, c = 0$ **65.** $a = 2, b = 1, c = -\dfrac{\pi}{4}$

67.

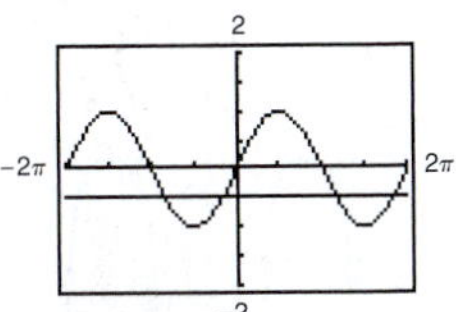

$x = -\dfrac{\pi}{6}, -\dfrac{5\pi}{6}, \dfrac{7\pi}{6}, \dfrac{11\pi}{6}$

69. (a) 6 seconds (b) 10 cycles per minute

(c)

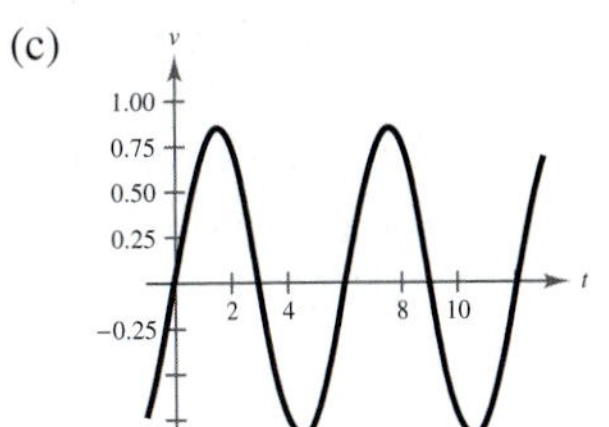

71. (a) $\frac{1}{440}$ second (b) 440 cycles per second

73. (a) $C(t) = 56.35 + 27.35 \cos\left(\dfrac{\pi t}{6} - 3.67\right)$

(b)

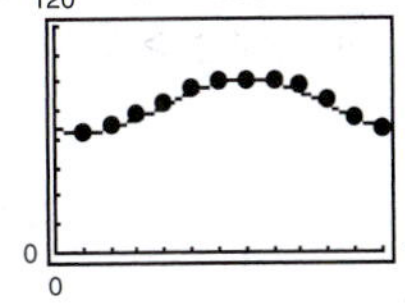

The model is a good fit.

(c)

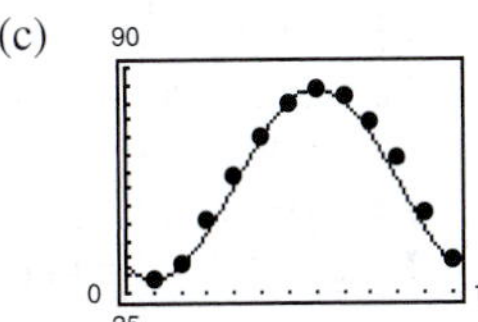

The model is a good fit.

(d) Tallahassee: 77.60°; Chicago: 56.35°

The constant term gives the annual average temperature.

(e) 12. Yes. One full period is 1 year.

(f) Chicago; amplitude; the greater the amplitude, the greater the variability in temperature.

75. (a) 365. Yes. One year is 365 days.

(b) 30.3 gallons; the constant term

(c)

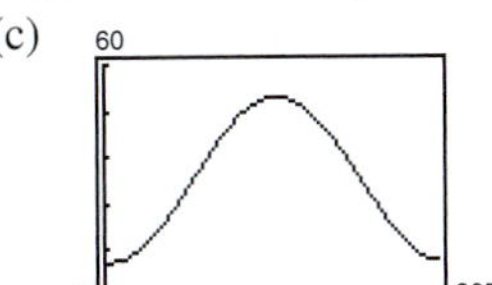

$124 < t < 252$

77. False. The function $y = \frac{1}{2} \cos 2x$ has an amplitude that is one-half that of $y = \cos x$. For $y = a \cos bx$, the amplitude is $|a|$.

79.

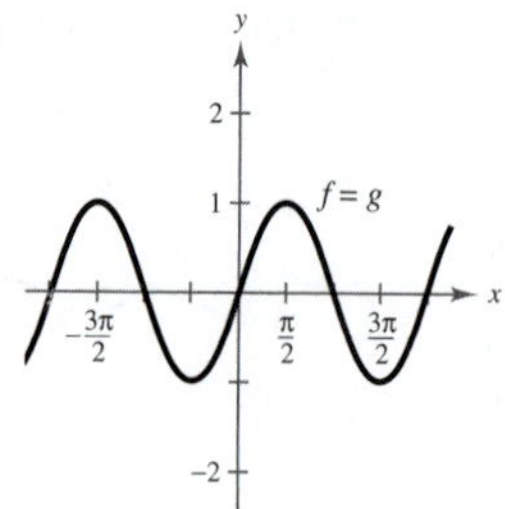

Conjecture:

$\sin x = \cos\left(x - \frac{\pi}{2}\right)$

81.

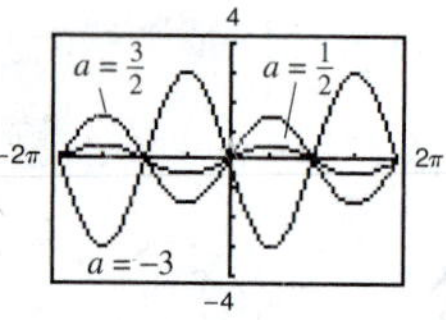

Amplitude changes

83.

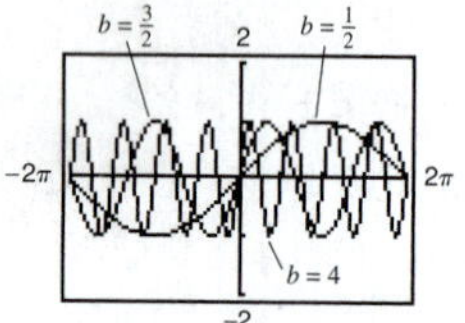

Period changes

85. (a)

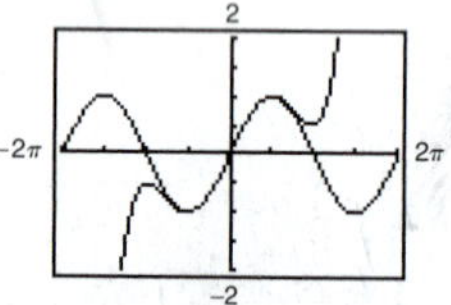

The graphs appear to coincide from $-\frac{\pi}{2}$ to $\frac{\pi}{2}$.

(b)

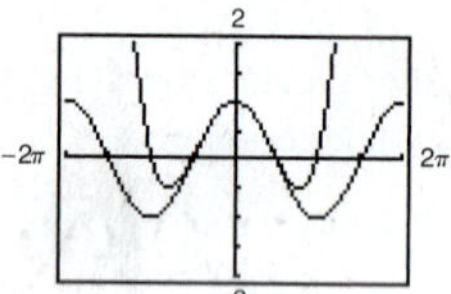

The graphs appear to coincide from $-\frac{\pi}{2}$ to $\frac{\pi}{2}$.

(c) $-\frac{x^7}{7!}, -\frac{x^6}{6!}$

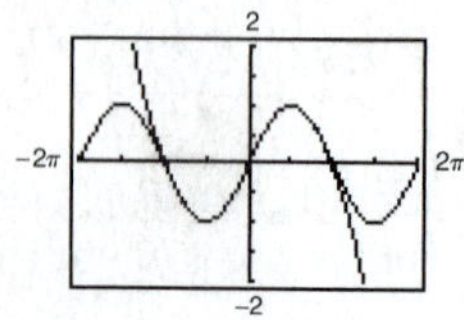

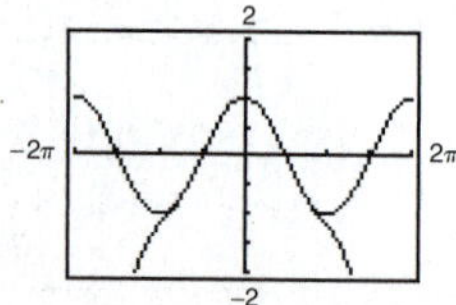

The interval of accuracy increased.

87. $\frac{1}{2}\log_{10}(x - 2)$ **89.** $3 \ln t - \ln(t - 1)$

91. $\log_{10} \sqrt{xy}$ **93.** $\ln \frac{3x}{y^4}$

Section 6.5 *(page 506)*

1. e, π **2.** c, 2π **3.** a, 1 **4.** d, 2π

5. f, 4 **6.** b, 4

7.

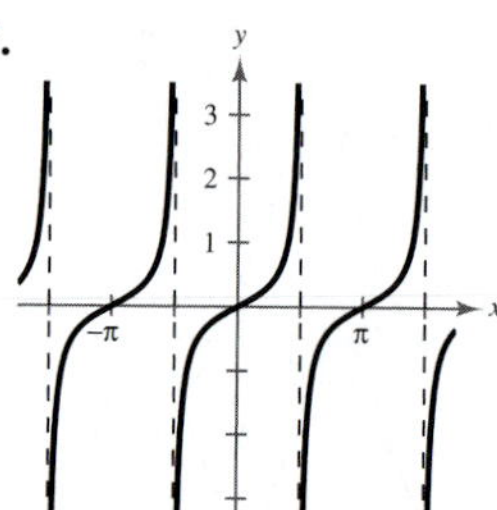

9.

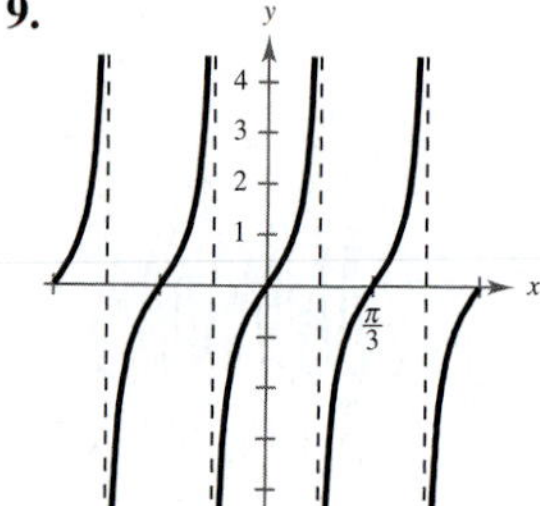

11.

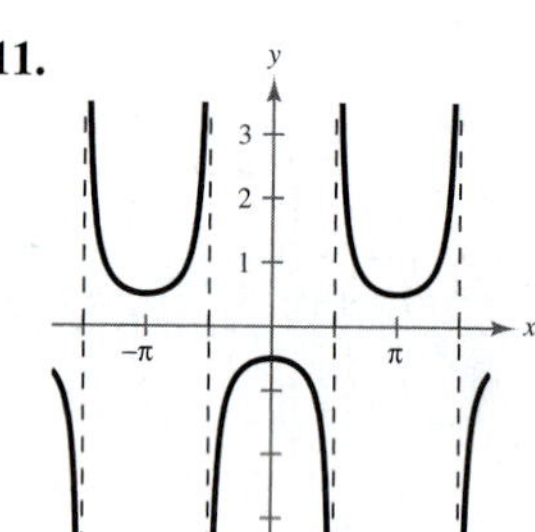

13.

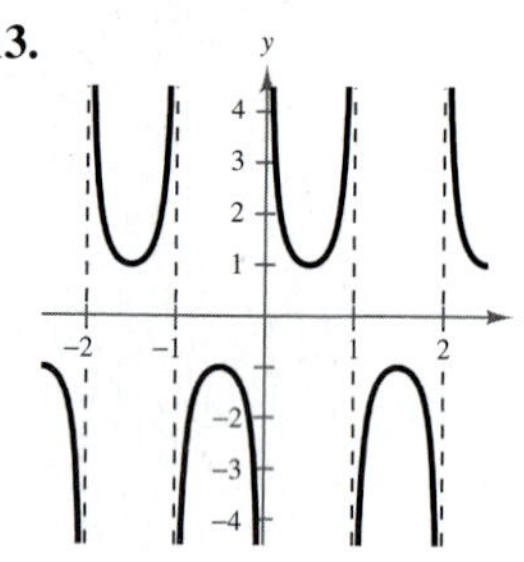

15.

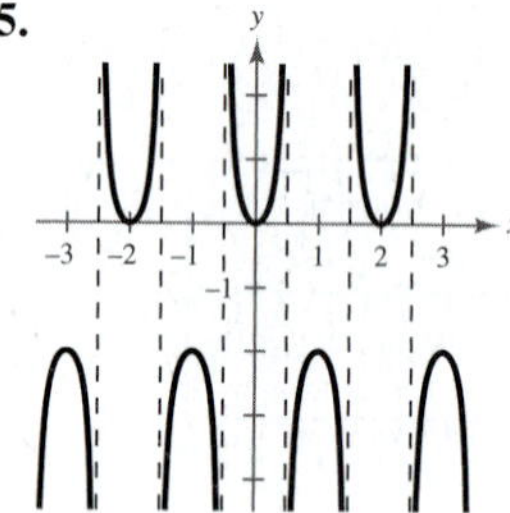

17.

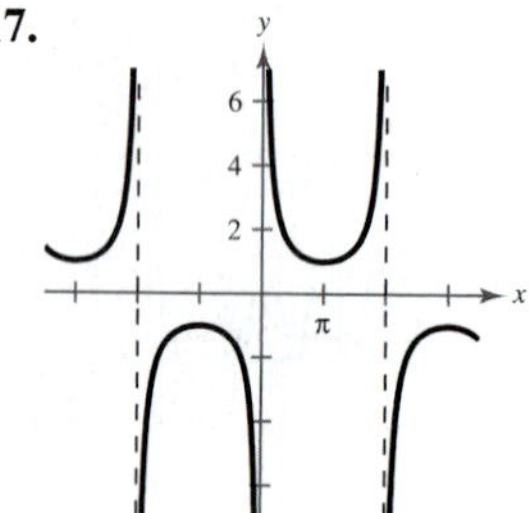

19.

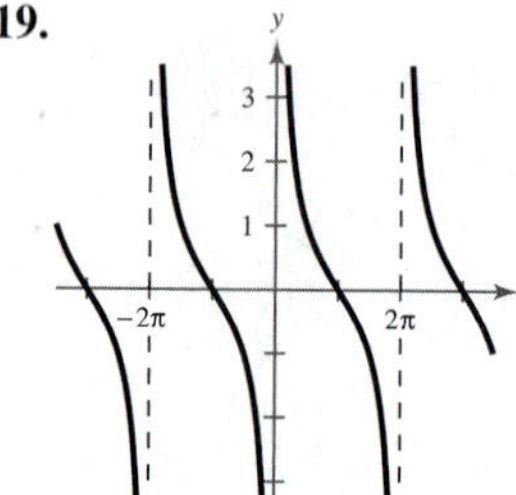

21.

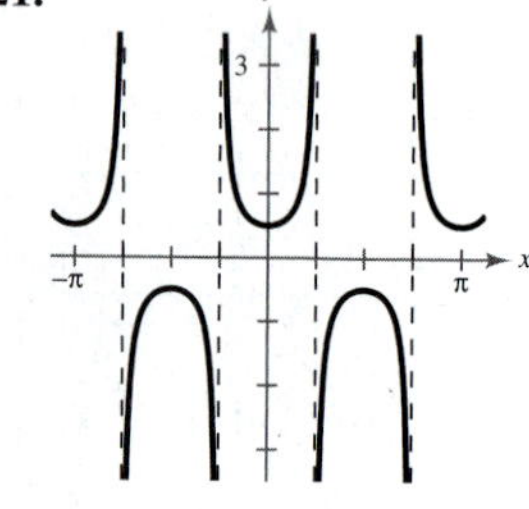

23.

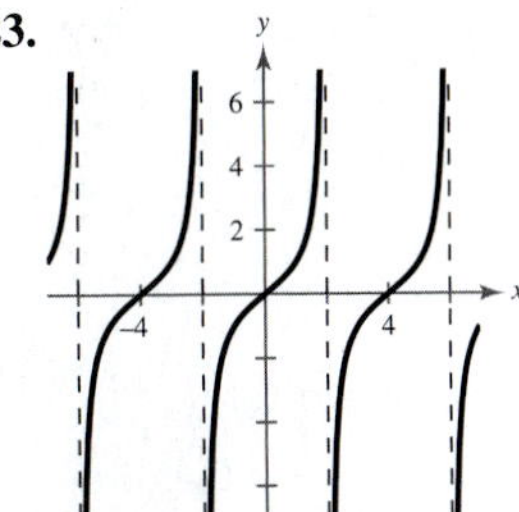

25.

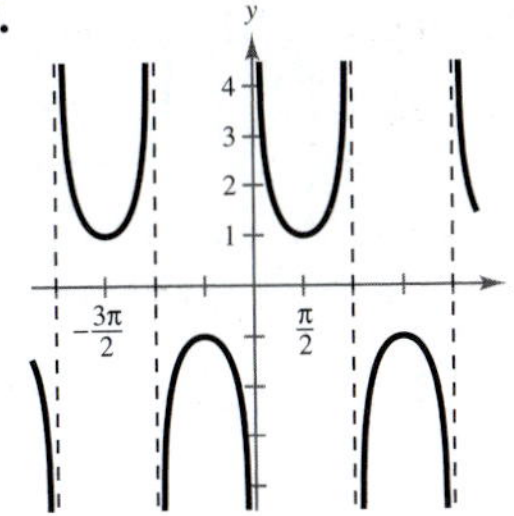

27.

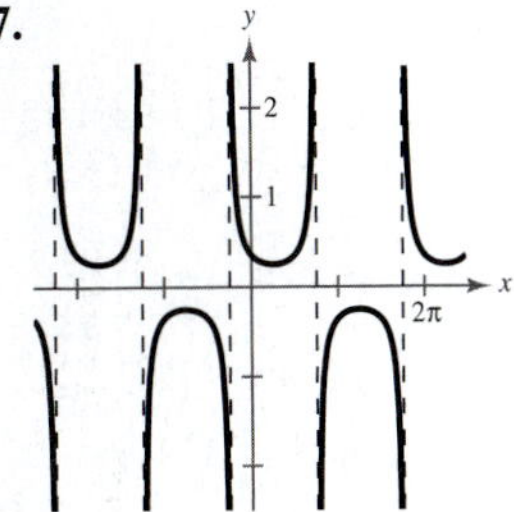

29.

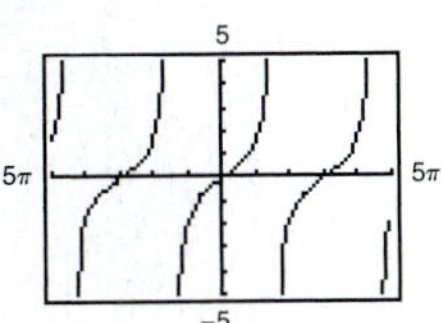

31.

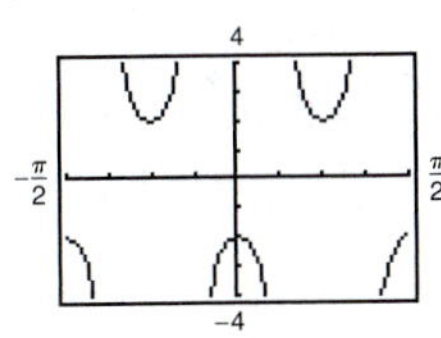

33.

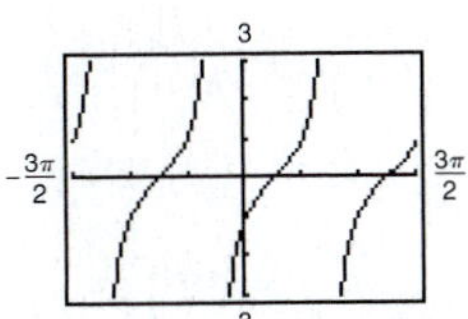

35.

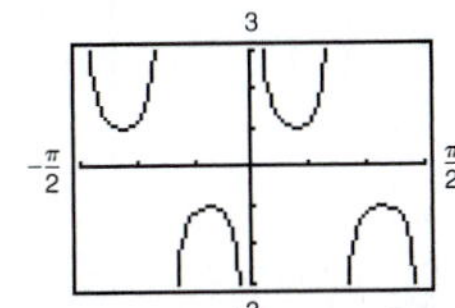

37.

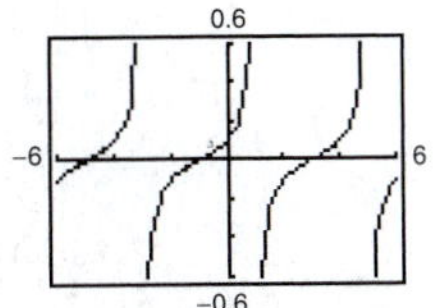

39. $-\frac{7\pi}{4}, -\frac{3\pi}{4}, \frac{\pi}{4}, \frac{5\pi}{4}$

41. $-\frac{4\pi}{3}, -\frac{\pi}{3}, \frac{2\pi}{3}, \frac{5\pi}{3}$

43. $-\frac{4\pi}{3}, -\frac{2\pi}{3}, \frac{2\pi}{3}, \frac{4\pi}{3}$

45. $-\frac{7\pi}{4}, -\frac{5\pi}{4}, \frac{\pi}{4}, \frac{3\pi}{4}$

47. Even

49. (a)

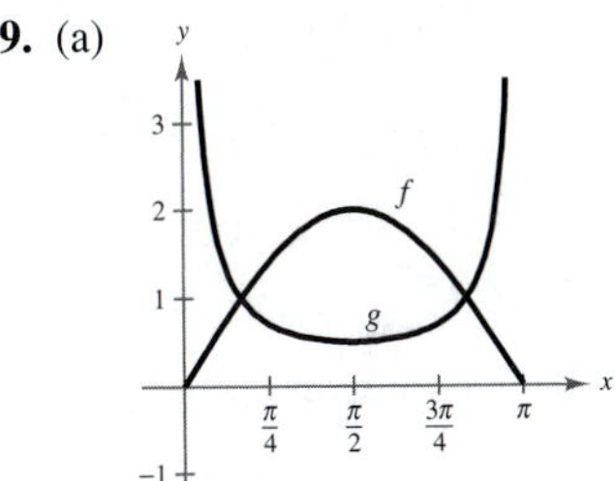

(b) $\frac{\pi}{6} < x < \frac{5\pi}{6}$

(c) f approaches 0 and g approaches $+\infty$ because the cosecant is the reciprocal of the sine.

51.

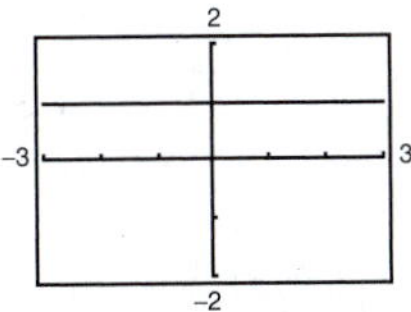

The expressions are equivalent except that when $\sin x = 0$, y_1 is undefined.

53.

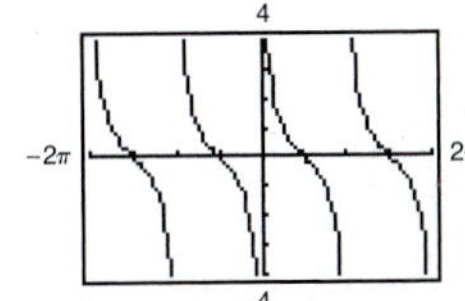

The expressions are equivalent.

55. d, $f \to 0$ as $x \to 0$.

56. a, $f \to 0$ as $x \to 0$.

57. b, $g \to 0$ as $x \to 0$.

58. c, $g \to 0$ as $x \to 0$.

59.

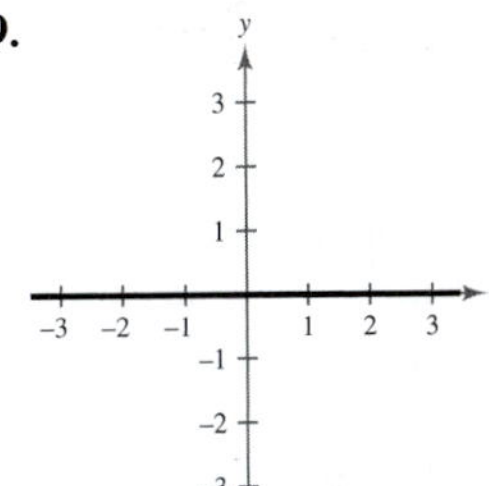

The functions are equal.

61.

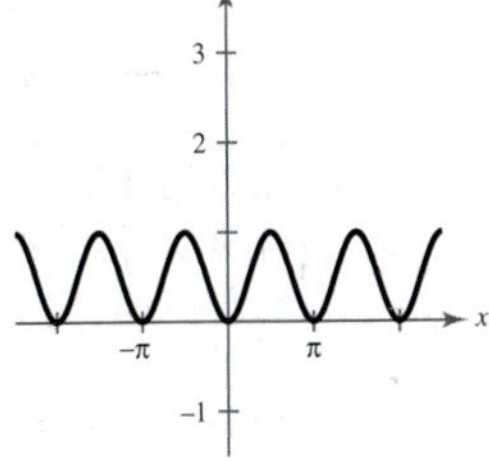

The functions are equal.

63.

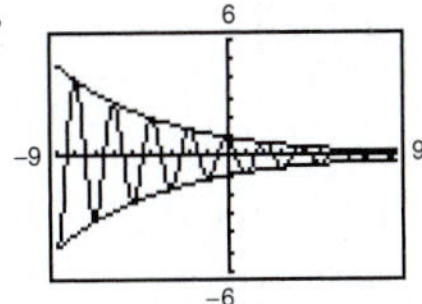

As $x \to \infty$, $f(x) \to 0$.

65.

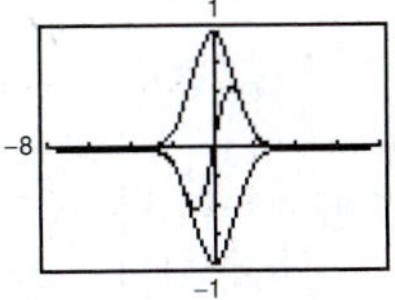

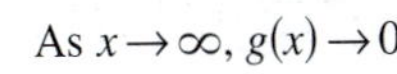

As $x \to \infty$, $g(x) \to 0$.

67.

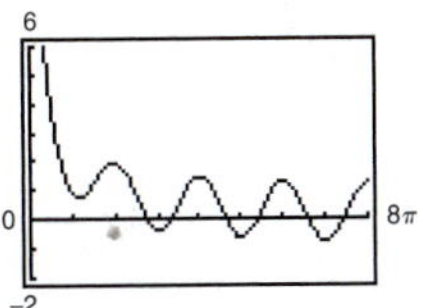

As $x \to 0$, $y \to \infty$.

69.

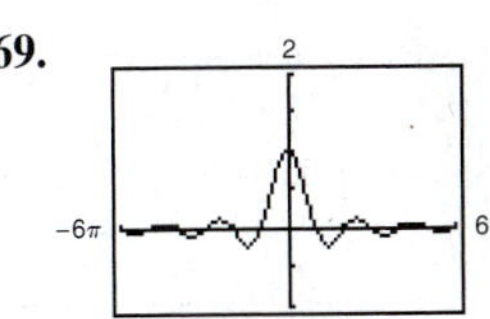

As $x \to 0$, $g(x) \to 1$.

71. 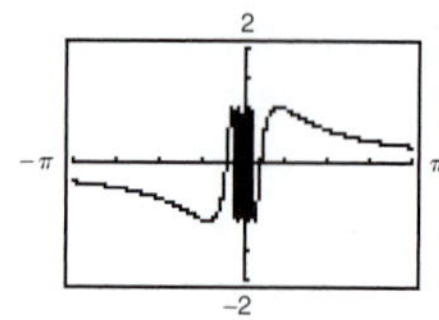

As $x \to 0$, $f(x)$ oscillates between 1 and -1.

73. $d = 7 \cot x$

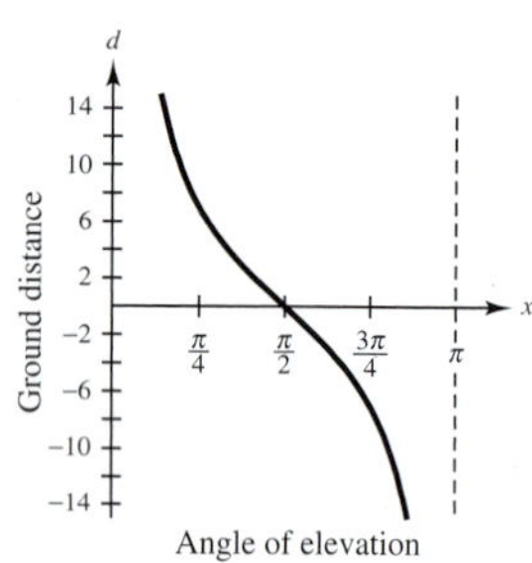

75. (a) 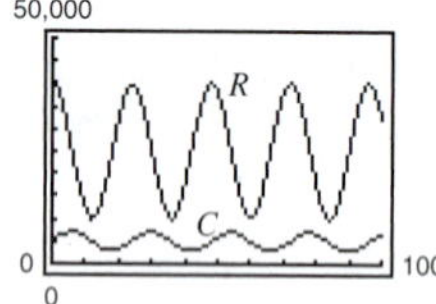

(b) As the predator population increases, the number of prey decreases. When the number of prey is small, the number of predators decreases.

(c) C: 24 months; R: 24 months

77. (a) 12

(b) Summer; Winter

(c) 1 month

79. True. For a given value of x, the y-coordinate of $\csc x$ is the reciprocal of the y-coordinate of $\sin x$.

81. As x approaches $\pi/2$ from the left, f approaches ∞. As x approaches $\pi/2$ from the right, f approaches $-\infty$.

83. (a) 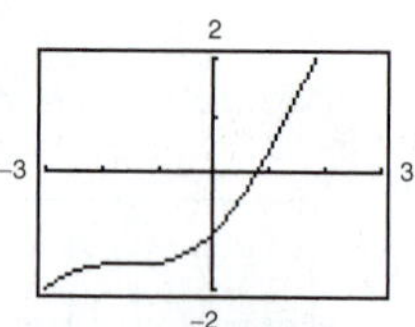

0.7391

(b) 1, 0.5403, 0.8576, 0.6543, 0.7935, 0.7014, 0.7640, 0.7221, 0.7504, 0.7314, . . . ; 0.7391

85. 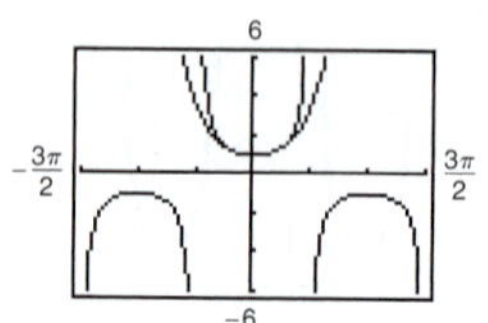

The graphs appear to coincide on the interval $-1.1 \le x \le 1.1$.

87. $\dfrac{\ln 54}{2} \approx 1.994$ **89.** $-\ln 2 \approx -0.693$

91. $\dfrac{2 + e^{73}}{3} \approx 1.684 \times 10^{31}$ **93.** $\pm\sqrt{e^{3.2} - 1} \approx \pm 4.851$

95. 2

Section 6.6 *(page 516)*

1. $\dfrac{\pi}{6}$ **3.** $\dfrac{\pi}{3}$ **5.** $\dfrac{\pi}{6}$ **7.** $\dfrac{5\pi}{6}$ **9.** $-\dfrac{\pi}{3}$ **11.** $\dfrac{2\pi}{3}$

13. $\dfrac{\pi}{3}$ **15.** 0 **17.** 1.29 **19.** -0.85 **21.** -1.25

23. 0.32 **25.** 1.99 **27.** 0.74 **29.** 0.85

31. 1.29 **33.** $-\dfrac{\pi}{3}, -\dfrac{\sqrt{3}}{3}, 1$

35. 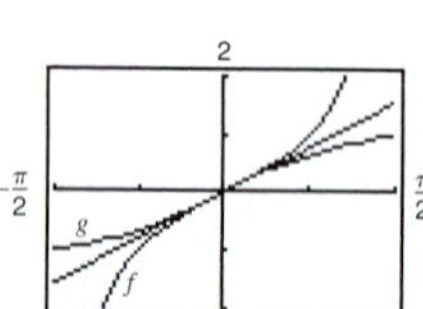

37. $\theta = \arctan \dfrac{x}{4}$

39. $\theta = \arcsin \dfrac{x + 2}{5}$ **41.** $\theta = \arccos \dfrac{x + 3}{2x}$ **43.** 0.3

45. -0.1 **47.** 0 **49.** $\dfrac{3}{5}$ **51.** $\dfrac{\sqrt{5}}{5}$ **53.** $\dfrac{12}{13}$

55. $\dfrac{\sqrt{34}}{5}$ **57.** $\dfrac{\sqrt{5}}{3}$ **59.** $\dfrac{1}{x}$ **61.** $\sqrt{1 - 4x^2}$

63. $\sqrt{1 - x^2}$ **65.** $\dfrac{\sqrt{9 - x^2}}{x}$ **67.** $\dfrac{\sqrt{x^2 + 2}}{x}$

69. 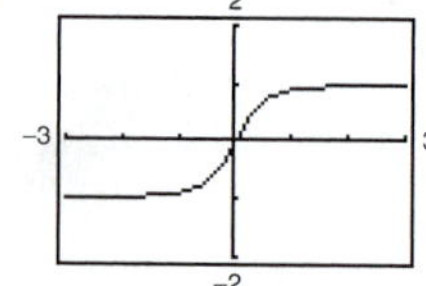

Asymptotes: $y = \pm 1$

71. $\dfrac{9}{\sqrt{x^2 + 81}}, x > 0; \dfrac{-9}{\sqrt{x^2 + 81}}, x < 0$

73. $\dfrac{|x - 1|}{\sqrt{x^2 - 2x + 10}}$

75.

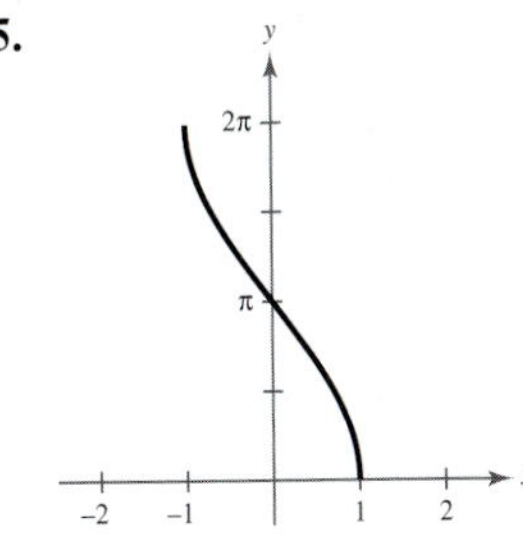

77.

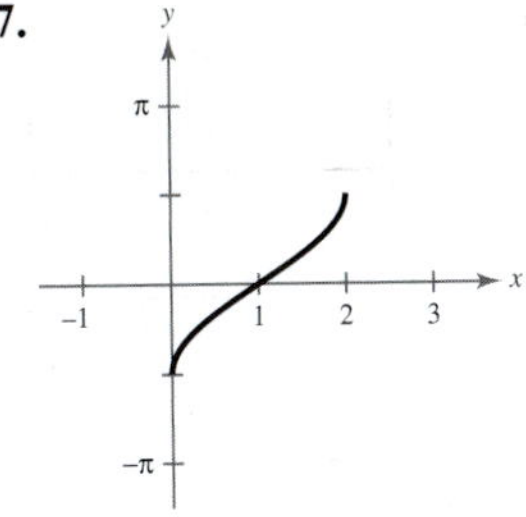

79.

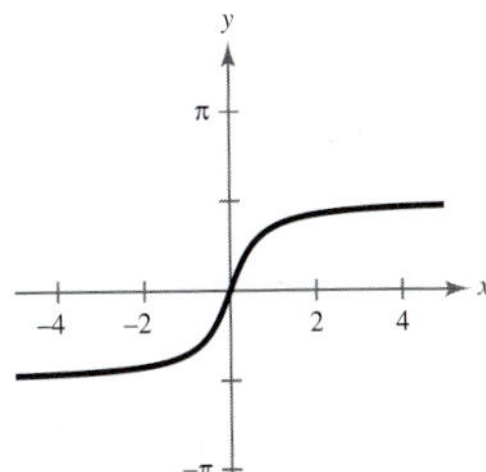

80.

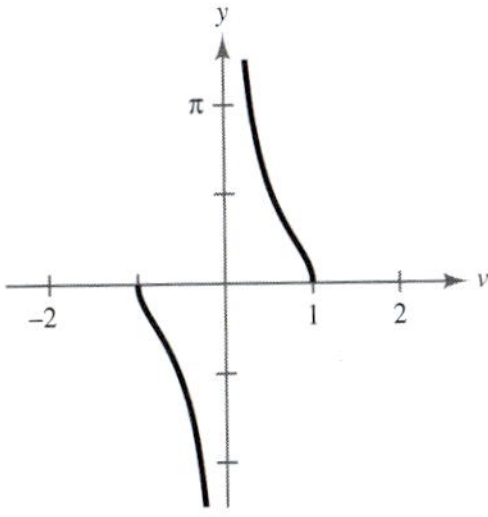

83.

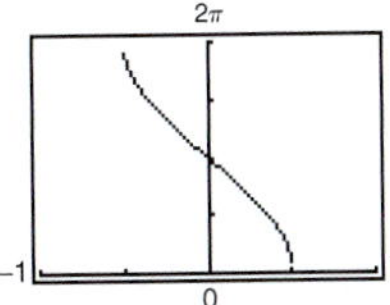

85.

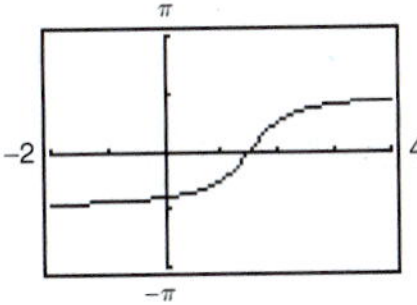

87.

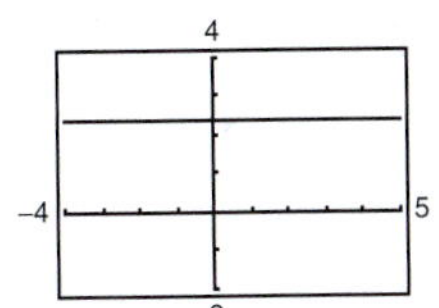

89. $3\sqrt{2}\sin\left(2t + \frac{\pi}{4}\right)$

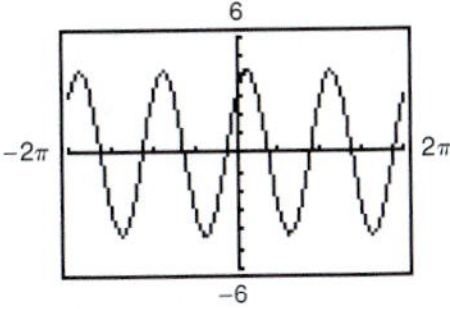

The graph implies that the identity is true.

91. (a) $\theta = \arcsin \frac{5}{s}$ (b) 0.13, 0.25

93. (a)

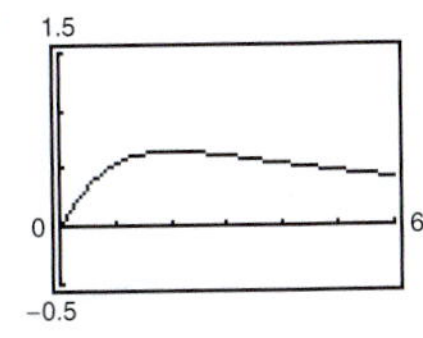

(b) 2 feet

(c) $\beta = 0$; As x increases, β approaches 0.

95. (a) $\theta \approx 26.0°$ (b) 24.4 feet

97. (a) $\theta = \arctan \frac{x}{20}$ (b) 14.0°, 31.0°

99. False. $\frac{5\pi}{4}$ is not in the range of the arctangent.

101. Domain: $(-\infty, -1] \cup [1, \infty)$

Range: $[0, \pi/2) \cup (\pi/2, \pi]$

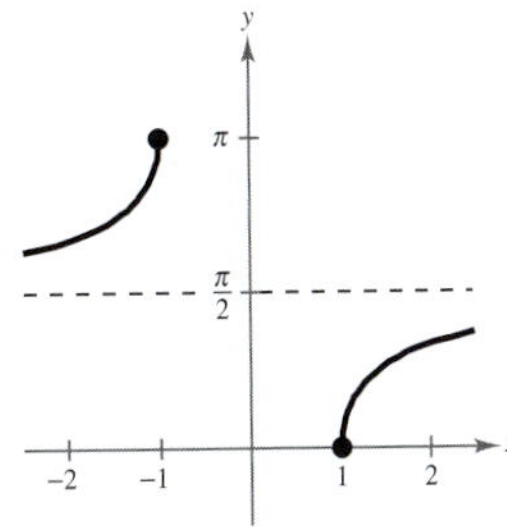

103. (a) $\frac{\pi}{4}$ (b) 0 (c) $\frac{5\pi}{6}$ (d) $\frac{\pi}{6}$

105.

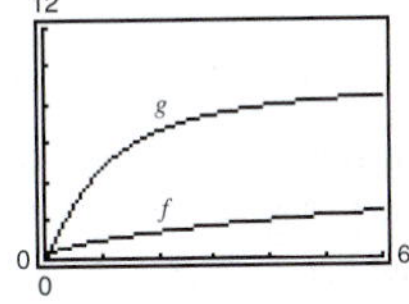

As x increases to infinity, g approaches 3π, but f has no maximum.

$a \approx 87.54$

107–111. Answers will vary.

113.

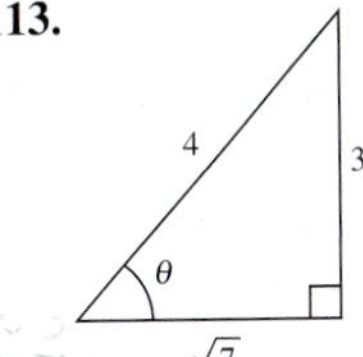

115.

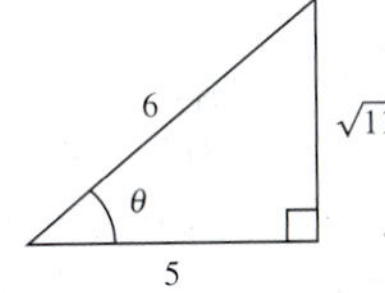

117. 1279.284 **119.** 117.391 **121.** Eight people

Section 6.7 *(page 526)*

1. $a \approx 3.64$
$c \approx 10.64$
$B = 70°$

3. $a \approx 8.26$
$c \approx 25.38$
$A = 19°$

5. $c \approx 11.66$
$A \approx 30.96°$
$B \approx 59.04°$

7. $a \approx 49.48$
$A \approx 72.08°$
$B \approx 17.92°$

9. $a \approx 91.34$
$b \approx 420.70$
$B = 77°45'$

11. 2.56 inches

13. 19.99 inches **15.** 107.2 feet **17.** 19.7 feet

19. (a)

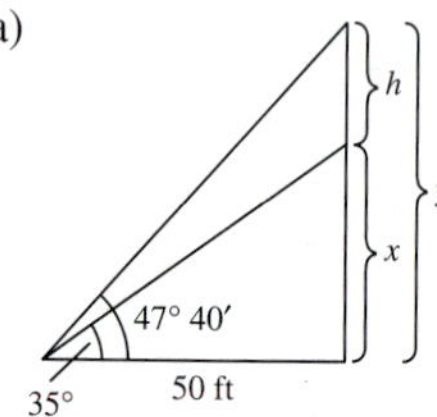

(b) $h = 50(\tan 47° 40' - \tan 35°)$ (c) 19.9 feet

21. 2236.8 feet

23. (a)

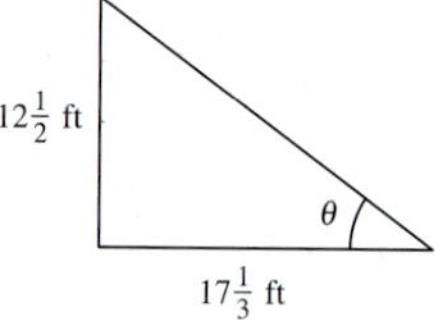

(b) $\tan \theta = \dfrac{12\frac{1}{2}}{17\frac{1}{3}}$

(c) 35.8°

25. 2.06° **27.** 0.73 mile

29. 554 miles north; 709 miles east **31.** 5.46 kilometers

33. 1933.3 feet **35.** $\approx$ 3.23 miles or $\approx$ 17,054 feet

37. 78.7° **39.** 35.3° **41.** 29.4 inches

43. $y = \sqrt{3}r$ **45.** $a \approx 12.2, b \approx 7$

47. (a) 4 (b) 4 (c) $\frac{1}{16}$

49. (a) $\frac{1}{16}$ (b) 60 (c) $\frac{1}{120}$ **51.** $d = 4\sin(\pi t)$

53. $d = 3\cos\left(\dfrac{4\pi t}{3}\right)$ **55.** $\omega = 528\pi$

57. (a)

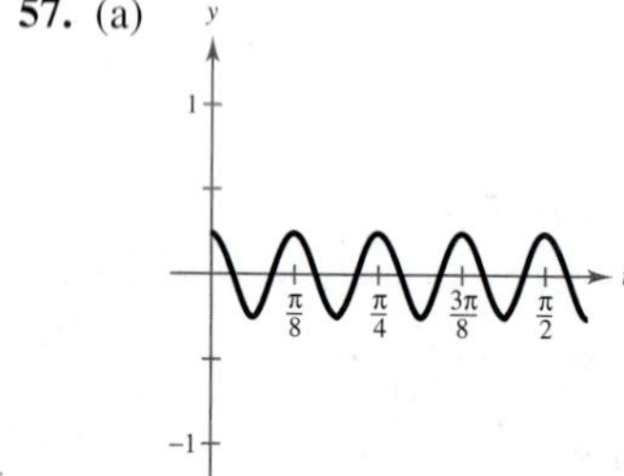

(b) $\dfrac{\pi}{8}$

(c) $\dfrac{\pi}{32}$

59. False. One period is the time for one complete cycle of the motion.

61. (a) and (b)

Base 1	Base 2	Altitude	Area
8	8 + 16 cos 10°	8 sin 10°	22.1
8	8 + 16 cos 20°	8 sin 20°	42.5
8	8 + 16 cos 30°	8 sin 30°	59.7
8	8 + 16 cos 40°	8 sin 40°	72.7
8	8 + 16 cos 50°	8 sin 50°	80.5
8	8 + 16 cos 60°	8 sin 60°	83.1
8	8 + 16 cos 70°	8 sin 70°	80.7

$\approx$83.1 square feet when $\theta = 60°$

(c) $A = 64(1 + \cos\theta)(\sin\theta)$

(d)

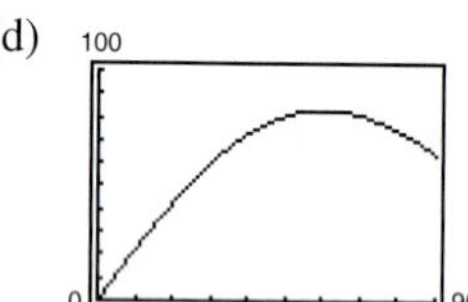

$\approx$83.1 square feet when $\theta = 60°$

63. (a)

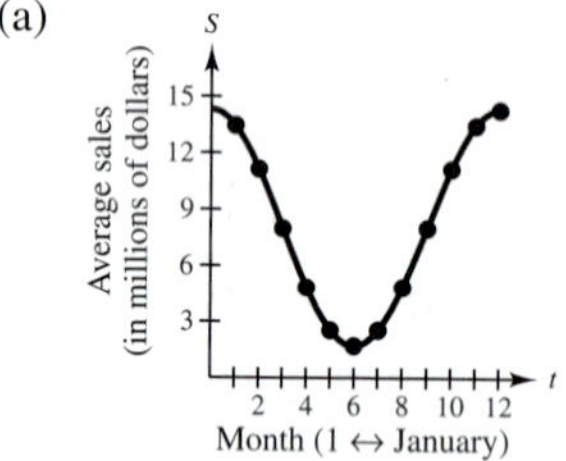

(b) $S = 8 + 6.3\cos\left(\dfrac{\pi}{6}t\right)$ or $S = 8 + 6.3\sin\left(\dfrac{\pi}{6}t + \dfrac{\pi}{2}\right)$

The model is a good fit.

(c) 12. Yes, sales of outerwear are seasonal.

(d) Maximum displacement from average sales of \$8 million

65.

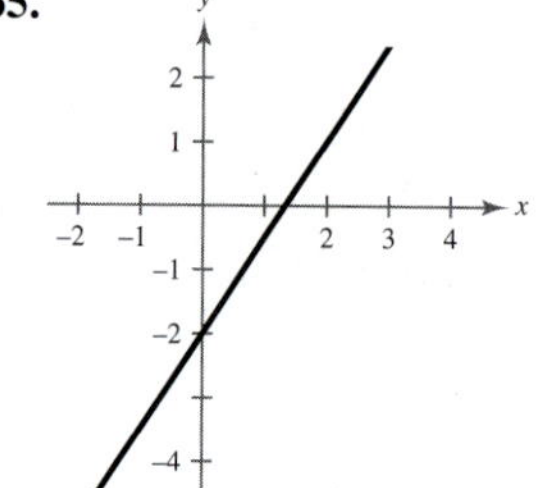

67.

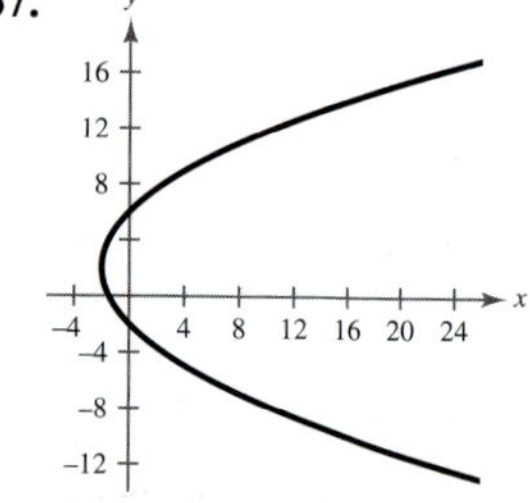

69.

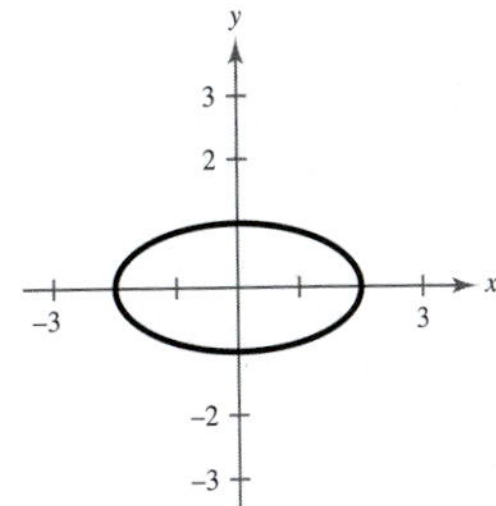

71.

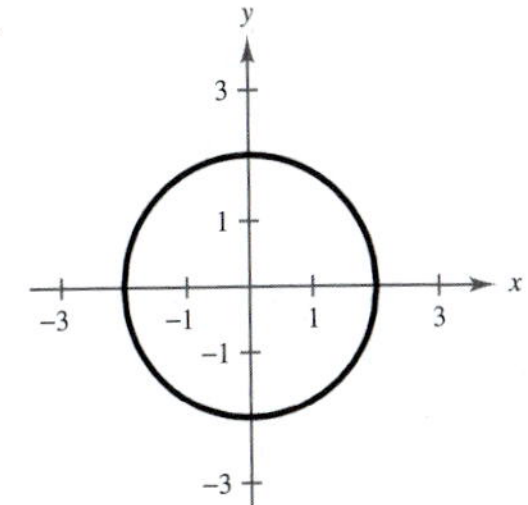

Review Exercises (page 532)

1. 40° **3.** 269°

5.

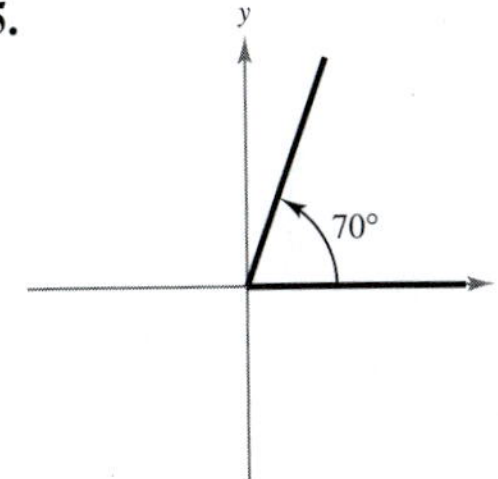

430°, −290°

7.

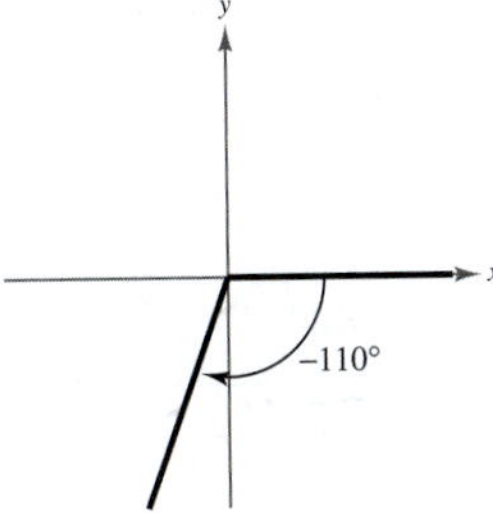

250°, −470°

9.

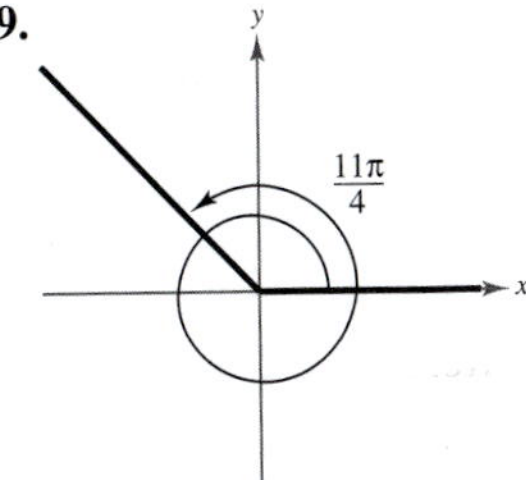

$\frac{3\pi}{4}, -\frac{5\pi}{4}$

11.

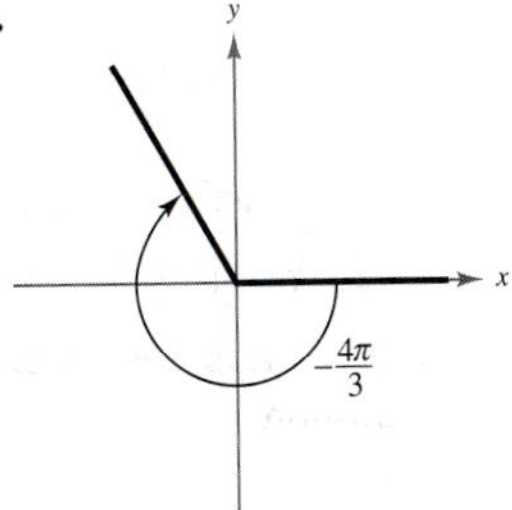

$\frac{2\pi}{3}, -\frac{10\pi}{3}$

13. 128.57° **15.** −108° **17.** −200.54°

19. 100.27° **21.** 8.3776 **23.** −0.2880

25. −0.5890 **27.** 1.4704

29. (a) $66\frac{2}{3}\pi$ radians per minute

(b) 400π inches per minute

31. $\sin\theta = \frac{4\sqrt{41}}{41}$, $\cos\theta = \frac{5\sqrt{41}}{41}$, $\tan\theta = \frac{4}{5}$, $\csc\theta = \frac{\sqrt{41}}{4}$, $\sec\theta = \frac{\sqrt{41}}{5}$, $\cot\theta = \frac{5}{4}$

33. $\sin\theta = \frac{\sqrt{3}}{2}$, $\cos\theta = \frac{1}{2}$, $\tan\theta = \sqrt{3}$, $\csc\theta = \frac{2\sqrt{3}}{3}$, $\sec\theta = 2$, $\cot\theta = \frac{\sqrt{3}}{3}$

35. (a) 3 (b) $\frac{2\sqrt{2}}{3}$ (c) $\frac{3\sqrt{2}}{4}$ (d) $\frac{\sqrt{2}}{4}$

37. (a) $\frac{1}{4}$ (b) $\frac{\sqrt{15}}{4}$ (c) $\frac{4\sqrt{15}}{15}$ (d) $\frac{\sqrt{15}}{15}$

39. 0.65 **41.** 0.56 **43.** 3.67 **45.** 0.98

47. 71.3 meters

49. $\sin\theta = \frac{4}{5}$, $\csc\theta = \frac{5}{4}$, $\cos\theta = \frac{3}{5}$, $\sec\theta = \frac{5}{3}$, $\tan\theta = \frac{4}{3}$, $\cot\theta = \frac{3}{4}$

51. $\sin\theta = \frac{15\sqrt{241}}{241}$, $\csc\theta = \frac{\sqrt{241}}{15}$, $\cos\theta = \frac{4\sqrt{241}}{241}$, $\sec\theta = \frac{\sqrt{241}}{4}$, $\tan\theta = \frac{15}{4}$, $\cot\theta = \frac{4}{15}$

53. $\sin\theta \approx 1$, $\csc\theta \approx 1$, $\cos\theta \approx -0.1$, $\sec\theta \approx -9$, $\tan\theta \approx -9$, $\cot\theta \approx -0.1$

55. $\sin\theta = \frac{4\sqrt{17}}{17}$, $\csc\theta = \frac{\sqrt{17}}{4}$, $\cos\theta = \frac{\sqrt{17}}{17}$, $\sec\theta = \sqrt{17}$, $\tan\theta = 4$, $\cot\theta = \frac{1}{4}$

57. $\sin\theta = -\frac{\sqrt{11}}{6}$, $\cos\theta = \frac{5}{6}$, $\tan\theta = -\frac{\sqrt{11}}{5}$, $\csc\theta = -\frac{6\sqrt{11}}{11}$, $\cot\theta = -\frac{5\sqrt{11}}{11}$

59. $\sin\theta = -\frac{5\sqrt{41}}{41}$, $\cos\theta = -\frac{4\sqrt{41}}{41}$, $\csc\theta = -\frac{\sqrt{41}}{5}$, $\sec\theta = -\frac{\sqrt{41}}{4}$, $\cot\theta = \frac{4}{5}$

61. $\sin \theta = \frac{12}{13}$

$\cos \theta = -\frac{5}{13}$

$\csc \theta = \frac{13}{12}$

$\sec \theta = -\frac{13}{5}$

$\cot \theta = -\frac{5}{12}$

63. $\sin \dfrac{\pi}{3} = \dfrac{\sqrt{3}}{2}$; $\cos \dfrac{\pi}{3} = \dfrac{1}{2}$; $\tan \dfrac{\pi}{3} = \sqrt{3}$

65. $\sin \dfrac{5\pi}{6} = \dfrac{1}{2}$; $\cos \dfrac{5\pi}{6} = -\dfrac{\sqrt{3}}{2}$; $\tan \dfrac{5\pi}{6} = -\dfrac{\sqrt{3}}{3}$

67. $\sin\left(-\dfrac{7\pi}{3}\right) = -\dfrac{\sqrt{3}}{2}$; $\cos\left(-\dfrac{7\pi}{3}\right) = \dfrac{1}{2}$;

$\tan\left(-\dfrac{7\pi}{3}\right) = -\sqrt{3}$

69. $\sin 495° = \dfrac{\sqrt{2}}{2}$; $\cos 495° = -\dfrac{\sqrt{2}}{2}$; $\tan 495° = -1$

71. $\sin(-150°) = -\dfrac{1}{2}$; $\cos(-150°) = -\dfrac{\sqrt{3}}{2}$;

$\tan(-150°) = \dfrac{\sqrt{3}}{3}$

73. -0.76 **75.** -1.06 **77.** 0.14

79. 0 **81.** 3.24 **83.** 0.41

85.

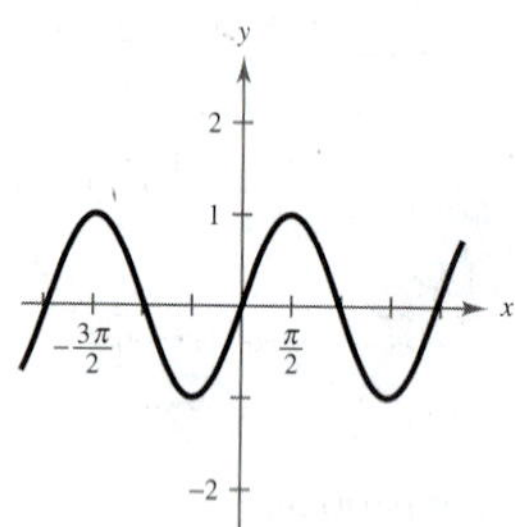

87.

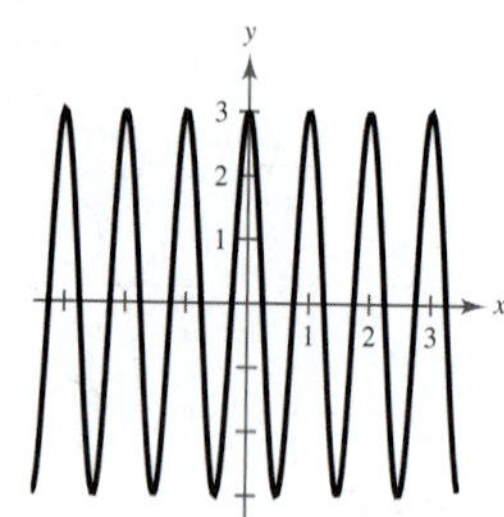

89.

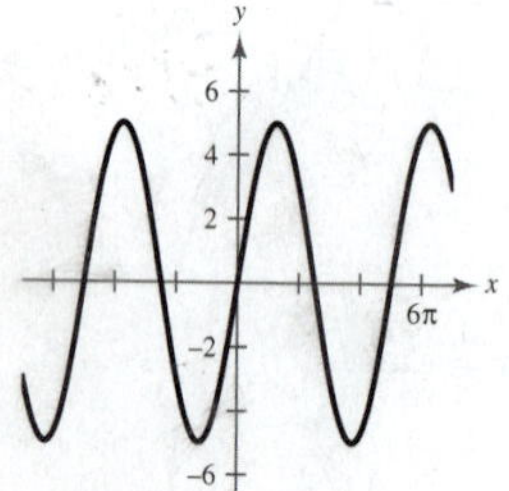

91.

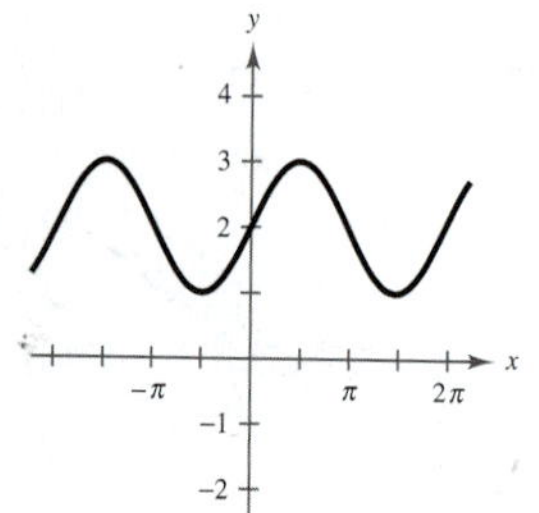

93.

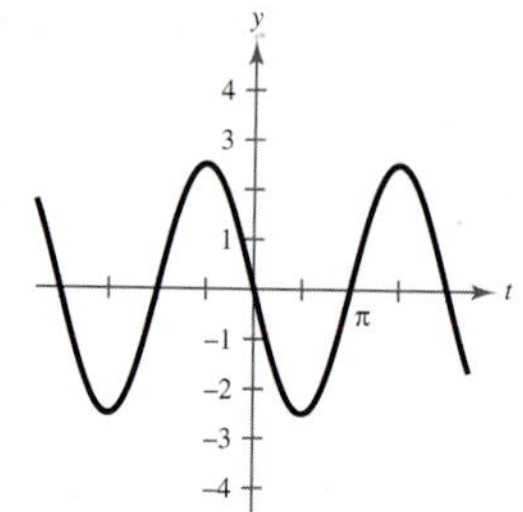

95. (a) $y = 2 \sin 528\pi x$

(b) 264 cycles per second

97.

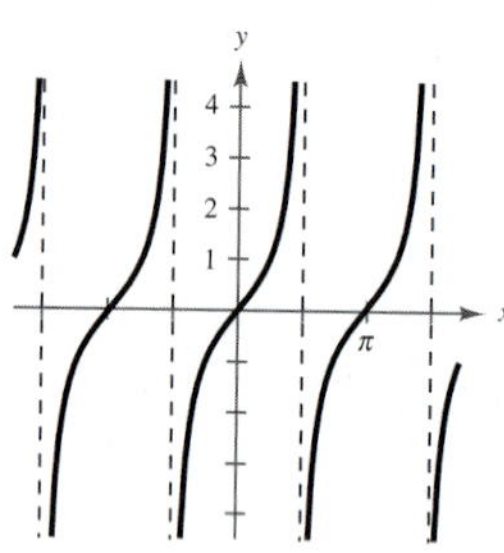

99.

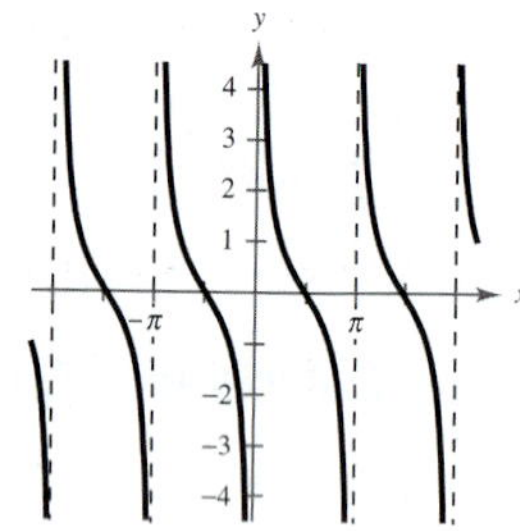

101.

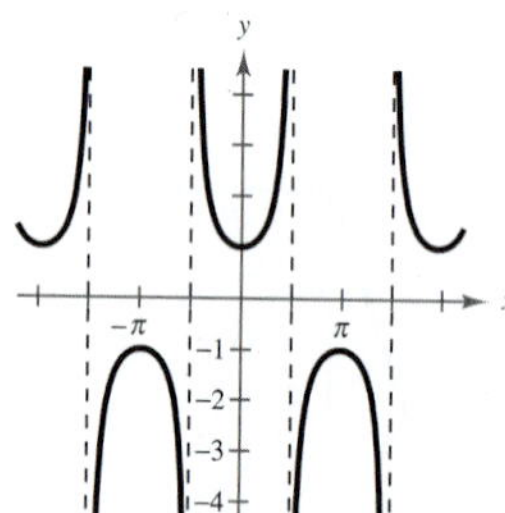

103.

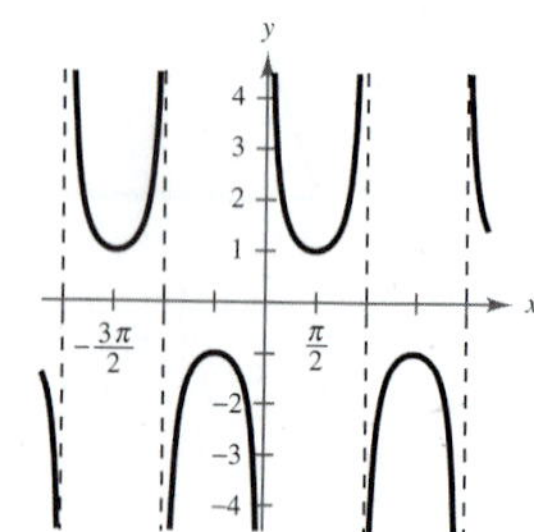

105.

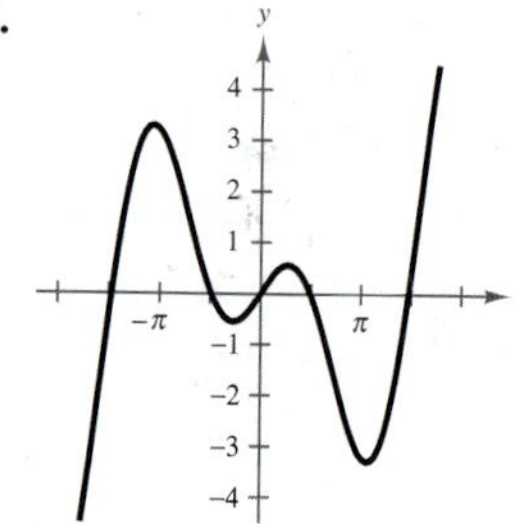

107. $-\dfrac{\pi}{6}$ **109.** 0.41

111. -0.46 **113.** $\dfrac{\pi}{6}$ **115.** π **117.** 1.24

119. 0.12 **121.** 1.40 **123.** -0.98 **125.** 0.72

127. 0 **129.** $\frac{4}{5}$ **131.** $\frac{13}{5}$ **133.** 66.8°

135. 1221 miles, 85.6°

137. False. The sine or cosine function is often useful for modeling simple harmonic motion.

139. False. For each θ there corresponds exactly one value of y.

141. d; The period is 2π and the amplitude is 3.

142. a; The period is 2π and, because $a < 0$, the graph is reflected in the x-axis.

143. b; The period is 2 and the amplitude is 2.

144. c; The period is 4π and the amplitude is 2.

145. The function is undefined because $\sec\theta = 1/\cos\theta$.

147. The ranges of the other four trigonometric functions are $(-\infty, \infty)$ or $(-\infty, -1] \cup [1, \infty)$.

149. (a) $A = 72(\tan\theta - \theta)$

(b)

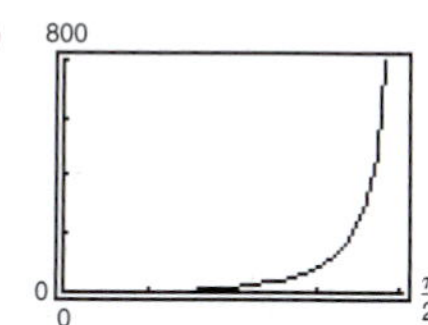

Area increases without bound as θ approaches $\pi/2$.

151. Answers will vary.

Chapter Test *(page 536)*

1. (a)

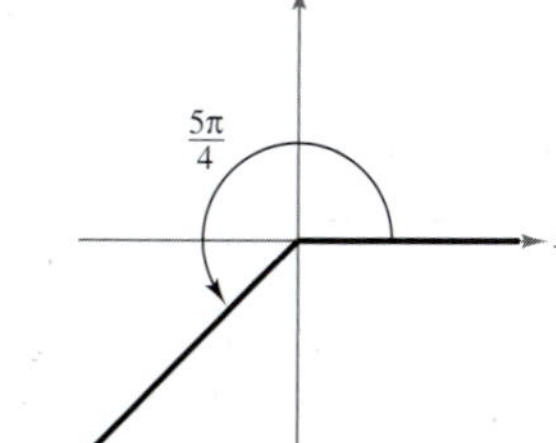

(b) 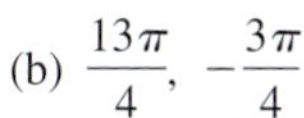 $\frac{13\pi}{4}, -\frac{3\pi}{4}$

(c) $225°$

2. 3000 radians per minute

3. $\sin\theta = \frac{3\sqrt{10}}{10}$ $\quad\csc\theta = \frac{\sqrt{10}}{3}$

$\cos\theta = -\frac{\sqrt{10}}{10}$ $\quad\sec\theta = -\sqrt{10}$

$\tan\theta = -3$ $\quad\cot\theta = -\frac{1}{3}$

4. For $0 \le \theta < \frac{\pi}{2}$: $\quad$ For $\pi \le \theta < \frac{3\pi}{2}$:

$\sin\theta = \frac{3\sqrt{13}}{13}$ $\quad\sin\theta = -\frac{3\sqrt{13}}{13}$

$\cos\theta = \frac{2\sqrt{13}}{13}$ $\quad\cos\theta = -\frac{2\sqrt{13}}{13}$

$\csc\theta = \frac{\sqrt{13}}{3}$ $\quad\csc\theta = -\frac{\sqrt{13}}{3}$

$\sec\theta = \frac{\sqrt{13}}{2}$ $\quad\sec\theta = -\frac{\sqrt{13}}{2}$

$\cot\theta = \frac{2}{3}$ $\quad\cot\theta = \frac{2}{3}$

5. $\theta' = 70°$

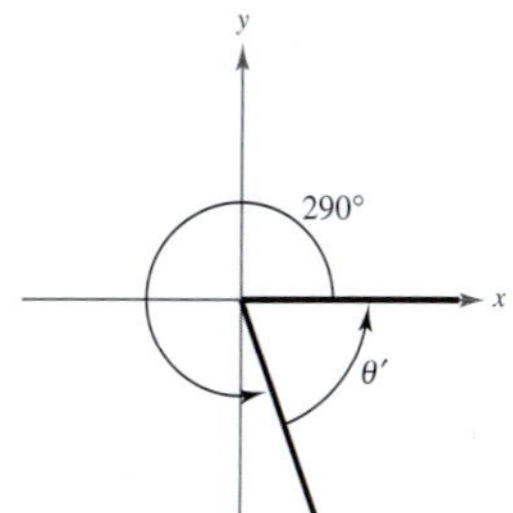

6. Quadrant III $\quad$ **7.** $150°, 210°$ $\quad$ **8.** $1.33, 1.81$

9. $\sin\theta = -\frac{4}{5}$ $\quad$ **10.** $\sin\theta = \frac{15}{17}$

$\tan\theta = -\frac{4}{3}$ $\quad\cos\theta = -\frac{8}{17}$

$\csc\theta = -\frac{5}{4}$ $\quad\tan\theta = -\frac{15}{8}$

$\sec\theta = \frac{5}{3}$ $\quad\csc\theta = \frac{17}{15}$

$\cot\theta = -\frac{3}{4}$ $\quad\cot\theta = -\frac{8}{15}$

11.

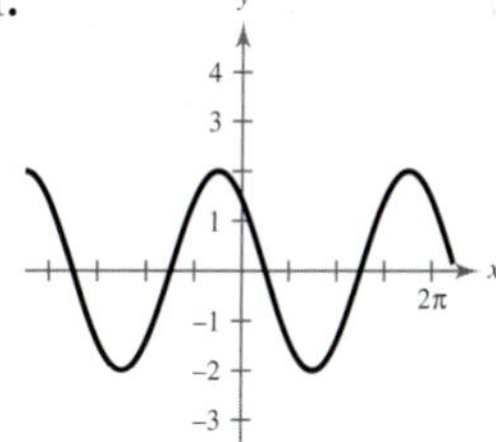

12.

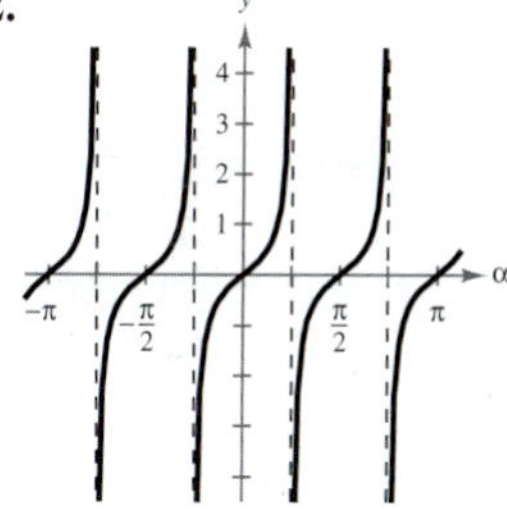

13.

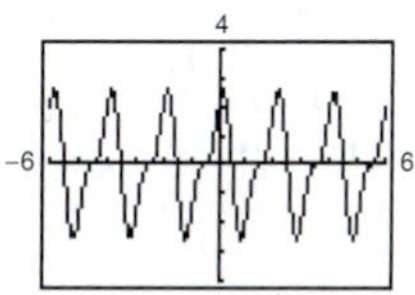

Period: 2

14.

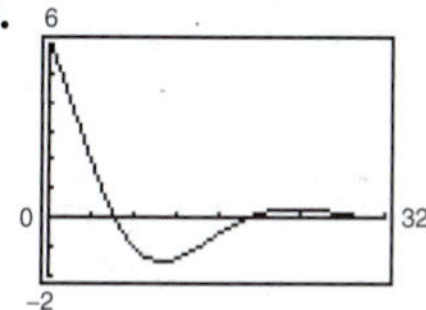

Not periodic

15. $a = -2, b = \frac{1}{2}, c = -\frac{\pi}{4}$ $\quad$ **16.** 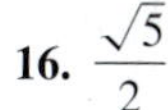 $\frac{\sqrt{5}}{2}$

17.

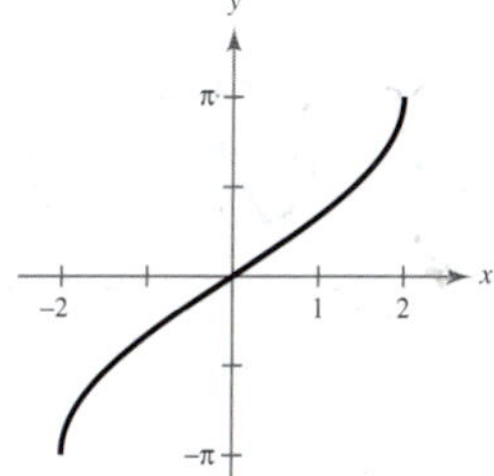

18. $310.1°$ $\quad$ **19.** $d = -6\cos\pi t$

Problem Solving (page 538)

1. (a) $\frac{11\pi}{2}$ radians or 990° (b) ≈ 816.42 feet

3. (a) 4767 feet (b) 3705 feet

(c) $\tan 63° = \frac{w + 3705}{3000}$,

$w = 2183$ feet

5. (a)

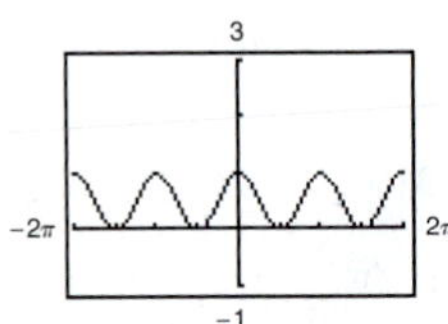

Even

(b)

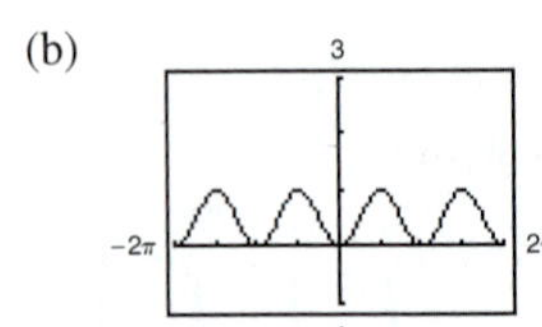

Even

7. $h = 51 - 50 \sin\left(8\pi t + \frac{\pi}{2}\right)$

9. (a)

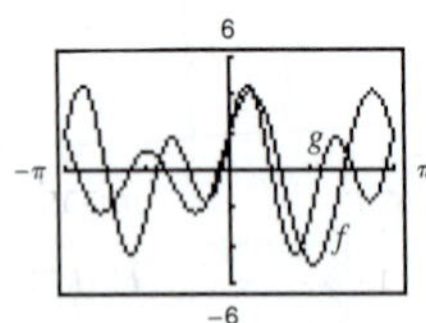

(b) Period of f: 2π

Period of g: π

(c) Yes, because the sine and cosine functions are periodic.

11. (a) Equal; two-period shift

(b) Not equal; $f\left(t + \frac{1}{2}c\right)$ is a horizontal translation and $f\left(\frac{1}{2}t\right)$ is a period change.

(c) Not equal; For example, $\sin\left[\frac{1}{2}(\pi + 2\pi)\right] \neq \sin\left(\frac{1}{2}\pi\right)$.

13. (a)

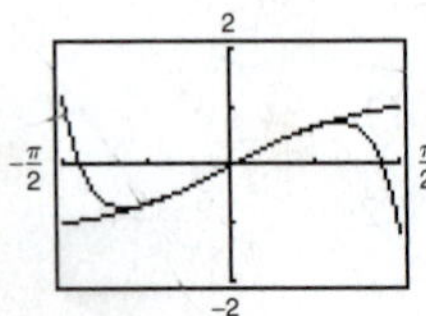

The approximation is accurate over the interval $-1 \leq x \leq 1$.

(b) $\frac{x^9}{9}$

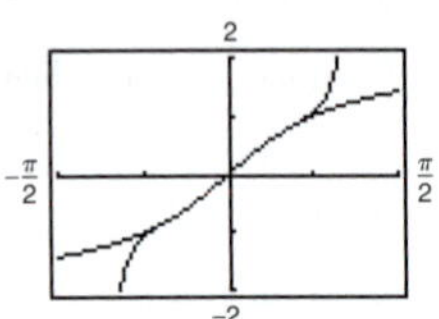

The accuracy improved.

Chapter 7

Section 7.1 (page 547)

1. $\tan x = -\sqrt{3}$

$\csc x = \frac{2\sqrt{3}}{3}$

$\sec x = -2$

$\cot x = -\frac{\sqrt{3}}{3}$

3. $\cos \theta = \frac{\sqrt{2}}{2}$

$\tan \theta = -1$

$\csc \theta = -\sqrt{2}$

$\cot \theta = -1$

5. $\sin x = -\frac{5}{13}$

$\cos x = -\frac{12}{13}$

$\csc x = -\frac{13}{5}$

$\cot x = \frac{12}{5}$

7. $\sin \phi = -\frac{\sqrt{5}}{3}$

$\cos \phi = \frac{2}{3}$

$\tan \phi = -\frac{\sqrt{5}}{2}$

$\cot \phi = -\frac{2\sqrt{5}}{5}$

9. $\sin x = \frac{1}{3}$

$\cos x = -\frac{2\sqrt{2}}{3}$

$\csc x = 3$

$\sec x = -\frac{3\sqrt{2}}{4}$

$\cot x = -2\sqrt{2}$

11. $\sin \theta = -\frac{2\sqrt{5}}{5}$

$\cos \theta = -\frac{\sqrt{5}}{5}$

$\csc \theta = -\frac{\sqrt{5}}{2}$

$\sec \theta = -\sqrt{5}$

$\cot \theta = \frac{1}{2}$

13. $\cos \theta = 0$

$\tan \theta$ is undefined.

$\csc \theta = -1$

$\sec \theta$ is undefined.

15. d **16.** a **17.** b **18.** f **19.** e **20.** c

21. b **22.** c **23.** f **24.** a **25.** e **26.** d

27. $\csc\theta$ **29.** $\cos^2\phi$ **31.** $\cos x$ **33.** $\sin^2 x$

35. 1 **37.** $\tan x$ **39.** $1 + \sin y$ **41.** $\sec\beta$

43. $\cos u + \sin u$ **45.** $\sin^2 x$ **47.** $\sin^2 x \tan^2 x$

49. $\sec x + 1$ **51.** $\sec^4 x$ **53.** $\sin^2 x - \cos^2 x$

55. $\cot^2 x(\csc x - 1)$ **57.** $1 + 2\sin x\cos x$

59. $4\cot^2 x$ **61.** $2\csc^2 x$ **63.** $2\sec x$

65. $1 + \cos y$ **67.** $3(\sec x + \tan x)$

69.

x	0.2	0.4	0.6	0.8	1.0
y_1	0.1987	0.3894	0.5646	0.7174	0.8415
y_2	0.1987	0.3894	0.5646	0.7174	0.8415

x	1.2	1.4
y_1	0.9320	0.9854
y_2	0.9320	0.9854

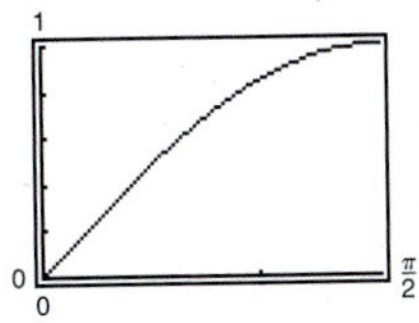

$y_1 = y_2$

71.

x	0.2	0.4	0.6	0.8	1.0
y_1	1.2230	1.5085	1.8958	2.4650	3.4082
y_2	1.2230	1.5085	1.8958	2.4650	3.4082

x	1.2	1.4
y_1	5.3319	11.6814
y_2	5.3319	11.6814

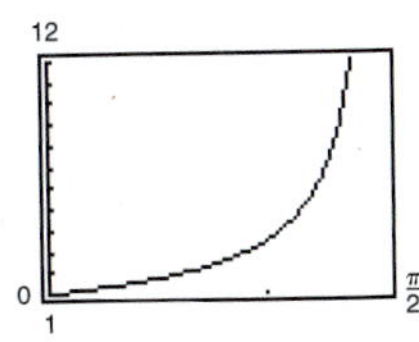

$y_1 = y_2$

73. $\csc x$ **75.** $\tan x$ **77.** $3\sin\theta$ **79.** $3\tan\theta$

81. $5\sec\theta$ **83.** $3\cos\theta = 3$; $\sin\theta = 0$; $\cos\theta = 1$

85. $4\sin\theta = 2\sqrt{2}$; $\sin\theta = \dfrac{\sqrt{2}}{2}$; $\cos\theta = \dfrac{\sqrt{2}}{2}$

87. $0 \le \theta \le \pi$ **89.** $0 \le \theta < \dfrac{\pi}{2}, \dfrac{3\pi}{2} < \theta < 2\pi$

91. $\ln|\cot x|$ **93.** $\ln|\csc t \sec t|$

95. (a) $\csc^2 132° - \cot^2 132° \approx 1.8107 - 0.8107 = 1$

(b) $\csc^2 \dfrac{2\pi}{7} - \cot^2 \dfrac{2\pi}{7} \approx 1.6360 - 0.6360 = 1$

97. (a) $\cos(90° - 80°) = \sin 80° \approx 0.9848$

(b) $\cos\left(\dfrac{\pi}{2} - 0.8\right) = \sin 0.8 \approx 0.7174$

99. $\mu = \tan\theta$

101. True. For example, $\sin(-x) = -\sin x$.

103. 1, 1 **105.** ∞, 0

107. Not an identity because $\cos\theta = \pm\sqrt{1 - \sin^2\theta}$

109. Not an identity because $\dfrac{\sin k\theta}{\cos k\theta} = \tan k\theta$

111. Identity because $\sin\theta \cdot \dfrac{1}{\sin\theta} = 1$

113. Answers will vary. For example:

$$\sin^2\theta + \cos^2\theta = \frac{y^2}{r^2} + \frac{x^2}{r^2} = \frac{x^2 + y^2}{r^2} = \frac{r^2}{r^2} = 1$$

115. $x - 25$

117. $\dfrac{x^2 + 6x - 8}{(x + 5)(x - 8)}$ **119.** $\dfrac{-5x^2 + 8x + 28}{(x^2 - 4)(x + 4)}$

Section 7.2 *(page 555)*

1–39. Answers will vary.

41. Identity **43.** Not an identity **45.** Identity

47. Identity **49.** Not an identity **51.** Identity

53. 1 **55.** 2 **57.** Answers will vary.

59. False. An identity is an equation that is true for all real values of θ.

61. The equation is not an identity because $\sin\theta = \pm\sqrt{1 - \cos^2\theta}$.

Possible answer: $\dfrac{7\pi}{4}$

63. $2 + \left(3 - \sqrt{26}\right)i$

65. $-8 + 4i$ **67.** $3 \pm \sqrt{3}i$ **69.** $-1 \pm \sqrt{3}i$

Section 7.3 *(page 564)*

1–5. Answers will vary. **7.** $\dfrac{2\pi}{3} + 2n\pi, \dfrac{4\pi}{3} + 2n\pi$

9. $\dfrac{\pi}{3} + 2n\pi, \dfrac{2\pi}{3} + 2n\pi$ **11.** $\dfrac{\pi}{6} + n\pi, \dfrac{5\pi}{6} + n\pi$

13. $n\pi, \dfrac{3\pi}{2} + 2n\pi$ **15.** $\dfrac{\pi}{3} + n\pi, \dfrac{2\pi}{3} + n\pi$

17. $\dfrac{\pi}{8} + \dfrac{n\pi}{2}, \dfrac{3\pi}{8} + \dfrac{n\pi}{2}$ **19.** $\dfrac{n\pi}{3}, \dfrac{\pi}{4} + n\pi$

21. $0, \frac{\pi}{2}, \pi, \frac{3\pi}{2}$ **23.** $0, \pi, \frac{\pi}{6}, \frac{5\pi}{6}, \frac{7\pi}{6}, \frac{11\pi}{6}$

25. $\frac{\pi}{3}, \frac{5\pi}{3}, \pi$ **27.** No solution **29.** $\pi, \frac{\pi}{3}, \frac{5\pi}{3}$

31. $\frac{\pi}{6}, \frac{5\pi}{6}, \frac{7\pi}{6}, \frac{11\pi}{6}$ **33.** $\frac{\pi}{6} + n\pi, \frac{5\pi}{6} + n\pi$

35. $\frac{\pi}{12} + \frac{n\pi}{3}$ **37.** $\frac{\pi}{2} + 4n\pi, \frac{7\pi}{2} + 4n\pi$

39. $-1 + 4n$ **41.** $-2 + 6n, 2 + 6n$

43. $\frac{2}{3}, \frac{3}{2}$; $0.8411 + 2n\pi, 5.4421 + 2n\pi$

45. 2.6779, 5.8195 **47.** 1.0472, 5.2360

49. 0.8603, 3.4256 **51.** 0, 2.6779, 3.1416, 5.8195

53. 0.9828, 1.7682, 4.1244, 4.9098

55. 0.3398, 0.8481, 2.2935, 2.8018

57. 1.9357, 2.7767, 5.0773, 5.9183

59. $\frac{\pi}{4}, \frac{5\pi}{4}$, arctan 5, arctan $5 + \pi$ **61.** $\frac{\pi}{3}, \frac{5\pi}{3}$

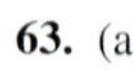

63. (a)

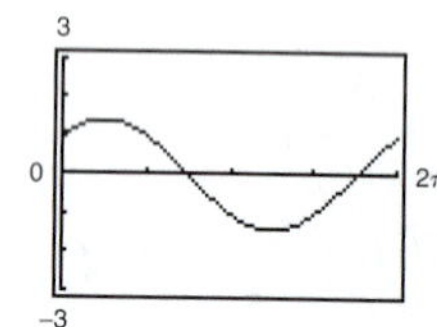

(b) $\frac{\pi}{4} \approx 0.7854$

$\frac{5\pi}{4} \approx 3.9270$

Maximum: (0.7854, 1.4142)

Minimum: (3.9270, −1.4142)

65. 1

67. (a) All real numbers $x \neq 0$

(b) y-axis symmetry; Horizontal asymptote: $y = 1$

(c) Oscillates

(d) Infinitely many solutions

(e) Yes, 0.6366

69. 0.04 second, 0.43 second, 0.83 second

71. February, March, and April **73.** 1.9°

75. (a)

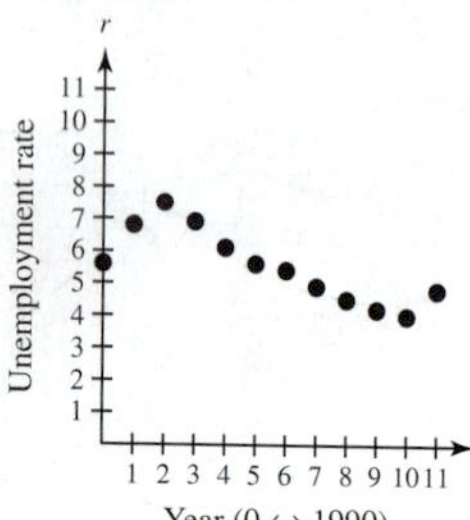

(b) (1)

(c) The constant term, 5.51

(d) ≈13 years

(e) 2004

77. True. The first equation has a smaller period than the second equation, so it will have more solutions in the interval $[0, 2\pi)$.

79. 1 **81.** $C = 24°$

$a \approx 54.8$

$b \approx 50.1$

83. $\sin 390° = \frac{1}{2}$

$\cos 390° = \frac{\sqrt{3}}{2}$

$\tan 390° = \frac{\sqrt{3}}{3}$

85. $\sin(-1845°) = -\frac{\sqrt{2}}{2}$

$\cos(-1845°) = \frac{\sqrt{2}}{2}$

$\tan(-1845°) = -1$

87. 1.36°

Section 7.4 *(page 572)*

1. (a) $\frac{\sqrt{2} - \sqrt{6}}{4}$ (b) $\frac{\sqrt{2} + 1}{2}$

3. (a) $\frac{1}{2}$ (b) $\frac{-\sqrt{3} - 1}{2}$

5. (a) $\frac{-\sqrt{2} - \sqrt{6}}{4}$ (b) $\frac{-1 + \sqrt{2}}{2}$

7. $\sin 105° = \frac{\sqrt{2}}{4}(\sqrt{3} + 1)$

$\cos 105° = \frac{\sqrt{2}}{4}(1 - \sqrt{3})$

$\tan 105° = -2 - \sqrt{3}$

9. $\sin 195° = \frac{\sqrt{2}}{4}(1 - \sqrt{3})$

$\cos 195° = -\frac{\sqrt{2}}{4}(\sqrt{3} + 1)$

$\tan 195° = 2 - \sqrt{3}$

11. $\sin \frac{11\pi}{12} = \frac{\sqrt{2}}{4}(\sqrt{3} - 1)$

$\cos \frac{11\pi}{12} = -\frac{\sqrt{2}}{4}(\sqrt{3} + 1)$

$\tan \frac{11\pi}{12} = -2 + \sqrt{3}$

13. $\sin \frac{17\pi}{12} = -\frac{\sqrt{2}}{4}(\sqrt{3} + 1)$

$\cos \frac{17\pi}{12} = \frac{\sqrt{2}}{4}(1 - \sqrt{3})$

$\tan \frac{17\pi}{12} = 2 + \sqrt{3}$

15. $\sin 285° = -\frac{\sqrt{2}}{4}(\sqrt{3} + 1)$

$\cos 285° = \frac{\sqrt{2}}{4}(\sqrt{3} - 1)$

$\tan 285° = -(2 + \sqrt{3})$

17. $\sin(-165^\circ) = -\frac{\sqrt{2}}{4}(\sqrt{3} - 1)$

$\cos(-165^\circ) = -\frac{\sqrt{2}}{4}(1 + \sqrt{3})$

$\tan(-165^\circ) = 2 - \sqrt{3}$

19. $\sin\frac{13\pi}{12} = \frac{\sqrt{2}}{4}(1 - \sqrt{3})$

$\cos\frac{13\pi}{12} = -\frac{\sqrt{2}}{4}(1 + \sqrt{3})$

$\tan\frac{13\pi}{12} = 2 - \sqrt{3}$

21. $\sin\left(-\frac{13\pi}{12}\right) = \frac{\sqrt{2}}{4}(\sqrt{3} - 1)$

$\cos\left(-\frac{13\pi}{12}\right) = -\frac{\sqrt{2}}{4}(\sqrt{3} + 1)$

$\tan\left(-\frac{13\pi}{12}\right) = -2 + \sqrt{3}$

23. $\cos 40^\circ$ **25.** $\tan 239^\circ$ **27.** $\sin 1.8$ **29.** $\tan 3x$

31. $-\frac{\sqrt{3}}{2}$ **33.** $\frac{\sqrt{3}}{2}$ **35.** -1 **37.** $-\frac{63}{65}$

39. $\frac{16}{65}$ **41.** $-\frac{63}{16}$ **43.** $\frac{65}{56}$ **45.** $\frac{3}{5}$ **47.** $-\frac{44}{117}$

49. $\frac{5}{3}$ **51.** 1 **53.** 0 **55–63.** Answers will vary.

65. $-\sin x$ **67.** $-\cos\theta$ **69.** $\frac{\pi}{2}$ **71.** $\frac{5\pi}{4}, \frac{7\pi}{4}$

73. $\frac{\pi}{4}, \frac{7\pi}{4}$ **75.** (a) $y = \frac{5}{12}\sin(2t + 0.6435)$

(b) $\frac{5}{12}$ feet (c) $\frac{1}{\pi}$ cycle per second

77. False. $\sin(u \pm v) = \sin u \cos v \pm \cos u \sin v$

79. False.

$\cos\left(x - \frac{\pi}{2}\right) = \cos x \cos\frac{\pi}{2} + \sin x \sin\frac{\pi}{2} = \sin x$

81. Answers will vary. **83.** Answers will vary.

85. (a) $\sqrt{2}\sin\left(\theta + \frac{\pi}{4}\right)$ (b) $\sqrt{2}\cos\left(\theta - \frac{\pi}{4}\right)$

87. (a) $13\sin(3\theta + 0.3948)$ (b) $13\cos(3\theta - 1.1760)$

89. $2\cos\theta$ **91.** 15°

93.

3

-2π 2π

-3

$\sin^2\left(\theta + \frac{\pi}{4}\right) + \sin^2\left(\theta - \frac{\pi}{4}\right) = 1$

95. Answers will vary. **97.** $f^{-1}(x) = \frac{x + 15}{5}$

99. Because f is not one-to-one, f^{-1} does not exist.

101. $4x - 3$ **103.** $6x - 3$

Section 7.5 *(page 582)*

1. $\frac{\sqrt{17}}{17}$ **3.** $\frac{15}{17}$ **5.** $\frac{8}{15}$ **7.** $\frac{17}{8}$

9. $0, \frac{\pi}{3}, \pi, \frac{5\pi}{3}$ **11.** $\frac{\pi}{12}, \frac{5\pi}{12}, \frac{13\pi}{12}, \frac{17\pi}{12}$

13. $0, \frac{2\pi}{3}, \frac{4\pi}{3}$ **15.** $\frac{\pi}{2}, \frac{\pi}{6}, \frac{5\pi}{6}, \frac{7\pi}{6}, \frac{3\pi}{2}, \frac{11\pi}{6}$

17. $0, \frac{\pi}{2}, \pi, \frac{3\pi}{2}$ **19.** $3\sin 2x$ **21.** $4\cos 2x$

23. $\sin 2u = \frac{24}{25}$
$\cos 2u = -\frac{7}{25}$
$\tan 2u = -\frac{24}{7}$

25. $\sin 2u = \frac{24}{25}$
$\cos 2u = \frac{7}{25}$
$\tan 2u = \frac{24}{7}$

27. $\sin 2u = -\frac{4\sqrt{21}}{25}$

$\cos 2u = -\frac{17}{25}$

$\tan 2u = \frac{4\sqrt{21}}{17}$

29. $\frac{1}{8}(3 + 4\cos 2x + \cos 4x)$ **31.** $\frac{1}{8}(1 - \cos 4x)$

33. $\frac{1}{16}(1 + \cos 2x - \cos 4x - \cos 2x \cos 4x)$

35. $\frac{4\sqrt{17}}{17}$ **37.** $\frac{1}{4}$ **39.** $\sqrt{17}$

41. $\sin 75^\circ = \frac{1}{2}\sqrt{2 + \sqrt{3}}$
$\cos 75^\circ = \frac{1}{2}\sqrt{2 - \sqrt{3}}$
$\tan 75^\circ = 2 + \sqrt{3}$

43. $\sin 112^\circ 30' = \frac{1}{2}\sqrt{2 + \sqrt{2}}$
$\cos 112^\circ 30' = -\frac{1}{2}\sqrt{2 - \sqrt{2}}$
$\tan 112^\circ 30' = -1 - \sqrt{2}$

45. $\sin\frac{\pi}{8} = \frac{1}{2}\sqrt{2 - \sqrt{2}}$

$\cos\frac{\pi}{8} = \frac{1}{2}\sqrt{2 + \sqrt{2}}$

$\tan\frac{\pi}{8} = \sqrt{2} - 1$

47. $\sin\frac{3\pi}{8} = \frac{1}{2}\sqrt{2 + \sqrt{2}}$

$\cos\frac{3\pi}{8} = \frac{1}{2}\sqrt{2 - \sqrt{2}}$

$\tan\frac{3\pi}{8} = \sqrt{2} + 1$

49. $\sin\frac{u}{2} = \frac{5\sqrt{26}}{26}$

$\cos\frac{u}{2} = \frac{\sqrt{26}}{26}$

$\tan\frac{u}{2} = 5$

51. $\sin\frac{u}{2} = \sqrt{\frac{89 - 8\sqrt{89}}{178}}$

$\cos\frac{u}{2} = -\sqrt{\frac{89 + 8\sqrt{89}}{178}}$

$\tan\frac{u}{2} = \frac{8 - \sqrt{89}}{5}$

53. $\sin \dfrac{u}{2} = \dfrac{3\sqrt{10}}{10}$

$\cos \dfrac{u}{2} = -\dfrac{\sqrt{10}}{10}$

$\tan \dfrac{u}{2} = -3$

55. $|\sin 3x|$ **57.** $-|\tan 4x|$

59. π

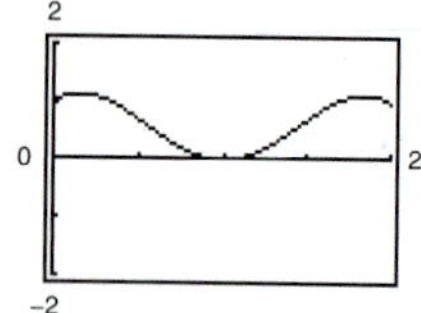

61. $\dfrac{\pi}{3}, \pi, \dfrac{5\pi}{3}$

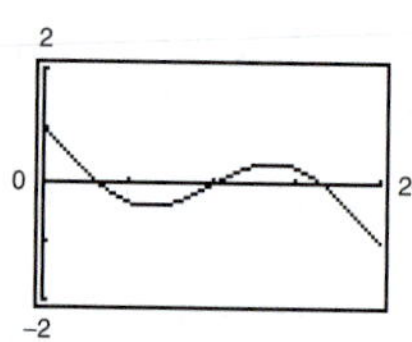

63. $3\left(\sin \dfrac{\pi}{2} + \sin 0\right)$

65. $\frac{1}{2}(\sin 10\theta + \sin 2\theta)$ **67.** $\frac{5}{2}(\cos 8\beta + \cos 2\beta)$

69. $\frac{1}{2}(\cos 2y - \cos 2x)$ **71.** $\frac{1}{2}(\sin 2\theta + \sin 2\pi)$

73. $5(\cos 60° + \cos 90°)$ **75.** $2 \sin 45° \cos 15°$

77. $-2 \sin \dfrac{\pi}{2} \sin \dfrac{\pi}{4}$ **79.** $2 \cos 4\theta \sin \theta$

81. $2 \cos 4x \cos 2x$ **83.** $2 \cos \alpha \sin \beta$

85. $-2 \sin \theta \sin \dfrac{\pi}{2}$

87. $0, \dfrac{\pi}{4}, \dfrac{\pi}{2}, \dfrac{3\pi}{4}, \pi, \dfrac{5\pi}{4}, \dfrac{3\pi}{2}, \dfrac{7\pi}{4}$

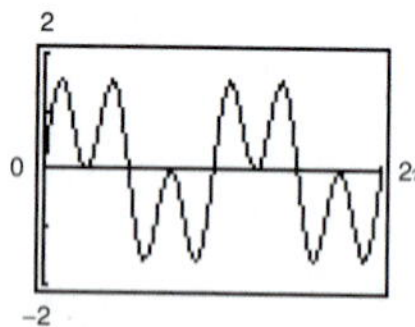

89. $\dfrac{\pi}{6}, \dfrac{5\pi}{6}$

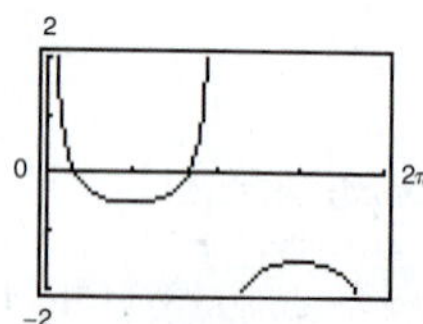

91. $\frac{25}{169}$ **93.** $\frac{4}{13}$ **95–109.** Answers will vary.

111.

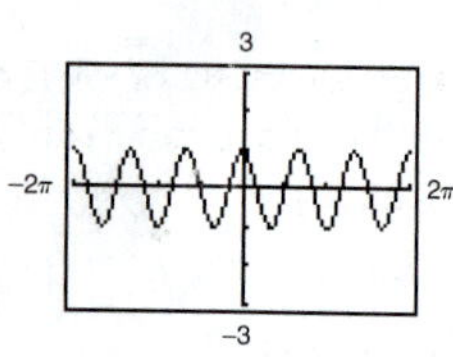

113.

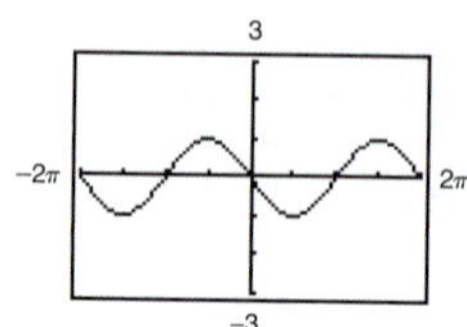

115.

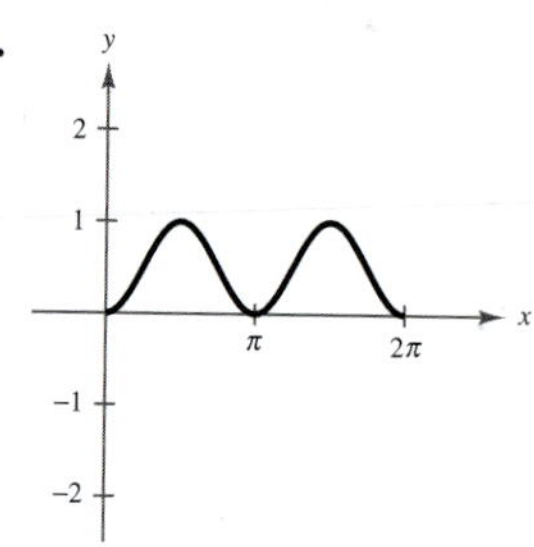

117. $2x\sqrt{1 - x^2}$

119. (a) $A = 100 \sin \dfrac{\theta}{2} \cos \dfrac{\theta}{2}$

(b) $A = 50 \sin \theta$

The area is maximum when $\theta = \pi/2$.

121. (a) π

(b) 0.4482

(c) 760 miles per hour; 3420 miles per hour

(d) $\theta = 2 \sin^{-1}\left(\dfrac{1}{m}\right)$

123. False. For $u < 0$,

$$\begin{aligned} \sin 2u &= -\sin(-2u) \\ &= -2 \sin(-u) \cos(-u) \\ &= -2(-\sin u) \cos u \\ &= 2 \sin u \cos u. \end{aligned}$$

125. (a)

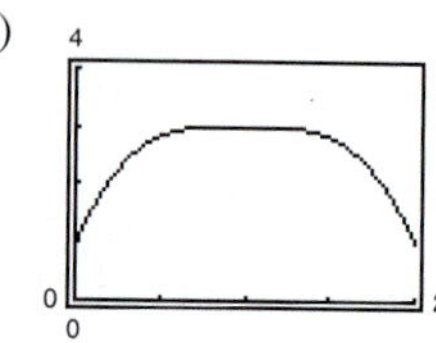

Maximum: $(\pi, 3)$

(b) π

127. (a) $\frac{1}{4}(3 + \cos 4x)$

(b) $2 \cos^4 x - 2 \cos^2 x + 1$

(c) $1 - 2 \sin^2 x \cos^2 x$

(d) $1 - \frac{1}{2} \sin^2 2x$

(e) No. There is often more than one way to rewrite a trigonometric expression.

129. $(x - 3)^2 = (y - 4)$

Vertex: $(3, 4)$

131. $(x - 1)^2 = \frac{1}{2}(y - 1)$

Vertex: $(1, 1)$

133. Vertex: $(5, -8)$

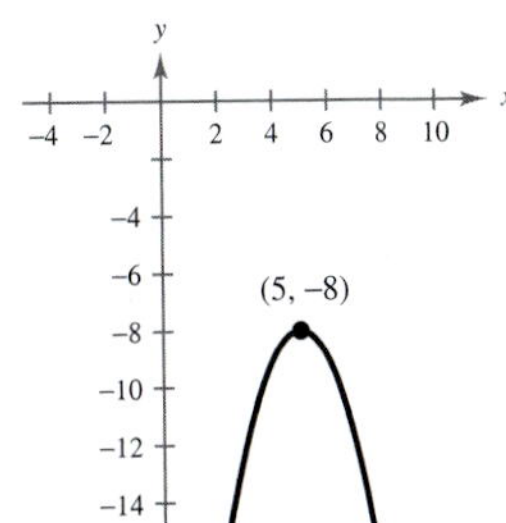

135. September: \$235,000 **137.** ≈ 127 feet
October: \$272,600

Review Exercises *(page 587)*

1. $\sec x$ **3.** $\cos x$ **5.** $\cot x$

7. $\tan x = \frac{3}{4}$, $\csc x = \frac{5}{3}$, $\sec x = \frac{5}{4}$, $\cot x = \frac{4}{3}$

9. $\cos x = \frac{\sqrt{2}}{2}$, $\tan x = -1$, $\csc x = -\sqrt{2}$, $\sec x = \sqrt{2}$, $\cot x = -1$

11. $\sin^2 x$ **13.** 1 **15.** $\cot \theta$ **17.** $\cot^2 x$

19. $\sec x + 2 \sin x$ **21.** $-2 \tan^2 \theta$

23–31. Answers will vary.

33. $\frac{\pi}{3} + 2n\pi, \frac{2\pi}{3} + 2n\pi$ **35.** $\frac{\pi}{6} + n\pi$

37. $\frac{\pi}{3} + n\pi, \frac{2\pi}{3} + n\pi$ **39.** $0, \frac{2\pi}{3}, \frac{4\pi}{3}$

41. $0, \frac{\pi}{2}, \pi$ **43.** $\frac{\pi}{8}, \frac{3\pi}{8}, \frac{9\pi}{8}, \frac{11\pi}{8}$

45. $0, \frac{\pi}{8}, \frac{3\pi}{8}, \frac{5\pi}{8}, \frac{7\pi}{8}, \frac{9\pi}{8}, \frac{11\pi}{8}, \frac{13\pi}{8}, \frac{15\pi}{8}$ **47.** $0, \pi$

49. $\arctan(-4) + \pi, \arctan(-4) + 2\pi, \arctan 3, \pi + \arctan 3$

51. $\sin 285° = -\frac{\sqrt{2}}{4}(\sqrt{3} + 1)$
$\cos 285° = \frac{\sqrt{2}}{4}(\sqrt{3} - 1)$
$\tan 285° = -2 - \sqrt{3}$

53. $\sin \frac{25\pi}{12} = \frac{\sqrt{2}}{4}(\sqrt{3} - 1)$
$\cos \frac{25\pi}{12} = \frac{\sqrt{2}}{4}(\sqrt{3} + 1)$
$\tan \frac{25\pi}{12} = 2 - \sqrt{3}$

55. $\sin 15°$ **57.** $\tan 35°$ **59.** $-\frac{3}{52}(5 + 4\sqrt{7})$

61. $\frac{1}{52}(5\sqrt{7} + 36)$ **63.** $\frac{1}{52}(5\sqrt{7} - 36)$

65. $\frac{\pi}{4}, \frac{7\pi}{4}$ **67.** $\frac{\pi}{6}, \frac{11\pi}{6}$

69. [graph]

71. $\sin 2u = \frac{24}{25}$
$\cos 2u = -\frac{7}{25}$
$\tan 2u = -\frac{24}{7}$

73. $\theta = 15°$ or $\frac{\pi}{12}$ **75.** $\frac{1 - \cos 4x}{1 + \cos 4x}$

77. $\frac{3 - 4 \cos 2x + \cos 4x}{4(1 + \cos 2x)}$

79. $\sin(-75°) = -\frac{1}{2}\sqrt{2 + \sqrt{3}}$
$\cos(-75°) = \frac{1}{2}\sqrt{2 - \sqrt{3}}$
$\tan(-75°) = -2 - \sqrt{3}$

81. $\sin \frac{19\pi}{12} = -\frac{1}{2}\sqrt{2 + \sqrt{3}}$
$\cos \frac{19\pi}{12} = \frac{1}{2}\sqrt{2 - \sqrt{3}}$
$\tan \frac{19\pi}{12} = -2 - \sqrt{3}$

83. $-|\cos 5x|$

85. $\sin \frac{u}{2} = \frac{\sqrt{10}}{10}$
$\cos \frac{u}{2} = \frac{3\sqrt{10}}{10}$
$\tan \frac{u}{2} = \frac{1}{3}$

87. $\frac{1}{2} \sin \frac{\pi}{3}$ **89.** $\frac{1}{2}(\cos 2\theta + \cos 8\theta)$

91. $2 \sin 75° \cos 15°$ **93.** $-2 \sin x \sin \frac{\pi}{6}$

95. (a) $y = \frac{1}{2}\sqrt{10} \sin\left(8t - \arctan \frac{1}{3}\right)$
(b) $\frac{1}{2}\sqrt{10}$ feet (c) $\frac{4}{\pi}$ cycles per second

97. False. Using the sum and difference formula, $\sin(x + y) = \sin x \cos y + \cos x \sin y$.

99. True by the product-to-sum formula

101. No. For an equation to be an identity, the equation must be true for all real numbers x. $\sin \theta = \frac{1}{2}$ has an infinite number of solutions but is not an identity.

103. $y_1 = y_2 + 1$

105. $-1.8431, 2.1758, 3.9903, 8.8935, 9.8820$

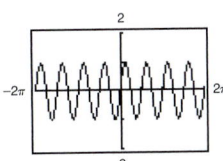

Chapter Test (page 590)

1. $\sin\theta = -\dfrac{3\sqrt{13}}{13}$ **2.** 1 **3.** 1 **4.** $\csc\theta\sec\theta$

$\cos\theta = -\dfrac{2\sqrt{13}}{13}$

$\csc\theta = -\dfrac{\sqrt{13}}{3}$

$\sec\theta = -\dfrac{\sqrt{13}}{2}$

$\cot\theta = \dfrac{2}{3}$

5. $\theta = 0, \dfrac{\pi}{2} < \theta \le \pi, \dfrac{3\pi}{2} < \theta < 2\pi$

6.

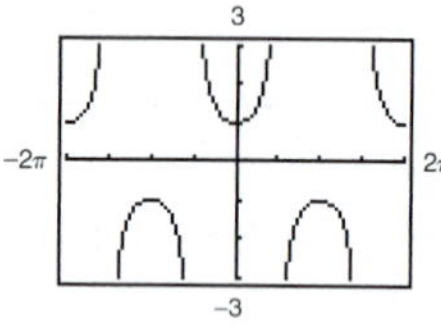

$y_1 = y_2$

7–12. Answers will vary.

13. $\dfrac{1}{16}\left(\dfrac{10 - 15\cos 2x + 6\cos 4x - \cos 6x}{1 + \cos 2x}\right)$ **14.** $\tan 2\theta$

15. $2(\sin 6\theta + \sin 2\theta)$ **16.** $-2\cos\dfrac{7\theta}{2}\sin\dfrac{\theta}{2}$

17. $0, \dfrac{3\pi}{4}, \pi, \dfrac{7\pi}{4}$ **18.** $\dfrac{\pi}{6}, \dfrac{\pi}{2}, \dfrac{5\pi}{6}, \dfrac{3\pi}{2}$

19. $\dfrac{\pi}{6}, \dfrac{5\pi}{6}, \dfrac{7\pi}{6}, \dfrac{11\pi}{6}$ **20.** $\dfrac{\pi}{6}, \dfrac{5\pi}{6}, \dfrac{3\pi}{2}$

21. $-2.938, -2.663, 1.170$

22. $\dfrac{\sqrt{2} - \sqrt{6}}{4}$ **23.** $\sin 2u = \dfrac{4}{5}, \tan 2u = -\dfrac{4}{3}$

24. Day 123 to day 223

Problem Solving (page 594)

1. (a) $\cos\theta = \pm\sqrt{1 - \sin^2\theta}$

$\tan\theta = \pm\dfrac{\sin\theta}{\sqrt{1 - \sin^2\theta}}$

$\cot\theta = \pm\dfrac{\sqrt{1 - \sin^2\theta}}{\sin\theta}$

$\sec\theta = \pm\dfrac{1}{\sqrt{1 - \sin^2\theta}}$

$\csc\theta = \dfrac{1}{\sin\theta}$

(b) $\sin\theta = \pm\sqrt{1 - \cos^2\theta}$

$\tan\theta = \pm\dfrac{\sqrt{1 - \cos^2\theta}}{\cos\theta}$

$\csc\theta = \pm\dfrac{1}{\sqrt{1 - \cos^2\theta}}$

$\sec\theta = \dfrac{1}{\cos\theta}$

$\cot\theta = \pm\dfrac{\cos\theta}{\sqrt{1 - \cos^2\theta}}$

3. Answers will vary. **5.** $y = \frac{1}{64}v^2\sin^2\theta$

7. $\sin\dfrac{\theta}{2} = \sqrt{\dfrac{1 - \cos\theta}{2}}$

$\cos\dfrac{\theta}{2} = \sqrt{\dfrac{1 + \cos\theta}{2}}$

$\tan\dfrac{\theta}{2} = \dfrac{\sin\theta}{1 + \cos\theta}$

9. (a)

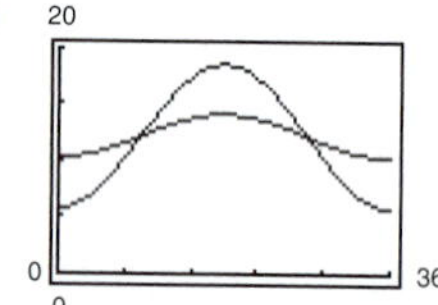

(b) $t = 91, t = 274$; Spring Equinox and Fall Equinox

(c) Seward; The amplitudes: 6.4 and 1.9

(d) 365.2 days

11. (a) $\dfrac{\pi}{6} \le x \le \dfrac{5\pi}{6}$

(b) $\dfrac{2\pi}{3} \le x \le \dfrac{4\pi}{3}$

(c) $\dfrac{\pi}{2} < x < \pi, \dfrac{3\pi}{2} < x < 2\pi$

(d) $0 \le x \le \dfrac{\pi}{4}, \dfrac{5\pi}{4} \le x \le 2\pi$

13. (a) $\sin(u + v + w)$

$= \sin u\cos v\ \cos w - \sin u\sin v\sin w$

$+ \cos u\sin v\cos w + \cos u\cos v\sin w$

(b) $\tan(u + v + w)$

$= \dfrac{\tan u + \tan v + \tan w - \tan u\tan v\tan w}{1 - \tan u\tan v - \tan u\tan w - \tan v\tan w}$

15. (a)

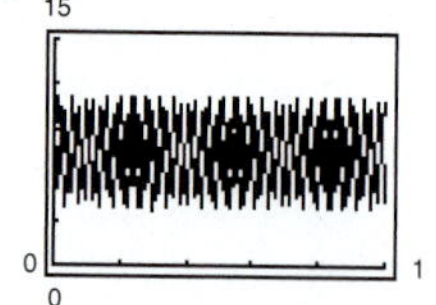

(b) 233.3 times per second

Chapter 8

Section 8.1 *(page 604)*

1. $C = 105°, b \approx 28.28, c \approx 38.64$

3. $C = 120°, b \approx 4.75, c \approx 7.17$

5. $B \approx 21.55°, C \approx 122.45°, c \approx 11.49$

7. $B = 60.9°, b \approx 19.32, c \approx 6.36$

9. $B = 42°4', a \approx 22.05, b \approx 14.88$

11. $A \approx 10°11', C \approx 154°19', c \approx 11.03$

13. $A \approx 25.57°, B \approx 9.43°, a \approx 10.53$

15. $B \approx 18°13', C \approx 51°32', c \approx 40.06$

17. $C = 83°, a \approx 0.62, b \approx 0.51$ **19.** No solution

21. No solution **23.** No solution

25. (a) $b \le 5,\ b = \dfrac{5}{\sin 36°}$

(b) $5 < b < \dfrac{5}{\sin 36°}$

(c) $b > \dfrac{5}{\sin 36°}$

27. (a) $b \le 10.8,\ b = \dfrac{10.8}{\sin 10°}$

(b) $10.8 < b < \dfrac{10.8}{\sin 10°}$

(c) $b > \dfrac{10.8}{\sin 10°}$

29. 10.4 **31.** 1675.2 **33.** 3204.5 **35.** 15.3 meters

37. 16.1° **39.** 77 meters

41. (a)

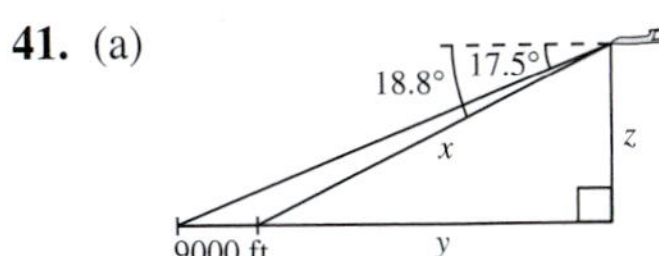

(b) 22.6 miles

(c) 21.4 miles

(d) 7.3 miles

43. 3.2 miles

45. True. If an angle of a triangle is obtuse (greater than 90°), than the other two angles must be acute and therefore less than 90°. The triangle is oblique.

47. (a) $\alpha = \arcsin(0.5 \sin \beta)$

(b)

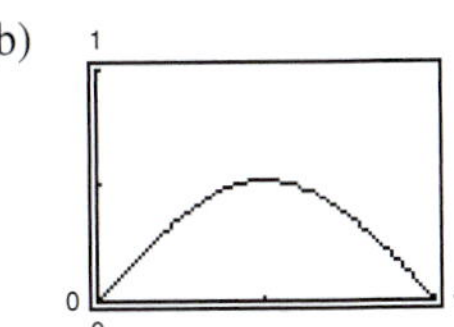

Domain: $0 < \beta < \pi$

Range: $0 < \alpha < \dfrac{\pi}{6}$

(c) $c = \dfrac{18 \sin[\pi - \beta - \arcsin(0.5 \sin \beta)]}{\sin \beta}$

(d) Domain: $0 < \beta < \pi$

Range: $9 < c < 27$

(e)

β	0.4	0.8	1.2	1.6
α	0.1960	0.3669	0.4848	0.5234
c	25.95	23.07	19.19	15.33

β	2.0	2.4	2.8
α	0.4720	0.3445	0.1683
c	12.29	10.31	9.27

As β increases from 0 to π, α increases and then decreases, and c decreases from 27 to 9.

49. $\cos x$ **51.** $\sin^2 x$ **53.** $3(\sin 11\theta + \sin 5\theta)$

Section 8.2 *(page 611)*

1. $A \approx 23.07°, B \approx 34.05°, C \approx 122.88°$

3. $B \approx 23.79°, C \approx 126.21°, a \approx 18.59$

5. $A \approx 31.99°, B \approx 42.38°, C \approx 105.63°$

7. $A \approx 92.94°, B \approx 43.53°, C \approx 43.53°$

9. $B \approx 13.45°, C \approx 31.55°, a \approx 12.16$

11. $A \approx 141°45', C \approx 27°40', b \approx 11.87$

13. $A = 27°10', C = 27°10', b \approx 56.94$

15. $A \approx 33.80°, B \approx 103.20°, c \approx 0.54$

	a	b	c	d	θ	ϕ
17.	5	8	12.07	5.69	45°	135°
19.	10	14	20	13.86	68.2°	111.8°
21.	15	16.96	25	20	77.2°	102.8°

23. 16.25 **25.** 10.44 **27.** 52.11

29. N 37.1° E, S 63.1° E

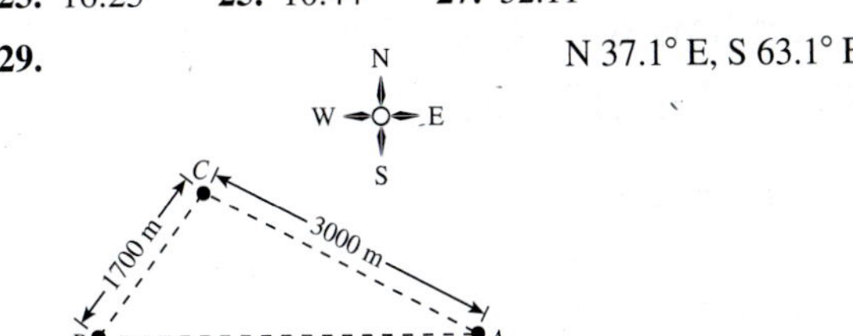

31. 373.3 meters **33.** 72.3° **35.** 43.3 miles

37. (a) N 58.4° W (b) S 81.5° W **39.** 63.7 feet

41. 24.2 miles **43.** $\overline{PQ} \approx 9.4, \overline{QS} = 5, \overline{RS} \approx 12.8$

45.

d (inches)	9	10	12	13	14
θ (degrees)	60.9°	69.5°	88.0°	98.2°	109.6°
s (inches)	20.88	20.28	18.99	18.28	17.48

d (inches)	15	16
θ (degrees)	122.9°	139.8°
s (inches)	16.55	15.37

47. 46,837.5 square feet

49. False. For s to be the average of the lengths of the three sides of the triangle, s would be equal to $(a + b + c)/3$.

51. False. The three side lengths do not form a triangle.

53. (a) 570.60 (b) 5909.2 (c) 177.09

55. Answers will vary. **57.** $-\frac{\pi}{2}$ **59.** $\frac{\pi}{3}$

61. $-\frac{\pi}{3}$ **63.** $\frac{1}{\sqrt{1 - 4x^2}}$ **65.** $\frac{1}{x - 2}$

67. $\cos \theta = 1$

$\sec \theta = 1$

$\csc \theta$ is undefined.

69. $\tan \theta = -\frac{\sqrt{3}}{3}$

$\sec \theta = \frac{2\sqrt{3}}{3}$

$\csc \theta = -2$

71. $-2 \sin \frac{7\pi}{12} \sin \frac{\pi}{4}$

Section 8.3 *(page 624)*

1. $\mathbf{v} = \langle 3, 2 \rangle$; $\|\mathbf{v}\| = \sqrt{13}$ **3.** $\mathbf{v} = \langle -3, 2 \rangle$; $\|\mathbf{v}\| = \sqrt{13}$

5. $\mathbf{v} = \langle 0, 5 \rangle$; $\|\mathbf{v}\| = 5$ **7.** $\mathbf{v} = \langle 16, 7 \rangle$; $\|\mathbf{v}\| = \sqrt{305}$

9. $\mathbf{v} = \langle 8, 6 \rangle$; $\|\mathbf{v}\| = 10$ **11.** $\mathbf{v} = \langle -9, -12 \rangle$; $\|\mathbf{v}\| = 15$

13.

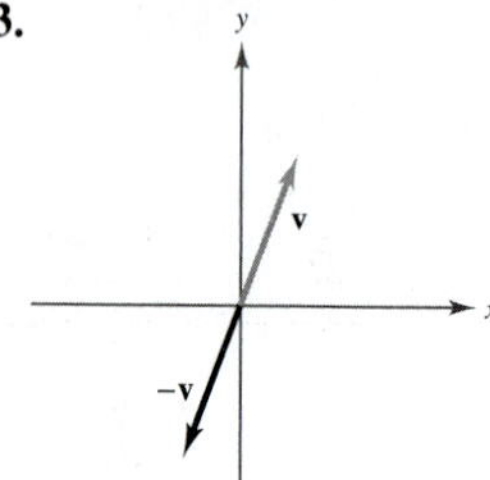

15.

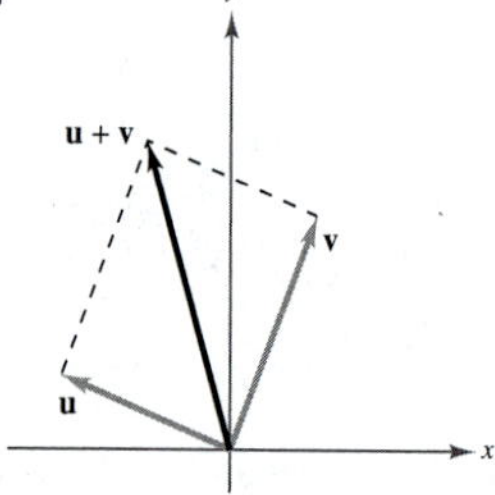

17.

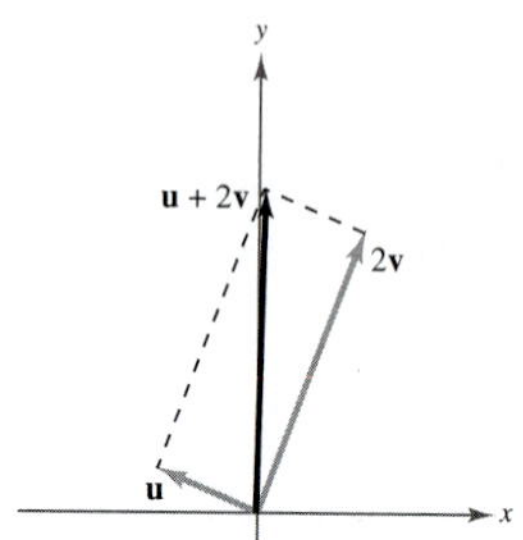

19. (a) $\langle 3, 4 \rangle$

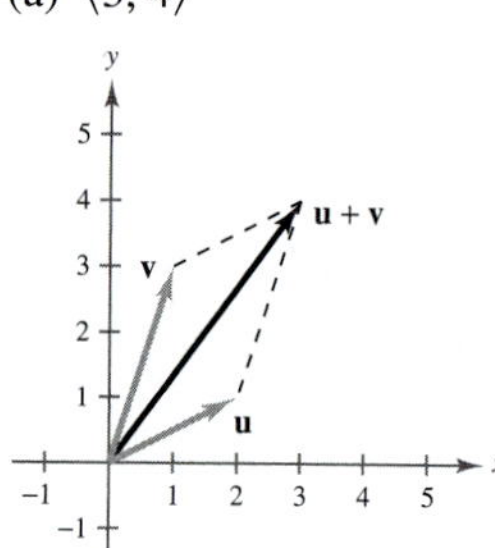

(b) $\langle 1, -2 \rangle$

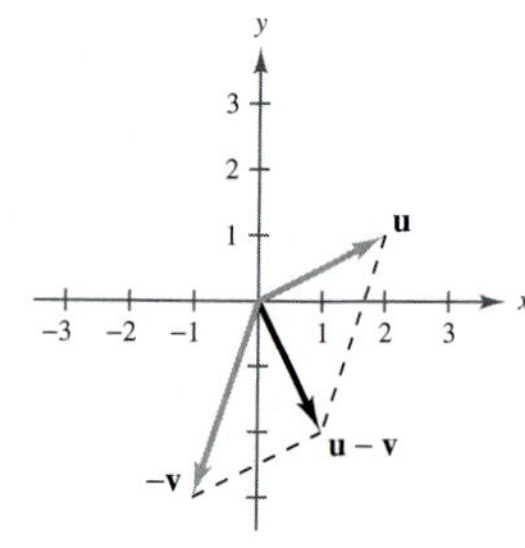

(c) $\langle 1, -7 \rangle$

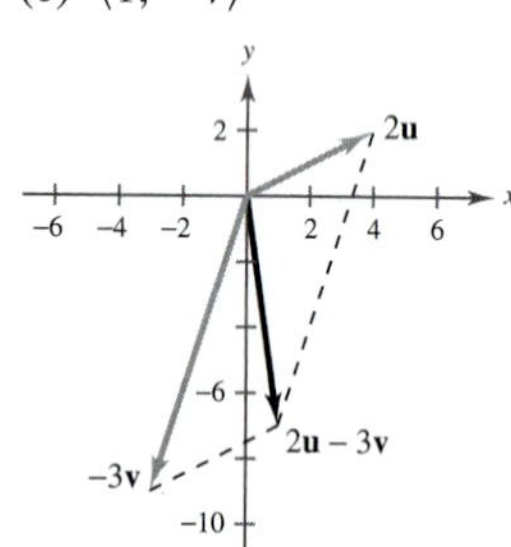

21. (a) $\langle -5, 3 \rangle$ (b) $\langle -5, 3 \rangle$

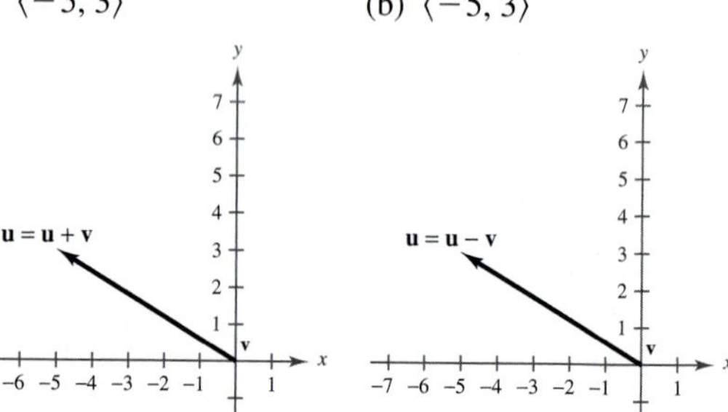

(c) $\langle -10, 6 \rangle$

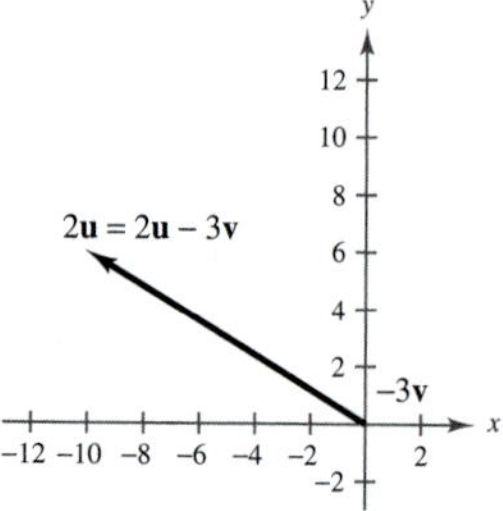

23. (a) $3\mathbf{i} - 2\mathbf{j}$

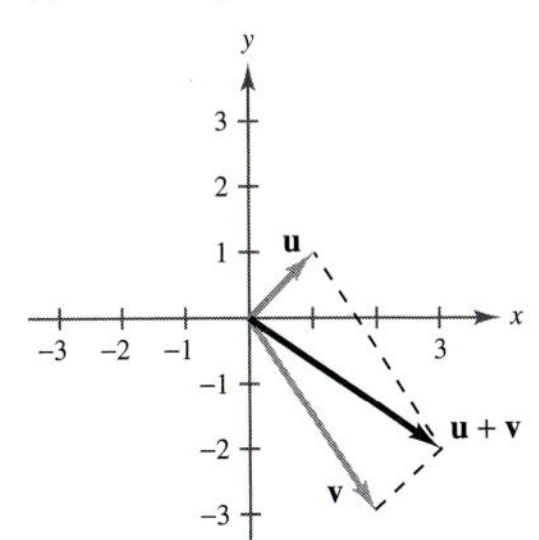

(b) $-\mathbf{i} + 4\mathbf{j}$

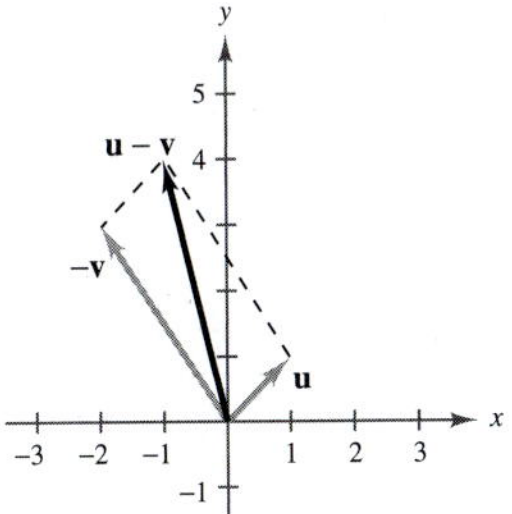

(c) $-4\mathbf{i} + 11\mathbf{j}$

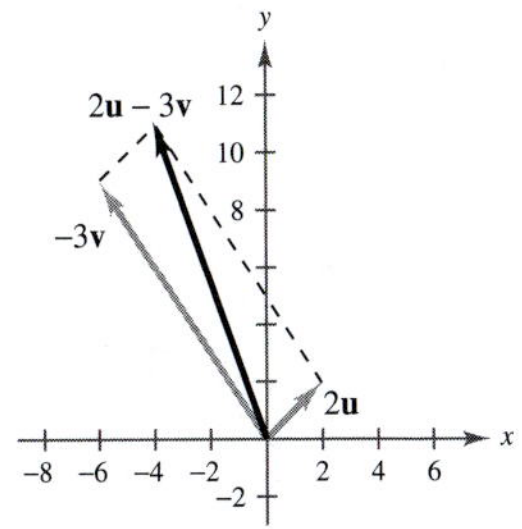

25. (a) $2\mathbf{i} + \mathbf{j}$

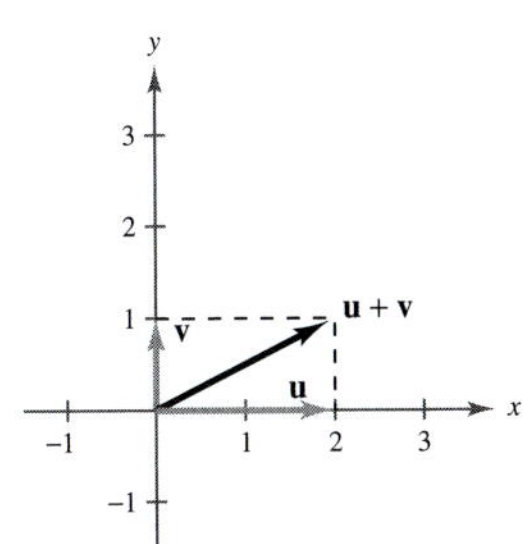

(b) $2\mathbf{i} - \mathbf{j}$

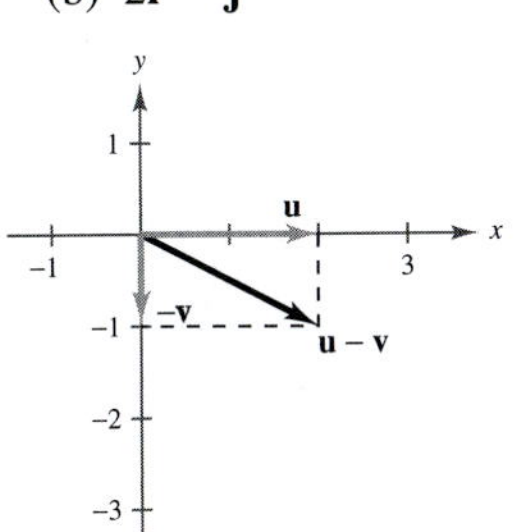

(c) $4\mathbf{i} - 3\mathbf{j}$

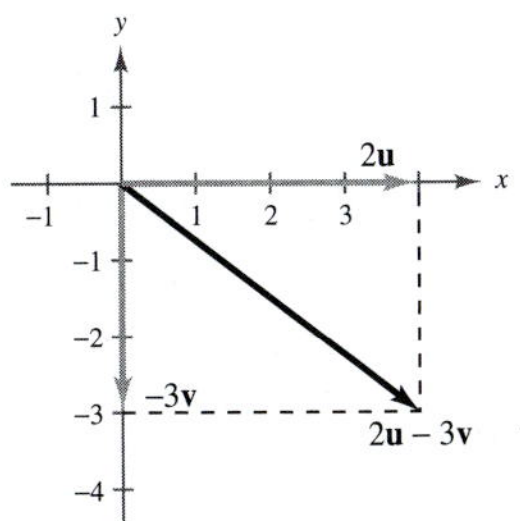

27. $\langle 1, 0 \rangle$ **29.** $\left\langle -\frac{\sqrt{2}}{2}, \frac{\sqrt{2}}{2} \right\rangle$ **31.** $\frac{3\sqrt{10}}{10}\mathbf{i} - \frac{\sqrt{10}}{10}\mathbf{j}$

33. $\mathbf{j}$ **35.** $\frac{\sqrt{5}}{5}\mathbf{i} - \frac{2\sqrt{5}}{5}\mathbf{j}$ **37.** $\left\langle \frac{5\sqrt{2}}{2}, \frac{5\sqrt{2}}{2} \right\rangle$

39. $\left\langle \frac{18\sqrt{29}}{29}, \frac{45\sqrt{29}}{29} \right\rangle$

41. $\mathbf{v} = \left\langle 3, -\frac{3}{2} \right\rangle$

43. $\mathbf{v} = \langle 4, 3 \rangle$

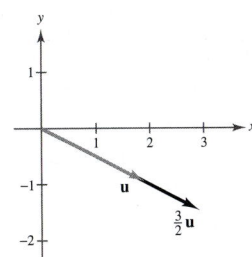

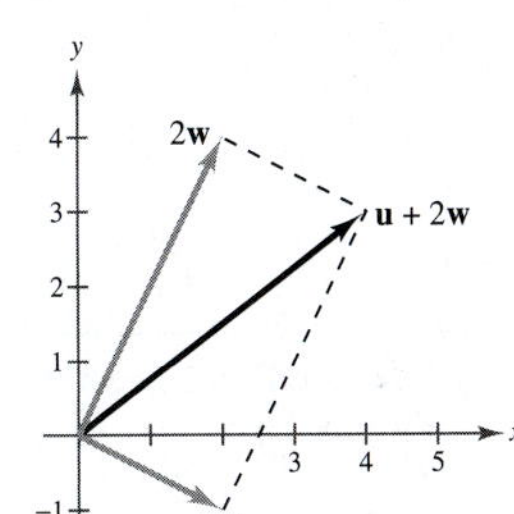

45. $\mathbf{v} = \left\langle \frac{7}{2}, -\frac{1}{2} \right\rangle$

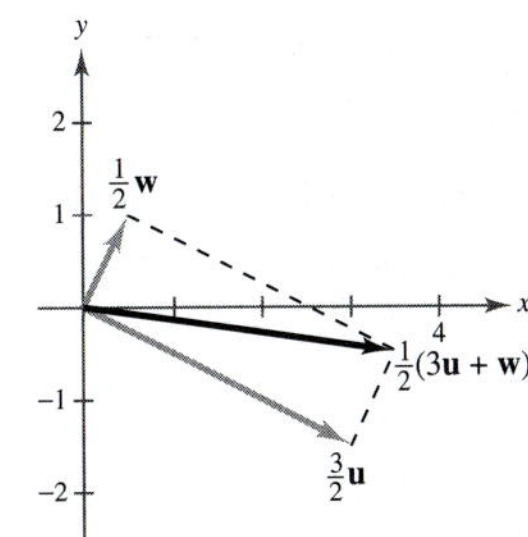

47. $\|\mathbf{v}\| = 3;\ \theta = 60°$ **49.** $\|\mathbf{v}\| = 6\sqrt{2};\ \theta = 315°$

51. $\mathbf{v} = \langle 3, 0 \rangle$

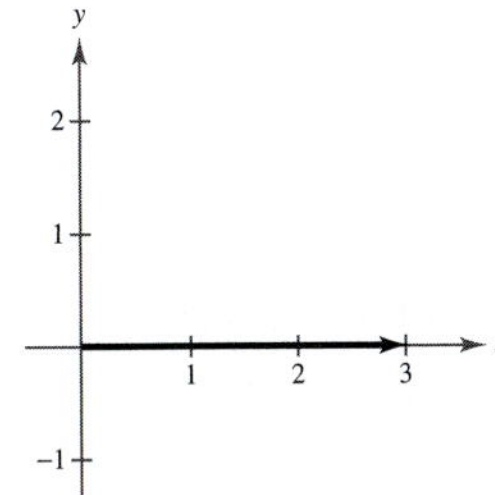

53. $\mathbf{v} = \left\langle -\frac{7\sqrt{3}}{4}, \frac{7}{4} \right\rangle$

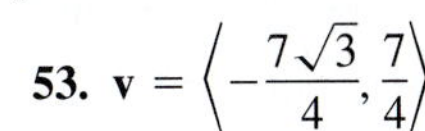

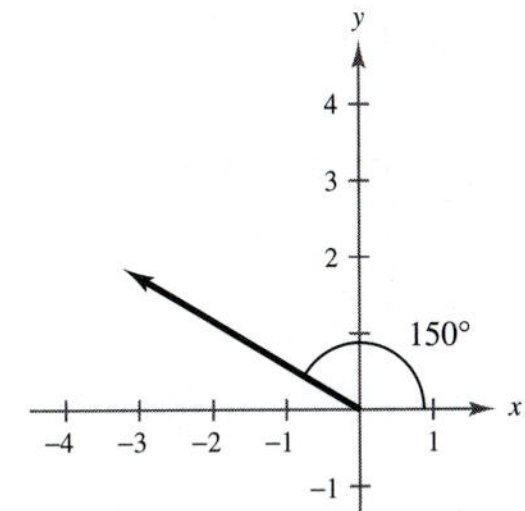

55. $\mathbf{v} = \left\langle -\frac{3\sqrt{6}}{2}, \frac{3\sqrt{2}}{2} \right\rangle$

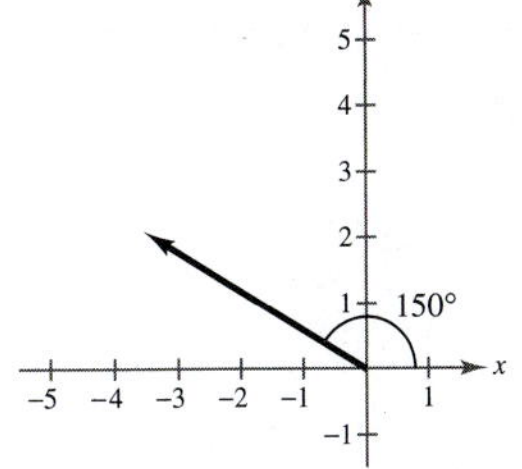

57. $\mathbf{v} = \left\langle \frac{\sqrt{10}}{5}, \frac{3\sqrt{10}}{5} \right\rangle$

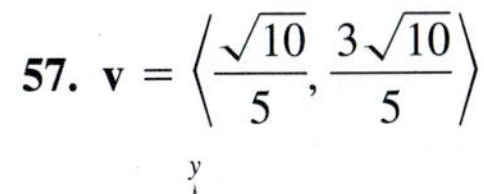

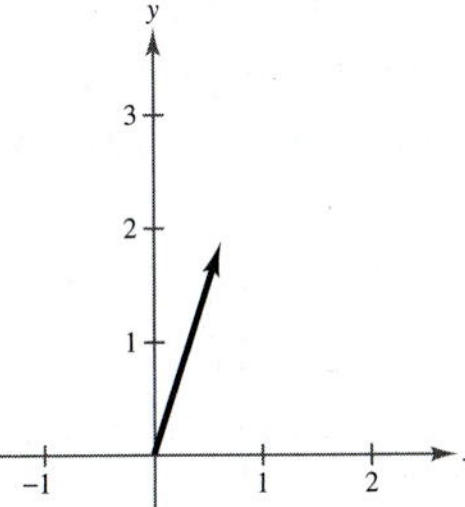

59. $\langle 5, 5 \rangle$ **61.** $\langle 10\sqrt{2} - 50, 10\sqrt{2} \rangle$ **63.** 90°

65. 63.4° **67.** 62.7° **69.** 12.8°; 398.32 newtons

71. 71.3°; 228.5 pounds

73. Vertical component: $70 \sin 35° \approx 40.15$ feet per second

Horizontal component: $70 \cos 35° \approx 57.34$ feet per second

75. $T_{AC} \approx 1758.8$ pounds **77.** 3154.4 pounds
$T_{BC} \approx 1305.4$ pounds

79. N 21.4° E; 138.7 kilometers per hour

81. 1928.4 foot-pounds **83.** True. See Example 1.

85. (a) 0° (b) 180°
(c) No. The magnitude is at most equal to the sum when the angle between the vectors is 0°.

87. Answers will vary. **89.** $\langle 1, 3\rangle$ or $\langle -1, -3\rangle$

91. $8 \tan \theta$ **93.** $6 \sec \theta$

95. $\frac{\pi}{2} + n\pi, \pi + 2n\pi$ **97.** $n\pi, \frac{\pi}{6} + 2n\pi, \frac{11\pi}{6} + 2n\pi$

Section 8.4 *(page 635)*

1. -9 **3.** 6 **5.** 8; scalar **7.** $\langle -6, 8\rangle$; vector

9. 13 **11.** $5\sqrt{41}$ **13.** 6 **15.** 90° **17.** 143.13°

19. 60.26° **21.** 90° **23.** $\frac{5\pi}{12}$

25. 26.57°, 63.43°, 90° **27.** 41.63°, 53.13°, 85.24°

29. -20 **31.** Parallel **33.** Neither **35.** Orthogonal

37. $\frac{1}{37}\langle 84, 14\rangle, \frac{1}{37}\langle -10, 60\rangle$ **39.** $\frac{45}{229}\langle 2, 15\rangle, \frac{6}{229}\langle -15, 2\rangle$

41. $\langle -5, 3\rangle, \langle 5, -3\rangle$ **43.** $\frac{2}{3}\mathbf{i} + \frac{1}{2}\mathbf{j}, -\frac{2}{3}\mathbf{i} - \frac{1}{2}\mathbf{j}$ **45.** 32

47. (a) \$58,762.50
This value gives the total revenue that can be earned by selling all of the units.
(b) $1.05\mathbf{v}$

49. 735 newton-meters **51.** 779.4 foot-pounds

53. False. Work is represented by a scalar.

55. (a) $\theta = \frac{\pi}{2}$ (b) $0 \le \theta < \frac{\pi}{2}$ (c) $\frac{\pi}{2} < \theta \le \pi$

57. Answers will vary. **59.** $12\sqrt{7}$ **61.** $-2\sqrt{6}$

63. $0, \frac{\pi}{6}, \pi, \frac{11\pi}{6}$ **65.** $0, \pi$ **67.** $-\frac{253}{325}$ **69.** $\frac{204}{325}$

Section 8.5 *(page 645)*

1.

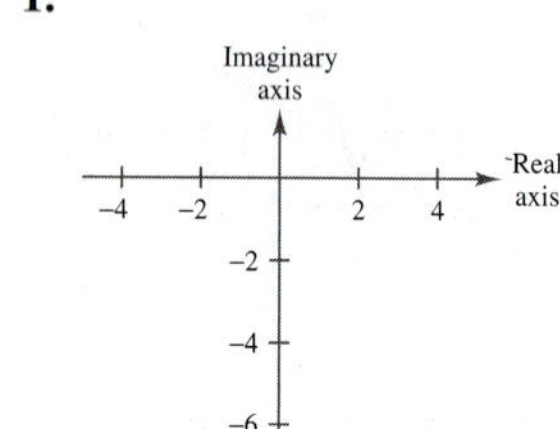

7

3.

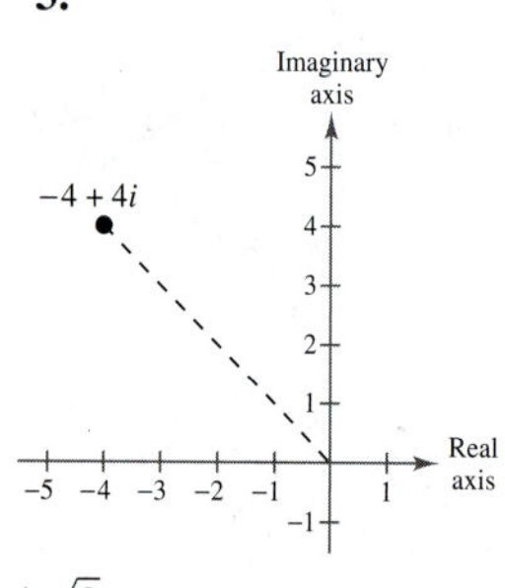

$4\sqrt{2}$

5.

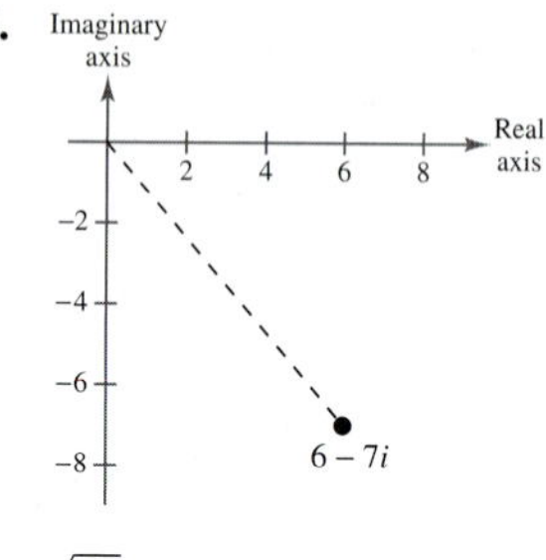

$\sqrt{85}$

7. $3\left(\cos \frac{\pi}{2} + i \sin \frac{\pi}{2}\right)$ **9.** $\sqrt{10}(\cos 5.96 + i \sin 5.96)$

11.

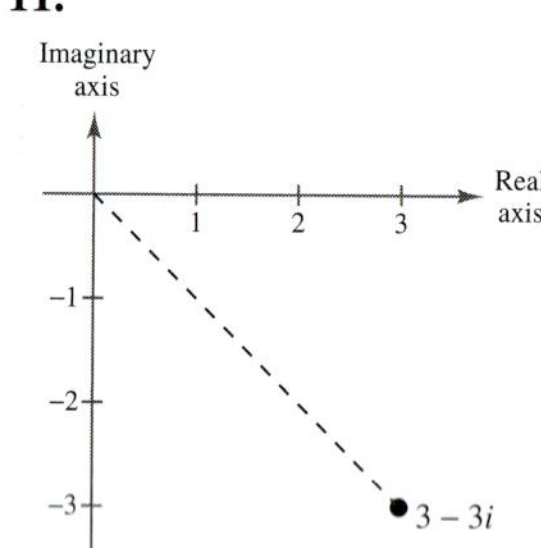

$3\sqrt{2}\left(\cos \frac{7\pi}{4} + i \sin \frac{7\pi}{4}\right)$

13.

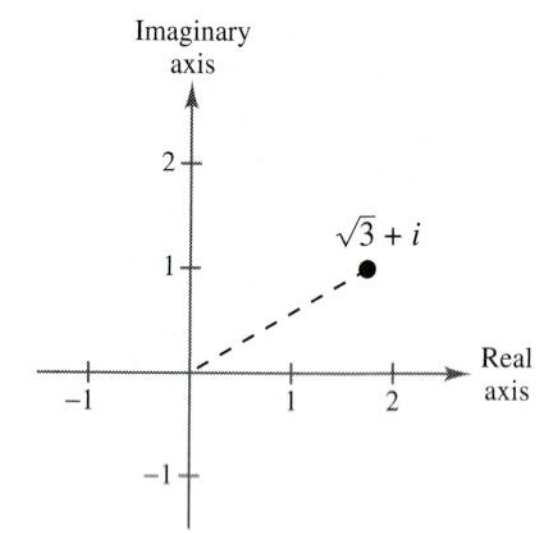

$2\left(\cos \frac{\pi}{6} + i \sin \frac{\pi}{6}\right)$

15.

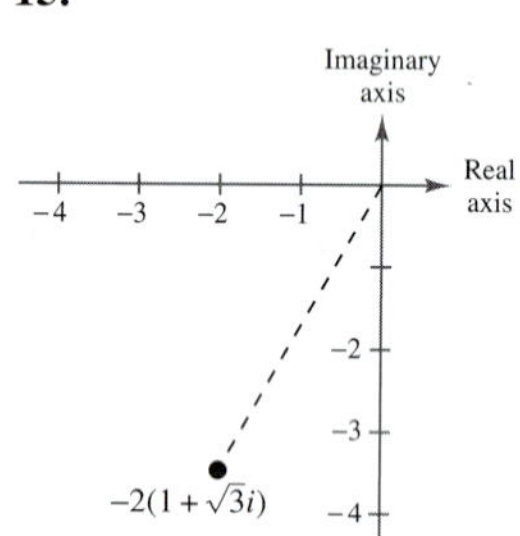

$4\left(\cos \frac{4\pi}{3} + i \sin \frac{4\pi}{3}\right)$

17.

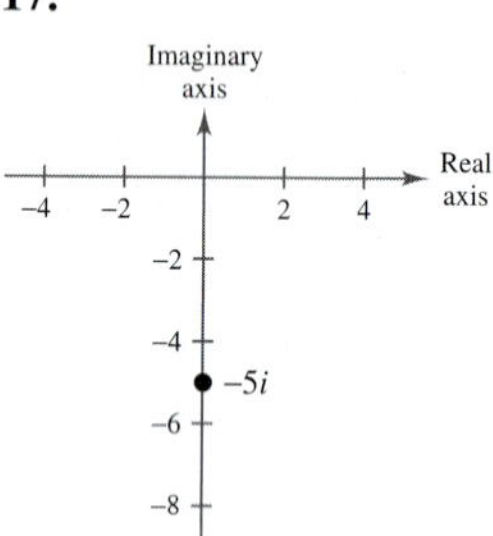

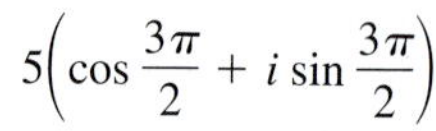
$5\left(\cos \frac{3\pi}{2} + i \sin \frac{3\pi}{2}\right)$

19.

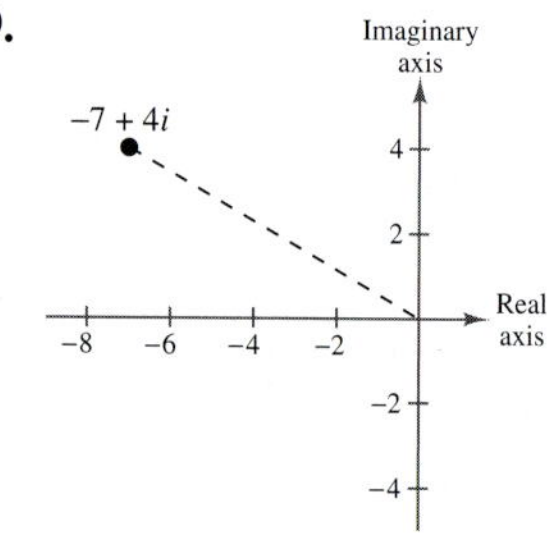

$\sqrt{65}(\cos 2.62 + i \sin 2.62)$

21.

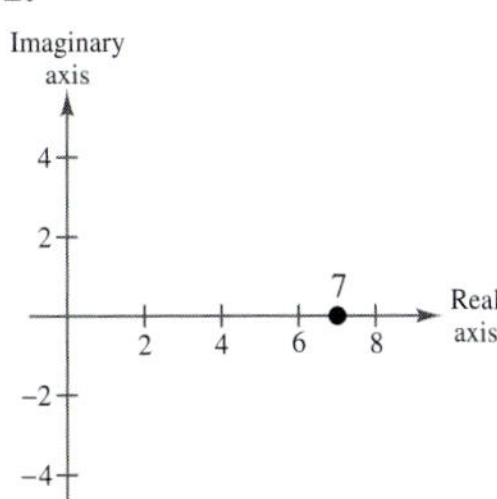

$7(\cos 0 + i \sin 0)$

23.

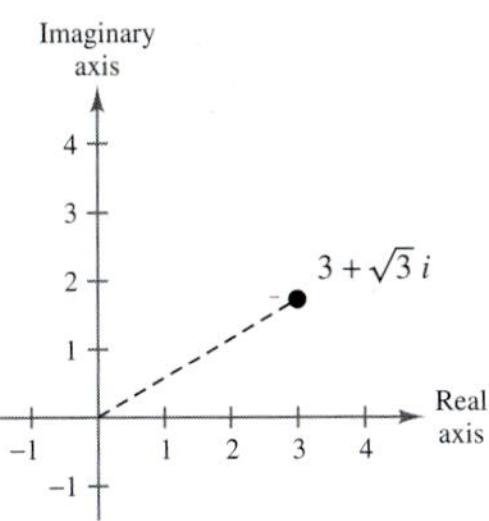

$2\sqrt{3}\left(\cos \frac{\pi}{6} + i \sin \frac{\pi}{6}\right)$

25.

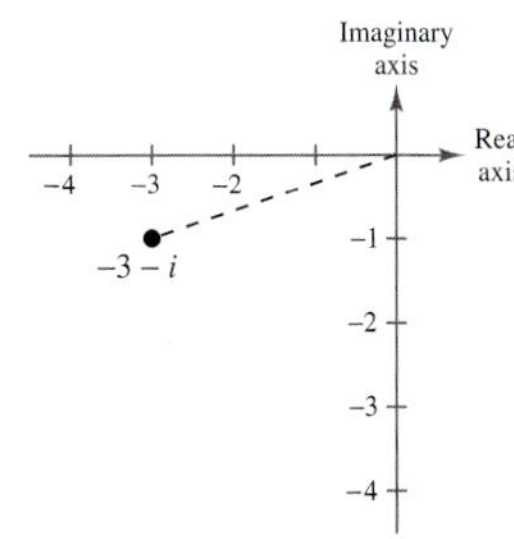

$\sqrt{10}(\cos 3.46 + i \sin 3.46)$

27. $5.39(\cos 0.38 + i \sin 0.38)$

29. $3.16(\cos 2.82 + i \sin 2.82)$

31. $8.19(\cos 5.26 + i \sin 5.26)$

33. $11.79(\cos 3.97 + i \sin 3.97)$

35.

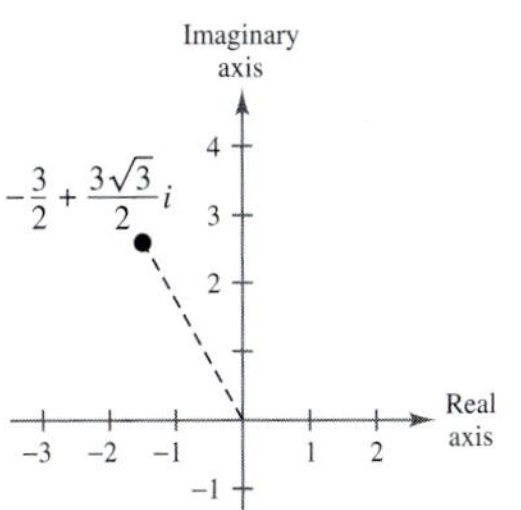

$-\frac{3}{2} + \frac{3\sqrt{3}}{2}i$

37.

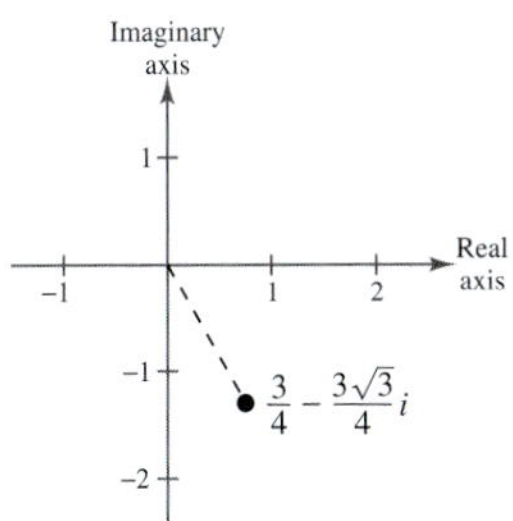

$\frac{3}{4} - \frac{3\sqrt{3}}{4}i$

39.

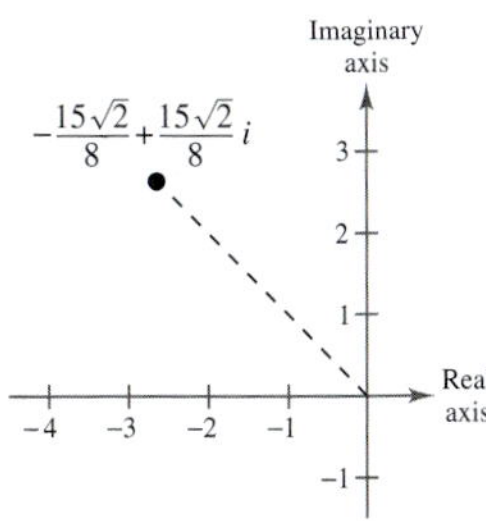

$-\frac{15\sqrt{2}}{8} + \frac{15\sqrt{2}}{8}i$

41.

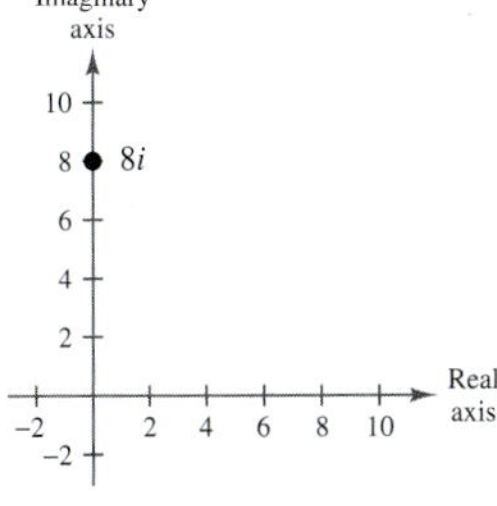

$8i$

43.

Imaginary axis, Real axis; $2.8408 + 0.9643i$

$2.8408 + 0.9643i$

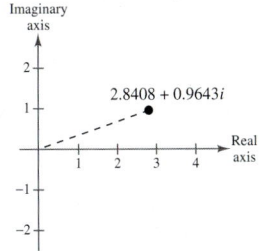

45. $4.6985 + 1.7101i$ **47.** $-2.9044 + 0.7511i$

49. $12\left(\cos \frac{\pi}{3} + i \sin \frac{\pi}{3}\right)$ **51.** $\frac{10}{9}(\cos 200° + i \sin 200°)$

53. $0.27(\cos 150° + i \sin 150°)$ **55.** $\cos 30° + i \sin 30°$

57. $\cos \frac{2\pi}{3} + i \sin \frac{2\pi}{3}$ **59.** $4(\cos 302° + i \sin 302°)$

61. (a) $\left[2\sqrt{2}\left(\cos \frac{\pi}{4} + i \sin \frac{\pi}{4}\right)\right]\left[\sqrt{2}\left(\cos \frac{7\pi}{4} + i \sin \frac{7\pi}{4}\right)\right]$

(b) $4(\cos 0 + i \sin 0) = 4$

(c) 4

63. (a) $\left[2\left(\cos \frac{3\pi}{2} + i \sin \frac{3\pi}{2}\right)\right]\left[\sqrt{2}\left(\cos \frac{\pi}{4} + i \sin \frac{\pi}{4}\right)\right]$

(b) $2\sqrt{2}\left(\cos \frac{7\pi}{4} + i \sin \frac{7\pi}{4}\right) = 2 - 2i$

(c) $-2i - 2i^2 = -2i + 2 = 2 - 2i$

65. (a) $[5(\cos 0.93 + i \sin 0.93)] \div \left[2\left(\cos \frac{5\pi}{3} + i \sin \frac{5\pi}{3}\right)\right]$

(b) $\frac{5}{2}(\cos 1.97 + i \sin 1.97) \approx -0.982 + 2.299i$

(c) $\approx -0.982 + 2.299i$

67. (a) $[5(\cos 0 + i \sin 0)] \div \left[\sqrt{13}(\cos 0.98 + i \sin 0.98)\right]$

(b) $\frac{5}{\sqrt{13}}(\cos 5.30 + i \sin 5.30) \approx 0.769 - 1.154i$

(c) $\frac{10}{13} - \frac{15}{13}i \approx 0.769 - 1.154i$

69.

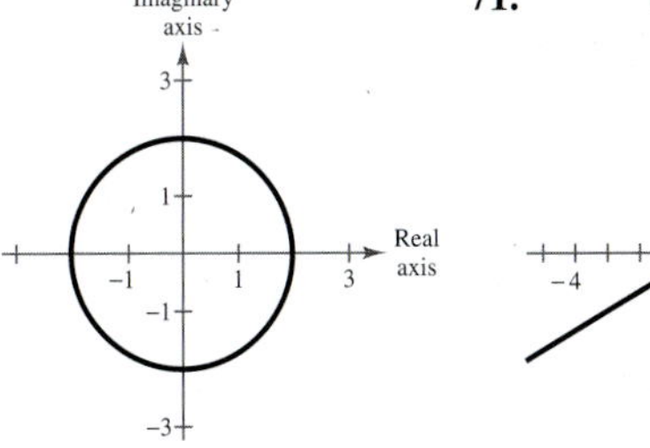

71.

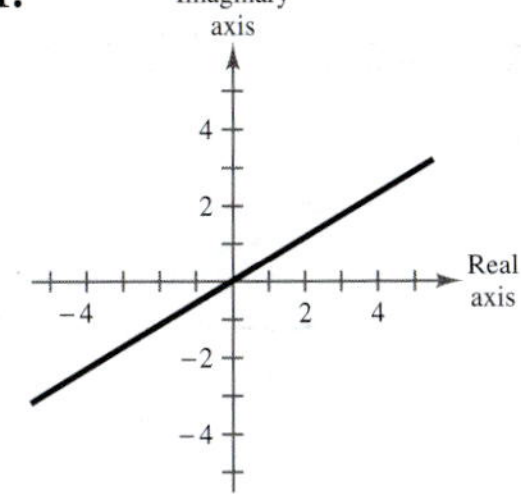

73. $-4 - 4i$ **75.** $-32i$ **77.** $-128\sqrt{3} - 128i$

79. $\frac{125}{2} + \frac{125\sqrt{3}}{2}i$ **81.** -1

83. $608.0204 + 144.6936i$

85. $-597 - 122i$ **87.** $\frac{81}{2} + \frac{81\sqrt{3}}{2}i$ **89.** $32i$

91. (a) $\sqrt{5}(\cos 60° + i \sin 60°)$

$\sqrt{5}(\cos 240° + i \sin 240°)$

(b)

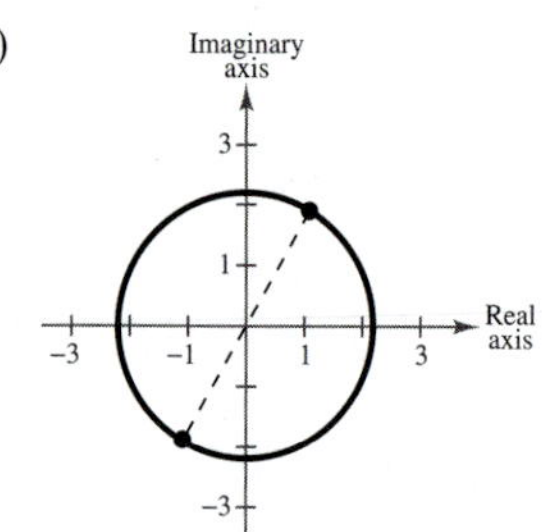

(c) $\frac{\sqrt{5}}{2} + \frac{\sqrt{15}}{2}i,\ -\frac{\sqrt{5}}{2} - \frac{\sqrt{15}}{2}i$

93. (a) $2\left(\cos \frac{2\pi}{9} + i \sin \frac{2\pi}{9}\right)$

$2\left(\cos \frac{8\pi}{9} + i \sin \frac{8\pi}{9}\right)$

$2\left(\cos \frac{14\pi}{9} + i \sin \frac{14\pi}{9}\right)$

(b)

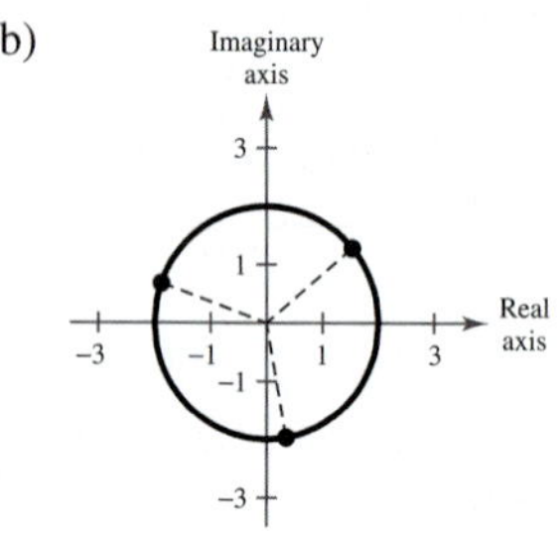

(c) $1.5321 + 1.2856i,\ -1.8794 + 0.6840i,$

$0.3473 - 1.9696i$

95. (a) $5\left(\cos \frac{3\pi}{4} + i \sin \frac{3\pi}{4}\right)$

$5\left(\cos \frac{7\pi}{4} + i \sin \frac{7\pi}{4}\right)$

(b)

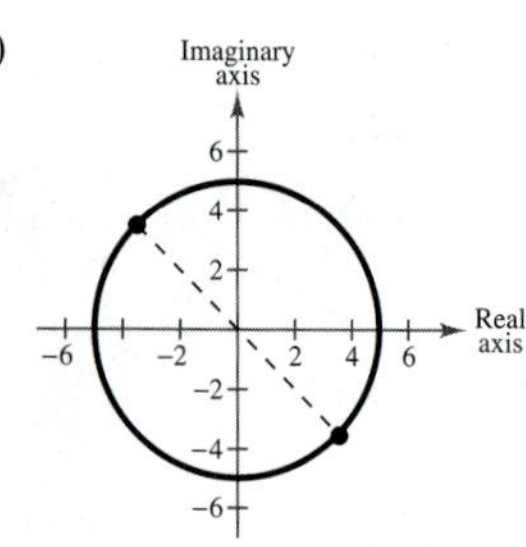

(c) $-\frac{5\sqrt{2}}{2} + \frac{5\sqrt{2}}{2}i,\ \frac{5\sqrt{2}}{2} - \frac{5\sqrt{2}}{2}i$

97. (a) $5\left(\cos \frac{4\pi}{9} + i \sin \frac{4\pi}{9}\right)$

$5\left(\cos \frac{10\pi}{9} + i \sin \frac{10\pi}{9}\right)$

$5\left(\cos \frac{16\pi}{9} + i \sin \frac{16\pi}{9}\right)$

(b)

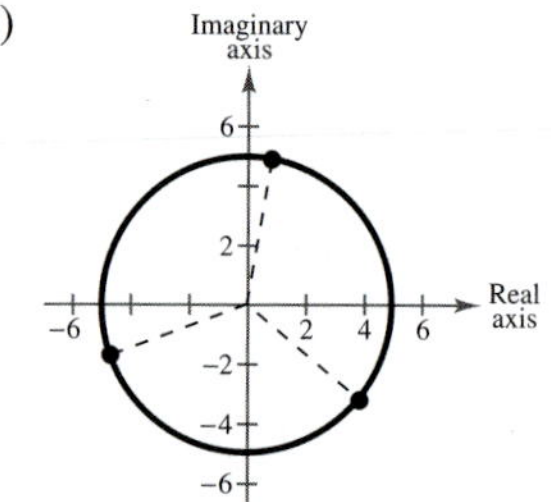

(c) $0.8682 + 4.9240i,\ -4.6985 - 1.7101i,$

$3.8302 - 3.2140i$

99. (a) $2(\cos 0 + i \sin 0)$

$2\left(\cos \frac{\pi}{2} + i \sin \frac{\pi}{2}\right)$

$2(\cos \pi + i \sin \pi)$

$2\left(\cos \frac{3\pi}{2} + i \sin \frac{3\pi}{2}\right)$

(b)

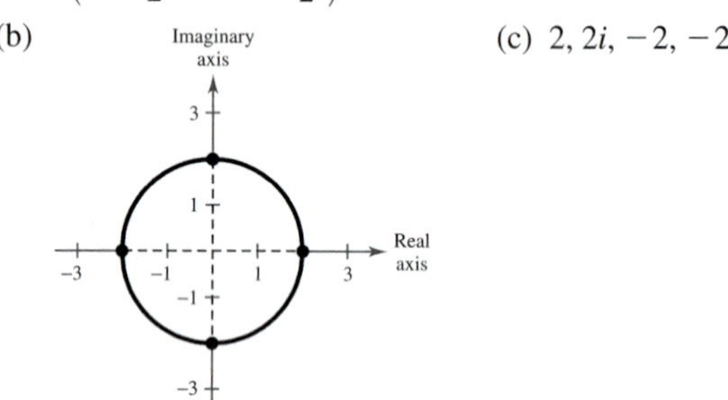

(c) $2, 2i, -2, -2i$

101. (a) $\cos 0 + i \sin 0$

$\cos \frac{2\pi}{5} + i \sin \frac{2\pi}{5}$

$\cos \frac{4\pi}{5} + i \sin \frac{4\pi}{5}$

$\cos \frac{6\pi}{5} + i \sin \frac{6\pi}{5}$

$\cos \frac{8\pi}{5} + i \sin \frac{8\pi}{5}$

(b)

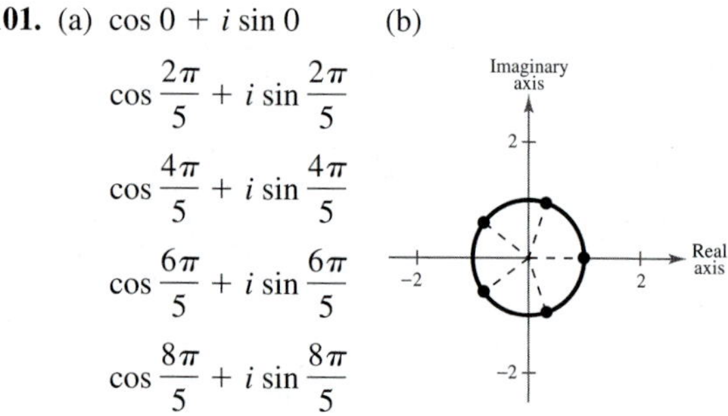

(c) $1, 0.3090 + 0.9511i,\ -0.8090 + 0.5878i,$

$-0.8090 - 0.5878i, 0.3090 - 0.9511i$

103. (a) $5\left(\cos \frac{\pi}{3} + i \sin \frac{\pi}{3}\right)$

$5(\cos \pi + i \sin \pi)$

$5\left(\cos \frac{5\pi}{3} + i \sin \frac{5\pi}{3}\right)$

(b)

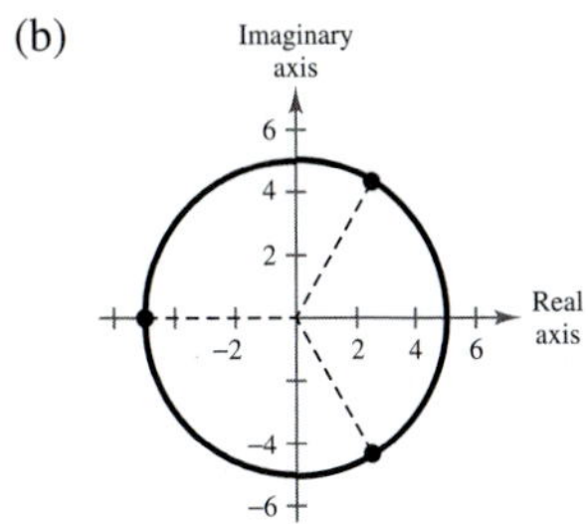

(c) $\frac{5}{2} + \frac{5\sqrt{3}}{2}i,\ -5,\ \frac{5}{2} - \frac{5\sqrt{3}}{2}i$

105. (a) $2\sqrt[5]{4\sqrt{2}}\left(\cos\frac{3\pi}{20} + i\sin\frac{3\pi}{20}\right)$

$2\sqrt[5]{4\sqrt{2}}\left(\cos\frac{11\pi}{20} + i\sin\frac{11\pi}{20}\right)$

$2\sqrt[5]{4\sqrt{2}}\left(\cos\frac{19\pi}{20} + i\sin\frac{19\pi}{20}\right)$

$2\sqrt[5]{4\sqrt{2}}\left(\cos\frac{27\pi}{20} + i\sin\frac{27\pi}{20}\right)$

$2\sqrt[5]{4\sqrt{2}}\left(\cos\frac{7\pi}{4} + i\sin\frac{7\pi}{4}\right)$

(b)

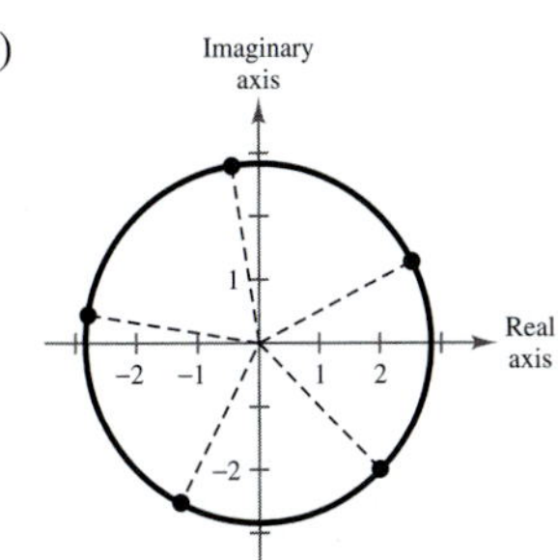

(c) $2.5201 + 1.2841i,\ -0.4425 + 2.7936i,$
$-2.7936 + 0.4425i,\ -1.2841 - 2.5201i,\ 2 - 2i$

107. $\cos\frac{3\pi}{8} + i\sin\frac{3\pi}{8}$

$\cos\frac{7\pi}{8} + i\sin\frac{7\pi}{8}$

$\cos\frac{11\pi}{8} + i\sin\frac{11\pi}{8}$

$\cos\frac{15\pi}{8} + i\sin\frac{15\pi}{8}$

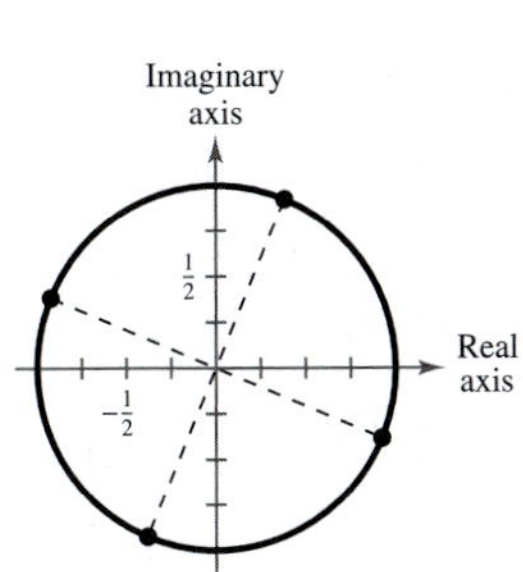

109. $3\left(\cos\frac{\pi}{5} + i\sin\frac{\pi}{5}\right)$

$3\left(\cos\frac{3\pi}{5} + i\sin\frac{3\pi}{5}\right)$

$3(\cos\pi + i\sin\pi)$

$3\left(\cos\frac{7\pi}{5} + i\sin\frac{7\pi}{5}\right)$

$3\left(\cos\frac{9\pi}{5} + i\sin\frac{9\pi}{5}\right)$

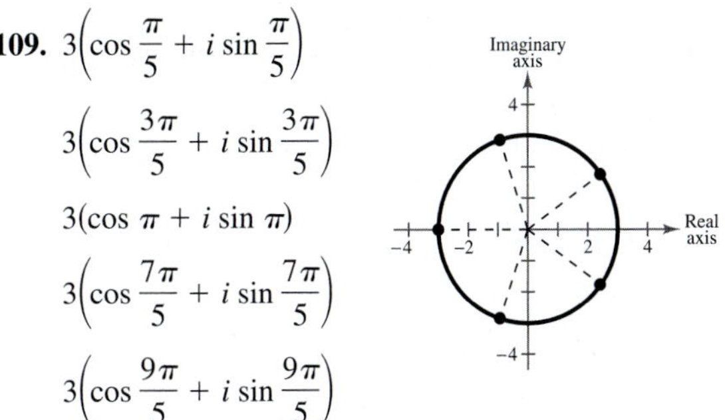

111. $2\left(\cos\frac{3\pi}{8} + i\sin\frac{3\pi}{8}\right)$

$2\left(\cos\frac{7\pi}{8} + i\sin\frac{7\pi}{8}\right)$

$2\left(\cos\frac{11\pi}{8} + i\sin\frac{11\pi}{8}\right)$

$2\left(\cos\frac{15\pi}{8} + i\sin\frac{15\pi}{8}\right)$

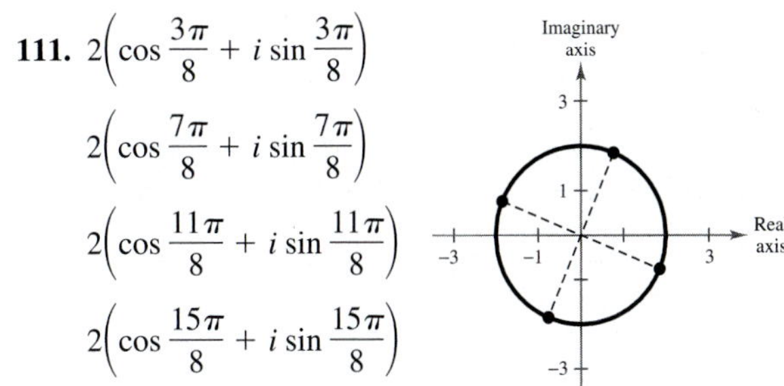

113. $\sqrt[6]{2}\left(\cos\frac{7\pi}{12} + i\sin\frac{7\pi}{12}\right)$

$\sqrt[6]{2}\left(\cos\frac{5\pi}{4} + i\sin\frac{5\pi}{4}\right)$

$\sqrt[6]{2}\left(\cos\frac{23\pi}{12} + i\sin\frac{23\pi}{12}\right)$

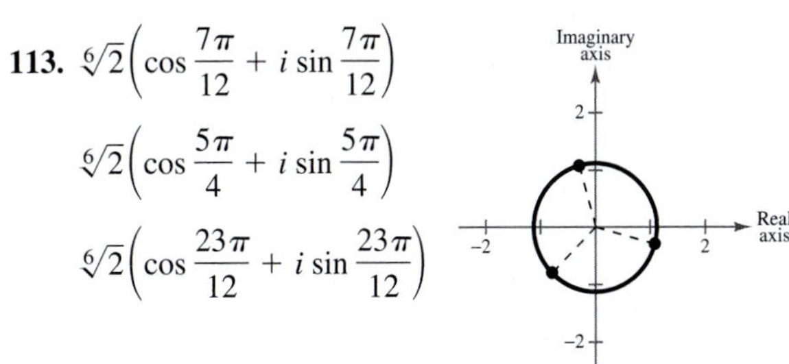

115. True, by the definition of the absolute value of a complex number.

117. True. $z_1z_2 = r_1r_2[\cos(\theta_1 + \theta_2) + i\sin(\theta_1 + \theta_2)] = 0$ if and only if $r_1 = 0$ and/or $r_2 = 0$.

119. Answers will vary. **121.** (a) r^2 (b) $\cos 2\theta + i\sin 2\theta$

123. Answers will vary.

125. (a) $2(\cos 30° + i\sin 30°)$ (b) $8i$
$2(\cos 150° + i\sin 150°)$
$2(\cos 270° + i\sin 270°)$

127. $B = 68°, b \approx 19.80, c \approx 21.36$

129. $B = 60°, a \approx 65.01, c \approx 130.02$

131. $B = 47°45', a \approx 7.53, b \approx 8.29$

133. 16; 2 **135.** $\frac{1}{16}$; 0

Review Exercises *(page 649)*

1. $C = 74°, b \approx 13.19, c \approx 13.41$

3. $A = 26°, a \approx 24.89, c \approx 56.23$

5. $C = 66°, a \approx 2.53, b \approx 9.11$

7. $B = 108°, a \approx 11.76, c \approx 21.49$

9. $A \approx 20.41°, C \approx 9.59°, a \approx 20.92$

11. $B \approx 39.48°, C \approx 65.52°, c \approx 48.24$

13. 7.9 **15.** 33.5 **17.** 31.1 meters **19.** 31.01 feet

21. $A \approx 29.69°, B \approx 52.41°, C \approx 97.90°$

23. $A \approx 29.92°, B \approx 86.18°, C \approx 63.90°$

25. $A = 35°, C = 35°, b \approx 6.55$

27. $A \approx 45.76°, B \approx 91.24°, c \approx 21.42$

29. 615.1 meters **31.** 9.80 **33.** 8.36

35.

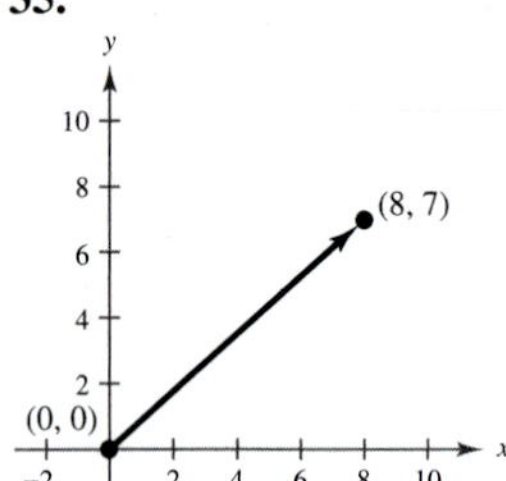

37.

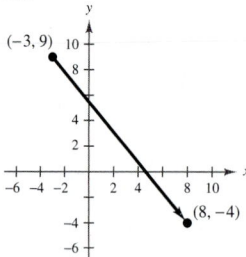

39. $\langle 7, -5\rangle$ **41.** $\langle 7, -7\rangle$ **43.** 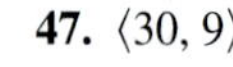 $\langle -4, 4\sqrt{3}\rangle$

45. $\langle 22, -7\rangle$

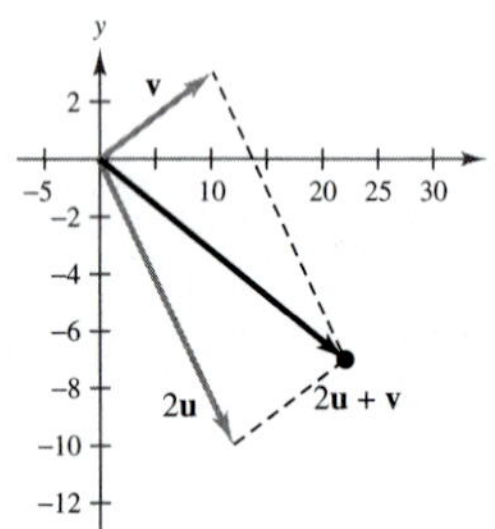

47. $\langle 30, 9\rangle$

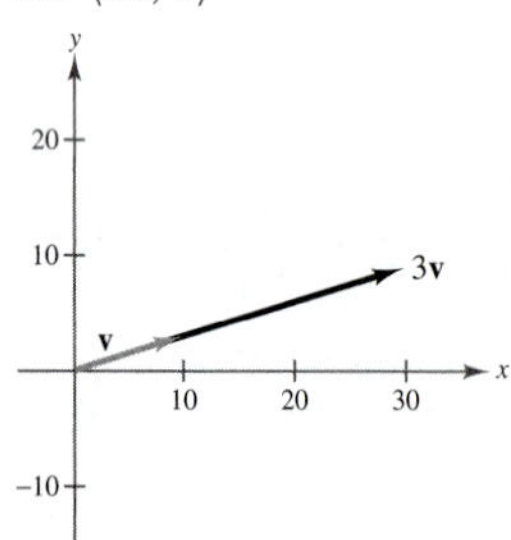

49. $-3\mathbf{i} + 4\mathbf{j}$ **51.** $6\mathbf{i} + 4\mathbf{j}$

53. $10\sqrt{2}(\cos 135°\,\mathbf{i} + \sin 135°\,\mathbf{j})$ **55.** $\|\mathbf{v}\| = 7;\ \theta = 60°$

57. $\|\mathbf{v}\| = \sqrt{41};\ \theta = 38.7°$ **59.** $\|\mathbf{v}\| = 3\sqrt{2};\ \theta = 225°$

61. The resultant force is 133.92 pounds and 5.6° from the 85-pound force.

63. 422.30 miles per hour; 130.4° **65.** 45 **67.** -2

69. 50; scalar **71.** $\langle 6, -8\rangle$; vector **73.** $\frac{11\pi}{12}$

75. 160.5° **77.** Orthogonal **79.** Neither

81. $-\frac{13}{17}\langle 4, 1\rangle, \frac{16}{17}\langle -1, 4\rangle$ **83.** $\frac{5}{2}\langle -1, 1\rangle, \frac{9}{2}\langle 1, 1\rangle$ **85.** 48

87.

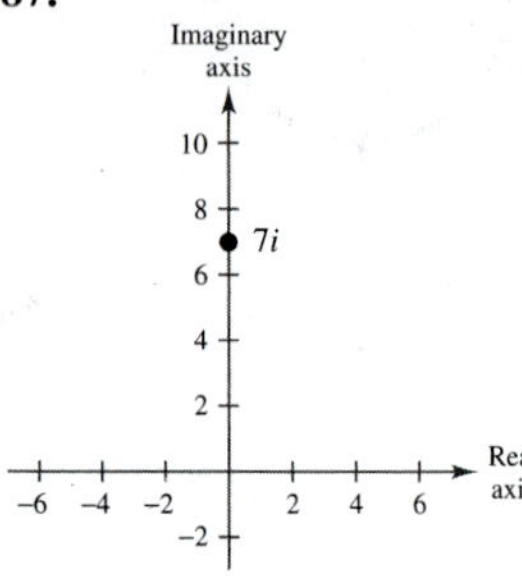

7

89.

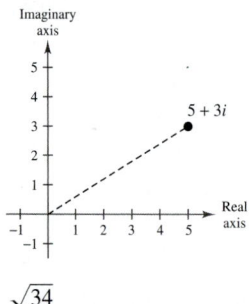

$\sqrt{34}$

91. $5\sqrt{2}\left(\cos\frac{7\pi}{4} + i\sin\frac{7\pi}{4}\right)$

93. $6\left(\cos\frac{5\pi}{6} + i\sin\frac{5\pi}{6}\right)$

95. (a) $z_1 = 4\left(\cos\frac{11\pi}{6} + i\sin\frac{11\pi}{6}\right)$

$z_2 = 10\left(\cos\frac{3\pi}{2} + i\sin\frac{3\pi}{2}\right)$

(b) $z_1z_2 = 40\left(\cos\frac{10\pi}{3} + i\sin\frac{10\pi}{3}\right)$

$\frac{z_1}{z_2} = \frac{2}{5}\left(\cos\frac{\pi}{3} + i\sin\frac{\pi}{3}\right)$

97. $\frac{625}{2} + \frac{625\sqrt{3}}{2}i$ **99.** $2035 - 828i$

101. (a) $4(\cos 60° + i\sin 60°)$ (b) -64

$4(\cos 180° + i\sin 180°)$

$4(\cos 300° + i\sin 300°)$

103. $3\left(\cos\frac{\pi}{4} + i\sin\frac{\pi}{4}\right)$

$3\left(\cos\frac{7\pi}{12} + i\sin\frac{7\pi}{12}\right)$

$3\left(\cos\frac{11\pi}{12} + i\sin\frac{11\pi}{12}\right)$

$3\left(\cos\frac{5\pi}{4} + i\sin\frac{5\pi}{4}\right)$

$3\left(\cos\frac{19\pi}{12} + i\sin\frac{19\pi}{12}\right)$

$3\left(\cos\frac{23\pi}{12} + i\sin\frac{23\pi}{12}\right)$

105. $3\left(\cos\frac{\pi}{4} + i\sin\frac{\pi}{4}\right) = \frac{3\sqrt{2}}{2} + \frac{3\sqrt{2}}{2}i$

$3\left(\cos\frac{3\pi}{4} + i\sin\frac{3\pi}{4}\right) = -\frac{3\sqrt{2}}{2} + \frac{3\sqrt{2}}{2}i$

$3\left(\cos\frac{5\pi}{4} + i\sin\frac{5\pi}{4}\right) = -\frac{3\sqrt{2}}{2} - \frac{3\sqrt{2}}{2}i$

$3\left(\cos\frac{7\pi}{4} + i\sin\frac{7\pi}{4}\right) = \frac{3\sqrt{2}}{2} - \frac{3\sqrt{2}}{2}i$

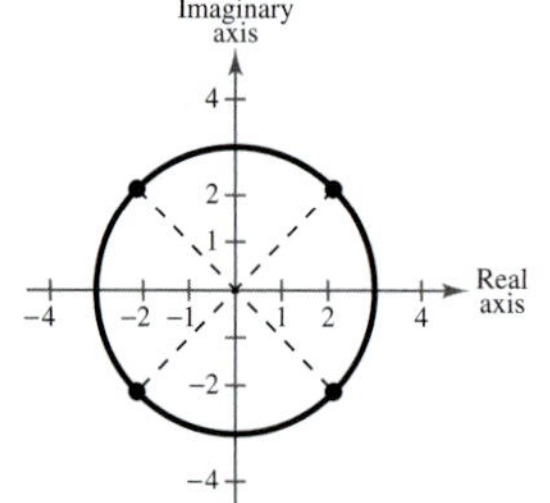

107. $2\left(\cos\frac{\pi}{2} + i\sin\frac{\pi}{2}\right) = 2i$

$2\left(\cos\frac{7\pi}{6} + i\sin\frac{7\pi}{6}\right) = -\sqrt{3} - i$

$2\left(\cos\frac{11\pi}{6} + i\sin\frac{11\pi}{6}\right) = \sqrt{3} - i$

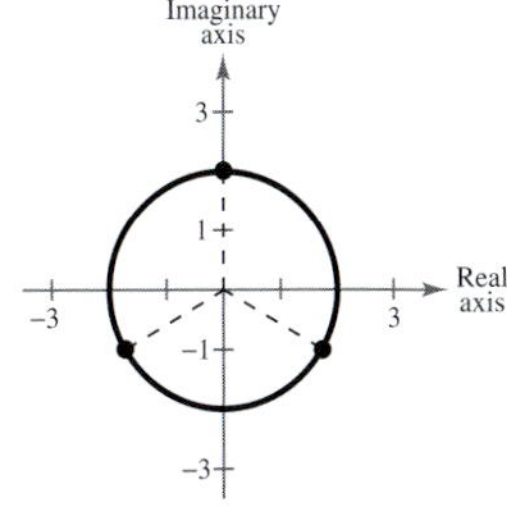

109. True. sin 90° is defined in the Law of Sines.

111. True. By definition, $\mathbf{u} = \frac{\mathbf{v}}{\|\mathbf{v}\|}$, so $\mathbf{v} = \|\mathbf{v}\|\mathbf{u}$.

113. False. The solutions to $x^2 - 8i = 0$ are $x = 2 + 2i$ and $x = -2 - 2i$.

115. $a^2 = b^2 + c^2 - 2bc\cos A$,
$b^2 = a^2 + c^2 - 2ac\cos B$,
$c^2 = a^2 + b^2 - 2ab\cos C$

117. **A** and **C**

119. If $k > 0$, the direction is the same and the magnitude is k times as great.

If $k < 0$, the result is a vector in the opposite direction and the magnitude is $|k|$ times as great.

121. $z_1 z_2 = -4$; $\frac{z_1}{z_2} = \cos(2\theta - \pi) + i\sin(2\theta - \pi)$
$= -\cos 2\theta - i\sin 2\theta$

Chapter Test *(page 653)*

1. $C = 88°$, $b \approx 27.81$, $c \approx 29.98$

2. $A = 43°$, $b \approx 25.75$, $c \approx 14.45$

3. Two solutions:

$B \approx 29.12°$, $C \approx 126.88°$, $c \approx 22.03$

$B \approx 150.88°$, $C \approx 5.12°$, $c \approx 2.46$

4. No solution **5.** $A \approx 39.96°$, $C \approx 40.04°$, $c \approx 15.02$

6. $A \approx 23.43°$, $B \approx 33.57°$, $c \approx 86.46$

7. 2052.5 square meters **8.** 606.3 miles; 29.1°

9. $\langle 14, -23\rangle$ **10.** $\left\langle \frac{18\sqrt{34}}{17}, -\frac{30\sqrt{34}}{17}\right\rangle$

11. $\langle -4, 6\rangle$

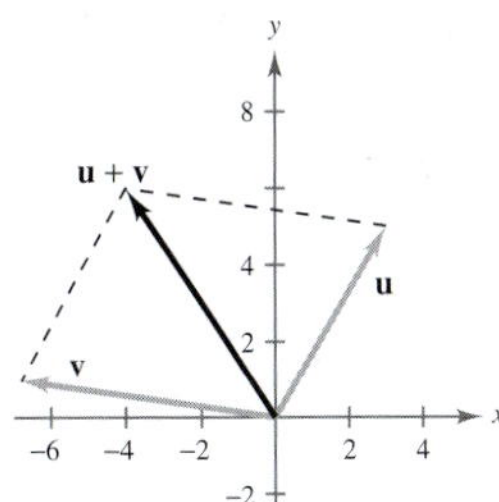

12. $\langle 10, 4\rangle$

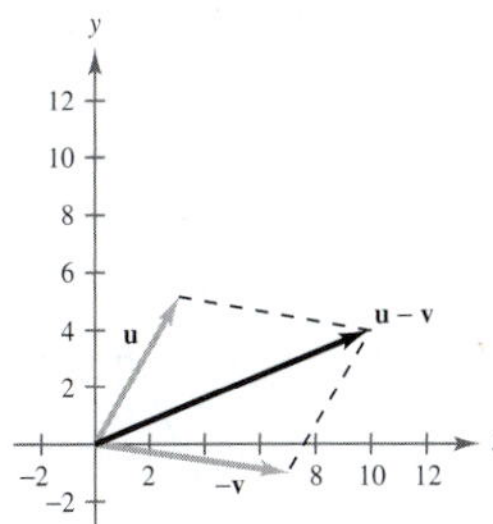

13. $\langle 36, 22\rangle$

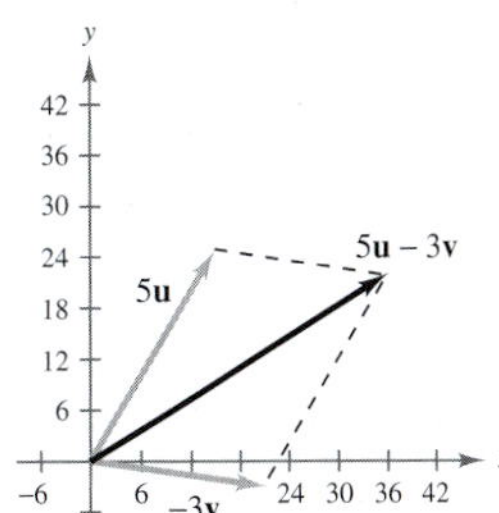

14. $\left\langle \frac{4}{5}, -\frac{3}{5}\right\rangle$

15. 14.9°; 250.15 pounds **16.** 135° **17.** No

18. $\frac{37}{26}\langle 5, 1\rangle$; $\frac{29}{26}\langle -1, 5\rangle$ **19.** $5\sqrt{2}\left(\cos\frac{7\pi}{4} + i\sin\frac{7\pi}{4}\right)$

20. $-3 + 3\sqrt{3}i$ **21.** $-\frac{6561}{2} - \frac{6561\sqrt{3}}{2}i$ **22.** $5832i$

23. $4\sqrt[4]{2}\left(\cos\frac{\pi}{12} + i\sin\frac{\pi}{12}\right)$

$4\sqrt[4]{2}\left(\cos\frac{7\pi}{12} + i\sin\frac{7\pi}{12}\right)$

$4\sqrt[4]{2}\left(\cos\frac{13\pi}{12} + i\sin\frac{13\pi}{12}\right)$

$4\sqrt[4]{2}\left(\cos\frac{19\pi}{12} + i\sin\frac{19\pi}{12}\right)$

24. $3\left(\cos\frac{\pi}{6} + i\sin\frac{\pi}{6}\right)$

$3\left(\cos\frac{5\pi}{6} + i\sin\frac{5\pi}{6}\right)$

$3\left(\cos\frac{3\pi}{2} + i\sin\frac{3\pi}{2}\right)$

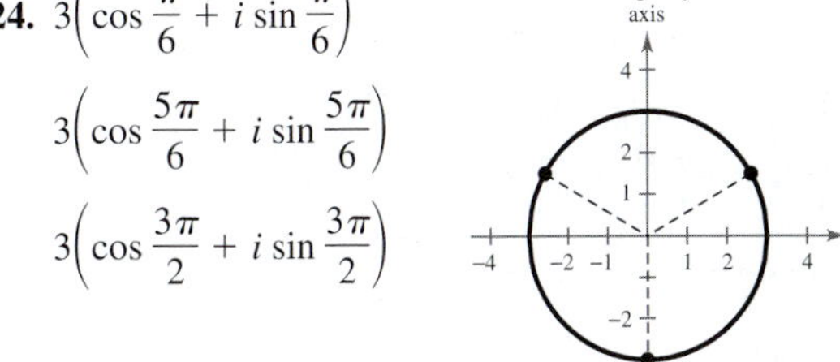

Cumulative Test for Chapters 6–8 *(page 654)*

1. (a) 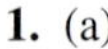

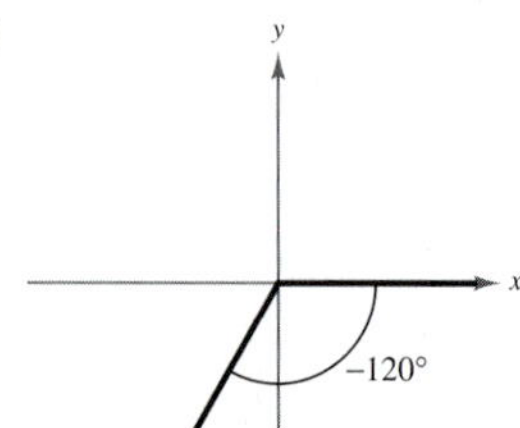

(b) 240°

(c) $-\dfrac{2\pi}{3}$

(d) 60°

(e) $\sin(-120°) = -\dfrac{\sqrt{3}}{2}$ $\quad \csc(-120°) = -\dfrac{2\sqrt{3}}{3}$

$\cos(-120°) = -\dfrac{1}{2}$ $\quad \sec(-120°) = -2$

$\tan(-120°) = \sqrt{3}$ $\quad \cot(-120°) = \dfrac{\sqrt{3}}{3}$

2. 134.6° **3.** $\frac{3}{5}$

4. Period: 2; Amplitude: 2

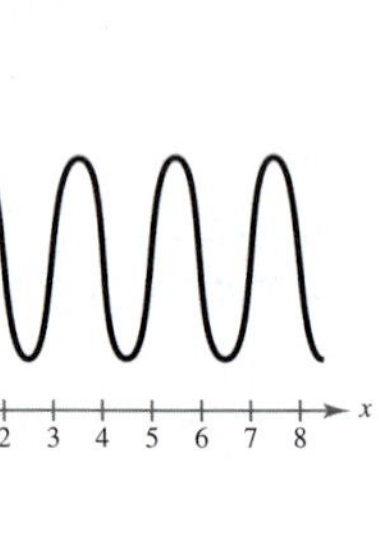

5. Period: π 

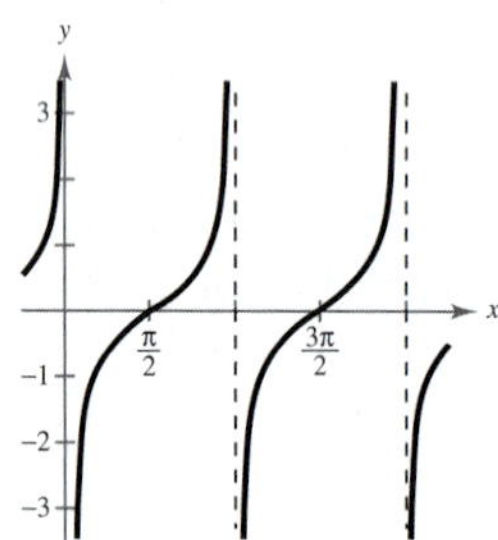

6. $a = -3, b = \pi, c = 0$

7.

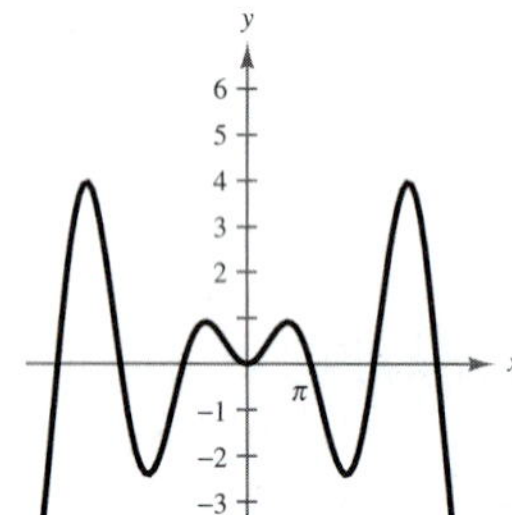

8. 6.7 **9.** $\frac{3}{4}$ **10.** $\sqrt{1 - 4x^2}$ **11.** 1

12. $2 \tan \theta$ **13–15.** Answers will vary.

16. $\dfrac{\pi}{3}, \dfrac{\pi}{2}, \dfrac{3\pi}{2}, \dfrac{5\pi}{3}$ **17.** $\dfrac{\pi}{6}, \dfrac{5\pi}{6}, \dfrac{7\pi}{6}, \dfrac{11\pi}{6}$ **18.** $\dfrac{3\pi}{2}$

19. $\dfrac{16}{63}$ **20.** $\dfrac{4}{3}$ **21.** $\dfrac{\sqrt{5}}{5}, \dfrac{2\sqrt{5}}{5}$ **22.** $\dfrac{5}{2}\left(\sin \dfrac{5\pi}{2} - \sin \pi\right)$

23. $B \approx 26.39°, C \approx 123.61°, c \approx 15.0$

24. $B \approx 52.48°, C \approx 97.52°, a \approx 5.04$

25. $B = 60°, a \approx 5.77, c \approx 11.55$

26. $A = 26.38°, B \approx 62.72°, C \approx 90.90°$

27. 36.4 square inches **28.** 85.2 square inches

29. $3\mathbf{i} + 5\mathbf{j}$ **30.** -5 **31.** $-\frac{1}{13}\langle 1, 5\rangle; \frac{21}{13}\langle 5, -1\rangle$

32. $2\sqrt{2}\left(\cos \dfrac{3\pi}{4} + i \sin \dfrac{3\pi}{4}\right)$ **33.** $-12\sqrt{3} + 12i$

34. $\cos 0 + i \sin 0 = 1$

$\cos \dfrac{2\pi}{3} + i \sin \dfrac{2\pi}{3} = -\dfrac{1}{2} + \dfrac{\sqrt{3}}{2}i$

$\cos \dfrac{4\pi}{3} + i \sin \dfrac{4\pi}{3} = -\dfrac{1}{2} - \dfrac{\sqrt{3}}{2}i$

35. $4\left(\cos \dfrac{\pi}{8} + i \sin \dfrac{\pi}{8}\right)$

$4\left(\cos \dfrac{5\pi}{8} + i \sin \dfrac{5\pi}{8}\right)$

$4\left(\cos \dfrac{9\pi}{8} + i \sin \dfrac{9\pi}{8}\right)$

$4\left(\cos \dfrac{13\pi}{8} + i \sin \dfrac{13\pi}{8}\right)$

36. 5 feet

37. ≈ 500 revolutions per minute; ≈ 20 minutes

38. 22.6° **39.** $d = 4 \cos \dfrac{\pi}{4}t$

40. 32.6°; 543.9 kilometers per hour

Problem Solving *(page 660)*

1. 2.01 feet

3. (a) Station A: 27.45 miles; Station B: 53.03 miles

(b) 11.03 miles; S 21.7° E

5. (a) (i) $\sqrt{2}$ (ii) $\sqrt{5}$ (iii) 1
(iv) 1 (v) 1 (vi) 1

(b) (i) 1 (ii) $3\sqrt{2}$ (iii) $\sqrt{13}$
(iv) 1 (v) 1 (vi) 1

(c) (i) $\dfrac{\sqrt{5}}{2}$ (ii) $\sqrt{13}$ (iii) $\dfrac{\sqrt{85}}{2}$
(iv) 1 (v) 1 (vi) 1

(d) (i) $2\sqrt{5}$ (ii) $5\sqrt{2}$ (iii) $5\sqrt{2}$
(iv) 1 (v) 1 (vi) 1

7. $\mathbf{w} = \frac{1}{2}(\mathbf{u} + \mathbf{v}); \mathbf{w} = \frac{1}{2}(\mathbf{v} - \mathbf{u})$

9. (a)

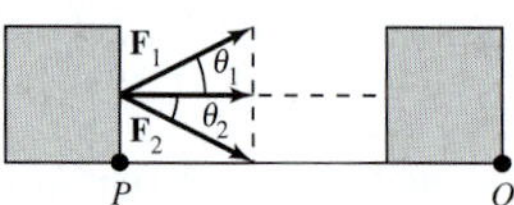

The amount of work done by $\mathbf{F}_1$ is equal to the amount of work done by $\mathbf{F}_2$.

(b)

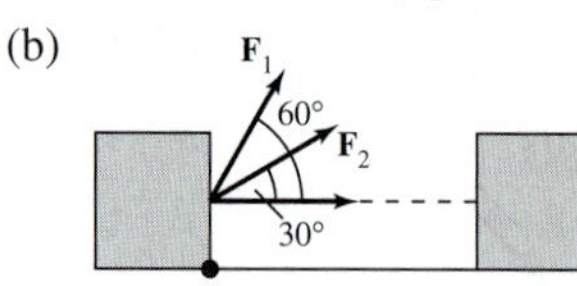

The amount of work done by $\mathbf{F}_2$ is $\sqrt{3}$ times as great as the amount of work done by $\mathbf{F}_1$.

Chapter 9

Section 9.1 *(page 671)*

1. d **3.** b **5.** (2, 2) **7.** (2, 6), (−1, 3)

9. (0, −5), (4, 3) **11.** (0, 0), (2, −4)

13. (0, 1), (1, −1), (3, 1) **15.** (5, 5) **17.** $\left(\frac{1}{2}, 3\right)$

19. (1, 1) **21.** $\left(\frac{20}{3}, \frac{40}{3}\right)$ **23.** No solution

25. (−2, 4), (0, 0) **27.** (0, 0), (−1, −1), (1, 1)

29. (4, 3) **31.** $\left(\frac{5}{2}, \frac{3}{2}\right)$ **33.** (2, 2), (4, 0)

35. (1, 4), (4, 7) **37.** $\left(4, -\frac{1}{2}\right)$

39. No solution **41.** (4, 3), (−4, 3)

43.

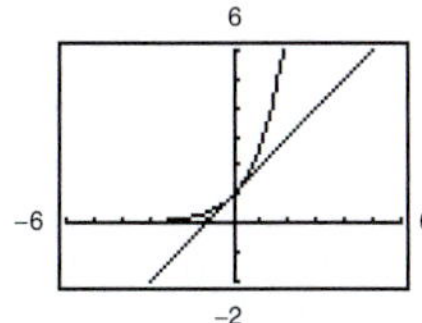

(0, 1)

45.

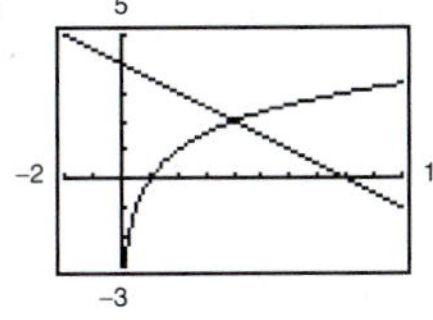

(4, 2)

47.

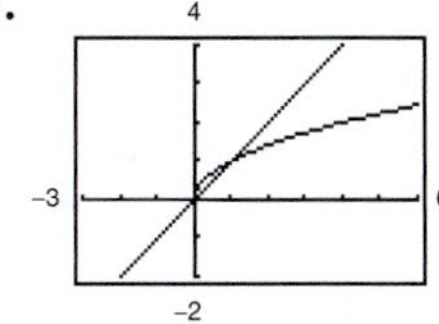

(0, 0), (1, 1)

49.

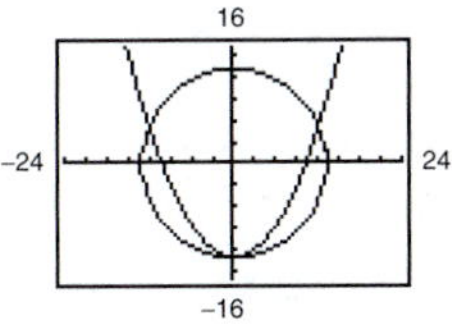

(0, −13), (±12, 5)

51. (1, 2) **53.** (−2, 0), $\left(\frac{29}{10}, \frac{21}{10}\right)$ **55.** No solution

57. (0.287, 1.751) **59.** (−1, 0), (0, 1), (1, 0)

61. $\left(\frac{1}{2}, 2\right), \left(-4, -\frac{1}{4}\right)$ **63.** 192 units **65.** 3133 units

67. (a) 781 units (b) 3708 units

69. (a) $\begin{cases} x + y = 25{,}000 \\ 0.06x + 0.085y = 2{,}000 \end{cases}$

(b)

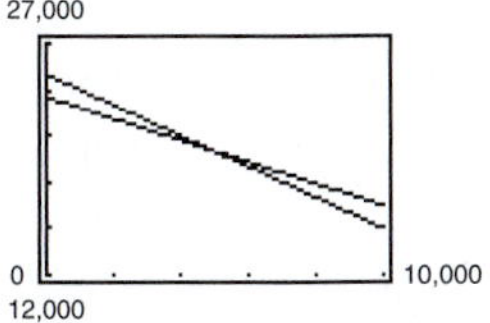

Decreases; Interest is fixed.

(c) $5000

71. More than $11,666.67

73. (a)

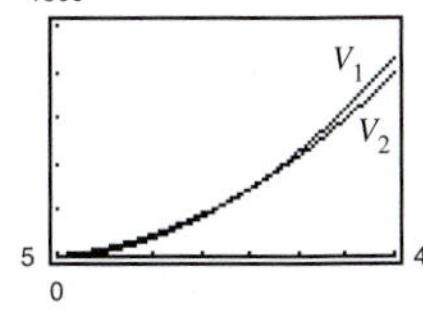

(b) 24.7 inches

(c) Doyle Log Rule

75. (a) $f(t) = 4270.2t + 65{,}082$

$g(t) = -552.00t^2 + 12{,}550.2t + 34{,}722$

(b)

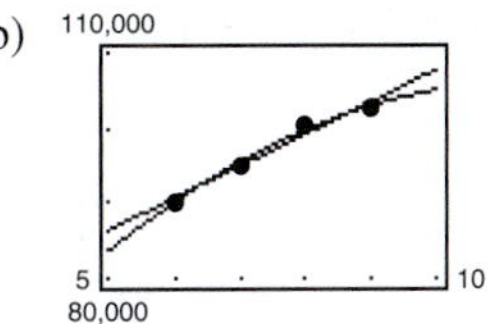

(c) (8.62, 101,882.73), (6.38, 92,334.27)

(d) (8.62, 101,888.70), (6.38, 92,323.45)

(e) $f = 129{,}135{,}000$, $g = 98{,}775{,}000$

(f) f is more accurate.

77. 6 × 9 meters **79.** 9 × 12 inches

81. 8 × 12 kilometers

83. False. To solve a system of equations by substitution, you can solve for either variable in one of the two equations and then back-substitute.

85. 1. *Solve* one of the equations for one variable in terms of the other.

2. *Substitute* the expression found in Step 1 into the other equation to obtain an equation in one variable.

3. *Solve* the equation obtained in Step 2.

4. *Back-substitute* the value obtained in Step 3 into the expression obtained in Step 1 to find the value of the other variable.

5. *Check* that the solution satisfies *each* of the original equations.

87. (a) $y = 2x$ (b) $y = 0$ (c) $y = x - 2$

89. $2x + 7y - 45 = 0$ **91.** $y - 3 = 0$

93. $30x - 17y - 18 = 0$

95. Domain: All real numbers $x \neq 6$

Horizontal asymptote: $y = 0$

Vertical asymptote: $x = 6$

97. Domain: All real numbers $x \neq \pm 4$

Horizontal asymptote: $y = 1$

Vertical asymptotes: $x = \pm 4$

Section 9.2 *(page 683)*

1. (2, 1)

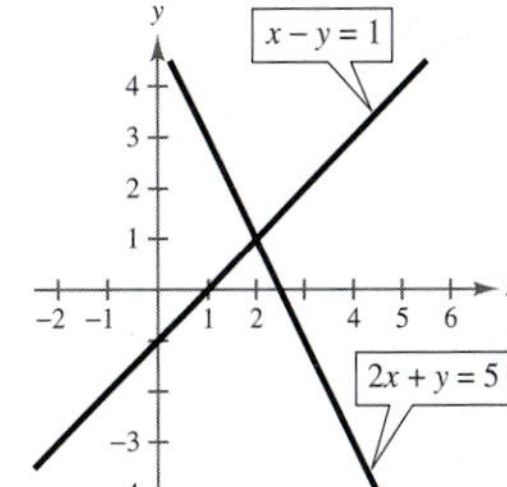

3. (1, −1)

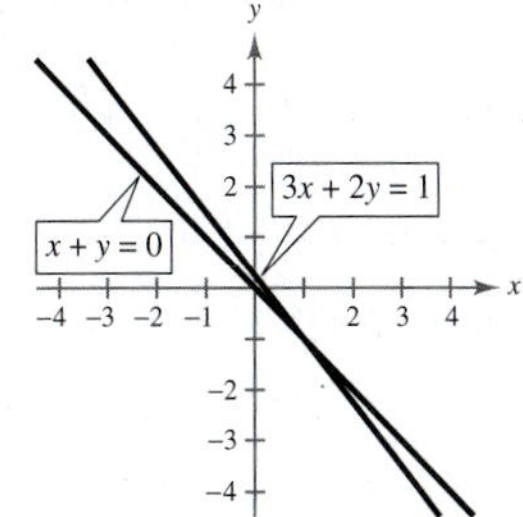

5. No solution

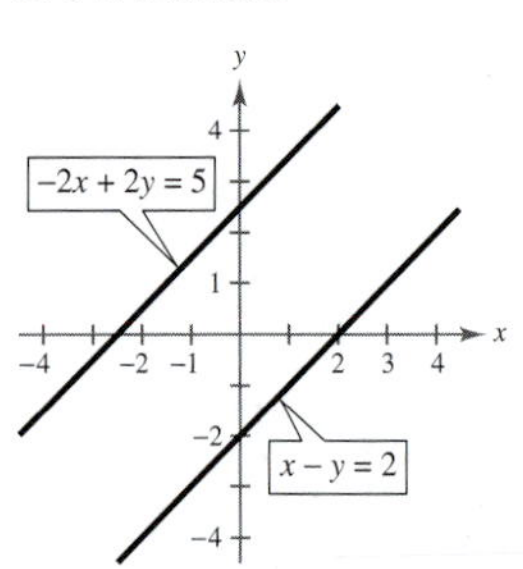

7. $\left(a, \frac{3}{2}a - \frac{5}{2}\right)$

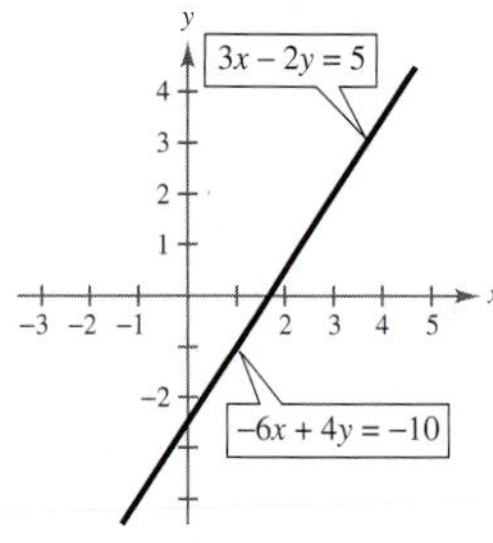

9. $\left(\frac{1}{3}, -\frac{2}{3}\right)$

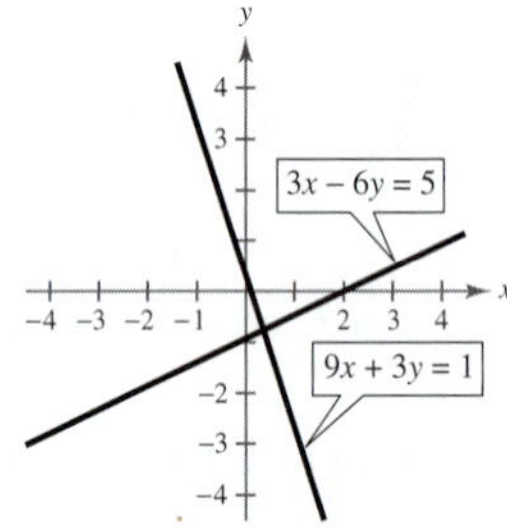

11. $\left(\frac{5}{2}, \frac{3}{4}\right)$

13. $(3, 4)$ **15.** $(4, -1)$

17. $\left(\frac{12}{7}, \frac{18}{7}\right)$ **19.** No solution **21.** $\left(\frac{18}{5}, \frac{3}{5}\right)$

23. $\left(a, \frac{5}{6}a - \frac{1}{2}\right)$ **25.** $\left(\frac{90}{31}, -\frac{67}{31}\right)$

27. $\left(-\frac{6}{35}, \frac{43}{35}\right)$ **29.** $(5, -2)$

31. b **32.** a **33.** c **34.** d

35. $(4, 1)$ **37.** $(2, -1)$ **39.** $(6, -3)$

41. $\left(\frac{43}{6}, \frac{25}{6}\right)$ **43.** $(80, 10)$ **45.** $(2{,}000{,}000, 100)$

47. 550 miles per hour, 50 miles per hour

49. (a) $\begin{cases} x + y = 10 \\ 0.2x + 0.5y = 3 \end{cases}$

(b)

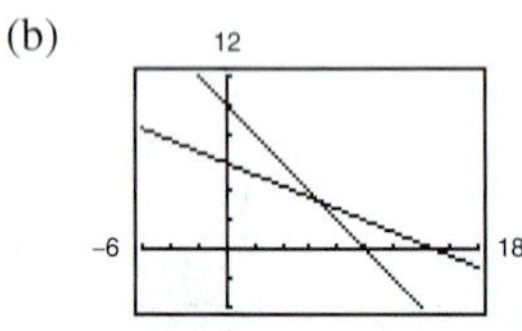

Decreases

(c) 20% solution: $6\frac{2}{3}$ liters

50% solution: $3\frac{1}{3}$ liters

51. \$6000 **53.** 400 adult, 1035 student

55. $y = 0.97x + 2.1$

57. $y = 0.32x + 4.1$

59. $y = -2x + 4$

61. (a) and (b) $y = 3.705t + 48.63$

(c)

Year	1995	1996	1997	1998	1999
y	\$67.16	\$70.86	\$74.57	\$78.27	\$81.98

(d) \$104.21 (e) 2003

63. False. Two lines that coincide have infinitely many points of intersection.

65. (39,600, 398). It is necessary to change the scale on the axes to see the point of intersection.

67. No. Two lines will intersect only once or will coincide, and if they coincide the system will have infinitely many solutions.

69. $k = -4$

71. $x \le -\frac{22}{3}$

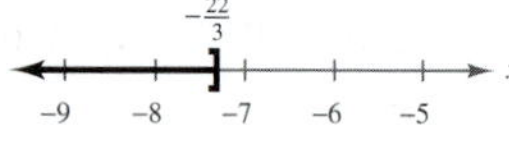

73. $x \le \frac{19}{16}$

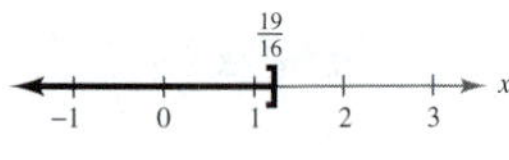

75. $-2 < x < 18$

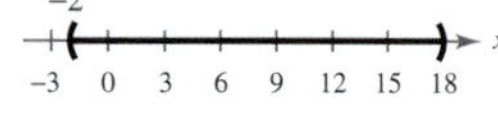

77. $-5 < x < \frac{7}{2}$

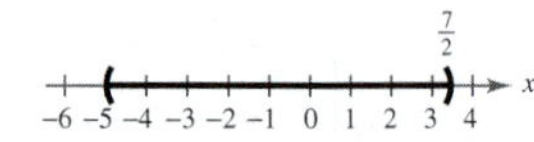

79. $-\dfrac{6}{x+5} + \dfrac{7}{x+6}$ **81.** $\ln 6x$ **83.** $\log_9 \dfrac{12}{x}$

85. No solution

Section 9.3 *(page 695)*

1. d **3.** c **5.** $(1, -2, 4)$ **7.** $(3, 10, 2)$

9. $\left(\frac{1}{2}, -2, 2\right)$

11. $\begin{cases} x - 2y + 3z = 5 \\ \quad\; y - 2z = 9 \\ 2x \quad\; - 3z = 0 \end{cases}$

First step in putting the system in row-echelon form

13. $(1, 2, 3)$ **15.** $(-4, 8, 5)$ **17.** $(5, -2, 0)$

19. No solution **21.** $\left(-\frac{1}{2}, 1, \frac{3}{2}\right)$

23. $(-3a + 10, 5a - 7, a)$ **25.** $(-a + 3, a + 1, a)$

27. $(2a, 21a - 1, 8a)$ **29.** $\left(-\frac{3}{2}a + \frac{1}{2}, -\frac{2}{3}a + 1, a\right)$

31. $(1, 1, 1, 1)$ **33.** No solution **35.** $(0, 0, 0)$

37. $(9a, -35a, 67a)$

39. $y = \frac{1}{2}x^2 - 2x$

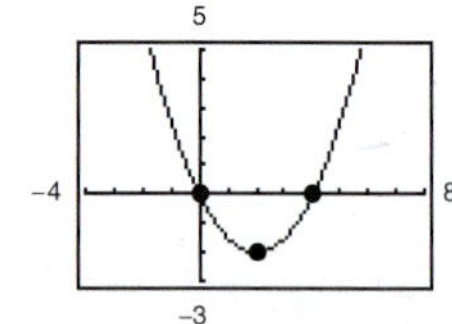

41. $y = x^2 - 6x + 8$

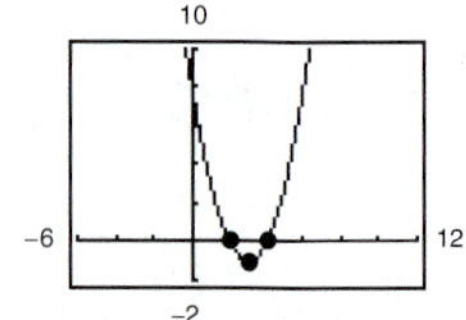

43. $x^2 + y^2 - 4x = 0$

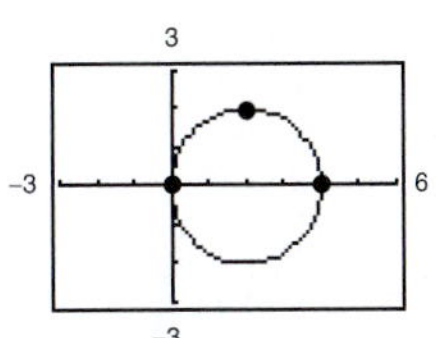

45. $x^2 + y^2 + 6x - 8y = 0$

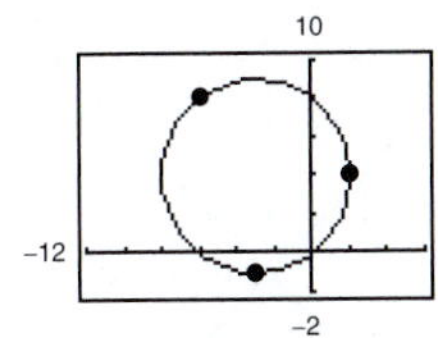

47. $s = -16t^2 + 144$ **49.** $s = -16t^2 - 32t + 500$

51. 8 touchdowns, 8 extra-point kicks, 2 field goals

53. \$300,000 at 8%
\$400,000 at 9%
\$75,000 at 10%

55. $250{,}000 - \frac{1}{2}s$ in certificates of deposit,
$125{,}000 + \frac{1}{2}s$ in municipal bonds,
$125{,}000 - s$ in blue-chip stocks,
s in growth stocks

57. Use four medium trucks or use two large, one medium, and two small trucks. Other answers are possible.

59. No 10% solution, $8\frac{1}{3}$ liters of 20% solution, $1\frac{2}{3}$ liters of 50% solution

61. $t_1 = 96$ pounds
$t_2 = 48$ pounds
$a = -16$ feet per second squared

63. $\frac{1}{2}\left(-\frac{2}{x} + \frac{1}{x-1} + \frac{1}{x+1}\right)$

65. $\frac{1}{2}\left(\frac{1}{x} - \frac{1}{x-2} + \frac{2}{x+3}\right)$

67. $y = -\frac{5}{24}x^2 - \frac{3}{10}x + \frac{41}{6}$ **69.** $y = x^2 - x$

71. (a) $y = -0.0075x^2 + 1.3x + 20$

(b)

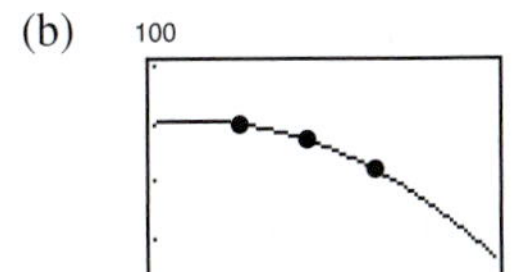

(c)

x	100	120	140
y	75	68	55

The values are the same.

(d) 24.25% (e) 156 females

73. $x = 5$
$y = 5$
$\lambda = -5$

75. $x = \pm\sqrt{2}/2$ or $x = 0$
$y = \frac{1}{2}$ $y = 0$
$\lambda = 1$ $\lambda = 0$

77. False. Equation 2 does not have a leading coefficient of 1.

79. No. Answers will vary.

81. $\begin{cases} 3x + y - z = 9 \\ x + 2y - z = 0 \\ -x + y + 3z = 1 \end{cases}$ $\begin{cases} x + y + z = 5 \\ x - 2z = 0 \\ 2y + z = 0 \end{cases}$

83. $\begin{cases} x + 2y - 4z = -5 \\ -x - 4y + 8z = 13 \\ x + 6y + 4z = 7 \end{cases}$ $\begin{cases} x + 2y + 4z = 9 \\ y + 2z = 3 \\ x - 4z = -4 \end{cases}$

85. 6.375 **87.** 80,000 **89.** $11 + i$

91. $22 + 3i$ **93.** $\frac{7}{2} + \frac{7}{2}i$

95. (a) $-4, 0, 3$ (b)

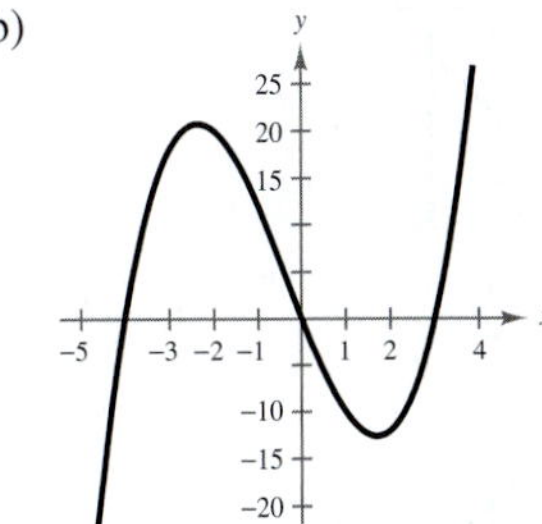

97. (a) $-4, -\frac{3}{2}, 3$

(b)

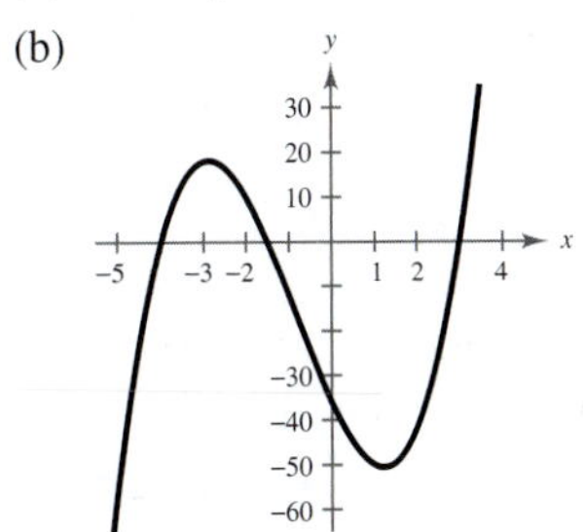

99.

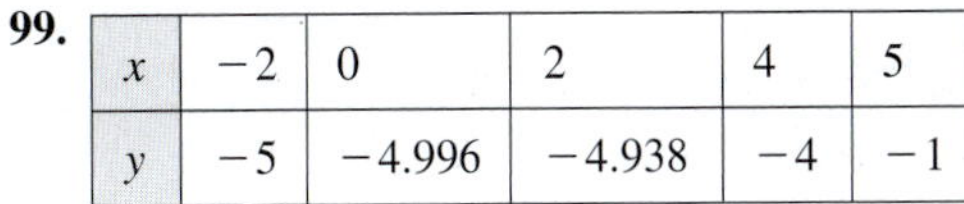

x	-2	0	2	4	5
y	-5	-4.996	-4.938	-4	-1

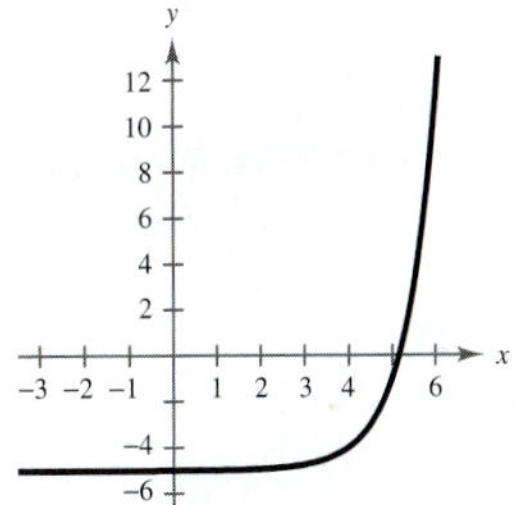

101.

x	-2	-1	0	1	2
y	5.793	4.671	4	3.598	3.358

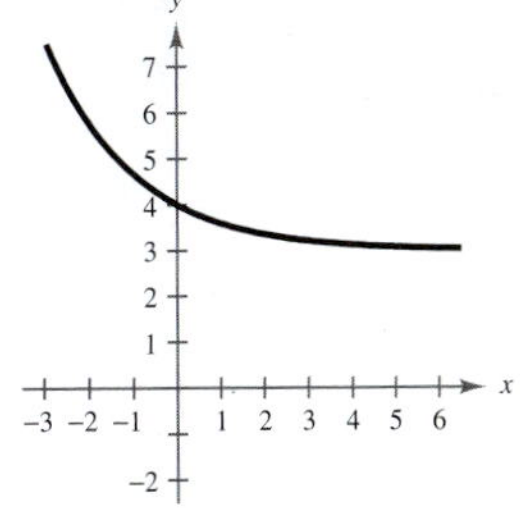

103. (40, 40)

Section 9.4 *(page 707)*

1.
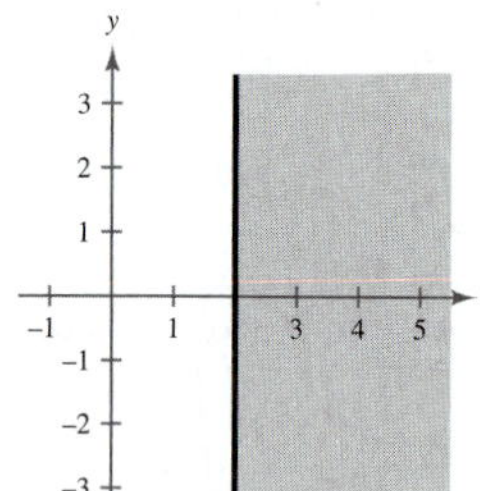

3.
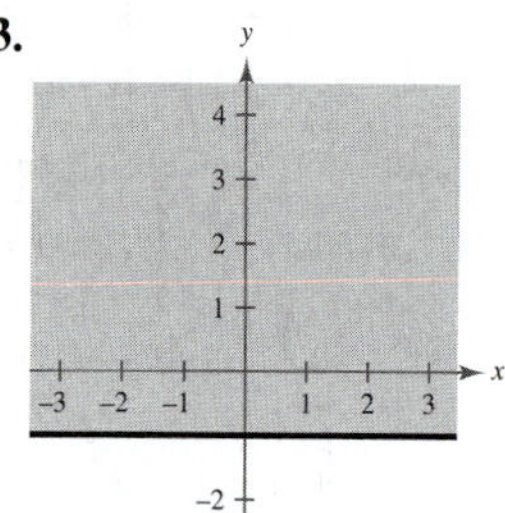

5.
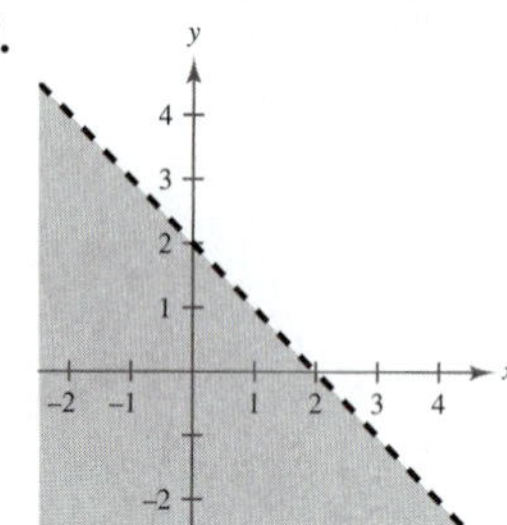

7.
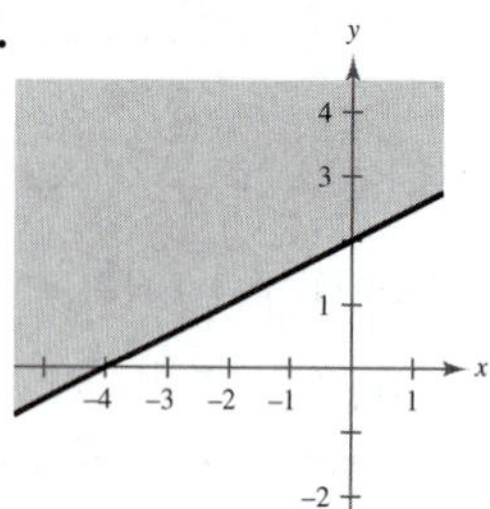

9.
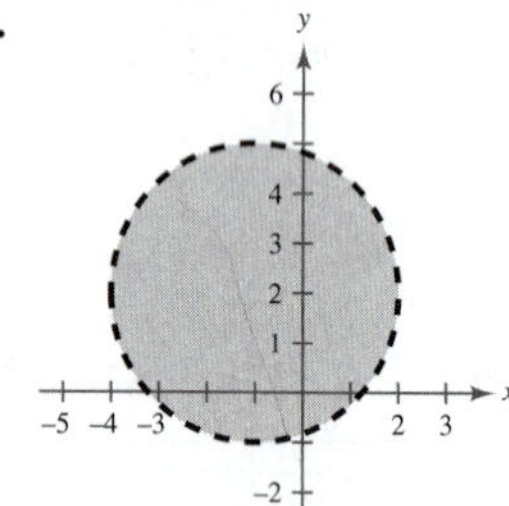

11.
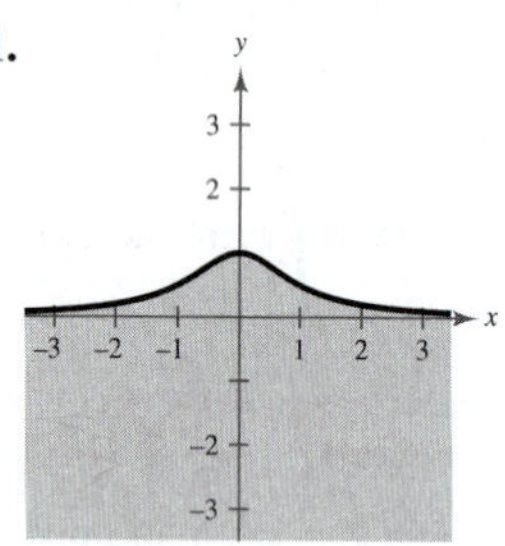

13.
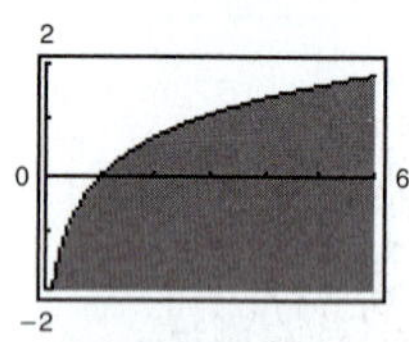

15.
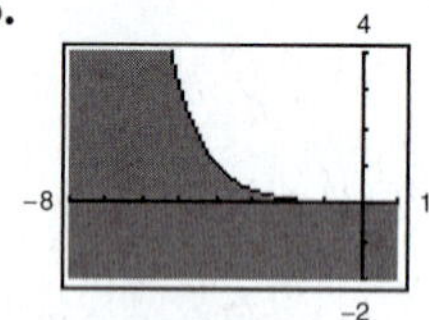

17.
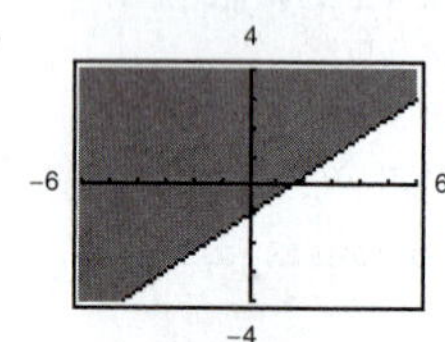

19.
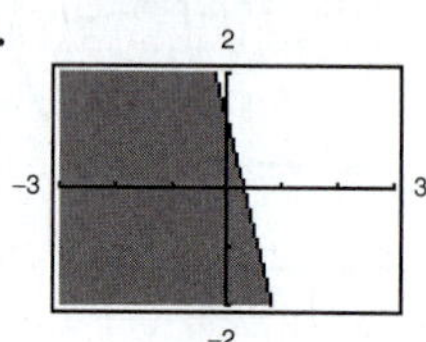

21.
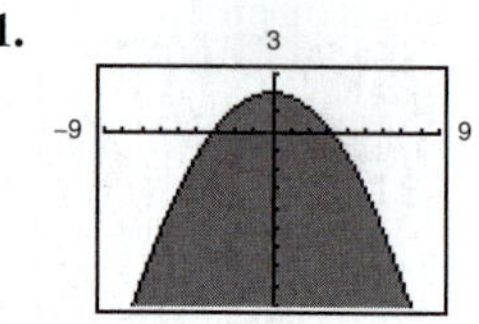

23.
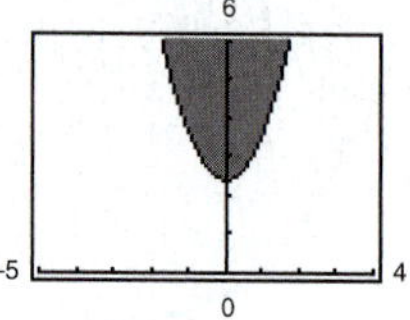

25. $y \le \frac{1}{2}x + 2$ **27.** $y \ge -\frac{2}{3}x + 2$

29. c and d **31.** a, c, and d

33.
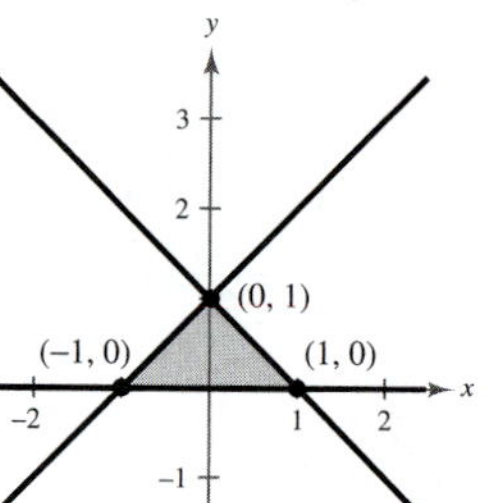

35.
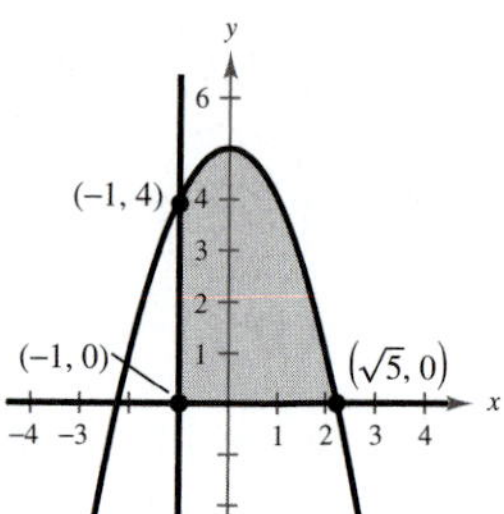

37.
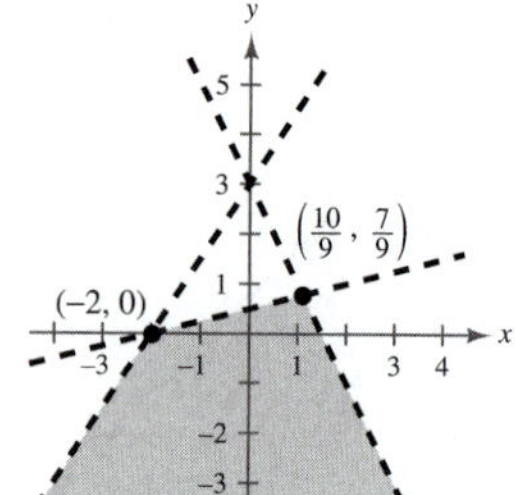

39. No solution

41.
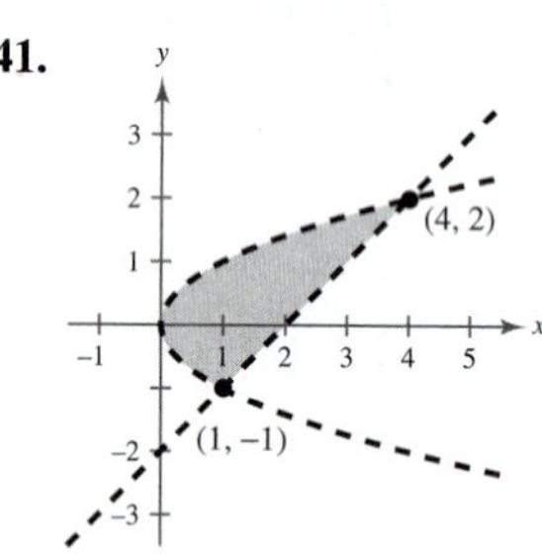

43.
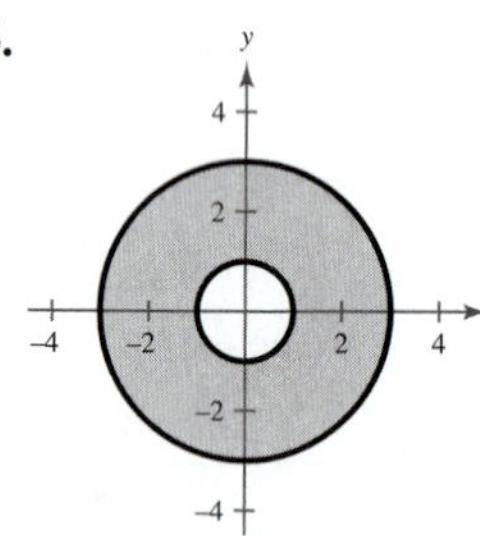

45.
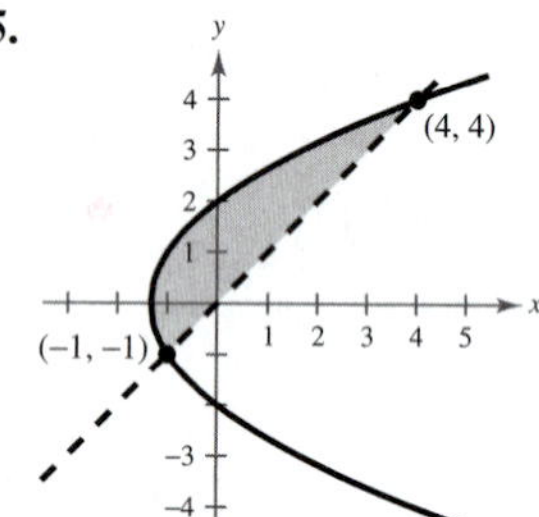

47.
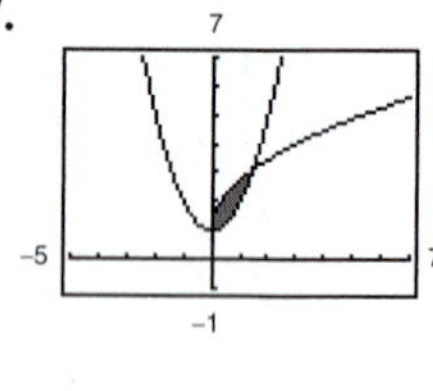

49.
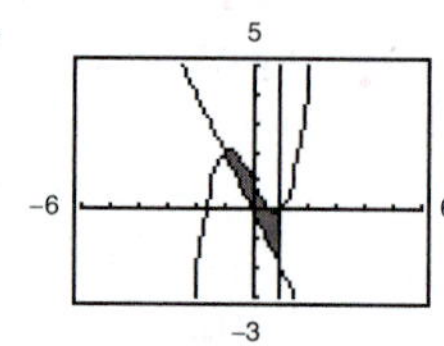

51.
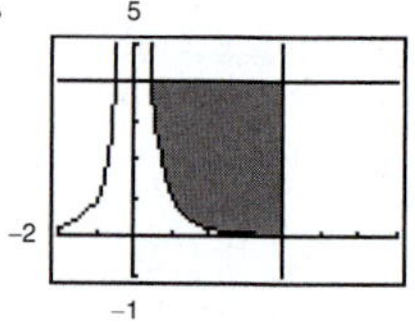

53. $\begin{cases} y \le 4 - x \\ x \ge 0 \\ y \ge 0 \end{cases}$ **55.** $\begin{cases} y \ge 4 - x \\ y \ge 2 - \frac{1}{4}x \\ x \ge 0,\ y \ge 0 \end{cases}$

57. $\begin{cases} x^2 + y^2 \le 16 \\ x \ge 0 \\ y \ge 0 \end{cases}$ **59.** $\begin{cases} 2 \le x \le 5 \\ 1 \le y \le 7 \end{cases}$ **61.** $\begin{cases} y \le \frac{3}{2}x \\ y \le -x + 5 \\ y \ge 0 \end{cases}$

63. Consumer surplus: \$1600

Producer surplus: \$400

65. Consumer surplus: $40,000,000

Producer surplus: $20,000,000

67. $\begin{cases} x + \frac{3}{2}y \le 12 \\ \frac{4}{3}x + \frac{3}{2}y \le 15 \\ x \ge 0 \\ y \ge 0 \end{cases}$

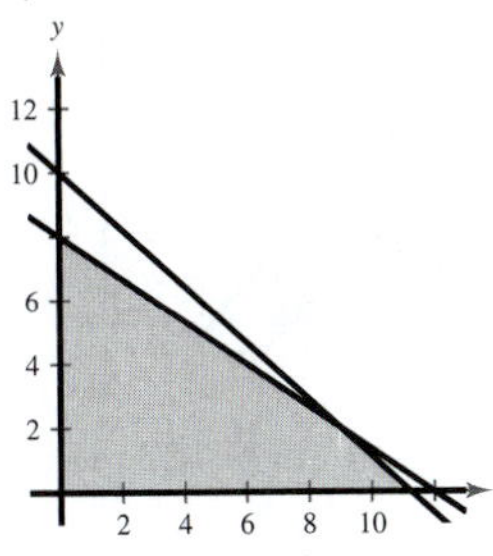

69. $\begin{cases} x + y \le 20{,}000 \\ y \ge 2x \\ x \ge 5{,}000 \\ y \ge 5{,}000 \end{cases}$

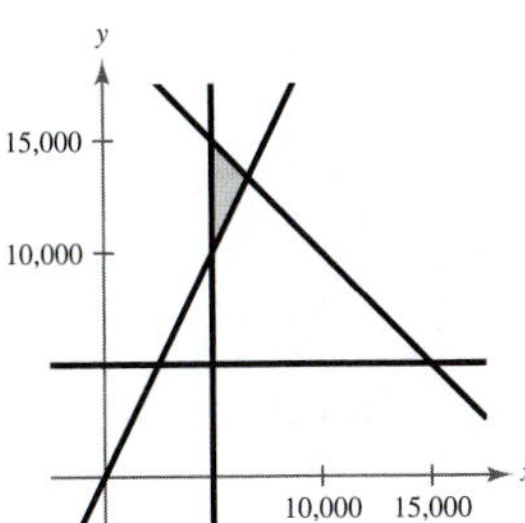

71. $\begin{cases} 55x + 70y \le 7500 \\ x \ge 50 \\ y \ge 40 \end{cases}$

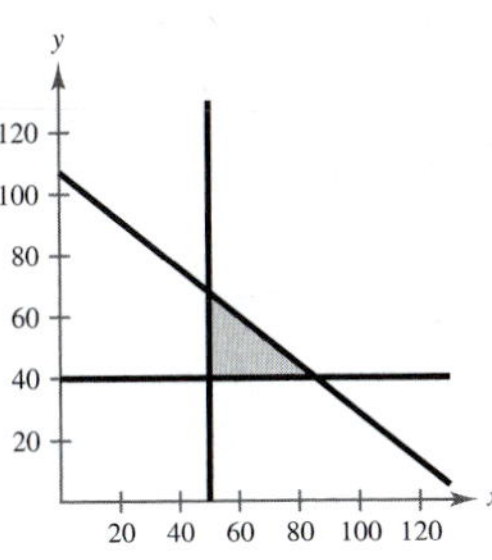

73. (a) $y = 1099.7t - 3484$

(b)

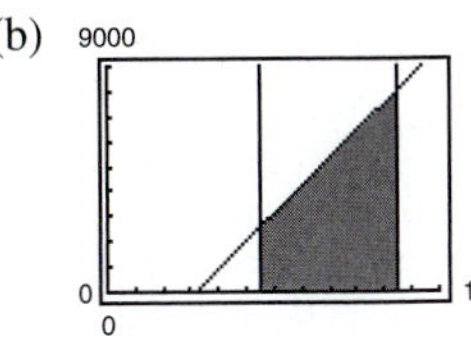

(c) 26,568 electric-powered vehicles

75. True. The figure is a rectangle with a length of 9 units and a width of 11 units.

77. The graph is a half-line on the real number line; on the rectangular coordinate system, the graph is a half-plane.

79. (a) $\begin{cases} \pi y^2 - \pi x^2 \ge 10 \\ y > x \\ x > 0 \end{cases}$

(b)

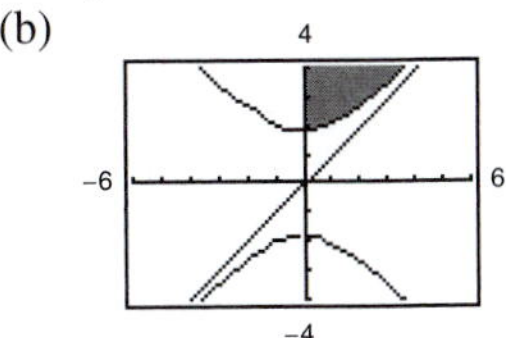

(c) The line is an asymptote to the boundary. The larger the circles, the closer the radii can be and the constraint will still be satisfied.

81. d **82.** b **83.** c **84.** a

85. $5x + 3y - 8 = 0$ **87.** $28x + 17y + 13 = 0$

89. $x + y + 1.8 = 0$

91. (a) $y_1 = 53.32t + 612.8$

$y_2 = 3.400t^2 + 9.12t + 746.5$

(b)

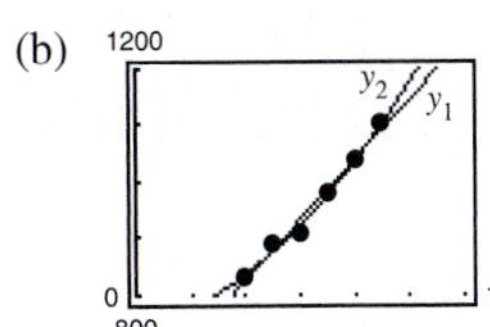

(c) The models are accurate.

93. 52.619 **95.** 0.064

Section 9.5 *(page 717)*

1. Minimum at (0, 0): 0

Maximum at (5, 0): 20

3. Minimum at (0, 0): 0

Maximum at (0, 5): 40

5. Minimum at (0, 0): 0

Maximum at (3, 4): 17

7. Minimum at (0, 0): 0

Maximum at (4, 0): 20

9. Minimum at (0, 0): 0

Maximum at (60, 20): 740

11. Minimum at (0, 0): 0

Maximum at any point on the line segment connecting (60, 20) and (30, 45): 2100

13.

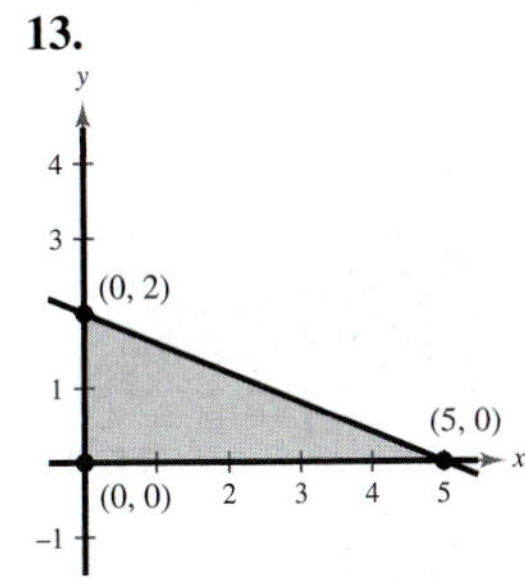

Minimum at (0, 0): 0

Maximum at (5, 0): 30

15.

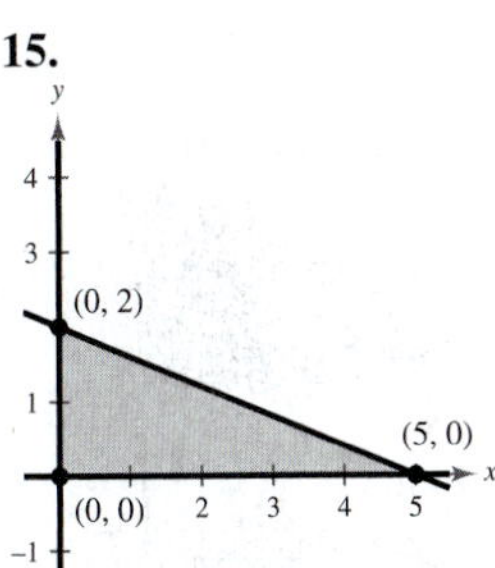

Minimum at (0, 0): 0

Maximum at (0, 2): 48

17.

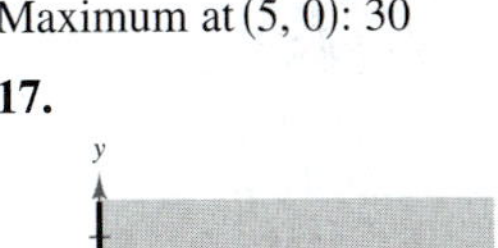

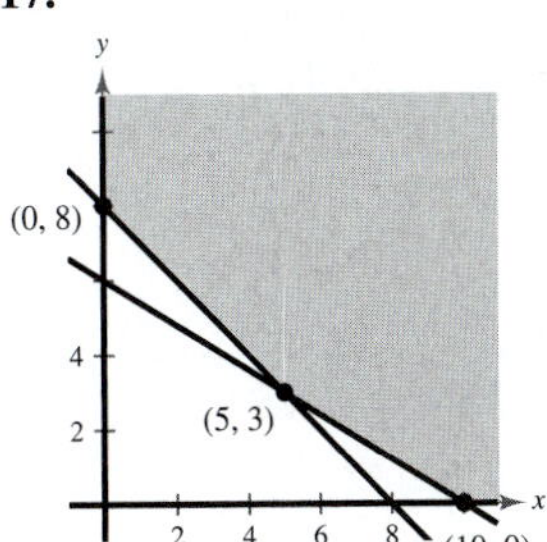

Minimum at (5, 3): 35

No maximum

19.

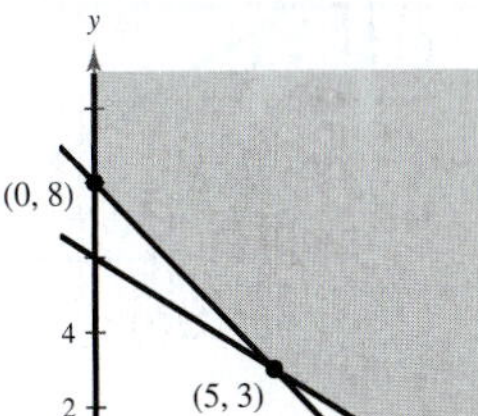

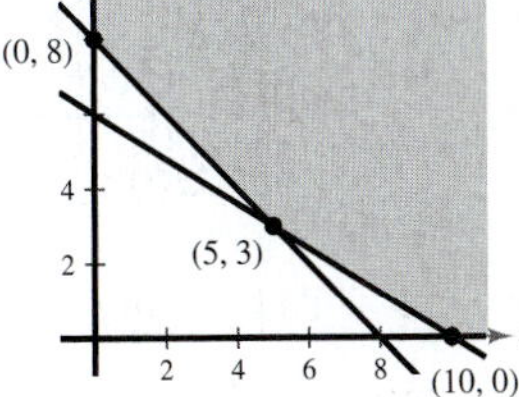

Minimum at (10, 0): 20

No maximum

21.
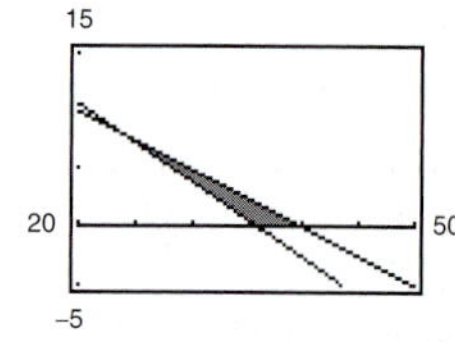

Minimum at (24, 8): 104
Maximum at (40, 0): 160

23. 15, 20, 50, −5

Minimum at (36, 0): 36
Maximum at (24, 8): 56

25. Maximum at (3, 6): 12 **27.** Maximum at (0, 10): 10

29. Maximum at (0, 5): 25 **31.** Maximum at $\left(\frac{22}{3}, \frac{19}{6}\right)$: $\frac{271}{6}$

33. 750 units of model A
1000 units of model B
Maximum profit: $83,750

35. (a) $P = 25x + 40y$ (b) $150x + 200y \le 40{,}000$
$x + y \le 250$
$x \ge 0$
$y \ge 0$

(c)
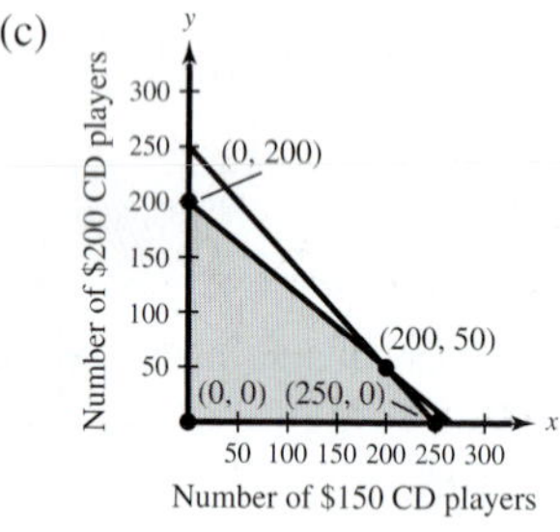

(d) 0 units of the $150 model
200 units of the $200 model

(e) Maximum profit: $8000

37. Three bags of brand X
Six bags of brand Y
Minimum cost: $195

39. Four audits
32 tax returns
Maximum revenue: $17,600

41. $62,500 to type A
$187,500 to type B

43.
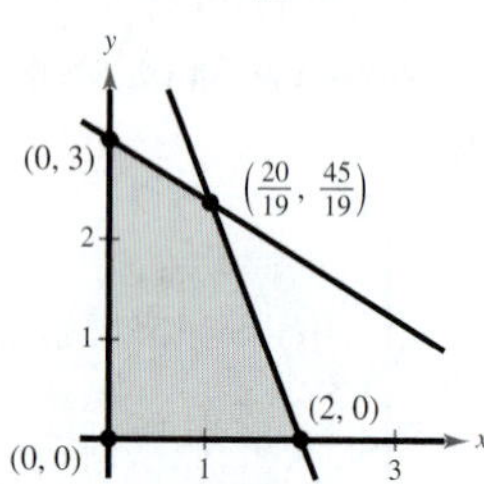

The maximum, 5, occurs at any point on the line segment connecting (2, 0) and $\left(\frac{20}{19}, \frac{45}{19}\right)$.

45.
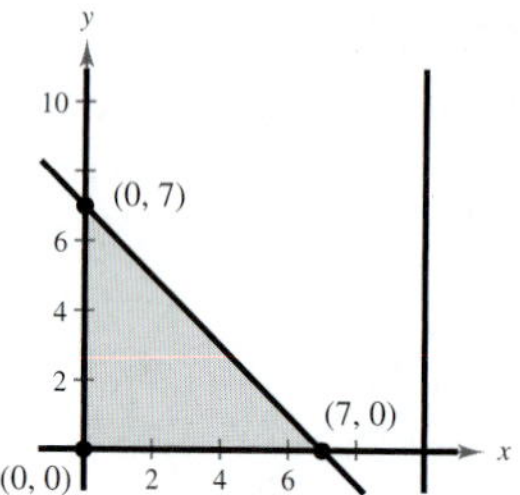

The constraint $x \le 10$ is extraneous. Maximum at (0, 7): 14

47.
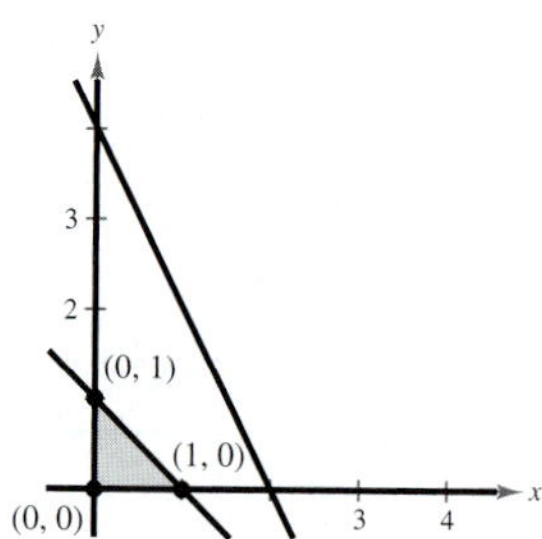

The constraint $2x + y \le 4$ is extraneous. Maximum at (0, 1): 4

49. True. The objective function has a maximum value at any point on the line segment connecting the two vertices.

51. (a) $t \ge 9$ (b) $\frac{3}{4} \le t \le 9$ **53.** $z = x + 5y$

55. $z = 4x + y$

57. $\dfrac{9}{2(x + 3)}, \ x \ne 0$ **59.** $\dfrac{x^2 + 2x - 13}{x(x - 2)}, \ x \ne \pm 3$

61. Parabola

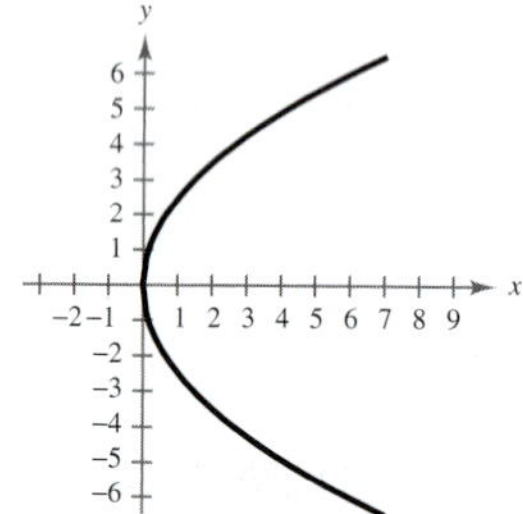

63. Ellipse

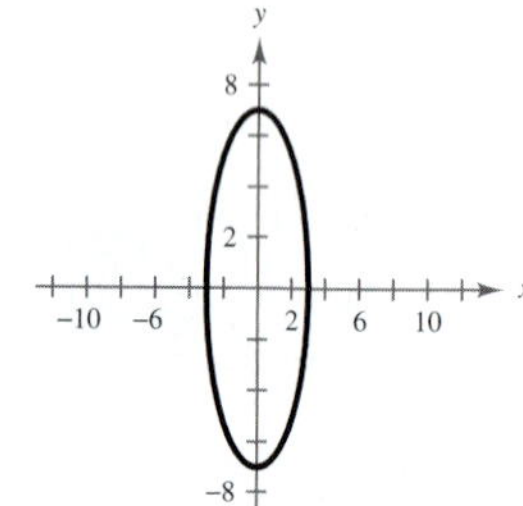

65. Hyperbola

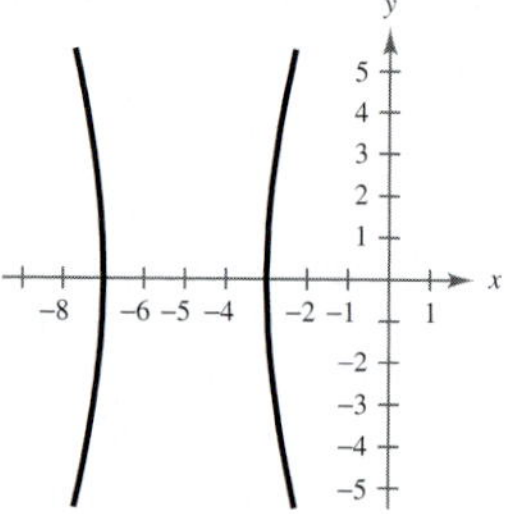

67. $\ln 3 \approx 1.099$ **69.** $4 \ln 38 \approx 14.550$

71. $\frac{1}{3}e^{12/7} \approx 1.851$ **73.** $(-4, 3, -7)$

Review Exercises (page 722)

1. (5, 4) **3.** (0, 0), (2, 8), (−2, 8) **5.** (4, −2)

7. (1.41, −0.66), (−1.41, 10.66)

9.

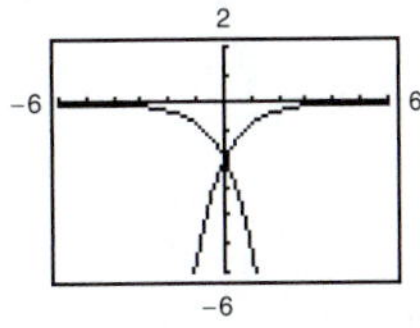

(0, −2)

11. 10,417 units **13.** 96 × 144 meters **15.** $\left(\frac{5}{2}, 3\right)$

17. (−0.5, 0.8) **19.** (0, 0) **21.** $\left(\frac{8}{5}a + \frac{14}{5}, a\right)$

23. d **24.** c **25.** b **26.** a

27. $\left(\dfrac{500{,}000}{7}, \dfrac{159}{7}\right)$ **29.** (2, −4, −5) **31.** $\left(\frac{24}{5}, \frac{22}{5}, -\frac{8}{5}\right)$

33. $(3a + 4, 2a + 5, a)$ **35.** $(a - 4, a - 3, a)$

37. $y = 2x^2 + x - 5$ **39.** $x^2 + y^2 - 4x + 4y - 1 = 0$

41. $-\dfrac{4}{x+4} + \dfrac{3}{x+2}$ **43.** $-\dfrac{1}{x+1} + \dfrac{2}{x+2}$

45. (a) $y = 596.50x^2 - 7147.5x + 20{,}083$

(b)

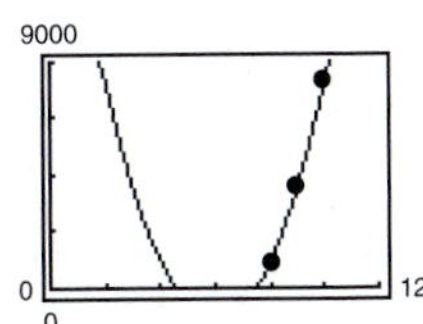

The model is a "good fit" for the data.

(c) 47,083. No.

47. \$16,000 at 7%

\$13,000 at 9%

\$11,000 at 11%

49.

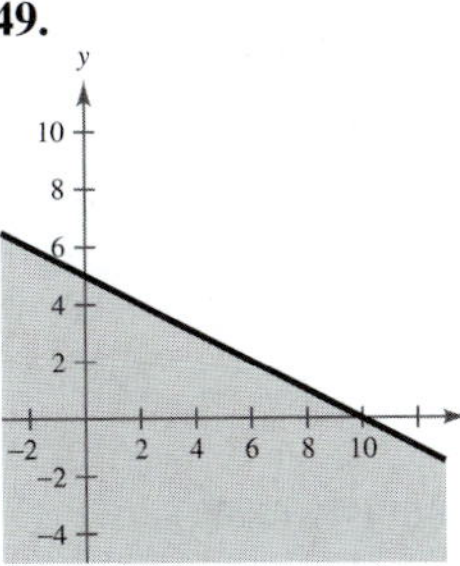

51.

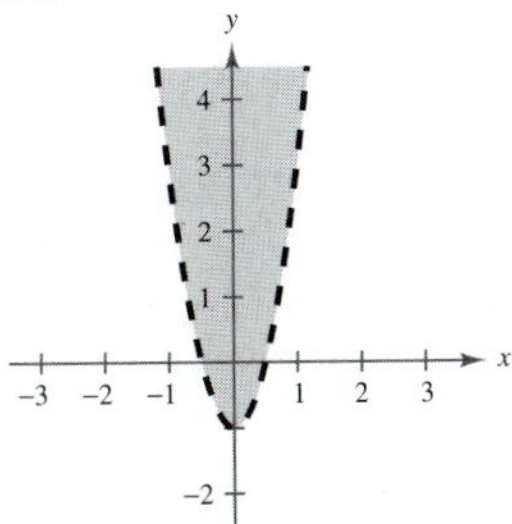

53.

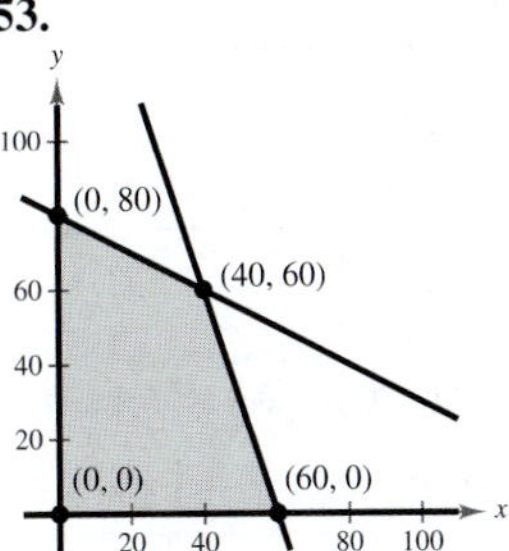

55.

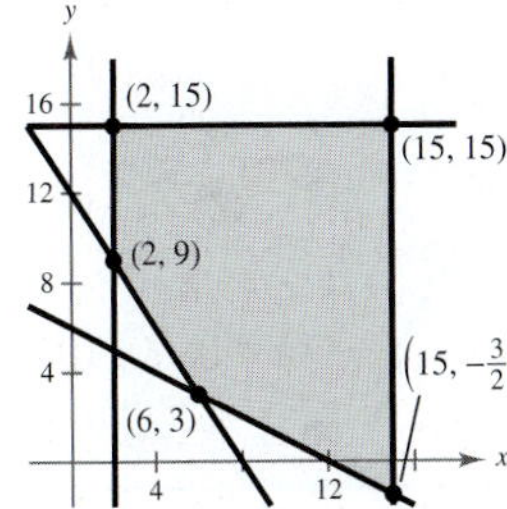

57.

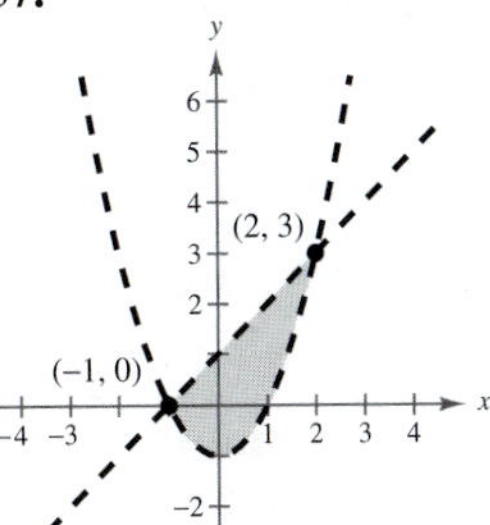

59.

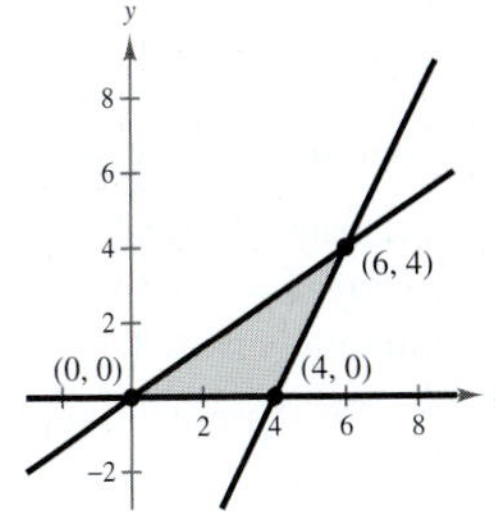

61. $\begin{cases} x + y \le 1500 \\ x \ge 400 \\ y \ge 600 \end{cases}$

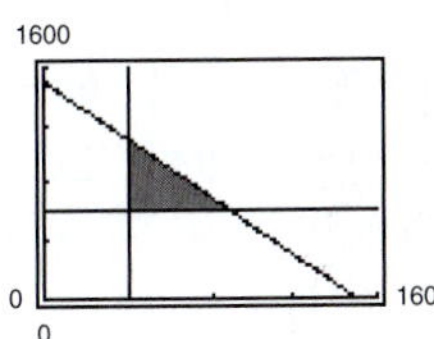

63. Consumer surplus: \$4,500,000

Producer surplus: \$9,000,000

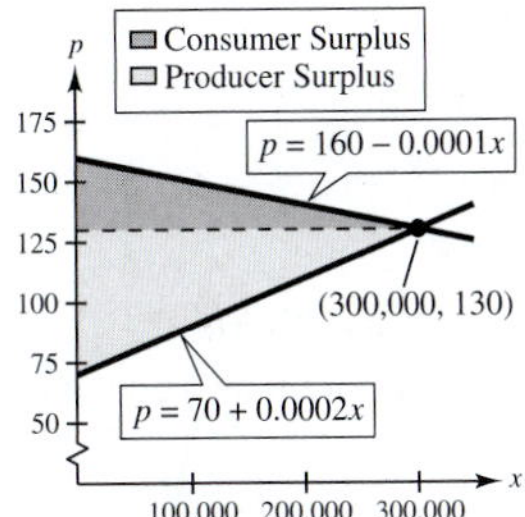

65.

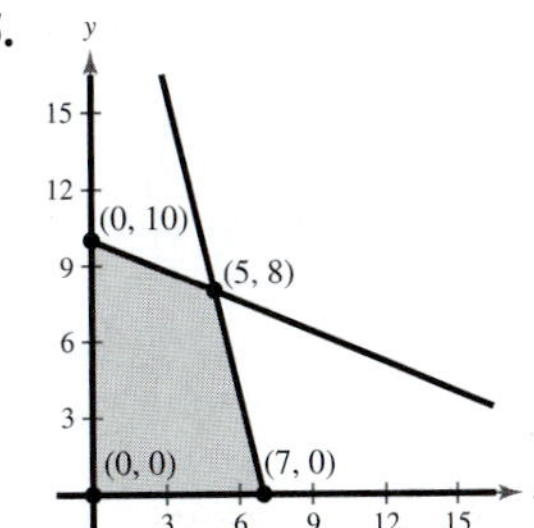

Minimum at (0, 0): 0

Maximum at (5, 8): 47

67.

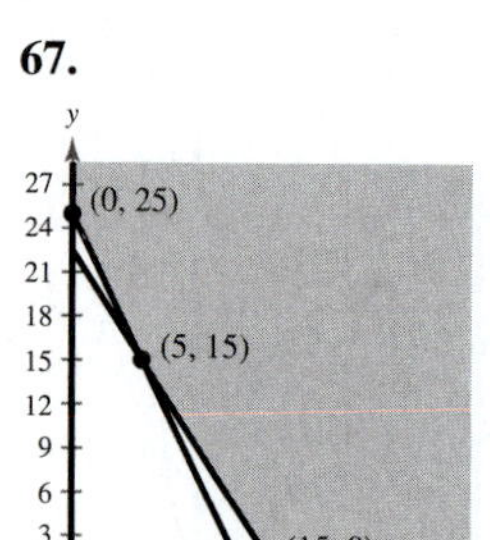

Minimum at (15, 0): 26.25

No maximum

69.

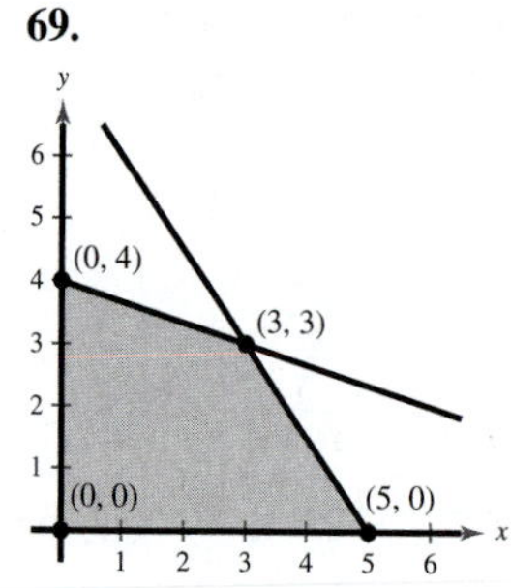

Minimum at (0, 0): 0

Maximum at (3, 3): 48

71. 72 haircuts, 0 permanents; Maximum revenue: $1800

73. Three bags of brand X
Two bags of brand Y
Minimum cost: $105

75. False. To represent a region covered by an isosceles trapezoid, the last two inequality signs should be $\leq$.

77. $\begin{cases} x + y = 2 \\ x - y = -14 \end{cases}$ **79.** $\begin{cases} 3x + y = 7 \\ -6x + 3y = 1 \end{cases}$

81. $\begin{cases} x + y + z = 6 \\ x + y - z = 0 \\ x - y - z = 2 \end{cases}$ **83.** $\begin{cases} 2x + 2y - 3z = 7 \\ x - 2y + z = 4 \\ -x + 4y - z = -1 \end{cases}$

85. An inconsistent system of linear equations has no solution.

87. Answers will vary.

Chapter Test *(page 726)*

1. $(-3, 4)$ **2.** $(0, -1), (1, 0), (2, 1)$

3. $(8, 4), (2, -2)$

4.

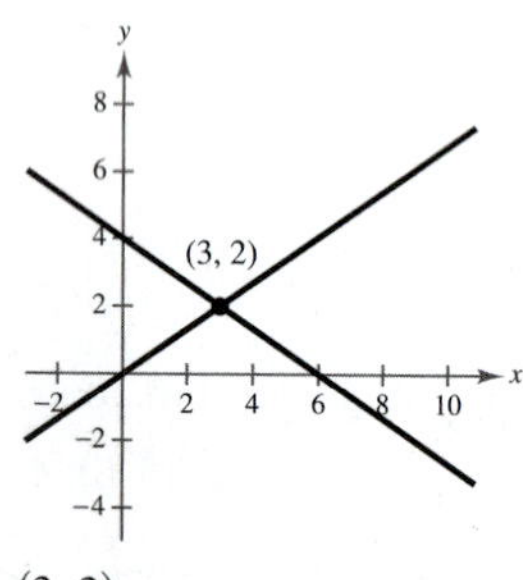

(3, 2)

5.

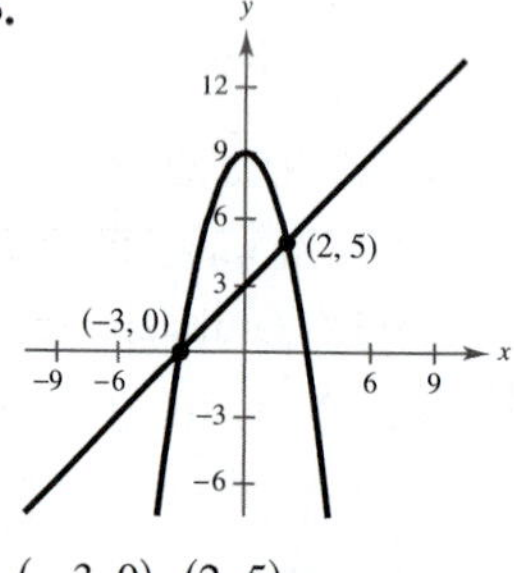

$(-3, 0), (2, 5)$

6.

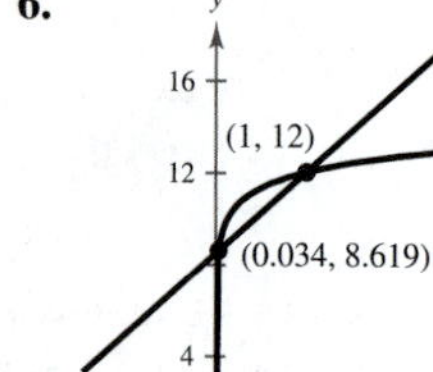

(1, 12), (0.034, 8.619)

7. (1, 5) **8.** $(2, -1)$ **9.** $(2, -3, 1)$ **10.** No solution

11.

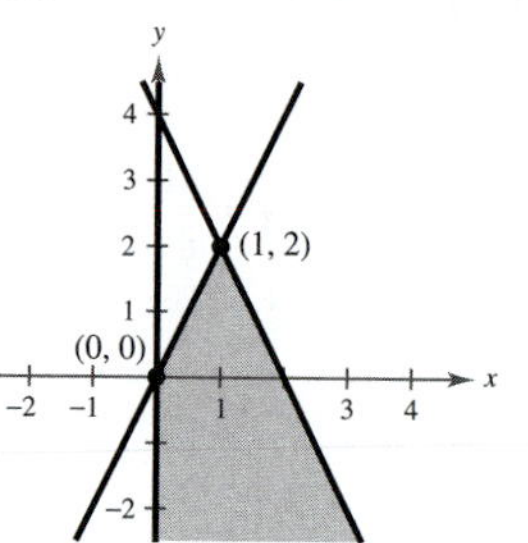

12.

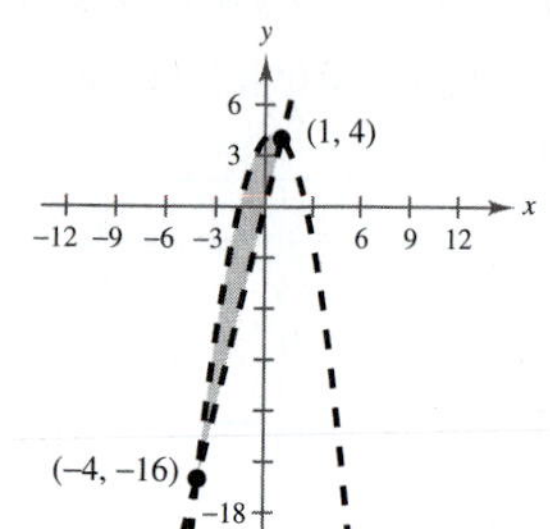

13.

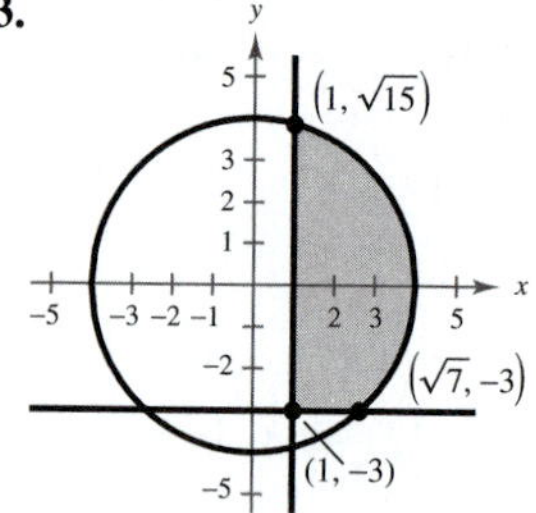

14. Maximum at (12, 0): 240

15. 8%: $20,000
8.5%: $30,000

16. $y = -\frac{1}{2}x^2 + x + 6$

17. 0 units of model I
5300 units of model II
Maximum profit: $212,000

Problem Solving *(page 728)*

1.

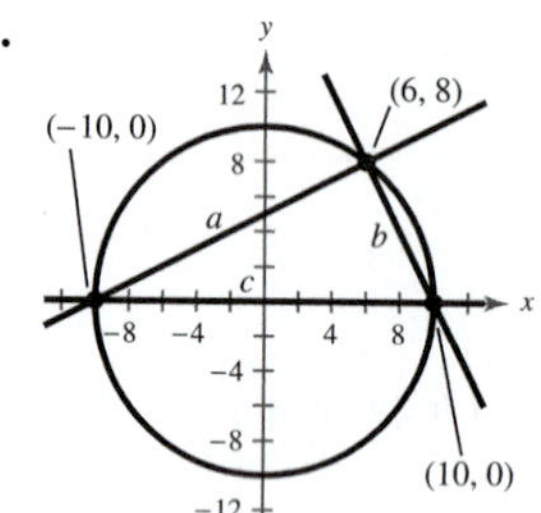

$a = 8\sqrt{5}, b = 4\sqrt{5}, c = 20$

$(8\sqrt{5})^2 + (4\sqrt{5})^2 = 20^2$

Therefore, the triangle is a right triangle.

3. $ad \neq cd$ **5.** (a) One (b) Two (c) Four

7. 10.1 feet deep; $\approx$ 252.7 feet long **9.** $12.00

11. (a) $(3, -4)$ (b) $\left(\frac{2}{-a + 5}, \frac{1}{4a - 1}, \frac{1}{a}\right)$

13. (a) $(-5a + 16, 5a - 16, 6a)$
(b) $(-11a + 36, 13a - 40, 14a)$
(c) $(-a + 3, a - 3, a)$ (d) Infinitely many

15. (a) $\begin{cases} y \geq 91 + 3.7x \\ y \leq 119 + 4.9x \\ x \geq 0 \\ y \geq 0 \end{cases}$ (b)

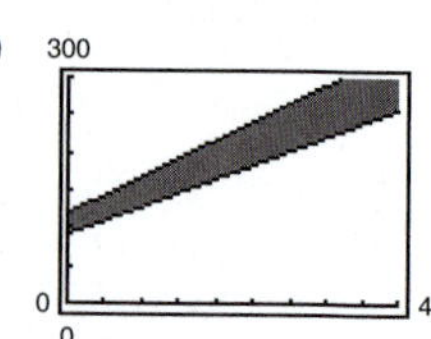

(c) 142.8 pounds $\leq y \leq$ 187.6 pounds

Chapter 10

Section 10.1 *(page 742)*

1. 1×2 **3.** 3×1 **5.** 2×2

7. $\begin{bmatrix} 4 & -3 & \vdots & -5 \\ -1 & 3 & \vdots & 12 \end{bmatrix}$ **9.** $\begin{bmatrix} 1 & 10 & -2 & \vdots & 2 \\ 5 & -3 & 4 & \vdots & 0 \\ 2 & 1 & 0 & \vdots & 6 \end{bmatrix}$

11. $\begin{bmatrix} 7 & -5 & 1 & \vdots & 13 \\ 19 & 0 & -8 & \vdots & 10 \end{bmatrix}$

13. $\begin{cases} x + 2y = 7 \\ 2x - 3y = 4 \end{cases}$ **15.** $\begin{cases} 2x \qquad + 5z = -12 \\ \quad y - 2z = 7 \\ 6x + 3y \qquad = 2 \end{cases}$

17. $\begin{cases} 9x + 12y + 3z \qquad = 0 \\ -2x + 18y + 5z + 2w = 10 \\ x + 7y - 8z \qquad = -4 \\ 3x \qquad + 2z \qquad = -10 \end{cases}$

19. $\begin{bmatrix} 1 & 4 & 3 \\ 0 & 2 & -1 \end{bmatrix}$ **21.** $\begin{bmatrix} 1 & 1 & 4 & -1 \\ 0 & 5 & -2 & 6 \\ 0 & 3 & 20 & 4 \end{bmatrix}$

$\begin{bmatrix} 1 & 1 & 4 & -1 \\ 0 & 1 & -\frac{2}{5} & \frac{6}{5} \\ 0 & 3 & 20 & 4 \end{bmatrix}$

23. Add 5 times Row 2 to Row 1.

25. Interchange Row 1 and Row 2. Add 4 times new Row 1 to Row 3.

27. Reduced row-echelon form

29. Not in row-echelon form

31. (a) $\begin{bmatrix} 1 & 2 & 3 \\ 0 & -5 & -10 \\ 3 & 1 & -1 \end{bmatrix}$ (b) $\begin{bmatrix} 1 & 2 & 3 \\ 0 & -5 & -10 \\ 0 & -5 & -10 \end{bmatrix}$

(c) $\begin{bmatrix} 1 & 2 & 3 \\ 0 & -5 & -10 \\ 0 & 0 & 0 \end{bmatrix}$ (d) $\begin{bmatrix} 1 & 2 & 3 \\ 0 & 1 & 2 \\ 0 & 0 & 0 \end{bmatrix}$

(e) $\begin{bmatrix} 1 & 0 & -1 \\ 0 & 1 & 2 \\ 0 & 0 & 0 \end{bmatrix}$

The matrix is in reduced row-echelon form.

33. $\begin{bmatrix} 1 & 1 & 0 & 5 \\ 0 & 1 & 2 & 0 \\ 0 & 0 & 1 & -1 \end{bmatrix}$ **35.** $\begin{bmatrix} 1 & -1 & -1 & 1 \\ 0 & 1 & 6 & 3 \\ 0 & 0 & 0 & 0 \end{bmatrix}$

37. $\begin{bmatrix} 1 & 0 & 0 \\ 0 & 1 & 0 \\ 0 & 0 & 1 \end{bmatrix}$ **39.** $\begin{bmatrix} 1 & 2 & 0 & 0 \\ 0 & 0 & 1 & 0 \\ 0 & 0 & 0 & 1 \\ 0 & 0 & 0 & 0 \end{bmatrix}$

41. $\begin{bmatrix} 1 & 0 & 3 & 16 \\ 0 & 1 & 2 & 12 \end{bmatrix}$

43. $\begin{cases} x - 2y = 4 \\ \quad y = -3 \end{cases}$ **45.** $\begin{cases} x - y + 2z = 4 \\ \quad y - z = 2 \\ \qquad z = -2 \end{cases}$

$(-2, -3)$ $(8, 0, -2)$

47. $(3, -4)$ **49.** $(-4, -10, 4)$ **51.** $(3, 2)$

53. $(-5, 6)$ **55.** $(-1, -4)$ **57.** Inconsistent

59. $(4, -3, 2)$ **61.** $(7, -3, 4)$ **63.** $(-4, -3, 6)$

65. $(2a + 1, 3a + 2, a)$

67. $(4 + 5b + 4a, 2 - 3b - 3a, b, a)$ **69.** Inconsistent

71. $(0, 2 - 4a, a)$ **73.** $(1, 0, 4, -2)$

75. $(-2a, a, a, 0)$ **77.** Yes; $(-1, 1, -3)$ **79.** No

81. $\begin{bmatrix} 1 & 3 & \frac{3}{2} & \vdots & 4 \\ 0 & 1 & \frac{7}{4} & \vdots & -\frac{3}{2} \\ 0 & 0 & 1 & \vdots & 2 \end{bmatrix}, \begin{bmatrix} 1 & 3 & 1 & \vdots & 3 \\ 0 & 1 & 2 & \vdots & -1 \\ 0 & 0 & 1 & \vdots & 2 \end{bmatrix}$

83. $\dfrac{4x^2}{(x + 1)^2(x - 1)} = \dfrac{1}{x - 1} + \dfrac{3}{x + 1} - \dfrac{2}{(x + 1)^2}$

85. \$150,000 at 7%
\$750,000 at 8%
\$600,000 at 10%

87. $y = x^2 + 2x + 5$

89. (a) $y = -0.004x^2 + 0.367x + 5$

(b)

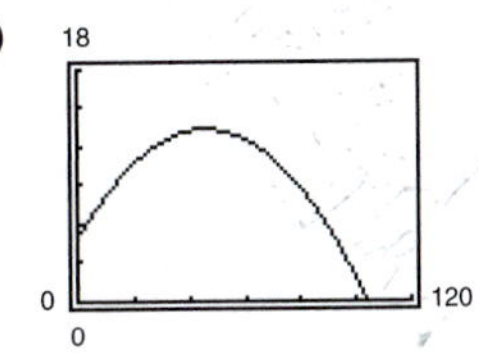

(c) 13 feet, 104 feet

(d) 13.418 feet, 103.793 feet

(e) The results are similar.

91. (a) $x_1 = s, x_2 = t, x_3 = 600 - s, x_4 = s - t,$
$x_5 = 500 - t, x_6 = s, x_7 = t$

(b) $x_1 = 0, x_2 = 0, x_3 = 600, x_4 = 0, x_5 = 500,$
$x_6 = 0, x_7 = 0$

(c) $x_1 = 0, x_2 = -500, x_3 = 600, x_4 = 500,$
$x_5 = 1000, x_6 = 0, x_7 = -500$

93. False. It is a 2×4 matrix.

95. False. Gaussian elimination reduces a matrix until a row-echelon form is obtained; Gauss-Jordan elimination reduces a matrix until a reduced row-echelon form is obtained.

97. (a) There exists a row with all zeros except for the entry in the last column.

(b) There are fewer rows with nonzero entries than there are variables and no rows as in (a).

99. They are the same.

101. Circle **103.** Line **105.** Ellipse

107.

109.

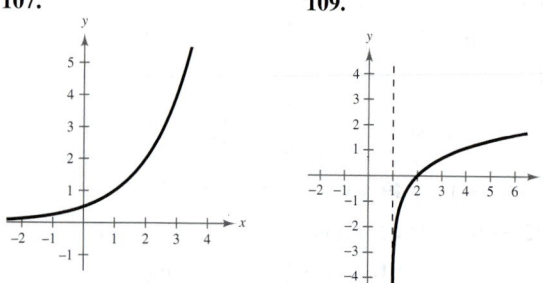

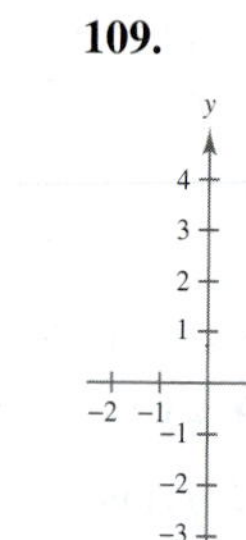

Section 10.2 *(page 757)*

1. $x = -4, y = 22$ **3.** $x = 2, y = 3$

5. (a) $\begin{bmatrix} 3 & -2 \\ 1 & 7 \end{bmatrix}$ (b) $\begin{bmatrix} -1 & 0 \\ 3 & -9 \end{bmatrix}$ (c) $\begin{bmatrix} 3 & -3 \\ 6 & -3 \end{bmatrix}$

(d) $\begin{bmatrix} -1 & -1 \\ 8 & -19 \end{bmatrix}$

7. (a) $\begin{bmatrix} 7 & 3 \\ 1 & 9 \\ -2 & 15 \end{bmatrix}$ (b) $\begin{bmatrix} 5 & -5 \\ 3 & -1 \\ -4 & -5 \end{bmatrix}$ (c) $\begin{bmatrix} 18 & -3 \\ 6 & 12 \\ -9 & 15 \end{bmatrix}$

(d) $\begin{bmatrix} 16 & -11 \\ 8 & 2 \\ -11 & -5 \end{bmatrix}$

9. (a) $\begin{bmatrix} 3 & 3 & -2 & 1 & 1 \\ -2 & 5 & 7 & -6 & -8 \end{bmatrix}$

(b) $\begin{bmatrix} 1 & 1 & 0 & -1 & 1 \\ 4 & -3 & -11 & 6 & 6 \end{bmatrix}$

(c) $\begin{bmatrix} 6 & 6 & -3 & 0 & 3 \\ 3 & 3 & -6 & 0 & -3 \end{bmatrix}$

(d) $\begin{bmatrix} 4 & 4 & -1 & -2 & 3 \\ 9 & -5 & -24 & 12 & 11 \end{bmatrix}$

11. (a), (b), and (d) not possible (c) $\begin{bmatrix} 18 & 0 & 9 \\ -3 & -12 & 0 \end{bmatrix}$

13. $\begin{bmatrix} -8 & -7 \\ 15 & -1 \end{bmatrix}$ **15.** $\begin{bmatrix} -24 & -4 & 12 \\ -12 & 32 & 12 \end{bmatrix}$

17. $\begin{bmatrix} 10 & 8 \\ -59 & 9 \end{bmatrix}$ **19.** $\begin{bmatrix} -17.143 & 2.143 \\ 11.571 & 10.286 \end{bmatrix}$

21. $\begin{bmatrix} -1.581 & -3.739 \\ -4.252 & -13.249 \\ 9.713 & -0.362 \end{bmatrix}$ **23.** $\begin{bmatrix} -6 & -9 \\ -1 & 0 \\ 17 & -10 \end{bmatrix}$

25. $\begin{bmatrix} 3 & 3 \\ -\frac{1}{2} & 0 \\ -\frac{13}{2} & \frac{11}{2} \end{bmatrix}$ **27.** Not possible **29.** $\begin{bmatrix} 3 & -4 \\ 10 & 16 \\ 26 & 46 \end{bmatrix}$

31. $\begin{bmatrix} 3 & 0 & 0 \\ 0 & -4 & 0 \\ 0 & 0 & -10 \end{bmatrix}$ **33.** $\begin{bmatrix} 0 & 0 & 0 \\ 0 & 0 & 0 \\ 0 & 0 & 0 \end{bmatrix}$

35. $\begin{bmatrix} 41 & 7 & 7 \\ 42 & 5 & 25 \\ -10 & -25 & 45 \end{bmatrix}$ **37.** $\begin{bmatrix} 151 & 25 & 48 \\ 516 & 279 & 387 \\ 47 & -20 & 87 \end{bmatrix}$

39. Not possible

41. (a) $\begin{bmatrix} 0 & 15 \\ 6 & 12 \end{bmatrix}$ (b) $\begin{bmatrix} -2 & 2 \\ 31 & 14 \end{bmatrix}$ (c) $\begin{bmatrix} 9 & 6 \\ 12 & 12 \end{bmatrix}$

43. (a) $\begin{bmatrix} 0 & -10 \\ 10 & 0 \end{bmatrix}$ (b) $\begin{bmatrix} 0 & -10 \\ 10 & 0 \end{bmatrix}$ (c) $\begin{bmatrix} 8 & -6 \\ 6 & 8 \end{bmatrix}$

45. (a) $\begin{bmatrix} 7 & 7 & 14 \\ 8 & 8 & 16 \\ -1 & -1 & -2 \end{bmatrix}$ (b) $[13]$ (c) Not possible

47. $\begin{bmatrix} 5 & 8 \\ -4 & -16 \end{bmatrix}$ **49.** $\begin{bmatrix} -4 & 10 \\ 3 & 14 \end{bmatrix}$

51. (a) $\begin{bmatrix} -1 & 1 \\ -2 & 1 \end{bmatrix}\begin{bmatrix} x_1 \\ x_2 \end{bmatrix} = \begin{bmatrix} 4 \\ 0 \end{bmatrix}$ (b) $\begin{bmatrix} 4 \\ 8 \end{bmatrix}$

53. (a) $\begin{bmatrix} -2 & -3 \\ 6 & 1 \end{bmatrix}\begin{bmatrix} x_1 \\ x_2 \end{bmatrix} = \begin{bmatrix} -4 \\ -36 \end{bmatrix}$ (b) $\begin{bmatrix} -7 \\ 6 \end{bmatrix}$

55. (a) $\begin{bmatrix} 1 & -2 & 3 \\ -1 & 3 & -1 \\ 2 & -5 & 5 \end{bmatrix}\begin{bmatrix} x_1 \\ x_2 \\ x_3 \end{bmatrix} = \begin{bmatrix} 9 \\ -6 \\ 17 \end{bmatrix}$ (b) $\begin{bmatrix} 1 \\ -1 \\ 2 \end{bmatrix}$

57. (a) $\begin{bmatrix} 1 & -5 & 2 \\ -3 & 1 & -1 \\ 0 & -2 & 5 \end{bmatrix}\begin{bmatrix} x_1 \\ x_2 \\ x_3 \end{bmatrix} = \begin{bmatrix} -20 \\ 8 \\ -16 \end{bmatrix}$ (b) $\begin{bmatrix} -1 \\ 3 \\ -2 \end{bmatrix}$

59. $\begin{bmatrix} 84 & 60 & 30 \\ 42 & 120 & 84 \end{bmatrix}$

61. (a) $A = \begin{bmatrix} 125 & 100 & 75 \\ 100 & 175 & 125 \end{bmatrix}$

The entries represent the numbers of bushels of each crop that are shipped to each outlet.

(b) $B = [\$3.50 \quad \$6.00]$

The entries represent the profits per bushel of each crop.

(c) $BA = [\$1037.50 \quad \$1400 \quad \$1012.50]$

The entries represent the profits from both crops at each of the three outlets.

63. $\begin{bmatrix} \$15{,}770 & \$18{,}300 \\ \$26{,}500 & \$29{,}250 \\ \$21{,}260 & \$24{,}150 \end{bmatrix}$

The entries represent the wholesale and retail values of the inventories at the three outlets.

65. $P^3 = \begin{bmatrix} 0.300 & 0.175 & 0.175 \\ 0.308 & 0.433 & 0.217 \\ 0.392 & 0.392 & 0.608 \end{bmatrix}$

$P^4 = \begin{bmatrix} 0.250 & 0.188 & 0.188 \\ 0.315 & 0.377 & 0.248 \\ 0.435 & 0.435 & 0.565 \end{bmatrix}$

$P^5 = \begin{bmatrix} 0.225 & 0.194 & 0.194 \\ 0.314 & 0.345 & 0.267 \\ 0.461 & 0.461 & 0.539 \end{bmatrix}$

$P^6 = \begin{bmatrix} 0.213 & 0.197 & 0.197 \\ 0.311 & 0.326 & 0.280 \\ 0.477 & 0.477 & 0.523 \end{bmatrix}$

$P^7 = \begin{bmatrix} 0.206 & 0.198 & 0.198 \\ 0.308 & 0.316 & 0.288 \\ 0.486 & 0.486 & 0.514 \end{bmatrix}$

$P^8 = \begin{bmatrix} 0.203 & 0.199 & 0.199 \\ 0.305 & 0.309 & 0.292 \\ 0.492 & 0.492 & 0.508 \end{bmatrix}$

Approaches the matrix

$\begin{bmatrix} 0.2 & 0.2 & 0.2 \\ 0.3 & 0.3 & 0.3 \\ 0.5 & 0.5 & 0.5 \end{bmatrix}$

67. True. The sum of two matrices of different orders is undefined.

69. Not possible **71.** Not possible **73.** 2×2

75. 2×3 **77.** $AC = BC = \begin{bmatrix} 2 & 3 \\ 2 & 3 \end{bmatrix}$

79. AB is a diagonal matrix whose entries are the products of the corresponding entries of A and B.

81. $-8, \dfrac{4}{3}$ **83.** $0, \dfrac{-5 \pm \sqrt{37}}{4}$ **85.** $4, \pm\dfrac{\sqrt{15}}{3}i$

87. $\left(7, -\frac{1}{2}\right)$ **89.** $(3, -1)$

Section 10.3 *(page 767)*

1–9. $AB = I$ and $BA = I$

11. $\begin{bmatrix} \frac{1}{2} & 0 \\ 0 & \frac{1}{3} \end{bmatrix}$ **13.** $\begin{bmatrix} -3 & 2 \\ -2 & 1 \end{bmatrix}$ **15.** $\begin{bmatrix} 1 & -1 \\ 2 & -1 \end{bmatrix}$

17. Does not exist **19.** Does not exist

21. $\begin{bmatrix} 1 & 1 & -1 \\ -3 & 2 & -1 \\ 3 & -3 & 2 \end{bmatrix}$ **23.** $\begin{bmatrix} 1 & 0 & 0 \\ -\frac{3}{4} & \frac{1}{4} & 0 \\ \frac{7}{20} & -\frac{1}{4} & \frac{1}{5} \end{bmatrix}$

25. $\begin{bmatrix} -\frac{1}{8} & 0 & 0 & 0 \\ 0 & 1 & 0 & 0 \\ 0 & 0 & \frac{1}{4} & 0 \\ 0 & 0 & 0 & -\frac{1}{5} \end{bmatrix}$ **27.** $\begin{bmatrix} -175 & 37 & -13 \\ 95 & -20 & 7 \\ 14 & -3 & 1 \end{bmatrix}$

29. $\begin{bmatrix} -1.5 & 1.5 & 1 \\ 4.5 & -3.5 & -3 \\ -1 & 1 & 1 \end{bmatrix}$ **31.** $\begin{bmatrix} -12 & -5 & -9 \\ -4 & -2 & -4 \\ -8 & -4 & -6 \end{bmatrix}$

33. $\begin{bmatrix} 0 & -1.\overline{81} & 0.\overline{90} \\ -10 & 5 & 5 \\ 10 & -2.\overline{72} & -3.\overline{63} \end{bmatrix}$ **35.** Does not exist

37. $\begin{bmatrix} 1 & 0 & 1 & 0 \\ 0 & 1 & 0 & 1 \\ 2 & 0 & 1 & 0 \\ 0 & 1 & 0 & 2 \end{bmatrix}$ **39.** $\begin{bmatrix} \frac{3}{19} & \frac{2}{19} \\ -\frac{2}{19} & \frac{5}{19} \end{bmatrix}$

41. Does not exist **43.** $\begin{bmatrix} \frac{16}{59} & \frac{15}{59} \\ -\frac{4}{59} & \frac{70}{59} \end{bmatrix}$ **45.** $(5, 0)$

47. $(-8, -6)$ **49.** $(3, 8, -11)$ **51.** $(2, 1, 0, 0)$

53. $(2, -2)$ **55.** No solution **57.** $(-4, -8)$

59. $(-1, 3, 2)$ **61.** $\left(\frac{5}{16}a + \frac{13}{16}, \frac{19}{16}a + \frac{11}{16}, a\right)$

63. $(-7, 3, -2)$ **65.** $(5, 0, -2, 3)$

67. \$7000 in AAA-rated bonds
\$1000 in A-rated bonds
\$2000 in B-rated bonds

69. \$9000 in AAA-rated bonds
\$1000 in A-rated bonds
\$2000 in B-rated bonds

71. (a) $I_1 = -3$ amperes, $I_2 = 8$ amperes, $I_3 = 5$ amperes
(b) $I_1 = 2$ amperes, $I_2 = 3$ amperes, $I_3 = 5$ amperes

73. True. If B is the inverse of A, then $AB = I = BA$.

75. Answers will vary.

77. $x \geq -5$ or $x \leq -9$

79. $\dfrac{2 \ln 315}{\ln 3} \approx 10.47$ **81.** $2^{6.5} \approx 90.51$

Section 10.4 *(page 775)*

1. 5 **3.** 5 **5.** 27 **7.** 0 **9.** 6 **11.** -9

13. 72 **15.** $\frac{11}{6}$ **17.** -0.002 **19.** -4.842 **21.** 0

23. (a) $M_{11} = -5, M_{12} = 2, M_{21} = 4, M_{22} = 3$
(b) $C_{11} = -5, C_{12} = -2, C_{21} = -4, C_{22} = 3$

25. (a) $M_{11} = -4, M_{12} = -2, M_{21} = 1, M_{22} = 3$
(b) $C_{11} = -4, C_{12} = 2, C_{21} = -1, C_{22} = 3$

27. (a) $M_{11} = 3, M_{12} = -4, M_{13} = 1, M_{21} = 2, M_{22} = 2, M_{23} = -4, M_{31} = -4, M_{32} = 10, M_{33} = 8$
(b) $C_{11} = 3, C_{12} = 4, C_{13} = 1, C_{21} = -2, C_{22} = 2, C_{23} = 4, C_{31} = -4, C_{32} = -10, C_{33} = 8$

29. (a) $M_{11} = 30, M_{12} = 12, M_{13} = 11, M_{21} = -36,$
$M_{22} = 26, M_{23} = 7, M_{31} = -4, M_{32} = -42, M_{33} = 12$
(b) $C_{11} = 30, C_{12} = -12, C_{13} = 11, C_{21} = 36, C_{22} = 26,$
$C_{23} = -7, C_{31} = -4, C_{32} = 42, C_{33} = 12$

31. (a) -75 (b) -75 **33.** (a) 96 (b) 96

35. (a) 170 (b) 170 **37.** 0 **39.** 0 **41.** -9

43. -58 **45.** -30 **47.** -168 **49.** 0 **51.** 412

53. -126 **55.** 0 **57.** -336 **59.** 410

61. (a) -3 (b) -2 (c) $\begin{bmatrix} -2 & 0 \\ 0 & -3 \end{bmatrix}$ (d) 6

63. (a) -8 (b) 0 (c) $\begin{bmatrix} -4 & 4 \\ 1 & -1 \end{bmatrix}$ (d) 0

65. (a) -21 (b) -19 (c) $\begin{bmatrix} 7 & 1 & 4 \\ -8 & 9 & -3 \\ 7 & -3 & 9 \end{bmatrix}$ (d) 399

67. (a) 2 (b) -6 (c) $\begin{bmatrix} 1 & 4 & 3 \\ -1 & 0 & 3 \\ 0 & 2 & 0 \end{bmatrix}$ (d) -12

69–73. Answers will vary. **75.** $-1, 4$ **77.** $-1, -4$

79. $8uv - 1$ **81.** e^{5x} **83.** $1 - \ln x$

85. True. If an entire row is zero, then each cofactor in the expansion is multiplied by zero.

87. Answers will vary.

89. A square matrix is a square array of numbers. The determinant of a square matrix is a real number.

91. (a) Columns 2 and 3 of A were interchanged.
$|A| = -115 = -|B|$
(b) Rows 1 and 3 of A were interchanged.
$|A| = -40 = -|B|$

93. All real numbers **95.** $-4 \le x \le 4$

97. All real numbers $t > 1$ **99.** $(y - 3)^2 = 8x$

101. $\dfrac{x^2}{64} + \dfrac{y^2}{28} = 1$

103.

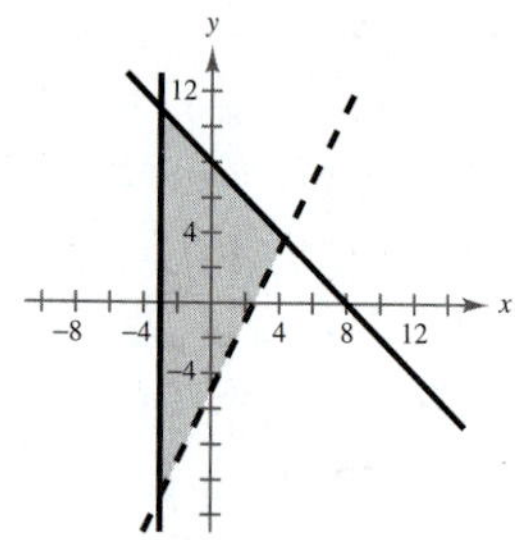

105. $\begin{bmatrix} \frac{1}{4} & \frac{1}{4} \\ 2 & 1 \end{bmatrix}$ **107.** Does not exist

Section 10.5 *(page 787)*

1. $(2, -2)$ **3.** $\left(\frac{32}{7}, \frac{30}{7}\right)$ **5.** $(-1, 3, 2)$

7. $(-2, 1, -1)$ **9.** $\left(0, -\frac{1}{2}, \frac{1}{2}\right)$ **11.** $(1, 2, 1)$ **13.** 7

15. 14 **17.** $\frac{33}{8}$ **19.** $\frac{5}{2}$ **21.** 28 **23.** $\frac{16}{5}$ or 0

25. -3 or -11 **27.** 250 square miles **29.** Collinear

31. Not collinear **33.** Collinear **35.** $x = -3$

37. $3x - 5y = 0$ **39.** $x + 3y - 5 = 0$

41. $2x + 3y - 8 = 0$

43. Uncoded: [20 18 15], [21 2 12], [5 0 9], [14 0 18], [9 22 5], [18 0 3], [9 20 25]
Encoded: -52 10 27 -49 3 34 -49 13 27 -94 22 54 1 1 -7 0 -12 9 -121 41 55

45. -6 -35 -69 11 20 17 6 -16 -58 46 79 67

47. -5 -41 -87 91 207 257 11 -5 -41 40 80 84 76 177 227

49. HAPPY NEW YEAR **51.** CLASS IS CANCELED

53. SEND PLANES **55.** MEET ME TONIGHT RON

57. False. The denominator is the determinant of the coefficient matrix.

59. False. If the determinant of the coefficient matrix is zero, the system has either no solution or infinitely many solutions.

61. $(-6, 4)$ **63.** $(-1, 0, -3)$

65.

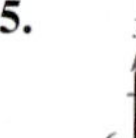

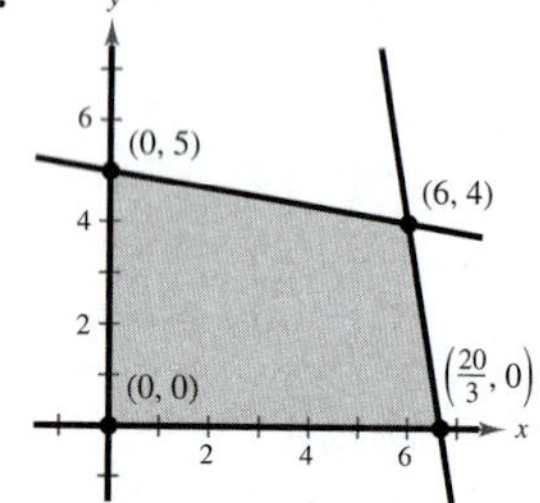

Minimum at (0, 0): 0

Maximum at (6, 4): 52

Review Exercises *(page 791)*

1. 3×1 **3.** 1×1 **5.** $\left[\begin{array}{cc|c} 3 & -10 & 15 \\ 5 & 4 & 22 \end{array}\right]$

7. $\begin{cases} 5x + y + 7z = -9 \\ 4x + 2y = 10 \\ 9x + 4y + 2z = 3 \end{cases}$ **9.** $\begin{bmatrix} 1 & 2 & 3 \\ 0 & 1 & 1 \\ 0 & 0 & 1 \end{bmatrix}$

11. $\begin{cases} x + 2y + 3z = 9 \\ y - 2z = 2 \\ z = 0 \end{cases}$

$(5, 2, 0)$

13. $\begin{cases} x - 5y + 4z = 1 \\ y + 2z = 3 \\ z = 4 \end{cases}$

$(-40, -5, 4)$

15. $(10, -12)$ **17.** $\left(-\frac{1}{5}, \frac{7}{10}\right)$ **19.** $(5, 2, -6)$

21. $\left(-2a + \frac{3}{2}, 2a + 1, a\right)$ **23.** $(1, 0, 4, 3)$

25. $(2, -3, 3)$ **27.** $(2, 3, -1)$ **29.** $(2, 6, -10, -3)$

31. $x = 12, y = -7$ **33.** $x = 1, y = 11$

35. (a) $\begin{bmatrix} -1 & 8 \\ 15 & 13 \end{bmatrix}$ (b) $\begin{bmatrix} 5 & -12 \\ -9 & -3 \end{bmatrix}$

(c) $\begin{bmatrix} 8 & -8 \\ 12 & 20 \end{bmatrix}$ (d) $\begin{bmatrix} -7 & 28 \\ 39 & 29 \end{bmatrix}$

37. (a) $\begin{bmatrix} 5 & 7 \\ -3 & 14 \\ 31 & 42 \end{bmatrix}$ (b) $\begin{bmatrix} 5 & 1 \\ -11 & -10 \\ -9 & -38 \end{bmatrix}$

(c) $\begin{bmatrix} 20 & 16 \\ -28 & 8 \\ 44 & 8 \end{bmatrix}$ (d) $\begin{bmatrix} 5 & 13 \\ 5 & 38 \\ 71 & 122 \end{bmatrix}$

39. $\begin{bmatrix} 17 & -17 \\ 13 & 2 \end{bmatrix}$ **41.** $\begin{bmatrix} 54 & 4 \\ -2 & 24 \\ -4 & 32 \end{bmatrix}$ **43.** $\begin{bmatrix} 48 & -18 & -3 \\ 15 & 51 & 33 \end{bmatrix}$

45. $\begin{bmatrix} -14 & -4 \\ 7 & -17 \\ -17 & -2 \end{bmatrix}$ **47.** $\begin{bmatrix} 3 & \frac{2}{3} \\ -\frac{4}{3} & \frac{11}{3} \\ \frac{10}{3} & 0 \end{bmatrix}$ **49.** $\begin{bmatrix} -30 & 4 \\ 51 & 70 \end{bmatrix}$

51. $\begin{bmatrix} 100 & 220 \\ 12 & -4 \\ 84 & 212 \end{bmatrix}$ **53.** $\begin{bmatrix} 14 & -2 & 8 \\ 14 & -10 & 40 \\ 36 & -12 & 48 \end{bmatrix}$ **55.** $\begin{bmatrix} 44 & 4 \\ 20 & 8 \end{bmatrix}$

57. $\begin{bmatrix} 24 & -8 \\ 36 & -12 \end{bmatrix}$ **59.** $\begin{bmatrix} 1 & 17 \\ 12 & 36 \end{bmatrix}$

61. $\begin{bmatrix} 14 & -22 & 22 \\ 19 & -41 & 80 \\ 42 & -66 & 66 \end{bmatrix}$ **63.** $\begin{bmatrix} 96 & 84 & 108 & 48 \\ 60 & 36 & 96 & 24 \\ 108 & 72 & 120 & 60 \end{bmatrix}$

65 and 67. $AB = I$ and $BA = I$

69. $\begin{bmatrix} 4 & -5 \\ 5 & -6 \end{bmatrix}$ **71.** $\begin{bmatrix} 13 & 6 & -4 \\ -12 & -5 & 3 \\ 5 & 2 & -1 \end{bmatrix}$

73. $\begin{bmatrix} \frac{1}{2} & -1 & -\frac{1}{2} \\ \frac{1}{2} & -\frac{2}{3} & -\frac{5}{6} \\ 0 & \frac{2}{3} & \frac{1}{3} \end{bmatrix}$ **75.** $\begin{bmatrix} -3 & 6 & -5.5 & 3.5 \\ 1 & -2 & 2 & -1 \\ 7 & -15 & 14.5 & -9.5 \\ -1 & 2.5 & -2.5 & 1.5 \end{bmatrix}$

77. $\begin{bmatrix} 1 & -1 \\ 4 & -\frac{7}{2} \end{bmatrix}$ **79.** $\begin{bmatrix} 2 & \frac{20}{3} \\ \frac{1}{10} & \frac{1}{6} \end{bmatrix}$

81. $(36, 11)$ **83.** $(-6, -1)$ **85.** $(2, -1, -2)$

87. $(6, 1, -1)$ **89.** $(-3, 1)$ **91.** $(1, 1, -2)$

93. -42 **95.** 550

97. (a) $M_{11} = 4, M_{12} = 7, M_{21} = -1, M_{22} = 2$

(b) $C_{11} = 4, C_{12} = -7, C_{21} = 1, C_{22} = 2$

99. (a) $M_{11} = 30, M_{12} = -12, M_{13} = -21,$
$M_{21} = 20, M_{22} = 19, M_{23} = 22, M_{31} = 5,$
$M_{32} = -2, M_{33} = 19$

(b) $C_{11} = 30, C_{12} = 12, C_{13} = -21,$
$C_{21} = -20, C_{22} = 19, C_{23} = -22,$
$C_{31} = 5, C_{32} = 2, C_{33} = 19$

101. 130 **103.** 279 **105.** $(4, 7)$ **107.** $(-1, 4, 5)$

109. 16 **111.** 10 **113.** Collinear

115. $x - 2y + 4 = 0$ **117.** $2x + 6y - 13 = 0$

119. Uncoded: [12 15 15], [11 0 15], [21 20 0], [2 5 12], [15 23 0]

Encoded: -21 6 0 -68 8 45 102 -42 -60 -53 20 21 99 -30 -69

121. SEE YOU FRIDAY

123. False. The matrix must be square.

125. The matrix must be square and its determinant nonzero.

127. No. The first two matrices describe a system of equations with one solution. The third matrix describes a system with infinitely many solutions.

129. $\lambda = \pm 2\sqrt{10} - 3$

Chapter Test *(page 796)*

1. $\begin{bmatrix} 1 & 0 & 0 \\ 0 & 1 & 0 \\ 0 & 0 & 1 \end{bmatrix}$ **2.** $\begin{bmatrix} 1 & 0 & -1 & 2 \\ 0 & 1 & 0 & -1 \\ 0 & 0 & 0 & 0 \\ 0 & 0 & 0 & 0 \end{bmatrix}$

3. $\left[\begin{array}{ccc|c} 4 & 3 & -2 & 14 \\ -1 & -1 & 2 & -5 \\ 3 & 1 & -4 & 8 \end{array}\right], \left(1, 3, -\frac{1}{2}\right)$

4. (a) $\begin{bmatrix} 1 & 5 \\ 0 & -4 \end{bmatrix}$ (b) $\begin{bmatrix} 15 & 12 \\ -12 & -12 \end{bmatrix}$

(c) $\begin{bmatrix} 7 & 14 \\ -4 & -12 \end{bmatrix}$ (d) $\begin{bmatrix} 4 & -5 \\ 0 & 4 \end{bmatrix}$

5. $\begin{bmatrix} \frac{1}{2} & \frac{2}{5} \\ 1 & \frac{3}{5} \end{bmatrix}$ **6.** $\begin{bmatrix} -\frac{5}{2} & 4 & -3 \\ 5 & -7 & 6 \\ 4 & -6 & 5 \end{bmatrix}$

7. $(13, 22)$ **8.** -196 **9.** 29 **10.** 43

11. $(-3, 5)$ **12.** $(-2, 4, 6)$ **13.** 7

14. Uncoded: [11 14 15], [3 11 0], [15 14 0], [23 15 15], [4 0 0]

Encoded: 115 -41 -59 14 -3 -11 29 -15 -14 128 -53 -60 4 -4 0

15. 75 liters of 60% solution

25 liters of 20% solution

Problem Solving (*page 798*)

1. (a) $AT = \begin{bmatrix} -1 & -4 & -2 \\ 1 & 2 & 3 \end{bmatrix}$

$AAT = \begin{bmatrix} -1 & -2 & -3 \\ -1 & -4 & -2 \end{bmatrix}$

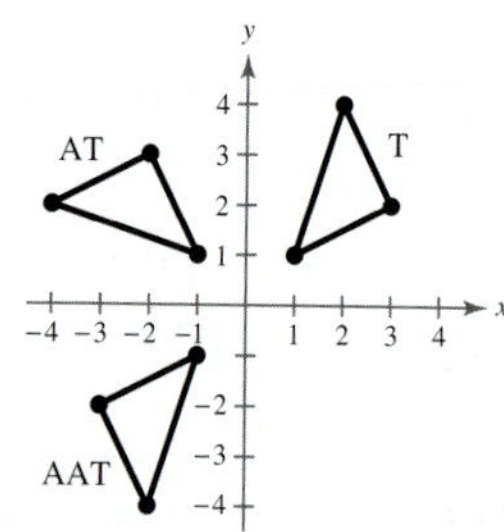

A represents reflections in the origin and in the *x*-axis.

(b) *AAT* is rotated clockwise 90° to obtain *AT*. *AT* is then rotated clockwise 90° to obtain *T*.

3. (a) Yes (b) No (c) No (d) No **5.** $x = 4$

7. Answers will vary. For example: $A = \begin{bmatrix} 1 & 1 \\ 0 & 0 \end{bmatrix}$

9. Answers will vary.

11. $\begin{vmatrix} x & 0 & 0 & d \\ -1 & x & 0 & c \\ 0 & -1 & x & b \\ 0 & 0 & -1 & a \end{vmatrix}$

13. Transformer: \$10.00

Foot of wire: \$0.20

Light: \$1.00

15. $A^T = \begin{bmatrix} -1 & 2 \\ 1 & 0 \\ -2 & 1 \end{bmatrix}$ $B^T = \begin{bmatrix} -3 & 1 & 1 \\ 0 & 2 & -1 \end{bmatrix}$

$(AB)^T = \begin{bmatrix} 2 & -5 \\ 4 & -1 \end{bmatrix} = B^TA^T$

17. $A^{-1} = \begin{bmatrix} 0.0625 & -0.4375 & 0.625 \\ 0.1875 & 0.6875 & -1.125 \\ -0.125 & -0.125 & 0.75 \end{bmatrix}$

$|A^{-1}| = \frac{1}{16}, |A| = 16$

$|A^{-1}| = \frac{1}{|A|}$

19. (a) Answers will vary.

(b) Squaring the 2×2 matrix yields the zero matrix. Cubing the 3×3 matrix yields the zero matrix.

(c) A^4 is the zero matrix.

(d) A^n is the zero matrix.

Chapter 11

Section 11.1 (*page 809*)

1. 4, 7, 10, 13, 16 **3.** 2, 4, 8, 16, 32

5. $-2, 4, -8, 16, -32$ **7.** $3, 2, \frac{5}{3}, \frac{3}{2}, \frac{7}{5}$

9. $3, \frac{12}{11}, \frac{9}{13}, \frac{24}{47}, \frac{15}{37}$ **11.** $0, 1, 0, \frac{1}{2}, 0$ **13.** $\frac{5}{3}, \frac{17}{9}, \frac{53}{27}, \frac{161}{81}, \frac{485}{243}$

15. $1, \frac{1}{2^{3/2}}, \frac{1}{3^{3/2}}, \frac{1}{8}, \frac{1}{5^{3/2}}$ **17.** $3, \frac{9}{2}, \frac{9}{2}, \frac{27}{8}, \frac{81}{40}$

19. $-1, \frac{1}{4}, -\frac{1}{9}, \frac{1}{16}, -\frac{1}{25}$ **21.** $\frac{2}{3}, \frac{2}{3}, \frac{2}{3}, \frac{2}{3}, \frac{2}{3}$

23. 0, 0, 6, 24, 60 **25.** -73 **27.** 0.000282 **29.** $\frac{44}{239}$

31.

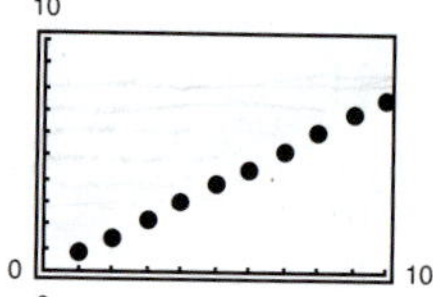

33. 18, 0, 10, −10

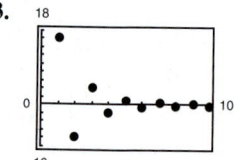

35.

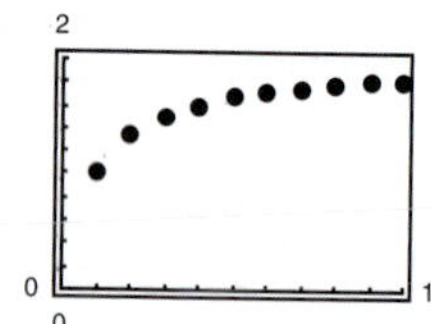

37. c **38.** b **39.** d **40.** a

41. $a_n = 3n - 2$ **43.** $a_n = n^2 - 1$

45. $a_n = \frac{(-1)^n(n+1)}{n+2}$ **47.** $a_n = \frac{n+1}{2n-1}$

49. $a_n = \frac{1}{n^2}$ **51.** $a_n = (-1)^{n+1}$ **53.** $a_n = 1 + \frac{1}{n}$

55. 28, 24, 20, 16, 12 **57.** 3, 4, 6, 10, 18

59. 6, 8, 10, 12, 14 **61.** 81, 27, 9, 3, 1

$a_n = 2n + 4$ $a_n = \frac{243}{3^n}$

63. $\frac{1}{30}$ **65.** 90 **67.** $n + 1$ **69.** $\frac{1}{2n(2n+1)}$

71. 35 **73.** 40 **75.** 30 **77.** $\frac{9}{5}$ **79.** 88

81. 30 **83.** 81 **85.** $\frac{47}{60}$

87. $\sum_{i=1}^{9} \frac{1}{3i}$ **89.** $\sum_{i=1}^{8} \left[2\left(\frac{i}{8}\right) + 3\right]$ **91.** $\sum_{i=1}^{6} (-1)^{i+1}3^i$

93. $\sum_{i=1}^{20} \frac{(-1)^{i+1}}{i^2}$ **95.** $\sum_{i=1}^{5} \frac{2^i - 1}{2^{i+1}}$ **97.** $\frac{75}{16}$ **99.** $-\frac{3}{2}$

101. $\frac{2}{3}$ **103.** $\frac{7}{9}$

105. (a) $A_1 = \$5100.00, A_2 = \$5202.00, A_3 = \$5306.04,$ $A_4 = \$5412.16, A_5 = \$5520.40, A_6 = \$5630.81,$ $A_7 = \$5743.43, A_8 = \$5858.30$

(b) $A_{40} = \$11{,}040.20$

107. (a) $b_n = 50.0n + 168$

(b) $c_n = -1.36n^2 + 65.0n + 138$

(c)

n	1	2	3	4	5
a_n	228	260	294	352	419
b_n	218	268	318	368	418
c_n	202	263	321	376	429

n	6	7	8	9	10
a_n	493	556	587	616	629
b_n	468	518	568	618	668
c_n	479	526	571	613	652

The quadratic model is a better fit.

(d) The quadratic model; 807

109. $a_0 = 3183.1, a_1 = 3615.4, a_2 = 4006.5,$ $a_3 = 4356.3, a_4 = 4665.0, a_5 = 4932.4,$ $a_6 = 5158.6, a_7 = 5343.5, a_8 = 5487.2,$ $a_9 = 5589.7, a_{10} = 5651.0$

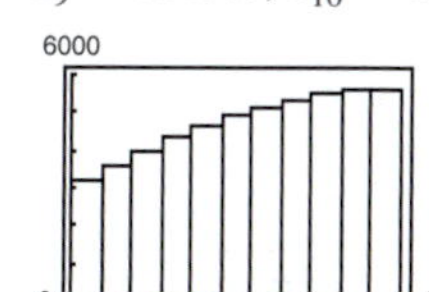

111. True by the Properties of Sums

113. 1, 1, 2, 3, 5, 8, 13, 21, 34, 55, 89, 144

$1, 2, \frac{3}{2}, \frac{5}{3}, \frac{8}{5}, \frac{13}{8}, \frac{21}{13}, \frac{34}{21}, \frac{55}{34}, \frac{89}{55}$

115. \$500.95 **117.** Answers will vary.

119. $x, \frac{x^2}{2}, \frac{x^3}{6}, \frac{x^4}{24}, \frac{x^5}{120}$

121. $-\frac{x^2}{2}, \frac{x^4}{24}, -\frac{x^6}{720}, \frac{x^8}{40{,}320}, -\frac{x^{10}}{3{,}628{,}800}$

123. $f^{-1}(x) = \frac{x+3}{4}$ **125.** $h^{-1}(x) = \frac{x^2-1}{5}, x > 0$

127. (a) $\begin{bmatrix} 8 & 1 \\ -3 & 7 \end{bmatrix}$ (b) $\begin{bmatrix} -26 & 1 \\ 15 & -24 \end{bmatrix}$

(c) $\begin{bmatrix} 18 & 9 \\ 18 & 0 \end{bmatrix}$ (d) $\begin{bmatrix} 0 & 6 \\ 27 & 18 \end{bmatrix}$

129. (a) $\begin{bmatrix} -3 & -7 & 4 \\ 4 & 4 & 1 \\ 1 & 4 & 3 \end{bmatrix}$ (b) $\begin{bmatrix} 10 & 25 & -10 \\ -12 & -11 & 3 \\ -3 & -9 & -8 \end{bmatrix}$

(c) $\begin{bmatrix} -2 & 7 & -16 \\ 4 & 42 & 45 \\ 1 & 23 & 48 \end{bmatrix}$ (d) $\begin{bmatrix} 16 & 31 & 42 \\ 10 & 47 & 31 \\ 13 & 22 & 25 \end{bmatrix}$

131. 26 **133.** -194

Section 11.2 *(page 819)*

1. Arithmetic sequence, $d = -2$

3. Not an arithmetic sequence

5. Arithmetic sequence, $d = -\frac{1}{4}$

7. Not an arithmetic sequence

9. Not an arithmetic sequence

11. 8, 11, 14, 17, 20

Arithmetic sequence, $d = 3$

13. $7, 3, -1, -5, -9$

Arithmetic sequence, $d = -4$

15. $-1, 1, -1, 1, -1,$

Not an arithmetic sequence

17. $-3, \frac{3}{2}, -1, \frac{3}{4}, -\frac{3}{5}$

Not an arithmetic sequence

19. 15, 19, 23, 27, 31; $d = 4$; $a_n = 4n + 11$

21. 200, 190, 180, 170, 160; $d = -10$; $a_n = -10n + 210$

23. $\frac{5}{8}, \frac{1}{2}, \frac{3}{8}, \frac{1}{4}, \frac{1}{8}$; $d = -\frac{1}{8}$; $a_n = -\frac{1}{8}n + \frac{3}{4}$

25. 5, 11, 17, 23, 29 **27.** $-2.6, -3.0, -3.4, -3.8, -4.2$

29. 2, 6, 10, 14, 18 **31.** $-2, 2, 6, 10, 14$

33. $a_n = 3n - 2$ **35.** $a_n = -8n + 108$

37. $a_n = 2xn - x$ **39.** $a_n = -\frac{5}{2}n + \frac{13}{2}$

41. $a_n = \frac{10}{3}n + \frac{5}{3}$ **43.** $a_n = -3n + 103$

45. b **46.** d **47.** c **48.** a

49.

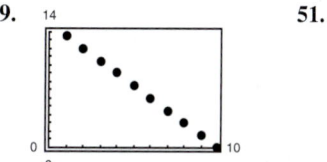

51.

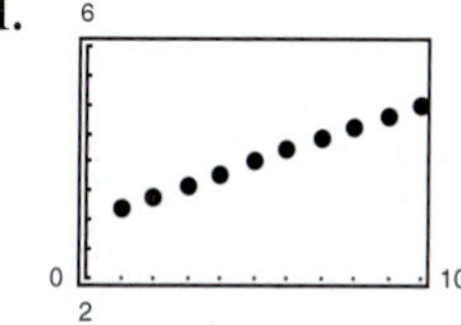

53. 620 **55.** 17.4 **57.** 265 **59.** 4000

61. 1275 **63.** 30,030 **65.** 355 **67.** 160,000

69. 520 **71.** 2725 **73.** 10,120 **75.** 10,000

77. (a) \$40,000 (b) \$217,500

79. 2340 seats **81.** 405 bricks **83.** 490 meters

85. (a) $a_n = 1048.8n + 19{,}188$ (b) $y = 1089.5n + 18{,}397$

This model is quite close to the arithmetic sequence.

(c)

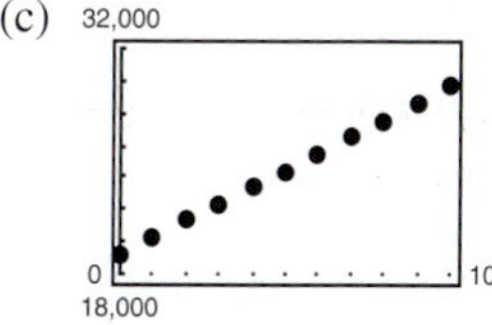

(d) 2001: \$30,725; 2003: \$32,822

(e) Answers will vary.

87. True. Given a_1 and a_2, $d = a_2 - a_1$ and $a_n = a_1 + (n - 1)d$.

89. (a) (b)

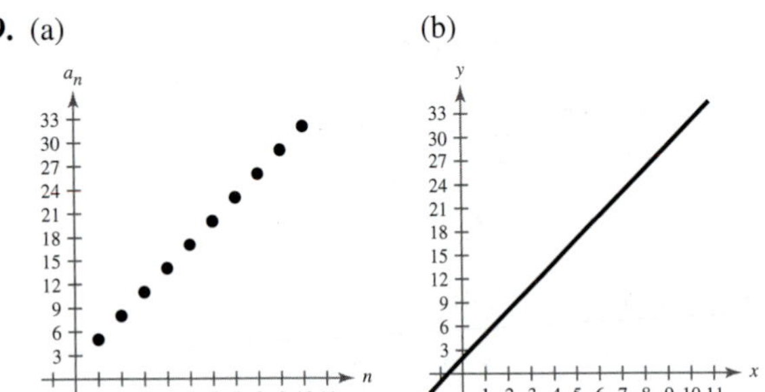

(c) The graph of $y = 3x + 2$ contains all points on the line. The graph of $a_n = 2 + 3n$ contains only points at the positive integers.

(d) The slope of the line and the common difference of the arithmetic sequence are equal.

91. 4

93. Slope: $\frac{1}{2}$; y-intercept: $\left(0, -\frac{3}{4}\right)$

95. Slope: undefined; No y-intercept

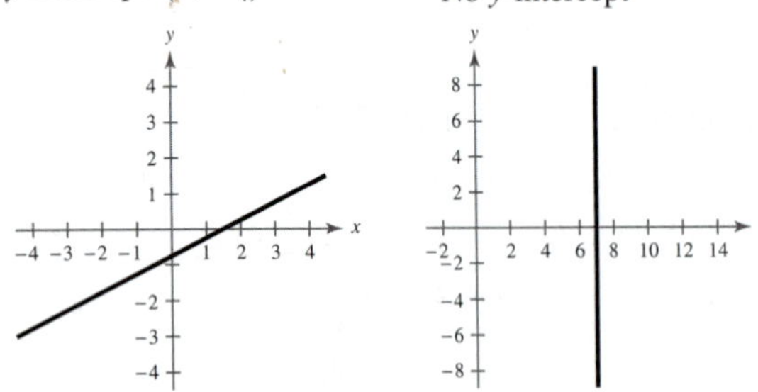

Section 11.3 *(page 828)*

1. Geometric sequence, $r = 3$

3. Not a geometric sequence

5. Geometric sequence, $r = -\frac{1}{2}$

7. Geometric sequence, $r = 2$

9. Not a geometric sequence

11. 2, 6, 18, 54, 162 **13.** $1, \frac{1}{2}, \frac{1}{4}, \frac{1}{8}, \frac{1}{16}$

15. $5, -\frac{1}{2}, \frac{1}{20}, -\frac{1}{200}, \frac{1}{2000}$ **17.** $1, e, e^2, e^3, e^4$

19. $2, \dfrac{x}{2}, \dfrac{x^2}{8}, \dfrac{x^3}{32}, \dfrac{x^4}{128}$

21. 64, 32, 16, 8, 4; $r = \frac{1}{2}$; $a_n = 128\left(\frac{1}{2}\right)^n$

23. 7, 14, 28, 56, 112; $r = 2$; $a_n = \frac{7}{2}(2)^n$

25. $6, -9, \frac{27}{2}, -\frac{81}{4}, \frac{243}{8}$; $r = -\frac{3}{2}$; $a_n = -4\left(-\frac{3}{2}\right)^n$

27. $\dfrac{1}{128}$ **29.** $-\dfrac{2}{3^{10}}$ **31.** $100e^{8x}$ **33.** 1082.372

35. 9 **37.** -2 **39.** a **40.** c **41.** b **42.** d

43. **45.**

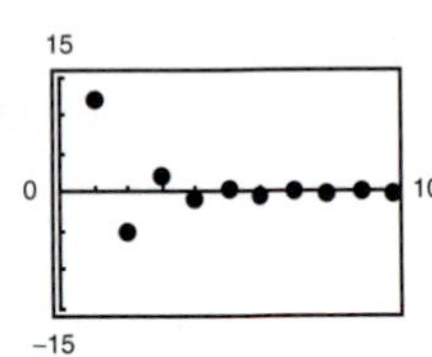

47. **49.**

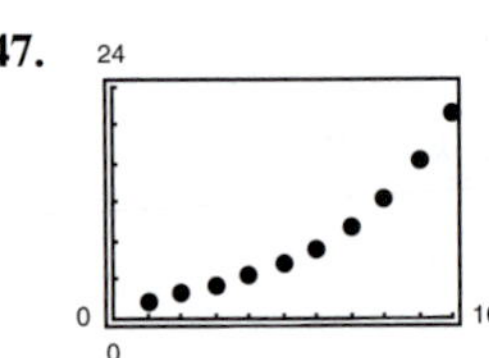

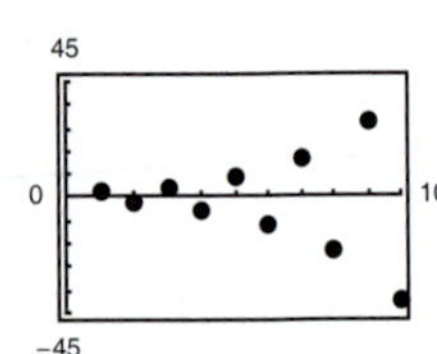

51. 511 **53.** 171 **55.** 43 **57.** $\frac{1365}{32}$

59. 29,921.311 **61.** 592.647 **63.** 2092.596

65. $\frac{8}{5}$ **67.** 6.400 **69.** 3.750

71. $\sum_{n=1}^{7} 5(3)^{n-1}$ **73.** $\sum_{n=1}^{7} 2\left(-\frac{1}{4}\right)^{n-1}$ **75.** $\sum_{n=1}^{6} 0.1(4)^{n-1}$

77. 2 **79.** $\frac{2}{3}$ **81.** $\frac{16}{3}$ **83.** $\frac{5}{3}$ **85.** -30

87. 32 **89.** Undefined **91.** $\frac{4}{11}$ **93.** $\frac{7}{22}$

95.

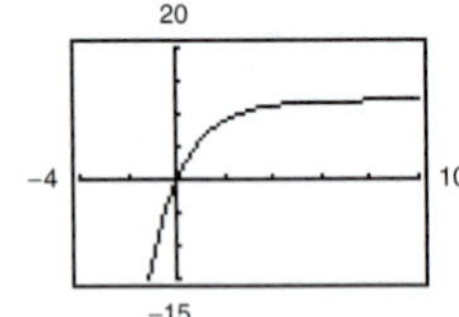

Horizontal asymptote: $y = 12$

Corresponds to the sum of the series

97. (a) $a_n = 1142.90(1.01)^n$

(b) The population is growing at a rate of 1% per year.

(c) 1394.6 million

(d) 2003

99. (a) \$11,652.39 (b) \$12,002.55

(c) \$12,188.60 (d) \$12,317.01

(e) \$12,380.41

101. \$7011.89 **103.** Answers will vary.

105. (a) \$26,198.27 (b) \$26,263.88

107. (a) \$637,678.02 (b) \$645,861.43

109. Answers will vary. **111.** 126 square inches

113. \$3,623,993.23

115. False. A sequence is geometric if the ratios of consecutive terms are the same.

117. The value of a real number between -1 and 1, raised to a power, approaches zero.

119. $x^2 + 2x$ **121.** $3x^2 + 6x + 1$

123. $x(3x + 8)(3x - 8)$ **125.** $(3x + 1)(2x - 5)$

127. $\dfrac{3x}{x - 3}, x \neq -3$ **129.** $\dfrac{2x + 1}{3}, x \neq 0, -\dfrac{1}{2}$

131. $\dfrac{5x^2 + 9x - 30}{(x + 2)(x - 2)}$

Section 11.4 *(page 840)*

1. $\dfrac{5}{(k + 1)(k + 2)}$ **3.** $\dfrac{(k + 1)^2(k + 2)^2}{4}$

5–17. Answers will vary.

19. 120 **21.** 91 **23.** 979 **25.** 70 **27.** -3402

29. $S_n = n(2n - 1)$ **31.** $S_n = 10 - 10\left(\frac{9}{10}\right)^n$

33. $S_n = \dfrac{n}{2(n + 1)}$ **35–47.** Answers will vary.

49. 0, 3, 6, 9, 12, 15

First differences: 3, 3, 3, 3, 3

Second differences: 0, 0, 0, 0

Linear

51. 3, 1, -2, -6, -11, -17

First differences: -2, -3, -4, -5, -6

Second differences: -1, -1, -1, -1

Quadratic

53. 2, 4, 16, 256, 65,536, 4,294,967,296

First differences: 2, 12, 240, 65,280, 4,294,901,760

Second differences: 10, 228, 65,040, 4,294,836,480

Neither

55. $a_n = n^2 - n + 3$ **57.** $a_n = \frac{1}{2}n^2 + n - 3$

59. (a) 1.9, 1.6, 1.9, 1.2

(b) A linear model can be used.

$a_n = 1.67n + 16.3$

(c) \$39.68 million

61. True. P_7 may be false.

63. True. If the second differences are all zero, then the first differences are all the same and the sequence is arithmetic.

65. $4x^4 - 4x^2 + 1$ **67.** $-64x^3 + 240x^2 - 300x + 125$

69. $x^2 + 2x - 8$ **71.** $4x^2 - 9x + 2$

73. (a) Intercept: $(0, 0)$

(b) Vertical asymptote: $x = -3$

Horizontal asymptote: $y = 1$

(c) No symmetry

(d)

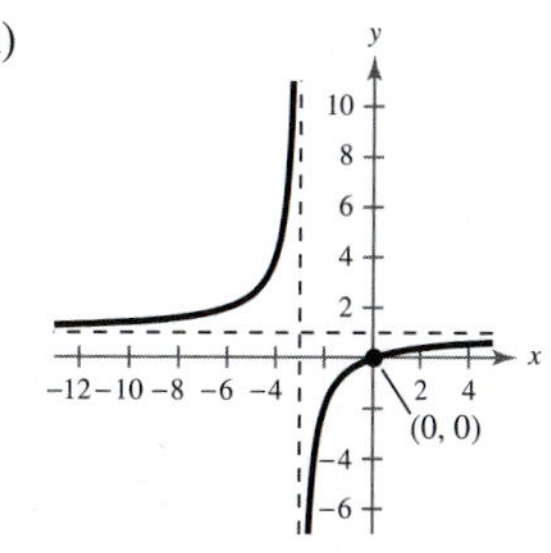

75. (a) t-intercept: $(7, 0)$

(b) Vertical asymptote: $t = 0$

Horizontal asymptote: $y = 1$

(c) No symmetry

(d)

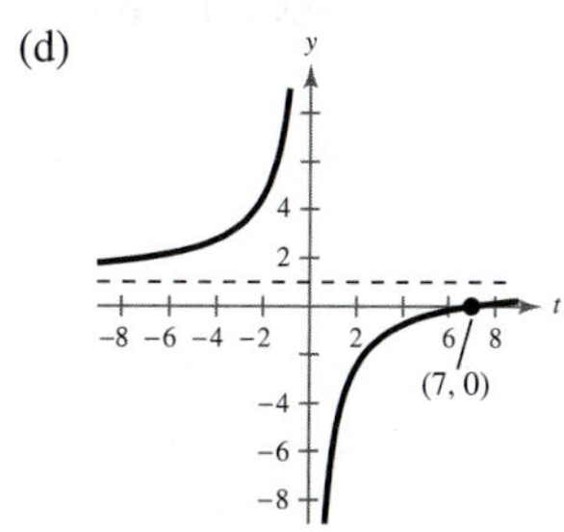

Section 11.5 *(page 847)*

1. 10 **3.** 1 **5.** 15,504 **7.** 210 **9.** 4950

11. 56 **13.** 35 **15.** $x^4 + 4x^3 + 6x^2 + 4x + 1$

17. $a^4 + 24a^3 + 216a^2 + 864a + 1296$

19. $y^3 - 12y^2 + 48y - 64$

21. $x^5 + 5x^4y + 10x^3y^2 + 10x^2y^3 + 5xy^4 + y^5$

23. $r^6 + 18r^5s + 135r^4s^2 + 540r^3s^3 + 1215r^2s^4 + 1458rs^5 + 729s^6$

25. $243a^5 - 405a^4b + 270a^3b^2 - 90a^2b^3 + 15ab^4 - b^5$

27. $1 - 6x + 12x^2 - 8x^3$

29. $x^8 + 20x^6 + 150x^4 + 500x^2 + 625$

31. $\dfrac{1}{x^5} + \dfrac{5y}{x^4} + \dfrac{10y^2}{x^3} + \dfrac{10y^3}{x^2} + \dfrac{5y^4}{x} + y^5$

33. $2x^4 - 24x^3 + 113x^2 - 246x + 207$

35. $32t^5 - 80t^4s + 80t^3s^2 - 40t^2s^3 + 10ts^4 - s^5$

37. $x^5 + 10x^4y + 40x^3y^2 + 80x^2y^3 + 80xy^4 + 32y^5$

39. $120x^7y^3$ **41.** $360x^3y^2$ **43.** $1,259,712x^2y^7$

45. $32,476,950,000x^4y^8$ **47.** 1,732,104

49. 180 **51.** $-326,592$ **53.** 210

55. $x^2 + 12x^{3/2} + 54x + 108x^{1/2} + 81$

57. $x^2 - 3x^{4/3}y^{1/3} + 3x^{2/3}y^{2/3} - y$

59. $3x^2 + 3xh + h^2, h \neq 0$

61. $\dfrac{1}{\sqrt{x+h}+\sqrt{x}},\ h \neq 0$ **63.** -4 **65.** $2035 + 828i$

67. 1 **69.** 1.172 **71.** 510,568.785

73.

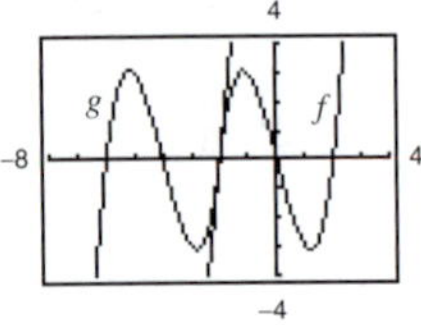

g is shifted four units to the left of f.

$g(x) = x^3 + 12x^2 + 44x + 48$

75. 0.273 **77.** 0.171

79. (a) $f(t) = 0.0071t^3 - 0.209t^2 + 2.55t - 4.1$

(b)

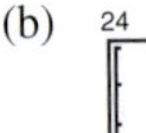

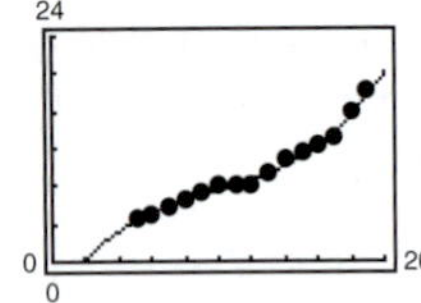

(c) $g(t) = 0.0071t^3 + 0.004t^2 + 0.50t + 7.6$

(d)

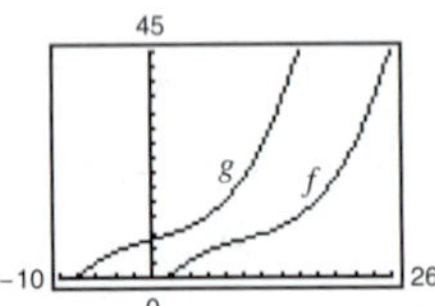

(e) 34.9 gallons; Yes

(f) The per capita consumption of bottled water is increasing. Answers will vary.

81. True. The coefficients from the Binomial Theorem can be used to find the numbers in Pascal's Triangle.

83. False. The coefficient of the x^{10}-term is 1,732,104 and the coefficient of the x^{14}-term is 192,456.

85.

1 8 28 56 70 56 28 8 1

1 9 36 84 126 126 84 36 9 1

1 10 45 120 210 252 210 120 45 10 1

87. The signs of the terms in the expansion of $(x - y)^n$ alternate between positive and negative.

89. Answers will vary. **91.** Answers will vary.

93.

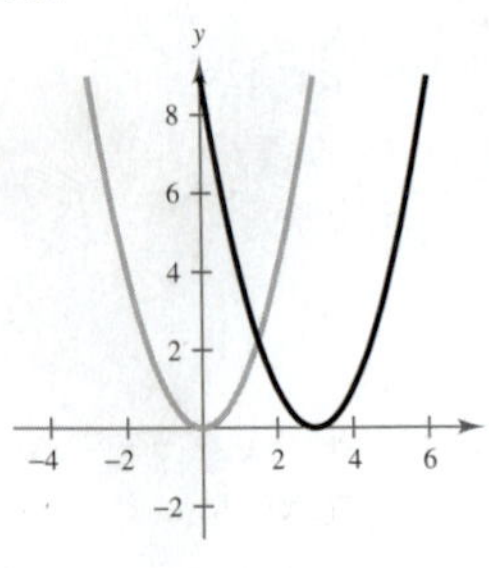

$g(x) = (x - 3)^2$

95.

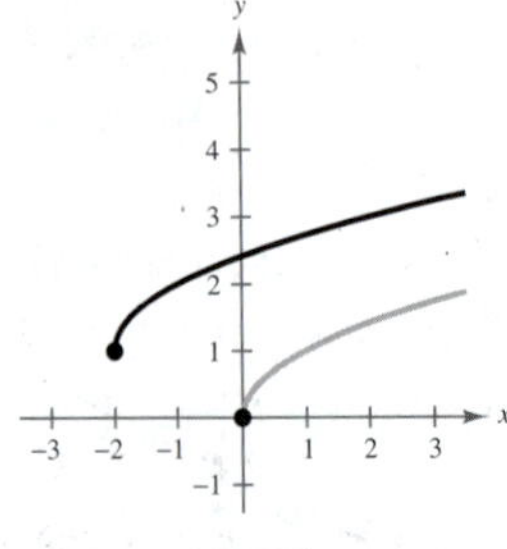

$g(x) = \sqrt{x+2} + 1$

97. $\begin{bmatrix} 4 & -5 \\ 5 & -6 \end{bmatrix}$

Section 11.6 *(page 857)*

1. 6 **3.** 5 **5.** 3 **7.** 7 **9.** 30 **11.** 15

13. 64 **15.** 4 **17.** 175,760,000

19. (a) 900 (b) 648 (c) 180 (d) 600 **21.** 64,000

23. (a) 40,320 (b) 384 **25.** 24 **27.** 336

29. 120 **31.** $n = 5$ or $n = 6$ **33.** 1,860,480

35. 970,200 **37.** 15,504 **39.** 420 **41.** 2520

43. ABCD, ABDC, ACBD, ACDB, ADBC, ADCB, BACD, BADC, CABD, CADB, DABC, DACB, BCAD, BDAC, CBAD, CDAB, DBAC, DCAB, BCDA, BDCA, CBDA, CDBA, DBCA, DCBA

45. AB, AC, AD, AE, AF, BC, BD, BE, BF, CD, CE, CF, DE, DF, EF

47. 120 **49.** 11,880 **51.** 15,504 **53.** 324,632

55. 3,921,225 **57.** 21 **59.** (a) 70 (b) 30

61. (a) 70 (b) 54 (c) 16 **63.** 5 **65.** 20

67. (a) 120, 526, 770

(b) If the jackpot is won, there is only one winning number.

(c) There are 22,957,480 possible winning numbers in the state lottery, which is considerably less than the possible number of winning Powerball numbers.

69. True by the definition of the Fundamental Counting Principle.

71. They are equal.

73. Answers will vary. **75.** Answers will vary.

77. No. For some calculators the number is too great.

79. (a) 35 (b) 8 (c) 83

81. (a) -4 (b) 0 (c) 0 **83.** 8.30 **85.** 35.00

87. $x^5 + 5x^4 + 10x^3 + 10x^2 + 5x + 1$

89. $x^{10} + 10x^8y + 40x^6y^2 + 80x^4y^3 + 80x^2y^4 + 32y^5$

Section 11.7 *(page 868)*

1. $\{(H, 1), (H, 2), (H, 3), (H, 4), (H, 5), (H, 6),$ $(T, 1), (T, 2), (T, 3), (T, 4), (T, 5), (T, 6)\}$

3. {ABC, ACB, BAC, BCA, CAB, CBA}

5. {AB, AC, AD, AE, BC, BD, BE, CD, CE, DE}

7. $\frac{3}{8}$ **9.** $\frac{7}{8}$ **11.** $\frac{3}{13}$ **13.** $\frac{3}{26}$ **15.** $\frac{1}{12}$

17. $\frac{11}{12}$ **19.** $\frac{1}{3}$ **21.** $\frac{1}{5}$ **23.** $\frac{2}{5}$ **25.** 0.3

27. $\frac{2}{3}$ **29.** 0.85 **31.** $\frac{7}{20}$

33. (a) 58% (b) 95.6% (c) 0.4%

35. (a) 38,570,000 (b) 23% (c) 58%

37. (a) $\frac{112}{209}$ (b) $\frac{97}{209}$ (c) $\frac{274}{627}$

39. $P(\{\text{Taylor wins}\}) = \frac{1}{2}$

$P(\{\text{Moore wins}\}) = P(\{\text{Jenkins wins}\}) = \frac{1}{4}$

41. (a) $\frac{21}{1292}$ (b) $\frac{225}{646}$ (c) $\frac{49}{323}$

43. (a) $\frac{1}{3}$ (b) $\frac{5}{8}$

45. (a) $\frac{1}{120}$ (b) $\frac{1}{24}$

47. (a) $\frac{1}{169}$ (b) $\frac{1}{221}$

49. (a) $\frac{14}{55}$ (b) $\frac{12}{55}$ (c) $\frac{54}{55}$

51. (a) $\frac{1}{4}$ (b) $\frac{1}{2}$ (c) $\frac{9}{100}$ (d) $\frac{1}{30}$

53. (a) 0.9702 (b) 0.9998 (c) 0.0002

55. (a) $\frac{1}{1024}$ (b) $\frac{243}{1024}$ (c) $\frac{781}{1024}$

57. 0.4746

59. $\frac{7}{16}$

61. True. Two events are independent if the occurrence of one has no effect on the occurrence of the other.

63. (a) As you consider successive people with distinct birthdays, the probabilities must decrease to take into account the birth dates already used. Because the birth dates of people are independent events, multiply the respective probabilities of distinct birthdays.

(b) $\frac{365}{365} \cdot \frac{364}{365} \cdot \frac{363}{365} \cdot \frac{362}{365}$ (c) Answers will vary.

(d) Q_n is the probability that the birthdays are *not* distinct, which is equivalent to at least two people having the same birthday.

(e)

n	10	15	20	23	30	40	50
P_n	0.88	0.75	0.59	0.49	0.29	0.11	0.03
Q_n	0.12	0.25	0.41	0.51	0.71	0.89	0.97

(f) 23

65. No real solution **67.** $0, \frac{1 \pm \sqrt{13}}{2}$ **69.** -4

71. $\frac{11}{2}$ **73.** -10 **75.** $\ln 27 \approx 3.296$

77. $\ln 1 = 0, \ln 3 \approx 1.099$ **79.** $\ln \frac{8}{3} \approx 0.981$

81. $e^8 \approx 2980.958$ **83.** $\frac{e^4}{6} \approx 9.100_3$

85.

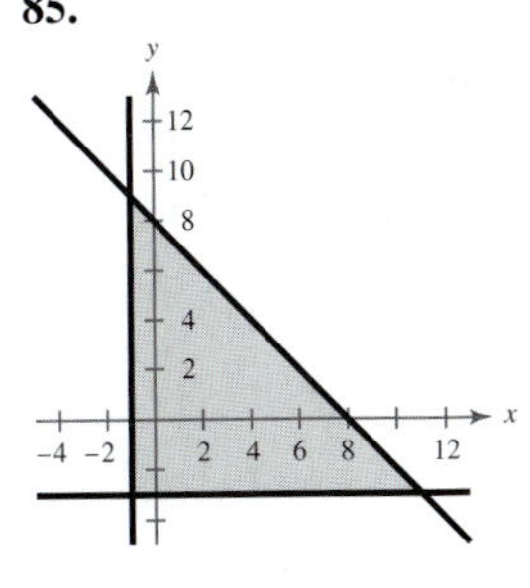

87.

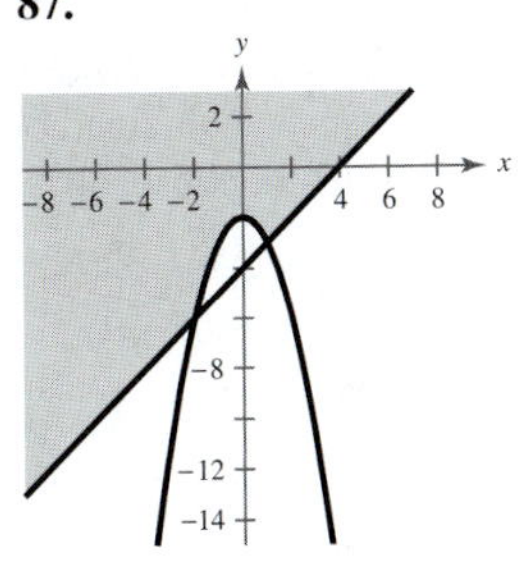

89. 15 **91.** 165

Review Exercises *(page 874)*

1. $8, 5, 4, \frac{7}{2}, \frac{16}{5}$ **3.** $72, 36, 12, 3, \frac{3}{5}$ **5.** $a_n = 2(-1)^n$

7. $a_n = \frac{4}{n}$ **9.** 120 **11.** 1 **13.** 30 **15.** $\frac{205}{24}$

17. 6050 **19.** $\sum_{k=1}^{20} \frac{1}{2k}$ **21.** $\frac{5}{9}$ **23.** $\frac{2}{99}$

25. (a) \$43,000 (b) \$192,500

27. Arithmetic sequence, $d = -2$

29. Arithmetic sequence, $d = \frac{1}{2}$

31. 4, 7, 10, 13, 16 **33.** 25, 28, 31, 34, 37

35. $a_n = 12n - 5$ **37.** $a_n = 3ny - 2y$

39. $a_n = -7n + 107$ **41.** 80 **43.** 88 **45.** 25,250

47. 43 minutes **49.** Geometric sequence, $r = 2$

51. Geometric sequence, $r = -2$ **53.** $4, -1, \frac{1}{4}, -\frac{1}{16}, \frac{1}{64}$

55. $9, 6, 4, \frac{8}{3}, \frac{16}{9}$ or $9, -6, 4, -\frac{8}{3}, \frac{16}{9}$

57. $a_n = 16\left(-\frac{1}{2}\right)^{n-1}$, 10.67

59. $a_n = 100(1.05)^{n-1}$, 3306.60 **61.** 127 **63.** $\frac{15}{16}$

65. 31 **67.** 24.85 **69.** 5486.45 **71.** 8 **73.** $\frac{10}{9}$

75. 12 **77.** (a) $a_t = 120{,}000(0.7)^t$ (b) \$20,168.40

79. Answers will vary. **81.** Answers will vary. **83.** 465

85. 4648 **87.** $S_n = n(2n + 7)$ **89.** $S_n = \frac{5}{2}\left[1 - \left(\frac{3}{5}\right)^n\right]$

91. 5, 10, 15, 20, 25
First differences: 5, 5, 5, 5
Second differences: 0, 0, 0
Linear

93. 16, 15, 14, 13, 12
First differences: $-1, -1, -1, -1$
Second differences: 0, 0, 0
Linear

95. 15 **97.** 56 **99.** 35 **101.** 28

103. $\frac{x^4}{16} + \frac{x^3y}{2} + \frac{3x^2y^2}{2} + 2xy^3 + y^4$

105. $a^5 - 15a^4b + 90a^3b^2 - 270a^2b^3 + 405ab^4 - 243b^5$

107. $41 + 840i$ **109.** 3 **111.** 10,000

113. 3,628,800 **115.** 56 **117.** $\frac{1}{9}$

119. (a) 43% (b) 82% **121.** $\frac{1}{216}$ **123.** $\frac{3}{4}$

125. True. $\frac{(n+2)!}{n!} = \frac{(n+2)(n+1)n!}{n!} = (n+2)(n+1)$

127. True by Properties of Sums

129. The set of natural numbers

131. (a) Each term is obtained by adding the same constant (common difference) to the preceding term.

(b) Each term is obtained by multiplying the same constant (common ratio) by the preceding term.

133. Each term of the sequence is defined in terms of preceding terms.

135. d **136.** a **137.** b **138.** c

139. 240, 440, 810, 1490, 2740

Chapter Test *(page 878)*

1. $-\frac{1}{5}, \frac{1}{8}, -\frac{1}{11}, \frac{1}{14}, -\frac{1}{17}$ **2.** $a_n = \frac{n+2}{n!}$

3. 50, 61, 72; 140 **4.** $a_n = 0.8n + 1.4$

5. 5, 10, 20, 40, 80 **6.** 86,100 **7.** 4

8. Answers will vary.

9. $x^4 + 8x^3y + 24x^2y^2 + 32xy^3 + 16y^4$ **10.** 180

11. (a) 72 (b) 328,440 **12.** (a) 330 (b) 720,720

13. 720 **14.** $\frac{1}{15}$ **15.** 3.908×10^{-10} **16.** 25%

Cumulative Test for Chapters 9–11 *(page 879)*

1. $(1, 2), \left(-\frac{3}{2}, \frac{3}{4}\right)$ **2.** $(2, -1)$

3. $(4, 2, -3)$ **4.** $(1, -2, 1)$

5.

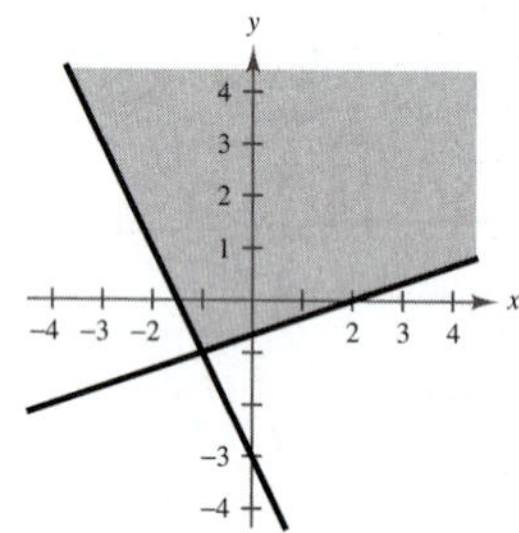

6.

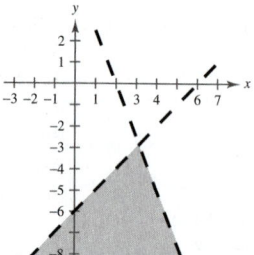

7.

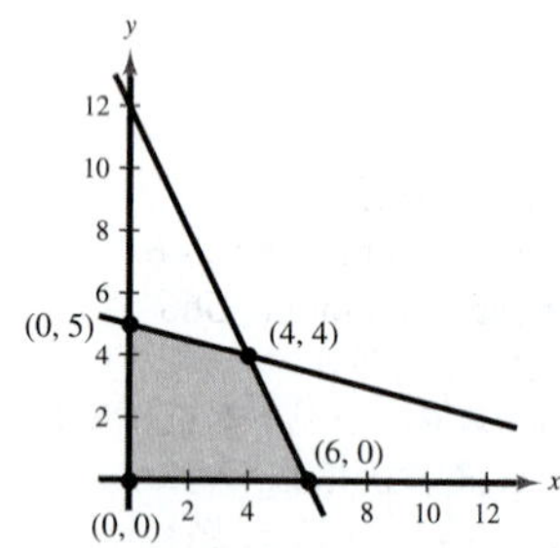

Maximum at (4, 4): $z = 20$

8. \$0.75 mixture: 120 pounds;
\$1.25 mixture: 80 pounds

9. $y = \frac{1}{3}x^2 - 2x + 4$

10. $\begin{bmatrix} -1 & 2 & -1 & \vdots & 9 \\ 2 & -1 & 2 & \vdots & -9 \\ 3 & 3 & -4 & \vdots & 7 \end{bmatrix}$ **11.** $(-2, 3, -1)$

12. $\begin{bmatrix} 5 & -3 \\ -2 & 2 \end{bmatrix}$ **13.** $\begin{bmatrix} 2 & -6 \\ -2 & 0 \end{bmatrix}$ **14.** $\begin{bmatrix} 6 & -6 \\ -3 & 2 \end{bmatrix}$

15. $\begin{bmatrix} -4 & 12 \\ 3 & -3 \end{bmatrix}$ **16.** 84 **17.** $\begin{bmatrix} -175 & 37 & -13 \\ 95 & -20 & 7 \\ 14 & -3 & 1 \end{bmatrix}$

18. Gym shoes: \$1936 million
Jogging shoes: \$1502 million
Walking shoes: \$3099 million

19. $(-5, 4)$ **20.** $(-3, 4, 2)$ **21.** 9

22. $\frac{1}{5}, -\frac{1}{7}, \frac{1}{9}, -\frac{1}{11}, \frac{1}{13}$ **23.** $a_n = \frac{(n+1)!}{n+3}$

24. 920 **25.** (a) 65.4 (b) $a_n = 3.2n + 1.4$

26. 3, 6, 12, 24, 48 **27.** $\frac{130}{9}$ **28.** Answers will vary.

29. $z^4 - 12z^3 + 54z^2 - 108z + 81$ **30.** 210 **31.** 600

32. 70 **33.** 120 **34.** 720 **35.** $\frac{1}{4}$

Problem Solving *(page 884)*

1. 1, 1.5, $1.41\overline{6}$, 1.414215686, 1.414213562, 1.414213562, . . .
x_n approaches $\sqrt{2}$.

3. (a)

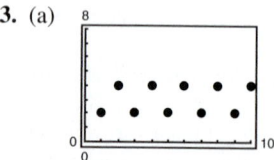

(b) If n is odd, $a_n = 2$, and if n is even, $a_n = 4$.

(c)

n	1	10	101	1000	10,001
a_n	2	4	2	4	2

(d) It is not possible to find the value of a_n as n approaches infinity.

5. (a) 3, 5, 7, 9, 11, 13, 15, 17
$a_n = 2n + 1$

(b) To obtain the arithmetic sequence, find the differences of consecutive terms of the sequence of perfect cubes. Then find the differences of consecutive terms of this sequence.

(c) 12, 18, 24, 30, 36, 42, 48
$a_n = 6n + 6$

(d) To obtain the arithmetic sequence, find the third sequence obtained by taking differences of consecutive terms in consecutive sequences.

(e) 60, 84, 108, 132, 156, 180
$a_n = 24n + 36$

7. $s_n = \left(\frac{1}{2}\right)^{n-1}$
$a_n = \frac{\sqrt{3}}{4}s_n^2$

9. Answers will vary.

11. (a) Answers will vary. (b) 17,710 **13.** $\frac{1}{3}$

Index of Applications

Biology and Life Sciences

Business

Chemistry and Physics

Construction

Consumer

Geometry

Interest Rate

Miscellaneous

Time and Distance

U.S. Demographics

Index

D

N

O

P

T

Definition of the Six Trigonometric Functions

Right triangle definitions, where $0 < \theta < \pi/2$.

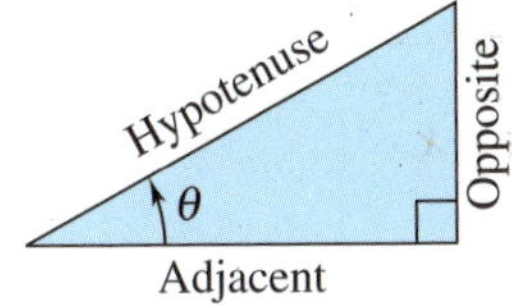

$\sin\theta = \frac{\text{opp.}}{\text{hyp.}}$ $\csc\theta = \frac{\text{hyp.}}{\text{opp.}}$

$\cos\theta = \frac{\text{adj.}}{\text{hyp.}}$ $\sec\theta = \frac{\text{hyp.}}{\text{adj.}}$

$\tan\theta = \frac{\text{opp.}}{\text{adj.}}$ $\cot\theta = \frac{\text{adj.}}{\text{opp.}}$

Circular function definitions, where θ *is any angle.*

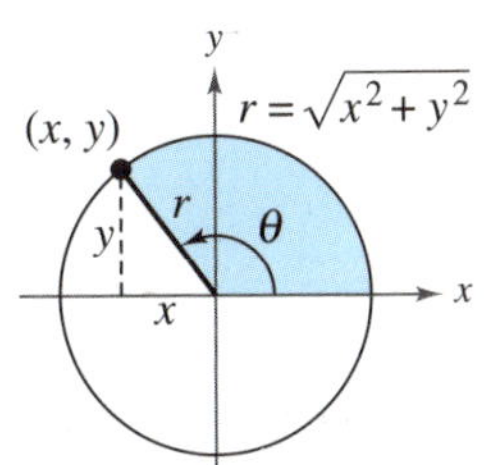

$\sin\theta = \frac{y}{r}$ $\csc\theta = \frac{r}{y}$

$\cos\theta = \frac{x}{r}$ $\sec\theta = \frac{r}{x}$

$\tan\theta = \frac{y}{x}$ $\cot\theta = \frac{x}{y}$

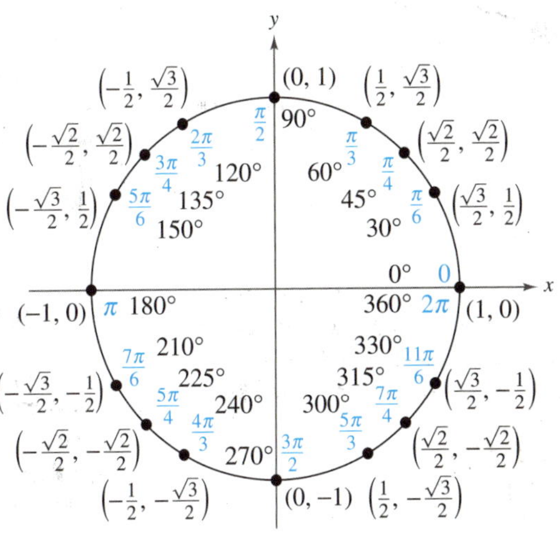

Reciprocal Identities

$\sin u = \frac{1}{\csc u}$ $\cos u = \frac{1}{\sec u}$ $\tan u = \frac{1}{\cot u}$

$\csc u = \frac{1}{\sin u}$ $\sec u = \frac{1}{\cos u}$ $\cot u = \frac{1}{\tan u}$

Quotient Identities

$\tan u = \frac{\sin u}{\cos u}$ $\cot u = \frac{\cos u}{\sin u}$

Pythagorean Identities

$\sin^2 u + \cos^2 u = 1$

$1 + \tan^2 u = \sec^2 u$ $1 + \cot^2 u = \csc^2 u$

Cofunction Identities

$\sin\left(\frac{\pi}{2} - u\right) = \cos u$ $\cot\left(\frac{\pi}{2} - u\right) = \tan u$

$\cos\left(\frac{\pi}{2} - u\right) = \sin u$ $\sec\left(\frac{\pi}{2} - u\right) = \csc u$

$\tan\left(\frac{\pi}{2} - u\right) = \cot u$ $\csc\left(\frac{\pi}{2} - u\right) = \sec u$

Even/Odd Identities

$\sin(-u) = -\sin u$ $\cot(-u) = -\cot u$

$\cos(-u) = \cos u$ $\sec(-u) = \sec u$

$\tan(-u) = -\tan u$ $\csc(-u) = -\csc u$

Sum and Difference Formulas

$\sin(u \pm v) = \sin u \cos v \pm \cos u \sin v$

$\cos(u \pm v) = \cos u \cos v \mp \sin u \sin v$

$\tan(u \pm v) = \frac{\tan u \pm \tan v}{1 \mp \tan u \tan v}$

Double-Angle Formulas

$\sin 2u = 2 \sin u \cos u$

$\cos 2u = \cos^2 u - \sin^2 u = 2\cos^2 u - 1 = 1 - 2\sin^2 u$

$\tan 2u = \frac{2\tan u}{1 - \tan^2 u}$

Power-Reducing Formulas

$\sin^2 u = \frac{1 - \cos 2u}{2}$

$\cos^2 u = \frac{1 + \cos 2u}{2}$

$\tan^2 u = \frac{1 - \cos 2u}{1 + \cos 2u}$

Sum-to-Product Formulas

$\sin u + \sin v = 2\sin\left(\frac{u+v}{2}\right)\cos\left(\frac{u-v}{2}\right)$

$\sin u - \sin v = 2\cos\left(\frac{u+v}{2}\right)\sin\left(\frac{u-v}{2}\right)$

$\cos u + \cos v = 2\cos\left(\frac{u+v}{2}\right)\cos\left(\frac{u-v}{2}\right)$

$\cos u - \cos v = -2\sin\left(\frac{u+v}{2}\right)\sin\left(\frac{u-v}{2}\right)$

Product-to-Sum Formulas

$\sin u \sin v = \frac{1}{2}[\cos(u - v) - \cos(u + v)]$

$\cos u \cos v = \frac{1}{2}[\cos(u - v) + \cos(u + v)]$

$\sin u \cos v = \frac{1}{2}[\sin(u + v) + \sin(u - v)]$

$\cos u \sin v = \frac{1}{2}[\sin(u + v) - \sin(u - v)]$

FORMULAS FROM GEOMETRY

Triangle:

$h = a \sin \theta$

$\text{Area} = \frac{1}{2}bh$

(Laws of Cosines)

$c^2 = a^2 + b^2 - 2ab \cos \theta$

Right Triangle:

(Pythagorean Theorem)

$c^2 = a^2 + b^2$

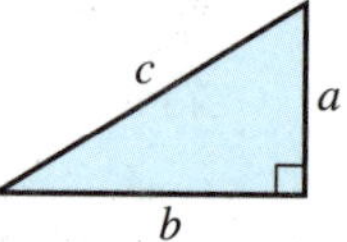

Equilateral Triangle:

$h = \frac{\sqrt{3}s}{2}$

$\text{Area} = \frac{\sqrt{3}s^2}{4}$

Parallelogram:

$\text{Area} = bh$

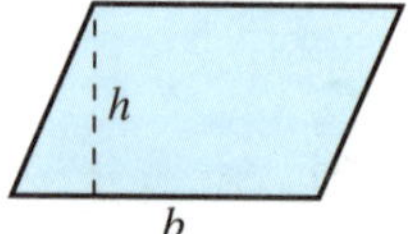

Trapezoid:

$\text{Area} = \frac{h}{2}(a + b)$

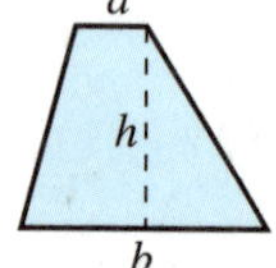

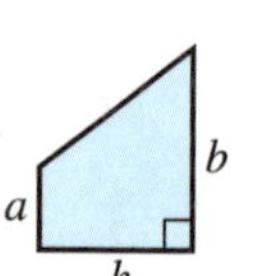

Circle:

$\text{Area} = \pi r^2$

$\text{Circumference} = 2\pi r$

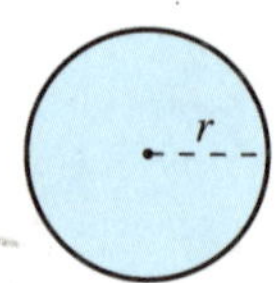

Sector of Circle:

(θ in radians)

$\text{Area} = \frac{\theta r^2}{2}$

$s = r\theta$

Circular Ring:

(p = average radius,

w = width of ring)

$\text{Area} = \pi(R^2 - r^2)$

$= 2\pi pw$

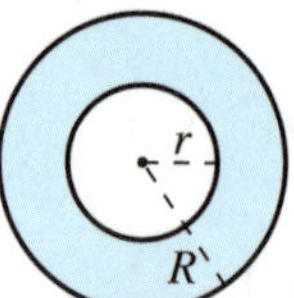

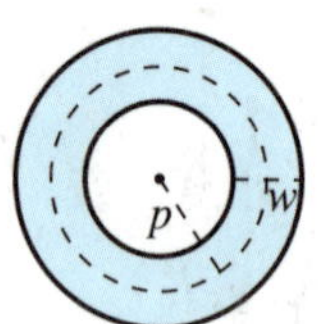

Sector of Circular Ring:

(p = average radius,

w = width of ring,

θ in radians)

$\text{Area} = \theta pw$

Ellipse:

$\text{Area} = \pi ab$

$\text{Circumference} \approx 2\pi \sqrt{\frac{a^2 + b^2}{2}}$

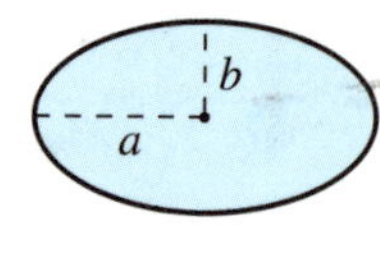

Cone:

(A = area of base)

$\text{Volume} = \frac{Ah}{3}$

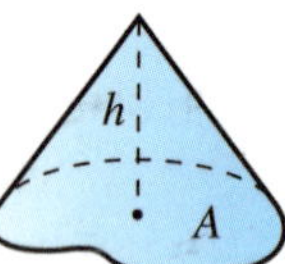

Right Circular Cone:

$\text{Volume} = \frac{\pi r^2 h}{3}$

$\text{Lateral Surface Area} = \pi r \sqrt{r^2 + h^2}$

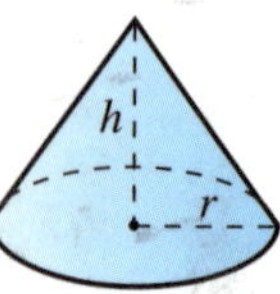

Frustum of Right Circular Cone:

$\text{Volume} = \frac{\pi(r^2 + rR + R^2)h}{3}$

$\text{Lateral Surface Area} = \pi s(R + r)$

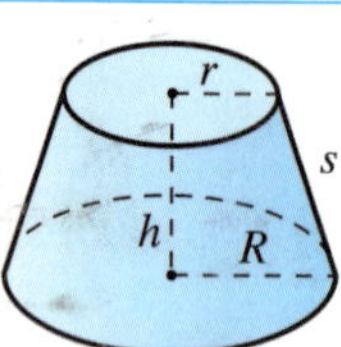

Right Circular Cylinder:

$\text{Volume} = \pi r^2 h$

$\text{Lateral Surface Area} = 2\pi rh$

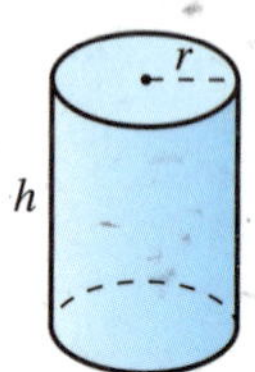

Sphere:

$\text{Volume} = \frac{4}{3}\pi r^3$

$\text{Surface Area} = 4\pi r^2$

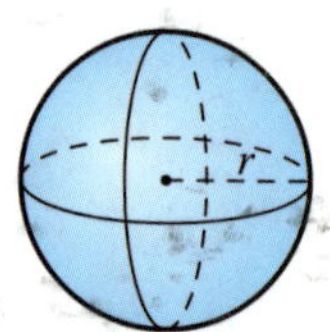

Wedge:

(A = area of upper face,

B = area of base)

$A = B \sec \theta$

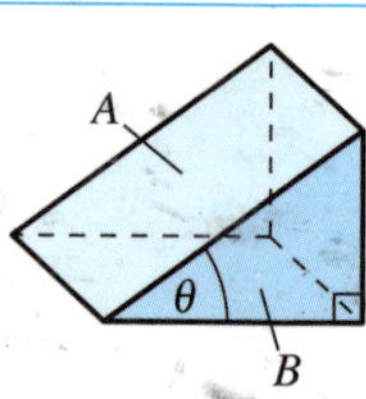